tenth edition

Prescott's Microbiology

Joanne M. Willey
HOFSTRA UNIVERSITY

Linda M. Sherwood
MONTANA STATE UNIVERSITY

Christopher J. Woolverton
KENT STATE UNIVERSITY

Mc
Graw
Hill
Education

PRESCOTT'S MICROBIOLOGY, TENTH EDITION

Published by McGraw-Hill Education, 2 Penn Plaza, New York, NY 10121. Copyright © 2017 by McGraw-Hill
Education. All rights reserved. Printed in the United States of America. Previous editions © 2014, 2011, and 2008. No
part of this publication may be reproduced or distributed in any form or by any means, or stored in a database or retrieval
system, without the prior written consent of McGraw-Hill Education, including, but not limited to, in any network or other
electronic storage or transmission, or broadcast for distance learning.

Some ancillaries, including electronic and print components, may not be available to customers outside the United States.

This book is printed on acid-free paper.

2 3 4 5 6 7 8 9 DOW 21 20 19 18 17 16

ISBN 978-1-259-28159-4
MHID 1-259-28159-0

Senior Vice President, Products & Markets: *Kurt L. Strand*
Vice President, General Manager, Products & Markets: *Marty Lange*
Vice President, Content Design & Delivery: *Kimberly Meriwether David*
Managing Director: *Michael S. Hackett*
Brand Manager: *Marija Magner*
Director, Product Development: *Rose Koos*
Product Developer: *Darlene M. Schueller*
Marketing Manager: *Kristine Rellihan*
Digital Product Developer: *Jake Theobald*
Director, Content Design & Delivery: *Linda Avenarius*
Program Manager: *Angela R. FitzPatrick*
Content Project Managers: *Jayne Klein/Christina Nelson*
Buyer: *Sandy Ludovissy*
Design: *Tara McDermott*
Content Licensing Specialists: *Carrie Burger/Lorraine Buczek*
Cover Image: © *Kevin Kemmerer/Getty Images/RF (background salt ponds);* © *Power and Syred/Science Source
(inset of purple halo bacterium archaea);* © *Dennis Kunkel Microscopy, Inc. (inset of gold halobacterium spp.-rod,
halophilic Archaea)*
Compositor: *Aptara®, Inc.*
Printer: *R. R. Donnelley*

All credits appearing on page or at the end of the book are considered to be an extension of the copyright page.

Library of Congress Cataloging-in-Publication Data
Willey, Joanne M.
  Prescott's microbiology / Joanne M. Willey, Hofstra University, Linda M. Sherwood, Montana State University,
Christopher J. Woolverton, Kent State University.—Tenth edition.
    pages cm
  ISBN 978-1-259-28159-4 (alk. paper)—ISBN 1-259-28159-0 (alk. paper) 1. Microbiology—Textbooks. I. Sherwood,
Linda. II. Woolverton, Christopher J. III. Title. IV. Title: Microbiology.
  QR41.2.P74 2015
  579—dc23
                                                                                2015027055

The Internet addresses listed in the text were accurate at the time of publication. The inclusion of a website does not
indicate an endorsement by the authors or McGraw-Hill Education, and McGraw-Hill Education does not guarantee the
accuracy of the information presented at these sites.

mheducation.com/highered

Brief Contents

About the Authors

Joanne M. Willey has been a professor at Hofstra University on Long Island, New York, since 1993, where she is Professor and Chair of the Department of Science Education at the Hofstra North Shore-Long Island Jewish School of Medicine. Dr. Willey received her B.A. in Biology from the University of Pennsylvania, where her interest in microbiology began with work on cyanobacterial growth in eutrophic streams. She earned her Ph.D. in biological oceanography (specializing in marine microbiology) from the Massachusetts Institute of Technology–Woods Hole Oceanographic Institution Joint Program in 1987. She then went to Harvard University, where she spent her postdoctoral fellowship studying the filamentous soil bacterium *Streptomyces coelicolor*. Dr. Willey continues to investigate this fascinating microbe and has coauthored a number of publications that focus on its complex developmental cycle. She is an active member of the American Society for Microbiology (ASM), and served on the editorial board of the journal *Applied and Environmental Microbiology* for nine years and as Chair of the Division of General Microbiology. Dr. Willey taught microbiology to biology majors for 20 years and now teaches microbiology and infectious disease to medical students. She has taught courses in cell biology, marine microbiology, and laboratory techniques in molecular genetics. Dr. Willey lives on the north shore of Long Island with her husband; she has two grown sons. She is an avid runner and enjoys skiing, hiking, sailing, and reading. She can be reached at joanne.m.willey@hofstra.edu.

Linda M. Sherwood recently retired from the Department of Microbiology at Montana State University after over 20 years of service to the department. Her interest in microbiology was sparked by the last course she took to complete a B.S. degree in Psychology at Western Illinois University. She went on to complete an M.S. degree in Microbiology at the University of Alabama, where she studied histidine utilization by *Pseudomonas acidovorans*. She subsequently earned a Ph.D. in Genetics at Michigan State University, where she studied sporulation in *Saccharomyces cerevisiae*. She briefly left the microbial world to study the molecular biology of *dunce* fruit flies at Michigan State University before moving to Montana State University. Dr. Sherwood has always had a keen interest in teaching, and her psychology training helped her to understand current models of cognition and learning and their implications for teaching. She taught courses in general microbiology, genetics, biology, microbial genetics, and microbial physiology. She served as the editor for ASM's *Focus on Microbiology Education* and participated in and contributed to numerous ASM Conferences for Undergraduate Educators (ASMCUE). She also worked with K–12 teachers to develop a kit-based unit to introduce microbiology into the elementary school curriculum and coauthored with Barbara Hudson a general microbiology laboratory manual, *Explorations in Microbiology: A Discovery Approach*. Her association with McGraw-Hill began when she prepared the study guides for the fifth and sixth editions of *Microbiology*. Her non-academic interests focus primarily on her family. She also enjoys reading, hiking, gardening, and traveling. She can be reached at lsherwood@montana.edu.

Christopher J. Woolverton is founding professor of Environmental Health Sciences, College of Public Health at Kent State University (KSU), and is the Director of the KSU Center for Public Health Preparedness, overseeing its BSL-3 Training Facility. He earned his B.S. in Biology from Wilkes College (PA), and his M.S. and Ph.D. in Medical Microbiology from West Virginia University, School of Medicine. He spent two years as a postdoctoral fellow at UNC-Chapel Hill. Dr. Woolverton's research is focused on the detection and control of pathogens. Dr. Woolverton has published and lectured widely on the mechanisms by which liquid crystals (LCs) act as microbial biosensors and on the LC characteristics of microbial proteins. Professor Woolverton has taught zombie preparedness, general microbiology, communicable diseases, immunology, prevention and control of disease, and microbial physiology. On faculty of the National Institutes of Health National Biosafety and Biocontainment Training Program, he teaches laboratory safety, risk assessment, decontamination strategies, and bioterrorism readiness. An active member of the American Society for Microbiology, Woolverton has served on its Board of Education, its distinguished lecturer program, and as a Conference for Undergraduate Educators co-chair, and is the immediate past editor-in-chief of its *Journal of Microbiology and Biology Education*. Woolverton and his wife, Nancy, have three grown daughters, two sons (in-law), and two grandchildren. He enjoys family time, hiking, camping, and cycling. His e-mail address is cwoolver@kent.edu.

Digital Tools for Your Success

Save time with auto-graded assessments. Gather powerful performance data.

McGraw-Hill Connect for Prescott's Microbiology provides online presentation, assignment, and assessment solutions, connecting your students with the tools and resources they'll need to achieve success.

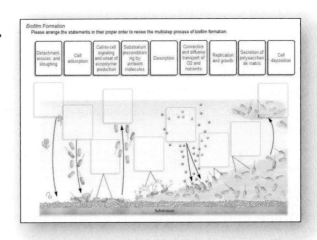

Homework and Assessment

With **Connect for Prescott's Microbiology,** you can deliver auto-graded assignments, quizzes, and tests online. Choose from a robust set of interactive questions and activities using high-quality art from the textbook and animations. Assignable content is available for every Learning Outcome in the book and is categorized according to the **ASM Curriculum Guidelines.** As an instructor, you can edit existing questions and author entirely new ones.

Significant faculty demand for content at higher Bloom's levels led us to examine assessment quality and consistency of our Connect content, to develop a scientific approach to systemically increase critical-thinking levels, and develop balanced digital assessments that promote student learning. The increased challenge at higher Bloom's levels will help the student grow intellectually and be better prepared to contribute to society.

Instructor Resources

Customize your lecture with tools such as PowerPoint® presentations, animations, and editable art from the textbook. An instructor's manual for the text saves you time in developing your course.

Learn more at connect.mheducation.com.

Detailed Reports

Track individual student performance—by question, by assignment, or in relation to the class overall—with detailed grade reports. Integrate grade reports easily with your Learning Management Systems (LMS).

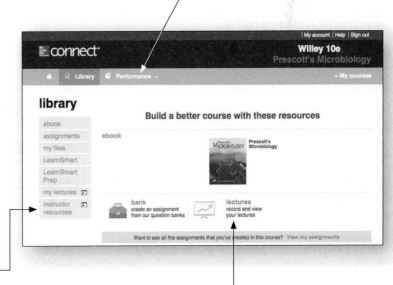

Lecture Capture

McGraw-Hill Tegrity® records and distributes your class lecture with just a click of a button. Students can view anytime, anywhere via computer or mobile device. Indexed as you record, students can use keywords to find exactly what they want to study.

McGraw-Hill Connect®
Learn Without Limits

Connect is a teaching and learning platform that is proven to deliver better results for students and instructors.

Connect empowers students by continually adapting to deliver precisely what they need, when they need it, and how they need it, so your class time is more engaging and effective.

Course outcomes improve with Connect.

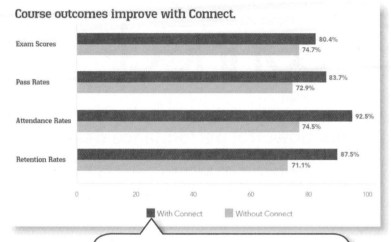

Using **Connect** improves passing rates by **10.8%** and retention by **16.4%**.

88% of instructors who use **Connect** require it; instructor satisfaction **increases** by 38% when **Connect** is required.

Analytics

Connect Insight®

Connect Insight is Connect's new one-of-a-kind visual analytics dashboard—now available for both instructors and students—that provides at-a-glance information regarding student performance, which is immediately actionable. By presenting assignment, assessment, and topical performance results together with a time metric that is easily visible for aggregate or individual results, Connect Insight gives the user the ability to take a just-in-time approach to teaching and learning, which was never before available. Connect Insight presents data that empowers students and helps instructors improve class performance in a way that is efficient and effective.

Connect helps students achieve better grades

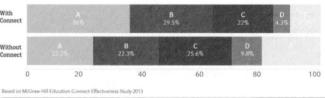

Based on McGraw-Hill Education Connect Effectiveness Study 2013

Students can view their results for any **Connect** course.

Mobile

Connect's new, intuitive mobile interface gives students and instructors flexible and convenient, anytime–anywhere access to all components of the Connect platform.

Adaptive

THE FIRST AND ONLY **ADAPTIVE READING EXPERIENCE** DESIGNED TO TRANSFORM THE WAY STUDENTS READ

> More students earn **A's** and **B's** when they use McGraw-Hill Education **Adaptive** products.

SmartBook®

Proven to help students improve grades and study more efficiently, SmartBook contains the same content within the print book, but actively tailors that content to the needs of the individual. SmartBook's adaptive technology provides precise, personalized instruction on what the student should do next, guiding the student to master and remember key concepts, targeting gaps in knowledge and offering customized feedback, and driving the student toward comprehension and retention of the subject matter. Available on smartphones and tablets, SmartBook puts learning at the student's fingertips—anywhere, anytime.

> Over **4 billion questions** have been answered, making McGraw-Hill Education products more intelligent, reliable, and precise.

STUDENTS WANT

SMARTBOOK®

95% of students reported **SmartBook** to be a more effective way of reading material

100% of students want to use the Practice Quiz feature available within **SmartBook** to help them study

100% of students reported having reliable access to off-campus wifi

90% of students say they would purchase **SmartBook** over print alone

95% reported that **SmartBook** would impact their study skills in a positive way

McGraw Hill Education

*Findings based on a 2015 focus group survey at Pellissippi State Community College administered by McGraw-Hill Education

www.learnsmartadvantage.com

A Modern Approach to Microbiology

Evolution as a Framework

Introduced immediately in chapter 1 and used as an overarching theme throughout, evolution helps unite microbiological concepts and provides a framework upon which students can build their knowledge.

An Introduction to the Entire Microbial World

Covered in chapters 3–6, the separate chapters on the structure and function of bacteria and archaea are followed by the discussion of eukaryotic cells preceding viruses.

Broad Coverage of Microbial Ecology

The importance and multidisciplinary nature of microbial ecology is demonstrated by content that ranges from global climate change to the human microbiome.

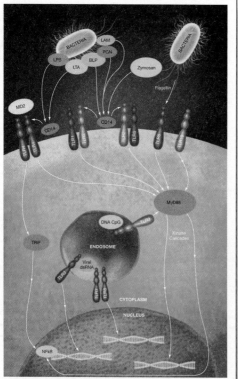

Figure 33.17 Recognition of Microbe-Associated Molecular Patterns (MAMPs) by Pattern Recognition Molecules (PRMs). MAMPs bind PRMs, especially toll-like receptors (TLRs), C-type lectin receptors (CLRs), and other pattern recognition receptors. PRM binding results in signaling that upregulates cytokine gene expression through common signal transduction pathways, like NF$_\kappa$B.

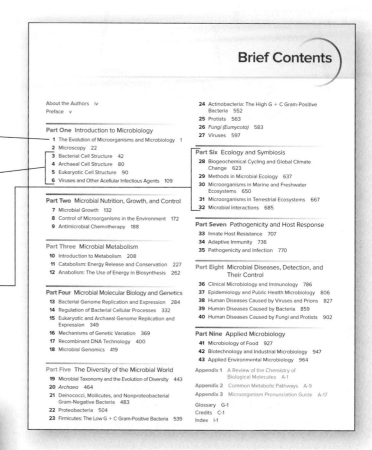

Molecular Microbiology and Immunology

The tenth edition includes updates on genetics, biotechnology, genomics and metagenomics, and immunology. The discussion of eukaryotic and archaeal genetics has been expanded, strengthening our understanding of the relatedness of genetic information flow observed in these organisms. A streamlined discussion of immunity, with enhanced detail between innate and adaptive linkages, helps students grasp the complexity and specificity of immune responses.

A Modern Approach to Microbiology

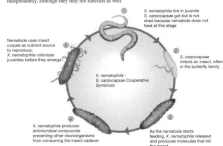

[Sample textbook page — Chapter 32 Microbial Interactions]

694 CHAPTER 32 | Microbial Interactions

Figure 32.8 The *Xenorhabdus nematophila* - *Steinernema carpocapsae* Is an Intriguing Model Bacterial-Nematode System.

21st-Century Microbiology

Prescott's Microbiology leads the way with text devoted to global climate change, biofuels, and microbial fuel cells. For more, see chapters 28, 30, 42, and 43.

Metagenomics and the Human Microbiome

The importance of metagenomics in understanding the role of microbes in all environments and in exploring symbionts of invertebrates and humans is threaded throughout the text. Special emphasis on the power of metagenomics is found in chapters 1, 18, 28, 32, and 42.

Laboratory Safety

Reflecting recommendations from the Centers for Disease Control and Prevention, along with the American Society for Microbiology, chapter 37 provides specific guidance for laboratory best practices to help instructors provide safe conditions during the teaching of laboratory exercises.

Special Interest Essays

Organized into four themes—Microbial Diversity & Ecology, Techniques & Applications, Historical Highlights, and Disease—these focused and interesting essays provide additional insight to relevant topics.

MICROBIAL DIVERSITY & ECOLOGY

4.1 What's in a Name?

Each day soon-to-be parents around the world agonize over what to name their babies. Is the name too popular or too unusual? Will it lead to undesirable nicknames? Was it the name of an unsavory historical figure? Though scientists probably don't agonize over what to call new organisms, cellular structures, or other natural phenomena, they do try to choose names carefully. Often the names have Greek or Latin roots that provide some information about the object being named. For instance, the archaeon *Pyrococcus furiosus*, a name that means rushing fireball, was so named because it is spherical, moves rapidly, and loves heat. Scientists also take care in naming things so that the names don't lead to misconceptions. Unfortunately, sometimes scientists get it wrong, and new names are suggested. Suggesting new names can lead to considerable debate and confusion about which terminology to use. Such is the case with the term *flagella*.

For decades, long, hairlike structures have been called flagella, and flagella have been identified in members of all three domains of life. In fact, the presence of flagella was long used as a criterion for distinguishing certain protists from others. Recall that protists and other eukaryotic organisms have another motility organelle, the cilium. As the ultrastructure of eukaryotic flagella and cilia were determined, it was found that they are the same. Both are very complex and make use of microtubules arranged in a characteristic 9 + 2 fashion. Furthermore, they move cells in a similar way: by whipping back and forth. Thus eukaryotic flagella are simply long cilia. Despite this, use of the term "flagella" when referring to long cilia persisted. When bacterial flagella discovered, they too were named flagella. Eventually ultrastructure and function were discovered and show distinct. As we describe in chapter 3, their structure is simpler, with the helical filament composed of a single of protein. It propels the cell by rotating.

With this knowledge, scientists began debating names for these structures. One suggestion was to re the term *flagella* for the bacterial organelle and to c the name of eukaryotic flagella to undilapodia, essentially means "waving feet." Undilapodia did n acceptance and finally scientists decided to use the cilia for both cilia and flagella. More recently, stud archaeal flagella led to the discovery that these are ve ferent from bacterial flagella and eukaryotic cilia, new debate has begun. Over the last few years some tists have suggested that three different terms be flagellum for the bacterial organelle, cilia for th eukaryotic organelles, and archaellum for the ar version of this motility organelle. Will this new name stick? Will the next edition of this text use the term? Will the discovery that archaeal flagella are evolutionarily related to bacterial type IV pili lead to a different name? Time will tell.

Source: Jarrell, K. F., and Albers, S.-V. 2012. The archaellum: An old motility structure with a new name. Trends Microbiol. 20(7):307–12.

DISEASE

26.1 White-Nose Syndrome Is Decimating North American Bat Populations

Bats evoke all kinds of images. Some people immediately think of vampire bats and are repulsed. Others think of large fruit bats often called flying foxes. If you have spent a summer evening outdoors on the east coast of North America, mosquitoes and the small bats that eat them may come to mind. A new scene can now be added to these: bats with white fungal hyphae growing around their muzzles (box figure). This is the hallmark of white-nose syndrome (WNS), and if its rate of infection continues unchecked, it is projected to eliminate the most common bat species in eastern North America (*Myotis lucifugus*) by 2026.

WNS was first spotted in 2006 among bats hibernating in a cave near Albany, NY. Scientists quickly became alarmed for two reasons. First, it spreads rapidly—it's known to occur in at least 11 bat species and is now found in 25 states in the United States and three Canadian provinces. Second, it is deadly. The population of bats declines from 30 to 99% in any given infected hibernacula (the place where bats hibernate, which unfortunately rhymes with Dracula).

WNS is caused by the ascomycete *Pseudogymnoascus destructans* (formerly *Geomyces destructans*). It colonizes a bat's

Geomyces destructans causes WNS. A little brown bat (*Myotis lucifugus*) with the white fungal hyphae (arrow) for which WNS is named.

wings, muzzle, and ears where it erodes the epidermis before invading the underlying skin and connective tissue. Despite the name WNS, the primary site of infection (and the anatomical site harmed most) is the wing. Wings provide a large surface area for colonization, and once infected, the thin layer of skin is easily damaged, leading to adverse physiological changes during hibernation. These in turn result in premature awakening, loss of essential fat reserves, and strange behavior.

Where did this pathogen come from and why does it infect bats? The best hypothesis regarding its origin is that humans inadvertently brought it from Europe, where it causes mild infection in at least one hibernating bat species. This makes *P. destructans* an apparent case of pathogen pollution—the human introduction of invasive pathogens of wildlife and domestic animal populations that threaten biodiversity and ecosystem function.

The capacity of *P. destructans* to sweep through bat populations results from a "perfect storm" of host- and pathogen-associated factors. *P. destructans* is psychrophilic, with a growth optimum around 12°C; it does not grow above 20°C. All infected bat species hibernate in cold and humid environments such as caves and mines. Because their metabolic rate is drastically reduced during hibernation, their body temperature reaches that of their surroundings, between 2 and 7°C. Thus WNS is only seen in hibernating bats or those that have just emerged from hibernation. When metabolically active, the bat's body temperature is too warm to support pathogen growth.

While it is too late to save the estimated 6 million bats that have already succumbed to WNS, microbiologists, conservationists, and government agencies are trying to limit the continued decline in bat populations. Caves have been closed to human traffic, and protocols for decontamination after visiting hibernacula have been developed to limit the spread from cave to cave. Although we cannot cure sick bats, it is our responsibility to stop the continued spread of this pathogen.

Read more: Langwig, K.E., et al. 2014. Invasion dynamics of white-nose syndrome fungus, midwestern United States, 2012–2014. Emerging Infectious Diseases. 21: 1023–1026.

Student-Friendly Organization

Micro Focus—Each chapter begins with a real-life story illustrating the relevance of the content covered in the upcoming text.

Readiness Check—The introduction to each chapter includes a skills checklist that defines the prior knowledge a student needs to understand the material that follows.

Learning Outcomes—Every section in each chapter begins with a list of content-based activities students should be able to perform after reading.

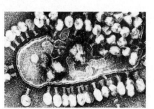

Animation Icon—This symbol indicates that material presented in the text is accompanied by an animation within Instructor Resources in Connect. Create a file attachment assignment in Connect to have your students view the animation, or post it to your Learning Management System for students.

Micro Inquiry—Selected figures in every chapter contain probing questions, adding another assessment opportunity for the student.

Cross-Referenced Notes—In-text references refer students to other parts of the book to review.

Retrieve, Infer, Apply—Questions within the narrative of each chapter help students master section concepts before moving on to other topics.

Student-Friendly Organization

Vivid Instructional Art Program—Three-dimensional renditions and bright, attractive colors enhance learning.

Annotated Figures—All key metabolic pathways and molecular processes are annotated, so that each step is clearly illustrated and explained.

Bacteria often have more than one transport system for a nutrient, as can be seen with *E. coli*. This bacterium has at least five transport systems for the sugar galactose, three systems each for the amino acids glutamate and leucine, and two potassium transport complexes. When several transport systems exist for the same substance, the systems differ in such properties as their energy source, their affinity for the solute transported, and the nature of their regulation. This diversity gives the bacterium an added competitive advantage in a variable environment.

Group Translocation

The distinguishing characteristic of **group translocation** is that a molecule is chemically modified as it is brought into the cell. The best-known group translocation system is the **phosphoenolpyruvate: sugar phosphotransferase system (PTS)**, which is observed in many bacteria. The PTS transports a variety of sugars while phosphorylating them, using phosphoenolpyruvate (PEP) as the phosphate donor. PEP is an important intermediate of a biochemical pathway used by many bacteria to extract energy from organic energy sources. PEP is a high-energy molecule that can be used to synthesize ATP, the cell's energy currency. However, when it is used in PTS reactions, the energy present in PEP is used to energize

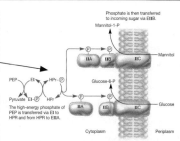

Figure 3.14 Group Translocation: Bacterial PTS Transport. Two examples of the phosphoenolpyruvate: sugar phosphotransferase system (PTS) are illustrated. The following components are involved in the system: phosphoenolpyruvate (PEP), enzyme I (EI), the low molecular weight heat-stable protein (HPr), and enzyme II (EII). EIIA is attached to EIIB in the mannitol transport system and is separate from EIIB in the glucose system.

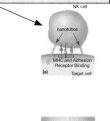

Figure 33.12 The System Used by Natural Killer (NK) Cells to Recognize and Destroy Abnormal Cells. (a) NK cell evaluates target cell using membrane ... es toward the target cell. (c) A lytic cleft forms and (d) granules exit the NK cell into the cleft. (e) Lytic perforins and ... eases the dead target cell and migrates toward another cell.

Key Concepts

17.1 Key Discoveries Led to the Development of Recombinant DNA Technology

- Genetic engineering became possible after the discovery of restriction enzymes and reverse transcriptase, and the development of essential methods in nucleic acid chemistry such as the Southern blotting technique.
- Restriction enzymes are important because they cut DNA at specific sequences, thereby releasing fragments of DNA that can be cloned or otherwise manipulated (**figure 17.3** and **table 17.1**).
- Gel electrophoresis is used to separate molecules according to charge and size.
- DNA fragments are separated on agarose and acrylamide gels. Because DNA is acidic, it migrates from the negative to the positive end of a gel (**figure 17.6**).

17.2 Polymerase Chain Reaction Amplifies Targeted DNA

- The polymerase chain reaction (PCR) allows small amounts of specific DNA sequences to be amplified, or increased in concentration thousands of times (**figure 17.8**).
- PCR consists of multiple cycles of 3 steps each: DNA denaturation, primer annealing, and DNA synthesis.
- PCR has numerous applications. It often is used to obtain genes for cloning and in diagnostic and forensic science.

17.3 Cloning Vectors Are Needed to Create Recombinant DNA

- There are four types of cloning vectors: plasmids, viruses, cosmids, and artificial chromosomes. Cloning vectors generally have at least three components: an origin of replication, a selectable marker, and a multicloning site or polylinker (**table 17.2; figures 17.10** and **17.12**).
- The most common approach to cloning is to digest both vector and DNA to be inserted with the same restriction enzyme or enzymes so that compatible sticky ends are generated. The vector and DNA to be cloned are then incubated in the presence of DNA ligase, which catalyzes the formation of phosphodiester bonds once the DNA fragment inserts into the vector.

- Once the recombinant plasmid has been introduced into host cells, cells carrying vector must be selected. This is often accomplished by allowing the growth of only antibiotic-resistant cells because the vector bears an antibiotic-resistance gene. Cells that took up vector with inserted DNA must then be distinguished from those that contain only vector. Often a blue-versus-white colony phenotype is used; this is based on the presence or absence, respectively, of a functional *lacZ* gene (**figure 17.11**).

17.4 Introducing Recombinant DNA into Host Cells

- The bacterium *E. coli* and the yeast *S. cerevisiae* are the most common host species.
- DNA can be introduced into microbes by transformation or electroporation.

17.5 Genomic Libraries: Cloning Genomes in Pieces

- It is sometimes necessary to find a gene without the knowledge of the gene's DNA sequence. A genomic library is constructed by cleaving an organism's genome into many fragments, each of which is cloned into a vector to make a unique recombinant plasmid.
- Genomic libraries are often screened for the gene of interest by either phenotypic rescue (genetic complementation) or DNA hybridization with an oligonucleotide probe (**figure 17.13**).

17.6 Expressing Foreign Genes in Host Cells

- An expression vector has the necessary features to express in high levels any recombinant gene it carries.
- If a eukaryotic gene is to be expressed in a bacterium, cDNA is used because it lacks introns; a bacterial leader must also be added to the 5′ end of the gene.
- Purification of recombinant proteins is often accomplished by fusing the coding sequence of a protein to six histidine residue codons found on some expression vectors. When introduced and expressed in bacteria, the His-tagged protein can be selectively purified (**figure 17.14**).
- Green fluorescent protein can be used to study the regulation of gene expression (transcriptional fusions) and protein localization (translational fusions) (**figure 17.15**).

Key Concepts—At the end of each chapter and organized by numbered headings, this feature distills the content to its essential components with cross-references to figures and tables.

Compare, Hypothesize, Invent

1. You are performing a PCR to amplify a gene encoding a tRNA from a bacterium that has only recently been grown in pure culture. You are expecting a product of 954 bp. However, you generate three different products; only one is the expected size. List at least two possible explanations (excluding experimental error).

2. You have cloned a structural gene required for riboflavin synthesis in *E. coli*. You find that an *E. coli* riboflavin auxotroph carrying the cloned gene on a vector makes less riboflavin than does the wild-type strain. Why might this be the case?

Compare, Hypothesize, Invent—Includes questions taken from current literature; designed to stimulate analytical problem-solving skills.

List of Content Changes

Each chapter has been thoroughly reviewed.

Part One

Chapter 1—Evolution is the driving force of all biological systems; this is made clear by introducing essential concepts of microbial evolution first. Advances in the discipline of microbiology and the increasing contributions of genomics and metagenomics are discussed.

Chapter 2—Microscopy was and is critical to the study of microorganisms and this chapter considers the most commonly used methods, including expanded coverage of phase-contrast microscopy.

Chapter 3—Coverage of bacterial cellular structure and function. A new chapter-opening story clearly establishes the importance of the material covered in this chapter.

Chapter 4—Growing understanding of the distinctive characteristics of archaea has warranted the creation of a new Microbial Diversity & Ecology box on the nature of motility organelles in the three domains of life. Comparisons to bacteria are made throughout the chapter.

Chapter 5—An introduction to eukaryotic cell structure and function, with emphasis on eukaryotic microbes. More detailed information on protist and fungal cells is presented in chapters 25 (*Protists*) and 26 (*Fungi*), which also focus on the diversity of these microbes. The current thought on the evolution of mitochondria and mitochondria-like organelles is considered. Comparisons between bacteria, archaea, and eukaryotes are included throughout the chapter.

Chapter 6—This chapter surveys the essential morphological, physiological, and genetic elements of viruses as well as viroids, satellites, and prions. Updated discussion of the role of viruses in causing cancer. This chapter completes our four-chapter introduction to microbial life.

Part Two

Chapter 7—Discussion of the growth of microbes outside the laboratory, including expanded and updated coverage of the "persister cell" phenomenon, is followed by topics related to laboratory culture of microbes.

Chapter 8—Updated to reflect emphasis on interruption of normal growth and reproduction functions to control microorganisms.

Chapter 9—Content focuses on the mechanism of action of each antimicrobial agent and stresses usage to limit drug resistance.

Part Three

Chapter 10—This introduction to metabolism includes a section outlining the nature of biochemical pathways. The concept of metabolic flux is presented by discussing the interconnected biochemical pathways used by cells.

Chapter 11—An introduction to metabolic diversity and nutritional types is followed by an exploration of the energy-conserving process of each nutritional type. The coverage of oxygenic photosynthesis is expanded and updated.

Chapter 12—New coverage of pathways used to synthesize porphyrins, lipopolysaccharides, sterols, and isoprenoid lipids.

Part Four

Chapter 13—Updated coverage of protein splicing, folding, and secretion.

Chapter 14—The regulation of bacterial cellular processes, with updated coverage of regulation by small RNAs.

Chapter 15—Eukaryal and archaeal genome replication and expression are considered together. In both cases, the discussion has been updated and expanded, and reflects the similarity of information flow as carried out by members of *Archaea* and *Eukarya*.

Chapter 16—Covers mutation, repair, and recombination in the context of processes that introduce genetic variation into populations. A new chapter-opening story introduces the complexity of understanding the growing problem of antibiotic resistance.

Chapter 17—Students are guided through the steps of cloning a microbial gene—from DNA purification through protein purification.

Chapter 18—Next-generation nucleotide sequencing and single-cell genome sequencing are covered in the context of metagenomics as it relates to the microbial ecology of natural systems, including the human microbiome.

Part Five

Chapter 19—This overview of microbial evolution includes discussion of the concepts of ecotype, microbial species, and superphylum.

Chapter 20—Expanded coverage of archaeal physiology includes archaeal-specific catabolic and anabolic pathways with particular attention CO_2 fixation. The evolutionary advantage of each pathway is discussed in the context of archaeal ecology.

List of Content Changes

Chapter 21—In addition to the ecology and physiology of photosynthetic bacteria, the recently described *Planctomycetes, Verrucomicrobia, Chlamydia* (PVC) superphylum is introduced with an updated review of each of these genera.

Chapter 22—This chapter's coverage includes newly recognized genera, an expanded discussion of sulfur metabolism, and an updated discussion of gliding motility that reflects recent advances.

Chapter 24—This overview of actinobacteria incorporates new figures illustrating the mycobacterial cell wall and a new photo program.

Chapter 25—This chapter introduces protist morphology and diversity, with an emphasis on physiological adaptation and ecology.

Chapter 26—Fungal diversity is presented within a phylogenetic framework. Morphology, ecology, and reproductive strategies are stressed.

Chapter 27—Updated discussion of virus taxonomy and phylogeny.

Part Six

Chapter 28—The description of each nutrient cycle is accompanied by a "student-friendly" figure that distinguishes between reductive and oxidative reactions. Updated coverage of the role of biogeochemical cycling in global climate change.

Chapter 29—This chapter continues to emphasize culture-based techniques as the "gold standard" and reviews some new, innovative approaches such as mass spectrometry in the identification of microbial taxa as well as metatranscriptomics and metaproteomics in the study of community activity.

Chapter 30—Updated and expanded discussion of the role of marine microbes in the global carbon budget as well as an update on subsurface microbes.

Chapter 31—New and updated coverage of the microbial ecology of the phyllosphere, rhizoplane, and rhizosphere. Expanded discussion of fungal plant pathogens.

Chapter 32—Important model systems for the exploration of microbial symbiosis are presented, along with increased coverage of the human microbiome.

Part Seven

Chapter 33—Updated to reflect the increasing overlap with the acquired immune functions, this chapter on innate host resistance provides in-depth coverage of physical and chemical components of the nonspecific host response, followed by an overview of cells, tissues, and organs of the immune system. Uniting these components are the ever-expanding methods for recognition of microorganisms by immune cells. Barriers, chemical mediators, and immune cells define the molecular mechanisms that drive phagocytosis and inflammation.

Chapter 34—Updated to enhance linkages between innate and adaptive immune activities. Discussions integrate concepts of cell biology, physiology, and genetics to present the immune system as a unified response having various components. Implications of dysfunctional immune actions are also discussed.

Chapter 35—This chapter has been reorganized to reflect the host-microorganism interaction that can lead to human disease. The essential elements required for a pathogen to establish infection are introduced, and virulence mechanisms are highlighted. This chapter is placed after the immunology chapters to stress that the host-parasite relationship is dynamic, with adaptations and responses offered by both host and parasite.

Part Eight

Chapter 36—This chapter has been updated to reflect the technological advances of a modern clinical laboratory. Emphasis is on modern diagnostic testing to identify infectious disease.

Chapter 37—Expanded focus on the important role of laboratory safety, especially in the teaching laboratory. Discussion emphasizes modern epidemiology as an investigative science and its role in preventative medicine. Disease prevention strategies are highlighted.

Chapter 38—Updated and expanded coverage includes viral pathogenesis, common viral infections, and prion-mediated diseases.

Chapter 39—Updated coverage of bacterial organisms and the ways in which they commonly lead to human disease.

Chapter 40—Updated and expanded coverage of fungal and protozoal diseases.

Part Nine

Chapter 41—Discussion of milk fermentation processes, including an updated description of cheese making.

Chapter 42—Includes updated coverage of biofuel production (first introduced in chapter 21) and an introduction to synthetic biology.

Chapter 43—This chapter complements our 21st-century approach to microbiology by emphasizing the importance of clean water and the power of microbial environmental remediation.

Lab Tools for Your Success

LearnSmart Labs® is an adaptive simulated lab experience that brings meaningful scientific exploration to students. Through a series of adaptive questions, LearnSmart Labs identifies a student's knowledge gaps and provides resources to quickly and efficiently close those gaps. Once students have mastered the necessary basic skills and concepts, they engage in a highly realistic simulated lab experience that allows for mistakes and the execution of the scientific method.

LearnSmart® Prep is an adaptive learning tool that prepares students for college-level work in Microbiology. LearnSmart Prep individually identifies concepts the student does not fully understand and provides learning resources to teach essential concepts so he or she enters the classroom prepared. Data-driven reports highlight areas where students are struggling, helping to accurately identify weak areas.

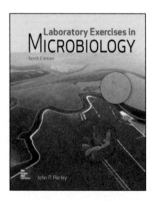

Laboratory Exercises in Microbiology, Tenth Edition

John P. Harley has revised this laboratory manual to accompany the tenth edition of *Prescott's Microbiology*. The class-tested exercises are modular which allows instructors to easily incorporate them into their course. This balanced introduction to each area of microbiology also has accompanying Connect content for additional homework and assessment opportunities. In addition, all artwork from the lab manual is available through the Instructor Resources in Connect for incorporation into lectures.

Acknowledgments

In the preparation of each edition, we have been guided by the collective wisdom of reviewers who are expert microbiologists and excellent teachers. They represent experience in community colleges, liberal arts colleges, comprehensive institutions, and research universities. We have followed their recommendations, while remaining true to our overriding goal of writing readable, student-centered content. Each feature incorporated into this edition has been carefully considered in terms of how it may be used to support student learning in both the traditional and the flipped learning environment.

Also in this edition, we are very excited to incorporate real student data points and input, derived from thousands of our LearnSmart users, to help guide our revision. With this information, we were able to hone both book and digital content.

The authors wish to extend their gratitude to our team at McGraw-Hill, including Marija Magner, Darlene Schueller, Kristine Rellihan, Jayne Klein, Tara McDermott, Christina Nelson, Lorraine Buczek, Carrie Burger, and David Tietz. Finally, we thank our spouses and children, who provided support and tolerated our absences (mental, if not physical) while we completed this demanding project.

Contents

Contents

Contents

Contents

Contents

Contents

Contents

Contents

The Evolution of Microorganisms and Microbiology

Artist's rendition of the six planets orbiting a star called Kepler-11. The drawing is based on observations made of the system by the Kepler spacecraft on August 26, 2010. Some are Earth-sized and may be habitable by life.

Over 4,000 Potential Planets Discovered

As of July 2015, the National Aeronautics and Space Administration (NASA) reported that over 4,000 potential planets and almost 1,000 confirmed planets had been discovered by the 2009 *Kepler* mission. Using a telescope in space, the light emanating from stars as far as 3,000 light-years away had been monitored every half-hour. The *Kepler* telescope identified planets as they circulated their star and caused a brief decrease in emitted light; just as an object is detected as a blip by radar, a blip of "darkness" indicates a possible planet.

Unless you are a science fiction fan, you might wonder why NASA is interested in finding planets. By finding other planets, scientists can gather evidence to support or refute current models of planet formation. These models predict a process that is chaotic and violent. Planets are thought to begin as dust particles circling around newly formed stars. As these particles collide, they grow in size, forming larger chunks. Eventually a series of such collisions results in planet-sized bodies. NASA astrobiologists are interested in identifying characteristics of a planet that may allow it to support life. Using Earth as a model, they hypothesize that life-supporting planets will share many features with Earth. But how will life be recognized? Scientists look to life on Earth to answer this question, and increasingly they are turning to micro-biologists for help.

Earth formed 4.5 billion years ago. Within the next billion years, the first cellular life forms—microbes—appeared. Since that time, microorganisms have evolved and diversified to occupy virtually every habitat on Earth: from oceanic geothermal vents to the coldest Arctic ice. The diversity of cellular microorganisms is best exemplified by their metabolic capabilities. Some carry out respiration, just as animals do. Others perform photosynthesis, rivaling plants in the amount of carbon dioxide they capture, forming organic matter and releasing oxygen into the atmosphere. Still other microbes are able to use inorganic molecules as sources of energy in both

oxic (oxygen available) and anoxic (no oxygen) conditions. Microbes also are diverse in terms of environmental conditions. Some withstand extremes of temperature, pressure, and pH. Indeed, studies have shown that some Earth microbes tolerate conditions that simulate those on Mars. These microbes are important for understanding what life might be like on other worlds.

Our goal in this chapter is to introduce you to the amazing world of microorganisms and to outline the history of their evolution and discovery. Microbiology is a biological science, and as such, much of what you will learn in this text is similar to what you have learned in high school and college biology classes that focus on large organisms. But microbes have unique properties, so microbiology has unique approaches to understanding them. These too will be introduced. But before you delve into this chapter, check to see if you have the background needed to get the most from it.

Readiness Check:
Based on what you have learned previously, you should be able to:
✔ List the features of eukaryotic cells that distinguish them from other cell types
✔ List the attributes that scientists use to determine if an object is alive

1.1 Members of the Microbial World

After reading this section, you should be able to:
- Differentiate the biological entities studied by microbiologists from those studied by other biologists
- Explain Carl Woese's contributions in establishing the three-domain system for classifying cellular life
- Provide an example of the importance to humans of each of the major types of microbes
- Determine the type of microbe (e.g., bacterium, fungus, etc.) when given a description of a newly discovered microbe

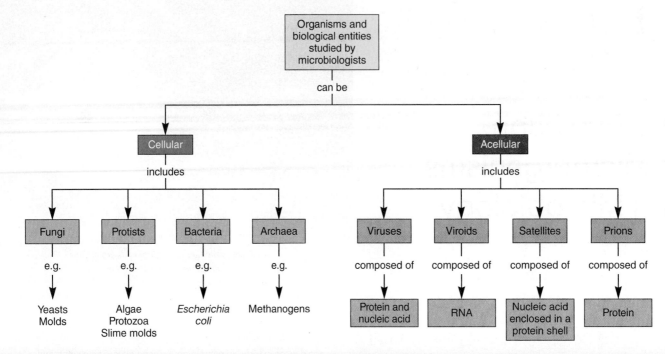

Figure 1.1 Concept Map Showing the Types of Biological Entities Studied by Microbiologists.

MICRO INQUIRY *How would you alter this concept map so that it also distinguishes cellular organisms from each other?*

Microorganisms are defined as those organisms too small to be seen clearly by the unaided eye (**figure 1.1**). They are generally 1 millimeter or less in diameter. Although small size is an important characteristic of microbes, it alone is not sufficient to define them. Some microbes, such as bread molds and filamentous photosynthetic microbes, are actually visible without microscopes. These macroscopic microbes are often colonial, consisting of small aggregations of cells. Some macroscopic microorganisms are multicellular. They are distinguished from other multicellular life forms such as plants and animals by their lack of highly differentiated tissues. Most unicellular microbes are microscopic. However, there are interesting exceptions, as we describe in chapter 3. In summary, cellular microbes are usually smaller than 1 millimeter in diameter, often unicellular and, if multicellular, lack differentiated tissues.

In addition to microorganisms, microbiologists study a variety of acellular biological entities (figure 1.1). These include viruses and subviral agents. Although the term "microorganism" is often applied only to cellular microbes, some texts use both "microorganism" and "microbe" when referring to these acellular agents.

The diversity of microorganisms has always presented a challenge to microbial taxonomists. The early descriptions of cellular microbes as either plants or animals were too simple. For instance, some microbes are motile like animals but also have cell walls and are photosynthetic like plants. Such microbes cannot be placed easily into either kingdom. An important breakthrough in microbial taxonomy arose from studies of their cellular architecture, when it was discovered that cells

exhibited one of two possible "floor plans." Cells that came to be called **prokaryotic cells** (Greek *pro,* before, and *karyon,* nut or kernel; organisms with a primordial nucleus) have an open floor plan. That is, their contents are not divided into compartments ("rooms") by membranes. The most obvious characteristic of these cells is that they lack the membrane-delimited nucleus observed in **eukaryotic cells** (Greek *eu,* true, and *karyon,* nut or kernel). Eukaryotic cells not only have a nucleus but also many other membrane-bound organelles that separate some cellular materials and processes from others.

These observations eventually led to the development of a classification scheme that divided organisms into five kingdoms: *Monera, Protista, Fungi, Animalia,* and *Plantae.* Microorganisms (except for viruses and other acellular infectious agents, which have their own classification system) were placed in the first three kingdoms. In this scheme, all organisms with prokaryotic cell structure were placed in *Monera.* The five-kingdom system was an important development in microbial taxonomy, but it is no longer accepted by microbiologists. This is because not all "prokaryotes" are the same and therefore should not be grouped together in a single kingdom. Furthermore, it is currently argued that the term *prokaryote* is not meaningful and should be abandoned. As we describe next, this discovery required several advances in the tools used to study microbes. ▶▶▎ *Use of the term "prokaryote" is controversial (section 3.1)*

Great progress has been made in three areas that profoundly affect microbial classification. First, much has been learned about the detailed structure of microbial cells from the use of electron microscopy. Second, microbiologists have determined the biochemical and physiological characteristics of many different microorganisms. Third, the sequences of nucleic acids and proteins from a wide variety of organisms have been compared. The comparison of ribosomal RNA (rRNA), begun by Carl Woese (1928–2012) in the 1970s, was instrumental in demonstrating that there are two very different groups of organisms with prokaryotic cell architecture: *Bacteria* and *Archaea*. Later studies based on rRNA comparisons showed that *Protista* is not a cohesive taxonomic unit (i.e., taxon) and that it should be divided into three or more kingdoms. These studies and others have led many taxonomists to reject the five-kingdom system in favor of one that divides cellular organisms into three domains: *Bacteria, Archaea,* and *Eukarya* (all eukaryotic organisms) (**figure 1.2**).

▶▶ *Nucleic acids (appendix I); Proteins (appendix I)*

Although the three-domain tree is widely accepted, other trees are also possible. Perhaps the leading alternate tree is a two-domain tree consisting of *Archaea* and *Bacteria*. In this tree eukaryotes are simply a lineage within the archaeal domain. Both trees have proponents who continue to debate which tree best represents the evolutionary history of life on Earth. Until the debate is settled, we will use the three-domain system throughout this text. A brief description of the three domains and the microorganisms in them follows.

Members of domain ***Bacteria*** are usually single-celled organisms.[1] Most have cell walls that contain the structural molecule peptidoglycan. Although most bacteria exhibit typical prokaryotic cell structure (i.e., they lack a membrane-bound nucleus), a few members of the unusual phylum *Planctomycetes* have their genetic material surrounded by a membrane. This inconsistency is another argument made for abandoning the term "prokaryote." Bacteria are abundant in soil, water, and air, including sites that have extreme temperatures, pH, or salinity. Bacteria are also major inhabitants of our bodies, forming the human **microbiome.** Indeed, more microbial cells are found in and on the human body than there are human cells. These microbes begin to colonize humans shortly after birth. As the microbes establish themselves, they contribute to the development of the body's immune system. Those microbes that inhabit the large intestine help the body digest food and produce vitamins. In these and other ways, the human microbiome helps maintain our health and well-being.

▶▶ *Phylum* Planctomycetes *(section 21.5)*

Unfortunately, some bacteria cause disease, and some of these diseases have had a huge impact on human history. In 1347 the plague (Black Death), an arthropod-borne disease, struck Europe with brutal force, killing one-third of the population (about 25 million people) within four years. Over the next 80 years, the disease struck repeatedly, eventually wiping out 75% of the European population. The plague's effect was so great that some historians believe it changed European culture and prepared the way for the Renaissance. Because of such plagues, it is easy for people to conclude that all bacteria are pathogens, but in fact, relatively few are. Most play beneficial

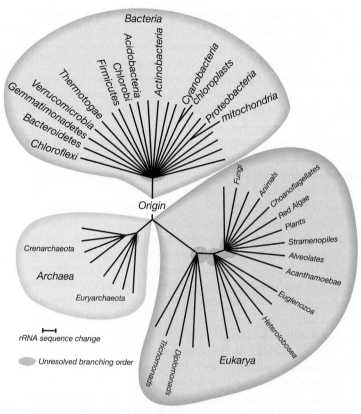

Figure 1.2 Universal Phylogenetic Tree. These evolutionary relationships are based on rRNA sequence comparisons. To save space, many lineages have not been identified.

MICRO INQUIRY *How many of the taxa listed in the figure include microbes?*

roles. In addition to maintaining human health by forming our microbiomes, they break down dead plant and animal material and, in doing so, cycle elements in the biosphere. Furthermore, they are used extensively in industry to make bread, cheese, antibiotics, vitamins, enzymes, and other products.

Members of domain ***Archaea*** are distinguished from bacteria by many features, most notably their distinctive rRNA sequences, lack of peptidoglycan in their cell walls, and unique membrane lipids. Some have unusual metabolic characteristics, such as the methanogens, which generate methane (natural) gas. Many archaea are found in extreme environments, including those with high temperatures (thermophiles) and high concentrations of salt (extreme halophiles). Although some archaea are members of a community of microbes involved in gum disease in humans, their role in causing disease has not been clearly established.

Domain ***Eukarya*** includes microorganisms classified as protists or fungi. Animals and plants are also placed in this domain. **Protists** are generally unicellular but larger than most bacteria and archaea. They have traditionally been divided into

---

1 In this text, the term *bacteria* (s., *bacterium*) is used to refer to those microbes belonging to domain *Bacteria*, and the term *archaea* (s., *archaeon*) is used to refer to those that belong to domain *Archaea*. In some publications, the term *bacteria* is used to refer to all cells having prokaryotic cell structure. That is not the case in this text.

protozoa and algae. Despite their use, none of these terms has taxonomic value as protists, algae, and protozoa do not form cohesive taxa. However, for convenience, we use them here.

The major types of protists are algae, protozoa, slime molds, and water molds. **Algae** are photosynthetic. They, together with cyanobacteria, produce about 75% of the planet's oxygen and are the foundation of aquatic food chains. **Protozoa** are unicellular, animal-like protists that are usually motile. Many free-living protozoa function as the principal hunters and grazers of the microbial world. They obtain nutrients by ingesting organic matter and other microbes. They can be found in many different environments, and some are normal inhabitants of the intestinal tracts of animals, where they aid in digestion of complex materials such as cellulose. A few cause disease in humans and other animals. **Slime molds** are protists that behave like protozoa in one stage of their life cycle but like fungi in another. In the protozoan phase, they hunt for and engulf food particles, consuming decaying vegetation and other microbes. **Water molds** are protists that grow on the surface of freshwater and moist soil. They feed on decaying vegetation such as logs and mulch. Some water molds have produced devastating plant infections, including the Great Potato Famine of 1846–1847 in Ireland, which led to the mass exodus of Irish to the United States and other countries. ▶▶| *Protists (chapter 25)*

Fungi are a diverse group of microorganisms that range from unicellular forms (yeasts) to molds and mushrooms. Molds and mushrooms are multicellular fungi that form thin, threadlike structures called hyphae. They absorb nutrients from their environment, including the organic molecules they use as sources of carbon and energy. Because of their metabolic capabilities, many fungi play beneficial roles, including making bread dough rise, producing antibiotics, and decomposing dead organisms. Some fungi associate with plant roots to form mycorrhizae. Mycorrhizal fungi transfer nutrients to the roots, improving growth of the plants, especially in poor soils. Other fungi cause plant diseases (e.g., rusts, powdery mildews, and smuts) and diseases in humans and other animals. ▶▶| Fungi *(chapter 26)*

The microbial world also includes numerous acellular infectious agents. **Viruses** are acellular entities that must invade a host cell to multiply. The simplest virus particles (also called virions) are composed only of proteins and a nucleic acid, and can be extremely small (the smallest is 10,000 times smaller than a typical bacterium). However, their small size belies their power: they cause many animal and plant diseases and have caused epidemics that have shaped human history. Viral diseases include smallpox, rabies, influenza, AIDS, the common cold, and some cancers. Viruses also play important roles in aquatic environments, and their role in shaping aquatic microbial communities is currently being explored. **Viroids** are infectious agents composed only of ribonucleic acid (RNA). They cause numerous plant diseases. **Satellites** are composed of a nucleic acid enclosed in a protein shell. They cause plant diseases and some important animal diseases such as hepatitis. Finally, **prions,** infectious agents composed only of protein, are responsible for causing a variety of spongiform encephalopathies such as scrapie and "mad cow disease." ▶▶| *Viruses and other acellular infectious agents (chapter 6)*

Retrieve, Infer, Apply

1. How did the methods used to classify microbes change, particularly in the last half of the twentieth century? What was the result of these technological advances?
2. Identify one characteristic for each of these types of microbes that distinguishes it from the other types: bacteria, archaea, protists, fungi, viruses, viroids, satellites, and prions.

1.2 Microbes Have Evolved and Diversified for Billions of Years

After reading this section, you should be able to:

- Propose a time line of the origin and history of microbial life and integrate supporting evidence into it
- Design a set of experiments that could be used to place a newly discovered cellular microbe on a phylogenetic tree based on small subunit (SSU) rRNA sequences
- Compare and contrast the definitions of plant and animal species, microbial species, and microbial strains

A review of figure 1.2 reminds us that in terms of the number of taxa, microbes are the dominant organisms on Earth. How has microbial life been able to radiate to such an astonishing level of diversity? To answer this question, we must consider microbial evolution. The field of microbial evolution, like any other scientific endeavor, is based on the formulation of hypotheses, the gathering and analysis of data, and the reformation of hypotheses based on newly acquired evidence. That is to say, the study of microbial evolution is based on the scientific method. To be sure, it is sometimes more difficult to amass evidence when considering events that occurred millions, and often billions, of years ago, but the advent of molecular methods has offered scientists a living record of life's ancient history. This section describes the outcome of this scientific research.

Theories of the Origin of Life Depend Primarily on Indirect Evidence

Dating meteorites through the use of radioisotopes places our planet at an estimated 4.5 to 4.6 billion years old. However, conditions on Earth for the first 100 million years or so were far too harsh to sustain any type of life. Eventually bombardment by meteorites decreased, water appeared on the planet in liquid form, and gases were released by geological activity to form Earth's atmosphere. These conditions were amenable to the origin of the first life forms. But how did this occur, and what did these life forms look like?

In order to find evidence of life and to develop hypotheses about its origin and subsequent evolution, scientists must be able to define life. Although even very young children can examine an object and correctly determine whether it is living or not, defining life succinctly has proven elusive for scientists.

Thus most definitions of life consist of a set of attributes. The attributes of particular importance to paleobiologists are an orderly structure, the ability to obtain and use energy (i.e., metabolism), and the ability to reproduce. Just as NASA scientists are using the characteristics of microbes on Earth today to search for life elsewhere (p. 1), so too are scientists examining **extant organisms,** those organisms present today, to explore the origin of life. Some extant organisms have structures and molecules that represent "relics" of ancient life forms. Furthermore, they can provide scientists with ideas about the type of evidence to seek when testing hypotheses.

The best direct evidence for the nature of primitive life would be a fossil record. There have been reports of microbial fossil discoveries since 1977 (**figure 1.3**). These have always met with skepticism because finding them involves preparing thin slices of ancient rocks and examining the slices for objects that look like cells. Unfortunately, some things that look like cells can be formed by geological forces that occurred as the rock was formed. The result is that the fossil record for microbes is sparse and always open to reinterpretation. Despite these problems, most scientists agree that life was present on Earth about 3.5 billion years ago (**figure 1.4**). To reach this conclusion, biologists have had to rely primarily on indirect evidence. Among the indirect evidence used are molecular fossils. These are chemicals found in rock or sediment that are chemically related to molecules found in cells. For instance, the presence in a rock of molecules called hopanes is an indication that when the rock was formed, bacteria were present. This conclusion is reached because hopanes are formed from hopanoids, which are found in the plasma membranes of bacteria. As you can see, no single piece of evidence can stand alone. Instead many pieces of evidence are put together in an attempt to get a coherent picture to emerge, much as for a jigsaw puzzle.

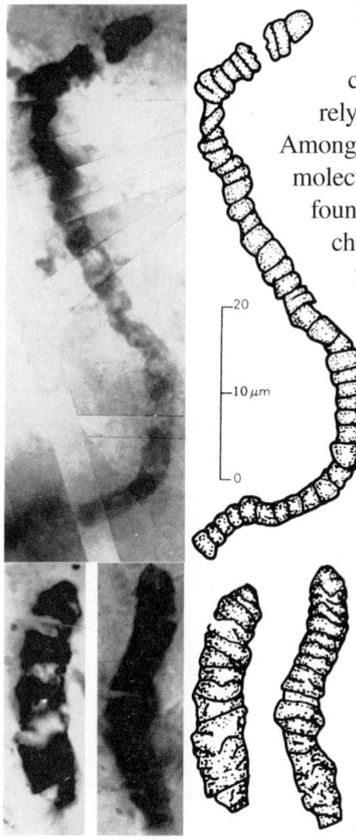

Figure 1.3 Possible Microfossils Found in the Archaeon Apex Chert of Australia. Chert is a type of granular sedimentary rock rich in silica. These structures were discovered in 1977. Because of their similarity to filamentous cyanobacteria they were proposed to be microfossils. In 2011 scientists reported that similar structures from the same chert were not biological in origin. They used spectrometry and microscopy techniques not available in 1977 to show that the structures were fractures in the rock filled with quartz and hematite. Scientists are still debating whether or not these truly are microfossils.

Early Life Was Probably RNA Based

The origin of life rests on a single question: How did early cells arise? At a minimum, modern cells consist of a plasma membrane enclosing water in which numerous chemicals are dissolved and subcellular structures float. It seems likely that the first self-replicating entity was much simpler than even the most primitive modern living cells. Before there was life, most evidence suggests that Earth was a very different place: hot and anoxic, with an atmosphere rich in water vapor, carbon dioxide, and nitrogen. In the oceans, hydrogen, methane, and carboxylic acids were formed by geological and chemical processes. Areas near hydrothermal vents or in shallow pools may have provided the conditions that allowed chemicals to react with one another, randomly "testing" the usefulness of the reaction and the stability of its products. Some reactions released energy and would eventually become the basis of modern cellular metabolism. Other reactions generated molecules that could function as catalysts, some aggregated with other molecules to form the predecessors of modern cell structures, and others were able to replicate and act as units of hereditary information.

In modern cells, three different molecules fulfill the roles of catalysts, structural molecules, and hereditary molecules (**figure 1.5**). Proteins have two major roles in modern cells: structural and catalytic. Catalytic proteins are called **enzymes,** and they speed up the myriad of chemical reactions that occur in cells. DNA stores hereditary information and can be replicated to pass the information on to the next generation. RNA is involved in converting the information stored in DNA into protein. Any hypothesis about the origin of life must account for the evolution of these molecules, but the very nature of their relationships to each other in modern cells complicates attempts to imagine how they evolved. As demonstrated in figure 1.5, proteins can do cellular work, but their synthesis involves other proteins and RNA, and uses information stored in DNA. DNA can't do cellular work. It stores genetic information and serves as the template for its own replication, a process that requires proteins. RNA is synthesized using DNA as the template and proteins as the catalysts for the reaction.

Based on these considerations, it is hypothesized that at some time in the evolution of life, there must have been a single molecule that could do both cellular work and replicate itself. This idea was supported in 1981 when Thomas Cech discovered a catalytic RNA molecule in a protist (*Tetrahymena* sp.) that could cut out an internal section of itself and splice the remaining sections back together. Since then, other catalytic RNA molecules have been discovered, including an RNA found in ribosomes that is responsible for forming peptide bonds—the bonds that hold together amino acids, the building blocks of proteins. Catalytic RNA molecules are now called **ribozymes.**

The discovery of ribozymes suggested that RNA at some time had the ability to catalyze its own replication, using itself as the template. In 1986 Walter Gilbert coined the term **RNA world** to describe a precellular stage in the evolution of life in which RNA was capable of storing, copying, and expressing genetic information, as well as catalyzing other chemical reactions.

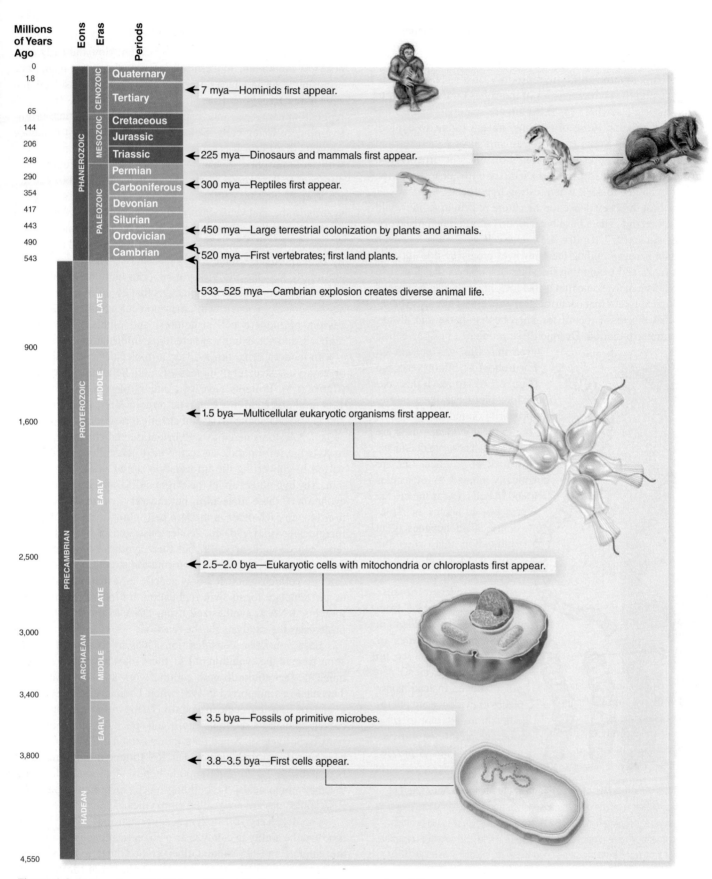

Figure 1.4 An Overview of the History of Life on Earth. mya = million years ago; bya = billion years ago.

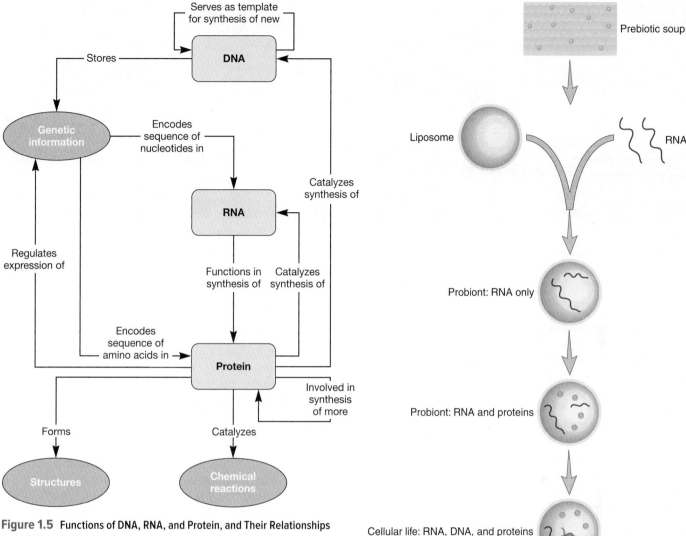

Figure 1.5 Functions of DNA, RNA, and Protein, and Their Relationships to Each Other in Modern Cells.

Figure 1.6 The RNA World Hypothesis for the Origin of Life.

MICRO INQUIRY *Why are the probionts pictured above not considered cellular life?*

However, for this precellular stage to proceed to the evolution of cellular life forms, a lipid membrane must have formed around the RNA (**figure 1.6**). This important evolutionary step is easier to imagine than other events in the origin of cellular life forms because lipids, major structural components of the membranes of modern organisms, spontaneously form liposomes—vesicles bounded by a lipid bilayer. ▶▶| *Lipids (appendix I)*

Jack Szostak is a leader in exploring how RNA-containing cells, so-called protocells, may have formed. His group has created liposomes using simpler fatty acids than those found in membranes today. These simpler fatty acids formed leaky liposomes that allowed single RNA nucleotides to move into the liposome, but prevented large RNA chains from moving out. Furthermore, they could prod the liposomes into growing and dividing. Szostak's group has also been able to create conditions in which an RNA molecule could serve as a template for synthesis of a complementary RNA strand. Thus, their experiments have recapitulated what may have happened in the early steps of the evolution of cells. As seen in figure 1.6, several other processes

would need to occur to reach the level of complexity found in extant cells.

Apart from its ability to perform catalytic activities, the function of RNA suggests its ancient origin. Consider that much of the cellular pool of RNA in modern cells exists in the ribosome, a structure that consists largely of rRNA and uses messenger RNA (mRNA) and transfer RNA (tRNA) to construct proteins. Also recall that rRNA itself catalyzes peptide bond formation during protein synthesis. Thus RNA seems to be well poised for its importance in the development of proteins. Because RNA and DNA are structurally similar, RNA could have given rise to double-stranded DNA. It is suggested that once DNA evolved, it became the storage facility for genetic information

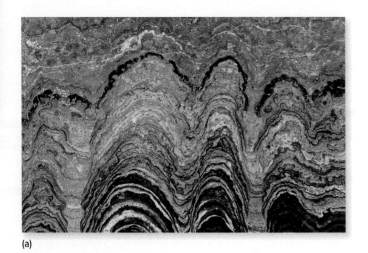

(a)

(b)

Figure 1.7 Stromatolites. (a) Section of a fossilized stromatolite. Evolutionary biologists think the layers of material were formed when mats of cyanobacteria, layered one on top of the other, became mineralized. (b) Modern stromatolites from Western Australia. Each stromatolite is a rocklike structure, typically 1 m in diameter, containing layers of cyanobacteria.

because it provided a more chemically stable structure. Two other pieces of evidence support the RNA world hypothesis: the fact that the energy currency of cells, ATP, is a ribonucleotide and the more recent discovery that RNA can regulate gene expression. So it would seem that proteins, DNA, and cellular energy can be traced back to RNA. ▶▶ *ATP (section 10.2); Riboswitches (section 14.3); Translational riboswitches (section 14.4)*

Despite the evidence supporting the RNA world hypothesis, it is not without problems, and many argue against it. Another area of research also fraught with considerable debate is the evolution of metabolism, in particular the evolution of energy-conserving metabolic processes. The early Earth was a hot environment that lacked oxygen. Thus the cells that arose there must have been able to use the available energy sources under these harsh conditions. Today there are heat-loving archaea capable of using inorganic molecules such as FeS as a source of energy. Some suggest that this interesting metabolic capability is a remnant of the first form of energy metabolism. Another metabolic strategy, oxygen-releasing photosynthesis (oxygenic photosynthesis), appears to have evolved perhaps as early as 2.7 billion years ago. Fossils of cyanobacteria-like cells found in rocks dating to that time support this hypothesis, as does the discovery of ancient stromatolites (**figure 1.7a**). Stromatolites are layered rocks, often domed, that are formed by the incorporation of mineral sediments into layers of microorganisms growing as thick mats on surfaces (figure 1.7*b*). Furthermore, chemical evidence, such as the presence of certain isotopes and oxidized minerals in rocks of this age, also support the antiquity of oxygenic photosynthesis. The appearance of cyanobacteria-like cells was an important step in the evolution of life on Earth. The oxygen they released is thought to have altered Earth's atmosphere to its current oxygen-rich state, allowing the evolution of additional energy-capturing strategies such as aerobic respiration, the oxygen-consuming metabolic process that is used by many microbes and animals.

Evolution of the Three Domains of Life

As noted in section 1.1, rRNA comparisons were an important breakthrough in the classification of microbes; this analysis also provides insights into the evolutionary history of all life. What began with the examination of rRNA from relatively few organisms has been expanded by the work of many others, including Norman Pace. Dr. Pace has developed a **universal phylogenetic tree** (figure 1.2) based on comparisons of small subunit rRNA molecules (SSU rRNA), the rRNA found in the small subunit of the ribosome. Here we examine how these comparisons are made and what the universal phylogenetic tree tells us. ▶▶ *Bacterial ribosomes (section 3.6); Microbial taxonomy and phylogeny are largely based on molecular characterization (section 19.3)*

Comparing SSU rRNA Molecules

The details of phylogenetic tree construction are discussed in chapter 19. However, the general concept is not difficult to understand. In one approach, the sequences of nucleotides in the genes that encode SSU rRNAs from diverse organisms are aligned, and pair-wise comparisons of the sequences are made. For each pair of SSU rRNA gene sequences, the number of differences in the nucleotide sequences is counted (**figure 1.8**). This value serves as a measure of the evolutionary distance between the organisms; the more differences counted, the greater the evolutionary distance. The evolutionary distances from many comparisons are used by sophisticated computer programs to construct the tree. The tip of each branch in the tree represents one of the organisms used in the comparison. The distance from the tip of one branch to the tip of another is the evolutionary distance between the two organisms.

Two things should be kept in mind when examining phylogenetic trees developed in this way. The first is that they are molecular trees, not organismal trees. In other words, they represent, as accurately as possible, the evolutionary history of a molecule

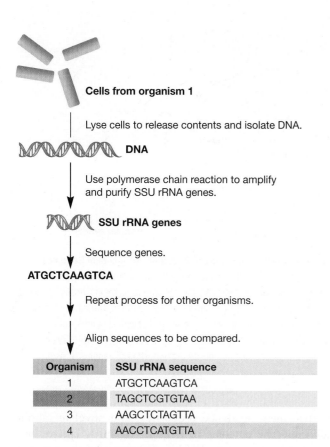

Cells from organism 1

Lyse cells to release contents and isolate DNA.

DNA

Use polymerase chain reaction to amplify and purify SSU rRNA genes.

SSU rRNA genes

Sequence genes.

ATGCTCAAGTCA

Repeat process for other organisms.

Align sequences to be compared.

Organism	SSU rRNA sequence
1	ATGCTCAAGTCA
2	TAGCTCGTGTAA
3	AAGCTCTAGTTA
4	AACCTCATGTTA

Count the number of nucleotide differences between each pair of sequences and calculate evolutionary distance (E_D).

Pair compared	E_D	Corrected E_D
1 → 2	0.42	0.61
1 → 3	0.25	0.30
1 → 4	0.33	0.44
2 → 3	0.33	0.44
2 → 4	0.33	0.44
3 → 4	0.25	0.30

For organisms 1 and 2, 5 of the 12 nucleotides are different:
$E_D = 5/12 = 0.42$.

The initial ED calculated is corrected using a statistical method that considers for each site the probability of a mutation back to the original nucleotide or of additional forward mutations.

Feed data into computer and use appropriate software to construct phylogenetic tree.

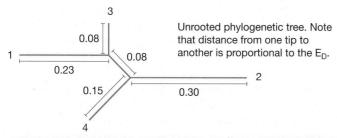

Unrooted phylogenetic tree. Note that distance from one tip to another is proportional to the E_D.

Figure 1.8 The Construction of Phylogenetic Trees Using a Distance Method. The polymerase chain reaction is described in chapter 17.

MICRO INQUIRY *Why does the branch length indicate amount of evolutionary change but not the time it took for that change to occur?*

and the gene that encodes it. Second, the distance between branch tips is a measure of relatedness, not of time. If the distance along the lines is very long, then the two organisms are more evolutionarily diverged (i.e., less related). However, we do not know when they diverged from each other. This concept is analogous to a map that accurately shows the distance between two cities but because of many factors (traffic, road conditions, etc.) cannot show the time needed to travel that distance.

LUCA

What does the universal phylogenetic tree tell us about the evolution of life? At the center of the tree is a line labeled "Origin" (figure 1.2). This is where data indicate the *last universal common ancestor* (LUCA) to all three domains should be placed. LUCA is on the bacterial branch, which means that *Archaea* and *Eukarya* evolved independently, separate from *Bacteria*. Thus the universal phylogenetic tree presents a picture in which all life, regardless of eventual domain, arose from a single common ancestor. One can envision the universal tree of life as a real tree that grows from a single seed.

The evolutionary relationship of *Archaea* and *Eukarya* is still the matter of considerable debate. According to the universal phylogenetic tree we show here, *Archaea* and *Eukarya* shared common ancestry but diverged and became separate domains. Other versions suggest that *Eukarya* evolved out of *Archaea*. The close evolutionary relationship of these two forms of life is still evident in the manner in which they process genetic information. For instance, certain protein subunits of archaeal and eukaryotic RNA polymerases, the enzymes that catalyze RNA synthesis, resemble each other to the exclusion of those of bacteria. However, archaea have other features that are most similar to their counterparts in bacteria (e.g., mechanisms for conserving energy). This has further complicated and fueled the debate. The evolution of the nucleus and endoplasmic reticulum is also at the center of many controversies. However, hypotheses regarding the evolution of other membrane-bound organelles are more widely accepted and are considered next.

Mitochondria, Mitochondria-Like Organelles, and Chloroplasts Evolved from Endosymbionts

The **endosymbiotic hypothesis** is generally accepted as the origin of several eukaryotic organelles, including mitochondria, chloroplasts, and hydrogenosomes. **Endosymbiosis** is an interaction between two organisms in which one organism lives inside the other. The initial statement of the endosymbiotic hypothesis proposed that over time a bacterial endosymbiont of an ancestral cell in the eukaryotic lineage lost its ability to live independently, becoming either a mitochondrion, if the intracellular bacterium used aerobic respiration, or a chloroplast, if the endosymbiont was a photosynthetic bacterium (*see figure 19.10*).

Although the mechanism by which the endosymbiotic relationship was established is unknown, there is considerable evidence to support the hypothesis. Mitochondria and chloroplasts contain DNA and ribosomes; both are similar to bacterial DNA and ribosomes. Peptidoglycan, the unique bacterial cell wall molecule (p. 3) has even been found between the two membranes that enclose the chloroplasts of some algae. Indeed, inspection of figure 1.2 shows that both organelles belong to the bacterial lineage based on SSU rRNA analysis. More specifically, mitochondria are most closely related to bacteria called alphaproteobacteria. The chloroplasts of plants and green algae are thought to have descended from an ancestor of the cyanobacterial genus *Prochloron*, which contains species that live within marine invertebrates. ▶▶| *Phylum* Cyanobacteria *(section 21.4)*

Recently the endosymbiotic hypothesis for mitochondria has been modified by the **hydrogen hypothesis.** This asserts that the endosymbiont was an anaerobic bacterium that produced H_2 and CO_2 as end products of its metabolism. Over time, the host became dependent on the H_2 produced by the endosymbiont. Ultimately the endosymbiont evolved into one of several organelles (*see figure 5.14*). If the endosymbiont developed the capacity to perform aerobic respiration, it evolved into a mitochondrion. Other endosymbionts evolved into other organelles such as a hydrogenosome—an organelle that produces ATP by a process called fermentation (*see figure 5.16*) and is found in some extant protists.

Evolution of Cellular Microbes

Although the history of early cellular life forms may never be known, we know that once they arose, they were subjected to the same evolutionary processes as modern organisms. The ancestral bacteria, archaea, and eukaryotes possessed genetic information that could be duplicated, lost, or mutated in other ways. These mutations could have many outcomes. Some led to the death of the mutant microbe, but others allowed new functions and characteristics to evolve. Those mutations that allowed the organism to increase its reproductive ability were selected for and passed on to subsequent generations. In addition to selective forces, isolation of populations allowed some groups to evolve separately from others. Thus selection and isolation led to the eventual development of new collections of genes (i.e., genotypes) and many new species.

In addition to mutation, other mechanisms exist for reconfiguring the genotypes of a species and therefore creating genetic diversity. Most eukaryotic species increase their genetic diversity by reproducing sexually. Thus each offspring of the two parents has a mixture of parental genes and a unique genotype. Bacterial and archaeal species do not reproduce sexually. They increase their genetic diversity by horizontal (lateral) gene transfer (HGT). During HGT, genetic information from a donor organism is transferred to a recipient, creating a new genotype. Thus genetic information can be passed from one generation to the next as well as between individuals of the same generation and even between species found in different domains of life. Genome sequencing has revealed that HGT has played an important role in the evolution of all species. Importantly, HGT still occurs in bacteria and archaea leading to the evolution of species with antibiotic

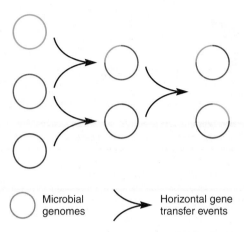

Figure 1.9 **The Mosaic Nature of Bacterial and Archaeal Genomes.** Horizontal gene transfer (HGT) events move pieces of the genome of one organism to another. Over time, HGT creates organisms having mosaic genomes composed of portions of the genomes of other microbes. Though not illustrated here, the genomes of eukaryotes are also littered with sequences of bacterial, archaeal, and viral origin. The length of segments drawn is arbitrary and is not meant to represent the actual size of the portion of genome transferred.

resistance, new virulence properties, and novel metabolic capabilities. The outcome of HGT is that many species have mosaic genomes composed of bits and pieces of the genomes of other organisms (**figure 1.9**). ▶▶| *Horizontal gene transfer (section 16.4)*

Microbial Species

All students of biology are introduced early in their careers to the concept of a species. But the term has different meanings, depending on whether the organism is sexual or not. Taxonomists working with plants and animals define a **species** as a group of interbreeding or potentially interbreeding natural populations that is reproductively isolated from other groups. This definition also is appropriate for the many eukaryotic microbes that reproduce sexually. However, bacterial and archaeal species cannot be defined by this criterion, since they do not reproduce sexually. Increasingly, comparisons of genome sequences are being used to distinguish one species from another. An appropriate definition is currently the topic of considerable discussion. A common definition is that bacterial and archaeal species are a collection of strains that share many stable properties and differ significantly from other groups of strains. A **strain** consists of the descendants of a single, pure microbial culture. Strains within a species may be described in a number of different ways. Biovars are variant strains characterized by biochemical or physiological differences, morphovars differ morphologically, serovars have distinctive properties that can be detected by antibodies (p. 16), and pathovars are pathogenic strains distinguished by the plants in which they cause disease. ▶▶| *What is a microbial species? (section 19.5)*

Microbiologists name microbes using the binomial system of the eighteenth-century biologist and physician Carl Linnaeus. The Latinized, italicized name consists of two parts. The first part, which is capitalized, is the generic name (i.e., the name of the genus to which the microbe belongs), and the second is the uncapitalized species epithet. For example, the bacterium that causes plague is called *Yersinia pestis*. Often the name of an organism will be shortened by abbreviating the genus name with a single capital letter (e.g., *Y. pestis*).

Retrieve, Infer, Apply

1. List two reasons RNA is thought to be the first self-replicating biomolecule.
2. Explain the endosymbiotic hypothesis of the origin of mitochondria, hydrogenosomes, and chloroplasts. List two pieces of evidence that support this hypothesis.
3. What is the difference between a microbial species and a strain?
4. What is the correct way to write this microbe's name: *bacillus subtilis*, Bacillus subtilis, *Bacillus Subtilis*, or *Bacillus subtilis*? Identify the genus name and the species epithet.

1.3 Microbiology Advanced as New Tools for Studying Microbes Were Developed

After reading this section, you should be able to:

- Evaluate the importance of the contributions to microbiology made by Hooke, Leeuwenhoek, Pasteur, Koch, Cohn, Beijerinck, von Behring, Kitasato, Metchnikoff, and Winogradsky
- Outline a set of experiments that might be used to decide if a particular microbe is the causative agent of a disease
- Predict the difficulties that might arise when using Koch's postulates to determine if a microbe causes a disease unique to humans

Even before microorganisms were seen, some investigators suspected their existence and role in disease. Among others, the Roman philosopher Lucretius (about 98–55 BCE) and the physician Girolamo Fracastoro (1478–1553) suggested that disease was caused by invisible living creatures. However, until microbes could actually be seen or studied in some other way, their existence remained a matter of conjecture. Therefore **microbiology** is defined not only by the organisms it studies but also by the tools used to study them. The development of microscopes was the critical first step in the evolution of the discipline. However, microscopy alone is unable to answer the many questions microbiologists ask about microbes. A distinct feature of microbiology is that microbiologists often remove microorganisms from their normal habitats and culture them isolated from other microbes. This is called a **pure** or **axenic culture.** The development of techniques for isolating microbes in pure culture was another critical step in microbiology's history. However, it is now recognized as having limitations. Microbes in pure culture are in some ways like animals in a zoo; just as a zoologist cannot fully understand the ecology of animals by studying them in zoos, microbiologists

cannot fully understand the ecology of microbes by studying them in pure culture. Today molecular genetic techniques and genomic analyses are providing new insights into the lives of microbes. ▶▶| *Methods in microbial ecology (chapter 29); Microbial genomics (chapter 18)*

Here we describe how the tools used by microbiologists have influenced the development of the field. As microbiology evolved as a science, it contributed greatly to the well-being of humans. This is exemplified by the number of microbiologists who have won the Nobel Prize. The historical context of some of the important discoveries in microbiology is shown in **figure 1.10**.

Microscopy Led to the Discovery of Microorganisms

The earliest microscopic observations of organisms appear to have been made between 1625 and 1630 on bees and weevils by the Italian Francesco Stelluti (1577–1652), using a microscope probably supplied by Galileo (1564–1642). Robert Hooke (1635–1703) is credited with publishing the first drawings of microorganisms in the scientific literature. In 1665 he published a highly detailed drawing of the fungus *Mucor* in his book *Micrographia. Micrographia* is important not only for its exquisite drawings but also for the information it provided on building microscopes. One design discussed in *Micrographia* was probably a prototype for the microscopes built and used by the amateur microscopist Antony van Leeuwenhoek (1632–1723) of Delft, the Netherlands (**figure 1.11a**). Leeuwenhoek earned his living as a draper and haberdasher (a dealer in men's clothing and accessories) but spent much of his spare time constructing simple microscopes composed of double convex glass lenses held between two silver plates (figure 1.11b). His microscopes could magnify about 50 to 300 times, and he may have illuminated his liquid specimens by placing them between two pieces of glass and shining light on them at a 45° angle to the specimen plane. This would have provided a form of dark-field illumination whereby organisms appeared as bright objects against a dark background. Beginning in 1673, Leeuwenhoek sent detailed letters describing his discoveries to the Royal Society of London. It is clear from his descriptions that he saw both bacteria and protists (figure 1.11c). ▶▶| *Dark-field microscope (section 2.2)*

Culture-Based Methods for Studying Microorganisms Were a Major Development

As important as Leeuwenhoek's observations were, the development of microbiology essentially languished for the next 200 years until techniques for isolating and culturing microbes in the laboratory were formulated. Many of these techniques began to be developed as scientists grappled with the conflict over the theory of spontaneous generation. This conflict and the subsequent studies on the role played by microorganisms in causing disease ultimately led to what is now called the golden age of microbiology.

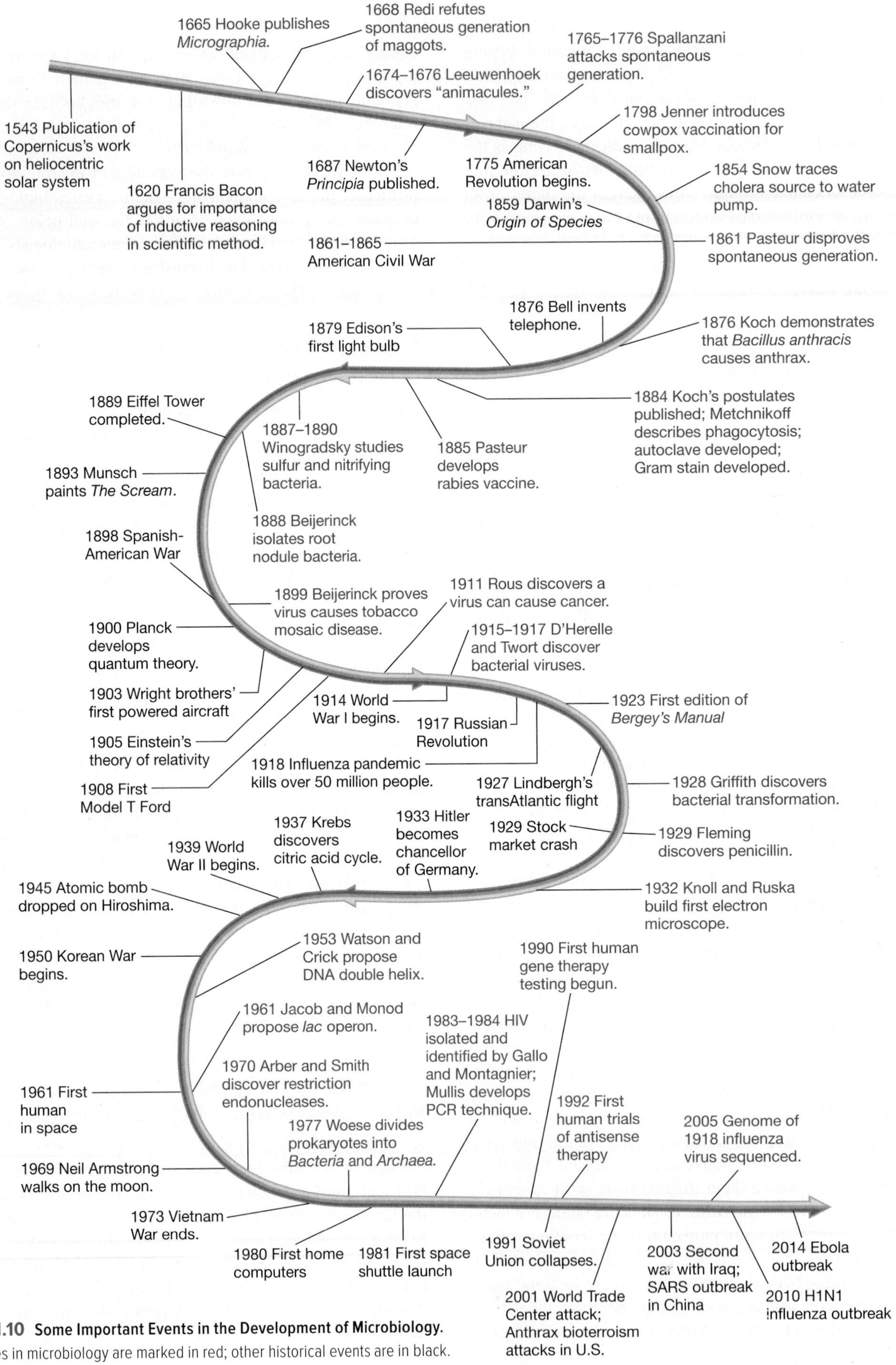

1665 Hooke publishes *Micrographia*.

1668 Redi refutes spontaneous generation of maggots.

1765–1776 Spallanzani attacks spontaneous generation.

1674–1676 Leeuwenhoek discovers "animacules."

1543 Publication of Copernicus's work on heliocentric solar system

1620 Francis Bacon argues for importance of inductive reasoning in scientific method.

1687 Newton's *Principia* published.

1775 American Revolution begins.

1798 Jenner introduces cowpox vaccination for smallpox.

1854 Snow traces cholera source to water pump.

1859 Darwin's *Origin of Species*

1861–1865 American Civil War

1861 Pasteur disproves spontaneous generation.

1876 Bell invents telephone.

1879 Edison's first light bulb

1876 Koch demonstrates that *Bacillus anthracis* causes anthrax.

1889 Eiffel Tower completed.

1887–1890 Winogradsky studies sulfur and nitrifying bacteria.

1885 Pasteur develops rabies vaccine.

1884 Koch's postulates published; Metchnikoff describes phagocytosis; autoclave developed; Gram stain developed.

1893 Munsch paints *The Scream*.

1888 Beijerinck isolates root nodule bacteria.

1898 Spanish-American War

1899 Beijerinck proves virus causes tobacco mosaic disease.

1911 Rous discovers a virus can cause cancer.

1915–1917 D'Herelle and Twort discover bacterial viruses.

1923 First edition of *Bergey's Manual*

1900 Planck develops quantum theory.

1903 Wright brothers' first powered aircraft

1914 World War I begins.

1917 Russian Revolution

1928 Griffith discovers bacterial transformation.

1905 Einstein's theory of relativity

1918 Influenza pandemic kills over 50 million people.

1927 Lindbergh's transAtlantic flight

1929 Fleming discovers penicillin.

1908 First Model T Ford

1937 Krebs discovers citric acid cycle.

1933 Hitler becomes chancellor of Germany.

1929 Stock market crash

1939 World War II begins.

1932 Knoll and Ruska build first electron microscope.

1945 Atomic bomb dropped on Hiroshima.

1953 Watson and Crick propose DNA double helix.

1990 First human gene therapy testing begun.

1950 Korean War begins.

1961 Jacob and Monod propose *lac* operon.

1983–1984 HIV isolated and identified by Gallo and Montagnier; Mullis develops PCR technique.

1970 Arber and Smith discover restriction endonucleases.

1992 First human trials of antisense therapy

1961 First human in space

1977 Woese divides prokaryotes into *Bacteria* and *Archaea*.

2005 Genome of 1918 influenza virus sequenced.

1969 Neil Armstrong walks on the moon.

1973 Vietnam War ends.

1991 Soviet Union collapses.

2014 Ebola outbreak

1980 First home computers

1981 First space shuttle launch

2003 Second war with Iraq; SARS outbreak in China

2001 World Trade Center attack; Anthrax bioterrorism attacks in U.S.

2010 H1N1 influenza outbreak

Figure 1.10 Some Important Events in the Development of Microbiology.
Milestones in microbiology are marked in red; other historical events are in black.

(a)

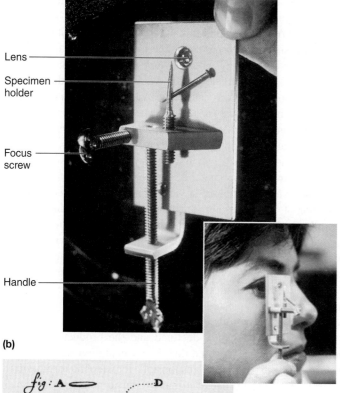

Lens

Specimen holder

Focus screw

Handle

(b)

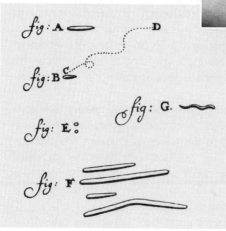

(c)

Figure 1.11 Antony van Leeuwenhoek.
(a) An oil painting of Leeuwenhoek. (b) A brass replica of the Leeuwenhoek microscope. Inset photo shows how it is held. (c) Leeuwenhoek's drawings of bacteria from the human mouth.

Spontaneous Generation

From earliest times, people had believed in **spontaneous generation**—that living organisms could develop from nonliving matter. This view finally was challenged by the Italian physician Francesco Redi (1626–1697), who carried out a series of experiments on decaying meat and its ability to produce maggots spontaneously. Using covered and uncovered containers of meat, Redi clearly demonstrated that the generation of maggots by decaying meat resulted from the presence of fly eggs, and meat did not spontaneously generate maggots, as previously believed. Other experiments helped discredit the theory for larger organisms. However, Leeuwenhoek's communications on microorganisms renewed the controversy. Some proposed that microbes arose by spontaneous generation even though larger organisms did not. They pointed out that boiled extracts of hay or meat gave rise to microorganisms after sitting for a while. In 1748 the English priest John Needham (1713–1781) suggested that the organic matter in these extracts contained a vital force that could confer the properties of life on nonliving matter. Such extracts were the forerunners of the culture media still used today in many microbiology laboratories.

A few years after Needham's experiments, the Italian priest and naturalist Lazzaro Spallanzani (1729–1799) sealed glass flasks that contained water and seeds and then placed the flasks in boiling water for about 45 minutes. He found that no growth took place as long as the flasks remained sealed. He proposed that air carried germs to the culture medium but also commented that external air might be required for growth of animals already in the medium. The supporters of spontaneous generation responded that heating the air in sealed flasks destroyed its ability to support life, and therefore did not discredit the theory of spontaneous generation.

Several investigators attempted to counter such arguments. Theodore Schwann (1810–1882) allowed air to enter a flask containing a sterile nutrient solution after the air had passed through a red-hot tube. The flask remained sterile. Subsequently Georg Friedrich Schroder (1810–1885) and Theodor von Dusch (1824–1890) allowed air to enter a flask of heat-sterilized medium after it had passed through sterile cotton wool. No growth occurred in the medium even though the air had not been heated. Despite these experiments, the French naturalist Felix Pouchet (1800–1872) claimed in 1859 to have carried out experiments conclusively proving that microbial growth could occur without air contamination.

Pouchet's claim provoked Louis Pasteur (1822–1895) to settle the matter of spontaneous generation. Pasteur (**figure 1.12**) first filtered air

Figure 1.12 Louis Pasteur.

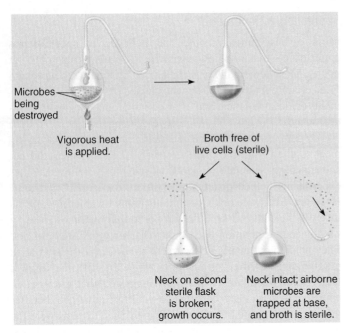

Figure 1.13 Pasteur's Experiments with Swan-Neck Flasks.

Microorganisms and Disease

For hundreds of years, most people believed that disease was caused by supernatural forces, poisonous vapors, and imbalances among the four humors thought to be present in the body. The role of the four humors (blood, phlegm, yellow bile [choler], and black bile [melancholy]) in disease had been widely accepted since the time of the Greek physician Galen (129–199). Support for the idea that microorganisms cause disease—that is, the germ theory of disease—began to accumulate in the early nineteenth century from diverse fields. Agostino Bassi (1773–1856) demonstrated in 1835 that a silkworm disease was due to a fungal infection. He also suggested that many diseases were due to microbial infections. In 1845 M. J. Berkeley (1803–1889) proved that the great potato blight of Ireland was caused by a water mold (then thought to be a fungus), and in 1853 Heinrich de Bary (1831–1888) showed that smut and rust fungi caused cereal crop diseases.

Pasteur also contributed to this area of research in several ways. His contributions began in what may seem an unlikely way. Pasteur was trained as a chemist and spent many years studying the alcoholic fermentations that yield ethanol and are used in the production of wine and other alcoholic beverages. When he began his work, the leading chemists were convinced that fermentation was due to a chemical instability that degraded the sugars in grape juice and other substances to alcohol. Pasteur did not agree; he believed that fermentations were carried out by living organisms.

In 1856 M. Bigo, an industrialist in Lille, France, where Pasteur worked, requested Pasteur's assistance. His business produced ethanol from the fermentation of beet sugars, and the alcohol yields had recently declined and the product had become sour. Pasteur discovered that the fermentation was failing because the yeast normally responsible for alcohol formation had been replaced by bacteria that produced acid rather than ethanol. In solving this practical problem, Pasteur demonstrated that all fermentations were due to the activities of specific yeasts and bacteria, and he published several papers on fermentation between 1857 and 1860.

Pasteur was also called upon by the wine industry in France for help. For several years, poor-quality wines had been produced. Pasteur referred to the wines as diseased and demonstrated that particular wine diseases were linked to particular microbes contaminating the wine. He eventually suggested a method for heating the wines to destroy the undesirable microbes. The process is now called pasteurization.

Indirect evidence for the germ theory of disease came from the work of the English surgeon Joseph Lister (1827–1912) on the

through cotton and found that objects resembling plant spores had been trapped. If a piece of the cotton was placed in sterile medium after air had been filtered through it, microbial growth occurred. Next he placed nutrient solutions in flasks, heated their necks in a flame, and drew them out into a variety of curves. The swan-neck flasks that he produced in this way had necks open to the atmosphere. Pasteur then boiled the solutions for a few minutes and allowed them to cool. No growth took place even though the contents of the flasks were exposed to the air (**figure 1.13**). Pasteur pointed out that growth did not occur because dust and germs had been trapped on the walls of the curved necks. If the necks were broken, growth commenced immediately. Pasteur had not only resolved the controversy by 1861 but also had shown how to keep solutions sterile.

The English physicist John Tyndall (1820–1893) and the German botanist Ferdinand Cohn (1828–1898) dealt a final blow to spontaneous generation. In 1877 Tyndall demonstrated that dust did indeed carry germs and that if dust was absent, broth remained sterile even if directly exposed to air. During the course of his studies, Tyndall provided evidence for the existence of exceptionally heat-resistant forms of bacteria. Working independently, Cohn discovered that the heat-resistant bacteria recognized by Tyndall were species capable of producing bacterial endospores. Cohn later played an instrumental role in establishing a classification system for bacteria based on their morphology and physiology. ▶▶| *Bacterial endospores are a survival strategy (section 3.9)*

These early microbiologists not only disproved spontaneous generation but also contributed to the rebirth of microbiology. They developed liquid media for culturing microbes. They also developed methods for sterilizing media and maintaining their sterility. These techniques were next applied to understanding the role of microorganisms in disease.

Figure 1.14 Robert Koch. Koch examining a specimen in his laboratory.

prevention of wound infections. Lister, impressed with Pasteur's studies on fermentation and putrefaction, developed a system of antiseptic surgery designed to prevent microorganisms from entering wounds. Instruments were heat sterilized, and phenol was used on surgical dressings and at times sprayed over the surgical area. The approach was remarkably successful and transformed surgery. It also provided strong indirect evidence for the role of microorganisms in disease because phenol, which kills bacteria, also prevented wound infections.

Koch's Postulates

The first direct demonstration that bacteria cause disease came from the study of anthrax by the German physician Robert Koch (1843–1910). Koch (**figure 1.14**) used the criteria proposed by his former teacher Jacob Henle (1809–1885) and others to establish the relationship between *Bacillus anthracis* and anthrax; he published his findings in 1876. Koch injected healthy mice with material from diseased animals, and the mice became ill. After transferring anthrax by inoculation through a series of 20 mice, he incubated a piece of spleen containing the anthrax bacillus in beef serum. The bacteria grew, reproduced, and produced endospores. When isolated bacteria or their spores were injected into healthy mice, anthrax developed. His criteria for proving the causal relationship between a microorganism and a specific disease are known as **Koch's postulates.**

After completing his anthrax studies, Koch fully outlined his postulates in his work on the cause of tuberculosis (**figure 1.15**). In 1884 he reported that this disease was caused by the rod-shaped bacterium *Mycobacterium tuberculosis*, and in 1905 he was awarded the Nobel Prize in Physiology or Medicine. Koch's postulates were quickly adopted by others and used to connect many diseases to their causative agent.

While Koch's postulates are still widely used, their application is at times not feasible. For instance, organisms such as *Mycobacterium leprae,* the causative agent of leprosy, cannot be isolated in pure culture. Some human diseases are so deadly (e.g., Ebola virus disease) that it would be unethical to use humans as the experimental organism; if an appropriate animal model does not exist, the postulates cannot be fully met. To avoid some of these difficulties, microbiologists sometimes use molecular and genetic evidence. For instance, molecular methods might be used to detect the nucleic acid of a virus in body tissues, rather than isolating the virus, or the genes thought to be associated with the virulence of a pathogen might be mutated. In this case, the mutant organism should have decreased ability to cause disease. Introduction of the normal gene back into the mutant should restore the pathogen's virulence.

Pure Culture Methods

During Koch's studies on bacterial diseases, it became necessary to isolate suspected bacterial pathogens in pure culture (p. 11). At first Koch cultured bacteria on the sterile surfaces of cut, boiled potatoes, but the bacteria did not always grow well. Eventually he developed culture media using meat extracts and protein digests, reasoning these were similar to body fluids. Initially he tried to solidify the media by adding gelatin. Separate bacterial colonies developed after the surface of the solidified medium had been streaked with a bacterial sample. The sample could also be mixed with liquefied gelatin medium. When the medium hardened, individual bacteria produced separate colonies.

Despite its advantages, gelatin was not an ideal solidifying agent because it can be digested by many microbes and melts at temperatures above 28°C. A better alternative was provided by Fanny Eilshemius Hesse (1850–1934), the wife of Walther Hesse (1846–1911), one of Koch's students and a physician. She often assisted her husband with his experiments and suggested using agar, which she used to make jellies, as a solidifying agent. Agar was not attacked by most bacteria. Furthermore, it did not melt until reaching a temperature of 100°C and, once melted, did not solidify until reaching a temperature of 50°C; this eliminated the need to handle boiling liquid. Some of the media developed by Koch and his associates, such as nutrient broth and nutrient agar, are still widely used. Another important tool developed in Koch's laboratory was a container for holding solidified media—the Petri dish (plate), named after Richard Petri (1852–1921), who devised it. These developments directly stimulated progress in all areas of microbiology. ▶▶◀ *Culture media (section 7.6); Enrichment and isolation of pure cultures (section 7.6)*

Our focus thus far has been on the development of methods for culturing bacteria. But viral pathogens were also being studied during this time, and methods for culturing them were also being developed. The discovery of viruses and their role in disease was made possible when Charles Chamberland (1851–1908), one of Pasteur's associates, constructed a porcelain bacterial filter in 1884. Dimitri Ivanowski (1864–1920) and Martinus Beijerinck (pronounced "by-a-rink"; 1851–1931) used the filter to study tobacco mosaic disease. They found that plant extracts and sap from diseased

Postulate	Experimentation
1. The microorganism must be present in every case of the disease but absent from healthy organisms.	Koch developed a staining technique to examine human tissue. *Mycobacterium tuberculosis* could be identified in diseased tissue.
2. The suspected microorganisms must be isolated and grown in a pure culture.	Koch grew *M. tuberculosis* in pure culture on coagulated blood serum.
3. The same disease must result when the isolated microorganism is inoculated into a healthy host.	Koch injected cells from the pure culture of *M. tuberculosis* into guinea pigs. The guinea pigs subsequently died of tuberculosis.
4. The same microorganisms must be isolated again from the diseased host.	Koch isolated *M. tuberculosis* in pure culture on coagulated blood serum from the dead guinea pigs.

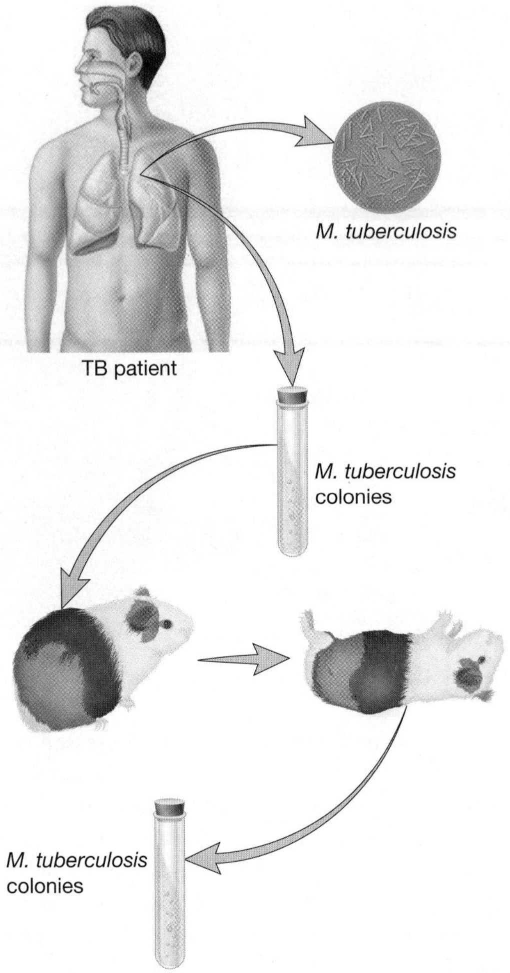

TB patient

M. tuberculosis

M. tuberculosis colonies

M. tuberculosis colonies

Figure 1.15 Koch's Postulates Applied to Tuberculosis.

plants were infectious, even after being filtered with Chamberland's filter. Because the infectious agent passed through a filter that was designed to trap bacterial cells, they reasoned that the agent must be something smaller than a bacterium. Beijerinck proposed that the agent was a "filterable virus." Eventually viruses were shown to be tiny, acellular infectious agents.

Retrieve, Infer, Apply

1. Discuss the contributions of Lister, Pasteur, and Koch to the germ theory of disease and the treatment or prevention of diseases. What other contributions did Koch make to microbiology?
2. Describe Koch's postulates. What is a pure culture? Why are pure cultures important to Koch's postulates?

Immunology

The ability to culture microbes also played an important role in early immunological studies. During studies on the bacterium that causes chicken cholera, Pasteur and Pierre Roux (1853–1933) discovered that incubating the cultures for long intervals between transfers caused the cultures to lose their ability to cause disease. These cultures were said to be attenuated. When chickens were injected with attenuated cultures, they not only remained healthy but also were able to resist the disease when exposed to virulent cultures. Pasteur called the attenuated culture a vaccine (Latin *vacca,* cow) in honor of Edward Jenner (1749–1823) because, many years earlier, Jenner had used material from cowpox lesions to protect people against smallpox (*see Historical Highlights 37.5*). Shortly after this, Pasteur and Chamberland developed an attenuated anthrax vaccine. ▶▶| *Vaccines immunize susceptible populations (section 37.6)*

Pasteur also prepared a rabies vaccine using an attenuated strain of rabies virus. During the course of these studies, Joseph Meister, a nine-year-old boy who had been bitten by a rabid dog, was brought to Pasteur. Since the boy's death was certain in the absence of treatment, Pasteur agreed to try vaccination. Joseph was injected 13 times over the next 10 days with increasingly virulent preparations of the attenuated virus. He survived. In gratitude for Pasteur's development of vaccines, people from around the world contributed to the construction of the Pasteur Institute in Paris, France. One of the initial tasks of the institute was vaccine production.

These early advances in immunology were made without any concrete knowledge about how the immune system works. Immunologists now know that the immune system uses chemicals and several types of blood cells to provide protection. Among the chemicals are soluble proteins called antibodies, which can be found in blood, lymph, and other body fluids. The role of soluble substances in preventing disease was recognized by Emil von Behring (1854–1917) and Shibasaburo Kitasato (1852–1931). After the discovery that diphtheria was caused by a bacterial toxin, they injected inactivated diphtheria toxin into rabbits. The inactivated toxin induced rabbits to produce an antitoxin, which protected against the disease. Antitoxins are now known to be antibodies that specifically bind and neutralize toxins. The first immune system cells were discovered when Elie Metchnikoff (1845–1916) found that some white blood cells could engulf disease-causing bacteria. He called these cells phagocytes and the process phagocytosis (Greek *phagein*, eating).

Microbial Ecology

Culture-based techniques were also applied to the study of microbes in soil and aquatic habitats. Early microbial ecologists studied microbial involvement in the carbon, nitrogen, and sulfur cycles. The Russian microbiologist Sergei Winogradsky (1856–1953) made many contributions to soil microbiology, including the discovery that soil bacteria could oxidize iron, sulfur, and ammonia to obtain energy and that many of these bacteria could incorporate CO_2 into organic matter much as photosynthetic organisms do. Winogradsky also isolated anaerobic nitrogen-fixing soil bacteria and studied the decomposition of cellulose. Martinus Beijerinck was one of the great general microbiologists who made fundamental contributions not only to virology (p. 15) but to microbial ecology as well. He isolated aerobic nitrogen-fixing bacteria (*Azotobacter* spp.), a root nodule bacterium also capable of fixing nitrogen (genus *Rhizobium*), and sulfate-reducing bacteria. Beijerinck and Winogradsky also developed enrichment culture techniques and selective media, which have been of great importance in microbiology. ▶▶| *Biogeochemical cycling sustains life on Earth (section 28.1); Culture media (section 7.6); Enrichment cultures (section 7.6)*

Retrieve, Infer, Apply

1. How did Jenner, Pasteur, von Behring, Kitasato, and Metchnikoff contribute to the development of immunology? How was the ability to culture microbes important to their studies?
2. How did Winogradsky and Beijerinck contribute to the study of microbial ecology? What new culturing techniques did they develop in their studies?
3. How might the work of Winogradsky and Beijerinck have contributed to research on bacterial pathogens? Conversely, how might Koch and Pasteur have influenced Winogradsky's and Beijerinck's study of microbial ecology?

1.4 Microbiology Encompasses Many Subdisciplines

After reading this section, you should be able to:

- Construct a concept map, table, or drawing that illustrates the diverse nature of microbiology and how it has improved human conditions
- Support the belief held by many microbiologists that microbiology is experiencing its second golden age

Microbiology today is as diverse as the organisms it studies. It has both basic and applied aspects. The basic aspects are concerned with the biology of microorganisms themselves. The applied aspects are concerned with practical problems such as disease, water and wastewater treatment, food spoilage and food production, and industrial uses of microbes. Despite this apparent dichotomy, the basic and applied aspects of microbiology are intertwined. Basic research is often conducted in applied fields, and applications often arise out of basic research.

An important recent development in microbiology is the increasing use of molecular and genomic methods to study microbes and their interactions with other organisms. These methods have led to a time of rapid advancement that rivals the golden age of microbiology. Indeed, many feel that microbiology is in its second golden age. Here we describe some of the important advances that have enabled microbiologists to use molecular and genomic techniques. We then discuss some of the important research being done in the numerous subdisciplines of microbiology.

Molecular and Genomic Methods for Studying Microbes

Molecular and genomic methods for studying microbes rely on the ability of scientists to manipulate the genes and genomes of the organisms being studied. An organism's **genome** is all the genetic information that organism contains. To study single genes or the entire genome, microbiologists must be able to isolate DNA and RNA, cut DNA into smaller pieces, insert one piece of DNA into another, and determine the sequence of nucleotides in DNA.

Cutting double-stranded DNA into smaller pieces was first accomplished using bacterial enzymes now known as restriction endonucleases, or simply, restriction enzymes. These enzymes were discovered by Werner Arber and Hamilton Smith in the 1960s. Their discovery was followed by the report in 1972 that David Jackson, Robert Symons, and Paul Berg had successfully generated recombinant DNA molecules—molecules made by combining two or more different DNA molecules together. They did this by cutting DNA from two different organisms with the same restriction enzyme, mixing the two DNA molecules together, and linking them together with an enzyme called DNA ligase. ▶▶| *Key discoveries led to the development of recombinant DNA technology (section 17.1)*

The next major breakthrough was the development of methods to determine the sequence of nucleotides in DNA. In the late 1970s, Frederick Sanger introduced a method that has since been modified and adapted for use in automated systems. Today entire genomes of organisms can be sequenced in a matter of days. In addition, newer, even more rapid sequencing methods have been devised. ▶▶| *Genome sequencing (section 18.2)*

Genome sequencing is the first step in **genomic analysis.** Once the genome sequence is in hand, microbiologists must decipher the information found in the genome. This involves identifying potential protein-coding genes, determining what they code for, and identifying other regions of the genome that may have other important functions (e.g., genes encoding tRNA and rRNA or sequences playing a role in regulating the function of genes). This work requires the use of computers, which has given rise to the scientific discipline **bioinformatics.** Bioinformaticists manage the ever-increasing amount of genetic information available for analysis. They also determine the function of genes and generate hypotheses that can be tested either in silico (i.e., in the computer) or in the laboratory. ▶▶| *Bioinformatics (section 18.4)*

DNA sequencing and related methods have led to a new and exciting phase in the evolution of microbiology: the use of these tools to understand microbes in their environment. This new approach, called **metagenomics** has allowed microbiologists to determine which organisms are present in an environment and what they are doing without first having to culture them. This is important because only a small fraction (1 to 2%) of microbes are easily cultivated. The environments being examined include numerous soil and aquatic habitats. One particularly interesting use of metagenomics is the study of the human microbiome—that is, those microbes occupying various habitats provided by the human body. The amount of knowledge gained about the human microbiome is expanding daily, and we describe some of the discoveries in chapter 32. ▶▶| *Metagenomics provides access to uncultured microbes (section 18.3)*

Major Fields in Microbiology

Microbiology is often divided into subdisciplines based on the type of microbe studied. Thus, microbiology encompasses bacteriology and virology as well as other microbe-specific fields. Microbiology can also be divided based on the activities of microbes—for instance, environmental microbiology and agricultural microbiology. Finally, microbiologists may study only one aspect of the biology of microbes, leading to subdisciplines such as microbial genetics and microbial physiology.

One of the most active and important fields in microbiology is medical microbiology, which deals with diseases of humans and animals. Medical microbiologists identify the agents causing infectious diseases and help plan measures for their control and elimination. Frequently they are involved in tracking down new, unidentified pathogens such as those causing variant Creutzfeldt-Jakob disease (the human version of "mad cow disease"), hantavirus pulmonary syndrome, and West Nile encephalitis. These microbiologists also study the ways microorganisms cause disease. As described in section 1.3, our understanding of the role of microbes in disease began to crystallize when we were able to isolate them in pure culture. Today, clinical laboratory scientists, the microbiologists who work in hospital and other clinical laboratories, use a variety of techniques to provide information needed by physicians to diagnose infectious disease. Increasingly, molecular genetic techniques are also being used.

Major epidemics have regularly affected human history. The 1918 influenza pandemic is of particular note; it killed more than 50 million people in about a year. Public health microbiology is concerned with the control and spread of such communicable diseases. Public health microbiologists and epidemiologists monitor the amount of disease in populations. Based on their observations, they can detect outbreaks and developing epidemics, and implement appropriate control measures. They also conduct surveillance for new diseases as well as bioterrorism events. Public health microbiologists working for local governments monitor community food establishments and water supplies to ensure they are safe and free from pathogens.

To understand, treat, and control infectious disease, it is important to understand how the immune system protects the body from pathogens; this question is the concern of immunology. Immunology is one of the fastest growing areas in science. Much of the growth began with the discovery of the human immunodeficiency virus (HIV), which specifically targets cells of the immune system. Immunology also deals with the nature and treatment of allergies and autoimmune diseases such as rheumatoid arthritis. ▶▶| *Innate host resistance (chapter 33); Adaptive immunity (chapter 34)*

Microbial ecology is another important field in microbiology. Microbial ecology developed when early microbiologists such as Winogradsky and Beijerinck chose to investigate the ecological role of microorganisms rather than their role in disease. Today, a variety of approaches, including non-culture-based techniques, are used to describe the vast diversity of microbes in terms of their morphology, physiology, and relationships with organisms and the components of their habitats. The importance of microbes in global and local cycling of carbon, nitrogen, and sulfur is well documented; however, many questions are still unanswered. Of particular interest is the role of microbes in both the production and removal of greenhouse gases such as carbon dioxide and methane. Microbial ecologists also are employing microorganisms in bioremediation to reduce pollution. A new frontier in microbial ecology is the study of the microbes normally associated with the human body—so-called human microbiota. Scientists are currently trying to identify all members of the human microbiota using molecular techniques that grew out of Woese's pioneering work to establish the phylogeny of microbes. ▶▶| *Global climate change (section 28.2); Biodegradation and bioremediation harness microbes to clean the environment (section 43.4)*

Agricultural microbiology is a field related to both medical microbiology and microbial ecology. Agricultural microbiology is concerned with the impact of microorganisms on agriculture. Microbes such as nitrogen-fixing bacteria play critical roles in the nitrogen cycle and affect soil fertility. Other microbes live in the digestive tracts of ruminants such as cattle and break down the plant materials these animals ingest. There are also plant and animal pathogens that have significant economic impact if not controlled. Furthermore, some pathogens of domestic animals also cause human disease. Agricultural microbiologists work on methods to increase soil fertility and crop yields, study rumen microorganisms in order to increase meat and milk production, and try to combat plant and animal diseases. Currently many agricultural microbiologists are studying the use of bacterial and viral insect pathogens as substitutes for chemical pesticides.

Agricultural microbiology has contributed to the ready supply of high-quality foods, as has the discipline of food and dairy microbiology. Numerous foods are made using microorganisms. On the other hand, some microbes cause food spoilage or are pathogens that are spread through food. Excellent examples of the latter are the rare *Escherichia coli* O104:H4, which in 2011 caused a widespread outbreak of disease in Europe thought to have been spread by bean sprouts. Also in 2011, contaminated ground turkey was implicated in a *Salmonella* outbreak in the United States. Food and dairy microbiologists explore the use of microbes in food production. They also work to prevent microbial spoilage of food and transmission of food-borne diseases. This involves monitoring the food industry for the presence of pathogens. Increasingly, molecular methods are being used to detect pathogens in meat and other foods. Food and dairy microbiologists also conduct research on the use of microorganisms as nutrient sources for livestock and humans. ▶▶| *Microbiology of food (chapter 41)*

Humans unknowingly exploited microbes for thousands of years. However, the systematic and conscious use of microbes in industrial microbiology did not begin until the 1800s. Industrial microbiology developed in large part from Pasteur's work on alcoholic fermentations, as described in section 1.3. His success led to the development of pasteurization to preserve wine during storage. Pasteur's studies on fermentation continued for almost 20 years. One of his most important discoveries was that some fermentative microorganisms were anaerobic and could live only in the absence of oxygen, whereas others were able to live either aerobically or anaerobically.

Another important advance in industrial microbiology occurred in 1929 when Alexander Fleming rediscovered that the fungus *Penicillium* sp. produced what he called penicillin, the first antibiotic that could successfully control bacterial infections. Although it took World War II for scientists to learn how to mass-produce penicillin, scientists soon found other microorganisms capable of producing additional antibiotics. Today industrial microbiologists also use microorganisms to make products such as vaccines, steroids, alcohols and other solvents, vitamins, amino acids, and enzymes. Microbes are also being used to produce biofuels such as ethanol. These alternative fuels are renewable and may help decrease pollution associated with burning fossil fuels. ▶▶| *Microbes are the source of many products of industrial importance (section 42.1); Biofuel production is a dynamic field (section 42.2)*

Industrial microbiologists identify or genetically engineer microbes of use to industrial processes, medicine, agriculture, and other commercial enterprises. They also utilize techniques to improve production by microbes and devise systems for culturing them and isolating the products they make.

The advances in medical microbiology, agricultural microbiology, food and dairy microbiology, and industrial microbiology are in many ways outgrowths of the labor of many microbiologists doing basic research in areas such as microbial physiology, microbial genetics, molecular biology, and bioinformatics. Microbes are metabolically diverse and can employ a wide variety of energy sources, including organic matter, inorganic molecules (e.g., H_2 and NH_3), and sunlight. Microbial physiologists study many aspects of the biology of microorganisms, including their metabolic capabilities. They also study the synthesis of antibiotics and toxins, the ways in which microorganisms survive harsh environmental conditions, and the effects of chemical and physical agents on microbial growth and survival. Microbial geneticists, molecular biologists, and bioinformaticists study the nature of genetic information and how it regulates the development and function of cells and organisms. The bacteria *E. coli* and *Bacillus subtilis*, the yeast *Saccharomyces cerevisiae* (baker's yeast), and bacterial viruses such as T4 and lambda continue to be important model organisms used to understand biological phenomena.

Clearly, the future of microbiology is bright. Genomics in particular is revolutionizing microbiology, as scientists are now beginning to understand organisms in toto, rather than in a reductionist, piecemeal manner. How the genomes of microbes evolve, the nature of host-pathogen interactions, the minimum set of genes required for an organism to survive, and many more topics are aggressively being examined by molecular and genomic analyses. This is an exciting time to be a microbiologist. Enjoy the journey.

Retrieve, Infer, Apply

1. Since the 1970s, microbiologists have been able to study individual genes and whole genomes at the molecular level. What advances made this possible?
2. What is metagenomics? How does it differ from genomic analysis? List one area where metagenomic analysis has been important.
3. Briefly describe the major subdisciplines in microbiology. Which do you consider to be applied fields? Which are basic?
4. Log all the microbial products you use in a week. Be sure to consider all foods and medications (including vitamins).
5. List all the activities or businesses you can think of in your community that directly depend on microbiology.

Key Concepts

1.1 Members of the Microbial World

- Microbiology studies microscopic cellular organisms that are often unicellular or, if multicellular, do not have highly differentiated tissues. Microbiology also focuses on biological entities that are acellular (**figure 1.1**).

- Microbiologists divide cellular organisms into three domains: *Bacteria, Archaea,* and *Eukarya* (**figure 1.2**).
- Domains *Bacteria* and *Archaea* consist of prokaryotic microorganisms. Eukaryotic microbes (protists and fungi) are placed in *Eukarya*. Viruses, viroids, satellites, and

prions are acellular entities that are not placed in any of the domains but are classified by a separate system.

1.2 Microbes Have Evolved and Diversified for Billions of Years

- Evolutionary biologists and others interested in the origin of life must rely on many types of evidence.
- Earth is approximately 4.5 billion years old. Within the first 1 billion years of its existence, life arose (**figure 1.4**).
- The RNA world hypothesis posits that the earliest self-replicating entity on the planet used RNA both to store genetic information and to conduct cellular processes (**figure 1.6**).
- Comparisons of small subunit (SSU) rRNA genes have been useful in creating universal phylogenetic trees. These trees provide information about the evolution of life after it arose (**figure 1.8**).
- The last universal common ancestor (LUCA) is placed on the bacterial branch of the universal phylogenetic tree. Thus, *Bacteria* are thought to have diverged first, and *Archaea* and *Eukarya* arose later.
- Mitochondria, chloroplasts, and hydrogenosomes are thought to have evolved from bacterial endosymbionts of ancestral cells in the eukaryotic lineage.
- The concept of a bacterial or archaeal species is difficult to define and is the source of considerable debate because these microbes do not reproduce sexually. Species are named using the binomial system of Linnaeus.

1.3 Microbiology Advanced as New Tools for Studying Microbes Were Developed

- Microbiology is defined not only by the organisms it studies but also by the tools it uses. Microscopy and culture-based techniques have played and continue to play important roles in the evolution of the discipline.
- Antony van Leeuwenhoek used simple microscopes and was the first person to extensively describe microorganisms (**figure 1.11**).
- Culture-based techniques for studying microbes began to develop as scientists debated the theory of spontaneous generation. Experiments by Redi and others disproved the theory of spontaneous generation of larger organisms. The

spontaneous generation of microorganisms was disproved by Spallanzani, Pasteur, Tyndall, Cohn, and others (**figure 1.13**).
- The availability of culture-based techniques played an important role in the study of microbes as the causative agents of disease. Support for the germ theory of disease came from the work of Bassi, Pasteur, Koch, Lister, and others.
- Koch's postulates are used to prove a direct relationship between a suspected pathogen and a disease. Koch and his coworkers developed the techniques required to grow bacteria on solid media and to isolate pure cultures of pathogens (**figure 1.15**).
- Viruses were discovered following the invention of a bacterial filter by Chamberland. Dimitri Ivanowski and Martinus Beijerinck were important contributors to the field of virology.
- The field of immunology developed as early microbiologists created vaccines and discovered antibodies and phagocytic cells. Pasteur, von Behring, Kitasato, and Mechnikof made important contributions to this field.
- Microbial ecology grew out of the work of Winogradsky and Beijerinck. They studied the role of microorganisms in carbon, nitrogen, and sulfur cycles and developed enrichment culture techniques and selective media.

1.4 Microbiology Encompasses Many Subdisciplines

- Today molecular, genomic, and metagenomic analyses have paved the way for understanding microbes as biological systems. These methods include recombinant DNA techniques, nucleic acid sequencing methods, and genome sequencing.
- Metagenomic analysis is being used to study microbes in their habitats without first culturing them. One interesting application of metagenomic analysis is to study the human microbiome.
- There are many fields in microbiology. These include medical, public health, industrial, and food and dairy microbiology. Microbial ecology, physiology, and genetics are important subdisciplines of microbiology.

Compare, Hypothesize, Invent

1. Microscopic organisms such as rotifers are not studied by microbiologists. Why is this so?

2. Why aren't viruses, viroids, satellites, and prions included in the three-domain system?

3. Why was the belief in spontaneous generation an obstacle to the development of microbiology as a scientific discipline?

4. Would microbiology have developed more slowly if Fanny Hesse had not suggested the use of agar? Give your reasoning.

5. Some individuals can be infected by a pathogen yet not develop disease. In fact, some become chronic carriers of the pathogen. How does this observation affect Koch's postulates? How might the postulates be modified to account for the existence of chronic carriers?

6. Develop a list of justifications for the usefulness of microorganisms as experimental models.

7. History is full of examples in which one group of people lost a struggle against another.

    a. Choose an example of a battle or other human activity such as exploration of new territory and determine the impact of microorganisms, either indigenous or transported to the region, on that activity.

    b. Discuss the effect that the microbe(s) had on the outcome in your example.

    c. Suggest whether the advent of antibiotics, food storage and preparation technology, or sterilization technology would have made a difference in the outcome.

8. Antony van Leeuwenhoek is often referred to as the father of microbiology. However, many historians feel that Louis Pasteur, Robert Koch, or perhaps both, deserve that honor. Decide who should be considered the father of microbiology and justify your decision.

9. Consider the discoveries described in sections 1.3 and 1.4. Which do you think were the most important to the development of microbiology? Why?

10. Support this statement: "Vaccinations against various childhood diseases have contributed to the entry of women, particularly mothers, into the full-time workplace."

11. Scientists are very interested in understanding when cyanobacteria first emerged because, as the first organisms capable of oxygenic photosynthesis, it is thought that they triggered a sharp rise in atmospheric oxygen. For many years, certain lipid biomarkers have served as "molecular fossils" to date the first appearance of cyanobacteria. However, a 2010 study questioned whether these lipids, called 2-methylhopanoids, provide accurate information in light of a 2007 discovery that they also are produced by an anoxygenic phototrophic bacterium—a bacterium that does not produce oxygen as it uses light energy. The authors of the 2010 study identified genes in extant bacteria involved in synthesis of the lipid biomarkers and then constructed phylogenetic trees based on comparisons of these genes. They also identified the phyla to which the bacteria belonged (based on SSU rRNA analysis) and noted the habitats and metabolic capabilities of the bacteria used in the study. Discuss the specific challenges encountered in the study of microbial evolution. What results from the phylogenetic analysis would support their claim that 2-methylhopanoids are not reliable biomarkers? Why were habitat and metabolic characteristics also part of their analysis?

    Read the original paper: Welander, P. V., et al. 2010. Identification of a methylase required for 2-methylhopanoid production and implications for the interpretation of sedimentary hopanes. *Proc. Natl. Acad. Sci. USA.* 107(19):8537.

12. It is possible to artificially create bacterial cells that completely lack a cell wall. The resulting cells, called L-forms, can be maintained and cultured. L-forms have been known for many years and have been of interest because of their unusual physiological characteristics. One of their most unusual features is that they reproduce by becoming deformed and then "falling apart," producing two or more additional L-forms. The progeny L-forms then grow larger and reproduce by a similar method. This is done without the aid of the normal division machinery. In 2013 scientists in the United Kingdom reported their attempts to understand how L-forms "divide." They generated mutant strains of *Bacillus subtilis* L-forms that promoted L-form proliferation. The only mutations found with this effect were those that increased production of membrane lipids. They argued that these lipids were inserted into the L-form membrane and that this increased the surface area of the L-forms without increasing their volume. The increased surface-to-volume ratio caused the cells to assume unusual shapes and experience torsional stress. The torsional stress served as the biophysical force causing the L-forms to split apart. They also suggested that the study of L-form division provided a good model for how primitive cells divided before the evolution of the complex division machinery that exists in modern cells. Why did they suggest this? Do you agree? Explain your answer.

    Read the original paper: Mercier, R., et al. 2013. Excess membrane synthesis drives a primitive mode of cell proliferation. *Cell* 152:997.

Learn More

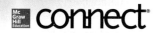

2

Microscopy

Hazardous materials (hazmat) personnel outside the U.S. Capitol building during the 2001 anthrax bioterrorism attack.

Anthrax Bioterrorism Attack 2001

While on a trip to North Carolina in late September 2001, a sixty-three-year-old photojournalist and outdoor enthusiast from Florida began feeling ill. His muscles ached, he felt nauseous, and he had a fever. His symptoms were annoying and varied in severity during his trip. After returning home, his symptoms worsened considerably. On October 2, he awoke with vomiting. Even more alarming was his mental confusion. When his wife took him to a local emergency room, doctors found that he was unable to answer simple questions not only about his illness but also about where he was and what time it was. His doctors performed a spinal tap and collected cerebral spinal fluid (CSF). The fluid was sent to the hospital laboratory where it was stained using the Gram-staining technique. Much to the surprise of the clinical lab scientists, the CSF contained long, Gram-positive rods that formed chains, a morphology unlike that of typical meningitis-causing bacteria. Based on the Gram-stain results, doctors made an initial diagnosis of inhalation anthrax. What made this diagnosis so surprising was that inhalation anthrax is extremely rare in the United States. Indeed, the last diagnosed case was in 1976. Doctors immediately began treatment with antibiotics, but the photojournalist's condition worsened, and he died October 5.

Thus began the first anthrax bioterrorism attack in the United States. Over the next several weeks, 16 other people developed inhalation anthrax; four of these individuals died. Much was learned about the disease from this attack. Much was also learned about the nation's readiness to deal with bioterrorism. This event is also a reminder of the continuing importance of microscopy in microbiology and in diagnosing disease.

As we describe in chapter 1, the invention of microscopes in the 1600s was a critical step in the evolution of microbiology. By the 1800s, microscopes had become relatively sophisticated, and methods for staining specimens were developed so that they could be seen more easily. Indeed, the Gram-staining technique is over 150 years old.

In this chapter, we introduce some of the most commonly used types of microscopy. We begin with light microscopes and then describe other common types of microscopes, as well as how specimens are prepared for examination by microscopes.

Readiness Check:
Based on what you have learned previously, you should be able to:

✔ Explain why microscopy is a major tool used by microbiologists (sections 1.1 and 1.3)
✔ Define magnification
✔ Express the size of organisms using the metric system (**table 2.1**)

Table 2.1	Common Units of Measurement	
Unit	**Abbreviation**	**Value**
1 centimeter	cm	10^{-2} meter or 0.394 inches
1 millimeter	mm	10^{-3} meter
1 micrometer	μm	10^{-6} meter
1 nanometer	nm	10^{-9} meter
1 Angstrom	Å	10^{-10} meter

2.1 Lenses Create Images by Bending Light

After reading this section, you should be able to:

■ Relate the refractive indices of glass and air to the path light takes when it passes through a prism or convex lens
■ Correlate lens strength and focal length

Light microscopes were the first microscopes invented and they continue to be the most commonly used type. To understand light microscopy, we must consider the way lenses bend and focus light to form images. When a ray of light passes from one medium to another, refraction occurs; that is, the ray is bent at the interface. The **refractive index** is a measure of how greatly a substance slows the velocity of light; the direction and magnitude of bending are determined by the refractive indices of the two media forming the interface. For example, when light passes from air into glass, a medium with a greater refractive index, it is slowed and bent toward the normal, a line perpendicular to the surface (**figure 2.1**). As light leaves glass and returns to air, a medium with a lower refractive index, it accelerates and is bent away from the normal. Thus a prism bends light because glass has a different refractive index from air and the light strikes its surface at an angle.

Lenses act like a collection of prisms operating as a unit. When the light source is distant so that parallel rays of light strike the lens, a convex lens focuses the rays at a specific point, the focal point (*F* in **figure 2.2**). The distance between the center of the lens and the focal point is called the focal length (*f* in figure 2.2).

Our eyes cannot focus on objects nearer than about 25 cm (i.e., about 10 inches). This limitation may be overcome by using

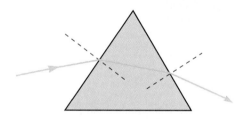

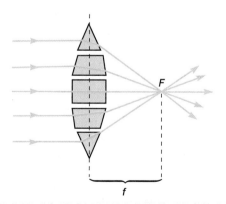

Figure 2.1 The Bending of Light by a Prism. Lines perpendicular to the surface of the prism are called normal; they are indicated by dashed lines. As light enters the glass, it is bent toward the first normal. When light leaves the glass and returns to air, it is bent away from the second normal. As a result, the prism bends light passing through it.

Figure 2.2 Lens Function. A lens functions somewhat like a collection of prisms. Light rays from a distant source are focused at the focal point *F*. The focal point lies a distance *f*, the focal length, from the lens center.

MICRO INQUIRY *How would the focal length change if the lens shown here were thicker?*

a convex lens as a simple magnifier (or microscope) and holding it close to an object. A magnifying glass provides a clear image at much closer range, and the object appears larger. Lens strength is related to focal length; a lens with a short focal length magnifies an object more than a lens having a longer focal length.

Retrieve, Infer, Apply

1. Define refraction, refractive index, focal point, and focal length.
2. Describe the path of a light ray through a prism or lens.
3. How is lens strength related to focal length? How is this principle applied to corrective eyeglasses?

2.2 There Are Several Types of Light Microscopes

After reading this section, you should be able to:

- Evaluate the parts of a light microscope in terms of their contributions to image production and use of the microscope
- Predict the relative degree of resolution based on light wavelength and numerical aperture of the lens used to examine a specimen
- Create a table that compares and contrasts the various types of light microscopes in terms of their uses, how images are created, and the quality of images produced

Microbiologists currently employ a variety of light microscopes in their work; bright-field, dark-field, phase-contrast, fluorescence, and confocal microscopes are most commonly used. Each is useful for certain applications. Modern microscopes are all compound microscopes; that is, they have two sets of lenses. The **objective lens** is the lens closest to the specimen. It forms a magnified image that is further enlarged by one or more additional lenses.

Bright-Field Microscope: Dark Object, Bright Background

The **bright-field microscope** is routinely used in microbiology labs to examine both stained and unstained specimens. It is called a bright-field microscope because it forms a dark image against a brighter background. It consists of a sturdy metal stand composed of a base and an arm to which the remaining parts are attached (**figure 2.3**). A light source, either a mirror or an electric illuminator, is located in the base. Two focusing knobs, the fine and coarse adjustment knobs, are located on the arm and can move either the stage or the nosepiece vertically to focus the image.

The stage is positioned about halfway up the arm. It holds microscope slides either by simple slide clips or by a mechanical stage clip. A mechanical stage uses stage control knobs to smoothly move a slide during viewing. The substage **condenser** (or simply, condenser) is mounted within or beneath the stage and focuses a cone of light on the slide. Its position often is fixed in simpler microscopes but can be adjusted vertically in more advanced models.

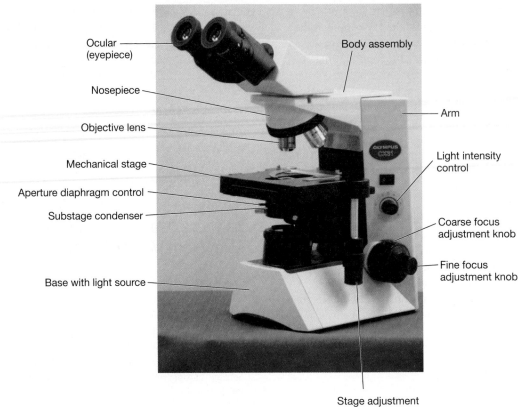

Figure 2.3 A Bright-Field Microscope.

The curved upper part of the arm holds the body assembly, to which a nosepiece and one or more **ocular lenses** (also called **eyepieces**) are attached. More advanced microscopes have eyepieces for both eyes and are called binocular microscopes. The body assembly contains a series of mirrors and prisms so that the barrel holding the eyepiece may be tilted for ease in viewing. The nosepiece holds three to five objective lenses of differing magnifying power and can be rotated to position any objective lens beneath the body assembly. Ideally a microscope should be **parfocal;** that is, the image should remain in focus when objective lenses are changed.

The image seen when viewing a specimen with a compound microscope is created by the objective and ocular lenses working together. Light from the illuminated specimen is focused by the objective lens, creating an enlarged image within the microscope. The ocular lens further magnifies this primary image. The total magnification is calculated by multiplying the objective and eyepiece magnifications together. For example, if a 45× objective lens is used with a 10× eyepiece, the overall magnification of the specimen is 450×.

Better Microscope Resolution Means a Clearer Image

The most important part of the microscope is the objective lens, which must produce a clear image, not just a magnified one. Thus resolution is extremely important. **Resolution** is the ability of a lens to separate or distinguish between small objects that are close together.

Resolution is described mathematically by an equation developed in the 1870s by Ernst Abbé (1840–1905), a German physicist responsible for much of the optical theory underlying microscope design. The Abbé equation states that the minimal distance (d) between two objects that reveals them as separate entities depends on the wavelength of light (λ) used to illuminate the specimen and on the **numerical aperture** of the lens ($n \sin \theta$), which is the ability of the lens to gather light.

$$d = \frac{0.5 \, \lambda}{n \sin \theta}$$

The smaller d is, the better the resolution, and finer detail can be discerned in a specimen; d becomes smaller as the wavelength of light used decreases and as the numerical aperture increases. Thus the greatest resolution is obtained using a lens with the largest possible numerical aperture and light of the shortest wavelength, light at the blue end of the visible spectrum (in the range of 450 to 500 nm).

The numerical aperture ($n \sin \theta$) of a lens is defined by two components: n is the refractive index of the medium in which the lens works (e.g., air) and θ is 1/2 the angle of the cone of light entering an objective (**figure 2.4**). A cone with a narrow angle does not adequately separate the rays of light emanating from closely packed objects, and the images are not resolved. A cone of light with a very wide angle does separate the rays, and the closely packed objects appear widely separated and resolved. Recall that the refractive indices of all materials through which

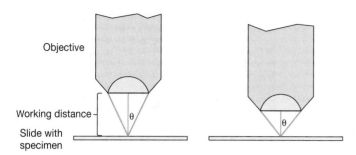

Figure 2.4 **Numerical Aperture in Microscopy.**   The numerical aperture of a lens is related to a value called the angular aperture (symbolized by θ), which is 1/2 the angle of the cone of light that enters a lens from a specimen. The equation for numerical aperture is *n* sin θ. In the right-hand illustration, the lens has larger angular and numerical apertures; its resolution is greater and its working distance smaller.

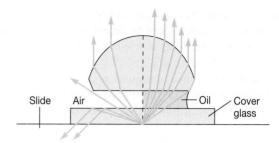

Figure 2.5 **The Oil Immersion Objective.**   An oil immersion objective operating in air and with immersion oil.

light waves pass determine the direction of the light rays emanating from the specimen. Some objective lenses work in air, which has a refractive index of 1.00. Since sin θ cannot be greater than 1 (the maximum θ is 90° and sin 90° is 1.00), no lens working in air can have a numerical aperture greater than 1.00. The only practical way to raise the numerical aperture above 1.00, and therefore achieve higher resolution, is to increase the refractive index with immersion oil, a colorless liquid with the same refractive index as glass (**table 2.2**). If air is replaced with immersion oil, many light rays that did not enter the objective due to reflection and refraction at the surfaces of the objective lens and slide will now do so (**figure 2.5**). This results in an increase in numerical aperture and resolution.

Numerical aperture is related to another characteristic of an objective lens, the working distance. The working distance of an objective is the distance between the surface of the lens and the surface of the cover glass (if one is used) or the specimen when it is in sharp focus (figure 2.4). Objectives with large numerical apertures and great resolving power have short working distances (table 2.2).

Numerical aperture is also an important feature of a microscope's condenser. The condenser is a large, light-gathering lens used to project a wide cone of light through the slide and into the objective lens. Most microscopes have a condenser with a

numerical aperture between 1.2 and 1.4. However, the condenser's numerical aperture will not be much above about 0.9 unless the top of the condenser is oiled to the bottom of the slide. During routine microscope operation, the condenser usually is not oiled, and this limits the overall resolution, even with an oil immersion objective.

The most accurate calculation of a microscope's resolving power considers both the numerical aperture of the objective lens and that of the condenser, as is evident from the following equation, where NA is the numerical aperture.

$$d_{microscope} = \frac{\lambda}{(NA_{objective} + NA_{condenser})}$$

However, in most cases the limit of resolution of a light microscope is calculated using the Abbé equation, which considers the objective lens only. The maximum theoretical resolving power of a microscope when viewing a specimen using an oil immersion objective (numerical aperture of 1.25) and blue-green light is approximately 0.2 μm.

$$d = \frac{(0.5)(530 \text{ nm})}{1.25} = 212 \text{ nm or } 0.2 \text{ μm}$$

At best, a bright-field microscope can distinguish between two dots about 0.2 μm apart (the same size as a very small bacterium). Thus the vast majority of viruses cannot be examined with a light microscope.

Given the limit of resolution of a light microscope, the largest useful magnification—the level of magnification needed to

Table 2.2	Properties of Objective Lenses				
		OBJECTIVE			
Property	**Scanning**	**Low Power**	**High Power**	**Oil Immersion**
Magnification	4×	10×	40–45×	90–100×
Numerical aperture	0.10	0.25	0.55–0.65	1.25–1.4
Approximate focal length (*f*)	40 mm	16 mm	4 mm	1.8–2.0 mm
Working distance	17–20 mm	4–8 mm	0.5–0.7 mm	0.1 mm
Approximate resolving power with light of 450 nm (blue light)	2.3 μm	0.9 μm	0.35 μm	0.18 μm

increase the size of the smallest resolvable object to be visible with the light microscope—can be determined. Our eye can just detect a speck 0.2 mm in diameter. When the acuity of the eye and the resolution of the microscope are considered together, it is calculated that the useful limit of magnification is about 1,000 times the numerical aperture of the objective lens. Most standard microscopes come with 10× eyepieces and have an upper limit of about 1,000× with oil immersion. A 15× eyepiece may be used with good objective lenses to achieve a useful magnification of 1,500×. Any further magnification does not enable a person to see more detail. Indeed, a light microscope can be built to yield a final magnification of 10,000×, but it would simply be magnifying a blur. Only the electron microscope provides sufficient resolution to make higher magnifications useful (p. 34).

Visualizing Living, Unstained Microbes

Bright-field microscopes are probably the most common microscope found in research and clinical laboratories and certainly in teaching laboratories. However, many microbes are unpigmented and are not clearly visible because there is little difference in contrast between the cells, subcellular structures, and water. As we discuss in section 2.3, one solution to this problem is to stain cells before observation to increase contrast and create variations in color between cell structures. Unfortunately, staining procedures usually kill cells. But what if an investigator must view living cells to observe a dynamic process such as movement or phagocytosis? Three types of light microscopes create detailed, clear images of living specimens: dark-field microscopes, phase-contrast microscopes, and differential interference contrast microscopes.

Dark-Field Microscope: Bright Object, Dark Background

The **dark-field microscope** produces detailed images of living, unstained cells and organisms by simply changing the way

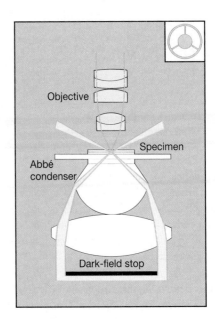

Figure 2.6 **Dark-Field Microscopy.** In dark-field microscopy, a dark-field stop (inset) is placed underneath the condenser lens system. The condenser then produces a hollow cone of light so that the only light entering the objective comes from the specimen.

in which they are illuminated. A hollow cone of light is focused on the specimen in such a way that unreflected and unrefracted rays do not enter the objective. Only light that has been reflected or refracted by the specimen forms an image (**figure 2.6**). The field surrounding a specimen appears black, while the object itself is brightly illuminated (**figure 2.7**). The dark-field microscope can reveal considerable internal structure in larger eukaryotic microorganisms (figure 2.7*b*). It also is used to identify certain bacteria such as the thin and distinctively shaped *Treponema pallidum*, the causative agent of syphilis (figure 2.7*a*).

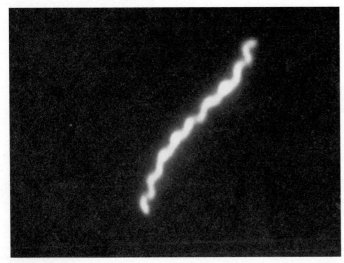

(a) *T. pallidum*

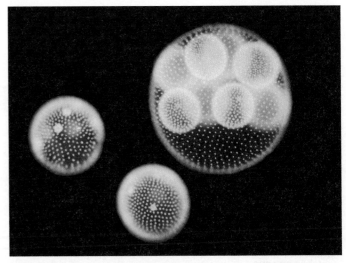

(b) *Volvox*

Figure 2.7 **Examples of Dark-Field Microscopy.** (a) *Treponema pallidum*, the spirochete that causes syphilis (×400). (b) *Volvox*. Note daughter colonies within the mature *Volvox* colony.

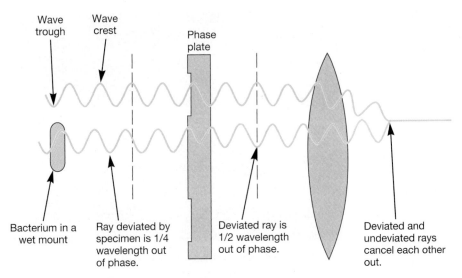

Figure 2.8 The Production of Contrast in Phase-Contrast Microscopy.   The behavior of deviated and undeviated (i.e., undiffracted) light rays in the positive-phase-contrast microscope. Because the light rays tend to cancel each other out, the image of the specimen will be dark against a brighter background.

Phase-Contrast Microscope

To understand phase-contrast microscopy, consider a bacterium in a wet mount as illustrated in **figure 2.8**. The refractive indices of many bacterial cell structures are greater than that of water. Therefore, light waves passing through a cell structure will be diffracted and slowed more than light waves passing through the surrounding medium (in this case water inside and outside the cell). Thus two kinds of light waves pass through a wet mount: deviated light waves that interact with bacterial cell structures and undeviated light waves that pass around and through the cell without interacting with the cell. Because the deviated light waves are slowed relative to the undeviated light waves, they are said to be out of phase. That is, the crests and troughs of the deviated and undeviated waves no longer align. Typically the deviated light waves are retarded by about ¼ wavelength compared to the undeviated light (figure 2.8).

Phase-contrast microscopes take advantage of this phenomenon and use it to create differences in light intensity that provide contrast to allow the viewer to see a clearer, more detailed image of the specimen (**figure 2.9**). They do so by separating the two types of light so that the undeviated light (primarily light from the surroundings) can be manipulated and then recombined with the deviated light (light from the bacterium) to form an image. Two components allow this to occur: a condenser annulus and a phase plate (**figure 2.10**). The condenser annulus is an opaque disk with a thin transparent ring. A ring of light is directed by the condenser annulus to the condenser, which focuses the light on the specimen as shown in figure 2.10. Deviated and undeviated light then pass through the objective toward the phase plate. The phase plate has a thin ring through which the undeviated light (i.e., from the surroundings) is focused (figure 2.8). In a common type of phase-contrast microscopy (positive phase contrast), the ring is coated with a substance that advances the phase of the undeviated light by ¼ wavelength. The deviated light is focused on the rest of the

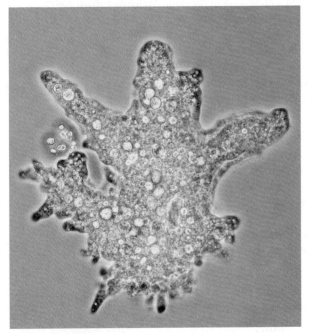

(a) An amoeba

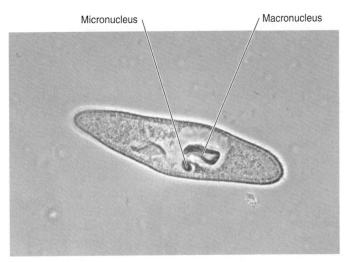

(b) *Paramecium* sp.

Figure 2.9 Examples of Phase-Contrast Microscopy.   (a) An amoeba, a eukaryotic microbe that moves by means of pseudopodia, which extend out from the main part of the cell body. (b) *Paramecium* sp. stained to show a large central macronucleus with a small spherical micronucleus at its side (×400).

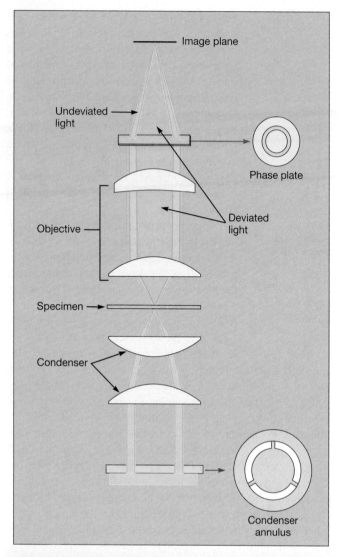

Figure 2.10 **Phase-Contrast Microscopy.** The optics of a positive-phase-contrast microscope.

MICRO INQUIRY *What is the purpose of the annular stop in a phase-contrast microscope? Is it found in any other kinds of light microscopes?*

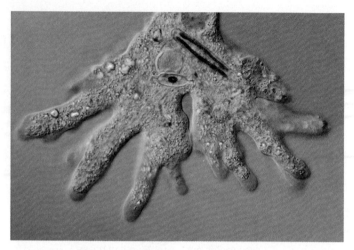

Figure 2.11 **Differential Interference Contrast Microscopy.** A micrograph of the protozoan *Amoeba proteus*. The image appears three-dimensional and contains considerable detail.

widely used to study eukaryotic cells. ▶▶| *Bacterial endospores are a survival strategy (section 3.9); Inclusions (section 3.6)*

Differential Interference Contrast Microscope

The **differential interference contrast (DIC) microscope** is similar to the phase-contrast microscope in that it creates an image by detecting differences in refractive indices and thickness. Two beams of plane-polarized light at right angles to each other are generated by prisms. In one design, the object beam passes through the specimen, while the reference beam passes through a clear area of the slide. After passing through the specimen, the two beams are combined and interfere with each other to form an image. A live, unstained specimen appears brightly colored and seems to pop out from the background, giving the viewer the sense that a three-dimensional image is being viewed (**figure 2.11**). Structures such as cell walls, endospores, granules, vacuoles, and nuclei are clearly visible.

Retrieve, Infer, Apply

1. List the parts of a light microscope and describe their functions.
2. If a specimen is viewed using a 5× objective in a microscope with a 15× eyepiece, how many times has the image been magnified?
3. How does resolution depend on the wavelength of light, refractive index, and numerical aperture? How are resolution and magnification related?
4. What is the function of immersion oil?
5. Why don't most light microscopes use 30× ocular lenses for greater magnification?

Fluorescence Microscopes Use Emitted Light to Create Images

The light microscopes thus far considered produce an image from light that passes through a specimen. An object also can be seen because it emits light: this is the basis of fluorescence microscopy. When some molecules absorb radiant energy, they become excited

phase plate, which lets the deviated light pass through unchanged. After leaving the phase plate, the deviated and undeviated light are now out of phase by ½ wavelength (figure 2.8). When the two rays of light recombine to form an image, they cancel each other out, a phenomenon called destructive interference. Destructive interference is also seen if the crest of a wave of water meets the trough of another wave—the two cancel each other out and the surface of the water remains calm at the point where they meet. In our example, the resulting image consists of a darker bacterium against a lighter background.

Phase-contrast microscopy is especially useful for studying microbial motility, determining the shape of living cells, and detecting bacterial structures such as endospores and inclusions. These are clearly visible because they have refractive indices markedly different from that of water. Phase-contrast microscopes also are

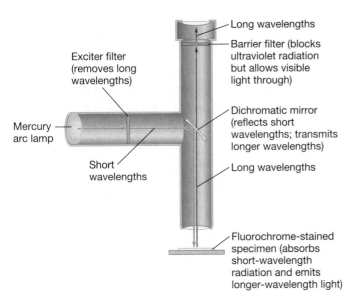

Figure 2.12 Epifluorescence Microscopy. The principles of operation of an epifluorescence microscope.

and release much of their trapped energy as light. Any light emitted by an excited molecule has a longer wavelength (i.e., has lower energy) than the radiation originally absorbed. **Fluorescent light** is emitted very quickly by the excited molecule as it gives up its trapped energy and returns to a more stable state.

The **fluorescence microscope** excites a specimen with a specific wavelength of light and forms an image with the fluorescent light emitted by the object. The most commonly used fluorescence microscopy is epifluorescence microscopy, also called incident light or reflected light fluorescence microscopy. Epifluorescence microscopes employ an objective lens that also acts as a condenser so the specimen is illuminated from above rather than below (**figure 2.12**). A mercury vapor arc lamp or other source produces an intense beam of light that passes through an exciter filter. The exciter filter transmits only the desired wavelength of light. The excitation light is directed down the microscope by the dichromatic mirror. This mirror reflects light of shorter wavelengths (i.e., the excitation light) but allows light of longer wavelengths to pass through. The excitation light continues down, passing through the objective lens to the specimen, which is usually stained with molecules called **fluorochromes** (**table 2.3**). The fluorochrome absorbs light energy from the excitation light and fluoresces brightly. The emitted fluorescent light travels up through the objective lens into the microscope. Because the emitted fluorescent light has a longer wavelength, it passes through the dichromatic mirror to a barrier filter, which blocks out any residual excitation light.

Table 2.3	Commonly Used Fluorochromes
Fluorochrome	**Uses**
Acridine orange	Stains DNA
Diamidino-2-phenyl indole (DAPI)	Stains DNA
Fluorescein isothiocyanate (FITC)	Often attached to DNA probes or to antibodies that bind specific cellular components
Tetramethyl rhodamine isothiocyanate (TRITC or rhodamine)	Often attached to antibodies that bind specific cellular components

Finally, the emitted light passes through the barrier filter to the eyepieces.

The fluorescence microscope has become an essential tool in microbiology. Bacterial pathogens can be identified after staining with fluorochromes or specifically tagging them with fluorescently labeled antibodies using immunofluorescence procedures. In ecological studies, fluorescence microscopy is used to observe microorganisms stained with fluorochrome-labeled probes or fluorochromes that bind specific cell constituents (table 2.3). In addition, microbial ecologists use epifluorescence microscopy to visualize photosynthetic microbes, as their pigments naturally fluoresce when excited by light of specific wavelengths. It is even possible to distinguish live bacteria from dead bacteria by the color they fluoresce after treatment with a specific mixture of stains (**figure 2.13a**). Thus the microorganisms can be viewed and directly counted in a relatively undisturbed

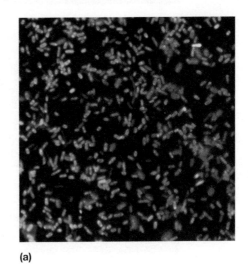

(a)

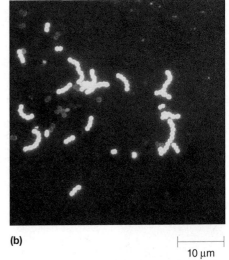

(b)

10 μm

Figure 2.13 Fluorescent Dyes and Tags. (a) Dyes that cause live cells to fluoresce green and dead ones red. (b) Fluorescent antibodies tag specific molecules. In this case, the antibody binds to a molecule that is unique to *Streptococcus pyogenes*, the bacterium that causes strep throat.

MICRO INQUIRY *How might the fluorescently labeled antibody used in figure 2.13b be used to diagnose strep throat?*

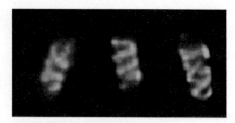

Figure 2.14 **Green Fluorescent Protein.** Visualization of Mbl, a cytoskeletal protein of *Bacillus subtilis*. The Mbl protein has been fused with green fluorescent protein and therefore fluoresces green.

ecological niche. ▶▶ *Genetic methods are used to assess microbial diversity (section 29.2); Identification of microorganisms from specimens (section 36.3)*

Another important use of fluorescence microscopy is the localization of specific proteins within cells. One approach is

to use genetic engineering techniques that fuse the gene for the protein of interest to a gene isolated from jellyfish belonging to the genus *Aequorea*. This jellyfish gene encodes a protein that naturally fluoresces green when exposed to light of a particular wavelength and is called green fluorescent protein (GFP). Thus when the protein is made by the cell, it is fluorescent. GFP has been used extensively in studies on bacterial cell division and related phenomena (**figure 2.14**). In fact, the 2008 Nobel Prize in Chemistry was awarded to Osamu Shimomura of Japan and Americans Martin Chalfie and Roger Tsien for their development of this important tool. ▶▶ *Fluorescence labeling (section 17.6)*

Confocal Microscopy

Like the large and small beads illustrated in **figure 2.15a**, biological specimens are three-dimensional. When three-dimensional objects

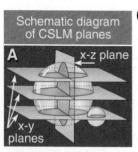

(a) All three-dimensional objects are defined by three axes: x, y, and z. The confocal scanning laser microscope (CSLM) is able to create images of planes formed by the x and y axes (x-y planes) and planes formed by the x and z axes (x-z planes).

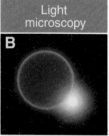

(b) The light microscope image of the two beads shown in **(a)**. Note that neither bead is clear and that the smaller bead is difficult to recognize as a bead. This is because the image is generated from light emanating from multiple planes of focus.

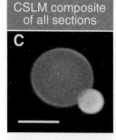

(c) A CSLM uses light from a single plane of focus to generate an image. A computer connected to a CSLM can make a composite image of the two beads using digitized information collected from multiple planes within the beads. The result is a much clearer and more detailed image.

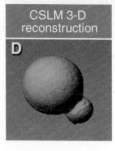

(d) The computer can also use digitized information collected from multiple planes within the beads to generate a three-dimensional reconstruction of the beads.

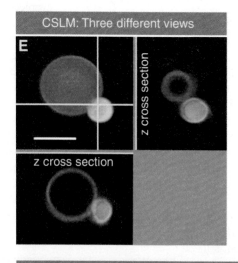

(e) The computer can also generate views of the specimen using different planes. The top left panel is the image of a single x-y plane (i.e., looking down from the top of the beads). The two lines represent the two x-z planes imaged in the other two panels. The vertical line indicates the x-z plane shown in the top right panel (i.e., a view from the right side of the beads) and the horizontal line indicates the x-z plane shown in the bottom panel (i.e., a view from the front face of the beads).

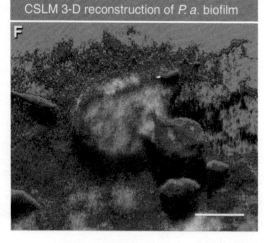

(f) A three-dimensional reconstruction of a *Pseudomonas aeruginosa* biofilm. The biofilm was exposed to an antibacterial agent and then stained with dyes that distinguish living (green) from dead (red) cells. The cells on the surface of the biofilm have been killed, but those in the lower layers are still alive. This image clearly demonstrates the difficulty of killing all the cells in a biofilm.

Figure 2.15 **Light and Confocal Microscopy.** (a–e) Two beads examined by light and confocal microscopy. (f) Three-dimensional reconstruction of a biofilm.

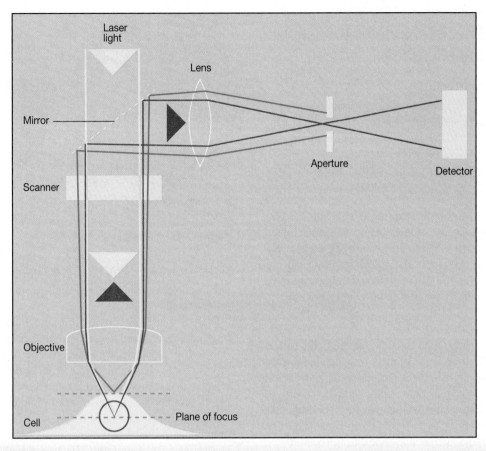

Figure 2.16 **Ray Diagram of a Confocal Microscope.** The yellow lines represent laser light used for illumination. Red lines symbolize the light arising from the plane of focus, and the blue lines stand for light from parts of the specimen above and below the focal plane.

MICRO INQUIRY *How does the light source differ between a confocal light microscope and other light microscopes?*

are viewed with traditional light microscopes, light from all areas of the object, not just the plane of focus, enters the microscope and is used to create an image. The resulting image is murky and fuzzy (figure 2.15*b*). This problem has been solved by the confocal scanning laser microscope (CSLM), or simply, **confocal microscope.** The confocal microscope uses a laser beam to illuminate a specimen, usually one that has been fluorescently stained. A major component of the confocal microscope is an opening (that is, an aperture) placed above the objective lens. The aperture eliminates stray light from parts of the specimen that lie above and below the plane of focus (**figure 2.16**). Thus the only light used to create the image is from the plane of focus, and a much sharper image is formed.

Computers are integral to the process of creating confocal images. A computer interfaced with the confocal microscope receives digitized information from each plane in the specimen that is examined. This information can be used to create a composite image that is very clear and detailed (figure 2.15*c*) or to create a three-dimensional reconstruction of the specimen (figure 2.15*d*). Images of x-z plane cross sections of the specimen can also be

generated, giving the observer views of the specimen from three perspectives (figure 2.15*e*). Confocal microscopy has numerous applications. One is the study of biofilms, which can form on many different types of surfaces, including indwelling medical devices such as hip joint replacements. As shown in figure 2.15*f*, it is difficult to kill all cells in a biofilm. This makes biofilms a particular concern to the medical field because their formation on medical devices can result in infections that are difficult to treat. ▶▶❘ *Biofilms are common in nature (section 7.5)*

Retrieve, Infer, Apply

1. Briefly describe how dark-field, phase-contrast, differential interference contrast, epifluorescence, and confocal microscopes work and the kind of image or images provided by each. Give a specific use for each type.

2. GFP is used much like a fluorochrome is used. However, the two are not the same. Explain why this is so.

3. Do all specimens examined by fluorescence microscopy need to be stained with a fluorochrome? Explain.

2.3 Staining Specimens Helps to Visualize and Identify Microbes

After reading this section, you should be able to:

- Recommend a fixation process to use when the microbe is a bacterium or archaeon and when the microbe is a protist
- Plan a series of appropriate staining procedures to describe an unknown bacterium as fully as possible
- Compare what happens to Gram-positive and Gram-negative bacterial cells at each step of the Gram-staining procedure

As noted in section 2.2, specimens examined by bright-field microscopy are often fixed and stained before being examined. Such preparation serves to increase the visibility of the microorganisms, accentuate specific morphological features, and preserve them for future study. Importantly, some staining procedures help microbiologists identify the organism being examined.

Fixation

Stained cells seen in a microscope should resemble living cells as closely as possible. **Fixation** is the process by which the internal and external structures of specimens are preserved and fixed in position. It inactivates enzymes that might disrupt cell morphology and toughens cell structures so that they do not change during staining and observation. A microorganism usually is killed and attached firmly to the microscope slide during fixation.

There are two fundamentally different types of fixation: heat fixation and chemical fixation. **Heat fixation** is routinely used to observe bacteria and archaea. Typically, a film of cells (a smear) is gently heated. Heat fixation preserves overall morphology and inactivates enzymes. However, it also destroys proteins in subcellular structures, which may distort their appearance. **Chemical fixation** is used to protect fine cellular substructure as well as morphology. It is often used when examining larger, more delicate microorganisms such as protists. Chemical fixatives penetrate cells and react with cellular components, usually proteins and lipids, to render them inactive, insoluble, and immobile. Common fixative mixtures contain ethanol, acetic acid, mercuric chloride, formaldehyde, and glutaraldehyde.

Dyes and Simple Staining

The many types of dyes used to stain microorganisms have two features in common. First, they have **chromophore groups**—

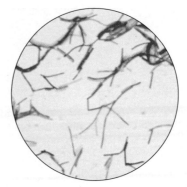
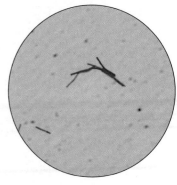

(a) Crystal violet stain of *Escherichia coli*

(b) Methylene blue stain of *Corynebacterium* sp.

Figure 2.17 Simple Stains.

chemical moieties with conjugated double bonds that give the dye its color. Second, they bind cells by ionic, covalent, or hydrophobic bonds.

Dyes that bind cells by ionic interactions are probably the most commonly used dyes. These ionizable dyes may be divided into two general classes based on the nature of their charged group: basic dyes and acidic dyes (**table 2.4**). The staining effectiveness of ionizable dyes may be altered by pH, since the nature and number of the charged moieties on cell components change with pH. Thus acidic dyes stain best under acidic conditions when proteins and many other molecules carry a positive charge; basic dyes are most effective at higher pH values.

Dyes that bind through covalent bonds or that have certain solubility characteristics are also useful. For instance, DNA can be stained by the Feulgen procedure in which the staining compound (Schiff's reagent) is covalently attached to the deoxyribose sugars of DNA. Sudan III (Sudan Black) selectively stains lipids because it is lipid soluble but does not dissolve in aqueous portions of the cell.

Microorganisms can be stained by **simple staining,** in which a single dye is used (**figure 2.17**). Simple staining's value lies in its ease of use. The fixed smear is covered with stain for a short time, excess stain is washed off with water, and the slide is blotted dry. Basic dyes such as crystal violet, methylene blue, and carbolfuchsin are frequently used in simple staining to determine the size, shape, and arrangement of bacterial and archaeal cells.

While most dyes directly stain the cell or object of interest, some dyes (e.g., India ink and nigrosin) are used in **negative staining.** In negative staining, the background is stained, not the

Table 2.4	Ionizable Dyes	
Type of Dye	**Examples**	**Characteristics**
Basic dyes	Methylene blue, basic fuchsin, crystal violet, safranin, malachite green	Have positively charged groups; bind to negatively charged molecules such as nucleic acids, many proteins, and the surfaces of bacterial and archaeal cells
Acidic dyes	Eosin, rose bengal, and acid fuchsin—possess groups such as carboxyls (—COOH) and phenolic hydroxyls (—OH).	In their ionized form, have a negative charge and bind to positively charged cell structures

Steps in Staining	State of Bacteria
Step 1: Crystal violet (primary stain) for 1 minute. Water rinse.	Cells stain purple.
Step 2: Iodine (mordant) for 1 minute. Water rinse.	Cells remain purple.
Step 3: Alcohol (decolorizer) for 10–30 seconds. Water rinse.	Gram-positive cells remain purple. Gram-negative cells become colorless.
Step 4: Safranin (counterstain) for 30–60 seconds. Water rinse. Blot dry.	Gram-positive cells remain purple. Gram-negative cells appear red.

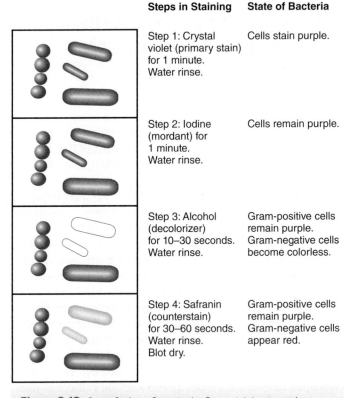

Figure 2.18 Gram Stain. Steps in the Gram-staining procedure.

MICRO INQUIRY *Why is the decolorization step considered the most critical in the Gram-staining procedure?*

cell; instead, the unstained cells appear as bright objects against a dark background.

Differential Staining

The **Gram stain,** developed in 1884 by the Danish physician Christian Gram, is the most widely employed staining method in bacteriology. It is routinely used in clinical laboratories and is an important first step in laboratory diagnosis of infectious disease, as we described in the chapter opening story. The Gram stain is an example of **differential staining**—procedures that distinguish organisms based on their staining properties. Use of the Gram stain divides most bacteria into two groups—those that stain Gram negative and those that stain Gram positive.

The Gram-staining procedure is illustrated in **figure 2.18.** In the first step, the smear is stained with a primary stain (crystal violet). This is followed by treatment with an iodine solution. Iodine functions as a **mordant,** a substance that helps bind the dye to a target molecule. Iodine increases the interaction between the cell and the dye so that the cell is stained more strongly. The smear is then decolorized by washing with alcohol or acetone. This step generates the differential aspect of the Gram stain; Gram-positive bacteria retain the crystal violet, whereas Gram-negative bacteria lose the crystal violet and become colorless. Finally, the smear is counterstained, usually

with safranin, which colors Gram-negative bacteria pink to red and leaves Gram-positive bacteria dark purple (**figure 2.19***a*).
▶▶ *There are two main types of bacterial cell walls (section 3.4)*

Acid-fast staining is another important differential staining procedure. It can be used to identify *Mycobacterium tuberculosis* and *M. leprae* (figure 2.19*b*), the pathogens responsible for tuberculosis and leprosy, respectively. These bacteria, as well as other mycobacteria, have cell walls containing lipids constructed from mycolic acids, a group of branched-chain hydroxy fatty acids, which prevent dyes from readily binding to the cells (*see figure 24.11*). A commonly used staining procedure, the cold Ziehl-Neelsen method, uses high concentrations of phenol and carbol fuchsin, as well as a wetting agent, to drive carbol fuchsin into mycobacterial cells. Once this dye has penetrated, the cells are not easily decolorized by acidified alcohol (acid-alcohol) and thus are said to be acid-fast. Non-acid-fast bacteria are decolorized by acid-alcohol and thus are stained another color by a counterstain.

Many differential staining procedures are based on detecting the presence of specific structures. **Endospore staining** uses heat to drive dye into an endospore, an exceptionally resistant structure produced by members of some Gram-positive bacterial genera (e.g., *Bacillus* and *Clostridium*). It can survive for long periods in an unfavorable environment and is called an endospore because it develops within the parent (mother) bacterial cell. Endospore morphology and location vary with species and often are valuable in identification. Endospores are not stained well by most dyes, but once stained, they strongly resist decolorization. This property is the basis of most endospore staining methods (figure 2.19*c*). ▶▶ *Bacterial endospores are a survival strategy (section 3.9); Class* Clostridia *(section 23.1); Class* Bacilli *(section 23.3)*

One of the simplest staining procedures is **capsule staining** (figure 2.19*d*), a technique that reveals the presence of capsules, a network usually made of polysaccharides that surrounds many bacteria and some fungi. Cells are mixed with India ink or nigrosin dye and spread out in a thin film on a slide. After air-drying, the cells appear as lighter bodies in the midst of a blue-black background because ink and dye particles cannot penetrate either the cell or its capsule. Thus capsule staining is an example of negative staining. The extent of the light region is determined by the size of the capsule and of the cell itself. There is little distortion of cell shape, and the cell can be counterstained for even greater visibility. ▶▶ *Capsules and slime layers (section 3.5)*

Flagella staining provides taxonomically valuable information about the presence and distribution pattern of flagella on bacterial and archaeal cells (figure 2.19*e; see also figure 3.41*). Their flagella are fine, threadlike organelles of locomotion that are so slender (about 10 to 30 nm in diameter), they can only be seen directly using an electron microscope (although bundles of flagella can be visualized by dark-field microscopy). To observe bacterial flagella with a light microscope, their thickness is increased by coating them with mordants such as tannic acid and potassium alum and then staining with a dye such as basic fuchsin. ▶▶ *Bacterial flagella (section 3.7)*

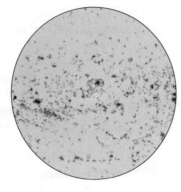

(a) Gram stain
Purple cells are Gram positive.
Red cells are Gram negative.

(b) Acid-fast stain
Red cells are acid-fast.
Blue cells are non-acid-fast.

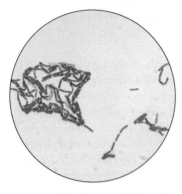

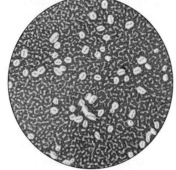

(c) Endospore stain of *Bacillus* sp., showing endospores (green) and vegetative cells (red)

(d) Capsule stain of bacteria

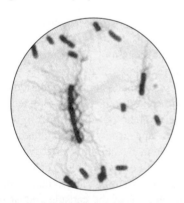

(e) Flagellar stain of *Proteus vulgaris.* A mordant was used to coat the flagella, which were then stained with a basic dye.

Figure 2.19 Differential Stains.

Retrieve, Infer, Apply

1. Describe the two general types of fixation. Which would you use when Gram staining a bacterium? Which would you use before observing the organelles of a protist?

2. Why would basic dyes be more effective under alkaline conditions?

3. Explain what happens to Gram-positive and Gram-negative bacterial cells at each step of the Gram-staining procedure.

4. What other differential staining procedures can be useful in identifying a bacterium? What information do they provide?

2.4 Electron Microscopes Use Beams of Electrons to Create Highly Magnified Images

After reading this section, you should be able to:

- Create a concept map, illustration, or table that compares transmission electron microscopes (TEMs) to light microscopes
- Decide when it would be best to examine a microbe by TEM, scanning electron microscopy (SEM), and electron cryotomography

For centuries the light microscope has been the most important instrument for studying microorganisms. However, even the best light microscopes have a resolution limit of about 0.2 μm, which greatly compromises their usefulness for detailed studies of many microorganisms (**figure 2.20**). Viruses, for example, are too small to be seen with light microscopes. Bacteria and archaea can be observed, but because they are usually only 1 to 2 μm in diameter, only their general shape and major morphological features are visible. The detailed internal structure of larger microorganisms also cannot be effectively studied by light microscopy. These limitations arise from the nature of visible light waves, not from any inadequacy of the light microscope itself. Electron microscopes have much greater resolution. Their use has transformed microbiology and added immeasurably to our knowledge. Here we briefly review the nature of the electron microscope and the ways in which specimens are prepared for observation.

Transmission Electron Microscope

Recall that the resolution of a light microscope increases with a decrease in the wavelength of the light it uses for illumination. In electron microscopes, electrons replace light as the illuminating beam. The electron beam can be focused, much as light is in a light microscope, but its wavelength is about 100,000 times shorter than that of visible light. Therefore electron microscopes have a practical resolution roughly 1,000 times better than the light microscope; with many electron microscopes, points closer than 0.5 nm can be distinguished, and the useful magnification is well over 100,000× (figure 2.20). The value of the electron microscope is evident on comparison of the photographs in **figure 2.21**: microbial morphology can now be studied in great detail.

A modern **transmission electron microscope** (**TEM**) is complex, but the basic principles behind its operation can be readily understood (**figure 2.22**). A heated tungsten filament in the electron gun generates a beam of electrons that is focused on the specimen by the condenser (**figure 2.23**). Since electrons cannot pass through a glass lens, doughnut-shaped electromagnets called magnetic lenses are used to focus the beam. The column containing the lenses and specimen must be under vacuum to obtain a clear image because electrons are deflected by collisions with air molecules. The specimen scatters some electrons, but those that pass through are used to form an enlarged image of the specimen on a fluorescent screen. A denser region

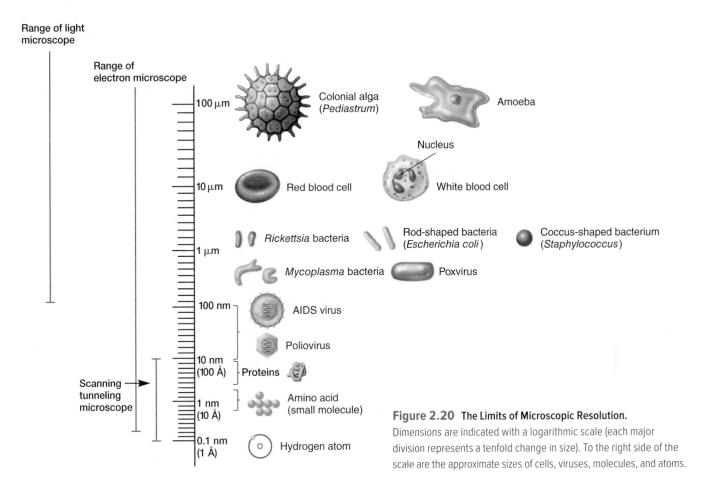

Figure 2.20 **The Limits of Microscopic Resolution.**
Dimensions are indicated with a logarithmic scale (each major division represents a tenfold change in size). To the right side of the scale are the approximate sizes of cells, viruses, molecules, and atoms.

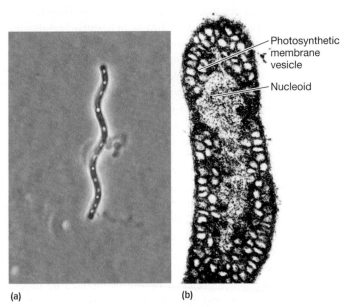

(a)   (b)

Figure 2.21 **Light and Electron Microscopy.**   A comparison of light and electron microscopic resolution. (a) The proteobacterium *Spirillum volutans* in phase-contrast light microscope (×1,000). (b) A thin section of another spiral-shaped proteobacterium *Rhodospirillum rubrum* in transmission electron microscope (×100,000).

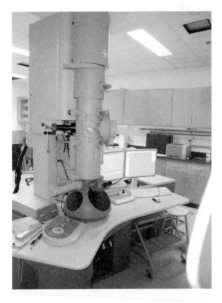

Figure 2.22 **A Transmission Electron Microscope.**   The electron gun is at the top of the central column, and the magnetic lenses are within the column. The image on the fluorescent screen may be viewed through a magnifier positioned over the viewing window. The camera is in a compartment below the screen.

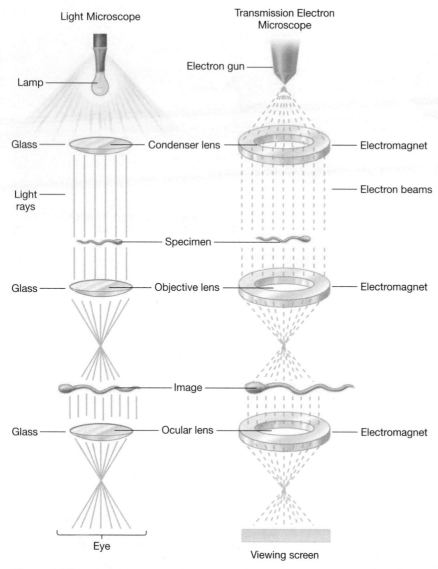

Figure 2.23 A Comparison of Light and Transmission Electron Microscopes.

in the specimen scatters more electrons and therefore appears darker in the image since fewer electrons strike that area of the screen; these regions are said to be "electron dense." In contrast, electron-transparent regions are brighter. The image can be captured as a photograph for a permanent record.

Table 2.5 compares some of the important features of light and transmission electron microscopes. The TEM has distinctive features that place harsh restrictions on the nature of samples that can be viewed and the means by which those samples must be prepared. Specimens must be viewed in a vacuum because electrons are deflected by air molecules. Only extremely thin slices (20 to 100 nm) of a specimen can be viewed because electron beams are easily absorbed and scattered by solid matter. To cut such a thin slice, the specimen must be embedded in a supportive matrix; the necessary support is provided by plastic. Specimens are prepared in the following way. After fixation with chemicals such as glutaraldehyde and osmium tetroxide to stabilize cell

structure, the specimen is dehydrated with organic solvents (e.g., acetone or ethanol). Complete dehydration is essential because most plastics used for embedding are not water soluble. Next the specimen is soaked in unpolymerized, liquid epoxy plastic until it is completely permeated, and then the plastic is hardened to form a solid block. Thin sections are cut from the block with a glass or diamond knife using a device called an ultramicrotome.

As with bright-field light microscopy, cells usually must be stained before they can be seen clearly with a TEM. The probability of electron scattering is determined by the density (atomic number) of atoms in the specimen. Biological molecules are composed primarily of atoms with low atomic numbers (H, C, N, and O), and electron scattering is fairly constant throughout an unstained cell or virus. Therefore specimens are prepared for observation by soaking thin sections with solutions of heavy metal salts such as lead citrate and uranyl acetate. The lead and uranium ions bind to structures in the specimen and make them more electron opaque, thus increasing contrast in the material. Heavy osmium atoms from the osmium tetroxide fixative also stain specimens and increase their contrast. The stained thin sections are then mounted on tiny copper grids and viewed.

Two other important techniques for preparing specimens are negative staining and shadowing. In negative staining, the specimen is spread out in a thin film with either phosphotungstic acid or uranyl acetate. Just as in negative staining for light microscopy, heavy metals do not penetrate the specimen but render the background dark, whereas the specimen appears bright. Negative staining is an excellent way to study the structure of virus particles, bacterial gas vacuoles, and other similar objects (**figure 2.24a**). In shadowing, a specimen is coated with a thin film of platinum or other heavy metal by evaporation at an angle of about 45° from horizontal so that the metal strikes the microorganism on only one side. In one commonly used imaging method, the area coated with metal appears dark in photographs, whereas the uncoated side and the shadow region created by the object are light (figure 2.24b). This technique is particularly useful in studying virus particle morphology, bacterial and archaeal flagella, and DNA.

The shapes of organelles within cells can be observed by TEM if specimens are prepared by the freeze-etching procedure. When cells are rapidly frozen in liquid nitrogen, they become very brittle and can be broken along lines of greatest weakness, usually down the middle of internal membranes (**figure 2.25**). The exposed surfaces are then shadowed and coated with layers of platinum and carbon to form a replica of

Table 2.5	Characteristics of Light and Transmission Electron Microscopes	
Feature	**Light Microscope**	**Transmission Electron Microscope**
Highest practical magnification	About 1,000–1,500	Over 100,000
Best resolution[1]	0.2 μm	0.5 nm
Radiation source	Visible light	Electron beam
Medium of travel	Air	High vacuum
Type of lens	Glass	Electromagnet
Source of contrast	Differential light absorption	Scattering of electrons
Focusing mechanism	Adjust lens position mechanically	Adjust current to the magnetic lens
Method of changing magnification	Switch the objective lens or eyepiece	Adjust current to the magnetic lens
Specimen mount	Glass slide	Metal grid (usually copper)

1 The resolution limit of a human eye is about 0.2 mm.

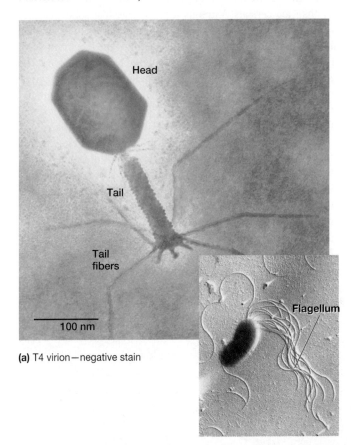

(a) T4 virion—negative stain

(b) *P. fluorescens*—after shadowing

Figure 2.24 Staining Microorganisms for the TEM. (a) T4 is a virus that infects *Escherichia coli*. (b) *Pseudomonas fluorescens* with its polar flagella.

MICRO INQUIRY *Why are all electron micrographs black and white (although they are sometimes artificially colorized after printing)?*

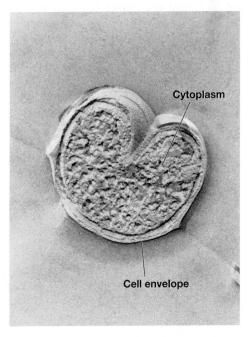

Figure 2.25 Example of Freeze-Etching. A freeze-etched preparation of the bacterium *Nitrospira* sp.

the surface. After the specimen has been removed chemically, this replica is studied in the TEM, providing a detailed view of intracellular structure. An advantage of freeze-etching is that it minimizes the danger of artifacts because the cells are frozen quickly, rather than being subjected to chemical fixation, dehydration, and embedding in plastic.

Scanning Electron Microscope

Transmission electron microscopes form an image from radiation that has passed through a specimen. The **scanning electron microscope (SEM)** produces an image from electrons released from atoms on an object's surface. The SEM has been used to examine the surfaces of microorganisms in great detail; many SEMs have a resolution of 7 nm or less.

Specimen preparation for SEM is relatively easy, and in some cases, air-dried material can be examined directly. However, microorganisms usually must first be fixed, dehydrated, and dried to preserve surface structure and prevent collapse of the cells when they are exposed to the SEM's vacuum. Before viewing, dried samples are mounted and coated with a thin layer of metal to prevent the buildup of an electrical charge on the surface and to give a better image.

To create an image, the SEM scans a narrow, tapered electron beam back and forth over the specimen (**figure 2.26**). When the beam strikes a particular area, surface atoms discharge a tiny shower of electrons called secondary electrons, and these are trapped by a detector. Secondary electrons strike a material in the detector that emits light when struck by electrons (the material is called a scintillator). The flashes of light are converted to an electrical current and amplified by a photomultiplier. The signal is digitized and sent to a computer, where it can be viewed.

The number of secondary electrons reaching the detector depends on the nature of the specimen's surface. When the electron beam strikes a raised area, a large number of secondary electrons enter the detector; in contrast, fewer electrons escape a depression in the surface and reach the detector. Thus raised areas appear lighter on the screen and depressions are darker. A realistic three-dimensional image of the microorganism's surface results (**figure 2.27**). The actual in situ location of microorganisms in ecological niches such as the human skin and the lining of the gut also can be examined.

Electron Cryotomography

Beginning approximately 40 years ago, a series of advances in electron microscopy paved the way for the development of **electron cryotomography,** a technique that since the 1990s has been providing exciting insights into the structure and function of cells and viruses. *Cryo-* refers to sample preparation and visualization. Samples are prepared by rapidly plunging the specimen into an extremely cold liquid (e.g., ethane); the sample is kept frozen while being examined. Rapid freezing of the sample forms vitreous ice rather than ice crystals. Vitreous ice is a glasslike solid that preserves the native state of structures and immobilizes the specimen so that it can be viewed in the high vacuum of the electron microscope. *Tomography* refers to the method used to create images. The object is viewed from many directions, referred to as a tilt series. The individual images are recorded and processed by computer programs, and finally merged to form a three-dimensional reconstruction of the object. Three-dimensional views, slices, and other types of representations of the object can be derived from the reconstruction (**figure 2.28**). The ultrastructure of bacterial and archaeal cells has been the focus of numerous studies using electron cryotomography. Some of these studies have revealed new cytoskeletal elements, such as those associated with magnetosomes, the inclusions used by some bacteria to orient themselves in magnetic fields (*see figure 3.37*). ▶▶▎ *Inclusions (section 3.6); Bacterial flagella (section 3.7)*

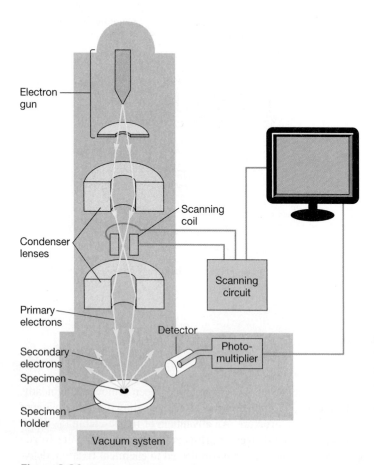

Figure 2.26 The Scanning Electron Microscope.

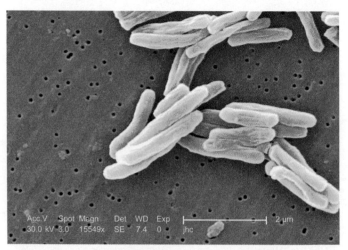

Figure 2.27 Scanning Electron Micrograph of *Mycobacterium tuberculosis.* Colorized image (×15,549).

(a)

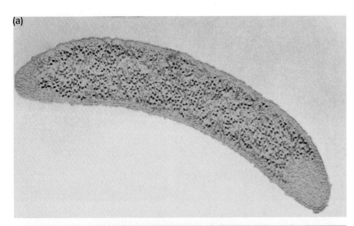

(b)

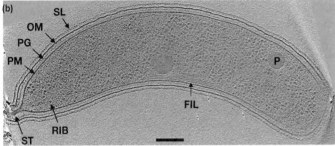

Figure 2.28 TEM and Electron Cryotomography. A comparison of
(a) a thin section of a *Caulobacter crescentus* cell that has been prepared
using conventional TEM procedures and (b) a central slice from a three-
dimensional reconstruction of an intact *C. crescentus* cell. PM, plasma
membrane; FIL, a bundle of filaments; OM, outer membrane; P, a presumed
inclusion; PG, peptidoglycan; RIB, ribosome; SL, S-layer; ST, stalk.

Retrieve, Infer, Apply

1. Why does the transmission electron microscope have much greater
resolution than the light microscope?
2. Describe in general terms how a TEM functions. Why must a TEM use
high vacuum and very thin sections?
3. Under what circumstances would it be desirable to prepare specimens
for a TEM by negative staining? Shadowing? Freeze-etching?
4. How does a scanning electron microscope operate, and in what way
does its function differ from that of a TEM? Which aspects of
morphology are studied using an SEM?

2.5 Scanning Probe Microscopy Can Visualize Molecules and Atoms

After reading this section, you should be able to:

■ Distinguish scanning tunneling from atomic force microscopes in
terms of how they create images and their uses
■ Evaluate light microscopy, electron microscopy, and scanning probe
microscopy in terms of their uses, resolution, and the quality of the
images created

Among the most powerful microscopes are **scanning probe
microscopes (SPMs).** These microscopes measure surface features

of an object by moving a sharp probe over the object's surface. One
type of SPM is the **scanning tunneling microscope,** which was
invented in 1980. It can achieve magnifications of 100 million
times, and it allows scientists to view atoms on the surface of a
solid. The scanning tunneling microscope has a needlelike probe
with a point so sharp that often there is only one atom at its tip. The
probe is lowered toward the specimen surface until its electron
cloud just touches that of the surface atoms. If a small voltage is
applied between the tip and specimen, electrons flow through a
narrow channel in the electron clouds. This tunneling current, as it
is called, is extraordinarily sensitive to distance and will decrease
about a thousandfold if the probe is moved away from the surface
by a distance equivalent to the diameter of an atom.

The arrangement of atoms on the specimen surface is determined by moving the probe tip back and forth over the surface
while keeping the probe at a constant height above the specimen.
As the tip moves up and down while following the surface con-
tours, its motion is recorded and analyzed by a computer to create
an accurate three-dimensional image of the surface atoms. The
surface map can be displayed on a computer screen or plotted on
paper. The resolution is so great that individual atoms are observed easily. Even more exciting is that the microscope can examine objects when they are immersed in water. Therefore it can
be used to study biological molecules such as DNA (**figure 2.29**).
The microscope's inventors, Gerd Binnig and Heinrich Rohrer
(1933–2013), shared the 1986 Nobel Prize in Physics for their
work, together with Ernst Ruska, the designer of the first transmission electron microscope.

A second type of SPM is the **atomic force microscope,** which
moves a sharp probe over the specimen surface while keeping the
distance between the probe tip and the surface constant. It does this
by exerting a very small amount of force on the tip, just enough to
maintain a constant distance but not enough force to damage the
surface. The vertical motion of the tip usually is followed by measuring the deflection of a laser beam that strikes the lever holding

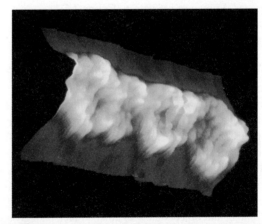

Figure 2.29 Scanning Tunneling Microscopy of DNA. The DNA
double helix with approximately three turns shown (false color; ×2,000,000).

MICRO INQUIRY *Compare the magnification of this image with that
shown in figure 2.27 (SEM). Which has the highest magnification?*

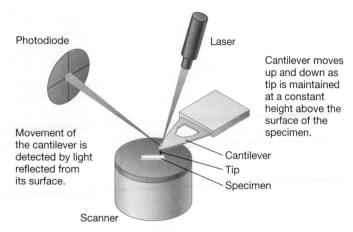

Figure 2.30 **The Basic Elements of an Atomic Force Microscope.** The tip used to probe the specimen is attached to a cantilever. As the probe passes over the "hills and valleys" of the specimen's surface, the cantilever is deflected vertically. A laser beam directed at the cantilever is used to monitor these vertical movements. Light reflected from the cantilever is detected by the photodiode and used to generate an image of the specimen.

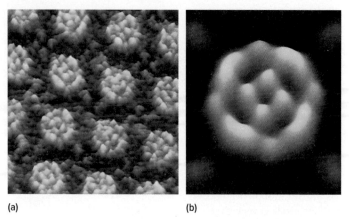

(a)   (b)

Figure 2.31 **The Membrane Protein Aquaporin Visualized by Atomic Force Microscopy.** Aquaporin is a membrane-spanning protein that allows water to move across the membrane. (a) Each circular structure represents the surface view of a single aquaporin protein. (b) A single aquaporin molecule observed in more detail and at higher magnification.

the probe (**figure 2.30**). Unlike the scanning tunneling microscope, the atomic force microscope can be used to study surfaces that do not conduct electricity well. The atomic force microscope has been used to study the interactions of proteins, to follow the behavior of living bacteria and other cells, and to visualize membrane proteins such as aquaporins (**figure 2.31**).

Retrieve, Infer, Apply

1. Briefly describe each of the most popular versions of the scanning probe microscope. What are these microscopes used for?
2. Scanning probe microscopy can be applied to samples that have not been dehydrated, while electron microscopy cannot. Why is this an important advantage?

Key Concepts

2.1 Lenses Create Images by Bending Light

- A light ray moving from air to glass or vice versa is bent in a process known as refraction (**figure 2.1**).
- Lenses focus light rays at a focal point and magnify images (**figure 2.2**).

2.2 There Are Several Types of Light Microscopes

- In a compound microscope such as the bright-field microscope, the primary image is an enlarged image formed by the objective lens. The primary image is further enlarged by the ocular lens to yield the final image (**figure 2.3**).
- Microscope resolution increases as the wavelength of radiation used to illuminate the specimen decreases. The maximum resolution of a light microscope is about 0.2 μm.
- The dark-field microscope uses only refracted light to form an image, and objects glow against a black background (**figure 2.6**).
- The phase-contrast microscope converts variations in the refractive index into changes in light intensity and thus makes colorless, unstained cells visible (**figures 2.8–2.10**).

- The differential interference contrast microscope uses two beams of light to create high-contrast images of live specimens (**figure 2.11**).
- The fluorescence microscope illuminates a fluorochrome-labeled specimen and forms an image from its fluorescence (**figure 2.12**).
- The confocal microscope is used to study thick, complex specimens. It creates an image by using only the light emanating from the plane of focus, while blocking out light from above and below the plane of focus (**figures 2.15** and **2.16**).

2.3 Staining Specimens Helps to Visualize and Identify Microbes

- Specimens are often fixed and stained before viewing them in the bright-field microscope. There are two fixation methods: heat fixation and chemical fixation.
- Most dyes are either positively charged basic dyes or negatively charged acidic dyes that bind to ionized parts of cells.
- In simple staining, a single dye is used to stain microorganisms (**figure 2.17**).

- Differential staining procedures such as Gram and acid-fast staining distinguish between microbial groups by staining them differently (**figures 2.18** and **2.19a,b**). Other differential staining techniques are specific for particular structures such as endospores, bacterial capsules, and flagella (**figure 2.19c–e**).

2.4 Electron Microscopes Use Beams of Electrons to Create Highly Magnified Images

- The transmission electron microscope (TEM) uses magnetic lenses to form an image from electrons that have passed through a very thin section of a specimen (**figure 2.23**). Resolution is high because the wavelength of a beam of electrons is very short.
- Specimens for TEM are usually prepared by methods that increase contrast. Specimens can be stained by treatment with solutions of heavy metals such as osmium tetroxide, uranium, and lead. They can also be prepared for TEM by negative staining, shadowing with metal, or freeze-etching (**figures 2.24** and **2.25**).

- The scanning electron microscope is used to study external surface features of microorganisms (**figure 2.26**).
- Electron cryotomography freezes specimens rapidly, keeps them frozen while being examined, and creates images from a series of directions that are combined and processed to form a three-dimensional reconstruction of the object (**figure 2.28**).

2.5 Scanning Probe Microscopy Can Visualize Molecules and Atoms

- Scanning probe microscopes reach very high magnifications that allow scientists to observe biological molecules (**figures 2.29** and **2.31**).
- Scanning tunneling microscopy enables the visualization of molecular surfaces using electron interaction between the probe and the specimen, whereas atomic force microscopy can scan the surface of molecules that do not conduct electricity well (**figure 2.30**).

Compare, Hypothesize, Invent

1. If you prepared a sample of a specimen for light microscopy, stained it with the Gram stain, and failed to see anything when you looked through your light microscope, list the things that you may have done incorrectly.

2. In a journal article, find an example of a light micrograph, a scanning or transmission electron micrograph, or a confocal image. Discuss why the figure was included in the article and why that particular type of microscopy was the method of choice for the research. What other figures would you like to see used in this study? Outline the steps that the investigators would take to obtain such photographs or figures.

3. How are freeze etching and electron cryotomography similar? How do they differ? What is the advantage of electron cryotomography?

4. STED microscopes were first developed about 10 years ago. They are modifications of confocal microscopes that use two highly focused beams of light rather than one as in typical confocal microscopy. The first beam uses a wavelength that excites the fluorophore with which a specimen is labeled. This creates a fluorescent spot where the fluorophore is located. The second beam is a hollow cone. It strikes the outermost edges of the fluorescent spot illuminated by the first beam. The wavelength used by the second beam returns the fluorophore to its ground state, thus only the center of the fluorescent spot is visible. The result is that the resolution of STED microscopes is greater and objects smaller than 200 nm can be resolved. In 2010 a STED microscope was used to observe the mechanism by which human immunodeficiency virus (HIV) particles are transferred from dendritic cells to T cells. These are important immune system cells and play critical roles in the development of HIV infection. Why was STED microscopy used rather than TEM or electron cryotomography?

Read the original paper: Felts, R. L., et al. 2010. 3D visualization of HIV transfer at the virological synapse between dendritic cells and T cells. *Proc. Natl. Acad. Sci. USA.* 107:13336.

Learn More

shop.mheducation.com   Enhance your study of this chapter with interactive study tools and practice tests. Also ask your instructor about the resources available through Connect, including adaptive learning tools and animations.

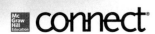

3

Bacterial Cell Structure

Hooking Up

Each year over 100 million people around the world become infected with *Neisseria gonorrhoeae*, the bacterium that causes gonorrhea. This troubling statistic is made even more disturbing by the increasing resistance of the bacterium to the antibiotics used to treat the disease. In males, infection is usually readily detected, but for females, infection is often asymptomatic and can lead to serious consequences such as pelvic inflammatory disease (PID) and sterility. These concerns have led scientists to consider methods for preventing infection. One method is to block transmission. Unfortunately, relatively little is known about the transmission process except that it occurs during sexual intercourse and that numerous hairlike structures (called pili) covering the surface of the bacterium play a role in establishing infection. The bacterium uses pili for a type of movement called twitching motility and for adherence to surfaces such as the sperm and epithelial cells of its human host. It has long been thought that by attaching to sperm cells the bacterium could hitch a ride to the female during sexual intercourse. This explained transmission from male to female. However, it did not clarify how transmission from female to male occurs.

In 2014 a study reported that exposure of *N. gonorrhoeae* to seminal fluid increases its twitching motility and enhances formation of small aggregates of bacteria. These changes promote infection of host epithelial cells and, in turn, increase the likelihood that the bacterium will encounter epithelial tissue of either partner during sexual intercourse. Importantly, this report helps explain how transmission from female to male might occur. The study also determined that seminal fluid proteins caused these changes and suggested that seminal fluid proteins alter the morphology and function of pili. In particular these proteins cause bundles of pili to separate into single filaments, enhancing the interaction of bacterial cells with each other and with host surfaces. Thus, the bacterium sensed the presence of seminal fluid proteins and responded to them so that it could better effect transmission and colonization.

As this story illustrates, even small, seemingly simple organisms such as bacteria can exhibit complex behaviors. To understand these amazing microbes, we must first examine their cell structure and begin to relate it to the functions they carry out. As we consider bacterial cell structure, it is important to remember that only about 1% of bacterial species have been cultured. Of the cultivated species, only a few have been studied in great detail. From this small sample, many generalizations are made, and it is presumed that most other bacteria are like the well-studied model organisms. However, part of the wonder and fun of science is that nature is full of surprises. As the biology of more and more bacteria is analyzed, our understanding of them may change in interesting and exciting ways.

Readiness Check:
Based on what you have learned previously, you should be able to:

✔ Describe the application of small subunit (SSU) rRNA analysis to the establishment of the three domain classification system proposed by Carl Woese (section 1.2)
✔ Identify the following structures or regions of a plant or animal cell and describe their functions: cell wall, plasma membrane, cytoplasm, mitochondria, chloroplasts, and ribosomes
✔ Define and give examples of essential nutrients; describe how they are used by cells

3.1 Use of the Term "Prokaryote" Is Controversial

After reading this section, you should be able to:

■ List the characteristics originally used to describe prokaryotic cells
■ Form an opinion on the "prokaryote" controversy using current evidence about bacterial cells

Bacteria and archaea have long been lumped together and referred to as prokaryotes. Although the term was first introduced early in the twentieth century, the concept of a prokaryote was not fully outlined until 1962, when R. Stanier and C. B. van Niel described prokaryotes in terms of what they lacked in comparison to eukaryotic cells. For instance, Stanier and van Niel pointed out that prokaryotes lack a membrane-bound nucleus, a cytoskeleton, membrane-bound organelles, and internal membranous structures such as the endoplasmic reticulum and Golgi apparatus.

Since the 1960s, biochemical, genetic, and genomic analyses, coupled with improved methods for imaging bacterial and archaeal cells, have shown that *Bacteria* and *Archaea* are distinct taxa. Because of this and other discoveries, some microbiologists question our traditional ways of thinking about prokaryotes. For instance, some members of the bacterial phylum *Planctomycetes* have their genetic material enclosed in a membrane; other members of this interesting taxon have a membrane-bound organelle called the anammoxosome. This organelle is the site of anoxic ammonia oxidation, an unusual metabolic process that is important in the nitrogen cycle. Other discoveries include the finding that cytoskeletal elements are widespread in members of both domains and that some bacteria have extensive intracytoplasmic membranes. Because of these discoveries, some microbiologists think the term prokaryote should be abandoned. ▶▶⃒ *Phylum* Planctomycetes *(section 21.5); Nitrogen-cycle (section 28.1)*

The current controversy creates a dilemma for both the textbook author and the student of microbiology—to use or not to use the term prokaryote. However, it also illustrates that microbiology is an exciting, dynamic, and rapidly changing field of study. All of us are waiting to see the outcome of this debate. Until then, we will try to limit our use of the term prokaryote and to be as explicit as possible about which characteristics are associated with members of *Bacteria*, which with members of *Archaea*, and which with members of both taxa.

3.2 Bacteria Are Diverse but Share Some Common Features

After reading this section, you should be able to:

■ Distinguish a typical bacterial cell from a typical plant or animal cell in terms of cell shapes and arrangements, size, and cell structures
■ Discuss the factors that determine the size and shape of a bacterial cell

Much of this chapter is devoted to a discussion of individual cell components. Therefore a preliminary overview of the features common to many bacterial cells is in order. We begin by considering overall cell morphology and then move to cell structures.

Shape, Arrangement, and Size

It might be expected that bacterial cells, being small and relatively simple, would be uniform in shape and size. This is

not the case, as the microbial world offers considerable variety in terms of morphology. However, the two most common shapes are cocci and rods (**figure 3.1**). **Cocci** (s., **coccus**) are roughly spherical cells. They can exist singly or can be associated in characteristic arrangements that can be useful in their identification. **Diplococci** (s., **diplococcus**) arise when cocci divide and remain together to form pairs. Long chains of cocci result when cells adhere after repeated divisions in one plane; this pattern is seen in the genera *Streptococcus, Enterococcus,* and *Lactococcus* (figure 3.1*a*). Members of the genus *Staphylococcus* divide in random planes to generate irregular, grapelike clusters (figure 3.1*b*). Divisions in two or three planes can produce symmetrical groupings of cocci. Bacteria in the genus *Micrococcus* often divide in two planes to form square groups of four cells called tetrads. In the genus *Sarcina,* cocci divide in three planes, producing cubical packets of eight cells.

Bacillus megaterium is an example of a bacterium with a **rod** shape (figure 3.1*c*). Rods, sometimes called **bacilli** (s., **bacillus**), differ considerably in their length-to-width ratio, the coccobacilli being so short and wide that they resemble cocci. The shape of the rod's end often varies between species and may be flat, rounded, cigar-shaped, or bifurcated. Although many rods occur singly, some remain together after division to form pairs or chains (e.g., *Bacillus megaterium* is found in long chains).

There are several less common cell shapes and arrangements. **Vibrios** are comma-shaped (**figure 3.2***a*). **Spirilla** are rigid, spiral-shaped cells (figure 3.2*b*). Many have tufts of flagella at one or both ends. **Spirochetes** are flexible, spiral-shaped bacteria that have a unique, internal flagellar arrangement (figure 3.2*c*). These bacteria are distinctive in other ways, and all belong to a single phylum, *Spirochaetes.* Some bacteria form stalks (e.g., *Caulobacter crescentus*) (figure 3.2*d*). Other bacteria are **pleomorphic,** being variable in shape and lacking a single, characteristic form. ▶▶⃒ *Phylum* Spirochaetes *(section 21.8);* Caulobacteraceae *and* Hyphomicrobiaceae *bacteria reproduce in unusual ways (section 22.1)*

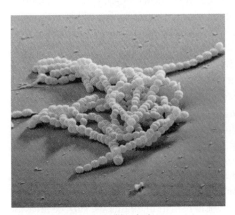

(a) *S. agalactiae*—cocci in chains

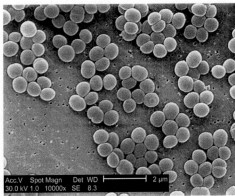

(b) *S. aureus*—cocci in clusters

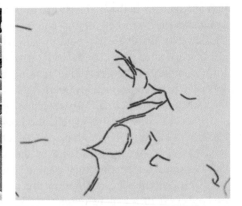

(c) *B. megaterium*—rods in chains

Figure 3.1 Cocci and Rods Are the Most Common Bacterial Shapes. (a) *Streptococcus agalactiae*, the cause of Group B streptococcal infections; color-enhanced scanning electron micrograph (×4,800). (b) *Staphylococcus aureus*; color-enhanced scanning electron micrograph. (c) *Bacillus megaterium*, Gram stain (×1,000).

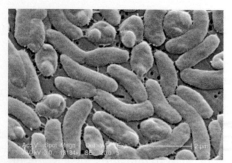

(a) *V. vulnificus*—comma-shaped vibrios

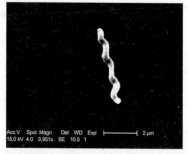

(b) *C. jejuni*—Spiral shaped

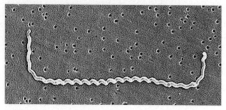

(c) *Leptospira interrogans*—a spirochete

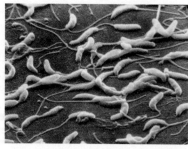

(d) *C. Crescentus*—a stalked bacterium

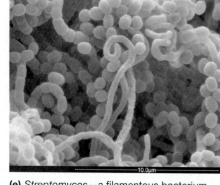

(e) *Streptomyces*—a filamentous bacterium

(f) *C. crocatus* fruiting body

Figure 3.2 Other Cell Shapes and Aggregations. (a) *Vibrio vulnificus*, scanning electron micrograph (SEM, X13,184). (b) *Campylobacter jejuni*, SEM. (c) *Leptospira interrogans*, the spirochete that causes the waterborne disease leptospirosis. (d) *Caulobacter crescentus*, SEM. (e) *Streptomyces* sp., SEM. (f) Fruiting body of the myxobacterium *Chondromyces crocatus*. The fruiting body is composed of thousands of cells.

Some bacteria can be thought of as multicellular. Many actinobacteria form long filaments called hyphae. The hyphae form a network called a **mycelium** (figure 3.2*e*), and in this sense, they are similar to eukaryotic filamentous fungi. Many cyanobacteria, a group of photosynthetic bacteria, are also filamentous. Being filamentous allows some degree of differentiation among cells in the filament. For instance, some filamentous cyanobacteria form heterocysts within the filament; these are specialized cells that carry out nitrogen fixation. Myxobacteria are of particular note. These bacteria sometimes aggregate to form complex structures called fruiting bodies (figure 3.2*f*). ▶▶| *Order* Streptomycetales *(section 24.1); Phylum* Cyanobacteria *(section 21.4); Order* Myxococcales *(section 22.4)*

Escherichia coli is an excellent representative of the average size of bacteria. This rod-shaped bacterium is 1.1 to 1.5 μm wide by 2.0 to 6.0 μm long. However, the size range of bacterial cells extends far beyond this average (**figure 3.3**). Near the small end of the size continuum are members of the genus *Mycoplasma* (0.3 μm in diameter). At the other end of the continuum are bacteria such as some spirochetes, which can reach 500 μm in length, and the cyanobacterium *Oscillatoria*,

which is about 7 μm in diameter (the same diameter as a red blood cell). Some bacteria are huge by "bacterial standards." For instance, *Epulopiscium fishelsoni* lives in the intestine of the brown surgeonfish *(Acanthurus nigrofuscus)* and grows as large as 600 by 80 μm, a little smaller than a printed hyphen and clearly larger than the well-known eukaryote *Paramecium* (**figure 3.4**). An even larger bacterium, *Thiomargarita namibiensis*, has been discovered in ocean sediment *(see figure 22.25)*. Thus a few bacteria are much larger than the average eukaryotic cell (typical plant and animal cells are around 10 to 50 μm in diameter).

The variety of sizes and shapes exhibited by bacteria raises a fundamental question. What causes a bacterial species to have a particular size and shape? Although far from being answered, recent discoveries have fueled a renewed interest in this question, and it is clear that size and shape determination are related and have been selected for during the evolutionary history of each bacterial species. For many years it was thought that microbes had to be small because being small increases the surface area-to-volume ratio (S/V ratio; **figure 3.5**). As this ratio increases, the uptake of nutrients and the diffusion of these and other molecules

Specimen	Approximate diameter or width × length in nm
Oscillatoria Red blood cell	7,000
E. coli	1,300 × 4,000
Streptococcus	800–1,000
Poxvirus	230 × 320
Influenza virus	85
T2 *E. coli* bacteriophage	65 × 95
Tobacco mosaic virus	15 × 300
Poliomyelitis virus	27

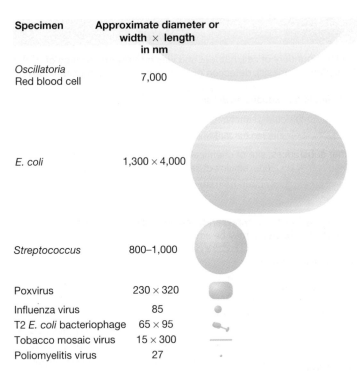

Figure 3.3 Sizes of Bacteria Relative to a Red Blood Cell and Viruses. Recall that 1,000 nm = 1 µm. Thus *E. coli* is 1.3 × 4 µm.

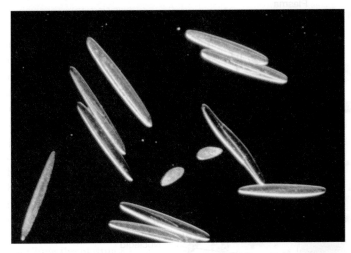

Figure 3.4 **A Giant Bacterium.** This photograph shows *Epulopiscium fishelsoni* dwarfing the paramecia, which are protozoa. *E. fishelsoni* cells are about 0.53 mm long. Phase-contrast microscopy.

within the cell become more efficient, which in turn facilitates a rapid growth rate. Shape affects the S/V ratio. A rod with the same volume as a coccus has a higher S/V ratio than does the coccus. This means that a rod can have greater nutrient flux across its plasma membrane. However, the discovery of *E. fishelsoni* demonstrates that bacteria can be very large. For bacteria to be large, they must have other characteristics that maximize their S/V ratio, or their size must be beneficial in some way. For instance,

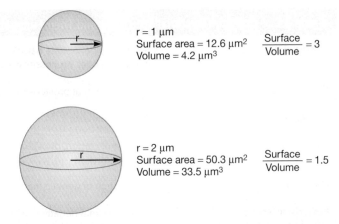

$r = 1$ µm
Surface area = 12.6 µm^2
Volume = 4.2 µm^3
$\dfrac{\text{Surface}}{\text{Volume}} = 3$

$r = 2$ µm
Surface area = 50.3 µm^2
Volume = 33.5 µm^3
$\dfrac{\text{Surface}}{\text{Volume}} = 1.5$

Figure 3.5 **The Surface-to-Volume Ratio Is an Important Determinant of Cell Size.** Surface area is calculated by the formula $4\pi r^2$. Volume is calculated by the formula $4/3\pi r^3$. Shape also affects the S/V ratio; rods with the same diameter as a coccus have a greater S/V ratio.

E. fishelsoni has a highly convoluted plasma membrane, which increases its S/V ratio. In addition, large cells are less likely to be eaten by predatory protists. Cells that are filamentous, have stalks, or are oddly shaped are also less susceptible to predation. ⟳ *Prokaryotic Cell Shapes*

Cell Organization

Structures often observed in bacterial cells are summarized and illustrated in **table 3.1** and **figure 3.6**. Note that no single bacterium possesses all of these structures at all times. Some are found only in certain cells in certain conditions or in certain phases of the life cycle. However, there are several common features of bacterial cell structure.

Bacterial cells are often surrounded by several layers, which are collectively called the cell envelope. The most common cell envelope layers are the plasma membrane, cell wall, and capsule or slime layer. The innermost layer of the cell envelope is the plasma membrane, which surrounds the cytoplasm. Most bacteria have a chemically complex cell wall, which covers the plasma membrane. Many bacteria are surrounded by a capsule or slime layer external to the cell wall. Because most bacteria do not contain internal, membrane-bound organelles, their interior appears morphologically simple. The genetic material is localized in a discrete region called the nucleoid and usually is not separated from the surrounding cytoplasm by membranes. Ribosomes and larger masses called inclusions are scattered about the cytoplasm. Finally, many bacteria use flagella for locomotion.

In the remaining sections of this chapter, we describe the major structures observed in bacterial cells in more detail. We begin with the cell envelope. We then proceed inward to consider structures located within the cytoplasm. Next the discussion moves outward, to a variety of appendages that are involved in attachment to surfaces, motility, or both. Finally, we consider the bacterial endospore.

proteins lying within the membrane lipid bilayer. ◄◄ *Electron microscopes use beams of electrons to create highly magnified images (section 2.4); Scanning probe microscopy can visualize molecules and atoms (section 2.5)*

The chemical nature of membrane lipids is critical to their ability to form bilayers. Most membrane-associated lipids (e.g., the phospholipids shown in figure 3.7) are **amphipathic:** they are structurally asymmetric, with polar and nonpolar ends (**figure 3.8**). The polar ends interact with water and are **hydro-philic;** the nonpolar **hydrophobic** ends are insoluble in water and tend to associate with one another. In aqueous environments, amphipathic lipids can interact to form a bilayer. The outer surfaces of the bilayer are hydrophilic, whereas hydrophobic ends are buried in the interior away from the surrounding water (figure 3.7). ►► *Lipids (appendix I)*

Two types of membrane proteins have been identified based on their ability to be separated from the membrane. **Peripheral**

membrane proteins are loosely connected to the membrane and can be easily removed (figure 3.7). They are soluble in aqueous solutions and make up about 20 to 30% of total membrane protein. The remaining proteins are **integral membrane proteins.** These are not easily extracted from membranes and are insoluble in aqueous solutions when freed of lipids. Integral membrane proteins, like membrane lipids, are amphipathic; their hydrophobic regions are buried in the lipid while the hydrophilic portions project from the membrane surface (figure 3.7).

Integral membrane proteins carry out some of the most important functions of the membrane. Many are transport proteins used to move materials either into or out of the cell. Others are involved in energy-conserving processes, such as the proteins found in electron transport chains. Those integral membrane proteins with regions exposed to the outside of the cell enable the cell to interact with its environment. ►► *Proteins (appendix I)*

Although most aspects of the fluid mosaic model are well supported by experimentation, some are being questioned. The fluid mosaic model suggests that membrane lipids are homogeneously distributed and that integral membrane proteins are free to move laterally within the membrane. However, the presence of microdomains enriched for certain lipids and the observation that some integral proteins are present at only certain sites do not support this view. Although the term mosaic initially referred to the clusters of proteins embedded in a homogeneous lipid bilayer, it may be more accurate to use the term to refer to the patchwork of lipid microdomains found in membranes. Research is ongoing to determine the physiological role of these microdomains.

Bacterial Plasma Membranes Contain Lipids Not Found in Eukaryotic Membranes

Bacterial membranes are similar to eukaryotic membranes in that they are lipid bilayers and many of their amphipathic lipids are phospholipids (figure 3.8). The plasma membrane is very dynamic: the lipid composition varies with environmental temperature in such a way that the membrane remains fluid during growth. For example, bacteria growing at lower temperatures have more unsaturated fatty acids in their membrane phospholipids; that is, there are one or more double covalent bonds in the long hydrocarbon chain. At higher temperatures, their phospholipids have more saturated fatty acids—those in which the carbon atoms are connected only with single covalent bonds. Other factors also affect the lipid composition of membranes. For instance, some pathogens change the lipids in their plasma membranes to protect themselves from antimicrobial peptides produced by the immune system. ►► *Environmental factors affect microbial growth (section 7.4); Antimicrobial peptides (section 33.3)*

Bacterial membranes usually differ from eukaryotic membranes in lacking sterols (steroid-containing lipids) such as cholesterol (**figure 3.9a**). However, many bacterial membranes contain hopanoids, which are similar to, but distinct from, steroids (figure 3.9b). **Hopanoids** are synthesized from the same precursors

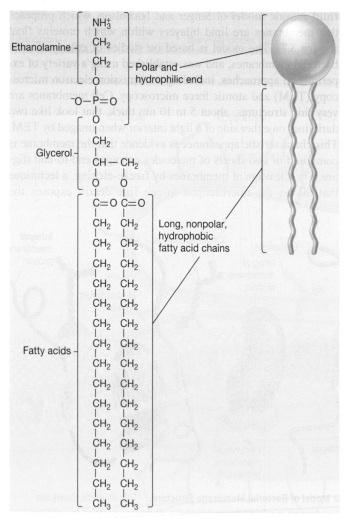

Figure 3.8 The Structure of a Phospholipid. Phosphatidylethanolamine, a phospholipid often found in bacterial membranes.

1. Why is the term prokaryote considered an inadequate descriptor by some microbiologists?
2. What characteristic shapes can bacteria assume? Describe the ways in which bacterial cells cluster together.
3. What advantages might a bacterial species that forms multicellular arrangements (e.g., clusters or chains) have that are not afforded unicellular bacteria?
4. What is the relevance of the surface area-to-volume ratio?

3.3 Bacterial Plasma Membranes Control What Enters and Leaves the Cell

After reading this section, you should be able to:

- Describe the fluid mosaic model of membrane structure and identify the types of lipids typically found in bacterial membranes
- Distinguish macroelements (macronutrients) from micronutrients (trace elements) and provide examples of each
- Provide examples of growth factors needed by some microorganisms
- Compare and contrast passive diffusion, facilitated diffusion, active transport, and group translocation, and provide examples of each
- Discuss the difficulty of iron uptake and describe how bacteria overcome this difficulty

The **cell envelope** is defined as the plasma membrane and all the surrounding layers external to it. The cell envelopes of many bacteria consist of the plasma membrane, cell wall, and at least one additional layer (e.g., capsule or slime layer). Of all these layers, the **plasma membrane** is the most important because it encompasses the cytoplasm and defines the cell. If it is removed, the cell's contents spill into the environment and the cell no longer exists. Furthermore, despite being the innermost layer of the cell envelope, the plasma membrane is responsible for much of the cell's relationship with the outside world. Thus we begin our consideration of bacterial cell structure by describing the plasma membrane.

First let's consider what cells do to survive. Cells must interact in a selective fashion with their environment, acquire nutrients, and eliminate waste. They also have to maintain their interior in a constant, highly organized state in the face of external changes. Plasma membranes are an absolute requirement for all living organisms because they are involved in carrying out these cellular tasks.

A primary role of all plasma membranes is that they are selectively permeable barriers: they allow particular ions and molecules to pass either into or out of the cell, while preventing the movement of others. Thus the plasma membrane prevents the loss of essential components through leakage while allowing the movement of other molecules. Bacterial plasma membranes play additional critical roles. They are the location of several crucial metabolic processes: respiration, photosynthesis, and the synthesis of lipids and cell wall constituents.

In addition to the plasma membrane, some bacteria have extensive intracytoplasmic membrane systems (p. 64). These internal membranes and the plasma membrane share a basic design. However, they can differ significantly in terms of the lipids and proteins they contain. To understand these chemical differences and the many functions of the plasma membrane and other membranes, it is necessary to become familiar with membrane structure.

Fluid Mosaic Model of Membrane Structure

The most widely accepted model for membrane structure is the **fluid mosaic model** of Singer and Nicholson, which proposes that membranes are lipid bilayers within which proteins float **(figure 3.7)**. The model is based on studies of eukaryotic and bacterial membranes, and was established using a variety of experimental approaches, including transmission electron microscopy (TEM) and atomic force microscopy. Cell membranes are very thin structures, about 5 to 10 nm thick, that look like two dark lines on either side of a light interior when imaged by TEM. This characteristic appearance is evidence that the membrane is composed of two sheets of molecules arranged end-to-end (figure 3.7). Cleavage of membranes by freeze-etching, a technique that allows the microscopist to see fine detail, exposes the

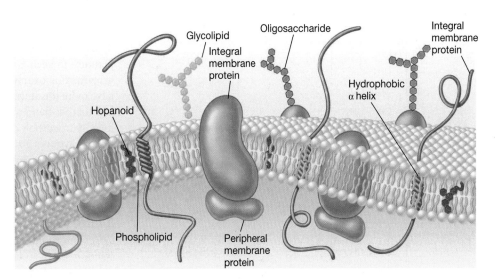

Figure 3.7 The Fluid Mosaic Model of Bacterial Membrane Structure. This diagram shows the integral membrane proteins (blue) floating in a lipid bilayer. Peripheral membrane proteins (purple) are associated loosely with the inner membrane surface. Small spheres represent the hydrophilic ends of membrane phospholipids, and wiggly tails are the hydrophobic fatty acid chains. Other membrane lipids such as hopanoids (red) may be present. Phospholipids are drawn much larger than their actual size.

proteins lying within the membrane lipid bilayer. ◄◄ *Electron microscopes use beams of electrons to create highly magnified images (section 2.4); Scanning probe microscopy can visualize molecules and atoms (section 2.5)*

The chemical nature of membrane lipids is critical to their ability to form bilayers. Most membrane-associated lipids (e.g., the phospholipids shown in figure 3.7) are **amphipathic:** they are structurally asymmetric, with polar and nonpolar ends (**figure 3.8**). The polar ends interact with water and are **hydrophilic;** the nonpolar **hydrophobic** ends are insoluble in water and tend to associate with one another. In aqueous environments, amphipathic lipids can interact to form a bilayer. The outer surfaces of the bilayer are hydrophilic, whereas hydrophobic ends are buried in the interior away from the surrounding water (figure 3.7). ►► *Lipids (appendix I)*

Two types of membrane proteins have been identified based on their ability to be separated from the membrane. **Peripheral membrane proteins** are loosely connected to the membrane and can be easily removed (figure 3.7). They are soluble in aqueous solutions and make up about 20 to 30% of total membrane protein. The remaining proteins are **integral membrane proteins.** These are not easily extracted from membranes and are insoluble in aqueous solutions when freed of lipids. Integral membrane proteins, like membrane lipids, are amphipathic; their hydrophobic regions are buried in the lipid while the hydrophilic portions project from the membrane surface (figure 3.7).

Integral membrane proteins carry out some of the most important functions of the membrane. Many are transport proteins used to move materials either into or out of the cell. Others are involved in energy-conserving processes, such as the proteins found in electron transport chains. Those integral membrane proteins with regions exposed to the outside of the cell enable the cell to interact with its environment. ►► *Proteins (appendix I)*

Although most aspects of the fluid mosaic model are well supported by experimentation, some are being questioned. The fluid mosaic model suggests that membrane lipids are homogeneously distributed and that integral membrane proteins are free to move laterally within the membrane. However, the presence of microdomains enriched for certain lipids and the observation that some integral proteins are present at only certain sites do not support this view. Although the term mosaic initially referred to the clusters of proteins embedded in a homogeneous lipid bilayer, it may be more accurate to use the term to refer to the patchwork of lipid microdomains found in membranes. Research is ongoing to determine the physiological role of these microdomains.

Bacterial Plasma Membranes Contain Lipids Not Found in Eukaryotic Membranes

Bacterial membranes are similar to eukaryotic membranes in that they are lipid bilayers and many of their amphipathic lipids are phospholipids (figure 3.8). The plasma membrane is very dynamic: the lipid composition varies with environmental temperature in such a way that the membrane remains fluid during growth. For example, bacteria growing at lower temperatures have more unsaturated fatty acids in their membrane phospholipids; that is, there are one or more double covalent bonds in the long hydrocarbon chain. At higher temperatures, their phospholipids have more saturated fatty acids—those in which the carbon atoms are connected only with single covalent bonds. Other factors also affect the lipid composition of membranes. For instance, some pathogens change the lipids in their plasma membranes to protect themselves from antimicrobial peptides produced by the immune system. ►► *Environmental factors affect microbial growth (section 7.4); Antimicrobial peptides (section 33.3)*

Bacterial membranes usually differ from eukaryotic membranes in lacking sterols (steroid-containing lipids) such as cholesterol (**figure 3.9a**). However, many bacterial membranes contain hopanoids, which are similar to, but distinct from, steroids (figure 3.9b). **Hopanoids** are synthesized from the same precursors

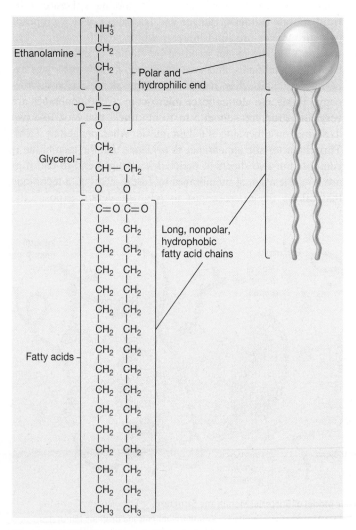

Figure 3.8 The Structure of a Phospholipid.
Phosphatidylethanolamine, a phospholipid often found in bacterial membranes.

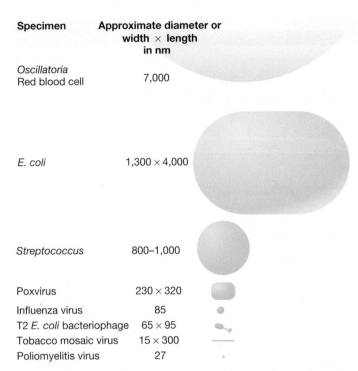

Specimen	Approximate diameter or width × length in nm
Oscillatoria Red blood cell	7,000
E. coli	1,300 × 4,000
Streptococcus	800–1,000
Poxvirus	230 × 320
Influenza virus	85
T2 *E. coli* bacteriophage	65 × 95
Tobacco mosaic virus	15 × 300
Poliomyelitis virus	27

Figure 3.3 Sizes of Bacteria Relative to a Red Blood Cell and Viruses. Recall that 1,000 nm = 1 μm. Thus *E. coli* is 1.3 × 4 μm.

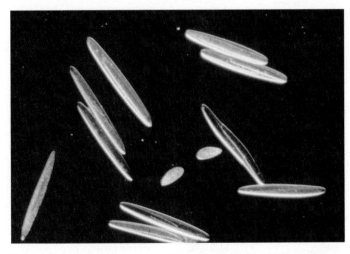

Figure 3.4 A Giant Bacterium. This photograph shows *Epulopiscium fishelsoni* dwarfing the paramecia, which are protozoa. *E. fishelsoni* cells are about 0.53 mm long. Phase-contrast microscopy.

within the cell become more efficient, which in turn facilitates a rapid growth rate. Shape affects the S/V ratio. A rod with the same volume as a coccus has a higher S/V ratio than does the coccus. This means that a rod can have greater nutrient flux across its plasma membrane. However, the discovery of *E. fishelsoni* demonstrates that bacteria can be very large. For bacteria to be large, they must have other characteristics that maximize their S/V ratio, or their size must be beneficial in some way. For instance,

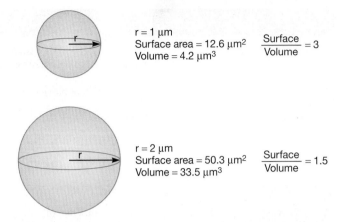

$r = 1\ \mu m$
Surface area = 12.6 μm²
Volume = 4.2 μm³
$\dfrac{\text{Surface}}{\text{Volume}} = 3$

$r = 2\ \mu m$
Surface area = 50.3 μm²
Volume = 33.5 μm³
$\dfrac{\text{Surface}}{\text{Volume}} = 1.5$

Figure 3.5 The Surface-to-Volume Ratio Is an Important Determinant of Cell Size. Surface area is calculated by the formula $4\pi r^2$. Volume is calculated by the formula $4/3\pi r^3$. Shape also affects the S/V ratio; rods with the same diameter as a coccus have a greater S/V ratio.

E. fishelsoni has a highly convoluted plasma membrane, which increases its S/V ratio. In addition, large cells are less likely to be eaten by predatory protists. Cells that are filamentous, have stalks, or are oddly shaped are also less susceptible to predation. ↻ *Prokaryotic Cell Shapes*

Cell Organization

Structures often observed in bacterial cells are summarized and illustrated in **table 3.1** and **figure 3.6**. Note that no single bacterium possesses all of these structures at all times. Some are found only in certain cells in certain conditions or in certain phases of the life cycle. However, there are several common features of bacterial cell structure.

Bacterial cells are often surrounded by several layers, which are collectively called the cell envelope. The most common cell envelope layers are the plasma membrane, cell wall, and capsule or slime layer. The innermost layer of the cell envelope is the plasma membrane, which surrounds the cytoplasm. Most bacteria have a chemically complex cell wall, which covers the plasma membrane. Many bacteria are surrounded by a capsule or slime layer external to the cell wall. Because most bacteria do not contain internal, membrane-bound organelles, their interior appears morphologically simple. The genetic material is localized in a discrete region called the nucleoid and usually is not separated from the surrounding cytoplasm by membranes. Ribosomes and larger masses called inclusions are scattered about the cytoplasm. Finally, many bacteria use flagella for locomotion.

In the remaining sections of this chapter, we describe the major structures observed in bacterial cells in more detail. We begin with the cell envelope. We then proceed inward to consider structures located within the cytoplasm. Next the discussion moves outward, to a variety of appendages that are involved in attachment to surfaces, motility, or both. Finally, we consider the bacterial endospore.

Table 3.1	Common Bacterial Structures and Their Functions
Plasma membrane	Selectively permeable barrier, mechanical boundary of cell, nutrient and waste transport, location of many metabolic processes (respiration, photosynthesis), detection of environmental cues for chemotaxis
Gas vacuole	An inclusion that provides buoyancy for floating in aquatic environments
Ribosomes	Protein synthesis
Inclusions	Storage of carbon, phosphate, and other substances; site of chemical reactions (microcompartments); movement
Nucleoid	Localization of genetic material (DNA)
Periplasmic space	In typical Gram-negative bacteria, contains hydrolytic enzymes and binding proteins for nutrient processing and uptake; in typical Gram-positive bacteria, may be smaller or absent
Cell wall	Protection from osmotic stress, helps maintain cell shape
Capsules and slime layers	Resistance to phagocytosis, adherence to surfaces
Fimbriae and pili	Attachment to surfaces, bacterial conjugation and transformation, twitching
Flagella	Swimming and swarming motility
Endospore	Survival under harsh environmental conditions

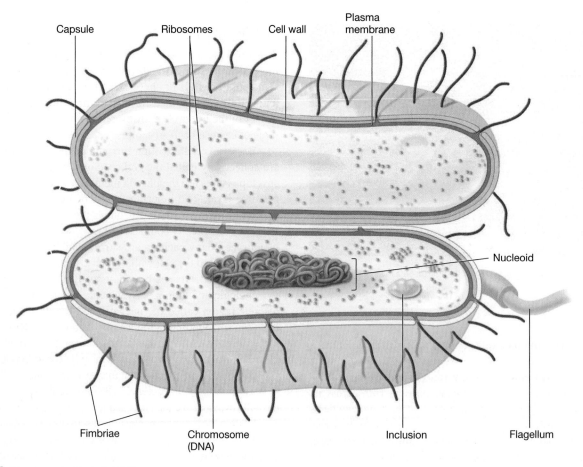

Figure 3.6 Structure of a Bacterial Cell.

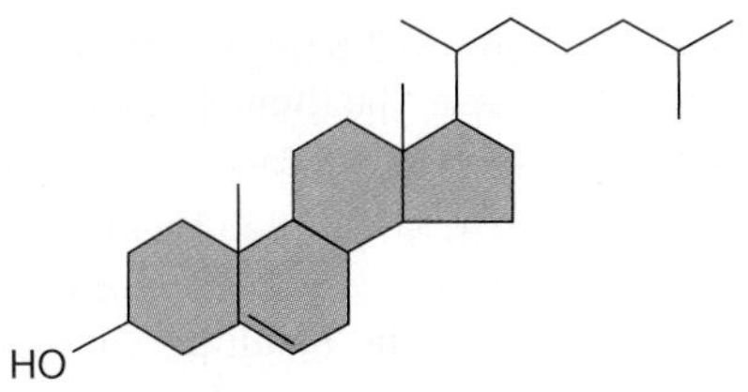

(a) Cholesterol (a steroid) is found in the membranes of eukaryotes.

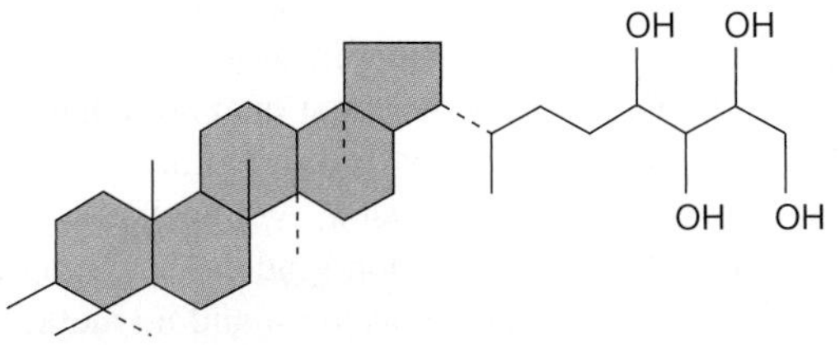

(b) Bacteriohopanetetrol (a hopanoid) is found in many bacterial membranes.

Figure 3.9 **Membrane Steroids and Hopanoids.**

as steroids, and like the sterols in eukaryotic membranes, they probably stabilize the membrane. Hopanoids are also of interest to ecologists and geologists: the total mass of hopanoids stored in sediments is estimated to be around 10^{11-12} tons—about as much as the total mass of organic carbon in all living organisms (10^{12} tons)—and evidence exists that hopanoids have contributed significantly to the formation of petroleum.

Bacteria Use Many Mechanisms to Bring Nutrients into the Cell

All plasma membranes function as barriers. Yet they must also allow movement of nutrients into the cell. If a microbe does not obtain nutrients from its environment, it will quickly exhaust its supply of amino acids, nucleotides, and other molecules needed to survive. In addition, if a microbe is to thrive and reproduce, it must have a source of energy. The energy source is used to generate the cell's major energy currency: the high-energy molecule ATP. Clearly, obtaining energy and nutrient sources is one of the most important jobs an organism has, and it is primarily a function of the plasma membrane. Here we discuss nutrient uptake, but first let's define some terms used to describe the nutrients needed by cells.

Microbiologists refer to carbon, oxygen, hydrogen, nitrogen, sulfur, and phosphorus as **macroelements** or macronutrients because they are required in relatively large amounts. They are found in organic molecules such as proteins, lipids, nucleic acids, and carbohydrates. Other macroelements are potassium, calcium, magnesium, and iron. They exist as cations and generally are associated with and contribute to the activity and stability of molecules and cell structures such as enzymes and ribosomes. Thus they are important in many cellular processes, including

protein synthesis and energy conservation. ▶▶❙ *Enzymes and ribozymes speed up cellular chemical reactions (section 10.6); Translation in bacteria (section 13.7); Electron transport chains (section 10.4)*

Other elements are required in small amounts—amounts so small that in the lab they are often obtained as contaminants in water, glassware, and growth media. Likewise in nature, they are ubiquitous and usually present in adequate amounts to support the growth of microbes. Microbiologists call these elements **micronutrients** or **trace elements.** The micronutrients—manganese, zinc, cobalt, molybdenum, nickel, and copper—are needed by most cells. Micronutrients are part of certain enzymes, and they aid in catalysis of reactions and maintenance of protein structure.

Some microbes are able to synthesize all the organic molecules they need from macroelements. However, some microbes are unable to synthesize certain molecules needed for survival. These molecules are called **growth factors,** and they must be obtained from the environment. There are three types of growth factors: amino acids, purines and pyrimidines, and vitamins.

What are the common features of nutrient uptake by bacteria? Bacteria can only take in dissolved molecules. Uptake mechanisms are specific; that is, the necessary substances, and not others, are acquired. It does a cell no good to take in a substance that it cannot use. Bacteria are able to transport nutrients into the cell even when the concentration of a nutrient inside the cell is higher than the concentration outside. Thus they are able to move nutrients up a concentration gradient. This is important because bacteria often live in nutrient-poor habitats. In view of the enormous variety of nutrients and the complexity of the task, it is not surprising that bacteria use several different transport mechanisms: passive diffusion, facilitated diffusion, primary and secondary active transport, and group translocation.

Passive Diffusion

Passive diffusion, often called diffusion or simple diffusion, is the process by which molecules move from a region of higher concentration to one of lower concentration; that is, the molecules move down the concentration gradient. The rate of passive diffusion depends on the size of the concentration gradient between a cell's exterior and its interior (**figure 3.10**). A large concentration gradient is required for adequate nutrient uptake by passive diffusion (i.e., the external nutrient concentration must be high while the internal concentration is low). The rate of diffusion decreases as more nutrient accumulates in the cell, unless the nutrient is used immediately upon entry.

Most substances cannot freely diffuse into a cell. However, some gases, including O_2 and CO_2, easily cross the plasma membrane by passive diffusion. H_2O also moves across membranes by passive diffusion. This is important because it allows the cell to adjust to differences in solute concentrations. Larger molecules, ions, and polar substances must enter the cell by other

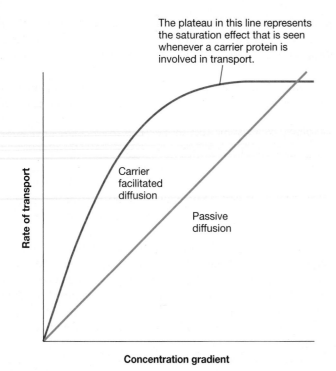

The plateau in this line represents the saturation effect that is seen whenever a carrier protein is involved in transport.

Carrier facilitated diffusion

Passive diffusion

Rate of transport

Concentration gradient

Figure 3.10 Passive and Facilitated Diffusion. The rate of diffusion depends on the size of the solute's concentration gradient (the ratio of the extracellular concentration to the intracellular concentration). This example of facilitated diffusion involves a carrier protein that can be saturated. Sometimes facilitated diffusion is mediated by a channel. Channels often do not exhibit a saturation effect.

mechanisms, all of which involve specialized proteins that are referred to as transport proteins. ⟳ *How Diffusion Works*

Transport Proteins

Bacterial cells employ a variety of transport proteins in their uptake mechanisms. These important proteins are embedded in membranes and are classified into several types. The two major types are channels and carriers. Channels, as their name indicates, are proteins that form pores in membranes through which substances can pass; they are often involved in facilitated diffusion. Channels show some specificity for the substances that pass through them, but this is considerably less than that shown by carriers, which are far more substrate specific. Carriers are so named because they carry nutrients across the membrane.

Facilitated Diffusion

During **facilitated diffusion,** substances move across the plasma membrane with the assistance of transport proteins that are either channels or carriers. The rate of facilitated diffusion increases

with the concentration gradient much more rapidly and at lower concentrations of the diffusing molecule than that of passive diffusion (figure 3.10). When the transporter is a carrier, the diffusion rate reaches a plateau above a specific gradient value because the carrier protein is saturated; that is, it is transporting as many solute molecules as possible. The resulting curve resembles an enzyme-substrate curve (*see figure 10.17*) and is different from the linear response seen with passive diffusion. An example of channel-mediated facilitated diffusion is that involving aquaporins (*see figure 2.31*), which transport water. Aquaporins are members of the major intrinsic protein (MIP) family of proteins. MIPs facilitate diffusion of small polar molecules, and they are observed in virtually all organisms.

Facilitated diffusion is truly diffusion, even though a transport protein is involved. A concentration gradient spanning the membrane drives the movement of molecules, and no metabolic energy input is required. If the concentration gradient disappears, net inward movement ceases. The gradient can be maintained by transforming the transported nutrient to another compound, as occurs when a nutrient is metabolized.

Considerable work has been done on the mechanism of carrier-mediated facilitated diffusion. After the solute molecule binds to the outside, the carrier is thought to change conformation and release the molecule on the cell interior (**figure 3.11**). The carrier subsequently changes back to its original shape and is ready to pick up another molecule. The net effect is that a hydrophilic molecule can enter the cell in response to its concentration gradient.

Facilitated diffusion has been documented in some bacteria. However, it does not seem to be the major uptake mechanism for these microbes. Recall that many bacteria live in environments where nutrient concentrations are low, and facilitated diffusion cannot concentrate nutrients inside cells. Therefore energy-dependent transport mechanisms that do concentrate nutrients are significantly more important uptake mechanisms for bacterial cells.

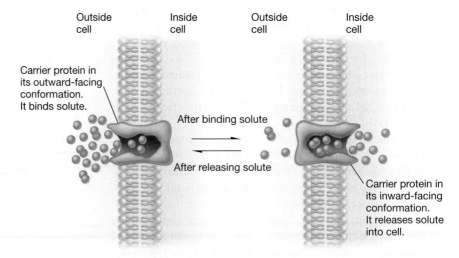

Outside cell   Inside cell   Outside cell   Inside cell

Carrier protein in its outward-facing conformation. It binds solute.

After binding solute

After releasing solute

Carrier protein in its inward-facing conformation. It releases solute into cell.

Figure 3.11 A Model of Facilitated Diffusion. Because there is no energy input, molecules continue to enter only as long as their concentration is greater on the outside.

Primary and Secondary Active Transport

Active transport is the transport of solute molecules to higher concentrations (i.e., against a concentration gradient) with the input of metabolic energy. Three types of active transport are observed in bacteria: primary active transport, secondary active transport, and group translocation. They differ in terms of the energy used to drive transport and on whether or not the transported molecule is modified as it enters.

Active transport resembles facilitated diffusion in that it involves carrier proteins. Recall that carrier proteins bind particular solutes with great specificity. Active transport is also characterized by the carrier saturation effect at high solute concentrations (figure 3.10). Nevertheless, active transport differs from facilitated diffusion because it uses metabolic energy and can concentrate substances. Metabolic inhibitors that block energy production inhibit active transport but do not immediately affect facilitated diffusion.

Primary active transport is mediated by carriers called primary active transporters. They use energy provided by ATP hydrolysis to move substances against a concentration gradient without modifying them. Primary active transporters are **uniporters;** that is, they move a single molecule across the membrane (**figure 3.12**). **A*TP-*b*inding *c*assette transporters** (**ABC transporters**) are important primary active transporters. Our focus here is on those ABC transporters that are used for import of substances. Other ABC transporters are used for export of substances, in particular proteins; these exporters are described in chapter 13.

▶▶| *Protein translocation and secretion in bacteria (section 13.8)*

Most ABC transporters consist of two hydrophobic membrane-spanning domains associated on their cytoplasmic surfaces with two ATP-binding domains (**figure 3.13**). The membrane-spanning domains form a pore in the membrane, and the ATP-binding domains bind and hydrolyze ATP to drive uptake. Most ABC transporters employ substrate-binding proteins to deliver the molecule to be transported to the transporter.

Secondary active transport couples the potential energy of ion gradients to transport of substances without modifying them. Secondary active transporters are cotransporters (figure 3.12). They

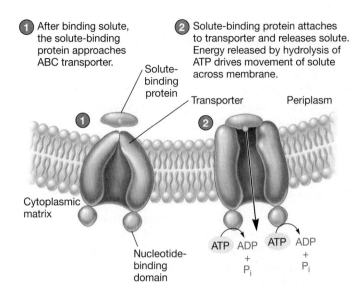

① After binding solute, the solute-binding protein approaches ABC transporter.

② Solute-binding protein attaches to transporter and releases solute. Energy released by hydrolysis of ATP drives movement of solute across membrane.

Solute-binding protein

Transporter

Periplasm

Cytoplasmic matrix

Nucleotide-binding domain

ATP ADP + P$_i$ ATP ADP + P$_i$

Figure 3.13 ABC Transporter Function.   Shown here is a transporter that works with a substrate-binding protein free in the periplasm. Other substrate-binding proteins are associated with the plasma membrane, always associated with the transporter, or even fused to the transporter.

move two substances simultaneously: the ion whose gradient powers transport and the substance being moved across the membrane. When the ion and other substance both move in the same direction, it is called **symport.** When they move in opposite directions, it is called **antiport.** ⟳ *Cotransport (Symport and Antiport)*

The ion gradients used by secondary active transporters arise primarily in three ways. The first results from bacterial metabolic activity. During energy-conserving processes, electron transport generates a proton gradient in which protons are at a higher concentration outside the cell than inside. The proton gradient is used to do cellular work, including secondary active transport. Some bacteria use the second method, in which an enzyme called a V-type ATPase hydrolyzes ATP and uses the energy released to create either a proton gradient or a sodium gradient across the plasma membrane. Finally, a proton gradient can be used to create another ion gradient such as a sodium gradient. This is accomplished by an antiporter that brings protons in as sodium ions are moved out of the cell. The sodium gradient can then be used to drive uptake of nutrients by a symport mechanism. ▶▶| *Electron transport and oxidative phosphorylation (step 3) generate the most ATP (section 11.6)*

The lactose permease of *E. coli* is a well-studied secondary active transporter. It is a single protein that transports a lactose molecule inward as a proton simultaneously enters the cell. The proton is moving down a proton gradient, and the energy released drives solute transport. X-ray diffraction studies show that the carrier protein exists in outward- and inward-facing conformations. When lactose and a proton bind to separate sites on the outward-facing conformation, the protein changes to its inward-facing conformation, and the sugar and proton are released into the cytoplasm. Thus this is an example of symport. *E. coli* also uses proton symport to take up amino acids and some organic acids.

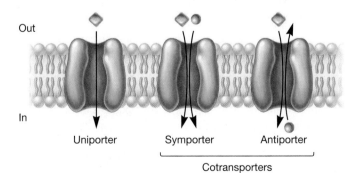

Out

In

Uniporter

Symporter

Antiporter

Cotransporters

Figure 3.12 Carrier Proteins Can Be Uniporters or Cotransporters.
Uniporters move a single substance into the cell. Cotransporters simultaneously move two substances across the membrane. When both substances move in the same direction, the carrier is a symporter. When the two substances move in opposite directions, the carrier is an antiporter.

Bacteria often have more than one transport system for a nutrient, as can be seen with *E. coli.* This bacterium has at least five transport systems for the sugar galactose, three systems each for the amino acids glutamate and leucine, and two potassium transport complexes. When several transport systems exist for the same substance, the systems differ in such properties as their energy source, their affinity for the solute transported, and the nature of their regulation. This diversity gives the bacterium an added competitive advantage in a variable environment.

Group Translocation

The distinguishing characteristic of **group translocation** is that a molecule is chemically modified as it is brought into the cell. The best-known group translocation system is the **phosphoenolpyruvate: sugar phosphotransferase system** (**PTS**), which is observed in many bacteria. The PTS transports a variety of sugars while phosphorylating them, using phosphoenolpyruvate (PEP) as the phosphate donor. PEP is an important intermediate of a biochemical pathway used by many bacteria to extract energy from organic energy sources. PEP is a high-energy molecule that can be used to synthesize ATP, the cell's energy currency. However, when it is used in PTS reactions, the energy present in PEP is used to energize sugar uptake rather than ATP synthesis. ▶▶ *ATP (section 10.2); Embden-Meyerhof pathway (section 11.4)*

The transfer of phosphate from PEP to the incoming molecule involves several proteins and is an example of a **phosphorelay system.** In *E. coli* and *Salmonella,* the PTS consists of two enzymes and a low molecular weight heat-stable protein (HPr). A phosphate is transferred from PEP to enzyme II with the aid of enzyme I and HPr (**figure 3.14**). Enzyme II then phosphorylates the sugar molecule as it is carried across the membrane. Many different PTSs exist, and they vary in terms of the sugars they transport. The specificity lies with the type of Enzyme II used in the PTS. Enzyme I and HPr are the same in all PTSs used by a bacterium. ▶▶ *Enzymes and ribozymes speed up cellular chemical reactions (section 10.6)*

PTSs are widely distributed in bacteria, being found primarily among facultatively anaerobic bacteria (bacteria that grow in either the presence or absence of O_2); some obligately anaerobic bacteria (e.g., *Clostridium* spp.) also have PTSs. However, most aerobic bacteria lack PTSs. Many carbohydrates are transported by PTSs. *E. coli* takes up glucose, fructose, mannitol, sucrose, *N*-acetylglucosamine, cellobiose, and other carbohydrates by group translocation. Besides their role in transport, PTSs function in numerous regulatory processes, including the regulation of carbon metabolism. One example is the role of PTSs in catabolite repression, a phenomenon in which the cell inhibits synthesis of degradative enzymes for some sugars so that it can catabolize a preferred sugar. We describe this in more detail in chapter 14. PTS proteins also can bind chemical attractants, toward which bacteria move by the process of chemotaxis (p. 74). ▶▶ *Oxygen concentration (section 7.4)* 🔁 *Active Transport by Group Translocation*

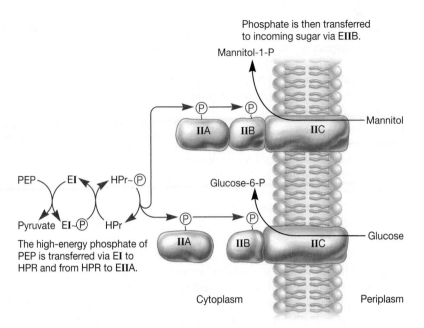

Figure 3.14 Group Translocation: Bacterial PTS Transport. Two examples of the phosphoenolpyruvate: sugar phosphotransferase system (PTS) are illustrated. The following components are involved in the system: phosphoenolpyruvate (PEP), enzyme I (EI), the low molecular weight heat-stable protein (HPr), and enzyme II (EII). EIIA is attached to EIIB in the mannitol transport system and is separate from EIIB in the glucose system.

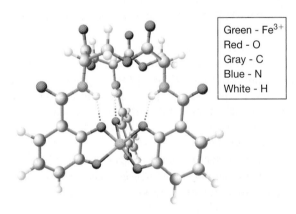

Green - Fe^{3+}
Red - O
Gray - C
Blue - N
White - H

Figure 3.15 Enterobactin: A Siderophore Produced by *E. coli.* Ball-and-stick model of enterobactin complexed with Fe^{3+}.

Iron Uptake

Almost all microorganisms require iron for building molecules important in energy-conserving processes (e.g., cytochromes), as well as for the function of many enzymes. Iron uptake is made difficult by the extreme insolubility of ferric iron (Fe^{3+}) and its derivatives, which leaves little free iron available for transport. Many bacteria have overcome this difficulty by secreting **siderophores** (Greek for iron bearers). Siderophores are low molecular weight organic molecules that bind ferric iron and supply it to the cell (**figure 3.15**). ▶▶ *Electron transport chains (section 10.4); Enzymes and ribozymes speed up cellular chemical reactions (section 10.6)*

Microorganisms secrete siderophores when iron is scarce in the medium. Once the iron-siderophore complex has reached the cell surface, it binds to a siderophore-receptor protein. Then either the iron is released to enter the cell directly or the whole iron-siderophore complex is transported inside by an ABC transporter. Iron is so crucial to microorganisms that they may use more than one route of iron uptake to ensure an adequate supply.

Retrieve, Infer, Apply

1. List the functions of bacterial plasma membranes. Why must their plasma membranes carry out more functions than the plasma membranes of eukaryotic cells?
2. Describe in words and with a labeled diagram the fluid mosaic model for cell membranes. What aspects of this model are currently being challenged? Why?
3. On what basis are elements divided into macroelements and trace elements?
4. Describe facilitated diffusion, primary and secondary active transport, and group translocation in terms of their distinctive characteristics and mechanisms. What advantage does a bacterium gain by using active transport rather than facilitated diffusion?
5. What are uniport, symport, and antiport?
6. What are siderophores? Why are they important?

3.4 There Are Two Main Types of Bacterial Cell Walls

After reading this section, you should be able to:

- Describe peptidoglycan structure
- Compare and contrast the cell walls of typical Gram-positive and Gram-negative bacteria
- Relate bacterial cell wall structure to the Gram-staining reaction

The **cell wall** is the layer that lies just outside the plasma membrane. It is one of the most important structures for several reasons: it helps maintain cell shape and protect the cell from osmotic lysis; it can protect the cell from toxic substances; and in pathogens, it can contribute to pathogenicity. Cell walls are so important that most bacteria have them. Those that do not have other features that fulfill cell wall function. Bacterial cell wall synthesis is targeted by several antibiotics. Therefore it is important to understand cell wall structure. ▶▶| *Antibacterial drugs (section 9.4)*

Overview of Bacterial Cell Wall Structure

After Christian Gram developed the Gram stain in 1884, it soon became evident that most bacteria could be divided into two major groups based on their response to the Gram-staining procedure. Gram-positive bacteria stained purple, whereas Gram-negative bacteria were pink or red. The true structural difference between these two groups did not become clear until the advent of the transmission electron microscope. Here we describe the long-held models of Gram-positive and Gram-negative cell walls developed from these studies. More recent studies of diverse groups of bacteria have shown that these models do not hold true for all bacteria

(**Microbial Diversity & Ecology 3.1**). Because of the ongoing discussions related to these new studies, we will refer to bacteria that fit the models as being typical Gram-positive or typical Gram-negative bacteria. |◀◀ *Differential staining (section 2.3)*

The cell walls of *Bacillus subtilis* and many other typical Gram-positive bacteria consist of a single, 20- to 80-nm-thick homogeneous layer of **peptidoglycan (murein)** lying outside the plasma membrane (**figure 3.16**). In contrast, the cell walls of

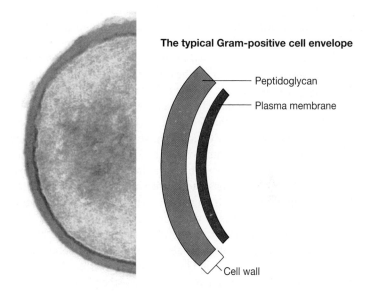

The typical Gram-positive cell envelope
— Peptidoglycan
— Plasma membrane
Cell wall

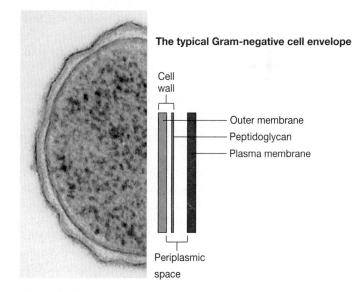

The typical Gram-negative cell envelope
Cell wall
— Outer membrane
— Peptidoglycan
— Plasma membrane
Periplasmic space

Figure 3.16 Cell Envelopes of Typical Gram-Positive and Gram-Negative Bacteria.   Cell envelopes consist of the plasma membrane and any layers (e.g., cell wall) exterior to it. For simplicity, we show only the plasma membrane and cell wall. Slime layers, capsules, and S-layers are common additional layers that lie outside the cell wall. *Staphylococcus aureus* (top) has a typical Gram-positive cell wall that consists primarily of peptidoglycan (also called murein). *Myxococcus xanthus* (bottom) has a typical Gram-negative cell wall consisting of a thin layer of peptidoglycan, an outer membrane, and the periplasmic space.

MICROBIAL DIVERSITY & ECOLOGY

3.1 Gram Positive and Gram Negative or Monoderms and Diderms?

The importance of the Gram stain in the history of microbiology cannot be overstated. The Gram stain reaction was for many years one of the critical pieces of information used by bacterial taxonomists to construct taxa, and it is still useful in identifying bacteria in clinical settings. The initial studies done to differentiate bacteria that stained Gram positive from those that stain Gram negative were done using model organisms such as *Bacillus subtilis* (Gram positive) and *Escherichia coli* (Gram negative). At the time, it was thought that all other bacteria would have similar cell wall structures. However, as the cell walls of more bacteria have been characterized, it has become apparent that it may be misleading to refer to bacteria as Gram positive or Gram negative. In other words, the long-held models of Gram-positive and Gram-negative cell walls do not hold true for all bacteria. Iain Sutcliffe has proposed that microbiologists stop referring to bacteria as either Gram positive or Gram negative. He suggests that instead we should more precisely describe bacterial cell envelope architectures by focusing on the observation that some bacteria have envelopes with a single membrane—the plasma membrane as seen in typical Gram-positive bacteria—while others have envelopes with two membranes—the plasma membrane and an outer membrane as seen in typical Gram-negative bacteria. He proposed calling the former monoderms and the latter diderms.

But why make this change? Sutcliffe begins by pointing out that some bacteria staining Gram positive are actually diderms and some staining Gram negative are actually monoderms. By referring to Gram-positive-staining diderms as Gram-positive bacteria, it is too easy to mislead scientists and

many a budding microbiologist into thinking that the bacterium has a typical Gram-positive envelope. He also argues that by relating cell envelope architecture to the phylogenies of various bacterial taxa, we may gain insight into the evolution of these architectures. He notes that the phyla *Firmicutes* and *Actinobacteria* are composed almost completely of monoderm bacteria, whereas almost all other bacterial phyla consist of diderms.

There are interesting exceptions to the relationship of phylogeny and cell envelope structure. For instance, members of the genus *Mycobacterium* (e.g., *M. tuberculosis*) belong to the predominantly monoderm phylum *Actinobacteria*. Mycobacteria have cell walls that consist of peptidoglycan and an outer membrane. The outer membrane is composed of mycolic acids rather than the phospholipids and lipopolysaccharides (LPSs) found in the typical Gram-negative cells' outer membrane. ▶▶❘ *Order* Corynebacteriales *includes important human pathogens (section 24.1)*

Members of the genus *Deinococcus* are another interesting exception. These bacteria stain Gram positive but are diderms. Their cell envelopes consist of the plasma membrane, what appears to be a typical Gram-negative cell wall, and an S-layer (p. 61). Their outer membrane is distinctive because it lacks LPS. Deinococci are not unique in this respect, however. It is now known that members of several taxa have outer membranes that lack LPS.

Source: Sutcliffe, I. C. 2010. A phylum level perspective on bacterial cell envelope architecture. Trends Microbiol. *18(10):464–70.*

E. coli and many other typical Gram-negative bacteria have two distinct layers: a 2- to 7-nm-thick peptidoglycan layer covered by a 7- to 8-nm-thick **outer membrane.**

One important feature seen in typical Gram-negative bacteria is a space between the plasma membrane and the outer membrane. It also is sometimes observed between the plasma membrane and cell wall in typical Gram-positive bacteria. This space is called the **periplasmic space.** The substance that occupies the periplasmic space is the **periplasm.**

Peptidoglycan Structure

The feature common to nearly all bacterial cell walls is the presence of peptidoglycan, which forms an enormous meshlike structure often referred to as the peptidoglycan sacculus. Peptidoglycan is composed of many identical subunits. Each subunit within the sacculus contains two sugar derivatives, *N*-acetylglucosamine (NAG) and

N-acetylmuramic acid (NAM), and several different amino acids. The amino acids form a short peptide, sometimes called the stem peptide, consisting of four alternating D- and L-amino acids; the peptide is connected to the carboxyl group of NAM (**figure 3.17**). Three of the amino acids are not found in proteins: D-glutamic acid, D-alanine, and *meso*-diaminopimelic acid. The presence of D-amino acids in the stem peptide protects against degradation by most peptidases, which recognize only the L-isomers of amino acid residues. The peptidoglycan subunit of many bacteria is shown in figure 3.17. ▶▶❘ *Carbohydrates (appendix I); Proteins (appendix I); Proteins are polymers of amino acids (section 13.2)*

The peptidoglycan sacculus is formed by linking the sugars of the peptidoglycan subunits together to form a strand; the strands are then cross-linked to each other by covalent bonds formed between the stem peptides extending from each strand. As seen in **figure 3.18**, the backbone of each strand is composed of alternating NAG and NAM residues. The strand is helical, and

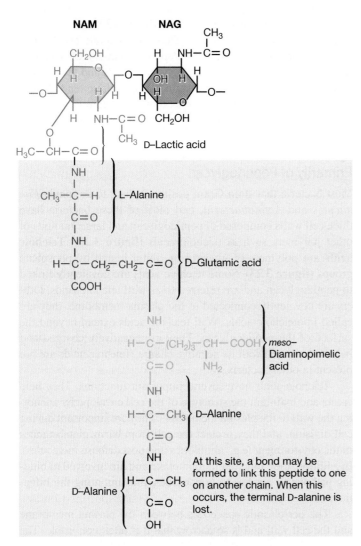

Figure 3.17 Peptidoglycan Subunit Composition. Shown is the peptidoglycan subunit of *E. coli,* many other typical Gram-negative bacteria, and many typical Gram-positive bacteria. This illustration shows the subunit before it has been inserted into the existing peptidoglycan polymer. NAG is *N*-acetylglucosamine. NAM is *N*-acetylmuramic acid. The stem peptide is composed of alternating D- and L-amino acids; it terminates with two D-alanines. The amino acids are shown in different shades of color for clarity.

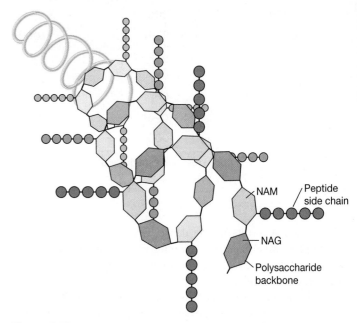

Figure 3.18 A Helical Peptidoglycan Strand. Because of the strand's helical nature, the stem peptides project out in different directions from the NAM-NAG backbone. Here the stem peptides are shown projecting out at 90° angles. Some studies suggest that the angle is actually 120°.

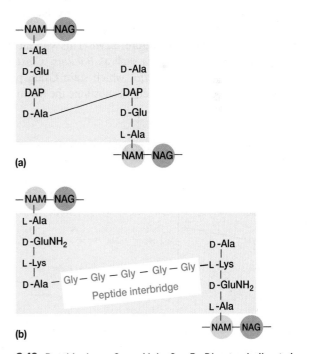

Figure 3.19 Peptidoglycan Cross-Links Can Be Direct or Indirect via a Peptide Interbridge. (a) *E. coli* peptidoglycan with direct cross-linking, typical of many Gram-negative bacteria. (b) *Staphylococcus aureus* peptidoglycan with an interbridge. *S. aureus* stains Gram positive. NAM is *N*-acetylmuramic acid. NAG is *N*-acetylglucosamine. Gly is glycine. D-GluNH₂ is D-glutamic acid with an NH₂ group attached to the α carbon (the carbon next to the carboxyl group).

the stem peptides extend out from the backbone in different directions. There are two types of cross-links: direct and indirect via a peptide interbridge. A direct cross-link is characterized by connecting the carboxyl group of an amino acid in one stem peptide to the amino group of an amino acid in another stem peptide. For instance, many bacteria cross-link the strands by connecting the carboxyl group of the D-alanine at position 4 of the stem peptide directly to the amino group of diaminopimelic acid (position 3) of the other peptidoglycan strand's stem peptide (the position 5 D-alanine is removed as the cross-link is formed). Bacteria that have indirect linkage use a **peptide interbridge** (also called an interpeptide bridge), a short chain of amino acids that links the stem peptide of one peptidoglycan strand to that of another (**figure 3.19**).

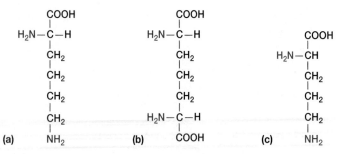

Figure 3.20 **Diamino Acids Present in Peptidoglycan.** (a) L-lysine. (b) *meso*-diaminopimelic acid, (c) D-ornithine.

Cross-linking results in one dense, interconnected network of peptidoglycan strands ▶▶| *Synthesis of peptidoglycan occurs in the cytoplasm, at the plasma membrane, and in the periplasmic space (section 12.4)*

The peptidoglycan sacculus is strong but elastic. It is able to stretch and contract in response to osmotic pressure. This is due to the rigidity of the backbone coupled with the flexibility of the cross-links. Peptidoglycan sacculi are also rather porous, allowing globular proteins having a molecular weight as large as 50,000 to pass through, depending on whether the sacculus is relaxed or stretched; thus only extremely large proteins are unable to pass through peptidoglycan.

Variants of peptidoglycan are often characteristic of particular groups and are therefore of some taxonomic value. Most typical Gram-negative bacteria have the amino acid composition and cross-linking shown in figure 3.19a. This variant is also observed in members of genera such as *Bacillus, Clostridium, Corynebacterium,* and *Nocardia,* which stain Gram positive. Other typical Gram-positive bacteria substitute the diamino acid

lysine for *meso*-diaminopimelic acid (**figure 3.20**) and cross-link chains via interpeptide bridges. These interpeptide bridges can vary considerably (**figure 3.21**). Peptidoglycan can also vary in terms of the length of the peptidoglycan strands and the amount of cross-linking. Bacteria that stain Gram positive tend to have much more cross-linking, whereas those that stain Gram negative have considerably less.

Typical Gram-Positive Cell Walls Consist Primarily of Peptidoglycan

Most bacteria that stain Gram positive belong to the phyla *Firmicutes* and *Actinobacteria*, and most of these bacteria have thick cell walls composed of peptidoglycan and large amounts of other polymers such as teichoic acids (**figure 3.22**). **Teichoic acids** are polymers of glycerol or ribitol joined by phosphate groups (**figure 3.23**). Some teichoic acids are covalently linked to peptidoglycan and are referred to as wall teichoic acids. Others are covalently connected to the plasma membrane; they are called lipoteichoic acids. Wall teichoic acids extend beyond the surface of the peptidoglycan. They are negatively charged and help give the cell wall its negative charge. Teichoic acids are not present in other bacteria.

Teichoic acids have several important functions. They help create and maintain the structure of the cell envelope by anchoring the wall to the plasma membrane. They are important during cell division, and they protect the cell from harmful substances in the environment (e.g., antibiotics and host defense molecules). In addition, they function in ion uptake and are involved in binding pathogenic species to host tissues, thus initiating the infectious disease process.

The periplasmic space lies between the plasma membrane and the cell wall and is so narrow that it is often not visible. The

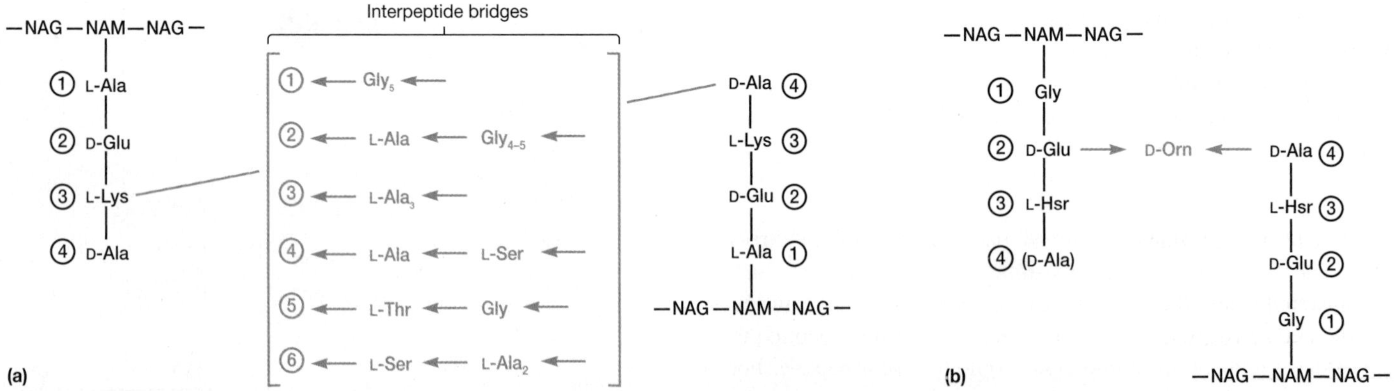

Interpeptide bridges

(a)

| —NAG —NAM—NAG— |
| ① L-Ala |
| ② D-Glu |
| ③ L-Lys |
| ④ D-Ala |

① ⟵ Gly₅ ⟵
② ⟵ L-Ala ⟵ Gly₄₋₅ ⟵
③ ⟵ L-Ala₃ ⟵
④ ⟵ L-Ala ⟵ L-Ser ⟵
⑤ ⟵ L-Thr ⟵ Gly ⟵
⑥ ⟵ L-Ser ⟵ L-Ala₂ ⟵

D-Ala ④
L-Lys ③
D-Glu ②
L-Ala ①
—NAG —NAM—NAG—

(b)

—NAG —NAM—NAG—
① Gly
② D-Glu → D-Orn ⟵ D-Ala ④
③ L-Hsr   L-Hsr ③
④ (D-Ala)   D-Glu ②
                     Gly ①
                —NAG —NAM—NAG—

Figure 3.21 **Examples of Peptidoglycan Cross-Links.** Most variation in peptidoglycan structure occurs in the composition of the stem peptide and the method by which peptidoglycan strands are cross-linked. (a) Examples of interpeptide bridges that link D-alanine in position 4 with L-lysine in position 3. The bracket contains six typical bridges: (1) *Staphylococcus aureus*, (2) *S. epidermidis*, (3) *Micrococcus roseus* and *Streptococcus thermophilus*, (4) *Lactobacillus viridescens*, (5) *Streptococcus salvarius,* and (6) *Leuconostoc cremoris*. The arrows indicate the polarity of peptide bonds running in the C to N direction. (b) An interpeptide bridge observed in *Corynebacterium poinsettiae.* The bridge extends between positions 2 and 4 and consists of the D-diamino acid ornithine (figure 3.20c). Note that L-homoserine (L-Hsr) is in position 3 rather than *meso*-diaminopimelic acid or L-lysine.

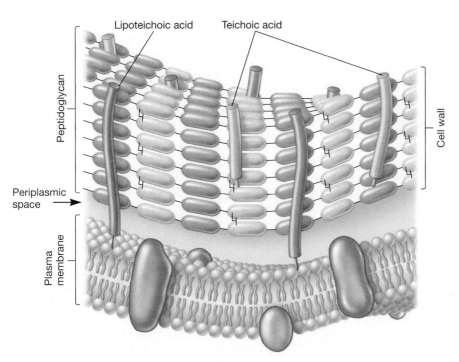

Figure 3.22 Typical Gram-Positive Cell Wall. This component of the cell envelope lies just outside the plasma membrane.

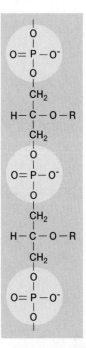

Figure 3.23 Teichoic Acid Structure. The segment of a teichoic acid made of phosphate, glycerol, and a side chain, R. R may represent D-alanine, glucose, or other molecules.

periplasm has relatively few proteins; this is probably because the peptidoglycan sacculus is porous and many proteins translocated across the plasma membrane pass through the sacculus. Some secreted proteins are enzymes called **exoenzymes.** Exoenzymes often serve to degrade polymers such as proteins and polysaccharides that would otherwise be too large for transport across the plasma membrane; the degradation products, the monomer building blocks, are then taken up by the cell. Those proteins that remain in the periplasmic space are usually attached to the plasma membrane.

In addition to the polymers embedded in the peptidoglycan sacculus, there are often proteins associated with its surface. These proteins are involved in interactions of the cell with its environment. Some are noncovalently bound to teichoic acids or other cell wall polymers. Other surface proteins are covalently attached to the peptidoglycan. Membrane-bound enzymes called sortases catalyze the formation of covalent bonds that join these proteins to the peptidoglycan. Many covalently attached proteins have roles in virulence. For example, the M protein of pathogenic streptococci aids in adhesion to host tissues and interferes with host defenses.

Typical Gram-Negative Cell Walls Include Additional Layers Besides Peptidoglycan

As just noted, most bacteria that stain Gram positive belong to the phyla *Firmicutes* and *Actinobacteria*. With a few exceptions,

bacteria belonging to the remaining phyla stain Gram negative (Microbial Diversity & Ecology 3.1). Even a brief inspection of figure 3.16 shows that typical Gram-negative cell walls are more complex than typical Gram-positive walls. One of the most striking differences is the paucity of peptidoglycan. The peptidoglycan layer is very thin (2 to 7 nm, depending on the bacterium) and sits within the periplasmic space.

The periplasmic space is usually 30 to 70 nm wide. Some studies indicate that it may constitute about 20 to 40% of the total cell volume. Thus it is much larger than that observed in typical Gram-positive cells. When cell walls are disrupted carefully or removed without disturbing the underlying plasma membrane, periplasmic enzymes and other proteins are released and may be easily studied. Some periplasmic proteins participate in nutrient acquisition—for example, hydrolytic enzymes and transport proteins. Some periplasmic proteins are involved in energy conservation. For instance, some bacteria have electron transport proteins in their periplasm (e.g., denitrifying bacteria, which convert nitrate to nitrogen gas). Other periplasmic proteins are involved in peptidoglycan synthesis and modification of toxic compounds that could harm the cell. ▶▶│ *Anaerobic respiration uses the same three steps as aerobic respiration (section 11.7), Nitrogen cycle (section 28.1)*

The outer membrane lies outside the thin peptidoglycan layer. It is linked to the cell by Braun's lipoprotein, the most

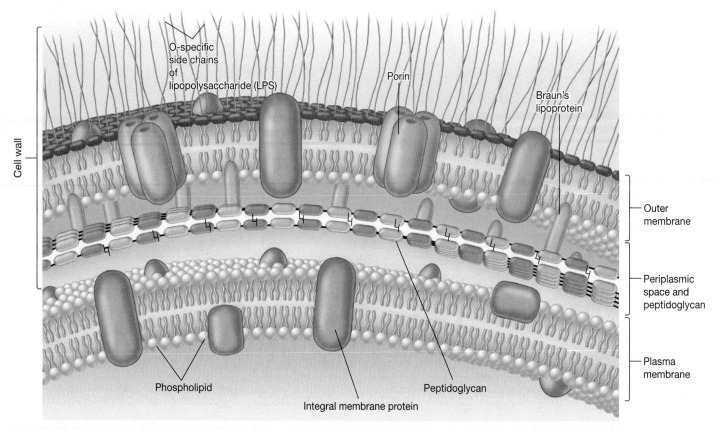

Figure 3.24 Typical Gram-Negative Cell Wall.   Notice that these bacteria are bounded by two membranes, the plasma membrane and the outer membrane of the cell wall.

MICRO INQUIRY *How does the outer membrane of the cell wall differ from the plasma membrane?*

abundant protein in the outer membrane (**figure 3.24**). This small lipoprotein is covalently joined to the underlying peptidoglycan and is embedded in the outer membrane by its hydrophobic end.

Possibly the most unusual constituents of the outer membrane are its **lipopolysaccharides** (**LPSs**). These large, complex molecules contain both lipid and carbohydrate, and consist of three parts: (1) lipid A, (2) the core polysaccharide, and (3) the O side chain. The LPS from *Salmonella* spp. has been studied most, and its general structure is described here (**figure 3.25**). **Lipid A** contains two glucosamine sugar derivatives, each with fatty acids and phosphate attached. The fatty acids of lipid A are embedded in the outer membrane, while the remainder of the LPS molecule projects from the surface. The **core polysaccharide** is joined to lipid A and is constructed of 10 sugars, many of them unusual in structure. The **O side chain** or **O antigen** is a polysaccharide chain extending outward from the core. It has several peculiar sugars and varies in composition between bacterial strains.

LPS has many important functions. (1) It contributes to the negative charge on the bacterial surface because the core polysaccharide usually contains charged sugars and phosphate (figure 3.25). (2) It helps stabilize outer membrane structure because lipid A is a major constituent of the exterior leaflet of the outer membrane. (3) It helps create a permeability barrier. The geometry of LPS (figure 3.25*b*) and interactions between neighboring LPS molecules are thought to restrict the entry of bile salts, antibiotics, detergents, and other toxic substances that might kill or injure the bacterium. (4) LPS helps protect pathogenic bacteria from host defenses. The O side chain of LPS is also called the O antigen because it elicits an immune response by an infected host. This response involves the production of antibodies that bind the strain-specific form of LPS that elicited the response. For example, microbiologists refer to specific strains of Gram-negative bacteria using the O antigen, such as *E. coli* O157; here the O side chain is the antigenic type number 157. Unfortunately, many bacteria can rapidly change the antigenic nature of their O side chains, thus thwarting host defenses. (5) Importantly, the lipid A portion of LPS can act as a toxin and is called endotoxin;

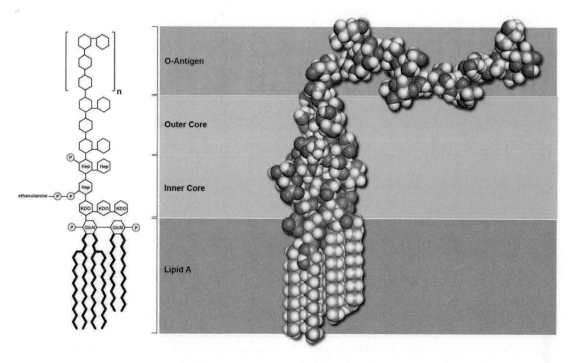

Figure 3.25 Lipopolysaccharide Structure. (a) A simplified diagram of LPS. Abbreviations: GlcN, glucosamine; Hep, heptulose; KDO, 2-keto-3-deoxyoctonate; P, phosphate. Lipid A is buried in the outer membrane. (b) Molecular model of a lipopolysaccharide. The lipid A and core polysaccharide are straight; the O side chain is bent at an angle in this model.

it causes some of the symptoms that arise in infections by Gram-negative pathogens. If LPS or lipid A enters the bloodstream, a form of septic shock develops for which there is no direct treatment. ▶▶❘ *Endotoxins (section 35.2); Antibodies are proteins that bind to specific 3-D molecules (section 34.7)*

As you have just seen, the makeup of the outer membrane differs from that of the plasma membrane. The two membranes also differ in terms of permeability. Even though LPS helps create a permeability barrier, the outer membrane is more permeable than the plasma membrane and permits the passage of small molecules such as glucose and other monosaccharides. This greater permeability is due to the presence of **porin proteins.** Most porin proteins cluster together to form a trimer in the outer membrane (figure 3.24 and **figure 3.26**). Each porin protein spans the outer membrane and is more or less tube-shaped; its narrow, water-filled channel allows passage of molecules smaller than about 600 daltons. However, larger molecules such as vitamin B_{12} cross the outer membrane through the action of specific outer membrane carrier proteins that deliver the molecule to ABC transporters in the plasma membrane.

Mechanism of Gram Staining

The difference between typical Gram-positive bacteria and typical Gram-negative bacteria is due to the physical nature of their

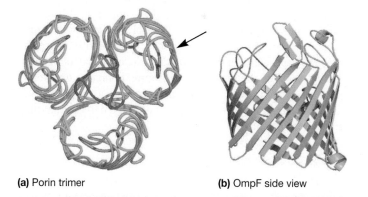

(a) Porin trimer   **(b)** OmpF side view

Figure 3.26 Porin Proteins. Two views of the OmpF porin of *E. coli*. (a) Structure of a trimeric porin observed when looking down at the outer surface of the outer membrane (i.e., top view). The center of each porin monomer is a water-filled channel. Each OmpF monomer can be divided into three loops: the green loop forms the channel, the blue loop interacts with other porin proteins to help form the trimer, and the orange loop narrows the channel. The arrow indicates the area of a porin molecule viewed from the side in panel (b). Side view of a porin monomer showing the β-barrel structure characteristic of porin proteins.

MICRO INQUIRY *Are these transporter proteins categorized as channels or carriers?*

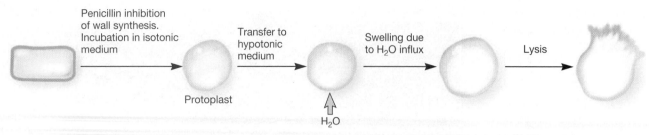

Figure 3.27 Protoplast Formation and Lysis.   Protoplast formation induced by incubation with penicillin in an isotonic medium. Transfer to a hypotonic medium will result in lysis.

cell walls. If the cell wall is removed, typical Gram-positive bacteria stain Gram negative. Furthermore, bacteria that never make cell walls, such as mycoplasmas, also stain Gram negative. ◀◀ *Differential staining (section 2.3)*

During the Gram-staining procedure, bacteria are first stained with crystal violet, a dye with a positive charge that is attracted to the bacterial cell's net negative charge, and next treated with iodine. Iodine is a mordant that interacts with the crystal violet, forming an insoluble complex and thus promoting dye retention. When bacteria are treated with ethanol in the decolorization step, the alcohol is thought to shrink the pores of the thick peptidoglycan found in the cell walls of typical Gram-positive bacteria, causing the peptidoglycan to act as a permeability barrier that prevents loss of crystal violet. Thus the dye-iodine complex is retained during the decolorization step and the bacteria remain purple, even after the addition of a second dye. In contrast, the peptidoglycan in typical Gram-negative cell walls is very thin, not as highly cross-linked, and has larger pores. Alcohol treatment also may extract enough lipid from the outer membrane to increase the cell wall's porosity further. For these reasons, alcohol more readily removes the crystal violet-iodine complex, decolorizing the cells. The counterstain safranin, also a dye with a net negative charge, easily stains the decolorized cells so that they appear red or pink.

Cell Walls and Osmotic Protection

Microbes have several mechanisms for responding to changes in osmotic pressure. This pressure arises when the concentration of solutes inside the cell differs from that outside, and the responses work to equalize the solute concentrations. However, in certain situations, osmotic pressure can exceed the cell's ability to acclimate. In these cases, additional protection is provided by the cell wall. When cells are in hypotonic solutions—ones in which the solute concentration is less than that in the cytoplasm—water diffuses into the cell, causing it to swell. Without the peptidoglycan layer of the cell wall, the pressure on the plasma membrane would become so great that the membrane would be disrupted and the cell would burst—a process called **lysis.** Conversely, in hypertonic solutions, water flows out and the cytoplasm shrivels up—a process called **plasmolysis.**

The protective nature of peptidoglycan is most clearly demonstrated when bacterial cells are treated with lysozyme or penicillin. The enzyme **lysozyme** attacks peptidoglycan by hydrolyzing the bond that connects *N*-acetylmuramic acid with *N*-acetylglucosamine (figure 3.17; *see also figure 33.4*). Penicillin works by a different mechanism. It inhibits the enzyme transpeptidase, which is responsible for making the cross-links between peptidoglycan chains. If bacteria are treated with either of these substances while in a hypotonic solution, they lyse. However, if they are in an isotonic solution, they can survive and grow normally. Treatment of typical Gram-positive bacteria with lysozyme or penicillin results in the complete loss of the cell wall, and the cell becomes a protoplast. When typical Gram-negative bacteria are exposed to lysozyme or penicillin, the peptidoglycan sacculus is destroyed, but the outer membrane remains. These cells are called **spheroplasts.** Both protoplasts and spheroplasts are osmotically sensitive. If they are transferred to a hypotonic solution, they lyse due to uncontrolled water influx (**figure 3.27**). ▶▶ *Antibacterial drugs (section 9.4)*

Bacteria That Lack Peptidoglycan or Cell Walls

Peptidoglycan is a signature molecule for bacteria. No other organisms have been found to synthesize it. However, members of the phyla *Chlamydiae* and *Planctomycetes,* which stain Gram negative, do not have peptidoglycan in their cell walls. These bacteria have an outer membrane like other Gram-negative bacteria, but there is no peptidoglycan layer between the plasma membrane and outer membrane. Chlamydia are intracellular parasites that cause disease in a variety of organisms, including humans. They also have complex life cycles in which they change from replicative forms to infective forms. Planctomycetes are of interest because some have cytoplasmic compartments surrounded by lipid membranes. Why these bacteria lack peptidoglycan is still an unanswered question. ▶▶ *Phylum* Planctomycetes *(section 21.5); Phylum* Chlamydiae *(section 21.6)*

Despite their lack of peptidoglycan, chlamydia and planctomycetes maintain a characteristic shape. This is not true of mycoplasmas, which lack a cell wall altogether; they tend to be pleomorphic (*see figure 21.4*). Mycoplasmas also are osmotically sensitive. Despite this, they often can grow in dilute media or terrestrial environments because their plasma membranes are more resistant to osmotic pressure than those of walled bacteria. The precise reason for this is not clear, although the presence of sterols in the membranes of many species may provide added strength.

Retrieve, Infer, Apply

1. Describe in detail the composition and structure of peptidoglycan. Why does peptidoglycan contain the unusual D-isomers of alanine and glutamic acid rather than the L-isomers observed in proteins?
2. List the major molecules that make up typical Gram-positive and Gram-negative cell walls and note which molecules are unique to each type of cell wall. How do these molecules contribute to the functions of the cell wall?
3. When protoplasts and spheroplasts are made, the shape of the cell becomes spherical regardless of the original cell shape. Why does this occur?
4. The cell walls of most members of the phyla *Firmicutes* and *Actinobacteria* lack porins. Why is this the case?
5. What two mechanisms allow the passage of nutrients across the outer membrane of typical Gram-negative bacteria?

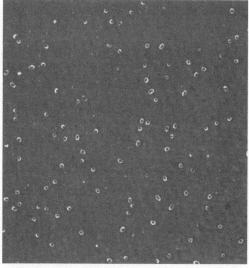

K. pneumoniae

Figure 3.28 Bacterial Capsules.   Capsule stain (a negative stain) of *Klebsiella pneumoniae*, bright-field light microscopy (×1,000).

3.5 The Cell Envelope Often Includes Layers Outside the Cell Wall

After reading this section, you should be able to:

■ Compile a list of the structures found in all the layers of bacterial cell envelopes, noting the functions and the major component molecules of each

Figures 3.22 and 3.24 show the bacterial envelope as consisting solely of the plasma membrane and cell wall. However, many bacteria have another layer in their cell envelopes that lies outside the cell wall. This layer is given different names depending on its makeup and how it is organized.

Capsules and Slime Layers

Capsules are layers that are well organized and not easily washed off (**figure 3.28**). They are most often composed of polysaccharides, but some are constructed of other materials. For example, *Bacillus anthracis* (the anthrax bacterium) has a proteinaceous capsule composed of poly-D-glutamic acid. Capsules are clearly visible in the light microscope when negative stains or specific capsule stains are employed; they also can be studied with the electron microscope.

Capsules are not required for growth and reproduction in laboratory cultures. However, they confer several advantages when bacteria grow in their normal habitats. They help pathogenic bacteria resist phagocytosis by host phagocytes. *Streptococcus pneumoniae*, which causes ear infections, pneumonia, and other diseases, provides a dramatic example. When it lacks a capsule, it is phagocytosed easily and does not cause disease. On the other

hand, the capsulated variant quickly kills mice. Capsules can also protect against desiccation because they contain a great deal of water. They exclude viruses and most hydrophobic toxic materials such as detergents.

A **slime layer** is a zone of diffuse, unorganized material that is removed easily. It is usually composed of polysaccharides but is not as easily observed by light microscopy. Gliding bacteria often produce slime, which in some cases has been shown to facilitate motility (section 3.8).

A bacterial **glycocalyx** is a layer consisting of a network of polysaccharides extending from the surface of the cell. The term can encompass both capsules and slime layers because they usually are composed of polysaccharides. The glycocalyx aids in attachment to solid surfaces, including tissue surfaces in plant and animal hosts. ▶▶◀ *Virulence factors (section 35.2)*

S-Layers

Many bacteria have a regularly structured layer called an **S-layer** on their surface. The S-layer has a pattern something like floor tiles and is composed of protein or glycoprotein (**figure 3.29**). In typical Gram-negative bacteria, the S-layer adheres directly to the outer membrane; it is associated with the peptidoglycan surface of typical Gram-positive cell walls.

S-layers are of considerable interest not only for their biological roles but also in the growing field of nanotechnology. Their biological roles include protecting the cell against ion and pH fluctuations, osmotic stress, enzymes, or predatory bacteria. The S-layer also helps maintain the shape and envelope rigidity of some cells, and it can promote cell adhesion to surfaces. Finally, the S-layer seems to protect some bacterial pathogens against host defenses, thus contributing to their virulence. The

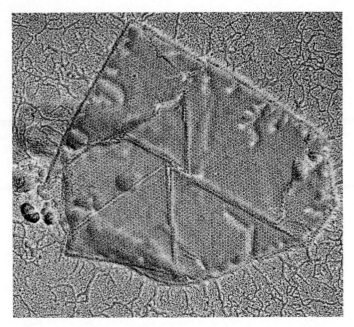

Figure 3.29 The S-Layer. An electron micrograph of the S-layer of the bacterium *Deinococcus radiodurans* after shadowing.

potential use of S-layers in nanotechnology is due to the ability of S-layer proteins to self-assemble; that is, S-layer proteins contain the information required to spontaneously associate and form the S-layer without the aid of any additional enzymes or other factors. Thus S-layer proteins could be used as building blocks for the creation of technologies such as drug-delivery systems and novel detection systems for toxic chemicals or bioterrorism agents.

Retrieve, Infer, Apply

1. What is the difference between a capsule and a slime layer? Why does the term glycocalyx usually encompass both?
2. S-layers and some capsules are composed of proteins. How does an S-layer differ from a proteinaceous capsule?

3.6 The Bacterial Cytoplasm Is More Complex than Once Thought

After reading this section, you should be able to:

- Create a table or concept map that identifies the components of the bacterial cytoplasm and describes their structure, molecular makeup, and functions

The plasma membrane and everything within is called the **protoplast.** The **cytoplasm** is the material bounded by the plasma membrane; thus the cytoplasm is a major part of the protoplast. The cytoplasm is further divided into the liquid component, called the **cytosol;** those structures such as inclusions, ribosomes,

and plasmids that float in the cytosol; and the many molecules dissolved in the cytosol. For many years, bacterial cells were thought of as bags of water, but the exciting discovery of cytoskeletal proteins in bacterial cells has forever changed that view. The bacterial cytoskeleton is less complex than that of eukaryotes. However, it is now clear that the bacterial cytoskeleton helps organize the cytoplasm. In this section, we consider the bacterial cytoskeleton and then a variety of other structures in the cytoplasm.

Bacterial Cytoskeleton

Eukaryotes possess three major cytoskeletal elements: actin filaments, microtubules, and intermediate filaments. Actin filaments are made from actin, and microtubules are made from tubulin. Intermediate filaments are composed of a mixture of one or more members of different classes of proteins. Homologues of all three types of eukaryotic proteins have been identified in bacteria (**table 3.2**). The bacterial cytoskeletal proteins are structurally similar to their eukaryotic counterparts and carry out similar functions: they participate in cell division, localize proteins to certain sites in the cell, and determine cell shape (table 3.2). In addition, some bacterial cytoskeletal proteins appear to be unique. Thus it is likely that the evolution of the cytoskeleton was an early event in the history of life on Earth.

▶▶| *The eukaryotic cytoplasm contains a complex cytoskeleton and many membranous organelles (section 5.3); Bacterial cell cycles can be divided into three phases (section 7.2)*

The **cytoskeletons** of *Escherichia coli, Bacillus subtilis,* and *Caulobacter crescentus* are the best studied and are the focus of our discussion. These three organisms are important bacterial model systems for several reasons. *E. coli* is a Gramnegative rod that has been extensively studied and can be easily manipulated. *B. subtilis* is a Gram-positive rod found in soil. It is an endospore-forming bacterium, making it a good model for cellular differentiation (section 3.9). Both were models for cell wall structure as described in section 3.4. *C. crescentus* is a curved rod found in aquatic habitats. It is interesting in part because it exhibits a complex life cycle that includes two different stages: a motile swarmer cell and a sessile, stalked cell that attaches to surfaces by a holdfast (*see figure 22.10*). ▶▶| Caulobacteraceae *and* Hyphomicrobiaceae *bacteria reproduce in unusual ways (section 22.1)*

The best studied bacterial cytoskeletal proteins are FtsZ, MreB, and CreS (also known as crescentin). FtsZ was one of the first bacterial cytoskeletal proteins identified and has since been found in most bacteria. FtsZ is a homologue of the eukaryotic protein tubulin. It forms a ring at the center of a dividing cell and is required for the formation of the septum that will separate the daughter cells (**figure 3.30***a*). MreB and its relative Mbl are actin homologues. Their major function is to determine cell shape in rod-shaped cells; MreB is not found in cocci. MreB and Mbl determine cell shape by properly positioning the machinery needed for peptidoglycan synthesis (figure 3.30*b,c*). CreS (crescentin) was discovered in *C. crescentus*

Table 3.2	Some Bacterial Cytoskeletal Proteins	
Type	**Function**	**Comments**
Tubulin Homologues		
FtsZ	Cell division	Widely observed in bacteria and archaea
BtubA/BtubB	Unknown	Observed only in *Prosthecobacter* spp.; thought to be encoded by eukaryotic tubulin genes obtained by horizontal gene transfer. In some species, Btub proteins assemble to form microtubule-like structures.
TubZ	Possibly plasmid segregation	Encoded by large plasmids observed in members of the genus *Bacillus*
Actin Homologues		
MamK	Positioning magnetosomes	Observed in magnetotactic species
MreB/Mbl	Helps determine cell shape; may be involved in chromosome segregation, localizing proteins, motility, and establishing cell polarity	Most rod-shaped bacteria
ParM	Plasmid segregation	Plasmid encoded
Intermediate Filament Homologues		
CreS (crescentin)	Induces curvature in curved rods	*Caulobacter crescentus*
Unique Bacterial Cytoskeletal Proteins		
MinD	Prevents polymerization of FtsZ at cell poles	Many rod-shaped bacteria
ParA	Segregates chromosomes and plasmids, helps localize chemotaxis proteins and type IV pili to one pole of certain rod-shaped bacteria	Observed in many species, including *Vibrio cholerae, C. crescentus,* and *Thermus thermophilus*

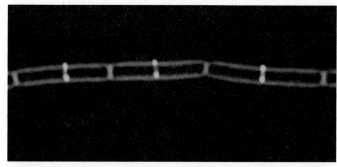

(a) FtsZ

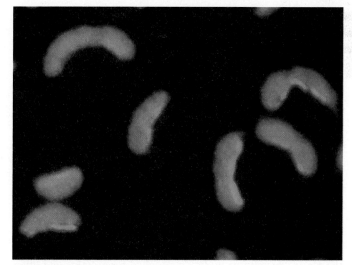

(d) Crescentin

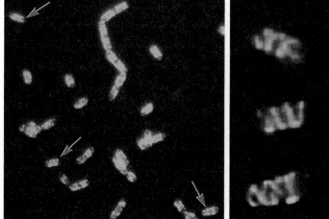

(b) Mbl

(c) Mbl

Figure 3.30 The Bacterial Cytoskeleton. (a) FtsZ protein in a chain of *Bacillus subtilis* cells; FtsZ-green fluorescent (GFP) fusion protein viewed by fluorescence microscopy. (b) The MreB-like cytoskeletal protein (Mbl) of *B. subtilis;* Mbl-GFP in live cells was examined by fluorescence microscopy. With this method, Mbl seems to form helices (arrows). More sensitive methods show that Mbl forms patches that move perpendicular to the long axis of the cell. (c) Three of the cells from (b) are shown at a higher magnification. (d) CreS (crescentin), in red, of *Caulobacter crescentus.* The DNA in the cells was stained blue with DAPI.

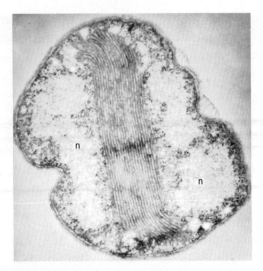

Figure 3.31 Internal Bacterial Membranes. The nitrifying bacterium *Nitrocystis oceanus* has parallel membranes traversing the whole cell. Note the nucleoid (n) with fibrillar structure.

and is responsible for its curved shape (figure 3.30*d*). CreS is a homologue of lamin and keratin, two intermediate filament proteins. ▶▶| *Cellular growth and determination of cell shape (section 7.2)*

Intracytoplasmic Membranes

Although members of *Bacteria* do not contain complex membranous organelles like mitochondria or chloroplasts, internal membranous structures are observed in some bacteria (**figure 3.31**). These can be extensive and complex in photosynthetic bacteria and in bacteria with very high respiratory activity, such as nitrifying bacteria. The internal membranes of the photosynthetic cyanobacteria are called thylakoids and are analogous to the thylakoids of chloroplasts. ▶▶| *Photosynthetic bacteria are diverse (section 21.4); Nitrifiying bacteria oxidize ammonium or nitrite to gain energy and electrons (section 22.1); Mitochondria, related organelles, and chloroplasts are involved in energy conservation (section 5.6)*

The internal membranous structures observed in bacteria may be aggregates of spherical vesicles, flattened vesicles, or tubular membranes (*see figure 22.3*). They are often connected to the plasma membrane and are thought to arise from it by invagination. However, these internal membranes differ from the plasma membrane by being enriched for proteins and other molecules involved in energy conservation. For instance, the thylakoids of cyanobacteria contain the chlorophyll and photosynthetic reaction centers responsible for converting light energy into ATP. Thus the function of internal membranes may be to provide a larger membrane surface for greater metabolic activity.

The membrane-bound organelle called the anammoxosome is observed in some members of the phylum *Planctomycetes* and deserves additional comment (*see figure 21.15*). This organelle is the site of anaerobic ammonia oxidation and

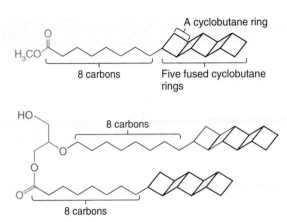

Figure 3.32 Ladderane Lipids Are Found in Anammoxosome Membranes. These lipids are unique to planctomycetes.

is unique to these bacteria. The anammoxosome membrane contains an unusual group of lipids called ladderane lipids (**figure 3.32**). Ladderane lipids are defined by the presence of two or more fused cyclobutane rings. They pack together very tightly, and this characteristic is thought to be important to anammoxosome function. ▶▶| *Phylum Planctomycetes (section 21.5)*

Inclusions

Inclusions are common in all cells. They are formed by the aggregation of substances that may be either organic or inorganic. The first bacterial inclusions were discovered in the late 1800s. Since then much has been learned about their structure and function. Inclusions can take the form of granules, crystals, or globules; some are amorphous. Some inclusions lie free in the cytoplasm. Other inclusions are enclosed by a shell that is single-layered and may consist of proteins or of both proteins and phospholipids. Some inclusions are surrounded by invaginations of the plasma membrane. Many inclusions are used for storage (e.g., of carbon compounds, inorganic substances, and energy) or to reduce osmotic pressure by tying up molecules in particulate form. The quantity of inclusions used for storage varies with the nutritional status of the cell. Some inclusions are so distinctive that they are increasingly being referred to as microcompartments. A brief description of several important inclusions follows.

Storage Inclusions

Cells have a wide variety of storage inclusions. Many are formed when one nutrient is in ready supply but another nutrient is not. Some store end products of metabolic processes. In some cases, these end products are used by the microbe when it is in different environmental conditions. The most common storage inclusions are glycogen inclusions, polyhydroxyalkonate granules, sulfur globules, and polyphosphate granules. Some storage inclusions are observed only in certain organisms, such as the cyanophycin granules in cyanobacteria. ▶▶| *Phylum Cyanobacteria (section 21.4)*

Carbon is often stored as polyhydroxyalkonate (PHA) granules. Several types of PHA granules have been identified, but the most common contain **poly-β-hydroxybutyrate (PHB)** (**figure 3.33**). The structure of PHB inclusions has been well studied, and PHB granules are now known to be surrounded by a single-layered shell composed of proteins and a small amount of phospholipids. Much of the interest in PHB and other PHA granules is due to their industrial use in making biodegradeable plastics. ▶▶| *Biopolymers (section 42.1)*

Polyphosphate granules and sulfur globules are inorganic inclusions observed in many organisms. **Polyphosphate granules** store the phosphate needed for synthesis of important cell constituents such as nucleic acids. In some cells, they act as an energy reserve, and polyphosphate also can serve as an energy source in some reactions, when the bond linking the final phosphate in the polyphosphate chain is hydrolyzed. Sulfur globules are formed by bacteria that use reduced sulfur-containing compounds as a source of electrons during their energy-conserving metabolic processes (**figure 3.34**). For example, some photosynthetic bacteria use hydrogen sulfide (rather than water) as an electron donor and accumulate the resulting sulfur either externally or internally. ▶▶| *Light reactions in anoxygenic photosynthesis (section 11.11); Class* Gammaproteobacteria *is the largest bacterial class (section 22.3)*

Microcompartments

Some bacterial inclusions are unique and serve functions other than simply storing substances for later use by the cell. These inclusions are called microcompartments. Microcompartments share several characteristics. They are relatively large polyhedrons formed by one

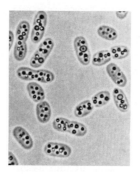

Figure 3.34 Sulfur Globules. *Chromatium vinosum*, a purple sulfur bacterium, with intracellular sulfur globules, bright-field microscopy (×2,000).

or more different proteins. Enclosed within the protein shell are one or more enzymes. Microcompartments include the ethanolamine utilization (Eut) microcompartment, the propandiol utilization (Pdu) microcompartment, and carboxysomes. We focus here on carboxysomes as they are the best studied.

Carboxysomes are present in many cyanobacteria and other CO_2-fixing bacteria (**figure 3.35**). Their polyhedral coat is composed of about six different proteins and is about 100 nm in diameter. Associated with the shell is the enzyme carbonic anhydrase, which converts carbonic acid into CO_2. Recall that biological membranes allow the free diffusion of CO_2. However, the carboxysome shell prevents CO_2 from escaping so it can accumulate within. Enclosed within the polyhedron is the enzyme ribulose-1, 5-bisphosphate carboxylase/oxygenase (RubisCO). RubisCO is the critical enzyme for CO_2 fixation, the process of converting CO_2 into sugar. Thus the carboxysome serves as a site for CO_2 fixation. As such, it is critical that carboxysomes be distributed to both daughter cells during cell division. A recent study has demonstrated that ParA, a cytoskeletal protein (table 3.2), helps ensure appropriate segregation of carboxysomes. ▶▶| *CO_2 fixation (section 12.3)*

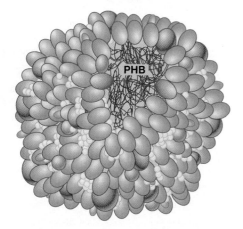

Figure 3.33 PHB Inclusions in Bacteria. PHB is a polymer of β-hydroxybutyrate molecules joined by ester bonds between the carboxyl and hydroxyl groups of adjacent molecules. PHB inclusions are around 0.2 to 0.7 μm in diameter. PHB is enclosed by a shell composed of several different proteins, including the PHB-synthesizing enzyme (red sphere) and the PHB-degrading enzyme (green sphere). Yellow spheres represent the phospholipids that are also found in the shell. Note that these phospholipids do not form a phospholipid bilayer.

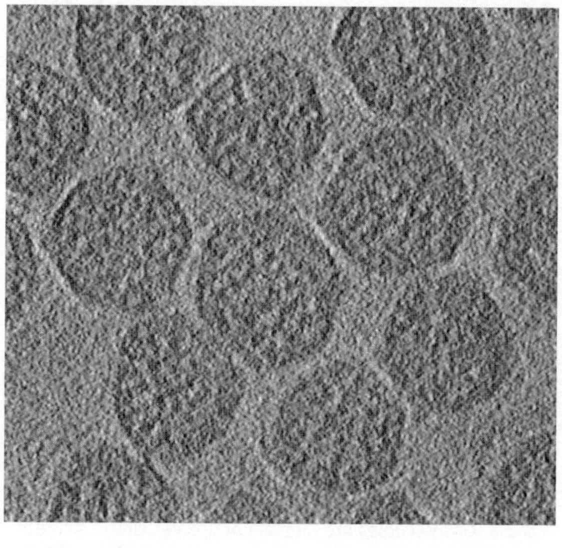

Figure 3.35 Carboxysomes. Carboxysomes in the bacterium *Halothiobacillus neapolitanus*. This is one image of a tilt series taken for electron cryotomography. Each carboxysome is approximately 100 nm in diameter.

Other Inclusions

Inclusions can be used for functions other than storage or as microcompartments. Two of the most remarkable inclusions are gas vacuoles and magnetosomes. Both are involved in bacterial movement.

The **gas vacuole** provides buoyancy to some aquatic bacteria, many of which are photosynthetic. Gas vacuoles are aggregates of enormous numbers of small, hollow, cylindrical structures called **gas vesicles** (figure 3.36). Gas vesicle walls are composed entirely of a single small protein. These protein subunits assemble to form a rigid cylinder that is impermeable to water but freely permeable to atmospheric gases. Cells with gas vacuoles can regulate their buoyancy to float at the depth necessary for proper light intensity, oxygen concentration, and nutrient levels. They descend by simply collapsing vesicles and float upward when new ones are constructed.

Aquatic magnetotactic bacteria use **magnetosomes** to orient themselves in Earth's magnetic field. Magnetosomes are intracellular chains of magnetite (Fe_3O_4) or greigite (Fe_3S_4) particles (figure 3.37). They are around 35 to 125 nm in diameter and enclosed within invaginations of the plasma membrane. The invaginations have been shown to contain distinctive proteins that are not found in the rest of the plasma membrane. Each iron particle is a tiny magnet: the Northern Hemisphere bacteria use their magnetosome chain to determine northward and downward directions, and swim down to nutrient-rich sediments or locate the optimum depth in freshwater and marine habitats. Magnetotactic bacteria in the Southern Hemisphere generally orient southward and downward, with the same result. For the cell to move properly within a magnetic field, magnetosomes must be arranged in a chain. A cytoskeletal protein called MamK is

currently thought to be responsible for establishing a framework upon which the chain can form (figure 3.37b).

Bacterial Ribosomes

Ribosomes are the site of protein synthesis, and large numbers of them are found in nearly all cells. The cytoplasm of bacterial cells is often packed with ribosomes, and other ribosomes may be loosely attached to the plasma membrane. The cytoplasmic ribosomes synthesize proteins destined to remain within the cell, whereas plasma membrane–associated ribosomes make proteins that will reside in the cell envelope or are transported to the outside.

Translation, the process of protein synthesis, is amazingly complex and is discussed in detail in chapter 13. This complexity is evidenced in part by the structure of ribosomes, which are made of numerous proteins and several ribonucleic acid (RNA) molecules. The ribosomes of all three domains of life share similarities. However, there are important differences. Here we focus strictly on bacterial ribosomes. We compare bacterial and

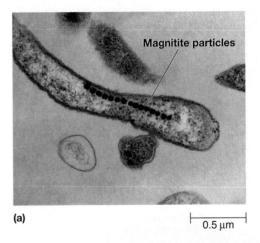

Magnitite particles

(a)

0.5 µm

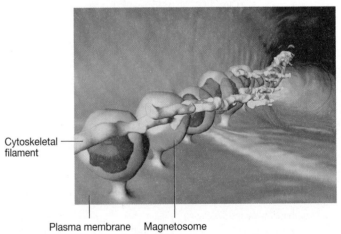

Cytoskeletal filament

Plasma membrane   Magnetosome

(b)

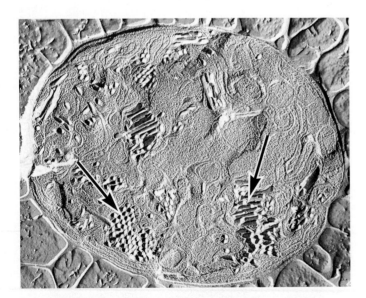

Figure 3.36 Gas Vacuoles Are Clusters of Gas Vesicles. A freeze-fracture preparation of *Anabaena flosaquae* (×89,000) showing gas vesicles and gas vacuoles. Both longitudinal and cross-sectional views of gas vesicles are indicated by arrows.

Figure 3.37 Magnetosomes.   (a) Transmission electron micrograph of the magnetotactic bacterium *Magnetospirillum magnetotacticum*. (b) Three-dimensional reconstruction of the magnetosomes of *M. magneticum* (electron cryotomograph).

archaeal ribosomes in chapter 4 and compare all three types of ribosomes in chapter 5.

Bacterial ribosomes are called 70S ribosomes and are constructed of a 50S and a 30S subunit (**figure 3.38**). The S in 70S and similar values stands for **Svedberg unit.** This is the unit of the sedimentation coefficient, a measure of sedimentation velocity in a centrifuge; the faster a particle travels when centrifuged, the greater its Svedberg value or sedimentation coefficient. The sedimentation coefficient is a function of a particle's molecular weight, volume, and shape. Heavier and more compact particles normally have larger Svedberg numbers and sediment faster.

Bacterial ribosomes are composed primarily of ribosomal RNA (rRNA) molecules. The small subunit contains 16S rRNA, whereas the large subunit consists of 23S and 5S rRNA molecules. Approximately 55 proteins make up the rest of the mass of the ribosome: 21 in the small subunit, and 34 in the large subunit.

Nucleoid

The **nucleoid** is an irregularly shaped region that contains the cell's chromosome and numerous proteins (**figure 3.39**). The chromosomes of most bacteria are a single circle of double-stranded **deoxyribonucleic acid (DNA)**, but some bacteria have a linear chromosome, and some bacteria, such as *Vibrio cholerae* and *Borrelia burgdorferi* (causative agents of cholera and Lyme disease, respectively), have more than one chromosome. A few bacteria (e.g., the very large bacteria *Epulopiscium* spp. and *Thiomargarita* sp.; p. 45) are polyploid.

Bacterial chromosomes are longer than the length of the cell. An important and still unanswered question is how these microbes manage to fit their chromosomes into the relatively small space occupied by the nucleoid. For instance, *E. coli*'s circular chromosome measures approximately 1,400 μm, or about 230–700 times longer than the cell (figure 3.39*b*). Thus the chromosome must be organized and packaged in a manner that decreases its overall size. Supercoiling is thought to be important. It produces a dense, central core of DNA with loops of DNA extending out from the core. Several nucleoid-associated proteins (NAPs) help compact the chromosome by causing the chromosome to bend and fold. Some NAPs form bridges between one section of the chromosome and another. NAPs are particularly important during cell division, when they further compact the chromosomes. This extra level of packing is important for proper segregation of daughter chromosomes during cell division. ▶▶ *DNA is a polymer of deoxyribonucleotides (section 13.2)*

For most bacteria, the nucleoid is simply a region in the cytoplasm; it is not separated from other components of the cytoplasm by a membrane. However, there are a few exceptions. Membrane-bound DNA-containing regions are present in at least two genera of the unusual bacterial phylum *Planctomycetes* (see figure 21.15). *Pirellula* has a single membrane that surrounds a region called the pirellulosome, which contains a fibrillar nucleoid and ribosome-like particles. The nuclear body of *Gemmata obscuriglobus* is bounded by two membranes. More work is required to determine the functions of these membranes and how widespread this phenomenon is. ▶▶ *Phylum* Planctomycetes *(section 21.5)*

Plasmids

In addition to the genetic material present in the nucleoid, many bacteria contain extrachromosomal DNA molecules called plasmids. Indeed, most of the bacterial genomes sequenced

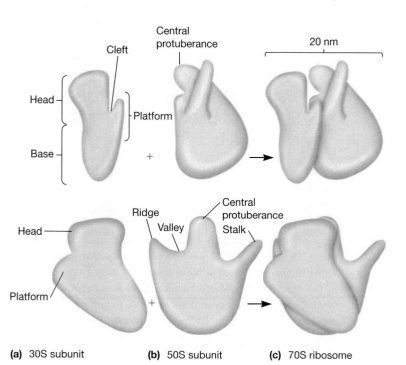

(a) 30S subunit   **(b)** 50S subunit   **(c)** 70S ribosome

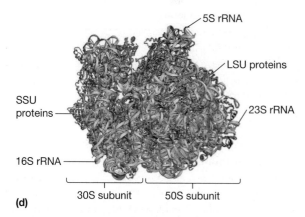

(d)

Figure 3.38 Bacterial Ribosomes. (a–c) Schematic representation of the two subunits and the complete 70S ribosome of *E. coli*. (d) Molecular structure of the 70S ribosome of *Thermus thermophilus*. The 50S subunit (LSU) includes 23S rRNA (gray) and 5S rRNA (lavender), while 16S rRNA (turquoise) is found in the 30S subunit (SSU). A molecule of tRNA (gold) is shown in the A site. To generate this ribbon diagram, crystals of purified bacterial ribosomes were prepared and exposed to X rays, and the resulting diffraction pattern analyzed.

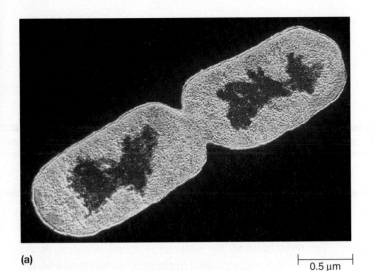

(a)

0.5 μm

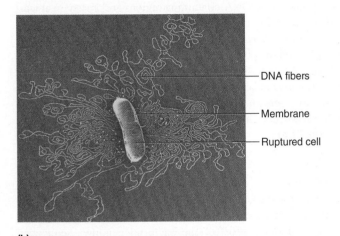

DNA fibers

Membrane

Ruptured cell

(b)

Figure 3.39 *E. coli* **Nucleoids and Chromosomes.** Bacterial chromosomes are located in the nucleoid, an area in the cytoplasm. (a) A color-enhanced transmission electron micrograph of a thin section of a dividing *E. coli* cell. The red areas are the nucleoids present in the two daughter cells. (b) Chromosome released from a gently lysed *E. coli* cell. Note how tightly packaged the DNA must be inside the cell.

thus far include plasmids. In some cases, numerous different plasmids within a single species have been identified. For instance, *B. burgdorferi* carries 12 linear and nine circular plasmids. Plasmids play many important roles in the lives of the organisms that have them. They also have proved invaluable to microbiologists and molecular geneticists in constructing and transferring new genetic combinations and in cloning genes, as described in chapter 17.

Plasmids are small, double-stranded DNA molecules that can exist independently of the chromosome. Both circular and linear plasmids have been documented, but most known plasmids are circular. Plasmids have relatively few genes, generally less than 30. Their genetic information is not essential to the bacterium, and cells that lack them usually function normally. However,

many plasmids carry genes that confer a selective advantage to the bacterium in certain environments.

Plasmids use the cell's DNA-synthesizing machinery to replicate, but their replication is not linked to any particular stage of the cell cycle. Thus regulation of plasmid and chromosomal replication are independent. However, some plasmids are able to integrate into the chromosome. Such plasmids are called **episomes** and when integrated are replicated as part of the chromosome. Plasmids are inherited stably during cell division, but they are not always equally apportioned into daughter cells and sometimes are lost. The loss of a plasmid is called **curing.** It can occur spontaneously or be induced by treatments that inhibit plasmid replication but not host cell reproduction. Some commonly used curing treatments are acridine mutagens, ultraviolet and ionizing radiation, thymine starvation, antibiotics, and growth above optimal temperatures.

Plasmids may be classified in terms of their mode of existence, spread, and function. A brief summary of the types of bacterial plasmids and their properties is given in **table 3.3**. These various types of plasmids can also differ in terms of the number of copies found within the cell. Single-copy plasmids produce only one copy per host cell. Multicopy plasmids may be present at concentrations of 40 or more per cell.

Retrieve, Infer, Apply

1. Briefly describe the nature and function of the cytoplasm, and the regions and structures within it. How is the cytosol different from the cytoplasm?
2. List the most common kinds of inclusions. How are they similar to eukaryotic organelles such as mitochondria and chloroplasts? How do they differ?
3. How do plasmids differ from chromosomes? What is an episome?
4. Explain the importance of each of the following plasmids: conjugative plasmid, R plasmid, Col plasmid, virulence plasmid, and metabolic plasmid.

3.7 Many Bacteria Have External Structures Used for Attachment and Motility

After reading this section, you should be able to:

- Distinguish pili (fimbriae) and flagella
- Illustrate the various patterns of flagella distribution

Bacteria must constantly respond to their changing environments. Sometimes, it is advantageous to attach to a surface. Other times, it is better to move toward or away from something in the environment. Many bacteria have structures that extend beyond the cell envelope and are involved in either attachment to surfaces or motility. In addition, these external structures can function in protection and horizontal gene transfer. Several are discussed in this section.

Table 3.3	Major Types of Bacterial Plasmids				
Type	**Function**	**Example**	**Size (kbp)**	**Hosts**	**Phenotypic Features[1]**
Conjugative Plasmids[2]	Transfer of DNA from one cell to another	F factor	95–100	*E. coli, Salmonella, Citrobacter*	Sex pilus, conjugation
R Plasmids	Carry antibiotic-resistance genes	RP4	54	*Pseudomonas* and many other Gram-negative bacteria	Sex pilus, conjugation, resistance to Amp, Km, Nm, Tet
Col Plasmids	Produce bacteriocins, substances that destroy closely related species	ColE1	9	*E. coli*	Colicin E1 production
Virulence Plasmids	Carry virulence genes	Ti	200	*Agrobacterium tumefaciens*	Tumor induction in plants
Metabolic Plasmids	Carry genes for enzymes	CAM	230	*Pseudomonas*	Camphor degradation

[1] Abbreviations used for resistance to antibiotics: Amp, ampicillin; Gm, gentamycin; Km, kanamycin; Nm, neomycin; Tet, tetracycline.
[2] Many R plasmids, metabolic plasmids, and others are also conjugative.

Bacterial Pili and Fimbriae

Many bacteria have fine, hairlike appendages that are thinner and typically shorter than flagella. These are usually called **fimbriae** (s., **fimbria**) or **pili** (s., **pilus**). The terms are synonymous, although certain structures are always called pilus (e.g., sex pilus), while others are always called fimbriae. We will use the terms interchangeably, except in those instances. A cell may be covered with up to 1,000 fimbriae, but they are only visible in an electron microscope due to their small size (**figure 3.40**). They are slender tubes composed of helically arranged protein subunits and are about 3 to 10 nm in diameter and up to several micrometers long. Pili grow by adding protein subunits to the base of the pilus. Several different types of fimbriae have been identified. Most function to attach cells to solid surfaces such as rocks in streams and host tissues. The pili of some bacteria function as nanowires that transport electrons from a cell to solid metals during their metabolic activities. Type IV pili are involved in motility (section 3.8) and two gene transfer mechanisms: bacterial transformation and bacterial conjugation. ▶▶| *Bacterial conjugation requires cell–cell contact (section 16.6); Bacterial transformation is the uptake of free DNA from the environment (section 16.7)*

Many bacteria have up to 10 **sex pili** (s., **sex pilus**) per cell. These hairlike structures differ from other pili in the following ways. Sex pili often are larger than other pili (around 9 to 10 nm in diameter). They are genetically determined by conjugative plasmids and are required for conjugation. Some bacterial viruses attach specifically to sex pili at the start of their multiplication cycle

Bacterial Flagella

Many motile bacteria move by use of **flagella** (s., **flagellum**), threadlike locomotor appendages extending outward from the plasma membrane and cell wall. Although the main function of

flagella is motility, they can have other roles. They can be involved in attachment to surfaces, and in some bacteria, they are virulence factors.

Bacterial flagella are slender, rigid structures about 20 nm across and up to 20 μm long. Flagella are so thin they cannot be observed directly with a bright-field microscope but must be stained with techniques designed to increase their thickness. The detailed structure of a flagellum can only be seen in the electron microscope.

Bacterial species often differ in their patterns of flagella distribution, and these patterns are useful in identifying bacteria. **Monotrichous** bacteria (*trichous* means hair) have one flagellum; if it is located at an end, it is said to be a **polar flagellum** (**figure 3.41***a*). **Amphitrichous** bacteria (*amphi* means on both

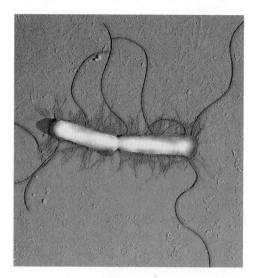

Figure 3.40 Flagella and Fimbriae. The long flagella and numerous shorter fimbriae are evident in this SEM of the bacterium *Proteus vulgaris*.

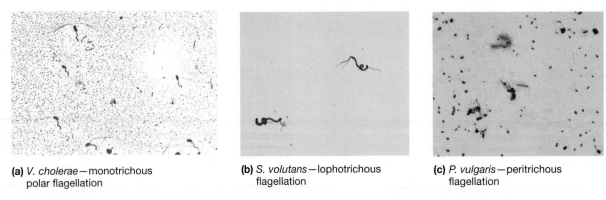

(a) *V. cholerae*—monotrichous polar flagellation

(b) *S. volutans*—lophotrichous flagellation

(c) *P. vulgaris*—peritrichous flagellation

Figure 3.41 Flagellar Distribution. Examples of various patterns of flagellation as seen in the light microscope. (a) Flagella stain of *Vibrio cholerae*, digitally colorized. (b) *Spirillum volutans* (×1,000). Only the tuft of flagella at one end is clearly visible. (c) *Proteus vulgaris* (×1,000).

sides) have a single flagellum at each pole. In contrast, **lophotrichous** bacteria (*lopho* means tuft) have a cluster of flagella at one or both ends (figure 3.41*b*). Flagella are spread evenly over the whole surface of **peritrichous** (*peri* means around) bacteria (figure 3.41*c*).

Transmission electron microscope studies have shown that the bacterial flagellum is composed of three parts (**figure 3.42**). (1) The **filament** is the longest and most obvious portion. It extends from the cell surface to the tip. (2) The **basal body** is embedded in the cell envelope; and (3) a short, curved segment, the **hook,** links the filament to its basal body and acts as a flexible coupling.

The filament is a hollow, rigid cylinder constructed of subunits of the protein **flagellin,** which ranges in molecular mass from 30,000 to 60,000 daltons, depending on the bacterial species. The filament ends with a capping protein. Some bacteria have sheaths surrounding their flagella. For example, *Vibrio cholerae* flagella have lipopolysaccharide sheaths.

The hook and basal body are quite different from the filament (figure 3.42). Slightly wider than the filament, the hook is made of different protein subunits. The basal body is the most complex part of a flagellum. The basal bodies of *E. coli* and most other typical Gram-negative bacteria have four rings: L, P, MS, and C, which are connected to a central rod (figure 3.42*a*). The L, P, and MS rings are embedded in the cell envelope, and the C ring is on the cytoplasmic side of the MS ring. Typical Gram-positive bacteria have only two rings: an inner ring connected to the plasma membrane and an outer one probably attached to the peptidoglycan (figure 3.42*b*).

The synthesis of bacterial flagella is complex and involves at least 20 to 30 genes. Besides the gene for flagellin, 10 or more genes code for hook and basal body proteins; other genes are concerned with control of flagellar construction or function. Many components of the flagellum lie outside the cell envelope and are transported out of the cell for assembly. Interestingly, the basal body is a specialized version of the type III protein secretion system observed in typical Gram-negative bacteria. Type III secretion systems have a needlelike structure through which proteins are secreted. In the flagellar type III secretion system, the filament replaces the needle. Individual flagellin subunits are transported through the hollow filament. When the subunits reach the tip, they spontaneously aggregate under the direction of a protein called the filament cap; thus the filament grows at its tip rather than at the base (**figure 3.43**). Filament synthesis, like S-layer formation, is an example of **self-assembly.** ▶▶❙ *Protein maturation and secretion (section 13.8)*

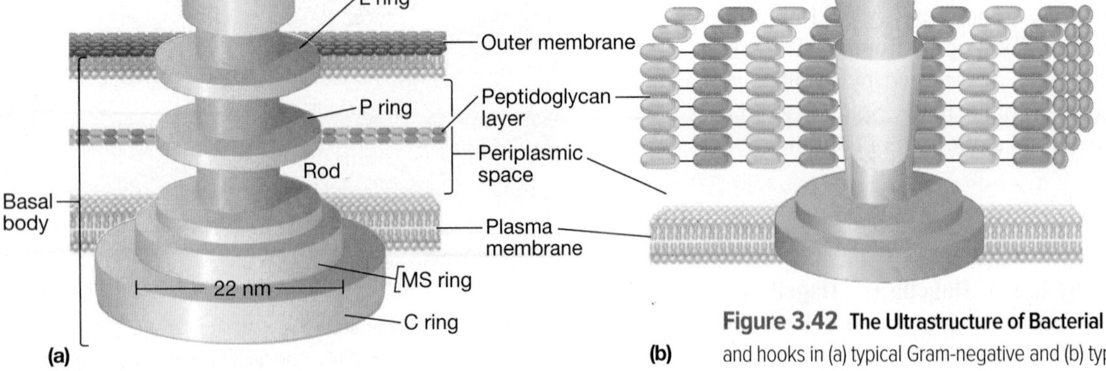

Filament

Hook

L ring

Outer membrane

P ring

Peptidoglycan layer

Rod

Periplasmic space

Basal body

Plasma membrane

22 nm

MS ring

C ring

(a)

(b)

Figure 3.42 The Ultrastructure of Bacterial Flagella. Flagellar basal bodies and hooks in (a) typical Gram-negative and (b) typical Gram-positive bacteria.

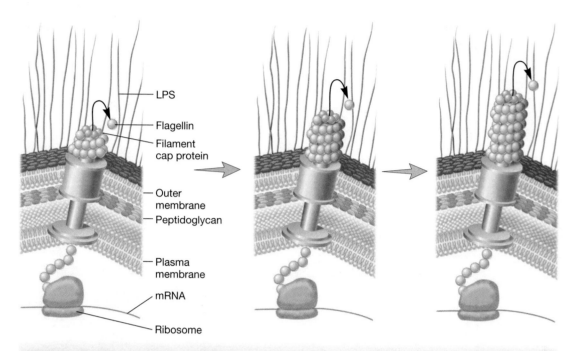

Figure 3.43 Flagellar Filaments Grow at the Tip. Flagellin subunits travel through the flagellar core and attach to the growing tip. Their attachment is directed by the filament cap protein.

MICRO INQUIRY *How does flagellum growth compare to pili growth?*

Labels: LPS, Flagellin, Filament cap protein, Outer membrane, Peptidoglycan, Plasma membrane, mRNA, Ribosome

---

Retrieve, Infer, Apply

1. What are the functions of fimbriae (pili) and sex pili?
2. What terms are used to describe the different flagella distribution patterns observed among bacteria?
3. What is self-assembly? Why does it make sense that the filament of a flagellum is assembled in this way?

3.8 Bacteria Move in Response to Environmental Conditions

After reading this section, you should be able to:

- Compare and contrast flagellar swimming motility, swarming, spirochete flagellar motility, and twitching and gliding motility
- State the source of energy that powers flagellar motility
- Explain why bacterial chemotaxis is referred to as a "biased random walk"

As we note in section 3.7, several structures extending beyond bacterial cell envelopes contribute to motility. Five major methods of movement have been observed: swimming movement conferred by flagella, flagella-mediated swarming, corkscrew movement of spirochetes, twitching motility associated with type IV pili, and gliding motility.

Motile bacteria do not move aimlessly. Rather, motility is used to move toward nutrients such as sugars and amino acids and away from many harmful substances and bacterial waste products. Movement toward chemical attractants and away from repellents is known as chemotaxis. Motile bacteria also can move in response to environmental cues such as temperature (thermotaxis), light (phototaxis), oxygen (aerotaxis), osmotic pressure (osmotaxis), and gravity.

Flagellar Movement

Swimming

The filament of a bacterial flagellum is in the shape of a rigid helix, and the cell moves when this helix rotates like a propeller on a boat. The flagellar motor can rotate very rapidly. The *E. coli* motor rotates 270 revolutions per second (rps); *Vibrio alginolyticus* averages 1,100 rps. For many bacteria in an aquatic environment, flagellar rotation results in two types of movement: a smooth swimming movement often called a **run,** which actually moves the cell from one spot to another, and a **tumble,** which serves to reorient the cell. As we shall see in our discussion of chemotaxis (p. 74), alternating between smooth swims and changes in direction is important for responding to environmental conditions. Often, the direction of flagellar rotation determines whether a run or a tumble occurs. For example, many bacteria with monotrichous, polar flagella use a counterclockwise rotation for a run (**figure 3.44**). When rotation is reversed, the cell tumbles. Many peritrichously flagellated bacteria operate in a somewhat similar way. To move forward in a run, the

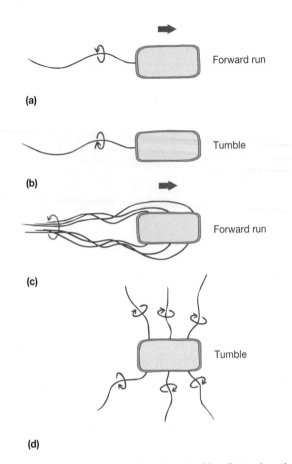

(a) Forward run

(b) Tumble

(c) Forward run

(d) Tumble

Figure 3.44 The Direction of Flagellar Rotation Often Determines the Way a Bacterium Moves. Parts (a) and (b) describe the motion of many monotrichous, polar bacteria. Parts (c) and (d) illustrate the movements of many peritrichous organisms.

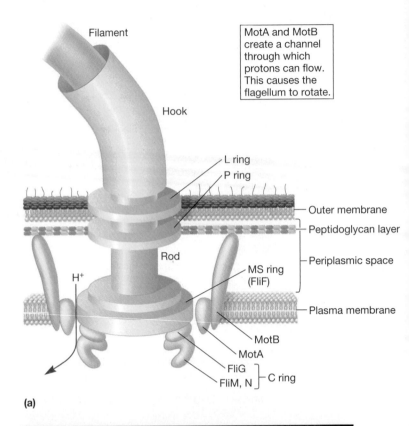

MotA and MotB create a channel through which protons can flow. This causes the flagellum to rotate.

Filament

Hook

L ring
P ring
Outer membrane
Peptidoglycan layer
Periplasmic space
Rod
MS ring (FliF)
H^+
Plasma membrane
MotB
MotA
FliG
FliM, N — C ring

(a)

(b)

Figure 3.45 Mechanism of Flagellar Movement. (a) This diagram of a flagellum shows some of the more important components and the flow of protons that drives rotation. Some of the many flagellar proteins are labeled. (b) A three-dimensional electron cryotomographic reconstruction of the flagellar motor.

MICRO INQUIRY *Would this flagellum be found in a typical Gram-negative or Gram-positive bacterium?*

flagella rotate counterclockwise. As they do so, the flagella bend at their hooks to form a rotating bundle that propels the cell forward. Clockwise rotation of the flagella disrupts the bundle and the cell tumbles.

Not all bacteria use runs and tumbles for swimming motility. For instance, *Rhodobacter sphaeroides* cells alternate between rotating their single flagellum in one direction (run) and no rotation, a so-called run-stop motility. When the cells are stopped, molecules in the environment bombard the cells and cause them to make small changes in orientation. When they resume a run, they move in a new direction. Another type of swimming motility is seen with the monotrichous bacterium *Vibrio alginolyticus*. It uses a run-reverse-flick pattern. It swims forward (run) when the flagellum rotates in one direction. When rotation reverses, the cell moves backward (reverse). Just as the rotation switches again for a run, the flagellum flicks, causing the cell to change its orientation and move in a new direction.

The motor that drives flagellar rotation is located at the base of the flagellum. Torque generated by the motor is transmitted to the hook and filament. The motor is composed of two components: the rotor and the stator. It is thought to function like an electrical motor, where the rotor turns in the center of a ring

of electromagnets, the stator. In typical Gram-negative bacteria, the rotor is composed of the MS ring and the C ring (**figure 3.45**). The C ring protein, FliG, is particularly important because it is thought to interact with the stator. The stator is composed of the proteins MotA and MotB, which form a channel through the plasma membrane. MotB also anchors MotA to cell wall peptidoglycan.

As with all motors, the flagellar motor must have a power source that allows it to generate torque and cause flagellar rotation.

The power used by most flagellar motors is a difference in charge and pH across the plasma membrane. This difference is called the proton motive force (PMF). PMF is largely created by the metabolic activities of organisms, as described in chapter 11. One important metabolic process is the transfer of electrons from an electron donor to a terminal electron acceptor via a chain of electron carriers called the electron transport chain (ETC). In bacterial cells, most components of the ETC are located in the plasma membrane. As electrons are transported down the ETC, protons are transported from the cytoplasm to the outside of the cell. Because there are more protons outside the cell than inside, the outside has more positively charged ions (the protons) and a lower pH. PMF is a type of potential energy that can be used to do work: mechanical work, as in the case of flagellar rotation; transport work, the movement of materials into or out of the cell (section 3.3); or chemical work such as the synthesis of ATP, the cell's major energy currency.

How can PMF be used to power the flagellar motor? The channels created by the MotA and MotB proteins allow protons to move across the plasma membrane from the outside to the inside (figure 3.45). Thus the protons move down the charge and pH gradient. This movement releases energy that is used to rotate the flagellum. In essence, the entry of a proton into the channel is like the entry of a person into a revolving door. The "power" of the proton generates torque, rather like a person pushing the revolving door. Indeed, the speed of flagellar rotation is proportional to the magnitude of the PMF.

The flagellum is a very effective swimming device. From the bacterium's point of view, swimming is quite a difficult task because the surrounding water seems as viscous as molasses. The cell must bore through the water with its corkscrew-shaped flagella, and if flagellar activity ceases, it stops almost instantly. Despite such environmental resistance to movement, bacteria can swim from 10 to 100 μm per second. This is equivalent to traveling from 2 to over 150 cell lengths per second. In contrast, an exceptionally fast human might be able to run around 5 to 6 body lengths per second.

Swarming

An increasing number of bacterial species has been found to exhibit an interesting type of motility called swarming. This motility occurs on moist surfaces and is a type of group behavior in which cells move in unison across the surface. Most bacteria that swarm have peritrichous flagella. Many also produce and secrete molecules that help them move across the substrate. When bacteria that swarm are cultured in the laboratory on appropriate solid media, they produce characteristic colony morphologies (**figure 3.46**).

Spirochete Motility

Spirochetes have flagella that work in a distinctive manner. In many spirochetes, multiple flagella arise from each end of the cell and wind around the cell (**figure 3.47**). The flagella do not extend outside the cell wall but rather remain in the periplasmic space and are covered by the outer membrane. They are called

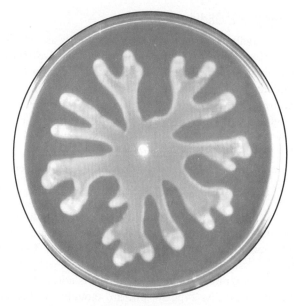

Figure 3.46 Swarming Bacteria Often Produce Distinctive Patterns on a Solid Growth Medium. These bacterial cells swarmed out from the center of the plate and produced a branching pattern called dendrites.

periplasmic flagella and are thought to rotate like the external flagella of other bacteria, causing the corkscrew-shaped outer membrane to rotate and move the cell through the surrounding liquid, even very viscous liquids. Flagellar rotation may also flex or bend the cell and account for the creeping or crawling movement observed when spirochetes are in contact with a solid surface. ▶▶❙ *Phylum* Spirochaetes *(section 21.8)*

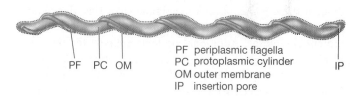

PF PC OM

PF periplasmic flagella
PC protoplasmic cylinder
OM outer membrane
IP insertion pore

IP

(a)

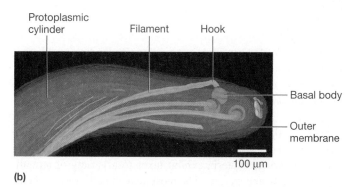

Protoplasmic cylinder   Filament   Hook

Basal body

Outer membrane

100 μm

(b)

Figure 3.47 Spirochete Flagella. (a) Numerous flagella arise from each end of the spirochete. These intertwine and wind around the cell, usually overlapping in the middle. (b) Electron cryotomographic image of the spirochete *Treponema denticola* showing three flagella arising from the tip of the cell.

Twitching and Gliding Motility

Twitching and gliding motility occur when cells are on a solid surface. However, unlike swarming, neither involves flagella. Gliding motility varies greatly in rate (from 2 to over 600 μm per minute) and in the nature of the motion. Although first observed over 100 years ago, the mechanism by which many bacteria glide remains a mystery. We describe the gliding motility of *Flavobacterium* spp. and *Mycoplasma* spp. in detail in chapter 21. ▶▶| *Phylum* Bacteroidetes *includes important gut microbiota (section 21.9); Class* Mollicutes, *Phylum* Tenericutes *(section 21.3)*

Here we focus on *Myxococcus xanthus* because it exhibits both twitching and gliding motility. Its twitching motility is called social (S) motility because it occurs when large groups of cells move together in a coordinated fashion. **Twitching motility** is characterized by short, intermittent, jerky motions of up to several micrometers in length and is normally seen on very moist surfaces. Type IV pili are thought to alternately extend and retract to move cells during twitching motility. The extended pilus contacts the surface at a point some distance from the cell body. When the pilus retracts, the cell is pulled forward. Hydrolysis of ATP likely powers the extension/retraction process. Although we have focused on *Myxococcus xanthus,* numerous other bacteria use type IV pili for twitching motility.

In contrast to the jerky movement of twitching motility, **gliding motility** is smooth. The gliding motility exhibited by *Myxococcus xanthus* is called adventurous (A) motility; it is observed when single cells move independently. In *Myxococcus xanthus,* proteins similar to MotA and MotB of the flagellar motor function as the motors for gliding motility. They are thought to be in the plasma membrane (like flagellar motors) and are associated with other proteins, forming relatively large protein complexes. Additional proteins connect the motors to MreB, a cytoskeletal protein (section 3.6). Other proteins connect the motors to the substrate along which the cell glides. The motors are powered by PMF. How the action of the motor complexes are translated into gliding motility is still unknown. ▶▶| *Order* Myxococcales *(section 22.4)*

Chemotaxis

Imagine a motile bacterium in an ocean. Depending on where it is in the water column, light, oxygen, and nutrient levels will vary. These also vary over time due to the activities of all the organisms present. In order to position itself in the most beneficial location in the water column, this bacterium needs to sense changes in the environment and move accordingly. The movement toward or away from attractants or repellents is called taxis. As noted earlier, bacteria exhibit taxes to a variety of stimuli, including light and oxygen. The movement of cells toward chemical attractants or away from chemical repellents (**chemotaxis**) is the best-studied type of taxis. Chemotaxis is readily observed in Petri dish cultures. If bacteria are placed in the center of a dish of semisolid agar containing an attractant, the bacteria will exhaust the local supply of the nutrient and swim outward following the

attractant gradient they have created. The result is an expanding ring of bacteria. When a disk of repellent is placed in a Petri dish of semisolid agar and bacteria, the bacteria will swim away from the repellent, creating a clear zone around the disk.

Attractants and repellents are detected by **chemoreceptors,** proteins that bind chemicals and transmit signals to other components of the chemosensing system. Chemosensing systems are very sensitive and allow the cell to respond to very low levels of attractants (about 10^{-8} M for some sugars). In typical Gram-negative bacteria, the chemoreceptor proteins are located in the plasma membrane. Some receptors also participate in the initial stages of sugar transport into the cell.

The chemotactic behavior of *E. coli* has been extensively studied and is our focus here. Its movements can be followed using a tracking microscope, a microscope with a moving stage that automatically keeps an individual bacterium in view. In the absence of a chemical gradient, *E. coli* cells move randomly, switching back and forth between a run and a tumble. During a run, the bacterium swims in a straight or slightly curved line. After a few seconds, the bacterium stops and tumbles. The tumble randomly reorients the cell so that it is facing in a different direction. Therefore when it begins the next run, it usually goes in a different direction (**figure 3.48***a*). In contrast, when *E. coli* is exposed to an attractant, it tumbles less frequently (or has longer runs) when traveling toward the attractant. Although the tumbles can still orient the bacterium away from the attractant, over time the cell gets closer and closer to the attractant (figure 3.48*b*). The opposite response occurs with a repellent. Tumbling frequency decreases (the run time lengthens) when the bacterium moves away from the repellent.

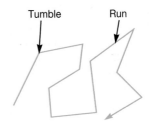

(a)

(b)

Figure 3.48 Tumbles and Runs Are Used to Direct the Movement of Many Bacteria Toward a Chemical Attractant. (a) Random movement of a bacterium in the absence of a concentration gradient. Tumbling frequency is fairly constant. (b) Movement in an attractant gradient. Tumbling frequency is reduced when the bacterium is moving up the gradient. Therefore runs in the direction of increasing attractant are longer.

E. coli must have some mechanism for sensing that it is getting closer to the attractant (or moving away from the repellent). The behavior of the bacterium is shaped by temporal changes in chemical concentration. The cell is able to compare the current concentration with the concentration a few seconds earlier. If the concentration of the attractant is increasing, tumbling is suppressed. Likewise, *E. coli* moves away from a repellent because it senses that the concentration of the repellent is decreasing. The bacterium's chemoreceptors play a critical role in this process. The molecular events that enable *E. coli* cells to sense a chemical gradient and respond appropriately are presented in chapter 14.

Retrieve, Infer, Apply

1. Describe the way many flagella operate to move a bacterium.
2. How does swimming differ from swarming?
3. Explain in a general way how bacteria move toward substances such as nutrients and away from toxic materials.
4. Suggest why chemotaxis is sometimes called a "biased random walk."

3.9 Bacterial Endospores Are a Survival Strategy

After reading this section, you should be able to:

- Describe the structure of a bacterial endospore
- Explain why bacterial endospores are of particular concern to the food industry and why endospore-forming bacteria are important model organisms
- Describe in general terms the process of sporulation
- Describe those properties of endospores that are thought to contribute to its resistance to environmental stresses
- Describe the three stages that transform an endospore into an active vegetative cell

Endospores, dormant cells formed within a so-called mother cell, are fascinating bacterial structures only produced by certain members of the genera *Bacillus* and *Clostridium* (rods), and *Sporosarcina* (cocci) within the phylum *Firmicutes*. There are several reasons why endospores have long held the interest of microbiologists. The primary reason is that endospores are extraordinarily resistant to environmental stresses such as heat, ultraviolet radiation, gamma radiation, chemical disinfectants, and desiccation. In fact, some endospores have remained viable for around 100,000 years. Another important reason is that several species of endospore-forming bacteria are dangerous pathogens. For example, *Clostridium botulinum* causes botulism, a food-borne disease that results from ingestion of botulinum toxin, the deadliest toxin known. In order to prevent botulism, food must be prepared and stored properly. The extreme heat resistance of *C. botulinum*'s endospores is a major concern of the food industry. *Bacillus anthracis* causes the deadly disease inhalational anthrax, which occurs when spores are inhaled and germinate in the lungs. *B. anthracis* spores can be produced in a laboratory and used as a bioterrorism agent, as was done in 2001 in the United States (*see chapter 2 opening story*). Other

medically important endospore-forming bacteria include *C. tetani* (causes tetanus) and *C. perfringens* (causes gas gangrene and food poisoning). Clearly, endospores are of great practical importance in food, industrial, and medical microbiology. Endospores also are of considerable interest to scientists because of their complex structures and the equally complex process that generates them. Over the decades, microbiologists have asked many questions about endospores and endospore formation. (1) How are endospores structurally different from vegetative cells? (2) What makes them so resistant to harsh environmental conditions? (3) What triggers endospore formation? (4) What are the steps in endospore formation and how is the process regulated? (5) How do spores convert back to vegetative cells? In this section, we relate some of the answers to these questions. We save the consideration of regulating spore formation for chapter 14. ▶▶❙ *Sporulation in* Bacillus subtilis *(section 14.5)*

Endospore Structure and Resistance

Electron microscopy has been an important tool for dissecting endospore structure. Electron micrographs show that the spore consists of a core surrounded by several layers that vary significantly in composition. The core has normal cell structures such as ribosomes and a nucleoid but has very low water content (**figure 3.49**). The core is surrounded by an inner membrane, which is in turn covered by the core wall (also called the germ cell wall, as it contains the peptidoglycan that will form the wall of the vegetative cell that grows out of the spore following germination). Next is the cortex, which may occupy as much as half the spore volume. It is made of peptidoglycan that is less cross-linked than that in vegetative cells. The cortex is surrounded by a phospholipid bilayer called the outer membrane. Outside of the outer membrane is the coat, a complex structure composed of several layers. It is composed of more than 70 different proteins, which are highly cross-linked to each other. Finally, many endospore-forming bacteria produce spores enclosed by a thin, delicate covering called the exosporium, which is made up of glycoproteins.

The ability of the spore to survive heat, radiation, and damaging chemicals requires that its enzymes and DNA be protected. The various layers of the spore contribute to this resistance in several ways. The spore coat protects the spore from chemicals and various lytic enzymes such as lysozyme. The inner membrane is extremely impermeable to various chemicals, including those that cause DNA damage. The core has very low water content, high amounts of dipicolinic acid complexed with calcium ions (Ca-DPA), and a slightly lower pH, all of which contribute to the spore's resistance to harsh conditions. The low water content seems to be especially important for endospore resistance. Evidence exists that the water content of the core is low enough to prevent rotation of enzymes and other proteins present. However, it is not low enough to prevent denaturation of the proteins. It has been suggested that the immobilization of the proteins prevents them from interacting with each other and becoming entangled. Thus, even though they may become denatured, they are able to refold to their proper active structure as the spore germinates. The spore's DNA is protected by two main mechanisms. Ca-DPA

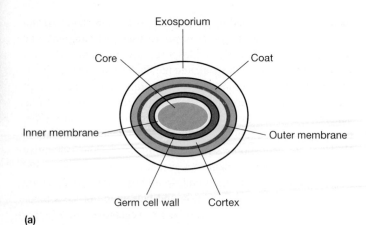

(a)

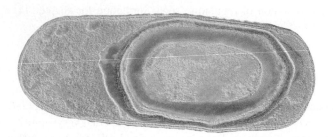

(b)

Figure 3.49 **Bacterial Endospores.** (a) A schematic showing the various layers of an endospore. They are not drawn to scale. (b) A colorized cross section of a *Bacillus subtilis* cell undergoing sporulation. The oval in the center is an endospore that is almost mature; when it reaches maturity, the mother cell will lyse to release it.

complexes are inserted between the nitrogenous bases of DNA, which helps stabilize it. The DNA is further stabilized by small, acid-soluble DNA-binding proteins (SASPs), which saturate spore DNA. There are several types of SASPs; the α/β type plays a major role in environmental resistance.

Sporulation: Making Endospores

Endospore-forming bacteria are common in soil, where they must be able to withstand fluctuating levels of nutrients. **Sporulation** normally commences when growth ceases due to lack of nutrients. Thus it is a survival mechanism that allows the bacterium to produce a dormant cell that can survive until nutrients are again available and vegetative growth can resume. These bacteria cycle between two states: vegetative growth and survival as an endospore. Vegetative growth is the normal, continuous cycle of growth and division. By contrast, sporulation is a complex process that occurs in a highly organized fashion over several hours. The mature endospore occupies a characteristic location in the mother cell (referred to as the sporangium), depending on the species of bacteria. Endospores may be centrally located, close to one end (subterminal), or terminal (**figure 3.50**). Sometimes an endospore is so large that it swells the sporangium. One of the best-studied endospore formers is *Bacillus subtilis*, which is an important model organism.

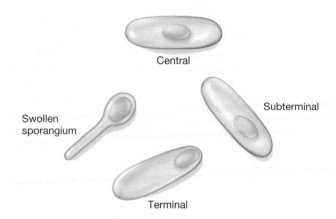

Figure 3.50 **Examples of Endospore Location and Size.**

Sporulation may be divided into seven stages (**figure 3.51**). The cell's DNA is replicated (stage I), followed by an inward folding of the cell membrane to enclose part of the DNA and produce the forespore septum (stage II). The mother cell membrane continues to grow and engulfs the immature endospore in a second membrane (stage III). Next, cortex is laid down in the space between the two membranes, and both calcium and dipicolinic acid are accumulated (stage IV). Protein coats are formed around the cortex (stage V), and maturation of the endospore occurs (stage VI). Finally, lytic enzymes destroy the sporangium, releasing the spore (stage VII). Sporulation requires about 8 to 10 hours. ⟳ *Bacterial Spore Formation*

Endospore to Vegetative Cell

The transformation of dormant spores into active vegetative cells is almost as complex as sporulation. It occurs in three stages: (1) activation, (2) germination, and (3) outgrowth. Activation is a process that prepares spores for germination and can result from treatments such as heating. This is followed by **germination,** the breaking of the spore's dormant state. It begins when proteins called germinant receptors, located in the inner membrane, detect small molecules such as sugars and amino acids. Upon detection of these molecules by the germinant receptors, a series of events occur. These include release of the Ca-DPA complexes, breakdown of the peptidoglycan in the cortex, and water uptake. Eventually water levels inside the germinating spore reach those characteristic of vegetative cells and enzymes in the core become active. This allows the spore to begin synthesizing various molecules needed to initiate spore outgrowth and return to a vegetative state.

Retrieve, Infer, Apply

1. Describe the structure of the bacterial endospore using a labeled diagram.
2. Briefly describe endospore formation and germination. What is the importance of the endospore?
3. What features of the endospore contribute to its resistance to harsh conditions?

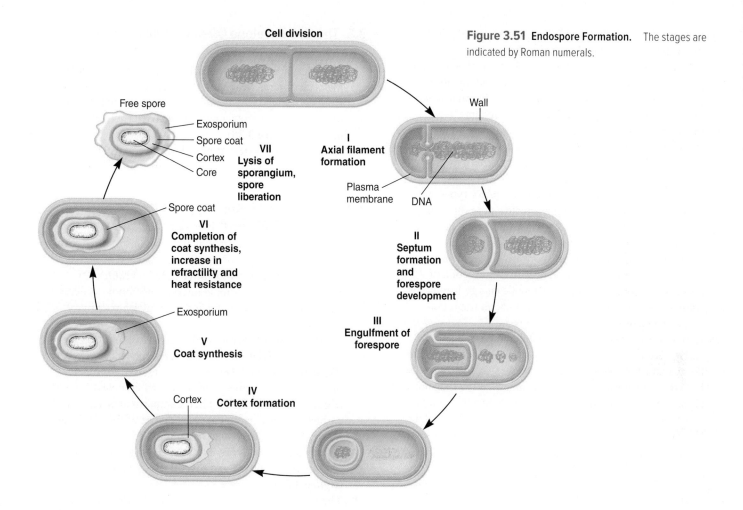

Figure 3.51 Endospore Formation. The stages are indicated by Roman numerals.

Key Concepts

3.1 Use of the Term "Prokaryote" Is Controversial

- Bacteria and archaea are often considered to be related because they share a common cell architecture. However, they are now known to have distinct characteristics and evolutionary lineages.
- Recent discoveries that challenge traditional notions of what a prokaryote is include the discovery of membrane-bound organelles in some bacteria and cytoskeletal elements in both bacteria and archaea.

3.2 Bacteria Are Diverse but Share Some Common Features

- Rods (bacilli) and cocci (spheres) are common bacterial shapes. In addition, bacteria may be comma-shaped (vibrio), spiral (spirillum and spirochetes), or filamentous; they may form buds and stalks; or they may have no characteristic shape (pleomorphic) (**figures 3.1** and **3.2**).
- Some cells remain together after division to form pairs, chains, and clusters of various sizes and shapes.

- Frequently observed bacterial cell structures include a cell wall, plasma membrane, cytoplasm, nucleoid, fimbriae, inclusions, capsule, ribosomes, and flagella (**figure 3.6**). **Table 3.1** summarizes the major functions of these structures.

3.3 Bacterial Plasma Membranes Control What Enters and Leaves the Cell

- The cell envelope consists of the plasma membrane and all external coverings, including the cell wall and other layers (e.g., capsules).
- The bacterial plasma membrane fulfills many roles, including acting as a semipermeable barrier, carrying out respiration and photosynthesis, and detecting and responding to chemicals in the environment.
- The fluid mosaic model proposes that cell membranes are lipid bilayers in which integral membrane proteins are buried. Peripheral membrane proteins are loosely associated with the membrane (**figure 3.7**).
- Bacterial membranes are bilayers composed of phospholipids constructed of fatty acids connected to glycerol by ester

linkages (**figure 3.8**). Bacterial membranes usually lack sterols but often contain hopanoids (**figure 3.9**).

■ Microorganisms require nutrients, materials that are used in energy conservation and biosynthesis. Macronutrients are needed in relatively large quantities. Micronutrients (trace elements) are used in very small amounts.

■ In passive diffusion, a substance moves through the membrane on its own. Movement is down a concentration gradient and does not require an input of energy. Only a few substances enter bacteria by passive diffusion.

■ Transport proteins are divided into two major types: channels and carriers. Channels create a pore through which a substance moves. Carriers are so called because they carry the molecule across the membrane.

■ In facilitated diffusion, a transport protein (either a channel or a carrier) helps move substances in the direction of decreasing concentration; no metabolic energy is required (**figure 3.11**). Facilitated diffusion is not an important uptake mechanism for bacterial cells.

■ Active transport systems use metabolic energy and carrier proteins to concentrate substances by transporting them against a gradient. ATP is used as an energy source by ABC transporters, a type of primary active transporter (**figure 3.13**). ABC transporters are uniporters (one substance is transported). Gradients of ions drive solute uptake in secondary active transport systems. Secondary active transporters are either symporters (two substances are transported in the same direction) or antiporters (two substances are transported in opposite directions) (**figure 3.12**).

■ Group translocation is a type of active transport in which bacterial cells transport organic molecules while modifying them (**figure 3.14**).

■ The secretion of siderophores, which bind ferric iron, enables the accumulation of iron within bacterial cells (**figure 3.15**).

3.4 There Are Two Main Types of Bacterial Cell Walls

■ The vast majority of bacteria have a cell wall outside the plasma membrane to help maintain their shape and protect them from osmotic stress.

■ Bacterial walls are chemically complex and usually contain peptidoglycan (**figures 3.16–3.21**). Typical Gram-positive walls have thick, homogeneous layers of peptidoglycan and teichoic acids (**figures 3.22 and 3.23**). Typical Gram-negative bacteria have a thin peptidoglycan layer surrounded by a complex outer membrane containing lipopolysaccharides (LPSs) and other components (**figures 3.24 and 3.25**). The lipid A portion of LPS is also called endotoxin. Its release into the human body can lead to septic shock.

■ The mechanism of the Gram stain is thought to depend on the thickness of peptidoglycan. The thick peptidoglycan of typical Gram-positive bacteria binds crystal violet tightly, preventing its loss during the ethanol wash.

3.5 The Cell Envelope Often Includes Layers Outside the Cell Wall

■ Capsules, slime layers, and glycocalyxes are layers of material lying outside the cell wall. They can protect cells from certain environmental conditions, allow cells to attach to surfaces, and protect pathogenic bacteria from host defenses (**figure 3.28**).

■ S-layers are the external-most layer in some bacteria. They are composed of proteins or glycoprotein and have a characteristic geometric shape (**figure 3.29**).

3.6 The Bacterial Cytoplasm Is More Complex than Once Thought

■ The bacterial cytoplasm contains proteins that are similar in structure and function to the cytoskeletal proteins observed in eukaryotes (**figure 3.30 and table 3.2**).

■ Some bacteria have simple internal membrane systems containing photosynthetic and respiratory machinery (**figure 3.31**).

■ Inclusions are observed in all cells. Most are used for storage (e.g., PHB inclusions and polyphosphate granules) (**figures 3.33 and 3.34**), but some are used for other purposes (e.g., magnetosomes and gas vacuoles) (**figures 3.36 and 3.37**). Microcompartments such as carboxysomes contain enzymes that catalyze important reactions (e.g., CO_2 fixation; **figure 3.35**).

■ Bacterial ribosomes are 70S in size. They are composed of numerous proteins and several rRNA molecules (**figure 3.38**).

■ The genetic material of bacterial cells is located in an area within the cytoplasm called the nucleoid. The nucleoid is not usually enclosed by a membrane (**figure 3.39**). In most bacteria, the nucleoid contains a single chromosome. The chromosome usually consists of a double-stranded, covalently closed, circular DNA molecule.

■ Plasmids are extrachromosomal DNA molecules found in many bacteria. Some are episomes—plasmids that are able to exist freely in the cytoplasm or can be integrated into the chromosome. Although plasmids are not required for survival in most conditions, they can encode traits that confer selective advantage in some environments. Many types of plasmids have been identified (**table 3.3**).

3.7 Many Bacteria Have External Structures Used for Attachment and Motility

■ Many bacteria have hairlike appendages called fimbriae or pili. Fimbriae function primarily in attachment to surfaces, but type IV pili are involved in twitching motility. Sex pili participate in the transfer of DNA from one bacterium to another (**figure 3.40**).

■ Many bacteria are motile, often by means of threadlike, locomotory organelles called flagella. Bacterial species differ in the number and distribution of their flagella (**figure 3.41**). Each bacterial flagellum is composed of a filament, hook, and basal body (**figure 3.42**).

3.8 Bacteria Move in Response to Environmental Conditions

- Several types of bacterial motility have been observed: swimming by flagella, swarming by flagella, spirochete motility, twitching motility, and gliding motility.
- The bacterial flagellar filament is a rigid helix that rotates like a propeller to push the bacterium through water (**figures 3.44** and **3.45**). When many bacteria swim, they alternate between two types of movement: a smooth swimming motion called a run and tumbling.
- Some bacterial species move as a group in a behavior called swarming. Swarming is mediated by flagella and occurs on moist surfaces.
- Spirochete motility is brought about by flagella that are wound around the cell and remain within the periplasmic space. When they rotate, the outer membrane of the spirochete is thought to rotate, thus moving the cell (**figure 3.47**).

- *Myxococcus* spp. exhibit both twitching and gliding motility. Twitching motility is a jerky movement brought about by type IV pili, whereas gliding motility is smooth.
- Motile cells can respond to gradients of attractants and repellents, a phenomenon known as chemotaxis. *E. coli* and many other peritrichously flagellated bacteria accomplish movement toward an attractant by increasing the length of time spent moving toward the attractant and shortening the time spent tumbling (**figure 3.48**). Conversely, they increase their run time when they move away from a repellent.

3.9 Bacterial Endospores Are a Survival Strategy

- Some bacteria survive adverse environmental conditions by forming endospores, dormant structures resistant to heat, desiccation, and many chemicals (**figure 3.49**).
- Both endospore formation and germination are complex processes made in response to certain environmental signals (**figure 3.51**).

Compare, Hypothesize, Invent

1. Propose a model for the assembly of a flagellum in a typical Gram-positive cell envelope.
2. The peptidoglycan of bacteria has been compared with the chain mail worn beneath a medieval knight's suit of armor. It provides both protection and flexibility. Describe other structures in biology that have an analogous function. How are they replaced or modified to accommodate the growth of the inhabitant?
3. Why might a microbe have more than one uptake system for certain substrates?
4. Design an experiment that illustrates the cell wall's role in protecting against lysis.
5. What would you expect to observe if you were able to "transplant" CreS into a rod-shaped bacterium such as *Bacillus subtilis*?
6. Develop a hypothesis to explain why gas vacuoles are bounded by proteins rather than a lipid bilayer membrane.
7. In 2009 it was reported that a member of the genus *Mycobacterium* was able to form endospores. This was greeted with surprise and skepticism because mycobacteria had always been thought to be non-endospore formers.

Design a set of experiments that would determine if a bacterium produces true endospores. Explain your experimental design.

Read the original paper: Ghosh, J., et al. 2009. Sporulation in mycobacteria. *Proc. Natl. Acad. Sci. USA* 106:10781.

8. LPS is synthesized in the cytoplasm and then transported to and inserted into the outer membrane. This is complicated by the amphipathic nature of LPS. Recently two different laboratories reported their studies on seven LPS transport proteins (Lpt proteins) that move LPS from the cytoplasm into the outer membrane. Several of the Lpt proteins fold to form a structure called a β-jelly roll, which can be described as being "greasy." Other Lpt proteins fold to form a β-barrel. Hypothesize how each fold might interact with an LPS molecule. Why would two different types of proteins be needed?

Read the original papers. Dong, H., et al. 2014. Structural basis for outer membrane lipopolysaccharide insertion. *Nature* 511:52; Qiao, S., 2014. Structural basis for lipopolysaccharide insertion in the outer membrane. *Nature* 511:108.

Learn More

shop.mheducation.com   Enhance your study of this chapter with interactive study tools and practice tests. Also ask your instructor about the resources available through Connect, including adaptive learning tools and animations.

4

Archaeal Cell Structure

Cows and Buffaloes and Sheep, Oh My!

Dairy cows are one group of ruminant animals that contribute methane to Earth's atmosphere. The methane is produced by methanogens living in the rumen of the animal and is released when the animal breathes. Methane has about 20 times the heat-retaining capacity of carbon dioxide.

The next time you drive or walk through the countryside near your home, count the number of domesticated ruminants you see—dairy cows, beef cattle, sheep, and goats, among others. It is estimated that there are about 3 billion head worldwide. These animals are important because they are a substantial protein source. But they do something else that is noteworthy: each year they exhale about 200 million metric tons of methane, a potent greenhouse gas (GHG). The manure these and other domesticated animals produce is thought to release another 18 million metric tons of methane. As detailed in chapter 28, carbon dioxide is an important GHG and is considered to be a significant contributor to global warming. Methane and nitrous oxide are also GHGs. Carbon dioxide is the dominant GHG in the atmosphere. However, methane has about 23 times the heat-retaining capacity of carbon dioxide. Thus, although the contribution of methane by domesticated ruminants and their manure represents only about 5% of the total GHG emissions worldwide each year, the impact of these emissions is of concern. But don't blame the animals; they are not directly responsible for generating methane. Rather, it is a unique group of archaea—the methanogens—that make methane from hydrogen gas, carbon dioxide, and acetate. Furthermore, these substrates arise in the animals as their food is digested with the help of a complex community of microbes living in the rumen. For the animal, methane is a waste product; for the farmer or rancher, methane represents a loss in production. There is considerable interest in figuring out how to decrease methane production so that GHG emissions are lowered and the farmer or rancher can earn more from each animal.

Methanogens are only one physiological group in domain *Archaea*. Other archaea have attracted attention because of their ability to live in extreme environments. All are of interest because of their chimeric natures: they have some features that are similar to bacteria and some that are similar to eukaryotes. Their cell structure is like that of bacteria; that is, they look like the canonical prokaryotic cell. However, the molecules used to construct their cell structures are often unique or are similar to those found in eukaryotes. As we describe in chapters 15 and 20, their physiologies exhibit this dichotomy. In general, the processes and molecules used to conserve energy are like those of bacteria, whereas the processes and molecules used to replicate and express their genomes are like those of eukaryotes. In this chapter, we describe some of the molecular and structural aspects of archaeal cells that set them apart.

4.1 Archaea Are Diverse but Share Some Common Features

After reading this section, you should be able to:

■ Describe a typical archaeal cell

Recall that for many years, **archaea** and bacteria had been lumped together and referred to as prokaryotes. Today *Archaea* and *Bacteria* are recognized as distinct taxa. The separation of bacteria and archaea into distinct taxa correlates with the observation that they each have unique and distinguishing characteristics. Some of these are summarized in **table 4.1**. Archaeal taxonomy is currently in a state of flux, but three phyla are well established: *Crenarchaeota, Euryarchaeota,* and *Thaumarchaeota*. Here we begin our discussion of archaeal cells by considering overall cell morphology and organization and then specific cell structures.

Shape, Arrangement, and Size

Archaeal cells, like bacterial cells, exhibit a variety of shapes. **Cocci** and **rods** are common (**figure 4.1a**). Both usually exist singly, but some cocci form clusters and some rods form chains. Curved rods, spiral shapes, and pleomorphic (many shaped) archaea have also been observed. To date, no spirochete-like and mycelial archaea have been discovered. However, some archaea exhibit unique shapes, such as the branched form of *Thermoproteus tenax* (figure 4.1b) and the flat, postage-stamp-shaped *Haloquadratum walsbyi,* an archaeon that lives in salt ponds and measures about 2 μm by 2 to 4 μm and only 0.25 μm thick. This shape has the advantage of greatly increasing the surface area-to-volume (S/V) ratio (*see figure 3.5*), which in turn increases efficiency of nutrient uptake, diffusion of

Table 4.1	Comparison of Bacterial and Archaeal Cells	
Property	**Bacteria**	**Archaea**
Plasma membrane lipids	Ester-linked phospholipids and hopanoids form a lipid bilayer; some have sterols	Glycerol diethers form lipid bilayers; glycerol tetraethers form lipid monolayers
Cell wall constituents	Peptidoglycan is present in nearly all; some lack cell walls	Very diverse but peptidoglycan is always absent: some consist of S-layer only, others combine S-layer with polysaccharides or proteins or both; some lack cell walls
Inclusions present	Yes, including gas vacuoles	Yes, including gas vacuoles
Ribosome size	70S	70S
Chromosome structure	Most are circular, double-stranded (ds) DNA; usually a single chromosome	All known are circular, dsDNA
Plasmids present	Yes; circular and linear dsDNA	Yes; circular dsDNA
External structures	Flagella, fimbriae (pili) common	Flagella, pili, and piluslike structures common
Capsules or slime layers	Common	Rare

molecules, and growth rate. ◀◀ *Bacteria are diverse but share some common features (section 3.2)*

Archaeal cells also vary in size as much as in shape. Typical rods are 1 to 2 μm wide by 1 to 5 μm long; cocci are typically 1 to 3 μm in diameter. However, extremely small and extremely large archaea have been identified. Several free-living, acid-loving (acidophilic), mine-dwelling microbes measure a mere 0.2 to 0.4 μm in diameter. Also at the small end of the size continuum is the parasitic archaeon *Nanoarchaeum equitans* (0.4 μm in diameter). At the other extreme are two recently observed giant archaea that form long filaments up to 30 mm in length. For one archaeon, the filament is composed of numerous cells, each measuring 8 to 10 μm wide by 20 to 24 μm long. An interesting characteristic of this archaeon is that its filaments are coated with a biofilm formed by a bacterium (**figure 4.2**). The nature of this interaction is not known, but it is thought that the archaeon is the host and the bacteria are symbionts.

Cell Organization

Structures observed in members of both *Bacteria* and *Archaea* are summarized and their differences noted in table 4.1. The archaeal plasma membrane is composed of strikingly different lipids than those found in bacterial membranes; in fact, the unusual lipids were one of the first pieces of evidence to suggest that these microbes are phylogenetically distinct from bacteria. Most archaea have a cell wall, but their walls are considerably more diverse than bacterial

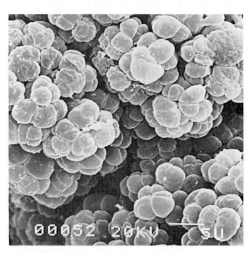

(a) *Methanosarcina mazei*—a coccus that forms clusters

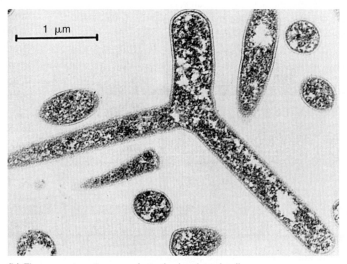

(b) *Thermoproteus tenax*—a branched archaeal cell

Figure 4.1 Archaeal Cell Morphology. (a) *Methanosarcina mazei;* SEM. Bar = 5 μm. (b) *Thermoproteus tenax,* electron micrograph.

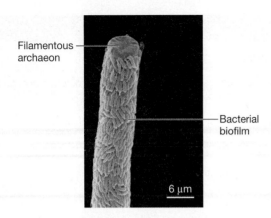

Filamentous archaeon

Bacterial biofilm

6 μm

Figure 4.2 A Giant Archaeon. *Candidatus* Giganthauma karukerense forms long filaments that are covered with a bacterial biofilm. The bacteria may be symbionts of the host archaeon.

walls. Notably, archaeal cell walls lack peptidoglycan. Capsules are not widespread among those archaea examined thus far. Within the archaeal cytoplasm, a nucleoid, ribosomes, and inclusions can be found. Finally, many archaea use flagella for locomotion.

In the remaining sections of this chapter, we describe the major structures observed in archaea in more detail, noting the aspects that distinguish them from the analogous bacterial structures. Understanding archaeal structures at the molecular level is an area of intense research because many archaea are found in extreme habitats. We note those structural adaptations that contribute to this lifestyle in the discussions. However, the topic of survival in extreme habitats is considered more thoroughly in chapter 7. ▶▶| *Environmental factors affect microbial growth (section 7.4)*

Retrieve, Infer, Apply

1. Which cell shapes are observed in members of both *Bacteria* and *Archaea*? Which are unique to bacteria? Which to archaea?
2. *Archaea* was first defined as a distinct taxon by comparisons of ribosomal RNA sequences. Identify two other molecules that could be used to determine if a microbe having a typical prokaryotic architecture is a bacterium or an archaeon.

4.2 Six Major Types of Archaeal Cell Envelopes Have Been Identified

After reading this section, you should be able to:

- Draw an archaeal cell envelope and identify the component layers
- Compare and contrast archaeal and bacterial cell envelopes in terms of their structure, molecular makeup, and functions
- Compare and contrast nutrient uptake mechanisms observed in bacteria and archaea

We define the **cell envelope** as the plasma membrane and any layers external to it. For bacteria, these extra layers include cell walls, S-layers, capsules, and slime layers. One of the most

distinctive features of archaeal cells is the nature of their cell envelopes. Their uniqueness begins with their plasma membranes. Recall from chapter 3 that the fluid mosaic model describes membranes as lipid bilayers within which proteins float (*see figure 3.7*). The model is based on studies of eukaryotic and bacterial membranes and is well established. Imagine the surprise of microbiologists when they found that some archaea have monolayer membranes that function like bilayers. This structural difference is due to the presence of unique lipids in archaeal plasma membranes. Archaeal cell envelopes also differ in terms of organization. For many archaea, an S-layer is the major, and sometimes only, component of the cell wall. Finally, capsules and slime layers are relatively rare among archaea and thus are not discussed.

Archaeal Plasma Membranes Are Composed of Unique Lipids but Function Like Bacterial Membranes

Archaeal membranes are composed primarily of lipids that differ from bacterial and eukaryotic lipids in two ways. First, they contain hydrocarbons derived from isoprene units—five-carbon, branched molecules (**figure 4.3**). Thus the hydrocarbons are branched as shown in **figure 4.4**. This affects the way the lipids pack together, which in turn affects the fluidity of the membrane and its permeability. This is especially important for extremophilic archaea for which membrane fluidity and permeability could be compromised by extreme conditions. Second, the hydrocarbons are attached to glycerol by ether links rather than ester links (figure 4.4*a*). Ether linkages are more resistant to chemical attack and heat than are ester links.

Two major types of archaeal lipids have been identified: glycerol diethers and diglycerol tetraethers. **Glycerol diether lipids** are formed when two hydrocarbons are attached to glycerol (figure 4.4*b*, lipid 3). Usually, the hydrocarbon chains in glycerol diethers are 20 carbons in length. **Diglycerol tetraether lipids** are formed when two glycerol residues are linked by two long hydrocarbons that are 40 carbons in length (figure 4.4*b*, lipids 1 and 2). Tetraethers are more rigid lipids than diethers. Cells can adjust the overall length of the tetraethers by cyclizing the chains to form pentacyclic rings (figure 4.4*b*, lipid 2). Phosphorus-, sulfur-, and sugar-containing groups can be attached to the glycerol moieties in the diethers and tetraethers, just like the phospholipids observed in bacterial and eukaryotic membranes.

Despite the significant differences in membrane lipids, the basic design of archaeal membranes is similar to that of bacterial and eukaryotic membranes: there are two hydrophilic surfaces and a hydrophobic core. When C_{20} diethers are used, a typical bilayer membrane is formed (**figure 4.5*a***). When the membrane

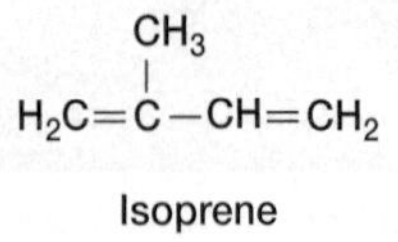

$$H_2C=\overset{\overset{\displaystyle CH_3}{|}}{C}-CH=CH_2$$

Isoprene

Figure 4.3 Isoprene. This five-carbon, branched molecule is the building block of archaeal lipids.

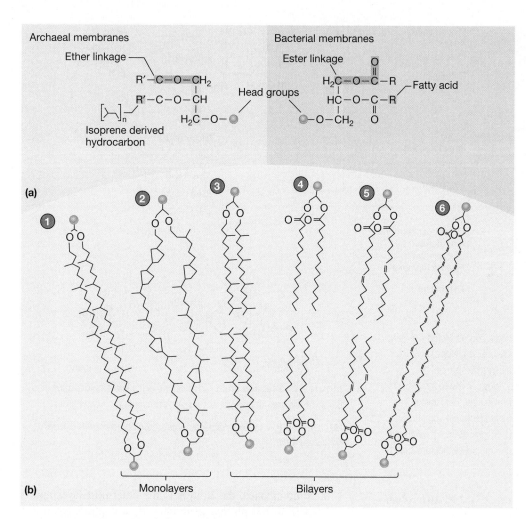

(a)

(b) Monolayers   Bilayers

Figure 4.4 **Comparison of Archaeal and Bacterial Membranes.** (a) Archaeal membrane lipids are attached to glycerol by ether linkages instead of ester linkages, as found in bacteria and eukaryotes. The stereochemistry also differs. In archaeal lipids, the stereoisomer of glycerol is *sn*-glycerol-1-phosphate; in bacterial lipids, the stereoisomer is *sn*-glycerol-3-phosphate. Thus in archaeal lipids, the side chains are attached to carbons 2 and 3 of glycerol, and in bacterial lipids, the side chains are attached to carbons 1 and 2. (b) Examples of archaeal lipids are lipids 1, 2, and 3. Lipids 4, 5, and 6 are bacterial lipids. Note that some archaeal lipids form monolayers (figure 4.5), whereas all bacterial lipids form bilayers.

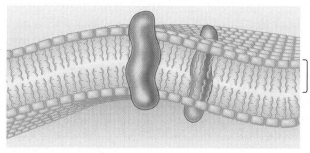

(a) Bilayer of C_{20} diethers

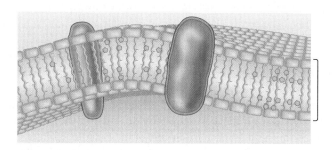

(b) Monolayer of C_{40} tetraethers

Figure 4.5 **Examples of Archaeal Plasma Membranes.**

is constructed of C_{40} tetraethers, a monolayer membrane with much more rigidity is formed (figure 4.5*b*). The addition of pentacyclic rings further increases this rigidity. As might be expected from their need for stability, the membranes of extreme thermophiles such as *Thermoplasma* and *Sulfolobus* species, which grow best at temperatures over 85°C, are almost completely tetraether monolayers. Archaea that live in moderately hot environments have membranes containing some regions with monolayers and some with bilayers. ▶▶| *Phylum* Crenarchaeota *(section 20.2)*

As we discuss in chapter 3, organisms have several options for obtaining the nutrients they need from their environment. Importantly, microbes often find themselves in nutrient-poor environments and therefore must be able to accumulate nutrients in their cytoplasm at concentrations higher than the external milieu. Thus, although passive and facilitated diffusion have been observed in archaea, they primarily use active transport for nutrient uptake. Both primary (e.g., ABC transport) and secondary active transport systems (e.g., symport and antiport) have been identified in archaeal cells, often by searching an archaeon's genome for the genes encoding components of these systems (i.e., genomic analysis). These systems tend to be similar to those seen in bacteria. The group translocation system phosphoenolpyruvate:sugar phosphotransferase system (PTS) also functions in some archaea. |◀◀ *Bacteria use many mechanisms*

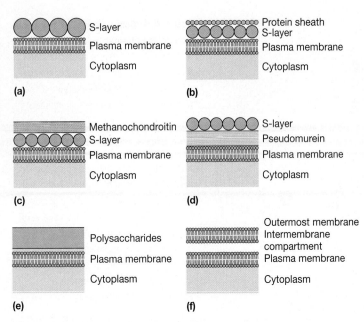

Figure 4.6 **Archaeal Cell Envelopes.** (a) *Methanococcus, Halobacterium, Pyrodictium, Sulfolobus,* and *Thermoproteus* species cell envelopes. (b) *Methanospirillum* spp. cell envelope. (c) *Methanosarcina* spp. cell envelope. (d) *Methanothermus* and *Methanopyrus* species cell envelopes. (e) *Methanobacterium, Methanosphaera, Methanobrevibacter, Halococcus,* and *Natronococcus* species cell envelopes. For *Methanosphaera* spp., the polysaccharide layer is composed of pseudomurein. (f) *Ignicoccus hospitalis* cell envelope. The outermost membrane contains protein complexes that form pores.

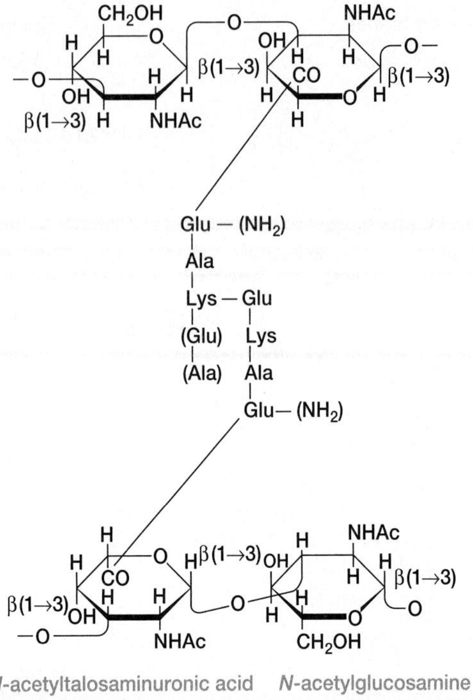

N-acetyltalosaminuronic acid   *N*-acetylglucosamine

Figure 4.7 **Pseudomurein.** The amino acids and amino groups in parentheses are not always present. Ac represents the acetyl group.

MICRO INQUIRY *How is pseudomurein similar to peptidoglycan? How does it differ?*

to bring nutrients into the cell (section 3.3) ▶▶ *Microbial genomics (chapter 18)*

There Are Many Different Types of Archaeal Cell Walls

Before they were distinguished as a unique domain of life, archaeal species were characterized as being either Gram positive or Gram negative based on their response to Gram staining. Thus, like all cells (even eukaryotic cells), they will stain either purple or pink when Gram stained. However, their staining reaction does not correlate reliably with a particular cell wall structure as it does for bacteria. Archaeal cell walls exhibit considerable variety in terms of their chemical makeup. Furthermore, their cell walls lack peptidoglycan.

The most common type of archaeal cell wall is an **S-layer** composed of either glycoprotein or protein (**figure 4.6a**). The S-layer may be as thick as 20 to 40 nm. Some methanogens (*Methanolobus* and *Methanococcus* species), salt-loving archaea (*Halobacterium* spp.), and extreme thermophiles (*Sulfolobus, Thermoproteus,* and *Pyrodictium* species) have S-layer cell walls.

Other archaea have additional layers of material outside the S-layer. For instance, *Methanospirillum* spp. have a protein sheath external to the S-layer (figure 4.6b). Other methanogens (*Methanosarcina* spp.) have a polysaccharide layer covering the S-layer (figure 4.6c). This material, called methanochondroitin, is similar to the chondroitin sulfate of animal connective tissue.

In some archaea, the S-layer is the outermost layer and is separated from the plasma membrane by a peptidoglycan-like molecule called pseudomurein (figure 4.6d). Pseudomurein differs from peptidoglycan in that it has L-amino acids instead of D-amino acids in its cross-links, *N*-acetyltalosaminuronic acid instead of *N*-acetylmuramic acid, and β(1→3) glycosidic bonds instead of β(1→4) glycosidic bonds (**figure 4.7**). These differences mean that lysozyme, penicillin, and other chemicals that affect peptidoglycan structure and synthesis in bacterial cell walls have no affect on pseudomurein-containing archaeal cell walls. ◀◀ *Cell walls and osmotic protection (section 3.4)*

The last type of archaeal cell wall consists of a single, thick homogeneous layer resembling that in Gram-positive bacteria (figure 4.6e). These archaea lack an S-layer and often stain Gram positive. Their wall chemistry varies from species to species but usually consists of complex polysaccharides such as pseudomurein.

Some archaea lack any layer resembling a cell wall. For instance, members of the acidophilic genera *Ferroplasma* and *Thermoplasma* have envelopes consisting only of a plasma membrane covered by a layer of slime. The slime, which is referred to as a glycocalyx, may provide some of the protection needed for these archaea to survive in their acidic habitats. The most unique wall-less archaeon is *Ignicoccus hospitalis*. Its envelope consists only of the plasma membrane and an outermost membrane, with an intermembrane compartment between them

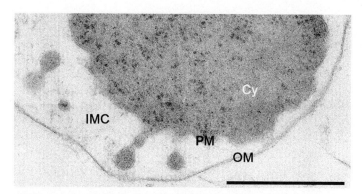

Figure 4.8 Cell Envelope of the Wall-less Archaeon *Ignicoccus hospitalis.* Cy is cytoplasm, PM is plasma membrane, IMC is intermembrane compartment, and OM is outermost membrane. TEM, Bar = 200 nm.

(figure 4.6*f* and **figure 4.8**). The outermost membrane contains protein complexes that form pores, much like bacterial porin proteins create pores in the outer membrane of typical Gram-negative bacteria.

Retrieve, Infer, Apply

1. Identify three features that distinguish archaeal plasma membranes from those of bacteria.
2. Both bacteria and archaea can have S-layers. How does their use as components of the cell envelope differ?

4.3 Archaeal Cytoplasm Is Similar to Bacterial Cytoplasm

After reading this section, you should be able to:
- Compare and contrast the cytoplasm of bacterial and archaeal cells

Overall, the cytoplasm of archaeal cells is very similar to that of bacteria. Within it can be found inclusions—polyhydroxyalkonates, polyphosphate granules, glycogen granules, and gas vacuoles; ribosomes; a nucleoid; and plasmids. Proteins that might form a **cytoskeleton** have also been identified, including FtsZ (tubulin homologue), MreB (actin homologue), and crenactin, an actin homologue unique to certain members of the phylum *Crenarchaeota*. In those archaea having FtsZ, the protein participates in cell division, as it does in bacteria. Evidence suggests that the actin homologues MreB and crenactin are involved in conferring a rod shape. Interestingly, a tubulin homologue called CetZ is observed in some members of the archaeal phylum *Euryarchaeota*. CetZ confers a rod shape to cells rather than functioning in cytokinesis as FtsZ does. Some structures found in the cytoplasm of archaeal cells have distinctive molecular makeup. In this section, we focus on those structures. ▶▶| *Phylum* Crenarchaeota *(section 20.2); Phylum* Euryarchaeota *(section 20.4)*

Archaeal Ribosomes Are the Same Size as Bacterial Ribosomes but Are Composed of Different Molecules

Like bacterial cells, the **ribosomes** of archaeal cells are 70S in size and are constructed of a 50S and a 30S subunit. However, their shape is somewhat different and their component molecules are not the same. Both have **ribosomal RNA (rRNA)** molecules of similar size: 16S in the small subunit, and 23S and 5S in the large subunit. However, as we discuss in chapter 1, the differences in nucleotide sequences of these molecules were the initial basis for establishing the taxon *Archaea*. Furthermore, at least one archaeon has an additional rRNA, a 5.8S rRNA, in the large subunit. This is of interest because the large subunit of eukaryotic ribosomes contains both 5S and 5.8S rRNA molecules.

The protein composition of bacterial and archaeal ribosomes also differs. Archaeal ribosomes have more proteins: about 68 rather than about 55. Archaeal ribosomal proteins can be sorted into three groups based on their similarity to proteins found in bacterial and eukaryotic ribosomes: (1) those observed in all three domains of life, (2) those unique to archaea, (3) those observed in both archaea and eukaryotes. There are only a few ribosomal proteins unique to archaea; the rest are about evenly divided between the other two groups. Importantly, all the ribosomal proteins present in both archaeal and bacterial ribosomes are also seen in eukaryotic ribosomes. Thus there are no ribosomal proteins that might be referred to as prokaryotic (i.e., present only in the ribosomes of archaea and bacteria). The difference in bacterial and archaeal ribosomes correlates with the fact that archaeal ribosomes are impervious to antibiotics that bind and inhibit bacterial ribosomes ▶▶| *Protein synthesis inhibitors (section 9.4).*

Nucleoid

The **nucleoid** provides another example of the difference between archaea and bacteria. This irregularly shaped region in the cytoplasm contains the cell's **chromosome** and numerous proteins. The chromosomes of all known archaea are circular, double-stranded **deoxyribonucleic acid (DNA)**. Evidence exists that some archaea are polyploid; that is, they have multiple copies of their chromosomes throughout their life cycles. Archaeal chromosomes, like bacterial chromosomes, are longer than the length of the cell and must be compacted to fit in the cell. In bacteria, super-coiling and nucleoid-associated proteins (NAPs) are important in compacting the chromosome. Considerably less is known about the compaction of archaeal chromosomes. However, many members of the phylum *Euryarchaeota* have histones associated with their chromosomes. These **histones** form **nucleosomes** that are similar to the nucleosomes observed in eukaryotes (*see figure 5.10*). However, they differ from eukaryotic nucleosomes in that they consist of four histones rather than eight and a smaller amount of DNA is part of the nucleosomes. It is hypothesized that in thermophilic archaea, histones help prevent denaturation of the chromosomes. Many euryarchaeotes also use NAPs for chromosome compaction. Remarkably, members of phylum *Crenarchaeota* do not have histones or nucleosomes (with rare exceptions). Instead crenarchaeotes rely solely on supercoiling and NAPs to compact their chromosomes. ▶▶| *Nucleoid (section 3.6); Nucleus (section 5.5)*

Retrieve, Infer, Apply

1. Thus far, homologues of intermediate filaments have not been identified in archaea. Is it likely that they will be identified eventually? Explain your answer.

2. Archaea are often described as hybrid organisms having some features that are similar to bacteria and some that are similar to eukaryotes. Based on what you have just read in this section, provide two examples that illustrate the similarity of archaea to bacteria; list two examples of their similarity to eukaryotes.

4.4 Many Archaea Have External Structures Used for Attachment and Motility

After reading this section, you should be able to:

- Describe cannulae and hami
- Compare and contrast bacterial and archaeal pili
- Compare and contrast bacterial and archaeal flagella in terms of their structure and function

Like bacteria, many archaea have structures that extend beyond the cell envelope. Some will be familiar: pili and flagella. However, some are unique to archaea. These external structures and their functions are discussed in this section.

Pili, Cannulae, and Hami

Many archaea have **pili,** but relatively little is known about their structural proteins. Some archaeal pili are composed of pilin proteins that are homologues of bacterial type IV pili proteins. However, the pilus formed from these proteins is unique in several ways, including the presence of a central lumen. Central lumens are observed in bacterial flagella but not in bacterial type IV pili. The function of archaeal pili has not been thoroughly investigated. However, some have been shown to attach archaeal cells to surfaces. |◀◀ *Bacterial pili and fimbriae (section 3.7)*

Two structures appear to be unique to archaea: cannulae and hami. **Cannulae** are hollow, tubelike structures observed on the surface of thermophilic archaea belonging to the genus *Pyrodictium* (**figure 4.9**). Members of this genus also have flagella, but in electron micrographs the cannulae are distinguishable by their larger diameter. The function of cannulae is unknown, but it is known that daughter cells arising from a single round of cell division remain connected to each other by cannulae. After many rounds of cell division, a network of cells is formed. **Hami** are of particular interest because of their shape (**figure 4.10**). They look like tiny grappling hooks, which suggests they might function to attach cells to surfaces. Indeed, archaeal cells that produce hami are members of biofilm communities generally consisting of a hami-producing archaeon and a bacterium.

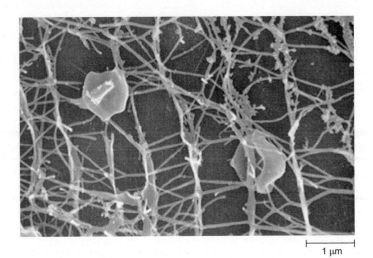

Figure 4.9 Cannulae. Cannulae are tubular structures about 25 nm in diameter. They have only been observed on *Pyrodictium* spp. They connect daughter cells, ultimately forming a dense network of cells. SEM, bar = 1 μm.

Archaeal Flagella and Motility

Archaeal **flagella** have been studied in detail in only a few model archaea. They are superficially similar to their bacterial counterparts, but important differences have been identified. These differences, plus the differences between bacterial and eukaryotic flagella led to the reexamination of the terminology to describe these motility organelles in all organisms (**Microbial Diversity & Ecology 4.1**). Archaeal flagella are thinner than bacterial flagella (10 to 14 nm rather than 18 to 22 nm) and some are composed of more than one type of flagellin subunit

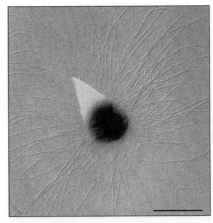

(a) Hami radiating from cell

(b) "Grappling hooks" at distal ends of hami

Figure 4.10 Hami. (a) A coccus-shaped cell is in the center of the electron micrograph and about 100 hami are radiating out from the platinum-shadowed cell. TEM, bar = 500 nm (b) At the ends of the hami are grappling hooklike structures that are thought to allow cells to adhere to surfaces, including other cells. TEM, bar = 50 nm.

MICROBIAL DIVERSITY & ECOLOGY

4.1 What's in a Name?

Each day soon-to-be parents around the world agonize over what to name their babies. Is the name too popular or too unusual? Will it lead to undesirable nicknames? Was it the name of an unsavory historical figure? Though scientists probably don't agonize over what to call new organisms, cellular structures, or other natural phenomena, they do try to choose names carefully. Often the names have Greek or Latin roots that provide some information about the object being named. For instance, the archaeon *Pyrococcus furiosus,* a name that means rushing fireball, was so named because it is spherical, moves rapidly, and loves heat. Scientists also take care in naming things so that the names don't lead to misconceptions. Unfortunately, sometimes scientists get it wrong, and new names are suggested. Suggesting new names can lead to considerable debate and confusion about which terminology to use. Such is the case with the term *flagella.*

For decades, long, hairlike structures have been called flagella, and flagella have been identified in members of all three domains of life. In fact, the presence of flagella was long used as a criterion for distinguishing certain protists from others. Recall that protists and other eukaryotic organisms have another motility organelle, the cilium. As the ultrastructure of eukaryotic flagella and cilia were determined, it was found that they are the same. Both are very complex and make use of microtubules arranged in a characteristic 9 + 2 fashion. Furthermore, they move cells in a similar way: by whipping back and forth. Thus eukaryotic flagella are simply long cilia. Despite this, use of the term "flagella" when referring to long cilia persisted. When bacterial flagella were discovered, they too were named flagella. Eventually their ultrastructure and function were discovered and shown to be distinct. As we describe in chapter 3, their structure is much simpler, with the helical filament composed of a single type of protein. It propels the cell by rotating.

With this knowledge, scientists began debating new names for these structures. One suggestion was to reserve the term *flagella* for the bacterial organelle and to change the name of eukaryotic flagella to undilapodia, which essentially means "waving feet." Undilapodia did not gain acceptance and finally scientists decided to use the term cilia for both cilia and flagella. More recently, studies of archaeal flagella led to the discovery that these are very different from bacterial flagella and eukaryotic cilia, and a new debate has begun. Over the last few years some scientists have suggested that three different terms be used: flagellum for the bacterial organelle, cilia for the two eukaryotic organelles, and archaellum for the archaeal version of this motility organelle. Will this new name stick? Will the next edition of this text use the term? Will the discovery that archaeal flagella are evolutionarily related to bacterial type IV pili lead to a different name? Time will tell.

Source: Jarrell, K. F., and Albers, S.-V. 2012. The archaellum: An old motility structure with a new name. Trends Microbiol. 20(7):307–12.

(figure 4.11). The flagellum is not hollow. Hooks have been observed for some archaeal flagella but not for others. These differences have prompted some to suggest a new name for archaeal flagella—archaella (Microbial Diversity & Ecology 4.1).

One interesting feature of the characterized flagellar proteins is that they are similar to the proteins in bacterial type IV pili. Recall that type IV pili are responsible for the twitching motility of some bacteria. Thus far, no homologues of bacterial flagellar proteins have been identified in archaeal genomes. Like type IV pili, the filament of the archaeal flagellum increases in length as flagellin subunits are added at the base of the filament rather than the tip. As we noted in our discussion of archaeal pili, some archaea construct pili from proteins similar to bacterial type IV pilins. Scientists think archaeal type-IV-like pili have the same functions as bacterial type IV pili, except that thus far twitching motility has not been reported in any archaea. The use of similar proteins to make two structures having distinct functions has raised many questions. It has been suggested that the machinery used to assemble archaeal flagella differs from that used to assemble pili but that some components may be shared. ◄◄ *Bacterial flagella (section 3.7)*

Despite their similarity to bacterial type IV pili, archaeal flagella work in a manner similar to bacterial flagella: rotation propels the cell. However, there are some important differences. First, flagellar rotation is powered by ATP hydrolysis rather than by proton motive force. Second, when the direction of rotation switches, it causes the cell to move in either the forward or reverse direction, just as is seen for some bacteria having polar flagella. Thus far, alternation between runs and tumbles has not been observed.

Halobacterium salinarum flagellar movement is the best studied. In this archaeon, clockwise rotation of the flagella pushes the cell forward, and counterclockwise rotation pulls the cell (i.e., the flagella are in front of the cell as it moves). *H. salinarum* exhibits both chemotaxis and phototaxis. Phototaxis is used to position the archaeon properly to absorb light when

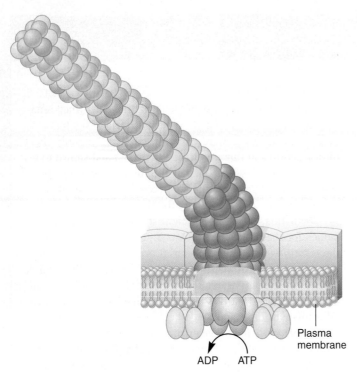

Plasma
membrane

ADP   ATP

Figure 4.11 Archaeal Flagellum. The different shades of blue in the filament illustrate that the filament may be composed of more than one type of flagellin.

carrying out rhodopsin-based phototrophy. Surprisingly, considering the difference in flagellar architecture, much of the machinery controlling chemotaxis and phototaxis in *H. salinarum* is homologous to that found in bacterial systems. ▶▶◀ *Phototrophy (section 11.11)*

Retrieve, Infer, Apply

1. At least some archaeal pili are similar to bacterial type IV pili. However, the archaeal pili are unique in terms of their structure. What is that distinctive feature?
2. What observations about cannulae and hami suggest that they allow archaeal cells to adhere to surfaces, including other cells?
3. List three aspects of archaeal flagella and flagellar motility that are similar to bacterial flagella and flagellar motility.

4.5 Comparison of *Bacteria* and *Archaea*

After reading this section, you should be able to:

■ Compare and contrast bacterial and archaeal cells in terms of the structures observed and their chemical makeup

Although members of *Bacteria* and *Archaea* share a common cell architecture, this chapter has outlined the significant differences between the two taxa. Here we review those differences.

The cell envelopes of bacteria and archaea are one of the distinctive features. Bacterial membranes contain lipids built by attaching fatty acids to glycerol by ester linkages. Archaeal membranes link isoprenoid hydrocarbons to glycerol by ether bonds. In some archaea, plasma membranes are monolayers rather than bilayers. Nearly all bacterial cell walls contain peptidoglycan. This molecule is unique to bacteria. Archaeal cell walls are much more varied than those of bacteria; many are composed solely of S-layers.

The ribosomes of bacteria and archaea are the same size but differ at the molecular level. The rRNA molecules are similar size, but the nucleotide sequences are distinguishable. Furthermore, the ribosomes of at least one archaeon have an additional rRNA that is not observed in bacterial ribosomes. A few ribosomal proteins are unique to archaea. However, the rest either have homologues in eukaryotic ribosomes or have homologues in both bacterial and eukaryotic ribosomes.

Most bacteria and archaea have chromosomes composed of double-stranded, circular DNA. These molecules are compacted using similar mechanisms. In addition, some archaea have histones that form nucleosomes similar to those observed in eukaryotic chromosomes.

Clearly, in terms of cell structure, the placement of archaea and bacteria into two separate domains of life is warranted. As you will see in subsequent chapters, archaea also differ from bacteria in their metabolic capabilities and gene expression. Considerably more is known about archaea now than a decade ago. As microbiologists have learned more, there have been many surprises. However, many questions about archaea remain unanswered. As more of these interesting microbes are isolated and studied, many of these questions will be answered and even more questions generated; expect more surprises.

Key Concepts

4.1 Archaea Are Diverse but Share Some Common Features

■ Members of *Bacteria* and *Archaea* share a common cell architecture. However, they have characteristics that define them as distinct taxa. **Table 4.1** summarizes some of the differences.

■ Rods and cocci are the most common archaeal shapes. Archaea also can be curved rods, spirals, branched, square, and pleomorphic (**figures 4.1**).

■ Most archaea are similar in size to bacteria. Archaea that are extremely small or are extremely large have also been identified (**figure 4.2**).

4.2 Six Major Types of Archaeal Cell Envelopes Have Been Identified

■ The cell envelope consists of the plasma membrane and all external coverings, including cell walls and other layers. Archaeal cell envelopes usually consist of only the plasma membrane and the cell wall.

- Archaeal membranes are composed of glycerol diether and diglycerol tetraether lipids (**figure 4.4**). Membranes composed of glycerol diether are lipid bilayers. Membranes composed of diglycerol tetraethers are lipid monolayers (**figure 4.5**). The overall structure of a monolayer membrane is similar to that of the bilayer membrane in that the membrane has a hydrophobic core and its surfaces are hydrophilic.
- Archaea typically use active transport systems to obtain nutrients from their environment.
- Archaeal cell walls do not contain peptidoglycan, and they exhibit great diversity in their makeup. The most common type of cell wall is one consisting of an S-layer only (**figure 4.6**).

4.3 Archaeal Cytoplasm Is Similar to Bacterial Cytoplasm

- Cytoskeletal proteins have been identified in archaeal cells. They include FtsZ (tubulin homologue), MreB (actin homologue), and crenactin (an unique archaeal actin homologue).
- Numerous inclusions are observed in archaeal cells, including gas vesicles.
- Bacterial and archaeal ribosomes are 70S in size but differ slightly in their morphology. They also differ in terms of their protein content, with many archaeal ribosomal proteins being more similar to those in eukaryotic ribosomes than to those in bacterial ribosomes.
- The genetic material of archaeal cells is located in the nucleoid, which is not enclosed by a membrane. All known archaeal chromosomes consist of a double-stranded, covalently closed, circular DNA molecule. In many archaea, the nucleoid contains a single chromosome. However, some archaea are polyploid, having more than one copy of their chromosome. Like eukaryotes, some archaea use histone proteins to organize their chromosomes.

4.4 Many Archaea Have External Structures Used for Attachment and Motility

- Many archaea have pili. Some archaeal pili are similar to bacterial type IV pili.
- Cannulae (**figure 4.9**) and hami (**figure 4.10**) are external structures that are unique to archaea. Their function is not known, but it is likely that they help cells attach to surfaces, including other cells.
- Many archaea are motile by means of flagella (**figure 4.11**). Archaeal flagella are structurally related to bacterial type IV pili. They are rigid helices that rotate, and the direction of rotation determines if the cell moves forward or backward. Rotation is powered by ATP hydrolysis.
- Motile archaea exhibit chemotaxis. Some are also phototactic. The archaeal taxis machinery is similar to that of bacteria.

4.5 Comparison of *Bacteria* and *Archaea*

- One major difference between bacterial and archaeal cells is that they have distinctive cell envelopes.
- There are also important differences between bacterial and archaeal ribosomes, chromosomes, flagella, and pili.

Compare, Hypothesize, Invent

1. Archaea with cell walls consisting of a thick, homogeneous layer of complex polysaccharides often retain crystal violet dye when stained using the Gram-staining procedure. Suggest a reason for this staining reaction.

2. Isoprene serves as a building block not only for the hydrocarbons observed in archaeal membranes but also for sterols, carotenoids, retinal, and quinones. Use any resources necessary to identify the function of these other isoprene-based molecules and to determine their distribution in nature. What does the use of isoprene to make this diverse array of molecules suggest about the nature of the last universal common ancestor (LUCA)?

3. As we note in section 4.3, archaeal ribosomal proteins can be divided into three classes: (1) those with homologues in all three domains of life, (2) those unique to archaea, and (3) those with homologues in eukaryotic ribosomes. Predict a class of archaeal ribosomal proteins that would refute the argument made by some microbiologists that the term *prokaryote* should be abandoned.

4. Archaea exhibit a wide variety of cell shapes, including some that are unique. Suggest why this diversity exists and what advantages the unique shapes might confer.

5. In the chapter opening case study, we focus on the production of methane by archaea living in the rumens of domesticated ruminant animals. However, there are many other sources of methane besides livestock. These include wild ruminants, the anoxic soils of rice paddies, and deforestation. There is considerable interest in more precisely measuring the output of methane by all these sources, yet current methods have many drawbacks. Some are laborious; some have a high degree of variability; others can only be used in highly controlled environments. In 2011 a group of scientists reported their attempts to correlate methane production with the amount of archaeol in the feces of cattle. They also determined archaeol levels resulting from different feeding regimens. Archaeol is a glycerol diether that can be measured by chemical procedures such as gas chromatography. What assumption did the scientists make when they chose to measure archaeol? Why did they choose to measure a glycerol diether rather than a diglycerol tetraether in their study? Suggest one other application of this method to better understand methane release into the atmosphere.

Read the original paper: Gill, F. L., et al. 2011. Analysis of archaeal ether lipids in bovine faeces. *Ani. Feed Sci. Tech.* 166–167:87.

5

Eukaryotic Cell Structure

Many Rocky Mountain forests have been ravaged by the mountain pine beetle and a fungus.

Red Means Dead

Anyone driving down a highway in the Rocky Mountains of Canada and the United States will observe an astonishing sight: large stands of rust-colored, dead pine trees in once verdant mountain forests. In British Columbia alone, over 40 million acres of trees have been killed. In 2008, about 4 million acres of trees were estimated killed in the northern Rocky Mountain states.

Two culprits working together have caused this destruction: the mountain pine beetle and its symbiont, the blue stain fungus *Grosmannia clavigera*. How they kill is an interesting example of a mutualistic relationship—one in which both benefit. The fungus is helped by the beetle, which carries the fungus on its mouth parts and transports it from one tree to another. The beetle is helped by the fungus, which protects the beetle from the tree's defense mechanisms and nourishes the beetle's young.

The killing process begins when the beetle chews through the bark of a pine tree to reach underlying tissues. The tree fights back by producing a resin toxic to the beetle. However, the fungus is able to degrade the toxic resin and use it as a carbon and energy source. Thus, with the fungus' help, the beetle is able to continue its attack, ultimately laying its eggs in the tree and carrying the fungus deeper into the tree. The fungus, now deep inside the tree, produces masses of filaments that help feed the beetle's progeny. These filaments are lethal as they literally clog the tree's circulatory system. Nutrients and water can no longer flow through the tree, and it dries up as it starves to death.

Much of this text is devoted to discussions of bacteria and archaea. But as just demonstrated, eukaryotic microbes also deserve our attention. They are prominent members of ecosystems, important model organisms, and exceptionally useful in industrial microbiology. A number are also major human pathogens.

Eukaryotic microbes can be divided into two major groups: protists and fungi. Protists are very diverse. Some use organic molecules as energy sources just as animals do; others are photosynthetic like plants. Many protist lineages have been identified in *Eukarya*. Currently, protists cannot be grouped together in a single monophyletic taxon—one that includes all descendants of the most recent common ancestor. Thus the term protist is a common name for these microbes rather than a valid taxon. Fungi, on the other hand, all share a common evolutionary history, so they are placed in one taxon (*Fungi*).

The division of eukaryotes into appropriate taxa is currently being debated, as is their evolution. The taxonomy and evolution of eukaryotes are discussed more completely in chapters 19, 25, and 26. In this chapter, we introduce eukaryotic cell structures and their functions. As we do so, comparisons are made to plants and animals. Thus it is important that you review what you already know about them.

Readiness Check:

Based on what you have learned previously, you should be able to:

✔ List the three domains of life (section 1.1)
✔ Describe the application of small subunit (SSU) rRNA analysis to the establishment of the three domain classification system proposed by Carl Woese (section 1.2)
✔ Identify the following structures or regions of a plant or animal cell and describe their functions: cell wall, plasma membrane, cytoplasm, mitochondria, chloroplasts, and ribosomes
✔ Identify the major cell structures observed in bacterial and archaeal cells, and describe the function and molecular makeup of each (chapters 3 and 4)

5.1 Eukaryotic Cells Are Diverse but Share Some Common Features

After reading this section, you should be able to:

■ Compare and contrast eukaryotic, bacterial, and archaeal cells in terms of their use of membranes, size, morphological diversity, and organelles

Before we begin our discussion of the numerous structures and organelles observed in eukaryotic cells, it is wise to begin with an overview. Here we make broad generalizations about eukaryotic cells. As we proceed in this chapter, we will give examples of deviations from the common features.

Shape and Size

One hallmark of eukaryotic cells is their morphological diversity. This is especially true for eukaryotic microbes, which are found in numerous habitats and have evolved morphological adaptations accordingly. Some of the diverse shapes of eukaryotic microbes are shown in **figure 5.1**.

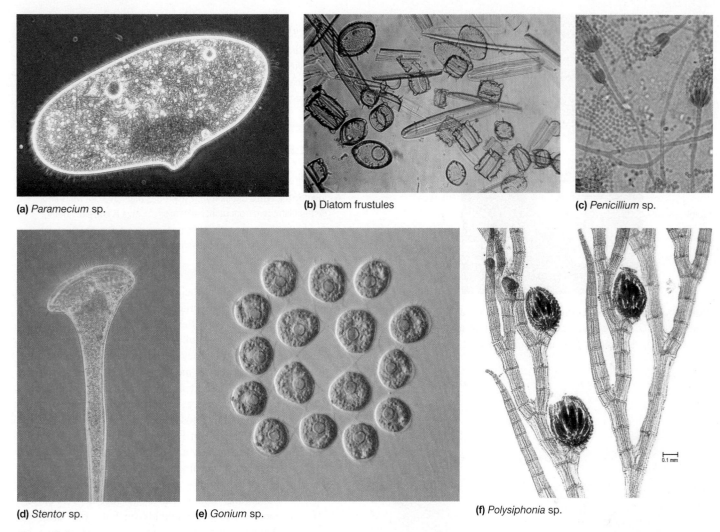

(a) *Paramecium* sp.

(b) Diatom frustules

(c) *Penicillium* sp.

(d) *Stentor* sp.

(e) *Gonium* sp.

(f) *Polysiphonia* sp.

Figure 5.1 Some Eukaryotic Microorganisms.   (a) *Paramecium* sp., a protist; interference-contrast microscopy (×115). (b) Mixed diatom frustules. (c) *Penicillium* sp., the mold's hyphae and conidia. (d) *Stentor* sp., a ciliated protist in its extended, actively feeding form; phase-contrast microscopy (×100). (e) *Gonium* sp., a freshwater, colonial alga (×40). (f) *Polysiphonia* sp., a filamentous, red alga (photosynthetic protist).

In general, eukaryotic cells are larger than bacterial and archaeal cells. This is true for eukaryotic microbes as well (*see figure 2.20*). However, just as some bacteria and archaea are very large, even larger than some of the largest protists (*see figure 3.4*), so too are there some protists smaller than many bacteria and archaea. These small eukaryotic microbes include *Ostreococcus tauri,* the smallest free-living alga known. Many other picoeukaryotes have been discovered in marine environments, as many as 10^3 to 10^4 per milliliter.

Cell Organization

Eukaryotic cells are distinctive because of their use of membranes (**figures 5.2** and **5.3**; compare these figures with figure 3.6). They have membrane-delimited nuclei, and membranes play a prominent part in the structure of many other organelles. **Organelles** are intracellular structures that perform specific functions in cells analogous to the functions of organs in the body of a multicellular organism.

The partitioning of the eukaryotic cell interior by membranes makes possible the placement of different cellular functions in separate compartments so that they can more easily take place simultaneously under independent control and proper coordination. Large membrane surfaces make possible greater respiratory and photosynthetic activity because these processes are located exclusively in membranes. The intracytoplasmic membrane complex also serves as a transport system to move materials between different cell locations. Thus abundant membrane systems probably are necessary in eukaryotic cells because of their large volume and the need for adequate regulation, metabolic activity, and transport.

Figures 5.2 and 5.3 illustrate most of the organelles discussed in this chapter. **Table 5.1** briefly summarizes the functions of the major organelles observed in most eukaryotes, including plants and animals. Our detailed discussion of eukaryotic cell structure begins with eukaryotic cell envelopes. We then proceed to the cytoplasm, organelles within the cytoplasm, and finally to external structures.

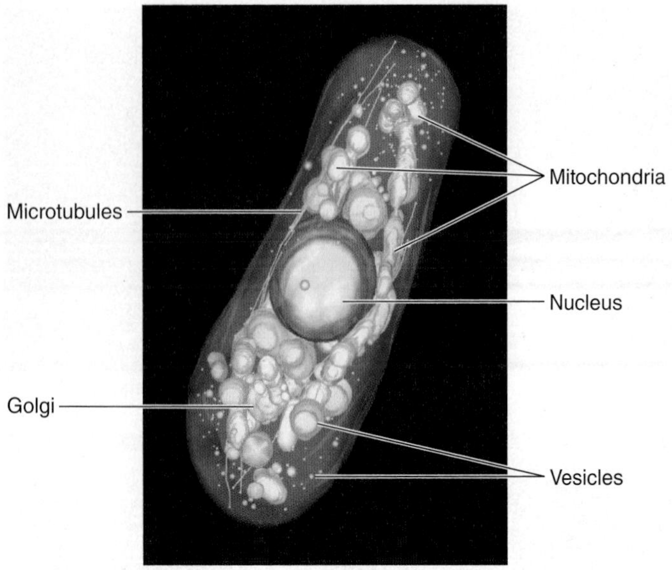

Figure 5.2 **Eukaryotic Cell Ultrastructure.** An electron cryotomogram of the yeast *Schizosaccharomyces pombe*. This image is the first high-resolution, three-dimensional reconstruction of a complete eukaryotic cell.

5.2 Eukaryotic Cell Envelopes

After reading this section, you should be able to:

- Identify the types of eukaryotic microbes that have cell walls and distinguish them from plant cell walls
- Compare and contrast the cell envelopes of members of *Bacteria*, *Archaea*, and *Eukarya* in terms of their component layers, molecular makeup, and function

The cell envelope consists of the plasma membrane and all coverings external to it. Eukaryotic microorganisms differ greatly from bacterial and archaeal cells in the structures they have external to the plasma membrane. Many eukaryotic microbes lack a cell wall. When cell walls are present, they are chemically distinctive.

The plasma membrane of eukaryotes is a lipid bilayer composed of a high proportion of sphingolipids and sterols (e.g., cholesterol and ergosterol) in addition to the phospholipids observed in bacterial membranes (**figure 5.4**). In phospholipids and sphingolipids, the fatty acids are attached to glycerol with ester links. The amounts of sphingolipids and sterols affect the fluidity and permeability of the plasma membrane; they are able to pack together very closely, more so than phospholipids can. Eukaryotic cells alter the quantity of sterols and sphingolipids in response to stresses such as high or low temperatures.

The distribution of lipids in the plasma membrane is asymmetric. This means that the lipids in the outer leaflet differ from those of the inner leaflet. Although most lipids in each leaflet mix freely with each other, there are small regions in the membrane called microdomains that differ in lipid and protein composition. These microdomains are sometimes referred to as lipid rafts. They have been implicated in a variety of cellular processes (e.g., signal transduction, assembly and release of virus particles, and endocytosis). As shown in figure 5.3*b*, many eukaryotic microbes as well as the cells of multicellular eukaryotes have integral membrane proteins and lipids decorated with carbohydrates. This often forms a carbohydrate-rich layer on the surface of the cell called the **glycocalyx.**
▶▶ *Viral life cycles have five steps (section 6.3)*

As with the plasma membranes of bacterial and archaeal cells, eukaryotic cells must be able to move materials across their membranes. Facilitated diffusion plays a more prominent role in eukaryotic cells than it does in bacterial and archaeal cells (*see figure 3.10*). Eukaryotes also use primary and

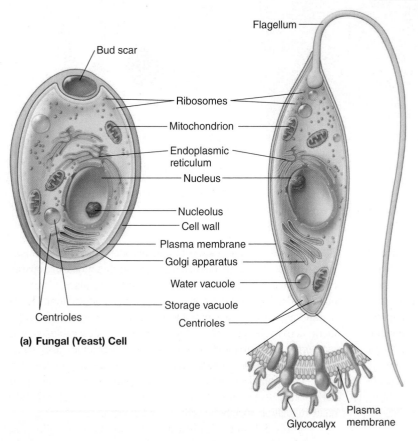

(a) Fungal (Yeast) Cell

(b) Protozoan Cell

Figure 5.3 **The Structure of Two Representative Eukaryotic Microbes.** Illustrations of a yeast cell (fungus) (a) and the flagellated protist *Peranema* (b).

MICRO INQUIRY *In addition to separating each organelle from the rest of the cytoplasm, what other functions do internal membranes serve?*

Table 5.1	Functions of Eukaryotic Organelles
Plasma membrane	Mechanical cell boundary; selectively permeable barrier with transport systems; mediates cell-cell interactions and adhesion to surfaces; secretion; signal transduction
Cytoplasm	Composed of cytosol (liquid portion) and organelles; location of many metabolic processes
Cytoskeleton	Composed of actin filaments, intermediate filaments, and microtubules; provides cell structure and movements
Endoplasmic reticulum	Transport of materials; lipid synthesis
Ribosomes	Protein synthesis
Golgi apparatus	Packaging and secretion of materials for various purposes; lysosome formation
Lysosomes	Intracellular digestion
Mitochondria	Energy production through use of the tricarboxylic acid cycle, electron transport, oxidative phosphorylation, and other pathways
Chloroplasts	Photosynthesis—trapping light energy and forming carbohydrate from CO_2 and water
Nucleus	Repository for genetic information
Nucleolus	Ribosomal RNA synthesis; ribosome construction
Cell wall and pellicle	Strengthen and give shape to the cell
Cilia and flagella	Cell movement
Vacuole	Temporary storage and transport; digestion (food vacuoles); water balance (contractile vacuole)

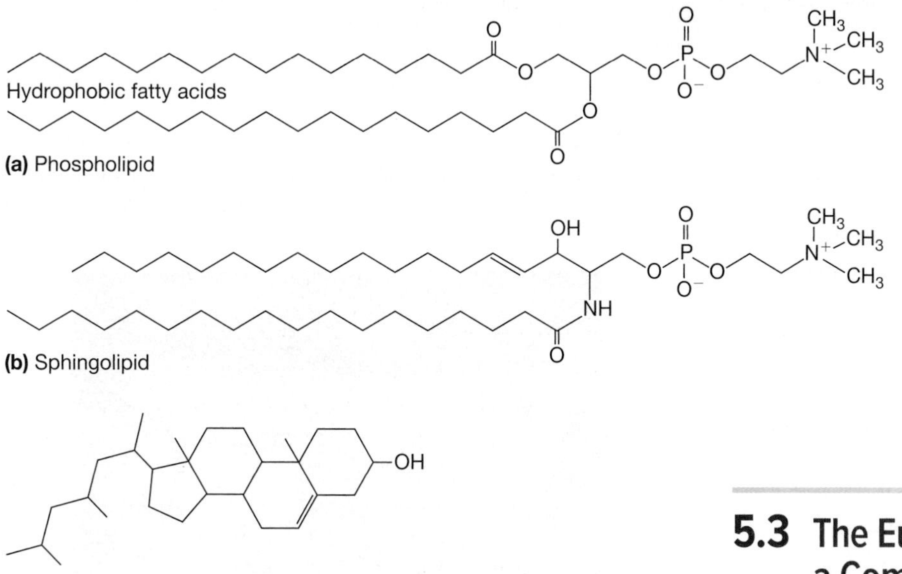

(a) Phospholipid

Hydrophobic fatty acids

(b) Sphingolipid

(c) Sterol

Figure 5.4 Examples of Eukaryotic Membrane Lipids.
(a) Phosphatidylcholine, a phospholipid. (b) Sphingomyelin, a sphingolipid. (c) Cholesterol, a sterol.

(commonly called algae) usually have a layered appearance and contain large quantities of polysaccharides such as cellulose and pectin. In addition, inorganic substances such as silica (in diatoms) or calcium carbonate may be present. Fungal cell walls normally are rigid. Their exact composition varies with the organism; usually cellulose, chitin, or glucan (a glucose polymer different from cellulose) is present. Despite their nature, the rigid materials in eukaryotic cell walls are chemically simpler than bacterial peptidoglycan. ◄◄ *There are two main types of bacterial cell walls (section 3.4)*

5.3 The Eukaryotic Cytoplasm Contains a Complex Cytoskeleton and Many Membranous Organelles

After reading this section, you should be able to:

- Describe the functions of the cytoplasm
- Identify the three filaments that make up the cytoskeleton of eukaryotic cells and describe their functions

secondary active transport and passive diffusion. In addition, eukaryotes have another option for bringing materials into cells: endocytosis. We focus on this unique process in section 5.4.

The chemical composition of the cell walls of eukaryotic microbes varies considerably. Those of photosynthetic protists

The **cytoplasm** is one of the most important and complex parts of a eukaryotic cell. It consists of a liquid component, the **cytosol,** in which many organelles are located. It is the location of many

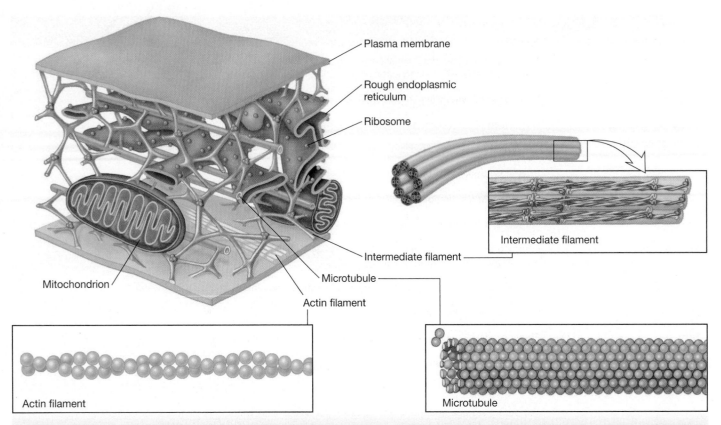

Plasma membrane

Rough endoplasmic reticulum

Ribosome

Intermediate filament

Intermediate filament

Microtubule

Mitochondrion

Actin filament

Actin filament

Microtubule

Figure 5.5 The Eukaryotic Cytoplasm and Cytoskeleton.   The cytoplasm of eukaryotic cells contains many important organelles. The cytoskeleton helps form a framework within which the organelles lie. The cytoskeleton is composed of three elements: actin filaments, microtubules, and intermediate filaments.

MICRO INQUIRY   *Which cytoskeletal filament is made of the proteins α- and β-tubulin?*

important biochemical processes, and several physical changes (e.g., viscosity changes and cytoplasmic streaming) are due to cytoplasmic activity. Because so many different kinds of organelles are observed in the eukaryotic cytoplasm, we address them separately in sections 5.4–5.6. Here we focus on the cytoskeleton, which helps organize the contents of the cytoplasm.

The **cytoskeleton** is a vast, dynamic network composed of three types of interconnected filaments: actin filaments (also called microfilaments), intermediate filaments, and microtubules (**figure 5.5**). Motor proteins (myosin, kinesin, and dynein) are often associated with actin filaments and microtubules. These cellular motors track along the cytoskeletal filaments, helping to move cell structures from one location to another.

Actin filaments are minute protein filaments, 4 to 7 nm in diameter, that are organized to form a variety of structures depending on the cell type (**figure 5.6**). They are composed of an actin protein that is similar to the actin contractile protein of muscle tissue. Actin filaments are involved in amoeboid movement, endocytosis, cytokinesis, and the movement of some structures within the cell. Interestingly, some bacterial pathogens use the actin proteins of their eukaryotic hosts to move rapidly through the host cell and to propel themselves into new host cells (*see figure 35.10*).

Intermediate filaments are heterogeneous elements of the cytoskeleton that play structural roles. They are about 10 nm in diameter and are assembled from a group of proteins that can be divided

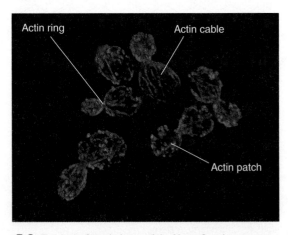

Actin ring

Actin cable

Actin patch

Figure 5.6 The Actin Cytoskeleton of the Yeast *Saccharomyces cerevisiae*.   Three kinds of structures made from actin filaments can be observed: actin patches, actin rings, and actin cables. Actin patches are sites where clathrin-dependent endocytosis occurs. Actin rings participate in cytokinesis and actin cables function in transport of material through the cytoplasm. The actin structures appear red in this micrograph.

into several classes (e.g., keratin and vimentin). Intermediate filaments having different functions are assembled from one or more of these classes of proteins. The role of intermediate filaments in eukaryotic microorganisms is unclear. Thus far, they have been identified and studied only in animals: some intermediate filaments form the nuclear lamina, a structure that provides support for the nuclear envelope (p. 99); some help position organelles within the cell; and others help link cells together to form tissues.

Microtubules are shaped like thin cylinders about 25 nm in diameter. They are constructed of two spherical protein subunits—α-tubulin and β-tubulin (figure 5.5). The two proteins are the same molecular weight and differ only slightly in terms of their amino acid sequence and tertiary structure. Each tubulin is approximately 4 to 5 nm in diameter. These subunits are assembled in a helical arrangement to form a cylinder with an average of 13 subunits in one turn.

Microtubules have several important functions. They form the spindle apparatus that separates chromosomes during mitosis and meiosis. They form tracks along which numerous types of organelles and vesicles are moved about the cell, and they are found in cilia and flagella, two organelles that confer motility (p. 104). Microtubules also are found in cell structures requiring support such as long, slender, rigid pseudopodia (called axopodia) observed in some protists.

Retrieve, Infer, Apply

1. What are the major lipids in eukaryotic membranes? Which other domain of life possesses similar lipids?

2. Which types of eukaryotic microbes have cell walls? How do their walls differ from those of bacterial and archaeal cells?

3. What are the three cytoskeletal elements observed in eukaryotic cells? What are their functions? Do eukaryotic microbes have all three elements?

5.4 Several Cytoplasmic Membranous Organelles Function in the Secretory and Endocytic Pathways

After reading this section, you should be able to:

- Differentiate the two types of endoplasmic reticulum in terms of structure and function
- Outline the pathway of molecules through transport and secretory pathways, noting the structures involved and their roles
- List the endocytic pathways observed in mammalian cells, noting the structures involved and their role in the process, and noting those pathways that have been observed in eukaryotic microbes

The cytoplasm of eukaryotic cells is permeated with an intricate complex of membranous organelles and vesicles that move materials into the cell from the outside (endocytic pathway) and from the inside of the cell out, as well as from location to location within the cell (secretory pathway). In this section, some of these organelles are described. This is followed by a summary of how the organelles function in the secretory and endocytic pathways.

Endoplasmic Reticulum

The **endoplasmic reticulum (ER)** (figure 5.3) is an irregular network of branching and fusing membranous tubules, around 40 to 70 nm in diameter, and many flattened sacs called cisternae (s., cisterna). The nature of the ER varies with the functional and physiological status of the cell. In cells synthesizing a great deal of protein to be secreted, a large part of the ER is studded on its outer surface with ribosomes and is called **rough endoplasmic reticulum (RER)**. Other cells, such as those producing large quantities of lipids, have ER that lacks ribosomes. This is **smooth endoplasmic reticulum (SER)**.

The endoplasmic reticulum has many important functions. Not only does it transport proteins, lipids, and other materials through the cell, it is also involved in the synthesis of many of the materials it transports. Lipids and proteins are synthesized by ER-associated enzymes and ribosomes. Polypeptide chains synthesized on RER-bound ribosomes may be inserted either into the ER membrane or into its lumen for transport elsewhere. The ER is also a major site of cell membrane synthesis.

The Golgi Apparatus Prepares Materials for Secretion

The **Golgi apparatus** is composed of flattened, saclike cisternae (**figure 5.7**). A complex network of tubules and vesicles is located at the edges of the cisternae. In many eukaryotes, the cisternae are stacked on each other, forming a structure called a dictyosome. There are usually around four to eight cisternae in a stack. In many cells, multiple stacks are linked to form a Golgi ribbon. Most eukaryotic microbes have just a single stack of cisternae, and in some, the cisternae are not in stacks (e.g., the yeast *Saccharomyces cerevisiae*).

A stack of cisternae has two faces that are quite different from one another. The sacs on the cis or forming face are closest to the ER and often associated with it. The sacs on the trans or maturing face are farthest from the ER. The two faces of the Golgi differ in thickness, enzyme content, and degree of vesicle formation.

The Golgi apparatus packages materials and prepares them for secretion, the exact nature of its role varying with the organism. For instance, the surface scales of some flagellated photosynthetic and radiolarian protists are constructed within the Golgi apparatus and then transported to the surface in vesicles. The Golgi often participates in the development of cell membranes and the packaging of cell products. The growth of some fungal hyphae occurs when Golgi vesicles contribute their contents to the wall at the hyphal tip. ▶▶ *Protists (chapter 25);* Fungi *(chapter 26)*

Lysosomes Are Degradative Organelles That Function in Endocytic Pathways

Lysosomes are found in animal cells. They are roughly spherical, are enclosed in a single membrane, and average about 500 nm in diameter but range from 50 nm to several μm in size. They are

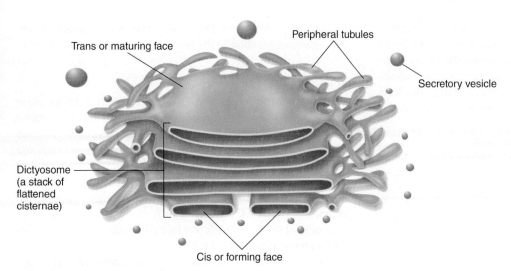

Figure 5.7 **Golgi Apparatus.** Drawing of the Golgi apparatus. Each cisterna is 15 to 20 nm thick and separated from other cisternae by 20 to 30 nm.

involved in intracellular digestion and contain the enzymes needed to digest all types of macromolecules. These enzymes, called hydrolases, catalyze the hydrolysis of molecules and function best under slightly acidic conditions (usually around pH 3.5 to 5.0). Lysosomes maintain an acidic environment by pumping protons into their interior. 🔁 *Lysosomes*

Organelles with the same function as lysosomes are found in fungal and protist cells, where they may be called lysosomes, vacuoles, phagocytic vacuoles, or food vacuoles. In addition to functioning in intracellular digestion, they may have other functions, including storage of calcium ions, phosphate, and amino acids. They also are components of the endocytic pathways observed in protists and fungi.

Transport and Secretory Pathways Move Materials to Other Locations, Including Outside the Cell

Various pathways are used to move materials about a cell. These share early steps with the pathways used to move materials into the plasma membrane or outside the cell. Thus, they are often described together and referred to as the biosynthetic-secretory pathway or simply the **secretory pathway.** The process of moving materials such as proteins and lipids around or outside the cell is complex and not fully understood. However, it involves numerous molecules that help target the protein or lipid to its proper location. The movement of proteins is of particular importance and is our focus.

Proteins destined for a cellular target or secretion follow a similar path to their final location. It begins when the proteins are synthesized by ribosomes attached to the rough endoplasmic reticulum (RER). These proteins have sequences of amino acids that target them to the lumen of the RER through which they move until released in small vesicles that bud from the ER. As the proteins pass through the ER, they are often modified by the addition of sugars—a process known as glycosylation. ▶▶ *Protein localization and secretion in eukaryotes (section 15.4)*

Vesicles released from the ER travel to the cis face of the Golgi apparatus (the side closest to the ER; figure 5.7). One

popular model of the secretory pathway posits that these vesicles fuse to form the cis face of the Golgi. How this occurs is still unresolved. As the proteins proceed from the cis to the trans side of the Golgi (the side farthest from the ER), they are further modified. Some of these modifications target the proteins for their final location.

Transport vesicles are released from the trans face of the Golgi and deliver their contents to various locations in the cell. Two types of secretory pathways transport materials to the plasma membrane. One type constitutively delivers proteins in an unregulated manner, releasing them to the outside of the cell as the transport vesicle (sometimes called a constitutive secretory vesicle) fuses with the plasma membrane. The other pathway, called the regulated secretory pathway, requires a signal before its vesicles (regulated secretory vesicles) fuse with the plasma membrane and release their contents.

One important feature of the secretory pathway is its quality-assurance mechanism. Proteins that fail to fold or have misfolded are not transported to their intended destination. Instead they are released into the cytosol, where they are targeted for destruction by the attachment of several small polypeptides known as ubiquitin, as detailed in **figure 5.8**. The ubiquitin-tagged proteins are degraded by a huge, cylindrical complex called a 26S **proteasome.** The protein is broken down to smaller peptides using energy supplied by adenosine triphosphate (ATP) when the terminal phosphate is removed. As the protein is broken down, ubiquitins are released. In animal cells, the proteasome is involved in producing peptides for antigen presentation during immunological responses, as described in chapters 33 and 34. Proteasomes and proteasome-like structures have also been identified in bacteria and archaea. However, they are simpler in structure and the proteins to be degraded are targeted to the degradation machinery by mechanisms other than ubiquitin. ▶▶ *ATP (section 10.2); Phagocytosis (section 33.5); Recognition of foreignness is critical for a strong defense (section 34.4)*

There Are Several Endocytic Pathways

Endocytosis is observed in all eukaryotic cells. It is used to bring materials into the cell from the outside. During endocytosis, a cell takes up solutes (pinocytosis) or particles (phagocytosis) by enclosing them in vesicles pinched off from the plasma membrane. Endocytosis occurs regularly in all eukaryotic cells as a mechanism for recycling molecules in the membrane. In addition, some eukaryotic cells have specialized **endocytic pathways** that allow them to concentrate materials outside the cell before bringing them in. Others use endocytic pathways as a feeding mechanism. Many viruses and other intracellular pathogens enter their eukaryotic host cells via the cell's endocytic pathways.

Figure labels: Trans or maturing face; Peripheral tubules; Secretory vesicle; Dictyosome (a stack of flattened cisternae); Cis or forming face

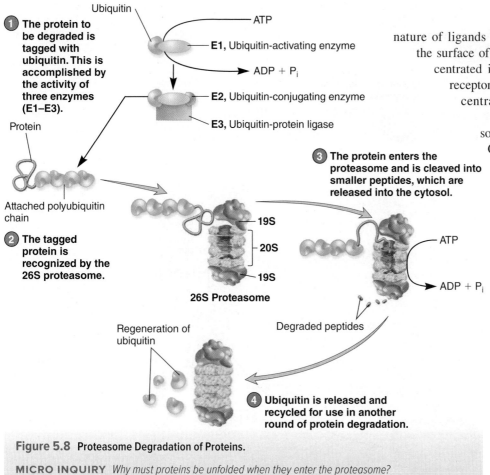

1 The protein to be degraded is tagged with ubiquitin. This is accomplished by the activity of three enzymes (E1–E3).

Ubiquitin

ATP

E1, Ubiquitin-activating enzyme

ADP + P$_i$

E2, Ubiquitin-conjugating enzyme

E3, Ubiquitin-protein ligase

Protein

Attached polyubiquitin chain

2 The tagged protein is recognized by the 26S proteasome.

19S

20S

19S

26S Proteasome

3 The protein enters the proteasome and is cleaved into smaller peptides, which are released into the cytosol.

ATP

ADP + P$_i$

Regeneration of ubiquitin

Degraded peptides

4 Ubiquitin is released and recycled for use in another round of protein degradation.

Figure 5.8   Proteasome Degradation of Proteins.

MICRO INQUIRY   *Why must proteins be unfolded when they enter the proteasome?*

nature of ligands varies (e.g., hormones or molecules on the surface of a virus). Sometimes receptors are concentrated in certain areas of the membrane, and receptor-mediated endocytosis serves to concentrate the ligand before it is endocytosed.

How materials are delivered to lysosomes depends on the endocytic pathway. Clathrin-coated vesicles and caveolin-coated vesicles first deliver their contents to small organelles containing hydrolytic enzymes. These organelles are called early **endosomes** (figure 5.9). Early endosomes mature into late endosomes, which fuse with lysosomes. The development of early endosomes into late endosomes is not well understood. It is thought that maturation involves the selective retrieval of membrane proteins. Phagosomes take a different route to lysosomes. They fuse directly with lysosomes (figure 5.9) to form a phagolysosome.

Endocytosis in eukaryotic microbes has not been studied as extensively as it has in mammals. It is known that some eukaryotic microbes carry out phagocytosis and probably macropinocytosis (figure 5.9). Clathrin-dependent endocytosis has been observed in fungi and is best characterized in the yeast *Saccharomyces cerevisiae*. Although some aspects of the endocytic process differ between this yeast and mammals, considerable similarity has been discovered, especially in terms of the mechanisms used to sort materials among early endosomes, late endosomes, lysosomes, and the plasma membrane.

Lysosomes are involved in another process that is increasingly attracting the attention of scientists: autophagy. Several types of autophagy have been identified. Some selectively sequester and destroy bacteria, viruses, aggregated proteins, and mitochondria. However, these selective processes are less well studied than **macroautophagy,** which nonselectively digests and recycles cytoplasmic components (including organelles such as mitochondria). This is a normal process that helps maintain cellular homeostasis. It is also sometimes used to destroy pathogens that have entered a host cell by endocytosis.

During macroautophagy, materials to be digested are surrounded by a double membrane to form a structure called an **autophagosome.** Autophagosome formation begins with creation of an isolation membrane. Evidence exists that the isolation membrane arises from the endoplasmic reticulum. Numerous autophagy-related proteins are involved in the formation of the isolation membrane and its subsequent maturation into the autophagosome. The additional lipids needed to complete the autophagosome are contributed by other membranous structures, including the Golgi apparatus, plasma membrane, and mitochondria. The mature autophagosome ultimately fuses with a lysosome.

Mammalian cells have several types of endocytic pathways (**figure 5.9**). All generate an endocytic vesicle of some type, and the substances in the vesicle are eventually delivered to a lysosome for degradation. **Phagocytosis** involves the use of protrusions from the cell surface to surround and engulf particulates. The endocytic vesicles formed by phagocytosis are called **phagosomes. Clathrin-dependent endocytosis** begins with coated pits, which are specialized membrane regions coated on the cytoplasmic side with the protein clathrin. The endocytic vesicles formed when these regions invaginate are called coated vesicles. Clathrin-dependent endocytosis is used to internalize hormones, growth factors, iron, and cholesterol. **Caveolin-dependent endocytosis** involves caveolae ("little caves"), tiny, cup-shaped invaginations of the plasma membrane (about 50 to 80 nm in diameter) that are enriched in cholesterol and the membrane protein caveolin. The vesicles formed when caveolae bud into the cytoplasm are called caveolin-coated vesicles. Caveolin-dependent endocytosis has been implicated in signal transduction, transport of small molecules such as folic acid, as well as transport of macromolecules. Evidence exists that toxins such as cholera toxin enter their target cells via caveolae. Caveolae also are used by many viruses, bacteria, and protozoa to enter host cells.

Endocytosis is often triggered by the binding of a molecule called a **ligand** to a **receptor** located in the cell membrane. This is referred to as **receptor-mediated endocytosis.** The

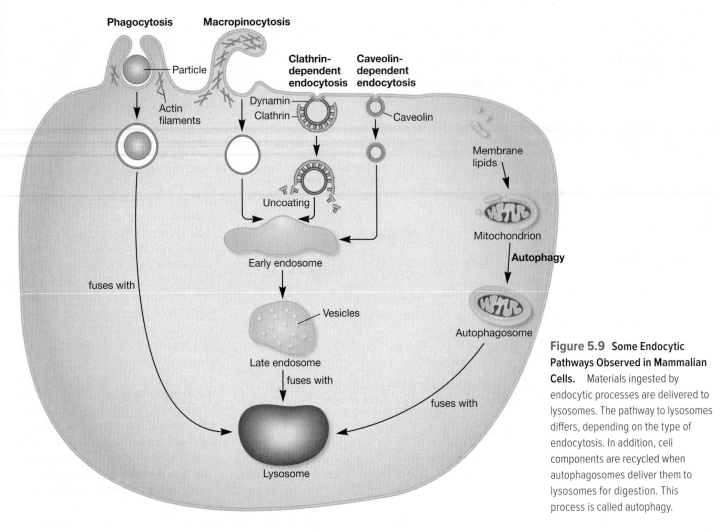

Figure 5.9 Some Endocytic Pathways Observed in Mammalian Cells.   Materials ingested by endocytic processes are delivered to lysosomes. The pathway to lysosomes differs, depending on the type of endocytosis. In addition, cell components are recycled when autophagosomes deliver them to lysosomes for digestion. This process is called autophagy.

No matter the route taken, digestion occurs once the targeted materials have been delivered to a lysosome. As the contents of the lysosome are digested, small products of digestion leave the lysosome and are used as nutrients or for other purposes. The resulting lysosome containing undigested material is sometimes called a **residual body.** In some cases, the residual body can release its contents to the cell exterior.

Retrieve, Infer, Apply

1. How do RER and SER differ from one another in terms of structure and function? List the processes in which the ER is involved.
2. What is a proteasome? Why is it important to the proper functioning of the ER?
3. Describe the structure of a Golgi apparatus. How do the cis and trans faces of the Golgi apparatus differ? List the major functions of the Golgi.
4. What are lysosomes? How do they participate in intracellular digestion? What might happen if lysosomes released their enzymatic contents into the cytoplasm?
5. Describe the secretory pathway. To what destinations does this pathway deliver proteins and other materials?
6. Define endocytosis. Describe the routes that deliver materials to lysosomes for digestion.

5.5 The Nucleus and Ribosomes Are Involved in Genetic Control of the Cell

After reading this section, you should be able to:

- Describe the structure and function of the nucleus, chromosomes, nucleolus, and eukaryotic ribosomes
- Compare and contrast the chromosomes and ribosomes of bacterial, archaeal, and eukaryotic cells

DNA is the molecule that houses the genetic blueprint of the cell. Eukaryotic cells differ dramatically from bacterial and archaeal cells in the way DNA is stored and used. In this section, the organelles involved with these important cellular functions are introduced.

Nucleus

The nucleus is by far the most visually prominent organelle in eukaryotic cells. It was discovered early in the study of cell structure and was shown by Robert Brown (1773–1858) in 1831 to be a constant feature of eukaryotic cells. The **nucleus** is the repository for the cell's genetic information.

Figure 5.10 Nucleosomes. DNA is wrapped around an ellipsoid composed of eight histone proteins (two each of H2A, H2B, H3, and H4). In most eukaryotes, the histone H1 associates with the nucleosome.

Nuclei are membrane-delimited spherical bodies about 5 to 7 μm in diameter (figures 5.2 and 5.3). They contain more than one chromosome; the exact number depends on the organism, cell type, and stage in the life cycle. Each eukaryotic **chromosome** is composed of chromatin. **Chromatin** is a complex of DNA and proteins, including histones. Histones are small basic proteins rich in the amino acids lysine, arginine, or both. There are five types of histones in most eukaryotic cells: H1, H2A, H2B, H3, and H4. Eight histone molecules form an ellipsoid about 11 nm long and 6.5 to 7 nm in diameter around which the DNA wraps to form a "beads-on-a-string" formation (**figure 5.10**). Each bead is called a **nucleosome.**

Chromosomes are very dynamic and vary in terms of their degree of compaction. When the cell is not dividing, there is less compaction. The highest degree of compaction occurs during cell division. Compaction is brought about in part by histones and other proteins. Some of these other proteins are similar to the nucleoid-associated proteins used by bacterial and archaeal cells to compact their chromosomes. Although it is clear that histones are important for compacting chromosomes, how they do so is still a matter of considerable research and debate. In vitro studies suggest that nucleosomes close to each other interact to form 30-nm fibers. Nucleosomes far from each other (and in distinct areas of the 30-nm fiber) interact to form even thicker fibers. However, it is unclear if these higher orders of chromatin structure exist in living cells. Tethering of the chromosome to nuclear structures also contributes to chromosome folding and compaction. ◄◄ *Nucleoid (sections 3.6 and 4.3)*

The nucleus is surrounded by the **nuclear envelope** (figure 5.3), a complex structure consisting of two lipid bilayer membranes, the inner and outer membranes, separated by a perinuclear space. The nuclear envelope is continuous with the ER at several points, and its outer membrane is covered with ribosomes. A network of intermediate filaments, called the nuclear lamina, is observed in animal cells. It lies against the inner surface of the nuclear envelope and supports it. Many nuclear pores penetrate the envelope, and each pore is formed by about 30 proteins; each pore plus the associated proteins is called a **nuclear pore complex** (**figure 5.11**). Pores are about 70 nm in diameter and collectively occupy about 10 to 25% of the nuclear surface. The nuclear pore complexes serve as transport routes between the nucleus and surrounding cytoplasm. Small molecules move through the nuclear pore complex unaided. However, large molecules are transported through the nuclear pore complex. Some nuclear pore complex proteins are involved in these transport processes.

Often the most noticeable structure within the nucleus is the **nucleolus** (**figure 5.12**). A nucleus may contain from one to many nucleoli. Although the nucleolus is not membrane-enclosed, it is a

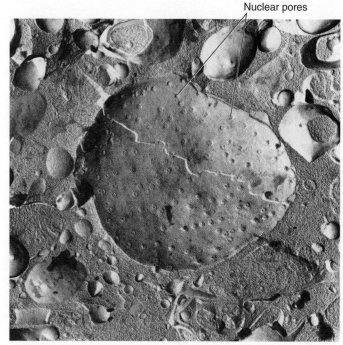

(a) Nucleus with nuclear pores

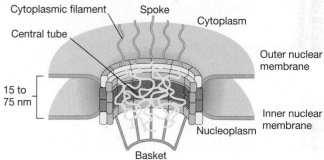

(b) Nuclear pore complex

Figure 5.11 The Nucleus and Nuclear Pore Complex. (a) A Transmission electron micrograph (TEM) of a freeze-etch preparation of a cell, showing the large, convex nuclear surface with pores scattered over it. (b) Schematic of the nuclear pore complex. Note that the nuclear envelope consists of two lipid bilayer membranes: outer nuclear membrane and inner nuclear membrane.

complex organelle with separate granular and fibrillar regions. It is present in nondividing cells but frequently disappears during mitosis. After mitosis, the nucleolus reforms around the nucleolar organizer, a particular part of a specific chromosome.

The nucleolus plays a major role in ribosome synthesis. The DNA of the nucleolar organizer directs the production of ribosomal RNA (rRNA). This RNA is synthesized in a single long piece that is cut to form the final rRNA molecules. The processed rRNAs combine with ribosomal proteins (which have been synthesized in the cytoplasm) to form partially completed ribosomal subunits. The granules seen in the nucleolus are probably these subunits. Immature ribosomal subunits then leave the nucleus, presumably by way of the nuclear pore complexes, and mature in the cytoplasm.

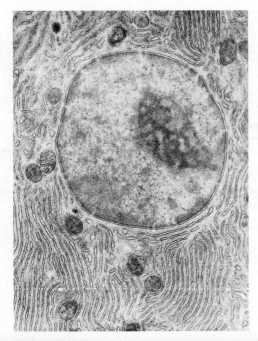

Figure 5.12 The Nucleolus. The nucleolus is a prominent feature of the nucleus, as seen in this electron micrograph. The nucleolus functions in rRNA synthesis and the assembly of ribosomal subunits.

MICRO INQUIRY *What are the granules within the nucleolus thought to be?*

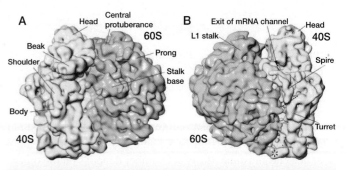

Figure 5.13 Eukaryotic Ribosome. This image showing two views of the 80S ribosome of *Trypanosoma cruzi* was generated using electron cryomicroscopy. The image is a density map with a resolution to 12 Å. Landmarks of the ribosome are shown; compare these to bacterial 70S ribosomes shown in figure 3.38.

MICRO INQUIRY *Which subunit attaches to the rough ER?*

Retrieve, Infer, Apply

1. What mechanisms contribute to the folding of eukaryotic chromosomes?
2. Suggest an explanation for the disappearance of the nucleolus during mitosis.
3. Describe the structure of the eukaryotic 80S ribosome and contrast it with bacterial and archaeal ribosomes.
4. How do free ribosomes and those bound to the ER differ in function?

Eukaryotic Ribosomes Are Larger and Contain Different Molecules than Bacterial and Archaeal Ribosomes

Eukaryotic ribosomes (i.e., those not found in mitochondria and chloroplasts) are larger than bacterial and archaeal 70S ribosomes. Each ribosome is a dimer of a 60S and a 40S subunit. The 60S subunit is composed of three rRNA molecules (5S, 28S, and 5.8S rRNAs) and about 50 proteins. The 40S subunit is composed of an 18S rRNA and about 30 proteins. Each ribosome is about 22 nm in diameter and has a sedimentation coefficient of 80S and a molecular weight of 4 million (**figure 5.13**). Eukaryotic ribosomes are either associated with the endoplasmic reticulum or free in the cytoplasm. When bound to the endoplasmic reticulum to form rough ER, they are attached through their 60S subunits.

Both free and ER-bound ribosomes synthesize proteins. Proteins made on the ribosomes of the RER are often secreted or are inserted into the ER membrane as integral membrane proteins. Free ribosomes are the sites of synthesis for nonsecreted and nonmembrane proteins. Some proteins synthesized by free ribosomes are inserted into organelles such as the nucleus, mitochondrion, and chloroplast. As we discuss in chapters 13 and 15, proteins called molecular chaperones aid the proper folding of proteins after synthesis. They also assist the transport of proteins into eukaryotic organelles such as mitochondria.

5.6 Mitochondria, Related Organelles, and Chloroplasts Are Involved in Energy Conservation

After reading this section, you should be able to:

- Draw a mitochondrion and identify its component parts
- Compare and contrast mitochondria and hydrogenosomes in terms of their structure and the chemical processes they carry out
- Draw a chloroplast and identify its component parts

Three important energy-conserving organelles are the focus of this section: mitochondria, hydrogenosomes, and chloroplasts. They are of scientific interest not only for this role but also because of their evolutionary history. All are thought to be derived from bacterial cells that invaded or were ingested by early ancestors of eukaryotic cells (*see figure 19.11*). ▶▶| *Evolution of the three domains of life (section 19.5)*

The Classic Mitochondrion Is Only One Member of the Mitochondrial Family of Organelles

In the last several years it has become increasingly apparent that eukaryotic microbes contain numerous organelles related to mitochondria. It is thought that they arose from a single endosymbiotic

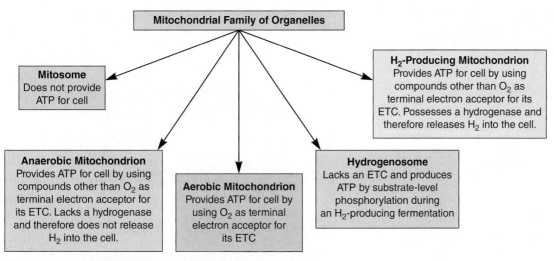

Figure 5.14 The Mitochondrial Family of Organelles. It is thought that the ancestral endosymbiont evolved in different cell lines to give rise to these five different organelles. Descriptions of the metabolic processes noted are described in chapter 11. ETC is electron transport chain.

other words, they are about the same size as bacterial cells.) Some cells possess 1,000 or more mitochondria; others (some yeasts, unicellular algae, and trypanosome protists) have a single, giant, tubular mitochondrion twisted into a continuous network permeating the cytoplasm. ▶▶◀ *Pyruvate to carbon dioxide (step 2) is accomplished by the tricarboxylic acid cycle (section 11.5); Electron transport and oxidative phosphorylation (step 3) generate the most ATP (section 11.6)*

event, as we describe in chapter 1, but then took different evolutionary paths to yield five different members of what is now referred to as the mitochondrial family of organelles (**figure 5.14**). In this section we describe two members of this family: the classic mitochondrion (aerobic mitochondrion) and the hydrogenosome.

The aerobic mitochondrion is the organelle in this family that is most familiar to biology students. For our discussion, we will simply refer to it as the mitochondrion. Found in most eukaryotic cells, these **mitochondria** (s., **mitochondrion**) frequently are called the "powerhouses" of the cell (**figure 5.15**). Metabolic processes such as the tricarboxylic acid cycle and generation of ATP, the major energy currency of all life forms, take place here, but only when oxygen is available. When viewed with a transmission electron microscope, many mitochondria are cylindrical structures and measure approximately 0.3 to 1.0 μm by 5 to 10 μm. (In

Mitochondria are bounded by two membranes: an outer mitochondrial membrane separated from an inner mitochondrial membrane by a 6 to 8 nm intermembrane space (figure 5.15). The outer mitochondrial membrane contains porins and thus is similar to the outer membrane of Gram-negative bacteria. The inner membrane has infoldings called **cristae** (s., **crista**), which greatly increase its surface area. The shape of cristae differs in mitochondria from various species. Platelike (laminar) cristae, cristae shaped like disks, tubular cristae, and cristae in the shape of vesicles have all been observed. The inner membrane encloses the mitochondrial matrix, a dense material containing ribosomes, DNA, and often large calcium phosphate granules. In many organisms, mitochondrial DNA is a closed circle, like most bacterial DNA. However, in some protists, mitochondrial DNA is linear.

Each mitochondrial compartment has a characteristic chemical and enzymatic composition. For example, the outer and inner

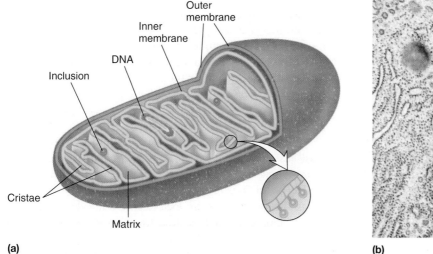

(a)

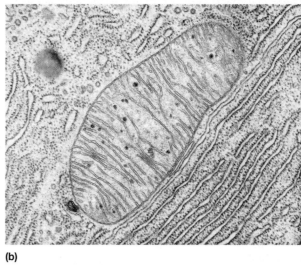

(b)

Figure 5.15 Mitochondrial Structure. (a) A diagram of mitochondrial structure. The insert shows the ATP-synthesizing enzyme ATP synthase lining the inner surface of the cristae. (b) Transmission electron micrograph (×85,000) of a mitochondrion showing the cristae.

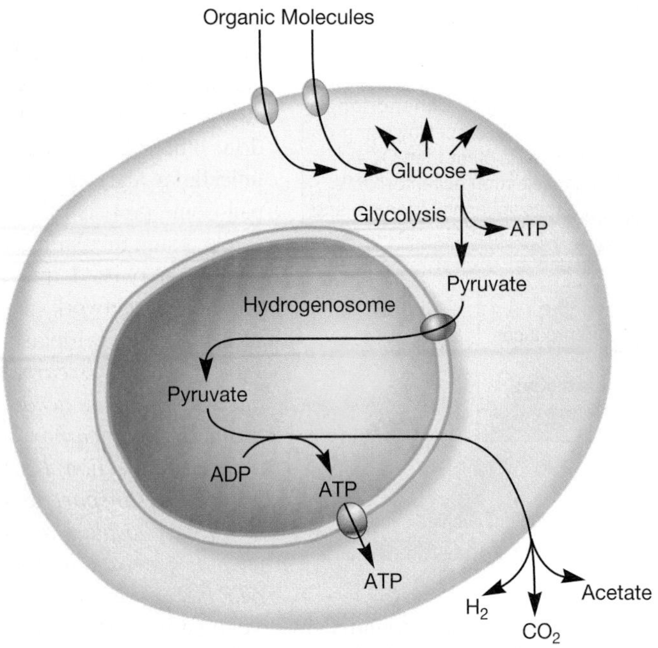

Figure 5.16 The Metabolic Activities of Hydrogenosomes.

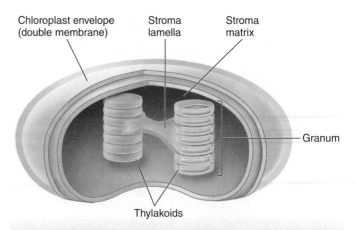

Figure 5.17 Diagram of Chloroplast Structure.

MICRO INQUIRY *In what regions of the chloroplast do the light reactions occur? The dark reactions?*

mitochondrial membranes possess different lipids. Enzymes and electron carriers involved in electron transport and oxidative phosphorylation (the formation of ATP during respiration) are located only in the inner membrane. Enzymes of the tricarboxylic acid cycle and those involved with the catabolism (breaking down) of fatty acids are located in the matrix. ▶▶| *Lipid catabolism (section 11.9)*

Mitochondria use their DNA and ribosomes to synthesize some mitochondrial proteins. In fact, mutations in mitochondrial DNA often lead to serious diseases in humans. However, most mitochondrial proteins are manufactured under the direction of the nucleus and must be transported into the mitochondrion. Mitochondria reproduce by binary fission, a reproductive process used by many bacteria. ▶▶| *Bacterial cell cycles can be divided into three phases (section 7.2); Protein localization and secretion in eukaryotes (section 15.4)*

Hydrogenosomes are small organelles involved in energy-conservation processes in some anaerobic protists (**figure 5.16**). Like mitochondria, hydrogenosomes are bound by a double membrane. However, they often lack cristae and usually lack DNA. They also differ from mitochondria in terms of the method used to generate ATP. Within hydrogenosomes, pyruvate is catabolized by a fermentative process rather than respiration, and CO_2, H_2, and acetate are formed. In some hydrogenosome-bearing protists, these metabolic products are consumed by symbiotic bacteria and archaea living within the protist. The symbiotic archaea include methanogens that consume the CO_2 and H_2 and generate methane (CH_4). ▶▶| *Fermentation does not involve an electron transport chain (section 11.8); Mutualism (section 32.1)*

Chloroplasts

Plastids are cytoplasmic organelles of photosynthetic protists and plants. They often possess pigments such as chlorophylls and

carotenoids, and are the sites of synthesis and storage of food reserves. The most important type of plastid is the chloroplast. **Chloroplasts** contain chlorophyll and use light energy to convert CO_2 and water to carbohydrates and O_2; that is, they are the site of photosynthesis. Two major types of chloroplasts have been identified: those that evolved from a primary endosymbiotic event and those that evolved from a secondary (or tertiary) event (**Microbial Diversity & Ecology 5.1**). The chloroplasts of plants and some photosynthetic protists are primary plastids and are the focus of this discussion.

Chloroplasts are quite variable in size and shape, but they share many structural features. Most are oval with dimensions of 2 to 4 μm by 5 to 10 μm, but some photosynthetic protists possess one huge chloroplast that fills much of the cell. Like mitochondria, chloroplasts are encompassed by two membranes (**figure 5.17**). The inner membrane surrounds a matrix called the stroma. The stroma contains DNA, ribosomes, lipid droplets, and starch granules. Also located within the stroma is a complex internal membrane system whose most prominent components are flattened, membrane delimited sacs called **thylakoids.** Clusters of two or more thylakoids are dispersed within the stroma of most algal chloroplasts (figure 5.17). In some photosynthetic protists, several disklike thylakoids are stacked on each other like coins to form grana (s., granum).

Photosynthetic reactions are separated structurally in chloroplasts just as electron transport and the tricarboxylic acid cycle are in mitochondria. The trapping of light energy to generate ATP, NADPH, and O_2 is referred to as the light reactions of photosynthesis. These reactions are located in the thylakoid membranes, where chlorophyll and electron transport components are also found. The ATP and NADPH formed by the light reactions are used to form carbohydrates from CO_2 and water in the dark reactions. The dark reactions of photosynthesis take place in the stroma. ▶▶| *Phototrophy (section 11.11)*

MICROBIAL DIVERSITY & ECOLOGY

5.1 There Was an Old Woman Who Swallowed a Fly

The children's song "There Was an Old Woman Who Swallowed a Fly" describes the unusual eating habits of an old woman. Her story began when she ate a fly but continued as she ate a series of other animals, all in an attempt to get rid of the fly. The evolutionary history of chloroplasts also appears to involve a series of meals. Molecular studies of the chloroplasts of a wide variety of photosynthetic protists have revealed that some chloroplasts arose when a phagocytic ancestor of a eukaryotic cell engulfed the ancestor of a cyanobacterium (**box figure**). This evolutionary event eventually gave rise to plants and the photosynthetic protists commonly called green algae and red algae. These organisms are now classified together in the supergroup *Archaeplastida* (*see chapter 25*). The chloroplasts that evolved from this endosymbiosis are the typical chloroplasts that biology students are familiar with. These primary plastids have two membranes. It is thought that these primary plastids gave rise

to at least two different plastids when two phagocytic plastid-free eukaryotes ingested an archaeplastid; one ingested a green alga, the other, a red alga. The ingested archaeplastid evolved into a chloroplast surrounded by three membranes. These chloroplasts are known as secondary plastids and are found in photosynthetic protists called stramenopiles. They include diatoms and brown algae. The final meal in the series was eaten by some dinoflagellates. These interesting protists can be either chemoorganotrophic or phototrophic. It appears that some photosynthetic dinoflagellates ingested a protist containing secondary plastids. The ingested protist evolved into a tertiary plastid surrounded by four membranes.

Source: Reyes-Prieto, A., et al. 2007. The origin and establishment of the plastid in algae and plants. Annu. Rev. Genet. 41:147; Dorrell, R. G., and Smith, A. G. 2011. Do red and green make brown?: Perspectives on plastid acquisitions within chromalveolates. Euk. Cell 10:856.

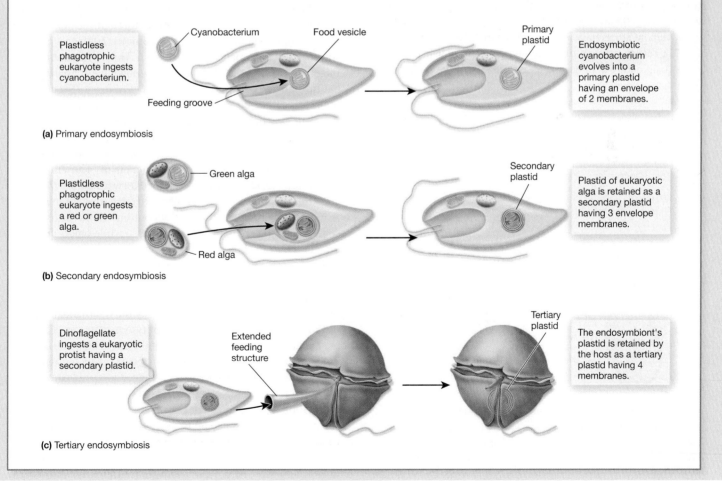

(a) Primary endosymbiosis

Plastidless phagotrophic eukaryote ingests cyanobacterium.

Cyanobacterium

Food vesicle

Feeding groove

Primary plastid

Endosymbiotic cyanobacterium evolves into a primary plastid having an envelope of 2 membranes.

(b) Secondary endosymbiosis

Plastidless phagotrophic eukaryote ingests a red or green alga.

Green alga

Red alga

Secondary plastid

Plastid of eukaryotic alga is retained as a secondary plastid having 3 envelope membranes.

(c) Tertiary endosymbiosis

Dinoflagellate ingests a eukaryotic protist having a secondary plastid.

Extended feeding structure

Tertiary plastid

The endosymbiont's plastid is retained by the host as a tertiary plastid having 4 membranes.

1. Describe the structure of aerobic mitochondria, hydrogenosomes, and chloroplasts. Where are the different components of mitochondria and chloroplast energy-trapping systems located?
2. What processes are used by aerobic mitochondria, hydrogenosomes, and chloroplasts to conserve energy?
3. What is the role of mitochondrial DNA?

5.7 Many Eukaryotic Microbes Have External Structures Used for Motility

After reading this section, you should be able to:

- Describe the structure of eukaryotic flagella and cilia
- Compare and contrast bacterial, archaeal, and eukaryotic flagella
- List the types of motility observed in eukaryotic microbes

Cilia (s., **cilium**) and **flagella** (s., **flagellum**) are the most prominent external structures observed on eukaryotic cells; they are associated with motility. Although both are whiplike and beat to move the microorganism along, they differ from one another in two ways. First, cilia are typically only 5 to 20 μm in length, whereas flagella are 100 to 200 μm long. Second, their patterns of movement are usually distinctive (**figure 5.18**). The beating of flagella causes them to undulate and generate planar or helical waves originating at either the base or the tip. If the wave moves

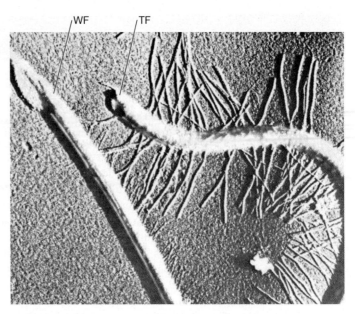

Figure 5.19 Whiplash and Tinsel Flagella. Transmission electron micrograph of a shadowed whiplash flagellum, WF, and a tinsel flagellum, TF, with mastigonemes.

from base to tip, the cell is pushed along; a beat traveling from the tip toward the base pulls the cell through the water. Some flagella have lateral hairs called flimmer filaments (thicker, stiffer hairs are called mastigonemes). These filaments change flagellar action so that a wave moving down the filament toward the tip pulls the cell along instead of pushing it. Such flagella often are called tinsel flagella, whereas naked flagella are referred to as whiplash flagella (**figure 5.19**). Cilia, on the other hand, normally have a beat with two distinctive phases. In the effective stroke, the cilium moves through the surrounding fluid like an oar, thereby propelling the organism. The cilium next bends along its length while it is pulled forward during the recovery stroke in preparation for another effective stroke (figure 5.18). A ciliated microorganism actually coordinates the beats so that some of its cilia are in the recovery phase while others are carrying out their effective stroke. This coordination allows the organism to move smoothly through the water.

Despite their differences, cilia and flagella are the same in ultrastructure. Many think that flagella are simply long cilia, and the classification of protists proposed by the International Society of Protistologists refers to both motility organelles as cilia. They are membrane-bound cylinders about 0.2 μm in diameter. Located in the matrix of the organelle is the axoneme, which usually consists of nine microtubule doublets arranged in a circle around two central microtubules (**figure 5.20**). This is called the 9 + 2 pattern of microtubules. Each doublet has pairs of arms projecting from subtubule A toward a neighboring doublet. A radial spoke extends from subtubule A toward the internal pair of microtubules with their central sheath. These microtubules are similar to those found in the cytoplasm. ▶▶ *What's in a name? (Microbial diversity & ecology 4.1); Protist taxonomy is controversial (section 25.1)*

A basal body lies in the cytoplasm at the base of each cilium or flagellum. It is a short cylinder with nine microtubule triplets around

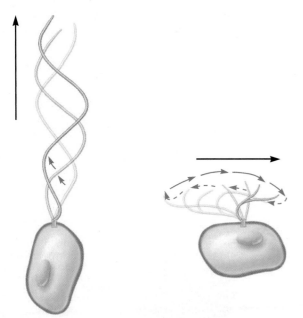

Figure 5.18 Patterns of Flagellar and Ciliary Movement. Flagellar and ciliary movement takes the form of waves. Flagella (left illustration) move either from the base of the flagellum to its tip or in the opposite direction. The motion of these waves propels the organism along. The beat of a cilium (right illustration) may be divided into two phases. In the effective stroke, the cilium remains fairly stiff as it swings through the water. This is followed by a recovery stroke in which the cilium bends and returns to its initial position. The black arrows indicate the direction of water movement in these examples.

its periphery (a 9 + 0 pattern) and is separated from the rest of the organelle by a basal plate. The basal body directs the construction of these organelles. Cilia and flagella appear to grow through the addition of preformed microtubule subunits at their tips.

Cilia and flagella bend because adjacent microtubule doublets slide along one another while maintaining their individual lengths. The doublet arms, about 15 nm long, are made of the protein dynein (figure 5.20). Dynein arms interact with the B subtubules of adjacent doublets to cause the sliding. The radial spokes also participate in this sliding motion. ATP powers the movement of cilia and flagella.

Cilia and flagella beat at a rate of about 10 to 40 strokes or waves per second and propel microorganisms rapidly. The flagellate protist *Monas stigmatica* swims at a rate of 260 μm/second (approximately 40 cell lengths per second); the euglenoid flagellate *Euglena gracilis* travels at around 170 μm or 3 cell lengths per second. The ciliate protist *Paramecium caudatum* swims about 2,700 μm/second (12 lengths per second). Such speeds are equivalent to or much faster than those seen in higher animals but not as fast as those in bacteria. ◄◄ *Bacterial flagella (section 3.7)*

Retrieve, Infer, Apply

1. Prepare and label a diagram showing the detailed structure of a cilium or flagellum.
2. How do cilia and flagella move, and what is dynein's role in the process? Contrast the ways in which flagella and cilia propel eukaryotic microorganisms through water.
3. How do the structure and mechanism of action of bacterial, archaeal, and eukaryotic flagella differ?

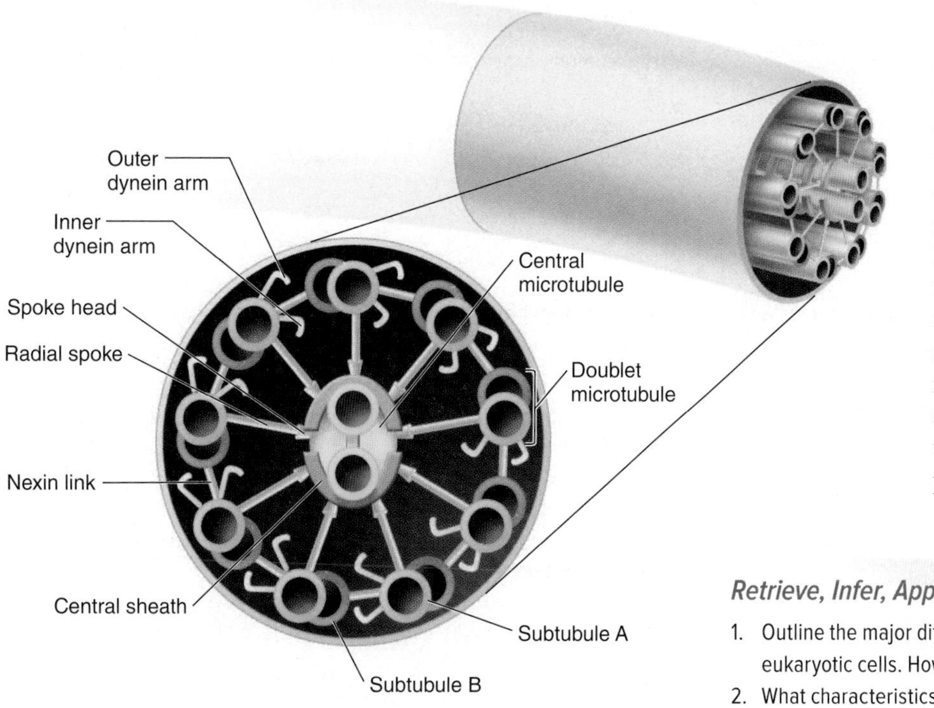

5.8 Comparison of Bacterial, Archaeal, and Eukaryotic Cells

After reading this section, you should be able to:

- Create a Venn diagram or concept map that clearly distinguishes bacterial, archaeal, and eukaryotic cells in terms of their genome organization, organelles, cell envelopes, ribosome size and component molecules, and cytoskeleton
- Determine the type of microbe when given a description of a newly discovered microbe

A comparison of the bacterial and eukaryotic cells in **figure 5.21** demonstrates that there are many fundamental differences between these cells. These differences are also observed between archaeal and eukaryotic cells, because archaea are similar to bacteria at the gross structural level. Eukaryotic cells have a membrane-enclosed nucleus. In contrast, bacterial and archaeal cells lack a true, membrane-delimited nucleus. Most bacteria and archaea are smaller than eukaryotic cells, often about the size of eukaryotic mitochondria and chloroplasts.

Many other major distinctions between these groups exist. It is clear from **table 5.2** that bacterial and archaeal cells are much simpler structurally. In particular, an extensive and diverse collection of membrane-delimited organelles is missing. Furthermore, bacterial and archaeal cells are functionally simpler in several ways. They lack mitosis and meiosis, and have a simpler genetic organization. Many complex eukaryotic processes are absent in bacteria and archaea: endocytosis, intracellular digestion, directed cytoplasmic streaming, and ameboid movement are just a few.

Despite the many significant differences, all cells are remarkably similar on the biochemical level, as we discuss in succeeding chapters. With a few exceptions, the genetic code is the same in all, as is the overall process by which genetic information is expressed. The principles underlying metabolic processes and many important metabolic pathways are identical. Thus beneath the profound structural and functional differences between bacterial, archaeal, and eukaryotic cells, there is an even more fundamental unity: a molecular unity that is basic to all known life processes.

Figure 5.20 Diagram of Cilia and Flagella Structure.

Labels: Outer dynein arm, Inner dynein arm, Spoke head, Radial spoke, Nexin link, Central sheath, Subtubule B, Subtubule A, Doublet microtubule, Central microtubule

Retrieve, Infer, Apply

1. Outline the major differences between bacterial, archaeal, and eukaryotic cells. How are they similar?
2. What characteristics make members of *Archaea* more like eukaryotes? What features make them more like bacteria?

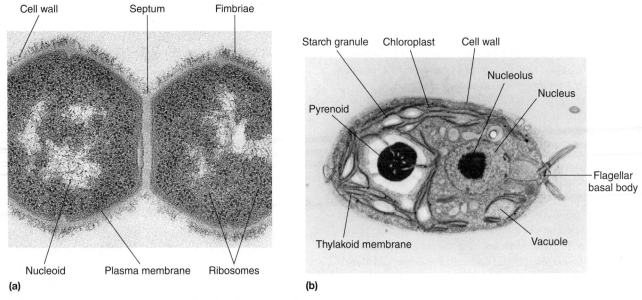

(a)

(b)

Figure 5.21 Comparison of Bacterial and Eukaryotic Cell Structure. (a) TEM of the bacterium *Streptococcus pyogenes* undergoing cell division. The septum separates the two daughter cells (×20,000). (b) TEM of the eukaryotic alga *Chlamydomonas reinhardtii*.

Table 5.2	Comparison of Bacterial, Archaeal, and Eukaryotic Cells		
Property	**Bacteria**	**Archaea**	**Eukaryotes**
Organization of Genetic Material			
True membrane-bound nucleus	No	No	Yes
DNA complexed with histones	No	Some	Yes
Chromosomes	Usually one circular chromosome; chromosomes have single origin of replication	One circular chromosome; some have chromosomes with multiple origins of replication; some are polyploid	More than one; chromosomes are linear; all have chromosomes with multiple origins of replication
Plasmids	Very common	Very common	Rare
Introns in genes	Rare	Rare	Yes
Nucleolus	No	No	Yes
Mitochondria, Chloroplasts, Endoplasmic Reticulum, Golgi, and Lysosomes Observed	No	No	Yes
Plasma Membrane Lipids	Ester-linked phospholipids and hopanoids; some have sterols	Glycerol diethers and diglycerol tetraethers	Ester-linked phospholipids and sterols
Flagella	Submicroscopic in size; filament composed of single type of flagellin	Submicroscopic in size; some filaments composed of more than one type of flagellin	Microscopic in size; membrane bound; usually 20 microtubules in 9 + 2 pattern
Peptidoglycan in Cell Walls	Yes	No	No
Ribosome Size and Structure	70S; 3 rRNAs; ~55 ribosomal proteins	70S; most have 3 rRNAs; ~68 ribosomal proteins	80S; 4 rRNAs and ~80 ribosomal proteins
Cytoskeleton	Rudimentary	Rudimentary	Yes
Gas Vesicles	Yes	Yes	No

Key Concepts

5.1 Eukaryotic Cells Are Diverse but Share Some Common Features

- Eukaryotic cells are generally larger than bacterial and archaeal cells, and they exhibit considerably more morphological diversity.
- The eukaryotic cell has a true, membrane-delimited nucleus and many membranous organelles (**table 5.1; figures 5.2** and **5.3**).
- The membranous organelles compartmentalize the cytoplasm of the cell. This allows the cell to carry out a variety of biochemical reactions simultaneously. It also provides more surface area for membrane-associated activities such as respiration.

5.2 Eukaryotic Cell Envelopes

- Eukaryotic membranes are similar in structure and function to bacterial membranes. The two differ in terms of their lipid composition.
- Eukaryotic membranes contain microdomains that are enriched for certain lipids and proteins, and participate in a variety of cellular processes.
- When a cell wall is present, it is constructed from polysaccharides (e.g., cellulose) that are chemically simpler than peptidoglycan, the molecule found in bacterial cell walls.

5.3 The Eukaryotic Cytoplasm Contains a Complex Cytoskeleton and Many Membranous Organelles

- The cytoplasm consists of a cytosol in which are found the organelles and cytoskeleton.
- The cytoskeleton is organized from actin filaments, intermediate filaments, and microtubules. These cytoskeletal elements are partly responsible for cell structure, endocytosis, movement of cell structures, cell division, and motility (**figure 5.5**).
- Actin filaments and microtubules have been observed in eukaryotic microbes. Actin filaments are composed of actin proteins; microtubules are composed of α-tubulin and β-tubulin.
- Intermediate filaments are assembled from a heterogeneous family of proteins. They are found in animal cells but have not been identified or studied in eukaryotic microbes.
- The uniquely eukaryotic organelles are numerous and include the endoplasmic reticulum, Golgi, mitochondria, and chloroplasts.

5.4 Several Cytoplasmic Membranous Organelles Function in the Secretory and Endocytic Pathways

- The cytoplasm is permeated with a complex of membranous organelles and vesicles. Some are involved in the synthesis and secretion of materials (secretory pathways). Some are involved in the uptake of materials from the extracellular milieu (endocytic pathways).
- The endoplasmic reticulum (ER) is an irregular network of tubules and flattened sacs (cisternae). Rough ER has ribosomes attached to it and is active in protein synthesis. Smooth ER lacks ribosomes.
- The ER donates some materials to the Golgi apparatus, an organelle usually composed of one or more stacks of cisternae (**figure 5.7**). The Golgi prepares and packages cell products for secretion.
- The Golgi apparatus buds off vesicles that deliver hydrolytic enzymes and other proteins to lysosomes and lysosome-like organelles. Lysosomes contain digestive enzymes and aid in intracellular digestion of extracellular materials delivered to them by endocytosis and macroautophagy.
- Mammalian cells use several kinds of endocytosis (**figure 5.9**). These include phagocytosis, clathrin-dependent endocytosis, and caveolin-dependent endocytosis. Some macromolecules are bound to receptors prior to endocytosis in a process called receptor-mediated endocytosis.
- Endocytic pathways in eukaryotic microbes are not as well studied as are mammalian pathways. However, similar mechanisms are employed.

5.5 The Nucleus and Ribosomes Are Involved in Genetic Control of the Cell

- The nucleus is a large organelle containing the cell's chromosomes. It is surrounded by a double-membrane envelope perforated by pores through which materials can move (**figure 5.11**).
- The nucleolus lies within the nucleus and participates in the synthesis of ribosomal RNA and ribosomal subunits (**figure 5.12**).
- Eukaryotic ribosomes are found either free in the cytoplasm or bound to the ER. They are 80S in size and also differ from bacterial and archaeal ribosomes in their molecular make up (**figure 5.13**).

5.6 Mitochondria, Related Organelles, and Chloroplasts Are Involved in Energy Conservation

- Canonical mitochondria are evolutionarily related to other organelles including hydrogenosomes (**figure 5.14**).
- Aerobic mitochondria are organelles bounded by two membranes, with the inner membrane folded into cristae (**figure 5.15**). They are responsible for energy conservation by cell respiration, which involves the tricarboxylic acid cycle, electron transport, and oxidative phosphorylation.
- Hydrogenosomes have a double membrane and carry out metabolic processes (fermentation) that generate hydrogen gas (**figure 5.16**).

- Chloroplasts are pigment-containing organelles that serve as the site of photosynthesis (**figure 5.17**). The trapping of light energy takes place in the thylakoid membranes of the chloroplast, whereas CO_2 fixation occurs in the stroma.

5.7 Many Eukaryotic Microbes Have External Structures Used for Motility

- Many eukaryotic cells are motile because of cilia and flagella, membrane-enclosed organelles with nine microtubule doublets surrounding two central microtubules (**figure 5.20**).
- Cilia and flagella exhibit a whiplike motion. This occurs when the microtubule doublets slide along each other, causing the cilium or flagellum to bend.

5.8 Comparison of Bacterial, Archaeal, and Eukaryotic Cells

- Bacterial and archaeal cells are similar in appearance but differ at the molecular levels. A major difference between these cells and eukaryotic cells is that eukaryotic cells contain a nucleus and numerous membranous structures.
- Despite the fact that eukaryotic, bacterial, and archaeal cells differ structurally in many ways (**table 5.2**), they are quite similar biochemically.

Compare, Hypothesize, Invent

1. Discuss the statement: "The most obvious difference between eukaryotic, bacterial, and archaeal cells is in their use of membranes." Specifically, compare the roles of bacterial and archaeal plasma membranes with those of the organelle membranes in eukaryotes.

2. Bacterial and archaeal cell size is limited by the rate at which nutrients diffuse from one place in the cell to another. Are eukaryotic cells similarly limited? Why or why not?

3. In addition to their role in packaging chromosomes, histones are involved in the regulation of transcription. How might this be the case?

4. Compare the mechanisms by which most eukaryotic cells acquire nutrients with those used by bacteria and archaea. What are the implications in terms of survival in different habitats and competition between eukaryotic microbes such as protists and bacteria and archaea?

5. In recent years, there have been advances in understanding longevity through research on three model organisms: mice, nematode worms, and the yeast *Saccharomyces cerevisiae*. It is clear that caloric restriction is linked to increased life span in multicellular organisms (mice and worms). This phenomenon can be studied at the molecular level in yeast, where loss of the protein Sch9p increases the yeast life span by increasing the rate of mitochondrial activity. In this way, caloric restriction is mimicked. List the pros and cons of the use of eukaryotic microorganisms as models for complex physiological and genetic questions that are relevant to higher organisms, including humans. What is the most significant advantage and the most unfortunate disadvantage?

Read the original paper: Lovoie, L., and Whiteway, M. 2008. Increased respiration in the *sch9*Δ mutant is required for increasing chronological life span but not replicative life span. *Euk. Cell* 7:1127.

Learn More

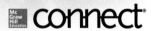

6

Viruses and Other Acellular Infectious Agents

This young man is clearly enjoying his hot dog. Hot dogs and other ready-to-eat meats are a favorite at picnics, baseball games, and other sporting events. However, they can also transmit food-borne pathogens.

Mustard, Catsup, and Viruses?

During the summer of 2010, over 21 million hot dogs were sold to fans attending games at major league baseball parks in the United States. Hot dogs and lunch meats are popular at outings such as baseball games and in lunches carried to work or school. Yet each year in the United States, approximately 1,600 people are sickened by a bacterium that can contaminate the meat and, even worse, survive and grow when the meat is properly refrigerated.

The disease culprit is *Listeria monocytogenes*, a Gram-positive rod found in soil and many other environmental sites. It is not only cold tolerant but salt and acid tolerant as well. Although it is in the minor leagues when compared to some of the big hitters of food-borne disease (e.g., *Salmonella enterica*), it is of concern for two reasons: who it kills and how many it kills. *L. monocytogenes* targets the young and old, pregnant women, and immunocompromised individuals; about 15% of those infected die.

Its effect on pregnant women is particularly heartbreaking. The woman usually only suffers mild, flulike symptoms; however, these innocuous symptoms belie the fact that the child she carries is in serious danger. Her pregnancy often ends in miscarriage or stillbirth. Newborns infected with the bacterium are likely to develop meningitis. Many will die as a result. Those who survive often have neurological disorders.

Currently, pregnant women are counseled against eating ready-to-eat foods unless they have been cooked prior to consumption. However, *L. monocytogenes* is known to contaminate many foods other than hot dogs and these can't always be heated. In 2006 the U.S. Food and Drug Administration (FDA) approved a new approach to prevent listeriosis: spraying viruses that attack and destroy the bacterium on ready-to-eat cold cuts and luncheon meats. In other words, the viruses are a food additive! The method is safe because the viruses only attack *L. monocytogenes,* not human cells.

Since approval, the use of viruses to control the transmission of listeriosis by other foods has been studied. Unfortunately, those studies did not include foods such as fresh fruit. In 2011 *L. monocytogenes*–contaminated

cantaloupe caused an outbreak of listeriosis in 20 states in the United States, which infected over 100 and killed over 20.

Viruses as agents of good will come as a surprise to many. Typically we think of them as major causes of disease. However, viruses are significant for other reasons. They are vital members of aquatic ecosystems. There they interact with cellular microbes and contribute to the movement of organic matter from particulate forms to dissolved forms. Bacterial viruses are being used in some European countries to treat infections caused by bacteria, and a number of animal viruses are being used to target and destroy cancer cells. Finally, they are important model organisms. In this chapter, we introduce viruses and other acellular infectious agents. ▶▶◀ *Aquatic viruses (section 30.2); Microorganisms can be controlled by biological methods (section 8.7)*

Readiness Check:
Based on what you have learned previously, you should be able to:
✔ Define the term acellular
✔ Compare and contrast in general terms viruses, viroids, satellites, and prions (section 1.1)

6.1 Viruses Are Acellular

After reading this section, you should be able to:
■ Define the terms virology, bacteriophages, and phages
■ List organisms that are hosts to viruses

The discipline of **virology** studies **viruses,** a unique group of infectious agents whose distinctiveness resides in their simple, acellular organization and pattern of multiplication. Despite this simplicity, viruses are major causes of disease. For instance, many human diseases are caused by viruses, and more are discovered every year, as demonstrated by the appearance of SARS in 2003, the H1N1 (swine) influenza virus in 2009, the H7N9

MICROBIAL DIVERSITY & ECOLOGY

6.1 Host-Independent Growth of an Archaeal Virus

Viruses cannot multiply without first infecting a host cell, and when outside a host cell, they have been thought to be inert. Thus they are often described as "entities" rather than organisms. So it was quite a surprise when an archaeal virus was discovered that develops long tails only when outside its host (**box figure**). This archaeal virus was found in acidic hot springs (pH 1.5, 85–93°C) in Italy, where it infects the hyperthermophilic archaeon *Acidianus convivator.* When the virus infects its host, lemon-shaped virions are assembled. Following host lysis, two "tails" begin to form at either end of the virion. These projections continue to assemble until they reach a length at least that of the capsid. Curiously, tails are only produced if virions are incubated at high temperature. The virus is called ATV for *Acidianus* two-tailed virus.

To learn more about the structure of the tails, the ATV genome was sequenced. ATV is a double-stranded DNA virus that encodes only nine structural proteins. The tail protein is 800 amino acids long and is homologous to eukaryotic intermediate filament proteins. Both intermediate filaments and purified ATV tail proteins assemble into filamentous structures without additional energy or cofactors. ◄◄ *The eukaryotic cytoplasm contains a complex cytoskeleton and many membranous organelles (section 5.3)*

It is suspected that the development of tails only at high temperatures may be a survival strategy for the virus when host cell density is low. ATV induces lysis rather than lysogeny in its archaeal host. Most other viruses that infect archaea living in acidic hot springs have evolved lysogeny as a means to survive these harsh conditions. Why this virus has

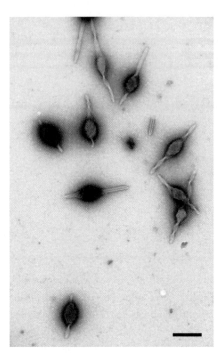

***Acidianus* Two-Tailed Virus, ATV.** Negative contrast electron micrograph of ATV virions at various steps of extracellular tail development. Scale bar = 200 nm.

evolved lysis and tail development is unknown, but it suggests that viruses may be more complicated than simple "entities."

Source: Häring, M.; Vestergaard, G.; Rachel, R.; Chen, L.; Garret, R. A.; and Prangishvili, D. 2005. Independent virus development outside a host. Nature 436:1101–02.

avian influenza virus in 2013, and the Middle East respiratory syndrome coronavirus (MERS-CoV) also in 2013. However, their simplicity also has made them attractive model organisms. They served as models for understanding DNA replication, RNA synthesis, and protein synthesis. Therefore the study of viruses has contributed significantly to the discipline of molecular biology. In fact, the field of genetic engineering is based in large part on the use of viruses and viral enzymes such as the retroviral enzyme reverse transcriptase. ►►I *Recombinant DNA technology (chapter 17)*

Viruses can exist either extracellularly or intracellularly. When extracellular, they are inactive (with one known interesting exception: **Microbial Diversity & Ecology 6.1**) because they possess few, if any, enzymes and cannot reproduce outside of living cells. When intracellular, viruses exist primarily as nucleic acids that can, at some point in the viral life cycle, commandeer

host cells and use them to synthesize viral components from which progeny virions are assembled and eventually released.

Viruses can infect all cell types. Numerous viruses infect bacteria. They are called **bacteriophages,** or **phages** for short. Fewer archaeal viruses have been identified. Most known viruses infect eukaryotic organisms, including plants, animals, protists, and fungi. Viruses have been classified into numerous families based primarily on genome structure, life cycle, morphology, and genetic relatedness. These families have been designated by the International Committee on Taxonomy of Viruses (ICTV), the agency responsible for standardizing the classification of viruses. ►►I *Virus phylogeny is difficult to establish (section 27.1)*

In this chapter, we introduce the structure and multiplication strategies of viruses. Their taxonomic diversity is presented in chapter 27, and their ecological relevance is discussed in chapter 30.

6.2 Virion Structure Is Defined by Capsid Symmetry and Presence or Absence of an Envelope

After reading this section, you should be able to:

- State the size range of virions
- Identify the parts of a virion and describe their function
- Distinguish enveloped viruses from nonenveloped viruses
- Describe the types of capsid symmetry

A complete virus particle is called a **virion.** Virion morphology has been intensely studied over the past decades because of the importance of viruses and the realization that virion structure is simple enough to be understood in detail. Progress has come from the use of several different techniques: electron microscopy, X-ray diffraction, biochemical analysis, and immunology. Although our knowledge is incomplete due to the large number of different viruses, we can discuss the general nature of virion structure.

General Structural Properties

Virions range in size from about 10 to 400 nm in diameter (**figure 6.1**). The smallest are a little larger than ribosomes, whereas mimiviruses, the largest viruses known, have virions larger than some of the smallest bacteria and can be seen in the light microscope. However, most virus particles must be viewed with electron microscopes. ◄◄ *Electron microscopes use beams of electrons to create highly magnified images (section 2.4)*

The simplest virions consist only of a **nucleocapsid,** which is composed of a nucleic acid, either DNA or RNA, and a protein coat called a **capsid** (**figure 6.2***a*). The capsid surrounds the viral nucleic acid, protects the viral genome, and often aids in its transfer between host cells. Among the proteins encoded by the viral genome are the capsid proteins, which are called **protomers.** Capsids self-assemble by a process that is not fully understood. Some viruses use noncapsid proteins as scaffolding upon which the capsids are assembled.

Probably the most important advantage of this design strategy is that the viral genome is used with maximum efficiency. For example, the *tobacco mosaic virus* (TMV) capsid is

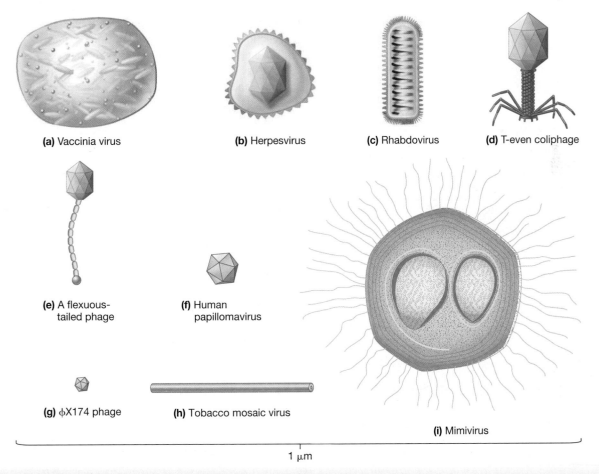

(a) Vaccinia virus **(b)** Herpesvirus **(c)** Rhabdovirus **(d)** T-even coliphage

(e) A flexuous-tailed phage **(f)** Human papillomavirus

(g) ϕX174 phage **(h)** Tobacco mosaic virus

(i) Mimivirus

1 μm

Figure 6.1 The Size and Virion Morphology of Selected Viruses. The virions are drawn to scale.

MICRO INQUIRY *Which capsids are icosahedral? Which are helical? Which have complex symmetry?*

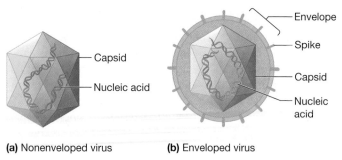

(a) Nonenveloped virus   **(b)** Enveloped virus

Figure 6.2 Generalized Structure of Virions. (a) The simplest virion is that of a nonenveloped virus (nucleocapsid), consisting of a capsid assembled around its nucleic acid. (b) Virions of enveloped viruses are composed of a nucleocapsid surrounded by a membrane called an envelope. The envelope usually has viral proteins called spikes inserted into it.

constructed using a single type of protomer (**figure 6.3**). Recall that the building blocks of proteins are amino acids and that each amino acid is encoded by three nucleotides, the building blocks of nucleic acids. The TMV protomer is 158 amino acids in length. Therefore only about 474 nucleotides are required to code for the coat protein. The TMV genome consists of 6,400 nucleotides. Thus only a small fraction of the genome is used to code for the capsid. Suppose, however, that the TMV capsid were composed of six different protomers, all about 150 amino acids in length. If this were the case, about 2,900 of the 6,400 nucleotides in the TMV genome would be

required just for capsid construction, and much less genetic material would be available for other purposes.

The various morphological types of virions primarily result from the combination of a particular type of capsid symmetry with the presence or absence of an envelope—a lipid layer external to the nucleocapsid. There are three types of capsid symmetry: helical, icosahedral, and complex. Viruses with virions having an envelope are called **enveloped viruses,** whereas those lacking an envelope are called **nonenveloped** or **naked viruses** (figure 6.2).

Helical Capsids

Helical capsids are shaped like hollow tubes with protein walls. Tobacco mosaic virus is a well-studied example of helical capsid structure (figure 6.3). The self-assembly of TMV protomers into a helical arrangement produces a rigid tube. The capsid encloses an RNA genome, which is wound in a spiral and lies within a groove formed by the protein subunits. Not all helical capsids are as rigid as the TMV capsid. The nucleocapsids of influenza viruses are thin and flexible and are enclosed within an envelope (**figure 6.4**). ▶▶ *Tobacco mosaic virus (section 27.5)*

The size of a helical capsid is influenced by both its protomers and the viral genome. The diameter of the capsid is a function of the size, shape, and interactions of the protomers. The length of the capsid appears to be determined by the nucleic acid because a helical capsid does not extend much beyond the end of the viral genome.

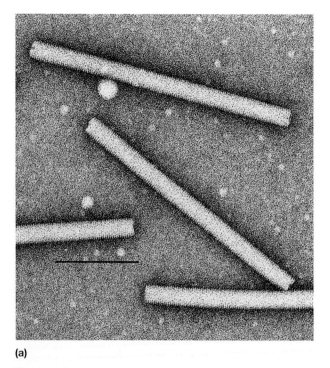

(a)

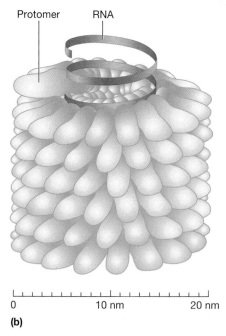

Protomer   RNA

0   10 nm   20 nm

(b)

(c)

Figure 6.3 Tobacco Mosaic Virus (TMV) Virions. (a) An electron micrograph of negatively stained helical capsids (×400,000). The virions are usually 15 to 18 nm in diameter and 300 nm long. (b) Illustration of a TMV nucleocapsid. Note that it is composed of a helical array of protomers with the RNA spiraling on the inside. (c) A model of a TMV virion.

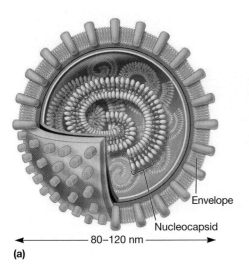

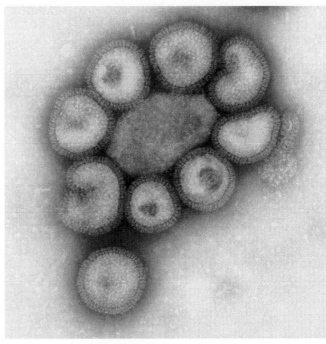

Envelope

Nucleocapsid

◄—— 80–120 nm ——►

(a)

Figure 6.4 Influenza Virus Virions: An Enveloped Virus with Helical Nucleocapsids. (a) Schematic view. Influenza viruses have segmented genomes consisting of seven to eight different RNA molecules. Each is coated by capsid proteins to create seven to eight flexible nucleocapsids. The envelope has spikes that project 10 nm from the surface at 7 to 8 nm intervals. (b) Because the nucleocapsids are flexible, the virions are pleomorphic. Electron micrograph (×350,000).

(b)

Icosahedral Capsids

An icosahedron is a regular polyhedron with 20 equilateral triangular faces and 12 vertices (figure 6.1*b,f,g*). **Icosahedral capsids** are the most efficient way to enclose a space. They are constructed from ring- or knob-shaped assemblages of five or six protomers; the assemblages are called **capsomers** (**figure 6.5**). Capsomers composed of five protomers are called pentamers (pentons); hexamers (hexons) are capsomers that possess six protomers. Pentamers are usually at the vertices of the icosahedron, whereas hexamers generally form its edges and triangular faces.

The virions of some RNA viruses have pentamers and hexamers constructed with only one type of subunit. Other virions have pentamers and hexamers composed of different proteins. Although many icosahedral capsids contain both pentamers and hexamers, some have only pentamers.

Capsids of Complex Symmetry

Most viruses have either icosahedral or helical capsids, but some viruses do not fit into either category. Poxviruses and large bacteriophages are two important examples.

(a)

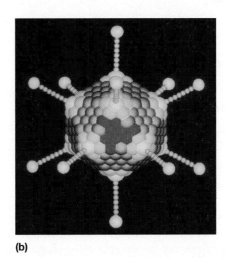

(b)

Figure 6.5 The Icosahedral Capsid of an Adenovirus. (a) Electron micrograph of adenovirus virions, 252 capsomers (×300,000). (b) Computer-simulated model of an adenovirus virion.

MICRO INQUIRY *Are the capsomers at the vertices of an adenovirus virion pentamers or hexamers? What is the difference between a pentamer and hexamer?*

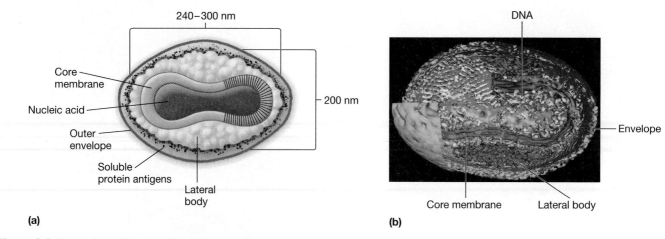

(a)

(b)

Figure 6.6 Morphology of Vaccinia Virus Virions.   (a) Diagram of virion structure. (b) Electron cryotomogram of the virion.

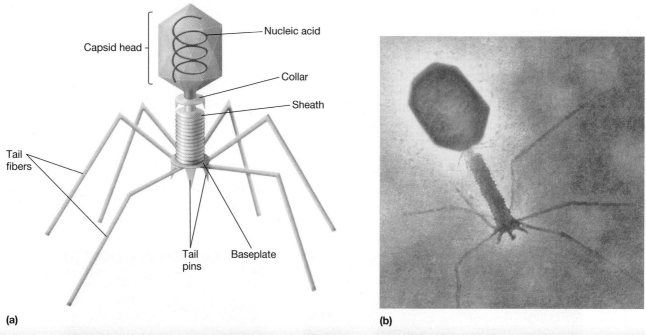

(a)

(b)

Figure 6.7 T4 Phage of *E. coli.*   (a) The structure of T4 bacteriophage virion. (b) Electron micrograph of the virion.

MICRO INQUIRY *Why is T4 said to have binal symmetry?*

Poxvirus virions are among the largest of the animal viruses (about 400 by 240 by 200 nm in size) and can be seen with a light microscope. They possess an exceptionally complex internal structure with an ovoid- to brick-shaped exterior. **Figure 6.6** shows the virion morphology of a vaccinia virus. Its double-stranded DNA genome is associated with proteins and contained in the core, a central structure shaped like a biconcave disk and surrounded by a membrane. Two lateral bodies lie between the core and the virion's outer envelope.

Some large bacteriophages have virions that are even more elaborate than those of poxviruses. The virions of T2, T4, and T6 phages (T-even phages) that infect *Escherichia coli* are said to have **binal symmetry** because they have a head that resembles an icosahedron and a tail that is helical. The icosahedral head is elongated by one or two rows of hexamers in the middle and contains the DNA genome (**figure 6.7**). The tail is composed of a collar joining it to the head, a central hollow tube, a sheath surrounding the tube, and a complex baseplate. In T-even phages,

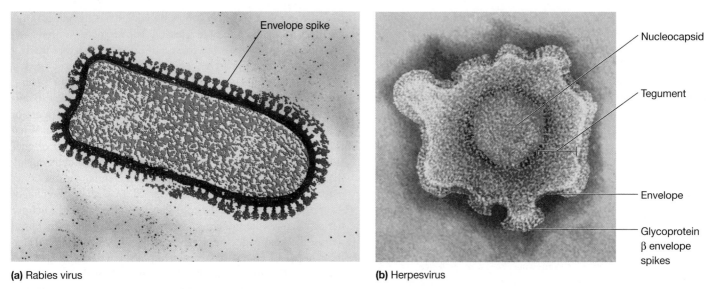

(a) Rabies virus　　　　　　　　　　　　　　　　**(b)** Herpesvirus

Figure 6.8 **Examples of Enveloped Viruses.** (a) Negatively stained virion of a rabies virus. (b) Herpesvirus virion. Images are artificially colorized.

the baseplate is hexagonal and has a pin and a jointed tail fiber at each corner. ▶▶◀ *Bacteriophage T4 (section 27.2)*

Viral Envelopes and Enzymes

The nucleocapsids of many animal viruses, some plant viruses, and at least one bacterial virus are surrounded by an outer membranous layer called an **envelope** (**figure 6.8**). Animal virus envelopes usually arise from the plasma or nuclear membranes of the host cell. Envelope lipids and carbohydrates are therefore acquired from the host. In contrast, envelope proteins are coded for by viral genes and may even project from the envelope surface as **spikes,** which are also called **peplomers** (figure 6.8). In many cases, spikes are involved in virion attachment to the host cell surface. Because spikes differ among viruses, they also can be used to identify some viruses. Many enveloped viruses have virions with a somewhat variable shape and are called pleomorphic. However, the envelopes of viruses such as the bullet-shaped rabies viruses are firmly attached to the underlying nucleocapsid and endow the virion with a constant, characteristic shape (figure 6.8*a*).

Influenza virus (figure 6.4) is a well-studied enveloped virus with two types of spikes. Some spikes consist of the enzyme **neuraminidase,** which functions in the release of mature virions from the host cell. Other spikes are **hemagglutinin** proteins, so named because they bind virions to red blood cells and cause the cells to clump together—a process called hemagglutination. Influenza virus's hemagglutinins participate in virion attachment to host cells. Most of its envelope proteins are glycoproteins—proteins that have carbohydrate attached to them. A nonglycosylated protein, the M (matrix) protein, is found on the inner surface of the envelope and helps stabilize it.

In addition to enzymes associated with the envelope or capsid (e.g., influenza neuraminidase), some viruses have enzymes within their capsids. Such enzymes are usually involved in nucleic acid replication. For example, influenza virus virions have an RNA genome and carry an enzyme that synthesizes RNA using an RNA template (*see figure 27.28*). Thus although viruses lack true metabolism and cannot reproduce independently of living cells, their virions may carry one or more enzymes essential to the completion of their life cycles.

Viral Genomes Are Structurally Diverse

One clear distinction between cellular organisms and viruses is the nature of their genomes. Cellular genomes are always double-stranded (ds) DNA. Viruses, on the other hand, employ all four possible nucleic acid types: dsDNA, single-stranded (ss) DNA, ssRNA, and dsRNA. All four types are used by animal viruses. Most plant viruses have ssRNA genomes, and most bacterial viruses have dsDNA. The size of viral genomes also varies greatly. Very small genomes are around 4,000 nucleotides—just large enough to code for three or four proteins. Some viruses save additional space by using overlapping genes. At the other extreme are the genomes of mimiviruses, which infect protists. They are about 1.2×10^6 nucleotides long, rivaling some bacteria and archaea in coding capacity. ▶▶◀ *What is a virus? (Microbial Diversity & Ecology 27.1)*

Most DNA viruses use dsDNA as their genetic material. However, some have ssDNA genomes (e.g., φX174 and M13). In both cases, the genomes may be either linear or circular. Some DNA genomes can switch from one form to the other. For instance, the *E. coli* phage lambda has a linear genome in its capsid, but it becomes circular once it enters the host cell. ▶▶◀ *Bacteriophages φX174 and fd (section 27.3); Bacteriophage lambda (section 27.2)*

Relatively few RNA viruses have dsRNA genomes. More common are viruses with ssRNA genomes. Polio, tobacco mosaic, SARS, rabies, mumps, measles, influenza, human immunodeficiency, and brome mosaic viruses are all ssRNA viruses.

Some RNA viruses have **segmented genomes**—genomes that consist of more than one piece (segment) of RNA. In many cases, each segment codes for one protein and there may be as many as 10 to 12 segments (figure 6.4). Usually all segments are enclosed in the same capsid; however, this is not always the case. For example, the genome of brome mosaic virus, a virus that infects certain grasses, is composed of three segments distributed among three different virions.

Retrieve, Infer, Apply

1. How are viruses similar to cellular organisms? In what fundamental way do they differ?
2. What is the difference between a nucleocapsid and a capsid?
3. Compare the structure of an icosahedral capsid with that of a helical capsid. How do pentamers and hexamers associate to form a complete icosahedron? Which virus would have a longer helical capsid: a virus with a 7,200 base pair DNA genome or one with an 11,000 base ssRNA genome?
4. What is an envelope? What are spikes (peplomers)? Why do some enveloped viruses have pleomorphic virions? Give two functions spikes might serve in the viral life cycle and the proteins that the influenza virus uses in these processes.
5. All four nucleic acid forms can serve as viral genomes. Describe each. What is a segmented RNA genome?
6. The RNA genomes of some RNA viruses resemble the messenger RNA (mRNA) of their eukaryotic hosts. What advantage would an RNA virus gain by having this type of genome?

6.3 Viral Life Cycles Have Five Steps

After reading this section, you should be able to:

- Describe the five steps common to the life cycles of viruses
- Discuss the role of receptors, capsid proteins, and envelope proteins in the life cycles of viruses
- Describe the two most common methods for virion release from a host cell

In the early years of virology, one of the fundamental questions about the biology of these unique entities was how they made progeny viruses. The one-step growth experiment devised in 1939 by Max Delbrück (1906–1981) and Emory Ellis (1906–2003) served as an experimental approach to answering this question. Delbrück and Ellis worked with bacteriophage T4. They knew that T4 killed its host, *Escherichia coli*, and released progeny phages by lysing the cells it infected. In their experiment, *E. coli* cells were mixed with T4. After a short interval, the mixture was greatly diluted so that any virions released upon host cell lysis would not be able to encounter and infect other cells. The diluted culture was then incubated, and over time samples were removed to determine the number of infectious phage particles in the culture. This was determined using a plaque assay as we describe in section 6.5. A plot of phage particles versus time shows several distinct periods in the resulting growth curve (**figure 6.9**, red line). The latent

period occurs immediately following addition of the phage. During this period, no virions are released. The rise period follows and is characterized by the rapid release of infective phages. Finally a plateau is reached and no more virions are produced.

The one-step growth experiment is important for several reasons. It employed procedures that are still used today to culture and enumerate viruses, and it ushered in the modern era of phage biology. It also led to another fundamental question: What is occurring during the latent period? A subsequent set of one-step growth experiments artificially lysed infected cells during the latent period and discovered that intracellular virions could not be detected early in the latent period. In essence, the phages disappeared once inside the cell. This period is called the eclipse period because the infecting virions were concealed or eclipsed within the host cell. These experiments also demonstrated that the number of completed, infective phages within the host cell increases after the end of the eclipse period. Still other experiments eventually showed that a carefully orchestrated series of events occurs during the latent period. These events are the focus of this section.

If a virus is to multiply and give rise to new progeny viruses, it must find and use (and in many cases abuse) a host cell. To accomplish this, a virus must use guile and subterfuge to access an

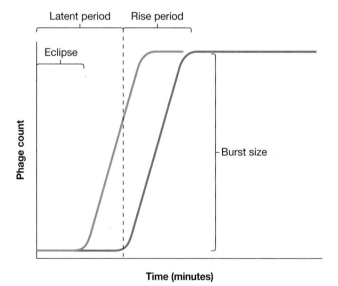

Figure 6.9 The One-Step Growth Curve. The eclipse period is the initial portion of the latent period in which host cells do not contain any complete, infective virions. During the remainder of the latent period, an increasing number of infective virions are present, but none are released. The latent period ends with host cell lysis and rapid release of virions (rise period). In this figure the blue line represents the total number of complete virions , both intracellular and free in the medium. The red line represents the number of free viruses (virions that were free in the culture and had not infected a host cell plus those virions that were released from infected cells). The burst size is the number of progeny viruses produced per infected cell.

appropriate host, enter the host, and avoid any defenses the host might employ to rid itself of the virus or prevent its multiplication. Once inside a host cell, a virus uses a repertoire of clever tricks to take control of cellular functions, thereby ensuring that viral genomes, mRNAs, and proteins are synthesized. The diversity of tricks viruses use has led to a plethora of distinctive viral life cycles (also called viral replicative cycles). The tricks used by a virus are often related to its virion structure, in particular the nature of its genome. Thus, viruses with a similar type of genome (e.g., dsDNA, ssRNA) often employ similar tricks, as we explore in chapter 27.

Despite the diversity of viral life cycles, a general pattern of viral replicative cycles can be discerned; it can be divided into five steps. Because viruses need a host cell in which to multiply, the first step is usually attachment (often called adsorption) to a host (**figure 6.10**). This is followed by entry of either the nucleocapsid or the viral nucleic acid into the host. If the nucleocapsid enters, uncoating of the genome usually occurs before the life cycle continues. Once inside the host cell, the synthesis stage begins. During this stage, viral genes are expressed. That is, the virus's genes are transcribed and translated. This allows the virus to take control of the host cell, forcing it to manufacture viral genomes and viral proteins. Then follows the assembly stage, during which new nucleocapsids are constructed by self-assembly of coat proteins with the nucleic acids. Finally, during the release step, mature virions escape the host.

Attachment (Adsorption)

All viruses, with the exception of plant viruses, must attach to a potential host cell long enough to gain entry into the cell. Attachment to the host is accomplished by specific interactions between molecules on the surface of the virion (ligands) and molecules on the surface of the host cell called **receptors.** For instance, some bacteriophages use cell wall lipopolysaccharides and proteins as receptors, while others use teichoic acids, flagella, or pili. Binding of an animal virus particle to its receptor often causes conformational changes in virion proteins that facilitate interaction with secondary receptors, entry into the host, and uncoating.

Receptor specificity is at least partly responsible for the preferences viruses have for a particular host. Bacteriophages not only infect a particular bacterial species but often infect only certain strains within a given species. Likewise, animal viruses infect specific animals and, in some cases, only particular tissues within that host. However, if the receptor recognized by a virus is present in numerous animals, then the virus will infect more than one animal species. Such is the case with rabies viruses.

Viruses have evolved such that they use host receptors that are always present on the surface of the host cell and are important for normal host cell function. Because the cell surface proteins are vital for cellular function, mutations that change them significantly are not tolerated, and this ensures that the virus can infect the host. In some cases, two or more host cell receptors are involved in attachment. For instance, *human immunodeficiency virus* (HIV) particles bind to two

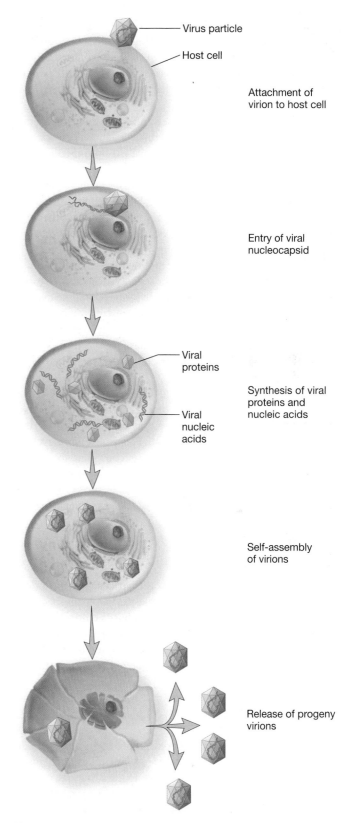

Figure 6.10 **Generalized Illustration of Virus Multiplication.** The life cycle of a virus that infects a eukaryotic cell is shown. Figure 6.16 illustrates the generalized life cycle of many bacterial and archaeal viruses. Virions are not drawn to scale.

Labels in figure:
- Virus particle
- Host cell
- Attachment of virion to host cell
- Entry of viral nucleocapsid
- Viral proteins
- Viral nucleic acids
- Synthesis of viral proteins and nucleic acids
- Self-assembly of virions
- Release of progeny virions

different proteins on human cells (e.g., CD4 and CCR5). Both of these host molecules normally bind cytokines—signaling molecules used by the immune system. ▶▶| *Retroviruses (section 27.7); Cytokines are chemical messages between cells (section 33.3); Acquired immune deficiency syndrome (AIDS; section 38.3)*

The distribution of the host cell receptors to which animal viruses attach also varies at the cellular level. Eukaryotic cell membranes have microdomains often called lipid rafts that are thought to be involved in both virion entrance and assembly. For example, the receptors for enveloped viruses such as HIV and Ebola are concentrated in lipid rafts. Distribution at the tissue level plays a crucial role in determining the **tropism** of the virus and the outcome of infection. For example, poliovirus receptors are found only in the human nasopharynx, gut, and anterior horn cells of the spinal cord. Therefore polioviruses infect these tissues, causing disease that ranges in severity from milder forms such as gastrointestinal disease to more serious paralytic disease. In contrast, measles virus receptors are present in most tissues and disease is disseminated throughout the body, resulting in the widespread rash characteristic of measles. |◀◀ *Eukaryotic cell envelopes (section 5.2)* ▶▶| *Human diseases caused by viruses and prions (chapter 38)*

Plant viruses are a notable exception to attachment based on receptor binding as no receptors have been identified for plant viruses. Rather, damage of host cells is required for the virus particles to access and enter the host. This is often achieved by plant-eating insects that carry virions from one plant to another. The virions are deposited in plant tissues as the insect devours the plant. Interestingly, evidence suggests that some viruses alter their plant hosts to promote activity of the insects and thereby foster transmission to new plants.

Entry into the Host

After attachment to the host cell, the virus's genome or the entire nucleocapsid enters the cytoplasm. For many bacteriophages only their nucleic acid enters the host's cytoplasm, leaving the capsid outside and attached to the cell. In contrast to phages, the nucleocapsid of many viruses of eukaryotes enters the cytoplasm with the genome still enclosed. Once inside the cytoplasm, some shed their capsid proteins in a process called uncoating, whereas others remain encapsidated. Because penetration and uncoating are often coupled, we consider them together.

The mechanisms of penetration and uncoating vary with the type of virus, and for many animal viruses, detailed mechanisms of penetration are unclear. However, it appears that one of three different modes of entry is usually employed by animal viruses: fusion of the viral envelope with the host cell's plasma membrane, entry by endocytosis, and release of viral nucleic acid into the cytoplasm of the host cell (**figure 6.11**). ℘ *Entry of Animal Viruses into Host Cells*

Fusion of viral envelopes with the host cell's plasma membrane often involves viral envelope glycoproteins that interact with proteins in the plasma membrane of the host cell (figure 6.11*a*). This interaction sets into motion events that allow the nucleocapsid to enter. For example, after attachment of paramyxovirus virions (single-stranded RNA viruses such as measles virus), membrane lipids rearrange, the adjacent halves of the contacting membranes merge, and a proteinaceous fusion pore forms. The nucleocapsid then enters the host cell cytoplasm, where a viral enzyme carried within the nucleocapsid begins synthesizing viral mRNA while it is still within the capsid.

Virions of nonenveloped viruses and some enveloped viruses enter cells by one of the endocytic pathways, including clathrin-dependent endocytosis and macropinocytosis (figure 6.11*b*). The resulting endocytic vesicle contains the virion and fuses with an endosome; depending on the virus, escape of the nucleocapsid or its genome from the endocytic vesicle may occur either before or after fusion with an endosome. Endosomal enzymes can aid in virion uncoating, and low pH often triggers the uncoating process. For some enveloped viruses, the viral envelope fuses with the endosomal membrane, and the nucleocapsid is released into the cytosol (the capsid proteins may have been partially removed by endosomal enzymes). Once in the cytosol, the viral nucleic acid may be released from the capsid upon completion of uncoating or may function while still attached to capsid components. Nonenveloped animal viruses cannot employ the membrane fusion mechanism for release from the endosome (figure 6.11*c*). In this case, it is thought that the low pH of the endosome causes a conformational change in the capsid. The altered capsid contacts the endosome membrane and either releases the viral nucleic acid into the cytosol or ruptures the membrane to release the intact nucleocapsid. |◀◀ *There are several endocytic pathways (section 5.4)*

Synthesis Stage

This stage of the viral life cycle differs dramatically among viruses because the genome of a virus dictates the events that occur. For dsDNA viruses, the synthesis stage can be very similar to the typical flow of information in cells. That is, the genetic information is stored in DNA and replicated by enzymes called DNA polymerases, recoded as mRNA (transcription), and decoded during protein synthesis (translation). Because of this similarity, some dsDNA viruses have the luxury of depending solely on their host cells' biosynthetic machinery to replicate their genomes and synthesize their proteins.

The same is not true for RNA viruses. Cellular organisms (except for plants) lack the enzymes needed to replicate RNA or to synthesize mRNA from an RNA genome. Therefore RNA viruses must carry in their nucleocapsids the enzymes needed to complete the synthesis stage, or the enzymes must be synthesized during the infection process. This is discussed in more detail in chapter 27.

Some animal and plant viruses carry out the synthesis stage and subsequent assembly step within the host's cytoplasm. To protect these processes from host defenses, some viruses bring about the reorganization of host cell membranes (e.g., membranes of the endoplasmic reticulum, Golgi apparatus, and lysosomes) to

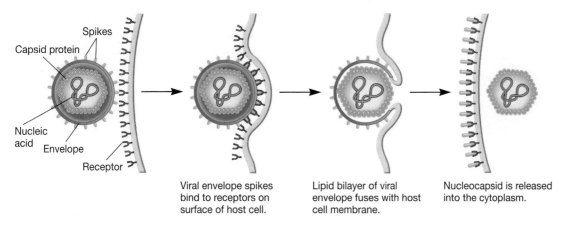

Viral envelope spikes bind to receptors on surface of host cell.

Lipid bilayer of viral envelope fuses with host cell membrane.

Nucleocapsid is released into the cytoplasm.

(a) Entry of enveloped virus by fusing with plasma membrane

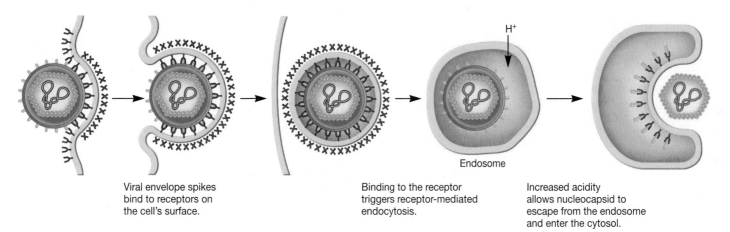

Viral envelope spikes bind to receptors on the cell's surface.

Binding to the receptor triggers receptor-mediated endocytosis.

Increased acidity allows nucleocapsid to escape from the endosome and enter the cytosol.

(b) Entry of enveloped virus by endocytosis

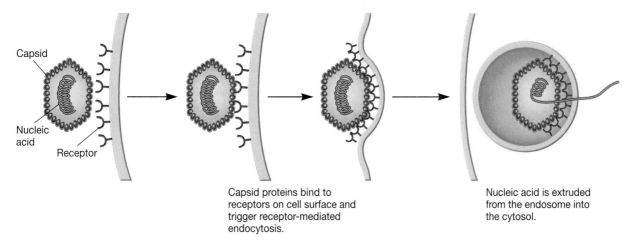

Capsid proteins bind to receptors on cell surface and trigger receptor-mediated endocytosis.

Nucleic acid is extruded from the endosome into the cytosol.

(c) Entry of nonenveloped virus by endocytosis

Figure 6.11 Animal Virus Entry. Examples of animal virus attachment and entry into host cells. Enveloped viruses can (a) enter after fusion of the envelope with the plasma membrane or (b) escape from an endosome after endocytosis. (c) Nonenveloped viruses may be taken up by endocytosis and then insert their nucleic acid into the cytosol through the vesicle membrane. It also is possible that they insert the nucleic acid directly through the plasma membrane.

MICRO INQUIRY *Which of these mechanisms involves the production of a vesicle within the host cell? Which involves the interaction between viral proteins and host cell membrane proteins?*

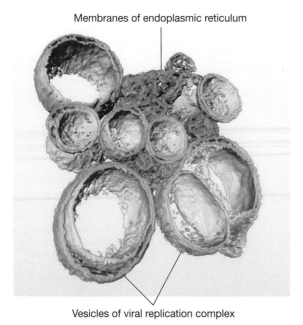

Membranes of endoplasmic reticulum

Vesicles of viral replication complex

Figure 6.12   A Viral Replication Complex.   An electron tomogram of cells infected with severe acute respiratory syndrome coronavirus (SARS-CoV). The vesicles of the viral replication complex have a double membrane derived from endoplasmic reticulum (ER) membranes. The membranes of vesicles are connected to the ER.

form membranous structures that enclose the machineries needed for genome replication, transcription, and protein synthesis (**figure 6.12**). The structures are called viral replication complexes, and they appear as vesicles, tubular structures, and other forms in electron micrographs of infected cells. Other viruses carry out synthesis and assembly in defined areas within the cytoplasm that are not enclosed by membranes. These areas of concentrated viral genomes, mRNAs and proteins are called viroplasms, and they are also visible in electron micrographs of infected cells. Both viral replication complexes and viroplasms are sometimes referred to as virus factories.

One important feature of the synthesis stage is the tight regulation of gene expression and protein synthesis. Genes are often referred to as early, middle, or late genes based on when they are expressed. The proteins they encode are likewise referred to as early, middle, or late proteins. Many early proteins are involved in taking over the host cell. Middle proteins often participate in replication of the viral genome or activation of expression of late genes. Late proteins usually include capsid proteins and other proteins involved in self-assembly and release.

Assembly

Several kinds of late proteins are involved in the assembly of mature virions. Some are nucleocapsid proteins, some are not incorporated into the nucleocapsid but participate in its assembly, and still other late proteins are involved in virion release. In addition, proteins and other factors synthesized by the host may be involved in assembling mature virions.

The assembly process can be quite complex with multiple subassembly lines functioning independently and converging in later steps to complete nucleocapsid construction. As shown in **figure 6.13**, the baseplate, tail fibers, and head components of bacteriophage T4 are assembled separately. Once the baseplate is finished, the tail tube is built on it and the sheath is assembled around the tube. The phage prohead (procapsid) is constructed with the aid of scaffolding proteins that are degraded or removed after assembly is completed. DNA is incorporated into the prohead by a complex of proteins sometimes called the "packasome." The packasome consists of a protein called the portal protein, which is located at the base of the prohead, and an enzyme called terminase, which moves DNA into the prohead. The movement of DNA consumes energy in the form of ATP, which is supplied by the metabolic activity of the host bacterium. After the head is completed, it spontaneously combines with the tail assembly.

Virion Release

Several release mechanisms have been identified. The two most common are release by lysing the host cell and release by budding. Release by lysis is especially common for bacterial

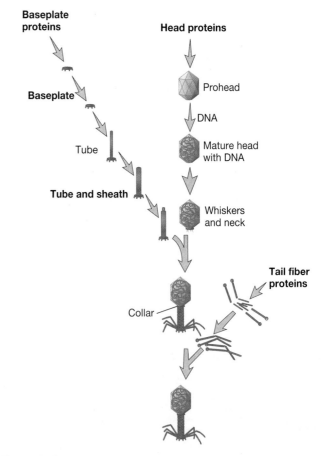

Baseplate proteins

Baseplate

Tube

Tube and sheath

Head proteins

Prohead

DNA

Mature head with DNA

Whiskers and neck

Tail fiber proteins

Collar

Figure 6.13   The Assembly of T4 Virions.   Note the subassembly lines for the baseplate, tail tube and sheath, tail fibers, and head.

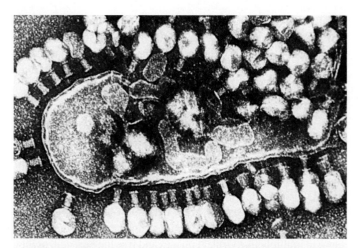

Figure 6.14 Release of T4 Virions by Lysis of the Host Cell. The host cell has been lysed (upper right portion of the cell) and virions have been released into the surroundings. Progeny virions also can be seen in the cytoplasm. In addition, empty capsids of the infecting virus particles coat the outside of the cell (×36,500).

MICRO INQUIRY *Why do the empty capsids remain attached to the cell after the viral genome enters the host cell?*

plasma membrane, enabling T4 lysozyme to move from the cytoplasm to the peptidoglycan.

Budding is frequently observed for enveloped viruses; in fact, envelope formation and virion release are usually concurrent processes. When virions are released by budding, the host cell may survive and continue releasing virions for some time. All envelopes of animal viruses are derived from host cell membranes by a multistep process. First, virus-encoded proteins are incorporated into the membrane. Then the nucleocapsid is simultaneously released and the envelope formed by membrane budding (**figure 6.15**). In several virus families, a matrix (M) protein attaches to the plasma membrane and aids in budding. Most envelopes arise from the plasma membrane. The endoplasmic reticulum, Golgi apparatus, and other internal membranes also can be used to form envelopes. ⚙ *Mechanism for Releasing Enveloped Virions*

Interestingly, some viruses are not released from their host cell into the surrounding environment. Rather, their virions move from one host cell directly to another host cell. Most fungal viruses lack an extracellular phase in their replicative cycles. Instead they are transmitted by cell division, spore formation, or during mating. Vaccinia viruses elicit the formation of long actin tails that propel nucleocapsids through the plasma membrane, directly into an adjacent cell. In this way, the virus avoids detection by the host immune system. The genomes or nucleocapsids of many plant viruses also move directly from cell to cell through small connections called plasmodesmata that link adjacent cells. This spread of the virus typically involves virus-encoded movement proteins. ◄◄ *The eukaryotic cytoplasm contains a complex cytoskeleton and many membranous organelles (section 5.3)*

viruses and some nonenveloped animal viruses. This process involves the activity of viral proteins. For instance, lysis of *E. coli* by T4 requires two specific proteins (**figure 6.14**). One is lysozyme, an enzyme that attacks peptidoglycan in the host's cell wall. The other, called holin, creates holes in *E. coli's*

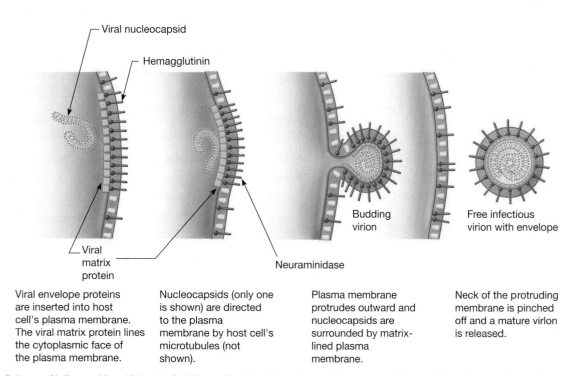

Figure 6.15 Release of Influenza Virus Virions by Budding. For simplicity, only one of the seven to eight possible nucleocapsids are shown.

1. Explain why the receptors that viruses have evolved to use are host surface proteins that serve very important, and sometimes essential, functions for the host cell.
2. What probably plays the most important role in determining the tissue and host specificity of viruses? Give some specific examples.
3. How do you think the complexity of the viral assembly process correlates with viral genome size?
4. In general, DNA viruses can be much more dependent on their host cells than can RNA viruses. Why is this so?
5. Consider the origin of viral envelopes and suggest why enveloped viruses that infect plants and bacteria are rare.
6. Why are the proteins involved in virion assembly synthesized late in the viral life cycle?

6.4 There Are Several Types of Viral Infections

After reading this section, you should be able to:

- Compare and contrast the major steps of the life cycles of virulent phages and temperate phages
- List examples of lysogenic conversion
- Differentiate among the types of viral infections of eukaryotic cells
- Summarize the current understanding of how oncoviruses cause cancer

So far, our discussion has focused on viral structures and modes of multiplication with little mention of the cost to host cells. The dependence of viruses on their host cells has many consequences—as anyone who has ever had a cold or the flu well understands. Next we discuss the interaction between virus and host cell.

Lytic and Lysogenic Infections Are Common for Bacterial and Archaeal Cells

Most bacteriophages are either virulent or temperate. A **virulent phage** is one that has only one option: to begin multiplying immediately upon entering its bacterial host, followed by release from the host by lysis. T4 is an example of a virulent phage. **Temperate phages** have two options: upon entry into the host, they can multiply like virulent phages and lyse the host cell, or they can remain within the host without destroying it (**figure 6.16**). Bacteriophage lambda is an example of this type of phage. The life cycles of both T4 and lambda are described in more detail in chapter 27. ↻ *Steps in the Replication of T4 Phage in* E. coli; *Lambda Phage Replication Cycle*

The relationship between a temperate phage and its host is called **lysogeny.** The form of the virus that remains within its host is called a **prophage.** A prophage is simply the viral nucleic acid either integrated into the bacterial chromosome or free in the cytoplasm. The infected bacteria are called **lysogens** or **lysogenic bacteria.** Lysogenic bacteria reproduce and in most other ways appear to be perfectly normal. However, they have two distinct characteristics. The first is that they cannot be

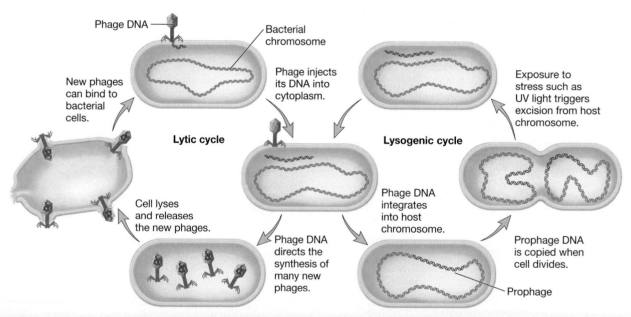

Figure 6.16 Lytic and Lysogenic Cycles of Temperate Phages. Temperate phages have two phases to their life cycles. The lysogenic cycle allows the genome of the virus to be replicated passively as the host cell's genome is replicated. Certain environmental factors such as UV light can cause a switch from the lysogenic cycle to the lytic cycle. In the lytic cycle, new virus particles are made and released when the host cell lyses. Virulent phages are limited to just the lytic cycle.

MICRO INQUIRY *Why is a lysogen considered a new or different strain of a given bacterial species?*

reinfected by the same virus; that is, they have immunity to superinfection. The second is that as they reproduce, the prophage is replicated and inherited by progeny cells. This can continue for many generations until conditions arise that cause the prophage to initiate synthesis of phage proteins and to assemble new virions, a process called **induction.** Induction is commonly caused by changes in growth conditions or ultraviolet irradiation of the host cell. As a result of induction, the **lysogenic cycle** ends and the **lytic cycle** commences; the host cell lyses and progeny phage particles are released.

Another important outcome of lysogeny is **lysogenic conversion.** This occurs when a temperate phage changes the phenotype of its host. Lysogenic conversion often involves alteration in surface characteristics of the host. For example, when a member of the genus *Salmonella* is infected by epsilon phage, the phage changes the activities of several enzymes involved in construction of the carbohydrate component of the bacterium's lipopolysaccharide. This eliminates the receptor for epsilon phage, so the bacterium becomes immune to infection by another epsilon phage. Other lysogenic conversions give the host pathogenic properties. This is the case when *Corynebacterium diphtheriae,* the cause of diphtheria, is infected with phage β. The phage genome encodes diphtheria toxin, which is responsible for the disease. Thus only those strains of *C. diphtheriae* that are infected by the phage (i.e., lysogens) cause disease. ▶▶| *Diphtheria (section 39.1)*

Clearly the infection of a bacterium by a temperate phage has significant impact on the host, but why would viruses evolve this alternate cycle? Two advantages of lysogeny have been recognized. The first is that lysogeny allows the viral nucleic acid to be maintained within a dormant host. Bacteria often become dormant due to nutrient deprivation, and while in this state, they do not synthesize nucleic acids or proteins. In such situations, a prophage would survive but most virulent bacteriophages would not be replicated, as they require active cellular biosynthetic machinery. Furthermore, their genome would be degraded as the host cell entered dormancy. The second advantage arises when there are many more phages in an environment than there are host cells, a situation virologists refer to as a high multiplicity of infection (MOI). In these conditions, lysogeny enables the survival of infected host cells within a population that has few uninfected cells. When MOI is high, a virulent phage would rapidly destroy the available host cells in its environment. However, a prophage will be replicated as the host cell reproduces.

Archaeal viruses can also be virulent or temperate. In addition, many archaeal viruses establish chronic infections. Unfortunately, little is known about the mechanisms they use to regulate their replicative cycles. ▶▶| *Archaeal viruses (section 27.2)*

Retrieve, Infer, Apply

1. Define the terms lysogeny, temperate phage, lysogen, prophage, immunity, and induction.
2. What advantages might a phage gain by being capable of lysogeny?
3. Describe lysogenic conversion and its significance.

Infections of Eukaryotic Cells

Viruses can harm their eukaryotic host cells in many ways. An infection that results in cell death is a cytocidal infection. As with bacterial and archaeal viruses, this can occur by lysis of the host (**figure 6.17a**). Infection does not always result in lysis of host cells. Some viruses (e.g., herpesviruses) can establish persistent infections lasting many years (figure 6.17b,c). Eukaryotic viruses can cause microscopic or macroscopic degenerative changes or abnormalities in host cells and in tissues that are distinct from lysis. These are called **cytopathic effects (CPEs).** Viruses use a variety of mechanisms to cause cytopathic and cytocidal effects. Many of these are noted in chapter 38. One mechanism of particular note is that some viruses cause the host cell to be transformed into a malignant cell (figure 6.17d). This is discussed next.

Viruses and Cancer

Cancer is one of the most serious medical problems in developed nations, and it is the focus of an immense amount of research. A tumor is a growth or lump of tissue resulting from **neoplasia**—unregulated abnormal new cell growth and reproduction. Tumor cells have aberrant shapes and altered plasma membranes that may contain distinctive molecules (tumor antigens). These changes result from the tumor cells becoming less differentiated. Their unregulated proliferation and loss of differentiation result in invasive growth that forms unorganized cell masses. This reversion to a more primitive or less differentiated state is called **anaplasia.**

Two major types of tumor growth patterns exist. If the tumor cells remain in place to form a compact mass, the tumor is benign. In contrast, cells from malignant or cancerous tumors actively spread throughout the body in a process known as metastasis. Some cancers are not solid but cell suspensions. For example, leukemias are composed of undifferentiated malignant white blood cells that circulate throughout the body. Indeed, dozens of kinds of cancers arise from a variety of cell types and afflict all kinds of organisms.

Some viruses have been shown to cause cancer in animals, including humans; it is estimated that about 10 to 20% of human cancers have a viral etiology. To understand the role viruses play in cancer, we must begin by considering carcinogenesis when viruses are not involved. Carcinogenesis is a complex, multistep process caused by mutations in multiple genes. Some mutations lead to the unregulated proliferation that is a major characteristic of a cancer cell. Other mutations are needed to allow that cell to grow into a cancerous tumor. For instance, one type of mutation promotes the growth of blood vessels (called angiogenesis) in the developing tumor so that it can obtain needed nutrients and oxygen to support its growth. In addition, mutations that promote metastasis must occur, so that the tumor can invade other tissues. ▶▶| *Mutations (section 16.1)*

Considerable research into the causes of cancer has focused on the mutations that allow cancerous cells to grow uncontrollably.

Figure 6.17 Types of Viral Infections and Their Effects on Animal Cells. (a) Lytic infections of host cells can lead to disease states in animal hosts called acute infections. There are two types of persistent infections: (b) latent infections and (c) chronic infections. (d) Some infections of animal cells cause the cell to be transformed into a malignant cell, which can cause cancer in the animal host.

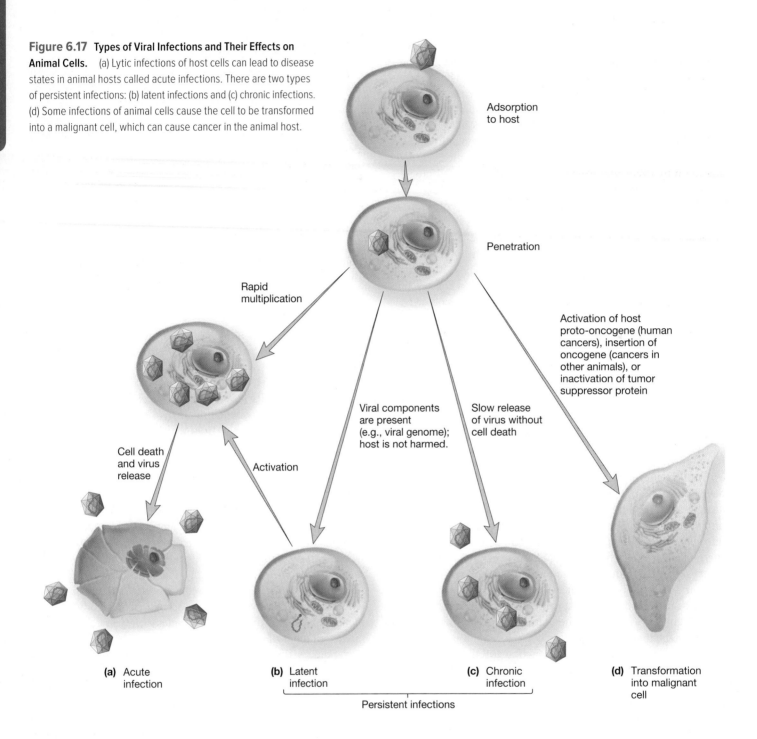

Adsorption to host

Penetration

Rapid multiplication

Activation of host proto-oncogene (human cancers), insertion of oncogene (cancers in other animals), or inactivation of tumor suppressor protein

Viral components are present (e.g., viral genome); host is not harmed.

Slow release of virus without cell death

Cell death and virus release

Activation

(a) Acute infection

(b) Latent infection

(c) Chronic infection

(d) Transformation into malignant cell

Persistent infections

Those studies show that there are two types of genes that must be mutated: proto-oncogenes and tumor suppressor genes. Their names suggest that these genes are somehow abnormal. But this is not the case. Proto-oncogenes and tumor suppressor genes are normal cellular genes. **Proto-oncogenes** must be expressed if cell division is to occur. However, they normally are expressed only if the cell receives an appropriate signal, such as the binding of a growth factor to a receptor on the cell surface. When such a signal is received, the cell initiates the cell cycle, but it cannot continue past a checkpoint controlled by the activity of proteins encoded by **tumor suppressor genes**.

These proteins are called **tumor suppressor proteins**. When tumor suppressor proteins are active, they prevent progression through the cell cycle. Thus, for cell division to occur, proto-oncogene proteins (sometimes called oncoproteins) must be active, and tumor suppressor proteins must be inactive. *Tumor suppressor genes*

During carcinogenesis, mutations arise that disrupt the normal activity of these two types of genes. The first type of mutation allows proto-oncogenes or their protein products to function without an appropriate signal. Thus the proto-oncogene or its oncoprotein is active even when it should not be. The

Table 6.1 Some Viruses Associated with Human Cancers

Virus	Genome Type	Cancer
Human herpesvirus 8 (HHV8)	Double-stranded (ds) DNA	Several, including Kaposi's sarcoma
Epstein-Barr virus (EBV)	dsDNA	Several, including Burkitt's lymphoma and nasopharyngeal carcinoma
Hepatitis B virus	dsDNA	Hepatocellular carcinoma
Hepatitis C virus	Single-stranded (ss) RNA	Liver cancer
Human papillomaviruses (HPV) strains 6, 11, 16, and 18	dsDNA	Cervical cancer and oral pharyngeal cancer
Human T-cell lymphotropic virus1 (HTLV-1)	ssRNA (retrovirus)	T-cell leukemia

second type of mutation prevents expression of tumor suppressor genes, yields nonfunctional tumor suppressor proteins, or in some other way leads to inactivation of a tumor suppressor protein. Thus the cell can continue cell division even when it shouldn't. Numerous proto-oncogenes and tumor suppressor genes have been identified. An important proto-oncogene encodes a protein called Myc. Myc controls transcription of many genes, including some involved in DNA replication and other cell-cycle-related functions. Two of the most important tumor suppressor proteins are Rb and p53. The protein p53 is often referred to as "the guardian of the genome" because not only does it cause cell cycle arrest, but it also initiates programmed cell death in response to DNA damage. Programmed cell death, also called apoptosis, is a process that causes cells to kill themselves. Thus, when p53 is inactivated, the cell not only can reproduce in an unregulated fashion, but it can avoid programmed cell death even when it should destroy itself. This is a catastrophic event for the cell because it can lead to the accumulation of additional mutations that contribute to carcinogenesis.

We now turn out attention to viruses and their role in causing cancer. These viruses are called **oncoviruses**. As you can see in **table 6.1**, most human oncoviruses have dsDNA genomes, and most of these viruses cause cancer in a similar way—by interacting with and inactivating either Rb or p53. Oncoviruses that are retroviruses (e.g., HTLV-1) exert their oncogenic powers in a different manner. Some carry genes called **oncogenes** that stimulate the activity of cellular proto-oncogenes. For example, HTLV-1 infects immune system cells called T cells. During the infection HTLV-1 produces a regulatory protein that activates expression of numerous cellular proto-oncogenes. Other retroviruses transform cells into cancerous cells when the viral genome integrates into the host chromosome such that strong, viral regulatory elements are near a cellular proto-oncogene. These elements cause the nearby proto-oncogene to be transcribed at a high level, causing the cellular gene to be considered an oncogene.

Retrieve, Infer, Apply

1. How does a latent infection differ from a chronic infection?
2. What is a cytocidal infection? What is a cytopathic effect?
3. Define the following terms: tumor, neoplasia, anaplasia, metastasis, proto-oncogene, oncogene, and tumor suppressor gene.
4. Distinguish the mechanism by which dsDNA viruses cause cancer from that of retroviruses. Why do you think there is an environmental component to many kinds of cancer?

6.5 Cultivation and Enumeration of Viruses

After reading this section, you should be able to:

- List the approaches used to cultivate viruses, noting which types of viruses are cultivated by each method
- Describe three direct counting methods and two indirect counting methods used to enumerate viruses
- Outline the events that lead to formation of a plaque in a lawn of bacterial cells
- Distinguish lethal dose from infectious dose

Because they are unable to reproduce outside of living cells, viruses cannot be cultured in the same way as cellular microorganisms. It is relatively simple to culture lytic bacterial and archaeal viruses as long as the host cell is easily grown in culture. Cells and virus particles are simply mixed together in a broth culture. Over time, more and more cells are infected and lysed, releasing virions into the broth. Culturing temperate viruses requires an additional step of inducing the lytic phase of their life cycles. When cultured with cells on a solid medium, lysis of the host cells is observed as **plaques** on the bacterial lawn (**figure 6.18**) Animal viruses are often cultivated by inoculating suitable host animals or embryonated eggs—fertilized chicken eggs incubated about 6 to 8 days after laying. Animal viruses also

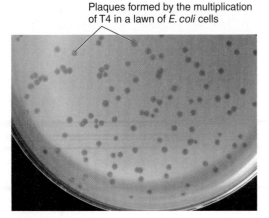

Plaques formed by the multiplication of T4 in a lawn of *E. coli* cells

Figure 6.18 Viral Plaques. The plaques are formed in the bacterial cell lawn by the process illustrated in figure 6.20.

are grown in tissue (cell) culture on monolayers of animal cells. If an animal virus causes its host cell to lyse, plaques often are formed in the monolayer of cells. Viruses that cause cytopathic effects are also grown and detected in tissue culture.

Plant viruses are cultivated in a variety of ways. Plant tissue cultures, cultures of separated cells, or cultures of protoplasts (cells lacking cell walls) may be used. Viruses also can be grown in whole plants, usually after the leaves are mechanically inoculated by rubbing them with a mixture of virus particles and an abrasive. When the cell walls are broken by the abrasive, the virions directly contact the plasma membrane and infect the exposed host cells. Some plant viruses can be transmitted only if a diseased part is grafted onto a healthy plant. A localized necrotic lesion often develops in infected plants due to the rapid death of cells in the infected area. Even when lesions do not occur, the infected plant may show symptoms such as changes in pigmentation or leaf shape (**figure 6.19**).

It is often necessary to determine the number of virions in a preparation of viruses or an environmental sample.

Figure 6.19 Necrotic Lesions on Plant Leaves. A tobacco plant leaf showing leaf color changes caused by TMV.

One direct approach is to count virions using an electron microscope. However, this is a time-consuming process that requires a high concentration of virus particles. Another direct counting method uses epifluorescence microscopy. This approach is often used to enumerate virus particles in aquatic habitats. The procedure stains the virions with fluorescent dyes prior to microscopic examination. Finally, sometimes rather than counting virions, the nucleic acid of a virus can be quantified by quantitative-polymerase chain reaction (qPCR). This is possible because PCR amplifies specific nucleic acids in a mixture of nucleic acids. ▶▶ *Aquatic viruses (section 30.2); Polymerase chain reaction amplifies targeted DNA (section 17.2)*

An indirect method of counting animal viruses is the **hemagglutination assay.** Many animal virus particles bind to the surface of red blood cells (*see figure 36.11*). If the ratio of virions to cells is large enough, virions will join the red blood cells together; that is, they agglutinate, forming a network that settles out of suspension. In practice, red blood cells are mixed with diluted samples of the virus particles, and each mixture is examined. The hemagglutination titer is the highest dilution of the virus preparation (or the reciprocal of the dilution) that still causes hemagglutination. ▶▶ *Agglutination (section 36.4)*

Other indirect assays determine virion numbers based on their infectivity, and many of these are based on the same techniques used for virus cultivation. For example, in the **plaque assay,** several dilutions of a sample containing virus particles are plated with appropriate host cells. Theoretically, when the multiplicity of infection (p. 123) is very low, then each plaque in a layer of host cells should have arisen from the multiplication of a single virion (**figure 6.20**). Therefore a count of the plaques produced at a particular dilution can be used to estimate the number of virions in the original sample. The resulting value is expressed as **plaque-forming units (PFU)** rather than as virions, for several reasons. First, not all virions may be infective. Furthermore, even though there are far fewer virus particles than host cells, it is still possible for more than one virion to infect the same cell (figure 6.14). Finally, it is possible for two infected cells to be plated to the same area, which would then give rise to a single plaque. Despite these caveats, the number of PFU is proportional to the number of virus particles: a preparation with twice as many virions will have twice the plaque-forming units.

The same approach employed in the plaque assay may be used with embryos and plants. Chicken embryos and plant leaves can be inoculated with a diluted preparation of virus particles. The number of pocks on embryonic membranes or necrotic lesions on leaves is used to calculate the concentration of infectious units.

When biological effects are not readily quantified by plaque and hemagglutination assays, the amount of virions required to cause disease or death can be determined. Organisms or cell cultures are inoculated with serial dilutions of a preparation of virus particles. The **lethal dose (LD$_{50}$)** is

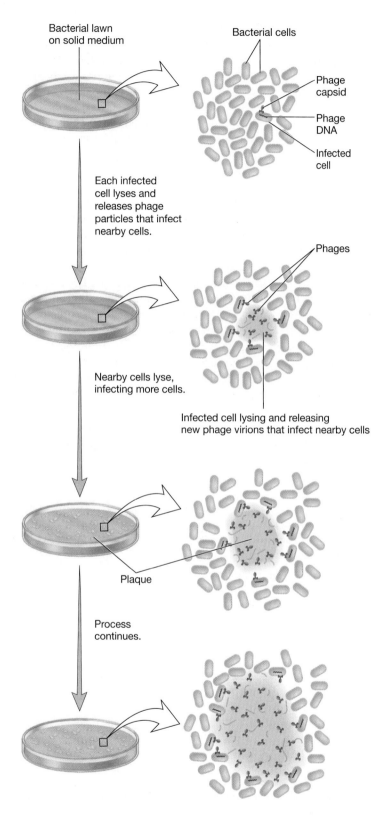

Bacterial lawn on solid medium

Bacterial cells

Phage capsid

Phage DNA

Infected cell

Each infected cell lyses and releases phage particles that infect nearby cells.

Phages

Nearby cells lyse, infecting more cells.

Infected cell lysing and releasing new phage virions that infect nearby cells

Plaque

Process continues.

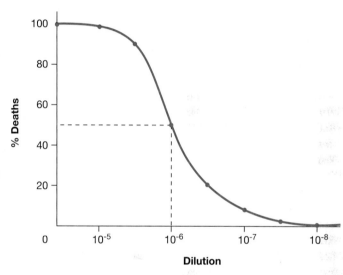

Figure 6.20 **Formation of Phage Plaques.** When phage particles and host bacterial cells are mixed at an appropriate ratio, only a portion of the cells are initially infected. When this mixture is plated, the infected cells will be separated from each other. The infected cells eventually lyse, releasing progeny phage particles. They infect nearby cells, which eventually lyse, releasing more phage particles. This continues and ultimately gives rise to a clear area within a lawn of bacteria. The clear area is a plaque.

MICRO INQUIRY *What would happen to a plate with individual plaques if it were returned to an incubator so that the viral infection could continue?*

Figure 6.21 **Determination of LD$_{50}$.** The LD$_{50}$ is indicated by the dashed line.

Retrieve, Infer, Apply

1. Discuss the ways that viruses can be cultivated. Define the terms plaque, cytopathic effect, and necrotic lesion.
2. Given that viruses must be cultivated to make vaccines against viral diseases, discuss the advantages and disadvantages that might occur with each approach (embryonated egg versus cell culture) to growing animal viruses.

6.6 Viroids and Satellites: Nucleic Acid-Based Subviral Agents

After reading this section, you should be able to:
- Describe the structure of a viroid and discuss the practical importance of viroids
- Distinguish satellite viruses from satellite nucleic acids

the dilution that contains a concentration (dose) of virions large enough to destroy 50% of the host cells or organisms. In a similar sense, the **infectious dose** (**ID$_{50}$**) is the dose that causes 50% of the host organisms to become infected (**figure 6.21**).

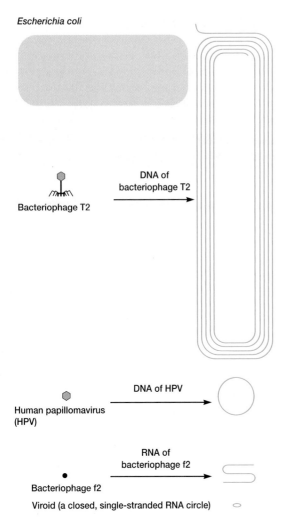

Figure 6.22 **Viroids, Viruses, and Bacteria.** A comparison of *E. coli*, several viruses, and the potato spindle-tuber viroid with respect to size and the amount of nucleic acid possessed. (All dimensions are enlarged approximately ×40,000.)

Although viruses are exceedingly simple, even simpler infectious agents exist. **Viroids** are infectious agents that consist only of RNA. They cause over 20 different plant diseases, including potato spindle-tuber disease, exocortis disease of citrus trees, and chrysanthemum stunt disease. Viroids are covalently closed, circular ssRNAs, about 250 to 370 nucleotides long (**figure 6.22**). The International Committee on Taxonomy of Viruses (ICTV) currently divides viroids into two families. Viroids in family *Pospiviroidae* have circular RNA that exists as a rodlike shape due to intrastrand base pairing, which forms double-stranded regions with single-stranded loops (**figure 6.23a**). The circular RNA of viroids in the family *Avsunviroidae* is shaped like a rod with a highly branched structure at one end, rather like a tree trunk with its roots. Each branch is formed by intramolecular base pairing of the RNA that creates a stem-loop structure. The two types of viroids replicate in different locations within the infected plant cell. Pospiviroids replicate in the nucleus, whereas avsunviroids replicate in plant plastids such as chloroplasts.

However, they replicate in similar ways. The RNA of viroids does not encode any gene products, so they cannot replicate themselves. Rather, they use a host cell enzyme called a DNA-dependent RNA polymerase. This enzyme normally functions in the host to synthesize RNA using DNA as the template during transcription. However, when infected by a viroid, the host polymerase uses the viroid RNA as a template for RNA synthesis. The host polymerase synthesizes a complementary RNA molecule, which then serves as the template for synthesis of new viroid RNAs.

A plant may be infected with a viroid without showing symptoms; that is, it may have a latent infection. However, the same viroid in another host species may cause severe disease. The pathogenicity of viroids is not well understood, but it is known that particular regions of the RNA are required; studies have shown that removing these regions blocks the development of disease (figure 6.23*a*). Some data suggest that viroids cause disease by triggering a eukaryotic response called **RNA silencing,** which normally functions to protect against infection by RNA viruses. During RNA silencing, the cell detects the presence of dsRNA and cuts it into small fragments. These are used by the RNA silencing machinery to destroy target mRNA molecules or prevent their translation. Viroids may usurp this response by hybridizing to specific host mRNA molecules to which they have a complementary nucleotide sequence. Formation of the viroid-host hybrid dsRNA molecule is thought to elicit RNA silencing. This results in destruction of the host mRNA and therefore silencing of the host gene. Failure to express a required host gene leads to disease in the host plant.

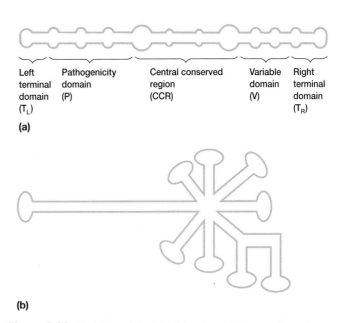

Figure 6.23 **Viroid Structure.** (a) This schematic diagram shows the general organization of a viroid belonging to the family *Pospiviroidae*. The closed single-stranded RNA circle has extensive intrastrand base pairing and interspersed unpaired loops. Also shown are the five domains identified in the molecule. Most changes in viroid pathogenicity arise from variations in the P and T$_L$ domains. (b) Schematic diagram of a viroid belonging to *Avsunviroidae*. These viroids lack the central conserved region observed in pospiviroids.

Satellites are similar to viroids in that they also consist of a nucleic acid (either DNA or RNA). They differ from viroids in that their nucleic acid gets enclosed in a capsid and they need a helper virus to replicate. There is little or no homology between the nucleic acid of the satellite and that of its helper virus (i.e., a satellite is not a defective version of its helper virus). Satellites are further divided into three types: satellite viruses, satellite RNAs, and satellite DNAs. Satellite viruses encode their own capsid proteins, whereas satellite RNAs and DNAs do not. Most satellites use plant viruses as their helper viruses.

6.7 Prions Are Composed Only of Protein

After reading this section, you should be able to:

- Describe prion structure and how prions are thought to replicate
- List characteristics common to all animal diseases caused by prions
- Name at least two human diseases caused by prions
- Describe the mechanisms by which a prion protein might first appear in a brain cell

Prions (for *proteinaceous infectious particle) cause a variety of neurodegenerative diseases in humans and other animals, including scrapie in sheep, bovine spongiform encephalopathy (BSE or "mad cow disease") and the human diseases kuru, fatal familial insomnia, Creutzfeldt-Jakob disease (CJD), and Gerstmann-Strässler-Scheinker syndrome (GSS). All result in progressive degeneration of the brain and eventual death. At present, no effective treatment exists. ▶▶┃ *Prion proteins transmit disease (section 38.6)*

The best-studied prion is the scrapie prion. Researchers have shown that scrapie is caused by an abnormal form of a cellular protein. The normal cellular protein is called PrP^C. Although its exact function is unknown, it has been shown to play a role in brain development as synaptic connections are formed. The abnormal form is called PrP^{Sc} (for *scrapie-associated prion protein*). Evidence supports a model in which entry of PrP^{Sc} into the brain of an animal causes the PrP^C protein to change from its normal conformation to the abnormal form (**figure 6.24**). The newly produced PrP^{Sc} molecules then convert more PrP^C molecules into the abnormal PrP^{Sc} form. How the PrP^{Sc} causes this conformational change is unclear. However, the best-supported model is that the PrP^{Sc} directly interacts with PrP^C, causing the change. It is noteworthy that mice lacking the PrP gene cannot be infected with PrP^{Sc}.

An important characteristic of PrP^{Sc} is its ability to oligomerize—that is, to form short chains of PrP^{Sc} proteins. This is important because oligomers may play a central role in continued conversion of PrP^C to PrP^{Sc}, as well as in the spread of PrP^{Sc} from cell to cell and from one tissue to another. Furthermore, evidence suggests that the toxic form of PrP^{Sc} is a trimer. However, how the PrP^{Sc} trimer causes disease is not understood.

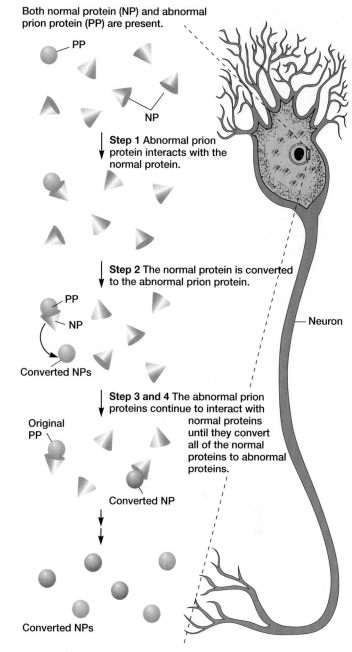

Both normal protein (NP) and abnormal prion protein (PP) are present.

PP

NP

Step 1 Abnormal prion protein interacts with the normal protein.

Step 2 The normal protein is converted to the abnormal prion protein.

PP

NP

Converted NPs

— Neuron

Step 3 and 4 The abnormal prion proteins continue to interact with normal proteins until they convert all of the normal proteins to abnormal proteins.

Original PP

Converted NP

Converted NPs

Figure 6.24 One Proposed Model for Prion Replication. The normal and prion proteins differ in their tertiary structures. Recent studies suggest that cofactors may function in prion replication. One proposed cofactor is RNA, which is thought to act as a catalyst for the conversion of PrP^C to PrP^{Sc}.

Retrieve, Infer, Apply

1. What are viroids and why are they of great interest?
2. How does a viroid differ from a virus? From a satellite?
3. What is a prion? In what way does a prion differ fundamentally from viruses and viroids?
4. Prions are difficult to detect in host tissues. Why do you think this is so? Why do you think we have not been able to develop effective treatments for prion-caused diseases?

Key Concepts

6.1 Viruses Are Acellular

- Virology is the study of viruses and other acellular infectious agents.
- Viruses cannot reproduce independently of living cells.

6.2 Virion Structure Is Defined by Capsid Symmetry and Presence or Absence of an Envelope

- All virions have a nucleocapsid composed of a nucleic acid, either DNA or RNA, held within a protein capsid made of one or more types of protein subunits called protomers (**figure 6.2**).
- Helical capsids resemble hollow protein tubes and may be either rigid or flexible (**figures 6.3** and **6.4**).
- Icosahedral capsids are usually constructed from two types of capsomers: pentamers (pentons) at the vertices and hexamers (hexons) on the edges and faces of the icosahedron (**figure 6.5**).
- Complex virions (e.g., poxviruses and large phages) have complicated morphology not characterized by icosahedral and helical capsid symmetry (**figure 6.6**). Virions of large phages often have binal symmetry: their heads are icosahedral and their tails are helical (**figure 6.7**).
- Some viruses have a membranous envelope surrounding their nucleocapsid. The envelope lipids usually come from the host cell; in contrast, many envelope proteins are viral and may project from the envelope surface as spikes (**figure 6.8**).
- Viral nucleic acids can be either single stranded (ss) or double stranded (ds), DNA or RNA. Most DNA viruses have dsDNA genomes that may be linear or closed circles. RNA viruses usually have ssRNA. Some RNA genomes are segmented.
- Although viruses lack true metabolism, some contain a few enzymes necessary for their multiplication.

6.3 Viral Life Cycles Have Five Steps

- The one-step growth experiment devised by Delbruck and Ellis was an important first step in understanding how viruses multiply (**figure 6.9**). This type of experiment showed that there is an initial latent period, during which no virions are released, followed by a rise period where virions are released. Early in the latent period is an eclipse period, in which no virus particles are present in infected cells. Viral life cycles can be divided into five steps: (1) attachment to host; (2) entry into host; (3) synthesis of viral nucleic acids and proteins; (4) self-assembly of nucleocapsids; and (5) release from host (**figures 6.10–6.15**).
- The synthesis stage of a virus's life cycle depends on the nature of its genome. DNA viruses may use enzymes that are similar to host enzymes. In some cases, they rely solely on their hosts for synthesis of their nucleic acids and proteins.

- RNA viruses must either encode the enzymes they need to make mRNA and replicate their genomes or carry these enzymes in their capsids.

6.4 There Are Several Types of Viral Infections

- Virulent bacteriophages and archaeal viruses lyse their host. In addition to host cell lysis, temperate bacterial and archaeal viruses can enter the lysogenic cycle in which they remain dormant in the host cell. This is often accomplished by integrating the viral genome into that of the host (**figure 6.16**).
- Viruses that infect eukaryotic cells can cause either host cell lysis or a more insidious cellular death; such viruses cause cytocidal infection. One relatively rare outcome of animal virus infection is the transformation of normal host cells into malignant or cancerous cells (**figure 6.17**).

6.5 Cultivation and Enumeration of Viruses

- Viruses are cultivated using tissue cultures, embryonated eggs, bacterial cultures, and other living hosts.
- Phages produce plaques in bacterial lawns. Sites of animal viral infection may be characterized by cytopathic effects such as pocks and plaques. Plant viruses can cause localized necrotic lesions in plant tissues (**figures 6.18–6.20**).
- Virions can be counted directly by transmission electron microscopy or indirectly by hemagglutination and plaque assays.
- Infectivity assays can be used to estimate virion numbers in terms of plaque-forming units, lethal dose (LD_{50}), or infectious dose (ID_{50}) (**figure 6.21**).

6.6 Viroids and Satellites: Nucleic Acid-Based Subviral Agents

- Viroids are infectious agents of plants that consist only of circular ssRNA molecules (**figures 6.22** and **6.23**). They do not encode any proteins.
- Satellites are subviral infectious agents that may encode some proteins but require a helper virus for replication. Satellite viruses encode their own capsid proteins, whereas satellite RNAs and satellite DNAs do not.

6.7 Prions Are Composed Only of Proteins

- Prions are small proteinaceous agents associated with at least six degenerative nervous system disorders: scrapie, bovine spongiform encephalopathy, kuru, fatal familial insomnia, Gerstmann-Strässler-Scheinker syndrome, and Creutzfeldt-Jakob disease.
- Most evidence supports the hypothesis that prion proteins exist in two forms: the infectious, abnormally folded form and a normal cellular form. The interaction between the abnormal form and the cellular form converts the cellular form into the abnormal form (**figure 6.24**).

Compare, Hypothesize, Invent

1. Many classification schemes are used to identify bacteria. These start with Gram staining, progress to morphology and arrangement characteristics, and include a battery of metabolic tests. Build an analogous scheme that could be used to identify viruses. You might start by considering the host, or you might start with viruses found in a particular environment, such as a marine filtrate.

2. The origin and evolution of viruses is controversial. Discuss whether you think viruses evolved before the first cell or whether they have coevolved and are perhaps still coevolving with their hosts.

3. Consider the separate stages of an animal virus life cycle. Assemble a short list of structures and processes that are unique to the virus and would make good drug targets for an antiviral agent. Explain your rationale for each choice.

4. Chronic wasting disease (CWD) is a neurodegenerative disease of cervid animals (e.g., deer, elk, moose) caused by prions. Unlike other prion diseases, CWD is readily transmitted from one animal to another in a herd. How this occurs in not understood. However, evidence suggests the mode of transmission is by animal excreta: saliva, urine, and feces. In this mechanism, a healthy animal would be exposed to the prion present in excreta from a diseased animal. Since many cervids congregate, the prion would easily spread from one animal to another. One problem facing scientists studying CWD is measuring the amount of infectious prions in excreta and determining their source (i.e., whether they are produced solely in nervous system tissue or in organs of excretion). Why is CWD of concern to fish and wildlife biologists and game managers in states where CWD is prevalent (e.g., Colorado, Wyoming, Wisconsin)? Why would the presence of CWD prions in tissues other than those of the central nervous system be important to know? (Hint: Learn about the variant Creutzfeldt-Jakob epidemic that occurred in the 1990s.)

Read the original article: Haley, N. J., et al. 2011. Detection of chronic wasting disease prions in salivary, urinary, and intestinal tissues of deer: potential mechanisms of prion shedding and transmission. *J. Virol.* 85:6309.

5. Syn5 is a virus that infects photosynthetic bacteria belonging to the genus *Synechococcus*. The Syn5 virion is icosahedral (660 Å in diameter) with a short tail and an appendage called a horn. The horn is located on a vertex directly opposite the tail. Syn5 follows the typical five steps observed for most viruses when they multiply, and progeny viruses are released when the host lyses. The assembly of Syn5 virions involves several proteins, including capsid proteins, scaffolding proteins, a portal protein, a terminase, tail proteins, and horn proteins. To better understand the assembly of Syn5 virions, scientists used a type of microscopy called Zernike phase contrast electron cryotomography. This type of microscopy improves the contrast typically seen for conventional cryotomography. In their studies, the scientists observed several different developing virus particles in infected cells. The particles were: (1) 660 Å icosahedron containing DNA and having a tail; (2) 660 Å icosahedron lacking DNA, tail, and horn; (3) 590 Å spherical particle lacking DNA, tail, and horn; (4) 660 Å icosahedron containing DNA and having a tail and horn; (5) 660 Å icosahedron containing DNA but lacking tail and horn. Based on these observations the scientists suggested a pathway for assembly of Syn5 virions. What are the functions of the portal protein and terminase? Organize the particles into an assembly pathway and indicate what you think has happened at each step. (*Hint: It will help if you draw the particles.*)

Read the original article: Dai, W., et al. 2013. Visualizing virus assembly intermediates inside marine cyanobacteria. *Nature* 502:707.

Learn More

shop.mheducation.com Enhance your study of this chapter with interactive study tools and practice tests. Also ask your instructor about the resources available through Connect, including adaptive learning tools and animations.

7

Microbial Growth

Body piercings are popular among teens and young adults but they can also serve as a habitat for microorganisms.

Metal or Plastic?

Each day in grocery stores, as their bill is tallied and their groceries readied for bagging, customers are asked, "Paper or plastic?" People who are about to pierce their tongues face a similar question: "Metal or plastic?" The desire is to have a tongue stud with "flash," which is thought to brighten the person's smile. However, there is more at stake than flash: recent studies suggest that pathogenic bacteria are more likely to accumulate on the surfaces of metal tongue studs (stainless steel and titanium) than on other materials (polypropylene, a plastic, and polytetrafluoroethylene, better known as Teflon). Even more worrisome is the observation that these bacteria are forming biofilms.

What are biofilms and why should they be of concern to those wanting a tongue piercing? Biofilms are complex, slime-encased communities of sessile microorganisms (not free-floating) growing on surfaces such as rocks in a stream, the hulls of ships, medical devices, and your teeth. The microbes in biofilms are different from planktonic cells, those that swim freely in their environment. Biofilm cells are more resistant to antimicrobial agents, including antibiotics, than are their planktonic relatives. Furthermore, they can escape the slime and travel to other sites: another rock, a different tooth, a heart valve.

Biofilm formation is just one aspect of microbial growth that is of concern to microbiologists. Even when microbes are not growing in a biofilm, they can create problems. Thus there are many situations in which it is important to inhibit their growth—in hospital rooms, kitchen counters, food-handling facilities, to name a few. In other situations, microbiologists want to promote microbial growth—from a clinical specimen so that a diagnosis can be made, in brewing vats at a brewery, or a fermentation vessel at a pharmaceutical plant.

To control the growth of microbes, it is imperative that microbiologists be able to understand how they reproduce, determine the environmental conditions that promote or hinder their growth, and measure the growth of microbial populations. This requires that microbiologists not only study the growth of microbes in natural settings but in the laboratory as well. Microbial growth in both settings is the focus of this chapter.

Readiness Check:

Based on what you have learned previously, you should be able to:

✔ Describe the functions of eukaryotic cytoskeletal proteins (section 5.3)
✔ Define and list examples of essential nutrients, and describe how they are used by cells
✔ Distinguish macroelements (macronutrients) from micronutrients (trace elements), list examples of each, and describe how they are used (section 3.3)
✔ Provide examples of growth factors needed by some microorganisms (section 3.3)
✔ Describe the eukaryotic cell cycle
✔ Describe the major events of mitosis and meiosis

7.1 Most Bacteria and Archaea Reproduce by Binary Fission

After reading this section, you should be able to:

■ Describe binary fission as observed in bacteria and archaea
■ Compare other reproductive strategies used by bacteria with binary fission

Eukaryotic microbes differ dramatically from bacteria and archaea in their reproductive strategies. Many eukaryotic microbes exhibit both asexual reproduction, involving mitosis, and sexual reproduction, involving meiosis to produce gametes or gamete-like cells. Furthermore, eukaryotic microbes often alternate between haploid and diploid stages in their life cycles. Some eukaryotic life cycles are described in chapters 25 and 26. Here our focus is on the types of cell division observed in bacterial and archaeal cells.

Most bacterial and archaeal cells reproduce by **binary fission (figure 7.1)**. Binary fission is a relatively simple type of cell division: the cell elongates as new material is synthesized, replicates its chromosome, and separates the newly formed DNA molecules so there is one chromosome in each

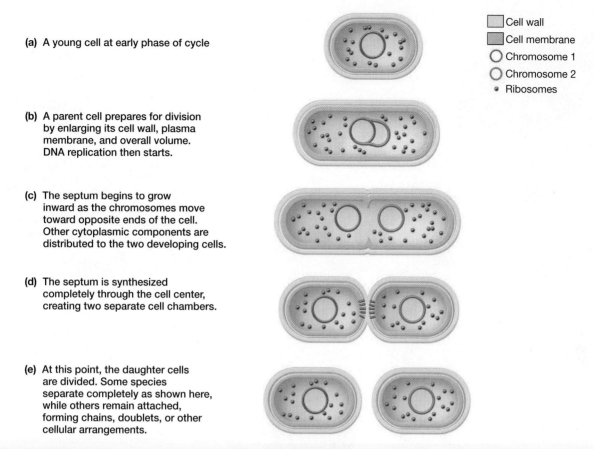

(a) A young cell at early phase of cycle

(b) A parent cell prepares for division by enlarging its cell wall, plasma membrane, and overall volume. DNA replication then starts.

(c) The septum begins to grow inward as the chromosomes move toward opposite ends of the cell. Other cytoplasmic components are distributed to the two developing cells.

(d) The septum is synthesized completely through the cell center, creating two separate cell chambers.

(e) At this point, the daughter cells are divided. Some species separate completely as shown here, while others remain attached, forming chains, doublets, or other cellular arrangements.

☐ Cell wall
■ Cell membrane
◯ Chromosome 1
◯ Chromosome 2
• Ribosomes

Figure 7.1 Binary Fission.

MICRO INQUIRY *In addition to chromosomes, what other cytoplasmic contents must be equally distributed between daughter cells?*

half of the cell. Finally, a septum (cross wall) is formed at midcell, dividing the parent cell into two progeny cells, each having its own chromosome and a complement of other cellular constituents.

Several other reproductive strategies have been identified in bacteria (**figure 7.2**). Some bacteria reproduce by forming a bud. Certain cyanobacteria undergo multiple fission. The progeny cells, called baeocytes, are held within the cell wall of the parent cell until they are released. Other bacteria, such as members of the genus *Streptomyces,* form multinucleoid filaments that eventually divide to form uninucleoid spores. These spores are readily dispersed, much like the dispersal spores formed by filamentous fungi. ▶▶ *Fungal reproduction (section 26.1)*

Despite the diversity of bacterial reproductive strategies, they share certain features. In all cases, the genome of the cell must be replicated and segregated to form distinct nucleoids. At some point during reproduction, each nucleoid and its surrounding cytoplasm becomes enclosed within its own plasma membrane. These processes are the major steps of the cell cycle. In section 7.2, we examine the bacterial cell cycle in more detail.

7.2 Bacterial Cell Cycles Can Be Divided into Three Phases

After reading this section, you should be able to:

- Summarize the three phases in a typical bacterial cell cycle
- Summarize current models for chromosome partitioning
- State the functions of cytoskeletal proteins during cytokinesis and in determining cell shape

The **cell cycle** is the complete sequence of events extending from formation of a new cell through the next division. It is of intrinsic interest as a fundamental biological process. However, understanding the cell cycle has practical importance as well. For instance, synthesis of peptidoglycan during the cell cycle is the target of numerous antibiotics used to treat bacterial infections. ▶▶ *Inhibitors of cell wall synthesis (section 9.4)*

The cell cycles of several bacteria—*Escherichia coli, Bacillus subtilis,* and the aquatic bacterium *Caulobacter crescentus*—have been examined extensively, and our understanding of the bacterial cell cycle is based largely on these studies. The bacterial

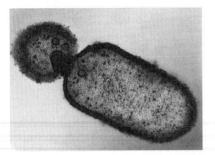

(a) *Listeria monocytogenes* mother cell and bud

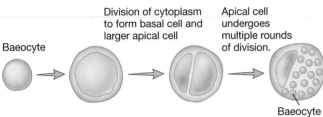

Baeocyte

Division of cytoplasm to form basal cell and larger apical cell

Apical cell undergoes multiple rounds of division.

Baeocyte

(b) Baeocyte formation

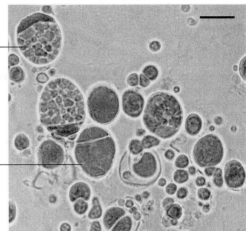

Cyanobacterium with one basal cell and an apical cell that has undergone multiple fission.

Cyanobacterium that has undergone one round of division, forming one basal cell and one large apical cell.

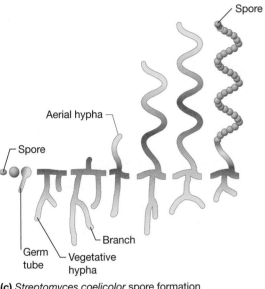

Spore

Aerial hypha

Spore

Germ tube

Branch

Vegetative hypha

(c) *Streptomyces coelicolor* spore formation

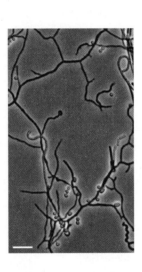

Figure 7.2 Some Bacteria Reproduce by Methods Other than Binary Fission. (a) Transmission electron micrograph (TEM) of a budding *Listeria monocytogenes* cell. (b) On the left is a drawing illustrating baeocyte formation. On the right is a photo of baeocytes produced by the cyanobacterium *Dermocarpella*. Bar = 10 μm. (c) Spore formation by *Streptomyces coelicolor*. The cells shown in the micrograph on the right are expressing a spore-formation related protein fused to the fluorescent molecule mCherry. Bar = 10 μm.

cell cycle consists of three phases: (1) a period of growth after the cell is born, which is similar to the G1 phase of the eukaryotic cell cycle; (2) chromosome replication and partitioning period, which functionally corresponds to the S and mitosis events of the M phase of the eukaryotic cycle; and (3) cytokinesis, during which a septum and daughter cells are formed (**figure 7.3**). Recall that in the eukaryotic cell cycle, the S phase is separated from the M phase by another period called G2. In G2, chromosome replication is completed and some time passes before chromosome segregation occurs. This is not the case for bacteria. As you will see in the discussion that follows, bacterial chromosome replication and partitioning occur concurrently. Furthermore, the initial events of cytokinesis actually occur before chromosome replication and partitioning are complete. Finally, some rapidly dividing bacteria are able to initiate new rounds of replication before the first round of replication and cytokinesis is finished.

Although chromosome replication and partitioning overlaps with cytokinesis, it is easiest to consider them separately.

Chromosome Replication and Partitioning

Most bacteria have a single circular chromosome. Each circular chromosome has a single site at which replication starts called the **origin of replication,** or simply the origin (figure 7.3). Replication is completed at the terminus, which is located directly opposite the origin. In a newly formed *E. coli* cell, the chromosome is compacted and organized so that the origin and terminus are in opposite halves of the cell. Early in the cell cycle, the origin and terminus move to midcell, and a group of proteins needed for chromosome replication assemble at the origin. This DNA synthesizing machinery is called the **replisome,** and DNA replication proceeds in both directions from the origin. As progeny

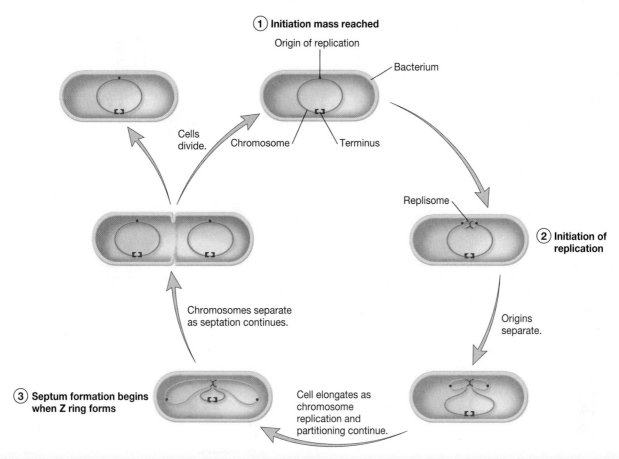

① **Initiation mass reached**

Origin of replication

Bacterium

Cells divide.

Chromosome

Terminus

Replisome

② **Initiation of replication**

Origins separate.

Chromosomes separate as septation continues.

③ **Septum formation begins when Z ring forms**

Cell elongates as chromosome replication and partitioning continue.

Figure 7.3 Cell Cycle of *E. coli.*   Initiation of the cell cycle results from an increase in the mass of a cell, which also results in accumulation of the DnaA protein, which initiates DNA replication. As the cell readies for DNA replication, the origin of replication migrates to the center of the cell and proteins that make up the replisome assemble. In this illustration, one round of DNA replication is completed before the cell divides. In rapidly growing cultures (generation times of about 20 minutes), second and third rounds of DNA replication are initiated before division of the original cell is completed. Thus the daughter cells inherit partially replicated DNA.

MICRO INQUIRY *Why is it important that the origin of replication migrate to the center of the cell prior to replication?*

chromosomes are synthesized, the two newly formed origins move toward opposite ends of the cell, and the rest of the chromosome follows in an orderly fashion.

Although the process of DNA synthesis and movement seems rather straightforward, the mechanism by which chromosomes are partitioned to each daughter cell has not been fully elucidated. Evidence suggests that the mechanism varies with bacterial species. Studies of *C. crescentus* have provided the most complete understanding of the partitioning process. This bacterium has an interesting life cycle in which a sessile stalked cell divides to give rise to a slightly smaller, flagellated swarmer cell; the flagellum forms at the pole opposite the stalk (*see* figure 22.10). ParA and ParB proteins work together to initiate partitioning. ParA is able to polymerize to form filaments, and ParB is a DNA-binding protein that binds a site on the chromosome called *parS*. This site is located near the origin of replication,

and ParB binds to each copy of the *parS* site on the two daughter chromosomes formed during replication (**figure 7.4**). One of the ParB/*parS* complexes remains at the stalk pole of the dividing cell. Interaction between a ParA filament and the other ParB/*parS* complex is thought to cause the ParA filament to depolymerize

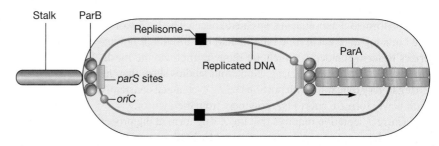

Stalk   ParB

Replisome

ParA

Replicated DNA

parS sites

oriC

Figure 7.4 The ParAB/*parS* Partitioning System of *Caulobacter crescentus.*   ParB binds each *parS* site on the two daughter chromosomes. ParA is a cytoskeletal protein that pulls one daughter chromosome to the pole opposite the stalk pole of the cell. Other proteins are involved in localizing ParA and ParB to their respective poles. These are not shown for simplicity.

and pull that daughter chromosome toward the opposite pole of the cell. Once the origin region of each daughter chromosome has been positioned at its pole, chromosome segregation continues as replication continues. It is thought that nucleoid-associated proteins may continue to pull the daughter chromosomes apart as they condense the chromosomes.

Despite the elegance of the *C. crescentus* ParAB/*parS* model, it is still unclear how widespread this chromosome segregation mechanism is among bacteria. Almost 70% of the about 400 bacterial and archaeal genomes studied have a recognizable *parS* site. *E. coli* is among the remaining 30%. Most of those having a *parS* site also have genes encoding ParA and ParB. Thus, the ParAB/*parS* system seems to be widespread among bacteria and is also present in at least a few archaea. Unfortunately, evidence for a role of ParA and ParB in chromosome segregation is lacking for many of those organisms encoding these proteins. Indeed, some studies of bacteria show that mutations in the genes encoding ParA or ParB do not significantly disrupt chromosome partitioning. Further studies are needed to discern other methods of chromosome partitioning among bacteria and archaea. ◄◄ *Bacterial cytoskeleton (section 3.6);* ►► Caulobacteraceae and Hyphomicrobiaceae *bacteria reproduce in unusual ways (section 22.1)* ⟳ *Binary Fission; Bidirectional DNA Replication*

Cytokinesis

Septation is the process of forming a cross wall between two daughter cells. **Cytokinesis,** a term that has traditionally been used to describe the formation of two eukaryotic daughter cells, is now used to describe this process in all cells. In bacteria, septation is divided into several steps: (1) selection of the site where the septum will be formed; (2) assembly of the Z ring, which is composed of the cytoskeletal protein FtsZ; (3) assembly of the cell wall–synthesizing machinery (i.e., for synthesis of peptidoglycan and other cell wall constituents); and (4) constriction of the cell and septum formation. Despite the widespread use of FtsZ in cytokinesis, there are bacteria that lack this protein. How these organisms complete cytokinesis is of considerable interest. ►► *Synthesis of peptidoglycan occurs in the cytoplasm, at the plasma membrane, and in the periplasmic space (section 12.4)*

Assembly of the Z ring is a critical step in septation, as it must be formed if subsequent steps are to occur. The FtsZ protein is a homologue of eukaryotic tubulin, and like tubulin, it polymerizes to form filaments, which are thought to create the meshwork that constitutes the Z ring. Numerous studies show that the Z ring is very dynamic, with portions being exchanged constantly with newly formed, short FtsZ polymers from the cytosol.

Correct septation requires that the Z ring form at the proper place at the proper time. In *E. coli,* the MinCDE system limits Z-ring formation to the center of the cell. Three proteins compose the system (MinC, MinD, and MinE). These proteins oscillate from one end of the cell to the other (**figure 7.5**). This oscillation creates high concentrations of MinC at the poles,

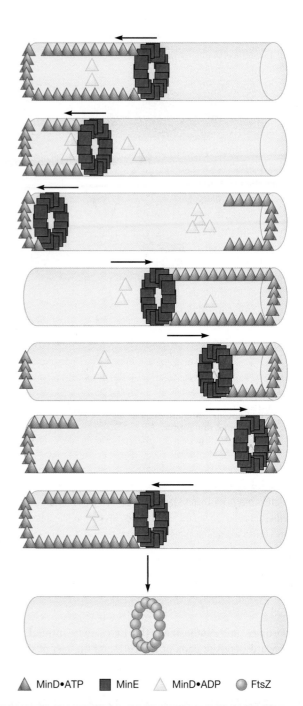

▲ MinD•ATP ■ MinE △ MinD•ADP ◯ FtsZ

Figure 7.5 MinCDE Proteins Help Establish the Site of Septum Formation. MinC (not shown) blocks septum formation. It oscillates with MinD. MinD's oscillations are controlled by MinE, which also follows MinD movements. As shown, MinD-ATP interacts with the plasma membrane and forms filaments. MinE binding to MinD-ATP causes hydrolysis of ATP, yielding MinD-ADP. MinD-ADP is released from the filament into the cytosol. There MinD-ADP is converted back to MinD-ATP, which then forms filaments at the opposite end of the cell. MinC and MinE follow and the process begins again. As a result of these oscillations, MinC concentrations are highest at the poles and septum formation is forced to occur at the center of the cell.

MICRO INQUIRY *What would be the outcome if FtsZ formed a Z ring that was not in the center of the cell? Consider both the morphology of the daughter cells and the partitioning of chromosomes.*

where it prevents formation of the Z ring; thus Z-ring formation can occur only at midcell, which lacks MinCDE. A mechanism called **nucleoid occlusion** helps ensure that the Z ring forms only after most of the daughter chromosomes have segregated from each other. This is important because the newly formed septum might otherwise guillotine the chromosome. It is currently thought that a protein called SlmA is critical to this process. SlmA binds to sites near the origin of replication. Recall that at the start of chromosome replication, the origin is located at midcell. As the daughter molecules move to the poles, SlmA moves with them. In vitro studies have shown that SlmA inhibits FtsZ polymerization. Thus, when SlmA is at midcell early in replication, the Z ring is prevented from forming. Once enough of the daughter chromosomes and SlmA molecules have moved away from midcell, the Z ring can form.

Once the Z ring forms, the rest of the division machinery, sometimes called the **divisome,** is constructed, as illustrated in **figure 7.6**. First, one or more anchoring proteins link the Z ring to the plasma membrane. Then the cell wall–synthesizing machinery is assembled (**table 7.1**). The final steps in division involve constriction of the Z ring, accompanied by invagination of the plasma membrane and synthesis of the septal wall.

The preceding discussion of the cell cycle describes what occurs in slowly growing *E. coli* cells. In these cells, the cell cycle takes approximately 60 minutes to complete. However, *E. coli* can reproduce at a much more rapid rate, completing the entire cell cycle in about 20 minutes, despite the fact that DNA replication always requires at least 40 minutes. *E. coli* accomplishes this by beginning a second round of DNA replication (and sometimes even a third or fourth round) before the first round of replication is completed. Thus the progeny cells receive

Table 7.1	Some *E. coli* Divisome Proteins and Their Functions
Divisome Protein	**Function**
FtsZ	Forms Z ring
FtsA, ZipA	Anchor Z ring to plasma membrane
FtsK	Chromosome segregation and separation of chromosome dimers
FtsQLB	May provide a scaffold for assembly of proteins involved in peptidoglycan synthesis
FtsI[1], FtsW	Peptidoglycan synthesis
FtsN	Thought to trigger constriction initiation

1 FtsI is also known as penicillin-binding protein 3 (PBP3).

a chromosome with two or more replication forks, and replication is continuous because the cells are always copying their DNA. *Bacterial Cell Cycle*

Cellular Growth and Determination of Cell Shape

As we have seen, bacterial and archaeal cells have defined shapes that are species specific. These shapes are neither accidental nor random, as demonstrated by the faithful propagation of shape from one generation to the next. In addition, some microbes change their shape under certain circumstances. For instance, *Sinorhizobium meliloti* switches from rod-shaped to Y-shaped cells when living symbiotically with plants. Likewise, *Helicobacter pylori,* the causative agent of gastric ulcers and stomach cancer, changes from its characteristic helical shape to a sphere in stomach infections and in prolonged culture.

To consider the shape of a cell, we must first review the function of the cell wall. The cell wall constrains the turgor pressure exerted by the cytoplasm, thereby preventing the cell from swelling and bursting. Turgor pressure is a term used to describe the force pushing against the cell wall as determined by the osmolarity of the cytoplasmic contents. The component of the bacterial cell wall that is responsible for protecting the cell from lysis is peptidoglycan. The problem faced by bacterial cells is that the strength of the existing peptidoglycan in the cell wall must be maintained as new peptidoglycan subunits are added. Thus understanding peptidoglycan synthesis is critical to understanding the determination of cell shape. ◀◀ *Peptidoglycan structure (section 3.4)*

Peptidoglycan synthesis involves many proteins, including a group of enzymes called **penicillin-binding proteins** (**PBPs**). They bear this name because they were first noted for their capacity to bind penicillin. While this property is important, their function is to link strands of peptidoglycan together or hydrolyze bonds in existing strands so that new units can be inserted

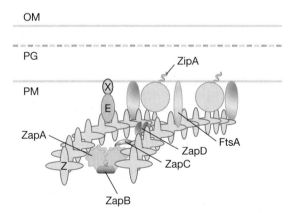

Figure 7.6 The *E. coli* Divisome. The cell division apparatus is composed of numerous proteins. The first step in divisome formation is the polymerization of FtsZ to form the Z ring. FtsA and ZipA proteins anchor the Z ring to the plasma membrane (PM), and then other proteins in the divisome assemble along the Z ring. FtsA and ZipA also stabilize the Z ring, along with ZapA, ZapB, ZapC, and ZapD. FtsE and FtsX help recruit and activate enzymes called amidases to the divisome. Amidases function to split the newly synthesized wall so that daughter cells can separate. Other divisome proteins and their functions are described in table 7.1. PG is peptidoglycan; OM is outer membrane.

1. **Peptidoglycan synthesis starts in the cytoplasm with the attachment of uridine diphosphate (UDP) to the sugar *N*-acetylglucosamine (NAG). Some of the UDP-NAG molecules are converted to UDP-NAM. Amino acid addition to NAM is not shown for simplicity.**

2. **NAM is transferred from UDP to bactoprenol, a carrier embedded in the plasma membrane. NAG is then attached to bactoprenol-NAM generating bactoprenol-NAM-NAG, called lipid II. The divisome protein FtsW (not shown) "flips" lipid II across the plasma membrane so that the NAM-NAG units are available for insertion into the sacculus.**

3. **Autolysins (cyan balls labeled "A") located at the divisome degrade bonds in the existing peptidoglycan sacculus. This permits the insertion of new NAM-NAG units into the sacculus.**

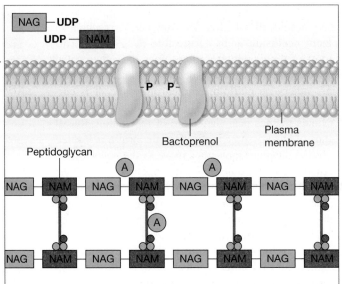

Figure 7.7 Components of the Peptidoglycan Synthesizing Machinery. See figure 12.9 for a detailed diagram of peptidoglycan synthesis.

during cell growth. The PBP enzymes that hydrolyze bonds of peptidoglycan are also called **autolysins;** other autolysins besides PBPs also help sculpt the peptidoglycan sacculus during cell growth and cell division. **Figure 7.7** illustrates some of the components of the peptidoglycan-synthesizing machinery and outlines a general scheme of peptidoglycan synthesis (*see figure 12.9* for a more detailed diagram). Synthesis of the NAG-NAM-pentapeptide building block is completed while it is attached to a lipid soluble carrier, bactoprenol, located in the plasma membrane. The carrier-bound building block is then flipped across the membrane by the divisome protein FtsW. Upon release of the NAG-NAM-pentapeptide into the periplasmic space, it is inserted into a peptidoglycan strand. The cellular location of autolysin activity and peptidoglycan export is not random and plays an important role in determining cell shape.

We begin our discussion with the simplest shape, the coccus. Although it has been stated that this is the "default" cellular shape, the growth of a spherical cell is more complicated than once thought. Studies of model cocci (e.g., *Enterococcus faecalis* and *Staphylococcus aureus*) show that new peptidoglycan forms only at the central septum (**figure 7.8a**). When daughter cells separate, each has one new and one old hemisphere. As in most bacteria, proper placement of the septum depends on FtsZ localization. In mutant cells without this tubulin homologue, peptidoglycan synthesis occurs in a random pattern around the cell, leading to bloated cells that lyse. Thus FtsZ placement determines the site of cell wall growth by recruiting PBPs and other enzymes needed for peptidoglycan synthesis to the divisome.

The peptidoglycan of rod-shaped cells is synthesized by two molecular machines that share many proteins but differ in terms of placement and time of function (figure 7.8b). The first

is responsible for elongation of the cell that occurs prior to septum formation. It is sometimes called the elongasome. The second is the divisome, which synthesizes peptidoglycan during cytokinesis.

During elongation, proteins in the actin homologue MreB family play an essential role. MreB proteins polymerize, creating patches of filaments along the cytoplasmic face of the plasma membrane. Although it is not clear how MreB controls elongation, much information has been gained by studying *Bacillus subtilis,* which has three MreB proteins—MreB, Mbl, and MreBH. During elongation, cell wall growth occurs in numerous bands around the circumference of the cell. The bands occur along the length of the cell but not at the poles. It has been suggested that MreB functions much like a scaffold inside the cytoplasm upon which the cell wall synthesizing machinery assembles. As the time for commencement of cytokinesis approaches, the FtsZ ring forms at the midcell, specifying peptidoglycan synthesis in this region. In some cells, MreB proteins and other elongasome proteins have been shown to redeploy from the sidewalls to the midcell, where they are thought to contribute to cell wall synthesis during cytokinesis. Thus, cell wall growth switches from the side wall to the septum at this time. Despite the uncertain mechanisms by which MreB proteins function, their importance in determining cell shape is demonstrated by two observations. First, rod-shaped cells in which MreB has been depleted assume a spherical shape. In addition, while almost all rod-shaped bacteria and archaea synthesize at least one MreB homologue, coccoid-shaped cells lack proteins in the MreB family.

The last cell shape we consider is that of comma-shaped cells, as studied in the aquatic bacterium *Caulobacter crescentus.* In addition to the actin homologue MreB and the tubulin-like

AI. Spherical cells build new peptidoglycan only at midcell, where the septum will form during division. This leads to daughter cells that have one old and one new cell wall hemisphere.

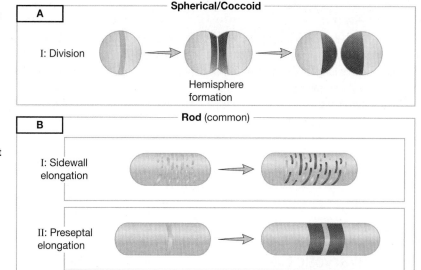

BI. During growth, prior to division, new cell wall is made along the side of the cell but not at the poles. This placement is thought to be determined by the position of MreB homologues.

BII. As division begins, FtsZ polymerization forms a Z ring and new cell wall growth is confined to the midcell.

BIII. Rod-shaped daughter cells are formed with one new pole and one old pole.

Figure 7.8 Cell Wall Biosynthesis and Determination of Cell Shape in Spherical and Rod-Shaped Cells.

MICRO INQUIRY *Which step in the development of rod-shaped cells is essential in determining cell morphology?*

protein FtsZ, these cells (and other vibrioid-shaped cells) produce a cytoskeletal protein called crescentin, a homologue of eukaryotic intermediate filaments (*see figure 3.30*). This protein localizes to one side of the cell, where it slows the insertion of new peptidoglycan units into the peptidoglycan sacculus (**figure 7.9**). The resulting asymmetric cell wall growth gives rise to the inner curvature that characterizes this cell shape. Interestingly, a metabolic enzyme called CTP synthase has recently been shown to regulate crescentin polymerization, and therefore, cell shape. The enzyme normally functions in synthesis of the nucleotide CTP. Why it would have the second function of regulating crescentin is unclear. ▶▶▏ Caulobacteraceae *and* Hyphomicrobiaceae *bacteria reproduce in unusual ways (section 22.1)*

It is clear from these examples that common microbial cell shapes require cytoskeletal proteins that bear startling similarity with those of eukaryotes. However, some bacterial shapes appear to be determined by other mechanisms. For example, the spirochete *Borrelia burgdorferi,* which causes Lyme disease, relies on its periplasmic flagella to confer its spiral shape. As another example, *Spiroplasma,* a citrus pathogen, has no cell wall or flagella. It relies on contractile cytoplasmic fibrils to give its spiral shape. ▶▶▏ *Phylum* Spirochaetes *(section 21.8); Class* Mollicutes, *Phylum* Tenericutes *(section 21.3)*

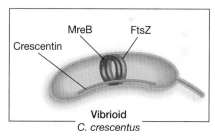

Figure 7.9 Vibrioid Shape Is Determined by Crescentin in *Caulobacter crescentus.* The intermediate filament homologue crescentin polymerizes along the length of the inner curvature of the cell. Cells depleted of crescentin are rod-shaped. Note that *C. crescentus* forms all three types of cytoskeleton protein homologues.

While the study of bacterial cytoskeletal elements and cell shape is inherently interesting, it also has practical aspects. Bacterial cytoskeletal homologues provide a tractable model for the study of more complicated eukaryotic cytoskeletal proteins and their assembly. In addition, these proteins may prove to be valuable targets for drug development. Indeed, molecules that block FtsZ function inhibit growth of the pathogen *Staphylococcus aureus* and might be effective treatments for staphylococcal diseases. ▶▶▏ *Antimicrobial chemotherapy (chapter 9)*

Retrieve, Infer, Apply

1. Describe the three phases of a bacterial cell cycle. The overlapping of cytokinesis and chromosome partitioning could potentially create problems for a cell during the cell cycle. What mechanisms does the cell use to prevent any problems?

2. How does the bacterial cell cycle compare with the eukaryotic cell cycle? List two ways they are similar and two ways they differ.

3. Do you think MinCDE functions in coccoid-shaped cells? Explain your answer.

4. Do you think *Spiroplasma* produces FtsZ? What about MreB? Explain your reasoning.

7.3 Some Archaeal Cell Cycles Resemble the Eukaryotic Cell Cycle

After reading this section, you should be able to:

- Compare and contrast the *Sulfolobus* spp. cell cycle and the typical eukaryotic cell cycle
- Compare and contrast the *Sulfolobus* spp. cell cycle and a bacterial cell cycle
- Identify the evidence suggesting that the *Sulfolobus* spp. cell cycle may not be universal among archaea

As is true with many other aspects of archaeal biology, understanding of the archaeal cell cycle lags behind that of bacterial cell cycles. However, studies of model archaea such as *Sulfolobus* spp. have yielded intriguing results. *Sulfolobus* spp. are members of the phylum *Crenarchaeota* and grow best in environments that are hot (80°C) and acidic (pH 3). Their cell cycle is reminiscent of a mitotic cell cycle. After a growth phase (G1), their DNA is replicated (S) using replicative machinery similar to that of eukaryotes. Of note is the fact that *Sulfolobus* spp. have three origins of replication. Furthermore, once replicated, the daughter chromosomes remain unsegregated for some time (G2); the *Sulfolobus* G2 phase consumes more than 50% of the cell cycle. G2 is followed by segregation of the chromosome and then cytokinesis occurs.

▶▶| *Archaea use homologues of eukaryotic replisome proteins (section 15.2)*

Advances have recently been made in elucidating how daughter chromosomes are segregated. Two proteins have been identified: SegA and SegB. SegA is somewhat similar to ParA proteins found in bacteria such as *Caulobacter crescentus* (p. 135). SegB is unique to archaea, but is thought to be the functional equivalent of bacterial ParB. The genes encoding the two proteins are located near one of the origins of replication, just as the genes for ParA and ParB are located near the bacterial origin. SegA polymerizes and is thought to exert the force, either pulling or pushing, that segregates the chromosomes. The role of SegB is unclear, although it is known to bind DNA and interact with SegA, enhancing SegA polymerization. However, *parS*-like sequences have not been clearly identified.

Unlike chromosome segregation, which appears to be bacteria-like, cytokinesis in *Sulfolobus* spp. is most similar to eukaryotic processes. These archaea undergo cytokinesis using proteins related to the ESCRT proteins of eukaryotes. ESCRT proteins are so named because of their roles in endocytosis: *e*ndosomal *s*orting *c*omplex *r*equired for *t*ransport. ESCRT proteins form four distinct complexes that are involved in formation of multivesicular bodies (MVB), a type of endosome (*see figure 5.9*). During MVB formation, some ESCRT proteins cause the endosome membrane to invaginate (**figure 7.10**). Then a complex of ESCRT proteins called ESCRT-III circles the neck formed by the invaginating membrane and brings about membrane scission, releasing a small vesicle into the developing MVB. Three division proteins have been identified in *Sulfolobus* spp.: CdvA, CdvB, and CdvC. The CdvB and CdvC proteins are homologues of ESCRT-III proteins. It is currently thought that CdvA first associates with the plasma membrane at the site where division will occur. It then recruits CdvB and CdvC proteins to the division site. Together the proteins form a ring, which eventually constricts, dividing the cell.

As exciting as these findings are, care must be taken not to assume that they apply to all archaea. For instance, some archaea do not have a G2-like phase in their cell cycle. It is thought that they either segregate their daughter chromosomes immediately following chromosome replication or that segregation and replication

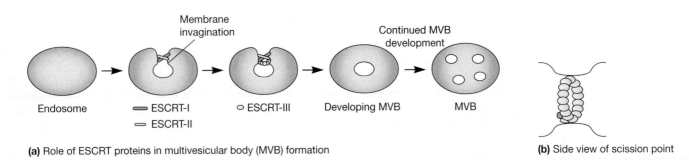

(a) Role of ESCRT proteins in multivesicular body (MVB) formation

(b) Side view of scission point

Figure 7.10 **Function of ESCRT Proteins in Eukaryotes.** (a) ESCRT proteins play a critical role in the maturation of an early endosome into a multivesicular body (MVB). ESCRT-III proteins create a scission point (b) during this process. Homologues of ESCRT-III proteins have been identified in *Sulfolobus* spp., where they function in cytokinesis.

occur at the same time, as in bacteria. In addition, the SegAB system seems to be limited to archaea belonging to the order *Sulfolobales* and a few members of the archaeal phylum *Euryarchaeota*. There is evidence that some archaea use homologues of the bacterial protein MinD (p. 136) to segregate chromosomes. Other evidence suggests that archaeal nucleoid-associated proteins are involved. Finally, many archaea have FtsZ proteins and appear to use a division process similar to that of bacteria. It is likely, then, that archaeal cell cycles are as diverse as bacterial cycles (perhaps even more diverse) and that even more interesting findings will arise in the coming years.

Retrieve, Infer, Apply

1. What elements of the *Sulfolobus* spp. cell cycle are similar to that of *Caulobacter crescentus*? What elements are similar to the eukaryotic cell cycle?
2. Many archaea have genes encoding an FtsZ homologue. Describe how FtsZ might function in an archaeal life cycle.

7.4 Environmental Factors Affect Microbial Growth

After reading this section, you should be able to:

- Use the terms that describe a microbe's growth range or requirement for each of the factors that influence microbial growth
- Summarize the adaptations of extremophiles to their natural habitats
- Summarize the strategies used by nonextremophiles to acclimate to changes in their environment
- Describe the enzymes observed in microbes that protect them against toxic O_2 products

Almost every year at Yellowstone National Park, a tourist is seriously burned or killed after falling into one of the park's hot springs. Yet, a rich microbial community lives in these same springs, as well as in the hot pots, fumaroles, and other thermal features of the park. Clearly, the adaptations of some microorganisms to inhospitable, **extreme environments** are truly remarkable. Indeed, microbes are thought to be present nearly everywhere on Earth. Bacteria such as *Bacillus infernus* are able to live over 2.4 km below Earth's surface, without oxygen and at temperatures above 60°C. Other microbes live at great ocean depths or in lakes such as the Great Salt Lake in Utah (USA) that have high sodium chloride concentrations. Microorganisms that grow in such harsh conditions are called **extremophiles.** To extremophiles, the harsh conditions of their habitats are "normal." Many other microbes are not extremophiles; they live in more moderate conditions. Whether an extremophile or not, all microbes must respond to changes in their environment. However, if the conditions exceed their ability to respond, they will not grow and eventually they may die. Thus all microbes have a characteristic range at which growth occurs for each environmental parameter. The range is defined by high and low

values beyond which the microbe cannot survive. Within the range is an optimal value at which growth is best. ▶▶◀ *Microorganisms in marine ecosystems (section 30.2); The subsurface biosphere is vast (section 31.4)*

To study the ecological distribution of microbes, it is important to understand the strategies they use to survive. An understanding of the environmental influences on microbes and their activity also aids in the control of microbial growth. In this section, we briefly review the effects of the most important environmental factors on microbial growth. Major emphasis is given to solutes and water activity, pH, temperature, oxygen level, pressure, and radiation. The adaptations that microbes have evolved in order to live in their natural environments also are discussed. **Table 7.2** summarizes how microorganisms are categorized in terms of their response to these factors.

Solutes Affect Osmosis and Water Activity

Water is critical to the survival of all organisms, but the behavior of water can also be destructive. Solutes in an aqueous solution alter the behavior of water. One way this occurs is the phenomenon of osmosis, which is observed when two solutions are separated by a semipermeable membrane that allows movement of water but not solutes. If the solute concentration of one solution is higher than the other, water moves to equalize the concentrations. In other words, water moves from solutions with lower solute concentrations to those with higher solute concentrations. Because a selectively permeable plasma membrane separates microorganisms from their environment, they can be affected by changes in the solute concentration of their surroundings. If a microorganism is placed in a hypotonic solution (one with a lower solute concentration; solute concentration is also referred to as osmotic concentration or osmolarity), water will enter the cell and cause it to burst unless something is done to prevent the influx of water or inhibit plasma membrane expansion. Conversely, if the microbe is placed in a hypertonic solution (one with a higher osmotic concentration), water will flow out of the cell. In microbes that have cell walls, the membrane shrinks away from the cell wall. Dehydration of the cell in hypertonic environments may damage the plasma membrane and cause the cell to become metabolically inactive.

Because of the potential damaging effects of uncontrolled osmosis, it is important that microbes be able to respond to changes in the solute concentrations of their environment. Microbes in hypotonic environments are protected in part by their cell wall, which prevents over-expansion of the plasma membrane. However, not all microbes have cell walls. Wall-less microbes can be protected by reducing the osmotic concentration of their cytoplasm; this protective measure is also used by many walled microbes to provide protection in addition to their cell walls. Microbes use several mechanisms to lower the solute concentration of their cytoplasm. For example, some bacteria have mechanosensitive (MS) channels in their plasma membrane. In a hypotonic environment, the membrane stretches due to an increase in hydrostatic pressure and cellular swelling. MS channels

Table 7.2	Microbial Responses to Environmental Factors	
Descriptive Term	**Definition**	**Representative Genera and Species**
Solute Concentration and Water Activity		
Halophile	Requires high levels of sodium chloride, usually above about 0.2 M, to grow	*Halobacterium, Dunaliella, Ectothiorhodospira*
Osmotolerant	Able to grow over wide ranges of water activity or osmotic concentration	*Staphylococcus aureus, Saccharomyces rouxii*
Xerophile	Organisms that grow best at low water activity, typically with optima at 0.85 or below	*Xeromyces bisporus*
pH		
Acidophile	Growth optimum between pH 0 and 5.5	*Sulfolobus, Picrophilus, Ferroplasma, Acontium*
Neutrophile	Growth optimum between pH 5.5 and 8.0	*Escherichia, Euglena, Paramecium*
Alkaliphile	Growth optimum between pH 8.0 and 11.5	*Bacillus alcalophilus, Natronobacterium*
Temperature		
Psychrophile	Grows at 0°C and has an optimum growth temperature of 15°C or lower	*Bacillus psychrophilus, Chlamydomonas nivalis*
Psychrotolerant	Can grow at 0–7°C; has an optimum between 20 and 30°C and a maximum around 35°C	*Listeria monocytogenes, Pseudomonas fluorescens*
Mesophile	Has growth optimum between 20 and 45°C	*Escherichia coli, Trichomonas vaginalis*
Thermophile	Can grow at 55°C or higher; optimum often between 55 and 65°C	*Geobacillus stearothermophilus, Thermus aquaticus, Cyanidium caldarium, Chaetomium thermophile*
Hyperthermophile	Has an optimum between 85 and about 113°C	*Sulfolobus, Pyrococcus, Pyrodictium*
Oxygen Concentration		
Obligate aerobe	Completely dependent on atmospheric O_2 for growth	*Micrococcus luteus,* most protists and fungi
Facultative anaerobe	Does not require O_2 for growth but grows better in its presence	*Escherichia, Enterococcus, Saccharomyces cerevisiae*
Aerotolerant anaerobe	Grows equally well in presence or absence of O_2	*Streptococcus pyogenes*
Obligate anaerobe	Does not tolerate O_2 and dies in its presence	*Clostridium, Bacteroides, Methanobacterium*
Microaerophile	Requires O_2 levels between 2–10% for growth and is damaged by atmospheric O_2 levels (20%)	*Campylobacter, Spirillum volutans, Treponema pallidum*
Pressure		
Piezophile (barophile)	Growth more rapid at high hydrostatic pressures	*Photobacterium profundum, Shewanella benthica*

then open and allow solutes to leave. Thus MS channels act as escape valves to protect cells from bursting. Many protists use contractile vacuoles to expel excess water.

Some microbes are adapted to extreme hypertonic environments, and can be called osmophiles. By definition, osmophiles should include **halophiles,** which require the presence of NaCl at a concentration above about 0.2 M (**figure 7.11**). However, usually **osmophile** is used to refer to organisms that require high concentrations of sugars. Here we focus on halophiles. Extreme halophiles have adapted so completely

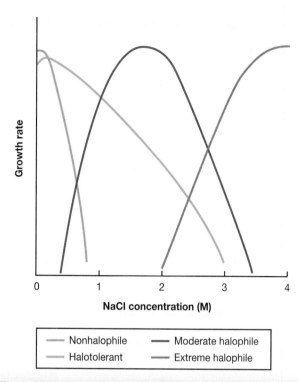

Figure 7.11 The Effects of Sodium Chloride on Microbial Growth. Four different patterns of microbial dependence on NaCl concentration are depicted. The curves are only illustrative and are not meant to provide precise shapes or salt concentrations required for growth.

MICRO INQUIRY *What is the difference between halophilic and halotolerant?*

solutes sucrose and polyols (e.g., arabitol, glycerol, and mannitol) for this purpose.

Some of the best-studied halophiles belong to the archaeal genus *Halobacterium*. These extremely halophilic archaea accumulate enormous quantities of potassium and chloride ions in order to remain hypertonic to their environment; the internal potassium concentration may reach 4 to 7 M. To do so, their enzymes, ribosomes, and transport proteins require high potassium levels for stability and activity. In addition, their plasma membranes and cell walls are stabilized by high concentrations of sodium ion. If the sodium concentration decreases too much, the wall and plasma membrane disintegrate. Such extreme halophiles have successfully adapted to environmental conditions that would destroy most organisms. However, in the process, they have become so specialized that they have lost ecological flexibility and can prosper only in a few extreme habitats. ▶▶| *Haloarchaea (section 20.4)*

Another way solutes change the behavior of water is by decreasing the availability of water to microbes. When solutes such as salts and sugars are present, the water is "tied up" by its interaction with the solutes. Microbiologists express quantitatively the degree of water availability by determining **water activity** (a_w). The water activity of a solution is 1/100 the relative humidity of the solution (when expressed as a percent). It is also equivalent to the ratio of the solution's vapor pressure (P_{soln}) to that of pure water (P_{water}). Distilled water has an a_w of 1, milk has an a_w of 0.97, a saturated salt solution has an a_w of 0.75, and the a_w of dried fruits is only about 0.5.

To survive in a habitat with a low a_w value microorganisms must maintain a high internal solute concentration to retain water. Some microorganisms can do this and are **osmotolerant;** they grow over wide ranges of water activity but optimally at higher levels. Osmotolerant organisms can be found in all domains of life. For example, *Staphylococcus aureus* is halotolerant, can be cultured in media containing sodium chloride concentration up to about 3 M, and is well adapted for growth on the skin. The yeast *Saccharomyces rouxii* grows in sugar solutions with a_w values as low as about 0.65. The photosynthetic protist *Dunaliella viridis* tolerates sodium chloride concentrations from 1.7 M to a saturated solution. In contrast to osmotolerant microbes, **xerophiles** grow best at low a_w. However, most microorganisms only grow well at water activities around 0.98 (the approximate a_w for seawater) or higher. This is why drying food or adding large quantities of salt and sugar effectively prevents food spoilage. ▶▶| *Various methods are used to control food spoilage (section 41.2)*

to hypertonic, saline conditions that they require NaCl concentrations between about 3 M and saturation (about 6.2 M). Members of the archaeal genus *Halobacterium* can be isolated from the Dead Sea (a salt lake between Israel and Jordan), the Great Salt Lake in Utah, and other aquatic habitats with salt concentrations approaching saturation.

Halophiles generally are able to live in their high-salt habitats because they synthesize or obtain from their environment molecules called **compatible solutes.** Compatible solutes can be kept at high intracellular concentrations without interfering with metabolism and growth. Some compatible solutes are inorganic molecules such as potassium chloride (KCl). Others are organic molecules such as choline, betaines (neutral molecules having both negatively charged and positively charged functional groups), and amino acids such as proline and glutamic acid. The use of compatible solutes extends beyond halophiles. This is also a strategy utilized by many osmophiles. Furthermore, many microorganisms, whether in hypotonic or hypertonic environments, use compatible solutes to keep the osmotic concentration of their cytoplasm somewhat above that of the habitat so that the plasma membrane is always pressed firmly against the cell wall. For instance, fungi and photosynthetic protists employ the compatible

Retrieve, Infer, Apply

1. How do microorganisms adapt to hypotonic and hypertonic environments?

2. Define water activity and briefly describe how it can be determined. Why is it difficult for microorganisms to grow at low a_w values?

3. What are halophiles and why do *Halobacterium* spp. require sodium and potassium ions?

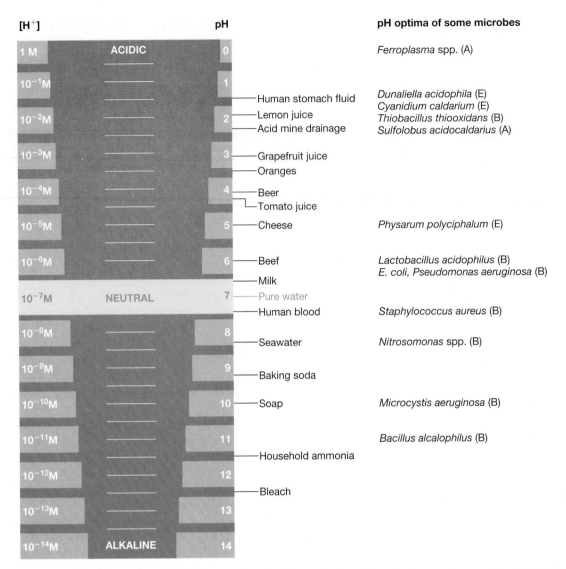

Figure 7.12 The pH Scale. The pH scale and examples of substances with different pH values. Several microorganisms are placed at their growth optima. A is archaeon; E, eukaryote; and B, bacterium.

pH

pH is a measure of the relative acidity of a solution and is defined as the negative logarithm of the hydrogen ion concentration (expressed in terms of molarity).

$$pH = -\log[H^+] = \log(1/[H^+])$$

The pH scale extends from pH 0.0 (1.0 M H^+) to pH 14.0 (1.0×10^{-14} M H^+), and each pH unit represents a tenfold change in hydrogen ion concentration. **Figure 7.12** shows that microbial habitats vary widely in pH—from pH 0 to 2 at the acidic end to alkaline lakes and soil with pH values between 9 and 10.

Each species has a definite pH growth range and pH growth optimum. **Acidophiles** have their growth optimum between pH 0 and 5.5; **neutrophiles,** between pH 5.5 and 8.0; and **alkaliphiles (alkalophiles)**, between pH 8.0 and 11.5. In general, different microbial groups have characteristic pH preferences. Most

known bacteria and protists are neutrophiles. Most fungi prefer more acidic surroundings, about pH 4 to 6; photosynthetic protists also seem to favor slight acidity. Many archaea are acidophiles. For example, the archaeon *Sulfolobus acidocaldarius* is a common inhabitant of acidic hot springs; it grows well from pH 1 to 3 and at high temperatures. Thus it is more accurately described as an acidophilic thermophile. The archaea *Ferroplasma acidarmanus* and *Picrophilus oshimae* can actually grow very close to pH 0. Alkaliphiles are distributed among all three domains of life. They include bacteria belonging to the genera *Bacillus, Micrococcus, Pseudomonas,* and *Streptomyces;* yeasts and filamentous fungi; and numerous archaea. Because seawater has a pH of about 8.3, marine microorganisms are alkaliphilic.

Although microorganisms often grow over wide ranges of pH and far from their optima, there are limits to their tolerance. When the external pH is low, the concentration of H^+ is much greater outside than inside, and H^+ will move into the cytoplasm and lower the cytoplasmic pH. Drastic variations in cytoplasmic pH can harm microorganisms by disrupting the plasma membrane or inhibiting the activity of enzymes and membrane transport proteins. Most microbes die if the internal pH drops much below 5.0 to 5.5. Changes in the external pH also can alter the ionization of nutrient molecules and thus reduce their availability to the organism.

Microorganisms respond to external pH changes using mechanisms that maintain a neutral cytoplasmic pH. Several responses to small changes in external pH have been identified. Neutrophiles appear to exchange potassium for protons using an antiport transport system. Internal buffering also contributes to pH homeostasis. However, if the external pH becomes too acidic, other mechanisms come into play. When the pH drops below about 5.5, *Salmonella enterica* serovar Typhimurium and *E. coli* synthesize an array of new proteins as part of what has been called their acidic tolerance response. An ATPase enzyme contributes to this protective response by pumping protons out of the cell, at the expense of ATP. If the external pH decreases to 4.5 or lower, acid shock proteins and heat shock proteins are synthesized. These prevent the denaturation of proteins and aid in refolding denatured proteins in acidic conditions. ◄◄ *Primary and secondary active transport (section 3.3)* ►► *Molecular chaperones (section 13.8)*

What about microbes that live at pH extremes? Extreme alkaliphiles such as *Bacillus alcalophilus* maintain their internal pH close to neutrality by exchanging internal sodium ions for external protons. Acidophiles use a variety of measures to maintain a neutral internal pH. These include the transport of cations (e.g., potassium ions) into the cell, thus decreasing the movement of H^+ into the cell; proton transporters that pump H^+ out if they get in; and highly impermeable cell membranes.

Retrieve, Infer, Apply

1. Define pH, acidophile, neutrophile, and alkaliphile.
2. Classify each of the following organisms as an alkaliphile, a neutrophile, or an acidophile: *Staphylococcus aureus, Microcystis aeruginosa, Sulfolobus acidocaldarius,* and *Pseudomonas aeruginosa* (see table 7.2). Which might be pathogens? Explain your choices.
3. Describe the mechanisms microbes use to maintain an internal neutral pH. Explain how extreme pH values might harm microbes.

Temperature

Microorganisms are particularly susceptible to external temperatures because they cannot regulate their internal temperature. An important factor influencing the effect of temperature on growth is the temperature sensitivity of enzyme-catalyzed reactions. Each enzyme has a temperature at which it functions optimally. At some temperature below the optimum, it ceases to be catalytic. As the temperature rises from this low point, the rate of catalysis increases to that observed for the optimal temperature. The velocity of the reaction roughly doubles for every 10°C rise in temperature. When all enzymes in a microbe are considered together, as the rate of each reaction increases, metabolism as a

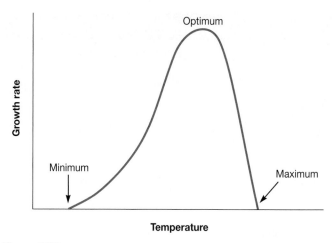

Figure 7.13 Temperature and Growth. The effect of temperature on growth rate.

whole becomes more active and the microorganism grows faster. However, beyond a certain point, further temperature increases actually slow growth, and sufficiently high temperatures are lethal. High temperatures denature enzymes, as well as transport and other proteins. Temperature also has a significant effect on microbial membranes. At very low temperatures, membranes solidify. At high temperatures, the lipid bilayer simply melts and disintegrates. Thus when organisms are above or below their optimum temperature, both function and cell structure are affected.

Because of these opposing temperature influences, microbial growth has a characteristic temperature dependence with distinct **cardinal temperatures**—minimum, optimum, and maximum growth temperatures (**figure 7.13**). Although the shape of temperature dependence curves varies, the temperature optimum is always closer to the maximum than to the minimum. The cardinal temperatures are not rigidly fixed. Instead, they depend to some extent on other environmental factors such as pH and available nutrients. For example, *Crithidia fasciculata*, a flagellated protist living in the gut of mosquitoes, grows in a simple medium at 22 to 27°C. However, if its medium is supplemented with extra metals, amino acids, vitamins, and lipids, it can grow at 33 to 34°C.

The cardinal temperatures vary greatly among microorganisms (**table 7.3**). Optima usually range from 0°C to 75°C, whereas microbial growth occurs at temperatures extending from less than −20°C to over 120°C. Some archaea even grow at 121°C (250°F), the temperature normally used in autoclaves. A major factor determining growth range is the availability of water. Even at the most extreme temperatures, microorganisms need liquid water to grow. The growth temperature range for a particular microorganism usually spans about 30 degrees. Some species (e.g., *Neisseria gonorrhoeae*) have a small range; others grow over a wide range of temperatures. The major microbial groups differ from one another regarding their maximum growth temperatures. The upper limit for protists is around 50°C. Some fungi grow at temperatures as high as 55 to 60°C. Bacteria and

Table 7.3	Temperature Ranges for Microbial Growth		
	CARDINAL TEMPERATURES (°C)		
Microorganism	**Minimum**	**Optimum**	**Maximum**
Bacteria and Archaea			
Planococcus halocryophilus	−15	25	37
Bacillus psychrophilum	0–3	25	30
Pseudomonas fluorescens	4	25–30	40
Escherichia coli	10	37	45
Neisseria gonorrhoeae	30	35–36	38
Thermus aquaticus	40	70–72	79
Pyrolobus fumarii	90	106	113
Protists			
Chlamydomonas nivalis	−36	0	4
Amoeba proteus	4–6	22	35
Trichomonas vaginalis	25	32–39	42
Cyclidium citrullis	18	43	47
Fungi			
Candida scotti	0	4–15	15
Saccharomyces cerevisiae	1–3	28	40
Mucor pusillus	21–23	45–50	50–58

have adapted to their environment in several ways. Their enzymes, transport systems, and protein synthetic machinery function well at low temperatures. The membranes of psychrophilic microorganisms have high levels of unsaturated fatty acids and remain semifluid when cold. Indeed, many psychrophiles begin to leak cellular constituents at temperatures higher than 20°C because of membrane disruption. Many psychrophiles accumulate compatible solutes. Rather than protecting against osmotic stress, in this case the compatible solutes decrease the freezing point of the cytosol. Still other psychrophiles use antifreeze proteins to decrease the freezing point of the cytosol. Psychrophilic bacteria and fungi are major causes of refrigerated food spoilage.

Mesophiles are microorganisms that grow in moderate temperatures. They have growth optima around 20 to 45°C and often have a temperature minimum of 15 to 20°C and a maximum of about 45°C. Almost all human pathogens are mesophiles, as might be expected because the human body is a fairly constant 37°C.

Microbes that grow best at high temperatures are thermophiles and hyperthermophiles. **Thermophiles** grow at temperatures between 45 and 85°C, and they often have optima between 55 and 65°C. The vast majority are members of *Bacteria* or *Archaea*, although a few photosynthetic protists and fungi are thermophilic. Thermophiles flourish in many habitats including composts, self-heating hay stacks, hot water lines, and hot springs. **Hyperthermophiles** have growth optima between 85°C and about 113°C. They usually do not grow below 55°C. *Pyrococcus abyssi* and *Pyrodictium occultum* are examples of marine hyperthermophiles found in hot areas of the seafloor.

archaea can grow at much higher temperatures than eukaryotes. It has been suggested that eukaryotes are not able to manufacture stable and functional organellar membranes at temperatures above 60°C.

Several terms are used to describe microbes based on their temperature ranges for growth (**figure 7.14**). Microbes that grow in cold environments are either psychrotolerants or psychrophiles. **Psychrotolerants** (sometimes called **psychrotrophs**) grow at 0°C or higher and typically have maxima at about 35°C. **Psychrophiles** (sometimes called **cryophiles**) grow well at 0°C and have an optimum growth temperature of 15°C; the maximum is around 20°C. They are readily isolated from Arctic and Antarctic habitats. Oceans constitute an enormous habitat for psychrophiles because 90% of ocean water is 5°C or colder. The psychrophilic protist *Chlamydomonas nivalis* can actually turn a snowfield or glacier pink with its bright red spores. Psychrophiles are widespread among bacterial taxa and are found in such genera as *Pseudomonas, Vibrio, Alcaligenes, Bacillus, Photobacterium,* and *Shewanella*. Psychrophilic microorganisms

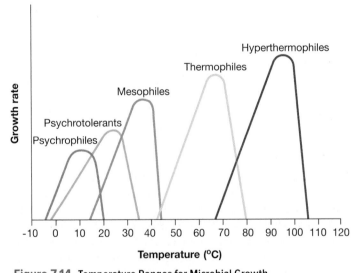

Figure 7.14 Temperature Ranges for Microbial Growth.
Microorganisms are placed in different classes based on their temperature ranges for growth. They are ranked in order of increasing growth temperature range as psychrophiles, psychrotolerants, mesophiles, thermophiles, and hyperthermophiles. Representative ranges and optima for these five types are illustrated.

Thermophiles and hyperthermophiles differ from mesophiles in many ways. They have heat-stable enzymes and protein synthesis systems that function properly at high temperatures. These proteins are stable for a variety of reasons. Heat-stable proteins have highly organized hydrophobic interiors and more hydrogen and other noncovalent bonds to stabilize their structure. Larger quantities of amino acids such as proline also make polypeptide chains less flexible and more heat stable. In addition, the proteins are stabilized and aided in folding by proteins called chaperones. Nucleoid-associated proteins appear to stabilize the DNA of thermophilic bacteria. Hyperthermophiles have an enzyme called reverse DNA gyrase that changes the topology of their DNA and enhances its stability. The membrane lipids of thermophiles and hyperthermophiles are also quite temperature stable. They tend to be more saturated, more branched, and of higher molecular weight. This increases the melting points of membrane lipids. Recall that archaea have membrane lipids with ether linkages. Such lipids are resistant to hydrolysis at high temperatures. Furthermore, the diglycerol tetraethers observed in the membranes of some archaeal thermophiles span the membrane to form a rigid, stable monolayer. ◀◀ *Six major types of archaeal cell envelopes have been identified (section 4.2)* ▶▶ *Proteins (appendix I)*

Retrieve, Infer, Apply

1. What are cardinal temperatures?
2. Why does the growth rate rise with increasing temperature and then fall again at higher temperatures?
3. Define psychrophile, psychrotroph, mesophile, thermophile, and hyperthermophile.
4. What metabolic and structural adaptations for extreme temperatures do psychrophiles and thermophiles have? How do these provide protection?

Oxygen Concentration

The importance of oxygen to the growth of an organism correlates with the processes it uses to conserve energy. Almost all energy-conserving metabolic processes involve the movement of electrons through a series of membrane-associated electron carriers called an electron transport chain (ETC). For chemotrophs, organisms that use chemical energy sources, an externally supplied terminal electron acceptor is critical to the functioning of the ETC. In many cases, the terminal electron acceptor is oxygen. ▶▶ *Electron transport chains (section 10.4)*

Traditionally, five types of relationships to oxygen have been described. Almost all multicellular organisms are completely dependent on atmospheric O_2 for growth; that is, they are **obligate aerobes** (table 7.2). Oxygen serves as the terminal electron acceptor for the ETC in the metabolic process called aerobic respiration. **Microaerophiles** are damaged by the atmospheric level of O_2 (20%) and require O_2 levels in the range of 2 to 10% for growth. **Facultative anaerobes** do not require O_2 for growth but grow better in its presence. In the presence of oxygen, they

use O_2 as the terminal electron acceptor during aerobic respiration. **Aerotolerant anaerobes** grow equally well whether O_2 is present or not; they can tolerate O_2, but they do not make use of it. Many have strictly fermentative metabolism and thus do not use O_2 in their energy-conserving processes. The ability to grow in both oxic and anoxic environments provides considerable flexibility and is an ecological advantage for facultative and aerotolerant anaerobes. In contrast, for strict or **obligate anaerobes,** O_2 is toxic, and they are usually killed by prolonged exposure to O_2. Strict anaerobes cannot generate energy through aerobic respiration and employ other metabolic strategies such as fermentation or anaerobic respiration, neither of which requires O_2. Although obligate anaerobes are killed by O_2, they may be recovered from habitats that appear to be oxic. In such cases, they associate with facultative anaerobes that use up the available O_2 and thus make the growth of strict anaerobes possible. For example, the strict anaerobe *Porphyromonas gingivalis* lives in the mouth, where it grows in the anoxic crevices around the teeth. ▶▶ *Aerobic respiration can be divided into three steps (section 11.3); Anaerobic respiration uses the same three steps as aerobic respiration (section 11.7); Fermentation does not involve an electron transport chain (section 11.8)*

The nature of bacterial O_2 responses can be readily determined by growing the microbe in a solid culture medium or a medium such as thioglycollate broth, which contains a reducing agent to lower O_2 levels (**figure 7.15**). Using these and other methods it is known that a microbial group may show more than one type of relationship to O_2. All five types are found among bacteria, archaea, and protists. Most fungi are aerobic, but a number of species—particularly among yeasts—are facultative anaerobes. Photosynthetic protists are usually obligate aerobes.

The different relationships with O_2 are due to several factors, including the inactivation of proteins and the effect of toxic O_2 derivatives. Enzymes can be inactivated when sensitive groups such as sulfhydryls are oxidized. However, even molecules that have evolved to function aerobically can be damaged by O_2. This is because the unpaired electrons in the outer shell of oxygen make it inherently unstable (*see figure AI.2*). Toxic O_2 derivatives are formed when cellular proteins such as flavoproteins transfer electrons to O_2. These toxic O_2 derivatives are called **reactive oxygen species** (**ROS**), and they can damage proteins, lipids, and nucleic acids. ROS include the superoxide radical, hydrogen peroxide, and the most dangerous hydroxyl radical.

$$O_2 + e^- \rightarrow O_2\bar{\bullet} \text{ (superoxide radical)}$$

$$O_2\bar{\bullet} + e^- + 2H^+ \rightarrow H_2O_2 \text{ (hydrogen peroxide)}$$

$$H_2O_2 + e^- + H^+ \rightarrow H_2O + OH\bullet \text{ (hydroxyl radical)}$$

A microorganism must be able to protect itself against ROS or it will be killed. Indeed, neutrophils and macrophages, two important immune system cells, use ROS to destroy invading pathogens. ▶▶ *Redox reactions (section 10.3); Phagocytosis (section 33.5)*

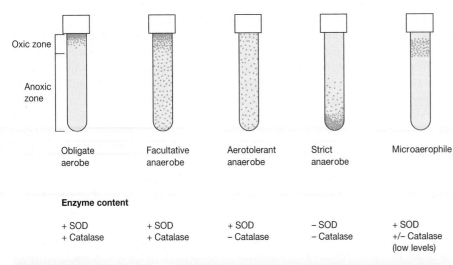

Enzyme content

Obligate aerobe	Facultative anaerobe	Aerotolerant anaerobe	Strict anaerobe	Microaerophile
+ SOD	+ SOD	+ SOD	– SOD	+ SOD
+ Catalase	+ Catalase	– Catalase	– Catalase	+/– Catalase (low levels)

Figure 7.15 Oxygen and Bacterial Growth. Each dot represents an individual bacterial colony within the agar or on its surface. The surface, which is directly exposed to atmospheric oxygen, is oxic. The oxygen content of the medium decreases with depth until the medium becomes anoxic toward the bottom of the tube. The presence and absence of the enzymes superoxide dismutase (SOD) and catalase for each type are shown.

MICRO INQUIRY *Why do facultative anaerobes grow best at the surface of the tube while aerotolerant anaerobes demonstrate a uniform growth pattern throughout the tube?*

Many microorganisms possess enzymes that protect against toxic O_2 products (figure 7.15). Obligate aerobes and facultative anaerobes usually contain the enzymes **superoxide dismutase (SOD)** and **catalase,** which catalyze the destruction of superoxide radical and hydrogen peroxide, respectively. Peroxidase also can be used to destroy hydrogen peroxide.

$$2O_2^{\overline{\bullet}} + 2H^+ \xrightarrow{\text{superoxide dismutase}} O_2 + H_2O_2$$

$$2H_2O_2 \xrightarrow{\text{catalase}} 2H_2O + O_2$$

$$H_2O_2 + NADH + H^+ \xrightarrow{\text{peroxidase}} 2H_2O + NAD^+$$

Strict anaerobes lack these enzymes or have them in very low concentrations and therefore cannot tolerate O_2.

Retrieve, Infer, Apply

1. Describe the five types of O_2 relationships seen in microorganisms.
2. What are the toxic effects of O_2? How do aerobes and other oxygen-tolerant microbes protect themselves from these effects?

Pressure

Organisms that spend their lives on land or the surface of water are always subjected to a pressure of 1 atmosphere (atm; 1 atm is ~0.1 megapascal, or MPa for short) and are never affected significantly by pressure. Other organisms, including many bacteria and archaea, live in the deep sea (ocean depths of 1,000 m or more), where the hydrostatic pressure can reach 600 to 1,100 atm and the temperature is about 2 to 3°C. These high hydrostatic pressures affect membrane fluidity and membrane-associated function. How do microbes survive under such conditions?

Many microbes found at great ocean depths are **barotolerant:** increased pressure adversely affects them but not as much as it does nontolerant microbes. Some are truly **piezophilic** (**barophilic**). A piezophile is defined as an organism that has a maximal growth rate at pressures greater than 1 atm. For instance, a piezophile recovered from the Mariana trench near the Philippines (depth about 10,500 m) grows only at pressures between about 400 to 500 atm when incubated at 2°C.

An important adaptation observed in piezophiles is that they change their membrane lipids in response to increasing pressure. For instance, bacterial piezophiles increase the amount of unsaturated fatty acids in their membrane lipids as pressure increases. They may also shorten the length of their fatty acids. Piezophiles are thought to play important roles in nutrient cycling in the deep sea. Thus far, they have been found among several bacterial genera (e.g., *Photobacterium, Shewanella, Colwellia*). ▶▶ *Microbial communities in benthic marine sediments are enormous and mysterious (section 30.2)*

Radiation

Our world is bombarded with electromagnetic radiation of various types (**figure 7.16**). Radiation behaves as if it were composed of waves like those traveling on the surface of water. The distance between two wave crests or troughs is the wavelength. As the wavelength of electromagnetic radiation decreases, the energy of the radiation increases; gamma rays and X rays are much more energetic than visible light or infrared waves. Electromagnetic radiation also acts like a stream of energy packets called photons, each photon having a quantum of energy whose value depends on the wavelength of the radiation.

Sunlight is the major source of radiation on Earth. It includes visible light, ultraviolet (UV) radiation, infrared rays, and radio waves. Visible light is a most conspicuous and important aspect of our environment because most life on Earth depends on the ability of photosynthetic organisms to trap the energy it holds. The visible spectrum consists of the ROYGBV wavelengths most of us memorized for early science classes. Sunlight is evenly distributed across these wavelengths and therefore appears white. Despite the importance of visible light, it is not the

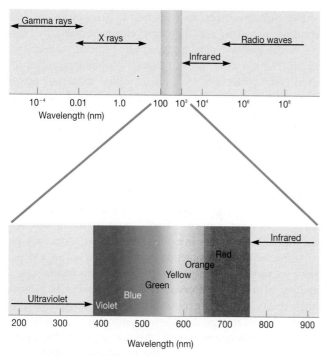

Figure 7.16 The Electromagnetic Spectrum. A portion of the spectrum showing the visible light range and the nearby ultraviolet and infrared wavelengths is expanded at the bottom of the figure.

predominant component of sunlight. Infrared rays hold that honor, making up almost 60% of the sun's radiation. Infrared radiation is the major source of Earth's heat. Only about 3% of the light reaching Earth's surface is UV radiation, and this includes only certain types of UV light. There are three major types of UV radiation: UVA, UVB, and UVC, which range from longest (UVA) to shortest (UVC) wavelengths. At sea level, there is very little UV radiation at wavelengths below about 290 nm (UVC and some UVB). This is because wavelengths shorter than this are absorbed by O_2 in Earth's atmosphere; this process forms a layer of ozone (O_3) between 40 and 48 km above Earth's surface. The ozone layer absorbs somewhat longer UV rays and reforms O_2. Thus, the major form of UV radiation reaching Earth's surface is UVA radiation. ▶▶| *Phototrophy (section 11.11)*

Many forms of electromagnetic radiation are very harmful to microorganisms. One of the most damaging is **ionizing radiation,** radiation of very short wavelength and high energy, which causes atoms to lose electrons (ionize). Two major forms of ionizing radiation are X rays, which are artificially produced, and gamma rays, which are emitted during natural radioisotope decay. Low levels of ionizing radiation may produce mutations that indirectly result in death, whereas higher levels are directly lethal. Ionizing radiation causes a variety of changes in cells. It breaks hydrogen bonds, oxidizes double bonds, destroys ring structures, and polymerizes some molecules.

Although microorganisms are more resistant to ionizing radiation than larger organisms, they are still destroyed by a

sufficiently large dose. Indeed, ionizing radiation can be used to sterilize items. However, bacterial endospores and bacteria such as *Deinococcus radiodurans* are extremely resistant to large doses of ionizing radiation. *D. radiodurans* is of particular note. This amazing microbe is able to piece together its genome after it is blasted apart by massive doses of radiation. How it does this is a matter of intense interest to microbiologists. ▶▶| *Radiation (section 8.4); Deinococcus-Thermus includes radiation-resistant bacteria (section 21.2)*

Ultraviolet (UV) radiation is another very damaging form of radiation. It can kill microorganisms due to its short wavelength (approximately from 10 to 400 nm) and high energy. The most lethal UV radiation has a wavelength of 260 nm, the wavelength most effectively absorbed by and damaging to DNA. The damage caused by UV light can be repaired by several DNA repair mechanisms, as we discuss in chapter 16. However, excessive exposure to UV light outstrips the organism's ability to repair the damage and death results. Longer wavelengths of UV light (near-UV radiation; 325 to 400 nm) can also harm microorganisms because they induce the breakdown of the amino acid tryptophan to toxic photoproducts. These toxic photoproducts plus the near-UV radiation itself produce breaks in DNA strands. ▶▶| *Mutations (section 16.1)*

Even visible light, when present in sufficient intensity, can damage or kill microbial cells. Usually pigments called photosensitizers and O_2 are involved. Photosensitizers include pigments such as chlorophyll, bacteriochlorophyll, cytochromes, and flavins, which can absorb light energy and become excited or activated. The excited photosensitizer (P) transfers its energy to O_2, generating singlet oxygen ($^1O_2$).

$$P \xrightarrow{\text{light}} P(\text{activated})$$
$$P(\text{activated}) + O_2 \rightarrow P + {}^1O_2$$

Singlet oxygen is a very reactive, powerful oxidizing agent that quickly destroys a cell.

Many microorganisms that are airborne or live on exposed surfaces use carotenoid pigments (*see figure 11.30*) for protection against photooxidation. Carotenoids effectively quench singlet oxygen; that is, they absorb energy from singlet oxygen and convert it back into the unexcited ground state. Both phototrophic and nonphototrophic microorganisms employ pigments in this way.

Retrieve, Infer, Apply

1. Where would you expect to find barotolerant and piezophilic bacteria? Explain your answer.

2. List the types of electromagnetic radiation in the order of decreasing energy or increasing wavelength.

3. What is the importance of ozone formation?

4. How do ionizing radiation, ultraviolet radiation, and visible light harm microorganisms? How do microorganisms protect themselves against damage from UV and visible light?

7.5 Microbial Growth in Natural Environments

After reading this section, you should be able to:

- Discuss the mechanisms used by microbes to survive starvation
- Distinguish sessile and planktonic microbial life styles
- Describe the formation of biofilms and summarize their importance in natural environments, industrial settings, and medicine
- Define quorum sensing and provide examples of cellular processes regulated by quorum sensing
- Discuss in general terms the communication that occurs between rhizobia and their plant hosts

Natural microbial environments—that is, those that are not created by humans—are complex and constantly changing. They expose microbes to many overlapping gradients of nutrients and other environmental factors. These include the environmental parameters we described in section 7.4, as well as inhibitory substances that limit microbial growth. Furthermore, microbes do not exist alone. Microbial habitats contain both micro- and macroorganisms. Microbes often associate with each other, sometimes forming biofilms, and they frequently form relationships with plants and animals. In this section, we explore three aspects of life in nature: the scarcity of nutrients, biofilms, and the mechanisms used by microbes to communicate among and between species.

Many Microbes Live in Oligotrophic Environments

Relatively few microbes are lucky enough to live in nutrient-rich **eutrophic** environments. Rather, most inhabit **oligotrophic** environments—ones in which nutrient levels are low—and some live in "feast or famine" environments. For microbes in oligotrophic or nutrient-labile environments, the ability to survive starvation conditions is paramount. Fortunately, microbes have evolved numerous responses to starvation.

Some bacteria respond to starvation with obvious morphological changes such as endospore formation. Recall that endospores are not only highly resistant to a variety of stressors (e.g., high temperatures and radiation) but are also metabolically dormant. Other microbes become dormant without such a drastic reorganization of the cell. Many eukaryotic microbes and some bacteria form cysts (*see figures 22.5 and 25.2*). Many become dormant while only decreasing somewhat in overall size. This is often accompanied by protoplast shrinkage and, in bacteria, nucleoid condensation.

In many bacteria, the action of a protein called RpoS is central to starvation survival strategies. RpoS is a component of RNA polymerase holoenzyme, the enzyme responsible for binding DNA and initiating RNA synthesis (transcription). RpoS is a transcription factor that directs the other subunits of the enzyme (collectively called RNA polymerase core enzyme) to appropriate locations along the chromosome, so that transcription can begin. Specifically, RpoS directs the core enzyme to genes encoding proteins that help the bacterium survive starvation, as well as proteins involved in a general stress response.
▶▶▶ *Transcription in bacteria (section 13.5)*

The proteins made in response to starvation are called **starvation proteins,** and they help ensure survival in several ways. Some increase peptidoglycan cross-linking and cell wall strength. The starvation protein Dps (*D*NA-binding *p*rotein from *s*tarved cells) protects DNA. Proteins called chaperone proteins prevent protein denaturation and renature damaged proteins. Because of these and many other mechanisms, starved cells become harder to kill and more resistant to starvation, damaging temperature changes, oxidative and osmotic damage, and toxic chemicals such as chlorine. These changes are so effective that some members of the bacterial population survive starvation for years. There is even evidence that *Salmonella enterica* serovar Typhimurium (*S.* Typhimurium) and some other bacterial pathogens become more virulent when starved. Clearly, these considerations are of great practical importance in medical and industrial microbiology.

An interesting phenomenon is often observed in a population of bacterial cells responding to starvation conditions: the formation of persisters. **Persisters** are bacterial cells identified by their ability to survive exposure to an antibiotic (often multiple antibiotics) even though they do not harbor antibiotic-resistance genes. They are nongrowing cells that are thought to be dormant. Indeed, dormancy is central to the hypotheses regarding their antibiotic-tolerant phenotype; in dormant cells, the target of an antibiotic is unavailable or inactive and would be unaffected by the antibiotic. Persisters are of considerable interest for several reasons, including their potential to cause recurrent infections after antibiotic treatment.

Two major questions about persister formation are when and how it occurs. Evidence suggests that in some cases, small subsets of cells in a population of growing cells spontaneously become dormant persister cells, even when nutrients are readily available. These persisters can be thought of as the population's attempt to prepare for future possible starvation conditions. There is also increasing evidence that persisters arise in response to various triggers, with starvation probably being the most important. Whether spontaneous or induced, toxin-antitoxin (TA) modules and the small molecule ppGpp are currently thought to play a role in persister formation. TA modules consist of an internal, nonsecreted toxin and a cognate antitoxin that prevents the toxin from exerting its effects on the cell. The toxin, when functioning, disrupts normal cellular functions, causing growth arrest and the persister phenotype. ppGpp is a signaling molecule that functions in a regulatory network called the stringent response, which we describe in more detail in chapter 14. ppGpp is often referred to as an alarmone because it is synthesized in response to many stressors and regulates the activity of numerous genes and proteins so that the cell can survive. The production of ppGpp is thought to set into motion destruction of the antitoxin of a TA module, thus releasing the toxin to attack its cellular target, cause growth to cease, and convert the cell into a persister.

Despite the attractiveness of this model of persistence, it may not hold true for all bacteria. For instance, at least one bacterium produces persisters that actively divide. Their antibiotic tolerance is mediated by efflux pumps that pump out the antibiotic as soon as it enters the cell. Another study demonstrated that in response to starvation, a population of nonreproducing cells

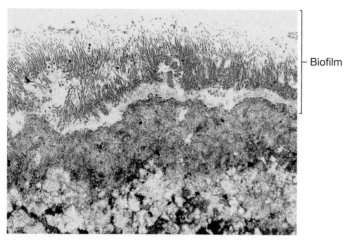

(a) Biofilm on surface of a stromatolite

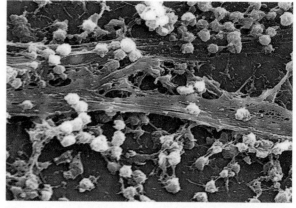

(b) Biofilm on internal surface of an indwelling catheter

Figure 7.17 Examples of Biofilms. Biofilms form on almost any surface exposed to microorganisms. (a) Biofilm on the surface of a stromatolite in Walker Lake (Nevada, USA), an alkaline lake. The biofilm consists primarily of the cyanobacterium *Calothrix* sp. (b) *Staphylococcus aureus* biofilm on the luminal surface of an indwelling catheter; colorized SEM X2,363.

was formed and that these cells could be divided into two groups: metabolically active and metabolically inactive. Only the metabolically active cells exhibited the persister phenotype. Thus, the story of persisters is still incomplete.

Biofilms Are Common in Nature

Many introductory microbiology students have scraped their teeth and examined the microbes in the scrapings, thus replicating the way Leeuwenhoek (*see figure 1.11*) often entertained dinner guests. What was sampled by all was dental plaque, an important example of a biofilm. Ecologists observed as early as the 1940s that more microbes in aquatic environments were found in biofilms on surfaces (sessile) than were free-floating (planktonic). However, only relatively recently has this fact gained the attention of microbiologists. **Biofilms** are complex, slime-encased communities of microbes. They are ubiquitous in nature, where they are most often seen as layers of slime on rocks or other objects in water or at water-air interfaces (**figure 7.17a**). When they form on the hulls of boats and ships, they cause corrosion, which limits the life of the ships and results in economic losses. Of major concern is the formation of biofilms on medical devices such as hip and knee implants and indwelling catheters (figure 7.17b). These biofilms often cause serious illness and failure of the medical device. Biofilms can also form on wound scabs and delay wound healing. Biofilm formation is apparently an ancient ability among microbes, as evidence for biofilms can be found in the fossil record from about 3.4 billion years ago.

Biofilms can form on virtually any surface, once it has been conditioned by proteins and other molecules present in the environment (**figure 7.18**). Initially microbes attach to the conditioned surface but can readily detach. Eventually they form a slimy matrix made up of various polymers, depending on the microbes in the biofilm. The polymers are collectively called extracellular polymeric substances (EPS) or extracellular matrix (ECM), and they include polysaccharides, proteins, glycoproteins, glycolipids, and DNA. The EPS allows the microbes to stick more stably to the surface. As the biofilm thickens and matures, the microbes

reproduce and secrete additional polymers. Eventually, conditions at locations in the biofilm can become detrimental to the cells and it becomes beneficial for cells to detach and escape the biofilm. This is of considerable importance for medical device-associated biofilms because the escaping cells can seed sites of infection elsewhere in the body.

A mature biofilm is a complex, dynamic community of microorganisms. It exhibits considerable heterogeneity due to differences in the metabolic activity of microbes at various locations within the biofilm; some are persister cells (**figure 7.19**). Biofilm microbes interact in a variety of ways. For instance, the waste products of one microbe may be the energy source for another microbe. The cells also use molecules to communicate with each other, as we describe next. Finally, DNA present in the EPS can be taken up by members of the biofilm community. Thus genes can be transferred from one cell (or species) to another.

While in the biofilm, microbes are protected from numerous harmful agents such as UV light and antibiotics. This is due in part to the EPS in which they are embedded (figure 7.19), but it also is due to physiological changes. Indeed, numerous proteins found in biofilm cells are not observed when these cells are free-living, planktonic cells, and vice versa. The resistance of biofilm cells to antimicrobial agents has important consequences. When biofilms form on a medical device such as a hip implant, treatment with antibiotics often fails, which can lead to serious systemic infections. The treatment failure is in part due to the presence of persisters in the biofilm. The persisters survive antibiotic treatment and then repopulate the biofilm once treatment ceases. Often the only way to manage patients in this situation is by removing the implant. Another problem with biofilms is that cells are regularly sloughed off (figure 7.18). This can have many consequences. For instance, biofilms in a city's water distribution pipes can clog or foul the pipes and serve as a source of contamination. ♻ *Biofilms*

Cell-Cell Communication Within Microbial Populations

For decades, microbiologists tended to think of bacterial populations as collections of individual cells growing and behaving

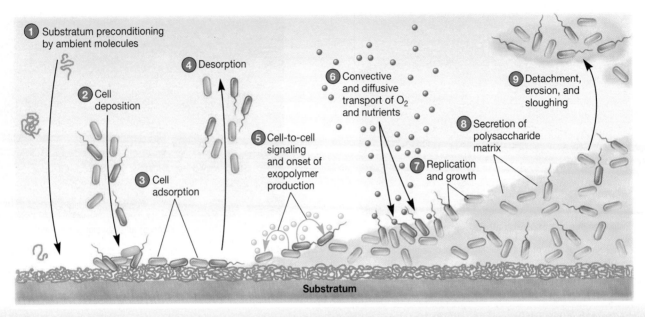

Figure 7.18 Biofilm Formation.

MICRO INQUIRY *What biomolecules make up the extracellular polymeric matrix, and what functions does the matrix serve?*

independently. But about 40 years ago, two examples of bacterial cells using molecular signals to communicate with each other in a density-dependent manner were discovered. This is now referred to as **quorum sensing;** a quorum usually refers to the minimum number of members in an organization, such as a legislative body, needed to conduct business.

The first example of quorum sensing was observed in the Gram-positive, pathogenic bacterium *Streptococcus pneumoniae*. *S. pneumoniae* cells produce and release a small protein (i.e., a peptide) into the environment. As population size increases, the amount of peptide increases. Eventually the concentration of the peptide is high enough to convert some cells in the population from a noncompetent state to a competent state. In the competent state, cells are able to take up DNA, a process called transformation. In addition, the competent cells release a chemical called bacteriocin that lyses the cells in the population that did not become competent. When the noncompetent cells lyse, they release DNA, which can be taken up by the competent cells. In addition, the lysed cells release virulence factors that help the competent cells invade tissues in a host organism, causing serious diseases such as pneumonia and meningitis. ▶▶┤ *Bacterial transformation is the uptake of free DNA from the environment (section 16.7)*

Although the quorum-sensing system of *S. pneumoniae* was the first discovered, the term was not coined until years later when the quorum-sensing system of a very different bacterium

was discovered. The marine luminescent bacterium *Vibrio fischeri* lives within the light organ of certain fish and squid. *V. fischeri* regulates its luminescence by producing a small, diffusible molecule called autoinducer. The autoinducer molecule is an **N-acylhomoserine lactone** (**AHL**). It is now known that many Gram-negative bacteria make AHL molecular signals that vary in

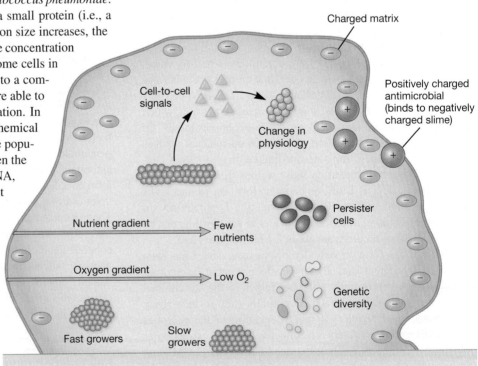

Figure 7.19 Biofilm Heterogeneity.

Signal and Structure	Representative Organism	Function Regulated
N-acylhomoserine lactone (AHL) 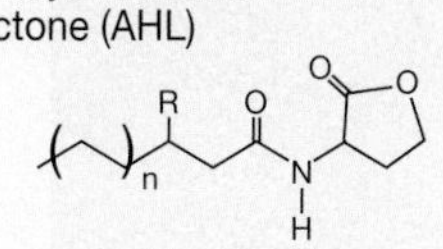	Vibrio fischeri	Bioluminescence
	Agrobacterium tumefaciens	Plasmid transfer
	Erwinia carotovora	Virulence and antibiotic production
	Pseudomonas aeruginosa	Virulence and biofilm formation
	Burkholderia cepacia	Virulence
Furanosylborate (AI-2) 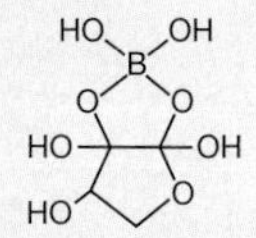	Vibrio harveyi [a]	Bioluminescence
Cyclic thiolactone (AIP-II) 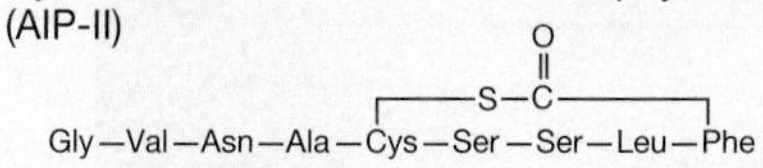	Staphylococcus aureus	Virulence
Gly—Val—Asn—Ala—Cys—Ser—Ser—Leu—Phe		
Hydroxy-palmitic acid methyl ester (PAME) 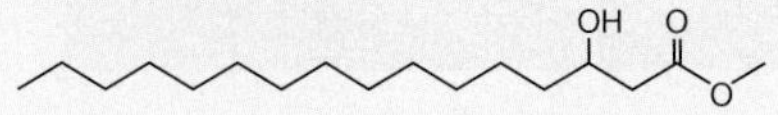	Ralstonia solanacearum	Virulence
Methyl dodecenoic acid 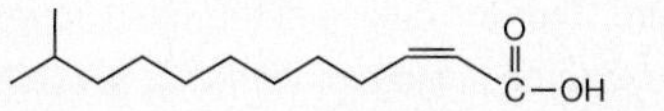	Xanthomonas campestris	Virulence
Farnesoic acid 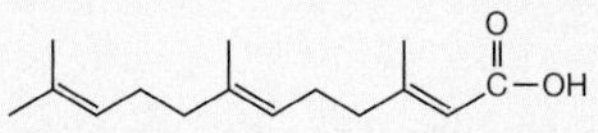	Candida albicans	Dimorphic transition and virulence
3-hydroxytridecan-4-one 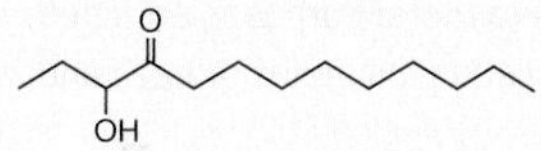	Vibrio cholerae	Virulence

[a] Other bacteria make a form of AI-2 that lacks boron.

Figure 7.20 Representative Cell-Cell Communication Molecules.

length and substitution at the third position of the acyl side chain (**figure 7.20**). In many of these species, AHL is freely diffusible across the plasma membrane. Thus at a low cell density, the diffusion gradient favors movement of AHL from the cytoplasm to the outside of the cell. However, when the cell population increases, the diffusion gradient is reversed so that movement of AHL into the cell is favored. Because the influx of AHL is cell density dependent, it enables individual cells to assess population density. When AHL reaches a threshold level inside the cell, it induces the expression of target genes that regulate a number of functions, depending on the microbe. These functions are most effective only if a large number of microbes are present. For instance, the light produced by one *V. fischeri* cell would be invisible and energetically costly, so bioluminescence is inhibited. However, cell densities within the light organ of marine fish and squid reach 10^{10} cells per milliliter. At such densities, light is clearly visible. Thus quorum sensing works to promote bioluminescence at high densities. This provides fish with a flashlight effect. In squid, bioluminescence provides countershading, which protects the squid from predation as it hunts at night. In return, the microbes have a safe and nutrient-enriched habitat (**figure 7.21**). ▶▶ *Quorum sensing by* Vibrio *spp. (section 14.5)*

Scientists have learned that many of the processes regulated by quorum sensing involve host-microbe interactions, including

pathogenicity. For instance, the Gram-negative, opportunistic bacterial pathogens *Burkholderia cepacia* and *Pseudomonas aeruginosa* use AHLs to regulate biofilm formation and the expression of virulence factors (figure 7.20). These bacteria cause debilitating pneumonia in people who are immunocompromised and are important pathogens in cystic fibrosis patients.

Like *S. pneumoniae,* other Gram-positive bacteria communicate using short peptides called oligopeptides. Examples include *Enterococcus faecalis,* whose oligopeptide signal is used to determine the best time to conjugate (transfer genes). Oligopeptide communication by *Staphylococcus aureus* and *B. subtilis* is also used to trigger the uptake of DNA from the environment.

The discovery of additional molecular signals made by a variety of microbes underscores the importance of cell-cell communication in regulating cellular processes. For instance, while only Gram-negative bacteria are known to make AHLs, both Gram-negative and Gram-positive bacteria make autoinducer-2 (AI-2). The soil microbe *Streptomyces griseus* produces a γ-butyrolactone known as A-factor. This small molecule regulates both morphological differentiation and the production of the antibiotic streptomycin. Eukaryotic microbes also rely on cell-cell communication to coordinate key activities within a population. For example, the pathogenic fungus *Candida albicans* secretes farnesoic acid to govern morphology and virulence (figure 7.20).

These examples of cell-cell communication demonstrate what might be called multicellular behavior in that many individual cells communicate and coordinate their activities to act as a unit. Other examples of such complex behavior are pattern formation in colonies (p. 159) and fruiting body formation in the myxobacteria. ▶▶ *Order* Myxococcales *(section 22.4)*

Interdomain Communication

The quorum-sensing phenomenon allows members of the same species to communicate with each other. As we describe in chapter 14, this system also allows different *Vibrio* species to communicate. Thus the *Vibrio* system is an example of interspecies communication. Even more dramatic examples of interspecies communication exist, including bacteria and plants "talking" to each other. ▶▶ *Quorum sensing by* Vibrio *spp. (section 14.5)*

One of the best-studied examples of interdomain communication is observed when members of the genus *Rhizobium* interact with a leguminous plant. A successful interaction ends with the invasion of a plant root by the rhizobial cells. Once inside, the rhizobia differentiate, becoming nitrogen-fixing forms called bacteroids. Nitrogen-fixation requires considerable energy, and the plant assists the bacterium by helping it meet its energy needs. In return, the bacterium supplies the plant with nitrogen in the form of amino acids.

(a) The bobtail squid, *E. scolopes*

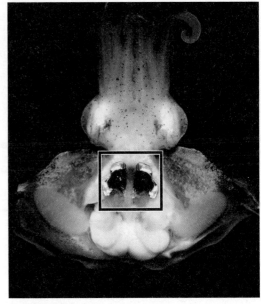

(b) Light organ

Figure 7.21 Squid—*Vibrio* Symbiosis. (a) The bobtail squid is a warm-water squid that remains buried in sand during the day and feeds at night. (b) When feeding, it uses its light organ (boxed, located on its ventral surface) to provide camouflage by projecting light downward. Thus the outline of the squid appears as bright as the water's surface to potential predators looking up through the water column. The light organ is colonized by a large number of *Vibrio fischeri* cells, so autoinducer accumulates to a threshold concentration, triggering light production. In addition to being a model for quorum sensing, this relationship is an important model for the most common type of animal-bacteria association, colonization of epithelial surfaces.

A successful interaction is initiated when the bacterium and plant "talk" to each other using chemicals they secrete. The plant releases molecules called flavonoids, which are taken up by the bacterium and bind to a bacterial protein called NodD. This activates the production of chemicals by the bacterium called Nod factors. Some Nod factors affect the outer cells of plant root hairs. Ultimately this causes the root hairs to curl, entrapping the bacteria. Other Nod factors trigger formation of an infection tube, through which the bacteria enter the plant. Thus by "speaking" to each other, both the plant and bacterium form a relationship that is beneficial to both. ▶▶| *Rhizobia are symbionts of leguminous plants (section 31.3)*

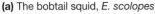

Retrieve, Infer, Apply

1. What is a biofilm? List two ways life in a biofilm is advantageous for microbes.
2. What medical challenges do biofilms present?
3. What is quorum sensing? Describe how it occurs, and briefly discuss its importance to microorganisms.
4. How is the communication that occurs between a rhizobium and a leguminous plant similar to that occurring between *Vibrio* species? How does it differ?

7.6 Laboratory Culture of Cellular Microbes Requires Media and Conditions That Mimic the Normal Habitat of a Microbe

After reading this section, you should be able to:

- Describe the importance of culturing microbes to the study of microorganisms
- Distinguish defined (synthetic) media from complex media and the uses of liquid from solid growth media
- List the characteristics of agar that make it a particularly useful solidifying agent
- Compare and contrast supportive (general purpose), enriched, selective, and differential media, listing examples of each and describing how each is used
- Discuss the use of enrichment cultures in isolating microbes
- Differentiate the streak-plate, spread-plate, and pour-plate methods for isolating pure cultures
- Use the terms commonly used by microbiologists to describe colony morphology

For decades studying microbes in their natural habitats has proven to be a significant hurdle for microbiologists to leap. For that reason, microbiology research has depended largely on the ability to grow and maintain microorganisms in the laboratory. Unfortunately, it is estimated that only 1 to 5% of all microbes are currently culturable. Recently nonculture-based alternatives have become available. Yet even with these techniques, an important goal is to eventually isolate microbes from their normal habitat. Thus culturing microbes continues to be an important tool used by microbiologists. ▶▶| *Microbial biology relies on cultures (section 29.1)*

Culture Media

Culturing microbes is possible only if suitable culture media are available. A **culture medium** is a solid or liquid preparation used to grow, transport, and store microorganisms. To be effective, the medium must contain all the nutrients the microorganism requires for growth. Specialized media are essential in the isolation and identification of microorganisms, the testing of antibiotic sensitivities, water and food analysis, industrial microbiology, and other activities. Although all microorganisms need sources of energy and macro- and micro-nutrients, the precise composition

Table 7.4	Types of Media
Basis for Classification	**Types**
Chemical composition	Defined (synthetic), complex
Physical nature	Liquid, semisolid, solid
Function	Supportive (general purpose), enriched, selective, differential

of a satisfactory medium depends on the species being cultivated due to the great variety of nutritional requirements. Knowledge of a microorganism's normal habitat often is useful in selecting an appropriate culture medium because its nutrient requirements reflect its natural surroundings. Frequently a medium is used to select and grow specific microorganisms or to help identify a particular species. Media can also be specifically designed to facilitate the growth of one type of microbe present in a sample from nature (p. 158). ◄◄ *Bacteria use many mechanisms to bring nutrients into the cell (section 3.3)*

Culture media can be classified based on several parameters: the chemical constituents from which they are made, their physical nature, and their function (**table 7.4**). The types of media defined by these parameters are described here.

Chemical and Physical Types of Culture Media

A medium in which all chemical components are known is a **defined** or **synthetic medium.** It can be in a liquid form (broth) or solidified by an agent such as agar. Defined media are often used to culture photoautotrophs such as cyanobacteria and photosynthetic protists. These microbes use CO_2 as a carbon source and light as an energy source. Thus they can be grown on media containing sodium carbonate or bicarbonate (sources of CO_2), nitrate or

ammonia as a nitrogen source, sulfate, phosphate, and other minerals (**table 7.5**). Many chemoorganoheterotrophs also can be grown in defined media. These organisms use reduced organic molecules as carbon and energy sources. Thus a medium with glucose as a carbon source and an ammonium salt as a nitrogen source will support their growth. Not all defined media are as simple as the examples in table 7.5; some are constructed from dozens of components. Defined media are used widely in research, as it is often desirable to know exactly what the microorganism is metabolizing.

Media that contain some ingredients of unknown chemical composition are **complex media.** They are very useful because a single complex medium may be able to meet all the nutritional requirements of many different microorganisms. In addition, complex media often are needed because the nutritional requirements of a particular microorganism are unknown, and thus a defined medium cannot be constructed. Complex media are also used to culture fastidious microbes, microbes with complicated nutritional or cultural requirements.

Most complex media contain undefined components such as peptones, meat extract, and yeast extract. Peptones are protein hydrolysates prepared by partial proteolytic digestion of meat, casein, soya meal, gelatin, and other protein sources. They serve as sources of carbon, energy, and nitrogen. Beef extract and yeast extract are aqueous extracts of lean beef and brewer's yeast, respectively. Beef extract contains amino acids, peptides, nucleotides, organic acids, vitamins, and minerals. Yeast extract is an excellent source of B vitamins as well as nitrogen and carbon compounds. Three commonly used complex media are nutrient broth, tryptic soy broth, and MacConkey agar (**table 7.6**).

Both liquid and solidified media are routinely used in laboratories. However, solidified media are particularly important because they can be used to isolate different microbes from each other to establish pure cultures. As we discuss in chapter 1, this is a critical step when using Koch's postulates to demonstrate the

Table 7.5	Examples of Defined Media		
BG–11 Medium for Cyanobacteria	**Amount (g/liter)**	**Medium for *Escherichia coli***	**Amount (g/liter)**
$NaNO_3$	1.5	Glucose	1.0
$K_2HPO_4 \cdot 3H_2O$	0.04	Na_2HPO_4	16.4
$MgSO_4 \cdot 7H_2O$	0.075	KH_2PO_4	1.5
$CaCl_2 \cdot 2H_2O$	0.036	$(NH_4)_2SO_4$	2.0
Citric acid	0.006	$MgSO_4 \cdot 7H_2O$	200.0 mg
Ferric ammonium citrate	0.006	$CaCl_2$	10.0 mg
EDTA (Na_2Mg salt)	0.001	$FeSO_4 \cdot 7H_2O$	0.5 mg
Na_2CO_3	0.02	Final pH 6.8–7.0	
Trace metal solution[1]	1.0 ml/liter		
Final pH 7.4			

Sources: Data from Rippka, R., et al. 1979. Journal of General Microbiology, 111:1–61; and Cohen, S. S., and Arbogast, R. 1950. Journal of Experimental Medicine, 91:619.
1 The trace metal solution contains H_3BO_3, $MnCl_2 \cdot 4H_2O$, $ZnSO_4 \cdot 7H_2O$, $Na_2Mo_4 \cdot 2H_2O$, $CuSO_4 \cdot 5H_2O$, and $Co(NO_3)_2 \cdot 6H_2O$.

Table 7.6   Some Common Complex Media

Nutrient Broth	Amount (g/liter)	Tryptic Soy Broth	Amount (g/liter)	MacConkey Agar	Amount (g/liter)
Peptone (gelatin hydrolysate)	5	Tryptone (pancreatic digest of casein)	17	Pancreatic digest of gelatin	17.0
Beef extract	3	Peptone (soybean digest)	3	Pancreatic digest of casein	1.5
Final pH 6.8		Glucose	2.5	Peptic digest of animal tissue	1.5
		Sodium chloride	5	Lactose	10.0
		Dipotassium phosphate	2.5	Bile salts	1.5
		Final pH 7.3		Sodium chloride	5.0
				Neutral red	0.03
				Crystal violet	0.001
				Agar	13.5
MacConkey Agar				Final pH 7.4	

relationship between a microbe and a disease. **Agar** is the most commonly used solidifying agent. It is a sulfated polymer composed mainly of D-galactose, 3,6-anhydro-L-galactose, and D-glucuronic acid. It usually is extracted from red algae. Agar is well suited as a solidifying agent for several reasons. One is that it melts at about 90°C but, once melted, does not harden until it reaches about 45°C. Thus after being melted in boiling water, it can be cooled to a temperature that is tolerated by human hands as well as microbes. Furthermore, microbes growing on agar medium can be incubated at a wide range of temperatures. Finally, agar is an excellent hardening agent because most microorganisms cannot degrade it.

Functional Types of Media

Media such as tryptic soy broth and tryptic soy agar are called general purpose or **supportive media** because they sustain the growth of many microorganisms. Blood and other nutrients may be added to supportive media to encourage the growth of fastidious microbes. These fortified media (e.g., blood agar) are called **enriched media (figure 7.22)**.

Selective media allow the growth of particular microorganisms, while inhibiting the growth of others (**table 7.7**). For instance, Gram-negative bacteria will grow on media containing bile salts or dyes such as basic fuchsin and crystal violet; however, the growth of Gram-positive bacteria is inhibited. Eosin methylene blue agar and MacConkey agar (table 7.7) are widely used for the detection of *E. coli* and related bacteria in water supplies and elsewhere. These media suppress the growth of Gram-positive bacteria.

Differential media are media that distinguish among different groups of microbes and even permit tentative identification of microorganisms based on their biological characteristics. Blood

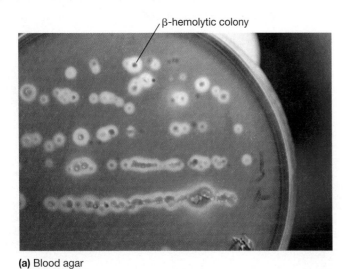

(a) Blood agar

(b) Chocolate agar

Figure 7.22   Enriched Media.   (a) Blood agar culture of bacteria from the human throat. (b) Chocolate agar is used to grow fastidious organisms such as *Neisseria gonorrhoeae*. The brown color is the result of heating red blood cells and lysing them before adding them to the medium. It is called chocolate agar because of its chocolate brown color.

Table 7.7	Mechanisms of Action of Selective and Differential Media	
Medium	**Functional Type**	**Mechanism of Action**
Blood agar	Enriched and differential	Blood agar supports the growth of many fastidious bacteria. These can be differentiated based on their ability to produce hemolysins—proteins that lyse red blood cells. Hemolysis appears as a clear zone (β-hemolysis) or greenish halo around the colony (α-hemolysis) (e.g., *Streptococcus pyogenes,* a β-hemolytic streptococcus).
Eosin methylene blue (EMB) agar	Selective and differential	Two dyes, eosin Y and methylene blue, inhibit the growth of Gram-positive bacteria. They also react with acidic products released by certain Gram-negative bacteria when they use lactose or sucrose as carbon and energy sources. Colonies of Gram-negative bacteria that produce large amounts of acidic products have a green, metallic sheen (e.g., fecal bacteria such as *E. coli*).
MacConkey (MAC) agar	Selective and differential	The selective components in MAC are bile salts and crystal violet, which inhibit the growth of Gram-positive bacteria. The presence of lactose and neutral red, a pH indicator, allows the differentiation of Gram-negative bacteria based on the products released when they use lactose as a carbon and energy source. The colonies of those that release acidic products are red (e.g., *E. coli*).
Mannitol salt agar	Selective and differential	A concentration of 7.5% NaCl selects for the growth of staphylococci. Pathogenic staphylococci can be differentiated based on the release of acidic products when they use mannitol as a carbon and energy source. The acidic products cause a pH indicator (phenol red) in the medium to turn yellow (e.g., *Staphylococcus aureus*).

agar is both a differential medium and an enriched one. It distinguishes between hemolytic and nonhemolytic bacteria. Some hemolytic bacteria (e.g., many streptococci and staphylococci isolated from throats) produce clear zones around their colonies because of red blood cell destruction (figure 7.22a). Blood agar is an enriched growth medium in that blood (usually sheep blood) provides protein, carbohydrate, lipid, iron, and a number of growth factors and vitamins necessary for the cultivation of fastidious organisms. MacConkey agar is both differential and selective. Because it contains lactose and neutral red dye, bacteria that catabolize lactose by fermenting it release acidic waste products that make colonies appear pink to red in color. These are easily distinguished from colonies of bacteria that do not ferment lactose.

Cultivation of Aerobes and Anaerobes

Because aerobes need O_2 and anaerobes are killed by it, radically different approaches must be used when they are cultivated. When large volumes of aerobic microorganisms are cultured, either they must be shaken to aerate the culture medium or sterile air must be pumped through the culture vessel. Without aeration, the low solubility of O_2 in liquid would prevent them from obtaining an adequate supply.

Precisely the opposite problem arises with anaerobes: all O_2 must be excluded. This is accomplished in several ways. (1) Anaerobic media containing reducing agents such as thioglycollate or cysteine may be used. The medium is boiled during preparation to dissolve its components and drive off oxygen. The reducing agents eliminate any residual dissolved O_2 in the medium so that anaerobes can grow beneath its surface. (2) Oxygen also may be eliminated from an enclosed work area, often called an anaerobic chamber or anaerobic workstation. Most of the air is removed with a vacuum pump followed by purges with nitrogen gas (**figure 7.23**). A gas mix containing hydrogen is then introduced into the workstation. In the presence

of a palladium catalyst, the hydrogen and last remaining molecules of O_2 react to form water, creating an anoxic environment. Often CO_2 is added to the chamber because many anaerobes require a small amount of CO_2 for best growth. (3) A method for culturing small numbers of anaerobes is the GasPak system, which also uses hydrogen and a palladium catalyst to remove O_2 (**figure 7.24**). (4) A similar approach uses plastic bags or pouches containing calcium carbonate and a catalyst, which produce an anoxic, carbon dioxide–rich atmosphere. (5) Increasingly in clinical laboratories, the GasPak system is being replaced by a bacterial enzyme that when added to a broth removes oxygen from the broth and the headspace of the container.

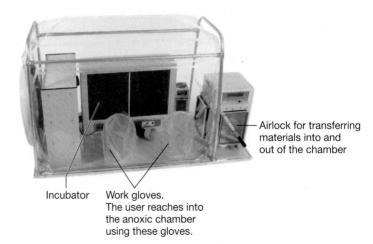

Airlock for transferring materials into and out of the chamber

Incubator   Work gloves.
The user reaches into
the anoxic chamber
using these gloves.

Figure 7.23 An Anaerobic Workstation and Incubator.   This system contains an oxygen-free work area and an incubator. The interchange compartment on the right of the work area allows materials to be transferred inside without exposing the interior to oxygen. The anoxic atmosphere is maintained largely with a vacuum pump and nitrogen purges. The remaining oxygen is removed by a palladium catalyst and hydrogen. The oxygen reacts with hydrogen to form water, which is absorbed by a desiccant.

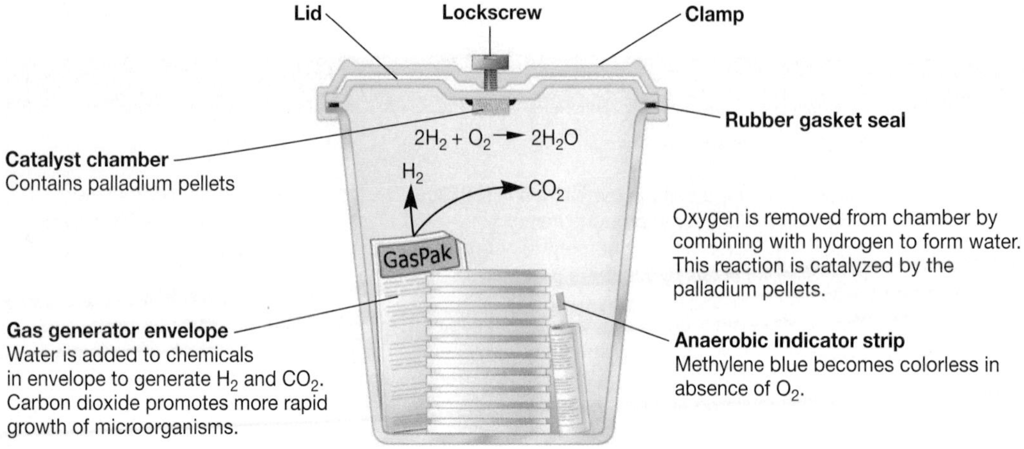

Figure 7.24 **The GasPak Anaerobic System.** Hydrogen and carbon dioxide are generated by a GasPak envelope. The palladium catalyst in the chamber lid catalyzes the formation of water from hydrogen and oxygen, thereby removing oxygen from the sealed chamber.

Enrichment and Isolation of Pure Cultures

In natural habitats, microorganisms often grow in complex, microbial communities containing many different species. This presents a problem for microbiologists because a single type of microorganism cannot be studied adequately in a mixed culture. One needs a **pure** or **axenic culture,** a population of cells arising from a single cell, to characterize an individual species. When isolating microbes from soil or aquatic habitats, microbiologists often begin by setting up conditions that favor growth of the desired microbes; that is, they enrich for the microbes of interest. This is followed by using other methods to obtain an axenic culture arising from a single cell.

Enrichment Cultures

The **enrichment culture** technique is a powerful tool used to encourage the growth of microbes having particular characteristics, while at the same time inhibiting the growth of other microbes. It has been used by microbiologists since it was first developed by Sergei Winogradsky and Martinus Beijerinck in the late 1800s. They used enrichment cultures to isolate numerous bacteria with interesting metabolic capabilities such as nitrogen fixation and the use of inorganic molecules (e.g., ammonia) as energy sources. To enrich for these organisms, they considered three factors: (1) a suitable source of the microbes, (2) nutrients that should and should not be included in the culture medium, and (3) environmental conditions provided during the incubation period.

Consider the following. Suppose you wanted to isolate an oil-degrading bacterium and study it in pure culture. A logical source of the bacterium (i.e., an inoculum) would be soil contaminated with oil. The assumption is that such a source would have exposed the indigenous bacteria to oil and selected for those able to use oil as a sole source of energy and carbon. To select for the oil-degrading bacteria, a medium would be used that contained inorganic nitrogen, sulfur, and phosphorus sources but no source of carbon and energy other than the types of molecules typically found in oil. Finally, a temperature, pH, and oxygen level similar to that of the soil environment would be used during incubation.

Despite its utility, the enrichment culture technique is not without problems. It is sometimes difficult to create the proper selective conditions to successfully isolate the desired organisms. Furthermore, even if microbes are successfully cultivated, they may not adequately represent the population as a whole at a sampling site. Generally, enrichment cultures yield the organisms that grow fastest,

and these may not be the predominant organisms in the habitat. Finally, enrichment cultures are not pure cultures. They usually contain more than one species having similar characteristics. To obtain a pure culture, one of the methods described next must be used.

Streak Plate

The streak-plate method and similar methods for isolating pure cultures were developed by the German bacteriologist Robert Koch. He used these techniques in a set of steps that now bear his name: Koch's postulates. As we describe in chapter 1, Koch's postulates are used to establish that a microbe is the causative agent of a particular disease. The use of these methods transformed microbiology, and within 20 years of their development, most pathogens responsible for major bacterial diseases had been isolated and identified. Koch reasoned that if cells from a mixture of microbes could be spatially isolated from each other, each cell would give rise to a completely separate **colony**—a macroscopically visible cluster of microorganisms in or on a solid medium. Because each colony arises from a single cell, each colony represents a pure culture.

One method for separating cells is the **streak plate.** In this technique, cells are transferred to the edge of an agar plate with an inoculating loop or swab and then streaked across the surface in one of several patterns (**figure 7.25**). After the first sector is streaked, the inoculating loop is sterilized and an inoculum for the second sector is obtained from the first sector. A similar process is followed for streaking the third sector, except that the inoculum is from the second sector. Thus this is essentially a dilution process. Eventually very few cells will be on the loop, and single cells will drop from it as it moves across the agar surface. These develop into separate colonies.

Spread Plate and Pour Plate

Spread-plate and pour-plate techniques are similar in that they both dilute a sample of cells before separating them spatially. They differ in that the spread plate spreads the cells on the surface of the agar, whereas the pour plate embeds the cells within the agar. Both methods can be used to determine the number of microorganisms in a sample (p. 166).

For the **spread plate,** a small volume of a diluted mixture containing around 25 to 250 cells is transferred to the center of an agar plate and spread evenly over the surface with a sterile bent rod (**figure 7.26**). The dispersed cells develop into isolated colonies.

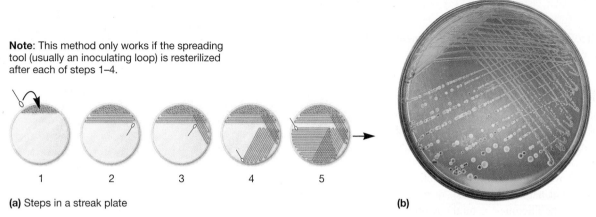

Note: This method only works if the spreading tool (usually an inoculating loop) is resterilized after each of steps 1–4.

1   2   3   4   5

(a) Steps in a streak plate

(b)

Figure 7.25 Streak-Plate Technique.   A typical streaking pattern is shown (a), as well as an example of a streak plate (b).

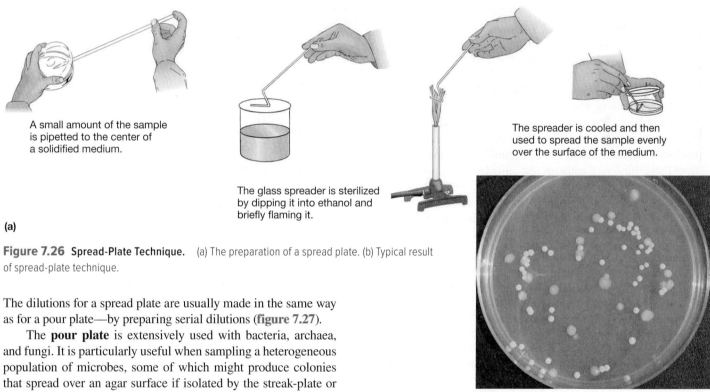

A small amount of the sample is pipetted to the center of a solidified medium.

The glass spreader is sterilized by dipping it into ethanol and briefly flaming it.

The spreader is cooled and then used to spread the sample evenly over the surface of the medium.

(a)

(b)

Figure 7.26 Spread-Plate Technique.   (a) The preparation of a spread plate. (b) Typical result of spread-plate technique.

The dilutions for a spread plate are usually made in the same way as for a pour plate—by preparing serial dilutions (**figure 7.27**).

The **pour plate** is extensively used with bacteria, archaea, and fungi. It is particularly useful when sampling a heterogeneous population of microbes, some of which might produce colonies that spread over an agar surface if isolated by the streak-plate or spread-plate methods. In the pour-plate method, the original sample is serially diluted to reduce the microbial population sufficiently to obtain separate colonies when plating (figure 7.27). Then small volumes of several diluted samples are mixed with liquid agar that has been cooled to about 45°C, and the mixtures are poured immediately into sterile culture dishes. Most microbes survive a brief exposure to the warm agar. Each cell becomes fixed in place to form an individual colony after the agar hardens.

Although the preparation of serial dilutions is the bane of many microbiology students, it has many applications other than the spread-plate and pour-plate methods. The numbers of cells in a solution can be diluted to the point where no cells are present in the dilution tube. If replicate dilutions are prepared and the number of tubes yielding growth determined, then the number of viable cells in the sample can be estimated by the most probable number method (MPN; *see figure 29.3*). Similarly, the number of viruses in a solution or the number of antibodies in a blood sample can be determined by first diluting the sample and then testing for the presence of viruses or antibodies. ◀◀ *Cultivation and enumeration of viruses (section 6.5)* ▶▶ *Antibodies are proteins that bind to specific 3-D molecules (section 34.7)*

Microbial Growth on Solid Media

Colony development on agar surfaces aids microbiologists in identifying microorganisms because individual species often form colonies of characteristic size and appearance (**figure 7.28**; also figure 7.26*b*). When a mixed population has been plated properly, it sometimes is possible to identify the desired colony based on its overall appearance and use it to obtain a pure culture.

It is obvious from the colonies pictured in figure 7.28 that bacteria growing on solid surfaces such as agar can form quite complex colony shapes. These patterns and colony size depend on many factors, including nutrient diffusion and availability, bacterial chemotaxis, the presence of liquid on the surface, and hardness of the agar. Cell-cell communication is important as well. Much research is currently focused on understanding the formation of bacterial colonies.

New Approaches to Culturing Microbes

The importance of culturing even those microbes that are the most recalcitrant to growth in the lab has led microbiologists to devise new techniques. As we have noted, traditional methods for culturing microbes have depended on being able to construct artificial habitats in a flask or petri dish that mimic the microbes' natural environments. However, it is often difficult to know all the environmental conditions required for growth. One way to solve this problem is to bring part of the natural environment into the lab. For instance, an aquarium containing seawater, beach sand, marine flora, and macroscopic organisms collected from a natural habitat can be used to culture marine microbes. This works very well for maintaining a mixed population in the lab, but it still leaves the problem of isolating particular microbes from the mixture. One solution is to create an enclosure within the man-made "natural" habitat that allows free diffusion of nutrients and other factors from the surroundings but keeps the microbes in the enclosure. For our aquarium example, a very dilute suspension of the seawater theoretically containing a single cell would be used to inoculate the enclosure. Thus an axenic culture would form in the enclosure.

Another approach is useful for microbes that require the presence of a different species in order to survive. In this case the microbe of interest is co-cultured with the required organism. Sometimes the required organism is an animal, animal tissue, a plant, or plant tissue. The microbe must be maintained growing in or on that organism. Another example of co-culturing occurs in a petri dish on which a dilute suspension of the desired microbe has been spread and then the helper microbe spotted onto the medium

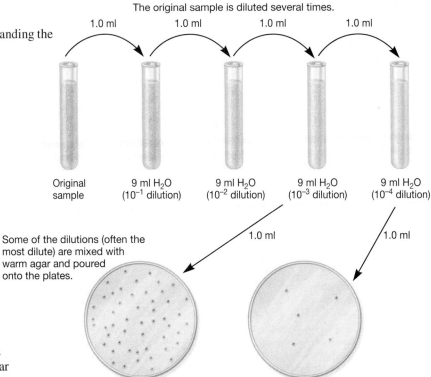

The original sample is diluted several times.

1.0 ml   1.0 ml   1.0 ml   1.0 ml

Original sample   9 ml H_2O (10^{-1} dilution)   9 ml H_2O (10^{-2} dilution)   9 ml H_2O (10^{-3} dilution)   9 ml H_2O (10^{-4} dilution)

Some of the dilutions (often the most dilute) are mixed with warm agar and poured onto the plates.

1.0 ml   1.0 ml

Isolated cells grow into colonies on the surface (appear round) and within the medium (appear lens-shaped). The isolated colonies can be counted or used to establish pure cultures.

Figure 7.27 Pour-Plate Technique.

at one location. The factors produced by the helper microbe diffuse into the medium and support the growth of the microbe of interest. This method is particularly useful for long-term maintenance of cultures once the microbe has been isolated in axenic culture.

Microbiologists are exploring even more methods for culturing the unculturable. In the future it is expected that with these new techniques microbiologists will gain a better understanding of the diversity of microbes and their roles in many environments. ▶▶| *Microbial biology relies on cultures (section 29.1)*

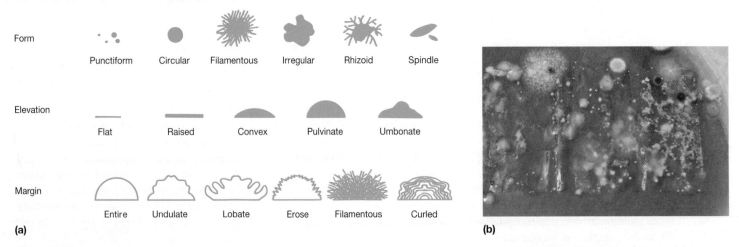

Form

Punctiform   Circular   Filamentous   Irregular   Rhizoid   Spindle

Elevation

Flat   Raised   Convex   Pulvinate   Umbonate

Margin

Entire   Undulate   Lobate   Erose   Filamentous   Curled

(a)   **(b)**

Figure 7.28 Bacterial Colony Morphology. (a) Variations in bacterial colony morphology seen with the naked eye. The general form of the colony and the shape of the edge or margin can be determined by looking down at the top of the colony. The nature of colony elevation is apparent when viewed from the side as the plate is held at eye level. (b) Portion of a Petri dish showing some commonly observed colony morphologies.

1. Describe the following kinds of media and their uses: defined media, complex media, supportive (general purpose) media, enriched media, selective media, and differential media. Give an example of each.
2. What are peptones, yeast extract, beef extract, thioglycollate, and agar? Why are they used in media?
3. Describe four ways in which anaerobes may be cultured.
4. What are pure cultures and why are they important? How are spread plates, streak plates, and pour plates prepared?
5. It is known that microbial growth varies within a colony. What factors might cause these variations?
6. How might an enrichment culture be used to isolate bacteria capable of growing photoautotrophically from a mixed microbial assemblage?

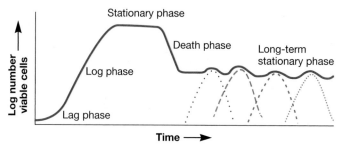

Figure 7.29 Microbial Growth Curve in a Closed System. The five phases of the growth curve are identified. The dotted lines shown during the long-term stationary phase represent successive waves of genetic variants that evolve during this phase of the growth curve.

MICRO INQUIRY *Identify the regions of the growth curve in which (1) nutrients are rapidly declining and (2) wastes accumulate.*

7.7 Growth Curves Consist of Five Phases

After reading this section, you should be able to:

- Describe the five phases of a microbial growth curve observed when microbes are grown in a batch culture
- Describe three hypotheses proposed to account for the decline in cell numbers during the death phase of a growth curve
- Predict how the presence of viable but nonculturable cells in food or water systems might impact public health
- Correlate changes in nutrient concentrations in natural environments with the five phases of a microbial growth curve
- Relate growth rate constant to generation (doubling) time and suggest how these values might be used by microbiologists doing basic research or working in industrial settings

In section 7.2, we commented on the changes in bacterial cell size that accompany its preparation for cell division. This is one type of growth of concern to microbiologists. However, microbiologists are more frequently concerned with the increase in population size that follows cell division. Therefore the term growth is also used to refer to growth in the size of a population.

Population growth is often studied by analyzing the growth of microbes in liquid (broth) culture. When microorganisms are cultivated in broth, they usually are grown in a **batch culture;** that is, they are incubated in a closed culture vessel with a single batch of medium. Fresh medium is not provided during incubation, so nutrient concentrations decline and concentrations of wastes increase over time. Population growth of microbes reproducing by binary fission in a batch culture can be plotted as the logarithm of the number of viable cells versus the incubation time. The resulting curve has five distinct phases (**figure 7.29**), which we examine in this section. Although this is "life in the lab," microbes do encounter conditions in their natural environments that mimic what occurs in a batch culture. Furthermore, humans routinely create artificial environments for microbes (e.g., the fermentation vessel in a pharmaceutical plant) that are batch cultures. Therefore understanding the growth curve is of paramount importance.

Lag Phase

When microorganisms are introduced into fresh culture medium, usually no immediate increase in cell number occurs. This period is called the **lag phase.** It is not a time of inactivity; rather cells are synthesizing new components. This can be necessary for a variety of reasons. The cells may be old and depleted of ATP, essential cofactors, and ribosomes; these must be synthesized before growth can begin. The medium may be different from the one the microorganism was growing in previously. In this case, new enzymes are needed to use different nutrients. Possibly the microorganisms have been injured and require time to recover. Eventually however, the cells begin to replicate their DNA, increase in mass, and divide. As a result, the number of cells in the population begins to increase.

Exponential Phase

During the **exponential (log) phase,** microorganisms are growing and dividing at the maximal rate possible given their genetic potential, the nature of the medium, and the environmental conditions. Their rate of growth is constant during the exponential phase; that is, they are completing the cell cycle and doubling in number at regular intervals (figure 7.29). The population is most uniform in terms of chemical and physiological properties during this phase; therefore exponential phase cultures are usually used in biochemical and physiological studies. The growth rate during log phase depends on several factors, including nutrient availability. When microbial growth is limited by the low concentration of a required nutrient, the final net growth or yield of cells increases with the initial amount of the limiting nutrient present (**figure 7.30a**). The rate of growth also increases with nutrient concentration (figure 7.30b) but in a hyperbolic manner much like that seen with many enzymes (*see figure 10.17*). The shape of the curve is thought to reflect the rate of nutrient uptake by microbial transport proteins. At sufficiently high nutrient levels, the transport systems are saturated, and the growth rate does not rise further with increasing nutrient concentration. ◄◄ *Bacteria use many mechanisms to bring nutrients into the cell (section 3.3)*

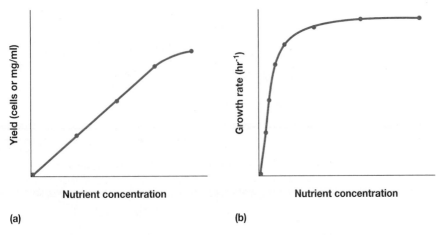

(a)

(b)

Figure 7.30 Nutrient Concentration and Growth. (a) The effect of changes in limiting nutrient concentration on total microbial yield. At sufficiently high concentrations, total growth will plateau. (b) The effect on growth rate.

Stationary Phase

In a closed system such as a batch culture, population growth eventually ceases and the growth curve becomes horizontal (figure 7.29). This **stationary phase** is attained by most bacteria at a population level of around 10^9 cells per milliliter. Protist cultures often have maximum concentrations of about 10^6 cells per milliliter. Final population size depends on nutrient availability and other factors, as well as the type of microorganism. In stationary phase, the total number of viable microorganisms remains constant. This may result from a balance between cell division and cell death, or the population may simply cease to divide but remain metabolically active.

One reason microorganisms enter the stationary phase is nutrient limitation; if an essential nutrient is severely depleted, population growth will slow and eventually stop. Thus the stationary phase is similar to growth in oligotrophic environments, as described in section 7.5. Interestingly, many of the survival strategies used in oligotrophic environments are also used by populations in the stationary phase of the growth cycle.

Microbes enter the stationary phase for other reasons besides nutrient limitation. Aerobic organisms often are limited by O_2 availability. Oxygen is not very soluble and may be depleted so quickly that only the surface of a culture will have an O_2 concentration adequate for growth. Population growth also may cease due to the accumulation of toxic waste products. This seems to limit the growth of many cultures growing in the absence of O_2. For example, streptococci can produce so much lactic acid and other organic acids from sugar fermentation that their medium becomes acidic and growth is inhibited. Finally, some evidence exists that growth may cease when a critical population level is reached. Thus entrance into the stationary phase may result from several factors operating in concert.

Death Phase

Cells growing in batch culture cannot remain in stationary phase indefinitely. Eventually they enter a phase known as the death phase (figure 7.29). During this phase, the number of viable cells declines exponentially, with cells dying at a constant rate. It was assumed that detrimental environmental changes such as nutrient deprivation and the buildup of toxic wastes caused irreparable harm to the cells. That is, even when bacterial cells were transferred to fresh medium, no cellular growth was observed. Because loss of viability was often not accompanied by a loss in total cell number, it was assumed that cells died but did not lyse.

This view is currently being debated. There are two alternative hypotheses that invoke an active process during the death phase. Some microbiologists think some cells are only temporarily unable to grow, at least under the laboratory conditions used. This phenomenon, in which the cells are called **viable but nonculturable** (**VBNC**), is thought to be the result of a genetic response triggered in starving, stationary phase cells. Once the appropriate conditions are available (e.g., a change in temperature or passage through an animal), VBNC microbes resume growth. VBNC microorganisms could pose a public health threat, as many assays that test for food and drinking water safety are culture based.

The second hypothesis is **programmed cell death.** In contrast to the VBNC hypothesis whereby cells are genetically programmed to survive, programmed cell death predicts that a fraction of the microbial population is genetically programmed to die after growth ceases. In this case, some cells die and the nutrients they leak enable the eventual growth of those cells in the population that did not initiate cell death. The dying cells are thus "altruistic"—they sacrifice themselves for the benefit of the larger population.

Long-Term Stationary Phase

Long-term growth experiments reveal that after a period of exponential death some microbes have a long period where the population size remains more or less constant. This **long-term stationary phase** (also called extended stationary phase) can last months to years (figure 7.29). During this time, the bacterial population continually evolves so that actively reproducing cells are those best able to use the nutrients released by their dying brethren and best able to tolerate the accumulated toxins. This dynamic process is marked by successive waves of genetically distinct variants. Thus natural selection can be witnessed within a single culture vessel.

Mathematics of Growth

Knowledge of microbial growth rates during the exponential phase is indispensable to microbiologists. Growth rate studies contribute to basic physiological and ecological research, and are applied in industry. The quantitative aspects of exponential phase growth discussed here apply to microorganisms that divide by binary fission.

Table 7.8 An Example of Exponential Growth

Time[1]	Division Number	2^n	Population[2] $(N_0 \times 2^n)$	$\log_{10}N_t$
0	0	$2^0 = 1$	1	0.000
20	1	$2^1 = 2$	2	0.301
40	2	$2^2 = 4$	4	0.602
60	3	$2^3 = 8$	8	0.903
80	4	$2^4 = 16$	16	1.204

1 The hypothetical culture begins with one cell having a 20-minute generation time.
2 Number of cells in the culture.

During the exponential phase, each microorganism is dividing at constant intervals. Thus the population doubles in number during a specific length of time called the **generation (doubling) time** (g). This can be illustrated with a simple example. Suppose that a culture tube is inoculated with one cell that divides every 20 minutes (**table 7.8**). The population will be 2 cells after 20 minutes, 4 cells after 40 minutes, and so forth. Because the population is doubling every generation, the increase in population is always 2^n where n is the number of generations. The resulting population increase is exponential; that is, logarithmic (**figure 7.31**).

The mathematics of growth during the exponential phase are illustrated in **figure 7.32**, which shows the calculation of two important values. The **growth rate constant (k)** is the number of

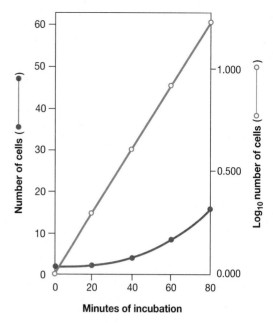

Figure 7.31 Exponential Microbial Growth. Four generations of growth are plotted directly (—) and in the logarithmic form (∘—∘). The growth curve is exponential, as shown by the linearity of the log plot.

generations per unit time and is often expressed as generations per hour. It can be used to calculate the generation time. As can be seen in figure 7.32, the generation time is simply the reciprocal of the growth rate constant. The generation time can also be determined directly from a semilogarithmic plot of growth curve data (**figure 7.33**). Once this is done, it can be used to calculate the growth rate constant.

Calculation of the growth rate constant

Let N_0 = the initial population number

N_t = the population at time t

n = the number of generations in time t

For populations reproducing by binary fission

$$N_t = N_0 \times 2^n$$

Solving for n, the number of generations, where all logarithms are to the base 10,

$$\log N_t = \log N_0 + n \cdot \log 2, \text{ and}$$

$$n = \frac{\log N_t - \log N_0}{\log 2} = \frac{\log N_t - \log N_0}{0.301}$$

The growth rate constant (k) is the number of generations per unit time $\left(\frac{n}{t}\right)$. Thus

$$k = \frac{n}{t} = \frac{\log N_t - \log N_0}{0.301t}$$

Calculation of generation (doubling) time

If a population doubles, then

$$N_t = 2N_0$$

Substitute $2N_0$ into the growth rate constant equation and solve for

$$k = \frac{\log (2N_0) - \log N_0}{0.301g} = \frac{\log 2 + \log N_0 - \log N_0}{0.301g}$$

$$k = \frac{1}{g}$$

The generation time is the reciprocal of the growth rate constant.

$$g = \frac{1}{k}$$

Figure 7.32 Calculation of the Growth Rate Constant and Generation Time. The calculations are only valid for the exponential phase of growth, when the growth rate is constant.

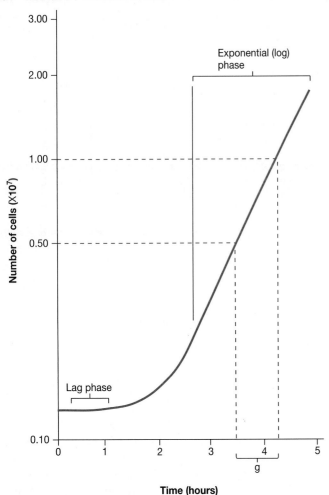

Figure 7.33 Generation Time Determination. The generation time can be determined from a microbial growth curve. The population data are plotted with the logarithmic axis used for the number of cells. The time to double the population number is then read directly from the plot. The log of the population number can also be plotted against time on regular axes.

Generation times vary markedly with the microbial species and environmental conditions. They range from less than 10 minutes (0.17 hours) to several days (**table 7.9**). Generation times in nature are usually much longer than in culture.

Retrieve, Infer, Apply

1. Define microbial growth.
2. Describe the phases of the growth curve and discuss the causes of each.
3. Why would cells that are vigorously growing when inoculated into fresh culture medium have a shorter lag phase than those that have been stored in a refrigerator?
4. Contrast and compare the viable but nonculturable status of microbes with that of programmed cell death as a means of responding to starvation.
5. Calculate the growth rate constant and generation time of a culture that increases in the exponential phase from 5×10^2 to 1×10^8 in 12 hours.
6. Suppose the generation time of a bacterium is 90 minutes and the number of cells in a culture is 10^3 cells at the start of the log phase. How many bacteria will there be after 8 hours of exponential growth?

Table 7.9	Examples of Generation Times[1]		
Microorganism		**Incubation Temperature (°C)**	**Generation Time (Hours)**
Bacteria			
Escherichia coli		40	0.35
Bacillus subtilis		40	0.43
Staphylococcus aureus		37	0.47
Pseudomonas aeruginosa		37	0.58
Clostridium botulinum		37	0.58
Mycobacterium tuberculosis		37	≈12
Treponema pallidum		37	33
Protists			
Tetrahymena geleii		24	2.2–4.2
Chlorella pyrenoidosa		25	7.75
Paramecium caudatum		26	10.4
Euglena gracilis		25	10.9
Giardia lamblia		37	18
Ceratium tripos		20	82.8
Fungi			
Saccharomyces cerevisiae		30	2
Monilinia fructicola		25	30

1 Generation times differ depending on the growth medium and environmental conditions used.

7.8 Microbial Population Size Can Be Measured Directly or Indirectly

After reading this section, you should be able to:

- Evaluate direct cell counts, viable counting methods, and cell mass measurements for determining population size
- Explain why plate count results are expressed in terms of colony forming units (CFU)
- Design appropriate approaches for measuring the population size of different types of samples

There are many ways to measure microbial growth to determine growth rate constants and generation times. Either population number or mass may be followed because growth leads to increases in both. Here the most commonly employed techniques for determining population size are examined and the advantages and disadvantages of each noted. No single technique is always best; the most appropriate approach depends on the experimental situation.

Direct Measurement of Cell Numbers

The most obvious way to determine microbial numbers is by **direct counts** using a counting chamber (e.g., Petroff-Hausser counting chamber). This approach is easy, inexpensive, and relatively quick. It also gives information about the size and morphology of microorganisms. Counting chambers consist of specially designed slides and coverslips; the space between the slide and coverslip creates a chamber of known depth. On the bottom of the chamber is an etched grid that facilitates counting the cells (**figure 7.34**). The number of microorganisms in a sample can be calculated by taking into account the chamber's volume and any dilutions made of the sample before counting. One disadvantage of using counting chambers is that to determine population size accurately, the microbial population must be relatively large and evenly dispersed because only a small volume of the population is sampled.

The number of bacteria in aquatic samples is frequently determined from direct counts after the bacteria have been trapped on membrane filters. In the membrane filter technique, the sample is first filtered through a black polycarbonate membrane filter. Then the bacteria are stained with nucleic acid fluorescent stains such as acridine orange or DAPI and observed microscopically (*see figure 29.5*). Alternatively, fluorescently labeled dyes that are specific for members of a given taxon may be used. The stained cells are easily observed against the black background of the membrane filter and can be counted when viewed with an epifluorescence microscope. Direct cell counts of environmental samples almost invariably result in higher cell densities than do methods that rely on culturing. This is because only a small percentage (about 1%) of cells growing in nature can be cultivated in the laboratory. ◄◄ *Fluorescence microscopes use emitted light to create images (section 2.2)*

Flow cytometry is increasingly being used to directly count microbes and to gain detailed information about them. A flow cytometer creates a stream of cells so narrow that one cell at a time passes through a beam of laser light. As each cell passes through the beam, the light is scattered. Scattered light is detected by the flow cytometer. Because cells are separated in space, each light-scattering event is detected independently. Thus the number of light-scattering events represents the number of cells in the sample (*see figure 29.4*). Cells of differing size, internal complexity, and other characteristics within a population can also be counted. This usually involves the use of fluorescent dyes or fluorescently labeled antibodies. These more sophisticated uses of flow cytometry can provide valuable information about characteristics of the population of cells. ►► *Flow cytometry (section 36.4)*

Microorganisms also can be directly counted with electronic counters such as the Coulter counter. In the Coulter counter, a microbial suspension is forced through a small hole. Electrical current flows through the hole, and electrodes placed

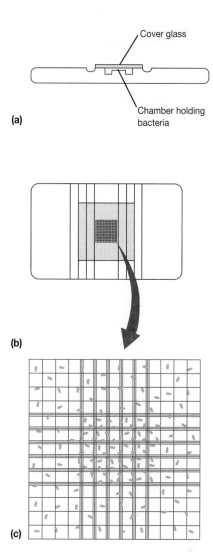

(a)

(b)

(c)

Figure 7.34 The Petroff-Hausser Counting Chamber. (a) Side view of the chamber showing the cover glass and the space beneath it that holds a bacterial suspension. (b) A top view of the chamber. The grid is located in the center of the slide. (c) An enlarged view of the grid. The bacteria in several of the central squares are counted, usually at ✕400 to ✕500 magnification. The average number of bacteria in these squares is used to calculate the concentration of cells in the original sample.

on both sides of the hole measure electrical resistance. Every time a microbial cell passes through the hole, electrical resistance increases (i.e., the conductivity drops), and the cell is counted.

Traditional methods for directly counting microbes in a sample usually yield cell densities that are much higher than the plating methods described next in part because direct counting procedures do not distinguish dead cells from culturable cells. Newer methods for direct counts help alleviate this problem. Commercially prepared fluorescent dyes can differentiate between live and dead cells, making it possible to count directly the number of live and dead microorganisms in a sample (*see figures 2.13a and 29.2*). Unfortunately, the method is not completely accurate. Just as scientists have difficulty

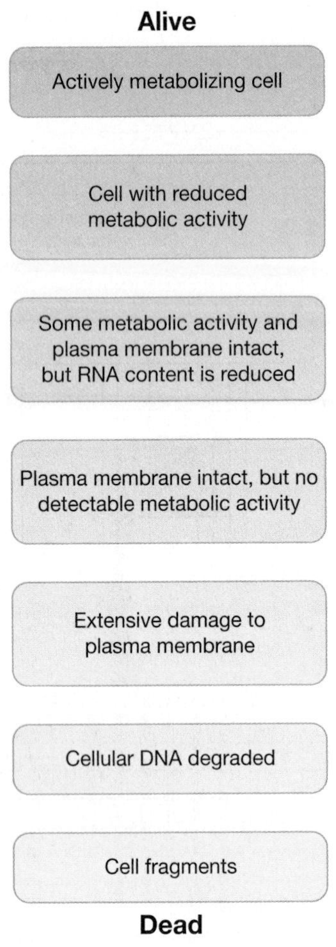

Alive

Actively metabolizing cell

Cell with reduced metabolic activity

Some metabolic activity and plasma membrane intact, but RNA content is reduced

Plasma membrane intact, but no detectable metabolic activity

Extensive damage to plasma membrane

Cellular DNA degraded

Cell fragments

Dead

Figure 7.35 **Alive or Dead?** Microbiologists have traditionally defined microbes as being dead when they could not be cultured. However, they have come to realize that cells may be inactive or damaged and therefore unable to reproduce temporarily. With time and appropriate conditions, the cells may recover and begin to reproduce. Various criteria have been used to establish these states; however, microbiologists are still unable to define the point at which a microbe is truly dead and cannot be resuscitated.

defining life, so too do they have difficulty deciding when a cell is truly dead (**figure 7.35**).

Viable Counting Methods

Several plating methods can be used to determine the number of viable microbes in a sample. These are referred to as either **viable counting methods** or **standard plate counts** because they count only those cells that are able to reproduce when cultured. Two commonly used procedures are the spread-plate and the pour-plate techniques, which are described in section 7.6. When using these methods to determine population size, the samples should yield between 25 and 250 colonies per plate for accurate counting. Once the number of colonies is known, the original number of viable microorganisms in the sample can be calculated from that number and the sample dilution. For example, if 1.0 milliliter of a solution diluted by a factor of 1×10^6 yielded 150 colonies (i.e., 1.5×10^2 colonies), then the original sample contained around 1.5×10^8 cells per milliliter. However, because it is not possible to be certain that each colony arose from an individual cell, the results are often expressed in terms of **colony forming units** (**CFU**), rather than the number of microorganisms.

Another commonly used plating method first traps bacteria in aquatic samples on a membrane filter. The filter is then placed on an agar medium or on a pad soaked with liquid media (**figures 7.36** and **7.37**) and incubated until each cell forms a separate colony. A colony count gives the number of microorganisms in the filtered sample, and selective media can be used to select for specific microorganisms. This technique is especially useful in analyzing water purity. ▶▶◀ *Sanitary analysis of waters (section 43.1)*

Plating techniques are simple, sensitive, and widely used for viable counts of bacteria and other microorganisms in samples of food, water, and soil. However, several problems can lead to inaccurate counts. The population will be underestimated if clumps of cells are not broken up and microorganisms well dispersed.

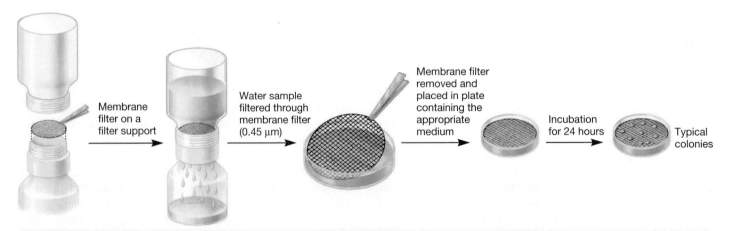

Figure 7.36 **The Membrane Filtration Procedure.** Membranes with different pore sizes are used to trap different microorganisms. Incubation times for membranes also vary with the medium and microorganism.

MICRO INQUIRY *Why is it important to have no more than about 250 colonies on the plate when counting colonies to determine population size?*

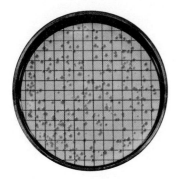

Figure 7.37 Colonies on a Membrane Filter. A sample was filtered using a membrane filter that did not let bacterial cells pass through. The filter was then placed on a culture medium and incubated. The resulting colonies are easily counted using the grid. The resulting value is used to determine colony-forming units in the sample.

The hot agar used in the pour-plate technique may injure or kill sensitive cells; thus spread plates sometimes give higher counts than pour plates. A major problem is that plate counts will be artificially low if the medium employed cannot support growth of all the viable microorganisms present. This is a common problem encountered by microbial ecologists trying to understand the makeup of microbial communities. In such studies, the number of cells observed by microscopy is often much higher than the population size determined by plate counts. This discrepancy is sometimes referred to as "the great plate count anomaly" and is discussed in detail in chapter 29. ▶▶| *Genetic methods are used to assess microbial diversity (section 29.2)*

Sometimes plate counts cannot be used to measure population size. For instance, plate counts are not helpful if the microbe cannot be cultured on solid media or if large colonies overgrow the surface of the plate, making it impossible to get an accurate count. In these cases, another approach is used: **most probable number** (**MPN**) determination. In this method, numerous replicates of several dilutions of a culture are made and added to tubes containing a suitable liquid growth medium. After incubation, each tube is examined to determine if growth occurred. It is assumed that the last tube in the dilution series that demonstrates growth was inoculated with between one and 10 cells, while the next tube had between 11 and 100 cells, and so on. If no growth is observed, then the tube is assumed not to have received any cells. In this way, the number of cells in the original sample is estimated (*for an example, see figure 29.3*). MPN values are most commonly used when a selective medium can be employed that supports the growth of a specific type of microbe. It is often used in assessing microbial density in water samples.

Measurement of Cell Mass

Techniques for measuring changes in cell mass also can be used to determine population size. One approach is the determination of microbial dry weight. Cells growing in liquid medium are collected by centrifugation, washed, dried in an oven, and weighed. This is an especially useful technique for measuring the growth of filamentous fungi. However, it is time-consuming and not very sensitive. Because bacteria weigh so little, it may be necessary

to centrifuge several hundred milliliters of culture to collect a sufficient quantity.

A more rapid and sensitive method for measuring cell mass is spectrophotometry (**figure 7.38**). Spectrophotometry depends on the fact that microbial cells scatter light that strikes them. Because microbial cells in a population are of roughly constant size, the amount of scattering is directly proportional to the biomass of cells present and indirectly related to cell number. When the concentration of bacteria reaches about a million (10^6) cells per milliliter, the medium appears slightly cloudy or turbid. Further increases in concentration result in greater turbidity, and less light is transmitted through the medium. The extent of light scattering (i.e., decrease in transmitted light) can be measured by a spectrophotometer and is called the absorbance (optical density) of the medium. Absorbance is almost linearly related to cell concentration at absorbance levels less than about 0.5. If the sample exceeds this value, it must first be diluted and then absorbance measured. Thus population size can be easily measured as long as the population is high enough to give detectable turbidity.

Cell mass can also be estimated by measuring the concentration of some cellular substance, as long as its concentration is constant in each cell. For example, a sample of cells can be analyzed for total protein or nitrogen. An increase in the microbial population will be reflected in higher total protein levels. Similarly, chlorophyll determinations can be used to measure phototrophic protist and cyanobacterial populations, and the quantity of ATP can be used to estimate the amount of living microbial mass.

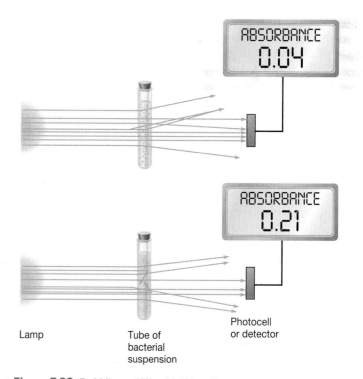

Lamp   Tube of bacterial suspension   Photocell or detector

Figure 7.38 Turbidity and Microbial Mass Measurement. Determination of microbial mass by measurement of light absorption. As the population and turbidity increase, more light is scattered and the absorbance reading given by the spectrophotometer increases.

Retrieve, Infer, Apply

1. Briefly describe each technique by which microbial population numbers may be determined and give its advantages and disadvantages.
2. When using direct cell counts to follow the growth of a culture, it may be difficult to tell when the culture enters the death phase. Why?
3. Why are plate count results expressed as colony forming units?
4. For each of the following, which enumeration technique would you use? Explain your choice. (a) A pure culture of *Staphylococcus aureus;* (b) a water sample that needs to be checked for *E. coli* contamination; (c) a sample of yogurt.

7.9 Chemostats and Turbidostats Are Used for Continuous Culture of Microorganisms

After reading this section, you should be able to:

- Distinguish batch culture and continuous culture
- Differentiate chemostats and turbidostats
- Discuss the relationship between the dilution rate of a chemostat and population size and growth rate

In section 7.7, our focus is on closed systems called batch cultures in which nutrients are not renewed nor wastes removed. Exponential (logarithmic) growth lasts for only a few generations and soon stationary phase is reached. However, it is possible to grow microorganisms in a system with constant environmental conditions maintained through continual provision of nutrients and removal of wastes. Such a system is called a **continuous culture system.** These systems can maintain a microbial population in exponential growth, growing at a known rate and at a constant biomass concentration for extended periods. Continuous culture systems make possible the study of microbial growth at very low nutrient levels, concentrations close to those present in natural environments. These systems are essential for research in many areas, particularly microbial ecology. For example, interactions between microbial species in environmental conditions resembling those in a freshwater lake or pond can be modeled. Continuous culture systems also are used in food and industrial microbiology. Two major types of continuous culture systems commonly are used: chemostats and turbidostats. ▶▶| *Microbiology of food (chapter 41); Biotechnology and industrial microbiology (chapter 42)*

Chemostats

A **chemostat** is constructed so that the rate at which a sterile medium is fed into a culture vessel is the same as the rate at which the medium containing microorganisms is removed (**figure 7.39**). The culture medium for a chemostat has a limited quantity of an essential nutrient (e.g., a vitamin). Because one nutrient is limiting, growth rate is determined by the rate at which sterile medium is fed into the growth chamber; the final

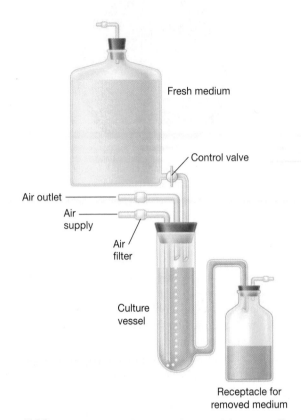

Figure 7.39 A Chemostat. Schematic diagram of the system. The fresh medium contains a limiting amount of an essential nutrient. Although not shown, usually the incoming medium is mechanically mixed with that already in the culture vessel. Growth rate is determined by the flow rate of medium through the culture vessel.

cell density depends on the concentration of the limiting nutrient. The rate of nutrient exchange is expressed as the dilution rate (*D*), the rate at which medium flows through the culture vessel relative to the vessel volume, where *f* is the flow rate (milliliter/hr) and *V* is the vessel volume (milliliter).

$$D = f/V$$

For example, if *f* is 30 milliliters/hr and *V* is 100 milliliters, the dilution rate is 0.30 hr^{-1}.

Both population size and generation time are related to the dilution rate (**figure 7.40**). When dilution rates are very low, only a limited supply of nutrient is available and the microbes can conserve only a limited amount of energy. Much of that energy must be used for cell maintenance, not for growth and reproduction. Slightly higher dilution rates make more nutrients available to the microbes. When the dilution rate provides enough nutrients for both maintenance and reproduction, the cell density will begin to rise. In other words, the growth rate can increase (and the generation time can decrease) when the total available energy provided by the nutrient supply exceeds the **maintenance energy.** Notice in figure 7.40 that for a wide range of dilution rates, the cell density remains stable. This is because the limiting nutrient is almost completely depleted by the rapid reproductive rate of the cells (as seen by the decreasing generation times). However,

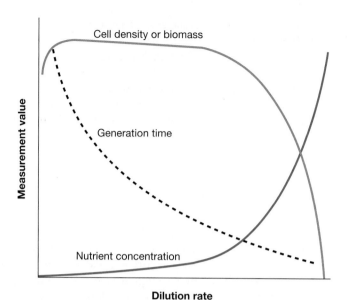

Figure 7.40 Chemostat Dilution Rate and Microbial Growth.
The effects of changing the dilution rate in a chemostat.

MICRO INQUIRY *Why does the cell density stay relatively constant over such a large range of dilution rates?*

if the dilution rate is too high, microorganisms can actually be washed out of the culture vessel before reproducing because the dilution rate is greater than the maximum growth rate.

Turbidostats

The second type of continuous culture system, the **turbidostat,** has a photocell that measures the turbidity (defined as the amount of light scattered) of the culture in the growth vessel. The flow rate of media through the vessel is automatically regulated to maintain a predetermined turbidity. Because turbidity is related to cell density, the turbidostat maintains a desired cell density. Turbidostats differ from chemostats in several ways. The dilution rate in a turbidostat varies, rather than remaining constant, and a turbidostat's culture medium contains all nutrients in excess. That is, none of the nutrients is limiting. A turbidostat operates best at high dilution rates; a chemostat is most stable and effective at lower dilution rates.

Retrieve, Infer, Apply

1. How does a continuous culture system differ from a closed culture system (i.e., a batch culture)?
2. Describe how chemostats and turbidostats operate. How do they differ?
3. What is the dilution rate? What is maintenance energy? How are they related?

Key Concepts

7.1 Most Bacteria and Archaea Reproduce by Binary Fission

■ Many eukaryotic microbes are capable of carrying out both sexual and asexual reproduction. The former involves meiosis and the latter, mitosis.
■ The most common type of cell division observed among bacteria and archaea is binary fission. During binary fission, the cell elongates, the chromosome is replicated, and then it segregates to opposite poles of the cell prior to completing the septum, which divides the cell into two progeny cells (**figure 7.1**).
■ Other bacterial reproductive strategies include budding, and baeocyte and spore formation (**figure 7.2**).

7.2 Bacterial Cell Cycles Can Be Divided into Three Phases

■ Bacterial cell cycles consist of an initial growth phase followed by chromosome replication and segregation, and cytokinesis. The latter two phases overlap (**figure 7.3**).
■ In bacterial cells, chromosome replication begins at a single site called the origin of replication. Many models for partitioning of the progeny chromosomes have been developed. The best-studied partitioning system is the ParAB/*parS* system of *Caulobacter crescentus* (**figure 7.4**). It is unclear how widespread this mechanism is among other bacteria.
■ Septum formation in most bacteria is accomplished by Z ring formation and assembly of cell wall–synthesizing

machinery at the center of the cell once the chromosome has been partially segregated. This is brought about by Min proteins and nucleoid exclusion (**figures 7.5** and **7.6**).
■ In rapidly dividing cells, initiation of DNA synthesis may occur before the previous round of synthesis is completed. This shortens the time needed for completing the cell cycle.
■ The shape of bacterial cells is maintained by the cell wall. In rod-shaped bacteria, the cytoskeletal proteins FtsZ and MreB are involved in determining where peptidoglycan will be synthesized. When rods are lengthening prior to cell division, MreB localizes along the length of the cell, and peptidoglycan synthesis occurs there as a result. During cell division, FtsZ localization confines peptidoglycan synthesis to the midcell (**figure 7.8**). In cocci, FtsZ and peptidoglycan synthesis localize to the midcell.

7.3 Some Archaeal Cell Cycles Resemble the Eukaryotic Cell Cycle

■ Members of the order *Sulfolobales* have a cell cycle that consists of G1-, S-, and G2-like phases followed by partitioning and cytokinesis. The partitioning process is similar to the ParAB/*parS* system of *C. crescentus*. However, these archaea use homologues of eukaryotic ESCRT proteins for cytokinesis (**figure 7.10**).
■ The cell cycles of other archaea are less well studied. However, many have homologues of FtsZ, which are thought to function in cytokinesis.

7.4 Environmental Factors Affect Microbial Growth

■ Solute concentrations can adversely affect microbes by causing lysis or shrinkage of the cytoplasm. Most bacteria, photosynthetic protists, and fungi have rigid cell walls that help protect them from osmotic stress. By synthesizing cytoplasmic compatible solutes, they can remain hypertonic relative to their habitats. Solutes also affect microbes because they interact with water, making it less available for microbial activities. The amount of water available to microorganisms is expressed as the water activity (a_w). Most microorganisms do not grow well at water activities below 0.98; osmotolerant organisms survive and even flourish at low a_w values. Halophiles actually require high sodium chloride concentrations for growth (**figure 7.11** and **table 7.2**).

■ Each microbial species has an optimum pH for growth and can be classified as an acidophile, neutrophile, or alkalophile (**figure 7.12** and **table 7.2**).

■ Microorganisms have distinct temperature ranges for growth with minima, maxima, and optima—the cardinal temperatures. These ranges are determined by the effects of temperature on the rates of catalysis, protein denaturation, and membrane disruption. There are five major classes of microorganisms with respect to temperature preferences: (1) psychrophiles, (2) psychrotolerants, (3) mesophiles, (4) thermophiles, and (5) hyperthermophiles (**figures 7.13** and **7.14** and **tables 7.2** and **7.3**).

■ Microorganisms can be placed into at least five different categories based on their response to the presence of O_2: obligate aerobes, microaerophiles, facultative anaerobes, aerotolerant anaerobes, and strict or obligate anaerobes (**figure 7.15** and **table 7.2**). Oxygen can become toxic because of the production of hydrogen peroxide, superoxide radical, and hydroxyl radical. These are destroyed by the enzymes superoxide dismutase, catalase, and peroxidase.

■ Most deep-sea microorganisms are barotolerant, but some are piezophilic and require high pressure for optimal growth.

■ High-energy or short-wavelength radiation harms organisms in several ways. Ionizing radiation—X rays and gamma rays—ionizes molecules and destroys DNA and other cell components. Ultraviolet (UV) radiation induces DNA damage. Visible light can provide energy for the formation of reactive singlet oxygen, which will destroy cells.

7.5 Microbial Growth in Natural Environments

■ Microbial growth in natural environments is profoundly affected by nutrient limitations and other adverse factors. Microbes are often found in oligotrophic environments. They have evolved strategies to survive in these habitats, including spore formation and other types of dormancy, persister formation, and the production of starvation proteins.

■ Many microbes form biofilms—aggregations of microbes growing on surfaces and held together by extracellular polymeric substances (EPS) (**figures 7.18** and **7.19**). Life in a biofilm has several advantages, including protection from harmful agents.

■ Bacteria often use chemical signals to communicate with one another in a density-dependent way and carry out a particular activity only when a certain population density is reached. This phenomenon is called quorum sensing (**figure 7.20**).

■ Some bacteria are able to communicate via chemicals with large organisms such as plants. A well-studied example is the rhizobium-plant interaction.

7.6 Laboratory Culture of Cellular Microbes Requires Media and Conditions That Mimic the Normal Habitat of a Microbe

■ Culture media can be constructed completely from chemically defined components (defined media or synthetic media) or constituents, such as peptones and yeast extract, whose precise composition is unknown (complex media). Culture media can be solidified by the addition of agar, a complex polysaccharide from red algae.

■ Culture media can be classified based on function. Supportive media are used to culture a wide variety of microbes. Enriched media are supportive media that contain additional nutrients needed by fastidious microbes. Selective media contain components that select for the growth of some microbes. Differential media contain components that allow microbes to be differentiated from each other, usually based on some metabolic capability.

■ Aerobes are cultivated by supplying oxygen. Anaerobes are cultivated using media and methods that exclude oxygen (**figures 7.23** and **7.24**).

■ Enrichment cultures are usually the first step in isolating pure cultures. They use selective conditions (media, environmental conditions) to promote the growth of certain microbes. After enrichment, pure cultures usually are obtained by isolating individual cells with any of three plating techniques: the streak-plate, spread-plate, and pour-plate methods. The streak-plate technique uses an inoculating loop to spread cells across an agar surface. The spread-plate and pour-plate methods usually involve diluting a culture or sample and then plating the dilutions. In the spread-plate technique, cells are spread on an agar surface using a bent rod; in the pour-plate technique, the cells are first mixed with cooled agar-containing media before being poured into a Petri dish (**figures 7.25–7.27**).

■ Microorganisms growing on solid surfaces tend to form colonies with distinctive morphology (**figure 7.28**).

■ New approaches are being developed to culture microbes that in the past have not been culturable. These include creating natural habitats within the lab setting and co-culturing microbes with organisms that help them survive.

7.7 Growth Curves Consist of Five Phases

■ Growth is an increase in cellular constituents and results in an increase in cell size, cell number, or both.

■ When microorganisms are grown in a batch culture, the resulting growth curve usually has five phases: lag, exponential (log), stationary, death, and long-term stationary phase (**figure 7.29**).

- In the exponential phase, the population number of cells undergoing binary fission doubles at a constant interval called the generation or doubling time (**figure 7.31**). The growth rate constant (k) is the reciprocal of the generation time (**figure 7.32**).

7.8 Microbial Population Size Can Be Measured Directly or Indirectly

- Microbial populations can be counted directly with counting chambers, flow cytometers, electronic counters, or fluorescence microscopy. However, it is sometimes difficult to discern if cells are living or dead (**figures 7.34** and **7.35**).

- Indirect measurements include viable counting techniques (standard plate counts) such as the spread-plate, the pour-plate, or the membrane filter method (**figure 7.36**).
- Population size also can be determined indirectly by measuring microbial mass through the measurement of dry weight, turbidity, or the amount of a cell component (**figure 7.38**).

7.9 Chemostats and Turbidostats Are Used for Continuous Culture of Microorganisms

- Microorganisms can be grown in an open system in which nutrients are constantly provided and wastes removed.
- A continuous culture system can maintain a microbial population in log phase. There are two types of these systems: chemostats and turbidostats (**figure 7.39**).

Hypothesize, Compare, Invent

1. As an alternative to diffusible signals, suggest another mechanism by which bacteria might quorum sense.

2. If you wished to obtain a pure culture of bacteria that could degrade benzene and use it as a carbon and energy source, how would you proceed?

3. Design an experiment to determine if a slow-growing microbial culture is in lag phase or exponential phase.

4. Suggest one specific mechanism underlying the observation that the cardinal temperatures of some microbes change depending on other environmental conditions (e.g., pH).

5. Consider cell-cell communication: bacteria that "subvert" and "cheat" have been described. Describe a situation in which it would be advantageous for one species to subvert another, that is, degrade an intercellular signal made by another species. Also, describe a scenario whereby bacterial cheaters—defined as bacteria that do not make a molecular signal but profit by the uptake and processing of signal made by another microbe—might have a growth advantage.

6. Suppose you discovered a new bacterial strain from the permafrost of Alaska. You are able to culture the bacterium, but it grows very slowly. What characteristics of the bacterium would you determine to see if you could optimize its growth rate? Explain your choices.

7. Because many persistent bacterial infections involve biofilms, there is great interest in developing strategies to prevent their formation and eliminate them once they have formed. The red alga *Delisea pulchra* produces a class of compounds known as halogenated furanones that inhibit biofilm formation by *Pseudomonas aeruginosa, E. coli, B. subtilis, Staphylococcus epidermidis,* and *Streptococcus* spp. In these microbes, the algal furanones inhibit cell-cell signaling. While furanones also inhibit biofilm formation in the food-borne pathogen *Salmonella enterica* serovar Typhimurium, they do not alter cell-cell signaling. How does the inhibition of intercellular communication inhibit biofilm formation in *P. aeruginosa* and others? Do you think the fact that the furanones fail to interrupt cell-cell communication in *Salmonella* means that these bacteria do not use intercellular signaling when constructing biofilms? What alternative explanations might explain the loss of biofilm formation without influencing cell-cell communication? Suggest a use for the algal furanones in a food-processing facility.

Read the original paper: Janssens, J. C., et al. 2008. Brominated furanones inhibit biofilm formation by *Salmonella enterica* serovar Typhimurium. *Appl. Environ. Microbiol.* 74:6639.

Learn More

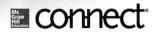 

8

Control of Microorganisms in the Environment

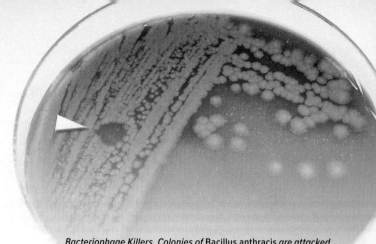

Bacteriophage Killers. Colonies of Bacillus anthracis *are attacked and killed by a highly specific gamma-phage. A single drop of gamma-phage applied to the 24-hour old culture of* B. anthracis *resulted in the large clearing, or plaque (arrow), where the bacteriophage lysed the host bacteria. The bacteriophage will continue to infect and replicate until no host cells remain.*

Bacterial Kamikazes Seek Out and Destroy Pathogens

There is no hiding anymore for the evasive pathogen *Pseudomonas aeruginosa*. Researchers have created a predator drone that can detect *P. aeruginosa,* a leading cause of hospital-acquired infections, and kill it. Using the tools of synthetic biology, scientists at Nanyang Technical University in Singapore engineered the common *Escherichia coli* bacterium to detect chemical communication signals—acylhomoserine lactones (AHLs)—produced by *P. aeruginosa* and, in turn, produce a novel protein antibiotic, pyocin S5. Sensing the AHLs, the engineered *E. coli* mass-produce the pyocin S5, resulting in its explosive release. The result is a lot of dead *P. aeruginosa*, along with the destruction of the *E. coli* drone. This innovative use of synthetic biology introduces an exciting extension of "biological" control. It paves a new way for eliminating bacteria that are difficult to control by conventional disinfection techniques.

P. aeruginosa is notorious for hiding in water supplies, bathrooms, and even soaps used for hand washing. It can then infect humans (especially those who are immunocompromised), colonizing the respiratory and gastrointestinal tracts where it rapidly develops resistance to conventional antibiotics. This use of engineered bacteria enables bacterial eradication with surgical precision and no effect on the remaining host microbiota. The predator drone grows and reproduces until it finds its pathogen targets, and then it is destroyed along with the pathogens. Perhaps someday you will see "Kamikaze Soap" in the grocery store.

In this chapter, we address the control and destruction of microorganisms, a topic of immense practical importance. Although most microorganisms are beneficial, some microbial activities have undesirable consequences, such as food spoilage and disease. Therefore it is essential to be able to destroy a wide variety of microorganisms or inhibit their replication to minimize their destructive effects. The goal is twofold: (1) to destroy pathogens and prevent their transmission, and (2) to reduce or eliminate microorganisms responsible for the contamination of water, food, and other substances.

Microorganism control continues to be a hot topic as microorganisms evolve to resist current strategies. Control efforts have a substantial role in public health to prevent disease as well as in therapeutic use to treat disease. Thus this chapter focuses on the control of microorganisms by physical, chemical, and biological agents (such as engineered bacteria). In general, any chemical, physical, or biological product that controls microorganisms is referred to as an **antimicrobial agent.** Chemotherapeutic agents that are used inside the body to treat human disease are discussed in chapter 9.

Readiness Check:

Based on what you have learned previously, you should be able to:

✔ Identify the structures and their functions of bacteria, protists, fungi, viruses, and prions, including their replication processes and energy requirements (sections 3.2–3.8, 5.1–5.7, 6.1–6.7, 7.1–7.7)

8.1 Microbial Growth and Replication Pathways: Targets for Control

After reading this section, you should be able to:

- Compare and contrast actions of disinfection, antisepsis, chemotherapy, and sterilization
- Distinguish between cidal (killing) and static (inhibitory) agents

The principles of microbial control are rooted in microbial nutrition, growth, and development, which are discussed in chapter 7. If you can starve or poison, or inhibit or prevent growth or replication, you can control microorganisms. Of course, it is not as simple as this sounds. Subtle differences in population density, degree of killing, and even what is used to remove microorganisms result in a somewhat complex vocabulary. Terminology is especially important when the control of microorganisms is discussed because words such as *disinfectant* and *antiseptic* often are used loosely, and as we shall see, they have different meanings. The situation is even more confusing because a particular treatment can either inhibit growth,

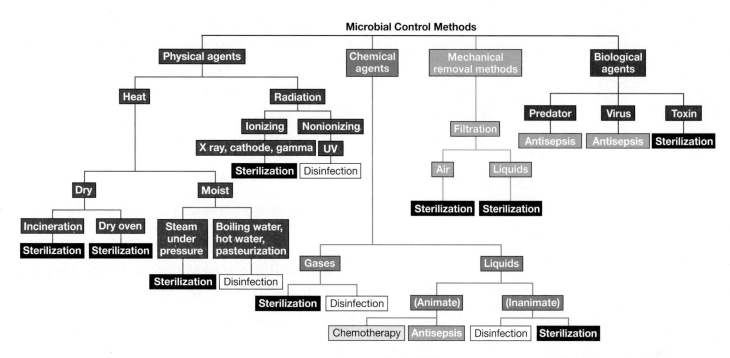

Disinfection: The destruction or removal of vegetative pathogens but not bacterial endospores. Usually used only on inanimate objects.
Sterilization: The complete removal or destruction of all viable microorganisms. Used on inanimate objects.
Antisepsis: Chemicals applied to body surfaces to destroy or inhibit vegetative pathogens.
Chemotherapy: Chemicals used internally to kill or inhibit growth of microorganisms within host tissues.

Figure 8.1 Microbial Control Methods.

MICRO INQUIRY *Which types of agents can be used for sterilization? Which can be used for antisepsis? What is the difference?*

inactivate replication, or kill, depending on the conditions. The types of control agents and their uses are outlined in **figure 8.1**. Note that not all agents are chemical. Rather, antimicrobial agents can be chemical, physical, mechanical, and biological. To simplify the terminology, we use the term **biocide** to mean all antimicrobial agents that can be used to control microorganisms. In general, to control microorganisms a biocide must be evaluated so as to determine the specific parameters under which it will be effective.

Sterilization (Latin *sterilis,* unable to produce offspring or barren), in our context, is the process by which all living cells, spores, and acellular entities (e.g., viruses, viroids, and prions) are either destroyed or removed from an object or habitat. A sterile object is totally free of viable microorganisms, spores, and other infectious agents. When sterilization is achieved by a chemical agent, the chemical is called a sterilant. In contrast, **disinfection** is the killing, inhibition, or removal of microorganisms that may cause disease; disinfection is the substantial reduction of the total microbial population and the destruction of potential pathogens. **Disinfectants** are agents, usually chemical, used to carry out disinfection and normally used only on inanimate objects. A disinfectant does not necessarily sterilize an object because viable spores and a few microorganisms may

remain. **Sanitization** is closely related to disinfection. In sanitization, the microbial population is reduced to levels that are considered safe by public health standards. The inanimate object is usually cleaned as well as partially disinfected. For example, sanitizers are used to clean eating utensils in restaurants. ◄◄ *Viroids and satellites (section 6.6); Prions are composed only of protein (section 6.7)*

It also is frequently necessary to control microorganisms on or in living tissue. **Antisepsis** (Greek *anti,* against, and *sepsis,* putrefaction) is the destruction or inhibition of microorganisms on living tissue; it is the prevention of infection or sepsis. **Antiseptics** are chemical agents applied to tissue to prevent infection by killing or inhibiting pathogen growth; they also reduce the total microbial population. Because they must not cause too much harm to the host, antiseptics are generally not as toxic as disinfectants. The exposure of microorganisms to increasing biocide concentrations decreases the number of viable organisms. **Figure 8.2** shows three possible population reduction curves resulting from three different biocides. The shape of the curve reflects various conditions that influence biocide effectiveness. Note that in each case, the eventual decline in viable microorganisms can occur as a staged interval of viability from antisepsis to sterilization. **Chemotherapy** is the use of

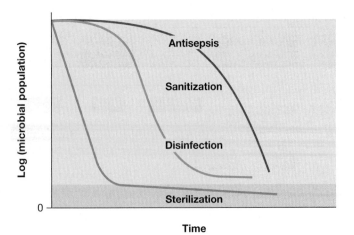

Figure 8.2 Impact of Biocide Exposure. Three exponential plots of survivors versus time of biocide exposure, indicating the potential kinetics of biocide action. Note how the general terms referencing microbial control are reflected by decreasing numbers of microbes. For example, sterilization refers to the absence of viable organisms regardless of biocide kinetics.

chemical agents to kill or inhibit the growth of microorganisms within host tissue and is the topic of chapter 9.

A suffix can be employed to denote the type of antimicrobial agent. Substances that kill organisms often have the suffix –*cide* (Latin *cida,* to kill); a cidal agent kills pathogens (and many nonpathogens) but not necessarily endospores. A disinfectant or antiseptic can be particularly effective against a specific group, in which case it may be called a **bactericide, fungicide,** or **viricide.** Other chemicals do not kill but rather prevent growth. If these agents are removed, growth will resume. Their names end in –*static* (Greek *statikos,* causing to stand or stopping); for example, **bacteriostatic** and **fungistatic.**

It is worth noting here that resistance to antimicrobial biocides has been increasing nearly as rapidly as resistance to antibiotics. Similar to antibiotic resistance mechanisms (discussed in Chapter 16), biocide resistance mechanisms include the induction of

efflux pumps, modified membrane permeability, and target modification. In fact, a number of studies have linked the acquisition of biocide resistance mechanisms to resistance to antibiotics. In many cases they are shared on plasmids and upregulated in response to ubiquitously low environmental biocide concentrations. It is clear that microorganisms adapt well to new situations, resulting in the sharing of successful resistance strategies.

Retrieve, Infer, Apply

1. Define the following terms: sterilization, sterilant, disinfection, disinfectant, sanitization, antisepsis, antiseptic, chemotherapy, biocide.
2. What is the difference between bactericidal and bacteriostatic? To which category do you think most household cleaners belong? Why?

8.2 The Pattern of Microbial Death Mirrors the Pattern of Microbial Growth

After reading this section, you should be able to:

- Calculate the decimal reduction time (D value)
- Correlate antisepsis, sanitization, disinfection, and sterilization with agent effectiveness

A microbial population is not killed instantly when exposed to a lethal agent. Population death is generally exponential (logarithmic); that is, the population is reduced by the same fraction at constant intervals (**table 8.1**). If the logarithm of the population number remaining is plotted against the time of exposure of the microorganism to the agent, a straight-line plot will result (**figure 8.3**). When the population has been greatly reduced, the rate of killing may slow due to the survival of a more resistant strain of the microorganism.

It is essential to have a precise measure of an agent's killing efficiency. One such measure is the **decimal reduction time** (D) or **D value.** The decimal reduction time is the time required to kill 90% of the microorganisms or spores in a sample under specified

Table 8.1	A Theoretical Microbial Heat-Killing Experiment			
Minute	Microbial Number at Start of Minute	Microorganisms Killed in 1 Minute (90% of Total)[1]	Microorganisms at End of 1 Minute	Log_{10} of Survivors
1	10^6	9×10^5	10^5	5
2	10^5	9×10^4	10^4	4
3	10^4	9×10^3	10^3	3
4	10^3	9×10^2	10^2	2
5	10^2	9×10^1	10	1
6	10^1	9	1	0
7	1	0.9	0.1	−1

1 Assume that the initial sample contains 10^6 vegetative microorganisms per milliliter and that 90% of the organisms are killed during each minute of exposure. The temperature is 121°C.

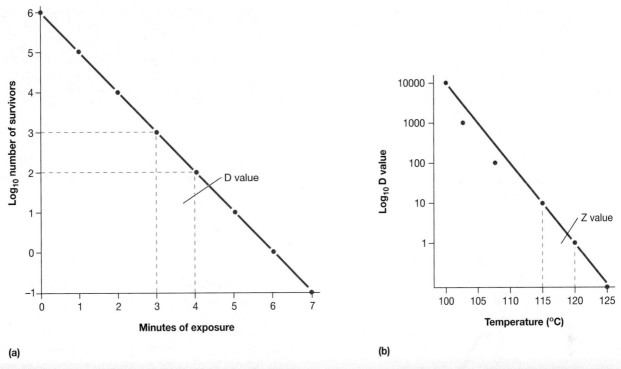

Figure 8.3 The Pattern of Microbial Death.   (a) An exponential plot of the survivors versus the minutes of exposure to heating at 121°C. In this example, the D value is 1 minute. The data are from table 8.1. (b) An exponential plot of D values versus temperature. The temperature change at a given D value that reduces the population by one log unit is known as the Z value.

MICRO INQUIRY *Examine graph (a). How long would it take to kill one-half of the original population of microorganisms? Keep in mind that the Y axis is exponential.*

conditions. For example, in a semilogarithmic plot of the population remaining versus the time of heating, the D value is the time required for the line to drop by one log cycle or tenfold (figure 8.3a). It is also possible to determine the temperature change at a given D value that decreases the microbial population by one log cycle (90%). This temperature change is referred to as the Z value and is predicted from a semilogarithmic plot of D values versus temperature (figure 8.3b).

To study the effectiveness of a lethal agent, one must be able to decide when microorganisms are dead, which may present some challenges. A microbial cell is often defined as dead if it does not grow when inoculated into culture medium that would normally support its growth. In like manner, an inactive virus cannot infect a suitable host. This definition has flaws, however. It has been demonstrated that when bacteria are exposed to certain conditions, they can remain alive but are temporarily unable to reproduce (*see figure 7.35*). When in this state, these cells are often referred to as viable but nonculturable (VBNC). In conventional tests to demonstrate killing by an antimicrobial agent, VBNC bacteria would be thought to be dead. This is a serious problem because the bacteria may regain their ability to reproduce and cause infection after a period of recovery. ◀◀ *Death phase (section 7.7)*

8.3 Mechanical Removal Methods Rely on Barriers

After reading this section, you should be able to:

- Explain the mechanism by which filtration removes microorganisms
- Propose filtration methods to selectively remove one type of microbe from a mixed population

Filtration is an excellent way to reduce the microbial population in solutions of heat-sensitive material and can be used to sterilize various liquids and gases (including air). Rather than directly destroying contaminating microorganisms, the filter simply acts as a barrier to remove them. There are two types of filters. **Depth filters** consist of fibrous or granular materials that have been bonded into a thick layer filled with twisting channels of small diameter. The solution containing microorganisms is sucked through this layer under vacuum, and microbial cells are removed by physical screening or entrapment and by adsorption to the surface of the filter material. Depth filters are made of diatomaceous earth (Berkefield filters), unglazed porcelain (Chamberland filters), asbestos, or other similar materials.

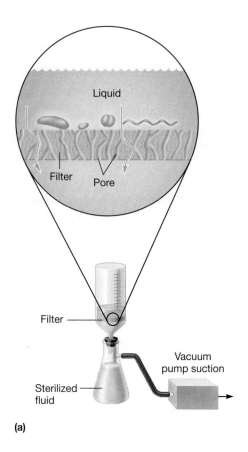

Liquid

Filter   Pore

Filter

Vacuum
pump suction

Sterilized
fluid

(a)

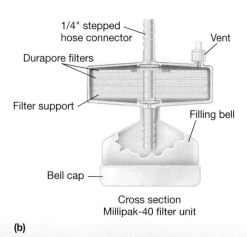

1/4" stepped
hose connector

Vent

Durapore filters

Filter support

Filling bell

Bell cap

Cross section
Millipak-40 filter unit

(b)

Figure 8.4 Membrane Filter Sterilization. The liquid to be sterilized is pumped through a membrane filter and into a sterile container. (a) Schematic representation of a membrane filtration setup that uses a vacuum pump to force liquid through the filter. The inset shows a cross section of the filter and its pores, which are too small for microbes to pass through. (b) Cross section of a membrane filtration unit. Several membranes are used to increase its capacity.

MICRO INQUIRY *How might one verify that filtration removed all microorganisms?*

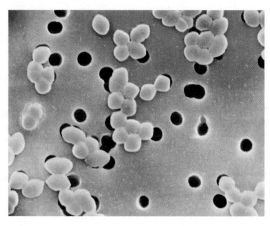

Figure 8.5 Membrane Filter. *Enterococcus faecalis* resting on a polycarbonate membrane filter with 0.4 μm pores (×5,900).

Membrane filters have replaced depth filters for many purposes. These filters are porous membranes, a little over 0.1 mm thick, made of cellulose acetate, cellulose nitrate, polycarbonate, polyvinylidene fluoride, or other synthetic materials. Although a wide variety of pore sizes are available, membranes with pores about 0.2 μm in diameter are used to remove most vegetative cells, but not viruses, from liquids ranging in volume from less than 1 milliliter to many liters. The membranes can be held in special holders (**figure 8.4**), and their use is often preceded by the use of depth filters made of glass fibers to remove larger particles that might clog the membrane filter. The liquid is pulled or forced through the filter with a vacuum or with pressure from a syringe, peristaltic pump, or nitrogen gas and collected in previously sterilized containers. Membrane filters remove microorganisms by screening them out much as a sieve separates large sand particles from small ones (**figure 8.5**). These filters are used to sterilize pharmaceuticals, ophthalmic solutions, culture media, oils, antibiotics, and other heat-sensitive solutions.

Air also can be filtered to remove microorganisms. Two common examples are N-95 disposable masks used in hospitals and labs, and **high-efficiency particulate air (HEPA) filters** in biosafety cabinets that let air move freely but restrict microorganisms. N-95 masks exclude 95% of particles that are larger than 0.3 μm. HEPA filters (a type of depth filter made from fiberglass) remove 99.97% of particles 0.3 μm or larger by both physical retention and electrostatic interactions. HEPA filters can also be made to adsorb viruses. HEPA filters are available that remove viruses that are 0.1 μm and smaller; they are used to sterilize air. Laminar flow biological safety cabinets or hoods force air through HEPA filters, then project a vertical curtain of sterile air across the cabinet opening. This protects a worker from microorganisms being handled within the cabinet and prevents contamination of the room (**figure 8.6**). A person uses these cabinets when working with dangerous agents such as *Mycobacterium tuberculosis*, pathogenic fungi, or tumor viruses. They are also employed in research labs and industries, such as the pharmaceutical industry, when a sterile working environment is needed.

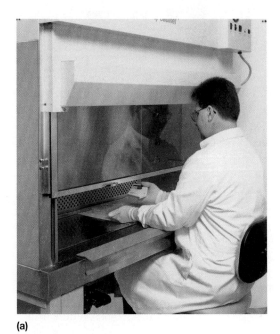

(a)

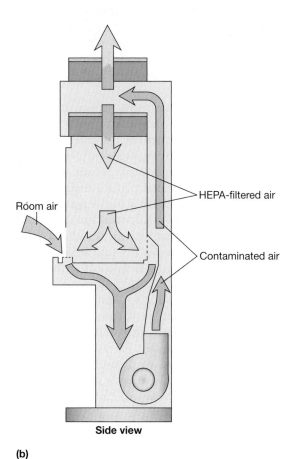

Room air

HEPA-filtered air

Contaminated air

Side view

(b)

Figure 8.6 A Biological Safety Cabinet. (a) A technician pipetting potentially hazardous material in a safety cabinet. (b) A schematic diagram showing the airflow pattern within a class II safety cabinet.

Retrieve, Infer, Apply

1. What are depth filters and membrane filters, and how are they used to sterilize liquids?
2. Describe the operation of a biological safety cabinet.
3. During the 2009–2010 pandemic flu outbreak, debate occurred regarding the use of N-95 masks to prevent virus spread. Should they have been used? What information would help you make a scientifically sound recommendation?

8.4 Physical Control Methods Alter Microorganisms to Make Them Nonviable

After reading this section, you should be able to:

- Describe the application of heat and radiation to control microorganisms
- Explain the mechanisms by which heat and radiation kill microbes
- Design novel antimicrobial control applications using heat and radiation to sterilize

Heat and other physical agents are normally used to control microbial growth and sterilize objects, as can be seen from the operation of the autoclave. The most frequently employed physical agents are heat and radiation.

Heat

Most microorganisms require specific temperatures for normal growth and replication; temperatures that exceed those damage structures and alter chemical reactions. Moist and dry heat readily destroy viruses, bacteria, and fungi in this way (**table 8.2**).

Moist heat destroys cells and viruses by degrading nucleic acids and denaturing enzymes and other essential proteins. It also disrupts cell membranes. Exposure to boiling water for 10 minutes is sufficient to destroy vegetative cells

Table 8.2	Approximate Conditions for Moist Heat Killing	
Organism	**Vegetative Cells**	**Spores**
Yeasts	5 minutes at 50–60°C	5 minutes at 70–80°C
Molds	30 minutes at 62°C	30 minutes at 80°C
Bacteria[1]	10 minutes at 60–70°C	2 to over 800 minutes at 100°C 0.5–12 minutes at 121°C
Viruses	30 minutes at 60°C	

1 Conditions for mesophilic bacteria.

(a)

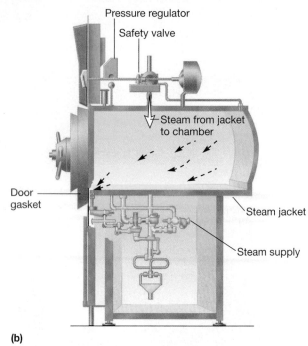

(b)

Figure 8.7 The Autoclave. (a) A modern, automatically controlled autoclave or sterilizer. (b) Longitudinal cross section of a typical autoclave showing some of its parts and the pathway of steam.

and eukaryotic spores. Unfortunately, the temperature of boiling water (100°C or 212°F at sea level) is not sufficient to destroy bacterial spores, which may survive hours of boiling. Therefore boiling can be used for disinfection of drinking water and objects not harmed by water, but boiling does not sterilize.

To destroy bacterial endospores, moist heat sterilization must be carried out at temperatures above 100°C, and this requires the use of saturated steam under pressure. Steam sterilization is carried out with an **autoclave** (figure 8.7), a device somewhat like a fancy pressure cooker. The development of the autoclave by Charles Chamberland in 1884 tremendously stimulated the growth of microbiology as a science. Water is boiled to produce steam, which is released into the autoclave's chamber (figure 8.7b). The air initially present in the chamber is forced out until the chamber is filled with saturated steam and the outlets are closed. Hot, saturated steam continues to enter until the chamber reaches the desired temperature and pressure, usually 121°C and 15 pounds per sq. in. (psi) of pressure.

Autoclaving must be carried out properly so that the saturated steam can destroy all vegetative cells and spores within the processed materials. If all air has not been flushed out of the chamber, it will not reach 121°C, even though it may reach a pressure of 15 psi. The chamber should not be packed too tightly because the steam needs to circulate freely and contact everything in the autoclave. Bacterial spores will be killed only if they are kept at 121°C for 10 to 12 minutes. When a large volume of liquid must be sterilized, an extended sterilization time is needed because it takes longer for the center of the liquid to reach 121°C; 5 liters of liquid may require about 70 minutes. In view of these potential difficulties, a biological indicator is often autoclaved along with other material. This indicator commonly consists of a culture tube containing a sterile ampule of medium and a paper strip covered with spores of *Geobacillus stearothermophilus*. After autoclaving, the ampule is aseptically broken and the culture incubated for several days. If the test bacterium

does not grow in the medium, the sterilization run has been successful. Sometimes indicator tape or paper that changes color upon sufficient heating is autoclaved with a load of material. These approaches that indicate heating has occurred are convenient and save time but are not as reliable as techniques that kill bacterial spores.

Many heat-sensitive substances, such as milk, are treated with controlled heating at temperatures well below boiling, a process known as **pasteurization** in honor of its developer, Louis Pasteur (1822–1895). In the 1860s the French wine industry was plagued by the problem of wine spoilage, which made wine storage and shipping difficult. Pasteur examined spoiled wine under the microscope and detected microorganisms that looked like the bacteria responsible for lactic acid and acetic acid fermentations; these bacteria do not form spores. He then discovered that a brief heating at 55 to 60°C would destroy these microorganisms and preserve wine for long periods. In 1886 the German chemists V. H. Soxhlet and F. Soxhlet adapted the technique for preserving milk and reducing milk-transmissible diseases. Milk pasteurization was introduced in the United States in 1889. Milk, beer, and many other beverages are now pasteurized. Pasteurization does not sterilize a beverage, but it does kill any pathogens present and drastically slows spoilage by reducing the level of nonpathogenic spoilage microorganisms (*see figure 41.1*).

Some materials cannot withstand the high temperature of the autoclave, and spore contamination precludes the use of other methods to sterilize them. For these materials, a process of intermittent sterilization, also known as **tyndallization** (for John Tyndall (1820–1893), the British physicist who used the technique to destroy heat-resistant microorganisms in dust) is used. The process also uses steam (30–60 minutes) to destroy vegetative bacteria. However, steam exposure is repeated for a total of three times with 23- to 24-hour incubations between steam exposures. The incubations permit remaining spores to germinate into heat-sensitive vegetative cells that are then destroyed upon subsequent steam exposures.

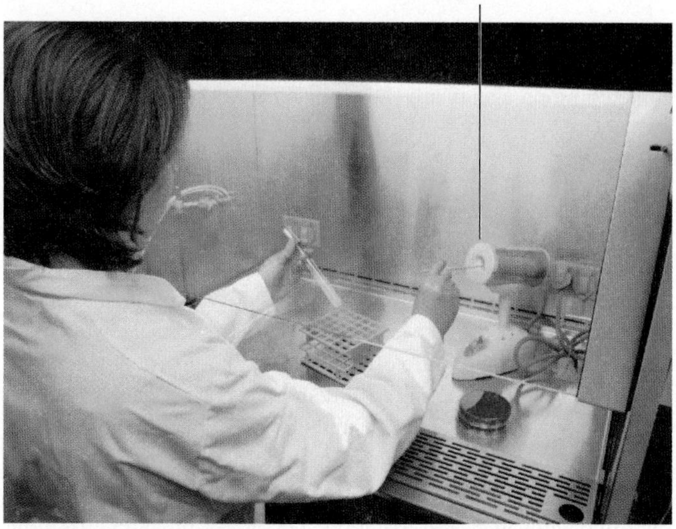

Incinerator

![Figure 8.8](bench-top incinerator photo)

Figure 8.8 Dry Heat Incineration. Bench-top incinerators are routinely used to sterilize inoculating loops used in microbiology laboratories.

Many objects are best sterilized by dry heat. Some items are sterilized by incineration. For instance, inoculating loops, which are used routinely in the laboratory, can be sterilized in a small, bench-top incinerator (**figure 8.8**). Other items are sterilized in an oven at 160 to 170°C for 2 to 3 hours. Microbial death results from the oxidation of cell constituents and denaturation of proteins. Dry air heat is less effective than moist heat. The spores of *Clostridium botulinum*, the cause of botulism, are killed in 5 minutes at 121°C by moist heat but only after 2 hours at 160°C by dry heat. However, dry heat has some definite advantages. It does not corrode glassware and metal instruments as moist heat does, and it can be used to sterilize powders, oils, and similar items. Despite these advantages, dry heat sterilization is slow and not suitable for heat-sensitive materials such as many plastic and rubber items.

Retrieve, Infer, Apply

1. Describe how an autoclave works. What conditions are required for sterilization by moist heat? What three things must one do when operating an autoclave to help ensure success?
2. In the past, spoiled milk was responsible for a significant proportion of infant deaths. Why is untreated milk easily spoiled?

Radiation

Ultraviolet (UV) radiation around 260 nm (*see figure 7.16*) is quite lethal. It causes thymine-thymine dimerization of DNA, preventing replication and transcription (*see figure 16.5*). However, UV radiation does not penetrate glass, dirt films, water, and other substances very effectively. Because of this disadvantage, UV radiation is used as a sterilizing agent only in a few specific situations. UV lamps are sometimes placed on the ceilings of rooms or in biological safety cabinets to sterilize the air and any exposed surfaces. Because UV radiation burns the skin and damages eyes, the UV lamps are off when the areas are in use. Commercial UV units are available for water treatment. Pathogens and other microorganisms are destroyed

when a thin layer of water is passed under the lamps. ▶▶| *Purification and sanitary analysis ensure safe drinking water (section 43.1)*

Ionizing radiation is an excellent sterilizing agent and penetrates deep into objects. Ionizing radiation has sufficient energy to dislodge electrons from atoms or molecules, producing chemically reactive free radicals. The free radicals react with nearby matter to weaken or destroy it. Ionizing radiation destroys bacterial spores and all microbial cells; however, ionizing radiation is not always effective against viruses. Gamma radiation from a cobalt 60 source and accelerated electrons from high-voltage electricity are used in the cold sterilization of antibiotics, hormones, sutures, and plastic disposable supplies such as syringes. Gamma radiation and electron beams have also been used to sterilize and "pasteurize" meat and other foods (**figure 8.9**). Irradiation can eliminate the threat of such pathogens as *E. coli* O157:H7, which causes a life-threatening intestinal disease; *Staphylococcus aureus,* which causes skin and blood infections, and readily colonizes medical devices used on patients; and *Campylobacter jejuni,* which contaminates poultry, causing human intestinal disease when undercooked meat is eaten. Based on the results of numerous studies, both the U.S. Food and Drug Administration and the World Health Organization have approved irradiated food and declared it safe for human consumption. Currently irradiation is being used to treat poultry, beef, pork, veal, lamb, fruits, vegetables, and spices. ▶▶| *Human diseases caused by bacteria (chapter 39); Various methods are used to control food spoilage (section 41.2)*

Retrieve, Infer, Apply

1. List the advantages and disadvantages of ultraviolet light and ionizing radiation as sterilizing agents. Provide a few examples of how each is used for this purpose.
2. What is the correlation between radiation "energy" and the mechanisms of sterilization?

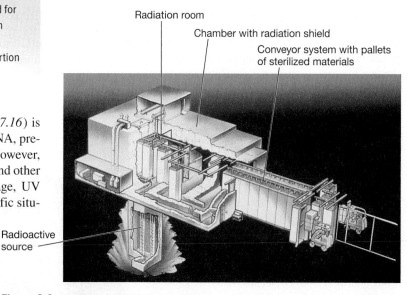

Radiation room

Chamber with radiation shield

Conveyor system with pallets of sterilized materials

Radioactive source

Figure 8.9 Sterilization with Ionizing Radiation. An irradiation machine that uses radioactive cobalt 60 as a gamma radiation source to sterilize fruits, vegetables, meats, fish, and spices.

8.5 Microorganisms Are Controlled with Chemical Agents

After reading this section, you should be able to:

- Describe the use of phenolics, alcohols, halogens, heavy metals, quaternary ammonium chlorides, aldehydes, and oxides to control microorganisms
- Explain the mechanisms of action for phenolics, alcohols, halogens, heavy metals, quaternary ammonium chlorides, aldehydes, and oxides
- Design novel antimicrobial control applications using phenolics, alcohols, halogens, heavy metals, quaternary ammonium chlorides, aldehydes, and oxides

Physical agents are generally used to sterilize objects. Chemicals can be employed for sterilization, disinfection, and antisepsis. The proper use of chemical agents is essential for personal safety. Chemicals also are employed to prevent microbial growth in food, and certain chemicals are used to treat infectious disease. The use of chemical agents for chemotherapy in humans is covered in chapter 9. Here we discuss chemicals used outside the body.

Many different chemicals are available for use as biocides, and many have been specifically formulated as disinfectants, each with its own advantages and disadvantages. Ideally the biocide must be effective against a wide variety of infectious agents (bacteria, bacterial spores, fungi, viruses, and prions) at low concentrations and in the presence of organic matter. Although the chemical must be toxic for infectious agents, it should not be toxic to people or corrosive for common materials. In practice, this balance between effectiveness and low toxicity for animals is hard to achieve. Some chemicals are used despite their low effectiveness because they are relatively nontoxic. The ideal disinfectant should be stable upon storage, odorless or with a pleasant odor, and soluble in water and lipids for penetration into microorganisms; have a low surface tension so that it can enter cracks in surfaces; and be relatively inexpensive.

Other chemical biocides are used as antiseptics. Recall that antiseptics are less toxic to humans than disinfectants and as such may be less effective at killing all the microorganisms that disinfectants can kill. In general, antiseptics should reduce the number of pathogens on human tissue to prevent infection. Examples of antiseptics include hand sanitizers, silver threads woven into clothing, and dilute iodine solutions that can be sprayed onto wounds. Antiseptics are meant to be used whenever thorough cleaning with soap and water will be delayed. One potentially serious problem is the overuse of antiseptics. For instance, resistance to the antibacterial agent triclosan (found in products such as deodorants, mouthwashes, soaps, cutting boards, and baby toys) has become a problem. There is now evidence that extensive use of triclosan also increases the frequency of bacterial resistance to antibiotics. Thus overuse of antiseptics can have unintended harmful consequences. ▶▶| *There are several mechanisms of drug resistance (section 16.9)*

The properties and uses of several groups of common disinfectants and antiseptics are surveyed next. Many of the characteristics of disinfectants and antiseptics are summarized in **tables 8.3** and **8.4**. Structures of some common agents are shown in **figure 8.10**.

Phenolics

Phenol was the first widely used antiseptic and disinfectant. In 1867 Joseph Lister employed it to reduce the risk of infection during surgery. Today phenol and phenol derivatives (phenolics) such as cresols, xylenols, and orthophenylphenol are used as disinfectants in laboratories and hospitals. The commercial disinfectant Lysol is made of a mixture of phenolics. Phenolics act by denaturing proteins and disrupting cell membranes. They have some real advantages as disinfectants: phenolics are tuberculocidal, effective in the presence of organic material, and remain active on surfaces long after application. However, they have a disagreeable odor and can cause skin irritation. The newer phenolic, triclosan (figure 8.10), is often used in hand sanitizers due to its effective blockage of bacterial fatty acid synthesis.

Alcohols

Alcohols are among the most widely used disinfectants, antiseptics, and sanitizers. They are bactericidal and fungicidal but not sporicidal; some enveloped viruses are also destroyed. The two most popular alcohol germicides are ethanol and isopropanol, usually used in about 60 to 80% concentration. They act by denaturing proteins and possibly by dissolving membrane lipids. A 10- to 15-minute soaking is sufficient to disinfect small instruments, while rubbing hands with specially formulated alcohol products sanitizes them by killing many pathogens.

Halogens

A halogen is any of the five elements (fluorine, chlorine, bromine, iodine, and astatine) in group VIIA of the periodic table. They exist as diatomic molecules in the free state and form saltlike compounds with sodium and most other metals. The halogens iodine and chlorine are important antimicrobial agents. Iodine is used as a skin antiseptic and kills by oxidizing cell constituents and iodinating cell proteins. At higher concentrations, it may even kill some spores. Iodine often is applied as tincture of iodine, 2% or more iodine in a water-ethanol solution of potassium iodide. Although it is an effective antiseptic, the skin may be damaged, a stain is left, and iodine allergies can result. Iodine can be complexed with an organic carrier to form an **iodophor.** Iodophors are water soluble, stable, and nonstaining, and release iodine slowly to minimize skin burns and irritation. They are used in hospitals for cleansing preoperative skin and in hospitals and laboratories for disinfecting. Some popular brands are Wescodyne for skin and laboratory disinfection, and Betadine for wounds.

Chlorine is the usual disinfectant for municipal water supplies and swimming pools, and is also employed in the dairy and food industries. It may be

Table 8.3	Activity Levels of Selected Biocides	
Class	**Use Concentration of Active Ingredient**	**Activity Level[1]**
Gas		
Ethylene oxide	450–500 mg/L[2]	High
Liquid		
Glutaraldehyde, aqueous	2–8%	High to intermediate
Formaldehyde + alcohol	8 + 70%	High to intermediate
Stabilized hydrogen peroxide	6–30%	High to intermediate
Formaldehyde, aqueous	6–8%	High to intermediate
Iodophors, high concentration	5,000–10,000 mg/L[3]	High to intermediate
Iodophors, low concentration	75–150 mg/L[3]	Intermediate to low
Iodine + alcohol	0.5 + 70%	Intermediate
Chlorine compounds	500–5,000 mg/L[4]	Intermediate
Phenolic compounds, aqueous	0.5–3%	Intermediate to low
Iodine, aqueous	1%	Intermediate
Alcohols (ethyl, isopropyl)	62–70%	Intermediate
Quaternary ammonium compounds	0.1–0.2% aqueous	Low
Chlorhexidine	0.75–4%	Low
Hexachlorophene	1–3%	Low
Mercurial compounds	0.1–0.2%	Low

1 High-level disinfectants destroy vegetative bacterial cells including *M. tuberculosis,* bacterial endospores, fungi, and viruses. Intermediate-level disinfectants destroy all of the above except spores. Low-level agents kill bacterial vegetative cells except for *M. tuberculosis,* fungi, and medium-sized lipid-containing viruses (but not bacterial endospores or small, nonlipid viruses).
2 In autoclave-type equipment at 55 to 60°C.
3 Available iodine.
4 Free chlorine.

applied as chlorine gas (Cl_2), sodium hypochlorite (bleach, NaOCl), or calcium hypochlorite [$Ca(OCl)_2$], all of which yield hypochlorous acid (HOCl):

$$Cl_2 + H_2O \rightarrow HCl + HOCl$$
$$NaOCl + H_2O \rightarrow NaOH + HOCl$$
$$Ca(OCl)_2 + 2H_2O \rightarrow Ca(OH)_2 + 2HOCl$$

The result is oxidation of cellular materials and destruction of vegetative bacteria and fungi. Death of almost all microorganisms usually occurs within 30 minutes. One potential problem is that chlorine reacts with organic compounds to form carcinogenic trihalomethanes, which must be monitored in drinking water. Some wastewater treatment facilities recover the chlorine from the water prior to discharge to prevent the formation of trihalomethanes. ▶▶❙ *Wastewater treatment processes (section 43.2)*

Chlorine is also an excellent disinfectant for individual use because it is effective, inexpensive, and easy to employ. Small quantities of drinking water can be disinfected with halazone tablets. Halazone (parasulfone dichloramidobenzoic acid) slowly releases chloride when added to water and disinfects it in about a half hour. It is frequently used by campers lacking access to uncontaminated drinking water. Of note is the fact that household bleach (diluted to 10% in water, 10-minute contact time) can be used to disinfect surfaces contaminated by human body fluids and that it is made more effective by the addition of household vinegar.

Heavy Metals

For many years the ions of heavy metals such as mercury, silver, arsenic, zinc, and copper were used as germicides. These have now been superseded by other less toxic and more effective germicides (many heavy metals are more bacteriostatic than bactericidal). However, there are a few exceptions. In some hospitals, a 1% solution of silver nitrate is added to the eyes of infants to prevent ophthalmic gonorrhea. Silver sulfadiazine is used on burns. Copper sulfate is an effective algicide in lakes and swimming pools. Heavy metals combine with proteins, often with their sulfhydryl groups, and inactivate them. They may also precipitate cell proteins. ▶▶❙ *Gonorrhea (section 39.3)*

Table 8.4	Relative Efficacy of Commonly Used Disinfectants and Antiseptics		
Class	Disinfectant[1]	Antiseptic	Comment
Gas			
Ethylene oxide	3–4	0	Sporicidal; toxic; good penetration; requires relative humidity of 30% or more; microbicidal activity varies with apparatus used; absorbed by porous material; dry spores highly resistant; presoaking is most desirable
Liquid			
Glutaraldehyde, aqueous	3	0	Sporicidal; active solution unstable; toxic
Stabilized hydrogen peroxide	3	0	Sporicidal; solution stable up to 6 weeks; toxic orally and to eyes; mildly skin toxic; little inactivation by organic matter
Formaldehyde + alcohol	3	0	Sporicidal; noxious fumes; toxic; volatile
Formaldehyde, aqueous	1–2	0	Sporicidal; noxious fumes; toxic
Phenolic compounds	3	0	Stable; corrosive; little inactivation by organic matter; irritates skin
Chlorine compounds	1–2	0	Fast action; inactivation by organic matter; corrosive; irritates skin
Alcohol	1	3	Rapidly microbicidal except for bacterial spores and some viruses; volatile; flammable; dries and irritates skin
Iodine + alcohol	0	4	Corrosive; very rapidly microbicidal; causes staining; irritates skin; flammable
Iodophors	1–2	3	Somewhat unstable; relatively bland; staining temporary; corrosive
Iodine, aqueous	0	2	Rapidly microbicidal; corrosive; stains fabrics; stains and irritates skin
Quaternary ammonium compounds	1	0	Inactivated by soap and anionics; compounds absorbed by fabrics; old or dilute solution can support growth of Gram-negative bacteria
Hexachlorophene	0	2	Insoluble in water, soluble in alcohol; not inactivated by soap; weakly bactericidal
Chlorhexidine	0	3	Soluble in water and alcohol; weakly bactericidal
Mercurial compounds	0	±	Greatly inactivated by organic matter; weakly bactericidal

1 Subjective ratings of practical usefulness in a hospital environment: 4 is maximal usefulness; 0 is little or no usefulness; ± signifies that the substance is sometimes useful.

Quaternary Ammonium Compounds

Quaternary ammonium compounds are detergents that have broad spectrum antimicrobial activity and are effective disinfectants used for decontamination purposes. **Detergents** (Latin *detergere,* to wipe away) are organic cleansing agents that are amphipathic, having both polar hydrophilic and nonpolar hydrophobic components. The hydrophilic portion of a quaternary ammonium compound is a positively charged quaternary nitrogen; thus quaternary ammonium compounds are cationic detergents. Their antimicrobial activity is the result of their ability to disrupt microbial membranes; they may also denature proteins.

Cationic detergents such as benzalkonium chloride and cetylpyridinium chloride kill most bacteria but not *M. tuberculosis* or spores. They have the advantages of being stable and nontoxic, but they are inactivated by hard water and soap. Cationic detergents are often used as disinfectants for food utensils and small instruments, and as skin antiseptics.

Aldehydes

Both of the commonly used aldehydes, formaldehyde and glutaraldehyde (figure 8.10), are highly reactive molecules that combine with nucleic acids and proteins, and inactivate them, probably by cross-linking and alkylating molecules (**figure 8.11**). They are sporicidal and can be used as chemical sterilants. Formaldehyde is usually dissolved in water or alcohol before use. A 2% buffered solution of glutaraldehyde is an effective disinfectant. It is less irritating than formaldehyde and is used to disinfect hospital and laboratory equipment. Glutaraldehyde usually disinfects objects within about 10 minutes but may require as long as 12 hours to destroy all spores.

Sterilizing Gases

Many heat-sensitive items such as disposable plastic Petri dishes and syringes, heart-lung machine components, sutures, and catheters are sterilized with ethylene oxide gas (figure 8.10). Ethylene oxide (EtO) is both microbicidal and sporicidal. It is a very

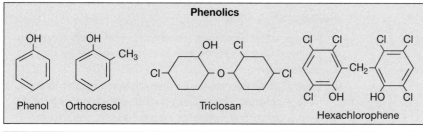

Phenolics

Phenol Orthocresol Triclosan Hexachlorophene

Alcohols

CH_3-CH_2-OH $CH_3-CH-CH_3$ with OH

Ethanol   Isopropanol

Halogenated compound

Halazone

Aldehydes

$H-C-H$ $H-C-CH_2-CH_2-CH_2-C-H$

Formaldehyde   Glutaraldehyde

Quaternary ammonium compounds

Cetylpyridinium   Benzalkonium

Gases

Ethylene oxide   Betapropiolactone   Hydrogen peroxide

Figure 8.10 Disinfectants and Antiseptics. The structures of some frequently used disinfectants and antiseptics.

MICRO INQUIRY *Why is it important that all of these compounds are relatively hydrophobic?*

strong alkylating agent that kills by reacting with functional groups of DNA and proteins to block replication and enzymatic activity. It is a particularly effective sterilizing agent because it rapidly penetrates packing materials, even plastic wraps.

Sterilization is carried out in an ethylene oxide sterilizer, which resembles an autoclave in appearance. It controls the EtO concentration, temperature, and humidity (**figure 8.12**). Because pure EtO is explosive, it is usually supplied in a 10 to 20% concentration mixed with either CO_2 or dichlorodifluoromethane. The EtO concentration, humidity, and temperature influence the rate of sterilization. A clean object can be sterilized if treated for 5 to 8 hours at 38°C or 3 to 4 hours at 54°C when the relative humidity is maintained at 40 to 50% and the EtO concentration at 700 mg/L. Because it is so toxic

to humans, extensive aeration of the sterilized materials is necessary to remove residual EtO.

Betapropiolactone (BPL) is occasionally employed as a sterilizing gas. In the liquid form, it has been used to sterilize vaccines and blood products. BPL decomposes to an inactive form after several hours and is therefore not as difficult to eliminate as EtO. It also destroys microorganisms more readily than EtO but does not penetrate materials well and may be carcinogenic. For these reasons, BPL has not been used as extensively as EtO.

Chlorine dioxide (ClO_2) gas is also used as a disinfectant. Of note is the fact that the chemistry of chlorine dioxide is very different from that of chlorine gas. ClO_2 is typically aerosolized in a humidified environment, 1 mg/liter (approximately 400 ppm-hour/liter) of air in 60% relative humidity, for at least 4 hours. At this concentration, ClO_2 provides a greater than 6 log reduction of spores and vegetative bacteria. It is effective between pH 4 and pH 10. It has been used to sterilize hospital operating and patient rooms. ClO_2 fumigation is used in the food industry to sanitize fruits and vegetables of contaminating yeasts and molds. As a disinfecting gas, it is best known for its use in sterilizing the U.S. Senate and Postal facilities after the 2001 anthrax attacks, and its use to kill molds that contaminated Gulf Coast homes in the aftermath of hurricane Katrina in 2005. ClO_2 has a broad killing spectrum, controlling both Gram-negative and Gram-positive bacteria, spores, fungi, and protozoa. It also penetrates and destroys biofilms. ClO_2 appears to have several mechanisms of action: it reacts readily with amino acids cysteine, tryptophan, and tyrosine, denaturing proteins; with free fatty acids, lysing membranes; and with nucleic acids, inhibiting replication. To date, no resistance to ClO_2 has been identified. ClO_2 is highly water soluble, especially in cold water, where it does not hydrolyze but remains as a dissolved gas in solution. Thus, a number of municipal water treatment facilities use it for water disinfection.

Vaporized hydrogen peroxide (VHP) can also be used to decontaminate biological safety cabinets, operating rooms, and other large facilities. VHP is produced from a solution of hydrogen peroxide in water that is passed over a vaporizer to achieve a vapor concentration between 140 and 1,400 parts per million (ppm), depending on the agent to be destroyed. VHP is then introduced as a sterilizing vapor into the enclosure for some time, depending on the size of the enclosure and the materials within. Hydrogen peroxide and its oxy-radical by-products are toxic (75 ppm are dangerous to human health) and kill a wide variety of microorganisms. During the course of the decontamination process, VHP breaks down to water and oxygen, both of which are harmless. Other advantages of VHP are that it can be used at a wide range of temperatures (4 to 80°C) and does not damage most nonliving materials.

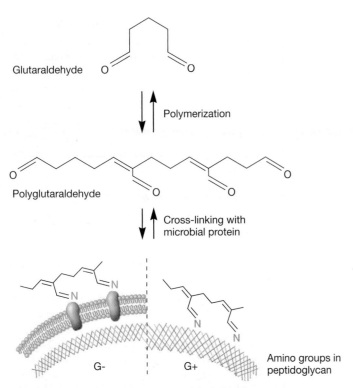

Figure 8.11 Effects of Glutaraldehyde. Glutaraldehyde polymerizes and then interacts with amino acids in proteins (left) or in peptidoglycan (right). As a result, the proteins are alkylated and cross-linked to other proteins, which inactivates them. The amino groups in peptidoglycan are also alkylated and cross-linked, which prevents them from participating in other chemical reactions such as those involved in peptidoglycan synthesis.

MICRO INQUIRY *Why are cross-linking agents such as glutaraldehyde often called "fixatives" or are said to "fix the cells"?*

(a)

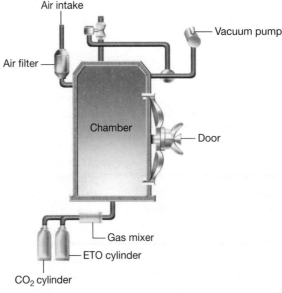

(b)

Figure 8.12 An Ethylene Oxide Sterilizer.
(a) An automatic ethylene oxide (EtO) sterilizer.
(b) Schematic of an EtO sterilizer. Items to be sterilized are placed in the chamber, and EtO and carbon dioxide are introduced. After the sterilization procedure is completed, the EtO and carbon dioxide are pumped out of the chamber and air enters.

8.6 Antimicrobial Agents Must Be Evaluated for Effectiveness

After reading this section, you should be able to:

- Predict the effects of (1) microbial population size and composition, (2) temperature, (3) exposure time, and (4) local environmental conditions on antimicrobial agent effectiveness
- Describe the processes used to measure microbial killing rates, dilution testing, and in-use testing of antimicrobial agents

The assessment of antimicrobial agent effectiveness is a complex process regulated by two different federal agencies. The Environmental Protection Agency regulates disinfectants, whereas agents used on humans and animals are under the control of the Food and Drug Administration. They establish the guidelines under which these agents are used and agent effectiveness is measured. Importantly, there are a number of variables that must also be considered when evaluating antimicrobial agent effectiveness.

Destruction of microorganisms and inhibition of microbial growth are not simple matters because the effectiveness of an antimicrobial agent is affected by at least six factors.

1. **Population size.** Because an equal fraction of a microbial population is killed during each interval, a larger population requires a longer time to die than does a smaller one (table 8.1 and figure 8.3).

2. **Population composition.** The effectiveness of an agent varies greatly with the nature of the organisms being treated because microorganisms differ markedly in susceptibility. Bacterial spores are much more resistant to most antimicrobial agents than are vegetative forms, and younger cells are usually more readily destroyed than mature organisms. Some species are able to withstand adverse conditions better than others. For instance, *M. tuberculosis,* which causes tuberculosis, is much more resistant to antimicrobial agents than most other bacteria.

3. **Concentration or intensity of an antimicrobial agent.** Often, but not always, the more concentrated a chemical agent or intense a physical agent, the more rapidly microorganisms are destroyed. However, agent effectiveness usually is not directly related to concentration or intensity. Over a short range, a small increase in concentration leads to an exponential rise in effectiveness; beyond a certain point, increases may not raise the killing rate much at all. Sometimes an agent is more effective at lower concentrations. For example, 70% ethanol is more bacteriocidal than 95% ethanol because the activity of ethanol is enhanced by the presence of water.

4. **Contact time.** The longer a population is exposed to a microbicidal agent, the more organisms are killed (figures 8.2 and 8.3). To achieve sterilization, contact time should be long enough to reduce the probability of survival by at least 6 logs.

5. **Temperature.** An increase in the temperature at which a chemical acts often enhances its activity. Frequently a lower concentration of disinfectant or sterilizing agent can be used at a higher temperature.

6. **Local environment.** The population to be controlled is not isolated but surrounded by environmental factors that may either offer protection or aid in its destruction. For example, because heat kills more readily at an acidic pH, acidic foods and beverages such as fruits and tomatoes are easier to pasteurize than more alkaline foods such as milk. A second important environmental factor is organic matter, which can protect microorganisms against physical and chemical disinfecting agents. Biofilms are a good example. The organic matter in a biofilm protects the biofilm's microorganisms. Furthermore, it has been clearly documented that bacteria in biofilms are altered physiologically, and this makes them less susceptible to many antimicrobial agents. Because of the impact of organic matter, it may be necessary to clean objects, especially medical and dental equipment, before they are disinfected or sterilized.

◀◀ *Biofilms are common in nature (section 7.5)*

The actual testing of antimicrobial agents often begins with an initial screening to see if they are effective and at what concentrations. This may be followed by more realistic in-use testing. The best-known disinfectant screening test is the **phenol coefficient test** in which the potency of a disinfectant is compared with that of phenol. A series of dilutions of phenol and the disinfectant being tested are prepared. Standard amounts of *Salmonella enterica* serovar Typhi and *Staphylococcus aureus* are added to each dilution; the dilutions are then placed in a 20 or 37°C water bath. At 5-minute intervals,

samples are withdrawn from each dilution and used to inoculate a growth medium, which is incubated and examined for growth. Growth in the medium indicates that the dilution at that particular time of sampling did not kill the bacteria. The highest dilution (i.e., the lowest concentration) that kills the bacteria after a 10-minute exposure but not after 5 minutes is used to calculate the phenol coefficient. The higher the phenol coefficient value, the more effective the disinfectant under these test conditions. A value greater than 1 means that the disinfectant is more effective than phenol.

The phenol coefficient test is a useful initial screening procedure, but the phenol coefficient can be misleading if taken as a direct indication of disinfectant potency during normal use. This is because the phenol coefficient is determined under carefully controlled conditions with pure bacterial cultures, whereas disinfectants are normally used on complex populations in the presence of organic matter and with significant variations in environmental factors such as pH, temperature, and presence of salts.

To more realistically estimate disinfectant effectiveness, other tests are often used. The rates at which selected bacteria are destroyed with various chemical agents may be experimentally determined and compared. A **use dilution test** can also be carried out. Stainless steel carriers are contaminated with one of three specific bacterial species under carefully controlled conditions. The carriers are dried briefly, immersed in the test disinfectants for 10 minutes, transferred to culture media, and incubated for 2 days. The disinfectant concentration that kills the bacteria on at least 59 out of 60 carriers (a 95% level of confidence) is determined. Disinfectants also can be tested under conditions designed to simulate normal in-use situations. In-use testing techniques allow a more accurate determination of the proper disinfectant concentration for a particular situation.

Retrieve, Infer, Apply

1. Briefly explain how the effectiveness of antimicrobial agents varies with population size, population composition, concentration or intensity of the agent, contact time, temperature, and local environmental conditions.

2. How does being in a biofilm affect an organism's susceptibility to antimicrobial agents?

3. Suppose hospital custodians have been assigned the task of cleaning all showerheads in patient rooms to prevent the spread of infectious disease. What two factors would have the greatest impact on the effectiveness of the disinfectant the custodians use? Explain what that impact would be.

4. Briefly describe the phenol coefficient test.

5. Why might it be necessary to employ procedures such as the use dilution and in-use tests?

8.7 Microorganisms Can Be Controlled by Biological Methods

After reading this section, you should be able to:

■ Propose predation, competition, and other methods for biological control of microorganisms

■ Suggest alternative decontamination and medical therapies using viruses of bacteria, fungi, and protozoa

The emerging field of biological control of microorganisms demonstrates great promise. Scientists are learning to exploit natural control processes such as predation of one microorganism on another, viral-mediated lysis, and toxin-mediated killing. While these control mechanisms occur in nature, their approval and use by humans is relatively new. Studies evaluating control of the human intestinal pathogens *Salmonella* spp., *Shigella* spp., and *E. coli* by Gram-negative predators such as *Bdellovibrio* spp. suggest that poultry farms may be sprayed with a predatory bacterium to reduce potential contamination. Another biological control method had its start in the early 1900s at the Pasteur Institute in France. Felix d'Herelle isolated bacteriophage from patients recovering from bacillary dysentery. After numerous tests in vitro, d'Herelle concluded that the bacteriophage participated in the destruction of dysentery-causing bacteria. Bacteriophage therapies were well on the way of development when penicillin ushered in the age of antibiotics. The control of human pathogens using bacteriophage is regaining wide support and appears to be effective in the eradication of a number of bacterial species by lysing the pathogenic host.

In fact, the U.S. FDA has now approved the use of a bacteriophage spray to eradicate *Listeria, Salmonella,* and *E. coli* in foods. Yet to be approved are several designs for treating human infectious diseases by similar methods. This seems intuitive, knowing that the virus lyses its specific bacterial host, yet unnerving when one thinks about maybe swallowing, injecting, or applying a virus (albeit a bacteriophage) to the human body. The use of microbial toxins (such as bacteriocins) to control susceptible populations suggests yet another method for potential control of other microorganisms. ◀◀ *Viruses and other acellular infectious agents (chapter 6)* ▶▶ *Bacteriocins (section 33.3); Various methods are used to control food spoilage (section 41.2)*

Retrieve, Infer, Apply

1. How would you explain to a patient that a virus can be used to eliminate a bone infection caused by bacteria that do not respond to antibiotics?
2. Propose the use of specific bacterial, viral, or fungal products that might be used to kill other, more virulent bacteria, viruses, or fungi.

Key Concepts

8.1 Microbial Growth and Replication Pathways: Targets for Control

■ Sterilization is the process by which all living cells, viable spores, viruses, viroids, and prions are either destroyed or removed from an object or habitat. Disinfection is the killing, inhibition, or removal of microorganisms (but not necessarily endospores) that can cause disease.

■ The main goal of disinfection and antisepsis is the removal, inhibition, or killing of pathogenic microbes. Both processes also reduce the total number of microbes. Disinfectants are chemicals used to disinfect inanimate objects; antiseptics are used on living tissue.

■ Antimicrobial agents that kill organisms often have the suffix *-cide,* whereas agents that prevent growth and reproduction have the suffix *-static.*

8.2 The Pattern of Microbial Death Mirrors the Pattern of Microbial Growth

■ Microbial death is usually exponential or logarithmic (**figure 8.3**).

■ The decimal reduction time measures an agent's killing efficiency. It represents the time needed to kill 90% of the microbes under specified conditions.

8.3 Mechanical Removal Methods Rely on Barriers

■ Microorganisms can be efficiently removed by filtration with either depth filters or membrane filters (**figure 8.4**).

■ Biological safety cabinets with high-efficiency particulate filters sterilize air by filtration (**figure 8.6**).

8.4 Physical Control Methods Alter Microorganisms to Make Them Nonviable

■ Moist heat kills by degrading nucleic acids, denaturing enzymes and other proteins, and disrupting cell membranes.

■ Although treatment with boiling water for 10 minutes kills vegetative forms, an autoclave must be used to destroy endospores by heating at 121°C and 15 pounds of pressure (**figure 8.7**).

■ Glassware and other heat-stable items may be sterilized by dry heat at 160 to 170°C for 2 to 3 hours.

■ Radiation of short-wavelength or high-energy ultraviolet and ionizing radiation can be used to sterilize objects (**figure 8.9**).

8.5 Microorganisms Are Controlled with Chemical Agents

■ Chemical agents usually act as disinfectants or antiseptics because they cannot readily destroy bacterial spores. Disinfectant effectiveness depends on concentration, treatment duration, temperature, and presence of organic material (**tables 8.3** and **8.4**).

■ Phenolics and alcohols are popular disinfectants and antiseptics that act by denaturing proteins and disrupting cell membranes (**figure 8.10**).

■ Halogens (iodine and chlorine) kill by oxidizing cellular constituents; cell proteins may also be iodinated. Iodine is applied as a tincture or iodophor. Chlorine may be added to water as a gas, hypochlorite, or an organic chlorine derivative.

- Heavy metals tend to be bacteriostatic agents. They are employed in specialized situations such as the use of silver nitrate in the eyes of newborn infants and copper sulfate in lakes and pools.
- Cationic detergents (e.g., quaternary ammonium compounds) are often used as disinfectants and antiseptics; they disrupt membranes and denature proteins.
- Aldehydes such as formaldehyde and glutaraldehyde can sterilize as well as disinfect because they kill spores.
- Ethylene oxide gas penetrates plastic wrapping material and destroys all life forms by reacting with proteins. It is used to sterilize packaged, heat-sensitive materials.
- Chlorine dioxide is aerosolized at low concentrations to kill bacteria, spores, fungi, and protozoa that contaminate living spaces and agriculture. It has multiple killing mechanisms.
- Vaporized hydrogen peroxide is used to decontaminate enclosed spaces (e.g., safety cabinets and small rooms). The vaporized hydrogen peroxide is a vapor that can be circulated throughout the space. The peroxide and its oxy-radical by-products are toxic to most microorganisms.

8.6 Antimicrobial Agents Must Be Evaluated for Effectiveness

- The effectiveness of a disinfectant or sterilizing agent is influenced by population size, population composition, concentration or intensity of the agent, exposure duration, temperature, and nature of the local environment.
- The presence of a biofilm can dramatically alter the effectiveness of an antimicrobial agent.
- The phenol coefficient test is frequently used to evaluate the effectiveness of antimicrobial agents. However, it does so using conditions that do not replicate real-life use.
- Other procedures used to determine the effectiveness of disinfectants include measurement of killing rates with germicides, use dilution testing, and in-use testing.

8.7 Microorganisms Can Be Controlled by Biological Methods

- Control of microorganisms by natural means such as through predation, viral lysis, and toxins is emerging as a promising field.

Compare, Hypothesize, Invent

1. Throughout history, spices have been used as preservatives and to cover up the smell or taste of food that is slightly spoiled. The success of some spices led to a magical, ritualized use, and possession of spices was often limited to priests or other powerful members of the community.

   a. Choose a spice and trace its use geographically and historically. What is its common use today?

   b. Spices grow and tend to be used predominantly in warmer climates. Explain.

2. Design an experiment to determine whether an antimicrobial agent is acting as a cidal or static agent. How would you determine whether an agent is suitable for use as an antiseptic rather than as a disinfectant?

3. Suppose you tested the effectiveness of disinfectants with the phenol coefficient test and obtained the results shown in the "Bacterial Growth After Treatment" table. What disinfectant can you safely say is the most effective? Can you determine its phenol coefficient from these results?

4. The death of *Geobacillus stearothermophilus* spores is routinely used to assess the effectiveness of sterilization procedures. In this procedure, endospores are exposed to the sterilization process (e.g., autoclaving) and then plated on germination medium. If sterilization has been successful, no growth occurs. This requires about 2 days. Yung and Ponce have developed a new approach that yields results within 15 minutes after sterilization. Their technique is based on the release of dipicolinic acid (DPA) from the spore coat by germinating cells. DPA release is monitored microscopically after endospores are placed in media containing alanine and terbium ion (Tb^{3+}). Alanine triggers germination and Tb^{3+} binds DPA, which then fluoresces green when illuminated with UV light. Design an experiment in which this new method could be shown to be as reliable as the culture-based approach. What other sterilization procedures besides autoclaving could this technique be used to monitor?

Read the original paper: Yung, P. T., and Ponce, A. 2008. Fast sterility by germinable-endospore biodensimetry. *Appl. Environ. Microbiol.* 74: 7669.

Bacterial Growth After Treatment			
Dilution	Disinfectant A	Disinfectant B	Disinfectant C
1/20	−	−	−
1/40	+	−	−
1/80	+	−	+
1/160	+	+	+
1/320	+	−	+

Antimicrobial Chemotherapy

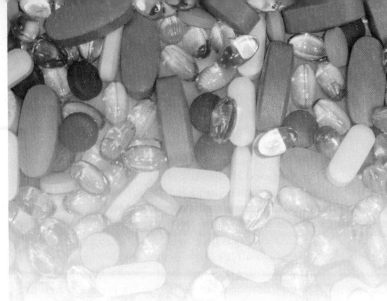

Many antimicrobial medications are available to combat infections. Nonetheless, they fall into a limited number of classes based on their modes of action.

A Teaspoon of Sugar Helps the Bacteria Go Down

A song in the musical *Mary Poppins* suggests sweetening off-tasting medicine to make it a bit more palatable. Certainly, no one would argue that adding sugar to medications helps to hide the taste. However, adding sugar to antibiotics may also offer a new treatment approach for those who suffer from chronic and recurrent infectious disease. Studies at Boston University have demonstrated that including sugars that activate bacterial metabolism makes biofilm bacteria (such as those found on intravenous and urinary catheters) more susceptible to the action of antibiotics. The studies evaluated persister bacteria from biofilms whose metabolic activity is substantially reduced, making them appear to be dormant, as they are not culturable. Many bacteria, including staphylococci, streptococci, mycobacteria, and *Escherichia coli,* become persisters and unresponsive to antibiotics during biofilm development. Persister cells are not necessarily antibiotic-resistant; they haven't mutated, nor have they acquired antibiotic resistance genes. Persisters are less sensitive to antibiotics in part because their metabolism is so slow. Using fructose or mannitol, investigators were able to coax persister *E. coli* and *Staphylococcus aureus* cells to increase the activity of their electron transport systems and trick them into importing a lethal antibiotic. The scientists demonstrated that *E. coli* and *S. aureus* persisters were unaffected by gentamicin alone. However, when these biofilm persister cells were treated with sugar and gentamicin together, their numbers were reduced by 99.9%. This novel approach suggests that biofilm persister cells, even if dormant, can be stimulated with sugars to increase their metabolism and respiration, and ultimately their membrane transport systems, to take in a common antibiotic that otherwise has no effect. ▶▶▎ *Microbial growth in natural environments (section 7.5)*

As evidenced by the vexing problem of bacterial antibiotic resistance, very few people living in the industrialized West reach adulthood without having taken an antibiotic. But what does the term antibiotic really mean? **Antibiotic** (Greek *anti,* against, and *bios,* life) refers to microbial products and their derivatives that kill susceptible microorganisms or inhibit their

growth. Drugs such as the sulfonamides are sometimes called antibiotics although they are synthetic chemotherapeutic agents. The rise in antibiotic resistance among microbes (discussed in chapter 16) has forced modern medicine to develop new drugs and rethink antimicrobial strategies (discussed in chapter 42). This chapter introduces the principles of **antimicrobial** chemotherapy and briefly reviews the characteristics of selected antibacterial, antifungal, antiprotozoan, and antiviral drugs.

Readiness Check:

Based on what you have learned previously, you should be able to:

✔ Identify the major structures and their functions of bacteria, protists, fungi, and viruses (sections 3.2–3.8, 5.1–5.7, 6.1–6.2, and 7.1–7.7)

✔ Compare the various growth and replication processes of viruses and bacteria (sections 6.3, 6.5, 7.1–7.2, and 7.4–7.5)

9.1 Antimicrobial Chemotherapy Evolved from Antisepsis Efforts

After reading this section, you should be able to:

■ Trace the general history of antimicrobial chemotherapy
■ Propose natural sources of new antimicrobial agents

Pasteur's ideas regarding "low forms of life" as the cause of some human diseases encouraged Sir Joseph Lister (1827–1912) to pursue methods to inhibit sepsis after surgery. His use of carbolic acid to fumigate the air above the surgery site is cited as one of the first uses of antiseptic technique. Lister's success stimulated subsequent scientific discussions that led to hospital practices of antiseptic handwashing, instrument disinfection and antiseptic wound dressings. Yet others longed for additional methods to control sepsis, especially outside of the hospital environment.

The modern era of chemotherapy began with the work of the German physician Paul Ehrlich (1854–1915). Ehrlich was fascinated with dyes that specifically bind to microbial cells. He reasoned that one of the dyes could be a chemical that would selectively destroy pathogens without harming human cells—a "magic bullet." By 1904 Ehrlich found that the dye trypan red was active against the trypanosome that causes African sleeping sickness (*see figure 25.6*) and could be used therapeutically. Subsequently Ehrlich and a young Japanese scientist named Sahachiro Hata (1873–1938) tested a variety of arsenic-based chemicals on syphilis-infected rabbits and found that arsphenamine was active against the syphilis spirochete. Arsphenamine was made available in 1910 under the trade name Salvarsan. Soon hundreds of compounds were being tested for their selective toxicity and therapeutic potential. ▶▶| *Microbiology advanced as new tools for studying microbes were developed (section 1.3)*

In 1927 Gerhard Domagk (1895–1964) screened a vast number of chemicals for other "magic bullets" and discovered that Prontosil red, a new dye for staining leather, protected mice completely against pathogenic streptococci and staphylococci without apparent toxicity. Jacques (1897–1977) and Therese (1892–1978) Trefouel later showed that the body metabolized the dye to sulfanilamide. Domagk received the 1939 Nobel Prize in Physiology or Medicine for his discovery of sulfonamides, or sulfa drugs.

Penicillin, the first true antibiotic because it is a natural microbial product, was first discovered in 1896 by a twenty-one-year-old French medical student named Ernest Duchesne (1874–1912). His work was forgotten until Alexander Fleming (1881–1955) accidentally rediscovered penicillin in 1928. After returning from a weekend vacation, Fleming noticed that a Petri plate of staphylococci also had mold growing on it and there were no bacterial colonies surrounding it (**figure 9.1**). Although the precise events are still unclear, it has been suggested that a *Penicillium notatum* spore had contaminated the Petri dish before it had been inoculated with the staphylococci. The mold apparently grew before the bacteria and produced penicillin. The bacteria nearest the fungus were lysed. Fleming correctly deduced that the mold produced a diffusible substance, which he called penicillin. Unfortunately, Fleming could not demonstrate that penicillin remained active in vivo long enough to destroy pathogens and he dropped the research.

In 1939 Howard Florey (1898–1968) and Ernst Chain (1906–1979), professors at Oxford University, obtained the *Penicillium* culture from Fleming and set about purifying the antibiotic. Norman Heatley (1911–2004), a biochemist, was enlisted to help. He devised the original assay, culture, and purification techniques needed to produce crude penicillin for further experimentation. When purified penicillin was injected into mice infected with streptococci or staphylococci, almost all the mice survived. Florey and Chain's success was reported in 1940, and subsequent human trials were equally successful. Fleming, Florey, and Chain received the Nobel Prize in 1945 for the discovery and production of penicillin.

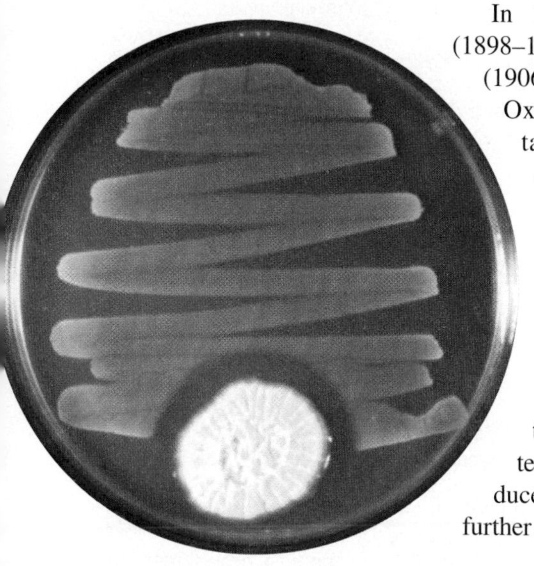

The discovery of penicillin stimulated the search for other antibiotics. Selman Waksman (1888–1973), while at Rutgers University, announced in 1944 that he and his associates had found a new antibiotic, streptomycin, produced by *Streptomyces griseus* (*see figure 24.15*). This discovery arose from the careful screening of about 10,000 strains of soil bacteria and fungi. The importance of streptomycin cannot be overstated, as it was the first drug to successfully treat tuberculosis. Waksman received the Nobel Prize in 1952, and his success led to a worldwide search for other antibiotic-producing soil microorganisms. Chloramphenicol, neomycin, oxytetracycline, and tetracycline were isolated from other *Streptomyces* species by 1953 (*see table 24.3*).

Retrieve, Infer, Apply

1. Why would a mold secrete an antibiotic?
2. Louis Pasteur is often credited with saying, "Chance favors the prepared mind." How does this apply to Fleming?

9.2 Antimicrobial Drugs Need to be Selectively Toxic over a Range of Effectiveness

After reading this section, you should be able to:

- Explain the difference between a narrow and broad-spectrum drug
- Correlate drug action with cidal and static effects

As Ehrlich so clearly saw, a successful chemotherapeutic agent has **selective toxicity:** it kills or inhibits the microbial pathogen while damaging the host as little as possible. The degree of selective toxicity may be expressed in terms of (1) the therapeutic dose—the drug level required for clinical treatment of a particular infection, and (2) the toxic dose—the drug level at which the agent becomes too toxic for the host. The **therapeutic index** is the ratio of the toxic dose to the therapeutic dose. The larger the therapeutic index, the better the chemotherapeutic agent in general.

A drug that disrupts a microbial structure or function not found in host cells often has greater selective toxicity and a higher therapeutic index. For example, penicillin inhibits bacterial peptidoglycan synthesis but has little effect on host cells because they lack cell walls; therefore penicillin's therapeutic index is high. A drug may have a low therapeutic index because it inhibits the same process in host cells or damages the host in other ways. This can lead to a diverse range of side effects that may involve almost any organ system. Because side effects can be severe, chemotherapeutic agents must be administered with great care.

Some bacteria and fungi naturally produce many of the commonly employed antibiotics (**table 9.1**). In contrast, several

Figure 9.1 Bacteriocidal Action of Penicillin. The *Penicillium* mold colony secretes penicillin that kills the *Staphylococcus aureus* streaked nearby.

Table 9.1   Properties of Some Common Antibacterial Drugs

Antibiotic Group	Primary Effect	Mechanism of Action	Example Members	Spectrum	Common Side Effects
Cell Wall Synthesis Inhibition					
Penicillins	Cidal	Inhibit transpeptidation enzymes involved in cross-linking the polysaccharide chains of peptidoglycan Activate cell wall lytic enzymes	Penicillin G, penicillin V, methicillin Ampicillin, carbenicillin	Narrow (Gram-positive) Broad (Gram-positive, some Gram-negative)	Allergic reactions (diarrhea, anemia, hives, nausea, renal toxicity)
Cephalosporins	Cidal	Same as above	Cephalothin, cefoxitin, cefaperazone, ceftriaxone	Broad (Gram-positive, some Gram-negative)	Allergic reactions, thrombophlebitis, renal injury
Vancomycin	Cidal	Prevents transpeptidation of peptidoglycan subunits by binding to D-Ala-D-Ala amino acids at the end of peptide side chains. Thus it has a different binding site than that of the penicillins.	Vancomycin	Narrow (Gram-positive)	Ototoxic (tinnitus and deafness), nephrotoxic, allergic reactions
Protein Synthesis Inhibition					
Aminoglycosides	Cidal	Bind to small ribosomal subunit (30S) and interfere with protein synthesis by directly causing misreading of mRNA	Neomycin, kanamycin, gentamicin Streptomycin	Broad (Gram-negative, mycobacteria) Narrow (aerobic Gram-negative)	Ototoxic, renal damage, loss of balance, nausea, allergic reactions
Tetracyclines	Static	Same as aminoglycosides	Oxytetracycline, chlortetracycline	Broad (including rickettsia and chlamydia)	Gastrointestinal upset, teeth discoloration, renal and hepatic injury
Macrolides	Static	Bind 23S rRNA of large ribosomal subunit (50S) to inhibit peptide chain elongation during protein synthesis	Erythromycin	Broad (aerobic and anaerobic Gram-positive, some Gram-negative)	Gastrointestinal upset, hepatic injury, anemia, allergic reactions
Lincosamines	Static	Act on the 50S ribosomal subunit, preventing transpeptidation by inhibiting peptidyl transferase activity	Clindamycin	Broad (most Gram-positive except enterococci and most anaerobes)	Gastrointestinal upset, diarrhea, hepatic injury, overgrowth of *C. difficile*
Chloramphenicol	Static	Same as macrolides	Chloramphenicol	Broad (Gram-positive, Gram-negative, rickettsia, and chlamydia	Depressed bone marrow function, allergic reactions
Nucleic Acid Synthesis Inhibition					
Quinolones and Fluoroquinolones	Cidal	Inhibit DNA gyrase and topoisomerase II, thereby blocking DNA replication	Norfloxacin, ciprofloxacin, Levofloxacin	Narrow (Gram-negatives better than Gram-positives) Broad spectrum	Tendonitis, headache, light-headedness, convulsions, allergic reactions
Rifampin	Cidal	Inhibits bacterial DNA-dependent RNA polymerase	R-Cin, rifacilin, rifamycin, rimactane, rimpin, siticox	*Mycobacterium* infections and some Gram-negatives (e.g., *Neisseria meningitidis* and *Haemophilus influenzae* b)	Nausea, vomiting, diarrhea, fatigue, anemia, drowsiness, headache, mouth ulceration, liver damage

Continued

Table 9.1	**Properties of Some Common Antibacterial Drugs** (*continued*)				
Cell Membrane Disruption					
Polymyxin B	Cidal	Binds to plasma membrane and disrupts its structure and permeability properties	Polymyxin B, polymyxin topical ointment	Narrow—mycobacterial infections, principally leprosy	Can cause severe kidney damage, drowsiness, dizziness
Antimetabolites					
Sulfonamides	Static	Inhibit folic acid synthesis by competing with *p*-aminobenzoic acid (PABA)	Silver sulfadiazine, sulfamethoxazole, sulfanilamide, sulfasalazine	Broad spectrum	Nausea, vomiting, and diarrhea; hypersensitivity reactions such as rashes, photosensitivity
Trimethoprim	Static	Blocks folic acid synthesis by inhibiting the enzyme tetrahydrofolate reductase	Trimethoprim (in combination with a sulfamethoxazole)	Broad spectrum	Same as sulfonamides but less frequent
Dapsone	Static	Thought to interfere with folic acid synthesis	Dapsone	Narrow—mycobacterial infections, principally leprosy	Back, leg, or stomach pains; discolored fingernails, lips, or skin; breathing difficulties, fever, loss of appetite, skin rash, fatigue
Isoniazid	Cidal if bacteria are actively growing, static if bacteria are dormant	Exact mechanism is unclear but thought to inhibit lipid synthesis (especially mycolic acid); putative enoyl-reductase inhibitor	Isoniazid	Narrow—mycobacterial infections, principally tuberculosis	Nausea, vomiting, liver damage, seizures, "pins and needles" in extremities (peripheral neuropathy)

important chemotherapeutic agents, such as sulfonamides, trimethoprim, ciprofloxacin, isoniazid, and dapsone, are synthetic—manufactured by chemical procedures independent of microbial activity. Some antibiotics are semisynthetic—natural antibiotics that have been structurally modified by the addition of chemical groups to make them less susceptible to stomach acids and inactivation by pathogens (e.g., ampicillin and methicillin).

Antimicrobial agents are often classified as **narrow-spectrum drugs**—that is, they are effective only against a limited variety of pathogens—or **broad-spectrum drugs** that attack many different kinds of bacteria. In addition, many semisynthetic drugs have a broader spectrum of antibiotic activity than do their parent molecules. This is particularly true of the semisynthetic penicillins (e.g., ampicillin, amoxicillin) as compared to the naturally produced penicillin G and penicillin V. Drugs may also be classified based on the general microbial group they act against: antibacterial, antifungal, antiprotozoan, and antiviral. A few agents can be used against more than one group; for example, sulfonamides are active against bacteria and some protozoa.

Finally, chemotherapeutic agents can be either **cidal** or **static.** Static agents reversibly inhibit growth; if the agent is removed, the microorganisms will recover and grow again. Although a cidal agent kills the target pathogen, it may be static at low levels. The effect of an agent also varies with the target species: an agent may be cidal for one species and static for another. Because static agents do not directly destroy the pathogen, elimination of the infection depends on the host's own immunity mechanisms. A static agent may not be effective if the host is immunosuppressed. Some idea of the effectiveness of a chemotherapeutic agent against a pathogen can be obtained from the **minimal inhibitory concentration (MIC)**. The MIC is the lowest concentration of a drug that prevents growth of a particular pathogen. On the other hand, the **minimal lethal concentration (MLC)** is the lowest drug concentration that kills the pathogen. A cidal drug generally kills pathogens at levels only two to four times more than the MIC, whereas a static agent kills at much higher concentrations, if at all. ▶▶| *Death phase (section 7.7)*

Retrieve, Infer, Apply

1. Define the following: selective toxicity, therapeutic index, side effect, narrow-spectrum drug, broad-spectrum drug, synthetic and semisynthetic antibiotics, cidal and static agents, minimal inhibitory concentration, and minimal lethal concentration.

2. How do semisynthetic antibiotics commonly differ from their parent molecules?

3. How would one use the MIC and MLC concepts to distinguish between cidal and static agents?

9.3 Antimicrobial Activity Can Be Measured by Specific Tests

After reading this section, you should be able to:

■ Explain how to determine the level of antibacterial drug activity using the dilution susceptibility test, the disk diffusion test, and the Etest®

■ Predict antimicrobial drug levels in vivo from in vitro data

Determination of antimicrobial drug effectiveness against specific pathogens is essential for proper therapy. Testing can show which agents are most effective against a pathogen and give an estimate of the proper therapeutic dose.

Dilution Susceptibility Tests

Dilution susceptibility tests can be used to determine MIC and MLC values. Antibiotic dilution tests can be done in both agar and broth. In the broth dilution test, a series of broth tubes (usually Mueller-Hinton broth) containing antibiotic concentrations in the range of 0.1 to 128 μg per milliliter (two-fold dilutions) is inoculated with a standard density of the test organism. The lowest concentration of the antibiotic resulting in no growth after 16 to 20 hours of incubation is the MIC. The MLC can be ascertained if the tubes showing no growth are then cultured into fresh medium lacking antibiotic. The lowest antibiotic concentration from which the microorganisms do not subsequently grow is the MLC. The agar dilution test is very similar to the broth dilution test. Plates containing Mueller-Hinton agar and various amounts of antibiotic are inoculated and examined for growth. Several automated systems for susceptibility testing and MIC determination with broth or agar cultures have been developed.

A number of automated antimicrobial microdilution tests have been recently commercialized. These products, along with specialized instruments to interpret the tests, miniaturize the broth dilution method to add convenience, increase time to results, and reduce technical errors. The robotic instruments inoculate the microdilution plates, incubate the cultures, collect data, and interpret the results.

Disk Diffusion Tests

If a rapidly growing microbe such as the Gram-positive *Staphylococcus aureus* or the Gram-negative *Pseudomonas aeruginosa* is being tested, a disk diffusion technique may be used. The principle behind this assay is fairly simple. When an antibiotic-impregnated, paper disk is placed on agar previously inoculated with the test bacterium, the antibiotic diffuses radially outward through the agar, producing a concentration gradient. The antibiotic is present at high concentrations near the disk and affects even minimally susceptible microorganisms; whereas, resistant organisms will grow close to the disk. As the distance from the disk increases, the antibiotic concentration decreases and only more susceptible pathogens are harmed. A clear zone or ring forms around an antibiotic disk after incubation if the agent inhibits bacterial growth. The wider the zone surrounding a disk, the more susceptible the pathogen is. Zone width also is a function of the antibiotic's initial concentration, its solubility, and its diffusion rate through agar. Thus zone width cannot be used to compare directly the effectiveness of different antibiotics.

The **Kirby-Bauer method** is the disk diffusion test most often used. It was developed in the early 1960s by William Kirby, A. W. Bauer, and their colleagues. Freshly grown bacteria are used to inoculate the entire surface of a Mueller-Hinton agar plate. After the agar surface has dried for about 5 minutes, the appropriate antibiotic test disks are placed on it, either with sterilized forceps or with a multiple applicator device (**figure 9.2**). After 16 to 18 hours of incubation at 35°C, the diameters of the zones of inhibition are measured to the nearest millimeter.

Kirby-Bauer test results are interpreted using a table that relates zone diameter to the degree of microbial resistance

(a)

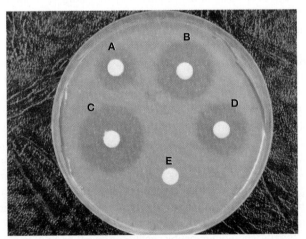

(b)

Figure 9.2 The Kirby-Bauer Method. (a) A multiple antibiotic disk dispenser and (b) disk diffusion test results.

MICRO INQUIRY *To which antibiotic (A, B, C, D, or E) is the plated bacterium resistant?*

Table 9.2 Inhibition Zone Diameter of Selected Chemotherapeutic Drugs

Chemotherapeutic Drug	Disk Content	ZONE DIAMETER (NEAREST MM)		
		Resistant	Intermediate	Susceptible
Carbenicillin (with *Proteus* spp. and *E. coli*)	100 μg	≤17	18–22	≥23
Carbenicillin (with *Pseudomonas aeruginosa*)	100 μg	≤13	14–16	≥17
Erythromycin	15 μg	≤13	14–17	≥18
Penicillin G (with staphylococci)	10 U[1]	≤20	21–28	≥29
Penicillin G (with other bacteria)	10 U	≤11	12–21	≥22
Streptomycin	10 μg	≤11	12–14	≥15
Sulfonamides	250 or 300 μg	≤12	13–16	≥17

1 One milligram of penicillin G sodium = 1,600 units (U).

(**table 9.2**). The values in table 9.2 were derived by finding the MIC values and zone diameters for many different microbial species. A plot of MIC (on a logarithmic scale) versus zone inhibition diameter (arithmetic scale) is prepared for each antibiotic (**figure 9.3**). These plots are then used to find the zone diameters corresponding to the drug concentrations actually reached in the body. If the zone diameter for the lowest level reached in the body is smaller than that seen with the test pathogen, the pathogen should have an MIC value low enough to be destroyed by the drug. A pathogen with too high an MIC value (too small a zone diameter) is resistant to the agent at concentrations that can normally be achieved in the body.

The Etest®

The Etest® from bioMérieux S.A. may be used in sensitivity testing under a majority of conditions. It is particularly convenient for use with anaerobic pathogens that grow relatively poorly in broth culture but quite well on agar. Thus agar-grown anaerobes that would not be evaluated by the Kirby-Bauer method (broth culture transferred to plates) can be tested for antimicrobial sensitivity by directly applying the Etest® strips to anaerobe-inoculated agar plates. In general, bacteria to be tested for antimicrobial sensitivities are individually inoculated on the surface of an agar medium and then Etest® strips are placed on the surface (**figure 9.4**). Each strip contains a gradient of an antibiotic and is labeled with a scale of MIC values. The lowest concentration in the strip lies at the center of the plate. After 24 to 48 hours of incubation, an elliptical zone of inhibition appears. As shown in figure 9.4, MICs are determined from the point of intersection between the inhibition zone and the strip's scale of MIC values.

Detecting Drug Resistance

One method designed to address antimicrobial drug efficacy is to determine if the infectious agent is expressing resistance factors and/or has the genetic potential to do so. Recall that resistance factors are genetically encoded. As such, it is possible to test for the gene product or for the gene itself, and a number of newer test systems have been commercialized to do so. Gene expression systems are designed to identify the production of a specific resistance factor, such as a target-modifying enzyme. Three test systems are available that measure the color change

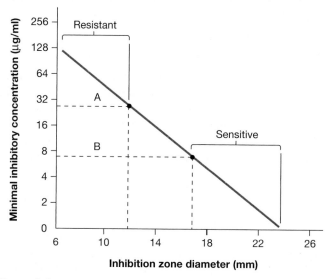

Figure 9.3 Interpretation of Kirby-Bauer Test Results. The relationship between the MICs of a hypothetical drug and the size of the zone around a disk in which microbial growth is inhibited is plotted to predict drug effectiveness. As the sensitivity of microorganisms to the drug increases, the MIC value decreases and the inhibition zone grows larger. Suppose that this drug varies from 7 to 28 μg/mL in the body during treatment. Dashed line A shows that any pathogen with a zone of inhibition less than 12 mm in diameter will have an MIC value greater than 28 μg/mL and will be resistant to drug treatment. A pathogen with a zone diameter greater than 17 mm will have an MIC less than 7 μg/mL and will be sensitive to the drug (see line B). Zone diameters between 12 and 17 mm indicate intermediate sensitivity and usually signify evolving resistance.

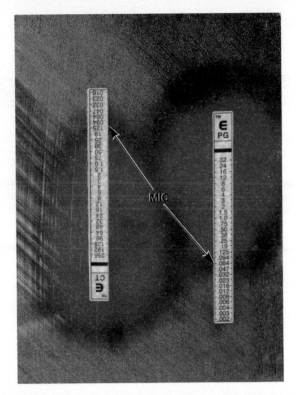

Figure 9.4 Etest®. An example of a bacterial culture plate with Etest® strips. The MIC value is read at the point where the inhibition ellipse intersects the scale on the strip, as shown by the arrows in this example. Etest® is a registered trademark of bioMérieux S.A. or one of its subsidiaries.

induced when a chromophore is acted upon by either a beta-lactamase (Cefinase disk or the DrySlide Nitrocefin) or chloramphenicol acetytransferase (CAT reagent kit). The resulting color changes are measured spectrophotometrically and protein concentration is extrapolated from a standardized curve. Thus, direct evidence of induced resistance can be made.

Gene, or nucleic-acid-based, detection systems are now commercially available to identify a chromosomally encoded or plasmid-encoded drug resistance factor. Two standard methods have been modified to achieve this. Polymerase chain reaction (PCR), with the aid of gel electrophoresis and a DNA-intercalating probe specific for the resistance gene, will reveal the resistance gene. The second system used is that of DNA hybridization. In this method, labeled DNA probes specific for the resistance gene bind, or hybridize, revealing the presence of the resistance gene. In some cases DNA chip technology can be used to screen large numbers of microbial isolates for resistance genes. ▶▶❙ *Polymerase chain reaction amplifies targeted DNA (section 17.2); DNA sequencing methods (section 18.1)*

Retrieve, Infer, Apply

1. How can dilution susceptibility tests and disk diffusion tests be used to determine microbial drug sensitivity?
2. Briefly describe the Kirby-Bauer test and its purpose.
3. What would you surmise if you examined a Kirby-Bauer assay and found individual colonies of the plated microbe growing within the zone of inhibition?
4. How is the Etest® carried out? When might it be used instead of the Kirby-Bauer test?

9.4 Antibacterial Drugs

After reading this section, you should be able to:

- Compare antibacterial drug mechanisms of action
- Correlate lack of microbial growth with selective toxicity
- Relate side effects of antibacterial drugs to mechanism of action
- Explain the relative effectiveness of various antibacterial agents based on drug target

Since Fleming's discovery of penicillin, many antibiotics have been discovered. However, most antibiotics fall into a limited number of classes. Part of the challenge in developing new antibiotics is to find bacterial structures or processes not already targeted. These novel targets have the best chance of beating antimicrobial resistance, at least for a while. Antimicrobial drug development is an active research area discussed in chapter 42. Here we present some of the more common antibiotics according to functional/structural class.

Inhibitors of Cell Wall Synthesis

The most selective antibiotics are those that interfere with bacterial cell wall synthesis. Drugs such as penicillins, cephalosporins, vancomycin, and bacitracin have a high therapeutic index because they target structures and functions not found in eukaryotic cells. ❙◀◀ *There are two main types of bacterial cell walls (section 3.4)*

Penicillins

Most penicillins (e.g., penicillin G or benzylpenicillin) are derivatives of 6-aminopenicillanic acid and differ from one another with respect to the side chain attached to the amino group (**figure 9.5**). The most crucial feature of the molecule is the **β-lactam ring,** which is essential for bioactivity. Many penicillin-resistant bacteria produce **penicillinases** (also called **β-lactamases**), enzymes that inactivate the antibiotic by hydrolyzing a bond in the β-lactam ring.

The structure of the penicillins resembles the terminal D-alanyl-D-alanine found on the peptide side chain of the peptidoglycan subunit. It is thought that this structural similarity blocks the enzyme catalyzing the transpeptidation reaction that forms the peptidoglycan cross-links (*see figure 12.10*). Thus formation of a complete cell wall is blocked, leading to osmotic lysis. This mechanism is consistent with the observation that penicillins act only on growing bacteria actively synthesizing new peptidoglycan. However, the mechanism of penicillin action is actually more complex. Penicillins may also destroy bacteria by activating their own autolytic enzymes. Penicillin may also stimulate proteins called bacterial holins to form holes or lesions in the plasma membrane, leading directly to membrane leakage and cell death.

Penicillins differ from each other in several ways. The two naturally occurring penicillins, penicillin G and penicillin V, are

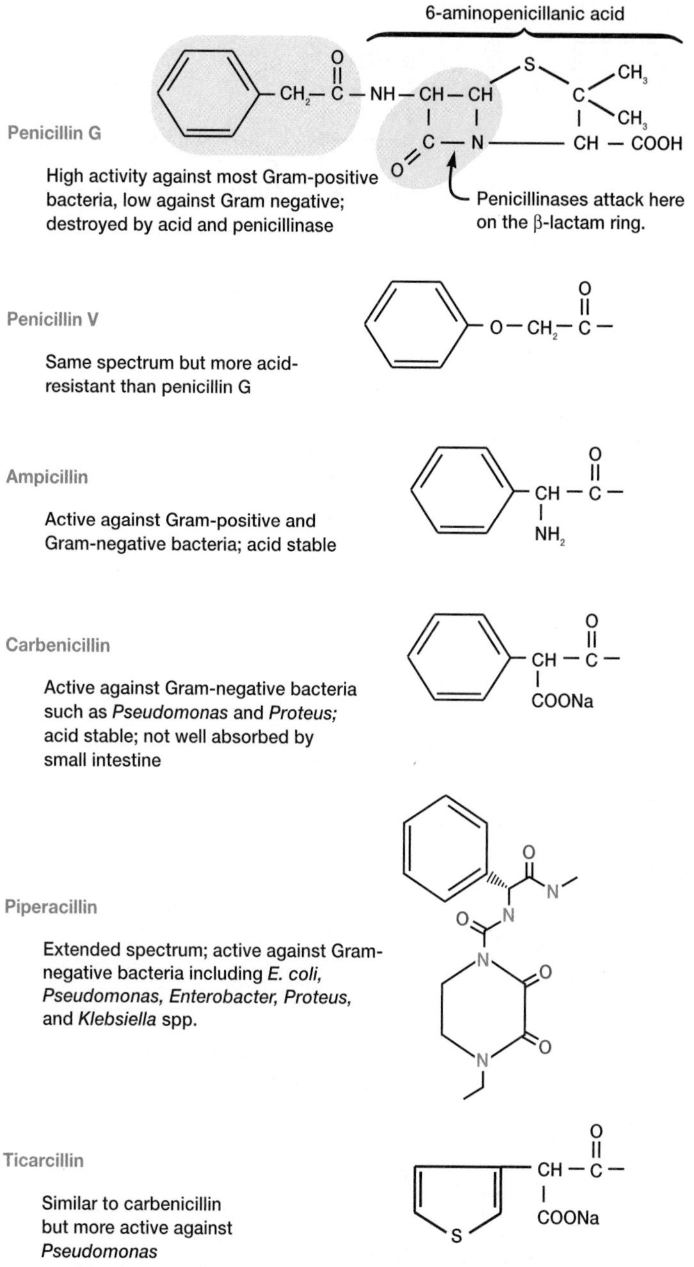

Figure 9.5 — Penicillins

6-aminopenicillanic acid

Penicillin G

High activity against most Gram-positive bacteria, low against Gram negative; destroyed by acid and penicillinase

Penicillinases attack here on the β-lactam ring.

Penicillin V

Same spectrum but more acid-resistant than penicillin G

Ampicillin

Active against Gram-positive and Gram-negative bacteria; acid stable

Carbenicillin

Active against Gram-negative bacteria such as *Pseudomonas* and *Proteus*; acid stable; not well absorbed by small intestine

Piperacillin

Extended spectrum; active against Gram-negative bacteria including *E. coli*, *Pseudomonas*, *Enterobacter*, *Proteus*, and *Klebsiella* spp.

Ticarcillin

Similar to carbenicillin but more active against *Pseudomonas*

Figure 9.5 Penicillins. The structures and characteristics of representative penicillins. All are derivatives of 6-aminopenicillanic acid; in each case, the shaded portion of penicillin G (purple) is replaced by the side chain indicated. The β-lactam ring is also shaded (blue).

MICRO INQUIRY *What is the difference between penicillin G and penicillin V? How do the semisynthetic penicillins differ from their parent compounds?*

narrow-spectrum drugs (figure 9.5). Penicillin G is effective against many Gram-positive pathogens. Penicillin G must be administered by injection (parenterally) because it is destroyed by stomach acid. Penicillin V is similar to penicillin G in spectrum of activity but can be given orally because it is more resistant to stomach acid. The semisynthetic penicillins have a broader spectrum of activity. Ampicillin can be administered orally and is effective against Gram-negative bacteria such as *Haemophilus influenzae* (middle-ear infections), *Salmonella* spp. (gastroenteritis), and *Shigella dysentariae* (dysentery). Carbenicillin and ticarcillin are potent against *Pseudomonas* spp. and *Proteus* spp. (wound and respiratory infections). ▶▶ *Human diseases caused by bacteria (chapter 39)*

An increasing number of bacteria have become resistant to natural and many semisynthetic penicillins. Physicians sometimes prescribe specific semisynthetic penicillins that are not destroyed by β-lactamases to combat antibiotic-resistant pathogens. These include nafcillin and oxacillin. However, this practice has been confounded by the emergence of nafcillin- and oxacillin-resistant bacteria. To combat this problem, extended-spectrum penicillins, such as piperacillin (figure 9.5) and the penem—penicillin (penam) and cephalosporin (cephem) hybrid—class of drugs, have been synthesized to specifically resist lactamases, and additionally have activity against Gram-negative bacteria.

Although penicillins are the least toxic of the antibiotics, about 1 to 5% of the adults in the United States develop allergies to them. Occasionally, a person will die of a violent allergic reaction. ▶▶ *Hypersensitivities (section 34.10)*

Cephalosporins

Cephalosporins are a family of antibiotics originally isolated in 1948 from the fungus *Cephalosporium*. They contain a β-lactam structure that is very similar to that of the penicillins (**figure 9.6**). As might be expected from their structural similarities to penicillins, cephalosporins also inhibit the transpeptidation reaction during peptidoglycan synthesis. They are broad-spectrum drugs frequently given to patients with penicillin allergies; however, about 10% of patients allergic to penicillin are also allergic to cephalosporins.

Vancomycin and Teicoplanin

Vancomycin is a glycopeptide antibiotic produced by the bacterium *Streptomyces orientalis*. It is a cup-shaped molecule composed of a peptide linked to a disaccharide. The peptide portion blocks the transpeptidation reaction by binding specifically to the D-alanyl-D-alanine terminal sequence on the pentapeptide portion of peptidoglycan (**figure 9.7**). The antibiotic is bactericidal for Gram-positive bacteria of the genus *Staphylococcus* and some members of the genera *Clostridium* (gangrene and severe diarrhea), *Bacillus* (food poisoning), *Streptococcus* ("strep" throat), and *Enterococcus* (urinary tract infections). It is given both orally (for *C. difficile* infections only) and intravenously, and has been particularly important in the treatment of antibiotic-resistant staphylococcal and enterococcal infections. However, vancomycin-resistant strains of *Enterococcus* have become widespread, and cases of resistant *Staphylococcus aureus* have appeared. In these bacteria, resistance is conferred when the bacteria produce a resistance factor that causes the terminal D-alanine to be replaced by either a D-lactate or D-serine residue, thus altering

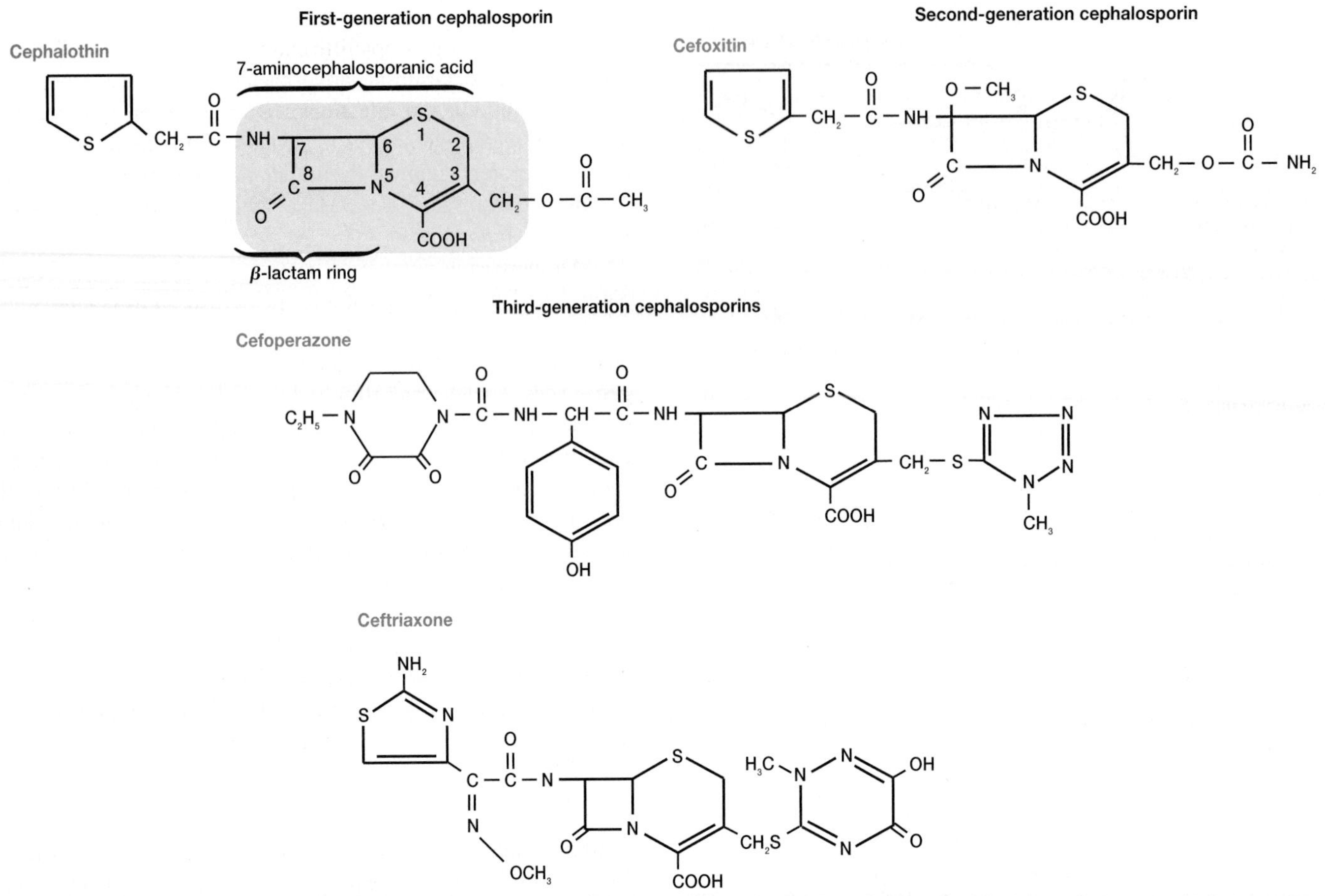

First-generation cephalosporin

Cephalothin

7-aminocephalosporanic acid

β-lactam ring

Second-generation cephalosporin

Cefoxitin

Third-generation cephalosporins

Cefoperazone

Ceftriaxone

Figure 9.6 Cephalosporin Antibiotics. These drugs are derivatives of 7-aminocephalosporanic acid and contain a β-lactam ring. Similar to electronic devices, as newer or next-generation antibiotics are developed, they are numbered first-generation, second-generation, etc.

MICRO INQUIRY *Do you think cephalosporin antibiotics are susceptible to degradation by β-lactamase enzymes? Explain.*

the target of the antibiotic. Vancomycin resistance poses a serious public health threat; vancomycin has been considered the "drug of last resort" in cases of antibiotic-resistant *S. aureus*. The mechanisms by which microbes acquire antibiotic resistance are discussed in chapter 16.

Teicoplanin, another glycopeptide antibiotic, is produced by *Actinoplanes teichomyceticus*. It is similar in structure and mechanism of action to vancomycin but has fewer side effects. It is active against staphylococci, enterococci, streptococci, clostridia, *Listeria* spp., and many Gram-positive pathogens. The diseases caused by these microbes are covered in chapter 39.

Protein Synthesis Inhibitors

Many antibiotics inhibit protein synthesis by binding the bacterial ribosome and other components of protein synthesis. Because these drugs discriminate between bacterial and eukaryotic ribosomes, their therapeutic index is fairly high but not as high as that of cell wall synthesis inhibitors. Several different steps in protein synthesis can be affected by drugs in this category. ▶▶ *Translation in bacteria (section 13.7)*

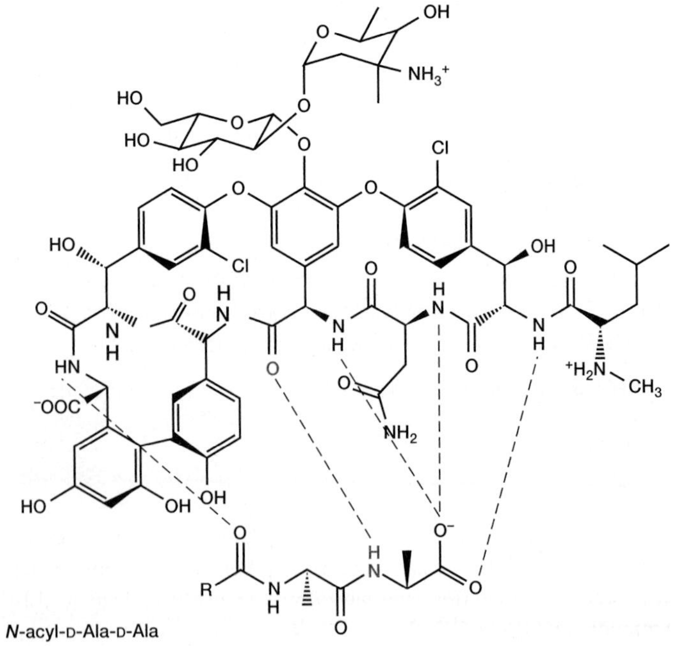

N-acyl-D-Ala-D-Ala

Figure 9.7 Vancomycin. The cup-shaped vancomycin molecule binds to the D-alanyl-D-alanine terminal sequence of peptidoglycan.

Aminoglycosides

Although considerable variation in structure occurs among several important **aminoglycoside antibiotics,** all contain a cyclohexane ring and amino sugars (**figure 9.8**). Streptomycin, kanamycin, neomycin, and tobramycin are synthesized by different species of the bacterial genus *Streptomyces,* whereas gentamicin comes from a related bacterium, *Micromonospora purpurea.* Streptomycin's usefulness has decreased greatly due to widespread drug resistance, but it may still be effective when other aminoglycosides should not be used (e.g., due to interactions with other drugs). Gentamicin is used to treat Gram-negative *Proteus, Escherichia, Klebsiella,* and *Serratia* infections. Aminoglycosides can be quite toxic, however, and can cause hearing and renal damage, loss of balance, nausea, and allergic reactions.

Although each aminoglycoside side chain binds slightly differently, the primary mechanism of action is the disruption of peptide elongation during translation. This occurs as aminoglycosides bind to ribosomal RNA of the bacterial 30S ribosomal subunit, interfering with mRNA reading and/or causing early termination of peptide synthesis. These antibiotics are bactericidal and tend to be most effective against Gram-negative pathogens. ▶▶ *Translation in bacteria (section 13.7)*

Tetracyclines

The **tetracyclines** are a family of antibiotics with a common four-ring structure to which a variety of side chains are attached (**figure 9.9**). Oxytetracycline and chlortetracycline are produced naturally by *Streptomyces* species, whereas other tetracyclines are semisynthetic. These antibiotics are similar to the aminoglycosides in that they can combine with the 30S subunit of the

Tetracycline (chlortetracycline, doxycycline)

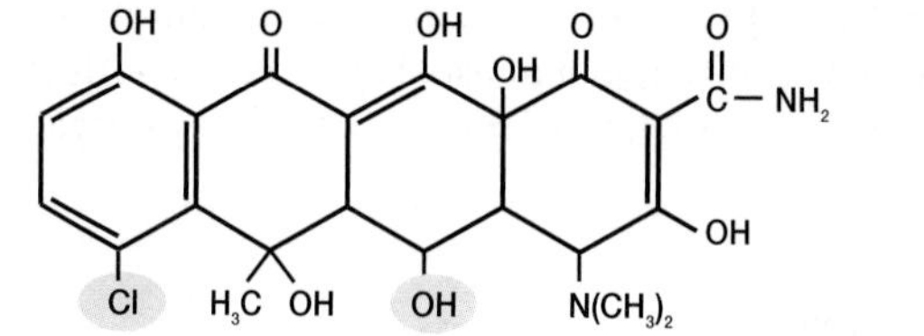

Figure 9.9 Tetracyclines. Three members of the tetracycline family. Tetracycline lacks both of the groups that are shaded. Chlortetracycline (aureomycin) differs from tetracycline in having a chlorine atom (blue); doxycycline consists of tetracycline with an extra hydroxyl (purple).

ribosome, inhibiting protein synthesis. Their action is only bacteriostatic, though. Tetracyclines are broad-spectrum antibiotics that are active against most bacteria, including the intracellular pathogens rickettsias, chlamydiae, and mycoplasmas.

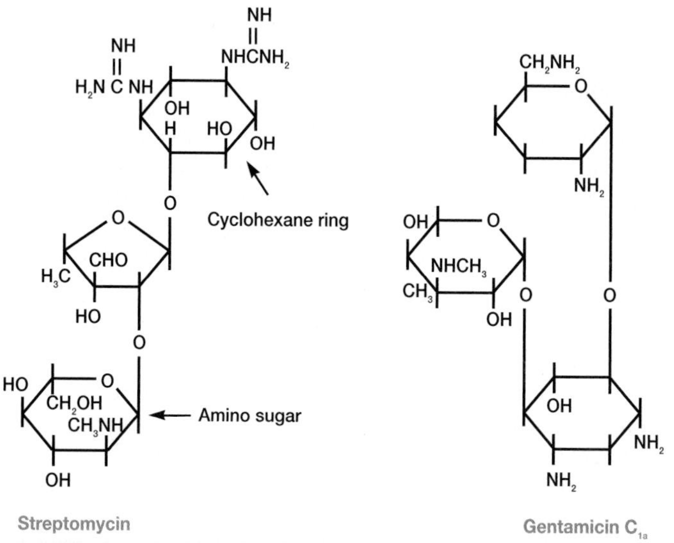

Macrolides

The **macrolide antibiotics** contain a ring structure consisting of 12 to 22 carbons called a lactone ring. The lactone ring is linked to one or more sugars (**figure 9.10**). Erythromycin binds to the 50S ribosomal subunit to inhibit bacterial protein synthesis. Erythromycin is a relatively broad-spectrum antibiotic effective against Gram-positive bacteria, mycoplasmas, and a few Gram-negative bacteria, but it is usually only bacteriostatic. It is used with patients who are allergic to penicillins and in the treatment of whooping cough, diphtheria, diarrhea

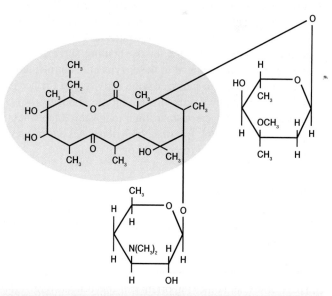

Figure 9.8 Representative Aminoglycoside Antibiotics. The cyclohexane ring and amino sugar are identified.

MICRO INQUIRY *How do these drugs inhibit protein synthesis?*

Figure 9.10 Erythromycin, a Macrolide Antibiotic. Erythromycin is composed of a 14-member lactone ring (shaded), so named because it contains a lactone or cyclic ester group, that is connected to two sugars.

MICRO INQUIRY *How is the mechanism by which macrolides block protein synthesis similar to that of the tetracyclines? How is it different?*

caused by *Campylobacter,* and pneumonia from *Legionella* or *Mycoplasma* infections. Azithromycin (Zithromax), which has surpassed erythromycin in use, is particularly effective against many bacteria, including the sexually transmitted *Chlamydia trachomatis.*

Lincosamines

Lincosamine antibiotics are produced by *Streptomyces* bacteria. They have a broad spectrum of activity against anaerobic bacteria and a more limited activity against aerobes. Most Gram-positive cocci are inhibited. Lincosamines are used sparingly, as they can support the overgrowth of *C. difficile,* leading to pseudomembraneous colitis and toxic megacolon, diseases resulting from the *C. difficile* toxins. Clindamycin is used to treat infections caused by *Bacteroides fragilis,* along with some staphylococcal and streptococcal infections.

Chloramphenicol

Chloramphenicol was first produced from cultures of *Streptomyces venezuelae* but is now synthesized chemically. Like erythromycin, this antibiotic binds the 50S ribosomal subunit to inhibit bacterial protein synthesis. It has a very broad spectrum of activity but, unfortunately, is quite toxic. Consequently this antibiotic is used only in life-threatening situations when no other drug is adequate.

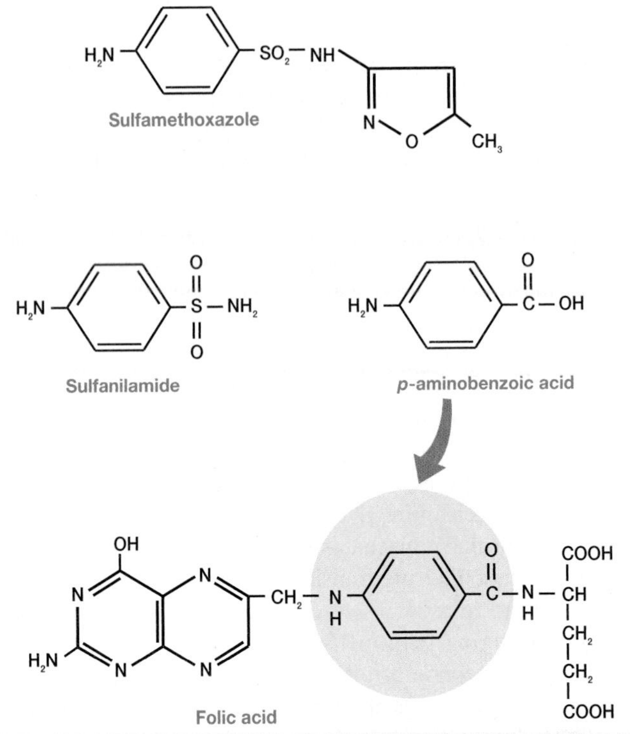

Metabolic Antagonists

Several valuable drugs act as **antimetabolites:** they antagonize, or block, the functioning of metabolic pathways. Antimetabolites are structurally similar to the substrates of key enzymes and compete with the metabolites for the binding site of these enzymes. However, once bound to the enzyme, the antimetabolites are different enough to block enzyme activity and further progression of the pathway. By preventing metabolism, they are broad spectrum but bacteriostatic; their removal reestablishes the metabolic activity. ▶▶ *Enzymes and ribozymes speed up cellular chemical reactions (section 10.6)*

Sulfonamides or Sulfa Drugs

Sulfonamides, or sulfa drugs, are structurally related to sulfanilamide, an analogue of *p*-aminobenzoic acid, or PABA (**figure 9.11**). PABA is an important component (cofactor) of many enzymes and is needed for folic acid (folate) synthesis. Folic acid contributes to the synthesis of purines, the bases used in the construction of DNA, RNA, and other important cell constituents (e.g., ATP). When sulfanilamide or another sulfonamide enters a bacterial cell, it competes with PABA for the active site of an enzyme involved in folic acid synthesis, causing a decline in folate concentration (*see figure 10.17*). The resulting inhibition of purine and pyrimidine synthesis leads to cessation of protein synthesis and DNA replication. Sulfonamides are selectively toxic for many bacteria and protozoa because these microbes manufacture their

Figure 9.11 Sulfa Drugs. Both sulfanilamide and sulfamethoxazole compete with *p*-aminobenzoic acid to block folic acid synthesis.

MICRO INQUIRY *Why do sulfa drugs have a high therapeutic index?*

own folate and cannot effectively take up this cofactor, whereas humans do not synthesize folate; instead, we must obtain it in our diet. Sulfonamides thus have a high therapeutic index. However, the increasing resistance of many bacteria to sulfa drugs limits their effectiveness.

Trimethoprim

Trimethoprim is a synthetic antibiotic that also interferes with the production of folic acid. It does so by binding to dihydrofolate reductase (DHFR), the enzyme responsible for converting dihydrofolic acid to tetrahydrofolic acid, competing against the dihydrofolic acid substrate (**figure 9.12**). Trimethoprim is a broad-spectrum antibiotic often used to treat respiratory and middle ear infections, urinary tract infections, and traveler's diarrhea. It is often combined with sulfa drugs to increase efficacy of treatment by blocking two key steps in the folic acid pathway. The inhibition of two successive steps in a single biochemical pathway means that less of each drug is needed in combination than when used alone. This is termed a synergistic drug interaction.

Nucleic Acid Synthesis Inhibition

The antibacterial drugs that inhibit nucleic acid synthesis function by inhibiting (1) DNA polymerase and topoisomerases or

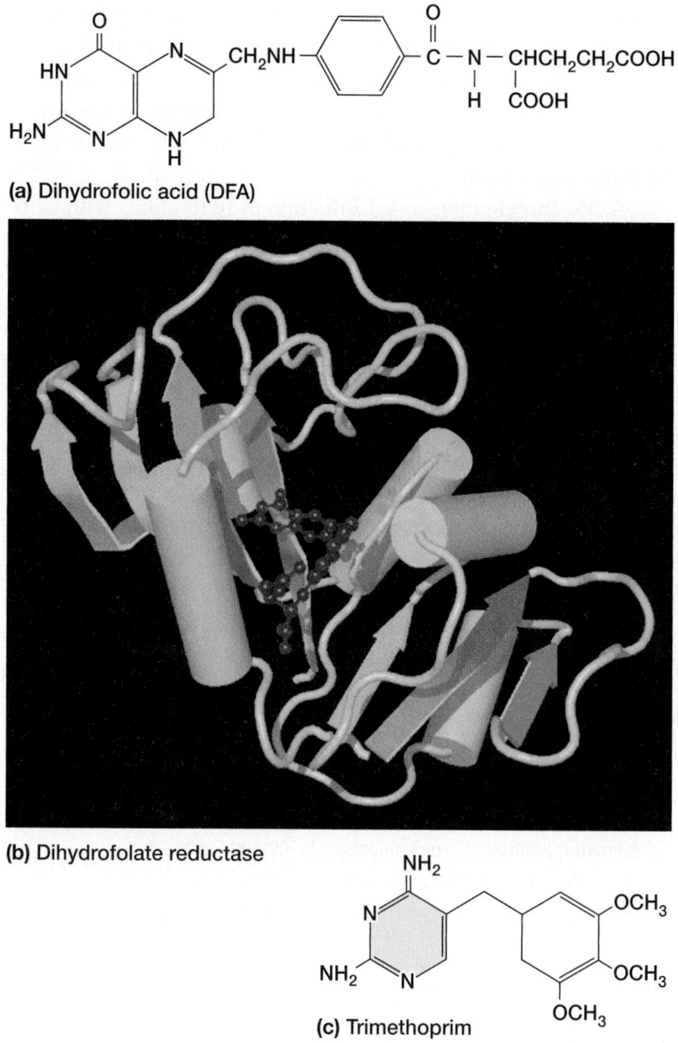

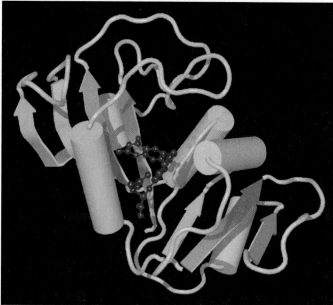

(a) Dihydrofolic acid (DFA)

(b) Dihydrofolate reductase

(c) Trimethoprim

Figure 9.12 **Competitive Inhibition of Dihydrofolate Reductase (DHFR) by Trimethoprim.** (a) Dihydrofolic acid (DFA) is the natural substrate for the DHFR enzyme of the folic acid pathway. (b) DHFR structure and its interaction with DFA (red), its natural substrate. Note the chemical structure and how it fits into the active site of the enzyme. (c) Trimethoprim mimics the structural orientation of the DFA and thus competes for the active site of the enzyme. This causes delayed or absent folic acid synthesis because the DFA cannot be converted to tetrahydrofolic acid when trimethoprim occupies the DHFR active site.

(2) RNA polymerase, to block replication or transcription, respectively. These drugs are not as selectively toxic as other antibiotics because bacteria and eukaryotes do not differ greatly with respect to nucleic acid synthesis. The most commonly used drugs in this category are the quinolones.

Quinolones

The **quinolones** are synthetic drugs that contain the 4-quinolone ring. They are increasingly used to treat a wide variety of infections. The first quinolone, nalidixic acid (**figure 9.13**), was synthesized in 1962. Since that time, generations of fluoroqui-

Nalidixic acid

Norfloxacin

Ciprofloxacin

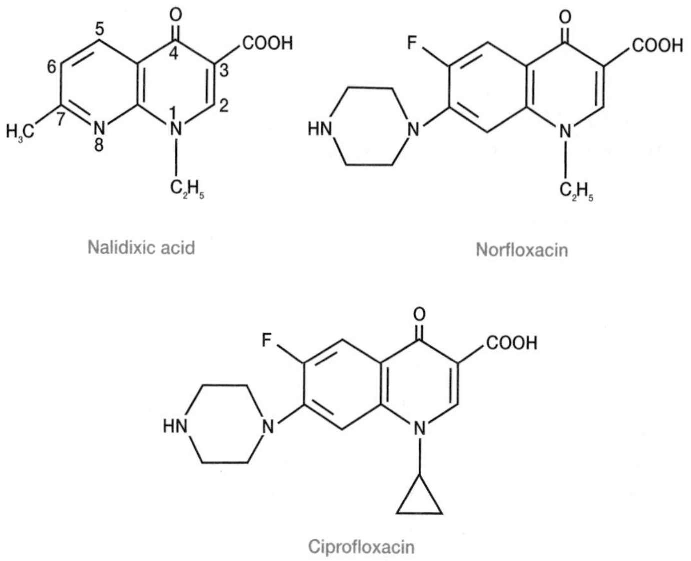

Figure 9.13 **Quinolone Antimicrobial Agents.** Ciprofloxacin and norfloxacin are newer-generation fluoroquinolones. The 4-quinolone ring in nalidixic acid has been numbered.

nolones have been produced. Three of these—ciprofloxacin, norfloxacin, and ofloxacin—are currently used in the United States.

Quinolones act by inhibiting the bacterial topoisomerases DNA gyrase and topoisomerase II. DNA gyrase introduces negative twist in DNA and helps separate its strands. Inhibition of DNA gyrase disrupts DNA replication and repair, bacterial chromosome separation during division, and other processes involving DNA. Fluoroquinolones also inhibit topoisomerase II, another enzyme that untangles DNA during replication. It is not surprising that quinolones are bactericidal. ▶▶| *DNA replication in bacteria (section 13.3)*

The quinolones are broad-spectrum antibiotics. They are highly effective against enteric bacteria such as *E. coli* and *Klebsiella pneumoniae*. They can be used with *Haemophilus* spp., *Neisseria* spp., *P. aeruginosa*, and other Gram-negative pathogens. The quinolones also are active against Gram-positive bacteria such as *S. aureus*, *Streptococcus pyogenes*, and *Mycobacterium tuberculosis*. Thus they are used in treating a wide range of infections.

Retrieve, Infer, Apply

1. Explain five ways in which chemotherapeutic agents kill or damage bacterial pathogens.
2. Why do penicillins and cephalosporins have a higher therapeutic index than most other antibiotics?
3. Would there be any advantage to administering a bacteriostatic agent along with penicillins? Any disadvantage?
4. What are antimetabolites? Why are these effective against protozoan pathogens while the other drugs presented here generally are not?

9.5 Antifungal Drugs

After reading this section, you should be able to:

- Compare antifungal drug mechanisms of action
- Explain why there are far fewer antifungal agents than there are antibacterial agents

Treatment of fungal infections generally has been less successful than that of bacterial infections largely because as eukaryotes, fungal cells are much more similar to human cells than are bacterial cells. Many drugs that inhibit or kill fungi are therefore quite toxic for humans and thus have a low therapeutic index. In addition, most fungi have a detoxification system that modifies many antifungal agents, limiting drug effectiveness. Most antifungals are fungistatic only as long as repeated application maintains high levels of unmodified drug. Nonetheless, a few drugs are useful in treating many major fungal diseases.

There are two major classes of antifungal drugs: polyenes and azoles; both block fungal cell membrane synthesis. This is a good drug target because fungal membranes require the sterol ergosterol, not found in human membranes. Polyenes bind directly to ergosterol, and azoles block the last step in ergosterol biosynthesis. Thus, both drugs prevent fungal membrane formation.

Fungal infections are often subdivided into infections called superficial mycoses, subcutaneous mycoses, and systemic mycoses. Treatment for these types of disease is very different. Several drugs are used to treat superficial mycoses. Three azole drugs—miconazole, ketoconazole (**figure 9.14**), and clotrimazole—are available as creams and solutions for the treatment of infections such as athlete's foot and oral and vaginal candidiasis. Nystatin (figure 9.14*a*), a polyene antibiotic from *Streptomyces,* is used to control *Candida* infections of the skin, vagina, or alimentary tract. It is too toxic to be taken systemically. Griseofulvin (figure 9.14*c*), an antimycotic formed by *Penicillium,* is given orally to treat chronic superficial mycoses. It is thought to disrupt the mitotic spindle and inhibit cell division; it also may inhibit protein and nucleic acid synthesis. ▶▶| *Human diseases caused by fungi and protists (chapter 40)*

Systemic fungal infections are very difficult to control and can be fatal. Drugs commonly used against systemic mycoses include amphotericin B, 5-flucytosine (figure 9.14*a* and *d*), and azoles such as fluconazole, itraconazole, and voriconazole. Amphotericin B from *Streptomyces* spp. is quite toxic to humans and used only for serious, life-threatening infections. The synthetic oral antimycotic agent 5-flucytosine (5-fluorocytosine) is effective against most systemic fungi, although drug resistance often develops rapidly. The drug is converted to 5-fluorouracil by fungi, which is an analogue of uracil, one of the building blocks of RNA. During transcription, 5-fluorouracil is incorporated into RNA in place of uracil and disrupts RNA function. Fluconazole is used in the treatment of candidiasis, cryptococcal meningitis, and coccidioidal meningitis. It

also disrupts fungal membrane sterols and ranges from fungistatic to fungicidal, depending on dose and organism. Because adverse effects of fluconazole are relatively uncommon, it is used prophylactically to prevent life-threatening fungal infections in AIDS patients and other individuals who are severely immunosuppressed.

A life-threatening fungal infection of individuals with compromised immune systems is caused by *Pneumocystis jiroveci* (formerly *P. carinii*). Interestingly, infections caused by this fungus are treated with the combination of antimetabolite drugs trimethoprim and sulfisoxazole (p. 198). This drug is effective because *P. jiroveci* synthesizes folic acid using enzymes that are very similar to those used by bacteria.

Subcutaneous mycoses are typically treated with combinations of drugs that would be used for superficial and systemic mycoses. The goal of this combination therapy is to provide less toxic drugs over longer time periods, with shorter term exposure to those that have greater toxic side effects. In this way, subcutaneous fungi are continuously targeted with antifungal drugs, while allowing patients respites from the side effects of the more toxic drugs. ▶▶| *Human diseases caused by fungi and protists (chapter 40)*

Retrieve, Infer, Apply

1. Summarize the mechanism of action and the therapeutic use of the following antifungal drugs: miconazole, nystatin, griseofulvin, amphotericin B, and 5-flucytosine.
2. Why are immunosuppressed individuals given antifungal agents?

9.6 Antiviral Drugs

After reading this section, you should be able to:

- Compare antiviral drug mechanisms of action
- Provide a rationale for combination drug therapy, using an anti-HIV model
- Explain why there are far fewer antiviral agents than there are antibacterial agents

Because viruses enter host cells and make use of host cell enzymes and constituents, it was long thought that a drug that blocked virus multiplication would be toxic for the host. However, the discovery of inhibitors of virus-specific enzymes and replication cycle processes has led to the development of antiviral drugs. Some important examples are shown in **figure 9.15**. |◀◀ *Viral life cycles have five steps (section 6.3)*

Most antiviral drugs disrupt critical stages in a virus's multiplication cycle. Probably the most publicized antiviral agent is **Tamiflu** (generically, oseltamivir phosphate). Tamiflu (figure 9.15*f*) inhibits the viral molecule neuraminidase, which is essential for release of newly synthesized influenza A virus particles from host cells. However, widespread use of Tamiflu initiated with the 2009–2010 H1N1 influenza pandemic has resulted in significant Tamiflu resistance in influenza virus, driving the

a. Drugs that bind to sterols, resulting in membrane damage

Amphotericin B

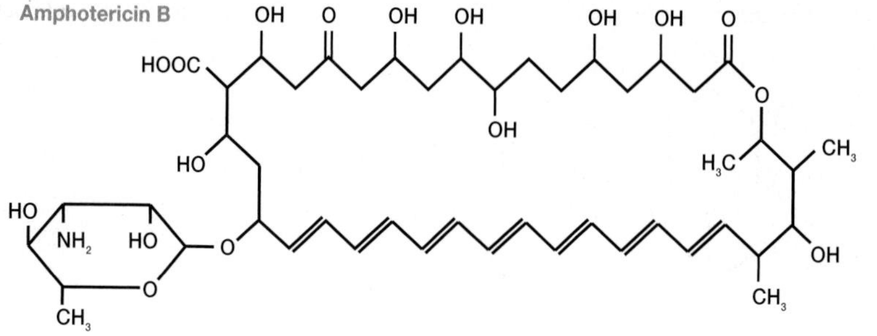

Nystatin

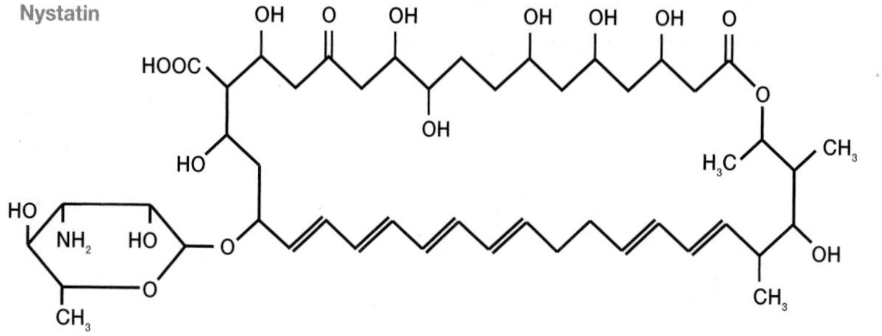

b. Drugs that inhibit sterol synthesis, resulting in altered membrane permeability

Miconazole

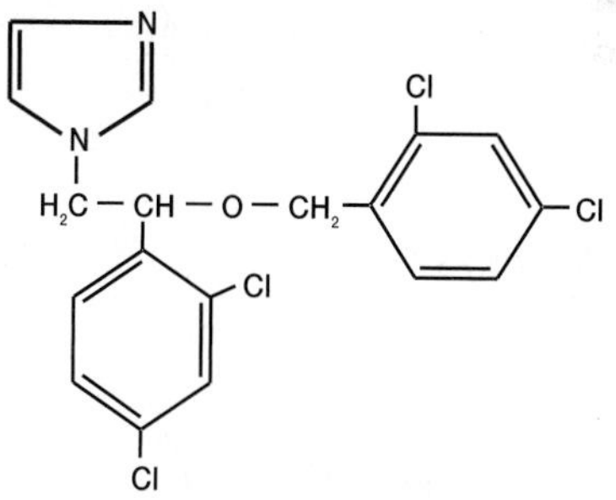

Ketoconazole

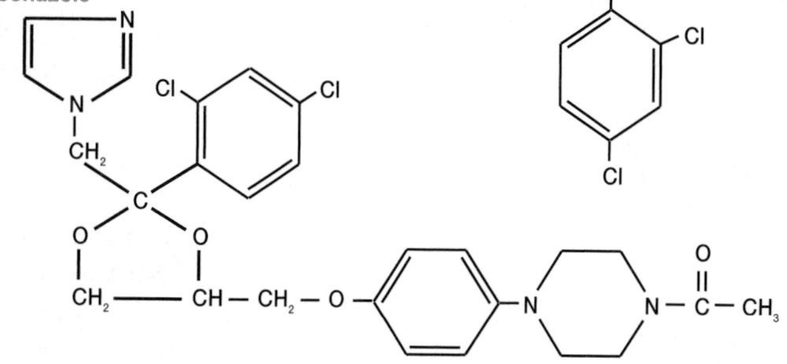

c. Drug that inhibits nucleic acid synthesis, protein synthesis, or cell division

Griseofulvin

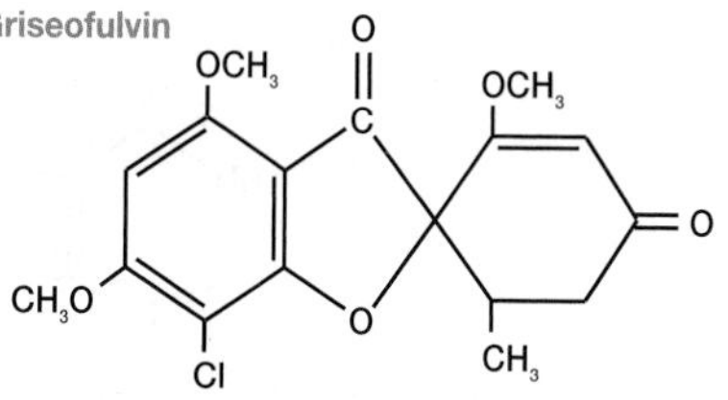

d. Drug that disrupts RNA function

5-flucytosine

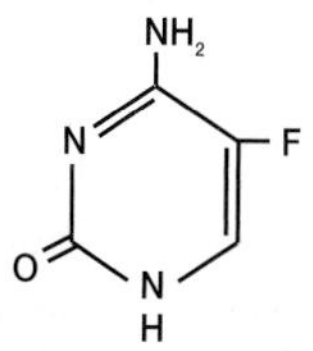

Figure 9.14 Antifungal Drugs.   Drugs are categorized by mechanism of action. Examples within each category are shown. Drug use is determined by the type of fungal infection, drug toxicity, and fungus sensitivity to the drug.

MICRO INQUIRY *What is the mechanism by which nystatin inhibits growth? How does this compare to that of amphotericin B? Do you think nystatin is less toxic than amphotericin? (Hint: Think about how the two drugs are delivered.)*

a. Nucleoside reverse transcriptase inhibitor

Azidothymidine
(AZT) or zidovudine

b. Viral protease inhibitor

Ritonavir

c. Viral fusion inhibitor

Foscarnet

d. Inhibitors of viral DNA polymerase

Cidofovir
(HPMPC)

Acyclovir

e. Blocks viral penetration and uncoating

Amantadine

f. Inhibits neuraminidase

Oseltamivir

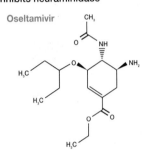

Figure 9.15 **Antiviral Drugs Target the Viral Replication Cycle.** Drugs
are categorized by mechanism of action. Their use is determined by type of
viral disease, sensitivity of virus to the drug, and drug toxicity. Viral disease of
the skin may permit use of a more toxic drug, for example, than viral disease
of the liver.

influenza A have developed resistance to these agents, and in
2014 the U.S. Food and Drug Administration approved another
neurominidase inhibitor, peramivir, to treat influenza infection in
adults. ▶▶| *Influenza (flu) (section 38.1)*

Several drugs are commonly used to treat illnesses caused by
viruses with DNA genomes. Adenine arabinoside (vida-
rabine) disrupts the activity of viral DNA polymerase
and several other enzymes involved in DNA and
RNA synthesis and function. It is given intravenously or applied as an ointment to treat herpes
infections (e.g., cold sores and genital herpes).
Another drug, acyclovir (figure 9.15*d*), is also used in
the treatment of herpes infections. Upon phosphorylation, acyclovir also inhibits viral DNA polymerase.
Unfortunately, acyclovir-resistant strains of herpes have
developed. Effective acyclovir derivatives are now available. Valacyclovir is an orally administered prodrug form of acyclovir.
Prodrugs are inactive until metabolized. Another kind of drug,
foscarnet (figure 9.15*c*), also inhibits the virus's DNA polymerase, and is very effective at treating illnesses caused by her-
pes simplex viruses and cytomegalovirus. ▶▶| *Direct contact
diseases can be caused by viruses (section 38.3)*

Several broad-spectrum anti-DNA virus drugs have been
developed. A good example is the drug HPMPC, also known as
cidofovir (figure 9.15*d*). It is effective against papillomaviruses

2014 CDC recommendation for limiting its use. While Tamiflu is
not a cure for neuraminidase-expressing viruses, patients who take
Tamiflu within 48 hours of influenza infection are relieved of flu
symptoms about a day and a half sooner than patients who do not
take Tamiflu. It is important to recognize that Tamiflu is not a
substitute for yearly flu vaccination and frequent hand washing.
Previously amantadine (figure 9.15*e*) and rimantadine were also
used to prevent influenza A illness. However, nearly all strains of

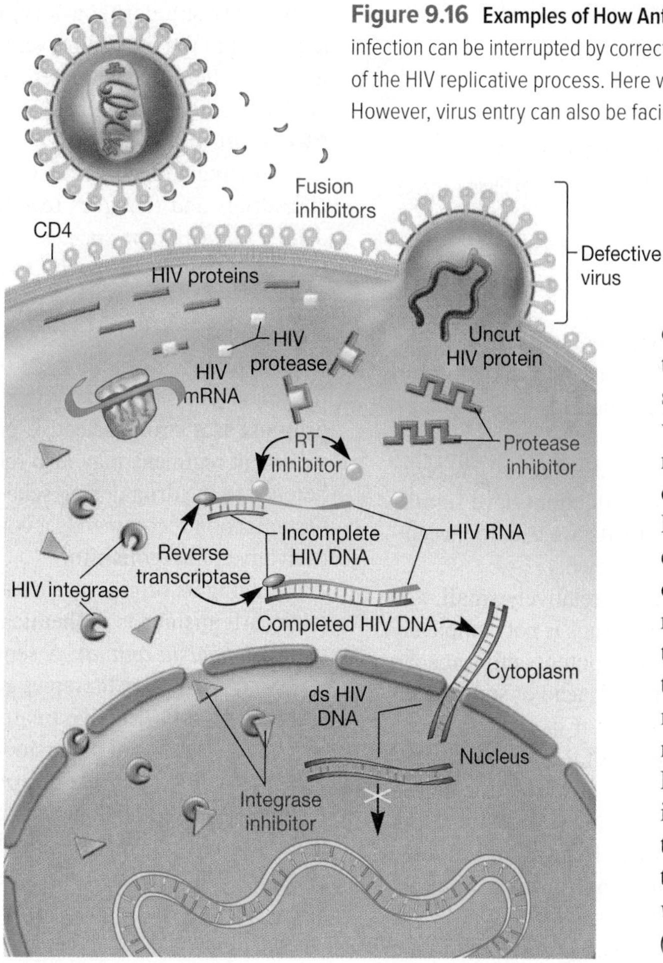

Figure 9.16 **Examples of How Anti-HIV Agents Block HIV Replication.** Viral infection can be interrupted by correct use of drugs that target specific components of the HIV replicative process. Here we show virus entry by membrane fusion. However, virus entry can also be facilitated by receptor-mediated endocytosis.

Infection begins with HIV fusion. Fusion inhibitors block this step.

Once inside a host cell, HIV uncoats and its reverse transcriptase (RT) forces the host to make DNA from the viral RNA. RT inhibitors block this step.

The viral DNA is transcribed and translated into polyproteins that are cut to release viral proteins. Protease inhibitors block this step.

Viral DNA is added to the host DNA by the action of a viral integrase. Integrase inhibitors block this step.

cycle of the virus—the conversion of the virus's RNA genome into double-stranded DNA (figure 9.16), blocks viral DNA synthesis and halts HIV replication. Protease inhibitors are effective because HIV, like many RNA viruses, synthesizes polyproteins that must be cleaved into the individual proteins required for virus replication. Protease inhibitors mimic the peptide bond that is normally attacked by the protease. Three of the most used PIs are saquinavir, indinavir, and ritonavir (figure 9.15*b*). Fusion inhibitors are particularly interesting as an effective blockade to viral entry into host cells, essentially preventing disease. ▶▶| *RNA viruses (section 27.4); Retroviruses (section 27.8)*

The most successful treatment approach to date in combating HIV/AIDS is to use drug combinations. Most effective is a cocktail of agents given at high dosages to prevent the development of drug resistance. For example, the combination of the two reverse transcriptase inhibitors AZT and 3TC, and the protease inhibitor ritonavir reduces HIV concentrations in plasma almost to zero. However, the treatment does not eliminate proviral HIV DNA that still resides in certain cells of the immune system (e.g., memory T cells) and possibly other cells. Thus with proper treatment, the virus often disappears from a patient's blood and drug-resistant strains do not seem to arise. But because HIV can remain dormant in memory T cells, it can survive drug cocktails and reactivate. Thus patients are not completely cured with drug treatment, requiring drug therapy for life. Unfortunately, side effects can be very severe, and treatment is prohibitively expensive for those without medical insurance. Globally, the vast majority of HIV-positive individuals do not have access to effective combination therapy. Recent studies have led the CDC to recommend a pre-HIV exposure prophylaxis (prevention) strategy to men and women who have sex with HIV-infected partners. The daily oral dose of two NRTIs is strongly correlated with reduced HIV transmission among uninfected individuals exposed to the virus through sexual contact. ▶▶| *Acquired immunodeficiency syndrome (AIDS) (section 38.3)*

(warts), adenoviruses (respiratory diseases), herpesviruses (oral and genital sores), cytomegalovirus, and poxviruses (chickenpox). The drug acts on the viral DNA polymerase as a competitive inhibitor. It is structurally similar to deoxycytosine triphosphate (dCTP), a substrate of DNA polymerase. Thus it blocks viral DNA synthesis.

Since the early days of human immunodeficiency virus (HIV) treatment, much effort has been focused on developing new drugs. These drugs target and interfere with critical steps in the viral replicative processes (**figure 9.16**). There are now five categories of drugs used to manage HIV infection: (1) nucleoside reverse transcriptase inhibitors (NRTIs), which are nucleoside analogues that produce faulty viral DNA (e.g., **azidothymidine** or AZT); (2) nonnucleoside reverse transcriptase inhibitors (NNRTIs), which prevent HIV DNA synthesis by selectively binding to and inhibiting the reverse transcriptase enzyme; (3) protease inhibitors (PIs), which block the activity of the HIV protease that is needed for the production of all viral proteins; (4) integrase inhibitors that prevent the incorporation of the HIV genome into the host's chromosomes; and (5) fusion inhibitors (FIs), a relatively new category of drugs that prevent HIV entry into cells (figure 9.16). Inhibition of reverse transcription, which catalyzes an early step in the multiplication

9.7 Antiprotozoan Drugs

After reading this section, you should be able to:

- Compare the mechanisms of action of antiprotozoan drugs
- Relate side effects and toxicity of antiprotozoan drugs to their mechanism of action
- Explain why there are far fewer antiprotozoan agents than there are antibacterial agents

As with other antimicrobial therapies, antiprotozoan drug effectiveness starts by identifying a unique target to which a drug can bind and thus prevent some vital function. However, because protozoa are eukaryotes, the potential for drug action on host cells and tissues is greater than it is when targeting bacteria. Most of the drugs used to treat protozoan infections have significant side effects; nonetheless, the side effects are usually acceptable when weighed against the parasitic burden.

The number of antiprotozoan drugs is relatively small, and the mechanism of action for most of these drugs is not completely understood. The drugs described here have potent antiprotozoan action and appear to act on protozoan nucleic acid or some metabolic process.

Malaria, caused by any of five species of the genus *Plasmodium,* kills about a million people annually—most of them children. Drugs to treat and prevent malaria include everything from ancient Chinese remedies to quinine water. In fact, gin and tonic is said to have been the most popular drink among imperial British in South Asia due to its antimalarial activity. We now use more purified forms of **quinine drugs,** including chloroquine and qualaquin to treat malaria (**figure 9.17**). These drugs suppress *Plasmodium* reproduction and are effective in eradicating asexual stages of the protozoan's life cycle that occur in red blood cells. Several mechanisms of action have been reported. They can raise the internal pH, clump the hemoglobin waste product hemazoin (often called the plasmodial pigment), and intercalate into plasmodial DNA. Chloroquine also inhibits malarial heme polymerase, an enzyme that converts toxic heme metabolites into the nontoxic hemazoin. Inhibition of this enzyme leads to a buildup of toxic heme metabolites. Another drug, mefloquine, has been found to swell the *Plasmodium falciparum* food vacuoles, where it may act by forming toxic complexes that damage membranes and other plasmodial components. Primaquine is

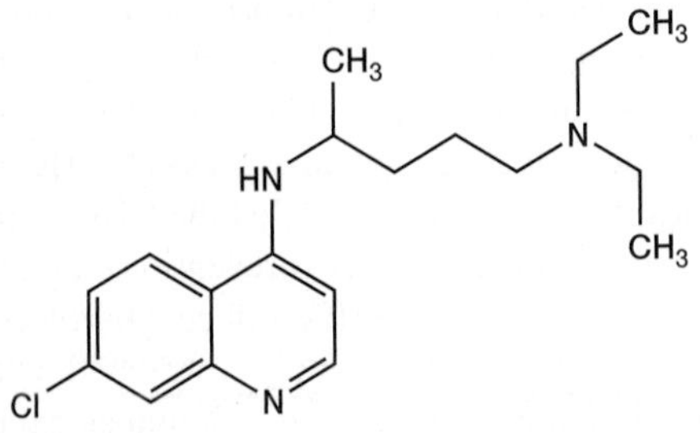

Figure 9.17 Chloroquine, a Quinine Drug. Quinine drugs have a long history of use as antimalarial agents.

active against a dormant form of the protists (hypnozoites) that are found in the liver; it prevents relapses. Individuals who travel to areas where malaria is endemic typically receive chemoprophylactic treatment with chloroquine or other antimalarial agents. ▶▶| *Malaria (section 40.3)*

Treatment of malaria has become a complex task, with specific drugs and therapies recommended according to (1) the endemic *Plasmodium* species, (2) drug sensitivity and resistance patterns, (3) drug cost, and (4) patient access to therapy. Because of these factors, a single drug therapy may be recommended in one area, while a different drug or even a combination therapy may be recommended in another. As mentioned, many factors impact treatment decisions, including patient compliance. Intermittent drug use is a strong selective pressure for drug resistance. Thus, treatment regimens may also rely on the likelihood of patient adherence to the drug dosing schedule. In general, the World Health Organization recommends a combination of a quinine drug and a derivative of **artemisinin.**

Having its origins as a treatment in traditional Chinese medicine, artemisinin is a chemical found in the sweet wormwood plant (*Artemisia annua*). A semisynthetic artemisinin derivative used in combination therapies is an effective treatment of malaria and a truly cost-effective treatment. The mechanism of artemisinin action is not well understood; it appears to form reactive oxygen intermediates inside *Plasmodium*-infected red blood cells, leading to altered hemoglobin catabolism and damage to the protist's electron transport chain. |◀◀ *Oxygen concentration (section 7.4)* ▶▶| *Electron transport chains (section 11.6)*

There are many other diseases caused by protozoa, as discussed in chapter 40. Amoebic dysentery is usually treated with **metronidazole.** Anaerobic organisms, such as the causative agent *Entamoeba* spp., readily reduce metronidazole to the active metabolite within the cytoplasm. A number of antibiotics that inhibit bacterial protein synthesis (e.g., the aminoglycosides clindamycin and paromomycin) are also used to treat protozoan infection. These agents are used to prevent or treat infections caused by protozoa known as apicomplexans. These microbes possess an organelle derived from plastids, the apicoplast, that is essential for viability. These protein synthesis inhibitors bind to the ribosomes in these organelles. ▶▶| *Supergroup* Amoebozoa *includes protists with pseudopodia (section 25.3); Supergroup SAR (section 25.4)*

Toxoplasmosis is a life-threatening infection in immunocompromised individuals and can cause severe birth defects in human fetuses. In the United States it is thought to be a leading cause of death associated with foodborne illness. The CDC reports that more than 60 million men, women, and children are infected with *Toxoplasma gondii*, but the immune system usually keeps the parasite from causing symptoms or illness. Toxoplasmosis is typically treated with combinations of pyrimethamine, dapsone, and sulfadiazine, plus folinic acid. Pyrimethamine and dapsone are thought to act in the same way as trimethoprim—interfering with folic acid synthesis by inhibition of dihydrofolate reductase. A drug that interferes with electron transport, nitazoxanide, is used to treat cryptosporidiosis, a food- and waterborne disease. ▶▶| *Toxoplasmosis (section 40.5); Cryptosporidiosis (section 40.5)*

9.8 Several Factors Influence Antimicrobial Drug Effectiveness

After reading this section, you should be able to:

- Predict the effects of (1) delivery route, (2) metabolism, and (3) local concentration on the effectiveness of an antimicrobial drug
- Correlate the sensitivity of a microorganism to an antimicrobial agent with microbial growth in the presence of that agent
- Identify practices that lead to antimicrobial drug resistance and suggest countermeasures

Drug Effectiveness

It is crucial to recognize that effective drug therapy is not a simple matter. Drugs can be administered in several different ways, and they do not always spread rapidly throughout the body or immediately kill all invading pathogens. A complex array of factors influences the effectiveness of antimicrobial drugs.

First, the drug must be able to reach the site of infection. Understanding the factors that control drug activity, stability, and metabolism in vivo are essential in drug formulation. For example, the mode of administration plays an important role. A drug such as penicillin G is not suitable for oral administration because it is relatively unstable in stomach acid. Some antibiotics—for example, gentamicin and other aminoglycosides—are not well absorbed from the intestinal tract and must be injected intramuscularly or given intravenously. Other antibiotics (neomycin, bacitracin) are so toxic that they can only be applied topically to skin lesions. Nonoral routes of administration are called **parenteral routes.** Even when an agent is administered properly, it may be excluded from the site of infection. For example, blood clots, necrotic tissue, or biofilms can protect bacteria from a drug, either because body fluids containing the agent may not easily reach the pathogens or because the agent is absorbed by materials surrounding them.

Second, the pathogen must be susceptible to the drug. Bacteria in biofilms or abscesses may be replicating very slowly and are therefore resistant to chemotherapy. Many antibiotic agents affect pathogens only if they are actively growing and dividing. A pathogen, even though growing, may simply not be susceptible to a particular agent. To control resistance, drug cocktails can be used to treat some infections. Notable examples of this are the use of clavulanic acid (to inactivate penicillinase) combined with ampicillin to treat penicillin-resistant bacteria and the combination of a replication inhibitor ledipasvir with a nucleoside analog inhibitor sofosbuvir to control hepatitis C virus.

Third, the chemotherapeutic agent must reach levels in the body that exceed the pathogen's MIC value if it is going to be effective. The concentration reached will depend on the amount of drug administered, the route of administration and speed of uptake, and the rate at which the drug is cleared or eliminated from the body. It makes sense that a drug will remain at high concentrations longer if it is absorbed over an extended period and excreted slowly.

Finally, chemotherapy has been rendered less effective and much more complex by the spread of drug-resistance genes and prevention of drug access by biofilm components. ▶▶ *Drug resistance can be transmitted from one bacterium to another by HGT (section 16.9)*

Overcoming Drug Resistance

As we discuss in chapter 16, bacteria are very adept at spreading resistance genes around, both within a population and to other members of a microbial community. However, what keeps the resistance genes in a population and contributes to the generation of drug-resistant bacterial strains is exposure to the drug. Clearly, it is important that physicians be able to treat infectious disease in their patients. The problem is that drugs are being overused, misused, and in many cases abused. Put simply, the more drugs are used, the more likely it is that bacteria will become resistant to them. That raises the question: Is there a way to overcome drug resistance?

Several strategies can be employed to discourage the emergence of drug resistance. Tighter controls over indiscriminate drug use and use of antibiotics in animal feed must be enacted. Drugs can be given in a high enough concentration to destroy susceptible microbes and most spontaneous mutants that might arise during treatment. Sometimes two or even three different drugs can be administered simultaneously with the hope that each drug will prevent the emergence of resistance to the other. This approach is used in treating tuberculosis, HIV, and malaria, for example. When treating tuberculosis (TB), several drugs are administered simultaneously (e.g., isoniazid [INH] plus rifampin, ethambutol, and pyrazinamide). These drugs are administered for 6 to 9 months as a way of decreasing the possibility that the bacterium develops drug resistance. If a patient fails to take prescribed antibiotics as directed (e.g., does not complete the course of treatment), resistant mutants survive and flourish because of their competitive advantage over nonresistant strains. Preventing the growth of such mutants is the rationale behind "directly observed therapy" (DOT). During DOT, a health-care worker observes TB patients take each antibiotic dose, thereby ensuring full antibiotic regimen compliance. ▶▶ *Human diseases caused by bacteria (section 39.1)*

Other strategies used to prevent drug resistance include the strict control on use of chemotherapeutic drugs, especially newly developed drugs to which bacteria have not become widely resistant. The use of broad-spectrum drugs must also be curtailed. These ideas and others are now incorporated into "antibiotic stewardship" programs at many health-care facilities. If possible, the pathogen should be identified, drug sensitivity tests performed, and the proper narrow-spectrum drug employed. Patient compliance is just as important because completing a full course of antimicrobial therapy often prevents full mutation to resistant phenotypes.

Despite efforts to control the emergence and spread of drug resistance, the situation continues to worsen. Thus an urgent need exists for new antibiotics that microorganisms have never encountered. Pharmaceutical and biotechnology companies collect and analyze samples from around the world in a search for completely new antimicrobial agents. Both culture-based and metagenomics approaches are used. Structure-based or rational drug design is another option. If the three-dimensional structure of a susceptible target molecule such as an enzyme essential to microbial function is known, computer programs can be used to design drugs that precisely fit the target molecule. These drugs might be able to bind to the target and disrupt its function sufficiently to destroy the pathogen. Pharmaceutical companies are using these approaches to develop drugs for the treatment of AIDS, cancer, septicemia caused by lipopolysaccharide, and the common cold. Information derived from the sequencing and analysis of pathogen genomes is also useful in identifying new targets for antimicrobial drugs. For example, genomics studies are providing data for research on inhibitors of both aminoacyl-tRNA synthetases and the enzyme that removes the formyl group from the N-terminal methionine during bacterial protein synthesis. The drug susceptibility of enzymes required for fatty acid synthesis is also being analyzed. A most interesting response to the current crisis is the renewed interest in an idea first proposed early in the twentieth century by Felix d'Herelle, one of the discoverers of bacterial viruses (bacteriophages). d'Herelle proposed that bacteriophages could be used to treat bacterial diseases. Although most microbiologists did not pursue his proposal actively due to technical difficulties and the advent of antibiotics, Russian scientists developed the medical use of bacteriophages. Currently Russian physicians use bacteriophages to treat many bacterial infections. Bandages are saturated with phage solutions, phage mixtures are administered orally, and phage preparations are given intravenously to treat *Staphylococcus* infections. Several American companies are actively conducting research on phage therapy and preparing to carry out clinical trials. ◀◀ *Viruses and other acellular infectious agents (chapter 6)* ▶▶| *Microbial genomics (chapter 18); Biotechnology and industrial microbiology (chapter 42)*

Retrieve, Infer, Apply

1. What factors do you think must be considered when treating an infection present in a biofilm on a medical implant (e.g., an artificial hip) versus a skin infection caused by the same microbe?
2. What is parenteral administration of a drug? Why is it used?
3. How might drug resistance be detected?
4. What are the primary medical practices that result in antimicrobial drug resistance? How can these be overcome?

Key Concepts

9.1 Antimicrobial Chemotherapy Evolved from Antisepsis Efforts

- The modern era of chemotherapy began with Paul Ehrlich's work on drugs against African sleeping sickness and syphilis.
- Other early pioneers were Gerhard Domagk, Alexander Fleming, Howard Florey, Ernst Chain, Norman Heatley, and Selman Waksman.

9.2 Antimicrobial Drugs Need to be Selectively Toxic over a Range of Effectiveness

- An effective chemotherapeutic agent must have selective toxicity. A drug with great selective toxicity has a high therapeutic index and usually disrupts a structure or process unique to the pathogen. It has fewer side effects.
- Antibiotics can be classified in terms of the range of target microorganisms (narrow spectrum versus broad spectrum); their source (natural, semisynthetic, or synthetic); and their general effect (static versus cidal) (**table 9.1**).

9.3 Antimicrobial Activity Can be Measured by Specific Tests

- Antibiotic effectiveness can be estimated through the determination of the minimal inhibitory concentration and the minimal lethal concentration with dilution susceptibility tests.

- Tests such as the Kirby-Bauer test (a disk diffusion test) and the Etest® are often used to estimate a pathogen's susceptibility to drugs quickly (**figures 9.2–9.4**).
- Antibiotic resistance can also be determined using protein and gene detection strategies.

9.4 Antibacterial Drugs

- Members of the penicillin family contain a β-lactam ring and disrupt bacterial cell wall synthesis, resulting in cell lysis (**figure 9.5**). Some, such as penicillin G, are usually administered by injection and are most effective against Gram-positive bacteria. Others can be given orally (penicillin V), are broad spectrum (ampicillin, carbenicillin), or are usually penicillinase-resistant (piperacillin).
- Cephalosporins are similar to penicillins but can be given to patients with penicillin allergies (**figure 9.6**).
- Vancomycin is a glycopeptide antibiotic that inhibits transpeptidation during peptidoglycan synthesis (**figure 9.7**). It is used against drug-resistant staphylococci, enterococci, and clostridia.
- Aminoglycoside antibiotics such as streptomycin and gentamicin bind to the small ribosomal subunit, inhibit protein synthesis, and are bactericidal (**figure 9.8**).

- Tetracyclines are broad-spectrum antibiotics having a four-ring nucleus with attached groups (**figure 9.9**). They bind to the small ribosomal subunit and inhibit protein synthesis.
- Erythromycin is a bacteriostatic macrolide antibiotic that binds to the large ribosomal subunit and inhibits protein synthesis (**figure 9.10**).
- Chloramphenicol is a broad-spectrum, bacteriostatic antibiotic that inhibits protein synthesis. It is quite toxic and used only for very serious infections.
- Sulfonamides or sulfa drugs resemble *p*-aminobenzoic acid and competitively inhibit folic acid synthesis (**figure 9.11**).
- Trimethoprim is a synthetic antibiotic that inhibits dihydrofolate reductase, which is required by organisms in the manufacture of folic acid (**figure 9.12**).
- Quinolones are a family of bactericidal synthetic drugs that inhibit DNA gyrase and thus inhibit DNA replication (**figure 9.13**).

9.5 Antifungal Drugs

- Because fungi are more similar to human cells than bacteria, antifungal drugs generally have lower therapeutic indexes than antibacterial agents and produce more side effects.
- Superficial mycoses can be treated with miconazole, ketoconazole, clotrimazole, tolnaftate, nystatin, and griseofulvin (**figure 9.14**). Amphotericin B, 5-flucytosine, and fluconazole are used for systemic mycoses.

9.6 Antiviral Drugs

- Antiviral drugs interfere with critical stages in the virus life cycle (amantadine, rimantadine, ritonavir) or inhibit the synthesis of virus-specific nucleic acids (zidovudine, adenine arabinoside, acyclovir) (**figure 9.15**).
- Drug combinations (cocktails) appear to be more effective than monotherapies.

9.7 Antiprotozoan Drugs

- The mechanisms of action of most drugs used to treat protozoan infection are unknown.
- Some antiprotozoan drugs interfere with critical steps in nucleic acid synthesis, protein synthesis, electron transport, or folic acid synthesis.

9.8 Several Factors Influence Antimicrobial Drug Effectiveness

- A variety of factors can greatly influence the effectiveness of antimicrobial drugs during use. These include route, effective concentration, and pathogen sensitivity.
- Drug resistance can often limit drug effectiveness. However, it can be overcome through the judicious use of antimicrobial agents (decreasing selective pressure), use of drug cocktails (attacking more than one microbial structure or function), and use of specific, narrow-spectrum drugs to target infecting microbes.

Compare, Hypothesize, Invent

1. What advantage might soil bacteria and fungi gain from the synthesis of antibiotics?

2. Some advocate stockpiling the drug Tamiflu in the event of an influenza pandemic. Others point out that wealthy, Western nations would have an unfair advantage because developing nations (where the pandemic is most likely to start) would not have access to this expensive antiviral. Furthermore, some fear that indiscriminate use of the drug would promote the evolution of resistant flu strains. Given these caveats, do you think developed nations should stockpile Tamiflu for the protection and treatment of their citizens? Explain your answer.

3. You are a pediatrician treating a child with an upper respiratory infection that is clearly caused by a virus. The child's mother insists that you prescribe antibiotics—she's not leaving without them! How do you convince the child's mother that antibiotics will do more harm than good?

4. Antibiotics that target bacterial molecules not previously exploited are desperately needed. One such target is the protein FtsZ. The small molecule 3-methoxybenzamide (3-MBA) is known to inhibit FtsZ in *Bacillus subtilis* but is not bacteriocidal. Nonetheless, researchers reasoned that 3-MBA offered a good starting point for the synthesis of a molecule that might be a potential drug candidate. Over 500 3-MBA analogues were synthesized and screened; one called PA190723 was extremely potent in its capacity to bind FtsZ and inhibit bacterial growth. In fact, when used in a mouse model, PA190723 was bacteriocidal against methicillin- and multidrug-resistant *Staphylococcus aureus*. What makes FtsZ a good drug target (*see chapters 3 and 7*)? What preliminary information about 3-MBA would be helpful if you were designing the 3-MBA analogues? As these researchers move forward with clinical (human) testing, what other parameters and outcomes must be assessed besides the bacteriocidal activity of PA190723?

Read the original paper: Haydon, D. J., et al. 2008. An inhibitor of FtsZ with potent and selective anti-staphylococcal activity. *Science* 321:1673.

10

Introduction to Metabolism

Each day, toilets send billions of gallons of wastewater to sewage treatment plants. The treatment of wastewater is a significant cost to municipalities.

Flushed Away

For one day, count the number of times you flush the toilet in your home. Is your toilet a low-volume toilet that sends 1.6 gallons or less out of your house with each flush? Or is it an older, water-greedy toilet? Did you ever wonder where the wastes you flush away go?

If you live in a city, your wastes travel through a labyrinth of pipes to a wastewater treatment plant. Now do some math: multiply the number of gallons you flushed away by the number of people in your town. That's how much wastewater is sent each day to your town's wastewater treatment plant. If you live in a large city, the amount of wastewater generated is astronomical: in New York City it's about 1.3 billion gallons per day!

The cost of treating wastewater so that it is safe to release into the environment is also astronomical. New York City spends an estimated $400 million each year. Thus wastewater treatment is a big item in any city's budget. But can these wastes be used to decrease the cost of treatment or, better yet, generate funds for municipalities? The answer to that question is yes, and it is the metabolic diversity of microbes that makes it possible.

As we describe in chapter 43, wastewater treatment plants are home to a complex microbial community. Working together, these microbes decrease the organic matter in wastewater and generate several by-products using their diverse metabolic abilities. One of the most important is methane, which is made when organic matter is degraded under anoxic conditions. For many years, wastewater treatment facilities simply burned the methane. Increasingly, wastewater treatment plants are using the methane to heat their buildings, and some municipalities, such as New York City, are exploring the possibility of selling methane to natural gas providers, which then would supply it to homes for heating.

Another by-product of wastewater treatment is sludge, the solid material composed of undigested material and biomass created as the organic material in wastewater is removed. Many towns truck the sludge to landfills. However, many landfills are at or nearing capacity and will no longer accept solid waste. Increasingly cities are selling sludge to farmers and ranchers to use as a soil amendment. Some are exploring the use of sludge for paving roads and as building materials. Of course the ultimate product of wastewater treatment is clean water. Indeed, some plants are capable of generating water so clean it could be used as a source of drinking water.

The treatment of wastewater is only one example of how humans exploit the metabolic prowess of microbes. Microbes generate numerous metabolites that humans have found desirable. As we describe in chapter 41, ethanol is important in the production of beer, ale, and wine. Propionic acid flavors Swiss cheese. Carbon dioxide leavens bread, and antibiotics save the lives of humans every day.

The impact of microbial metabolites goes beyond their use in industry and food production. It is now recognized that the microbes living in and on the human body contribute compounds that are needed for human metabolism. Indeed, it is thought that the metabolic capabilities of both the human normal microbiota and their human hosts are highly integrated as the result of coevolution. ▶▶ *Biotechnology and industrial microbiology (chapter 42); The human-microbe ecosystem (section 32.2)*

The importance of microbial metabolism is evidenced in less obvious but equally significant ways. Microbes are not visible to us, but they constitute the largest component of biomass on Earth and therefore contribute to the cycling of elements in ecosystems. The nitrogen cycle is of particular note. Four of the eight transformations of nitrogen illustrated in **figure 10.1** are done only by microbes, and microbes contribute to the remaining four. ▶▶ *Biogeochemical cycling sustains life on Earth (section 28.1)*

Clearly, microbial metabolism is extremely important to the well-being of humans, and much of the study of microbes has focused on this aspect of their biology. Chapters 10 through 12 focus on the metabolic processes that conserve the energy supplied by an organism's energy source and how that energy is used to synthesize the building blocks from which an organism is constructed. In this chapter, we lay the foundation for that discussion.

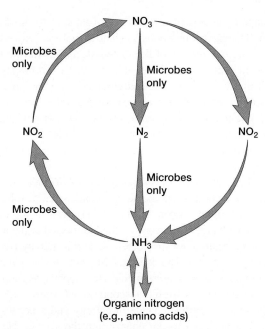

Figure 10.1 Contributions of Microbes to the Nitrogen Cycle. This is a simplified representation of the nitrogen cycle. A more complete cycle is presented in chapter 28.

Readiness Check:

Based on what you have learned previously, you should be able to:

✔ Distinguish metabolism, catabolism, and anabolism

✔ Describe the structure of bacterial and archaeal cell envelopes (sections 3.3–3.5 and 4.2)

✔ Describe mitochondrial structure and function (section 5.6)

10.1 Metabolism: Important Principles and Concepts

After reading this section, you should be able to:

■ List the features common to all types of metabolism

■ Describe the three types of work carried out by cells

■ State the relationship between cellular work and energy

■ Paraphrase the first and second laws of thermodynamics

■ Use the equation for free energy change and the equation that links standard free energy change to a reaction's equilibrium constant and explain their importance

■ Infer from a standard free energy change whether a reaction is exergonic or endergonic

It would be nice to think that microbes existed for the benefit of humans and that their metabolisms evolved for us to exploit, but of course this is not the case. Rather, they use their vast repertoire of chemical reactions to survive and reproduce, just as all organisms do. Thus **metabolism** is central to all life. The forces of evolution have shaped the metabolic process used by organisms for billions of years. Despite the diversity of chemical reactions that have evolved, there are several aspects of metabolism that are

common to all organisms. Here we list the common features, noting where they are discussed in this chapter.

■ **Life obeys the laws of thermodynamics** (this section). Therefore to maintain their structural complexity, organisms must obtain energy from their environment.

■ **The energy cells obtain from their environment is most often conserved as a molecule called ATP** (section 10.2). ATP is used to supply the energy needed for certain chemical reactions to proceed.

■ **Oxidation-reduction (redox) reactions play a critical role in energy conservation** (section 10.3). Many of these redox reactions occur in electron transport chains (section 10.4).

■ **The chemical reactions that occur in cells are organized into pathways.** The product of one reaction is the substrate for the next reaction in the pathway (section 10.5).

■ **Each reaction of a pathway is catalyzed by an enzyme or a ribozyme.** Enzymes and ribozymes are critical to all life as they speed up the reactions that enable life to exist (section 10.6).

■ **The functioning of biochemical pathways is regulated** so that their products are made at the correct time (section 10.7).

Cellular Work and Energy Transfers

Examine the common features of metabolism just described. It should be clear that cells must do work in order to survive and reproduce. Cells carry out three major types of work. **Chemical work** involves the synthesis of complex biological molecules from much simpler precursors (i.e., anabolism); energy is needed to increase the molecular complexity of a cell. **Transport work** requires energy to take up nutrients, eliminate wastes, and maintain ion balances. Energy input is needed because molecules and ions often must be transported across cell membranes against an electrochemical gradient. The third type of work is **mechanical work.** Energy is required for cell motility and the movement of structures within cells, such as partitioning chromosomes during cell division.

As just indicated, cells need energy to do work. Indeed, **energy** may be defined most simply as the capacity to do work. This is because all physical and chemical processes are the result of the application or movement of energy. Organisms obtain the energy they need from an energy source present in their environment. They convert the energy it provides into a useful form. The most commonly used form of cellular energy is the nucleoside triphosphate ATP. In addition, other nucleoside triphosphates and other high-energy molecules are required for specific processes.

To understand how energy is conserved in ATP and how ATP is used to do cellular work, some knowledge of the basic principles of thermodynamics is required. The science of **thermodynamics** analyzes energy changes in a collection of matter (e.g., a cell or a plant) called a system. All other matter in the universe is called the surroundings. Thermodynamics focuses on the energy differences between the initial state and the final state of a system. It is not concerned with the rate of the process. For instance, if a pan of water is heated to boiling, only the condition of the water at the start and at boiling is important in thermodynamics, not how fast it is heated.

Laws of Thermodynamics

Two important laws of thermodynamics are critical to understanding energy transfers. The **first law of thermodynamics** says that energy can be neither created nor destroyed. The total energy in the universe remains constant, although it can be redistributed, as it is during the many energy exchanges that occur during chemical reactions. For example, heat is given off by exothermic reactions and absorbed during endothermic reactions. However, the first law alone cannot explain why heat is released by one chemical reaction and absorbed by another. Explanations for this require the **second law of thermodynamics** and a condition of matter called entropy. **Entropy** is a measure of the randomness or disorder of a system. The greater the disorder of a system, the greater its entropy. The second law states that physical and chemical processes proceed in such a way that the randomness or disorder of the universe (the system and its surroundings) increases. However, even though the entropy of the universe increases, the entropy of any given system within the universe can increase, decrease, or remain unchanged.

Two types of energy units are employed to specify the amount of energy used in or evolving from a particular process. A **calorie** (cal) is the amount of heat energy needed to raise 1 gram of water from 14.5 to 15.5°C. The amount of energy also may be expressed in terms of **joules** (J), the units of work capable of being done. One cal of heat is equivalent to 4.1840 J of work. One thousand calories, or a kilocalorie (kcal), is enough energy to boil 1.9 milliliters of water. A kilojoule is enough energy to boil about 0.44 milliliters of water or enable a person weighing 70 kilograms to climb 35 steps.

Free Energy Change Predicts the Nature of a Chemical Reaction

The first and second laws of thermodynamics can be combined in a useful equation, relating the changes in energy that can occur in chemical reactions and other processes.

$$\Delta G = \Delta H - T\,\Delta S$$

ΔG is the change in free energy, ΔH is the change in enthalpy, T is the temperature in Kelvin (°C + 273), and ΔS is the change in entropy occurring during the reaction. The change in **enthalpy** is the change in heat content. Cellular reactions occur under conditions of constant pressure and volume. Thus the change in enthalpy is about the same as the change in total energy during the reaction. The **free energy change** is the amount of energy in a system (or cell) available to do useful work at constant temperature and pressure. Therefore the change in entropy (ΔS) is a measure of the proportion of the total energy change that the system cannot use in performing work. Free energy and entropy changes do not depend on how the system gets from start to finish. A reaction will occur spontaneously—that is, without any external cause—if the free energy of the system decreases during the reaction or, in other words, if ΔG is negative. It follows from the equation that a reaction with a large positive change in entropy will normally tend to have a negative ΔG value and therefore occur spontaneously. A decrease in entropy will tend to make ΔG more positive and the reaction less favorable.

It is helpful to think of the relationship between entropy (ΔS) and change in free energy (ΔG) in terms that are more concrete. Consider the Greek myth of Sisyphus, king of Corinth. For his assorted crimes against the gods, he was condemned to roll a large boulder to the top of a steep hill for all eternity. This represents a very negative change in entropy—a boulder poised at the top of a hill is neither random nor disordered—and this activity (reaction) has a very positive ΔG. That is to say, Sisyphus had to put a lot of energy into the system. Unfortunately for Sisyphus, as soon as the boulder was at the top of the hill, it spontaneously rolled back down. This represents a positive change in entropy and a negative ΔG. Sisyphus did not need to put energy into the system. He could simply stand at the top of the hill and watch the reaction proceed.

The change in free energy has a definite, concrete relationship to the direction of chemical reactions. Consider this simple reaction.

$$A + B \rightleftharpoons C + D$$

If molecules A and B are mixed, they will combine to form the products C and D. Eventually C and D will become concentrated enough to combine and produce A and B at the same rate as C and D are formed. The reaction is now at **equilibrium:** the rates in both directions are equal and no further net change occurs in the concentrations of reactants and products. This situation is described by the **equilibrium constant (K_{eq})**, relating the equilibrium concentrations of products and substrates to one another.

$$K_{eq} = \frac{[C]\,[D]}{[A]\,[B]}$$

If the equilibrium constant is greater than one, the products are in greater concentration than the reactants at equilibrium; that is, the reaction tends to go to completion as written.

The equilibrium constant of a reaction is directly related to its change in free energy. When the free energy change for a process is determined at carefully defined standard conditions of concentration, pressure, pH, and temperature, it is called the **standard free energy change** ($\Delta G°$). If the pH is set at 7.0 (which is close to the pH of living cells), the standard free energy change is indicated by the symbol $\Delta G°'$. The change in standard free energy may be thought of as the maximum amount of energy available from the system for useful work under standard conditions. Using $\Delta G°'$ values allows comparisons of reactions without considering variations in ΔG due to differences in environmental conditions. The relationship between $\Delta G°'$ and K_{eq} is given by this equation.

$$\Delta G°' = -2.303RT \cdot \log K_{eq}$$

R is the gas constant (1.9872 cal/mole-degree or 8.3145 J/mole-degree), and T is the absolute temperature. Inspection of this

Exergonic reactions

$$A + B \rightleftharpoons C + D$$

$$K_{eq} = \frac{[C][D]}{[A][B]} > 1.0$$

$\Delta G^{\circ\prime}$ is negative.

Endergonic reactions

$$A + B \rightleftharpoons C + D$$

$$K_{eq} = \frac{[C][D]}{[A][B]} < 1.0$$

$\Delta G^{\circ\prime}$ is positive.

Figure 10.2 The Relationship of $\Delta G^{\circ\prime}$ to the Equilibrium of Reactions. Note the differences between exergonic and endergonic reactions.

MICRO INQUIRY *Which reaction would release heat? Explain your answer.*

equation shows that when $\Delta G^{\circ\prime}$ is negative, the equilibrium constant is greater than one and the reaction goes to completion as written. It is said to be an **exergonic reaction (figure 10.2)**. In an **endergonic reaction,** $\Delta G^{\circ\prime}$ is positive and the equilibrium constant is less than one. That is, the reaction is not favorable, and little product will be formed at equilibrium under standard conditions. Keep in mind that the $\Delta G^{\circ\prime}$ value shows only where the reaction lies at equilibrium, not how fast the reaction reaches equilibrium.

Retrieve, Infer, Apply

1. What kinds of work are carried out in a cell? Suppose a bacterium was doing the following: synthesizing peptidoglycan, rotating its flagellum and swimming, and secreting siderophores. What type of work is the bacterium doing in each case?
2. What is thermodynamics? Summarize the first and second laws of thermodynamics.
3. Define entropy and enthalpy. Do living cells increase entropy within themselves? Do they increase entropy in the environment?
4. Define free energy. What are exergonic and endergonic reactions?
5. Suppose that a chemical reaction had a large negative $\Delta G^{\circ\prime}$ value. Is the reaction endergonic or exergonic? What would this indicate about its equilibrium constant? If displaced from equilibrium, would the reaction proceed rapidly to completion? Would much or little free energy be made available?

10.2 ATP: The Major Energy Currency of Cells

After reading this section, you should be able to:

- Draw a simple schematic that illustrates the structure of ATP
- Describe ATP's role as a coupling agent that links exergonic and endergonic reactions
- Compare ATP's phosphate transfer potential to that of glucose 6-phosphate and phosphoenolpyruvate (PEP) and relate this to ATP's function in cells
- Describe the energy cycle observed in all organisms
- Describe the metabolic functions of three nucleoside triphosphates (other than ATP)

Energy is released from a cell's energy source in exergonic reactions (i.e., those reactions with a negative ΔG). Rather than wasting this energy, much of it is trapped in a practical form that allows its transfer to the cellular systems doing work. These systems carry out endergonic reactions (e.g., anabolism), and the energy captured by the cell is used to drive these reactions to completion. In living organisms, the most commonly used practical form of energy is the nucleotide **adenosine 5′-triphosphate (ATP; figure 10.3)**. In a sense, cells carry out certain processes so that they can "earn"

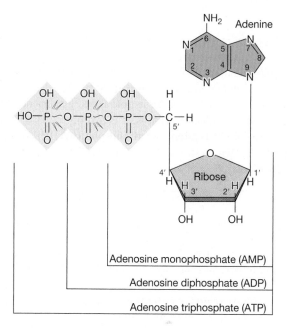

(a) ⌣ Bond that releases energy when broken

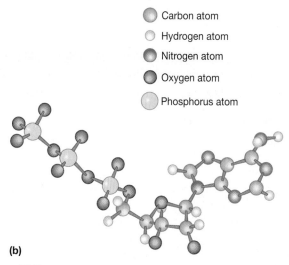

(b)

Figure 10.3 Adenosine Triphosphate. (a) Structure of ATP, ADP, and AMP. The two red bonds (~) are more easily broken and release considerable energy that can be used in endergonic reactions. The pyrimidine ring atoms have been numbered, as have the carbon atoms in ribose. (b) A stick-and-ball model of ATP.

Endergonic reaction alone

$$A + B \rightleftharpoons C + D$$

Endergonic reaction coupled to ATP breakdown

ATP → ADP + P$_i$

$$A + B \rightleftharpoons C + D$$

Figure 10.4 ATP as a Coupling Agent. ATP is formed by exergonic reactions and then used to drive endergonic reactions.

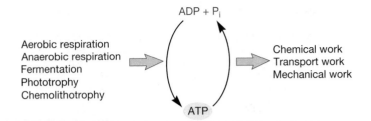

Aerobic respiration
Anaerobic respiration
Fermentation
Phototrophy
Chemolithotrophy

ADP + P$_i$

ATP

Chemical work
Transport work
Mechanical work

Figure 10.5 The Cell's Energy Cycle. ATP is formed from energy made available during aerobic respiration, anaerobic respiration, fermentation, chemolithotrophy, and phototrophy. Its breakdown to ADP and phosphate (P$_i$) makes chemical, transport, and mechanical work possible.

ATP and carry out other processes in which they "spend" their ATP. Thus ATP is often referred to as the cell's energy currency. In the cell's economy, ATP serves as the link between exergonic reactions and endergonic reactions (**figure 10.4**).

What makes ATP suited for its role as energy currency? ATP is a high-energy molecule. That is, it is hydrolyzed almost completely to the products **adenosine diphosphate (ADP)** and orthophosphate (P$_i$), with a strongly exergonic $\Delta G^{\circ\prime}$ of -7.3 kcal/mole (-30.5 kJ/mole).

$$ATP + H_2O \rightleftharpoons ADP + P_i + H^+$$

The energy released is used to power endergonic reactions (figure 10.4).

The very negative $\Delta G^{\circ\prime}$ of hydrolysis of ATP is related to another important characteristic of ATP: its ability to transfer a phosphoryl group to another molecule. ATP is said to have a **high phosphate transfer potential** because it readily donates a phosphoryl group to other molecules. Note in **table 10.1** that there are other phosphorylated compounds found in cells. These molecules are generated during catabolism of organic molecules such as glucose. Some of these molecules have even higher phosphate transfer potentials than ATP. The fact that

ATP does not have the highest phosphate transfer potential means that it can easily be made by cells from ADP, using molecules such as phosphoenolpyruvate (PEP) as the source of the phosphoryl group. This mechanism for making ATP is called **substrate-level phosphorylation.** ▶▶ *Glucose to pyruvate (section 11.4)*

ATP, ADP, and P$_i$ form an energy cycle. As seen in **figure 10.5**, the energy released from an energy source is used to synthesize ATP from ADP and P$_i$. When ATP is hydrolyzed, the energy released drives endergonic processes such as anabolism, transport, and mechanical work. The mechanisms for synthesizing ATP are described in more detail in chapter 11.

ATP is the major energy currency for cells, but it is not the only energy currency. Other nucleoside triphosphates (NTPs) have major roles in metabolism. Guanosine 5′-triphosphate (GTP) supplies some of the energy used during protein synthesis. Cytidine 5′-triphosphate (CTP) is used during lipid synthesis, and uridine 5′-triphosphate (UTP) is used for the synthesis of peptidoglycan and other polysaccharides. ▶▶ *Translation in bacteria (section 13.7); Lipid synthesis (section 12.7); Synthesis of carbohydrates (section 12.4)*

Table 10.1	Phosphate Transfer Potential of Common Phosphorylated Compounds[1]	
Phosphorylated Molecule	**$\Delta G^{\circ\prime}$ of Hydrolytic Removal of Phosphate (KJ/mol)**	**Phosphate Transfer Potential**
High-Energy Phosphorylated Compounds		
Phosphoenolpyruvate[2]	-61.9	61.9
1, 3-bisphosphoglycerate[2]	-49.3	49.3
ATP (hydrolysis to AMP)	-45.6	45.6
ATP (hydrolysis to ADP)	-30.5	30.5
Low-Energy Phosphorylated Compounds		
Glucose 6-phosphate	-13.8	13.8
Glycerol 1-phosphate	-9.2	9.2

1 Phosphate transfer potential is defined as the negative of the standard free energy change ($\Delta G^{\circ\prime}$) for the hydrolytic removal of phosphate from the phosphorylated molecule.
2 Phosphoenolpyruvate and 1,3-bisphosphoglycerate are generated during the catabolism of organic molecules such as glucose.

10.3 Redox Reactions: Reactions of Central Importance in Metabolism

After reading this section, you should be able to:

- Describe a redox reaction, noting the role of the two half reactions and identifying the electron donor, electron acceptor, and conjugate redox pairs of the reaction
- Relate the standard reduction potential of a conjugate redox pair to its tendency to act as an electron-donating half reaction
- State the equation that links standard free energy change to the difference in standard reduction potentials of the two half reactions of a redox reaction
- Predict which molecule will act as an electron donor, which molecule will act as an electron acceptor, and the relative amount of energy released by a redox reaction, using the standard reduction potentials of the reaction's conjugate redox pairs

Free energy changes are related to the equilibria of all chemical reactions, including the equilibria of oxidation-reduction reactions. The release of energy from an energy source normally involves oxidation-reduction reactions. **Oxidation-reduction (redox) reactions** are those in which electrons move from an **electron donor** to an **electron acceptor**.[1] As electrons move from the donor to acceptor, the donor becomes less energy rich and the acceptor becomes more energy rich. Thus electrons can be thought of as packets of energy. The more electrons a molecule has and is able to donate in a redox reaction, the more energy rich the molecule is. This explains why molecules such as glucose, which can donate up to 24 electrons in redox reactions, are such excellent sources of energy.

Each redox reaction consists of two half reactions. One half reaction functions as the electron-donating half reaction (i.e., an oxidation reaction), and the other functions as the electron-accepting half reaction (i.e., the reduction). By convention, half reactions are written as reductions. Thus each half reaction consists of a molecule that can accept electrons (on the left side of the chemical equation), the number (n) of electrons (e^-) it accepts, and the molecule it becomes after accepting the electrons. The latter is placed on the right side of the chemical equation and is referred to as a donor, because it has electrons it can give up. The acceptor and donor of a half reaction are referred to as a **conjugate redox pair.**

$$\text{Acceptor} + ne^- \rightleftharpoons \text{donor}$$

The equilibrium constant for a redox half reaction is called the **standard reduction potential** (E_0) and is a measure of the tendency of the donor of a half reaction to lose electrons. By convention, the standard reduction potentials for half reactions, such as those in **table 10.2**, are determined at pH 7 and are represented by E_0'. Standard reduction potentials are measured in volts, a unit of electrical potential or electromotive force. Therefore conjugate redox pairs are a potential source of energy.

The reduction potential has a concrete meaning. Conjugate redox pairs with more negative reduction potentials will spontaneously donate electrons to pairs with more positive potentials and greater affinity for electrons. Thus electrons tend to move from donors at the top of the list in table 10.2 to acceptors at the

Table 10.2	Selected Biologically Important Half Reactions	
Half Reaction		E_0' **(Volts)**[1]
$2H^+ + 2e^- \rightarrow H_2$		−0.42
Ferredoxin (Fe^{3+}) + $e^- \rightarrow$ ferredoxin (Fe^{2+})		−0.42
$NAD(P)^+ + H^+ + 2e^- \rightarrow NAD(P)H$		−0.32
$S + 2H^+ + 2e^- \rightarrow H_2S$		−0.27
Acetaldehyde + $2H^+ + 2e^- \rightarrow$ ethanol		−0.20
Pyruvate$^-$ + $2H^+ + 2e^- \rightarrow$ lactate^{2-}		−0.19
FAD + $2H^+ + 2e^- \rightarrow FADH_2$		−0.18[2]
Oxaloacetate^{2-} + $2H^+ + 2e^- \rightarrow$ malate^{2-}		−0.17
Fumarate^{2-} + $2H^+ + 2e^- \rightarrow$ succinate^{2-}		0.03
Cytochrome b (Fe^{3+}) + $e^- \rightarrow$ cytochrome b (Fe^{2+})		0.08
Ubiquinone + $2H^+ + 2e^- \rightarrow$ ubiquinone H_2		0.10
Cytochrome c (Fe^{3+}) + $e^- \rightarrow$ cytochrome c (Fe^{2+})		0.25
Cytochrome a (Fe^{3+}) + $e^- \rightarrow$ cytochrome a (Fe^{2+})		0.29
Cytochrome a_3 (Fe^{3+}) + $e^- \rightarrow$ cytochrome a_3 (Fe^{2+})		0.35
$NO_3^- + 2H^+ + 2e^- \rightarrow NO_2^- + H_2O$		0.42
$NO_2^- + 8H^+ + 6e^- \rightarrow NH_4^+ + 2H_2O$		0.44
$Fe^{3+} + e^- \rightarrow Fe^{2+}$		0.77[3]
$\frac{1}{2}O_2 + 2H^+ + 2e^- \rightarrow H_2O$		0.82

1 E_0' is the standard reduction potential at pH 7.0.
2 The value for FAD/FADH$_2$ applies to the free cofactor because it can vary considerably when bound to an apoenzyme.
3 The value for free Fe, not Fe complexed with proteins (e.g., cytochromes).

MICRO INQUIRY *For the $Fe^{3+} + e^- \rightarrow Fe^{2+}$ half reaction, which form of iron is the acceptor and which is the donor?*

---

1 In redox reactions, the electron donor is often called the reducing agent or reductant because it is donating electrons to the acceptor and thus reducing it. The electron acceptor is called the oxidizing agent or oxidant because it is removing electrons from the donor and oxidizing it.

bottom because the latter have more positive potentials. This may be expressed visually in the form of an electron tower in which the most negative reduction potentials are at the top (**figure 10.6**). Electrons "fall down" the tower from donors higher in the tower (i.e., those having more negative potentials) to acceptors lower in the tower (i.e., those having more positive potentials).

Consider the case of the electron acceptor **nicotinamide adenine dinucleotide** (**NAD⁺**). The NAD^+/NADH conjugate redox pair has a very negative E'_0, and NADH can therefore give electrons to many acceptors, including O_2.

$$NAD^+ + 2H^+ + 2e^- \rightleftharpoons NADH + H^+ \quad E'_0 = -0.32 \text{ volts}$$

$$\tfrac{1}{2}O_2 + 2H^+ + 2e^- \rightleftharpoons H_2O \quad E'_0 = +0.82 \text{ volts}$$

Because the reduction potential of NAD^+/NADH is more negative than that of $\tfrac{1}{2}O_2/H_2O$, electrons flow from NADH (the donor) to O_2 (the acceptor), as shown in figure 10.6. The redox reaction is shown here.

$$NADH + H^+ + \tfrac{1}{2}O_2 \rightarrow H_2O + NAD^+$$

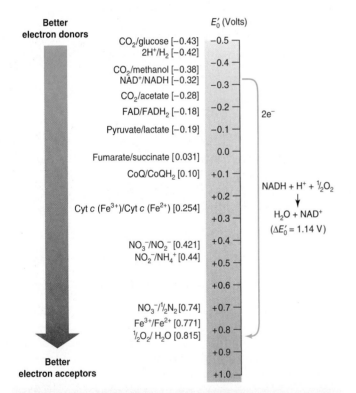

Figure 10.6 Electron Movement and Reduction Potentials. Electrons spontaneously move from donors higher on the tower (more negative potentials) to acceptors lower on the tower (more positive potentials). That is, the donor is always higher on the tower than the acceptor. For example, NADH will donate electrons to oxygen and form water in the process. Some typical conjugate redox pairs are shown on the left, and their reduction potentials are given in brackets.

MICRO INQUIRY *Why would energy be required to move electrons "up" the tower, from water to nitrate, for example?*

The relatively negative E'_0 of the NAD^+/NADH pair also means that the pair stores more potential energy than redox pairs with less negative (or more positive) E'_0 values. It follows that when electrons move from a donor to an acceptor with a more positive redox potential, free energy is released. The $\Delta G^{\circ\prime}$ of the reaction is directly related to the magnitude of the difference between the reduction potentials of the two couples ($\Delta E'_0$). The larger the $\Delta E'_0$, the greater the amount of free energy made available, as is evident from the equation

$$\Delta G^{\circ\prime} = -nF \cdot \Delta E'_0$$

in which n is the number of electrons transferred, F is the Faraday constant (23,062 cal/mole-volt; 96,480 J/mole-volt), and $\Delta E'_0$ is the E'_0 of the acceptor minus the E'_0 of the donor. For every 0.1 volt change in $\Delta E'_0$, there is a corresponding 4.6 kcal (19.3 kJ) change in $\Delta G^{\circ\prime}$ when a two-electron transfer takes place. This is similar to the relationship of $\Delta G^{\circ\prime}$ and K_{eq} in other chemical reactions: the larger the equilibrium constant, the greater the $\Delta G^{\circ\prime}$. The difference in reduction potentials between NAD^+/NADH and $\tfrac{1}{2}O_2/H_2O$ is 1.14 volts, a large $\Delta E'_0$ value. When electrons move from NADH to O_2, a large amount of free energy is made available and can be used to synthesize ATP and do other work.

🌀 *How NAD⁺ Works*

10.4 Electron Transport Chains: Sets of Sequential Redox Reactions

After reading this section, you should be able to:

- List the molecules that are commonly found in electron transport chains (ETCs) and indicate if they transfer electrons and protons or just electrons
- Arrange electron carriers in an ETC using their standard reduction potentials
- Indicate the location of ETCs in bacterial, archaeal, and eukaryotic cells

We have focused our attention on the reduction of O_2 by NADH because NADH plays a central role in the metabolism of many organisms. For instance, glucose is a common organic energy source. As glucose is catabolized, it is oxidized. Many of the electrons released from glucose are accepted by NAD^+, reducing it to NADH, which transfers the electrons to O_2. However, it does not do so directly. Instead, the electrons are transferred to O_2 via a series of electron carriers that are organized into a system called an **electron transport chain** (**ETC**). An ETC is similar to a bucket brigade. Each carrier represents a person receiving a pail of water (electrons) that are destined for the fire (terminal electron acceptor). The pail of water is passed down the line, just as electrons are passed from carrier to carrier. As soon as one person hands off a pail of water to the person after him in the line, he can receive a new bucket from the person before him in the line. Likewise, as electrons flow through the ETC, each carrier is

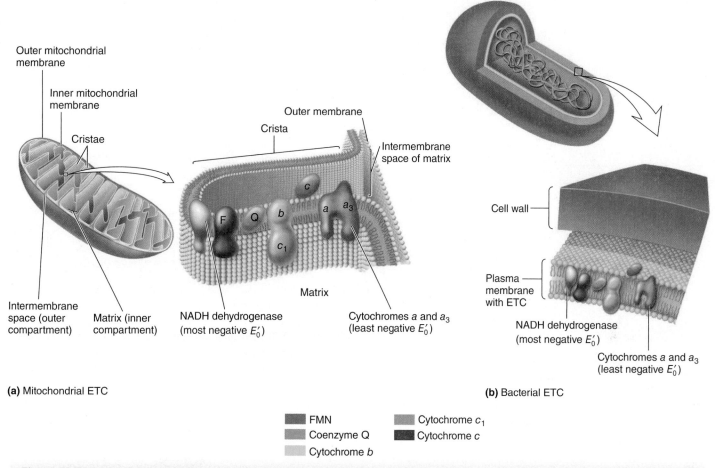

(a) Mitochondrial ETC

(b) Bacterial ETC

■ FMN	■ Cytochrome c_1
■ Coenzyme Q	■ Cytochrome c
■ Cytochrome b	

Figure 10.7 Electron Transport Chains. Electron transport chains (ETCs) are associated with membranes. Electrons flow from the electron carrier having the most negative reduction potential to the carrier having the most positive reduction potential. During respiratory processes (e.g., aerobic respiration; *see figure 11.2*), the source of electrons is a reduced molecule such as glucose and an exogenous molecule such as oxygen serves as the terminal electron acceptor (*see figures 11.11–11.13*). (a) The mitochondrial ETC. (b) A typical bacterial ETC.

MICRO INQUIRY *Refer to figure 10.6 and determine the E_0' for NAD$^+$/NADH and coenzyme Q/CoQH$_2$. Suggest a plausible E_0' value for FMN.*

sequentially reduced (given the pail full of water) and then reoxidized (passes the pail on to the next person in the brigade), and is ready to accept more electrons as catabolism continues.

The first electron carrier in an ETC has the most negative E_0', each successive carrier is slightly less negative (**figure 10.7**). Thus electrons are transferred spontaneously from one carrier to the next. The carriers direct the electrons to the terminal electron acceptor (in this case, O_2). This protects the cells from random, nonproductive reductions of other molecules in the cell. The use of several carriers in a chain also releases energy from an energy source such as glucose in a controlled manner so that more can be conserved and used to form ATP.

Electron transport chains are associated with or are in the plasma membranes and intracytoplasmic membranes of bacterial and archaeal cells (figure 10.7). In eukaryotes they are localized to the internal membranes of mitochondria and chloroplasts.

They are important to almost all types of energy-conserving processes, as we discuss in chapter 11.

The electron carriers associated with ETCs differ in terms of their chemical nature and the way they carry electrons. NADH and its chemical relative **nicotinamide adenine dinucleotide phosphate** (**NADPH**), which donate electrons to ETCs, contain a nicotinamide ring (**figure 10.8**). This ring accepts two electrons and one proton from a donor (e.g., an intermediate formed during the catabolism of glucose), and a second proton is released. **Flavin adenine dinucleotide** (**FAD**) and **flavin mononucleotide** (**FMN**) bear two electrons and two protons on the complex ring system shown in **figure 10.9**. Proteins bearing FAD and FMN are often called flavoproteins. **Coenzyme Q** (**CoQ**) or **ubiquinone** is a quinone that transports two electrons and two protons (**figure 10.10**). **Cytochromes** and several other carriers use iron atoms to transport one electron at a time. In cytochromes, the iron atoms are part of a heme group or other similar iron-porphyrin rings

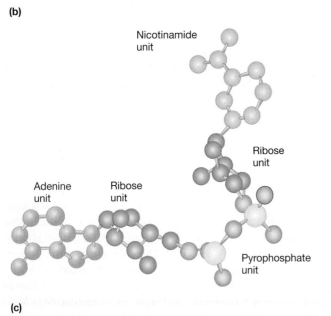

(a)

(b)

(c)

Figure 10.8 **The Structure and Function of NAD.** (a) The structure of NAD$^+$ and NADP$^+$. NADP$^+$ differs from NAD$^+$ in having an extra phosphate on one of its ribose sugar units. (b) NAD$^+$ can accept two electrons and one proton from a reduced substrate (S). In this example, the electrons and proton are supplied by the hydrogen atoms of the reduced substrate; recall that each hydrogen atom consists of one proton and one electron. (c) Stick-and-ball model of NAD$^+$.

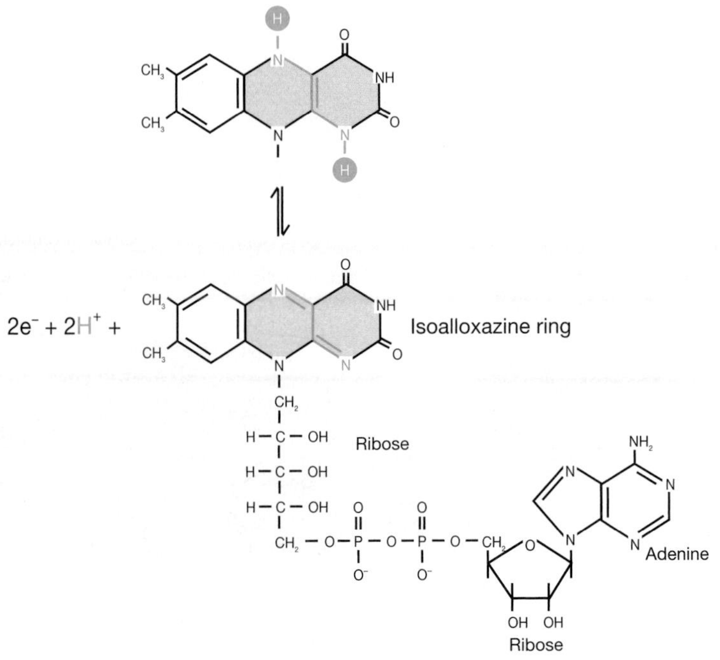

Figure 10.9 **The Structure and Function of FAD.** The vitamin riboflavin is composed of the isoalloxazine ring and its attached ribose sugar. FMN is riboflavin phosphate. The portion of the ring directly involved in oxidation-reduction reactions is in color.

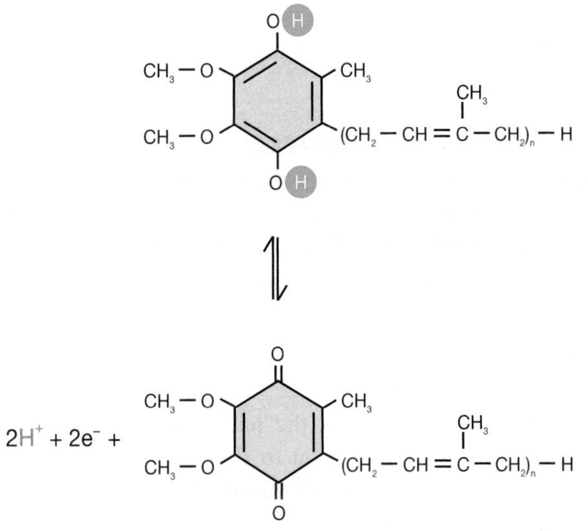

Figure 10.10 **The Structure and Function of Coenzyme Q (Ubiquinone).** The length of the side chain varies among organisms from n = 6 to n = 10.

(figure 10.11). There are several different cytochromes, each consisting of a protein and an iron-porphyrin ring. Some iron-containing electron-carrying proteins lack a heme group and are called **nonheme iron proteins.** They are often referred to as **iron-sulfur (Fe-S) proteins** because the iron is associated with sulfur atoms (figure 10.12). **Ferredoxin** is an Fe-S protein active in photosynthesis-related electron transport. Like cytochromes, Fe-S proteins carry only one electron at a time. The differences in the number of electrons and protons transported by electron carriers are of great importance to the operation of ETCs, and we discuss them further in chapter 11. 🌀 *Proton Pump*

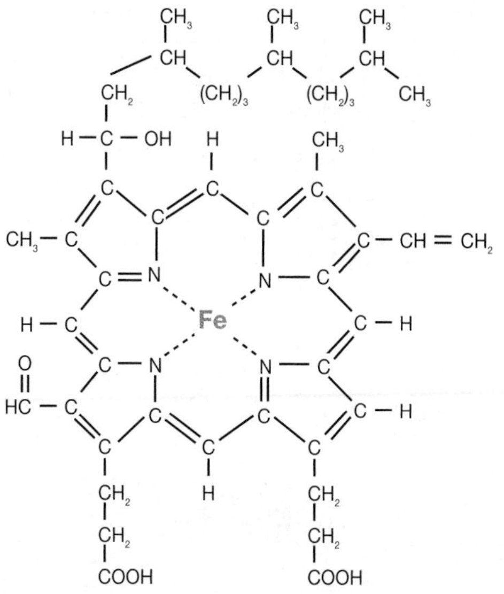

Figure 10.11 **The Structure of Heme.** Heme is composed of a porphyrin ring and an attached iron atom. It is the nonprotein component of many cytochromes. The iron atom alternatively accepts and releases an electron.

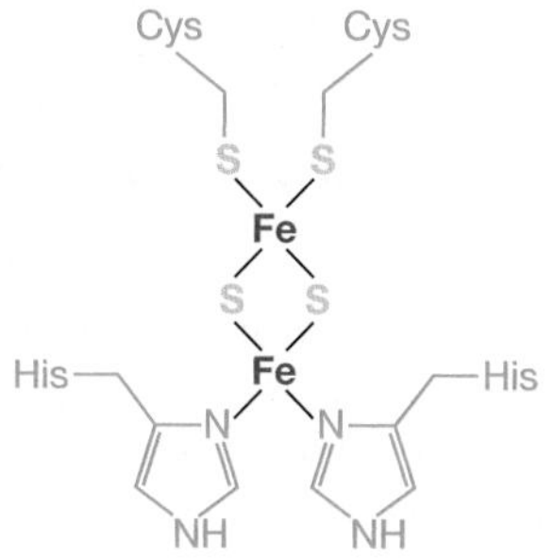

Figure 10.12 **The Iron-Sulfur Cluster of an Iron-Sulfur Protein.** This Fe-S protein consists of four S atoms and 2 Fe atoms. The cluster is bound to the polypeptide portion of the protein (green) by bonds between one Fe and the two S atoms provided by cysteine (Cys) residues and by bonds between the other Fe and two N atoms provided by histidine (His) residues. The iron atoms are responsible for transferring one electron at a time to the next carrier in an ETC.

Retrieve, Infer, Apply

1. How is the direction of electron flow between conjugate redox pairs related to the standard reduction potential and the release of free energy?

2. When electrons flow from the NAD^+/NADH conjugate redox pair to the $\frac{1}{2}O_2$/H_2O redox pair, does the reaction begin with NAD^+ or with NADH? What is produced—O_2 or H_2O?

3. Which among the following would be the best electron donor? Which would be the worst? Ubiquinone/ubiquinone H_2, NAD^+/NADH, FAD/$FADH_2$, NO_3^-/NO_2^-. Explain your answers.

4. In general terms, how is $\Delta G^{\circ\prime}$ related to $\Delta E_0'$? What is the $\Delta E_0'$ when electrons flow from the NAD^+/NADH redox pair to the Fe^{3+}/Fe^{2+} redox pair? How does this compare to the $\Delta E_0'$ when electrons flow from the Fe^{3+}/Fe^{2+} conjugate redox pair to the $\frac{1}{2}O_2$/H_2O pair? Which will yield the largest amount of free energy to the cell?

5. Name and briefly describe the major electron carriers found in cells. Why is NADH a good electron donor? Why is ferredoxin an even better donor?

10.5 Biochemical Pathways: Sets of Linked Chemical Reactions

After reading this section, you should be able to:

- Describe the components of a biochemical pathway and how they are organized

Organisms carry out a myriad of chemical reactions, and the products of these reactions are called **metabolites.** The reactions are organized into **biochemical pathways,** which can take various forms. Some are linear (**figure 10.13a**). In linear pathways, the first molecule in the pathway is often termed the starting molecule or the substrate of the pathway. The last molecule is called the end product. Those metabolites in between are pathway intermediates. As seen in figure 10.13b, some linear pathways are branched and can yield more than one product. Biochemical pathways can also be cycles (figure 10.13c). In this case, all molecules in the pathway can be thought of as intermediates. For the cycle to continue running, there must be inputs into it. In both types of pathways, each reaction is represented by an arrow and is catalyzed by an enzyme or ribozyme, as discussed in section 10.6.

Biochemical pathways are often illustrated as though they exist in isolation of each other. Unfortunately, this is very misleading (especially to students). In reality, biochemical pathways are connected and form a complex network (figure 10.13d). Thus, an intermediate of a pathway may be diverted from one pathway to another pathway. This is important to recognize because biochemical pathways are dynamic. As long as the starting molecules or inputs of the pathway are available and some end product is needed, these molecules will flow into and out of the many pathways that function in cells. Microbiologists are often concerned with metabolite flux. **Metabolite flux** is the rate of turnover of a metabolite; that is, the rate at which a metabolite is formed and then used. Metabolite flux can be used as a measure of pathway activity and to understand metabolic networks.

10.6 Enzymes and Ribozymes Speed Up Cellular Chemical Reactions

After reading this section, you should be able to:

- State the function and chemical makeup of enzymes
- Distinguish apoenzyme from holoenzyme and prosthetic group from coenzyme
- Draw a diagram that shows the effect of an enzyme on the activation energy of a chemical reaction
- Describe the effects of substrate concentration, pH, and temperature on enzyme activity
- Differentiate competitive and noncompetitive inhibitors of enzymes
- Discuss the importance of the discovery of ribozymes
- Compare and contrast ribozymes and enzymes
- List two examples of ribozymes and the reactions they catalyze

As we discussed in section 10.1, an exergonic reaction is one with a negative $\Delta G^{\circ\prime}$ and an equilibrium constant greater

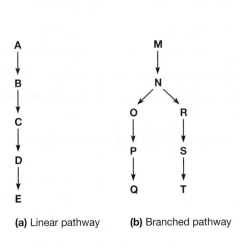

(a) Linear pathway   **(b)** Branched pathway

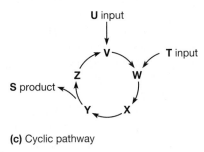

(c) Cyclic pathway

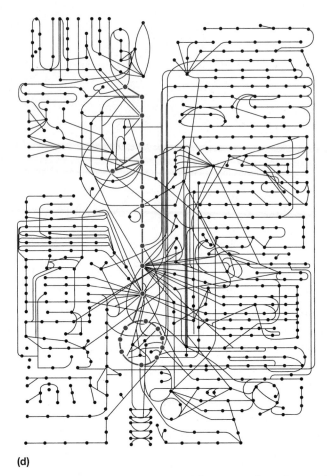

(d)

Figure 10.13 Chemical Reactions Carried Out by Cells Are Organized into Biochemical Pathways. Biochemical pathways take two main forms: linear and cyclic. (a) Molecules in the linear forms are distinguished based on their placement in the pathway. The first molecule (A) is the starting molecule (substrate) of the pathway. Those in the middle of the pathway are called intermediates (molecules B, C, and D). The last molecule of the pathway (E) is the product of the pathway. (b) Some linear biochemical pathways are branched and thus have more than one end product (molecules Q and T). (c) The molecules present in biochemical cycles are intermediates of the pathway (molecules V, W, X, Y, and Z). Biochemical cycles also generate products (molecule S); however, to keep the cycle running, there must be inputs into the cycle (molecules U and T). (d) Metabolic pathways are interconnected and form a complex network. In this representation of a metabolic network, each ball represents a metabolite formed by reactions in the network. Each line represents the enzyme-catalyzed reaction that converts one metabolite into another.

than one. An exergonic reaction proceeds to completion in the direction written (i.e., toward the right of the equation). Nevertheless, reactants for an exergonic reaction often can be combined with no obvious result. For instance, if a polysaccharide such as starch is mixed in water, the hydrolysis of the starch into its component monosaccharides (glucose) is exergonic and will occur spontaneously; that is, it will occur on its own, given enough time. However, the time needed is very long. Even if an organic chemist carried out this reaction in 6 moles/liter (M) HCl and at 100°C, it would still take several hours to go to completion. A cell, on the other hand, can accomplish the same reaction at neutral pH, at a much lower temperature, and in just fractions of a second. Cells can do this because they manufacture biological catalysts that speed up chemical reactions. Most of these catalysts are proteins called enzymes. Other catalysts are RNA molecules termed ribozymes. Enzymes and ribozymes are critically important to cells, since most biological reactions occur very slowly without them. Indeed, enzymes and ribozymes make life possible.

Enzyme Structure

Enzymes are protein catalysts that have great specificity for the reaction catalyzed, the molecules acted on, and the products they yield. A **catalyst** is a substance that increases the rate of a chemical reaction without being permanently altered itself. The reacting molecules are called **substrates,** and the substances formed are the **products.** Enzymes may be placed in one of six general classes and usually are named in terms of the substrates they act on and the type of reaction catalyzed (**table 10.3**). ▶▶ *Proteins (appendix I)*

Many enzymes are composed only of proteins. However, some are composed of two parts: a protein component called the **apoenzyme** and a nonprotein component called a **cofactor.** Cofactors include metal ions and a variety of organic molecules. The complete enzyme consisting of the apoenzyme and its cofactor is called the **holoenzyme.** If the cofactor is firmly attached to the apoenzyme, it is a **prosthetic group.** If the cofactor is loosely attached and can dissociate from the apoenzyme after products have been formed, it is called a **coenzyme.** Many vitamins required by humans serve as coenzymes or as their precursors (e.g., riboflavin is incorporated into FAD). Coenzymes often carry one of the products of a chemical reaction to another enzyme or transfer chemical groups from one substrate to another (**figure 10.14**).

How Enzymes Speed Up Reactions

It is important to keep in mind that enzymes increase the rates of reactions but do not alter their equilibrium constants. If a reaction is endergonic, the presence of an enzyme will not shift its equilibrium so that more products are formed. Enzymes simply speed up the rate at which a reaction proceeds toward its final equilibrium. ⟳ *How Enzymes Work*

How do enzymes catalyze reactions? Some understanding of the mechanism can be gained by considering the course of a simple exergonic chemical reaction.

$$A + B \rightleftharpoons C + D$$

When molecules A and B approach each other to react, they form a transition-state complex, which resembles both the substrates and the products (**figure 10.15**). **Activation energy** is required to bring the reacting molecules together in the correct way to reach the transition state. The transition-state complex can then resolve to yield the products C and D. The difference in free energy level between reactants and products is $\Delta G^{\circ\prime}$. Thus the equilibrium in our example lies toward the products because $\Delta G^{\circ\prime}$ is negative (i.e., the products are at a lower energy level than the substrates).

As seen in figure 10.15, A and B will not be converted to C and D if they are not supplied with an amount of energy equivalent to the activation energy. Enzymes accelerate reactions by lowering the activation energy; therefore more substrate molecules will have sufficient energy to come together and form products. Even though the equilibrium constant (or $\Delta G^{\circ\prime}$) is

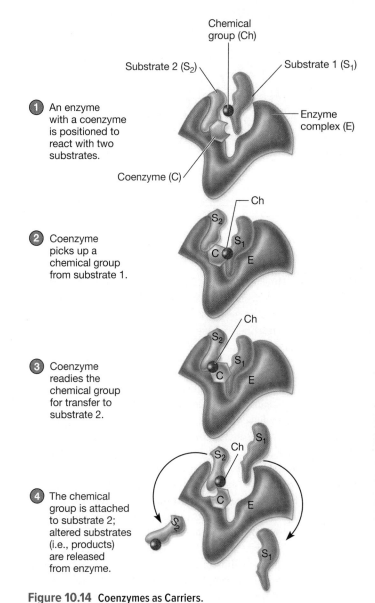

1. An enzyme with a coenzyme is positioned to react with two substrates.

2. Coenzyme picks up a chemical group from substrate 1.

3. Coenzyme readies the chemical group for transfer to substrate 2.

4. The chemical group is attached to substrate 2; altered substrates (i.e., products) are released from enzyme.

Figure 10.14 Coenzymes as Carriers.

Table 10.3	Enzyme Classification	
Type of Enzyme	**Reaction Catalyzed by Enzyme**	**Example of Reaction**
Oxidoreductase	Oxidation-reduction reactions	Lactate dehydrogenase: Pyruvate $+$ NADH $+$ H$^+$ $\rightleftharpoons$ lactate $+$ NAD$^+$
Transferase	Reactions involving the transfer of chemical groups between molecules	Aspartate carbamoyltransferase: Aspartate $+$ carbamoylphosphate $\rightleftharpoons$ carbamoylaspartate $+$ phosphate
Hydrolase	Hydrolysis of molecules	Glucose 6-phosphatase: Glucose 6-phosphate $+$ H$_2$O $\rightarrow$ glucose $+$ P$_i^1$
Lyase	Breaking of C—C, C—O, C—N and other bonds by a means other than hydrolysis	Fumarase: L-malate $\rightleftharpoons$ fumarate $+$ H$_2$O
Isomerase	Reactions involving isomerizations	Alanine racemase: L-alanine $\rightleftharpoons$ D-alanine
Ligase	Joining of two molecules using ATP (or the energy of other nucleoside triphosphates)	Glutamine synthetase: Glutamate $+$ NH$_3$ $+$ ATP $\rightarrow$ glutamine $+$ ADP $+$ P$_i$

1 P$_i$ is inorganic phosphate.

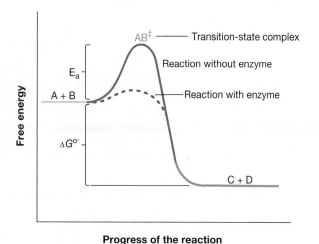

Figure 10.15 Enzymes Lower the Energy of Activation. This figure traces the course of a chemical reaction in which A and B are converted to C and D. The transition-state complex is represented by AB‡, and the activation energy required to reach it, by E$_a$. The red line represents the course of the reaction in the presence of an enzyme. Note that the activation energy is much lower in the enzyme-catalyzed reaction.

unchanged, equilibrium is reached more rapidly in the presence of an enzyme because of this decrease in activation energy.

Researchers have worked hard to discover how enzymes lower the activation energy of reactions. Enzymes bring substrates together at a specific location in the enzyme called the **active** or **catalytic site** to form an enzyme-substrate complex. How the enzyme and substrate interact still has not been fully elucidated. However, the induced fit model describes the interaction for many enzymes. In the induced fit model, the enzyme changes shape when it binds the substrate so that the active site surrounds and precisely fits the substrate. This mechanism is used by hexokinase, as illustrated in **figure 10.16**. The formation of an enzyme-substrate complex can lower the activation energy in many ways. For example, by bringing the substrates together at the active site, the enzyme is, in effect, concentrating them and speeding up the reaction. An enzyme does not simply concentrate its substrates, however. It also binds them so that they are

correctly oriented with respect to each other. Such an orientation lowers the amount of energy that the substrates require to reach the transition state. These and other catalytic site activities speed up a reaction by hundreds of thousands of times.

Substrate Concentration Affects Enzyme Activity

Enzyme activity varies in response to substrate concentration. Examination of **figure 10.17** shows that the curve describing the change in enzyme activity (usually expressed as the rate of product formation) in response to substrate concentration is a hyperbola. Why is this so? At very low substrate concentrations, an enzyme makes product slowly because it seldom contacts a substrate molecule. If more substrate molecules are present, an enzyme binds substrate more often, and the velocity of the reaction is greater than at a lower substrate concentration. Eventually further increases in substrate concentration do not result in a greater reaction velocity because the available enzyme molecules are binding substrate and converting it to product as rapidly as possible. That is, the enzyme is saturated with substrate and operating at maximal velocity (V_{max}).

The substrate concentrations in cells are often low. Therefore it is useful to know the substrate concentration an enzyme needs to function adequately. Usually the **Michaelis constant** (K_m), the substrate concentration required for the enzyme to achieve half-maximal velocity, is used as a measure of the apparent affinity of an enzyme for its substrate. The lower the K_m value, the lower the substrate concentration at which an enzyme catalyzes its reaction. Enzymes with a low K_m value are said to have a high affinity for their substrates. Since the concentrations of substrates in cells are often low, enzymes with lower K_m values are able to function better.

Enzyme Denaturation Destroys Enzyme Activity

Enzyme activity is changed not only by substrate concentration but also by alterations in pH and temperature. Each enzyme functions most rapidly at a specific pH optimum. When the pH deviates too greatly from an enzyme's optimum, activity slows and the enzyme may be damaged. Enzymes likewise have temperature

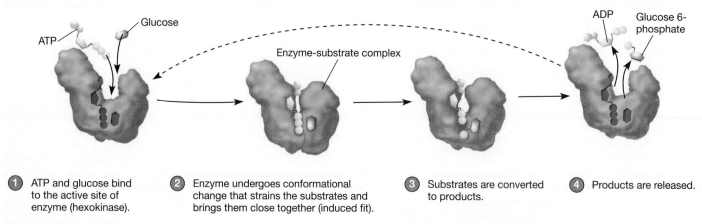

1. ATP and glucose bind to the active site of enzyme (hexokinase).

2. Enzyme undergoes conformational change that strains the substrates and brings them close together (induced fit).

3. Substrates are converted to products.

4. Products are released.

Figure 10.16 The Induced Fit Model of Enzyme Function. Shown is the reaction between glucose and ATP catalyzed by the enzyme hexokinase.

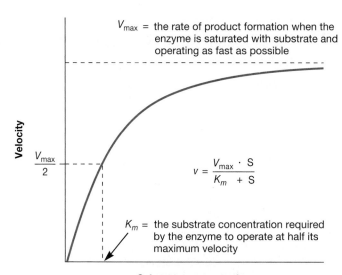

Figure 10.17 **The Effect of Substrate Concentration on Enzyme Activity.** Enzyme activity increases as substrate concentration increases until a maximum velocity is reached (V_{max}). The resulting curve is described by the Michaelis-Menten equation, which was derived by biochemists Leonor Michaelis (1875–1949) and Maud Menten (1879–1960). It relates reaction velocity (v) to substrate concentration (S) using V_{max} and the Michaelis constant (K_m).

MICRO INQUIRY *Will an enzyme with a relatively high K_m have a high or low affinity for its substrate? Explain.*

optima for maximum activity. If the temperature rises too much above the optimum, an enzyme's structure will be disrupted and its activity lost. This phenomenon is known as **denaturation.** The pH and temperature optima of a microorganism's enzymes often reflect the pH and temperature of its habitat. Not surprisingly, bacteria and archaea that grow best at high temperatures often have enzymes with high temperature optima and great heat stability. ◄◄ *Environmental factors affect microbial growth (section 7.4)*

Enzyme Inhibition

Microorganisms can be poisoned by a variety of chemicals (e.g., cyanide), and many of the most potent poisons are enzyme inhibitors. A **competitive inhibitor** directly competes with the substrate at an enzyme's catalytic site and prevents the enzyme from forming product (**figure 10.18**). Competitive inhibitors usually resemble normal substrates, but they cannot be converted to products.

Competitive inhibitors are important in the treatment of many microbial diseases. Sulfa drugs such as sulfanilamide (figure 10.18*b*) resemble *p*-aminobenzoate (PABA), a molecule used in the formation of the coenzyme folic acid. Sulfa drugs compete with PABA for the catalytic site of an enzyme involved in folic acid synthesis. This blocks the production of folic acid and inhibits the growth of organisms that require its synthesis. Humans are not harmed because they do not synthesize folic acid but rather obtain it in their diet. ◄◄ *Metabolic antagonists (section 9.4)*

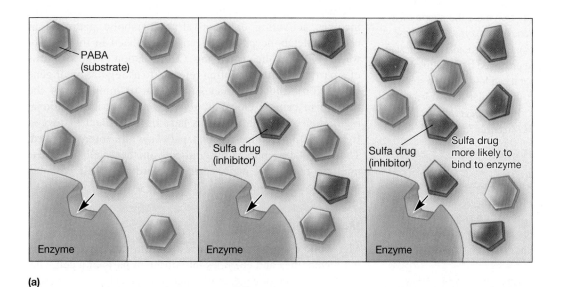

(a)   **(b)**

Figure 10.18 **Competitive Inhibition of Enzyme Activity.** (a) A competitive inhibitor is usually similar in shape to the normal substrate of the enzyme and therefore can bind the active site of the enzyme. This prevents the substrate from binding, and the reaction is blocked. (b) Structure of sulfanilamide, a structural analogue of PABA. PABA is the substrate of an enzyme involved in folic acid biosynthesis. When sulfanilamide binds the enzyme, activity of the enzyme is inhibited and synthesis of folic acid is stopped.

MICRO INQUIRY *Based on this figure, explain the notion of a minimum inhibitory concentration of sulfanilamide.*

Table 10.4	Examples of Self-Splicing Ribozymes
Function	**Where observed**
Splicing of pre-rRNA	*Tetrahymena* spp.
Splicing of mitochondrial rRNA and mRNA	Numerous fungi
Splicing of chloroplast tRNA, rRNA, and mRNA	Plants and algae
Splicing of viral mRNA	Viruses (e.g., T4, hepatitis delta virus)

Noncompetitive inhibitors affect enzyme activity by binding to the enzyme at some location other than the active site. This alters the enzyme's shape, rendering it inactive or less active. These inhibitors are called noncompetitive because they do not directly compete with the substrate. Heavy metals such as mercury frequently are noncompetitive inhibitors of enzymes.

Ribozymes: Catalytic RNA Molecules

Biologists once thought that all cellular reactions were catalyzed by proteins. However, in the early 1980s Thomas Cech and Sidney Altman discovered that some RNA molecules also can catalyze reactions. Catalytic RNA molecules are called **ribozymes.** One important ribozyme is located in ribosomes and is responsible for catalyzing peptide bond formation between amino acids during protein synthesis. However, the best-studied ribozymes catalyze self-splicing, in which they cut themselves and then join segments of themselves back together (**table 10.4**). Just as with enzymes, the shape of a ribozyme is essential to catalytic efficiency. Ribozymes even have Michaelis-Menten kinetics (figure 10.17). One particularly interesting ribozyme is a self-splicing ribozyme produced by hepatitis delta virus, a satellite virus. It is unusual in that it can fold into two shapes with quite different catalytic activities: the regular RNA cleavage activity and an RNA ligation reaction. ◀◀ *Viroids and satellites (section 6.6)* ▶▶ *Translation in bacteria (section 13.7)*

Retrieve, Infer, Apply

1. What is an apoenzyme? A holoenzyme? What are the two types of cofactors? What roles do cofactors play in enzyme function?
2. Illustrate the effect enzymes have on the activation energy of the reactions they catalyze. What is a transition state complex? Use your drawing to explain why enzymes do not change the equilibria of the reactions they catalyze.
3. How does enzyme activity change with substrate concentration? How does the Michaelis constant (K_m) relate to enzyme function?
4. What special properties might an enzyme isolated from a psychrophilic bacterium have? Will enzymes need to lower the activation energy more or less in thermophiles or psychrophiles?
5. What are competitive and noncompetitive inhibitors, and how do they inhibit enzymes?
6. How are enzymes and ribozymes similar? How do they differ?

10.7 Metabolism Must Be Regulated to Maintain Homeostasis and Prevent Waste

After reading this section, you should be able to:

- List the three general approaches cells use to regulate metabolism
- Describe metabolic channeling and one example of how it is accomplished; predict whether it is an important regulatory mechanism in bacterial, archaeal, and eukaryotic cells
- Distinguish allosteric regulation and covalent modification
- Describe the structure of an allosteric enzyme
- Create a concept map that illustrates how positive allosteric and negative allosteric effectors regulate the activity of an enzyme
- List the three chemical groups commonly used to covalently modify an enzyme and its activity
- Explain how feedback inhibition is used to control the functioning of biosynthetic pathways
- Predict which enzymes of a biochemical pathway are likely to be regulated by either allosteric control or covalent modification
- Develop a model illustrating how feedback inhibition can be used to regulate a multiply branched biosynthetic pathway

Microorganisms must regulate their metabolism to conserve raw materials and energy, and to maintain a balance among various cell components. Microbes often live in environments where the nutrients, energy sources, and physical conditions change rapidly. Therefore microbes must continuously monitor internal and external conditions, and respond accordingly. This involves activating or inactivating metabolic pathways as needed. Metabolic pathways can be regulated in three major ways: (1) metabolic channeling, (2) regulation of the synthesis of a particular enzyme (often referred to as regulation of gene expression), and (3) direct stimulation or inhibition of the activity of critical enzymes, referred to as posttranslational regulation. Although we focus here on individual pathways, figure 10.13*d* reminds us that pathways are interconnected. Thus, regulating one pathway affects another. Cells often use multiple regulatory approaches to coordinate their complex metabolic activities.

Metabolic Channeling

Metabolic channeling influences pathway activity by localizing metabolites and enzymes into different parts of a cell. One of the most common metabolic channeling mechanisms is **compartmentation,** the differential distribution of enzymes and metabolites among separate cell structures or organelles. Compartmentation is particularly important in eukaryotic microorganisms with their many membrane-bound organelles. For example, fatty acid catabolism is located within the mitochondrion, whereas fatty acid synthesis occurs in the cytosol. Microcompartments and the periplasm in Gram-negative bacteria are also examples of compartmentation. Compartmentation makes possible the simultaneous but separate operation and regulation of similar pathways. Furthermore, pathway activities can be coordinated through regulation of the transport of metabolites and coenzymes between cell compartments. ▶▶◀ *Microcompartments (section 3.6)*

Regulation of Gene Expression

In the second regulatory mechanism—regulation of the synthesis of a particular enzyme—transcription and translation rates are altered to control the amount of an enzyme present in the cell. When the enzyme is not needed, very little, if any, of the enzyme is synthesized. Conversely, when the enzyme is needed, more enzyme molecules are synthesized. Regulation of gene expression is a relatively slow response to changes in the cell's state, but it ultimately saves the cell considerable energy and raw materials. Cells have evolved numerous mechanisms for regulating gene expression. We save the discussion of these mechanisms for chapters 14 and 15.

Posttranslational Regulation of Enzyme Activity

In contrast to regulation of gene expression, the direct stimulation or inhibition of the activity of critical enzymes rapidly alters pathway activity. It is often called **posttranslational regulation** because it occurs after the enzyme has been synthesized. A number of posttranslational regulatory mechanisms are known. Some are irreversible; for instance, cleavage of a protein can either increase or decrease its activity. Others, such as allosteric regulation and covalent modification, are reversible.

Allosteric Regulation

Most regulatory enzymes are **allosteric enzymes.** The activity of an allosteric enzyme is altered by a small molecule known as an **allosteric effector.** The effector binds reversibly by noncovalent forces to a **regulatory site** separate from the catalytic site and causes a change in the shape (conformation) of the enzyme (**figure 10.19**). The activity of the catalytic site is altered as a result. A positive effector increases enzyme activity, whereas a negative effector decreases activity (i.e., inhibits the enzyme). These changes in activity often result from alterations in the apparent affinity of the enzyme for its substrate, but changes in maximum velocity also can occur.

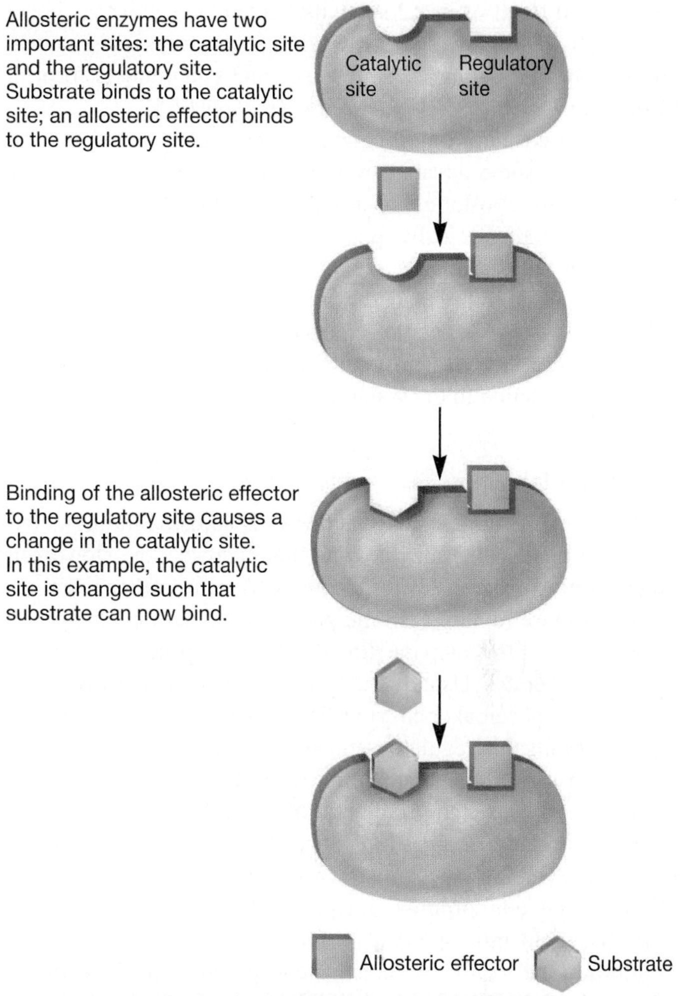

Covalent Modification of Enzymes

Regulatory enzymes also can be switched on and off by **reversible covalent modification.** Usually this occurs through the addition and removal of a particular chemical group, typically a phosphoryl, methyl, or adenylyl group.

One of the most intensively studied regulatory enzymes is *Escherichia coli*'s glutamine synthetase, an enzyme involved in nitrogen assimilation. It is a large, complex enzyme consisting of 12 subunits, each of which can be covalently modified by an adenylic acid residue (**figure 10.20**). When an adenylic acid residue

Allosteric enzymes have two important sites: the catalytic site and the regulatory site. Substrate binds to the catalytic site; an allosteric effector binds to the regulatory site.

Catalytic site   Regulatory site

Binding of the allosteric effector to the regulatory site causes a change in the catalytic site. In this example, the catalytic site is changed such that substrate can now bind.

■ Allosteric effector   ⬡ Substrate

Figure 10.19 Allosteric Regulation.   The structure and function of an allosteric enzyme.

MICRO INQUIRY *Is this allosteric effector a positive or negative effector? Is this true of all allosteric effectors?*

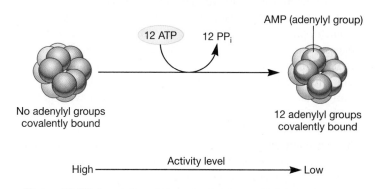

12 ATP   12 PP$_i$

AMP (adenylyl group)

No adenylyl groups covalently bound

12 adenylyl groups covalently bound

High ———————— Activity level ————————▶ Low

Figure 10.20 Regulation of Glutamine Synthetase Activity by Covalent Modification.   Glutamine synthetase consists of 12 subunits. Six subunits form a ring, and one ring (darker color) is above the other (lighter color). Not all of the lower ring subunits are shown. Each subunit can be adenylylated. Only the adenylyl groups on the upper ring are shown. As the number of adenylyl groups increases, the activity of the enzyme decreases.

is attached to all of its 12 subunits, glutamine synthetase is not very active. Removal of AMP groups produces more active deadenylylated glutamine synthetase, and glutamine is formed. ▶▶| *Inorganic nitrogen assimilation (section 12.5)*

Using covalent modification for the regulation of enzyme activity has some advantages. These interconvertible enzymes often are also allosteric. For instance, glutamine synthetase also is regulated allosterically. Because each form can respond differently to allosteric effectors, systems of covalently modified enzymes are able to respond to more stimuli in varied and sophisticated ways. Regulation can also be exerted on the enzymes that catalyze the covalent modifications, which adds a second level of regulation to the system.

Feedback Inhibition

The rate of many metabolic pathways is adjusted through control of a pacemaker enzyme, one that catalyzes the slowest or rate-limiting reaction in the pathway. Because other reactions proceed more rapidly than the pacemaker reaction, changes in the activity of this enzyme directly alter the speed with which a pathway operates. Usually the first reaction in a pathway is catalyzed by a pacemaker enzyme. The end product of the pathway often inhibits this regulatory enzyme, a process known as **feedback inhibition** or **end product inhibition.** Feedback inhibition ensures balanced production of a pathway end product. If the end product becomes too concentrated, it inhibits the regulatory enzyme and slows its own synthesis. As the concentration of the end product decreases, pathway activity again increases and more product is formed. In this way, feedback inhibition automatically matches end product supply with the demand. 🔁 *A Biochemical Pathway, Feedback Inhibition of Biochemical Pathways*

Frequently a biosynthetic pathway branches to form more than one end product. In such a situation, the synthesis of all end products must be coordinated precisely. It would not do to have one end product present in excess while another is lacking. Branching biosynthetic pathways usually achieve a balance among end products by using regulatory enzymes at branch points (**figure 10.21**). If an end product is present in excess, it often inhibits the first enzyme on the branch leading to its formation, in this way regulating its own formation without affecting the synthesis of other products. In figure 10.21, notice that both products also inhibit the initial enzyme in the pathway. An excess of one product slows the flow of carbon into the whole pathway and also inhibits the appropriate branch-point enzyme. Because less carbon is required when only one branch is functioning, feedback inhibition of the initial pacemaker enzyme helps match the supply with the demand in branching pathways. The regulation of branched pathways is often made even more sophisticated by the presence of **isoenzymes** (**isozymes**), different forms of an enzyme that catalyze the same reaction. In the pathway shown in figure 10.21, end product P would control one isoenzyme and end product Q would control a different isoenzyme of the first

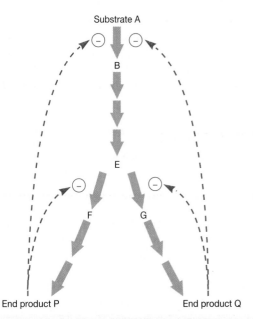

Figure 10.21 Feedback Inhibition. Feedback inhibition in a branching pathway with two end products. The branch-point enzymes, those catalyzing the conversion of intermediate E to F and G, are regulated by feedback inhibition. Products P and Q also inhibit the initial reaction in the pathway. A colored line with a minus sign at one end indicates that an end product, P or Q, is inhibiting the enzyme catalyzing the step next to the minus.

MICRO INQUIRY *What would happen to product formation if P and Q inhibited the formation of intermediate E independently of one another?*

enzyme in the pathway. If Q were present in excess, it would affect only its isoenzyme. The P isoenzyme would still function, and the pathway would still operate, but at a decreased level. Thus an initial pacemaker step may be catalyzed by several isoenzymes, each under separate control. Therefore, an excess of a single end product reduces pathway activity but does not completely block pathway function because some isoenzymes are still active.

Retrieve, Infer, Apply

1. Briefly describe the three ways a metabolic pathway may be regulated.
2. Define the terms metabolic channeling and compartmentation. How are they involved in the regulation of metabolism?
3. Define allosteric enzyme and allosteric effector.
4. How can regulatory enzymes be influenced by reversible covalent modification? What group is used for this purpose with glutamine synthetase, and which form of this enzyme is active?
5. What is a pacemaker enzyme? Feedback inhibition? How does feedback inhibition automatically adjust the concentration of a pathway end product?
6. What is the significance of the fact that regulatory enzymes often are located at pathway branch points? What are isoenzymes, and why are they important in pathway regulation?

Key Concepts

10.1 Metabolism: Important Principles and Concepts

■ Despite the tremendous metabolic diversity observed in the microbial world, there are common features to all types of metabolisms. These are (1) life obeys the laws of thermodynamics, (2) ATP is the major form of energy currency for cells, (3) redox reactions are important in energy-conserving processes, (4) cellular chemical reactions are organized into biochemical pathways, (5) cellular chemical reactions are catalyzed by enzymes or ribozymes, and (6) the functioning of biochemical pathways is regulated.

■ Living cells carry out three major kinds of work: chemical work of biosynthesis, transport work, and mechanical work. Energy is required to do cellular work and can be defined as the capacity to do work.

■ Thermodynamics is the field that analyzes energy changes in systems such as cells. The first law of thermodynamics states that energy is neither created nor destroyed. The second law of thermodynamics states that changes occur in such a way that the randomness or disorder of the universe increases to the maximum possible. That is, entropy always increases during spontaneous processes.

■ The first and second laws of thermodynamics can be combined to determine the amount of energy made available for useful work.

$$\Delta G = \Delta H - T \cdot \Delta S$$

In this equation the change in free energy (ΔG) is the energy made available for useful work, the change in enthalpy (ΔH) is the change in heat content, and the change in entropy is ΔS.

■ The standard free energy change ($\Delta G^{\circ\prime}$) for a chemical reaction is directly related to the equilibrium constant.

■ In exergonic reactions, $\Delta G^{\circ\prime}$ is negative and the equilibrium constant is greater than one; the reaction goes to completion as written. Endergonic reactions have a positive $\Delta G^{\circ\prime}$ and an equilibrium constant less than one (**figure 10.2**).

10.2 ATP: The Major Energy Currency of Cells

■ ATP is a high-energy molecule that transports energy in a useful form from one reaction or location in a cell to another (**figure 10.3**).

■ ATP is readily synthesized from ADP and P_i using energy released from exergonic reactions; when hydrolyzed back to ADP and P_i, it releases the energy, which is used to drive endergonic reactions. This cycling of ATP with ADP and P_i is called the cell's energy cycle (**figure 10.5**).

10.3 Redox Reactions: Reactions of Central Importance in Metabolism

■ In oxidation-reduction (redox) reactions, electrons move from an electron donor to an electron acceptor. Each redox reaction consists of two half reactions: one is the electron-donating half reaction; the other is the electron-accepting half reaction.

■ Each half reaction of a redox reaction consists of a molecule that can accept electrons and one that can donate electrons. These two molecules are called a conjugate redox pair.

■ The standard reduction potential measures the tendency of the donor of a conjugate redox pair to give up electrons.

■ Conjugate redox pairs with more negative reduction potentials donate electrons to those with more positive potentials, and energy is made available during the transfer (**figure 10.6** and **table 10.2**).

10.4 Electron Transport Chains: Sets of Sequential Redox Reactions

■ Many metabolic processes use a series of electron carriers to transport electrons from the primary electron donor (e.g., an organism's energy source) to a final electron acceptor (e.g., oxygen). A series of electron carriers is called an electron transport chain (ETC). ETCs are associated with membranes (**figure 10.7**).

■ Some of the most important electron carriers in cells are NAD, NADP, FAD, FMN, coenzyme Q, cytochromes, and nonheme iron (Fe-S) proteins (**figures 10.8–10.12**). They differ in several ways, including how many electrons and protons they transfer.

10.5 Biochemical Pathways: Sets of Linked Chemical Reactions

■ The chemical reactions carried out by cells are organized to form biochemical pathways. The intermediates and products of pathways are called metabolites (**figure 10.13**).

■ Metabolic pathways are interconnected such that the intermediates and end product of one pathway may be the starting molecules of others (**figure 10.13d**). Metabolite flux refers to the turnover of metabolites as they are used in various biochemical pathways.

10.6 Enzymes and Ribozymes Speed Up Cellular Chemical Reactions

■ Enzymes are protein catalysts that catalyze specific reactions.

■ Many enzymes consist of a protein component, the apoenzyme, and a cofactor. Cofactors may be metals or organic molecules. Some cofactors are prosthetic groups, which are tightly bound to the apoenzyme. Other cofactors are loosely bound and are called coenzymes.

■ Enzymes and ribozymes speed chemical reactions by binding substrates at their active sites and lowering the activation energy (**figures 10.15** and **10.16**).

■ The rate of an enzyme- or ribozyme-catalyzed reaction increases with substrate concentration at low substrate levels and reaches a plateau (the maximum velocity) at saturating substrate concentrations. The Michaelis constant is the substrate concentration that the enzyme or ribozyme requires to achieve half maximal velocity (**figure 10.17**).

- Enzymes have pH and temperature optima for activity. Their activity can be slowed by competitive and noncompetitive inhibitors (**figure 10.18**).
- Some ribozymes alter either their own structure or that of other RNAs. An important ribozyme is located in the ribosome, where it links amino acids together during protein synthesis.

10.7 Metabolism Must Be Regulated to Maintain Homeostasis and Prevent Waste

- The regulation of metabolism keeps cell components in proper balance and conserves metabolic energy and material. There are three major approaches to regulating metabolism: metabolic channeling, regulation of gene expression, and posttranslational regulation of enzyme activity.
- Metabolic channeling localizes metabolites and enzymes in different parts of the cell and influences pathway activity. A common channeling mechanism is compartmentation.

- Regulation of gene expression is brought about by controlling the rate of transcription and translation. This is discussed in chapters 14 and 15.
- Posttranslational regulation is the direct control of enzyme activity. There are two major types of reversible posttranslational regulation: allosteric control and covalent modification. In allosteric control, certain enzymes in a pathway are regulatory enzymes to which an allosteric effector binds noncovalently to a regulatory site separate from the catalytic site. This causes a conformational change in the enzyme and alters its activity (**figure 10.19**). In covalent modification, enzyme activity is regulated by reversible attachment of a chemical group, such as a phosphoryl, methyl, or adenylyl group (**figure 10.20**).
- The first enzyme in a pathway and enzymes at branch points often are subject to feedback inhibition by one or more end products. Excess end product slows its own synthesis (**figure 10.21**).

Compare, Hypothesize, Invent

1. Examine the structures of macromolecules in appendix I. Which type has the most electrons to donate? Suggest why carbohydrates are usually the primary source of energy for bacteria that use organic molecules as energy sources?

2. Most enzymes do not operate at their biochemical optima inside cells. Suggest why this is the case.

3. Examine the branched pathway shown here for the synthesis of the amino acids aspartate, methionine, lysine, threonine, and isoleucine.

   Scenario 1: The microbe is cultured in a medium containing aspartate and lysine but lacking methionine, threonine, and isoleucine.

   Scenario 2: The microbe is cultured in a medium containing a rich supply of all five amino acids.

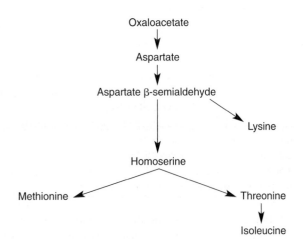

For each of these two scenarios, answer the following questions:

a. Which portion(s) of the pathway would need to be shut down in this situation?

b. How might allosteric control be used to accomplish this?

4. Chlamydia are pathogenic bacteria that must be grown within a eukaryotic host cell. They rely on the host cell for many nutrients, including nucleotides, lipids, and amino acids. It is thus no surprise that when grown in coculture with a host depleted of any of these metabolites, chlamydial growth declines. However, investigators were surprised to find that high levels of certain amino acids also inhibit chlamydial growth. Further investigation found that the amino acids leucine, isoleucine, methionine, and phenylalanine could retard growth, but this effect could be reversed if valine was added at the same time. The authors discovered that the amino acid transporter BrnQ was blocked by the four growth inhibitory amino acids.

   Develop a model whereby BrnQ transports valine and the observed growth inhibition is due to competitive inhibition by high levels of these four amino acids. Your model should take into account the similarity between valine, leucine, and isoleucine (*see appendix I*) and the fact that BrnQ transports methionine under normal conditions.

   Read the original paper: Braun, P. R., et al. 2008. Competitive inhibition of amino acid uptake suppresses chlamydial growth: Involvement of the chlamydial amino acid transporter BrnQ. *J. Bacteriol.* 190:1822.

11

Catabolism: Energy Release and Conservation

The Richest Hill On Earth

The history of Butte, Montana, is a colorful one filled with interesting characters, great tragedies, and the amazing resilience of its people. The city has gone through boom-and-bust cycles related to its location—it sits on a mountain rich in copper and other metals. From the 1860s until 1982, mining shaped life in Butte. It began with underground mining, which was accompanied by one of the worst mining disasters in U.S. history. In 1917 the Granite Mountain fire took the lives of 168 miners.

The advent of open-pit mining of the mountain in 1955 raised the hope that earlier mining tragedies would be averted. That hope came to fruition. Over the next 20 years, the strip mine devoured the side of the mountain and parts of the city, creating the Berkeley Pit, a deep hole a mile and a half across. In the process, it yielded over 1 billion tons of copper, gold, lead, and other metals, and was a boon to Butte's economy.

As the mine operated, groundwater and runoff entered the pit, requiring that the water be continuously pumped out. When the mine closed in 1982, the pumps were shut down and water began to accumulate. Now the north edge of Butte borders on the Berkeley Pit Lake, a lake containing water contaminated with toxic metals and a pH of about 2.5. The danger of the water was tragically demonstrated in 1995 when 342 migrating snow geese died after landing in the lake.

But the Berkeley Pit Lake is not devoid of life. In 1998 a search for microbial life in the lake began. The study, led by Montana Tech University scientist Grant Mitman, led to the discovery of algae, fungi, and bacteria living in the lake. This led scientists to suggest that microbes might be used to bioremediate (clean up) the water. Later studies by Don and Andrea Stierle of the University of Montana-Missoula led to the discovery of microbes that produce interesting chemicals that may be useful for treating cancer, migraines, and other illnesses. Thus, the Berkeley Pit Lake may be able to restore the title of "the richest hill on Earth" to the Butte area due to the metabolic activity of its resident microbes.

The story of the Berkeley Pit clearly demonstrates that microbes are the most successful organisms on Earth as witnessed by their growth under almost every conceivable condition. In large part, their success results from the diversity of their fueling reactions—those reactions that convert energy from an organism's energy source into ATP, provide reducing power, and generate precursor metabolites. The fueling reactions are the focus of this chapter.

Readiness Check:

Based on what you have learned previously, you should be able to:

✔ Distinguish metabolism, catabolism, and anabolism
✔ Discuss the function of ATP and how that function is related to its phosphate transfer potential (section 10.2)
✔ Relate the standard reduction potential of a conjugate redox pair to its tendency to act as an electron-donating half reaction (section 10.3)
✔ List the molecules that are commonly found in electron transport chains (ETCs) and indicate if they transfer electrons and protons or just electrons (section 10.4)
✔ Describe the organization of cellular chemical reactions into interconnected biochemical pathways (section 10.5)
✔ State the function of enzymes and ribozymes in cellular chemical reactions (section 10.6)

11.1 Metabolic Diversity and Nutritional Types

After reading this section, you should be able to:

- Use the terms that describe a microbe's carbon source, energy source, and electron source
- State the carbon, energy, and electron sources of photolithoautotrophs, photoorganoheterotrophs, chemolithoautotrophs, chemolithoheterotrophs, and chemoorganoheterotrophs
- Describe the products of the fueling reactions
- Discuss the metabolic flexibility of microorganisms

As we discuss in chapter 10, all organisms require an energy source to maintain their order and complexity. Furthermore, the importance of redox reactions in cellular metabolism dictates that organisms must also have an electron source. Finally, life on Earth is carbon-based. Thus all organisms must obtain carbon from their environments to synthesize the organic molecules they need to survive. The central importance of energy, electrons, and carbon has led biologists to describe organisms in terms of how these requirements are met. The terms devised can be combined to characterize an organism's nutritional type. The energy, electrons, and carbon that an organism obtains from its environment are used in chemical reactions that are referred to as fueling reactions. The nutritional types and fueling reactions are the focus of this section. In addition, we explore how the fueling reactions relate to other metabolic processes carried out by microbes.

Nutritional Types Are Defined by the Source of an Organism's Energy, Electrons, and Carbon

For each requirement—energy, electrons, and carbon—a dichotomy exists for how an organism fulfills that need. Only two sources of energy are available to organisms: light and certain chemical molecules. **Phototrophs** use light as their energy source (**table 11.1**); **chemotrophs** obtain energy from the oxidation of chemical compounds (either organic or inorganic). Likewise, organisms have only two sources for electrons. **Lithotrophs** (i.e., "rock-eaters") use reduced inorganic substances as their electron source, whereas **organotrophs** extract electrons from reduced organic compounds. Finally, organisms may be **heterotrophs**—organisms that use reduced, preformed organic molecules as their carbon source for growth, or they may be **autotrophs**—organisms that use carbon dioxide (CO_2) as their sole or principal source of carbon.

By combining the roots of these terms, most organisms can be placed in one of five nutritional types based on their primary sources of carbon, energy, and electrons (**table 11.2**). The majority

of microorganisms thus far studied are either photolithoautotrophic, chemolithoautotrophic, or chemoorganoheterotrophic. **Photolithoautotrophs** (often called simply **photoautotrophs**) use light energy and have CO_2 as their carbon source. Photosynthetic protists and cyanobacteria employ water as the electron donor and release oxygen. Other photolithoautotrophs, such as the purple sulfur bacteria and the green sulfur bacteria, cannot oxidize water but extract electrons from inorganic donors such as hydrogen, hydrogen sulfide, and elemental sulfur. **Chemolithoautotrophs** oxidize reduced inorganic compounds such as iron-, nitrogen-, or sulfur-containing molecules to derive both energy and electrons for biosynthesis, using carbon dioxide as the carbon source. Chemolithoautotrophs contribute greatly to the chemical transformations of elements (e.g., the conversion of ammonia to nitrate or sulfur to sulfate) that continually occur in ecosystems. Photoautotrophs and chemolithoautotrophs also are important primary producers in ecosystems. That is, they fix

Table 11.1	Sources of Carbon, Energy, and Electrons
Carbon Sources	
Autotrophs	CO_2 sole or principal biosynthetic carbon source
Heterotrophs	Reduced, preformed, organic molecules from other organisms
Energy Sources	
Phototrophs	Light
Chemotrophs	Oxidation of organic or inorganic compounds
Electron Sources	
Lithotrophs	Reduced inorganic molecules
Organotrophs	Organic molecules

Table 11.2	Major Nutritional Types of Microorganisms			
Nutritional Type	**Carbon Source**	**Energy Source**	**Electron Source**	**Representative Microorganisms**
Photolithoautotroph	CO_2	Light	Inorganic e^- donor	Purple and green sulfur bacteria, cyanobacteria, diatoms
Photoorganoheterotroph	Organic carbon	Light	Organic e^- donor	Purple nonsulfur bacteria, green nonsulfur bacteria
Chemolithoautotroph	CO_2	Inorganic chemicals	Inorganic e^- donor	Sulfur-oxidizing bacteria, hydrogen-oxidizing bacteria, methanogens, nitrifying bacteria, iron-oxidizing bacteria
Chemolithoheterotroph	Organic carbon	Inorganic chemicals	Inorganic e^- donor	Some sulfur-oxidizing bacteria (e.g., *Beggiatoa* spp.)
Chemoorganoheterotroph	Organic carbon	Organic chemicals, often same as C source	Organic e^- donor, often same as C source	Most nonphotosynthetic microbes, including most pathogens, fungi, and many protists and archaea

CO_2, making reduced organic molecules that sustain the chemoorganoheterotrophs that share their habitats. ▶▶ *Nitrifying bacteria oxidize ammonium or nitrite to gain energy and electrons (section 22.1); Photosynthetic bacteria are diverse (section 21.4); Purple sulfur bacteria perform anoxygenic photosynthesis (section 22.3)*

Chemoorganoheterotrophs (sometimes called **chemoheterotrophs** or chemoorganotrophs) use the reduced organic compounds made by autotrophs as sources of energy, electrons, and carbon. Frequently the same organic nutrient will satisfy all these requirements. Chemoorganotrophs contribute to biogeochemical cycles, such as the carbon cycle and nitrogen cycle, in which elements are converted into different forms. In addition, they are of considerable practical importance. Many chemoorganotrophs are used industrially to make foods (e.g., yogurt, pickles, cheese), medical products (e.g., antibiotics), and beverages (e.g., beer and wine). Nearly all pathogenic microorganisms are chemoorganoheterotrophs. ▶▶ *Biogeochemical cycling sustains life on Earth (section 28.1)*

The other nutritional types have fewer known microorganisms but are very important ecologically. Some phototrophic bacteria (e.g., purple nonsulfur and green bacteria) use organic matter as their electron donor and carbon source. These **photoorganoheterotrophs** are common inhabitants of polluted lakes and streams. **Chemolithoheterotrophs** use reduced inorganic molecules as their energy and electron source but derive their carbon from organic sources. Like chemolithoautotrophs, they contribute to numerous biogeochemical cycles. ▶▶ *Purple nonsulfur bacteria perform anoxygenic photosynthesis (section 22.1)*

Although we have sorted microbes into a particular nutritional type, microbes are not always so easily categorized. Some show great metabolic flexibility and alter their metabolism in response to environmental changes. For example, many purple nonsulfur bacteria act as photoorganoheterotrophs in the absence of oxygen but oxidize organic molecules and function chemoorganotrophically at normal oxygen levels (*see figure 22.4*). When oxygen is low, phototrophic and chemoorganotrophic metabolism may function simultaneously. This allows the bacterium to gain energy from both light and organic molecules, and still supply the carbon it needs for biosynthesis. Some of these bacteria also can grow as photolithoautotrophs with molecular hydrogen as an electron donor. This sort of flexibility gives these microbes a distinct advantage if environmental conditions frequently change.

Fueling Reactions Convert Energy, Electron, and Carbon Sources into ATP, Reducing Power, and Precursor Metabolites

Despite the diversity of energy, electron, and carbon sources used by organisms, all are used to generate three main products: **ATP,** the primary molecule used to conserve the energy supplied by an energy source; **reducing power,** molecules that serve as a ready supply of electrons for a variety of chemical reactions; and **precursor**

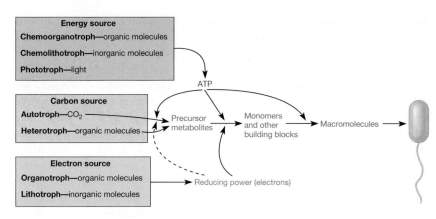

Figure 11.1 The Fueling Reactions Convert an Organism's Carbon, Energy, and Electron Sources into Precursor Metabolites, ATP, and Reducing Power.   In autotrophs, the precursor metabolites arise from CO_2-fixation and related pathways, which are discussed in chapter 12. In heterotrophs, the precursor metabolites arise from reactions of the central metabolic pathways (*see figure 12.2*). ATP and reducing power are used in anabolic reactions that convert precursor metabolites into monomers and other building blocks. Monomers are used to generate macromolecules such as proteins, which are assembled into various cell structures.

metabolites, small organic molecules that provide the carbon skeletons needed for biosynthesis of important chemical building blocks (monomers) such as amino acids (**figure 11.1**). The processes that supply ATP, reducing power, and precursor metabolites are called **fueling reactions.** With the exception of CO_2-fixation, the fueling reactions are part of catabolism. As seen in figure 11.1, the generation of monomers using the products of the fueling reactions paves the way for anabolic reactions that synthesize macromolecules (e.g., polymers such as proteins). The macromolecules are used to construct cellular structures such as ribosomes and flagella.

Retrieve, Infer, Apply

1. Discuss the ways in which organisms are classified based on their requirements for energy, carbon, and electrons.
2. Describe the nutritional requirements of the major nutritional groups and give some microbial examples of each.
3. Compare photolithoautotrophy with chemolithoautotrophy. Do you think it is possible for an ecosystem to exist solely on the organic carbon generated by chemolithotrophy (i.e., without any contribution by photosynthetic organisms)? Explain your reasoning.
4. What are the three major products generated by the fueling reactions? Summarize how they are used in anabolic processes.

11.2 There Are Three Chemoorganotrophic Fueling Processes

After reading this section, you should be able to:

- List the three types of chemoorganotrophic metabolisms
- List the pathways of major importance to chemoorganotrophs and explain their importance
- Propose an explanation that accounts for the existence of amphibolic pathways

As illustrated in figure 11.1, chemoorganotrophs use reduced organic compounds for catabolic and anabolic processes. When the energy source is oxidized to release the energy from the energy source (catabolism), it also provides the carbon and electrons needed for anabolism. There are two general approaches a chemoorganotroph may use to catabolize its energy source: respiration and fermentation. When the organic energy source is oxidized, the electrons released are accepted by electron carriers such as NAD$^+$ and FAD. When these reduced electron carriers (e.g., NADH, FADH$_2$) in turn donate the electrons to an electron transport chain, the metabolic process is called **respiration** and may be divided into two different types (**figure 11.2**). In aerobic respiration, the final electron acceptor is oxygen, whereas the terminal acceptor in anaerobic respiration is a different oxidized molecule such as NO$_3^-$, SO$_4^{2-}$, CO$_2$, Fe^{3+}, or SeO$_4^{2-}$. Organic acceptors such as fumarate and humic acids also may be used. As just noted, respiration involves the activity of an electron transport chain. As electrons pass through the chain to the final electron acceptor, a type of potential energy called the proton motive force (PMF) is generated and used to synthesize ATP from ADP and phosphate (P$_i$). In contrast, **fermentation** (Latin *fermentare*, to cause to rise) uses an electron acceptor that is endogenous (from within the cell) and does not involve an electron transport chain. The endogenous electron acceptor is usually an intermediate (e.g., pyruvate) of the catabolic pathway used to degrade and oxidize the organic energy source. During fermentation, ATP is synthesized almost exclusively by substrate-level phosphorylation, a process in which a phosphate is transferred to ADP from a high-energy molecule (e.g., phosphoenolpyruvate) generated by catabolism of the energy source.

By convention, aerobic respiration, anaerobic respiration, and fermentation are usually described with glucose as the energy source. This is done for several reasons. One is that glucose is used by many chemoorganotrophs as an energy source. But perhaps more important is the way catabolic pathways are organized.

Most chemoorganotrophs use a wide variety of organic molecules as energy sources (**figure 11.3**). They are degraded by pathways that either generate glucose or intermediates of the pathways used in glucose catabolism. Thus nutrient molecules are funneled into ever fewer metabolic intermediates. Indeed, a common pathway often degrades many similar molecules (e.g., several different sugars). The existence of a few metabolic pathways that each degrade many nutrients greatly increases metabolic efficiency by avoiding the need for a large number of less metabolically flexible pathways.

The diversity of organic molecules used as energy sources by chemoorganotrophic microbes contributes to their ecology. They are found in habitats rich in organic molecules, particularly those molecules they can use as energy sources. For instance, lactose catabolizers are common in the intestinal tracts of mammals, where they may have a ready supply of this sugar. Clostridia and other soil bacteria that use amino acids as energy sources are important decomposers of proteins in protein-rich habitats. One-carbon molecules such as methane can also be catabolized by microbes, some doing so aerobically, others anaerobically. Furthermore, some bacteria such as the soil-dwelling pseudomonads are able to degrade unusual and often complex organic molecules, including pesticides, polystyrene, polychlorinated biphenyls (PCBs), and even the phenolic resins used as glue in the production of plywood and fiberboard. These microbes are useful in bioremediation. ▶▶ *Biodegradation and bioremediation (section 43.4)*

The catabolic pathways of greatest importance to chemoorganotrophs are the glycolytic pathways (section 11.4) and the tricarboxylic acid (TCA) cycle (section 11.5). They are important not only for their role in catabolism but also for their roles in anabolism. They supply material needed for biosynthesis, including

Chemoorganotrophic Fueling Processes

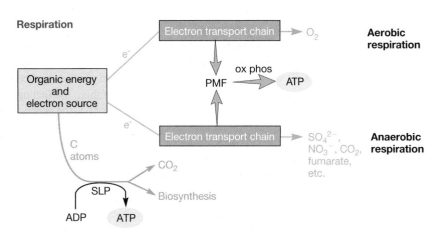

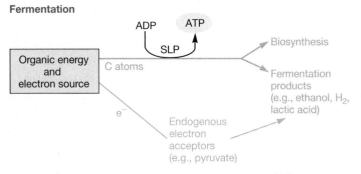

Figure 11.2 Chemoorganotrophic Fueling Processes. Organic molecules serve as energy and electron sources for all three fueling processes used by chemoorganotrophs. In aerobic respiration and anaerobic respiration, the electrons pass through an electron transport chain. This generates a proton motive force (PMF), which is used to synthesize most of the cellular ATP by a mechanism called oxidative phosphorylation (ox phos); a small amount of ATP is made by a process called substrate-level phosphorylation (SLP). In aerobic respiration, O$_2$ is the terminal electron acceptor, whereas in anaerobic respiration, exogenous molecules other than O$_2$ serve as electron acceptors. During fermentation, endogenous organic molecules act as electron acceptors, electron transport chains do not function, and most organisms synthesize ATP only by substrate-level phosphorylation.

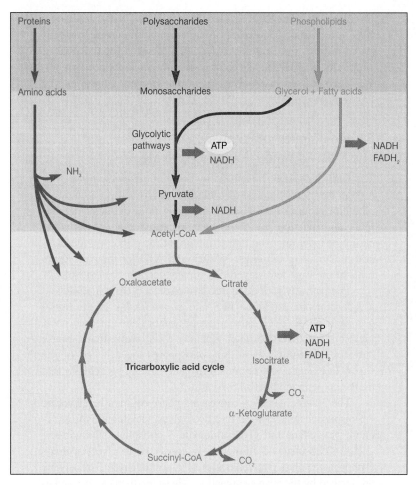

Figure 11.3 Pathways Used by Chemoorganotrophs to Catabolize Organic Energy Sources.   Notice that these pathways funnel metabolites into the glycolytic pathways and the tricarboxylic acid cycle, thus increasing metabolic efficiency and flexibility.

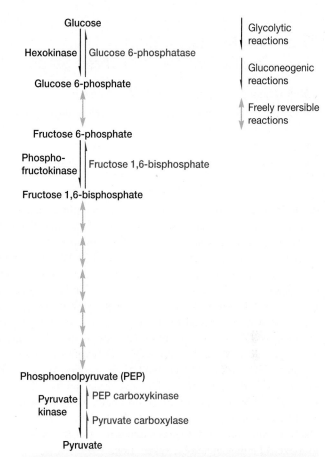

Figure 11.4 The Embden-Meyerhof Pathway Is an Example of an Amphibolic Pathway.   Many reactions of this pathway are catalyzed by enzymes that function in glycolysis and in an anabolic pathway called gluconeogenesis. Gluconeogenesis reverses the glycolytic process and allows cells to synthesize glucose from smaller molecules such as pyruvate. Note that some glycolytic reactions are catalyzed by enzymes unique to that pathway. Likewise, some gluconeogenic reactions are catalyzed by enzymes unique to gluconeogenesis. Therefore, although they share several enzymes, the two pathways are distinct.

MICRO INQUIRY *Is NAD$^+$ reduced to NADH in the catabolic or anabolic direction of this pathway?*

precursor metabolites and reducing power. Furthermore, some enzymes of these pathways are reversible and can function either catabolically or anabolically. ▶▶ *Precursor metabolites (section 12.2)*

Pathways with enzymes that function both catabolically and anabolically are often called **amphibolic pathways** (Greek *amphi,* on both sides). For instance, many of the enzymes of the Embden-Meyerhof pathway (a glycolytic pathway) are freely reversible. They function catabolically during glycolysis but anabolically during gluconeogenesis, a pathway that generates glucose from pyruvate and other small molecules (**figure 11.4**). Other reactions in glycolysis are not reversible. During glycolysis, an enzyme catalyzes the reaction in a catabolic direction, and during gluconeogenesis, a different enzyme catalyzes the reverse, anabolic reaction. Thus, although glycolysis and gluconeogenesis share many enzymes, they are two distinct pathways.

In the following sections, we explore the metabolism of chemoorganotrophs in more detail. The discussion begins with aerobic respiration and introduces the glycolytic pathways, TCA cycle, and other important processes. Many of these also occur during anaerobic respiration. Thus in many cases, anaerobic respiration differs from aerobic respiration mainly in terms of the terminal electron acceptor. Fermentation is quite distinct from respiration. It involves only a subset of the reactions that function during respiration and only partially catabolizes the energy source. However, it is widely used by microbes and has important practical applications, including the production of many foods and the identification of pathogenic microbes. ▶▶ *Microbiology of fermented foods (section 41.5; Identification of microorganisms from specimens (section 36.3)*

Retrieve, Infer, Apply

1. Give examples of the types of electron acceptors used by fermentation and respiration. What is the difference between aerobic respiration and anaerobic respiration?

2. Why is it to a cell's advantage to catabolize diverse organic energy sources by funneling them into a few common pathways?

3. What are amphibolic pathways? Why are they important?

11.3 Aerobic Respiration Can Be Divided into Three Steps

After reading this section, you should be able to:

- Describe in general terms what happens to a molecule of glucose during aerobic respiration
- List the end products made during aerobic respiration
- Identify the process that generates the most ATP during aerobic respiration

Aerobic respiration is a process that can completely catabolize a reduced organic energy source to CO_2 using the glycolytic pathways and TCA cycle with O_2 as the terminal electron acceptor for an electron transport chain (figure 11.2). We focus in this section and the following three sections on the aerobic respiration of glucose. The catabolism of glucose can be divided into three steps. It begins with the formation of pyruvate, using one or more pathways described in section 11.4. These pathways also produce NADH, $FADH_2$, or both. Next pyruvate is fed into the TCA cycle and oxidized completely to CO_2 with the production of some GTP or ATP, NADH, and $FADH_2$ (section 11.5). Finally, the NADH and $FADH_2$ formed by glycolysis and the TCA cycle are oxidized by an electron transport chain, using O_2 as the terminal electron acceptor (section 11.6). It is the activity of the electron transport chain that conserves most of the energy used to make ATP during aerobic respiration.

11.4 Glucose to Pyruvate: The First Step

After reading this section, you should be able to:

- List the three major pathways that catabolize glucose to pyruvate
- Describe substrate-level phosphorylation
- Diagram the major changes made to glucose as it is catabolized by the Embden-Meyerhof, Entner-Doudoroff, and pentose phosphate pathways
- Identify those reactions of the Embden-Meyerhof, Entner-Doudoroff, and pentose phosphate pathways that consume ATP, produce ATP and NAD(P)H, generate precursor metabolites, or are redox reactions
- Calculate the yields of ATP and NAD(P)H by the Embden-Meyerhof, Entner-Doudoroff, and pentose phosphate pathways
- Summarize the function of the Embden-Meyerhof, Entner-Doudoroff, and pentose phosphate pathways
- Draw a simple diagram that shows the connection between the Entner-Doudoroff pathway and the Embden-Meyerhof pathway and the connection between the pentose phosphate pathway and the Embden-Meyerhof pathway
- Create a table that shows which types of organisms use each of the glycolytic pathways

Microorganisms employ several metabolic pathways to catabolize glucose to pyruvate, including (1) the Embden-Meyerhof pathway, (2) the Entner-Doudoroff pathway, and (3) the pentose phosphate pathway. We refer to these pathways collectively as

glycolytic pathways or as glycolysis (Greek *glyco,* sweet, and *lysis,* a loosening). However, in some texts, the term glycolysis refers only to the Embden-Meyerhof pathway. For the sake of simplicity, the detailed structures of some metabolic intermediates are not used in pathway diagrams. However, these can be found in appendix II.

Embden-Meyerhof Pathway: The Most Common Route to Pyruvate

The **Embden-Meyerhof pathway** is undoubtedly the most common pathway for glucose degradation to pyruvate. It is found in all major groups of microorganisms, as well as in plants and animals, and functions in the presence or absence of O_2. As noted earlier, it is an important amphibolic pathway and provides several precursor metabolites, NADH, and ATP for the cell. Reactions of the Embden-Meyerhof pathway occur in the cytosol.

The pathway is divided into two parts: a 6-carbon phase and a 3-carbon phase (**figure 11.5** and *appendix II*). In the initial 6-carbon phase, ATP is used to phosphorylate glucose twice, yielding fructose 1,6-bisphosphate. This preliminary phase "primes the pump" by adding phosphates to each end of the sugar. In essence, the organism invests some of its ATP so that more can be made later in the pathway.

The 3-carbon, energy-conserving phase begins when fructose 1,6-bisphosphate is cleaved into two halves, yielding dihydroxyacetone phosphate and glyceraldehyde 3-phosphate. Dihydroxyacetone phosphate, is immediately converted to glyceraldehyde 3-phosphate. Thus two molecules of glyceraldehyde 3-phosphate are formed by the cleavage reaction. These are then catabolized to pyruvate in a five-step process. Because dihydroxyacetone phosphate can be easily converted to glyceraldehyde 3-phosphate, both halves of fructose 1,6-bisphosphate are used in the 3-carbon phase.

NADH and ATP are produced in the 3-carbon phase of the pathway. NADH is formed when glyceraldehyde 3-phosphate is oxidized with NAD^+ as the electron acceptor, and a phosphate is simultaneously incorporated to give a high-energy molecule called 1,3-bisphosphoglycerate. This reaction sets the stage for ATP production. The phosphate on the first carbon of 1,3-bisphosphoglycerate is donated to ADP to produce ATP. This is an example of **substrate-level phosphorylation** because ADP phosphorylation is coupled with the exergonic hydrolysis of a high-energy molecule having a higher phosphate transfer potential than ATP (*see table 10.1*). A second ATP is made by substrate-level phosphorylation when the phosphate on phosphoenolpyruvate (the last intermediate of the pathway) is donated to ADP. This reaction also yields pyruvate, the final product of the pathway. ◄◄ *ATP (section 10.2)* ↻ *How NAD+ Works*

The yields of ATP and NADH by the Embden-Meyerhof pathway may be calculated. In the 6-carbon phase, two ATP are used to form fructose 1,6-bisphosphate. For each glyceraldehyde 3-phosphate transformed into pyruvate, one NADH and two ATP are formed. Because two glyceraldehyde 3-phosphates arise from a single glucose (one by way of dihydroxyacetone phosphate), the 3-carbon phase generates four ATP and two NADH per glucose. Subtraction of the ATP used in the 6-carbon

Glucose is phosphorylated at the expense of one ATP, generating glucose 6-phosphate, a precursor metabolite and the starting molecule for the pentose phosphate pathway.

Isomerization of glucose 6-phosphate (an aldehyde) to fructose 6-phosphate (a ketone and a precursor metabolite)

ATP is consumed to phosphorylate C1 of fructose. The cell is spending some of its energy currency in order to earn more in the next part of the pathway.

Fructose 1, 6-bisphosphate is split into two 3-carbon molecules, one of which is a precursor metabolite. DHAP is readily converted to glyceraldehyde 3-phosphate.

Glyceraldehyde 3-phosphate is oxidized and simultaneously phosphorylated, generating a high-energy molecule. The electrons released reduce NAD^+ to NADH.

ATP is made by substrate-level phosphorylation. Another precursor metabolite is made.

Another precursor metabolite is made.

The oxidative breakdown of one glucose results in the formation of two pyruvate molecules. Pyruvate is one of the most important precursor metabolites.

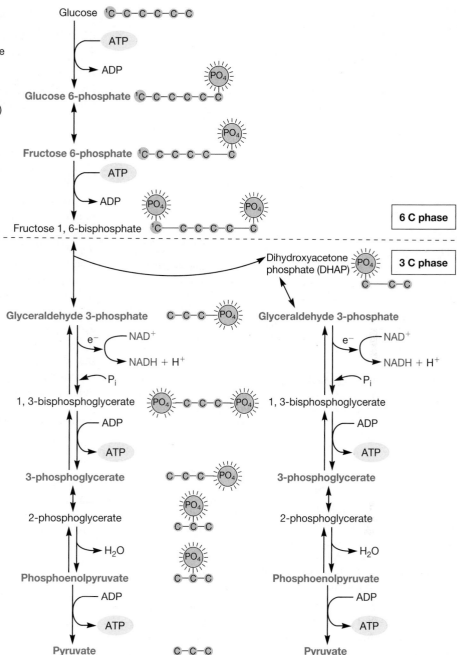

Figure 11.5 Embden-Meyerhof Pathway. This is one of three pathways used to catabolize glucose to pyruvate, and it can function during aerobic respiration, anaerobic respiration, and fermentation. When used during respiration, the electrons accepted by NAD^+ are transferred to an electron transport chain and are ultimately accepted by an exogenous electron acceptor. When used during fermentation (section 11.8), the electrons accepted by NAD^+ are donated to an endogenous electron acceptor (e.g., pyruvate). The Embden-Meyerhof pathway also generates several precursor metabolites (shown in blue).

MICRO INQUIRY *Which reactions are examples of substrate-level phosphorylation?*

phase from that produced by substrate-level phosphorylation in the 3-carbon phase gives a net yield of two ATP per glucose. Thus the catabolism of glucose to pyruvate can be represented by this equation.

$$Glucose + 2ADP + 2P_i + 2NAD^+ \rightarrow$$
$$2 \text{ pyruvate} + 2ATP + 2NADH + 2H^+$$

The NADH is used during aerobic respiration to transport electrons to an electron transport chain (ETC), as we describe in section 11.6. *How Glycolysis Works*

Entner-Doudoroff Pathway

The **Entner-Doudoroff pathway** is used by some Gram-negative bacteria, especially those found in soil. To date, very

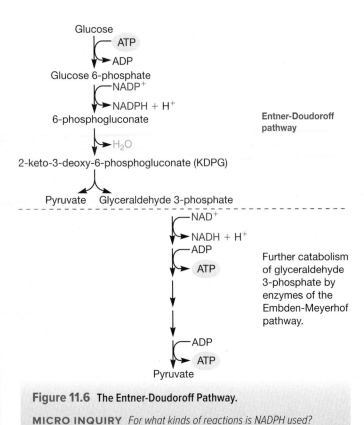

Figure 11.6 **The Entner-Doudoroff Pathway.**

MICRO INQUIRY *For what kinds of reactions is NADPH used?*

few Gram-positive bacteria have been found to use this pathway, with the intestinal bacterium *Enterococcus faecalis* being a rare exception. It is not used by eukaryotes.

The Entner-Doudoroff pathway essentially replaces the first phase of the Embden-Meyerhof pathway and yields pyruvate and glyceraldehyde 3-phosphate. (Compare this to the two molecules of glyceraldehyde 3-phosphate produced by the Embden-Meyerhof pathway.) A key intermediate of the Entner-Doudoroff pathway is 2-keto-3-deoxy-6-phosphogluconate (KDPG), which is formed from glucose by three reactions that consume one ATP and produce one NADPH (**figure 11.6** and *appendix II*). KDPG is then cleaved to pyruvate and glyceraldehyde 3-phosphate. Bacteria that use this pathway also have the enzymes that function in the second phase of the Embden-Meyerhof pathway. These enzymes may be used to catabolize glyceraldehyde 3-phosphate to form a second pyruvate molecule. If this occurs, two ATP and one NADH are formed. Thus the catabolism of one glucose molecule to two pyruvates by way of the Entner-Doudoroff pathway coupled with the second half of the Embden-Meyerhof pathway has a net yield of one ATP, one NADH, and one NADPH. During aerobic respiration, the NADH is used to transport electrons to an ETC; the NADPH is used as reducing power for anabolic reactions.

Pentose Phosphate Pathway: A Major Producer of Reducing Power for Anabolic Reactions

The **pentose phosphate pathway** (also called the **hexose monophosphate pathway**) may be used at the same time as either the Embden-Meyerhof or the Entner-Doudoroff pathways. It can operate either aerobically or anaerobically and is important in both biosynthesis and catabolism. It is thought to be used by all organisms because of its role in providing reducing power and important precursor metabolites. However, its role in archaea has not been fully established.

The pentose phosphate pathway begins with the oxidation of glucose 6-phosphate to 6-phosphogluconate followed by the oxidation of 6-phosphogluconate to the 5-carbon sugar (i.e., a pentose sugar) ribulose 5-phosphate and CO_2 (**figure 11.7** and *appendix II*). NADPH is produced during these oxidations. Ribulose 5-phosphate is then converted to a mixture of 3- to 7-carbon sugar phosphates. Two enzymes play a central role in these transformations: (1) transketolase catalyzes the transfer of 2-carbon groups, and (2) transaldolase transfers a 3-carbon group from sedoheptulose 7-phosphate (a 7-carbon molecule) to glyceraldehyde 3-phosphate (a 3-carbon molecule). The overall result is that three glucose 6-phosphates are converted to two fructose 6-phosphates, glyceraldehyde 3-phosphate, and three CO_2 molecules, as shown in this equation.

$$3 \text{ glucose 6-phosphate} + 6NADP^+ + 3H_2O \rightarrow$$
$$2 \text{ fructose 6-phosphate} + \text{glyceraldehyde 3-phosphate} +$$
$$3CO_2 + 6NADPH + 6H^+$$

These intermediates are used in two ways. Fructose 6-phosphate can be changed back to glucose 6-phosphate while glyceraldehyde 3-phosphate is converted to pyruvate by enzymes of the Embden-Meyerhof pathway. Alternatively two glyceraldehyde 3-phosphates may combine to form fructose 1,6-bisphosphate, which is eventually converted back into glucose 6-phosphate. This results in the complete degradation of glucose 6-phosphate to CO_2 and the production of a great deal of NADPH.

$$\text{Glucose 6-phosphate} + 12NADP^+ + 7H_2O \rightarrow$$
$$6CO_2 + 12NADPH + 12H^+ + P_i$$

The pentose phosphate pathway is an important amphibolic pathway because: (1) NADPH produced by the pathway serves as a source of electrons for the reduction of molecules during biosynthesis. Indeed, the pentose phosphate pathway is the major source of reducing power for cells, yielding two NADPH molecules for each glucose metabolized to pyruvate in this way. (2) The pathway produces two important precursor metabolites: erythrose 4-phosphate and ribose 5-phosphate. Erythrose 4-phosphate is used to synthesize aromatic amino acids and vitamin B_6; ribose 5-phosphate is a major component of nucleic acids. Furthermore, when a microorganism is growing on a 5-carbon sugar, the pathway can function biosynthetically to supply 6-carbon sugars (hexose sugars, such as the glucose needed for peptidoglycan synthesis). (3) Intermediates

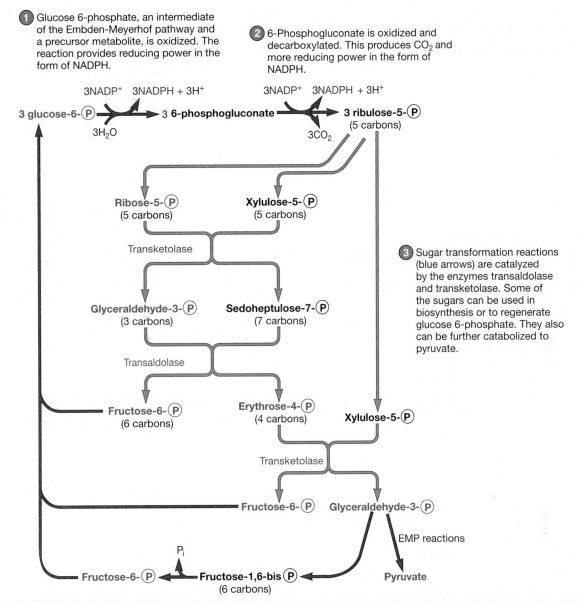

① Glucose 6-phosphate, an intermediate of the Embden-Meyerhof pathway and a precursor metabolite, is oxidized. The reaction provides reducing power in the form of NADPH.

② 6-Phosphogluconate is oxidized and decarboxylated. This produces CO_2 and more reducing power in the form of NADPH.

$3NADP^+$ $3NADPH + 3H^+$

$3NADP^+$ $3NADPH + 3H^+$

3 glucose-6-(P) → **3 6-phosphogluconate** → **3 ribulose-5-(P)**
(5 carbons)

$3H_2O$

$3CO_2$

Ribose-5-(P) (5 carbons)   **Xylulose-5-(P)** (5 carbons)

Transketolase

③ Sugar transformation reactions (blue arrows) are catalyzed by the enzymes transaldolase and transketolase. Some of the sugars can be used in biosynthesis or to regenerate glucose 6-phosphate. They also can be further catabolized to pyruvate.

Glyceraldehyde-3-(P) (3 carbons)   **Sedoheptulose-7-(P)** (7 carbons)

Transaldolase

Fructose-6-(P) (6 carbons)   **Erythrose-4-(P)** (4 carbons)   **Xylulose-5-(P)**

Transketolase

Fructose-6-(P)   **Glyceraldehyde-3-(P)**

EMP reactions

P_i

Fructose-6-(P) ← **Fructose-1,6-bis (P)** ← (6 carbons)   **Pyruvate**

Figure 11.7 The Pentose Phosphate Pathway. The catabolism of three glucose 6-phosphate molecules to two fructose 6-phosphates, a glyceraldehyde 3-phosphate, and three CO_2 molecules is traced. Note that the pentose phosphate pathway generates several intermediates that are also intermediates of the Embden-Meyerhof pathway (EMP). These intermediates can be fed into the EMP with two results: (1) continued degradation to pyruvate or (2) regeneration of glucose 6-phosphate by gluconeogenesis. The pentose phosphate pathway also plays a major role in producing reducing power (NADPH) and several precursor metabolites (shown in blue) for biosynthesis. The sugar transformations are indicated with blue arrows.

MICRO INQUIRY *For what macromolecule is ribose 5-phosphate a precursor?*

in the pathway may be used to produce ATP. For instance, glyceraldehyde 3-phosphate can enter the 3-carbon phase of the Embden-Meyerhof pathway. As it is degraded to pyruvate, two ATP are formed by substrate-level phosphorylation. ▶▶ *Synthesis of amino acids consumes many precursor metabolites (section 12.5); Synthesis of purines, pyrimidines, and nucleotides (section 12.6)*

Retrieve, Infer, Apply

1. Summarize the major features of the Embden-Meyerhof, Entner-Doudoroff, and pentose phosphate pathways. Include the starting points, the products of the pathways, the ATP yields, and the metabolic roles each pathway has.

2. What is substrate-level phosphorylation?

11.5 Pyruvate to Carbon Dioxide (Step 2) Is Accomplished by the Tricarboxylic Acid Cycle

After reading this section, you should be able to:

- State the alternate names for the tricarboxylic acid (TCA) cycle
- Diagram the major changes made to pyruvate as it is catabolized by the TCA cycle
- Identify those reactions of the TCA cycle that produce ATP (or GTP) and NAD(P)H, generate precursor metabolites, or are redox reactions
- Calculate the yields of ATP (or GTP), NAD(P)H, and $FADH_2$ by the TCA cycle
- Summarize the function of the TCA cycle
- Diagram the connections between the various glycolytic pathways and the TCA cycle
- Locate the TCA cycle enzymes in bacterial, archaeal, and eukaryotic cells

In the glycolytic pathways, glucose is oxidized to pyruvate. During aerobic respiration, the catabolic process continues by oxidizing pyruvate to three CO_2. The first step of this process employs a multienzyme system called the pyruvate dehydrogenase complex. It oxidizes and cleaves pyruvate to form one CO_2 and the 2-carbon molecule **acetyl-coenzyme A (acetyl-CoA)** (**figure 11.8**). Acetyl-CoA is energy rich because hydrolysis of the bond that links acetic acid to coenzyme A (a thioester bond) has a large negative change in free energy, just as hydrolysis of the bond in many phosphate-containing molecules does. As shown in figure 11.3, carbohydrates as well as fatty acids and amino acids can be converted to acetyl-CoA (section 11.9).

Acetyl-CoA then enters the **tricarboxylic acid (TCA) cycle**, which is also called the **citric acid cycle** or the **Krebs cycle** (figure 11.8 and *appendix II*). In the first reaction, acetyl-CoA is condensed with (i.e., added to) the 4-carbon intermediate oxaloacetate to form citrate, a molecule with six carbons. Citrate is rearranged to give isocitrate, a more readily oxidized alcohol. Isocitrate is subsequently oxidized and decarboxylated twice to yield α-ketoglutarate (five carbons) and then succinyl-CoA (four carbons), another high-energy molecule containing a thioester bond. At this point, two NADH molecules have been formed and two carbons lost from the cycle as CO_2. The cycle continues when succinyl-CoA is converted to succinate. This involves hydrolysis of the thioester bond in succinyl-CoA and using the large amount of energy released to form either one ATP or one GTP by substrate-level phosphorylation. GTP is also a high-energy molecule, and it is functionally equivalent to ATP. It is used in protein synthesis and to make other nucleoside triphosphates, including ATP. Two oxidation steps follow, yielding one $FADH_2$ and one NADH. The last oxidation step regenerates oxaloacetate, and as long as there is a supply of acetyl-CoA, the cycle can repeat itself. Inspection of figure 11.8 shows that the TCA cycle generates two CO_2 molecules, three NADH molecules, one $FADH_2$, and either one ATP or GTP for each acetyl-CoA molecule oxidized. ♻ *How the Krebs Cycle Works*

TCA cycle enzymes are widely distributed among microorganisms. In bacteria and archaea, they are located in the cytoplasm.

In eukaryotes, they are found in the mitochondrial matrix. The complete cycle appears to be functional in many aerobic bacteria, free-living protists, and fungi. This is not surprising because the cycle plays an important role in energy conservation by producing numerous NADH and $FADH_2$ (section 11.6). Even those microorganisms that lack the complete TCA cycle usually have most of its enzymes, because the TCA cycle is also a key source of precursor metabolites for use in biosynthesis. ▶▶ *Anaplerotic reactions replace the precursor metabolites used for amino acid biosynthesis (section 12.5)*

Retrieve, Infer, Apply

1. Identify the substrate and products of the TCA cycle. Describe its organization in general terms. What are its major functions?
2. What chemical intermediate links pyruvate to the TCA cycle?
3. How many times must the TCA cycle be performed to oxidize one molecule of glucose completely to six molecules of CO_2? Why?
4. In what eukaryotic organelle is the TCA cycle found? Where is the cycle located in bacterial and archaeal cells?
5. Why is it desirable for a microbe with the Embden-Meyerhof pathway and the TCA cycle also to have the pentose phosphate pathway?

11.6 Electron Transport and Oxidative Phosphorylation (Step 3) Generate the Most ATP

After reading this section, you should be able to:

- Compare and contrast the mitochondrial electron transport chain (ETC) and bacterial ETCs
- Describe the chemiosmotic hypothesis
- Correlate length of an ETC and the carriers in it with the magnitude of the proton motive force (PMF) it generates
- Explain how ATP synthase uses PMF to generate ATP
- Draw a simple diagram that shows the connections between the glycolytic pathways, TCA cycle, ETC, and ATP synthesis
- List the ways the PMF is used by bacterial cells in addition to ATP synthesis
- Calculate the maximum possible ATP yields when glucose is completely catabolized to six molecules of CO_2 during aerobic respiration

During the oxidation of glucose to six CO_2 molecules by glycolysis and the TCA cycle, as many as four ATP molecules are generated by substrate-level phosphorylation. Thus at this point, the work done by the cell has yielded relatively little ATP. However, in oxidizing glucose, the cell has also generated numerous molecules of NADH and $FADH_2$. Both of these molecules have a relatively negative E'_0 and can be used to conserve energy. In fact, most of the ATP generated during respiration comes from the energy conserved when these electron carriers are oxidized by an ETC. We examine the mitochondrial electron transport chain first because it has been well studied and because it functions in fungi and many protists. We then turn to bacterial chains and a discussion of ATP synthesis.

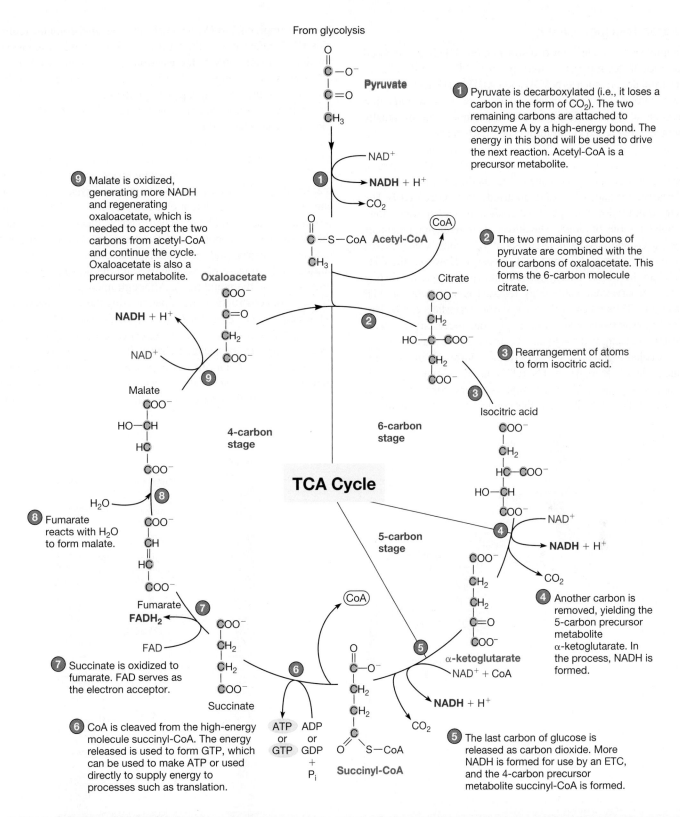

Figure 11.8 The Tricarboxylic Acid Cycle. The TCA cycle is linked to glycolysis by a connecting reaction catalyzed by the pyruvate dehydrogenase complex. The reaction decarboxylates pyruvate and generates acetyl-CoA. The cycle may be divided into three stages based on the size of its intermediates. The three stages are separated from one another by two decarboxylation reactions. Precursor metabolites are shown in blue. NADH and FADH$_2$ are shown in purple; they can transfer electrons to an electron transport chain (ETC).

MICRO INQUIRY *What reaction provides the energy to fuel the condensation (joining) of the 4-carbon oxaloacetate with the 2-carbon acetyl-CoA to form citrate?*

Electron Transport Chains

The mitochondrial **electron transport chain** (**ETC**) is composed of a series of electron carriers that operate together to transfer electrons from donors, such as NADH and $FADH_2$, to O_2 (**figure 11.9**). The electrons flow from carriers with more negative reduction potentials to those with more positive potentials and eventually combine with O_2 and H^+ to form water. Each carrier is in turn reduced and then reoxidized. Thus the carriers are constantly recycled as electrons are transported through the chain. The pattern of electron flow is exactly the same as seen in the electron tower that is described in chapter 10 and reproduced here (**figure 11.10**). The electrons move down this potential gradient much like water flowing down a series of rapids. The difference in reduction potentials between O_2 and NADH is large, about 1.14 volts, which makes possible the release of a great deal of energy. However, the ETC breaks up the overall energy release into small steps, facilitating energy conservation. The energy released is sufficient for ATP production. Thus much as the energy from waterfalls can be harnessed by waterwheels and used to generate electricity, so too is the flow of electrons through an ETC harnessed to make ATP.

In eukaryotes, ETC carriers reside within the inner membrane of the mitochondrion. The ETC consists of four large, multiprotein complexes and two additional carriers, coenzyme Q and the small protein cytochrome c, that connect complex I to complex III and complex III to IV, respectively. The final complex (complex IV) is referred to as the terminal oxidase as it is responsible for the final step transferring electrons to O_2 (**figure 11.11**). As electrons move through the ETC, protons are moved across the inner membrane, forming a proton gradient.

Although some bacterial and archaeal ETCs resemble the mitochondrial chain, they are frequently very different. First, bacterial and archaeal ETCs are located primarily within the plasma membrane. Furthermore, some Gram-negative bacteria have ETC carriers in the periplasmic space and even the outer membrane (*see figure 22.29*). Bacterial and archaeal ETCs also can be composed of different electron carriers, employ different terminal oxidases, and be branched. That is, electrons may enter the chain at several points and leave through several terminal oxidases. Bacterial and archaeal ETCs also may be shorter, resulting in the release of less energy (and the transport of fewer protons across the membrane). Although microbial ETCs differ in details of construction, they operate using the same fundamental principles. ⌬ *Electron Transport System and ATP Synthesis*

The ETCs of *Paracoccus denitrificans* and *Escherichia coli* will serve as examples of bacterial chains. *P. denitrificans* is a Gram-negative, soil bacterium that is a facultative anaerobe and is extremely versatile metabolically. Under oxic conditions, it carries out aerobic respiration using the ETC shown in **figure 11.12**. This

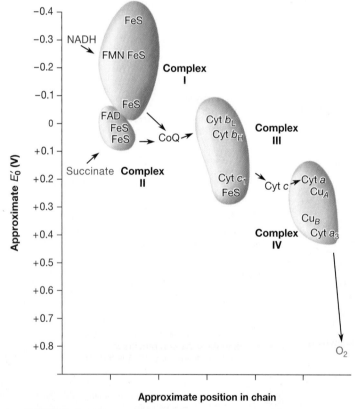

Figure 11.9 The Reduction Potentials of Carriers in the Mitochondrial Electron Transport Chain. In mitochondria, the electron carriers are organized into four complexes that are linked by the electron carriers coenzyme Q (CoQ) and cytochrome c (Cyt c). Electrons flow from NADH and succinate down the reduction potential gradient to oxygen.

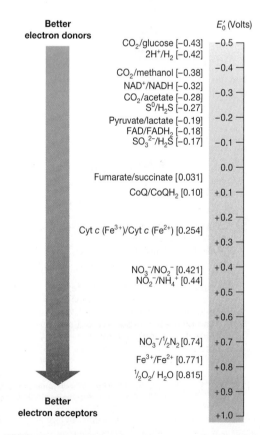

Figure 11.10 The Electron Tower. Electrons move spontaneously from donors with more negative reduction potentials to acceptors with more positive reduction potentials.

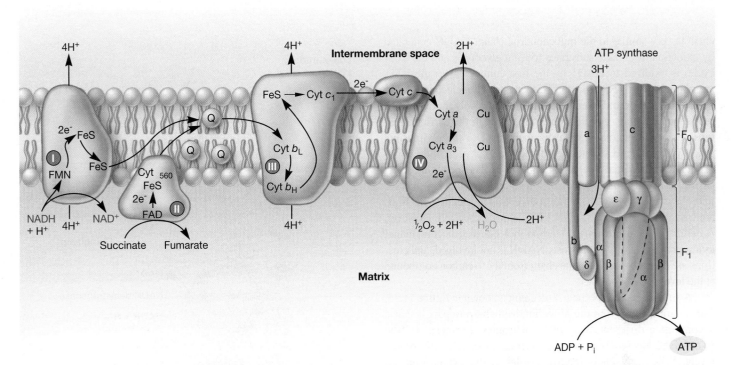

Figure 11.11 Creation of a Proton Gradient by the Mitochondrial Electron Transport Chain and Its Use in Synthesis of ATP. As electrons move along the ETC, protons are transported across the membrane creating a proton gradient. Proton release occurs when electrons are transferred from carriers, such as FMN and coenzyme Q (Q), that carry both electrons and protons to components such as nonheme iron proteins (FeS) and cytochromes (Cyt) that transport only electrons. Complex IV is the terminal oxidase that transfers electrons to oxygen. The number of protons moved across the membrane at each site per pair of electrons transported is still somewhat uncertain and may vary among species; the current consensus is that 10 protons move outward during NADH oxidation. One molecule of ATP is synthesized and released from the enzyme ATP synthase for every three protons that move back into the matrix by passing through the enzyme.

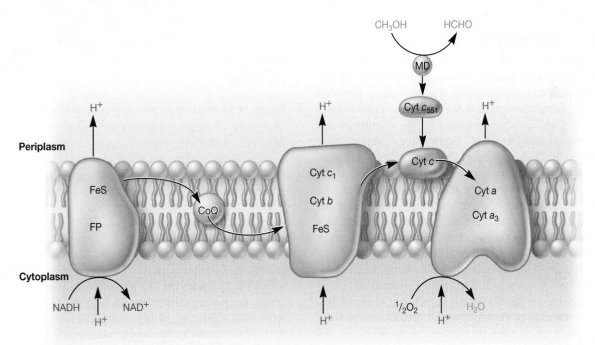

Figure 11.12 *Paracoccus denitrificans* Electron Transport Chain Used During Aerobic Respiration. *P. denitrificans* can respire both aerobically and anaerobically (figure 11.18). The aerobic ETC resembles a mitochondrial ETC and uses oxygen as the terminal electron acceptor. This metabolically versatile bacterium can also use methanol and methylamine as electron donors. They donate electrons to the chain at the level of cytochrome *c*. Locations of proton movement are shown, but not the number of protons involved. However, 8 to 10 protons are thought to be translocated. Abbreviations used: Flavoprotein (FP), methanol dehydrogenase (MD).

MICRO INQUIRY *How does the number of H$^+$ moved across the membrane as a result of NADH oxidation compare with the number of H$^+$ moved following methanol oxidation?*

chain is very similar to the mitochondrial chain. NADH generated by oxidation of organic substrates (during glycolysis and the TCA cycle) is oxidized to NAD^+ by the first component in the ETC, membrane-bound NADH dehydrogenase. The electrons from NADH are transferred to carriers with progressively more positive reduction potentials. As electrons move through the carriers, protons are moved across the plasma membrane to the periplasmic space, rather than to an intermembrane space, as seen in the mitochondrion (compare figures 11.11 and 11.12). The metabolic diversity of the bacterium is also demonstrated in figure 11.12. When the microbe is growing on an energy source such as glucose, electrons are transferred to the ETC by NADH, as just described. However, *P. denitrificans* also can use 1-carbon molecules such as methanol as a source of energy. In this case, NADH is not involved, and electrons are donated directly to the chain from the 1-carbon compound at the level of cytochrome *c*.

A simplified view of the *E. coli* chain is shown in **figure 11.13**. This chain differs from the mitochondrial chain in part because it contains a different array of cytochromes. Furthermore, the *E. coli* ETC has two branches that operate under different oxygen levels. When oxygen is readily available, the cytochrome *bo* branch is used (lower half of figure 11.13). When oxygen is less plentiful, the cytochrome *bd* branch is used because it has a higher affinity for oxygen (upper half of figure 11.13). However, it is less efficient than the *bo* branch because the *bd* branch moves fewer protons into the periplasmic space. ♻ *Electron Transport Systems and Formation of ATP*

Oxidative Phosphorylation

Oxidative phosphorylation is the process by which ATP is synthesized as the result of electron transport driven by the oxidation of a chemical energy source. The mechanism by which oxidative phosphorylation takes place has been studied intensively for years and is best explained by the **chemiosmotic hypothesis,** which was formulated by British biochemist Peter Mitchell (1920–1992). According to the chemiosmotic hypothesis, mitochondrial ETCs are organized so that protons move across the inner membrane from the mitochondrial matrix to the intermembrane space as electrons are transported down the chain (figure 11.11). In bacteria and archaea, the protons usually are moved across the plasma membrane from the cytoplasm to the periplasmic space (figures 11.12 and 11.13).

The movement of protons across the membrane is not completely understood. Some protons are actively pumped across the membrane (e.g., by complex IV of the mitochondrial chain; figure 11.11). In other cases, translocation of protons results from the juxtaposition of carriers that accept both electrons and protons with carriers that accept only electrons. For instance, coenzyme Q, a lipid-soluble molecule that moves freely in the membrane, accepts two electrons and two protons but delivers only two electrons to complex III of the mitochondrial chain. This transfer of electrons is complicated by the fact that cytochrome *b* and the FeS protein of complex III accept only one electron at a time and do not accept protons. This difference in

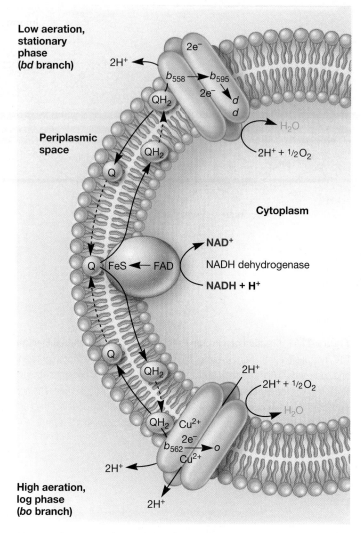

Figure 11.13 The Electron Transport Chain of *E. coli*.   NADH transfers electrons from an organic energy and electron source to the ETC. Ubiquinone-8 (Q) connects the NADH dehydrogenase with two terminal oxidase systems. The upper (*bd* branch) branch operates when the bacterium is in stationary phase and there is little oxygen. It is called the *bd* branch because it uses the cytochromes b_{558}, b_{595}, and *d*. The lower branch (*bo* branch) functions when *E. coli* is growing rapidly with good aeration. It is so named because it uses two cytochromes, b_{562} and *o*.

protons and electrons carried sets the stage for a phenomenon called the **Q cycle;** the Q cycle ultimately moves four protons across the membrane (**figure 11.14**). The Q cycle involves the sequential oxidation of two reduced coenzyme Q molecules (QH_2), the transfer of two of the four electrons released by these oxidations to one molecule of the oxidized form of coenzyme Q (located nearby in the membrane) via the cytochrome *b* located in complex III, and the transfer of the remaining two electrons to the FeS protein of complex III.

The result of proton expulsion during electron transport is the formation of a concentration gradient of protons (Δ pH; chemical potential energy) and a charge gradient ($\Delta\psi$; electrical

Oxidation of first QH₂

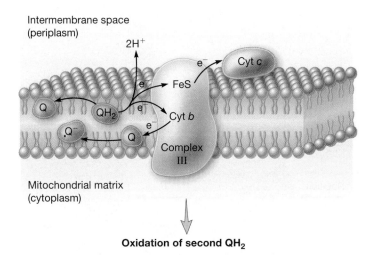

Oxidation of second QH₂

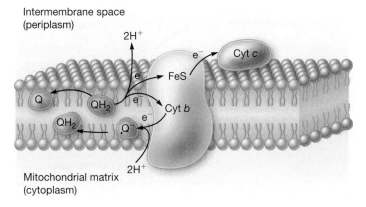

Net reaction:

QH₂ + 2Cyt *c* (oxidized) + 2H⁺ (matrix side)

↓

Q + 2Cyt *c* (reduced) + 4H⁺ (intermembrane
space side)

Figure 11.14 **The Q Cycle.** This simplified illustration shows the major events in the Q cycle.

potential energy). Thus as long as the membrane is intact, the mitochondrial matrix is more alkaline and more negative than the intermembrane space. Likewise with bacterial and archaeal cells, the cytoplasm is more alkaline and more negative than the periplasmic space. The combined chemical and electrical potential differences make up the **proton motive force** (**PMF**). The PMF is used to perform work when protons flow back across the membrane, down the concentration and charge gradients, and into the mitochondrial matrix (or the cytoplasm of bacterial and archaeal cells). This flow is exergonic and is often used to phosphorylate ADP to ATP. PMF is also used by many secondary active transport systems to move nutrients into the cell and to rotate the bacterial flagellar motor. Thus the PMF plays a central role in a cell's physiology (**figure 11.15**). ◄◄ *Primary and secondary active transport (section 3.3); Bacterial flagella (section 3.7)*

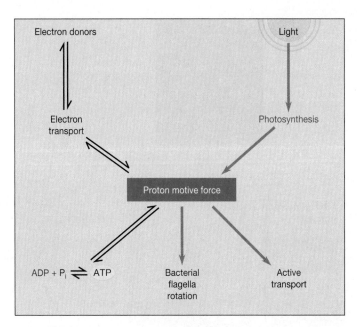

Figure 11.15 **The Central Role of Proton Motive Force.** Active transport is not always driven by PMF.

The use of PMF for ATP synthesis is catalyzed by **ATP synthase.** The best-studied ATP synthases are the F_1F_0 ATP synthases found in mitochondria, chloroplasts, and bacteria. They are also known as F_1F_0 ATPases because they can catalyze ATP hydrolysis. The mitochondrial F_1 component appears as a spherical structure attached to the mitochondrial inner membrane surface by a stalk (**figure 11.16**). The F_0 component is embedded in the membrane. ATP synthase is on the inner surface of the plasma membrane in bacterial cells. F_0 participates in proton movement across the membrane. F_1 is a large complex in which three α subunits alternate with three β subunits. The catalytic sites for ATP synthesis are located on the β subunits. At the center of F_1 is the γ subunit. The γ subunit extends through F_1 and interacts with F_0.

It is now known that ATP synthase functions like a rotary engine, much like the rotary motor of bacterial flagella. It is thought that the flow of protons down the proton gradient through the F_0 subunit causes F_0 and the γ subunit to rotate. As the γ subunit rotates rapidly within the F_1 (much like a car's crankshaft), conformation changes occur in the β subunits (figure 11.16*b*). One conformation change ($β_E$ to $β_{HC}$) allows entry of ADP and P_i into the catalytic site. Another conformation change ($β_{HC}$ to $β_{DP}$) loosely binds ADP and P_i in the catalytic site. ATP is synthesized when the $β_{DP}$ conformation is changed to the $β_{TP}$ conformation, and ATP is released when $β_{TP}$ changes to the $β_E$ conformation, to start the synthesis cycle anew.

ATP Yield During Aerobic Respiration

It is possible to estimate the number of ATP molecules synthesized per NADH or FADH₂ oxidized by an ETC. During aerobic respiration, a pair of electrons from NADH and FADH₂ is donated

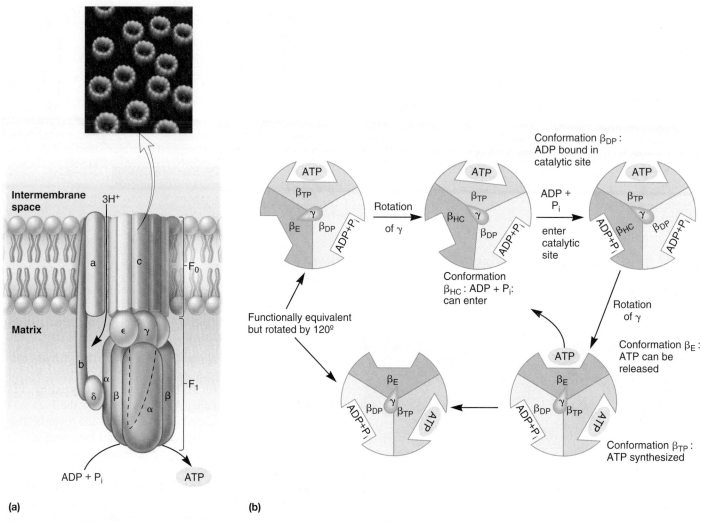

Figure 11.16 ATP Synthase Structure and Function.   (a) The major structural features of ATP synthase deduced from X-ray crystallography and other studies. F_1 is a spherical structure composed largely of alternating α and β subunits; the three active sites are on the β subunits. The γ subunit extends through the center of the sphere and can rotate. The stalk (γ and ϵ subunits) connects the sphere to F_0, the membrane embedded complex that serves as a proton channel. F_0 contains one a subunit, two b subunits, and nine to 12 c subunits. The stator arm is composed of subunit a, two b subunits, and the δ subunit; it is embedded in the membrane and attached to F_1. A ring of c subunits in F_0 is connected to the stalk and may act as a rotor and move past the a subunit of the stator. As the c subunit ring turns, it rotates the shaft ($\gamma\epsilon$ subunits). The inset is an atomic force micrograph of the F_0 rotor. (b) The binding change mechanism is a widely accepted model of ATP synthesis. This simplified drawing of the model shows the three catalytic β subunits and the γ subunit, which is located at the center of the F_1 complex. As the γ subunit rotates, it causes conformational changes in each subunit. The β_E (empty) conformation is an open conformation, which does not bind nucleotides. When the γ subunit rotates, β_E is converted to the β_{HC} (half closed) conformation. P_i and ADP can enter the catalytic site when it is in this conformation. The next rotation by the γ subunit is critical because it brings about three significant conformational changes: (1) β_{HC} to β_{DP} (ADP bound), (2) β_{DP} to β_{TP} (ATP bound), and (3) β_{TP} to β_E. Change from β_{DP} to β_{TP} is accompanied by the formation of ATP; change from β_{TP} to β_E allows for release of ATP from ATP synthase.

to the ETC and ultimately used to reduce an atom of oxygen to H_2O. The energy released drives the synthesis of ATP. Prior to the acceptance of the chemiosmotic hypothesis, the concept of the phosphorus to oxygen (P/O) ratio was used as a measure of the number of ATP molecules (phosphorus) generated per oxygen (O) reduced as NADH and $FADH_2$ were oxidized. With the acceptance of the chemiosmotic hypothesis, it was recognized that the important measurement was the number of protons transported across the membrane by NADH oxidation and the number of protons consumed during the synthesis of ATP.

Despite this change, the expression of the amount of ATP synthesized as the result of the oxidation of NADH is still often referred to as the P/O ratio. The best estimates of the number of protons moved across the membrane due to NADH oxidation is 10; six electrons are thought to be transferred when $FADH_2$ oxidation initiates electron flow to oxygen. It is currently thought that four protons are consumed during ATP synthesis; three are used by ATP synthase and one is used for transport of ATP, ADP, and P_i. Thus the P/O ratio is estimated to be 2.5 for NADH and 1.5 for $FADH_2$.

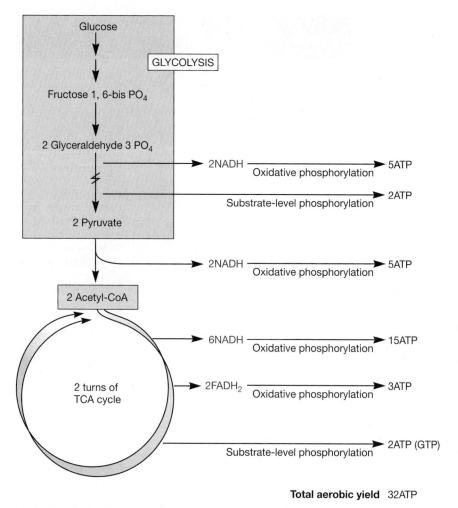

Figure 11.17 Maximum Theoretic ATP Yield from Aerobic Respiration. To calculate the theoretic maximum yield of ATP, P/O ratios of 2.5 for the oxidation of NADH and 1.5 for $FADH_2$ are assumed.

Bacterial ETCs are often shorter and therefore transport fewer protons across the plasma membrane. Thus their ETCs have lower P/O ratios than eukaryotic chains, and their ATP yields are smaller. For example, *E. coli,* with its truncated ETC, has a P/O ratio around 1.3 when using the cytochrome *bo* path at high oxygen levels and a ratio of only about 0.67 when employing the cytochrome *bd* branch (figure 11.13) at low oxygen concentrations. In this case, ATP production varies with environmental conditions. Perhaps because *E. coli* normally grows in habitats such as the intestinal tract that are very rich in nutrients, it does not have to be particularly efficient in ATP synthesis. Presumably the ETC functions when *E. coli* is in an oxic freshwater environment between hosts.

Two other factors affect the yield of ATP from the catabolism of glucose by aerobic respiration, thus making the theoretical maximum a value that is rarely reached. One is that the PMF generated by electron transport is used for functions other than ATP synthesis (e.g., bacterial flagella rotation). The second is related to the amphibolic nature of the pathways used to catabolize glucose. Recall that an important function of these pathways is the generation of precursor metabolites for anabolism. For each molecule of glucose degraded, numerous precursor metabolites are made, and as each is made, a microbe must "decide" if that metabolite is needed for anabolism or if it can continue the catabolic process. If the precursor metabolite is used for biosynthesis, fewer NADH and $FADH_2$ molecules may be made and fewer ATP molecules generated. Cells must carefully monitor and regulate the flow of the carbons from glucose such that ATP production is appropriately balanced with biosynthesis. The mechanisms for this regulation are described in chapters 10 and 14.

Given this information, the maximum ATP yield of aerobic respiration in a eukaryote can be calculated (**figure 11.17**). Substrate-level phosphorylation yields at most two ATP molecules per glucose converted to pyruvate (figures 11.5 and 11.7). Two ATP are generated by substrate-level phosphorylation during the two turns of the TCA cycle needed to oxidize two acetyl-CoA molecules (figure 11.8). Thus during aerobic respiration of glucose, at most four ATP molecules are made by substrate-level phosphorylation. Most of the ATP made during aerobic respiration is generated by oxidative phosphorylation. Up to 10 NADH (2 from glycolysis, 2 from pyruvate conversion to acetyl-CoA, and 6 from the TCA cycle) and 2 $FADH_2$ (from the TCA cycle) are generated when glucose is oxidized completely to 6 CO_2. Assuming a P/O ratio of 2.5 for NADH oxidation and 1.5 for $FADH_2$ oxidation, the 10 NADH could theoretically drive the synthesis of 25 ATP, while oxidation of the 2 $FADH_2$ molecules would add another 3 ATP for a maximum of 28 ATP generated via oxidative phosphorylation. Thus oxidative phosphorylation accounts for seven times more ATP than does substrate-level phosphorylation. The maximum total yield of ATP during aerobic respiration by eukaryotes is 32 ATP.

11.7 Anaerobic Respiration Uses the Same Three Steps as Aerobic Respiration

After reading this section, you should be able to:

- Compare and contrast aerobic respiration and anaerobic respiration using glucose as carbon source
- List examples of terminal electron acceptors used during anaerobic respiration
- Defend this statement: "The use of nitrate (NO_3^-) as a terminal electron acceptor is dissimilatory nitrate reduction."
- Predict the relative amount of energy released for each of the common terminal electron acceptors used during anaerobic respiration, as compared to energy released during aerobic respiration
- List three examples of the importance of anaerobic respiration

As we have seen, during aerobic respiration, sugars and other organic molecules are oxidized and their electrons transferred to NAD^+ and FAD to generate NADH and $FADH_2$, respectively. These carriers then donate the electrons to an ETC that uses O_2 as the terminal electron acceptor. However, it is also possible for other terminal electron acceptors to be used for electron transport. **Anaerobic respiration** is the chemoorganotrophic process whereby an exogenous terminal electron acceptor other than O_2 is used for electron transport. It is carried out by many bacteria and archaea, and some eukaryotic microbes. The most common terminal electron acceptors used during anaerobic respiration are nitrate, sulfate, and CO_2, but metals and a few organic molecules can also be reduced (**table 11.3**). Interestingly, when some metals are used, the metals remain outside the cell. Bacteria then use either nanowires or specialized shuttling molecules to transfer electrons from the cell surface to the metal. Some nanowires are specialized types of pili, and others are thin extensions of the outer membrane and periplasm along which cytochromes localize; shuttling molecules include flavin mononucleotides. ▶▶| *Bacterial pili and fimbriae (section 3.7); Order Alteromonadales includes anaerobes that use a range of electron acceptors (section 22.3)*

The diversity of alternate electron acceptors has important ecological and practical consequences. For instance, numerous soil microbes use nitrate (NO_3^-) as an electron acceptor in **anoxic** soils. When this occurs in agricultural soils, it depletes the soil nitrogen and decreases the yield from that field. It also forces farmers to use nitrogen-containing fertilizers that can have detrimental consequences such as the contamination of nearby wells, streams, and rivers. The contamination of wells is of particular concern because NO_3^- is toxic to humans (e.g., "blue baby syndrome") and other animals. On the other hand, if NO_3^- is being used as an electron acceptor by bacteria in a sewage treatment plant, NO_3^- levels in the plant effluent are decreased, as is the likelihood that the effluent will cause pollution when it is discharged. Finally, the use of NO_3^- and other electron acceptors by chemoorganotrophs links the carbon cycle to other biogeochemical cycles such as the nitrogen cycle and the sulfur cycle. This is the basis for many of the interactions

Table 11.3	Some Electron Acceptors Used in Respiration		
	Electron Acceptor	**Reduced Products**	**Examples of Microorganisms**
Aerobic	O_2	H_2O	All aerobic bacteria, fungi, and protists
Anaerobic	NO_3^-	NO_2^-	Enteric bacteria
	NO_3^-	NO_2^-, N_2O, N_2	*Pseudomonas, Bacillus,* and *Paracoccus species*
	SO_4^{2-}	H_2S	*Desulfovibrio* and *Desulfotomaculum*
	CO_2	CH_4	Methanogens
	CO_2	Acetate	Acetogens
	S^0	H_2S	*Desulfuromonas* and *Thermoproteus species*
	Fe^{3+}	Fe^{2+}	*Pseudomonas, Bacillus,* and *Geobacter species*
	$HAsO_4^{2-}$	$HAsO_2$	*Bacillus, Desulfotomaculum, Sulfurospirillum species*
	SeO_4^{2-}	Se, $HSeO_3^-$	*Aeromonas, Bacillus, Thauera species*
	Fumarate	Succinate	*Wolinella species*

that occur between microbes in their habitats. ▶▶| *Biogeochemical cycling sustains life on Earth (section 28.1)*

Many microbes that carry out anaerobic respiration will perform aerobic respiration instead if oxygen is available. In fact, some bacteria anaerobically respire sugars and other organic molecules using the exact same pathways used for aerobic respiration. Thus, depending on the organism, glycolysis and the TCA cycle function in the same way during both aerobic respiration and anaerobic respiration. Furthermore, most of the ATP generated during anaerobic respiration is made by oxidative phosphorylation, as is the case during aerobic respiration.

An example of a bacterium that can carry out both anaerobic respiration and aerobic respiration is *Paracoccus denitrificans*. We have already described the ETC used by this microbe during aerobic respiration (figure 11.12). Under anoxic conditions, *P. denitrificans* uses NO_3^- as its electron acceptor, reducing it to gaseous dinitrogen (N_2). As shown in **figure 11.18**, the anaerobic ETC is more complex than the aerobic chain. The anaerobic ETC is branched and uses different electron carriers, some of which are located in the periplasmic

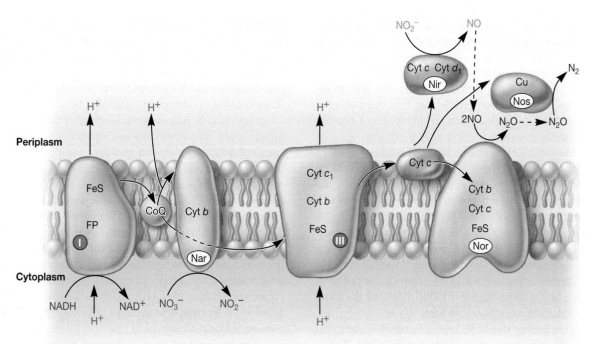

Figure 11.18 *Paracoccus denitrificans* **Electron Transport Chain Used During Anaerobic Respiration.** This branched ETC is made of both membrane and periplasmic proteins. Nitrate is reduced to diatomic nitrogen (N_2) by the collective action of four different reductases that receive electrons from CoQ and cytochrome *c*. Locations of proton movement are indicated. Four protons are pumped into the periplasm by complex I, two by nitrate reductase (Nar), and two by complex III. However, two protons are used by nitric oxide reductase (Nor) to reduce nitric oxide to nitrous oxide. Thus six protons, net, are used to create a PMF. Abbreviations used: flavoprotein (FP), nitrite reductase (Nir), and nitrous oxide reductase (Nos).

space. Not as many protons are pumped across the membrane during anaerobic growth, but nonetheless, a PMF is established.

The anaerobic reduction of NO_3^- makes it unavailable for assimilation into the cell. That is, it can't be used to construct N-containing molecules such as amino acids and nucleotides. Therefore this process is called **dissimilatory nitrate reduction.** When dissimilatory nitrate reduction leads to the production of gaseous compounds such as N_2 that are released into the atmosphere, the process is also called **denitrification.** As can be seen in figure 11.18, denitrification carried out by *P. denitrificans* is a multistep process with four enzymes participating. These enzymes function sequentially to reduce NO_3^- to nitrite (NO_2^-), then nitrite to nitric oxide (NO), followed by NO to nitrous oxide (N_2O), and finally N_2O to N_2.

In addition to *P. denitrificans,* some members of the genera *Pseudomonas* and *Bacillus* carry out denitrification. All use denitrification as an alternative to aerobic respiration and are considered facultative anaerobes. Indeed, if O_2 is present, the synthesis of nitrate reductase is repressed and these bacteria use aerobic respiration.

Anaerobic respiration does not produce as much ATP as aerobic respiration. The lower ATP yield is due to the fact that alternate electron acceptors such as NO_3^- have less positive reduction potentials than O_2. The difference in standard reduction potentials between NADH and NO_3^- is smaller than the difference between NADH and O_2 (figure 11.10). Less energy is available to make ATP in anaerobic respiration because energy yield is directly related to the magnitude of the reduction potential difference. Nevertheless, anaerobic respiration is useful because

it allows ATP synthesis by electron transport and oxidative phosphorylation in the absence of O_2.

Retrieve, Infer, Apply

1. Describe the process of anaerobic respiration. Does anaerobic respiration yield as much ATP as aerobic respiration? Why or why not?
2. What is denitrification? Why do farmers dislike this process?
3. *E. coli* can use O_2, fumarate, or nitrate as a terminal electron acceptor under different conditions. What is the order of energy yield from highest to lowest for these electron acceptors? Explain your answer in thermodynamic terms.

11.8 Fermentation Does Not Involve an Electron Transport Chain

After reading this section, you should be able to:

- Compare and contrast aerobic respiration, anaerobic respiration, and fermentation of glucose
- List the pathways that may function during fermentation if glucose is the organism's carbon and energy source
- Create a table that lists some of the common fermentation pathways and their products, and gives examples of their importance
- Compare the use of ATP synthase during respiration and fermentation

Despite the tremendous ATP yield obtained by oxidative phosphorylation, some chemoorganotrophic microbes never respire and others may be unable to respire under certain conditions. Those microbes incapable of respiration lack ETCs. Those that carry out aerobic respiration may repress the synthesis of ETC components under anoxic conditions, making anaerobic respiration impossible. Finally, some facultative anaerobes able to respire both aerobically and anaerobically may find themselves in an environment that lacks both oxygen and the terminal electron acceptor they use for anaerobic respiration (e.g., nitrate). For each of these microbes, NADH produced by the Embden-Meyerhof pathway reactions during glycolysis (figure 11.5) must still be oxidized back to NAD$^+$. If NAD$^+$ is not regenerated, the oxidation of glyceraldehyde 3-phosphate will cease and glycolysis will stop. Many microorganisms solve this problem by slowing or stopping pyruvate dehydrogenase activity and using pyruvate or one of its derivatives as an electron acceptor for the reoxidation of NADH in a fermentation process (**figure 11.19**). There are many kinds of fermentations, and they often are characteristic of particular microbial groups (**figure 11.20**). A few of the more common fermentations are introduced here.

Four unifying themes should be kept in mind when microbial fermentations are examined: (1) NADH is oxidized to NAD$^+$; (2) O$_2$ is not needed; (3) the electron acceptor is often either pyruvate or a pyruvate derivative; and (4) an ETC is not used to reoxidize NADH, and this reduces the ATP yield per glucose significantly. Thus in fermentation, the substrate (e.g., glucose) is only partially catabolized and ATP is formed in most organisms exclusively by substrate-level phosphorylation. However, not having a functioning ETC creates a problem for fermenting microbes: they still need a PMF to do work, in particular to drive active transport. To solve

this dilemma, they use their ATP synthase in the reverse direction. That is, the ATP synthase pumps protons out of the cell, using the energy released when ATP is hydrolyzed to ADP and P$_i$. Although fermentation yields considerably less ATP, it is an important component of the metabolic repertoire of many microbes, allowing them to adjust to changes in their habitats.

Fermentation pathways are named after the major acid or alcohol produced. The most common fermentation is lactic acid (lactate) fermentation, the reduction of pyruvate to lactate (figure 11.20, number 1). It occurs in bacteria (lactic acid bacteria, *Bacillus* spp.), protists (*Chlorella* spp. and some water molds), and animal skeletal muscle. Lactic acid fermenters can be separated into two groups. **Homolactic fermenters** use the Embden-Meyerhof pathway and directly reduce almost all their pyruvate to lactate with the enzyme lactate dehydrogenase. **Heterolactic fermenters** use the pentose phosphate pathway to form substantial amounts of products other than lactate; many also produce ethanol and CO$_2$. Lactic acid bacteria are important to the food industry, where they are used to make a variety of fermented foods (e.g., yogurt). ▶▶ *Microbiology of fermented foods (section 41.5)*

Another common fermentation is carried out by many fungi, protists, and some bacteria. These microbes ferment sugars to ethanol and CO$_2$ in a process called **alcoholic fermentation.** Pyruvate is decarboxylated to acetaldehyde, which is then reduced to ethanol by alcohol dehydrogenase with NADH as the electron donor (figure 11.20, number 2). This fermentation produces the ethanol desired by brewers making beer and ale and by vintners making wine.

Many bacteria, especially members of the family *Enterobacteriaceae,* metabolize pyruvate to numerous products using several pathways simultaneously. One such complex fermentation is the **mixed acid fermentation,** which results in the excretion of ethanol and a mixture of acids, particularly acetic, lactic, succinic, and formic acids (**table 11.4**). Members of the genera *Escherichia, Salmonella,* and *Proteus* carry out mixed acid fermentation. They use pathways numbered 1, 5, 8, and 9 in

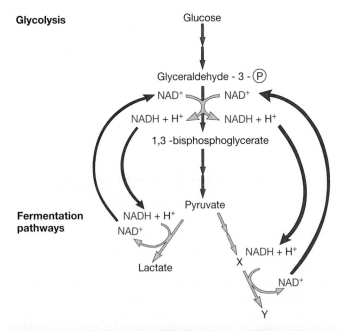

Figure 11.19 Reoxidation of NADH During Fermentation. NADH from glycolysis is reoxidized by being used to reduce pyruvate or a pyruvate derivative (X). Either lactate or reduced product Y result.

MICRO INQUIRY *How many NADH are reoxidized to NAD$^+$ for each glucose catabolized?*

Table 11.4	Mixed Acid Fermentation Products of *Escherichia coli*	
	FERMENTATION BALANCE (µM PRODUCT/100 µM GLUCOSE)	
	Acid Growth (pH 6.0)	**Alkaline Growth (pH 8.0)**
Ethanol	50	50
Formic acid	2	86
Acetic acid	36	39
Lactic acid	80	70
Succinic acid	11	15
Carbon dioxide	88	2
Hydrogen gas	75	0.5

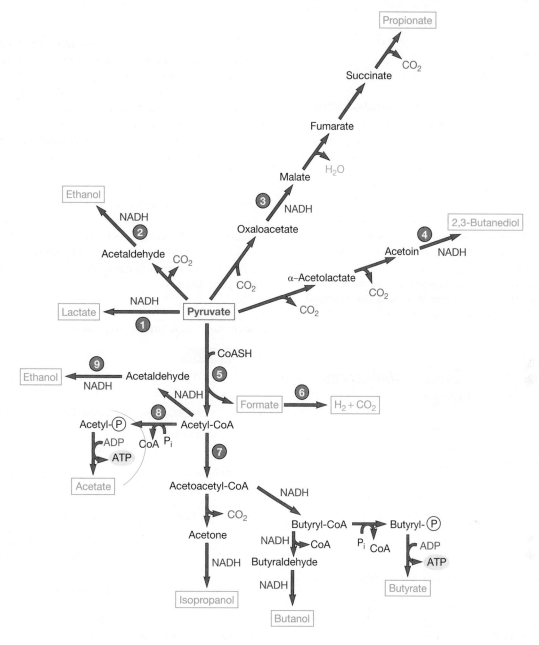

Figure 11.20 Some Common Microbial Fermentations. Only pyruvate fermentations are shown for the sake of simplicity; many other organic molecules can be fermented. Most of these pathways have been simplified by deletion of one or more steps and intermediates. Pyruvate and major end products are shown in color. Enteric bacteria include members of the genera *Escherichia, Enterobacter, Salmonella,* and *Proteus.*

1. Lactic acid bacteria (*Streptococcus, Lactobacillus* species), *Bacillus* spp., enteric bacteria
2. Yeast, *Zymomonas* spp.
3. Propionic acid bacteria (*Propionibacterium* spp.)
4. *Enterobacter, Serratia, Bacillus* species

5. Enteric bacteria
6. Enteric bacteria
7. *Clostridium* spp.
8. Enteric bacteria
9. Enteric bacteria

figure 11.20 to make all the fermentation products, except succinate. If the enzyme formic hydrogenlyase is present (figure 11.20, number 6), formate will be degraded to H_2 and CO_2 (*see figure 22.31*). Another complex fermentation is the **butanediol fermentation,** which is characteristic of members of the genera *Enterobacter, Serratia, Erwinia,* and some species of *Bacillus.* The predominant pathway used during this fermentation yields butanediol (figure 11.20, number 4). However, large amounts of ethanol also are produced (figure 11.20, number 9), as are smaller

amounts of lactic acid (figure 11.20, number 1) and formic acid (figure 11.20, number 5); in some bacteria, the formic acid is further catabolized to H_2 and CO_2 (figure 11.20, number 6). The use of either mixed acid fermentation or butanediol fermentation is important in differentiating members of the family *Enterobacteriaceae,* many of which are clinically important. ▶▶| *Order* Enterobacteriales *includes pathogens and mutualists (section 22.3); Class* Bacilli *(section 23.3); Identification of microorganisms from specimens (section 36.3)*

11.9 Catabolism of Organic Molecules Other Than Glucose

After reading this section, you should be able to:

- Discuss the role of ATP and UTP in the catabolism of monosaccharides other than glucose
- Differentiate the catabolism of disaccharides and polysaccharides by hydrolysis from their catabolism by phosphorolysis
- Discuss the fate of the fatty acid and glycerol components of triglycerides when triglycerides are catabolized
- State the name of the enzymes responsible for hydrolyzing proteins into amino acids
- Distinguish deamination from transamination and explain how the two are related
- Draw a simple diagram that illustrates how the pathways used to catabolize reduced organic molecules other than glucose connect to the glycolytic pathways and the TCA cycle

Thus far our main focus has been on the catabolism of glucose. However, microorganisms can catabolize many other organic molecules, including, surprisingly, antibiotics. Other interesting and sometimes useful catabolic processes include fermentations of acetate, lactate, and propionate. For instance, citrate can be fermented to diacetyl, which is used by the dairy industry to give flavor to fermented milk. Also common is the catabolism of a variety of other carbohydrates, lipids, and amino acids. The use of these energy sources is discussed next. ▶▶│ *Microbiology of fermented foods (section 41.5)*

Carbohydrates

The other carbohydrates used by microbes may come either from outside the cell or from internal sources generated during normal metabolism. Often the initial steps in catabolism of external

carbohydrate polymers differ from those employed with internal reserves. **Figure 11.21** outlines some catabolic pathways for the monosaccharides (single sugars) glucose, fructose, mannose, and galactose. The first three are phosphorylated using ATP and easily enter the Embden-Meyerhof pathway. In contrast, galactose must be converted to uridine diphosphate galactose (UDP-gal, *see figure 12.7*) after initial phosphorylation, then converted to glucose 6-phosphate in a three-step process.

The common disaccharides are cleaved to monosaccharides by at least two mechanisms: hydrolysis and phosphorolysis (figure 11.21). Maltose, sucrose, and lactose can be directly hydrolyzed to their constituent sugars. Many disaccharides (e.g., maltose, cellobiose, and sucrose) are split by phosphorolysis, in which a phosphate attacks the bond joining the two sugars. This yields the two constituent monosaccharides, one of which is phosphorylated.

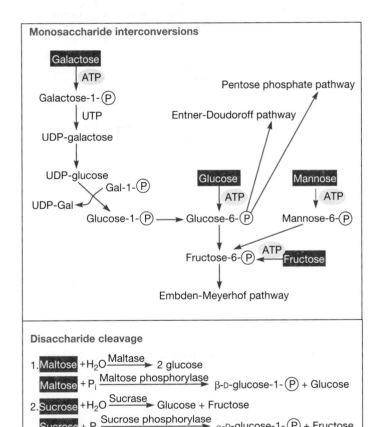

Figure 11.21 Carbohydrate Catabolism. Examples of enzymes and pathways used in disaccharide and monosaccharide catabolism. UDP is an abbreviation for uridine diphosphate. See figure AI.9 to learn how monosaccharides are named.

MICRO INQUIRY *What is the difference between a hydrolase and phosphorylase?*

Polysaccharides, like disaccharides, are cleaved by both hydrolysis and phosphorolysis depending on whether they are extracellular or intracellular. Because of their large size, polysaccharides in the environment cannot be transported into the cell. Instead, bacteria, archaea, and fungi degrade external polysaccharides by secreting hydrolytic enzymes. These exoenzymes cleave polysaccharides into smaller molecules that can then be taken up and catabolized. Microbes can catabolize numerous polysaccharides in this way. For instance, starch and glycogen are hydrolyzed by amylases to glucose, maltose, and other products. Cellulose, which is more difficult to digest, is hydrolyzed by the cellulases produced by many fungi and a few bacteria, yielding cellobiose and glucose. Some bacteria excrete an agarase that degrades agar (e.g., *Cytophaga* spp.). Many soil bacteria and bacterial plant pathogens degrade pectin, a polymer of galacturonic acid (a galactose derivative) that is an important constituent of plant cell walls and tissues. Recall that in times of plenty, microorganisms store excess carbon by synthesizing intracellular reserves such as glycogen and starch. They use these to survive for long periods in the absence of exogenous nutrients. Under such circumstances, they catabolize their intracellular stores. Glycogen and starch are degraded by phosphorylases that shorten the polysaccharide chain by one glucose, yielding glucose 1-phosphate.

$$(\text{Glucose})_n + P_1 \rightarrow (\text{glucose})_{n-1} + \text{glucose-1-P}$$

Glucose 1-phosphate can enter glycolytic pathways after conversion to glucose 6-phosphate (figure 11.21). ▶▶ *Soils are an important microbial habitat (section 31.1)*

Lipid Catabolism

Chemoorganotrophic microorganisms frequently use lipids as energy sources. Triglycerides (also called triacylglycerols) are composed of a glycerol core to which are attached three fatty acids. They are common energy sources and will serve as our examples (**figure 11.22**). Triglycerides can be hydrolyzed to glycerol and fatty acids by enzymes called lipases. The glycerol is then phosphorylated and oxidized to dihydroxyacetone phosphate, an intermediate of the Embden-Meyerhof pathway (EMP), which allows for its continued degradation as shown in figure 11.5.

Fatty acids from triacylglycerols and other lipids are often oxidized in the **β-oxidation pathway** after being linked to coenzyme A (**figure 11.23**). In this pathway, fatty acids are shortened by two carbons with each turn of the cycle. The two carbon units are released as acetyl-CoA, which can be fed into the TCA cycle or used in biosynthesis. One turn of the cycle produces acetyl-CoA, NADH, and $FADH_2$; NADH and $FADH_2$ can be oxidized by an ETC to provide more ATP. Thus, fatty acids are a rich source of energy for microbial growth. In a similar fashion, some microorganisms grow well on petroleum hydrocarbons.

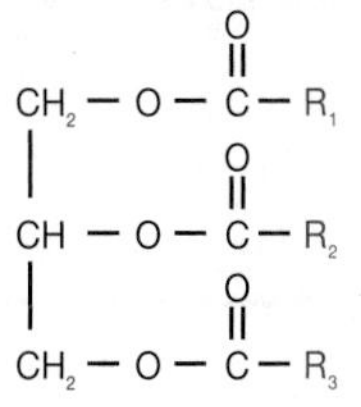

Protein and Amino Acid Catabolism

Some bacteria and fungi—particularly pathogenic, food spoilage, and soil microorganisms—use proteins as their source of carbon and energy. They secrete enzymes called **proteases** that hydrolyze proteins to amino acids, which are transported into the cell and catabolized.

The first step in amino acid catabolism is **deamination,** the removal of the amino group from an amino acid. This is often accomplished by **transamination.** The amino group is transferred from an amino acid to an α-keto acid acceptor (**figure 11.24**). The organic acid resulting from deamination can be converted to pyruvate, acetyl-CoA, or a TCA cycle intermediate. Therefore, depending on the organic acid produced, it may be used in several ways. Depending on the microbe, the organic acid may be fermented to release energy. For example, some members of the genus *Clostridium* ferment mixtures of amino acids using the Stickland reaction in which one amino acid is oxidized and a second amino acid acts as the electron acceptor (*see figure 23.4*). Most microbes catabolize the organic acid in the TCA cycle or use it as a source of carbon for the synthesis of cell constituents. Excess nitrogen from deamination may be excreted as ammonium ion. ▶▶ *Class Clostridia (section 23.1)*

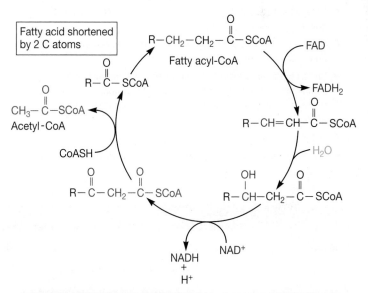

Figure 11.23 Fatty Acid β-Oxidation. The portions of the fatty acid being modified are shown in red.

MICRO INQUIRY *Why are fatty acids a rich source of energy even though no ATP is generated when they are degraded by the β-oxidation pathway?*

Figure 11.22 A Triacylglycerol (Triglyceride). The R groups represent the fatty acid side chains.

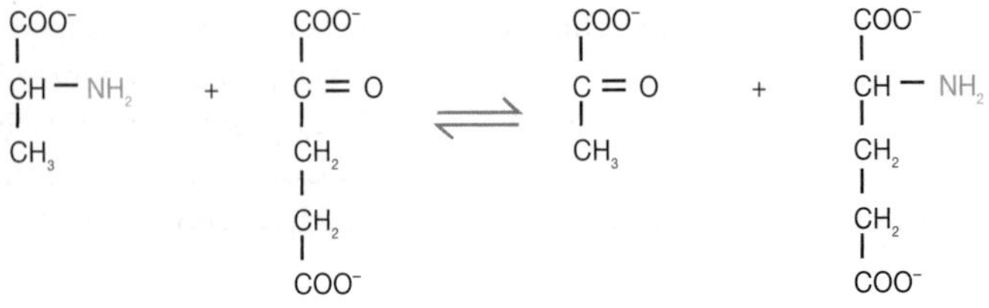

Figure 11.24 An Example of Transamination.   The α-amino group (blue) of alanine is transferred to the acceptor α-ketoglutarate, forming pyruvate and glutamate. The pyruvate can be fermented, catabolized in the tricarboxylic acid cycle, or used in biosynthesis.

Retrieve, Infer, Apply

1. Briefly discuss the ways in which microorganisms degrade and use common monosaccharides, disaccharides, and polysaccharides from both external and internal sources.
2. Can members of the genus *Cytophaga* be cultured on standard solidified media? Why or why not?
3. Describe how a microorganism might derive carbon and energy from the lipids and proteins in its diet. What is β-oxidation? Deamination? Transamination?

11.10 Chemolithotrophy: "Eating Rocks"

After reading this section, you should be able to:

- Describe in general terms the fueling reactions of chemolithotrophs
- List the molecules commonly used as energy sources and electron donors by chemolithotrophs
- Discuss the use of electron transport chains and oxidative phosphorylation by chemolithotrophs
- Predict the relative amount of energy released for each of the commonly used energy sources of chemolithotrophs, as compared to energy released during aerobic and anaerobic respiration of glucose
- Differentiate nitrification from denitrification
- List three examples of important chemolithotrophic processes

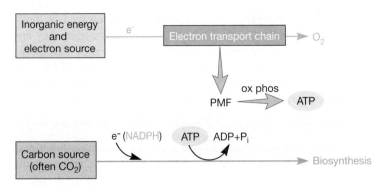

Figure 11.25 Chemolithotrophic Fueling Processes. Chemolithotrophic bacteria and archaea oxidize inorganic molecules (e.g., H_2S and NH_3), which serve as energy and electron sources. The electrons released pass through an electron transport chain, generating a proton motive force (PMF). ATP is synthesized by oxidative phosphorylation (ox phos). Most chemolithotrophs use O_2 as the terminal electron acceptor. However, some can use other exogenous molecules as terminal electron acceptors. Note that a molecule other than the energy source provides carbon for biosynthesis. Many chemolithotrophs are autotrophs.

So far, we have considered microbes that synthesize ATP with the energy liberated by the oxidation of organic substrates such as carbohydrates, lipids, and proteins. Certain bacteria and archaea are **chemolithotrophs.** These microbes donate electrons to their ETCs by oxidizing inorganic molecules rather than organic nutrients (**figure 11.25**). The most common energy sources (electron donors) are hydrogen, reduced nitrogen compounds, reduced sulfur compounds, and ferrous iron (Fe^{2+}). The acceptor is usually O_2, but sulfate and nitrate may also be used (**table 11.5**).

Much less energy is available from the oxidation of inorganic molecules than from the complete oxidation of glucose to CO_2 (**table 11.6**). This is because the reduction potentials of most inorganic substrates used by chemolithotrophs are much more positive than the reduction potentials of organic molecules such as glucose (figure 11.10). The low yield of ATP means that chemolithotrophs must oxidize a large quantity of inorganic material to grow and reproduce. This is particularly true of chemolithoautotrophs, those chemolithotrophs that fix CO_2 into reduced organic molecules. CO_2-fixation pathways consume considerable amounts of ATP and reducing power (often NADPH). Because

Table 11.5	Representative Chemolithotrophs and Their Energy Sources			
Bacteria	**Electron Donor**	**Electron Acceptor**	**Products**	
Alcaligenes, Hydrogenophaga, and *Pseudomonas* species	H_2	O_2	H_2O	
Nitrobacter spp.	NO_2^-	O_2	NO_3^-, H_2O	
Nitrosomonas spp.	NH_4^+	O_2	NO_2^-, H_2O	
Thiobacillus denitrificans	S^0, H_2S	NO_3^-	SO_4^{2-}, N_2	
Acidithiobacillus ferrooxidans	Fe^{2+}, S^0, H_2S	O_2	Fe^{3+}, H_2O, H_2SO_4	

Table 11.6 Energy Yields from Oxidations Used by Chemolithotrophs

Reaction	$\Delta G^{\circ\prime}$ (kcal/mole)[1]
$H_2 + \frac{1}{2}O_2 \rightarrow H_2O$	−56.6
$NO_2^- + \frac{1}{2}O_2 \rightarrow NO_3^-$	−17.4
$NH_4^+ + 1\frac{1}{2}O_2 \rightarrow NO_2^- + H_2O + 2H^+$	−65.0
$S^0 + 1\frac{1}{2}O_2 + H_2O \rightarrow H_2SO_4$	−118.5
$S_2O_3^{2-} + 2O_2 + H_2O \rightarrow 2SO_4^{2-} + 2H^+$	−223.7
$2Fe^{2+} + 2H^+ + \frac{1}{2}O_2 \rightarrow 2Fe^{3+} + H_2O$	−11.2

[1] The $\Delta G^{\circ\prime}$ for complete oxidation of glucose to CO_2 is −686 kcal/mole. A kcal is equivalent to 4.184 kJ.

they must consume a large amount of inorganic material, chemolithotrophs have significant ecological impact. They make important contributions to several biogeochemical cycles, including the nitrogen, sulfur, and iron cycles. ▶▶ *CO_2 fixation (section 12.3); Biogeochemical cycling sustains life on Earth (section 28.1)*

Members of several bacterial and archaeal genera can oxidize hydrogen gas using a hydrogenase enzyme (table 11.5).

$$H_2 \rightarrow 2H^+ + 2e^-$$

The $2H^+/H_2$ conjugate redox pair has a very negative standard reduction potential, and its electrons can be donated to either an ETC or NAD^+, depending on the hydrogenase. If NADH is produced, it can be used in ATP synthesis by electron transport and oxidative phosphorylation, with O_2, Fe^{3+}, S^0, and carbon monoxide (CO) as the terminal electron acceptors. Often these hydrogen-oxidizing microorganisms use organic compounds as energy sources when available.

Some bacteria and archaea use the oxidation of nitrogenous compounds as a source of electrons. Among these chemolithotrophs, the **nitrifying bacteria** are best understood. They are soil and aquatic bacteria of considerable ecological significance that carry out **nitrification**—the oxidation of ammonia to nitrate

(*see figure 22.15*). Nitrification is a two-step process that depends on the activity of at least two different microbes. In the first step, ammonia is oxidized to nitrite by members of a number of bacterial genera (e.g., *Nitrosomonas*) as well as some mesophilic archaea. In the second step, the nitrite is oxidized to nitrate by members of bacterial genera such as *Nitrobacter*.

Care should be taken not to confuse nitrification and denitrification. Nitrification is a chemolithotrophic process in which ammonia is oxidized in a two-step process to yield nitrate. The electrons released are donated to an ETC that uses oxygen as the terminal electron acceptor. By contrast, denitrification is a form of anaerobic respiration, and as such it involves the oxidation of an organic compound under anoxic conditions. The electrons are donated to an ETC that uses an oxidized nitrogenous compound (e.g., nitrate) as the terminal electron acceptor; it is further reduced to gaseous forms of nitrogen such as nitrogen gas (figure 11.18).

A unique metabolic process that combines aspects of both nitrification and denitrification is carried out by anammox bacteria (phylum *Planctomycetes*). These bacteria carry out *an*aerobic *amm*onia *ox*idation (anammox) using a membrane-bound organelle called the anammoxosome. In the anammoxosome, ammonia oxidation is coupled with the reduction of nitrite, generating N_2 and H_2O (*see figure 21.16*). The anammox reaction is unusual in that it features both the oxidation and reduction of nitrogenous compounds.

Sulfur-oxidizing microbes are the third major group of chemolithotrophs. The metabolism of *Thiobacillus* and *Acidithiobacillus* species has been best studied. These bacteria oxidize sulfur (S^0), hydrogen sulfide (H_2S), thiosulfate ($S_2O_3^{2-}$), and other reduced sulfur compounds to sulfuric acid; therefore they have a significant ecological impact. Interestingly, they generate ATP by both oxidative phosphorylation and substrate-level phosphorylation (*see figure 22.20*). Substrate-level phosphorylation involves adenosine 5′-phosphosulfate (APS), a high-energy molecule formed from sulfite and adenosine monophosphate (**figure 11.26**).

Some sulfur-oxidizing microbes are extraordinarily flexible metabolically. For example, *Sulfolobus brierleyi*, an archaeon, and some bacteria can grow aerobically by oxidizing sulfur with oxygen as the electron acceptor (*see figure 20.11*). However, in the absence of O_2, they carry out anaerobic respiration and

(a) Direct oxidation of sulfite

$$SO_3^{2-} \xrightarrow{\text{sulfite oxidase}} SO_4^{2-} + 2e^-$$

(b) Formation of adenosine 5′-phosphosulfate

$$2SO_3^{2-} + 2AMP \longrightarrow 2APS + 4e^-$$
$$2APS + 2P_i \longrightarrow 2ADP + 2SO_4^{2-}$$
$$2ADP \longrightarrow AMP + ATP$$

$$2SO_3^{2-} + AMP + 2P_i \longrightarrow 2SO_4^{2-} + ATP + 4e^-$$

(c) Adenosine 5′-phosphosulfate

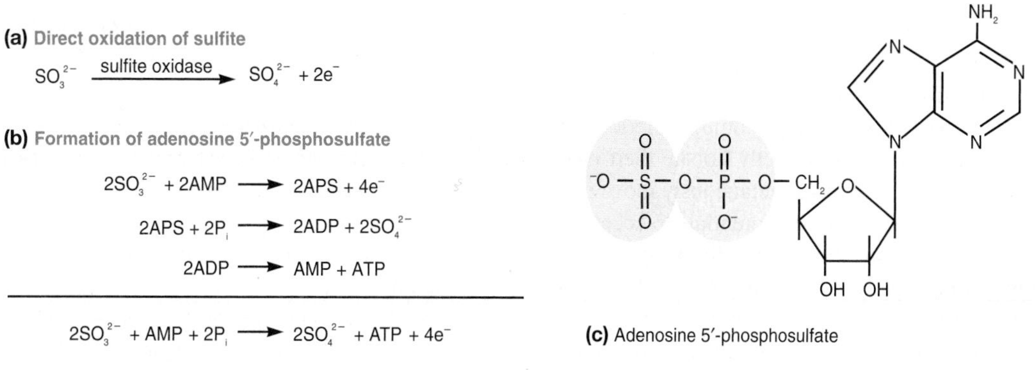

Figure 11.26 Energy Generation by Sulfur Oxidation. (a) Sulfite can be directly oxidized to provide electrons for electron transport and oxidative phosphorylation. (b) Sulfite can also be oxidized and converted to adenosine 5′-phosphosulfate (APS). This route produces electrons for use in electron transport and ATP by substrate-level phosphorylation with APS. (c) The structure of APS.

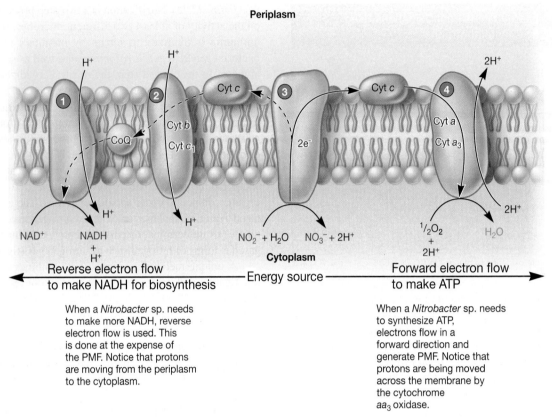

Figure 11.27 Electron Flow in a _Nitrobacter_ spp. Electron Transport Chain.   _Nitrobacter_ spp. oxidize nitrite and carry out normal electron transport to generate proton motive force for ATP synthesis. This is the right-hand branch of the diagram. Some of the proton motive force also is used to force electrons to flow up the reduction potential gradient from nitrite to NAD^+ (left-hand branch). Cytochrome _c_ (Cyt c), Coenzyme Q (CoQ), and four protein complexes make up this ETC. The four complexes are NADH-ubiquinone oxidoreductase (1), ubiquinol-cytochrome _c_ oxidoreductase (2), nitrite oxidase (3), and cytochrome aa_3 oxidase (4).

oxidize organic material with sulfur as the electron acceptor. Furthermore, many sulfur-oxidizing chemolithotrophs use CO_2 as their carbon source but will grow heterotrophically if they are supplied with reduced organic carbon sources such as glucose or amino acids.

Energy released by the oxidation of ammonia, nitrite, and sulfur-containing compounds is used to make ATP by oxidative phosphorylation. As we have already noted, the ATP yields for these chemolithotrophs are low. Complicating this already difficult lifestyle is the fact that many of these microbes are also autotrophs, and autotrophs need NAD(P)H (reducing power) as well as ATP to reduce CO_2 and other molecules (figure 11.25). Molecules such as ammonia, nitrite, and H_2S have more positive reduction potentials than the $NAD(P)^+/NAD(P)H$ conjugate redox pair (figure 11.10). Therefore, they cannot directly donate their electrons to $NAD(P)^+$. Recall that electrons spontaneously move only from donors with more negative reduction potentials to acceptors with more positive potentials. Thus these chemolithotrophs face a major dilemma: how to make the NAD(P)H they need. They solve this problem by using a process called **reverse electron flow.** During reverse electron flow, electrons derived from oxidation of inorganic substrates (reduced nitrogen or sulfur compounds) are moved up their ETCs to reduce $NAD(P)^+$ to NAD(P)H

(figure 11.27). Of course, this is not thermodynamically favorable, so energy in the form of the PMF must be diverted from performing other cellular work (e.g., ATP synthesis, transport, motility) to "push" the electrons from molecules of relatively positive reduction potentials to those that are more negative. Chemolithotrophs can afford this inefficiency, as they have no serious competitors for their unique energy sources.

Retrieve, Infer, Apply

1. How do chemolithotrophs obtain their ATP and NAD(P)H? What is their most common source of carbon?
2. Describe energy production by hydrogen-oxidizing bacteria, nitrifying bacteria, and sulfur-oxidizing bacteria.
3. Why can hydrogen-oxidizing bacteria and archaea donate electrons to NAD^+, whereas sulfur- and ammonia-oxidizing bacteria and archaea cannot?
4. What is reverse electron flow and why do many chemolithotrophs perform it?
5. Arsenate is a compound that inhibits substrate-level phosphorylation. Compare the effect of this compound on an H_2-oxidizing chemolithotroph, on a sulfite-oxidizing chemolithotroph, and on a chemoorganotroph carrying out fermentation.

11.11 Phototrophy

After reading this section, you should be able to:

- Describe in general terms the fueling reactions of phototrophs
- Differentiate phototrophy from photosynthesis
- Describe the light and dark reactions that occur during photosynthesis
- Summarize the structure and function of the light-absorbing pigments used by oxygenic and anoxygenic phototrophs
- Defend this statement: "Oxidative phosphorylation and photophosphorylation by chlorophyll-based phototrophs differ primarily in the energy source driving the process."
- Distinguish cyclic photophosphorylation from noncyclic photophosphorylation
- Compare and contrast oxygenic photosynthesis, anoxygenic phototrophy, and rhodopsin-based phototrophy
- List two examples of the importance of chlorophyll-based phototrophy

Many microbes capture the energy in light and use it to synthesize ATP and reducing power (e.g., NADPH). When the ATP and reducing power are used to reduce and incorporate CO_2 (CO_2 fixation), the process is called **photosynthesis.** Photosynthesis is one of the most significant metabolic processes on Earth because almost all our energy is ultimately derived from solar energy. Photosynthetic organisms serve as the base of most food webs in the biosphere. One type of photosynthesis is also responsible for replenishing the supply of O_2 in Earth's atmosphere. Although most people associate photosynthesis with plants, over half the photosynthesis on Earth is carried out by microorganisms (**table 11.7**). Indeed, *Prochlorococcus*, a cyanobacterium, is thought to be the most abundant photosynthetic organism on Earth and thus a major contributor to the functioning of the biosphere.

Photosynthesis is divided into two parts. In the **light reactions,** light energy is trapped and converted to chemical energy (ATP) and reducing power. ATP synthesis is accomplished by formation of a PMF, initiated by the absorption of light—a process called **photophosphorylation.** The ATP and reducing power are used to fix CO_2 and synthesize cell constituents in the **dark reactions.** The term **phototrophy** refers to the use of light energy to fuel a variety of cellular activities but not necessarily CO_2 fixation. In this sense, photoheterotrophs, which use light to drive ATP synthesis but not carbon fixation, are considered phototrophs but are not photosynthetic. In this section, three types of phototrophy are discussed: oxygenic photosynthesis, anoxygenic photosynthesis, and rhodopsin-based phototrophy (**figure 11.28**). The dark reactions of photosynthesis are reviewed in chapter 12. ▶▶▎ *CO_2 fixation (section 12.3)*

Light Reactions in Oxygenic Photosynthesis

Photosynthetic eukaryotes and cyanobacteria carry out **oxygenic photosynthesis,** so named because oxygen is generated and released into the environment when light energy is converted to

Table 11.7	Diversity of Phototrophic Microorganisms
Eukaryotes	Multicellular green, brown, and red algae; unicellular protists (e.g., euglenoids, dinoflagellates, diatoms)
Bacteria	Cyanobacteria, green sulfur bacteria, green nonsulfur bacteria, purple sulfur bacteria, purple nonsulfur bacteria, heliobacteria, acidobacteria
Archaea	Halophiles

chemical energy. Central to this process, and to all other phototrophic processes, are light-absorbing pigments (**table 11.8**). In oxygenic phototrophs, the most important pigments are **chlorophylls.** Chlorophylls are large planar molecules composed of four rings (pyrrole rings) with a central magnesium atom coordinated to the four nitrogen atoms of the pyrrole rings

Chlorophyll-based phototrophy

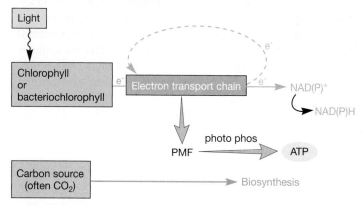

Rhodopsin-based phototrophy

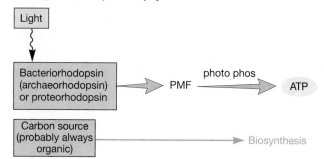

Figure 11.28 Phototrophic Fueling Reactions. Phototrophs use light to generate a proton motive force (PMF), which is then used to synthesize ATP by a process called photophosphorylation (photo phos). The process requires light-absorbing pigments. When the pigments are chlorophyll or bacteriochlorophyll, the absorption of light triggers electron flow through an electron transport chain, accompanied by the pumping of protons across a membrane. The electron flow can be either cyclic (dashed line) or noncyclic (solid line), depending on the organism and its needs. Rhodopsin-based phototrophy differs in that the PMF is formed directly by the light-absorbing pigment, which is a light-driven proton pump. Many phototrophs are autotrophs and must use much of the ATP and reducing power they make to fix CO_2.

Table 11.8   Properties of Chlorophyll-Based Photosynthetic Systems

Property	Eukaryotes	Cyanobacteria	Green Bacteria, Purple Bacteria, Heliobacteria, and Acidobacteria
Photosynthetic pigment	Chlorophyll a	Chlorophyll a[1]	Bacteriochlorophyll
Number of photosystems	2	2[2]	1
Photosynthetic electron donors	H_2O	H_2O	H_2, H_2S, S, organic matter
O_2 production pattern	Oxygenic	Oxygenic[3]	Anoxygenic
Primary products of energy conversion	ATP + NADPH	ATP + NADPH	ATP
Carbon source	CO_2	CO_2	Organic or CO_2

1 Members of the cyanobacterial genus *Prochlorococcus* have divinyl chlorophyll *a* and *b*.
2 A recently discovered cyanobacterium lacks photosystem II.
3 Some cyanobacteria can function anoxygenically under certain conditions. For example, *Oscillatoria* can use H_2S as an electron donor instead of H_2O.

(**figure 11.29**). The pyrrole rings are decorated with specific chemical groups that distinguish one type of chlorophyll from another. A long hydrophobic tail (the R side chain at the bottom left of figure 11.29) aids in chlorophyll's attachment to membranes, the site of the light reactions. Several chlorophylls are found in eukaryotes; the two most important are chlorophyll *a* and chlorophyll *b*. These two molecules differ slightly in their structure and

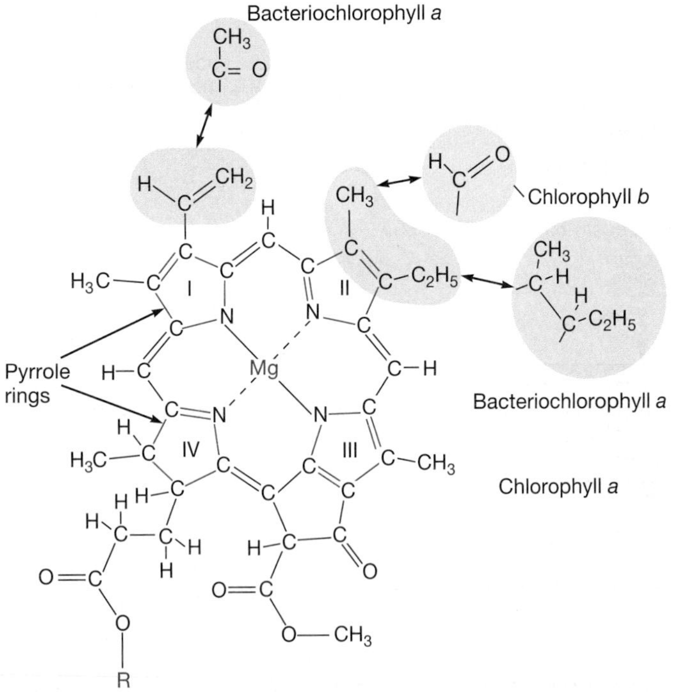

Figure 11.29 Chlorophyll Structure. The structures of chlorophyll *a*, chlorophyll *b*, and bacteriochlorophyll *a*. The complete structure of chlorophyll *a* is given. Only one group is altered to produce chlorophyll *b*, and two modifications in the ring system are required to change chlorophyll *a* to bacteriochlorophyll *a*. The side chain (R) of bacteriochlorophyll *a* may be either phytyl (a 20-carbon chain also found in chlorophylls *a* and *b*) or geranylgeranyl (a 20-carbon side chain similar to phytyl but with three more double bonds).

spectral properties. They both absorb light in the red range of the visible light spectrum (*see figure 7.16*), but chlorophyll *a* has a light absorption peak at a slightly longer wavelength than chlorophyll *b* (665 nm versus 645 nm). In addition to absorbing red light, chlorophylls also absorb blue light strongly. Because chlorophylls absorb primarily in the red and blue ranges, green light is transmitted, and these organisms appear green.

Other photosynthetic pigments are used to trap light energy. The most widespread of these are carotenoids, long molecules, usually yellowish in color, that possess an extensive conjugated double bond system (**figure 11.30**). β-Carotene is present in cyanobacteria belonging to the genus *Prochloron* and most photosynthetic protists; fucoxanthin is found in protists such as diatoms and dinoflagellates. Red algae and cyanobacteria have photosynthetic pigments called **phycobiliproteins,** consisting of a protein with a linear tetrapyrrole attached (figure 11.30). Phycoerythrin and phycocyanin are phycobiliproteins that give these microbes their characteristic colors.

Carotenoids and phycobiliproteins are often called **accessory pigments** because of their role in photosynthesis. They are important because they absorb light in the range not absorbed by chlorophylls (the blue-green through yellow range; about 470 to 630 nm). This light energy is transferred to chlorophyll. In this way, accessory pigments make photosynthesis possible over a broader range of wavelengths. In addition, this allows organisms to use light not used by other phototrophs in their habitat. Accessory pigments also protect microorganisms from intense sunlight, which could oxidize and damage the photosynthetic apparatus.

About 300 chlorophyll molecules and the accessory pigments are assembled in highly organized arrays called antennas whose purpose is to create a large surface area to trap as many photons as possible. Light energy captured in an antenna is transferred from accessory pigment to chlorophyll and from chlorophyll to chlorophyll until it reaches a **reaction-center chlorophyll pair** that is directly involved in photosynthetic electron transport (*see figure 21.11*). In oxygenic phototrophs, there are two kinds of antennas associated with two different photosystems (**figure 11.31**).

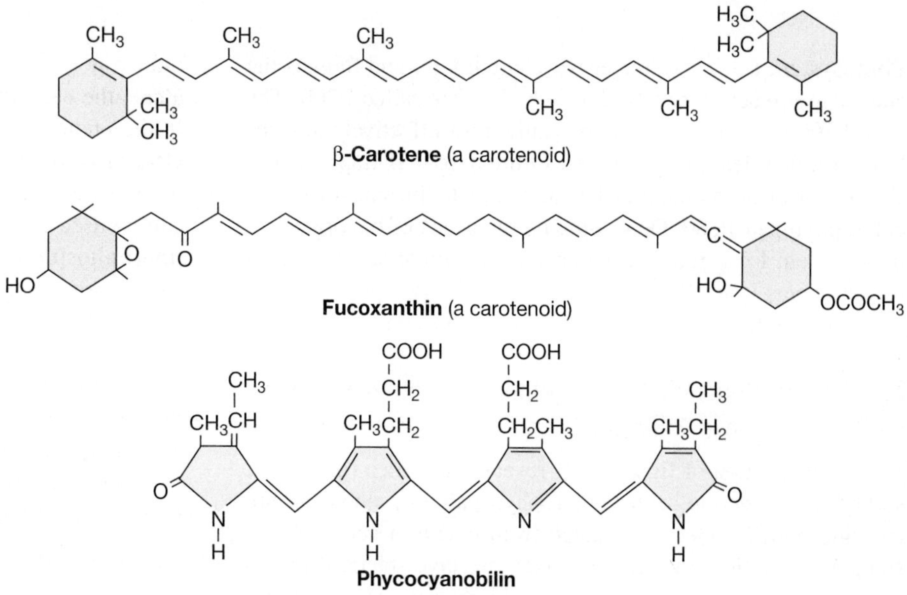

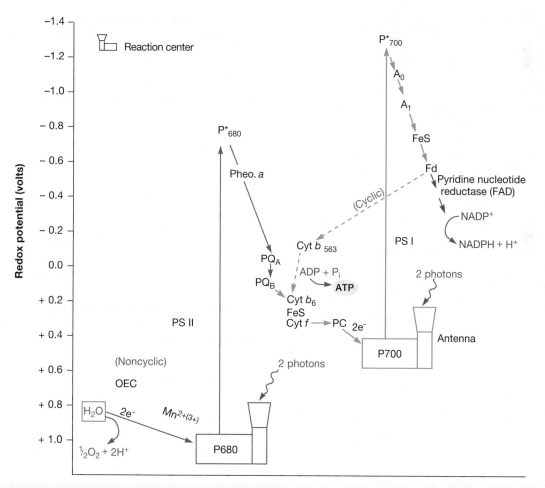

Figure 11.30 Representative Accessory Pigments. Beta-carotene is a carotenoid found in photosynthetic protists and plants. Note that it has a long chain of alternating double and single bonds called conjugated double bonds. Fucoxanthin is a carotenoid accessory pigment in several types of algae. Phycocyanobilin is a linear molecule consisting of four pyrrole rings (tetrapyrrole; see figure 11.29); it is attached to a protein to form a phycobiliprotein.

β-**Carotene** (a carotenoid)

Fucoxanthin (a carotenoid)

Phycocyanobilin

Figure 11.31 Oxygenic Photosynthesis. Cyanobacteria and eukaryotic algae are similar in having two photosystems, although they may differ in some details. The carriers involved in electron transport are ferredoxin (Fd) and other FeS proteins; cytochromes b_6, b_{563}, and f; plastiquinones A and B (PQ_A and PQ_B); copper-containing plastocyanin (PC); pheophytin a (Pheo. a); possibly a specialized form of chlorophyll (A_0); and phylloquinone (A_1). Both photosystem I (PS I) and photosystem II (PS II) are involved in noncyclic photophosphorylation; only PS I participates in cyclic photophosphorylation. The oxygen-evolving complex (OEC) that extracts electrons from water contains manganese ions that enable electron transfer to the PS II reaction center.

MICRO INQUIRY *When electrons from P700 are used to reduce NADP⁺, what compound supplies electrons for the re-reduction of P700?*

Photosystem I absorbs longer wavelength light and funnels the energy to a reaction center chlorophyll *a* pair called P700. The term P700 signifies that this molecule most effectively absorbs light at a wavelength of 700 nm. **Photosystem II** traps light at shorter wavelengths and transfers its energy to the reaction center chlorophyll pair P680. Depending on the needs of the organism, photosystem I can function independently, while at other times photosystems I and II are used together. The two photosystems working together is where we begin our discussion.

Two Photosystems Working Together Generate ATP and NADPH; Oxygen Is Released

A close inspection of figure 11.31 reveals that when light is absorbed by a reaction-center chlorophyll pair in a photosystem, the chlorophyll molecules change from having a very positive standard reduction potential to a very negative standard reduction potential. That is, the absorption of light causes the reaction-center chlorophylls to become excellent electron donors. With that in mind, let's first examine photosystem II. When it absorbs light, the energized P680 (P*$_{680}$ in figure 11.31) donates electrons to a series of electron carriers that ultimately donate the electrons to P700 of photosystem I. The flow of electrons through the P680-associated ETC generates a PMF, which can be used by ATP synthase to make ATP from ADP and P$_i$. In the meantime, photosystem I has also absorbed light and its energized reaction-center chlorophylls (P*$_{700}$ in figure 11.31) donate electrons to a short ETC that ultimately transfers electrons to NADP$^+$, yielding NADPH. Thus when both photosystems work together, both ATP and NADPH are produced. Because the electrons flow in a noncyclic pattern, this is called noncyclic electron flow, and the ATP produced is said to be made by **noncyclic photophosphorylation.**

At this point the cell has a problem. When P700 was oxidized, its electrons were replaced by electrons arising from the oxidation of P680. But if P680 is to continue to absorb light, its electrons must be replaced. Notice in figure 11.31 that the standard reduction potential of P680 is more positive than that of the ½O$_2$/H$_2$O conjugate redox pair. This means that H$_2$O can be used to donate electrons to P680, resulting in the release of oxygen. This is accomplished by a metal-containing enzyme called the oxygen-evolving complex (OEC) or the water-splitting complex, which is closely associated with photosystem II. As photosystem II absorbs light, Mn atoms in the OEC become increasingly oxidized, enabling them to extract electrons from H$_2$O and pass them to a protein in the photosystem, which then passes them to P680. In the process, protons are moved across the membrane in which the photosynthetic machinery is located. Thus, the evolution of oxygen from water also contributes to the PMF.

Photosystem I Working Alone Yields Only ATP

Now let us return our attention to photosystem I functioning alone. Notice in figure 11.31 that when electrons are donated from P*$_{700}$ to the photosystem I-associated ETC, they reach a branch point at ferredoxin (Fd). When photosystem I functions alone, the electrons follow the dashed line to an electron carrier in the photosystem II-associated ETC (cyt b_6). From there, the electrons eventually return to P700. In this case the electrons have flowed in a cyclic manner. This still generates a PMF, which can be used to synthesize ATP, and this is referred to as **cyclic photophosphorylation.** Also note that because the electrons were diverted at Fd, no electrons were available to reduce NADP$^+$ to NADPH. Thus when photosystem I functions alone, only ATP is made available to the cell.

You might be asking why an organism needs to use both photosystems at certain times, and only photosystem I at others. The answer is related to the requirements of the dark reactions. The dark reactions of oxygenic phototrophs use three ATP and two NADPH to reduce one CO$_2$ to carbohydrate (CH$_2$O).

$$CO_2 + 3ATP + 2NADPH + 2H^+ + H_2O \rightarrow$$
$$(CH_2O) + 3ADP + 3P_i + 2NADP^+$$

Oxygenic phototrophs use noncyclic and cyclic electron flow to supply these needs. The noncyclic system generates one NADPH and one ATP per pair of electrons; therefore four electrons passing through the system produce two NADPH and two ATP. A total of 8 quanta of light energy (4 quanta for each photosystem) is needed to propel the four electrons from water to NADP$^+$. Because the ratio of ATP to NADPH required for CO$_2$ fixation is 3:2, at least one more ATP must be supplied. Cyclic photophosphorylation probably operates independently to generate the extra ATP. This requires absorption of another 2 to 4 quanta. It follows that around 10 to 12 quanta of light energy are needed to reduce and incorporate one molecule of CO$_2$ during photosynthesis. ▶▶︎ *CO$_2$ fixation (section 12.3)*

It is worth reemphasizing that although light is the source of energy for chlorophyll-based phototrophy, the process used to make ATP is virtually the same as seen for chemotrophs: oxidation-reduction reactions occurring in ETCs generate a PMF that is used by ATP synthase to make ATP. Furthermore, just as is true of mitochondrial electron transport, photosynthetic electron transport takes place within a membrane. Chloroplast granal membranes contain both photosystems and their antennas. **Figure 11.32** shows a thylakoid membrane carrying out noncyclic photophosphorylation. Protons move to the thylakoid interior during photosynthetic electron transport and return to the stroma when ATP is formed. It is thought that stromal lamellae possess only photosystem I and are involved in cyclic photophosphorylation alone. In cyanobacteria, photosynthetic light reactions are located in thylakoid membranes within the cell. ↻ *Photosynthetic Electron Transport and ATP Synthesis*

Light Reactions in Anoxygenic Photosynthesis

Certain bacteria carry out a second type of photosynthesis called **anoxygenic photosynthesis.** Although the term photosynthesis is used, many of these bacteria can function as photoheterotrophs and don't always use the energy gained from light to fix CO$_2$. Anoxygenic phototrophy is a better term; it

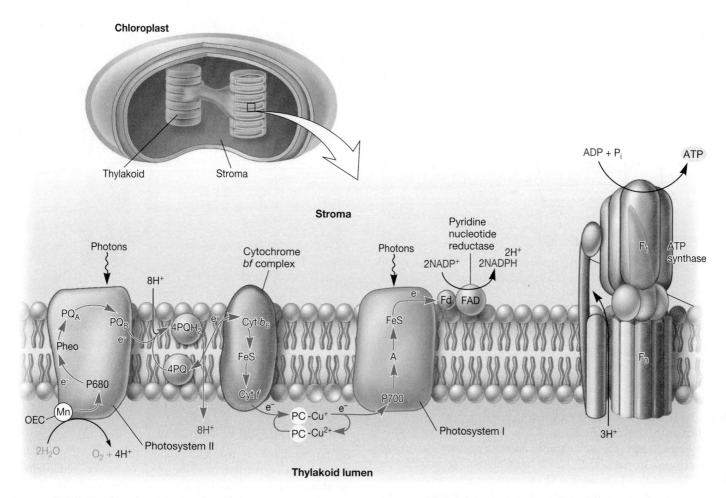

Figure 11.32 The Light Reactions of Photosynthesis. An illustration of the chloroplast thylakoid membrane showing photosynthetic ETC function and noncyclic photophosphorylation. The chain is composed of three complexes: photosystem (PS) I, the cytochrome *bf* complex, and PS II. Two diffusible electron carriers connect the three complexes. Plastoquinone (PQ) connects PS II with the cytochrome *bf* complex, and plastocyanin (PC) connects the cytochrome *bf* complex with PS I. The light-driven electron flow pumps protons across the thylakoid membrane and generates a PMF, which can then be used to make ATP. Water is the source of electrons and the oxygen-evolving complex (OEC) produces oxygen. A in PS I represents both A_0 and A_1, as shown in figure 11.31.

derives its name from the fact that molecules other than water are used as an electron source and therefore O_2 is not produced. The process also differs in terms of the pigments used, the participation of just one photosystem, and the mechanisms used to generate reducing power. Members of five bacterial phyla carry out anoxygenic phototrophy: *Proteobacteria* (purple sulfur and purple nonsulfur bacteria), *Chlorobi* (green sulfur bacteria) *Chloroflexi* (green nonsulfur bacteria), *Firmicutes* (heliobacteria), and *Acidobacteria*. The biology and ecology of these organisms are described in more detail in chapters 21, 22, and 23.

Anoxygenic phototrophs have light-absorbing pigments called **bacteriochlorophylls** (figure 11.29). Recall that bacteriochlorophylls are similar to the chlorophylls used in oxygenic photosynthesis, but differ in terms of the modifications made to the core of the pigment formed by the four pyrrole rings. In some bacteria, bacteriochlorophylls are located in membranous vesicles called chlorosomes (*see figure 21.9*).

The absorption maxima of bacteriochlorophylls are at longer wavelengths than those of chlorophylls, reaching into the infrared wavelengths. This better adapts anoxygenic phototrophic bacteria to their ecological niches. ▶▶ *Photosynthetic bacteria are diverse (section 21.4)*

The differences between anoxygenic and oxygenic phototrophs are because anoxygenic phototrophs have a single photosystem. This means that they are restricted to cyclic photophosphorylation and are unable to produce O_2 from H_2O. Indeed, almost all anoxygenic phototrophs are strict anaerobes. A tentative scheme for the phototrophic ETC of a purple nonsulfur bacterium is given in **figures 11.33** and **11.34**. When the reaction-center bacteriochlorophyll P870 is excited, it donates an electron to an ETC that transfers the electron back to P870 while generating sufficient PMF to drive ATP synthesis by ATP synthase. Note that although both green and purple bacteria lack two photosystems, the purple bacteria have a photosynthetic apparatus similar to photosystem II of oxygenic phototrophs,

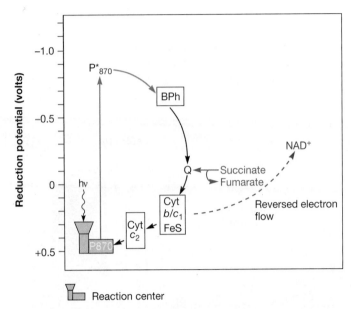

Figure 11.33 Purple Nonsulfur Bacterial Phototrophy. The ETC of the purple nonsulfur bacterium *Rhodobacter sphaeroides*. Ubiquinone (Q) is very similar to coenzyme Q. BPh stands for bacteriopheophytin. The electron source succinate is in blue.

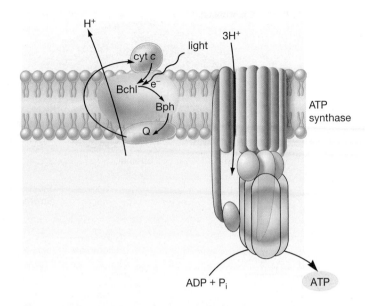

Figure 11.34 Cyclic Electron Flow During Anoxygenic Phototrophy. The photosystem of *Rhodopseudomonas viridis* is illustrated. Bchl, bacteriochlorophyll; Bph, bacteriophaeophytin; Q, quinones; and cyt *c*, cytochrome *c*.

whereas the green sulfur bacteria have a system similar to photosystem I. ▶▶ *Purple nonsulfur bacteria perform anoxygenic photosynthesis (section 22.1); Phylum Chlorobi (section 21.4)*

In addition to making ATP, anoxygenic phototrophs must also generate reducing power (NAD[P]H or reduced ferredoxin) for CO_2 fixation (photoautotrophs) and other biosynthetic processes (photoautotrophs and photoheterotrophs). They are able to generate reducing power in at least three ways, depending on the bacterium. (1) Some have hydrogenases that are used to produce NAD(P)H directly from the oxidation of hydrogen gas. This is possible because hydrogen gas has a more negative reduction potential than NAD^+ (figure 11.10). (2) Others, such as the photosynthetic purple bacteria, use reverse electron flow to generate NAD(P)H (figure 11.33). In this mechanism, electrons are drawn off the photosynthetic ETC and "pushed" to $NAD(P)^+$ using PMF. This process is similar to that seen for chemolithotrophs having inorganic energy sources with more positive reduction potentials than that of $NAD(P)^+/NAD(P)H$. (3) Phototrophic green bacteria and heliobacteria also draw off electrons from their ETCs. However, because the reduction potential of the carrier where this occurs is more negative than NAD^+, the electrons flow spontaneously to these electron acceptors. Thus these bacteria exhibit a simple form of noncyclic photosynthetic electron flow (**figure 11.35**). The diversion of electrons from the photosynthetic ETC means that continued generation of PMF will cease unless the electrons are replaced. Electrons from electron donors such as hydrogen sulfide, elemental sulfur, and organic compounds replace the electrons removed from the ETC in this way. ◔ *Cyclic and Noncyclic Photophosphorylation*

Rhodopsin-Based Phototrophy

So far, we have discussed chlorophyll-based types of phototrophy; that is, the use of chlorophyll or bacteriochlorophyll to absorb light and initiate the conversion of light energy to chemical energy. Within the last decade, it has become apparent that many bacteria and archaea are capable of

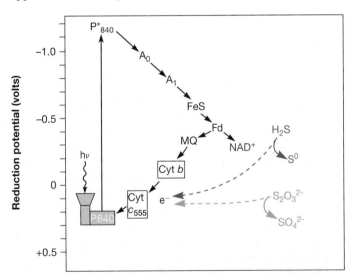

Figure 11.35 Green Sulfur Bacterial Light Reactions. Light energy is used to make ATP by cyclic photophosphorylation and to move electrons from thiosulfate ($S_2O_3^{2-}$) and H_2S (green and blue) to NAD^+. The ETC consists of A_0 (bacteriochlorophyll 663); A_1, a quinone-like molecule; FeS (iron sulfur molecules within the reaction center); Fd (ferredoxin); MQ (menaquinone) and two different cytochromes (Cyt).

chlorophyll-independent phototrophy. These microbes rely on a form of microbial rhodopsin, a molecule similar to that found in the eyes of many multicellular organisms. The first microbial rhodopsin was discovered several decades ago in certain archaea. In fact, it was first described so long ago that it was called bacteriorhodopsin; we now know that it is more correctly called **archaerhodopsin.** While a variety of rhodopsins have since been found in members of *Proteobacteria* (and thus called proteorhodospin), *Flavobacteria,* extremely halophilic bacteria (xanthorhodopsin), and some eukaryotes, the archaerhodopsin of the halophilic archaeon *Halobacterium salinarum* remains the best studied and is the focus of our discussion here. ▶▶◀ *Haloarchaea (section 20.4); Microorganisms in the open ocean are adapted to nutrient limitation (section 30.2)*

H. salinarum normally depends on aerobic respiration for the release of energy from an organic energy source. It cannot grow anaerobically by anaerobic respiration or fermentation. However, under conditions of low oxygen and high light intensity, it synthesizes archaerhodopsin, a deep-purple pigment. Archaerhodopsin's chromophore is retinal, a type of carotenoid. The chromophore is covalently attached to the pigment protein, which is embedded in the plasma membrane in such a way that retinal is in the center of the membrane.

Archaerhodopsin functions as a light-driven proton pump. When retinal absorbs light, a proton is released and the archaerhodopsin undergoes a sequence of conformation changes that translocate the proton into the periplasmic space (*see figure 20.18*). The light-driven proton pumping generates a pH gradient that can be used to power the synthesis of ATP by chemiosmosis. This phototrophic capacity is particularly useful to *H. salinarum* because oxygen is not very soluble in concentrated salt solutions and may decrease to an extremely low level in its habitat. When the surroundings become temporarily anoxic, the archaeon uses light energy to synthesize sufficient ATP to survive until oxygen levels rise again. Note that this type of phototrophy does not involve electron transport.

Retrieve, Infer, Apply

1. Define the following terms: light reactions, dark reactions, chlorophyll, carotenoid, phycobiliprotein, antenna, and photosystems I and II.
2. What happens to a reaction center chlorophyll pair, such as P700, when it absorbs light?
3. What is the function of accessory pigments?
4. What is photophosphorylation? What is the difference between cyclic and noncyclic photophosphorylation?
5. Compare and contrast anoxygenic phototrophy and oxygenic photosynthesis. How do these two types of phototrophy differ from rhodopsin-based phototrophy?
6. Suppose you isolated a bacterial strain that carried out oxygenic photosynthesis. What photosystems would it possess, and what group of bacteria would it most likely belong to?

Key Concepts

11.1 Metabolic Diversity and Nutritional Types

- All organisms require a source of carbon, energy, and electrons. Biologists have devised terms to describe how these needs are fulfilled (**table 11.1**).
- Heterotrophs use reduced organic molecules as their source of carbon. Autotrophs use CO_2 as their primary or sole carbon source.
- Phototrophs use light energy, and chemotrophs obtain energy from the oxidation of chemical compounds.
- Electrons are extracted from reduced inorganic substances by lithotrophs and from reduced organic compounds by organotrophs.
- Most organisms can be placed into one of five nutritional types: photolithoautotroph, photoorganoheterotroph, chemolithoautotroph, chemolithoheterotroph, and chemoorganoheterotroph (**table 11.2**).

11.2 There Are Three Chemoorganotrophic Fueling Processes

- Chemoorganotrophic microorganisms can use three kinds of electron acceptors during energy metabolism (**figure 11.2**). Electrons from the oxidized nutrient can be accepted by an endogenous electron acceptor (fermentation), by oxygen (aerobic respiration), or by another external electron acceptor (anaerobic respiration).
- Chemoorganotrophs can use a variety of molecules as their energy, electron, and carbon source. These molecules are degraded by pathways that funnel into pathways used for glucose catabolism (**figure 11.3**).
- Some of the pathways used by chemoorganotrophs are amphibolic, having both catabolic and anabolic functions (**figure 11.4**).

11.3 Aerobic Respiration Can Be Divided into Three Steps

- Aerobic respiration of glucose begins with glycolytic pathways, which catabolize glucose to pyruvate. Pyruvate is fed into the TCA cycle for completion of catabolism.
- The glycolytic pathways and the TCA cycle generate numerous NADH and $FADH_2$ molecules. These are oxidized by an electron transport chain, using oxygen as the terminal electron acceptor. Electron flow generates a proton motive force (PMF), which is used to synthesize ATP by oxidative phosphorylation.

11.4 Glucose to Pyruvate: The First Step

- Glycolysis, used in its broadest sense, refers to all pathways used to break down glucose to pyruvate.
- The Embden-Meyerhof pathway has a net production of two NADH and two ATP, the latter being produced by substrate-level phosphorylation. It also produces several precursor metabolites (**figure 11.5**).
- In the Entner-Doudoroff pathway, glucose is catabolized to pyruvate and glyceraldehyde 3-phosphate (**figure 11.6**). The latter product can be oxidized by enzymes of the Embden-Meyerhof pathway to provide ATP, NADH, and another molecule of pyruvate.
- In the pentose phosphate pathway, glucose 6-phosphate is oxidized twice and converted to pentoses and other sugars. It is a source of NADPH, ATP, and several precursor metabolites (**figure 11.7**).

11.5 Pyruvate to Carbon Dioxide (Step 2) Is Accomplished by the Tricarboxylic Acid Cycle

- Pyruvate from the glycolytic pathways is fed into the tricarboxylic acid cycle by a reaction that converts pyruvate to acetyl-CoA. In the process, one of pyruvate's carbons is released in the form of carbon dioxide.
- The tricarboxylic acid cycle oxidizes acetyl-CoA to CO_2 and forms one GTP or ATP, three NADH, and one $FADH_2$ per acetyl-CoA (**figure 11.8**). It also generates several precursor metabolites.

11.6 Electron Transport and Oxidative Phosphorylation (Step 3) Generate the Most ATP

- The NADH and $FADH_2$ produced from the oxidation of reduced organic molecules can be oxidized by an electron transport chain (ETC). Electrons flow from carriers with more negative reduction potentials to those with more positive potentials (**figures 11.9–11.11**). As electrons flow through an ETC, protons are transported across the membrane, generating proton motive force (PMF). PMF is used for ATP synthesis by oxidative phosphorylation.
- Bacterial and archaeal ETCs are often different from eukaryotic chains with respect to such aspects as carriers and branching (**figures 11.12** and **11.13**).
- ATP synthase catalyzes the synthesis of ATP (**figure 11.16**). In eukaryotes, it is located on the inner surface of the inner mitochondrial membrane. Bacterial and archaeal ATP synthases are on the inner surface of the plasma membrane. ATP synthase uses the PMF for ATP synthesis (**figure 11.11**).
- In eukaryotes, the P/O ratio for NADH is about 2.5 and that for $FADH_2$ is around 1.5; P/O ratios are usually much lower in bacterial and archaeal chains. Aerobic respiration in eukaryotes can theoretically yield a maximum of 32 ATP (**figure 11.17**).

11.7 Anaerobic Respiration Uses the Same Three Steps as Aerobic Respiration

- Anaerobic respiration uses an exogenous molecule other than O_2 as the terminal electron acceptor for electron transport. The most common acceptors are nitrate, sulfate, and CO_2 (**figure 11.18** and **table 11.3**).
- For some microorganisms, the same pathways used to aerobically respire an organic energy source are also used for anaerobic respiration. However, less energy is provided by anaerobic respiration than aerobic respiration because the alternate electron acceptors have reduction potentials that are less positive than the reduction potential of oxygen.

11.8 Fermentation Does Not Involve an Electron Transport Chain

- During fermentation, an endogenous electron acceptor is used to reoxidize any NADH generated by the catabolism of glucose to pyruvate (**figure 11.19**). Flow of electrons from the electron donor to the electron acceptor does not involve an ETC, and in most organisms, ATP is synthesized only by substrate-level phosphorylation.
- There are many different fermentation pathways. These are of practical importance in clinical and industrial settings (**figure 11.20**).

11.9 Catabolism of Organic Molecules Other Than Glucose

- Microorganisms catabolize many extracellular carbohydrates. Monosaccharides are taken in and phosphorylated; disaccharides may be cleaved to monosaccharides by either hydrolysis or phosphorolysis. External polysaccharides are degraded by hydrolysis and the products are absorbed. Intracellular glycogen and starch are converted to glucose 1-phosphate by phosphorolysis (**figure 11.21**).
- Triglycerides are hydrolyzed to glycerol and fatty acids by enzymes called lipases. Fatty acids are usually oxidized to acetyl-CoA in the β-oxidation pathway (**figure 11.23**).
- Proteins are hydrolyzed to amino acids that are then deaminated (**figure 11.24**). The carbon skeletons produced by deamination can be fermented or fed into the TCA cycle (**figure 11.3**).

11.10 Chemolithotrophy: "Eating Rocks"

- Chemolithotrophs synthesize ATP by oxidizing inorganic compounds—usually hydrogen, reduced nitrogen and sulfur compounds, or ferrous iron—with an ETC. O_2 is the usual electron acceptor (**figure 11.25** and **table 11.5**). The PMF produced is used by ATP synthase to make ATP.
- Many of the energy sources used by chemolithotrophs have a more positive standard reduction potential than the NAD^+/NADH conjugate redox pair. These chemolithotrophs must expend energy (PMF) to drive reverse electron flow and produce the NADH they need for CO_2 fixation and other processes (**figure 11.27**).

11.11 Phototrophy

- In oxygenic photosynthesis, eukaryotes and cyanobacteria trap light energy with chlorophyll and accessory pigments, and move electrons through photosystems I and II to make ATP and NADPH (the light reactions). The ATP and NADPH are used in the dark reactions to fix CO_2.

- Cyclic photophosphorylation involves the activity of photosystem I alone and generates ATP only. In noncyclic photophosphorylation, photosystems I and II operate together to move electrons from water to $NADP^+$, producing ATP, NADPH, and O_2 (**figure 11.31**). In both cases, electron flow generates PMF, which is used by ATP synthase to make ATP.

- Anoxygenic phototrophs differ from oxygenic phototrophs in possessing bacteriochlorophyll and having only one photosystem (**figures 11.33–11.35**). Cyclic electron flow generates a PMF, which is used by ATP synthase to make ATP (i.e., cyclic photophosphorylation). They are anoxygenic because they use molecules other than water as an electron donor for electron flow and the production of reducing power.

- Some bacteria and archaea use a type of phototrophy that involves a rhodopsin, a proton-pumping pigment. This type of phototrophy generates PMF but does not involve an ETC.

Compare, Hypothesize, Invent

1. Without looking in chapters 21 and 22, predict some characteristics that would describe niches occupied by green and purple photosynthetic bacteria.

2. From an evolutionary perspective, discuss why most microorganisms use aerobic respiration to generate ATP.

3. How would you isolate a thermophilic chemolithotroph that uses sulfur compounds as a source of energy and electrons? What changes in the incubation system would be needed to isolate bacteria using sulfur compounds in anaerobic respiration? How would you tell which process is taking place through an analysis of the sulfur molecules present in the spent medium?

4. Certain chemicals block ATP synthesis by allowing protons and other ions to "leak across membranes," disrupting the charge and proton gradients established by electron flow through an ETC. Does this observation support the chemiosmosis hypothesis? Explain your reasoning.

5. Two flasks of *E. coli* are grown in batch culture in the same medium (2% glucose and amino acids; no nitrate) and at the same temperature (37°C). Culture #1 is well aerated. Culture #2 is anoxic. After 16 hours the following observations are made:

   - Culture #1 has a high cell density; the cells appear to be in stationary phase, and the glucose level in the medium is reduced to 1.2%.

   - Culture #2 has a low cell density; the cells appear to be in logarithmic phase, although their doubling time is prolonged (over 1 hour). The glucose level is reduced to 0.2%.

   Why does culture #2 have so little glucose remaining relative to culture #1, even though culture #2 displayed slower growth and has less biomass?

6. A cyanobacterium having photosystem I but not photosystem II has been discovered. It is unable to fix CO_2. How do you think this microbe makes ATP? Reducing power? Why is it to this microbe's advantage to be a heterotroph?

7. Review the description of the Berkeley Pit Lake in this chapter's opening case study. In addition to heavy metals, the water has a high concentration of sulfate. Predict the nutritional types that might be isolated from the lake. What might they use as energy and electron sources? What might any chemotrophs in the lake use as terminal electron acceptors for their ETCs?

8. The archaeon *Metallosphaera sedula* is of great interest to microbiologists in the field of biomining because it might be used to recover base and precious metals. This aerobic thermoacidophile can use organic carbon, ferrous iron (Fe^{2+}), and reduced inorganic sulfur compounds including elemental sulfur (S^0) and tetrathionate ($S_4O_6^{2-}$) as a source of electrons. Thus *M. sedula* can grow chemoorganotrophically and chemolithotrophically. Genome sequencing reveals that *M. sedula* has five major terminal oxidase complexes. Refer to table 10.2 and figure 11.10 to propose the flow of electrons through the electron transport chain when using each of these compounds as the electron source. Which will conserve the most energy and which the least? Why do you think this microbe requires five different terminal oxidases if it always uses O_2 as the terminal electron acceptor?

   Read the original paper: Auernik, K. S., and Kelly, R. M. 2008. Identification of components of electron transport chains in the extremely thermoacidophilic crenarchaeon *Metallosphaera sedula* through iron and sulfur oxidation transcriptomes. *App. Environ. Microbiol.* 74:7723.

9. Nitrite-oxidizing bacteria have been thought to be restricted to using nitrite (NO_2^-) as an energy and electron source. However, the bacterium *Nitrospira moscoviensis* recently was shown to be capable of oxidizing hydrogen (H_2) and using the energy released for growth. In fact, it can carry out both H_2 and NO_2^- oxidation simultaneously. The authors of the study suggest that simultaneous H_2 and NO_2^- might be particularly useful in low oxygen environments. Explain why this would be so.

   Read the original paper: Koch, H., et al. 2014. Growth of nitrite-oxidizing bacteria by aerobic hydrogen oxidation, *Science* 345:1052.

12

Anabolism: The Use of Energy in Biosynthesis

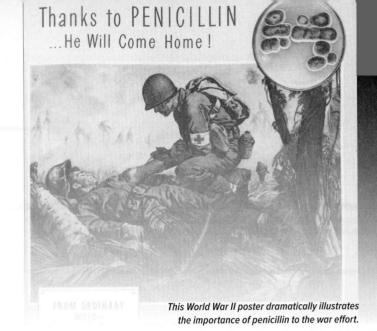

Thanks to PENICILLIN
...He Will Come Home!

This World War II poster dramatically illustrates the importance of penicillin to the war effort.

An Author's Life Saved

When one of the authors of this text (LS) was an infant, she developed pneumonia. If she had been born a few years earlier, this story would not be told, and a different author would be writing this textbook. Her life was saved by antibiotics.

The story of Alexander Fleming's rediscovery of penicillin in 1929 is well known. However, Fleming had little to do with developing penicillin as a treatment for infectious disease because after some initial experiments, he abandoned the study of its activity. Fortunately, Ernst Chain developed an interest in the chemical and eventually purified it. In 1940, as many nations battled each other in World War II, Howard Florey and his assistants began clinical trials with penicillin. By that time, Selman Waksman had discovered several other antibacterial agents and had coined the term antibiotic. The penicillin trials were successful, and the need for antibiotics during World War II to treat wound infections was the impetus for finding and developing more antibiotics for clinical use. A few years later a ready supply was available and an author's life was saved. ▶▶▌ *Antimicrobial chemotherapy evolved from antisepsis efforts (section 9.1)*

Antibiotics are interesting molecules for several reasons. One is that they are made at all. Why would a microbe expend materials and energy to make an antibiotic? What selective advantage do they provide? How do they contribute to the ecology of the microbe? Do they serve a purpose other than killing other microbes (*see chapter 24 opening story*)? These questions are still not fully answered. Another reason these molecules are interesting is that they are structurally complex. Despite this, they are built using the same precursor metabolites and many of the same biochemical processes used in the anabolic pathways that sustain life.

During anabolism, an organism begins with simple inorganic molecules and a carbon source, and constructs ever more complex molecules until new organelles and cells arise (**figure 12.1**). Thus anabolism is the creation of order. Because a cell is highly ordered and immensely complex, much energy is required for biosynthesis. This is readily apparent from estimates of the biosynthetic capacity of rapidly growing *Escherichia*

coli cells (**table 12.1**). Furthermore, even nongrowing cells need energy for the biosynthetic processes they carry out. This is because nongrowing cells continuously degrade and resynthesize cellular molecules during a process known as turnover. Thus cells are never the same from one instant to the next. Although there is considerably less diversity in anabolic processes as compared to catabolic processes, anabolism is amazing in its own right. Using just 12 precursor metabolites, a cell is able to manufacture the myriad of molecules from which it is constructed. In this chapter, we discuss the synthesis of some of the most important types of cell constituents.

Readiness Check:

Based on what you have learned previously, you should be able to:

✔ Distinguish metabolism, catabolism, and anabolism
✔ List the products of the fueling reactions (section 11.1)
✔ Describe how carbon, hydrogen, oxygen, nitrogen, phosphorus, and sulfur are used by cells
✔ Differentiate between heterotrophy and autotrophy (section 11.1)
✔ Describe the components of a biochemical pathway and how they are organized (section 10.5)
✔ Discuss the function of ATP and how that function is related to its phosphate transfer potential (section 10.2)
✔ State the function of enzymes and ribozymes (section 10.6)

12.1 Principles Governing Biosynthesis

After reading this section, you should be able to:

■ Describe in general terms the steps organisms use to convert a carbon source and inorganic molecules to cells
■ Discuss the principles that govern biosynthesis

The problem faced by all cells is how to make the many molecules they need as efficiently as possible. They have solved this problem by carrying out biosynthesis using a few basic principles. These are now briefly discussed.

1. **Large molecules are made from small molecules.** The construction of large **macromolecules** (complex molecules) from a few simple structural units (monomers) linked together by a single type of covalent bond saves much genetic storage capacity, biosynthetic raw material, and energy, making their synthesis highly efficient.

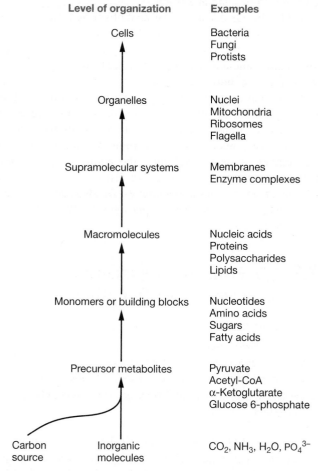

2. **Many enzymes do double duty. However, some enzymes function in one direction only.** As discussed in chapter 11, many enzymes are used for both catabolic and anabolic processes in pathways that are said to be amphibolic (*see figure 11.4*), saving additional materials and energy. Although many steps of amphibolic pathways are catalyzed by enzymes that act reversibly, some are not. These steps require the use of separate enzymes: one to catalyze the catabolic reaction, the other to catalyze the anabolic reaction. The use of two enzymes allows independent regulation of catabolism and anabolism. Thus catabolic and anabolic pathways are never identical, even though they share many enzymes. Although both types of pathways can be regulated by their end products as well as by the concentrations of ATP, ADP, AMP, and $NAD^+/NADH$, end-product regulation generally assumes more importance in anabolic pathways.

3. **Anabolism consumes energy.** Many reactions in anabolic pathways are endergonic and will not proceed in the direction of biosynthesis. Cells solve this problem by connecting endergonic biosynthetic reactions to the hydrolysis of ATP and other nucleoside triphosphates. When these two processes are coupled, the free energy made available during nucleoside triphosphate breakdown drives the biosynthetic reaction to completion. ◄◄ *ATP (section 10.2)*

4. **Catabolism and anabolism can be physically separated.** Catabolic and anabolic pathways can be localized into distinct cellular compartments—a process called compartmentation. In eukaryotes, this is easily done because eukaryotic cells have numerous membrane-bound organelles that carry out specific functions. Compartmentation also occurs in bacterial and archaeal cells. For instance, carboxysomes separate CO_2 fixation from other processes in bacteria. Compartmentation makes it easier for catabolic and anabolic pathways to operate simultaneously yet independently.

5. **Catabolism and anabolism often use different cofactors.** Usually catabolic oxidations produce NADH, a substrate for electron transport. In contrast, when an electron donor is needed during biosynthesis, NADPH often serves as the donor.

Figure 12.1 The Construction of Cells. The biosynthesis of cells and their constituents is organized in levels of ever greater complexity.

Table 12.1	Biosynthesis in *Escherichia coli*		
Cell Constituent	**Number of Molecules per Cell[1]**	**Molecules Synthesized per Second**	**Molecules of ATP Required per Second for Synthesis**
DNA	1[2]	0.00083	60,000
RNA	15,000	12.5	75,000
Polysaccharides	39,000	32.5	65,000
Lipids	15,000,000	12,500.0	87,000
Proteins	1,700,000	1,400.0	2,120,000

1 Estimates for a cell with a volume of 2.25 µm³, a total weight of 1×10^{-12} g, a dry weight of 2.5×10^{-13} g, and a 20-minute cell division cycle.
2 It should be noted that bacteria can contain multiple copies of their genomic DNA.

After macromolecules have been constructed from simpler precursors, they are assembled into larger, more complex structures such as supramolecular systems and organelles by a process called **self-assembly** (figure 12.1). This is possible because many macromolecules contain the necessary information to build multimolecular complexes without enzymatic help.

Retrieve, Infer, Apply

1. Summarize the principles by which biosynthetic pathways are organized.
2. What is self-assembly? What biological entity also exhibits self-assembly?

12.2 Precursor Metabolites: Starting Molecules for Biosynthesis

After reading this section, you should be able to:

- List the central metabolic pathways, noting which precursor metabolites are generated by each pathway
- Draw a simple diagram that lists all the precursor metabolites and illustrates how they are used in biosynthesis

Generation of **precursor metabolites** is critical to anabolism because they give rise to all other molecules made by the cell. Precursor metabolites are carbon skeletons (i.e., carbon chains) used as the starting substrates for the synthesis of monomers and other building blocks needed to make macromolecules (*see figure 11.1*). During biosynthesis, functional moieties such as amino and sulfhydryl groups are added to the carbon chains. The precursor metabolites and how they are used in biosynthesis are shown in **figures 12.2** and **12.3**. Several things should be noted in figure 12.2. First, all the precursor metabolites are intermediates of the glycolytic pathways (Embden-Meyerhof, Entner-Doudoroff, and the pentose phosphate pathways) and the tricarboxylic acid (TCA) cycle. Therefore these pathways play a central role in metabolism and are often referred to as the **central metabolic pathways.** Note, too, that most of the precursor metabolites are used for synthesis of amino acids and nucleotides. ◀◀ *Glucose to pyruvate (section 11.4); Pyruvate to carbon dioxide (step 2) is accomplished by the tricarboxylic acid cycle (section 11.5)* ▶▶ *Common metabolic pathways (appendix II)*

If an organism is a chemoorganotroph using glucose as its energy, electron, and carbon source (either aerobically or anaerobically), it generates the precursor metabolites as it generates ATP and reducing power. But what if the chemoorganotroph is using an amino acid as its sole source of carbon, electrons, and energy? And what about autotrophs? How do they generate precursor metabolites from CO_2, their carbon source? Heterotrophs growing on something other than glucose convert that carbon source into one or more intermediates of the central metabolic pathways. From there, they can generate the remaining precursor metabolites. Autotrophs must first convert CO_2 into reduced organic carbon from which they can generate the precursor metabolites. Many of the reactions that autotrophs use to synthesize the precursor metabolites are reactions of the central metabolic pathways, operating in either the catabolic direction or the anabolic direction. Thus the central metabolic pathways are important to the anabolism of both heterotrophs and autotrophs.

We begin our discussion of anabolism by first considering CO_2 fixation by autotrophs. Once CO_2 is converted to organic carbon, the synthesis of other precursor metabolites, amino acids, nucleotides, and additional building blocks is very similar in both autotrophs and heterotrophs. Recall that the precursor metabolites provide the carbon skeletons for the synthesis of other important organic molecules. In the process of transforming a precursor metabolite into an amino acid or a nucleotide, the carbon skeleton is modified in a number of ways, including the addition of nitrogen, phosphorus, and sulfur. Thus as we discuss the synthesis of monomers from precursor metabolites, we also address the assimilation of nitrogen, sulfur, and phosphorus. ♻ *A Biochemical Pathway*

12.3 CO₂ Fixation: Reduction and Assimilation of CO₂ Carbon

After reading this section, you should be able to:

- List the pathways used by microbes to fix CO_2, noting the types of organisms (eukaryotes, bacteria, archaea) that use each pathway
- Describe in general terms the three phases of the Calvin-Benson cycle
- Identify the steps in the Calvin-Benson cycle that consume ATP and NADPH

Autotrophs use CO_2 as their sole or principal carbon source, and the reduction and incorporation of CO_2 requires much energy. Photoautotrophs obtain energy by trapping light during the light reactions of photosynthesis, and chemolithoautotrophs derive energy from the oxidation of inorganic electron donors. Autotrophic CO_2 fixation is crucial to life on Earth because it provides the organic matter on which heterotrophs depend. ◀◀ *Chemolithotrophy (section 11.10); Phototrophy (section 11.11)* ▶▶ *Carbon cycle (section 28.1)*

Six different CO_2-fixation pathways have been identified in microorganisms. Most autotrophs use the **Calvin-Benson cycle,** which is also called the Calvin cycle. The Calvin-Benson cycle is found in photosynthetic eukaryotes and most photosynthetic bacteria. Other pathways are used by some obligatory anaerobic and microaerophilic bacteria and archaea. We consider the Calvin-Benson cycle first and then briefly introduce the other CO_2-fixation pathways.

Calvin-Benson Cycle

The Calvin-Benson cycle is also called the reductive pentose phosphate cycle because it is essentially the reverse of the pentose

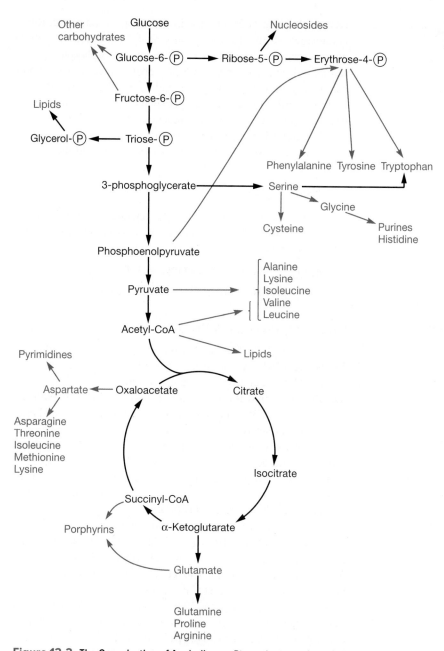

Figure 12.2 The Organization of Anabolism. Biosynthetic products (in blue) are derived from precursor metabolites. Porphyrin synthesis in non-plant eukaryotes and alphabacteria begins with succinyl-CoA. For simplicity, the role of glycine in the succinyl-CoA pathway is not illustrated. In plants and all other bacteria, porphyrin synthesis begins with glutamate (section 12.5).

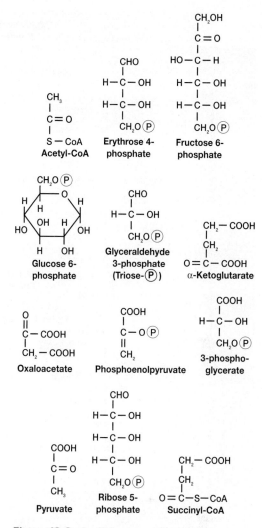

Figure 12.3 The 12 Precursor Metabolites. In chemoorganoheterotrophs, these are produced by the central metabolic pathways. Autotrophs use CO₂-fixation pathways and other pathways (e.g., gluconeogenesis) to make precursor metabolites.

phosphate pathway. Thus many of the reactions are similar, in particular the sugar transformations. The reactions of the Calvin-Benson cycle occur in the chloroplast stroma of eukaryotic autotrophs. In cyanobacteria, some nitrifying bacteria, and thiobacilli (sulfur-oxidizing chemolithotrophs), the cycle is associated with inclusions called **carboxysomes.** These polyhedral structures contain the enzyme critical to the Calvin-Benson cycle and are the site of CO₂ fixation. ◀◀ *Pentose phosphate pathway (section 11.4); Inclusions (section 3.6)*

The Calvin-Benson cycle is divided into three phases: carboxylation phase, reduction phase, and regeneration phase (**figure 12.4** and *appendix II*). During the carboxylation phase, the enzyme **ribulose 1,5-bisphosphate carboxylase/oxygenase** (RuBisCO) catalyzes the addition of CO₂ to the 5-carbon molecule ribulose 1,5-bisphosphate (RuBP), forming a 6-carbon intermediate that rapidly and spontaneously splits into two molecules of 3-phosphoglycerate (PGA). PGA is an intermediate of the Embden-Meyerhof pathway (EMP), and in the reduction

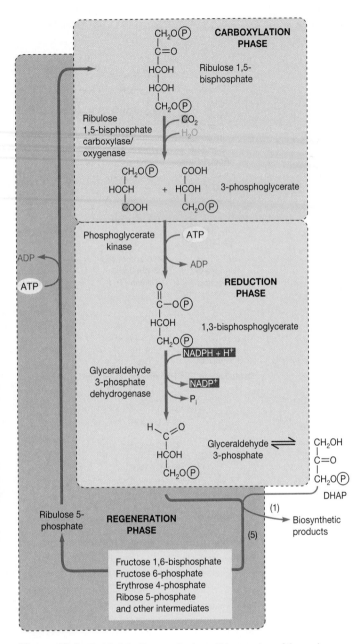

Figure 12.4 The Calvin-Benson Cycle. This overview of the cycle shows only the carboxylation and reduction phases in detail. Three ribulose 1,5-bisphosphates are carboxylated (the incorporated carbon atom is shaded in pink) to give six 3-phosphoglycerates in the carboxylation phase. These are converted to six glyceraldehyde 3-phosphates, which can be converted to dihydroxyacetone phosphate (DHAP). Five of the six trioses (glyceraldehyde phosphate and dihydroxyacetone phosphate) are used to reform three ribulose 1,5-bisphosphates in the regeneration phase. The remaining triose is used in biosynthesis. The numbers in parentheses at the lower right indicate this carbon flow.

phase of the Calvin-Benson cycle, PGA is reduced to glyceraldehyde 3-phosphate by two reactions that reverse two EMP reactions. The EMP reactions differ from the Calvin-Benson cycle reactions in that the Calvin cycle enzyme glyceraldehyde 3-phosphate dehydrogenase uses NADPH rather than NADH (compare

figures 12.4 and 11.5). Finally, in the regeneration phase, RuBP is reformed, so that the cycle can repeat. In addition, this phase produces carbohydrates such as glyceraldehyde 3-phosphate, fructose 6-phosphate, and glucose 6-phosphate, all of which are precursor metabolites (figures 12.2 and 12.3). This portion of the cycle is similar to the pentose phosphate pathway and involves transketolase and transaldolase reactions. 🐾 *The Calvin Cycle*

To synthesize fructose 6-phosphate or glucose 6-phosphate from CO_2, the cycle must operate six times to yield the desired hexose and reform the six RuBP molecules.

$$6RuBP + 6CO_2 \rightarrow 12PGA \rightarrow 6RuBP + \text{fructose-6-P}$$

The incorporation of one CO_2 into organic material requires three ATP and two NADPH. The formation of glucose from CO_2 may be summarized by the following equation:

$$6CO_2 + 18ATP + 12NADPH + 12H^+ + 12H_2O \rightarrow$$
$$\text{glucose} + 18ADP + 18P_i + 12NADP^+$$

Other CO₂-Fixation Pathways

The reactions of the Calvin-Benson cycle were determined in the 1940s and 1950s. At the time, it was thought that this pathway was used by all autotrophic organisms. This was shown not to be the case in the mid-1960s, when a second CO_2-fixation pathway was discovered. Over the years, and at an increasingly rapid rate, four more pathways were discovered. Many microbiologists think that there are still others. Why so many pathways have evolved is still an unanswered question. However, evidence suggests that certain pathways correlate with certain lifestyles. In other words, the ecological niche occupied by a microbe has shaped the evolution of its CO_2-fixation pathway.

The first alternative CO_2-fixation pathway discovered was the **reductive TCA cycle** (figure 12.5), which is used by some autotrophs in the bacterial phyla *Aquificae, Proteobacteria, Nitrospirae,* and *Chlorobi.* The reductive TCA cycle is so named because it runs in the reverse direction of the normal, oxidative TCA cycle (compare figures 12.5 and 11.8). ▶▶| Aquificae *and* Thermotogae *are ancient bacterial lineages (section 21.1); Phylum* Chlorobi *(section 21.4); Class* Epsilonproteobacteria *ranges from pathogens to deep-sea bacteria (section 22.5)*

The remaining CO_2-fixation pathways are used by members of the bacterial phylum *Chloroflexi* (3-hydroxypropionate bicycle; *see figure 21.10*): the reductive acetyl-CoA pathway (*see figure 20.3*) is used by some methanogens; the 3-hydroxypropionate/4-hydroxybutyrate pathway (*see figure 20.4*) is used by members of the order *Sulfolobales;* and the dicarboxylate/4-hydroxybutyrate cycle (*see figure 20.4*) was discovered in members of orders *Thermoproteales* and *Desulfurococcales.* These pathways are discussed in chapters 20 and 21.

Retrieve, Infer, Apply

1. Briefly describe the three phases of the Calvin-Benson cycle. What other pathways are used to fix CO_2?

2. Which two enzymes are specific to the Calvin cycle?

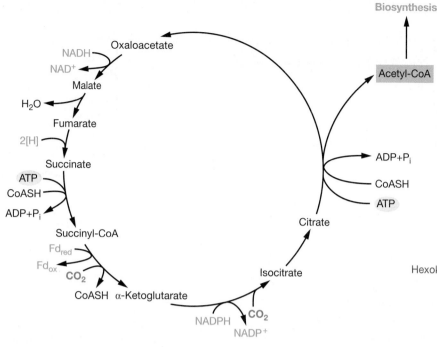

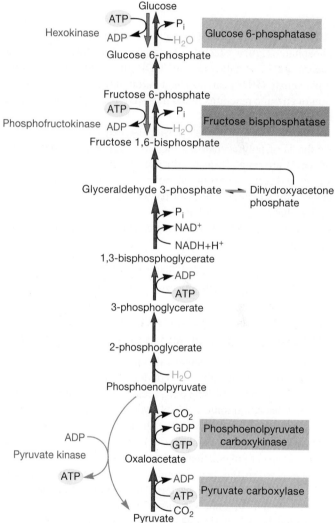

Figure 12.5 The Reductive TCA Cycle. This cycle is used by green sulfur bacteria and some chemolithotrophic bacteria to fix CO_2. The cycle runs in the opposite direction as the TCA cycle. ATP and reducing power (NADH, NADPH, and [H]) drive the reversal. The product of the pathway is acetyl-CoA.

MICRO INQUIRY *How might acetyl-CoA be used by the cell?*

12.4 Synthesis of Carbohydrates

After reading this section, you should be able to:

- Compare and contrast gluconeogenesis and the Embden-Meyerhof pathway
- Describe the role of ATP and UTP in the synthesis of monosaccharides (other than glucose) and polysaccharides
- Outline the major steps in peptidoglycan synthesis
- Evaluate the effectiveness of targeting antibiotics to peptidoglycan synthesis

Many microbes need to synthesize glucose for anabolic purposes. The synthesis of glucose from noncarbohydrate precursors is called **gluconeogenesis.** The gluconeogenic pathway shares six enzymes with the Embden-Meyerhof pathway. However, the two pathways are not identical (**figure 12.6**). Four reactions are catalyzed by enzymes that are specific for gluconeogenesis. Two of these enzymes are involved in the conversion of pyruvate to phosphoenolpyruvate. The third enzyme catalyzes the formation of fructose 6-phosphate from fructose 1,6-bisphosphate, and the fourth removes the phosphate from glucose 6-phosphate to generate glucose.

Synthesis of Monosaccharides and Polysaccharides Often Involves Nucleoside Diphosphate Carriers

As can be seen in figure 12.6, gluconeogenesis synthesizes fructose 6-phosphate, glucose 6-phosphate, and glucose. Once these sugars have been formed, other common sugars can be manufactured. For example, mannose comes directly from fructose 6-phosphate by a simple rearrangement of a hydroxyl group (*see figure A1.9*). Several sugars are synthesized while attached to a **nucleoside diphosphate.** One of the most important nucleoside diphosphate sugars is uridine diphosphate glucose

Figure 12.6 Gluconeogenesis Pathway Used by Many Microorganisms. The names of the four enzymes catalyzing irreversible reactions that are different from those found in the Embden-Meyerhof pathway (EMP) are in shaded boxes. EMP steps are shown in blue for comparison.

MICRO INQUIRY *Why is it important that the enzymes dedicated to gluconeogenesis (rather than those used in both catabolic and anabolic reactions) are found in both the 3-carbon and 6-carbon stages?*

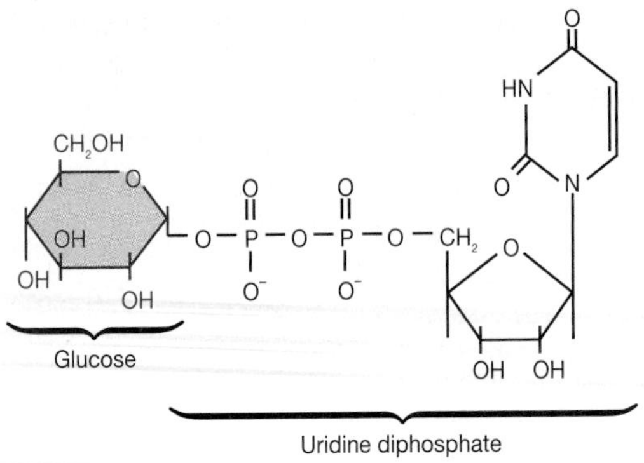

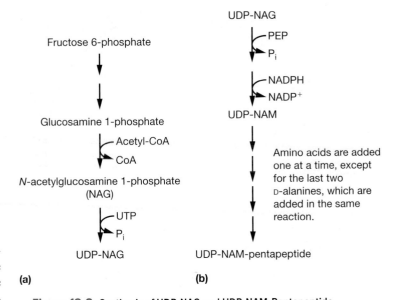

Figure 12.7 Uridine Diphosphate Glucose (UDPG).

(UDPG), which is formed when glucose reacts with uridine triphosphate (**figure 12.7**). Other important uridine diphosphate sugars are UDP-galactose and UDP-glucuronic acid. Uridine diphosphate (UDP) carries sugars around the cell for participation in enzyme reactions much like ADP bears phosphate in the form of ATP.

Nucleoside diphosphate sugars also play a central role in the synthesis of polysaccharides such as starch and glycogen, both of which are long chains of glucose. Again, biosynthesis is not simply a direct reversal of catabolism. For instance, during glycogen and starch synthesis in bacteria and protists, adenosine diphosphate glucose (ADP-glucose) is formed from glucose 1-phosphate and ATP. It then donates glucose to the end of growing glycogen and starch chains.

$$\text{ATP} + \text{glucose 1-phosphate} \rightarrow \text{ADP-glucose} + \text{PP}_i$$
$$(\text{Glucose})_n + \text{ADP-glucose} \rightarrow (\text{glucose})_{n+1} + \text{ADP}$$

Synthesis of Peptidoglycan Occurs in the Cytoplasm, at the Plasma Membrane, and in the Periplasmic Space

Nucleoside diphosphate sugars also participate in the synthesis of peptidoglycan. Recall that peptidoglycan is a large, complex molecule consisting of long polysaccharide chains made of alternating *N*-acetylmuramic acid (NAM) and *N*-acetylglucosamine (NAG) residues. The NAM groups bear stem peptides consisting of five amino acids (a pentapeptide). During peptidoglycan synthesis, adjacent polysaccharide chains are cross-linked by bonds formed between the stem peptides. In Gram-negative bacteria and many Gram-positive bacteria, the pentapeptide stems are directly linked. In some Gram-positive bacteria, the stems are connected by an interbridge consisting of one or more amino acids (*see figures 3.19 and 3.21*). ◀◀ *There are two main types of bacterial cell walls (section 3.4)*

Not surprisingly, such an intricate structure requires an equally intricate biosynthetic process, especially because some reactions occur in the cytoplasm, others at the plasma membrane, and others in the periplasmic space. Peptidoglycan synthesis involves two carriers. The first, uridine diphosphate, functions in the cytoplasmic reactions. The second carrier, bactoprenol phosphate, is needed for reactions that occur first on the cytoplasmic and then the periplasmic side of the plasma membrane.

Figure 12.8 Synthesis of UDP-NAG and UDP-NAM-Pentapeptide. These are the initial steps in peptidoglycan synthesis. NAG, *N*-acetylglucosamine; PEP, phosphoenolpyruvate; NAM, *N*-acetylmuramic acid.

In the first step of peptidoglycan synthesis, UDP derivatives of NAM and NAG are formed (**figure 12.8**). Amino acids are then added sequentially to UDP-NAM to form the pentapeptide stem. NAM-pentapeptide is then transferred to bactoprenol phosphate (also called undecaprenyl phosphate), which is located at the cytoplasmic side of the plasma membrane (**figure 12.9**). The resulting intermediate is called Lipid I. Bactoprenol is linked to NAM by two phosphates (pyrophosphate; **figure 12.10**). Next UDP transfers NAG to Lipid I, generating Lipid II. This creates the peptidoglycan repeat unit. If the peptidoglycan unit requires an interbridge, it is thought to be added while the repeat unit is at the cytoplasmic face of the plasma membrane. Once Lipid II is formed, the repeat unit is flipped across the plasma membrane so that it is now on the periplasmic side of the plasma membrane, while still attached to bactoprenol. This is catalyzed by an enzyme often referred to as a "flippase." The identity of flippase has been a topic of considerable speculation, but the most recent evidence suggests that it is a protein called MurJ. Enzymes called glycotransferases then attach the repeat unit to a growing nascent peptidoglycan strand, which is still attached to bactoprenol. Glycotransferases are one type of the penicillin-binding proteins (PBPs) that function during peptidoglycan synthesis. The growing chain is removed from bactoprenol during the final step in peptidoglycan synthesis: **transpeptidation** (**figure 12.11**), which creates the cross-links between the peptidoglycan chains. Enzymes called transpeptidases catalyze this reaction. They too are PBPs. During transpeptidation the terminal D-alanine is removed as the cross-link is formed. ↺ *Peptidoglycan Biosynthesis*

The result of the process we have just described is a peptidoglycan sacculus that is thicker by one layer. However, the peptidoglycan of any bacterium usually has a typical thickness that does not increase or decrease dramatically. Furthermore, for rod-shaped cells to elongate, and for any bacterial cell to divide, the

1 UDP derivatives of NAM and NAG are synthesized (figure 12.8).

2 Sequential addition of amino acids to UDP-NAM to form the NAM-pentapeptide (figure 12.8b).

3 NAM-pentapeptide is transferred to bactoprenol phosphate (figure 12.10), generating Lipid I. They are joined by a pyrophosphate bond.

4 UDP transfers NAG to the bactoprenol-NAM-pentapeptide, generating Lipid II. If a pentaglycine interbridge is required, it is created using special glycyl-tRNA molecules but not ribosomes. Interbridge formation occurs at the cytoplasmic side of the membrane.

5 Lipid II is flipped across the membrane by the flippase enzyme (not shown). The bactoprenol remains embedded in the membrane while the repeat unit now extends into the periplasmic space.

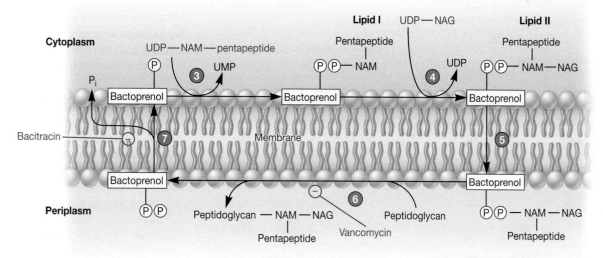

8 Peptide cross-links between peptidoglycan chains are formed by transpeptidation (figure 12.11).

7 The bactoprenol donor moves back across the membrane. As it does, it loses one phosphate, becoming bactoprenol phosphate. It is now ready to begin a new cycle.

6 The NAG-NAM-pentapeptide is attached to the growing end of a nascent peptidoglycan chain that is held at the membrane by a bactoprenol molecule. This increases the chain's length by one repeat unit.

Figure 12.9 Peptidoglycan Synthesis. NAM is *N*-acetylmuramic acid and NAG is *N*-acetylglucosamine. Inhibition by bacitracin and vancomycin also is shown.

MICRO INQUIRY *What is the difference between Lipid I and Lipid II?*

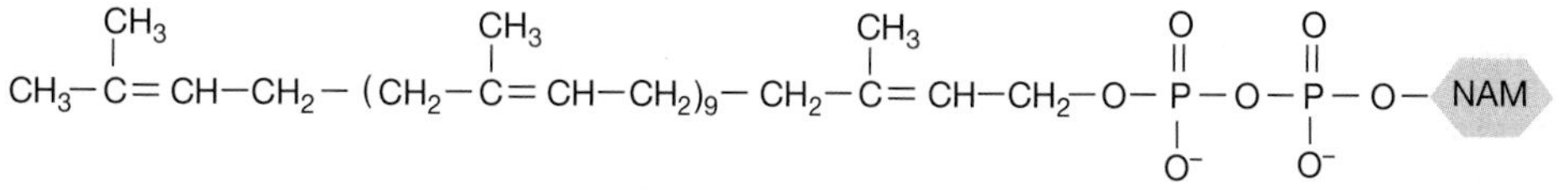

$$CH_3-C=CH-CH_2-(CH_2-C=CH-CH_2)_9-CH_2-C=CH-CH_2-O-P-O-P-O-\boxed{NAM}$$

Figure 12.10 Bactoprenol-NAM. Bactoprenol is a 55-carbon, isoprene-derived alcohol (section 12.7) that is connected to *N*-acetylmuramic acid (NAM) by pyrophosphate.

structure of the sacculus must be sculpted and modified to allow for maintenance of the proper thickness, elongation, and separation during cytokinesis. This is accomplished by hydrolytic enzymes often called **autolysins.** These enzymes are also PBPs and they function in a precise and well-regulated way while maintaining wall shape and integrity in the presence of high osmotic pressure (*see figure 7.8*). Some autolysins attack the polysaccharide chains, while others hydrolyze the peptide cross-links. Autolysin activity is tightly controlled because unregulated activity would lead to weakening of the peptidoglycan sacculus. How autolysin activity is controlled is still poorly understood. ◀◀ *Cellular growth and determination of cell shape (section 7.2)*

Peptidoglycan synthesis is a particularly effective target for antimicrobial agents because of its importance to bacterial cell

wall structure and function and because it is absent in animal cells. Inhibition of any stage of synthesis weakens the cell wall and can lead to lysis. For example, penicillin inhibits the transpeptidation reaction (figure 12.11), and bacitracin blocks the dephosphorylation of bactoprenol pyrophosphate (figure 12.9). ◀◀ *Inhibitors of cell wall synthesis (section 9.4)*

Retrieve, Infer, Apply

1. What is gluconeogenesis? Why is it important?
2. Describe the formation of mannose, galactose, starch, and glycogen. What are nucleoside diphosphate sugars? How do microorganisms use them?
3. Suppose a microorganism is growing on a medium that contains amino acids but no sugars. In general terms, how would it synthesize the pentose and hexose sugars it needs?
4. Diagram the steps involved in the synthesis of peptidoglycan and show where they occur in the cell. What are the roles of bactoprenol and UDP?
5. What would happen to a cell if it did not produce any autolysins? What if it produced too many in an unregulated fashion?

E. coli transpeptidation

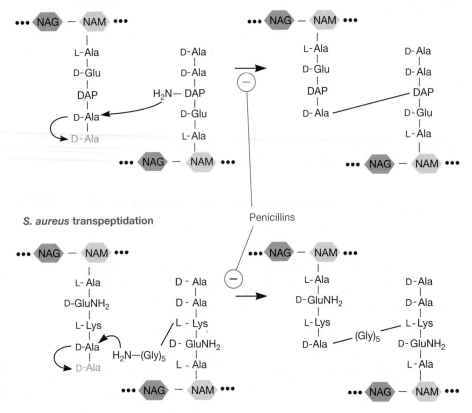

Figure 12.11 Transpeptidation. The transpeptidation reactions in the formation of the peptidoglycans of *E. coli* and *Staphylococcus aureus*. Note that the terminal D-alanine is removed during transpeptidation. The disruption of transpeptidation by penicillins is indicated. Other β-lactam antibiotics (e.g., cephalosporins) also inhibit this reaction. DAP is diaminopimelic acid.

12.5 Synthesis of Amino Acids Consumes Many Precursor Metabolites

After reading this section, you should be able to:

- Discuss the three mechanisms microorganisms use to assimilate inorganic nitrogen and the role of transaminases in them
- Describe the two methods microbes use to assimilate sulfur
- Differentiate assimilatory nitrate reduction from dissimilatory nitrate reduction and assimilatory sulfate reduction from dissimilatory sulfate reduction
- Evaluate the efficiency of using branched pathways for synthesizing amino acids
- List the major anaplerotic reactions and explain their importance
- Describe the two pathways for porphyrin synthesis and identify the types of organisms that use each

Many of the precursor metabolites serve as starting substrates for the synthesis of amino acids (figure 12.2). In the amino acid biosynthetic pathways, the carbon skeleton is remodeled and an amino group, and sometimes sulfur, is added. In this section, we first examine the mechanisms by which nitrogen and sulfur are assimilated and incorporated into amino acids. This is followed by a brief consideration of the organization of amino acid biosynthetic pathways.

Inorganic Nitrogen Assimilation

Nitrogen is a major component not only of proteins but also of nucleic acids, coenzymes, and many other cell constituents. Thus the cell's ability to assimilate inorganic nitrogen is exceptionally important. Although nitrogen gas is abundant in the atmosphere, only a few bacteria and archaea can reduce the gas and use it as a nitrogen source. Most incorporate either ammonia or nitrate. We examine ammonia and nitrate assimilation first and then discuss nitrogen assimilation in microbes that fix N_2.

Ammonia Incorporation Occurs in Two Ways

Ammonia can be incorporated into organic material relatively easily and directly because it is more reduced than other forms of inorganic nitrogen. Ammonia is initially incorporated into carbon skeletons by one of two mechanisms: reductive amination or the glutamine synthetase–glutamate synthase system. Once incorporated, the nitrogen can be transferred to other carbon skeletons.

The **reductive amination pathway** forms glutamate from α-ketoglutarate. It is catalyzed in many bacteria and fungi by glutamate dehydrogenase when the ammonia concentration is high (**figure 12.12**). Once glutamate has been synthesized, the

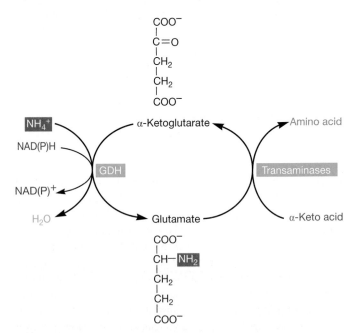

Figure 12.12 Ammonia Assimilation by Reductive Amination and Transaminases. Either NADPH- or NADH-dependent glutamate dehydrogenases may be involved. This route is most active at high ammonia concentrations. GDH is glutamate dehydrogenase.

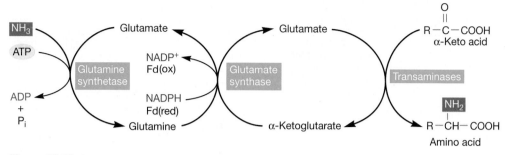

Figure 12.13 Ammonia Incorporation Using Glutamine Synthetase, Glutamate Synthase, and Transaminases. This route is effective at low ammonia concentrations. Fd is ferredoxin in either its reduced (Fd$_{red}$) state or oxidized (Fd$_{ox}$) form.

newly formed α-amino group can be transferred by enzymes called **transaminases** to other carbon skeletons. Microorganisms have a number of transaminases, each of which catalyzes the formation of several amino acids.

Many bacteria use the **glutamine synthetase–glutamate synthase (GS-GOGAT) system** to assimilate ammonia when its levels are low (**figure 12.13**). Incorporation of ammonia begins with synthesis of glutamine from glutamate in a reaction catalyzed by glutamine synthetase (**figure 12.14**). In this reaction, the carboxyl group found in the side chain of glutamate reacts with ammonia, and the incorporated nitrogen exists as an amide in glutamine. Glutamate synthase then transfers the amide nitrogen to

α-ketoglutarate to generate a new glutamate molecule (figure 12.14). Because glutamate acts as an amino donor in transaminase reactions, ammonia may be used to synthesize all common amino acids when suitable transaminases are present.

Assimilatory Nitrate Reduction: NO$_3$ to NH$_3$

The nitrogen in nitrate (NO$_3^-$) is much more oxidized than that in ammonia (NH$_3$). Therefore nitrate must first be reduced to ammonia before the nitrogen can be converted to an organic form. This reduction of nitrate is called **assimilatory nitrate reduction.** It is important to clearly distinguish assimilatory nitrate reduction from the type of nitrate reduction discussed in chapter 11—dissimilatory nitrate reduction. In dissimilatory nitrate reduction, nitrate serves as the terminal electron acceptor for anaerobic respiration. The reduced forms of nitrogen produced (e.g., N$_2$, N$_2$O) are released into the environment and therefore are not incorporated into cell components. In assimilatory nitrate reduction, nitrate is incorporated into cell material (biomass) and does not participate in energy conservation. The process is widespread among bacteria, fungi, and photosynthetic protists, and it is an important step in the nitrogen cycle. ◄◄ *Anaerobic respiration uses the same three steps as aerobic respiration (section 11.7)* ►► *Nitrogen cycle (section 28.1)*

Assimilatory nitrate reduction takes place in the cytoplasm in bacteria. The first step in nitrate assimilation is its reduction to nitrite by **nitrate reductase,** an enzyme that contains both FAD and molybdenum (**figure 12.15**). Nitrite is next reduced to

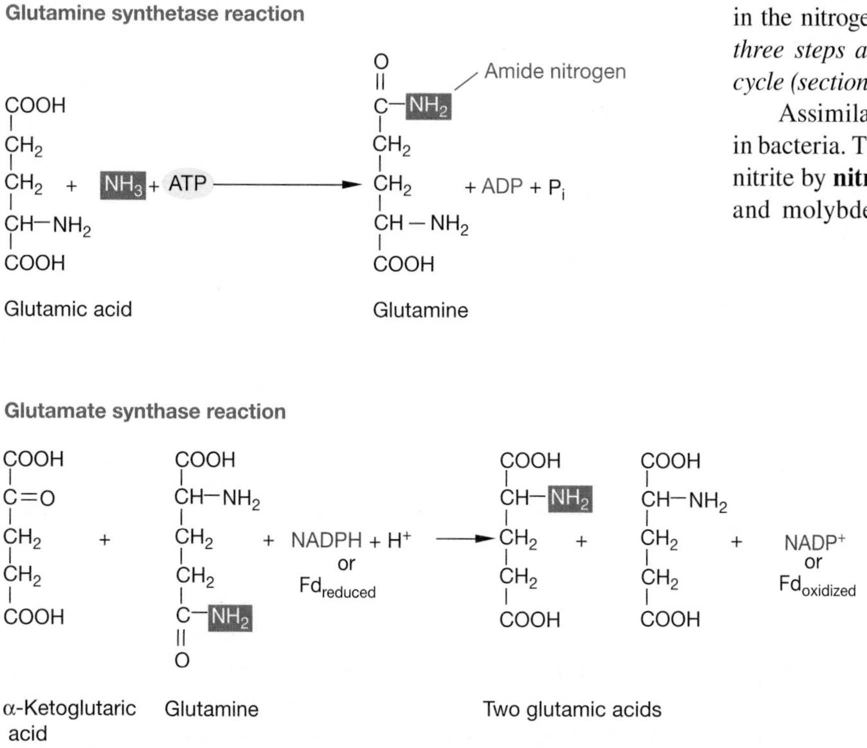

Glutamine synthetase reaction

COOH
|
CH$_2$
|
CH$_2$ +   NH$_3$ + ATP
|
CH—NH$_2$
|
COOH

Glutamic acid

→

O
||
C—NH$_2$ ← Amide nitrogen
|
CH$_2$
|
CH$_2$ + ADP + P$_i$
|
CH — NH$_2$
|
COOH

Glutamine

Glutamate synthase reaction

COOH
|
C=O
|
CH$_2$ +
|
CH$_2$
|
COOH

α-Ketoglutaric acid

COOH
|
CH—NH$_2$
|
CH$_2$ + NADPH + H$^+$
| or
CH$_2$ Fd$_{reduced}$
|
C—NH$_2$
||
O

Glutamine

→

COOH
|
CH—NH$_2$
|
CH$_2$ +
|
CH$_2$
|
COOH

COOH
|
CH—NH$_2$
|
CH$_2$ + NADP$^+$
| or
CH$_2$ Fd$_{oxidized}$
|
COOH

Two glutamic acids

Figure 12.14 Glutamine Synthetase and Glutamate Synthase Reactions Involved in Ammonia Assimilation. Some glutamate synthases use NADPH as an electron source; others use reduced ferredoxin (Fd).

MICRO INQUIRY *What purpose is served for the cell when these reactions are used to produce two glutamate rather than a single glutamine?*

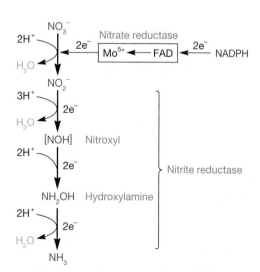

Figure 12.15 Assimilatory Nitrate Reduction. This sequence is thought to operate in bacteria that can reduce and assimilate nitrate nitrogen. Here we show the electron donor as NADPH. However, some bacteria use other electron donors such as ferredoxin or NADH.

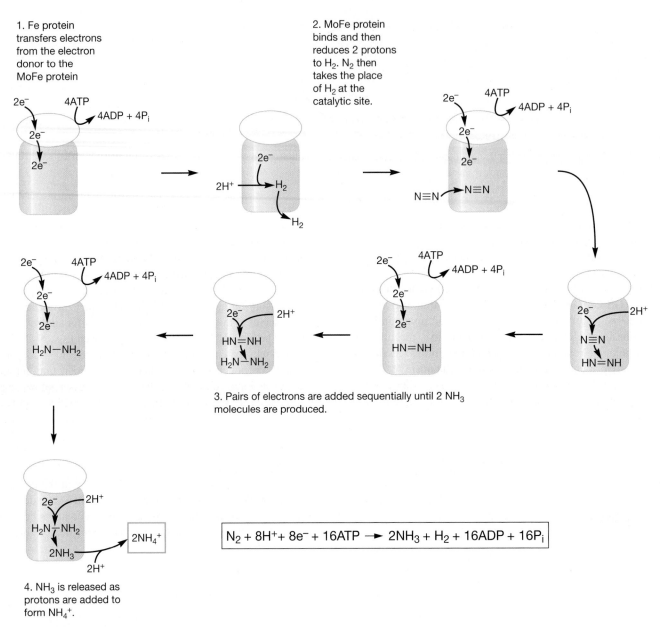

1. Fe protein transfers electrons from the electron donor to the MoFe protein

2. MoFe protein binds and then reduces 2 protons to H_2. N_2 then takes the place of H_2 at the catalytic site.

3. Pairs of electrons are added sequentially until 2 NH_3 molecules are produced.

$$N_2 + 8H^+ + 8e^- + 16ATP \longrightarrow 2NH_3 + H_2 + 16ADP + 16P_i$$

4. NH_3 is released as protons are added to form NH_4^+.

Figure 12.16 **The Proposed Sequence of Nitrogen Reduction by Nitrogenase.** Ferredoxin is often the electron donor for nitrogen fixation.

ammonia through a series of two electron additions catalyzed by nitrite reductase. The ammonia is then incorporated into amino acids by reductive amination or the GS-GOGAT system.

Nitrogen Fixation: N_2 to NH_3

The reduction of atmospheric gaseous nitrogen to ammonia is called **nitrogen fixation.** Only a few bacteria and archaea carry out nitrogen fixation: (1) free-living chemotrophic bacteria and archaea (e.g., *Azotobacter, Klebsiella, Clostridium,* and *Methanococcus* species), (2) bacteria living in symbiotic associations with plants such as legumes (e.g., *Rhizobium* spp., *see figure 31.9*), and (3) cyanobacteria (e.g., *Nostoc, Anabaena,* and *Trichodesmium* species). These microbes play a critical role in the nitrogen cycle. They complete the cycle from NO_3^- (the most oxidized form of nitrogen) to ammonia (the most reduced form of nitrogen) via N_2,

which has an intermediate oxidation state. The biological aspects of nitrogen fixation are discussed in chapters 28 and 31. The biochemistry of nitrogen fixation is the focus of this section.

During nitrogen fixation, nitrogen is reduced by two-electron additions in a way similar to that illustrated in **figure 12.16**. The reduction of molecular nitrogen to ammonia is quite exergonic, but the reaction has a high activation energy because molecular nitrogen is an unreactive gas with a triple bond connecting the two nitrogen atoms. Therefore nitrogen reduction is expensive, requiring a large ATP expenditure—at least 8 electrons and 16 ATP molecules (4 ATP per pair of electrons); symbiotic nitrogen-fixing bacteria can consume almost 20% of the ATP supplied by their host plants. Note in figure 12.16 that two of the electrons and four of the ATP molecules are expended to reduce 2 protons to hydrogen gas at the start of the process. This is often thought of as a wasteful expenditure of electrons and energy. However, some bacteria have evolved mechanisms for capturing the hydrogen gas and using it as a source of electrons for nitrogen reduction. Interestingly, some scientists are exploring the use of nitrogen fixation as a means to obtain a hydrogen biofuel source.

The reduction of nitrogen to ammonia is catalyzed by the enzyme **nitrogenase** (figure 12.16). The best studied nitrogenase is a complex enzyme consisting of two major protein components, a MoFe protein (MW 220,000) joined with a Fe protein (MW 64,000). The MoFe protein has an organic cofactor that contains molybdenum, iron, and sulfur atoms, whereas the Fe protein associates with iron atoms. Fe protein is reduced by ferredoxin and then binds ATP. ATP binding changes the conformation of the Fe protein and lowers its reduction potential (-0.29 to -0.40 V), enabling it to reduce the MoFe protein. ATP is hydrolyzed when this electron transfer occurs. Finally, reduced MoFe protein donates electrons to protons (the first two reactions of the process) or sequentially to atomic nitrogen (the series of reactions outlined by number 3 in figure 12.16). Nitrogenase is quite sensitive to O_2 and must be protected from O_2 inactivation within the cell. Microbes use a variety of strategies to protect nitrogenase, as we discuss more fully in chapters 22 and 31.

Sulfur Assimilation: SO₄ to S-Bearing Molecules

Sulfur is needed for the synthesis of the amino acids cysteine and methionine. It is also needed for synthesis of several coenzymes (e.g., coenzyme A and biotin). Sulfur is obtained from two sources: (1) the amino acids cysteine and methionine, and (2) sulfate. Many microorganisms use cysteine and methionine obtained from either external sources or intracellular amino acid reserves. Because the sulfur is already in a reduced organic form, it is readily incorporated into other organic molecules. In contrast, the sulfur atom in sulfate is more oxidized than it is in cysteine and methionine; thus sulfate must be reduced before it can be assimilated. This process is known as **assimilatory sulfate reduction** to distinguish it from the dissimilatory sulfate reduction that takes place when sulfate acts as an electron acceptor during anaerobic respiration. ◄◄ *Anaerobic respiration uses the same three steps as aerobic respiration (section 11.7)* ►►| *Sulfur cycle (section 28.1)*

Assimilatory sulfate reduction involves sulfate activation through the formation of phosphoadenosine 5′-phosphosulfate (**figure 12.17**). Sulfate activation prepares the sulfate for

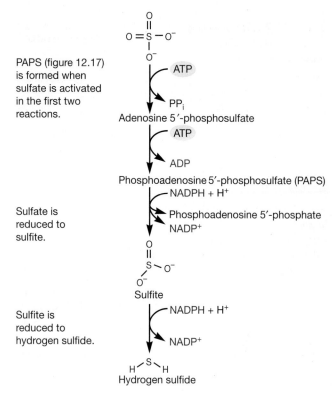

PAPS (figure 12.17) is formed when sulfate is activated in the first two reactions.

Sulfate is reduced to sulfite.

Sulfite is reduced to hydrogen sulfide.

Figure 12.18 The Sulfate Reduction Pathway.

reduction, which follows. As shown in **figure 12.18**, sulfate is first reduced to sulfite (SO_3^{2-}), then to hydrogen sulfide. Hydrogen sulfide is then used to synthesize cysteine in two ways (**figure 12.19**). Fungi combine hydrogen sulfide with serine to form cysteine (figure 12.19*a*), whereas many bacteria join hydrogen sulfide with O-acetylserine instead (figure 12.19*b*). Once formed, cysteine can be used in the synthesis of other sulfur-containing organic compounds, including the amino acid methionine.

Amino Acid Biosynthetic Pathways

Some amino acids are made directly by transamination of a precursor metabolite. For example, alanine and aspartate are

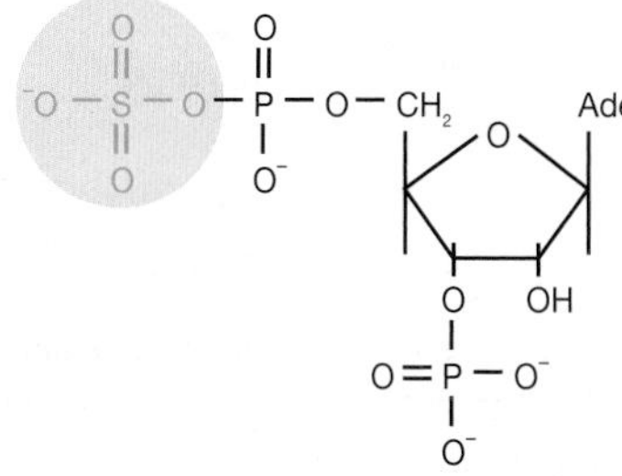

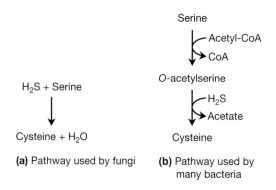

(a) Pathway used by fungi

(b) Pathway used by many bacteria

Figure 12.17 Phosphoadenosine 5′-phosphosulfate (PAPS). The sulfate group is in color.

Figure 12.19 Incorporation of Sulfur from H₂S. H₂S is the product of sulfate reduction, as shown in figure 12.18.

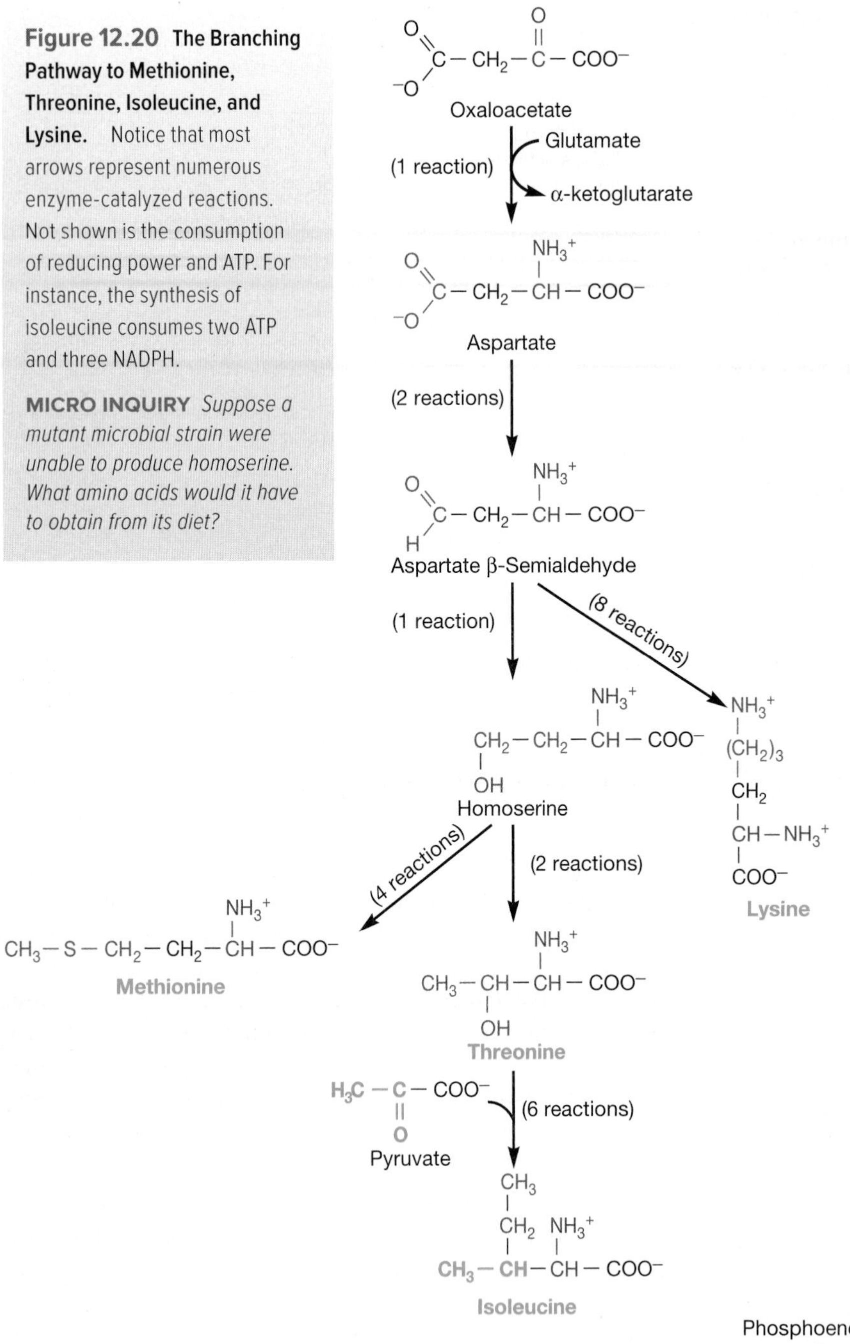

Figure 12.20 The Branching Pathway to Methionine, Threonine, Isoleucine, and Lysine.   Notice that most arrows represent numerous enzyme-catalyzed reactions. Not shown is the consumption of reducing power and ATP. For instance, the synthesis of isoleucine consumes two ATP and three NADPH.

MICRO INQUIRY *Suppose a mutant microbial strain were unable to produce homoserine. What amino acids would it have to obtain from its diet?*

Oxaloacetate

(1 reaction) — Glutamate → α-ketoglutarate

Aspartate

(2 reactions)

Aspartate β-Semialdehyde

(1 reaction) / (8 reactions)

$CH_2-CH_2-CH-COO^-$, NH_3^+, OH
Homoserine

Lysine

(4 reactions) / (2 reactions)

$CH_3-S-CH_2-CH_2-CH-COO^-$, NH_3^+
Methionine

$CH_3-CH-CH-COO^-$, NH_3^+, OH
Threonine

Pyruvate

(6 reactions)

Isoleucine

made directly from pyruvate and oxaloacetate, respectively, using glutamate as the amino group donor. However, most precursor metabolites must be altered by more than just the addition of an amino group. In many cases, the carbon skeleton must be reconfigured, and for cysteine and methionine, the carbon skeleton must be amended by the addition of sulfur. These biosynthetic pathways are more complex, and they often involve many steps and are branched. By using branched pathways, a single precursor metabolite can be used for the synthesis of a family of related amino acids. For example, the amino acids lysine, threonine, isoleucine, and methionine are synthesized from oxaloacetate by a branching route (**figure 12.20**).

Because of the need to conserve nitrogen, carbon, and energy, amino acid synthetic pathways are usually tightly regulated by feedback mechanisms. ▶▶ *Posttranslational regulation of enzyme activity (section 10.7)*

Anaplerotic Reactions Replace the Precursor Metabolites Used for Amino Acid Biosynthesis

When an organism is actively synthesizing amino acids, a heavy demand for precursor metabolites is placed on the central metabolic pathways, especially the TCA cycle. Therefore it is critical that TCA cycle intermediates be readily available. This is especially true for organisms carrying out fermentation, where the TCA cycle does not function to catabolize glucose. To ensure an adequate supply of TCA cycle–generated precursor metabolites, microorganisms use **anaplerotic reactions** (Greek *anaplerotic,* filling up) to replenish TCA cycle intermediates.

Most microorganisms replace TCA cycle intermediates by generating oxaloacetate; this is done in two ways. One adds a CO_2 as a carboxyl group to pyruvate (catalyzed by pyruvate carboxylase), and the other adds CO_2 to phosphoenolpyruvate (catalyzed by phosphoenolpyruvate carboxylase) (**figure 12.21**). The

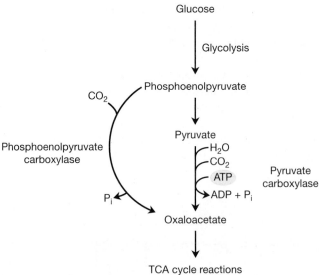

Figure 12.21 Anaplerotic Reactions That Replace Oxaloacetate, a TCA Cycle Intermediate.   Pyruvate and phosphoenolpyruvate are formed by the Embden-Meyerhof glycolytic pathway. Biotin is a cofactor for pyruvate carboxylase and many other carboxylases, and therefore biotin is a required growth factor for many organisms.

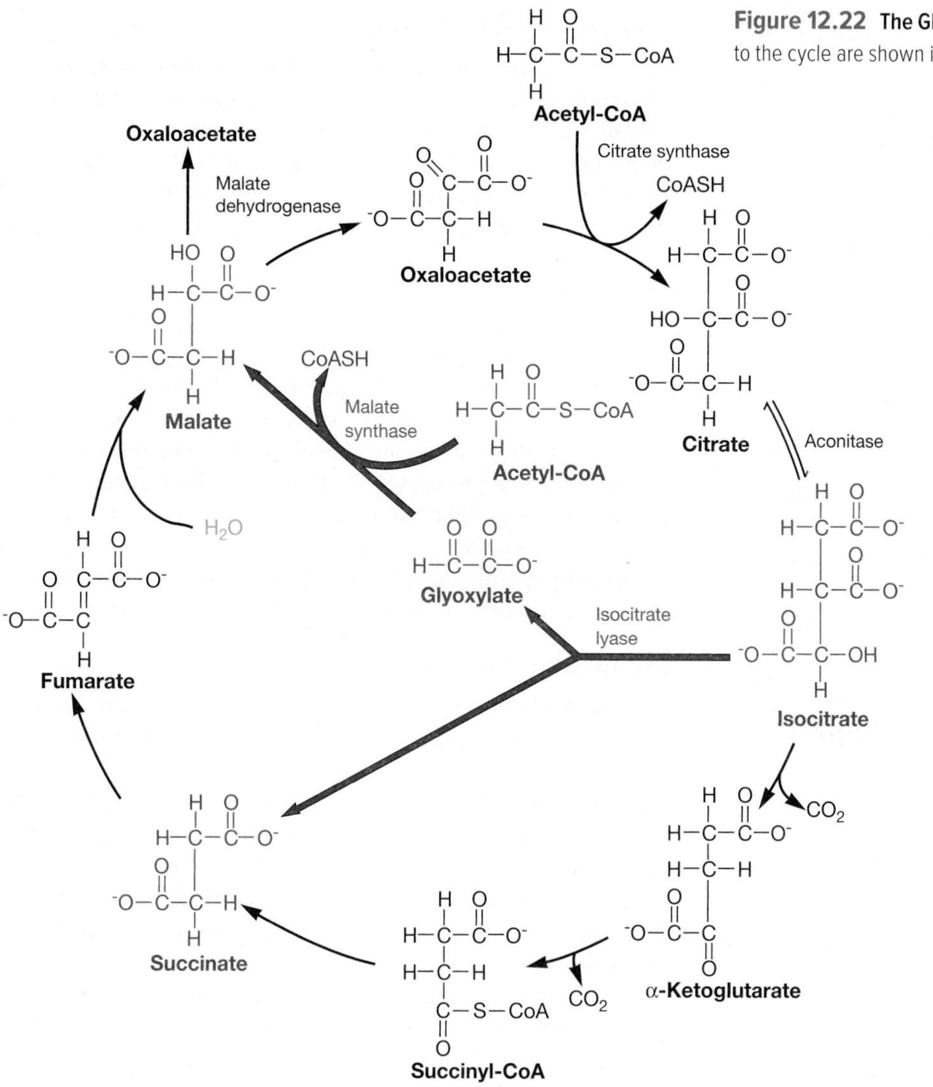

Figure 12.22 The Glyoxylate Cycle. The reactions and enzymes unique to the cycle are shown in red and blue, respectively.

the acetyl-CoA carbons as CO_2. ◄◄ *Pyruvate to carbon dioxide (step two) is accomplished by the tricarboxylic acid cycle (section 11.5)*

Synthesis of Porphyrins

As we demonstrate in section 12.6 and in figure 12.2, in addition to their role in protein synthesis, amino acids are also required for the synthesis of other molecules. Here we consider their role in the synthesis of porphyrins, molecules that contain four pyrrole rings (*see figures 10.11 and 11.29*), Porphyrins are found in a number of critically important molecules, including hemoglobin, cytochromes, phycobilins, chlorophylls, and vitamin B_{12}. As can be seen in **figure 12.23**, there are two pathways leading to these important products: the succinyl-CoA/glycine pathway, used by all nonphotosynthetic eukaryotes and α-proteobacteria; and the glutamate pathway, used by all other bacteria, photosynthetic eukaryotes, and archaea. Both pathways converge at the intermediate 5-aminolevulinic acid. The remaining reactions are highly conserved in all organisms studied thus far.

pyruvate carboxylase reaction is observed in yeasts and some bacteria. The phosphoenolpyruvate carboxylase reaction is used by other microorganisms (e.g., *E. coli* and *Salmonella* spp.).

Other anaplerotic reactions are part of the **glyoxylate cycle,** which functions in some bacteria, fungi, and protists (**figure 12.22**). This cycle is made possible by two unique enzymes, isocitrate lyase and malate synthase. The glyoxylate cycle is actually a modified TCA cycle. The two decarboxylations of the TCA cycle are bypassed, making possible the conversion of acetyl-CoA to oxaloacetate without the loss of

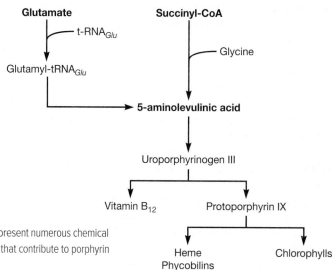

Figure 12.23 A Simplified Illustration of Porphyrin Biosynthesis. Most arrows represent numerous chemical reactions. Not shown are the many electron donors, ATP molecules, and other molecules that contribute to porphyrin synthesis.

Retrieve, Infer, Apply

1. Describe the roles of glutamate dehydrogenase, glutamine synthetase, glutamate synthase, and transaminases in ammonia assimilation.
2. How is nitrate assimilated? How does assimilatory nitrate reduction differ from dissimilatory nitrate reduction? What is the fate of nitrate following assimilatory nitrate reduction versus its fate following denitrification?
3. What is nitrogen fixation? Briefly describe the structure and mechanism of action of nitrogenase.
4. How do organisms assimilate sulfur? How does assimilatory sulfate reduction differ from dissimilatory sulfate reduction?
5. Why is using branched pathways an efficient mechanism for synthesizing amino acids?
6. Define an anaplerotic reaction. Give three examples of anaplerotic reactions.
7. Describe the glyoxylate cycle. How is it similar to the TCA cycle? How does it differ?
8. What are porphyrins? Which amino acids and precursor metabolites are important in their synthesis?
9. List the types of organisms that use each porphyrin biosynthetic pathway.

12.6 Synthesis of Purines, Pyrimidines, and Nucleotides

After reading this section, you should be able to:

- Distinguish a purine from a pyrimidine, and identify which nitrogenous bases are purines and which are pyrimidines
- Draw a simple diagram that illustrates the chemical moieties found in nucleosides and nucleotides
- Discuss in general terms how phosphorus is assimilated
- Compare and contrast purine and pyrimidine biosynthesis
- Discuss the methods used to convert ribonucleotides to deoxyribonucleotides

Purine and pyrimidine biosynthesis is critical for all cells because these molecules are used to synthesize ATP, several cofactors, ribonucleic acid (RNA), deoxyribonucleic acid (DNA), and other important cell components. Nearly all microorganisms can synthesize their own purines and pyrimidines as these are crucial to cell function. ▶▶| *DNA replication in bacteria (section 13.3); Transcription in bacteria (section 13.5)*

Purines and **pyrimidines** are cyclic nitrogenous bases with several double bonds. **Adenine** and **guanine** are purines, consisting of two joined rings, whereas pyrimidines (**uracil, cytosine, and thymine**) have only one ring. A purine or pyrimidine base joined with a pentose sugar, either ribose or deoxyribose, is a **nucleoside**. A **nucleotide** is a nucleoside with one or more phosphate groups attached to the sugar.

Amino acids participate in the synthesis of nitrogenous bases and nucleotides in a number of ways, including providing the nitrogen that is part of all purines and pyrimidines. The phosphorus present in nucleotides is provided by other mechanisms. We begin this section by examining phosphorus assimilation. We then examine the pathways for synthesis of nitrogenous bases and nucleotides.

Phosphorus Assimilation

In addition to nucleic acids, phosphorus is found in proteins (i.e., phosphorylated proteins), phospholipids, and coenzymes such as $NADP^+$. The most common phosphorus sources are inorganic phosphate and organic molecules containing a phosphoryl group. Inorganic phosphate is incorporated through the formation of ATP in one of three ways: (1) photophosphorylation, (2) oxidative phosphorylation, and (3) substrate-level phosphorylation. ◄◄ *Glucose to pyruvate (section 11.4); Electron transport and oxidative phosphorylation (step 3) generate the most ATP (section 11.6); Phototrophy (section 11.11)*

Microorganisms may obtain organic phosphates from their surroundings in dissolved or particulate form. **Phosphatases** very often hydrolyze organic phosphate-containing molecules to release inorganic phosphate. Gram-negative bacteria have phosphatases in the periplasmic space, which allows phosphate to be taken up immediately after release. On the other hand, protists can directly use organic phosphates after ingestion or hydrolyze them in lysosomes and then incorporate the phosphate.

Purine Biosynthesis

The biosynthetic pathway for purines is a complex, 11-step sequence (*see appendix II*) in which seven different molecules contribute parts to the final purine skeleton (**figure 12.24**). The pathway begins with ribose 5-phosphate, and the purine skeleton is constructed on this sugar. Therefore the first purine product of the pathway is the nucleotide inosinic acid, not a free purine base. The cofactor folic acid is very important in purine biosynthesis because folic acid derivatives contribute two carbons to the purine skeleton.

Once inosinic acid has been formed, relatively short pathways synthesize adenosine monophosphate and guanosine monophosphate (**figure 12.25**), and produce nucleoside diphosphates and triphosphates by phosphate transfers from ATP. At this point, the cell has synthesized ribonucleotides. However, it needs deoxynucleotides for DNA synthesis. In deoxyribonucleotides there is an H atom rather than a hydroxyl group on carbon two of the sugar. Thus, to convert a ribonucleotide

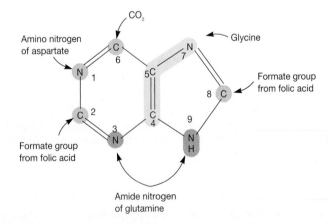

Figure 12.24 Sources of the Nitrogen and Carbon in Purines. The nitrogen numbered 9 is linked to the sugar ribose 5-phosphate, and the base is built while attached to the sugar.

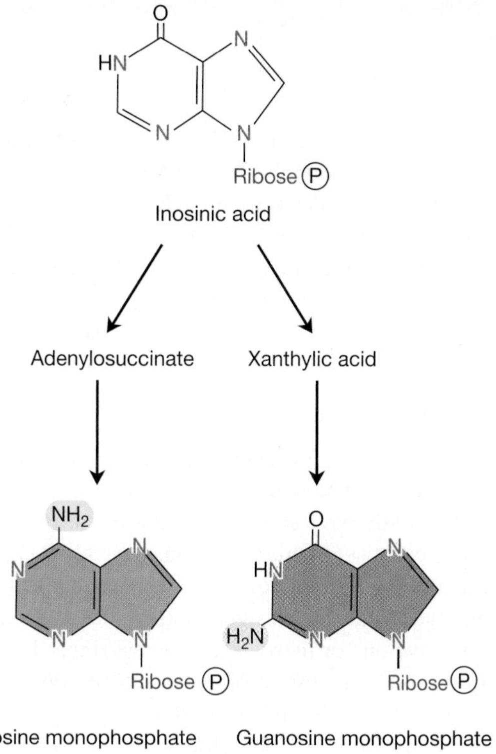

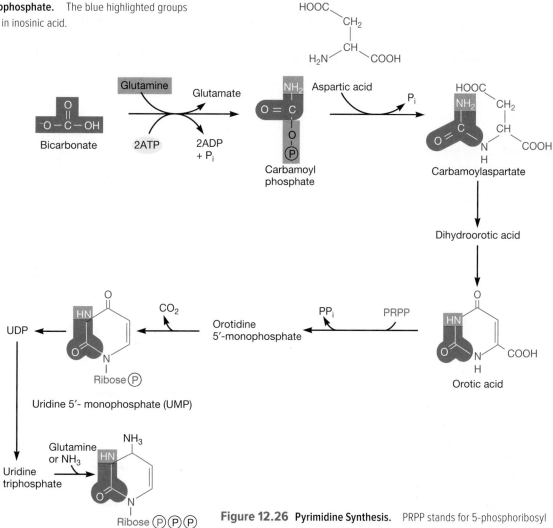

Figure 12.25 Synthesis of Adenosine Monophosphate and Guanosine Monophosphate.   The blue highlighted groups differ from those in inosinic acid.

into a deoxyribonucleotide, that carbon must be reduced. Deoxyribonucleotides arise from the reduction of nucleoside diphosphates or nucleoside triphosphates by two different routes. Some microorganisms reduce the ribose in nucleoside triphosphates with a system requiring vitamin B_{12} as a cofactor. Others reduce the ribose in nucleoside diphosphates. Both systems employ a small, sulfur-containing protein called thioredoxin as their reducing agent.

Pyrimidine Biosynthesis

Pyrimidine biosynthesis is equally as complex as purine synthesis. However, the two processes differ in an important way: purines are synthesized by adding chemical moieties to ribose 5-phosphate and never exist as free bases; construction of pyrimidine rings is completed before ribose is attached.

Pyrimidine biosynthesis begins with aspartic acid and carbamoyl phosphate, a high-energy molecule synthesized from bicarbonate and ammonia provided by the amino acid glutamine (**figure 12.26**). Aspartate carbamoyltransferase catalyzes the

Figure 12.26 Pyrimidine Synthesis.   PRPP stands for 5-phosphoribosyl 1-pyrophosphate, which provides the ribose 5-phosphate. The parts derived from bicarbonate and glutamine are shaded.

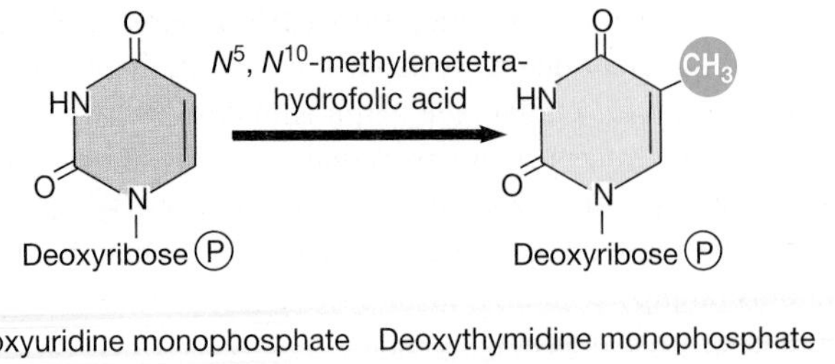

Figure 12.27 Deoxythymidine Monophosphate Synthesis.
Deoxythymidine differs from deoxyuridine in having the shaded
methyl group.

joining (condensation) of these two substrates to form carbamo-
ylaspartate, which is then converted to the initial pyrimidine
product, orotic acid. The high-energy molecule 5-phosphoribosyl
1-pyrophosphate (PRPP) provides the ribose 5-phosphate that is
attached to orotic acid to form a nucleotide.

Orotidine monophosphate is sequentially converted to
uridine monophosphate, uridine triphosphate, and cytidine
triphosphate. These ribonucleotides are reduced to deoxy-
forms in the same way that purine ribonucleotides are. That
leaves the cell with only one deoxynucleotide yet to be made:
deoxythimidine monophosphate. It is made when deoxyuridine monophosphate is methylated with a folic acid derivative
(**figure 12.27**).

Retrieve, Infer, Apply

1. How is phosphorus assimilated? What roles do phosphatases
   play in phosphorus assimilation? Why can phosphate be directly
   incorporated into cell constituents, whereas nitrate, nitrogen gas,
   and sulfate cannot?
2. Explain the difference between a purine and a pyrimidine, and
   between a nucleoside and a nucleotide.
3. Outline the way in which purines and pyrimidines are
   synthesized. How is the deoxyribose component of
   deoxyribonucleotides made?

12.7 Lipid Synthesis

After reading this section, you should be able to:

- Describe the role of acetyl-CoA, acyl carrier protein (ACP), and
  fatty acid synthase in fatty acid synthesis
- Distinguish saturated fatty acids from unsaturated fatty acids
- Describe the roles of dihydroxyacetone phosphate and fatty acids
  in the synthesis of triacylglycerol
- Describe the roles of diacylglycerol and CTP in the synthesis of
  phospholipids
- Summarize lipopolysaccharide synthesis
- Describe the Lpt pathway
- Compare sterol synthesis to archaeal isoprenoid lipid synthesis

Lipids are absolutely required by cells, as they are the major
components of cell membranes. Most bacterial and eukaryal
lipids contain fatty acids or their derivatives. Archaeal lipids are
built using an isoprene building block. In this section we consider the pathways leading to these molecules. We also consider
the synthesis of lipopolysaccharides, major components of the
outer membrane of typical Gram-negative bacteria.

Fatty Acids and Phospholipids

Fatty acids consist of a carboxyl group (i.e., they are monocarboxylic acids) attached to a long chain of carbons (alkyl chains);
the chains usually have an even number of carbons (the average
length is 18 carbons). Some fatty acids have alkyl chains that are
unsaturated; that is, they have one or more double bonds. Most
microbial fatty acids are straight chained, but some are branched
and some have one or more cyclopropane rings. Their synthesis
is complex and is the target of some antimicrobial agents used
to treat infectious disease caused by eukaryotic pathogens.
▶▶| *Lipids (appendix I)* |◀◀ *Antifungal drugs (section 9.5)*

Synthesis of saturated fatty acids is catalyzed by the
fatty acid synthase complex with acetyl-CoA and malonyl-
CoA as the substrates and NADPH as the electron donor.
Malonyl-CoA arises from the ATP-driven carboxylation of
acetyl-CoA (**figure 12.28**; top right portion of the figure).

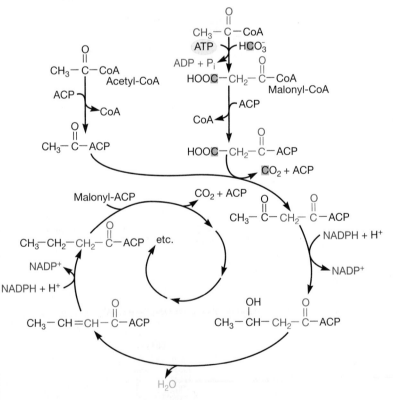

Figure 12.28 Fatty Acid Synthesis. The cycle is repeated until the proper
chain length has been reached. Carbon dioxide carbon and the remainder of
malonyl-CoA are shown in red. ACP stands for acyl carrier protein.

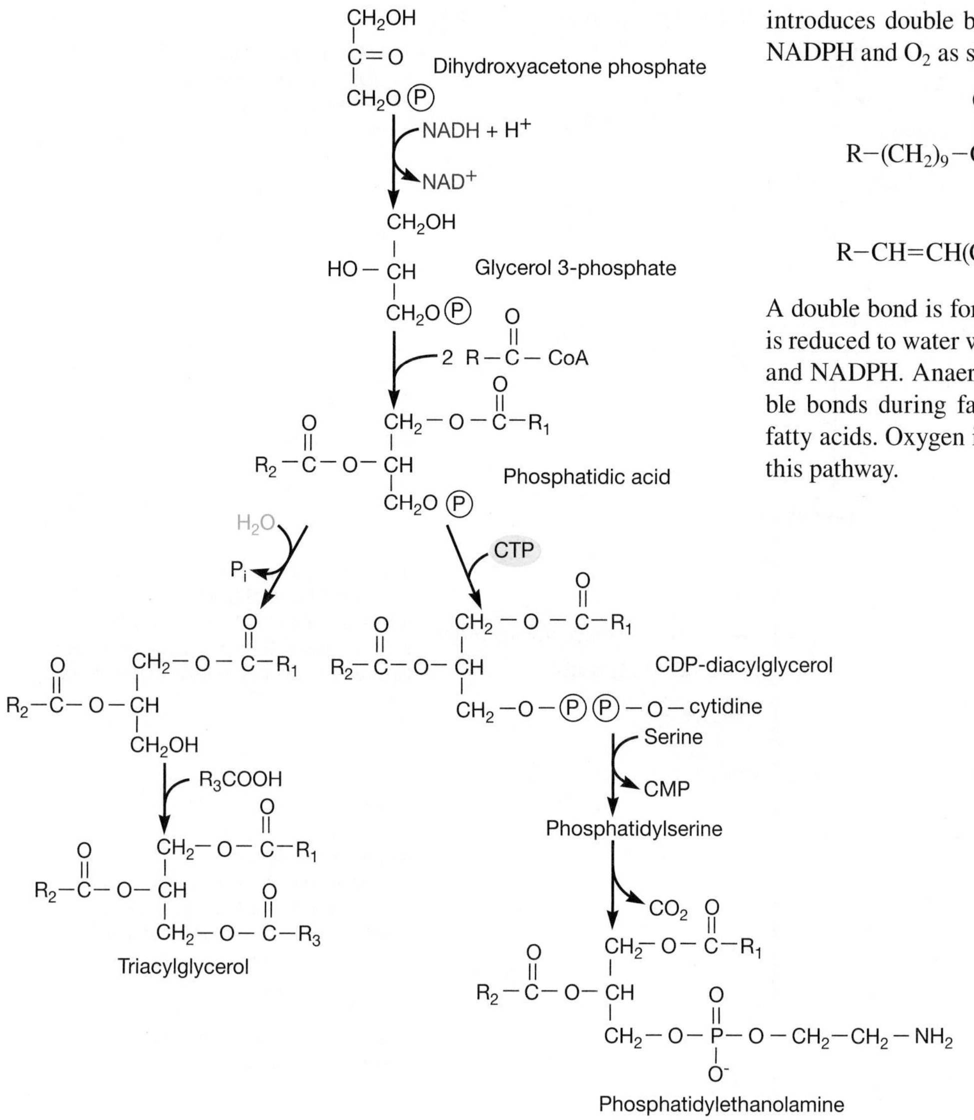

introduces double bonds into a saturated fatty acid using both NADPH and O_2 as shown in the following equation.

$$R-(CH_2)_9-\overset{\overset{\displaystyle O}{\|}}{C}-SCoA + NADPH + H^+ + O_2 \rightarrow$$

$$R-CH=CH(CH_2)_7-\overset{\overset{\displaystyle O}{\|}}{C}-SCoA + NADP^+ + 2H_2O$$

A double bond is formed between carbons nine and ten, and O_2 is reduced to water with electrons supplied by both the fatty acid and NADPH. Anaerobic bacteria and some aerobes create double bonds during fatty acid synthesis by dehydrating hydroxy fatty acids. Oxygen is not required for double-bond synthesis by this pathway.

Fatty acids are used to make other lipids such as triacylglycerol and phospholipids. Both of these molecules have glycerol linked to fatty acids by ester bonds and both are synthesized using the same branched pathway (**figure 12.29**). Glycerol arises from the reduction of dihydroxyacetone phosphate, an intermediate of the Embden-Meyerhoff pathway, to glycerol 3-phosphate. Glycerol 3-phosphate is then esterified with two fatty acids to give phosphatidic acid. Phosphatidic acid is the branchpoint of the pathway. The branch to triacylglycerol begins with the hydrolysis of phosphate from phosphatidic acid, giving a diacylglycerol; the third fatty acid is then attached to yield a triacylglycerol. The branch to a phospholipid begins when phosphatidic acid is attached to a cytidine diphosphate (CDP) carrier that plays a role similar to that of uridine and adenosine diphosphate carriers in carbohydrate biosynthesis. For example, bacteria synthesize phosphatidylethanolamine, a major cell membrane component, through the initial formation of CDP-diacylglycerol (figure 12.29). This CDP derivative then reacts with serine to form the phospholipid phosphatidylserine, and decarboxylation yields phosphatidylethanolamine. In this way, a complex membrane lipid is constructed from the products of glycolysis, fatty acid biosynthesis, and amino acid biosynthesis.

Figure 12.29 Triacylglycerol and Phospholipid Synthesis.
CTP is cytidine triphosphate; CDP is cytidine diphosphate.

Synthesis takes place after acetate (a 2-carbon molecule) and malonate (a 3-carbon molecule) have been transferred from coenzyme A (CoA) to acyl carrier protein (ACP), a small protein that carries the growing fatty acid chain during synthesis. Fatty acid synthase adds two carbons at a time to the carboxyl end of the growing fatty acid chain in a two-stage process (figure 12.28). First, malonyl-ACP reacts with the fatty acyl-ACP to form a fatty acyl-ACP two carbons longer. Malonate's third carbon is released as CO_2, and this loss drives the reaction to completion. In the second stage of synthesis, the β-keto group arising from the initial condensation reaction is removed in a three-step process involving two reductions and a dehydration. The fatty acid is then ready for the addition of two more carbon atoms.

Unsaturated fatty acids are synthesized in two ways. Eukaryotes and many aerobic bacteria use a pathway that

Lipopolysaccharides

Lipopolysaccharides are important components of the outer membrane of typical Gram-negative bacteria. Recall from chapter 3 that these molecules consist of a lipid component (lipid A), an oligosaccharide core, and the O polysaccharide (O antigen). We consider the synthesis of LPS here because its dual nature (lipid and carbohydrate) requires an understanding of the processes used to synthesize both types of molecules. Furthermore,

Figure 12.30 Lipopolysaccharide Biosynthetic Pathway. Most arrows represent multiple reactions. Note the use of carrier molecules such as UDP and ACP. N-acetylglucosamine is abbreviated NAG in figure 12.8. We use GlcNAc as the abbreviation for N-acetylglucosamine GlcNAc here so that the fate of acetate can be discerned.

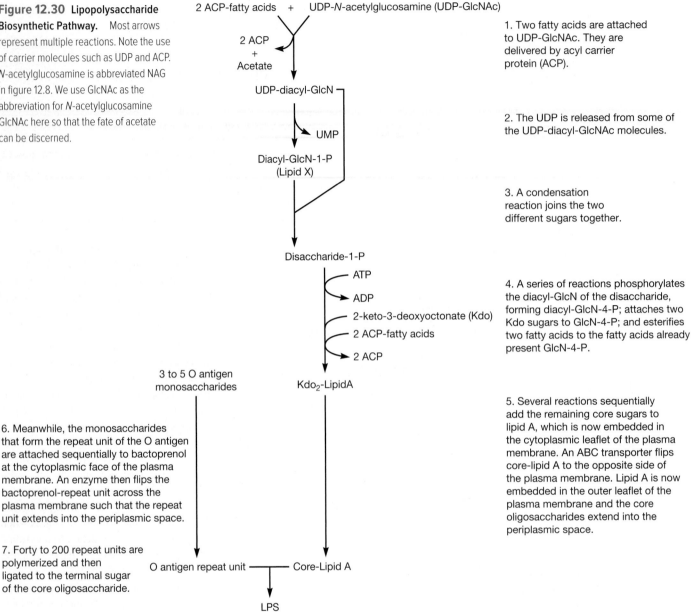

1. Two fatty acids are attached to UDP-GlcNAc. They are delivered by acyl carrier protein (ACP).

2. The UDP is released from some of the UDP-diacyl-GlcNAc molecules.

3. A condensation reaction joins the two different sugars together.

4. A series of reactions phosphorylates the diacyl-GlcN of the disaccharide, forming diacyl-GlcN-4-P; attaches two Kdo sugars to GlcN-4-P; and esterifies two fatty acids to the fatty acids already present GlcN-4-P.

5. Several reactions sequentially add the remaining core sugars to lipid A, which is now embedded in the cytoplasmic leaflet of the plasma membrane. An ABC transporter flips core-lipid A to the opposite side of the plasma membrane. Lipid A is now embedded in the outer leaflet of the plasma membrane and the core oligosaccharides extend into the periplasmic space.

6. Meanwhile, the monosaccharides that form the repeat unit of the O antigen are attached sequentially to bactoprenol at the cytoplasmic face of the plasma membrane. An enzyme then flips the bactoprenol-repeat unit across the plasma membrane such that the repeat unit extends into the periplasmic space.

7. Forty to 200 repeat units are polymerized and then ligated to the terminal sugar of the core oligosaccharide.

many of the biosynthetic strategies we have described earlier are apparent in the LPS biosynthetic pathway (**figure 12.30**). This is a complex pathway that involves many steps; most arrows in our rendition of the pathway represent multiple reactions, each catalyzed by a different enzyme. Briefly summarized, the pathway consists of two branches: one for the synthesis of lipid A attached to the core oligosaccharide (lipid A-core), and one for the synthesis of the O-antigen repeat unit and its polymerization into the final full-length O antigen.

The lipid A-core branch begins with UDP-N-acetylglucosamine (UDP-GlcNAc; abbreviated UDP-NAG in figure 12.8) and, using numerous reactions, generates the acyldisaccharide (i.e., a disaccharide linked to fatty acids) platform on which sugars that make up the core oligosaccharide are added. The fatty acids are delivered during the process by acyl carrier protein

(ACP), which was introduced in our discussion of fatty acid biosynthesis (figure 12.28). The process occurs mainly at the cytoplasmic face of the plasma membrane, where the fatty acids are inserted into the plasma membrane; the developing core oligosaccharide extends into the cytoplasm. Once synthesis of the lipid A-core portion of LPS is finished, it is flipped to the periplasmic face of the plasma membrane, with the core oligosaccharides in the periplasmic space.

The O antigen branch of the pathway begins with the three to five sugars (each carried by a nucleoside diphosphate) that form the repeat unit of the O antigen. These are added sequentially to bactoprenol on the cytoplasmic face of the plasma membrane; the repeat-unit sugars are in the cytoplasm (recall the use of bactoprenol in peptidoglycan synthesis; figure 12.9). The bactoprenol-repeat unit is next flipped to the periplasmic side of

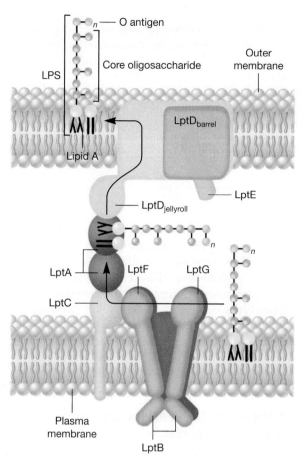

Figure 12.31 The Lpt Pathway Moves LPS from the Plasma Membrane to the Outer Membrane. This schematic of the pathway is based on studies of *Salmonella enterica* serovar Typhimurium and *Shigella flexneri*.

the plasma membrane. The repeat units, which are now extending into the periplasmic space, are polymerized as a block, yielding the O antigen. Finally, the O antigen is ligated to the lipidA-core, generating the final LPS molecule.

Figure 12.30 illustrates synthesis of LPS, but does not show how the mature molecule is inserted into the outer leaflet of the outer membrane. The problem faced by the cell is twofold. First, the carbohydrate portion of LPS is hydrophilic, making its transport across the outer membrane difficult. Second, the lipid A portion of LPS is hydrophobic, making its transport across the periplasmic space difficult. Current evidence supports a model in which seven proteins form a "walking path" along which the LPS moves (**figure 12.31**). The proteins are called Lpt proteins (short for *l*ipo*p*olysaccharide *t*ransport proteins) and the path they create is called the Lpt pathway. Some of the Lpt proteins are embedded in the plasma membrane. Their function is to remove the LPS from the plasma membrane and hand it off to a protein that forms a bridge between the plasma membrane and the outer membrane. The bridge hides lipid A from the periplasmic water-filled environment, using regions that fold into what is called a β-jelly roll. The β-jelly roll is a hydrophobic portion of the

protein (often described as "greasy"), and thus readily interacts with lipid A. The core oligosaccharides and O antigen extend out from the bridge. This positions the O antigen near the next part of the pathway, a complex of two proteins. One of the proteins (LptD) folds to form a β-barrel. Recall from chapter 3 that β-barrels are common in transport proteins (e.g., porin proteins) that translocate molecules across membranes. The second protein, LptE, resides at the periplasmic side of the LptD β-barrel. It is thought that LptE interacts with the O antigen and reorients it so that it moves vertically into the β-barrel. Lipid A is briefly exposed to the periplasm as this occurs, but then is quickly inserted into the outer leaflet of the outer membrane.

Sterols and Isoprenoid Lipids

Sterols are important lipids found in eukaryotic membranes, and isoprenoid lipids (also called isoprenes) are the type of lipid found in archaeal membranes. We consider them together because isoprenes are synthesized from two intermediates of the sterol biosynthetic pathway. The two intermediates are isopentenyl diphosphate and its isomer dimethylallyl diphosphate (**figure 12.32a,b**). These two molecules are most commonly synthesized by a pathway that starts with acetyl-CoA, an important precursor metabolite. Notice that both isopentenyl diphosphate and dimethylallyl diphosphate have two distinct ends. The diphosphate end is called the head and the opposite end is called a tail. The two molecules participate in a variety of condensation reactions that link molecules head to tail (figure 12.32c), head to head, or tail to tail, leading to the production of a variety of different molecules, including intermediates that give rise to cholesterol and steroid hormones, carotenoids, and archaeal lipids. The pathways to the various sterols found in eukaryotic cells are well understood and involve numerous reactions and enzymes. The archaeal isoprene synthesis pathways are less understood. Still unresolved is when the lipids are linked to glycerol to form glycerol diether lipids, when the hydrophilic head group is added, and how glycerol tetraethers are formed.

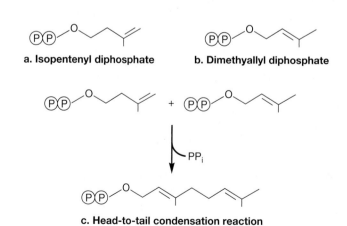

Figure 12.32 Isoprenoid Lipid Building Blocks. (a) Isopentenyl diphosphate and (b) dimethylallyl diphosphate are shown in (c) undergoing a head-to-tail condensation reaction.

Retrieve, Infer, Apply

1. What is a fatty acid? Describe in general terms how fatty acid synthase manufactures a fatty acid.
2. How are unsaturated fatty acids made?
3. Briefly describe the pathways for triacylglycerol and phospholipid synthesis. Of what importance are phosphatidic acid and CDP-diacylglycerol?

4. Activated carriers participate in carbohydrate, peptidoglycan, lipid, and LPS synthesis. Briefly describe these carriers and their roles. Are there any features common to all the carriers? Explain your answer.
5. What two building blocks are important for both sterol and isoprene biosynthesis. What pathway is most commonly used to provide them?

Key Concepts

12.1 Principles Governing Biosynthesis

- Many important cell constituents are macromolecules, large polymers constructed of simple monomers.
- Although many catabolic and anabolic pathways share enzymes, making metabolism more efficient, some of their enzymes are separate and independently regulated.
- Macromolecular components often undergo self-assembly to form the final supramolecular system.

12.2 Precursor Metabolites: Starting Molecules for Biosynthesis

- Precursor metabolites are carbon skeletons used as the starting substrates for biosynthetic pathways. They are intermediates of glycolytic pathways and the TCA cycle (i.e., the central metabolic pathways) (**figures 12.2** and **12.3**).
- Most precursor metabolites are used for amino acid biosynthesis; others are used for synthesis of porphyrins, purines, pyrimidines, and lipids.

12.3 CO₂ Fixation: Reduction and Assimilation of CO₂ Carbon

- Six different CO_2-fixation pathways have been identified in autotrophic microorganisms: the Calvin-Benson cycle, the reductive TCA cycle, the acetyl-CoA pathway, the 3-hydroxypropionate bi-cycle, the 3-hydroxypropionate/4-hydroxybutyrate pathway, and the dicarboxylate/4-hydroxybutyrate cycle.
- The Calvin-Benson cycle is used by most autotrophs to fix CO_2. It can be divided into three phases: the carboxylation phase, the reduction phase, and the regeneration phase (**figure 12.4**). Three ATP and two NADPH are used during the incorporation of one CO_2.
- The reductive TCA cycle is used by a number of autotrophs belonging to the bacterial phyla *Aquificae*, *Chlorobi*, *Proteobacteria*, and *Nitrospirae* (**figure 12.5**).

12.4 Synthesis of Carbohydrates

- Gluconeogenesis is the synthesis of glucose and related sugars from nonglucose precursors.

- Glucose, fructose, and mannose are gluconeogenic intermediates or are made directly from them (**figure 12.6**); galactose is synthesized with nucleoside diphosphate derivatives (**figure 12.7**). Bacteria and protists synthesize glycogen and starch from adenosine diphosphate glucose.
- Peptidoglycan synthesis is a complex process involving both UDP derivatives and the lipid carrier bactoprenol, which helps transport NAG-NAM-pentapeptide units across the plasma membrane. Cross-links are formed by transpeptidation (**figures 12.8–12.11**).

12.5 Synthesis of Amino Acids Consumes Many Precursor Metabolites

- The addition of nitrogen to the carbon chain provided by a precursor metabolite is an important step in amino acid biosynthesis. Ammonia, nitrate, or N_2 can serve as the source of nitrogen.
- Ammonia can be directly assimilated by the activity of transaminases and either glutamate dehydrogenase or the glutamine synthetase–glutamate synthase system (**figures 12.12–12.14**).
- Nitrate is incorporated through assimilatory nitrate reduction catalyzed by the enzymes nitrate reductase and nitrite reductase (**figure 12.15**).
- Nitrogen fixation is catalyzed by nitrogenase. Atmospheric molecular nitrogen is reduced to ammonia, which is then incorporated into amino acids (**figure 12.16**).
- Microorganisms can use cysteine, methionine, and inorganic sulfate as sulfur sources. Sulfate must be reduced to sulfide before it is assimilated. This occurs during assimilatory sulfate reduction (**figures 12.18** and **12.19**).
- Some amino acids are made directly by the addition of an amino group to a precursor metabolite, but most amino acids are made by pathways that are more complex. Many amino acid biosynthetic pathways are branched. Thus a single precursor metabolite can give rise to several amino acids (**figure 12.20**).
- Anaplerotic reactions replace TCA cycle intermediates to keep the cycle in balance while it supplies precursor metabolites. The anaplerotic reactions include the glyoxylate cycle (**figures 12.21** and **12.22**).

- Amino acids are also used in the synthesis of porphyrins, molecules consisting of four pyrrole rings. Two pathways have been identified for porphyrin synthesis: glutamate pathway and succinyl-CoA pathway (**figure 12.23**).

12.6 Synthesis of Purines, Pyrimidines, and Nucleotides

- Purines and pyrimidines are nitrogenous bases found in DNA, RNA, and other molecules. The nitrogen is supplied by certain amino acids. Phosphorus is provided by either inorganic phosphate or organic phosphate.
- Phosphorus can be assimilated directly by phosphorylation reactions that form ATP from ADP and P_i. Organic phosphorus sources are the substrates of phosphatases that release phosphate from the organic molecule.
- The purine skeleton is synthesized beginning with ribose 5-phosphate and initially produces inosinic acid (**figures 12.24** and **12.25**). Pyrimidine biosynthesis starts with carbamoyl phosphate and aspartate, and ribose is added after the skeleton has been constructed (**figures 12.26** and **12.27**).

12.7 Lipid Synthesis

- Fatty acids are synthesized from acetyl-CoA, malonyl-CoA, and NADPH by fatty acid synthase. During synthesis, the intermediates are attached to the acyl carrier protein (**figure 12.28**). Double bonds can be added in two different ways.
- Triacylglycerols are made from fatty acids and glycerol phosphate. Phosphatidic acid is an important intermediate in this pathway (**figure 12.29**).
- Phospholipids can be synthesized from phosphatidic acid by forming CDP-diacylglycerol, then adding an amino acid.
- Lipopolysaccharides are synthesized by a complex branched pathway that uses one branch to make the lipidA-core portion and the other to synthesize the O antigen (**figure 12.30**). These two are joined together in the periplasmic space. LPS must then move from the plasma membrane to the outer leaflet of the outer membrane. This is accomplished using the Lpt pathway (**figure 12.31**).
- Sterols and isoprenes are synthesized from two building blocks called isopentenyl diphosphate and dimethylallyl diphosphate. Numerous condensation reactions using these molecules in various ways build the ring structures of sterols and the glycerol diethers and diglycerol tetraethers found in archaeal membranes (**figure 12.32**).

Compare, Hypothesize, Invent

1. What would happen if a microorganism that depended on the glyoxylate cycle no longer produced the enzyme isocitrate lyase? Could this defect be corrected by nutrients in the media? If so, which nutrients?

2. Intermediary carriers are in a limited supply: When they cannot be recycled because of a metabolic block, serious consequences ensue. Think of some examples of these consequences.

3. Magnetotactic bacteria are morphologically and metabolically complex microbes that produce intracellular, membrane-bound magnetic crystals (*see figure 3.37*). The magnetotactic bacterial isolates called MV-1 and MC-1 both grow chemolithoautotrophically. However, MV-1 uses the Calvin-Benson cycle to fix carbon, whereas MC-1 uses the reductive TCA cycle.

Refer to figures 12.4 and 12.5 to identify key enzymatic steps in each pathway. How do you think the microbiologists studying these fascinating microbes were able to conclude that these two bacterial isolates use two different pathways for carbon fixation?

Read the original paper: Williams, T. J., et al. 2006. Evidence for autotrophy via the reverse tricarboxylic acid cycle in the marine magnetotactic coccus strain MC-1. *Appl. Environ. Microbiol.* 72:1322.

Learn More

shop.mheducation.com   Enhance your study of this chapter with interactive study tools and practice tests. Also ask your instructor about the resources available through Connect, including adaptive learning tools and animations.

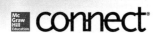 

13

Bacterial Genome Replication and Expression

Humans have long used coded messages such as this telegram sent in 1917 suggesting that Germany and Mexico form an alliance against the United States. As clever as humans have been at coding and decoding messages, nature devised highly successful coding and decoding mechanisms billions of years before.

Making Code

Since ancient times, humans have sent important messages in code. Julius Caesar sent coded messages to his generals. Mary Queen of Scots used code to detail her plans to overthrow Queen Elizabeth the First. George Washington regularly communicated to his troops in code, as have commanders in all wars. Today work continues to create codes that are harder to break and more content rich.

Code writing is related to another field: storage of information, which often involves converting the information using a code—a process termed encryption. The history of information storage parallels the history of communicating secrets in code. Initially coded messages were spoken; likewise, information was stored in song and dance. Later, coded messages were hand-written on paper; manuscripts and books stored valuable data. As codes became more sophisticated and their content larger, the physical form of coded messages evolved to include radio communications, recordings, and binary digits in computers. Similar methods were developed to store data sets.

In the early years of this century, scientists began exploring a new method of encrypting and storing valuable data. This new method is called biostorage or bioencryption. The earliest approaches stored coded data in pieces of DNA inserted into the chromosome of a bacterium. Scientists turned to DNA because they estimated that the information in about 1,000 books having 500 pages of small print is the equivalent of the information stored in one strand of human DNA. One of the earliest breakthroughs came in 2007 when a group of Japanese scientists used a soil bacterium to store Einstein's famous equation $E = MC^2$. Another advance came in 2011, when a group of students attending Hong Kong's Chinese University developed a method of encrypting data that takes up less space in the bacterial

genome. More recent approaches have forgone the use of cells, instead taking advantage of new methods for synthesizing and sequencing DNA, as we describe in chapter 18. The data are stored as small chains of deoxyribonucleotides (oligonucleotides) that are kept in wells of specially designed plates, each plate holding 1,536 wells. Scientists predict that about 50 zettabytes (a zettabyte is 2^{70} or about 10^{21} bytes) of information could be stored in approximately 1,000 of these plates.

For millennia cellular organisms have used DNA to store genetic information, and they have devised methods for replicating, recoding, and decoding genetic information—processes called DNA replication, transcription, and translation, respectively. All three construct polymers (DNA, RNA, protein) from monomers and are required for survival. In this chapter, we explore the metabolic reactions called polymerization reactions that generate these polymers (*see figure 11.1*). These reactions are central to the inheritance of traits of organisms.

The current understanding of inheritance of traits began in the mid-1800s when Gregor Mendel established how various traits in pea plants were inherited. At that time, chemists had not yet discovered DNA. Unfortunately, Mendel's work went largely unnoticed, but was rediscovered in the early twentieth century and furthered by scientists working with fruit flies and plants such as corn. The use of microorganisms as models for genetic studies soon followed. Microorganisms, especially bacteria, have significant advantages as model organisms, in part because of their unique characteristics. One important feature of bacteria is the nature of their genomes, which are usually haploid (1N). In addition, they often carry plasmids, which are useful to cells because they allow genetic information to move from one cell to another. Plasmids are also useful to scientists for genetic analyses and recombinant DNA technologies. Using bacteria (and often their viruses) as model systems was important in the elucidation of the nature of genetic information, gene structure, the genetic code, and mutation. Thus we begin by considering bacterial genetics in this chapter and the next. Because of the distinctive characteristics of eukaryotes and archaea, the replication and expression of their genomes are discussed in chapter 15.

Readiness Check:

Based on what you have learned previously, you should be able to:

✔ Define the terms genome, chromosome, plasmid, haploid, diploid, genotype, and phenotype

✔ Outline the flow of genetic information, identifying the processes used

✔ Draw a simple diagram that illustrates the chemical moieties found in nucleotides (section 12.6)

✔ State the base-pairing rules

✔ Describe the structure of bacterial genomes (section 3.6)

13.1 Experiments Using Bacteria and Viruses Demonstrated that DNA Is the Genetic Material

After reading this section, you should be able to:

■ Summarize Griffith's experiments on transformation

■ Relate how the contributions of Avery, MacLeod, McCarty, Hershey, and Chase confirmed that DNA stores genetic material

Although it is now hard to imagine, it was once thought that DNA was too simple a molecule to store genetic information. It is composed of only four different nucleotides, and it seemed that a molecule of much greater complexity must house the genetic information of a cell. It was argued that proteins, being composed of 20 different amino acids, were the better candidate for this important cellular function.

The early work of Fred Griffith in 1928 on the transfer of virulence (that is, the ability to cause disease) in the pathogen *Streptococcus pneumoniae*, commonly called pneumococcus, set the stage for research showing that DNA was indeed the genetic material. Griffith found that if he boiled virulent bacteria and injected them into mice, the mice were not affected and no pneumococci could be recovered from the animals (**figure 13.1**). When he injected a combination of killed virulent bacteria and a living nonvirulent strain, the mice died; moreover, he could recover living virulent bacteria from the dead mice. Griffith called this change of nonvirulent bacteria into virulent pathogens transformation.

Oswald Avery and his colleagues then set out to discover which constituent in the heat-killed virulent pneumococci was responsible for transformation. These investigators used enzymes to selectively destroy DNA, RNA, or protein in purified extracts of virulent pneumococci (S strain). They then exposed nonvirulent pneumococcal strains (R strains) to the treated extracts. Transformation of the nonvirulent bacteria was blocked only if DNA was destroyed, suggesting that DNA was carrying the information required for transformation (**figure 13.2**). The publication of these studies by Avery, C. M. MacLeod, and M. J. McCarty in 1944 provided the first evidence that DNA carried genetic information.

Eight years later, Alfred Hershey and Martha Chase wanted to know if protein or DNA carried the genetic information of a bacterial virus called T2 bacteriophage. They performed experiments in which they made the virus's DNA radioactive with ^{32}P, or they labeled its protein coat with ^{35}S. They mixed radioactive virions (either ^{32}P-labeled or ^{35}S-labeled) with *Escherichia coli* and incubated the mixture for a few minutes. This allowed the virions to attach to *E. coli* and initiate the infection process (*see figure 6.16*). The culture was then centrifuged to separate the infected cells from any unadsorbed phage particles. The infected cells in the resulting pellet were resuspended and the suspension was agitated violently in a blender. The blender treatment sheared off the bacteriophage particles adsorbed to the *E. coli* cells (**figure 13.3**). Some of the suspension was used in a plaque assay (*see figure 6.20*) to determine if the blender treatment affected the ability of the phage to multiply within the host cells. The remaining suspension was centrifuged to separate cells from the sheared-off phage particles. After centrifugation, radioactivity in the supernatant (where the phage particles remained) versus that in the bacterial cells in the pellet was determined.

The results of their experiments showed that blender treatment did not disrupt the infection process, because progeny phages were produced. They also demonstrated that when ^{35}S-labeled T2 was used in the experiment, the majority of the radioactive protein was in the supernatant; whereas when ^{32}P-labeled T2 was used, the radioactive DNA was in the bacterial cells that formed the pellet. Because DNA entered the cells and the protein did not, the phage DNA must have been carrying the genetic information needed to complete the infection process. Some luck was involved in their discovery, for the genetic material of many viruses is RNA and the researchers happened to select a DNA virus for their studies. Imagine the confusion if T2 had been an RNA virus! The controversy surrounding the nature of genetic information might have lasted considerably longer than it did.

Subsequent studies on the genetics of viruses and bacteria were largely responsible for the rapid development of molecular genetics. Furthermore, much of the recombinant DNA technology described in chapter 17 has arisen from studies of bacterial

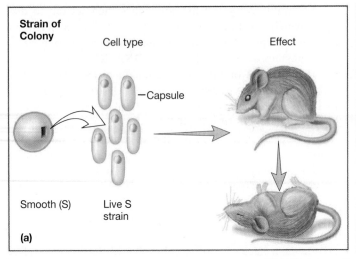

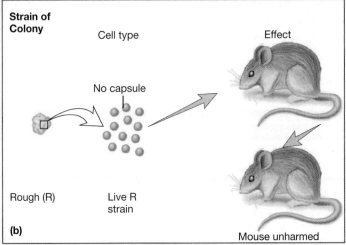

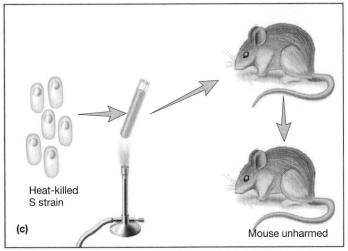

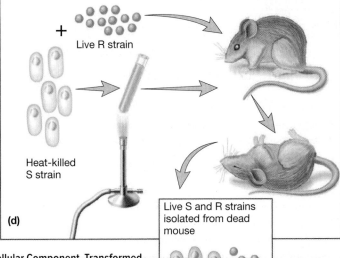

Figure 13.1 Griffith's Transformation Experiments Demonstrated that a Cellular Component Transformed Nonpathogenic Bacteria into Pathogenic Bacteria.   (a) Mice died of pneumonia when injected with pathogenic strains of pneumococci, which have a capsule and form smooth-looking colonies (S strains). (b) Mice survived when injected with nonpathogenic strains of pneumococci, which lack a capsule and form rough colonies (R strains). (c) Injection with heat-killed S strains had no effect. (d) Injection with a live R strain and a heat-killed S strain gave the mice pneumonia, and live S strain pneumococci could be isolated from the dead mice.

MICRO INQUIRY *Based on what we now know about proteins, why can we conclude from this experiment that genetic information was unlikely to be carried by proteins?*

and viral genetics. Research in microbial genetics has had a profound impact on biology as a science and on technology that affects everyday life.

Retrieve, Infer, Apply

1. Briefly summarize the experiments of Griffith; Avery, MacLeod, and McCarty; and Hershey and Chase.
2. Explain how protein was ruled out as the molecule of genetic information storage in each of the experiments performed by these important microbiologists.

13.2 Nucleic Acid and Protein Structure

After reading this section, you should be able to:

- Draw schematic representations of DNA, RNA, and amino acids that show their major features
- Compare and contrast the structure of DNA and RNA
- Identify the covalent bonds that are used to link nucleotides together to form a nucleic acid and amino acids together to form a polypeptide

DNA, RNA, and proteins are often called informational molecules. The information exists as the sequence of monomers

The experiments:

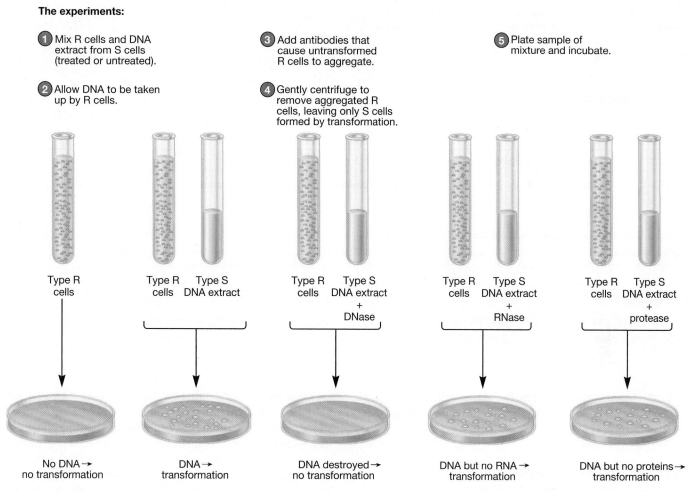

① Mix R cells and DNA extract from S cells (treated or untreated).

② Allow DNA to be taken up by R cells.

③ Add antibodies that cause untransformed R cells to aggregate.

④ Gently centrifuge to remove aggregated R cells, leaving only S cells formed by transformation.

⑤ Plate sample of mixture and incubate.

Type R cells

Type R cells | Type S DNA extract

Type R cells | Type S DNA extract + DNase

Type R cells | Type S DNA extract + RNase

Type R cells | Type S DNA extract + protease

No DNA → no transformation

DNA → transformation

DNA destroyed → no transformation

DNA but no RNA → transformation

DNA but no proteins → transformation

Figure 13.2 Some Experiments on the Transforming Principle. Earlier experiments done by Avery, MacLeod, and McCarty had shown that only DNA extracts from S cells caused transformation of R cells to S cells. To demonstrate that contaminating molecules in the DNA extract were not responsible for transformation, the DNA extract from S cells was treated with RNase, DNase, or protease, and then mixed with R cells. Only treatment of the DNA extract from S cells with DNase destroyed the ability of the extract to transform the R cells.

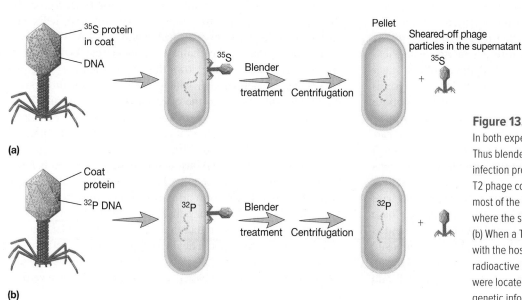

Pellet

^{35}S protein in coat

DNA

^{35}S

Blender treatment

Centrifugation

Sheared-off phage particles in the supernatant

^{35}S

(a)

Coat protein

^{32}P DNA

^{32}P

Blender treatment

Centrifugation

^{32}P

(b)

Figure 13.3 The Hershey-Chase Experiment. In both experiments, phage progeny were produced. Thus blender treatment did not interfere with the infection process. (a) When *E. coli* was infected with a T2 phage containing ^{35}S protein, after centrifugation most of the radioactivity remained in the supernatant where the sheared-off phage particles were located. (b) When a T2 phage containing ^{32}P DNA was mixed with the host bacterium, after centrifugation the radioactive DNA was in the pellet where the cells were located. Thus DNA was carrying the virus's genetic information.

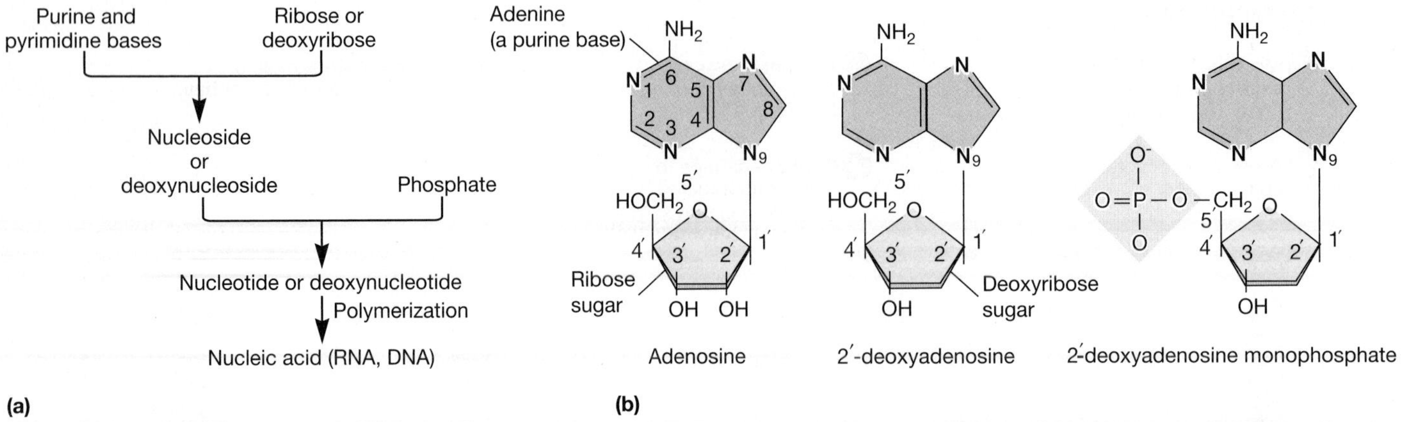

Figure 13.4 The Composition of Nucleic Acids.   (a) A diagram showing the relationships of various nucleic acid components. Combination of a purine or pyrimidine base with ribose or deoxyribose gives a nucleoside (a ribonucleoside or deoxyribonucleoside). A nucleotide contains a nucleoside and one or more phosphates. Nucleic acids polynucleotide chains formed when nucleotides are connected by phosphodiester bonds (figure 13.5a). (b) Examples of nucleosides—adenosine and 2′-deoxyadenosine—and the nucleotide 2′-deoxyadenosine monophosphate. The carbons of nucleoside and nucleotide sugars are indicated by numbers with primes to distinguish them from the carbons in the bases.

MICRO INQUIRY   *To which carbon of ribose (deoxyribose) is each of the following bonded: adenine and the hydroxyls of adenosine and deoxyadenosine?*

from which they are built. Here we describe the monomers and how they are linked together to form these important macromolecules.

DNA Is a Polymer of Deoxyribonucleotides

Deoxyribonucleic acid (DNA) is a polymer of deoxyribonucleotides (**figure 13.4**) linked together by phosphodiester bonds (**figure 13.5a**). It contains the bases adenine, guanine, cytosine, and thymine. DNA molecules are very large and are usually composed of two polynucleotide chains coiled together to form a double helix 2.0 nm in diameter (figure 13.5). The monomers of DNA are called deoxyribonucleotides because the sugar found in them is deoxyribose (figure 13.4b). The bond that links the monomers together to form the polymer is called a phosphodiester bond because it consists of a phosphate that forms a bridge between the 3′-hydroxyl of one sugar and the 5′-hydroxyl of an adjacent sugar. Purine and pyrimidine bases are attached to the 1′-carbon of the deoxyribose sugars, and the bases extend toward the middle of the cylinder formed by the two chains. (The numbers designating the carbons in the sugars are given a prime notation to distinguish them from the numbers designating the carbons and nitrogens in the nitrogenous bases.) The bases from each strand interact with those of the other strand, forming base pairs. The base pairs are stacked on top of each other in the center, one base pair every 0.34 nm. The purine adenine (A) of one strand is always paired with the pyrimidine thymine (T) of the opposite strand by two hydrogen bonds. The purine guanine (G) pairs with cytosine (C) by three hydrogen bonds. This AT and

GC base pairing means that the two strands in a DNA double helix are **complementary.** In other words, the bases in one strand match up with those of the other according to specific base-pairing rules. Because the sequences of bases in these strands encode genetic information, considerable effort has been devoted to determining the base sequences of DNA and RNA from many organisms, including hundreds of microbes. ▶▶| *Microbial genomics (chapter 18)*

The two polynucleotide strands of DNA fit together much like the pieces in a jigsaw puzzle. Inspection of figure 13.5b,c shows that the two strands are not positioned directly opposite one another. Therefore when the strands twist about one another, a wide major groove and narrower minor groove are formed by the backbone. There are 10.5 base pairs per turn of the helix, and each turn of the helix has a vertical length of 3.4 Å. The helix is right-handed; that is, the chains turn counterclockwise as they approach a viewer looking down the longitudinal axis. The two backbones are antiparallel, which means they run in opposite directions with respect to the orientation of their sugars. Thus, the 5′ end of one strand is paired with the 3′ end of the complementary strand; or stated another way, one strand is oriented 5′ to 3′ and the other, 3′ to 5′ (figure 13.5b).

The structure of DNA just described is that of the B form, the most common form in cells. Two other forms of DNA have been identified. The A form primarily differs from the B form in that it has 11 base pairs per helical turn, rather than 10.5, and a vertical length of 2.6 Å, rather than 3.4. Thus it is wider than the B form. The Z form is dramatically different, having a left-handed helical structure, rather than right-handed as seen in the B and A forms. The Z form has 12 base pairs per helical turn and a vertical rise of 3.7 Å. Thus it is more slender than the B form. At this time, it is unclear whether the A form is found in cells. However, evidence exists that small portions of chromosomes can be in the Z form. It is thought that they may relieve torsional stress placed on the DNA molecule when it is being actively transcribed.

DNA Structure

Another property of DNA needs to be addressed: supercoiling. DNA is helical; that is, it is a coil. Whenever the rotation of a coil is restrained in some way, it causes the coil to coil on itself. The coiling of a coil is supercoiling. Recall that most bacterial chromosomes are closed, circular double-stranded DNA molecules. In this state, the two strands are unable to rotate freely relative to each other, and the molecule is said to be strained. The strain is relieved by supercoiling. There are two

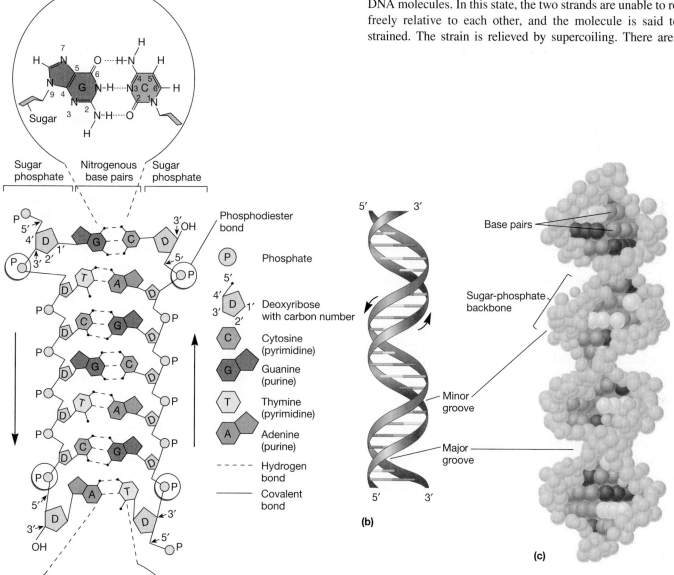

Figure 13.5 DNA Structure—B Form of DNA. DNA Is Usually a Double-Stranded Molecule.

(a) A schematic, nonhelical model. In each strand, phosphates are esterified to the 3′-carbon of one deoxyribose sugar (blue) and the 5′-carbon of the adjacent sugar. The two strands are held together by hydrogen bonds (dashed lines). Because of the specific base pairing, the base sequence of one strand determines the sequence of the other. The two strands are antiparallel; that is, the backbones run in opposite directions, as indicated by the two arrows, which point in the 5′ to 3′ direction. (b) A simplified model that highlights the antiparallel arrangement and the major and minor grooves. (c) A space-filling model of the B form of DNA. Note that the sugar-phosphate backbone spirals around the outside of the helix and the base pairs are embedded inside.

MICRO INQUIRY *How many H bonds are there between adenine and thymine, and between guanine and cytosine?*

types of supercoiling: positive and negative. For DNA, these are defined by the change in number of base pairs per turn in the double helix. As just discussed, the B form of DNA has 10.5 base pairs per turn of the helix. Supercoiling that decreases the number of base pairs per turn is said to be negative supercoiling. Likewise, supercoiling that increases the number of base pairs per turn is called positive supercoiling. Bacterial chromosomes are generally negatively supercoiled.

What is the importance of supercoiling? Supercoiling helps compact DNA so that it fits into the cell. Importantly for this chapter, supercoiling also "loosens" up the DNA, making it easier to separate the two strands from each other. Separation of the two strands is an important early step in both DNA replication and transcription, as we discuss in sections 13.3 and 13.5, respectively. Furthermore, positive supercoiling is often introduced into DNA during DNA replication. This can interfere with DNA replication and must be removed.

RNA Is a Polymer of Ribonucleotides

Ribonucleic acid (**RNA**) is a polymer of ribonucleotides (figure 13.4) that contains the sugar ribose and the bases adenine, guanine, cytosine, and uracil (instead of thymine). The nucleotides are joined by a phosphodiester bond, just as they are in DNA. In cells, RNA molecules are single stranded. However, an RNA strand can coil back on itself to form secondary structures such as hairpins with complementary base pairing and helical organization (p. 309). The formation of double-stranded regions in RNA is often critical to its function. ▶▶| *Attenuation and riboswitches can stop transcription prematurely (section 14.3); Translational riboswitches (section 14.4)*

Proteins Are Polymers of Amino Acids

Proteins are polymers of amino acids linked by peptide bonds; thus they are also called polypeptides. An amino acid is defined by the presence of a central carbon (the α carbon) to which are attached a carboxyl group, an amino group, and a side chain (**figure 13.6**). Twenty amino acids are normally used to form proteins. However, two unusual amino acids are found in some

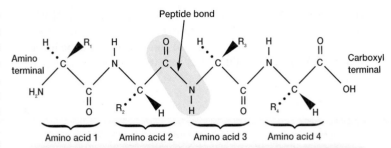

Peptide bond

Figure 13.7 **Peptide Bonds Link Amino Acids Together in Peptide Chains.** A tetrapeptide chain is shown. One of the peptide bonds linking the four amino acids together is highlighted in blue. At one end of the peptide is an amino group (amino or N terminal); at the other end is a carboxyl group (carboxyl or C terminal).

MICRO INQUIRY *Identify the two other peptide bonds of the tetrapeptide chain.*

proteins (section 13.6). Amino acids differ in terms of their side chains. Depending on the structure of the side chain, the amino acid is described as nonpolar, polar, or charged. The peptide bonds linking the amino acids together are formed by a reaction between the carboxyl group of one amino acid and the amino group of the next amino acid in the protein (**figure 13.7**). A polypeptide has polarity just as DNA and RNA do. At one end of the chain is an amino group, and at the other end is a carboxyl group. Thus a polypeptide has an amino or N terminus and a carboxyl or C terminus.

Proteins do not typically exist as extended chains of amino acids. Rather, they fold back on themselves to form three-dimensional structures, often more or less spherical in shape. The final shape is determined to a large extent by the sequence of amino acids in the polypeptide. This sequence is called the primary structure. Secondary and tertiary structures result from the folding of the chain. Finally, two or more polypeptide strands can interact to form the final, functional protein. This level of structure is called quaternary structure. These higher levels of structure are stabilized by intra- (and inter-) chain bonds. Protein structure is described in more detail in appendix I.

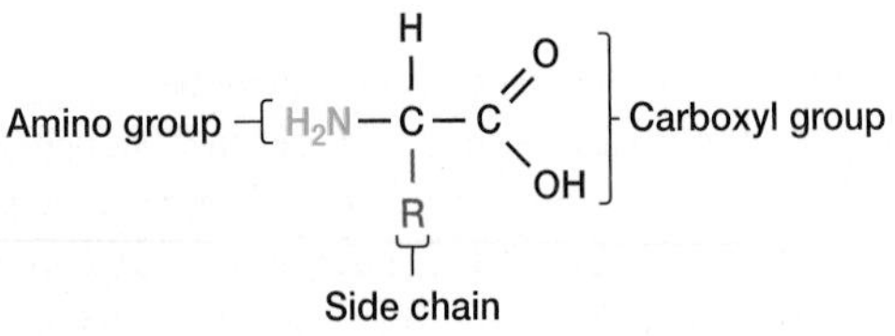

Figure 13.6 **The General Structure of an Amino Acid.** All amino acids have a central carbon (the α carbon) to which are attached a carboxyl group (COOH), an amino group (NH_2), and a side chain (R). Amino acids differ in terms of their side chains, which may be nonpolar, polar, negatively charged (acid), or positively charged (basic).

Retrieve, Infer, Apply

1. What are nucleic acids? How do DNA and RNA differ in structure?
2. What does it mean to say that the two strands of the DNA double helix are complementary and antiparallel? Examine figure 13.5*b* and explain the differences between the minor and major grooves.
3. Amino acids are described as nonpolar, polar, and charged, depending on the molecular makeup of their side chains. Which type of amino acid might be found in the transmembrane portion of a polypeptide located in a cell's plasma membrane? Explain your answer.

13.3 DNA Replication in Bacteria

After reading this section, you should be able to:

- Describe a bacterial replicon
- Summarize the events that occur during the three phases of DNA replication
- Create a table or concept map that illustrates the function of the major proteins found in a bacterial replisome
- List the enzymatic and structural elements needed by DNA polymerases for DNA synthesis
- Outline the major events that occur at the replication fork

DNA replication is an extraordinarily important and complex process upon which all life depends. During DNA replication, the two strands of the double helix are separated; each then serves as a template for the synthesis of a complementary strand according to the base-pairing rules. Each of the two progeny DNA molecules consists of one new strand and one old strand, and DNA replication is said to be semiconservative (**figure 13.8**). DNA replication is also extremely accurate; *E. coli* makes errors with a frequency of only 10^{-9} or 10^{-10} per base pair replicated (or about one in a million [10^{-6}] per gene per generation). Despite its complexity and accuracy, replication is very rapid. In bacteria, replication rates approach 750 to 1,000 base pairs per second. Most of our discussion in this section is based on studies of chromosomal DNA replication in *E. coli*.

Bacterial DNA Replication Initiates from a Single Origin of Replication

The replication of chromosomal DNA begins at a single point, the **origin of replication.** Synthesis of DNA occurs at the **replication fork,** the place at which the DNA helix is unwound and individual strands are replicated. Two replication forks move outward from the origin until they have copied the whole **replicon**—the portion of the genome that contains an origin that is replicated as a unit. When the replication forks move around the circular chromosomes observed in most bacteria, a structure shaped like the Greek letter theta (θ) is formed (**figure 13.9**). Because the bacterial chromosome is a single replicon, the forks meet on the other side and two separate chromosomes are released. Less is known about the replication of linear bacterial chromosomes. One well-studied organism with linear chromosomes belongs to the spirochete genus *Borrelia*. The replication of its chromosomes is considered on p. 298. 🔁 *DNA Replication Fork; Bidirectional DNA Replication*

Replisomes Contain DNA Polymerases and Other Proteins Needed for DNA Replication

DNA replication is essential to organisms, and a great deal of effort has been devoted to understanding its mechanism. The

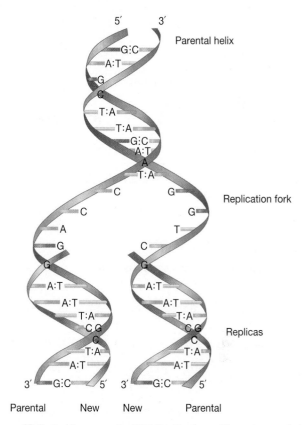

Figure 13.8 Semiconservative DNA Replication. The replication fork of DNA showing the synthesis of two progeny strands. Newly synthesized strands are purple. Each copy contains one new and one old strand.

replication of *E. coli* DNA requires at least 30 proteins that form a huge complex called the **replisome** (**table 13.1**). As we describe in chapter 15, many of the archaeal and eukaryotic replication enzymes and proteins differ from those used by bacterial cells. In general, the archaeal replication machinery is more similar to the eukaryotic machinery than it is to the bacterial system. Despite these differences, the overall process of DNA replication is similar in all organisms.

Replisome enzymes called **DNA polymerases** catalyze DNA synthesis. All known DNA polymerases catalyze DNA synthesis in the 5′ to 3′ direction, and the nucleotide to be added is a deoxyribonucleoside triphosphate (dNTP; often referred to simply as deoxynucleoside triphosphates or deoxynucleotides). Deoxynucleotides are linked by phosphodiester bonds formed by a reaction between the hydroxyl group at the 3′ end of the growing DNA strand and the phosphate closest to the 5′ carbon (the α-phosphate) of the incoming deoxynucleotide (**figure 13.10**). The energy needed to form the phosphodiester bond is provided by release of the terminal two phosphates as pyrophosphate (PP_i) from the nucleotide that is added. The PP_i is subsequently hydrolyzed to two separate phosphates (P_i). Thus the deoxynucleoside triphosphates dATP, dTTP, dCTP, and dGTP serve as DNA polymerase substrates while

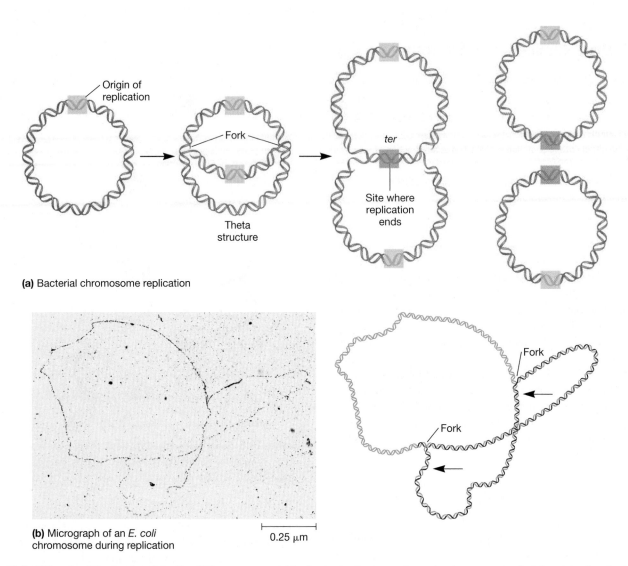

(a) Bacterial chromosome replication

(b) Micrograph of an *E. coli* chromosome during replication

0.25 μm

Figure 13.9 **Bidirectional Replication of the *E. coli* Chromosome.** (a) Replication begins at one site on the chromosome, called the origin of replication. Two replication forks proceed in opposite directions from the origin until they meet at a site called the replication termination site (*ter*). A theta structure is a commonly observed intermediate of the process. (b) An autoradiograph of a replicating *E. coli* chromosome; about one-third of the chromosome has been replicated. To the right is a schematic representation of the chromosome. Parental DNA is blue; new DNA strands are purple; arrow represents direction of fork movement.

deoxynucleoside monophosphates (dNMPs: dAMP, dTMP, dCMP, dGMP) are incorporated into the growing chain.

🔖 *How Nucleotides Are Added in DNA Replication*

DNA polymerase needs three things to catalyze DNA synthesis. The first is a template, which is read in the 3′ to 5′ direction and is used to direct synthesis of a complementary DNA strand. The second is a primer (e.g., an RNA strand or a DNA strand) to provide a free 3′-hydroxyl group to which nucleotides can be added (figure 13.10). The third is a set of dNTPs. *E. coli* has five different DNA polymerases (DNA polymerase I-V). DNA polymerase III plays the major role in replication, although it is assisted by DNA polymerase I.

DNA polymerase III holoenzyme is a multifunctional enzyme composed of 10 different proteins. For many years it was thought

that the complex contained two core enzymes (**figure 13.11**). However, most recent evidence supports a model consisting of three core polymerases. Each core enzyme binds a strand of DNA and is responsible for catalyzing DNA synthesis and proofreading the product to ensure fidelity of replication. Associated with each core enzyme is a subunit called the β clamp. The β clamp tethers a core enzyme to the DNA. At the center of the holoenzyme, and represented by an octopus-like structure in figure 13.11, is a complex of proteins called the τ (tau) complex clamp loader, which includes proteins responsible for loading the β clamp onto DNA. The tau proteins in the complex hold the holoenzyme together. Two of the core enzymes replicate one of the DNA strands, and the third core enzyme replicates the other strand. Thus, both strands of DNA are bound by a single DNA polymerase III holoenzyme.

Table 13.1	Components of the *E. coli* Replication Machinery
Protein	**Function**
DnaA (initiator protein)	Initiation of replication; binds origin of replication (*oriC*)
DnaB	Helicase (5'→3'); breaks hydrogen bonds holding two strands of double helix together; promotes DNA primase activity; involved in primosome assembly
DNA gyrase	Relieves supercoiling of DNA produced as DNA strands are separated by helicases; separates daughter molecules in final stages of replication
SSB proteins	Bind single-stranded DNA after strands are separated by helicases
DnaC	Helicase loader; working with DnaA, directs DnaB (helicase) to DNA template
DNA primase	Synthesis of RNA primer; component of primosome
DNA polymerase III holoenzyme	Catalyzes most of the DNA synthesis that occurs during DNA replication; has 3'→5' exonuclease (proofreading) activity
DNA polymerase I	Removes RNA primers; fills gaps in DNA formed by removal of RNA primer
Ribonuclease H	Removes RNA primers
DNA ligase	Seals nicked DNA, joining DNA fragments together
Tus	Termination of replication
Topoisomerase IV	Separation of chromosomes upon completion of DNA replication

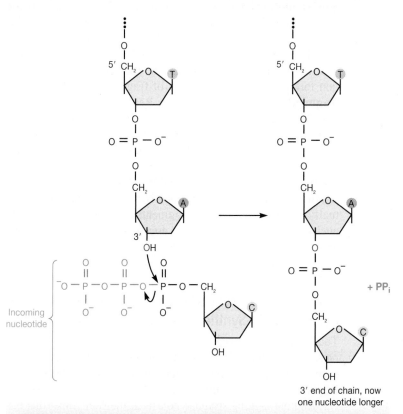

Figure 13.10 The DNA Polymerase Reaction. The hydroxyl of the 3' terminal deoxyribose makes a nucleophilic attack on the α-phosphate of the nucleotide substrate (in this example, adenosine attacks cytidine triphosphate).

MICRO INQUIRY *What provides the energy to fuel this reaction?*

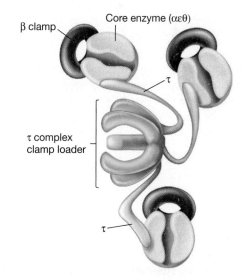

Figure 13.11 DNA Polymerase III Holoenzyme.   The holoenzyme consists of three core enzymes (each composed of three different proteins; α, ε, θ, not shown) and several other proteins. Each core enzyme is associated with a β clamp, which tethers a DNA template to the core enzyme. The tau (τ) proteins are part of a large complex called the τ complex clamp loader, so named because some of the proteins in the complex load β clamps onto DNA.

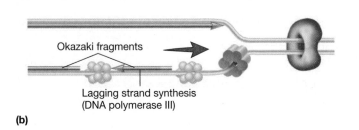

(a)

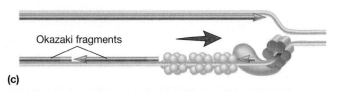

(b)

(c)

Figure 13.12 Other Replisome Proteins. A single replication fork showing the activity of replisome proteins other than DNA polymerase III holoenzyme (not shown) is illustrated. DnaB helicase is responsible for separating the two strands of parental DNA. The strands are kept apart by single-stranded DNA binding proteins (SSB), which allows for synthesis of an RNA primer by DNA primase. DNA gyrase eases the strain introduced into the DNA double helix by helicase activity. Both leading strand and lagging strand synthesis are illustrated. The lagging strand is synthesized in short fragments called Okazaki fragments. A new RNA primer is required for the synthesis of each Okazaki fragment.

MICRO INQUIRY *What is the difference between helicase and gyrase? Which is a topoisomerase?*

Other proteins found in the replisome include helicases, single-stranded DNA binding proteins, and topoisomerases (**figure 13.12**). **Helicases** are responsible for separating (unwinding) the DNA strands just ahead of the replication fork, using energy from ATP hydrolysis. **Single-stranded DNA binding proteins** (SSBs) keep the strands apart once they have been separated, and **topoisomerases** relieve the tension generated by the rapid unwinding of the double helix (the replication fork may rotate as rapidly as 75 to 100 revolutions per second). This is important because rapid unwinding can lead to the formation of positive supercoils in the helix ahead of

the replication fork, and these can impede replication if not removed. Topoisomerases change the structure of DNA by transiently breaking one or two strands without altering the nucleotide sequence of the DNA (e.g., a topoisomerase might tie or untie a knot in a DNA strand). **DNA gyrase** is an important topoisomerase in *E. coli*. It is not only important during DNA replication but also for introducing negative supercoiling in the bacterial chromosome that helps compact it. ◀◀ *Nucleoid (section 3.6)*

Once the template is prepared, the primer needed by DNA polymerase III is synthesized by an enzyme called **primase.** The primer is a short RNA strand, usually around 10 nucleotides long and complementary to the DNA (figure 13.12). RNA is used as the primer because unlike DNA polymerase, RNA polymerases (such as primase) can initiate RNA synthesis without an existing 3′-OH. Primase and several accessory proteins form a complex called the **primosome** (table 13.1). The primosome is another important component of the replisome.

As noted, DNA polymerase enzymes synthesize DNA in the 5′ to 3′ direction. Therefore one of the DNA polymerase core enzymes is able to move in the same direction as the replication fork and synthesize DNA continuously as the DNA unwinds ahead of the core enzyme. This strand is called the **leading strand** (figure 13.12). The other strand, called the **lagging strand,** cannot be extended in the same direction as the movement of the replication fork because there is no free 3′-OH to which a nucleotide can be added. As a result, the lagging strand is synthesized discontinuously in the 5′ to 3′ direction (i.e., in the direction opposite of the movement of the replication fork) and produces a series of fragments called **Okazaki fragments,** after their discoverer, Reiji Okazaki (1930–1975). Discontinuous synthesis occurs as primase makes many RNA primers along the template strand. DNA polymerase III then extends these primers with DNA, and eventually the Okazaki fragments are joined to form a complete strand. Thus while the leading strand requires only one RNA primer to initiate synthesis, the lagging strand has many RNA primers that must eventually be removed. Okazaki fragments are about 1,000 to 2,000 nucleotides long in bacteria and approximately 100 nucleotides long in eukaryotic cells. ⟳ *DNA Replication; Structural Basis of DNA Replication*

The Replication Fork Is Where the Complex Events of DNA Synthesis Occur

The details of DNA replication are outlined in **figure 13.13**. We present replication as a series of discrete steps, but in the cell these events occur quickly and simultaneously on both the leading and lagging strands. Although we show two core DNA polymerases carrying out lagging strand synthesis at the same time, this has yet to be definitively demonstrated.

In *E. coli*, DNA replication is initiated at specific nucleotides called the *oriC* locus (for *ori*gin of *c*hromosomal replication). This site is AT rich. Recall that adenines pair with thymines using only two hydrogen bonds, so AT-rich segments of DNA

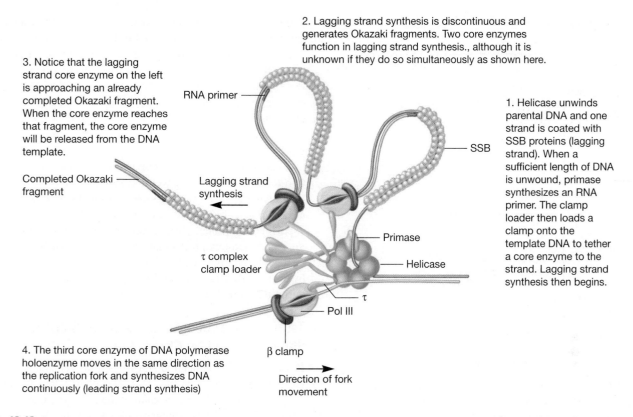

3. Notice that the lagging strand core enzyme on the left is approaching an already completed Okazaki fragment. When the core enzyme reaches that fragment, the core enzyme will be released from the DNA template.

2. Lagging strand synthesis is discontinuous and generates Okazaki fragments. Two core enzymes function in lagging strand synthesis., although it is unknown if they do so simultaneously as shown here.

RNA primer

1. Helicase unwinds parental DNA and one strand is coated with SSB proteins (lagging strand). When a sufficient length of DNA is unwound, primase synthesizes an RNA primer. The clamp loader then loads a clamp onto the template DNA to tether a core enzyme to the strand. Lagging strand synthesis then begins.

SSB

Completed Okazaki fragment

Lagging strand synthesis

Primase

τ complex clamp loader

Helicase

τ

Pol III

4. The third core enzyme of DNA polymerase holoenzyme moves in the same direction as the replication fork and synthesizes DNA continuously (leading strand synthesis)

β clamp

Direction of fork movement

Figure 13.13 Events at the Replication Fork.   Events are numbered and move in a counterclockwise direction. DNA polymerase III holoenzyme and other components of the replisome are responsible for synthesis of both leading and lagging strands. The arrows show the direction of movement of lagging strand synthesis and the replication fork. Okazaki fragments are eventually joined together after removal of the RNA primer and synthesis of DNA to fill the gap, both catalyzed by DNA polymerase I (figure 13.14); DNA ligase then seals the nick and joins the two fragments (figure 13.15).

become single stranded more readily than do GC-rich regions. This is important for initiation of replication.

1. The bacterial initiator protein DnaA is responsible for triggering DNA replication. DnaA proteins bind regions in *oriC* throughout the cell cycle, but to initiate replication, DnaA proteins must bind a few particular *oriC* sequences. The presence of DnaA at these sites recruits a helicase (usually DnaB helicase) to the origin.

2. The helicase unwinds the helix with the aid of topoisomerases such as DNA gyrase (figure 13.13, step 1). The single strands are kept separate by SSBs.

3. Primase synthesizes RNA primers as needed (figure 13.13, step 1). DNA polymerase III holoenzyme catalyzes both leading strand and lagging strand synthesis from the RNA primers. Lagging strand synthesis is particularly amazing because of the "gymnastic" feats performed by the holoenzyme. It must discard old β clamps, load new β clamps, and tether the template to the core enzyme with each new round of Okazaki fragment synthesis. All of this occurs as DNA polymerase III is synthesizing DNA.

4. After most of the lagging strand has been synthesized by the formation of Okazaki fragments, DNA polymerase

I removes the RNA primers using its unique ability to snip off nucleotides one at a time starting at the 5′ end while moving toward the 3′ end of the RNA primer. This ability is referred to as 5′ to 3′ exonuclease activity. DNA polymerase I begins its exonuclease activity at the free 5′ end of each RNA primer. With the removal of each ribonucleotide, the adjacent 3′-OH from the deoxynucleotide is used by DNA polymerase I to fill the gap between Okazaki fragments (**figure 13.14**).

5. Finally, the Okazaki fragments are joined by the enzyme **DNA ligase,** which forms a phosphodiester bond between the 3′-OH of the growing strand and the 5′-phosphate of an Okazaki fragment (**figure 13.15**).

Amazingly, DNA polymerase III, like all DNA polymerases (except DNA polymerases called reverse transcriptases), has an additional function that is critically important: **proofreading.** Proofreading is the removal of a mismatched base immediately after it has been added; its removal must occur before the next base is incorporated. One of the protein subunits of the DNA polymerase III core enzyme (the ε subunit) has 3′ to 5′ exonuclease activity. This activity enables the polymerase core enzyme to check each newly incorporated base to see that it forms stable hydrogen bonds. In this way, mismatched bases can be detected.

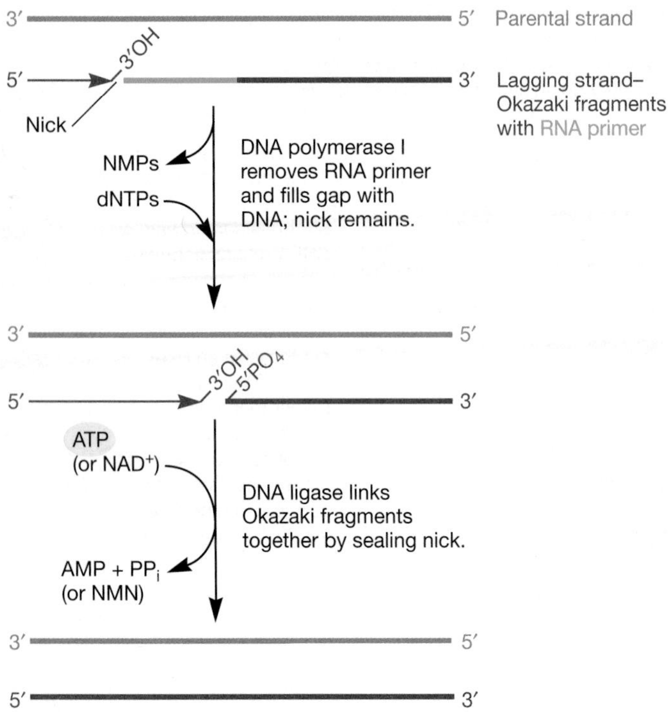

Figure 13.14 **DNA Polymerase I Completes Lagging Strand Synthesis.** NMPs, nucleoside monophosphates; dNTPs, deoxynucleoside triphosphates; NAD⁺, nicotinamide adenine dinucleotide; NMN, nicotinamide mononucleotide.

If the wrong base has been mistakenly added, the exonuclease activity is used to remove it, but only as long as it is still at the 3′ end of the growing strand (**figure 13.16**). Once removed, holoenzyme backs up and adds the proper nucleotide in its place. DNA proofreading is not 100% efficient, and as discussed in chapter 16, the mismatch repair system is the cell's second line of defense against the potential harm caused by the incorporation of the incorrect nucleotide. ⟳ *Proofreading Function of DNA Polymerase*

As we have seen, DNA polymerase III is a remarkable multiprotein complex, with several enzymatic activities. In *E. coli,* the polymerase component is encoded by the *dnaE* gene. *Bacillus subtilis,* a Gram-positive bacterium that is another important experimental model, has a second polymerase gene called *dnaE_{Bs}.* Its protein product appears to be responsible for synthesizing the lagging strand. Thus while the overall mechanism by which DNA is replicated is highly conserved, there can be variations in replisome components.

Termination of Replication Requires Steps to Separate Daughter Chromosomes

In *E. coli,* DNA replication stops when the replisome reaches a termination site (*ter*) on the DNA. A protein called Tus binds to the *ter* sites and halts progression of the forks. In many other bacteria, replication stops spontaneously when the forks meet. Regardless of how fork movement is stopped, there are two problems that often must be solved by the replisome. One is the formation of interlocked chromosomes called **catenanes** (**figure 13.17a**). The other is a dimerized chromosome—two chromosomes joined together to form a single chromosome twice as long (figure 13.17b). Catenanes are produced when topoisomerases break and rejoin DNA strands to ease supercoiling ahead of the replication fork. The two daughter DNA molecules are separated by the action of other topoisomerases that break both strands of one molecule, pass the other DNA molecule through the break, and then rejoin the strands (figure 13.17a). Dimerized chromosomes result from DNA recombination that sometimes occurs between two daughter molecules during DNA replication.

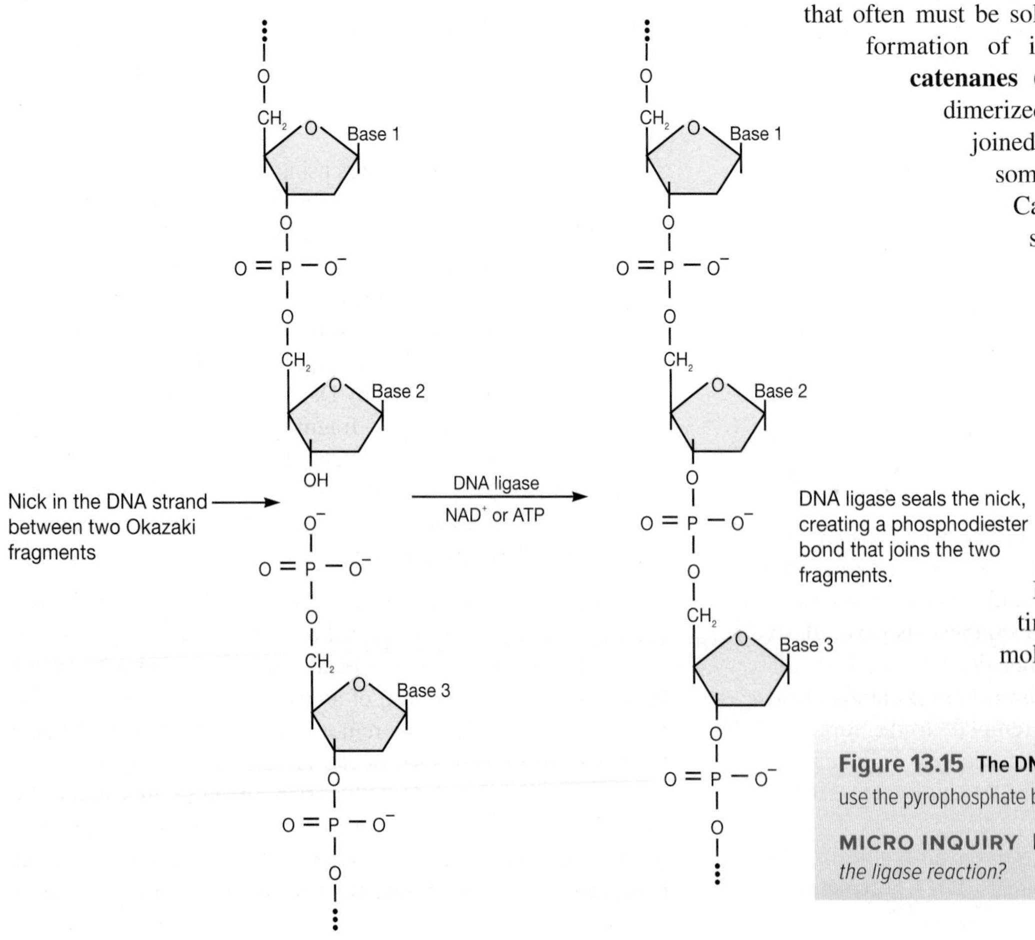

DNA ligase seals the nick, creating a phosphodiester bond that joins the two fragments.

Nick in the DNA strand between two Okazaki fragments

DNA ligase
NAD⁺ or ATP

Figure 13.15 **The DNA Ligase Reaction.** Bacterial ligases use the pyrophosphate bond of NAD⁺ or ATP as an energy source.

MICRO INQUIRY *Why can't DNA polymerase I perform the ligase reaction?*

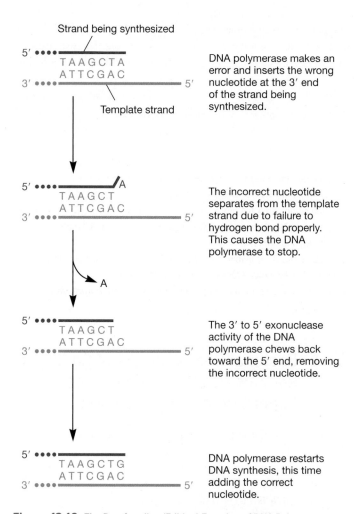

Figure 13.16 The Proofreading (Editing) Function of DNA Polymerase.

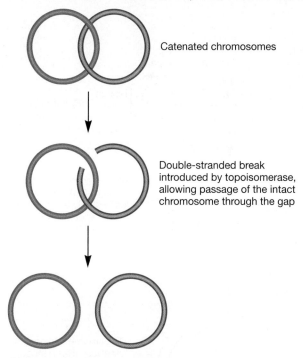

(a) Resolution of catenated daughter chromosomes by a topoisomerase

Catenated chromosomes

Double-stranded break introduced by topoisomerase, allowing passage of the intact chromosome through the gap

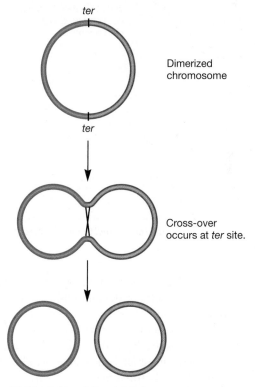

ter

Dimerized chromosome

ter

Cross-over occurs at *ter* site.

(b) Resolution of dimerized chromosome by the XerCD recombinase

Figure 13.17 Resolving Catenated and Dimerized Chromosomes.
(a) Catenated chromosomes arise from the activity of topoisomerases during chromosome replication. Accordingly, they are resolved by topoisomerases. (b) Dimerized chromosomes arise from recombination events that can occur during chromosome replication between the daughter chromosomes. They are resolved by recombinase enzymes such as XerCD of *E. coli*. These enzymes catalyze a cross-over that separates the two chromosomes.

Recombinase enzymes (e.g., XerCD in *E. coli*) catalyze an intramolecular cross-over that separates the two chromosomes (figure 13.17*b*).

Linear Chromosome Ends Must Be Protected During Replication

All eukaryotic chromosomes and some bacterial chromosomes are linear. In both cases, this poses a problem during replication because all DNA polymerases synthesize DNA in the 5′ to 3′ direction from a primer that provides a free 3′-OH. When the RNA primer for the Okazaki fragment at the 5′ end of the daughter strand is removed, the daughter molecule is shorter than the parent molecule. Over numerous rounds of DNA replication and cell division, this leads to a progressively shortened chromosome. Ultimately the chromosome loses critical genetic information, which is lethal to the cell. This is called the "end replication problem," and a cell must solve it if it is to survive. Eukaryotic cells have solved the end replication problem with an enzyme called telomerase, as we describe in chapter 15. Bacteria have taken a different approach. ▶▶ *Telomeres and telomerases (section 15.2).*

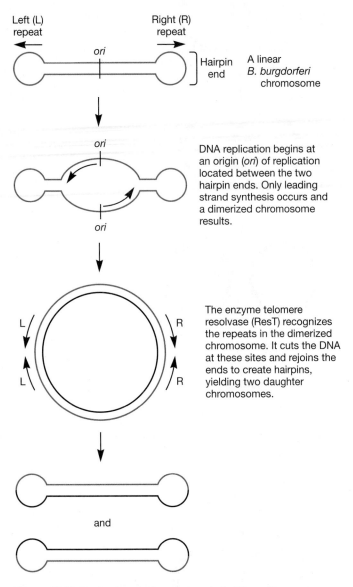

Figure 13.18 Replication of a *Borrelia burgdorferi* Linear Chromosome.

Of those bacteria having linear chromosomes, the best understood mechanism for solving the end replication problem is that used by *Borrelia burgdorferi*. The approach *B. burgdorferi* cells use is to disguise the ends so well that they aren't really ends. How are the ends disguised? Consider a typical linear, double-stranded DNA molecule. At each terminus of the double helix is a strand having a 3′-OH and the complementary strand having a 5′-phosphate. Now examine **figure 13.18**, which shows that *B. burgdorferi* chromosomes do not have these free ends. Rather, a phosphodiester bond links the two complementary strands together. This is made possible by inverted repeats at each terminus and the formation of a hairpin. The origin of replication is located between the hairpins. Interestingly, only leading strand synthesis occurs at each replication fork. When replication is complete, a circular molecule has been formed that is twice the length of

the parent chromosome. Thus it is a dimerized chromosome. An enzyme called telomere resolvase (ResT) cuts the two chromosomes apart as it forms hairpin ends for each daughter molecule.

Retrieve, Infer, Apply

1. How many replicons do typical bacterial cells have (i.e., those having a single chromosome)? How many replication forks are used to replicate a circular chromosome? Is a primosome part of the replisome, or is the replisome part of the primosome? What is the function of each?
2. Describe the nature and functions of the following replication components and intermediates: DNA polymerases I and III, topoisomerase, DNA gyrase, helicase, single-stranded DNA binding proteins, Okazaki fragment, DNA ligase, leading strand, lagging strand, and primase.
3. Outline the steps involved in DNA synthesis at the replication fork. How do DNA polymerases correct their mistakes?
4. What is the end replication problem? How does *B. burgdorferi* solve it?

13.4 Bacterial Genes Consist of Coding Regions and Other Sequences Important for Gene Function

After reading this section, you should be able to:

- Draw a typical bacterial protein-coding gene; label the important portions of the gene and the conventions for numbering base pairs in the gene
- Draw typical tRNA- and rRNA-coding genes

DNA replication allows genetic information to be passed from one generation to the next. But how is the information used? To answer that question, we must first look at how genetic information is organized. The basic unit of genetic information is the gene. Genes have been regarded in several ways. At first, it was thought that a gene contained information for the synthesis of one enzyme—the one gene–one enzyme hypothesis. This was modified to the one gene–one polypeptide hypothesis because of the existence of enzymes and other proteins composed of two or more different polypeptide chains (subunits) coded for by separate genes. A segment of DNA that encodes a single polypeptide is sometimes termed a cistron. However, not all genes encode proteins; some code instead for ribosomal RNA (rRNA) and transfer RNA (tRNA), both of which function in protein synthesis (**figure 13.19**). In addition, it is now known that some eukaryotic genes encode more than one protein. Thus a **gene** might be defined as a polynucleotide sequence that codes for one or more functional products (i.e., a polypeptide, tRNA, or rRNA). In this section, we consider the structure of each of these three types of genes.

Protein-Coding Genes

Most of the genes found in bacterial genomes encode proteins. However, DNA does not serve directly as the template for protein

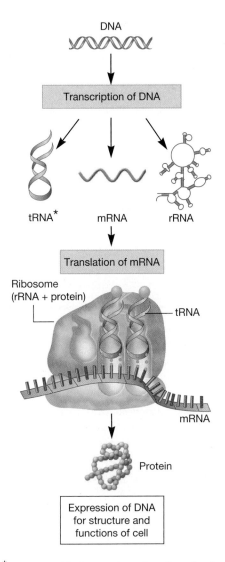

DNA

↓

Transcription of DNA

tRNA*   mRNA   rRNA

↓

Translation of mRNA

Ribosome
(rRNA + protein)

tRNA

mRNA

↓

Protein

Expression of DNA
for structure and
functions of cell

*The sizes of RNA are enlarged to show details.

Figure 13.19 **Transcription Yields Three Major Types of RNA Molecules.** Messenger RNA (mRNA) molecules arise from transcription of protein-coding genes. They are translated into protein with the aid of the other two major types of RNA: transfer RNA (tRNA) molecules carry amino acids to the ribosome during translation; ribosomal RNA (rRNA) molecules have several functions, including catalyzing peptide bond formation.

synthesis. Rather, the genetic information in the gene is transcribed to give rise to a messenger RNA (mRNA), which is translated (section 13.7) into a protein (figure 13.19). For this to occur, protein-coding genes must contain signals that indicate where transcription should start and stop, and signals in the resulting mRNA that indicate where translation should start and stop. As we describe in more detail in section 13.5, during transcription only one strand of a gene directs mRNA synthesis. This strand is called the **template strand,** and the complementary DNA strand is known as the sense strand because it is the same nucleotide sequence as the mRNA, except in DNA bases (**figure 13.20**). Messenger RNA is synthesized from the 5′ to the 3′ end in a manner similar to DNA synthesis.

Therefore the polarity of the DNA template strand is 3′ to 5′. In other words, the beginning of the gene is at the 3′ end of the template strand.

An important site called the **promoter** is located at the start of the gene. The promoter is the binding site for RNA polymerase, the enzyme that synthesizes RNA. The promoter is neither transcribed nor translated; it functions strictly to orient RNA polymerase so it is a specific distance from the first DNA nucleotide that will serve as a template for RNA synthesis. The promoter thus specifies which strand is to be transcribed and where transcription should begin. As we discuss in chapter 14, the sequences near the promoter often are very important in regulating when and at what rate a gene is transcribed. ▶▶| *Regulation of transcription initiation saves considerable energy and materials (section 14.2)*

The transcription start site (labeled +1 in figure 13.20) represents the first nucleotide in the mRNA synthesized from the gene. However, the initially transcribed portion of the gene does not necessarily code for amino acids. Instead, it is a **leader** that is transcribed into mRNA but is not translated into amino acids. In bacteria, the leader includes a region called the **Shine-Dalgarno sequence,** which is important in the initiation of translation. The leader sometimes is also involved in regulation of transcription and translation. ▶▶| *Attenuation and riboswitches can stop transcription prematurely (section 14.3); Riboswitches and small RNAs can control translation (section 14.4)*

Immediately next to (and downstream of) the leader is the most important part of the gene, the **coding region** (figure 13.20). The coding region typically begins with the template DNA sequence 3′-TAC-5′. This is transcribed into the start codon, 5′-AUG-3′, which codes for the first amino acid of the polypeptide encoded by the gene. The remainder of the coding region is transcribed into a sequence of codons that specifies the sequence of amino acids for the rest of the protein. The coding region ends with a sequence that, when transcribed, is a stop codon. It signals the end of the protein and stops the ribosome during translation. The stop codon is immediately followed by the **trailer,** which is transcribed but not translated. The trailer contains sequences that prepare RNA polymerase for release from the template strand. Indeed, just beyond the trailer (and sometimes slightly overlapping it) is the **terminator.** The terminator is a sequence that signals RNA polymerase to stop transcription.

In bacteria, the coding region is usually continuous, unlike the coding regions of eukaryotic genes, which are often interrupted by noncoding sequences called **introns.** Those rare bacterial genes that do contain introns are transcribed into an intron-containing mRNA. The introns are eventually removed but by a mechanism different than that used to remove introns from eukaryotic mRNAs. ▶▶| *From pre-mRNA to mRNA (section 15.3)*

tRNA and rRNA Genes

Actively growing cells need a ready supply of tRNA and rRNA molecules so that protein synthesis can occur. To ensure this, bacterial cells often have more than one gene for each of these molecules. Furthermore, it is important that the number of each

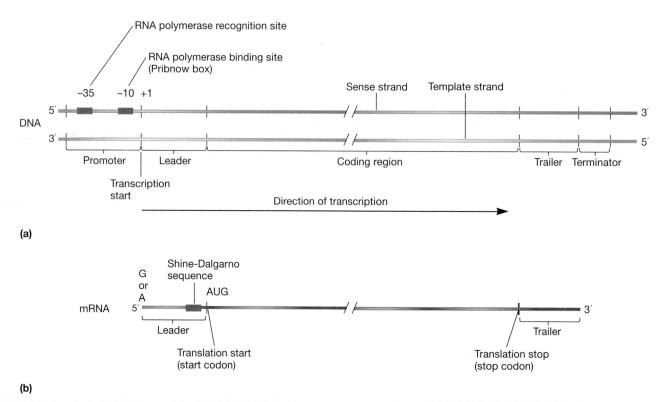

Figure 13.20   A Bacterial Structural Gene and Its mRNA Product.   (a) The organization of a typical structural gene in a bacterial cell. Some genes lack leaders or trailers or both. Transcription begins at the +1 position in DNA and proceeds to the right, as shown. The numbering of nucleotides to the left of this spot is in a negative direction, while the numbering to the right is in a positive direction. For example, the nucleotide that is immediately to the left of the +1 nucleotide is numbered −1, and the nucleotide to the right of the +1 nucleotide is numbered +2. There is no zero nucleotide in this numbering system. In many bacterial promoters, sequence elements at the −35 and −10 regions play a key role in promoting transcription. During transcription, the template is read in the 3′ to 5′ direction. Regulatory sites are not shown but are usually upstream of the coding region and may overlap with the promoter. (b) Messenger RNA product of the gene shown in part a. The first nucleotide incorporated into mRNA is usually GMP or AMP. Translation of the mRNA begins with the AUG start codon.

MICRO INQUIRY   *Why is the nontemplate strand called the "sense strand"?*

tRNA or rRNA relative to other tRNAs or rRNAs be controlled. This is accomplished in part by having several tRNA or rRNA genes transcribed together from a single promoter.

In bacteria, genes for tRNA consist of a promoter, tRNA coding region, leader, and trailer. When more than one tRNA is transcribed from the promoter, the coding regions are separated by short spacer sequences (**figure 13.21a**). Whether the gene encodes a single tRNA or multiple tRNAs, the initial transcript must be processed to remove the noncoding sequences (i.e., leader, trailer, and spacers, if present). This is called posttranscriptional modification, and it is accomplished by ribonucleases—enzymes (and in some cases ribozymes) that cut RNA. In addition, many bacterial tRNA genes contain introns that must be removed during tRNA maturation. ◀◀ *Ribozymes (section 10.6)*

Bacterial cells usually contain more than one rRNA gene. Each gene has a promoter, trailer, and terminator, and encodes all three types of rRNA (figure 13.21b). Thus, as seen for tRNA genes, the transcript from an rRNA gene is a single, large precursor molecule that is cut up by ribonucleases to yield the final

rRNA products. Interestingly, in many bacteria, the trailer regions and the spacers often contain tRNA genes. Thus the precursor rRNA encodes for both tRNA and rRNA. Recently the 16S rRNA genes of several giant bacteria (e.g., *Thiomargarita namibiensis*) were shown to contain several introns. This was surprising because introns had never been observed before in bacterial 16S rRNA genes. Like other bacterial introns, they are removed by self-splicing rather than by a spliceosome as in eukaryotes.

Retrieve, Infer, Apply

1. The coding region of a gene is said to be "downstream" from the leader. Conversely, the leader is said to be "upstream" of the coding region. For each of the following portions of a gene, indicate whether it is downstream or upstream of the coding region: promoter, +1 nucleotide, trailer, terminator. Which portions of a gene are transcribed but not translated?

2. Which strand of a gene has sequences that correspond to the start (i.e., ATG rather than AUG) and stop codons in the mRNA product of the gene?

3. Briefly discuss the general organization of tRNA and rRNA genes. How does their expression differ from that of protein-coding genes with respect to posttranscriptional modification of the gene product?

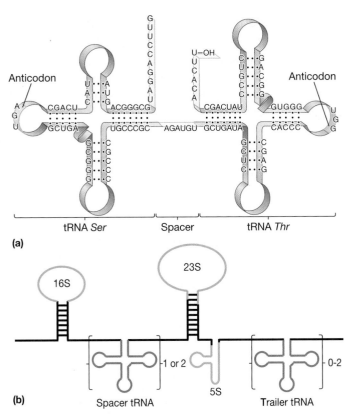

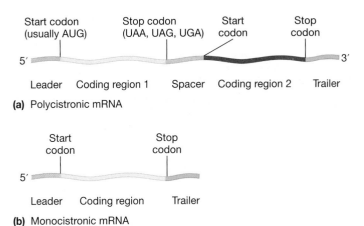

(a) Polycistronic mRNA

(b) Monocistronic mRNA

Figure 13.22 Polycistronic and Monocistronic mRNAs. Polycistronic mRNAs are commonly observed in bacteria and archaea. Eukaryotic genes usually give rise to monocistronic mRNAs.

Figure 13.21 tRNA and rRNA Genes. (a) A tRNA precursor from *E. coli* that contains two tRNA molecules. The spacer and extra nucleotides at both ends are removed during processing. (b) The *E. coli* rRNA gene codes for a large transcription product that is cleaved into three rRNAs and one to three tRNAs. The 16S, 23S, and 5S rRNA segments are represented by blue lines, and tRNA sequences are placed in brackets. The seven copies of this gene vary in the number and kind of tRNA sequences.

13.5 Transcription in Bacteria

After reading this section, you should be able to:

- Illustrate the organization of bacterial genes in a typical operon
- Describe the structure of a typical bacterial RNA polymerase holoenzyme
- Outline the events that occur during the three phases of transcription
- Discuss the role of bacterial promoters and sigma factors in transcription initiation
- Distinguish factor-independent termination of transcription from rho-dependent termination of transcription

Synthesis of RNA under the direction of DNA is called **transcription,** and the RNA product has a sequence complementary to the DNA template directing its synthesis. Although adenine directs the incorporation of thymine during DNA replication, it usually codes for uracil during RNA synthesis. Transcription generates three major kinds of RNA. **Transfer RNA (tRNA)** carries amino acids during protein synthesis, and **ribosomal**

RNA (**rRNA**) molecules are components of ribosomes. **Messenger RNA (mRNA)** bears the message for protein synthesis. Bacterial genes encoding proteins involved in a related process (e.g., encoding enzymes for synthesis of an amino acid) are often located close to each other and are transcribed from a single promoter. Such a transcriptional unit is termed an **operon.** Transcription of an operon yields an mRNA consisting of a leader followed by one coding region, which is separated from the second coding region by a spacer, and so on, with the final sequence of nucleotides being the trailer. Such mRNAs are said to be **polycistronic mRNAs** (**figure 13.22a**). Each coding region in the polycistronic mRNA is defined by a start and stop codon. Thus each coding region is translated separately to give rise to a single polypeptide. Many archaeal mRNAs are also polycistronic. However, polycistronic mRNAs are rare in eukaryotes. Instead, their mRNAs are usually **monocistronic mRNAs** (figure 13.22b), containing information of a single gene. ▶▶◀ *Transcription (section 15.3)*

Bacterial RNA Polymerases Consist of Five Different Proteins

RNA is synthesized by enzymes called **RNA polymerases.** In bacteria, a single RNA polymerase transcribes all genes. Most bacterial RNA polymerases contain five types of polypeptide chains: α, β, β′, ω and σ (**figure 13.23**). The **RNA polymerase core enzyme** is composed of five polypeptides (two α subunits, β, β′, and ω) and catalyzes RNA synthesis. The **sigma factor (σ)** has no catalytic activity but instead functions as a transcription factor by helping the core enzyme recognize the promoter. When sigma is bound to the core enzyme, the six-subunit complex is termed **RNA polymerase holoenzyme.** Only holoenzyme can begin transcription, but the core enzyme completes RNA synthesis once it has been initiated.

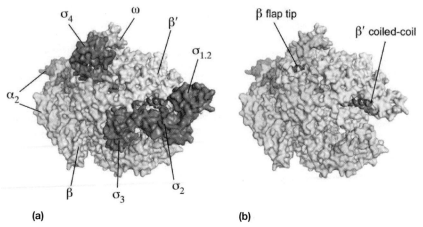

(a) **(b)**

Figure 13.23 RNA Polymerase Structure. (a) The holoenzyme of the bacterium *Thermus aquaticus*. Note that several portions of σ factor have been identified using subscripts. (b) Core enzyme showing two interactions sites (β flap tip and β′ coiled-coil) for σ with the β and β′ proteins, respectively.

Stages of Transcription: Initiation, Elongation, and Termination

Transcription involves three separate processes: initiation, elongation, and termination, which together are often referred to as the transcription cycle (**figure 13.24**). The transcription factor sigma is critical to the initiation process. As part of the RNA polymerase holoenzyme, it helps position the core enzyme at the promoter. Bacterial promoters have several characteristic features. Two of particular note are a sequence of six bases (often TTGACA) about 35 base pairs before (upstream) the transcription starting point and a TATAAT sequence called the **Pribnow box,** usually about 10 base pairs upstream of the transcriptional start site (**figure 13.25**; also figure 13.20). These regions are called the −35 and −10 sites, respectively, because these are their

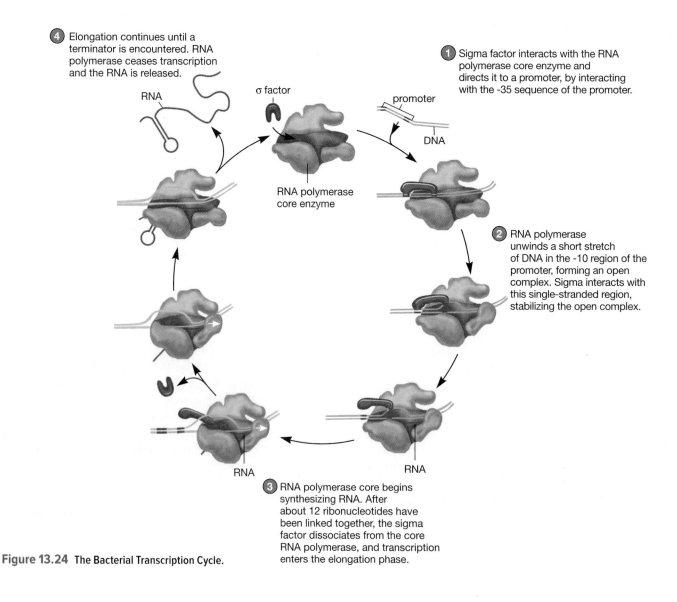

4 Elongation continues until a terminator is encountered. RNA polymerase ceases transcription and the RNA is released.

1 Sigma factor interacts with the RNA polymerase core enzyme and directs it to a promoter, by interacting with the -35 sequence of the promoter.

RNA

σ factor

promoter

DNA

RNA polymerase core enzyme

2 RNA polymerase unwinds a short stretch of DNA in the -10 region of the promoter, forming an open complex. Sigma interacts with this single-stranded region, stabilizing the open complex.

RNA

RNA

3 RNA polymerase core begins synthesizing RNA. After about 12 ribonucleotides have been linked together, the sigma factor dissociates from the core RNA polymerase, and transcription enters the elongation phase.

Figure 13.24 The Bacterial Transcription Cycle.

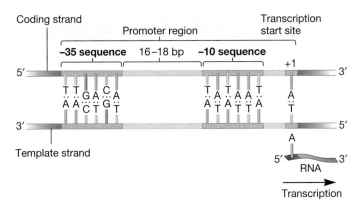

Figure 13.25 Promoter region, –35 sequence 16–18 bp –10 sequence, Coding strand 5′ ... 3′, Template strand 3′ ... 5′, Transcription start site +1, RNA 5′ 3′, Transcription →

T T G A C A / A A C T G T ... T A T A A T / A T A T T A ... A / T ... A

Figure 13.25 A σ⁷⁰ Promoter. Many bacterial promoters are recognized by the sigma factor σ⁷⁰. Its promoters have a characteristic set of nucleotides centered at about −10 and −35. Shown are the consensus sequences for these two important promoter sites. Bacterial cells produce additional sigma factors that recognize different promoter sequences (table 13.2).

MICRO INQUIRY *Are the −35 and −10 regions considered "upstream" or "downstream" of the +1 nucleotide?*

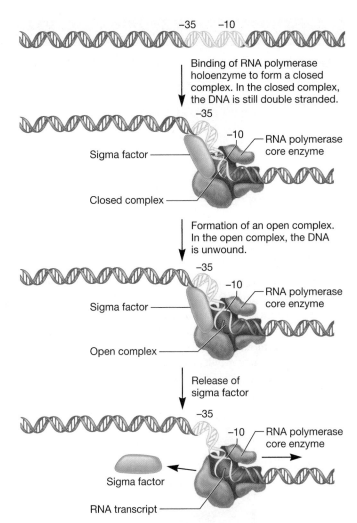

Figure 13.26 Initiation of Transcription in Bacteria. The sigma factor of the RNA polymerase holoenzyme is responsible for positioning the core enzyme properly at the promoter. Sigma factor recognizes two regions in the promoter, one centered at −35 and the other centered at −10. Once positioned properly, the DNA at the −10 region unwinds to form an open complex. The sigma factor dissociates from the core enzyme after transcription is initiated.

approximate distances in nucleotides upstream of the first nucleotide to be transcribed (i.e., the +1 site). Sigma factor first recognizes the −35 sequence, allowing the holoenzyme to "settle down" on that region of the promoter. Sigma and proteins in the core enzyme undergo conformational changes that cause the DNA strands in the AT-rich −10 region to separate, forming what will become the transcription bubble. Sigma then interacts with one of the strands and stabilizes the interaction of RNA polymerase with the unwound DNA. The resulting complex of RNA polymerase holoenzyme and unwound DNA is called the open complex (**figure 13.26**). *Stages of Transcription*

At this point in our discussion, it is worth noting that bacterial cells produce more than one type of sigma factor. Each sigma factor preferentially directs RNA polymerase to a distinct set of promoters. For instance, in *E. coli*, most genes have promoters recognized by a sigma factor called σ⁷⁰. This sigma factor recognizes promoters having the −10 and −35 sequences shown in figure 13.25 and **table 13.2**. These sequences are the **consensus sequences** for σ⁷⁰-recognized promoters. Promoters recognized by other sigma factors have different consensus sequences. The use of different sigma factors to initiate transcription is a common bacterial regulatory mechanism, as we describe in chapter 14. Our focus here is on transcription of genes recognized by σ⁷⁰.

Once the open complex is formed, transcription can begin. Within the open complex is a region of unwound DNA equivalent to about 16 to 20 base pairs. This is the "transcription bubble," and it moves with the RNA polymerase as it synthesizes mRNA from the template DNA strand during elongation (**figure 13.27**). Within the transcription bubble, a temporary RNA:DNA hybrid is formed. As RNA polymerase holoenzyme progresses along the DNA template, the sigma factor dissociates

from the other subunits and can help another RNA polymerase core enzyme initiate transcription (figure 13.24).

The reaction catalyzed by RNA polymerase is quite similar to that catalyzed by DNA polymerase (figure 13.10). ATP, GTP, CTP, and UTP are used to produce RNA complementary to the DNA template, and pyrophosphate is produced as ribonucleoside monophosphates are incorporated into the growing RNA chain. Pyrophosphate is hydrolyzed to fuel the process. RNA synthesis also proceeds in a 5′ to 3′ direction with new ribonucleotides being added to the 3′ end of the growing chain, making the RNA complementary and antiparallel to the template DNA. As elongation of the mRNA continues, single-stranded mRNA is released, and the two strands of DNA behind the transcription bubble resume their double helical structure. As shown in figure 13.24, RNA polymerase is a remarkable enzyme capable of several

Table 13.2 *E. coli* Sigma Factors and the Sequences They Recognize

Sigma Factor	Consensus Promoter Sequences[1]		Genes Transcribed from Promoter
σ^{70}	TTGACAT	TATAAT	Most genes
σ^{54}	CTGGNA[2]	TTGCA	Genes for nitrogen metabolism
σ^{38}	TTGACA	TCTATACTT	Genes for stationary phase and stress responses
σ^{32}	TCTCNCCCTTGAA	CCCCATNTA	Genes for heat-shock response
σ^{28}	CTAA	CCGATAT	Genes for chemotaxis and motility

1 With the exception of the σ^{54} promoters, all consensus sequences are located at −35 and −10, respectively. The σ^{54} consensus sequences are located at −24 and −12.
2 N indicates any nucleotide.

activities, including unwinding the DNA, moving along the template, and synthesizing RNA.

Termination of transcription occurs when the core RNA polymerase dissociates from the template DNA. This is brought about by the terminator. There are two kinds of terminators. The first type causes factor-independent termination (**figure 13.28**). This terminator consists of an inverted repeat followed by an A-rich nucleotide sequence. RNA polymerase transcribes the inverted repeat, but it pauses within the A-rich region. This allows the inverted repeat to fold back on itself, forming a hairpin-shaped stem-loop structure. The A-U base pairs holding the DNA and RNA together in the transcription bubble are too weak to hold the RNA:DNA duplex together and RNA polymerase falls off.

The second kind of terminator is termed factor-dependent terminator because it requires the aid of a protein. The best-studied termination factor is rho factor (ρ). Rho factor can be involved in transcription termination of all types of genes, but its action is best studied for protein-coding genes. Current models propose that rho

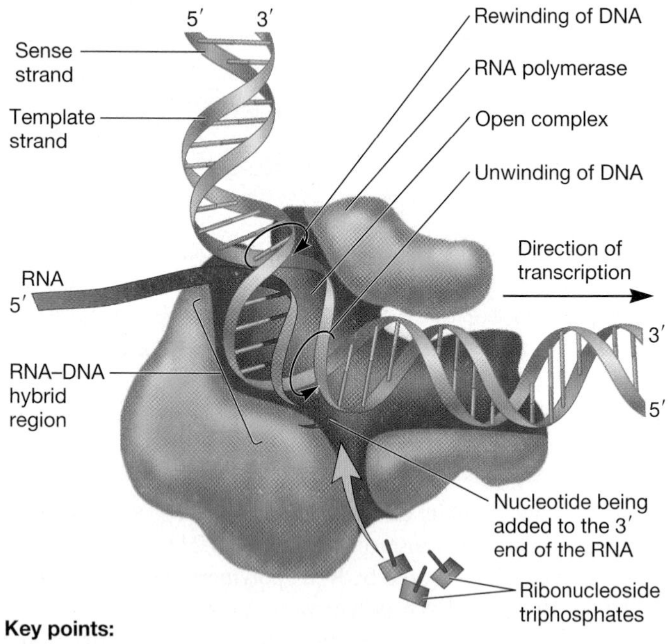

Key points:

• RNA polymerase slides along the DNA, creating an open complex as it moves.

• The template strand is used to make a complementary copy of RNA as an RNA–DNA hybrid.

• The RNA is synthesized in a 5′ to 3′ direction using ribonucleoside triphosphates as precursors. Pyrophosphate is released (not shown).

• The complementarity rule is the same as the AT/GC rule except that U is substituted for T in the RNA.

Figure 13.27 The Transcription Bubble.

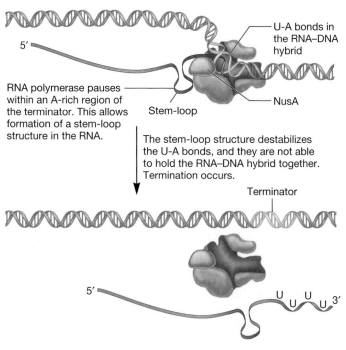

Figure 13.28 **Factor-Independent Termination of Transcription.**
This type of terminator contains an inverted repeat and an A-rich sequence downstream from the repeat. A protein called NusA stimulates termination.

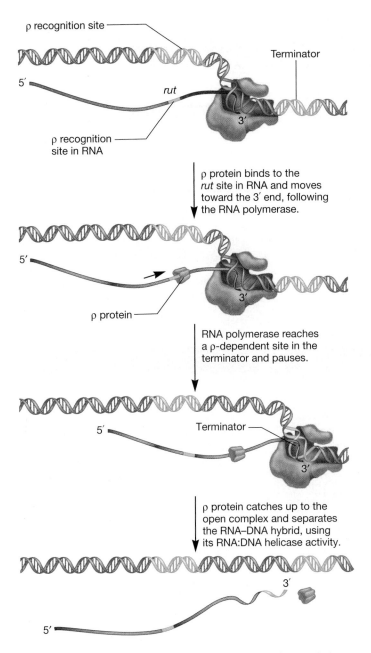

ρ recognition site

Terminator

5′

rut

3′

ρ recognition site in RNA

ρ protein binds to the *rut* site in RNA and moves toward the 3′ end, following the RNA polymerase.

5′

3′

ρ protein

RNA polymerase reaches a ρ-dependent site in the terminator and pauses.

Terminator

5′

3′

ρ protein catches up to the open complex and separates the RNA–DNA hybrid, using its RNA:DNA helicase activity.

3′

5′

Figure 13.29 **Rho-Factor (ρ)-Dependent Termination of Transcription.** The *rut* site stands for *rho-utilization* site.

binds to mRNA at a site called *rut* for *rho-ut*ilization site. For rho to bind, *rut* must be free of ribosomes, as shown in **figure 13.29.** Rho uses energy supplied by ATP hydrolysis to move along the mRNA, as it tries to catch up with RNA polymerase. However, rho's rate of movement is slower than that of RNA polymerase. Thus rho can only catch up with RNA polymerase if the polymerase pauses at a rho-dependent pause site. If this occurs, rho catches up with RNA polymerase and causes RNA polymerase to dissociate from DNA. Rho is known to have hybrid RNA:DNA helicase activity, and this activity is thought to cause unwinding of the mRNA-DNA complex and release of RNA polymerase from the template DNA strand. ❧ *mRNA Synthesis*

13.6 The Genetic Code Consists of Three-Letter "Words"

After reading this section, you should be able to:

- Explain the importance of the reading frame of a protein-coding gene
- Describe the universal genetic code
- List deviations from the universal genetic code that have been identified in some microorganisms
- Explain how the wobble hypothesis enables organisms to encode fewer tRNA molecules

The final step in expression of protein-coding genes is translation. Protein synthesis is called translation because it is a decoding process. The information encoded in the language of nucleic acids must be rewritten in the language of proteins. During translation, the sequence of nucleotides is "read" in discrete sets of three nucleotides, each set being a **codon.** Each codon codes for a single amino acid. The sequence of codons is "read" in only one way—the **reading frame (figure 13.30)**—to give rise to the amino acid sequence of a polypeptide. Deciphering the genetic code was one of the great achievements of the twentieth century. Here we examine the nature of the genetic code. Translation is considered in section 13.7.

The genetic code, presented in RNA form, is summarized in **table 13.3.** Close inspection of the code reveals several features that are related not only to the way cells use DNA to store information but also to why it is valuable for storing data, as described in the chapter opening story. One feature is that the code words (codons) are three letters (bases) long; thus one small "word" conveys a significant amount of information. Each codon is recognized by an anticodon present on a tRNA molecule. Another feature is that the code has "punctuation." One codon, AUG, is almost always the first codon in the protein-coding portion of mRNA molecules. It is called the **start codon** because it serves as the start site for translation by coding for the initiator tRNA. Three other codons (UGA, UAG, and UAA) terminate translation and are called **stop** or **nonsense codons.** These codons do not encode an amino acid and therefore do not have a tRNA bearing their anticodon. Thus only 61 of the 64 codons in the code, the **sense codons,** direct amino acid incorporation into

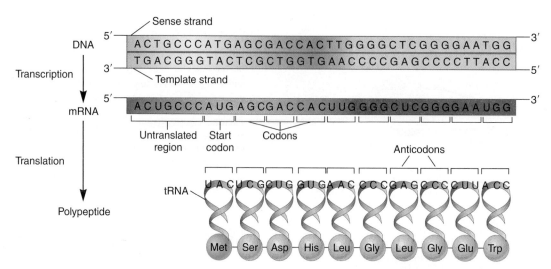

Figure 13.30 Reading Frame. During transcription, an mRNA complementary to the template strand of DNA is synthesized. The nucleotides in the mRNA are organized into groups of three, each group being a codon. The first codon translated into protein is the start codon. It establishes the reading frame and therefore the sequence of amino acids in the polypeptide chain that is made from the mRNA.

Table 13.3	The Genetic Code[1]			

SECOND POSITION

First Position (5′ End)[2]	U	C	A	G	Third Position (3′ End)
U	UUU UUC } Phe F UUA UUG } Leu L	UCU UCC UCA UCG } Ser S	UAU UAC } Tyr Y UAA UAG } STOP	UGU UGC } Cys C UGA STOP UGG Trp W	U C A G
C	CUU CUC CUA CUG } Leu L	CCU CCC CCA CCG } Pro[3] P	CAU CAC } His H CAA CAG } Gln Q	CGU CGC CGA CGG } Arg R	U C A G
A	AUU AUC AUA } Ile I AUG Met M	ACU ACC ACA ACG } Thr T	AAU AAC } Asn N AAA AAG } Lys K	AGU AGC } Ser S AGA AGG } Arg R	U C A G
G	GUU GUC GUA GUG } Val V	GCU GCC GCA GCG } Ala A	GAU GAC } Asp D GAA GAG } Glu E	GGU GGC GGA GGG } Gly G	U C A G

[1] Next to each codon or set of codons is the R group characteristic of that amino acid.
[2] The code is presented in RNA form. Codons run in the 5′ to 3′ direction. The 3- and 1-letter abbreviations for the amino acids designated by the codon are provided, as is the structure of the R chain of that amino acid.
[3] Proline is an imino acid, rather than an amino acid.

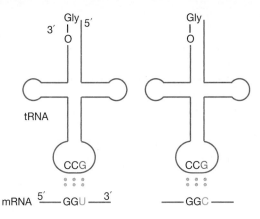

(a) Base pairing of one glycine tRNA with two codons due to wobble

Glycine mRNA codons: GGU, GGC, GGA, GGG (5′ —→ 3′)

Glycine tRNA anticodons: CCG, CCU, CCC (3′ —→ 5′)

(b) Glycine codons and anticodons

Figure 13.31 Wobble Decreases the Number of tRNAs Needed to Specify a Particular Amino Acid. The use of wobble in coding for the amino acid glycine. (a) Because of wobble, G in the 5′ position of the anticodon can pair with either C or U in the 3′ position of the codon. Thus two codons can be recognized by the same tRNA. (b) Because of wobble, only three tRNA anticodons are needed to translate the four glycine (Gly) codons.

protein. Finally, the genetic code exhibits **code degeneracy** (also called redundancy); that is, there are up to six different codons for a given amino acid.

Despite the existence of 61 sense codons, there are fewer than 61 different tRNAs. It follows that not all codons have a corresponding tRNA. Cells can successfully translate mRNA using fewer tRNAs because loose pairing between the 5′ base in the anticodon and the 3′ base of the codon is tolerated. Thus as long as the first and second bases in the codon correctly base pair with an anticodon, the tRNA bearing the correct amino acid will bind to the mRNA during translation. This is evident on inspection of the code. Note that the codons for a particular amino acid most often differ at the third position (table 13.3). This somewhat loose base pairing is known as **wobble,** and it relieves cells of the need to synthesize so many tRNAs (**figure 13.31**). Wobble also decreases the effects of some mutations. ▶▶ *Mutations (section 16.1)*

The description of the genetic code just provided is of the universal genetic code. However, there are exceptions to the code (**table 13.4**). The first exceptions discovered were stop codons that encoded one of the 20 amino acids. For instance, mycoplasma bacteria use the stop codon UGA to code for glutamine (Gln). More dramatic deviations from the code have also been discovered. Members of all three domains of life encode proteins containing the amino acid selenocysteine, the twenty-first amino acid (**figure 13.32a**). Most selenocysteine-containing enzymes catalyze redox reactions. Pyrrolysine, the twenty-second amino acid, can be found in the proteins of several methanogenic archaea and at least one bacterium (figure 13.32b). In the methanogens, pyrrolysine is found in

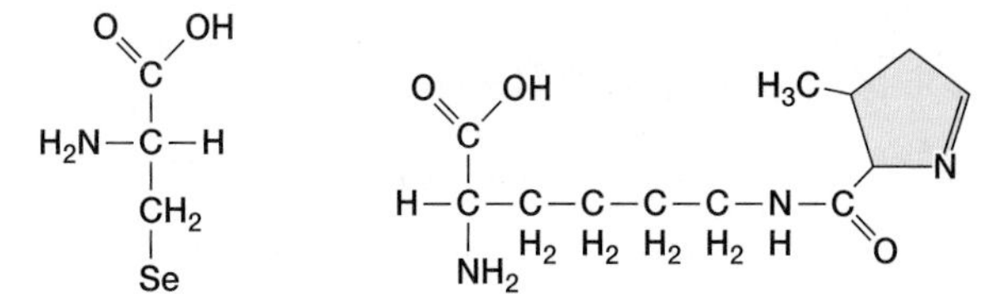

(a) Selenocysteine   **(b)** Pyrrolysine

Figure 13.32 Selenocysteine and Pyrrolysine. These unusual amino acids are not coded for by DNA but are inserted in response to specific stop codons and stem-loop structures in mRNA.

methyltransferase enzymes, which function in methanogenesis. Genomic analysis indicates that pyrrolysine might also exist in many other bacteria and some eukaryotes. Selenocysteine is inserted at certain UGA codons, whereas pyrrolysine is inserted at UAG codons (p. 314).

Retrieve, Infer, Apply

1. List the "punctuation" codons of the genetic code and indicate what each signifies.
2. What is the difference between a codon and an anticodon?
3. What is meant by code degeneracy (redundancy)? How does wobble help alleviate the energy cost exacted on a cell due to code degeneracy?
4. Is the genetic code truly universal? Explain your answer.

Table 13.4	Some Exceptions to the Universal Genetic Code	
Codon	**Amino Acid Inserted**	**Where Observed**
AGA and AGG	Stop	Mammalian mitochondria
AGA and AGG	Serine (Ser)	Invertebrate mitochondria
AUA	Methionine (Met)	Mammalian, invertebrate, and yeast mitochondria
CUA	Threonine (Thr)	Yeast mitochondria
CUG	Serine	Some fungi
UAA and UAG	Glutamine (Gln)	Some protists
UAG	Pyrrolysine	Some methanogens and bacteria
UGA	Selenocysteine	Members of all three domains
UGA	Tryptophan (Trp)	Mammalian, invertebrate, and yeast mitochondria
UGA	Glutamine	Mycoplasma bacteria

MICRO INQUIRY *For each of the above codons, determine what it normally encodes. Choose one of the variants and suggest how it might have evolved.*

13.7 Translation in Bacteria

After reading this section, you should be able to:

- Relate the general structure of a tRNA molecule to its role in amino acid activation and translation
- Summarize the formation of a translation initiation complex
- Describe the structure of bacterial ribosomes
- State the initiator tRNA used by bacteria
- Outline the events that occur at the A, P, and E sites of the bacterial ribosome during the elongation phase of translation

Translation involves decoding mRNA and covalently linking amino acids together to form a polypeptide; this occurs at the ribosome. Translation begins when a ribosome binds mRNA and is positioned properly so that translation will yield the correct amino acid sequence in the polypeptide chain. Transfer RNA molecules carry amino acids to the ribosome so that they can be added to the polypeptide chain as the ribosome moves down the mRNA molecule. Just as DNA and RNA synthesis proceeds in one direction, so too does protein synthesis. Polypeptide synthesis begins with the amino acid at the end of the chain with a free amino group (the N-terminal) and moves in the C-terminal

direction. Thus translation is said to occur in the amino terminus to carboxyl terminus direction. Protein synthesis is accurate and very rapid. In *E. coli*, synthesis occurs at a rate of at least 900 amino acids added per minute.

Cells that grow quickly must use each mRNA with great efficiency to synthesize proteins at a sufficiently rapid rate. To achieve rapid rates of protein synthesis, mRNAs often are simultaneously complexed with several ribosomes, each ribosome reading the mRNA message and synthesizing a polypeptide. At maximal rates of mRNA use, there may be a ribosome every 80 nucleotides along the mRNA or as many as 20 ribosomes simultaneously reading an mRNA that codes for a 50,000 dalton polypeptide. A complex of mRNA with several ribosomes is called a **polyribosome** or polysome (**figure 13.33**). Polysomes are present in all organisms. Bacteria can further increase the efficiency of gene expression by coupling transcription and translation (figure 13.33*b*). While RNA polymerase is synthesizing an mRNA, ribosomes can already be attached to the mRNA so that transcription and translation occur simultaneously. Coupled transcription and translation is possible in bacterial cells because a nuclear envelope does not separate the translation machinery from DNA, as it does in eukaryotes. ↻ *How Translation Works; Protein Synthesis*

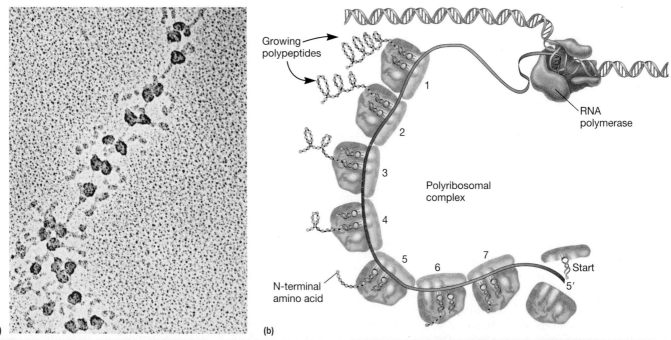

Figure 13.33 Coupled Transcription and Translation in Bacteria. (a) A transmission electron micrograph showing a polyribosome. (b) A schematic representation of coupled transcription and translation. As the DNA is transcribed, ribosomes bind the free 5′ end of the mRNA. Thus translation is started before transcription is completed. Note that there are multiple ribosomes bound to the mRNA, forming a polyribosome. The ribosomes are shown at different points in the translation process. Ribosomes 1, 2, 5, and 7 have completed the transpeptidation reaction, but translocation has not yet occurred. Ribosomes 3 and 4 have an A site containing an incoming aminoacyl-tRNA. Transpeptidation has not occurred. Ribosome 6 shows elongation upon completion of both transpeptidation and translocation. The tRNA bearing the growing polypeptide is in the P site and the empty tRNA is in the E site.

MICRO INQUIRY *Why is simultaneous transcription and translation impossible in eukaryotes?*

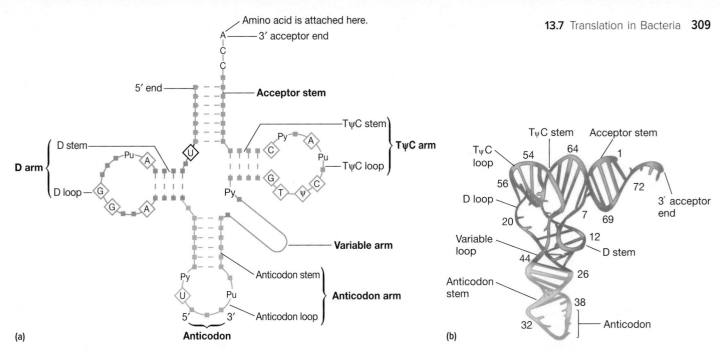

Figure 13.34 tRNA Structure. (a) The two-dimensional cloverleaf structure for tRNA. In addition to the anticodon arm, three other arms are readily observed: the D or DHU arm, the T or TψC arm, and the variable arm. The D and T arms are named because of the presence of unusual nucleotides. The variable arm is of different lengths depending on the tRNA; the other arms are fairly constant in size. Bases found in all tRNAs are in diamonds; purine and pyrimidine positions in all tRNAs are labeled Pu and Py, respectively. (b) The three-dimensional structure of tRNA. The various regions are distinguished with different colors.

Amino Acid Activation: Attachment of an Amino Acid to a Transfer RNA

For translation to occur, a ready supply of tRNA molecules bearing the correct amino acid must be available. Thus a preparatory step for protein synthesis is **amino acid activation,** the process in which amino acids are attached to tRNA molecules. Before we discuss this process, we need to examine the structure of tRNA molecules.

Transfer RNA molecules are about 70 to 95 nucleotides long and possess several characteristic structural features. These features become apparent when the tRNA is folded so that base pairing within the tRNA strand is maximized. When represented two-dimensionally, this base pairing causes the tRNA to assume a cloverleaf conformation (**figure 13.34a**). However, the three-dimensional structure looks like the letter L (figure 13.34b). One important feature of tRNAs is the acceptor stem, which holds the activated amino acid. The 3′ end of all tRNAs has the same CCA sequence, and in all cases the amino acid is attached to the A nucleotide. Another important feature of a tRNA is the **anticodon.** The anticodon is complementary to an mRNA codon and is located on the anticodon arm (figure 13.34a).

Enzymes called **aminoacyl-tRNA synthetases** catalyze amino acid activation (**figure 13.35**). As is true of DNA and RNA synthesis, the reaction is driven to completion when ATP is hydrolyzed to release pyrophosphate. The amino acid is attached to the tRNA by a high-energy bond. The storage of energy in this bond provides the fuel needed to generate the peptide bond when the amino acid is added to the growing peptide chain.

There are at least 20 aminoacyl-tRNA synthetases, each specific for a single amino acid and its tRNAs (cognate tRNAs). It is critical that each tRNA attach the corresponding amino acid because if an incorrect amino acid is attached to a tRNA, it will be incorporated

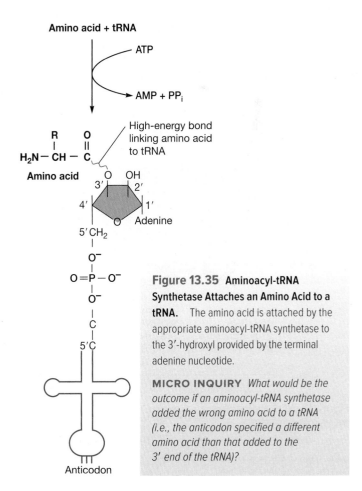

Figure 13.35 Aminoacyl-tRNA Synthetase Attaches an Amino Acid to a tRNA. The amino acid is attached by the appropriate aminoacyl-tRNA synthetase to the 3′-hydroxyl provided by the terminal adenine nucleotide.

MICRO INQUIRY *What would be the outcome if an aminoacyl-tRNA synthetase added the wrong amino acid to a tRNA (i.e., the anticodon specified a different amino acid than that added to the 3′ end of the tRNA)?*

into a polypeptide in place of the correct amino acid. The protein synthetic machinery recognizes only the anticodon of the aminoacyl-tRNA and cannot tell whether the correct amino acid is attached. Some aminoacyl-tRNA synthetases proofread just like DNA polymerases do. If the wrong amino acid is attached to tRNA, the enzyme hydrolyzes the amino acid from the tRNA, rather than release the incorrect product. ⮐ *Aminoacyl-tRNA Structure*

Ribosomes Have Three tRNA Binding Sites

Protein synthesis takes place on ribosomes that serve as workbenches, with mRNA acting as the blueprint. Recall that ribosomes are formed from two subunits, the large subunit and the small subunit, and each contains one or more rRNA molecules and numerous polypeptide chains. A bacterial ribosome and its components are shown in **figure 13.36**. The ribosome can be divided into two functional domains, the translational domain and the exit domain. Both subunits contribute to the formation of the translational domain, which interacts with tRNAs and is responsible for forming peptide bonds. The exit domain is located solely in the large subunit. Three sites are found within the translational domain for binding tRNAs: A, P, and E sites. The **A** (**aminoacyl** or **acceptor**) site receives tRNAs carrying an amino acid to be added to the protein being synthesized. The **P** (**peptidyl** or **donor**) site holds a tRNA attached to the growing polypeptide. The **E** (**exit**) site is the location from which empty tRNAs leave the ribosome. The growing peptide chain emerges from the large subunit at the exit domain.

Ribosomal RNA is thought to have three roles. (1) All three rRNA molecules contribute to ribosome structure. (2) The 16S rRNA of the 30S subunit is needed for initiation of protein synthesis because its 3′ end binds to a site on the leader of the mRNA called the Shine-Dalgarno sequence; thus the Shine-Dalgarno sequence is part of the **ribosome-binding site** (**RBS**). This helps position the mRNA on the ribosome. The 16S rRNA also binds a protein needed to initiate translation (initiation factor 3) and the 3′ CCA end of amino-acyl-tRNA. (3) The 23S rRNA is a ribozyme that catalyzes peptide bond formation.

Protein Synthesis Begins with Formation of the 70S Initiation Complex

Like transcription and DNA replication, protein synthesis is divided into three stages: initiation, elongation, and termination. The initiation of protein synthesis is very elaborate. Apparently the complexity is necessary to ensure that the ribosome does not start synthesizing a polypeptide chain in the middle of a gene—a disastrous error.

Bacteria begin protein synthesis with a modified aminoacyl-tRNA, *N*-formylmethionyl-tRNA^{fMet} (fMet-tRNA), which is called the **initiator tRNA** and is coded for by the start codon AUG (**figure 13.37**). The amino acid of the initiator tRNA has a formyl group covalently bound to the amino group and can be used only for initiation because of the presence of the formyl group. When methionine is to be added to a growing polypeptide chain (i.e., at an AUG codon in the middle of the mRNA), a normal methionyl-tRNA^{Met} is employed. Although bacteria start protein synthesis with *N*-formylmethionine, the formyl group is not retained but is hydrolytically removed. In fact, one to three amino acids may be removed from the amino terminus of the polypeptide after synthesis.

Protein synthesis in bacteria begins with formation of the 30S initiation complex, consisting of the initiator tRNA, the mRNA to be translated, and the 30S ribosomal subunit; two **initiation factors** (IF-1 and IF-2) are involved (**figure 13.38**). Positioning of the initiator fMet-tRNA on the mRNA is crucial for proper translation of the mRNA. This is accomplished with the help of the 16S rRNA within the 30S subunit, which is complementary to and binds the Shine-Dalgarno sequence in the leader sequence of the mRNA. By aligning the Shine-Dalgarno sequence with the 16S rRNA, the start codon (AUG or sometimes GUG) specifically binds with the fMet-tRNA anticodon. This ensures that the start codon will be translated first.

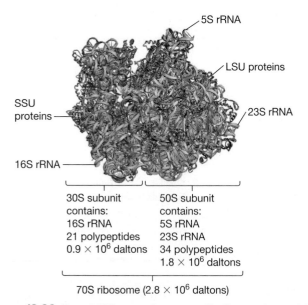

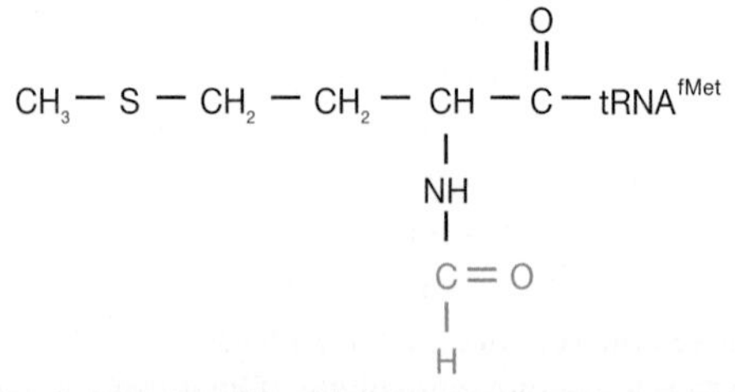

Figure 13.36 Bacterial Ribosome Structure. The *Thermus thermophilus* 70S ribosome viewed from the right-hand side with the 30S subunit (SSU) on the left and the 50S subunit (LSU) on the right. The A site tRNA (gold) is visible in the interface cavity. 16S rRNA, cyan; 23S rRNA, gray; 5S rRNA, light blue; 30S proteins, dark blue; 50S proteins, magenta.

Figure 13.37 *N***-formylmethionyl-tRNA^{fMet} Is the Initiator tRNA Used by Bacteria.** The formyl group is in color. Archaea and eukaryotes use methionyl-tRNA for initiation.

MICRO INQUIRY *Why would it be impossible for fMet-tRNA to initiate peptide bond formation with another amino acid? (Hint: Examine figure 13.40 closely.)*

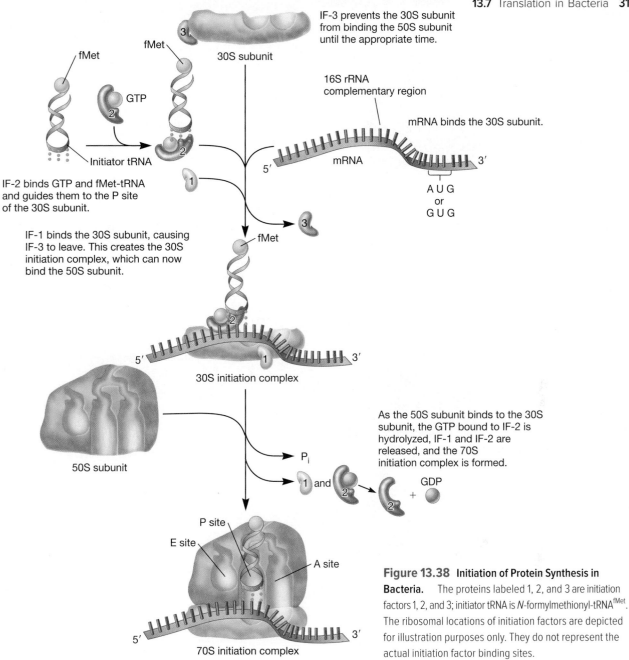

IF-3 prevents the 30S subunit from binding the 50S subunit until the appropriate time.

fMet

fMet

30S subunit

GTP

16S rRNA complementary region

mRNA binds the 30S subunit.

Initiator tRNA

mRNA

5′ 3′

IF-2 binds GTP and fMet-tRNA and guides them to the P site of the 30S subunit.

A U G or G U G

IF-1 binds the 30S subunit, causing IF-3 to leave. This creates the 30S initiation complex, which can now bind the 50S subunit.

fMet

5′ 3′

30S initiation complex

As the 50S subunit binds to the 30S subunit, the GTP bound to IF-2 is hydrolyzed, IF-1 and IF-2 are released, and the 70S initiation complex is formed.

50S subunit

P$_i$

1 and

GDP

+

P site

E site

A site

Figure 13.38 **Initiation of Protein Synthesis in Bacteria.** The proteins labeled 1, 2, and 3 are initiation factors 1, 2, and 3; initiator tRNA is *N*-formylmethionyl-tRNA^{fMet}. The ribosomal locations of initiation factors are depicted for illustration purposes only. They do not represent the actual initiation factor binding sites.

5′ 3′

70S initiation complex

Once the 30S initiation complex is formed, it binds the 50S ribosomal subunit, forming the 70S initiation complex. The fMet-tRNA is positioned at the peptidyl or P site. At this juncture, you may be wondering what kept the 30S and 50S subunits from binding each other earlier in the initiation stage. The answer is the third initiation factor (IF-3), as illustrated in figure 13.38. Also revealed in this figure is the energy cost of initiation. GTP, like ATP, is a high-energy molecule. Hydrolysis of GTP to GDP provides the energy needed to accomplish initiation.

The mechanism of translation initiation we have just described holds for mRNAs having a leader (i.e., a 5′ untranslated region) with a Shine-Dalgarno sequence. In recent years, mRNAs lacking a Shine-Dalgarno sequence in the leader or lacking leaders altogether have been identified. Current evidence suggests that translation initiation of leaderless mRNAs is accomplished

by intact 70S ribosomes. How translation initiation of Shine-Dalgarno-lacking mRNAs occurs is still unclear.

Elongation of the Polypeptide Chain

Every addition of an amino acid to a growing polypeptide chain is the result of an elongation cycle composed of three phases: aminoacyl-tRNA binding, the transpeptidation reaction, and translocation. The process is aided by proteins called **elongation factors (EF).** In each turn of the cycle, an amino acid corresponding to the proper mRNA codon is added to the C-terminal end of the polypeptide chain as the ribosome moves down the mRNA in the 5′ to 3′ direction.

At the beginning of an elongation cycle, the P site is filled with either the initiator fMet-tRNA or a tRNA bearing a growing polypeptide chain (peptidyl-tRNA), and the A and E sites are empty (**figure 13.39**). Messenger RNA is bound to the ribosome

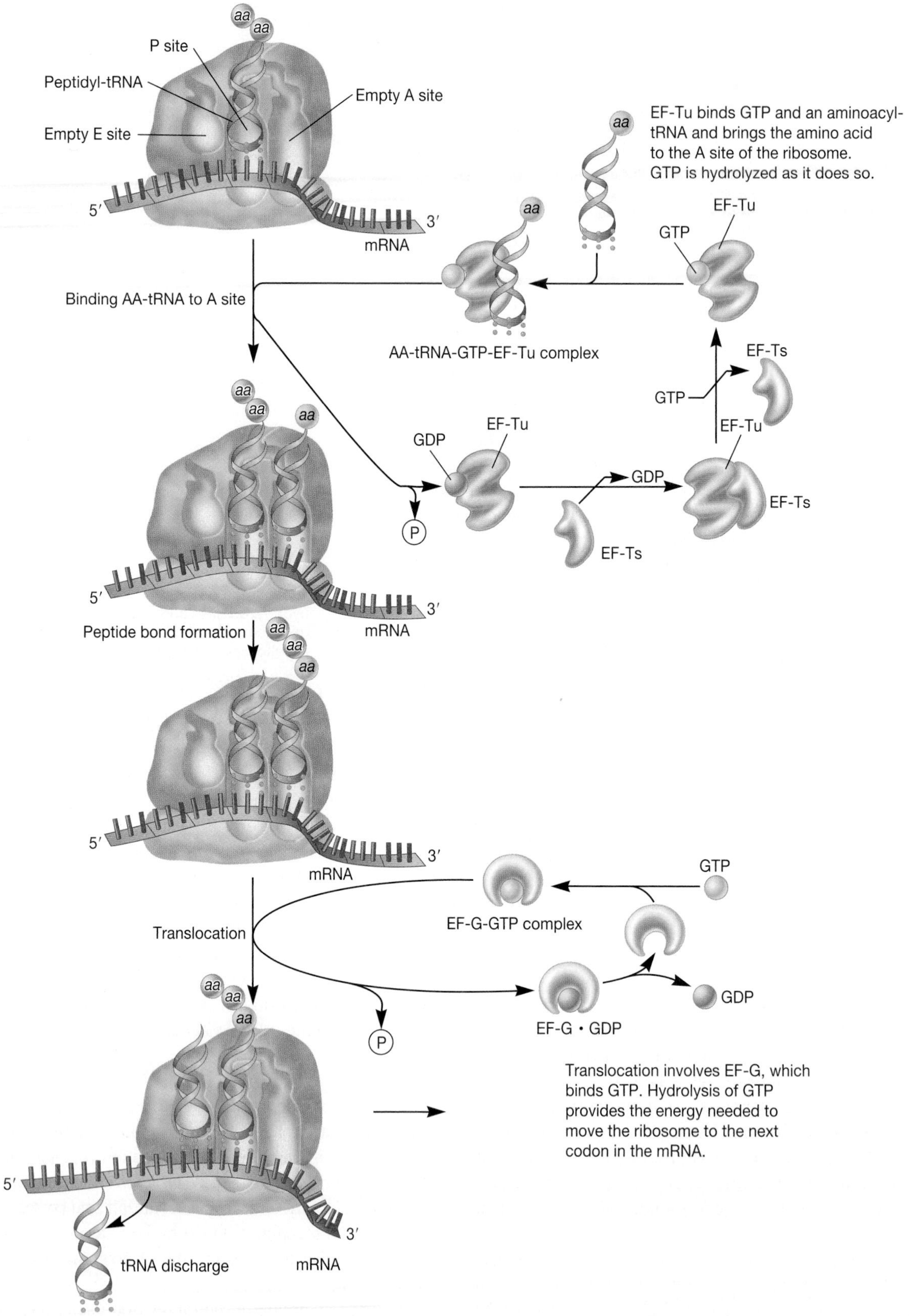

Figure 13.39 The Elongation Cycle of Protein Synthesis. The ribosome possesses three sites, a peptidyl or donor site (P site), an aminoacyl or acceptor site (A site), and an exit site (E site). The arrow to the right of the ribosome in the translocation step shows the direction of ribosome movement.

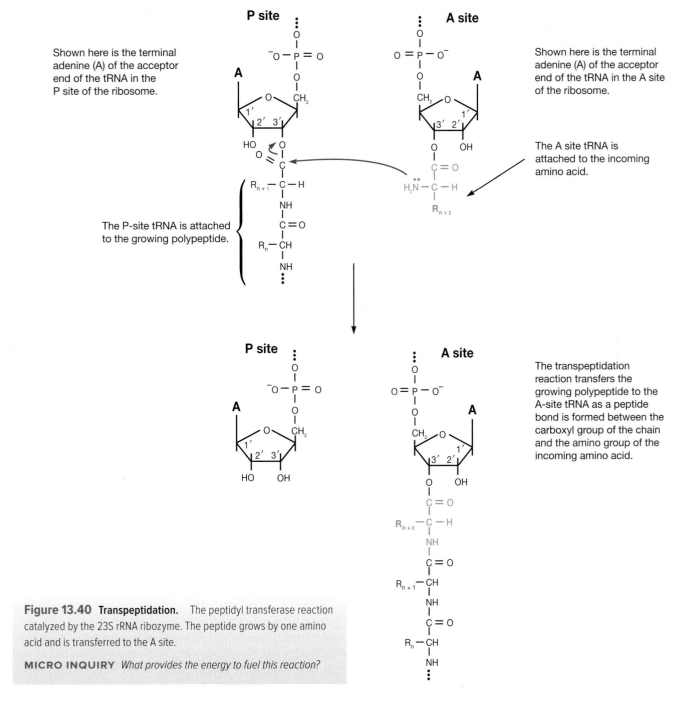

Shown here is the terminal adenine (A) of the acceptor end of the tRNA in the P site of the ribosome.

Shown here is the terminal adenine (A) of the acceptor end of the tRNA in the A site of the ribosome.

The A site tRNA is attached to the incoming amino acid.

The P-site tRNA is attached to the growing polypeptide.

The transpeptidation reaction transfers the growing polypeptide to the A-site tRNA as a peptide bond is formed between the carboxyl group of the chain and the amino group of the incoming amino acid.

Figure 13.40 Transpeptidation. The peptidyl transferase reaction catalyzed by the 23S rRNA ribozyme. The peptide grows by one amino acid and is transferred to the A site.

MICRO INQUIRY *What provides the energy to fuel this reaction?*

in such a way that the proper codon interacts with the P site tRNA (e.g., an AUG codon for fMet-tRNA). The next codon is located within the A site and is ready to accept an aminoacyl-tRNA.

In the aminoacyl-tRNA binding phase, the first phase of the cycle, the aminoacyl-tRNA corresponding to the codon in the A site is inserted so its anticodon is aligned with the codon on the mRNA. In bacterial cells, this is aided by two elongation factors and requires the expenditure of one GTP (figure 13.39). Once the proper aminoacyl tRNA is in the A site, the second phase of the elongation cycle, the transpeptidation reaction, occurs (figure 13.39 and **figure 13.40**).

Transpeptidation is catalyzed by the **peptidyl transferase** activity of the 23S rRNA ribozyme, which is part of the 50S ribosomal subunit. In this reaction, the amino group of the A site amino acid reacts with the carboxyl group of the C-terminal amino acid on the P site tRNA (figure 13.40). This results in the transfer of the peptide chain from the tRNA in the P site to the tRNA in the A site, as a peptide bond is formed between the peptide chain and the incoming amino acid. No extra energy source is required for peptide bond formation because the bond linking an amino acid to tRNA is high in energy (figure 13.35).

The final phase in the elongation cycle is **translocation.** Three things happen simultaneously: (1) the peptidyl-tRNA moves from the A site to the P site; (2) the ribosome moves one codon along mRNA so that a new codon is positioned in the A site; and (3) the empty tRNA moves from the P site to the E site and subsequently leaves the ribosome. Translocation involves rotations of the 30S and 50S subunits relative to each other. In addition, the head portion of the 30S subunit (*see figure 3.38*) swivels. These changes in ribosome structure move the tRNAs into their new locations; the codon-anticodon interactions between the tRNAs and the mRNA move the mRNA as the tRNAs move. One elongation factor participates and one GTP is hydrolyzed during this intricate process. ➋ *Translation Elongation*

Insertion of Selenocysteine and Pyrrolysine

Insertion of the unusual amino acids selenocysteine and pyrrolysine during translation occurs by two distinctive mechanisms. Selenocysteine is synthesized from serine after it has been attached to certain tRNAs. The enzyme catalyzing the conversion is selenocysteine synthase. Once formed, the amino acid is recognized by a specific elongation factor and is incorporated when a UGA stop codon is encountered in association with nucleotide sequences called cis-acting *sele*no*c*ysteine *i*nsertion *s*equence elements (SECIS). In bacteria, SECIS are found immediately after the UGA stop codon.

Pyrrolysine insertion differs from that of selenocysteine in several ways. Pyrrolysine is synthesized from lysine before being attached to a tRNA. Organisms that use pyrrolysine make an unusual tRNA with a CUA anticodon; the pyrrolysine is attached by a specific aminoacyl-tRNA synthetase. Pyrrolysine is inserted at UAG stop codons located near a sequence element called *p*yrroly*s*ine *i*nsertion *s*equence (PYLIS). Both SECIS and PYLIS form stem-loop structures that prevent cessation of translation.

Protein Synthesis Ends When the Ribosome Reaches a Stop Codon

Protein synthesis stops when the ribosome reaches one of three stop codons: UAA, UAG, and UGA (**figure 13.41**). The stop codon is found on the mRNA immediately before the trailer. Three release factors (RF-1, RF-2, and RF-3) aid the ribosome in recognizing these codons. Because there is no cognate tRNA for a stop codon, the ribosome halts. Peptidyl transferase hydrolyzes the bond linking the polypeptide to the tRNA in the P site, and the polypeptide and the empty tRNA are released. GTP hydrolysis occurs during this sequence of events. Next the ribosome dissociates from its mRNA and separates into 30S and 50S subunits. IF-3 binds the 30S subunit, which prepares it for the next round of protein synthesis. ➋ *Translation Termination*

Ensuring Accuracy During Translation

Protein synthesis is a very expensive process. Two ATP high-energy bonds are required for amino acid activation, initiation consumes one GTP, two GTP molecules are used during each elongation cycle, and another GTP is hydrolyzed when protein

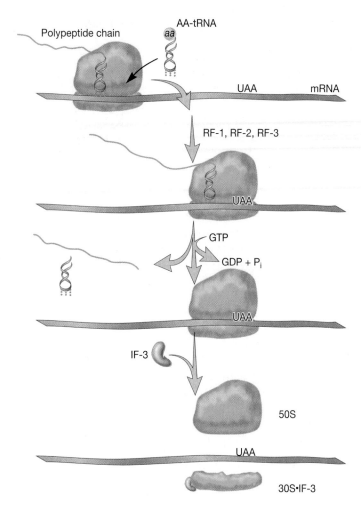

Figure 13.41 Termination of Protein Synthesis in Bacteria. Although three different nonsense codons can terminate chain elongation, UAA is most often used. Three release factors (RF) assist the ribosome in recognizing nonsense codons and terminating translation and GTP is hydrolyzed. Binding of IF-3 to the 30S subunit prepares it for the next round of translation initiation. Transfer RNAs are in pink.

synthesis terminates (figures 13.35, 13.38, 13.39, and 13.41). Presumably this large energy expenditure is required to ensure the fidelity of protein synthesis. Fidelity is assessed both before and after formation of the peptide bond. When an aminoacyl-tRNA enters the A site, correct pairing of the anticodon and codon causes conformational changes in components of the ribosome such that the aminoacyl-tRNA is "locked into place" in a manner that facilitates peptide bond formation. These conformational changes do not occur for an incorrect aminoacyl-tRNA and the tRNA is ejected. However, on rare occasions the incorrect aminoacyl-tRNA is selected and a peptide bond is formed between the growing polypeptide and the wrong amino acid. The ribosome is able to detect its error, but how it does so is not clear. The presence of an incorrect amino acid is also recognized by release factors. This leads to hydrolysis of the aberrant polypeptide from the tRNA, its release from the ribosome, and termination of translation.

1. In which direction are polypeptides synthesized? What is a polyribosome and why is it useful?
2. Briefly describe the structure of transfer RNA and relate this to its function. How are amino acids activated for protein synthesis, and why is the specificity of the aminoacyl-tRNA synthetase reaction so important?
3. What are the translational and exit domains of the ribosome? What roles do ribosomal RNAs have?
4. Tabulate the nature and function of the following: fMet-tRNA, start codon, initiation factors, elongation cycle, elongation factors, peptidyl and aminoacyl sites, transpeptidation reaction, peptidyl transferase, translocation, stop codon, and release factors.
5. How many ATP and GTP molecules would be hydrolyzed in the synthesis of a 125 amino acid protein? Explain why this is a good argument for careful regulation of gene expression (especially considering that most proteins are larger than 125 amino acids).

13.8 Protein Maturation and Secretion

After reading this section, you should be able to:

- Describe the role of protein splicing in protein maturation
- Discuss the role of molecular chaperones in protein folding, and list some important examples of chaperones
- Distinguish translocation of proteins from protein secretion
- List bacterial translocation and secretion systems, and indicate whether they function in Gram-positive, Gram-negative, or both types of bacteria

As a polypeptide emerges from a ribosome, it is not yet ready to assume its cellular functions. Protein function depends on its three-dimensional shape. Some proteins have extra amino acids that must be removed. They then must be properly folded and in some cases associated with other protein subunits to generate a functional enzyme (e.g., DNA and RNA polymerases are multimeric proteins). In addition, proteins must be delivered to the proper subcellular or extracellular site. We now discuss these posttranslational events.

Protein Splicing

One complexity in the formation of proteins is observed in microbes belonging to all three domains of life: **protein splicing.** One type of protein splicing is called cis-splicing. In this process, a part of the polypeptide is removed before the polypeptide folds into its final shape. Cis-splicing proteins begin as larger precursor proteins composed of an internal intervening sequence called an **intein** (about 130 to 600 amino acids in length) flanked by external sequences called **exteins** (**figure 13.42**). Inteins remove themselves from the precursor protein. When the splicing is completed, two products have been formed: the intein and the protein formed by splicing the two exteins together. Another type of splicing is called trans-splicing. It differs from cis-splicing in that the intein is divided between two different polypeptides, such that the N-terminal portion of the intein is part of one polypeptide, and the C-terminal portion of the intein is part of another. The two parts of

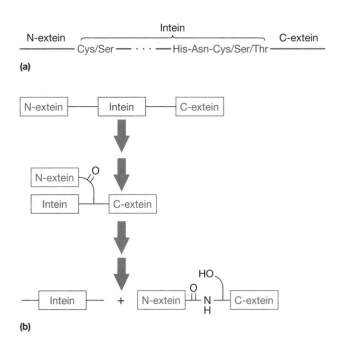

Figure 13.42 Protein Splicing. A cis-splicing protein is shown. (a) A generalized illustration of intein structure. The amino acids that are commonly present at each end of the inteins are shown. Note that many are thiol- or hydroxyl-containing amino acids. (b) A simplified overview of cis-splicing.

the intein must first associate with each other before splicing continues. The final protein produced arises from the exteins present on the two precursor polypeptides.

Molecular Chaperones: Proteins That Help Proteins Fold

The cytoplasm is very crowded, filled with many proteins, including those that are newly synthesized. These crowded conditions can cause proteins to fold improperly or to aggregate forming nonfunctional protein complexes. To ensure that this does not happen, cells use proteins called **molecular chaperones,** or simply chaperones, to suppress incorrect folding and in some cases to reverse any incorrect folding that has already taken place. They are so important that chaperones are present in cells in all domains of life.

Several chaperones and cooperating proteins aid proper protein folding in bacteria. As can be seen in **figure 13.43**, most proteins are successfully folded with the aid of the chaperone called trigger factor (TF). TF binds the growing polypeptide chain as it leaves the ribosome, and is thought to mask hydrophobic regions of the protein so they don't interact with each other or with other proteins. Thus, TF functions cotranslationally. However, about 30% of proteins need additional help folding, either as they are synthesized or posttranslationally. For these proteins, other chaperones, including DnaK, DnaJ, GroEL, and GroES, complete the folding process. DnaJ and DnaK fold many of the proteins that TF alone cannot fold. This requires the expenditure of ATP. Sometimes the polypeptide still does not reach its native conformation and the partially folded protein may be transferred to chaperones GroEL and GroES, which complete the folding.

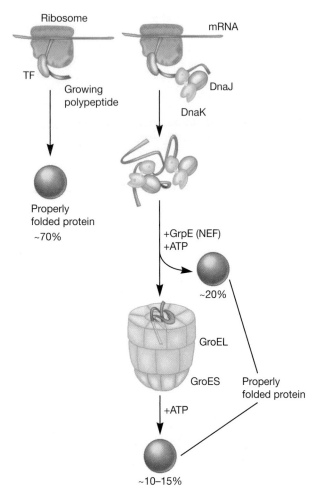

Figure 13.43 Chaperones Assist Polypeptide Folding. Most bacterial proteins are folded with the aid of a ribosome-associated chaperone called trigger factor (TF). TF binds nascent polypeptides as they leave the ribosome. Those proteins that can't be folded by the TF alone are assisted in folding by other chaperones. GrpE is a nucleotide exchange factor (NEF) that regulates the activity of DnaK.

This chaperone system also expends ATP as it folds the protein into its proper conformation.

Chaperones were first discovered because they dramatically increase in concentration when cells are exposed to stressful conditions that cause protein denaturation. For example, when an *E. coli* culture is switched from 30 to 42°C, the concentrations of about 20 different proteins increase greatly within about 5 minutes. Some of these proteins are chaperones that protect proteins from denaturing at the higher temperatures. Thus many chaperones are called **heat-shock proteins (HSPs)**. Chaperones are often categorized in part based on their size. For instance, DnaK is an HSP70 protein, DnaJ an HSP40 protein, and GroEL an HSP60 protein (also called chaperonins), where the 70, 40, and 60, indicate the protein's size in kilodaltons.

Protein Translocation and Secretion in Bacteria

To survive, cells must obtain energy from their environment and convert it to a usable form (e.g., ATP). They also obtain nutrients from their environment, avoid harmful chemicals, elaborate external structures such as fimbriae and flagella, and if pathogenic, may release toxins. All of these processes require that some of the proteins synthesized in the cytoplasm be moved to membranes, the periplasmic space, or the external milieu. It has been estimated that almost one-third of the proteins synthesized by cells leave the cytoplasm. Therefore it is not surprising that over 15 different systems for moving proteins out of the cytoplasm have evolved. Some of these systems are found in all domains of life. Others are unique to bacterial cells, and others are observed only in Gram-negative or Gram-positive bacteria. When proteins are moved from the cytoplasm to the membrane or to the periplasmic space, the movement is called translocation. Protein secretion refers to the movement of proteins from the cytoplasm to the external environment. Many of the secretion pathways are designated with numbers (e.g., type I secretion system, type II secretion system, etc.). All protein translocation and secretion systems described here require the expenditure of energy at some step in the process. The energy is usually supplied by hydrolysis of high-energy molecules such as ATP and GTP. However, the proton motive force also sometimes plays a role. ◄◄ *ATP (section 10.2); Electron transport and oxidative phosphorylation (step 3) generate the most ATP (section 11.6)*

The variations in cell envelope structure pose different challenges for protein secretion. For most Gram-positive bacteria to secrete proteins, the proteins must be translocated across the plasma membrane. Once across the plasma membrane, the protein either passes through the relatively porous peptidoglycan into the external environment or becomes embedded in or attached to the peptidoglycan. Gram-negative bacteria have more hurdles to jump when they secrete proteins. They, too, must transport the proteins across the plasma membrane, but to complete secretion, the proteins must be transported across the outer membrane. Interestingly, mycobacteria, which are Gram-positive but have a complex cell wall that includes an outer membrane-like component (*see figure 24.11*), face the same difficulty as Gram-negative bacteria. They have evolved a unique secretion system (type VII secretion system) that spans their entire complex cell envelope.

Some Translocation and Secretion Systems Are Found in Both Gram-Positive and Gram-Negative Bacteria

Two translocation systems (Sec system and Tat system) and two secretion systems (Type I and Type IV) are observed in both Gram-positive and Gram-negative bacteria. The **Sec system,** sometimes called the general secretion pathway, is highly conserved, having been identified in all three domains of life (**figure 13.44**). It translocates unfolded proteins across the plasma membrane or integrates them into the membrane itself. It does so either posttranslationally or cotranslationally. Generally, posttranslational translocation moves proteins across the membrane, whereas cotranslational translocation is used to insert proteins into the plasma membrane.

In posttranslational translocation, the protein is synthesized and released from the ribosome as a preprotein. Its folding is prevented by chaperone proteins. An amino acid sequence at the amino terminus of the preprotein, called the **signal peptide,** is recognized and bound by the protein SecA. SecA "delivers" the

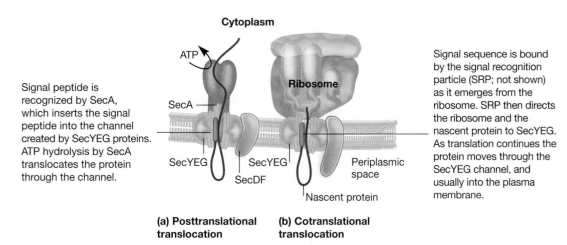

Cytoplasm

Signal peptide is recognized by SecA, which inserts the signal peptide into the channel created by SecYEG proteins. ATP hydrolysis by SecA translocates the protein through the channel.

Signal sequence is bound by the signal recognition particle (SRP; not shown) as it emerges from the ribosome. SRP then directs the ribosome and the nascent protein to SecYEG. As translation continues the protein moves through the SecYEG channel, and usually into the plasma membrane.

ATP

SecA

Ribosome

SecYEG   SecYEG

SecDF

Periplasmic space

Nascent protein

(a) Posttranslational translocation

(b) Cotranslational translocation

Figure 13.44 Posttranslational and Cotranslational Translocation by the Sec System Move Unfolded Proteins Across or Into the Plasma Membrane. (a) Posttranslational translocation uses chaperones to keep the protein unfolded until it is recognized by SecA. Here we show SecA existing as a dimer. There is also evidence that a monomer of SecA is sufficient for translocation. (b) In cotranslational translocation the protein has not had a chance to fold before it is bound by SRP and delivered, along with the ribosome, to SecYEG.

preprotein to the Sec system. Certain Sec proteins (SecY, SecE, and SecG) form a channel in the membrane through which the preprotein passes. SecA is thought to insert the peptide signal into the YEG channel and then to act as a motor, using the energy released from ATP hydrolysis to translocate the preprotein. Two other proteins (SecDF) use the proton motive force to help fuel translocation through the plasma membrane. When the preprotein emerges from the plasma membrane, an enzyme called signal peptidase removes the signal peptide. The protein then folds into the proper shape. In Gram-negative bacteria, proteins delivered to the periplasm by the Sec system may be transported across the outer membrane by type II, IV, or V secretion systems.

Cotranslational movement of proteins by the Sec system is mediated by a complex of RNA and protein called the **signal recognition particle (SRP)**. It is thought that SRP binds a signal sequence (not the same as the signal peptide) in the protein as it leaves the ribosome and directs the protein together with the translating ribosome to SecYEG. As translation continues, the protein is threaded into the SecYEG channel and inserted into the plasma membrane, often with the aid of a protein called YidC. This process is similar to the cotranslational translocation that occurs in the eukaryotic Sec system that moves proteins across the membranes of the endoplasmic reticulum. ▶▶ *Protein localization and secretion in eukaryotes (section 15.4)*

The **Tat system (figure 13.45)** is distinguished from the Sec system by the nature of the protein transported. The Sec system translocates unfolded proteins; the Tat system translocates folded proteins. Furthermore, the Tat system only moves proteins that feature two, or "twin," arginine residues in their signal sequence—in fact, *Tat* stands for *t*win *a*rginine *t*ranslocase. In Gram-negative bacteria, proteins translocated by the Tat system are delivered to a type II or type V secretion system for transport across the outer membrane.

Type I secretion systems are ubiquitous in Gram-positive and Gram-negative bacteria (figure 13.45). Type I systems are members of a protein superfamily defined by the ABC transporters

described in chapter 3. In Gram-negative bacteria, type 1 secretion systems (T1SSs) consist of three components: the transporter that resides in the plasma membrane; membrane fusion proteins present in the periplasmic space; and outer membrane factors, which form a channel through which the secreted protein passes. Gram-positive T1SSs consist only of the transporter and the membrane fusion proteins. In pathogenic Gram-negative bacteria, T1SSs are involved in the secretion of toxins (e.g., α-hemolysin), as well as other proteins (e.g., proteases).

Type IV secretion systems are unique in that they are used to secrete proteins as well as to transfer DNA from a donor bacterium to a recipient during a process called bacterial conjugation. These systems are observed in both Gram-positive and Gram-negative bacteria; however, in Gram-positive bacteria, they only function in DNA transfer. The type IV systems of Gram-negative bacteria are best studied. They are composed of many different proteins and are described in more detail in chapter 16.

The Presence of an Outer Membrane Means that Gram-Negative Bacteria Must Have Distinctive Secretion Systems

Currently six protein secretion systems (types I to VI) have been identified in Gram-negative bacteria (figure 13.45; the type VI system is not shown). Some of these systems have already been described as they are present in both Gram-negative and Gram-positive bacteria (type I and type IV secretion systems). All others are unique to Gram-negative bacteria. Most are used to secrete virulence factors produced by plant and animal pathogens. Gram-negative bacteria use the **type II** and type V systems to transport proteins across the outer membrane after the protein has first been translocated across the plasma membrane by the Sec or Tat system. The type I, III, and VI systems do not transport proteins with the help of the Sec system, so they are said to be Sec-independent. The type IV pathway sometimes is linked to the Sec pathway but usually functions on its own.

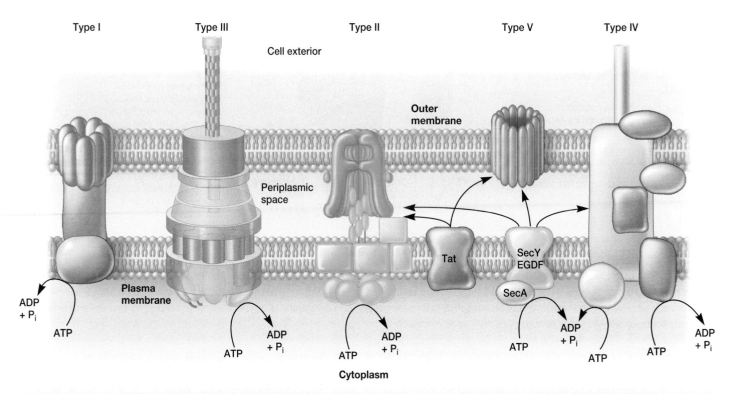

Figure 13.45 Some Protein Secretion Systems of Gram-Negative Bacteria. Secretion systems type I through V of Gram-negative bacteria are shown. The Sec and Tat systems deliver proteins from the cytoplasm to the periplasmic space and also are found in Gram-positive bacteria. Type II, type V, and sometimes type IV secretion systems complete the secretion process begun by the Sec system. Type I and type III secretion systems bypass Sec and Tat, moving proteins directly from the cytoplasm to the extracellular space. The type IV secretion system can work either alone or with the Sec system to transport proteins to the extracellular space. Type IV systems are also observed in Gram-positive bacteria, where they function in DNA transfer only. Not all type IV systems have a needle as shown here.

MICRO INQUIRY *What are two distinguishing features of protein translocation via the Tat system as compared to that using the Sec system?*

Type III and VI systems, and some type IV systems, form a needlelike structure that extends beyond the outer membrane and can make contact with other cells (figure 13.45; *also see figure 39.9*). The **type III secretion system** (T3SS) is best studied because some pathogens use it to inject virulence factors directly into their plant and animal host cells. Its needle, called the injectisome, delivers a variety of proteins, including virulence factors such as toxins, phagocytosis inhibitors, stimulators of cytoskeleton reorganization in the host cell, and promoters of host cell suicide (apoptosis). The proteins secreted are delivered to the type III apparatus by chaperone proteins. T3SSs are also referred to as contact-dependent; the translocon portion at the end of the injectisome is formed only upon contact with the eukaryotic target of the pathogen. Once formed, the translocon delivers virulence factors into the target cell. The participation of T3SSs in bacterial virulence is further discussed in chapter 39.

Type V and **VI secretion systems** also warrant comment. Some type V systems employ proteins called autotransporters because after being translocated across the plasma membrane by the Sec system, the proteins are able to transport themselves across the outer membrane. Autotransporters have three domains. One is recognized by the Sec system, and another forms a pore in the outer membrane through which the third domain (a virulence factor) is transported. Type VI systems (T6SSs) are of interest because they are similar to the delivery systems used by some bacteriophages to release their genomes into the cytoplasm of their bacterial hosts. The first T6SSs discovered were used by bacteria to deliver virulence factors into the cytoplasm of eukaryotic target cells. However, it is now thought that most T6SSs are used by bacteria to deliver toxic effector molecules to other bacteria with which they compete. More recent studies of T6SSs has shown that not all of their components are phage related. Rather, some are similar to components of type IV secretion systems. ▶▶| *Bacteriophage T4 (section 27.2)*

Retrieve, Infer, Apply

1. What are molecular chaperones and heat-shock proteins? Describe their functions.
2. Would an intein-containing protein be successfully folded by trigger factor alone? Explain your answer.
3. Give the major characteristics and functions of the protein secretion systems described in this section.
4. Which translocation or secretion system is most widespread?
5. What is a signal peptide? Suggest why a protein's signal peptide is not removed until after the protein is translocated across the plasma membrane?

Key Concepts

13.1 Experiments Using Bacteria and Viruses Demonstrated that DNA Is the Genetic Material

- DNA is composed of only four different building blocks (nucleotides). Therefore it was originally thought to be too simple to function as an organism's genetic material.
- The knowledge that DNA is the genetic material for cells came from studies on transformation by Griffith and Avery and from experiments on T2 phage by Hershey and Chase (**figures 13.1–13.3**).

13.2 Nucleic Acid and Protein Structure

- DNA and RNA are polymers of deoxyribonucleotides and ribonucleotides, respectively. DNA differs in composition from RNA in having deoxyribose and thymine, rather than ribose and uracil.
- DNA is double stranded, with complementary AT and GC base pairing between the strands. The strands run antiparallel and are twisted into a right-handed double helix (**figure 13.5**).
- RNA is normally single stranded, although it can coil upon itself and base pair to form hairpin structures.
- Proteins are polymers of amino acids (**figure 13.6**) linked by peptide bonds (**figure 13.7**).

13.3 DNA Replication in Bacteria

- Circular bacterial DNAs have a single origin of replication. They are copied by two replication forks moving around the circle to form a theta-shaped (θ) figure (**figure 13.9**).
- The replisome is a huge complex of proteins and is responsible for DNA replication (**table 13.1** and **figures 13.11** and **13.12**).
- DNA polymerase enzymes catalyze the synthesis of DNA in the 5′ to 3′ direction while reading the DNA template in the 3′ to 5′ direction. The double helix is unwound by helicases with the aid of topoisomerases such as DNA gyrase. DNA binding proteins keep the strands separate (**figures 13.10, 13.12,** and **13.13**).
- DNA polymerase III holoenzyme, the replicative DNA polymerase in bacteria, synthesizes a complementary DNA copy beginning with a short RNA primer made by the enzyme primase. The leading strand is replicated continuously, whereas DNA synthesis on the lagging strand is discontinuous and forms Okazaki fragments (**figures 13.13** and **13.14**). DNA polymerase I excises the RNA primers and fills in the resulting gap. DNA ligase then joins the fragments together (**figures 13.14** and **13.15**).
- The linear chromosomes of *Borrelia burgdorferi* have hairpin ends and an origin of replication in the middle of the chromosome. DNA replication proceeds bidirectionally from the origin, but only leading strand synthesis occurs at each fork. The resulting molecule is a dimer that is then cut into the single linear molecules as hairpin ends are formed (**figure 13.18**).

13.4 Bacterial Genes Consist of Coding Regions and Other Sequences Important for Gene Function

- A gene may be defined as a nucleic acid sequence that codes for a polypeptide, tRNA, or rRNA.
- The template strand of DNA carries genetic information and directs the synthesis of an RNA transcript.
- A gene also contains a promoter, a coding region, and a terminator; it may have a leader and a trailer (**figure 13.20**).
- The genes for tRNA and rRNA often code for a precursor that is subsequently processed to yield several products (**figure 13.21**).

13.5 Transcription in Bacteria

- Genes of related function are often organized into a transcriptional unit called an operon. A single promoter directs transcription of the operon, yielding a polycistronic mRNA. Each coding region in the polycistronic mRNA has its own start and stop codons and gives rise to a single polypeptide (**figure 13.22**).
- RNA synthesis is catalyzed by enzymes called RNA polymerases. Bacterial RNA polymerases are composed of several protein subunits (**figure 13.23**).
- The transcription cycle (**figure 13.24**) includes initiation, elongation, and termination.
- A protein called sigma factor is a transcription factor that directs RNA polymerase to a promoter during transcription initiation. Sigma recognizes certain sequences in the promoter (**figure 13.25**). Once bound, RNA polymerase unwinds a short stretch of DNA (**figure 13.26**). It then begins moving down the template strand (**figure 13.27**), synthesizing RNA in a 5′ to 3′ direction (elongation). Termination of transcription is brought about by two mechanisms: factor-independent and factor-dependent termination (**figures 13.28** and **13.29**).

13.6 The Genetic Code Consists of Three-Letter "Words"

- Genetic information is carried in the form of 64 nucleotide triplets called codons (**table 13.3**); 61 sense codons direct amino acid incorporation, and three stop codons terminate translation. The code is degenerate (redundant); that is, there is more than one codon for most amino acids.
- Some proteins contain two rare amino acids: selenocysteine and pyrrolysine (**figure 13.32**).

13.7 Translation in Bacteria

- During translation, ribosomes attach to mRNA and synthesize a polypeptide beginning at the N-terminal end. A polysome or polyribosome is a complex of mRNA with several ribosomes (**figure 13.33**).
- Amino acids are activated for protein synthesis by attachment to the 3′ end of transfer RNAs. Activation requires ATP, and the reaction is catalyzed by aminoacyl-tRNA synthetases (**figure 13.35**).

- Ribosomes are large, complex organelles composed of rRNAs and many polypeptides (**figure 13.36**).
- Protein synthesis begins with the formation of the 30S initiation complex made up of the mRNA, initiator fMet-tRNA, two initiation factors, and the 30S subunit. The initiator tRNA is bound to the start codon on the mRNA. The 50S subunit then binds to form the 70S initiation complex (**figures 13.37** and **13.38**).
- In the elongation cycle, the proper aminoacyl-tRNA binds to the A site (**figure 13.39**). Then the transpeptidation reaction is catalyzed by peptidyl transferase (**figure 13.40**). Finally, during translocation, the peptidyl-tRNA moves to the P site and the ribosome moves down the mRNA by one codon. The empty tRNA leaves the ribosome by way of the exit site.
- Protein synthesis stops when a stop codon is reached. Bacteria require three release factors for codon recognition and ribosome dissociation from the mRNA (**figure 13.41**).

13.8 Protein Maturation and Secretion

- Some proteins are self-splicing and excise portions of themselves before folding into their final shape (**figure 13.42**).
- Chaperone proteins help ensure that proteins fold properly or are delivered to their destination site (**figure 13.43**).
- Many proteins must be transported across the plasma membrane of cells. The most commonly used mechanism is the Sec system, which is found in all organisms (**figure 13.44**). Type I secretion systems are observed in many bacteria. Type IV secretion systems are observed in both Gram-positive and Gram-negative bacteria (**figure 13.45**).
- Gram-negative bacteria have evolved additional systems for moving proteins across the outer membrane of the cell wall. These include type II, III, V, and VI secretion systems (**figure 13.45**).

Compare, Hypothesize, Invent

1. *Streptomyces coelicolor* has a linear chromosome. Interestingly, there are no genes that encode essential proteins near the ends of the chromosome in this bacterium. Why do you think this is the case?

2. You have isolated several *E. coli* mutants:

   Mutant #1 has a point mutation (a single base-pair change) in the −10 region of the promoter of a gene encoding an enzyme needed for synthesis of the amino acid serine.

   Mutant #2 has a mutation in the −35 region in the promoter of the same gene.

   Mutant #3 is a double mutant with mutations in both the −10 and −35 region of the promoter of the same gene.

   Only Mutant #3 is unable to make serine. Why do you think this is so?

3. DNA polymerase I (Pol I) of *E. coli* consists of three functional parts (domains): an N-terminal domain with 5′ to 3′ exonuclease activity required for removal of the RNA primer, a central domain responsible for 3′ to 5′ exonuclease proofreading, and a C-terminal domain with polymerase activity. Pol I is thought to simultaneously remove RNA primers and fill in the gaps that result (figure 13.14). A group of proteins known as RNaseH also have 5′ to 3′ exonuclease activity and can thus remove RNA primers. However, they lack the other two functions observed for Pol I. Predict the ability of the following mutants to replicate DNA: (1) a strain with a mutant gene encoding Pol I such that it no longer has polymerase activity (but retains both types of nuclease activities); (2) a strain without RNaseH proteins; (3) a strain with a mutant gene encoding Pol I such that it no longer has 5′ to 3′ exonuclease activity (but retains 3′ to 5′ nuclease and polymerase activities); (4) a strain with the mutant Pol I described in (3) and a strain lacking all RNaseH proteins. Explain your reasoning for each.

   Read the original paper: Fukushima, S., et al. 2008. Reassessment of the in vivo functions of DNA polymerase I and RNaseH in bacterial cell growth. *J. Bacteriol.* 189:8575.

4. When bacteria enter stationary phase they inactivate their ribosomes and block protein synthesis in two ways. One mechanism involves two proteins called ribosome modulation factor (RMF) and hibernation promoting factor (HPF). RMF and HPF work together to not only inhibit ribosome activity, but also to cause two ribosomes to interact and form a ribosome dimer (100S dimer). The other method is accomplished by a protein called YfiA. This protein interacts with the ribosome and stops protein synthesis, but the ribosome remains a 70S ribosome monomer. The exact nature of both inactivation processes was unclear until scientists obtained high resolution crystal structures of 70S ribosomes bound to each of these inactivating proteins. Suggest sites on the ribosome where a factor might bind. For each site you suggest, offer a hypothesis explaining how factor binding would disrupt protein synthesis.

   Read the original paper: Polikanov, Y. S., et al. 2012. How hibernation factors RMF, HPF, and YfiA turn off protein synthesis. *Science* 336:915.

14

Regulation of Bacterial Cellular Processes

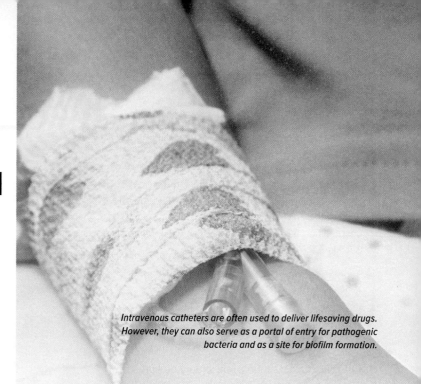

Intravenous catheters are often used to deliver lifesaving drugs. However, they can also serve as a portal of entry for pathogenic bacteria and as a site for biofilm formation.

Letting Go

Treatment of diseases such as cancer often requires the placement of intravenous catheters for the long-term delivery of medications through the bloodstream. The catheters stay in the vein between chemotherapy sessions and are rinsed with anticoagulant solutions after each use. Unfortunately, the catheters and the solutions used to rinse them can be a source of danger, rather than a critical part of the treatment plan. Such was the case in December 2004 when an anticoagulant solution contaminated with the bacterium *Pseudomonas fluorescens* was used to rinse the intravenous catheters of patients. Thirty-six patients in four different states developed blood infections from the contaminated solution. The Centers for Disease Control and Prevention (CDC) investigated the outbreak, and by the end of January 2005 the U.S. Food and Drug Administration had alerted the nation and all existing bottles of contaminated solution were recalled. Problem solved? Unfortunately not.

In March 2005 the CDC was again alerted that 28 new patients developed bloodstream infections caused by *P. fluorescens*. Investigations determined that the infections were caused by the same *P. fluorescens* strain isolated from the original patients identified in 2004. These new patients had been exposed to the contaminated solution many months earlier but did not become ill. Thus they were not identified in the 2004 outbreak. Why was their illness so delayed? Subsequent CDC investigations showed that the catheters in the new patients had a *P. fluorescens* biofilm. When the biofilm-coated catheters were flushed with sterile solutions, the dislodged cells caused the 2005 blood infections.

For these patients, physical forces released the cells from the biofilms, but it has long been known that bacteria have mechanisms for "deciding" when to establish a biofilm and when to "let go" and escape the biofilm. This involves sensing their environment and then altering cellular processes to make the switch. Switching back and forth between free-swimming (planktonic) and biofilm (sessile) growth is an example of a regulated behavior. Importantly, understanding how the change is made

is critical to developing strategies for controlling biofilm formation and dissolution.

A series of experiments culminating in a 2011 report have begun to reveal the mechanism for letting go, at least in one situation. The model includes an adhesion molecule located in the outer membrane (LapA), another protein (LapG), and describes how they work to control *P. fluorescens* cells leaving a biofilm when phosphate levels are low.

Why phosphate? Phosphate is needed for synthesis of nucleic acids. No phosphate, no DNA replication or transcription, dead cell. When phosphate is abundant, *P. fluorescens* cells want to stay put. They do so by sticking to a surface with LapA, the adhesion protein. Biofilm building then commences. However, when phosphate levels wane, if the cells are to survive, they must escape the biofilm and hunt for a location with more phosphate. They accomplish this with LapG, a periplasmic protease that cleaves LapA, removing it from the outer membrane. Without LapA holding them to the surface, the bacterial cells are free to leave the biofilm.

The capacity of *P. fluorescens* to stay or leave a biofilm is just one example of the many "decisions" microbes make. Bacteria are constantly detecting changes in their environment and responding accordingly. Cells use two general approaches to regulate cellular processes. (1) They can alter the activity of enzymes and other proteins. This type of regulation occurs after the protein is synthesized and is called posttranslational control. (2) They can change the rate of synthesis of enzymes and other proteins. This type of regulation is often called regulation of gene expression. Posttranslational control acts rapidly to adjust metabolic activity and other processes from moment to moment. The control of gene expression occurs over longer intervals. However, it conserves considerable energy and raw materials, maintains the balance between the amounts of various cell proteins, and enables microbes to acclimate to long-term environmental change. Thus control of gene expression complements posttranslational control. Both types of regulatory mechanisms are illustrated in this chapter. ◄◄ *Posttranslational regulation of enzyme activity (section 10.7)*

Readiness Check:

Based on what you have learned previously, you should be able to:

✔ Describe the mechanisms often used to regulate enzyme activity (section 10.7)

✔ Discuss genome structure and organization of genetic information in bacterial cells (sections 13.3 and 13.4)

✔ Summarize the events that occur during the three phases of transcription and translation (sections 13.5 and 13.7)

✔ Describe the phosphoenolpyruvate: sugar phosphotransferase system (section 3.3)

14.1 Bacteria Use Many Regulatory Options

After reading this section, you should be able to:

■ List when, during the flow of genetic information, bacterial cells can regulate gene expression

■ Speculate why microbial geneticists for many years focused almost exclusively on the regulation of transcription initiation

Figure 14.1 summarizes the steps leading from the information coded in DNA to a functional protein. As we discuss in chapter 13, the process begins with transcription of the gene, followed by translation of the mRNA to yield a protein. Regulation that occurs by controlling the processes of transcription and translation is often called regulation of gene expression. When gene expression is regulated by controlling mRNA synthesis, it is said to be "at the level of transcription" or at the "transcriptional level." Likewise, when gene expression is governed by controlling protein synthesis, it is said to be "at the level of translation" or at the "translational level." Translation yields a protein that may or may not be functional. The activity level of proteins may be altered by posttranslational modification.

Although the overall processes of transcription and translation are similar for all organisms, gene expression is somewhat different for members in each domain. Our focus in this chapter is on well-understood bacterial regulatory processes at the levels of transcription, translation, and posttranslation.

Regulation at the level of transcription has been the focus of microbial geneticists for many years and is well understood. Transcription can be regulated by governing its initiation or its elongation, and these topics are covered in sections 14.2 and 14.3, respectively. We begin our discussion by introducing two phenomena: induction of enzyme synthesis and repression of enzyme synthesis. Induction and repression provided the first models for gene regulation. These models involve the action of regulatory proteins, and the notion that gene expression is regulated solely by proteins persisted for many years. Eventually it was demonstrated that RNA molecules also have regulatory functions.

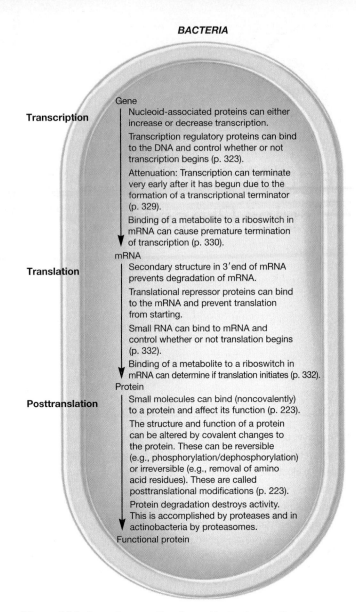

Figure 14.1 Some Common Regulatory Mechanisms in Bacteria.

14.2 Regulation of Transcription Initiation Saves Considerable Energy and Materials

After reading this section, you should be able to:

■ Compare and contrast housekeeping, constitutive, inducible, and repressible genes

■ Describe two common motifs in DNA-binding proteins

■ Summarize how negative transcriptional control and positive transcriptional control can be used to regulate both inducible and repressible genes

■ Outline the regulatory "decisions" made by cells

Induction and repression are historically important, as they were the first regulatory processes for gene expression to be understood in any detail. In this section, we first describe these phenomena and then examine the underlying regulatory mechanisms.

Induction and Repression of Enzyme Synthesis

Just as an automobile or a refrigerator does not work forever and eventually must be replaced, so too enzymes remain functional only for a certain time. The "old" enzymes are eventually degraded by proteasomes and other protein-degrading systems. This degradation not only removes enzymes that may not be functioning properly but also recycles amino acids so that they can be used to synthesize new proteins. However, this valuable process creates a problem for a cell because many enzymes catalyze reactions that are needed almost all the time (e.g., enzymes of the central metabolic pathways). Their functions are often referred to as "housekeeping functions," and the genes that encode them are often called **housekeeping genes.** To maintain the proper amount of housekeeping enzymes, many housekeeping genes are expressed continuously by the cell. Genes expressed continuously are said to be **constitutive genes.** Other enzymes are only needed at certain times and in certain environments. To conserve energy and cellular materials such as amino acids and nucleotides, genes encoding these enzymes are expressed only when needed; therefore their expression is regulated. The β-galactosidase gene is an example of a regulated gene.

The enzyme β-galactosidase catalyzes hydrolysis of the disaccharide sugar lactose to glucose and galactose (**figure 14.2**). When *E. coli* grows with lactose as its only carbon source, each cell contains about 3,000 β-galactosidase molecules, but it has less than three molecules in the absence of lactose. β-galactosidase is an inducible enzyme; that is, its level rises in the presence of a small effector molecule called an **inducer** (in this case, the lactose derivative allolactose). Likewise, the genes that encode inducible enzymes such as β-galactosidase are referred to as **inducible genes.**

β-galactosidase is an enzyme that functions in a catabolic pathway, and many catabolic enzymes are inducible enzymes. On the other hand, the genes for enzymes involved in biosynthetic pathways are often called **repressible genes,** and their products are called repressible enzymes. For instance, an amino acid present in the surroundings may inhibit the formation of enzymes responsible for its biosynthesis. This makes sense because the microorganism does not need the biosynthetic enzymes for a particular substance if it is already available. Generally, repressible enzymes are necessary for synthesis and are present unless the end product of their pathway is available. Inducible enzymes, in contrast, are required only when their substrate is available; they are missing in the absence of the inducer.

Regulatory Proteins Often Control Transcription Initiation

Proteins are often used to regulate transcription initiation, and therefore induction and repression. Many of these **transcriptional regulatory proteins** are DNA-binding proteins that form dimers and attach to short, inverted sequences of bases in the DNA called palindromes. Only a small portion of the proteins actually interacts with the palindromes, and these are referred to as DNA-binding domains. Examination of numerous transcriptional regulatory proteins has led to the recognition of two common motifs in DNA-binding domains: helix-turn-helix and zinc fingers.

Helix-turn-helix DNA-binding domains are often observed in transcriptional regulatory proteins (e.g., *lac* repressor, *trp* repressor, and CAP; pp. 326, 327, and 337, respectively). The helix-turn-helix domains are generally about 20 amino acids in length and are folded into two α-helices (*see figure A1.14*) separated by a β-turn. One of the helices protrudes from the surface of the protein and is often positioned in the major groove of the DNA. Thus when the regulatory protein forms a dimer, two α-helices, one from each protein subunit, interact with the major groove of DNA.

Zinc finger DNA-binding domains have been observed in some bacterial regulatory proteins. Zinc fingers are long loops formed by about 30 amino acids. The loop is stabilized by an interaction between certain amino acids in the loop and a Zn^{2+} ion. Regulatory proteins often have more than one zinc finger, each of which interacts with DNA. However, the function of each finger can differ. For instance, one finger may bind DNA nonspecifically, whereas another is important for recognition of a particular sequence in the DNA.

Regulatory proteins can exert either negative or positive control. **Negative transcriptional control** occurs when the binding of the

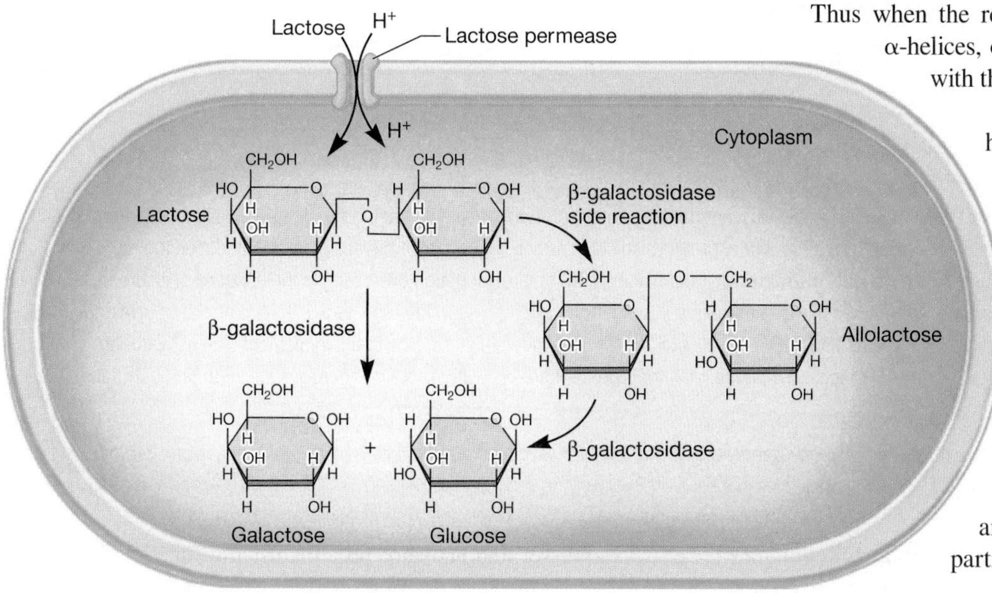

Figure 14.2 The Reactions of β-Galactosidase. The main reaction catalyzed by β-galactosidase is the hydrolysis of lactose, a disaccharide, into the monosaccharides galactose and glucose. The enzyme also catalyzes a minor reaction that converts lactose to allolactose. Allolactose acts as the inducer of β-galactosidase synthesis.

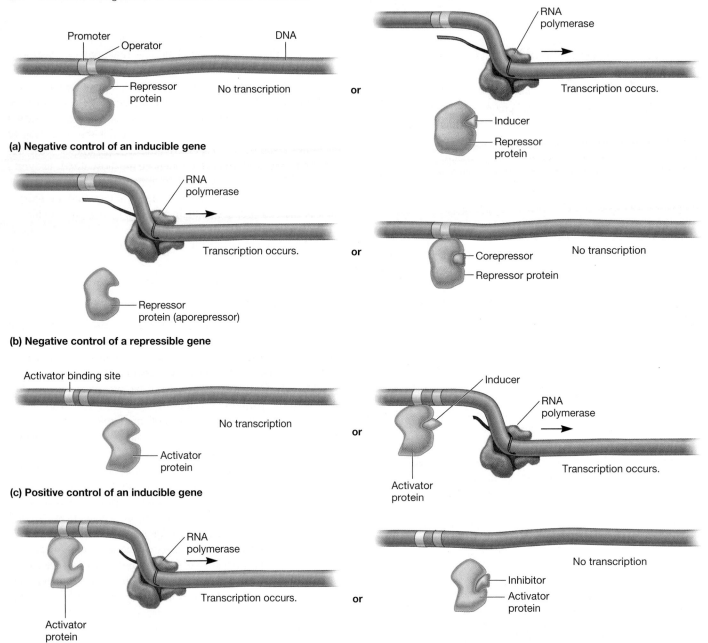

(a) Negative control of an inducible gene

(b) Negative control of a repressible gene

(c) Positive control of an inducible gene

(d) Positive control of a repressible gene

Figure 14.3 Action of Bacterial Regulatory Proteins. Bacterial transcriptional regulatory proteins have two binding sites, one for a small effector molecule and one for DNA. The binding of the effector molecule changes the regulatory protein's ability to attach to DNA. (a) In the absence of an inducer, the repressor protein blocks transcription. The presence of an inducer prevents the repressor from binding DNA and transcription occurs. (b) In the absence of a corepressor, the repressor is unable to bind DNA and transcription occurs. When the corepressor interacts with the repressor, the repressor is able to attach to DNA and transcription is blocked. (c) The activator protein is only able to bind DNA and activate transcription when it interacts with the inducer. (d) The activator protein binds DNA and promotes transcription unless the inhibitor is present. When the inhibitor is present, the activator protein undergoes a conformational change that prevents it from binding DNA; this inhibits transcription.

MICRO INQUIRY *In what way is an inducer molecule that binds a repressor protein similar to an inhibitor molecule that binds an activator protein?*

protein to DNA inhibits initiation of transcription. Regulatory proteins that act in this fashion are called **repressor proteins. Positive transcriptional control** occurs when the binding of the protein to DNA promotes transcription initiation. These proteins are called **activator proteins.** ⮐ *Regulatory Proteins*

Repressor and activator proteins usually function by binding DNA at specific sites. In bacteria, repressor proteins attach to a region called the **operator,** which usually overlaps or is downstream of the promoter (i.e., closer to the coding region) (**figure 14.3a,b**). When bound, the repressor protein either blocks binding of RNA polymerase to the promoter or prevents its movement. Activator proteins attach to **activator-binding sites** (figure 14.3*c,d*). These are often upstream of the promoter (i.e., farther away from the coding region). Binding of an activator to its regulatory site generally facilitates RNA polymerase binding. This is accomplished in two main ways. In some cases, binding of the activator alters the structure of the promoter, making it a better target for RNA polymerase. More often, the activator protein interacts with RNA polymerase's α or σ subunits to facilitate binding.

Repressor and activator proteins must exist in both active and inactive forms if transcription initiation is to be controlled appropriately. Their activity is usually modified by small effector molecules, most of which bind the repressor or activator protein noncovalently (i.e., allosteric regulation). Figure 14.3 shows the four basic ways in which the interactions of an effector and a regulatory protein can affect transcription. (1) For negatively controlled inducible genes (e.g., those encoding enzymes needed for catabolism of a sugar), the repressor protein is active and prevents transcription when the substrate of the pathway is not available (figure 14.3*a*). It is inactivated by the inducer (e.g., the substrate of the pathway). (2) For negatively controlled repressible genes (e.g., those encoding enzymes needed for the synthesis of an amino acid), an inactive repressor protein, called the **aporepressor,** is initially synthesized. It is activated by the **corepressor** (figure 14.3*b*). For repressible enzymes that function in a biosynthetic pathway, the corepressor is often the product of the pathway (e.g., an amino acid). (3) The activator of a positively regulated inducible gene is activated by the inducer (figure 14.3*c*), whereas (4) the activator protein of a positively regulated

repressible gene is inactivated by an inhibitor (figure 14.3*d*). ◀◀ *Allosteric regulation (section 10.7)*

Before we continue our discussion, we must consider two general aspects of regulation. The first is that gene expression is rarely an all-or-nothing phenomenon; it is a continuum. Inhibition of transcription usually does not mean that genes are "turned off" (though this terminology is frequently used). Rather it means the level of mRNA synthesis is decreased significantly and in most cases is occurring at very low levels. In other words, many promoters of regulated genes and operons are considered "leaky" in that there is always some low, basal level of transcription. The second aspect of regulation is the "decision-making" process used by microbial cells. Although cells do not have thought processes, it is convenient to think of regulation in this way. Consider the regulatory "decisions" made by an *E. coli* cell. It need only synthesize the enzymes of a specific catabolic pathway if the substrate of the pathway is present in the environment and a preferred carbon and energy source (e.g., glucose) is not (**figure 14.4**; *also see figure 22.4*). Preferred carbon and energy sources usually are more easily catabolized or yield more energy. Thus it is to the cell's advantage to use the preferred source before another source. Conversely, synthesis of enzymes for biosynthetic pathways is inhibited when the end product of the pathway is present.

Recall that functionally related bacterial genes are often transcribed from a single promoter. The **structural genes**—the genes coding for nonregulatory polypeptides (e.g., enzymes)—are simply lined up together on the DNA, and a single, polycistronic mRNA carries all the messages (*see figure 13.22*). The structural genes, together with the promoter and operator or activator-binding sites, is called an **operon.** Many operons have been discovered and studied. Three well-studied operons are discussed next. They demonstrate different ways that regulatory proteins can be used to control gene expression at the level of transcription initiation.

Lactose Operon: Negative Transcriptional Control of Inducible Genes

In the late 1930s Jacques Monod (1910–1976) began examining bacterial growth and its regulation. He chose *E. coli* as a model bacterium and eventually focused his attention on genes involved in the growth of *E. coli* on lactose. He was joined in his studies about 15 years later by François Jacob (1920–2013). Together they developed the concept of an operon, and as a result of their studies, the lactose (*lac*) operon of *E. coli* is probably the best-studied negative control system.

The *lac* operon contains three structural genes controlled by the *lac* repressor (LacI). LacI is encoded by *lacI*, which is located upstream of the *lac*

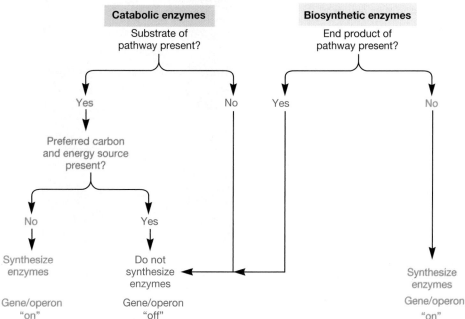

Figure 14.4 Examples of Regulatory "Decisions" Made by Cells.

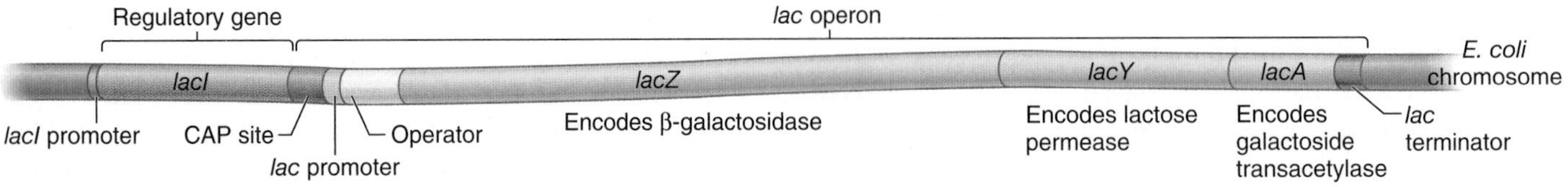

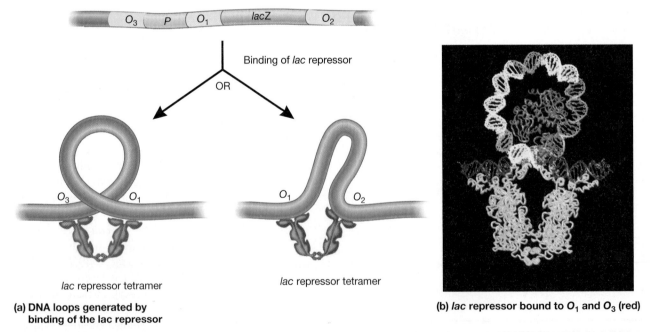

Regulatory gene

lac operon

lacI

lacZ

lacY

lacA

E. coli chromosome

lacI promoter

CAP site

Operator

lac promoter

Encodes β-galactosidase

Encodes lactose permease

Encodes galactoside transacetylase

lac terminator

Figure 14.5 **The *lac* Operon.** The *lac* operon consists of three structural genes—*lacZ, lacY,* and *lacA*—that are transcribed as a single unit from the *lac* promoter. The operon is regulated both negatively and positively. Negative control is brought about by the *lac* repressor, which is the product of the *lacI* gene. The operator is the site of *lac* repressor binding. Positive control results from the action of CAP. CAP binds the CAP site located just upstream from the *lac* promoter. CAP is partly responsible for a phenomenon called catabolite repression, an example of a global control network, in which numerous operons are controlled by a single protein (section 14.5). For simplicity, the operator is represented as a single region. In reality, the *lac* operator consists of three distinct sites, as shown in figure 14.6.

operon (**figure 14.5**). One gene in the operon codes for β-galactosidase; a second gene directs the synthesis of β-galactoside permease, the protein responsible for lactose uptake. The third gene codes for the enzyme β-galactoside transacetylase, whose function still is uncertain. The presence of the first two genes in the same operon ensures that the rates of lactose uptake and breakdown will vary together.

Lactose is one of many organic molecules *E. coli* can use as a carbon and energy source. It is wasteful to synthesize enzymes of the *lac* operon when lactose is not available. Therefore the cell expresses this operon at high levels when lactose is available and a preferred carbon and energy source is not; the *lac* repressor is responsible for inhibiting transcription when there is no lactose.

The *lac* repressor is composed of four identical subunits (i.e., it is a tetramer), each with a helix-turn-helix DNA-binding domain. The tetramer is formed when two dimers interact. When

lactose catabolism is not required, each dimer of the tetramer recognizes and tightly binds one of three different *lac* operator sites: O_1, O_2, and O_3 (**figure 14.6a**). O_1 is the main operator site and must be bound by the repressor if transcription is to be blocked. When one dimer is at O_1 and another is at one of the two other operator sites, the dimers bring the two operator sites close together, with a loop of DNA forming between them. Attachment of the *lac* repressor is a two-step process. First, it binds nonspecifically to DNA. Then it rapidly slides along the DNA until it reaches an operator site. Two α-helices of the repressor fit into the major groove of operator-site DNA (figure 14.6b).

How does the repressor inhibit transcription? The *lac* promoter is located near the *lac* operator sites. When there is no lactose, the repressor binds O_1 and one of the other operator sites, bending the DNA in the promoter region. This prevents initiation of transcription either because RNA polymerase cannot access the

O_3 | P | O_1 | lacZ | O_2

Binding of *lac* repressor

OR

O_3 O_1

lac repressor tetramer

O_1 O_2

lac repressor tetramer

(a) DNA loops generated by binding of the lac repressor

(b) *lac* repressor bound to O_1 and O_3 (red)

Figure 14.6 **The *lac* Operator Sites.** The *lac* operon has three operator sites: O_1, O_2, and O_3 (a). As shown in (a) and (b), the *lac* repressor (violet) binds O_1 and one of the other operator sites, forming a DNA loop. The DNA loop contains the −35 and −10 binding sites (green) recognized by RNA polymerase holoenzyme. Thus these sites are inaccessible and transcription is blocked. The DNA loop also contains the CAP binding site, and CAP (blue) is shown bound to the DNA (b). When the *lac* repressor is bound to the operator, CAP is unable to activate transcription.

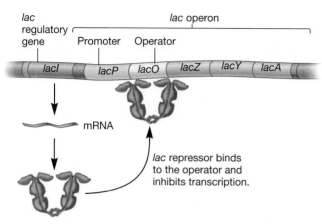

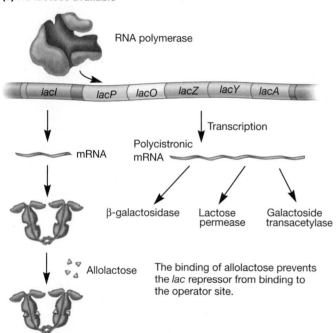

(a) No lactose available

(b) Lactose present

Figure 14.7 **Regulation of the *lac* Operon by the *lac* Repressor.** (a) The *lac* repressor is active and can bind the operator when allolactose is not present. Binding of the repressor to the operator inhibits transcription of the operon by RNA polymerase. (b) When lactose is available, some of it is converted to allolactose by β-galactosidase. When sufficient amounts of allolactose are present, it binds and inactivates the *lac* repressor. The repressor leaves the operator and RNA polymerase is free to initiate transcription.

MICRO INQUIRY *Is allolactose a corepressor or inducer molecule? Explain your answer.*

promoter or because it is blocked from moving into the coding region (**figure 14.7a**). When lactose is available, it is brought into the cell by lactose permease. Once inside the cell, β-galactosidase converts lactose to allolactose, the inducer of the operon (figure 14.2).

This occurs because there is always a low level of permease and β-galactosidase synthesis. Allolactose noncovalently interacts with the *lac* repressor protein and causes it to change to an inactive shape that is unable to bind any operator sites. The inactivated repressor leaves the DNA and transcription of the operon occurs (figure 14.7b).

Close examination of figures 14.5 and 14.6 clearly shows that regulation of the *lac* operon is not as simple as has just been described. That is because the *lac* operon is regulated by a second regulatory protein called *catabolite activator protein* (CAP). CAP functions in a global regulatory network that allows *E. coli* to use glucose preferentially over all other carbon and energy sources by a mechanism called catabolite repression. The use of two different regulatory proteins to control the expression of an operon illustrates another important point about regulatory processes—there are often layers of regulation. In the case of the lactose operon, the lactose repressor regulates gene expression in response to the presence or absence of lactose. CAP regulates the operon in response to the presence or absence of glucose. As described in section 14.5, the use of two regulatory proteins generates a continuum of expression levels. The highest levels of transcription occur when lactose is available and glucose is not; the lowest levels occur when lactose is not available and glucose is. For almost all of the examples described in this chapter, regulation occurs by more than one mechanism. 🐾 *The* lac *Operon*

Tryptophan Operon: Negative Transcriptional Control of Repressible Genes

The tryptophan (*trp*) operon of *E. coli* consists of five structural genes that encode enzymes needed for synthesis of the amino acid tryptophan (**figure 14.8**). It is regulated by the *trp* repressor, which is encoded by the *trpR* gene. Because the enzymes encoded by the *trp* operon function in a biosynthetic pathway, it is wasteful to make the enzymes needed for tryptophan synthesis when tryptophan is readily available. Therefore the operon functions only when tryptophan is not present and must be made de novo from precursor molecules (figure 14.4). To accomplish this regulatory goal, the *trp* repressor protein is synthesized in an inactive form that cannot bind the *trp* operator as long as tryptophan levels are low (figure 14.8a). When tryptophan levels increase, tryptophan acts as a corepressor, binding the repressor and activating it. The repressor-corepressor complex then attaches to the operator, blocking transcription initiation (figure 14.8b).

Like the *lac* operon, the *trp* operon is subject to another layer of regulation. In addition to being controlled at the level of transcription initiation by the *trp* repressor, expression of the *trp* operon is also controlled at the level of transcription elongation by a process called attenuation. This mode of regulation is discussed in section 14.3. 🐾 *Tryptophan Repressor*

Arabinose Operon: Transcriptional Control by a Protein That Acts Both Positively and Negatively

Regulation of the *E. coli* arabinose (*ara*) operon illustrates how the same protein can function either positively or negatively,

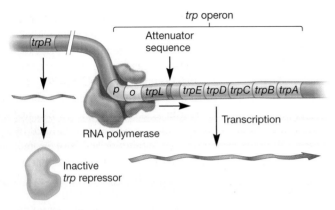

(a) Low tryptophan levels, transcription of the entire *trp* operon occurs

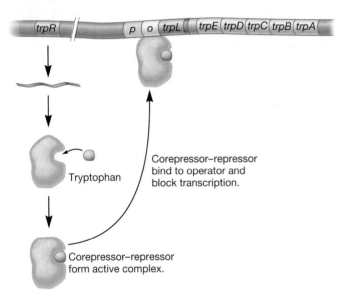

(b) High tryptophan levels, repression occurs

Figure 14.8 Regulation of the *trp* Operon by Tryptophan and the *trp* Repressor. The *trp* repressor is inactive when first synthesized and therefore is unable to bind the operator. It is activated by the binding of tryptophan, which serves as the corepressor. (a) When tryptophan levels are low, the repressor is inactive and transcription occurs. The enzymes encoded by the operon catalyze the reactions needed for tryptophan biosynthesis. (b) When tryptophan levels are sufficiently high, it binds the repressor. The repressor-corepressor complex binds the operator, and transcription of the operon is inhibited.

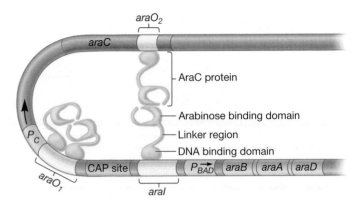

(a) Operon inhibited in the absence of arabinose

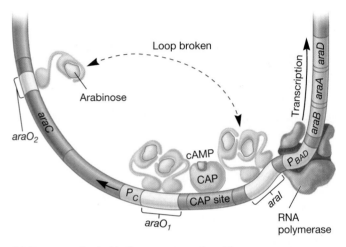

(b) Operon activated in the presence of arabinose

Figure 14.9 Regulation of the *ara* Operon by the AraC Protein. The AraC protein can act both as a repressor and as an activator, depending on the presence or absence of arabinose. (a) When arabinose is not available, the protein acts as a repressor. Two AraC proteins are involved. One binds *araI* and the other binds *araO₂*. The two proteins interact in such a way that the DNA between the two sites is bent, making it inaccessible to RNA polymerase. (b) When arabinose is present, it binds AraC, disrupting the interaction between the two AraC proteins. Subsequently, two AraC proteins, each bound to arabinose, form a dimer, which binds to *araI*. The AraC dimer functions as an activator; it interacts with sigma factor and facilitates both binding the RNA polymerase to the promoter and transition to the open complex.

depending on the environmental conditions. The *ara* operon encodes enzymes needed for catabolism of arabinose to xylulose 5-phosphate, an intermediate of the pentose phosphate pathway. The *ara* operon is regulated by AraC, which can interact with three different regulatory sequences: *araO₂*, *araO₁*, and *araI* (**figure 14.9**). When arabinose is not present, one molecule of AraC binds *araI* and another binds *araO₂*. The two AraC proteins interact, causing the DNA to bend. This prevents RNA polymerase from binding to the *ara* promoter, thereby

blocking transcription. In these conditions, AraC acts as a repressor (figure 14.9*a*). However, when arabinose is present, it binds AraC and prevents AraC molecules from interacting. This breaks the DNA loop. Furthermore, binding of two AraC-arabinose complexes to *araI* promotes transcription. Thus when arabinose is available, AraC acts as an activator (figure 14.9*b*). The *ara* operon, like the *lac* operon, is also subject to catabolite repression (section 14.5). ◀◀ *Pentose phosphate pathway (section 11.4)*

1. Many genes and operons are regulated at the level of transcription initiation. Why is it advantageous to regulate gene expression at the transcriptional level, rather than during or after translation?
2. What are induction and repression? How do bacteria use them to respond to changing nutrient supplies?
3. Using figure 14.4 as a guide, trace the "decision-making" pathway of an *E. coli* cell that is growing in a medium containing arabinose but lacking tryptophan.
4. Suppose synthesis of an amino acid was accomplished by four enzymes encoded by four genes that formed an operon. Draw a model of the operon showing how it would be controlled by positive regulation of transcription initiation.

14.3 Attenuation and Riboswitches Can Stop Transcription Prematurely

After reading this section, you should be able to:

- Describe in general terms how the *trp* operon of *E. coli* is regulated by attenuation
- Compare and contrast regulation of transcription elongation by attenuation and riboswitches
- Explain why the coupling of transcription and translation in bacterial cells is important to the regulatory mechanisms they use

In addition to regulating transcription initiation, bacteria can also regulate transcription termination. In this type of regulation, transcription is initiated but prematurely stopped depending on the environmental conditions and the needs of the organism. Attenuation was the first example of this kind of regulation. It was discovered by Charles Yanofsky in his 1970s studies of the *trp* operon. More recently riboswitches have been discovered. These regulatory sequences in the leader of an mRNA both sense and respond to environmental conditions by either prematurely terminating transcription or blocking translation. Both attenuation and riboswitches are described in this section.

Attenuation: Ribosome Behavior Influences Transcription

As noted in section 14.2, the tryptophan (*trp*) operon of *E. coli* is under the control of a repressor protein, and excess tryptophan inhibits transcription of operon genes by acting as a corepressor and activating the repressor protein. Although the operon is regulated mainly by repression, the continuation of transcription also is controlled. That is, there are two decision points involved in transcriptional control: the initiation of transcription and the continuation of transcription into the operon's structural genes. This additional level of control serves to adjust transcription in a more subtle fashion, such that the two systems of control can decrease transcription more than either one alone. When the repressor protein is not active, RNA polymerase begins

transcription of the leader. However, it often does not progress to the first structural gene in the operon. Instead, transcription is terminated within the leader; this is called **attenuation.** ◄◄ *Protein-coding genes (section 13.4)*

The ability to attenuate transcription is based on two things: the nature of the leader and the fact that transcription and translation are coupled in bacteria *(see figure 13.33)*. The *trp* operon leader contains a sequence of nucleotides called *trpL*. Interestingly, a small portion of *trpL* is transcribed and translated, giving rise to a small peptide called the leader peptide. The leader peptide has never been isolated, presumably because it is rapidly degraded. In addition to encoding the leader peptide, *trpL* contains **attenuator** sequences (**figure 14.10**). When transcribed, these sequences form stem-loop secondary structures in the newly formed mRNA. We define these sequences numerically (regions 1, 2, 3, and 4). When regions 1 and 2 pair with one another (1:2; figure 14.10*a*), they form a secondary structure called the pause loop, which causes RNA polymerase to slow down. The pause loop forms just prior to the formation of the terminator loop, which is made when regions 3 and 4 base-pair (3:4; figure 14.10*a*). A poly-U sequence follows the 3:4 terminator loop, just as it does in factor-independent transcriptional termination *(see figure 13.28)*. However, in this case, the terminator is in the leader, rather than at the end of the operon. Another stem-loop structure can be formed in the mRNA by pairing regions 2 and 3 (2:3, figure 14.10*b*). The formation of this antiterminator loop prevents generation of both the 1:2 pause and 3:4 terminator loops.

How do these various loops control transcription termination? Three scenarios describe the process. In the first, translation is not coupled to transcription because protein synthesis is not occurring. In other words, no ribosome is associated with the mRNA. In this scenario, the pause and terminator loops form, stopping transcription before RNA polymerase reaches the *trpE* gene (figure 14.10*a*).

In the next two scenarios, translation and transcription are coupled; that is, a ribosome initiates synthesis of the leader peptide. The interaction between RNA polymerase and the nearest ribosome determines which stem-loop structures are formed. As a ribosome translates the mRNA, it follows the RNA polymerase. Among the first several nucleotides of region 1 are two tryptophan codons; this is unusual because normally there is only one tryptophan residue per 100 amino acids in *E. coli* proteins. If tryptophan levels are low, there will not be enough charged tRNA^{trp} to fill the A site of the ribosome when the ribosome encounters the two tryptophan codons, and it will stall (figure 14.10*b*). Meanwhile RNA polymerase continues to transcribe mRNA, moving away from the stalled ribosome. The presence of the ribosome on region 1 prevents region 1 from base-pairing with region 2. As RNA polymerase continues, region 3 is transcribed, enabling the generation of the 2:3 antiterminator loop. This prevents the formation of the 3:4 terminator loop. Because the terminator loop is not present, RNA polymerase is not ejected from the DNA and transcription continues into the *trp* structural genes. If, on the other hand, there is plenty of tryptophan in the

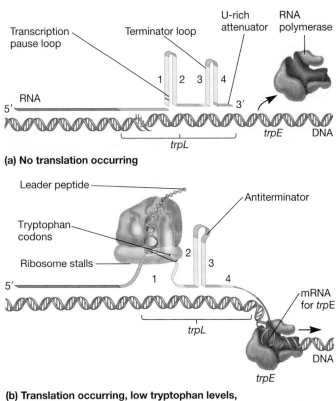

(a) No translation occurring

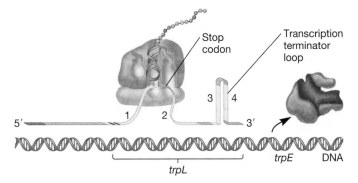

(b) Translation occurring, low tryptophan levels,
2:3 forms → transcription continues

(c) Translation occurring, high tryptophan levels,
3:4 forms → transcription is terminated

Figure 14.10 Attenuation of the *trp* Operon. (a) When protein synthesis has slowed, transcription and translation are not tightly coupled. Under these conditions, the most stable form of the mRNA occurs when region 1 hydrogen bonds to region 2 (RNA polymerase pause loop) and region 3 hydrogen bonds to region 4 (transcription terminator or attenuator loop). The formation of the transcription terminator causes transcription to stop within *trpL*. (b) When protein synthesis is occurring, transcription and translation are coupled, and the behavior of the ribosome on *trpL* influences transcription. If tryptophan levels are low, the ribosome pauses at the tryptophan codons in *trpL* because of insufficient amounts of charged tRNA^trp. This blocks region 1 of the mRNA, so that region 2 can hydrogen bond only with region 3. Because region 3 is already hydrogen bonded to region 2, the 3:4 terminator loop cannot form. Transcription proceeds and the *trp*-encoded biosynthetic enzymes are made. (c) If tryptophan levels are high, translation of *trpL* progresses to the stop codon, blocking region 2. Regions 3 and 4 can hydrogen bond and transcription terminates.

MICRO INQUIRY *How does this attenuation respond to rates of protein synthesis in addition to the intracellular level of tryptophan?*

the bacterium begins to synthesize protein rapidly, tryptophan may be scarce and the concentration of trp-tRNA may be low. This would reduce attenuation activity and stimulate operon transcription, resulting in larger quantities of the tryptophan biosynthetic enzymes. Acting together, repression and attenuation can coordinate the rate of synthesis of amino acid biosynthetic enzymes with the availability of amino acid end products and with the overall rate of protein synthesis. When tryptophan is present at high concentrations, any RNA polymerases not blocked by the activated repressor protein probably will not get past the attenuator sequence. Repression decreases transcription about 70-fold and attenuation slows it another eight-to ten-fold; when both mechanisms operate together, transcription can be slowed about 600-fold.

Attenuation is important in regulating at least five other operons that encode amino acid biosynthetic enzymes. In all cases, the leader peptide sequences resemble the tryptophan system in organization. For example, the leader peptide sequence of the histidine operon codes for seven histidines in a row and is followed by an attenuator that is a terminator sequence.

Riboswitches: Effector–mRNA Interaction Regulates Transcription

Riboswitches (also called sensory RNAs) are regions in the 5′ untranslated leader of an mRNA to which a small effector molecule (i.e., not a protein) can bind. Binding of the effector determines whether or not its target mRNA continues to be synthesized. Other riboswitches function at the level of translation. Our focus here is on transcriptional regulation; translational regulation is considered in section 14.4. For most transcriptional riboswitches, if the leader of an mRNA is folded one way, transcription continues; if folded another, transcription is terminated. The leader is called a riboswitch because it is analogous to a light switch turning lights on or

cell, there will be an abundance of charged tRNA^trp, and the ribosome will translate the two tryptophan codons in the leader peptide sequence without hesitation. Thus the ribosome remains close to the RNA polymerase. As RNA polymerase and the ribosome continue through *trpL*, regions 1 and 2 are transcribed and readily form a pause loop. Then regions 3 and 4 are transcribed, the terminator loop forms, and RNA polymerase is ejected from the DNA template. Finally, the presence of a UGA stop codon between regions 1 and 2 (i.e., the end of the leader peptide-coding region) causes termination of translation (figure 14.10c). ◄◄ *The genetic code consists of three-letter "words" (section 13.6)*

How is attenuation useful to *E. coli*? If the bacterium is deficient in an amino acid other than tryptophan, protein synthesis will slow and tryptophanyl-tRNA^Trp (trp-tRNA) will accumulate. Transcription of the *trp* operon will be inhibited by attenuation. When

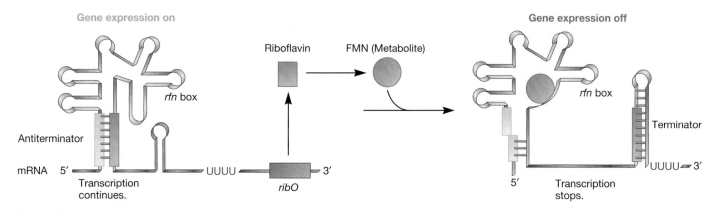

Figure 14.11 **Riboswitch Control of the Riboflavin (*rib*) Operon of *Bacillus subtilis*.** The *rib* operon produces enzymes needed for the synthesis of riboflavin, a component of flavin mononucleotide (FMN). Binding of FMN to the rfn (ribo*flavin*) box in the leader of *rib* mRNA causes the formation of a transcription terminator and cessation of transcription.

off, depending on whether the switch is in the on position (one folding pattern of the mRNA) or the off position (another folding pattern of the mRNA). What makes riboswitches unique and exciting is that the mRNA folding is determined by binding of the effector directly to the mRNA—a capability previously thought confined to proteins.

The riboswitch that regulates the riboflavin (*rib*) biosynthetic operon of *Bacillus subtilis* serves as an example (**figure 14.11**). The production of riboflavin biosynthetic enzymes is repressed by flavin mononucleotide (FMN), which is derived from riboflavin. When transcription of the *rib* operon begins, sequences in the leader of the mRNA fold into a structure called the RFN-element. This element binds FMN and in doing so alters the folding of the leader, creating a terminator that stops transcription. Transcription is stopped with the aid of rho protein, the RNA helicase that functions in factor-mediated transcription termination. ◄◄ *Stages of transcription (section 13.5)*

Since the initial discovery of riboswitches, many other novel riboswitches have been discovered; examples are listed in **table 14.1**. Examination of table 14.1 reveals that some riboswitches are coordinated with other types of regulatory

Table 14.1	Regulation of Gene Expression by Riboswitches	
System	**Target genes encode:**	**Effector and Regulatory Response**
T box	Amino acid biosynthetic enzymes	Uncharged tRNA; anticodon base-pairs to 5′ end of mRNA, preventing formation of transcriptional terminator.
Vitamin B$_{12}$ element	Cobalamine biosynthetic enzymes	Adenosylcobalamine (AdoCbl) binds to *btuB* mRNA and blocks translation.
THI box	Thiamine (vitamin B$_1$) biosynthetic and transport proteins	Thiamine pyrophosphate (TPP) causes either premature transcriptional termination or blocks ribosome binding.
RFN-element	Riboflavin biosynthetic enzymes	Flavin mononucleotide (FMN) cases premature transcriptional termination.
S box	Methionine biosynthetic enzymes	S-adenosylmethionine (SAM) causes premature transcriptional termination.
Cyclic-di-GMP	Self-splicing ribozyme	Binding of c-di-GMP alters splicing site.
Metal-sensing riboswitches (e.g., M box)	Metal transporter proteins	Intracellular levels of metals such as Mg^{2+} regulate availability of transporters for the metal either by prematurely stopping transcription or by triggering degradation of the transporter mRNA.
glmS riboswitch	GlmS enzyme	*glmS* mRNA, a self-cleaving ribozyme, encodes an enzyme that catalyzes synthesis of glucosamine 6-phosphate (GlcN6P). Binding of GlcN6P to riboswitch promotes self-cleavage of *glmS* mRNA, yielding a smaller RNA that is degraded by an RNase enzyme.
EutX riboswitch	*eut* operon	Coordinates availability of vitamin B$_{12}$, a cofactor for ethanolamine utilization (eut), with availability of ethanolamine. Control of the *eut* operon by a two-component signal transduction system, which senses eut, is linked to control of the availability of the response regulator by a small, noncoding RNA (EutX), which senses B$_{12}$.

mechanisms. For instance, some respond to second messengers such as cyclic dimeric GMP (c-di-GMP; p. 335). Others function in conjunction with small RNAs (section 14.4) and two-component signal transduction systems (p. 334). This is another example of how bacteria use multiple regulatory mechanisms to control their activities.

14.4 Riboswitches and Small RNAs Can Control Translation

After reading this section, you should be able to:

- Distinguish translational riboswitches from transcriptional riboswitches
- Discuss regulation of translation by small RNAs

It appears that in general, the riboswitches found in Gram-positive bacteria function by transcriptional termination, whereas the riboswitches discovered in Gram-negative bacteria regulate the translation of mRNA. Translation is usually regulated by blocking its initiation. In addition, some small RNA molecules can control translation initiation of target mRNAs. Both riboswitch and small RNA regulatory mechanisms are described now.

Translational Riboswitches

Similar to the riboswitches described in section 14.3, riboswitches that function at the translational level contain effector-binding elements at the 5′ end of the mRNA. Binding of the effector molecule affects the folding pattern of the mRNA leader, which often results in occlusion of the Shine-Dalgarno sequence and other elements of the ribosome-binding site. This inhibits ribosome binding and initiation of translation (**figure 14.12**). An example of this type of regulation is observed for the thiamine biosynthetic operons of numerous bacteria. The leaders of thiamine operons contain a structure called the THI-element, which can bind thiamine pyrophosphate. Association of thiamine pyrophosphate with the THI-element causes a conformational change in the leader that sequesters the Shine-Dalgarno sequence and blocks translation initiation.

Small RNA Molecules

A large number of RNA molecules have been discovered that do not function as mRNAs, tRNAs, or rRNAs. Microbiologists often refer to them as **small RNAs (sRNAs)** or as noncoding RNAs (ncRNAs); sRNAs range in size from around 25 to 500 nucleotides. Although some sRNAs have been implicated in the regulation of DNA replication and transcription, many function at the level of translation. Those that regulate translation do so by targeting mRNA molecules. These translation-controlling sRNAs are considered in this section.

Two main types of mRNA-targeting sRNAs can be described based on where they are encoded on the genome: cis-encoded

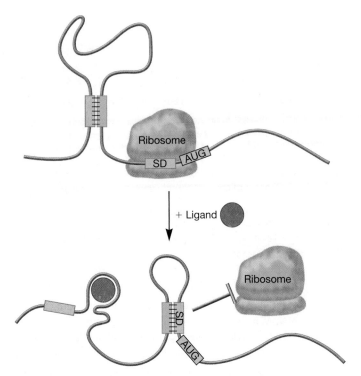

Figure 14.12 Inhibition of Translation by a Riboswitch. In the absence of a relevant metabolite, an effector binding site is formed in the leader of the mRNA (red) when complementary sequences (orange box and green box) hydrogen bond. This folding pattern exposes important sequences in the ribosome-binding site (e.g., the Shine-Dalgarno sequence; blue box) and translation occurs. When the appropriate effector molecule is present, it binds the leader, creating a structure containing the ribosome-binding site. Thus the ribosome-binding site becomes inaccessible and translation is blocked.

and trans-encoded. Cis-encoded sRNAs are synthesized from the nontemplate strand (i.e., the sense strand) of the gene that gives rise to their mRNA target. Thus they are complementary to the target mRNA and are called **antisense RNAs.** Trans-encoded sRNAs are synthesized from genes located at sites distinct from the genes encoding their target mRNAs. Despite this, they have some regions of complementarity to their targets and are sometimes also referred to as antisense RNAs. Both usually work in a similar way, by binding to the mRNA. As you might imagine, binding of an sRNA to an mRNA target can interfere with either ribosome binding or the ability of the ribosome to move down the mRNA. Thus it seems logical that many sRNAs would inhibit translation. However, some sRNAs actually promote translation upon binding to the mRNA. Whether inhibiting or activating, most trans-encoded sRNAs work with the help of an RNA chaperone, a protein that interacts with RNA to promote changes in its structure, stability, and interactions with other RNA molecules. Trans-encoded sRNAs are the best studied and are the focus of our discussion.

The regulation of synthesis of the OmpF porin protein provides an example of translation control by a trans-encoded sRNA.

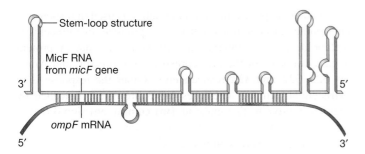

Figure 14.13 Inhibition of Translation by a Trans-Encoded sRNA. The *ompF* mRNA encodes the porin OmpF. Translation of this mRNA is regulated by the trans-encoded sRNA MicF, product of the *micF* gene. MicF is partially complementary to *ompF* mRNA and, when bound to it, prevents translation.

MICRO INQUIRY *How does inhibition of translation by sRNA differ from translational inhibition by riboswitches?*

Recall that the outer membrane of Gram-negative bacteria contains channels made of porin proteins (*see figure 3.26*). The two most important *E. coli* porins are OmpF and OmpC (Omp for *outer membrane protein*). OmpC pores are slightly smaller and are made when the bacterium grows at high osmotic pressures. It is the dominant porin when *E. coli* is in the intestinal tract. The larger OmpF pores are favored when *E. coli* grows in a dilute environment; OmpF allows solutes (e.g., nutrients) to diffuse into the cell more readily. The cell must maintain a constant level of porin protein in the membrane, with the relative levels of the two porins corresponding to the osmolarity of the medium. Expression of the *ompF* gene is regulated by MicF RNA, the product of the *micF* gene (*mic* for *mRNA-interfering complementary RNA*). The MicF RNA is complementary to *ompF* mRNA at the translation initiation site (**figure 14.13**). When MicF RNA base-pairs with *ompF* mRNA, translation is repressed. MicF RNA is produced under conditions such as high osmotic pressure or the presence of some toxic material, both of which favor *ompC* expression. Production of MicF RNA helps ensure that OmpF protein is not produced at high levels at the same time as OmpC protein. Some other sRNAs are listed in **table 14.2**. In addition to being regulated by MicF RNA, *ompF* (and *ompC*) is regulated by a two-component signal transduction system, as we describe in section 14.5.

Retrieve, Infer, Apply

1. What are the functions of *trpL* and the ribosome in attenuation?
2. Describe how attenuation would change in *E. coli* from its inoculation into a tryptophan-rich medium to one that is tryptophan deficient.
3. What are translational riboswitches? Compare them to transcriptional riboswitches.
4. How are translational riboswitches similar to attenuation as described for the *trp* operon? How do they differ? How do translational riboswitches differ from sRNAs?
5. Suggest a mechanism by which *micF* transcription might be regulated.

Table 14.2	Regulation of Gene Expression by Small Regulatory RNAs		
Small RNA	**Size**	**Bacterium**	**Function**
RhyB	90 nt[1]	*E. coli*	Represses translation of mRNAs encoding iron-containing proteins when iron availability is low
Spot 42	109 nt	*E. coli*	Inhibits translation of *galK* mRNA (encodes galactokinase)
RprA	105 nt	*E. coli*	Promotes translation of *rpoS* mRNA (encodes σ^s, a stationary phase sigma factor); antisense repressor of global negative regulator H-NS (involved in stress responses)
MicF	109 nt	*E. coli*	Inhibits *ompF* mRNA translation
OxyS	109 nt	*E. coli*	Inhibits translation of transcriptional regulator *fhlA* mRNA and *rpoS* mRNA
DsrA	85 nt	*E. coli*	Increases translation of *rpoS* mRNA
CsrB	366 nt	*E. coli*	Inhibits CsrA, a translational regulatory protein that positively regulates flagella synthesis, acetate metabolism, and glycolysis
RNAIII	512 nt	*Staphylococcus aureus*	Activates genes encoding secreted proteins (e.g., α-hemolysin); represses genes encoding surface proteins
RNA α	650 nt	*Vibrio anguillarum*	Decreased expression of fat, an iron-uptake protein
RsmB'	259 nt	*Erwinia carotovora* subsp. *carotovora*	Stabilizes mRNA of virulence proteins (e.g., cellulases, proteases, pectinolytic enzymes)

1 nt: nucleotides

C-di-GMP was discovered in the 1980s but drew relatively little interest until the annotation of many bacterial genomes led to the discovery that an unexpectedly large number of proteins contained the characteristic GGDEF and EAL sequences. When the expression of these proteins was altered by genetic manipulation, a host of important and complex behaviors were affected. These included biofilm formation, motility, spore formation, cell cycle events, and production of virulence factors. Interestingly, if a GGDEF-containing protein was overexpressed and affected a particular behavior, then overexpression of an EAL-containing protein had the opposite effect on the same behavior.

Only recently have scientists begun to unravel the mechanism of action of c-di-GMP. Both DGCs and PDEs have at least one sensory domain that can detect environmental signals such as oxygen levels, decreasing nutrients, and light. When a DGC detects its signal, it catalyzes formation of c-di-GMP. When PDE receives its signal, it degrades c-di-GMP. When c-di-GMP is available, it interacts with an effector. Most effectors thus far identified are proteins, but as we note in section 14.3, some are riboswitches. Upon interacting with c-di-GMP, the effector is able to influence the activity of its target. Some effectors are transcriptional regulatory proteins and thus control transcription. Others act posttranslationally, directly changing the activity of enzymes.

Recall from this chapter's opening story that *P. fluorescens* is able to "decide" whether to stay in or flee a biofilm. How is this accomplished? The answer is an "inside-out" signaling pathway where c-di-GMP plays a central role as a second messenger communicating phosphate levels to an effector called LapD. The DGC and PDE involved sense phosphate levels (the signal) and respond accordingly. When phosphate levels are high, DGC activity is high and PDE activity is low. Thus, c-di-GMP levels are high. It enters a pocket in a cytoplasmic domain in LapD, a protein embedded in the plasma membrane. LapD (the effector) also has a domain that extends into the periplasmic space. When c-di-GMP is in the pocket, LapD assumes a shape that promotes interaction with the protease LapG (the target), sequestering it. LapG is unable to cleave LapA, and cells form or remain in a biofilm. When phosphate levels decrease, PDE activity increases, and the

enzyme destroys c-di-GMP. With no c-di-GMP in the LapD pocket, LapD undergoes a conformational change and LapG is released. It cleaves LapA, and the cell is free to leave the biofilm.

The final mechanism we introduce is unique to bacteria: the use of **alternate sigma factors,** which can immediately change expression of many genes as they direct RNA polymerase to specific subsets of bacterial genes. This is possible because RNA polymerase core enzyme needs the assistance of a sigma factor to bind a promoter and initiate transcription. Each sigma factor recognizes promoters that differ in sequence (*see table 13.2*). ◄◄ *Transcription in bacteria (section 13.5)*

E. coli synthesizes several sigma factors (**table 14.3**). Under normal conditions, a sigma factor called σ^{70} directs RNA polymerase activity. (The superscript number or letter indicates the size or function of the sigma factor; 70 stands for 70,000 daltons.) When flagella and chemotactic proteins are needed, *E. coli* produces σ^F (σ^{28}). σ^F then binds its consensus sequences in promoters of genes whose products are needed for flagella biosynthesis and chemotaxis. If the temperature rises too high, σ^H (σ^{32}) is produced and stimulates the formation of about 17 heat-shock proteins that protect the cell from thermal destruction. Importantly, each sigma factor has its own set of promoter consensus sequences to which it binds (*see table 13.2*).

Sigma (σ) factors known as *extracytoplasmic function* (ECF) sigmas deserve mention because they function in signal transduction. Numerous mechanisms for their regulatory activities have been described. The best-studied system uses a plasma membrane-bound protein called an anti-sigma factor to sense an extracellular stimulus and thereby determine the activity of the ECF σ. In the absence of the stimulus, the anti-σ binds the ECF σ, sequestering it so that it is not able to bind its cognate promoters. Upon detecting the appropriate signal, the anti-σ is destroyed by proteolysis, and the ECF σ is released and activates transcription of its target genes. Other ECF sigmas are phosphorylated by plasma membrane-bound kinases (serine/threonine kinases) that detect the signal and phosphorylate and activate the ECF σ, which then initiates transcription.

In the remainder of this section, we explore several examples of global control networks. As you examine these examples, notice that many involve multiple mechanisms of control that function at

Table 14.3	*E. coli* Sigma Factors
Sigma Factor	**Genes Transcribed**
σ^{70}	Genes needed during exponential growth
σ^S	Genes needed during the general stress response and during stationary phase
σ^E	Genes needed to restore membrane integrity and the proper folding of membrane proteins
σ^H (σ^{32})	Genes needed to protect against heat shock and other stresses, including genes encoding chaperones that help maintain or restore proper folding of cytoplasmic proteins and proteases that degrade damaged proteins
FecI σ	Genes that encode the iron citrate transport machinery in response to iron starvation and the availability of iron citrate
σ^F (σ^{28})	Genes involved in flagellum assembly
σ^{60}	Genes involved in nitrogen metabolism

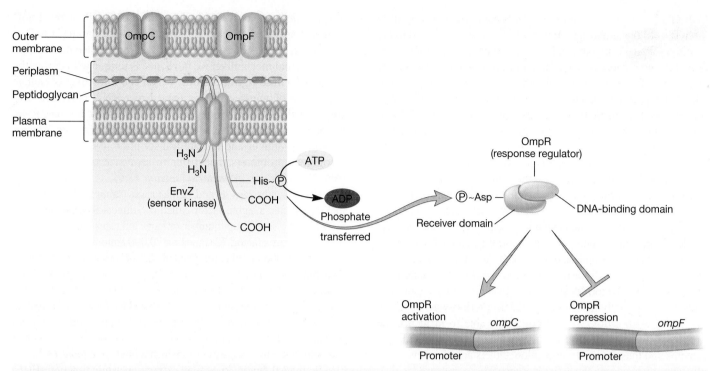

Figure 14.14 Two-Component Signal Transduction System and the Regulation of Porin Proteins. In this system, the sensor kinase protein EnvZ loops through the cytoplasmic membrane so that both its C- and N-termini are in the cytoplasm. When EnvZ senses an increase in osmolarity, it autophosphorylates a histidine residue at its C-terminus. EnvZ then passes the phosphoryl group to the response regulator OmpR, which accepts it on an aspartic acid residue located in its N-terminus. This activates OmpR so that it is able to bind DNA and repress *ompF* expression and enhance that of *ompC*.

MICRO INQUIRY *Relative to each promoter, where would you predict phosphorylated OmpR would bind the* ompC *and* ompF *genes?*

Two-component signal transduction systems involve a simple relay where the sensor kinase transfers a phosphoryl group directly to the response regulator. However, there are instances when more proteins participate in the transfer of phosphoryl groups. These more complex pathways are called **phosphorelay systems.** Phosphorelays are important control measures observed in quorum-sensing by *Vibrio harveyi* (p. 342) and endospore formation by

B. subtilis (p. 345). The effectiveness of two-component systems is illustrated by their abundance. Most bacterial cells use a variety of these systems to respond to an array of environmental stresses. For example, members of the soil bacterial genus *Streptomyces* have at least 80 such systems.

Many bacteria use **second messengers** in their global control systems. Second messengers are small molecules produced within a cell in response to a signal (i.e., the first messenger) that is outside the cell. Two bacterial second messengers were discovered many years ago: cyclic adenosine monophosphate (cAMP) and guanosine tetraphosphate (ppGpp). Newer additions to the bacterial second messenger family include **cyclic dimeric GMP (c-di-GMP; figure 14.15a)**, c-di-AMP, and cGMP. The role of cAMP in catabolite repression and that of ppGpp in the stringent response are discussed on pages 337 and 344, respectively. Here we consider c-di-GMP.

The second messenger c-di-GMP is synthesized by enzymes called diguanylate cyclases (DGCs). Once formed, c-di-GMP functions until it is destroyed by enzymes called phophodiesterases (PDEs; figure 14.15b). Each type of enzyme includes a characteristic sequence of amino acids: GGDEF for DGCs, and EAL (or HD-GYP) for PDEs (*see table 13.3*).

(a) Cyclic dimeric GMP

(b)

2 GTP

↓ Diguanylate cyclase

c - di - GMP

↓ Phosphodiesterase

pGpG

↓

2 GMP

Figure 14.15 Cyclic Dimeric GMP. (a) Structure of cyclic dimeric GMP (c-di-GMP). This second messenger consists of two GMP molecules linked together by their phosphates. (b) The two enzymes that control levels of c-di-GMP in cells. Diguanylate cyclases form c-di-GMP and phosphodiesterases degrade c-di-GMP.

C-di-GMP was discovered in the 1980s but drew relatively little interest until the annotation of many bacterial genomes led to the discovery that an unexpectedly large number of proteins contained the characteristic GGDEF and EAL sequences. When the expression of these proteins was altered by genetic manipulation, a host of important and complex behaviors were affected. These included biofilm formation, motility, spore formation, cell cycle events, and production of virulence factors. Interestingly, if a GGDEF-containing protein was overexpressed and affected a particular behavior, then overexpression of an EAL-containing protein had the opposite effect on the same behavior.

Only recently have scientists begun to unravel the mechanism of action of c-di-GMP. Both DGCs and PDEs have at least one sensory domain that can detect environmental signals such as oxygen levels, decreasing nutrients, and light. When a DGC detects its signal, it catalyzes formation of c-di-GMP. When PDE receives its signal, it degrades c-di-GMP. When c-di-GMP is available, it interacts with an effector. Most effectors thus far identified are proteins, but as we note in section 14.3, some are riboswitches. Upon interacting with c-di-GMP, the effector is able to influence the activity of its target. Some effectors are transcriptional regulatory proteins and thus control transcription. Others act posttranslationally, directly changing the activity of enzymes.

Recall from this chapter's opening story that *P. fluorescens* is able to "decide" whether to stay in or flee a biofilm. How is this accomplished? The answer is an "inside-out" signaling pathway where c-di-GMP plays a central role as a second messenger communicating phosphate levels to an effector called LapD. The DGC and PDE involved sense phosphate levels (the signal) and respond accordingly. When phosphate levels are high, DGC activity is high and PDE activity is low. Thus, c-di-GMP levels are high. It enters a pocket in a cytoplasmic domain in LapD, a protein embedded in the plasma membrane. LapD (the effector) also has a domain that extends into the periplasmic space. When c-di-GMP is in the pocket, LapD assumes a shape that promotes interaction with the protease LapG (the target), sequestering it. LapG is unable to cleave LapA, and cells form or remain in a biofilm. When phosphate levels decrease, PDE activity increases, and the

enzyme destroys c-di-GMP. With no c-di-GMP in the LapD pocket, LapD undergoes a conformational change and LapG is released. It cleaves LapA, and the cell is free to leave the biofilm.

The final mechanism we introduce is unique to bacteria: the use of **alternate sigma factors,** which can immediately change expression of many genes as they direct RNA polymerase to specific subsets of bacterial genes. This is possible because RNA polymerase core enzyme needs the assistance of a sigma factor to bind a promoter and initiate transcription. Each sigma factor recognizes promoters that differ in sequence (*see table 13.2*). ◀◀ *Transcription in bacteria (section 13.5)*

E. coli synthesizes several sigma factors (**table 14.3**). Under normal conditions, a sigma factor called σ^{70} directs RNA polymerase activity. (The superscript number or letter indicates the size or function of the sigma factor; 70 stands for 70,000 daltons.) When flagella and chemotactic proteins are needed, *E. coli* produces σ^F (σ^{28}). σ^F then binds its consensus sequences in promoters of genes whose products are needed for flagella biosynthesis and chemotaxis. If the temperature rises too high, σ^H (σ^{32}) is produced and stimulates the formation of about 17 heat-shock proteins that protect the cell from thermal destruction. Importantly, each sigma factor has its own set of promoter consensus sequences to which it binds (*see table 13.2*).

Sigma (σ) factors known as *extracytoplasmic function* (ECF) sigmas deserve mention because they function in signal transduction. Numerous mechanisms for their regulatory activities have been described. The best-studied system uses a plasma membrane-bound protein called an anti-sigma factor to sense an extracellular stimulus and thereby determine the activity of the ECF σ. In the absence of the stimulus, the anti-σ binds the ECF σ, sequestering it so that it is not able to bind its cognate promoters. Upon detecting the appropriate signal, the anti-σ is destroyed by proteolysis, and the ECF σ is released and activates transcription of its target genes. Other ECF sigmas are phosphorylated by plasma membrane-bound kinases (serine/threonine kinases) that detect the signal and phosphorylate and activate the ECF σ, which then initiates transcription.

In the remainder of this section, we explore several examples of global control networks. As you examine these examples, notice that many involve multiple mechanisms of control that function at

Table 14.3	*E. coli* Sigma Factors
Sigma Factor	**Genes Transcribed**
σ^{70}	Genes needed during exponential growth
σ^S	Genes needed during the general stress response and during stationary phase
σ^E	Genes needed to restore membrane integrity and the proper folding of membrane proteins
σ^H (σ^{32})	Genes needed to protect against heat shock and other stresses, including genes encoding chaperones that help maintain or restore proper folding of cytoplasmic proteins and proteases that degrade damaged proteins
FecI σ	Genes that encode the iron citrate transport machinery in response to iron starvation and the availability of iron citrate
σ^F (σ^{28})	Genes involved in flagellum assembly
σ^{60}	Genes involved in nitrogen metabolism

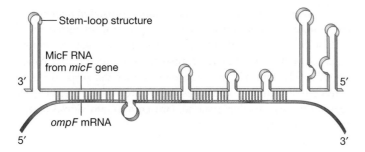

Figure 14.13 Inhibition of Translation by a Trans-Encoded sRNA.
The *ompF* mRNA encodes the porin OmpF. Translation of this mRNA is regulated by the trans-encoded sRNA MicF, product of the *micF* gene. MicF is partially complementary to *ompF* mRNA and, when bound to it, prevents translation.

MICRO INQUIRY *How does inhibition of translation by sRNA differ from translational inhibition by riboswitches?*

Recall that the outer membrane of Gram-negative bacteria contains channels made of porin proteins (*see figure 3.26*). The two most important *E. coli* porins are OmpF and OmpC (Omp for *outer membrane protein*). OmpC pores are slightly smaller and are made when the bacterium grows at high osmotic pressures. It is the dominant porin when *E. coli* is in the intestinal tract. The larger OmpF pores are favored when *E. coli* grows in a dilute environment; OmpF allows solutes (e.g., nutrients) to diffuse into the cell more readily. The cell must maintain a constant level of porin protein in the membrane, with the relative levels of the two porins corresponding to the osmolarity of the medium. Expression of the *ompF* gene is regulated by MicF RNA, the product of the *micF* gene (*mic* for *m*RNA-*i*nterfering *c*omplementary RNA). The MicF RNA is complementary to *ompF* mRNA at the translation initiation site (**figure 14.13**). When MicF RNA base-pairs with *ompF* mRNA, translation is repressed. MicF RNA is produced under conditions such as high osmotic pressure or the presence of some toxic material, both of which favor *ompC* expression. Production of MicF RNA helps ensure that OmpF protein is not produced at high levels at the same time as OmpC protein. Some other sRNAs are listed in **table 14.2**. In addition to being regulated by MicF RNA, *ompF* (and *ompC*) is regulated by a two-component signal transduction system, as we describe in section 14.5.

Retrieve, Infer, Apply

1. What are the functions of *trpL* and the ribosome in attenuation?
2. Describe how attenuation would change in *E. coli* from its inoculation into a tryptophan-rich medium to one that is tryptophan deficient.
3. What are translational riboswitches? Compare them to transcriptional riboswitches.
4. How are translational riboswitches similar to attenuation as described for the *trp* operon? How do they differ? How do translational riboswitches differ from sRNAs?
5. Suggest a mechanism by which *micF* transcription might be regulated.

Table 14.2	Regulation of Gene Expression by Small Regulatory RNAs		
Small RNA	**Size**	**Bacterium**	**Function**
RhyB	90 nt[1]	*E. coli*	Represses translation of mRNAs encoding iron-containing proteins when iron availability is low
Spot 42	109 nt	*E. coli*	Inhibits translation of *galK* mRNA (encodes galactokinase)
RprA	105 nt	*E. coli*	Promotes translation of *rpoS* mRNA (encodes σ^s, a stationary phase sigma factor); antisense repressor of global negative regulator H-NS (involved in stress responses)
MicF	109 nt	*E. coli*	Inhibits *ompF* mRNA translation
OxyS	109 nt	*E. coli*	Inhibits translation of transcriptional regulator *fhlA* mRNA and *rpoS* mRNA
DsrA	85 nt	*E. coli*	Increases translation of *rpoS* mRNA
CsrB	366 nt	*E. coli*	Inhibits CsrA, a translational regulatory protein that positively regulates flagella synthesis, acetate metabolism, and glycolysis
RNAIII	512 nt	*Staphylococcus aureus*	Activates genes encoding secreted proteins (e.g., α-hemolysin); represses genes encoding surface proteins
RNA α	650 nt	*Vibrio anguillarum*	Decreased expression of fat, an iron-uptake protein
RsmB′	259 nt	*Erwinia carotovora* subsp. *carotovora*	Stabilizes mRNA of virulence proteins (e.g., cellulases, proteases, pectinolytic enzymes)

1 nt: nucleotides

14.5 Bacteria Combine Several Regulatory Mechanisms to Control Complex Cellular Processes

After reading this section, you should be able to:

- Compare and contrast two-component signal transduction systems with phosphorelay signal transduction systems
- List the nucleotides that are used by bacteria as second messengers
- Explain how alternate sigma factors can be used to initiate transcription of distinct sets of operons
- Summarize how CAP, cAMP, and the *lac* repressor work together to cause diauxic growth as well as other outcomes related to regulation of the lactose operon
- Hypothesize how other operons encoding catabolic enzymes might be regulated by catabolite repression
- Outline how the proteins of the *E. coli* chemotaxis system function to control flagellar rotation in the presence of a chemoattractant
- Predict the events that would occur if *E. coli* cells were in a gradient of a chemorepellent
- Differentiate quorum sensing observed in *Vibrio fischeri* from that in *Vibrio harveyi*
- Describe the stringent response
- Outline the regulatory mechanisms used by *Bacillus subtilis* to control endospore formation

Consider the bacterium *Neisseria meningitidis,* more commonly known as meningococcus. This Gram-negative coccus is a normal inhabitant of the human nasopharynx, where it does no harm to its host. However, if the bacterium somehow enters the bloodstream, it experiences an increase in temperature. Meningococcus responds to the higher temperature by activating many genes in numerous operons that convert the harmless bacterium to a deadly pathogen. How does meningococcus sense a higher temperature and how does that lead to such a dramatic change in its nature and relationship to its host? Although the answer has not been fully elucidated, it is known that it does so using one or more **global regulatory systems:** regulatory systems that affect many genes, operons, and pathways simultaneously. In this section, we consider some ways bacteria control complex processes using global regulation. We begin by introducing some common "themes" in global regulation. We then provide several examples that illustrate the use of these common mechanisms.

Mechanisms Used for Global Regulation

Bacteria have several tools in their regulatory tool belts that enable them to control processes such as chemotaxis, translation, quorum sensing, and endospore formation. Perhaps the simplest approach is to use a single **global regulatory protein** to alter the expression of multiple operons (or genes). The collection of genes and operons controlled by the global regulator is called a **regulon.** A good example of global regulatory protein is CAP, which is involved in the phenomenon called catabolite repression (p. 337).

Another approach to global regulation is used in response to environmental conditions that do not produce an intracellular metabolite that can interact directly with a regulatory protein (e.g., as we saw for the *lac* repressor, *trp* repressor, and AraC; pp. 325, 327, and 327, respectively). Rather the response is to extracellular stimuli. These responses are often controlled by regulatory proteins that make up **two-component signal transduction systems.** These systems link events occurring outside the cell to gene expression inside the cell.

Two-component signal transduction systems are named after the two proteins that govern the regulatory pathway. The first is an enzyme called a **sensor kinase** that spans the plasma membrane so that part of it is exposed to the extracellular environment (periplasm in Gram-negative bacteria), while another part is exposed to the cytoplasm. In this way, the sensor kinase can detect specific changes in the environment and communicate information to the cell's interior. It does so by first phosphorylating itself and then transferring the phosphate to the second component, the **response regulator.** There are two types of response regulators. Most are DNA-binding proteins that, when activated by the sensor kinase, can act as either activators or repressors of transcription initiation. Other response regulators function posttranslationally to control the activity of enzymes or other proteins in the global network. An example of this type of system is chemotaxis, which is described on page 339.

The EnvZ/OmpR system is an excellent example of a two-component signal transduction system that functions at the level of transcription; it is used to control porin production in *E. coli.* As we introduce in section 14.4, *E. coli* uses sRNA to control OmpF production. Thus two regulatory mechanisms are used to control the response to osmolarity. In fact, the two mechanisms are intertwined, with the two-component system also controlling the synthesis of the MicF sRNA. Increasing evidence suggests that two-component systems and sRNAs often operate together to form complex regulatory networks. In the EnvZ/OmpR system, EnvZ (Env for cell *env*elope) is the sensor kinase and OmpR is the response regulator. EnvZ is a homodimeric protein (that is, two identical polypeptides interact to form the functional protein) that is an integral membrane protein anchored to the membrane by two membrane-spanning domains. EnvZ is looped through the membrane such that a central domain protrudes into the periplasm, while the amino and carboxyl termini are exposed to the cytoplasm (**figure 14.14**).

The response regulator OmpR is a soluble, cytoplasmic protein that controls transcription of the *ompF* and *ompC* structural genes. The N-terminal end of OmpR is called the receiver domain because it possesses a specific aspartic acid residue that accepts the signal (a phosphoryl group) from the sensor kinase. Upon receipt of the signal, the C-terminal end of OmpR is able to regulate transcription by binding DNA. When EnvZ senses high osmolarity, it phosphorylates itself (autophosphorylation) on a specific histidine residue. This phosphoryl group is quickly transferred to the N-terminus of OmpR. Once OmpR is phosphorylated, it is able to regulate transcription of the porin genes so that *ompF* transcription is repressed and *ompC* transcription is activated.

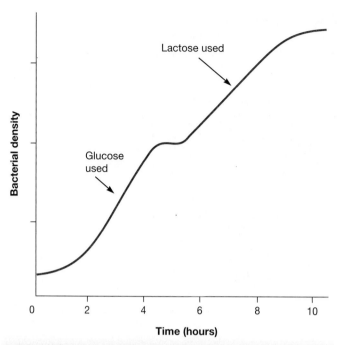

Figure 14.16 Diauxic Growth. The diauxic growth curve of *E. coli* grown with a mixture of glucose and lactose. Glucose is first used, then lactose. A short lag in growth is present while the bacteria synthesize the enzymes needed for lactose use.

MICRO INQUIRY *For what other compounds would you also see this kind of diauxic growth?*

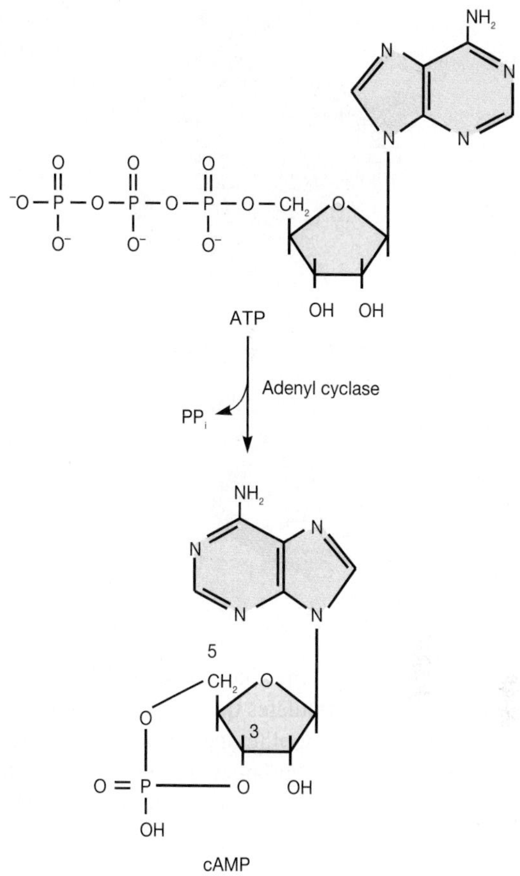

Figure 14.17 Cyclic Adenosine Monophosphate (cAMP). The phosphate extends between the 3′ and 5′ hydroxyls of the ribose sugar. The enzyme adenyl cyclase forms cAMP from ATP.

different levels (e.g., transcriptionally or posttranslationally). Remember that these examples were chosen because they are relatively well studied and are important model systems for understanding the regulation of even more complex networks. We encourage you to imagine how these models might help elucidate the control of the cell cycle, developmental processes such as fruiting body formation by *Myxococcus* spp., or the expression of virulence factors by pathogenic bacteria. We begin these discussions by returning to the "grand daddy" of operons, the lactose operon and its regulation by catabolite repression. ▶▶ *Order* Myxococcales *(section 22.4)*

Catabolite Repression in *E. coli*

If *E. coli* grows in a medium containing both glucose and lactose, it uses glucose preferentially until the sugar is exhausted. Then after a short lag, growth resumes at a slower rate with lactose as the carbon source (**figure 14.16**). This allows glucose, the more easily catabolized and more energy-yielding molecule, to be used first, followed by other energy sources that may be more difficult to degrade or yield less energy. This biphasic growth pattern is called **diauxic growth** and it is brought about by a global regulatory process called **catabolite repression.** In *E. coli,* two regulatory networks bring about catabolite repression. One network directly inhibits the transporters that take up the less preferred sugars resulting in inducer exclusion. The other network dramatically decreases expression of the genes encoding the enzymes that catabolize the non-preferred

sugars. Interestingly, the two networks are linked, as we describe in a moment. We begin by looking at the latter network.

The enzymes for glucose catabolism are constitutive. However, operons that encode enzymes required for the catabolism of carbon sources that must first be modified before entering glycolysis, such as the *lac, ara, mal* (maltose), and *gal* (galactose) operons, are called catabolite operons, and their expression is coordinately repressed when glucose is plentiful. This is accomplished by **catabolite activator protein (CAP)**, which is also called *c*yclic AMP *r*eceptor *p*rotein (CRP). CAP exists in two states. It is active when the second messenger **3′, 5′-cyclic adenosine monophosphate (cAMP; figure 14.17)** is bound, and it is inactive when it is free of cAMP. The enzyme adenyl cyclase synthesizes cAMP from ATP. Adenyl cyclase catalyzes cAMP synthesis only when little or no glucose is available. Thus the level of cAMP varies inversely with that of glucose. When glucose is unavailable and the catabolism of another sugar might be needed, the amount of cAMP in the cell increases, allowing cAMP to bind to and activate CAP.

All catabolite operons contain a CAP recognition site, and CAP must be bound to this site before RNA polymerase can begin transcription efficiently. Upon binding, CAP interacts with the carboxyl terminal domain of the α subunits of RNA

Figure 14.18 CAP Structure and DNA Binding.   (a) The CAP dimer binding to DNA at the *lac* operon promoter. The recognition helices fit into two adjacent major grooves on the double helix. (b) A model of the *E. coli* CAP-DNA complex.

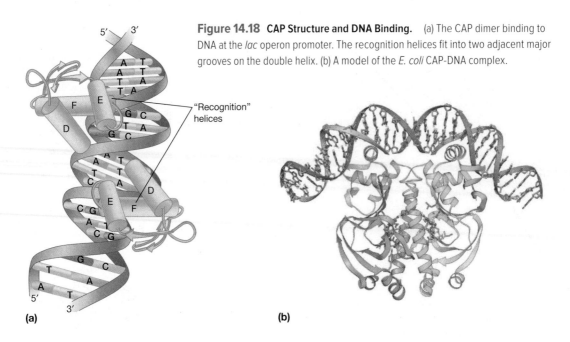

polymerase and stabilizes the interaction of RNA polymerase with the promoter; this stimulates transcription (figure 14.6*b* and **figure 14.18**). Thus all catabolite operons are controlled by two regulatory proteins: the regulatory protein specific to each operon (e.g., *lac* repressor and AraC protein) and CAP. In the case of the *lac* operon, if glucose is absent and lactose is present, the inducer allolactose will bind to and inactivate the *lac* repressor protein, CAP will be in the active form (with cAMP attached), and transcription will proceed (**figure 14.19*a***). However, if glucose and lactose are both in short supply, even though CAP binds

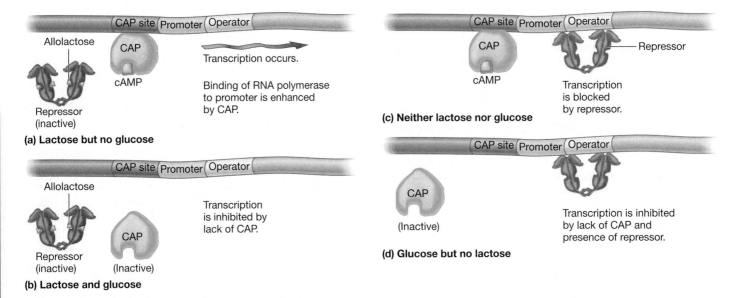

Figure 14.19 Regulation of the *lac* Operon by the *lac* Repressor and CAP.   A continuum of *lac* mRNA synthesis is brought about by the action of CAP, an activator protein, and the *lac* repressor. (a) When lactose is available and glucose is not, the repressor is inactivated and cAMP levels increase. Cyclic AMP binds CAP, activating it. CAP binds the CAP binding site near the *lac* promoter and facilitates binding of RNA polymerase. Under these conditions, transcription occurs at maximal levels. (b) When both lactose and glucose are available, both CAP and the *lac* repressor are inactive. Because RNA polymerase cannot bind the promoter efficiently without the aid of CAP, transcription levels are low. (c) When neither glucose nor lactose is available, both CAP and the *lac* repressor are active. In this situation, both proteins are bound to their regulatory sites. CAP binding enhances the binding of RNA polymerase to the promoter. However, the repressor blocks transcription. Transcription levels are low. (d) When glucose is available and lactose is not, CAP is inactive and the *lac* repressor is active. Thus RNA polymerase binds inefficiently, and those polymerase molecules that do bind are blocked by the repressor. This condition results in the lowest levels of transcription observed for the *lac* operon.

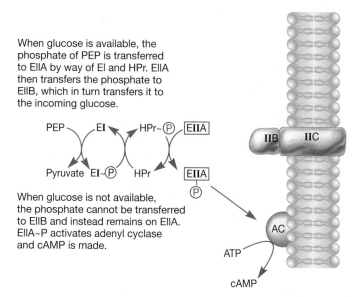

When glucose is available, the phosphate of PEP is transferred to EIIA by way of EI and HPr. EIIA then transfers the phosphate to EIIB, which in turn transfers it to the incoming glucose.

When glucose is not available, the phosphate cannot be transferred to EIIB and instead remains on EIIA. EIIA~P activates adenyl cyclase and cAMP is made.

Figure 14.20 **Activation of Adenyl Cyclase by the Phosphoenolpyruate: Sugar Phosphotransferase System (PTS).** PEP is phosphoenolpyruvate. EI and EII are enzymes I and II of the PTS, respectively. EII is composed of three subunits: A, B, and C. HPr stands for heat-stable protein. This model does not account for all experimental results, and therefore a factor called factor x has also been postulated to be involved in activating adenyl cyclase.

to its site on the DNA, transcription will be inhibited by the presence of the repressor protein, which remains bound to the operator in the absence of inducer (figure 14.19c). Dual control ensures that the *lac* operon is expressed only when lactose catabolic genes are needed.

We have seen how CAP controls catabolite operons; now let's turn our attention to the regulation of the levels of cAMP and how inducer exclusion occurs. Both depend on the state of a protein that functions in the phosphoenolpyruvate:sugar phosphotransferase system (PTS) **(figure 14.20)**. Recall from chapter 3 that in the PTS, a phosphoryl group is transferred by a series of proteins from phosphoenolpyruvate (PEP) to glucose, which then enters the cell as glucose 6-phosphate *(see figure 3.14)*. When glucose is present, enzyme IIA transfers the phosphoryl group to enzyme IIB, which phosphorylates glucose. The dephosphorylated enzyme IIA directly interacts with the transporters for the nonpreferred carbon sources, causing inducer inclusion. Furthermore, it inhibits adenyl cyclase, so that even if some inducer is present, CAP will be inactive and transcription of the catabolite operons will be repressed. However, when glucose is absent, the phosphoryl groups from PEP are transferred to enzyme IIA but are not transferred to enzyme IIB. The phosphorylated form of enzyme IIA accumulates and is thought to be responsible for activating adenyl cyclase. This then stimulates cAMP production. Thus the transporters for alternate carbon sources are functional, and transcription of the catabolite operons is enhanced if the appropriate carbon source is available. ◀◀ *Group translocation (section 3.3)* ◈ *Combination of Switches: The* lac *Operon*

Chemotaxis in *E. coli*

Recall from chapter 3 that bacteria are able to sense chemicals in their environment and move either toward them or away from them, depending on whether the chemical is an attractant or a repellent. For simplicity, we only concern ourselves with movement toward an attractant. The best-studied chemotactic system is that of *E. coli,* which, like many other peritrichously flagellated bacteria, exhibits two movement modalities: a smooth swimming motion called a run, which is interrupted by the second type of movement called a tumble. A run occurs when the flagellum rotates in a counterclockwise direction (CCW), and a tumble occurs when the flagellum rotates clockwise (CW) *(see figures 3.44 and 3.48)*. The cell alternates between these two types of movements, with the tumble establishing the direction of movement in the run that follows.

When *E. coli* is in an environment that is homogenous—that is, the concentration of all chemicals in the environment is the same throughout its habitat—the cell moves about randomly, with no apparent direction or purpose; this is called a random walk. However, if a chemical gradient exists in its environment, the frequency of tumbles decreases as long as the cell is moving toward the attractant. In other words, the length of time spent moving toward the attractant is increased, and eventually the cell gets closer to the attractant. The process is not perfect, and the cell must continually readjust its direction through a trial-and-error process that is mediated by tumbling. When one examines the path taken by the cell, it is similar to a random walk but is biased toward the attractant.

For many decades, scientists have been dissecting this complex behavior in order to understand how *E. coli* senses the presence of an attractant, how it switches from a run to a tumble and back again, and how it "knows" it is heading in the correct direction. These studies reveal that the chemotactic response of *E. coli* involves a two-component signal transduction system and covalent modification of chemotaxis receptors. As we noted previously, most two-component systems are used to regulate transcription initiation (p. 334). However, some regulate protein activity. This is the case for chemotaxis.

For chemotaxis to occur, *E. coli* must determine if an attractant is present and then modulate the activity of the two-component system that dictates the rotational direction of the flagellum (i.e., either run or tumble). *E. coli* senses chemicals in its environment when they bind to chemoreceptors **(figure 14.21)**. Numerous chemoreceptors have been identified. Here we focus on one class of receptors called methyl-accepting chemotaxis proteins (MCPs). The signal transduction system that controls direction of flagellar rotation consists of the sensor kinase CheA and the response regulator CheY. When activated, CheA phosphorylates itself using ATP (figure 14.21c). This phosphoryl group is then quickly transferred to CheY. Phosphorylated CheY diffuses through the cytoplasm to the flagellar motor. Upon interacting with the motor, the direction of rotation is switched from CCW to CW, and a tumble ensues. When CheA is inactive,

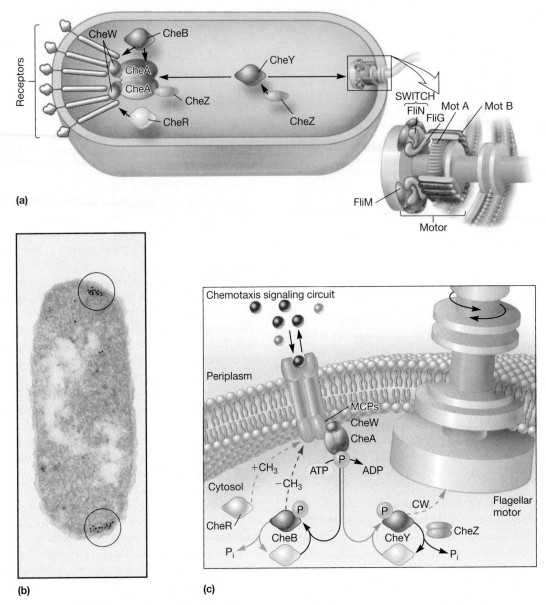

Figure 14.21 Proteins and Signaling Pathways of the Chemotaxis Response in *E. coli.* (a) The methyl-accepting chemotaxis proteins (MCPs) form clusters associated with the CheA and CheW proteins. CheA is a sensor kinase that, when activated, phosphorylates two different response regulators: CheB, a methylesterase, or CheY. Phosphorylated CheY interacts with the FliM protein of the flagellar motor, causing rotation of the flagellum to switch from counterclockwise (CCW) to clockwise (CW). This results in a switch from a run (CCW rotation) to a tumble (CW rotation). (b) MCPs, CheW, and CheA complexes form large clusters of receptors at either end of the cell, as shown in this electron micrograph of *E. coli*. Gold-tagged antibodies were used to label the receptor clusters, which appear as black dots (encircled). (c) The chemotactic signaling pathways of *E. coli*. The pathways that increase the probability of CCW rotation are shown in red. CCW rotation is the default rotation. It is periodically interrupted by CW rotation, which causes tumbling. The pathways that lead to CW rotation are shown in green. Molecules shown in gray are unphosphorylated and inactive. Note that MCP, CheA, and CheZ are homodimers. CheW, CheB, CheY, and CheR are monomers.

the flagellum rotates in its default mode (CCW), and the cell moves forward in a smooth run.

As implied by the preceding discussion, the state of the MCPs must be communicated to the CheA/CheY system. How is this accomplished? The MCPs are embedded in the plasma membrane with different parts exposed on each side of the membrane (figure 14.21*c*). The periplasmic side of each MCP

has a binding site for one or more attractant molecules. The cytoplasmic side of an MCP interacts with two proteins, CheW and CheA. The CheW protein associates with the MCP and helps attach the CheA protein to an MCP. Together with CheW and CheA, the MCP receptors form large receptor clusters at one or both poles of the cell (figure 14.21*b*). It is thought that smaller aggregations of the MCPs, CheA, and CheW function

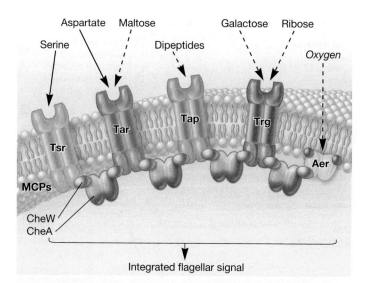

Figure 14.22 The Methyl-Accepting Chemotaxis Proteins of E. coli. The attractants sensed by each methyl-accepting chemotaxis protein (MCP) are shown. Some are sensed directly, when the attractant binds the MCP (solid lines). Others are sensed indirectly (dashed lines). The attractants maltose, dipeptides, galactose, and ribose are detected by their interaction with periplasmic binding proteins. Oxygen is detected indirectly by the Aer chemoreceptor, which differs from other MCPs in that it lacks a periplasmic sensing domain. Instead, the cytoplasmic domain has a binding site for FAD. FAD is an important electron carrier found in many electron transport systems. The redox state of the MCP-bound FAD molecule is used to monitor the functioning of the electron transport system. This in turn mediates a tactic response to oxygen.

MICRO INQUIRY *Why doesn't the Aer receptor need a periplasmic domain?*

But how does *E. coli* measure the concentration of attractant in its environment, and how does it know when it is moving toward the attractant? *E. coli* measures the concentration of an attractant every few seconds and determines if the concentration is increasing or decreasing over time. As long as the concentration increases, the cell continues a run. If the concentration decreases, a tumble is triggered. To compare concentrations of the attractant over time, *E. coli* must have a mechanism for "remembering" the previous concentration. *E. coli* accomplishes this by using covalent modification, specifically methylation of the MCPs. *E. coli* compares the overall methylation level of the MCPs (on the cytoplasmic side) with the overall amount of attractant bound (on the periplasmic face). The cytoplasmic portion of each MCP has four to six sites that can be methylated. As long as the concentration of the attractant keeps increasing, the number of MCPs bound to attractant remains high, and the MCP methylation level remains high (i.e., most or all of the sites are methylated). However, if the attractant concentration decreases, the level of methylation will exceed the level of attractant bound. This disparity in methylation level and MCP-bound attractant stimulates CheA to autophosphorylate. As a result, the signal for CW flagellar rotation is initiated, and the cell tumbles in an attempt to reorient itself in the gradient so that it is moving up the gradient (toward the attractant), rather than down the gradient (away from the attractant). At the same time, some of the methyl groups are removed from the MCPs by the methylesterase CheB, establishing a methylation level that is commensurate with the number of MCPs bound to the attractant. A few seconds later, the number of MCPs bound to attractant will be compared to this new methylation level. Based on the correspondence of the two, the cell will determine if it is again moving up the gradient. If it is, tumbling will be suppressed and the run will continue.

Retrieve, Infer, Apply

1. What are global regulatory systems and why are they necessary? What are ECF sigma factors? How do they differ from other alternate sigma factors?

2. What is diauxic growth? Explain how catabolite repression causes diauxic growth.

3. Describe the events that occur with *E. coli* in each of the following growth conditions: in a medium containing glucose but not lactose; in a medium containing both sugars; in a medium containing lactose but no glucose; and in a medium containing neither sugar.

4. *E. coli* has two phosphate uptake systems (one with a high affinity for phosphate, the other with a low affinity). Describe how a two-component regulatory system might be used by *E. coli* to regulate phosphate transport.

5. Describe the MCP-CheW-CheA receptor complex. What two proteins are phosphorylated by CheA? What is the role of each?

6. How does an MCP regulate the rate of CheA autophosphorylation? How does this mediate chemotaxis?

as signaling teams and are the building blocks of the receptor clusters (**figure 14.22**). The signaling teams become interconnected by an unknown mechanism to form the receptor clusters visible at the poles of the cell.

Evidence exists that the MCPs in each signaling team work cooperatively to modulate CheA activity. When any one of the MCPs in the signaling team is bound to an attractant, CheA autophosphorylation is inhibited, the flagellum continues rotating CCW, and the cell continues in its run. Because of this cooperation, the cell can respond to very low concentrations of attractant. Furthermore, it can integrate signals from all receptors in the team (figure 14.22). On the other hand, if attractant levels decrease, so that the level of attractant bound to the MCPs in a signaling team decreases, CheA is stimulated to autophosphorylate, the phosphate is then transferred to CheY, and the cell begins to tumble. However, tumbling does not continue indefinitely. About 10 seconds after the switch to CW rotation occurs, the phosphoryl group is removed from CheY by the CheZ protein, and CCW rotation is resumed.

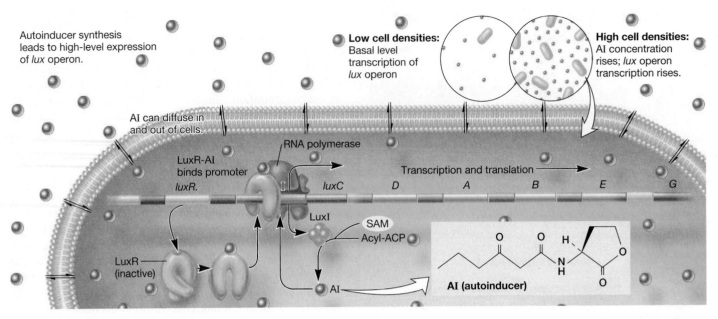

Figure 14.23 Quorum Sensing in *V. fischeri.* The AHL signaling molecule (AI) diffuses out of the cell; when cell density is high, AHL diffuses back into the cell, where it binds to and activates the transcriptional regulator LuxR. Active LuxR then stimulates transcription of the gene coding for AHL synthase (*luxI*), as well as the genes encoding proteins needed for light production.

Quorum Sensing by *Vibrio* spp.

Cell-to-cell communication among bacterial cells often occurs by the exchange of chemicals often termed signals or signaling molecules. The exchange of signaling molecules is essential in the coordination of gene expression in microbial populations. It has become clear that cell-cell communication plays an essential role in the regulation of genes whose products are needed for the establishment of virulence, symbiosis, biofilm production, and morphological differentiation in a wide range of bacteria. Quorum-sensing mechanisms vary among microbes. Here we focus on two of the best-studied systems.

The marine bioluminescent bacterium *Vibrio fischeri* uses quorum sensing to regulate light production within the light organ of its squid host; cells produce light only if cells are at a high density. Quorum sensing is also used by the bacterium to control genes whose products are needed for maintenance of the symbiotic relationship between *V. fisheri* and its host. As a result, the squid/*V. fisheri* symbiosis has become an important model for understanding animal-bacterial associations. Our focus is on the regulation of a single operon, that involved with bioluminescence. However, it should be kept in mind that **quorum sensing** regulates multiple genes and operons. ◄◄ *Cell-cell communication within microbial populations (section 7.5)*

Quorum sensing in *V. fischeri* and many other Gram-negative bacteria uses an **N-acylhomoserine lactone (AHL)** signal (**figure 14.23**). Synthesis of this small molecule is catalyzed by an enzyme called AHL synthase, the product of the *luxI* gene. The *luxI* gene is subject to positive autoregulation; that is, transcription of *luxI* increases as AHL accumulates in the cell. This is accomplished through the transcriptional activator

LuxR, which is active only when AHL binds to it. Thus a simple feedback loop is created. Without AHL-activated LuxR, the *luxI* gene is transcribed only at basal levels. When *V. fisheri* cell density within the squid light organ is low, the small amounts of AHL produced by the bacterial cells freely diffuse out of each cell and accumulate in the environment. As cell density increases, the concentration of AHL outside each cell eventually exceeds that inside the cell, and the concentration gradient is reversed. As AHL flows back into a cell, it binds and activates LuxR. LuxR can now increase transcription of *luxI* and the genes whose products are needed for bioluminescence (*luxCDABEG*). Quorum sensing is often called **autoinduction,** and the AHL signal is termed the **autoinducer (AI)** to reflect the autoregulatory nature of this system.

🐾 *Quorum Sensing*

Another kind of quorum sensing depends on an elaborate phosphorelay signal transduction system. It is found in both Gram-negative and Gram-positive bacteria, including *Staphylococcus aureus, Ralstonia solanacearum, Salmonella enterica, Vibrio cholerae,* and *E. coli.* It has been best studied in the bioluminescent bacterium *Vibrio harveyi.*

Unlike *V. fischeri, V. harveyi* responds to three autoinducer molecules: HAI-1, AI-2, and CAI-1. HAI-1 (harveyi autoinducer-1) is a homoserine lactone, and its synthesis depends on the *luxM* gene. AI-2 (autoinducer-2) is furanosylborate, a small molecule that contains a boron atom—quite an unusual component in an organic molecule (*see figure 7.20*). Its synthesis relies on the product of the *luxS* gene. CAI-1 (cholerae autoinducer-1) is the product of the enzyme CqsA. As shown in **figure 14.24,** HAI-1, AI-2, and CAI-1 are secreted by cells, which then use separate proteins called LuxN, LuxPQ, and

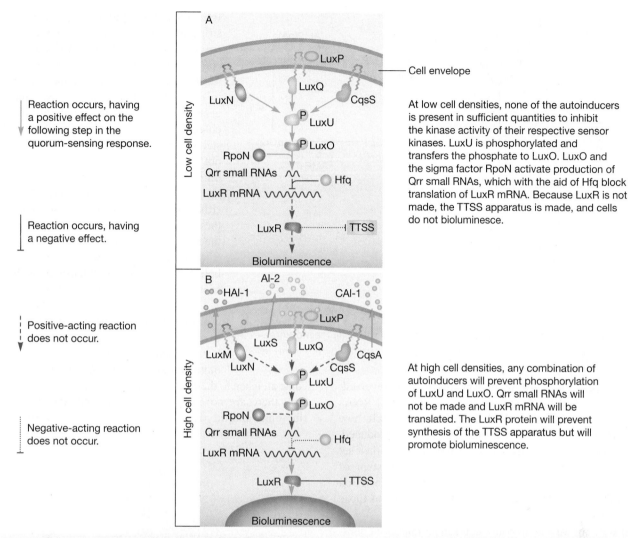

Reaction occurs, having a positive effect on the following step in the quorum-sensing response.

Reaction occurs, having a negative effect.

Positive-acting reaction does not occur.

Negative-acting reaction does not occur.

At low cell densities, none of the autoinducers is present in sufficient quantities to inhibit the kinase activity of their respective sensor kinases. LuxU is phosphorylated and transfers the phosphate to LuxO. LuxO and the sigma factor RpoN activate production of Qrr small RNAs, which with the aid of Hfq block translation of LuxR mRNA. Because LuxR is not made, the TTSS apparatus is made, and cells do not bioluminesce.

At high cell densities, any combination of autoinducers will prevent phosphorylation of LuxU and LuxO. Qrr small RNAs will not be made and LuxR mRNA will be translated. The LuxR protein will prevent synthesis of the TTSS apparatus but will promote bioluminescence.

Figure 14.24 Quorum Sensing in *V. harveyi.* Three autoinducing signals—HAI-1, AI-2, and CAI-1—are detected. HAI-1 is only produced by *V. harveyi* and one other closely related species. AI-2 is produced by many Gram-negative and Gram-positive bacteria. CAI-1 (3-hydroxytridecan-4-one) is produced by many members of the genus *Vibrio*. Thus *V. harveyi* not only "measures" the density of its own population but that of other bacteria as well. LuxN, LuxA, and CqsA are sensor kinases involved in detecting HAI-1, AI-2, and CAI-1, respectively. Their kinase activity is inhibited by the presence of their respective autoinducer. All three quorum-sensing signals converge at LuxU, which when phosphorylated transfers the phosphate to LuxO. LuxO working with the sigma factor RpoN activates transcription of the *qrr* (quorum regulatory RNA) genes. The products of these genes are small RNAs that block translation of the mRNA encoding the protein LuxR. This is accomplished in conjunction with a small protein, Hfq, which is an RNA chaperone that helps RNA molecules bind each other or aids in their folding. Lux R is a regulatory protein that represses transcription of genes encoding components of a type III secretion system (TTSS) and activates transcription of genes required for bioluminescence.

MICRO INQUIRY *Why does* V. harveyi *make three separate signaling molecules?*

CqsS, respectively, to detect their presence. LuxN, LuxQ, and CqsS are sensor kinases. At low cell density in the absence of any autoinducer, the three sensor kinases autophosphorylate and converge on a single phosphotransferase protein called LuxU. LuxU accepts phosphates from each sensor kinase and then phosphorylates the response regulator LuxO. Phosphorylated LuxO in turn activates the transcription of genes encoding several small RNAs that destabilize *luxR* mRNA. This is accomplished with the aid of an RNA chaperone (p. 332).

LuxR is a transcriptional activator of the operon *luxCDABE*, which encodes proteins needed for bioluminescence. Because *luxR* mRNA is not translated, LuxR protein is not made. Therefore cells do not make light at low cell density.

An interesting thing happens as the density of any one of these autoinducers increases: LuxN binds HAI-1, LuxPQ binds AI-2, and CqsS binds CAI-1. When this happens, the proteins switch from functioning as kinases to phosphatases, proteins that dephosphorylate, rather than phosphorylate,

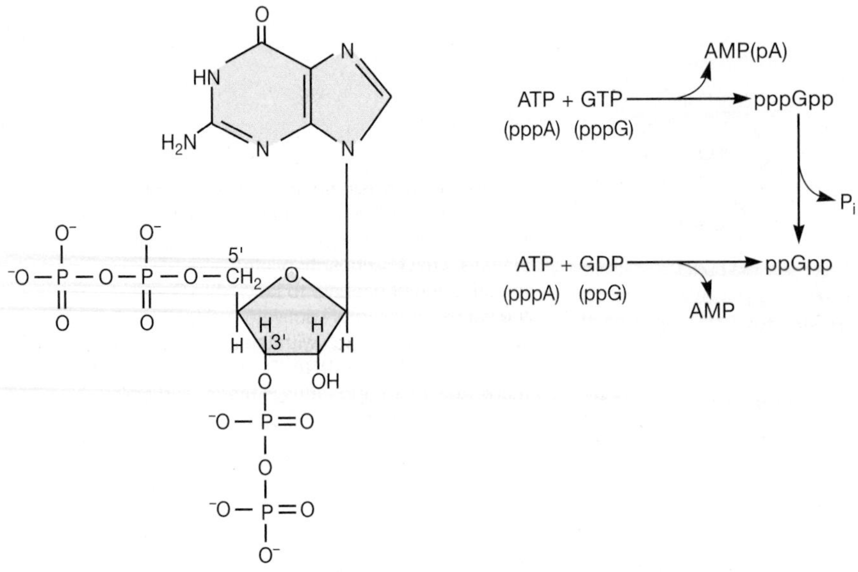

(a) ppGpp

(b) Synthesis of ppGpp

ATP + GTP → pppGpp
(pppA) (pppG)

AMP(pA)

P_i

ATP + GDP → ppGpp
(pppA) (ppG)

AMP

Figure 14.25 **Guanosine Tetraphosphate.** (a) The structure of ppGpp. (b) The two processes that generate ppGpp. In the stringent response, these reactions are catalyzed by the enzyme RelA.

tetraphosphate (**ppGpp; figure 14.25**). The stringent response is observed in many bacterial species but is best studied in *E. coli*. When *E. coli* cells are growing in a nutrient-rich environment, many of the RNA polymerase molecules in the cell are transcribing tRNA and rRNA genes, providing a ready supply of these molecules for protein synthesis. When cells are starved for amino acids, protein synthesis cannot proceed as it does in nutrient-rich conditions. Therefore it is wasteful for the cell to keep synthesizing large quantities of tRNA and rRNA. Instead, the cell decreases synthesis of these molecules and increases transcription of the appropriate amino acid biosynthetic genes. Eventually the cell "resets" its level of metabolic activity to one commensurate with the nutrients available in its environment.

The mechanism by which these changes are brought about has become increasingly clear over the 40-plus years this phenomenon has been studied. When *E. coli* is starved for one or more amino acids, tRNA molecules cannot become attached to their cognate amino acids. Thus during translation, there are no charged tRNAs to enter the A site of the ribosome, and the ribosome stalls (**figure 14.26**). However, an

their substrates. The flow of phosphates is now reversed; LuxO is inactivated by dephosphorylation, the small RNAs are not made, and *luxR* mRNA is translated. LuxR now activates transcription of *luxCDABE* and light is produced. Careful inspection of figure 14.24 reveals that another set of genes is controlled by the quorum-sensing system of *V. harveyi*. In this microbe, genes for a type III protein secretion system (TTSS) are controlled in the opposite manner as those for bioluminescence.

At this point, we should pause and ask an important question: Why does *V. harveyi* need three different autoinducers? It appears that these molecules allow the bacterium to carry out three different kinds of conversations. HAI-1 is specific to *V. harveyi* (and one other closely related species). Thus it is thought to allow *V. harveyi* to communicate with members of its own species. In essence, this autoinducer conveys the message, "There are many *V. harveyi* nearby." AI-2 is made by many Gram-negative and Gram-positive bacteria. Thus its message is thought to be, "There are many bacteria nearby." Finally, CAI-1 is produced by other members of the genus *Vibrio*, including *V. cholerae*, for which it is named. Its message is thought to be, "There are many *Vibrio* species nearby." As might be expected, each autoinducer has a different signal strength. HAI-1 is the strongest and CAI-1 is the weakest. The presence of all three autoinducers maximizes expression of the bioluminescence genes and fully represses expression of the type III secretion system genes.

Stringent Response in *E. coli*

The **stringent response** occurs when bacterial cells are starved for amino acids and is regulated by the second messenger **guanosine**

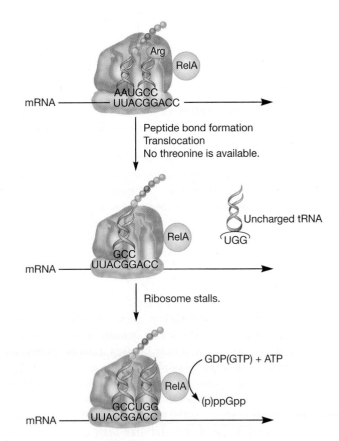

Figure 14.26 **Activation of RelA in Response to Amino Acid Starvation.**

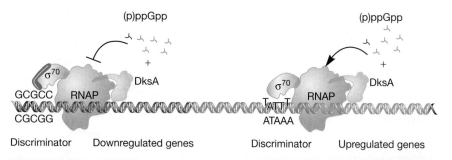

Figure 14.27 Effects of ppGpp and DksA on Transcription. Genes encoding tRNA and rRNA have a region between the −10 site of the promoter and the +1 nucleotide of the coding region that is GC rich. This region interacts with sigma factor as it positions the RNA polymerase core enzyme at the promoter. The presence of the GC region destabilizes open complexes formed during initiation. During the stringent response, ppGpp working with DksA further destabilizes the open complexes, thus lowering the rate of transcription. Elsewhere in the genome are genes encoding enzymes needed for amino acid biosynthesis. The promoters for these genes are rich for AT base-pairs in the same area of the promoter. These regions do not destabilize open complex formation. The interaction of ppGpp and DksA promotes open complex formation, and transcription is increased for these genes.

MICRO INQUIRY *Why would having many GC pairs in the −10 to +1 region of the promoter make it more difficult to maintain an open complex?*

uncharged tRNA eventually enters the A site of the stalled ribosome. This triggers an enzyme called RelA to catalyze the formation of ppGpp from GDP and ATP (figure 14.25b). RelA also catalyzes a reaction between GTP and ATP, generating guanosine pentaphosphate (pppGpp), which is quickly converted to ppGpp. ppGpp then interacts with RNA polymerase and downregulates synthesis of tRNA and rRNA, while simultaneously upregulating transcription of amino acid biosynthetic genes. However, ppGpp does not act alone. Instead, it exerts its effects with the help of the protein DksA. Together they destabilize open complexes (*see figure 13.26*) formed during transcription initiation of the downregulated genes and enhance the rate of open complex formation of the upregulated genes (**figure 14.27**).

An important question for understanding the stringent response is how the cell knows which promoters to downregulate and which to upregulate. Comparisons of the promoters of rRNA genes and amino acid synthetic genes have answered this question. Recall that sigma factor (σ) positions RNA polymerase core enzyme properly on the promoter. The enzyme then melts the two strands of DNA, forming an open complex. In ppGpp-regulated genes, there is a region between the −10 site of the promoter and the +1 nucleotide that is important for controlling transcription initiation. In downregulated genes (i.e., rRNA and tRNA genes), this region is GC rich, whereas in amino acid biosynthetic genes, it is AT rich (figure 14.27). The GC-rich regions form unstable open complexes during transcription initiation. DksA and ppGpp further destabilize the open complexes, lowering transcription levels. The AT-rich regions of the upregulated genes are thought to form much more stable open complexes that are not destabilized by DksA and ppGpp.

More recently, ppGpp has been shown to function as an "alarm," made in response to a number of stresses in addition to amino acid starvation. In some cases ppGpp regulates transcription levels of relevant genes. In other cases, ppGpp affects protein stability or directly interacts with proteins and alters their activity. In these situations, production of ppGpp is catalyzed by a different enzyme, called SpoT. Despite being an area of intensive research for many years, many of the details of this global response to stress are still being explored.

Sporulation in *B. subtilis*

As we discuss in chapter 3, endospore formation is a complex process that involves asymmetric division of the cytoplasm to yield a large mother cell and a smaller forespore, engulfment of the forespore by the mother cell, and construction of additional layers of spore coverings (**figures 14.28a** and 3.51). Sporulation takes approximately 8 hours. Numerous bacterial species form endospores, including *B. subtilis*, which has served for many years as the model organism for understanding sporulation.

In *B. subtilis*, sporulation is controlled by phosphorelay, posttranslational modification of proteins, numerous transcription initiation regulatory proteins, and alternate sigma factors. The latter are particularly important. When growing vegetatively, *B. subtilis* RNA polymerase uses sigma factors σ^A and σ^H to recognize genes for normal survival. However, when cells sense a starvation signal, a cascade of events is initiated that results in the production of other sigma factors that are differentially expressed in the developing endospore and mother cell.

Initiation of sporulation is controlled by the protein Spo0A (Spo is short for sporulation, the number 0 represents the stage of sporulation at which the protein functions), a response regulator protein that is part of a phosphorelay system (figure 14.28b). Sensor kinases associated with this system detect environmental stimuli that trigger sporulation. One of the most important sensor kinases is KinA, which senses nutrient starvation. When *B.*

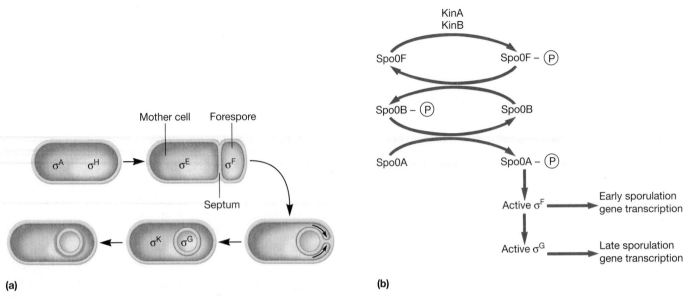

Figure 14.28 Regulation of Sporulation in B. subtilis. (a) The initiation of sporulation is governed in part by the activities of two spatially separated sigma factors. σ^F is located in the forespore, while σ^E is confined to the mother cell. These sigma factors direct the initiation of transcription of genes whose products are needed for early events in sporulation. Later, σ^G and σ^K are localized to the developing endospore and mother cell, respectively. They control the expression of genes whose products are involved in later steps of sporulation. (b) The activation of σ^F is accomplished through a phosphorelay system that is triggered by the activation of the sensor kinase protein KinA. When KinA senses starvation, it autophosphorylates a specific histidine residue. The phosphoryl group is then passed in relay fashion from SpoOF to SpoOB and finally to SpoOA.

MICRO INQUIRY *Compartmentation—the spatial localization of proteins as a means of regulation—is usually considered in the context of eukaryotic cells. How is the process of sporulation an example of compartmentation?*

subtilis finds its nutrients are depleted, KinA autophosphorylates a specific histidine residue. The phosphoryl group is then transferred to an aspartic acid residue on SpoOF, the second member of the phosphorelay. SpoOF donates the phosphoryl group to a histidine on SpoOB. SpoOB in turn relays the phosphoryl group to SpoOA. Phosphorylated SpoOA positively controls genes needed for sporulation and negatively controls genes that are not needed. In response to SpoOA, the expression of over 500 genes is altered. This has earned SpoOA the name "master regulator." Among the genes whose expression is stimulated by SpoOA is *sigF,* the gene encoding σ^F, and *spoIIGB,* the gene encoding an inactive form of σ^E (pro-σ^E).

When sporulation starts, the chromosome has replicated, with one copy remaining in the mother cell and another to be partitioned in the forespore. Shortly after the formation of the spore septum, σ^F is found in the forespore, and pro-σ^E is localized to the mother cell. Pro-σ^E is cleaved by a protease to form active σ^E. The two sigma factors, σ^F and σ^E, bind to the promoters of genes needed in the forespore and mother cell, respectively. There they direct the expression of genes encoding products needed for the early steps of endospore formation. These genes are primarily responsible for the engulfment process. Another gene regulated by σ^F is one that encodes σ^G, which will replace σ^F in the developing endospore. Likewise, σ^E directs the transcription of a mother-cell–specific

sigma factor, σ^K. σ^K ensures that genes encoding late-stage sporulation products are transcribed. These include genes for synthesis of the cortex and coat layers of the endospore. Overall, temporal regulation is achieved because σ^F and σ^E direct transcription of genes that are needed early in the sporulation process, whereas σ^G and σ^K are needed for the transcription of genes whose products function later. In addition, spatial control of gene expression is accomplished because σ^F and σ^G are located in the forespore and developing endospore, whereas σ^E and σ^K are found only in the mother cell.

Retrieve, Infer, Apply

1. What would be the phenotype of a *V. fischeri* mutant that could not regulate *luxI,* so that it was constantly producing autoinducer at high levels?

2. Why might bacteria use quorum sensing to regulate genes needed for virulence? How might this reason be related to the rationale behind using quorum sensing to establish a symbiotic relationship?

3. Why is the stringent response typically active during nitrogen starvation but not in response to carbon sources?

4. Briefly describe how a phosphorelay system and sigma factors are used to control sporulation in *B. subtilis.* Give one example of posttranslational modification as a means to regulate this process.

Key Concepts

14.1 Bacteria Use Many Regulatory Options

- Regulation of gene expression can be controlled at many levels, including transcription initiation, transcription elongation, translation, and posttranslation (**figure 14.1**).
- The three domains of life differ in terms of their genome structure and the steps required to complete gene expression. These differences affect the regulatory mechanisms they use.

14.2 Regulation of Transcription Initiation Saves Considerable Energy and Materials

- Induction and repression of enzyme levels are two important regulatory phenomena. They usually occur because of the action of regulatory proteins.
- Transcriptional regulatory proteins are DNA-binding proteins. When bound to DNA, they can either inhibit transcription (negative control) or promote transcription (positive control). Their activity is modulated by small effector molecules called inducers, corepressors, and inhibitors (**figure 14.3**).
- Repressors are responsible for negative control. They block transcription by binding an operator and interfering with the binding of RNA polymerase to its promoter or by blocking the movement of RNA polymerase after it binds DNA.
- Activator proteins are responsible for positive control. They bind DNA sequences called activator-binding sites and, in doing so, promote binding of RNA polymerase to its promoter.
- The *lac* operon of *E. coli* is an example of a negatively controlled inducible operon. When there is no lactose in the surroundings, the *lac* repressor is active and transcription is blocked. When lactose is available, it is converted to allolactose by the enzyme β-galactosidase. Allolactose acts as the inducer of the *lac* operon by binding the repressor and inactivating it. The inactive repressor cannot bind the operator and transcription occurs (**figure 14.7**).
- The *trp* operon of *E. coli* is an example of a negatively controlled repressible operon. When tryptophan is not available, the *trp* repressor is inactive and transcription occurs. When tryptophan levels are high, tryptophan acts as a corepressor and binds the *trp* repressor, activating it. The *trp* repressor binds the operator and blocks transcription (**figure 14.8**).
- The *ara* operon of *E. coli* is an example of an inducible operon that is regulated by the dual-function regulatory protein AraC. AraC functions as a repressor when arabinose is not available. It functions as an activator when arabinose, the inducer, is available (**figure 14.9**).

14.3 Attenuation and Riboswitches Can Stop Transcription Prematurely

- In the tryptophan operon, a region called *trpL* lies between the operator and the first structural gene (**figure 14.10**). It codes for synthesis of a leader peptide and contains an

attenuator, a factor-independent termination site. Synthesis of the leader peptide by a ribosome while RNA polymerase is transcribing *trpL* regulates transcription. Therefore the tryptophan operon is expressed only when insufficient tryptophan is available. This mechanism of transcription control is called attenuation.
- The 5′ ends (leaders) of some mRNA molecules can bind metabolites and other effectors. Binding of the effector to the mRNA causes a change in the leader structure, which can terminate transcription. This regulatory mechanism is called a riboswitch (**figure 14.11**).

14.4 Riboswitches and Small RNAs Can Control Translation

- Some riboswitches regulate gene expression at the level of translation. For these riboswitches, the binding of an effector to specific sequences in the leader region of the mRNA alters leader structure and affects ribosome binding either negatively or positively (**figure 14.12**).
- Translation can also be controlled by small RNAs. These small RNA molecules are noncoding. They base-pair to the mRNA and usually inhibit translation (**figure 14.13**).

14.5 Bacteria Combine Several Regulatory Mechanisms to Control Complex Cellular Processes

- Some cellular processes are so complex they require the action of multiple operons that must be simultaneously controlled. Such operon networks are global regulatory systems.
- Global regulatory systems often involve many layers of regulation. Mechanisms such as global regulatory proteins, two-component signal transduction systems, phosphorelay systems, second messengers, and alternate sigma factors are often used.
- Expression of the porin genes *ompF* and *ompC* is controlled by a two-component signal transduction system. These systems have a sensor kinase that detects an environmental change. The sensor kinase transduces the environmental signal to the response-regulator protein by transferring a phosphoryl group to it. The response regulator then activates genes needed to adapt to the new environmental conditions and inhibits expression of those genes that are not needed (**figure 14.14**).
- Cyclic dimeric GMP (c-di-GMP) is a recently discovered second messenger. Its level in bacterial cells depends on the activity of enzymes called diguanylate cyclases (DGCs), which synthesize c-di-GMP, and phosphodiesterases (PDEs), which degrade c-di-GMP (**figure 14.15**). The DGCs and PDEs have sensory domains that detect environmental signals. Their response to the signal determines the level of c-di-GMP. C-di-GMP interacts with effector molecules that affect a target, bringing about the appropriate response to the environmental signal.

■ Diauxic growth is observed when *E. coli* is cultured in the presence of glucose and another sugar such as lactose (**figure 14.16**). This growth pattern is the result of catabolite repression, where glucose is used preferentially over other sugars. Operons that are part of the catabolite repression system are regulated by the activator protein CAP. CAP activity is modulated by the second messenger cAMP, which is produced only when glucose is not available (**figure 14.20**). Thus when there is no glucose, CAP is active and promotes transcription of operons needed for the catabolism of other sugars (**figure 14.19**).

■ Chemotaxis in *E. coli* is regulated by covalent modification of chemoreceptor proteins and a two-component signal transduction system that controls the direction of flagellar rotation (**figure 14.21**). Methylation is used to measure the amount of a chemoattractant encountered over time. In this way, *E. coli* can determine if it is moving toward or away from an attractant.

■ Quorum sensing is a type of cell-to-cell communication mediated by small signaling molecules such as *N*-acylhomoserine lactone (AHL). Quorum sensing couples cell density to regulation of transcription. Well-studied quorum-sensing systems include the regulation of bioluminescence in *Vibrio* spp. (**figures 14.23** and **14.24**). *V. harveyi* quorum sensing is an example of a phosphorelay signal transduction pathway.

■ The stringent response occurs when bacteria are starved for amino acids. It is mediated by the second messenger guanosine tetraphosphate (ppGpp). The stringent response results in a decrease in transcription of tRNA and rRNA genes and increased transcription of amino acid biosynthetic genes (**figures 14.25–14.27**). ppGpp also serves as a general "alarm" for other stress conditions.

■ Endospore formation in *B. subtilis* is regulated by a phosphorelay system that is important in initiation of sporulation and the use of alternate sigma factors (**figure 14.28**).

Compare, Hypothesize, Invent

1. Attenuation affects anabolic pathways, whereas repression affects either anabolic or catabolic pathways. Provide an explanation for this.

2. Describe the phenotype of the following *E. coli* mutants when grown in two different media: glucose only and lactose only. Explain the reasoning behind your answer.
   a. A strain with a mutation in the gene encoding the *lac* repressor; the mutant repressor cannot bind allolactose.
   b. A strain with a mutation in the gene encoding CAP; the mutant form of CAP binds but cannot release cAMP.
   c. A strain in which the Shine-Dalgarno sequence has been deleted from the gene encoding adenyl cyclase.

3. What would be the phenotype of an *E. coli* strain in which the tandem tryptophan codons in the leader region were mutated so that they coded for serine instead?

4. What would be the phenotype of a *B. subtilis* strain whose gene for σ^G has been deleted? Consider the ability of the mutant to survive in nutrient-rich versus nutrient-depleted conditions.

5. Propose a mechanism by which a cell might sense and respond to levels of Na^+ in its environment.

6. *Neisseria meningitidis,* commonly called meningococcus, causes meningitis, a serious infection of the membranes surrounding the brain and spinal cord (meninges); these infections often lead to death. Surprisingly, meningococcus is a normal inhabitant of the nasopharynx in about a third of humans, where it causes no harm. It was known that meningococcal meningitis often follows an inflammatory response such as one caused by influenza. Inflammatory responses are often accompanied by an increase in temperature, especially in the tissues where the response occurs. Thus, in the case of influenza, inflammation of the nasopharynx would expose the meningococcus bacteria to higher temperatures and contribute to the development of meningitis. Knowing this, scientists wondered how the bacterium sensed the temperature increase and how this contributed to its ability to escape human defenses, cross the blood-brain barrier, and cause disease. Scientists in the United Kingdom tried to answer this question by isolating mutants resistant to immune system defenses present in human serum when exposed to the serum at body temperature. They then characterized the mutated genes and found that each had a small change in the 5′ untranslated region (5′ UTR) that altered its response to temperature. The normal, wild-type gene was expressed at high levels only when the bacterium was exposed to higher temperatures, whereas the mutant genes were expressed at high levels even when the bacteria were exposed to lower temperatures (e.g., body temperature). Thus the 5′ UTR was involved in regulating expression of these genes in response to temperature, and the scientists referred to it as a thermosensor. One of the genes (*cssA*) encodes an enzyme involved in capsule formation, and its higher expression leads to bacteria forming more capsular material, thus rendering them resistant to a variety of host defenses. Suggest a mechanism by which the 5′ UTR might function as a thermosensor. (HINT: Remember that the 5′ UTR is transcribed into mRNA but is not translated.)

Read the original paper: Loh, E., et al. 2013. Temperature triggers immune evasion by *Neisseria meningitidis. Nature* 502:237.

15

Eukaryotic and Archaeal Genome Replication and Expression

Approximately 140 million tons of plastics are purchased each year. Most are discarded and very slowly, if ever, degrade.

Plastics: Brought to You by Microbes

The next time you discard a plastic water bottle, consider this: that plastic bottle will take thousands of years to be converted back into the smaller chemical molecules from which it was made, assuming it ever degrades. Also consider that each year 140 million tons of equally recalcitrant plastics are purchased worldwide (and usually discarded). And what is often forgotten is that to make all that plastic, 150 million tons of fossil fuels must be processed each year, contributing additional waste, diversion of the fuels from refineries and power plants, and exacerbating geopolitical problems centered on oil and other fossil fuels. Could microbes solve this problem? Microbiologists have been trying for years to answer that question.

You may be asking why microbiologists would suggest microbes as a solution to the plastics problem. This is why. Plastics are organic polymers, such as polyurethane, polystyrene, polyethylene, and polyester. Microbiologists have known for decades that many bacteria and archaea store carbon as intracellular inclusions of poly-β-hydroxybutyrate (PHB) and other polyhydroxyalkanoates (PHAs). PHAs are polyesters; thus bacteria and archaea have evolved to synthesize a type of plastic. And even more important, because PHA-producing bacteria and archaea have evolved a means to use the PHA when carbon is in short supply, they produce the enzymes needed for PHA degradation.

For many years, microbiologists have worked to optimize production of PHAs by bacteria such as *Ralstonia eutropha,* and they have succeeded—up to 80% of cellular dry weight is produced by some strains. However, the cost of feeding these chemoorganoheterotrophic bacteria so that they produce PHAs makes the plastics they produce very expensive. Therefore, although some companies manufacture bioplastics, their share of the market is minuscule relative to that produced each year by chemical means.

"If expense arises from feeding a chemoorganotroph," scientists thought, "why not move the bacterial genes encoding PHB-synthesizing enzymes into an organism that feeds itself?" The logical autotrophs were plants such as thale cress (*Arabidopsis thaliana*) and tobacco (*Nicotiana tabacum*). Though successful, scientists were then faced with another dilemma: the use of land to "grow" plastics rather than to grow food.

Once again, microbes may provide a solution. Several German microbiologists have introduced the genes for PHB synthesis into a microalga, the diatom *Phaeodactylum tricornutum*. The advantages of an alga-based system are that these photoautotrophs need only be provided with sunlight and CO_2, and they can be grown in areas (e.g., a desert) that are not used for food production. This greatly decreases cost and avoids the ethical problems associated with plant-based systems.

Although the German scientists did nothing to maximize PHB production by the diatom, it was a "proof of concept" experiment. It showed that an alga-based system for plastic production may be successful. The next step is to maximize PHB production in a cost-effective manner. This will require that microbiologists take into account the differences between gene expression in bacteria and eukaryotes. They will need to modify the PHB-synthesis genes such that they are transcribed at a high rate in the diatom, producing mRNA molecules that are recognized and translated by eukaryotic ribosomes. Fortunately, microbiologists have a wealth of information provided by geneticists and molecular biologists studying plant systems and eukaryotic microbes such as the yeast *Saccharomyces cerevisiae* to guide them. What these geneticists and molecular biologists have learned is the focus of this chapter.

Readiness Check:

Based on what you have learned previously, you should be able to:

✔ Discuss bacterial genome structure and the organization of genetic information (sections 13.3 and 13.4)

✔ Describe the events that occur during the three phases of DNA replication and how the components of the bacterial replisome participate (section 13.3)

the chromosome to be replicated much faster than it could be if there were only one origin per chromosome. Two replication forks move outward from each origin until they encounter replication forks that formed at adjacent origins. Thus eukaryotic chromosomes consist of multiple **replicons,** rather than the single replicon (i.e., the entire chromosome) observed in bacteria.

In eukaryotes, origins of replication are "marked" by a complex of proteins called the **origin recognition complex (ORC)**, which remains bound to the origins throughout much of the cell cycle (table 15.1). All ORCs consist of six different ORC proteins (Orc1 to Orc6), but the identity of these proteins varies among eukaryotic species. ORC serves as a platform on which other proteins assemble in a cell cycle-dependent fashion. The first protein to associate with ORC is the helicase loader, either Cdc6 (*cell-division-cycle* protein 6) or Cdt1 (*Cdc10-dependent-transcript* 1 protein), depending on the organism. This occurs in the late M/early G1 phases of the cell cycle. Together, ORC and the helicase loader recruit the helicase enzyme, a set of proteins called the MCM (*minichromosome maintenance*) complex, to the origin, thereby forming the pre-replicative complex (pre-RC). Pre-RCs are activated as the cell cycle transitions from the G1 phase to the S phase by phosphorylation of the helicase loader, which is then replaced by numerous other proteins, forming the complete replisome (table 15.1). Within the replisome, the MCM helicase is associated with two co-activators and begins unwinding DNA to initiate replication.

Table 15.2	DNA Polymerase Families	
Family	**Example(s)**	**Functions/Comments**
A	Bacterial DNA polymerase I	Replacement of RNA primers present in Okazaki fragments
B	Archaeal DNA polymerase B (Pol B) Eukaryotic DNA polymerases α, δ, and ε	Replicative DNA polymerases
C	Bacterial DNA polymerase III	Replicative DNA polymerase
D	Archaeal DNA polymerase D (Pol D)	Replicative DNA polymerase in some archaea; unique to archaea
X	Eukaryotic DNA polymerase β	DNA repair
Y	Bacterial DNA polymerase IV (DinB)	DNA repair
Reverse transcriptase (RT)	Retroviral reverse transcriptases Telomerase RT	RNA-dependent DNA polymerase

Eukaryotic Primases and DNA Polymerases

DNA polymerases from many organisms have been analyzed and sorted into seven families based on their amino acid sequence and structure. As shown in **table 15.2**, eukaryotic DNA polymerases are found in different families than are the bacterial enzymes. In eukaryotes, three DNA polymerases function in DNA replication: DNA polymerases α-primase, ε, and δ. Primer synthesis is accomplished with the enzyme DNA polymerase α-primase (often called simply Pol α-primase). Pol α-primase is so named because it has two activities: DNA synthesis (Pol α) and RNA synthesis (primase). This interesting enzyme consists of four subunits: two for DNA synthesis (Pol 1 and Pol 2), and two for RNA synthesis (Pri1 and Pri2). The Pri1 subunit contains the catalytic site. The primer is made in two steps: the primase component of the enzyme makes a short RNA strand (~10 nucleotides), which is then transferred to the active site of Pol α. Pol α adds an additional 20 or so deoxyribonucleotides to the RNA strand. Thus in eukaryotes, the single-stranded primer for DNA replication is an RNA-DNA hybrid molecule. Once the primer is formed, the other two DNA polymerases take over. DNA polymerase ε (Pol ε) is responsible for leading-strand synthesis, whereas DNA polymerase δ (Pol δ) carries out lagging-strand synthesis. Thus unlike most bacteria where DNA polymerase III synthesizes both leading and lagging strands, two distinct DNA polymerases carry out these functions in eukaryotes.

Telomeres and Telomerases: Protecting the Ends of Linear DNA Molecules

The fact that the DNA in eukaryotic chromosomes is linear poses several problems. Without protection, the ends are susceptible to degradation by enzymes called DNases. They are also able to fuse with the ends of other DNA molecules, thus generating aberrant chromosomes. Finally, linear chromosomes present a problem during replication because of DNA polymerase's need for a primer that provides a free 3′-OH. When the primer for the Okazaki fragment at the end of the daughter strand is removed, the daughter molecule is shorter than the parent molecule. Over numerous rounds of DNA replication and cell division, this leads to a progressively shortened chromosome. Ultimately the chromosome loses critical genetic information, which could be lethal to the cell. This is called the "end replication problem," and a cell must solve it if it is to survive.

Eukaryotic cells have solved the difficulties related to having linear chromosomes by forming complex structures called **telomeres (figure 15.3)** at the ends of their chromosomes and by using an enzyme called **telomerase (figure 15.4)**. Telomeres are protein-DNA complexes that protect the linear DNA within them from degradation and end fusion. The protein component of a telomere varies from species to species, as does the length of DNA present. Telomeric DNA contains many copies of a particular sequence of nucleotides, placed one after the other (tandem

Table 15.1	Major Replisome Proteins in Bacteria, Eukaryotes, and Archaea		
Function	**Bacteria**	**Eukaryotes**	**Archaea**
Initiator; binds origin[1]	DnaA	ORC	Orc1/Cdc6
Helicase loader	DnaC	Cdc6/Cdt1	Orc1/Cdc6
Helicase	DnaB	MCM[2]	MCM[2]
Single-strand DNA binding	SSB proteins	Replication protein factor A (RPA)	SSB or RPA, depending on the species
Primer synthesis	DnaG	Pol α-primase	PriS/PriL[3]
Replicative DNA polymerase	DNA polymerase III (C-family polymerase)	DNA polymerase (Pol) δ and DNA polymerase ε (B-family polymerases)	B-family polymerase (both crenarchaeotes and euryarchaeotes) and D-family polymerase (euryarchaeotes)
Clamp loader	τ complex	Replication factor C (RFC)-L and RFC-S (1-4)[4]	RFC-L and RFC-S[4]
Clamp	β clamp	Proliferating-cell nuclear antigen (PCNA)	PCNA
Primer removal	Ribonuclease (RNase) H[5]; DNA polymerase I	RNaseH[5]; Dna2; Flap endonuclease (FEN)-1	RNaseH[5]; FEN-1
Ligase	NAD[+]-dependent DNA ligase; ATP-dependent DNA ligase[6]	ATP-dependent DNA ligase[6]	ATP-dependent DNA ligase; NAD[+]-dependent DNA ligase[6]

1 Although we have included the initiators in this table, they are involved only in the formation of the replisome and do not remain in the replisome as it advances from the origin of replication.
2 In eukaryotes, MCM is heteromeric, composed of six different Mcm proteins (Mcm2-7). It is also part of the CMG complex consisting of Cdc45, MCM, and GINS. GINS is short for the first letters of the Japanese numbers representing the proteins composing GINS. In most archaea, MCM is homohexameric. The archaeal MCM is associated with GINS homologues. A protein called RecJ may be the archaeal version of Cdc45.
3 PriS and PriL are homologues of the primase subunits of Pol α-primase.
4 The eukaryotic clamp loader is a pentameric structure consisting of one large subunit (RFC-L) and four small subunits encoded by four different genes. The archaeal clamp loader is also pentameric consisting of RFC-L and four small subunits encoded by one or two genes, depending on the archaeon.
5 The bacterial RNase H is distinct, whereas the eukaryotic and archaeal enzymes are homologues.
6 All NAD[+]-dependent DNA ligases belong to the same family of ligases and are evolutionarily related. Likewise ATP-dependent DNA ligases are related. NAD[+]-dependent DNA ligases are primarily responsible for ligating Okazaki fragments in bacteria, whereas ATP-dependent DNA ligases play this role in eukaryotes and archaea.

DNA Replication in Eukaryotes

The distinctive features of eukaryotic DNA replication arise from differences in the replication machinery and genome structure. Most eukaryotes use a similar set of proteins to replicate their DNA (**table 15.1** and **figure 15.1**). However, eukaryotic replisome proteins are given different names in part to reflect that they are generally unrelated to bacterial replisome proteins. The major features of eukaryotic genome structure relevant to DNA replication include having multiple chromosomes, each of which is usually much larger than a typical bacterial chromosome. Furthermore, the DNA in eukaryotic chromosomes is linear, which means that a mechanism for replicating chromosome ends is needed. Finally, eukaryotic DNA is wound around histones in nucleosomes. Nucleosomes must be deconstructed ahead of the replication fork (i.e., histones removed) and then formed on the parental and newly synthesized strands behind the replication fork.

DNA Replication Initiates from Multiple Origins

Replication of the large, eukaryotic chromosomes is initiated at multiple origins of replication (**figure 15.2**). This allows

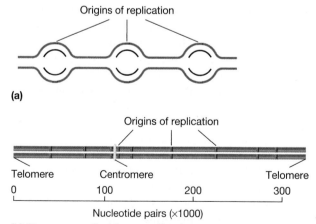

(a)

(b) Chromosome III of the yeast *Saccharomyces cerevisiae*

Figure 15.2 Eukaryotic Chromosomes Are Replicated from Multiple Origins of Replication. (a) Drawing of a eukaryotic chromosome being replicated. (b) Schematic of chromosome III from the yeast *Saccharomyces cerevisiae*. It has 180 genes and is replicated from nine origins of replication.

the chromosome to be replicated much faster than it could be if there were only one origin per chromosome. Two replication forks move outward from each origin until they encounter replication forks that formed at adjacent origins. Thus eukaryotic chromosomes consist of multiple **replicons,** rather than the single replicon (i.e., the entire chromosome) observed in bacteria.

In eukaryotes, origins of replication are "marked" by a complex of proteins called the **origin recognition complex (ORC)**, which remains bound to the origins throughout much of the cell cycle (table 15.1). All ORCs consist of six different ORC proteins (Orc1 to Orc6), but the identity of these proteins varies among eukaryotic species. ORC serves as a platform on which other proteins assemble in a cell cycle-dependent fashion. The first protein to associate with ORC is the helicase loader, either Cdc6 (*cell-division-cycle* protein 6) or Cdt1 (*Cdc10-dependent-transcript* 1 protein), depending on the organism. This occurs in the late M/early G1 phases of the cell cycle. Together, ORC and the helicase loader recruit the helicase enzyme, a set of proteins called the MCM (*minichromosome maintenance*) complex, to the origin, thereby forming the pre-replicative complex (pre-RC). Pre-RCs are activated as the cell cycle transitions from the G1 phase to the S phase by phosphorylation of the helicase loader, which is then replaced by numerous other proteins, forming the complete replisome (table 15.1). Within the replisome, the MCM helicase is associated with two co-activators and begins unwinding DNA to initiate replication.

Eukaryotic Primases and DNA Polymerases

DNA polymerases from many organisms have been analyzed and sorted into seven families based on their amino acid sequence and structure. As shown in **table 15.2**, eukaryotic DNA polymerases are found in different families than are the bacterial enzymes. In eukaryotes, three DNA polymerases function in DNA replication: DNA polymerases α-primase, ε, and δ. Primer synthesis is accomplished with the enzyme DNA polymerase α-primase (often called simply Pol α-primase). Pol α-primase is so named because it has two activities: DNA synthesis (Pol α) and RNA synthesis (primase). This interesting enzyme consists of four subunits: two for DNA synthesis (Pol 1 and Pol 2), and two for RNA synthesis (Pri1 and Pri2). The Pri1 subunit contains the catalytic site. The primer is made in two steps: the primase component of the enzyme makes a short RNA strand (~10 nucleotides), which is then transferred to the active site of Pol α. Pol α adds an additional 20 or so deoxyribonucleotides to the RNA strand. Thus in eukaryotes, the single-stranded primer for DNA replication is an RNA-DNA hybrid molecule. Once the primer is formed, the other two DNA polymerases take over. DNA polymerase ε (Pol ε) is responsible for leading-strand synthesis, whereas DNA polymerase δ (Pol δ) carries out lagging-strand synthesis. Thus unlike most bacteria where DNA polymerase III synthesizes both leading and lagging strands, two distinct DNA polymerases carry out these functions in eukaryotes.

Table 15.2	DNA Polymerase Families	
Family	**Example(s)**	**Functions/Comments**
A	Bacterial DNA polymerase I	Replacement of RNA primers present in Okazaki fragments
B	Archaeal DNA polymerase B (Pol B) Eukaryotic DNA polymerases α, δ, and ε	Replicative DNA polymerases
C	Bacterial DNA polymerase III	Replicative DNA polymerase
D	Archaeal DNA polymerase D (Pol D)	Replicative DNA polymerase in some archaea; unique to archaea
X	Eukaryotic DNA polymerase β	DNA repair
Y	Bacterial DNA polymerase IV (DinB)	DNA repair
Reverse transcriptase (RT)	Retroviral reverse transcriptases Telomerase RT	RNA-dependent DNA polymerase

Telomeres and Telomerases: Protecting the Ends of Linear DNA Molecules

The fact that the DNA in eukaryotic chromosomes is linear poses several problems. Without protection, the ends are susceptible to degradation by enzymes called DNases. They are also able to fuse with the ends of other DNA molecules, thus generating aberrant chromosomes. Finally, linear chromosomes present a problem during replication because of DNA polymerase's need for a primer that provides a free 3'-OH. When the primer for the Okazaki fragment at the end of the daughter strand is removed, the daughter molecule is shorter than the parent molecule. Over numerous rounds of DNA replication and cell division, this leads to a progressively shortened chromosome. Ultimately the chromosome loses critical genetic information, which could be lethal to the cell. This is called the "end replication problem," and a cell must solve it if it is to survive.

Eukaryotic cells have solved the difficulties related to having linear chromosomes by forming complex structures called **telomeres (figure 15.3)** at the ends of their chromosomes and by using an enzyme called **telomerase (figure 15.4)**. Telomeres are protein-DNA complexes that protect the linear DNA within them from degradation and end fusion. The protein component of a telomere varies from species to species, as does the length of DNA present. Telomeric DNA contains many copies of a particular sequence of nucleotides, placed one after the other (tandem

15

Eukaryotic and Archaeal Genome Replication and Expression

Approximately 140 million tons of plastics are purchased each year. Most are discarded and very slowly, if ever, degrade.

Plastics: Brought to You by Microbes

The next time you discard a plastic water bottle, consider this: that plastic bottle will take thousands of years to be converted back into the smaller chemical molecules from which it was made, assuming it ever degrades. Also consider that each year 140 million tons of equally recalcitrant plastics are purchased worldwide (and usually discarded). And what is often forgotten is that to make all that plastic, 150 million tons of fossil fuels must be processed each year, contributing additional waste, diversion of the fuels from refineries and power plants, and exacerbating geopolitical problems centered on oil and other fossil fuels. Could microbes solve this problem? Microbiologists have been trying for years to answer that question.

You may be asking why microbiologists would suggest microbes as a solution to the plastics problem. This is why. Plastics are organic polymers, such as polyurethane, polystyrene, polyethylene, and polyester. Microbiologists have known for decades that many bacteria and archaea store carbon as intracellular inclusions of poly-β-hydroxybutyrate (PHB) and other polyhydroxyalkanoates (PHAs). PHAs are polyesters; thus bacteria and archaea have evolved to synthesize a type of plastic. And even more important, because PHA-producing bacteria and archaea have evolved a means to use the PHA when carbon is in short supply, they produce the enzymes needed for PHA degradation.

For many years, microbiologists have worked to optimize production of PHAs by bacteria such as *Ralstonia eutropha,* and they have succeeded— up to 80% of cellular dry weight is produced by some strains. However, the cost of feeding these chemoorganoheterotrophic bacteria so that they produce PHAs makes the plastics they produce very expensive. Therefore, although some companies manufacture bioplastics, their share of the market is minuscule relative to that produced each year by chemical means.

"If expense arises from feeding a chemoorganotroph," scientists thought, "why not move the bacterial genes encoding PHB-synthesizing enzymes into an organism that feeds itself?" The logical autotrophs were plants such as thale cress (*Arabidopsis thaliana*) and tobacco (*Nicotiana tabacum*). Though successful, scientists were then faced with another dilemma: the use of land to "grow" plastics rather than to grow food.

Once again, microbes may provide a solution. Several German microbiologists have introduced the genes for PHB synthesis into a microalga, the diatom *Phaeodactylum tricornutum.* The advantages of an alga-based system are that these photoautotrophs need only be provided with sunlight and CO_2, and they can be grown in areas (e.g., a desert) that are not used for food production. This greatly decreases cost and avoids the ethical problems associated with plant-based systems.

Although the German scientists did nothing to maximize PHB production by the diatom, it was a "proof of concept" experiment. It showed that an alga-based system for plastic production may be successful. The next step is to maximize PHB production in a cost-effective manner. This will require that microbiologists take into account the differences between gene expression in bacteria and eukaryotes. They will need to modify the PHB-synthesis genes such that they are transcribed at a high rate in the diatom, producing mRNA molecules that are recognized and translated by eukaryotic ribosomes. Fortunately, microbiologists have a wealth of information provided by geneticists and molecular biologists studying plant systems and eukaryotic microbes such as the yeast *Saccharomyces cerevisiae* to guide them. What these geneticists and molecular biologists have learned is the focus of this chapter.

Readiness Check:

Based on what you have learned previously, you should be able to:

✔ Discuss bacterial genome structure and the organization of genetic information (sections 13.3 and 13.4)

✔ Describe the events that occur during the three phases of DNA replication and how the components of the bacterial replisome participate (section 13.3)

✔ Describe the structure of a typical bacterial RNA polymerase holoenzyme and how it functions during the three phases of transcription (section 13.5)

✔ List the major features of the universal genetic code and comment on any deviations from it (section 13.6)

✔ Describe the events that occur during the three phases of translation and the structure of bacterial ribosomes, and identify the initiator tRNA used by bacteria (section 13.7)

✔ Summarize the posttranslational events that result in a properly folded protein in its appropriate location within or outside a bacterial cell (section 13.8)

✔ Trace the movement of materials via the endocytic and secretory pathways of eukaryotic cells (section 5.4)

✔ Describe the structure of mitochondria and chloroplasts (section 5.6)

✔ Outline the steps of the eukaryotic cell cycle

15.1 Why Consider Eukaryotic and Archaeal Genetics Together?

If it looks like a duck, walks like a duck, and quacks like a duck, it must be a duck. Unfortunately, the logic of this saying fails when applied to archaea. For many years, microbiologists thought they looked like bacteria, did the things bacteria do, and therefore were just another group of bacteria. And why not? Archaea look like bacteria, and archaeal energy-conserving and biosynthetic processes and enzymes are similar to those of bacteria. However, archaea are not bacteria, as demonstrated by differences in cell walls membrane lipids, and rRNA molecules, as well as the unique archaeal ability to carry out methanogenesis. Following Carl Woese's discovery that bacterial and archaeal rRNA molecules differ, other components of their genome replication and expression systems were also found to be distinct. These studies led to the realization that the archaeal machinery for genetic information processing is more similar to that of eukaryotes than of bacteria.

In this chapter, our goal is to accomplish three things: (1) to illustrate the similarities of archaea and eukaryotes with respect to how they replicate and express the information in their genomes, (2) to point out those aspects of the archaeal machinery that are distinct, and (3) to draw comparisons among the members of the three domains of life. To that end, we organize this chapter with sections that follow the steps of genetic information flow in cells. We begin each section with a brief summary of the most salient information about each process as it occurs in bacteria. We then describe the process in eukaryotes and finally consider archaea. As you will see, archaea are an interesting mix of bacterial and eukaryotic characteristics. It is this mix that has prompted further studies to elucidate the evolutionary history of microorganisms.

15.2 DNA Replication: Similar Overall, but with Different Replisome Proteins

After reading this section, you should be able to:

- Create a table or concept map that illustrates the differences in eukaryotic, archaeal, and bacterial genome structure and organization
- Distinguish the replicons of eukaryotes and archaeal cells from bacterial replicons
- Compare and contrast the replisomes of eukaryotes, archaeal cells, and bacterial cells
- Construct a model to demonstrate the role of telomerase in the replication of eukaryotic chromosomes

DNA replication is similar in all cellular organisms. It is accomplished by a huge complex of many different proteins called the **replisome.** Recall from chapter 13 that the DNA of most bacteria is circular and replication begins at a single origin of replication (*see figure 13.9*). The bacterial replicative DNA polymerase (DNA polymerase III; *see figure 13.11*) is recruited to the origin only after the initiator protein DnaA begins assembly of the bacterial replisome, which is composed of at least 30 proteins. The replication forks move bidirectionally from the origin until the entire circular chromosome is replicated. We now compare this to what occurs in eukaryotes.

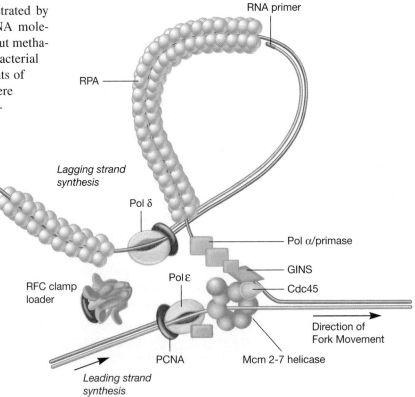

Figure 15.1 A Typical Eukaryotic Replication Fork.   The replisome proteins and their functions are defined in table 15.1. Compare this figure to figure 13.13.

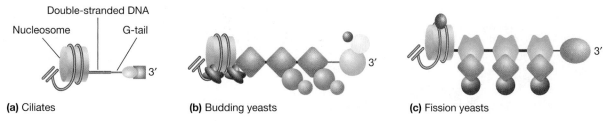

(a) Ciliates **(b)** Budding yeasts **(c)** Fission yeasts

Figure 15.3 Telomere Structure. Telomeres consist of the end of the DNA molecule complexed with proteins. The location, number, and nature of the proteins differ from organism to organism. For each telomere illustrated, identical proteins are shown having the same shape. Different shapes indicate that the protein is not the same. Note that some proteins bind the single-stranded G-tail, whereas others bind upstream where the DNA is double stranded. (a) Telomere of the ciliate protists *Stylonychia* and *Oxytricha* spp. (b) Telomeres of budding yeasts such as *Saccharomyces cerevisiae*. (c) Telomere of fission yeasts such as *Schizosaccharomyces pombe*.

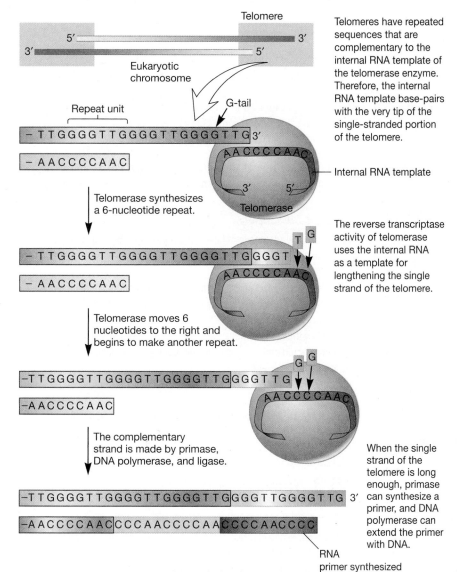

Telomeres have repeated sequences that are complementary to the internal RNA template of the telomerase enzyme. Therefore, the internal RNA template base-pairs with the very tip of the single-stranded portion of the telomere.

The reverse transcriptase activity of telomerase uses the internal RNA as a template for lengthening the single strand of the telomere.

When the single strand of the telomere is long enough, primase can synthesize a primer, and DNA polymerase can extend the primer with DNA.

Figure 15.4 Telomerase Replicates the Telomeric DNA of Eukaryotic Chromosomes. Telomerase contains an RNA molecule that can base-pair with a small portion of the 3' overhang (the G-tail). The RNA serves as a template for DNA synthesis catalyzed by the reverse transcriptase activity of the enzyme. The 3'-OH of the telomere DNA serves as the primer and is lengthened. The process shown is repeated many times until the 3' overhang is long enough to serve as the template for synthesis of the complementary telomere DNA strand by the replisome.

repeats). Importantly, the DNA is single-stranded at the very end and is called the G-tail because it is rich in guanosine bases; telomerase uses the G-tail to maintain chromosome length.

Telomerase has two important components: an internal RNA template and an enzyme called telomerase reverse transcriptase (RT). The internal RNA template is complementary to a portion of the G-tail and base-pairs with it (figure 15.4). The internal RNA template provides the template for DNA synthesis, which is catalyzed by telomerase RT (i.e., the 3'-OH of the G-tail serves as the primer for DNA synthesis). After being lengthened sufficiently, there is room for synthesis of an RNA primer, and the single strand of telomere DNA can serve as the template for synthesis of the complementary strand. Thus the length of the chromosome is maintained.

Telomerase RT deserves additional comment. As described in the preceding paragraph, telomerase RT synthesizes DNA using an RNA template. Enzymes with this capability are defined as RNA-dependent DNA polymerases. RNA-dependent DNA polymerases are not unique to telomerases; certain viruses use RNA-dependent DNA polymerases to complete their life cycles (e.g., human immunodeficiency viruses and hepatitis B viruses). These interesting viruses are described more in chapters 27 and 38.

Archaea Use Homologues of Eukaryotic Replisome Proteins

The mixture of bacterial and archaeal features is obvious when archaeal genomes are examined. Archaeal chromosomes are similar in size to those of bacteria. In addition, all known archaeal chromosomes are circular, like most bacterial chromosomes. However, some archaea have histones associated with their chromosomes just like eukaryotes.

DNA replication in archaea also illustrates their "hybrid" nature. The circular chromosomes of archaea are replicated by replisomal proteins that

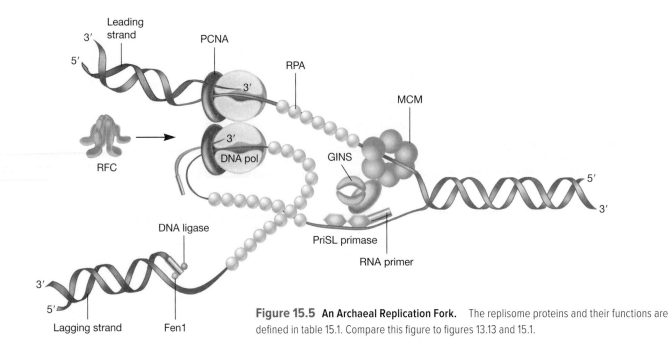

Figure 15.5 An Archaeal Replication Fork. The replisome proteins and their functions are defined in table 15.1. Compare this figure to figures 13.13 and 15.1.

are similar to those of eukaryotes (**figure 15.5**; table 15.1). Most archaeal cells have a single origin of replication, like bacterial chromosomes; however, a few have more than one. The reason for this is unknown, as archaeal chromosomes are not especially large. However, having more than one origin may be related to the speed at which the replisome travels around the DNA. Those with faster speeds have single origins, whereas those with slower speeds have more than one origin. Thus having multiple origins may serve the same purpose as in eukaryotes: decreasing the time needed to replicate the chromosome.

Analysis of archaeal primases and replicative DNA polymerases has led to some intriguing discoveries. Nearly all archaeal primases are heterodimeric; that is, they are composed of two different protein subunits, PriS and PriL, just like their eukaryotic counterparts. One (PriS) is the catalytic subunit and the other (PriL) is thought to function as a regulatory subunit. Archaeal primases vary dramatically in terms of their activity. Some are thought to make RNA-DNA hybrid primers as in eukaryotes. Others appear to make RNA primers. Furthermore, some archaeal primases have functions other than priming DNA synthesis, including filling gaps as part of DNA repair processes. Other archaea have monomeric primases consisting of both PriS and PriL sequences fused into a single polypeptide.

Archaeal DNA polymerases fall into two families. All archaea studied to date have DNA polymerases in the same family as the replicative eukaryotic DNA polymerases (B family polymerases; table 15.2) and are called Pol B. However, members of one archaeal phylum, *Euryarchaeota*, have in addition to Pol B a unique archaeal DNA polymerase called Pol D. Pol D enzymes are so distinct, they are the sole members of the D family of DNA polymerases (table 15.2). Current models of DNA replication in euryarchaeotes propose that Pol D has two functions: lagging-strand synthesis and initiation of leading-strand synthe-

sis. Pol B is thought to eventually replace Pol D and complete synthesis of the leading strand.

Retrieve, Infer, Apply

1. What is the advantage of having multiple origins of replication for a eukaryote? What is the most plausible reason some archaeal species have multiple origins?
2. Refer to figure 15.1 and draw the last replication fork present near the end of a chromosome. Use your drawing to illustrate the end replication problem.
3. Review the steps in mitosis and meiosis. Why would having unprotected ends of linear DNA disrupt these two important cellular processes?
4. Outline the process used by telomerase to maintain chromosome length.
5. The discovery of telomerase activity in an archaeon would be a surprise. Why?

15.3 Transcription

After reading this section, you should be able to:

- Compare and contrast gene structure of eukaryotic, archaeal, and bacterial cells
- Create a table to distinguish eukaryotic, archaeal, and bacterial promoters, RNA polymerases, and transcription factors
- Discuss the modifications made to eukaryotic mRNA molecules during and after synthesis

Transcription is essentially the same in all cellular organisms: the template strand directs RNA synthesis, and RNA is made from the 5′ to the 3′ end by enzymes called RNA polymerases. Although this unifies the process of transcription across all domains of life, several key aspects differ. These arise in part from differences in cell and chromosomal structure, the nature of RNA polymerase, and the organization of protein-coding genes. Bacteria use a single RNA polymerase that consists of the catalytic core responsible for synthesizing mRNA and the intrinsic transcription factor subunit, sigma (σ; *see figure 13.23*). Bacterial genes with related function are often organized into operons, and are thereby transcribed from the same promoter. This results in a polycistronic mRNA (*see figure 13.22*) from which individual proteins are translated. Some regions of a bacterial gene (e.g., leader and trailer) are transcribed but not translated, and for almost all protein-coding genes, the coding information within a gene is continuous. Now let's compare this to transcription in eukaryotes.

Transcription in Eukaryotes

In eukaryotes, transcription occurs in the nucleus, and the RNA products must move to the cytoplasm, where they function in translation. Just as noted for DNA replication, chromatin structure must be altered for RNA polymerases to be recruited to promoters and ahead of the RNA polymerases as transcription proceeds. This is accomplished by chromatin-remodeling and chromatin-modifying enzymes (p. 364). Gene structure and

organization in eukaryotes also is strikingly different from that seen in bacteria. For the vast majority of eukaryotes, each protein-coding gene has its own promoter, and the transcript synthesized from the gene is **monocistronic.** Furthermore, many eukaryotic protein-coding genes are composed of sequences called exons and introns. An **exon** contains sequences that code for part of the polypeptide, whereas an **intron** is a stretch of noncoding sequences. Thus the coding portion of a eukaryotic gene is not continuous. Both exons and introns are transcribed, yielding an exon- and intron-containing transcript called a **primary** or **pre-mRNA.** The introns are spliced out of the primary transcript before the protein is made (p. 356).

Eukaryotic RNA Polymerases Include Homologues of the Bacterial RNA Polymerase Core Subunits

Eukaryotes have three major RNA polymerases. RNA polymerase I catalyzes rRNA synthesis, RNA polymerase II is responsible for mRNA synthesis, and RNA polymerase III synthesizes all the tRNAs and one rRNA (5S rRNA). All five of the bacterial RNA polymerase core enzyme subunits are conserved in eukaryotic RNA polymerases (**table 15.3**). However, eukaryotic RNA polymerases have several additional subunits.

Table 15.3	Subunits of RNA Polymerases (RNAP) and Their Homologues			
Bacteria Core RNAP Subunits[1]	Archaeal RNAP Subunits[2]	Eukaryotic RNAP II Subunits	Eukaryotic RNAP III Subunits	Eukaryotic RNAP I Subunits
β	Rpo1 (RpoA)[3]	RPB1	C160	A190
β	Rpo2 (RpoB)	RPB2	C128	A135
α	Rpo3 (RpoD)	RPB3	AC40	AC40
α	Rpo11 (RpoL)	RPB11	AC19	AC19
ω	Rpo6 (RpoK)	RPB6	RPB6	RPB6
	Rpo5 (RpoH)	RPB5	RPB5	RPB5
	Rpo10 (RpoN)	RPB10	RPB10	RPB10
	Rpo12 (RpoP)	RPB12	RPB12	RPB12
	Rpo4 (RpoF)	RPB4	C17	A14
	Rpo7 (RpoE)	RPB7	C25	A43
		RPB8[4]	RPB8	RPB8
		RPB9	C11	A12

1 Sigma (σ) factors found in the bacterial holoenzyme are transcription factors and are not included in this table. The numerous archaeal and eukaryal transcription factors are shown in table 15.4. Most archaeal transcription factors have homologues in eukaryotic organisms.
2 Only those archaeal subunits observed in all archaea studied thus far are listed.
3 Alternate names for subunits are in parentheses.
4 Some archaea have a subunit homologous to RPB8.

Table 15.4	Eukaryotic and Archaeal Basal Transcription Factors	
Eukaryotic Transcription Factor	**Homologous Archaeal Transcription Factor**
TATA-binding protein (TBP)[1]	TBP
TFIIB[2]	TFB[3]
TFIIE	TFE
TFIIF	
TFIIB	
TFIIH	

1 Eukaryotic TBP is often part of a complex of proteins called TFIID.
2 TFII indicates that the protein is a *transcription factor* for RNA polymerase *II*. TF indicates that the protein is a *transcription factor* for archaeal RNA polymerase.

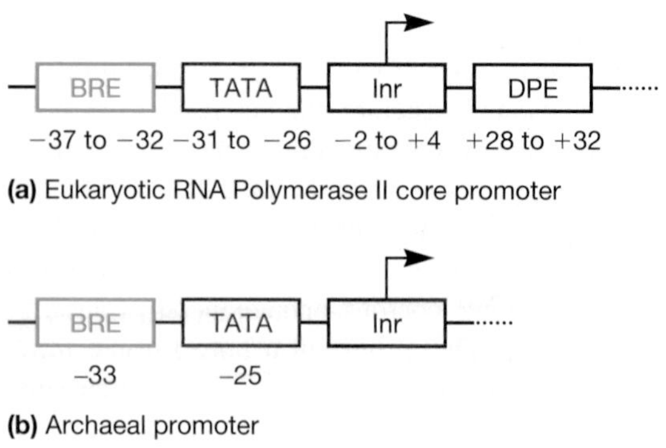

(a) Eukaryotic RNA Polymerase II core promoter

(b) Archaeal promoter

Figure 15.6 Eukaryotic and Archaeal Promoters. (a) Eukaryotic RNA polymerase II core promoters contain one or more of the elements shown. However, it is very rare for all of the elements to be present. BRE is the TFII*B* recognition element. Inr is the *ini*tiator element; transcription initiates within this site (indicated by the arrow). DPE is the *d*ownstream *p*romoter *e*lement. The approximate positions of each promoter element are indicated, relative to the transcriptional start site. (b) An archaeal promoter. Promoter element designations are the same as those in eukaryotic promoters.

Transcription Initiation in Eukaryotes

Most research on transcription in eukaryotes has been on the transcription of protein-coding genes, and that is our focus here. In eukaryotes, RNA polymerase II (often referred to as RNAPII) is responsible for transcribing protein-coding genes. This enzyme is a large aggregate, at least 500,000 daltons in size, with about 10 subunits (table 15.3).

Transcription initiation requires that RNAPII be recruited to a promoter. This is accomplished by transcription factors called basal or general transcription factors, some of which bind the core promoter and then interact with other transcription factors to attract RNAPII to the promoter (**table 15.4**). RNAPII-recognized promoters usually have two regions, the core promoter and a regulatory region. The core promoter has been defined by in vitro studies to be the minimal region needed for transcription to occur. One of the first features discovered in eukaryotic RNAPII core promoters was the conserved nucleotide sequence TATA; thus the region is named the TATA box (**figure 15.6**). Although it was initially thought that all RNAPII core promoters contain a TATA box, it is now known that many RNAPII core promoters lack a TATA box. TATA-less promoters are generally associated with housekeeping genes. Those that do contain a TATA box tend to be part of highly regulated genes.

Eukaryotic transcription factors are referred to as extrinsic factors because they are not part of the RNA polymerase. The TATA-binding protein (TBP) is one of the most important for transcription initiation. It is often part of a complex of proteins called TFIID (short for *t*ranscription *f*actor *D* for RNA polymerase *II*). Despite its name, TBP functions in transcription initiation for both TATA box-containing promoters and those that lack the TATA box. One sequence of events to recruit RNAPII to

the promoter is outlined in **figure 15.7**. As seen there, initiation requires the formation of the pre-initiation complex (PIC). Once formed, TFIIH acts as a helicase to unwind the DNA, forming an open complex (*see figure 13.26*). RNAPII can then begin synthesizing mRNA. However, transcription cannot continue to the elongation phase until the contacts between TFIIB and RNAPII are broken. This is accomplished by TFIIH, which phosphorylates the C-terminal domain of one of the RNAPII subunits. Once this occurs, RNAPII is released from the PIC and elongation proceeds.

Studies of transcription initiation in numerous eukaryotes have shown that this is not the only sequence of events nor are these the only proteins involved. This is especially true for highly regulated genes. For these genes, three protein complexes that function as co-activators are important. One is a complex of proteins called **mediator.** Another co-activator is called SAGA. Both mediator and SAGA are sometimes involved in recruiting certain basal transcription factors to the promoter during transcription initiation. Interestingly, the basal transcription factor TFIID is also a co-activator and plays important roles in regulating gene expression, as we discuss in section 15.5.

From Pre-mRNA to mRNA: A Unique Eukaryotic Process

Unlike nearly all bacterial mRNAs, the initial transcript synthesized from a eukaryotic protein-coding gene must be modified before giving rise to a translatable mRNA. Modification occurs in the nucleus and begins during the elongation phase of transcription; it is completed after transcription ends. Modification starts with the addition of an unusual nucleotide to the 5′ end of the transcript when the transcript is only about 25 ribonucleotides long (**figure 15.8**). The **5′ cap,** as it is called, is 7-methylguanosine. After the 5′ cap is added and a sufficiently long transcript is generated, removal of introns begins. After synthesis is completed, any remaining introns are removed, and the pre-mRNA is modified further by the addition of a **3′ poly-A tail,** about 200 nucleotides long (figure 15.8). Both the 5′ cap and poly-A tail help protect the mRNA from enzymatic attack. Furthermore, they enable the cell to recognize that the mRNA is intact and ready to be transported from the nucleus to the cytoplasm. The 5′ cap has the additional important function of serving as a recognition signal for binding of ribosomes to the mRNA so that it can be translated.

Introns are removed from the pre-mRNA by a large complex of proteins and RNA molecules that is unique to eukaryotic cells. This large complex, called a **spliceosome,** is composed of several small nuclear ribonucleoproteins (snRNPs, pronounced "snurps") and several non-snRNP splicing factors. The snRNPs consist of small nuclear RNA (snRNA) molecules (about 60 to 300 nucleotides long) associated with proteins. Some snRNPs recognize and bind exon-intron junctions (**figure 15.9**).

Sometimes a pre-mRNA is spliced so that different patterns of exons remain or the junction of two exons varies. This **alternative splicing** allows a single gene to code for more than one protein. The splice pattern determines which protein is synthesized. Splice patterns can be cell-type specific or determined by the needs of the cell. The importance of alternative splicing in multicellular eukaryotes was clearly demonstrated when it was discovered that the human genome has only about 20,000 genes, rather than the anticipated 100,000. It is thought that alternative splicing is one mechanism by which human cells produce a vast array of proteins using fewer genes. This mechanism for expanding the coding capacity of a genome is not available to bacteria and archaea because introns in their protein-coding genes are very rare. ↻ *Processing of Genetic Information: Prokaryotes Versus Eukaryotes*

Transcription in Archaea

If transcription in archaea were described in a newspaper article, the headline might be something like this: "Eukaryotic-like RNA polymerase functions in bacteria-like environment." This is because archaeal transcription occurs in the cytoplasm, as it does for bacteria. Many (although perhaps not quite as many) archaeal genes are organized into operons, as are bacterial genes. Thus many archaeal mRNAs are polycistronic. Introns are rare in protein-coding archaeal genes (as they are in bacteria), and they are removed by a process

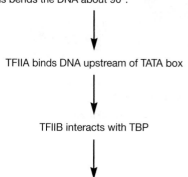

TATA-binding protein (TBP) binds the TATA box. This bends the DNA about 90°.

TFIIA binds DNA upstream of TATA box

TFIIB interacts with TBP

RNA polymerase II and TFIIF join the other proteins at the promoter. TFIIB interacts extensively with RNA polymerase II, acting as a bridge between TBP and RNA polymerase II.

TFIIE then TFIIH bind forming the pre-initiation complex.

(a) Formation of the Pre-initiation Complex

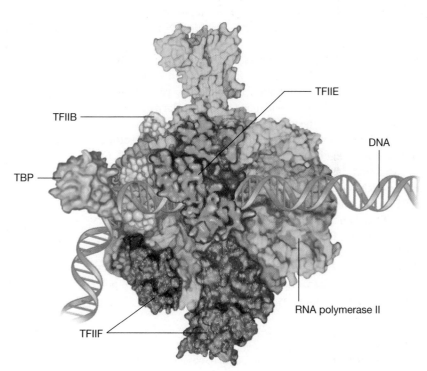

(b) Pre-initiation Complex

Figure 15.7 Initiation of Transcription in Eukaryotes. (a) Outline of the formation of the pre-initiation complex. (b) Schematic showing the location of some of the transcription factors in the pre-initiation complex (PIC). RNA polymerase II is brown. The two red proteins are the two proteins of TFIIF. TBP is turquoise. The orange shape represents the two proteins of TFIIE. TFIIB is yellow. TFIIH and TFIIA are not visible in this view of the PIC. Note how closely the transcription factors are associated with RNA polymerase II. The arrow shows the direction of transcription, once it initiates.

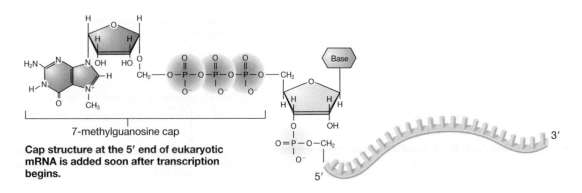

Cap structure at the 5′ end of eukaryotic mRNA is added soon after transcription begins.

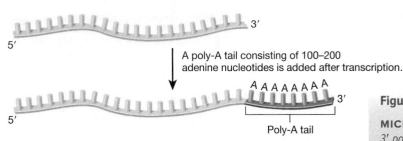

A poly-A tail consisting of 100–200 adenine nucleotides is added after transcription.

A A A A A A A A

Poly-A tail

Addition of a poly-A tail at the 3′ end of eukaryotic mRNA

Figure 15.8 **Modifications Made to the Ends of Eukaryotic mRNA.**

MICRO INQUIRY *What functions are served by the 5′ cap and the 3′ poly-A tail?*

different than that used by eukaryotes (again as they are in bacteria). Finally, archaeal cells use a single RNA polymerase to catalyze transcription.

The archaeal RNA polymerase is large and most similar to RNA polymerase II of eukaryotes (table 15.3 and **figure 15.10**). Likewise, transcription initiation in archaea is very similar to that seen in eukaryotes, as evidenced by the similarity in promoters (figure 15.6) and the use of similar basal transcription factors (table 15.4). The archaeal process can be thought of as a simplified version of that in eukaryotes. As shown in **figure 15.11**, two archaeal transcription factors are important for recruiting the archaeal RNA polymerase to a promoter: TATA-binding protein (TBP) and transcription factor B (TFB). A third transcription factor TFE then associates with RNA polymerase and promotes unwinding of the DNA to form the open complex (*see figure 13.26*). RNA polymerase then moves down the gene, transcribing it. TFB remains behind bound to the promoter, ready to initiate another round of transcription. TFE remains with RNA polymerase until transcription is terminated.

Retrieve, Infer, Apply

1. What elements in archaeal promoters are also observed in eukaryotic RNA polymerase II (RNAPII) core promoters? What additional elements are observed in RNAPII core promoters? How do bacterial promoters differ from those seen in eukaryotic and archaeal cells?

2. Compare the role of bacterial sigma factors in transcription initiation to that of eukaryotic and archaeal transcription factors. Why are bacterial sigma factors referred to as intrinsic transcription factors, whereas eukaryotic and archaeal transcription factors are extrinsic?

3. Outline the events that generate a mature, translatable mRNA in eukaryotes. Where does this processing occur?

15.4 Translation and Protein Maturation and Localization

After reading this section, you should be able to:

- Differentiate the structure of eukaryotic, archaeal, and bacterial ribosomes
- Identify the initiator tRNA used by eukaryotic and archaeal cells
- Outline the steps in translation initiation observed in eukaryotes and compare them to those observed in bacteria and archaea
- Discuss the role of molecular chaperones in eukaryotic and archaeal cells
- Create a concept map or table that distinguishes the translocation and secretion systems observed in the three domains of life

For organisms to survive, they must be able to synthesize the proteins they need (translation), fold them into a functional conformation, and localize the proteins to the correct site within (or outside) the cell. In bacteria, translation initiation requires three protein initiation factors (IFs), the initiator tRNA (fMet-tRNA$_i$^{Met}), and the two ribosomal subunits (50S and 30S subunits) (*see figure 13.38*). The 16S rRNA and the Shine-Dalgarno sequence interact to position the ribosome properly at the start codon. Elongation is assisted by three elongation factors (EFs; *see figure 13.39*). Translation is tightly coupled with transcription, and polysomes are often observed (*see figure 13.33*). Translation termination involves three release factors (RFs). **Chaperones** help the protein fold (*see figure 13.43*), and some chaperones deliver proteins to protein translocation or secretion systems.

In this section, we focus on translation, and protein folding and localization as they occur in eukaryotes and archaea. As

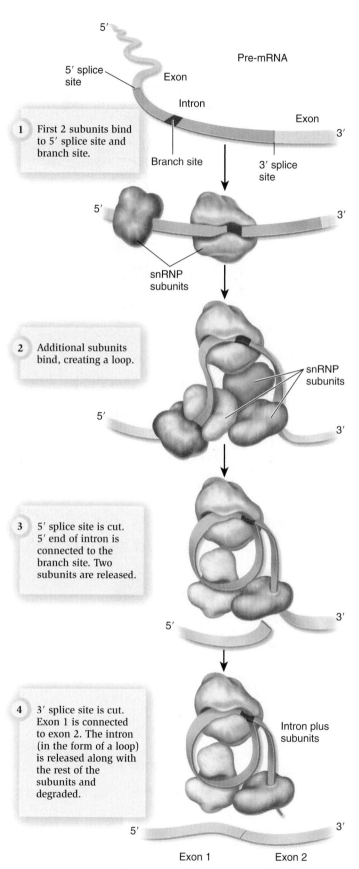

Figure 15.9 Removal of an Intron by a Spliceosome.

1. First 2 subunits bind to 5′ splice site and branch site.

2. Additional subunits bind, creating a loop.

3. 5′ splice site is cut. 5′ end of intron is connected to the branch site. Two subunits are released.

4. 3′ splice site is cut. Exon 1 is connected to exon 2. The intron (in the form of a loop) is released along with the rest of the subunits and degraded.

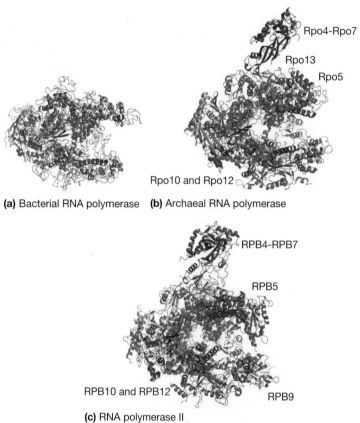

(a) Bacterial RNA polymerase   (b) Archaeal RNA polymerase

(c) RNA polymerase II

Figure 15.10 Bacterial, Archaeal, and Eukaryotic RNA Polymerases. Ribbon diagrams of the RNA polymerases of (a) the bacterium *Thermus aquaticus*, (b) the archaeon *Sulfolobus shibatae*, and (c) RNA polymerase II of eukaryotes. The subunits found in all three RNA polymerases are shown in blue. Those that are unique to archaea and RNA polymerase II are shown in magenta.

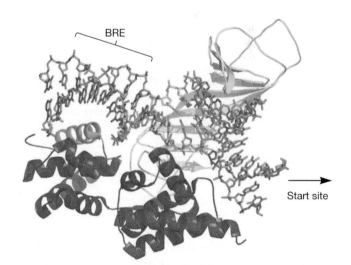

Figure 15.11 Binding of TBP and TFB to an Archaeal Promoter. The crystal structure of the ternary complex between TBP, the carboxyl terminus of TFB, and the region of an archaeal promoter containing the TATA box and BRE. DNA is shown in gray; TBP is the yellow ribbon structure; and TFB is magenta and turquoise. The turquoise portion of TFB is an α-helix that recognizes BRE. TBP is TATA-box binding protein; TFB is transcription factor B; BRE is B recognition element.

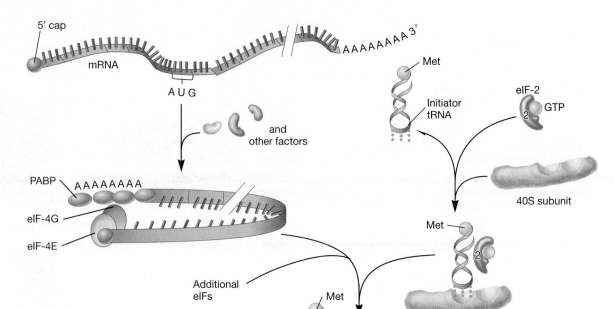

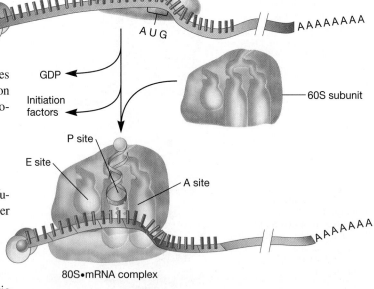

Figure 15.12 **Translation Initiation in Eukaryotes.** To simplify the drawing, many additional eukaryotic initiation factors (eIFs) are not shown. The upper left portion of the figure shows mRNA activation; the upper right shows the formation of the 40S complex. The 40S complex binds to the 5′ end of the activated mRNA and then moves down the mRNA to find the first AUG. This positions the initiator tRNA carrying methionine in what will become the peptidyl (P) site of the ribosome after the addition of the 60S subunit. The acceptor (A) site is vacant and ready for the arrival of the tRNA carrying the second amino acid encoded by the mRNA. Met is methionine. PABP is poly-A binding protein. E site is exit site.

with DNA replication and transcription, the overall processes are similar in all organisms. Therefore we focus our attention primarily on those aspects of the eukaryotic and archaeal processes that differ from those observed in bacteria.

Translation in Eukaryotes Initiates in a Unique Manner but Proceeds Like in Bacteria

There are many obvious differences between bacterial and eukaryotic translation. One is that eukaryotic ribosomes are larger than those in bacteria and require more initiation factors to be positioned properly on the mRNA. The 5′ cap and the 3′ poly-A tail of eukaryotic mRNAs are integral to translation initiation. The process begins with mRNA activation. mRNA activation is brought about by the binding of several eukaryotic initiation factors (eIFs) to the 5′ end of the mRNA and attachment of poly-A binding proteins (PABPs) to the 3′ end. The eIFs and PABPs interact, causing the mRNA to fold back on itself with the eIFs and PABPs forming a bridge between the two ends (**figure 15.12**). Meanwhile, the initiator tRNA (Met-tRNA$_i^{Met}$) is

loaded onto the 40S ribosomal subunit (the small subunit, SSU) to form a structure termed the 43S complex. Formation of the 43S complex is facilitated by several additional eIFs, one of which delivers the initiator tRNA to the 40S subunit. The 43S complex then binds the activated mRNA, generating the 43S·mRNA complex.

As seen in figure 15.12, at this point in translation initiation, the SSU is not positioned at the start codon of the mRNA as it is in bacteria. Rather, it is at the 5′ end of the mRNA. The 43S complex slides down the mRNA scanning for the start codon. Once the start codon is encountered, scanning ceases and the 60S ribosomal subunit (large subunit) joins with the 43S·mRNA complex, forming the complete 80S ribosome. Translation can now commence.

Our description of translation initiation holds for mRNAs having a leader (i.e., a 5′ untranslated region). In recent years mRNAs lacking a leader have been identified. For instance, all mRNAs produced by the protist *Giardia intestinalis* lack leaders. How these leaderless mRNAs are translated is still unclear. However, evidence suggests that intact 80S ribosomes bind the start codon and initiate translation.

Translation elongation and termination proceed in a fashion similar to that seen in bacteria. Three elongation factors function in both bacteria and eukaryotes, and the bacterial and eukaryotic factors are very similar structurally and functionally. Polysomes are used during translation of eukaryotic mRNAs. However, because of the circularization of the mRNA during translation initiation, the polysomes have a characteristic structure (**figure 15.13**). Another difference is seen in the termination step. In eukaryotes, a single release factor (eRF) functions as opposed to three RFs in bacteria.

Protein Folding in Eukaryotes Is Similar to That Seen in Bacteria

Eukaryotes, like bacteria, use proteins called chaperones to help many proteins fold properly and restore proper folding if the protein is denatured by environmental conditions. As we introduce in chapter 13, chaperones were first identified by their role in protecting the proteins of heat-stressed cells and are often called heat-shock proteins (HSPs). There are numerous HSPs, and these are often sorted into types of HSPs based on their size. For instance, the bacterial chaperones GroEL and GroES have a molecular weight of about 60,000 and are included with other HSPs of similar size. All are referred to as HSP60s. Representatives of all types of chaperones are found in eukaryotes.

Just as with bacteria, eukaryotes use several pathways to fold their proteins properly (**figure 15.14**). Some of the chaperones involved perform their tasks cotranslationally, while others function posttranslationally. Most proteins are successfully folded by *nascent-chain-associated chaperone* (NAC) working with an Hsp70 chaperone (similar to the bacterial DnaK protein) and an Hsp40 cochaperone (similar to the bacterial DnaJ protein). Other proteins are folded by pathways that use Hsp60 chaperones similar to the GroEL system of bacteria. In eukaryotes this is called TRiC/CTT [*t*ailless *c*omplex polypeptide-1 (TCP-1) *r*ing *c*omplex/*c*haperonin-*c*ontaining *T*CP-1].

Protein Localization and Secretion in Eukaryotes

Proteins produced by eukaryotic cells have numerous possible destinations because these cells are highly compartmentalized,

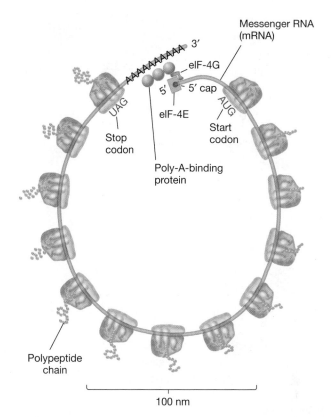

Figure 15.13 Eukaryotic Polysome. Schematic representation of a polysome. The distinctive circular form results from the interaction of the two ends of the mRNA brought about by the poly-A-binding proteins, eIF-4E, eIF-4G, and other initiation factors.

having many membrane-bound organelles such as the endoplasmic reticulum (ER), mitochondria, chloroplasts, and lysosomes. Although mitochondria and chloroplasts have genomes that encode some of their proteins, most are encoded in the nucleus and translated in the cytosol (i.e., the liquid portion of the cytoplasm). Thus some proteins synthesized in the cytosol stay there, but many are moved to specialized cellular compartments or are secreted. As described in chapter 5, proteins that are secreted or that move between the endoplasmic reticulum, Golgi apparatus, endosomes, and other membranous components of the endocytic and secretory pathways are delivered to these organelles or the plasma membrane by small vesicles that pinch off from the membrane of one organelle (e.g., the ER) and then fuse with a target membrane. This mechanism for localizing and secreting proteins is often called **vesicular transport,** and it does not involve the translocation of proteins across membranes. However, in order for proteins to move from the cytosol to the ER, mitochondria, and chloroplasts, the proteins must be translocated across membranes. This transmembrane transport involves translocator systems that are similar to certain translocation and protein secretion systems used by bacteria, as well as other proteins found in bacterial membranes. ◄◄ *Several cytoplasmic membranous organelles function in the secretory and endocytic pathways (section 5.4); Protein translocation and secretion in bacteria (section 13.8)*

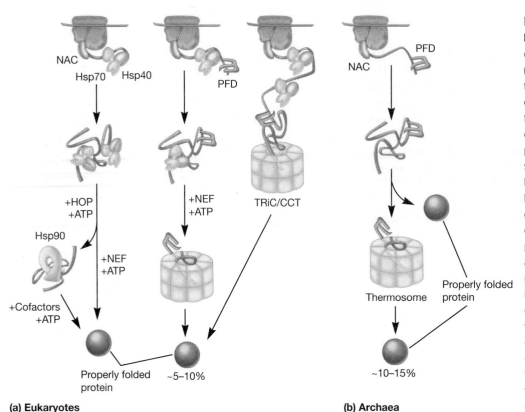

(a) Eukaryotes

(b) Archaea

Figure 15.14 **Protein Folding in Eukaryotes and Archaea.** (a) The eukaryotic pathways all employ NAC (nascent-chain-associated chaperone), the cochaperone Hsp40 and an Hsp70 chaperone. Some proteins are properly folded by these chaperones and the cochaperone alone. However, other proteins require additional chaperones such as Hsp90, PFD, and TRiC/CCT. The latter is an Hsp60 chaperone as is GroEL of bacteria. PFD is prefoldin, HOP is Hsp90-organizing protein, NEF is nucleotide exchange factor. (b) The archaeal folding pathways also employ NAC and PFD. Most archaea lack Hsp70 chaperones. Most proteins are folded using only NAC and PFD. Others require additional folding steps conferred by an Hsp60 chaperone called the thermosome. Although all three domains of life use Hsp60 chaperones, those of eukaryotes and archaea are more similar to each other than they are to the bacterial versions. Although not shown, all Hsp60 chaperones hydrolyze ATP.

The best-studied protein translocation system in eukaryotes is the **Sec system** (**figure 15.15**). Central to the functioning of this pathway in eukaryotes is the Sec61αβγ translocon through which unfolded proteins move. Several proteins that make up the Sec system in eukaryotes are homologous to Sec proteins in bacteria (**table 15.5**). In eukaryotes, the Sec system functions to move proteins from the cytosol into the membrane or lumen of the ER. It also functions in chloroplasts to move proteins from the stroma into the thylakoid membrane or the thylakoid lumen. Like the Sec system of bacteria, cotranslational translocation involves a signal recognition particle, which helps deliver the

translating ribosome and nascent polypeptide to the Sec system. Likewise, posttranslationally translocated proteins are targeted to the Sec translocon by a **signal peptide** at their amino-terminus (*see figure 13.44*).

In addition to the Sec system, the twin-arginine translocation (Tat) system is thought to function in the chloroplasts and mitochondria of some eukaryotes. For instance, Tat systems have been found in the thylakoid membranes of algal chloroplasts, and a homologue of one bacterial Tat system protein (TatC) is encoded by the mitochondrial genomes of many protists. On the other hand, fungi lack Tat systems. The localization of most mitochondrial proteins

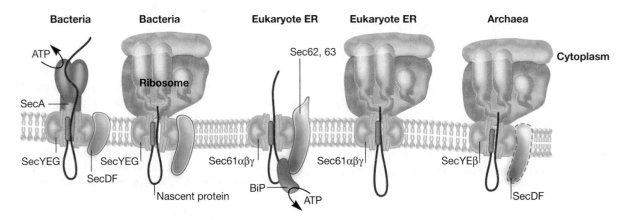

Figure 15.15 **Comparison of the Sec Systems Found in Bacteria, Eukaryotes, and Archaea.** In bacteria, translocation can occur posttranslationally or cotranslationally. In most cases, cotranslational translocation inserts the protein into the plasma membrane. In eukaryotes, posttranslational and cotranslational translocation are also observed. In archaea, cotranslational translocation occurs and is relatively well studied. More recently posttranslational translocation has been discovered. Its mechanism is unknown and so is not included in this figure. SecDF proteins are common in euryarchaeotes but not crenarchaeotes.

Table 15.5	Sec Protein Homologues	
Bacteria	**Eukaryotes**	**Archaea**
SecY	Sec61α	SecY[1]
SecE	Sec61γ	SecE[2]
SecG	—	—
—	Sec61β[2]	Sec61β

1 Although given the same name as the bacterial protein, this archaeal protein's structure is more like the eukaryotic homologue.
2 All Sec systems have three proteins that form the pore through which proteins move. However, the Sec61β protein differs from SecG both in structure and in function.

in all eukaryotes is mediated by very distinctive systems, which we describe next.

Localization of proteins in the membranes, intermembrane spaces, and matrix of mitochondria is illustrated in **figure 15.16**. Proteins are moved across the outer mitochondrial membrane by a system of proteins called the TOM complex (*t*ranslocase of the *o*uter mitochondrial *m*embrane). Once across the outer membrane, they may remain in the intermembrane space, be inserted into the outer membrane by another complex of proteins called SAM (*s*orting and *a*ssembly *m*achinery), or cross or be inserted into the inner membrane. Two distinct TIM complexes (*t*ranslocase of the *i*nner mitochondrial *m*embrane) move proteins across or into the inner membrane, TIM23 and TIM22, respectively. ◄◄ *Mitochondria, related organelles, and chloroplasts are involved in energy conservation (section 5.6)*

Translation, Protein Maturation, and Protein Localization in Archaea

Most studies of archaeal translation have focused on translation initiation. These studies have yielded some surprising results. First, transcription and translation are tightly coupled, as in bacteria. Thus translation of mRNAs can begin before synthesis of the mRNA is finished. Some archaeal mRNAs have a Shine-Dalgarno sequence in the leader. Translation initiation of these mRNAs proceeds by a mechanism similar to that seen for bacteria. However, the initiator tRNA is Met-tRNA$_i^{Met}$, as in eukaryotes. Furthermore, more initiation factors (called archaeal IFs or aIFs) are involved than in bacteria. Some are similar to eIFs, some to bacterial IFs, and other aIFs are unique. Many archaeal mRNAs are either leaderless or lack a Shine-Dalgarno sequence. Leaderless mRNAs are translated using an intact 70S ribosome as observed for leaderless bacterial and eukaryal mRNAs. For those archaeal mRNAs that are polycistronic, translation is initiated and proceeds from initiation sites for each coding-sequence carried by the mRNA. Thus each coding-region in the polycistronic mRNA yields a separate protein, as occurs in bacteria.

Translation elongation and termination have been less well studied than translation initiation. However, several archaeal elongation factors have been characterized. They are similar to those in both bacteria and eukaryotes. Thus elongation factors, as opposed to initiation factors, are highly conserved in all cellular organisms. Finally, archaea use a single release factor during translation termination, as do eukaryotes.

Most archaeal proteins are folded using two chaperones called NAC and PFD, both of which are homologues of like-named chaperones in eukaryotes (figure 15.14*b*). Some proteins are folded with the additional help of an Hsp60 chaperone called the thermosome. As you might imagine, archaeal chaperones have been of particular interest to microbiologists because many archaea are thermophilic. One well-studied archaeal chaperone is the thermosome of *Pyrodictium occultum,* an archaeon that grows at temperatures as high as 110°C. This Hsp60-type chaperone, like other HSP60s, hydrolyzes ATP as it helps proteins fold (*see figure 13.43*). The *P. occultum* thermosome hydrolyzes ATP most rapidly at 100°C and makes up almost three-quarters of the cell's soluble protein when *P. occultum* grows at 108°C.

Archaeal protein translocation systems are less well studied than those of bacteria and eukaryotes. Thus far, two protein translocation systems have been identified: the Sec and the Tat systems (*see figure 13.45*). The archaeal Sec system is best studied and presents a clear example of the mixing of bacterial and

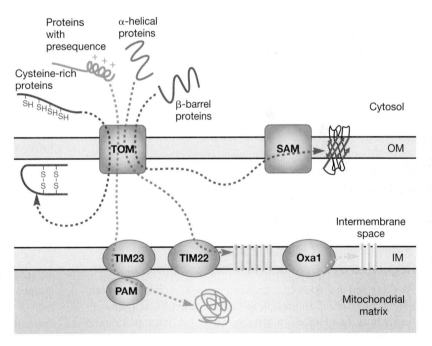

Figure 15.16 Translocation of Proteins from Cytosol to Locations in the Mitochondrion. A simplified diagram of the translocation systems used to localize proteins in mitochondria. The type of protein dictates its final location. TOM moves proteins across the outer membrane (OM) of the mitochondrion. Some of these proteins will be moved across the inner membrane (IM) by the TIM23 system or into the IM by the TIM22 system. Some eukaryotic microbes use TIM23 for both functions. Other proteins translocated across the OM by TOM are inserted into the OM by SAM (*s*orting and *a*ssembly *m*achinery). A complex of proteins similar to SAM exists in Gram-negative bacteria (BAM). Oxa1 inserts proteins made in the mitochondrial matrix into the IM. Oxa1 is related to a bacterial translocase called YidC, which inserts proteins into the plasma membrane.

eukaryotic characteristics (figure 15.15 and table 15.5). In archaea, the Sec translocon is called the SecYEβ translocon because it consists of SecY and Sec E, which have homologues in both bacteria and eukaryotes but are most similar to those of eukaryotes (table 15.5), and Sec61β, which is found in eukaryotic cells. ◄◄ *Protein translocation and secretion in bacteria (section 13.8)*

15.5 Regulation of Cellular Processes

After reading this section, you should be able to:

- List the steps at which eukaryotic and archaeal cells can regulate cellular processes
- Explain why the separation in time and space of transcription and translation in eukaryotic cells is important to the possible regulatory mechanisms used
- Write a paragraph comparing the mechanisms by which eukaryotic and bacterial regulatory proteins control transcription
- Discuss the importance of small RNA molecules in regulating gene expression in eukaryotes

As is the case in bacteria, the regulation of cellular processes in eukaryotes and archaea can occur at transcriptional, translational, and posttranslational levels (**figure 15.17**; and *see figure 14.1*). Much of the research on regulation in eukaryotes and archaea has focused on transcription initiation. That is our focus here.

Transcription Initiation in Eukaryotes Is Often Regulated by Activator and Repressor Proteins

As described in section 15.3, transcription initiation in eukaryotes involves numerous transcription factors called basal transcription factors. In addition, co-activators (e.g., mediator; p. 356) can play important roles, depending on the organism and the gene being transcribed. Regulation of transcription initiation in eukaryotes is often brought about by DNA-binding proteins that can act either positively (i.e., transcription activators that increase transcription levels) or negatively (i.e., transcription repressors that decrease transcription levels).

The binding sites of numerous eukaryotic activator and repressor proteins have been identified. Two different transcription activator binding sites are sometimes distinguished: **upstream activating sequences** (**UASs**) and enhancers. UASs are near the promoter of the gene they regulate. **Enhancers** can be either upstream or downstream of the promoter, and importantly, they can exert their effects from a great distance (**figure 15.18**). Despite the differences between UASs and enhancers, they function in a similar way. Both are binding sites for transcription activators which then interact with other proteins to promote transcription, including mediator and TFIID. Transcription repressors bind DNA sites called **silencers.** Silencers, like enhancers, can regulate transcription even when located far from the gene they regulate. Eukaryotic repressor proteins interact with other proteins, including mediator, to bring about repression. Thus, although mediator is often called a co-activator, it can also function in repression. The existence of regulatory sites that are far removed from the gene being regulated presents a problem: how does the cell prevent an enhancer or silencer from affecting transcription of the wrong gene? Eukaryotic cells have solved this problem by having sites called insulators. **Insulators** are so named because they "insulate" genes by preventing the "spread" of the effects of activators and repressor proteins beyond a certain area of the chromosome.

Regulation of the *GAL* genes in the yeast *Saccharomyces cerevisiae* has served as an important model for regulation of transcription initiation in eukaryotes (**figure 15.19**). These genes encode proteins needed for the uptake and catabolism of the sugar galactose. *GAL* genes are controlled by an activator protein (Gal4) that binds a UAS near each *GAL* gene. The Gal4 protein's ability to activate transcription of *GAL* genes is controlled by another protein (Gal80), which inhibits Gal4 when no galactose is present. When galactose is available, Gal80 is prevented from inhibiting Gal4, and Gal4 activates transcription by recruiting several proteins to the core promoter, including the basal transcription factors. Finally, RNA polymerase II binds, setting the stage for transcription to start. The regulation of yeast *GAL* genes has been an important model because it exemplifies many of the features of eukaryotic regulation. These include chromatin remodeling, chromatin modification, formation of activation complexes, and the indirect influence of activators and repressors on genes.

Chromatin remodeling and chromatin modification deserve additional comment. Both are important in regulation of gene expression because they facilitate changes in chromatin structure such that promoters can become either more or less accessible to the transcription machinery. Both chromatin remodeling and chromatin modification are catalyzed by enzymes that alter nucleosomes (figure 15.19). When **chromatin remodeling** occurs, the histones present in a nucleosome are changed. Recall from chapter 5 that the nucleosome core is composed of DNA wound around two copies each of the major histone proteins H2A, H2B, H3, and H4. In addition to the major histone proteins, which are produced in great abundance, eukaryotic cells also produce small quantities of variant forms of histones H2A, H2B, and H3. These variant forms can replace a major histone, thereby altering chromatin structure. **Chromatin modification** refers to the covalent modification of the major histone proteins present in a nucleosome—often by addition or removal of acetyl, methyl, or phosphoryl groups. Variant histones and covalent modification of histones are used by cells as signals; some indicate that a gene should be expressed, whereas others repress transcription. For instance, acetylation of histones promotes transcription; methylation of histones often inhibits transcription.

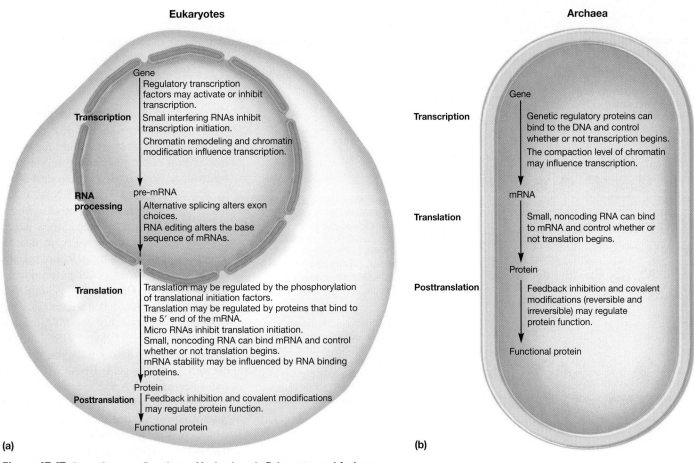

Eukaryotes

Gene
| Regulatory transcription factors may activate or inhibit transcription.

Transcription
Small interfering RNAs inhibit transcription initiation.

Chromatin remodeling and chromatin modification influence transcription.

pre-mRNA

RNA processing
Alternative splicing alters exon choices.

RNA editing alters the base sequence of mRNAs.

Translation
Translation may be regulated by the phosphorylation of translational initiation factors.

Translation may be regulated by proteins that bind to the 5′ end of the mRNA.

Micro RNAs inhibit translation initiation.

Small, noncoding RNA can bind mRNA and control whether or not translation begins.

mRNA stability may be influenced by RNA binding proteins.

Protein

Posttranslation
Feedback inhibition and covalent modifications may regulate protein function.

Functional protein

(a)

Archaea

Gene
Genetic regulatory proteins can bind to the DNA and control whether or not transcription begins.

The compaction level of chromatin may influence transcription.

Transcription

mRNA

Translation
Small, noncoding RNA can bind to mRNA and control whether or not translation begins.

Protein

Posttranslation
Feedback inhibition and covalent modifications (reversible and irreversible) may regulate protein function.

Functional protein

(b)

Figure 15.17 Some Common Regulatory Mechanisms in Eukaryotes and Archaea.

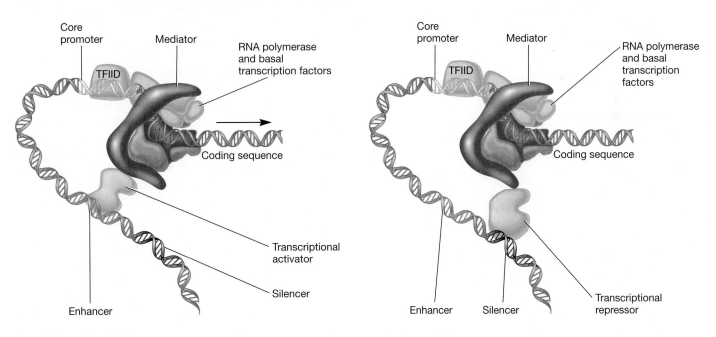

Core promoter   Mediator   RNA polymerase and basal transcription factors
TFIID
Coding sequence
Transcriptional activator
Silencer
Enhancer

The transcriptional activator interacts with mediator. This enables RNA polymerase to form a preinitiation complex that can proceed to the elongation phase of transcription.

Core promoter   Mediator   RNA polymerase and basal transcription factors
TFIID
Coding sequence
Transcriptional repressor
Enhancer   Silencer

The transcriptional repressor interacts with mediator so that transcription is repressed.

Figure 15.18 Enhancers and Silencers Exert Their Effects from a Distance.   Enhancers and silencers are regulatory sites on DNA bound by activator proteins, which increase the rate of transcription, and repressor proteins, which inhibit transcription, respectively. Both sites can be quite far from the genes they regulate. Most models propose that when the regulatory protein binds its recognition site, it subsequently interacts with mediator or a basal transcription factor (e.g., TFIID), causing a DNA loop to form. Evidence exists that when mediator functions in repression, additional mediator subunits called the CDK8 (*c*yclin-*d*ependent *k*inase 8) module are involved.

Gal4 protein is an activator that is part of the regulatory system controlling transcription of *GAL* genes. It recognizes and binds to an upstream activating sequence specific for GAL genes (UAS$_{GAL}$).

In the absence of galactose, the protein Gal80 binds Gal4 and prevents it from activating transcription.

Figure 15.19 Regulation of *GAL* genes in *Saccharomyces cerevisiae*. TBP is TATA-binding protein, one of the basal transcription factors (BTFs) recruited to the promoter when galactose is available.

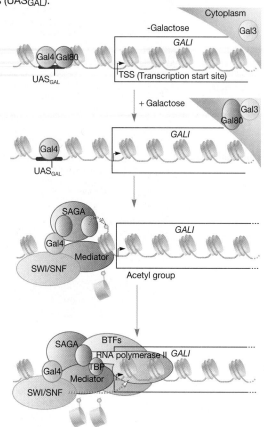

When galactose is available, the protein Gal3 moves into the nucleus and binds Gal80. The Gal80/Gal3 complex moves to the cytoplasm. This allows Gal4 to begin to activate transcription of *GAL* genes.

Gal4 recruits a number of protein complexes, including mediator, SAGA, and SWI/SNF. SAGA transfers an acetyl group to a histone in a nucleosome. This facilitates the removal of the nucleosome by SWI/SNF.

Removal of the nucleosome enables SAGA and mediator to recruit the basal transcription factors (BTFs) and RNA polymerase II to the core promoter.

Eukaryotes Use Several Other Regulatory Mechanisms

In addition to controlling transcription initiation with regulatory proteins, eukaryotes use a variety of other methods to regulate gene expression (figure 15.17*a*). Two-component signal transduction pathways have been identified in a number of eukaryotic microbes. The small nucleotide cAMP serves as a second messenger in eukaryotes, and c-di-GMP signaling has recently been shown to function in the complex process of fruiting body formation by *Dictyostelium discoideum*. Eukaryotes produce cis-encoded small RNAs (sRNAs), often called antisense RNAs, that function at the level of translation. Riboswitches have been identified in some eukaryotes. However, they work by controlling whether or not RNA splicing occurs, rather than affecting transcription elongation or translation. Eukaryotes also produce RNA molecules that are important components of the spliceosome, where they contribute to the selection of splice sites used during mRNA processing (alternative splicing; p. 357). In doing so, different proteins can be made at certain times in the life cycle of the organism by combining different exons. ◀◀ *Riboswitches (section 14.3); Riboswitches and small RNAs can control translation (section 14.4); Mechanisms used for global regulation*

(section 14.5): Supergroup Amoebozoa *includes protists with pseudopodia (section 25.3)*

One of the most interesting regulatory mechanisms observed in eukaryotes is RNA interference (RNAi). RNAi is brought about by two types of very small RNAs: micro RNAs and small interfering RNAs. **Micro RNAs (miRNAs)** are formed when larger, cell-encoded RNA molecules with double-stranded regions are cut by an RNase (usually an RNase called Dicer), yielding a tiny double-stranded miRNA. Once formed, the miRNAs interact with a protein called argonaute (AGO) to form a complex called the *RNA-induced silencing complex* (RISC). RISC surveys mRNAs present in the cell using one strand of the miRNA as the guide RNA. When RISC encounters an mRNA with sequences complementary to the guide RNA, the guide strand base-pairs with a target mRNA. This leads to inhibition of gene expression at the level of translation by several mechanisms. These include preventing translation initiation, inhibiting translation elongation, and degradation of mRNA. RNAi caused by miRNAs has been shown to play important roles in developmental processes observed in plants, animals, fungi, and some protists.

Small interfering RNAs (siRNAs) are formed in a similar manner as miRNAs. The difference lies in their source. Many naturally occurring siRNAs are formed from double-stranded RNAs (dsRNAs) that arise in cells during a viral infection. Once detected by the cell, an RNase (e.g., dicer) cuts the dsRNA, forming the mature siRNA. This can interact with argonaute to form a RISC, with one strand of the siRNA serving as the guide RNA as seen for miRNAs. siRNAs also differ from miRNAs in that they can function at the level of transcription. They do so by interacting with several proteins, including argonaute, to form a complex termed *RNA-induced transcriptional silencing* (RITS) complex. The siRNA directs the RITS complex to a target area on the chromosome by a mechanism that is not yet understood. Once bound to the chromosome, the RITS complex initiates chromatin remodeling that results in a highly compacted form of chromatin called heterochromatin. Heterochromatin is so compacted that any promoters in it cannot be recognized by the basal transcription factors and RNA polymerase. Therefore transcription is prevented.

🔁 *Control of Gene Expression in Eukaryotes*

Regulation of Gene Expression in Archaea

The regulation of gene expression in archaea is of interest because archaeal transcription and translation machinery is most similar to that of eukaryotes, yet it functions in anucleate cells with circular chromosomes. The question being asked is whether archaeal regulation of gene expression is more like bacterial regulation or more like eukaryotic regulation. Unfortunately, studying regulation of gene expression has proven difficult in archaea, and relatively little is known. For instance, evidence exists that some archaea use sRNAs and riboswitches to regulate gene expression, but little is known about how they function. Better success has been obtained for other regulatory methods.

Some archaea use transcriptional regulatory proteins that are bacteria-like in their action. For example, repressor proteins have been identified that attach to operators in a fashion similar to bacterial repressors (see figure 14.3). However, other repressor proteins function in a unique manner: instead of binding an operator, they associate with the TATA box in the promoter, thereby preventing the TATA-binding protein (TBP) from attaching. A few archaeal activator proteins have been studied. They bind near the promoter and recruit transcription factors (e.g., TATA-binding protein) to the promoter, thus facilitating RNA polymerase binding.

One interesting observation is that many archaea contain different forms of transcription factor B (TFB) and TBP (figure 15.11). Evidence suggests that certain combinations of TFB and TBP are used to initiate transcription of distinct subsets of genes and operons in the genome. This is reminiscent of the use of different sigma factors by bacteria to control transcription of suites of genes in response to certain environmental conditions.

Finally, the association of histones with the chromosomes of some archaea opens the possibility that they play a role in regulating transcription. Some evidence suggests that the presence of histones at or near a promoter can inhibit transcription. Furthermore, one transcription activator protein has been shown to counteract this inhibition. One explanation for this observation is that the activator protein competes with histones for promoter binding. Once the activator attaches to the promoter, it allows TFB, TBP, and RNA polymerase to bind and initiate transcription.

Retrieve, Infer, Apply

1. List two similarities and two differences between bacterial and eukaryotic regulatory proteins.
2. What regulatory sequences in bacterial genomes are analogous to the enhancers and silencers observed in eukaryotic genomes?
3. How are cis-encoded RNAs, micro RNAs, and small interfering RNAs alike? How do they differ? What other non-coding RNAs can affect protein synthesis in eukaryotes?

Key Concepts

15.1 Why Consider Eukaryotic and Archaeal Genetics Together?

- Despite having prokaryotic cell architecture, archaea exhibit unique characteristics that distinguish them from bacteria.
- Examination of the machinery involved in processing and expressing archaeal genomes has revealed that it is more like that of eukaryotes than that of bacteria.

15.2 DNA Replication: Similar Overall, but with Different Replisome Proteins

- Eukaryotic DNA has many replicons and replication origins (**figure 15.2**). DNA replication is initiated when certain proteins are recruited to origins of replication by a set of proteins called the origin recognition complex (ORC; **table 15.1**).
- In eukaryotes, three different DNA polymerases are part of the replisome (**figure 15.1**). DNA polymerase α-primase synthesizes an RNA-DNA hybrid molecule that serves as the primer for DNA synthesis. DNA polymerase ε carries out leading strand synthesis, and DNA polymerase δ is responsible for lagging strand synthesis.
- Telomerase is responsible for replicating the ends of eukaryotic chromosomes (**figure 15.4**).
- All known archaeal cells have circular DNA. Their chromosomes are thought to be replicated in a manner similar to circular bacterial chromosomes. However, some archaeal species have more than one origin of replication.
- In archaea, the replisomes of all species examined thus far contain a primase and a DNA polymerase (Pol B) similar to those in eukaryotic replisomes (**figure 15.5**). Some

archaea have a second DNA polymerase (Pol D) that is unique to archaea (**table 15.2**).

15.3 Transcription

- Most eukaryotic protein-coding genes are not organized into operons. Therefore the mRNAs that arise from them are monocistronic.
- Three RNA polymerases transcribe genes in eukaryotes. RNA polymerase II transcribes protein-coding genes. RNA polymerase I transcribes rRNA genes, and RNA polymerase III transcribes the 5S rRNA and all tRNA genes. All three eukaryotic RNA polymerases have more subunits than bacterial RNA polymerases (**table 15.3**).
- The RNA polymerase II core promoters of eukaryotes are distinctive, as is the mechanism of transcription initiation (**figure 15.6**). Several basal transcription factors are involved in recruiting RNA polymerase II to a promoter (**figure 15.7**).
- RNA polymerase II synthesizes pre-mRNA, which is modified by addition of a 3′ poly-A sequence and a 5′ cap (**figure 15.8**) and by removal of introns by spliceosomes (**figure 15.9**).
- Many archaeal genes are organized into operons (though perhaps not as many as in bacteria). The single archaeal RNA polymerase is similar to eukaryotic RNA polymerase II. The archaeal RNA polymerase requires two transcription factors to be positioned properly at a promoter (**figure 15.11**). Archaeal promoters have elements that are similar to those in the RNA polymerase II core promoters of eukaryotes, including a TATA box (**figures 15.6** and **15.11**).

15.4 Translation and Protein Maturation and Localization

- Translation initiation in eukaryotes requires more initiation factors than it does in bacteria (**figure 15.12**). Initiation involves an interaction of the 5′ and 3′ ends of the mRNA, followed by binding of the 43S complex (SSU of the ribosome and initiator tRNA) to the 5′ end of the message. The 43S complex then scans the mRNA until it encounters a start codon. Translation elongation is similar in eukaryotes and bacteria. Translation termination requires a single release factor, rather than the three used in bacteria.

- Protein folding in eukaryotes is aided by chaperone proteins (**figure 15.14**) and can occur cotranslationally or posttranslationally. Homologues of all types of chaperones are found in eukaryotes.

- In eukaryotes, proteins are translocated into the endoplasmic reticulum (ER) membrane by a Sec system (**figure 15.15**). The proteins are then trafficked between the ER, plasma membrane, Golgi, and other components of the endocytic and secretory pathways by vesicular transport. The Sec system is also used to move some proteins from the chloroplast stroma to the thylakoid membrane or the thylakoid lumen. The Tat system functions in some eukaryotes to translocate proteins into mitochondria and chloroplasts. In addition, mitochondrial proteins synthesized in the cytosol are translocated into mitochondria by the TOM and TIM systems (**figure 15.16**).

- In archaea, transcription and translation are coupled, as in bacteria. Some archaeal mRNAs have a Shine-Dalgarno sequence that is thought to function in positioning the ribosome as it does in bacteria. However, other archaea lack a Shine-Dalgarno sequence or are leaderless. It is thought that intact 70S ribosomes initiate translation of these mRNAs. Other aspects of archaeal translation are similar to those seen in eukaryotes, including the protein factors and initiator tRNA used.

- Archaea use chaperones to help fold proteins. They resemble those of eukaryotes.

- Archaea use the Sec and Tat pathways to translocate proteins into or across the plasma membrane. The archaeal Sec system proteins are similar to the eukaryotic Sec system proteins (**figure 15.15**).

15.5 Regulation of Cellular Processes

- Eukaryotes have regulatory sites for the binding of transcription regulatory proteins. Activators bind either enhancers or upstream activating sites, and repressors bind sites called silencers (**figure 15.18**). The regulatory proteins exert their actions indirectly by affecting the activity of transcription factors, chromatin-modifying enzymes, and chromatin-remodeling enzymes (**figure 15.19**).

- Micro RNAs are used by eukaryotic cells to regulate translation. Micro RNAs are part of a complex that targets a particular mRNA and inhibits its translation.

- Small interfering RNAs are also used to control translation, in a manner similar to that used by micro RNAs. However, some small interfering RNAs regulate transcription. This is accomplished by causing the formation of heterochromatin, a highly compacted form of chromatin, in the region targeted by the small interfering RNA.

- The mechanism of action of archaeal proteins that regulate transcription initiation is best understood for repressor proteins. Some repressor proteins bind an operator sequence and are thought to work in a manner similar to that seen in bacteria. Other repressor proteins inhibit transcription by localizing to the TATA box, thereby preventing the archaeal transcription factors from attaching to the DNA and initiating transcription.

Compare, Hypothesize, Invent

1. Budding yeasts such as *S. cerevisiae* exhibit telomerase activity throughout their life cycles, whereas human somatic cells do not exhibit telomerase activity. Suggest why this is so.

2. All of the subunits in bacterial RNA polymerases have homologues in both archaeal and eukaryotic RNA polymerases. What does this suggest about the evolution of these enzymes?

3. Would you expect that one day microbiologists might discover small interfering RNAs in archaea? If so, how might they function? Explain your answer.

4. In the chapter opening story, it was stated that the next step in using algae to produce plastics is to maximize expression of the bacterial polyhydroxyalkanoate-synthesizing gene in algal cells. Suggest one genetic change that might be made to increase its expression.

5. The yeast *S. cerevisiae* continues to be an important model organism for understanding regulation of gene expression. In 2011 scientists from the University of Wisconsin-Madison reported their studies to determine the mechanism of action of a yeast repressor protein called PUF. In previous studies, it had been shown that PUF binds elements in the 3′ untranslated region (UTR) of mRNA molecules. This observation raised the possibility that PUF might interact with poly-A binding proteins (PABPs) to exert its effect. The scientists examined translation in three different experimental conditions: translation of mRNAs that lacked poly-A tails, translation in mutants that lacked PABP, and translation in mutant forms of PABP. What information would be provided by examining translation in each of the experimental conditions used? Suggest a mechanism by which PUF interaction with PABP might repress translation.

Read the original paper: Chritton, J. J., and Wickens, M. 2011. A role for the poly(A)-binding protein Pab1p in PUF protein-mediated repression. *J. Biol. Chem.* 286(38):33268.

16

Mechanisms of Genetic Variation

Manure represents a loss in production unless it is used in other ways. It is common to use raw or composted manure as fertilizer for crops.

Manure Happens

Consider this: a milk-producing dairy cow generates on average 150 pounds (68 kilograms) of manure each day and a beef cow will have produced about 9,800 pounds (4,445 kilograms) of manure during the finishing process that prepares it for slaughter. Multiply this by the number of dairy cows and beef cattle around the world and that's a lot of manure!

This much manure creates many problems: odor, release of volatile compounds into the atmosphere, water pollution, and the possible dissemination of pathogens. The latter is of considerable concern because it is coupled with another problem: the potential spread of antibiotic-resistant bacteria, pathogens and nonpathogens alike. Antibiotic resistance (AR) has been increasing since the beginning of antibiotic use. Currently, more than 30,000 people in Europe and the United States die each year from infections caused by microbes resistant to antimicrobial agents.

You may be wondering why there might be antibiotic-resistant bacteria in manure. The answer is simple. Each year food animals are exposed to antibiotics either through therapeutic use or at lower amounts (subtherapeutic levels) in their feed. These antibiotics are a selective pressure that acts on the resident microbiota of the animal, leading to the evolution of antibiotic-resistant bacteria. These bacteria pass their AR genes on to their progeny. But more worrisome is their ability to share AR genes with other, unrelated bacteria in their habitat by horizontal gene transfer (HGT). Bacteria use HGT to create genetic variability within a species, which in turn helps at least some members of a species survive when encountering new environmental conditions (e.g., exposure to an antibiotic).

The therapeutic use of antibiotics to cure infection is understandable, but what may be less obvious is why food animals are given subtherapeutic levels of antibiotics in their feed. Indeed, many argue the practice should be stopped because of the risk of selecting for antibiotic-resistant bacteria. However, considerable evidence shows there is a clear benefit to the practice. The major advantage is that animal feed is used more efficiently. When feed is used more efficiently, more meat or milk is produced from each animal. Therefore fewer animals are needed, less animal feed needs to be raised, more cropland is available for producing human food, and less manure is generated. This translates to greater profits for the farmer or rancher and lower costs for the consumer.

Do the benefits of subtherapeutic antibiotic use outweigh the risk of spreading AR genes from manure to other organisms in the habitats where the manure is stored, handled, or dispensed? This important question is difficult to answer, in part because insufficient information is available regarding the amount and type of antibiotic-resistant bacteria and their AR genes in natural habitats. Importantly, this inquiry also encompasses many other questions. Are there antibiotic-resistant bacteria in manure? If there are, can these bacteria contaminate crop plants and enter the food chain if manure is used as a fertilizer? Can AR genes be transferred from manure bacteria to soil or aquatic bacteria and enter the food chain or contaminate drinking water sources when manure is used as a fertilizer? Finally, even if AR genes are transferred from manure to other microbes, are they likely to end up in organisms of clinical importance?

Studies using molecular approaches such as those described in chapter 18 have clearly shown the presence of AR genes in manure. Many of the genes are only distantly related to clinically relevant AR genes, and some define an entirely new family of AR genes. This is either good news or bad news. It could be that most AR genes present in manure are not spreading to humans and causing disease. However, it could also mean that the future looks grim, with manure being a reservoir of many novel AR genes waiting to make their way into clinically important bacteria. Further complicating the issue is a recent report that "pristine" manure, that is, manure from cows never exposed to antibiotics, increases the number of soil bacteria with AR genes encoding β-lactamases (penicillin-destroying enzymes). Why this occurs is not clear. But what is clear is that bacteria are very adept at altering their genomes, thereby generating more genetic diversity in a population.

The increase of antibiotic resistance in bacteria is only one example of microbial evolution. All bacteria and archaea use a variety of mechanisms to help ensure their survival and to evolve into new species. HGT among and between bacteria and archaea plays a particularly important role. As more and more bacterial and archaeal genomes have been sequenced and annotated, it has become increasingly clear that HGT has occurred many times. Furthermore, genes have also been transferred from bacteria and archaea to eukaryotes and from eukaryotes to bacteria and archaea.

369

How the genomes of microbes are altered to increase genetic diversity is the topic of this chapter. Our main focus is on how this is accomplished in the absence of sexual reproduction. Some of the mechanisms we discuss can create changes in the genome that are actually detrimental. Thus we also consider processes that help balance the stability of the genome with the capacity to introduce genetic variation.

Readiness Check:

Based on what you have learned previously, you should be able to:

✔ Draw a simple diagram that illustrates the three chemical moieties that define a nucleotide (section 13.2)

✔ State the base-pairing rules

✔ Explain the importance of the reading frame of a protein-coding gene (section 13.6)

✔ Describe the major events of meiosis

✔ Describe the major types of plasmids (section 3.6)

✔ Create a concept map or table that distinguishes the secretion systems observed in Gram-negative bacteria (section 13.8)

16.1 Mutations: Heritable Changes in a Genome

After reading this section, you should be able to:

■ Distinguish spontaneous from induced mutations, and list the most common ways each arises

■ Construct a table, concept map, or picture to summarize how base analogues, DNA-modifying agents, and intercalating agents cause mutations

■ Discuss the possible effects of mutations

Perhaps the most obvious way genetic diversity can be created is by **mutations** (Latin *mutare,* to change), which, in cells, are heritable changes in DNA sequence. Several types of mutations exist. Some arise from the alteration of single pairs of nucleotides and from the addition or deletion of one nucleotide pair in the coding regions of a gene. Such small changes in DNA are called **point mutations** because they affect only one base pair in a given location. Larger mutations are less common. These include large insertions, deletions, inversions, duplications, and translocations of nucleotide sequences.

Mutations occur in one of two ways. (1) **Spontaneous mutations** arise occasionally in all cells and in the absence of any added agent. (2) **Induced mutations** are the result of exposure to a **mutagen,** which can be either a physical or a chemical agent. Mutations are characterized according to either the kind of genotypic change that has occurred or their phenotypic consequences. In this section, the molecular basis of mutations and mutagenesis is first considered. Then the phenotypic effects of mutations are discussed.

Spontaneous Mutations

Spontaneous mutations result from errors in DNA replication, spontaneously occurring lesions in DNA, or the action of mobile genetic elements such as transposons (section 16.5). A few of the more prevalent mechanisms are described here.

Replication errors can occur when the nitrogenous base of a nucleotide shifts to a different form (isomer) called a tautomeric form. Nitrogenous bases typically exist in the keto form (*see figure 13.5*) but are in equilibrium with the imino and enol forms (**figure 16.1a**). The shift from one form to another

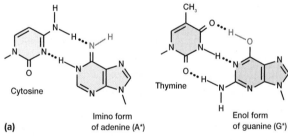

(a)

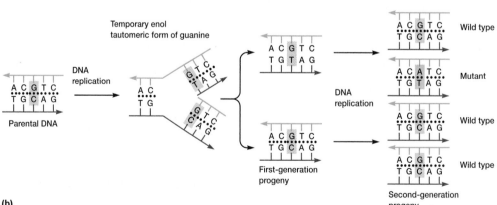

(b)

Figure 16.1 Tautomerization and Transition Mutations.   Errors in replication due to base tautomerization. (a) Normally AT and GC pairs are formed when keto groups participate in hydrogen bonds. In contrast, enol tautomers produce AC and GT base pairs. The alteration in the base is shown in blue. (b) Mutation as a consequence of tautomerization during DNA replication. The temporary enolization of guanine leads to the formation of a GT base pair in one of the first-generation progeny molecules. Thus, the pyrimidine cytosine is replaced by the pyrimidine thymine. If this abnormal pairing is not detected by repair mechanisms, then when that progeny DNA molecule is replicated, one of the second-generation progeny molecules will contain an AT base pair and a GC–to–AT transition mutation will have occurred. Note that the process requires two replication cycles. Wild type is the form of the gene before mutation occurred.

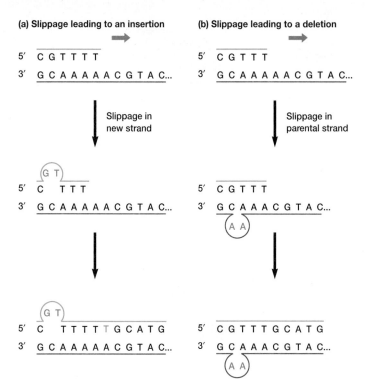

(a) Slippage leading to an insertion

(b) Slippage leading to a deletion

Figure 16.2 Insertions and Deletions. A mechanism for the generation of insertions and deletions during replication. The direction of replication is indicated by the blue arrow. In each case, there is strand slippage resulting in the formation of a small loop that is stabilized by hydrogen bonding in the repetitive sequence, the AT stretch in this example. (a) If the new strand slips, an addition of one T results. (b) Slippage of the parental strand yields a deletion (in this case, a loss of two Ts).

changes the hydrogen-bonding characteristics of the bases, allowing purine for purine (e.g., an A for a G) or pyrimidine for pyrimidine (e.g., a T for a C) substitutions that can eventually lead to a stable alteration of the nucleotide sequence (figure 16.1b). Such substitutions are known as **transition mutations** and are relatively common. On the other hand, **transversion mutations**—mutations where a purine is substituted for a pyrimidine (e.g., an A for a T) or a pyrimidine for a purine—are rarer due to the steric problems of pairing purines with purines and pyrimidines with pyrimidines.

Replication errors can also result in the insertion and deletion of nucleotides. These mutations generally occur where there is a short stretch of repeated nucleotides. In such a location, the pairing of template and new strand can be displaced, leading to insertions or deletions of bases in the new strand (**figure 16.2**).

Spontaneous mutations can originate from lesions in DNA as well as from replication errors. For example, it is possible for purine nucleotides (A and G) to be depurinated; that is, to lose their base. This results in the formation of an apurinic site, which does not base pair normally and may cause a mutation after the next round of replication (**figure 16.3**). Likewise, pyrimidine bases (T and C) can be lost, forming an apyrimidinic site. Other lesions are caused by reactive forms of oxygen such as oxygen free radicals and peroxides produced during aerobic metabolism. For example, guanine can be converted to 8-oxo-7,8-dihydrodeoxyguanine, which often pairs with adenine, rather than cytosine, during replication.

Induced Mutations Are Caused by Mutagens

Any agent that damages DNA, alters its chemistry, or in some way interferes with its functioning will probably induce mutations. Mutagens can be conveniently classified according to their mode of action. Three common types of chemical mutagens are base analogues, DNA-modifying agents, and intercalating agents. A number of physical agents (e.g., radiation) are mutagens that damage DNA.

Base analogues are structurally similar to normal nitrogenous bases and can be incorporated into the growing polynucleotide chain during replication (**table 16.1**). Once in place, these compounds typically exhibit base-pairing properties different from the bases they replace and can eventually cause a stable mutation. A widely used base analogue is 5-bromouracil, an analog of thymine. It undergoes a tautomeric shift from the normal keto form to an enol much more frequently than does a normal

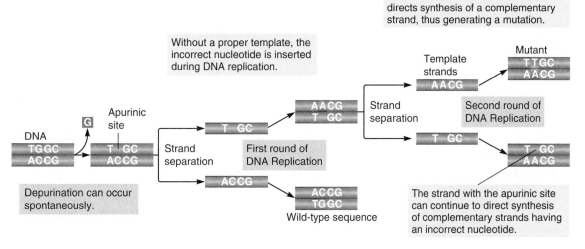

Figure 16.3 Mutation Due to Formation of an Apurinic Site. Apurinic sites can arise spontaneously (depurination). They do not provide a template for DNA replication, and as a result, an incorrect nucleotide can be inserted. A similar process can introduce mutations at sites where apyrimidinic sites arise.

Table 16.1	Examples of Mutagens
Mutagen	**Effect(s) on DNA Structure**
Chemical	
5-Bromouracil	Base analogue
2-Aminopurine	Base analogue
Ethyl methanesulfonate	Alkylating agent that adds an ethyl group to guanine
Hydroxylamine	Hydroxylates cytosine
Nitrogen mustard	Alkylating agent that adds a methyl group to guanine
Nitrous oxide	Deaminates bases
Proflavin	Intercalating agent
Acridine orange	Intercalating agent
Physical	
UV light	Promotes pyrimidine dimer formation
X rays	Cause base deletions, single-strand nicks, cross-linking, and chromosomal breaks

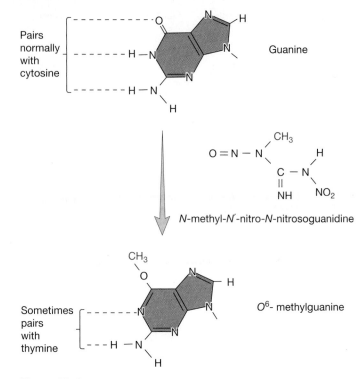

Figure 16.4 Methyl-Nitrosoguanidine Mutagenesis. Methyl-nitrosoguanidine methylates guanine.

base. The enol tautomer forms hydrogen bonds like cytosine, pairing with guanine rather than adenine. The mechanism of action of other base analogues is similar to that of 5-bromouracil.

There are many **DNA-modifying agents**—mutagens that change a base's structure and therefore alter its base-pairing specificity. Some of these mutagens preferentially react with certain bases and produce a particular kind of DNA damage. For example, methyl-nitrosoguanidine adds methyl groups to guanine, causing it to mispair with thymine (**figure 16.4**). A subsequent round of replication can then result in a GC-AT transition.

Hydroxylamine is another example of a DNA-modifying agent. It attaches a hydroxyl group (i.e., hydroxylates) to cytosine causing it to base pair like thymine.

Intercalating agents distort DNA to induce single nucleotide pair insertions and deletions. These mutagens are planar and insert themselves (intercalate) between the stacked bases of the helix. This results in a mutation, possibly through the formation of a loop in DNA.

Many mutagens, and indeed many carcinogens, damage bases so severely that hydrogen bonding between base pairs is impaired or prevented and the damaged DNA can no longer act as a template for replication. For instance, ultraviolet (UV) radiation often generates thymine dimers between adjacent thymines (**figure 16.5**). Other examples are ionizing radiation and carcinogens such as the fungal toxin aflatoxin B1.

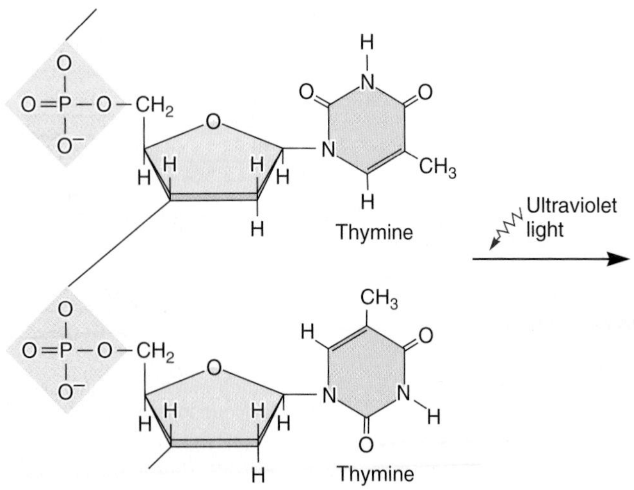

Figure 16.5 Thymine Dimers in a DNA Strand Are Formed by Ultraviolet Radiation. Thymine dimers are one type of cyclobutane dimer (i.e., a dimer formed between adjacent pyrimidines) generated by UV radiation.

Effects of Mutations

The effects of mutations can be described at the protein level and in terms of observed phenotypes. In all cases, the impact is readily noticed only if it produces a change in phenotype. In general, the more prevalent form of a gene and its associated phenotype is called the **wild type.** A mutation from wild type to a mutant form is a **forward mutation (table 16.2).** It is possible to restore a wild-type phenotype with a second mutation. If the second mutation is at the same site as the original mutation (e.g., the same base pair in a codon), it is called a **reversion mutation.**

Table 16.2	Types of Point Mutations	
Type of Mutation	**Change in DNA**	**Example**
Forward Mutations		
None	None	5′-A-T-G-A-C-C-T-C-C-C-C-G-A-A-A-G-G-G-3′ Met - Thr - Ser - Pro - Lys - Gly
Silent	Base substitution	5′-A-T-G-A-C-A-T-C-C-C-C-G-A-A-A-G-G-G-3′ Met - Thr - Ser - Pro - Lys - Gly
Missense	Base substitution	5′-A-T-G-A-C-C-T-G-C-C-C-G-A-A-A-G-G-G-3′ Met - Thr - Cys - Pro - Lys - Gly
Nonsense		5′-A-T-G-A-C-C-T-C-C-C-C-G-T-A-A-G-G-G-3′ Met - Thr - Ser - Pro - STOP!
Frameshift	Insertion/deletion	5′-A-T-G-A-C-C-T-C-C-G-C-C-G-A-A-A-G-G-G-3′ Met - Thr - Ser - Ala - Glu - Arg
Reverse Mutations		
	Base substitution	5′-A-T-G-A-C-C-T-C-C $\xrightarrow{\text{forward}}$ A-T-G-C-C-C-T-C-C $\xrightarrow{\text{reverse}}$ A-T-G-A-C-C-T-C-C Met - Thr - Ser Met - Pro - Ser Met - Thr - Ser
	Base substitution	5′-A-T-G-A-C-C-T-C-C $\xrightarrow{\text{forward}}$ A-T-G-A-C-C-T-G-C $\xrightarrow{\text{reverse}}$ A-T-G-A-C-C-A-G-C Met - Thr - Ser Met - Thr - Cys Met - Thr - Ser
	Base substitution	5′-A-T-G-A-C-C-T-C-C $\xrightarrow{\text{forward}}$ A-T-G-C-C-C-T-C-C $\xrightarrow{\text{reverse}}$ A-T-G-C-T-C-T-C-C Met - Thr - Ser Met - Pro - Ser Met - Leu - Ser (polar amino acid) (nonpolar amino acid) (polar amino acid) pseudo-wild type
Suppressor Mutations		

Frameshift of opposite sign (intragenic suppressor)	Insertion/deletion	5′-A-T-G-A-C-C-T-C-C-C-C-G-A-A-A-G-G-G-3′
		Met - Thr - Ser - Pro - Lys - Gly
		
		↓ Forward
		  mutation
		5′-A-T-G-A-C-C-T-C-C-G-C-C-G-A-A-A-G-G-G-3′
		Met - Thr - Ser - Ala - Glu - Arg
		
		↓ Suppressor
		  mutation (deletion)
		5′-A-T-G-A-C-C-C-C-G-C-C-G-A-A-A-G-G-G-3′
		Met - Thr - Pro - Pro - Lys - Gly

Extragenic suppressor Nonsense suppressor		Gene (e.g., for tyrosine tRNA) undergoes a mutation in its anticodon region that enables it to recognize and align with a nonsense mutation (e.g., UAG). Thus an amino acid (tyrosine) is inserted at the mutant stop codon and translation continues.
Physiological suppressor		A defect in one chemical pathway is circumvented by another mutation; for example, one that opens up another chemical pathway to the same product or one that permits more efficient uptake of a compound produced in small quantities because of the original mutation.

Some reversion mutations re-establish the original wild-type sequence. Others create a new codon that codes for the same amino acid. Reversions can also restore the wild-type phenotype by creating a codon that replaces the wild-type amino acid with a similar amino acid (e.g., both amino acids are nonpolar). If the wild-type phenotype is restored by a second mutation at a different site than the original mutation, it is called a **suppressor mutation.** Suppressor mutations may be within the same gene (intragenic suppressor mutation) or in a different gene (extragenic suppressor mutation). Because point mutations are the most common types of mutations, their effects are our focus.

Mutations in Protein-Coding Genes

Point mutations in protein-coding genes can affect protein structure in a variety of ways. Point mutations are named according to if and how they change the encoded protein. The most common types of point mutations are silent mutations, missense mutations, nonsense mutations, and frameshift mutations. Examples of each are shown in table 16.2.

Silent mutations change the nucleotide sequence of a codon but do not change the amino acid encoded by that codon. This is possible because the genetic code exhibits degeneracy. Therefore when there is more than one codon for a given amino acid, a single base substitution may result in the formation of a new codon for the same amino acid. For example, if the codon CGU were changed to CGC, it would still code for arginine, even though a mutation had occurred. When there is no change in the protein, there is no change in the phenotype of the organism. ◄◄ *The genetic code consists of three-letter "words" (section 13.6)*

Missense mutations involve a single base substitution that changes a codon for one amino acid into a codon for another. For example, the codon GAG, which specifies glutamic acid, could be changed to GUG, which codes for valine. The effects of missense mutations vary. They alter the primary structure of a protein, but the effect of this change may range from complete loss of activity to no change at all. This is because the effect of missense mutations on protein function depends on the type and location of the amino acid substitution. For instance, replacement of a nonpolar amino acid in the protein's interior with a polar amino acid can drastically alter the protein's three-dimensional structure and therefore its function. Similarly the replacement of a critical amino acid at the active site of an enzyme often destroys its activity. However, the replacement of one polar amino acid with another at the protein surface may have little or no effect. Such mutations are called **neutral mutations.** Missense mutations play a very important role in providing new variability to drive evolution because they often are not lethal and therefore remain in the gene pool. ►►| *Proteins (appendix I)* ↻ *Mutation by Base Substitution*

Nonsense mutations convert a sense codon (i.e., one that codes for an amino acid) to a nonsense codon (i.e., a stop codon: one that does not code for an amino acid). This causes the early termination of translation and therefore results in a shortened polypeptide. Depending on the location of the mutation, the phenotype may be more or less severely affected. Most proteins retain some

function if they are shortened by only one or two amino acids; complete loss of normal function usually results if the mutation occurs closer to the beginning or middle of the gene.

Frameshift mutations arise from the insertion or deletion of base pairs within the coding region of the gene. Since the code consists of a precise sequence of triplet codons, the addition or deletion of fewer than three base pairs causes the reading frame to be shifted for all subsequent codons downstream. Frameshift mutations usually are very deleterious and yield mutant phenotypes resulting from the synthesis of nonfunctional proteins. In addition, frameshift mutations often produce a stop codon so that the peptide product is shorter as well as different in sequence. Of course, if the frameshift occurs near the end of the gene or if there is a second frameshift shortly downstream from the first that restores the reading frame, the phenotypic effect might not be as drastic. A second, nearby frameshift that restores the proper reading frame is an example of an intragenic suppressor mutation (table 16.2). ↻ *Addition and Deletion Mutations*

Changes in protein structure can alter the phenotype of an organism in many ways. They may change the microorganism's colonial or cellular morphology. Lethal mutations, when expressed, result in the death of the microorganism. Because a microbe must be able to grow to be isolated and studied, lethal mutations are recovered only if they are conditional mutations. **Conditional mutations** are those that are expressed only under certain environmental conditions. For example, a conditional lethal mutation might not be expressed when a bacterium is cultured at a low temperature but would be expressed at a high temperature. Thus the mutant would grow normally at cooler temperatures but would die at high temperatures.

Other common mutations inactivate a biosynthetic pathway, frequently eliminating the capacity of the mutant to make an essential molecule such as an amino acid or nucleotide. A strain bearing such a mutation has a conditional phenotype: it is unable to grow on medium lacking that molecule but grows when the molecule is provided. Such mutants are called **auxotrophs,** and they are said to be auxotrophic for the molecule they cannot synthesize. The wild-type strain from which the mutant arose is called a **prototroph.** Another interesting mutant is the resistance mutant. These mutants have acquired resistance to some pathogen (e.g., bacteriophage), chemical (e.g., antibiotic), or physical agent. Auxotrophic and resistance mutants are quite important in microbial genetics due to the ease of their detection and their relative abundance.

Mutations in Regulatory Sequences

Some of the most interesting and informative mutations studied by microbial geneticists are those that occur in the regulatory sequences responsible for controlling gene expression. Constitutive lactose operon mutants in *E. coli* are excellent examples. Many of these mutations map in the operator site and produce altered operator sequences that are not recognized by the *lac* repressor protein. Therefore the operon is continuously transcribed, and β-galactosidase is always synthesized. Mutations in promoters also have been identified. If the mutation renders the promoter sequence nonfunctional, the mutant will be unable to synthesize the product, even though the coding region of the structural gene is completely normal. Without a fully functional promoter, RNA polymerase cannot transcribe a gene as well as wild type. ◄◄ *Regulation of transcription initiation saves considerable energy and materials (section 14.2)*

Mutations in tRNA and rRNA Genes

Mutations in tRNA and rRNA alter the phenotype of an organism through disruption of protein synthesis. In fact, these mutants often are initially identified because of their slow growth. On the other hand, a suppressor mutation involving tRNA restores normal (or near normal) growth rates. In these mutations, a base substitution in the anticodon region of a tRNA allows the insertion of the correct amino acid at a mutant codon (table 16.2).

Retrieve, Infer, Apply

1. List three ways in which spontaneous mutations might arise.
2. Compare and contrast the means by which the mutagens 5-bromouracil, methyl-nitrosoguanidine, proflavin, and UV radiation induce mutations.
3. Give examples of intragenic and extragenic suppressor mutations.
4. Sometimes a point mutation does not change the phenotype. List all the reasons why this is so.
5. Why might a missense mutation at a protein's surface not affect the phenotype of an organism, whereas the substitution of an internal amino acid does?

16.2 Detection and Isolation of Mutants

After reading this section, you should be able to:

- Differentiate mutant detection from mutant selection
- Design an experiment to isolate mutant bacteria that are threonine auxotrophs
- Propose an experiment to isolate revertants of a threonine auxotroph and predict the types of mutations that might lead to the revertant phenotype
- Explain how the Ames test is used to screen for potential carcinogens and evaluate its effectiveness

Mutations often arise spontaneously and provide genetic diversity, which enhances survival during changing environmental conditions; thus mutations are of value to microbes. Mutations are also of practical importance to microbial geneticists. Mutant strains have been used to reveal mechanisms of complex processes such as DNA replication, endospore formation, and regulation of transcription. They are also useful as selective markers in recombinant DNA procedures. ►►I *Recombinant DNA technology (chapter 17)*

To study microbial mutants, they must be readily detected, even when they are rare, and then efficiently isolated from wild-type organisms and other mutants that are not of interest. Microbial geneticists typically increase the likelihood of obtaining mutants by using mutagens to increase the rate of mutation. The rate can increase from the usual one mutant per 10^7 to 10^{11} cells to about one per 10^3 to 10^6 cells. Even at this rate, carefully devised means for detecting or selecting a desired mutation must be used. This section describes some techniques used in mutant detection, selection, and isolation.

Mutant Detection: Screening for Mutants with an Observable Phenotype

When collecting mutants of a particular organism, the wild-type characteristics must be known so that an altered phenotype can be recognized. A suitable detection system for the mutant phenotype also is needed. The use of detection systems is called screening. Screening for mutant phenotypes in haploid organisms is straightforward because the effects of most mutations can be seen immediately as long as the mutated gene is expressed under the growth conditions used. Some screening procedures require only examination of colony morphology. For instance, if colorless-colony mutants of a bacterium that normally produces red colonies are sought, detection simply requires visual observation of colony color. Other screening methods are more complex. For example, the **replica plating** technique is used to screen for auxotrophic mutants. It distinguishes between mutants and the wild-type strain based on their ability to grow in the absence of a particular biosynthetic end product (**figure 16.6**). A lysine auxotroph, for instance, grows on lysine-supplemented media but when the colonies that grow on this medium are transferred to a medium lacking lysine, they will not grow, because they cannot synthesize this amino acid.

Once a screening method is devised, mutants are collected. However, mutant collection can present practical problems. Consider a search for the colorless-colony mutants mentioned previously. If the mutation rate were around one in a million, on average a million or more organisms would have to be tested to find one colorless-colony mutant. This probably would require several thousand plates. The task of isolating auxotrophic mutants in this way would be even more taxing with the added labor of replica plating. Thus, if possible, it is more efficient to use a selection system employing some environmental factor to inhibit the growth of wild-type microorganisms. Examples of selection systems are described next.

Mutant Selection: Using Conditions to Inhibit Growth of Wild-Type Organisms

Mutant selection techniques use incubation conditions under which the mutant grows because of properties conferred by the mutation, whereas the wild type does not. Selection methods often involve reversion or suppressor mutations that restore the wild-type phenotype. For example, if the intent is to isolate a prototroph from a lysine auxotroph (Lys⁻), a large population of lysine auxotrophs is plated on minimal medium lacking lysine, incubated, and examined for colony formation. Only cells that

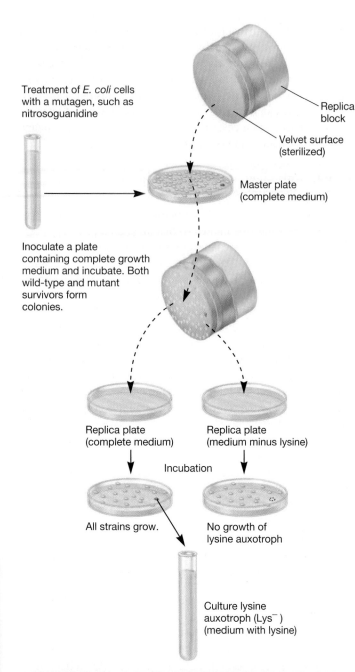

Treatment of *E. coli* cells with a mutagen, such as nitrosoguanidine

Replica block

Velvet surface (sterilized)

Master plate (complete medium)

Inoculate a plate containing complete growth medium and incubate. Both wild-type and mutant survivors form colonies.

Replica plate (complete medium)

Replica plate (medium minus lysine)

Incubation

All strains grow.

No growth of lysine auxotroph

Culture lysine auxotroph (Lys⁻) (medium with lysine)

Figure 16.6 Replica Plating. The use of replica plating to isolate a lysine auxotroph. After growth of a mutagenized culture on a complete medium, a piece of sterile velvet is pressed on the plate surface to pick up bacteria from each colony. Then the velvet is pressed to the surface of other plates, and organisms are transferred to the same position as on the master plate. After the location of Lys⁻ colonies growing on the replica with complete medium is determined, the auxotrophs can be isolated and cultured.

MICRO INQUIRY *How would you screen for a tryptophan auxotroph? How would you select for a mutant that is resistant to the antibiotic ampicillin but sensitive to tetracycline (assume the parental stain is resistant to both antibiotics)?*

have mutated to restore the ability to manufacture lysine (i.e., prototrophy) will grow on minimal medium. Several million cells can be plated on a single Petri dish, but only the rare revertant cells will grow. Thus many cells can be tested for mutations by scanning a few Petri dishes for growth.

Other selection methods are used to identify mutants resistant to a particular environmental stress such as virus attack, antibiotic treatment, or specific temperatures. Therefore it is possible to grow the microbe in the presence of the stress and look for surviving organisms. Consider the example of an antibiotic-sensitive wild-type bacterium. When it is cultured in medium lacking the antibiotic and then plated on selective medium containing the antibiotic, any colonies that form are resistant to the antibiotic and carry a mutated gene that confers antibiotic resistance.

Mutant screening and selection methods are used for purposes other than understanding more about the nature of genes or the biochemistry of a particular microorganism. One very important role of mutant selection and screening techniques is in the study of carcinogens. One of the first and perhaps best known of the carcinogen testing systems is the Ames Test, named after Bruce Ames. The Ames test is based on the observation that many carcinogens also are mutagens. The test determines if a substance increases the rate of mutation; that is, if it is a mutagen. If the substance is a mutagen, then it is also likely to be carcinogenic if an animal is exposed to it at sufficient levels. Note that the test does not directly test for carcinogenicity. This is because it uses a bacterium as the test organism. Carcinogenicity can only be directly demonstrated with animals, and testing is extremely expensive and takes much longer to complete than does the Ames test. Thus the Ames test serves as an inexpensive procedure to identify chemicals that may be carcinogenic and thus deserve further testing.

The Ames test is called a mutational reversion assay because it looks for histidine prototrophs that arise from histidine auxotrophs. The assay employs several "tester" strains of *Salmonella enterica* serovar Typhimurium. Each tester strain has a different mutation in the histidine biosynthesis operon. They also have mutations that make their cell walls more permeable to test substances and impair their ability to repair DNA.

In the Ames test, tester strains of *S.* Typhimurium are plated with the substance being tested and the number of visible colonies that form are determined (**figure 16.7**). To ensure that DNA replication can take place and allow for the development of a mutation (figure 16.1) in the presence of the potential mutagen, the bacteria and test substance are mixed in dilute molten top agar to which a trace of histidine has been added. This molten mix is then poured on top of minimal agar plates and incubated for 2 to 3 days at 37°C. All of the histidine auxotrophs grow for the first few hours in the presence of the test compound until the histidine is depleted. This initial growth does not produce a visible colony. Once the histidine supply is exhausted, only revertants that have mutated to regain the ability to synthesize histidine continue to grow and produce visible colonies. These colonies need only be counted and compared to controls to estimate the relative mutagenicity of the compound: the more colonies, the greater the mutagenicity.

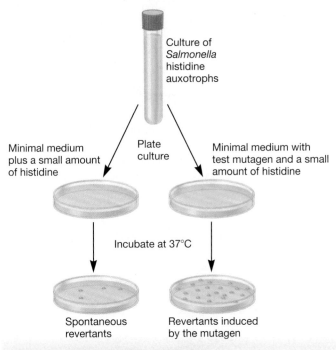

Figure 16.7 The Ames Test for Mutagenicity.

MICRO INQUIRY *Why is a small amount of histidine added to the test medium?*

Retrieve, Infer, Apply

1. Describe how replica plating is used to detect and isolate auxotrophic mutants.
2. Why are mutant selection techniques generally preferable to screening methods?
3. Briefly discuss how reversion mutations and resistance to an environmental factor (e.g., presence of an antibiotic) can be employed in mutant selection.
4. Design an experiment that selects for mutants of a bacterium that have gained the ability to catabolize sucrose (i.e., the wild type can't, but the mutant can).
5. Describe how you would isolate a mutant that required histidine for growth and was resistant to penicillin. The wild type is a prototroph.
6. What is the Ames test and how is it carried out? What assumption concerning mutagenicity and carcinogenicity is it based upon?

16.3 DNA Repair Maintains Genome Stability

After reading this section, you should be able to:

- Compare and contrast proofreading, excision repair, direct repair, mismatch repair, and recombinational repair
- Propose a scenario that would elicit the SOS response and describe the response to those conditions

If there is a microbial equivalent of an extreme sport, *Deinococcus radiodurans*'s ability to repair its genome after it has been blasted apart by a high dose of radiation might be a contender. Surprisingly, this ability is primarily related to the resistance of *D. radiodurans*

proteins and the structure of its genome, rather than to its DNA repair mechanisms. Its radiation-resistant proteins are able to begin repairing the genome quickly. Repair is aided by the genome consisting of two chromosomes, each having numerous areas of homology. This allows the DNA fragments to anneal to each other, facilitating the piecing together of the shattered genome. Other than this, *D. radiodurans* uses pretty much the same DNA repair mechanisms as other organisms. Obviously mutations can have disastrous effects. Therefore it is imperative that a microorganism be able to repair changes. There are numerous repair mechanisms and these are well conserved across all domains of life. Repair in *E. coli* is best understood and is our focus in this section. ▶▶| Deinococcus-Thermus *includes radiation-resistant bacteria (section 21.2)*

Proofreading: The First Line of Defense

As we discuss in chapter 13, replicative DNA polymerases sometimes insert the incorrect nucleotide during DNA replication. However, these DNA polymerases have the ability to evaluate the hydrogen bonds formed between the newly added nucleotide and the template nucleotide, and correct any errors immediately; that is, before the next nucleotide is added. This ability is called **proofreading.** When a DNA polymerase detects that a mistake has been made, it backs up, removing the incorrect nucleotide with its 3′ to 5′ exonuclease activity (*see figure 13.16*). It then restarts DNA replication, this time inserting the correct nucleotide. Proofreading is very efficient, but it does not always correct errors in replication. Furthermore, it is not useful for correcting induced mutations. *E. coli* uses other repair mechanisms to help ensure the stability of its genome. |◀◀ *DNA replication in bacteria (section 13.3)*

Mismatch Repair

When proofreading by replicative DNA polymerases fails, mismatched bases are usually detected and repaired by the **mismatch repair** system (**figure 16.8**). In *E. coli* the enzyme MutS scans the newly replicated DNA for mismatched pairs. Another enzyme, MutH, removes a stretch of newly synthesized DNA around the mismatch. A DNA polymerase then replaces the excised nucleotides, and the resulting nick is sealed by DNA ligase.

Successful mismatch repair depends on the ability of mismatch repair enzymes to distinguish between parental and newly synthesized DNA strands. This distinction is possible because newly synthesized DNA strands temporarily lack methyl groups on their bases, whereas the parental DNA has methyl groups on the bases. Thus for a short time after the replication fork has passed, the DNA is hemimethylated. The repair system cuts out the mismatch from the unmethylated strand. Eventually adenine bases in the GATC sequences of the newly synthesized strand are methylated by enzyme DNA adenine methyltransferase (DAM).

Excision Repair

Excision repair corrects damage that causes distortions in the DNA double helix. Two types of excision repair systems have been described: nucleotide excision repair and base excision repair. Both use the same approach to repair; they remove the

The MutS protein slides along the DNA until it finds a mismatch. MutL binds MutS and the MutS/MutL complex binds to MutH, which is already bound to a hemimethylated sequence.

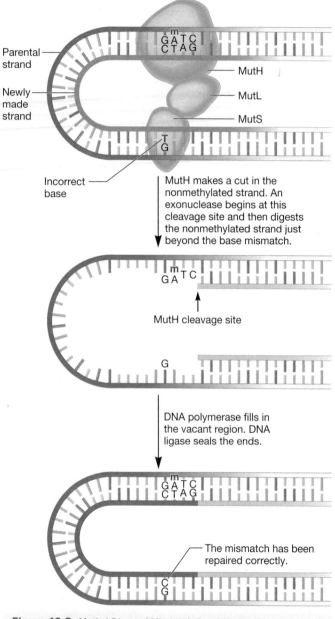

Figure 16.8 **Methyl-Directed Mismatch Repair in *E. coli*.** The role of MutH is to identify the methylated strand of DNA, which is the nonmutated parental strand. The methylated adenine is designated by an m.

MICRO INQUIRY *How is mismatch repair similar to DNA polymerase proofreading? How is it different?*

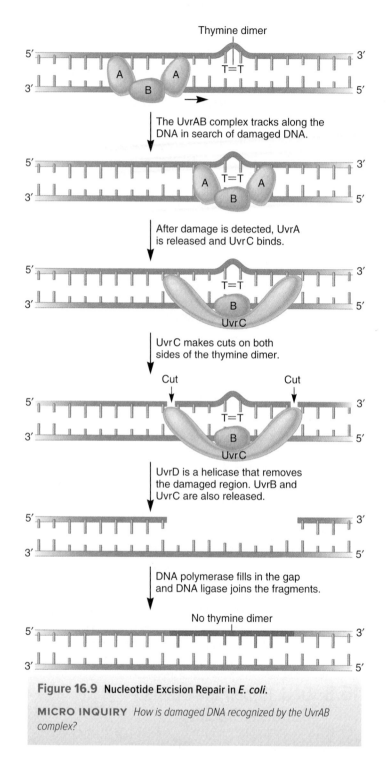

Figure 16.9 **Nucleotide Excision Repair in *E. coli*.**

MICRO INQUIRY *How is damaged DNA recognized by the UvrAB complex?*

damaged portion of a DNA strand and use the intact complementary strand as the template for synthesis of new DNA. The two repair systems are distinguished by the enzymes used to correct the damage.

In **nucleotide excision repair,** an *E. coli* enzyme called UvrABC endonuclease removes damaged nucleotides and a few nucleotides on either side of the lesion. The resulting single-stranded gap is filled by DNA polymerase I, and DNA ligase joins the fragments (**figure 16.9**). This system can remove thymine dimers (figure 16.5) and repair almost any other injury that produces a detectable distortion in DNA.

Base excision repair employs enzymes called DNA glycosylases. These enzymes remove damaged or unnatural bases yielding apurinic or apyrimidinic (AP) sites. Enzymes called AP

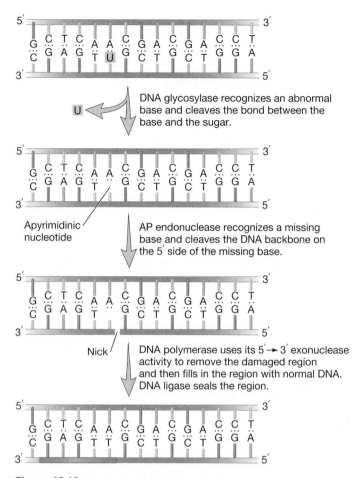

Figure 16.10 Base Excision Repair.

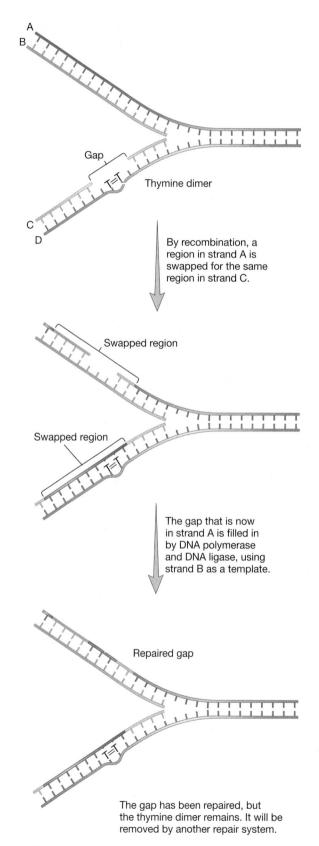

Figure 16.11 Recombinational Repair.

endonucleases recognize the damaged DNA and nick the backbone at the AP site (**figure 16.10**). DNA polymerase I removes the damaged region, using its 5′ to 3′ exonuclease activity. It then fills in the gap, and DNA ligase joins the DNA fragments.

Direct Repair

Thymine dimers and alkylated bases (e.g., those with methyl or ethyl groups attached) often are corrected by **direct repair.** Thymine dimers (figure 16.5) are split apart by a process called **photoreactivation,** which requires visible light and is catalyzed by the enzyme photolyase. Methyl and some other alkyl groups that have been added to guanine can be removed with the help of an enzyme known as alkyltransferase or methylguanine methyltransferase. Thus damage to guanine from mutagens such as methyl-nitrosoguanidine (figure 16.4) can be repaired directly. ↻ *Direct Repair*

Recombinational Repair

Recombinational repair corrects damaged DNA in which both bases of a pair are missing or damaged, or where there is a gap opposite a lesion. In this type of repair, the protein **RecA** cuts a piece of template DNA from a sister molecule and puts it into the gap or uses it to replace a damaged strand (**figure 16.11**). Although most bacterial cells are haploid, another copy of the damaged segment often is available because either it has recently been replicated or the cell is growing rapidly and contains more than one copy of its chromosome. Once the template is in place, the remaining damage can be corrected by another repair system.

SOS Response

Despite having multiple repair systems, sometimes the damage to an organism's DNA is so great that the normal repair mechanisms just described cannot repair all the damage. As a result, DNA synthesis stops completely. In such situations, a global control network called the **SOS response** is activated. In this response, over 50 genes are activated when a transcriptional repressor protein called LexA is destroyed. Once LexA is destroyed, these genes are transcribed and the SOS response ensues.

The SOS response, like recombinational repair, depends on the activity of RecA. RecA binds to single- or double-stranded DNA breaks and gaps generated by cessation of DNA synthesis. RecA binding initiates recombinational repair. Simultaneously RecA takes on coprotease function. It interacts with LexA, causing LexA to destroy itself (autoproteolysis). Destruction of LexA increases transcription of genes for excision repair and recombinational repair, in particular.

The first genes transcribed in the SOS response are those that encode the Uvr proteins needed for nucleotide excision repair (figure 16.9). Then expression of genes involved in recombinational repair is further increased. To give the cell time to repair its DNA, the protein SfiA is produced; SfiA blocks cell division. Finally, if the DNA has not been fully repaired after about 40 minutes, a process called **translesion DNA synthesis** is triggered. In this process, DNA polymerase IV (also known as DinB; Din is short for *d*amage *in*ducible) and DNA polymerase V (UmuCD; Umu is short for *UV mu*tagenesis) synthesize DNA across gaps and other lesions (e.g., thymine dimers) that had stopped DNA polymerase III. However, because an intact template does not exist, these DNA polymerases often insert incorrect bases. Furthermore, they lack proofreading activity. Therefore even though DNA synthesis continues, it is highly error prone and results in the generation of numerous mutations.

The SOS response is so named because it is made in a life-or-death situation. The response increases the likelihood that some cells will survive by allowing DNA synthesis to continue. For the cell, the risk of dying because of failure to replicate DNA is greater than the risk posed by the mutations generated by this error-prone process.

Retrieve, Infer, Apply

1. Compare and contrast the two types of excision repair.
2. What role does DNA methylation play in mismatch repair?
3. When *E. coli* cells are growing rapidly, they may contain up to four copies of their chromosomes. Why is this important if *E. coli* needs to carry out recombinational repair?
4. Explain how the following DNA alterations and replication errors would be corrected (there may be more than one way): base addition errors by DNA polymerase III during replication, thymine dimers, AP sites, methylated guanines, and gaps produced during replication.

16.4 Microbes Use Mechanisms Other than Mutation to Create Genetic Variability

After reading this section, you should be able to:

- Describe in general terms how recombinant eukaryotic organisms arise
- Distinguish vertical gene transfer from horizontal gene transfer
- Summarize the four possible outcomes of horizontal gene transfer
- Compare and contrast homologous recombination and site-specific recombination

In section 16.1, we discuss the consequences of mutations in terms of their effect on a protein and the phenotype of the organisms bearing the mutation. However, the effect also depends on the environment in which the organism lives. In other words, those mutations that don't immediately kill the organism are subject to selective pressure. Selective pressure determines if a mutation will persist in a population to become an alternate form of the gene. An alternate form of a gene is termed an **allele.** The existence of alleles for each gene in a genome means each organism in a population can have a distinctive (and probably a unique) set of alleles making up its genome; that is, its genotype. Each genotype in a population can be selected for or selected against. Organisms with genotypes, and therefore phenotypes, that are best suited to the environment survive and are most likely to pass on their genes. Shifts in environmental pressures can lead to changes in the population and ultimately result in the evolution of new species. The mechanisms by which new combinations of genes are generated are the topic of this section. All involve **recombination,** the process in which one or more nucleic acid molecules are rearranged or combined to produce a new genotype. Geneticists refer to organisms produced following a recombination event as recombinant organisms or simply **recombinants.**

Sexual Reproduction and Genetic Variability

The transfer of genes from parents to progeny is called **vertical gene transfer.** This type of gene transfer is observed in all organisms. In eukaryotes capable of sexual reproduction, vertical gene transfer is accompanied by genetic recombination, which occurs in two ways. The first is from crossing-over between sister chromosomes during meiosis. This creates new mixtures of alleles on homologous chromosomes. The second type of recombination results from fusion of gametes. Since each gamete contains alleles from the parent, when the gametes fuse, the parental alleles are combined in the zygote. As a result of these two types of recombination events, progeny organisms are not identical to their parents or to each other.

Horizontal Gene Transfer: Creating Variability the Asexual Way

Bacteria and archaea do not reproduce sexually. This suggests that genetic variation in populations of these microbes should be

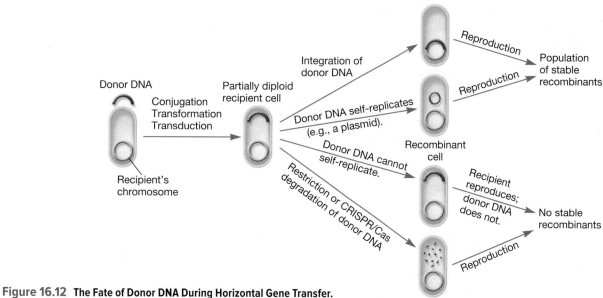

Figure 16.12 The Fate of Donor DNA During Horizontal Gene Transfer.

relatively limited, only occurring with the advent of a new mutation and its transfer to the next generation by vertical gene transfer. However, this is not the case. Bacteria and archaea have evolved three major mechanisms for creating recombinants. These mechanisms are referred to collectively as **horizontal (lateral) gene transfer** (**HGT**). HGT is distinctive from vertical gene transfer because genes from one independent, mature organism are transferred to another mature organism, often creating a stable recombinant having characteristics of both the donor and the recipient. ⮑ *Horizontal Gene Transfer*

The importance of HGT cannot be overstated. HGT has been important in the evolution of many species, and it continues to be commonplace in many environments. Indeed, studies reveal that HGT is the main mechanism by which bacteria and archaea evolve. Many examples are known of DNA transfer between distantly related genera and across domains of life, particularly among microbes sharing a habitat (e.g., the human gut or an antarctic deep lake). HGT may also help bacteria and archaea survive environmental stresses. For instance, some archaea respond to UV radiation by increasing their ability to carry out HGT so that they can obtain and use DNA from other cells to repair any UV-damaged DNA in their genomes. Another important example is the transfer of groups of genes encoding virulence factors from pathogenic bacteria to other bacteria. These groups of genes are called pathogenicity islands, and their discovery has provided valuable insights into the evolution of pathogenic bacteria. Finally, in most cases, the acquisition of new traits by HGT enables microbes to rapidly expand their ecological niche. This is clearly seen when HGT results in the spread of antibiotic-resistance genes among pathogenic bacteria. ▶▶ *Evolutionary processes and the concept of a microbial species inspire debate (section 19.5); Pathogenicity islands (section 35.2)*

During HGT, a piece of donor DNA, sometimes called the exogenote, enters a recipient cell. The transfer can occur in three ways: direct transfer between two cells temporarily in physical contact (conjugation), transfer of a naked DNA fragment (transformation), and transport of DNA by viruses (transduction). If the donor DNA contains genes already present in the recipient, the recipient will become temporarily diploid for those genes. This partially diploid cell is sometimes called a merozygote (*meros,* Greek meaning "part"; **figure 16.12**). The donor DNA has four possible fates in the recipient. First, when the donor DNA has a sequence homologous to one in the recipient's chromosome (sometimes called an endogenote), integration may occur. That is, the donor's DNA may pair with the recipient's DNA and be incorporated to yield a recombinant genome. The recombinant then reproduces, yielding a population of stable recombinants. Second, if the donor DNA is able to replicate itself (e.g., it is a plasmid), it may persist separate from the recipient's chromosome. When the recipient reproduces, the donor DNA replicates and a population of stable recombinants is formed. Third, the donor DNA remains in the cytoplasm but is unable to replicate. When the recipient reproduces, the progeny cells lack the donor DNA and it is eventually lost from the population. Finally, host restriction or CRISPR/Cas degradation of donor DNA may occur, thereby preventing the formation of a recombinant cell. ▶▶ *Restriction enzymes (section 17.1); The CRISPR/Cas system (section 27.2)*

Molecular Recombination: Joining DNA Molecules Together

Crossing-over during meiosis and integration of donor DNA into the recipient's chromosome during HGT occur by similar mechanisms. **Homologous recombination** is the most common mechanism, and many of the enzymes used are similar for all organisms. Homologous recombination occurs wherever there are long regions of the same or similar nucleotide sequence in two DNA molecules (e.g., sister chromosomes in diploid eukaryotes or similar sequences on both a chromosome and plasmid).

Homologous recombination results from DNA strand breakage and reunion leading to crossing-over. It is carried out by enzymes, many of which are also important for DNA repair (e.g., RecA; section 16.3).

The other major type of recombination is **site-specific recombination.** Site-specific recombination differs from homologous recombination in three significant ways. First, it does not require long regions of sequence homology. Second, recombination occurs at specific target sites in DNA molecules, and third, it is catalyzed by enzymes collectively called **recombinases.** Site-specific recombination is used by some plasmids and viral genomes to integrate into host chromosomes. It is also the type of recombination used by transposable elements to move from one place to another in DNA molecules, as we discuss in section 16.5.

Retrieve, Infer, Apply

1. An antibiotic-resistance gene located on a bacterium's chromosome is transferred from the bacterium to its progeny. What type of gene transfer is this?
2. What are the three mechanisms of horizontal gene transfer?
3. What four fates can DNA have after entering a bacterium?
4. How does homologous recombination differ from site-specific recombination?

16.5 Transposable Elements Move Genes Within and Between DNA Molecules

After reading this section, you should be able to:

- Differentiate insertion sequences from transposons
- Distinguish simple transposition from replicative transposition
- Defend this statement: "Transposable elements are important factors in the evolution of bacteria and archaea"

As more and more genomes have been sequenced and annotated, it has become increasingly apparent that the genomes of all organisms are rife with genetic elements that are able to move into and out of the genomes where they reside. These elements are often referred to as "jumping genes," mobile genetic elements, and transposable elements. **Transposition** refers to the movement of a mobile genetic element.

Mobile genetic elements were first discovered in the 1940s by Barbara McClintock (1902–1992) during her studies on maize genetics (a discovery for which she was awarded the Nobel Prize in 1983). As more and more types of mobile genetic elements were discovered, new nomenclature was developed to distinguish one type from another. As is often the case when a scientific field expands and its language evolves, the nomenclature became unwieldy and confusing. In 2008 a group of prominent scientists studying bacterial and archaeal mobile genetic elements proposed a new definition for **transposable elements:** "specific DNA segments that can repeatedly insert into one or more sites or into one or more genomes." This new definition encompasses elements that differ in structure, mechanisms of integration and excision, target sites, and ability to be transferred from one cell to another by HGT (**table 16.3**). ⟐ *Transposition: Shifting Segments of the Genome*

The enzymes that function in transposition are collectively termed recombinases. However, the recombinase used by a specific mobile genetic element may be called an integrase, resolvase, or transposase. There is tremendous interest in these enzymes because their mechanisms of action are similar to those used to rearrange the gene segments that encode important immune system proteins such as antibodies and T-cell receptors (*see figures 34.16 and 34.17*). They are also similar to those that move so-called gene cassettes (p. 396) into and out of sites that control their expression.

The simplest mobile genetic elements in bacteria are **insertion sequences,** or IS elements for short (**figure 16.13a**). An IS element is a short sequence of DNA (around 750 to 1,600 base pairs [bp] in length). It contains only the gene for the enzyme transposase, and it is bounded at both ends by inverted repeats—identical or very similar sequences of nucleotides in reversed orientation. Inverted repeats are usually about 15 to 25 base pairs long and vary among IS elements so that each type of IS has a specific nucleotide sequence in its inverted repeats. **Transposase** is required for transposition and accurately recognizes the ends of the IS. Each IS element is named by giving it the prefix IS

Table 16.3	Some Types of Transposable Elements
Transposable Element	**Description**
Insertion sequences (IS)	Small genetic elements consisting of a transposase gene bracketed by inverted repeats
Composite transposons	Contain genes unrelated to transposition bounded at each end by IS elements, which supply the transposase activity needed for transposition
Unit transposons	Contain one or more genes encoding enzymes needed for transposition, as well as other genes unrelated to transposition (e.g., antibiotic-resistance genes). They are not associated with an IS element.
Integrative conjugative elements	In addition to transposition functions, contain genes for conjugative transfer to a new host cell and other genes (e.g., antibiotic-resistance genes)

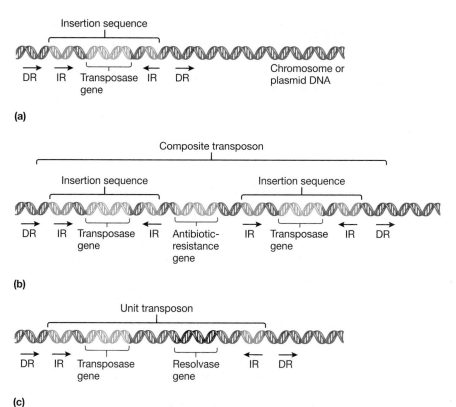

(a)

(b)

(c)

Figure 16.13 Transposable Elements. All transposable elements contain common features. These include inverted repeats (IRs) at the ends of the element and a recombinase (e.g., transposase) gene. (a) Insertion sequences (IS) consist only of IRs on either side of the transposase gene. (b) Composite transposons and (c) unit transposons contain additional genes (e.g., antibiotic-resistance genes) in addition to the recombinases that enable them to transpose. In composite transposons, the additional genes are flanked by insertion sequences, which supply the transposase. Unit transposons are not associated with insertion sequences. DRs, direct repeats in host DNA, flank a transposable element.

MICRO INQUIRY *What features are common to all types of transposable elements?*

replicate autonomously, it can transfer itself from *E. faecalis* to a variety of recipients and integrate into their chromosomes. ICEs are primarily observed in Gram-positive bacteria.

Two major transposition methods have been identified: simple transposition and replicative transposition (**figure 16.14**). **Simple transposition** is also called **cut-and-paste transposition.** In this method, transposase catalyzes excision of the transposable element, followed by cleavage of a new target site and ligation of the element into this site (**figure 16.15**). Target sites are specific sequences about five to nine base pairs long. When a mobile genetic element inserts at a target site, the target sequence is duplicated so that short, direct-sequence repeats flank the element's terminal inverted repeats. In **replicative transposition,** the original transposon remains at the parental site on the chromosome and a copy is inserted at the target DNA site (figure 16.14*b*). ↪ *Simple Transposition*

Transposable elements are of interest for many reasons. Their movement into a chromosome can alter gene function either by causing mutations in the genes into which they move or by providing promoters that are regulated differently than the normal promoter. They are also important evolutionary forces. They contribute to the evolution of an organism's chromosome, plasmids, and other mobile genetic elements. Of particular note is the role of mobile genetic elements in the spread of antibiotic resistance, as we discuss in section 16.9.

followed by a number. IS elements have been observed in a variety of bacteria and some archaea.

Transposons are more complex in structure than IS elements. Some transposons (composite transposons; table 16.3) consist of a central region containing genes unrelated to transposition (e.g., antibiotic-resistance genes) flanked on both sides by IS elements that are identical or very similar in sequence (figure 16.13*b*). The flanking IS elements encode the transposase used by the transposon to move. Other transposons lack IS elements and encode their own transposition enzymes (figure 16.13*c*). Most transposon names begin with the prefix Tn.

Some transposons bear transfer genes and can move between bacteria through the process of conjugation, as we discuss in section 16.6. They are called **conjugative transposons** or *integrative conjugative elements* (**ICEs**). A well-studied example of an ICE is Tn*916* from *Enterococcus faecalis.* Although Tn*916* cannot

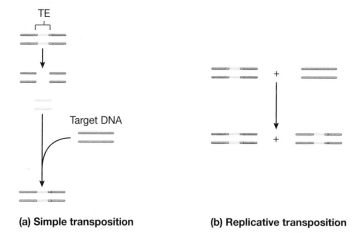

(a) Simple transposition

(b) Replicative transposition

Figure 16.14 Comparison of Simple and Replicative Transposition. TE, transposable element.

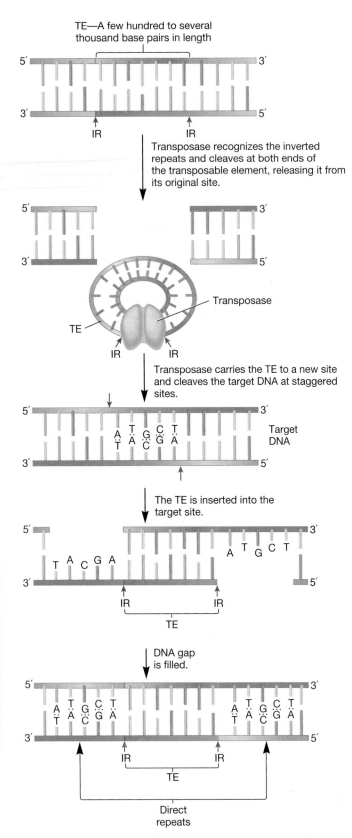

Figure 16.15 Simple Transposition. TE, transposable element; IR, inverted repeat.

1. How does a transposon differ from an insertion sequence?
2. What is simple (cut-and-paste) transposition? What is replicative transposition? How do the two mechanisms of transposition differ? What happens to the target site during transposition?
3. What effect would you expect the existence of transposable elements to have on the rate of microbial evolution? Give your reasoning.

16.6 Bacterial Conjugation Requires Cell-Cell Contact

After reading this section, you should be able to:

- Identify the type of plasmids that are important creators of genetic variation
- Describe the features of the F factor that allow it to (1) transfer itself to a new host cell and (2) integrate into a host cell's chromosome
- Outline the events that occur when an F$^+$ cell encounters an F$^-$ cell
- Distinguish F$^+$, Hfr, and F′ cells from each other
- Explain how Hfr cells arise
- Outline the events that occur when an Hfr cell encounters an F$^-$ cell

Conjugation, the transfer of DNA by direct cell-to-cell contact, depends on the presence of a **conjugative plasmid.** Recall from chapter 3 that **plasmids** are small, double-stranded DNA molecules that can exist independently of host chromosomes. They have their own replication origins, replicate autonomously, and are stably inherited. Some plasmids are **episomes,** plasmids that can exist either with or without being integrated into host chromosomes.

Perhaps the best-studied conjugative plasmid is **F factor.** It plays a major role in conjugation in *E. coli,* and it was the first conjugative plasmid to be described (**figure 16.16**). The F factor is about 100,000 bases long and bears genes responsible for cell attachment and plasmid transfer between specific *E. coli* cells. Most of the information required for plasmid transfer is located in the *tra* operon, which contains at least 28 genes. Many of these direct the formation of sex pili that attach the F$^+$ cell (the donor cell containing an F plasmid) to an F$^-$ cell (**figure 16.17**). Other gene products aid DNA transfer. In addition, the F factor has several IS elements that assist plasmid integration into the host cell's chromosome. Thus the F factor is an episome that can exist outside the bacterial chromosome or be integrated into it (**figure 16.18**).

The initial evidence for bacterial conjugation came from an elegant experiment performed in 1946 by Joshua Lederberg (1925–2008) and Edward Tatum (1909–1975). They mixed two auxotrophic strains, incubated the culture for several hours in nutrient medium, and then plated it on minimal medium. To reduce the chance that their results were due to a reversion or suppressor mutation, they used double and triple auxotrophs on the assumption that two or three simultaneous reversion or suppressor mutations would be extremely rare. When recombinant prototrophic colonies appeared on the minimal medium after incubation, they concluded that the two auxotrophs were able to associate and undergo recombination.

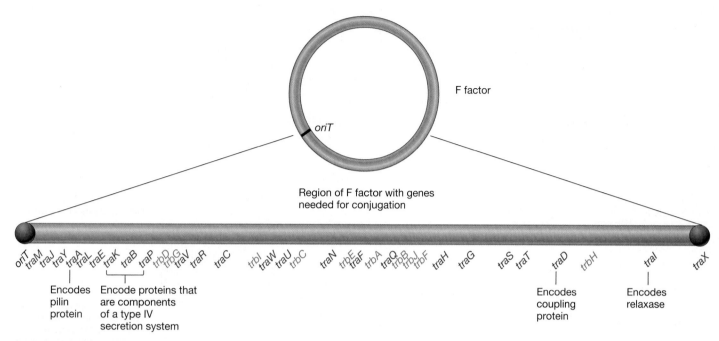

Figure 16.16 The F plasmid.   Transfer (*tra*) genes are shown in red, and some of their functions are indicated. The plasmid also contains three insertion sequences and a transposon. The site for initiation of rolling-circle replication and gene transfer during conjugation is *oriT*.

Lederberg and Tatum did not directly prove that physical contact of the cells was necessary for gene transfer. This evidence was provided several years later by Bernard Davis (1919–1994), who constructed a U-tube consisting of two pieces of curved glass tubing fused at the base to form a U shape with a glass filter between the halves. The filter allowed passage of media but not bacteria. The U-tube was filled with a growth medium and each side inoculated with a different auxotrophic strain of *E. coli.*

During incubation, the medium was pumped back and forth through the filter to ensure medium exchange between the halves. When the bacteria were later plated on minimal medium, Davis discovered that if the two auxotrophic strains were separated

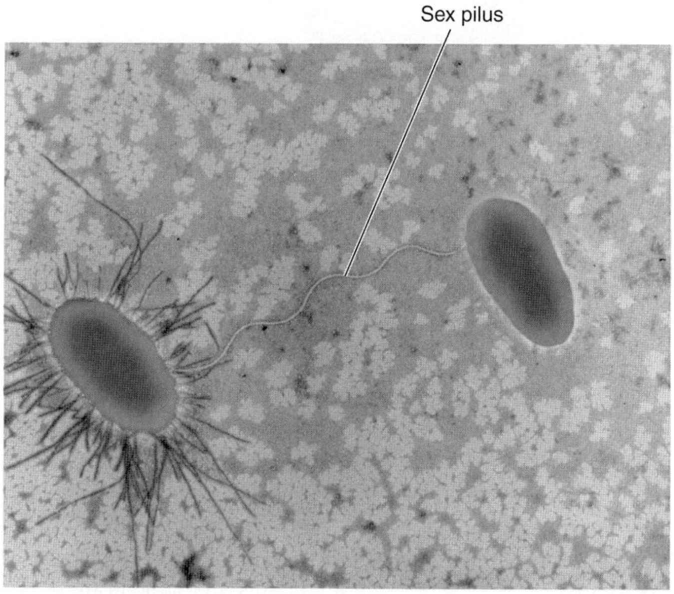

Sex pilus

Figure 16.17 Bacterial Conjugation.   An electron micrograph of two *E. coli* cells in an early stage of conjugation. The F⁺ cell to the left is covered with fimbriae, and a sex pilus connects the two cells.

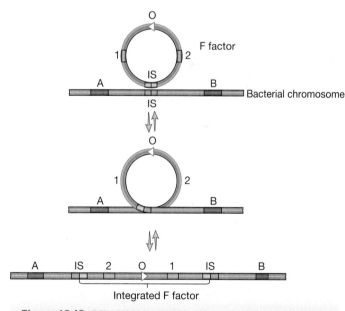

Figure 16.18 F Plasmid Integration.   The reversible integration of an F plasmid or factor into a host bacterial chromosome. The process begins with association between plasmid and bacterial insertion sequences. The O arrowhead (white) indicates the site at which oriented transfer of chromosome to the recipient cell begins. A, B, 1, and 2 represent genetic markers.

MICRO INQUIRY *What does the term episome mean?*

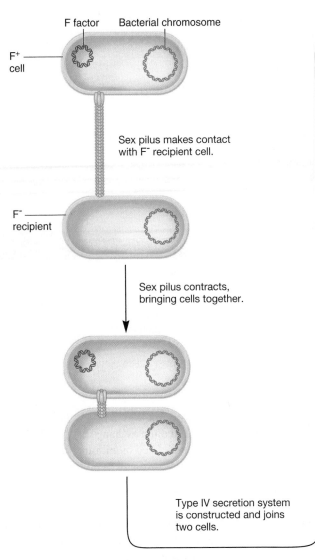

Figure 16.19 F Factor–Mediated Conjugation. The F factor encodes proteins for building the sex pilus and the type IV secretion system that transfers DNA from the donor to the F⁻ recipient. One protein, the coupling factor, brings the DNA to the secretion system. During F⁺ × F⁻ conjugation, only the F factor is transferred because the plasmid is extrachromosomal. The recipient cell becomes F⁺.

from each other by the filter, gene transfer did not take place. Therefore direct contact was required for the recombination that Lederberg and Tatum had observed. ⚙ *Bacterial Conjugation*

F⁺ × F⁻ Mating: Plasmid Only Transfer

In 1952 William Hayes (1913–1994) demonstrated that the gene transfer observed by Lederberg and Tatum was unidirectional. That is, there were definite donor (F⁺, or fertile) and recipient (F⁻, or nonfertile) strains, and gene transfer was nonreciprocal. He also found that in F⁺ × F⁻ mating, the progeny were only rarely changed with regard to auxotrophy (i.e., chromosomal genes usually were not transferred). However, F⁻ strains frequently became F⁺.

These results are now understood and readily explained in the following way. An F⁺ strain contains an extrachromosomal F factor carrying the genes for sex pilus formation and plasmid transfer. The **sex pilus** is used to establish contact between the F⁺ and F⁻ cells (**figure 16.19**). Once contact is made, the pilus

Figure 16.20 The Conjugation Machinery Encoded by F Factor Is a Modified Type IV Secretion System. The F factor–encoded type IV secretion system is composed of numerous Tra proteins, including TraA proteins, which form the sex pilus, and TraD, which is the coupling factor. TraI, TraM, TraY, and the host protein IHF form the relaxosome. Here we show the relaxosome interacting with TraD. Also note that both ends of the nicked DNA strand are attached to the relaxosome. The 5′ end is attached to a tyrosine residue (Y) in TraI. Plasma membrane (PM); periplasm (P); peptidoglycan (PG); outer membrane (OM); lipopolysaccharide (LPS). Only the 5′ and 3′ ends of the DNA are shown. The rest of the DNA would extend below the figure.

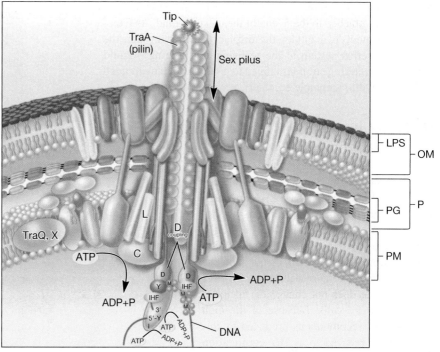

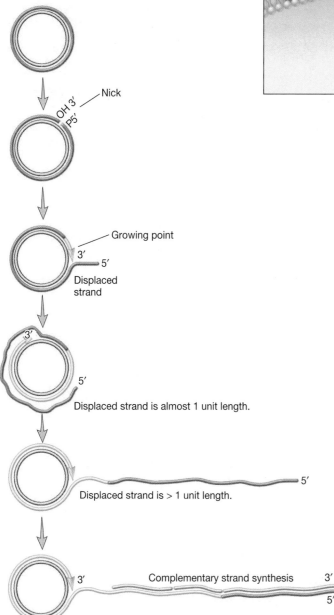

retracts, bringing the cells into close physical contact. The F⁺ cell prepares for DNA transfer by assembling a type IV secretion system, using many of the same genes used for sex pilus biogenesis (**figure 16.20**). Once built, the secretion system begins to move the plasmid to the recipient. ⮑ *Conjugation: Transfer of the F Plasmid*

As the F plasmid is being transferred, it is also copied using a process called **rolling-circle replication,** one strand of the circular DNA is nicked, and the free 3′-hydroxyl end is extended by replication enzymes (**figure 16.21**). The 3′ end is lengthened while the growing point rolls around the circular template and the 5′ end of the strand is displaced to form an ever-lengthening tail, much like the peel of an apple is displaced by a knife as an apple is pared. We are concerned here with rolling-circle replication of a plasmid. However, rolling-circle replication is also observed during the replication of some viral genomes (e.g., phage lambda). ⮑ *Rolling-Circle Replication*

During conjugation, rolling-circle replication is initiated by the relaxosome, a complex of proteins encoded by the F factor (figure 16.16). The relaxosome nicks one strand of the F factor at a site called *oriT* (for origin of transfer). The major component of

Figure 16.21 Rolling-Circle Replication. A single-stranded tail, often composed of more than one genome copy, is generated and can be converted to the double-stranded form by synthesis of a complementary strand. The "free end" of the rolling-circle strand is probably bound to the primosome. OH 3′ is the 3′-hydroxyl and P 5′ is the 5′-phosphate created when the DNA strand is nicked.

the relaxosome is a protein called TraI. It has relaxase activity and remains attached to the 5′ end of the nicked strand. As F factor is replicated, TraI guides the displaced strand through the type IV secretion system to the recipient cell. During plasmid transfer, the entering strand is copied to produce double-stranded DNA. When this is completed, the F⁻ recipient cell becomes F⁺.

Hfr Conjugation Transfers Chromosomal DNA

By definition, an F⁺ cell has the F factor free from the chromosome, so in an F⁺ × F⁻ mating, chromosomal DNA is not transferred. However, within this population, a few cells have the F plasmid integrated (i.e., recombined) into their chromosomes. This explains why not long after the discovery of F⁺ × F⁻ mating, a second type of F factor–mediated conjugation was discovered. In this type of conjugation, the donor transfers chromosomal genes with great efficiency but does not change the recipient bacteria into F⁺ cells. Because of the *high frequency* of *recombinants* produced by this mating, it is referred to as **Hfr conjugation** and the donor is called an **Hfr strain.**

Hfr strains contain the F factor integrated into their chromosome, rather than free in the cytoplasm (**figure 16.22**). When integrated, the F plasmid's *tra* operon is still functional; the plasmid can direct the synthesis of pili, carry out rolling-circle replication, and transfer genetic material to an F⁻ recipient cell. However, rather than transferring just itself, the F factor also directs the transfer of the host chromosome. DNA transfer begins when the integrated F factor is nicked at *oriT*. As it is replicated, the F factor begins to move into the recipient (**figure 16.23**). Initially only part of the F factor is transferred, followed by the donor's chromosome. If the cells remain connected, the entire chromosome with the rest of the integrated F factor will be transferred; this takes about 100 minutes to accomplish. However, the connection between the cells usually breaks before this process is finished. Thus a complete F factor is rarely transferred, and the recipient remains F⁻.

When an Hfr strain participates in conjugation, bacterial genes are transferred to the recipient in either a clockwise or a counterclockwise direction around a circular chromosome, depending on the orientation of the integrated F factor. After the replicated donor chromosome enters the recipient cell, it may be degraded or incorporated into the F⁻ genome by recombination. ♻ *Transfer of Chromosomal DNA*

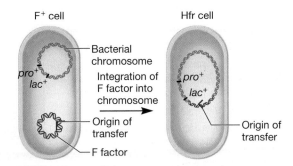

Figure 16.22 Insertion of an F Factor into the Donor Cell's Chromosome Creates an Hfr Cell.

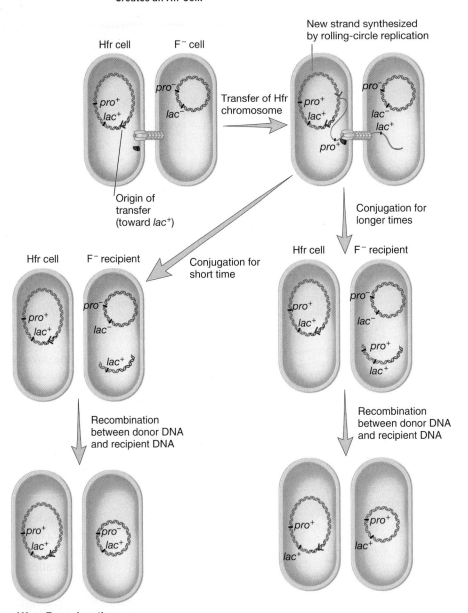

Hfr × F⁻ conjugation

Figure 16.23 Hfr × F⁻ Conjugation. Shown are the two cells after initial contact and elaboration of the type IV secretion system. As can be seen, during Hfr × F⁻ conjugation, some plasmid genes and some chromosomal genes are transferred to the recipient. Note that only a portion of the F factor moves into the recipient. Because the entire plasmid is not transferred, the recipient remains F⁻. In addition, the incoming DNA must recombine into the recipient's chromosome if it is to be stably maintained.

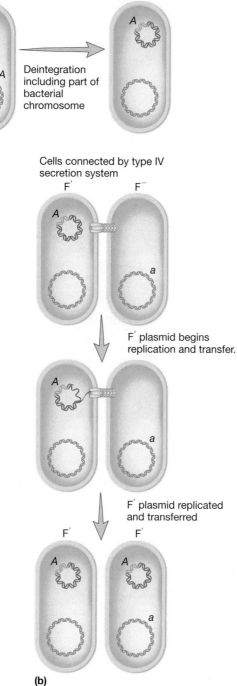

(a)

Deintegration including part of bacterial chromosome

Cells connected by type IV secretion system

F′ plasmid begins replication and transfer.

F′ plasmid replicated and transferred

(b)

Figure 16.24 F′ Conjugation. (a) Due to an error in excision, the *A* gene of an Hfr cell is picked up by the F factor. (b) During conjugation, the *A* gene is transferred to a recipient, which becomes diploid for that gene (i.e., Aa).

F′ Conjugation

Because the F plasmid is an episome, it can leave the bacterial chromosome and resume status as an autonomous F factor. Sometimes during excision an error occurs and a portion of the chromosome is excised, becoming part of the F plasmid. Because this erroneously excised plasmid is larger and genotypically distinct from the original F factor, it is called an **F′ plasmid** (**figure 16.24a**). A cell containing an F′ plasmid retains all of its genes, although some of them are on the plasmid. It mates only with an F⁻ recipient, and F′ × F⁻

conjugation is similar to an F⁺ × F⁻ mating. Once again, the plasmid is transferred as it is copied by rolling-circle replication. Bacterial genes on the chromosome are not transferred (figure 16.24b), but bacterial genes on the F′ plasmid are transferred. These genes need not be incorporated into the recipient chromosome to be expressed. The recipient becomes F′ and is partially diploid because the same bacterial genes present on the F′ plasmid are also found on the recipient's chromosome. In this way, specific bacterial genes may spread rapidly throughout a bacterial population.

Bacterial Conjugation in Gram-Positive Bacteria

Although most research on plasmids and conjugation has been done using *E. coli* and other Gram-negative bacteria, conjugative plasmids are present in numerous Gram-positive bacteria. With the one exception (*Streptomyces* spp.), Gram-positive conjugative systems share similarities with those of Gram-negative bacteria. For instance, they have homologues of TraI (relaxase) and TraD (coupling factor), as well as other type IV secretion system proteins. Unlike Gram-negative bacteria, which establish contact by way of the sex pilus, cell-to-cell contact in these Gram-positive bacteria is established by surface substances that enable cells to directly adhere to one another. One of the best-studied Gram-positive systems is that of *Enterococcus faecalis*. The recipient cells of this bacterium release short peptide signals that activate transfer genes in donor cells containing the proper plasmid. The conjugation systems of *Streptomyces* spp. differ from those of other Gram-positive bacteria in that TraI homologues as well as a specific nick site have not been identified. 🌀 *Transfer of a Plasmid*

Retrieve, Infer, Apply

1. What is bacterial conjugation and how was it discovered?
2. For F⁺, Hfr, and F⁻ strains of *E. coli,* indicate which acts as a donor during conjugation, which acts as a recipient, and which transfers chromosomal DNA.
3. Describe how F⁺ × F⁻ and Hfr conjugation processes proceed, and distinguish between the two in terms of mechanism and the final results.
4. Compare and contrast F⁺ × F⁻ and F′ × F⁻ conjugation.

16.7 Bacterial Transformation Is the Uptake of Free DNA from the Environment

After reading this section, you should be able to:

- Describe the factors that contribute to a bacterium being naturally transformation competent
- Predict the outcomes of transformation using a DNA fragment versus using a plasmid
- Design an experiment to transform bacteria that are not naturally competent with a plasmid that carries genes encoding ampicillin resistance and the protein that generates green fluorescence

Another HGT mechanism is transformation, discovered by Fred Griffith in 1928. **Transformation** is the uptake by a cell of DNA, either a plasmid or a fragment of linear DNA, from the surroundings and maintenance of the DNA in the recipient in a heritable form. Natural transformation has been discovered in some archaea and in members of several bacterial phyla. Natural transformation occurs in soil and aquatic ecosystems and in vivo during infection; it also is an important route of HGT in biofilm and other microbial communities.

Natural transformation is best studied in *Streptococcus pneumoniae, Bacillus subtilis,* and *Neisseria* spp. It occurs when donor bacteria lyse and release their DNA into the surrounding environment. These fragments may be relatively large and contain several genes. If a fragment contacts a **competent cell**—a cell that is able to take up DNA and be transformed—the DNA can be bound to the cell and taken inside (**figure 16.25a**). The transformation frequency of very competent cells is around 10^{-3} for most genera when an excess of DNA is used. That is, about one cell in every thousand will take up and integrate the gene.

Competency is a complex phenomenon that is induced in most bacteria during a certain stage of growth. For instance *S. pneumoniae* becomes competent during early exponential phase, whereas *B. subtilis* becomes competent in stationary phase. Regulation of competence induction can involve alternative sigma factors, two-component regulatory systems, and transcription activator proteins. A few species are constitutively competent (e.g., *Neisseria* spp. and *Thermus* spp.) Most bacteria are able to take up DNA from any source (i.e., other bacteria, archaea, and eukaryotes). Some take up DNA only from closely related species. For instance, *Haemophilus influenzae* only takes up DNA containing a specific 9-base-pair sequence that is repeated more than 1,400 times in the *H. influenzae* genome.

Natural transformation is a multistep process (**figure 16.26**). Double-stranded (ds) DNA must first bind a competent cell. One strand is then hydrolyzed as the remaining intact strand is brought into the cell. The incoming single-stranded (ss) fragment then interacts with proteins that integrate the DNA into the recipient's chromosome by homologous recombination.

Despite the difference in cell envelope architecture, Gram-positive and Gram-negative bacteria (with one known exception) use similar proteins as the main components of their DNA uptake machinery. These proteins are related to those used in type II secretion systems and type IV pili. The major differences arise from the need for Gram-negative bacteria to move dsDNA across the outer membrane prior to its translocation as ssDNA across the plasma membrane (**figure 16.27**). Despite knowing the basic machinery, scientists still have not determined the precise mechanisms by which DNA enters the cell or how each component interacts with the DNA. How this energy-expending process is powered is also unclear for many bacteria. In Gram-positive bacteria belonging to *Firmicutes* (e.g., *Bacillus, Streptococcus,* and *Enterococcus* species), an ATP-dependent translocase (ComFA) has been identified. ↻ *Bacterial Transformation*

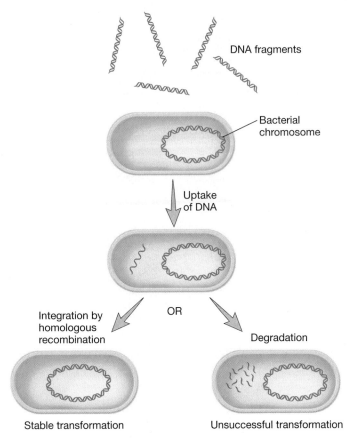

(a) Transformation with DNA fragments

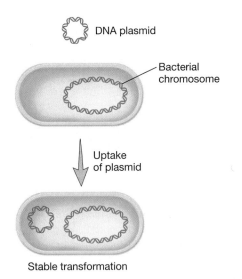

(b) Transformation with a plasmid

Figure 16.25 Bacterial Transformation. Transformation with (a) DNA fragments and (b) plasmids. Transformation with a plasmid often is induced artificially in the laboratory. The transforming DNA is in purple, and integration is at a homologous region of the genome.

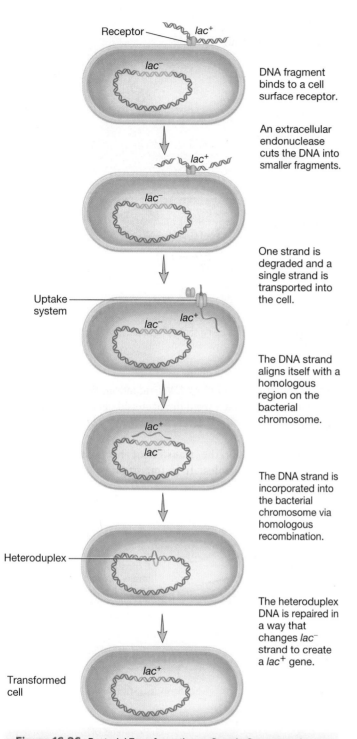

Figure 16.26 **Bacterial Transformation as Seen in *S. pneumoniae.***

MICRO INQUIRY *According to this model, what would happen if DNA that lacked homology to the* S. pneumoniae *chromosome were taken into the cell?*

Microbial geneticists exploit transformation to move DNA (usually recombinant DNA) into cells. Because many species, including *E. coli,* are not naturally transformation competent, these bacteria must be made artificially competent by certain

treatments. Two common techniques are electrical shock and exposure to calcium chloride. Both approaches render the cell membrane temporarily more permeable to DNA, and both are used with *E. coli.* To increase the transformation frequency with *E. coli,* strains that lack one or more nucleases are used. These strains are especially important when transforming the cells with linear DNA, which is vulnerable to attack by nucleases. It is easier to transform bacteria with plasmid DNA since plasmids can replicate within the host and are not as easily degraded as are linear fragments (figure 16.25b). ▶▶| *Introducing recombinant DNA into host cells (section 17.4)*

Retrieve, Infer, Apply

1. Define transformation and competence.
2. Describe how transformation occurs in *S. pneumoniae.* How does the process differ in *H. influenzae* and *B. subtilis?*
3. Discuss two ways in which artificial transformation can be used to place functional genes within bacterial cells.

16.8 Transduction Is Virus-Mediated DNA Transfer

After reading this section, you should be able to:

- Differentiate generalized transduction from specialized transduction
- Correlate a phage's life cycle to its capacity to mediate generalized or specialized transduction
- Draw a figure, create a concept map, or construct a table that distinguishes conjugation, transformation, and transduction

Transduction is mediated by viruses. It is a frequent mode of HGT in nature. Indeed evidence suggests that the number of genes moved by marine viruses from one host cell to another is huge (perhaps 10^{24} per year). Furthermore, viruses in marine environments and hot springs move genes between organisms in all three domains of life. ▶▶| *Aquatic viruses (section 30.2)*

Recall from chapter 6 that virus particles are structurally simple, often composed of just a nucleic acid genome protected by a protein coat called the capsid. Viruses are unable to multiply autonomously. Instead, they infect and take control of a host cell, forcing the host to make many copies of the virus. Viruses that infect bacteria are called bacteriophages, or phages for short. **Virulent bacteriophages** multiply in their bacterial host immediately after entry. After the progeny phage particles reach a certain number, they cause the host to lyse, so they can be released and infect new host cells (**figure 16.28**). Thus this process is called the **lytic cycle. Temperate bacteriophages,** on the other hand, do not immediately kill their host. Instead, the phage establishes a relationship with their host called **lysogeny,** and bacteria that have been lysogenized are called **lysogens.** Many temperate phages establish lysogeny

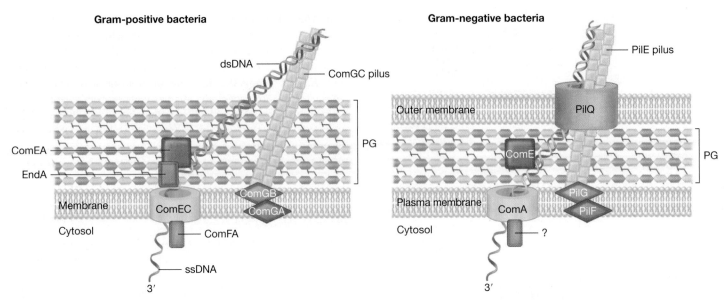

Gram-positive bacteria

dsDNA

ComGC pilus

PG

ComEA

EndA

Membrane

Cytosol

ComEC

ComGB
ComGA

ComFA

ssDNA

3′

Gram-negative bacteria

PilE pilus

Outer membrane

PilQ

ComE

PG

Plasma membrane

Cytosol

ComA

PilG
PilF

?

3′

Figure 16.27 DNA Uptake Systems of Gram-Positive and Gram-Negative Bacteria. Com proteins, Com being short for competence, are type II secretion-related proteins. Pil proteins, short for pilin, are type IV pilus-related proteins.

by inserting their genomes into the bacterial chromosome. The inserted viral genome is called a **prophage.** The host bacterium is unharmed by this, and the phage genome is passively replicated as the host cell's genome is replicated. Temperate phages can remain inactive in their hosts for many generations. However, they can be induced to switch to a lytic cycle under certain conditions, including UV irradiation. When this occurs, the prophage is excised from the bacterial genome and the lytic cycle proceeds. ◄◄ *There are several types of viral infections (section 6.4)*

Transduction is the transfer of bacterial or archaeal genes by virus particles. It is important to understand that

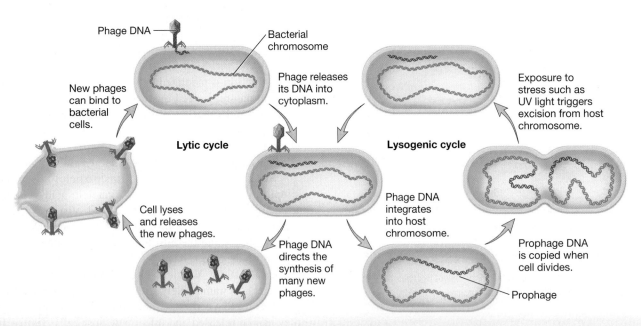

Phage DNA

Bacterial chromosome

New phages can bind to bacterial cells.

Phage releases its DNA into cytoplasm.

Exposure to stress such as UV light triggers excision from host chromosome.

Lytic cycle

Lysogenic cycle

Cell lyses and releases the new phages.

Phage DNA directs the synthesis of many new phages.

Phage DNA integrates into host chromosome.

Prophage DNA is copied when cell divides.

Prophage

Figure 16.28 Lytic and Lysogenic Cycles of Temperate Phages. Virulent phages undergo only the lytic cycle. Temperate phages have two phases to their life cycles. The lysogenic cycle allows the genome of the virus to be replicated passively as the host cell's genome is replicated. Certain environmental factors such as UV light can cause a switch from the lysogenic cycle to the lytic cycle. In the lytic cycle, new virions are made and released when the host cell lyses.

MICRO INQUIRY *What is the term used to describe a temperate phage genome when it is integrated into the host genome?*

host genes are packaged in the virus particle because of errors made during the virus's life cycle. The virion containing these genes then transfers them to a recipient cell. Two kinds of bacterial transduction have been described: generalized and specialized.

Generalized Transduction Transfers Any Donor Gene

Generalized transduction most often occurs during the lytic cycle of virulent phages but sometimes happens during the lytic cycle of temperate phages. Any part of the bacterial genome can be transferred after being partially degraded as the virus takes

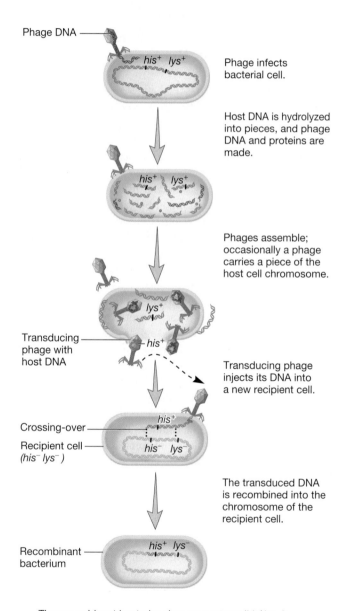

Phage DNA

his⁺ lys⁺

Phage infects bacterial cell.

Host DNA is hydrolyzed into pieces, and phage DNA and proteins are made.

his⁺ lys⁺

Phages assemble; occasionally a phage carries a piece of the host cell chromosome.

lys⁺

Transducing phage with host DNA

his⁺

Transducing phage injects its DNA into a new recipient cell.

his⁺

Crossing-over

Recipient cell (his⁻ lys⁻)

his⁻ lys

The transduced DNA is recombined into the chromosome of the recipient cell.

Recombinant bacterium

his⁺ lys⁻

The recombinant bacterium has a genotype (*his⁺lys⁻*) that is different from recipient bacterial cell (*his⁻ lys⁻*).

Figure 16.29 **Generalized Transduction in Bacteria.**

control of its host (**figure 16.29**). During the assembly stage, viral genomes are packaged by the "headful mechanism;" that is, only genomes of a certain size (i.e., number of nucleotides in length) are packaged. During general transformation, a fragment of the host genome that happens to be about the same size as the phage genome is mistakenly packaged. Such a phage is called a generalized transducing particle, because once it is released, it may encounter a susceptible host cell and eject the bacterial DNA it carries into that cell. However, because it lacks viral genes this does not initiate a lytic cycle. As in transformation, once the DNA fragment has been released into the recipient cell, it must be incorporated into the recipient cell's chromosome to preserve the transferred genes. The DNA remains double stranded during transfer, and both strands are integrated into the recipient's chromosome. About 70 to 90% of the transferred DNA is not integrated but often is able to remain intact temporarily and be expressed. Abortive transductants are bacteria that contain this nonintegrated, transduced DNA and are partial diploids.

Specialized Transduction Transfers Specific Donor Genes

In **specialized transduction,** only specific portions of the bacterial genome are carried by transducing particles. Specialized transduction is made possible by an error in the lysogenic life cycle of temperate phages that insert their genomes into a specific site in the host chromosome. When a prophage is induced to leave the host chromosome, excision is sometimes carried out improperly. The resulting phage genome contains portions of the bacterial chromosome (about 5 to 10% of the bacterial DNA) next to the integration site, much like the situation with F′ plasmids (**figure 16.30**). However, the transducing particle is defective because it lacks some viral genes and cannot reproduce without assistance. In spite of this, it will inject the remaining viral genome and any bacterial genes it carries into another bacterium. The bacterial genes may become stably incorporated under the proper circumstances.

The best-studied example of specialized transduction is carried out by the *E. coli* phage lambda. The lambda genome inserts into the host chromosome at specific locations known as attachment or *att* sites (**figure 16.31**). The phage *att* sites and bacterial *att* sites are similar and can complex with each other. The *att* site for lambda is between the *gal* and *bio* genes on the *E. coli* chromosome; consequently when lambda excises incorrectly to generate a specialized transducing particle, these bacterial genes are most often present. The product of cell lysis (lysate) resulting from the induction of a population of lysogenized *E. coli* cells contains normal phage and a few defective transducing particles. These particles are called lambda *dgal* if they carry the galactose utilization genes or lambda *dbio* if they carry the *bio* from the other side of the *att* site (figure 16.31). ▶▶ *Bacteriophage lambda (section 27.2)*
🔄 *Specialized Transduction*

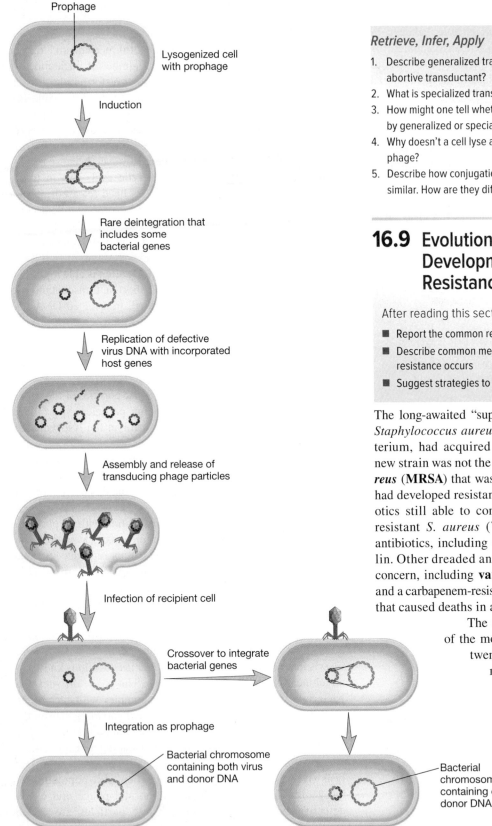

Figure 16.30 **Specialized Transduction by a Temperate Bacteriophage.** Recombination can produce two types of transductants.

MICRO INQUIRY *Compare the number of transducing particles that arise during generalized (figure 16.29) and specialized transduction. Why is there such a big difference?*

Retrieve, Infer, Apply

1. Describe generalized transduction and how it occurs. What is an abortive transductant?
2. What is specialized transduction and how does it come about?
3. How might one tell whether horizontal gene transfer was mediated by generalized or specialized transduction?
4. Why doesn't a cell lyse after successful transduction with a temperate phage?
5. Describe how conjugation, transformation, and transduction are similar. How are they different?

16.9 Evolution in Action: The Development of Antibiotic Resistance in Bacteria

After reading this section, you should be able to:

- Report the common reasons for increasing drug resistance
- Describe common mechanisms by which antimicrobial drug resistance occurs
- Suggest strategies to overcome drug resistance

The long-awaited "superbug" arrived in the summer of 2002. *Staphylococcus aureus*, a common but sometimes deadly bacterium, had acquired a new antibiotic-resistance gene. This new strain was not the well-known **methicillin-resistant *S. aureus* (MRSA)** that was the bane of hospitals. This newer strain had developed resistance to vancomycin, one of the few antibiotics still able to control *S. aureus*. This new vancomycin-resistant *S. aureus* (VRSA) strain also resisted most other antibiotics, including ciprofloxacin, methicillin, and penicillin. Other dreaded antibiotic-resistant strains are also of special concern, including **vancomycin-resistant enterococci (VRE)** and a carbapenem-resistant *Enterobacteriaceae* (CRE) bacterium that caused deaths in a Los Angeles hospital in 2015.

The spread of drug-resistant pathogens is one of the most serious threats to public health in the twenty-first century. There are two types of resistance: inherent and acquired. An example of inherent resistance is that of the cell wall-less mycoplasma's resistance to penicillin, which interferes with peptidoglycan synthesis. Similarly, many Gram-negative bacteria are unaffected by penicillin because it cannot penetrate the bacterial outer membrane. Acquired resistance occurs when there is a change in the genome of a bacterium that converts it from one that is sensitive to an antibiotic to one that is now resistant. This section describes the ways in which bacteria acquire drug resistance and

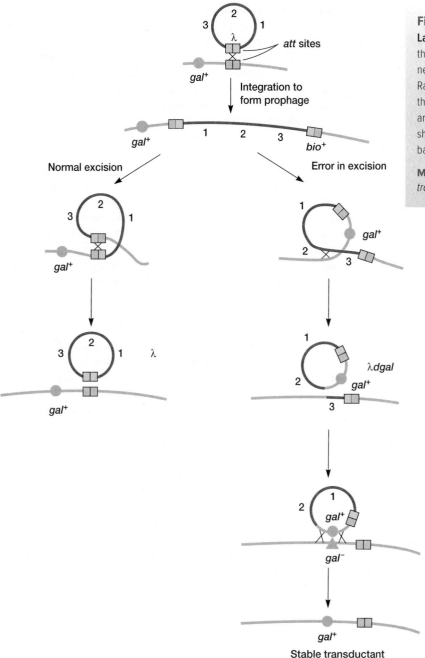

Figure 16.31 The Mechanism of Transduction for Phage Lambda and *E. coli.* Integrated lambda phage lies between the *gal* and *bio* genes. When it excises normally (top left), the new phage is complete and contains no bacterial genes. Rarely, excision occurs asymmetrically (top right), and either the *gal* or *bio* genes are picked up and some phage genes are lost (only aberrant excision involving the *gal* genes is shown). The result is a defective lambda phage that carries bacterial genes and can transfer them to a new recipient.

MICRO INQUIRY *Why can't the* gal *and* bio *genes be transduced by the same transducing particle?*

mutating a gene that functions in the synthesis of the target or by acquiring by HGT a gene that either encodes an alternative version of the target or encodes an enzyme that modifies the target. This resistance mechanism is possible because each chemotherapeutic agent acts on a specific target enzyme or cellular structure. For instance, resistance to vancomycin arises when bacteria "pick up" the *vanA* gene that encodes a protein that changes the terminal D-alanine in the pentapeptide of peptidoglycan to either D-lactate or D-serine (*see figure 12.9*). The affinity of ribosomes for erythromycin and chloramphenicol can be decreased by mutating the 23S rRNA to which they bind. Antimetabolite action may be resisted through alteration of susceptible enzymes. For example, in sulfonamide-resistant bacteria, the enzyme that uses *p*-aminobenzoic acid during folic acid synthesis often has a much lower affinity for sulfonamides. ◄◄ *Antibacterial drugs (section 9.4)*

A second resistance strategy is **drug inactivation.** The best-known example is hydrolysis of the β-lactam ring of penicillins by penicillinase and other β-lactamase enzymes. Drugs also are inactivated by the addition of chemical groups. For example, chloramphenicol can be modified by the addition of acetyl-CoA to either of two hydroxyl groups. Aminoglycosides (*see figure 9.8*) are modified and inactivated by acetylation of amino groups and phosphorylation or adenylylation of hydroxyl groups present on the aminoglycoside. Because these resistance strategies involve gaining enzymes that catalyze the inactivating reaction, the new function is most often acquired by HGT.

The third resistance strategy minimizes the concentration of the antibiotic in the cell. This can be accomplished by altering membrane structure, particularly the outer membrane for Gram-negative bacteria, so that less antibiotic enters the cell. Another approach is to pump the drug out of the cell after it has entered, using translocases, often called **efflux pumps,** that

how resistance spreads within a bacterial population. ◄◄ *Synthesis of peptidoglycan occurs in the cytoplasm, at the plasma membrane, and in the periplasmic space (section 12.4); Typical Gram-negative cell walls include additional layers besides peptidoglycan (section 3.4)* ►► *Class* Mollicutes, *phylum* Tenericutes *(section 21.3)*

There Are Several Mechanisms of Drug Resistance

Bacteria have evolved several resistance mechanisms (**figure 16.32**). One is to modify the target of the antibiotic. This occurs by

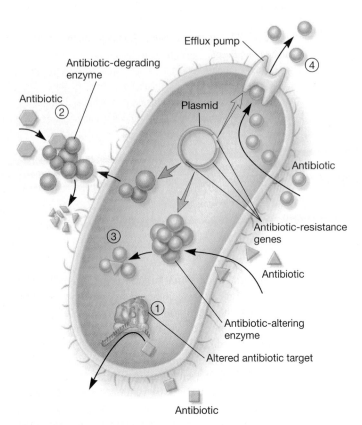

Figure 16.32 Antibiotic-Resistance Mechanisms. Bacteria can resist the action of antibiotics by (1) preventing access to (or altering) the target of the antibiotic, (2) degrading the antibiotic, (3) altering the antibiotic, or (4) rapidly extruding the antibiotic.

expel drugs. Efflux pumps are relatively nonspecific and pump many different drugs; therefore, they often confer multidrug-resistance. Many efflux pumps are drug/proton antiporters; that is, protons enter the cell as the drug leaves.

Finally, resistant bacteria may either use an **alternate pathway** to bypass the biochemical reaction inhibited by the agent or increase the production of the target metabolite. For example, some bacteria are resistant to sulfonamides simply because they use preformed folic acid from their surroundings, rather than synthesize it themselves. Other strains increase their rate of folic acid production and thus counteract sulfonamide inhibition. Again, these changes are most often mediated by HGT.

Drug Resistance Can Be Transmitted from One Bacterium to Another by HGT

Within three years after the widespread use of penicillin began, penicillin-resistant bacteria were found in clinical specimens. To understand the origin of drug resistance, it is important to recall that in nature, antibiotic-producing microbes must also protect themselves from the antibiotics they secrete. In other words, gaining the function of antibiotic production requires that the microbe also gain resistance—otherwise they would "commit

suicide" by producing the antibiotic. In antibiotic-producing microorganisms, the genes that encode resistance proteins are often referred to as immunity genes. Immunity genes are usually coordinately regulated with genes that code for antibiotic biosynthetic enzymes. It is believed that many genes encoding antibiotic resistance in bacteria were "captured" from antibiotic-producing bacteria and moved by HGT to nonproducers, giving rise to a large pool of resistance-encoding genes outside the producing microorganisms. It is therefore not surprising that in nonproducing bacteria, genes for drug resistance may be present on bacterial chromosomes, plasmids, transposons, and other mobile genetic elements. Because they are often found on mobile genetic elements, they can freely exchange between bacteria. In addition to the resistance genes transferred to nonproducers, some nonproducers can become resistant due to spontaneous chromosomal mutations. Usually such mutations result in a change in the drug target; therefore the antibiotic cannot bind and inhibit growth. This is exemplified by the chloramphenicol and sulfonamide resistance discussed earlier (p. 395).

Frequently a bacterial pathogen is drug resistant because it has a plasmid bearing one or more resistance genes; such plasmids are called **R plasmids** (resistance plasmids; **figure 16.33**). Plasmid-borne resistance genes often code for enzymes that destroy or modify drugs. Plasmid-associated genes have been implicated in resistance to aminoglycosides, chloramphenicol, penicillins, cephalosporins, erythromycin, tetracyclines, sulfonamides, and others. Once a bacterial cell possesses an R plasmid, the plasmid (or its genes) may be transferred to other cells quite rapidly through HGT. Because a single plasmid may carry genes for resistance to several drugs, a pathogen population can become resistant to several antibiotics simultaneously, even though the infected patient is being treated with only one drug.

Antibiotic-resistance genes can be located on genetic elements other than plasmids. Many transposons contain genes for antibiotic resistance and can move rapidly between plasmids and through a bacterial population. As seen in figure 16.33, some R1 resistance genes are provided by a transposon. Transposons are found in both Gram-negative and Gram-positive bacteria. Some examples and their resistance markers are Tn*5* (kanamycin, bleomycin, streptomycin), Tn*21* (streptomycin, spectinomycin, sulfonamide), Tn*551* (erythromycin), and Tn*4001* (gentamicin, tobramycin, kanamycin). Tn*5* is of particular note because it is a replicative transposon. Thus it not only moves but also leaves a copy of itself in its original location. Recall that some transposons are conjugative transposons (or ICEs; table 16.3). Because they are capable of moving between bacteria by conjugation, they are also effective in spreading resistance.

Antibiotic-resistance genes are sometimes part of a DNA segment called an integron. An **integron** consists of a gene encoding an integrase (*intI*), a recombination site (*attI*), and a promoter that drives transcription of one or more "gene cassettes" that have been captured by the integron. Gene cassettes typically encode a single protein and have a recombination site called *attC*. The integrase enzyme catalyzes excision of the gene cassette from the

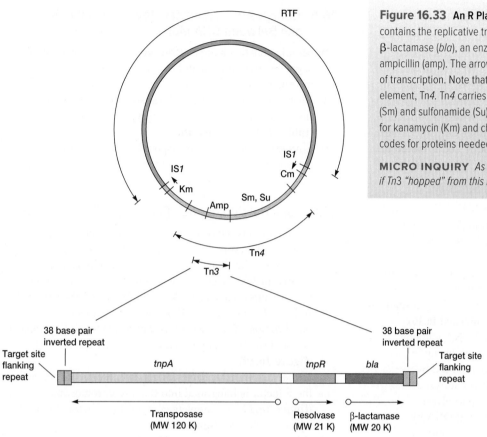

Figure 16.33 An R Plasmid.   Plasmid R1 is an R plasmid that contains the replicative transposon Tn*3*. Tn*3* contains the gene for β-lactamase (*bla*), an enzyme that confers resistance to the antibiotic ampicillin (amp). The arrows below the Tn*3* genes indicate the direction of transcription. Note that Tn*3* is inserted into another transposable element, Tn*4*. Tn*4* carries genes that provide resistance to streptomycin (Sm) and sulfonamide (Su). The R1 plasmid also carries resistance genes for kanamycin (Km) and chloramphenicol (Cm). The RTF region of R1 codes for proteins needed for plasmid replication and transfer.

MICRO INQUIRY *As a replicative transposon, what would happen if Tn*3 *"hopped" from this R1 plasmid into a different plasmid?*

integron. The cassette can be "captured" by another integron using its integrase enzyme; this involves a site-specific recombination event occurring between *attI* and *attC*. Thus gene cassettes can be moved from one integron to another. If a gene cassette is captured by an integron in a conjugative plasmid, the gene cassette can be transferred to other cells by conjugation (section 16.6).

Our focus here has been on the origin and transmission of resistance genes from antibiotic producers to nonproducers. What is important to recognize is that expression of these genes by nonproducers can be costly, expending considerable energy and metabolites. Resistance genes can be lost if the bacteria are not consistently exposed to the antibiotic, and in some cases, to biocides and heavy metals. Thus, exposure to the antibiotic is a selective force that helps maintain the resistance gene in a population of bacteria. Of concern is the relatively low concentrations of antibiotics and heavy metals needed to maintain the carriage of resistance genes. In most cases, sublethal levels of 1% of the minimum inhibitory concentration (MIC) of a single antibiotic is sufficient to maintain the resistant phenotype. Unfortunately, the levels of antibiotic and heavy metal contamination in the environment far exceed this amount. As we discuss in chapter 9, measures are being developed to decrease the use (especially misuse) of antibiotics to help alleviate the antibiotic-resistance problem, including a multinational effort to reduce antibiotic use in agriculture.

Key Concepts

16.1 Mutations: Heritable Changes in a Genome

- A mutation is a stable, heritable change in the nucleotide sequence of the genetic material.
- Spontaneous mutations can arise from replication errors (transition, transversion, and insertion and deletion of nucleotides), and from DNA lesions (apurinic sites, apyrimidinic sites, oxidation of DNA) (**figures 16.1–16.3**).
- Induced mutations are caused by mutagens. Mutations may result from the incorporation of base analogs, specific mispairing due to alterations of a base caused by DNA-modifying agents, the presence of intercalating agents, and severe damage to the DNA caused by exposure to radiation (**table 16.1; figures 16.4** and **16.5**).

- Mutations are usually recognized when they cause a change from the more prevalent wild-type phenotype. A mutant phenotype can be restored to wild type by reversion mutations or suppressor mutations (**table 16.2**).
- There are four important types of point mutations: silent mutations, missense mutations, nonsense mutations, and frameshift mutations (**table 16.2**).
- Mutations can affect phenotype in numerous ways. Some mutations affect colony or cell morphology, viability (i.e., lethal mutations), biochemical pathways, and susceptibility to environmental stresses (e.g., resistance mutations). Some mutations are conditional, being expressed under some conditions but not others (e.g., high temperature but not low temperatures).

16.2 Detection and Isolation of Mutants

■ A sensitive and specific screening method is needed for detecting and isolating mutants. An example is replica plating for the detection of auxotrophs (**figure 16.6**).

■ One of the most effective techniques for isolating mutants is to select for mutants by adjusting environmental conditions so that the mutant will grow while the wild type does not.

■ Because many carcinogens are also mutagenic, the Ames test can be used to measure mutagenicity and is therefore an indirect indication of carcinogenicity (**figure 16.7**).

16.3 DNA Repair Maintains Genome Stability

■ Proofreading during DNA replication helps correct errors in replication. However, cells have multiple mechanisms for correcting mispaired and damaged DNA when proofreading is not effective.

■ Mismatch repair replaces incorrect base pairs (**figure 16.8**).

■ Excision repair systems remove damaged portions from a single strand of DNA (e.g., thymine dimers) and use the other strand as a template for filling in the gap (**figures 16.9** and **16.10**).

■ Direct repair systems correct damaged DNA without removing damaged regions. For instance, during photoreactivation, thymine dimers are repaired by splitting the two thymines apart. This is catalyzed in the presence of light by the enzyme photolyase.

■ Recombinational repair removes damaged DNA by recombination of the damaged DNA with a normal DNA strand elsewhere in the cell (**figure 16.11**).

■ When DNA damage is severe, DNA replication is halted, and this triggers the SOS response. During the SOS response, genes of DNA repair systems are transcribed at a higher rate. In addition, special DNA polymerases are produced, which are able to replicate damaged DNA. However, they do so without a proper template and therefore create mutations.

16.4 Microbes Use Mechanisms Other than Mutation to Create Genetic Variability

■ In sexual organisms, the mixing of parental DNA molecules in the progeny, as well as crossing-over that occurs during gamete formation generates genetic diversity.

■ Horizontal gene transfer (HGT) is an important mechanism for creating genetic diversity, especially in bacteria and archaea. It is a one-way process in which donor DNA is transferred from the donor to a recipient. In many transfers, the donor DNA must be integrated into the recipient's chromosome to be stably maintained (**figure 16.12**).

■ The formation of recombinant microbes by HGT involves recombination of DNA molecules. Often this is accomplished by homologous recombination, in which DNA molecules pair due to large areas of sequence similarity. Some viral genomes and all known transposable elements integrate into other DNA molecules by site-specific recombination. This type of recombination does not require extensive sequence homology.

16.5 Transposable Elements Move Genes Within and Between DNA Molecules

■ Mobile genetic elements (transposable elements) are DNA segments that move about the genome in a process known as transposition.

■ There are numerous types of transposable elements, including insertion sequences and several types of transposons (**table 16.3** and **figure 16.13**).

■ Simple (cut-and-paste) transposition and replicative transposition are two distinct mechanisms of transposition (**figures 16.14** and **16.15**).

16.6 Bacterial Conjugation Requires Cell-Cell Contact

■ Conjugation is the transfer of genes between bacteria that depends upon direct cell-to-cell contact and is mediated by a plasmid. The F factor is one type of conjugative plasmid (**figure 16.16**). F factor conjugation in *E. coli* is accomplished by a sex pilus and a type IV secretion system (**figure 16.20**).

■ In F⁺ × F⁻, mating, the F factor remains independent of the chromosome and a copy is transferred to the F⁻ recipient; donor chromosomal genes are not usually transferred (**figure 16.19**).

■ Hfr strains transfer bacterial genes to recipients because the F factor is integrated into the host chromosome (**figure 16.22**). A complete copy of the F factor is not often transferred (**figure 16.23**).

■ When the F factor leaves an Hfr chromosome, it occasionally picks up some bacterial genes to become an F′ plasmid, which readily transfers these genes to other bacteria (**figure 16.24**).

16.7 Bacterial Transformation Is the Uptake of Free DNA from the Environment

■ Transformation is the uptake of naked DNA by a competent cell and its incorporation into the genome (**figures 16.25** and **16.26**).

■ Only certain bacterial species are naturally transformation competent. With one exception, naturally competent bacteria have uptake machinery composed of proteins similar to those of type II secretion systems and type IV pili (**figure 16.27**). Other species can be made competent by artificial means.

16.8 Transduction Is Virus-Mediated DNA Transfer

■ Bacterial viruses (bacteriophages) can multiply and destroy the host cell (lytic cycle) or become a latent prophage that remains within the host (lysogenic cycle) (**figure 16.28**).

■ Transduction is the transfer of genes by viruses.

■ In generalized transduction, any host DNA fragment can be packaged in a virus capsid and transferred to a recipient (**figure 16.29**).

■ Certain temperate phages carry out specialized transduction by incorporating bacterial genes during prophage induction and then donating those genes to another bacterium (**figures 16.30** and **16.31**).

16.9 Evolution in Action: The Development of Antibiotic Resistance in Bacteria

■ Bacteria can become resistant to a drug by modifying the target of the drug; altering or degrading the drug; or by pumping the drug out as soon as it enters, using an efflux pump (**figure 16.32**).

■ Genes encoding resistance can be found on chromosomes, R plasmids (**figure 16.33**), and transposable elements. Thus resistance genes are readily moved from one cell to another by horizontal gene transfer.

Compare, Hypothesize, Invent

1. Mutations are often considered harmful. Give an example of a mutation that would be beneficial to a microorganism. What gene would bear the mutation? How would the mutation alter the gene's role in the cell, and what conditions would select for this mutant allele?

2. Mistakes made during transcription affect the cell but are not considered "mutations." Why not?

3. Suppose that transduction took place when a U-tube experiment was conducted. How would you confirm that a virus was passed through the filter and transduced the recipient?

4. Suppose that you carried out a U-tube experiment with two auxotrophs and discovered that recombination was not blocked by the filter but was stopped by treatment with deoxyribonuclease. What gene transfer process is responsible? Why would it be best to use double or triple auxotrophs in this experiment?

5. What would be the evolutionary advantage of having a period of natural "competence" in a bacterial life cycle? What would be possible disadvantages?

6. Studies of phage therapy to treat bacterial infections were largely abandoned in the United States when antibiotics were discovered. However, the increasing number of important bacterial pathogens resistant to antibiotics has reinvigorated research in this area. One approach being examined is the use of phage to deliver genes that reverse an antibiotic-resistant phenotype. This is then coupled with treatment using the antibiotic to which the bacterium is once again susceptible. To determine if this was feasible, scientists transduced streptomycin-resistant *E. coli* cells with one of two different temperate phages. One phage (φ1) carried a wild-type *rpsL* gene; this gene encodes one of the proteins found in the 30S ribosomal subunit, the target of streptomycin (*see section 9.4*). The second phage (φ2) carried two *rpsL* genes: the wild-type gene and one containing numerous silent mutations. Both lysogenized bacteria were tested for their sensitivity to streptomycin by determining the minimum inhibitory concentration (MIC; *see section 9.3*) and comparing these MIC values to that of

the parent resistant strain. *E. coli* cells lysogenized with φ1 had an MIC one-tenth that of the resistant strain. Those bacteria lysogenized with φ2 had an MIC level nearly identical to that of streptomycin-sensitive *E. coli* strains. Suggest why scientists used this approach rather than simply using a virulent strain to kill streptomycin-resistant strains. Why do you think scientists included the *rpsL* gene having many silent mutations? Would this approach work with penicillin-resistant bacteria? Explain your answer.

Read the original paper: Edgar, R., et al. 2012. Reversing bacterial resistance to antibiotics by phage-mediated delivery of dominant sensitive genes. *Appl. Environ. Microbiol.* 78(3):744.

7. *Enterococcus faecalis* is a major cause of hospital-acquired infections. It contains a pathogenicity island (PAI) that can be transferred from one *E. faecalis* strain to another. The PAI encodes a number of virulence factors, including toxins and aggregation factors. Examination of the genes encoded by PAI revealed elements suggesting that it transfers by an integrative conjugative element (ICE)-like mechanism. However, this had not been clearly established and it was possible that the PAI transfers by an Hfr conjugative mechanism. This possibility was strengthened by the knowledge that *E. faecalis* strains contain resident conjugative plasmids known to mediate transfer of chromosomal genes. To determine which process moves PAI from one strain to another, scientists mutated the ICE-related elements in the PAI and then characterized the recombinant bacteria to see what had been transferred. How would transfer of PAI as an ICE differ from transfer mediated by one of the resident plasmids? What ICE-like elements were likely mutated in the study? Explain why scientists would have chosen these elements to mutate.

Read the original paper: Manson, J. M., et al. 2010. Mechanism of chromosomal transfer of *Enterococcus faecalis* pathogenicity island, capsule, antimicrobial resistance, and other traits. *Proc. Natl. Acad. Sci.* 107(27):12269.

17

Recombinant DNA Technology

The polymerase chain reaction (PCR) has revolutionized forensic biology. Whether scientists are solving a crime or exploring ancient Egyptian civilizations, PCR allows them to answer questions they previously would not have asked.

Archeological Digs Reveal Source of Ancient Pathogen

A 9,000-year-old grave holding a mother and her infant, and 4,000-year-old Egyptian mummies have helped unravel the mystery of the origin of tuberculosis (TB). Probably the world's most widespread disease, TB currently affects 2 billion people, about a third of the world's population. Most are infected with the Gram-positive bacterium *Mycobacterium tuberculosis,* but *M. canettii* and *M. africanum* are also known to cause disease in humans.

Animals, including cows, also can contract TB. Before pasteurization, *M. bovis* was responsible for about 6% of human TB deaths. Because people can contract TB from cattle, it was long believed that the first human TB case was caused by *M. bovis*. A *M. bovis* strain was thought to have slowly evolved to the current *M. tuberculosis*. However, this notion was questioned in 2003, when it was reported that a third of 85 mummies examined from Thebes West, Upper Egypt, were infected with an ancient strain of *M. tuberculosis*. (Note that the incidence of TB has not improved in over 4,000 years!) Then in 2008, a team of scientists from Britain and Israel tested the remains of a mother and her baby buried together prior to the advent of dairy farming. The burial site, Atlit-Yam, Israel, was later flooded and covered with a thick layer of mud and clay, and eventually seawater. This kept the bodies in an anoxic state, slowing the rate of decay. In both Egypt and Israel, preservation made it possible for scientists to recover snippets of DNA from bones and use the polymerase chain reaction (PCR; section 17.2) to verify the presence of nucleotide sequences unique to *M. tuberculosis,* but not *M. bovis*.

So where did this pathogen come from? The best hypothesis is that TB arose in Africa from an ancestor of *M. africanum*, perhaps as many as 3 million years ago. Subsequently an *M. africanum*-like strain left Africa with the migration of ancient populations. Although this microbe most likely evolved into the successful pathogen that still dominates human infectious disease, there is a surprising twist to the story. Genome sequences of TB bacteria from Peruvian mummies suggest that the disease was carried to the New World by seals or sea lions.

The success of such amazing archeological detective work was made possible by PCR. In this chapter, we introduce some of the techniques used in recombinant DNA technology—a field revolutionized by PCR.

PCR enables the production of massive quantities of targeted DNA and often represents one of the first steps in genetic cloning—the generation of a large number of genetically identical DNA molecules (**figure 17.1**). Here we review the essential methods used in biotechnology; this requires an understanding of basic genetic principles.

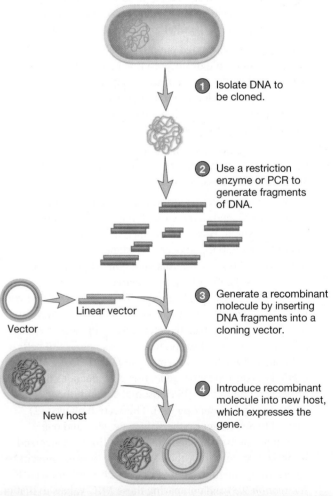

1. Isolate DNA to be cloned.

2. Use a restriction enzyme or PCR to generate fragments of DNA.

3. Generate a recombinant molecule by inserting DNA fragments into a cloning vector.

Vector

Linear vector

4. Introduce recombinant molecule into new host, which expresses the gene.

New host

Figure 17.1 Steps in Cloning a Gene. Each step shown in this overview is discussed in more detail in this chapter.

MICRO INQUIRY *Which of the DNA molecules shown are recombinant?*

Readiness Check:

Based on what you have learned previously, you should be able to:

✔ Identify the key structural elements of DNA, RNA, and protein (section 13.2)

✔ Diagram the flow of cellular information from gene to protein as well as protein localization (e.g., subcellular localization or export) (section 13.8)

✔ Explain the general structural differences in a eukaryotic gene and a bacterial gene, including regulatory, coding, and noncoding sequences (sections 13.3, 13.5, and 15.3)

✔ Describe the capacity of proteins to recognize and bind to DNA (sections 13.2 and 14.2)

✔ List three differences between a plasmid and a chromosome (sections 3.6 and 16.6)

17.1 Key Discoveries Led to the Development of Recombinant DNA Technology

After reading this section, you should be able to:

■ Explain how restriction enzymes recognize and digest DNA to create either blunt or sticky ends

■ Explain the general principles by which molecules are electrophoretically separated

■ Draw an agarose gel in which molecular weight markers and digested DNA can be visualized

■ Discuss the use of single-stranded oligonucleotide probes to identify specific fragments of DNA (or RNA)

■ Diagram the reaction catalyzed by reverse transcriptase and describe its application to biotechnology

The pace of research and discovery in the last half century has been remarkable thanks to the development of recombinant DNA technologies. Many early discoveries are essential to biotechnology as it is practiced today; these are now discussed.

Restriction Enzymes

Recombinant DNA is DNA with a new nucleotide sequence. Such DNA is formed by joining fragments from two or more different sources. One of the first breakthroughs leading to recombinant DNA technology was the discovery by Werner Arber and Hamilton Smith in the late 1960s of bacterial enzymes that cut double-stranded DNA. These enzymes, known as **restriction enzymes** or restriction endonucleases, recognize and cleave specific sequences about four to eight base pairs long (**figure 17.2**). Restriction enzymes identify specific DNA sequences called recognition sites. Each restriction enzyme has its own recognition site. Hundreds of different restriction enzymes have been purified and are commercially available. Type I and type III endonucleases identify their unique recognition sites and then cleave DNA at a defined distance from it. The more common type II endonucleases cut

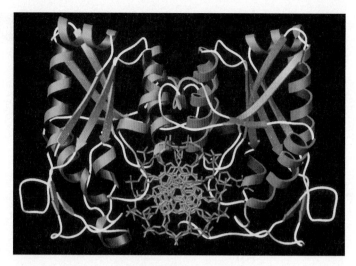

Figure 17.2 Restriction Endonuclease Bound to DNA. The structure of BamHI bound to DNA viewed down the DNA axis. The enzyme's two subunits lie on each side of the DNA double helix. The α-helices are in green, the β conformations in purple, and the DNA is in orange.

DNA directly at their recognition sites. These enzymes can be used to prepare DNA fragments containing specific genes or portions of genes. For example, the restriction enzyme EcoRI, isolated by Herbert Boyer in 1969 from *Escherichia coli*, cleaves DNA between G and A in the base sequence 5′-GAATTC-3′ (**figure 17.3**). Because DNA is antiparallel, this sequence is reversed on the complementary strand of DNA. When EcoRI cleaves between the G and A residues, unpaired 5′-AATTC-3′ remains at the end of each strand. The complementary bases on two EcoRI-cut fragments can hydrogen

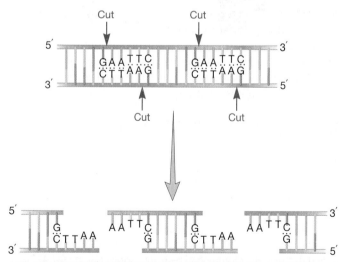

Figure 17.3 Restriction Endonuclease Action. The cleavage catalyzed by the restriction endonuclease EcoRI. The enzyme makes staggered cuts on the two DNA strands to form sticky ends.

MICRO INQUIRY *Examine the uncut piece of DNA shown in the upper half of this figure. Where would an exonuclease cleave?*

Table 17.1   Some Type II Restriction Endonucleases and Their Recognition Sequences

Enzyme	Microbial Source	Recognition Sequence[1]	Ends Produced
AluI	*Arthrobacter luteus*	↓ 5′AGCT 3′ 3′ TCGA 5′ ↑	5′ AG CT 3′ 3′ TC GA 5′
BamHI	*Bacillus amyloliquefaciens* H	↓ 5′ GGATCC 3′ 3′ CCTAGG5′ ↑	5′ G GATCC 3′ 3′ CCTAG G 5′
EcoRI	*Escherichia coli*	↓ 5′ GAATTC 3′ 3′ CTTAAG 5′ ↑	5′ G AATTC 3′ 3′ CTTAA G 5′
HindIII	*Haemophilus influenzae* d	↓ 5′ AAGCTT 3′ 3′ TTCGAA 5′ ↑	5′ A AGCTT 3′ 3′ TTCGA A 5′
NotI	*Nocardia otitidis-caviarum*	↓ 5′ GCGGCCGC 3′ 3′ CGCCGGCG 5′ ↑	5′ GC GGCCGC 3′ 3′ CGCCGG CG 5′
SalI	*Streptomyces albus*	↓ 5′ GTCGAC 3′ 3′ CAGCTG 5′ ↑	5′ G TCGAC 3′ 3′ CAGCT G 5′

1 The arrows indicate the sites of cleavage on each strand.

MICRO INQUIRY: *Which of the above enzymes yield blunt ends? Which yield sticky ends with a 5′ overhang? What about a 3′ overhang?*

bond, thus EcoRI and other endonucleases like it generate cohesive or **sticky ends.** In contrast, cleavage by restriction enzymes such as AluI and HaeIII leaves blunt ends. A few restriction enzymes and their recognition sites are listed in **table 17.1**. Note that each enzyme is named after the bacterium from which it is purified. ⮔ *Restriction Enzymes*

Genetic Cloning and cDNA Synthesis

An important advance in cloning DNA came in 1972, when David Jackson, Robert Symons, and Paul Berg reported that they had successfully generated recombinant DNA molecules. They allowed the sticky ends of fragments to anneal—that is, to base pair with one another (figure 17.3), and then covalently joined the fragments with the enzyme DNA ligase (*see figure 13.15*). Within a year, plasmid vectors into which foreign DNA fragments had been inserted during gene cloning had been developed (**figure 17.4**). Such recombinant plasmids replicate within a microbial host, thereby maintaining the cloned fragment of DNA. ⮔ *Early Genetic Engineering Experiment*

Once it was demonstrated that foreign genes could be recombined, or inserted, into cloning vectors, biologists sought to clone specific genes from various organisms. However, it was evident that cloning eukaryotic DNA into bacterial hosts would be problematic. This is because eukaryotic pre-mRNA must be processed (e.g., introns spliced out), and bacteria lack the molecular machinery to perform this task. In 1970 Howard Temin and David Baltimore independently discovered the enzyme that solved this dilemma. They isolated the enzyme **reverse transcriptase** (**RT**) from retroviruses. These viruses have an RNA genome that is copied into DNA prior to replication. The mechanism by which reverse transcriptase accomplishes this is outlined in **figure 17.5**. Processed mRNA can be used as a template for **complementary DNA (cDNA)** synthesis in vitro. The resulting cDNA can then be cloned, without the need for RNA processing. ▶▶ *Retroviruses (section 27.8)* ⮔ *cDNA*

Gel Electrophoresis

The notion that **gel electrophoresis** could be used to separate macromolecules (e.g., DNA and proteins) on the basis of size was first introduced in the 1930s with the development of sucrose gel electrophoresis. The following 40 years brought the development of a variety of different gel systems, but it was not until the 1970s that the precursors to present-day techniques appeared.

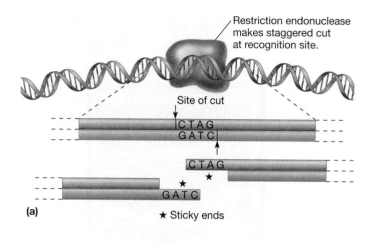

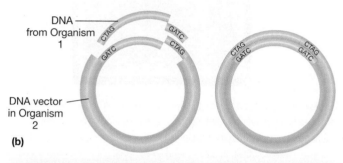

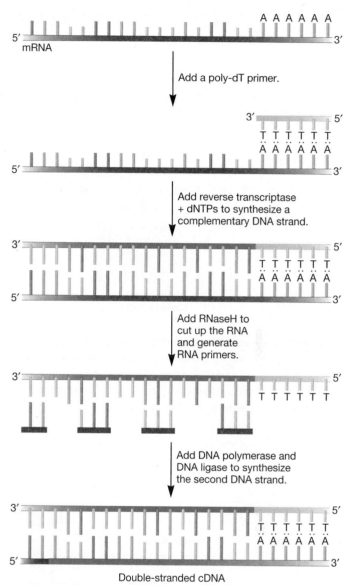

Figure 17.4 Recombinant Plasmid Construction. (a) A restriction endonuclease recognizes and cleaves DNA at its specific recognition site. Cleavage produces sticky ends that accept complementary tails for gene splicing. (b) The sticky ends can be used to join DNA from different organisms by cutting it with the same restriction enzyme, ensuring that all fragments have complementary ends.

MICRO INQUIRY *Why might two DNA fragments inadvertently be cloned into a single vector when using this cloning strategy?*

Figure 17.5 Synthesis of cDNA. A poly-dT primer anneals to the 3' end of mRNAs. Reverse transcriptase then catalyzes the synthesis of a complementary DNA strand (cDNA). RNaseH digests the mRNA into short pieces that are used as primers by DNA polymerase to synthesize the second DNA strand. The 5' to 3' exonuclease function removes all of the RNA primers except the one at the 5' end (because there is no primer upstream from this site). This RNA primer can be removed by the subsequent addition of another RNase. After the double-stranded cDNA is made, it can be inserted into vectors, as described in figure 17.4.

MICRO INQUIRY *Why must introns be removed from eukaryotic DNA before it can be expressed in a bacterium?*

Agarose and polyacrylamide gel electrophoresis is now routinely used to separate and visualize DNA fragments. When DNA molecules are placed at the negative end of an electrical field, they migrate toward the pole with a positive charge (**figure 17.6**). Each fragment's migration rate is determined by its molecular weight so that the smaller a fragment is, the faster it moves through the gel. Migration rate is also a function of gel density. In practice, this means that higher concentrations of gel material (agarose or acrylamide) provide better resolution of small fragments and vice versa. Plasmid and bacterial chromosomal DNA that has not been digested with restriction enzymes is usually supercoiled and does not migrate to a position corresponding to its molecular weight. For this and other reasons, DNA is usually cut with restriction enzymes prior to electrophoresis.

Southern Blotting

Another problem early biotechnologists faced was the inability to distinguish the fragment of DNA possessing the gene of interest from the numerous chromosomal fragments produced by restriction enzyme digestion of an organism's genome. In 1975 Edwin Southern solved this problem with his **Southern blotting technique.** This procedure enables the detection of specific DNA fragments from a mixture of DNA molecules. As outlined in **figure 17.7**, Southern blotting is a three-step

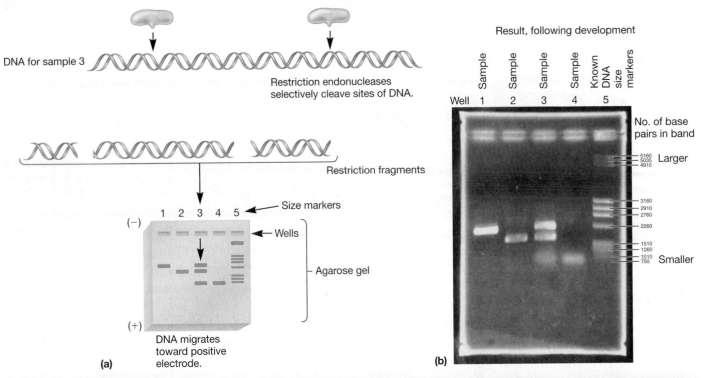

Figure 17.6 Gel Electrophoresis of DNA. (a) After cleavage into fragments, DNA is loaded into wells on one end of an agarose gel. When an electrical current is passed through the gel (from the negative pole to the positive pole), the DNA, being negatively charged, migrates toward the positive pole. The larger fragments, measured in numbers of base pairs, migrate more slowly and remain nearer the wells than the smaller (shorter) fragments. (b) Developed and stained gel revealing the separation pattern of the fragments of DNA. The size of a given DNA band can be determined by comparing it to a known set of molecular weight markers (lane 5) called a ladder.

MICRO INQUIRY *If a linear piece of DNA is cut with a restriction enzyme for which it has four recognition sites, how many bands would be visible on a gel if the products of digestion were electrophoresed? (Hint: Make a drawing of the DNA and its restriction sites.)*

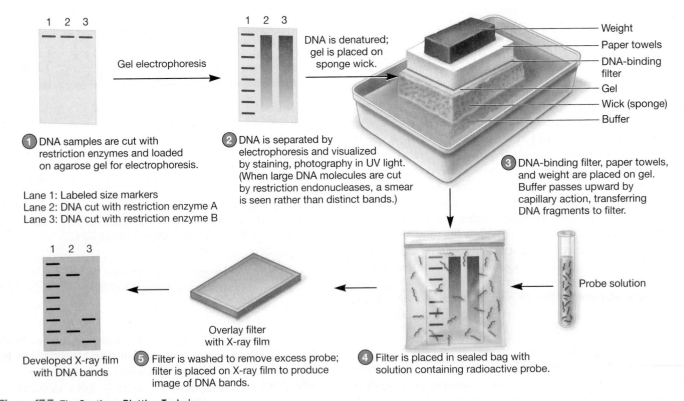

Figure 17.7 The Southern Blotting Technique.

17.1 Streptavidin-Biotin Binding and Biotechnology

Egg white contains many proteins and glycoproteins with unique properties. One of the most interesting, which binds tenaciously to biotin, was isolated in 1963. This glycoprotein, called avidin due to its "avid" binding of biotin, was suggested to play an important role: making egg white antimicrobial by "tying up" the biotin needed by many microorganisms. Avidin, which functions best under alkaline conditions, has the highest known binding affinity between a protein and a ligand. Several years later, scientists at Merck & Co., Inc., discovered a similar protein produced by the actinomycete *Streptomyces avidini,* which binds biotin at a neutral pH and does not contain carbohydrates. These characteristics make this protein, called streptavidin, an ideal

binding agent for biotin, and it has been used in an almost unlimited range of biotechnology applications. The streptavidin protein is joined to a probe. When a sample is incubated with the biotinylated binder, the binder attaches to any available target molecules. The presence and location of target molecules can be determined by treating the sample with a streptavidin probe because the streptavidin binds to the biotin on the biotinylated binder, and the probe is then visualized. This detection system is employed in a wide variety of biotechnological applications, including use as a nonradioactive probe in hybridization studies and as a critical component in biosensors for a wide range of environmental monitoring and clinical applications.

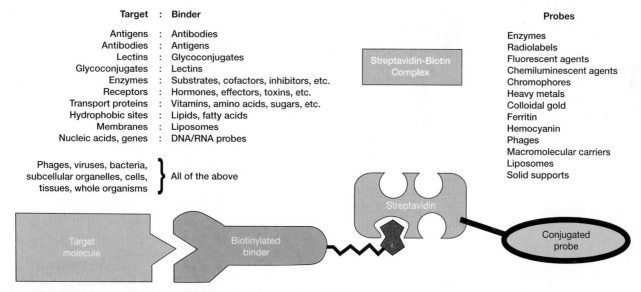

Target	:	Binder
Antigens	:	Antibodies
Antibodies	:	Antigens
Lectins	:	Glycoconjugates
Glycoconjugates	:	Lectins
Enzymes	:	Substrates, cofactors, inhibitors, etc.
Receptors	:	Hormones, effectors, toxins, etc.
Transport proteins	:	Vitamins, amino acids, sugars, etc.
Hydrophobic sites	:	Lipids, fatty acids
Membranes	:	Liposomes
Nucleic acids, genes	:	DNA/RNA probes

Phages, viruses, bacteria, subcellular organelles, cells, tissues, whole organisms } All of the above

Probes

Enzymes
Radiolabels
Fluorescent agents
Chemiluminescent agents
Chromophores
Heavy metals
Colloidal gold
Ferritin
Hemocyanin
Phages
Macromolecular carriers
Liposomes
Solid supports

Streptavidin-Biotin Complex

Target molecule

Biotinylated binder

Streptavidin

Conjugated probe

Streptavidin-Biotin Binding Systems Are Widely Used in Biotechnology, Medicine, and Environmental Studies.   Each molecule of streptavidin, a protein derived from an actinomycete, has four sites by which it can bind tenaciously to biotin (noted in red). By attaching a binder to the biotin and a probe, such as a fluorescent molecule, to the streptavidin, the target molecule can be detected at low concentrations. Target binders, probes, and applications are noted.

process: separate DNA molecules by electrophoresis, transfer the single-stranded (ss) DNA molecules to a membrane, and hybridize a labeled ssprobe specific for the gene of interest to its corresponding genomic complement. An **oligonucleotide** (Greek *oligo,* few or scant) **probe,** is a fragment of ssDNA complementary to the DNA of interest. It is labeled, often with radioactive phosphate, and is usually 15 to 30 nucleotide bases long. If the probe is radioactive, the DNA to which the probe hydrogen bonds becomes radioactive and is detected by **autoradiography.** In autoradiography, a sheet of photographic film is placed over the Southern blot membrane that has both the chromosomal DNA and the bound probe. When the film is developed, bands appear wherever the radioactive probe is bound

because the energy released by the isotope exposes the film. Probes can alternatively be labeled with a nonradioactive ("cold") molecule. A common label is the small molecule biotin, as discussed in **Techniques & Applications 17.1.** Nonradioactive labels often are more rapidly detected and are safer to use than radioisotopes. ⮩ *Southern Blotting*

Thanks to the advent of the polymerase chain reaction (PCR) and genomic sequencing, Southern blotting is not performed as often as it once was. However, it was the first technique to employ hybridization of a probe to target DNA, an approach that is the basis of a variety of important methods that continue to be used today. ▶▶ *Microbial genomics (chapter 18)*

1. Describe restriction enzymes, sticky ends, and blunt ends. Can you think of a cloning situation where blunt-ended DNA might be more useful than DNA with sticky ends?

2. What is cDNA? Why is it necessary to generate cDNA before cloning and expressing a eukaryotic gene in a bacterium?

3. You want to visualize a digested plasmid that yields fragments of 100 bp, 400 bp, and 3,000 bp. You have another plasmid digested with the same endonuclease that yields two fragments, 4,000 bp and 5,500 bp. How many recognition sites are there for this enzyme in each plasmid? For which plasmid would you use a 1.5% agarose gel? For which would a 0.8% gel be best? Explain your answer.

4. What is the purpose of Southern blotting? How is a probe selected? Why was the Southern blotting technique was an important breakthrough when it was introduced?

17.2 Polymerase Chain Reaction Amplifies Targeted DNA

After reading this section, you should be able to:

- Differentiate between a PCR cycle and step, and define the function of each of the three steps used in a PCR cycle
- Explain why PCR results in the amplification of a specific DNA sequence despite many competing sequences
- Explain why PCR generates billions of products that are all the same size
- Contrast real-time, quantitative PCR to end-point collection PCR and identify an application for each
- Summarize the importance of PCR in biology

The **polymerase chain reaction (PCR)**, invented by Kary Mullis in the early 1980s, exploded onto the biotechnology landscape. It changed the way genes are cloned, nucleic acids sequenced,

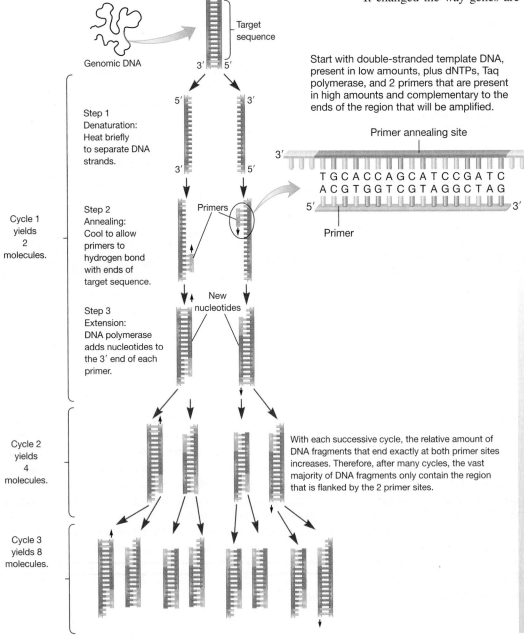

Start with double-stranded template DNA, present in low amounts, plus dNTPs, Taq polymerase, and 2 primers that are present in high amounts and complementary to the ends of the region that will be amplified.

Primer annealing site

T G C A C C A G C A T C C G A T C
A C G T G G T C G T A G G C T A G

Primer

With each successive cycle, the relative amount of DNA fragments that end exactly at both primer sites increases. Therefore, after many cycles, the vast majority of DNA fragments only contain the region that is flanked by the 2 primer sites.

Figure 17.8 The Polymerase Chain Reaction (PCR) Technique. During each cycle, oligonucleotides that are complementary to the ends of the targeted DNA sequence bind to the DNA and act as primers for the synthesis of this DNA region. The primers are usually 15 to 20 nucleotides in length. The region between the two primers is typically hundreds of nucleotides in length, not just several nucleotides as shown here. The net result of PCR is the synthesis of many copies of DNA in the region that is flanked by the two primers.

MICRO INQUIRY *Why, after three cycles, are the vast majority of amplified DNA molecules (i.e., PCR products) the size defined by the distance between the forward and reverse primers?*

diseases diagnosed, and crimes solved. Why is PCR so versatile and important? Quite simply, it enables the rapid synthesis of billions of copies of a specific DNA fragment from a complex mixture of DNA molecules. Researchers can thus obtain large quantities of specific pieces of DNA for experimental and diagnostic purposes.

Figure 17.8 outlines how PCR works. Suppose that one wishes to make large quantities of a particular gene or other DNA sequence, a process known as gene or **DNA amplification.** The first step is to synthesize **oligonucleotide primers.** These DNA fragments are designed to be complementary to the DNA just before (i.e., 5′) and after (3′) the DNA to be amplified. Primers provide the 3′-OH needed for DNA synthesis during PCR. Because primers are based on these flanking nucleotide sequences, only DNA with known (or nearly known) sequence can be amplified by PCR. Oligonucleotide primers are made with a DNA synthesizer and are generally between 15 and 30 nucleotides long. They are added to the reaction mixture, along with the template DNA (often copies of an entire genome), a thermostable DNA polymerase, and each of the four deoxyribonucleoside triphosphates (dNTPs).

PCR requires a series of repeated reactions, called cycles. Each cycle precisely executes three steps in a machine called a thermocycler. In the first step, the DNA containing the sequence to be amplified is denatured by raising the temperature to about 95°C. Next the temperature is lowered to about 50°C so that the primers can hydrogen bond (anneal) to complementary DNA on both sides of the target sequence. Finally, the temperature is raised, usually to 68 to 72°C, so that DNA polymerase can extend the primers and synthesize copies of the target DNA sequence using dNTPs. Only polymerases that function at the high temperatures can be used. The most commonly used thermostable enzyme is **Taq polymerase** from the thermophilic bacterium *Thermus aquaticus.*

At the end of one PCR cycle, the targeted sequences on both strands have been copied. When the three-step cycle is repeated (figure 17.8), the two strands from the first cycle are copied to produce four fragments. These are amplified in the third cycle to yield eight double-stranded products. Thus, each cycle increases the number of target DNA molecules exponentially. Depending on the initial concentration of the template DNA and other parameters such as the G + C content of the DNA to be amplified, it is theoretically possible to produce about 1 million copies of targeted DNA sequence after 20 cycles and over 1 billion after 30 cycles. Pieces ranging in size from less than 100 base pairs to several thousand base pairs in length can be amplified, and the initial concentration of target DNA can be as low as 10^{-20} to 10^{-15} M. ◀◀ *DNA replication in bacteria (section 13.3)* 🕭 *Polymerase Chain Reaction*

PCR is most frequently used in two ways. If large quantities of a specific piece of DNA are needed, the reaction products are collected and purified at the end of a designated number of cycles. This is sometimes called end-point PCR, and the final number of DNA fragments amplified is not quantitative. This means that the amount of final product does not reflect the amount of template DNA present. In contrast,

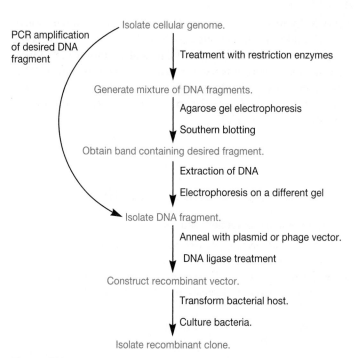

Figure 17.9 The Use of PCR in Cloning DNA Fragments. The use of PCR to synthesize DNA bypasses several steps traditionally used in cloning procedures. In addition, primers can be designed so that the restriction enzyme recognition site of choice is present at each end of the DNA fragment to be cloned.

real-time PCR is quantitative; in fact, it is referred to as qPCR. That is, the amount of DNA or RNA template (which is converted to DNA with reverse transcriptase prior to starting PCR) present in a given sample can be determined. This is accomplished by adding a fluorescently labeled probe to the reaction mixture and measuring its signal during the initial cycles. This is when the rate of DNA amplification is logarithmic. However, as the PCR cycles continue, substrates are consumed and polymerase efficiency declines. So although the amount of product increases, its rate of synthesis is no longer exponential (this is why end-point collection of PCR products is not quantitative). Specially designed thermocyclers record the amount of PCR product generated as it occurs, thus the term real-time PCR. Gene expression studies often rely on real-time PCR because mRNA transcripts can be copied by reverse transcriptase to cDNA, which is then quantified. Therefore the procedure monitors transcription of the gene targeted by the primers.

PCR is an essential tool in many areas of molecular biology, medicine, and biotechnology. As shown in **figure 17.9**, when PCR is used to obtain DNA for cloning, a number of steps in traditional cloning procedures are no longer required. PCR is also used to generate DNA for nucleotide sequencing. Because the primers used in PCR target specific DNA, PCR can isolate particular fragments of DNA (e.g., genes) from materials that contain many different genomes, such as soil, water, and blood. This explains why PCR has become an essential part of certain diagnostic tests, including those for

AIDS, Lyme disease, chlamydia, hepatitis, human papillomavirus, and other infectious agents and diseases. The tests are rapid, sensitive, and specific. PCR is also employed in forensic science, where it is used in criminal cases as part of DNA fingerprinting technology. ▶▶| *Metagenomics provides access to uncultured microbes (section 18.3); Genomic fingerprinting (section 19.3)*

17.3 Cloning Vectors Are Needed to Create Recombinant DNA

After reading this section, you should be able to:

- List the three features of a cloning vector and why these elements are needed
- Differentiate between plasmids, phage-based cloning vectors, cosmids, and artificial chromosomes in terms of structure and application
- Explain how a piece of foreign DNA is recombined in vitro into a cloning vector

Recombinant DNA technology depends on the propagation of many copies of the nucleotide sequence of choice. To accomplish this, genes or other genetic elements to be cloned are inserted into **cloning vectors** that replicate in a host organism. There are four major types of vectors: plasmids, bacteriophages and other viruses, cosmids, and artificial chromosomes (**table 17.2**). Each type has its own advantages and applications, so the selection of the proper cloning vector is critical to the success of any cloning experiment. Most cloning vectors share three important features: an origin of replication; a region of DNA that bears unique restriction sites, called a multicloning site or polylinker; and a selectable marker. These elements are described next in the context of plasmids, the

most frequently used cloning vectors. ↻ *Construction of a Plasmid Vector*

Plasmids

Plasmids make excellent cloning vectors because they replicate autonomously (i.e., independently of the chromosome) and are easy to purify. They can be taken up by microbes by conjugation or transformation. Many different plasmids are used in biotechnology, all derived from naturally occurring plasmids that have been genetically engineered (**figure 17.10**). |◀◀ *Plasmids (section 3.6); Bacterial conjugation requires cell-cell contact (section 16.6); Bacterial transformation is the uptake of free DNA from the environment (section 16.7)*

Origin of Replication

The **origin of replication** (*ori*) allows the plasmid to replicate in the microbial host independently of the chromosome. pUC19, an *E. coli* plasmid, has an *ori* that generates a "high copy number" because it directs about 50–100 plasmid replications in the course of one cell cycle. High copy number is often important because it facilitates plasmid purification and can dramatically increase the amount of cloned gene product produced by the cell. Some plasmids have two origins of replication, each recognized by different host organisms. These plasmids are called **shuttle vectors** because they can be transferred, or "shuttle," from one host to another. YEp24 is a shuttle vector that can replicate in yeast (*Saccharomyces cerevisiae*) and in *E. coli* because it has the 2μ circle yeast replication element and the *E. coli* origin of replication (figure 17.10).

Selectable Marker

Following the uptake of vector by host cells, it is necessary to discriminate between cells that successfully obtained vector

Table 17.2	Recombinant DNA Cloning Vectors		
Vector	**Insert Size (kb, 1 kb = 1,000 bp)**	**Example**	**Characteristics**
Plasmid	<20 kb	pBR322, pUC19	Replicates independently of microbial chromosome so many copies may be maintained in a single cell
Bacteriophage	9–25 kb	λ 1059, λ gt11, M13mp18, EMBL3	Packaged into lambda phage particles; single-stranded DNA viruses such as M13 have been modified (e.g., M13mp18) to generate either double- or single-stranded DNA in the host
Cosmids	30–47 kb	pJC720, pSupercos	Can be packaged into lambda phage particles for efficient introduction into bacteria, then replicates as a plasmid
BACs (bacterial artificial chromosomes)	75–300 kb	pBAC108L	Modified F plasmid that can carry large DNA inserts; very stable within the cell
YACs (yeast artificial chromosomes)	100–1,000 kb	pYAC	Can carry largest DNA inserts; replicates in *Saccharomyces cerevisiae*

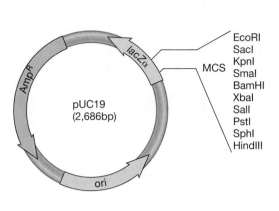

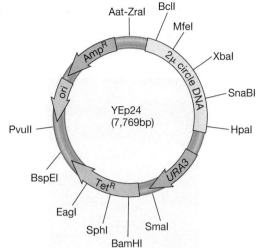

Figure 17.10 **The Cloning Vectors pUC19 and YEp24.** Restriction sites that are present only once in each vector are shown. pUC19 replicates only in *E. coli*, while YEp24 replicates in both *E. coli* and *S. cerevisiae*. MCS is multiple cloning site.

MICRO INQUIRY *Which plasmid is a shuttle vector? Why?*

(transformants) from those that did not (nontransformants). Furthermore, selective conditions for the presence of plasmid must be maintained, otherwise the host cell may stop replicating it. This is achieved by the presence of a gene encoding a protein needed for the cell to survive under certain conditions. Such a gene is called a **selectable marker.** In the case of pUC19, the selectable marker, amp^R, encodes the ampicillin-resistance enzyme. Because *E. coli* is normally susceptible to ampicillin, only those cells that have taken up plasmid (i.e., transformants) will grow when plated on agar containing ampicillin. The shuttle vector YEp24 bears both the amp^R gene to be used when selecting *E. coli* transformants and URA3, which encodes a protein essential for uracil biosynthesis in yeast. Therefore this plasmid must be used in strains of *S. cerevisiae* that are uracil auxotrophs, otherwise it would be impossible to select, or identify, transformants.

Multicloning Site or Polylinker

In order to clone a fragment of DNA into a plasmid, both the fragment and plasmid are cut with the same restriction enzyme (or enzymes) so that compatible sticky ends are generated. It is essential that these restriction enzymes cut at only one place in the plasmid. Cleavage at a unique restriction site generates a linear plasmid. Alternatively, two different, unique sites may be cleaved and the DNA sequence between the two sites replaced with cloned DNA. Plasmids used for cloning have been designed with many restriction sites clustered in a single region called the **multicloning site** (**MCS**) (figure 17.10). When the plasmid and the DNA to be inserted are cut with one of these enzymes and then incubated in the presence of DNA ligase, the compatible sticky ends hydrogen bond and phosphodiester bonds are generated between the cloned DNA fragment and the vector (**figure 17.11**).

Unfortunately, the ligation of foreign DNA into a vector is never 100% efficient. Thus when the ligation mixture is introduced into host cells, one must distinguish cells that carry plasmid without a DNA insert from those that carry plasmid into which DNA was successfully cloned. To accomplish this, the MCS in plasmids such as pUC19 is located within the 5′ end of the *lacZ* gene. *lacZ* encodes β-galactosidase (β-Gal), an enzyme that cleaves the disaccharide lactose into galactose and glucose (*see figure 14.2*). When DNA has been inserted into the MCS, the *lacZ* gene is interrupted and a functional β-galactosidase is not produced. This can be detected by the color of colonies: cells turn blue when β-Gal splits the alternative substrate, X-Gal (5-bromo-4-chloro-3-indolyl-β-D-galactopyranoside), which is included in the medium (figure 17.11c). This is important because all *E. coli* cells that take up plasmid (with or without insert) are selected by their resistance to ampicillin (Amp^R); that is to say, only pUC19 transformants grow. Among these, colonies with plasmid lacking the desired DNA insert will be blue (due to the presence of functional *lacZ* gene), while those with pUC19 into which DNA was successfully cloned will be white. There are a number of other clever ways in which cells with vector versus those with vector plus insert can be differentiated; the detection of blue versus white colonies, though, is a common approach.

Phage Vectors

Phage vectors are phage genomes that have been genetically modified to include useful restriction enzyme recognition sites for the insertion of foreign DNA. Once DNA has been inserted, the recombinant phage genome is packaged into viral capsids and used to infect host cells. The resulting phage lysate consists of thousands of phage particles that carry cloned DNA as well as the genes needed for host lysis. Two commonly used vectors are derived from the *E. coli* bacteriophages T7 and lambda (λ), both of which have double-stranded DNA genomes. In addition, phage vectors have been engineered for a number of different bacterial host species. ▶▶| *Bacteriophage lambda (section 27.2)*

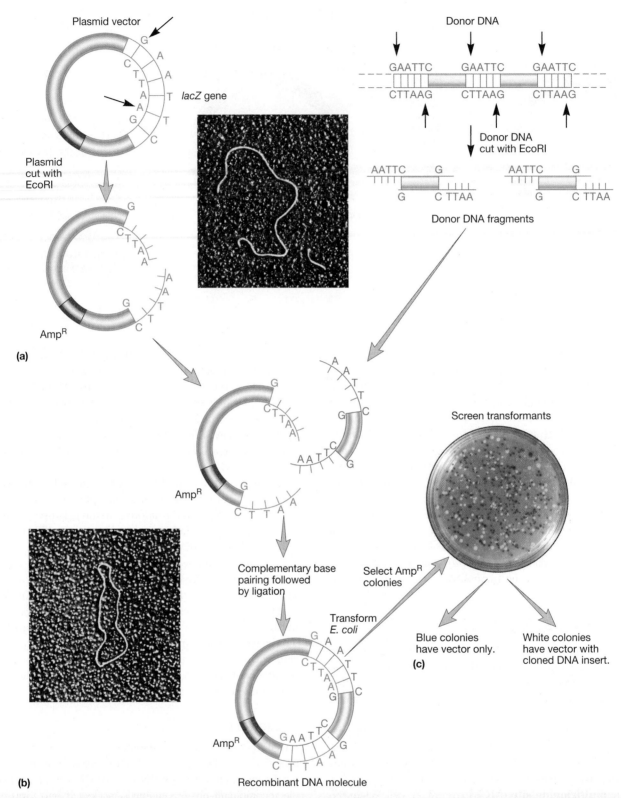

Figure 17.11 Recombinant Plasmid Construction and Cloning. The construction and cloning of a recombinant plasmid vector using an antibiotic-resistance gene to select for the presence of the plasmid. The interruption of the *lacZ* gene by cloned DNA is used to detect vectors with insert. The scale of the sticky ends of the fragments and plasmid has been enlarged to illustrate complementary base pairing. (a) The electron micrograph shows a plasmid that has been cut by a restriction enzyme and a donor DNA fragment. (b) The micrograph shows a recombinant plasmid. (c) After transformation, *E. coli* cells are plated on medium containing ampicillin and X-Gal so that only ampicillin-resistant transformants grow; X-Gal enables the visualization of colonies that were transformed with recombinant vector (vector + insert, white colonies).

MICRO INQUIRY *What would you conclude if you obtained only blue colonies after cloning into a vector that enables blue/white screening of transformants? Why might such a result occur?*

Cosmids

DNA inserts larger than about 15,000 bp to 25,000 bp cannot be stably maintained in plasmids and phage vectors, respectively. Instead, **cosmids** can be used to clone such large fragments of DNA (table 17.2). These engineered vectors have a *cos* site from phage and a selectable marker, origin of replication, and MCS from plasmids (thus the term "cos-mid"). These hybrid vectors replicate as plasmids within the host cell, but the presence of the *cos* sites means that the vector can be packaged into phage capsids and transferred to new host cells by transduction. ▶▶ *Transduction is virus-mediated DNA transfer (section 16.8)*

Artificial Chromosomes

Artificial chromosomes are special cloning vectors used when particularly large fragments of DNA must be cloned, as when constructing a genomic library. In fact, the genomic library containing the human genome was constructed in **bacterial artificial chromosomes** (**BACs**), which then enabled nucleotide sequencing. BACs were also used in the construction of a synthetic microbial genome (**Techniques & Applications 17.2**). Like natural chromosomes, artificial chromosomes replicate only once per cell cycle. **Yeast artificial chromosomes** (**YACs**) were developed first and consist of a yeast telomere (TEL) at each end, a centromere sequence (CEN), a yeast origin of replication (ARS, *a*utonomously *r*eplicating *s*equence), a selectable marker such as URA3, and an MCS to facilitate the insertion of foreign DNA (**figure 17.12a**). YACs are used when extraordinarily large DNA pieces (up to 1,000 kb; table 17.2) are to be cloned. BACs were developed, in part, because YACs tend to be unstable and may recombine with host chromosomes, causing mutations and rearrangement of the cloned DNA. Although BACs accept smaller DNA inserts than do YACs (up to 300 kb), they are generally more stable. BACs are based on the F fertility factor of *E. coli* (*see figure 16.16*). The example shown in figure 17.12b is typical in that it includes genes that ensure a replication complex will be formed (*repE*), as well as proper partitioning of one newly replicated BAC to each daughter cell (*sopA, sopB,* and *sopC*). It also includes features common to many plasmids such as an MCS within the *lacZ* gene for blue/white colony screening and a selectable marker, in this case for resistance to the antibiotic chloramphenicol (Cm^R).

Retrieve, Infer, Apply

1. Briefly describe the polymerase chain reaction. What is the difference between a step and a cycle?
2. Why is PCR used to detect infectious agents that are often hard to diagnose?
3. How would you use PCR to measure the concentration of a specific gene's mRNA?
4. Why is it possible to visualize a PCR product on an agarose gel even if the template genome is present at such a low concentration that it cannot be seen?
5. You want to clone a 6,000 bp DNA fragment in *E. coli*. Which cloning vectors would be appropriate? How will you select transformants?

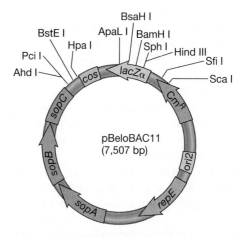

| TEL | TRP1 | ARS | CEN | MCS | URA3 | TEL |

(a) Yeast artificial chromosome (YAC)

(b) Bacterial artificial chromosome (BAC)

Figure 17.12 Artificial Chromosomes Can Be Used as Cloning Vectors. (a) A yeast artificial chromosome and (b) the bacterial artificial chromosome pBeloBAC11.

MICRO INQUIRY *In what ways does the BAC shown here differ from the plasmid pUC19 shown in figure 17.10?*

17.4 Introducing Recombinant DNA into Host Cells

After reading this section, you should be able to:

- Identify commonly used host cells
- Compare two common techniques by which recombinant DNA constructed in vitro is introduced into host cells

In cloning procedures, the selection of a host organism is as important as the choice of cloning vector. *E. coli* is the most frequent bacterial host and *S. cerevisiae* is most common among eukaryotes. Host microbes engineered to lack restriction enzymes and recombination enzymes (RecA in bacteria, and RecA homologues in *S. cerevisiae*) make better hosts because it is less likely that the newly acquired DNA will be degraded or recombined with the host chromosome. There are several ways to introduce recombinant DNA into a host microbe. Chemical transformation and electroporation are two commonly employed techniques when the host microbe does not have the capacity to be transformed naturally. This is the case with *E. coli* and most Gram-negative bacteria as well as many Gram-positive bacteria and archaea. In chemical transformation, the host cells are made competent by treatment with divalent cations and then transformed by heat-shocking the cells in a solution containing plasmid. ◀◀ *Bacterial transformation is the uptake of free DNA from the environment (section 16.7)*

TECHNIQUES & APPLICATIONS

17.2 How to Build a Microorganism

Biotechnology has repeatedly demonstrated how useful the genetic modification of an organism can be. From the production of recombinant human insulin to recombinant indigo (the dye used in blue jeans), it is clear that "gene juggling" can be used for the production of a wide range of products. But what if, rather than modifying one gene at a time, it were possible to design the ideal genome? The construction of such a genome would enable synthesis of the precise genotype best suited for the application at hand, whether that is generating biofuels or pharmaceuticals.

The first step in reaching this goal is to build a genome from scratch. Molecular biologists at the J. Craig Venter Institute (JCVI) accomplished this in 2008 with the one of the smallest bacterial genomes, that of *Mycoplasma genitalium*. The assembly of chemically synthesized DNA was performed in a series of BACs and then YACs to generate the entire bacterial chromosome in a yeast host. A year later, JCVI scientists showed it was possible to "transplant" a bacterial genome from a yeast host into a bacterium. So by 2010 the team was ready to go the distance and construct a synthetic genome and transplant it into a bacterial host.

The JCVI researchers used *M. mycoides* as their model genome. At 1.1 million base pairs (Mb), this microbe's genome is about twice the size of *M. genitalium*. The first step was to chemically synthesize 1,078 oligonucleotides, each 1080 bp long (**box figure**). If lined up end to end, these oligonucleotides would represent the entire genome. Importantly, 80 bp at each end of every oligonucleotide overlapped with its neighbor, so within an *E. coli* host, these could be assembled in the same order as that found in the native chromosome. Assembly in groups of 10 generated 109 chromosomal fragments, each about 10,000 bp. These were then transformed into yeast, where they were joined into 11 larger cassettes of 100,000 bp each. Finally, these were recombined to yield the entire genome. In addition to the native genes, four nucleotide sequences, called watermarks or barcodes, were included. Just as a barcode is used to identify an object, these nucleotide sequence barcodes were added to identify the genome as artificial.

The final step was to transplant the synthetic genome from the yeast host to the final bacterial host, *M. capricolum*.

Just as in any other cloning experiment, this depended on the presence of a selectable marker. In this case, a gene that conferred resistance to the antibiotic tetracycline was included in the synthetic genome. Upon selection of tetracycline-resistant *M. capricolum* cells bearing the *M. mycoides* genome, a series of verification tests were performed. These involved PCR and analysis of restriction endonuclease digestion patterns. Once it was confirmed that the host cell was carrying the synthetic genome, all that was left to do was to name the new organism. The JCVI scientists dubbed their new organism *M. mycoides* JCVI-syn1.0.

Source: Gibson, D. G., et al., 2010. Creation of a bacterial cell controlled by a chemically synthesized genome. Science. *329:52.*

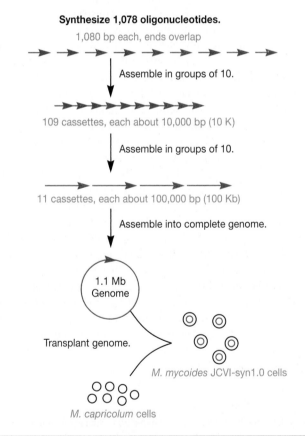

Synthesize 1,078 oligonucleotides.
1,080 bp each, ends overlap

Assemble in groups of 10.

109 cassettes, each about 10,000 bp (10 K)

Assemble in groups of 10.

11 cassettes, each about 100,000 bp (100 Kb)

Assemble into complete genome.

1.1 Mb Genome

Transplant genome.

M. mycoides JCVI-syn1.0 cells

M. capricolum cells

Electroporation is a simple technique for transforming bacteria, plant, and animal cells. In this procedure, cells are mixed with the recombinant DNA and exposed to a brief pulse of high-voltage electricity. The plasma membrane becomes temporarily permeable and DNA molecules are taken up by some of the cells. The cells are then grown on media that select for the presence of the cloning vector, as described in section 17.3.

17.5 Genomic Libraries: Cloning Genomes in Pieces

After reading this section, you should be able to:

- Explain why genomic libraries are useful
- Outline the construction of a genomic library and how the gene of interest might be selected

When cloning, the DNA to be inserted into a vector can be obtained in several ways. It can be synthesized by PCR using the microbe's chromosome as template, it can be located on the chromosome by Southern blotting, or it can be amplified from DNA extracted from a natural environment, like soil or water. PCR and Southern blotting require knowledge of the nucleotide sequence of the DNA to be cloned. However, what if researchers want to clone a gene based on the function of its product but have no idea what its DNA sequence might be? A genomic library must then be constructed and screened.

The goal of **genomic library** construction is to have all the DNA derived from a single organism or isolated from a single environment inserted into a series of cloning vectors. When cloning DNA directly from the environment in this way, a **metagenomic library** is produced. Genomic and metagenomic library construction starts by isolating and cutting the DNA into many fragments with restriction endonucleases. Each fragment is then cloned into a separate vector. Each vector now carries a separate piece of chromosome. When a collection of these recombinant vectors is introduced (e.g., by transformation) into a population of host cells, each transformant carries a vector that has a different piece of foreign chromosome. Ideally the entire organism's genome or all the DNA from the sample environment is represented in the collection of recombinant vectors. In this way, isolated groups of genes can be analyzed and isolated (**figure 17.13**). ▶▶ *Metagenomics provides access to uncultured microbes (section 18.3)*

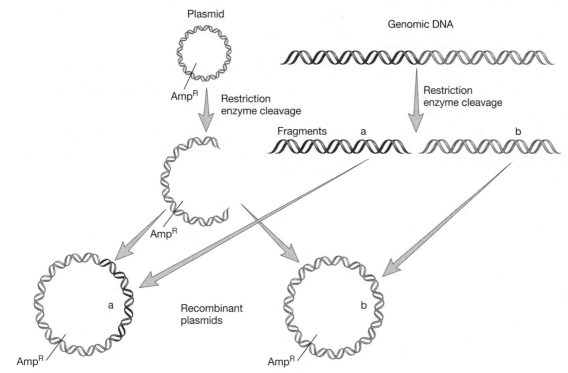

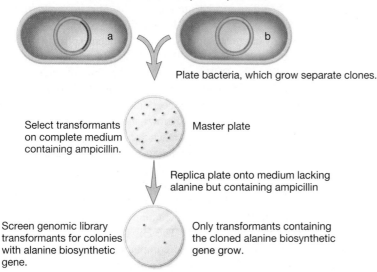

Figure 17.13 Construction of a Genomic Library and Screening by Phenotypic Rescue. A genomic library is made by cloning fragments of an organism's entire genome into a vector. For simplicity, only two genomic fragments and recombinant vectors are shown. In reality, a large mixture of vectors with inserts is generated. This mixture is then introduced into a suitable host. Phenotypic rescue is one way to screen the colonies for the gene of interest. It involves using a host with a genetic defect that can be complemented or "rescued" by the expression of a specific gene that has been cloned.

MICRO INQUIRY *Why are long fragments (e.g., 20,000 bp) of genomic DNA often desired when constructing a genomic library?*

Genomic and metagenomic libraries are often constructed so that the nucleotide sequence of the cloned DNA can be determined, as discussed in chapter 18. Alternatively, libraries are constructed to find a gene or genetic element with a specific function. In this case, it is necessary to know something about the function of the target gene or genetic element. If the genomic library has been introduced into a microbe that expresses the foreign gene, it may be possible to assay each clone for a specific protein or phenotype. For example, if one is studying a newly isolated soil bacterium and wants to find genes that encode enzymes needed for the biosynthesis of the amino acid alanine, the library could be expressed in an *E. coli* alanine auxotroph (figure 17.13). Recall that the growth of alanine auxotrophs requires the addition of this amino acid to the medium. Following introduction of the genomic library into host cells, those that now grow without alanine would be good candidates for the genomic library fragment that possesses the alanine biosynthetic genes. Success with this approach depends on the assumption that the activity of the cloned gene product is similar in both organisms. If this is not the case, the host must be the same species from which the library was prepared. In this example, a soil bacterial mutant lacking the trait in question (e.g., an alanine auxotroph) is used as the genomic library host. The **genetic complementation** of a deficiency in the host cell is sometimes called **phenotypic rescue** because the mutant phenotype is (at least partially) restored, or "rescued." ◀◀ *Detection and isolation of mutants (section 16.2)*

If a genomic library is prepared from a eukaryote, a cDNA library is usually constructed. This removes introns and only the protein-coding regions of the genome are cloned. cDNA is prepared (figure 17.5) and cloned into a suitable vector. After the library is introduced into the host microbe, host cells carrying the gene of interest are typically identified by phenotypic rescue or PCR amplification of the gene of interest after the vector has been isolated from candidate transformants. In some cases these approaches are not feasible. Then the researcher must develop a novel approach that suits the particular set of circumstances to screen the genomic library.

17.6 Expressing Foreign Genes in Host Cells

After reading this section, you should be able:

- Explain the utility of expression vectors
- Outline the procedure by which a His-tagged protein is generated in vivo and purified in vitro
- Summarize the role of GFP in protein analysis, and differentiate between a transcriptional and a translational GFP fusion

When a gene from one organism is cloned into another, it is said to be a **heterologous gene.** However, to be transcribed, a recombinant, heterologous gene must have a promoter that is recognized by the host RNA polymerase. Translation of its mRNA depends on proper ribosome binding. For instance, if the host is a bacterium

and the gene has been cloned from a eukaryote, a bacterial promoter and leader must be provided and introns removed.

The problems of expressing recombinant genes in host cells are largely overcome with the help of specially constructed cloning vectors called **expression vectors.** These vectors enable the high-level transcription of a gene because it is cloned adjacent to a particularly "strong" promoter. Often expression vectors contain regulatory regions of the *lac* operon (or another inducible promoter) so that expression of the cloned genes can be controlled by the addition of inducer molecules. This is important because sometimes high-level expression of a heterologous protein can be toxic to the host cell. The presence of an inducible promoter allows the biotechnologist to grow the cells to a certain density before inducing the expression of the cloned gene. ◀◀ *Regulation of transcription initiation saves considerable energy and materials (section 14.2)*

Purification and Study of Recombinant Proteins

It is often necessary to isolate the protein product of the cloned gene so that its structure and activity can be studied. In addition, it is sometimes desirable to determine the subcellular localization of a protein. Finally, as discussed in chapter 42, heterologous genes are often expressed for the production of commercially prepared recombinant proteins. Here we discuss some clever ways in which proteins can be purified and visualized in living cells.

Protein Purification

A common method by which proteins are purified is called **polyhistidine tagging,** or simply **His-tagging.** His-tagging involves adding a series of histidine amino acid residues to either the N- or C-terminus of the protein. Most often six residues are added; this is called a 6xHis-tag. Histidine is used because it has a high affinity for metal ions, so a protein bearing multiple histidine residues will bind preferentially to a solid-phase material, called a resin, which has exposed nickel or cobalt atoms. This permits the separation of the His-tagged protein from other cellular constituents.

The process of purifying a His-tagged protein begins by adding the histidine residues to the protein. This is most commonly accomplished by cloning the protein into an expression vector that already has the histidine-encoding sequence present (**figure 17.14**, step 1). Alternatively, the histidine-encoding sequence can be amplified by PCR and added to the gene of interest. Once the protein-coding nucleotide sequence is fused to the histidine codons, the vector is introduced into the microbe of choice, usually *E. coli* (step 2). As *E. coli* grows, the His-tagged protein is produced (step 3). When the cells reach a sufficiently high density, they are lysed by treatment with detergent or enzymes, and all the soluble proteins in the cell—including the His-tagged protein—are collected. These proteins are mixed with the metal resin that is bound to a column (step 4) so that only the His-tagged protein sticks to the resin. The remaining, unwanted material is washed from the resin (step 5). The His-tagged protein is then removed by passing the aromatic compound imidazole through the column (step 6). Depending on the purification system used, the

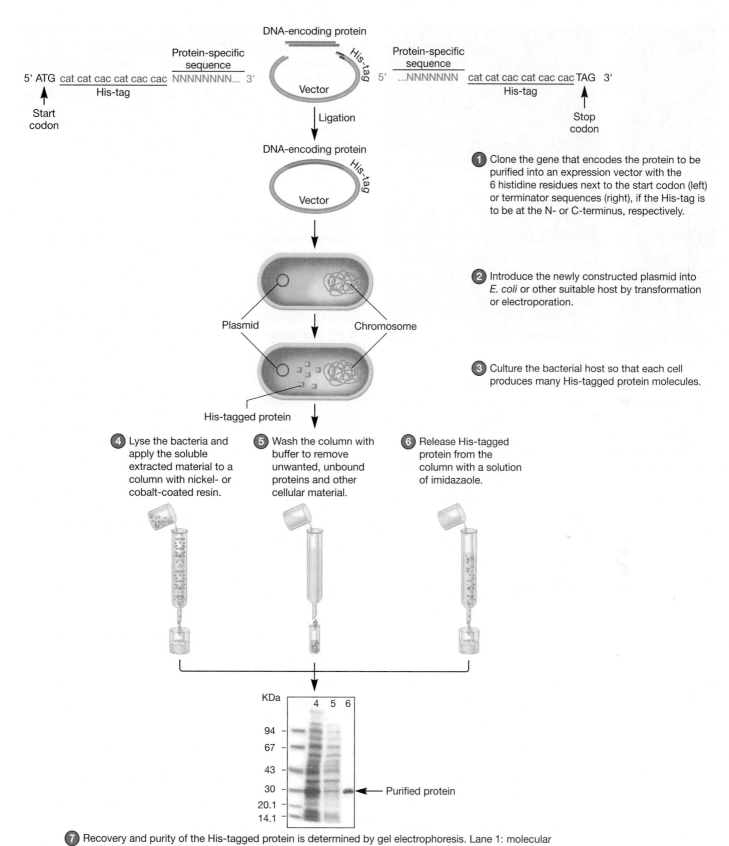

DNA-encoding protein

Protein-specific
sequence

Protein-specific
sequence

5' ATG cat cat cac cat cac cac NNNNNNNN... 3'   His-tag   5'   ...NNNNNNN cat cat cac cat cac cac TAG   3'

His-tag

His-tag

Start
codon

Stop
codon

Vector

Ligation

DNA-encoding protein

His-tag

Vector

Plasmid

Chromosome

His-tagged protein

① Clone the gene that encodes the protein to be
purified into an expression vector with the
6 histidine residues next to the start codon (left)
or terminator sequences (right), if the His-tag is
to be at the N- or C-terminus, respectively.

② Introduce the newly constructed plasmid into
E. coli or other suitable host by transformation
or electroporation.

③ Culture the bacterial host so that each cell
produces many His-tagged protein molecules.

④ Lyse the bacteria and
apply the soluble
extracted material to a
column with nickel- or
cobalt-coated resin.

⑤ Wash the column with
buffer to remove
unwanted, unbound
proteins and other
cellular material.

⑥ Release His-tagged
protein from the
column with a solution
of imidazaole.

KDa   4 5 6

94 –
67 –

43 –

30 –   ← Purified protein
20.1 –
14.1 –

⑦ Recovery and purity of the His-tagged protein is determined by gel electrophoresis. Lane 1: molecular
weight markers, Lane 2: soluble proteins obtained in step 4, Lane 3: proteins eluted in step 5, Lane 4:
His-tagged protein released in step 6.

Figure 17.14 **Construction and Purification of a Polyhistidine-Tagged Protein.**

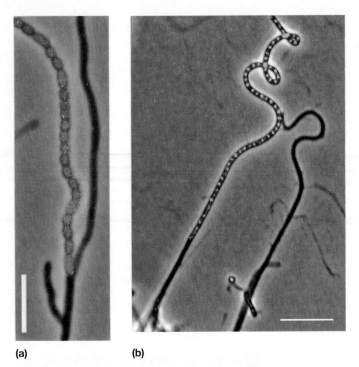

(a)   **(b)**

Figure 17.15 Fluorescent Protein Labeling. (a) A transcriptional fusion places the gene encoding a fluorescent peptide under the control of a promoter, replacing the coding sequence of the gene. In this case, the mCherry gene is controlled by the promoter that normally governs transcription of a gene in the actinomycete *Streptomyces coelicolor* that is expressed only when aerial hyphae form cross walls (septa) that give rise to a chain of spores. Thus red fluorescence is seen uniformly throughout the chain of spores but not in hyphae without septae. (b) A translational fusion generates a chimeric protein consisting of the normal protein attached to the GFP label. Here the FtsZ gene has been fused to GFP. This shows that FtsZ localizes at the septa. Notice that the filament to the right is not forming septae, so FtsZ is not present.

MICRO INQUIRY *What special considerations are necessary if one is constructing a translational fusion of a protein known to be exported?*

that the jellyfish *Aequorea victoria* produces a protein called **green fluorescent protein** (**GFP**). GFP is encoded by a single gene that, when translated, undergoes self-catalyzed modification to generate a strong, green fluorescence. This means that it is easily cloned and expressed in any organism. In fact, an entire palette of fluorescent labels is now available. These include mutant versions of the GFP protein that glow throughout the blue-green-yellow spectrum. In addition, other fluorescent peptides that likewise undergo self-catalyzed light production have been developed.

There are two ways in which the amino acid sequence that encodes fluorescent peptides can be attached to a gene. A transcriptional fusion is used to study gene regulation by replacing the entire protein coding sequence of the gene of interest with the fluorescence-encoding gene (**figure 17.15a**). This places the expression of the fluorescence gene under the control of the promoter of the replaced gene. Thus, when the promoter is turned "on," it directs the transcription of the fluorescence gene and the cells glow. If the transcriptional fusion has been made in a multicellular organism, then not only is the timing of promoter activity seen but the specific cell type in which the gene is normally expressed is also determined. On the other hand, a translational fusion is used to determine where a particular protein is localized in the cell. Translational fusions add the GFP-coding sequence to the end of the structural gene of interest. This creates a chimeric protein—a protein that consists of two parts: the protein being studied and GFP. Now whenever that protein is made, and wherever it goes in the cell, the protein will fluoresce. Of course, care must be taken to ensure that the protein fusion still functions like the original protein. One way to do this is to test for phenotypic rescue of a mutant lacking the structural gene of interest. Once it has been determined that the chimeric protein is fully functional, the cellular "address" of the protein can be determined (figure 17.15b).

His-tag can be cleaved from the protein while it is attached to the resin, or it can be removed following release of the protein. Often the protein is functional with the His-tag, so it may not be removed. His-tagging is used if the protein is soluble and is found in the cytoplasm. While His-tagging is certainly the most popular approach to purifying recombinant proteins, other approaches exist, especially for purifying membrane-associated proteins.

Fluorescence Labeling

What if the goal of cloning a gene is to study the regulation or function of the protein product in vivo? It is possible to visualize the activity of a specific promoter as well as observe the localization of a protein by fluorescence microscopy. The development of fluorescent labeling of living cells began when it was discovered

Retrieve, Infer, Apply

1. Describe how a genomic library might be screened for a gene that confers the production of an extracellular enzyme that degrades casein, a protein found in milk. (Hint: Agar with skim milk loses its opacity when casein is degraded.)
2. Explain the selection for antibiotic resistance followed by blue versus white screening of colonies containing recombinant plasmids. Why must both antibiotic selection and color screening be used?
3. How can one prevent recombinant DNA from undergoing recombination in a bacterial host cell?
4. List several reasons why a cloned gene might not be expressed in a heterologous host cell.
5. Think of a situation in which you might want to keep the His-tag on a recombinant protein and a situation in which it would be important to remove it.
6. You are studying chemotaxis proteins in a newly described bacterium. You have cloned a gene that encodes the CheA protein. You need to be sure that this protein localizes to the inside face of the plasma membrane (*see figure 14.21*). Will you use a transcriptional or translational fusion? Explain your choice.

Key Concepts

17.1 Key Discoveries Led to the Development of Recombinant DNA Technology

- Genetic engineering became possible after the discovery of restriction enzymes and reverse transcriptase, and the development of essential methods in nucleic acid chemistry such as the Southern blotting technique.
- Restriction enzymes are important because they cut DNA at specific sequences, thereby releasing fragments of DNA that can be cloned or otherwise manipulated (**figure 17.3** and **table 17.1**).
- Gel electrophoresis is used to separate molecules according to charge and size.
- DNA fragments are separated on agarose and acrylamide gels. Because DNA is acidic, it migrates from the negative to the positive end of a gel (**figure 17.6**).

17.2 Polymerase Chain Reaction Amplifies Targeted DNA

- The polymerase chain reaction (PCR) allows small amounts of specific DNA sequences to be amplified, or increased in concentration thousands of times (**figure 17.8**).
- PCR consists of multiple cycles of 3 steps each: DNA denaturation, primer annealing, and DNA synthesis.
- PCR has numerous applications. It often is used to obtain genes for cloning and in diagnostic and forensic science.

17.3 Cloning Vectors Are Needed to Create Recombinant DNA

- There are four types of cloning vectors: plasmids, viruses, cosmids, and artificial chromosomes. Cloning vectors generally have at least three components: an origin of replication, a selectable marker, and a multicloning site or polylinker (**table 17.2; figures 17.10** and **17.12**).
- The most common approach to cloning is to digest both vector and DNA to be inserted with the same restriction enzyme or enzymes so that compatible sticky ends are generated. The vector and DNA to be cloned are then incubated in the presence of DNA ligase, which catalyzes the formation of phosphodiester bonds once the DNA fragment inserts into the vector.

- Once the recombinant plasmid has been introduced into host cells, cells carrying vector must be selected. This is often accomplished by allowing the growth of only antibiotic-resistant cells because the vector bears an antibiotic-resistance gene. Cells that took up vector with inserted DNA must then be distinguished from those that contain only vector. Often a blue-versus-white colony phenotype is used; this is based on the presence or absence, respectively, of a functional *lacZ* gene (**figure 17.11**).

17.4 Introducing Recombinant DNA into Host Cells

- The bacterium *E. coli* and the yeast *S. cerevisiae* are the most common host species.
- DNA can be introduced into microbes by transformation or electroporation.

17.5 Genomic Libraries: Cloning Genomes in Pieces

- It is sometimes necessary to find a gene without the knowledge of the gene's DNA sequence. A genomic library is constructed by cleaving an organism's genome into many fragments, each of which is cloned into a vector to make a unique recombinant plasmid.
- Genomic libraries are often screened for the gene of interest by either phenotypic rescue (genetic complementation) or DNA hybridization with an oligonucleotide probe (**figure 17.13**).

17.6 Expressing Foreign Genes in Host Cells

- An expression vector has the necessary features to express in high levels any recombinant gene it carries.
- If a eukaryotic gene is to be expressed in a bacterium, cDNA is used because it lacks introns; a bacterial leader must also be added to the 5′ end of the gene.
- Purification of recombinant proteins is often accomplished by fusing the coding sequence of a protein to six histidine residue codons found on some expression vectors. When introduced and expressed in bacteria, the His-tagged protein can be selectively purified (**figure 17.14**).
- Green fluorescent protein can be used to study the regulation of gene expression (transcriptional fusions) and protein localization (translational fusions) (**figure 17.15**).

Compare, Hypothesize, Invent

1. You are performing a PCR to amplify a gene encoding a tRNA from a bacterium that has only recently been grown in pure culture. You are expecting a product of 954 bp. However, you generate three different products; only one is the expected size. List at least two possible explanations (excluding experimental error).

2. You have cloned a structural gene required for riboflavin synthesis in *E. coli*. You find that an *E. coli* riboflavin auxotroph carrying the cloned gene on a vector makes less riboflavin than does the wild-type strain. Why might this be the case?

3. Suppose you transformed a plasmid vector carrying a human interferon gene into *E. coli* but none of the transformed bacteria produced interferon. Give as many plausible reasons as possible for this result.

4. You are interested in the activity and regulation of a protease made by the Gram-positive bacterium *Geobacillus stearothermophilus.* What would be the purpose of constructing each of the following: a His-tagged protease, a transcriptional GFP fusion to the protease gene, and a translational GFP fusion to the protease gene?

5. Rapid and accurate identification of pathogens in clinical samples is paramount for effective patient care. To that end, the identification of microbes in human tissue and blood using PCR is used for some infectious agents. However, before a PCR approach can be adopted by clinical microbiologists, it must be shown that it is specific and accurate. Fungal infections are particularly difficult to treat and have become more widespread as the population of patients with compromised immune systems grows. A PCR protocol for the identification of several important fungal pathogens, including *Aspergillus, Candida,* and *Cryptococcus,* has been described. This assay is based on the amplification of the 28S large subunit ribosomal RNA gene from each microbe. Recall that as a eukaryote, the human host also has a homologous gene. Outline how you think these researchers performed the PCRs and what they did to control for false positive and false negatives. Why do you think real-time PCR was used, rather than end-point PCR?

Read the original paper: Vollmer, T., et al. 2008. Evaluation of novel broad-range real-time PCR assay for rapid detection of human pathogenic fungi in various clinical specimens. *J. Clin. Microbiol.* 46:1919.

6. *Zymomonas mobilis* is a Gram-negative bacterium that is used to make ethanol on an industrial scale. There is great interest in enhancing and expanding this microbe's industrial uses. This prompted a group at the University of Hong Kong to develop a new *E. coli–Z. mobilis* shuttle vector. To do so, they started with a plasmid naturally found in some *Z. mobilis* strains. What genes and genetic elements might the researchers need to add in order to construct a successful shuttle plasmid? What genetic elements do you think were added to obtain high-level expression of heterologous genes? Finally, how do you think they determined the copy number of the resulting shuttle vector in *E. coli* and in *Z. mobilis*?

Read the original paper: So, L. Y., et al., 2014. pZMO7-Derived shuttle vectors for heterologous protein expression and proteomic applications in the ethanol-producing bacterium *Zymomonas mobilis. BMC Microbiol.* 14:68.

Learn More

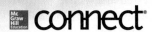

18

Microbial Genomics

The science of genomics began in the mid-1990s with the sequencing of viral, then bacterial, and finally mammalian genomes. The new frontier of genomics is the construction of completely synthetic genomes so that organisms can be designed to perform a particular suite of functions.

"Synthetic Life": Oxymoron or the Future?

"To live, to err, to fall, to triumph, to recreate life out of life!" When James Joyce wrote these words in 1916 in *A Portrait of the Artist as a Young Man,* he could never have imagined that they would be encrypted in the genetic code inserted in the synthetic genome of a novel bacterium. But 94 years later, Daniel Gibson and his co-workers at the J. Craig Venter Institute (JCVI) did just that. Building on 15 years of investigation, the JCVI researchers used a computer to design a genome and a DNA synthesizer to make short stretches of DNA. These were then stitched together to create a 1.08 million base pair genome based on the chromosome of *Mycoplasma mycoides* (*see Techniques & Applications 17.2*). The genome was then transplanted into *M. capricolum,* which after a few rounds of replication consisted entirely of molecules whose synthesis was directed by a chromosome that started as a computer program and four bottles of deoxynucleotides.

Clearly, this feat raises lots of questions. Is this a synthetic cell? Some say yes, some say no; it is a synthetic genome inside a naturally produced cell. Why construct this microbe? Craig Venter views it as a $30 million proof of concept that can (will?) lead to the construction of genomes specifically designed for the development of vaccines, pharmaceuticals, clean water and food products, and biofuels. In fact, Dr. Venter is working with Exxon Mobil Corporation to construct a biofuel-producing algal cell.

Is synthetic life ethical? Depends on whom you ask. Some environmental groups say synthetic genome research should stop until specific regulations are in place. Most scientists and bioethicists take a more measured view, envisioning two ways such a microbe might be released: bioterror and bioerror. Protection against bioterror relies on a host of security measures, many of which are already in place. Bioerror brings us back to James Joyce. The scientists who designed the genome inserted a "watermark"—sequences unique to this microbe—into the genome so that if it were to escape the lab, its identity could be determined. They

developed a cipher to convert the genetic code so that when "translated," the watermark spells out the Joyce quote, as well as "See things not as they are, but as they might be," from the book *American Prometheus* (about Robert Oppenheimer, who oversaw the Manhattan Project), and perhaps most fittingly, the words of the late physicist Richard Feynman, "What I cannot create, I do not understand."

The construction of this bacterium, *M. mycoides* JCVI-syn1.0, represents the culmination of decades of work. In order to understand how we have arrived in a world where a computer is the starting point for a new organism, we need to review the individual technologies required. This involves the integration of a number of genetic and biotechnology principles; be sure you succeed in the readiness check, because the world of genomics is a vast and amazing place well worth exploring.

Readiness Check:

Based on what you have learned previously, you should be able to:

✔ Identify the key structural elements of DNA, RNA, and protein (section 13.2)

✔ Discuss what is meant by the term genome

✔ Explain the process and utility of PCR (section 17.2)

✔ Diagram the fundamental principles of gel electrophoresis (section 17.1)

✔ Summarize the purpose, construction, and screening of a genomic library (sections 17.3–17.5)

18.1 DNA Sequencing Methods

After reading this section, you should be able to:

- Explain how DNA is sequenced by the Sanger chain termination method
- Contrast and compare the advantages and disadvantages of the Sanger method with next-generation sequencing

The genomic era is considered a 21st-century phenomenon. One of the most crucial steps leading to the genomic revolution was figuring out how to sequence DNA. This was accomplished in 1977 by Alan Maxam and Walter Gilbert (collaborating on one technique), and Frederick Sanger. Sanger's method became the most commonly used and is now discussed.

Sanger DNA Sequencing

The Sanger method involves the synthesis of a new strand of DNA using the DNA to be sequenced as a template. The reaction begins when single strands of template DNA are mixed with an oligonucleotide primer (a short piece of DNA complementary to the region to be sequenced), DNA polymerase, the four deoxynucleoside triphosphates (dNTPs), and dideoxynucleoside triphosphates (ddNTPs). ddNTPs differ from dNTPs in that the 3′ carbon lacks a hydroxyl group (**figure 18.1**). In such a reaction mixture, DNA synthesis will continue until a ddNTP, rather than a dNTP, is added to the growing chain. Without a 3′-OH group to

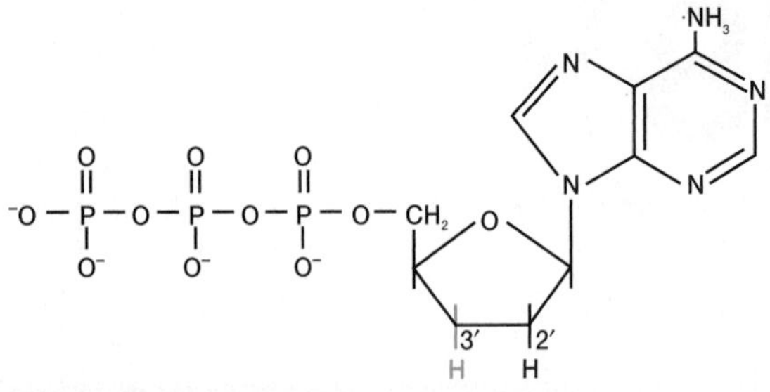

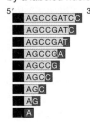

Figure 18.1 Dideoxyadenosine Triphosphate (ddATP). Note the lack of a hydroxyl group on the 3′ carbon, which prevents further chain elongation by DNA polymerase.

MICRO INQUIRY *What is the function of the 3′-OH during DNA synthesis?*

attack the 5′-PO₄ of the next dNTP to be incorporated, synthesis stops (*see figure 13.10*). Indeed, Sanger's technique is frequently referred to as the **chain-termination DNA sequencing** method.

To obtain sequence information, four separate synthesis reactions must be prepared, one for each ddNTP (**figure 18.2**). The

1 Isolated unknown DNA fragment

5′ CACTTAGCCGATCC 3′
3′ GTGAATCGGCTAGG 5′
Original DNA to be sequenced

2 DNA is denatured to produce single template strand.

3′ GTGAATCGGCTAGG 5′

3 Labeled specific primer molecule hybridizes to the DNA strand.

Primer
CACTT
3′ GTGAATCGGCTAGG 5′

4 DNA polymerase and regular nucleotide mixture (dATP, dCTP, dGTP, and dTTP) are added; ddGTP, ddATP, ddCTP, and ddTTP are placed in separate reaction tubes with the regular nucleotides. The dd nucleotides are labeled with some type of tracer, which allows them to be visualized.

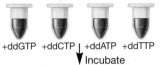

+ddGTP +ddCTP +ddATP +ddTTP
↓ Incubate

5 Newly replicated strands are terminated at the point of addition of a dd nucleotide.

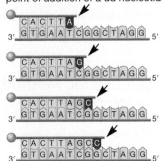
CACTTA
3′ GTGAATCGGCTAGG 5′

CACTTAG
3′ GTGAATCGGCTAGG 5′

CACTTAGC
3′ GTGAATCGGCTAGG 5′

CACTTAGCC
3′ GTGAATCGGCTAGG 5′

6 Schematic view of how all possible positions on the fragment are occupied by a labeled nucleotide

5′   3′
AGCCGATCC
AGCCGATC
AGCCGAT
AGCCGA
AGCCG
AGCC
AGC
AG
A

+ddGTP +ddCTP +ddATP +ddTTP

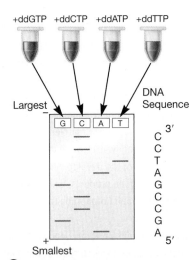

Largest

G C A T

DNA Sequence
C 3′
C
T
A
G
C
C
G
A 5′

+ Smallest

7 Running the reaction tubes in four separate gel lanes separates them by size and nucleotide type. Reading from bottom to top, one base at a time, provides the correct DNA sequence.

Figure 18.2 The Sanger Method of DNA Sequencing. Steps 1–6 are used for both manual and automated sequencing. Step (7) shows preparation of a gel for manual sequencing in which radiolabeled ddNTPs are used.

ddNTP is mixed with all four normal dNTPs, and as DNA synthesis proceeds, only sometimes will the ddNTP be incorporated into the growing DNA strand, rather than its dNTP cousin. This results in a collection of DNA fragments of varying lengths, each ending in the same ddNTP. For example, a reaction prepared with ddATP + dATP, dCTP, dGTP, and dCTP produces fragments ending with an A, those with ddTTP produce fragments with T termini, and so forth. After DNA synthesis is completed, the DNA is made single stranded, usually by heating. If the DNA is to be manually sequenced, radioactive dNTPs are used and each reaction is separated by electrophoresis in a separate lane of a polyacrylamide gel. Recall that each fragment's migration rate is inversely proportional to the log of its molecular weight. Simply put, the smaller a fragment is, the faster it moves through the gel. Because synthesis always adds a nucleotide to the 3′-OH of the growing strand, the ddNTP at the end of the shortest fragment is assigned to the 5′ end of the DNA sequence, while the largest fragment is the 3′ end. In this way, the DNA sequence can be read directly from the gel from the smallest to the largest fragment. ◄◄ *Gel electrophoresis (section 17.1)* ↻ *Sanger Sequencing*

DNA is usually prepared for automated sequencing (**figure 18.3**). Here each ddNTP is labeled with a different colored fluorescent dye. The resulting fragments are then separated by electrophoresis, and as DNA fragments exit the bottom of the gel from smallest to largest, a laser beam detects the identity of the ddNTP by its fluorescent color. This is recorded on a graph called a chromatogram in which the amplitude of each spike represents the fluorescent intensity of each particular fragment (figure 18.3*b*). The corresponding DNA sequence is listed above the chromatogram. Typically, an automated Sanger chain termination sequencing system can accurately read 500–800 bp in a single electrophoresis run.

Next-Generation DNA Sequencing

Sanger's chain termination method was used to complete the first human genome sequence in 2001. This cost about $300 million and took about a decade to finish. Although this was an amazing feat at the time, it illustrates two limitations of Sanger sequencing: it is expensive and time consuming. Because scientists want to sequence genomes faster and more cheaply, innovative, newer DNA sequencing techniques have been invented.

About a decade ago, several **next-generation sequencing** (NGS) technologies were developed. All commercially available NGS approaches start by shearing the DNA into short pieces to be sequenced and attaching oligonucleotides called adapters to the ends of the resulting fragments (**figure 18.4**). The adapter at one end of each DNA fragment attaches it to a solid substrate, and the adapter at the other end anneals to a primer used to initiate the polymerase chain reaction (PCR). PCR results in the production of many copies of the same fragment, which are sequenced simultaneously. Because thousands of identical DNA fragments are sequenced at the same time, these methods are sometimes called **massively parallel sequencing techniques.** In addition to making genomic sequencing faster and cheaper, as we discuss later in this chapter, NGS avoids the need to insert (i.e., clone) individual DNA fragments into vectors. This is important because it is almost impossible to clone every DNA fragment into any given genomic library. ◄◄ *Polymerase chain reaction amplifies target DNA (section 17.2); Genomic libraries (section 17.5)*

Although several NGS techniques have been marketed, currently **reversible chain termination sequencing** is most frequently used. Here, the DNA is sequenced as each nucleotide is incorporated. This is called **sequencing by synthesis.** Once adapter nucleotides have been attached to the genomic fragments, the fragments are introduced into a flow cell, which is a glass slide into which grooves have been cut. The surface of each groove is covered with oligonucleotides whose sequences complement the adapter sequence at the end of each DNA fragment (figure 18.4). Reagents needed for PCR are introduced, and the reaction proceeds using the DNA fragments as template. Each newly generated fragment then doubles over like a hairpin when the sequence at the other end of the new strand hybridizes to its complement oligonucleotide attached to the flow cell surface. This process is called "bridge amplification" and it creates clusters of double-stranded fragments of DNA scattered over the surface of the slide. As we will see, it is critical that fragments with identical nucleotide sequences are clustered together in the flow cell. In the next step, the

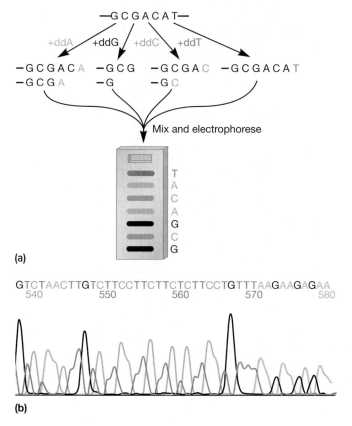

Figure 18.3 Automated Sanger DNA Sequencing. (a) Part of an automated DNA sequencing run. Here the ddNTPs are labeled with fluorescent dyes. (b) Data generated during an automated DNA sequencing run. Bases 538 to 580 are shown.

double-stranded fragments are denatured, and the flow cell is flushed. This leaves bundles of single-stranded linear fragments ready for sequencing.

Similar to Sanger sequencing, sequencing by synthesis uses a modified fluorescent nucleotide that, when introduced into the growing strand, stops the reaction because the 3′-OH is blocked (**figure 18.5**). However, rather than using a dideoxynucleotide as in Sanger sequencing (figure 18.1), here a small chemical group is added that can be enzymatically removed. Another big difference is that the modified nucleotide does not fluoresce until it is

incorporated. Because sequencing proceeds in a flow cell, reagents needed for DNA synthesis can be flushed from the flow cell once a modified dNTP is incorporated and detected. This is followed by infusion of the enzyme cocktail that removes the fluorescent tag and blocking group. Therefore each nucleotide determined by laser optics involves the following events: (1) The DNA polymerase, all four modified dNTPs, and other reagents needed for DNA synthesis, using the tethered fragments as template, are added to the flow cell. (2) The incorporation of a modified nucleotide as determined by the template strand occurs,

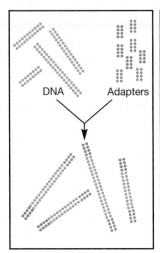

1. Genomic DNA is fragmented and adapters are attached to both ends of each fragment.

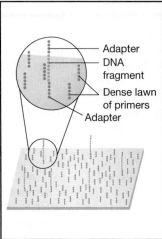

2. Single-stranded fragments are bound to flow cell surface.

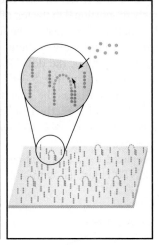

3. PCR reagents, including unlabeled nucleotides, are added to begin bridge PCR amplification of each fragment.

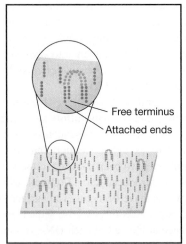

4. Bridge PCR amplification generates double-stranded fragments bound to the flow cell surface at both ends.

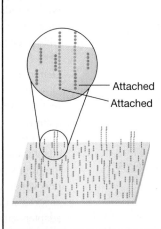

5. Fragments are denatured to become single stranded.

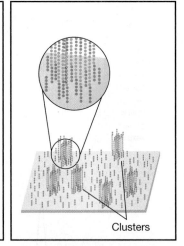

6. Each fragment serves as PCR template to generate millions of clusters of identical fragments.

Figure 18.4 Reversible Chain Termination Sequencing.

MICRO INQUIRY *Why is it important that identical fragments of DNA to be sequenced are clustered together?*

(Continued)

Figure 18.4 *(Continued)*

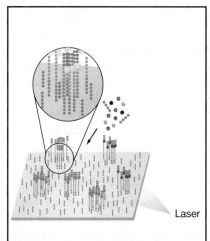

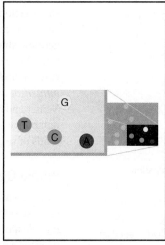

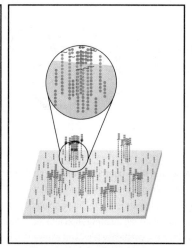

7. The first sequencing cycle begins when reagents, including primers and labeled nucleotides with reversible terminators, are added.

8. Laser excitation is followed by image capture of fluorescent signal from each cluster of fragments. This identifies the first base, which is recorded.

9. Enzymatic cleavage of the reversible terminator enables the second sequencing cycle to begin.

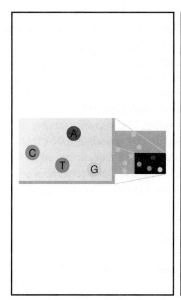

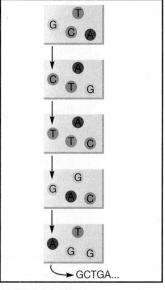

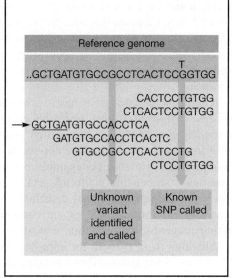

10. Laser excitation generates the signal that identifies the second base in each cluster of fragments.

11. Cleavage of terminator and sequencing cycles are repeated to determine base sequence of each cluster of fragments.

12. Sequences are aligned and compared with reference data base. SNP, single nucleotide polymorphism.

causing its fluorescent tag to emit light while at the same time stalling synthesis. This pause allows the identity of the incorporated base to be determined by the color of the fluorescent tag. (3) The reagents needed for synthesis are then flushed from the flow cell, and (4) the cleavage enzyme reagents are introduced. (5) The fluorescent label and the blocking molecule are removed, exposing the 3′OH. Finally, (6) the enzyme cocktail is flushed, and synthesis reagents are added once again and the cycle repeats. Because the light emitted by any single nucleotide is too dim to accurately record, each cluster of identical fragments must grow synchronously so that a large enough signal is generated. The incorporation of the same nucleotide at the same time into each identical fragment in a cluster generates a signal with sufficient amplitude for detection. This explains why the length of each read (the number of nucleotides determined in a single run) is much shorter than it is for Sanger sequencing: after about 75 to 150 bases have been incorporated, the synchronicity of nucleotide incorporation deteriorates. That is, the synthesis of

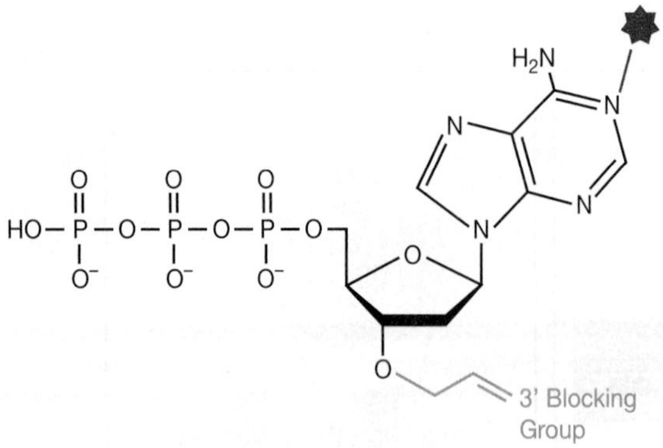

Figure 18.5 Modified base used in reversible chain termination sequencing. This example shows modified dATP, with a 3′OH blocking group and a fluorescent tag attached to the adenine. The base only fluoresces when the PP$_i$ is removed during incorporation into the growing strand. The 3′ blocking group is then enzymatically removed exposing the 3′OH needed for continued DNA synthesis.

complementary strands in a cluster of identical templates becomes "out of sync." Without a strong synchronous signal, the data becomes ambiguous. Nonetheless, short read length is compensated by the volume of data produced: at least 1.5 GB (billion bases) are read per flow cell in about a day, with newer systems reading up to 75 GB in about 2 days. Compare this with the output of automated Sanger sequencing, of about 1 million bases per day. With the average bacterial genome measuring about 4 million base pairs, one can begin to appreciate the enormous power of next-generation sequencing.

In addition to the method described here, various new approaches are being developed. In fact, reversible chain termination is now sometimes called "second generation" sequencing to distinguish it from technologies that are not yet (or are just becoming) commercially available. For example, newer, third- and fourth-generation sequencing platforms seek to streamline sample preparation, improve signal detection, optimize surfaces to more rapidly attain high reagent concentrations while preventing reagents from sticking, and attain longer read lengths. Although there is room for improvement, next-generation DNA sequencing technologies have already revolutionized genomics by drastically reducing time and cost, and increasing accuracy.

18.2 Genome Sequencing

After reading this section, you should be able to:

- List the steps used in whole-genome shotgun cloning
- Compare and contrast the assembly of genomes using Sanger and next-generation sequencing (NGS) approaches
- Describe the multiple strand displacement method and how this technique is used
- Define metagenomics and explain why NGS is the sequencing method of choice for such analyses

In 1995 J. Craig Venter, Hamilton Smith, and their collaborators were the first to sequence a bacterial genome. Prior to that, only the small genomes of viruses had been sequenced. In order to reduce bacterial genomes to manageable sizes, they developed **whole-genome shotgun sequencing** and the computer software needed to assemble sequence data into a complete genome. They used their new method to sequence the genomes of the bacteria *Haemophilus influenzae* and *Mycoplasma genitalium*. With this accomplishment, Venter and Smith ushered in the genomic era. Within 20 years the number of complete genomes published grew from two to thousands of sequenced genomes spanning all three domains of life.

Whole-Genome Shotgun Sequencing

Although revolutionary when it was introduced, sample preparation is now considered one of the biggest disadvantages of Sanger sequencing. Sanger sequencing requires whole-genome shotgun cloning, which is the construction of a genomic library. The entire sequencing process can be broken into four stages: library construction, random sequencing, fragment alignment and gap closure, and editing.

1. **Library construction.** The genome is randomly broken into fragments using ultrasonic waves and the fragments are then purified (**figure 18.6**). These fragments are next inserted into cosmid or bacterial artificial chromosome (BAC; *see figure 17.12*) vectors and isolated. *Escherichia coli* strains are next transformed with the cosmids or BACs to produce a library of clones whose inserts represent the entire genome to be sequenced. ◀◀ *Cloning vectors are needed to create recombinant DNA (section 17.3)*

2. **Random sequencing.** The vectors carrying the cloned DNA are purified and thousands of DNA fragments are sequenced with automated sequencers, employing primers that recognize the plasmid DNA sequences adjacent to the cloned, chromosomal insert. Up to 96 samples can be sequenced simultaneously, making it possible to sequence as many as 1 million bases per day, per sequencer. Usually all stretches of the genome are sequenced between eight and ten times to increase the accuracy of the final results.

3. **Fragment alignment and gap closure.** Using computer analysis, the DNA sequence information of each fragment is assembled into longer stretches of sequence. Two fragments are joined together to form a larger stretch of DNA if the sequences at their ends overlap and match. This comparison process results in a set of larger, contiguous nucleotide sequences called **contigs**. Sometimes an overlapping sequence is missing, generating gaps between contigs. There are several strategies to obtain the missing sequences. Ultimately, however, the contigs are aligned in the proper order to form the complete genome sequence. The term scaffold is used to describe sequence data with gaps that persist between contigs.

4. **Editing.** The sequence is then carefully proofread to resolve any ambiguities or frameshift mutations in the sequence. Proofreading is accomplished by ensuring that all reads of the same sequence are identical and the sequences of the two DNA strands are complementary.

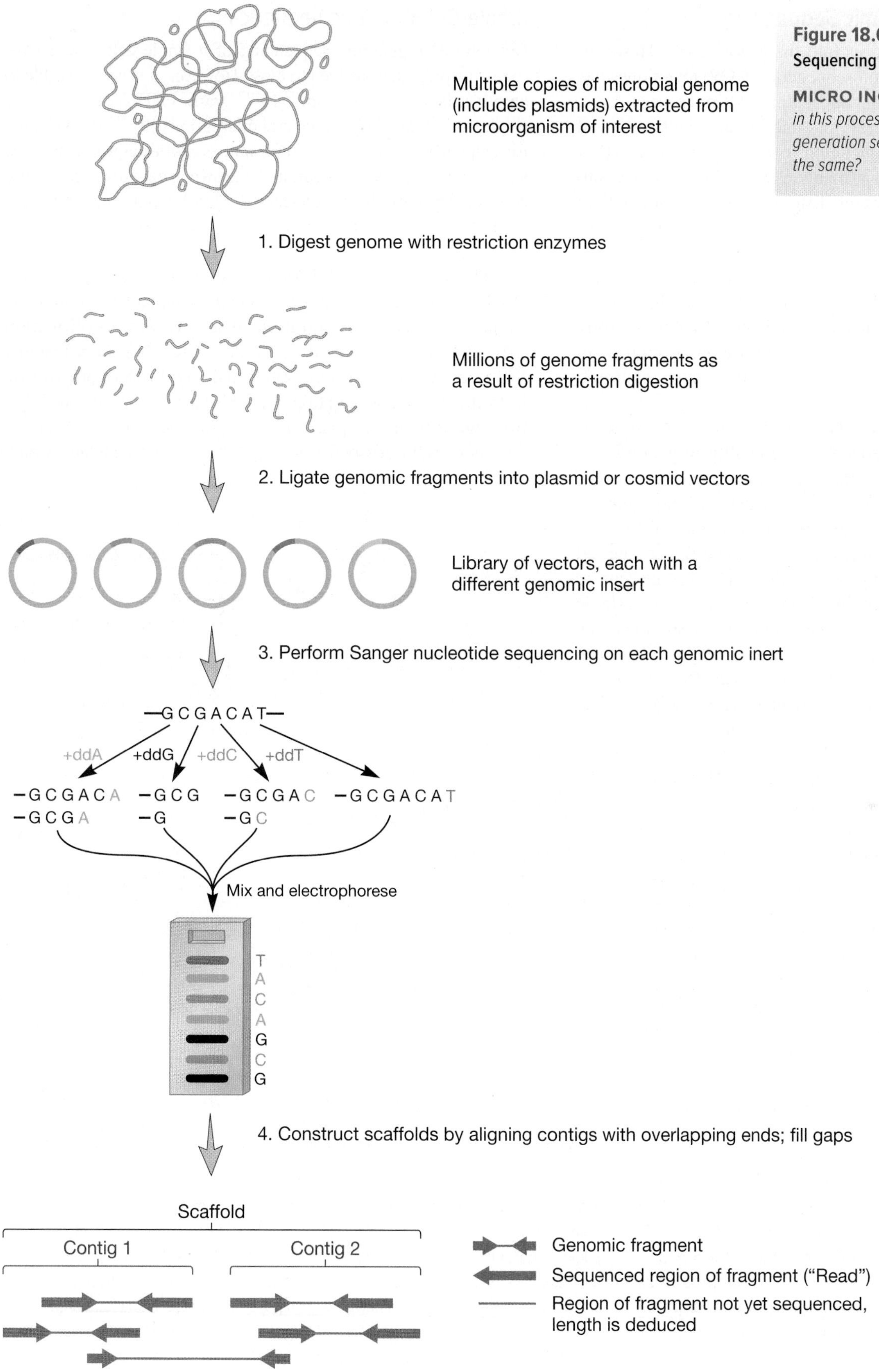

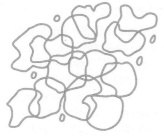

Multiple copies of microbial genome (includes plasmids) extracted from microorganism of interest

Figure 18.6 Whole-Genome Shotgun Sequencing Using Sanger Sequencing.

MICRO INQUIRY *Which step (or steps) in this process is (are) not used in next generation sequencing? Which are the same?*

1. Digest genome with restriction enzymes

Millions of genome fragments as a result of restriction digestion

2. Ligate genomic fragments into plasmid or cosmid vectors

Library of vectors, each with a different genomic insert

3. Perform Sanger nucleotide sequencing on each genomic inert

```
            —G C G A C A T—
  +ddA    +ddG    +ddC    +ddT
—G C G A C A  —G C G   —G C G A C   —G C G A C A T
—G C G A      —G       —G C
```

Mix and electrophorese

```
T
A
C
A
G
C
G
```

4. Construct scaffolds by aligning contigs with overlapping ends; fill gaps

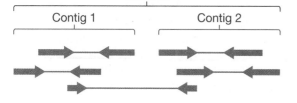

Scaffold

Contig 1   Contig 2

➤◄ Genomic fragment

◄ Sequenced region of fragment ("Read")

—— Region of fragment not yet sequenced, length is deduced

Next-Generation Genomic Sequencing

Sanger sequencing ushered us into the genomic era. However, the advent of next-generation sequencing (NGS) techniques has made genomic sequencing, particularly of microorganisms, much more practical in terms of time, money, and improved outcome. Like Sanger sequencing, preparing genomic DNA for sequencing by synthesis involves shearing the DNA into smaller pieces. The important difference is that, rather than inserting genomic fragments into cloning vectors as is the case with Sanger sequencing, adapters are added to the genomic DNA fragments, which are then attached to a solid substrate (figure 18.4). This is vastly more efficient; typically close to 100% of the genomic fragments with adapters bind to the solid substrate (e.g., flow cell surface), whereas cloning more than 80% of the genomic fragments into a cloning vector requires a lot of skill and luck.

The advent of NGS has had a significant impact on the quality of the final genomic sequence. The quality of genomic sequencing is assessed by two specific factors: depth of coverage and breadth of coverage. **Coverage** refers to the average number of times each nucleotide is sequenced in a genome (or other sequencing project). This redundancy of coverage is also referred to as the depth of sequencing, so "depth" and "coverage" can be used interchangeably. Depending on the technique used, a single nucleotide might be read (sequenced) 18 times or 80 times. The latter is described as deep sequencing; each nucleotide has been sequenced a very high average number of times. Breadth of coverage refers to how much of the entire genome was sequenced. A coverage of 100% means the genome sequence is complete, without gaps between contigs. Ideally one wants deep sequencing with 100% breadth of coverage. Because NGS does not involve the construction of a genomic library, it almost always yields higher breadth of coverage. The depth of coverage is also quite different. When Sanger genomic sequencing is used, any given region of the genome is typically sequenced no more than ten times. By contrast, the massively parallel sequencing that occurs in NGS results in the same genomic fragment being read 30 to 100 times, greatly increasing accuracy. This is important because the DNA polymerase used in sequencing reactions lacks proofreading capability, so it cannot correct mismatched bases. Deep NGS overcomes the limitations of the DNA polymerase reaction by re-reading the same stretch of genome many times. A base at a given position that is not in agreement with the majority of the reads is recognized as a mistake.

One feature that is not terribly different between the two types of approaches is the assembly of a complete genome. Recall that NGS generates much shorter reads (about 100 bases rather than 500 to 800 bases for Sanger); nonetheless, the reads are aligned by overlapping ends (figure 18.6). A great deal of computing power is required to align and assemble the thousands of short, overlapping sequences.

Single-Cell Genomic Sequencing

Not long ago, genomic sequencing of a single cell would have seemed a preposterous idea to most. However, it is now possible to amplify the few femtograms (10^{-15} gram) of DNA present in a single microbial cell to the several micrograms (10^{-6} gram) needed for sequencing. This is an important breakthrough because the vast majority of microbes cannot be grown axenically (i.e., in pure culture). The capacity to sequence the DNA from a single cell extracted from its natural environment is prompting new research strategies in microbial genetics, ecology, and infectious disease.

The process of **single-cell genomic sequencing** requires DNA amplification, but rather than PCR, a method called multiple strand displacement or **multiple displacement amplification** (**MDA**) is used (**figure 18.7**). Unlike PCR DNA amplification, MDA occurs at a single temperature and uses the DNA polymerase from the bacteriophage phi29 to synthesize new strands of DNA from the genome template. This polymerase is used because it does not readily dissociate from ("fall off") the template strand;

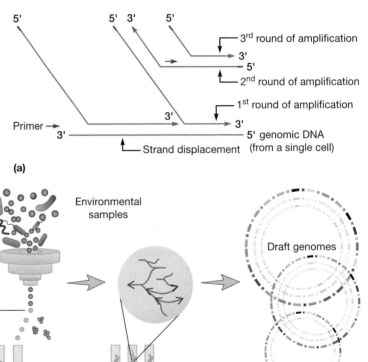

(a)

(b)

Figure 18.7 Single-Cell Genomic Sequencing. (a) Many copies of DNA extracted from a cell are generated by multiple displacement amplification. Random hexamer primers (red) bind to complementary template sequences, and DNA polymerase from bacteriophage phi29 is used to catalyze synthesis in the 5′ to 3′ direction (purple arrowheads). When the end of a newly synthesized strand meets double-stranded DNA, one strand is displaced by the growing DNA. (b) Single-cell genomic sequencing can be used to discover uncultivated bacteria and archaea from natural samples.

the importance of this feature will be made clear shortly. In addition, phi29 polymerase rarely incorporates the wrong base; that is to say, it has higher fidelity than most thermostable DNA polymerases used in PCR. Another difference between PCR and MDA is the use of a collection of primers with random sequences, six bases in length (hexamers), to initiate synthesis. The primers hydrogen bond to complementary sequences scattered throughout the genome. As DNA synthesis proceeds from each primer, the growing 3′ end of one newly made strand will eventually bump into and then displace the 5′ end of another newly growing strand. Recall that the phi29 polymerase does not easily dissociate once bound to the DNA, so both new strands will continue to grow. In this way, many new strands are rapidly synthesized. The new strands have an average length of about 12,000 bases (12 kb) but can be as long as 100 kb. This makes them suitable for DNA sequencing. NGS is then performed to avoid the requirement of genomic library construction and to ensure higher breadth and depth of coverage.

Recently an international group of microbiologists used single-cell genomics to sequence the genomes of 201 uncultivated microbes representing 29 undefined bacterial and archaeal taxa sampled from nine different environments (figure 18.7b). These microbes constitute what the researchers call "microbial dark matter," because we had no prior knowledge of their existence. Although the average breadth of coverage was only about 40%, reflecting a major limitation of single-cell genomics, a variety of discoveries were made, including about 20,000 new hypothetical protein families, evidence for two new superphyla, and the surprising presence of genes encoding sigma factors in archaea. ▶▶ *Microbial taxonomy and the evolution of diversity (chapter 19)*

Retrieve, Infer, Apply

1. Why is the Sanger technique of DNA sequencing also called the chain-termination method?
2. Explain the difference between a dideoxynucleotide used in Sanger sequencing and the modified bases used in reversible chain termination sequencing.
3. Why does reversible chain termination sequencing yield short reads?
4. How would one recognize a gap in the genome sequence following nucleotide sequencing?
5. Suggest a medical and an ecological application of single-cell genomic sequencing.

18.3 Metagenomics Provides Access to Uncultured Microbes

After reading this section, you should be able to:

- Differentiate between the construction and screening of a genomic library and a metagenomic library
- List two applications of metagenomics in any field of microbiology
- Define the human microbiome and explain the role metagenomics plays in its investigation

Metagenomics—the study of microbial genomes based on DNA extracted directly from the environment—has emerged as a key technology across a number of biological disciplines, including ecology, environmental microbiology, infectious disease, and immunology. The impact of metagenomics on such a large number of disciplines reflects the fact that so few microbial species can be cultured. Because metagenomics samples the entire pool of nucleic acids found in any given ecosystem, it is most frequently used to determine the members of the microbial community living there. In the past such analysis was performed using PCR to amplify small subunit rRNA genes (16S for bacteria and archaea, 18S for eukaryotes) or other target genes. This approach continues to be used, but metagenomics enables the sequencing of large portions of genomes, thereby revealing new microbial groups and novel genes (**figure 18.8**). ▶▶ *Microbial taxonomy and phylogeny are largely based on molecular characterization (section 19.3)*

To illustrate the impact of coupling direct DNA extraction with NGS, consider a microbial census based only on the microorganisms that can be cultured from an environmental sample (e.g., soil). Based on the average recovery of microbes in culture, this approach will miss about 98% of the microbial species present. Such an approach would be like taking a census of a big city by randomly sampling a few individuals. Next best would be to base a microbial census on genes cloned into bacterial vectors that are then sequenced by Sanger chain termination, as this would certainly improve representation, but by an unknown amount. Keeping with our city census analogy, this would be like taking a census by sending out a questionnaire to each household, hoping everyone would respond, but understanding that only an unknown fraction of residents will. Metagenomic analysis using next-generation sequencing, on the other hand, is the equivalent of sending a team of census takers into the community to ensure that each person is counted: the team probably won't reach every single resident, but depth and breadth of coverage will be very high. Similarly, without the need to construct a genomic library and by sequencing deeply, metagenomic analysis that employs NGS can detect organisms present in low numbers.

Once nucleotide sequences are obtained from DNA extracted directly from the environment, partial or full genomes can be detected in two ways. First, they can be assembled as previously discussed, by aligning overlapping sequences at the ends of reads (figure 18.6). However, one disadvantage of metagenomics is that the number of gapped sequences tends to be large (yielding scaffolds rather than genomes), so this approach is somewhat limited. Instead, one can attempt to align short reads to previously sequenced genomes in an effort to identify identical or related species. This too has the major drawback that many taxa obtained by metagenomics are uncultured and are therefore unrepresented in existing genomic databases; they too are microbial dark matter. This problem prompted the establishment of the Genomic Encyclopedia of Bacteria and Archaea Project. The goal of this program is to improve reference databases by sequencing the genomes of a wide diversity of cultured microorganisms. Metagenomic sampling of natural environments has so far led to the discovery of over 60 bacterial and archaeal phyla. Recall that a phylum is a group of related organisms above the level of class, order, and family, so this means that over 60 novel types of organisms have been discovered.

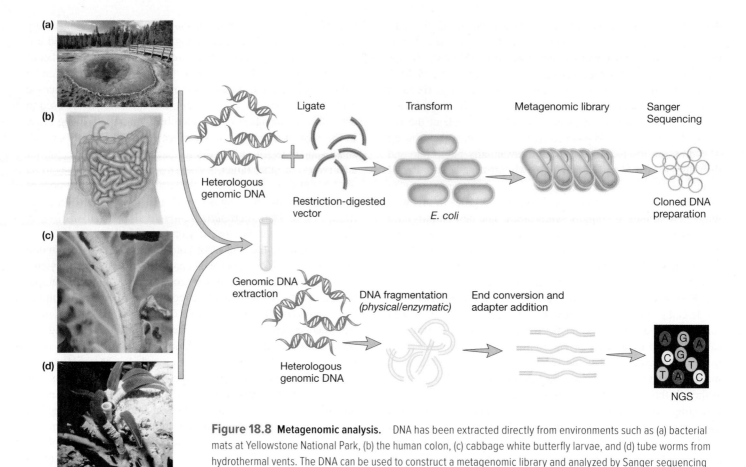

Figure 18.8 Metagenomic analysis. DNA has been extracted directly from environments such as (a) bacterial mats at Yellowstone National Park, (b) the human colon, (c) cabbage white butterfly larvae, and (d) tube worms from hydrothermal vents. The DNA can be used to construct a metagenomic library and analyzed by Sanger sequencing (top); however, NGS is more likely to be used (bottom), thereby avoiding the need for genomic library construction.

In addition to the microbial ecology of soils and waters, metagenomic analysis has had a huge impact on the microbial ecology of human beings. The census of microbes associated with human hosts has triggered a revolution in immunology, human physiology, and pathophysiology. A microbiome is the collection of microorganisms (and their genes) that are normally present in and on an organism. As discussed in further detail in chapter 32, we now know that there are about three times as many microbial cells in and on the human body than the somatic cells that comprise any given person. Furthermore, collectively these microbes have about 100-fold more genes than those encoded on the human genome. Metagenomics has helped identify who these microbes are and the many ways in which they influence human health. ▶▶◀ *The human-microbe ecosystem (section 32.2)*

Retrieve, Infer, Apply

1. What is a phylotype?
2. NGS techniques are considered well suited for metagenomics because a genomic library does not have to be constructed. Apart from convenience, explain why this is important for metagenomic analysis.
3. Examine figure 18.8. How might metagenomics be used to isolate genes encoding a potentially new antibiotic?

18.4 Bioinformatics: What Does the Sequence Mean?

After reading this section, you should be able to:

- Explain how a potential protein-coding gene is recognized within a genome sequence
- Compare the meaning of the terms orthologue and paralogue
- Differentiate between a conserved hypothetical protein and a putative protein of unknown function
- Describe the construction of a physical genome map

The analysis of entire genomes generates a huge volume of information regarding genome content, structure, and arrangement as well as data detailing protein structure and function. The field **bioinformatics** combines biology, mathematics, computer science, and statistics to determine the location and potential function of genes or presumed genes on sequenced genomes using a complex process called **genome annotation.** Once genes have been identified, bioinformaticists can perform computer or **in silico analysis** to further examine the genome.

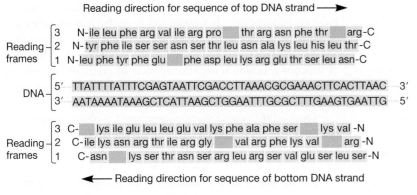

Figure 18.9 Finding Potential Protein-Coding Genes. Annotation of genomic sequence requires that both strands of DNA be translated from the 5′ to 3′ direction in each of three possible reading frames. Stop codons are shown in green.

Obviously, obtaining nucleotide sequences without any understanding of the location and function of individual genes would be a pointless exercise. The goal of genome annotation is to identify every potential (putative) protein-coding gene as well as each rRNA- and tRNA-coding gene. A protein-coding gene is usually first recognized as an **open reading frame** (**ORF**); to find all ORFs, both strands of DNA must be analyzed in all three reading frames (**figure 18.9**). A bacterial or archaeal ORF is generally defined as a sequence of at least 100 codons that it is not interrupted by a stop codon and has terminator sequences at the 3′ end. The 5′ end of the gene should also bear a ribosome-binding site. Only if these elements are present is an ORF considered a putative protein-coding gene. This process is performed by gene prediction programs designed to find genes that encode proteins or functional RNA

products. Ideally, computer-identified genes are then manually inspected by bioinformaticists to verify the computer-generated gene assignments. This process is called genome curation.

ORFs that appear to encode proteins are called **coding sequences** (**CDS**). Bioinformaticists have developed algorithms to compare the sequence of predicted CDS with those in large databases containing nucleotide and amino acid sequences of known proteins. The base-by-base comparison of two or more gene sequences is called **alignment.** Alignments can also be performed by comparing amino acid sequences between two proteins. Scientists most often use **BLAST** (***b**asic **l**ocal **a**lignment **s**earch **t**ool*) programs to perform this task. These programs compare the nucleotide (or amino acid) sequence of interest, called the query sequence, to all other sequences entered in the database. The results ("hits") are ranked in order of decreasing similarity. An E-value is assigned to each alignment; this value measures the possibility of obtaining the alignment by chance; thus highly homologous sequences have very low E-values.

Considerable information can often be inferred from translated amino acid sequences of potential genes. Often a short pattern of amino acids, called a motif or domain, will represent a functional unit within a protein, such as the active site of an enzyme. For instance, **figure 18.10** shows the C-terminal domain of the cell division protein MinD from a number of microbes. Because these amino acids are found in such a wide range of organisms, they are considered phylogenetically well conserved. In

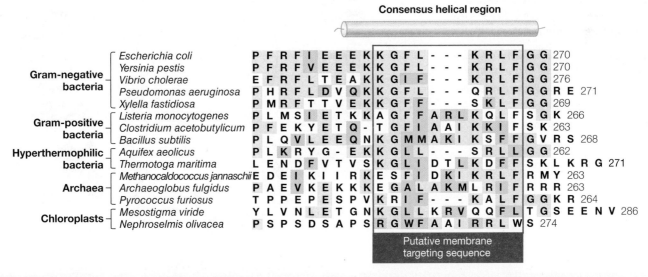

Figure 18.10 Analysis of Conserved Regions of Phylogenetically Well-Conserved Proteins. C-terminal amino acid residues of MinD from 15 organisms and chloroplasts are aligned to show strong similarities. Amino acid residues identical to *E. coli* are boxed in yellow, and conservative substitutions (e.g., one hydrophobic residue for another) are boxed in orange. Dashes indicate the absence of amino acids in those positions; such gaps may be included to maintain an alignment. The number of the last residue shown relative to the entire amino acid sequence is shown at the extreme right of each line.

MICRO INQUIRY *Which amino acids are most highly conserved?*

this case, the conserved region is predicted to form a coil needed for proper localization of the protein to the membrane. Finding this high level of conservation allows the genome curator to confidently assign a function to the domain. ◄◄ *Cytokinesis (section 7.2)*

Genes from different organisms with such similar ORFs are called **orthologues.** Sometimes there appear to be duplicated genes on the same genome. This is discovered when two or more genes have very similar nucleotide sequences. Such genes are called **paralogues.**

As the number of sequenced genomes has expanded, so has the need to carefully define the vocabulary by which new genes and proteins are named. The use of a structured vocabulary is called ontology, and a standard **gene ontology (GO)** has been adopted as the means by which proteins, or motifs within proteins, are commonly named. This is based on the similarities of amino acid sequences among orthologous proteins. A GO term not only reflects protein function but also defines the cellular process in which the protein participates (e.g., motility) and the cellular location of the protein (e.g., flagellum).

Proteins that do not align with known amino acid sequences fall into two classes: (1) **Conserved hypothetical proteins** are encoded by genes that have matches in the database but no function has yet been assigned to any of the sequences. (2) **Proteins of unknown function** are the products of genes unique to that organism. However, as more genomes are published, future comparisons may reveal a match in another organism.

Once all the genes have been annotated, a **physical map** representing the entire genome may be drawn (**figure 18.11**). It is helpful to color-code genes according to category. For instance, when the *H. influenzae* genome was first annotated, about one-third of the genes were of unknown function (denoted by white regions on the map) and about 65 genes were assigned regulatory functions (dark-blue regions). However, the *H. influenzae* genome was only the second bacterial genome sequenced. With thousands of bacterial and archaeal sequences now available for comparison, the number of genes with unknown function has fallen dramatically and will continue to decline as additional genome sequences are made publicly available.

Figure 18.11 Physical Map of the *Haemophilus influenzae* Genome. The predicted coding regions in the outer concentric circle are indicated with colors representing their functional roles. The outer perimeter shows the *Not*I, *Rsr*II, and *Sma*I restriction sites. The inner concentric circle shows regions of high G + C content (red and blue) and high A + T content (black and green). The third circle shows the coverage by λ clones (blue). The fourth circle shows the locations of rRNA operons (green), tRNAs (black), and the Mu-like prophage (blue). The fifth circle shows simple tandem repeats and the probable origin of replication (outward pointing green arrows). The red lines are potential termination sequences.

Science by MOSES KING, Reproduced with permission of AMERICAN ASSOCIATION FOR THE ADVANCEMENT OF SCIENCE in the format Educational/Instructional Program via Copyright Clearance Center.

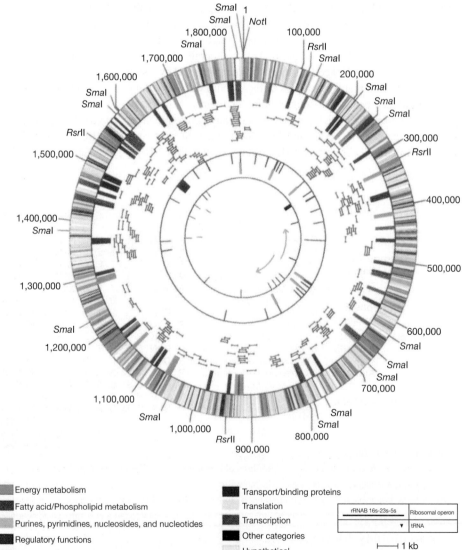

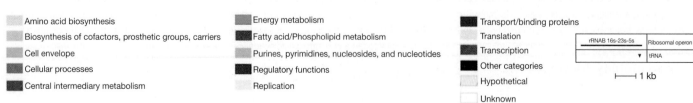

18.5 Functional Genomics Links Genes to Phenotype

After reading this section, you should be able to:

- Explain how genome annotation can be used to graphically represent the metabolism, transport, motility, and other key features of a microbe
- Contrast and compare microarray analysis with RNA-seq in the study of transcriptomes
- Explain how 2-D gel electrophoresis is able to separate proteins of identical molecular weight
- Summarize the importance of mass spectrometry in analyzing protein structure
- Explain why DNA-protein interactions are of interest and how they can be experimentally identified

Functional genomics seeks to place genomic information in a biological context. For instance, the careful annotation of a microbe's genome can be used to piece together metabolic pathways, transport systems, and potential regulatory and signal transduction mechanisms. A common outcome of a genome project is to use the identified genes to infer the functional metabolic pathways, transport mechanisms, and other physiological features of the microbe (**figure 18.12**). One of the first microbes for which such analysis was performed was the causative agent of syphilis, *Treponema pallidum*. Because *T. pallidum* has not yet been grown in pure culture, we know little about its metabolism or the way it avoids host defenses. The sequencing and annotation of the *T. pallidum* genome reveal that *T. pallidum* is metabolically crippled. It can use carbohydrates as an energy source but lacks TCA cycle and oxidative phosphorylation enzymes. *T. pallidum* also lacks many biosynthetic pathways (e.g., for enzyme cofactors, fatty acids, nucleotides, and some electron transport proteins) and must rely on molecules supplied by its host. In fact, about 5% of its genes code for transport proteins. The genes for surface proteins are of particular interest. *T. pallidum* has a family of surface protein genes characterized by many repetitive sequences. Some have speculated that these genes might undergo recombination to generate new surface proteins, enabling the organism to avoid attack by the immune system. The annotated

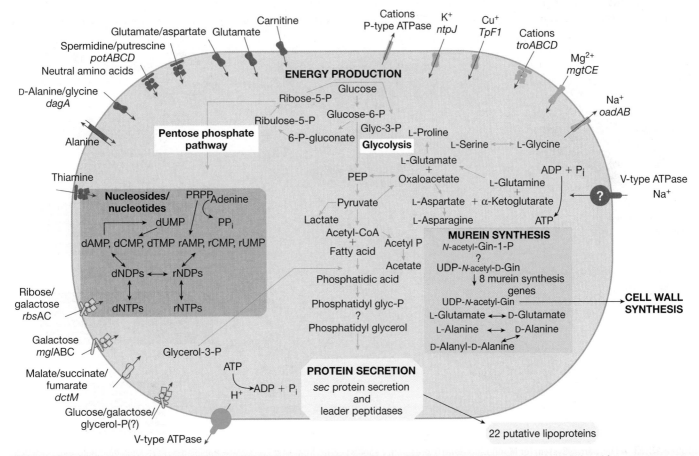

Figure 18.12 Metabolic Pathways and Transport Systems of *Treponema pallidum*. This depicts *T. pallidum* metabolism as deduced from genome annotation. Note the limited biosynthetic capabilities and extensive array of transporters. Although glycolysis is present, the TCA cycle and respiratory electron transport are lacking. Question marks indicate where uncertainties exist or expected activities have not been found.

MICRO INQUIRY *Based on this genomic reconstruction, can you determine if* T. pallidum *has a respiratory or fermentative metabolism?*

genome should ultimately help us understand how *T. pallidum* causes syphilis. ▶▶│ *Phylum* Spirochaetes *(section 21.8)*

Transcriptome Analysis: Microarrays and RNA-Seq

Once the identity and function of the genes that comprise a genome have been analyzed, the key question remains, "Which genes are expressed at any given time?" The entire collection of mRNAs that are produced at any one time by an organism is its **transcriptome,** which is studied in the field of transcriptomics. Prior to the genomic era, researchers could identify only a limited number of genes whose expression was altered under specific circumstances. There are now two technologies that allows scientists to look at the expression level of a vast collection of genes: DNA microarrays and RNA-Seq. The first to be developed, **DNA microarrays,** consists of solid supports, usually of glass or silicon, upon which DNA is attached in an organized grid fashion. Each spot of DNA, called a **probe,** represents a single gene. The probe may be a PCR product generated from genes of interest or complementary DNA (cDNA). The location and identity of each probe on the grid is carefully recorded. This enables the investigator to monitor the expression of all the genes represented on the grid.

The analysis of gene expression using microarray technology, like many other molecular genetic techniques, is based on hybridization between the single-stranded probe DNA (i.e., the genes attached to the microarray) and the target nucleotides from the microbe of interest. These are typically single-stranded cDNAs made from mRNA (**figure 18.13**). The target nucleotides are labeled with fluorescent dyes and incubated with the microarray under conditions that ensure proper binding of target cDNA to its complementary probe (gene). Unbound target is washed off and the microarray is then scanned with laser optics. Fluorescence at each spot or probe indicates that cDNA hybridized. Analysis of the color and intensity of each probe quantifies the level of expression of each gene represented by a probe. 🔎 *DNA Probe*

DNA microarray analysis can be used to determine which genes are expressed at specific times during growth or how a mutation in one gene influences the expression of other genes. One common application of microarray technology is to determine the genes whose expression is changed (either up- or down-regulated) in response to environmental changes (figure 18.13). For instance, the microbe *Deinococcus radiodurans* has the remarkable ability to survive intense desiccation and γ-radiation at doses many times above that needed to kill humans. In order to determine which *D. radiodurans* genes are up- or down-regulated upon exposure to ionizing radiation, total cellular mRNA from bacteria grown under normal conditions can be isolated and

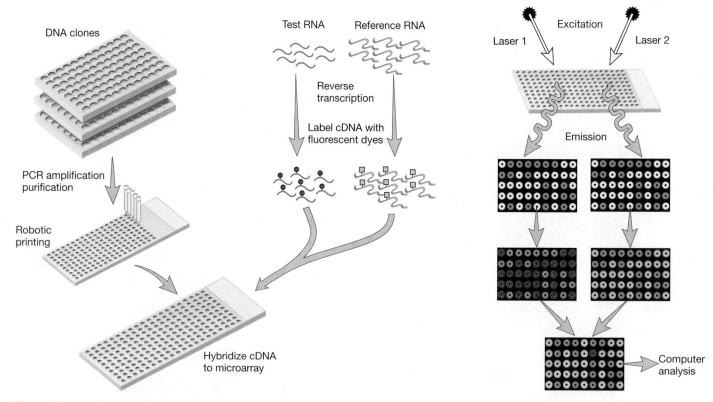

Figure 18.13 A Microarray System for Monitoring Gene Expression.   Cloned genes from an organism are amplified by PCR, and after purification, samples are applied by a robotic printer to generate a microarray. To monitor enzyme expression, mRNA from test and reference cultures is converted to cDNA by reverse transcriptase and labeled with two different fluorescent dyes. The labeled mixture is hybridized to the microarray and scanned using two lasers with different excitation wavelengths. The fluorescence responses are measured as normalized ratios that show whether the test gene response is higher or lower than that of the reference.

converted to cDNA by reverse transcription (*see figure 17.5*). This cDNA is tagged with a green fluorochrome to serve as a control or reference, while cDNA made from mRNA recovered from cells exposed to γ-radiation is labeled red. The green (reference) and red (experimental) cDNAs are then mixed and hybridized to the same microarray. During incubation, the red and green cDNAs for any given gene must compete with each other for binding to the same probe. After the unattached cDNA is washed off, the microarray is scanned and the image is computer analyzed. A yellow spot or probe indicates that roughly equal numbers of green and red cDNA molecules were bound, so there was no change in the level of gene expression for that gene. If a target is red, more mRNA from irradiated bacteria was present when the sample was taken, thus this gene was up-regulated. Conversely, a green target indicates that the gene was down-regulated upon exposure to radiation. Careful image analysis is used to determine the relative intensity of each spot so that the magnitude of change in expression of each gene can be approximated. ❷ *Microarrays*

Microarray experiments yield a vast amount of information that must be organized in some meaningful fashion. One common approach is to use **hierarchical cluster analysis,** which groups genes according to function or in terms of level of expression: up-regulated genes (red spots) are grouped separately from down-regulated genes (green spots); genes whose expression remains unaltered are shown in black. The hierarchical cluster analysis shown in **figure 18.14** was generated when *D. radiodurans* was exposed to γ-radiation. Ionizing radiation causes double-stranded breaks in DNA—the most lethal form of DNA damage. *D. radiodurans* is able to reassemble its genome after it has been fragmented into thousands of pieces. It was thought that sequencing this bacterium's genome would reveal that *D. radiodurans* is superb in executing DNA repair. But, surprisingly, *D. radiodurans* was discovered to have fewer DNA repair genes than *E. coli*.

To understand these unexpected findings, microbiologists used microarrays with about 94% of *D. radiodurans* genes represented to examine the transcriptome following radiation treatment. These results were then scanned for genes with similar functions and then grouped by relatedness (degree of relatedness is statistically quantified by a correlation coefficient, or "r value"). Such analysis confirmed that the DNA repair gene *recA* and genes involved in DNA replication and recombination are dramatically upregulated following irradiation. However, detection of a gene whose function was poorly understood led to studies in which it was found that unlike repair found in other cells in which one strand serves as a template for synthesis of a new strand, *Deinococcus* uses an unusual DNA polymerase to synthesize two new strands. The newly synthesized fragments are then pieced together in a process that depends on RecA. The newly repaired genome thus becomes a mosaic of old and new DNA fragments. ▶▶| Deinococcus-Thermus *includes radiation-resistant bacteria (section 21.2)*

Although microarrays gave birth to transcriptome analysis, they present three principal drawbacks. First,

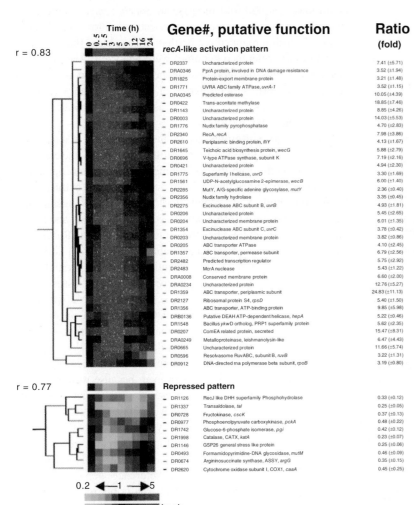

Figure 18.14 Hierarchical Cluster Analysis of Gene Expression of *D. radiodurans* Following Exposure to γ-Radiation. Each row of colored strips represents a single gene, and the color indicates the level of expression over nine time intervals. The far-left column is the control and thus is black (at control levels of expression). The level of induction or repression relative to the control value is indicated as the ratio (fold). Each group of genes has been scored for relatedness, and a "tree" has been generated on the far left of the clusters, with the indicated correlation coefficient (r value). Many genes encoding DNA repair, synthesis, and recombination proteins are induced upon radiation. These are grouped separately from repressed genes involved in other aspects of metabolism.

MICRO INQUIRY *Based on your interpretation of the data, identify two genes that are strongly induced and two genes that are strongly repressed.*

the mRNAs identified are only those that have a corresponding probe on the array. This means that noncoding RNAs and coding sequences that may have been missed during annotation go undetected. Second, nonspecific hybridization of some cDNAs to probe sequences can introduce a high level of background noise, making analysis difficult. Finally, microarrays have what is called "limited dynamic range." This means that small changes in gene expression are overlooked and the magnitude of very high gene expression is impossible to determine because the signal saturates at a fairly low level.

The advent of NGS techniques has made it possible to directly sequence total cellular mRNA, a process known as **RNA-Seq.** Quantification of mRNA is accomplished by measuring the amount of data ("reads") matching each gene. Because RNA-Seq

generates data without the limitations encountered in microarray analysis, it has replaced microarrays as the transcriptomic tool of choice.

The process of RNA-Seq, like that of microarrays, begins with converting all the cellular mRNA into cDNA by incubating the mRNA in the presence of the enzyme reverse transcriptase and dNTPs (**figure 18.15**; *also see figure 17.5*). Next, adapter sequences are added, at either one or both ends of each cDNA fragment. Each cDNA, representing an individual mRNA, is then sequenced using NGS. The resulting nucleotide sequences can be analyzed in two ways. If the microorganism being studied has a sequenced genome, the cDNA sequence is identified by alignment with the known genome sequence, called a reference genome. If no reference genome is available, the nucleotide sequences are converted to amino acid sequences and compared with databases that house millions of protein sequences.

RNA-Seq offers many advantages over microarrays. It is not limited to the identification of only those mRNA molecules with a corresponding gene on the array, or even a gene that has been previously sequenced. Indeed, many new genes and small RNAs have been discovered by RNA-Seq. Because the number of reads is not limited, there is no upper limit to the level of gene induction that can be detected. For instance, under specific growth conditions, a 9,000-fold increase in *Saccharomyces cerevisiae* (baker's yeast), gene expression was estimated for several genes when 16 million mapped reads (that is, sequences that could be aligned to the reference genome) were analyzed. Also, because RNA-Seq yields nucleotide sequence, information is revealed about the transcripts themselves. For instance, variations in transcribed sequences can be detected, as can the nucleotides where transcription starts and stops.

Metatranscriptomics, as its name suggests, is similar to metagenomics in that it is based on the analysis of nucleic acids obtained directly from the environment. In metatranscriptomics, however, RNA rather than DNA is extracted. The extraction of RNA from fluids, soils, blood, stool, and so forth is technically challenging because RNA is not very stable and tRNA and rRNA molecules vastly outnumber the more labile mRNA. Recent advances in extraction procedures enable the isolation of sufficient quantities of mRNA from diverse microorganisms to assess gene expression at the community level. Once obtained, the environmental mRNA is processed like that described for RNA-Seq from a single, cultured microorganism. However, because metatranscriptomics describes the transcriptome of an entire ecosystem, it will yield transcripts that map to multiple reference genomes as well as those that fail to align with any reference genomes. These novel transcripts represent newly discovered genes.

Figure 18.15 RNA-Seq.

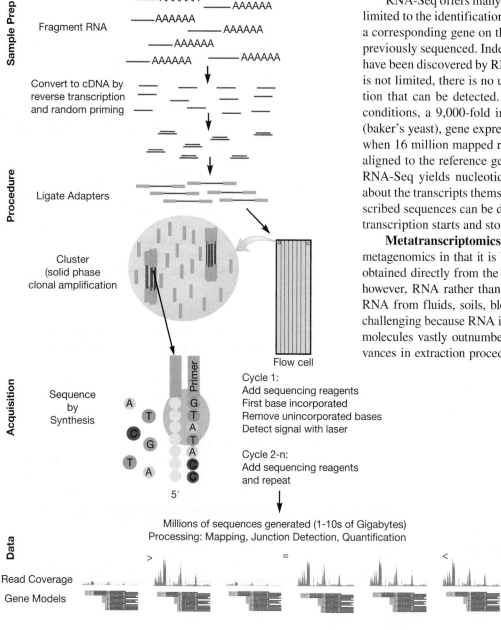

Retrieve, Infer, Apply

1. What is genome annotation? Why does it require knowledge of mathematics, statistics, biology, and computer science?
2. How might the following scientists use transcriptomics in their research?
   (a) A microbial ecologist who is interested in how the soil microbe *Rhodopseudomonas palustris* degrades the toxic compound 3-chlorobenzene
   (b) A medical microbiologist who wants to learn how the pathogen *Salmonella* survives within a host cell
3. What specific limitations of microarrays are overcome by RNA-seq?
4. Compare and contrast metagenomics with metatranscriptomics.

Proteomics Explores Total Cellular Protein

Genome function can be studied at the level of translation (i.e., proteins) as well as transcription (i.e., mRNA). The entire collection of proteins produced by an organism is called its **proteome.** Thus **proteomics** is the study of the proteome or the collection of proteins an organism produces. Proteomics provides information about genome function that mRNA studies cannot because a direct correlation between mRNA and the pool of cellular proteins does not always exist. Proteomics strives to determine the function of different cellular proteins, how they interact with one another and other molecules within the cell, and the ways in which they are regulated.

Although new techniques in proteomics are continuously being developed, we focus briefly only on the most common approaches, including **two-dimensional gel electrophoresis.** In this procedure, a mixture of proteins is separated using two different electrophoretic procedures, called dimensions. This permits the visualization of thousands of proteins that would not otherwise be separated based on molecular weight, because many proteins have roughly the same mass. As shown in **figure 18.16***a,* the first dimension makes use of **isoelectric focusing,** in which proteins move through a pH gradient (e.g., pH 3 to 10). In isoelectric focusing, a protein mixture is applied to an acrylamide gel in a tube with a pH gradient and electrophoresed. Each protein moves along the pH gradient until the protein's net charge is zero and the protein stops moving. The pH at this point is equal to the protein's **isoelectric point.** Thus the first dimension (or gel) separates proteins based on the content of ionizable amino acids. The second dimension (gel) is SDS *p*oly*a*crylamide *g*el *e*lectrophoresis (SDS-PAGE). SDS (sodium dodecyl sulfate) is an anionic detergent that denatures proteins and coats them so that now all the proteins have a negative charge. After the isoelectric gel has been completed, the tube gel is soaked in SDS buffer and then placed at the edge of an SDS-PAGE gel. A voltage is then applied. Because all the proteins are negatively charged thanks to SDS treatment, they migrate from the negative pole of the gel to the positive. In this way, polypeptides, which are already separated along the length of the tube gel by charge, are now separated by their molecular weight; that is, the smallest polypeptide will travel fastest and farthest. Two-dimensional gel electrophoresis can resolve thousands of proteins; each protein is visualized as a spot of varying

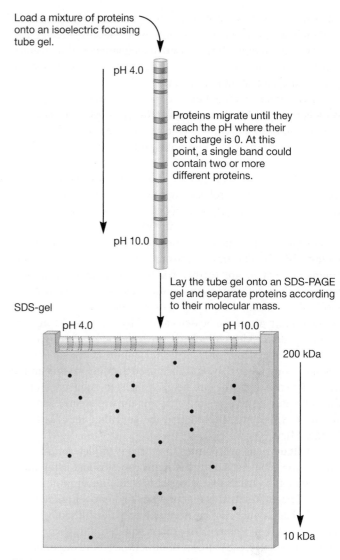

(a) The technique of two-dimensional gel electrophoresis

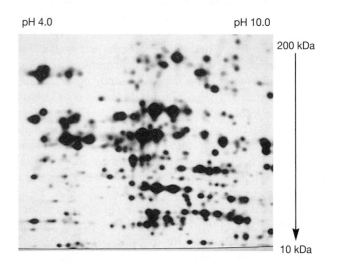

(b) An autoradiograph of a two-dimensional gel. Each protein is a discrete spot.

Figure 18.16 Two-Dimensional Gel Electrophoresis.

intensity, depending on its cellular abundance (figure 18.16*b*). Radiolabeled proteins are typically used, enabling greater sensitivity. Computer analysis is used to compare two-dimensional gels from microbes grown under different conditions or to compare wild-type and mutant strains.

Two-dimensional gel electrophoresis is even more powerful when coupled with **mass spectrometry** (**MS**). To perform MS, a protein spot is cut from the gel and the protein is cleaved into fragments by treatment with proteolytic enzymes. Then the fragments are analyzed by a mass spectrometer and the mass of the fragments is plotted. This mass fingerprint can be used to estimate the probable amino acid composition of each fragment and tentatively identify the protein. Sometimes proteins or collections of fragments are run through two mass spectrometers in sequence, a process known as **tandem MS** (**figure 18.17**). The first spectrometer separates proteins and fragments, which are further fragmented. The second spectrometer then determines the amino acid sequence (based on molecular weight) of each smaller fragment. The sequence of a whole protein often can be determined by analysis of fragment sequence data. Alternatively, if the genome of the organism has been sequenced, only a partial amino acid sequence is needed. Computer analysis is then used to compare this amino acid sequence with the predicted translated sequences of all the annotated protein-coding genes on the organism's genome. In this way, both the protein and the gene that encodes it can be identified.

In **structural proteomics** the focus is on determining the three-dimensional structures of many proteins and using these to predict the structures of other proteins and protein complexes. The assumption is that proteins fold into a limited number of shapes and can be grouped into families of similar structures. When a number of protein structures are determined for a given family, the patterns of protein structure organization or protein-folding rules will be known. Then computational biologists use this information to predict the most likely shape of a newly discovered protein; a process known as **protein modeling.**

Not surprisingly, a variety of other "-omics" are studied. For instance, **lipidomics** investigates a cell's lipid profile at a particular time, and **glycomics** is the systematic study of the cellular pool of carbohydrates. The goal of **metabolomics** is to identify all the small-molecule metabolites present in the cell at a given point in time. Because many metabolites are used in multiple pathways, the overview presented by a complete metabolome enables a researcher to discern which pathways are functional under a given set of conditions. Thus metabolomics provides a high-resolution means to evaluate physiological status. These "-omics" rely heavily on the use of chromatography and as such are beyond the scope of this chapter. Nonetheless, the development of these fields illustrates the dynamic and ever-changing nature of microbiology and the need to take a holistic view of cells and their communities if one seeks to understand the structure, function, and behavior of microbes.

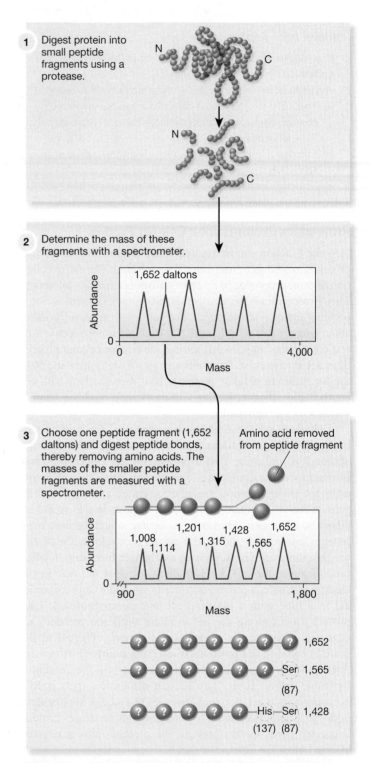

1 Digest protein into small peptide fragments using a protease.

2 Determine the mass of these fragments with a spectrometer.

1,652 daltons

Abundance

0
0   Mass   4,000

3 Choose one peptide fragment (1,652 daltons) and digest peptide bonds, thereby removing amino acids. The masses of the smaller peptide fragments are measured with a spectrometer.

Amino acid removed from peptide fragment

Abundance

1,008   1,114   1,201   1,315   1,428   1,565   1,652

0
900   1,800
Mass

? ? ? ? ? ? ?   1,652
? ? ? ? ? ?—Ser   1,565
                    (87)
? ? ? ? ?—His—Ser   1,428
              (137) (87)

Figure 18.17 The Use of Tandem Mass Spectrometry to Determine the Amino Acid Sequence of a Peptide.

MICRO INQUIRY *How would determining the amino acid sequence of a 30 amino acid residue peptide differ from determining the amino acid sequence of a 200,000 dalton protein?*

Probing DNA-Protein Interactions

As described in chapters 13 and 15, many proteins directly interact with DNA; for example, regulatory proteins and proteins involved in replication and transcription function by binding to DNA. Until recently, the most common way to study protein-DNA interactions was through a procedure known as electrophoretic mobility shift assays (EMSA). This technique is also called a gel shift assay because when a DNA-binding protein is added to a mixture of purified target DNA, the mobility of the DNA is slowed in an agarose gel. When compared to the same DNA without protein, the DNA-protein complex is shifted to a higher molecular weight. The magnitude of the shift is usually related to the protein:DNA ratio such that when sufficient protein is added, the largest shift is seen. To perform EMSA, the target protein must be purified and the DNA to which it binds must be amplified. ◀◀ *Gel electrophoresis (section 17.1)*

What if an investigator suspects a DNA-binding protein binds to more than one target DNA? He or she would turn to **chromatin immunoprecipitation (ChIP)** coupled with microarray technology or next-generation sequencing. In contrast to EMSA, which is entirely in vitro and uses one protein and one amplified fragment of DNA, ChIP surveys the living cell. First, cells are treated with a cross-linking agent such as formaldehyde, which "fixes" any DNA-binding proteins to their genomic targets (**figure 18.18**). The cells are then broken open and the DNA (with bound proteins) is cut into small pieces. Antibodies, which are molecules that bind to specified proteins, are used to tag the protein of interest bound to the DNA, and the antibody-protein-DNA complexes are precipitated. Next, the proteins are removed and the DNA to which the proteins were bound can be identified by hybridizing the DNA to a microarray (ChIP-chip) or by shearing the DNA so that a next-generation sequencing technique can be applied. This is called ChIP-Seq, and it is the preferred approach for the same reasons that RNA-Seq has displaced microarrays. ▶▶ *Antibodies are proteins that bind to specific molecules (section 34.7)*

Regardless of whether ChIP-chip or ChIP-Seq is performed, this technology enables the identification of DNA-protein interactions under different conditions. The same microbial strain can be treated with cross-linking agents when grown in different situations or at different phases of growth, so the DNA-binding profile of a protein can be followed.

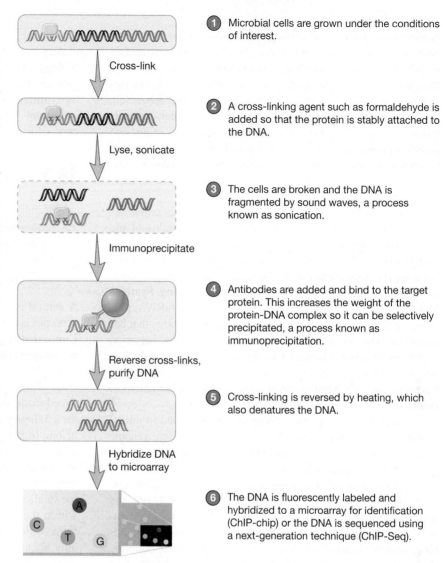

Figure 18.18   ChIP-Chip and ChIP-Seq Analysis.   This technology can be used to identify the DNA to which a protein binds.

1. Microbial cells are grown under the conditions of interest.

2. A cross-linking agent such as formaldehyde is added so that the protein is stably attached to the DNA.

3. The cells are broken and the DNA is fragmented by sound waves, a process known as sonication.

4. Antibodies are added and bind to the target protein. This increases the weight of the protein-DNA complex so it can be selectively precipitated, a process known as immunoprecipitation.

5. Cross-linking is reversed by heating, which also denatures the DNA.

6. The DNA is fluorescently labeled and hybridized to a microarray for identification (ChIP-chip) or the DNA is sequenced using a next-generation technique (ChIP-Seq).

Retrieve, Infer, Apply

1. Why does two-dimensional gel electrophoresis allow the visualization of many more cellular proteins than seen in single-dimension electrophoresis (i.e., SDS-PAGE)?

2. What is the difference between mass spectrometry and tandem mass spectrometry?

3. Describe a ChIP-Seq experiment that will enable the identification of promoters to which the stationary phase sigma factor RpoS binds in *E. coli*. How could you also determine when RpoS binds to DNA?

18.6 Systems Biology: Making and Testing Complex Predictions

After reading this section, you should be able to:

- Explain the role of genomics in shaping systems biology
- Compare and contrast systems biology and synthetic biology

Imagine the following experiment: You have created a microbial strain in which you have deleted, or "knocked out," a single gene;

it doesn't really matter what the gene encodes. You want to compare the transcriptome and proteome of the knockout mutant with its parent strain, which is identical to it in every way except the presence of this particular gene. To compare the transcriptomes of wild-type and mutant strains, you perform RNA-Seq, and to compare their proteomes, you run two-dimensional gels followed by mass spectroscopy. You find that the loss of just one gene has huge ramifications with regard to gene expression and protein synthesis. Not only is the pathway that your gene product was known to be involved in altered, but the expression of many more genes and proteins has also been affected.

This scenario has been repeated many times for numerous microorganisms. Life scientists now realize that most of the genes and proteins that have been widely investigated have a vastly broader role in cellular physiology than previously thought. Gone are the days of reductionist biology, when a single metabolic or regulatory pathway could be considered in isolation. **Systems biology** seeks to integrate the "parts list" of cells—mRNA, proteins, small molecules—with the molecular interactions that become pathways for catabolism, anabolism, regulation, behavior, responses to environmental signals, etc. To accomplish this, systems biology is interdisciplinary, requiring the expertise of physiologists, biochemists, bioinformaticists, mathematicians, geneticists, ecologists, and others.

One area where systems biology has been applied to microbiology is the study of metabolism (**figure 18.19**). Here genome sequencing and annotation leads to one set of predictions that can be tested by transcriptomics, proteomics, and metabolomics. The interactions between the individual products—RNA, protein, metabolites—can then be linked in a predictive network and used to construct testable hypotheses. Like most scientific endeavors, systems biology is an iterative process involving multiple rounds of data collection, analysis, and hypothesis building. What sets it apart, however, is the volume of data, the depth of sampling, and the use of sophisticated predictive models to guide the development of new hypotheses.

As systems biology matured, scientists realized that understanding cellular regulatory systems could lead to a new level of metabolic engineering such that a microbe could be constructed with novel genetic networks. The development of artificial regulatory pathways to yield specific products is called **synthetic biology.** Among the biggest successes of synthetic biology to date are biofuel production in *E. coli* by rerouting amino acid biosynthetic pathways, and the commercial production of the important antimalaria drug artemisinin from *Saccharomyces cerevisiae.* ▶▶❙ *Synthetic biology (section 42.4)*

18.7 Comparative Genomics

After reading this section, you should be able to:

- Discern the general relationship between genome size and organism complexity
- Describe the genomic differences that distinguish intracellular parasites from free-living microbes

A natural consequence of the tremendous number of sequenced genomes is the ability to make meaningful genome comparisons. **Comparative genomics** focuses on similar nucleotide and amino acid sequences among organisms to infer gene function and evolution. Perhaps the most fundamental comparison is the relative sizes of microbial genomes. Several generalizations can be deduced from such analysis. Inspection of **figure 18.20** reveals that the smallest genomes published to date, across all three domains, belong to parasitic microbes. These organisms depend on their hosts for many metabolites and as they have co-evolved with their hosts, have lost the genes encoding enzymes needed for their assembly. For example, one of the smallest genomes (if not the smallest) belongs to *Candidatus* Carsonella rudii, which lives within the sap-sucking insect *Pachypsylla venusta.* This bacterium, which cannot be grown without its host, has a genome of only about 160,000 (160 Kb) base pairs and 182 genes. This is roughly equivalent in size to the cytomegalovirus (CMV) with about 160 proteins encoded on 230-Kb genome.

Generalizations about genome size can also be made across the three domains of life. Bacterial and archaeal genomes tend to be smaller than eukaryotic genomes, in part because they have more open reading frames per nucleotide; that is, they have a higher gene density than eukaryotes. This is because bacteria and archaea lack introns and much of the other noncoding sequences found in eukaryotes. Indeed, some archaea and bacteria have reduced noncoding sequences to a minimum and have many overlapping genes. This is particularly true for those living in nutrient-limited environments, because DNA synthesis requires both phosphate and nitrogen and is energetically costly. Furthermore, archaea tend to have smaller genomes than many bacteria (figure 18.20). The largest known archaeal genome, that of the methanogen *Methanosarcina acetivorans,* falls

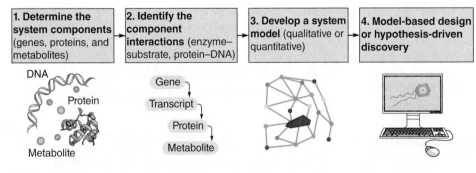

| 1. Determine the system components (genes, proteins, and metabolites) | 2. Identify the component interactions (enzyme–substrate, protein–DNA) | 3. Develop a system model (qualitative or quantitative) | 4. Model-based design or hypothesis-driven discovery |

DNA
Protein
Metabolite

Gene
Transcript
Protein
Metabolite

Figure 18.19 Systems-Based Biology. Microbial metabolism is an active area of systems biology research. It can involve (1) identifying the system components, (2) recognizing their interactions, and (3) developing a model that predicts how they interact, which then (4) informs the synthesis of hypotheses that can be tested either by further modeling or by experimentation.

in the middle of the genome size range for bacteria. The genomes of free-living bacteria range from the small (about a million base pairs) to over 9 million base pairs (Mb); genome size generally reflects metabolic and morphological complexity. Although the majority of eukaryotic microbes have genomes that are bigger than bacterial and archaeal genomes, figure 18.20 demonstrates that eukaryotic microbes have the widest range of genome sizes. The genome of the intracellular pathogenic microsporidian fungus *Encephalitozoon cuniculi* is smaller than many bacterial and archaeal genomes, yet the ciliated protist *Paramecium tetraurelia* has a tremendously large genome (possibly due to multiple rounds of genomic duplication). ▶▶❙ *Order Streptomycetales (section 24.1); Alveolata (section 25.4); Nitrogen-fixing bacteria are vital to agriculture (section 31.3)*

Comparing genome sequences between genera, species, and strains has revealed a high level of **horizontal gene transfer** (**HGT**). As detailed in chapter 16, HGT is the exchange of genetic material between organisms that need not be of similar evolutionary lineages. Surprisingly, many genes involved in a myriad of cellular functions reside on mobile genetic elements, including transposons, plasmids, and phages. Genome analysis has revealed that HGT is frequently mediated by viruses and that lysogeny may be the rule, rather than the exception. In fact, some bacteria and archaea carry multiple proviruses. Some temperate phages carry lysogenic conversion genes, so called because they change the phenotype of their hosts. For instance, the virulence genes for some important pathogens, including the bacteria that cause diphtheria and cholera, are encoded by phages. ❙◀◀ *Lytic and lysogenic infections are common for bacterial and archaeal cells (section 6.4)*

Figure 18.20 Microbial Genome Sizes. The distribution of genome sizes in millions of base pairs (Mb) among some archaea, bacteria, and eukaryotic microbes. Asterisk denotes that the microbe is parasitic.

MICRO INQUIRY *Two* E. coli *strains are shown in this figure. One is a pathogen and one is not. Explain why genome size cannot be used to determine which is pathogenic.*

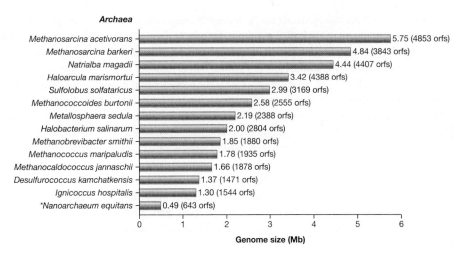

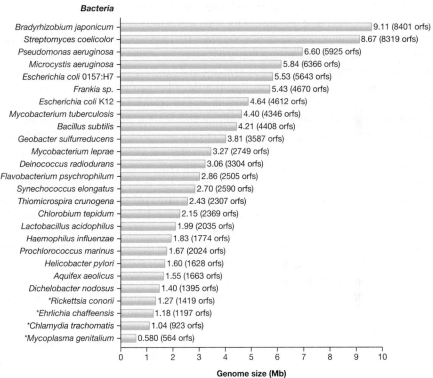

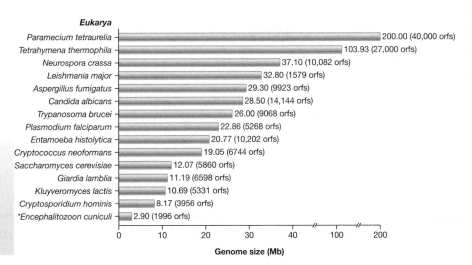

When mobile genetic elements are permanently integrated into a microbial genome, they are called **genomic islands.** If these new genes encode proteins that contribute to or confer virulence, they are known as **pathogenicity islands.** How is it known that certain genes were obtained by HGT? The most obvious clue is the presence of the same or similar genes in another distantly related microorganism. In addition, genomic islands often have a G + C content (measured as percentage) that differs from the remainder of the genome. Genomic islands also sometimes retain inverted repeat sequences or other elements needed for transfer or integration. It has become clear that HGT is a major evolutionary force in short-term microbial evolution and long-term speciation. ◄◄ *Microbes use mechanisms other than mutation to create genetic variability (section 16.4)* ►► *Evolutionary processes and the concept of a microbial species inspire debate (section 19.5)*

It follows that phylogenetic relationships between microbes can be explored by comparative genomics. One approach is to examine how similar the organization of orthologous genes (i.e., homologous genes found in different organisms) are in the genomes of the microbes being compared. This is called **synteny,** and just as syntax refers to the order of words in a sentence, synteny describes the order of genes on a given genome and compares this order between the genomes of two or more organisms. Organisms that share a more recent common evolutionary history are more closely related microbes and so generally have a high degree of synteny. It may be easiest to understand synteny by first considering the order of genes in similar operons found in two different microbes. For instance, one would not be surprised to learn that the order of tryptophan biosynthetic genes is the same (conserved) in the *E. coli* and *Salmonella* spp. genomes. Analysis of synteny takes this comparison to the level of whole genomes. As shown in **figure 18.21,** when the order of genes in the genome of the plant symbiont *Sinorhizobium meliloti* is compared with that of the plant pathogen *Agrobacterium tumefaciens,* it is clear that the gene order is very similar. This is in contrast to *E. coli* and *S. meliloti.* Although these two bacteria belong to the same phylum, *Proteobacteria,* they are not sufficiently related to show synteny. ►► *Class* Alphaproteobacteria *includes many oligotrophs (section 22.1); Microbe-plant interactions can be positive, negative, or neutral (section 31.3)*

Comparative genomics can also provide great insight into genes and gene products that are associated with virulence when pathogen genomes are analyzed. Such insights are particularly needed in understanding *Mycobacterium tuberculosis,* the causative agent of tuberculosis (TB). About one-third of the human population has TB. After establishing residence in immune cells in the lung, *M. tuberculosis* often remains in a dormant state until the host's immune system is compromised. It then goes on to cause active disease, killing about 2 million people annually. The *M. tuberculosis* genome has been compared to the genomes of two relatives—*M. leprae,* which causes leprosy, and *M. bovis,* the causative agent of TB in a wide range of animals, including cows and humans. The genomes of *M. bovis* and *M. tuberculosis* are most similar—about 99.5% identical at the sequence level. However, the *M. bovis* genome is missing 11 separate regions, making its genome slightly smaller (4.3 Mb versus 4.4. Mb). The sequence dissimilarities involve the inactivation of some genes, leading to major differences in the way the two bacteria respond to environmental conditions. This may account for the host range differences between these two closely related pathogens. ►► *Order* Corynebacterinales *includes important human pathogens (section 24.1); Mycobacterium infections (section 39.1)*

The divergence between *M. tuberculosis* and *M. leprae* is even more striking. The *M. leprae* genome is a third smaller than that of *M. tuberculosis.* About half the genome is devoid of functional genes. Instead, there are over 1,000 degraded, nonfunctional genes called **pseudogenes (figure 18.22).** In total, *M. leprae* seems to have lost more than 1,000 genes during its career as an intracellular parasite. *M. leprae* even lacks some of the enzymes required for energy production and DNA replication. This might explain why the bacterium has such a long doubling time, about 2 weeks in mice. One hope from genomic analysis is that critical surface

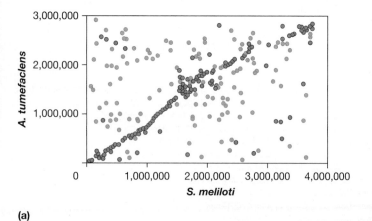

(a)

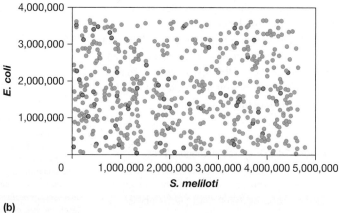

(b)

Figure 18.21 Synteny. Each axis on the graphs maps the genes on each organism's genome. For instance, each number on the X axis in both (a) and (b) indicates the position of genes (blue dots) on the 3.7 Mb *S. meliloti* genome. (a) When the genomes of *S. meliloti* and *A. tumefaciens* are compared, it is evident that the distribution of orthologous genes on the genome is quite similar, as shown by the red line indicating similar placement on the genome (red dots represent *A. tumefaciens* genes). (b) *E. coli* and *S. meliloti* are not closely related; thus their genomes show no synteny (red dots represent *E. coli* genes).

proteins can be discovered and used to develop a sensitive test for early detection of leprosy. This would allow immediate treatment of the disease before nerve damage occurs. ▶▶❘ *Leprosy (section 39.3)*

Retrieve, Infer, Apply

1. Cite an infectious disease for which you think a systems biology approach would be appropriate. List three specific questions you would like to explore regarding this host-pathogen interaction.
2. For what types of microorganisms is extensive gene loss common? What is the most likely explanation for this phenomenon?
3. How would you infer that a toxin gene was recently acquired by HGT in a food pathogen such as *E. coli* O157:H7?

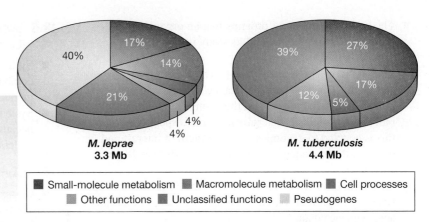

Figure 18.22 Comparative Genomic Analysis Between Two *Mycobacterium* Species. The genome of *M. leprae* is significantly smaller than that of *M. tuberculosis*. Over one-third of its genes are nonfunctional pseudogenes that will presumably be lost as genome reduction continues.

Key Concepts

18.1 DNA Sequencing Methods

- Genomics is the study of the molecular organization of genomes, their information content, and the gene products they encode.
- DNA fragments have traditionally been sequenced using dideoxynucleotides and the Sanger chain termination technique (**figures 18.2** and **18.3**).
- Next-generation sequencing (NGS) technologies are faster and cheaper than the Sanger method and are based on sequencing by synthesis, employing reversible chain termination (**figures 18.4** and **18.5**).

18.2 Genome Sequencing

- When using the Sanger sequencing technique, the whole-genome shotgun technique is used. Four stages are involved: library construction, sequencing of randomly produced fragments, fragment alignment and gap closure, and editing the final sequence (**figure 18.6**).
- Genome sequencing using NSG techniques has accelerated the volume and pace of microbial genome sequencing.
- It is now possible to use the multiple strand displacement technique to copy the genome of a single microbial cell so there is sufficient DNA for sequencing. In this way, the genomes of uncultivated microbes can be studied (**figure 18.7**).

18.3 Metagenomics Provides Access to Uncultured Microbes

- Metagenomics enables the study of the biodiversity and metabolic potential of microbial communities without culturing individual microbes (**figure 18.8**).

- Metagenomics involves sequencing DNA extracted directly from the environment. NGS approaches are used to avoid genomic library construction.
- Metagenomics has been applied to many natural microbial habitats, including the human body.

18.4 Bioinformatics: What Does the Sequence Mean?

- Analysis of vast amounts of genome data requires sophisticated computer software; these analytical procedures are part of the discipline of bioinformatics.
- An open reading frame is a putative protein-coding gene; the identity of the gene and its product are inferred by aligning the translated amino acid sequence with similar sequences in the database (**figures 18.9** and **18.10**).
- Bioinformatics enables the comparison of genes within genomes to identify paralogues and the comparison of genes between different organisms to identify orthologues.
- The physical map of a microbe's genome is often displayed so the location of genes with related functions are easily identified (**figure 18.11**).

18.5 Functional Genomics Links Genes to Phenotype

- The presence of genes whose products are known to function in specific metabolic processes such as catabolic pathways and transport enables the genomic reconstruction of the organism—a type of map that illustrates these processes within the cell (**figure 18.12**).
- Two techniques are used to study transcriptomes: DNA microarrays and RNA-seq. Both assess gene expression as a measure of individual gene transcripts (mRNA). Gene expression can be determined for mutant versus wild-type strains or for a given organism under specific environmental conditions (**figures 18.13** to **18.15**).

- The entire collection of proteins that an organism can produce is its proteome, and its study is called proteomics.
- The proteome is often analyzed by two-dimensional gel electrophoresis, in which the total cellular protein pool can be visualized. In many cases, the amino acid sequence of individual proteins is determined by mass spectrometry; if this is coupled to genomics, both a protein of interest and the gene that encodes it can be identified (**figures 18.16** and **18.17**).
- Structural proteomics seeks to model the three-dimensional structure of proteins based on computer analysis of amino acid sequence data.
- Protein-DNA interactions can be analyzed by chromatin immunoprecipitation, which enables the identification of both the protein and its target DNA (**figure 18.18**).

18.6 Systems Biology: Making and Testing Complex Predictions

- Systems biology is an interdisciplinary field of study that combines experimentally derived data sets with predictive modeling to an organismal process.
- Systems biology is the opposite of reductionist biology because it integrates molecular networks with the goal of understanding the function of the cell as a whole.

18.7 Comparative Genomics

- Comparing genomes reveals information about genome structure and evolution, including the importance of horizontal gene transfer.
- Comparative genomics is an important tool in discerning how microbes have adapted to particular ecological niches and in developing new therapeutic agents.

Compare, Hypothesize, Invent

1. Propose an experiment that can be done easily with a DNA microarray or RNA-seq that would have required years to do before these technologies were developed.

2. You have discovered a new DNA-binding protein. You show by EMSA that it binds near the promoter of an operon that encodes enzymes required for carotenoid biosynthesis. You wonder if any other genes are controlled by this new protein. How will you find other regions of DNA to which the protein might bind? How will you determine if the new protein increases or decreases transcription of the genes it regulates?

3. You are developing a new vaccine for a pathogen. You want your vaccine to recognize specific cell-surface proteins. Explain how you will use genome analysis to identify potential protein targets. What functional genomics approach will you use to determine which of these proteins is produced when the pathogen is in its host?

4. You are annotating the genomic sequence of a new soil bacterium. You believe you have found a gene that encodes nitrate reductase. How did you arrive at this conclusion, and how will you strengthen this annotation using only in silico analysis?

5. The bacterial genus *Borrelia* is the cause of several tick-borne diseases. In addition, *B. recurrentis* causes louse-borne relapsing fever. In 2008 a team of scientists reported that the genome of *B. recurrentis* is a smaller, degraded version of the genome of *B. duttonii*, the cause of tick-borne relapsing fever. What do you think these researchers discovered when they compared these two genomes that enabled them to make this conclusion? In addition, it was found that *B. recurrentis* lacks genes encoding *recA* and *mutS*; refer to section 16.3 and explain

why the loss of these genes may have accelerated the decay of this genome.

Read the original paper: Lescot, M., et al. 2008. The genome of *Borrelia recurrentis*, the agent of deadly louse-borne relapsing fever, is a degraded subset of tick-borne *Borrelia duttonii*. *PLoS Genetics* 4:e1000185.

6. The origin of tuberculosis (TB) in the Western Hemisphere, or New World, has long been a mystery. It is generally agreed that the TB-causing bacterium, *Mycobacterium tuberculosis,* radiated out of Africa with our ancestors (*see chapter 17 opener*). However, it has not been clear how preserved human remains with evidence of TB found in the New World acquired the disease prior to the entry of the first Europeans. In 2014 an international team of scientists sequenced the 1,000-year-old mycobacterial genomes from three Peruvian human skeletons. Surprisingly, these genomes are most similar to those mycobacterial genomes found in seals and sea lions rather than genomes of *M. tuberculosis* adapted to humans. The team hypothesizes that sea mammals transmitted TB to the ancient New World human population. Explain why next-generation nucleotide sequencing has made the results of this type of forensic epidemiology credible. How were the genomes from the Peruvian skeletons matched to seals and sea lions, and how was the ancestral role of human-adapted *M. tuberculosis* strains excluded? Your answer should explore the role of orthologs and synteny.

Read the original paper: Bos, K. I., et al., 2014. Pre-Columbian mycobacterial genomes reveal seals as a new source of New World human tuberculosis. *Nature.* doi:10.1038/nature13591.

19

Microbial Taxonomy and the Evolution of Diversity

There are more microbes on Earth than stars in the universe. This vast diversity reflects the roughly 3.5 billion years in which microbes have evolved to adapt to seemingly every environment on (and in) the planet, regardless of how inhospitable it may seem to a plant or animal.

Scientists Query: "Is the Microbial Universe Expanding?"

Should the audacious question "Does life exist that cannot be classified as archaeal, bacterial, or eukaryotic?" be taken seriously? After all, the elegant work of Carl Woese, who first described archaea, has survived the test of time to emerge stronger and more widely accepted. Yet this is the question that some microbiologists and evolutionary biologists ask when they consider "microbial dark matter." Just as astronomical dark matter represents the majority of mass in the observable universe, most of the biomass in microbial ecosystems (including the human body) is also difficult to define.

One way to view all the microbes on Earth is to place them into one of three categories: explored, unexplored, and undiscovered. Explored microbes are those we have in culture. Here we see an overrepresentation of pathogens and microbes with medical, industrial, or food applications. The unexplored are those we know exist based on gene sequences obtained from natural settings, but have not yet been grown in the lab. These are "environmental microbes" that contribute to nutrient cycling and host-microbe physiology in ways that are defined in aggregate but not at the level of a single microbial taxon (e.g., genus or species). The undiscovered, however, are most intriguing. These are the microbes yet to be found—the real microbial dark matter. Do they exist? If so, how much bigger is the microbial universe than previously thought?

The only way to answer these questions is to explore. As discussed in chapter 18, metagenomic and single-cell genome sequencing have the capacity to reveal the existence of novel microorganisms. These techniques harness the power of next-generation sequencing to probe genes and genomes that belong to uncultured microorganisms in natural systems. The results have been startling: more than 60 new archaeal and bacterial phyla have recently been proposed. Even more exciting is the fact that some newly discovered gene sequences don't match any previously identified archaeal, bacterial, or eukaryotic nucleotide sequences. For example, very large and unusual viruses have been found to possess genes whose evolutionary history places them

somewhere between archaeal and eukaryotic domains. If the cellular counterparts to these viruses exist, would they represent a new domain?

As we discuss in this chapter, within each domain, phyla (s. phylum) are higher-level taxa to which a group of related organisms belong. The recent, sudden increase in the number of phyla has raised concern among taxonomists. While the criteria used to define a particular bacterial or archaeal species have long been a subject of discussion, many taxonomists feel the explosion in metagenomic and single-cell genomic data warrants a discussion regarding what truly constitutes higher-level taxa, including the newly minted term "superphylum," as well as the traditionally held phylum, class, order, family, and perhaps even domain. It seems the expansion of the microbial universe is a messy business that may require not only new nomenclature, but a new way of thinking about life itself.

In this chapter, we explore microbial taxonomy and evolution—how and why microorganisms are classified using an evolutionary framework. This was not always the case, and the move to an evolution-based system resulted in the reclassification of some genera, to the consternation of some scientists and clinicians. Indeed, the fields of microbial taxonomy and evolution continue to present challenges and controversy. But ultimately, microbial taxonomy and phylogeny open the door to a degree of biological diversity unmatched by multicellular organisms.

Readiness Check:

Based on what you have learned previously, you should be able to:

✔ Describe the structure of DNA (section 13.2) and summarize methods used in DNA sequencing and metagenomics (sections 18.1 to 18.3)

✔ Outline in general terms how small subunit rRNA has been used to define the three domains of life (section 1.2)

✔ Summarize the mechanisms of horizontal gene transfer (chapter 16)

✔ Describe the rationale for the RNA world hypothesis of the origin of life (section 1.2)

✔ Summarize why the term prokaryote is controversial (section 3.1)

✔ Contrast and compare the structural differences between bacterial, archaeal, and eukaryotic microbes (chapters 3–5)

✔ Explain the principles of next-generation sequencing and metagenomics (sections 18.1–18.3)

19.1 Microbial Taxonomy Is Based on the Evolution of Multiple Traits

After reading this section, you should be able to:

- Explain the utility of taxonomy and systematics
- Draw a concept map illustrating the differences between phenetic, phylogenetic, and genotypic classification

Microbiologists are faced with the daunting task of understanding the diversity of life forms that cannot be seen with the naked eye but can live anywhere on Earth. Obviously a reliable classification system is paramount. The science of classifying living things is called **taxonomy** (Greek *taxis,* arrangement or order, and *nomos,* law, or *nemein,* to distribute or govern). In a broader sense, taxonomy consists of three separate but interrelated parts: classification, nomenclature, and identification. A taxonomic scheme is used to arrange organisms into groups called taxa (s., **taxon**) based on mutual similarity. **Nomenclature** is the branch of taxonomy concerned with the assignment of names to taxonomic groups in agreement with published rules. Identification is the practical side of taxonomy—the process of determining if a particular isolate belongs to a recognized taxon and, if so, which one. The term **systematics** is often used for taxonomy, although it sometimes infers a more general scientific study of organisms with the ultimate objective of arranging them in an orderly manner. Thus systematics encompasses disciplines such as morphology, ecology, epidemiology, biochemistry, genetics, molecular biology, and physiology.

One of the oldest classification systems, called **natural classification,** arranges organisms into groups whose members share many characteristics. The Swedish botanist Carl von Linné, or Carolus Linnaeus as he often is called, developed the first natural classification in the middle of the eighteenth century. It was based largely on anatomical characteristics and was a great improvement over previously employed systems because natural classification provided information about many biological properties. For example, classification of humans as mammals denotes that they have hair, self-regulating body temperature, and milk-producing mammary glands in the female.

When natural classification is applied to higher organisms, evolutionary relationships become apparent simply because the morphology of a given structure (e.g., wings) in a variety of organisms (ducks, songbirds, hawks) suggests how that structure might have been modified to adapt to specific environments or behaviors. However, the traditional taxonomic assignment of microbes was not rooted in evolutionary relatedness. For instance, bacterial pathogens and microbes of industrial importance were historically given names that described the diseases they cause or the processes they perform (e.g., *Vibrio cholerae, Clostridium tetani,* and *Lactococcus lactis*). Although these labels are of practical use, they do little to guide the taxonomist concerned with the vast majority of microbes that are neither pathogenic nor of industrial consequence. Our present understanding of the evolutionary relationships among microbes now serves as the theoretical underpinning for taxonomic classification.

In practice, determining the genus and species of a newly isolated microbe is based on **polyphasic taxonomy.** As the term "polyphasic" suggests, this encompasses many aspects that describe the microorganism. These include phenotypic, phylogenetic (i.e., the evolutionary history), and genotypic features. To understand how all of these data are incorporated into a coherent profile of taxonomic criteria, we must first consider the individual components.

Phenetic Classification

For a very long time, microbial taxonomists had to rely exclusively on a **phenetic system,** which classifies organisms according to their phenotypic similarity (**Table 19.1**). This system succeeded in bringing order to biological diversity and clarified the function of morphological structures. For example, because motility and flagella are always associated in particular microorganisms, it is reasonable to suppose that flagella are involved in at least some types of motility. Although phenetic studies can reveal possible evolutionary relationships, this is not always the case. For example, not all flagellated bacteria belong to the same phylum. This is why the best phenetic classification is one constructed by comparing as many attributes as possible.

Genotypic Classification

As the name suggests, **genotypic classification** seeks to compare the genetic similarity between organisms. Individual genes or

Table 19.1	Components of Polyphasic Taxonomy		
Classification System	**Basis of Classification**	**Traits**	**Microbial Example**
Phenetic	Phenotype	Morphology Physiology	*Clostridium* spp. (anaerobic, endospore-forming, Gram-positive rod) compared to *Bacillus* spp. (aerobic, endospore-forming, Gram-positive rod)
Genotypic	Genes and genomes	DNA:DNA hybridization Average nucleotide identity	*E. coli* strain K12 compared to *E. coli* O157:H7 *E. coli* compare to *Salmonella enterica*
Phylogenetic	Evolution	rRNA genes Conserved protein-coding gene	Archaea compared to Bacteria

whole genomes can be compared. Since the 1970s it has been widely accepted that bacteria and archaea whose genomes are at least 70% homologous belong to the same species (p. 449). However, there is now a consensus building to replace this metric with a genomics-based assay that measures the average nucleotide identity between organisms. The means by which microbes are genotypically classified is discussed further in section 19.3.

Phylogenetic Classification

With the publication in 1859 of Charles Darwin's *On the Origin of Species,* biologists began developing **phylogenetic** or **phyletic classification systems** that sought to compare organisms on the basis of evolutionary relationships. The term **phylogeny** (Greek *phylon,* tribe or race, and *genesis,* generation or origin) refers to the evolutionary development of a species. Scientists realized that when they observed differences and similarities between organisms as a result of evolutionary processes, they also gained insight into the history of life on Earth. However, for much of the twentieth century, microbiologists could not effectively employ phylogenetic classification systems, primarily because of the lack of a good fossil record. When Carl Woese and George Fox proposed using small subunit (SSU) rRNA nucleotide sequences to assess evolutionary relationships among microorganisms, the door opened to the resolution of long-standing inquiries regarding the origin and evolution of the majority of life forms on Earth—microbes. As discussed later (p. 449), the power of rRNA as a phylogenetic and taxonomic tool rests on the features of the rRNA molecule that make it a good indicator of evolutionary history and on the ever-increasing size of the rRNA sequence database.

Retrieve, Infer, Apply

1. What is a natural classification? What microbial features might have been considered when devising a natural classification scheme?
2. What is polyphasic taxonomy, and what three types of data does it consider? Should each type of data be of equal weight? Why or why not?
3. Consider the finding that bacteria capable of anoxygenic photosynthesis belong to several different phyla. How do you think these bacteria might have been originally classified, and what types of data do you think were key in making the most recent taxonomic assignments?

19.2 Taxonomic Ranks Provide an Organizational Framework

After reading this section, you should be able to:

- Outline the general scheme of taxonomic hierarchy
- Explain how the binomial system of Linnaeus is used in microbial taxonomy

The definition of a bacterial or archaeal species is widely debated, as discussed in section 19.5. Nonetheless, for practical reasons it is essential that the established rules of taxonomy are followed. Microbes are placed in taxonomic levels arranged in a nonoverlapping hierarchy so that each level includes not only the traits that define the rank above it but also a new set of more restrictive traits (**figure 19.1**). Thus within each domain—*Bacteria, Archaea,* or *Eukarya*—each organism is assigned (in descending order) to a phylum, class, order, family, genus, and species epithet or name. Some microbes are also given a subspecies designation. Microbial groups at each level have a specific suffix that indicates rank or level.

The application of Linnaeus's classification system to bacteria began in Pasteur and Koch's time, about 1872. However, within 20 years microbiologists became dissatisfied, considered it haphazard, and called for more uniform criteria in organizing phyla, classes, orders, families, genera, and species. With the ongoing explosion in metagenomic analysis, some would argue that history is repeating itself. Within the last decade or so, thousands of 16S rRNA genes and protein-coding genes have been sequenced that cannot convincingly be affiliated with any previously defined taxa. This data explosion has led to the recent development of the taxonomic classification "**superphylum,**" below domain and above phylum (e.g., in figure 19.1, superphylum would be placed between *Bacteria* and *Proteobacteria*). Ideally a superphylum includes organisms of several phyla that share a number of distinctive characteristics, such as unusual morphological or metabolic features. However, some feel that the term is being loosely applied based on insufficient data—for instance, to SSU rRNA sequences alone.

At the other end of the classification scheme is the most fundamental definition of a bacterial or archaeal **species.** A species is a collection of strains that share many stable properties and differ significantly from other groups of strains. A **strain** consists of the descendants of a single, pure microbial culture. Strains within a species may be described in a number of different ways. **Biovars** are variant strains characterized by biochemical or physiological differences, **morphovars** differ morphologically, and **serovars** have distinctive antigenic (immunologically reactive) properties. Because changes in species assignment are not uncommon, one strain is designated as the **type strain** for each species. The type strain is the holder of the species name. This ensures permanence of names when nomenclature revisions occur because the type species must remain within the original species. It is usually one of the first strains studied and often is more fully characterized than others; however, it does not have to be the most representative member. Only those strains very similar to the type strain or type species are included in a species. Each species is assigned to

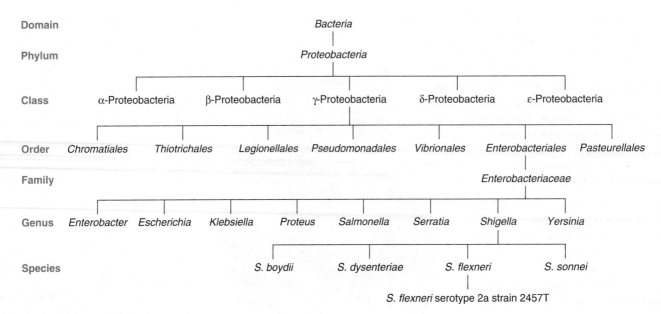

Figure 19.1 Hierarchical Arrangement in Taxonomy.   In this example, members of the genus *Shigella* are placed within higher taxonomic ranks. Not all classification possibilities are given for each rank to simplify the diagram. Note that *-ales* denotes order and *-ceae* indicates family.

a **genus,** the next rank in the taxonomic hierarchy. A genus is a well-defined group of one or more species that is clearly separate from other genera. In practice, considerable subjectivity occurs in assigning species to a genus, and taxonomists may disagree about the composition of genera.

Microbiologists name microorganisms using the **binomial system** of Linnaeus. The Latinized, italicized name consists of two parts. The first part is the generic name (i.e., the genus), and the second is the species name (e.g., *Yersinia pestis,* the causative agent of plague). The species name is stable; the oldest epithet for a particular organism takes precedence and must be used. In contrast, a generic name can change if the organism is assigned to another genus. For example, some members of the genus *Streptococcus* were placed into two new genera, *Enterococcus* and *Lactococcus,* based on rRNA analysis and other characteristics. Thus *Streptococcus faecalis* is now *Enterococcus faecalis.* To be recognized as a new species, genomic, metabolic, morphological, reproductive, and ecological data must be accepted and published in the *International Journal of Systematic and Evolutionary Microbiology;* until that time, the new species name will appear in quotation marks. Microbes that have not been grown in pure culture but for which there is sufficient genetic characterization may be given a provisional genus and species name preceded by the term *Candidatus,* meaning candidate. For instance, a novel aerobic phototrophic member of the phylum *Acidobacterium* has been grown only in coculture with another isolate. It has been given the provisional name *Candidatus* Chloracidobacterium thermophilum (note that the genus and species are not italicized). *Bergey's Manual of Systematic Bacteriology* contains only recognized bacterial and archaeal species as discussed in section 19.6. ▶▶| *Photosynthetic bacteria are diverse (section 21.4)*

Retrieve, Infer, Apply

1. What is the difference between a microbial species and a strain?
2. Why is it important to have a type strain for each species?
3. The genus *Salmonella* was once thought to contain five species. Most scientists now consider only two species valid: *S. bongori* and *S. enterica.* The latter contains six subspecies, and *S. enterica* subspecies *enterica* is further subdivided into eight serovars. Three of these, Enteritidis, Typhi, and Typhimurium, were once considered *Salmonella* species. (a) Inspect figure 19.1 and construct a similar lineage for *Salmonella.* (b) What criteria do you think were used to make these taxonomic changes?

19.3 Microbial Taxonomy and Phylogeny Are Largely Based on Molecular Characterization

After reading this section, you should be able to:

- Construct a concept map describing the approaches commonly used to determine taxonomic classification
- Assess the impact molecular methods have had on the field of microbial taxonomy and phylogeny
- Compare and contrast nucleotide sequencing and nonsequencing-based molecular approaches used in microbial taxonomy and phylogeny
- Select an appropriate technique to identify a microbial genus, species, and strain
- Predict the basic biological as well as public health implications of microbial taxonomic identification

Many different approaches are used in classifying and identifying microorganisms that have been isolated and grown in pure culture. For clarity, we divide them into two groups: classical and molecular. The most durable identifications are those that are based on a combination of approaches. Methods often employed in routine laboratory identification of pathogenic bacteria are covered in chapter 36, Clinical Microbiology and Immunology.

Classical Characteristics

Classical approaches to taxonomy make use of morphological, physiological, biochemical, and ecological characteristics. These characteristics have been employed in microbial taxonomy for many years and form the basis for phenetic (phenotypic) classification. When used in combination, they are quite useful in routine identification of well-characterized microbes.

Morphological Characteristics

Morphological features are important in microbial taxonomy for many reasons (**table 19.2**). Morphology is easy to study and analyze, particularly in eukaryotic microorganisms and more complex bacteria and archaea. In addition, morphological comparisons are valuable because structural features depend on the expression of many genes and are usually genetically stable. Thus morphological similarity often is a good indication of phylogenetic relatedness.

Physiological and Metabolic Characteristics

Physiological and metabolic characteristics are very useful because they are directly related to the nature and activity of microbial proteins. For instance, the detection of specific end products of fermentation in a newly discovered microorganism reveals the presence of

specific catabolic enzymes and the genes that encode them. Therefore analysis of characteristics, such as energy metabolism and nutrient transport, provides an indirect comparison of microbial genomes. **Table 19.3** lists some of the most important of these properties.

Biochemical Characteristics

Among the more useful biochemical characteristics used in microbial taxonomy are bacterial fatty acids, which can be analyzed using a technique called *f*atty *a*cid *m*ethyl *e*ster (FAME) analysis. A fatty acid profile reveals specific differences in chain length, degree of saturation, branched chains, and hydroxyl groups. Microbes of the same species will have identical fatty acid profiles, provided they are grown under the same conditions; this limits FAME analysis to only those microbes that can be grown in pure culture. Finally, because the identification of a species is done by comparing the results of the unknown microbe in question with the FAME profile of other, known microbes, identification is only possible if the species in question has been previously analyzed. Nonetheless, FAME analysis is particularly important in public health, food, and water microbiology. In these applications, microbiologists seek to identify specific microbial pathogens. ▶▶| *Microbiology of food (chapter 41); Purification and sanitary analysis ensures safe drinking water (section 43.1)*

Advances in mass spectrometry (MS) have resulted in the fast and accurate identification of bacteria based on the presence of specific, highly abundant proteins. The specific type of MS used

Table 19.2	Some Morphological Features Used in Classification and Identification
Feature	**Microbial Groups**
Cell shape	All major groups[1]
Cell size	All major groups
Colonial morphology	All major groups
Ultrastructural characteristics	All major groups
Staining behavior	Bacteria, some fungi
Cilia and flagella	All major groups
Mechanism of motility	Gliding bacteria, spirochetes, protists
Endospore shape and location	Some Gram-positive bacteria
Spore morphology and location	Bacteria, protists, fungi
Cellular inclusions	All major groups
Colony color	All major groups

1 Used in classifying and identifying at least some bacteria, archaea, fungi, and protists.

Table 19.3	Some Physiological and Metabolic Characteristics Used in Classification and Identification
Carbon and nitrogen sources	
Cell wall constituents	
Energy sources	
Fermentation products	
General nutritional type	
Growth temperature optimum and range	
Motility	
Osmotic tolerance	
Oxygen relationships	
pH optimum and growth range	
Photosynthetic pigments	
Salt requirements and tolerance	
Secondary metabolites formed	
Storage inclusions	

is called matrix-assisted laser desorption/ionization-time of flight (MALDI-ToF). MALDI-ToF enables the analysis of complex biomolecules that could not previously be studied by MS. The material to be analyzed is dried on a sample holder (called a target) and then mixed with a molecular film called a matrix. When a UV laser beam strikes the sample target, the matrix helps stimulate release of the sample from the surface; this is matrix-assisted desorption. Once the matrix-sample is released (desorbed), the matrix transfers protons to the sample, which then becomes ionized; this is matrix-assisted ionization. Once ionized, biomolecules are "flown" across a space within the instrument; the time taken for a molecule to fly from one side to the other is used to determine the mass of each molecule; this is time of flight.

In its simplest form, MALDI-ToF identification of bacteria involves the transfer of whole cells from a single colony grown under specific conditions to a sample target. Once dried, a matrix is deposited and the bacterial samples are analyzed. MALDI-ToF yields the masses of many highly abundant bacterial proteins. Like FAME analysis, each experimentally derived protein profile must be matched to a protein profile from a known bacterium. MALDI-ToF is becoming increasingly important in medical microbiology laboratories where the same strains of organisms are regularly encountered. Like FAME, microbes must be grown under very specific conditions and MALDI-ToF cannot be used to identify a newly discovered microbe. ◄◄ *Proteomics explores total cellular proteins (section 18.5)*

Ecological Characteristics

The ability of a microorganism to colonize a specific environment is of taxonomic value. Some microbes may be very similar in many other respects but inhabit different ecological niches, suggesting they may not be as closely related as first suspected. Some examples of taxonomically important ecological properties are life cycle patterns; the nature of symbiotic relationships; the ability to cause disease in a particular host; and habitat preferences such as requirements for temperature, pH, oxygen, and osmotic concentration. Many growth requirements are considered physiological characteristics as well (table 19.3). ◄◄ *Environmental factors affect microbial growth (section 7.4)* ►►| *Many types of microbial interactions exist (section 32.1)*

Molecular Characteristics

It is hard to overestimate how the study of DNA, RNA, and proteins has advanced our understanding of microbial evolution and taxonomy. Evolutionary biologists studying plants and animals draw from a rich fossil record to assemble a history of morphological changes; in these cases, molecular approaches supplement such data. In contrast, microorganisms have left almost no fossil record, so molecular analysis is the only feasible means of collecting a large and accurate data set that explores microbial evolution. When scientists are careful to make only valid comparisons, phylogenetic inferences based on molecular approaches provide the most robust analysis of microbial evolution.

Nucleic Acid Base Composition

Microbial genomes can be directly compared and taxonomic similarity can be estimated in many ways. The first, and possibly the simplest, technique to be employed is the determination of DNA base composition. Base-pairing rules dictate that the (G + C)/(A + T) ratio or **G + C content**—the percent of G + C in DNA—reflects the base sequence and varies with sequence changes as follows:

$$\text{mol\% G + C} = \frac{G + C}{G + C + A + T} \times 100$$

The base composition of DNA can be determined in several ways. The most obvious is genome sequencing. However, for microbes whose genomes have not yet been sequenced, the G + C content can be determined from the **melting temperature** (T_m) of DNA. In double-stranded DNA, three hydrogen bonds join GC base pairs and two bonds connect AT base pairs. As a result, strands of DNA with a greater G + C content separate at higher temperatures; that is, the DNA has a higher melting temperature. For example, compare *Mycoplasma hominis,* with a mol% G + C of about 29 and a T_m of 65°C, with *Micrococcus luteus,* which is almost 79% G + C with a T_m of 85°C.

DNA melting can be easily followed spectrophotometrically because the absorbance of DNA at 260 nm increases during strand separation. When a DNA sample is slowly heated, the absorbance increases as hydrogen bonds are broken and reaches a plateau when all the DNA has become single-stranded (**figure 19.2**). The midpoint of the rising curve gives the melting temperature, a direct measure of the G + C content. If a microbe's genome has been sequenced, the T_m is determined mathematically.

The G + C content of thousands of organisms has been determined. In animals and higher plants, it averages around 40% and ranges between 30 and 50%. In contrast, the G + C content of bacteria and archaea ranges from about 25 to almost 80% and is more variable than that of fungi and protists. Despite such a

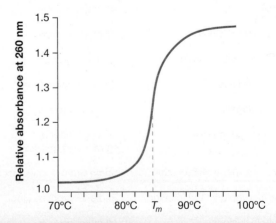

Figure 19.2 A DNA Melting Curve. The T_m is indicated.

MICRO INQUIRY *Would this curve be shifted to the left or the right for a microbe with an exceptionally low G + C composition? Explain your answer.*

wide range of variation, the G + C content of strains within a particular species is constant and varies very little within a genus. If two organisms differ in their G + C content by more than about 10%, their genomes have quite different base sequences, indicating that they are not closely related. On the other hand, it is not safe to assume that organisms with very similar G + C contents also have similar DNA base sequences because two very different base sequences can be constructed from the same proportions of A + T and G + C base pairs.

Nucleic Acid Hybridization

The similarity between genomes can be compared more directly by use of nucleic acid hybridization studies, also called **DNA-DNA hybridization** (**DDH**). If the genomes of two microbial isolates are heated to become single-stranded (ss) DNA and then cooled and held at a temperature about 25°C below the T_m, strands with complementary base sequences will reassociate to form stable double-stranded (ds) DNA. However, noncomplementary strands will remain unpaired. Because strands with similar but not identical sequences form less temperature-stable dsDNA hybrids, incubation of the mixture at 30 to 50°C below the T_m allows hybrids of more diverse ssDNAs to form. Incubation at 10 to 15°C below the T_m permits hybrid formation only with almost identical strands. Two strains whose DNAs show at least 70% relatedness under optimal experimental hybridization conditions and less than a 5% difference in T_m often, but not always, are considered members of the same species. 🔊 *DNA Probe (DNA-DNA Hybridization)*

The use of DDH has been considered a vital tool in microbial classification since its introduction in the late 1960s. However, DDH techniques are cumbersome and the results are relatively crude when compared to genomic data. For these and other reasons, an alternate approach has been sought. **Average nucleotide identity** (**ANI**), which uses pairwise alignment between sequenced DNA from two organisms, is a promising replacement for DDH. ANI was initially used to compare complete genome sequences but has since been used to compare short stretches of nucleotides generated using next-generation sequencing techniques. In general, an ANI value of 95 to 96% is comparable to the 70% value as a minimum value needed for species delineation. Several other nucleotide sequencing-based approaches are also under consideration; although it is unclear which will ultimately replace DDH as the part of the "gold standard," there is no doubt that DDH has become outdated. ◀◀ *DNA sequencing methods (section 18.1)*

Nucleic Acid Sequencing

The rRNAs from small ribosomal subunits (16S from bacterial and archaeal cells and 18S from eukaryotes) have become the molecules of choice for inferring microbial phylogenies and making taxonomic assignments at the genus level. The **small subunit rRNAs** (**SSU rRNAs**) are almost ideal for studies of microbial evolution, relatedness, and genus identification for several important reasons (**figure 19.3**). First, they play the same role in all microorganisms. In addition, because ribosomes are

absolutely necessary for survival and ribosomes cannot function without SSU rRNAs, the genes encoding these rRNAs cannot tolerate large mutations. Thus these genes change very slowly with time. Finally, rRNA genes do not appear to be subject to horizontal gene transfer, an important factor in comparing sequences for phylogenetic purposes. The utility of SSU rRNAs is extended by the presence of certain sequences within SSU rRNA genes that are variable among organisms as well as other regions that are quite similar. The variable regions enable comparison between closely related microbes, whereas the stable sequences allow the comparison of distantly related microorganisms.

Comparative analysis of SSU rRNA sequences from thousands of organisms has demonstrated the presence of **oligonucleotide signature sequences.** These are short, conserved nucleotide sequences that are specific for phylogenetically defined groups of organisms. Thus the signature sequences found in bacterial rRNAs are rarely or never found in archaeal rRNAs and vice versa. Likewise, the 18S rRNA of eukaryotes bears signature sequences that are specific to the domain *Eukarya*. The nucleotide sequences of genes encoding the complete rRNAs or, more often, specific rRNA fragments can be compared. The Ribosomal Database Project website is a repository of hundreds of thousands of rRNA sequences and facilitates accurate comparative analysis.

The ability to amplify regions of rRNA genes (rDNA) by the polymerase chain reaction (PCR) and sequence the DNA using next-generation sequencing technology has greatly increased the efficiency by which SSU rRNA sequences can be obtained. PCR can be used to amplify rDNA from the genomes of different organisms because conserved nucleotide sequences flank the nucleotides that can be used to reveal the microbe's identity. In practice, this means that PCR primers can be generated to amplify rDNA from both cultured and uncultured microbes. The validity of SSU rRNA sequences as a taxonomic marker has been evaluated in at least two ways. First, analysis of thousands of bacterial and archaeal 16S rRNA sequences reveals that no two microbes with 98.5% 16S rRNA sequence identity have been found with less than the 70% DNA-DNA hybridization cutoff value for the operational definition of a microbial species. Second, genome-based studies have demonstrated that 16S rRNA is a valid and robust measurement of taxonomic relatedness. ◀◀ *Polymerase chain reaction amplifies targeted DNA (section 17.2)*

Signature sequences are present in genes other than those encoding rRNA. Many genes have nucleotide insertions or deletions of specific lengths and sequences at fixed positions. A particular nucleotide sequence that is inserted or deleted may be found exclusively among all members of a phylum. These taxon specific insertions and deletions are called conserved **indels** (for *in*sertion/*del*etion). Indels are particularly useful in phylogenetic studies when they are flanked by conserved regions. In such cases, changes in the signature sequence cannot be due to sequence misalignments. The signature sequences located in some highly conserved housekeeping genes do not appear to be greatly affected by horizontal gene transfers and, like SSU rRNA, can be employed in phylogenetic analysis.

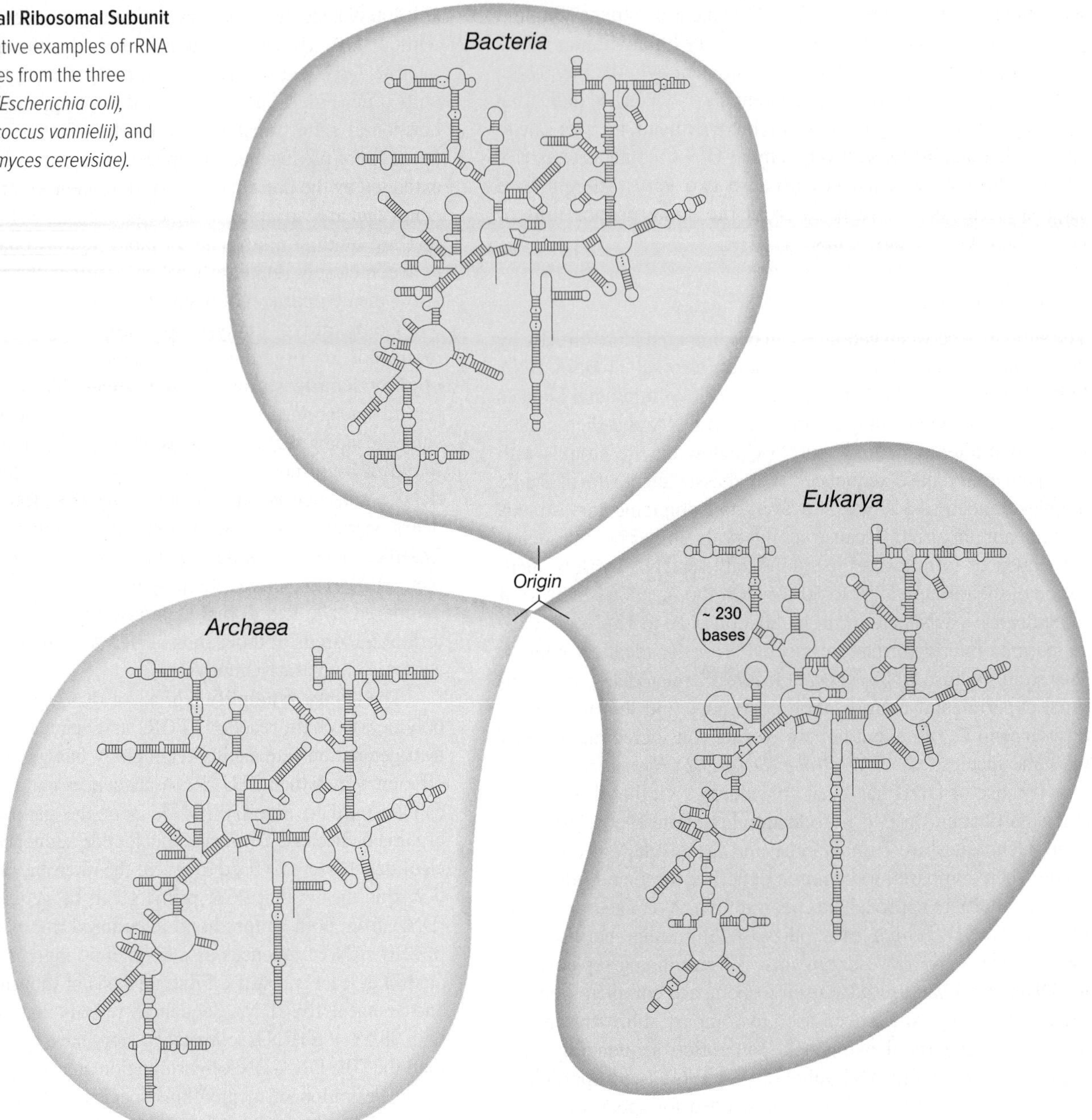

Figure 19.3 **Small Ribosomal Subunit RNA.** Representative examples of rRNA secondary structures from the three domains: *Bacteria (Escherichia coli)*, *Archaea (Methanococcus vannielii)*, and *Eukarya (Saccharomyces cerevisiae).*

Bacteria

Origin

Eukarya

~ 230 bases

Archaea

Genomic Fingerprinting

SSU rRNA can also be analyzed by methods that do not require nucleotide sequencing. Instead, the differences in rRNA gene sequences can be discerned by digestion with restriction enzymes. Recall that restriction enzymes cut only at their recognition sequences, so when these sequences vary (even by one nucleotide) in the SSU rRNA gene from different microbial isolates, bands of differing sizes (called restriction fragments) will be seen when the DNA is electrophoresed *(see table 17.1 and figure 17.6).* There are two ways to examine SSU rRNA genes using restriction enzymes. The first method, called **restriction fragment length polymorphism** (**RFLP**), requires PCR ampli-

fication of the gene encoding the rRNA to provide enough DNA for analysis. The DNA is then digested with restriction enzymes and run on a gel (**figure 19.4**). The second method, termed **ribotyping,** omits the need for PCR because the rRNA genes are detected by a labeled nucleotide probe. Specifically, the microbe's entire genome is cut with one or more restriction enzymes. The digested DNA is run on a gel and transferred to a nylon filter, and the rRNA encoding DNA fragments are visualized after hybridization with a labeled rRNA gene probe, much like a Southern blot. Importantly, each strain yields a reproducible pattern that can be deposited in a database; such databases are commonly used in clinical diagnostics and food and water analyses. Because both RFLP and ribotyping generate specific patterns that are used to reveal microbial identity, they are often referred to as **genomic fingerprinting,** an analogy to the specific patterns found on human fingerprints that are used to determine identity. Any uncultivated microorganism that is identified solely on its nucleic acid sequence (or other observable, quantifiable characteristic) is called a **phylotype.** ◄◄ *Restriction enzymes; Southern blotting (section 17.1)* ↻ *Restriction Fragment Length Polymorphisms*

So far we have discussed methods that generally resolve identity to the genus level. To identify a microbial species, genes that evolve more quickly than those that encode rRNA must be analyzed. In fact, rather than using a single gene, five to seven conserved housekeeping genes can be sequenced and compared in a technique called **multilocus sequence analysis** (**MLSA**). Multiple genes are examined to avoid misleading results that can arise through horizontal gene transfer. MLSA compares orthologous genes from a number of strains that may not belong to the same genus. Because many different versions, or alleles, of each gene can exist, the finding that two microbial isolates share the same alleles for multiple genes is very strong evidence that the two strains are closely related, perhaps even the same strain. MLSA was derived from **multilocus sequence typing** (**MLST**), which is used to discriminate among strains belonging to the same pathogenic species. MLSA differs from MLST because its broad application to microbial taxonomy often requires the comparison of genes from more than one microbial species.

Highly conserved and repetitive DNA sequences present in many copies in the genomes of most Gram-negative and some Gram-positive bacteria can also be used to help identify microbes. Three families of repetitive sequences are typically used for microbial identification: the 154 bp BOX elements, the 124–127 bp enterobacterial repetitive intergenic consensus (ERIC) sequence, and 35–40 bp repetitive extragenic palindromic (REP) sequences. These sequences are found at distinct sites between genes; that is, they are intergenic. Because they are conserved among genera, oligonucleotide primers can be used to specifically amplify the repetitive sequences by PCR. Different primers are used for each type of repetitive element, and the results are classified as arising from BOX-PCR, ERIC-PCR, or REP-PCR. In each case, the amplified fragments from many microbial samples can be resolved and visualized on an agarose gel. Each lane of the gel corresponds to a single bacterial isolate, and the pattern created by many samples resembles a UPC bar code. The "bar code" is then computer analyzed using pattern recognition software as well as software that calculates phylogenetic relationships. Because this enables identification to the level of species,

subspecies, and often strain, it is most often used in the identification of human, animal, and plant pathogens.

In comparing the methods used to define and identify a species, we see that with the exception of DNA-DNA hybridization and ANI, only a small region of the genome is sampled. As shown in **figure 19.5,** 16S rRNA analysis focuses on one specific gene, whereas MLSA and MLST sample multiple housekeeping genes. While much information is gained using these approaches,

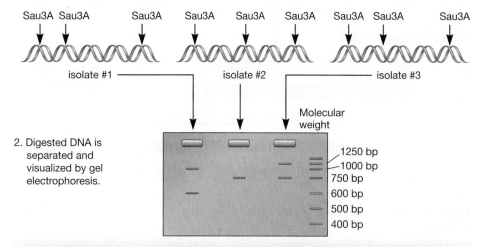

1. DNA amplified from different microbial isolates are digested with the same endonuclease (e.g., Sau3A).

2. Digested DNA is separated and visualized by gel electrophoresis.

Molecular weight

1250 bp
1000 bp
750 bp
600 bp
500 bp
400 bp

isolate #1 — isolate #2 — isolate #3

Figure 19.4 **Examination of SSU rRNA Genes from Three Different Microbial Isolates Using RFLP Analysis.**

MICRO INQUIRY *Why does isolate #2 yield only one DNA fragment (band)?*

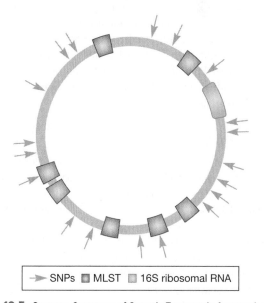

→ SNPs ▪ MLST ▪ 16S ribosomal RNA

Figure 19.5 **Genome Coverage of Genetic Taxonomic Approaches.**
A typical bacterial or archaeal genome contains at least one locus encoding 16S rRNA. MLSA samples multiple genes throughout the genome. SNP analysis compares more, but shorter, nucleotide sequences that span the entire genome.

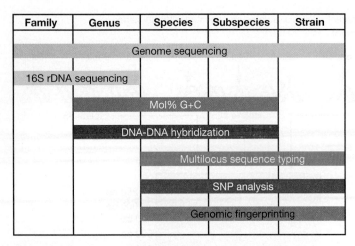

Family	Genus	Species	Subspecies	Strain
Genome sequencing				
16S rDNA sequencing				
Mol% G+C				
DNA-DNA hybridization				
Multilocus sequence typing				
SNP analysis				
Genomic fingerprinting				

Figure 19.6 Relative Taxonomic Resolution of Various Molecular Techniques.

several disadvantages have been noted. Novel species of bacteria and archaea can be missed when PCR is used to amplify 16S rRNA sequences from natural samples, because the sequences of their rRNA genes may not be complementary to the PCR primers. Evolutionary changes in housekeeping genes of certain bacteria, particularly pathogens and other microbes that live in association with other organisms, can be driven by interactions with the host. These changes are not necessarily correlated with housekeeping functions and therefore may not be good candidates for MLSA or MLST.

An alternate technique, **single nucleotide polymorphisms (SNP,** pronounced "snip"), samples a larger fraction of the genome than either 16S rRNA analysis or MLSA (figure 19.5). Originally designed for use in humans, SNP analysis looks at single nucleotide changes, or polymorphisms, in specific genes, intergenic regions, or other noncoding regions. These particular regions are targeted because they are normally conserved, so single base-pair differences reveal evolutionary change. **Figure 19.6** shows the taxonomic utility of the molecular methods discussed here with the caveat that a combination of approaches is always best.

Retrieve, Infer, Apply

1. What are the advantages of using each major group of characteristics (morphological, physiological and metabolic, biochemical, ecological, and molecular) in classification and identification? How is each group related to the nature and expression of the genome? Give examples of each type of characteristic.

2. Why is it not safe to assume that two microorganisms with the same G + C content belong to the same species? In what ways are G + C content data taxonomically valuable?

3. What are the drawbacks of DDH?

4. Why is rRNA so suitable for determining relatedness?

5. How is ribotyping similar to rRNA sequence analysis? How do the two techniques differ? Do you think one is more accurate than the other?

19.4 Phylogenetic Trees Illustrate Evolutionary Relationships

After reading this section, you should be able to:

- Paraphrase the rationale underpinning the construction of phylogenetic trees
- Compare and contrast rooted and unrooted trees
- Outline the general considerations used in building a phylogenetic tree
- Characterize the challenges horizontal gene transfer introduces in the study of microbial evolution

Microbial taxa within *Bacteria* and *Archaea* form discrete, genealogically clustered groups that can be illustrated in phylogenetic trees. **Phylogenetic trees** show inferred evolutionary relationships in the form of multiple branching lineages connected by nodes (**figure 19.7**). The organism whose nucleotide sequences have been analyzed is identified at the tip of each branch. Each node (branchpoint) represents a divergence event, and the length of the branches represents the number of molecular changes that have taken place between the two nodes.

Often sequences are obtained from well classified microbes that have been grown in pure culture; however, this is not always the case. SSU rRNA sequences have become particularly important in both identifying microbes in nature and constructing phylogenetic trees to describe their evolutionary relationships. An inclusive term for the organism at each branch tip, regardless of whether or not it has been cultured, is the **operational taxonomic unit** (**OTU**). We now briefly describe how phylogenetic trees can be built with the intention of increasing an understanding of what they represent.

There are five steps in building a phylogenetic tree. First, the nucleotide or amino acid sequence must be aligned. Alignment is usually done using an online resource such as CLUSTAL, although manual inspection of the alignment is also important (figure 19.7a). In addition to SSU rRNA gene sequences, protein-coding genes may also be analyzed, in which case the alignment of amino acids is preferred. This is because the genetic code is degenerate, so even if a nucleotide sequence is not conserved, the amino acid sequence may be. For instance, if a protein-coding gene were mutated so that a codon changed from AGA to AGG, arginine would still be added to the growing polypeptide during translation. Next the alignment must be examined for a phylogenetic signal; this will determine if it is appropriate to continue with tree building (figure 19.7b). There are two extremes in this regard: at one end of the spectrum, the sequences align perfectly; the other extreme is the absence of any matches whatsoever. Phylogenetic analysis can only be performed on those sequences that fall in the middle, with a mixture of random and matched positions. The third step is the hardest: one must choose which tree-building method to use. We briefly review some of the more popular methods next. The last two steps involve the application of the selected method, which is performed by a computer, followed by manual examination of the resulting

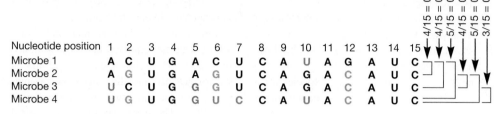

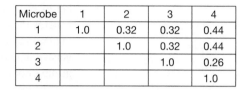

Nucleotide position	1	2	3	4	5	6	7	8	9	10	11	12	13	14	15
Microbe 1	A	C	U	G	A	C	U	C	A	U	A	G	A	U	C
Microbe 2	A	G	U	G	A	G	U	C	A	G	A	C	A	U	C
Microbe 3	U	C	U	G	G	G	U	C	A	G	A	C	A	U	C
Microbe 4	U	G	U	G	G	U	C	C	A	U	A	C	A	U	C

$4/15 = 0.27$
$4/15 = 0.27$
$5/15 = 0.33$
$4/15 = 0.27$
$5/15 = 0.33$
$3/15 = 0.20$

(a) Sequence alignment and analysis

Microbe	1	2	3	4
1	1.0	0.27	0.27	0.33
2		1.0	0.27	0.33
3			1.0	0.20
4				1.0

(b) Calculated evolutionary distance

Microbe	1	2	3	4
1	1.0	0.32	0.32	0.44
2		1.0	0.32	0.44
3			1.0	0.26
4				1.0

(c) Corrected evolutionary distance

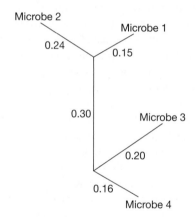

(d) Phylogenetic tree

Figure 19.7 Constructing a Phylogenetic Tree Using a Distance Method.   (a) Nucleotide sequences are aligned and pairwise comparison is made. The number of nonidentical nucleotides is then scored. For instance, when sequences from microbes 1 and 2 are compared, there are 4 mismatches out of 15 total nucleotides, yielding a calculated evolutionary distance (E_D) of 0.27. (b) The calculated E_D values are corrected to account for back mutation to the original genotype or other forward mutations that could have occurred at the same site before generating the observed genotype. (c) A tree-building method is then selected (in this case, a distance method is used), and computer analysis of the values generates a phylogenetic tree as shown in (d). E_D values are indicated for each branch.

Character-based methods for phylogenetic tree building are more complicated but generate more robust trees. These methods start with assumptions about the pathway of evolution, infer the ancestor at each node, and choose the best tree according to a specific model of evolutionary change. These methods include maximum parsimony, which assumes that the fewest number of changes occurred between ancestor and extant (living) organisms. Another approach is called maximum likelihood. This requires a large data set because for each possible tree that can be built, its probability (i.e., the likelihood) based on certain evolutionary and molecular information is determined so that the tree with the greatest probability based on these criteria is selected. All tree building methods have their advantages and disadvantages, so it is usually advisable to use several methods to analyze the same data set. Similar trees generated by different approaches is the best outcome.

Importantly, a tree may be unrooted or rooted. An unrooted tree (**figure 19.8a**) simply represents phylogenetic relationships but does not indicate which organisms are more primitive relative others. Figure 19.8a shows that A is more closely related to C than it is to either B or D, but it does not indicate which of the four species might be the oldest. In contrast, the rooted tree (figure 19.8b) gives a node that serves as the common ancestor and shows the development of the four species from this root. It is much more difficult to develop a rooted tree. For example, there are 15 possible rooted trees that connect four species but only three possible unrooted trees.

An unrooted tree can be rooted by adding data from an outgroup—a species known to be very distantly related to all the species in the tree (figure 19.8c). The root is determined by the point of the tree where the outgroup joins. This provides a point of reference to identify the oldest node on the tree, which is the node closest to the outgroup. So, for example, in figure 19.8c, organism Z is the outgroup and the oldest node on the tree is marked with an arrow.

Once a tree is constructed, it is important to get a sense of whether the placement of its branches and nodes is legitimate. There are a variety of methods to assess the "strength" of a tree, but the most common is bootstrapping. Bootstrapping involves phylogenetic analysis of a randomly selected subset of the data presented on the tree. A bootstrap value is the percent of analyses in which that

tree to make sure it makes sense. For example, a computer-generated tree that places a mammal and an archaeon on the same branch would certainly require manual correction.

The different approaches to building a phylogenetic tree can be divided into two broad categories: a distance-based (phenetic) approach and a character-based (cladistic) approach. Distance-based approaches are the most intuitive. Here the differences between the aligned sequences are counted for each pair and summarized into a single statistic, which is roughly the percent difference between the two sequences (figure 19.7b,c). A tree is then generated by serially linking pairs that are ever more distantly related (i.e., start with those with the least number of sequence differences and move to those with the most). This is called cluster analysis and should be carefully applied as it has the unattractive capability of generating trees even in the absence of evolutionary relationships. Neighbor joining is another distance-based method that uses a slightly different matrix that attempts to avoid this problem. It does this by modifying the distance between each pair of nodes based on the average divergence from all other nodes.

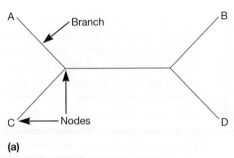

(a)

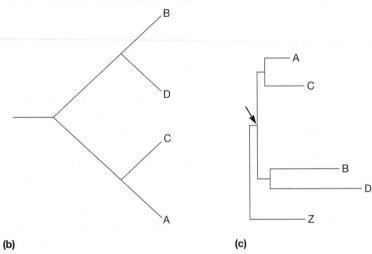

(b)

(c)

Figure 19.8 Phylogenetic Tree Topologies. (a) Unrooted tree joining four taxonomic units. (b) Rooted tree. (c) The tree shown in (a) can be rooted by adding an outgroup, represented by Z.

particular branch was found. Typically bootstrap values of 70% or greater are thought to support a tree. Another approach, called Bayesian inference, may also be used. Rather than looking at a single tree, Bayesian inference analyzes multiple potential trees and calculates the probability that each branch would appear based on this comparison. Although these values are also reported as percentages, they are not directly comparable to bootstrap values. Only values greater than 95% are acceptable when Bayesian inference is used.

An important feature in phylogenetic trees is the scale. Just as a scale bar on a road map indicates distance in number of centimeters per kilometer, the scale bar on a phylogenetic tree illustrates the evolutionary distance. This is usually measured in number of mutations per 1,000 nucleotides or amino acid substitutions per 1,000 amino acid residues. This may be expressed as a number without units, e.g., 0.02 (2 per 1,000). To continue our analogy with a road map, just as a map does not reveal how long it takes to get from one point to another (due to traffic, weather, variable road conditions), the branches on a phylogenetic tree do not indicate the length of time it took for an ancestral microbe to give rise to an extant form. As discussed in section 19.5, the theory of punctuated equilibria is one important reason evolutionary distance, as measured by the similarity of genes or proteins in living organisms, provides little or no information regarding how long ago evolutionary divergence occurred.

One of the biggest challenges in constructing a satisfactory tree is widespread, frequent horizontal gene transfer (HGT). Although microbiologists are careful to exclude from their analysis genes and proteins known to have been subject to HGT, the influence of HGT on phylogeny and evolution cannot be ignored. Indeed, eukaryotes possess genes from both bacteria and archaea, and there has been frequent gene swapping between the bacterial and archaeal

domains, particularly from bacteria to archaea. Some bacteria even have acquired eukaryotic genes. Clearly, the pattern of microbial evolution is not truly linear and treelike.

Assessing the impact HGT has had on microbial evolution has been guided by genome sequence analysis. Comparisons of the genomes of strains within a species and among species within a phylum have revealed that microbial genomes consist of an older core genome and a more recently acquired pan-genome (**figure 19.9**). The **core genome** is that set of genes found in all members of a species (or other monophyletic group). Thus it is thought to represent the minimal number of genes needed for the group of microbes to survive. In general, these genes encode "informational" proteins involved in DNA replication, transcription, and translation (e.g., rRNA genes). These genes are thought to have been present in the group's common ancestor. By contrast, the **pan-genome** consists of every gene in all strains of a species (or other taxonomic unit), so it includes the core genome plus every additional gene found only in some (or a single) strain. Genes outside the core genome are more recently acquired genes that enable microbial colonization of new niches. A comparison of the core genome size with the actual genome size of a particular strain thus indicates the evolution of new traits. For instance, current values of the core genome and pan-genome of *Bacillus anthracis* differ by only roughly 200 genes (about 3,600 versus 3,800, respectively), reflecting the limited genetic diversity within this species. By contrast, the *E. coli* core genome consists of about 2,800 genes, whereas some estimate that the pan-genome consists of roughly 16,000 genes. The broad genetic diversity among *E. coli* strains is illustrated by comparing the genomes of the nonpathogenic *E. coli* strain K12 and the pathogenic stain O157:H7. Their last common ancestor is estimated to have lived about 4.5 million years ago. During this time *E. coli* has mutated and exchanged genes, allowing the development of many strains. These have radiated to numerous habitats to which strains continue to adapt. Obviously, for any given species the estimated size of the core and pan-genome depends on the number of strains with sequenced genomes. Indeed, as more strains of each species are sequenced, the core genome tends to get smaller, as strains are discovered that lack genes once thought to common to all genomes. It follows that as core genomes shrink, pan-genomes expand. ◀◀ *Mechanisms of genetic variation (chapter 16)*

Retrieve, Infer, Apply

1. Could a phylotype be considered an OTU? What about a species?
2. List the differences between distance-based and character-based methods for constructing a phylogenetic tree. Which type is maximum parsimony? Explain your answer.
3. What is the difference between a rooted and unrooted tree? Which provides more information?
4. You are building a tree based on 16S rRNA sequence alignments of a group of spirochetes. Suggest a possible outgroup so that you can build a rooted tree. Refer to chapter 21 if you need more information about these bacteria.
5. Is HGT involved in movement of genes in the core or pan-genome? Explain.

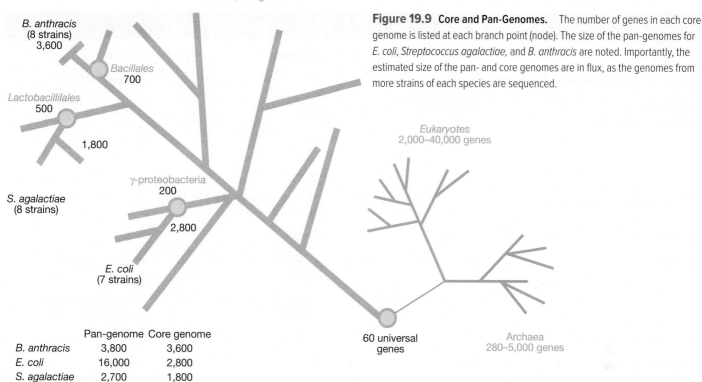

Figure 19.9 Core and Pan-Genomes. The number of genes in each core genome is listed at each branch point (node). The size of the pan-genomes for *E. coli*, *Streptococcus agalactiae*, and *B. anthracis* are noted. Importantly, the estimated size of the pan- and core genomes are in flux, as the genomes from more strains of each species are sequenced.

	Pan-genome	Core genome
B. anthracis	3,800	3,600
E. coli	16,000	2,800
S. agalactiae	2,700	1,800

19.5 Evolutionary Processes and the Concept of a Microbial Species Inspire Debate

After reading this section, you should be able to:

- Diagram the endosymbiotic theory of the origin of mitochondria and chloroplasts
- Compare and contrast the two theories that address the origin of the nucleus
- Explain why the concept of a microbial species is difficult to define
- List the "gold standard" taxonomic methods currently applied to species designation
- Explain the importance of adaptive mutations in giving rise to new ecotypes

It is our goal here to describe current models that seek to explain the evolution of new microbial species. Before we begin this discussion, however, it is helpful to consider the origin of all microbes—bacterial, archaeal, and eukaryotic. We then review the controversy that surrounds the word species as applied to bacteria and archaea. Only then will it be possible to understand and appreciate the evolutionary mechanisms that drive the development of new species of microorganisms.

Evolution of the Three Domains of Life

As we present in chapter 1, many scientists think that the first self-replicating entity was RNA. This is because RNA has the capacity to reproduce itself as well as catalyze chemical reactions. It is thought that when early RNA became enclosed in lipid spheres, the first primitive cell-like forms were generated (*see figures 1.5 and 1.6*). Considerable evidence indicates that by at least 3.5 billion years ago, such proto-cells (Greek, *protos,* meaning first) had evolved to form the ancestors of our extant microbes. Moreover, by 2.5 billion years ago, bacteria and archaea not only abounded, but each had evolved distinct taxonomic lineages. For instance, the Gram-positive bacterial phylum *Firmicutes* and Gram-negative phyla *Proteobacteria* and *Cyanobacteria* had developed. Indeed, the ancestors of modern cyanobacteria performed the oxygenic photosynthesis responsible for converting our anoxic planet to an oxygenated one.

A reexamination of the tree of life based on SSU rRNA (*see figure 1.2*) shows that the root of the tree is on the earliest region of the bacterial branch. As we discuss in chapter 1, the root is considered the *last universal common ancestor*, or LUCA. The placement of LUCA indicates that although bacteria and archaea share similar cellular construction, they are not phylogenetically linked. Because LUCA maps to the bacterial branch of the tree, it is thought that *Archaea* and *Eukarya* evolved independently of *Bacteria*. Recall that this was first suggested by Carl Woese and George Fox in the 1970s. Since that time, the distinction they inferred between bacterial and archaeal lineages has been confirmed by a number of biochemical differences. These include membrane lipids, cell wall construction, and enzymes involved in gene transcription (**table 19.4**).

Although *Archaea* and *Eukarya* share a recent common ancestor, eukaryotes possess both archaeal and bacterial traits. There are several hypotheses that account for genes of both archaeal and bacterial ancestry on the eukaryotic nuclear genome. One hypothesis asserts that the first eukaryotic cell arose upon the fusion of an archaeon and a bacterium that lived in close association. Over time, archaeal genes involved in metabolism were lost while bacterial genes involved in information processing were also degraded.

Table 19.4   Comparison of *Bacteria*, *Archaea*, and *Eukarya*

Property	Bacteria	Archaea	Eukarya
Membrane-Enclosed Nucleus with Nucleolus	Absent	Absent	Present
Complex Internal Membranous Organelles	Absent	Absent	Present
Cell Wall	Almost always have peptidoglycan containing muramic acid	Variety of types, no muramic acid; some have pseudomurein	No muramic acid
Membrane Lipid	Have ester-linked, straight-chained fatty acids	Have ether-linked, branched isoprene-derived chains	Have ester-linked, straight-chained fatty acids
Gas Vesicles	Present	Present	Absent
Transfer RNA	Thymine present in most tRNAs	No thymine in T or TΨC arm of tRNA	Thymine present
	N-formylmethionine carried by initiator tRNA	Methionine carried by initiator tRNA	Methionine carried by initiator tRNA
Polycistronic mRNA	Present	Present	Present in some protists
mRNA Introns	Rare	Rare	Present
mRNA Splicing, Capping, and Poly-A Tailing	Absent	Absent	Present
Ribosomes			
Size	70S	70S	80S
Elongation factor 2 reaction with diphtheria toxin	Does not react	Reacts	Reacts
Sensitivity to chloramphenicol and kanamycin	Sensitive	Insensitive	Insensitive
Sensitivity to anisomycin	Insensitive	Sensitive	Sensitive
DNA-Dependent RNA Polymerase			
Number of enzymes	One	One	Three
Structure	Simple subunit pattern (6 subunits)	Complex subunit pattern similar to eukaryotic enzymes (8–12 subunits)	Complex subunit pattern (12–14 subunits)
Rifampicin sensitivity	Sensitive	Insensitive	Insensitive
RNA Polymerase II Type Promoters	Absent	Present	Present
Metabolism			
Similar ATP synthase	No	Yes	Yes
Methanogenesis	Absent	Present	Absent
Nitrogen fixation	Present	Present	Absent
Chlorophyll-based photosynthesis	Present	Absent	Present[1]
Chemolithotrophy	Present	Present	Absent

1 Present in chloroplasts (of bacterial origin).

Contrary to this "single-step" hypothesis, others suggest a more multi-step scenario involving a series of endosymbioses. Here an archaeal cell is thought to have engulfed a bacterium, which then donated genes that would eventually become the nuclear genome of an ancestral eukaryote. Most recently, evidence supporting the hypothesis that the first eukaryotic cell arose from within the archaeal lineage was reported. In 2015, archaeal DNA sequences recovered from Loki's Castle, a sampling location in the mid-Atlantic ridge, were found to encode a number of eukaryotic-like proteins, such as membrane proteins involved in phagocytosis, small GTPases, and a homologue of a eukaryotic ribosome protein. These sequences were proposed to constitute a new archaeal phylum "Lokiarchaeota" that is more closely related to eukaryotes than to any other archaeal or bacterial lineage. Ironically, the name "Loki" refers to the shape-shifting Norse god who has given rise to numerous scholarly debates.

Unlike the uncertainty surrounding the origin of the first eukaryotic proto-cell, there is general agreement that mitochondria and chloroplasts arose by the incorporation of endosymbiotic bacteria. Like bacteria, most mitochondria and chloroplasts have a single, circular chromosome and undergo binary fission. In addition, mitochondria and chloroplasts have 70S, not 80S, ribosomes. These observations led to the development of the **endosymbiotic hypothesis (figure 19.10)**. This posits the following series of events. An ancestral eukaryotic cell lost its rigid cell wall but had evolved actin (or its precursor) that enabled amoeboid motility. This nucleated cell was thus able to develop endocytosis. These mobile proto-eukaryotes became predators of other cells, including bacteria. Predation imposed selection for cellular enlargement and increased motility. Engulfment without digestion of bacterial prey evolved because a smaller bacterial cell provided energy for the larger host cell, while the host protected and supplied nutrients to the bacterial cell. The energy supplied by the endosymbiont conferred a growth advantage to the proto-eukaryote, enabling its dominance over and eventually eliminating other cells that lacked both cell walls and endosymbionts.

As the endosymbiont became more dependent on its host for nutrients and protection, there was little selective pressure for the retention of genes involved in these processes. Conversely, because the endosymbiont was "permitted" to remain only if it captured and stored energy (presumably as ATP), there was strong selective pressure to retain the genes involved in energy conservation. Thus genes whose products were redundant to the host were eventually lost. In fact, such genome reduction is the rule, rather than the exception, among obligate intracellular microbes (*see figure 18.20*).

Finally, as genome reduction continued, the endosymbiont was so dependent on its host, it had evolved into an energy-providing organelle. It was originally postulated that the endosymbiont was capable of oxidative phosphorylation and so gave rise to the mitochondrion. However, more recent evidence suggests that the endosymbiont was an anaerobic bacterium with a fermentative metabolism. These endosymbionts then evolved into mitochondria or other mitochondria-like organelles (e.g., hydrogenosomes and mitosomes). Hydrogenosomes are found in some extant protists where, like mitochondria, they take up pyruvate that results from glycolysis within the host cytoplasm. Unlike mitochondria, however, pyruvate in the hydrogenosome is reduced to acetate, H_2, and CO_2 with an additional ATP generated (*see figure 5.16*). The simi-

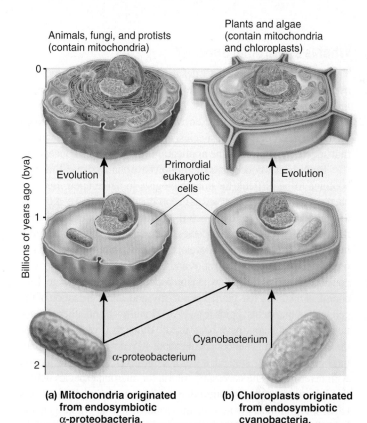

Figure 19.10   The Endosymbiotic Theory.   (a) According to this hypothesis, mitochondria were derived from an α-proteobacterium. (b) A similar phenomenon occurred for chloroplasts, which were derived from cyanobacteria.

MICRO INQUIRY   *On what evidence is this hypothesis based?*

larity between certain key genes (and thus their protein products) supports the notion that hydrogenosomes and mitochondria evolved from a single common ancestor, most likely an α-proteobacterium. Mitosomes have only recently been described, and like hydrogenosomes, are found in some protists. Of these three organelles, mitochondria appear to be most highly derived (i.e., continued to evolve), since these organelles are the site of oxidative phosphorylation. ▶▶ *The classic mitochondrion is only one member of the mitochondrial family of organelles (section 5.6)*

Chloroplasts arose when these new aerobic eukaryotes engulfed a cyanobacterium—probably an ancestor of *Prochlorococcus* (*see p. 494*). Again, this led to the development of a mutualistic relationship that evolved into our extant green plants and algae—organisms that possess both mitochondria and chloroplasts. Such endosymbioses exist today in certain protists that retain living cyanobacteria or the functional chloroplasts of their algal prey. Here it is thought a eukaryotic cell with a mitochondrion-like organelle engulfed a photosynthetic cell that possessed both a mitochondrion and chloroplast. Recent genomic data support this hypothesis, that is to say that plants arose from an ancestral protist that engulfed a cyanobacterium. The situation is more complicated for red algae, which are derived from a protist that engulfed this ancient photosynthetic eukaryote. Thus, the ancient photosynthetic organism is ancestral to all plants. ▶▶ *Protists (chapter 25)*

What Is a Microbial Species?

The term "species concept" describes a theoretical framework used to understand how and why certain organisms can be sorted into discrete taxonomic groups. We discussed "species definition" in sections 19.2 and 19.3, when we reviewed the criteria used to identify a microbial genus, species, or strain. From this, we can see that species definition is the application of the species concept.

Both species concept and definition have changed over time and continue to be difficult for microbiologists to agree upon. Because bacteria and archaea lack sexual reproduction, extensive morphological features, and a fossil record, microbiologists are at a distinct disadvantage when defining species, as compared to biologists studying other forms of life. The development of chromosome-based approaches, including G + C ratios and DNA-DNA hybridization, was considered a breakthrough when introduced about 60 years ago. Biochemical methods also gained popularity, and characterization of the cell wall components, lipids (e.g., FAME analysis), and other chemotaxonomic indicators have been used. Although Carl Woese introduced SSU rRNA gene sequencing about 40 years ago and its validity has been intensively studied, it was not until 2002 that those charged with determining the necessary criteria for microbial species definition, the International Committee on Systematics of Prokaryotes (ICSP), recommended its use. Importantly, unlike whole genome hybridization, SSU rRNA gene comparisons can be performed on microorganisms that are not in pure culture.

Historically, the application of different criteria in making species assignments has led to taxonomic confusion. In some cases, a single microbial species is so metabolically and genetically diverse that it seems probable that the group represents multiple species. On the other hand, some species are very narrowly defined, such that two species differ very little. For instance, *Bacillus anthracis* strains are so similar to *B. cereus,* many believe that all *B. anthracis* strains are really members of the *B. cereus* species. It is argued that only because *B. anthracis* causes anthrax does it have its own species designation.

In an effort to clarify and standardize microbial taxonomy, the ICSP recommends four criteria to meet a "gold standard" for species assignment: The microbe must be phenotypically similar to others in the group, whole genome similarity as determined by DNA-DNA hybridization must be at least 70%, the melting temperature of the DNA (a reflection of the G + C content) within 5°C, and less than 3% divergence in rRNA gene sequence. As noted previously, DNA-DNA hybridization, as well as melting temperature, may soon be replaced with a genomic metric, such as average nucleotide identity. Even with these proposed updates, some remain uncomfortable with these criteria. They point out that two microorganisms with, for instance, only 75% similarity in DNA and 98% rRNA gene sequence identity can be

considered the same species, but if these criteria were applied to eukaryotes, all primates (monkey, apes, you) would be lumped together as a single species! Indeed, it remains unresolved whether or not the species concept can be applied to microbes.

Microbial Evolutionary Processes

While the debate regarding the operational definition of a microbial species continues, microbiologists agree that the microbial species concept is grounded in natural selection and evolution. As the most ancient life forms on Earth, bacteria and archaea have had the opportunity to evolve and adapt to virtually every habitat. While their diverse metabolic strategies and ability to tolerate extreme conditions explain *why* microbes display such enormous diversity, natural selection explains *how* this diversity came to be.

Recall that genetic diversity in members of *Archaea* and *Bacteria* must occur asexually. Thus heritable genetic changes in these organisms are introduced principally by two mechanisms: mutation and HGT, both of which are subject to natural selection. Generally speaking, it is thought that mutation drives initial speciation events, and HGT permits more rapid radiation thereafter. In other words, a new species must arise from its single, ancestral species. This makes sense if one assumes that the ancestral population of microbes was genetically homogeneous. Genetic variation can only arise within a population possessing identical (or nearly identical) genomes by mutation, gene loss and gain, and intragenomic recombination (**figure 19.11a**). By definition, there is not enough genetic diversity to drive speciation among such a population by HGT.

Anagenesis, also known as **genetic drift,** refers to small, random genetic changes that occur over generations. It might seem that very small genetic differences within a microbial population would be of little evolutionary significance. However,

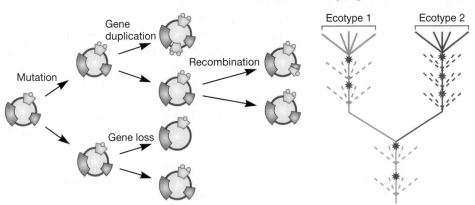

(a) Mechanisms of genetic variation within a homogeneous population

(b) Stable ecotype model

Figure 19.11 Evolution of Microbial Diversity. (a) Mechanisms by which a single, genetically homogeneous population of microbes can develop genetic variation include mutation, gene loss and duplication, and recombination. (b) Genetic changes within a population can lead to the development of new ecotypes. Each ecotype is subject to periodic selection events (indicated by stars) that enable cells with adaptive mutations to outcompete other lineages, which are eventually driven to extinction (dotted lines). The solid lines represent successful populations or lineages; those at the top are extant.

MICRO INQUIRY *Construct a scenario in which each of the following factors lead to the establishment of two ecotypes from a single common ancestor, as shown in (b): the availability of carbon and nitrogen sources; terminal electron acceptor; and mean local temperature.*

model studies designed to assess competition between microbial populations has led to some surprising observations. When selection is applied, very small genetic differences can result in one population overtaking another. How does this happen when individuals within a population have similar mutation rates and most of these mutations are neutral and have no phenotypic effect? Only those rare mutations that confer a growth advantage, called **adaptive mutations,** are retained and passed from one generation to the next, in which case we say the mutation is fixed. The descendants of that individual

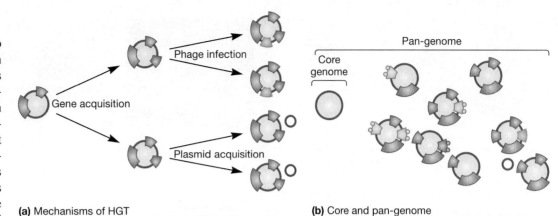

(a) Mechanisms of HGT **(b)** Core and pan-genome

Figure 19.12 Horizontal Gene Transfer (HGT) and the Pan-Genome. (a) The acquisition of genes from other populations of microbes is mediated by phage infection, conjugation, and transformation. During conjugation and transformation, transferred genes can be integrated directly into the chromosome or encoded on a plasmid, as shown here. (b) The prevalence of genes that have been horizontally transferred has given rise to the concept of a core genome—genes found in all members of a taxon, and a pan-genome—the combination of the core genome and all additional genes acquired principally by HGT within the taxon.

MICRO INQUIRY *Which do you think would have a pan-genome more closely related to its core genome: a microbial species whose strains are obligate intracellular symbionts, or a species whose strains are part of the normal flora of the mammalian gut? Explain your answer.*

continue to evolve through mutation and other intraspecific mechanisms.

Adaptive mutation is key to the ecotype model of microbial evolution. An **ecotype** is a population of microbes that is genetically very similar but ecologically distinct. Ecotypes arise when members of a microbial population living in a specific ecosystem undergo a genetic event (or series of events) that enables them to outcompete the remainder of the population. According to the ecotype model, the acquisition of adaptive mutations ultimately drives the remaining members of the population into extinction and reduces the amount of genetic diversity within the surviving population (figure 19.11*b*). The fossil record shows that the pace of evolution does not always occur at a constant rate but is periodically interrupted by rapid bursts of speciation driven by abrupt changes in the environment. Niles Eldredge and Steven Jay Gould coined the term **punctuated equilibria** to describe this phenomenon. Certainly, the 3.5-billion-year history of microbial life on Earth affords the accumulation of many, many mutations; that in turn has resulted in vast speciation. ◀◀ *Mechanisms of genetic variation (chapter 16); Comparative genomics (section 18.7)*

What about horizontal transfer of genetic material? Recall that the pan-genome is the complete gene repertoire of a taxon, so it includes the core genome plus housekeeping and dispensable genes. Housekeeping genes are generally defined as those genes whose products are required for normal metabolism and growth. In general, genes unique to the pan-genome are considered to have been acquired by HGT (**figure 19.12**).

Unlike the variation introduced in the ecotype model, HGT-driven genetic variation requires genetically diverse groups of microbes. This is because HGT does not rely on replication but rather on the exchange of genetic material between microbes. The rate of HGT is extremely variable. Some microbes have very reduced genomes with no evidence of HGT. These microbes are generally highly

adapted to a specific, stable ecological niche. The most extreme examples are obligate intracellular symbionts that can only grow within their host cells, where no other microbes exist with which to exchange genes. By contrast, some microbes appear to have high rates of HGT with more than half of their genome acquired from genetic exchange, as is the case for the members of the ancient phylum *Thermotogae*. In such bacteria and archaea, genes acquired by HGT frequently expand metabolic capabilities, thereby enabling rapid adaptation to new environmental challenges. ▶▶ *Aquaficae and Thermotogae are ancient bacterial lineages (section 21.1)*

Finally, we pose the question: How many bacterial and archaeal species are there? The easy answer is, unfortunately, not the correct answer: There are about 11,000 archaeal and bacterial species in culture. There are two major obstacles to formulating the correct answer. First, most microbial species resist growth in the laboratory, so they can only be detected by metagenomic or other culture-independent approaches. Second, as we have seen, microbiologists cannot agree on a biological species concept. So we must resort to the operational definition of a species in terms of G + C content and percent SSU rRNA and DNA homologies, as well as similarities in physiology, morphology, and ecology. Based on these criteria, estimates range from 100,000 to 1,000,000 species in nature, with about 10^{30} individual cells. These estimates reveal that there are probably about a billion more microbes on Earth than stars in the universe.

Retrieve, Infer, Apply

1. Define ecotype. Do you think it is necessary to obtain microbes in pure culture before assigning different ecotypes? Explain.
2. What is the difference between the core genome and pan-genome? What might you infer if you compare two genera, one in which the size of the core genome and pan-genome are very similar, and one in which the core genome is much smaller than the pan-genome?

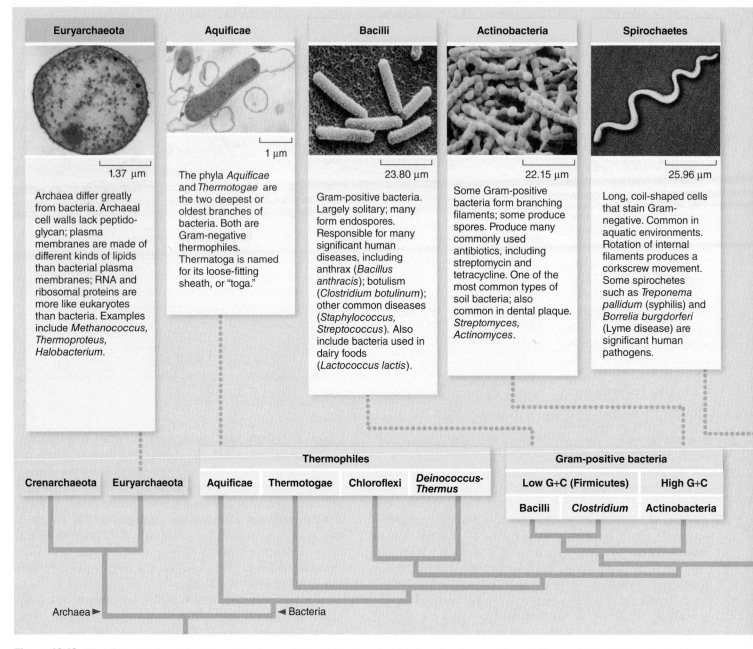

Euryarchaeota

1.37 μm

Archaea differ greatly from bacteria. Archaeal cell walls lack peptidoglycan; plasma membranes are made of different kinds of lipids than bacterial plasma membranes; RNA and ribosomal proteins are more like eukaryotes than bacteria. Examples include *Methanococcus*, *Thermoproteus*, *Halobacterium*.

Aquificae

1 μm

The phyla *Aquificae* and *Thermotogae* are the two deepest or oldest branches of bacteria. Both are Gram-negative thermophiles. Thermatoga is named for its loose-fitting sheath, or "toga."

Bacilli

23.80 μm

Gram-positive bacteria. Largely solitary; many form endospores. Responsible for many significant human diseases, including anthrax (*Bacillus anthracis*); botulism (*Clostridium botulinum*); other common diseases (*Staphylococcus*, *Streptococcus*). Also include bacteria used in dairy foods (*Lactococcus lactis*).

Actinobacteria

22.15 μm

Some Gram-positive bacteria form branching filaments; some produce spores. Produce many commonly used antibiotics, including streptomycin and tetracycline. One of the most common types of soil bacteria; also common in dental plaque. *Streptomyces*, *Actinomyces*.

Spirochaetes

25.96 μm

Long, coil-shaped cells that stain Gram-negative. Common in aquatic environments. Rotation of internal filaments produces a corkscrew movement. Some spirochetes such as *Treponema pallidum* (syphilis) and *Borrelia burgdorferi* (Lyme disease) are significant human pathogens.

Thermophiles				**Gram-positive bacteria**	
Crenarchaeota	Euryarchaeota	Aquificae	Thermotogae	Chloroflexi	*Deinococcus-Thermus*	Low G+C (Firmicutes)	High G+C
						Bacilli / *Clostridium*	Actinobacteria

Archaea ►   ◄ Bacteria

Figure 19.13 Some Major Clades of Bacteria and Archaea. This classification scheme is based on *Bergey's Manual of Systematic Bacteriology*, 2nd ed.

3. Of the following genes, which do you think are part of the pan-genome and which are part of the core genome: the genes for lactose catabolism in *E. coli*; the genes for heat-stable DNA polymerase in *Thermus aquaticus*; the genes for proteorhodopsin in marine bacteria; the genes for toxin production in *Vibrio cholerae*?

4. Would a protein encoded on the core genome or one encoded only on the pan-genome be best to use in constructing a phylogenetic tree? Explain your answer.

19.6 *Bergey's Manual of Systematic Bacteriology*

After reading this section, you should be able to:

■ Employ *Bergey's Manual* to investigate the defining taxonomic elements used for a bacterium or archaeon that is unfamiliar to you

In 1923 David Bergey (1860–1937), professor of bacteriology at the University of Pennsylvania, and four colleagues published *Bergey's Manual of Determinative Bacteriology*, a classification of bacteria that could be used for the identification of many bacterial species. The ninth edition of this single-volume manual was published in 1994. Despite its age, this text continues to serve as a relatively brief reference guide in the identification of bacteria based on physiological and morphological traits.

In 1984 the first edition of *Bergey's Manual of Systematic Bacteriology* was published. It contained descriptions of all bacterial and archaeal species then identified. A more recent second edition consists of five volumes published over a number of years, starting in 2001. Each volume covers a specific group of microbes and is written by experts in that particular field. The morphology, physiology, growth conditions, ecology, and other

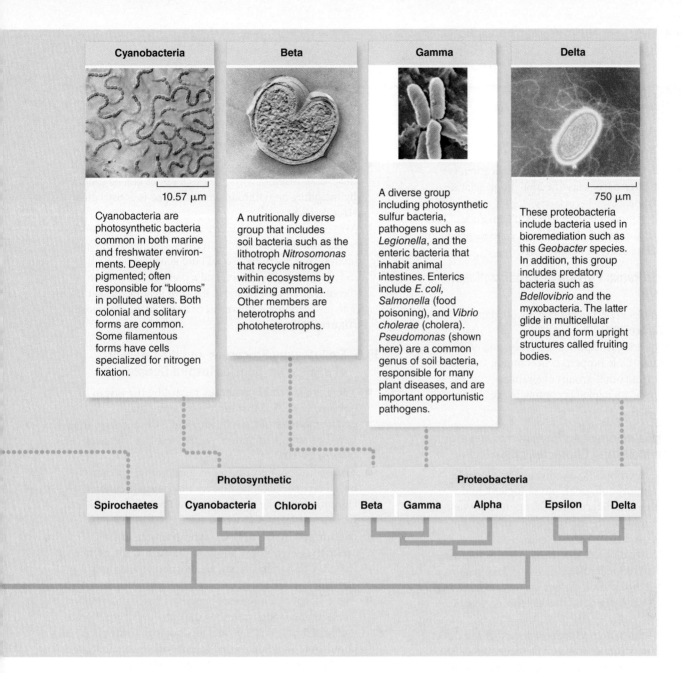

Cyanobacteria

Cyanobacteria are photosynthetic bacteria common in both marine and freshwater environments. Deeply pigmented; often responsible for "blooms" in polluted waters. Both colonial and solitary forms are common. Some filamentous forms have cells specialized for nitrogen fixation.

10.57 μm

Beta

A nutritionally diverse group that includes soil bacteria such as the lithotroph *Nitrosomonas* that recycle nitrogen within ecosystems by oxidizing ammonia. Other members are heterotrophs and photoheterotrophs.

Gamma

A diverse group including photosynthetic sulfur bacteria, pathogens such as *Legionella*, and the enteric bacteria that inhabit animal intestines. Enterics include *E. coli*, *Salmonella* (food poisoning), and *Vibrio cholerae* (cholera). *Pseudomonas* (shown here) are a common genus of soil bacteria, responsible for many plant diseases, and are important opportunistic pathogens.

Delta

These proteobacteria include bacteria used in bioremediation such as this *Geobacter* species. In addition, this group includes predatory bacteria such as *Bdellovibrio* and the myxobacteria. The latter glide in multicellular groups and form upright structures called fruiting bodies.

750 μm

Spirochaetes	**Photosynthetic**		**Proteobacteria**				
	Cyanobacteria	Chlorobi	Beta	Gamma	Alpha	Epsilon	Delta

information is provided, making this a valuable reference for microbiologists.

The second edition of *Bergey's Manual of Systematic Bacteriology* reflects the enormous progress that has been made in microbial taxonomy since the first edition was published. This is particularly true of the molecular approaches to phylogenetic analysis. Whereas microbial classification in the first edition was phenetic (based on phenotypic characterization), classification in the second edition of *Bergey's Manual* is largely phylogenetic. Although Gram-staining properties are generally considered phenetic characteristics, they also play a role in the phylogenetic classification of microbes.

In addition to the reorganization based on phylogeny, the second edition has more ecological information about individual taxa. It does not group all the clinically important bacteria together as the first edition did. Instead, pathogenic species are placed phylogenetically and thus scattered throughout the five volumes:

Volume 1, *The Archaea and the Deeply Branching and Phototrophic Bacteria*
Volume 2, *The Proteobacteria*
Volume 3, *The Firmicutes*
Volume 4, *The Bacteroidetes, Spirochaetes, Tenericutes (Mollicutes), Acidobacteria, Fibrobacteres, Fusobacteria, Dictyoglomi, Gemmatimonadetes, Lentisphaerae, Verrucomicrobia, Chlamydiae, and Planctomycetes*
Volume 5, *The Actinobacteria*

Figure 19.13 illustrates most of the groups covered in *Bergey's Manual* and in chapters 20–24.

Retrieve, Infer, Apply

1. Why is the second edition of *Bergey's Manual* no longer based on phenetic classification?
2. Describe two different situations in which it would be essential to identify the genus and species of a bacterium or archaeon.

Key Concepts

19.1 Microbial Taxonomy Is Based on the Evolution of Multiple Traits

■ Taxonomy, the science of biological classification, is composed of three parts: classification, nomenclature, and identification.

■ A polyphasic approach is used to classify microbes. This incorporates information gleaned from genetic, phenotypic, and phylogenetic analysis (**Table 19.1**).

19.2 Taxonomic Ranks Provide an Organizational Framework

■ Taxonomic ranks are arranged in a nonoverlapping hierarchy (**figure 19.1**).

■ A bacterial or archaeal species is a collection of strains that have many stable properties in common and differ significantly from other groups of strains.

■ Microorganisms are named according to the binomial system.

19.3 Microbial Taxonomy and Phylogeny Are Largely Based on Molecular Characterization

■ Historically, microbial taxonomic and phylogenic analysis used morphological, physiological, and ecological characteristics. These remain important in building a complete picture that also includes molecular information.

■ The G + C content of DNA is taxonomically valuable because it is an indirect reflection of the base sequence.

■ Nucleic acid hybridization studies are used to compare DNA or RNA sequences and thus determine genetic relatedness.

■ Nucleic acid sequencing is the most powerful and direct method for comparing genomes. The sequences of SSU rRNA are used most often in phylogenetic studies of microbes (**figure 19.3**). Alternatively, SSU rRNA genes can be analyzed by phylotyping, which does not require nucleotide sequencing (**figure 19.4**).

■ Additional techniques must be applied to identify a microbe at the species or strain level. They include restriction fragment length polymorphism (RFLP) analysis, multilocus sequence typing (MLST), the study of repetitive sequences, and single nucleotide polymorphism analysis (**figures 19.5** and **19.6**).

19.4 Phylogenetic Trees Illustrate Evolutionary Relationships

■ Phylogenetic relationships often are shown in the form of branched diagrams called phylogenetic trees. Trees are based on pairwise comparison of amino acid or nucleotide sequences, followed by computer analysis (**figure 19.7**).

■ Trees may be either rooted or unrooted and are created in several different ways. Unrooted trees can be rooted by including an outgroup when the tree is constructed (**figure 19.8**).

■ Microbes have a long history of horizontal gene transfer, which confuses taxonomic analysis. Complete genome analysis has revealed a set of core genes found in all members of a given taxon and a pan-genome, which is the sum of all genes outside the core genome of that taxon. The pan-genome is the result of horizontal gene transfer (**figure 19.9**).

19.5 Evolutionary Processes and the Concept of a Microbial Species Inspire Debate

■ There are several hypotheses regarding the origin of eukaryotic cells. Most biologists agree that endosymbioses of a bacterium and a cyanobacterium gave rise to mitochondria and chloroplasts, respectively (**figure 19.10**).

■ The operational definition of a microbial species is based on criteria approved by the International Committee on the Systematics of Prokaryotes. This includes at least 70% whole genome similarity as determined by DNA-DNA hybridization, at least 97% 16S rRNA homology, no more than a 5°C difference in % G + C, and physiological, morphological, and ecological similarity.

■ The concept of a microbial species is based on evolution. The ecotype model describes the outcome of a periodic natural selection on a genetically homogeneous microbial population. Individuals that acquire adaptive mutations are the source of microbial diversity (**figure 19.11**).

■ Horizontal gene transfer is also important in microbial evolution. However, speciation is thought to be the outcome of mutation, while rapid adaptation to new niches is mediated by horizontal gene transfer (**figure 19.12**).

19.6 *Bergey's Manual of Systematic Bacteriology*

■ *Bergey's Manual of Systematic Bacteriology* is based on the accepted system of prokaryotic taxonomy.

■ The second edition of *Bergey's Manual* provides phylogenetic classifications. Bacteria are classified among 26 phyla and archaea are divided between two phyla. Comparisons of nucleic acid sequences, particularly 16S rRNA sequences, are the foundation of this classification.

Compare, Hypothesize, Invent

1. Consider the fact that the use of 16S rRNA sequencing as a taxonomic and phylogenetic tool has resulted in tripling the number of bacterial phyla. Why has the advent of this genetic technique expanded the currently accepted number of microbial phyla?

2. *Bacteria* and *Archaea* were classified phenetically in the first edition of *Bergey's Manual of Systematic Bacteriology.* What are the advantages and disadvantages of the phylogenetic classification used in the second edition?

3. You have recently established a pure culture of a new archaeon from soil. Describe the approaches you would use to identify your new microbe to the species level.

4. Discuss the problems in developing an accurate phylogenetic tree. Do you think it is possible to create a completely accurate universal phylogenetic tree? Explain your answer.

5. Why is the current classification system for *Bacteria* and *Archaea* likely to change considerably? How would one select the best features to use in the identification of unknown microbes and determination of relatedness?

6. In 2007 a severe food-borne outbreak of enterohemorrhagic *E. coli* (EHEC) O157:H7 occurred in the United States. The strain that caused this outbreak seemed to be more virulent, causing more severe illness, than previous outbreaks. To test if new, more virulent strains of EHEC might be evolving, SNP analysis was performed on 500 clinical isolates of the bacterium that were collected before, during, and after the 2007 outbreak. SNPs were detected in 96 loci, and 39 discrete SNP genotypes were identified. These could be separated into nine different clades, as shown in this phylogenetic tree. The amount of toxin produced and the severity of disease differed among clades. It was discovered that members of clade 8 caused the 2007 outbreak and this clade was associated with high levels of toxin production. Discuss the evolution of this pathogen. Specifically consider each of these mechanisms of genetic variation: horizontal gene transfer, mutation, gene amplification, gene deletion, and intragenic recombination. Which do you think is (or are) the most important? Explain your answer.

Read the original paper: Manning, S. D., et al. 2008. Variation in virulence among clades of *Escherichia coli* O157:H7 associated with disease outbreaks. *Proc. Natl. Acad. Sci. USA.* 105:4868.

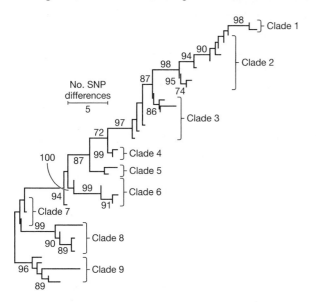

7. The enzymes responsible for DNA replication in archaea are often considered simplified versions of eukaryotic replication machinery. However, archaea also possess replication enzymes that are unique to this domain, and some archaeal replication proteins have eukaryotic homologs that are not involved in replication. It is thought that genes involved in "information processing" (i.e., replication, transcription, and translation) are typically part of the core genome, but replication proteins vary greatly among archaeal lineages. These observations prompted a group of French microbiologists to survey 22 genes that encode DNA replication-related proteins in 142 archaeal genomes. Suggest a means by which the authors were able to assign these genes to the core genome or the pan-genome for each archaeal phylum and genus analyzed. How do you think the authors used these data to infer the last common archaeal ancestor with respect to the replication apparatus? What would you predict regarding the presence of eukaryotic replication genes in archaea if the last common eukaryotic ancestor arose from an archaeon?

Read the original paper: Raymann, K., et al. 2014. Global phylogenomic analysis disentangles the complex evolutionary history of DNA replication in archaea. *Genome Biol. Evol.* 6:193.

Learn More

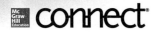

20

Archaea

The recovery of natural gas (methane, produced by archaea) is a boon to the U.S. energy economy, but the method by which it is extracted—hydraulic fracturing—is not without controversy.

Methanogenic Archaea Fuel Domestic Energy Debate

Long before gas prices surged and politicians began to argue about energy security, archaea were busy generating methane, otherwise known as natural gas. Although methane deposits are billions of years old, only recently have geologists recognized that at least 2,000 trillion cubic feet (Tcf) of natural gas could be recovered within the United States. This upward revision is largely due to the inclusion of 750 Tcf of methane trapped in underground shale. This domestic source of gas, much of it under Appalachian and Rocky Mountain states, could provide over 20% of the total U.S. gas supply by 2020. This should be good news as the United States consumes about 25 Tcf annually. But there is a hitch. The methane has to be extracted from the earth.

Before we tackle the extraction problem, let's discuss how all that natural gas got there in the first place. There are two sources of methane: abiotic and biotic. Abiotic methane results principally from thermal decomposition of sedimentary rocks and is almost negligible compared to that produced by microbes. All biologically produced methane is generated by methanogenic archaea from either $CO_2 + H_2$ or acetate (section 20.4). Evidently, shale beneath the United States has been home to a vast and productive archaeal community for eons.

Although the abundance of natural gas in shale has been known for a long time, only recently has methane recovery become financially worthwhile. The price of extracting the gas trapped within the soft, porous shale far exceeded any later profits. But the recent development of horizontal drilling and hydraulic fracturing ("fracking") technologies changed the economic equation. Now a typical well can be drilled vertically about 3,000 meters and then extended just as far horizontally. Fracking involves high-pressure pumping of up to 9 million liters of water, sand, and chemicals into the well. This cracks the shale, while a "propping agent" keeps the pores open so the gas can be recovered.

However, some citizens in communities where fracking takes place are concerned about groundwater contamination, fracking-induced earthquakes, the escape of contaminated drilling fluids at the well surface, and methane leaking from the wells, thereby increasing the carbon footprint of natural gas usage. Federal regulations mandate that recovered drilling fluids be treated before they are discharged back into the environment. In some cases, drillers have used municipal wastewater treatment facilities for this, creating a tremendous burden on local infrastructure. Escaping methane is considered problematic because methane is a harmful greenhouse gas (*see section 28.2*). If the United States is to continue to use this tremendous natural resource, these concerns must be addressed to the satisfaction of producers and consumers.

In this chapter, we review archaea, which were once considered almost exclusively extremophiles—organisms that thrive in extreme conditions, where they do unusual things, such as make methane. However, we know now that they are physiologically and ecologically diverse, found in both mundane and challenging environments. Indeed, the third domain of life deserves our respect, especially if they are to provide us with energy to fuel our power grid.

Readiness Check:
Based on what you have learned previously, you should be able to:

✔ Draw the basic structure of the phylogenetic tree of life based on SSU rRNA sequences and explain the placement of *Archaea* (section 1.2)
✔ Describe an archaeal cell wall and membrane, including component parts (section 4.2)
✔ List environmental factors that influence microbial growth, including oxygen (section 7.4)
✔ Describe the difference between heterotrophy and autotrophy (section 11.1)
✔ Define chemolithotrophy and identify an example of an electron donor and an acceptor (section 11.10)
✔ Review archaeal DNA replication, gene transcription, and protein synthesis and secretion (chapter 15)
✔ Describe the use of metagenomics in the discovery of new microbial taxa (section 18.3)

20.1 Overview of *Archaea*

After reading this section, you should be able to:

- List some common habitats in which archaea reside
- Describe the debate that surrounds archaeal taxonomy
- Compare at least three key metabolic pathways that are central to archaeal physiology with those used by bacteria

Archaea have many features in common with eukaryotes, others in common with bacteria, and still other elements that are uniquely archaeal. In general, archaeal genes that encode proteins involved in DNA replication, transcription, and translation, often called information genes, share homology with those of eukaryotes, whereas genes involved in metabolism are similar to bacterial genes. Some unique archaeal features include their tRNA structure and the production of methane. Like bacteria, archaea are quite diverse, both in morphology and physiology. They may be spherical, rod-shaped, spiral, lobed, cuboidal, triangular, plate-shaped, irregularly shaped, or pleomorphic. Some are single cells, whereas others form filaments or aggregates. They range in diameter from 0.3 to over 15 μm, and some filaments can grow up to 200 μm in length. Although they stain either Gram positive or Gram negative, they have unique cell walls—different from that of bacteria. Multiplication is usually by binary fission but may be by budding, fragmentation, or other mechanisms. Archaea are just as diverse physiologically. They can be aerobic, facultatively anaerobic, or strictly anaerobic. Nutritionally, they range from chemolithoautotrophs to organotrophs. They include psychrophiles, mesophiles, and hyperthermophiles that can grow above 100°C.

Archaea inhabit a wide variety of habitats, but for many years they were considered microbes of extreme environments, or extremophiles. Indeed, many, but by no means all, archaea inhabit niches that have very high or low temperatures or pH, concentrated salts, or are completely anoxic. For instance, in some hypersaline environments, their populations become so dense that the brine is red with archaeal pigments. By contrast, archaea also make up about 20% of the prokaryotic biomass of marine plankton and are important members of some soil communities—environments that cannot be considered extreme. In addition, some are symbionts in the digestive tracts of animals, but to date, no pathogenic archaea have been confirmed. Thus the notion that archaea are exclusively extremophiles is no longer valid as archaea are known to inhabit temperate and tropical soils and waters as well as the human body.

Archaeal Taxonomy

As shown in **table 20.1**, well-characterized archaea can be divided into six major groups based on physiological and morphological differences. However, we use a taxonomic organization to discuss archaea. Published in 2001, the first volume of *Bergey's Manual of Systematic Bacteriology* divides *Archaea* into the phyla *Euryarchaeota* (Greek *eurus,* wide, and *archaios,* ancient or primitive) and *Crenarchaeota* (Greek *crene,* spring or fount, and *archaios*) (**figure 20.1**). The euryarchaeotes are given this name because they occupy a wide variety of ecological niches and are therefore metabolically diverse. Methanogens, extreme halophiles, sulfate reducers, and many extreme thermophiles with sulfur-dependent metabolism are placed in *Euryarchaeota.* Crenarchaeotes are thought to resemble the ancestral archaea, and all are thermophiles or hyperthermophiles found in unusual environments such as acidic hot springs and submarine hydrothermal vents.

Crenarchaeota phylogeny has been the subject of considerable debate over the last decade or so. Metagenomic sequences suggested the presence of what were dubbed "mesophilic crenarchaeota," distinguishing them from thermophilic crenarchaeotes in pure culture. More recently, many of these organisms have been placed in the phylum *Thaumarchaeota* (Greek *thaumas,* wonder) (**figure 20.2**). The phylum was initially proposed based on 16S rRNA data from uncultured archaea and the genomic sequence of *Cenarchaeum symbiosum,* a marine sponge symbiont. More recently the genomes of *Nitrosopumilus maritimus* and *Candidatus* (*Ca.*) Nitrososphaera gargensis have been analyzed. In addition, there is now an abundance of metagenomic and single-cell genome studies that demonstrate the presence of these organisms in both marine and soil environments. Thaumarchaea are aerobic ammonia-oxidizing archaea; all such nitrifying archaea belong to this lineage.

Three additional phyla have been proposed. The phylum *Korarchaeota* (from the Greek word for "young man") was suggested over a decade ago, based solely on the SSU rRNA data of uncultured microbes. Since that time, there have been many metagenomic and single-cell genomic analyses, but as yet no members of this candidate phylum have been grown in pure culture. Perhaps this is because members of this lineage are hyperthermophilic anaerobes, residing in marine and terrestrial geothermal habitats. Recent genome analysis of (*Ca.*) Korarchaeum cryptofilum reveals tRNA maturation proteins that resemble those of euryarchaeotes. It is argued that this helps support the notion that *Korarchaeota* represents a unique lineage, separate from *Crenarchaeota,* where *Bergey's* places these microbes (figure 20.2). K. cryptofilum also lacks genes for many biosynthetic pathways, which may also help explain why it has not yet been grown in pure culture. Because this archaeon has not been grown axenically, it has candidate (*Candidatus*) species status and its name is not italicized.

The proposed phylum, *Aigarchaeota,* is more tenuous. It is based largely on the genome sequence of the archaeon Ca. Caldiarchaeum subterraneum, which was found in a subsurface geothermal stream. This is very controversial, with some taxonomists arguing that C. subterraneum and related taxa are members of *Thaumarchaeota,* while others feel that these microbes belong in *Crenarchaeota.* ◄◄ *Comparative genomics (section 18.7); Phylogenetic trees illustrate evolutionary relationships (section 19.4)*

The third proposed phylum is also based solely on genomic sequences recovered from the environment. The discovery of *Lokiarchaeota* is particularly exciting because these sequences encode "eukaryotic signature proteins." If further evidence continues to support this lineage, the lokiarchaeotes may represent the closest extant relatives of the first eukaryotic cell, lending support to the hypothesis that eukaryotes arose from an archaeal ancestor. ◄◄ *Evolution of the the three domains of life (section 19.5)*

Table 20.1 Characteristics of the Major Archaeal Physiological Groups

Group	General Characteristics	Representative Genera
Methanogenic archaea	Strict anaerobes. Methane is the major metabolic end product. S^0 may be reduced to H_2S without yielding energy. Cells possess coenzyme M, factors 420 and 430, and methanopterin.	*Methanobacterium* (E)[1] *Methanococcus* (E) *Methanomicrobium* (E) *Methanosarcina* (E)
Archaeal sulfate reducers	Regular and irregular cocci. H_2S formed from thiosulfate and sulfate. Autotrophic growth with thiosulfate and H_2. Can grow heterotrophically. Traces of methane also formed. Extremely thermophilic and strictly anaerobic. Possess factor 420 and methanopterin but not coenzyme M or factor 430.	*Archaeoglobus* (E)
Extremely halophilic archaea	Rods, cocci, or irregular shaped cells that may include pyramids or cubes. Primarily chemoorganoheterotrophs. Most species require sodium chloride ≥1.5 M, but some survive in as little as 0.5 M. Most produce characteristic bright-red colonies; some are unpigmented. Neutrophilic to alkalophilic. Generally mesophilic; however, at least one species is known to grow at 55°C. Possess either archaerhodopsin or halorhodopsin and can use light energy to produce ATP.	*Halobacterium* (E) *Halococcus* (E) *Natronobacterium* (E)
Cell wall-less archaea	Pleomorphic cells. Thermoacidophilic and chemoorganotrophic. Facultatively anaerobic. Plasma membrane contains mannose-rich glycoproteins and lipoglycans.	*Thermoplasma* (E)
Extremely thermophilic S^0-metabolizers	Rods, filaments or cocci. Obligately thermophilic (optimum growth temperature between 70–100°C). Usually strict anaerobes but may be aerobic or facultative. Acidophilic or neutrophilic. Autotrophic or heterotrophic. Most are sulfur metabolizers. S^0 reduced to H_2S anaerobically; H_2S or S^0 oxidized to H_2SO_4 aerobically.	*Desulfurococcus* (C) *Pyrodictium* (C) *Pyrococcus* (E) *Sulfolobus* (C) *Thermococcus* (E) *Thermoproteus* (C)
Mesophilic aerobic ammonia-oxidizers	Globally distributed in marine and soil environments, chemolithoautotrophic, using ammonia as electron donor and oxygen as terminal electron acceptor. Use HP/HB[2] pathway for carbon fixation. Evidence that some deep-ocean thaumarchaea may use urea as carbon and nitrogen source.	*Cenarchaeum symbiosum* (T) *Nitrosopumilus maritimus* (T)

1 Indicates phylum; *E, Euryarchaeota, C, Crenarchaeota, T, Thaumarchaeota*
2 HP/HB, 3-Hydroxypropionate/4-Hydroxybutyrate Cycle

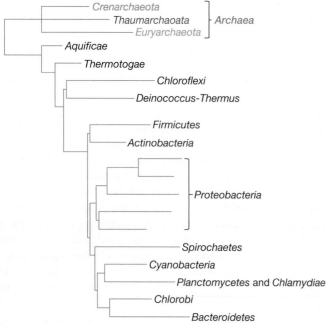

Figure 20.1 Phylogenetic Relationships Among Members of *Bacteria* and *Archaea*. Recognized archaeal lineages are highlighted.

Thus although *Bergey's* currently accepts only two archaeal phyla, one new phylum has been accepted, and *Korarchaeota* is fairly likely to gain full standing within the next 5 years. This state of flux in archaeal phylogeny demonstrates how dynamic microbial taxonomy can be. The use of molecular probes to dissect microbial communities, combined with innovative culture techniques, ensures that phylogenetic analysis will continue to evolve.

Metabolism

In view of the diversity of archaeal lifestyles, it is not surprising that archaeal metabolism varies greatly among members of different groups. Some archaea are heterotrophs; others are autotrophic. Here we review the pathways by which archaea fix carbon and catabolize carbohydrates.

CO₂-Fixation Pathways

Like all anabolic processes, carbon fixation requires reducing power and energy. Because inorganic carbon (CO_2 or HCO_3^-) has an oxidation state of +4, it requires four reducing equivalents for its assimilation into cellular carbon (oxidation state of 0). Most known autotrophic archaea are either anaerobic or live in low oxygen conditions. These microbes tend to use ferredoxin

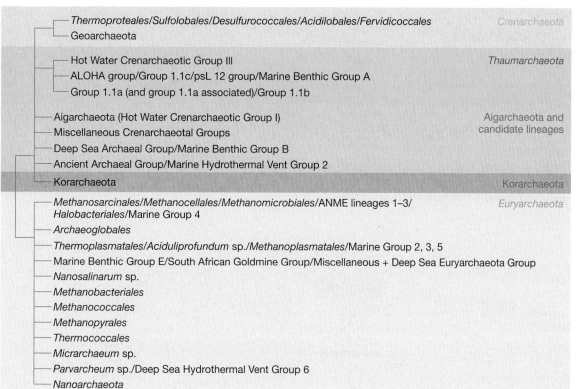

Thermoproteales/Sulfolobales/Desulfurococcales/Acidilobales/Fervidicoccales	*Crenarchaeota*
Geoarchaeota	
Hot Water Crenarchaeotic Group III	*Thaumarchaeota*
ALOHA group/Group 1.1c/psL 12 group/Marine Benthic Group A	
Group 1.1a (and group 1.1a associated)/Group 1.1b	

Figure 20.2 Phylogenetic Tree of *Archaea*. This schematic representation shows the phylogenetic diversity of archaea but also illustrates the dynamic nature of archaeal systematics. Phyla in gray font have not been accepted as valid. *Lokiarchaeota* sequences were not available for consideration.

(Fd) instead of NADPH as their source of electrons. Because reduced Fd has a standard reduction potential (E_0') of -400 mV, it bears more energy than NADPH (E_0' of -320 mV), and most anaerobic pathways require less ATP per CO_2 fixed than aerobic carbon fixation pathways (e.g., the Calvin cycle). This is important as some autotrophic archaea, particularly methanogens, live close to their thermodynamic limit and therefore have very little (if any) ATP to spare. ◄◄ *Redox reactions (section 10.3); CO_2 fixation (section 12.3)*

Three different carbon fixation pathways have been characterized in archaea: the reductive acetyl-CoA pathway, the 3-hydroxyproprionate/4-hydroxybutyrate (HP/HB) cycle, and the dicarboxylate/4-hydroxybutyrate (DC/HB) cycle (**table 20.2**). Recently it has been shown that the euryarchaeotes *Archaeoglobus* and *Ferroglobus* spp. use the reductive acetyl-CoA pathway (**figure 20.3**). By contrast, it has long been known that methanogens (also euryarchaeotes) use this pathway, as it is the most energeti-

cally favorable route. It requires only about one ATP per pyruvate synthesized, compared to seven for the Calvin cycle. As the pathway's name reflects, two CO_2 molecules are reduced and incorporated into one acetyl-CoA. One CO_2 is reduced to carbon monoxide (CO) by a CO dehydrogenase that also functions as an acetyl-CoA synthase. Another CO_2 is reduced to the level of CH_3 and bound to a tetrahydropterin coenzyme. Methanogens withdraw this from the pathway and reduce it to methane, as discussed in section 20.3. Thus methanogens use the acetyl-CoA pathway for both carbon assimilation and energy conservation. *Archaeoglobus* and *Ferroglobus* spp. capture energy using anaerobic respiration with sulfate as the terminal electron acceptor, so this pathway is used only for carbon fixation. Likewise, several anaerobic bacteria, including some planctomycetes, proteobacteria, and spirochetes, use the reductive acetyl-CoA pathway to fix CO_2.

The HP/HB cycle is found in members of the crenarchaeote order *Sulfolobales* and in *Thaumarchaeota* (**figure 20.4**). It

Table 20.2	Comparison of Archaeal CO_2-Fixation Pathways to the Calvin-Benson Cycle				
CO_2-Fixation Pathway	**Figure**	**Oxygen Tolerance**	**ATP Consumed per Pyruvate Produced**	**Archaeal Phyla Known to Use Pathway**
Reductive Acetyl-CoA	20.3	Anaerobic	1	*Euryarchaeaota*
3-Hydroxpropionate/ 4-Hydroxybutyrate	20.4	Aerobic	9	*Crenarchaeota* *Thaumarchaeota*
Dicarboxylate/ 4-Hydroxybutyrate	20.4	Anaerobic	5	*Crenarchaeota*
Calvin-Benson	12.4	Aerobic	7	None

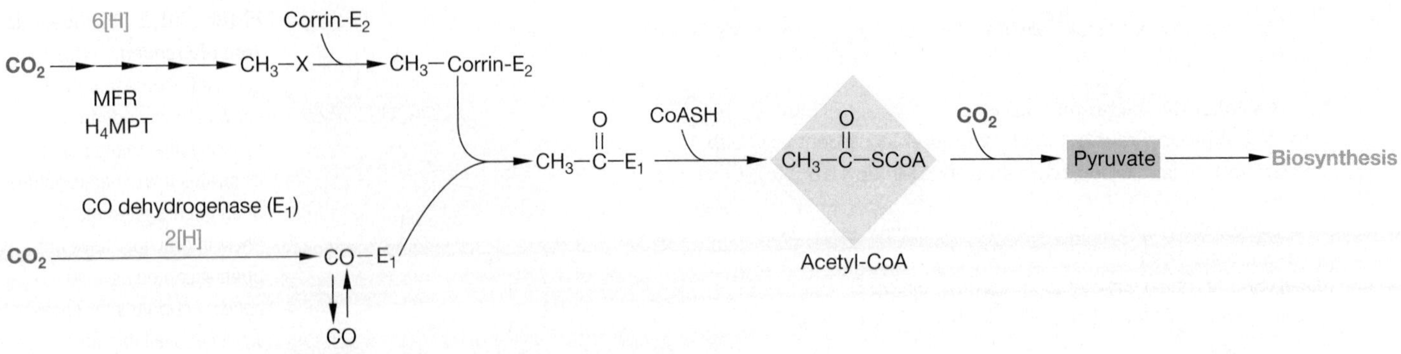

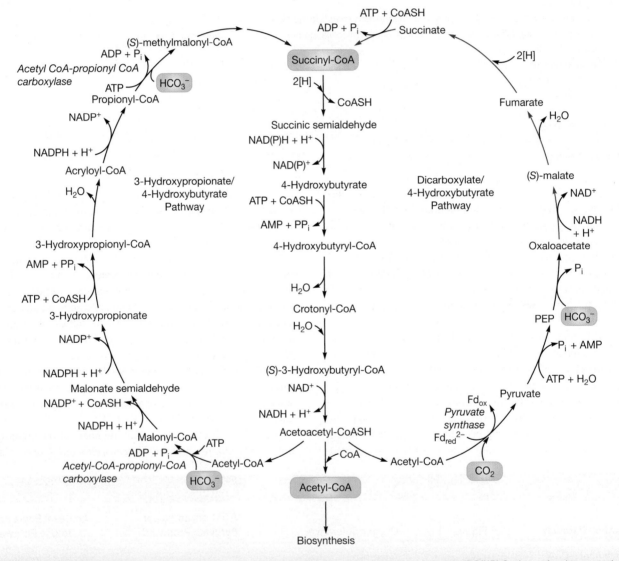

Figure 20.3 The Reductive Acetyl-CoA Pathway.   Methanogens reduce two molecules of CO_2, each by a different mechanism, and combine them to form acetyl and then acetyl-CoA.

Figure 20.4 The 3-Hydroxypropionate/4-Hydroxybutyrate (HP/HB) and Dicarboxylate/4-Hydroxybutyrate (DC/HB) Cycles.   Aerobic crenarchaea such as *Sulfolobus* and some thaumarchaea use the HP/HB pathway (left) to convert two bicarbonate ions into acetyl-CoA. Some crenarchaea fix one molecule of CO_2 and one HCO_3^- using the anaerobic DC/HB pathway (right). The reactions shown with red arrows are also used in the reductive tricarboxylic acid cycle.

MICRO INQUIRY   *Given that the DC/HB pathway uses far less ATP per pyruvate synthesized than the HP/HB pathway, why do some archaea use the HP/HB pathway instead?*

requires more energy than any other CO_2 fixation pathway (9 ATP/pyruvate synthesized) but has other advantages. It has a low demand for metal cofactors, and unlike either the acetyl-CoA pathway or the DC/HB cycle, it can operate under aerobic conditions (table 20.2). The key carboxylase enzyme, a biotin-dependent acetyl-CoA/propionyl-CoA carboxylase, accepts two bicarbonate ions (HCO_3^-) in a series of steps that yields succinyl-CoA and net synthesis of one acetyl-CoA. Interestingly, bacteria and eukaryotes use acetyl-CoA carboxylase as the first step in fatty acid biosynthesis, but because archaea do not make fatty acids, this enzyme can be dedicated to carbon assimilation.

The last autotrophic pathway known to occur in archaea is the anaerobic DC/HB pathway (figure 20.4). It is found in the anaerobic and microaerobic members of two crenarchaeote families, *Thermoproteales* and *Desulfurococcales*. The DC/HB pathway starts when an acetyl-CoA and a CO_2 molecule are condensed to form pyruvate, which is then converted to phosphoenolpyruvate (PEP). PEP carboxylase then adds HCO_3^- to PEP to form oxaloacetate (OAA). Through a series of reductive tricarboxylic acid cycle steps (i.e., backward TCA cycle steps), OAA is converted to succinyl-CoA, which is transformed through 4-hydroxybutyrate into two molecules of acetyl-CoA. As in the HP/HB pathway, one acetyl-CoA is then used for biosynthesis, and the other is used as a CO_2 acceptor in the next round of the cycle. The DC/HB cycle uses only five ATP per pyruvate made, but some of its enzymes are oxygen sensitive. Thus while the conversion of succinyl-CoA to acetyl-CoA is the same in both the HP/HB and DC/HB pathways, aerobes must perform the more expensive HP/HB pathway.

Inspection of figures 20.3 and 20.4 reveals that autotrophic pathways used by archaea result in the production of the 2-carbon compound (C2) acetyl-CoA that must be converted to the central metabolite pyruvate (C3) for anabolic purposes. The most common means by which this is accomplished is through the glyoxylate cycle (**figure 20.5**). Here the enzyme isocitrate lyase splits isocitrate (C6) into glyoxylate (C2) and succinate (C4); glyoxylate is then condensed with acetyl-CoA (C2) to form malate (C4). While autotrophic archaea and haloarchaea such as *Haloferax volcanii* use the glyoxylate cycle, other haloarchaea that must assimilate acetyl-CoA do not. These microbes use the methylaspartate pathway. This pathway also yields malate and succinate but requires many more steps. In the methylaspartate pathway (figure 20.5), α-ketoglutarate (C5) is converted to the amino acid glutamate, which is then rearranged to form the nonstandard amino acid methylaspartate. The ammonia is next removed, and several steps are required before the cleavage of a C5 intermediate into propionyl-CoA (C3) and glyoxylate (C2). Just as in the glyoxylate pathway, acetyl-CoA condenses with glyoxylate to form malate, which is used for biosynthesis. Unlike the glyoxylate pathway, another CO_2 (as HCO_3^-) is accepted by propionyl-CoA to form succinate, which is also converted to malate. All this begs the question, why use a more complicated (and therefore costly) mechanism? The answer may be linked to the fact that so far, the methylaspartate pathway is found only among haloarchaea. These microbes are

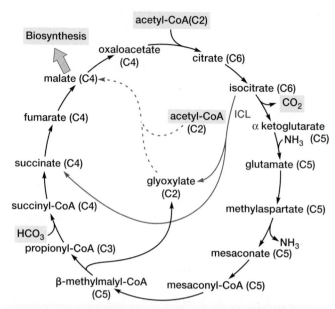

Figure 20.5 Glyoxylate and Methylaspartate Pathways Are Used for Acetyl-CoA Assimilation. Autotrophic archaea and some haloarchaea use the glyoxylate pathway (red and dashed arrows) to incorporate acetyl-CoA. Other haloarchaea lack the key enzyme isocitrate lyase (ICL) and assimilate acetyl-CoA via the longer methylaspartate pathway instead (black and dashed arrows).

MICRO INQUIRY *Why might the production of glutamate as an intermediate be beneficial for some haloarchaea?*

under constant osmotic stress in the high-salt environments they inhabit. Glutamate is used as an osmoprotectant, so it is hypothesized that haloarchaea may use the accumulation of intracellular glutamate as a signal to activate the methylaspartate pathway when growing on acetate or compounds that produce acetate as an intermediate. ◄◄ *Solutes affect osmosis and water activity (section 7.4)*

Very few details are known about other archaeal biosynthetic pathways. Evidence suggests that pathways for amino acid, purine, and pyrimidine biosynthesis are similar to those in other organisms. Some methanogens can fix atmospheric dinitrogen, and this pathway also seems to be well conserved among bacteria and archaea. ◄◄ *Anabolism: the use of energy in biosynthesis (chapter 12)*

Chemoorganotrophic Pathways

Among chemoorganotrophic archaea, a least three novel carbohydrate catabolic pathways have been identified: a modified Embden-Meyerhof (EM) pathway and two variations of the Entner-Doudoroff pathway. The modified EM pathway used by hyperthermophilic euryarchaeotes involves several novel enzymes, including an ADP-dependent glucokinase and phosphofructokinase (**figure 20.6**). These kinases each result in the production of AMP; in fact, unlike the classical EM pathway, there is no net ATP produced. Another unique step in

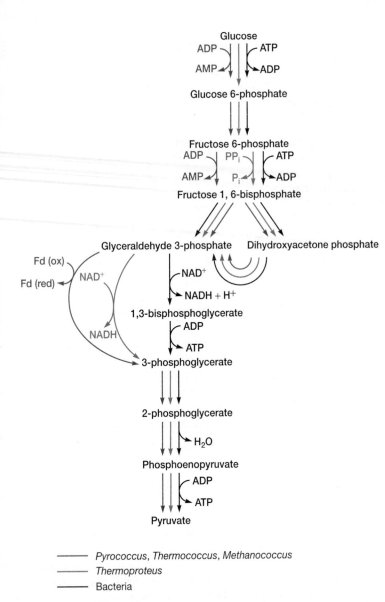

Figure 20.6 **Modified Embden-Meyerhof Pathway.** The pathway used by some archaea compared to that of bacteria. The only difference between the pathway used by *Thermoproteus* spp. and the other archaea shown is the use of NAD$^+$ as the electron acceptor rather than ferredoxin when glyceraldehyde 3-phosphate is oxidized to 3-phosphoglycerate. There is no net ATP produced in the archaeal pathway.

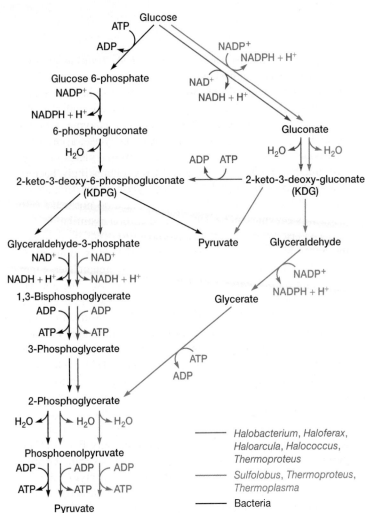

Figure 20.7 **Two Alternate Versions of the Entner-Doudoroff Pathway.** The phosphorylative pathway (red) used by halophilic archaea involves the ATP-dependent phosphorylation of KDG to KDPG; the remainder of the pathway is like that of bacteria and an ATP is gained. The nonphosphorylative pathway (purple) used by *Sulfolobus* spp. and others joins the bacterial ED pathway when glycerate is converted to 2-phosphoglycerate, so no net ATP is produced.

this modified EM pathway is the conversion of glyceraldehyde 3-phosphate directly to 3-phosphoglycerate, with the use of ferredoxin as the electron acceptor, rather than NAD$^+$. By contrast, members of the thermophilic crenarchaeote genus *Thermoproteus* employs NAD$^+$ as the electron acceptor in the modified EM pathway.

The two Entner-Doudoroff pathways include one called the phosphorylative pathway and another dubbed the nonphosphorylative pathway. Extreme halophiles and *Thermoproteus* spp. catabolize glucose using the phosphorylative Entner-Doudoroff (ED) pathway. As shown in **figure 20.7**, glucose is oxidized to gluconate and then dehydrated to 2-keto-3-deoxy-gluconate (KDG). KDG is

phosphorylated at the expense of ATP to KDPG, and the remainder of the ED pathway is like that of bacteria. This results in a net yield of one ATP. The nonphosphorylative ED pathway is found in *Sulfolobus, Thermoplasma,* and *Thermoproteus* spp. (note that thermoprotei can utilize all three archaeal-specific pathways). This pathway begins like that of the halophiles, but KDG is cleaved by KDG aldolase to pyruvate and glyceraldehyde. Glyceraldehyde is oxidized to glycerate, which is then phosphorylated to 2-phosphoglycerate. Pyruvate is then generated as in the classical and halophilic ED pathway. Because these microbes bypass the energy-conserving step in which 1,3-bisphosphoglycerate is converted to 3-phosphoglycerate, there is no net yield of ATP. ◄◄ *Glucose to pyruvate (section 11.4)*

All archaea that have been studied can oxidize pyruvate to acetyl-CoA. However, they lack the pyruvate dehydrogenase complex present in eukaryotes and respiratory bacteria, and use the enzyme pyruvate ferridoxin oxidoreductase instead. Halophiles and the extreme thermophiles *Thermoplasma* spp. appear to have a functional tricarboxylic acid cycle. Methanogens do not catabolize glucose to any significant extent, so it is not surprising that they lack a complete tricarboxylic acid cycle. However, some store glycogen, which is catabolized via a modified EM pathway (figure 20.6). ◄◄ *Pyruvate to carbon dioxide (step 2) is accomplished by the tricarboxylic acid cycle (section 11.5)*

Retrieve, Infer, Apply

1. How are euryarchaeotes, crenarchaeotes, and thaumarchaeotes distinguished?
2. Why do methanogens use the reductive acetyl-CoA pathway for carbon fixation?
3. Why do haloarchaea use the methylaspartate rather than the glyoxylate cycle for the incorporation of acetate?
4. Compare the ATP and NADH yield of each of the archaeal glycolytic pathways with that used by bacteria.

20.2 Phylum *Crenarchaeota*: Metabolically Diverse Thermophiles

After reading this section, you should be able to:

■ List the major physiological types among crenarchaea
■ Discuss hyperthermophilic and thermoacidophilic growth

Bergey's Manual lists only the well-characterized members of the phylum *Crenarchaeota,* all of which are thermophilic. They belong to a single class, *Thermoprotei,* which is divided into four

Figure 20.8 Habitat for Thermophilic Archaea. The Sulfur Cauldron in Yellowstone National Park. The water is at its boiling point and very rich in sulfur. *Sulfolobus* spp. grow well in such habitats.

orders. **Table 20.3** shows the further division of these orders and summarizes some of the characteristics of each family.

Many crenarchaeotes are sulfur dependent. The sulfur may be used either as an electron acceptor in anaerobic respiration or as an electron donor by lithotrophs. They are often found in geothermally heated water or soils that contain elemental sulfur. These environments are sometimes called solfatara and are scattered all over the world. Familiar terrestrial examples are the sulfur-rich hot springs in Yellowstone National Park (**figure 20.8**).

Table 20.3	Phylum *Crenarchaeota;* Class *Thermoprotei*				
Order	**Family or Families**	**Typical Morphology**	**Optimal Growth Temp Range**	**Growth Characteristics**	**Representative Genera**
Thermoproteales	*Thermoproteaceae*	Cocci, clubs, rods	75 to >100°C	Facultatively anaerobic or anaerobic, chemolithoautotropic reduction of S⁰ with H₂ or chemoorganotrophic	*Thermoproteus* *Pyrobaculum* *Thermocladium*
	Thermofilaceae	Thin rods, filamentous	80 to 90°C	Obligately anaerobic, acidOphilic, sulfur respiration	*Thermofilum*
Desulfurococcales	*Desulfurococcaceae*	Cocci, disc-shaped	85 to 95°C	Most anaerobic, few aerobic; chemolithoautotrophic or heterotrophic growth with sulfur reduction or fermentation	*Desulfurococcus* *Ignicoccus* *Thermodiscus* *Staphylothermus*
	Pyrodictiaceae	Cocci, disc-shaped	108 to 113°C	Facultatively anaerobic, or anaerobic; chemolithoautotrophic reduction of S⁰ or fermentative	*Pyrodictium* *Pyrolobus* *Hyperthermus*
Sulfolabales	*Sulfolobaceae*	Cocci, irregular	75 to 85°C	Aerobic, facultatively anaerobic, or anaerobic; Chemolithoautotrophic or chemoorganotrophic; Acidiphilic	*Sulfolobus,* *Metallosphaera* *Acidianus* *Sulfurisphaera* *Sulfurococcus*
Caldisphaerales	*Caldisphaeracae*	Cocci	70 to 75°C	Anaerobic, chemoorganotrophic, acidophilic	*Caldisphaera*

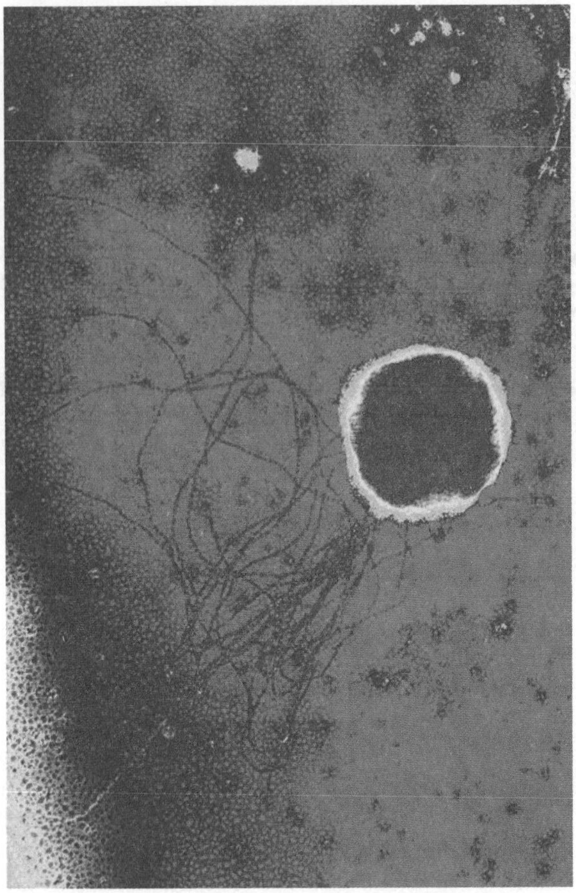

Figure 20.9 An Extremely Hyperthermophilic Crenarchaeote.   A member of the family *Pyrodictiaceae,* this archaeon grows following autoclaving at 121°C, and so was dubbed "strain 121." It reduces Fe(III) to magnetite when incubated anaerobically; colorized electron micrograph.

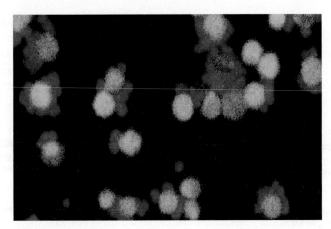

Figure 20.10 Parasitic Cells of *Nanoarchaeum equitans* Attached to the Surface of Its Host, *Ignicoccus hospitalis.*   The crenarchaeote *I. hospitalis* is the only archaeon known to have an outer membrane, which may be important for *N. equitans* attachment. In this confocal laser scanning micrograph, *I. hospitalis* is stained green and *N. equitans* is red.

Many hyperthermophiles have also been isolated from the waters surrounding submarine volcanoes (*see figure 32.4*). These archaea are classified as **hyperthermophiles** if their optimum growth temperature exceeds 85°C. The most extreme example is *Pyrolobus fumarii,* which was isolated from an active hydrothermal vent in the northeast Pacific Ocean. This member of the *Pyrodictiaceae* family grows best at about 105°C, but it can survive autoclaving at 121°C for 1 hour. It is strictly anaerobic, using Fe(III) as a terminal electron acceptor and H_2 or formate as electron donors and energy sources (**figure 20.9**).

Hydrothermal vents are also home to the only parasitic archaeon, *Nanoarchaeum equitans* (a euryarchaeote), which relies on its crenarchaeal host, *Ignicoccus hospitalis,* for replication (**figure 20.10**). *I. hospitalis* is an anaerobic, chemolithoautotrophic hyperthermophilic member of *Desulfurococcaceae* that uses H_2 as an electron donor and elemental sulfur as its electron acceptor. It uses the DC/HB pathway for carbon fixation (figure 20.4). Its cell envelope is unique as it lacks an S layer and is the only archaeon to possess a double membrane. The plasma membrane is composed of di- and tetraethers that frequently give rise to small, bleblike

vesicles. The outermost membrane surrounds a wide periplasm-like compartment between the two membranes. These lipids, which should not be confused with fatty acids, are called caldarchaeol lipids. They are extremely thermostable tetraethers found in many hyperthermophilic archaea. They consist of two C40 isoprenoid units linked to glycerol at both ends (i.e., it has four ether bonds), thus generating a monolayer rather than the more familiar lipid bilayer seen in bacteria and eukaryotes (*see figures 4.4 and 4.6*). The outermost membrane of *I. hospitalis* houses H_2:sulfur oxidoreductase and ATP synthase complexes, and is thereby energized. Its parasitic partner *N. equitans* is also quite interesting. It cannot be grown without its host and forms tiny cells of only 350 to 500 nm that attach to the *I. hospitalis* outermost membrane. Very little is known about the physiology of *N. equitans,* although genome analysis reveals that it lacks the capacity to synthesize most lipids, amino acids, nucleotides, and cofactors. At 490,000 bp, it seems *N. equitans* has taken genome reduction to the extreme. ◄◄ *Comparative genomics (section 18.7)*

Two of the best-studied members of *Crenarchaeota* belong to the genera *Sulfolobus* and *Thermoproteus* (table 20.3). Members of *Sulfolobus* are aerobic, and spherical, with a temperature optimum around 80°C and a pH optimum of only 2 to 3. For this reason, they are **thermoacidophiles.** Their cell walls contain carbohydrate and archaeal-specific lipoproteins. They grow chemoorganotrophically under oxic conditions, but they can also grow chemolithoautotrophically, using H_2, H_2S, and FeS_2 as electron donors and with oxygen as the terminal electron acceptor, although the use of ferric iron has been reported. When growing chemolithoautotrophically, they use the HP/HB cycle, and when growing chemoorganotrophically, they employ the nonphosphorylative version of the Entner-Doudoroff pathway (figure 20.7) and a complete TCA cycle. Unlike most organisms, *S. solfataricus* rarely uses NAD^+ as an electron acceptor; instead, it uses $NADP^+$ and ferredoxin-dependent oxidoreductases. In fact, archaea in general seem to lack NADH dehydrogenase, whereas it is integral to the function of bacterial and mitochondrial electron transport chains. The genomes of *Sulfolobus* spp. show a high level of plasticity; indeed, the *S. solfataricus* genome has 200 integrated insertion sequences, indicating a history of horizontal gene transfer. Although *S. solfataricus* grows at pH 2 to 4, it maintains a cytoplasmic pH of about 6.5, thereby generating a large pH gradient across the plasma membrane. This energy is conserved in the formation of ATP by membrane-bound ATP synthases and at least 15 secondary transport systems that couple the transport of organic solutes (e.g., sugars, amino acids) with the movement of protons (**figure 20.11**). The ABC transporters also used for nutrient uptake have high substrate affinity, an important adaptation in low nutrient environments. ◄◄ *Electron transport and oxidative phosphorylation (step 3) generate the most ATP (section 11.6)*

Thermoproteus spp. are long, thin rods that can be bent or branched, with cell walls composed of glycoprotein. Thermoprotei are anaerobic and grow at temperatures from 75 to 100°C. Some are acidophiles, with optimum pH values between 3 and 4, while others are neutrophiles. They are found in hot springs and other hot aquatic habitats rich in sulfur. *Thermoproteus* spp. can grow organotrophically and oxidize glucose, amino acids, alcohols, and organic acids with elemental sulfur as the electron acceptor during anaerobic respiration. Like *Sulfolobus* spp., they use the nonphosphorylative Entner-Doudoroff pathway to catabolize sugars. They also grow chemolithotrophically, oxidizing H_2 with S^0 as the electron acceptor. CO_2 is incorporated through the DC/HB cycle (figure 20.4).

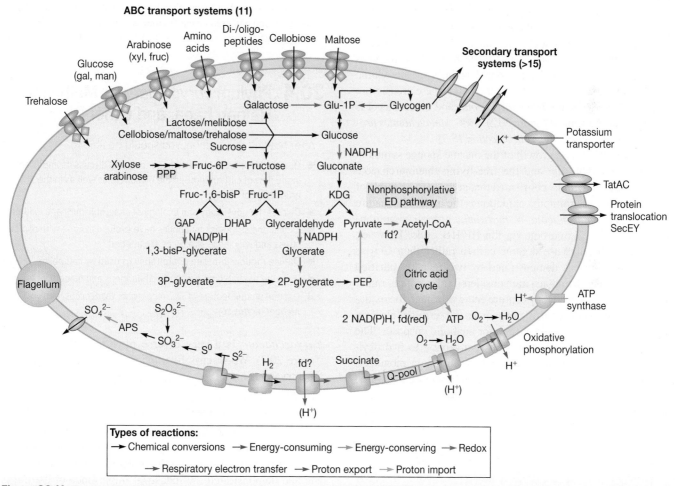

Figure 20.11 **Genomic Reconstruction of *Sulfolobus solfataricus*.** Each type of reaction is denoted by a different color arrow, as shown in the key. Components of the aerobic respiratory network include the reduction of a quinone (Q) pool, a putative ferredoxin dehydrogenase (Fd?), and succinate dehydrogenase. Alternative electron donors include hydrogen and sulfide, which reduce the quinone pool through hydrogenase and sulfide reductase, respectively. Elemental sulfur and thiosulfate are converted to sulfate. The ABC transport systems import xylose (xyl), fructose (fruc), glucose (glu), galactose (gal), and mannose (man). The flagellar apparatus more closely resembles type IV pili, as is the case with other archaea, and a gene encoding a methyl-accepting chemotaxis protein is present.

MICRO INQUIRY *Identify the potential electron donors to the electron transport chain. What is the source of energy for the potassium transporter shown on the right-hand side of the cell? Hint: Examine the arrow color code box.*

20.3 Phylum Thaumarchaeota: Mesophilic Ammonia Oxidizers

After reading this section, you should be able to:

- Describe the importance of archaeal-specific lipids in the discovery of thaumarchaeoetes
- Explain, in basic terms, the metabolism of thaumarchaeotes
- Predict where you might find members of this phylum

The first evidence for what would become the phylum *Thaumarchaeota* was the discovery of an archaeal-specific lipid called, at that time, crenarchaeol (**figure 20.12**). The detection of this lipid, coupled with metagenomic data, showed what appeared to be mesophilic crenarchaeotes in marine plankton from polar, temperate, and tropical waters. These organisms were also found in rice paddies, soils, and freshwater lake sediments, and at least two symbiotic species have been isolated, one from a cold-water sea cucumber and the other from a marine sponge. Collectively, these microbes were for many years called the group I archaea or mesophilic *Crenarchaeota*, but as discussed in section 20.1, these archaea are now placed in the phylum *Thaumarchaeota*, and crenarchaeol is now called **thaumarchaeol** (figure 20.2). ◀◀ *Metagenomics provides access to uncultured microbes (section 18.3)*

The purification and growth of the marine sponge symbiont *Cenarchaeum symbiosum* and the free-living thaumarchaeote *Nitrosopumilus maritimus* confirmed that archaea are capable of nitrification. Indeed, ammonia oxidation is the metabolic feature that defines *Thaumarchaeota*. *N. maritimus* grows chemolithoautotrophically, fixing carbon via the HP/HB cycle. However, some thaumarchaeotes use organic carbon rather than CO_2 as their carbon source, as demonstrated by two recently purified coastal strains. These archaea are considered mixotrophic rather than heterotropic because they capture energy through the oxidation of ammonia to nitrite using oxygen as the terminal electron acceptor while using organic carbon for anabolic purposes. The first step in ammonia oxidation is its conversion to hydroxylamine (NH_2OH), a reaction catalyzed by ammonia monooxygenase (AMO; *see figure 22.15*). AMO is composed of three subunits, AmoA, AmoB, and AmoC; the archaeal versions of the genes encoding these proteins are phylogenetically distinct from bacterial *amo* genes. This has made it possible to assess the presence and, through reverse transcription PCR studies, the activity of archaeal nitrification in soils and waters. While it had long been thought that α- and β-proteobacteria were solely responsible for nitrification, many metagenomic studies have shown that ammonia oxidation by thaumarchaeotes is very important. This is particularly true of marine environments. ▶▶ *Nitrifying bacteria oxidize ammonium or nitrate to gain energy and electrons (section 22.1)*

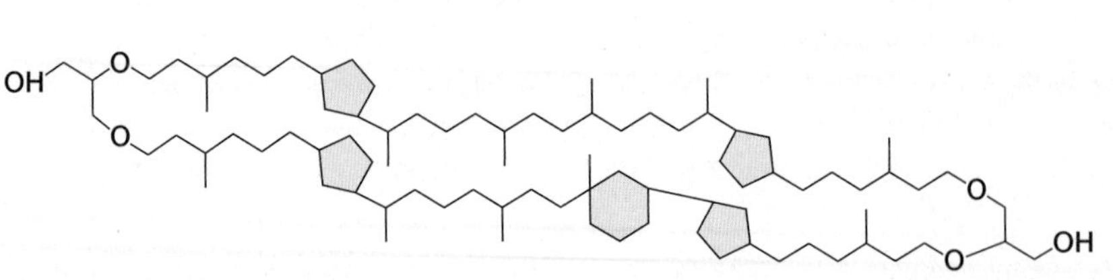

Retrieve, Infer, Apply

1. What are thermoacidophiles and where do they grow? In what ways do they use sulfur in their metabolism?
2. List three aspects of the *Ignicoccus-Nanoarchaeum* system that makes it unusual.
3. Compare the carbon and sulfur metabolism of *Sulfolobus* and *Thermoproteus* spp.
4. Discuss the role of external pH on the magnitude of the proton motive force generated by *Sulfolobus* spp.
5. You have discovered a new microorganism that oxidizes ammonia. How will you determine if this microbe is a thaumarchaeote or a bacterium?

20.4 Phylum *Euryarchaeota:* Methanogens, Haloarchaea, and Others

After reading this section, you should be able to:

- Outline the process of methanogenesis and discuss its importance in the flow of carbon through the biosphere as well as in the production of energy
- Discuss the physiology and ecology of anaerobic methane oxidation
- Explain the strategies halophiles have evolved to cope with osmotic stress and why these strategies are needed
- Outline rhodopsin-based phototrophy as used by halophiles
- Describe the habitats in which methanogens and halophiles reside
- List one unique feature of *Thermoplasma, Pyrococcus,* and *Archaeoglobus* spp.

Euryarchaeota is a very diverse phylum with many genera (figure 20.2). We discuss five major physiologic groups within the euryarchaeotes.

Methanogens and Methanotrophs

Late in the eighteenth century, Italian physicist Alessandra Volta discovered that he could ignite gas released from anoxic marshes. In this way, Volta discovered the biological production of methane and demonstrated the industrial importance of natural gas. The production of methane—**methanogenesis**—is the last step in the anaerobic degradation of organic compounds. Because the ΔG of methanogenesis compares unfavorably to other forms of respiration, methanogenesis occurs only when O_2 and most other electron acceptors are unavailable, making it a strictly anaerobic process (*see figure 28.1*). All methanogenic microbes belong to *Euryarchaeota* and are called **methanogens.** They

Figure 20.12 Thaumarchaeol. This membrane lipid is composed of dicyclic biphytane and a tricyclic biphytane.

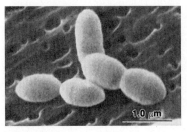
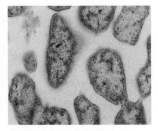

(a) **(b)**

Figure 20.13 Methanogens. (a) *Methanobrevibacter smithii*; SEM. (b) *Methanococcoides burtonii*; TEM

generate methane from H_2 and CO_2 or from short-chain organic compounds, such as formate, acetate, and methanol. These compounds are usually the fermentation products of other microbes that live in the same community. When using H_2 and CO_2, their growth is autotrophic. As discussed in section 20.1, these archaea use the reductive acetyl-CoA pathway to incorporate CO_2 (figure 20.3).

Methanogens are the largest group of cultured archaea. There are five orders (*Methanobacteriales, Methanococcales, Methanomicrobiales, Methanosarcinales,* and *Methanopyrales*) that differ greatly in overall shape, 16S rRNA sequence, cell wall chemistry and structure, membrane lipids, and other features. For example, methanogens construct three different types of cell walls. Several genera have walls with pseudomurein; other walls contain either proteins or heteropolysaccharides (*see figure 4.6*). The morphology of two representative methanogens is shown in **figure 20.13**, and selected properties are presented in **table 20.4**.

Methanogenesis

Methanogenesis, or the production of methane, is a complex process that requires several unique cofactors. As pictured in **figure 20.14**, these include *methanofuran* (MFR), tetrahydro*methanopterin* (H_4MPT), coenzyme F_{420}, *coenzyme M* (CoM; 2-mercaptoethane-sulfonic acid), and coenzyme F_{430}. These are used in the following series of reactions (**figure 20.15**): (1) Gaseous CO_2 is bound to MFR, and as such, is now activated and part of a formyl group (—HC=O). (2) The formyl group is transferred to H_4MPT and dehydrated. (3) Using F_{420} as the electron donor, the dehydrated formyl group (=HC—) is reduced to the methyl level (—CH_3) in a two-step process. (4) The methyl group is transferred to CoM. This reaction releases energy. (5) The methyl group is now ready for final reduction to methane (CH_4). This is catalyzed by the F_{430}-containing *methyl-coenzyme M reductase* (MCR). H_2 is the electron donor and protons are released.

It appears that ATP synthesis is linked with methanogenesis by electron transport, proton pumping, and a chemiosmotic mechanism. The precise means by which a proton motive force is generated is debated but may involve the oxidation of H_2 on the outer surface of the membrane (figure 20.15, step 5). In addition, the transfer of the methyl groups from methyl-H_4MPT to HS-CoM releases sufficient energy for the uptake of sodium ions (step 4). This results in a sodium motive force that could also drive ATP synthesis. This may explain why methanogens require about 1 mM Na^+ for growth. ◄◄ *Electron transport and oxidative phosphorylation (step 3) generate the most ATP (section 11.6)*

It can be useful to compare methanogens that possess cytochromes with those that lack them. Among the five orders of methanogens, only members of the youngest order,

Table 20.4 Selected Characteristics of Representative Methanogens

Order and Genera	Morphology	Wall Composition	Motility	Methanogenic Substrates Used	Capable of N_2 Fixation
Order *Methanobacteriales*					Yes
Methanobacterium	Long rods or filaments	Pseudomurein	−	$H_2 + CO_2$, formate	
Methanothermus	Straight to slightly curved rods	Pseudomurein with an outer protein S-layer	+	$H_2 + CO_2$	No
Order *Methanococcales*					Yes
Methanococcus	Irregular cocci	Protein	+	$H_2 + CO_2$, formate	
Order *Methanomicrobiales*					No
Methanomicrobium	Short curved rods	Protein	+	$H_2 + CO_2$, formate	
Methanogenium	Irregular cocci	Protein or glycoprotein	−	$H_2 + CO_2$, formate	No
Methanospirillum	Curved rods or spirilla	Protein	+	$H_2 + CO_2$, formate	Yes
Order *Methanosarcinales*					Yes
Methanosarcina	Irregular cocci, packets	Protein sometimes with polysaccharide	−	$H_2 + CO_2$, methanol, methylamines, acetate	

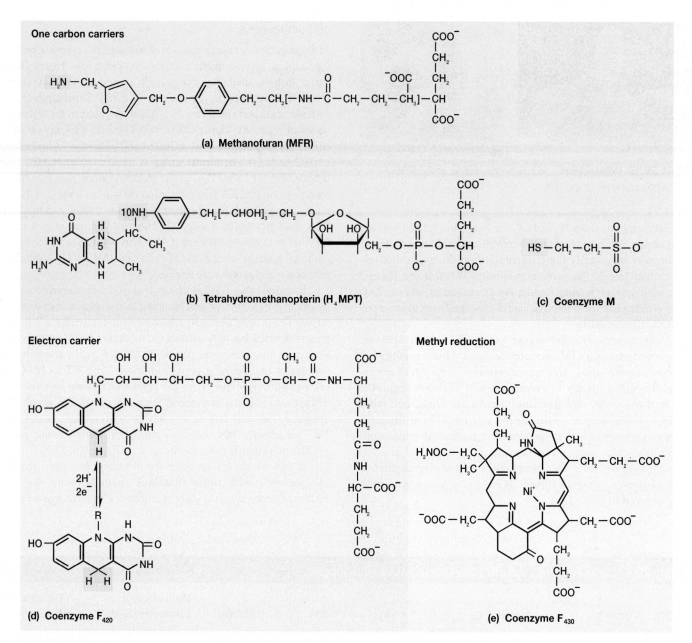

Figure 20.14 Methanogen Coenzymes. (a) Coenzyme MFR, (b) H₄MPT, and (c) coenzyme M are used to carry 1-carbon units during methanogenesis. MFR and a simpler form of H₄MPT called methanopterin (MPT; not shown) also participate in the synthesis of acetyl-CoA. The portions of the coenzymes that carry the 1-carbon units are highlighted. H₄MPT carries carbon units on nitrogens 5 and 10, like the more common enzyme tetrahydrofolate. (d) Coenzyme F_{420} participates in redox reactions. The part of the molecule that is reversibly oxidized and reduced is highlighted. (e) Coenzyme F_{430} participates in reactions catalyzed by the enzyme methyl-CoM methylreductase.

Methanosarcinales, have cytochromes and methanophenazine, a menaquinone-like, redox active molecule. These are the only methanogens with the capacity to grow on acetate, methanol, and methylamines, with a few species of *Methanosarcina* also growing on H_2 and CO_2. These archaea tend to have relatively high growth yields of around 7 grams of dried cellular material per mole of CH_4 produced. By contrast, most members of the other four orders convert H_2 and CO_2 to methane; a few can also use formate as a methanogenic substrate. These autotrophic methanogens have growth yields of between

1.5 to 3 grams per mole CH_4. Given these better growth yields, it is not surprising that members of *Methanosarcinales* account for about two-thirds of archaeal methanogenesis.

Methanogenic archaea are potentially of great practical importance, as discussed in the chapter opening story. Anaerobic digesters use fermentative bacteria to degrade particulate wastes such as sewage sludge to H_2, CO_2, and acetate (*see figure 43.7*). CO_2-reducing methanogens then form CH_4 from CO_2 and H_2, while acetoclastic methanogens cleave acetate to CO_2 and CH_4. The tight thermodynamic relationship between the H_2-producing

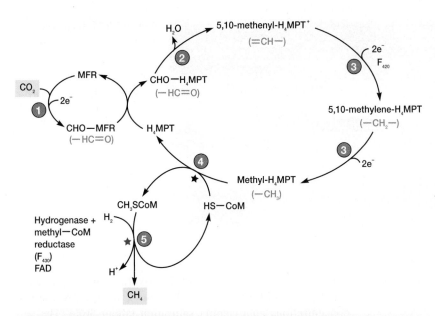

Figure 20.15 Methane Synthesis. Pathway for CH_4 synthesis from CO_2 in *M. thermoautotrophicus*. Cofactor abbreviations: methanopterin (MPT), methanofuran (MFR), and 2-mercaptoethanesulfonic acid or coenzyme M (CoM). The nature of the carbon-containing intermediates leading from CO_2 to CH_4 is indicated in parentheses. Stars indicate sites of energy conservation. The circled numbers correspond to steps described in text.

MICRO INQUIRY *What are the mechanisms by which methanogens are thought to couple CO_2 reduction to ATP generation? Why do you think this has been so hard to demonstrate?*

fermenting bacteria and methanogens is called interspecies hydrogen transfer and is discussed on page 693. A kilogram of organic matter can yield up to 600 liters of methane. It is quite likely that future innovation will greatly increase the efficiency of industrial methane production and make methanogenesis from organic waste an important source of energy. ▶▶▎ *Biofuel production is a dynamic field (section 42.2)*

Methanogenic archaea are estimated to produce about 1 billion tons of methane annually. In freshwater ecosystems, rates of methane production can be so great that bubbles of methane sometimes rise to the surface of lakes and ponds. Rumen methanogens are so active that a cow can belch 200 to 400 liters of methane a day. Methanogenesis is also an environmental problem. Methane absorbs infrared radiation and is a more potent greenhouse gas than CO_2. Atmospheric methane concentrations have been rising over the last 200 years. Methane production significantly promotes global warming; this subject is discussed in chapter 28. ▶▶▎ *The rumen ecosystem (section 32.1); Global climate change (section 28.2)*

Methanotrophy

No doubt the levels of archaeal-generated methane would render this planet uninhabitable were it not for the oxidation of methane by microbes capable of methanotrophy. For many decades, it was thought that this process—the capacity to oxidize methane and

use it as a carbon source—was restricted only to obligate aerobic proteobacteria. However, it is now recognized that about 90% of methane produced in marine sediments is oxidized anaerobically by archaea. The use of fluorescent probes for specific DNA sequences has revealed assemblages of archaea and sulfate-reducing bacteria inhabiting anoxic, methane-rich sediments. The best-studied assemblage is characterized by clusters of about 100 cells from the order *Methanosarcinales* surrounded by a layer of bacteria related to the proteobacterial genus *Desulfosarcina* (**figure 20.16**). Although no anaerobic methane-oxidizing archaeon has yet been grown in pure culture, additional assemblages have since been identified and grouped into three distinct lineages called anaerobic methanotrophic archaea (ANME)-1, ANME-2, and ANME-3. Those in ANME-1 are related to members in the order *Methanomicrobiales,* while those in ANME-2 and ANME-3 are related to *Methanosarcinales.* These microbes are most often found in regions of the seafloor where methane seeps into the surrounding cold water. They have also been detected in hydrothermal vents, anoxic seawater columns, soils, and freshwater habitats.

Because none of the ANME archaea have been cultivated, the mechanism of methane oxidation has been deciphered by metagenomic analysis. The ANME archaea appear to possess all the genes needed for methanogenesis, implying that anaerobic methane oxidation is accomplished by running methanogenesis backward. The demonstration that purified enzyme methyl-coenzyme M reductase can catalyze the energetically unfavorable conversion of methane to methyl-coenzyme M (denoted with a red star in figure 20.15) supports the notion that methane is the electron

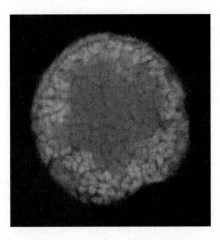

Figure 20.16 Methanotrophic Archaea Grow in Association with Sulfate-Reducing Bacteria. A cluster of methanotrophic archaea, stained red by an archaeal fluorescent 16S rRNA probe, surrounded by a layer of sulfate-reducing bacteria labeled by a green bacterial 16S rRNA fluorescent probe.

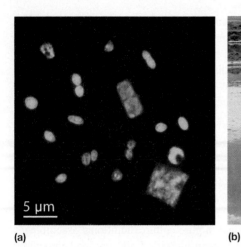

5 μm

(a)

(b)

Figure 20.17 Halophilic **Archaea.** (a) A sample taken from a saltern in Australia viewed by fluorescence microscopy. Note the range of cell shapes includes cocci, rods, and cubes. (b) A solar evaporation pond is extremely high in salt and mineral content. The archaea that dominate this hot, saline habitat produce brilliant red pigments.

donor. In the ANME consortia with sulfate-reducing bacteria, sulfate is the terminal electron acceptor:

$$CH_4 + SO_4^{2-} \rightarrow HCO_3^- + HS^- + H_2O \quad (\Delta G'^o = -17 \text{ kJ/mol})$$

It is not clear how electrons are transferred from the archaeon to the bacterium. The archaea provide the bacteria with much of their organic carbon. In addition, ANME archaea also fix N_2 gas and provide the bacteria with organic nitrogen.

The transfer of electrons from ANME archeae to sulfate-reducing *Deltaproteobacteria* appears to be the most common means by which methane is oxidized anaerobically. However, there have been some exceptions, including a freshwater ANME archaeon that couples the anaerobic oxidation of methane with bacterial denitrification—the reduction of nitrate to N_2 during anaerobic respiration. In addition, there have been at least two reports of ANME capable of anaerobic oxidation of methane without a bacterial partner. One involves an ANME-2 archaeon capable of independent methane oxidation because it possesses the genes for dissimilatory sulfate reduction. The other is also an ANME-2 archeaon, but it is capable of using nitrate as the terminal electron acceptor. It appears to have acquired the genes needed for nitrate reduction by horizontal gene transfer from bacteria.

Haloarchaea

The **extreme halophiles** include some bacteria, eukaryotic microorganisms, and **haloarchaea,** which belong to the order *Halobacteriales*. This euryarchaeal group currently has 17 genera in one family, *Halobacteriaceae*. These microbes received this unfortunate name before *Archaea* and *Bacteria* were recognized as different taxa. Most are aerobic chemoorganotrophs with respiratory metabolism. Extreme halophiles demonstrate a wide variety of nutritional capabilities. When grown in defined media, these archaea use carbohydrates or simple compounds such as glycerol, acetate, or pyruvate as their carbon source. Halophiles are found in a variety of cell shapes, including cubes and pyramids in addition to rods and cocci (**figure 20.17***a*). Some are motile.

The most obvious distinguishing trait of this family is its absolute dependence on a high concentration of NaCl, requir-

ing at least 1.5 M NaCl (about 8%, wt/vol). Growth optima are usually 3 to 4 M NaCl (17 to 23%), but some will grow at salt concentrations approaching saturation (about 36%). The cell walls of most halophiles are so dependent on the presence of NaCl that they disintegrate when the NaCl concentration drops below 1.5 M. Thus they only grow in high-salinity habitats such as marine salterns and salt lakes such as the Dead Sea between Israel and Jordan, and the Great Salt Lake in Utah. Haloarchaea often have red-to-yellow pigmentation from carotenoids that are probably used as protection against strong sunlight. They can reach such high population levels that salt lakes, called salterns turn red (figure 20.17*b*).

Halophiles use two strategies to cope with osmotic stress. The first approach is to increase cytoplasmic osmolarity by accumulating small organic molecules called **compatible solutes.** These include glycine, betaine, polyols, ectoine, and amino acids. When this strategy is used, salt adaptation of cytoplasmic proteins is not necessary. Compatible solutes are found in extremely halophilic methanogenic archaea, as well as halophilic bacteria. By contrast, members of the orders *Halobacteriales, Haloanaerobiales,* and *Salinibacter* employ a "salt-in" approach. These archaea use Na^+/H^+ antiporters and K^+ symporters to concentrate KCl and NaCl to levels equivalent to the external environment. Thus proteins need to be protected from salt denaturation and dehydration. Proteins from these microbes have evolved to possess a limited number of hydrophobic amino acids and a greater number of acidic residues. The acidic amino acids tend to be located on the surface of the folded protein, where they attract cations, which form a hydrated shell around the protein, thereby maintaining its solubility.

Probably the best-studied member of the family is *Halobacterium salinarum*. This archaeon is unusual because it produces a protein called **archaerhodopsin** (originally termed bacteriorhodopsin) that can harvest light energy without the presence of chlorophyll. Structurally similar to the rhodopsin found in the mammalian eye, archaerhodopsin functions as a light-driven proton pump. Like all members of the rhodopsin family, archaerhodopsin has two distinct features: (1) a chromophore that is a derivative of retinal (an aldehyde of vitamin A), which is covalently

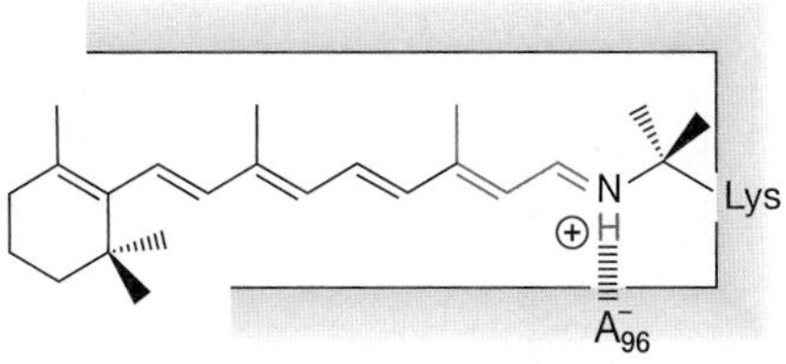

Figure 20.18 The Chromophore of Archaerhodopsin Is a Retinal Schiff Base.

attached to the protein by a Schiff base with the amino group of lysine (**figure 20.18**); and (2) seven membrane-spanning domains connected by loops on either side of the membrane with the retinal resting within the membrane. Archaerhodopsin molecules form aggregates in a modified region of the plasma membrane called the **purple membrane.** As shown in **figure 20.19,** this light-driven proton pump generates a pH gradient without an electron transport chain. Nonetheless, this gradient can be used to power the synthesis of ATP by a chemiosmotic mechanism. ◄◄ *Oxidative phosphorylation (section 11.6); Rhodopsin-based phototrophy (section 11.11)*

 H. salinarum has three additional rhodopsins, each with a different function. Halorhodopsin uses light energy to transport chloride ions into the cell and maintain a 4 to 5 M intracellular KCl concentration. The two additional rhodopsins are called sensory rhodopsin I (SRI) and SRII. **Sensory rhodopsins** act as photoreceptors; in this case, one for red light and one for blue. They control flagellar activity to position the organism optimally in the water column. *Halobacterium* cells move to a location of high light intensity that optimizes archaerhodopsin proton pumping but limits the lethal effects of ultraviolet light.

 Rhodopsin is widely distributed among bacteria as well as archaea. DNA sequence analysis of uncultivated marine bacterioplankton reveals the presence of rhodopsin genes among some α- and β-proteobacteria and bacteroidetes. This rhodopsin is called **proteorhodopsin.** Cyanobacteria also have sensory rhodopsin proteins, which like SRI and SRII of the haloarchaea, sense the spectral quality of light. All these bacterial and archaeal rhodopsins conserve the seven transmembrane helices through the cell membrane and the lysine residue that forms the Schiff base linkage with retinal. ►► *Microorganisms in the open ocean are adapted to nutrient limitation (section 30.2)*

 Genomic analysis of *Halobacterium* NRC-1 illustrates many of these physiologic strategies (figure 20.19). The use of both archaerhodopsin-mediated proton motive force and oxidative phosphorylation to drive ATP synthesis is shown, as is the import of cations to increase cytoplasmic osmolarity. The signals from sensory rhodopsins converge on the flagellar apparatus, along with input from 17 methyl-accepting chemotactic proteins. *Halobacterium* NRC-1 uses a variety of ATP transporters for the uptake of cationic amino acids, peptides, and sugars. Carbohydrates are oxidized via the phosphorylative variant of the Entner-Doudoroff pathway (figure 20.7) and the TCA cycle. Interestingly,

genes for exporting the toxic heavy metals arsenite and cadmium are present, as well as genes homologous to nonspecific multidrug-resistance exporters.

Thermoplasms

Archaea in the class *Thermoplasmata* are thermoacidophiles that lack cell walls. At present, three genera—*Thermoplasma, Picrophilus,* and *Ferroplasma*—are known. They are sufficiently different from one another to be placed in separate families: *Thermoplasmataceae, Picrophilaceae,* and *Ferroplasmataceae.*

 Members of the genus *Thermoplasma* grow in refuse piles of coal mines. These piles contain large amounts of pyrite (FeS), which is oxidized to sulfuric acid by chemolithotrophic bacteria. As a result, the piles become very hot and acidic. This is an ideal habitat for these archaea because they grow best at 55 to 59°C and pH 1 to 2. Although they lack cell walls, their plasma membranes are strengthened by large quantities of caldarchaeol, lipid-containing polysaccharides, and glycoproteins. At 59°C, *Thermoplasma* spp. take the form of irregular filaments, whereas at lower temperatures, they are spherical. The cells may be flagellated and motile.

 Picrophilus spp. are even more unusual than *Thermoplasma.* They have an S-layer outside the plasma membrane. The cells grow as irregularly shaped cocci, around 1 to 1.5 μm in diameter, and have large cytoplasmic cavities that are not surrounded by a membrane. They are aerobic and grow between 47 and 65°C with an optimum of 60°C. They have remarkable pH requirements, growing only below pH 3.5 with a growth optimum at pH 0.7. Growth even occurs at about pH 0.

Extremely Thermophilic S⁰-Reducers

This physiological group contains the class *Thermococci,* with one order, *Thermococcales.* These archaea are strictly anaerobic chemoorganoheterotrophs that use a variety of carbon substrates (e.g., peptides and carbohydrates) and reduce sulfur to sulfide. They are motile by flagella and have optimum growth temperatures around 88 to 100°C. The order contains one family and three genera, *Thermococcus, Paleococcus,* and *Pyrococcus.*

Sulfate-Reducing *Euryarchaeota*

Although many euryarchaea reduce sulfur during anaerobic respiration, only a few can use sulfate as a terminal electron acceptor. The sulfate reducers are found in the class *Archaeoglobi* and the order *Archaeoglobales.* This order has only one family and three genera. *Archaeoglobus* spp. are irregular coccoid cells with cell walls consisting of glycoprotein subunits. They can extract electrons from a variety of electron donors (e.g., H_2, lactate, glucose) and reduce sulfate, sulfite, or thiosulfate to sulfide. Elemental sulfur is not used as an acceptor. They are thermophilic (optimum growth is about 83°C) and have been isolated from marine hydrothermal vents. Although they are not considered methanogens, they use coenzymes F_{420} and methanopterin to produce tiny amounts of methane. Because they lack the genes encoding the critical enzyme methyl-CoM reductase (figure 20.15,

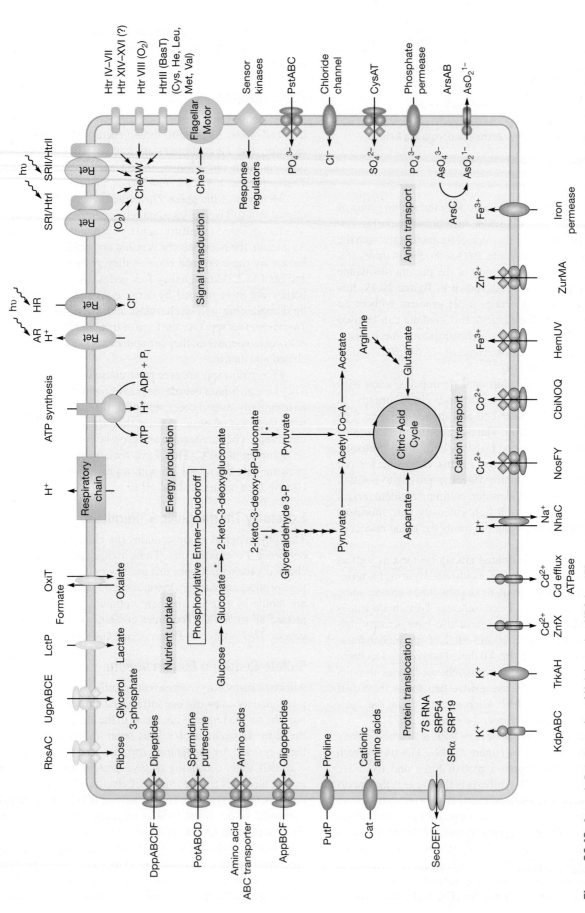

Figure 20.19 Genomic Reconstruction of *Halobacterium* NRC-1. ATP synthesis occurs through respiratory-dependent chemiosmosis and the ATP synthase, light-driven proton pumping by archaerhodopsin (AR, pink oval), and chloride transport by halorhodopsin (HR, blue oval). Nutrient uptake systems are shown in brown and yellow; names of transporters correspond with the gene(s) encoding them; the substrates transported are shown in the inside of the transporter. Cation transporters are shown in purple and anion transporters in red. Sensory rhodopsins (orange) and 17 light, oxygen, and amino acid receptors (Htr proteins) regulate chemotaxis and are thus shown linked to the flagellar motor through the chemotaxis protein CheY.

step 5), the mechanism by which methane is made remains unclear. Some species are autotrophic and, like methanogens, use the reductive acetyl-CoA pathway to assimilate CO_2.

A recently characterized thermophilic euryarchaeote that is not yet listed in *Bergey's Manual* bears mention. The isolated *Aciduliprofundum* strain, *Candidatus* Aciduliprofundum boonei is an acidophile that requires a pH of 3.3 to 5.8 and a temperature between 60 and 75°C for growth. Culture-independent analysis shows that members of this genus are often abundant in microbial communities growing at hydrothermal vents with actively venting sulfide deposits. It follows that A. boonei is a sulfur- and iron-reducing heterotroph. Its significance is twofold. It is the first thermoacidophile (as opposed to acid-tolerant) microbe to be isolated from environments rich in sulfide deposits. Also, its abundance at such sites implies that it is important in sulfur and iron cycling in these ecosystems. ▶▶◀ *Biogeochemical cycling sustains life on Earth (section 28.1)*

Retrieve, Infer, Apply

1. Why are the growth yields of most methanogens much lower than those of microbes that grow aerobically?
2. Why do you think methanogenesis requires so many cofactors? (Hint: What would happen to CO_2 and CH_3 if they were not bound to larger molecules?)
3. What is the ecological and practical importance of methanogens?
4. Where are extreme halophiles found, and what is unusual about their cell walls and growth requirements?
5. What is the difference between sensory rhodopsin and archaerhodopsin? Find examples of each in figure 20.19.
6. How is *Thermoplasma* able to live in acidic, very hot coal refuse piles when it lacks a cell wall?
7. In what way are members of the genus *Archaeoglobus* similar to methanogens, and how do they differ from other extreme thermophiles?

Key Concepts

20.1 Overview of *Archaea*

- *Archaea* contains a highly diverse group of microbes with respect to morphology, reproduction, physiology, and ecology. Although best known for their growth in anoxic, hypersaline, and high-temperature habitats, they also inhabit marine Arctic, temperate, and tropical waters.

- Archaea may be divided into six physiological groups: methanogenic archaea, sulfate reducers, extreme halophiles, cell wall-less archaea, extremely thermophilic S^0-metabolizers, and mesophilic ammonia oxidizers (**table 20.1**).

- The second edition of *Bergey's Manual* divides *Archaea* into two phyla, *Crenarchaeota* and *Euryarchaeota*; however, the phylum *Thaumarchaeota* was recently recognized (**figure 20.2**).

- Autotrophic archaea generally use one of three pathways for CO_2 assimilation: the reductive acetyl-CoA pathway, the 3-hydroxypropionate/4-hydroxybutyrate (HP/HB) cycle, or the dicarboxylate/4-hydroxybutyrate (DC/HB) cycle. Acetate is incorporated either by the glyoxylate or methylaspartate cycle. Most other anabolic pathways appear to be similar to those found in bacteria (**table 20.2; figures 20.3 to 20.5**).

- Although much of archaeal catabolism appears similar to that of bacteria, archaea differ with respect to glucose catabolism, using modified versions of the Embden-Meyerhof and Entner-Doudoroff pathways (**figures 20.6 and 20.7**).

20.2 Phylum *Crenarchaeota*: Metabolically Diverse Thermophiles

- The extremely thermophilic S^0-metabolizers in the phylum *Crenarchaeota* depend on sulfur for growth and are frequently acidophiles. The sulfur may be used as an electron acceptor in anaerobic respiration or as an electron donor by chemolithotrophs. Many are strict anaerobes and grow in geothermally heated soil and water that is rich in sulfur (**table 20.3**).

20.3 Phylum *Thaumarchaeota*: Mesophilic Ammonia Oxidizers

- The discovery of an archaeal-specific lipid, now called thaumarchaeol, led to the discovery of globally distributed marine, mesophilic archaea.

- These mesophiles are placed in the phylum *Thaumarchaeota* and are united by their capacity to oxidize ammonia. They are found in many marine and terrestrial habitats, where they can be important contributors to nitrification.

20.4 Phylum *Euryarchaeota*: Methanogens, Haloarchaea, and Others

- The phylum *Euryarchaeota* contains five major physiological groups: methanogens, haloarchaea, thermoplasms, extremely thermophilic S^0-reducers, and sulfate-reducing archaea.

- Methanogenic archaea are strict anaerobes that obtain energy through the synthesis of methane. They have several unusual cofactors that are involved in methanogenesis (**figures 20.14 and 20.15**).

- Anaerobic methane-oxidizing archaea belong to *Methanomicrobiales* and *Methanosarcinales,* and are commonly found in association with sulfate-reducing bacteria. These methanotrophs oxidize methane by reversing the methanogenic pathway.

- Extreme halophilic archaea are aerobic chemoorganotrophs that require at least 1.5 M NaCl for growth. They are found in habitats such as salterns and salt lakes.

- *Halobacterium salinarum* can carry out phototrophy without chlorophyll or bacteriochlorophyll by using archaerhodopsin, which employs retinal to pump protons across the plasma membrane (**figures 20.18** and **20.19**).

- The thermophilic *Thermoplasma* spp. grow in hot, acidic coal refuse piles and survive despite the lack of a cell wall. Another thermoplasm, *Picrophilus* spp. grow at pH 0.

- The class *Thermococci* contains extremely thermophilic organisms that can reduce sulfur to sulfide.

- The extreme thermophiles in the genus *Archaeoglobus* differ from other archaea in using a variety of electron donors to reduce sulfate. They also contain methanogen cofactors F_{420} and methanopterin.

Compare, Hypothesize, Invent

1. Some believe that archaea should not be separate from bacteria because both groups are prokaryotic. Do you agree or disagree? Explain your reasoning.

2. Explain why the fixation of CO_2 by *Thermoproteus* spp. using the DC/HB cycle is not photosynthesis.

3. When the temperature increases, some archaea change their shapes from elongated rods into spheres. Suggest one reason for this change.

4. Why would ether linkages be more stable in membranes than ester lipids? How would the presence of tetraether linkages stabilize a thermophile's membrane?

5. Suppose you wished to isolate an archaeon from a hot spring in Yellowstone National Park. How would you go about it?

6. Thermophilic archaea capable of ammonia oxidation to nitrite were discovered in enrichment cultures derived from microbial mats from the Siberian Garga hot spring. To demonstrate the presence of these microbes and their ability to oxidize ammonia, the authors did the following:

   - Screened for α- and β-proteobacterial 16S rRNA genes and the bacterial gene that encodes the key enzyme in ammonia oxidation (*see chapter 22*).

   - Screened for archaeal-specific 16S rRNA and archaeal-specific genes that encode the ammonia oxidation subunits AmoA and AmoB.

   - Constructed two phylogenetic trees—one based on the archaeal 16S rRNA genes and one based on the AmoA and AmoB amino acid sequences as compared to other archaeal AmoA and AmoB sequences.

   - Performed reverse-transcriptase PCR using mRNA extracted from the enrichment cultures. Primers were specifically targeted archaeal *amoA* nucleotide sequences.

   What was the rationale (i.e., what were the authors testing) for each procedure? The authors concluded that a thermophilic ammonia-oxidizing archaeote was responsible for the conversion of ammonia to nitrite in this system; what was the likely result of each test?

   Read the original paper: Hatzenpichler, R., et al. 2008. A moderately thermophilic ammonia-oxidizing crenarchaeote from a hot spring. *Proc. Natl. Acad. Sci. USA.* 105:2134–39.

7. The Mariana forearc, found in the western Pacific, is part of the deepest region of the world's oceans. Analysis of 16S rRNA sequences obtained from mud volcanoes located along the Mariana forearc shows an abundance of methane-oxidizing and sulfate-reducing phylotypes. Researchers report rRNA signature sequences from both *Euryarchaota* and *Crenarchaeota*. Using tables 20.1, 20.3, and 20.4, predict the genera to which these phylotypes might belong.

   Read the original paper: Curtis, A. C., et al. 2013. Mariana forearc serpentinite mud volcanoes harbor novel communities of extremophilic *Archaea*. *Geomicrobiol. J.* 30:430–441.

Learn More

shop.mheducation.com   Enhance your study of this chapter with interactive study tools and practice tests. Also ask your instructor about the resources available through Connect, including adaptive learning tools and animations.

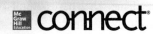

Deinococci, Mollicutes, and Nonproteobacterial Gram-Negative Bacteria

Biofuel production is an active area of microbiological research.

Cyanobacteria Stimulate Broad Appeal for Biofuel Production

Biofuels could be part of a comprehensive, global energy solution. They are renewable and can be produced domestically. But there is a downside. In the United States, most biofuel is made from corn, thereby co-opting thousands of acres of land for producing energy rather than food. This has the unintended consequence of driving food prices higher. Given that about 14% of the world's population is undernourished, this is not the best use of arable land.

Some scientists have turned from corn to cyanobacteria for an answer. Globally, there are a number of projects that seek to turn these Gram-negative bacteria, once known as blue-green algae, into a source of biofuel. One approach, used by researchers at Arizona State University's Biodesign Institute, involves harvesting lipids from the cyanobacteria that belong to the genus *Synechocystis*. These microbes can be genetically engineered to increase production of the kinds of lipids used in biodiesel. Their growth requirements are simple: CO_2, phosphorus, and trace elements. Because *Synechocystis* spp. fix atmospheric N_2, there is no need for petroleum-based nitrogenous fertilizer. And it gets even better: it may be possible to feed the cyanobacterium CO_2 generated by power plants, thereby reducing emissions of this greenhouse gas. Lastly, cyanobacterial lipid production is more than 100 times more efficient per acre than plants, and their growth need not eliminate land from food production.

Not far from Arizona, bioengineers at the University of Texas are using a different approach to capitalize on the benefits of cyanobacteria. Like others, they would like to replace corn as a biofuel source. The use of corn is expensive, involving enzymatic, chemical, and mechanical methods to break down cellulose into its glucose monomers, which is then fermented to ethanol (*see section 42.2*). They have developed a strain of the cyanobacterium *Oscillatoria* that secretes high levels of glucose, sucrose, and cellulose. Importantly, unlike plant cellulose, cyanobacterial cellulose is not mixed with lignin and other recalcitrant compounds, making it more easily fermented. Thus these cyanobacteria can easily provide the starting materials for biofuel fermentation and avoid the sacrifice of thousands of acres to the growth of a food product for biofuel generation.

These projects are only two of many under development around the world that aim to harness cyanobacteria for biofuel generation. In this chapter, we cover these microbes as well as nine other phyla of Gram-negative bacteria (**figure 21.1**). This includes a variety of morphologically and physiologically amazing microbes. From deinococci, which can withstand levels of ionizing radiation that would kill a human, to the spirochetes that cause syphilis, this chapter truly illustrates the depth of microbial diversity.

Readiness Check:
Based on what you have learned previously, you should be able to:

✔ Draw a Gram-negative bacterial cell wall and list its constituent parts (section 3.4)

✔ Differentiate between swimming and gliding motility (section 3.8)

✔ List environmental factors that influence microbial growth (section 7.4)

✔ Compare and contrast the growth and physiology of obligate aerobes, facultative anaerobes, and strict anaerobes (section 7.4)

✔ Describe the difference between heterotrophy and autotrophy (section 11.1)

✔ Define chemolithotrophy and identify an example of an electron donor and an acceptor (section 11.10)

✔ Compare and contrast oxygenic and anoxygenic phototrophy (section 11.11)

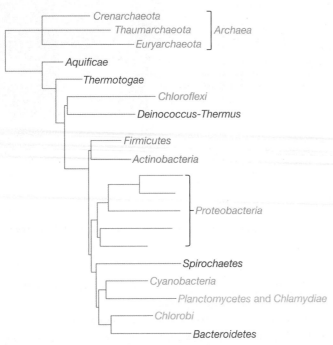

Figure 21.1 Phylogenetic Relationships Among *Bacteria* and *Archaea*. Nonproteobacterial Gram-negative bacteria are highlighted.

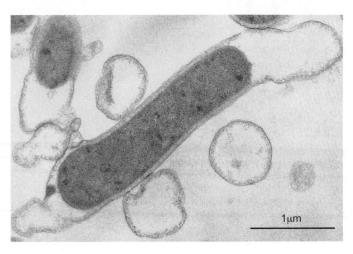

Figure 21.2 *Thermotoga maritima.* Note the loose sheath extending from each end of the cell.

21.1 *Aquificae* and *Thermotogae* Are Ancient Bacterial Lineages

After reading this section, you should be able to:

■ Compare and contrast the physiological and structural differences between members of the phyla *Aquificae* and *Thermotogae*

Thermophilic microbes are found in both the bacterial and archaeal domains, but to date, most hyperthermophilic microbes (those with optimum growth temperatures above 80°C) belong to *Archaea*. The phyla *Aquificiae* and *Thermotoga* are two examples of bacterial hyperthermophiles.

The phylum *Aquificae* is thought to represent the deepest, oldest branch of *Bacteria*. Two of the best-studied genera are *Aquifex* and *Hydrogenobacter. Aquifex* spp. are chemolithoautotrophs that capture energy by oxidizing hydrogen, thiosulfate, and sulfur with oxygen as the terminal electron acceptor. The reductive citric acid cycle is used to fix CO_2 (*see figure 12.5*). *Aquifex pyrophilus* is a microaerophilic rod that has been shown to grow with a temperature optimum of 85°C and a maximum of 95°C. Because the genera *Aquifex* and *Hydrogenobacter* contain thermophilic chemolithoautotrophic species, it has been suggested that the original bacterial ancestor was probably thermophilic and chemolithoautotrophic. ◄◄ *Chemolithotrophy (section 11.10)*

The second deepest branch is the phylum *Thermotogae*. Members of the genus *Thermotoga* (Greek *therme*, heat; Latin *toga*, outer garment), like *Aquifex*, are thermophiles with a growth optimum of 80°C and a maximum of 90°C. They are

Gram-negative rods with an outer sheathlike envelope (like a toga) that can extend or balloon out from the ends of the cell (**figure 21.2**). They grow in active geothermal areas found in marine hydrothermal systems and terrestrial solfataric springs. In contrast to *Aquifex, Thermotoga* species are chemoorganotrophs with a functional glycolytic pathway that can grow anaerobically using protons, elemental sulfur, or ferric iron as terminal electron acceptors. Like other members of the phylum *Thermotogae*, the genomes of *Thermotoga* spp. have been subject to horizontal gene transfer (HGT). Among this phylum, estimates of genes of archaeal origin range from 25% to 50%. ◄◄ *Horizontal gene transfer (section 16.4)*

21.2 *Deinococcus-Thermus* Includes Radiation-Resistant Bacteria

After reading this section, you should be able to:

■ Explain why members of *Deinococcus-Thermus* have erroneously been considered Gram positive

■ Describe habitats in which deinococci can be isolated

■ Discuss the unique capacity of deinococci to tolerate desiccation and high doses of radiation

The phylum *Deinococcus-Thermus* contains the orders *Deinococcales* (Greek, *deinos*, unusual) and *Thermales*. There are only three genera in the phylum. Most are mesophilic, but *Thermus aquaticus* was one of the first thermophilic organisms discovered. It grows optimally at 70°C, and thermostable "Taq" DNA polymerase is commonly used in polymerase chain reactions. Deinococci are spherical or rod-shaped and nonmotile, and they are often seen in pairs or tetrads (**figure 21.3a**). They are aerobic heterotrophs producing acid from only a few sugars. Although they stain Gram positive (and are sometimes referred to as such), their cell wall includes an outer membrane (figure 21.3b). They also differ from Gram-positive cocci in having

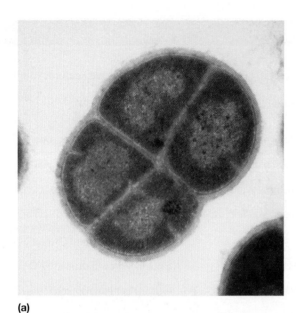

(a)

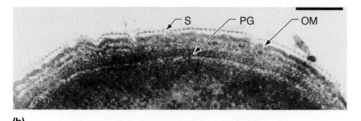

(b)

Figure 21.3 Deinococci. (a) A *Deinococcus radiodurans* tetracoccus, or cluster of four cells (average cell diameter 2.5 μm). (b) The cell wall of *D. radiodurans* with an S-layer (S), a peptidoglycan layer (PG), and an outer membrane (OM). Scale bar = 100 nm.

L-ornithine in their peptidoglycan, lacking teichoic acid, and having a plasma membrane with large amounts of palmitoleic acid rather than phosphatidylglycerol phospholipids. Their cell envelope also includes an S-layer. Almost all strains are extraordinarily resistant to all sources of oxidative stress, including desiccation and radiation; they can survive as much as 3 to 5 million rad of radiation. By contrast, exposure to 100 rad can be lethal to humans.

Deinococci can be isolated from ground meat, feces, air, freshwater, and other sources, but their natural habitat is not yet known. Their resistance to radiation results from their ability to repair a severely damaged genome, which consists of a circular chromosome, a megaplasmid, and a small plasmid. When exposed to high levels of radiation, radical oxygen species are generated and the genome is broken into hundreds of fragments. Amazingly, within 12 to 24 hours, the genome is pieced back together, ensuring viability. It is unclear how this is accomplished. Evidence suggests that while *Deinococcus radiodurans* DNA is not especially radiation resistant, its proteins are *(see figure 18.14)*. The ability to accumulate high levels of Mn(II) helps protect the microbe from damage caused by radiation-induced toxic oxygen species. Interestingly, manganese ions selectively protect proteins, rather than DNA. Thus, unlike other microbes, *D. radiodurans* DNA repair enzymes survive severe oxidative stress.

21.3 Class *Mollicutes*, Phylum *Tenericutes:* Bacteria That Lack Cell Walls

After reading this section, you should be able to:

- Relate the structure of the mollicute cell envelope to their morphology
- Outline how arginine can be catabolized to yield ATP by substrate-level phosphorylation
- Draw the flux of protons that enables *Ureaplasma urealyticum* to generate a proton motive force despite the absence of an electron transport chain
- Illustrate how gliding motility is accomplished by mycoplasmas
- List several habitats in which mollicutes reside and relate this to their capacity to be pathogenic

Until recently the class *Mollicutes* was placed in the phylum *Firmicutes.* Although it is generally agreed that members of this class are related to certain clostridia (low G + C Gram-positive bacteria), the fourth volume of *Bergey's Manual of Systematic Bacteriology* lists *Mollicutes* in the phylum *Tenericutes* (Latin *tener,* tender, and *cutis,* skin). *Tenericutes* includes all bacteria that lack walls and do not synthesize peptidoglycan precursors. Members of *Mollicutes* are commonly called **mycoplasmas;** these bacteria are characterized by the absence of cell walls, their small genomes, and simplified metabolic pathways. Their small genomes appear to be the result of genome reduction such that they now lack a variety of metabolic capabilities, including the ability to synthesize peptidoglycan precursors.

Some of the major characteristics of the best-studied mycoplasmas are summarized in **table 21.1.** Because they are bound only by a plasma membrane, they are pleomorphic, varying in shape from spherical or pear-shaped microbes from 0.3 to 0.8 μm in diameter, to slender branched or helical filaments (**figure 21.4**). Some mycoplasmas (e.g., *M. genitalium*) have a specialized terminal structure that projects from the cell and gives them a flask or pear shape. This structure aids in attachment to eukaryotic cells. They are among the smallest bacteria capable of self-reproduction. Most species are facultative anaerobes, but a few are obligate anaerobes. When grown on agar, most form colonies with a "fried egg" appearance because they grow into the agar surface at the center while spreading outward on the surface at the colony edges (**figure 21.5**).

Mollicute genomes are among the smallest found among bacteria, ranging from 0.7 to 1.7 Mb (table 21.1). The genomes of the human pathogens *Mycoplasma genitalium, M. pneumoniae,* and *Ureaplasma urealyticum* have fewer than 1,000 genes, suggesting a minimal genome size for a free-living existence. Their limited number of genes reflects their inability to synthesize a number of macromolecules. For instance, because *M. genitalium* relies so heavily on its host's biosynthetic capacity, it has lost genes that encode enzymes needed to synthesize amino acids, purines, pyrimidines, and fatty acids. In addition, most species require sterols for growth, which they obtain from the host as cholesterol. Sterols are an essential component of the mycoplasma plasma membrane,

Table 21.1	Properties of Some Members of the Class *Mollicutes*				
Genus	No. of Recognized Species	Genome Size (Mb)	Sterol Requirement	Habitat	Other Distinctive Features
Acholeplasma	13	1.50–1.65	No	Vertebrates, some plants and insects	Optimum growth 30–37°C
Anaeroplasma	4	1.50–1.60	Yes	Bovine or ovine rumen	Oxygen-sensitive anaerobes
Asteroleplasma	1	1.50	No	Bovine or ovine rumen	Oxygen-sensitive anaerobes
Entomoplasma	5	0.79–1.14	Yes	Insects, plants	Optimum growth 30°C
Mesoplasma	12	0.87–1.10	No	Insects, plants	Optimum growth 30°C; sustained growth in serum-free medium only with 0.04% detergent (Tween 80)
Mycoplasma	104	0.60–1.35	Yes	Humans, animals	Optimum growth usually 37°C
Spiroplasma	22	0.94–2.20	Yes	Insects, plants	Helical filaments; optimum growth at 30–37°C
Ureaplasma	6	0.75–1.20	Yes	Humans, animals	Urea hydrolysis

Adapted from J. G. Tully, et al., "Revised Taxonomy of the Class Mollicutes" in International Journal of Systematic Bacteriology, 43(2):378–85. Copyright © 1993 American Society for Microbiology, Washington, D.C.

Figure 21.4 Mycoplasmas. A scanning electron micrograph of *Mycoplasma pneumoniae* shows its pleomorphic nature (×26,000).

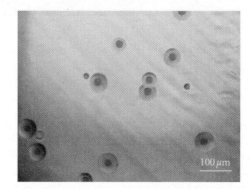

Figure 21.5 Mycoplasma Colonies. Note the "fried egg" appearance.

where they may facilitate osmotic stability in the absence of a cell wall. Some mycoplasmas produce ATP by the Embden-Meyerhof pathway and lactic acid fermentation. The pentose phosphate pathway seems to be functional in at least some mycoplasmas; none appears to have the complete tricarboxylic acid cycle. Arginine can be catabolized to generate ATP in the following series of reactions:

Arginine + H_2O → citrulline + NH_3
Citrulline + P_i → ornithine + carbamyl-P
Carbamyl-P + ADP → ATP + CO_2 + NH_3

The first reaction is catalyzed by the enzyme arginine deaminase, which is found in several firmicutes, including *Bacillus* and *Clostridium* species. ▶▶| Firmicutes: *Low G + C bacteria (chapter 23)*

The small size of the mycoplasma genome has made it the model organism for development of a "synthetic" bacterium. As discussed in chapter 17 (*see p. 412*), this has resulted in the construction of a bacterium with a genome that has been chemically synthesized. This work provides proof of principle that cells can be manufactured to encode only those products desired by the biotechnologist. |◀◀ *Techniques & Applications 17.2*

Ureaplasma urealyticum, a causative agent of urinary tract infections, has an interesting way of conserving energy: it hydrolyzes urea (found in urine) to generate an electrochemical gradient via the accumulation of ammonia/ammonium (NH_3/NH_4^+). This gradient is responsible for the chemiosmotic potential that drives ATP synthesis (**figure 21.6**).

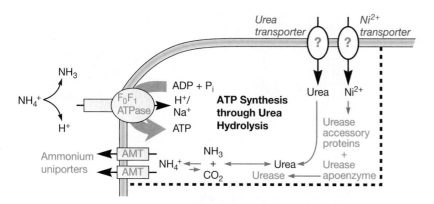

Figure 21.6 **Energy Conservation in *Ureaplasma urealyticum*.** To generate an electrochemical gradient through the accumulation of NH_3/NH_4^+, *U. urealyticum* is thought to import urea through specific transporters. Once inside the cell, urease (a nickel-binding enzyme) catalyzes the hydrolysis of urea to NH_3 and CO_2. Ammonia accepts a proton to become NH_4^+, which is exported through NH_4^+ uniporters (AMT). Once outside the cell, NH_4^+ is converted to NH_3 and H^+, providing the protons to drive ATP synthesis via the membrane-bound ATP synthase.

Some mycoplasmas are capable of gliding motility. Recall that gliding motility enables movement across solid surfaces. *Mycoplasma mobile* moves at a steady pace of 2 to 5 μm per second. This appears to be mediated by cell surface proteins that surround the "neck" of the cell (**figure 21.7**). In *M. mobile,* these proteins are thought to attach to the cytoskeleton and function like microscopic legs. Unlike flagellar motility, gliding in this microbe is powered by ATP hydrolysis. By contrast, most other motile mycoplasmas move more like inchworms, at a tenth of the speed of *M. mobile*. For instance the terminal structure of *M. pneumoniae* is covered with cytoskeletal proteins that may function as motors to cause the contraction and extension of the cell. ◄◄ *Twitching and gliding motility (section 3.8)*

Mycoplasmas are remarkably widespread and can be isolated from animals, plants, the soil, and even compost piles. Although their complex growth requirements can make their growth in pure (axenic) cultures difficult, at least 10% of the mammalian cell cultures in use are probably contaminated with mycoplasmas. This seriously interferes with tissue culture experiments and has resulted in the development of commercially available rapid tests for the detection of mycoplasmas in cell culture.

In animals, mycoplasmas colonize mucous membranes and joints, and often are associated with diseases of the respiratory and urogenital tracts. Mycoplasmas cause several major diseases in livestock; for example, contagious bovine pleuropneumonia in cattle (*M. mycoides*), chronic respiratory disease in chickens (*M. gallisepticum*), and pneumonia in swine (*M. hyopneumoniae*). Spiroplasmas have been isolated from insects, ticks, and a variety of plants. They cause disease in citrus plants, cabbage, broccoli, corn, honeybees, and other hosts. Arthropods often act as vectors and carry the spiroplasmas between plants. In humans, *U. urealyticum* and *M. hominis* are common parasitic microorganisms of the genital tract, and their transmission is related to sexual activity. Both mycoplasmas can opportunistically cause inflammation of the reproductive organs of males

and females. In addition, *Ureaplasma urealyticum* is associated with premature delivery of newborns, as well as neonatal meningitis and pneumonia. *M. pneumoniae* causes primary atypical pneumonia ("walking pneumonia") in humans. ►►I *Mycoplasmal pneumonia (section 39.1)*

Retrieve, Infer, Apply

1. In what ways are members of *Aquifex* and *Thermotoga* similar? How do they differ?
2. How do deinococci survive high doses of radiation?
3. Design an experiment designed to explore the role of sterols in mycoplasma structure and/or physiology.
4. Explain the relationship between genome size and growth requirements of mycoplasmas. Why is genome reduction common in intracellular microbes?

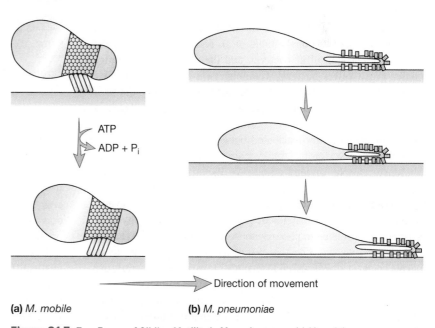

(a) *M. mobile* **(b)** *M. pneumoniae*

Figure 21.7 **Two Forms of Gliding Motility in Mycoplasmas.** (a) *M. mobile* uses cytoskeletal "leg" proteins to move the microbe at a constant speed. (b) *M. pneumoniae* has cytoplasmic motor proteins that cause the cell to expand and contract, so that it moves more slowly and at an uneven, inchwormlike pace.

21.4 Photosynthetic Bacteria Are Diverse

After reading this section, you should be able to:

- Assess the importance of photosynthetic pigments in the distribution of photosynthetic bacteria in nature
- Draw a generic chlorosome and identify the function of its structural elements
- Predict the habitat closest to your home where you might find members of the phyla *Chlorobi, Chloroflexi,* and *Cyanobacteria*
- Draw a generic cyanobacterial cell and label its intracellular structures
- List three types of specialized cells made by cyanobacteria and describe the function of each
- Compare and contrast the prochlorophytes with other cyanobacteria
- Draw a concept map describing the structure and physiology of the six types of photosynthetic bacteria listed in table 21.2

Historically, the term photosynthetic has been applied to both photoheterotrophic and photoautotrophic microbes. We continue to use this term in this way, but care should be taken not to assume that the capture of light energy (phototrophy) is always linked with CO_2 fixation (autotrophy). As we shall see, unlike plants, photosynthetic microbes often have more than one metabolic strategy and may uncouple phototrophy from autotrophy and grow photoheterotrophically.

Bacteria capable of chlorophyll-based photosynthesis have been identified in seven phyla, with the most recent discovered in 2014 when a chlorophyll-containing photoheterotrophic, Gram-negative bacterium belonging to the phylum *Gemmatimonadetes* was isolated from a freshwater lake in the Gobi Desert. However, because volume 1 of *Bergey's Manual* was published in 2001, it distributes distributes phototrophic bacteria between five bacterial phyla (table 21.2). The phylum *Chloroflexi* contains green nonsulfur bacteria, and the phylum *Chlorobi* includes green sulfur bacteria. Cyanobacteria are placed in their own phylum, *Cyanobacteria*. Purple bacteria are divided between three groups. Purple sulfur bacteria are placed in γ-proteobacteria (families *Chromatiaceae* and *Ectothiorhodospiraceae*). Purple nonsulfur bacteria are distributed between α-proteobacteria and β-proteobacteria. *Aerobic Anoxygenic Phototrophs* (AAnPs) are photoheterotrophs divided among α-, β-, and γ-proteobacteria. Finally, Gram-positive heliobacteria in the phylum *Firmicutes* are also photosynthetic. There appears to have been considerable horizontal transfer of photosynthetic genes among the five phyla. At least 50 genes related to photosynthesis are common to all five. In this chapter, we describe the cyanobacteria and green bacteria; purple bacteria are discussed in chapter 22, while heliobacteria are discussed in chapter 23.

Cyanobacteria differ fundamentally from the other photosynthetic bacteria because they perform **oxygenic photosynthesis:** they have photosystems I and II, use water as an electron donor, and generate oxygen during photosynthesis. In contrast, purple bacteria, green bacteria, and AAnPs have only one photosystem and use **anoxygenic photosynthesis.** Because they are unable to

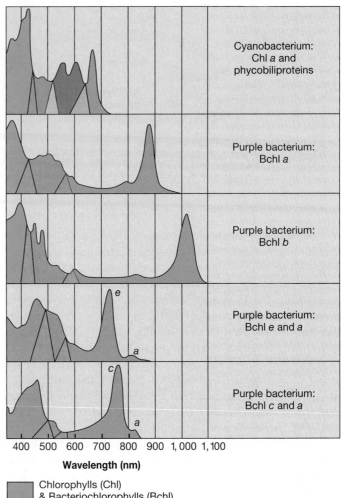

Figure 21.8 Photosynthetic Pigments. Absorption spectra of five photosynthetic bacteria showing the differences in absorption maxima and the contributions of various accessory pigments.

use water as an electron source, they employ other reduced molecules as electron donors. AAnPs and purple nonsulfur bacteria are photoorganoheterotrophs, so they use electrons obtained during the oxidation of organic substrates but make ATP from light energy. The purple and green sulfur bacteria use inorganic electron donors such as hydrogen sulfide, sulfur, and hydrogen, as well as organic matter, as an electron source for the reduction of $NAD(P)^+$ to $NAD(P)H$. Those that use sulfide as an electron source often form sulfur granules. Purple sulfur bacteria accumulate granules within their cells, whereas green sulfur bacteria deposit the sulfur granules outside their cells. ◄◄ *Phototrophy (section 11.11)*

Photosynthetic microbes are major contributors of fixed carbon, termed primary productivity, in a variety of habitats. The differences in photosynthetic pigments and oxygen requirements among photosynthetic bacteria have important ecological consequences. As shown in **figure 21.8**, chlorophylls

Table 21.2 Characteristics of the Major Groups of Gram-Negative Photosynthetic Bacteria

Characteristic	Green Sulfur	Green Nonsulfur	Purple Sulfur	Purple Nonsulfur	Aerobic Anoxygenic Phototrophs	Cyanobacteria
Major photosynthetic pigments	Bacteriochlorophyll a plus c, d, or e	Bacteriochlorophylls a and c	Bacteriochlorophyll a or b	Bacteriochlorophyll a or b	Bacteriochlorophyll a	Chlorophyll a plus phycobiliproteins. *Prochlorococcus* has divinyl derivatives of chlorophyll a and b
Morphology of photosynthetic membranes	Photosynthetic system partly in chlorosomes	Chlorosomes present when grown anaerobically	Photosynthetic system contained in spherical or lamellar membrane complexes that are continuous with the plasma membrane	Photosynthetic system contained in spherical or lamellar membrane complexes that are continuous with the plasma membrane	Few, if any, intracytoplasmic membranes	Thylakoid membranes lined with phycobilisomes
Photosynthetic electron donors	H_2, H_2S, S^0	Photoheterotrophic donors—a variety of sugars, amino acids, and organic acids; photoautotrophic donors—H_2S, H_2	H_2, H_2S, S^0	Usually organic molecules: sometimes reduced sulfur compounds or H_2	A variety of sugars, amino acids, and organic acids	H_2O
Sulfur deposition	Outside of the cell	N/A[1]	Inside the cell[2]	Outside of the cell in a few cases	N/A	N/A
Nature of phototrophy	Anoxygenic	Anoxygenic	Anoxygenic	Anoxygenic	Anoxygenic	Oxygenic (some are also facultatively anoxygenic)
General metabolic type	Obligate anaerobic photolithoautotrophs	Usually photoheterotrophic; sometimes photoautotrophic or chemoheterotrophic (when aerobic and in the dark)	Obligate anaerobic photolithoautotrophs	Usually anaerobic photoorgano-heterotrophs; some facultative photolithoautotrophs; some are chemo-organoheterotrophs in the dark	Aerobic photoorgano-heterotrophs	Aerobic photo-lithoautotrophs
Motility	Many nonmotile; some glide; some have gas vesicles	Gliding	Motile with polar or peritrichous flagella	Motile with polar flagella or nonmotile; some have gas vesicles	Some are motile with a single or a few polar or subpolar flagella	Nonmotile, swimming motility without flagella; some glide; some have gas vesicles
Taxonomic distribution	*Chlorobi*	*Chloroflexi*	γ-proteobacteria	α-proteobacteria, β-proteobacteria (*Rhodocyclus*)	α-proteobacteria, β-proteobacteria, γ-proteobacteria	*Cyanobacteria*

1 N/A: Not applicable.
2 With the exception of *Ectothiorhodospira*.

(Chl), bacteriochlorophylls (Bchl), and their associated accessory pigments have narrow absorption spectra. Microbes therefore use accessory pigments to broaden their energy input. These pigments collect light of different wavelengths and transfer it to Chl or Bchl, where it is converted to chemical energy. Oxygenic cyanobacteria and diatoms dominate the aerated upper layers of freshwater and marine microbial communities, where they absorb large amounts of red and blue light. Anoxygenic purple and green photosynthetic bacteria inhabit the deeper anoxic zones that are rich in hydrogen sulfide and other reduced compounds that can be used as electron donors. Their Bchl and accessory pigments enable their use of light in the far-red spectrum (i.e., > 750 nm) that is not used by other photosynthetic organisms (compare the Bchl a and Bchl b with Chl a in figure 21.8). Bchl absorption has additional peaks between 350 to 550 nm, allowing growth at greater depths because shorter-wavelength light can penetrate water farther.

Before we describe the established phyla, it is worth mentioning an interesting chlorophyll-containing thermophilic bacterium in the phylum *Acidobacteria* discovered in Yellowstone National Park hot springs in 2007. This Gram-negative bacterium contains bacteriochlorophylls a and c under oxic conditions and grows photoheterotrophically. The discovery of 16S rRNA sequences in Thailand and Tibet similar to those found at Yellowstone suggests a global distribution of this thermophile. The phylum *Acidobacteria* is small, but it may be far more diverse than previously imagined.

Indeed, the number of phototrophic microbes may exceed our long-held beliefs.

Phylum *Chlorobi*: Green Sulfur Bacteria

Members of the phylum *Chlorobi* are commonly called **green sulfur bacteria.** This is a small group of obligately anaerobic photolithoautotrophs that use hydrogen sulfide, elemental sulfur, and hydrogen as electron sources. When sulfide is oxidized, elemental sulfur is deposited outside the cell. Their photosynthetic pigments are located in ellipsoidal vesicles called **chlorosomes** or chlorobium vesicles, which are attached to the plasma membrane by a proteinaceous baseplate that contains Bchl a molecules (**figure 21.9**). The chlorosome membrane is an unusual lipid monolayer, rather than bilayer. Within chlorosomes, bacteriochlorophyll molecules (Bchl c, d, or e) are grouped into rodlike structures held together by carotenoids and lipids. The light harvested by these pigments is transferred by baseplate Bchl a to reaction centers in the plasma membrane, which also contain Bchl a. The ATP produced by photophosphorylation is used to fuel CO_2 fixation by the reductive citric acid cycle (*see figure 12.5*); the more familiar Calvin-Benson cycle is used only by aerobes.

These bacteria flourish in the anoxic, sulfide-rich zones of lakes. Although they lack flagella and most are nonmotile, members of at least one genus glides. Some species have gas vesicles to adjust their depth for optimal light and hydrogen sulfide (*see figure 3.36*). Those without gas vesicles are found in sulfide-rich muds at the bottom of lakes and ponds.

Green sulfur bacteria are very diverse morphologically. They may be rods, cocci, or vibrios; some grow singly, and others form chains and clusters. They are either grass-green or chocolate-brown in color. Representative genera are *Chlorobium, Prosthecochloris,* and *Pelodictyon.*

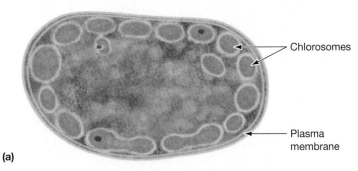

Chlorosomes

Plasma membrane

(a)

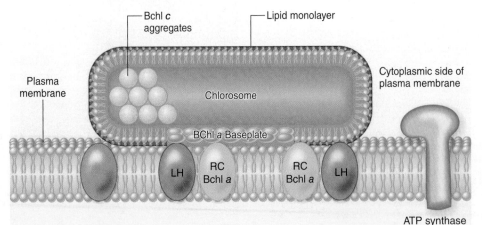

Bchl c aggregates

Lipid monolayer

Plasma membrane

Chlorosome

Cytoplasmic side of plasma membrane

BChl a Baseplate

LH

RC Bchl a

RC Bchl a

LH

ATP synthase

(b)

Figure 21.9 Green Sulfur Bacteria.
(a) Light-harvesting chlorosomes surround the inside perimeter of the cytoplasmic membrane. Antenna bacteriochlorophylls (Bchl c, d, or e) transfer light energy to reaction center Bchl a in the plasma membrane. (b) A chlorosome is attached to the inner surface of the plasma membrane by a baseplate composed of a lipid monolayer and Bchl a molecules. The chlorosome absorbs light by Bchl c, d, or e molecules aligned in rod-shaped complexes (Bchl c is shown here). The reaction center (RC) is located in the plasma membrane and contains only Bchl a. The light-harvesting complex (LH) and ATP synthase are also located in the plasma membrane.

Phylum *Chloroflexi*: Green Nonsulfur Bacteria

The phylum *Chloroflexi* is a deep and ancient branch of the bacterial tree. It has two orders: one consists of thermophilic, facultatively photosynthetic bacteria (*Chlorflexales*), and the other consists of nonphotosynthetic thermophiles and mesophiles (*Herpetosiphonales*). Members of the genus *Chloroflexus* make up the majority of photosynthetic **green nonsulfur bacteria.** However, the term green nonsulfur is a misnomer because not all members of this group are green, and some can use sulfide. *Chloroflexus* spp. are filamentous, gliding, bacteria often isolated from neutral to alkaline hot springs, where they grow in the form of orange-reddish mats, usually in association with cyanobacteria. They possess small chlorosomes with accessory Bchl *c*. The light-harvesting complexes contain Bchl *a* and are in the plasma membrane. *Chloroflexus* spp. are metabolically very diverse. Some are facultative anaerobes, others are anaerobic, and while some are chemoorganotrophic, most are photoheterotrophic. They can carry out anoxygenic photosynthesis using sulfide or hydrogen as a source of electrons. Heterotrophic growth is supported by pyruvate, acetate, glycerol, and glucose oxidation. When growing autotrophically, they use an unusual pathway, the 3-hydroxypropionate bi-cycle to fix CO_2 **(figure 21.10)**. An interesting aspect of this pathway is that the left half produces glyoxylate, which is then used in the right half to make pyruvate.

Members of the nonphotosynthetic, gliding, rod-shaped, or filamentous genus *Herpetosiphon* are aerobic chemoorganotrophs with respiratory metabolism. They can be isolated from freshwater and soil habitats.

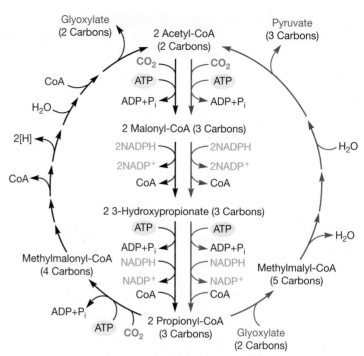

Figure 21.10 The 3-Hydroxypropionate Bi-Cycle. This pathway functions when photosynthetic members of the phylum *Chloroflexi* grow autotrophically; CO_2 exists as HCO_3^- when it is added to intermediates of the pathway.

Retrieve, Infer, Apply

1. Which of the phototrophs discussed perform anoxygenic photosynthesis? Which are also capable of chemoorganoheterotrophy?
2. What is the function of chlorosomes? How does their unique structure contribute to this function?
3. Compare the capacity of green nonsulfur (*Chloroflexi*) and green sulfur bacteria (*Chlorobi*) to position themselves in light of optimum wavelength.

Phylum *Cyanobacteria*: Oxygenic Photosynthetic Bacteria

Cyanobacteria are the largest and most diverse group of photosynthetic bacteria. Originally studied by botanists, they were classified as *Chlorophyceae,* erroneously known as blue-green algae. *Bergey's Manual* describes 56 genera and divides cyanobacteria into five subsections **(table 21.3)**. At present all taxonomic schemes must be considered tentative, and while many genera have been assigned, species names have not yet been designated in most cases.

Recall that endosymbiotic cyanobacteria are thought to have evolved into chloroplasts. It follows that the cyanobacterial photosynthetic apparatus closely resembles that of eukaryotes; cyanobacteria have chlorophyll *a* and photosystems I and II, and thereby perform oxygenic photosynthesis. Light-harvesting and electron transport chain components are located in thylakoid membranes to which particles called **phycobilisomes** are attached **(figure 21.11)**. Phycobilisomes are protein complexes made up of **phycobiliprotein** pigments, particularly **phycocyanin** and **phycoerythrin,** held together by linker proteins (figure 21.11*b*). Each phycobilisome consists of a core phycobiliprotein pigment, allophycocyanin, surrounded by other pigments arranged in stacklike structures. Each pigment absorbs light of different wavelengths, and this arrangement enables the efficient transfer energy to photosystem II. ◄◄ *Evolution of the three domains of life (section 19.5)*

Carbon dioxide is assimilated through the Calvin-Benson cycle (*see figure 12.4*). The enzymes needed for this process are found in internal structures (microcompartments) called **carboxysomes.** During daylight hours, when cyanobacteria fix carbon for anabolic purposes, they also synthesize **glycogen** as a reserve carbohydrate to supply electrons needed for respiration at night. Some cyanobacteria also store extra nitrogen as polymers of arginine and aspartic acid called **cyanophycin** granules. Phosphate is stored in polyphosphate granules. Because cyanobacteria lack the enzyme α-ketoglutarate dehydrogenase (α-KDH), it was long thought that they lack a fully functional TCA cycle. Recently two enzymes that bypass the need for α-KDH were identified in a unicellular cyanobacterium belonging to the genus *Synechococcus*. In this microbe, and most cyanobacteria, the first enzyme converts α-ketoglutarate to succinic semi-aldehyde, which is then oxidized to succinate by the second enzyme; the intermediate succinyl-CoA is not produced (*see figure 11.8*) ◄◄ *Microcompartments (section 3.6)*

Table 21.3   Characteristics of the Cyanobacterial Subsections

Subsection	General Shape	Reproduction and Growth	Heterocysts	Other Properties	Representative Genera
I	Unicellular rods or cocci; nonfilamentous aggregates	Binary fission, budding	−	Nonmotile or swim without flagella	*Chroococcus* *Gleocapsa* *Prochlorococcus* *Synechococcus*
II	Unicellular rods or cocci; may be held together in aggregates	Multiple fission to form baeocytes	−	Some baeocytes are motile.	*Pleurocapsa* *Dermocarpella* *Chroococcidiopsis*
III	Filamentous, unbranched trichome with only vegetative cells	Binary fission in a single plane, fragmentation	−	Usually motile	*Lyngbya* *Oscillatoria* *Prochlorothrix* *Spirulina* *Pseudanabaena*
IV	Filamentous, unbranched trichome may contain specialized cells	Binary fission in a single plane, fragmentation to form hormogonia	+	Often motile, may produce akinetes	*Anabaena* *Cylindrospermum* *Nostoc*
V	Filamentous trichomes either with branches or composed of more than one row of cells	Binary fission in more than one plane, hormogonia formed	+	May produce akinetes; greatest morphological complexity and differentiation in cyanobacteria	*Fischerella* *Stigonema* *Geitleria*

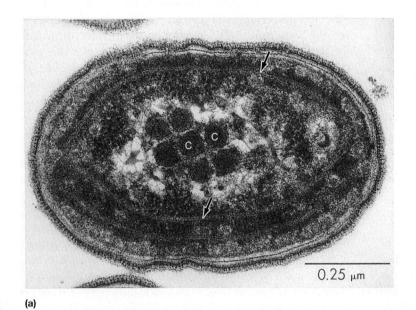

(a)

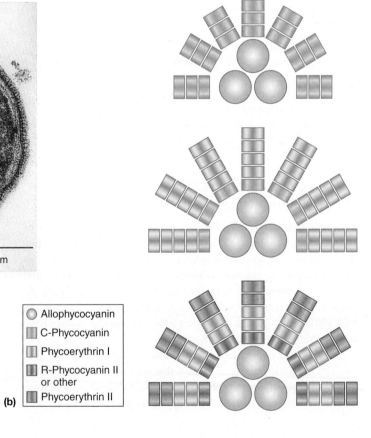

Legend:
- ○ Allophycocyanin
- ▦ C-Phycocyanin
- ▦ Phycoerythrin I
- ▦ R-Phycocyanin II or other
- ▦ Phycoerythrin II

(b)

Figure 21.11 Cyanobacterial Thylakoids and Phycobilisomes.
(a) Marine *Synechococcus* with thylakoids (arrows). Carboxysomes are labeled c.
(b) Examples of pigment arrangements in phycobilisomes. Phycobilisomes are found in all cyanobacteria except prochlorophytes.

(a) *Oscillatoria*

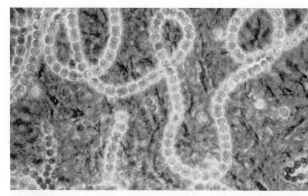

(c) *Nostoc*

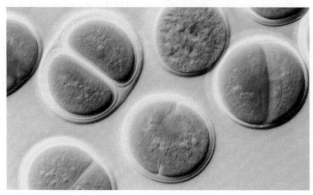

(b) *Chroococcus turgidus*

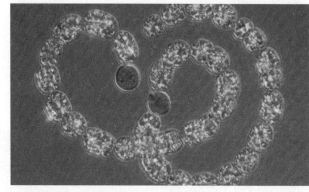

(d) *Anabaena spiroides*

Figure 21.12 Cyanobacteria Are Morphologically Diverse.   (a) *Oscillatoria* trichomes seen with Nomarski interference-contrast optics (×250). (b) *Chroococcus turgidus* produces a gelatinous sheath that can surround one, two, or four cells. (c) *Nostoc* sp. produces heterocysts in the absence of combined nitrogen; here it has only vegetative cells. (d) The cyanobacteria *Anabaena spiroides* with heterocysts. (×1,000).

Although most cyanobacteria are obligate photolithoauto-trophs, a few can grow in the dark as chemoheterotrophs by oxidizing glucose and a few other sugars. Under anoxic conditions, *Oscillatoria limnetica* oxidizes hydrogen sulfide instead of water and carries out anoxygenic photosynthesis much like green photosynthetic bacteria. *Oscillatoria* is one of several cyanobacterial genera whose members produce geosmins, volatile organic compounds that often have an earthy odor. When populations of these cyanobacteria become dominant in lakes and rivers, this can be a public nuisance, even changing the taste of drinking water. ▶▶❘ *Microorganisms in marine and freshwater ecosystems (chapter 30)*

Cyanobacteria also vary greatly in shape and appearance. They range in diameter from about 1 to 10 μm and may be unicellular, exist as colonies of many shapes, or form filaments called trichomes (**figure 21.12a**). A **trichome** is a row of bacterial cells that are in close contact with one another over a large area. Although many cyanobacteria appear blue-green because of phycocyanin, isolates from the open ocean are red or brown because their phycobilisomes contain the pigment phycoerythrin (figure 21.11*b*). Because each pigment evolved to absorb light of different wavelengths, cyanobacteria modulate the relative amounts of these pigments in response to available light in a process known as **chromatic adaptation.** When orange light is sensed, the production of phycocyanin (absorption

peak at 618 nm) is stimulated, whereas blue and blue-green light promote the production of phycoerythrin (absorption peaks at 490, 546, and 576 nm). Chromatic adaptation means that cyanobacteria must be able to sense light quality, and they use a variety of strategies to do so, including sensory rhodopsins (*see p. 479*), plantlike proteins called phytochromes, and membrane-associated receptors that sense ultraviolet and blue light. Many cyanobacterial species use gas vacuoles to position themselves in optimum illumination in the water column—a form of **phototaxis.** Gliding motility is used by many filamentous cyanobacteria, including those found in dense microbial assemblages in sediments called microbial mats (*see figure 29.10*). Although cyanobacteria lack flagella, about one-third of the marine *Synechococcus* strains swim at rates up to 25 μm per second. Swimming motility is not used for phototaxis; instead, it appears to be used for chemotaxis toward simple nitrogenous compounds such as urea. The mechanisms for both gliding and swimming motility in cyanobacteria remain unresolved. ◀◀ *Chemotaxis (section 3.8)*

Cyanobacteria show great diversity with respect to reproduction and employ a variety of mechanisms: binary fission, budding, fragmentation, and multiple fission. In the last process, a cell enlarges and then divides several times to produce many smaller progeny called **baeocytes** (*see figure 7.2*), which are released upon the rupture of the parental cell. Some baeocytes disperse through gliding motility. Fragmentation of filamentous cyanobacteria can generate small, motile filaments called **hormogonia.** Some filamentous species develop **akinetes,** specialized, dormant, thick-walled resting cells that are resistant

50 μm

Jason Oyadomari

Akinetes   Heterocyst

(a)

(b)

O₂ CO₂   N₂   CO₂ O₂

CH_2O NH_3 CH_2O

Glutamate

Vegetative   Heterocyst   Vegetative
cells   cells

(c)

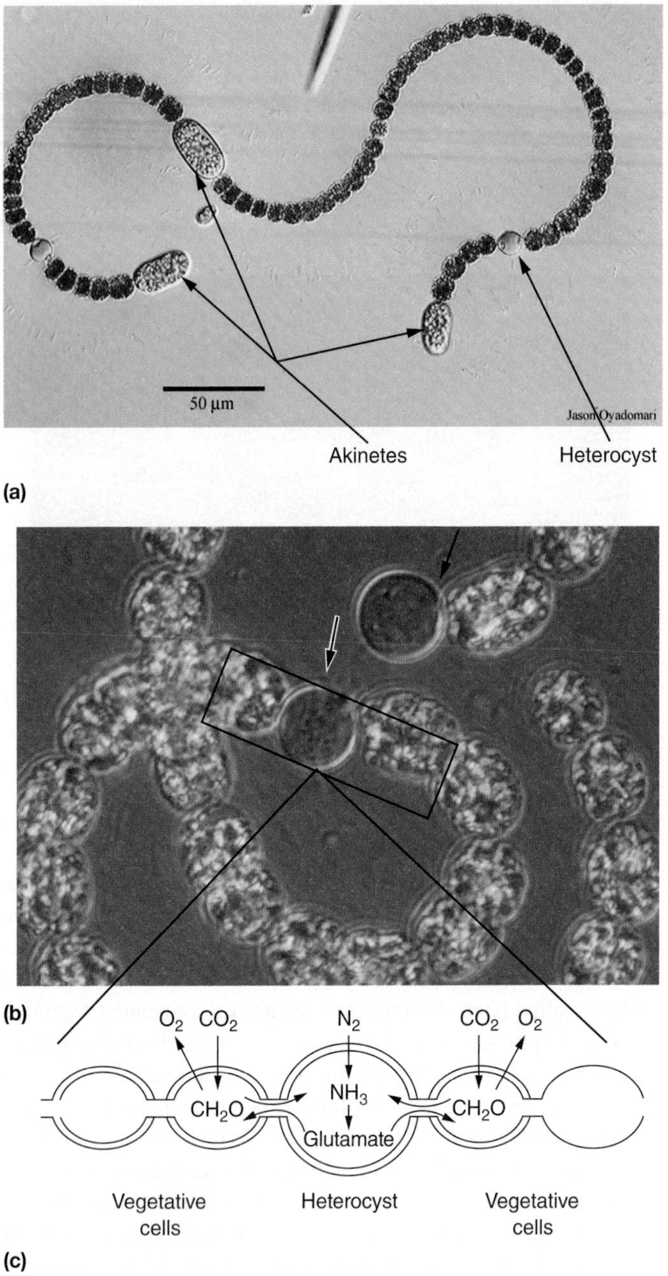

Figure 21.13 Cyanobacteria with Akinetes and Heterocysts.
(a) The filamentous cyanobacterium *Anabaena* with akinetes and heterocysts. (b) *Anabaena* heterocysts (arrow) and vegetative cells. (c) Heterocysts fix dinitrogen to form glutamate, which is exchanged with adjacent vegetative cells for carbohydrate. Heterocysts use only photosystem I, so photosynthesis is anoxygenic in these differentiated cells.

MICRO INQUIRY *Why do you think NH₃ is converted to the amino acid glutamate in heterocysts before its transfer to neighboring vegetative cells?*

to desiccation (**figure 21.13a**). These later germinate to form new filaments.

Many filamentous cyanobacteria fix atmospheric nitrogen by means of special cells called **heterocysts** (figure 21.13). Around 5 to 10% of the cells develop into heterocysts when these cyanobacteria are deprived of both nitrate and ammonia, their preferred nitrogen sources. Within these specialized cells, photosynthetic membranes are reorganized and the proteins that make up photosystem II and phycobilisomes are degraded. Photosystem I remains functional to produce ATP, but no oxygen is generated. Lack of O_2 production is critical because the N_2-fixing enzyme nitrogenase is extremely oxygen sensitive. Heterocysts develop thick cell walls, which slow or prevent O_2 diffusion into the cell, and any O_2 present is consumed during respiration. Heterocyst structure and physiology ensure that these specialized cells remain anaerobic; the heterocyst is dedicated to nitrogen fixation, not reproduction. Nutrients are obtained from adjacent vegetative cells, while the heterocysts contribute fixed nitrogen in the form of amino acids. Nitrogen fixation also is carried out by some cyanobacteria that lack heterocysts. Some fix nitrogen under dark, anoxic conditions in microbial mats. Planktonic forms such as those belonging to the genus *Trichodesmium* fix nitrogen and contribute significantly to the marine nitrogen budget *(see figure 30.9).* ◄◄ *Nitrogen fixation (section 12.5)* ►► *Nitrogen cycle (section 28.1)*

An important group of cyanobacteria is collectively referred to as **prochlorophytes** (genera *Prochloron, Prochlorococcus,* and *Prochlorothrix*). These microbes differ from other cyanobacteria by the presence of chlorophyll *b* as well as *a*, and the lack of phycobilins. This pigment arrangement imparts a grass-green color. As the only bacteria to possess chlorophyll *b*, the ancestors of unicellular prochlorophytes are considered the best candidates for the endosymbionts that gave rise to chloroplasts. ◄◄ *Evolution of the three domains of life (section 19.5)*

The three recognized prochlorophyte genera are quite different from one another. Members of the genus *Prochloron* were first discovered as symbionts growing either on the surface or within the cloacal cavity of marine colonial ascidian invertebrates (commonly called "sea squirts"). These bacteria are single-celled, spherical, and 8 to 30 μm in diameter. By contrast, those in the genus *Prochlorothrix* are free living, with cylindrical cells that form filaments growing in freshwater.

Prochlorococcus spp. are tiny coccobacilli less than 1 μm in diameter that flourish in marine plankton. They differ from other prochlorophytes in having divinyl chlorophyll *a* and *b*, and α-carotene, instead of chlorophyll *a* and β-carotene. Other cyanobacteria use monovinyl chlorophyll *a*; the difference between mono- and divinyl chlorophyll is the presence of a double bond between carbons attached to pyrrole ring II *(see figure 11.29)*. *Prochlorococcus marinus* is globally distributed in the ocean between the latitudes 40°N and 40°S and to 200 meters in depth, making it the most abundant photosynthetic organism on Earth.

Prochlorococci have a profound effect on the global carbon cycle. Together with members of the genus *Synechococcus,*

Figure 21.14 Bloom of Cyanobacteria and Algae in a Eutrophic Pond.

Prochlorococcus accounts for at least one-third of global CO_2 fixation. These microbes have coevolved to take maximum advantage of incident light. *Synechococcus* spp. occupy the upper 25 meters of the open ocean, where white and blue-green light can be harvested. In the waters beneath *Synechococcus,* two ecotypes of *Prochlorococcus* species exist: the high-light ecotype grows between about 25 to 100 meters, while the low-light ecotype lives from 80 to 200 meters, where the waters are penetrated by blue-violet light. This prevents the two genera of cyanobacteria from competing for the essential resource of light. ◄◄ *Microbial evolutionary processes (section 19.5)* ►► *Microorganisms in the open ocean are adapted to nutrient limitation (section 30.2)*

In addition to those in coastal and open ocean waters, many other cyanobacteria inhabit a wide range of other habitats. A variety of cyanobacteria live in symbiotic association with protists and fungi. Thermophilic species may grow at temperatures up to 75°C in neutral to alkaline hot springs. Because these photoautotrophs are so hardy, they are primary colonizers of soils and surfaces that are devoid of plant growth. Some unicellular forms even grow in the fissures of desert rocks. In nutrient-rich warm ponds and lakes, filamentous cyanobacteria such as *Anacystis* and *Anabaena* spp. can reproduce rapidly to form blooms (**figure 21.14**). The release of large amounts of organic matter upon the death of the bloom microorganisms stimulates the growth of chemoheterotrophic bacteria. These microbes subsequently deplete available oxygen. This kills fish and other organisms. Some species can produce toxins that can kill animals that drink the water. Other cyanobacteria (e.g., *Oscillatoria* spp.) are so pollution resistant and characteristic of freshwater with high organic matter content that they are used as water pollution indicators. ►► *Microorganisms in coastal ecosystems are adapted to a changeable environment (section 30.2)*

Retrieve, Infer, Apply

1. What are the major characteristics that distinguish cyanobacteria from other photosynthetic organisms?
2. What is a trichome and how does it differ from a simple chain of cells?
3. Briefly discuss the ways in which cyanobacteria reproduce.
4. Describe how a vegetative cell, a heterocyst, and an akinete are different. How are heterocysts modified to carry out nitrogen fixation?
5. Compare prochlorophytes with other cyanobacteria. Why do they occupy stratified layers in the ocean?
6. Describe some important positive and negative impacts cyanobacteria have on humans and the environment.

21.5 Phylum *Planctomycetes:* Bacteria with Intracellular Compartments

After reading this section, you should be able to:

- Draw the unusual structural features of planctomycetes
- Identify the electron donor and acceptor used in the anammox reaction
- Explain why the anammox reaction is important to the global flux of nitrogen

The phylum *Planctomycetes* was largely underappreciated until recently. Bacteria in this phylum are morphologically unique, having compartmentalized cells (**figure 21.15**). Although each species is unique, all follow a basic cellular organization that includes a plasma membrane closely surrounded by a proteinacious cell wall that lacks peptidoglycan. At least one candidate species is known to have an S layer. The intracytoplasmic membrane is separated from the plasma membrane by a peripheral, ribosome-free region called the paryphoplasm (sometimes referred to as the pirellulosome). Some planctomycetes, such as *Gemmata obscuriglobus,* enclose their genome in a compartment called the nuclear body. Unlike a eukaryotic nucleus, the nuclear body contains ribosomes as well as nucleic acids. Members of the genus *Planctomyces* attach to surfaces through a stalk and holdfast. Most of these bacteria have life cycles in which sessile cells bud to produce motile swarmer cells. The swarmer cells are flagellated and swim for a while before attaching to a substrate and beginning budding reproduction.

Members of the genera *Brocadia, Kuenenia, Scalindua,* and *Anammoxoglobus* lack a nuclear body but possess another compartment, the **anammoxosome.** This is the site of *ana*erobic *ammo*nia *ox*idation, called the **anammox reaction,** or simply anammox. This unique form of chemolithoautotrophy employs ammonium ion (NH_4^+) as the electron donor and nitrite (NO_2^-) as the terminal electron acceptor—it is reduced to nitrogen gas (N_2) (**figure 21.16**). Hydrazine (N_2H_4) is an important intermediate in anammox. It is normally a cellular toxin, but in this case its very negative reduction potential

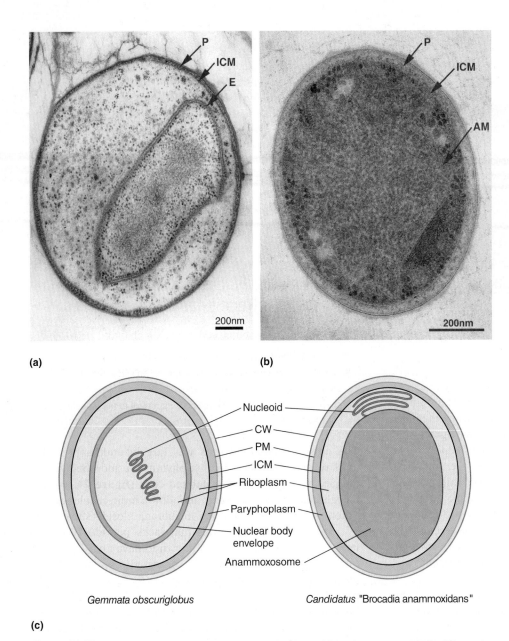

(a)

(b)

Gemmata obscuriglobus

Candidatus "Brocadia anammoxidans"

(c)

Figure 21.15 Planctomycete Cellular Compartmentalization. (a) An electron micrograph of *Gemmata obscuriglobus* showing the nuclear body envelope (E), the intracytoplasmic membrane (ICM), and the paryphoplasm (P). (b) An electron micrograph of the anaerobic ammonia-oxidizing planctomycete *Candidatus* Brocadia anammoxidans. The anammoxosome is labeled AM. (c) Schematic drawings corresponding to (a) and (b): cell wall (CW), plasma membrane (PM).

enables it to donate electrons to ferredoxin. The four high-energy electrons carried by reduced ferredoxin can then be used to produce a proton motive force (PMF) that drives ATP synthesis by an ATP synthase embedded in the anammoxosome membrane. Reduced ferredoxin is also used in CO_2 fixation.

One interesting feature of the physiology of anammox bacteria is that they use nitrite not only as the electron acceptor for ammonium oxidation and generation of a PMF but also as an electron donor for CO_2 fixation, which is accomplished using the reductive acetyl-CoA pathway (*see figure 20.3*). Therefore nitrite acts in both anabolic and catabolic processes, as shown in the following reactions:

An anabolic reaction:

$$2NO_2^- + CO_2 + H_2O \rightarrow CH_2O + 2NO_3^-$$

And a catabolic reaction:

$$NO_2^- + NH_4^+ \rightarrow N_2 + 2H_2O$$

The anammoxosome is bounded by a single bilayer membrane that contains ladderane lipids (*see figure 3.32*). These lipids are unique among planctomycetes and they are thought to form an impermeable barrier, preventing the escape of toxic anammox intermediates into the riboplasm. It is important to note that anammox is estimated to contribute as much as 70% to the cycling of nitrogen in the world's oceans.
▶▶| *Nitrogen cycle (section 28.1)*

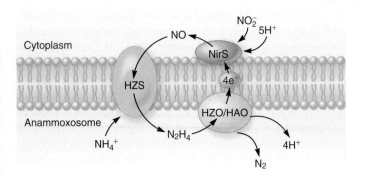

Figure 21.16 The Anammox Reaction. Ammonium (NH_4^+) combines with nitric oxide (NO) to form hydrazine (N_2H_4) in a reaction catalyzed by hydrazine synthase (HZS). Oxidation of hydrazine by a hydrazine/hydroxylamine oxidoreductase (HZO/HAO) results in the formation of one N_2 and four protons. The electrons are passed to NO_2^-, which is reduced by nitrite:nitric oxide reductase (NirS) to form another molecule of NO and the cycle continues.

MICRO INQUIRY *How does a planctomycete use the anammox reaction to generate a proton motive force?*

21.6 Phylum *Chlamydiae:* Obligate Intracellular Parasites

After reading this section, you should be able to:

- Explain the term obligate intracellular parasite
- Diagram the chlamydial life cycle
- Deduce why chlamydia can survive despite significant metabolic limitations

Chlamydiae are Gram-negative obligate intracellular parasites. This means they must grow and reproduce within host cells. Although their ability to cause disease is widely recognized, many species grow within protists and animal cells without adverse effects. These hosts represent a natural reservoir for chlamydiae. Within the single class *Chlamydiae,* there are four families, of which *Chlamydiales* is most well known. This family was founded with the single genus, *Chlamydia.* In 1999 several species were split from this genus and placed in a new genus, *Chlamydophila.* Although controversial, this was based on molecular data including 16S rRNA sequences and DNA-DNA hybridization. Since that time the complete genome sequences of many species in both genera have been published, fueling continued debate and recommendations to reunite the genera. Both genera include human and animal pathogens, including *Chlamydophila pneumoniae, Chlamydophila psittici,* and *Chlamydia trachomatis.* In the human host, *C. trachomatis* is responsible for the most common sexually transmitted disease, nongonococcal urethritis. This microbe is also the most common cause of preventable blindness in the world. These diseases are discussed in chapter 39. ◀◀ *Microbial taxonomy and phylogeny are largely based on molecular characteristics (section 19.3)*

Members of the genera *Chlamydia* and *Chlamydophilia* are nonmotile, coccoid bacteria, ranging in size from 0.2 to 1.5 μm. For many years the presence of peptidoglycan was debated, because although the genes for its synthesis are found on the genomes of sequenced species, peptidoglycan could not be detected. Recently a new approach to labeling D-amino acids demonstrated peptidoglycan in *Chlamydia trachomatis.* Chlamydiae are extremely limited metabolically, relying on their host cells for key metabolites, including ATP. Although they have an ATP synthase, it is run in reverse such that the hydrolysis of ATP generates a proton motive force used to drive the uptake of nutrients from the host. Like other intracellular parasites, its genome is small at 1.0 to 1.3 Mb. Comparative genomics reveals that some chlamydial genes were acquired by horizontal gene transfer (HGT) from free-living bacteria to a common chlamydial ancestor, and HGT accounts for the transfer of genes from chlamydiae to their protist hosts in the genera *Trypanosoma* and *Leishmania.* ▶▶| *Euglenozoa (section 25.2)*

Chlamydial reproduction is unique. It begins with the attachment of an **elementary body (EB)** to the host cell surface (**figure 21.17**). EBs are 0.2 to 0.6 μm in diameter, contain electron-dense nuclear material and a rigid cell wall, and are infectious. They achieve osmotic stability in the extracellular environment by cross-linking their outer membrane proteins, and possibly periplasmic proteins, with disulfide bonds. Host cells become infected with EBs by endocytosing them. An EB is held in an endosome that evades fusion with lysosomes. There the EB differentiates into a **reticulate body (RB)** and undergoes binary fission, which continues until the host cell dies. Reticulate bodies are 0.5 to 1.5 μm in diameter and have less dense nuclear material and more ribosomes than EBs; their walls are also more flexible. After 20 to 25 hours, RBs prepare to leave the host cell and seek new host cells by differentiating into infectious EBs. The host cell lyses and releases EBs 48 to 72 hours after infection.

Chlamydial metabolism is very different from that of other bacteria. It had been thought that chlamydiae cannot catabolize carbohydrates or synthesize ATP. *Chlamydophila psittaci,* one of the best-studied species, lacks both flavoprotein and cytochrome electron transport chain carriers but has a membrane translocase that acquires host ATP in exchange for ADP. The protist parasite *Chlamydia* UWE25 cannot synthesize NAD^+ but synthesizes a transport protein for the import of host NAD^+. Thus some chlamydiae may be energy parasites that are completely dependent on their hosts for ATP. However, in addition to two genes for ATP/ADP translocases, the *C. trachomatis* genome also has genes for substrate-level phosphorylation, electron transport, and oxidative phosphorylation. When supplied with precursors from the host, RBs can synthesize DNA, RNA, glycogen, lipids, and proteins. They also can synthesize at least some amino acids and coenzymes. The EBs have minimal metabolic activity and cannot take in ATP or synthesize proteins. They are designed exclusively for transmission.

21.7 Phylum *Verrucomicrobia* Includes Human Symbionts and Methylotrophs

After reading this section, you should be able to:

- Describe habitats where methane oxidation might occur
- List unusual features that unite some members of *Planctomycetes, Verrucomicrobia,* and *Chlamydiae*

Although the phylum *Verrucomicrobium* has just a few cultured representatives, it has attracted attention, as it is one of the phyla within the "PVC superphylum." This superphylum includes *Planctomycetes, Verrucomicrobia, Chlamydiae,* and several others. This grouping is supported by various phylogenetic markers and the observation that some genera within each of these phyla have complex internal membranes, synthesize sterols, and lack the cell division protein FtsZ. Because these features are not uniformly present, it is believed that the last common ancestor of these microbes had all these characteristics, which were subsequently lost by some. This has led some to the controversial suggestion that these organisms might be relatives of the ancestral microorganism that gave rise to both archaea and eukaryotes (*see figure 1.2*).

Recent investigations have revealed some interesting members of the phylum *Verrucomicrobia* for example, in 2007 a new member of this phylum was reported, *Acidimethylosilex fumarolicum*.

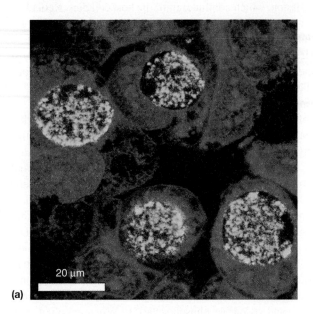

(a)

The discovery and cultivation of this bacterium is noteworthy because it is thermophilic, acidophilic, and methanotrophic; that is, it can grow using methane as a carbon and energy source.

To be used as a carbon source, methane must be oxidized. Methane oxidation is significant because methane is an important greenhouse gas. For many years, it was thought that only 13 genera of aerobic α- and β-proteobacteria could oxidize methane. We now know that certain archaea anaerobically oxidize methane, usually when growing in association with sulfate-reducing bacteria. In all cases, methane oxidation ceases in acid environments below pH 4.2. However, it had been noted that methane is produced abiotically in certain geochemically active areas. These sites are also rich in H_2S, which microbes convert to sulfuric acid (H_2SO_4), so the pH can be as low as 1.0. Researchers hypothesized that despite the low pH, aerobic methane oxidation might occur in such environments. They used PCR to directly amplify from soil a methane monooxygenase gene conserved in proteobacterial methanotrophs. Enrichment cultures were then used to isolate *A. fumarolicum*. This bacterium grows between pH 0.8 and 5.8 with a temperature optimum of 55°C. Since the discovery of *A. fumarolicum* in Italy, other *Verrucomicrobia* isolates have been found in similar sites across the globe. These bacteria demonstrate that methanotrophy is more widespread and that this phylum is more diverse than previously thought. Indeed, the discovery of additional novel *Verrucomicrobia*

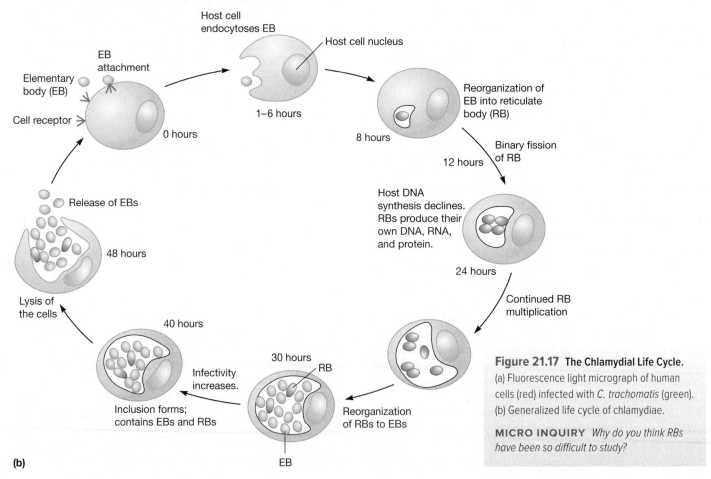

Figure 21.17 The Chlamydial Life Cycle.

(a) Fluorescence light micrograph of human cells (red) infected with *C. trachomatis* (green). (b) Generalized life cycle of chlamydiae.

MICRO INQUIRY *Why do you think RBs have been so difficult to study?*

(b)

strains is anticipated as environmental clone libraries from a variety of ecosystems contain SSU rRNA of microbes that may be assigned to this phylum. ◄◄ *Methanotrophy (section 20.4)* ►►│ *Order* Methylococcales *consists of C1-metabolizing bacteria (section 22.3)*

Another member of the phylum *Verrucomicrobia* has received a great deal of recent attention: *Akkermansia mucinophila.* This bacterium makes up 3 to 5% of the typical human gut microbial community and, as discussed in chapter 32, has been shown to diminish obesity-related systemic inflammation. As its name suggests, *A. mucinophila* consumes mucus and was initially grown in pure culture by providing gastric mucin as the sole carbon source. It is an obligate anaerobe with a fermentative metabolism, nonmotile and coccobacillus in shape. ►►│ *Cooperation (section 32.1)*

Retrieve, Infer, Apply

1. Describe planctomycetes and their distinctive morphological and metabolic properties.
2. Explain how some planctomycetes produce N_2.
3. Briefly describe the steps in the chlamydial life cycle. Why do you think chlamydiae differentiate into specialized cell types for infection and reproduction?
4. How does chlamydial metabolism differ from that of other bacteria?
5. What might have been some of the culture conditions used to isolate *A. fumarolicum?*

21.8 Phylum *Spirochaetes:* Bacteria with a Corkscrew Morphology

After reading this section, you should be able to:

- Draw a spirochete cell as it would be observed in a scanning electron micrograph and in cross section by transmission electron microscopy
- Describe how the unusual flagellar arrangement is well suited for motility in common spirochete habitats

The phylum *Spirochaetes* (Greek *spira,* a coil, and *chaete,* hair) contains Gram-negative, chemoorganotrophic bacteria distinguished by their structure and mechanism of motility. Carbohydrates, amino acids, long-chain fatty acids, and long-chain fatty alcohols may serve as carbon and energy sources. **Table 21.4** summarizes some of the more distinctive properties of selected genera.

Spirochetes are morphologically unique. They are slender, long bacteria (0.1 to 3.0 μm by 5 to 250 μm) with a flexible, helical shape (**figure 21.18**). The central protoplasmic cylinder, which contains cytoplasm and the nucleoid, is bounded by a plasma membrane and a Gram-negative cell wall (**figure 21.19**). Two to more than a hundred flagella, called **axial fibrils** or **periplasmic flagella,** extend from both ends of the cylinder and often overlap one another in the center third of the cell (*also see figure 3.47*). The whole complex of periplasmic flagella lies inside a flexible outer

Table 21.4	Characteristics of Spirochete Genera			
Genus	Dimensions (μm) and Flagella	Oxygen Relationship	Carbon + Energy Source	Habitats
Spirochaeta	0.2–0.75 × 5–250; 2–40 periplasmic flagella (almost always 2)	Facultatively anaerobic or anaerobic	Carbohydrates	Aquatic and free living
Cristispira	0.5–3.0 × 30–180; ≥100 periplasmic flagella	Thought to be facultatively anaerobic	N.A.	Mollusk digestive tract
Treponema	0.1–0.4 × 5–20; 2–16 periplasmic flagella	Anaerobic or microaerophilic	Carbohydrates or amino acids	Mouth, intestinal tract, and genital areas of animals; some are pathogenic (syphilis, yaws)
Borrelia	0.2–0.5 × 3–20; 14–60 periplasmic flagella	Anaerobic or microaerophilic	Carbohydrates	Mammals and arthropods; pathogens (relapsing fever, Lyme disease)
Leptospira	0.1 × 6–24; 2 periplasmic flagella	Aerobic	Fatty acids and alcohols	Free living or pathogens of mammals, usually located in the kidney (leptospirosis)
Leptonema	0.1 × 6–20; 2 periplasmic flagella	Aerobic	Fatty acids	Mammals
Brachyspira	0.2 × 1.7–6.0; 8 periplasmic flagella	Anaerobic	Carbohydrates	Mammalian intestinal tract
Serpulina	0.3–0.4 × 7–9; 16–18 periplasmic flagella	Anaerobic	Carbohydrates and amino acids	Mammalian intestinal tract

1 N.A., information not available.

Figure 21.18 *Treponema pallidum,* **the Causative Agent of Syphilis.** These spirochetes range from 6 to 20 μm in length.

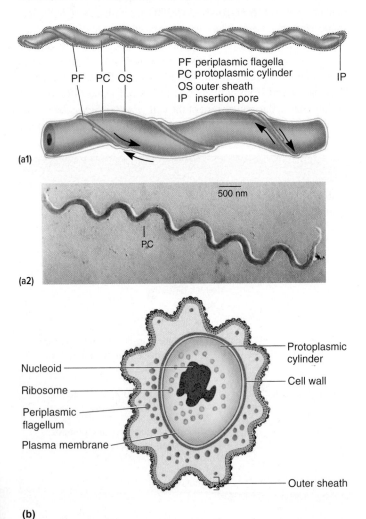

PF periplasmic flagella
PC protoplasmic cylinder
OS outer sheath
IP insertion pore

PF   PC   OS

IP

(a1)

500 nm

PC

(a2)

Nucleoid

Ribosome

Periplasmic flagellum

Plasma membrane

Protoplasmic cylinder

Cell wall

Outer sheath

(b)

Figure 21.19 **Spirochete Morphology.** (a1) A surface view of spirochete structure as interpreted from electron micrographs. (a2) A longitudinal view of *Treponema pallidum.* (b) A cross section of a typical spirochete showing morphological details.

membrane sheath. This outer sheath is made of lipid, protein, and carbohydrate, and varies in structure between different genera. Its precise function is unknown, but the sheath is essential because spirochetes die if it is damaged or removed. The outer membrane sheath of *Treponema pallidum,* the causative agent of syphilis, has few proteins exposed on its surface. This helps the spirochete avoid attack by host antibodies. *T. pallidum* is also discussed in chapter 18 (*see figure 18.12*). ▶▶❙ *Syphilis (section 39.3)*

Spirochete motility is uniquely adapted to movement through viscous solutions. Like the external flagella of other bacteria, the periplasmic flagella rotate. This results in the rotation of the outer membrane sheath in the opposite direction relative to the protoplasmic cylinder, and the cell moves in a corkscrewlike movement through the medium (**figure 21.20**). Flagellar rotation also enables flexing and crawling on solid substrates. Thus, periplasmic flagella mediate movement through liquids and on solid surfaces.

Spirochetes are exceptionally diverse ecologically and grow in habitats ranging from mud to the human mouth. Members of the genus *Spirochaeta* are free-living and often grow in anoxic and sulfide-rich freshwater and marine environments. Some species of the genus *Leptospira* grow in oxic water and moist soil. Some spirochetes form symbiotic associations with other organisms and are found in a variety of locations: the hindguts of termites and wood-eating roaches, the digestive tracts of mollusks (*Cristispira*) and mammals, and the oral cavities of animals (*Treponema denticola, T. oralis*). Spirochetes from termite hindguts and freshwater sediments can fix nitrogen. Spirochetes coat the surfaces of many protists that live in the guts of termites and wood-eating roaches. For example, the flagellate *Mixotricha paradoxa,* a termite symbiont, is covered with slender spirochetes (0.15 by 10 μm in length) that are firmly attached and help move the protist. ▶▶❙ *Microorganism-insect mutualisms (section 32.1)*

Borrelia burgdorferi is one of several *Borrelia* species that cause Lyme disease. It is an unusual bacterium in several respects. Like many other intracellular bacteria, *B. burgdorferi* lacks genes for many cellular biosynthetic pathways. For instance, *N*-acetyl glucosamine (NAG) is required for growth. Recall that NAG is a component of peptidoglycan; *B. burgdorferi* may also use it as an energy source. The microbe uses the Embden-Meyerhof pathway to oxidize glucose monomers to pyruvate. Consistent with the microaerophilic nature of the bacterium, *B. burgdorferi* lacks a TCA cycle and oxidative phosphorylation.

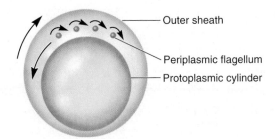

Outer sheath

Periplasmic flagellum

Protoplasmic cylinder

Figure 21.20 **Spirochete Motility.** A hypothetical mechanism for spirochete motility.

21.9 Phylum *Bacteroidetes* Includes Important Gut Microbiota

After reading this section, you should be able to:

- Explain how the human is host to members of this phylum
- Describe unusual carbon substrates used by some bacteroidetes
- Explain the proposed mechanism for *Flavobacterium johnsoniae* gliding motility, and relate gliding motility to the microbe's capacity to degrade complex organic substrates

The phylum *Bacteroidetes* is very diverse. All known members are chemoheterotrophs, and many contribute to the digestion of cellulose, agar (a component of seaweed), chitin (found in arthropod shells and fungal cell walls), proteins, and many other macromolecules. They are common in a variety of terrestrial and marine environments, as well as sewage treatment plants where they presumably contribute to the treatment process. ▶▶| *Wastewater treatment processes (section 43.2)*

The genus *Bacteroides* contains anaerobic, Gram-negative, nonspore-forming rods; motility is rare. They are important constituents of the human intestinal tract and the rumen of ruminants. *Bacteroides ruminicola* is a major component of the rumen flora; it ferments starch, pectin, and other carbohydrates. About 30% of the bacteria cultured from human feces are members of the genus *Bacteroides,* and these organisms may provide extra nutrition by degrading cellulose, pectins, and other complex carbohydrates. Although they often benefit their hosts, they are also involved in human disease. Some are associated with diseases of major organ systems, ranging from the central nervous system to the skeletal system. *B. fragilis* is a particularly common anaerobic pathogen found in abdominal, pelvic, pulmonary, and blood infections. Another important pathogen within the *Bacteroides* phylum is *Porphyromonas gingivalis,* which as its species name suggests, is associated with human gum disease. *P. gingivalis* is related to members of the genus *Bacteroides* and shares many features with them, such as anaerobic growth. ▶▶| *The rumen ecosystem (section 32.1); Normal microbiota of the human body adapt to many sites (section 32.3)*

Many genera within the phylum *Bacteroidetes* are yellow to orange because of carotenoid or flexirubin pigments. Some of the flexirubins are chlorinated, which is unusual for biological molecules. Members of the genera *Cytophaga* and *Sporocytophaga* are aerobes that actively degrade cellulose. *Sporocytophaga* species produce spherical resting cells called microcysts. Most members of the genus *Flavobacterium* are free-living soil or aquatic bacteria, but some are

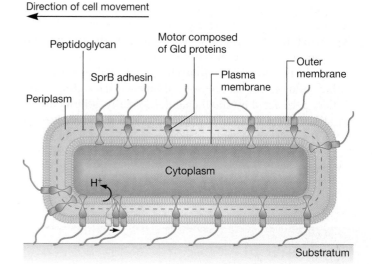

Direction of cell movement

Figure 21.21 Proposed Mechanism for *Flavobacterium johnsoniae* Gliding Motility. Gld proteins within the cell envelope are thought to form motors that propel adhesins such as SprB along the cell surface.

MICRO INQUIRY *How would you test whether or not Gld proteins are necessary for gliding motility? Explain the experiment you would perform and what your prediction would be.*

pathogens of vertebrate hosts. *Flavobacterium columnare* causes columnaris, a highly contagious fish infection that results in major fish kills in natural systems, commercial aquaculture, and home aquaria.

The **gliding motility** so characteristic of members of the genera *Cytophaga, Sporocytophaga,* and *Flavobacterium* can be very rapid; gliding members of the phylum *Bacteroidetes* typically travel 120 μm per minute, but some are considerably faster. Low nutrient levels usually stimulate gliding. Unlike flagellar-mediated swimming motility, the mechanisms that enable gliding motility are not well understood and are not highly conserved across bacterial phyla. Gliding members of the phylum *Bacteroidetes,* such as *Flavobacterium johnsoniae,* depend on motility proteins that are unique to this phylum. Some of these are called glide (Gld) proteins and are thought to assemble into motors that cross the plasma membrane and extend into the periplasm (**figure 21.21**). Like the motor proteins in flagellar basal bodies, the Gld proteins probably harvest the energy of the proton motive force. Gld proteins are thought to propel large, surface proteins called SprB, resulting in cell movement along helical tracks. |◀◀ *Twitching and gliding motility (section 3.8)*

Gliding motility gives a bacterium many advantages. Many aerobic chemoheterotrophic gliding bacteria actively digest insoluble macromolecular substrates such as cellulose and chitin, and gliding motility is ideal for searching these out. Because many of the digestive enzymes are cell wall-associated, the bacteria must be in contact with insoluble nutrient sources; gliding motility makes this possible. Gliding movement is well adapted to drier habitats and to movement within solid masses such as soil, sediments, and rotting wood that are permeated by small channels. Finally, gliding bacteria, like flagellated bacteria, can position themselves at optimal levels of oxygen, hydrogen sulfide, light intensity, temperature, and other factors that influence growth and survival.

Key Concepts

21.1 *Aquificae* and *Thermotogae* Are Ancient Bacterial Lineages

- *Aquifex* and *Thermotoga* contain hyperthermophilic, Gram-negative rods that are the two deepest or oldest phylogenetic branches of *Bacteria*.
- *Aquifex* spp. are chemolithoautotrophs. *Thermotoga* spp. are chemoorganotrophs.

21.2 *Deinococcus-Thermus* Includes Radiation-Resistant Bacteria

- Members of the order *Deinococcales* are aerobic cocci and rods that have an unusual cell wall.
- *Deinococcus radiodurans* is distinctive because of its tremendous resistance to desiccation and radiation.

21.3 Class *Mollicutes*, Phylum *Tenericutes:* Bacteria That Lack Cell Walls

- Mycoplasmas stain Gram negative because they lack cell walls and cannot synthesize peptidoglycan, but many require sterols for growth.
- Many mycoplasmas are plant and animal pathogens. They are some of the smallest bacteria capable of self-reproduction.

21.4 Photosynthetic Bacteria Are Diverse

- Cyanobacteria carry out oxygenic photosynthesis; purple and green bacteria use anoxygenic photosynthesis.
- The four most important groups of purple and green photosynthetic bacteria are the purple sulfur bacteria, the purple nonsulfur bacteria, both of which are proteobacteria; green sulfur bacteria (phylum *Chlorobi*), and green nonsulfur bacteria in the phylum *Chloroflexi* (**table 21.2**).
- The bacteriochlorophyll pigments of purple and green bacteria enable them to live in deeper, anoxic zones of aquatic habitats.
- Green sulfur bacteria are obligately anaerobic photolithoautotrophs that use hydrogen sulfide, elemental sulfur, and hydrogen as electron sources.

- Green nonsulfur bacteria include members of the genus *Chloroflexus*, gliding thermophilic filaments metabolically similar to purple nonsulfur bacteria. Both *Chlorobi* and *Chloroflexus* spp. have chlorosomes, which function to harvest light that is transferred to the reaction center in the plasma membrane (**figure 21.9**).
- Cyanobacteria carry out oxygenic photosynthesis by means of a photosynthetic apparatus similar to that of the eukaryotes. Phycobilisomes contain the light-harvesting pigments phycocyanin and phycoerythrin (**figure 21.11**).
- Cyanobacteria reproduce by binary fission, budding, multiple fission, and fragmentation by filaments to form hormogonia. Some produce dormant akinetes.
- Many nitrogen-fixing cyanobacteria form heterocysts, specialized cells in which nitrogen fixation occurs (**figure 21.13**).
- Members of the *Prochlorococcus* taxon are the most abundant photosynthetic organism on Earth. Together with other unicellular cyanobacteria in the genus *Synchcoccus*, they account for about one-third of global carbon fixation.

21.5 Phylum *Planctomycetes:* Bacteria with Intracellular Compartments

- Members of the phylum *Planctomycetes* lack peptidoglycan in their cell walls and undergo budding division.
- Planctomycetes have unusual cellular compartmentalization, and some perform the anammox reaction (**figures 21.15 and 21.16**).

21.6 Phylum *Chlamydiae:* Obligate Intracellular Parasites

- Chlamydiae are nonmotile, coccoid, Gram-negative bacteria that lack peptidoglycan and must reproduce within the cytoplasmic vacuoles of host cells by a life cycle involving elementary bodies (EBs) and reticulate bodies (RBs) (**figure 21.17**).

21.7 Phylum *Verrucomicrobia* Includes Human Symbionts and Methylotrophs

- A thermoacidophilic member of this phylum is the first microbe known to oxidize methane at very low pH. It is methanotrophic, meaning it can use methane as a carbon and energy source.

21.8 Phylum *Spirochaetes*: Bacteria with a Corkscrew Morphology

- Spirochetes are slender, long, helical, Gram-negative bacteria.
- Spirochetes are motile by means of periplasmic flagella underlying an outer membrane sheath or outer membrane (**figures 21.18–21.20**).

21.9 Phylum *Bacteroidetes* Includes Important Gut Microbiota

- Members of the phylum *Bacteroidetes* are obligately anaerobic, chemoorganotrophic, nonsporing, motile, or nonmotile rods of various shapes.
- Some of these bacteria are important rumen and intestinal symbionts; others can cause disease. Many are important in degrading cellulose, chitin, and other macromolecules in terrestrial and aquatic environments.
- Gliding motility is present in a diverse range of bacteria; in members of *Bacteroidetes,* gliding is mediated by a unique group of transmembranous and surface proteins (**figure 21.21**).

Compare, Hypothesize, Invent

1. The cyanobacterium *Anabaena* grows well in liquid medium that contains nitrate as the sole nitrogen source. Suppose you transfer some of these filaments to the same medium except it lacks nitrate. Describe the morphological and physiological changes you would observe.

2. Compare the structural and functional differences between chlorosomes and thylakoid membranes. Do you think this is an example of convergent evolution—two structures evolving separately to fulfill the same function, or divergent evolution—two structures that arose from a single ancestor but adapted to meet the particular needs of the organism that bears them? What other data would you need (apart from structure and function) to answer this question?

3. Many types of movement are employed by bacteria discussed in this chapter. Review them and propose mechanisms by which energy (ATP or proton gradients) might drive the locomotion.

4. Some believe that an ancient planctomycete evolved into the first eukaryotic cell. List the similarities and differences between these bacteria and a eukaryotic cell, and develop a hypothesis that would account for the cellular and molecular events needed for this transition.

5. Microbiologists using transmission electron microscopy have examined the cellular structure of a thermoacidophilic methanotroph belonging to the phylum *Verrucomicrobia* that was isolated from an acidic hot spring in Kamchatka, Russia. Surprisingly, this microbe has polyhedral microcompartments that look much like carboxysomes (figure 21.10). By contrast, proteobacterial methanotrophs possess intracytoplasmic membranes in which methane monooxygenase is embedded. This enzyme uses O_2 to oxidize CH_4. What do you think is the utility of such enzyme compartmentalization? How would you determine if methane monooxygenase is localized to the polyhedral inclusion bodies found in thermoacidiphilic methanotrophs? Assuming the monooxygenase is confined to these structures, why do you think enzyme compartmentalization evolved in both types of bacteria?

Read the original paper: Islam, R., et al. 2008. Methane oxidation at 55°C and pH 2 by a thermoacidophilic bacterium belonging to the *Verrucomicrobia* phylum. *Proc. Natl. Acad. Sci. USA.* 105:300.

6. In 2014 a team of Chinese and Czech microbiologists reported the discovery a new microbe capable of photoheterotrophic growth isolated from a freshwater lake in the Gobi Desert. The group performed many experiments to characterize this isolate, including 16S rRNA followed by whole-genome sequencing. Predict what they discovered in these analyses that supported the conclusions that the microbe is a member of the phylum *Gemmatimonadetes,* that it is a photoheterotroph, and that it acquired its genes for photosynthesis via horizontal gene transfer from purple sulfur bacteria.

Read the original paper: Zeng, Y., et al. 2014. Functional type 2 photosynthetic reaction centers found in the rare bacterial phylum *Gemmatimonadetes. Proc. Nat. Acad. Sci., USA.* 111:7795–7800.

Learn More

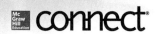

22

Proteobacteria

Bison in Yellowstone National Park, as shown here, have been the cause of much controversy. Ranchers surrounding the park fear that bison may infect their cattle with brucellosis when they forage outside park boundaries during the winter.

Bison and Brucellosis Spark Controversy

Winter in Yellowstone National Park is said to be a "magical" time for visitors. But for the roughly 4,900 Yellowstone bison, winter brings the challenge of finding enough to eat. And as these iconic animals forage, some find themselves overstepping the bounds of the park. That's when things get complicated.

Brucellosis is a disease of ruminant animals caused by the bacterium *Brucella abortus*. As its name implies, it causes miscarriages in pregnant animals. Because Yellowstone bison may be asymptomatic carriers of *B. abortus,* ranchers fear they will infect cattle when they wander out of the park. Unfortunately, the first sign of infection is when an animal aborts a highly contagious stillborn calf.

In 2000 federal, state, and tribal agencies sought to address fears of disease transmission with a bison management plan aimed at maintaining a sustainable wild population without negative impact on the cattle industry. The basics include hazing and transport, screening animals for *B. abortus* exposure, and slaughter for those testing positive. Hazing—a process that can include anything from noisemakers to helicopters to rubber bullets—is used to discourage bison from leaving the park and to capture those that have. Bison found outside the park are transported to designated capture areas where they are tested for brucellosis. Those with *B. abortus* antibodies (seropositive) are sent to slaughter. Seronegative calves and yearlings are vaccinated and sent to a separate holding area along with seronegative adults. These animals are tagged and released back into the park in the spring.

There is only one problem with this management plan: no one likes it. Many ranchers fear that wandering bison will decimate their cattle. They recall a time when brucellosis cost the cattle industry $400 million annually. Others rebut that there has never been a documented case of brucellosis transmission from Yellowstone bison to cattle. They remind the public of the vast herds of bison that were so grossly overhunted that by 1902 only 50 remained in Yellowstone. Meanwhile, each winter hundreds of out-of-bounds bison are transported, tested, and either vaccinated and

released or culled. At the same time, the public lands have been opened for regulated hunting of bison. It seems that the only thing the pro- and anti-bison factions can agree upon is that the current management plan has neither reduced the prevalence of brucellosis in the bison herd nor generated a sustainable population.

So what will be the fate of Yellowstone bison? The delicate balancing act between ranching, tourism, and environmental conservation will continue to drive any modifications to the bison management plan. It is in many ways odd to think that the future of these 1-ton animals rests upon a microbe that measures only 2 μm.

B. abortus is a member of the vast phylum **Proteobacteria.** This diverse group of microbes includes many pathogens as well as bacteria that play key roles in nutrient cycling. Comparison of 16S rRNA sequences reveals five lineages of descent within the phylum (**figure 22.1**). When reading the chapter, challenge yourself to find attributes of environmental microbes that are also found in proteobacterial pathogens.

Readiness Check:
Based on what you have learned previously, you should be able to:

✔ Draw a Gram-negative cell wall and list its constituent parts (section 3.4)
✔ Differentiate between swimming and gliding motility (section 3.8)
✔ List environmental factors that influence microbial growth with a particular emphasis on oxygen (section 7.4)
✔ Describe the difference between heterotrophy and autotrophy (section 11.1)
✔ Compare anaerobic and aerobic respiration and fermentation (sections 11.7 and 11.8)
✔ Draw an electron transport chain indicating the relative standard reduction potentials of the electron donor, acceptor, and electron transporters (sections 10.4 and 11.6)
✔ Define chemolithotrophy and identify an example electron donor and acceptor (section 11.10)
✔ Compare and contrast oxygenic and anoxygenic phototrophy (section 11.11)
✔ Summarize the function of the Calvin-Benson cycle and the enzyme nitrogenase (sections 12.3 and 12.5)

22.1 Class *Alphaproteobacteria* Includes Many Oligotrophs

After reading this section, you should be able to:

- Describe the oligotrophic environments inhabited by some α-proteobacteria
- Outline the metabolic flexibility of some purple photosynthetic bacteria
- Draw the photosynthetic apparatus found in purple photosynthetic bacteria and label its constituent parts
- List two unusual metabolic features of rickettsias and relate these to their intracellular lifestyle
- Diagram the life cycles of *Hyphomicrobium* and *Caulobacter* spp.
- Compare and contrast the relationship between *Rhizobium* and *Agrobacterium* spp. and plants
- Identify the electron donors and the electron acceptor used in nitrification
- List the physiological and taxonomic groups of microbes capable of nitrification

Since *Bergey's Manual of Systematic Bacteriology* was published in 2005, there have been several reassessments of the proposed phylogeny and taxonomy of *Alphaproteobacteria*. Some of these studies are based on rRNA sequences, while other analyses include additional vertically transmitted, slowly evolving essential genes. Here we present one recently developed phylogenetic tree based on rRNA homologies (**figure 22.2**). Important updates include the new orders *Magnetococcales,* made up

of marine magnetotactic, bacteria, and *Pelagibacterales,* which includes the world's most abundant microorganisms, belonging to the SAR11 clade. Also, note the branch just above *Pelagibacterales* in figure 22.2 labeled "Promitochondrion." This represents the extinct lineage of bacteria that, through endosymbiosis, gave rise to eukaryotic mitochondria and related organelles. The placement of this lineage is controversial, as is the placement of *Rickettsiales.* ◀◀ *Evolution of the three domains of life (section 19.5)*

A wide variety of bacteria belong to the class *Alphaproteobacteria,* including most of the oligotrophic proteobacteria (those capable of growing at low nutrient levels). Others have evolved to live within plant and animal cells as either mutualists, whereby both kinds of organisms benefit, or as pathogens. Some α-proteobacteria have unusual metabolic modes such as methylotrophy—the ability to grow using one-carbon compounds as their sole carbon source (*Methylobacterium*), chemolithotrophy (*Nitrobacter*), and the ability to fix nitrogen (*Rhizobium*). **Table 22.1** summarizes the general characteristics of many of the bacteria discussed in the following sections.

Purple Nonsulfur Bacteria Perform Anoxygenic Photosynthesis

All **purple nonsulfur bacteria** are α-proteobacteria, with the exception of *Rhodocyclus* (β-proteobacteria). These bacteria use anoxygenic photosynthesis and possess bacteriochlorophylls *a* or *b* (*see table 21.2*). Their photosynthetic apparatus is continuous with the plasma membrane. The plasma membrane is invaginated, making many intracellular folds (**figure 22.3a,b**). These intra*c*ytoplasmic *m*embranes (ICMs) increase the surface area, thereby providing more space for *p*hoto*s*ynthetic *u*nits (PSUs). Each PSU includes a *r*eaction *c*enter (RC), consisting of

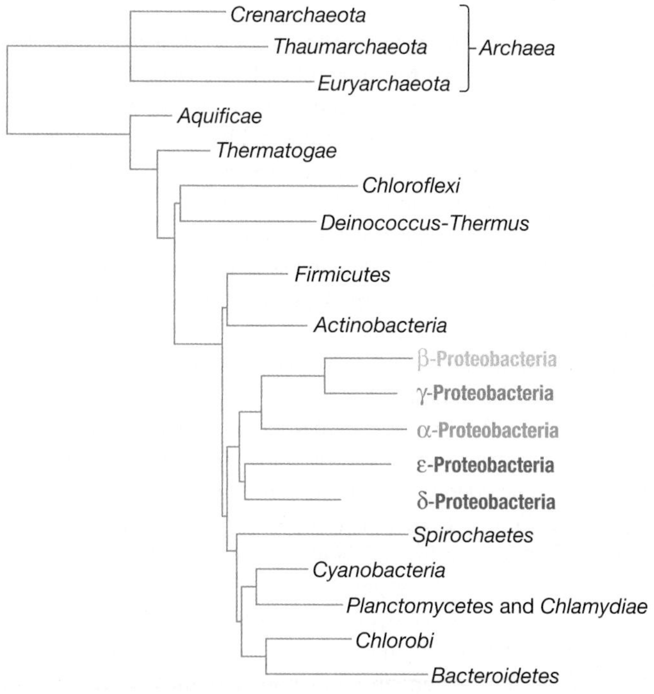

Figure 22.1 Phylogenetic Relationships Among *Bacteria* and *Archaea.* Proteobacterial groups are highlighted.

(Figure 22.1 tree — tips:) Crenarchaeota, Thaumarchaeota, Euryarchaeota [Archaea]; Aquificae; Thermatogae; Chloroflexi; Deinococcus-Thermus; Firmicutes; Actinobacteria; β-Proteobacteria; γ-Proteobacteria; α-Proteobacteria; ε-Proteobacteria; δ-Proteobacteria; Spirochaetes; Cyanobacteria; Planctomycetes and Chlamydiae; Chlorobi; Bacteroidetes

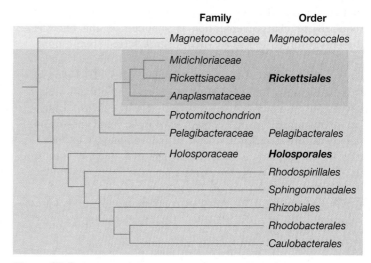

Family	Order
Magnetococcaceae	Magnetococcales
Midichloriaceae	
Rickettsiaceae	**Rickettsiales**
Anaplasmataceae	
Protomitochondrion	
Pelagibacteraceae	Pelagibacterales
Holosporaceae	**Holosporales**
	Rhodospirillales
	Sphingomonadales
	Rhizobiales
	Rhodobacterales
	Caulobacterales

Figure 22.2 Proposed Subclasses of Alphaproteobacteria. Three proposed subdivisions are *Magnetococcida, Rickettsidae,* and *Calobacteriridae.* Protomitochondrion is an extinct organism that gave rise to the mitochondrial family of organelles of eukaryotes.

Table 22.1   Characteristics of Selected α-Proteobacteria

Genus	Dimensions (μm) and Morphology	Genome Size (Mb)	Oxygen Requirement	Other Distinctive Characteristics
Agrobacterium	0.6–1.0 × 1.5–3.0; motile with peritrichous flagella, nonsporing rods	2.5	Aerobic	Chemoorganotrophs that can invade plants and cause tumors
Caulobacter	0.4–0.6 × 1–2; comma-shaped with polar flagellum and prostheca with holdfast	4.0	Aerobic	Chemoorganotrophic and oligotrophic; asymmetric cell division
Hyphomicrobium	0.3–1.2 × 1–3; rod-shaped or oval with polar prosthecae	3.6	Aerobic	Reproduce by budding; methylotrophic
Nitrobacter	0.5–0.9 × 1.0–2.0; rod- or pear-shaped, sometimes motile by flagella	3.4	Aerobic	Chemolithotrophs, oxidize nitrite to nitrate
Rhizobium	0.5–1.0 × 1.2–3.0; motile rods with flagella	5.1	Aerobic	Invade leguminous plants to produce nitrogen-fixing root nodules
Rhodospirillum	0.7–1.5 wide; spiral cells with polar flagella	4.4	Anaerobic, microaerophilic, aerobic	Anoxygenic photoheterotrophs under anoxic conditions
Rickettsia	0.3–0.5 × 0.8–2.0; short nonmotile rods	1.1–1.3	Aerobic	Obligate intracellular parasites
Pelagibacter	0.2 × 0.5; rod-shaped	1.3	Aerobic	Chemoorganotrophic; proteorhodopsin present
Magnetococcus	1–2 in diameter, roughly spherical; two bundles of flagella at single pole; single chain of cytoplasmic magnetite crystals	4.7	Microaerophilic	Chemoorganotrophic and chemolithotrophic; magnetotactic

a protein-bacteriochlorophyll (Bchl) complex that absorbs light in the infrared range. The RC-Bchl is responsible for donating electrons to the electron transport chain that results in a proton motive force sufficient to fuel ATP synthesis. The energy required for charge separation is supplied by the photons directly absorbed by the RC-Bchl or transferred from a light-harvesting complex. All purple bacteria (nonsulfur and sulfur) have a primary light-harvesting complex called LH-I closely associated with the RC. Most also have additional light-harvesting complexes, called LH-II (figure 22.3c). Light-harvesting complexes are composed of Bchl *a* and carotenoid pigments. LH-IIs encircle the RC/LH-I complex. This maximizes the efficiency of energy transfer. ◄◄ *Light reactions in anoxygenic photosynthesis (section 11.11); Photosynthetic bacteria are diverse (section 21.4)*

Purple nonsulfur bacteria vary considerably in morphology. Most are motile by polar flagella. They may be spirals (*Rhodospirillum*), rods (*Rhodopseudomonas*), half circles or circles (*Rhodocyclus*), or they may even form prosthecae and buds (*Rhodomicrobium*). They are most prevalent in the mud and water of lakes and ponds with abundant organic matter and low sulfide levels.

Purple nonsulfur bacteria show exceptional metabolic flexibility. Normally they grow anaerobically as photoorganoheterotrophs, trapping light energy and employing organic molecules as both electron and carbon sources (table 22.1). Although they are called nonsulfur bacteria, some species can oxidize very low, nontoxic levels of sulfide to sulfate, but they do not oxidize elemental sulfur. Some can use also use hydrogen as an electron source. In the absence of light, most purple nonsulfur bacteria grow aerobically as chemoorganoheterotrophs, but some species carry out fermentation anaerobically. Oxygen inhibits bacteriochlorophyll and carotenoid synthesis so that cultures growing aerobically in the dark are colorless.

Rhodospirillum rubrum is perhaps the best-studied purple nonsulfur bacterium. It exemplifies the metabolic flexibility of this group. The "decision-making" process this microbe uses is displayed in **figure 22.4**. When in oxic conditions, the photosynthetic apparatus is inhibited, and it grows chemoorganotrophically. However, when oxygen is not present, it can use one of three metabolic options, depending on the presence of light and fixed carbon. These include fermentation, photoheterotrophy, or photoautotrophy. When growing phototrophically, *R. rubrum* makes photopigments that give it a rich red color. *R. rubrum*

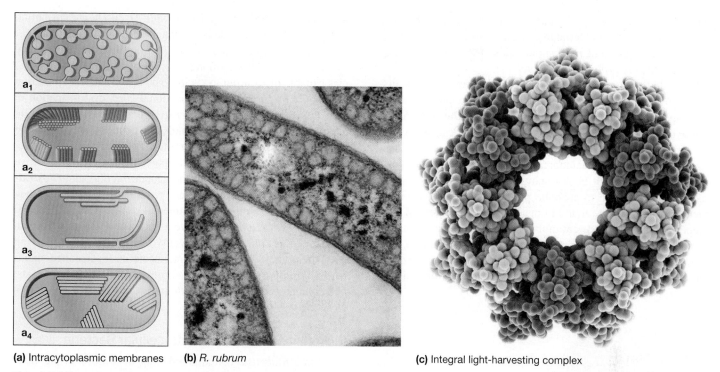

(a) Intracytoplasmic membranes   **(b)** *R. rubrum*   **(c)** Integral light-harvesting complex

Figure 22.3 The Photosynthetic Apparatus of Purple Bacteria.   (a) The arrangement of intracytoplasmic membranes (ICMs) varies, but in all cases, they are associated with the plasma membrane. (a$_1$) ICMs that resemble vesicles are found in the nonsulfur α-proteobacteria *Rhodospirillum rubrum* and *Rhodobacter capsulatus,* and the sulfur γ-proteobacteria *Chromatium vinusum* and *Thiocapsa roseopersicina.* (a$_2$) Tubelike ICMs are found in *T. pfennigii.* (a$_3$) ICMs arranged in small stacks are seen in *Rhodospirillum molischianum* and the γ-proteobacterium *Ectothiorhodospira mobilis.* (a$_4$) *Rhodopseudomonas palustris* has larger stacked ICMs. (b) A transmission electron micrograph of *Rhodospirlllium* sp. grown anaerobically in the light shows vesicle-like ICMs. (c) Molecular model of light-harvesting complex-II (LH-II) from *Rhodopseudomonas acidophila.* LH-II in this organism contains bacteriochlorophyll *a* and carotenoids that transfer electrons to the centrally located reaction center (RC) and closely associated light-harvesting complex I (LH-I); not shown.

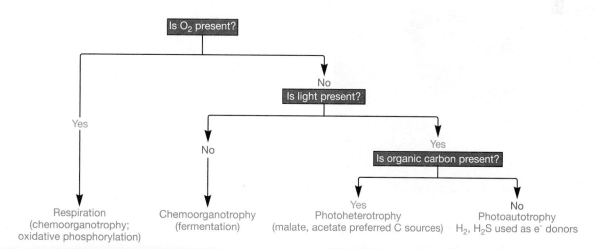

Figure 22.4 The Metabolic Flexibility of *Rhodospirillum rubrum*.   This microbe can switch from one type of metabolism to another, depending on the availability of light and oxygen.

MICRO INQUIRY *What would* Rhodospirillum rubrum *use for carbon and energy sources if grown (a) anaerobically in the dark in the presence of malate, (b) anaerobically in the light without organic carbon, and (c) aerobically in the light in the presence of malate?*

has sparked the interest of industrial microbiologists because it produces novel biodegradable plastics when grown on β-hydroxycarboxylic and *n*-alkanoic acids. ▶▶| *Biotechnology and industrial microbiology (chapter 42)*

In addition to the anaerobic anoxygenic photosynthesis displayed by the purple nonsulfur bacteria, some α-proteobacteria are capable of **aerobic anoxygenic photosynthesis (AAnP)**. These aerobic bacteria are photoheterotrophs that catabolize carbon and other organic materials using O_2-dependent respiration. They supplement this chemoorganotrophic source of energy using Bchl *a*–dependent phototrophy. This allows them to shunt more organic carbon to anabolic pathways, thereby enhancing their growth. They are found in marine and freshwater environments, and are often present in microbial mat communities. Important α-proteobacterial genera include the α-proteobacteria *Erythromonas, Roseococcus, Porphyrobacter,* and *Roseobacter,* although metagenomic analysis reveals that a wider diversity of

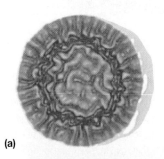

(a)

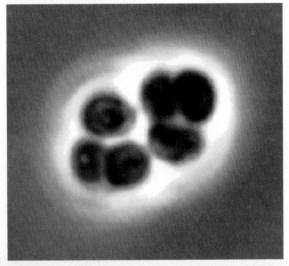

(b)

Figure 22.5 Cyst Formation in *Rhodospirillum centenum*.
(a) Colony morphology when grown on a medium that limits nutrients.
(b) The cells within the colony (a) have formed cysts. In this species, cysts are most commonly found in clusters of four surrounded by a thick outer coat consisting of lipopolysaccharides and lipoproteins. Cysts are about 1 μm in diameter, nonmotile, and more spherical than growing cells.

MICRO INQUIRY *How are bacterial cysts different from the endospores of certain Gram-positive bacteria?*

bacteria are capable of AAnP (*see table 21.2*). Their contribution to the global carbon cycle is an area of active investigation.

Rhodospirillum and *Azospirillum* (both in the family *Rhodospirillaceae*) are among several bacterial genera with species capable of forming **cysts** in response to nutrient limitation (**figure 22.5**). In addition to nutritional cues, cyst formation in both *R. centenum* and *A. brasilense* is regulated by the secretion and reuptake of cyclic GMP, a molecule better known for its role as a second messenger in eukaryotes. These resting cells differ from the well-characterized endospores made by Gram-positive bacteria, such as members of the genera *Bacillus* and *Clostridium*. Although cysts are also very resistant to desiccation, they are less tolerant of other environmental stresses such as heat and ultraviolet (UV) light. They have a thick outer coat and store an abundance of poly-β–hydroxybutyrate (PHB). Cyst-forming bacteria are not limited to α-proteobacteria; for instance, members of *Azotobacter*, which are γ-proteobacteria, also form cysts. |◀◀ *Inclusions (section 3.6); Bacterial endospores are a survival strategy (section 3.9)*

Rickettsias Are Obligate Intracellular Bacteria

The genus *Rickettsia* contains rod-shaped, coccoid, or pleomorphic cells with typical Gram-negative walls and no flagella. Although their size varies, they tend to be very small (e.g., members of the genus *Rickettsia* are 0.3 to 0.5 μm in diameter and 0.8 to 2.0 μm long). All species are parasitic or mutualistic. The parasitic forms grow in vertebrate erythrocytes, macrophages, and vascular endothelial cells. Some also live in blood-sucking arthropods such as fleas, ticks, mites, or lice, which serve as vectors or primary hosts (**figure 22.6**).

Rickettsias are believed to have descended from a free-living, aerobic bacterium that became an intracellular parasite of an ancestral eukaryotic cell lacking organelles. In this way, the bacterium was able to assimilate many host substrates, making genes for the biosynthesis of these substrates unnecessary. Thus, as these genes mutated, there was no selection to

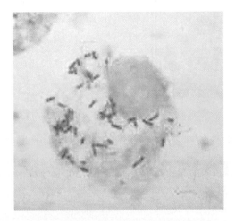

Figure 22.6 *Rickettsia rickettsii*.   A tick hemolymph cell filled with *Rickettsia rickettsii*, the causative agent of Rocky Mountain spotted fever. These bacteria are very small, ranging in size from 0.2 × 0.5 μm to 0.3 × 2.0 μm.

prevent their loss and conversion to pseudogenes (genes that have lost the capacity to encode products due to mutation of critical elements, e.g., promoter, placement of an early stop codon). To decrease cellular energy burden, many pseudogenes were eventually lost. This gene reduction gave rise to a small genome that prevented the bacterium from living outside the host cell (*see figure 18.20*).

Rickettsias are therefore very different from most other bacteria in physiology and metabolism. They have lost the genes encoding glycolytic pathway enzymes, so they cannot use glucose as an energy source. Instead, they oxidize glutamate and tricarboxylic acid cycle intermediates such as succinate. The rickettsial plasma membrane has carrier-mediated transport systems for the assimilation of host cell nutrients and coenzymes. Its membrane also has a carrier that exchanges ADP for external ATP. However, *R. prowazekii* possesses a complete TCA cycle and a membrane-bound ATP synthase.

This genus includes important human pathogens including *R. prowazekii* and *R. typhi,* the causative agents of louse-borne and murine typhus, respectively, as well as *R. rickettsii,* which causes Rocky Mountain spotted fever. These microbes enter the host cell by inducing phagocytosis but immediately escape the phagocytic vacuole to reproduce by binary fission in the cytoplasm. Eventually the host cell bursts, releasing new organisms. ▶▶| *Phagocytosis (section 33.5); Rocky mountain spotted fever (section 39.2)*

Caulobacteraceae and *Hyphomicrobiaceae* Bacteria Reproduce in Unusual Ways

These two groups of α-proteobacteria have interesting life cycles that feature a prostheca or reproduction by budding. A **prostheca** (pl., prosthecae), also called a **stalk,** is an extension of the cell, including the plasma membrane and cell wall, that is narrower than the mature cell. **Budding** is a form of cellular reproduction that is distinctly different from binary fission. The bud first appears as a small protrusion at a single point on the mother cell and enlarges to form a mature cell. This new cell pinches free of the mother cell and is noticeably smaller than its parent. Most or all of the bud's cell envelope is newly synthesized. |◀◀ *Bacterial cell cycles can be divided into three phases (section 7.2)* ↻ *Appendaged Bacteria*

The families *Caulobacteraceae* and *Hyphomicrobiaceae* contain two of the best-studied prosthecate genera: *Hyphomicrobium* and *Caulobacter*. The genus *Hyphomicrobium* contains chemoorganotrophic, aerobic, budding bacteria that frequently attach to solid objects in freshwater, marine, and terrestrial environments. The vegetative cell measures about 0.5 to 1.0 by 1 to 3 μm (**figure 22.7**). At the beginning of the reproductive cycle, the mature cell produces a prostheca (also called a hypha) 0.2 to 0.3 μm in diameter that grows to several μm in length (**figure 22.8**). The chromosome replicates, and a copy moves into the hypha while a bud forms at its

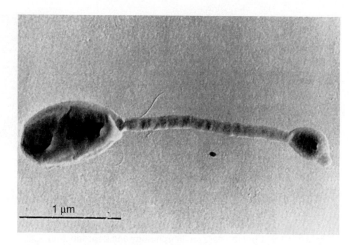

Figure 22.7 Prosthecate, Budding Bacteria. *Hyphomicrobium facilis* with hypha and young bud.

end. As the bud matures, it produces one to three flagella, and a septum divides the bud from the hypha. The bud is finally released as an oval- to pear-shaped swarmer cell, which swims off, then settles down and begins budding. The mother cell may bud several times at the tip of its hypha.

Members of the genus *Hyphomicrobium* also have distinctive physiology and nutrition. Sugars and most amino acids do not support abundant growth; instead, these bacteria grow on ethanol and acetate, as well as 1-carbon (C1) compounds such as methanol, formate, and formaldehyde. This makes them facultative **methylotrophs** because they gain energy as well as carbon for biosynthesis from reduced C1 compounds (p. 511), as well as other, larger organic substrates. In fact, they are so efficient at acquiring C1 molecules they can grow in medium without an added carbon source (presumably the medium absorbs sufficient atmospheric carbon compounds). *Hyphomicrobium* spp. may comprise up to 25% of the total bacterial population in oligotrophic (nutrient-poor) freshwater habitats.

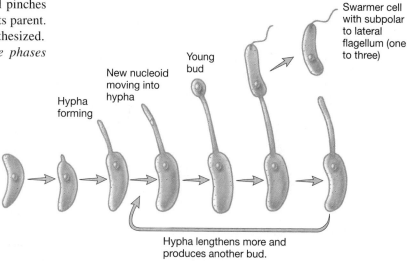

Hypha forming

New nucleoid moving into hypha

Young bud

Swarmer cell with subpolar to lateral flagellum (one to three)

Hypha lengthens more and produces another bud.

Figure 22.8 The Life Cycle of *Hyphomicrobium*.

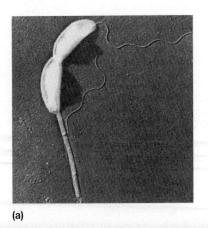
(a)

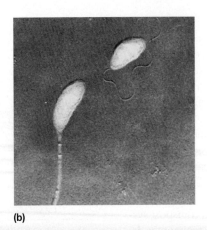
(b)

1 μm
(c)

Figure 22.9 *Caulobacter* **spp. Morphology and Reproduction.** (a) A cell dividing to produce a swarmer (×6,030). Note prostheca and flagellum. (b) A stalked cell and a flagellated swarmer cell (×6,030). (c) A rosette of cells as seen in the electron microscope.

MICRO INQUIRY *If* Caulobacter *bacteria were grown in a phosphate-limited broth, what would happen to the length of the stalks?*

Bacteria in the genus *Caulobacter* form vibroid-shaped cells (*see figure 7.9*), which alternate between polarly flagellated rods and cells that possess a prostheca and a **holdfast,** by which they attach to solid substrata (**figure 22.9**). Incredibly, the material secreted at the end of the *Caulobacter crescentus* holdfast is the strongest biological adhesion molecule known—a sort of bacterial superglue. These bacteria often adhere to other microorganisms and may absorb nutrients released by those microbes. The prostheca differs from that of *Hyphomicrobium* spp. in that it lacks cytoplasmic components and is composed almost totally of the plasma membrane and cell wall. It grows longer in nutrient-poor media and can reach more than 10 times the length of the cell body. The prostheca may improve the efficiency of nutrient uptake from dilute habitats by increasing surface area.

The life cycle of *Caulobacter crescentus* is a model for the investigation of the bacterial cell cycle (**figure 22.10**). When ready to reproduce, the cell elongates and a single polar flagellum forms at the end opposite the prostheca. The cell then undergoes asymmetric transverse binary fission to produce a flagellated swarmer cell that swims away. The swarmer, which cannot reproduce, comes to rest, ejects its flagellum, and forms a new prostheca on the formerly flagellated end. The new stalked cell then starts the cycle anew. This process takes about 2 hours to complete.

Order *Rhizobiales* Includes Important N₂-Fixing Bacteria

The order *Rhizobiales* of α-proteobacteria includes the family *Rhizobiaceae,* which includes aerobic bacteria in the genera *Rhizobium* and *Agrobacterium.* Members of the genus *Rhizobium* are motile rods that become pleomorphic (having many shapes) under adverse conditions. They live symbiotically within root nodules of legumes as specialized nitrogen-fixing cells called bacteroids. Bacteroid physiology is optimized for nitrogen fixation within the nodule (**figure 22.11**; *see also figure 31.9*). The presence of rhizobium nodules has led to the success of leguminous plants likes soybeans. Indeed, *Leguminosae* is the most successful plant family on Earth, with over 18,000 species. Within the nodules, the microbes fix atmospheric nitrogen to ammonia, making it available to the plant host. The process by which bacteria perform this fascinating and important symbiosis is discussed in chapter 31. ◄◄ *Nitrogen fixation (section 12.5)* ▶▶ *Rhizobia are symbionts of leguminous plants (section 31.3)*

The genus *Agrobacterium* differs from *Rhizobium* because it contains plant pathogens. Although genetic analysis supports assigning many species of *Agrobacterium* to the genus *Rhizibium,* like many others, we continue to refer to the genus as *Agrobacterium* as defined by *Bergey's Manual of Systematic Bacteriology.* Agrobacteria invade the crown, roots, and stems of many plants and transform plant cells into autonomously

Pili synthesis   Swarmer
cell

Flagellar rotation   Flagellum
                                       shedding

**Completion
of
cytokinesis**

              Stalked
              cell   Stalk
                                       formation

Figure 22.10 *Caulobacter* **Life Cycle.** Stalked cells attached to a substrate undergo asymmetric binary fission, producing a stalked and a flagellated cell, called a swarmer cell. The swarmer cell swims freely and makes pili until it settles, ejects its flagella, and forms a stalk. Only stalked cells can divide.

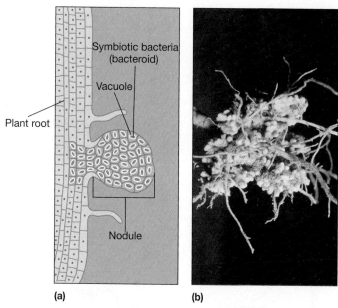

Figure 22.11 *Rhizobium.* (a) Diagram of differentiated bacteria, called bacteroids, in the nodule of a legume root. (b) Nodules on a legume root.

proliferating tumor cells (*see figures 31.12 and 31.13*). The best-studied species is *A. tumefaciens,* which enters many broad-leaved plants through wounds and causes crown gall disease (**figure 22.12**). The ability to produce tumors depends on a large Ti (for *t*umor *i*nducing) plasmid. Tumor production by *A. tumefaciens* is discussed in greater detail in chapter 31, while the use of the Ti plasmid in the genetic engineering of plants is covered in chapter 42 (*see figure 42.9*).

The order *Rhizobiales* also includes the family *Brucellaceae.* As described in the chapter-opening story, the genus *Brucella* contains important human and animal pathogens. Brucellosis, also called undulant fever, is a zoonotic disease—a disease transmitted from animals to humans. It is usually caused by tiny, faintly staining coccobacilli of the species *B. abortus, B. melitensis, B. suis,* or *B. canis.* ▶▶ *Brucellosis (section 39.5)*

Another group of microbes within the order *Rhizobiales* are obligate methylotrophs—bacteria whose growth is supported only by the assimilation of C1 compounds such as methanol and methylamine. These include members of the genera *Methylocystis* and *Methylosinus.* In addition to C1 compounds like formate, the microbes can also use methane as a carbon and energy source, so they are also considered **methanotrophs.** All methanotrophs have extensive intracellular membranes and form a resting cell; α-proteobacterial methanotrophs typically form exospores. **Exospores** are produced when a single cell differentiates into a dormant form with a thick cell wall. Exospores are not as environmentally resistant as endospores. Unlike endospores, which are made only by certain Gram-positive bacteria, exospores are produced by both Gram-positive and Gram-negative bacteria. Interestingly, like some other members of the order *Rhizobiales,* these methanotrophs can also fix nitrogen, but not in association with leguminous plants.

Figure 22.12 *Agrobacterium.* Crown gall tumor of a plant caused by *Agrobacterium tumefaciens.*

The use of a C1 compound as both a carbon and energy source can be simplified to a three-step process (**figure 22.13**). First, methane (or other single-carbon compound) is oxidized to formaldehyde. Next formaldehyde is either assimilated to produce biomass or oxidized to CO_2, and its electrons used to generate a proton motive force. The use of formaldehyde as the central intermediate is paradoxical because it is poisonous. The

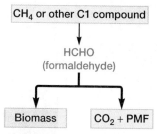

Figure 22.13 **Methanotrophy and Methylotrophy.** Methane or another reduced single-carbon compound (i.e., not CO_2) is converted to the toxic intermediate formaldehyde. This is tolerated only because it is rapidly incorporated into larger molecules to increase biomass or it is quickly oxidized to generate a proton motive force (PMF). Thick arrows indicate the conversion of formaldehyde must occur faster than it can accumulate in the cell.

key to survival is the speed at which formaldehyde is either oxidized to CO_2 or reduced and assimilated into cellular constituents. It is incorporated into cell material by the activity of one of two pathways; α-proteobacteria use the serine pathway (**figure 22.14**), which features the deamination of serine to generate acetyl-CoA. By contrast, γ-proteobacterial methylotrophs use the ribulose monophosphate pathway (p. 524).

The oxidation of methane (CH_4) to formaldehyde (HCHO) requires the formation of methanol (CH_3OH) as an intermediate. This was long thought to be exclusively an aerobic process because molecular oxygen is used in the process. Recently a methanotrophic proteobacterium was reported that oxidizes methane in anoxic environments by coupling methanotrophy with nitrite reduction (denitrification). In the process of reducing nitrite (NO_2^-) to nitrogen gas (N_2), molecular oxygen is released within the microbe, which it harnesses for methane oxidation. Oxidation of methane by bacteria is profoundly different than reverse methanogenesis used by some archaea (*see figure 28.3*). ◀◀ *Methanogens and methanotrophs (section 20.4)* ▶▶ *Carbon cycle (section 28.1)*

Nitrifying Bacteria Oxidize Ammonium or Nitrite to Gain Energy and Electrons

Nitrifying bacteria are a very diverse collection of microbes. Nitrifiers are chemolithoautotrophs that use either ammonium or nitrite as electron donors in a process called **nitrification.** Electrons are donated to an electron transport chain from ammonia, which is oxidized to nitrite, or from nitrite, which is then oxidized to nitrate (**figure 22.15**). Oxygen serves as the terminal electron acceptor. No microbe can perform both reactions, so some proteobacteria are classified as ammonia-oxidizing bacteria,

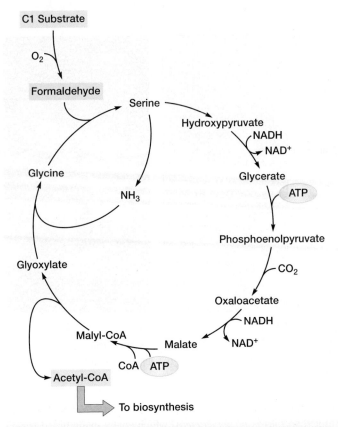

Figure 22.14 The Serine Pathway Is Used by α-Proteobacterial Methylotrophs to Assimilate C1 Compounds. The net yield of this pathway is an acetyl-CoA molecule that can be used in central metabolism. Note that two ATP are hydrolyzed and two NADH are oxidized.

MICRO INQUIRY *What steps are also found in the TCA cycle? What is the structural difference between serine and glycine?*

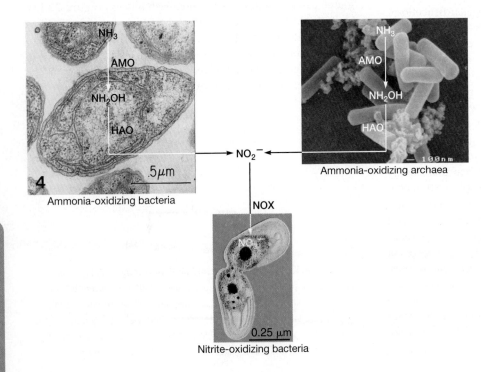

Ammonia-oxidizing bacteria

Ammonia-oxidizing archaea

Nitrite-oxidizing bacteria

Figure 22.15 Nitrification. The process of nitrification involves three different groups of microorganisms. Ammonia-oxidizing bacteria (e.g., *Nitrosomonas europaea*) and archaea (*Nitrosopumilus maritimus*) use the enzyme ammonia monooxygenase (AMO) to convert ammonia to the intermediate hydroxylamine. Hydroxylamine oxidoreductase (HAO) then catalyzes the production of nitrite, which is released into the environment. Nitrite-oxidizing bacteria such as *Nitrobacter* sp. use nitrite oxidoreductase (NOX) to catalyze the oxidation of nitrite to nitrate.

MICRO INQUIRY *Why do nitrifying microbes possess intracellular membranes?*

Table 22.2	Selected Characteristics of Representative Nitrifying Bacteria				
Species	Cell Morphology and Size (μm)	Reproduction	Motility	Cytomembranes	Habitat
Ammonia-Oxidizing Bacteria					
Nitrosomonas europaea (β-proteobacteria)	Rod; 0.8–1.1 × 1.0–1.7	Binary fission	−	Peripheral, lamellar	Soil, sewage, freshwater, marine
Nitrosococcus oceani (γ-proteobacteria)	Coccoid; 1.8–2.2 in diameter	Binary fission	+; 1 or more subpolar flagella	Centrally located parallel bundle, lamellar	Obligately marine
Nitrosospira briensis (β-proteobacteria)	Spiral; 0.3–0.4 in diameter	Binary fission	+ or −; 1 to 6 peritrichous flagella	Rare	Soil
Nitrite-Oxidizing Bacteria					
Nitrobacter winogradskyi (α-proteobacteria)	Rod, often pear-shaped; 0.5–0.9 × 1.0–2.0	Budding	+ or −; 1 polar flagellum	Polar cap of flattened vesicles in peripheral region of the cell	Soil, freshwater, marine
Nitrococcus mobilis (γ-proteobacteria)	Coccoid; 1.5–1.8 in diameter	Binary fission	+; 1 or 2 subpolar flagella	Tubular cytomembranes randomly arranged in cytoplasm	Marine

From Brenner, D. J., et al., eds. 2005. *Bergey's Manual of Systemic Bacteriology*, 2d ed., vol. 2: *The Proteobacteria*. Garrity, G. M. Ed-in-Chief. New York: Springer.

while others are listed as nitrite-oxidizing bacteria (**table 22.2**). Interestingly, in 2014 *Nitrospira moscoviensis* was found to be the first nitrifying microbe also capable of using H_2 as an electron donor. Although most nitrifying bacteria are chemolithoautotrophs, some can also use reduced organic carbon sources and are considered chemolithoheterotrophs. ◀◀ *Chemolithotrophy (section 11.10)*

Bergey's Manual places nitrifying bacteria in several families. All terrestrial ammonia-oxidizing bacteria are β-proteobacteria, including *Nitrosomonas* and *Nitrosospira* in the *Nitrosomonadaceae* family, while marine ammonia oxidizers include both β-proteobacteria and γ-proteobacteria such as *Nitrosococcus* in the *Chromatiaceae* family. Nitrite oxidation is carried out by bacteria in the α-proteobacterial class, *Nitrobacter* (*Bradyrhizobiaceae* family), and the γ-proteobacteria class, *Nitrococcus* (*Ectothiorhodospiraceae* family). These microbes differ considerably in morphology and may be rod-shaped, ellipsoidal, spherical, spirillar, or lobate. They may possess either polar or peritrichous flagella. Often they have extensive membrane complexes in their cytoplasm. These intracellular membranes increase the surface area available for electron transport. Chemolithotrophy captures very little free energy, so reduced inorganic compounds such as ammonia or nitrite must be oxidized at a great rate in order to conserve enough energy to support growth. This is analogous to an old car that gets bad gas

mileage; it must burn more gas (electrons) to cover the same distance (do work) than an efficient, newer car.

Nitrifying microorganisms make important contributions to the nitrogen cycle. Nitrification occurs rapidly in oxic soil treated with fertilizers containing ammonium salts. By converting ammonium to nitrite and then to nitrate, nitrification increases the availability of nitrogen to plants because nitrate is readily used by plants. The discovery of ammonia-oxidizing archaea has triggered a reassessment of the relative contribution of bacteria to global nitrification. ▶▶ *Nitrogen cycle (section 28.1)*

Magnetococcus: A Newly Discovered Genus

Magnetococcus marinus is the only spherical magnetotactic bacterium in pure culture (**figure 22.16**). The microbe has two sheathed bundles of flagella at one pole, a morphology called bilophotrichous. Magnetotaxis is mediated by a single row of magnetite crystals arranged within the cytoplasm (*see also figure 3.37*). It was isolated from the oxic/anoxic interface of an estuary and is obligately microaerophilic when grown in culture. It grows both chemoorganoheterotrophically, using acetate as a carbon source, or chemolithoautotrophically with thiosulfate or sulfide as electron donors. When growing autotrophically, it uses the reductive TCA cycle to fix CO_2 (*see figure 12.5*). ◀◀ *Inclusions (section 3.6)*

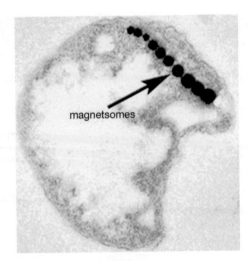

Figure 22.16 *Magnetococcus marinum.* Cells are roughly spherical with a single chain of magnetite crystals and two bundles of polar flagella (not shown) that mediate magnetotaxis through marine sediments. Average cell size is 1-2 μm.

SAR11 Clade, the Most Abundant Microorganisms on Earth

Candidatus Pelagibacter ubique is the sole cultured representative of the world's most abundant group of bacteria, called the SAR11 clade. "SAR" represents *Sar*gasso Sea, where these marine bacteria were first discovered. All members of the SAR11 clade are chemoorganoheterotrophs, and information about their physiology comes from combined analysis of cultures, the complete genome of *Ca.* P. ubique, and metagenomic sequences of environmental bacteria. These microbes appear to reverse the Embden-Meyerhof pathway for gluconeogenesis and use a modified version of the Enter-Douderoff pathway for glucose oxidation (**figure 22.17**). Culture experiments demonstrate that *Ca.* P. ubique can use glucose and low-molecular-weight organic acids as sole carbon sources, but metagenomic analysis suggests this is not true of all members of this clade.

The genome of *Ca.* P. ubique is 1.31 Mb, which is quite small for an independently growing microbe. As this bacterium evolved, it became optimized for a life with scant nutrients. The small size of its genome reflects the almost complete absence of intergenic nucleotides, with a median of only three nucleotides between genes. There are no pseudogenes, phage genes, or duplicated genes. It encodes a relatively large number of high-affinity transporters, as it is energetically cheaper to transport macromolecules than to make them *de novo*. Even its small cell size maximizes the surface-to-volume ratio (*see figure 3.5*). The discovery of genes for proteorhodopsin perplexed microbiologists, as initial experiments failed to show that light enhanced rates of growth, as predicted if this light-driven proton pump contributed to the production of ATP. However, when such experiments were later performed on cells starved for carbon, there was an observable growth advantage for cells cultivated in the light. ◄◄ *Rhodopsin-based phototrophy (section 11.11)* ▶▶ *Microorganisms in the open ocean are adapted to nutrient limitation (section 30.2)*

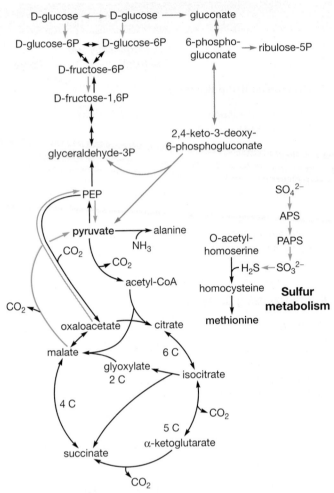

Figure 22.17 **Central Metabolism in *Ca.* P. Ubique as Determined by Genome Analysis.** The upper half of the figure shows the Embden-Meyerhof pathway for glucose oxidation as used by many bacteria (gray arrows); *Ca.* P. ubique runs this pathway in reverse for gluconeogenesis (black arrows); glucose is oxidized by the modified Entner-Douderoff pathway shown with blue arrows. The bottom half of the figure shows the glyoxylate cycle; unlike other microbes that convert oxaloacetate to PEP (gray arrow), *Ca.* P. ubique starts gluconeogenesis by converting malate to pyruvate (green arrow), *Ca.* P. ubique lacks genes for assimilatory sulfate reduction (gray arrows), and requires reduced organic sulfur compounds such as methionine for growth. G3-P, 3-phosphoglycerate; 3-HP, 3-hydroxypyruvate

MICRO INQUIRY *Why can oxaloacetate and glucose, but not acetate, replace the need for pyruvate for growth?*

Retrieve, Infer, Apply

1. What feature unites α-proteobacteria?
2. Why are purple nonsulfur bacteria called "nonsulfur" bacteria?
3. Why do rickettsias have very small genomes?
4. Compare and contrast the life cycle of members of the genera *Hyphomicrobium* and *Caulobacter*.
5. How do *Agrobacterium* and *Rhizobium* spp. differ in lifestyle?

6. What is a methylotroph? How do methane-oxidizing bacteria use methane as both an energy source and a carbon source? How do these microbes avoid poisoning themselves with the formaldehyde they produce as an intermediate in C1 metabolism?

7. What are the two steps of nitrification? Why do you think that to date, no single microbe has been identified that can perform both steps?

8. Contrast and compare chemolithoautotrophic growth of a nitrifying bacterium with that of *Magnetococcus marinum*.

9. Why does the presence of proteorhodopsin enable phototrophy but not photosynthesis by SAR11?

22.2 Class *Betaproteobacteria* Includes Chemoheterotrophs and Chemolithotrophs

After reading this section, you should be able to:

- List human diseases caused by *Neisseria* and *Bordetella* spp.
- Discuss the function of the sheath in the physiology of *Leptothrix* spp.
- Summarize the reactions used by *Nitrosomonas europaea* to oxidize ammonium to nitrite
- Differentiate between nitrification and denitrification, and identify the electron donors and acceptors for each
- Explain how sulfur-oxidizing bacteria obtain ATP from reduced sulfur species via substrate-level phosphorylation

β-proteobacteria are similar to α-proteobacteria metabolically but tend to use substances released from organic decomposition in anoxic habitats. Some of these bacteria use hydrogen,

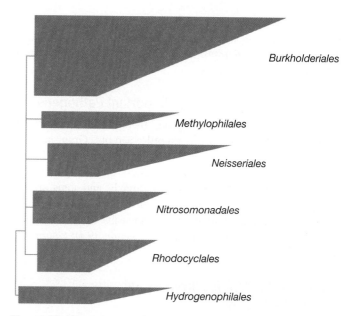

Figure 22.18 Phylogenetic Relationships Among Major Groups of β-Proteobacteria. The relationships are based on 16S rRNA sequence data.

ammonia, volatile fatty acids, and similar substances. As with α-proteobacteria, there is considerable metabolic diversity; β-proteobacteria may be chemoorganotrophs, photolithotrophs, and chemolithotrophs.

The class *Betaproteobacteria* includes 12 families. **Figure 22.18** shows the phylogenetic relationships among major groups of β-proteobacteria, and **table 22.3** summarizes the general characteristics of many of the bacteria discussed in this

Table 22.3	Characteristics of Selected β-Proteobacteria		
Genus	**Dimensions (μm) and Morphology**	**Oxygen Requirement**	**Other Distinctive Characteristics**
Bordetella	0.2–0.5 × 0.5–2.0; nonmotile coccobacillus	Aerobic	Require organic sulfur and nitrogen; mammalian parasite
Burkholderia	0.5–1.0 × 1.5–4; straight rods with single flagellum or a tuft at the pole	Aerobic, some capable of anaerobic respiration with NO_3^-	Poly-β-hydroxybutyrate as reserve; can be pathogenic
Leptothrix	0.6–1.5 × 2.5–15; straight rods in chains with sheath, free cells flagellated	Aerobic	Sheaths encrusted with iron and manganese oxides
Neisseria	0.6–1.9; cocci in pairs with flattened adjacent sides	Aerobic	Inhabitants of mammalian mucous membranes
Nitrosomonas	Size varies with strain; spherical to ellipsoidal cells with intracytoplasmic membranes	Aerobic	Chemolithotrophs that oxidize ammonia to nitrite
Sphaerotilus	1.2–2.5 × 2–10; single chains of cells with sheaths, may have holdfasts	Aerobic	Sheaths not encrusted with iron and manganese oxides
Thiobacillus	0.3–0.5 × 0.9–4; rods, often with polar flagella	Aerobic	All chemolithotrophic; oxidize reduced sulfur compounds to sulfate; some also capable of chemoorganotrophy

section. Because sequence data suggest that the separation between *Betaproteobacteria* and *Gammaproteobacteria* is less distinct than once thought, *Betaproteobacteria* may one day be merged with *Gammaproteobacteria*.

Order *Neisseriales* Includes Important Pathogens

Neisseria is the best-studied genus in the order *Neisseriales*. Members of this genus are nonmotile, aerobic, Gram-negative cocci that most often occur in pairs with adjacent sides flattened. They may have capsules and pili. These chemoorganotrophs inhabit the mucous membranes of mammals, and some are human pathogens. *Neisseria gonorrhoeae* is the causative agent of gonorrhea; *Neisseria meningitidis* is responsible for some cases of bacterial meningitis. ▶▶| *Meningitis (section 39.1); Gonorrhea (section 39.3)*

Order *Burkholderiales* Includes Recently Evolved Pathogens

The order *Burkholderiales* has three well-known families. *Burkholderia* is placed in the family *Burkholderiaceae*. Members of this genus are Gram-negative, aerobic, mesophilic straight rods. With the exception of one species, all are motile with a single polar flagellum or a tuft of polar flagella. Most species use poly-β-hydroxybutyrate as their carbon reserve. One of the most important species is *B. cepacia*, which can degrade over 100 different organic molecules and is very active in recycling organic materials in nature. Originally described as the plant pathogen that causes onion rot, it has emerged in the last 20 years as a major nosocomial (hospital-acquired) pathogen. It is a particular problem for cystic fibrosis patients. Two other species, *B. mallei* and *B. pseudomallei*, are pathogens that could be misused as bioterrorism agents. ▶▶| *Bioterrorism readiness is an integral component of public health microbiology (section 37.7)*

Surprisingly, two genera within the *Burkholderiaceae* family are capable of forming nitrogen-fixing symbioses with legumes much like the rhizobia (α-proteobacteria). Genome analysis of nitrogen-fixing *Burkholderia* and *Cupriavidus* (formerly *Ralstonia*) isolates reveals the presence of nodulation (*nod*) genes that are very similar to those of rhizobia. This suggests a common origin. It is thought these β-proteobacteria gained the capacity to form symbiotic, nitrogen-fixing nodules with legumes through horizontal gene transfer. |◀◀ *Horizontal gene transfer (section 16.4)*

Some members of the order *Burkholderiales* have a **sheath**—a hollow, tubelike structure surrounding a chain of cells. Sheaths often are close fitting, but they are never in intimate contact with the cells they enclose. Some contain ferric or manganic oxides. They have at least two functions. Sheaths help bacteria attach to solid surfaces and acquire nutrients from slowly running water as it flows past, even if it is nutrient-poor. Sheaths also protect against predators such as protozoa.

Two well-studied sheathed genera are in the family *Comamonadaceae*. *Sphaerotilus* bacteria form long, sheathed chains of rods attached to submerged plants, rocks, and other solid objects.

These chains release swarmer cells with a bundle of flagella into the water, where they attach to an object by means of a holdfast and form a new chain. They grow best in slowly running, polluted freshwater. In fact, they thrive in activated sewage sludge and sometimes form tangled masses of filaments that interfere with the proper settling of sludge (*see figure 43.5*). *Leptothrix* spp. characteristically deposit large amounts of iron and manganese oxides in the sheath.

The family *Alcaligenaceae* contains the genus *Bordetella*. This genus is composed of Gram-negative, aerobic coccobacilli. They are chemoorganotrophs with respiratory metabolism and require organic sulfur and nitrogen (amino acids) for growth. As mammalian parasites, they multiply in respiratory epithelial cells. *B. bronchiseptica* is the causative agent of the canine condition kennel cough, and *B. pertussis* is a nonmotile, encapsulated species that causes pertussis (whooping cough); *B. parapertussis* is a closely related species that causes a milder form of the disease. ▶▶| *Pertussis (section 39.1)*

Order *Nitrosomonadales* Consists of Ammonia-Oxidizing Bacteria

A number of chemolithotrophs are found in the order *Nitrosomonadales* including the genera of nitrifying bacteria *Nitrosomonas* and *Nitrosospira* (figure 22.15). Both of these bacteria oxidize ammonia to nitrite; a metabolic and transport reconstruction based on the genome of *Nitrosomonas europaea* is shown in **figure 22.19**. This microbe uses the enzyme ammonia monooxygenase (AMO) and hydroxylamine oxidoreductase to oxidize ammonia to nitrite in the reaction:

$$NH_3 + O_2 + 2H^+ + 2e^- \rightarrow NH_3OH + H_2O \rightarrow$$
$$NO_2 + 5H^+ + 4e^-$$

Of the four electrons released, two must return to AMO to continue ammonia oxidation, while the other two can be used either as reductants for biosynthesis or be donated to the electron transport chain to reduce the terminal electron acceptor, oxygen. Figure 22.19 also illustrates the autotrophic metabolism of this microbe, which uses the Calvin-Benson cycle to fix CO_2.

Order *Hydrogenophilales* Includes a Model Colorless Sulfur Bacterium

This small order contains the genus *Thiobacillus*, which are the best-studied **colorless sulfur bacteria**, so named because they use sulfur-containing compounds as energy sources but are not photosynthetic and therefore lack photosynthetic pigments. Many are unicellular, rod-shaped or spiral, sulfur-oxidizing bacteria that can be either nonmotile or flagellated (**table 22.4**). *Bergey's Manual* divides the colorless sulfur bacteria between two classes; for example, the genera *Thiobacillus* and *Macromonas* are β-proteobacteria, whereas the genera *Thiomicrospira, Thiobacterium, Thiospira, Thiothrix, Beggiatoa*, and others are γ-proteobacteria. Only some of these bacteria have been grown in pure culture.

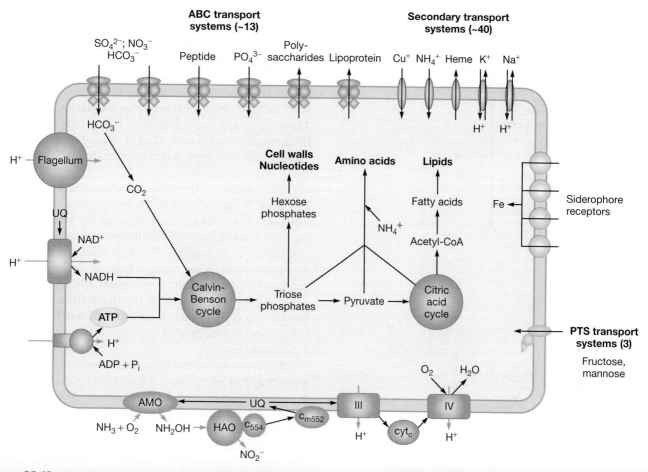

Figure 22.19 Genomic Reconstruction of *Nitrosomonas europaea*. This chemolithoautotroph uses the Calvin-Benson cycle to fix CO_2. Ammonia oxidation is catalyzed by a membrane-bound ammonia monooxygenase (AMO) and hydroxylamine oxidoreductase (HAO). Electrons from HAO flow through two cytochromes (c_{554} and cm_{552}) into an electron transport chain at the level of ubiquinone-cytochrome c oxidoreductase. O_2 is the terminal electron acceptor. Electrons from ammonia oxidation must also flow in reverse for NAD^+ and $NADP^+$ reduction. These enzymes, the flagellar motor, the ATP synthase, and secondary transport systems are fueled by the proton motive force established by ammonia oxidation.

MICRO INQUIRY *Suggest why* N. europaea *has PTS transport systems for the sugars fructose and mannose if it is autotrophic.*

Table 22.4   Colorless Sulfur-Oxidizing Genera

Genus	Cell Shape	Motility; Location of Flagella	Sulfur Deposit[1]	Nutritional Type
Thiobacillus	Rods	+; polar	Extracellular	Obligate or facultative chemolithotrophs
Thiomicrospira	Spirals, comma, or rod-shaped	− or +; polar	Extracellular	Obligate chemolithotrophs
Thiobacterium	Rods embedded in gelatinous masses	−	Intracellular[2]	Probably chemoorganoheterotrophs
Thiospira	Spiral rods, usually with pointed ends	+; polar (single or in tufts)	Intracellular	Unknown
Macromonas	Rods, cylindrical or bean-shaped	+; polar tuft	Intracellular[2]	Probably chemoorganoheterotrophs

1 When hydrogen sulfide is oxidized to elemental sulfur.
2 May use sulfur oxidation to detoxify H_2O_2.

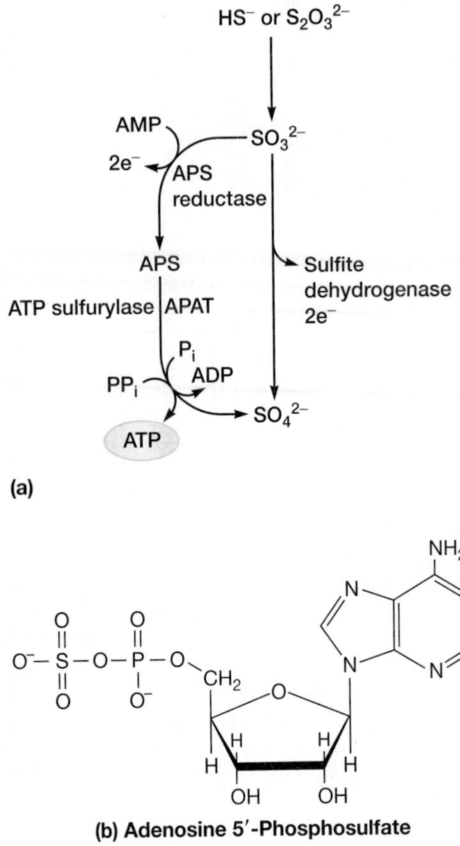

(a)

(b) Adenosine 5′-Phosphosulfate

Figure 22.20 Sulfur Oxidation in *Thiobacillus ferrooxidans.*
(a) Reduced sulfur compounds such as sulfide (HS^-) or thiosulfate ($S_2O_3^{2-}$) are oxidized to sulfite (SO_3^{2-}). Electrons from sulfite can be donated to an electron transport chain via sulfite dehydrogenase and used to generate a proton motive force. Alternatively, the enzyme APS reductase catalyzes the oxidation of sulfite to sulfate, which is used to convert adenosine monophosphate (AMP) to APS. ATP sulfurylase and APS:phosphate adenylyltransferase (APAT) then catalyze the exchange of the sulfate on APS with PP_i to generate ATP via substrate-level phosphorylation. (b) The structure of APS.

MICRO INQUIRY *Identify the reactions that contribute to substrate-level phosphorylation and those that might contribute to ATP synthesis by a membrane-bound ATP synthase.*

Thiobacillus spp. metabolism has been intensely studied. The genus includes obligate aerobes and facultative denitrifiers. They use a variety of reduced, inorganic sulfur compounds as energy and electron sources and oxygen as electron acceptor. ATP is generated in two ways, both of which require that the cell first oxidize sulfide or thiosulfate to sulfite (SO_3^{2-}; **figure 22.20a**). When sulfite donates its electrons

to an electron transport chain, ATP is made by oxidative phosphorylation and sulfate is the end product. Alternatively, sulfite can be bound to AMP to form the molecule adenosine 5′-phosphosulfate (APS, figure 22.20b). The sulfate in APS is next enzymatically replaced with PP_i, yielding sulfate and ATP though substrate-level phosphorylation. Although all *Thiobacillus* spp. are autotrophic, fixing CO_2 via the Calvin-Benson cycle (*see figure 12.4*), *T. novellus* and a few other species can also grow heterotrophically. ◀◀ *Chemolithotrophy (section 11.10)*

The metabolic flexibility of *Thiobacillus denitrificans* has made it the focus of much investigation. Remarkably it can use either oxygen or nitrate as its terminal electron acceptor when growing chemolithotrophically, using hydrogen sulfide or thiosulfate to donate electrons to an electron transport chain. When using nitrate as a terminal electron acceptor, *T. denitrificans* can sequentially reduce nitrate to nitrogen gas—a process called **denitrification.** It can also use ferrous iron (Fe^{2+}) as an electron donor; this form of chemolithotrophy has been well documented in another species, *T. ferrooxidans.*

Sulfur-oxidizing bacteria have a wide distribution and great practical importance. *Thiobacillus* spp. are ubiquitous in soil, freshwater, and marine habitats. Many demonstrate great acid tolerance (e.g., *T. thiooxidans* grows at pH 0.5 and cannot grow above pH 6), and these bacteria prosper in habitats they have acidified by sulfuric acid production. Drainage from mining operations is arguably the most important high-sulfide, low-pH environment in the United States. *T. ferrooxidans* and *Leptospirillum ferrooxidans* control the levels of sulfide (as sulfuric acid and iron disulfide) in these waters. Acid mine drainage is inhospitable to life other than acidophilic bacteria and archaea and can pollute nearby streams and corrode natural and man made structures (**Microbial Diversity & Ecology 22.1**). Conversely, sulfur-oxidizing bacteria can be beneficial. Other, neutrophilic species can increase soil fertility by oxidizing elemental sulfur to sulfate. Thiobacilli are used in processing low-grade metal ores because of their ability to leach metals from ore. There is much interest in using *T. denitrificans* to decontaminate water with high levels of nitrate. ▶▶ *Sulfur cycle (section 28.1)*

Retrieve, Infer, Apply

1. Members of which β-proteobacterial genera make important contributions to the environment? Which are human or animal pathogens?
2. What is the end product of ammonium oxidation?
3. Examine figure 22.19. How does *N. europaea* generate a proton motive force (PMF), and what processes does this PMF fuel?
4. How do colorless sulfur bacteria obtain energy by oxidizing sulfur compounds?
5. What role does iron play in the redox metabolism of *Thiobacillus ferrooxidans*?

MICROBIAL DIVERSITY & ECOLOGY

22.1 Acid Mine Drainage

Each year millions of tons of sulfuric acid flow to the Ohio River from the coal mines of the Appalachian Mountains. This sulfuric acid is of microbial origin and leaches enough metals from the mines to make the river reddish and acidic. The primary culprit is *Acidithiobacillus ferrooxidans,* a chemolithotrophic bacterium that derives its energy from oxidizing ferrous ion to ferric ion and sulfide ion to sulfate ion. The combination of these two energy sources is important because of the solubility properties of iron. Ferrous ion is somewhat soluble and can be formed at pH values of 3.0 or less in moderately reducing environments. However, when the pH is greater than 4.0 to 5.0, ferrous ion is spontaneously oxidized to ferric ion by O_2 in the water and precipitates as a hydroxide. If the pH drops below 2.0 to 3.0 because of sulfuric acid production by spontaneous oxidation of sulfur or sulfur oxidation by thiobacilli and other bacteria, the ferrous ion remains reduced, soluble, and available as an energy source. Remarkably, *A. ferrooxidans* grows well at such acidic pHs and actively oxidizes ferrous ion to an insoluble ferric precipitate. The water is rendered toxic for most aquatic life and unfit for human consumption.

The ecological consequences of this metabolic lifestyle arise from the common presence of pyrite (FeS_2) in coal mines. The bacteria oxidize both elemental components of pyrite for their growth and in the process form sulfuric acid, which leaches the remaining minerals.

Autoxidation or bacterial action

$$2FeS_2 + 7O_2 + 2H_2O \rightarrow 2Fe^{2+} + 4SO_4^{2-} + 4H^+$$

A. ferrooxidans

$$2Fe^{2+} + 1/2O_2 + 2H^+ \rightarrow 2Fe^{3+} + H_2O$$

Pyrite oxidation is further accelerated because the ferric ion generated by bacterial activity readily oxidizes more pyrite to sulfuric acid and ferrous ion. In turn, the ferrous ion supports further bacterial growth. It is difficult to prevent *A. ferrooxidans* growth as it requires only pyrite and common inorganic salts. Because *A. ferrooxidans* gets its O_2 and CO_2 from the air, the only feasible method of preventing its damaging growth is to seal the mines to render the habitat anoxic.

22.3 Class *Gammaproteobacteria* Is the Largest Bacterial Class

After reading this section, you should be able to:

- Relate the electron donors used by a variety of γ-proteobacteria to the habitats in which they thrive
- Describe the multicellular arrangements and unusual cellular morphologies displayed by some γ-proteobacteria
- Outline the oxidation of methane by methanotrophs
- Compare the intracellular life cycle of *Legionella* spp. to that of *Coxiella* spp.
- Describe four practical aspects of pseudomonads
- List at least five electron acceptors used by *Shewanella* cells and explain the term dissimilatory metal reduction
- Describe three strategies used by dissimilatory metal-reducing bacteria that enable their use of insoluble compounds as electron acceptors
- Compare and contrast the physiology and importance of members of *Vibrionales* with those in *Enterobacteriales* and *Pasteurellales*
- Summarize mixed acid fermentation
- Describe the use of metabolic assays to distinguish between enteric bacteria

As the largest class of *Proteobacteria,* **γ-proteobacteria** demonstrate an extraordinary variety of physiological types. According to some molecular analyses, *Gammaproteobacteria* is composed of several deeply branching groups. One consists of purple sulfur bacteria; a second includes intracellular parasites in the genera *Legionella* and *Coxiella*. The two largest groups contain a wide variety of nonphotosynthetic genera. These microbes can be grouped into rRNA superfamilies based on the degree of sequence conservation among rRNA (*see figure 19.3*). Ribosomal RNA superfamily I includes the families *Vibrionaceae, Enterobacteriaceae,* and *Pasteurellaceae*. These bacteria use the Embden-Meyerhof and pentose phosphate pathways to catabolize carbohydrates. Most are facultative anaerobes. Ribosomal RNA superfamily II contains mostly aerobes that often use the Entner-Doudoroff and pentose phosphate pathways to catabolize many different kinds of organic molecules. The genera *Pseudomonas, Azotobacter, Moraxella,* and *Acinetobacter* belong to this superfamily. ◄◄ *Microbial taxonomy and phylogeny are largely based on molecular characterization (section 19.3)*

The exceptional diversity of these bacteria is evident from the fact that *Bergey's Manual* divides γ-proteobacteria into 14 orders and 28 families. **Figure 22.21** illustrates the phylogenetic relationships among major groups and selected γ-proteobacteria,

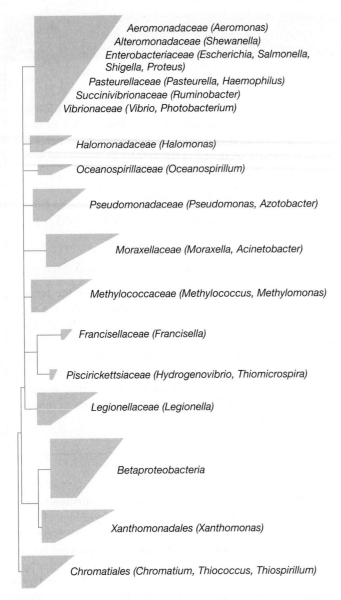

Figure 22.21 Phylogenetic Relationships Among γ-Proteobacteria.
The major phylogenetic groups based on 16S rRNA sequence comparisons.
Representative genera are given in parentheses. Each tetrahedron in the tree
represents a group of related organisms; its horizontal edges show the shortest
and longest branches in the group. Multiple branching at the same level
indicates that the relative branching order of the groups cannot be determined
from the data. The placement of *Betaproteobacteria* has given rise to the
suggestion that this class should be merged with *Gammaproteobacteria*.

and **table 22.5** outlines the general characteristics of some of the
bacteria discussed in this section.

Purple Sulfur Bacteria Perform Anoxygenic Photosynthesis

Purple sulfur bacteria are strict anaerobes and usually photolithoautotrophic. *Bergey's Manual* places **purple sulfur bacteria** in

the order *Chromatiales,* which has two families: *Chromatiaceae*
and *Ectothiorhodospiraceae. Ectothiorhodospira* is composed of
red, spiral-shaped, polarly flagellated cells that deposit sulfur
globules externally. Internal photosynthetic membranes are organized as lamellar stacks (figure 22.3).

The majority of purple sulfur bacteria are in the family
Chromatiaceae. Recall that during anoxygenic photosynthesis, electrons are required for the reduction of NADPH. These
microbes use hydrogen sulfide as an electron donor and deposit the resulting oxidized sulfur granules internally; often
they further oxidize the sulfur to sulfate. Hydrogen also may
serve as an electron donor. *Thiospirillum, Thiocapsa,* and
Chromatium spp. are typical purple sulfur bacterial genera,
although a *Thiocapsa* isolate is able to use nitrite as an electron donor during anoxygenic photosynthesis. These bacteria
are found in anoxic, sulfide-rich zones of lakes, marshes, and
lagoons where large blooms can occur under certain condi-
tions (**figure 22.22**). ▶▶| *Light reactions in anoxygenic pho-
tosynthesis (section 11.11)*

Order *Thiotrichales* Includes Large Filamentous Bacteria

The largest family in the order *Thiotrichales* is *Thiotrichaceae.*
This family has several genera whose members oxidize sulfur
compounds. Morphologically both rods and filamentous forms
are present, but the most striking members are filamentous.

Two genera in this family are *Beggiatoa* and *Leucothrix,*
which contain well-studied gliding species. *Beggiatoa* spp. are
microaerophilic and grow in sulfide-rich habitats such as sulfur
springs, freshwater with decaying plant material, rice paddies,
salt marshes, and marine sediments. Their filaments contain
short, disklike cells (**figure 22.23**). These bacteria oxidize hydrogen sulfide to form large sulfur grains located in pockets formed
by invaginations of the plasma membrane. *Beggiatoa* spp. also
oxidize the sulfur to sulfate when donating electrons to the electron transport chain for energy production. Many strains grow
heterotrophically with acetate as a carbon source, and some incorporate CO_2 autotrophically. Carbon fixation typically occurs
using the Calvin-Benson cycle, but there is metagenomic evidence that some *Beggiatoa* from marine sediments perform the
reductive TCA cycle as well (*see figure 12.5*). Freshwater species
are much smaller than marine.

Members of three genera of *Thiotrichales—Beggiatoa, Thioploca,* and *Thiomargarita*—are notable because they form large
mats on the seafloor near hydrothermal vents and adjacent to nutrient-rich coasts where methane seeps out of Earth's crust. These
filamentous microbes are among the largest known bacteria. For
example, *Thiomargarita* spp. can be over 100 μm in diameter
(about twice as thick as a fingernail) and hundreds of centimeters
long (**figure 22.24**). All three groups grow in bundles, making
them even more conspicuous. These bacteria appear hollow because they form large intracellular vacuoles where high concentrations of nitrate (up to 500 mM) are stored. Sulfur inclusions are
found in a thin layer of cytoplasm around the vacuoles.

Table 22.5   Characteristics of Selected γ-Proteobacteria

Genus	Dimensions (μm) and Morphology	Oxygen Requirement	Other Distinctive Characteristics
Azotobacter	1.5–2.0; ovoid cells, pleomorphic, peritrichous flagella or nonmotile	Aerobic	Can form cysts, fix nitrogen nonsymbiotically
Beggiatoa	1–200 × 2–10; colorless cells form filaments, either single or in bundles	Aerobic or microaerophilic	Gliding motility; can form sulfur inclusions with hydrogen sulfide present
Chromatium	1–6 × 1.5–16; rod-shaped or ovoid, straight or slightly curved, polar flagella	Anaerobic	Anoxygenic photolithoautotrophs that can use sulfide; sulfur stored within the cell
Ectothiorhodospira	0.7–1.5 in diameter; vibrioid- or rod-shaped, polar flagella	Anaerobic, some aerobic or microaerophilic	Internal lamellar stacks of membranes; deposit sulfur granules outside cells
Escherichia	1.1–1.5 × 2–6; straight rods, peritrichous flagella or nonmotile	Facultatively anaerobic	Mixed acid fermenters; formic acid converted to H_2 and CO_2, lactose fermented, citrate not used
Haemophilus	<1.0 in width, variable lengths; coccobacilli or rods, nonmotile	Aerobic or facultatively anaerobic	Fermentative; require growth factors present in blood; parasites on mucous membranes
Leucothrix	Long filaments of short cylindrical cells, usually holdfast is present	Aerobic	Dispersal by gonidia, filaments don't glide; rosettes formed; chemoorganotrophic
Methylococcus	0.8–1.5 × 1.0–1.5; cocci with capsules, nonmotile	Aerobic	Can form cysts; use methane, methanol, and formaldehyde as sole carbon and energy sources
Photobacterium	0.8–1.3 × 1.8–2.4; straight, plump rods with polar flagella	Facultatively anaerobic	Two species can emit blue-green light; Na^+ needed for growth
Pseudomonas	0.5–1.0 × 1.5–5.0; straight or slightly curved rods, polar flagella	Aerobic or facultatively anaerobic	Respiratory metabolism with oxygen or nitrate as acceptor; some use H_2 or CO as energy source
Vibrio	0.5–0.8 × 1.4–2.6; straight or curved rods with sheathed polar flagella	Facultatively anaerobic	Fermentative or respiratory metabolism; sodium ions stimulate or are needed for growth; oxidase positive

Figure 22.22 Purple Photosynthetic Sulfur Bacteria.   Purple photosynthetic sulfur bacteria growing in a salt marsh.

Figure 22.23 *Beggiatoa alba.*   A light micrograph showing part of a colony (×400). Note the dark sulfur granules within many of the filaments.

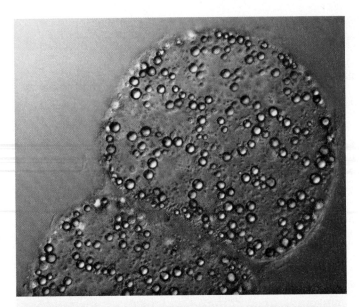

Figure 22.24 *Thiomargarita namibiensis,* **the World's Largest Known Bacterium.** This bacterium, usually 100–300 μm in diameter, occasionally reaches a size of 750 μm (larger than a period on this page), 100 times the size of a common bacterium. *T. namibiensis* uses sulfide from bottom sediments as an energy source and nitrate, which is found in the overlying waters, as an electron acceptor. Vacuoles that store high concentrations of nitrate appear beadlike in this image.

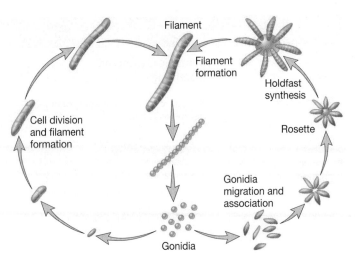

Figure 22.25 Life Cycle and Morphology of *Leucothrix mucor.*

Leucothrix spp. are aerobic chemoorganotrophs that form filaments or trichomes up to 400 μm long. Most are marine, and attach to solid substrates by a holdfast, a microscopic root-like structure. These filamentous bacteria have a complex life cycle in which they are dispersed by the formation of small spherical cells called gonidia (**figure 22.25**). *Thiothrix* is a related genus whose members form sheathed filaments and release gonidia from the open end of the sheath. In contrast to *Leucothrix, Thiothrix* includes species that are facultatively autotrophic, chemoorganotrophic or mixotrophic, oxidizing hydrogen sulfide and depositing sulfur granules internally. *Thiothrix* spp. grow in sulfide-rich flowing water and activated sludge sewage systems.

The genus *Thiomicrospira,* family *Piscirickettsiaceae,* has an interesting history. Originally it included all marine, spiral-shaped, sulfur-oxidizing bacteria. Later 16S rRNA analysis showed that this group was not monophyletic, and members were split between γ- and ε-proteobacterial classes. For instance *Thiomicrospira denitrificans* is now known as *Sulfurimonas denitrificans* and is a member of ε-proteobacteria.

Figure 22.26 shows a cell model based on the annotated genome of *Thiomicrospira crunogena* XCL-2. This deep-sea microbe was isolated from a hydrothermal vent (*see figure 32.4*). This unique environment is characterized by the mixture of hot, anoxic, highly reduced vent fluid that rapidly mixes with cold, oxic bottom water. Of importance to this autotroph is the difference in dissolved CO_2 concentrations that it encounters: vent

fluid is rich in CO_2, whereas bottom water is not. This has led to the evolution of a high affinity transporter for both HCO_3^- and CO_2. Also of note is the multienzyme complex capable of oxidizing a number of reduced sulfur compounds to sulfate. This "Sox" complex (for *s*ulfur *ox*idation) is conserved in a number of sulfur-dependent chemolithoautotrophs such as the β-proteobacteria *Thiobacillus* spp. Although *T. crunogena* can use a variety of reduced sulfur compounds as electron donors, it can use only oxygen as an electron acceptor. Interestingly, genome sequencing reveals the presence of what appears to be a complete prophage within the chromosome.

Order *Methylococcales* Consists of C1-Metabolizing Bacteria

The single family *Methylococcaceae* in the order *Methylococcales* includes the genera *Methylomonas, Methylobacter,* and *Methylococcus,* all of which are obligate methylotrophs that oxidize methane and other C1 compounds as their only carbon and energy source. Like the α-proteobacterial methanotrophs, these microbes also form extensive intracellular membranes; unlike α-proteobacteria, they form cysts instead of exospores (figure 22.5). They also differ in the use of the ribulose monophosphate pathway for the incorporation of C1 compounds (**figure 22.27**). ◀◀ *Methanogens and methanotrophs (section 20.4)*

Order *Legionellales* Includes Intracellular Pathogens

Two families make up the order *Legionellales.* The first is *Legionellaceae,* with its single genus, *Legionella.* The second family is *Coxiellaceae,* which has two genera, *Coxiella* and *Rickettsiella* (not to be confused with the α-proteobacterial genus *Rickettsia*). All these microbes are intracellular pathogens that display a dimorphic (i.e., two cell types) lifestyle that is reminiscent of the chlamydial life cycle. ◀◀ *Phylum Chlamydiae (section 21.6)*

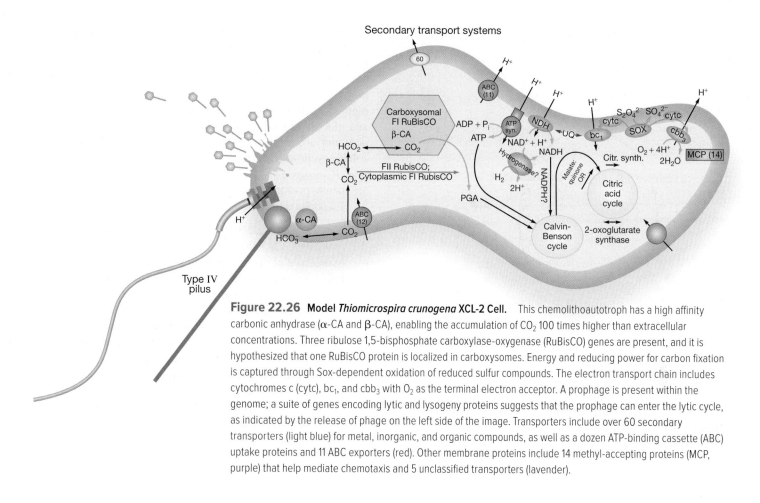

Figure 22.26 Model *Thiomicrospira crunogena* XCL-2 Cell. This chemolithoautotroph has a high affinity carbonic anhydrase (α-CA and β-CA), enabling the accumulation of CO_2 100 times higher than extracellular concentrations. Three ribulose 1,5-bisphosphate carboxylase-oxygenase (RuBisCO) genes are present, and it is hypothesized that one RuBisCO protein is localized in carboxysomes. Energy and reducing power for carbon fixation is captured through Sox-dependent oxidation of reduced sulfur compounds. The electron transport chain includes cytochromes c (cytc), bc_1, and cbb_3 with O_2 as the terminal electron acceptor. A prophage is present within the genome; a suite of genes encoding lytic and lysogeny proteins suggests that the prophage can enter the lytic cycle, as indicated by the release of phage on the left side of the image. Transporters include over 60 secondary transporters (light blue) for metal, inorganic, and organic compounds, as well as a dozen ATP-binding cassette (ABC) uptake proteins and 11 ABC exporters (red). Other membrane proteins include 14 methyl-accepting proteins (MCP, purple) that help mediate chemotaxis and 5 unclassified transporters (lavender).

Although there are over 40 species of *Legionella, L. pneumophila* has been most intensely studied because it causes a pneumonia in human hosts called Legionnaire's disease. In nature, *L. pneumophila* is an intracellular parasite of protozoa that live in moist soil and aquatic environments, including cooling towers, air conditioning units, and hot tubs. This explains why outbreaks of Legionnaire's disease occur in hotels, hospitals, and cruise ships—all of which have centralized cooling systems. In its protozoan hosts, as well as in some human cell cultures, *L. pneumophila* replicates by binary fission to produce slender, rod-shaped cells with a typical Gram-negative cell wall. The replicative forms (RFs) reside in a host vacuole called the replicative endosome. Late during infection, RFs differentiate into cystlike cells called *mature intracellular forms* (MIFs), which when released are nonreplicative; they exist only to infect the next host. MIFs are chubby rods that are metabolically dormant. They also have a distinct cell wall structure that includes a high concentration of the protein Hsp60. This is important because this protein acts as an invasin, that is, a protein required for host cell invasion. In addition, MIFs are heat tolerant and resistant to a variety of antibiotics. These features may help explain why *L. pneumophila* becomes more infective to human

cells after it has grown within a protist. ▶▶ *Legionnaire's disease (section 39.1)*

Coxiella spp. also display two cell types within the host, although they have a much more impressive range of hosts, including insects, fish, birds, rodents, sheep, goats, and humans. As with *L. pneumophila*, humans become infected by inhaling contaminated aerosols, although in the case of the pathogen *C. burnetti*, the resulting Q fever is a flulike illness. The developmental life cycle of *C. burnetti* is similar to that of *L. pneumophila*. A small, infectious form called the *small cell variant* (SCV, similar to MIFs and chlamydial elementary bodies) enters the host cell by phagocytosis. Once the phagosome has fused with a lysosome, the low pH triggers the SCV to become metabolically active. The resulting vacuole is called the parasitophorous (parasite-eating) vacuole. While they are in this vacuole, *C. burnetti* SCVs differentiate into the replicative form known as the *large cell variant* (LCV). Both LCVs and SCVs undergo binary fission and infect host cells, so one might wonder why the microbe evolved the capacity to form SCVs. It is thought that SCVs, like MIFs, are responsible for long-term survival outside the host. This makes these pathogens distinct from chlamydiae and rickettsiae, which cannot survive outside

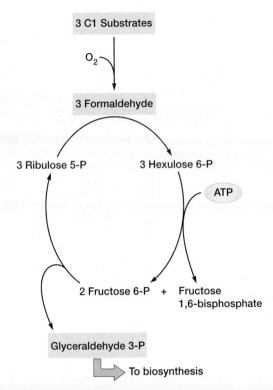

Figure 22.27 The Ribulose Monophosphate Pathway Is Used by γ-Proteobacterial Methylotrophs to Assimilate C1 Compounds. The net yield of this pathway is a glyceraldehyde 3-P molecule that can be used in central metabolism.

MICRO INQUIRY *How does the ribulose monophosphate pathway compare to the serine pathway (figure 22.14) in terms of energy and reducing power used? What might be the ecological consequence of this difference?*

the host. In addition, it has been suggested that the compact and protected genome of SCVs makes these forms better suited for the initial respiratory burst and release of degradative enzymes when phagosomes and lysosomes initially fuse. ◀◀ *Phylum* Chlamydiae *(section 21.6)* ▶▶ *Phagocytosis (section 33.5); Q fever (section 39.5)*

Order *Pseudomonadales* Includes Important and Metabolically Diverse Bacteria

Pseudomonas is the most important genus in the order *Pseudomonadales,* family *Pseudomonaceae.* These bacteria are straight or slightly curved rods, and are motile by one or several polar flagella. They are chemoorganotrophs that usually carry out aerobic respiration, although some use nitrate as the terminal electron acceptor in anaerobic respiration. Most hexoses are degraded by the Entner-Doudoroff pathway; all pseudomonads have a functional tricarboxylic acid cycle. ◀◀ *Glucose to pyruvate (section 11.4); Pyruvate to carbon dioxide (step 2) is accomplished by the tricarboxylic acid cycle (section 11.5)*

The genus *Pseudomonas* is an exceptionally heterogeneous taxon. Many of the 60 or so species can be placed in one of seven rRNA homology groups. The three best-characterized groups are subdivided according to properties such as the presence of poly-β-hydroxybutyrate (PHB), the production of a fluorescent pigment, pathogenicity, the presence of arginine dihydrolase, and glucose utilization. For example, the fluorescent subgroup does not accumulate PHB and produces a diffusible, water-soluble, yellow-green pigment that fluoresces under UV radiation. *P. aeruginosa, P. fluorescens, P. putida,* and *P. syringae* are members of this group.

Pseudomonads have a number of practical impacts, including:

1. Many can degrade an exceptionally wide variety of organic molecules. Thus they are very important in the microbial breakdown of organic materials to inorganic substances (mineralization) in nature and in sewage treatment. The fluorescent pseudomonads can use approximately 80 different substances as their carbon and energy sources.

2. Several species (e.g., *P. aeruginosa*) are important experimental subjects. Many advances in microbial physiology and biochemistry have come from their study. For example, the study of *P. aeruginosa* has significantly advanced our understanding of how bacteria form biofilms and the role of extracellular signaling in bacterial communities and pathogenesis. The genome of *P. aeruginosa* has an unusually large number of genes for catabolism, nutrient transport, the efflux of organic molecules, and metabolic regulation, reflecting its ability to grow in many environments and resist antibiotics. ◀◀ *Biofilms are common in nature (section 7.5); Cell-cell communication within microbial populations (section 7.5)*

3. Some pseudomonads are major animal and plant pathogens. *P. aeruginosa* is an opportunistic pathogen that infects cystic fibrosis patients. It also invades tissues that have been burned and causes urinary tract infections. *P. syringae* is an important plant pathogen.

4. Pseudomonads such as *P. fluorescens* are involved in the spoilage of refrigerated milk, meat, eggs, and seafood because they grow at 4°C and degrade lipids and proteins.

The genus *Azotobacter* also is in the family *Pseudomonaceae.* These bacteria are ovoid, and may be motile by peritrichous flagella. Cells are often pleomorphic, ranging from rods to coccoid shapes, and form cysts as the culture ages. They are aerobic and fix nitrogen nonsymbiotically. *Azotobacter* spp. are widespread in soil and water. ▶▶ *Nitrogen-fixing bacteria are vital to agriculture (section 31.3)*

Order *Alteromonadales* Includes Anaerobes That Use a Range of Electron Acceptors

Many members of order *Alteromonadales,* including members of the genera *Marinomonas, Pseudoalteromonas,* and *Alteromonas,* are found in marine environments. *Shewanella* spp. have also been isolated from seawater, as well as lake sediments and

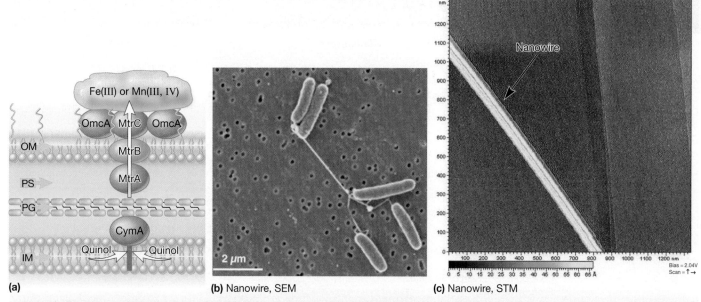

Figure 22.28 External Terminal Electron Acceptors. (a) *Shewanella oneidensis* MR1 transfers electrons from the plasma membrane, through the outer membrane, to an insoluble metal. CymA is a quinol dehydrogenase that oxidizes quinol and passes the electrons to the c-type cytochrome MtrA. MtrB is not a cytochrome but helps move the electrons through the outer membrane to the surface-exposed c-type cytochromes, MtrC and OmcA. These redox proteins donate electrons to the oxidized, insoluble metals iron and manganese, which thereby serve as external, terminal electron acceptors. (b) When electron acceptors are in short supply, *S. oneidensis* MR1 makes nanowires along which electrons can travel to a terminal electron acceptor. This scanning electron micrograph shows the production of nanowires that occurs when cells are oxygen limited. (c) A scanning tunneling micrograph of a *S. oneidensis* MR1 nanowire with a diameter of 100 nm. Note the ridges and troughs that run along the axis of the nanowire. IM, inner or plasma membrane; PG, peptidoglycan; PS, periplasmic space; OM, outer membrane.

MICRO INQUIRY *Could a Gram-positive microbe use an insoluble metal as a terminal electron acceptor in the same way* S. oneidensis *does in (a)?*

salted foods. Bacteria in the genus *Shewanella* have generated much recent interest and are the focus of our discussion. This genus includes 48 facultatively anaerobic species that form straight or curved rods. It was first noted that *S. oneidensis* could use over 10 different electron acceptors and reduce thiosulfate and elemental sulfur to sulfide, a property usually seen only in strict anaerobes. Later it was discovered that all known *Shewanella* spp. show similar metabolic flexibility and can use a variety of metal electron acceptors, including uranium, chromium, iodate, technetium, neptunium, plutonium, selnite, tellurite, and vanadate, as well as nitroaromatic compounds. This wide range of metal electron acceptors makes *Shewanella* spp. excellent candidates as agents for bioremediation of environments contaminated with radionuclides. In addition, these bacteria can use a number of carbon substrates, and some can use H_2 as an energy and electron source. ▶▶ *Biodegradation and bioremediation (section 43.4)*

The use of metals as terminal electron acceptors is considered dissimilatory metal reduction because the metals are not incorporated into biomass. Dissimilatory metal reduction presents a problem for the microbe because many metals, such as iron and manganese, are insoluble in their oxidized forms, which is the form needed by the microbe for use as an electron acceptor.

Dissimilatory metal-reducing bacteria have evolved several strategies that enable the use of solid metals such as Fe(III) or Mn(IV) as electron acceptors. The most direct approach is to localize cytochromes in the outer membrane, rather than the plasma membrane (**figure 22.28a**). This enables the direct transfer of electrons from cytochromes to the extracellular metal. This approach is used by *Shewanella,* as well as by a number of Gram-negative dissimilatory metal-reducing bacteria such as the δ-proteobacterium *Geobacter sulfurreducens* (p. 531). Another strategy used by members of the genera *Shewanella* and *Pseudomonas* is to transfer electrons to the metal via external, intermediary compounds such as humic acids found in the environment or their own secreted metabolites. These compounds are collectively known as **electron shuttles** because they pass (shuttle) electrons from the last point on the electron transport chain (ETC) through the growth substrate to the mineral surface. Species of *Shewanella, Geobacter,* and some cyanobacteria have evolved a third strategy: the production of electrically conductive **nanowires.** These appendages measure about 100 nm in diameter and tens of μm in length. In *Shewanella* spp., nanowires are known to be thread-like extensions of plasma membrane. They transfer electrons from the terminal point in the ETC to a metal surface (figure 20.28b,c). Nanowires enable the transport of electrons to solid-phase acceptors

Table 22.6   Characteristics of Families of Facultatively Anaerobic Gram-Negative Rods

Characteristics	*Enterobacteriaceae*	*Vibrionaceae*	*Pasteurellaceae*
Cell dimensions	0.3–1.0 × 1.0–6.0 μm	0.3–1.3 × 1.0–3.5 μm	0.2–0.4 × 0.4–2.0 μm
Morphology	Straight rods; peritrichous flagella or nonmotile	Straight or curved rods; polar flagella; lateral flagella may be produced on solid media	Coccoid to rod-shaped cells, sometimes pleomorphic; nonmotile
Physiology	Oxidase negative	Oxidase positive; all can use D-glucose as sole or principal carbon source	Oxidase positive; heme and/or NAD$^+$ often required for growth; organic nitrogen source required
Symbiotic relationships	Some are parasitic on mammals and birds; some species are plant pathogens	Most are not pathogens; several inhabit light organs of marine organisms	Parasites of mammals and birds
Representative genera	*Escherichia, Shigella, Salmonella, Citrobacter, Klebsiella, Enterobacter, Erwinia, Serratia, Proteus, Yersinia*	*Vibrio, Photobacterium*	*Pasteurella, Haemophilus*

Data from Garrity, G. M., editor-in-chief. Bergey's Manual of Systematic Bacteriology, vol. 2. Copyright © 2005 New York: Springer. Reprinted by permission.

that are physically distant from the microbe. Interestingly, when near a redox active metal, *Shewanella* cells display electrokinesis—behavior that is analogous to chemotaxis. ◄◄ *Chemotaxis (section 3.8)* ►►| *Microbial fuel cells (section 43.3)*

Order *Vibrionales* Includes Aquatic Bioluminescent Bacteria and Pathogens

Three closely related orders of *Gammaproteobacteria* contain a number of important bacterial genera. Each order has only one family of facultatively anaerobic Gram-negative rods. **Table 22.6** summarizes the distinguishing properties of the families *Enterobacteriaceae, Vibrionaceae,* and *Pasteurellaceae.*

The order *Vibrionales* contains the family *Vibrionaceae*. Members of this family are straight or curved, flagellated rods. All use D-glucose as their sole or primary carbon and energy source (table 22.6). The majority are aquatic microorganisms, widespread in freshwater and the sea. The family has eight genera: *Vibrio, Photobacterium, Salinivibrio, Listonella, Allomonas, Enterovibrio, Catencoccus,* and *Grimontia.*

Several vibrios are important pathogens. *Vibrio cholerae* causes cholera, and *V. parahaemolyticus* can cause gastroenteritis in humans following consumption of contaminated seafood. *V. anguillarum* and others are responsible for fish diseases, which can be especially problematic in fish farms. ►►| *Cholera (section 39.4)*

Some members of *Vibrionaceae* are unusual in being bioluminescent. *Vibrio fischeri, V. harveyi,* and at least two species of *Photobacterium* are marine bacteria capable of bioluminescence. They emit a blue-green light catalyzed by the enzyme luciferase (**Microbial Diversity & Ecology 22.2**). The light is usually blue-green in color (472 to 505 nm), but one strain of *V. fischeri* emits yellow light with a major peak at 545 nm. Although many of these bacteria are free living, *V. fischeri, V. harveyi, P. phosphoreum,* and *P. leiognathi* live symbiotically in the luminous organs of fish (**figure 22.29**) and squid (*see figure 7.21*).

(a)

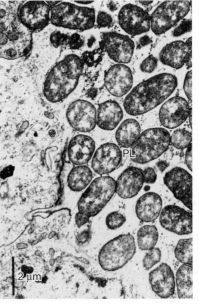

(b)

Figure 22.29   Bioluminescence.   (a) A photograph of the Atlantic flashlight fish *Kryptophanaron alfredi*. The light area under the eye is the fish's luminous organ, which can be covered by a lid of tissue. (b) Ultrathin section of the luminous organ of a fish, *Equulities novaehollandiae,* with the bioluminescent bacterium *Photobacterium leiognathi,* PL.

MICROBIAL DIVERSITY & ECOLOGY

22.2 Bacterial Bioluminescence

Several species in the genera *Vibrio* and *Photobacterium* can emit light of a blue-green color. The enzyme luciferase catalyzes the reaction and uses reduced flavin mononucleotide, molecular oxygen, and a long-chain aldehyde as substrates.

$$FMNH_2 + O_2 + RCHO \xrightarrow{\text{luciferase}} FMN + H_2O + RCOOH + \text{light}$$

Evidence suggests that an enzyme-bound, excited flavin intermediate is the direct source of luminescence. Because the electrons used in light generation are probably diverted from the electron transport chain and ATP synthesis, the bacteria expend considerable energy on luminescence. It follows that luminescence is regulated and can be turned off or on under the proper conditions.

There is much speculation about the role of bacterial luminescence and its value to bacteria, particularly because it is such an energetically expensive process. Luminescent bacteria do not emit light when they grow as free-living organisms in sea-water; the light emitted by a single cell would not be visible. When free-living luminescent bacteria infect the light organ of young fish or squid, they reproduce and reach densities where light emission is visible. Bacterial luminescence was the first example of density-dependent gene regulation by the quorum-sensing molecule autoinducer.

It remains unclear what advantage bacterial luminescence confers on some animal hosts. However, the mechanism by which autoinducer regulates light production in these marine bacteria is an important model for understanding quorum sensing in many Gram-negative bacteria, including a number of pathogens. ◄◄ *Cell-cell communication within microbial populations (section 7.5); Quorum sensing by Vibrio spp. (section 14.5)*

Order *Enterobacteriales* Includes Pathogens and Mutualists

The order *Enterobacteriales* has only one family, *Enterobacteriaceae,* which contains 44 genera. This family is the largest listed in table 22.6. It contains peritrichously flagellated or nonmotile, facultatively anaerobic, straight rods with simple nutritional requirements.

The metabolic properties (particularly during anaerobic growth) of members of *Enterobacteriaceae* are very useful in characterizing its constituent genera. These bacteria, often called **enterobacteria** or **enteric bacteria** (Greek *enterikos,* pertaining to the intestine), all degrade sugars by means of the Embden-Meyerhof pathway. Under microaerobic or anoxic conditions, the enzyme **pyruvate formate-lyase** (PFL) catalyzes the cleavage of pyruvate to formate and acetyl-CoA. Those enteric bacteria that produce large amounts of gas during sugar fermentation, such as *Escherichia* spp., also have the enzyme **formate dehydrogenase** (FDH), which splits formic acid to H_2 and CO_2 (**figure 22.30**). This enzyme complex is crucial to survival during fermentation, when one-third of the carbon derived from glucose is converted to formate. If this intermediate were allowed to accumulate within the cell, the internal pH would become far too acidic to support viability. In fact, *E. coli* has three different FDH enzymes. FDH-H (H for hydrogen production) is synthesized only during fermentative conditions. FDH-N is produced under anoxic conditions when nitrate serves as the terminal electron acceptor. FDH-O is produced under microaerobic conditions as well as when nitrate is the terminal electron acceptor.

The family can be divided into two groups based on their fermentation products. The majority (e.g., the genera *Escherichia, Proteus, Salmonella,* and *Shigella*) carry out mixed acid fermentation and produce mainly lactate, acetate, succinate, ethanol, formate, H_2, and CO_2. In contrast, *Enterobacter, Serratia, Erwinia,* and *Klebsiella* spp. are butanediol fermenters, producing butanediol, ethanol, and CO_2. ◄◄ *Fermentation does not involve an electron transport chain (section 11.8)*

Because enteric bacteria are so similar in morphology, biochemical tests are normally used to identify them after a preliminary examination of their morphology, motility, and growth responses (**figure 22.31** provides a simple example). Some more commonly used tests are those for the type of fermentation, lactose and citrate utilization, indole production from tryptophan, urea hydrolysis, and hydrogen sulfide production. For example, lactose fermentation occurs in some genera (e.g., *Escherichia, Klebsiella,* and *Enterobacter*) but not in others (e.g., *Shigella, Salmonella,* and *Proteus*). **Table 22.7** summarizes a few of the biochemical properties useful in distinguishing between genera of enteric bacteria. The mixed acid fermenters are located on the left in this table and the butanediol fermenters on the right. The two types of fermentations are distinguished by the methyl red and Voges-Proskauer tests, respectively. The usefulness of biochemical tests in identifying enteric bacteria is shown by the popularity of commercial identification systems, such as the Enterotube II and API 20E systems, that incorporate these tests. ►► *Identification of microorganisms from specimens (section 36.3)*

Members of *Enterobacteriaceae* are so widespread and important that they are probably seen more often in laboratories

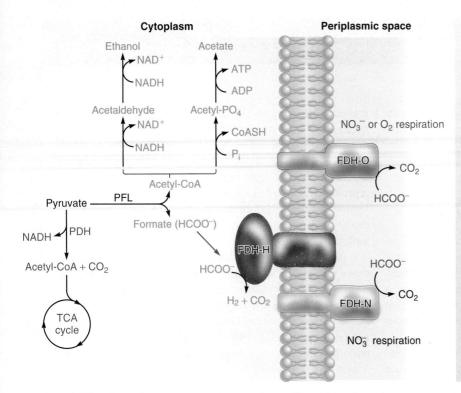

Cytoplasm

Ethanol   Acetate

Periplasmic space

NO_3^- or O_2 respiration

FDH-O → CO_2

$HCOO^-$

Acetyl-CoA

Pyruvate —PFL—

Formate (HCOO⁻)

NADH ↑ PDH

$HCOO^-$

Acetyl-CoA + CO_2

FDH-H

$H_2 + CO_2$

$HCOO^-$

FDH-N → CO_2

TCA cycle

NO_3^- respiration

Figure 22.30 Metabolism of Pyruvate and Formate in *E. coli*. During aerobic growth, pyruvate is converted to CO_2 and acetyl-CoA by the pyruvate dehydrogenase complex (PDH) so that acetyl-CoA can enter the TCA cycle. Under low oxygen conditions, some pyruvate is also cleaved to formate and acetyl-CoA by pyruvate formate-lyase (PFL). Formate must then be converted into CO_2 and H_2 by one of three formate dehydrogenase (FDH) enzymes. FDH-O is active during aerobic respiration when oxygen levels are low. If respiration continues under anoxic conditions with nitrate as the terminal electron acceptor, both FDH-N and FDH-O split formate. These enzymes are located on the periplasmic face of the plasma membrane. During fermentative conditions, FDH-H is made, resulting in mixed acid fermentation: formate is cleaved to yield CO_2 and H_2, whereas acetyl-CoA is converted to ethanol and acetate. FDH-H is a peripheral membrane protein, with its active site exposed to the cytoplasm. Aerobic metabolism is shown in pink, fermentative metabolism in green.

MICRO INQUIRY *Why do you think FDH-O and FDH-N are on the periplasmic face of the plasma membrane while FDH-H is on the cytoplasmic face?*

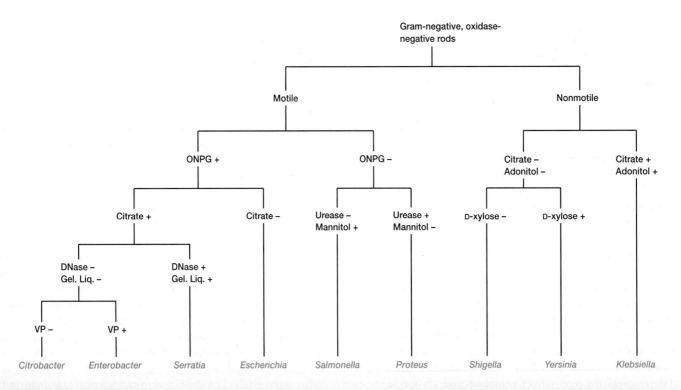

Figure 22.31 Identification of Enterobacterial Genera. A dichotomous key to selected genera of enteric bacteria based on motility and biochemical characteristics. The following abbreviations are used: ONPG, o-nitrophenyl-β-D-galactopyranoside (a test for β-galactosidase); DNase, deoxyribonuclease; Gel. Liq., gelatin liquefaction; and VP, Voges-Proskauer (a test for butanediol fermentation).

MICRO INQUIRY *What would be the identity of a motile microbe that synthesizes the enzyme β-galactosidase, assimilates citrate, produces DNase, and liquefies gelatin?*

than any other bacteria. *E. coli* is undoubtedly the best-studied bacterium and the experimental organism of choice for many microbiologists. It is an inhabitant of the colon of humans and other warm-blooded animals, and it is useful in the analysis of water for fecal contamination. Some strains cause gastroenteritis or urinary tract infections. Several other genera contain very important human pathogens responsible for a variety of diseases: *Salmonella*, typhoid fever and gastroenteritis; *Shigella*, bacillary dysentery; *Klebsiella*, pneumonia; and *Yersinia*, plague. Members of the genus *Erwinia* are major pathogens of crop plants and cause blights, wilts, and several other plant diseases.

Order *Pasteurellales* Includes Important Pathogens

Bacteria in the family *Pasteurellaceae* (order *Pasteurellales*) differs from those in *Vibrionales* and *Enterobacteriales* in several ways (table 22.6). Most notably, they are smaller, nonmotile, normally oxidase positive, have complex nutritional requirements, and are parasitic in vertebrates. The family contains seven genera: *Pasteurella, Haemophilus, Actinobacillus, Lonepinella, Mannheimia, Phocoenobacter,* and *Gallibacterium*.

Members of this family are best known for the diseases they cause in humans and many animals. *Pasteurella multocida* is responsible for fowl cholera, which kills many chickens, turkeys, ducks, and geese each year. *P. haemolytica* is at least partly responsible for pneumonia in cattle, sheep, and goats (e.g., "shipping fever" in cattle). *H. influenzae* serotype b is a major human pathogen that causes a variety of diseases, including meningitis, sinusitis, pneumonia, and bronchitis. Fortunately, a sharp reduction in the incidence of *H. influenzae* serotype b infections began in the mid-1980s due to administration of the *H. influenzae* type b ("Hib") vaccine. However, this bacterium still causes at least 3 million cases of serious disease and several hundreds of thousands of deaths each year globally. ▶▶◀ *Meningitis (section 39.1)*

Retrieve, Infer, Apply

1. What are the major characteristics of purple sulfur bacteria? Contrast the features that define the families *Chromatiaceae* and *Ectothiorhodospiraceae*.
2. Describe the metabolism of *Beggiatoa, Leucothrix,* and *Thiothrix* spp. Why do *Thiomargarita* cells appear hollow?
3. What are the environmental impacts of the bacteria in the genera *Pseudomonas* and *Azotobacter*?
4. List three attributes that account for the importance of pseudomonads.
5. Why is metal reduction by *Shewanella* spp. considered dissimilatory? Describe two morphological adaptations that enable oxidation of insoluble metals outside the cell.
6. List the major distinguishing traits of bacteria in the families *Vibrionaceae, Enterobacteriaceae,* and *Pasteurellaceae*.
7. What two cofactors are required for bioluminescence?
8. Into what two groups can enteric bacteria be placed based on their fermentation patterns?

22.4 Class *Deltaproteobacteria* Includes Chemoheterotrophic Anaerobes and Predators

After reading this section, you should be able to:

- Characterize the common physiological features of sulfate-reducing bacteria (SRB)
- Identify common carbon substrates used by SRB
- Draw an electron transport chain capable of dissimilatory sulfate reduction
- Relate the physiology of SRB to their common environmental niches
- Compare and contrast the physiology and ecology of *Shewanella* and *Geobacter* spp.
- Draw the life cycle of a bdellovibrio and explain why it is considered a bacterial predator
- Relate the life cycle of myxobacteria to their habitat and means of acquiring nutrients
- Describe the two types of gliding motility displayed by myxobacteria, and sketch their respective mechanistic models

Although **δ-proteobacteria** are not a large assemblage of genera, they show considerable morphological and physiological diversity. These bacteria can be divided into two general groups, all of them chemoorganoheterotrophs. Some are predators, such as bdellovibrios and myxobacteria. Others are anaerobes that use sulfate and sulfur as terminal electron acceptors while oxidizing organic nutrients. The class has eight orders and 20 families. **Figure 22.32** illustrates the phylogenetic relationships among major groups of δ-proteobacteria, and

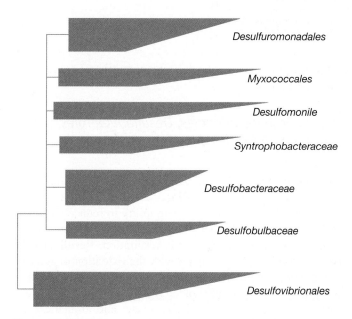

Desulfuromonadales

Myxococcales

Desulfomonile

Syntrophobacteraceae

Desulfobacteraceae

Desulfobulbaceae

Desulfovibrionales

Figure 22.32 Phylogenetic Relationships Among Major Groups Within the δ-Proteobacteria. The relationships are based on 16S rRNA sequence.

Table 22.7   Some Characteristics of Selected Genera in *Enterobacteriaceae*

Characteristics	*Escherichia*	*Shigella*	*Salmonella*	*Citrobacter*	*Proteus*
Methyl red	+	+	+	+	+
Voges-Proskauer	−	−	−	−	d
Indole production	(+)	d	−	d	d
Citrate use	−	−	(+)	+	d
H$_2$S production	−	−	(+)	d	(+)
Urease	−	−	−	(+)	+
β-galactosidase	(+)	d	d	+	−
Gas from glucose	+		(+)	+	+
Acid from lactose	+		(−)	d	−
Phenylalanine deaminase	−	−	−	−	+
Lysine decarboxylase	(+)	−	(+)	−	−
Ornithine decarboxylase	(+)	d	(+)	(+)	d
Motility	d	−	(+)	+	+
Gelatin liquefaction (22°C)	−	−	−	−	+
Genome size (Mb)	4.6–5.5	4.6	4.5–4.9	Nd^4	Nd
Other characteristics	1.1–1.5 × 2.0–6.0 μm; peritrichous flagella when motile	No gas from sugars	0.7–1.5 × 2–5 μm; peritrichous flagella	1.0 × 2.0–6.0 μm; peritrichous flagella	0.4–0.8 × 1.0–3.0 μm; peritrichous flagella

1 (+) Usually present.
2 (−) Usually absent.
3 d, Strains or species vary in possession of characteristic.
4 Nd: Not determined; genome not yet sequenced.

table 22.8 summarizes the defining properties of some representative genera.

Orders *Desulfovibrionales, Desulfobacterales,* and *Desulfuromonadales*: Anaerobic Sulfate/Sulfur Reducers

Desulfovibrionales, Desulfobacterales, and *Desulfuromonadales* are a diverse group of **sulfate-** or **sulfur-reducing bacteria (SRB)** that are united by their ability to reduce elemental sulfur or sulfate and other oxidized sulfur compounds to hydrogen sulfide during anaerobic respiration. Recall that anaerobic respiration is coupled with the oxidation of reduced organic compounds and in general the SRB can be grouped according to carbon metabolism. Many SRB oxidize lactate, formate, butyrate, propionate, pyruvate, and aromatic compounds to acetate, which is not further metabolized. Other

SRBs oxidize acetate completely to CO$_2$. ◄◄ *Anaerobic respiration uses the same three steps as aerobic respiration (section 11.7)*

The use of sulfate as a terminal electron acceptor is a multistep process. Sulfate cannot be directly reduced. Because sulfate is a stable molecule, it must first be activated by reacting with ATP to form adenosine phosphosulfate, or APS (**figure 22.33**; figure 22.20b). The fate of APS during anaerobic respiration differs from that seen in sulfide-oxidizing chemolithoautotrophs (figure 22.20*a*). During sulfate reduction, APS accepts two electrons, which results in the production of sulfite (SO$_3^{2-}$) and adenosine monophosphate (AMP). Six additional electrons are needed to reduce sulfite completely to hydrogen sulfide. These are captured by a hydrogenase enzyme located on the periplasmic face of the plasma membrane. The hydrogenase donates the electrons from molecular hydrogen to cytochrome c_3, which in turn transfers

Yersinia	Klebsiella	Enterobacter	Erwinia	Serratia
+	(+)[1]	(−)[2]	+	d[3]
− (37°C)	(+)	+	(+)	+
d	d	−	(−)	(−)
(−)	(+)	+	(+)	+
−	−	−	(+)	−
d	(+)	(−)	−	−
+	(+)	+	+	+
(−)	(+)	(+)	(−)	d
(−)	(+)	(+)	d	d
−	−	(−)	(−)	−
(−)	(+)	d	−	d
d	−	(+)	−	d
− (37°C)	−	+	+	+
(−)	−	d	d	(+)
4.6	Nd	Nd	5.1	5.1
0.5–0.8 × 1.0–3.0 μm; peritrichous flagella when motile	0.3–1.0 × 0.6–6.0 μm; capsulated	0.6–1.0 × 1.2–3.0 μm; peritrichous flagella	0.5–1.0 × 1.0–3.0 μm; peritrichous flagella; plant pathogens and saprophytes	0.5–0.8 × 0.9–2.0 μm; peritrichous flagella; colonies often pigmented

them to the integral membrane protein Hmc. Finally, an iron-sulfur protein delivers the electrons to sulfite reductase, and hydrogen sulfide is produced. Note that hydrogenase releases protons in the periplasm; the resulting proton motive force is sufficient to drive the synthesis of ATP.

SRB are very important in the cycling of sulfur within ecosystems. Because significant amounts of sulfate are present in almost all aquatic and terrestrial habitats, SRB are widespread and active in locations made anoxic by microbial digestion of organic materials. The SRB thrive in habitats such as muds and sediments of polluted lakes and streams, sewage lagoons and digesters, and waterlogged soils. *Desulfuromonas* spp. are most prevalent in anoxic marine and estuarine sediments. They also can be isolated from methane digesters and anoxic hydrogen sulfide–rich muds of freshwater habitats. Often sulfate and sulfur reduction are apparent from the smell of hydrogen sulfide (a "rotten egg" odor) and the blackening of water and sediment by iron sulfide. Hydrogen sulfide production in waterlogged soils can kill animals, plants, and microorganisms. SRB negatively

impact industry because of their primary role in the anaerobic corrosion of iron in pipelines, heating systems, and other structures. ▶▶❙ *Sulfur cycle (section 28.1)*

Order *Desulfuromonales* Includes Metabolically Flexible Anaerobes

The order *Desulfuromonales* encompasses three families, *Desulfuromonaceae*, *Geobacteraceae*, and *Pelobacteraceae*. All are strictly anaerobic with respiratory or fermentative metabolism. They can be chemolithoheterotrophs, obtaining electrons from reduced inorganic compounds, or chemoorganotrophs. They are mesophilic and have been isolated from anoxic marine and freshwater environments.

Geobacter metallireducens was the first microbe described to couple the oxidation of organic substrates such as acetate with the use of Fe(III) as a terminal electron acceptor. The capacity of *Geobacter* spp. to conserve energy from dissimilatory metal reduction is of great interest to environmental microbiologists.

Table 22.8 Characteristics of Selected δ- and ε-Proteobacteria

Class Genus	Dimensions (μm) and Morphology	Oxygen Requirement	Other Distinctive Characteristics
δ-Proteobacteria			
Bdellovibrio	0.2–0.5 × 0.5–1.4; comma-shaped rods with a sheathed polar flagellum; also capable of gliding motility	Aerobic	Prey on other Gram-negative bacteria and grow in the periplasm; alternate between predatory and intracellular reproductive phases
Desulfovibrio	0.5–1.5 × 2.5–10; curved or sometimes straight rods, motile by polar flagella	Anaerobic	Oxidize organic compounds to acetate and reduce sulfate or sulfur to H_2S
Desulfuromonas	0.4–0.9 × 1.0–4.0; straight or slightly curved or ovoid rods, lateral or subpolar flagella	Anaerobic	Reduce sulfur to H_2S, oxidize acetate to CO_2; form pink or peach-colored colonies
Myxococcus	0.4–0.7 × 2–8; slender rods with tapering ends, gliding motility	Aerobic	Form fruiting bodies with microcysts not enclosed in a sporangium
Stigmatella	0.7–0.8 × 4–8; straight rods with tapered ends, gliding motility	Aerobic	Stalked fruiting bodies with sporangioles containing myxospores (0.9–1.2 × 2–4 μm)
ε-Proteobacteria			
Campylobacter	0.2–0.8 × 0.5–5; spirally curved cells with a single polar flagellum at one or both ends	Microaerophilic	Carbohydrates not fermented or respired; oxidase positive and urease negative; found in intestinal tract, reproductive organs, and oral cavity of animals
Helicobacter	0.2–1.2 × 1.5–10; helical, curved, or straight cells with rounded ends; multiple, sheathed flagella	Microaerophilic	Catalase and oxidase positive; urea rapidly hydrolyzed; found in the gastric mucosa of humans and other animals

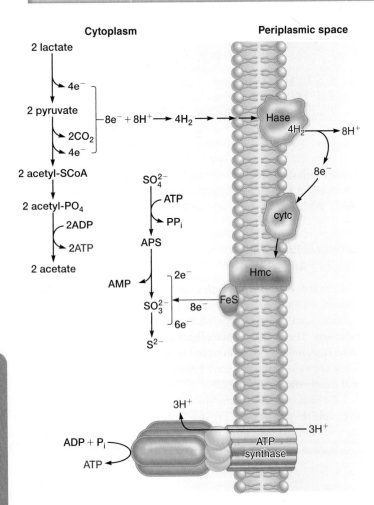

They have discovered that these bacteria, like γ-proteobacteria in the genus *Shewanella*, can reduce a number of toxic and radioactive metals. In addition, when *Geobacter* spp. oxidize organic substrates, they can transfer the electrons directly to an electrode. Like *Shewanella*, *Geobacter* spp. have an outer membrane reductase that transfers electrons they receive from the plasma membrane-bound ETC to an external, insoluble acceptor. They also synthesize nanowires (figure 22.28). However, they do not use external electron shuttles, as do *Shewanella* spp. There is great interest in using members of both genera in bioremediation and to generate electricity with microbial fuel cells. ▶▶ *Microbial fuel cells (section 43.3)*

Order *Bdellovibrionales* Includes Predators That Invade Other Gram-Negative Bacteria

The order *Bdellovibrionales* has only the family *Bdellovibrionaceae* and four genera. The genus *Bdellovibrio* (Greek *bdella*,

Figure 22.33 Dissimilatory Sulfate Reduction by *Desulfovibrio*. These microbes make ATP by substrate-level phosphorylation during acetate fermentation and by ATP synthase. Synthesis of ATP by ATP synthase is driven by a proton motive force generated by a periplasmic hydrogenase, which oxidizes H_2 to protons and electrons. The electrons are passed back into the cell, where they are used to reduce sulfate completely to sulfide. As detailed in the text, the oxidation of sulfate to sulfide requires eight electrons and the activation of sulfate by its covalent attachment to AMP to form APS.

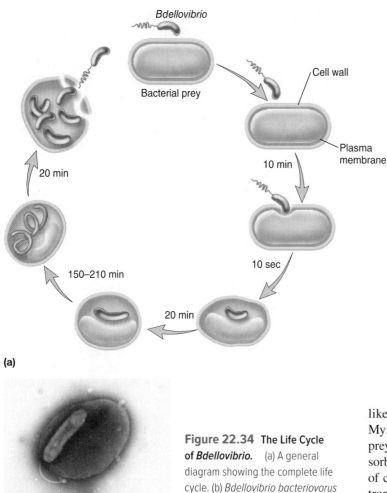

Bdellovibrio

Bacterial prey

Cell wall

Plasma membrane

10 min

10 sec

20 min

150–210 min

20 min

(a)

(b)

Figure 22.34 The Life Cycle of *Bdellovibrio*. (a) A general diagram showing the complete life cycle. (b) *Bdellovibrio bacteriovorus* encapsulated in the periplasm of *E. coli.*

leech) contains aerobic, Gram-negative, curved rods with polar flagella. The flagellum is unusually thick due to the presence of a sheath that is continuous with the cell wall. Bdellovibrios have a distinctive lifestyle: they prey on other Gram-negative bacteria and alternate between a nongrowing predatory phase and an intracellular reproductive phase.

The life cycle of a bdellovibrio is complex, although it requires only 1 to 3 hours for completion (**figure 22.34**). The free bacterium swims along very rapidly (about 100 cell lengths per second) until it collides violently with its prey. It attaches to the prey bacterium's surface, begins to rotate as fast as 100 revolutions per second, and bores a hole through the host cell wall in a process involving type IV pili and the release of several hydrolytic enzymes. Its flagellum is lost during penetration of the cell. These bacteria can also scout for bacterial prey by gliding slowly over solid substrates. ◄◄ *Bacterial pili and fimbriae (section 3.7)*

After entry, the bdellovibrio cell takes control of the host cell and grows in the periplasm while the host cell loses its shape and rounds up. The predator quickly inhibits host DNA,

RNA, and protein synthesis, and disrupts the host's plasma membrane so that cytoplasmic constituents leak out of the cell. The growing bdellovibrio uses host amino acids as its carbon, nitrogen, and energy source. It employs host fatty acids and nucleotides directly in biosynthesis, thus saving carbon and energy. The bdellovibrio rapidly grows into a long filament and then divides into many smaller, flagellated progeny, which escape upon host cell lysis. Such multiple fission is rare in bacteria and archaea. ◉ *Bdellovibrio*

Order *Myxococcales*: Bacteria with Morphological Complexity and Multicellularity

Myxobacteria are Gram-negative, aerobic soil bacteria characterized by gliding motility and a complex life cycle that includes production of multicellular structures called fruiting bodies and the formation of spores called myxospores. Myxobacterial cells are rods, which may be either slender with tapered ends or stout with rounded, blunt ends. The order *Myxococcales* is divided into six families based on the shape of vegetative cells, myxospores, and sporangia.

Most myxobacteria are predators of other microbes but, unlike bdellovibrios, do not depend on predation for reproduction. Myxobacteria secrete lytic enzymes and antibiotics to kill their prey. The digestion products, primarily small peptides, are absorbed. Most myxobacteria use amino acids as their major source of carbon, nitrogen, and energy. All are aerobic chemoorganotrophs with respiratory metabolism.

The myxobacterial life cycle is quite distinctive and in many ways resembles that of cellular slime molds (**figure 22.35**). In the presence of a food supply, myxobacteria glide along a solid surface, feeding and leaving slime trails. During this stage, the cells often form a swarm and move in a coordinated fashion. Some species congregate to produce a sheet of cells that moves rhythmically to generate waves or ripples. When their nutrient supply is exhausted, the myxobacteria aggregate and differentiate into a fruiting body, a structure that is largely made up of individual cells that have differentiated into exospores called myxospores. The life cycle of the species *Myxococcus xanthus* serves as a model system for bacterial differentiation and behavior. Development in this microbe is induced by nutrient limitation and requires at least five different extracellular signaling molecules that allow the cells to communicate with one another. ◄◄ *Bacteria move in response to environmental conditions (section 3.8)*

Fruiting bodies range in height from 50 to 500 μm and often are colored red, yellow, or brown by carotenoid pigments. Each species forms a characteristic fruiting body. They vary in complexity from simple globular objects made of about 100,000 cells (*Myxococcus*) to the elaborate, branching, treelike structures formed by members of the genera *Stigmatella* and *Chondromyces*. In some species, cells develop into dormant myxospores enclosed in walled structures called sporangioles or

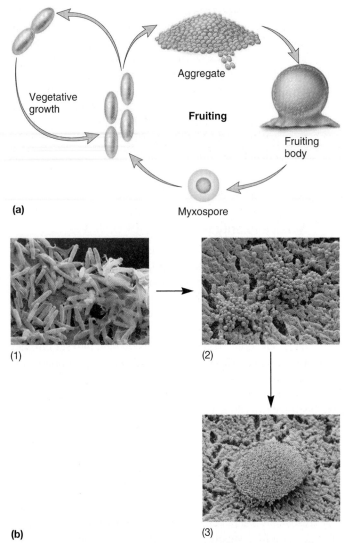

(a)

Vegetative growth

Aggregate

Fruiting

Fruiting body

Myxospore

(b)

(1)   (2)

(3)

Figure 22.35   Life Cycle of *Myxococcus xanthus*.   (a) When nutrients are plentiful, *M. xanthus* grows vegetatively. However, when nutrients are depleted, the exchange of extracellular signaling molecules triggers cells to aggregate and form fruiting bodies. Most of the cells within a fruiting body become resting myxospores that will not germinate until nutrients are available. (b) Scanning electron micrographs taken during (1) vegetative growth (2), aggregation, and (3) fruiting body formation.

sporangia. As might be expected for such complex structures, a variety of new proteins are synthesized during fruiting body formation, and *Myxococcus xanthus* has one of the largest bacterial genomes sequenced to date.

Myxospores are dormant and desiccation-resistant; they may survive up to 10 years under adverse conditions (compare this with thousands of years for Gram-positive endospores). The use of fruiting bodies provides further protection for the myxospores and assists in their dispersal. Because myxospores are kept together within the fruiting body, a colony of myxobacteria automatically develops when the myxospores are

released and germinate. This communal organization may be advantageous because myxobacteria obtain nutrients by secreting hydrolytic enzymes and absorbing soluble digestive products. A mass of myxobacteria can produce enzyme concentrations sufficient to digest their prey more easily than can an individual cell.

Fruiting body development requires gliding motility. Gliding motility in *M. xanthus* is much slower (2–4 mm/min) than in flavobacteria (*see figure 21.21*). Two types of motility have been characterized in *M. xanthus*. Social (S) motility is a kind of twitching that is governed by the production of retractable type IV pili from the front end of the cell (**figure 22.36***a*). When the pili retract, the cell creeps forward. This type of motility was originally called social motility because it is only observed when cells are in a group. It is now known that cell-to-cell contact is required for S motility because cells share outer membrane lipoproteins involved in pili secretion. The second type of motility is called adventurous (A) motility because it is exhibited by single cells that leave the group, perhaps scouting for prey.

The mechanism mediating adventurous motility in *M. xanthus* remained a mystery for many decades. It is now known that A motility requires a multiprotein motility "machine" (figure 22.36*b*). The molecular motor is made up of Agl proteins, which have several functions. Agl proteins couple the translocation of protons with gliding motility, much like Mot proteins in the flagellar basal body in swimming bacteria (*see figure 3.45*). When gliding, the Agl proteins move in a continuous looped path lengthwise along the cell surface. The cytoskeletal protein MreB may be involved with the helical movement of the Agl proteins. The mechanical work of the Agl motor proteins is converted to movement by the Glt (*gl*iding *t*ransducer) proteins located in the cell wall. Because the Glt proteins contact one of the Agl proteins, they too are moved along the cell surface. Critically, Glt proteins use this movement to produce thrust when the entire multiprotein complex comes into contact with the substrate at specific areas in the cell envelope called focal adhesions. Although it is not known how Glt proteins contact the substrate, it is clear that this is facilitated by slime secreted by the bacterium.

It is intriguing that a single microbe has evolved two motility systems, and while S motility appears to be similar to twitching motility in *Pseudomonas aeruginosa*, neither system seems to be similar to the mechanism by which *Bacteroidetes*, cyanobacteria, or *Mycoplasma* species glide. This suggests that gliding motility has evolved independently multiple times. ◄◄ *Bacterial cytoskeleton (section 3.6); Twitching and gliding motility (section 3.8); Class Mollicutes, Phylum Tenericutes (section 21.3); Phylum Bacteroidetes includes important gut microbiota (section 21.9)*

Myxobacteria are found in soils worldwide. They are most commonly isolated from neutral soils or decaying plant material such as leaves and tree bark, and from animal dung. Although they grow in habitats as diverse as tropical rain forests and the Arctic tundra, they are most abundant in warm areas.

Figure 22.36 Model for Gliding Motility in *Myxococcus xanthus*. *M. xanthus* displays two types of surface-associated motility. During social motility, type IV pili extend and retract from the leading pole, much like grappling hooks. When cells reverse direction, the pili disassemble and reassemble at the opposite pole. (b) Adventurous motility involves a motor (Agl proteins) bound to Glt proteins in the cell wall; motility is driven by the proton motive force. Glt proteins attach to the substrate in an unknown way. When the entire multiprotein complex moves in a helical fashion along the length of the cell, the Glt proteins produce the thrust needed to propel the cell.

MICRO INQUIRY *How could the analysis of motility mutants be used to determine the role of the Glt proteins?*

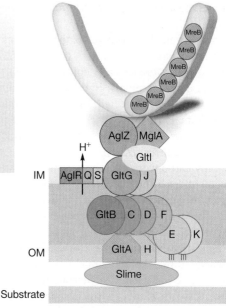

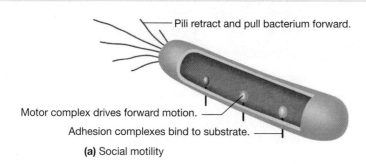

Pili retract and pull bacterium forward.

Motor complex drives forward motion.

Adhesion complexes bind to substrate.

(a) Social motility

(b) Adventurous motility

Retrieve, Infer, Apply

1. What general features are common to most δ-proteobacteria?
2. What is the role of fermentation in dissimilatory sulfate reduction? How do SRB make ATP?
3. What mechanisms do *Geobacter* spp. use to transfer electrons to extracellular insoluble metals? How does this compare with *Shewanella* spp?
4. Characterize *Bdellovibrio* spp. and outline the life cycle in detail. Why can they prey only on Gram-negative bacteria?
5. Compare the growth of a myxobacterium in nutrient-rich and nutrient-poor conditions. Why are extracellular signaling molecules required for fruiting body formation?

22.5 Class *Epsilonproteobacteria* Ranges from Pathogens to Deep-Sea Bacteria

After reading this section, you should be able to:

- List two common human illnesses caused by ε-proteobacteria
- Describe the unusual metabolism of *Helicobacter* spp. and explain how *H. pylori* can tolerate the acidic conditions within the stomach
- Describe some of the unusual habitats that recently discovered ε-proteobacteria inhabit

ε-proteobacteria make up the smallest of the five proteobacterial classes. They all are slender, Gram-negative rods, which can be straight, curved, or helical. Although *Bergey's Manual* places ε-proteobacteria in one order, *Campylobacterales* (families

Campylobacteraceae, Helicobacteraceae, and *Hydrogenimonaceae*), recent phylogenetic analysis that includes SSU rRNA sequences from newly cultured and uncultured organisms identifies an additional order called *Nautiliales*. Although ε-proteobacteria are physiologically and ecologically diverse, most inhabit extreme environments such as the acidic gastric mucosa, hydrothermal vents, and sulfidic caves. Two genera, *Campylobacter* and *Helicobacter,* contain members that are microaerophilic, motile, helical or vibrioid, Gram-negative rods. Table 22.8 summarizes some of the characteristics of these two genera.

The genus *Campylobacter* contains both nonpathogens and species pathogenic for humans and other animals. *C. foetus* causes reproductive disease and abortions in cattle and sheep. Other campylobacters are associated with conditions in humans ranging from septicemia (pathogens or their toxins in the blood) to enteritis (inflammation of the intestinal tract). Infection with some species frequently precedes the development of Guillain-Barré syndrome (GBS), the most common form of flaccid paralysis. It is thought that these bacteria trigger GBS by a phenomenon known as molecular mimicry. In this case, the structure of *Campylobacter* spp. lipopolysaccharide resembles components found on nerve cells. This causes the immune system to mistakenly attack the host's peripheral nervous system even after infection has cleared.

C. jejuni is a slender, Gram-negative, motile, curved rod found in the intestinal tract of animals. It causes an estimated 2 million human cases of *Campylobacter* gastroenteritis—inflammation of the stomach and intestine with subsequent diarrhea—in the United States each year. Studies with chickens, turkeys, and cattle have shown that as much as 50 to 100% of a flock or herd of these birds or other animals excrete *C. jejuni*. Campylobacteria are transmitted to humans by contaminated food and water, contact with infected animals, or anal-oral sexual activity.

The other ε-proteobacterial genus that contains pathogens is *Helicobacter.* There are at least 23 species of *Helicobacter,* all isolated from the stomachs and upper intestines of humans, dogs, cats, and other mammals. In developing countries, 70 to 90% of the population is infected; the rate in developed countries ranges from 25 to 50%. The major human pathogen is *Helicobacter pylori,* which causes gastritis, peptic ulcer disease, and gastric cancer. *H. pylori* is an obligate microaerophile. The only carbohydrate the bacterium can use is glucose, which it metabolizes by both respiratory and fermentative pathways. It uses the pentose phosphate pathway, which provides NADPH for biosynthesis, and the Entner-Doudoroff pathway to produce pyruvate. The microbe does not have a complete TCA cycle and appears to run the cycle "forward" (i.e., oxidatively) from oxaloacetate to α-ketoglutarate and "backward" (i.e., in the reductive direction) from oxaloacetate to succinate. Fumarate may act as the terminal electron acceptor during anaerobic respiration. It can also carry out aerobic respiration. Pyruvate is fermented to lactate, acetate, and formate during fermentation. As a resident of the gastric mucosa, it would seem that *H. pylori* is exposed to very acidic conditions, yet in the laboratory, it cannot be cultured below pH 4.5. The presence of urease enzymes in the cytoplasm and the cell surface helps explain this paradox. *H. pylori* burrows within the gastric mucosa, which is not as acidic as the lumen. There it depends on urease activity to convert urea to CO_2 and NH_3. The ammonia drives the local pH up while the microbe uses chemotaxis to maintain its position within the mucous membranes. Indeed, motility is required for colonization of the host.

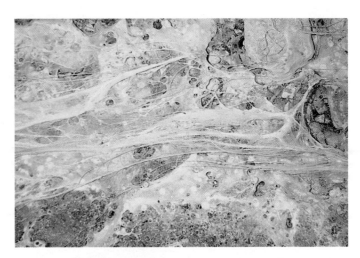

Figure 22.37 ε-Proteobacteria Dominate Filamentous Microbial Mats in a Wyoming Sulfidic Cave Spring.

ε-proteobacteria are also found in a variety of unusual ecosystems. For instance, filamentous microbial mats in anoxic, sulfide-rich cave springs are dominated by ε-proteobacteria (**figure 22.37**). Many have been cultured from marine hydrothermal vents; terrestrial forms have been isolated from groundwater, oil-field brines, limestone caves, and sulfidic springs. These include thermophilic, chemolithoautotrophic, chemoorganotrophic, and chemolithoheterotrophic species (**table 22.9**).

Chemolithoheterotrophic organisms use organic carbon for biosynthesis but inorganic compounds, such as sulfide and molecular hydrogen, as electron donors for energy conservation. The autotrophic species use the reductive TCA cycle for CO_2 fixation (*see figure 12.5*). Many ε-proteobacteria that

Table 22.9	Some Recently Isolated ε-Proteobacteria					
Species	**Isolation Site**	**Optimum Growth Temperature**	**Carbon Metabolism**	**Electron Donor**	**Electron Acceptor**	**Sulfur/Nitrogen Reduction Product**
Nautilia lithotrophica	Hydrothermal vent	52°C	Heterotroph	H_2, formate	SO_3^{2-}, S^0	H_2S
Caminibacter hydrogeniphilus	Hydrothermal vent	60°C	Heterotroph	H_2, complex organic compounds	NO_3^-, S^0	H_2S, NH_3
Nitratiruptor tergarcus	Hydrothermal vent	55°C	Autotroph	H_2	O_2, (microaerobic), NO_3^-, S^0	H_2S, N_2
Sulfurospirillum sp. str. Am-N	Hydrothermal vent	41°C	Heterotroph	Formate, fumarate	S^0	H_2S
Arcobacter sp. str. FWKO B	Oil-field production water	30°C	Autotroph	H_2, formate, HS^-	O_2, (microaerobic), NO_3^-, S^0	H_2S, N_2O^-
Sulfuricurvum kujiense	Oil-field production water	25°C	Autotroph	H_2, HS^-, $S_2O_3^{2-}$, S^0	O_2, (microaerobic), NO_3^-	NO_2^-

From Campbell, B. J., et al. 2006. The versatile ε-proteobacteria: Key players in sulphidic habitats. Nature Rev. Microbiol. 4:458–67.

inhabit hydrothermal vents are capable of nitrate respiration, that is, anaerobic respiration using nitrate as the terminal electron acceptor. This discovery suggests that microbial metabolism is more diverse in these interesting ecosystems usually thought of as sites of sulfur cycling. ▶▶| *Sulfide-based mutualisms (section 32.1)*

Key Concepts

22.1 Class *Alphaproteobacteria* Includes Many Oligotrophs

- The purple nonsulfur bacteria can grow anaerobically as photoorganoheterotrophs and aerobically as chemoorganoheterotrophs. When growing photosynthetically, purple bacteria localize the light-harvesting proteins and reaction centers on intracytoplasmic membranes (**figure 22.3**).
- Rickettsias are obligate intracellular parasites. They have numerous transport proteins in their plasma membranes and make extensive use of host cell nutrients, coenzymes, and ATP.
- Many proteobacteria have prosthecae or stalks or reproduce by budding. Two examples of budding or appendaged bacteria are *Hyphomicrobium* spp. (budding bacteria that produce swarmer cells) and *Caulobacter* spp. (bacteria with prosthecae and holdfasts) (**figures 22.7–22.10**).
- *Rhizobium* spp. carry out nitrogen fixation, whereas agrobacteria cause the development of plant tumors. Both are in the family *Rhizobiaceae* (**figures 22.11** and **22.12**).
- Proteobacteria that use 1-carbon compounds for carbon and energy are called methylotrophs; some of these microbes can use methane and are called methanotrophs. In bacteria, this is an aerobic process that produces the intermediate formaldehyde, which is either oxidized for energy or assimilated for anabolic purposes (**figures 22.13, 22.14, and 22.28**).
- Nitrifying bacteria are chemolithotrophic aerobes that oxidize either ammonia to nitrite or nitrite to nitrate and are responsible for nitrification (**figure 22.15** and **table 22.2**).
- *Magnetococcus marinus* is the only cultured representative of a new order of magnetotactic bacteria. These marine microbes are facultatively chemolithoautotrophic and microaerophilic.
- Members of the globally distributed marine SAR11 clade are the most abundant microorganisms on Earth. They are uniquely adapted to life in oligotrophic waters.

22.2 Class *Betaproteobacteria* Includes Chemoheterotrophs and Chemolithotrophs

- The genus *Neisseria* contains nonmotile, aerobic, Gram-negative cocci that usually occur in pairs. They colonize mucous membranes and cause several human diseases, including meningitis and gonorrhea.

- *Sphaerotilus, Leptothrix,* and several other genera contain species with sheaths, hollow tubelike structures that surround chains of cells without being in intimate contact with the cells.
- The order *Nitrosomonadales* includes two genera of nitrifying bacteria, including the ammonia-oxidizing microbe *Nitrosomonas europaea* (**table 22.2; figure 22.19**).
- Colorless sulfur bacteria such as *Thiobacillus* spp. oxidize elemental sulfur, hydrogen sulfide, and thiosulfate to sulfate while generating energy chemolithotrophically (**figure 22.20**).

22.3 Class *Gammaproteobacteria* Is the Largest Bacterial Class

- γ-proteobacteria are the largest subgroup of proteobacteria and have great variety in physiological types (**table 22.5**).
- Purple sulfur bacteria are anaerobes and usually photolithoautotrophs. They oxidize hydrogen sulfide to sulfur and deposit the granules internally (**figure 22.22**).
- Bacteria such as *Beggiatoa* and *Leucothrix* spp. grow in long filaments or trichomes. They have gliding motility. *Beggiatoa* spp. are primarily chemolithotrophs and *Leucothrix* spp. are chemoorganotrophs.
- Marine *Beggiatoa, Thioploca,* and *Thiomargarita* spp. are among the largest microbes. Members of the genus *Thiomargarita* form large intracellular vacuoles in which nitrate is stored (**figure 22.24**).
- The marine bacterium *Thiomicrospira crunogena* is a chemolithoautotroph that uses reduced sulfur compounds as a source of energy and electrons and O_2 as the terminal electron acceptor (**figure 22.26**).
- Members of the genus *Pseudomonas* are straight or slightly curved, Gram-negative, aerobic rods that are motile by one or several polar flagella and do not have prosthecae or sheaths.
- Pseudomonads participate in natural mineralization processes, are major experimental subjects, cause many diseases, and often spoil refrigerated food.
- The genus *Shewanella* is notable for species with metabolic flexibility and the ability to use extracellular, insoluble metals as terminal electron acceptors. They are among several microbes that have evolved strategies to accomplish this, including the use of electron shuttles and formation of nanowires (**figure 22.28**).

- The most important facultatively anaerobic, Gram-negative rods are found in three families: *Vibrionaceae, Enterobacteriaceae,* and *Pasteurellaceae* (**table 22.6**).
- *Enterobacteriaceae,* often called enterobacteria or enteric bacteria, are peritrichously flagellated or nonmotile, facultatively anaerobic, straight rods with simple nutritional requirements (**table 22.7**).
- Enteric bacteria are usually identified by a variety of physiological tests. *E. coli* performs a mixed acid fermentation, which depends on the enzymes pyruvate formate-lyase and formate dehydrogenase (**figure 22.30**).

22.4 Class *Deltaproteobacteria* Includes Chemoheterotrophic Anaerobes and Predators

- δ-proteobacteria are aerobic chemoorganotrophic bacteria that can use elemental sulfur and oxidized sulfur compounds as electron acceptors in anaerobic respiration. Other δ-proteobacteria are predatory aerobes (**table 22.8**).
- Sulfate-reducing bacteria are very important in sulfur cycling in ecosystems. *Desulfovibrio* spp. conserve energy by fermentation and a proton motive force (**figure 22.33**).

- *Bdellovibrio* spp. are aerobic curved rods with a sheathed polar flagellum that consume other Gram-negative bacteria and grow within the periplasmic space of their prey (**figure 22.34**).
- *Myxobacteria* spp. are Gram-negative, aerobic soil bacteria with gliding motility and a complex life cycle that leads to the production of dormant myxospores held within fruiting bodies (**figures 22.35** and **22.36**).

22.5 Class *Epsilonproteobacteria* Ranges from Pathogens to Deep-Sea Bacteria

- ε-proteobacteria are the smallest of the proteobacterial classes and include two important genera with pathogenic species: *Campylobacter* and *Helicobacter*. These bacteria are microaerophilic, motile, helical or vibrioid, Gram-negative rods (**table 22.8**).
- Recently the order *Nautiliaceae* has been added. Many of these bacteria are chemolithoautotrophs from deep-sea hydrothermal vent ecosystems (**table 22.9**).

Compare, Hypothesize, Invent

1. The advantages of metabolic flexibility in *Rhodospirillum rubrum* must offset the energetic burden of carrying all the structural and regulatory genes involved in these processes. What makes this trade-off possible?

2. Compare nitrification with denitrification: which is found in heterotrophs and which is found in autotrophs? Are they both dissimilatory processes?

3. Why might the ability to form dormant cysts be of great advantage to an agrobacterium but not as much to a rhizobium?

4. Examine figures 22.20 and 22.33. Note that in both cases APS is formed, but in the case of *T. ferrooxidans*, it is involved in sulfide oxidation, whereas for *Desulfovibrio*, it is used during sulfate reduction. What other similarities and differences are present? Why must APS be used in both cases?

5. How might electron shuttles facilitate growth of bacteria in a biofilm that has become anoxic in the interior?

6. Intercellular communication has been studied in detail in *Myxococcus xanthus*. Review the life cycle of this developmentally complex microbe and indicate at what

points during this process intercellular signaling molecules might be exchanged and the message they would convey.

7. The γ-proteobacterium *Congregibacter litoralis* is one of several cultured aerobic anoxygenic phototrophs (AAnPs). AAnPs are also found among α- and β-proteobacteria. These microbes are photoheterotrophs that contain Bchl *a*. They cannot grow photoautotrophically or anaerobically; *C. litoralis* is a microaerophile. It possesses the genes needed for the oxidation of reduced sulfur compounds, but the addition of reduced, inorganic sulfur to the media does not enhance growth. Compare the distribution of the AAnPs among the proteobacterial classes with that of the purple sulfur and nonsulfur bacteria. How would you determine if the AAnPs and the purple bacteria evolved from a single photosynthetic ancestor or have acquired the capacity to grow phototrophically by horizontal gene transfer? Finally, if *C. litoralis* does not use reduced sulfur compounds as a source of electrons, what do you think is the electron donor?

Read the original paper: Fuchs, B. M., et al. 2007. Characterization of a marine gammaproteobacterium capable of aerobic anoxygenic photosynthesis. *Proc. Natl. Acad. Sci. USA*. 104:2891.

Learn More

shop.mheducation.com   Enhance your study of this chapter with interactive study tools and practice tests. Also ask your instructor about the resources available through Connect, adaptive learning tools and animations.

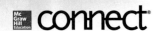

Firmicutes: The Low G + C Gram-Positive Bacteria

The father of the Muppets, Jim Henson, succumbed to an early death as the result of a Streptococcus pyogenes *infection.*

Invasive Strep Strikes Young, Old, and Famous

In 1990 the world mourned the untimely death of a genius. At the age of 53, Jim Henson, the creator of the Muppets, succumbed to pneumonia. Big Bird, Elmo, Bert, Ernie, children, and grown-ups were shocked to hear that a man in his prime could die from a disease thought to be fatal only among the very young and very old. How could this be?

You might be surprised to learn that the same bacterium that causes strep throat in schoolchildren caused Henson's death. In fact, up to 20% of kids carry *Streptococcus pyogenes* asymptomatically. How can a single bacterial species have no effect on children but kill an adult? The short answer is it depends on the strain, but the long answer is more interesting. *S. pyogenes* is a member of group A streptococci, called GAS for short. It was first described in 1874, when it was cultured from infected wounds. Later named *S. pyogenes* (pyogenic means pus producing), it has also been named *S. hemolyticus* (reflecting its ability to lyse red blood cells), *S. erysipelatos* (referring to reddened inflammation), and *S. scarlatinae*, as it is also the cause of scarlet fever. As this changeable but descriptive nomenclature suggests, *S. pyogenes* infection should not be ignored.

S. pyogenes is transmitted by direct contact with contaminated mucus or wounds. Strep throat is easily treated with antibiotics; but in cases where the microbe gains access to the blood, muscle, or lungs, this is not the case. Such "invasive strep" bacteria produce a suite of toxins and other virulence factors that make their treatment difficult. Some invasive streptococci have been dubbed "flesh-eating bacteria" because they cause necrotizing fasciitis, which results in the rapid destruction of muscle, skin, and fat. The pneumonia that killed Jim Henson was caused by an invasive strain that secretes a toxin that causes streptococcal toxic shock syndrome (STSS). This is not the same toxic shock syndrome associated with tampon use; that is caused by *Staphylococcus aureus*, a different firmicute. The *S. pyogenes* toxin is a superantigen (*see section 34.5*) that leads to a rapid plunge in blood pressure and organ failure. Once a lethal concentration of toxin has been released into the bloodstream, antibiotics cannot reverse the course of disease.

The firmicutes discussed in this chapter may sometimes seem like a hit parade of pathogens. However, we also rely on firmicutes for fermented dairy products, including yogurt and cheese. In fact, these so-called lactic acid bacteria are so similar to streptococci, they were once placed in a single genus. Both lactic acid bacteria and the present-day streptococci are fermenters. They lack oxidative phosphorylation, but despite their limited metabolic flexibility, they have remarkable, if sometimes very sad, capabilities.

Phylogenetic relationships among Gram-positive bacteria show that they can be divided into a low G + C group (**figure 23.1**) and a high G + C group, which is covered in chapter 24. The low G + C Gram-positive bacteria are placed in the phylum *Firmicutes* and divided into three classes: *Clostridia, Bacilli*, and the recently added *Negativicutes*. **Figure 23.2** shows the phylogenetic relationships among some of the bacteria reviewed in this chapter, but it does not illustrate the *Negativicutes*. This class includes many newly discovered species as well as some that were previously considered *Clostridia*.

Readiness Check:

Based on what you have learned previously, you should be able to:

✔ Draw a Gram-positive cell wall and list its constituent parts (section 3.4)
✔ Differentiate between swimming and gliding motility (section 3.8)
✔ Outline the process of endospore formation and germination (section 3.9)
✔ Distinguish heterotrophy from autotrophy (section 11.1)
✔ List environmental factors that influence microbial growth (section 7.4)
✔ Compare and contrast the growth and physiology of obligate aerobes, facultative anaerobes, microaerophiles, and strict anaerobes (section 7.4)
✔ Explain how biofilms form and why they are important (section 7.5)
✔ Compare and contrast anaerobic and aerobic respiration, oxygenic and anoxygenic phototrophy, and substrate-level phosphorylation and oxidative phosphorylation (sections 11.4, 11.5, and 11.11)
✔ Draw an electron transport chain indicating the relative standard reduction potentials of the electron donor, acceptor, and electron transporters (sections 10.4 and 11.6)
✔ Summarize the generation of ATP and the fate of pyruvate in fermentation (section 11.8)

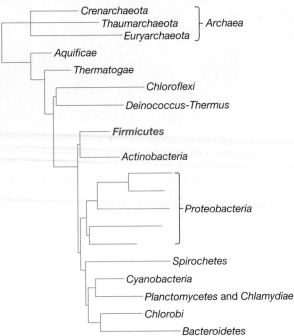

Figure 23.1 **Phylogenetic Relationships Among *Bacteria* and *Archaea*.** The phylum *Firmicutes* is highlighted.

23.1 Class *Clostridia*: Anaerobic Endospore-Forming Bacteria

After reading this section, you should be able to:

- Identify several *Clostridia* spp. that form endospores
- Outline the reaction by which some clostridia ferment amino acids, and relate this carbon substrate to their environmental distribution and pathogenic potential
- Identify the terminal electron acceptors used by *Desulfotomaculum* spp.
- Distinguish phototrophy in heliobacteria from that of other anoxygenic phototrophic bacteria
- Discuss the importance of *Veillonella* spp. to human disease

Figure 23.2 **Phylogenetic Relationships in the Phylum *Firmicutes* (Low G + C Gram Positives).** The major phylogenetic groups are shown. Each tetrahedron in the tree represents a group of related organisms; its horizontal edges show the shortest and longest branches in the group. Multiple branching at the same level indicates that the relative branching order of the groups cannot be determined from the data.

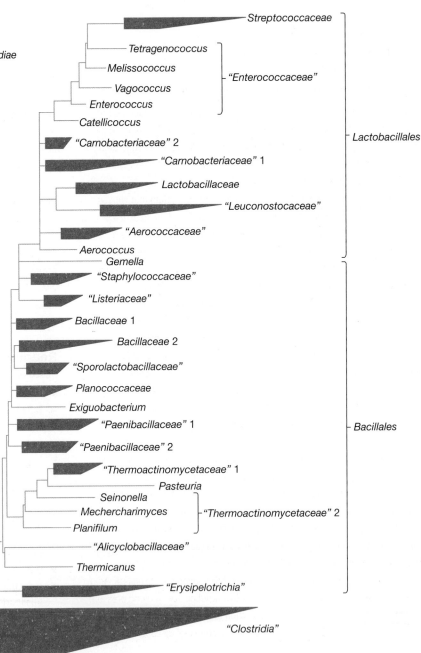

Table 23.1   Characteristics of Selected Members of the Class *Clostridia*

Genus	Dimensions (μm), Morphology, and Motility	Genome Size (Mb)	Oxygen Relationship	Other Distinctive Characteristics
Clostridium	0.3–2.0 × 1.5–20; rod-shaped, often pleomorphic, nonmotile or peritrichous flagella	Most species 3.0–4.0	Anaerobic	Usually chemoorganotrophic, fermentative, and catalase negative; form oval or spherical endospores
Desulfotomaculum	0.3–1.5 × 3–9; straight or curved rods, peritrichous or polar flagella	2.7–3.9	Anaerobic	Reduce sulfate to H_2S, form subterminal to terminal endospores; stain Gram negative but have Gram-positive wall; catalase negative
Heliobacterium	1.0 × 4–10; rods that are frequently bent, gliding motility	3.1	Anaerobic	Photoheterotrophic with bacteriochlorophyll *g;* stain Gram negative but have Gram-positive wall; some form endospores

There are two classes of low G + C bacteria that form endospores: *Clostridia* and *Bacilli.* Recall that endospores are made within the mother cell upon nutrient deprivation. Once released, these **spores** are remarkably resistant to environmental stresses, sometimes surviving thousands of years. ◀◀ *Bacterial endospores are a survival strategy (section 3.9)*

The class *Clostridia* contains a diverse group of bacteria. The characteristics of some of the more important genera are summarized in **table 23.1.** With over 100 species, *Clostridium* is the largest genus in the class. It includes obligately anaerobic, fermentative, Gram-positive bacteria that form endospores (**figure 23.3**). Members of this genus have great practical impact. Because they are anaerobic and form heat-resistant endospores, they are responsible for many cases of food spoilage, even in canned foods. Clostridia often can ferment amino acids via the **Stickland reaction** (**figure 23.4**). In this process, one amino acid is oxidized to produce pyruvate, which then is fermented to yield acetate and ATP. A different amino acid serves as an electron acceptor, so it is reduced, thereby reoxidizing NADH to NAD^+ and forming an additional two acetate molecules. This reaction also generates ammonia, hydrogen sulfide, and fatty acids during the anaerobic decomposition of proteins. These products are responsible for many unpleasant odors arising during putrefaction.

Although some clostridia are industrially valuable (e.g., *C. acetobutylicum* is used to manufacture butanol), the pathogenic

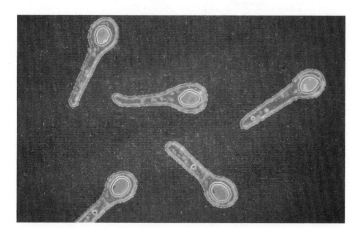

Figure 23.3 *Clostridium tetani.* This pathogen makes endospores that are round and terminal. When the mother cell surrounding each endospore lyses, the released spores remain viable for hundreds, if not thousands of years (×3,000)

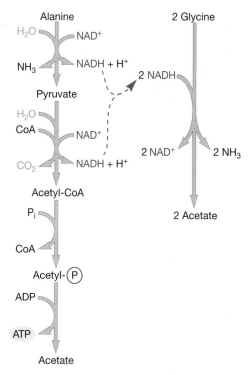

Figure 23.4 **The Stickland Reaction.** Alanine is oxidized to acetate, and glycine is used to reoxidize the NADH generated during alanine fermentation. The fermentation also produces some ATP.

species produce well-known toxins. For instance, *C. botulinum* produces the botulism toxin and *C. tetani* causes tetanus through the release of tetanus toxin. *C. perfringens* secretes a variety of protein toxins and, depending on the strain, causes gas gangrene and food poisoning. Although it lacks a TCA cycle and makes ATP only by fermentation, it has an extraordinary

doubling time of only 8 to 10 minutes when in the human host. This is presumably due to the ready supply and assimilation of host macromolecules. ▶▶| *Botulism (section 39.4); Tetanus (section 39.3)*

C. tetani ferments a number of amino acids to lactate, butyrate, and H_2 (**figure 23.5**). It lacks the genes to catabolize

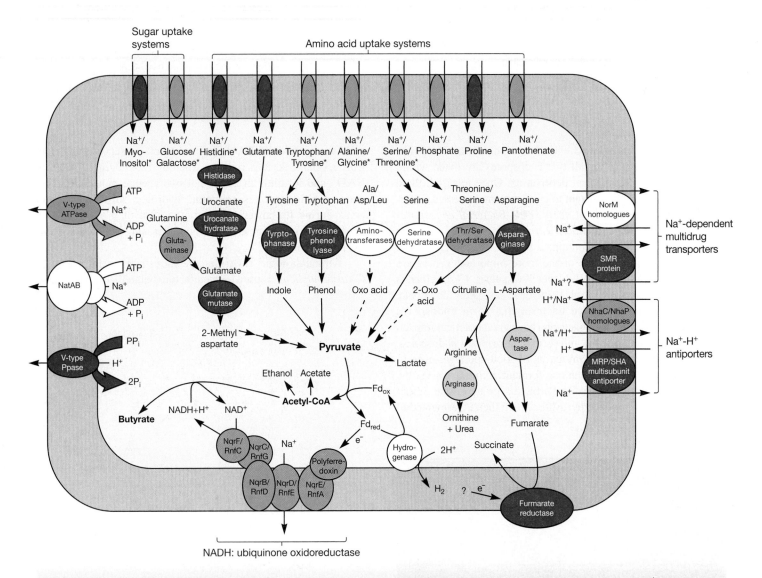

Figure 23.5 Amino Acid Fermentation and Na⁺ Flux in *Clostridium tetani*.   The fermentation of amino acids and the generation of a Na⁺ motive force is central to survival of this pathogen. Note that amino acids are transported into the cell by Na⁺-dependent uptake systems. Sugars are also transported in this manner but are used in anabolic reactions only. Several Na⁺-dependent transporters with unknown substrates are present. A V-type ATPase and an ATP-binding cassette transporter (NatAB) pump Na⁺ out of the cell at the expense of ATP hydrolysis. A proton translocating pyrophosphatase (V-type Ppase) is used to export H⁺. Components of a membrane-bound electron transport chain are present, but it is thought that the NADH dehydrogenase participates in amino acid fermentation. The lack of quinones suggests that the ETC is not used to establish a proton motive force.

MICRO INQUIRY   *The presence of an electron transport chain in a strictly fermentative microbe is unusual. Describe how this microbe uses its ETC and how this differs from a respiratory microbe.*

sugars. By contrast, *C. acetobutylicum* and *C. perfringens* can catabolize both carbohydrates and amino acids. *C. tetani* uses an ATP-binding cassette transporter and a V-type ATPase to couple ATP hydrolysis to the flow of sodium ions out of the cell. V-type ATPases pump protons, or in this case sodium ions, out of the cell at the expense of ATP made by substrate-level phosphorylation during fermentation. This is not a futile exercise, because it creates a sodium motive force, which in turn drives amino acid and sugar uptake systems. Although *C. tetani* also exports protons via a pyrophosphatase, external protons seem to serve only as the counter ion in Na^+-H^+ antiport. Surprisingly, *C. tetani* possesses genes for a membrane-bound electron transport chain; microbes that make ATP exclusively by fermentation do not generally possess electron transport chains. In this case, it appears that this chain links Na^+-extrusion with oxidation of NADH generated during fermentation. ◄◄ *Primary and secondary active transport (section 3.3)*

Desulfotomaculum is a genus that contains endospore-forming species that reduce sulfate and sulfite to hydrogen sulfide during anaerobic respiration (**figure 23.6**). Although they stain Gram negative, electron microscopy studies show that these bacteria have Gram-positive cell walls, and phylogenetic studies place the genus with the low G + C Gram positives.

Heliobacteria further illustrate the diversity of this class. The genera *Heliobacterium* and *Heliophilum* contain unusual anaerobic, phototrophic species characterized by the presence of bacteriochlorophyll *g*. They have a photosystem I–type reaction center like green sulfur bacteria but have no intracytoplasmic photosynthetic membranes; pigments are contained in the plasma membrane (**figure 23.7**). They also differ from many other anoxygenic phototrophic bacteria because they cannot grow autotrophically. Pyruvate is used during photoheterotrophic growth in light; it is fermented in dark, anoxic conditions. Like the clostridia, heliobacteria are capable of nitrogen fixation. Although they have a Gram-positive cell wall, they have a low peptidoglycan content, and they stain Gram negative. Some heliobacteria form endospores. ◄◄ *Light reactions in anoxygenic photosynthesis (section 11.11); Photosynthetic bacteria are diverse (section 21.4)*

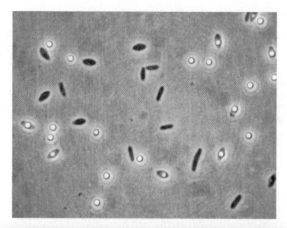

Figure 23.6 *Desulfotomaculum.* *Desulfotomaculum acetoxidans* with endospores (bright spheres); phase contrast.

MICRO INQUIRY *In addition to its Gram-staining characteristics, why do you think* Desulfotomaculum *spp. were originally thought to be Gram-negative microbes?*

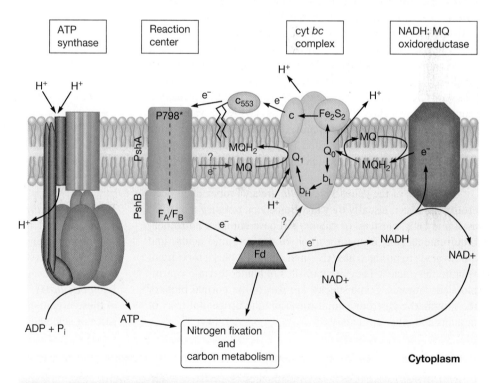

Figure 23.7 **Electron Transfer in *Heliobacterium modesticaldum.*** Genes encoding all subunits of NADH:quinone oxidoreductase are present on the *H. modesticaldum* genome. This suggests that electrons flow from NADH to the NADH:menaquinonone (MQ) oxidoreductase, to cytochrome (cyt) *bc* complex, which in turn reduces a 2Fe-2S subunit and cytochrome b_L. Electrons from cyt *bc* can be transferred via cyt c_{553} to the reaction center (RC) primary electron donor P798. Based on similarities between the heliobacterial RC and photosystem I in other anoxygenic photosynthetic bacteria, it is hypothesized that a proton motive force is generated by cyclic electron flow.

MICRO INQUIRY *What are the two sources of electrons for the* H. modesticaldum *electron transport chain?*

23.2 Class *Negativicutes*: Gram-Positive Bacteria with Outer Membranes

After reading this section, you should be able to:

- Explain why these bacteria are considered Gram positive despite possessing outer membranes.
- Discuss why new taxonomic ranks must be developed
- Identify the location where *Veillonella* sp. might be found on your body

In 2010 three bacterial isolates were cultured from human skin and soft-tissue wounds. What happened next illustrates the ever-changing world of microbial taxonomy and systematics. Further investigation of these bacteria revealed that although they had an inner and outer membrane, like Gram-negative bacteria, 16S rRNA sequences showed that they were closely related to microbes in the Gram-positive family *Veillonellaceae*. Because the three isolates were distinct from any other bacteria in this family, they were assigned to a new genus and species: *Negativicoccus succinicivorans*. The name reflects their unusual cell envelope and the fermentation of succinate.

The discovery of *N. succinicivorans* triggered a reexamination of the class *Clostridia*. It was determined that bacteria that have peptidoglycan between the inner and outer membranes and therefore stain Gram-negative, but are otherwise genetically *Firmicutes*, require their own class. Therefore the new class *Negativicutes* and order *Selenomonadales* were approved (**table 23.2**). Two existing families, *Acidaminococcaceae* and *Veillonellaceae*, were then assigned to this new order. Such changes are always controversial, and this one is no exception.

Members of the genus *Veillonella* are anaerobic, chemoheterotrophic cocci. Usually they are diplococci, but they may exist as single cells, clusters, or chains. All have complex nutritional requirements and ferment carbohydrates, organic acids, and amino acids to produce gas (CO_2 and often H_2) plus a mixture of volatile fatty acids. They are parasites of homeothermic (warm-blooded) animals. Some species are part of the normal biota of the mouth, the gastrointestinal tract, and the urogenital tract of humans and other animals.

Retrieve, Infer, Apply

1. List three species of *Clostridium* and why they are notable.
2. Suggest why *C. tetani* uses a sodium motive force rather than a proton motive force.
3. What is the evolutionary significance of the discovery of a phototrophic bacterium within the low G + C Gram-positive bacteria? Do you think the phototrophic apparatus in *Heliobacterium* spp. evolved independently of those found in Gram-negative anoxygenic phototrophs? Explain.
4. What kind of genetic evidence, in addition to 16S rRNA sequences, might have demonstrated that *Negativicutes* should be considered Gram positive?

23.3 Class *Bacilli*: Aerobic Endospore-Forming Bacteria

After reading this section, you should be able to:

- List terminal electron acceptors and fermentation products produced by *Bacillus* spp. and staphylococci
- Discuss reasons that make *Bacillus subtilis* an important model organism
- List three reasons *Bacillus* spp. are of practical importance
- Summarize the evolution and emergence of MRSA, and list at least two diseases caused by *Staphylococcus aureus*
- Describe the structure and medical importance of *Listeria monocytogenes*
- Identify the genera considered lactic acid bacteria, and discuss their importance in the food industry
- Distinguish between enterococci and streptococci
- Name bacterial genera with species capable of hemolysis, and differentiate between α-hemolysis and β-hemolysis

The class *Bacilli* is very large but includes only two orders, *Bacillales* and *Lactobacillales*. These include genera representing cocci, endospore-forming rods and cocci, and nonsporing rods. We first describe the biology of some members of the order *Bacillales* and then important representatives of the order *Lactobacillales*. The phylogenetic relationships between some of these organisms are shown in figure 23.2, and the characteristics of selected genera are summarized in **table 23.3**.

Table 23.2	**Characteristics of Selected Members of the Class *Negativicutes***			
Genus	**Dimension (μm), Morphology, Motility**	**Genome Size (Mb)**	**Oxygen Relationship**	**Other Distinctive Characteristics**
Veillonella	0.3–0.5; cocci in pairs, short chains, and masses; nonmotile	2.8	Anaerobic	Pyruvate and lactate fermented but not carbohydrates; parasitic in mouths, intestines, and respiratory tracts of animals.
Megasphaera	1.7–2.6, cocci, nonmotile	1.8–2.6	Anaerobic	Ferment a variety of carbohydrates, including lactate; isolated from mouths, intestines of animals, and spoiled beer
Negativicoccus	0.4, cocci; nonmotile	1.45	Anaerobic and microaerophilic	Isolated from human skin and soft-tissue wounds; ferments succinate to acetate and propionate

Genus	Dimensions (μm), Morphology, and Motility	Genome Size (Mb)	Oxygen Relationship	Other Distinctive Characteristics
Bacillus	0.5–2.5 × 1.2–10; straight rods, peritrichous flagella, spore-forming	4.2–5.4	Aerobic or facultative	Catalase positive; chemoorganotrophic
Caryophanon	1.5–3.0 × 10–20; multicellular rods with rounded ends, peritrichous flagella, nonsporing	Nd[1]	Aerobic	Acetate only major carbon source; catalase positive; trichome cells have greater width than length; trichomes can be in short chains
Enterococcus	0.6–2.0 × 0.6–2.5; spherical or ovoid cells in pairs or short chains, nonsporing, sometimes motile	3.2	Aerotolerant	Ferment carbohydrates to lactate with no gas; complex nutritional requirements; catalase negative; occur widely, particularly in fecal material
Lactobacillus	0.5–1.2 × 1.0–10; usually long, regular rods, nonsporing, rarely motile	1.9–3.3	Facultative or microaerophilic	Fermentative, at least half the end product is lactate; require rich, complex media; catalase and cytochrome negative
Lactococcus	0.5–1.2 × 0.5–1.5; spherical or ovoid cells in pairs or short chains, nonsporing, nonmotile	2.4	Aerotolerant	Chemoorganotrophic with fermentative metabolism; lactate without gas produced; catalase negative; complex nutritional requirements; in dairy and plant products
Leuconostoc	0.5–0.7 × 0.7–1.2; cells spherical or ovoid, in pairs or chains; nonmotile and nonsporing	1.7–3.2	Facultative	Require fermentable carbohydrate and nutritionally rich medium for growth; fermentation produces lactate, ethanol, and gas; catalase and cytochrome negative
Staphylococcus	0.9–1.3; spherical cells occurring singly and in irregular clusters, nonmotile and nonsporing	2.5–2.8	Facultative	Chemoorganotrophic with both respiratory and fermentative metabolism; usually catalase positive; associated with skin and mucous membranes of vertebrates
Streptococcus	0.5–2.0; spherical or ovoid cells in pairs or chains, nonmotile and nonsporing	1.8–2.2	Aerotolerant	Fermentative, producing mainly lactate and no gas; catalase negative; commonly attack red blood cells (α- or β-hemolysis); complex nutritional requirements; commensals or parasites on animals
Thermoactinomyces	0.4–1.0 in diameter; branched, septate mycelium resembles those of actinomycetes	3.4	Aerobic	Usually thermophilic; true endospores form singly on hyphae; numerous in decaying hay, vegetable matter, and compost

Table 23.3 Characteristics of Members of the Class *Bacilli*

1 Nd: Not determined; genome not yet sequenced.

Order *Bacillales*

The genus *Bacillus,* family *Bacillaceae,* is the largest in the order *Bacillales.* The genus contains endospore-forming, chemoheterotrophic rods that are usually motile with peritrichous flagella (**figure 23.8**). Members of this genus are aerobic or facultative, and catalase positive. *Bacillus subtilis,* the type species *(see page 445)* for the genus, is the most well-studied Gram-positive bacterium. It is a facultative anaerobe that can use nitrate as a terminal electron acceptor or perform mixed acid fermentation with lactate, acetate, and acetoin as major end products. It is nonpathogenic and a terrific model organism for the study of gene regulation, cell division, quorum sensing, and cellular differentiation. Its 4.2-Mb genome was one of the first genomes to be completely sequenced and

reveals a number of interesting elements. For instance, several families of genes have been expanded by gene duplication; the largest such family encodes ABC transporters—the most abundant class of protein in *B. subtilis.* There are 18 genes that encode sigma factors. Recall that the use of alternative sigma subunits of RNA polymerase is one way in which bacteria regulate gene expression. In this case, many of the sigma factors govern spore formation and other responses to stressful conditions. ◄◄ *Bacterial endospores are a survival strategy (section 3.9); Protein maturation and secretion (section 13.8); Sporulation in* Bacillus subtilis *(section 14.5)*

   B. subtilis is a soil-dwelling microbe whose spores are easily extracted from diverse soil types. Interestingly, it is often found in its vegetative state (i.e., not as a spore) when

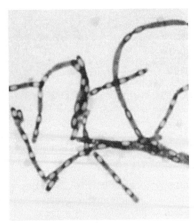

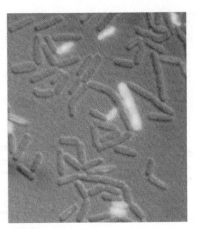

(a) *Bacillus anthracis*

(b) *B. cereus*

Figure 23.8 *Bacillus.* (a) *B. anthracis* endospores are elliptical and central (×1,600). (b) *B. cereus* stained with SYTOX Green nucleic acid stain and viewed by epifluorescence and differential interference contrast microscopy. The cells that glow yellow are dead.

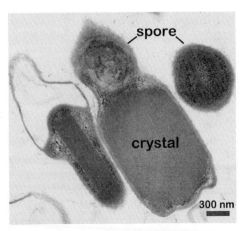

Figure 23.10 **The Parasporal Body.** An electron micrograph of a *B. thuringiensis* sporulating cell containing a parasporal crystal just beneath the endospore. A free spore is also shown.

(a) Biofilm formation

(b) Fruiting body formation

Figure 23.9 **Biofilm Formation by *Bacillus subtilis*.** (a) When grown for several days without agitation in nutrient-limiting medium, *B. subtilis* forms a hydrophobic biofilm at the aqueous-air interface. (b) *B. subtilis* forms multicellular structures called fruiting bodies, when grown on minimal-salts agar for several days. These cells are labeled so that spores turn blue, demonstrating that spores are abundant at the tips of these structures.

MICRO INQUIRY *What growth or survival advantage might this type of multicellular behavior confer to* B. subtilis?

growing in association with actively decaying material and in close association with plant roots. In addition to forming spores, *B. subtilis* also develops biofilms made up of vegetative cells at the base and differentiated structures containing spores at the tip (**figure 23.9**). ◀◀ *Biofilms are common in nature (section 7.5)*

Many species of *Bacillus* are of considerable importance. They produce the antibiotics bacitracin, gramicidin, and polymyxin. *B. cereus* (figure 23.8*b*) can cause food poisoning. Several species are used as insecticides. For example, *B. thuringiensis*

and *B. sphaericus* form a solid protein crystal, the **parasporal body,** next to their endospores during endospore formation (**figure 23.10**). As discussed in chapter 42, the *B. thuringiensis* parasporal body contains protein toxins that kill over 100 species of moths when activated in the alkaline gut of caterpillars thereby destroying the epithelium. The *B. sphaericus* parasporal body contains proteins toxic for mosquito larvae and may be useful in controlling the mosquitoes that carry the malaria parasite *Plasmodium*. *B. anthracis* is the causative agent of the disease anthrax, which can kill both farm animals and humans. ▶▶ *Anthrax (section 39.5); Agricultural biotechnology relies on a plant pathogen (section 42.5)*

Members of the genus *Thermoactinomyces* historically were classified as actinomycetes because, like these microbes, their soil-associated substrate hyphae (filamentous cells) differentiate into upwardly growing aerial hyphae. However, phylogenetic analysis places them with low G + C bacteria in the order *Bacillales,* family *Thermoactinomycetaceae*. Their G + C content (52–55 mol%) is considerably lower than that of actinobacteria. Also unlike actinobacteria, *Thermoactinomyces* species form true endospores within both the aerial and substrate hyphae. These bacteria are thermophilic and grow between 45 and 60°C. *Thermoactinomyces* spp. are commonly found in damp haystacks and compost piles. In fact, *T. vulgaris* from haystacks, grain storage silos, and compost piles is a causative agent of farmer's lung, an allergic disease of the respiratory system. Spores from *T. vulgaris* recovered from the mud of a Minnesota lake were found to be viable after about 7,500 years of dormancy.

One other genus within the order *Bacillales* bears mentioning: *Sporosarcina*. This genus is one of five genera in the family *Panococcaceae,* and it contains the only known endospore-forming bacterium that has a coccoid rather than rod shape. *S. ureae* is particularly interesting because it is tolerant of very alkaline conditions (up to pH 10), which it creates when it degrades urea to CO_2 and NH_3. Interestingly, it is easily isolated from agricultural soils on which animals frequently urinate.

Family *Staphylococcaceae*

The family *Staphylococcaceae* contains five genera, the most important of which is *Staphylococcus*. Members of this genus are facultatively anaerobic, nonmotile cocci, 0.5 to 1.5 μm in diameter, occurring singly, in pairs, and in tetrads, and characteristically dividing in more than one plane to form irregular clusters (**figure 23.11**). They are usually catalase positive and oxidase negative. They respire using oxygen as the terminal electron acceptor, although some can reduce nitrate to nitrite. They are also capable of fermentation and convert glucose principally to lactate.

Staphylococci are normally associated with the skin and mucous membranes of warm-blooded animals. They are responsible for many human diseases. *S. aureus* has an impressive collection of virulence factors to evade and subvert the host (figure 23.11*b*). These include Protein A, which binds host antibodies by the conserved Fc region, rendering them ineffective. Two types of proteins aid in colonization and dissemination; clumping factors enable bacteria to attach to platelets and wounded tissue, thereby promoting initial colonization. By contrast, the protein staphylokinase dissolves fibrin clots that would otherwise keep the bacterium from invading beyond the site of initial colonization. The enzyme coagulase, which causes blood plasma to clot, is used as a marker for the identification of *S. aureus* but has no clear role in pathogenesis. Growth and hemolysis patterns on blood agar are also useful in identifying these staphylococci. The staphylococcal β-hemolysin lyses cells by forming solvent-filled channels in target plasma membranes. ▶▶| *Staphylococcal diseases (section 39.3); Staphylococcal food poinsoning (section 39.4)*

Family *Listeriaceae*

The family *Listeriaceae* includes two genera: *Brochothrix* and *Listeria*. *Brochothrix* spp. are most commonly found in meat but are not pathogenic. *Listeria* is the medically important genus in this family. It contains short rods that are facultatively anaerobic, catalase positive, and motile by peritrichous flagella. In addition to aerobic respiration, these microbes ferment glucose mainly to lactate. *Listeria* spp. are widely distributed in nature, particularly in decaying matter. *L. monocytogenes* is a pathogen of humans and other animals, and causes listeriosis, an important food-borne infection. ▶▶| *Food-borne disease outbreaks (section 41.3)*

Order *Lactobacillales*

Streptococcus, Enterococcus, Lactococcus, Lactobacillus, and *Leuconostoc* are all genera in the order *Lactobacillales*. These bacteria do not form endospores and are usually nonmotile. They are strictly fermentative, depending on sugars for carbon and energy. They lack cytochromes and an electron transport chain so they obtain energy only by substrate-level phosphorylation. Those that produce lactic acid as their major or sole fermentation product are sometimes collectively called **lactic acid bacteria** (**LAB**). Nutritionally, they are fastidious, and many vitamins, amino acids, purines, and pyrimidines must be supplied because of their limited biosynthetic capabilities. Members of this order usually are categorized as facultative anaerobes but can be considered aerotolerant. |◀◀ *Oxygen concentration (section 7.4); Fermentation does not involve an electron transport chain (section 11.8)*

The largest genus in this order is *Lactobacillus*, with about 100 species. Lactobacilli include rods and some coccobacilli. All lack catalase and cytochromes, and produce lactic acid as their

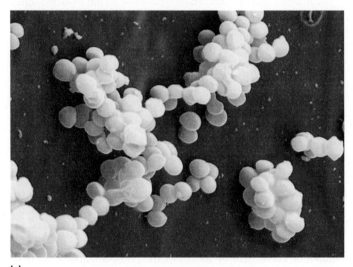

(a)

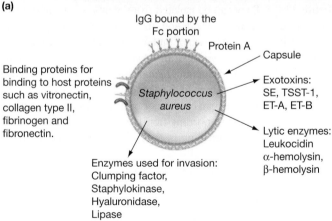

IgG bound by the Fc portion

Protein A

Capsule

Binding proteins for binding to host proteins such as vitronectin, collagen type II, fibrinogen and fibronectin.

Staphylococcus aureus

Exotoxins: SE, TSST-1, ET-A, ET-B

Lytic enzymes: Leukocidin α-hemolysin, β-hemolysin

Enzymes used for invasion: Clumping factor, Staphylokinase, Hyaluronidase, Lipase

(b)

Figure 23.11 *Staphylococcus.* (a) *S. aureus* cocci arranged in grape-like clusters; color-enhanced scanning electron micrograph. Each cell is about 1 μm in diameter. (b) *S. aureus* produces many virulence factors that account for its success as a pathogen. Staphylokinase catalyzes the degradation of host fibrin, facilitating microbial invasion. Hyaluronidase degrades hyaluronic acid, which is found in connective and epithelial tissues. See chapter 39 for further discussion of staphylococcal diseases.

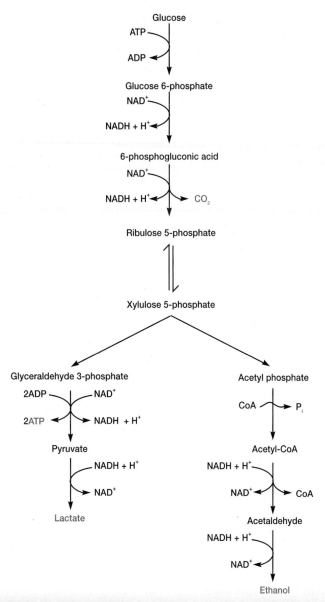

Figure 23.12 Heterolactic Fermentation and the Phosphoketolase Pathway. The phosphoketolase pathway converts glucose to lactate, ethanol, and CO_2.

MICRO INQUIRY *What is the net yield of ATP for each glucose that enters the phosphoketolase pathway?*

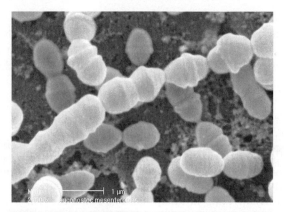

Figure 23.13 *Leuconostoc.* *Leuconostoc mesenteroides;* scanning electron micrograph.

Lactobacillus species are indispensable to the food and dairy industry. They are used in the production of fermented vegetable foods (e.g., sauerkraut, pickles), sourdough bread, Swiss and other hard cheeses, yogurt, and sausage. Yogurt is probably the most popular fermented milk product in the United States, and *Streptococcus thermophilus* and *Lactobacillus delbrueckii* subspecies *bulgaricus* are used in its production. At least one species, *L. acidophilus,* is sold commercially as a probiotic agent that may provide some health benefits for the consumer. On the other hand, some lactobacilli are problematic. They can be responsible for spoilage of beer, wine, milk, and meat. ▶▶ *Fermented milks (section 41.5); Probiotics (section 41.6)*

Leuconostoc, family *Leuconostocaceae,* contains facultative Gram-positive cocci, which may be elongated or elliptical and arranged in pairs or chains (**figure 23.13**). Leuconostocs carry out heterolactic fermentation by converting glucose to D-lactate and ethanol or acetic acid by means of the phosphoketolase pathway (figure 23.12). They can be isolated from plants, silage, and milk. They are used in wine production, in the fermentation of vegetables, and in the manufacture of buttermilk, butter, and cheese. *L. mesenteroides* synthesizes dextrans from sucrose and is important in industrial dextran production. *Leuconostoc* species are involved in food spoilage and tolerate high sugar concentrations so well that they grow in syrup and can be a major problem in sugar refineries.

Bacteria in the important families *Enterococcaceae* and *Streptococcaceae* occur in pairs or chains when grown in liquid media (**figure 23.14** and **table 23.4**). They are unable to respire and have a strictly fermentative metabolism. They ferment sugars to produce lactic acid but no gas; that is, they carry out homolactic fermentation. A few species are strict anaerobes rather than aerotolerant. Enterococci such as *E. faecalis* are normal residents of the intestinal tracts of humans and most other animals. *E. faecalis* is an opportunistic pathogen that can cause urinary tract infections and endocarditis. Enterococci are major agents in the horizontal transfer of antibiotic-resistance genes.

The family *Streptococcaceae* includes only two genera: *Lactococcus* and *Streptococcus. Lactococcus* spp. ferment sugars to lactic acid and can grow at 10°C but not at 45°C. *L. lactis* is widely used in the production of buttermilk and cheese because it can

main or sole fermentation product (*see figure 11.20*). They carry out either **homolactic fermentation** using the Embden-Meyerhof pathway or **heterolactic fermentation** with the phosphoketolase pathway (**figure 23.12**). They grow optimally under slightly acidic conditions, at a pH between 4.5 and 6.4, and are found in dairy products, meat, water, sewage, beer, fruits, and many other materials. Lactobacilli also are part of the normal flora of the human body; in particular they are important normal vaginal flora.

curdle milk and add flavor through the synthesis of diacetyl and other products.

The genus *Streptococcus* is large and complex. All species in the genus are facultatively anaerobic and catalase negative. Many characteristics are used to identify these cocci. One of the most important is the ability to lyse red blood cells when grown on blood agar, an agar medium containing 5% sheep or horse blood (**figure 23.15**). In **α-hemolysis,** a 1- to 3-mm greenish zone of incomplete hemolysis forms around the colony; **β-hemolysis** is characterized by a clear zone of complete lysis. Serological studies are also very important in identification because streptococci often have distinctive cell wall antigens. These bacteria produce a surface protein, the M protein, that is important in pathogenesis and species identification. The **Lancefield grouping system,** used for many years, was based on structural variations in M protein structure. This system has largely been replaced by sequence analysis of the *eem* gene, which encodes the M protein *(see figure 39.5).* Biochemical and physiological tests, and sensitivity to bacitracin, sulfa drugs, and optochin (a quinine analogue) also are used to identify particular species. ▶▶| *Clinical microbiology and immunology (chapter 36)*

The streptococci have been divided into three groups: pyogenic streptococci, oral streptococci, and other streptococci. Pyogenic streptococci are pathogens associated with pus formation.

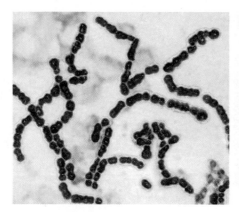
(a) *Streptococcus pyogenes*

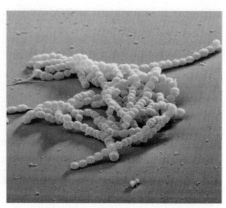
(b) *S. agalactiae*

Figure 23.14 *Streptococcus.* (a) *Streptococcus pyogenes.* (b) *Streptococcus agalactiae,* the cause of Group B streptococcal infections. Note the long chains of cells; color-enhanced scanning electron micrograph.

Most species are β-hemolytic on blood agar and form chains of cells. The major human pathogen in this group is *S. pyogenes,* which causes streptococcal sore throat, acute glomerulonephritis, and rheumatic fever. The normal habitat of oral streptococci is the oral cavity and upper respiratory tract of humans and other animals. In other respects, oral streptococci are not necessarily similar. *S. mutans* is associated with the formation of dental caries. *S. pneumoniae* is α-hemolytic and grows as pairs of cocci. It is associated with lobar pneumonia and otitis media (inflammation of the middle ear) in young children. ▶▶| *Streptococcal diseases (sections 39.1 and 39.3)*

Table 23.4	Classification of Streptococci, Entercocci, and Lactococci		
Characteristics	***Streptococcus***	***Enterococcus***	***Lactococcus***
Predominant arrangement (most common first)	Chains, pairs	Pairs, chains	Pairs, short chains
Capsule/slime layer	+	−	−
Habitat	Mouth, respiratory tract	Gastrointestinal tract	Dairy products
Growth at 45°C	Variable	+	−
Growth at 10°C	Variable	Usually +	+
Growth at 6.5% NaCl broth	Variable	+	−
Growth at pH 9.6	Variable	+	−
Hemolysis	Usually β (pyogenic) or α (oral)	α, β, −	Usually −
Serological group (Lancefield)	Variable (A–O)	Usually D	Usually N
Representative species	Pyogenic streptococci: *S. agalactiae, S. pyogenes, S. equi, S. dysgalactiae;* Oral streptococci: *S. gordonii, S. salivarius, S. sanguis, S. oralis, S. pneumoniae, S. mitis, S. mutans;* Other streptococci: *S. bovis, S. thermophilus*	*E. faecalis, E. faecium, E. avium, E. durans, E. galinarum*	*L lactis, L. raffinolactis, L. plantarum*

(a) Alpha-hemolysis   **(b)** Beta-hemolysis   **(c)** No hemolysis

Figure 23.15 Staphylococcal and Streptococcal Hemolytic Patterns. (a) *Staphylococcus aureus* on blood agar, illustrating α-hemolysis. (b) *Streptococcus pyogenes* on blood agar, illustrating β-hemolysis. (c) *Staphylococcus epidermidis* on blood agar with no hemolysis, also referred to as γ-hemolysis.

MICRO INQUIRY *Why is there a green zone around α-hemolytic colonies?*

Retrieve, Infer, Apply

1. You are studying what you think is a new *Bacillus* isolate. Without performing molecular analyses, what morphological and physiological traits would you assay?
2. What makes the genus *Thermoactinomyces* unique? What disease does *T. vulgaris* cause?
3. List three environments where you would find *Staphylococcus* spp. On what basis are pathogenic and nonpathogenic species and strains distinguished?

4. List the major properties that define the genus *Lactobacillus*. In what ways are species in this genus important in the food and dairy industries?
5. How are various species of *Streptococcus* identified?
6. Of what practical importance are leuconostocs? What are lactic acid bacteria?
7. What is the difference between α-hemolysis and β-hemolysis?

Key Concepts

23.1 Class *Clostridia*: Anaerobic Endospore-Forming Bacteria

- Members of the genus *Clostridium* are anaerobic, Gram-positive rods that form endospores and do not carry out dissimilatory sulfate reduction. They are responsible for food spoilage, botulism, tetanus, and gas gangrene. They have a fermentative metabolism (**figures 23.3–23.5**).
- *Desulfotomaculum* contains anaerobic, endospore-forming species that reduce sulfate to sulfide during anaerobic respiration.
- Heliobacteria are anaerobic, phototrophic bacteria with bacteriochlorophyll *g*. Some form endospores (**figure 23.7**).

23.2 Class *Negativicutes*: Gram-Positive Bacteria with Outer Membranes

- The discovery of several new Gram-negative staining bacteria that are phylogenetically *Firmicutes* led to the development of the class *Negativicutes*.

- The class has two families previously considered *Clostridia*: *Acidaminococcacaea* and *Veillonellaceae*. All members have an outer membrane and are fermentative; many are residents of warm-blooded animals (**table 23.2**).

23.3 Class *Bacilli*: Aerobic Endospore-Forming Bacteria

- The class *Bacilli* is divided into two orders: *Bacillales* and *Lactobacillales*.
- The genus *Bacillus* contains aerobic and facultative, catalase-positive, endospore-forming, chemoorganotrophic rods that are usually motile with peritrichous flagella. *Bacillus subtilis* is a model organism that forms endospores and multicellular structures such as biofilms and fruiting bodies (**figures 23.8** and **23.9**).
- *Thermoactinomyces* spp. are thermophilic and form a mycelium and true endospores. *T. vulgaris* causes allergic reactions that lead to farmer's lung.

- Members of the genus *Staphylococcus* are facultatively anaerobic, nonmotile, Gram-positive cocci that form irregular clusters. They carry out aerobic respiration; some also ferment and use nitrate as a terminal electron acceptor. *S. aureus* is a pathogen that possesses an array of virulence factors (**figure 23.11**).

- Several important genera such as *Lactobacillus* and *Listeria* contain regular, nonsporing rods. *Lactobacillus* spp. carry out lactic acid fermentation and are extensively used in the food and dairy industries. *Listeria monocytogenes* is an important agent in food poisoning. *Leuconostoc* spp. carry out heterolactic fermentation using the phosphoketolase pathway (**figure 23.12**); they are involved in the production of fermented vegetable and dairy products.

- The genera *Streptococcus*, *Enterococcus*, and *Lactococcus* contain cocci arranged in pairs and chains that are usually facultative anaerobes that carry out fermentation (**table 23.4**). Some important species are the pyogenic coccus *S. pyogenes*, *S. pneumoniae*, the enterococcus *E. faecalis*, and the lactococcus *L. lactis*.

Compare, Hypothesize, Invent

1. Some species of *Clostridium* demonstrate much faster doubling times when in the host than when grown in vitro. Design an experiment that would explore this phenomenon.

2. What physiological properties might account for the ease with which anaerobic clostridia can be isolated from soil and other generally aerobic niches.

3. *S. aureus* strains that are resistant to methicillin (MRSA) are also resistant to many other antibiotics and are more virulent than non-MRSA strains. Refer to chapters 16 and 18, and develop a testable hypothesis that predicts how these many traits came to be in a single group of strains.

4. Cells of *Bacillus subtilis* that are committed to sporulate secrete a signaling protein that prevents sister cells from entering sporulation and a killing factor that causes these sisters to lyse. The nutrients released by these cells are consumed by the "killer" cells, providing the nutrients needed to complete spore formation. It has been proposed that this provides evidence that sporulation is a "last resort" for survival by this microbe. Do you agree that the capacity to cannibalize sister cells is evidence that the *B. subtilis* uses spore formation as last resort? Explain your answer, being careful not to project human terms onto the microbes (i.e., anthropomorphize).

Read the original papers: Gonzalez-Paster, E., et al. 2003. Cannibalism by sporulating bacteria. *Science* 301:510; and Ellermeier, C. D., et al. 2006. A three-protein signaling pathway governing immunity to a bacterial cannibalism toxin. *Cell* 124:549.

Learn More

shop.mheducation.com   Enhance your study of this chapter with interactive study tools and practice tests. Also ask your instructor about the resources available through Connect, including adaptive learning tools and animations.

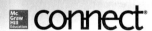

24

Actinobacteria: High G + C Gram-Positive Bacteria

If this pharmacist is dispensing an antibiotic, odds are, it was made by a streptomycete. Members of the genus Streptomyces *synthesize the majority of antibiotics used in medicine today. They also produce other bioactive compounds that have important medicinal and industrial uses.*

Antibiotic Production: Is It Actually Bacterial Chit-Chat?

A mong bacteria, it is hard to find a genus whose constituents have given more to humankind than *Streptomyces*. When growing in their natural soil habitat, streptomycetes are key players in a microbial community that eats decaying plant and animal material. And like diners in a crowded restaurant, these microbes are "talking" up a storm, using a vocabulary of small molecules.

In fact, some microbiologists hypothesize that antibiotics are the language of inter- and intraspecies communication. If this is the case, then streptomycetes are very talkative indeed. About two-thirds of all antibiotics prescribed are made by these bacteria, including tetracycline, vancomycin, and neomycin. Noting that streptomycetes produce and secrete such a wide range of natural products, pharmaceutical companies have screened countless isolates, discovering that they also make anticancer compounds (e.g., doxorubicin), antifungals (e.g., nystatin), and ivermectin, used to treat parasitic infections in humans and to prevent heartworm in dogs. All told, *Streptomyces* species produce about 10,000 bioactive products, but models suggest there may be 990,000 yet undiscovered!

How can it be that after 60 years of research, so few streptomycete compounds are known? And how do we know this? One stark piece of evidence was revealed in 2002 when the *S. coelicolor* genome was published. This species was known to make four antibiotics, yet the genes for the production of 20 additional compounds were found. These genes are tightly regulated; to find and test new antibiotics, their regulatory "off" switch must be unlocked. This is where the ability of streptomycetes to communicate gets interesting. About 60 years ago, streptomycin, the first drug to treat tuberculosis, was found to be made by *S. griseus*. Ten years later, researchers discovered that streptomycin production depended on the release of tiny amounts of a small molecule called A-factor, a gamma-butyrolactone. If *S. griseus* cells fail to exchange A-factor, no streptomycin is produced. Much later it was discovered that a different group of chemicals, 2-alkyl-4-hydroxymethylfuran-3-carboxylic acids (or AHFCAs), controls the release of other antibiotics. And studies with natural isolates show that when cultured alone, many *Streptomyces* spp. lack discernible antibiotic activity. But when grown in the company of other streptomycetes, and allowed to exchange secreted molecules, new antibiotics are released.

What does all this bacterial chitchat mean? To the microbes it probably provides a way to ensure that enough of their kind are present. After all, the amount of antibiotic secreted by an individual isn't a very effective defense. To the world of drug discovery, it means that screening for new natural products is more complicated than simply growing a *Streptomyces* isolate in pure culture. New and creative methods of culturing and screening must be developed.

By definition, high G + C Gram-positive bacteria have a DNA base composition above approximately 50 mol% G + C; these bacteria belong to the phylum *Actinobacteria* (**figure 24.1**). Based on 16S rRNA sequence data,

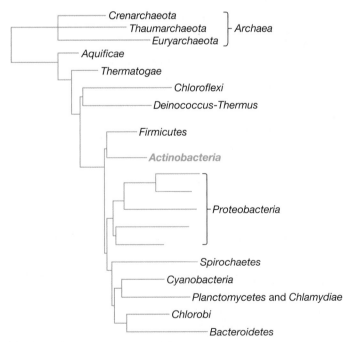

Figure 24.1 Phylogenetic Relationships Among *Bacteria* and *Archaea*. The phylum *Actinobacteria* is highlighted.

Table 24.1 Characteristics of Actinobacteria

Genus	Dimensions (μm), Morphology and Motility	Genome Size Range (Mb)	Oxygen Relationship	Other Distinctive Characteristics
Actinoplanes	Nonfragmenting, branching mycelium with little aerial growth; sporangia formed; motile spores with polar flagella	8.7–9.4	Aerobic	Hyphae often in palisade arrangement; highly colored; type II cell walls; found in soil and decaying plant material
Arthrobacter	0.8–1.2 × 1.0–8.0; young cells are irregular rods, older cells are small cocci; usually nonmotile	3.9–5.2	Aerobic	Have rod-coccus growth cycle; respiratory metabolism; catalase positive; mainly in soil
Bifidobacterium	0.5–1.3 × 1.5–8; rods of varied shape, usually curved; nonmotile	1.9–2.8	Anaerobic	Cells can be clubbed or branched, pairs often in V arrangement; ferment carbohydrates to acetate and lactate but no CO_2; catalase negative
Corynebacterium	0.3–0.8 × 1.5–8.0; straight or slightly curved rods with tapered or clubbed ends; nonmotile	2.3–3.4	Facultatively anaerobic	Cells often arranged in a V formation or in palisades of parallel cells; catalase positive and fermentative; polyphosphate granules
Frankia	0.5–2.0 in diameter; vegetative hyphae with limited-to-extensive branching and no aerial mycelium; multilocular sporangia formed	5.3–9.0	Aerobic to microaerophilic	Sporangiospores nonmotile; usually fix nitrogen; type III cell walls; most strains are symbiotic with angiosperm plants and induce nodules
Micrococcus	0.5–2.0 diameter; cocci in pairs, tetrads, or irregular clusters; usually nonmotile	2.3–2.6	Aerobic	Colonies usually yellow or red; catalase positive with respiratory metabolism; primarily on mammalian skin and in soil
Mycobacterium	0.2–0.6 × 1.0–10; straight or slightly curved rods, sometimes branched; acid-fast; nonmotile and nonsporing	3.3–7.0	Aerobic	Catalase positive; can form filaments that are readily fragmented; walls have high lipid content; in soil and water; some parasitic
Nocardia	0.5–1.2 in diameter; rudimentary to extensive vegetative hyphae that can fragment into rod-shaped and coccoid forms	6.2–9.4	Aerobic	Aerial hyphae formed; catalase positive; type IV cell walls; widely distributed in soil
Propionibacterium	0.5–0.8 × 1–5; pleomorphic nonmotile rods, may be forked or branched; nonsporing	2.5–2.6	Anaerobic to aerotolerant	Fermentation produces propionate and acetate, and often gas; catalase positive
Streptomyces	0.5–2.0 in diameter; vegetative mycelium extensively branched; aerial mycelium forms chains of three to many spores	8.5–10.1	Aerobic	Form discrete lichenoid or leathery colonies that often are pigmented; respiratory metabolism; use many organic compounds as nutrients; soil organisms

there are six classes in the phylum. In this chapter, we cover members of the class **Actinobacteria.** Many actinobacteria are also called actinomycetes. **Actinomycetes** are aerobic bacteria that are distinctive because most have filamentous cells called hyphae that differentiate to produce asexual spores. A summary of the actinobacteria discussed in this chapter is presented in **table 24.1.**

Readiness Check:

Based on what you have learned previously, you should be able to:

✔ Draw a Gram-positive cell wall and list its constituent parts (section 3.4)

✔ Compare and contrast the growth and physiology of obligate aerobes, facultative anaerobes, microaerophiles, and strict anaerobes (section 7.4)

✔ List environmental factors that influence microbial growth (section 7.4)

✔ Explain how biofilms form and why they are important (section 7.5)

✔ Discriminate between substrate-level phosphorylation and oxidative phosphorylation (sections 11.4 and 11.6)

✔ Draw an electron transport chain indicating the relative standard reduction potentials of the electron donor, acceptor, and electron transporters (sections 10.4 and 11.6)

✔ Compare anaerobic and aerobic respiration (sections 11.3 and 11.7)

✔ Summarize the generation of ATP and the fate of pyruvate in fermentation (section 11.8)

24.1 Class *Actinobacteria*

After reading this section, you should be able to:

- Draw the life cycle of at least two different actinobacteria
- Distinguish an exospore from an endospore
- Discuss the taxonomic importance of cell wall structure
- List common environments in which these microbes can be found
- List one veterinary and two human diseases caused by actinobacteria
- Explain snapping division and how this results in the palisade arrangement of cells
- Relate the structure of mycolic acids to the properties of mycobacterial cell walls
- Relate the unusual carbon and energy substrates used by bacteria in the family *Nocardiaceae* to their importance in bioremediation
- Describe the unusual morphology of actinoplanete spores
- Name two reasons propionibacteria are important to humans
- Explain how streptomycete morphology helps this group thrive in soils
- Discuss the origin of the antibiotic streptomycin and list five other medicinal agents made by streptomycetes
- Describe the morphology of *Frankia* spp., relate this to their colonization of plants, and explain the advantage these bacterial symbionts confer to plants such as alder trees
- Describe the morphology of members of *Bifidobacteriaceae*
- Differentiate between the human pathogenic and nonpathogenic members of class *Actinobacteria*

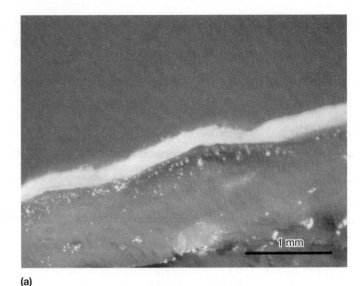

(a)

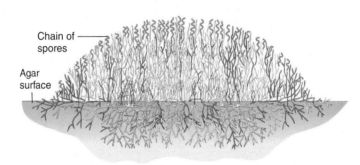

Chain of spores

Agar surface

(b)

Figure 24.2 Cross Section of an Actinobacterial Colony. (a) This photomicrograph shows *Streptomyces griseus* substrate hyphae (yellow) growing into the agar and the white aerial hyphae growing away from the colony surface. (b) The colony consists of actively growing aerial hyphae that cannibalize the substrate hyphae to obtain nutrients. Live hyphae are blue; dead are white.

Actinobacteria are fascinating microorganisms. Although all are chemoorganoheterotrophs, recently several have been discovered to use atmospheric H_2 as an electron donor to maintain electron flow into the electron transport chain when growing in nutrient-depleted aerobic conditions. As described in the chapter opening story, they also produce metabolites used as antibiotics, anticancer drugs, antihelminthics, and drugs that suppress the immune system in patients who have received organ transplants. This practical aspect of actinobacteria is linked very closely to their mode of growth, which in many cases involves a complex life cycle. Many actinobacteria make filaments, called **hyphae,** instead of more familiar, smaller rods or cocci. Septae (crosswalls) usually divide the hyphae into long cells (20 μm and longer), each containing several nucleoids. When growing on a solid substratum such as soil or agar, filamentous actinobacteria develop a branching network of hyphae (**figure 24.2**). Hyphae grow on and into the substratum to form a dense hyphal mat called a **substrate mycelium.** In nature, the substratum is soil or other sediments. In the laboratory, substrate hyphae grow into agar, where the tangle of substrate hyphae imparts a leathery texture to the colonies of many actinomycetes. In many actinobacteria, substrate hyphae go on to differentiate into upwardly growing hyphae to form an **aerial mycelium** that gives the colony a fuzzy appearance. It is at this time that medically useful compounds are formed. Because the physiology of the

actinobacteria has switched from actively growing vegetative cells into this special, secondary cell type, these compounds are often called **secondary metabolites.**

Over time, aerial hyphae septate to form chains of spores. When the septum completes constriction, individual exospores are released into the environment. These spores are considered **exospores** because they do not develop within a mother cell as do the endospores of *Bacillus* and *Clostridium* spp. If the spores are located in a sporangium (a sheathlike structure), they may be called **sporangiospores.** The spores can vary greatly in shape (**figure 24.3**). Like spore formation in other bacteria, actinobacterial sporulation is in response to nutrient deprivation and chemical signaling between microbes. In general, actinobacterial spores are not particularly heat resistant but withstand desiccation well, so they are well suited to the sediment environment. Most actinobacteria are nonmotile, and spores dispersed by wind or animals may find a new habitat to provide needed nutrients. Interestingly, among the few motile actinobacteria, movement is confined to flagellated spores.

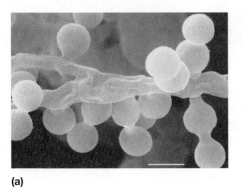
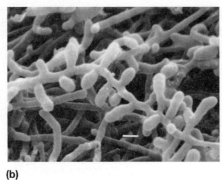
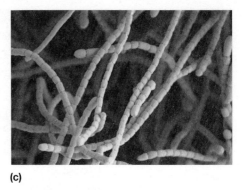

(a)   (b)   (c)

Figure 24.3 Examples of Actinomycete Spores as Seen in the Scanning Electron Microscope. (a) Spores of *Pseudonocardia yunnanensis* (bar = 1 μm). (b) Single spores of *Saccharomonospora viridis* are closely spaced along aerial hyphae. This organism causes farmer's lung. (c) Spore chain of *Streptomyces venezuelae*, which produces the antibiotic chloramphenicol.

Actinobacterial cell wall composition varies greatly among different groups and remains of taxonomic importance, despite the ubiquity of 16S rRNA analysis. Four major cell wall types can be distinguished based on the structure of the peptidoglycan and the cell wall sugar content (**figure 24.4** and **table 24.2**; *also see figure 3.17*). Peptidoglycan containing unusual other sugars can be used to categorize actinomycetes. Two aspects of amino acid content are also helpful: the amino acid in position 3 of the tetrapeptide side chain and the presence of glycine in interpeptide bridges. Some other taxonomically valuable properties include the color of the mycelium and sporangia, the surface features and arrangement of spores, the phospholipid composition of cell membranes, and spore heat resistance. ◄◄ *Peptidoglycan structure (section 3.4)*

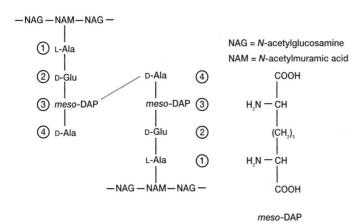

NAG = *N*-acetylglucosamine
NAM = *N*-acetylmuramic acid

meso-DAP

Figure 24.4 Actinobacterial Peptidoglycan Structure. Peptidoglycan with *meso*-diaminopimelic acid in position 3 and a direct cross-linkage between positions 3 and 4 of the peptide subunits is shown here. This is found in a variety of bacteria, including actinobacteria containing type III and IV cell walls (e.g., *Frankia* and *Nocardia*). Other actinobacteria have a glycine residue linking tetrapeptides.

MICRO INQUIRY *Refer to table 24.2 and determine how the cell walls of* Streptomyces *and* Actinoplanes *species differ from that shown here.*

Order *Actinomycetales*

The order *Actinomycetales* includes the genera *Actinomyces, Actinobaculum, Arcanobacterium,* and *Mobiluncus.* Most are irregularly shaped, nonsporing, Gram-positive rods with aerobic, facultative, or microaerophilic metabolism. The rods may be straight or slightly curved, and usually have swellings, club shapes, or other deviations from normal rod-shape morphology.

Members of the genus *Actinomyces* are either straight or slightly curved rods, or slender filaments with true branching (**figure 24.5**). The rods and filaments may have swollen or clubbed ends. They are either facultative or strict anaerobes that require CO_2 for optimal growth. *Actinomyces* species are normal inhabitants of mucosal surfaces of humans and other warm-blooded animals; the oral cavity is their preferred habitat. *A. bovis* causes lumpy jaw in cattle. *A. israelii* is most commonly responsible for actinomycoses, ocular infections, and lumpy jaw in humans.

Order *Micrococcales*

Of the wide variety of genera within the order *Micrococcales,* two of the best known are *Micrococcus* and *Arthrobacter.* The genus *Micrococcus* contains aerobic, catalase-positive cocci that occur mainly in pairs, tetrads, or irregular clusters and are usu-

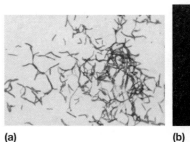

(a)   (b)

Figure 24.5 Representatives of the Genus *Actinomyces.* (a) *Actinomyces* sp.; Gram stain shows branching forms (×400). (b) *Actinomyces* sp.: color-enhanced scanning electron micrograph (×2,500).

Table 24.2	Actinobacterial Cell Wall Types and Whole Cell Sugar Patterns			
Cell Wall Type	**Diaminopimelic Acid Isomer**	**Glycine in Interpeptide Bridge**	**Characteristic Sugars**	**Representative Genera**
I	L, L	+	NA[1]	*Nocardioides, Streptomyces*
II	*meso*	+	NA	*Micromonospora, Pilimelia, Actinoplanes*
III	*meso*	–	NA	*Actinomadura, Frankia*
IV	*meso*	–	Arabinose, galactose	*Saccharomonospora, Nocardia*
Whole Cell Sugar Patterns[2]		**Characteristic Sugars**		**Representative Genera**
A		Arabinose, galactose		*Nocardia, Rhodococcus, Saccharomonospora*
B		Madurose[3]		*Actinomadura, Streptosporangium, Dermatophilus*
C		None		*Thermomonospora, Actinosynnema, Geodermatophilus*
D		Arabinose, xylose		*Micromonospora, Actinoplanes*

1 NA, either not applicable or no diagnostic sugars.
2 Characteristic sugar patterns are present only in wall types II–IV, those actinobacteria with *meso*-diaminopimelic acid.
3 Madurose is 3-*O*-methyl-D-galactose.

ally nonmotile (**figure 24.6**). Unlike many other actinobacteria, micrococci do not undergo morphological differentiation. Their colonies often are yellow, orange, or red. They are widespread in soil, water, and on mammalian skin.

The genus *Arthrobacter* contains aerobic, catalase-positive rods with respiratory metabolism. Their most distinctive feature is a rod-coccus growth cycle (**figure 24.7**). When in exponential phase, these bacteria are irregular, branched rods that undergo snapping division (p. 557). As they enter stationary phase, the cells change to a coccoid form. Upon transfer to fresh medium, the coccoid cells differentiate to form actively growing rods. Although arthrobacters often are isolated from fish, sewage, and plant surfaces, like many other actinobacteria, their most important habitat is soil, where they constitute a significant component of the culturable microbial community. They are well adapted to this niche because they are resistant to desiccation and nutrient deprivation, even though they do not form spores. Arthrobacters are unusually flexible nutritionally and can even degrade some herbicides and pesticides.

Retrieve, Infer, Apply

1. Describe the function of the substrate mycelium, aerial mycelium, and exospore. Explain how these structures confer a survival advantage.
2. How can cell wall structure and sugar content be used to classify actinobacteria?
3. Why are actinobacteria of such practical interest?
4. List two genera that are commonly associated with humans.

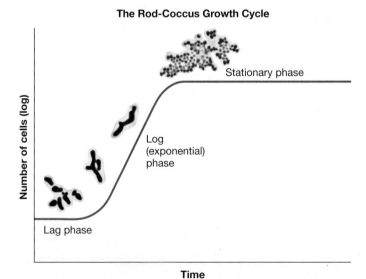

Figure 24.7 The Rod-Coccus Growth Cycle of *Arthrobacter* spp. When grown at 25°C, coccoid cells of about 1 μm in diameter (like those in stationary phase) swell and elongate to form branching rods of about 7–10 μm in length in exponential phase. As nutrients are depleted and cells enter stationary phase, the cells become spherical and form clusters.

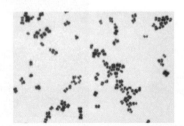

Figure 24.6 *Micrococcus.* *Micrococcus luteus*, Gram stain.

MICRO INQUIRY *What makes the morphology of* M. luteus *distinct from most other* actinobacteria *discussed in this chapter?*

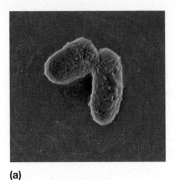
(a)

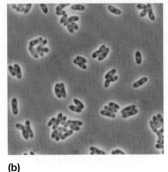
(b)

Figure 24.8 *Corynebacterium diphtheriae* **Undergoes Snapping Division, Giving Rise to Cells in Palisades.** (a) Following snapping division, daughter cells remain attached and at an angle (×28,000). (b) This angular association of pairs of cells results in a palisade arrangement (×1,000).

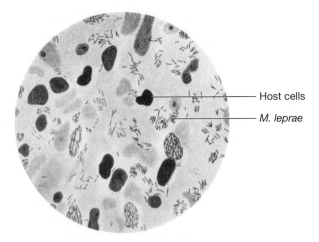

Host cells

M. leprae

Figure 24.9 **Mycobacteria.** *Mycobacterium leprae* from a leprosy skin lesion, acid-fast stain (×1,000).

Order *Corynebacteriales* Includes Important Human Pathogens

The order *Corynebacteriales* includes three very important genera: *Corynebacterium, Mycobacterium,* and *Nocardia*. Species in these genera produce long-chain fatty acids, called mycolic acids. These lipids are a virulence factor and are essential for viability only in mycobacteria.

The family *Corynebacteriaceae* has one genus, *Corynebacterium,* which includes aerobic and facultative, catalase-positive, straight to slightly curved rods. Club-shaped cells are common. Both corynebacteria and arthrobacters have a two-layered cell wall that leads to **snapping division** (**figure 24.8a**). This occurs because during cell division, only the inner layer grows inward to generate a septum dividing new cells. As the septum thickens, tension rises on the outer wall layer, which still holds the two cells together. Eventually increasing tension ruptures the outer layer at its weakest point, and a snapping movement tears the outer layer apart around most of its circumference. The new cells now rest at an angle to each other and are held together by the remaining portion of the outer layer, which acts as a hinge. This results in angular arrangements of the cells or a **palisade arrangement** in which rows of cells are lined up side by side (figure 24.8b). Although some species are harmless saprophytes, many corynebacteria are plant or animal pathogens. For example, *C. diphtheriae* is the causative agent of diphtheria in humans. ▶▶▶ *Diphtheria (section 39.1)*

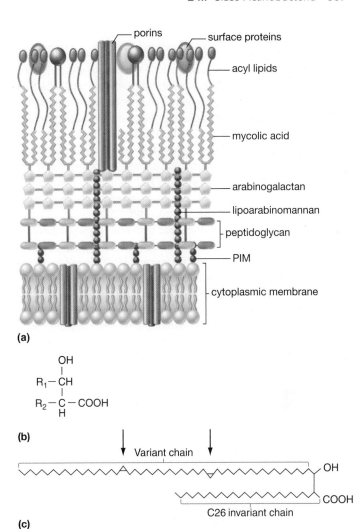

porins
surface proteins
acyl lipids
mycolic acid
arabinogalactan
lipoarabinomannan
peptidoglycan
PIM
cytoplasmic membrane

(a)

$$R_1 - \overset{\displaystyle OH}{\underset{\displaystyle }{CH}}$$
$$R_2 - \overset{\displaystyle }{\underset{\displaystyle H}{C}} - COOH$$

(b)

Variant chain

OH

COOH

C26 invariant chain

(c)

Figure 24.10 **Mycobacterial Cell Wall.** (a) The peptidoglycan layer is tethered to the plasma membrane by the glycolipids lipoarabinomannan and phosphatidyinositol mannosides (PIM). Arabinogalactan (D-arabinose and D-galactose), which is linked to mycolic acids, is also attached to the peptidoglycan, and is covered by a lipid bilayer. Like Gram-negative bacteria, mycobacteria have porins to regulate the passage of small molecules across the outer membrane. (b) The generic structure of mycolic acids, a family that includes over 500 different types. (c) An example mycolic acid with two cyclopropane rings.

MICRO INQUIRY *Why do* Mycobacterium *spp. have porins?*

The family *Mycobacteriaceae* contains the genus *Mycobacterium,* which is composed of slightly curved or straight rods that sometimes branch or form filaments (**figure 24.9**). Mycobacterial filaments readily fragment into rods and coccoid bodies. They are aerobic and catalase positive. Mycobacteria grow very slowly and must be incubated for 2 to 40 days after inoculation on a complex medium to form a visible colony.

Although mycobacteria are considered Gram positive, their cell walls possess an outer membrane, and their peptidoglycan is held within the periplasm (**figure 24.10a**). The outer membrane is constructed of tightly packed, covalently linked **mycolic acids** and noncovalently attached shorter lipids. Mycolic acids are complex fatty acids with an invariant C26 fatty acid that is attached to a longer, variable fatty acid (figure 24.10*b*). Mycolic acids in the outer membrane are covalently linked to the peptidoglycan by arabinogalactan, a polysaccharide made of arabinose and galactose subunits. The cell wall is extremely hydrophobic and impenetrable to most organic molecules, including antibiotics. In fact, mycobacteria have evolved porins that enable the entry of small molecules (*see figure 3.26*). In addition, the mycobacterial outer membrane possesses a recently discovered mechanism for protein secretion called the type VII secretion system. The lipid content makes the cell wall **acid-fast** (basic fuchsin dye cannot be removed from the cell by acid alcohol treatment). ◄◄ *Differential staining (section 2.3); Protein maturation and secretion (section 13.8)*

Although some mycobacteria are free-living saprophytes, they are best known as animal pathogens. *M. bovis* causes tuberculosis in cattle, other ruminants, and primates. Because this bacterium can produce tuberculosis (TB) in humans, dairy cattle are tested for the disease yearly; milk pasteurization kills the pathogen. Prior to widespread milk pasteurization, contaminated milk was a significant source of transmission. Today *M. tuberculosis* is the chief source of tuberculosis in humans. It is estimated that about one-third of the world's population is infected with *M. tuberculosis* (*see figure 39.4*). Leprosy is also a mycobacterial human disease; it is caused by *M. leprae* (figure 24.9). Finally, *M. avium* and the related species *M. intracellulare* comprise *M. avium* complex (MAC), which has become a common mycobacterial disease in developed countries, particularly among individuals with compromised immunity. ◄◄ *Comparative genomics (section 18.7)* ▶▶ Mycobacterium *infections (section 39.1); Mycobacterial skin infections (section 39.3)*

The family *Nocardiaceae* is composed of two genera, *Nocardia* and *Rhodococcus*. These bacteria develop a substrate mycelium that readily breaks into rods and coccoid fragments (**figure 24.11**). They also form an aerial mycelium that may produce conidia. Almost all are strict aerobes. These and related bacteria that resemble members of the genus *Nocardia* (named after Edmond Nocard [1850–1903], French bacteriologist and veterinary pathologist) are collectively called **nocardioforms.**

Nocardia bacteria are distributed worldwide in soil and aquatic habitats. They are involved in the degradation of hydrocarbons and waxes, and can contribute to the biodeterioration of rubber joints in water and sewage pipes. Although most are free-living saprophytes, some species, particularly *N. asteroides,* are opportunistic pathogens that cause nocardiosis in humans and other animals. People with low resistance due to other health problems, such as individuals with HIV-AIDS, are most at risk. The lungs are typically infected; other organs and the central nervous system may be invaded as well.

Rhodococcus contains species that are widely distributed in soils and aquatic habitats. They are of considerable interest because they can degrade an enormous variety

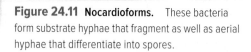

Nocardia

Figure 24.11 Nocardioforms. These bacteria form substrate hyphae that fragment as well as aerial hyphae that differentiate into spores.

of molecules such as petroleum hydrocarbons, detergents, benzene, polychlorinated biphenyls (PCBs), and various pesticides.

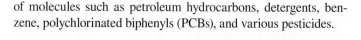

Order *Micromonosporales* Includes Bacteria That Produce Morphologically Diverse Spores

The order *Micromonosporales* contains the genera *Micromonospora, Dactylosporangium, Pilimelia,* and *Actinoplanes.* These bacteria are often called actinoplanetes (Greek *actinos,* a ray or beam, and *planes,* a wanderer). They have an extensive substrate mycelium but usually lack all but a rudimentary aerial mycelium. Often the hyphae are highly colored, and diffusible pigments may be produced. Spores are usually formed within a sporangium raised above the surface of the substratum at the end of a special hypha called a sporangiophore. The spores are quite unusual because in some species, they demonstrate flagellated motility. Flagella attach to a new substrate, where spores remain until triggered to germinate. The morphology and arrangement of spores is also variable. For instance, *Actinoplanes* and *Pilimelia* spp. have sporangia that may have only a few spores, while other species produce thousands of spores per sporangium (**figure 24.12**). The spores are arranged in coiled or parallel chains. *Dactylosporangium* spp. form club-shaped, fingerlike, or pyriform sporangia with one to six spores (figure 24.12*c*). *Micromonospora* bacteria bear single spores, which often occur in branched clusters of sporophores.

Actinoplanetes grow in almost all soil habitats, ranging from forest litter to beach sand. Some flourish in freshwater, particularly in streams and rivers (probably because of abundant oxygen and plant debris), others are marine. Actinoplanes is the source of the drug acarbose used to treat type 2 diabetes. The soil-dwelling species may have an important role in the decomposition of plant and animal material. *Pilimelia* bacteria grow in association with keratin. *Micromonospora* spp. actively degrade chitin and cellulose, and produce antibiotics such as gentamicin.

Order *Propionibacteriales*

The order *Propionibacteriales* contains the genus *Propionibacterium,* pleomorphic, nonmotile, nonsporing rods that are often club-shaped with one end tapered and the other end rounded. The most famous members of this genus are used to produce Swiss cheese, where its gaseous products of fermentation are responsible for the holes. Cells also may be coccoid or branched. They can be single, in short chains, or in clumps. They are facultatively anaerobic or aerotolerant; lactate and sugars are fermented to produce large quantities of propionic and acetic acids, and often carbon dioxide. In animals, these bacteria are found growing in

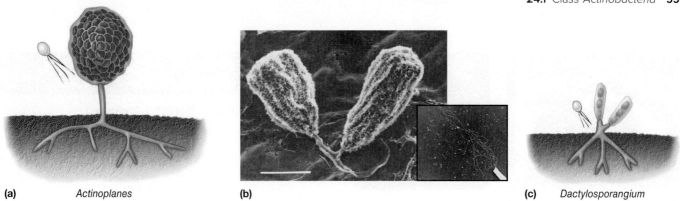

(a) *Actinoplanes* **(b)** **(c)** *Dactylosporangium*

Figure 24.12 *Micromonosporaceae.* Actinoplanete morphology. (a) *Actinoplanes* structure. (b) A scanning electron micrograph of mature *Actinoplanes capillaceus* sporangia; inset shows a motile spore. (c) *Dactylosporangium* structure.

the digestive tract and on the skin. *P. acnes* is involved with the development of body odor and is thought to cause acne vulgaris.

Order *Streptomycetales:* An Important Source of Antibiotics

The order *Streptomycetales* has only one family, *Streptomycetaceae,* and three genera, the most important of which is *Streptomyces.* This is a large genus; there are about 150 species. All are considered strict aerobes. Members of this family and similar bacteria are often called **streptomycetes** (Greek *streptos,* bent or twisted, and *myces,* fungus).

Like Gram-negative myxobacteria, streptomycetes are a good example of true multicellularity among bacteria. This is illustrated by their life cycle (**figure 24.13**). Single, nonmotile exospores germinate to produce long substrate hyphae. These form a dense substrate mycelium. Upon receipt of cues that indicate nutrient limitation and intercellular signals, the substrate hyphae grow up into the air to form aerial hyphae. Aerial hyphae obtain nutrients by cannibalizing the filaments within the substrate mycelium. It is at this point in the life cycle that the stressed but still surviving substrate hyphae produce antibiotics and other secondary metabolites. As aerial hyphae continue to grow, they curl and

septate in regular intervals. Ultimately the aerial hyphae become chains of spores that pinch off into single exospores to begin the cycle anew. Interestingly, the process of exospore formation is dependent on the cytoskeletal proteins FtsZ and MreB, but vegetative growth is not, because substrate hyphae grow by extending their tips rather than by binary fission. This results in long filaments with only occasional septae. Because multiple nucleoids are enclosed in the same cell, this is called syncytial growth. ◄◄ *Cytokinesis (section 7.2)* ►► *Antibiotics (section 42.1)*

Streptomycetes are very important both ecologically and medically. The natural habitat of most streptomycetes is the soil, where they may constitute from 1 to 20% of the culturable population. In fact, the odor of moist earth is largely the result of streptomycete production of volatile substances such as geosmin. Streptomycetes play a major role in mineralization—the degradation of organic matter to inorganic elements. As substrate hyphae grow into the earth, they secrete a variety of exoenzymes enabling the degradation of resistant substances such as pectin, lignin, chitin, keratin, latex, agar, and aromatic compounds. ►► *Diverse microorganisms inhabit soil (section 31.2)*

Selman Waksman's discovery that *S. griseus* (**figure 24.14a**) produces streptomycin, the first drug to cure TB, set off a massive search resulting in the isolation of new *Streptomyces* species

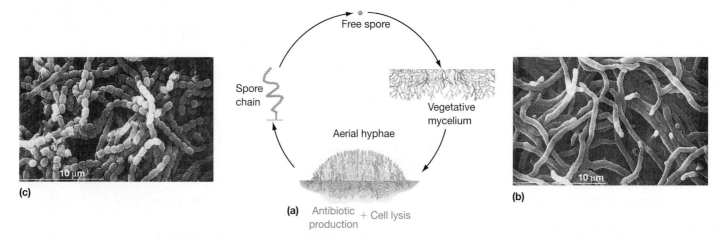

Figure 24.13 *Streptomyces* **Life Cycle.** (a) Streptomycetes form three types of cells: exospores germinate to produce vegetative (also called substrate) hyphae, which then differentiate into aerial hyphae. These give rise to chains of exospores. Scanning electron micrographs are shown of (b) *S. coelicolor* vegetative hyphae and (c) chains of exospores.

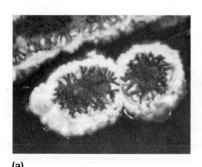

(a)   **(b)**

Figure 24.14 Streptomycetes of Practical Importance.   (a) *Streptomyces griseus.* Colonies of the actinomycete that produces streptomycin. Each colony is about 0.5 cm in diameter. (b) *S. scabies* growing on a potato.

Table 24.3	Examples of Natural Products Made by Streptomycetes	
Microbe	**Natural Product**	**Application**
Streptomyces orientals	Vancomycin	Antibiotic; cell wall inhibitor
S. mediterranei	Rifamycin	Antibiotic: transcription inhibitor
S. rimosus	Tetracycline	Antibiotic: protein synthesis inhibitor
S. venezuelae	Chloramphenicol	Antibiotic: protein synthesis inhibitor
S. clavuligerus	Clavulanic acid	β-lactamase inhibitor
S. nodosis	Amphotericin B	Antifungal
S. noursei	Nystatin	Antifungal
S. peucetius	Daunorubicin, doxorubicin epirubicin	Anticancer
S. verticillus	Bleomycin	Anticancer

that produce other compounds of medicinal importance. Since that time, streptomycetes have been found to produce over 10,000 bioactive compounds; a few are listed in **table 24.3**. Hundreds of these natural products are now used in medicine and industry; about two-thirds of the antimicrobial agents used in human and veterinary medicine are derived from streptomycetes. Antibiotic-producing bacteria have "immunity" genes that encode proteins (or small noncoding RNA) that make them resistant to the antibiotics they synthesize. ◀◀ *Antimicrobial chemotherapy (chapter 9)*

The genome of *Streptomyces coelicolor,* which produces four antibiotics, is one of the largest bacterial genomes (8.67 Mbp). Its large number of genes (7,825) no doubt reflects the number of proteins required to undergo a complex life cycle. Many genes are devoted to regulation, with an astonishing 65 predicted RNA polymerase sigma transcription factors and about 80 two-component regulatory systems. The ability to exploit a variety of soil nutrients is also demonstrated by the presence of a large number of ABC transporters, the Sec and two Tat protein translocation systems, and over 50 secreted degradative enzymes. ◀◀ *Protein maturation and secretion (section 13.8); Regulation of transcription initiation saves considerable energy and materials (section 14.2)*

Although most streptomycetes are nonpathogenic saprophytes, a few are associated with plant and animal diseases. *Streptomyces scabies* causes scab disease in potatoes and beets (figure 24.14*b*). *S. somaliensis* is the only streptomycete known to be pathogenic to humans. It is associated with actinomycetoma, an unusual infection of subcutaneous tissues that produces lesions that lead to swelling, abscesses, and even bone destruction if untreated. *S. albus* and other species have been isolated from patients with various ailments and may be pathogenic.

Order *Streptosporangiales*

The order *Streptosporangiales* includes maduromycetes, so named because they produce

the sugar derivative **madurose** (3-*O*-methyl-D-galactose). Aerial hyphae bear pairs or short chains of spores, and the substrate hyphae are branched (**figure 24.15**). Members of some genera form sporangia; spores are not heat resistant. *Thermomonospora* bacteria

(a) *Actinomadura madurae*

(b)   *Streptosporangium*

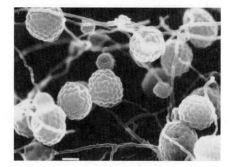

Figure 24.15 Maduromycetes.   (a) *Actinomadura madurae* morphology; illustration and electron micrograph of a spore chain. (b) *Streptosporangium roseum* morphology; illustration and micrograph of sporangia and hyphae; SEM.

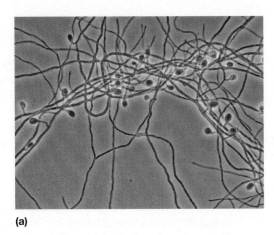

(a)

(b)

Figure 24.16 *Frankia.* (a) A phase contrast micrograph showing hyphae, multilocular sporangia, and spores. (b) Nodules of the alder *Alnus rubra*.

produce single spores on the aerial hyphae or on both the aerial and substrate hyphae. They have been isolated from moderately high-temperature habitats such as compost piles and hay and can grow at 40 to 48°C.

Order *Frankiales:* Bacteria That Fix N₂ in Trees

The order *Frankiales* includes the genera *Frankia* and *Geodermatophilus*. Species in both form multilocular sporangia, characterized by clusters of spores when hyphae divide both transversely and longitudinally. (Multilocular means having many cells or compartments.) *Geodermatophilus* spp. have motile spores and are aerobic soil organisms. *Frankia* spp. form nonmotile spores in a sporogenous body (**figure 24.16**). They grow in symbiotic association with the roots of at least eight families of higher nonleguminous plants (e.g., alder trees). They are microaerophilic and can fix atmospheric nitrogen.

Frankia bacteria fix nitrogen so efficiently that a plant such as an alder can grow in the absence of combined nitrogen (e.g., NO_3^-) (figure 24.16*b*). Within the nodule, these bacteria form branching hyphae with globular vesicles at their ends, which may be the sites of nitrogen fixation. Their nitrogenase resembles that of rhizobia in that it is oxygen sensitive and requires molybdenum and cobalt. ▶▶◀ *Nitrogen-fixing bacteria are vital to agriculture (section 31.3)*

Order *Bifidobacteriales:* Pathogens and Probiotics

The order *Bifidobacteriales* includes members of the genera *Falcivibrio* and *Gardnerella*, which are found in the human genital/urinary tract. *Gardnerella* spp. are thought to be a major cause of bacterial vaginitis. Bifidobacteria have gained recent attention as more is learned about the importance of human gut microflora. They are nonmotile, nonsporing, Gram-positive rods of varied shapes that are slightly curved and clubbed; often they are branched (**figure 24.17**). The rods can be single or in clusters and V-shaped pairs. They are anaerobic and actively ferment carbohydrates to produce acetic and lactic acids but not carbon dioxide. They are

found in the mouth and intestinal tract of warm-blooded vertebrates, in sewage, and in insects. *B. bifidum* is a pioneer colonizer of the human intestinal tract, especially when babies are breast fed. *Bifidobacterium* species are sold as probiotic agents (particularly in yogurt) because they are thought to impart a health benefit. ▶▶◀ *Normal microbiota of the human body adapt to many sites (section 32.3); Probiotics (section 41.6)*

Retrieve, Infer, Apply

1. List the distinguishing properties of actinoplanetes. What other actinomycetes form motile spores?
2. Where would you most likely find a member of the genus *Propionibacterium* in your home or dormitory?
3. List three types of natural products made by members of the genus *Streptomyces*.
4. Describe three ways in which *Streptomyces* spp. are of ecological importance.
5. Briefly describe the defining properties of genera in the order *Streptosporangiales*. What is madurose?
6. Why are *Frankia* spp. considered ecologically important?
7. Examine the ingredients list on a commercial yogurt cup; are bifidobacteria present?

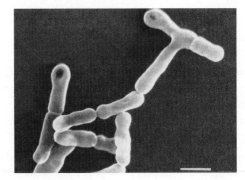

Figure 24.17 *Bifidobacterium. Bifidobacterium bifidum*; scanning electron photomicrograph.

Key Concepts

24.1 Class *Actinobacteria*

- Members of this class that form branching, usually nonfragmenting hyphae and asexual spores are collectively called actinobacteria or actinomycetes (**figure 24.2**).

- The asexual spores borne on aerial hyphae are called exospores because they do not develop within a mother cell. Some produce sporangiospores.

- Actinobacteria have several distinctively different types of cell walls and often also vary in terms of the sugars present in cell extracts. Properties such as color and morphology are also taxonomically useful (**tables 24.1 and 24.2; figure 24.4**).

- Members of the genus *Actinomyces* are irregularly shaped, nonsporing rods that can cause disease in cattle and humans.

- Members of the genus *Micrococcus* are unusual members of this class because they form simple coccoid cells. Some species are normal inhabitants on human skin.

- Arthrobacters have an unusual rod-coccus growth cycle and, like corynebacteria, carry out snapping division (**figures 24.7**).

- Corynebacteria form irregular and club-shaped cells that are often arranged in palisades. *C. diphtheriae* is the cause of diphtheria (**figure 24.8**).

- Mycobacteria form either rods or filaments that readily fragment. Their cell walls have a high lipid content and mycolic acids, which makes them acid-fast (**figures 24.9 and 24.10**).

- Nocardioform actinobacteria have hyphae that readily fragment into rods and coccoid elements, and often form aerial mycelia with spores (**figure 24.11**).

- *Micromonospora* and *Actinoplanes* spp. have an extensive substrate mycelium and form aerial sporangia and are present in soil, freshwater, and the ocean (**figure 24.12**).

- Members of the genus *Propionibacterium* are common skin and intestinal inhabitants, and are important in cheese manufacture and the development of acne vulgaris.

- The order *Streptomycetales* includes the genus *Streptomyces*. Members of this genus produce substrate hyphae that differentiate into aerial hyphae-bearing chains of 3 to 50 or more exospores (**figure 24.13**).

- Streptomycetes are important in the degradation of resistant organic material in the soil and produce many useful antibiotics. A few cause diseases in plants and animals. However, they are most notable for the variety of antibiotics and other medically and industrially important compounds they produce (**table 24.3**).

- The genera *Frankia* and *Geodermatophilus* are placed in the order *Frankiales*. *Frankia* spp. grow in symbiotic association with nonleguminous plants and fix nitrogen (**figure 24.16**).

- The genus *Bifidobacterium* is placed in the order *Bifidobacteriales*. Bifidobacteria are irregular, anaerobic rods and one of the first colonizers of the intestinal tract in nursing babies.

- Members of the genus *Bifidobacterium* are often sold for their probiotic activity.

Compare, Hypothesize, Invent

1. Even though actinobacteria are high G + C organisms, there are regions of the genome that are AT-rich. Suggest a few such regions and explain why they must be more AT-rich.

2. Given that mycolic acids are essential for mycobacterial viability, how would you test their importance in mycobacterial infection?

3. *Streptomyces coelicolor* is studied as a model system for cellular differentiation. Some of the genes involved in sporulation contain an AT-rich leucine codon not used in vegetative genes. Suggest how *Streptomyces* might use this rare codon to regulate sporulation.

4. Suppose that you discovered a nodulated plant that could fix atmospheric nitrogen. How might you show that a bacterial symbiont was involved and that a *Frankia* sp. rather than a *Rhizobium* sp. was responsible?

5. It has long been held that soil microbes such as the streptomycetes produce antibiotics so that they can kill neighboring microbes and use the nutrients released. However, it has been found that each of 480 soil bacterial isolates were resistant to 6 to 20 different antibiotics. Separate research demonstrates that hundreds of soil bacteria can use assorted antibiotics as their sole carbon source. These data, together with the ability of sublethal concentrations of antibiotics to trigger diverse responses in nontarget microbes, have stimulated the hypothesis that in nature, antibiotics are used as signaling molecules rather than weapons. Do you think this argument has any merit and how would you test it?

Read the original papers: D'Costa, V. M., et al. 2006. Sampling the antibiotic resistome. *Science* 311:374; and Dantas, G., et al. 2008. Bacteria subsisting on antibiotics. *Science* 320:100.

25

Protists

Fruiting bodies of the social protist Dictyostelium discoideum. *This amazing microbe farms its food and has three "sexes," or mating types.*

Sustainable Farming Practiced by Amoebae

The advent of farming changed the course of human evolution; without the need to hunt and gather, our ancestors could form larger, more cohesive social groups. However, we are not the first farmers. For decades, scientists have known of beetles, termites, and ants that farm fungi (*see figure 32.12*). Now an even more primitive kind of farming has been discovered. It is performed by an amoeba called *Dictyostelium discoideum* (Dicty for short).

Dicty lives in soil where it grows as solitary, single-celled microbes preying on bacteria. If Dicty runs short of bacterial food, cells join forces and form a multicellular slug that migrates to a new, food-rich area (section 25.3). Once there, the slug forms a fruiting body consisting of two parts: a spherical sorus that houses cells that have become spores and a supporting stalk, also made of Dicty cells. While it was well established that fruiting body formation is a means of dispersal to a site with a better food supply, researchers at Rice University in Texas have now shown that about a third of Dicty fruiting bodies bring their food with them.

When grown in the presence of bacterial food, the farming Dicty stop eating before those that do not farm. Dicty farmers then pack up the leftovers (no brown bags involved) in each sorus. The sorus of nonfarming Dicty contains only amoebae. When slugs arrive at a spot deemed suitable for germination, the farmers have a ready food supply consisting of the bacteria they packed in the sorus. So regardless of where they move, they are sure to have their "favorite" species for consumption.

But there are a couple of drawbacks with this system. First, if farmer and nonfarmer slugs are placed in a head-to-head migration race, the nonfarmers go farther (but not necessarily faster). This means the nonfarmer is more likely to escape the poor conditions that triggered slug formation in the first place. Furthermore, if both types of Dicty settle in regions of equal food supply, the farmers have no advantage; farmers and nonfarmers grow equally well. Nonetheless, bacterial husbandry can be a lifesaver, so why is it found in only a third of all isolates? The answer may rest with the fact that Dicty eats a varied diet, consuming a wide range of soil bacteria. This presumably has prevented the kind of co-evolution between prey and predator seen in other systems ranging from viruses and their cellular hosts to superpredators and their quarry on the African plains.

So far, our discussion of microbial diversity has focused on *Bacteria* and *Archaea.* We now survey the vast diversity of the eukaryotic microbial world, starting with protists and moving on to fungi in chapter 26. The general structure of eukaryotic cells is reviewed in chapter 5; here we focus on adaptations that make these microorganisms successful competitors.

Eukaryotic microorganisms suffer from a confused taxonomic history. The kingdom Protista, as presented in many introductory biology texts, is an artificial grouping of over 64,000 different single-celled life forms that lack common evolutionary heritage; that is, they are not monophyletic (*see figure 1.2*). In this text, we refer to these organisms simply as **protists.** Protists are unified only by what they lack: absent is the level of tissue organization found in the more evolved fungi, plants, and animals. The term **protozoa** (s., protozoan; Greek *protos,* first, and *zoon,* animal) refers to chemoorganotrophic protists. The term **algae** can be used to describe photosynthetic protists that, like plant cells, possess a cell wall. The study of photosynthetic protists (algae) is often referred to as **phycology** and is the realm of both botanists and protistologists. The study of all protists, regardless of their metabolic type, is called **protistology,** although the term **protozoology** is more commonly used.

Readiness Check:
Based on what you have learned previously, you should be able to:

✔ Draw a eukaryotic cell envelope and identify the structure and function of its component layers (section 5.2)

✔ List the three filaments that make up the cytoskeleton of eukaryotic cells and describe how they are used (section 5.3)

✔ Identify the structure and function of eukaryotic cellular organelles (sections 5.4–5.6)

✔ Describe the endocytic pathway in general terms (section 5.4)

✔ Explain the difference between a mitochondrion and a hydrogenosome (section 5.6)

✔ Describe the structure and function of eukaryotic flagella and cilia (section 5.7)

25.1 Protist Diversity Reflects Broad Phylogeny

After reading this section, you should be able to:

- Describe three nutritional strategies used by protists
- Draw a typical protist cell membrane and underlying features, labeling the plasmalemma, ecto- and endoplasm, and pellicle
- Explain how contractile vacuoles are used to maintain osmotic stability
- Trace the passage of a food particle as it is consumed by a protozoan
- Compare binary fission as it occurs in a protist with that in a bacterium or archaeon
- Outline the process of sexual reproduction and how it can vary in protists

Most eukaryotes are microbes. It is therefore not surprising that the vast diversity of protists is a function of their capacity to thrive in a wide variety of habitats. Their one common requirement is moisture because all are susceptible to desiccation. Most protists are free living and inhabit freshwater or marine environments. Many terrestrial chemoorganotrophic forms can be found in decaying organic matter and soil. Whether terrestrial or aquatic, protists play an important role in nutrient cycling.

▶▶| *Biogeochemical cycling sustains life on Earth (section 28.1)*

Some protozoa, or chemoorganoheterotrophic protists, are **saprophytes,** securing nutrients from dead organic material by releasing degradative enzymes into the environment. They then absorb the soluble products—a process sometimes called **osmotrophy.** Other protozoa employ **holozoic nutrition,** in which solid nutrients are acquired by phagocytosis. Photoautotrophic protists are strict aerobes and, like cyanobacteria, use photosystems I and II to perform oxygenic photosynthesis. It is difficult to classify the nutritional strategies of some protists because they simultaneously use reduced organic molecules and CO_2 as carbon sources. This strategy is called **mixotrophy.**

Protist Morphology

Despite their diversity, protists share many common features. In many respects, their morphology and physiology are the same as the cells of multicellular plants and animals. However, because many protists are unicellular, all of life's various functions must be performed within a single cell. Those that are multicellular lack highly differentiated tissues. Therefore the structural complexity observed in protists arises at the level of specialized organelles.

The protist cell membrane is called the **plasmalemma** and is identical to that of multicellular organisms (**figure 25.1**). In some protists, the cytoplasm immediately under the plasmalemma is divided into an outer gelatinous region called the **ecto-plasm** and an inner fluid region, the **endoplasm.** The ectoplasm imparts rigidity to the cell body. Many protists also have a supportive mechanism called the **pellicle.** The pellicle consists of the plasmalemma and a relatively rigid layer just beneath it. The pellicle may be simple in structure. For example, *Euglena* spp.

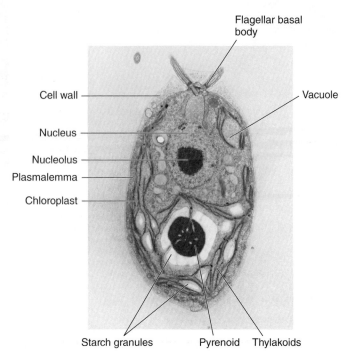

Figure 25.1 Protist Structure. A transmission electron micrograph of the photosynthetic protist *Chlamydomonas reinhardtii* is shown as an example.

are protists with a series of overlapping strips with a ridge at the edge of each strip fitting into a groove on the adjacent one, much like how the "tongue-and-groove" boards of a hardwood floor fit together. In contrast, the pellicles of ciliate protists are exceptionally complex with two membranes and a variety of associated structures. Although pellicles are not as strong and rigid as cell walls, for those that possess them, pellicles impart the characteristic shape associated with that particular species.

One or more vacuoles are usually present in the cytoplasm of protozoa. These are differentiated into contractile, secretory, and food or phagocytic vacuoles. **Contractile vacuoles** function as osmoregulatory organelles in protists that live in hypotonic environments, such as freshwater lakes. Contractile vacuoles maintain osmotic balance by continuous water expulsion. **Phago-cytic vacuoles** are conspicuous in protists that ingest food by phagocytosis (holozoic protists). Phagocytic vacuoles are the sites of food digestion. In some organisms, they may occur anywhere on the cell surface, while others have a specialized structure for phagocytosis called the **cytostome** (cell mouth). Once the cytostome has engulfed a food particle, digestion commences when the phagocytic vacuole becomes acidic. As digestion proceeds, the vacuole membrane forms small blebs that pinch off and carry nutrients throughout the cytoplasm. The undigested contents of the original phagocytic vacuole are expelled from the cell either at a random site on the cell membrane or at a designated position called the **cytoproct** (Greek, *proktos,* anus).

The diversity of energy-conserving organelles observed in protists reflects their evolutionary history and life style. Most aerobic chemoorganotrophic protists have mitochondria, while photosynthetic forms have chloroplasts. A dense proteinaceous area, the **pyrenoid,** which is associated with the synthesis and storage of starch, may be present in chloroplasts. The majority of anaerobic chemoorganotrophic protists lack mitochondria; some of these organisms have hydrogenosomes.

Many protists feature cilia or flagella at some point in their life cycle. Their formation is associated with a basal bodylike organelle called the kinetosome. In addition to aiding in motility, these organelles may be used to generate water currents for feeding and respiration.

Encystment and Excystment

Many protists are capable of **encystment.** During encystment, the organism becomes simpler in morphology and develops into a resting stage called a cyst. The cyst is a dormant form marked by the presence of a cell wall and very low metabolic activity. Cyst formation is particularly common among aquatic, free-living protists and parasitic forms. Cysts serve three major functions: (1) they protect against adverse changes in the environment, such as nutrient deficiency, desiccation, adverse pH, and low levels of O_2; (2) they are sites for nuclear reorganization and cell division (reproductive cysts); and (3) in parasitic species they are the infectious stage between hosts. Protists escape from cysts by a process called **excystment (figure 25.2).** Although the exact stimulus for excystment is unknown for most protists, it is generally triggered by a return to favorable environmental conditions. For example, cysts of parasitic species excyst after ingestion by the host.

Protist Reproductive Cells and Structures

Most protists have both asexual and sexual reproductive phases in their life cycles. The most common method of asexual reproduction is **binary fission.** During this process, the nucleus first undergoes mitosis and then the cytoplasm divides by cytokinesis to form two identical individuals (**figure 25.3**). Multiple fission is also common, as is budding. Some filamentous, photosynthetic protists undergo fragmentation so that each piece of the broken filament contains a nucleus and grows independently.

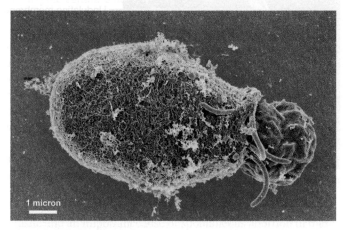

Figure 25.2 Excystment. This colorized scanning electron micrograph shows the protist *Giardia* sp. undergoing excystment. The flagellated vegetative cell, called a trophozoite, is emerging from the right of the cyst.

Sexual reproduction involves the formation of gametes. Protist cells that produce gametes are termed **gamonts.** The fusion of haploid gametes is called **syngamy.** Among protists, syngamy can involve the fusion of two morphologically similar gametes (isogamy) or two morphologically different types (anisogamy). Meiosis may occur before the formation and union of gametes, as in most animals, or just after fertilization, as is the case with lower plants. Furthermore, the exchange of nuclear material may occur in the familiar fashion—between two different individuals (**conjugation**)—or by the development of a genetically distinct nucleus within a single individual (autogamy).

With this level of reproductive complexity, perhaps it is not surprising that the nuclei among protists show considerable diversity. Most commonly, a vesicular nucleus is present. This is 1 to 10 μm in diameter, spherical, and has a distinct nucleolus and uncondensed chromosomes (figure 25.1). The uncondensed chromosomes give these nuclei a lacy or vesicular appearance when viewed microscopically. Ovular nuclei are up to 10 times this size and possess many peripheral nucleoli. Still others have chromosomal nuclei, in which the chromosomes remain condensed throughout the cell cycle. Finally, many ciliated forms have two types of nuclei: a large macronucleus with distinct nucleoli and condensed chromatin, and a smaller, diploid micronucleus with dispersed chromatin but lacking nucleoli. Genes in the macronucleus encode proteins involved in housekeeping functions and binary fission, whereas genes on the micronucleus function only in sexual reproduction (p. 574).

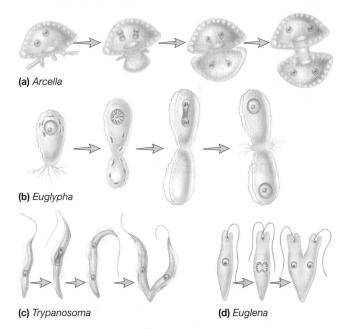

(a) *Arcella*

(b) *Euglypha*

(c) *Trypanosoma*   **(d)** *Euglena*

Figure 25.3 Binary Fission in Protists. (a) The two nuclei of a testate (shelled) amoeba in the genus *Arcella* divide as some of its cytoplasm is extruded and a new test for the daughter cell is secreted. (b) Members of another testate amoeba genus, *Euglypha,* begin to secrete new platelets before cytoplasm begins to move out of the aperture. The nucleus divides while the platelets are used to construct the test for the daughter cell. For many protists, all organelles must be replicated before the cell divides. This is also the case with species of (c) *Trypanosoma* and (d) *Euglena*.

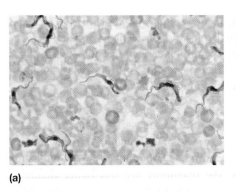

(a)

(b)

Figure 25.6 **The Euglenozoan *Trypanosoma* and Its Insect Host.** (a) *Trypanosoma* among red blood cells. Note the dark-staining nuclei, anterior flagella, and undulating, changeable shape. (b) The tsetse fly, shown here sucking blood from a human arm, is an important vector of the *Trypanosoma* species that causes African sleeping sickness.

to *40.13*). ▶▶| *Leishmaniasis (section 40.3); Trypanosomiasis (section 40.3)*

The trypanosomes *T. gambiense* and *T. rhodesiense* (often considered a subspecies of *T. brucei*) cause African sleeping sickness (**figure 25.6**). Ingestion of these parasites by the blood-sucking tsetse fly triggers a complex cycle of development and reproduction, first in the fly's gut and then in its salivary glands (*see figure 40.12*). From there the parasite is easily transferred to a vertebrate host, where it often causes a fatal infection. It is estimated that about 65,000 people die annually of sleeping sickness. The presence of this dangerous parasite prevents the use of about 11 million square kilometers of African grazing land.

Retrieve, Infer, Apply

1. List at least three features that distinguish *Parabasalia* from *Euglenozoa*.
2. What is the function of the plasmalemma and pellicle in euglenid cells? How are these structures similar to the bacterial cell wall?
3. Members of which *Euglenozoa* genera cause disease?

25.3 Supergroup *Amoebozoa* Includes Protists with Pseudopodia

After reading this section, you should be able to:

- Draw three different types of pseudopodia
- Differentiate between a naked and a testate amoeba
- Compare and contrast the structure, motility, and life cycles of acellular and cellular slime molds
- Explain the importance of molecular signaling in the morphological development of *Dictyostelium discoideum*
- Identify a nearby location where you might find members of the genera *Tubulinea*, *Entamoebida*, and *Eumycetozoa*

Because the amoeboid form arose independently numerous times from various flagellated ancestors, some amoebae are placed in

the supergroup *Amoebozoa*, while others are in *Rhizaria*. One of the morphological hallmarks of amoeboid motility is the use of **pseudopodia** (meaning "false feet") for both locomotion and feeding (**figure 25.7**). Pseudopodia can be rounded (**lobopodia**), long and narrow (**filopodia**), or form a netlike mesh (**reticulopodia**). Amoebae that lack a cell wall and are surrounded only by a plasma membrane are called **naked amoebae.** In contrast, the plasma membrane of a **testate amoeba** is covered by material that is either made by the protist itself or collected by the organism from the environment. Binary fission is the usual means of asexual division, although some amoebae form cysts that undergo multiple fission.

Tubulinea

Members of *Tubulinea* inhabit almost any environment where they will remain moist; this includes glacial meltwater, marine environments, tide pools, lakes, and streams. Free-living forms are known to dwell in ventilation ducts and cooling towers, where they feed on microbial biofilms. Others are endosymbionts, commensals, or parasites of invertebrates, fishes, and mammals. Some harbor intracellular symbionts, including algae,

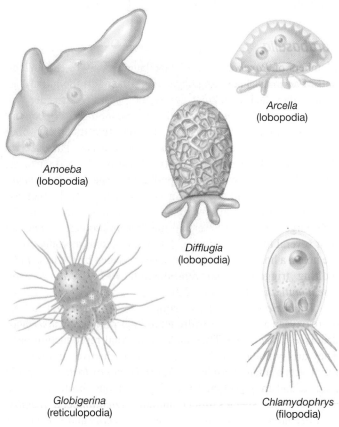

Amoeba (lobopodia)

Arcella (lobopodia)

Difflugia (lobopodia)

Globigerina (reticulopodia)

Chlamydophrys (filopodia)

Figure 25.7 Pseudopodia.

consume contaminated stream or river water. It is the most common cause of epidemic waterborne diarrheal disease in the United States (about 30,000 cases yearly). Members of *Fornicata* bear flagella and lack mitochondria. However, they possess mitochondria-like double-bounded membrane organelles called **mitosomes.** Although these primitive organelles share ancestry with mitochondria and hydrogenosomes, they do not appear to be involved in ATP synthesis. Instead the mitosomes of *G. intestinalis* are involved in producing FeS proteins. ▶▶| *The classic mitochondrion is only one member of the mitochondrial family of organelles (section 5.6)*

Cyst formation is key to *G. intestinalis* survival. Ingestion of only about 10 cysts is needed for infection, which involves its attachment to gut cells called enterocytes. When these enterocytes die and are sloughed from the gut, the microbe must "decide" whether to attach to another enterocyte or encyst and exit to find a new host. When a cyst enters a new host, the presence of bile and the slightly alkaline conditions of the small intestine trigger excystment (figure 25.2). ▶▶| *Giardiasis (section 40.5)*

Unlike *Giardia* spp. most members of *Fornicata* are harmless symbionts. The few free-living forms are most often found in waters that are heavily polluted with organic nutrients (a condition known as eutrophication). Only asexual reproduction by binary fission has been observed. Other pathogenic species include *Hexamida salmonis,* a fish parasite found in hatcheries and fish farms, and *H. meleagridis,* a turkey pathogen that is responsible for the annual loss of millions of dollars in poultry revenue.

Parabasalia

Members of *Parabasalia* are flagellated; most are endosymbionts of animals. They use phagocytosis to engulf food items, although they lack a distinct cytostome. Here we consider two subgroups: *Trichonymphida* and *Trichomonadida.* Trichonymphids are obligate mutualists in the digestive tracts of wood-eating insects such as termites and wood roaches. They secrete the enzyme cellulase needed for the digestion of wood particles, which they entrap with pseudopodia. One species, *Trichonympha campanula,* can account for up to one-third of the biomass of an individual termite (*see figure 32.2*). This species is particularly large for a protist (several hundred micrometers) and can bear several thousand flagella. Although asexual reproduction is the norm, a hormone called ecdysone produced by the host when molting triggers sexual reproduction. ▶▶| *Microorganism-insect mutualisms (section 32.1)*

Trichomonadida spp., or simply trichomonads, do not require oxygen and possess hydrogenosomes rather than mitochondria (*see figure 5.16*). They undergo asexual reproduction only. They are symbionts of the digestive, reproductive, and respiratory tracts of many vertebrates. *Tritrichomonas foetus* is a cattle parasite and an important cause of spontaneous abortion in these animals. Four species infect humans: *Dientamoeba fragilis, Pentatrichomonas hominis, Trichomonas tenax,* and *T. vaginalis* (figure 25.4*b*). *D. fragilis* has recently been recognized as a cause of diarrhea, while *T. vaginalis* has long been known to be

pathogenic. It is found in the genitourinary tract of both men and women, and the sexual transmission of pathogenic strains accounts for an estimated 7 million cases of trichomoniasis annually in the United States and 180 million cases annually worldwide. ▶▶| *Protist pathogens (section 40.4)*

Euglenozoa

Members of *Euglenozoa* are commonly found in freshwater, although a few species are marine. About one-third of euglenids are photoautotrophic; the remaining are free-living chemoorganotrophs. Most chemoorganotrophic forms are saprophytic, although a few parasitic species have been described. The representative genus is *Euglena,* which contains photoautotrophic euglenids. A typical *Euglena* spp. cell (**figure 25.5**) is elongated and bounded by a plasmalemma. The pellicle consists of proteinaceous strips and microtubules; it is elastic enough to enable turning and flexing of the cell, yet rigid enough to prevent excessive alterations in shape. *Euglena* spp. contain chlorophylls *a* and *b* together with carotenoids. The large nucleus contains a prominent nucleolus. Euglenids store carbon as paramylon (a polysaccharide composed of β(1→3) linked glucose molecules), which is unique to euglenids. A red eyespot called a **stigma** helps the organism orient to light and is located near an anterior reservoir. A contractile vacuole near the reservoir continuously collects water from the cell and empties it into the reservoir, thus regulating the osmotic pressure within the organism. Two flagella arise from the base of the reservoir, although only one emerges from the canal and actively beats to move the cell. Reproduction in euglenids is by longitudinal mitotic cell division.

Trypanosomes are important members of *Euglenozoa.* These microbes exist only as parasites of plants and animals, and have global significance. For example, members of the genus *Leishmania,* which cause a group of conditions collectively termed leishmaniasis, are typanosomes that affect some 12 million people. Chagas' disease is caused by *Trypanosoma cruzi,* which is transmitted by "kissing bugs" (*Triatominae*), so called because they bite the face of sleeping victims (*see figures 40.10*

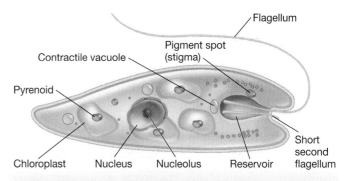

Figure 25.5 *Euglena:* **Principle Structures Found in This Euglenoid.** Notice that a short second flagellum does not emerge from the anterior invagination. In some euglenids, both flagella are emergent.

MICRO INQUIRY *What are the functions of the stigma and the reservoir?*

(a)

(b)

Figure 25.6 The Euglenozoan *Trypanosoma* and Its Insect Host. (a) *Trypanosoma* among red blood cells. Note the dark-staining nuclei, anterior flagella, and undulating, changeable shape. (b) The tsetse fly, shown here sucking blood from a human arm, is an important vector of the *Trypanosoma* species that causes African sleeping sickness.

to *40.13*). ▶▶| *Leishmaniasis (section 40.3); Trypanosomiasis (section 40.3)*

The trypanosomes *T. gambiense* and *T. rhodesiense* (often considered a subspecies of *T. brucei*) cause African sleeping sickness (**figure 25.6**). Ingestion of these parasites by the blood-sucking tsetse fly triggers a complex cycle of development and reproduction, first in the fly's gut and then in its salivary glands (*see figure 40.12*). From there the parasite is easily transferred to a vertebrate host, where it often causes a fatal infection. It is estimated that about 65,000 people die annually of sleeping sickness. The presence of this dangerous parasite prevents the use of about 11 million square kilometers of African grazing land.

Retrieve, Infer, Apply

1. List at least three features that distinguish *Parabasalia* from *Euglenozoa*.
2. What is the function of the plasmalemma and pellicle in euglenid cells? How are these structures similar to the bacterial cell wall?
3. Members of which *Euglenozoa* genera cause disease?

25.3 Supergroup *Amoebozoa* Includes Protists with Pseudopodia

After reading this section, you should be able to:

- Draw three different types of pseudopodia
- Differentiate between a naked and a testate amoeba
- Compare and contrast the structure, motility, and life cycles of acellular and cellular slime molds
- Explain the importance of molecular signaling in the morphological development of *Dictyostelium discoideum*
- Identify a nearby location where you might find members of the genera *Tubulinea*, *Entamoebida*, and *Eumycetozoa*

Because the amoeboid form arose independently numerous times from various flagellated ancestors, some amoebae are placed in

the supergroup *Amoebozoa*, while others are in *Rhizaria*. One of the morphological hallmarks of amoeboid motility is the use of **pseudopodia** (meaning "false feet") for both locomotion and feeding (**figure 25.7**). Pseudopodia can be rounded (**lobopodia**), long and narrow (**filopodia**), or form a netlike mesh (**reticulopodia**). Amoebae that lack a cell wall and are surrounded only by a plasma membrane are called **naked amoebae.** In contrast, the plasma membrane of a **testate amoeba** is covered by material that is either made by the protist itself or collected by the organism from the environment. Binary fission is the usual means of asexual division, although some amoebae form cysts that undergo multiple fission.

Tubulinea

Members of *Tubulinea* inhabit almost any environment where they will remain moist; this includes glacial meltwater, marine environments, tide pools, lakes, and streams. Free-living forms are known to dwell in ventilation ducts and cooling towers, where they feed on microbial biofilms. Others are endosymbionts, commensals, or parasites of invertebrates, fishes, and mammals. Some harbor intracellular symbionts, including algae,

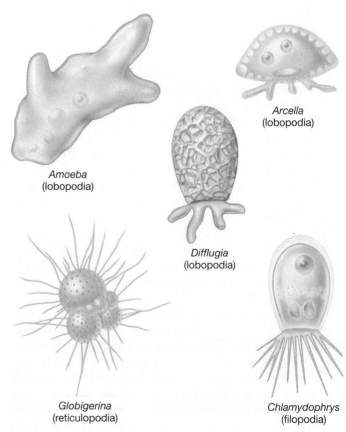

Amoeba (lobopodia)

Arcella (lobopodia)

Difflugia (lobopodia)

Globigerina (reticulopodia)

Chlamydophrys (filopodia)

Figure 25.7 Pseudopodia.

area, the **pyrenoid,** which is associated with the synthesis and storage of starch, may be present in chloroplasts. The majority of anaerobic chemoorganotrophic protists lack mitochondria; some of these organisms have hydrogenosomes.

Many protists feature cilia or flagella at some point in their life cycle. Their formation is associated with a basal bodylike organelle called the kinetosome. In addition to aiding in motility, these organelles may be used to generate water currents for feeding and respiration.

Encystment and Excystment

Many protists are capable of **encystment.** During encystment, the organism becomes simpler in morphology and develops into a resting stage called a cyst. The cyst is a dormant form marked by the presence of a cell wall and very low metabolic activity. Cyst formation is particularly common among aquatic, free-living protists and parasitic forms. Cysts serve three major functions: (1) they protect against adverse changes in the environment, such as nutrient deficiency, desiccation, adverse pH, and low levels of O_2; (2) they are sites for nuclear reorganization and cell division (reproductive cysts); and (3) in parasitic species they are the infectious stage between hosts. Protists escape from cysts by a process called **excystment (figure 25.2).** Although the exact stimulus for excystment is unknown for most protists, it is generally triggered by a return to favorable environmental conditions. For example, cysts of parasitic species excyst after ingestion by the host.

Protist Reproductive Cells and Structures

Most protists have both asexual and sexual reproductive phases in their life cycles. The most common method of asexual reproduction is **binary fission.** During this process, the nucleus first undergoes mitosis and then the cytoplasm divides by cytokinesis to form two identical individuals (**figure 25.3**). Multiple fission is also common, as is budding. Some filamentous, photosynthetic protists undergo fragmentation so that each piece of the broken filament contains a nucleus and grows independently.

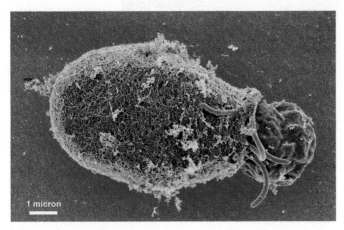

Figure 25.2 Excystment. This colorized scanning electron micrograph shows the protist *Giardia* sp. undergoing excystment. The flagellated vegetative cell, called a trophozoite, is emerging from the right of the cyst.

Sexual reproduction involves the formation of gametes. Protist cells that produce gametes are termed **gamonts.** The fusion of haploid gametes is called **syngamy.** Among protists, syngamy can involve the fusion of two morphologically similar gametes (isogamy) or two morphologically different types (anisogamy). Meiosis may occur before the formation and union of gametes, as in most animals, or just after fertilization, as is the case with lower plants. Furthermore, the exchange of nuclear material may occur in the familiar fashion—between two different individuals (**conjugation**)—or by the development of a genetically distinct nucleus within a single individual (autogamy).

With this level of reproductive complexity, perhaps it is not surprising that the nuclei among protists show considerable diversity. Most commonly, a vesicular nucleus is present. This is 1 to 10 μm in diameter, spherical, and has a distinct nucleolus and uncondensed chromosomes (figure 25.1). The uncondensed chromosomes give these nuclei a lacy or vesicular appearance when viewed microscopically. Ovular nuclei are up to 10 times this size and possess many peripheral nucleoli. Still others have chromosomal nuclei, in which the chromosomes remain condensed throughout the cell cycle. Finally, many ciliated forms have two types of nuclei: a large macronucleus with distinct nucleoli and condensed chromatin, and a smaller, diploid micronucleus with dispersed chromatin but lacking nucleoli. Genes in the macronucleus encode proteins involved in housekeeping functions and binary fission, whereas genes on the micronucleus function only in sexual reproduction (p. 574).

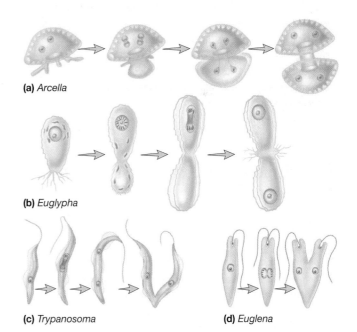

(a) *Arcella*

(b) *Euglypha*

(c) *Trypanosoma*   **(d)** *Euglena*

Figure 25.3 Binary Fission in Protists. (a) The two nuclei of a testate (shelled) amoeba in the genus *Arcella* divide as some of its cytoplasm is extruded and a new test for the daughter cell is secreted. (b) Members of another testate amoeba genus, *Euglypha,* begin to secrete new platelets before cytoplasm begins to move out of the aperture. The nucleus divides while the platelets are used to construct the test for the daughter cell. For many protists, all organelles must be replicated before the cell divides. This is also the case with species of (c) *Trypanosoma* and (d) *Euglena.*

Protist Taxonomy Is Controversial

Ever since Antony van Leeuwenhoek described the first protozoan "animalcule" in 1674, the taxonomic classification
of protists has remained in flux. During the twentieth century,
classification schemes were based on morphology and protists
were typically classified into four major groups based on their
means of locomotion: flagellates (*Mastigophora*), ciliates (*Infuso-
ria* or *Ciliophora*), amoebae (*Sarcodina*), and stationary forms
(*Sporozoa*). Although these terms may still be encountered, they
are without evolutionary context and should be avoided. While it is
now agreed that the old classification system is best abandoned,
little agreement exists on what should take its place. Here we use a
modified version of the higher-level classification system for the
eukaryotes based on morphological, biochemical, and phylogenetic analyses proposed by the International Society of Protistologists in 2005. This scheme does not use formal hierarchical rank
designations such as class and order, reflecting the fact that protist
taxonomy remains an area of active research. The *Classification of
the Protists* as proposed by the International Society of Protistologists is presented in a table posted to Connect®.

Retrieve, Infer, Apply

1. Describe the pellicle. What is its function?
2. Trace the path of a food item from the phagocytic vacuole to the
   cytoproct.
3. What functions do cysts serve for a typical protist? What causes
   excystment to occur?
4. What is a gamont? What is the difference between anisogamy,
   isogamy, and syngamy?
5. How does conjugation differ from autogamy?
6. Describe vesicular, ovular, and chromosomal nuclei. Which is most
   like the nuclei found in higher eukaryotes?
7. Why has taxonomic classification of protists been problematic?

25.2 Supergroup *Excavata:* Primitive Eukaryotes

After reading this section, you should be able to:

- Differentiate a mitosome from a mitochondrion
- Identify which members of *Excavata* have hydrogenosomes
- Identify a location near your home or school where you might find
  protists in the taxa *Fornicata, Parabasalia,* and *Euglena*
- List one human and one veterinary disease caused by species of
  Fornicata and *Parabasalia*
- Draw a typical euglenoid cell and explain the function of the stigma
  and pyrenoid
- Discuss the etiology of Chagas' disease and African sleeping sickness

Excavata includes some of the most primitive, or deeply branching, eukaryotes. Most possess a cytostome characterized by a
suspension-feeding groove with a posteriorly directed flagellum
used to generate a feeding current. This enables the capture of
small particles. Those that lack this morphological feature are

presumed to have had it at one time during their evolution; that
is, it is thought to have been secondarily lost.

Fornicata

Ever curious, van Leeuwenhoek described *Giardia lamblia*
(**figure 25.4a**) from his own diarrheic feces. Over 300 years
later, this species, now known as *G. intestinalis,* continues to be
a public health concern. Today this microaerophilic protist most
often infects campers and other individuals who unwittingly

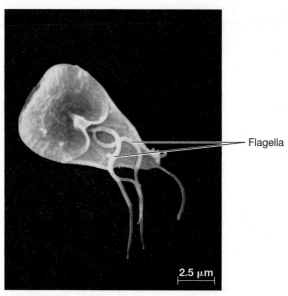

(a) *Giardia intestinalis*

2.5 μm — Flagella

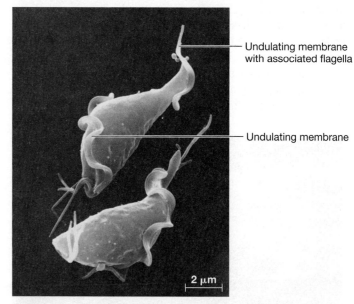

Undulating membrane
with associated flagella

Undulating membrane

2 μm

(b) *Trichomonas vaginalis*

Figure 25.4 Parasitic Members of the Supergroup *Excavata*.
(a) *Giardia intestinalis*. (b) *Trichomonas vaginalis*. These specialized
parasitic flagellates absorb nutrients from living hosts.

MICRO INQUIRY *How are these two protists morphologically
similar but physiologically different?*

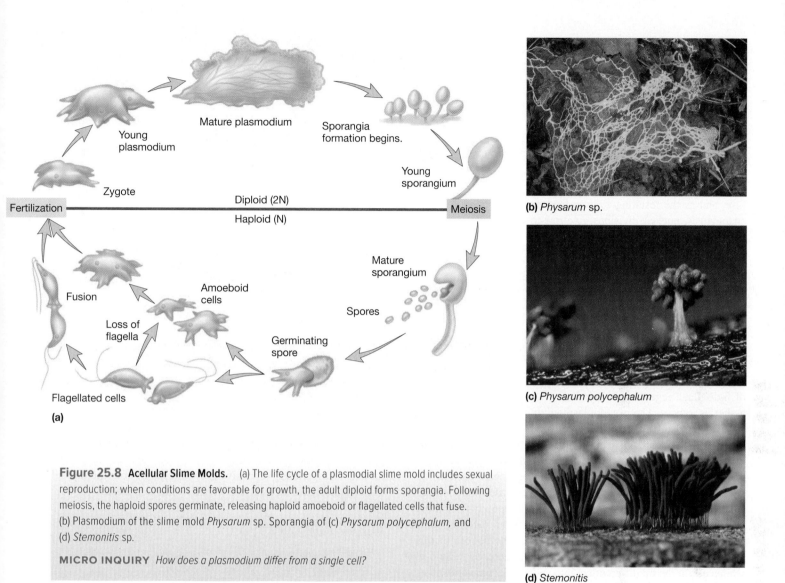

(b) *Physarum* sp.

(c) *Physarum polycephalum*

(d) *Stemonitis*

Figure 25.8   Acellular Slime Molds.   (a) The life cycle of a plasmodial slime mold includes sexual reproduction; when conditions are favorable for growth, the adult diploid forms sporangia. Following meiosis, the haploid spores germinate, releasing haploid amoeboid or flagellated cells that fuse. (b) Plasmodium of the slime mold *Physarum* sp. Sporangia of (c) *Physarum polycephalum,* and (d) *Stemonitis* sp.

MICRO INQUIRY   *How does a plasmodium differ from a single cell?*

bacteria, and viruses, but the nature of these relationships is not well understood. *Amoeba proteus,* a favorite among introductory biology laboratory instructors, is included in this group.

Entamoebida

Like *Excavata* protists, entamoebid protists lack mitochondria and hydrogenosomes but contain mitosomes. However, the identification of about 20 putative mitochondrial proteins in the free-living amoeba *Mastigamoeba balamuthi* suggests that these mitosomes are more advanced than those found in *Giardia* spp. *M. balamuthi* is a relative of *Entamoeba histolytica,* the cause of amoebic dysentery—the third-leading cause of parasitic death worldwide. Individuals acquire this pathogen by eating feces-contaminated food or by drinking water contaminated by *E. histolytica* cysts. These pass unharmed through the stomach and undergo multiple fission when introduced to the alkaline conditions in the intestines (*see figure 40.19*). There they graze on bacteria and produce a suite of digestive enzymes that degrade the gut epithelium. *E. histolytica* can cause intestinal ulceration and penetrate into the

bloodstream, and migrate to the liver, lungs, or skin. Cysts in feces remain viable for weeks but are killed by heat greater than 40°C. ▶▶❘ *Amebiasis (section 40.5)*

Eumycetozoa

First described in the 1880s, members of *Eumycetozoa* or "slime molds" have been classified as plants, animals, and fungi. As we examine their morphology and behavior, the source of this confusion should become apparent. Recent analysis of certain proteins (e.g., elongation factor EF-1, α-tubulin, and actin) as well as physiological, behavioral, biochemical, and developmental data point to a monophyletic group. *Eumycetozoa* includes *Myxogastria* and *Dictyostelia.*

The **acellular slime mold (*Myxogastria*)** life cycle includes a distinctive stage when the organisms exist as streaming masses of colorful protoplasm. The protoplasm creeps along in amoeboid fashion over moist, rotting logs, leaves, and other organic matter, which it degrades (**figure 25.8**). Acellular slime molds derive their name from the stage in their life cycle when a large, multinucleate mass called a plasmodium is formed; there can be as many as 10,000 synchronously dividing nuclei within a single plasmodium (figure 25.8*b*). When starved or dried, the plasmodium develops ornate fruiting bodies. As these mature, they form stalks with cellulose walls that are resistant to environmental stressors (figure 25.8*c,d*). When conditions improve, spores germinate and release haploid amoeboflagellates. These fuse, and as the resulting zygotes feed by endocytosis, nuclear divisions give rise to the multinucleate plasmodium.

Cellular slime molds (*Dictyostelia*) are strictly amoeboid and use endocytosis to feed on bacteria and yeasts. Their complex life cycle involves true multicellularity, despite their primitive evolutionary status (**figure 25.9**). The species *Dictyostelium discoideum* is an attractive model organism. During its life cycle, a pseudoplasmodium is formed. This consists of thousands of individual vegetative cells moving together as a mass. The pseudoplasmodium forms when starved cells release cyclic AMP (cAMP) and a specific glycoprotein, which serve as molecular signals. Other *D. discoideum* cells sense these compounds and respond by forming an aggregate around the signal-producing cells. In this way, large, motile, multicellular slugs develop and serve as precursors to fruiting body formation. Fruiting body morphogenesis commences when the slug stops and cells pile on top of each other. Cells at the bottom of this vertically oriented structure form a stalk by secreting cellulose, while cells at the tip differentiate into spores. Germinated spores become vegetative amoebae to start this asexual cycle anew.

Dictyostelium spp. display complex behaviors. In addition to farming, described in the chapter-opening story, they also differentiate to resemble primitive immune cells. During slug formation, some cells become "sentinel cells" and vanquish harmful bacteria. Sentinel cells accomplish this by producing proteins that are similar to those involved in immune responses in higher organisms. These cells roam within the slug as if patrolling for pathogenic bacteria such as *Legionella pneumophila,* which are known to infect *Dictyostelium* spp. Sentinel cells have been found in several species related to *Dictyostelium discoideum*; immunologists are not too surprised, noting that all multicellular organisms need protection against bacterial pathogens.

Sexual reproduction in *D. discoideum* involves the formation of special spores call macrocysts. These arise by a form of conjugation that has some unusual features. First, a group of amoebae become enclosed within a wall of cellulose. This brings cells into close contact so that conjugation can occur between members of different mating types. There are three mating types determined by the sequence of a single gene, as well as those capable of self-fertilization. Following conjugation, a large amoeba forms and cannibalizes the

remaining amoebae. The now-giant amoeba matures into a macrocyst. Macrocysts can remain dormant within their cellulose walls for extended periods. Vegetative growth resumes after the diploid nucleus undergoes meiosis to generate haploid amoebae.

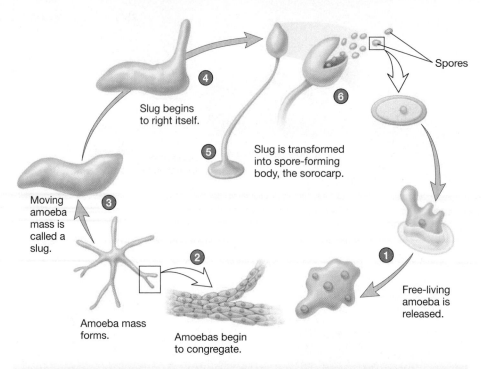

Figure 25.9 Life Cycle of *Dictyostelium discoideum,* a Cellular Slime Mold.

Slug begins to right itself.

Slug is transformed into spore-forming body, the sorocarp.

Spores

Moving amoeba mass is called a slug.

Amoeba mass forms.

Amoebas begin to congregate.

Free-living amoeba is released.

MICRO INQUIRY *How would a mutant strain unable to secrete cAMP behave if mixed with a wild-type strain? What if it were mixed with another cAMP mutant?*

Retrieve, Infer, Apply

1. Describe filopodia, lobopodia, and reticulopodia form and function.
2. Why have slime molds been so hard to classify?
3. What is a plasmodium? How does it differ from the pseudoplasmodium produced by cellular slime molds?
4. Describe the life cycle of *Dictyostelium discoideum.* Why is this organism a good model for the study of cellular differentiation and coordinated cell movement?

25.4 Supergroup SAR: Protists of Great Importance

After reading this section, you should be able to:

- Describe the morphology of the filopodia and axopodia found in members of *Rhizaria*
- Explain the morphology, feeding mechanisms, and reproduction of members of *Radiolaria* and *Foraminifera*
- Infer why the accretion of foraminiferan tests occurs and why it is important in oil exploration
- Describe the morphology, motility, and feeding behaviors of dinoflagellates

- Explain the importance of dinoflagellates to coral biology and the generation of "red tides"
- Diagram how a ciliate captures a food particle and consumes it
- Compare and contrast the morphology and function of the micro- and macronucleus and trace their fate during ciliate conjugation
- Identify the unique structures found in an apicomplexan cell and explain their function
- Outline the apicomplexan life cycle
- List at least two human diseases caused by apicomplexans
- Explain the importance of heterokont flagella in the taxonomy of the *Stramenopila*
- Review the structure, morphology, reproduction, and ecological importance of diatoms
- Deduce why the peronosporomycetes and *Labyrinthulids* were erroneously considered fungi

The supergroup SAR, which stands for stramenopiles, alveolates, and *Rhizaria,* has only recently been recognized. Phylogenomic evidence strongly supports the view that these organisms form a monophyletic clade. As such, it elevates stramenopiles and alveolates to the level of supergroup, and eliminates *Chromalveolata,* to which they were formerly assigned. The SAR supergroup encompasses some of the world's most important protists, including those that fix as much carbon as the rainforests (diatoms) and those that infect over 200 million people worldwide (*Plasmodium* spp., which cause malaria).

Rhizaria

Members of *Rhizaria* include those in *Radiolaria* and *Foraminifera.* These organisms are amoeboid in morphology and can be distinguished by their filopodia, which can be simple, branched, or connected. Filopodia supported by microtubules are known as **axopodia** (s., axopodium). Axopodia protrude from a central region of the cell called the axoplast and are primarily used in feeding (**figure 25.10***a*).

Radiolaria

Most radiolaria have an internal skeleton made of siliceous material; however, members of the subgroup *Acantharia* have endoskeletons consisting of strontium sulfate. A few genera contain species with an exoskeleton of siliceous spines or scales, while some lack a skeleton completely. Skeletal morphology is highly variable and often includes radiating spines that help the organisms float, as does the storage of oils and other low-density fluids (figure 25.10*b*).

Radiolaria feed by endocytosis, using mucus-coated filopodia to entrap prey including bacteria, other protists, and even small invertebrates. Large prey items are partially digested extracellularly by the secretion of degradative enzymes before becoming encased in a food vacuole. Many surface-dwelling radiolarians have algal symbionts thought to enhance their net carbon assimilation. Acantharians reproduce only sexually by consecutive mitotic and meiotic divisions that release hundreds of biciliated cells. Asexual reproduction (binary or multiple fission, or budding) is most common in other radiolaria, but sexual reproduction can be triggered by nutrient limitation or a heavy feeding. In this

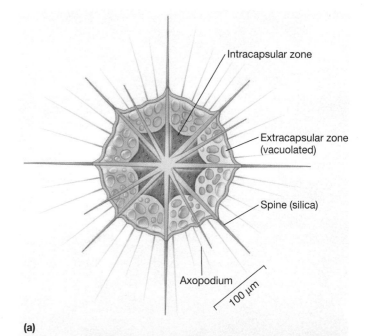

(a)

(b) Radiolarian shells

Figure 25.10 Radiolaria. (a) The radiolarian *Acanthometra elasticum* demonstrates the internal skeleton. (b) Radiolarian shells made of silica.

case, two haploid nuclei fuse to form a diploid zygote encased in a cyst from which it is released when survival conditions improve.

Foraminifera

Members of *Foraminifera* (also called forams) range in size from roughly 20 μm to several centimeters. Their filopodia are arranged in a branching network called reticulopodia. These bear vesicles at

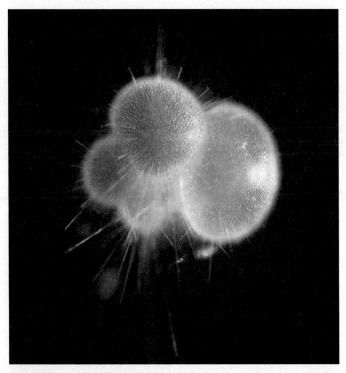

Figure 25.11 **A Foraminiferan.** The reticulopodia project through pores in the calcareous test, or shell, of this protist.

Figure 25.12 **White Cliffs of Dover.** The limestone that forms these cliffs is composed almost entirely of fossil shells of protists, including foraminifera.

their tips that secrete a sticky substance used to trap prey. Many species harbor endosymbiotic algae that migrate out of reticulopodia (without being eaten) to expose themselves to more sunlight. These algae contribute significantly to foram nutrition. Forams have characteristic tests (plates) arranged in multiple chambers that are sequentially added as the protist grows (**figure 25.11**).

Some forams are free-floating planktonic forms; others are benthic, living within the sediments. A few benthic forams inhabiting anoxic sediments perform denitrification, reducing nitrate completely to dinitrogen gas. Foraminifera were the first example of a denitrifying eukaryote; it is now known that a some fungi are also capable of this process. ◀◀ *Anaerobic respiration uses the same three steps as aerobic respiration (section 11.7)* ▶▶ *Nitrogen cycle (section 28.1)*

Foraminiferan life cycles can be complex. While some smaller species reproduce only asexually by budding or multiple fission, larger forms frequently alternate between sexual and asexual phases. During the sexual phase, flagellated gametes pair, fuse, and generate asexual individuals (agamonts). Meiotic division of the agamonts gives rise to haploid gamonts. There are several mechanisms by which gamonts return to the diploid condition. For instance, a variety of forams release flagellated gametes that become fertilized in the open water. In others, two or more gamonts attach to one another, enabling gametes to fuse within the chambers of the paired tests. When the shells separate, newly formed agamonts are released.

Forams are found in marine and estuarine habitats. Some are planktonic, but because most are benthic, their tests accumulate on the seafloor, where they create a fossil record dating back to the Early Cambrian (543 million years ago). This is helpful in oil

exploration. Their remains, or ooze, can be up to hundreds of meters deep in some tropical regions. Foram tests make up most modern-day chalk, limestone, and marble, and are familiar to most as the white cliffs of Dover in England (**figure 25.12**). They also formed the stones used to build the great pyramids of Egypt.

Alveolata

Alveolata is a large group that includes *Dinoflagellata* (dinoflagellates), *Ciliophora* (ciliates), and *Apicomplexa* (apicomplexans). We begin our discussion with **dinoflagellates**—a large group most commonly found in marine plankton, where some species are responsible for the phosphorescence sometimes seen in seawater. Their nutrition is complex; photoautotrophy, heterotrophy, and mixotrophy are all observed. Most of the heterotrophic forms are saprophytic (using osmotrophy to assimilate nutrients), but some also use endocytosis (i.e., they are holozoic). Each cell bears two distinctively placed flagella: One is wrapped around a transverse groove (the girdle), and the other is draped in a longitudinal groove (the sulcus; **figure 25.13**). The orientation and beating patterns of these flagella cause the cell to spin as it is propelled forward; the name dinoflagellate is derived from the Greek *dinein*, "to whirl." Many dinoflagellates are covered with cellulose plates that are secreted within the alveolar sacs that lie just under the plasma membrane. These forms are said to be thecate or armored; those with empty alveoli are called athecate or naked and include members of the luminescent genus *Noctiluca*. Like some other protists, dinoflagellates have specialized compressed proteins called **trichocysts.** When attacked, trichocysts shoot out from the cell as a means of defense.

Most dinoflagellates are free living, although some form important associations with other organisms. Endosymbiotic

Figure 25.13 Dinoflagellates. (a) The surface of *Peridiniopsis berolinensis* is smooth because the alveoli appear empty. Two types of flagella can be seen in this freshwater dinoflagellate. One flagellum coils around the girdle; as it moves, this flagellum causes the cell to spin. The straight flagellum that extends from the cell acts as a rudder to determine the direction of movement. (b) The alveoli of a marine dinoflagellate in the genus *Ornithocercus* contains cellulose cell-wall plates. The cell wall extends to form sail-like structures.

MICRO INQUIRY *What do you think is the function of the "sails" in* Ornithocercus *sp.?*

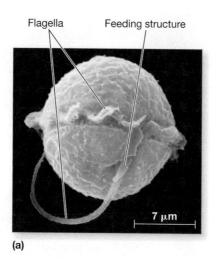

Flagella   Feeding structure

7 μm

(a)

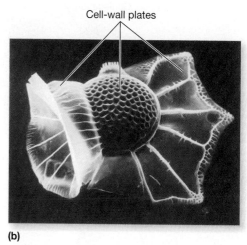

Cell-wall plates

(b)

dinoflagellates that live as undifferentiated cells occasionally send out motile cells called **zooxanthellae.** The most well-known zooxanthellae belong to the genus *Symbiodinium.* These are photosynthetic endosymbionts of reef-building coral (*see figure 32.3*). They provide fixed carbon to the coral animal and help maintain the internal chemical environment needed for the coral to secrete its calcium carbonate exoskeleton. Dinoflagellates are also responsible for toxic "red tides" that harm other organisms, including humans. Common red tide dinoflagellates include members of the genera *Alexandrium* and *Lingulodinium* (formerly called

Gonyaulax; see figure 30.3). ▶▶◀ *Microorganisms in coastal ecosystems are adapted to a changeable environment (section 30.2); Zooxanthellae (section 32.1)*

Ciliates (*Ciliophora*) include about 12,000 known species. All are chemoorganotrophic and range from about 10 μm to 4.5 mm long. They inhabit both benthic and planktonic communities in marine and freshwater systems, as well as moist soils. As their name implies, ciliates employ many cilia for locomotion and feeding. The cilia are generally arranged either in longitudinal rows (**figure 25.14**) or in spirals around the body of the organism.

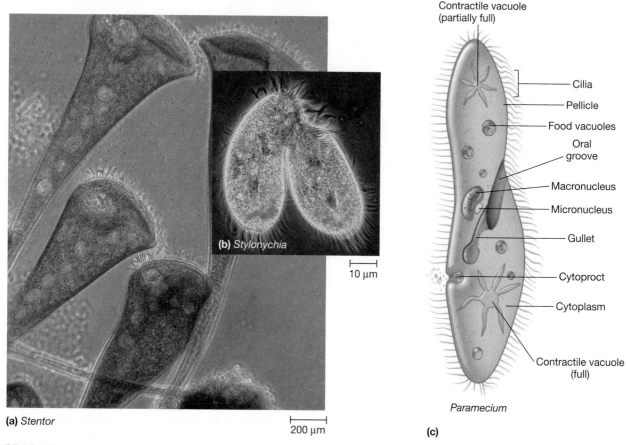

Contractile vacuole (partially full)

Cilia
Pellicle
Food vacuoles
Oral groove
Macronucleus
Micronucleus
Gullet
Cytoproct
Cytoplasm
Contractile vacuole (full)

Paramecium

(b) *Stylonychia*

10 μm

(a) *Stentor*

200 μm

(c)

Figure 25.14 *Ciliophora.* (a) *Stentor*, a large, vase-shaped, freshwater protozoan. (b) Two *Stylonychia* cells conjugating. (c) Structure of *Paramecium* spp.

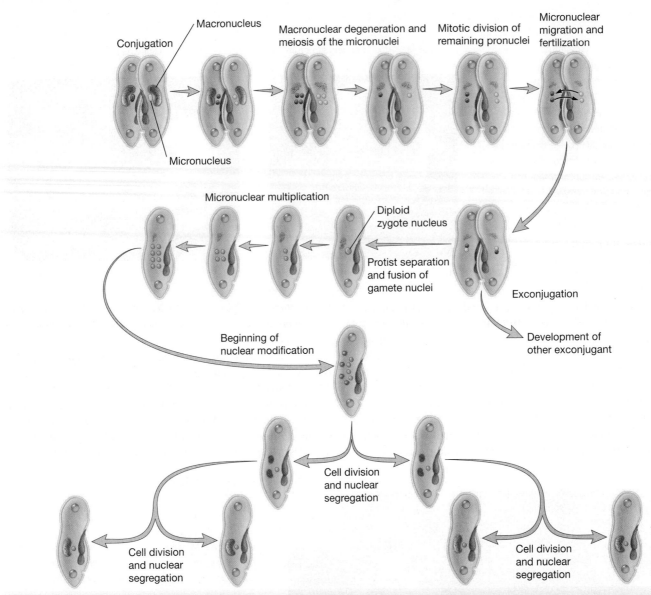

Figure 25.15 Conjugation in *Paramecium caudatum*. After the conjugants separate, only one of the exconjugants is shown; however, a total of eight new protists result from each conjugation.

MICRO INQUIRY *What are the functions of the micronucleus and macronucleus?*

They beat with an oblique stroke, causing the protist to revolve as it swims. Ciliary beating is so precisely coordinated that ciliates can go both forward and backward. There is great variation in shape, and most ciliates do not look like the slipper-shaped members of the genus *Paramecium*. Some species, including *Vorticella* spp., attach to substrates by a long stalk. *Stentor* spp. attach to substrates and stretch out in a trumpet shape to feed. A few species have tentacles for the capture of prey. Some can discharge toxic, threadlike darts called toxicysts, which are used in capturing prey. A striking feature of ciliates is their ability to quickly entrap many particles by the action of the cilia around the buccal cavity. Food first enters the cytostome and passes into phagocytic vacuoles that fuse with lysosomes after detachment from the cytostome and the vacuole's contents are digested. After the digested material has been absorbed into the cytoplasm, the vacuole fuses with the cytoproct and waste material is expelled.

Most ciliates have two types of nuclei: a large **macronucleus** and a smaller **micronucleus.** The micronucleus is diploid and contains the normal somatic chromosomes. It divides by mitosis and transmits genetic information through meiosis and sexual reproduction. Macronuclei gene expression is needed to maintain routine cellular functions and control normal cell metabolism. Macronuclei are derived from micronuclei by a complex series of steps. Within the macronucleus are many chromatin bodies, each containing many copies of only one or two genes. Macronuclei are thus polyploid and divide by elongating and then constricting.

The fate of the macro- and micronuclei can be followed during conjugation of *Paramecium caudatum* (**figure 25.15**). In this

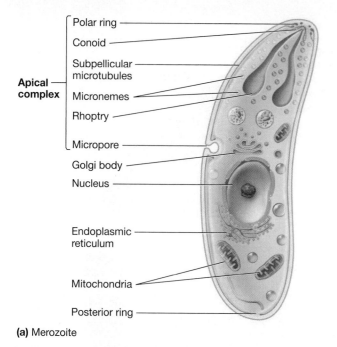

(a) Merozoite

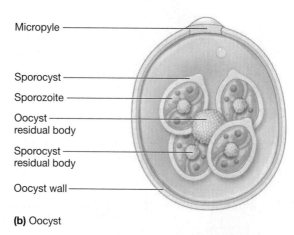

(b) Oocyst

Figure 25.16 **The Apicomplexan Cell.** (a) The vegetative cell, or merozoite, illustrating the apical complex, which consists of the polar ring, conoid, rhoptries, subpellicular microtubules, and micropore. (b) The infective oocyst of *Eimeria* sp. The oocyst is the resistant stage and has undergone multiple fission after zygote formation (sporogony).

process, there is an exchange of gametes between paired cells of complementary mating types (conjugants). At the beginning of conjugation, two ciliates fuse their pellicles at the site of contact. The macronucleus is no longer needed, so it is degraded in each cell. The individual micronuclei undergo meiosis to form four haploid pronuclei, three of which disintegrate. The remaining pronucleus divides again mitotically to form two gametic nuclei—a stationary one and a migratory one. The migratory nuclei pass into the respective conjugants. Then the ciliates separate, the gametic nuclei fuse, and the resulting diploid zygote nucleus undergoes three rounds of mitosis. The eight resulting nuclei have different fates: one nucleus is retained as a micronucleus; three others are destroyed; and the four remaining nuclei develop into macronuclei. Each separated conjugant now undergoes cell division. Eventually progeny with one macronucleus and one micronucleus are formed.

Although most ciliates are free living, symbiotic forms also exist. Some live as harmless commensals; for example, *Entodinium* spp. are found in the rumen of cattle, and *Nyctotherus* protists occur in the colon of frogs. Other ciliates are strict parasites; for example, *Balantidium coli* lives in the intestines of mammals, including humans, where it can cause dysentery. *Ichthyophthirius* spp. live in freshwater, where they can attack many species of fish, producing a disease known as "ick."

All **apicomplexans** are either intra- or intercellular parasites of animals and are distinguished by a unique arrangement of fibrils, microtubules, vacuoles, and other organelles, collectively called the **apical complex,** which is located at one end of the cell (**figure 25.16a**). There are three secretory organelles—the rhoptries, micronemes, and dense granules—that release enzymes and calcium needed to penetrate host cells. In addition, apicomplexans (with the exception of *Cryptosporidium* spp.) contain plastids, called **apicoplasts.** Although these arose by endosymbiosis of an ancient cyanobacterium, unlike chloroplasts, apicoplasts are not the site of photosynthesis. In the malaria-causing

pathogen *Plasmodium* spp., apicoplasts are known to be the site of isoprenoid biosynthesis. In all apicomplexans studied to date, the apicoplast is essential for viability. This makes them attractive targets for the development of drugs to inhibit the growth of pathogenic apicomplexans. ◀◀ *Sterols and isoprenoid lipids (section 12.7)*

Apicomplexans have complex life cycles in which certain stages sometimes occur in one host and other stages occur in a different host. The life cycle has both asexual (clonal) and sexual phases, and is characterized by an alternation of haploid and diploid generations. The clonal and sexual stages are haploid, except for the zygote. The motile, infective stage is called the **sporozoite.** When this haploid form infects a host, it differentiates into a gamont; male and female gamonts pair and undergo multiple fission, which produces many haploid cells called merozoites. Merozoites may continue to infect new host cells or pair and fuse to form zygotes. Each zygote secretes a protective covering and is then considered a spore. Within the spore, the nucleus undergoes meiosis (restoring the haploid condition) followed by mitosis to generate eight sporozoites ready to infect a new host (figure 25.16b). Motility (flagellated or amoeboid) is confined to the gametes and zygotes of a few species.

Among the most important human pathogens are the five species in the apicomplexan genus *Plasmodium,* which cause malaria: *P. falciparum, P. malariae, P. vivax, P. knowlesi,* and *P. ovale.* The life cycle of *P. vivax* is shown in figure 40.7; it infects both the liver and red blood cells. Prevention measures have resulted in about a 42% decline in malaria since 2000; however, the disease still causes about 600,000 deaths per year. *P. falciparum* is the most deadly of the five species, accounting for at least one-third of malaria deaths. The energy metabolism of this microbe is unusual. While it has the genes for a mitochondrial ATP synthase, it is fermentative. Like all apicomplexans, *P. falciparum* cannot convert pyruvate to acetyl-CoA in the mitochondrion. Instead, phosphoenolpyruvate

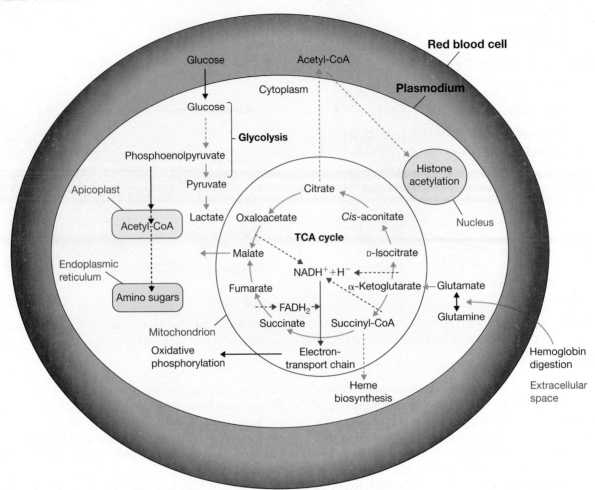

Figure 25.17 *Plasmodium falciparum* **Drives the TCA Cycle in Two Directions.** Glutamate is deaminated to yield α-ketoglutarate from which reactions proceed in both the oxidative and reductive direction. Pyruvate is fermented in the parasite's cytosol to lactate rather than oxidized to acetyl-CoA. Phosphoenolpyruvate is converted to acetyl-CoA in the apicoplast and used to synthesize amino sugars. The host may convert citrate produced by the protist to acetyl-CoA, which then re-enters the microbe, where it is thought to be used to acetylate histones. The uncertainty of these steps is denoted by a dashed line.

generated by the Embden-Meyerhof pathway in the cytosol is transported to the apicoplast where it is converted to acetyl-CoA. Without acetyl-CoA in the mitochondrion, the TCA cycle begins with the deamination of glutamate to α-ketoglutarate. Reactions then proceed in both the reductive and oxidative directions. Reductive TCA steps result in citrate and malate, which are exported from the mitochondrion, while in the oxidative direction, succinyl-CoA, a key precursor for heme biosynthesis, is generated (**figure 25.17**). ▶▶| *Malaria (section 40.3)*

A number of other apicomplexans are important infectious agents. Several species of *Eimera* are the causative agents of cecal coccidiosis in chickens, a condition that costs hundreds of millions of dollars in lost animals each year in the United States. *Theileria parva* and *T. annulata* are tick-borne parasites that cause a fatal disease in cattle called East Coast fever. This kills over a million animals annually in sub-Saharan Africa, costing over $200 million and affecting farmers who can least afford such losses. Toxoplasmosis, caused by members of the genus *Toxoplasma*, is transmitted either by consumption of undercooked meat or by fecal contamination from a cat's litter box. Cryptosporidia are responsible for cryptosporidiosis, an infection

that begins in the intestines but can disseminate to other parts of the body. Cryptosporidiosis and members of another apicomplexan genus, *Cyclospora*, have become problematic for AIDS patients and other immunocompromised individuals. ▶▶| *Toxoplasmosis (section 40.5); Cryptosporidiosis (section 40.5)*

Retrieve, Infer, Apply

1. What adaptations do planktonic radiolaria have to help them float?
2. Compare how radiolaria use axopodia with the way forams use reticulopodia.
3. What are alveoli? Describe the difference between a naked and an armored dinoflagellate.
4. What is the morphology of a typical ciliate? Why do you think ciliates are the fastest-moving protists?
5. Describe conjugation as it occurs in ciliates. What is the fate of the micronucleus and the macronucleus during this process?
6. What is the apical complex seen in apicomplexans?
7. Consider the complex life cycle of apicomplexans. Which stage would be most vulnerable to drug treatment in an effort to treat a human disease caused by an apicomplexan?

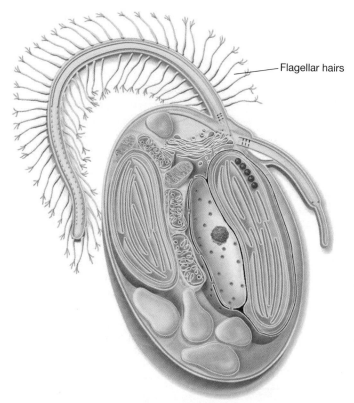

Flagellar hairs

Figure 25.18 A Stramenopile Cell. Note the two flagella, one with "straw hairlike" projections.

Stramenopila

The large and diverse *Stramenopila* group (informally known as stramenopiles) includes photosynthetic protists such as diatoms, brown and golden algae (*Chrysophyceae*), as well as chemoorganotrophic (saprophytic) genera such as öomycetes (*Peronosporomycetes*), labyrinthulids (slime nets), and members of *Hyphochytriales*. Stramenopiles also include brown seaweeds and kelp that form large, rigid structures and macroscopic forms that were once considered fungi and plants. One unifying feature of these very diverse protists is the possession of **heterokont flagella** at some point in the life cycle. This is characterized by two flagella—one extending anteriorly and the other posteriorly. These flagella bear small hairs with a unique, three-part morphology; stramenopila means "straw hair" (**figure 25.18**).

Diatoms (*Bacillariophyta*) possess chlorophylls a and c_1/c_2, and the carotenoid fucoxanthin. When fucoxanthin is the dominant pigment, the cells have a golden-brown color. Their major carbohydrate reserve is chrysolaminarin, a polysaccharide storage product composed principally of $\beta(1\rightarrow3)$–linked glucose residues. Diatoms have a distinctive, two-piece cell wall of silica called a **frustule** composed of crystallized silica [Si(OH)$_4$] with very fine markings. Diatom frustules are composed of two halves or thecae that overlap like those of a Petri dish (**figure 25.19**). The larger half is the epitheca, and the smaller half is the hypotheca. Frustule morphology is very useful in diatom identification, and diatom frustules have a number of practical applications. The fine detail and precise morphology of these frustules have made them attractive for nanotechnology. ▶▶❘ *Some microbes are products (section 42.6)*

Although the majority of diatoms are strictly photoautotrophic, some are facultative chemoorganotrophs, absorbing carbon-containing molecules through the holes in their walls. The vegetative cells of diatoms are diploid and can be unicellular, colonial, or filamentous. They lack flagella and have a single, large nucleus and smaller plastids. Reproduction consists of the organism dividing asexually, with each half then constructing a new hypotheca (**figure 25.20***a*). Because the epitheca and hypotheca are of different sizes, each time the parental hypotheca is used as an eptithecal template to construct a new hypotheca, the diatom gets smaller. This cannot go on forever without the diatom

(a) *Cyclotella meneghiniana*

(b)

Figure 25.19 Diatoms.
(a) The siliceous epitheca and hypotheca of the diatom *Cyclotella meneghiniana* fit together like halves of a Petri dish. (b) A variety of diatoms show the intricate structure of the silica cell wall.

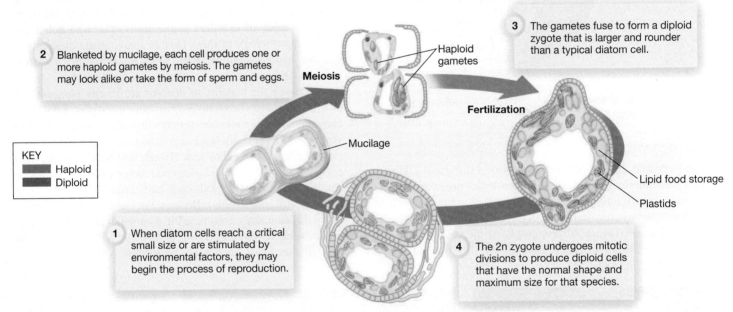

(a) Asexual reproduction in diatoms

After many cell divisions, some progeny cells are very small.

Figure 25.20 **Gametic Life Cycle as Illustrated by Diatoms.**
(a) Diatom asexual reproduction involves repeated mitotic divisions. Because a new lower cell-wall piece is always synthesized, asexual reproduction eventually causes the mean cell size to decline in a diatom population. (b) Small cell size triggers sexual reproduction, which regenerates maximal cell size.

MICRO INQUIRY *How does the trigger for sexual reproduction (i.e., reduced size) compare with more common stimuli that induce sexual reproduction? (Hint: Compare to radiolaria.)*

2 Blanketed by mucilage, each cell produces one or more haploid gametes by meiosis. The gametes may look alike or take the form of sperm and eggs.

Meiosis

Haploid gametes

3 The gametes fuse to form a diploid zygote that is larger and rounder than a typical diatom cell.

Fertilization

KEY
Haploid
Diploid

Mucilage

Lipid food storage

Plastids

1 When diatom cells reach a critical small size or are stimulated by environmental factors, they may begin the process of reproduction.

4 The 2n zygote undergoes mitotic divisions to produce diploid cells that have the normal shape and maximum size for that species.

(b) Sexual reproduction in diatoms

disappearing altogether, so when a cell has diminished to about 30% of its original size, sexual reproduction is usually triggered. The diploid vegetative cells undergo meiosis to form gametes, which then fuse to produce a zygote. The zygote develops into an auxospore, which increases in size again and forms a new wall. The mature auxospore eventually divides mitotically to produce vegetative cells with frustules of the original size (figure 25.20*b*).

Diatoms are found in freshwater lakes, ponds, streams, and throughout the world's oceans. Marine planktonic diatoms produce about 50% of the organic carbon in the ocean; they are therefore critical in global carbon cycling. In fact, marine diatoms are thought to contribute as much fixed carbon as all rain forests combined. ▶▶︎ *Carbon cycle (section 28.1); Microorganisms in the open ocean are adapted to nutrient limitation (section 30.2)*

A group of protists once considered fungi and traditionally called **öomycetes,** meaning "egg fungi," are now called **peronosporomycetes.** They differ from fungi in a number of ways, including their cell wall composition (cellulose and β-glucan instead of chitin) and the fact that they are diploid throughout their life cycle. When undergoing sexual reproduction, they form a relatively large egg cell (öogonium) that is fertilized by either a sperm cell or a smaller gametic cell (called an antheridium) to produce a zygote. When the zygote germinates, the asexual zoospores display heterokont flagellation.

Some peronosporomycetes help degrade plant and animal detritus, while others are pathogenic. For instance, species of *Saprolegnia* and *Achlya* are saprophytes that form cottony masses on dead algae and animals, mainly in freshwater environments. By contrast, *Peronospora hyoscyami* is responsible for "blue mold" on tobacco plants, and grape downy mildew is caused by *Plasmopara viticola.* Certainly the most famous peronosporomycete is *Phytophthora infestans,* which attacked the European potato crop in the mid-1840s, spawning the Irish famine. The original classification of *P. infestans* as a fungus was misleading, and for decades farmers attempted to control its growth with fungicides, to which it is (of course) resistant. This protist continues to take its toll; potato blight costs some \$6.7 billion annually worldwide. *P. sojae* is another agriculturally important species; it infects soybeans and costs this industry millions of dollars per year. Finally, *P. ramorum* is a plant pathogen that causes a disease called sudden oak death, which kills oak trees and a variety of woody shrubs.

Labyrinthulids also have a complex taxonomic history: like peronosporomycetes, they were formerly considered fungi. However, molecular phylogenetic evidence combined with the observation that they form heterokont flagellated zoospores places them in *Stramenopila.* The nonflagellated stage of the life cycle features spindle-shaped cells that form complex colonies that glide rapidly along an ectoplasmic net made by the organism. This net is actually an external network of calcium-dependent contractile fibers made up of actinlike proteins that facilitate the movement of cells. Their feeding mechanism is like that of fungi: osmotrophy aided by the production of extracellular degradative enzymes. In marine habitats, members of the genus *Labyrinthula* grow on plants and algae, and are thought to play a role in the wasting disease of eelgrass, an important intertidal plant.

Haptophyta

The phylogenetics of *Haptophyta* remains unsettled; it may be a sister group to the supergroup SAR. It includes the *Coccolithales,* photosynthetic protists that bear ornate calcite scales called coccoliths (**figure 25.21**). Together with forams, **coccolithophores** precipitate calcium carbonate ($CaCO_3$) in the ocean, thereby influencing Earth's carbon budget. Cells are usually biflagellated and possess a unique organelle called a haptonema, which is somewhat similar to a flagellum but differs in microtubule arrangement. One species, *Emiliania huxleyi,* has been studied extensively. Like all coccolithophores, it is planktonic and can cause massive blooms in the open ocean (*see figure 30.11*). These high concentrations or blooms of *E. huxleyi* can significantly alter nutrient flux as well as alter the weather by emitting sulfur (as dimethylsulfide or DMS) to the atmosphere. The release of DMS causes clouds to grow larger. Other coccolithophores are known to cause toxic blooms. ▶▶❙ *Water is the largest microbial habitat (section 30.1)*

Picozoa

The phylogenetic origin of another marine lineage, *Picozoa,* is very mysterious. Evidence for *Picozoa* (meaning "tiny animals") emerged in the early 2000s when 18S rRNA metagenomic data

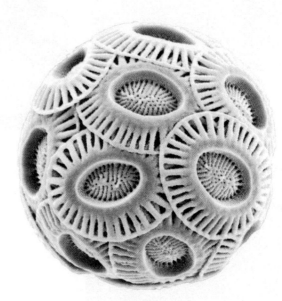

Figure 25.21 **The Haptophyte *Emiliania huxleyi.*** The ornamental scales are made of calcite.

revealed the presence of eukaryotic organisms small enough to fit through filters of 2–3 μm. Originally thought to be photosynthetic, they were given the tentative name "phycobiliphytes." Further analysis showed that these protists are not photosynthetic. They are broadly distributed in high numbers throughout the world's oceans.

In 2013 the first cultured picozoan was reported. These protists, named *Picomonas judraskeda,* were isolated from marine coastal waters and consist of two roughly spherical halves divided by a deep cleft. Total cell size ranges from 2×2.5 μm to 2.5×3.8 μm. They have two flagella and demonstrate an unusual kind of motility that has been described as cycles of "jump-drag-skedaddle." They consume colloidal organic materials by endocytosis, and it is thought that their unusual motility assists in feeding.

25.5 Supergroup *Archaeplastida* Includes "Green Algae"

After reading this section, you should be able to:

- List at least three different morphologies found among the *Chloroplastida*
- Describe the structure and life cycle of the model organism, *Chlamydomonas*

Archaeplastida includes all organisms with a photosynthetic plastid that arose through an ancient endosymbiosis with a cyanobacterium. It thus includes all higher plants as well as many protist species. ❙◀◀ *Mitochondria, mitochondria-like organelles, and chloroplasts evolved from endosymbionts (section 1.2)*

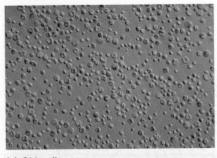

(a) *Chlorella*

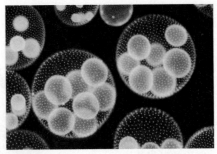

(b) *Volvox*

(c) *Spirogyra*

(d) *Acetabularia*

Figure 25.22 *Chlorophyta* **(Green Algae); Light Micrographs.** (a) *Chlorella*, a unicellular nonmotile chlorophyte (×160). (b) *Volvox*, which demonstrates colonial growth. (c) *Spirogyra;* each filament contains ribbonlike, spiral chloroplasts. (d) *Acetabularia,* the mermaid's wine goblet.

Chloroplastida

Members of *Chloroplastida* are often referred to as green algae (Greek *chloros,* green). These phototrophs grow in fresh- and saltwater, in soil, and on and within other organisms. They have chlorophylls *a* and *b* along with specific carotenoids, and they store carbohydrates such as starch. Many have cell walls made of cellulose. They exhibit a wide diversity of body forms, ranging from unicellular to colonial, filamentous, membranous or sheetlike, and tubular types (**figure 25.22**). The smallest known free-living eukaryote, *Ostreococcus tauri,* is a member of *Chloroplastida* (subgroup *Prasinophyceae*). This marine, planktonic microbe is smaller than most bacteria and archaea, with an average size of 0.8 μm in diameter. Some species have a holdfast structure that anchors them to a substratum. Both asexual and sexual reproduction are observed.

Chlamydomonas spp. are members of the subgroup *Chlorophyta* (**figure 25.23**). Individuals have two flagella of equal

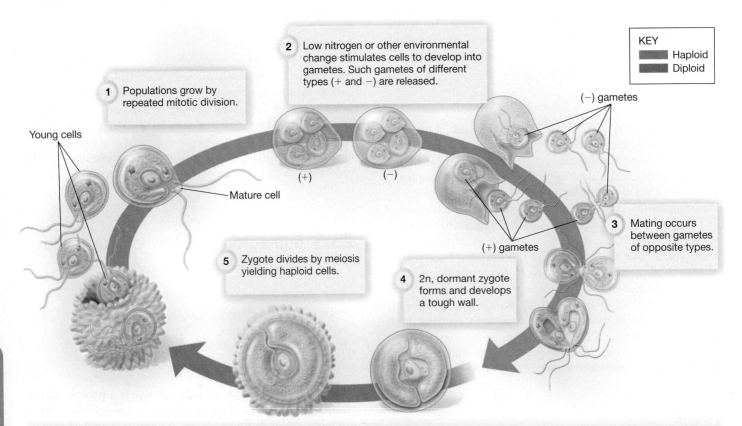

1 Populations grow by repeated mitotic division.

2 Low nitrogen or other environmental change stimulates cells to develop into gametes. Such gametes of different types (+ and −) are released.

3 Mating occurs between gametes of opposite types.

4 2n, dormant zygote forms and develops a tough wall.

5 Zygote divides by meiosis yielding haploid cells.

Young cells

Mature cell

(+) (−)

(−) gametes

(+) gametes

KEY
Haploid
Diploid

Figure 25.23 *Chlamydomonas reinhardtii:* **The Structure and Life Cycle of This Motile Green Alga.** During asexual reproduction, all structures are haploid; during reproduction, only the zygote is diploid.

MICRO INQUIRY *Which of the cells shown above are haploid and which are diploid?*

length at the anterior end by which they move rapidly in water. Each cell has a single haploid nucleus, a large chloroplast, a conspicuous pyrenoid, and a stigma that aids the cell in phototactic responses. Two small contractile vacuoles at the base of the flagella function as osmoregulatory organelles. *Chlamydomonas* spp. reproduce asexually by producing zoospores through mitotic division. Sexual reproduction occurs when some cells act as gametes and fuse to form a four-flagellated, diploid zygote that ultimately loses the flagella and enters a resting phase. Meiosis occurs at the end of this resting phase and produces four haploid cells that give rise to adults.

The chlorophyte *Prototheca moriformis* causes the disease protothecosis in humans and animals. *P. moriformis* cells are fairly common in the soil, and it is from this site that most infections occur. Severe systemic infections, such as massive invasion of the bloodstream, have been reported in animals. The subcutaneous type of infection is more common in humans. It starts as a small lesion and spreads slowly through the lymph glands, covering large areas of the body.

Retrieve, Infer, Apply

1. Explain the unique structural features of diatoms. How does their morphology play a role in the alternation between asexual and sexual reproduction?
2. Why do you think öomycetes and labyrinthulids were formerly considered fungi?
3. What is the ecological importance of coccolithophores?
4. What does the discovery of a large group of protists seemingly unrelated to any other organisms suggest about the state of our knowledge of the eukaryotic world?
5. Compare the morphology of members of the genus *Chlamydomonas* with those belonging to *Euglena* (figure 25.23).

Key Concepts

25.1 Protist Diversity Reflects Broad Phylogeny

- Protists are widespread in the environment, found wherever water, suitable nutrients, and an appropriate temperature occur.
- Protists are important components of many terrestrial, aquatic, and marine ecosystems, where they contribute to nutrient cycling. Many are parasitic in humans and animals, and some have become very useful in the study of molecular biology.
- The protistan cell membrane is called the plasmalemma, and the cytoplasm can be divided into the ectoplasm and the endoplasm. A pellicle lies beneath the plasma membrane of some protists. The cytoplasm contains contractile, secretory, and food (phagocytic) vacuoles.
- Energy metabolism occurs within mitochondria, hydrogenosomes, or chloroplasts.
- Some protists can secrete a resistant covering and go into a resting stage (encystment) called a cyst. Cysts protect the organism against adverse environments, function as a site for nuclear reorganization, and serve as a means of transmission in parasitic species (**figure 25.2**).
- Most protists reproduce asexually by binary or multiple fission or budding (**figure 25.3**).
- Protists also use a variety of sexual reproductive strategies, including syngamy and autogamy.
- Protist phylogeny is the subject of active research and debate. The classification scheme presented here is based on that of the International Society of Protistologists.

25.2 Supergroup *Excavata*: Primitive Eukaryotes

- The super group *Excavata* includes *Fornicata, Parabasalia,* and *Euglenozoa.* Most have a cytoproct and use a flagellum for suspension feeding.
- The human pathogen *Giardia lamblia* is a member of *Fornicata.* It is considered one of the most primitive eukaryotes (**figure 25.4**).
- Most parabasalids are flagellated endosymbionts of animals. They include the obligate mutualists of wood-eating insects such as *Trichonympha* spp. and the human pathogen *Trichomonas vaginalis.*
- Many members of *Euglenozoa* are photoautotrophic (**figure 25.5**). The remainder are chemoorganotrophs; most of these are saprophytic. Important human pathogens include members of this group cause leishmaniasis, Chagas' disease, and African sleeping sickness.

25.3 Supergroup *Amoebozoa* Includes Protists with Pseudopodia

- Amoeboid forms use pseudopodia, which can be lobopodia, filopodia, or reticulopodia (**figure 25.7**). Amoebae that bear external plates are called testate; those without plates are called naked or atestate amoebae.
- The *Amoebozoa* subgroup *Eumycetozoa* includes acellular and cellular slime molds. Acellular slime molds form a large, multinucleate mass of protoplasm (**figure 25.8**).
- Cellular slime molds produce a pseudoplasmodium, consisting of many individual cells moving together as a mass. *Dictyostelium discoideum* is a cellular slime mold that is used as a model organism in the study of chemotaxis, cellular development, and cellular behavior (**figure 25.9**).

25.4 Supergroup SAR: Protists of Great Importance

- *Rhizaria* encompasses amoeboid forms, including members of *Radiolaria,* which have a siliceous internal skeleton and filopodia that are used to entrap prey (**figure 25.10**).
- Members of *Foraminifera* bear netlike reticulopodia and tests. Most foraminifera are benthic, and their tests accumulate on the ocean floor, where they are useful in oil exploration (**figure 25.11**).

- *Alveolata* consists of dinoflagellates, ciliates, and apicomplexans.
- Dinoflagellates are a large group of nutritionally complex protists. Most are marine and planktonic. They are known for their phosphorescence and for causing toxic blooms (**figure 25.13**).
- Ciliates are chemoorganotrophic protists that use cilia for locomotion and feeding. In addition to asexual reproduction, conjugation is used in sexual reproduction (**figures 25.14 and 25.15**).
- Apicomplexans are parasitic with complex life cycles. The motile, infective stage is called the sporozoite. This group includes a number of pathogens, including five species of *Plasmodium*, which cause malaria (**figure 25.16**).
- Stramenopiles are extremely diverse and include diatoms, golden and brown algae, peronosporomycetes, and labyrinthulids. Diatoms are found in fresh- and saltwater,

and are important components of marine plankton (**figures 25.18 to 25.20**).
- Haptophytes include coccolithophores, planktonic photosynthetic protists that contribute to the global carbon budget by precipitating calcium carbonate for their ornate scales (**figure 25.21**).
- The *Picozoa* are new group that includes very small, heterotrophic, marine protists with a global distribution.

25.5 Supergroup *Archaeplastida* Includes "Green Algae"

- *Archaeplastida* includes *Chlorophyta,* also known as green algae. All are photosynthetic with chlorophylls *a* and *b* along with specific carotenoids. They exhibit a wide range of morphologies (**figure 25.22**).
- *Chlamydomonas* spp. are model protists, which undergo sexual and asexual reproduction (**figure 25.23**).

Compare, Hypothesize, Invent

1. Protist encystment is usually triggered by changes in the environment. Compare and contrast this with endospore formation in Gram-positive bacteria.

2. Which of the protists discussed in this chapter do you think are the most evolutionarily advanced or derived? Explain your reasoning.

3. Vaccine development for diseases caused by protists (e.g., malaria, Chagas' disease) has been much less successful than for bacterial or viral diseases. Discuss one biological reason and one geopolitical reason for this fact.

4. Some protists reproduce asexually when nutrients are plentiful and conditions are favorable for growth but reproduce sexually when environmental or nutrient conditions are not favorable. Why is this an evolutionarily important and successful strategy?

5. Most apicomplexans bear the unpigmented chloroplast remnant called the apicoplast (p. 575). Only the recently described apicomplexan *Chromera velia* has been shown

to use the apicoplastid as the site of photosynthesis. What genetic and physiological features do you think were characterized to show that this organism is truly photosynthetic? How would you determine if this protist is photoautotrophic, photoheterotrophic, or mixotrophic?

6. Benthic foraminifera inhabit marine sediments. It was discovered that the benthic foram *Globobulimina pseudospinescens* stores intracellular nitrate, which it reduces to nitrogen gas. This is the first and only case of eukaryotic denitrification. How would you go about designing a comparative analysis of eukaryotic and bacterial denitrification? What physical factors and physiological and enzymatic features would you characterize?

Read the original paper: Risgaard-Peterson, N., et al. 2006. Evidence for complete denitrification in a benthic foraminifera. *Nature.* 443:93.

Learn More

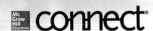

26

Fungi (Eumycota)

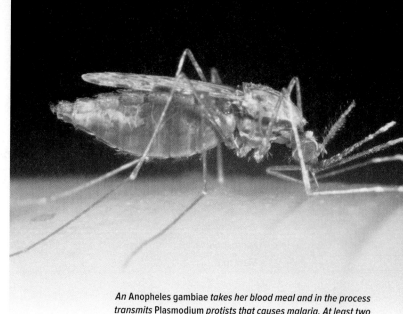

An Anopheles gambiae *takes her blood meal and in the process transmits* Plasmodium *protists that causes malaria. At least two fungal species show promising signs of becoming biocontrol agents to fight the transmission of this devastating disease.*

Fungi May Be Key to Quelling Malaria

Every minute someone dies of malaria. That someone is usually a child under the age of five living in sub-Saharan Africa. Sadly, these children could be protected by bed nets impregnated with insecticide. But even for those kids lucky enough to have bed nets, the insecticide may be losing its effectiveness because mosquitoes are acquiring resistance to these chemicals. Likewise, the protozoa that cause malaria—members of the genus *Plasmodium*—are losing susceptibility to the antimalarial drugs used to treat the disease. All things considered, it's a grim situation.

What's needed is a new approach. Some see entomopathogenic fungi as just that. Simply put, entomopathogenic fungi kill insects, in this case mosquitoes. There are two entomopathogenic fungal species of particular interest: *Metarhizium anisopliae* and *Beauveria bassiana*. Both have performed well in laboratory tests, where they are toxic to *Anopheles* mosquitoes carrying *Plasmodium* cells. But there is serious debate about how to harness these potential biocontrol agents.

Is it best to kill the adult mosquito or the larvae, or to focus on preventing transmission? Researchers targeting adult mosquitoes field-tested bed nets treated with *B. bassiana.* Although this killed more than twice as many insecticide-resistant mosquitoes than untreated nets, the viability of the fungus declined within days. In separate tests designed to prevent maturation of mosquito larvae, fungal spores were mixed with nontoxic oil and spread over the surface of *Anopheles*-infested water. Again there was a two-fold increase in mosquito death, and of those that survived, only about a fifth were able to pupate. However, some scientists argue that by killing mosquitoes susceptible to fungi, resistant mosquitoes will ultimately predominate. To address this, a genetically modified (GM) *M. anisopliae* strain that prevents the entrance of *Plasmodium* sporozoites into the mosquito's salivary gland when taking a blood meal was developed. Within two days of infection with GM fungi, 80% of the mosquitoes failed to transmit *Plasmodium* cells compared with 32% of those infected with non-GM fungi and 14% of fungal-free mosquitoes.

There are clear advantages and disadvantages in all three approaches. Some question the wisdom of releasing GM fungi. Others point to the cost of introducing any new technology. Perhaps there is no single correct approach. The only certainty is that too many lives continue to be lost to malaria.

In this chapter, we introduce **Fungi,** sometimes referred to as true fungi or *Eumycota* (Greek *eu,* true, and *mykes,* fungus). The term fungus (pl., fungi; Latin *fungus,* mushroom) describes eukaryotic organisms that are spore-bearing, have absorptive nutrition, lack chlorophyll, and reproduce both sexually and asexually. Scientists who study fungi are **mycologists,** and the scientific discipline devoted to fungi is called **mycology.** The study of fungal toxins is called **mycotoxicology,** and the diseases caused by fungi in animals are known as **mycoses** (s., mycosis).

Fungi consists of an enormous group of organisms; about 90,000 fungal species have been described, although some estimate the true number of species to be about 1.5 million. It is thus not surprising that the taxonomy of these organisms has been revised numerous times. The application of molecular techniques combined with morphological and ecological considerations has led to a better understanding of the phylogenetic relationships between various fungal groups (**figure 26.1**). For instance, neither *Zygomycota* nor *Chytridiomycota* are monophyletic. Most recently, metagenomic analysis from a wide range of soils revealed a large clade of environmental RNA sequences. This has led to the approval of a new phylum called *Cryptomycota,* meaning "hidden fungi" because they may constitute almost half the fungal biomass in some locations but until now have been unnoticed. Here we present the six major fungal groups that have representatives grown in culture: *Chytridiomycota, Zygomycota, Glomeromycota, Ascomycota, Basidiomycota,* and *Microsporidia,* shown in **figure 26.2** and summarized in **table 26.1**.

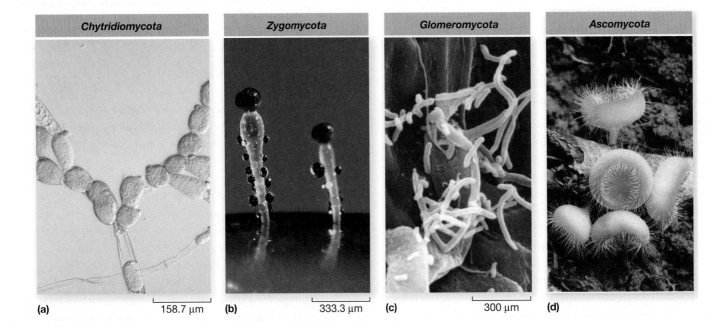

Ascomycota ⌉
Basidiomycota ⌋ Dikarya
Glomeromycota
Mucormycotina (*Zygomycota*)
Entomophthorales (*Zygomycota*)
Olpidium (*Chytridiomycota*)
Blastocladiales (*Chytridiomycota*)
Euchytrids (*Chytridiomycota*)
Cryptomycota
Microsporidia
Metazoa

Figure 26.1 The Major Branches of *Fungi*. Phylogenetic analysis of the seven fungal groups. Some suggest that the *Microsporidia* and *Cryptomycota* be combined into a single phylum.

MICRO INQUIRY *Compare the nodes and branches that lead to the monophyletic taxa* Basidiomycota *and* Ascomycota. *How do they differ from those leading to* Zygomycota *and* Chytridiomycota? *What can be concluded from this?*

Chytridiomycota

(a) 158.7 μm **(b)** 333.3 μm **(c)** 300 μm **(d)**

Zygomycota

Glomeromycota

Ascomycota

Basidiomycota

(e)

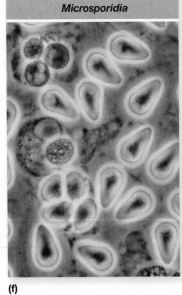

Microsporidia

(f)

Figure 26.2 Representatives of the Fungal Groups. (a) The free living chytrid *Allomyces macrogynus,* which lives in mud. (b) *Pilobolus* sp., a zygomycete, grows on animal dung and on culture medium. Stalks about 10 mm long contain dark, spore-bearing sacs. (c) Mycorrhizae are glomeromycetes associated with roots. (d) The cup fungus *Cookeina tricholoma* is an ascomycete from the rain forest of Costa Rica. In cup fungi, the spore-producing structures line the cup; in basidiomycetes that form mushrooms such as *Amanita* sp. they line the gills beneath the cap of the mushroom. (e) *Pleurotus ostreatus,* the oyster mushroom, is a basidiomycete. All visible structures of fungi arise from an extensive network of hyphae that penetrate and are interwoven with the substrate as they grow. (f) By contrast, the unusual fungi that belong to *Microsporidia* form spores that germinate in host cells. Shown here is the mosquito pathogen *Edhazardia aedis.*

Table 26.1   Abbreviated Classification of *Fungi*

Subclass	Characteristics	Examples
Chytridiomycota	Flagellated cells in at least one stage of life cycle; may have one or more flagella. Cell walls with chitin and β-1,3 / 1,6-glucan; glycogen is used as a storage carbohydrate. Sexual reproduction often results in a zygote that becomes a resting spore or sporangium; saprophytic or parasitic. Chytrid subdivisions include *Blastocladiales, Monoblepharidales, Neocallimastigaceae, Spizellomycetales,* and *Chytridiales.*	*Allomyces Blastocladiella Coelomomyces Physoderma Synchytrium*
Zygomycota	Thalli usually filamentous and nonseptate, without cilia; sexual reproduction gives rise to thick-walled zygospores that are often ornamented. Includes seven subdivisions: *Basidiobolus, Dimargaritales, Endogonales, Entomophthorales, Harpellales, Kickxellales, Mucorales,* and *Zoopagales.* Human pathogens found in *Mucorales* and *Entomophthorales.*	*Amoebophilus Mucor Phycomyces Rhizopus Thamnidium*
Ascomycota	Sexual reproduction involves meiosis of a diploid nucleus in an ascus, giving rise to haploid ascospores; most also undergo asexual reproduction with the formation of conidiospores with specialized aerial hyphae called conidiophores. Many produce asci within complex fruiting bodies called ascocarps. Includes saprophytic, parasitic forms; many form mutualisms with phototrophic microbes to form lichens. Four monophyletic subdivisions: *Saccharomycetes, Pezizomycotina, Taphrinomycotina,* and *Neolecta.*	*Ascobolus Aspergillis Candida Crinula Neurospora Penicillium Pneumocystis Saccharomyces*
Basidiomycota	Includes many common mushrooms and shelf fungi. Sexual reproduction involves formation of a basidium (small, club-shaped structure that typically forms spores at the ends of tiny projections) within which haploid basidiospores are formed. Usually four spores per basidium but can range from one to eight. Sexual reproduction involves fusion with opposite mating type resulting in a dikaryotic mycelium with parental nuclei paired but not initially fused. No subdivisions recognized. Also includes plant pathogens (rusts and smuts) belonging to *Urediniomycota* and *Ustilaginomycota.*	*Agaricus Boletes Dacrymyces Lycoperdon Polyporus Uromyces Ustilago*
Glomeromycota	Filamentous, most are endomycorrhizal, arbuscular; lack cilium; form asexual spores outside of host plant; lack centrioles, conidia, and aerial spores. No subdivisions recognized.	*Acaulospora Entrophospora Glomus*
Microsporidia	Obligate intracellular parasites usually of animals. Lack mitochondria, peroxisomes, kinetosomes, cilia, and centrioles; spores have an inner chitin wall and outer wall of protein; produce a tube for host penetration. Subdivisions currently uncertain.	*Amblyospora Encephalitozoon Enterocytozoon Nosema*
Cryptomycota	3–5 µm roughly spherical cells, flagellated, osmotrohpic. Cell wall does not possess chitin. Broadly distributed in soil, freshwater, and aquatic sediments.	*Rozella*

Adapted from: Adl, S. M., et al. 2005. The new higher level classification of eukaryotes with emphasis on the taxonomy of protists. J. Eukaryot. Microbiol. 52:399–451.

Readiness Check:

Based on what you have learned previously, you should be able to:

✔ Draw a eukaryotic cell envelope and identify the structure and function of its component layers (section 5.2)

✔ List the three filaments that make up the cytoskeleton of eukaryotic cells and describe how they are used (section 5.3)

✔ Identify the structure and function of eukaryotic cellular organelles (sections 5.4–5.6)

✔ Describe the endocytic pathway in general terms (section 5.4)

✔ Describe the structure and function of eukaryotic flagella and cilia (section 5.7)

26.1 Fungal Biology Reflects Vast Diversity

After reading this section, you should be able to:

■ Discuss the importance of fungi in the environment as well as their practical importance

■ Differentiate between a yeast and a mold, and between hypha and mycelium

■ Compare and contrast different spore morphologies

■ Explain how mating types govern sexual reproduction

■ Infer what is meant by a "dikaryotic fungus"

■ Assess the role of fungal decomposition on global carbon flux

Fungal Distribution and Importance

Unlike protists, fungi are primarily terrestrial organisms. They have a global distribution from polar to tropical regions. Fungi are **saprophytes,** securing nutrients from dead organic material by releasing degradative enzymes into the environment. This enables their absorption of the soluble products—a process sometimes called **osmotrophy** (Greek, *osmos,* pushing; perhaps a misnomer, as nutrients are transported rather than pushed into the cell). Fungi are important decomposers. They degrade complex organic materials in the environment to simple organic compounds and inorganic molecules. In this way, carbon, nitrogen, phosphorus, and other critical constituents of dead organisms are released and made available for living organisms. Many fungi are pathogenic, with over 5,000 species known to attack economically valuable crops and many other plants. About 20 new human fungal pathogens are documented each year. Conversely, fungi also form beneficial relationships with other organisms. For example, the vast majority of vascular plant roots form important associations with fungi called mycorrhizae. ▶▶| *Mycorrhizae facilitate the growth of most plants (section 31.3)*

Fungi, especially yeasts (single-celled fungi), are essential to many industrial processes involving fermentation. Examples include the making of bread, wine, beer, cheeses, and soy sauce. They are also important in the commercial production of many organic acids (citric, gallic) and certain drugs (ergometrine, cortisone), and in the manufacture of many antibiotics (penicillin, griseofulvin) and the immunosuppressive drug cyclosporine. In addition, fungi are important research tools in the study of fundamental biological processes. Cytologists, geneticists, biochemists, biophysicists, and microbiologists regularly use fungi in their research. The yeast *Saccharomyces cerevisiae* is the best understood eukaryotic cell. It has been a valuable model organism in the study of cell biology, genetics, and cancer. ▶▶| *Microbiology of fermented foods (section 41.5); Microbes are the source of many products of industrial importance (section 42.1)*

Fungal Structure

The body or vegetative structure of a fungus is called a **thallus** (pl., thalli). It varies in complexity and size. Single-cell microscopic fungi are referred to as yeasts, while multicellular masses are called molds. Fungi also include macroscopic puffballs and mushrooms. Like most bacteria, fungi possess cell walls; however, fungal cell walls are made of glucans, mannans, glycoproteins, and chitin. Chitin is a strong but flexible nitrogen containing polysaccharide consisting of *N*-acetylglucosamine residues.

A **yeast** is a unicellular fungus with a single nucleus that reproduces either asexually by budding and transverse division or sexually through spore formation. Each bud that separates can grow into a new cell, and some group together to form colonies. Generally yeast cells are larger than bacteria and are commonly spherical to egg-shaped. They lack flagella and cilia but have most other eukaryotic organelles (*see figures 5.2 and 5.3a*).

The thallus of a **mold** consists of long, branched, threadlike filaments called **hyphae** (s., hypha; Greek *hyphe,* web) that form a tangled mass called a **mycelium** (pl., mycelia). In some fungi, protoplasm streams lengthwise through the hyphae, uninterrupted by cross walls. These hyphae are called **coenocytic** or **aseptate hyphae (figure 26.3a)**. The hyphae of other fungi (figure 26.3b) have cross walls called **septa** (s., **septum**) with either a single pore (figure 26.3c) or multiple pores (figure 26.3d) that enable cytoplasmic streaming. These hyphae are termed **septate hyphae.**

Hyphae are composed of an outer cell wall and an inner lumen, which contains the cytosol and organelles. A plasma membrane surrounds the cytoplasm and lies next to the cell wall. The filamentous nature of hyphae results in a large surface area relative to the volume of cytoplasm. This makes nutrient absorption by osmotrophy possible.

Fungal Reproduction

Reproduction in fungi can be either asexual or sexual. Asexual reproduction is accomplished in several ways: (1) a parent cell undergoes mitosis and divides into two daughter cells by a central constriction and formation of a new cell wall (**figure 26.4a**), or (2) mitosis in vegetative cells may be concurrent with budding to produce a smaller daughter cell. This is very common in yeasts. The formation of asexual spores often accompanies asexual reproduction and is usually used as a means of dispersal. There are many types of asexual spores, each with its own name. **Arthroconidia** (**arthrospores**) are formed when hyphae fragment (figure 26.4b). **Sporangiospores** develop within a sac called a **sporangium** (pl., sporangia) at a hyphal tip (figure 26.4c).

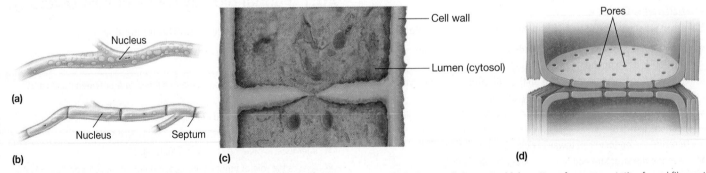

Figure 26.3 Hyphae. Drawings of (a) coenocytic hyphae (aseptate) and (b) hyphae divided into cells by septa. (c) A section of a representative fungal filament demonstrates that in some species, cellular compartments communicate by a single pore. (d) Drawing of a multiperforate septal structure.

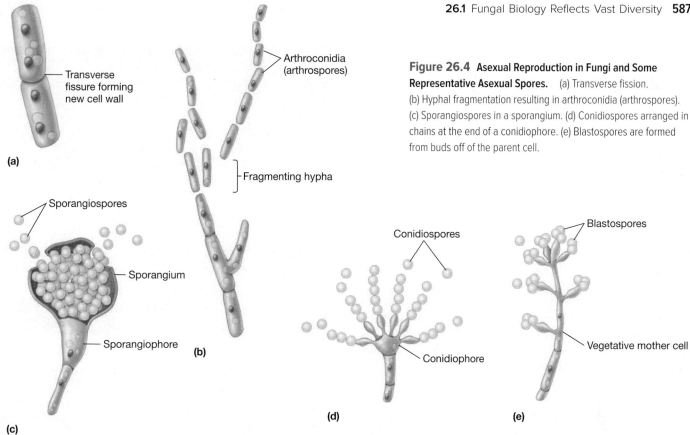

(a) Transverse fissure forming new cell wall

Arthroconidia (arthrospores)

Fragmenting hypha

(b)

Sporangiospores

Sporangium

Sporangiophore

(c)

Conidiospores

Conidiophore

(d)

Blastospores

Vegetative mother cell

(e)

Figure 26.4 **Asexual Reproduction in Fungi and Some Representative Asexual Spores.** (a) Transverse fission. (b) Hyphal fragmentation resulting in arthroconidia (arthrospores). (c) Sporangiospores in a sporangium. (d) Conidiospores arranged in chains at the end of a conidiophore. (e) Blastospores are formed from buds off of the parent cell.

Conidiospores are spores that are not enclosed in a sac but produced at the tips or sides of the hypha (figure 26.4*d*). **Blastospores** are produced from a vegetative mother cell budding (figure 26.4*e*).

Sexual reproduction in fungi involves the fusion of compatible nuclei. Homothallic fungal species are self-fertilizing and produce sexually compatible gametes on the same mycelium. Heterothallic species require outcrossing between different but sexually compatible mycelia called mating types. Depending on the species, sexual fusion may occur between haploid gametes, gamete-producing bodies called **gametangia,** or hyphae. Sometimes both the cytoplasm and haploid nuclei fuse immediately to produce the diploid zygote, as seen in higher eukaryotes. More often, cytoplasmic fusion occurs before nuclear fusion. This produces a **dikaryotic stage** in which a single cell contains two separate haploid nuclei (N + N), one from each parent **(figure 26.5).** After a period of dikaryotic existence, the two nuclei fuse and undergo meiosis to yield haploid spores. This is seen in both ascomycetes and basidiomycetes, so these are sometimes referred to as dikaryotic fungi.

Fungal spores are important for several reasons. They enable fungi to survive environmental stresses such as desiccation, nutrient limitation, and extreme temperatures, although they are not as stress resistant as bacterial endospores. They aid in fungal dissemination, which helps explain their wide distribution. Because spores are often small and light, they can remain suspended in air for long periods and are often spread by adhering to the bodies of insects and other animals. The bright colors and fluffy textures of many molds often are due to their aerial hyphae and spores. Finally, the size, shape, color, and number of spores are useful in the identification of fungal species.

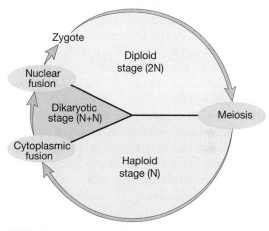

Zygote

Nuclear fusion

Diploid stage (2N)

Dikaryotic stage (N+N)

Meiosis

Cytoplasmic fusion

Haploid stage (N)

Figure 26.5 **Reproduction in Fungi.** A drawing of the generalized life cycle for fungi showing the alternation of haploid and diploid stages. Some fungal species do not pass through the dikaryotic stage indicated in this drawing. The asexual (haploid) stage is used to produce spores that aid in the dissemination of the species. The sexual (diploid) stage involves the formation of spores that survive adverse environmental conditions (e.g., cold, dryness, heat).

Retrieve, Infer, Apply

1. How can fungi be defined? What is the difference between a yeast and a mold?
2. What is the distribution of these microbes?
3. Why does the mycelial morphology of fungi make them especially effective saprophytes?
4. What organelles would you expect to find in the cytoplasm of a typical fungi?
5. Compare and contrast the following types of asexual fungal spores: sporangiospore, arthrospore, conidiospore, and blastospore.

26.2 *Chytridiomycota* Produce Motile Spores

After reading this section, you should be able to:

- Explain why chytrids are unique among fungi
- Identify several habitats where chytrids grow
- Describe general chytrid morphology

Chytridiomycota, commonly called **chytrids,** includes free-living and parasitic genera. Free-living chytrids are saprophytic, living on plant or animal matter in freshwater, mud, or soil. Parasitic forms infect aquatic plants and animals, including insects **(figure 26.6)**. A few are found in the anoxic rumen of herbivores. The chytrid *Batrachochytrium dendrobatidis* is responsible for large-scale mortality of amphibians worldwide. This includes over 100 species of harlequin frogs (*Atelopus*) native to tropical America, in which infection has caused rapid population declines and, in some cases, extinction.

Chytrids are unique among fungi in the production of a spore with a single, posterior, whiplash flagellum. It has been hypothesized that all fungi were once flagellated with at least four independent losses of flagella during the evolution of more advanced fungal groups. The loss of flagella is thought to have coincided with the development of other means of spore distribution, such as aerial dispersal from mycelial mats. Chytrids are not considered monophyletic (note three clades of *Chytridiomycota* in figure 26.1). Instead, they derive from early diverging lineages that bear spores. The largest clade, *Euchytrids,* unites several orders of chytrids.

Chytrids display a variety of life cycles involving both asexual and sexual reproduction. Members of this group are microscopic in size and may consist of a single cell, a small multinucleate mass, or a true mycelium. Many are capable of degrading cellulose and even keratin, which enables the degradation of crustacean exoskeletons. The model chytrid *Allomyces macrogynus* has a complex life cycle that includes four types of sporangia and five developmentally distinct spore types.

26.3 *Zygomycota:* Fungi with Coenocytic Hyphae

After reading this section, you should be able to:

- Recognize zygomycetes in nature
- Outline the life cycle of a *Rhizopus* species
- List at least two reasons zygomycetes benefit humans

Zygomycota contains fungi informally called **zygomycetes.** Most live on decaying plant and animal matter in the soil; a few are parasites of plants, insects, and other animals, including humans. The hyphae of zygomycetes are coenocytic (aseptate), with many haploid nuclei. Asexual spores develop in sporangia at the tips of aerial hyphae and are usually wind dispersed. Sexual reproduction produces

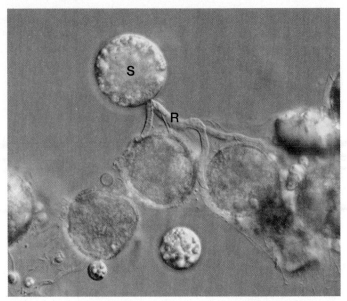

(a)

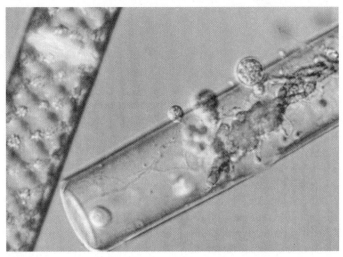

(b)

Figure 26.6 *Chytridiomycota.* (a) A chytrid with a single round sporangium (S) that is about 50 μm in diameter. In this image, the fungus is growing between brown pollen grains, which it will eventually penetrate with its branched rhizoids (R). (b) Parasitic chytrids attached to the surface of a photosynthetic protist (a green alga).

MICRO INQUIRY *What is the function of the chytrid rhizoids?*

tough, thick-walled zygotes called **zygospores** that can remain dormant when the environment is too harsh for growth of the fungus.

The mold *Rhizopus stolonifer* is a common member of this division. This fungus grows on the surface of moist, carbohydrate-rich foods, such as breads, fruits, and vegetables. Hyphae called rhizoids extend into the substrate and absorb nutrients. Other hyphae (stolons) become erect, then arch back into the substratum, forming new rhizoids. Still others remain erect and produce at their tips asexual sporangia filled with hundreds of black spores, giving the mold its characteristic color. Each spore, when liberated, can germinate to start a new mycelium.

Rhizopus spp. usually reproduce asexually, but if food becomes scarce or environmental conditions unfavorable, sexual reproduction occurs (**figure 26.7**). Sexual reproduction requires compatible strains of opposite mating types (called + and −). When + and − hyphae are close, each produces a different hormone, called a pheromone, that causes their hyphae to form projections called progametangia; these mature into gametangia. First the gametangia fuse, a process called plasmogamy. Then, the nuclei of the two gametes fuse (karyogamy), forming a zygote. The zygote develops a thick, rough, black coat and becomes a dormant zygospore. Meiosis often occurs at the time of germination; the zygospore then splits open and produces a hypha that bears an asexual sporangium filled with haploid spores that begin the cycle again.

One important member of the genus *Rhizopus* is involved in the rice disease known as seedling blight. If one considers that rice feeds more people on Earth than any other crop, the implications of this disease are obvious. It was thought that the fungus secreted a toxin that kills rice seedlings, so scientists set about isolating the toxin and the genes that produce it. Much to everyone's surprise, the toxin is produced by an α-proteobacterium, *Burkholderia* sp., growing within the fungus.

Zygomycetes also contribute to human welfare. For example, one species of *Rhizopus* is used in Indonesia to produce a food called tempeh from boiled, skinless soybeans. Other zygomycetes (*Mucor* spp.) are used with soybeans in Asia to make a curd called sufu. Others are employed in the commercial preparation of some anesthetics, birth control agents, industrial alcohols, meat tenderizers, and the yellow coloring used in margarine and butter substitutes. ▶▶| *Microbiology of fermented foods (section 41.5)*

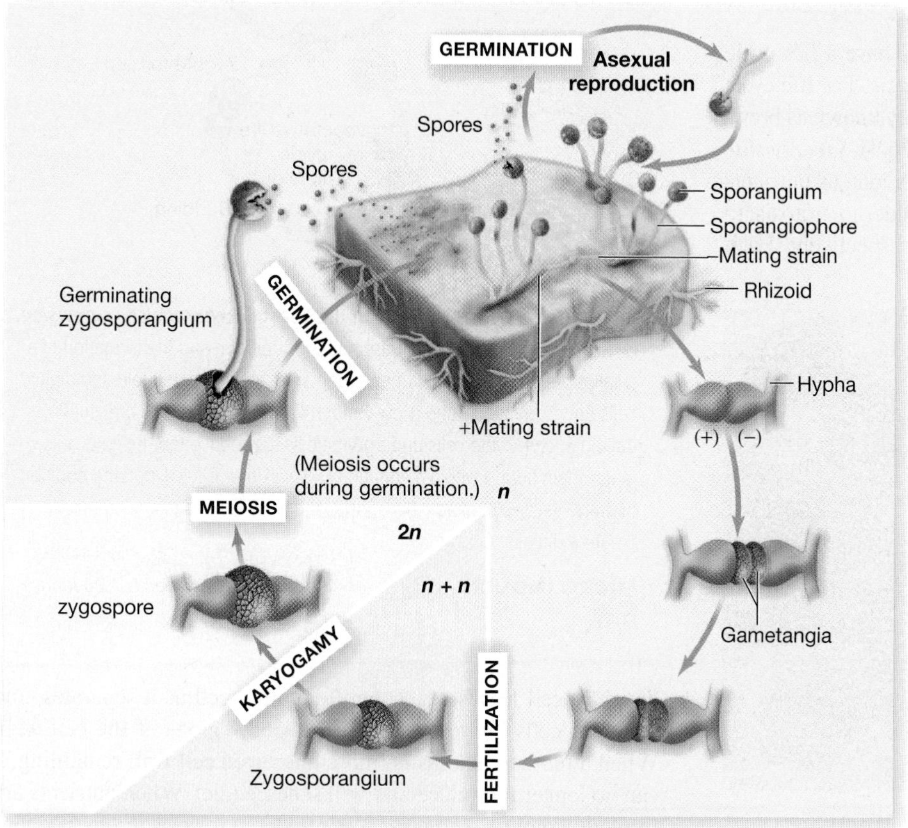

26.4 *Glomeromycota* Are Mycorrhizal Symbionts

After reading this section, you should be able to:

- Describe the functional importance of glomeromycetes
- Differentiate between ectomycorrhizae and arbuscular mycorrhizae

Glomeromycetes are of critical ecological importance because they are mycorrhizal symbionts of vascular plants. **Mycorrhizal fungi** form important associations with the roots of almost all herbaceous plants and tropical trees. This is considered a mutualistic relationship because both the host plant and the fungus benefit: the fungus delivers soil nutrients to the plant, and the plant in turn provides carbohydrate to the fungus. As discussed in chapter 31, most mycorrhizal fungi belong to one of two groups. Ectomycorrhizae do not penetrate host root cells; instead, the fungal hyphae grow between and around cells. By contrast, arbuscular mycorrhizal hyphae penetrate the host root cell wall but not the plasma membrane (*see figures 31.4–31.8*). ▶▶| *Mycorrhizae facilitate the growth of most plants (section 31.3)*

Glomeromycetes have aseptate hyphae and produce large, multinucleate spores. Only asexual reproduction is known to occur. This can occur by fragmentation of filaments in the soil or on a plant. Spores germinate when in contact with the roots of a suitable host plant. Unlike other filamentous fungi, glomeromycetes have evolved specialized flat hyphae called **appressoria** (s., appressorium) to penetrate their host plants.

Figure 26.7 *Rhizopus stolonifer*, a Zygomycete That Grows on Simple Sugars. This fungus is often found on moist bread or fruit. The life cycle of *Rhizopus* spp. involves sexual and asexual phases as shown. *Zygomycota* is named for the zygosporangia characteristic of members of the genus *Rhizopus*.

Retrieve, Infer, Apply

1. What morphological feature makes the chytrids unique among fungi?
2. What is the difference between hyphae called rhizoids and those called stolons?
3. How do different mating types of *Rhizopus stolonifer* recognize each other?
4. Given that mycorrhizal fungi obtain carbohydrates from their plant partners, what specific nutrients are the fungi likely to provide to plants?
5. Which fungal class would you most likely find in each of the following environments: two-week-old cake, mud, tree roots?

26.5 *Ascomycota* Includes Yeasts and Molds

After reading this section, you should be able to:

- Describe ecological, economic, and medical importance of ascomycetes
- Diagram the *Saccharomyces cerevisiae* life cycle
- Outline the mechanisms used by typical filamentous ascomycetes to reproduce, survive unfavorable conditions, and disperse
- Explain the function of sclerotia
- Identify the origin of St. Anthony's fire, LSD, and "sick building syndrome"

Members of *Ascomycota,* or **ascomycetes,** commonly known as **sac fungi,** are named for their characteristic reproductive structure, the saclike **ascus** (pl., asci; Greek *askos,* sac), which holds **ascospores.** Ascomycetes are ecologically important in freshwater, marine, and terrestrial habitats because they degrade many chemically stable organic compounds, including lignin, cellulose, and collagen. Many species are quite familiar and economically important (**figure 26.8**). For example, most of the red, brown, and blue-green molds that cause food spoilage are ascomycetes. The powdery mildews that attack plant leaves and the fungi that cause chestnut blight and Dutch elm disease are ascomycetes. Many, if not most, fungi that infect humans are ascomycetes. On the other hand, many yeasts as well as edible morels and truffles are also ascomycetes. The pink bread mold *Neurospora crassa* is an important research tool in genetics and biochemistry.

Some ascomycetes are yeasts, while others have a life cycle that alternates between yeast and filamentous forms. The life cycle of the yeast *Saccharomyces cerevisiae,* commonly known as brewer's or baker's yeast, is well understood (**figure 26.9**). *S. cerevisiae* alternates between haploid and diploid states. As long as nutrients remain plentiful, haploid and diploid cells undergo mitosis to produce haploid and diploid daughter cells, respectively. Each

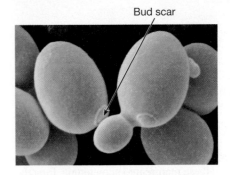

(a) *Saccharomyces cerevisiae:* budding division

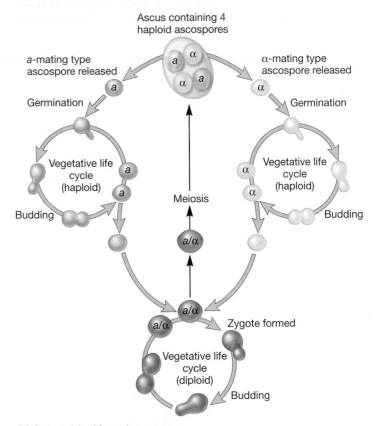

(b) *S. cerevisiae* life cycle

Figure 26.9 The Life Cycle of the Yeast *Saccharomyces cerevisiae.* (a) Budding division results in asymmetric septation and the formation of a smaller daughter cell. (b) When nutrients are abundant, haploid and diploid cells undergo mitosis and grow vegetatively. When nutrients are limited, diploid *S. cerevisiae* cells undergo meiosis to produce four haploid cells that remain bound within a common cell wall, the ascus. Upon the addition of nutrients, two haploid cells of opposite mating types (a and α) fuse to create a diploid cell.

MICRO INQUIRY *What determines when a yeast cell can no longer bud?*

(a) *Morchella esculenta*

(b) *Tuber brumale*

Figure 26.8 *Ascomycota.* (a) The common morel, *Morchella esculenta,* is one of the choicest edible fungi. It fruits in the spring. (b) The black truffle, *Tuber brumale,* is highly prized for its flavor by gourmet cooks.

daughter cell leaves a scar on the mother cell as it separates, and daughter cells bud only from unscarred regions of the cell wall. When a mother cell has no more unscarred cell wall remaining, it can no longer reproduce and will senesce (die). When nutrients are limited, diploid *S. cerevisiae* cells undergo meiosis to produce four haploid cells that remain bound within a common cell wall, the ascus. Upon the addition of nutrients, two haploid cells of opposite mating types (a and α) come into contact and fuse to create a

Figure 26.10 **Asexual Reproduction in Filamentous Ascomycetes.** Characteristic conidiospores of *Aspergillus* spp. as viewed with the scanning electron microscope.

diploid. Typically only cells of opposite mating types can fuse; this process is tightly regulated by the action of pheromones.

S. cerevisiae is a valuable model organism. Research on this organism has revealed the importance of many cellular processes. For instance, it is a favorite model system for studying cell cycling and the events during mitosis. This research is critical not only for

our understanding of normal cell division, but also for exploring the loss of cell cycle control that occurs in cancerous cells.

One of the hallmarks of filamentous ascomycetes is the formation of septate hyphae. Asexual reproduction is common and is associated with the production of externally produced asexual spores called **conidia** (**figure 26.10**). Sexual reproduction involves two strains of opposite mating types. Hyphae from one mating type form a specialized structure called an ascogonium, and hyphae of the other mating type form an antheridium (**figure 26.11**). The nuclei from the antheridium migrate to the ascogonium during the exchange of cytoplasm (plasmogamy). The nuclei do not yet fuse, rather nuclei from each mating type pair and replicate in unison during the ensuing dikaryotic phase. The ascogonium is said to be fertilized and the hyphae are called **ascogenous hyphae.** Ascogenous hyphae, supported by sterile hyphae, next develop into a cuplike structure called an **ascocarp.** This is where ascogenous

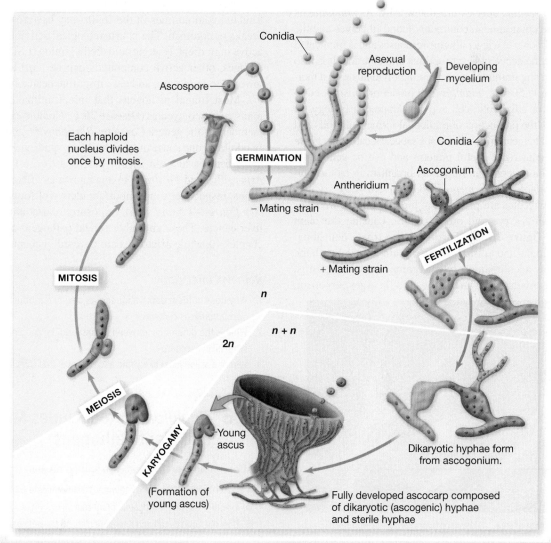

Figure 26.11 **The Typical Life Cycle of a Filamentous Ascomycete.** Sexual reproduction involves the formation of asci and ascospores. The events pictured on the left side of the figure (karyogamy, meiosis, mitosis, and release of ascospores) occur within the ascocarp.

MICRO INQUIRY *In what ways is the life cycle of filamentous ascomycetes similar to that of the zygomycete shown in figure 26.7? How do the two life cycles differ?*

hyphae mature, and nuclear fusion (karyogamy) takes place. The diploid zygote nucleus then undergoes meiosis, and the resulting four haploid nuclei divide mitotically again to produce a row of eight nuclei in each developing ascus. These nuclei are walled off from one another and each will develop into an **ascospore.** Thousands of asci may be packed together in an ascocarp. When the ascospores mature, they often are released from the asci with great force. If the mature ascocarp is jarred, it may appear to belch puffs of "smoke" consisting of thousands of ascospores. Upon reaching a suitable environment, the ascospores germinate and start the cycle anew.

Although conidia are the major form of dissemination, structures called **sclerotia** are compact masses of dormant hyphae that permit survival in stressful conditions. Sclerotia confer a competitive advantage to the fungi that produce them. For instance, some species of *Aspergillus* that infect plants form sclerotia to remain viable in the soil, where they can take advantage of nutrient resources when the temperature rises.

Several *Aspergillus* species are noteworthy. *A. fumigatus* is ubiquitous in the environment, commonly found in homes and the workplace. It is known to trigger allergic responses and is implicated in the increased incidence of severe asthma and sinusitis. It is also pathogenic, infecting immunocompromised individuals with a mortality rate of nearly 50%. *A. nidulans* is a model organism used to study questions of eukaryotic cell and developmental biology. *A. oryzae* is used in the production of traditional fermented foods and beverages in Japan, including saki and soy sauce. Because *A. oryzae* secretes many industrially useful proteins and can be genetically manipulated, it has become an important organism in biotechnology. ▶▶| *Microbiology of fermented foods (section 41.5)*

Many ascomycetes are parasites of higher plants. *Claviceps purpurea* parasitizes rye and other grasses, causing the plant disease **ergot** (**figure 26.12**). Ergotism, the toxic, delusional condition in humans and animals that eat grain infected with the fungus, is often accompanied by gangrene, nervous spasms, abortion, and convulsions. During the Middle Ages, ergotism,

Figure 26.13 Fungal "Sick Building Syndrome" Can Be Caused by an Ascomycete. Scanning electron micrograph of dense mycelia and conidiophores characteristic of *Stachybotrys chartarum.*

then known as St. Anthony's fire, killed thousands of people. For example, over 40,000 deaths from ergot poisoning were recorded in France in the year 943. It has been suggested that the widespread accusations of witchcraft in Salem Village and other New England communities in the 1690s may have resulted from outbreaks of ergotism. The pharmacological activities are due to an active ingredient, lysergic acid diethylamide (LSD). In controlled dosages, other active compounds can be used to induce labor, lower blood pressure, and ease migraine headaches.

Most fungal pathogens that infect animals, including humans, are ascomycetes (**Disease 26.1**). Notable examples include members of the genera *Candida, Blastomyces,* and *Histoplasma.* In addition, one cause of "sick building syndrome," *Stachybotrys chartarum* is also an ascomycete (**figure 26.13**). Finally, the *Aspergillus* and *Fusarium* toxins known as aflatoxins and fumonisins, respectively, are important causes of food contamination (*see figures 41.5 and 41.6*). Exposure to aflatoxins can result in liver cancer. These and other fungal pathogens are discussed in chapter 40, while aflatoxins are covered in chapters 40 and 41.

Retrieve, Infer, Apply

1. Why does nutrient deprivation trigger sexual reproduction in *Saccharomyces cerevisiae*?
2. What is the difference between an antheridium and an ascogonium? What is an ascocarp?
3. Where are you most likely find ascomycetes that form sclerotia?

26.6 *Basidiomycota* Includes Mushrooms and Plant Pathogens

After reading this section, you should be able to:

- Describe the life cycle of a typical basidiomycete
- Identify the fungi that cause plant rusts
- Describe one human and one plant disease caused by a basidiomycete

Fungi in *Basidiomycota* are commonly known as **basidiomycetes,** or **club fungi.** Examples include jelly fungi, rusts, shelf fungi, stinkhorns, puffballs, toadstools, mushrooms, and bird's nest fungi. Most are saprophytes that decompose plant debris,

Figure 26.12 Ergot of Rye. The ascomycete *Claviceps purpurea* infects rye and other grasses, producing hard masses of hyphae known as ergots in place of some of the grains (fruits). Ergots produce alkaloids that cause psychotic delusions in humans and animals when products made with infected grain are consumed. Ergots have been used to treat migraine headaches and hasten childbirth.

DISEASE

26.1 White-Nose Syndrome Is Decimating North American Bat Populations

Bats evoke all kinds of images. Some people immediately think of vampire bats and are repulsed. Others think of large fruit bats often called flying foxes. If you have spent a summer evening outdoors on the east coast of North America, mosquitoes and the small bats that eat them may come to mind. A new scene can now be added to these: bats with white fungal hyphae growing around their muzzles (**box figure**). This is the hallmark of white-nose syndrome (WNS), and if its rate of infection continues unchecked, it is projected to eliminate the most common bat species in eastern North America (*Myotis lucifugus*) by 2026.

WNS was first spotted in 2006 among bats hibernating in a cave near Albany, NY. Scientists quickly became alarmed for two reasons. First, it spreads rapidly—it's known to occur in at least 11 bat species and is now found in 25 states in the United States and three Canadian provinces. Second, it is deadly. The population of bats declines from 30 to 99% in any given infected hibernacula (the place where bats hibernate, which unfortunately rhymes with Dracula).

WNS is caused by the ascomycete *Pseudogymnoascus destructans* (formerly *Geomyces destructans*). It colonizes a bat's

wings, muzzle, and ears where it erodes the epidermis before invading the underlying skin and connective tissue. Despite the name WNS, the primary site of infection (and the anatomical site harmed most) is the wing. Wings provide a large surface area for colonization, and once infected, the thin layer of skin is easily damaged, leading to adverse physiological changes during hibernation. These in turn result in premature awakening, loss of essential fat reserves, and strange behavior.

Where did this pathogen come from and why does it infect bats? The best hypothesis regarding its origin is that humans inadvertently brought it from Europe, where it causes mild infection in at least one hibernating bat species. This makes *P. destructans* an apparent case of pathogen pollution—the human introduction of invasive pathogens of wildlife and domestic animal populations that threaten biodiversity and ecosystem function.

The capacity of *P. destructans* to sweep through bat populations results from a "perfect storm" of host- and pathogen-associated factors. *P. destructans* is psychrophilic, with a growth optimum around 12°C; it does not grow above 20°C. All infected bat species hibernate in cold and humid environments such as caves and mines. Because their metabolic rate is drastically reduced during hibernation, their body temperature reaches that of their surroundings, between 2 and 7°C. Thus WNS is only seen in hibernating bats or those that have just emerged from hibernation. When metabolically active, the bat's body temperature is too warm to support pathogen growth.

While it is too late to save the estimated 6 million bats that have already succumbed to WNS, microbiologists, conservationists, and government agencies are trying to limit the continued decline in bat populations. Caves have been closed to human traffic, and protocols for decontamination after visiting hibernacula have been developed to limit the spread from cave to cave. Although we cannot cure sick bats, it is our responsibility to stop the continued spread of this pathogen.

Geomyces destructans causes WNS. A little brown bat (*Myotis lucifugus*) with the white fungal hyphae (arrow) for which WNS is named.

Read more: Langwig, K.E., et al. 2014. Invasion dynamics of white-nose syndrome fungus, midwestern United States, 2012–2014. Emerging Infectious Diseases. 21: 1023–1026.

especially cellulose and lignin. For example, the common fungus *Polyporus squamosus* forms large, shelflike structures that project from the lower portion of dead trees, which they help decompose. Many mushrooms are used as food, and the cultivation of the mushroom *Agaricus campestris* is a multimillion-dollar business. Of course, not all mushrooms are edible. Many mushrooms produce alkaloids that act as either poisons or hallucinogens. One such example is the "death angel" mushroom, *Amanita phalloides* (**figure 26.14**). Two toxins isolated from

this species are phalloidin and α-amanitin. Phalloidin primarily attacks liver cells, where it binds to and ruptures plasma membranes. Alpha-amanitin attacks the cells lining the stomach and small intestine, causing severe gastrointestinal symptoms associated with mushroom poisoning. *A. muscaria* is both poisonous and hallucinogenic.

Basidiomycetes are named for their characteristic structure or cell, the **basidium,** which is involved in sexual reproduction (figure 24.15*b*). A basidium (Greek *basidion,* small base) is

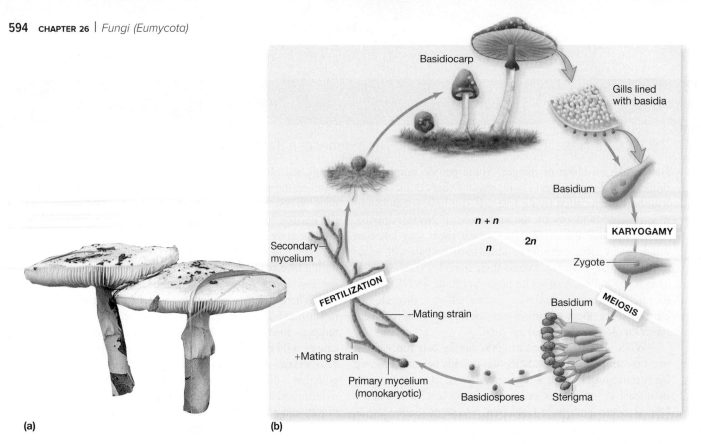

(a)

(b)

Figure 26.14 *Basidiomycota.* (a) The death cap mushroom *Amanita phalloides* is usually fatal when eaten. (b) The life cycle of a typical soil basidiomycete starts with a basidiospore germinating to produce a monokaryotic mycelium (a single nucleus in each septate cell). The mycelium grows and spreads throughout the soil. When this primary mycelium meets another monokaryotic mycelium of a different mating type, the two fuse to initiate a dikaryotic secondary mycelium. The secondary mycelium is divided by septa into cells, each of which contains two nuclei, one of each mating type. This dikaryotic mycelium is eventually stimulated to produce basidiocarps. A solid mass of hyphae forms a button that pushes through the soil, elongates, and develops a cap. The cap contains many platelike gills, each of which is coated with basidia. The two nuclei in the tip of each basidium fuse to form a diploid zygote nucleus, which immediately undergoes meiosis to form four haploid nuclei. These nuclei push their way into the developing basidiospores, which are then released at maturity.

produced at the tip of hyphae and normally is club-shaped. Two or more **basidiospores** are produced by the basidium, and basidia may be held within fruiting bodies called **basidiocarps.** Basidiocarps can reach 60 cm in diameter and have many pores, each lined with basidia that produce basidiospores. Thus a single fruiting body can produce millions of spores. Like ascomycete hyphae, the hyphae of basidiomycetes are septate.

The basidiomycete *Cryptococcus neoformans* is an important human and animal pathogen. It causes a disease called cryptococcosis, an infection primarily involving the lungs and central nervous system. In the human host, this fungus always grows as a large, budding yeast, and the production of an elaborate capsule is an important virulence factor for the microbe. In the environment, *C. neoformans* is a saprophyte with a worldwide distribution. ▶▶| *Cryptococcosis (section 40.2)*

Basidiomycetes known as urediniomycetes and ustilaginomycetes include important plant pathogens causing "rusts" and "smuts," respectively, that destroy millions of dollars worth of crops annually. These fungi do not form large basidiocarps. Instead, small basidia arise from hyphae at the surface of the host plant. The hyphae grow extracellularly in plant tissue. In addition, some urediniomycetes are human pathogens.

The ustilaginomycete *Ustilago maydis* is a common corn pathogen that has become a model organism for plant smuts (**figure 26.15**). It is dimorphic; plant-associated fungi grow in

the mycelial form but are yeastlike in the external environment. Yeastlike saprophytic *U. maydis* cells can be easily grown in the laboratory. In nature, the yeast (haploid sporidia) must mate to produce infectious, filamentous dikaryons that depend on the

Figure 26.15 *Ustilago maydis.* This pathogen causes tumor formation in corn. Note the enlarged tumors releasing dark teliospores in place of normal corn kernels.

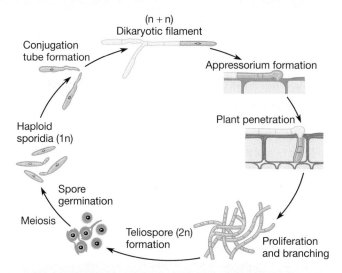

Figure 26.16 Life Cycle of the Plant Pathogen *Ustilago maydis*. Haploid sporidia must mate to form infectious dikaryons, which in turn depend on the host plant for continued development. The formation of fungal appressoria within the host induces tumor formation.

MICRO INQUIRY *What adaptations do you think* U. maydis *has evolved that has led to its pathogenicity? Contrast this with mycorrhizae.*

host plant for continued development (**figure 26.16**). Once a plant is infected, *U. maydis* forms appressoria within the host. This triggers the plant to form tumors in which the fungus proliferates and eventually produces diploid spores called **teliospores.** Upon germination, cells undergo meiosis and haploid sporidia are released. These sporidia can spread from plant to plant, where they start the growth and infection cycle again.

26.7 *Microsporidia* Are Intracellular Parasites

After reading this section, you should be able to:

- Explain why microsporidia are no longer considered protists
- Describe the microsporidian life cycle and relate its unique morphology to host invasion
- List at least two human diseases caused by microsporidia

Of all the fungi, members of **Microsporidia** have the most confused taxonomic history. These obligately endoparasitic microbes are sometimes still considered protists. However, analysis of ribosomal RNA and specific proteins such as α- and β-tubulin shows that they are fungi. However, unlike other fungi, they lack mitochondria, peroxisomes, and centrioles but possess

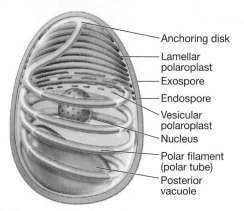

Figure 26.17 The Unique Structure of a Microsporidian Spore. Upon germination in a host cell, the polar tube is ejected with enough force to pierce the host cell membrane, allowing the fungus to gain entry.

mitosomes, like some protists. Their genomes are highly reduced in size and complexity. But like most fungi, microsporidian cell walls contain chitin and trehalose.

Microsporidial morphology is also unique among eukaryotes. It is distinct in the possession of an organelle called the polar tube that is essential for host invasion. Small spores of 1 to 40 μm are viable outside the host. Depending on the species, spores may be spherical, rodlike, or egg- or crescent-shaped. Spore germination occurs only in the host, where it expels the tightly packed polar tube or filament (**figure 26.17**). The polar tube is ejected with enough force to pierce the host cell membrane, permitting parasite entry. Once inside the host cell, the microsporidian undergoes a developmental cycle that differs among the various microsporidian species. However, in all cases, more spores are produced and eventually take over the host cell.

Microsporidia are important obligate intracellular parasites that infect insects, fish, and humans. They are particularly problematic for immunocompromised individuals, especially those with HIV/AIDS. Human pathogens include *Enterocytozoon bieneusi,* which causes diarrhea and pneumonia (depending on whether it was acquired through ingestion or inhalation, respectively), and *Encephalitozoon cuniculi,* which causes encephalitis and nephritis (kidney disease). Aquatic birds are an important source of pathogenic microsporidia in nature; indeed, a single flock of waterfowl can deposit over 10^8 microsporidian spores in surface waters after one visit. ▶▶| *Microsporidiosis (section 40.6)*

Retrieve, Infer, Apply

1. What is the difference between a basidiocarp and a basidium?
2. How do cryptococci differ from most basidiomycetes?
3. How does *Ustilago maydis* trigger tumor formation? How is tumor formation advantageous to the fungi?
4. Why do you think members of *Microsporidia* are still sometimes considered protists? What structure is essential for the germination of a microsporidian spore? Describe its function.

Key Concepts

26.1 Fungal Biology Reflects Vast Diversity

- Fungi are important decomposers that break down organic matter; live as parasites on animals, humans, and plants; play a role in many industrial processes;

and are used as research tools in the study of fundamental biological processes. Fungi secrete enzymes outside their body structure and absorb the digested food.

- A fungus is a eukaryotic, spore-bearing organism that has absorptive nutrition and lacks chlorophyll; reproduces asexually, sexually, or by both methods; and normally has a cell wall usually containing chitin.
- Hyphae are long, branched, threadlike filaments that form a tangled mass called a mycelium. Hyphae may be either septate or coenocytic (nonseptate). The mycelium can produce reproductive structures (**figure 26.3**).
- Asexual reproduction in the fungi often leads to the production of specific types of spores that are easily dispersed (**figure 26.4**).
- Sexual reproduction is initiated in fungi by the fusion of hyphae (or cells) of different mating strains. In some fungi, the nuclei in the fused hyphae immediately combine to form a zygote. In others, the two genetically distinct nuclei remain separate, forming pairs that divide synchronously. Eventually some nuclei fuse (**figure 26.5**).

26.2 *Chytridiomycota* Produce Motile Spores

- Fungi belonging to *Chytridiomycota* produce motile spores. Most are saprophytic; some reside in the rumen of herbivores (**figure 26.6**).
- These fungi are microscopic and display both sexual and asexual reproduction.

26.3 *Zygomycota:* Fungi with Coenocytic Hyphae

- Most zygomycetes are saprophytic. One example is the common bread mold, *Rhizopus stolonifer*. Sexual reproduction occurs through conjugation between strains of opposite mating type (**figure 26.7**).
- Zygomycetes are used in food and pharmaceutical manufacturing.

26.4 *Glomeromycota* Are Mycorrhizal Symbionts

- *Glomeromycota* includes mycorrhizal fungi that grow in association with plant roots. They serve to increase plant nutrient uptake.
- Most mycorrhizal fungi are either ectomycorrhizae, which do not penetrate the host plant cells, or arbuscular mycorrhizae, which grow between the plant cell wall and plasma membrane.

26.5 *Ascomycota* Includes Yeasts and Molds

- Members of *Ascomycota* are known as sac fungi because they form a sac-shaped reproductive structure called an ascus. Ascomycetes can have either a yeast morphology (**figure 26.9**) or a mold morphology (**figure 26.11**).
- Important ascomycetes include the yeast *Saccharomyces cerevisiae*, *Aspergillus* spp., *Claviceps purpurea* (the cause of ergotism), and the indoor mold *Stachybotrys chartarum*.

26.6 *Basidiomycota* Includes Mushrooms and Plant Pathogens

- Members of *Basidiomycota* are called club fungi. They are named after their basidium that produces basidiospores (**figure 26.14**).
- Genera from urediniomycetes and ustilaginomycetes include important plant pathogens (**figure 26.15**).

26.7 *Microsporidia* Are Intracellular Parasites

- Members of *Microsporidia* have a unique morphology and are still sometimes considered protists.
- They include virulent human pathogens; some infect other vertebrates and insects (**figure 26.17**).

Compare, Hypothesize, Invent

1. What are some physiological or morphological targets to exploit in treating animals or plants suffering from fungal infections? Compare these with the targets you would use when treating infections caused by bacteria.

2. Some fungi can be viewed as coenocytic organisms that exhibit little differentiation. When differentiation does occur, such as in the formation of reproductive structures, it is preceded by septum formation. Why does this occur?

3. Both bacteria and fungi are major environmental decomposers. Obviously competition exists in any given environment, but fungi usually have an advantage. What characteristics specific to fungi provide this advantage?

4. *Neurospora crassa* and several other filamentous fungi demonstrate a circadian rhythm such that conidiation occurs in roughly 21.5-hour cycles, depending on temperature and the light/dark regime. This can be observed in nature as bands of conidia alternating with bands of undifferentiated fungi. Suggest why these fungi evolved a circadian rhythm? If you were to study this phenomenon, what questions would you want to explore?

5. Analysis of sclerotia and conidia formation in the filamentous ascomycete *Aspergillus flavus* suggests that formation of these structures is cell density dependent. Specifically, high cell density cultures yield conidia, whereas low cell density triggers sclerotia formation. Review the structure and function of conidia and sclerotia, and formulate a hypothesis to explain why their development would be regulated in a density-dependent manner. List several specific testable predictions based on your hypothesis.

   Read the original paper: Horowitz-Brown, S., et al. 2008. Morphological transitions governed by density dependence and lipooxygenase activity in *Aspergillus flavus*. *Appl. Environ. Microbiol.* 74:5674.

27

Viruses

Sheep are just one of the targets of Schmallenberg virus.

Deadly New Virus Strikes European Farm Animals

Imagine the shock and dismay of a farmer who sees that his prized ewe just gave birth to a horribly deformed, stillborn lamb. This is exactly what some farmers in several European countries experienced in 2012 as a newly discovered virus spread from farm to farm.

Disease caused by Schmallenberg virus (SBV) was first detected in adult animals in November 2011, when dairy cows in Schmallenberg, Germany, became ill. The cows had a fever, but the most striking symptom was drastically reduced milk production, an obvious loss of income for farmers. Unfortunately, this was just the tip of the iceberg. Within about a month, the first deformed lambs were born, indicating that the virus not only had spread from cows to sheep, but also transmitted vertically from cows and ewes to their offspring. Goats were the next animals to be attacked. During 2012, increasing numbers of adult animals were infected, and more and more of their offspring affected. For farmers, this virus is a two-pronged threat: decreased milk production from the adult animals and loss of progeny from infected females.

The emergence of Schmallenberg virus should not come as a surprise. Over the last several decades, several new viruses or virus strains have been discovered—consider HIV, SARS and MERS coronaviruses, Ebola virus, and new influenza viruses. The ongoing discovery of new viruses is a reminder of the diversity of viruses.

As we outline in chapter 6, viral diversity is in part due to virion structure, especially the nature of the viral genome. A virus's genome dictates important aspects of its life cycle and is in part the basis of a scheme (the Baltimore system) that greatly simplifies the discussion of viral life cycles. The Baltimore system is not the classification system used by virologists to identify and name viruses. That system is developed by the International Committee on Taxonomy of Viruses; however, our discussion is organized by the Baltimore system due to its relative simplicity. For each Baltimore group, we present a brief overview of how viruses in the group multiply, as well as the life cycle of at least one virus.

Another aspect of viral diversity is the variety of strategies viruses use to regulate events in their life cycles and activities of their host cells. Like many viruses, SBV inhibits transcription of host cell genes. This in turn prevents the host from mounting an immune response against the virus. We examine some of these strategies in this chapter as well. In many ways, this discussion is a continuation of topics introduced in chapters 14 and 15, where we consider the regulation of cellular processes, and is continued in chapter 38, where we discuss diseases caused by human viruses. For a virus to multiply (and therefore cause disease), it must "know" its host intimately so that it can take advantage of cellular processes as needed. Indeed, viruses are masters of regulation, both of their own life cycles and of the behaviors of their hosts.

Readiness Check:

Based on what you have learned previously, you should be able to:

✔ Describe common virion structures and morphology (section 6.2)
✔ Summarize the five steps common to the life cycle of most viruses (section 6.3)
✔ Describe the events that occur during the three phases of DNA replication, transcription, and translation (sections 13.3, 13.5, 13.7 15.2, 15.3, and 15.4)
✔ Draw typical bacterial, archaeal, and eukaryotic protein-coding genes (sections 13.4, 13.5, and 15.3)

27.1 Virus Phylogeny Is Difficult to Establish

After reading this section, you should be able to:

- Distinguish between the Baltimore system of grouping viruses and the official taxonomy of viruses proposed by the International Committee on Taxonomy of Viruses
- Determine if a virus has a positive- or negative-strand genome
- Differentiate the Baltimore groups of viruses

In 1971 the International Committee on Taxonomy of Viruses (ICTV) developed a uniform classification system for viruses.

Since then the number of viruses and taxa has continued to expand. In its ninth report, the ICTV describes over 2,000 virus species and places them in 6 orders, 87 families, 19 subfamilies, and 349 genera. (See the table **Characteristics of Some Virus Families** posted to Connect.) The committee considers many viral characteristics but places greatest weight on the following properties to define families: nucleic acid type, presence or absence of an envelope, symmetry of the capsid, and dimensions of the virion and capsid. Virus order names end in *-virales;* virus

Order: *Mononegavirales*
(*mono* = single; *neg* = negative)

A group of related viruses having negative-strand RNA genomes

Contains four families, including:

Family: *Paramyxoviridae*
(paramyxo = Greek *para,* by the side of, and *myxo,* mucus)

Contains two subfamilies, including:

Subfamily: *Paramyxovirinae*

Contains five genera, including:

Genus: *Rubulavirus*
(rubula from *rubula infans,* an old name for mumps)

Genus: *Morbillivirus*
(morbilli from Latin *morbillus,* diminutive form of *morbus* = disease)

Type species: *Mumps virus*

Type species: *Measles virus*

Virus name: Mumps virus (MuV)

Virus name: Measles virus (MeV)

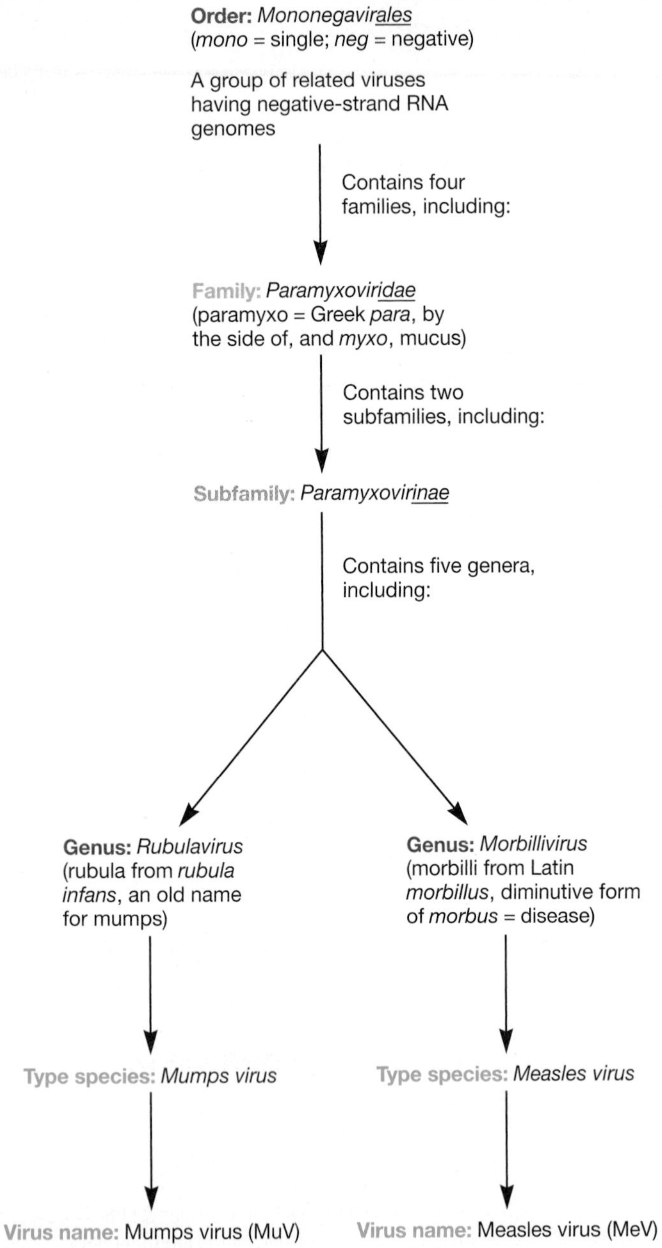

Figure 27.1 Virus Taxonomy. Because of the difficulty in establishing evolutionary relationships, most virus families have not been placed into an order. The taxa names are derived from various aspects of the biology and history of the members of the taxa, including features of their structure, diseases they cause, and locations where they were first identified or recognized. The suffixes used for each type of taxon (e.g., order, family) are underlined.

family names in *-viridae;* subfamily names in *-virinae;* and genus names in *-virus.* An example of this nomenclature scheme is shown in **figure 27.1**. ◄◄ *Virion structure is defined by capsid symmetry and presence or absence of an envelope (section 6.2)*

As we discuss in chapter 19, a goal of taxonomy is to classify organisms based on their evolutionary history. To piece together the phylogenetic relationships of viruses, virologists are increasingly using two major approaches: comparisons of genome sequences and comparisons of protein folds observed in their major capsid proteins. Both approaches are challenging, in part because horizontal gene transfer among unrelated viruses and between viruses and their host cells is clearly evident. Despite this, evidence suggests that retroviruses (section 27.7) and reverse transcribing DNA viruses (section 27.8) share a common evolutionary history. All of these viruses use an enzyme called reverse transcriptase in their life cycles. Evidence also suggests that double-stranded (ds) DNA viruses can be divided into two lineages based on folding of their capsid proteins. These viruses include nucleocytoplasmic large DNA (NCLD) viruses (section 27.2), which have garnered considerable interest in part because they seem to blur the distinctions between cells and viruses.

Although ICTV reports are the official authority on viral taxonomy, many virologists find it useful to group viruses using a scheme devised by Nobel laureate David Baltimore. The Baltimore system complements the ICTV system but focuses on the viral genome and the process used to synthesize viral mRNA. Recall from chapter 6 that all four nucleic acid types can be found in viruses: dsDNA, single-stranded (ss) DNA, dsRNA, and ssRNA. The characterization of the genome of a ssRNA virus is further differentiated by the sense of the ssRNA—that is, whether the genome is identical to or complementary to the mRNA produced by the virus. ssRNA viruses with an RNA genome that is identical in base sequence to that of the mRNA it produces are said to have **plus-strand** or **positive-strand RNA** genomes (**figure 27.2**). Other ssRNA viruses have genomes that are complementary to the mRNA they produce. These viruses are said to have **minus-strand** or **negative-strand RNA.** Baltimore's system organizes viruses into seven groups (**table 27.1**). This system helps virologists (and microbiology students) simplify the vast array of viral life cycles into a relatively small number of basic types; thus we use it here.

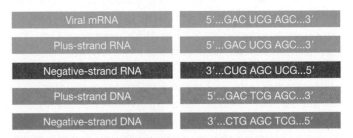

Viral mRNA	5′...GAC UCG AGC...3′
Plus-strand RNA	5′...GAC UCG AGC...3′
Negative-strand RNA	3′...CUG AGC UCG...5′
Plus-strand DNA	5′...GAC TCG AGC...3′
Negative-strand DNA	3′...CTG AGC TCG...5′

Figure 27.2 Plus-Strand and Negative-Strand Viral Genomes. The genomes and replication intermediates of viral genomes can be either plus strand or negative strand. This designation is relative to the sequence of nucleotides in the mRNA of the virus. Plus-strand genomes have the same sequence as the mRNA, either using DNA nucleotides if a DNA genome or RNA nucleotides if an RNA genome. Negative-strand genomes are complementary to the viral mRNA.

Table 27.1	The Baltimore System
Group	**Description**
Double-stranded (ds) DNA viruses	Genome replication: dsDNA → dsDNA mRNA synthesis: dsDNA → mRNA
Single-stranded (ss) DNA viruses	Genome replication: ssDNA → dsDNA → ssDNA mRNA synthesis: ssDNA → dsDNA → mRNA
Double-stranded RNA viruses	Genome replication: dsRNA → ssRNA → dsRNA mRNA synthesis: dsRNA → mRNA
Plus-strand RNA (+RNA) viruses	Genome replication: +RNA → −RNA → +RNA mRNA synthesis: +RNA = mRNA → −RNA → mRNA
Negative-strand RNA (−RNA) viruses	Genome replication: −RNA → +RNA → −RNA mRNA synthesis: −RNA → mRNA
Retroviruses	Genome replication: ssRNA → dsDNA → ssRNA mRNA synthesis: ssRNA → dsDNA → mRNA
Reverse transcribing DNA viruses	Genome replication: dsDNA → ssRNA → dsDNA mRNA synthesis: dsDNA → mRNA

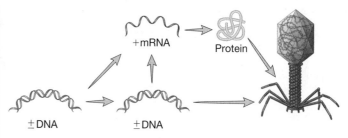

Figure 27.3 Multiplication Strategy of Double-Stranded DNA Viruses. Because the genome of dsDNA viruses is similar to that of the host, genome replication and mRNA synthesis closely resemble that of the host cell and can involve host polymerases, viral polymerases, or both. The DNA genome serves as the template for DNA replication and mRNA synthesis. Translation of the mRNA by the host cell's translation machinery yields viral proteins, which are assembled with the viral DNA to make mature virions. These are eventually released from the host.

Retrieve, Infer, Apply

1. List some characteristics used in classifying viruses. Which seem to be the most important?
2. Consider these terms: *Equine torovirus, Nidovirales, Torovirinae, Equine torovirus (EToV), Toroviridae,* and *Torovirus.* Construct a figure similar to figure 27.1 using these terms.

27.2 Double-Stranded DNA Viruses Infect All Cell Types

After reading this section, you should be able to:

- Describe in general terms the strategy used by double-stranded (ds) DNA viruses to synthesize their nucleic acids and proteins
- Describe in general terms how bacteriophage lambda regulates the switch between lytic and lysogenic cycles
- Choose one specific bacterial, archaeal, and eukaryal dsDNA virus and outline the major events in their life cycles, noting, when possible, the specific mechanisms used to accomplish each step

Perhaps the largest group of known viruses is the double-stranded (ds) DNA viruses; most bacteriophages have dsDNA genomes, as do several insect viruses and a number of important vertebrate viruses, including herpesviruses and poxviruses. The pattern of multiplication for dsDNA viruses is shown in **figure 27.3.** The synthesis of DNA and RNA is similar to what occurs in cellular organisms; therefore some dsDNA viruses can rely entirely on their host's DNA and RNA polymerases.

Bacteriophage T4: A Virulent Bacteriophage

The life cycle of T4 bacteriophage (family *Myoviridae,* species *Enterobacteria phage T4*) serves as our example of a virulent dsDNA

phage. Virulent (lytic) bacteriophages are capable only of the lytic cycle. That is, their infection of a host always ends with cell lysis. As with most viruses (except plant viruses), the first step of viral infection is attachment (adsorption) to the host cell surface. T4 attachment begins when a tail fiber contacts either the lipopolysaccharide or certain proteins in the outer membrane of its *Escherichia coli* host (**figure 27.4a**). As more tail fibers make contact, the baseplate settles down on the surface (figure 27.4b). After the baseplate is seated firmly on the cell surface, the baseplate and sheath change shape, and the tail sheath reorganizes so that it shortens from a cylinder 24 rings long to only 12 rings (figure 27.4c,d). As the sheath becomes shorter and wider, the central tube located within the sheath is pushed through the bacterial cell wall. The baseplate contains the protein gp5, which has lysozyme activity. Lysozyme is an enzyme that breaks the bonds linking the sugars of peptidoglycan together (*see figure 33.4*). Thus this protein aids in the penetration of the tube through the peptidoglycan layer. Finally, the linear DNA is extruded from the head, probably through the central tube, and into the host cell (figure 27.4e,f). The tube may interact with the plasma membrane to form a pore through which DNA passes.

Within 2 minutes after entry of T4 DNA into an *E. coli* cell, the *E. coli* RNA polymerase starts synthesizing T4 mRNA (**figure 27.5**). This mRNA is called early mRNA because it is made before viral DNA is made. Within 5 minutes, viral DNA synthesis commences, catalyzed by a virus-encoded **DNA-dependent DNA polymerase.** DNA replication is initiated from several origins of replication and proceeds bidirectionally from each. Viral DNA replication is followed by the synthesis of late mRNAs, which are important in later stages of the infection.

As with most viruses, expression of T4 genes is carefully timed, and the transcripts and protein products of the genes are named to reflect the time of appearance. Thus depending on the virus, terms such as immediate-early, middle, and late may be used. T4 controls the expression of its genes by regulating the activity of the *E. coli* RNA polymerase. Initially T4 genes are transcribed by the host RNA polymerase and the housekeeping sigma factor σ[70] (*see table 14.3*). After a short interval, a viral enzyme

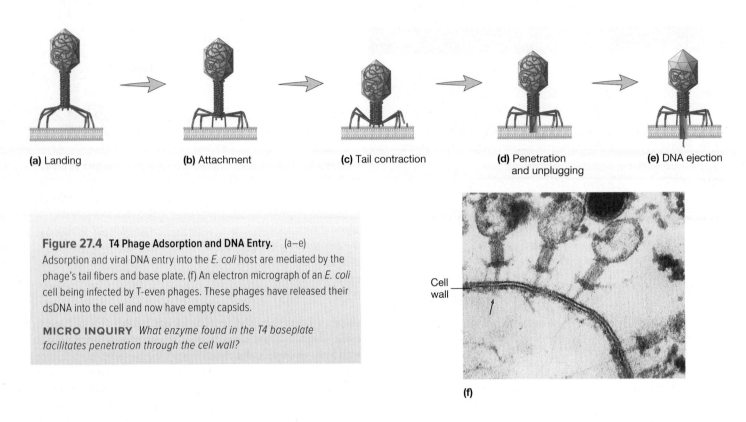

(a) Landing

(b) Attachment

(c) Tail contraction

(d) Penetration and unplugging

(e) DNA ejection

Figure 27.4 **T4 Phage Adsorption and DNA Entry.** (a–e) Adsorption and viral DNA entry into the *E. coli* host are mediated by the phage's tail fibers and base plate. (f) An electron micrograph of an *E. coli* cell being infected by T-even phages. These phages have released their dsDNA into the cell and now have empty capsids.

MICRO INQUIRY *What enzyme found in the T4 baseplate facilitates penetration through the cell wall?*

Cell wall

(f)

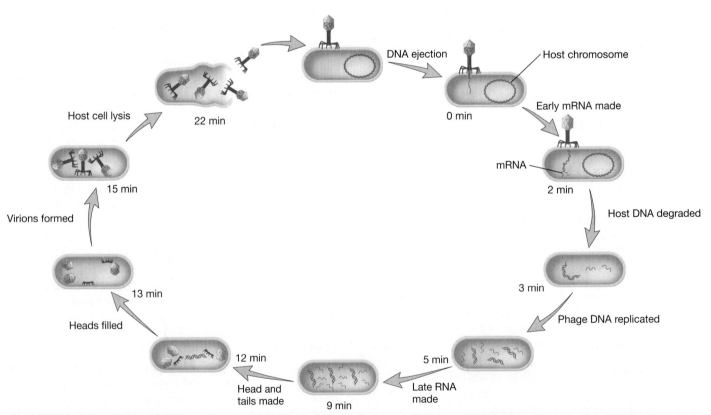

DNA ejection

Host chromosome

Host cell lysis

22 min

0 min

Early mRNA made

15 min

mRNA

Virions formed

2 min

Host DNA degraded

13 min

3 min

Heads filled

Phage DNA replicated

12 min

5 min

Head and tails made

Late RNA made

9 min

Figure 27.5 **The Life Cycle of Bacteriophage T4.** A diagram depicting the life cycle with the minutes after DNA ejection given for each stage.

MICRO INQUIRY *Why do you think T4 evolved to initiate DNA replication from multiple origins, rather than from a single origin of replication as seen in its host cell?*

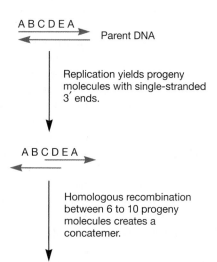

Figure 27.6 5-Hydroxymethylcytosine (HMC). In T4 DNA, the HMC often has glucose attached to its hydroxyl.

MICRO INQUIRY *What function does HMC glycosylation serve?*

catalyzes addition of the chemical group ADP-ribose to one of the α-subunits of RNA polymerase (*see figure 13.23*). This modification helps inhibit the transcription of host genes and promotes viral gene expression. Later the second α-subunit receives an ADP-ribosyl group. This halts transcription of some early T4 genes but not before the product of one early gene (*motA*) stimulates transcription of somewhat later genes. One of these later genes encodes the sigma factor gp55. This viral sigma factor helps the host cell's RNA polymerase core enzyme bind to viral late promoters and transcribe the late genes, around 10 to 12 minutes after infection. ◄◄ *Bacteria combine several regulatory mechanisms to control complex cellular processes (section 14.5)*

The tight regulation of T4 gene expression is aided by the organization of the T4 genome, in which genes with related functions—such as the genes for phage head or tail fiber construction—are usually clustered together. Early and late genes also are clustered separately on the genome; they are even transcribed in different directions. A considerable portion of the T4 genome encodes products needed for its replication, including all the protein subunits of its replisome and enzymes needed to prepare for DNA synthesis. Some of these enzymes synthesize an important component of T4 DNA, hydroxymethylcytosine (HMC) (**figure 27.6**). HMC is a modified nucleotide that replaces cytosine in T4 DNA. Once HMC is synthesized, replication ensues by a mechanism similar to that seen in bacteria. After T4 DNA has been synthesized, it is modified by the addition of glucose (a process called glycosylation) to the HMC residues. Glycosylated HMC residues protect T4 DNA from attack by *E. coli* endonucleases called **restriction enzymes,** which would otherwise cleave the viral DNA and destroy it. This bacterial defense mechanism is called **restriction.** Other phages also chemically modify their DNA to protect against host restriction. ◄◄ *Restriction enzymes (section 17.1)*

The linear dsDNA genome of T4 is generated in an interesting manner involving the formation of long DNA molecules called **concatemers,** which are composed of several genome units linked together in the same orientation (**figure 27.7**). How does this occur? As we discuss in chapters 13 and 15, the ends of linear DNA molecules cannot be replicated without mechanisms such as the enzyme telomerase. T4 and its *E. coli* host do not have telomerase enzyme activity. Therefore each progeny viral DNA molecule has single-stranded 3′ ends. These ends participate in homologous recombination with double-stranded regions of other progeny DNA molecules, generating concatemers. During

Figure 27.7 The Terminally Redundant, Circularly Permuted Genome of T4. The formation of concatemers during replication of the T4 genome is an important step in phage multiplication. During assembly of virions, the phage head is filled with DNA cleaved from the concatemer. Because slightly more than one set of T4 genes is packaged in each head, each virion contains a different DNA fragment (note that the ends of the fragments are different). However, if each genome were circularized, the sequence of genes would be the same.

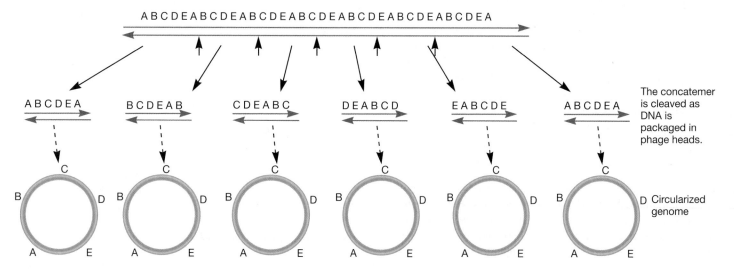

assembly, concatemers are cleaved such that the genome packaged in the capsid is slightly longer than the T4 gene set. Thus each progeny virus has a genome unit that begins with a different gene and ends with the same set of genes. If each genome of the progeny viruses were circularized, the sequence of genes in each virion would be the same. Therefore the T4 genome is said to be terminally redundant and circularly permuted, and the genetic map of T4 is drawn as a circular molecule.

The formation of new T4 phage particles is an exceptionally complex self-assembly process that involves viral proteins and some host cell factors (*see figure 6.13*). Late mRNA transcription begins about 9 minutes after entry of T4 DNA into *E. coli* (figure 27.5). Late mRNA products include phage structural proteins, proteins that help with phage assembly but do not become part of the virion, and proteins involved in cell lysis and phage release. These proteins are used in four fairly independent subassembly lines that ultimately converge to generate a mature T4 virion.

A critical step in T4 virion construction is filling the head portion of the virion with the T4 genome. This is no simple matter—the dsDNA genome is somewhat rigid and has many negatively charged moieties. Therefore the dsDNA must essentially be crammed into the developing head. This is accomplished by a complex of proteins sometimes called the "packasome." The T4 packasome has more power than an automobile engine.

The packasome includes a protein called terminase, which has two functions: to cut the concatemers formed during T4 genome replication and to push the DNA into the T4 head. For each head filled, it uses these functions in the order "cut, push, cut." The first cut generates a double-stranded end on a concatemer. This end is then threaded into the portal, as terminase pushes the DNA into the head, using energy supplied by ATP hydrolysis, until the phage head is filled with DNA—a DNA molecule roughly 3% longer than the length of one set of T4 genes. Terminase then makes its second cut, and the packaging process for that head is complete. Terminase then leaves the head, and several other viral proteins bind the head at the portal through which the DNA entered. This seals the head and prepares it for addition of the tail and tail fibers.

Finally, virions are released so that they can infect new cells and begin the cycle anew. T4 lyses *E. coli* when about 150 virus particles have accumulated in the host cell. T4 encodes two proteins to accomplish this. The first, called holin, creates holes in the *E. coli* plasma membrane. The second, an endolysin called T4 lysozyme, degrades peptidoglycan in the host's cell wall. Thus the activity of holin enables T4 lysozyme to move from the cytoplasm to the peptidoglycan so that both the plasma membrane and the cell wall are destroyed.

Bacteriophage Lambda: A Temperate Bacteriophage

Unlike bacteriophage T4, temperate bacteriophages, such as phage lambda (family *Siphoviridae*, species *Enterobacteria phage lambda*), can enter either the lytic or lysogenic cycle upon infecting a host cell. If a temperate virus enters the lysogenic cycle, its dsDNA genome often is integrated into the host's

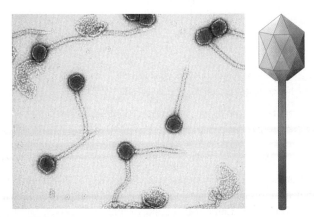

Figure 27.8 Bacteriophage Lambda (λ).   The icosahedral head illustrated in the drawing is not obvious in the photomicrograph at this resolution.

chromosome, where it resides as a prophage until conditions for induction occur (*see figure 6.16*). Upon induction, the viral genome is excised from the host genome and the lytic cycle is initiated. |◄◄ *Lytic and lysogenic infections are common for bacterial and archaeal cells (section 6.4)*

How does a temperate phage "decide" which cycle to follow? The process by which phages make this decision is best illustrated by bacteriophage lambda (λ), which infects *E. coli*. As shown in **figure 27.8**, λ has an icosahedral head 55 nm in diameter and a noncontractile tail with a thin tail fiber at its end. Its DNA genome is a linear molecule with cohesive ends—single-stranded stretches, 12 nucleotides long, that are complementary to each other and can base-pair.

Like most bacteriophages, λ attaches to its host and then releases its genome into the cytoplasm, leaving the capsid outside. Once inside the cell, the linear genome is circularized when the two cohesive ends base-pair with each other; the breaks in the strands are sealed by the host cell's DNA ligase (**figure 27.9**). The λ genome has been carefully mapped, and over 40 genes have been identified (**figure 27.10**). Most genes are clustered according to their function, with separate groups involved in head synthesis, tail synthesis, lysogeny, DNA replication, and cell lysis. This organization is important because once the genome is circularized, a cascade of regulatory events occurs that determine if the phage pursues a lytic cycle or establishes

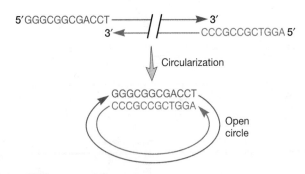

Figure 27.9 Lambda Phage DNA.   A diagram of λ phage DNA showing its 12-base, single-stranded cohesive ends.

lysogeny. Regulation of appropriate genes is facilitated by clustering and coordinated transcription from the same promoters.

The cascade of events leading to either lysogeny or the lytic cycle serves as a model for complex regulatory processes. It involves the action of several regulatory proteins that function as transcriptional repressors or activators or both, proteins that regulate transcription termination, and antisense RNA molecules (**figure 27.11**). The protein cII is an activator protein that plays a pivotal role in determining if λ will establish lysogeny or follow a lytic pathway. If the cII protein reaches high enough levels early in the infection, lysogeny will occur; if it does not reach a critical level, the lytic cycle will occur.

Transcription of the λ genome is catalyzed by the host cell's **DNA-dependent RNA polymerase** (an RNA polymerase that uses DNA as the template for RNA synthesis), and the cII protein is synthesized relatively early in the infection. If the cII protein levels are high enough, it will increase transcription of the *int* gene, which encodes the enzyme integrase. **Integrase** catalyzes integration of the λ genome into the host cell's chromosome, thus establishing lysogeny. The cII protein also increases transcription of the *cI* gene. This gene encodes a regulatory protein that is often called the λ repressor because it represses the transcription of all genes (except its own). This repression maintains the lysogenic state. As just noted, the cI protein (λ repressor) allows transcription of its own gene. This is because cI functions as an activator protein when it binds to the P_{RM} promoter from which the *cI* gene can be transcribed (**table 27.2**).

Integration of the λ genome into the host chromosome takes place at a site in the host chromosome called the attachment site (*att*). A corresponding site is present on the phage genome, and the λ integrase catalyzes site-specific recombination between the two sites. The bacterial site is located between the galactose (*gal*) and biotin (*bio*) operons, and as a result of integration, the circular λ genome becomes a linear stretch of DNA located between these two host operons (*see figure 16.31*). The prophage can remain integrated indefinitely, being replicated as the bacterial genome is replicated. ◄◄ *Molecular recombination (section 16.4)*

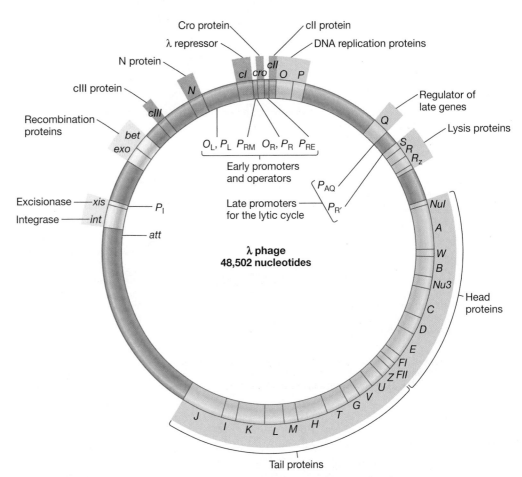

Figure 27.10 The Genome of Phage λ.   The genes are color-coded according to function. Those genes shaded in orange encode regulatory proteins that determine if the lytic or lysogenic cycle will be followed. Genes involved in establishing lysogeny are shaded in yellow. Those required for the lytic cycle are shaded in tan. Important promoters and operators are also noted.

Because the cII protein is made early in the infection, it might seem that it would accumulate quickly and ensure that lysogeny occurs. However, cII is degraded by a host enzyme (HflB) unless it is protected by a viral protein called cIII. The cIII protein is synthesized at the same time as cII, and as long as its levels remain high enough, cII will be protected. However, if cII is not protected sufficiently from the host degradative enzyme, the level of a protein called Cro (product of the *cro* gene) will increase. Cro protein is both a repressor protein and an activator protein. It inhibits transcription of the *cIII* and *cI* genes, further decreasing the amount of cII and λ repressor. However, it increases its own synthesis as well as the synthesis of another regulatory protein called Q. When Q protein accumulates at a high enough level, it activates transcription of genes required for the lytic cycle. When that occurs, the infection process is committed to the lytic cycle. Ultimately the host is lysed and new virions are released.

We have now considered the regulatory processes that dictate whether lysogeny is established or the lytic cycle is pursued. However, how does induction reverse lysogeny? Induction usually occurs in response to environmental factors such as

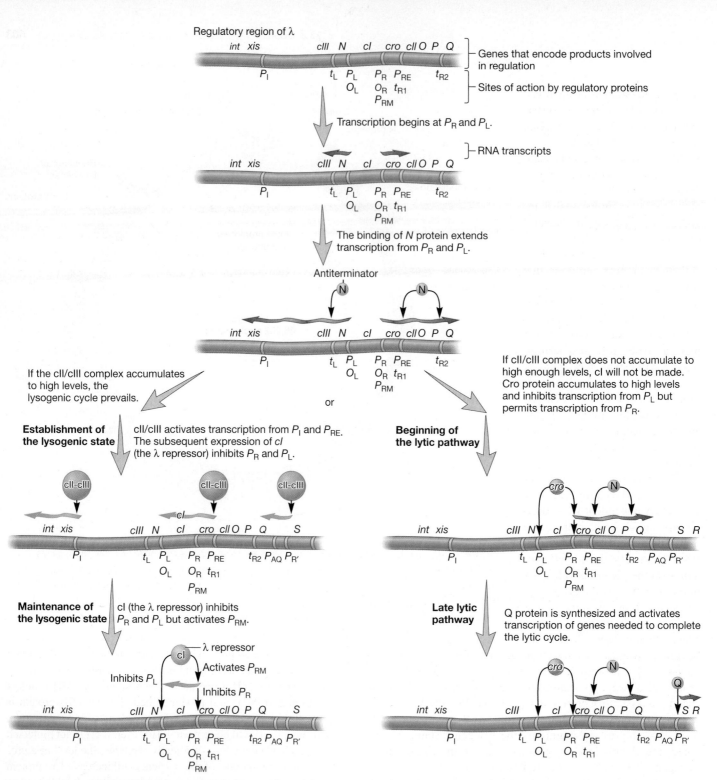

Figure 27.11 The Decision-Making Process for Establishing Lysogeny or the Lytic Pathway. The initial transcripts are synthesized by the host RNA polymerase. These encode the N protein and the Cro protein. The N protein is an antiterminator that allows transcription to proceed past the terminator sequences t_L, t_{R1}, and t_{R2}. This allows transcription of other regulatory genes, as well as the *xis* and *int* genes. The latter genes encode the enzymes excisionase and integrase, respectively. The left side of the figure illustrates what occurs if lysogeny is established. The right side of the figure shows the lytic pathway.

ultraviolet light or chemical mutagens that damage DNA. This damage alters the activity of the host cell's **RecA** protein. As we describe in chapter 16, RecA plays important roles in recombination and DNA repair. When activated by DNA damage, RecA interacts with λ repressor, causing the repressor to cleave itself.

As more and more λ repressor proteins destroy themselves, transcription of the *cI* gene is decreased, further lowering the amount of λ repressor in the cell. Eventually the level becomes so low that transcription of the *xis*, *int*, and *cro* genes begins. The *xis* gene encodes the protein excisionase. It binds integrase, causing

Table 27.2	Functions of Lambda Promoters and Operators	
Promoter or Operator	**Name Derivation**	**Function**
P_L	**P**romoter **L**eftward	Promoter for transcription of *N*, *cIII*, *xis*, and *int* genes; important in establishing lysogeny
O_L	**O**perator **L**eftward	Binding site for lambda repressor and Cro protein; binding by lambda repressor maintains lysogenic state; binding by Cro protein prevents establishment and maintenance of lysogeny
P_R	**P**romoter **R**ightward	Promoter for transcription of *cro*, *cII*, *O*, *P*, and *Q* genes; Cro, O, P, and Q proteins are needed for lytic cycle; CII protein helps establish lysogeny
O_R	**O**perator **R**ightward	Binding site for lambda repressor and Cro protein; binding by lambda repressor maintains lysogenic state; binding by Cro allows transcription to occur
P_{RE}	**P**romoter for Lambda **R**epressor **E**stablishment	Promoter for *cI* gene (lambda repressor gene); recognized by CII protein, a transcriptional activator; important in establishing lysogeny
P_I	**P**romoter for **I**ntegrase Gene	Transcription from P_I generates mRNA for integrase protein but not excisionase; recognized by the transcriptional activator CII; important for establishing lysogeny
P_{AQ}	**P**romoter for **A**nti-**Q** mRNA	Transcription from P_{AQ} generates an antisense RNA that binds *Q* mRNA, preventing its translation; recognized by the transcriptional activator CII; important for establishing lysogeny
P_{RM}	**P**romoter for **R**epressor **M**aintenance	Promoter for transcription of lambda repressor gene (*cI*); activated by lambda repressor; important in maintaining lysogeny
$P_{R'}$	**P**romoter **R**ightward'	Promoter for transcription of viral structural genes; activated by Q protein; important in lytic cycle

it to reverse the integration process, and the prophage is freed from the host chromosome. As λ repressor levels continue to decline, the Cro protein levels increase. Eventually synthesis of λ repressor is completely blocked, protein Q levels become high, and the lytic cycle proceeds to completion.

Our attention has been on λ phage, but there are many other temperate phages. Most, like λ, exist as integrated prophages in the lysogen. However, not all temperate phages integrate into the host chromosome at specific sites. Bacteriophage Mu integrates randomly into the genome. It then expresses a repressor protein that inhibits lytic growth. Furthermore, integration is not an absolute requirement for lysogeny. The *E. coli* phage P1 is similar to λ in that it circularizes after infection and begins to manufacture repressor. However, it remains as an independent circular DNA molecule in the lysogen and is replicated at the same time as the host chromosome. When *E. coli* divides, P1 DNA is apportioned between the daughter cells so that all lysogens contain one or two copies of the phage genome.

Retrieve, Infer, Apply

1. Explain why the T4 genome is circularly permuted.
2. Precisely how, in molecular terms, is a bacterial cell made lysogenic by a temperate phage such as λ?
3. How is a prophage induced to become active again?
4. Describe the roles of cII, CIII, λ repressor (CI), Cro, Q, RecA, integrase, and excisionase in lysogeny and induction.
5. How do the temperate phages Mu and P1 differ from λ?

Archaeal Viruses

Our understanding of archaeal viruses lags significantly behind that of bacterial viruses and animal viruses. However, as more archaeal viruses are isolated and studied, virologists are beginning to develop a clearer picture of the nature of these interesting biological entities. Nearly all known archaeal viruses have dsDNA genomes; the few exceptions have ssDNA genomes. However, metagenomic evidence suggests that RNA archaeal viruses exist; they have not yet been isolated. Archaeal virus genomes can be either linear or circular. Some archaeal viruses exhibit morphologies similar to that of other viruses. However, many others have unusual morphologies, such as bottle-shaped, droplet-shaped, and spindle-shaped (**figure 27.12**; *see also Microbial Diversity & Ecology 6.1 box figure*.) These unusually shaped viruses have defined several new virus families. Unlike most bacteriophages, many archaeal viruses are enveloped.

With a few exceptions, relatively little is known about archaeal virus life cycles. However, they do exhibit some interesting differences from those of bacteriophages. Thus far, most archaeal viruses studied establish chronic infections, whereby viruses are continuously produced and extruded from the host cell. This type of infection is similar to that of fd bacteriophage and other related ssDNA phages in the family *Inoviridae* (section 27.3). Some archaeal viruses have been shown to establish lysogeny by integrating their genomes into the host cell's chromosome.

Although relatively few archaeal viruses always cause lytic infections, one such virulent archaeal virus is of particular

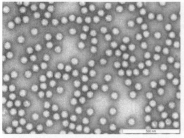

(a) STIV virions

(b) Virions of *Sulfolobus shibatae* virus 2 (SSV-2-Ss)

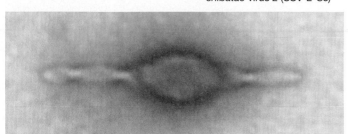

(c) *Acidianus* two-tailed virus virion

Figure 27.12 **Virion Morphology of Some Archaeal Viruses.**
(a) *Sulfolobus* turreted icosahedral virus (STIV) has a virion morphology similar to some bacterial and animal viruses. Scale bar = 500 nm. Transmission electron microscopy (TEM). (b) TEM image of *Sulfolobus shibatae* virus 2 (SSV-2-Ss) virions. These droplet-shaped virions have a short tail at one end. Virus particles often attach to membranes or each other by their tails. (c) TEM image of an *Acidianus* two-tailed virus virion.

interest. The *Sulfolobus* turreted icosahedral virus (STIV) is released from its host *Sulfolobus* spp. in an unusual way. After the genome enters the host cell (by an unknown mechanism), transcription of the viral genome begins. Some of the viral gene products regulate host cell genes, thus enabling the virus to control the replication process using the host's DNA replication, transcription, and translation machinery. Eventually the assembly phase begins. STIV virions have an icosahedral capsid that surrounds an internal lipid bilayer; the lipid bilayer encloses the virus's genome. Electron cryotomography studies suggest that during assembly, the capsid and internal membrane are assembled together, leading to the formation of an empty virion (procapsid) devoid of the viral genome. Once the procapsids (capsid plus membrane) are complete, the DNA is packaged by an unknown mechanism. Virion release results from the formation of pyramid-like structures on the surface of the host cell (**figure 27.13**). These virus-associated pyramids, as they are called, open up, much like the petals of a

Virus-associated pyramids

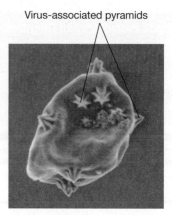

Figure 27.13 **Virus-Associated Pyramids.** Scanning electron micrograph of a *Sulfolobus solfataricus* cell infected with *Sulfolobus* turreted icosahedral virus (STIV). STIV virions are released from the cell by way of pyramid-like structures that form on the surface of the infected cell. The pyramids are seven-sided, and each face of the pyramid "peels back," providing an opening through which the virions escape.

flower bud open. This allows escape of the progeny virions, leaving the empty cell ghosts behind.

The CRISPR/Cas System: Bacteria and Archaea Fight Back

In our discussion of T4 bacteriophage, we introduce a defense mechanism—restriction—used by bacteria to protect themselves against attack from viruses. Many bacteria and archaea have another defense mechanism: the CRISPR/Cas system. This intriguing defense system has been likened to the adaptive immunity used by animals to protect themselves from microbial attack. The similarity is based on the observation that the CRISPR/Cas system uses parts of the invading virus's genome to "remember" an attack and prepare for a possible future attack, just as the adaptive immune system uses parts of an invading microbe to remember and prepare for the next invasion.

CRISPR/Cas was discovered when analyses of numerous bacterial and archaeal genomes identified sets of repeated nucleotide sequences separated by short spacers. These sets form a CRISPR system, CRISPR being short for *c*lustered, *r*egularly *i*nterspaced *s*hort *p*alindromic *r*epeats. The sequence of base pairs in the repeats are identical (or nearly so) within each CRISPR system. The spacers, however, differ considerably in sequence. Surprisingly, when the spacer sequences were used to probe nucleic acid databases, it was discovered that the spacers exhibit significant similarity to a variety of bacterial and archaeal virus genomes. CRISPR regions were later shown to be associated with a set of genes encoding proteins called Cas proteins (short for *CRISPR-a*ssociated *s*equences). Thus the new designation: a CRISPR/Cas system (**figure 27.14**).

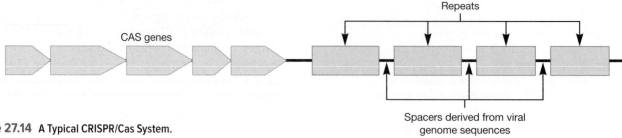

Repeats

CAS genes

Spacers derived from viral genome sequences

Figure 27.14 **A Typical CRISPR/Cas System.**

The surprising similarity between viral genes and CRISPR/Cas spacers led scientists to question how the spacers came to exist and what their role was in the host cell. The function of CRISPR/Cas systems is currently divided into three stages: acquisition, CRISPR RNA biogenesis, and interference. The acquisition stage occurs when a cell is infected by a virus. If the cell survives (e.g., because the virus is defective and can't successfully complete its life cycle), it adds portions of the viral genome to its CRISPR region. The addition is made at the end of the region closest to the CAS genes, and each new addition consists of a new repeat and a spacer consisting of DNA from the infecting virus. Thus the CRISPR sequences are akin to the growth rings on trees. Just as the history of the yearly growth of a tree can be ascertained by examining its growth rings, so too can the history of viral infections be determined. In any given bacterial or archaeal strain, a CRISPR sequence inserted during the oldest infection is located farthest from the Cas genes and the most recent is closest to the CAS genes.

During the CRISPR RNA (crRNA) biogenesis stage the CRISPR region is transcribed to yield a large RNA containing all repeats and spacers. Cas proteins associate with this full-length RNA and process it into mature crRNAs. Each of these consists of one repeat and one spacer. Cas proteins remain associated with the crRNAs, and during a subsequent infection, the Cas-crRNAs associate with either viral DNA or mRNA, leading to destruction of the molecule (interference). This prevents virus multiplication, and infection is thwarted. The interference stage bears significant similarity to silencing RNAs (siRNAs), as we describe in chapter 15.

Our focus has been on the use of CRISP/Cas systems by bacteria and archaea to defend against viral attacks. However, there is considerably more to these systems. It is now clear that bacteria and archaea use CRISPR/Cas for the more general process of protecting their genomes from foreign nucleic acids that might enter via conjugation or natural transformation. Furthermore, evidence suggests that crRNAs may regulate gene expression of host genes. Finally, viruses have been shown to evolve to evade their hosts' CRISPR/Cas defense, often by having a mutation in the target sequence. The host cell, in turn, responds by obtaining new spacers specific for that virus. It has been suggested that this back-and-forth interaction between virus and host plays an important role in the evolution of both. ◄◄ *Bacterial conjugation requires cell-cell contact (section 16.6); Bacterial transformation is the uptake of free DNA from the environment (section 16.7)*

Herpesviruses

So far our discussion of dsDNA viruses has focused on bacteriophages and archaeal viruses. We now turn our attention to some important dsDNA viruses of eukaryotes, beginning with herpesviruses. As you will see, the life cycle of these viruses shares many features with that of bacteriophages such as T4. However, important distinctions in the life cycles exist because of the nature of their respective host cells.

Herpesviruses are members of the order *Herpesvirales*. This order includes herpes simplex virus type 1 (HSV-1; also called human herpesvirus 1) and herpes simplex virus type 2 (HSV-2; also called human herpesvirus 2), which cause cold sores and genital herpes, respectively (*see figures 38.13 and 38.16*); varicella-zoster virus (also called human herpesvirus 3), which causes chickenpox and shingles (*see figures 38.1 and 38.2*); cytomegaloviruses (species *Human herpesvirus 5*); Epstein-Barr virus (also called human herpesvirus 4), which causes infectious mononucleosis and has been implicated in some human cancers; and HHV 8 (species *Human herpesvirus 8*), the cause of Kaposi's sarcoma in AIDS patients. ►►I *Cold sores (section 38.3); Genital herpes (section 38.3); Chickenpox (varicella) and shingles (herpes zoster) (section 38.1); Cytomegalovirus inclusion disease (section 38.3); Mononucleosis (infectious) (section 38.3); Acquired immune deficiency syndrome (AIDS) (section 38.3)*

Herpesvirus virions are 125 to 200 nm in diameter, somewhat pleomorphic, and enveloped (**figure 27.15**). The envelope contains distinct viral projections (called spikes) that are regularly dispersed over the surface. The envelope surrounds a layer of proteins called the tegument (Latin, *tegumentum,* to cover), which in turn surrounds the nucleocapsid. Herpesvirus genomes are linear, about 125 to 295 kilobase pairs (kb) long, and encode 70 to over 200 proteins. When herpesviruses target cells of vertebrate hosts, some bind to epithelial cells, others to neurons. Host cell selection is mediated by the binding of envelope spikes to specific host cell surface proteins termed receptors.

Herpesviruses cause both productive infections and latent infections. In a productive infection, the virus multiplies explosively; between 50,000 and 200,000 new virions are produced from each infected cell. As the virus multiplies, the host cell's metabolism is inhibited and the host's DNA is degraded. As a result, the cell dies. The first exposure to a herpesvirus usually causes this type of infection. Some of the cells infected in the initial infection develop a latent infection. During the latent infection, virions cannot be detected. However, the virus can be reactivated in the host cells, leading to a productive infection. The viral genome remains in the host cell after reactivation; thus once infected, the host can experience repeated productive infections.

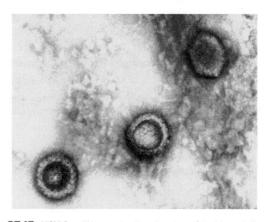

Figure 27.15 HSV-2. Herpes simplex virus type 2 inside an infected cell.

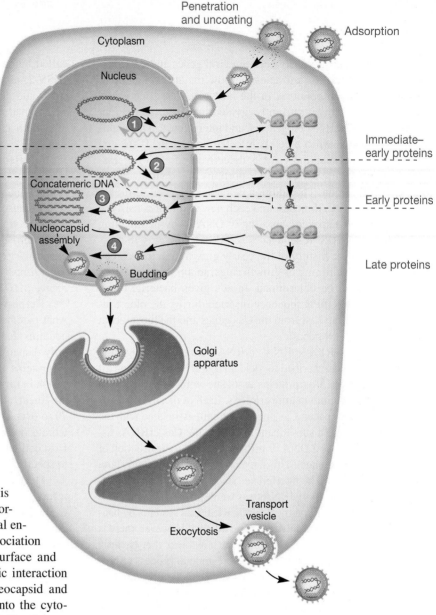

① Circularization of genome and transcription of immediate-early genes

② Immediate-early proteins (products of immediate-early genes) stimulate transcription of early genes.

③ Early proteins (products of early genes) function in DNA replication, yielding concatemeric DNA. Late genes are transcribed.

④ Late proteins (products of late genes) participate in virion assembly.

Figure 27.16 Life Cycle for Herpes Simplex Virus Type 1.

MICRO INQUIRY *How is the envelope of this virus formed? How does this differ from most enveloped viruses?*

A productive infection caused by HSV-1 is shown in **figure 27.16**. It begins with receptor-mediated attachment followed by fusion of the viral envelope with the host cell membrane. The initial association is between proteoglycans of the epithelial cell surface and viral glycoproteins. This is followed by a specific interaction with one of several cellular receptors. The nucleocapsid and some associated tegument proteins are released into the cytoplasm and transported by the host cell's microtubule/dyenin system to the nucleus. The linear dsDNA and some of the tegument proteins then enter the nucleus by way of a nuclear pore complex. Immediately upon entry of the viral DNA into the host nucleus, the DNA circularizes and is transcribed by host DNA-dependent RNA polymerase to form mRNAs, which are translated to yield several immediate early and early proteins. These are mostly regulatory proteins and enzymes required for replication of viral DNA (figure 27.16, steps 1 and 2). Replication of the genome with a virus-specific DNA-dependent DNA polymerase begins in the cell nucleus within 4 hours after infection (figure 27.16, step 3). Viral structural proteins are the products of the late genes. These proteins enter the nucleus, where they are assembled into nucleocapsids. Capsid formation and genome packaging are similar to that seen for bacteriophage T4.

The acquisition of the tegument and envelope, and exit from the host cell are interesting processes that require several steps to complete. Once the nucleocapsid is assembled, it makes contact with the inner membrane of the nucleus. It then buds into the space between the two nuclear membranes. This generates an envelope that is called the primary viral envelope. The primary envelope is lost when it fuses with the outer nuclear membrane, releasing the herpesvirus nucleocapsid into the cytoplasm. At this point in the life cycle, some tegument proteins associate with the nucleocapsid. Additional tegument proteins are added when the developing virion is enveloped by membranes of either the Golgi apparatus or endosomes. This yields a mature enveloped virion that is transported to the plasma membrane in a membrane vesicle. The membrane vesicle fuses with the plasma membrane and releases the mature virion from the host cell. Thus unlike many other enveloped viruses, the source of the HSV-1 envelope is an organelle membrane, rather than the plasma membrane.

Just as the life cycle of λ has been of interest as a model of regulatory processes, so too has the establishment of latency by herpesviruses. It appears that both viral and host proteins play a role in establishing a latent HSV-1 infection. Several studies have demonstrated that in epithelial cells, where productive infections occur, an HSV-1 protein called VP16 and a host protein called simply host cell factor (HCF) enter the nucleus with the viral genome. VP16 and HCF are needed for full expression of the

MICROBIAL DIVERSITY & ECOLOGY

27.1 What Is a Virus?

The discovery of mimivirus, a giant virus that infects the amoeba *Acanthamoeba polyphaga* and other megaviruses, has renewed a debate among biologists, chemists, virologists, and others: Are viruses alive? It has also introduced a new level of complexity to this debate: What makes a virus a virus? The confusion and debate stem from several sources. The first is that biologists have difficulty defining life. Instead, biologists have a list of attributes associated with life. Viruses share some of those attributes: they multiply, they evolve, they alter their gene expression in response to environmental stimuli (e.g., induction). However, there are others they lack: metabolic capabilities, especially energy-conserving processes, and an ability to carry out any of the processes associated with life outside a host cell. Another shift in thinking about viruses occurred when Wendell Stanley (1904–1971) and colleagues (all chemists) crystallized tobacco mosaic virus virions. The fact that they could be crystallized caused many to begin thinking of viruses simply as complexes of chemicals, rather than as biological entities.

The large size and coding capacity of megaviruses and the surprising discovery that they need not rely totally on their hosts for translation have fueled the debate. Indeed, the genomes of megaviruses are extensive enough to support the replication of their own viruses. In other words, some

megaviruses are parasitized by viruses called virophages. Virophages use megavirus-encoded enzymes to complete their multiplication cycles. Surprisingly, one virophage genome is integrated into its megavirus "host."

In 2008 Didier Raoult and Patrick Forterre formalized their thoughts about what makes a virus a virus in a paper they published in *Nature Reviews Microbiology*. They proposed, first of all, that viruses are living organisms. They then argued that organisms should be divided into two groups: those that synthesize translational machinery (i.e., cellular organisms) and those that synthesize capsid (i.e., viruses). As you might imagine, this has caused considerable controversy.

These giant viruses have generated other controversies. Some scientists suggest that megaviruses be included in the tree of life as the fourth domain. Others argue that megaviruses may have given rise to the nucleus during the evolution of eukaryotic cells. There is also much debate about when and how megaviruses arose. Don't be surprised if more controversies arise as more is learned about these viruses.

Sources: LaScola, B., et al. 2003. A giant virus in amoebae. Science 299:2033; Raoult, D., and Forterre, P. 2008. Redefining viruses: Lessons from mimivirus. Nature Rev. Microbiol. 6:315–19; LaScola, B., et al. 2008. The virophage as a unique parasite of the giant mimivirus. Nature 455:100–4; Forterre, P., et al. 2014. Cellular domains and viral lineages. Trends Microbiol. 22:554–8.

virus's immediate early genes. However, VP16 and HCF do not enter the nucleus of infected neurons, and the expression of many immediate early genes is decreased. In addition, small noncoding RNAs produced by the virus further decrease expression of immediate early genes needed for a lytic infection. The inhibition of early gene expression helps establish latency. Interestingly, the production of VP16 in virus-infected neurons is an early event in the reactivation of the virus.

Nucleocytoplasmic Large DNA Viruses (Megaviruses)

Nucleocytoplasmic large DNA (NCLD) viruses are viruses of eukaryotic cells that are thought to have arisen from a common ancestor; a new virus order, *Megavirales,* has been proposed to encompass them. NCLD viruses, now often referred to as megaviruses, have similar life cycles in which most, if not all, of the events occur in the cytoplasm of their hosts. Most have large icosahedral capsids that enclose a lipid membrane. At the center of their virions is a large dsDNA genome (compare the 10-kilobase genome of an influenza virus to the 100- to 1,200-kilobase genomes of megaviruses). Thus both the virion itself and the genome are large, often rivaling the sizes of small bacteria and their genomes. Megaviruses infect a variety of eukaryotic hosts, including animals, algae, and protozoa.

The large genomes of megaviruses enable them to encode many proteins. Thus they are very self-sufficient and rely on their hosts for much less than do other viruses. They generally encode all the proteins needed for DNA replication, enzymes involved in recombination, RNA polymerases and associated transcription factors, and chaperone proteins. In addition, they encode some components of the translational machinery (tRNAs and aminoacyl-tRNA synthetases). Indeed, their coding capacity and size have led many virologists to rethink the definition of viruses and cells (**Microbial Diversity & Ecology 27.1**).

Retrieve, Infer, Apply

1. The CRISPR/Cas system has been a boon to biotechnology because it allows targeted silencing of specific genes. Explain how the CRISPR/Cas system can be modified to target eukaryotic genes.

2. Why do cold sores recur throughout the lifetime of an HSV-1–infected individual?

3. In what part of the host cell does a herpesvirus genome replicate? Where does the viral genome reside during a latent infection?

4. Many small DNA viruses rely on host enzymes for replication and transcription. Why are megaviruses able to use their own DNA-dependent RNA polymerases and DNA polymerases?

27.3 Single-Stranded DNA Viruses Use a Double-Stranded Intermediate in Their Life Cycles

After reading this section, you should be able to:

- Describe in general terms the strategy used by single-stranded (ss) DNA viruses to synthesize their nucleic acids and proteins
- Choose one specific bacterial and eukaryal ssDNA virus and illustrate the major events in their life cycles, noting, when possible, the specific mechanisms used to accomplish each step

Most DNA viruses are double-stranded, but several important viruses with single-stranded (ss) DNA genomes have been described. The life cycles of ssDNA viruses are similar to those of dsDNA viruses with one major exception. An additional step must occur in the synthesis stage because the ssDNA genome needs to be converted to a dsDNA molecule. A few ssDNA viruses are discussed next.

Bacteriophages φX174 and fd

The life cycle of φX174 (family *Microviridae,* species *Enterobacteria phage phiX174*) begins with attachment to the cell wall of its *E. coli* host. The circular ssDNA genome enters the cell with the help of a pilot protein called protein H. Protein H is thought to form a tubelike structure through which the DNA passes; the protein capsid remains outside the cell. The φX174 ssDNA genome has the same base sequence as viral mRNA and is therefore **plus-strand DNA.** For either transcription or genome replication to occur, the phage DNA must be converted to a double-stranded **replicative form (RF)** (**figure 27.17**). This is catalyzed by the host's DNA polymerase. The RF directs the synthesis of more RF copies and plus-strand DNA, both by rolling-circle replication (*see figure 16.21*). Assembly of φX174 virions begins with formation of a procapsid. This involves scaffolding proteins that interact with capsid proteins, causing them to undergo conformational changes and form a procapsid with the proper morphology. Once the procapsid is generated, the ssDNA genome is inserted, creating the mature virion. After

assembly of virions, the host is lysed by a viral enzyme (enzyme E) that blocks peptidoglycan synthesis. Enzyme E inhibits the activity of the bacterial protein MraY, which catalyzes the transfer of peptidoglycan precursors to lipid carriers (*see figure 12.9*). Blocking cell wall synthesis weakens the host cell wall, causing the cell to lyse and release the progeny virions.

We include *Escherichia coli* phage fd (family *Inoviridae*) in our discussion because of one striking feature: its ability to be released from its host without lysis; it is also a good example of viral self-assembly. It, like other filamentous phages (e.g., *Pseudomonas* phage Pf1), is released by a secretory process. Extrusion begins when phage coat proteins are inserted into the host cell's plasma membrane. The coat then assembles around the viral DNA as it is secreted through the host plasma membrane (**figure 27.18**). Although the host cell is not lysed, it grows and divides at a slightly reduced rate. This mechanism is rare among bacterial viruses but is more common among archaeal viruses, as we note in section 27.2.

Parvoviruses

Parvoviruses (family *Parvoviridae*) infect numerous animal hosts, including crustaceans, dogs, cats, mice, and humans. Since its discovery in 1974, human parvovirus B19 (genus *Erythrovirus*) has emerged as a significant human pathogen. Parvovirus virions are uniform, icosahedral, nonenveloped particles approximately

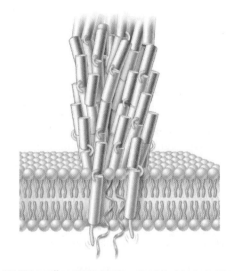

Figure 27.18 Release of Pf1 Phage. The filamentous Pf1 phage is released from *Pseudomonas aeruginosa* without lysis. In this illustration, the blue cylinders are hydrophobic α-helices that span the plasma membrane, and the red cylinders are amphipathic helices that lie on the membrane surface before virion assembly. In each protomer, the two helices are connected by a short, flexible peptide loop (yellow). It is thought that the blue helix binds with circular, single-stranded viral DNA (green) as it is extruded through the membrane. The red helix simultaneously attaches to the growing viral coat that projects from the membrane surface. Eventually the blue helix leaves the membrane and also becomes part of the capsid.

MICRO INQUIRY *What is the fate of the host cell infected with Pf1 phage?*

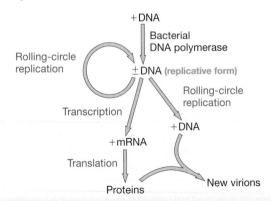

Figure 27.17 The Multiplication Strategy of φX174, a Plus-Strand DNA Phage.

MICRO INQUIRY *Why is the φX174 genome considered plus stranded?*

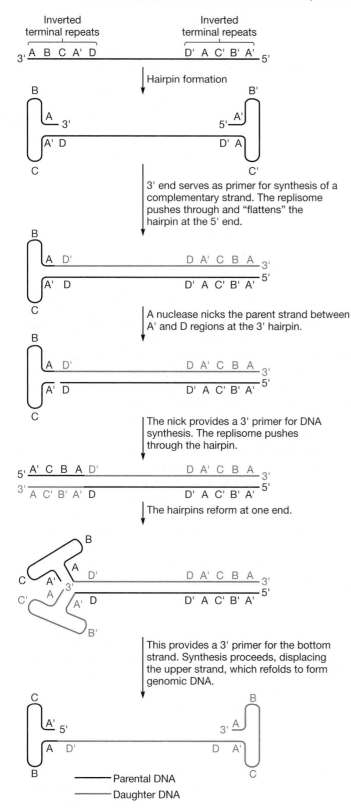

Figure 27.19 The Rolling-Hairpin Replication Used by Parvoviruses to Replicate Their DNA.

genome is so small that it directs the synthesis of only three proteins and some smaller polypeptides. None has enzymatic activity. ▶▶ *Human parvovirus B19 infection (section 38.3)*

Each parvovirus uses a particular host cell molecule for attachment. For instance, the B19 virus uses a molecule found only on the surface of red blood cell progenitors. The virion then enters by receptor-mediated endocytosis. The virion escapes the endosome and is thought to be transported to the nucleus by the host cell's microtubules. The nucleocapsid enters the nucleus, followed by release of viral DNA from the capsid.

Since the parvovirus genome does not code for any enzymes, the virus must use host cell enzymes for all biosynthetic processes. Thus viral DNA can only be replicated in the nucleus during the S phase of the cell cycle, when the host cell replicates its own DNA. Because the genome is negative-strand DNA, it serves as the template for mRNA synthesis. Some of the RNA products encode polypeptides required for the interesting way the virus's genome is replicated. The ends of the parvovirus genome are palindromic sequences that can fold back on themselves. Formation of a hairpin at the 3′ end of the genome provides the primer needed for replication (**figure 27.19**). This is recognized by the host DNA polymerase, and DNA replication ensues by a process that is somewhat similar to rolling-circle replication. The parvovirus version of this replication method is often called rolling-hairpin replication because DNA polymerase seems to shuttle back and forth as it synthesizes genomes. The process involves a dsDNA intermediate, much like ϕX174.

Retrieve, Infer, Apply

1. Why is it necessary for some ssDNA viruses to manufacture a replicative form?
2. From the point of view of the virus, compare the advantages and disadvantages of host cell lysis, as seen in ϕX174, with the continuous release of phage without lysis, as with fd phage and other filamentous phages.
3. How do parvoviruses "trick" the host DNA polymerase into replicating their genomes? Why must this occur in the nucleus only during the S phase of the host's cell cycle?

27.4 Double-Stranded RNA Viruses: RNA-Dependent RNA Polymerase Replicates the Genome and Synthesizes mRNA

After reading this section, you should be able to:

- Distinguish RNA-dependent RNA polymerases from DNA-dependent RNA polymerases
- Describe what an RNA-dependent RNA polymerase is doing when functioning as a replicase and when functioning as a transcriptase
- Describe in general terms the strategy used by dsRNA viruses to synthesize their nucleic acids and proteins
- Describe the major events in the life cycles of ϕ6 and rotaviruses, noting, when possible, the specific mechanisms used to accomplish each step

26 nm in diameter. Their genomes are composed of one ssDNA molecule of about 5,000 bases. Most of the genomes are **negative-strand DNA** molecules. That is, their sequence of nucleotides is complementary to that of the viral mRNA (figure 27.2). Parvoviruses are among the simplest of the DNA viruses. The

The Baltimore system divides viruses with RNA genomes into four groups (table 27.1). However, they all share the same dilemma: their host cells have dsDNA genomes and thus lack a polymerase that can make RNA from an RNA template. Therefore RNA viruses classified in the Baltimore System as double-stranded, plus-strand, and negative-strand RNA viruses must produce an enzyme called **RNA-dependent RNA polymerase** (RdRp; an RNA polymerase that uses an RNA molecule as the template for RNA synthesis). When the RdRp is used to replicate the viral RNA genome, it is often referred to as a **replicase.** When it is used to synthesize mRNA, the RdRp is often said to have **transcriptase** activity. In most cases, an identical RdRp carries out both functions. Retroviruses use a very different approach: they first convert their ssRNA genomes into dsDNA, which is then recognized by the host cell's DNA-dependent RNA polymerase.

Among viruses with RNA genomes, those with dsRNA appear to be least abundant. These viruses share a common multiplication strategy (**figure 27.20**). Here we discuss two representative dsRNA viruses: a bacteriophage and a vertebrate virus. As you will see, these two viruses use one or more strategies to ensure that all needed viral proteins are made. Our discussions of other RNA viruses will introduce additional strategies. These strategies function at the levels of genome structure and organization, transcription, and translation.

Bacteriophage φ6

Several dsRNA phages have been discovered and assigned to the family *Cystoviridae*. The best studied, *Pseudomonas* phage phi6 (φ6), is unusual for several reasons. One is that it is an enveloped phage. The envelope encloses the nucleocapsid, which consists of an outer protein shell, composed of a viral protein called P8, and an inner core sometimes called the polymerase complex. It is so named because it consists of the virus's RNA-dependent RNA polymerase (RdRp) and three other proteins, and encloses the genome. The φ6 genome illustrates one of the methods used by RNA viruses to ensure synthesis of all viral proteins: it is a **segmented genome,** composed of more than one distinct RNA molecule. For most viruses, each segment encodes a single protein. However, φ6 deviates from this because each of its three dsRNA molecules generates a polycistronic mRNA and several proteins. Many other viruses also generate polycistronic mRNAs, much like the polycistronic mRNAs observed in bacteria and archaea (*see figure 13.22*).

The life cycle of the virus also has unusual features. Like some other phages, φ6 attaches to the side of a pilus. However, φ6 uses an envelope spike to facilitate adsorption. Retraction of the pilus brings the phage into contact with the outer membrane of its Gram-negative host. The viral envelope then fuses with the cell's outer membrane, a process mediated by another envelope spike. Fusion of the two membranes delivers the nucleocapsid into the periplasmic space. A protein associated with the nucleocapsid digests the peptidoglycan, allowing the nucleocapsid to cross this layer of the cell wall. Finally, the intact nucleocapsid enters the host cell by a process that resembles endocytosis. Evidence suggests that an interaction between the shell protein P8 and plasma membrane phospholipids mediates this mechanism of entry.

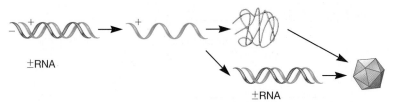

Figure 27.20 Multiplication Strategy of Double-Stranded RNA Viruses. Although we have designated plus and minus RNA strands, both strands may serve as templates for synthesis of mRNA by transcriptase, depending on the virus. For those viruses, each strand is said to be ambisense. The strand of the viral genome we have designated as the negative strand serves as the template for synthesis of positive strands by replicase. The positive strands are used by replicase to synthesize negative strands, thus replicating the dsRNA genome.

Once inside the host, the viral RNA polymerase acts as a transcriptase, catalyzing synthesis of viral mRNA from each dsRNA segment while they are still inside the capsid. The mRNAs leave the capsid and some are translated to early proteins. These assemble to form new, empty polymerase complexes (i.e., those lacking RNAs). Meanwhile, the virus's RNA polymerase acts as a replicase, synthesizing plus-strand RNA from each segment. The plus-strand RNAs become enclosed within the newly formed polymerase complexes, where they serve as templates for the synthesis of the complementary negative strand, regenerating the dsRNA genome. Proteins made later in the infection cycle include the proteins found in the outer shell. These associate with the polymerase complex, forming the nucleocapsid. Once the nucleocapsid is completed, a nonstructural viral protein called P12 surrounds the nucleocapsid with a plasma membrane-derived envelope while the nucleocapsid is within the host cytoplasm. Finally, additional viral proteins are added to the envelope and the host cell is lysed, releasing mature virions.

Rotaviruses

Human rotaviruses (family *Reoviridae*) are responsible for the deaths of over 600,000 children worldwide each year. They cause severe diarrhea, which rapidly causes dehydration and death if appropriate therapy is not provided. Because of their impact on humans, rotaviruses have been studied intensely to better understand their life cycles and pathogenesis. ▶▶| *Gastroenteritis (viral) (section 38.4)*

Viewed by electron microscopy, rotavirus virions have a characteristic wheel-like appearance (rotavirus is derived from the Latin *rota,* meaning wheel). Virions are nonenveloped and are composed of 11 segments of dsRNA surrounded by three concentric layers of proteins (*see figure 38.21*). The RNA segments code for six structural and six nonstructural proteins.

When a rotavirus virion enters a host cell, it loses the outermost protein layer and is then referred to as a double-layered particle (DLP; **figure 27.21**). The genome is transcribed by the viral transcriptase while still inside the DLP. The mRNA passes through channels in the DLP and is released into the cytosol of the host cell. There the mRNAs are translated by the host cell's

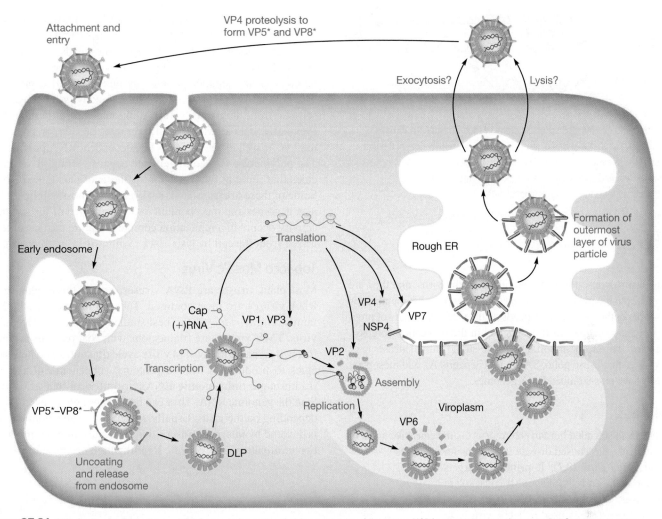

Figure 27.21 Life Cycle of a Rotavirus. VP1-8 are viral proteins that become part of the virion. NSP4 is a nonstructural protein that functions in virion assembly but is not incorporated into the virion. The method of release of progeny virions has not been definitively determined.

protein synthesis machinery. The newly formed proteins cluster together, forming an inclusion called a viroplasm. It is within the viroplasm that new DLPs are formed. Initially the DLPs contain plus-strand RNA, but this is soon used as a template for synthesis of the negative strand. Thus the dsRNA molecule is synthesized within the developing DLP.

DLPs containing dsRNA eventually leave the viroplasm and are enveloped by membranes of the endoplasmic reticulum. While in the endoplasmic reticulum, the outermost layer is added to the DLP, converting it to the mature triple-layered virion. The mature virion is released by an unknown mechanism.

Retrieve, Infer, Apply

1. The rotavirus genome encodes 12 proteins. Suggest one or more strategies it might use to ensure that all are synthesized.
2. Describe the life cycle of φ6 phage. What makes this phage unusual when compared with other bacteriophages?
3. How does the rotavirus structure give rise to its name?
4. In what ways are the life cycles of φ6 and rotaviruses similar? How do they differ?

27.5 Plus-Strand RNA Viruses: Genomes That Can Be Translated upon Entry

After reading this section, you should be able to:

- Describe in general terms the strategy used by plus-strand RNA viruses to synthesize their nucleic acids and proteins
- Outline the major events in the life cycles of poliovirus and tobacco mosaic virus, noting, when possible, the specific mechanisms used to accomplish each step

Plus-strand RNA viruses have genomes that can act as mRNA and be translated upon entry into the host cell. One of the first products is an RNA-dependent RNA polymerase, which catalyzes synthesis of negative-strand RNAs; these are then used to make more plus-strand RNAs (**figure 27.22**). In some cases, this occurs by way of a double-stranded replicative form (RF), just as seen with ssDNA viruses (section 27.3). In plant and animal viruses, viral genome replication occurs in a structure formed within the cytoplasm called a replication complex (*see figure 6.12*). These are compartments

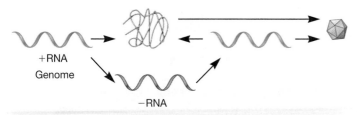

Figure 27.22 Multiplication Strategy of Plus-Strand RNA Viruses.
The plus-strand RNA genome can serve directly as mRNA. One of the first viral proteins synthesized is replicase, which replicates the genome, sometimes via a double-stranded replicative form. Transcriptase is responsible for synthesizing additional mRNA molecules.

MICRO INQUIRY *Where in the host does the plus-strand RNA genome replicate?*

formed in response to factors produced by the virus, and they are derived from the membranes of cell organelles (e.g., ER membrane); the source of the membrane depends on the virus. Assembly of progeny virions sometimes also occurs within the replication complex. There are a number of important positive-strand animal viruses (e.g., the viruses that cause polio, SARS, and hepatitis A), and most plant viruses have plus-stranded RNA genomes.

Poliovirus

Poliovirus (also called human enterovirus C), the causative agent of poliomyelitis, has caused disease in humans for centuries (*see Historical Highlights 38.1*). It primarily targets children, with paralysis being the tragic result in some cases. The life cycle of poliovirus illustrates another strategy used by RNA viruses to ensure synthesis of needed proteins: synthesis of a polyprotein. ▶▶ *Poliomyelitis (section 38.4)*

The biology of polioviruses (family *Picornaviridae*) has been intensely studied for decades. But many questions linger. This non-enveloped virus makes its way into a human host by ingestion. It attaches to a cell surface molecule called human PV receptor (PV for poliovirus). The nucleocapsid enters the host cell, and the plus-strand RNA genome is released into the cytosol while the virion is at the cell periphery and held within an endocytic vesicle (**figure 27.23**). The genome acts as mRNA and is translated by host cell ribosomes. This process has garnered considerable attention because the virus's RNA does not have the 5′ cap found on eukaryotic mRNAs, which is important for ribosome binding. Poliovirus "tricks" its host into translating its capless RNA using a 5′ region on the RNA called the internal ribosome entry site (IRES). In this region the ssRNA folds back on itself and forms extensive secondary structures

(regions of dsRNA and numerous ssRNA loops), which are important for recognition of the RNA by ribosomes (**figure 27.24**).

Translation of the poliovirus genome yields a single polyprotein (**figure 27.25**). A **polyprotein** is a large protein that can be cleaved into smaller proteins by enzymes called proteases. The poliovirus polyprotein has protease activity and cleaves itself into three smaller proteins. Additional cuts in these proteins eventually yield all the structural proteins needed for capsid formation, as well as the virus's RNA-dependent RNA polymerase. The polymerase is used to generate negative-strand RNA molecules that serve as templates for synthesis of plus-strand RNAs. Some of these are translated, but eventually most will be used as new genomes and incorporated into capsids. All of the synthetic activity occurs in a replication complex. Mature particles are released from the cell by lysis. ◀◀ *Synthesis stage (section 6.3)*

Tobacco Mosaic Virus

Most plant viruses are RNA viruses, and of these, plus-strand RNA viruses are most common. Tobacco mosaic virus (TMV; family *Virgaviridae*) is the best-studied plus-strand RNA plant virus. TMV virions are filamentous with coat proteins arranged in a helical pattern. The TMV life cycle illustrates two new strategies to produce its needed proteins: subgenomic RNAs and readthrough. **Subgenomic RNAs** are mRNAs that are shorter than the genomic RNA. They can be synthesized in several ways, depending on the virus. Usually, synthesis of subgenomic RNAs is dictated by internal transcription start sites, internal transcription termination sites, or both. **Readthrough** occurs at the level of

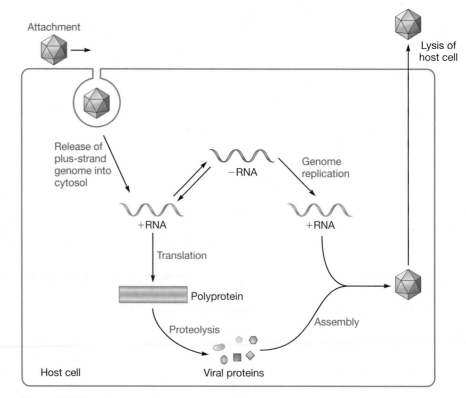

Figure 27.23 Poliovirus Life Cycle.

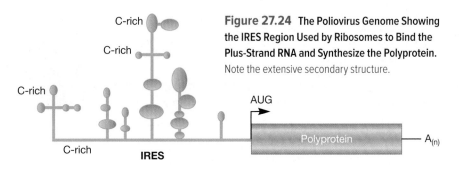

Figure 27.24 **The Poliovirus Genome Showing the IRES Region Used by Ribosomes to Bind the Plus-Strand RNA and Synthesize the Polyprotein.** Note the extensive secondary structure.

translation. It occurs when the ribosome reaches a stop codon, ignores the stop signal, and continues translation.

Plant viruses, including TMV, usually enter the host through an abrasion or wound on the plant; biting insects are often involved in transmission of the virus. Following entry into its host, the TMV RNA genome is translated. Two proteins are produced, one smaller (about 125 kilodaltons; kD) and one larger (about 185 kD). The larger protein is produced by readthrough of the stop codon at the end of the coding region for the 125 kD protein. The 185 kD protein has RNA-dependent RNA polymerase activity and functions as both a transcriptase and a replicase. Soon thereafter, TMV uses membranes of the endoplasmic reticulum (ER) to form a replication complex. There, the RNA-dependent RNA polymerase synthesizes negative-strand RNA using the plus-strand genome as the template. It is not clear whether a double-stranded RF is created in vivo, but RFs have been observed in vitro. The negative-strand RNA serves as a template for synthesis of new genomes. It also serves as the template for synthesis of subgenomic mRNAs that are translated into a variety of proteins needed by the virus.

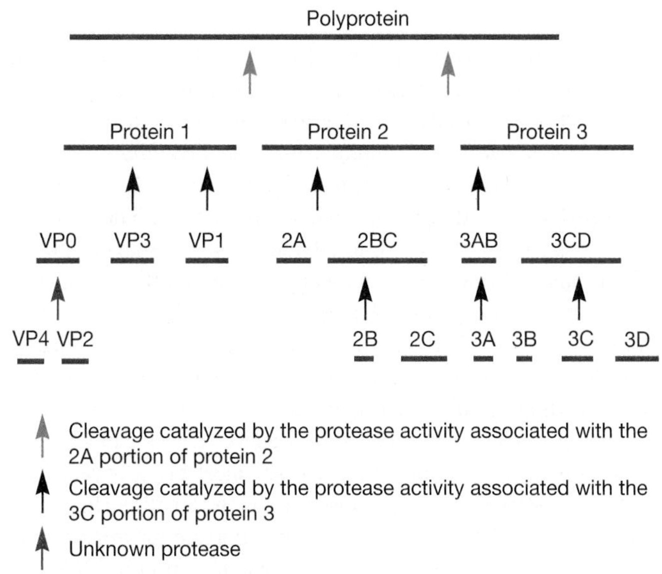

After the coat protein and RNA genome of TMV have been synthesized, they spontaneously assemble into complete TMV virions in a highly organized process (**figure 27.26**). The protomers come together to form disks composed of two layers of protomers arranged in a helical spiral. Association of coat protein with TMV RNA begins at a specific assembly initiation site close to the 3′ end of the genome. The helical capsid grows by the addition of protomers, probably as disks, to the end of the rod. As the rod lengthens, the RNA passes through a channel in its center

Polyprotein

Protein 1   **Protein 2**   **Protein 3**

VP0   VP3   VP1   2A   2BC   3AB   3CD

VP4 VP2   2B   2C   3A   3B   3C   3D

↑ Cleavage catalyzed by the protease activity associated with the 2A portion of protein 2

↑ Cleavage catalyzed by the protease activity associated with the 3C portion of protein 3

↑ Unknown protease

Figure 27.25 **Cleavage of the Poliovirus Polyprotein.** Proteins VP1-4 are structural proteins. Protein 3D is the RNA-dependent RNA polymerase.

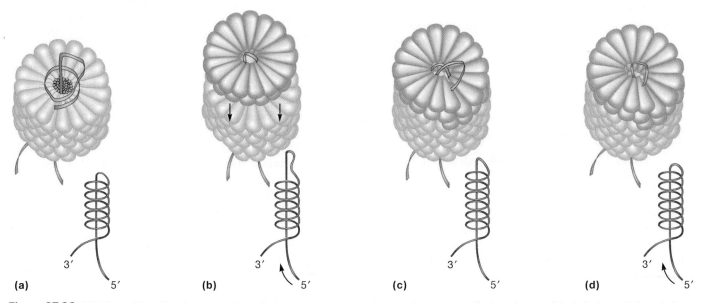

(a)   (b)   (c)   (d)

Figure 27.26 **TMV Assembly.** The elongation phase of tobacco mosaic virus nucleocapsid construction. The lengthening of the helical capsid through the addition of a protein disk to its end is shown in a sequence of four illustrations; line drawings depicting RNA behavior are included. The RNA genome inserts itself through the hole of an approaching disk and then binds to the groove in the disk as it locks into place at the end of the cylinder.

and forms a loop at the growing end. In this way, the RNA easily fits as a spiral into the interior of the helical capsid.

Multiplication of plant viruses within their host depends on the virus's ability to spread throughout the plant. Viruses can move to new sites in the plant through the plant vasculature; usually they travel in the phloem. They can also spread more locally in nonvascular tissue. Recall that plant cells have tough cell walls. Therefore, plant viruses move from cell to cell through plant cell plasmodesmata. These are slender bridges of cytoplasmic material, including extensions of the ER. Plasmodesmata extend through holes in cell walls to join adjacent plant cells. Viral "movement proteins" are required for transfer through the plasmodesmata. TMV movement proteins are closely associated with ER membranes and so are near the TMV replication complex where viral genomes are being synthesized. They bind the viral genomic RNA (vRNA) and also interact with components of the cytoskeleton. Therefore they facilitate the movement of vRNA from the replication complex to the plasmodesmata. TMV movement proteins also cause the plasmodesmata to increase in diameter, allowing the vRNA to pass through to the adjacent cell.

Several cytological changes can take place in TMV-infected cells. These include microscopically visible intracellular inclusions that are similar to the replication complexes observed in animal cells infected with plus-strand viruses. Hexagonal crystals of almost pure TMV virions sometimes develop in TMV-infected cells. In addition, host cell chloroplasts become abnormal and often degenerate, while new chloroplast synthesis is inhibited.

Retrieve, Infer, Apply

1. How do some plus-strand viruses use polyproteins to complete their life cycles? Describe two other strategies used by plus-strand viruses to generate all their proteins.
2. What is an IRES? Why is it important?
3. How does TMV spread from one host cell to another?

27.6 Minus-Strand RNA Viruses: RNA-Dependent RNA Polymerase Is Part of the Virion

After reading this section, you should be able to:

- Describe in general terms the strategy used by minus-strand RNA viruses to synthesize their nucleic acids and proteins
- Explain how having a segmented genome impacts synthesis of viral mRNA and proteins and the generation of new strains of a virus
- Create a flow chart that summarizes the life cycle of influenza virus, noting the specific mechanisms it uses to accomplish each step of its life cycle

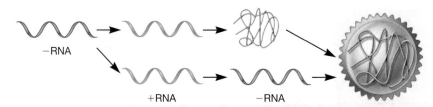

Figure 27.27 Multiplication Strategy of Negative-Strand RNA Viruses. An RNA-dependent RNA polymerase enters the host cell at the same time the negative-strand RNA genome enters. The genome serves as a template for synthesis of mRNA. Later in the infection, the negative-strand genome is used for plus-strand synthesis. These plus strands then act as templates for replication of the negative-strand genomes.

Negative-strand viruses are found in many families, including *Rhabdoviridae* (e.g., rabies virus), *Filoviridae* (e.g., Marburg and Ebola viruses), *Paramyxoviridae* (e.g., measles and mumps viruses), *Bunyaviridae* (e.g., hantaviruses), and *Orthomyxoviridae* (e.g., influenza viruses). Most negative-strand RNA viruses are enveloped viruses that infect plants and animals. They vary in morphology from spherical, to filamentous, rod-shaped, bullet-shaped, and pleomorphic. Members of four families have non-segmented linear genomes and are grouped into the order *Mononegavirales*. The remaining negative-strand RNA viruses have segmented genomes that range from two to eight segments, each encoding usually one protein.

The genomes of negative-strand RNA viruses cannot function as mRNA. Therefore these viruses must bring at least one RNA-dependent RNA polymerase into the host cell during entry. Initially the viral genome serves as the template for mRNA synthesis (**figure 27.27**). Later the virus switches from mRNA synthesis to genome replication, as the RNA-dependent RNA polymerase synthesizes a distinct plus-strand RNA for replication. During this phase of the life cycle, the plus-strand RNA molecules synthesized from the minus-strand genome serve as templates for the manufacture of new negative-strand RNA genomes.

Our focus in this section is on influenza viruses. There are three types of influenza virus: influenza viruses A, B, and C. Influenza virus virions are composed of an envelope enclosing seven or eight nucleocapsids, depending on the type of influenza virus. Each nucleocapsid consists of a single, negative-strand RNA that exists as a double helical hairpin (**figure 27.28a**). Formation and maintenance of the double helix does not involve base pairing. Rather, this is accomplished by numerous copies of the nucleocapsid protein (NP), which coat the RNA. The three proteins of the influenza polymerase (PA, PB1, PB2) are attached to the open end of the hairpin.

An influenza virion enters a host cell by receptor-mediated endocytosis. This encloses the virus in an endosome (figure 27.28b). When the endosomal pH decreases, a conformational change in the virion facilitates contact with the endosome membrane, allowing fusion of the endosome membrane and viral envelope to occur, releasing the nucleocapsids into the cytosol. ▶▶| *Influenza (flu; section 38.1)*

Once the RNA segments and associated RNA-dependent RNA polymerase enter the host cell nucleus, the genome segments serve as templates for mRNA synthesis (figure 27.28b,

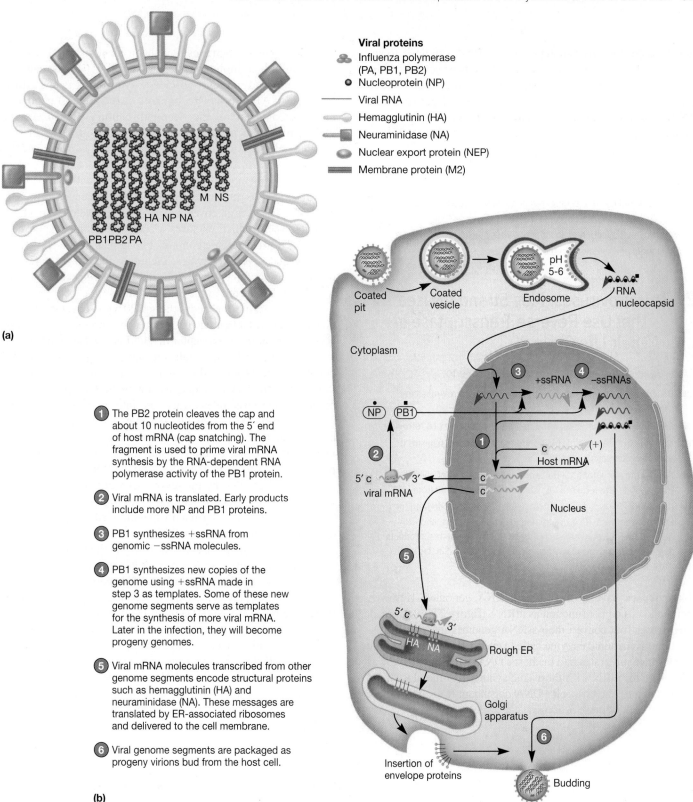

(a)

Viral proteins

Influenza polymerase (PA, PB1, PB2)
Nucleoprotein (NP)
Viral RNA
Hemagglutinin (HA)
Neuraminidase (NA)
Nuclear export protein (NEP)
Membrane protein (M2)

1 The PB2 protein cleaves the cap and about 10 nucleotides from the 5′ end of host mRNA (cap snatching). The fragment is used to prime viral mRNA synthesis by the RNA-dependent RNA polymerase activity of the PB1 protein.

2 Viral mRNA is translated. Early products include more NP and PB1 proteins.

3 PB1 synthesizes +ssRNA from genomic −ssRNA molecules.

4 PB1 synthesizes new copies of the genome using +ssRNA made in step 3 as templates. Some of these new genome segments serve as templates for the synthesis of more viral mRNA. Later in the infection, they will become progeny genomes.

5 Viral mRNA molecules transcribed from other genome segments encode structural proteins such as hemagglutinin (HA) and neuraminidase (NA). These messages are translated by ER-associated ribosomes and delivered to the cell membrane.

6 Viral genome segments are packaged as progeny virions bud from the host cell.

(b)

Figure 27.28 An Influenza Virus and Its Life Cycle.   (a) Schematic representation of an influenza A virus showing the double helical hairpin RNA genome. The protein or proteins encoded by each RNA segment are shown below each nucleocapsid. A protein called M1 (lavender; M stands for matrix) lines the lipid bilayer envelope (dark blue). Another matrix protein (M2) is embedded in the envelope. NS proteins (NS1 and NS2) are nonstructural proteins that function in the life cycle of the virus but are not part of the virion. (b) This simplified version of the influenza life cycle omits several steps and represents the viral RNA molecules as simple lines. Negative-strand RNA is red and positive-strand RNA is green. After entry by receptor-mediated endocytosis, the virus envelope fuses with the endosome membrane, releasing the nucleocapsids into the cytosol. The nucleocapsids enter the nucleus, where synthesis of viral mRNA and genomes occurs. The influenza virus polymerase is composed of three subunits. PB1 has RNA-dependent RNA polymerase activity. PB2 is an endonuclease responsible for the cap snatching described in step 1. PA is a protease. Steps 1 through 6 illustrate the remaining steps of the virus's life cycle.

step 1). Later the virus switches from mRNA synthesis to genome replication. During this phase of the life cycle, the plus-strand RNA molecules synthesized from the minus-strand genome segments serve as templates for new negative-strand RNA genomes (figure 27.28, step 4). Virions exit the host cell by budding and thus acquire their envelope (figure 27.28*b*, step 6; *see also figure 6.15*).

Retrieve, Infer, Apply

1. How does that use of a segmented genome by influenza viruses differ from that used by bacteriophage ɸ6?
2. Trace the multiplication of an influenza virus starting with host cell attachment and ending with the exit of virions.

27.7 Retroviruses: Plus-Strand Viruses That Use Reverse Transcriptase in Their Life Cycles

After reading this section, you should be able to:

- Describe in general terms the strategy used by retroviruses to synthesize their nucleic acids and proteins
- Differentiate a segmented genome from the genome of a retrovirus
- Describe the life cycle of HIV, noting the specific mechanisms it uses to accomplish each step
- Defend the statement "all three enzymatic activities of reverse transcriptase are required for dsDNA to be synthesized from RNA"
- Distinguish DNA-dependent DNA polymerases, DNA-dependent RNA polymerases, RNA-dependent RNA polymerases, and RNA-dependent DNA polymerases in terms of their templates, products synthesized, and proofreading activity, as discussed in this and previous sections

Retroviruses have positive-strand RNA genomes. However, their genomes do not function as mRNA (**figure 27.29**). Instead, retroviruses first convert their ssRNA genomes into dsDNA using a multifunctional enzyme called **reverse transcriptase.** The dsDNA then integrates into the host's DNA, where it can serve as a template for mRNA synthesis and synthesis of the plus-strand RNA genome. The host cell's DNA-dependent RNA polymerase catalyzes both of these processes.

Numerous retroviruses have been identified and studied. However, human immunodeficiency virus (HIV), the cause of AIDS (acquired immune deficiency syndrome), is of particular interest. AIDS is now recognized as the greatest pandemic of the second half of the twentieth century. Because of its global importance, we focus exclusively on HIV in this section. ▶▶❘ *Acquired immune deficiency syndrome (AIDS) (section 38.3)*

HIV is a member of the genus *Lentivirus* within the family *Retroviridae*. In the United States, AIDS is caused primarily by HIV-1. HIV-1 is an enveloped virus. The envelope surrounds an outer shell, which encloses a somewhat cone-shaped core (**figure 27.30**). The core contains two copies of the HIV RNA genome and several enzymes, including the enzymes reverse transcriptase and integrase. Thus far 10 virus-specific proteins have been discovered in the HIV virion.

After entering the body, the gp120 viral envelope protein binds to host cells that have a surface glycoprotein called CD4. The virus requires a coreceptor in addition to CD4, and this varies depending on the host cell infected (*see figure 38.10*). Still being debated is how HIV enters a host cell. Initially it was thought that the HIV envelope fused with the cell's plasma membrane and the virus released its core into the cytoplasm. However, evidence exists that virions enter by receptor-mediated endocytosis (figure 27.30). Some scientists suggest that virions may enter by either method. Inside the infected cell, the core protein dissociates from the RNA, and the RNA is copied into a single strand of DNA by the reverse transcriptase enzyme. The RNA is next degraded by reverse transcriptase, and the DNA strand is duplicated to form a double-stranded DNA copy of the original RNA genome (figure 27.29).

Reverse transcription is a critical step in the life cycle of HIV, and reverse transcriptase is a remarkable enzyme with multiple activities. It is an **RNA-dependent DNA polymerase,** a DNA-dependent DNA polymerase, and a ribonuclease. This latter function is referred to as RNaseH activity. Despite its versatility, reverse transcriptase lacks a function observed in other DNA polymerases: proofreading. Thus it makes many errors as it synthesizes DNA. Nonetheless, reverse transcription is an amazing process that involves the use of a host tRNA molecule as primer for initial steps in DNA synthesis (**figure 27.31**). In the process, a newly synthesized, small negative-strand DNA molecule is transferred

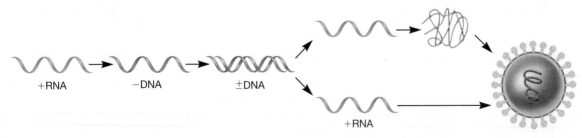

+RNA   −DNA   ±DNA   +RNA

Figure 27.29 Multiplication Strategy of Retroviruses. Retroviruses have a plus-strand RNA genome that is first converted into dsDNA by the enzyme reverse transcriptase. The viral dsDNA integrates into the host chromosome, where it serves as the template for synthesis of viral mRNA and viral genomes. Both are synthesized using the host cell's DNA-dependent RNA polymerase.

from one end of the RNA template to the other to prime the synthesis of the rest of the minus-strand DNA. Later the full-length minus-strand DNA circularizes to allow completion of the plus-strand DNA.

Once dsDNA is formed, a complex of dsDNA (the provirus), integrase enzyme, and other factors (including some host molecules) moves into the nucleus, where the proviral DNA is integrated into the cell's DNA by integrase. Once integrated, the provirus can force the cell to synthesize viral mRNA (figure 27.30). Several different mRNAs are transcribed using the host cells DNA-dependent RNA polymerase. Some are full-length mRNAs and others are subgenomic mRNAs. The subgenomic mRNAs are formed by alternative splicing that brings different coding regions together, eliminating any intervening coding regions (rather than introns, as typically occurs during alternative splicing). Translation of two of the mRNAs yields polyproteins that are cleaved to give rise to numerous proteins. One of the polyproteins is synthesized by ribosomal frameshifting followed by readthrough. **Ribosomal frameshifting** is often used when viral genes overlap but are in different reading frames. It occurs at specific sites at which the ribosome pauses and then either continues in the same reading frame or shifts to the alternative reading frame. In the case of HIV, the ribosome reaches a stop codon, pauses, and either stops to yield one protein or shifts reading frame and continues translating the mRNA to yield the larger polyprotein. Some of the earliest proteins synthesized are used to regulate cellular activities so that HIV genes are preferentially expressed. Proteins needed to form HIV virions are made later. Eventually viral proteins and the complete HIV-1 RNA genome are assembled into new nucleocapsids that bud from the infected host cell (figure 27.30). After some time, the host cell dies, in part from repeated budding but by other processes as well.

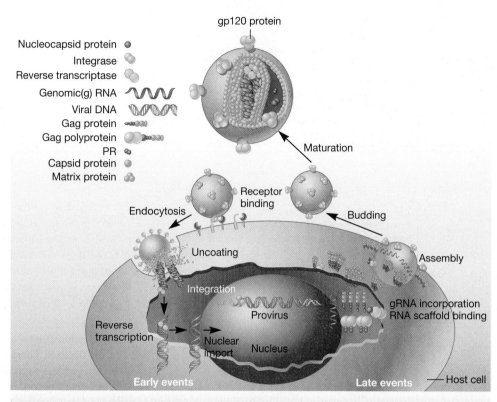

Figure 27.30 The HIV-1 Life Cycle.

MICRO INQUIRY *What three enzymatic activities does reverse transcriptase display? When during the HIV life cycle is each activity used?*

Retrieve, Infer, Apply

1. What is the function of each of the following HIV products: gp120, reverse transcriptase, and integrase?
2. CD4 is found on several different immune system cells. Consult chapter 34. Why is the presence of this molecule on the surface of many immune system cells important to the development of AIDS?
3. What role does alternative splicing play in the life cycle of HIV-1? For what purpose does the virus use ribosomal frameshifting and readthrough?

27.8 Reverse Transcribing DNA Viruses

After reading this section, you should be able to:

- Describe in general terms the strategy used by reverse transcribing DNA viruses to synthesize their nucleic acids and proteins
- Compare the role of reverse transcriptase in the life cycle of a retrovirus to that in the life cycle of a hepadnavirus

Reverse transcribing DNA viruses use reverse transcriptase to replicate their genomes. Members of four virus families fall into this category: *Caulimoviridae* (e.g., cauliflower mosaic virus and other plant viruses), *Metaviridae* (fungal, plant, and animal viruses), *Pseudoviridae* (fungal, plant, protist, and animal viruses), and *Hepadnaviridae* (e.g., hepatitis B virus and related animal viruses). Hepatitis B virus (HBV) is one of the best-studied reverse transcribing DNA viruses and is our focus here.

HBV virions are 42-nm spherical particles that contain the viral genome. The HBV genome is a circular dsDNA molecule that consists of one complete but nicked strand and a complementary strand that has a large gap; that is, it is incomplete (**figure 27.32**). The genome is 3.2 kb in length and consists of four partially overlapping, open reading frames that encode viral proteins. Production of new virions takes place predominantly in liver cells (hepatocytes).

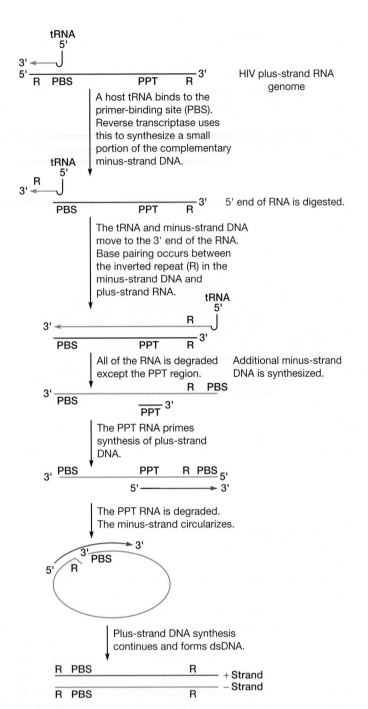

Figure 27.31 **Reverse Transcription.**

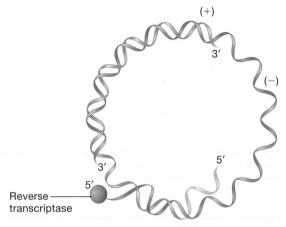

Figure 27.32 **The Gapped Genome of Hepadnaviruses.** The genomes of hepadnaviruses are unusual in several respects. The negative strand of the dsDNA molecule is complete but nicked. The enzyme reverse transcriptase is attached to its 5′ end. The positive strand is gapped; that is, it is incomplete (shown in purple). A short stretch of RNA is attached to its 5′ end (shown in green). Upon entry into the host nucleus, the nick in the negative strand is sealed and the gap in the positive strand is filled, yielding a covalently closed, circular dsDNA molecule.

After infecting a host cell, the viral-gapped DNA is released into the nucleus. There host repair enzymes fill the gap and seal the nick, yielding a covalently closed, circular DNA. Transcription of viral genes occurs in the nucleus using host RNA polymerase and yields several mRNAs, including a large 3.4-kb RNA known as the pregenome (a plus-strand RNA). The RNAs move to the cytoplasm, and the mRNAs are translated to produce viral proteins, including core protein and reverse transcriptase. Reverse transcriptase then associates with the plus-strand RNA pregenome and core protein to form an immature core particle. Reverse transcriptase subsequently reverse transcribes the RNA using a protein primer to form a minus-strand DNA from the pregenome RNA. After almost all the pregenome RNA has been degraded by the RNaseH activity of reverse transcriptase, the remaining RNA fragment serves as a primer for synthesis of the gapped dsDNA genome, using the minus-strand DNA as template. Finally, the nucleocapsid is completed and progeny virions are released.

Retrieve, Infer, Apply

1. Describe the HBV genome. How is it converted to covalently closed, circular DNA in the host?
2. Trace the HBV multiplication cycle, paying particular attention to localization within the host cell during the biosynthetic stage.

Key Concepts

27.1 Virus Phylogeny Is Difficult to Establish

- Currently viruses are classified with a taxonomic system placing primary emphasis on the type and strandedness of viral nucleic acids, and on the presence or absence of an envelope (**figure 27.1**).

- Increasingly, comparisons of viral genome sequences and the folding patterns of viral capsid proteins are being used to establish phylogenetic relationships among viruses.
- The Baltimore system is used by many virologists to organize viruses based on their genome type and the mechanisms used to synthesize mRNA and replicate their genomes (**table 27.1**).

27.2 Double-Stranded DNA Viruses Infect All Cell Types

- T4 is a virulent bacteriophage that causes lytic infections of *E. coli*. After attachment to a specific receptor site on the bacterial surface, T4 releases its dsDNA into the cell (**figures 27.4 and 27.5**). T4 DNA contains hydroxymethylcytosine (HMC) in place of cytosine, and glucose is often added to the HMC to protect the phage DNA from attack by host restriction enzymes (**figure 27.6**). T4 DNA replication produces concatemers, long strands of several genome copies linked together (**figure 27.7**).

- Lambda (λ) phage is a temperate bacteriophage. It can establish lysogeny, rather than pursuing a lytic infection. During lysogeny, the viral DNA, called a prophage, is replicated as the cell's genome is replicated. Lysogeny is reversible, and the prophage can be induced to become active again and lyse its host. This highly regulated process is an important model system for regulatory processes. The protein cII plays a central role in regulating the choice between lysogeny and a lytic cycle. If cII protein levels are high enough, lysogeny is established. If not, the lytic cycle is initiated (**figures 27.10 and 27.11**).

- Less is known about archaeal viruses. However, many of those discovered have interesting morphologies, which has led to the creation of several new viral families (**figure 27.12**). Many establish chronic infections, whereas others establish lysogenic and lytic infections (**figure 27.13**).

- The CRISPR/Cas system is a set of sequences found in many archaeal and bacterial genomes (**figure 27.14**). It is a defense system that protects the cells from viral attack. It functions in a manner similar to RNA silencing, an antiviral defense mechanism observed in eukaryotic cells.

- Herpesviruses are a large group of dsDNA viruses. They cause acute infections such as cold sores, genital herpes, chickenpox, and mononucleosis. This is followed by a lifelong latent infection in which the virus genome resides within neurons. The virus can be reactivated at later times to cause a productive infection.

- Nucleocytoplasmic large DNA (NCLD) viruses, also called megaviruses, are members of several virus families. They are thought to share an evolutionary history and are interesting due to their large genome and virion size. Furthermore, their genomes contain genes for translation-related functions that are not usually observed in viral genomes.

27.3 Single-Stranded DNA Viruses Use a Double-Stranded Intermediate in Their Life Cycles

- φX174 is an example of a ssDNA bacteriophage. Its replication involves the formation of a dsDNA replicative form (RF) (**figure 27.17**).

- fd phage is filamentous phage that upon infection is continuously released by the host without causing lysis.

- Parvoviruses cause a spectrum of diseases in a wide variety of animals. The parvovirus genome is replicated by host DNA polymerase in the host cell's nucleus using a process that is similar to rolling-circle replication (**figure 27.19**).

27.4 Double-Stranded RNA Viruses: RNA-Dependent RNA Polymerase Replicates the Genome and Synthesizes mRNA

- Double-stranded RNA viruses use a viral enzyme called RNA-dependent RNA polymerase to synthesize mRNA (transcriptase activity) and replicate their genomes (replicase activity) (**figure 27.20**).

- Bacteriophage φ6 is an unusual phage in that it is enveloped and enters the host bacterium through a process that resembles endocytosis. It has a segmented genome, in which each segment encodes a polycistronic mRNA that yields several proteins.

- Rotaviruses have segmented genomes, in which each segment codes for one or two proteins. They multiply in an inclusion called the viroplasm (**figure 27.21**).

27.5 Plus-Strand RNA Viruses: Genomes That Can Be Translated upon Entry

- The genomes of plus-strand RNA viruses serve directly as mRNA molecules. Among the first viral proteins synthesized is an RNA-dependent RNA polymerase that replicates the plus-strand RNA genome, sometimes by forming a dsRNA replicative intermediate. The negative RNA strands produced by the RNA-dependent RNA polymerase can be used to make either more genomes or mRNA (**figure 27.22**).

- Poliovirus genomic RNA is translated into a single polyprotein that is cleaved to form all the proteins needed by the virus during its life cycle (**figures 27.23–27.25**).

- TMV is like most known plant viruses in that it has a plus-strand RNA genome. Translation of TMV genomic RNA yields two proteins, the larger of which is produced by a readthrough mechanism. The RNA-dependent RNA polymerase encoded by the genome synthesizes minus-strand RNAs that serve as templates for synthesis of subgenomic mRNAs that are translated to yield other proteins needed by the virus. The minus-strand RNAs also serve as templates for synthesis of new plus-strand RNA genomes. The TMV nucleocapsid forms spontaneously when disks of coat protein protomers complex with the RNA (**figure 27.26**).

27.6 Minus-Strand RNA Viruses: RNA-Dependent RNA Polymerase Is Part of the Virion

- For minus-strand RNA viruses to synthesize mRNA and replicate their genomes, their virions must carry an RNA-dependent RNA polymerase, which enters the host cells as the genome does. The polymerase first synthesizes mRNA from the negative-strand genome. Later it is used to replicate the genome by way of a plus-strand intermediate (**figure 27.27**).

- Influenza viruses have segmented genomes; most segments encode a single protein. The viruses exit by budding (**figure 27.28**).

27.7 Retroviruses: Plus-Strand Viruses That Use Reverse Transcriptase in Their Life Cycles

■ Retroviruses replicate their genomes and synthesize mRNA via a dsDNA intermediate. The dsDNA is formed by a multifunctional enzyme called reverse transcriptase (**figure 27.29**).

■ HIV is an enveloped retrovirus with a cone-shaped core that contains two copies of its genome and several enzymes, including reverse transcriptase. Upon infection, its ssRNA genome is converted to dsDNA, which is then integrated into the host genome (**figure 27.30**). HIV uses alternative splicing to generate numerous different mRNA molecules. One encodes two polyproteins. One of the polyproteins arises by a combined ribosomal frameshifting, readthrough mechanism.

27.8 Reverse Transcribing DNA Viruses

■ There are four groups of DNA viruses that use reverse transcriptase in their life cycles. One group, hepadnaviruses, includes hepatitis B virus, which has a dsDNA genome that consist of one complete but nicked strand and an incomplete (i.e., gapped) complementary strand (**figure 27.32**).

■ Upon infection with hepatitis B virus, the host cell repairs the gap and seals the nick to generate a covalently closed, circular viral genome. This serves as the template for the synthesis of pregenome RNA, which is the template for reverse transcription. Reverse transcription produces the dsDNA, gapped genome.

Compare, Hypothesize, Invent

1. No temperate RNA phages have yet been discovered. How might this absence be explained?

2. The choice between lysogeny and lysis is influenced by many factors. How would external conditions such as starvation or crowding be "sensed" and communicated to the transcriptional machinery and influence this choice?

3. The most straightforward explanation as to why the endolysin (i.e., the T4 lysozyme) of T4 is expressed so late in infection is that its promoter is recognized by the gp55 alternative sigma factor. Propose a different explanation.

4. You are studying RNA viruses and have discovered a new one that grows well in a culture of eukaryotic cells. You know that the virus is a single-stranded RNA virus, but you don't know if it is plus or minus stranded. Your lab-mate says, "Well, just treat your cell culture with cyclohexamide and see if the virus replicates its genome." You know that cyclohexamide inhibits protein elongation by binding to eukaryotic ribosomes. What is the basis of your lab-mate's suggestion?

5. White spot syndrome virus (WSSV) is an enveloped dsDNA virus that infects crustaceans. It is particularly problematic for shrimp farmers as white spot disease can cause 100% shrimp mortality within a week. During infection, WSSV produces a large amount of a protein known as ICP11. This protein is a DNA mimic; that is, its folded structure resembles the acidic double helical structure of DNA. ICP11 binds directly to the DNA-binding site of host histones. How can a protein look like DNA, given that each is a different macromolecule composed of distinct monomers? Do you think ICP11 alone could cause host cell death? If so, explain the specific steps that would lead to host cell mortality. If not, what other virus-mediated activities would be necessary?

Read the original paper: Wang, H. C., et al. 2008. White spot syndrome virus protein ICP11: A histone-binding DNA mimic that disrupts nucleosome assembly. *Proc. Natl. Acad. Sci. USA.* 105:20758.

6. Upon infection of host epithelial cells, papillomavirus (family *Papillomaviridae*) genomes are stably maintained in the nuclei for many years. These viral genomes are extrachromosomal and lack the capacity to segregate during host cell division. To prevent their loss during host cell division, the viral genomes use a protein known as E2 to attach to host chromosomes. For instance, the E2 protein of bovine papillomavirus type 1 (BPV-1) binds the host protein Brd4, which helps the viral genome attach to host chromosomes during mitosis. By contrast, the E2 protein of human papillomavirus 8 (HPV-8) does not require other host factors and binds directly to specific regions of host DNA during mitosis. How do you think the viral genome is released from a host chromosome following cell division? How do you think it was established that the HPV-8 E2 functions independently of host proteins, whereas the E2 of BPV-1 requires Brd4? Do you think the E2 proteins could be exploited to develop antiviral drugs? If so, why? How would you determine that a drug that targets E2 is specific to the viral protein and does not interfere with host cell function?

Read the original paper: Poddar, A., et al. 2009. The human papillomavirus type 8 E2 tethering protein targets the ribosomal DNA loci of host mitotic chromosomes. *J. Virol.* 83:640.

7. Associated with the envelope of herpesviruses are numerous proteins called tegument proteins. Upon fusion of the viral envelope with the cell's plasma membrane, some of the tegument proteins are released into the cytosol. The fate of these proteins is unknown, although many are thought to be degraded. Herpes simplex virus 1 (HSV-1) encodes a tegument protein called vhs. This protein is released into the cytosol as HSV-1 enters the cell, but it is not degraded. The vhs protein has been shown to make cuts in RNA molecules (i.e., it is an endoribonuclease). Suggest a role for vhs in the life cycle of HSV-1. Justify your suggestion.

Read the original paper: Dauber, B., et al. 2011. The Herpes simplex virus 1 vhs protein enhances translation of viral true later mRNAs and virus production in a cell type-dependent manner. *J. Virol.* 85(11):5363.

28

Biogeochemical Cycling and Global Climate Change

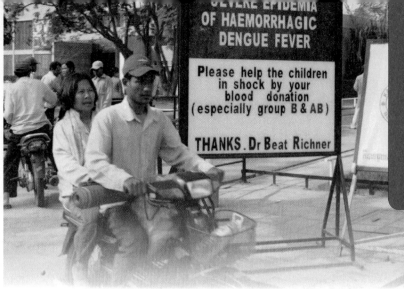

Global climate change has already increased the range of a number of insect vectors that cause disease. As the incidence of certain infectious diseases continues to shift and possibly increase, outbreaks of certain arthropod-borne diseases, such dengue fever, which is transmitted by mosquitoes, will become more common.

Global Climate Change; Global Infectious Disease Change?

When most people think of the bubonic plague, they think of medieval Europe. Malaria brings to mind sub-Saharan Africa. And if they have even heard of dengue (aka breakbone) fever, they think of a jungle somewhere in the tropics. However, these are just three of the diseases that may threaten residents living in temperate climates thanks to global climate change, according to scientists and the World Health Organization.

The links between climate and infectious disease are complex, so the exact shifts in disease patterns are difficult to predict. But consider the following facts. Insects are the vectors for many of the world's worst scourges: plague, malaria, Lyme disease, yellow and dengue fevers (just to name a few). Warmer temperatures not only expand the geographical range of many insects, they also lengthen their breeding season, boost their reproductive rates, and drive them to feed (i.e., bite) more frequently. In addition to diseases carried by insects, about 600 of the known 14,000 infectious agents, such as the viruses causing Ebola and avian flu, are shared between humans and animals. Historically, changes in the environment have been associated with the emergence of new diseases or the reemergence of old diseases.

With these facts in mind, the effects of global climate change on infectious disease can be broadly divided into three categories: altered disease patterns that are already being witnessed, those that can be predicted with confidence, and those that will surprise us. Let's start with a disease whose range has already increased—plague. This bacterial disease is endemic in human and rodent populations in parts of New Mexico, where its prevalence is correlated with the timing and amount of rainfall. Cases of human plague have recently been diagnosed farther north than ever before; plague is predicted to move as far north as Wyoming and Idaho. Similar temperature-dependent shifts in the number of rodents carrying plague bacteria have been documented in Central Asia.

The patterns of insect-borne diseases are also likely to change. Mosquitoes have brought malaria to Texas, Michigan, New Jersey, and New York. Dengue fever–bearing mosquitoes have spread as far north as

Chicago and the Netherlands. Lyme disease, a tick-borne disease, is expected to migrate into Canada. And as drenching rain and flooding becomes more common, the rates of cholera and other diarrheal diseases are expected to rise.

This leaves us with diseases that could surprise us. These could be completely new diseases or diseases that were thought to be well controlled. But whether the infectious diseases are new or old, their prevention and treatment will require the expertise of a new generation of epidemiologists, health care workers, and (of course) microbiologists.

It is obvious why microbiologists are involved in the study of the link between climate and infectious disease. What may be less obvious is the role of microbes in climate change. However, as this chapter illustrates, the cycling of all elements (not just those involved in climate change) is mediated by microorganisms. The sum of the microbial, physical, and chemical processes that drive the flow of elements between sediments, waters, and the atmosphere is known as **biogeochemical cycling.** We explore biogeochemical cycling by focusing on the flux of carbon, nitrogen, sulfur, phosphorus, iron, and magnesium. We then turn our attention to global climate change, which is driven by changes in the carbon and nitrogen cycles. Collectively, this is the realm of **environmental microbiology,** which is concerned with microbial processes that occur in ecosystems, rather than the individual microorganisms dwelling there. However, it is critical to understand that bacteria, archaea, fungi, and protists are the agents of change. With only about 1% of microbes currently in culture, the identities and specific physiological activities of most microbes await discovery. Chapter 29 describes some approaches microbial ecologists employ to make these discoveries.

Readiness Check:
Based on what you learned previously, you should be able to:

✔ Distinguish macroelements from micronutrients, list examples of each, and describe their cellular function (section 3.3)

✔ Explain how nutrient availability and environmental factors influence microbial growth (sections 3.3, 7.4, and 7.5)

✔ Compare and contrast the growth and physiology of obligate aerobes, facultative anaerobes, and strict anaerobes (section 7.4)

✔ Explain oxidation-reduction (redox) reactions (section 10.3)

✔ Diagram, in general terms, the standard reduction potentials of electron donors as compared to electron acceptors and relate this to standard free-energy change (sections 10.3 and 10.4)

✔ Describe the difference between heterotrophy and autotrophy (section 11.1)

✔ Compare anaerobic respiration, aerobic respiration, and fermentation (sections 11.3, 11.7, and 11.8)

✔ Define chemolithotrophy and identify an example electron donor and acceptor (section 11.10)

✔ Compare and contrast oxygenic and anoxygenic phototrophy (section 11.11)

28.1 Biogeochemical Cycling Sustains Life on Earth

After reading this section, you should be able to:

■ Explain the term redox potential and how it influences the flux of elements found in specific habitats

■ Sketch generalized carbon, nitrogen, phosphorus, sulfur, iron, and manganese cycles

■ Identify, by metabolic type, the microorganisms responsible for carbon and nitrogen fixation, and the mineralization of organic carbon and nitrogen under oxic and anoxic conditions

■ List environments in which methanogenesis and methane oxidation occur

■ Trace the fate of the N atom during dissimilatory nitrate reduction, nitrification, denitrification, and the anammox reaction, and indicate whether each must occur in an oxic or anoxic environment

■ Contrast and compare dissimilatory nitrate reduction with dissimilatory sulfate reduction, and assimilatory nitrate reduction with assimilatory sulfate reduction

■ Identify, by metabolic type, microorganisms responsible for dissimilatory nitrate, sulfate, and iron reduction; nitrification; and sulfur and iron oxidation

■ Contrast and compare the iron and manganese cycles

■ Provide at least two examples of how the flux of one element influences the cycling of another

■ Summarize the importance of biogeochemical cycling in maintaining life on Earth

The long success and vast diversity of microbes reflect the variety of habitats found on Earth. In addition to differences in temperature, pressure, salinity, pH, and nutrient availability, variation in redox potential is also extremely important in determining microbial activity. **Redox potential,** or more formally, oxidation-reduction potential, is a measure of the tendency of molecules in a system to accept or donate electrons. In chapter 10, we discuss redox potential as it relates to specific conjugate redox pairs *(see table 10.2)*. Here our focus is on redox potential in the context of microbial environments, such as soil, sludge, or sterile growth media. The redox potential of these environments can be quantified in volts or millivolts, but to do so, each environment must be compared to a standard. Because no "standard environment" exists, a hydrogen electrode set to zero volts is used. Thus the redox potential of an environment is determined by the electrical potential difference between the environment and a standard hydrogen electrode. The value is denoted as E_h. Environments composed principally of compounds with high redox potentials (more positive) will be more likely to accept electrons (i.e., become reduced) when a new compound is added. The newly added compound that donates electrons then becomes oxidized (**table 28.1**).

Table 28.1	Elements Important in Biogeochemical Cycling and Their Redox States					
		MAJOR FORMS AND VALENCES				
Cycle	**Significant Gaseous Component Present?**	**Reduced Form**	**Intermediate Oxidation State Forms**			**Oxidized Form**
C	Yes	Methane: CH_4 (−4)	Carbon monoxide: CO (+2)			CO_2 (+4)
N	Yes	Ammonium: NH_4^+; organic N (−3)	Nitrogen gas: N_2 (0)	Nitrous oxide N_2O (+1)	Nitrite: NO_2^- (+3)	Nitrate: NO_3^- (+5)
S	Yes	Hydrogen sulfide: H_2S; SH groups in organic matter (−2)	Elemental sulfur: S^0 (0)	Thiosulfate: $S_2O_3^{2-}$ (+2)	Sulfite: SO_3^{2-} (+4)	Sulfate: SO_4^{2-} (+6)
Fe	No	Ferrous iron: Fe^{2+} (+2)				Ferric Iron: Fe^{3+} (+3)
P	No	None	Always present in maximally oxidized state			Phosphate: PO_4^- (+5)
Mn	No	Manganese ion: Mn^{2+} (+2)	Manganese oxide MnO (+2)			Manganese dioxide MnO_2 (+4)

Relative Concentration of Element

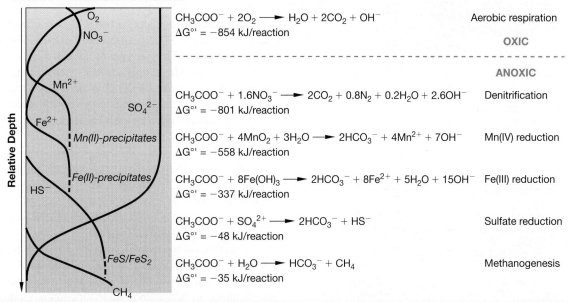

$$CH_3COO^- + 2O_2 \longrightarrow H_2O + 2CO_2 + OH^-$$
$$\Delta G^{\circ\prime} = -854 \text{ kJ/reaction}$$
Aerobic respiration

OXIC

- -

ANOXIC

$$CH_3COO^- + 1.6NO_3^- \longrightarrow 2CO_2 + 0.8N_2 + 0.2H_2O + 2.6OH^-$$
$$\Delta G^{\circ\prime} = -801 \text{ kJ/reaction}$$
Denitrification

$$CH_3COO^- + 4MnO_2 + 3H_2O \longrightarrow 2HCO_3^- + 4Mn^{2+} + 7OH^-$$
$$\Delta G^{\circ\prime} = -558 \text{ kJ/reaction}$$
Mn(IV) reduction

$$CH_3COO^- + 8Fe(OH)_3 \longrightarrow 2HCO_3^- + 8Fe^{2+} + 5H_2O + 15OH^-$$
$$\Delta G^{\circ\prime} = -337 \text{ kJ/reaction}$$
Fe(III) reduction

$$CH_3COO^- + SO_4^{2+} \longrightarrow 2HCO_3^- + HS^-$$
$$\Delta G^{\circ\prime} = -48 \text{ kJ/reaction}$$
Sulfate reduction

$$CH_3COO^- + H_2O \longrightarrow HCO_3^- + CH_4$$
$$\Delta G^{\circ\prime} = -35 \text{ kJ/reaction}$$
Methanogenesis

Figure 28.1 The Redox Potential, pH, and Oxygen Concentration Influence the Availability of Terminal Electron Acceptors.   As oxygen concentration decreases with depth, other terminal electron acceptors are used in anaerobic respiration. The distribution of microbes in sediments or soils is dictated by the free energy yield for each respiratory process. The relative concentration of elements with depth is shown for marine sediments; freshwater sediments and soil have less sulfate, thus methanogenic archaea may be more abundant.

MICRO INQUIRY *Why do sulfate reduction and methanogenesis rarely occur at the same depth or in the same microhabitat?*

Why do microbiologists care about redox potentials? The redox state of the environment plays a critical role in determining the types of microbes present because it determines which oxidized compounds are available as terminal electron acceptors for anaerobic respiration and what reduced molecules are present for use as electron donors in chemolithotrophy. In general, electron acceptors used in anaerobic respiration follow a continuum defined by the free energy yield of each reaction. The terminal electron acceptors and their vertical zonation are shown in **figure 28.1**; however, microbial metabolic versatility is far more amazing than illustrated here. Redox cycling of many other elements such as arsenic, vanadium, uranium, and chromium can also be used to support microbial growth. ◄◄ *Redox reactions (section 10.3)*

Carbon Cycle

Carbon is the element that defines life, and as such it is everywhere. The carbon cycle is often regarded in terms of sources, sinks, and reservoirs of CO_2. Sources, such as heterotrophic microbes, release carbon as CO_2, whereas sinks include plants and phytoplankton, which take up CO_2 from the atmosphere. Reservoirs store carbon for geological periods of time. The exchange of elements between sources and sinks is referred to as flux. Carbon is present in reduced forms, such as methane (CH_4) and other, more complex organic matter, and in the oxidized, inorganic forms, carbon monoxide (CO) and carbon dioxide (CO_2). Although carbon is continuously transformed from one form to another, for the

sake of clarity, we shall say that the cycle "begins" with carbon fixation—the conversion of CO_2 into organic matter (**figure 28.2**). Plants such as trees and crops are often regarded as the principal CO_2-fixing organisms, but at least half the carbon on Earth is fixed

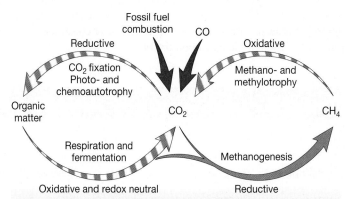

Figure 28.2 The Carbon Cycle.   Carbon fixation occurs through the activities of photoautotrophic and chemoautotrophic microorganisms. Methane is produced from inorganic substrates ($CO_2 + H_2$) or from organic matter. Carbon monoxide (CO)—produced by sources such as automobiles and industry—is returned to the carbon cycle by CO-oxidizing bacteria. Aerobic processes are indicated by red arrows, those that occur under both oxic and anoxic conditions are noted with striped arrows, and a purple arrow leads to methanogenesis, an anaerobic process.

MICRO INQUIRY *What microbes are capable of methanogenesis and methane oxidation?*

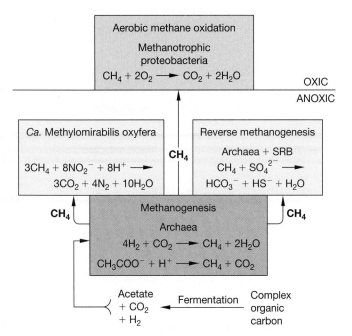

Figure 28.3 Methanogenesis and Methanotrophy. As methane is produced by archaea in terrestrial and marine anoxic sediments, it diffuses upward where it is oxidized by one of three general types of methanotrophs: the newly discovered proteobacterium *Candidatus* Methylomirabilis oxyfera, anaerobic archaea that reverse the methanogenic pathway in association with sulfate-reducing bacteria (SRB), and aerobic proteobacteria.

by microbes, particularly marine photosynthetic bacteria and protists (e.g., cyanobacteria in the genera *Prochlorococcus* and *Synechococcus,* and diatoms, respectively). Importantly, microbes also fix carbon in anoxic environments using anoxygenic photosynthesis as well as by chemolithoautotrophy in the absence of light. In fact, recent evidence suggests that bacterial chemolithoautotrophy in deep, dark subsurface sediments may constitute a significant fraction of global carbon fixation.

Alternatively, inorganic carbon (CO_2) can be reduced anaerobically to methane (CH_4). Recall that only archaea form methane from either $H_2 + CO_2$ or H_2 + acetate. As discussed in chapters 20 and 22, methane produced in sediments is then oxidized either aerobically by proteobacteria or anaerobically by archaea or newly discovered bacteria that couple methane oxidation with denitrification (**figure 28.3**). Large amounts of methane are also generated in the guts of ruminant animals. Globally, sediments found in rice paddies, coal mines, sewage treatment plants, landfills, marshes and mangrove swamps, and archaea found in the guts of ruminant animals and even termites are important sources of methane. ◄◄ *Methanogens and methanotrophs (section 20.4); Class* Alphaproteobacteria *includes many oligotrophs (section 22.1); Class* Gammaproteobacteria *is the largest bacterial class (section 22.3)*

All fixed carbon enters a common pool of organic matter that can then be oxidized back to CO_2 through aerobic or anaerobic respiration and fermentation (i.e., heterotrophy). In the carbon cycle depicted in figure 28.2, no distinction is made between the different types of organic matter formed and degraded. This is a marked oversimplification because organic matter varies widely in terms of elemental composition, structure of basic repeating units, and linkages between repeating units. Its degradation is also influenced by other factors, including (1) oxidation-reduction potential; (2) availability of competing nutrients; (3) abiotic conditions such as pH, temperature, O_2, and osmotic conditions; and (4) the microbial community present.

Many of the complex organic substrates used by microorganisms are summarized in **table 28.2**. Lignin, an important structural component in mature plant materials, is notoriously stable. Lignin is actually a family of complex amorphous polymers linked by carbon-carbon and carbon-ether bonds (*see figure 31.2*). Filamentous microorganisms—fungi and the streptomycetes—secrete hydrolytic enzymes that degrade lignin by oxidative depolymerization, a process that requires oxygen. A few microbes, such as the purple bacterium *Rhodopseudomonas palustris,* can

			ELEMENTS PRESENT IN LARGE QUANTITY					DEGRADATION	
Substrate	**Basic Subunit**	**Linkages (If Critical)**	**C**	**H**	**O**	**N**	**P**	**With O$_2$**	**Without O$_2$**
Starch	Glucose	$\alpha(1\rightarrow4)$ $\alpha(1\rightarrow6)$	+	+	+	–	–	+	+
Cellulose	Glucose	$\beta(1\rightarrow4)$	+	+	+	–	–	+	+
Hemicellulose	C6 and C5 monosaccharides	$\beta(1\rightarrow4)$, $\beta(1\rightarrow3)$, $\beta(1\rightarrow6)$	+	+	+	–	–	+	+
Lignin	Phenylpropene	C—C, C—O bonds	+	+	+	–	+	+	+/–
Chitin	*N*-acetylglucosamine	$\beta(1\rightarrow4)$	+	+	+	+	–	+	+
Hydrocarbon	Aliphatic, cyclic, aromatic		+	+	–	–	–	+	+/–

Table 28.2   Complex Organic Substrate Characteristics That Influence Decomposition and Degradability

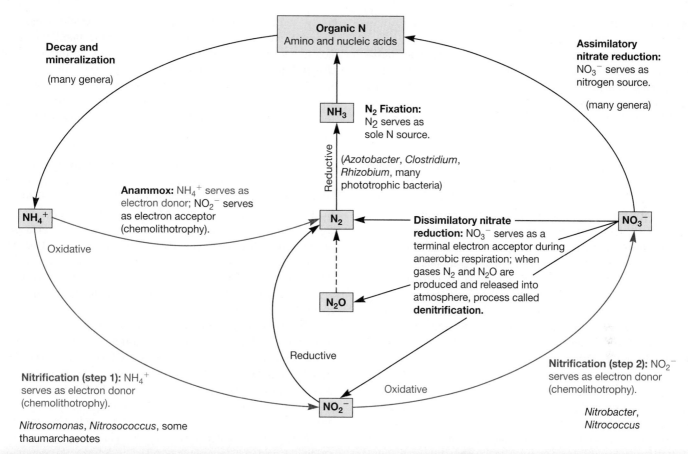

Figure 28.4 A Simplified Nitrogen Cycle. Depending on their oxidative state, some nitrogen species can function as either electron donors or electron acceptors and therefore play very different roles in the environment. NH_4^+ is fully reduced; therefore it can serve only as an electron donor. The fully oxidized species NO_3^- can function as a terminal electron acceptor, or plants and microbes can reduce it and incorporate it into biomass during assimilatory reduction.

MICRO INQUIRY *What are the products of denitrification, nitrification, dissimilatory nitrate reduction, and assimilatory nitrate reduction?*

degrade lignin anaerobically but very slowly. This diminished biodegradability under anoxic conditions results in accumulation of lignified materials, including the formation of peat bogs.

Two terms are helpful in considering the fate and availability of organic materials in nature: mineralization and immobilization. **Mineralization** describes the decomposition of organic matter to simpler, inorganic compounds (e.g., CO_2, NH_3, CH_4, H_2). These compounds may or may not be recycled within the same environment. By contrast, nutrients (including carbon) that are converted into biomass become temporarily unavailable for nutrient cycling; this is called nutrient **immobilization.** Saprophytes, which degrade dead organic material, as well as viruses, protists, and other predators are important in recycling immobilized organic compounds.

Many complex substrates contain only carbon, hydrogen, and oxygen. Growth using these substrates demands that microbes acquire the remaining nutrients (e.g., N, P, S, and Fe) elsewhere in the environment. This is often very difficult, as the concentration of nitrogen, phosphorus, and iron may be very low. When the supply of a macronutrient is insufficient to support maximal growth, that nutrient is said to be limiting. For instance, in open-ocean microbial communities, growth of many microbes is often nitrogen limited. In

other words, if higher concentrations of usable nitrogen (e.g., NO_3^-, NH_4^+) were available, the rate of microbial growth would increase. This phenomenon is called Leibig's law of the minimum, named after the nineteenth-century chemist Justus van Leibig.

Retrieve, Infer, Apply

1. What are the possible fates of methane in anoxic zones?
2. When is a nutrient mineralized and when is it immobilized?
3. Where will lignin degrade faster: in a shallow riverbed or in a rice paddy? Explain.
4. What C, N, and S forms will accumulate after anaerobic degradation of organic matter? Compare these with the forms that accumulate after aerobic degradation.
5. What kinds of microbes would be able to recycle the carbon atoms in CO_2 and CH_4?

Nitrogen Cycle

Nitrogen exists in redox states from -3 to $+5$, and this is reflected in the transformations that drive nitrogen cycling (table 28.1). As shown in **figure 28.4**, depending on its level of oxidation, nitrogen species can serve as electron acceptors in anaerobic respiration (e.g., NO_3^- and NO_2^-) or as electron donors (e.g., NO_2^- and NH_4^+) in chemolithotrophy. This can lead to some confusion, but the trick is to consider the oxidation state of the nitrogen species: is it fully oxidized (i.e., NO_3^-)? If so, it can only accept electrons. Conversely, if it is fully reduced (NH_4^+), its role is limited to electron donor. Nitrite (NO_2^-) is neither fully oxidized nor reduced, so it can function as either electron donor or acceptor.

We begin our discussion of this cycle with **nitrogen fixation**—the reduction of the inorganic gaseous molecule N_2 to its organic form (e.g., amino acids, purines, and pyrimidines). Nitrogen fixation is performed only by some bacteria and archaea. Although the nitrogenase enzyme is sensitive to oxygen, nitrogen fixation can be carried out under both oxic and anoxic conditions. Microbes such as *Azotobacter* spp. and cyanobacteria in the genus *Trichodesmium* fix nitrogen aerobically, while free-living anaerobes such as *Clostridium* spp. also fix nitrogen. Perhaps the best-studied nitrogen-fixing microbes are the bacterial symbionts of leguminous plants, including rhizobia, their α-proteobacterial relatives, and a few β-proteobacteria. Other bacterial symbionts fix nitrogen as well. For instance, actinomycete species in the genus *Frankia* fix nitrogen while colonizing many types of woody shrubs, and heterocyst-forming cyanobacteria *Anabaena* spp. fix nitrogen when in association with the water fern *Azolla* spp. ◄◄ *Nitrogen fixation (section 12.5); Phylum* Cyanobacteria *(section 21.4); Order* Frankiales *(section 24.1)* ►►| *Nitrogen-fixing bacteria are vital to agriculture (section 31.3)*

The product of N_2 fixation is ammonia (NH_3), which is immediately incorporated into organic matter as an amine. These amine N atoms are introduced into proteins, nucleic acids, and other biomolecules. Eventually these organic molecules are degraded (dissimilated) and mineralized, and the nitrogen is released as ammonium (NH_4^+). Many microbial genera are capable of dissimilation of organic nitrogen substrates, but complete mineralization requires an assemblage of microbes.

Because ammonium is fully reduced, it can only donate electrons, and this is what happens in the two-step chemolithotrophic process of **nitrification** (figure 28.4). In the first step, ammonium is oxidized to nitrite (NO_2^-), and in the second step, NO_2^- is oxidized to nitrate (NO_3^-). No single microbial genus can perform both steps of nitrification. For example, some mesophilic archaea and bacteria in the genera *Nitrosomonas* and *Nitrosococcus* play important roles in ammonia oxidation, while *Nitrobacter* spp. and related bacteria carry out nitrite oxidation. Both steps of nitrification are usually aerobic, with O_2 as the terminal electron acceptor. An exception is the β-proteobacterium *Nitrosomonas eutropha*, which oxidizes ammonium anaerobically to nitrite and nitric oxide (NO) using nitrogen dioxide (NO_2) as an acceptor in a denitrification-related reaction (*see figure 22.15*).

The production of nitrate is important because it can be reduced and incorporated into microbial and plant cell biomass in a process known as **assimilatory nitrate reduction.** Alterna-

tively, some microorganisms use nitrate as a terminal electron acceptor during anaerobic respiration. When respired, the nitrogen is not incorporated into cellular material, so this is called **dissimilatory nitrate reduction.** A variety of microbes, including *Geobacter metallireducens* and *Desulfovibrio* spp., are capable of dissimilatory nitrate reduction. When nitrate is fully reduced to dinitrogen gas (N_2), nitrogen is removed from the ecosystem and returned to the atmosphere through a series of reactions collectively known as **denitrification.** This dissimilatory process is performed by a variety of heterotrophic bacteria, such as *Pseudomonas denitrificans*. The major products of dissimilatory nitrate reduction include nitrogen gas (N_2) and nitrous oxide (N_2O), although nitrite (NO_2^-) also can accumulate (*see figure 11.18*). N_2O, an important greenhouse gas, is also produced during ammonia oxidation to nitrite; that is, the first step of nitrification. This is particularly true of marine ammonia oxidizing archaea. ◄◄ *Anaerobic respiration uses the same three steps as aerobic respiration (section 11.7); Inorganic nitrogen assimilation (section 12.5)*

The **anammox reaction** (*anoxic ammonium oxidation*) is an anaerobic reaction performed by chemolithotrophs in the phylum *Planctomycetes*. Here ammonium ion (NH_4^+) serves as the electron donor and nitrite (NO_2^-) as the terminal electron acceptor; it is reduced to nitrogen gas (N_2) (*see figure 21.16*). In effect, the anammox reaction is a shortcut to N_2, proceeding directly from ammonium and nitrite, without having to cycle first through nitrate (figure 28.4). The discovery that planctomycete bacteria oxidize considerable amounts of NH_4^+ to N_2, thereby removing ammonia from the environment, has sparked keen interest in the wastewater management field, where high levels of ammonia are undesirable. ◄◄ *Phylum* Planctomycetes *(section 21.5)*

Retrieve, Infer, Apply

1. Organic nitrogen that is introduced into an ecosystem directly by nitrogen fixation is sometimes called "new" nitrogen. Explain why.

2. Describe the two-step process that makes up nitrification. Could microbes performing nitrification and anammox live in the same environment? Explain your answer.

3. What is the difference between assimilatory nitrate reduction and denitrification? Which reaction is performed by many microbes, and which is a more specialized metabolic capability?

4. Refer to section 21.5 and figure 21.15 (*p. 496*). What nitrogenous intermediates are used in the anammox reaction? Why don't they accumulate in the environment?

Phosphorus Cycle

Biogeochemical cycling of phosphorus is important for a number of reasons. All living cells require phosphorus for nucleic acids, and some lipids and polysaccharides. However, unlike carbon and nitrogen, phosphorus exists in a single valence state of +5 and has no gaseous component (table 28.1). Indeed, all phosphorus is originally derived solely from the weathering of phosphate-containing rocks (**figure 28.5**). However, recent evidence suggests that phosphonates, which bear a C—P bond and are thus organic, may be a significant source of phosphorus for some marine microorganisms. Nonetheless, phosphorus is present in low concentrations and frequently limits growth.

In soil, phosphorus exists in both inorganic and organic forms. Organic phosphorus includes not only that found in biomass but also that in materials such as humus and other organic compounds. The phosphorus in such organic materials is readily recycled by microbial activity. Inorganic phosphorus is negatively charged, so it complexes with positively charged elements in the environment, such as iron, aluminum, and calcium. These compounds are relatively insoluble, and their dissolution is pH dependent such that phosphate is most available to plants and microbes between pH 6 and 7. Microbes transform simple orthophosphate (PO_4^{3-}) to more complex forms. These include the polyphosphates present as inclusions as well as more familiar macromolecules such as nucleotides and phospholipids. ◄◄ *Inclusions (section 3.6); Phosphorus assimilation (section 12.6)*

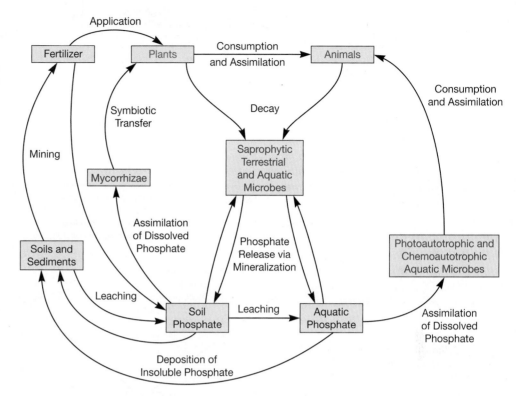

Figure 28.5 A Simplified Phosphorus Cycle. Phosphorus enters soil and water through the degradation of plants and animals, weathering of rocks, and fertilizer application. Phosphate in soil and waters is consumed by terrestrial and aquatic microbes and passed onto larger organisms. However, much of the soil phosphorus can leach great distances or, like phosphate in aquatic systems, complex with cations to form relatively insoluble compounds. Terrestrial chemolithotrophic microbes also take up soil phosphate but are not shown.

Sulfur Cycle

Microorganisms contribute greatly to the sulfur cycle; a simplified version is shown in **figure 28.6**. The sulfur cycle is similar to the nitrogen cycle in that, depending on the oxidation state of the sulfur species, it can serve as an electron acceptor, an electron donor, or both. Sulfate, the fully oxidized species, is reduced by plants and microbes for use in amino acid and protein biosynthesis; this is described as **assimilatory sulfate reduction.** By contrast, when sulfate diffuses into anoxic habitats, it provides an opportunity for microbial **dissimilatory sulfate reduction.** Here sulfate serves as a terminal electron acceptor during anaerobic respiration by a variety of microbes, including δ-proteobacteria such *Desulfovibrio* and *Desulfonema* spp. and archaea belonging to the genus *Archaeoglobus.* This results in sulfide accumulation. Sulfide, which is fully reduced, can then serve as an electron source for anoxygenic photosynthetic microorganisms and chemolithoautotrophs, including members of the phylum *Chlorobi* and the genus *Thiobacillus,* respectively. These microbes convert sulfide to elemental sulfur and sulfate. ◄◄ *Proteobacteria (chapter 22)*

Other microorganisms have been found to carry out dissimilatory elemental sulfur (S^0) reduction. These include members of the genus *Desulfuromonas,* thermophilic archaea, and cyanobacteria in hypersaline sediments. Sulfite (SO_3^{2-}) is another critical intermediate that can be reduced to sulfide by a wide variety of microorganisms, including members of the genera *Alteromonas, Clostridium, Desulfovibrio,* and *Desulfotomaculum.*

Dimethylsulfoniopropionate (DMSP) is an important organic sulfur compound. It is produced by marine phytoplankton as a compatible solute. When these cells die and release DMSP, bacterioplankton (floating bacteria) use it as a sulfur and carbon source. In the process, it is metabolized to dimethylsulfide (DMS) and released into the atmosphere. There DMS is rapidly converted into a variety of sulfur compounds that serve as nuclei for water droplet formation, contributing to the formation of clouds. Because clouds help keep the Earth's surface cool, it is hypothesized that increased DMS production could help mitigate the effects of global climate change.

Iron Cycle

The iron cycle features the interchange of ferrous iron (Fe^{2+}) to ferric iron (Fe^{3+}) (**figure 28.7**). In environments that are fully aerated with a neutral pH, iron is present primarily as insoluble minerals of either oxidation state. The solubility of both reduced, ferrous iron

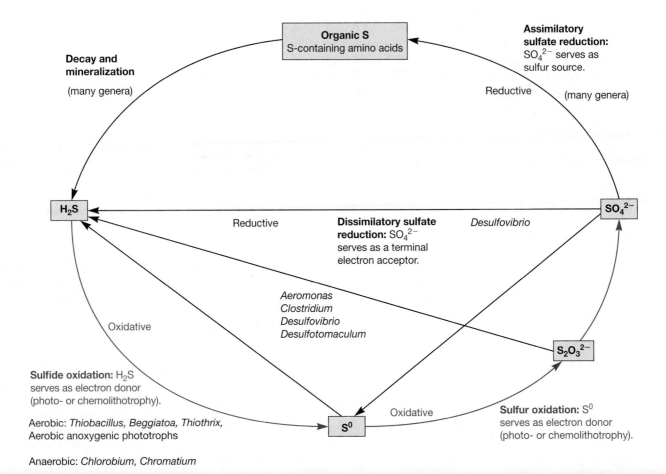

Figure 28.6 A Simplified Sulfur Cycle. Sulfate is fully oxidized, thus it serves as the terminal electron acceptor in anaerobic respiration, termed dissimilatory sulfate reduction. Alternatively, plants and microbes can donate the S atom to organic compounds during assimilatory sulfate reduction. H_2S is fully reduced, so it can serve as an electron donor during chemo- and photolithotrophy. Because elemental sulfur and thiosulfate are neither fully oxidized nor reduced, they can serve as either electron donors or acceptors.

MICRO INQUIRY *Which sulfur species is gaseous?*

Figure 28.7 The Iron Cycle. Dissimilatory reduction of iron takes place when Fe^{3+} is used as a terminal electron acceptor during anaerobic respiration. This occurs only in anoxic environments. By contrast, oxidation of Fe^{2+} can occur under both oxic and anoxic conditions. Most well-known is the use of Fe^{2+} as an electron donor and O_2 as electron acceptor. Fe^{2+} can also be oxidized by lithotrophs that use NO_3^- or even the environmental contaminants chlorate (ClO_3^-) and perchlorate (ClO_4^-) as electron acceptors.

MICRO INQUIRY *Why is Fe^{2+} used as an electron donor chiefly in acidic environments?*

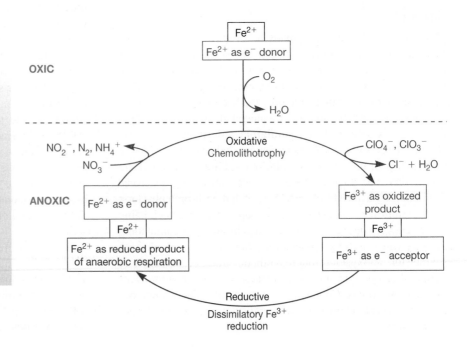

(Fe^{2+}) and oxidized, ferric iron (Fe^{3+}) increases with acidification, so that below pH 4, iron is found in the aqueous form. Its assimilation presents two challenges for most aerobic microorganisms. First, while iron is an essential element, free iron is usually present in very small quantities. Second, iron typically exists in the wrong oxidation state for uptake: Fe^{3+} dominates in oxic environments, but microbes generally incorporate Fe^{2+}. The use of siderophores solves both problems. Siderophores are low molecular weight organic molecules that bind Fe^{3+}, facilitating its transport into the cell, where it is reduced to Fe^{2+} (*see figure 3.15*). ◀◀ *Iron uptake (section 3.3)*

Once again, we see that dissimilatory reduction involves anaerobic respiration. In the case of dissimilatory iron reduction, ferric iron (Fe^{3+}) serves as a terminal electron acceptor during anaerobic respiration. In most environments, Fe^{3+} is found chiefly in a crystalline phase (e.g., hematite and magnetite) and as a component of sediment clays. As described more fully in chapter 22, some microbes donate electrons from the electron transport chain to these solid forms of Fe^{3+} outside the cell. Different microbes appear to use different strategies to transfer electrons to these external electron acceptors. For instance, δ-proteobacteria in the genus *Geobacter* and γ-proteobacteria in the genus *Shewanella* use electrically conductive threadlike structures called nanowires to transfer electrons from a reductase in the outer membrane to particulate Fe^{3+} (*see figure 22.28*). *Shewanella* spp. also produce redox-active flavins that shuttle electrons from the outer membrane to crystalline Fe^{3+} oxides. ◀◀ *Anaerobic respiration uses the same three steps as aerobic respiration (section 11.7); Order Alteromonadales includes anaerobes that use a range of electron acceptors (section 22.3)*

A wide range of archaea and bacteria are capable of dissimilatory Fe^{3+} reduction. This includes archaeal genera from *Euryarchaeota* and *Crenarchaeota*, as well as all five classes of *Proteobacteria*, *Firmicutes*, *Deferribacteraceae*, *Acidobacteria*, *Thermotoga*, and *Thermus*. This phylogenetic diversity may reflect the antiquity of Fe^{3+} reduction. It is believed that life began 3.5 billion to 3.8 billion years ago in an environment that was rich in Fe^{2+}. The photooxidation of Fe^{2+} to Fe^{3+} and H_2 would have provided an electron acceptor and energy source, respectively, to early cellular forms. Banded iron formation that occurred when atmospheric oxygen levels were beginning to increase at the end of the Precambrian era may be evidence of increased bacterial iron metabolism. ◀◀ *Microbes have evolved and diversified for billions of years (section 1.2)*

In addition to ferric iron (Fe^{3+}) as a terminal electron acceptor, some magnetotactic bacteria such as *Aquaspirillum magnetotacticum* transform extracellular iron to the mixed valence iron oxide mineral magnetite (Fe_3O_4) and construct intracellular magnetic compasses (*see figure 3.37*). Magnetotactic bacteria may be described as magnetoaerotactic bacteria because they are thought to use magnetic fields to migrate to the position in a bog or swamp where the oxygen level best meets their needs. Furthermore, some dissimilatory iron-reducing bacteria accumulate magnetite as an extracellular product. ◀◀ *Inclusions (section 3.6)*

Because it is in the reduced state, ferrous iron (Fe^{2+}) can be used as an electron donor by lithotrophic microbes in acidic, oxic

environments where it is soluble and oxygen can serve as the terminal electron acceptor. This has been well characterized in the α-proteobacterium *Acidithiobacillus ferrooxidans* and thermophilic crenarchaeotes in the genus *Sulfolobus*. In addition, it is now known that a number of microbes oxidize Fe^{2+} at neutral pH under oxic conditions. Best studied are γ-proteobacteria in the genus *Marinobacter* and β-proteobacteria in the genera *Leptothrix* and *Gallionella*. Ferrous iron can also be oxidized under anoxic conditions with nitrate as the electron acceptor. One interesting anaerobic microbe is *Dechlorosoma suillum*, which oxidizes Fe^{2+} using perchlorate (ClO_4^-) and chlorate (ClO_3^-) as electron acceptors. Because perchlorate is a major component of explosives and rocket propellants, it is a frequent contaminant at retired munitions facilities and military bases. Thus *D. suillum* may be used in the bioremediation (biological cleanup) of such sites. This process also occurs in aquatic sediments with depressed levels of oxygen and may be another route by which large zones of oxidized iron have accumulated in environments with lower oxygen levels. ▶▶ *Biodegradation and bioremediation harness microbes to clean the environment (section 43.4)*

Manganese and Mercury Cycles

Much like the iron cycle, the manganese cycle involves the transformation of the reduced form, manganous ion (Mn^{2+}), to the oxidized species, MnO_2 (equivalent to manganic ion [Mn^{4+}]). This occurs at the oxic-anoxic interface in hydrothermal vents, bogs, and stratified lakes (**figure 28.8**). Phylogenetically diverse bacteria are able to use Mn^{2+} as an electron donor with oxygen or

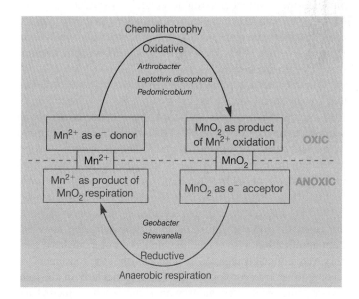

Figure 28.8 The Manganese Cycle, Illustrated in a Stratified Lake. Microorganisms make many important contributions to the manganese cycle. After diffusing from anoxic (pink) to oxic (blue) zones, manganous ion (Mn^{2+}) is oxidized chemically and by many morphologically distinct microorganisms in the oxic water column to manganic oxide—MnO_2(IV), valence equivalent to 4+. When the MnO_2(IV) diffuses into the anoxic zone, bacteria such as *Geobacter* spp. and *Shewanella* spp. carry out the complementary reduction process. Similar processes occur across oxic/anoxic transitions in soils, muds, and other environments.

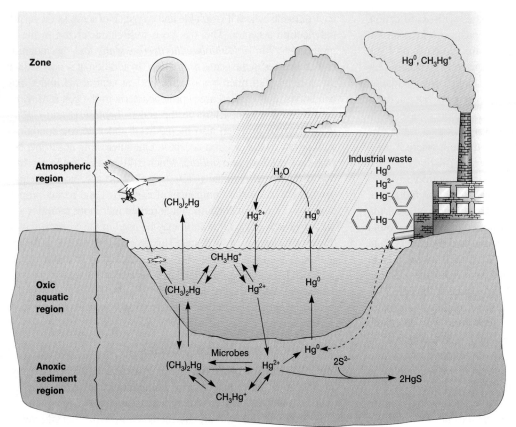

Figure 28.9 The Mercury Cycle. Interactions between the atmosphere, oxic water, and anoxic sediment are critical. Microorganisms in anoxic sediments, primarily *Desulfovibrio* spp., can transform mercury to methylated forms that can be transported to water and the atmosphere. These methylated forms also undergo biomagnification. The production of volatile elemental mercury (Hg^0) releases this metal to waters and the atmosphere. Sulfide, if present in the anoxic sediment, can react with ionic mercury to produce less soluble HgS.

nitrate at the terminal electron acceptor. *Shewanella* spp., *Geobacter* spp., and other chemoorganotrophs can carry out the complementary manganese reduction process, using MnO_2 as an electron acceptor. ▶▶ *Microorganisms in lakes can cause seasonal blooms (section 30.3)*

The mercury cycle illustrates many characteristics of metals that can be methylated. Mercury compounds were once widely used in industrial processes. A devastating situation developed in southwestern Japan when large-scale mercury poisoning occurred in the Minamata Bay region because industrial mercury was released into the marine environment. Inorganic mercury that accumulated in bottom muds of the bay was methylated by anaerobic bacteria of the genus *Desulfovibrio* (**figure 28.9**). Methylated mercury forms are volatile and lipid soluble, and mercury concentrations increase with each link of the food web in a process known as **biomagnification.** Fish containing high levels of methylated mercury were ultimately ingested by humans—the "top consumers"—leading to severe neurological disorders, particularly in children. Clearly, microbial interactions with metals such as methylated mercury can have far-reaching consequences.

Interaction Between Elemental Cycles

So far, we have presented biogeochemical cycling as discrete nutrient fluxes. However, it is important to understand that biogeochemical cycling involves dynamic, interconnected processes that over geologic time keep the biosphere in a self-sustaining steady state. Although links between cycles occur at a global

scale, it is easiest to consider such links at the level of a single type of microbe. Here we present two illustrative examples: chemolithotrophy and nitrogen fixation.

When a community of chemolithotrophic microbes uses ammonium (NH_4^+) and nitrite (NO_2^-) as its source of electrons and CO_2 as its sole source of carbon, it contributes to both the flux of carbon and nitrogen by producing organic carbon and nitrate, which can be taken up by assimilatory nitrate reduction or returned to the atmosphere following denitrification (**figure 28.10**). Because

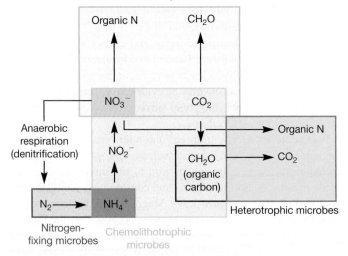

Figure 28.10 Examples of Interaction Between the N and C Cycles. The activities of microbes with different metabolic strategies result in linkage between elemental cycles. Common pools of organic carbon and nitrogen are typically created in an ecosystem; however, they are shown separately here for clarity.

chemolithoautotrophs fix CO_2 and use reduced inorganic compounds such as ammonium or ferrous iron as a source of electrons, they must oxidize large amounts of these compounds to generate enough ATP and reducing equivalents for carbon fixation. This reflects the relatively small amount of energy released during the oxidation of most inorganic chemicals. Thus chemolithotrophs not only link the carbon cycle to either the nitrogen, sulfur, or iron (or other metal) cycles, they also significantly impact the global flux of these elements.

In the case of nitrogen fixation, bacteria and archaea convert N_2 to ammonia, which is then incorporated into organic nitrogen molecules such as amino and nucleic acids. In many environments, the amount of carbon that is taken up by plants and microbes (either as CO_2 or as dissolved organic carbon) is dictated by how much nitrogen is available to support growth. Thus by introducing "new" nitrogen into the system, nitrogen-fixing microbes can also increase the rate of carbon flux.

Although we present only two examples, links between nutrient cycles abound, and new interactions are being discovered. For example, recent advances in microbial ecology have revealed that microbial methane oxidation links the carbon cycle to the nitrogen and sulfur cycles (figure 28.3).

Retrieve, Infer, Apply

1. Compare the products of dissimilatory sulfate reduction to those of dissimilatory nitrate reduction.
2. How might iron have been important in the evolution of early life?
3. What are some important microbial genera that contribute to manganese cycling?
4. How can microbial activity render some metals more or less toxic to warm-blooded animals?
5. Suggest how a chemolithoautotroph might link the carbon and sulfur cycles.

28.2 Global Climate Change: Biogeochemical Cycling Out of Balance

After reading this section, you should be able to:

- Differentiate between the terms "global warming" and "global climate change" and identify greenhouse gases
- Explain the link between greenhouse gas accumulation and microbial cycling of carbon and nitrogen
- Discuss the origin of the greenhouse gases CO_2, methane, and nitrogen oxide species

Microbial activity is critical in maintaining the dynamic equilibrium that defines our biosphere. The capacity of microbes to thrive in seemingly every niche on Earth is the result of 3.5 billion years of microbial evolution. Importantly, changes in the physical and chemical environment to which microbes have adapted generally occurred over geological time scales. However, since the beginning of the twentieth century, the rate at which CO_2 and other so-called greenhouse gases have entered the atmosphere

has been much faster than at any other time in the known history of life on Earth. Atmospheric gases such as CO_2, CH_4, and nitrogen oxides are called **greenhouse gases** because they trap the heat that is reflected from Earth's surface in the atmosphere, rather than allowing it to radiate into space. The rate at which these gases are entering the atmosphere exceeds the rate by which the natural carbon and nitrogen cycles can recycle them, thus they accumulate. It is the ever-increasing atmospheric concentration of these gases that has resulted in what is often called global warming. However, the term global climate change more accurately reflects the changes in patterns of wind, precipitation, and ocean temperatures we are now experiencing.

The most abundant greenhouse gas is CO_2. Since the onset of the industrial era, about 150 years ago, global CO_2 levels have risen from 278 ppm (parts per million) to a current level over 400 ppm (**figure 28.11**). Most of this can be attributed to the

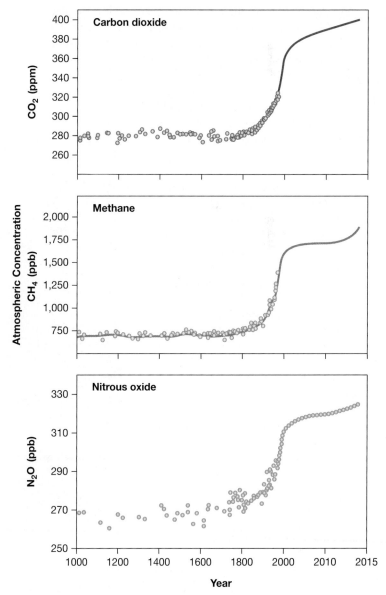

Figure 28.11 Global Atmospheric Concentrations of Greenhouse Gases.
ppm = parts per million, ppb = parts per billion.

burning of fossil fuel, which releases about 10 billion tons of CO_2 each year. Changes in land use management, principally through deforestation, account for another 1.2 billion tons of CO_2 annually. To understand why these activities result in the release of CO_2, consider that fossil fuel (e.g., oil) is formed from decaying organic matter. This organic matter arose through millions of years of CO_2 fixation (i.e., removal of CO_2 from the atmosphere). It has been locked deep within the Earth for almost as long. When fossil fuels are burned, the energy of combustion oxidizes this material back to CO_2 in a tiny fraction of the time it took to accumulate.

The situation is different with deforestation. Forests are generally regarded as CO_2 sinks because the abundance of photoautotrophic plants pulls a great deal of CO_2 out of the atmosphere and converts it to biomass. Indeed, during the 1990s terrestrial ecosystems sequestered about 3 gigatons of carbon per year (a gigaton is 1 billion metric tons). Thus when forests are cut down for firewood or agricultural reasons, the Earth loses a carbon sink. While some speculate that other, remaining plants will assimilate the extra CO_2 and simply grow faster, changes in the rate of soil microbial respiration complicate matters. Recall that as respiration rates increase, so does the quantity of CO_2, CH_4, and N_2O released. In fact, recent analysis shows that warming permafrost in subarctic regions has dramatically increased soil microbial respiration. Thus the balance of CO_2 and other greenhouse gases in terrestrial ecosystems remains unclear. The important role oceans play in maintaining atmospheric CO_2 is discussed in chapter 30.

Methane is a greenhouse gas of increasing concern because it has about 25 times the global warming potential of CO_2. This means that a single molecule of CH_4 released into the atmosphere has the same thermal retention capacity as 25 CO_2 molecules. Based on analyses of gas bubbles in glacier ice cores, the levels of methane in the atmosphere remained essentially constant until about 150 years ago. Since then, methane levels have increased 2.5 times to the present level of about 1,890 parts per billion (ppb) (figure 28.11). Considering these trends, there is a worldwide interest in understanding the factors that control methane production and oxidation by microorganisms.

To understand the chief source of the nitrogen oxide greenhouse gases NO and N_2O, collectively referred to as "Nox," we need to examine food production. The "green revolution" of the mid-twentieth century ensured that agricultural output, boosted by the application of human-made fertilizer, could keep pace with population growth. The manufacture of fertilizer is an energy-intensive process that uses hydrogen gas to reduce N_2 to NH_4^+ at high temperature and pressure. This process was first described by Fritz Haber and Karl Bosch in 1913, but it was not until the middle of the twentieth century that the Haber-Bosch reaction was used at an industrial level. Current fertilizer production is about 1.3×10^{14} g per year, or about 400 times greater now than it was in 1940. In contrast to the positive outcome of increased agricultural output, the addition of so much ammonium to the soil has altered the balance of the nitrogen cycle.

What becomes of the extra NH_4^+? That which is not taken up by plants generally has one of two fates: runoff or nitrification followed by denitrification (**figure 28.12**). Ammonium runoff leaches into lakes and streams, frequently causing **eutrophication**—an increase in nutrient levels that stimulates the growth of a limited number of organisms, thereby disturbing the ecology of these aquatic ecosystems. By contrast, microbial nitrification results in the oxidation of ammonium to more nitrate than can be immobilized by plants and microbes, because organisms need a specific ratio of C:N:P, *(see page 669).* The process of denitrification converts this extra nitrate to N_2 and the reactive greenhouse nitrogen oxides. This cycle of nitrification/denitrification

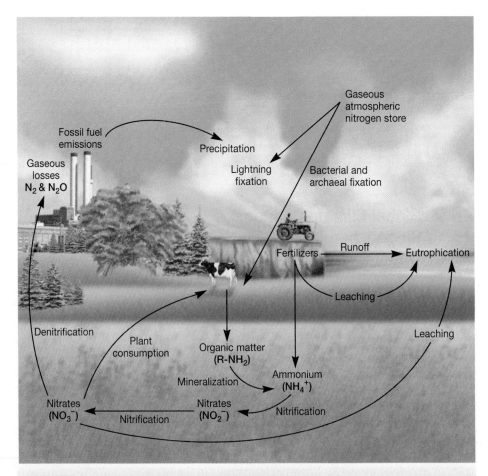

Figure 28.12 Natural and Human-Made Influences on the Nitrogen Cycle.

MICRO INQUIRY *What organisms benefit from nitrification?*

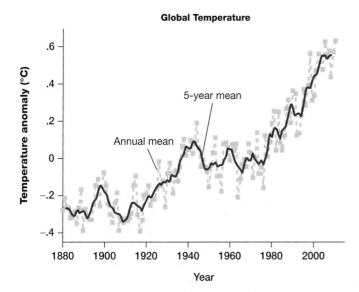

Figure 28.13 **Global Annual-Mean Surface Air Temperature Change.** Data is derived from the meteorological station network, Goddard Institute for Space Sciences, http://data.giss.nasa.gov/gistemp/graphs/.

fueled by NH_4^+ introduced as fertilizer is responsible for the highest N_2O levels in 650,000 years.

What are the consequences of disrupting the carbon and nitrogen cycles? Global climate change is the most obvious example. It is important to keep in mind that weather is not the same as climate. While North America has suffered some of the hottest summers on record in the past decade, a single day or week in July that is particularly hot is not, by itself, evidence of global climate change. Global climate change is measured over decades and includes many parameters such as surface temperature on land and sea, and in the atmosphere and troposphere; rates of precipitation; and frequency of extreme weather. Based on these analyses, the average global temperature has increased 0.74°C since the early twentieth century, and this rise is directly correlated with fossil fuel combustion to CO_2 (**figure 28.13**). Depending on the rate of continued increase in greenhouse gases, the average global surface temperature is predicted to rise between 1.1 and 6.4°C by 2100.

An important question is how will microbes respond to a changing world. Because for the vast majority of Earth's history, microorganisms have been the drivers of elemental cycling, changes in microbial activities will have a major impact on the rate and magnitude of greenhouse gas accumulation and global climate change. The role microbes play in balancing carbon and nitrogen fluxes has opened new avenues of research in microbial ecology.

Retrieve, Infer, Apply

1. List three greenhouse gases. Discuss their origins.
2. Discuss the possible role of forests in the control of CO_2.
3. How do changes in the nitrogen cycle caused by fertilization influence the carbon cycle?
4. Given that each microbial group has an optimum temperature range for growth, how might you predict changes to a soil microbial community living in your geographic area?

Key Concepts

28.1 Biogeochemical Cycling Sustains Life on Earth

- The redox potential of the environment has a major impact on the type of metabolic processes that can occur (**figure 28.1**).
- The cycling between inorganic (CO_2) and organic carbon occurs aerobically and anaerobically by all types of microbes. Only methanogenic archaea can produce methane, an anaerobic process. Methane can be oxidized by bacteria in the presence of oxygen, or anaerobically by archaea (**figure 28.3**).
- Assimilatory processes involve incorporation of nutrients into the organism's biomass during metabolism; dissimilatory processes involve the release of nutrients to the environment after metabolism. Dissimilatory reduction usually involves the use of the oxidized compound (e.g., NO_3^- or SO_4^{2-}) as the terminal electron acceptor during anaerobic respiration (**figures 28.4** and **28.6**).
- Chemolithotrophs use reduced elements as electron donors and generate the oxidized form, which can then be used for assimilatory or dissimilatory processes (**figures 28.4** and **28.6**).
- Phosphorus is cycled in its oxidized state, phosphate. Unlike other elements, there is no gaseous form. Phosphate must be leached from rocks and soils. (**figure 28.5**).

- Iron and manganese cycle between their oxidized and reduced forms without the accumulation of intermediates (**figures 28.7** and **28.8**).
- Mercury cycling is complex but can have significant effects on the health of higher organisms (**figure 28.9**).
- Although it is convenient to consider a single element cycling, in nature there are extensive and complex interactions between nutrient cycles (**figure 28.10**).

28.2 Global Climate Change: Biogeochemical Cycling Out of Balance

- Microbes have evolved slowly over time; this has given rise to the biogeochemical cycles that sustain life on Earth. However, since the beginning of the twentieth century, the rate at which CO_2, CH_4, and nitrogen oxides have been released into the atmosphere has outpaced the rate at which they can be recycled. Thus these greenhouse gases are accumulating in Earth's atmosphere (**figures 28.11** and **28.12**).
- The increase in greenhouse gases is correlated with an increase in the global annual-mean surface air temperature (**figure 28.13**).

Compare, Hypothesize, Invent

1. Compare the production of N_2 gas by denitrification and anammox. Are they both aerobic processes? What are the implications for a given pool of NO_2^- and NO_3^-?

2. Examine figure 28.12. Farmers in Argentina use a crop/cattle rotation system in which cattle graze on pastures for about 5 years, then are moved to a different area and crops are planted where they had grazed. Crops are grown for 3 years, and then cattle are rotated back and the field reverts to a pasture. This dramatically reduces the amount of nitrogen fertilizer that must be added when crops are grown. Why is this the case? Why is it argued that this practice also minimizes fossil fuel emissions?

3. A bacterium isolated from sewage sludge was recently found to be capable of anoxygenic photosynthesis using nitrite as the electron donor, converting it to nitrate. Compare this form of nitrification to that which is well characterized. By removing nitrite from the environment for reductant, what other processes within the nitrogen cycle might this bacterium be influencing?

4. It is well documented that aquatic ecosystems have been negatively impacted by the influx of nitrogen from anthropogenic sources, including fertilizer and fossil fuel combustion. Although some of this N is removed by nitrification and immobilization, a better understanding of removal processes is needed for appropriate management. To that end, globally distributed lakes were analyzed and surprisingly showed that the availability of phosphorus promotes nitrogen removal. This has real-world implications, as many communities have implemented management plans to reduce phosphorus influx to natural waters. Explain how an increase in one element, phosphorus, could lead to the removal of another, nitrogen. Think about the carbon cycle. How could an increase in another element in a given ecosystem alter the loss of carbon as CO_2?

Read the original paper: Finlay, J. C., et al., 2013. Human influences on nitrogen removal in lakes. *Science.* 342:247.

Learn More

 shop.mheducation.com   Enhance your study of this chapter with interactive study tools and practice tests. Also ask your instructor about the resources available through Connect, including adaptive learning tools and animations.

29

Methods in Microbial Ecology

Recent advances in the methods used by microbial ecologists have made it possible to explore the deep-sea sediment microbial community. Surprisingly, the benthos may be home to as many as half the microbes on the planet.

Scientists Search for Intraterrestrial Life—and Find It

In 1955 the "father of marine microbiology," Claude ZoBell, concluded that there was very little, if any, microbial life beneath the seabed. He based this assessment on the absence of bacterial growth in cultures from marine sediments deeper than about 10 meters. About a decade later, this notion was reinforced when the deep-sea submersible *Alvin* broke free of its tethers and sank. Although the scientists in the sub were rescued, a sandwich was not. Ten months later, *Alvin* was recovered and the sandwich was still intact. "Aha!" said scientists, "the ocean depths are too extreme to support microbial life!" Not surprisingly, this approach to studying deep-sea microbial ecology has not stood the test of time.

The modern age of subsurface microbiology was launched in the mid-1980s, when the U.S. Department of Energy drilled wells 200 m deep and a variety of microorganisms from subsurface aquifers were cultured. This was followed by a 1994 report of the growth of microorganisms from sediments collected 500 m below the seabed. In both cases, novel culture conditions were employed that more closely resembled the pressure, temperature, and redox potential of subsurface environments. Nonetheless, it was hard to convince the scientific community that these microbes weren't contaminants.

In the twenty-first century, advances in the tools available to microbial ecologists have propelled the "deep hot biosphere" to one of the most dynamic fields in microbiology and ecology. Technological advances including the advent of culture-independent methods have been crucial in probing what may be half of all microbial life on Earth. These techniques are designed to ask not just "Is there life?" but "Who is there?" "How fast are they growing?" and "What is their food?" Obtaining microorganisms in pure culture remains the most powerful tool available for understanding the taxonomy, genetics, and physiology of a single microbial strain. But just as it is impossible to learn everything about zebras by studying only those in zoos, microbiologists cannot fully understand microbes without exploring them in nature. In the past, investigations were often limited by technology; now this rarely seems to be the case. What would ZoBell think?

Recent estimates suggest that most microbial communities have between 10^{10} to 10^{17} individuals representing at least 10^7 different taxa. How can such huge groups of organisms exist and, moreover, survive together in a productive fashion? To answer this question, one must know which microbes are present and how they interact. In other words, one must study **microbial ecology**—an interdependent series of inquiries that involves traditional, laboratory-based analyses, metagenomics, and in situ (Latin, in place) biogeochemical assessments (**figure 29.1**). The application of these techniques has brought an explosion of recent advances and generated a new appreciation of the vast diversity of microbes and their role in biogeochemical cycling.

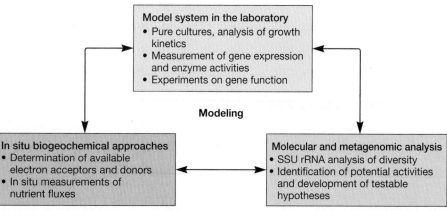

Figure 29.1 Microbial Ecology Is an Interdisciplinary Science. It relies on the interplay between a variety of approaches, which can be broadly classified as shown here.

Readiness Check:

Based on what you have learned previously, you should be able to:

✔ Differentiate between defined and complex culture media (section 7.6)

✔ Describe conventional techniques used to isolate single colonies and establish a pure culture (section 7.6)

✔ List environmental factors that influence microbial growth (section 7.4)

✔ Explain why and how microbes frequently grow in biofilms (section 7.5)

✔ Evaluate the importance of intercellular molecular signaling to microbial growth in nature (section 7.5)

✔ Explain why PCR results in the amplification of a specific DNA sequence despite many competing sequences (section 17.2)

✔ Summarize the principle and application of gel electrophoresis (section 17.1)

✔ Review the theory and practice of metagenomics and functional genomics (sections 18.3 and 18.5)

✔ Compare and contrast different molecular approaches to microbial taxonomy and phylogeny (section 19.3)

✔ Explain why SSU rRNA is used as a taxonomic marker (section 19.3)

✔ Explain how the oxidation-reduction (redox) state of an environment impacts the types of microbial metabolic processes that might be found there (section 28.1)

29.1 Microbial Biology Relies on Cultures

After reading this section, you should be able to:

■ Explain why, to the best of our knowledge, most microorganisms resist culturing in the laboratory

■ Describe the use of enrichment cultures

■ State when and how to apply the most probable number (MPN) method

■ Describe at least two innovative approaches used to coax growth in the laboratory of previously uncultured microbes

■ Explain the fundamentals of flow cytometry

It is now well understood that less than 5% of microbes have been grown under laboratory conditions. There are many reasons a particular microbe may resist culturing; a few are listed in **table 29.1**. The discrepancy between the number of microbial cells observed by microscopic examination and the number of colonies cultivated from the same natural sample has been called the **great plate count anomaly.** This has led to the description of such potentially viable microbes as being "nonculturable" (*see figure 7.35*). A microbe is deemed **viable but nonculturable** (**VBNC**) if it fails to be cultivated but is determined to be viable by one of several methods. A common technique uses dyes that discriminate between live and dead cells that can be visualized microscopically (**figure 29.2**). A molecular approach, called viability polymerase chain reaction (PCR) also uses a dye to differentiate between live and dead cells. In this case, the dye enters dead cells whose membranes have become damaged. Upon photoactivation, the dye crosslinks the DNA, thereby preventing subsequent PCR amplification of target genes (e.g., SSU rRNA). In this way, the DNA of only viable cells is present in the pool of PCR templates and is therefore detected as amplified DNA product.

Although the "gold standard" is an **axenic culture** (i.e., a pure culture) of the microbe of interest, one typically has to purify the microbe from a natural microbial assemblage. In such cases, it is often best to increase the number of microbes of interest relative to others in the same sample. **Enrichment culture** techniques are based on recreating a microenvironment in the laboratory to promote the growth of a microorganism formerly restricted to a small ecological niche. At the same time, growth of other microbial types is inhibited. This approach often plays a central role in finding new microbes. As described in chapter 7, the success of an enrichment culture depends on a thorough understanding of the specific niche the microbe of interest inhabits and the physiological features that set that microbe apart from others. ◀◀ *Enrichment cultures (section 7.6)*

Table 29.1	Reasons Microbes May Be "Unculturable"
Potential Problem	**Example Method Designed to Overcome**
Microbe is slow growing.	Incubation times of weeks to months
Microbe is present in very low abundance.	Extinction cultures, using many replicates
Different microbes in same habitat are physiologically very similar. OR Inhibition by other microbes in a mixed culture	Remove other microbes from the sample by physical methods such as filtration or density-gradient centrifugation, or use extinction cultures.
Fastidious growth requirement	Assess growth requirements of similar microbes, if known; use annotation of metagenomic sequences to infer nutritional capabilities and requirements; grow in diffusion chambers that allow influx of small molecules from natural environmental samples without microbial contamination.
Crossfeeding or communication signals from other microbes are needed.	Co-cultivate strains; use diffusion chambers; grow in conditioned spent medium of "helper" microbe.
Triggers for growth or exit from a dormant state are not present.	Add known growth triggers such as N-acetylmuramic acid.

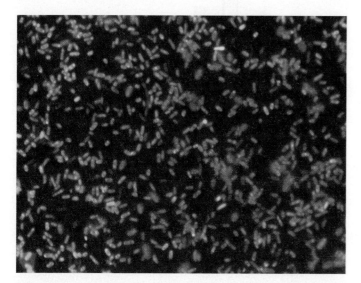

Figure 29.2 Assessment of Microbial Viability by Use of Direct Staining. By using differential staining methods, it is possible to estimate the portion of cells in a given population that are viable. In the LIVE/DEAD *Bac*Light Bacterial Viability procedure, two stains are used: a membrane-permeable green fluorescent nucleic acid stain and propidium iodide (red), which penetrates only cells with damaged membranes. In this *Escherichia coli* culture, living cells stain green, while dead and dying cells stain red.

If the microbes of interest can be cultured, the number of microbes in the sample can be estimated by using the **most probable number** (**MPN**) technique. This can be used to quantify specific types of microbes in natural samples, enrichment cultures, or, as is often the case, in food or water samples. The MPN technique involves establishing serial, ten-fold dilutions of the sample. Triplicate tubes of each dilution are usually established, although sometimes five or six replicates of each dilution are used for more accurate results. The theoretical basis of the MPN approach is that even if only one cell is inoculated into a tube, growth will occur. As shown in **figure 29.3**, after incubation at the appropriate temperature and duration, the tubes are examined for growth. Ideally, beyond a certain dilution, tubes fail to show growth because no cells were introduced into the medium. This is seen in the 10^{-5} and 10^{-6} triplicates in figure 29.3. In this case, the first dilution without growth would be marked as the last dilution in a set of three that includes the two dilutions that precede it. The number of tubes in the first set (10^{-3}, in this case) and the middle set is recorded. In our example, we see that this gives us a pattern of three tubes with growth in the 10^{-3} dilution tubes, one tube with growth in 10^{-4}, and zero tubes in 10^{-5}. A table based on statistical and theoretical considerations is consulted (figure 29.3*b*), and we see that the pattern 3-1-0 gives a MPN of 0.43. What does this mean? The most probable outcome is that an average of

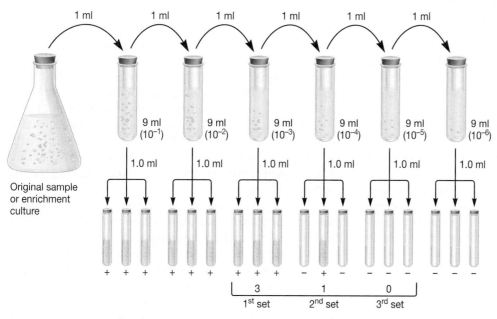

① A dilution series of the sample is made. Usually ten-fold dilutions are used.

② 1.0 ml of each dilution is used to inoculate triplicate tubes of growth medium.

③ The tubes are evaluated for growth. The first set of dilutions that fails to show growth (or the set with the least number of tubes without growth) is used to bracket a set of three dilutions.

④ The MPN is determined by consulting a table that has been established using statistical analysis. (Only a small part of the MPN table is shown here.)

(a) Procedure for MPN analysis

Number of positive tubes in:			MPN
1st set	2nd set	3rd set	
3	0	3	0.95
3	1	0	0.43
3	1	1	0.75

MPN equals
4.3 x 10^3 microbes per 1.0 ml of sample

(b) MPN table

Figure 29.3 Most Probable Number Technique.

0.43 organisms were inoculated into each of the tubes of the middle set (in this case, 10^{-4}). From this, we conclude that the most-probable number of organisms per 1 milliliter of the original, undiluted sample would be 0.43×10^4 or 4.3×10^3.

Inspection of the MPN table in figure 29.3 reveals that the results do not always show the expected lack of growth only in the lowest dilution tubes. This can happen if the microbes were not randomly distributed within the sample or if they grew in clumps or chains. When this occurs, less than ideal results are obtained, and it may be best to use more replicates or a plating technique so that individual colonies can be counted.

Clearly, the MPN technique is only valid if the microbe of interest can be grown in the laboratory. Although this is not true of most microorganisms found in nature, it would be premature to claim that microbes currently unculturable will never be cultivated. Traditional growth media are now giving way to novel media and incubation conditions. One approach is to use unusual electron donors and acceptors. For instance, a novel isolate from Mono Lake (California) sediments was cultivated only when anoxic conditions with arsenite as the electron donor and nitrate as the electron acceptor were provided. It has also become clear that growing "unculturable" microbes requires patience. Microbiologists are most familiar with doubling times in the order of hours or, at the most, a few days, when in fact, this is not always the case. For example, newly described anaerobic soil isolates were discovered only when solid cultures were incubated over a 3-month period. Indeed, 20% of the colonies did not appear until after the second month. Furthermore, microbiologists have been trained to equate turbidity in liquid culture with growth. This is not always true, as some bacteria may stop dividing (i.e., reach stationary phase) at cell densities too low to appear turbid.

A variety of new culture techniques have been applied that address specific needs of microorganisms growing in diverse environments. Efficiency can be greatly improved by using "high-throughput" methods. One popular approach is the **extinction culture technique,** so named because natural samples are first microscopically examined and then diluted to a density of one to 10 cells, that is, diluted to extinction, much like in the MPN assay. Multiple 1-milliliter cultures are established in 48 well microtiter dishes. After incubating for the desired time under appropriate conditions, the samples are stained and examined microscopically to screen for growth. In many cases, cultures that do not appear turbid will nonetheless show signs of viable microbial growth and can be transferred to fresh medium. It is important that the initial culture has fewer than 10 cells; additional transfers and dilutions can ensure that a culture arising from a single cell (i.e., clonal growth) eventually is obtained.

Recent progress in cataloging microbes that make up the human microbiome has impelled the development of a high-throughput cultivation technique called culturomics. **Culturomics** involves at least three steps: first, samples of a mixed assemblage of microbes are used as inoculum for hundreds of different culture media incubated under a variety of physical conditions (e.g., temperature, pH,

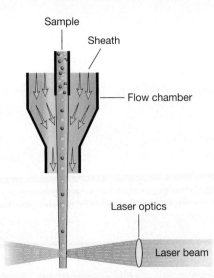

Figure 29.4 Flow Cytometry. Fluorescently labeled cell samples are injected into a fluid stream that narrows using a technique called hydrodynamic focusing. This forces only one cell at a time to pass a laser beam and scatter light in all directions. Forward-scattering light is a function of cell size, while the nature of laterally scattered light is determined by morphological complexity. This information is detected, measured, and converted to an electrical signal that reports the number and types of cells present.

MICRO INQUIRY *What happens to the flow stream in fluorescence activated cell sorting (FACS)?*

and oxygen concentration). Microorganisms from colonies that grow are then identified by mass spectrometry (*see page 447*) and small subunit (SSU; 16S for bacteria and archaea, 18S for eukaryotic microorganisms) rRNA gene sequencing. From these initial cultures, larger cultures can be started—that is, cultures are "scaled up," and in some cases organisms are grown in pure culture. Even if pure culture is not possible, these microbes can be the source for genome sequencing by single-cell genomics (*see page 426*). In this way, it is possible to grow and identify thousands of microorganisms and purify hundreds of microbial isolates, including bacteria, archaea, and fungi.

Two different techniques are commonly used to isolate single cells from a mixed population: flow cytometry and optical tweezers. In **flow cytometry,** cells are tagged with a fluorescent dye and injected into a flowing stream of fluid. As shown in **figure 29.4**, the diameter of the stream narrows, forcing one cell at a time through a laser beam that scatters its fluorescence. A detector collects and measures the light scattered in the forward direction and converts it to a voltage pulse that is proportional to the size of the cell. Light that scatters to the side is a function of morphological complexity of the cell, and this also is detected and quantified. Thus based on cell size and morphology, flow cytometry can detect different populations of cells. This enables an additional function, **fluorescence activated cell sorting (FACS)**. Here, as the stream leaves the detector, vibration causes the fluid to break into single droplets with no more than one fluorescent cell per drop. The detector software identifies the specific fluorescent signal that depends on the size and complexity of the cell. Cells become electrostatically charged as they pass the detector. The charge is used to guide the cell into a collection vessel. Thousands of cells per second pass through the laser beam, making "flow" an extremely efficient means to count and sort cells.

A less efficient but more visually rewarding approach of isolating single cells is accomplished by **optical tweezers**. This technique uses a highly focused laser beam to manipulate microbial cells. As the laser beam is focused, it passes through a very strong electric field gradient. Cells are dielectric (i.e., they can be polarized in an electric field), so they are attracted to the center of the light beam within the gradient. Because light is a collection of particles (photons) that interact with matter in its path, the traveling laser beam impinges on cells as they travel through the electric current, moving the cells toward the narrowest point of the focused beam. In this way, an optical trap is made that can be used to move microbial cells (and other small, dielectric particles, including DNA and proteins) in a directed fashion. When coupled with micromanipulation, a desired cell or cellular organelle can be drawn up into a micropipette. As with flow cytometry, once the microbe is isolated, it can be used as inoculum for the development of an axenic culture. Alternatively, PCR amplification of DNA from the individual cell can provide sequence data for phylogenetic analysis or single cell genomic sequencing. ◄◄ *Single-cell genomic sequencing (section 18.2)*

Retrieve, Infer, Apply

1. When might a microbe be considered VBNC?
2. What is an enrichment culture? Design an enrichment culture to isolate free-living nitrogen-fixing soil bacteria.
3. Describe the rationale upon which the extinction culture technique is based.
4. Would you use fluorescence activated cell sorting or optical tweezers to obtain cells for analysis from (a) lake water, (b) a biofilm, (c) soil? Explain your choices.

29.2 Genetic Methods Are Used to Assess Microbial Diversity

After reading this section, you should be able to:

- Explain why microbial communities are often investigated in situ
- Describe why, when, and how FISH and CARD-FISH are used
- Summarize how PCR is used to take a microbial census
- Explain why and how DGGE is used
- Assess the advantages and disadvantages of DNA association studies
- Compare and contrast the use of phylochips with other microarray-based technologies designed to assess microbial communities
- Predict which techniques would be appropriate for assessing the microbial community in a gram of forest soil and in a milliliter of seawater

Microbial ecologists study natural microbial communities that exist in soils or waters, or in association with other organisms, including humans. In all cases, scientists seek to assess microbial diversity.

Diversity can be a measure of the number of genera, species, or ecotypes in an ecosystem. Alternatively, diversity can reflect the various types of metabolic strategies used in a single habitat: What carbon sources, and electron donors and acceptors are used by the different microbes? A microbial **population** is defined as a group of microorganisms within an ecosystem that are the same species. When a group of microbes is defined by its physiological activity, it is called a **guild,** in the same way that guild of people often represents their occupation, for example, a carpenter's guild. A microbial example would be a guild of nitrifying microbes in a soil sample.

Staining Techniques

The most direct way to assess microbial diversity is to observe microbes in nature. This can be carried out in situ using immersed microscope slides or electron microscope grids placed in a location of interest, which are then recovered later for observation. Samples taken from the environment often are examined in the laboratory using classical cellular stains or fluorescent dyes that stain specific molecules. Individual cells can then be counted and their concentration calculated. The fluorescent stains DAPI (4′,6-*dia*mido-2-*p*henyl*i*ndole) and acridine orange are commonly used to visualize microbes in environmental, food, and clinical samples (**figure 29.5**). These dyes specifically label nucleic acids

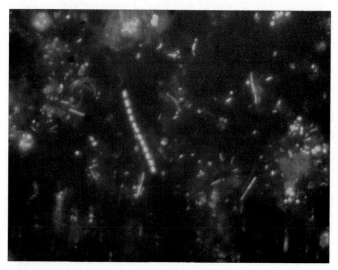

Figure 29.5 DAPI Stains Nucleic Acids. A microbial assemblage from the North Sea off the Belgian coast as visualized following DAPI staining.

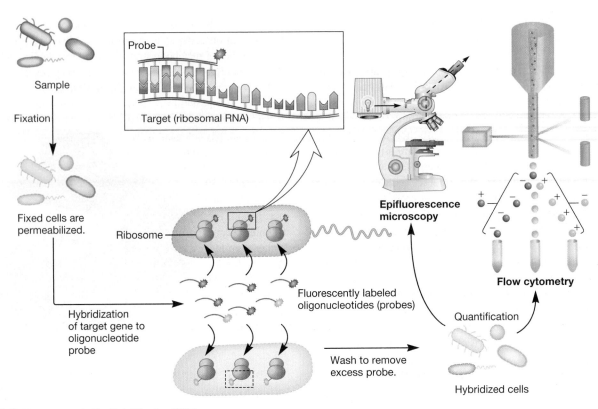

Figure 29.6 Fluorescence in Situ Hybridization (FISH).

(i.e., both DNA and RNA), and little sample preparation is required prior to their application. The nucleic acid dye SYBR Green is also used, in part because the signal is so bright virus particles can be seen.

Often microbiologists want to enumerate a specific genus or type of microbe. In these cases, a technique called *fluorescent in situ hybridization* (**FISH**) can be used. Here fluorescently labeled oligonucleotides (small lengths of DNA) that are known to be specific to the microbe of interest are used to label natural samples (**figure 29.6**). These oligonucleotides are called **probes.** Sequences within the gene encoding SSU rRNA are commonly used probes (*see figure 19.3*). To perform FISH, microbial cell membranes are made permeable so the probe can be internalized. Once inside the cell, it hybridizes to complementary nucleotide sequences. Probe that does not bind is washed off, and probe fluorescence reveals the organisms of interest. Probe fluorescence is visualized by **epifluorescence microscopy,** in which the microscope is fitted with light filters that enable the excitation of the specimen with a specific wavelength. The fluorescent tag then emits light of a longer wavelength (**figure 29.7**). Alternatively, individual cells can be separated and counted (but not visualized) by flow cytometry (figure 29.4). ◄◄ *Fluorescence microscopes use emitted light to create images (section 2.2)*

Depending on the probe used, FISH can identify a particular species, strain, or ecotype. This has made it a popular tool in clinical diagnostics and food microbiology, as well as microbial ecology. For example, FISH was used to discover the association

between the archaea *Nanoarchaeum equitans* and *Ignicoccus hospitalis (see figure 20.10)*. Two fluorescent tags—one green, the other red—were attached to 16S rRNA probes, each having a different genus-specific nucleotide sequence. In this way, it was found that

Figure 29.7 **The Use of FISH to Study Microbial Populations.** In this study, 16S rRNA probes tagged with different color fluorescent labels were used to study the microbial diversity within an earthworm gut.

MICRO INQUIRY *Under what circumstances would one use epifluorescence microscopy, rather than flow cytometry, and vice versa?*

N. equitans, which is only 0.4 μm in diameter, grows attached to the larger archae on as an obligate symbiont. Thus the combination of direct observation and molecular probes made it possible to document the lifestyle of this unusual archaeal symbiont and extend our understanding of archaeal diversity. ▶▶| *Identification of microorganisms from specimens (section 36.3)*

Sometimes when FISH is used in natural samples, the fluorescent signal is not bright enough to be detected microscopically. This is because microbes living in these environments often grow quite slowly, so each cell has relatively few ribosomes and little probe is bound. A clever modification of the FISH technique amplifies the signal produced by each cell. The trick is to label the fluorescent probe (i.e., label the label) with an enzyme that makes lots of fluorescent product when exposed to substrate. This is called *c*atalyzed *r*eported *d*eposition-FISH, or simply **CARD-FISH.** The enzyme attached to the oligonucleotide is usually horseradish peroxidase (HRP). After the oligonucleotides are hybridized to the sample, the HRP substrate tyramide is added. Tyramide is oxidized by HRP and the level of fluorescence is amplified. In this way, a single oligonucleotide bound to just one molecule of rRNA can generate an "enzyme boosted" fluorescent signal.

Molecular Techniques

Because so few microbes can be grown easily in culture, the importance of molecular techniques to identify and quantify microorganisms is paramount. Metagenomic analysis has become increasingly important in studying microbial diversity, particularly when coupled with next-generation sequencing. |◀◀ *Next-generation DNA sequencing (section 18.1); Metagenomics provides access to uncultured microbes (section 18.3)*

Microbial ecologists regularly use SSU rRNA analysis to identify the microbes that populate a community, although the **internal transcribed spacer region (ITS)** between the 16S and 23S rRNA genes may also be used. Sometimes other genes besides SSU rRNA analysis are used. When protein-coding genes are used to define phylotypes, the genes must meet three important criteria: they need to be found in all known microbes, they need to have only one copy in the genome, and they need to rarely show evidence of horizontal gene transfer.

Several approaches can be used to generate a molecular census. DNA from samples of soil, water, or other natural material (e.g., sputum or blood in a clinical setting) can be cloned into a plasmid vector to generate a metagenomic library. PCR is then used to amplify the gene or genes of interest. More often, PCR is used to amplify rRNA genes directly from natural samples without first constructing a metagenomic library. In this way, all the genomes recovered from the environment can be sequenced by a next-generation technique. Alternatively, single-cell genomic sequencing may be performed. |◀◀ *Polymerase chain reaction amplifies targeted DNA (section 17.2); Genomic libraries (section 17.5); Single-cell genomic sequencing (section 18.2)*

Despite the power of PCR, its use under these circumstances can be problematic because the rRNA genes of some microbes are more readily amplified than others. The most obvious reason

for this is relative abundance of a microbe—a microbe present in great numbers typically generates more PCR product than a rare microbe—but this is not the only reason. Other factors include (but are not limited to) the G + C content of a microbe's genome, the amount of nucleic acid degradation that might have occurred, and the relative ease with which some microbes can be lysed and thus their nucleic acids extracted (e.g., spores vs. vegetative cells). The observation that certain nucleotide templates are more readily amplified than others is termed **PCR bias.** PCR bias is typically more of a problem when nucleotide fragments are cloned into a metagenomic library prior to nucleotide sequencing.

Once SSU rRNA nucleotide sequences are obtained, they can be compared with sequences isolated from other microbes that have been deposited in databases such as the Ribosomal Database Project. In this way, microbial ecologists can get a reasonable idea of the identity of the microbes that occupy a specific niche. Because whole organisms are not isolated and studied, it is said that specific **phylotypes** have been identified.

As discussed in chapter 19, it is sometimes desirable to avoid nucleic acid sequencing and obtain a DNA fingerprint instead. DNA fingerprints generate a barcodelike image that represents community diversity most commonly by separating the amplified rRNA gene fragments on an agarose gel (*see figure 19.4*). One microbial community can then be compared with another (differing in either time or space). Often a single pair of primers is used to amplify the same region of a SSU rRNA gene from a mixture of genomes. Despite different nucleotide sequences, the PCR products will have very similar molecular weights and thus appear as a single band on an agarose gel. To be of any use, such PCR products must be separated from one another. This is accomplished through **denaturing gradient gel electrophoresis (DGGE) (figure 29.8).**

DGGE relies on the fact that although the PCR fragments are about the same size, they differ in nucleotide sequence. Recall that the rate at which double-stranded DNA becomes single stranded varies with the G + C content; that is, it varies with DNA sequence. DGGE takes advantage of the observation that DNA of different nucleotide sequences will denature at varying rates and uses a gradient of chemicals to denature the DNA (usually urea and formamide). In this technique, DNA fragments generated by PCR are loaded in a single well in a gradient gel. As electrophoresis proceeds, fragments will migrate until they become denatured. What appeared to be a single DNA fragment (band) on a nongradient gel will resolve into separate fragments (multiple bands) by DGGE. Because DGGE can be limited by poor gel-to-gel reproducibility, its popularity has declined in recent years. |◀◀ *Gel electrophoresis (section 17.1); DNA sequencing methods (section 18.1)*

Microarray technology has also been applied as a means to assess microbial diversity in natural samples without nucleotide sequencing. These specialized microarrays are called **phylochips (figure 29.9).** As described more fully in chapter 18, microarrays are prepared robotically with single-stranded DNA "probes" attached in a gridlike pattern on a solid matrix, often a glass slide. Each DNA probe is a single gene (or part of a gene), and its identity and location on the slide is recorded. Single-stranded

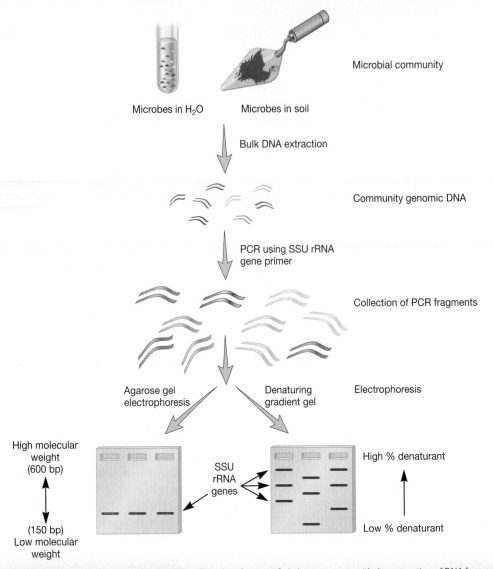

Microbial community

Microbes in H₂O   Microbes in soil

Bulk DNA extraction

Community genomic DNA

PCR using SSU rRNA
gene primer

Collection of PCR fragments

Agarose gel   Denaturing   Electrophoresis
electrophoresis   gradient gel

High molecular   High % denaturant
weight
(600 bp)   SSU
                                  rRNA
                                  genes

                                                           Low % denaturant
(150 bp)
Low molecular
weight

Figure 29.8 Denaturing Gradient Gel Electrophoresis (DGGE). The identification of phylotypes starts with the extraction of DNA from a microbial community and PCR amplification of the gene of choice, usually that which encodes SSU rRNA. Because the majority of amplified DNA fragments have about the same molecular weight, when visualized by agarose gel electrophoresis, they appear identical, as shown in the gel on the left. However, DGGE uses a gradient of DNA denaturing agents to separate the fragments based on the concentration at which they become single stranded. When a fragment is denatured, it stops migrating through the gel matrix (gel on right). Individual DNA fragments can then be cut out of the gel and cloned, and the nucleotide sequence determined.

MICRO INQUIRY *How many different phylotypes appear to be represented by this DGGE?*

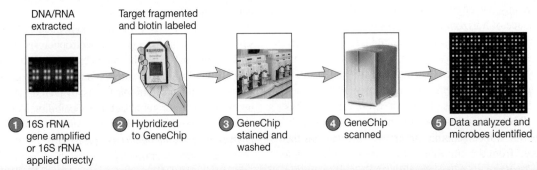

DNA/RNA   Target fragmented
extracted   and biotin labeled

1 16S rRNA   **2** Hybridized   **3** GeneChip   **4** GeneChip   **5** Data analyzed and
gene amplified   to GeneChip   stained and   scanned   microbes identified
or 16S rRNA   washed
applied directly

Figure 29.9 Phylochip Analysis. This form of microarray analysis enables the rapid evaluation of microbial diversity. (1) 16S rRNA genes or 16S rRNA molecules are obtained from the natural sample and labeled. (2) The 16S nucleic acids are hybridized to probes on a microarray called a phylochip. The probes consist of short regions of 16S rRNA genes obtained from preselected microbes. (3) The chip is washed to remove labeled 16S rRNA nucleic acids that could not hybridize because their complement is not found among the probes. (4) Laser beams scan the phylochip and (5) generate a series of spots of various colors and intensities that indicate the presence, absence, and, to a lesser extent, relative abundance of each microbe represented with an rRNA probe. This data is computer analyzed.

MICRO INQUIRY *How do phylochips differ from the microarrays discussed in chapter 18 (see p. 432)?*

nucleotides from an environmental sample are incubated with the phylochip so that complementary probe and sample sequences can hybridize. The level of hybridization to each probe (gene) is measured by laser beams. Analysis of DNA hybridization thus reveals the presence or absence of the specific genes in the environment. ◄◄ *Transcriptome analysis (section 18.5)*

29.3 Assessment of Microbial Community Activity Relies on Biochemistry and Genetics

After reading this section, you should be able to:

- Explain how microbial ecologists measure community activity
- Describe why microelectrode measurements are important tools in the study of microbial community activity
- Compare the application of traditional stable isotope analysis with stable isotope probing
- Assess the advantages and disadvantages of measuring in situ mRNA abundance
- List the type of data that can be generated by MAR-FISH and compare this with functional gene arrays
- Describe the construction and application of reporter microbes
- Differentiate metagenomics from metaproteomics

Microbial ecologists want to understand how microbes drive and respond to the biogeochemical cycles that are essential for life on Earth. Thus they seek to identify and measure the flux of nutrients, the interactions between microbes that make these fluxes possible,

and the influences of abiotic factors. Here we divide methods into biogeochemical and molecular techniques, but it should become clear that the results of one type of approach are often used to formulate hypotheses that will be tested by another (figure 29.1). This illustrates why microbial ecology might be considered the most multidisciplinary branch of microbiology. ◄◄ *Biogeochemical cycling sustains life on Earth (section 28.1)*

Biogeochemical Approaches

Microbial growth rates in complex systems can be measured directly. Changes in microbial numbers are followed over time, and the frequency of dividing cells is also used to estimate production. Alternatively, the incorporation of radiolabeled components such as thymidine (a DNA constituent, usually labeled with ^{3}H) into microbial biomass provides information about growth rates and microbial turnover. This approach is most easily applied to aquatic microorganisms; it is more difficult to directly observe terrestrial microbial communities.

Aquatic microorganisms can be recovered directly by filtration; the volume, dry weight, or chemical content of the microorganisms can then be measured. It is sometimes not possible to directly measure cell density. In that case, carbon, nitrogen, phosphorus, or an organic constituent of the cells (such as protein) can be determined. This will give a single-value estimate of the microbial community. Such chemical measurements are typically expressed as microbial biomass.

To determine community activity, it is necessary to have a good understanding of the microenvironment inhabited by the microbial community in question. In many ecosystems, these microhabitats are studied by **microelectrodes**—electrodes capable of measuring pH, oxygen tension, H_2, H_2S, or nitrogen species. The tips of these electrodes range from 2–5 μm wide and 100–200 μm long, enabling measurements to be taken as the tip penetrates the sample in increments of a millimeter or less. Because these parameters include electron donors (e.g., sulfide, ammonium) and acceptors (e.g., oxygen, nitrate, sulfate), microbial activities can be inferred. Microelectrodes have been widely used in the study of **microbial mats (figure 29.10)**. These highly

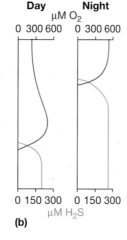

Day   Night
μM O_2
0 300 600   0 300 600

0 150 300   0 150 300
μM H_2S

(a)

(b)

Figure 29.10 Microelectrode Analysis of Microbial Mats. (a) Microorganisms, through their metabolic activities, can create environmental gradients resulting in layered ecosystems. A vertical section of a hot spring (55°C) microbial mat, showing the various layers of microorganisms. (b) The measurement of oxygen and H_2S with microelectrodes demonstrates the diurnal shift in gas production: during the day, oxygen produced by cyanobacteria in the upper layers inhibits anaerobic respiration, so sulfate reduction is confined to the lower (dark) strata. At night, atmospheric oxygen has a limited range of diffusion into the mat, and H_2S diffuses upward, where it can be used as an electron donor by purple photosynthetic bacteria.

striated communities of photoautotrophs and chemoheterotrophs develop in saline lakes, marine intertidal zones, and hot springs.

In determining community function, it is often instructive to explore the cycling of a particular elemental species. In aquatic and marine systems, the amount of carbon fixed by microbes is an important measurement. This is most commonly accomplished by measuring how much radioactive bicarbonate ($H^{14}CO_3$) is taken up by microbes. As we discuss in chapter 30, when bicarbonate dissolves in water, it dissociates to CO_2. Thus when added to a sample of microbes, its incorporation into particulate carbon (i.e., biomass) represents the gross primary production; that is, the total amount of carbon fixation without accounting for CO_2 lost through respiration. Specialized CO_2 traps can be added to the incubation vessel so that the amount of CO_2 lost by respiration is also measured. When CO_2 respired is subtracted from the amount of CO_2 fixed into biomass, net primary production has been assessed.

In addition to radioactive isotope incorporation, **stable isotope analysis** may be used. Isotopes are those forms of elements that differ in atomic weight because they bear different numbers of neutrons. Isotopes can be stable or unstable; only radioactive isotopes are unstable and decay. As shown in **table 29.2**, stable isotopes are rare when compared to their lighter counterparts. For example, both ^{14}N and ^{15}N occur in nature, but ^{14}N is found in much greater abundance. Organisms discriminate between stable isotopes, and for any given element, the lighter (in this case, ^{14}N) is preferentially incorporated into biomass, a phenomenon known as **isotope fractionation.** This discrimination can be used to follow the fate of a specific element by labeling the compound of interest with the heavier isotope. For instance, one might want to explore the fate of NO_3^- in soil. Samples of the soil would be incubated in the presence of $^{15}NO_3^-$. One would then sample the soil microorganisms and measure the amount of ^{15}N incorporated into biomass, released as $^{15}N_2$, and reduced to $^{15}NH_4^+$. However, because the absolute value of ^{15}N is so small, it cannot be directly measured. Instead, the ratio of ^{15}N to ^{14}N in the sample is compared to the standard ratio found in inorganic material. An organism that assimilates the $^{15}NO_3^-$ would contain more ^{15}N when compared to the standard. This difference, called delta (δ), is calculated as follows:

$$[(R_{sample} - R_{standard})/(R_{standard})] \times 1{,}000 = \delta_{sample\text{-}standard}$$

where R_{sample} is the ratio of heavy to light isotope in the sample and $R_{standard}$ is the ratio of heavy to light isotope in the standard. For instance, if our hypothetical soil community is found to have a $^{15}N/^{14}N$ ratio greater than the standard by 3 parts per thousand, this value is reported as $\delta\ ^{15}N = +3\ \delta\ \text{o/oo}$. The true value of stable isotope fractionation rests in the fact that different microbes discriminate between heavy and light isotopes differently. For instance, nitrate production can be discerned from N_2 formation because the isotopic enrichment factor for denitrification is less than that for nitrification (about $-26\ \text{o/oo}$ vs. $-34\ \text{o/oo}$, respectively).

Another technique, called **stable isotope probing,** can be used to examine nutrient cycling in microbial communities as well as identify the microbes taking up the element of interest. As an example, this technique was employed to explore

Table 29.2	Natural Abundances of Some Biologically Relevant Isotopes			
Hydrogen	**Carbon**	**Nitrogen**	**Oxygen**	**Sulfur**
^1H 99.984%	^{12}C 98.89%	^{14}N 99.64%	^{16}O 99.763%	^{32}S 95.02%
^2D 0.0156% (deuterium)	^{13}C 1.11%	^{15}N 0.36%	^{17}O 0.0375%	^{33}S 0.75%
			^{18}O 0.1995%	^{34}S 4.21%
				^{36}S 0.02%

methanogenesis in a rice paddy soil ecosystem. Researchers built **microcosms**—small incubation chambers with conditions that mimic the natural setting. In this case, they simulated natural rice paddies and introduced $^{13}CO_2$. Gaseous $^{13}CH_4$ was collected and RNA was isolated from the soil. The ^{12}C-containing RNA was separated from the ^{13}C-RNA on the basis of differing densities. The ^{13}C-containing rRNA, which could only be synthesized by those bacteria that assimilated the $^{13}CO_2$, was then used to identify these archaea; recall that all methanogenic microbes are archaea. This study demonstrates the value and importance of culture-independent approaches that combine biogeochemistry and molecular biology to understand microbial community ecology and physiology. ◄◄ *Methanogens and methanotrophs (section 20.4)*

Molecular Approaches

As introduced in section 18.3, the use of metagenomics to investigate natural populations has revolutionized the depth to which community dynamics can be assessed. Metagenomics applied to diverse natural habitats can have several outcomes, including the identification of new genes and gene products from uncultured microbes, the assembly of whole genomes, and comparisons of community gene content from microbial assemblages of different origin. Metagenomic data can also be used to validate the measurement of gene expression when mRNA is extracted directly from the environment. Researchers are finding large numbers of new genes from metagenomic analysis, and it is agreed that the discovery of new genes is far from complete. In fact, the number of metagenomic sequences obtained from uncultivated microbes in mixed assemblages outnumbers data sets obtained from microbes grown in pure culture. Here we discuss a few metagenomic approaches that are most commonly used by microbial ecologists.

Recall that metagenomics involves the nucleotide sequencing of community DNA fragments. These nucleotide sequences can be used to infer community function. Metagenomes from a variety of biomes (e.g., mines, hypersaline ponds, marine water,

Table 29.3	Example Genes Used to Assess Community Function and Metabolic Diversity	
Process	**Gene**	**Gene Product**
Ammonium oxidation	amoC, amoC, amoB	Ammonia monooxygenase
Anammox	hzf	Hydrazine hydrolase
Dissimilatory nitrate reduction	narG	Nitrate reductase
	nosZ	Nitrous oxide reductase
	nir	Nitrite reductase
Nitrogen fixation	nifH, nifD, nifK	Different subunits of nitrogenase
Methanogensis	mcrA	Methyl coenzyme M reductase
Methane oxdition	pmoA	Methane monooxygenase
Sulfate reduction	apsA	Adenosine phosphosulfate reductase

freshwater, etc.) can predict the biogeochemical conditions of a habitat. For instance, community metagenomes from a well-oxygenated coral reef are rich in genes encoding respiration proteins, while the metagenomes from the colons of terrestrial animals have a high number of genes coding for enzymes involved in fermentation.

mRNA present in the environment can also be monitored, a technique called **metatranscriptomics**. Here mRNA extracted from the natural environment is reverse transcribed to cDNA (*see figure 17.5*) and directly analyzed by next-generation sequencing. The presence of mRNA demonstrates that a gene is actively transcribed and the gene product is probably active. Thus, one can assess community activity at the time the sample was collected. However, rather than surveying the entire genetic landscape, a researcher may be interesting in a particular metabolic function. In this case, a specific gene or mRNA encoding enzymes of interest would be targeted by specific primers. Example genes are listed in **table 29.3**. Such studies have shown that both dissimilatory nitrate reduction and the anammox reaction account for loss of nitrogen in specific habitats.

Microarray analysis can also be used to assess community function. **Functional gene arrays (FGAs)** are prepared with sequences of genes known to be important in biogeochemical cycling processes (table 29.3). However, the use of microarrays to assess microbial communities is inherently problematic: only known genes can be used to make oligonucleotide probes. This generates an information bias such that no novel genes in the community can be detected. In addition, the interpretation of community-based microarrays is challenging because the mRNA pool is derived from multiple genera. This results in varying levels of homology to different probes. In turn, this can generate high background levels of fluorescence and obscure valid results. It is for these reasons that metatranscriptomics is becoming the

preferred method for evaluating microbial gene expression in environmental samples. ◀◀ *Transcriptome analysis (section 18.5)*

Unfortunately, most mRNAs are unstable, making recovery technically challenging. Instead of measuring mRNA, a **metaproteomics** approach can be used to identify each protein present at the time of sampling. There are two general approaches by which all the proteins in a community can be sampled and identified. The more labor-intensive technique uses 2D-polyacrylamide gel electrophoresis (2D-PAGE) so that each protein recovered from the environment is visualized (*see figure 18.16*). Individual proteins can be eluted from the gel and analyzed by mass spectroscopy (MS). Total peptide mass and amino acid sequence information is then used to search databases to identify the protein. This approach is best if the relative concentration of individual proteins is considered important. If a census of proteins is the goal, a more automated, high-throughput approach is taken. Here community proteins are extracted and digested with a protease to generate protein fragments called peptides. This mixture of peptides is separated by a technique known as 2D nano-liquid chromatography (nano-LC). Just as 2D gels separate proteins first by charge and then by molecular weight, 2D nano-LC separates peptides by two different properties: peptides are first separated by charge and then by hydrophobicity. The peptide masses are determined by advanced mass spectrometry, and finally, tandem MS is used to obtain amino acid sequence information. In both cases, the protein sequences obtained are compared to metagenomic data obtained from the same environment. Metaproteomics is limited by the complexity of the environment being assessed. Complex environments such as a pristine soil may have as many as a billion or more different proteins per gram. Even the most advanced and automated MS approaches are not yet able to resolve such a large and diverse group of proteins. ◀◀ *Proteomics explores total cellular proteins (section 18.5)*

Both metatranscriptomics and metaproteomics seek to assess activities that are actually taking place in the environment. Another method that also measures active processes uses FISH to identify specific mRNAs (figure 29.7). This approach is called **in situ reverse transcriptase (ISRT)-FISH**. Recall that in situ means "in place," so this technique allows the microbial ecologist to examine specimens and determine if a gene of interest is being expressed, as well as visualize the microbes expressing it. Let's say that one is interested in recently identified archaea capable of nitrification (*see figure 22.15*). These microbes must express the *amoA* (ammonia monooxygenase) gene to oxidize ammonium to nitrite (table 29.3). One could design an *amoA* DNA probe—a short piece of DNA that is complementary to the *amoA* mRNA. Once the DNA probe has hybridized to the *amoA* mRNA in the natural sample, reverse transcriptase is used to generate a cDNA, which is then amplified by PCR. One now has an abundance of nucleotide fragments that can be fluorescently tagged in FISH. In our example of the archaeal nitrifier, we might want to determine where these archaea are metabolically active in the water column of an oceanic sampling station. Water samples of varying depths would be collected and filtered on board a research vessel. The filters would most likely be fixed

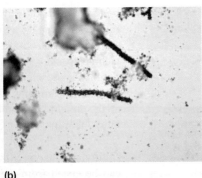

(a)   (b)

Figure 29.11 MAR-FISH Technique Combines Microautoradiography with FISH in a Single Sample. Cells of interest are triple-labeled for identification and measurement of metabolic activity even if present in a complex microbial community. (a) The filamentous microbe of interest was identified using a FISH probe that generates an orange signal under fluorescence microscopy. The green fluorescence is the result of a dye (YO-PRO-1) that binds to double-stranded nucleic acid such as DNA of the bacteria. (b) MAR resulted in black staining of the filamentous bacteria when viewed under bright field microscopy, indicating that the microbes took up (assimilated) ^{3}H-labeled amino acids.

FISH can be employed. This method is thus called MAR-FISH. To understand MAR-FISH, it is necessary to first explain MAR. In MAR, a radioactive substrate is mixed with the sample and incubated under specific conditions, usually closely resembling those in the field. At the completion of the incubation period, the sample is washed to remove any radioactive substrate not taken up by the cells. A sample attached to a microscope slide is treated with a photographic emulsion. The slide is carefully wrapped and stored in complete darkness. As the incorporated radioactive material decays, the silver grains in the photographic emulsion are exposed. When developed, this leaves a series of black dots around the cells that assimilated the radioactive substrate (**figure 29.11**). When MAR is combined with FISH, one can analyze the phylogenetic identity and the specific substrate uptake patterns of microbes of interest at the level of a single cell in complex microbial communities.

with preservative for later analysis (it is difficult to perform microscopy on a moving boat). The data collected by ISRT-FISH would indicate where archaeal nitrification was most active. This could then be correlated with physical parameters such as light penetration, oxygen concentration, and levels of nitrate, nitrite, ammonia, and dissolved organic nitrogen. In this way, the microbial ecologist can begin to build a hypothesis about the importance and distribution of these interesting microbes.

But what if one wanted to identify the microbe responsible for a specific metabolic activity? Recall that ISRT-FISH enables the visualization of the microbes expressing a specific gene but does not identify the microbe. To simultaneously assess metabolic activity and phylogenetic identity, a technique that combines an older method, **microautoradiography** (**MAR**), with

Retrieve, Infer, Apply

1. How do radioactive and stable isotopes differ? Design an experiment to determine whether or not denitrification is taking place in a soil ecosystem.
2. How might microelectrodes be used to gather data in preparation for a stable isotope study?
3. How do functional gene arrays differ from phylochips? When would one use a functional gene array, rather than a phylochip?
4. What is metaproteomics? How is it similar to metagenomics? How does it differ?
5. What is the difference between MAR and MAR-FISH? Suggest a microbial ecology question that could be answered by MAR and one that would require MAR-FISH.

Key Concepts

29.1 Microbial Biology Relies on Cultures

- Enrichment cultures are used to promote the growth of one type of microbe, while suppressing the growth of others. This is accomplished by designing media and culture conditions that favor the growth of the desired microorganisms but limit the growth of others.
- Although most microbes cannot be grown in pure culture, novel approaches have been developed that have resulted in the isolation of previously uncultured microorganisms.

29.2 Genetic Methods Are Used to Assess Microbial Diversity

- A variety of staining techniques are used to observe microbes in natural environments. Fluorescent in situ hybridization (FISH) labels specific microbes that possess

a specific nucleotide sequence, usually a region of the SSU rRNA gene (**figures 29.5–29.7**).
- A survey of SSU rRNA is the most common approach for assessing microbial diversity. There are several approaches by which SSU rRNA can be recovered; those that do not use PCR to amplify rRNA genes are gaining popularity because they avoid PCR bias.
- Microarrays that have SSU rRNAs as probes, called phylochips, can quickly evaluate population and community diversity (**figure 29.9**).

29.3 Assessment of Microbial Community Activity Relies on Biochemistry and Genetics

- Microelectrodes can be used to determine physical and biological parameters such as pH, O_2, and H_2S concentrations in microniches (**figure 29.10**).

- Stable isotope analysis is based on the fact that organisms discriminate between heavy and light naturally occurring isotopes. It is often used to monitor nutrient flux through ecosystems.
- Metatranscriptomics assesses community activity by measuring microbial gene expression.

- Metaproteomics can be used to identify the proteins that are expressed by a microbial community. Because the number of proteins is so large, traditional 2D gel analysis is usually not practical and high-throughput chromatographic techniques are used instead.

Compare, Hypothesize, Invent

1. How might you attempt to grow in the laboratory a chemolithoautotroph that uses ferrous iron as an electron donor and oxygen as an electron acceptor?

2. What genes would you use as probes on a functional array designed to assess microbial activity in the plankton of a nitrogen-limited aquatic ecosystem? Would direct sequencing of the mRNA from these genes provide a more accurate assessment? Why or why not?

3. Both acetate and CO_2 can be used by methanogenic archaea to generate methane (CH_4). How would you determine the source of carbon for methanogenesis in a waterlogged peat?

4. Kindaichi and colleagues used FISH and microelectrodes to investigate the spatial organization of nitrifying and anammox microbes in a biofilm. Recall that nitrification is

an aerobic process that involves ammonia oxidation to nitrite by both archaea and bacteria followed by nitrite oxidation to nitrate by bacteria. Gram-negative bacteria in the phylum *Planctomycetes* perform the anammox reaction—the anaerobic oxidation of ammonium with nitrite as the electron acceptor and N_2 as the major end product (*see figure 21.16*). Explain the purpose of each technique used by these investigators. How do you think these approaches would contribute to an understanding of the spatial distribution of each type of microbe as well as help document the metabolic process occurring in each microregion of the biofilm?

Read the original paper: Kindaichi, T., et al. 2007. In situ activity and spatial organization of anaerobic, ammonium-oxidizing (anammox) bacteria in biofilms. *Appl. Environ. Microbiol.* 73:4931.

Learn More

30

Microorganisms in Marine and Freshwater Ecosystems

The Mississippi Delta is home to a dead zone. Oxygen depletion and loss of macrofauna result in turbid, brown water, strikingly different from the well-oxygenated blue water.

Ocean Death Coming Soon to a Coast Near You

If you live near a coast, you probably live not far from an ocean dead zone. These regions, more technically called "oxygen minimum zones," have been increasing in size and distribution over the past decade and could occupy as much as 20% of the world's oceans by 2100.

Ocean dead zones are regions in which all plant and animal life dwelling below the surface has suffocated. The twenty-first century has brought the convergence of global climate change and water pollution (in the form of nitrogen and phosphorus fertilizer) to coastal ecosystems. Dead zones result when phytoplankton respond to increases in nutrients and temperature by growing (and growing and growing) and then sinking. Initially, phytoplankton on the ocean floor means more food for heterotrophic organisms. But as the number of oxygen-consuming animals and microbes increases, available oxygen dwindles. Invertebrates and fish die. Plants cannot grow because there is not enough light. The only organisms left are anaerobic chemolithoautotrophs—microbes that make a living fixing CO_2 and extracting energy from reduced inorganic compounds.

Of the more than 500 oxygen minimum zones globally covering over 245,000 square kilometers, the Baltic Sea is the largest, while the second largest dead zone is at the mouth of the Mississippi River. This mighty river dumps fertilizer collected from watersheds ranging from Montana to Louisiana. The addition of literally tons of extra nitrogen and phosphorus to the Gulf of Mexico has resulted in a dead zone that is 22,015 km², roughly the size of New Jersey. The presence of such a large dead zone has huge and obvious implications for fisheries, sea birds, tourism, and the overall health of the Gulf.

Are dead zones reversible? In some cases, yes. But like most problems, prevention is the best approach. Prudent use of fertilizer and limiting surface runoff are good places to start. Another area of concern is the deposition of animal waste into local waterways. Finally, proper sewage

treatment (*section 43.2*) and the restoration of natural wetlands to filter and consume much of the nitrogen and phosphorus before it flows to the coast are projects worth pursuing.

In this chapter, we turn our attention to the microbial ecology of aquatic environments. The two main disciplines of aquatic biology are oceanography and limnology. **Oceanography** is the study of marine systems and the biological, physical, geological, and chemical factors that impact biogeochemical cycling, water circulation, and climate. Microbial oceanography seeks to understand marine microorganisms at the level of the individual taxon and at the same time elucidate how they influence biogeochemical cycling at a global level. **Limnology** is the investigation of aquatic systems within continental boundaries, including glaciers, groundwater, rivers, streams, and wetlands. All terrestrial life depends on the sustained provision of clean freshwater from these ecosystems. This chapter provides a brief introduction to these dynamic and interdisciplinary fields.

Readiness Check:

Based on what you have learned previously, you should be able to:

✔ Explain the global impact that microbes have on carbon and nitrogen cycling (section 28.1)

✔ Diagram the viral life cycle (section 6.3)

✔ List environmental factors that influence microbial growth (section 7.4)

✔ Compare and contrast the growth and physiology of obligate aerobes, facultative anaerobes, and strict anaerobes (section 7.4)

✔ Explain why and how microbes frequently grow in biofilms (section 7.5)

✔ Explain how the oxidation-reduction (redox) state of an environment impacts the types of microbial metabolic processes that might be found there (section 28.1)

✔ Identify, by nutritional type, microorganisms responsible for carbon fixation, nitrogen fixation, nitrification, dissimilatory nitrate reduction, denitrification, nitrification, the anammox reaction, and sulfur reduction (sections 20.3, 21.4, 21.5, 22.1, 22.3, and 22.4)

✔ Review innovative methods for growing microbes isolated from nature in the laboratory (section 29.1)

30.1 Water Is the Largest Microbial Habitat

After reading this section, you should be able to:

- Explain the carbonate equilibrium system and why it is important in the study of marine microbial biology

Microbes drive marine and freshwater ecosystems. They are the most diverse and abundant organisms in these habitats. The importance of microbial activities in oceans, rivers, and lakes has been brought into sharp focus by global climate change. To understand the capacity of microbes to influence the worldwide flux of nutrients and greenhouse gases, we must first consider marine and freshwater habitats.

The nature of water as a microbial habitat depends on a number of physical factors such as temperature, pH, and light penetration. One of the most important factors is dissolved oxygen. The flux rate of oxygen through water is about 10,000 times less than its rate through air. However, in some aquatic habitats, the limits to oxygen diffusion can be offset by the increased solubility of oxygen at colder temperatures and increasing atmospheric pressures. For instance, in the very deep ocean, the dissolved oxygen concentration actually increases with depth, even though the air-water interface can be literally miles away. On the other hand, tropical lakes and summertime temperate lakes may become oxygen limited only meters below the surface.

As seen in the development of dead zones (chapter opening story), light is also critical for the health of marine and freshwater ecosystems. Like all life on Earth, all organisms in these environments depend on **primary producers**—autotrophic organisms—to fix CO_2, thereby providing organic carbon. In streams, lakes, and coastal marine systems, macroscopic algae and plants are the chief primary producers. Organic carbon also enters these systems in terrestrial runoff. The situation is very different in the open ocean, where all organic carbon is the product of microbial autotrophy. In fact, about half of all **primary production** (carbon fixation) on Earth is the result of this microbial carbon fixation.

Solar radiation warms the water, and this can lead to thermal stratification. Warm water is less dense than cool water, so as the sun heats the surface in tropical and temperate waters, a **thermocline** develops. A thermocline can be thought of as a mass of warmer water floating on top of cooler water. These two water masses remain separate until there is either a substantial mixing event, such as a severe storm, or in temperate climates, the onset of autumn. As the weather cools, the upper layer of warm water becomes cooled and the two water masses mix. This is often associated with a pulse of nutrients from the lower, darker waters to the surface. This pulse of nutrients can trigger a sudden and rapid increase in the population of certain microbes, and a bloom may develop. This is considered more fully in the discussion of microorganisms in coastal marine systems (p. 653) and lakes (p. 663).

Because CO_2 is critical in many chemical and biological processes, it is important to understand its fate in freshwater and marine ecosystems. The pH of unbuffered distilled water is determined by dissolved CO_2 in equilibrium with the air and is typically 5.0 to 5.5. The pH of freshwater systems such as lakes and streams, which are usually only weakly buffered, is therefore controlled by terrestrial input (e.g., minerals that may be either acidic or alkaline) and the rate at which CO_2 is removed by photoautotrophic organisms such as diatoms; CO_2 fixation can increase the pH of the water.

In contrast, seawater is strongly buffered by the balance of CO_2, bicarbonate (HCO_3^-), and carbonate (CO_3^{2-}). Atmospheric CO_2 enters the oceans and either is converted to organic carbon by photosynthesis or reacts with seawater to form carbonic acid (H_2CO_3), which quickly dissociates to form bicarbonate and carbonate (**figure 30.1**):

$$CO_2 + H_2O \rightleftharpoons H_2CO_3 \rightleftharpoons H^+ + HCO_3^- \rightleftharpoons 2H^+ + CO_3^{2-}$$

The oceans are effectively buffered between pH 7.6 and 8.2 by this **carbonate equilibrium system.** Much like the buffer one might use in a chemistry experiment, the pH of seawater is determined by the relative concentrations of bicarbonate and carbonate.

The reactions of the carbonate equilibrium system have taken on new importance as increases in atmospheric CO_2 result in higher rates of CO_2 dissolution in seawater. In fact, the ocean is already absorbing about half the anthropogenically produced CO_2. This has led to **ocean acidification;** the average pH of the ocean has dropped by 0.1 unit (to 8.1) since the pre-industrial age. It is expected to decline another 0.35 to 0.50 by 2100, unless

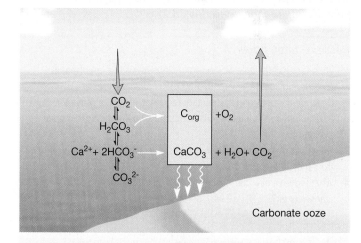

Carbonate ooze

Figure 30.1 The Carbonate Equilibrium System. Atmospheric CO_2 enters seawater and is converted to organic carbon (C_{org}) or to carbonic acid (H_2CO_3) that rapidly dissociates into the weak acids bicarbonate (HCO_3^-) and carbonate (CO_3^{2-}). Calcium carbonate ($CaCO_3$), a solid, precipitates to the seafloor, where it helps form a carbonate ooze. This system keeps seawater buffered at about pH 8.0.

MICRO INQUIRY *Which carbon species, CO_2 or CO_3^{2-}, acidifies seawater?*

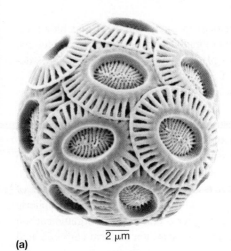

(a)

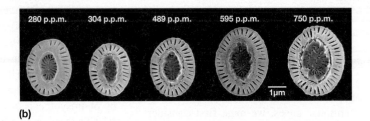

(b)

Figure 30.2 Coccolithophores Respond to Seawater Acidification. (a) *Emiliania huxleyi*, an abundant species in the open ocean. (b) *E. huxleyi* coccoliths increase in size when grown in high partial pressures of CO$_2$. In addition to an increase in size, the coccolith shape can become distorted above 400 ppm CO$_2$, the value of atmospheric CO$_2$ measured in 2014.

effective means of limiting CO$_2$ emissions are implemented. This is significant because as CO$_2$ increases, the carbonate equilibrium shifts away from CaCO$_3$, potentially limiting its availability for the formation of shells and skeletons of marine organisms. Because protists known as coccolithophores produce about a third of the marine CaCO$_3$, they have been the focus of investigations seeking to predict the effects of ocean acidification. While responses are species specific, it is clear that these seemingly slight shifts in pH have an impact (**figure 30.2**). Ocean acidification is also having a negative impact on corals. Although it is difficult to predict, some models estimate that when atmospheric CO$_2$ reaches about 550 ppm, many corals will begin to erode faster rather than grow. If left unchecked, this threshold will be crossed in about 50 years. ◄◄ Haptophyta *(section 25.4); Global climate change (section 28.2)*

Retrieve, Infer, Apply

1. What factors influence oxygen solubility? How is this important in considering aquatic environments?
2. Describe the buffering system that regulates the pH of seawater. What might be the implications of this stable buffering on microbial evolution?
3. How does primary production in lakes and coastal ecosystems differ from that in the open ocean?
4. What features of a thermocline make it similar to hot and cool masses of air?

30.2 Microorganisms in Marine Ecosystems

After reading this section, you should be able to:

- Explain the ecological and economic importance of estuaries and harmful algal blooms
- Diagram carbon and sulfur flux in a Winogradsky column
- Assess the importance of marine microbial primary productivity in the global carbon budget

- Evaluate the interplay between carbon, nitrogen, and iron in the open ocean and how the relative concentrations of these elements influence the productivity of these waters
- Illustrate the microbial loop and evaluate its importance in the open ocean
- Evaluate the importance of members of the SAR11 clade in marine ecosystems
- Describe at least three mechanisms microbes have evolved to adapt to the low nutrient conditions found in the open ocean
- Explain the role of viruses in maintaining and influencing marine microbial populations, as well as their contribution to nutrient cycling
- Convince a layperson that the vast majority of microbes reside below the surface of the Earth
- Explain the distribution of microbial types that inhabit a cross-section sediment

As terrestrial organisms, we must remind ourselves that about 96% of Earth's water is in marine environments. From a microbiological perspective, surface ocean waters have been most intensely studied; only recently have scientists begun to probe the benthos—the deep-sea sediments and the subsurface. Investigations of this kind are revealing many surprises. We begin our discussion of marine ecosystems with estuaries, then consider the microbial communities that inhabit the open ocean and, finally, the dark, cold, high-pressure benthos.

Microorganisms in Coastal Ecosystems Are Adapted to a Changeable Environment

An estuary is a semi-enclosed coastal region where a river meets the sea. Estuaries are defined by tidal mixing between freshwater and saltwater. Most estuaries undergo large-scale tidal flushing; this forces organisms to adapt to changing salt concentrations on a daily basis. Microbes that live under such conditions combat the resulting osmotic stress by adjusting their intracellular osmolarity to limit the difference with that of the surrounding water. Most protists and fungi produce osmotically active carbohydrates for this purpose, whereas bacteria and archaea regulate internal concentrations of potassium or amino acids such as proline and betaine. Salt is another stressor, and most microbes that inhabit estuaries are halotolerant, which is distinct from halophilic. **Halotolerant** microbes can withstand significant changes in salinity; halophilic microorganisms have an absolute requirement for high salt concentrations. ◄◄ *Solutes affect osmosis and water activity (section 7.4)*

Estuaries are unique in many respects. Their calm, nutrient-rich waters serve as nurseries for juvenile forms of many commercially important fish and invertebrates. However, despite their importance to the commercial fishing industry, estuaries are among the most polluted marine environments. Estuaries and the rivers that feed them are the receptacles of toxins discharged during industrial processes as well as agricultural and street runoff. These pollutants include organic materials that can be used as nutrients. If too much organic matter is consistently added, the system is said to be **eutrophic,** and the rate of respiration greatly exceeds that of photosynthesis. Thus the water may become anoxic (chapter opening story). By contrast, if nutrients are introduced in a pulselike fashion, a single photoautotrophic species, either algal or cyanobacterial, grows at the expense of all other organisms in the community. This phenomenon, called a **bloom,** often results from the introduction of nutrients combined with changes in light, temperature, and mixing of water masses. If the microbes produce a toxic product or are in themselves poisonous to other organisms such as shellfish or fish, the term **harmful algal bloom** (**HAB**) is used. Some HABs are called **red tides** because the microbial density is so great that the water becomes red or pink, the color of the algae (**figure 30.3**).

The number of HABs has dramatically increased in the last decade or so. It is predicted that global climate change may trigger more HABs, in part because warmer water forms a more stable thermocline that lasts longer into the autumn. HABs are sometimes responsible for killing large numbers of fish or marine mammals. For instance, off the coast of California, periodic HABs are responsible for sudden, large-scale deaths of sea lions. These blooms are often caused by diatoms in the genus *Pseudonitzschia*. Anchovies consume the diatoms, and a potent neurotoxin, domoic acid, accumulates in the fish. The mammals are poisoned after they ingest large quantities of the fish. ◄◄ Stramenopila *(section 25.4)*

In addition to diatoms, HABs are often caused by dinoflagellates. Some bloom-causing dinoflagellates produce potent neurotoxins. Dinoflagellates of the genus *Alexandrium* produce a toxin responsible for paralytic shellfish poisoning (PSP), which affects humans as well as other animals in coastal, temperate North America. The toxin produced by the dinoflagellate *Karenia brevis* killed a number of endangered manatees and bottlenose dolphins in 2002 in a bloom in Florida. Another dinoflagellate, *Pfiesteria piscicida,* has become a problem in the Chesapeake Bay and regions south. This protist produces lethal lesions in fish (**figure 30.4**) and has had a devastating effect on the local fishing industry. Exposure to this microbe may also cause neurological damage to humans. ◄◄ Alveolata *(section 25.4)*

Figure 30.3 Harmful Algal Blooms Are Also Called Red Tides.
Blooms often color the water with the pigment of the microbe responsible. In this case, a dinoflagellate has turned the water pink-orange.

Figure 30.4 *Pfiesteria piscicida* Lesions. Lesions on a fish resulting from parasitism by the dinoflagellate *Pfiesteria piscicida*.

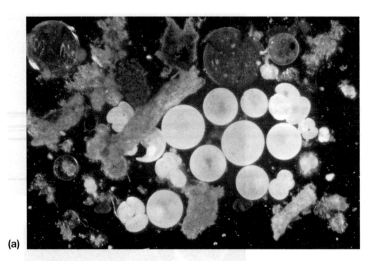

(a)

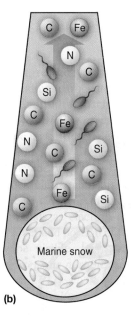

(b)

Figure 30.7 Marine Snow. The export of organic matter out of the photic zone to deeper water occurs through the sinking of marine snow. (a) Material collected in a sediment trap at 5,367 m on the Sohm Abyssal Plain in the Sargasso Sea. It includes cylindrical fecal pellets, planktonic tests (round white objects), transparent snail-like pteropod shells, radiolarians, and diatoms. (b) The hydrolytic activity of the exoenzymes secreted by the microbes that colonize marine snow generates a plume of nutrients during sinking. This in turn attracts more microbes.

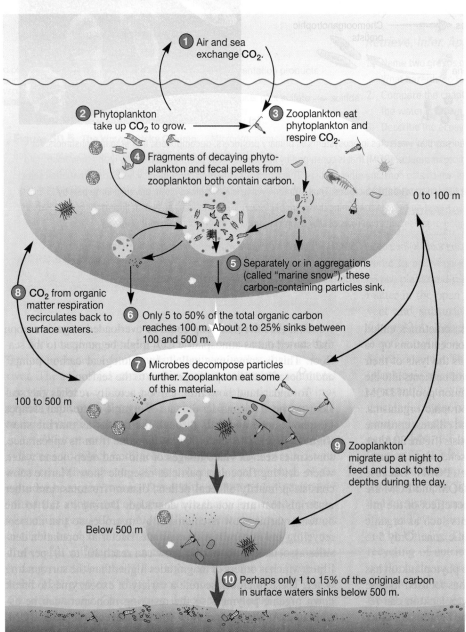

1. Air and sea exchange CO_2.

2. Phytoplankton take up CO_2 to grow.

3. Zooplankton eat phytoplankton and respire CO_2.

4. Fragments of decaying phytoplankton and fecal pellets from zooplankton both contain carbon.

5. Separately or in aggregations (called "marine snow"), these carbon-containing particles sink.

0 to 100 m

8. CO_2 from organic matter respiration recirculates back to surface waters.

6. Only 5 to 50% of the total organic carbon reaches 100 m. About 2 to 25% sinks between 100 and 500 m.

7. Microbes decompose particles further. Zooplankton eat some of this material.

100 to 500 m

9. Zooplankton migrate up at night to feed and back to the depths during the day.

Below 500 m

10. Perhaps only 1 to 15% of the original carbon in surface waters sinks below 500 m.

nitrogen, phosphorus, and iron. These nutrients attract chemotactic free-swimming microbes, which help to deplete the nutrient pool even further (figure 30.7*b*). By the time a piece of marine snow reaches the seafloor, less than 1% of photosynthetically derived organic carbon remains unaltered. Once there, only a tiny fraction will be buried in the sediments; most will return to the surface in upwelling regions (**figure 30.8**).

An urgent question is whether or not the oceans can draw down (remove and bury) more CO_2 as atmospheric concentrations continue to increase. Presently it is estimated that about 3 billion tons of carbon per year are buried (sequestered) in deep-ocean sediments. Can increases in atmospheric CO_2 be offset by increasing the rate and volume of carbon sequestration? Could this in turn slow global warming? Such a scenario has been pursued in projects that fertilize

Figure 30.8 Fate of Carbon from the Photic Zone to the Deep Benthos. The vast majority of the carbon fixed by microbial autotrophs at the surface of the open ocean remains in the photic zone. Although a small fraction reaches the seafloor, much of it returns to the surface in regions where deep and surface waters mix (upwelling regions; not shown). The flux of CO_2 in and out of the world's oceans is in equilibrium, but this equilibrium may be upset by the increase in atmospheric CO_2, the hallmark of global warming.

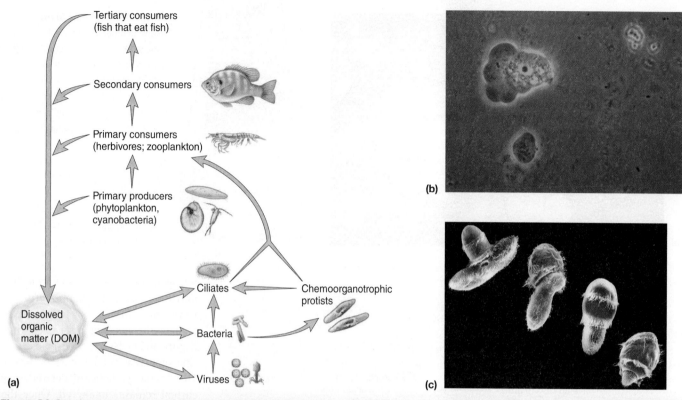

Figure 30.6 The Microbial Loop. (a) Microorganisms play vital roles in ecosystems as primary producers, decomposers, and primary consumers. All organisms contribute to a common pool of dissolved organic matter (DOM) that is consumed by microbes. Viruses contribute DOM by lysing their hosts, and bacterial and archaeal cells are consumed by protists, which also consume other protists. These microbes are then consumed by herbivores that often select food items by size, thereby ingesting both heterotrophic and autotrophic microbes. Thus nutrient cycling is a complex system driven in large part by microbes. (b) Protists consume bacteria; in this case, a naked amoeba is consuming the cyanobacterium *Synechococcus* sp. which fluoresces red. (c) Protists consume protists; here, the ciliate *Didinium* sp. (rounded organism with two rows of cilia) is preying upon another ciliate, *Paramecium* sp.

MICRO INQUIRY *How do heterotrophic microbes contribute to the pool of DOM? How does this compare with the manner by which microbial autotrophs contribute to the DOM pool?*

leaks from the photoautotrophic microbes is sometimes called **photosynthate.** Marine viruses, present at concentrations up to 10^8 per milliliter, are also a source of DOM as the lysis of their host cells contributes significantly to the return of nutrients into the microbial loop (p. 658). This generates a common pool of DOM that serves as a nutrient source for all osmotrophic organisms. Heterotrophic protists, including flagellates and ciliates, consume smaller microbes, which may include other protists (figure 30.6*b,c*). Microbial cells, insoluble detritus, and other solid organic material are called **particulate organic matter** (**POM**). Because protists are consumed by zooplankton, both DOM and POM are recycled at a number of trophic levels. The net effect of the microbial loop is the reuse of essential nutrients such as organic carbon, nitrate, and phosphate within the photic zone. Only 5 to 15% this material sinks to the waters below.

The fate of all the carbon fixed by marine phytoplankton has practical importance. Global climate change has focused intense scrutiny on determining how much CO_2 phytoplankton can remove from the atmosphere and sequester in the benthos. In other words, if the microbial loop could be overloaded, organic carbon that started out as atmospheric CO_2 might be pumped to the seafloor. This is sometimes called the "biological carbon pump," and it describes the export of carbon to the seafloor.

To understand how little carbon actually reaches the seafloor, we need to consider the fate of organic matter that escapes the photic zone and falls through the depths as **marine snow** (**figure 30.7***a*). This material gets its name from its appearance, sometimes seen in video images of mid- and deep-ocean water, where drifting flocculent particles resemble snow. Marine snow consists primarily of fecal pellets, diatom frustules, and other materials that are not easily degraded. During its fall to the bottom, marine snow is colonized by microbes so that nutrient recycling and mineralization continue. Bacterial population densities associated with marine snow can reach 10^8 to 10^9 per milliliter, which is orders of magnitudes higher than the surrounding water. These microbes secrete a variety of exoenzymes to break down organic polymers so the resulting monomers can be assimilated. As the marine snow sinks, it leaves a plume of DOM,

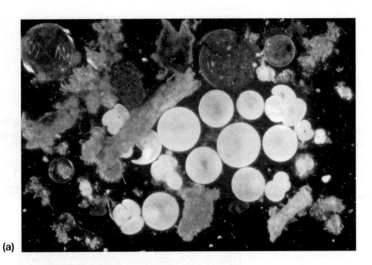

(a)

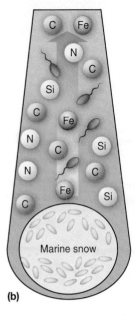

Marine snow

(b)

Figure 30.7 Marine Snow. The export of organic matter out of the photic zone to deeper water occurs through the sinking of marine snow. (a) Material collected in a sediment trap at 5,367 m on the Sohm Abyssal Plain in the Sargasso Sea. It includes cylindrical fecal pellets, planktonic tests (round white objects), transparent snail-like pteropod shells, radiolarians, and diatoms. (b) The hydrolytic activity of the exoenzymes secreted by the microbes that colonize marine snow generates a plume of nutrients during sinking. This in turn attracts more microbes.

nitrogen, phosphorus, and iron. These nutrients attract chemotactic free-swimming microbes, which help to deplete the nutrient pool even further (figure 30.7*b*). By the time a piece of marine snow reaches the seafloor, less than 1% of photosynthetically derived organic carbon remains unaltered. Once there, only a tiny fraction will be buried in the sediments; most will return to the surface in upwelling regions (**figure 30.8**).

An urgent question is whether or not the oceans can draw down (remove and bury) more CO_2 as atmospheric concentrations continue to increase. Presently it is estimated that about 3 billion tons of carbon per year are buried (sequestered) in deep-ocean sediments. Can increases in atmospheric CO_2 be offset by increasing the rate and volume of carbon sequestration? Could this in turn slow global warming? Such a scenario has been pursued in projects that fertilize

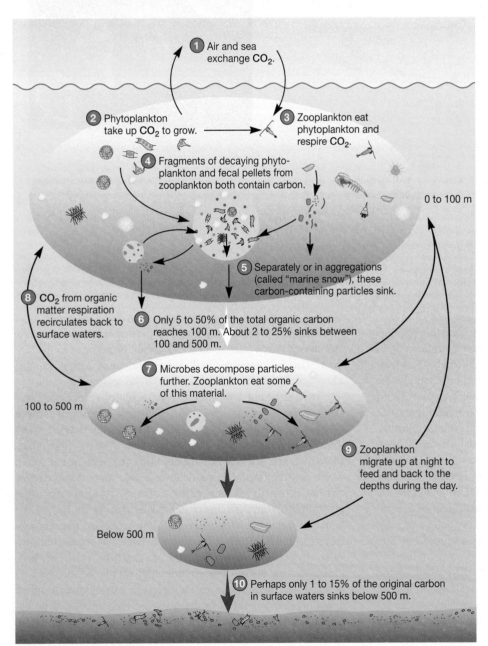

① Air and sea exchange CO_2.

② Phytoplankton take up CO_2 to grow.

③ Zooplankton eat phytoplankton and respire CO_2.

④ Fragments of decaying phytoplankton and fecal pellets from zooplankton both contain carbon.

0 to 100 m

⑤ Separately or in aggregations (called "marine snow"), these carbon-containing particles sink.

⑧ CO_2 from organic matter respiration recirculates back to surface waters.

⑥ Only 5 to 50% of the total organic carbon reaches 100 m. About 2 to 25% sinks between 100 and 500 m.

⑦ Microbes decompose particles further. Zooplankton eat some of this material.

100 to 500 m

⑨ Zooplankton migrate up at night to feed and back to the depths during the day.

Below 500 m

⑩ Perhaps only 1 to 15% of the original carbon in surface waters sinks below 500 m.

Figure 30.8 Fate of Carbon from the Photic Zone to the Deep Benthos. The vast majority of the carbon fixed by microbial autotrophs at the surface of the open ocean remains in the photic zone. Although a small fraction reaches the seafloor, much of it returns to the surface in regions where deep and surface waters mix (upwelling regions; not shown). The flux of CO_2 in and out of the world's oceans is in equilibrium, but this equilibrium may be upset by the increase in atmospheric CO_2, the hallmark of global warming.

internal concentrations of potassium or amino acids such as proline and betaine. Salt is another stressor, and most microbes that inhabit estuaries are halotolerant, which is distinct from halophilic. **Halotolerant** microbes can withstand significant changes in salinity; halophilic microorganisms have an absolute requirement for high salt concentrations. ◄◄ *Solutes affect osmosis and water activity (section 7.4)*

Estuaries are unique in many respects. Their calm, nutrient-rich waters serve as nurseries for juvenile forms of many commercially important fish and invertebrates. However, despite their importance to the commercial fishing industry, estuaries are among the most polluted marine environments. Estuaries and the rivers that feed them are the receptacles of toxins discharged during industrial processes as well as agricultural and street runoff. These pollutants include organic materials that can be used as nutrients. If too much organic matter is consistently added, the system is said to be **eutrophic,** and the rate of respiration greatly exceeds that of photosynthesis. Thus the water may become anoxic (chapter opening story). By contrast, if nutrients are introduced in a pulselike fashion, a single photoautotrophic species, either algal or cyanobacterial, grows at the expense of all other organisms in the community. This phenomenon, called a **bloom,** often results from the introduction of nutrients combined with changes in light, temperature, and mixing of water masses. If the microbes produce a toxic product or are in themselves poisonous to other organisms such as shellfish or fish, the term **harmful algal bloom (HAB)** is used. Some HABs are called **red tides** because the microbial density is so great that the water becomes red or pink, the color of the algae (**figure 30.3**).

The number of HABs has dramatically increased in the last decade or so. It is predicted that global climate change may trigger more HABs, in part because warmer water forms a more stable thermocline that lasts longer into the autumn. HABs are sometimes responsible for killing large numbers of fish or marine mammals. For instance, off the coast of California, periodic HABs are responsible for sudden, large-scale deaths of sea lions. These blooms are often caused by diatoms in the genus *Pseudonitzschia.* Anchovies consume the diatoms, and a potent neurotoxin, domoic acid, accumulates in the fish. The mammals are poisoned after they ingest large quantities of the fish. ◄◄ Stramenopila *(section 25.4)*

In addition to diatoms, HABs are often caused by dinoflagellates. Some bloom-causing dinoflagellates produce potent neurotoxins. Dinoflagellates of the genus *Alexandrium* produce a toxin responsible for paralytic shellfish poisoning (PSP), which affects humans as well as other animals in coastal, temperate North America. The toxin produced by the dinoflagellate *Karenia brevis* killed a number of endangered manatees and bottlenose dolphins in 2002 in a bloom in Florida. Another dinoflagellate, *Pfiesteria piscicida,* has become a problem in the Chesapeake Bay and regions south. This protist produces lethal lesions in fish (**figure 30.4**) and has had a devastating effect on the local fishing industry. Exposure to this microbe may also cause neurological damage to humans. ◄◄ Alveolata *(section 25.4)*

Figure 30.3 Harmful Algal Blooms Are Also Called Red Tides.
Blooms often color the water with the pigment of the microbe responsible. In this case, a dinoflagellate has turned the water pink-orange.

Figure 30.4 *Pfiesteria piscicida* **Lesions.** Lesions on a fish resulting from parasitism by the dinoflagellate *Pfiesteria piscicida.*

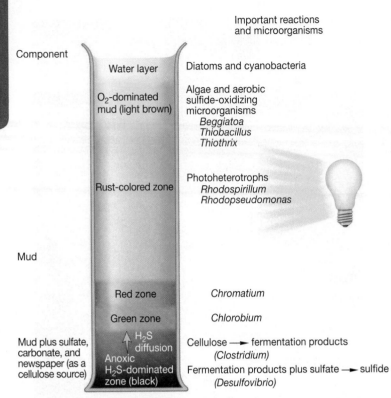

Figure 30.5 The Winogradsky Column. Microorganisms and nutrients interact over a vertical gradient in these microcosms. Fermentation products and sulfide migrate up from the reduced lower zone, and oxygen penetrates from the surface. This creates conditions similar to those in a lake or salt marsh with nutrient-rich sediments. Light simulates the penetration of sunlight into the anoxic lower region, which stimulates photosynthetic microorganism growth.

MICRO INQUIRY *How is sulfur cycled between the anoxygenic photosynthetic microbes belonging to the genera* Chlorobium *and* Chromatium, *and anaerobic heterotrophic* Desulfovibrio *spp.?*

Salt marshes generally differ from estuaries because their freshwater input is from multiple sources, rather than a single river. The microbial communities within salt marsh sediments are very dynamic. These ecosystems can be modeled in **Winogradsky columns (figure 30.5)**, named after the pioneering microbial ecologist Sergei Winogradsky (1856–1953).

A Winogradsky column is easily constructed using a glass cylinder into which sediment is placed and then overlaid with saltwater. When the top of the column is sealed, much of the cylinder becomes anoxic. The addition of shredded newspaper introduces a source of organic carbon that is fermented by members of the genus *Clostridium*. With these fermentation products available as electron donors, *Desulfovibrio* spp. and other microbes use sulfate as a terminal electron acceptor to produce hydrogen sulfide (H_2S). The H_2S diffuses upward, creating a stable H_2S gradient where photolithoautotrophs of the genera *Chlorobium* and *Chromatium* develop as visible olive green and purple zones, respectively. These green and purple sulfur

bacteria, respectively, use H_2S as an electron source and CO_2, from sodium carbonate, as a carbon source. Above this region, purple nonsulfur bacteria of the genera *Rhodospirillum* and *Rhodopseudomonas* can grow. These photoorganotrophs use organic matter as an electron donor under anoxic conditions and function in a zone where the sulfide level is lower. Both O_2 and H_2S may be present higher in the column, promoting the growth of other microorganisms. These include the chemolithotrophs in the genera *Beggiatoa* and *Thiothrix*, which use reduced sulfur compounds as electron donors and O_2 as an acceptor. In the upper, aerated portion of the column, diatoms and cyanobacteria may be visible. ◄◄ *Anaerobic respiration uses the same three steps as aerobic respiration (section 11.7); Phototrophy (section 11.11); Photosynthetic bacteria are diverse (section 21.4); Purple nonsulfur bacteria perform anoxygenic photosynthesis (section 22.1); Class* Gammaproteobacteria *is the largest bacterial class (section 22.3); Class* Clostridia *(section 23.1)*

Retrieve, Infer, Apply

1. Name two groups of protists known to cause HABs in marine ecosystems. What hazards do HABs pose for other marine organisms and humans?
2. Compare the chapter opening figure with figure 30.3 and explain why the water is brown in anoxic zones and may be red during an HAB.
3. Describe the ecosystem that develops within a Winogradsky column. Are any of these organisms chemolithotrophic? What would happen if the column were incubated in the dark for an extended period of time?

Microorganisms in the Open Ocean Are Adapted to Nutrient Limitation

Sometimes called "the invisible rain forest," the illuminated upper 200 to 300 meters of the open ocean, termed the **photic zone,** is home to a diverse collection of microbes. **Phytoplankton** (Greek *phyto,* plant, and *planktos,* wandering) are the source of all organic matter in the open ocean, with the exception of the hydrothermal vent and subsurface chemolithoautotrophic communities (*see figure 32.4*). Roughly half the carbon fixation in the open ocean is performed by cyanobacteria in the genera *Prochlorococcus* and *Synechococcus*. These microbes help constitute the **picoplankton** (planktonic microbes between 0.2 and 2.0 μm in size), which can represent 20 to 80% of the total phytoplankton biomass. Larger unicellular cyanobacteria (*Crocosphera watsonii*) and eukaryotic autotrophs, especially diatoms, contribute the other 50% of fixed carbon. ◄◄ *Phylum* Cyanobacteria *(section 21.4);* Stramenopila *(section 25.4)* ►► *Sulfide-based mutualisms (section 32.1)*

In contrast to coastal regions, nutrient levels are extremely low in the open ocean, which is unaffected by rivers, streams, and terrestrial runoff. In the open-ocean nitrogen, phosphorus, iron, and even the silica diatoms need to construct their frustules can be limiting. The key to survival in the photic zone is the tight recycling of nutrients, rather than allowing them to sink to the seafloor; this is the essence of the **microbial loop (figure 30.6)**. Within the microbial loop, heterotrophic microbes consume **dissolved organic matter (DOM)** released as exudates by other microbes as well as larger plants and animals. The DOM that

Figure 30.9 The Filamentous Cyanobacterium *Trichodesmium* sp. Fixes N₂ in the Open Ocean.

specific regions of the oceans. These regions, known as high-nitrate, low-chlorophyll (HNLC) areas, are limited by iron. A number of experiments have been performed wherein large transects of the southern Pacific have been fertilized with iron to trigger diatom blooms. However, as shown in figure 30.8, the amount of CO_2 sequestered is minimal; the vast majority of the extra carbon that falls from the ocean's surface is consumed (and converted back to CO_2) before it reaches the bottom.

Unlike the HNLC areas, most of the open ocean is limited by nitrogen, not iron. Because there is no terrestrial runoff, the only source of "new" organic nitrogen at sea is biological nitrogen fixation. There appear to be at least four significant sources of N₂ fixation—all of which are cyanobacterial. For many years, it was thought that the filamentous cyanobacterium *Richelia intracellularis* was the only microbe that fixed significant amount of N₂. This microbe lives in association with certain diatoms. However, filamentous cyanobacteria of the genus *Trichodesmium* are now known to be much more important in terms of the amount of N₂ fixed (**figure 30.9**). This microbe drifts on the surface in rafts that cover many square meters in tropical and subtropical oceans. It protects its oxygen-sensitive nitrogenase enzyme by partitioning it to regions of trichomes (filaments) called diazocytes. Because the region of a filament that functions as a diazocyte is not separated from the rest of the trichome by a septum, these structures differ from heterocysts seen on other filamentous cyanobacteria (*see figure 21.13*). Another source of "new" nitrogen is the N₂-fixing, unicellular cyanobacterium *Crocosphera watsonii*. It measures 3 to 8 μm in diameter, making it much bigger than the other unicellular cyanobacteria in the

genera *Synechococcus* or *Prochlorococcus*. *C. watsonii* is found in tropical surface waters.

The last source of "new" nitrogen is an unusual cyanobacterium first reported in 2008. This cyanobacterium fixes N₂ but lacks photosystem II and is photoheterotrophic. Organic carbon is fermented and because it uses only photosystem I no O_2 is produced, so, it can fix N₂ during the day. This microbe has not yet been cultured and is known only as UCYN-A. When observed in natural samples, it is a little less than 1 μm in diameter. It inhabits temperate latitudes, and unlike *Trichodesmium* spp., it is found in coastal waters as well as the open ocean. ◄◄ *Nitrogen fixation (section 12.5); Phylum* Cyanobacteria *(section 21.4)*

In addition to nitrogen fixation, other components of the nitrogen cycle occur in the open ocean. Much of this cycling occurs below the photic zone, where oxygen concentrations reach a minimum before increasing with depth. Here two types of microbial-mediated reactions result in the loss of nitrogen as N₂ gas. The first is denitrification—anaerobic respiration using nitrate (NO_3^-) as a terminal electron acceptor and reducing it to N₂. The second is the anammox reaction—the anaerobic oxidation of ammonium (NH_4^+) to N₂ using nitrite (NO_2^-) as the terminal electron acceptor (*see figure 21.16*). These bacteria are members of the interesting phylum *Planctomycetes*. Nitrogen that escapes denitrification or anammox in the oxygen minimum zone and sinks further is mostly NH_3/NH_4^+. However, most of the nitrogen that accumulates at depth is nitrate (NO_3^-). This is because the NH_3/NH_4^+ is oxidized by microbes performing nitrification—the oxidation of NH_3 first to NO_2^- and then to NO_3^-. Most marine nitrification is accomplished by archaea in the phylum *Thaumarchaeota*. These mesophilic archaea oxidize NH_3 to NO_2^- which is then oxidized to NO_3^- by bacteria. The urgency to understand oceanic nitrogen cycling reflects concern about release of the greenhouse gas N_2O (principally by ammonia oxidizing archaea) and increasing levels of atmospheric CO_2. Many question the capacity of the world's oceans to consume more atmospheric CO_2 while maintaining overall ecosystem equilibrium. ◄◄ *Phylum* Thaumarchaeota *(section 20.3); Phylum* Planctomycetes *(section 21.5); Nitrogen cycle (section 28.1)*

Given the enormous volume of the ocean, it is probably not surprising that the most abundant monophyletic organisms on Earth are marine α-proteobacteria. Members of the clade called **SAR11** (named after the *Sar*gasso Sea, where they were first discovered) have been detected by their rRNA genes from almost all open-ocean samples taken worldwide. In addition, they have been found at depths of 3,000 m, in coastal waters, and even in some freshwater lakes. Using fluorescence in situ hybridization (FISH; *see figure 29.6*), SAR11 bacteria have been found to constitute 25 to 50% of the total bacteria and archaea in the surface waters in both nearshore and open-ocean samples. Indeed, SAR11 bacteria are estimated to constitute about 25% of all microbial life on the planet. ◄◄ *SAR11 clade, the most abundant microorganisms on Earth (section 22.1); Microbial biology relies on cultures (section 29.1)*

The physiology of SAR11 illustrates a recurring metabolic theme that has emerged in the study of marine microorganisms: many have evolved strategies to supplement their energy reserves.

Since the first discovery of rhodopsin-containing γ-proteobacteria in 2000, similar genes have been found in other marine microbes, including α-proteobacteria (i.e., SAR11), β-proteobacteria, bacteroidetes, and new members of the archaeal phylum *Euryarchaeota*. In all cases, the rhodopsin is structurally distinct from the archaerhodopsin found in haloarchaea. None of these microorganisms is autotrophic; it appears that they use the light-driven rhodopsin pump to supplement ATP pools in nutrient-depleted waters. The importance of rhodopsin in marine microorganisms is revealed by metagenomic analysis, which indicates that 13 to 80% of heterotrophic marine bacteria and archaea (depending on sampling location) have the genes needed for proteorhodopsin biosynthesis. ◀◀ *Rhodopsin-based phototrophy (section 11.11); Haloarchaea (section 20.4)*

Another group of marine microbes that appear to be supplementing their energy reserves are aerobic anoxygenic phototrophs (AAnPs). Although these bacteria were discovered several decades ago, for a long time they were considered a physiological curiosity. Bacteria capable of **aerobic anoxygenic phototrophy** are known to possess bacteriochlorophyll *a* (Bchl *a*), but it was a great surprise when it was discovered that this photosynthetic pigment is widely distributed in surface ocean waters. A combination of cultivation and metagenomic techniques demonstrates that some α-proteobacteria (the *Roseobacter* clade), as well as β- and γ-proteobacteria, are capable of this form of phototrophy (*see table 21.2*). It would be incorrect to call these bacteria photosynthetic because they are not autotrophs. Like the rhodopsin-containing microbes, they use light to drive the synthesis of ATP (photophosphorylation) but meet their carbon needs by the assimilation of organic carbon; that is, they are photoheterotrophs. By meeting some of their ATP demands phototrophically, AAnPs and rhodopsin-containing microorganisms can divert more organic carbon to biosynthesis and less to catabolic processes. However, open-ocean bacteria that possess Bchl *a* are much less abundant than those that produce proteorhodopsin. This may reflect the fact that genes for proteorhodopsin have been widely subject to horizontal gene transfer, as only six genes are required for a functional light-driven proton pump. This is in contrast to the roughly 50 genes needed to assemble the anoxygenic light-harvesting complex, reaction center, Bchl *a*, and carotenoids found in AAnPs.

The α-proteobacterium *Silicobacter pomeroyi* appears to be capable of a third interesting metabolic strategy to increase its ATP yield: **lithoheterotrophy** (or chemolithoheterotrophy). Like photoheterotrophy, in which light is used for energy and electrons while organic carbon is consumed, lithoheterotrophy uses inorganic chemicals as a source of energy but requires organic carbon for anabolic purposes. A number of marine carbon monoxide (CO)-oxidizing heterotrophs have been cultured. These bacteria produce CO dehydrogenase, thereby oxidizing CO to CO_2, but unlike other CO-oxidizing bacteria, they do not have the enzymes necessary for the conversion of CO_2 into biomass. As has been shown with *S. pomeroyi*, they seem to use CO as a supplemental electron donor for their electron transport chain. Thus they need not oxidize as much organic carbon to generate NADH for this purpose. Instead, more organic carbon can be used for biosynthesis. This hypothesis is supported by thermodynamic considerations: the oxidation of one molecule of CO should release enough energy to drive the production of one ATP. Measurements of CO lithoheterotrophs show that they make up about 7% of the bacterioplankton in the open ocean and coastal waters sampled.

It is not difficult to hypothesize that these alternatives to "traditional" organotrophy would be beneficial to individual cells in a nutrient-stressed environment. What is more problematic is estimating how much these processes contribute to the carbon budget of marine habitats. Theoretically, the advantage these processes confer must more than offset the energetic cost of carrying the genes required for their expression. But it remains to be seen if these processes actually make these microbes better competitors. Likewise, the impact these metabolic strategies have on carbon flux at the community, ecosystem, and global levels awaits more investigation. It is questions like these that make microbial oceanography such a dynamic and exciting field.

Aquatic Viruses: Mortality at Sea

As mentioned in our discussion of the microbial loop (figure 30.6), viruses are important members of marine and freshwater microbial communities. In fact, **virioplankton** are the most numerous members of marine ecosystems. However, quantifying viruses is tricky: The traditional method of plaque formation requires knowledge of both the virus and its host, and the ability to grow the host in the laboratory. Because so few microbes have been cultured, this prevents the measurement of virus diversity by examining actual virus infection. Instead, virus particles may be visualized directly. There are at least two approaches for this. Electron microscopy is the most rigorous, but fluorescence microscopy is a common and convenient method used to look at viruses in aquatic systems. When seawater is filtered, viruses can be collected on filters with a pore size of 0.02 μm. Viral nucleic acids are stained with a fluorescent dye, such as YO-PRO or SYBR Green for fluorescence microscopy. Although microscopy reveals the presence of virions outside the host cell, it does not prove that they can infect a host cell. Therefore, viruses enumerated in this way are called **virus***like* **particles** (**VLPs**). Using this approach, the average VLP density in seawater is between 10^6 to 10^7 per milliliter (although in some cases, it may be closer to 10^8 per milliliter); their numbers decline to roughly 10^6 below about 250 m. Marine viruses are so abundant that VLPs are now recognized as the most abundant microbes on Earth (**figure 30.10**).

Viral diversity is vast, including single- and double-stranded RNA and DNA viruses that infect archaea, bacteria, and protists. Recent advances in metagenomic analysis of viral genomes have been used to explore viral diversity. In one such study, viral communities were sampled from the Arctic Ocean, North Pacific coastal waters, the Gulf of Mexico, and the Sargasso Sea. Of the roughly 1.8 million nucleotides obtained, 90% had no recognizable match on any database. Thus not only is viral genetic diversity immense, it is largely unexplored. ◀◀ *Viruses (chapter 27)*

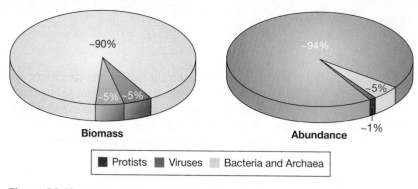

Figure 30.10 Relative Biomass and Abundance of Marine Microbes. (a) The biomass of bacteria and archaea far exceeds that of both protists and viruses. (b) When individuals are counted, however, the vast majority of marine microbes are viruses.

The abundance of marine viruses indicates that they must be major agents of mortality in the sea. Indeed, viruses are thought to kill on average about 20% of the marine microbial biomass daily. However, measurement of virus-induced microbial mortality shows that it is highly variable. Viral abundance frequently corresponds to the microbial host that is most active in the community. Models predict that as one microbial species (or strain) becomes numerically dominant, it soon will be targeted by lytic viruses and thus its population will decline. This permits another microbial species (or strain) to flourish, which then becomes subject to intense viral lysis, and so on. This "kill the winner" model has garnered much attention, but more experimental evidence is needed before it is widely accepted. Perhaps most compelling is the fate of blooms of the coccolithophore *Emiliania huxleyi*. Such blooms are so intense they can be imaged from space, yet their collapse is thought to be (at least in part) the result of viral lysis (**figure 30.11**). Although this phenomenon is interesting, in most cases, viral lysis does not result

in the complete collapse of a host population. This may be due partly to strains of the host species that are resistant to virus infection, as has been shown with members of the abundant cyanobacterial genus *Synechococcus*. Computer modeling and model experiments indicate that viruses contribute to nutrient cycling by accelerating the rate at which their microbial hosts are converted to POM and DOM, thereby feeding other microorganisms without first making them available for protists and other bacteriovores. This "short-circuits" the microbial loop (**figure 30.12**).

As might be predicted, viruses are important vectors for horizontal gene transfer in marine ecosystems. In fact, it has been calculated that in the oceans, phage-mediated gene transfer occurs at an astounding rate of 20 billion times per second. The importance of phage-mediated horizontal gene transfer is demonstrated by cyanophages that infect species of *Synechococcus* and *Prochlorococcus*. These phages carry the structural genes for photosynthetic reaction center proteins. By shuttling these genes between strains of cyanobacteria, these phages may play a critical role in the evolution of these important primary producers.

Microbial Communities in Benthic Marine Sediments Are Enormous and Mysterious

Sixty-seven percent of the Earth's surface lies under the sea, which means that of all the world's microbial ecosystems, we know the least about the largest. However, the combination of deep-ocean drilling projects and the exploration of geologically active sites, such as submarine volcanoes and hydrothermal vents, has revealed that the study of ocean sediments, or benthos, can be rewarding and surprising. Marine sediments range from the very shallow to the deepest trenches, from dimly illuminated to completely dark, and from the newest sediment on Earth to material that is millions of years old. The temperature and age of marine sediments depend on their proximity to geologically active areas. Hydrothermal vent communities with large and diverse invertebrates, some of which depend on endosymbiotic chemolithotrophic bacteria, have been intensely investigated since their exciting discovery in the late 1970s. These microbes are discussed in chapter 32. Because the vast majority of Earth's crust lies at great depth far from geothermally active regions, most benthic marine microbes live under high pressure, without light, and at temperatures between 1°C and 4°C.

Deep-ocean sediments were once thought to be devoid of all life and thus not worth the considerable effort it takes to study them. But sediment samples from water depths up to 8,200 m (at its deepest, the ocean is about 11,000 m) reveal the presence of a vast "piezosphere" (Greek, *piezein*, to squeeze or press). The piezosphere describes the biome at depths with pressures exceeding 100 atm (pressure increases about 1 atm per 10 m depth). Microbes that grow at such great pressures are termed **piezophilic** and their density is 10 to 10,000 greater per unit volume than in productive surface waters. Reasons for this difference are complicated but include energy

Figure 30.11 Viral Lysis and the Termination of a Massive Plankton Bloom. This satellite image shows the demise of a 500 km-long bloom (light blue region) of the coccolithophore *Emiliania huxleyi*. While the exhaustion of available nutrients also contributes to the death of such an enormous bloom, research has demonstrated that viruses specific to this protist are principle agents in the bloom's demise. The green land mass is Ireland.

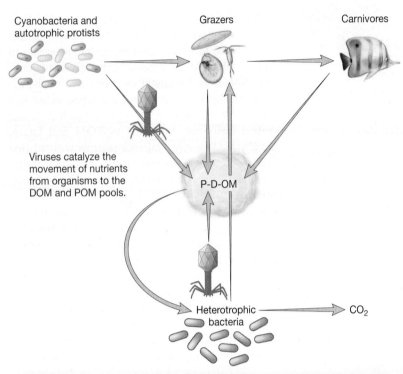

Cyanobacteria and autotrophic protists

Grazers

Carnivores

Viruses catalyze the movement of nutrients from organisms to the DOM and POM pools.

P-D-OM

Heterotrophic bacteria → CO_2

Figure 30.12 **The Role of Viruses in the Microbial Loop.** Viral lysis of autotrophic and heterotrophic microbes accelerates the rate at which these microbes are converted to particulate and dissolved organic matter (P-D-OM). This is thought to increase net community respiration and decrease the efficiency of nutrient transfer to higher trophic levels.

MICRO INQUIRY *When might viruses choose lysogeny rather than lysis and how would this affect carbon flow (also see figure 30.6)?*

availability and limited grazing pressure. This result is especially surprising because the majority of the ocean floor receives only about 1 gram of organic carbon per square meter per year.

It is now accepted that the biosphere below the Earth's crust represents the "hidden majority" of all microbial biomass. Estimates range from half to five-sixths of the Earth's total microbial biomass and up to one-third of Earth's total living biomass. As might be predicted, most of these microbes are not amenable to laboratory culture. In fact, these microbes "breathe" about a million times slower than surface microorganisms, making their metabolism so slow that doubling times are measured in hundreds, and sometimes thousands, of years. Subsurface microbes truly stretch the limits of what we have come to define as life.

Global surveys show that the deep subsurface biosphere within relatively nutrient-rich continental margin sediments varies from that in the organic-poor distant abyss. Seawater is rich in sulfate, which is used as an electron acceptor in anaerobic respiration. In the continental margin, sulfate in sediments is depleted within 100 m depth, at which point methanogenesis becomes dominant. However, in the abyss, sulfate penetrates much more deeply, and distinct zonation between sulfate-reducing bacteria and methanogenic archaea is absent. Despite biochemical evidence for the presence of sulfate reduction and methanogenesis, SSU rRNA sequence analysis of DNA has been surprising in the number of genes that lack homology to known sulfate reducers and methanogens.

Furthermore, the sediment surface and subsurface appear to be teeming with viruses; it is estimated that each year viral lysis in the benthos releases up to 630 million tons of carbon that had been sequestered by marine snow and other falling particulates. These exciting findings suggest that a significant fraction of Earth's biomass is amazingly active yet largely uncharacterized.

Although thermophilic and hyperthermophilic bacteria and archaea are found at hydrothermal vent sites, most deep-sea microbes are psychrophilic. The majority recovered so far are Gram-negative, facultative anaerobic bacteria. Common genera include γ-proteobacteria in the genera *Shewanella, Photobacterium, Colwellia, Moritella,* and *Psychromonas*. Interestingly, it appears that the subterranean depth limit of microbial growth is not governed by pressure; rather, it is determined by temperature. It seems unlikely that the maximum temperature at which life can be sustained has been identified. ◄◄ *Environmental factors affect microbial growth (section 7.4)*

Exploration of deep-sea subsurface microbiology has turned our understanding of bacterial energetics literally upside down. As discussed in chapters 11 and 28, it is generally understood that anaerobic respiration occurs such that there is preferential use of available terminal electron acceptors (*see figure 28.1*). Acceptors that yield the most energy (more negative ΔG) from the oxidation of NADH or a reduced inorganic compounds (e.g., H_2, H_2S) will be used before those that produce a less negative ΔG. Thus following oxygen depletion, nitrate will be reduced, then manganese, iron, sulfate, and, finally, carbon dioxide. This is demonstrated in sediment cores up to 420 m deep collected off the coast of Peru (**figure 30.13**). However, when these signature chemical compounds were measured at greater depth, the profile was reversed. This suggests the presence of unknown sources of these electron acceptors at subsurface depths of more than 420 m. In addition, contrary to our long-held notion of thermodynamic limits, methane formation (methanogenesis) and iron and manganese reduction co-occur in sediments with high sulfate concentrations. Although the identity of the microbes that make up these communities awaits further study, it is clear that with densities of 10^8 cells per gram of sediment at the seafloor surface and 10^4 cells per gram in deep subsurface sediments, these communities are important. ◄◄ *Redox reactions (section 10.3); Anaerobic respiration uses the same three steps as aerobic respiration (section 11.7); Chemolithotrophy (section 11.10)*

Another outcome of deep-ocean sediment exploration has been the discovery of microbial communities on continental margins fueled by the release of hydrocarbons at depths between 200 and 3,500 m. Depending on the surface topology and the rate at which hydrocarbons are emitted, they are called pockmarks, gas chimneys, mud volcanoes, brine ponds, and oil and asphalt seeps. A variety of microbes can use hydrocarbons as a sole carbon source, oxidizing them with oxygen or sulfate as the electron acceptor. In addition, methane-oxidizing archaea grow in consortia with sulfate-reducing bacteria of the *Desulfococcus* and *Desulfobulbus* groups, as discussed in chapter 20. Some regions are the

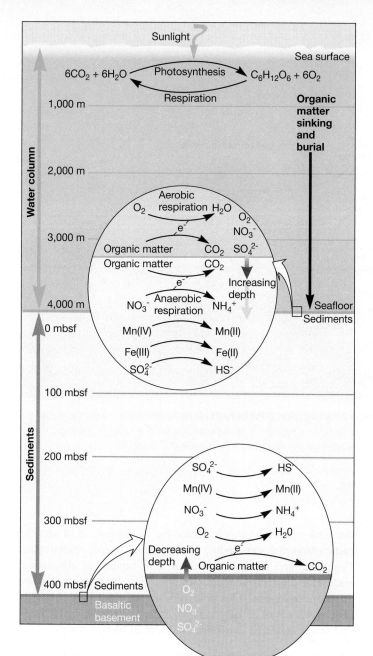

Figure 30.13 Microbial Activity in Deep-Ocean Sediments. At the surface of the seafloor, reduction of oxidized substrates that serve as electron acceptors in anaerobic respiration follows a predictable stratification based on thermodynamic considerations. This sequence reverses in very deep subsurface sediments, suggesting a source of electron acceptors from deep within Earth's crust. Meters below seafloor, mbsf.

sites of microbial mats, which can sometimes cover several hundreds of meters. These mats are dominated by the giant, vacuolated sulfur-oxidizing bacteria in the genera *Beggiatoa* and *Thiomargarita* (*see figures 22.23 and 22.24*). ◄◄ *Class Gammaproteobacteria is the largest bacterial class (section 22.3)*

Perhaps the most industrially important of these hydrocarbon-fueled communities are those found at **methane hydrates.** Methane hydrates are lattice-like ice structures in which methane is trapped. The methane is produced by both geothermal activity and methanogenic archaea. The lack of associated methane-oxidizing microbes has enabled this pool of natural gas to accumulate in latticelike cages of crystalline water 500 m or more below the

sediment surface in many regions of the world's oceans. This discovery is significant because there may be up to 10^{13} metric tons of methane hydrate worldwide. Thus 80,000 times the world's current known natural gas reserve awaits the development of the technology to access it. ◄◄ *Phylum* Euryarchaeota (*section 20.4*)

Retrieve, Infer, Apply

1. What is marine snow? Why is it important in CO_2 drawdown?
2. Describe two recently recognized sources of organic nitrogen in the open ocean.
3. Why do you think that, despite their great abundance, the SAR11 clade was not discovered until the late twentieth century?
4. List some metabolic strategies that have evolved to enable microbial survival in oligotrophic marine habitats. Which do you think is most successful in terms of numbers of microbes and metabolic flexibility? Explain your answer.
5. Describe the role of marine viruses in the microbial loop. How might community dynamics change in the absence of viral lysis?
6. Explain what is meant by "upside-down microbial energetics," as described for some deep-subsurface ocean sediments.
7. What are methane hydrates? Why are they important?

30.3 Microorganisms in Freshwater Ecosystems

After reading this section, you should be able to:

■ Appreciate the scarcity of freshwater on Earth
■ Contrast and compare the microbial communities found in lotic systems with those in lentic systems
■ Describe the differences between the microbial communities found in a deep, stratified lake with those of a shallow lake that lacks stratification
■ Describe how seasonal changes influence lake microbial communities and how these changes can lead to blooms

While the vast majority of water on Earth is in marine environments, freshwater is crucial to our terrestrial existence. As shown in **figure 30.14**, of the 3.9% of Earth's water that is not saline, only a tiny fraction represents the familiar streams, rivers, and wetlands

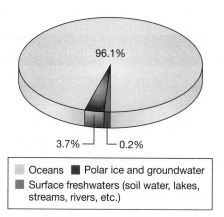

Figure 30.14 Relative Distribution of Freshwater on Earth.

that we associate with the term freshwater. We begin our discussion with the largest fraction of freshwater—that which is frozen at the Earth's polar regions. We then move to streams, rivers, and lakes.

Glaciers and Permanently Frozen Lakes Support Active Microbial Communities

Much of the ice on Earth has remained frozen for millions of years. Indeed, it is important to note that a majority of the Earth's surface never exceeds a temperature of 5°C. This includes polar regions, the deep ocean, and high-altitude terrestrial locations throughout the world. Surprisingly, microbes within glaciers are not dormant. Rather, evidence that active microbial communities exist in these environments has been growing over the last decade. In fact, this is an exciting time in glacial microbiology; determining the diversity in these systems and assessing the role of these microorganisms in biogeochemical cycling can involve novel and creative techniques. In addition, since the discovery of ice on Mars and on Jupiter's moon Europa, astrobiologists have become very interested in ice-dwelling psychrophilic microbes. ◄◄ *Temperature (section 7.4).*

The biology of lakes associated with glaciers has gained importance as it has been recognized that these ecosystems function as sentinels of climate change. Glacial lakes are constrained by low temperature and reduced solar radiation, and in the Antarctic, lakes are separated from macrofauna as well. Here phytoplankton, bacteria, and protists form simple food webs almost completely devoid of multicellular organisms (**figure 30.15**). Microbial growth rates are slow, so it not surprising that primary productivity within

these lakes must be supplemented by the deposition of organic carbon from the surrounding environment. Indeed, phototrophic flagellated and ciliated protists are also capable of consuming bacteria and dissolved organic carbon. These mixotrophs are not the only consumers of bacteria; most of the viruses are bacteriophages. Viral lysis of bacteria in Antarctic lakes is almost 10 times that found in similar temperate and tropical lakes.

Microbial Communities in Streams and Rivers Differ

Streams and rivers are much more dynamic than glacial lakes. Some of the physical and biological factors that help regulate the microbial communities of these freshwater systems are listed in **table 30.1**. Lakes, rivers, and streams can be broadly

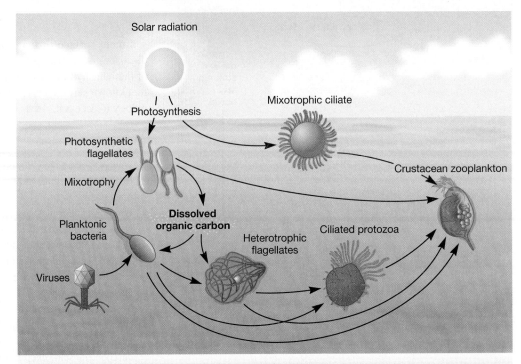

Figure 30.15 **Nutrient Cycling in Antarctic Lakes Is Performed Entirely by Microbes.** Antarctic glacial lakes lack multicellular organisms, so all nutrient cycling is between various types of microbes, including bacteriophages. The surrounding environment provides some organic carbon.

MICRO INQUIRY *List three major differences between this microbial loop and that found in the open ocean (figure 30.6).*

Table 30.1	Physical and Biological Factors Influencing Microbial Communities in Freshwater Habitats		
Habitat	**Major Microbial Community**	**Major Physical Characteristic**	**Major Biological Constraints**
Lake	Plankton	Stratification, wind-generated turbulence	Nutrient competition, grazing, parasitism
Floodplain	Plankton, biofilms attached to plants	Periodic desiccation	Competition between algae and macrophytes
Rivers and streams	Benthic microbes	Flow-generated turbulence	Colonization competition, biofilm grazing
Estuary—mudflats	Biofilms	Desiccation, high light, salt exposure	Competition and grazing at mud surface
Estuary—outflow	Plankton	Mixing with saltwater, turbidity	Grazing in water column

classified as either **lotic** systems with free-running waters (streams, most rivers, canals) or **lentic** systems with free-standing waters (e.g., ponds; lakes; marshes; very broad, slow-moving rivers). In large, lentic rivers and lakes, phytoplankton thrive and are responsible for most of the fixed carbon. The biologically available carbon that is produced within the system is called **autochthonous.** By contrast, the constant flow of water in streams and all but the largest rivers prevents the development of significant planktonic communities. In addition, phytoplankton growth can be further limited by the amount of available light due to overhanging foliage, turbidity, and rapid mixing of the water. In most streams and rivers, the source of nutrients therefore comes from the surrounding land. Such nutrients are called **allochthonous.** This distinction is important, as lentic systems often have a net autotrophic metabolism, whereas lotic systems are generally heterotrophic.

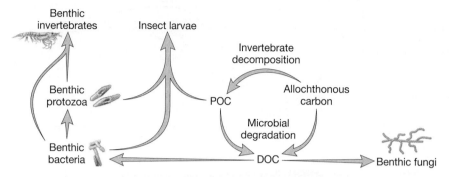

Figure 30.16 **A Benthic Microbial Loop in Lotic Systems.** In contrast to the microbial loop in planktonic systems (figure 30.6), the benthos of a lotic system, such as a fast-moving river or stream, features small eukaryotes such as invertebrates and insect larvae. These organisms play an important part in nutrient cycling. At the same time, macrofauna such as fish and other top carnivores may be absent or play relatively minor roles.

The allochthonous carbon that enters rivers and streams includes both dissolved and particulate organic carbon (DOC and POC, respectively). The amount of carbon that enters freshwater systems depends on surrounding vegetation, proximity to human activity, and local geology. In addition, hydrological factors such as how long water stays in the soil before entering the river or stream also influence the nature of the organic carbon. Thus DOC ranges from material that can be readily assimilated to refractory carbon that only a few microbial groups can slowly degrade. Much of the particulate organic carbon includes large items such as leaf litter, which provide substrates for fungi and bacteria to colonize. Initially, the activity of these microbes releases a pulse of DOC; this is followed by the slow leaching of nutrients that may not be as easy to assimilate.

Most rivers and streams have healthy benthic communities of attached microbes that form biofilms consumed by invertebrates and other grazing organisms. In shallow rivers and streams, some biofilms are dominated by diatoms and other photosynthetic protists, which contribute to the total primary production of the system. Biofilms that consist primarily of heterotrophic microbes are important sources of degradation and mineralization. The role of benthic microbes in rivers and streams can be generally characterized by the microbial loop shown in **figure 30.16**.

As in any microbial community, when river and stream chemoorganotrophic microorganisms metabolize the available organic material, they recycle nutrients within the ecosystem. Autotrophic microorganisms require minerals released from organic matter. This leads to the production of O_2 during daylight hours, when the rates of photosynthesis and respiration are in dynamic equilibrium. When the amount of organic matter added to streams and rivers does not upset this balance, so that heterotrophy and autotrophy are roughly equal, productive streams and rivers are maintained. However, as seen with ocean dead zones, when the supply of organic matter supports a vast heterotrophic population whose respiration rate outpaces oxygenic photosynthesis, the water may become anoxic. This can occur in streams and rivers adjacent to urban and agricultural areas. The release of inadequately treated municipal wastes and other materials from a specific location along a river or stream represents point source pollution. Such additions of organic matter can produce distinct and predictable changes in the microbial community and oxygen concentration, creating an oxygen sag curve in which oxygen is depleted just downstream of the pollution source (**figure 30.17**). Runoff from agriculturally active fields and feedlots is an example of nonpoint source pollution. This can also cause disequilibrium in the microbial community, leading to algal or cyanobacterial blooms.

Retrieve, Infer, Apply

1. Why is mixotrophy suited for survival in Antarctic glacial lakes?
2. What is an oxygen sag curve? What changes in a river cause this effect?
3. What are point and nonpoint source pollution? Can you think of examples in your community?

Microorganisms in Lakes Can Cause Seasonal Blooms

Lakes differ from rivers and streams because they are lentic systems, thus dominated by planktonic microbes and invertebrates. The topology and hydrology of lakes are key factors in the development of these communities. Lake hydrology incorporates all aspects of water flow: the water in the lake itself, plus its inflow and outflow. Because much of the water in a lake comes from the surrounding land, the geology and use of the land that feeds the lake largely determine the chemistry of the inflow water. For instance, mountain lakes are fed water from relatively infertile land, resulting in oligotrophic conditions within the lake. This explains why mountain lakes tend to be so clear: microbial growth is limited to those species that tolerate low nutrient levels. Such microbes tend to have long generation times. On the other hand, low-lying lakes generally catch water from fertile, cultivated soils and so are nutrient rich. This supports a high level of planktonic growth, and the lakes may appear turbid. The picoplankton often include freshwater species of *Synechococcus,* whereas filamentous cyanobacteria

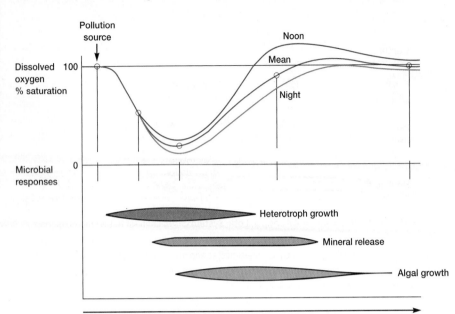

Figure 30.17 **The Dissolved Oxygen Sag Curve.** Microorganisms can create gradients over distance and time when nutrients are added to rivers. An excellent example is the dissolved oxygen sag curve, caused when organic wastes are added to a clean river system. During the later stages of self-purification, the phototrophic community will again become dominant, resulting in diurnal changes in river oxygen levels.

such as those in the genera *Anabaena* and *Microcystis* constitute the macroplankton (microbes larger than 200 μm).

Deep lakes can be divided into two main regions: the **littoral zone,** or shoreline, and the central, **pelagic zone** (these terms also apply to marine systems). As shown in **figure 30.18a**, the littoral region is also defined by the depth of the photic zone. The two zones have distinct microbial communities. In the littoral zone, allochthonous DOC introduces organic substrates and nutrients such as phosphate and nitrate. Because the water is shallow, light penetrates throughout the water column, enabling the development of larger plants. Biofilms frequently form on these plants and in the sediments.

The microbial community within the pelagic zone is partly determined by the depth of the lake. Light penetrates the entire water column in shallow lakes (less than about 10 m), so benthic autotrophs, including diatoms and cyanobacteria, may be the principle primary producers (figure 30.18b). In deeper lakes, the central, pelagic region can become thermally stratified. This is the case in temperate regions (**figure 30.19**). In the summer, the upper layer, called the **epilimnion,** becomes warmed and oxygen is released into the atmosphere. The thermocline spans a region called the **metalimnion;** this region acts as a barrier to mixing of upper and lower water masses. The lower, colder, dark region, known as the **hypolimnion,** can become anoxic because oxygenic photosynthesis is confined to the epilimnion. Because these two water masses do not mix (or mix very little), the upper layer supports a rich diversity of primary producers and consumers, which can deplete nutrients. Just the opposite is true for the hypolimnion, where dark, anoxic, cold conditions limit growth. Nutrients accumulate and the microbial community primarily consists of

benthic heterotrophs. When autumn weather cools the surface and storms physically mix the two layers, a short-lived bloom can occur as nutrients from the bottom of the lake are mixed into the photic zone. Much longer blooms may occur in the spring, when thermal stratification occurs but nutrients in the epilimnion have not yet been depleted by microbial growth.

Tropical lakes lack significant temperature changes during the year but have seasonal differences in rainfall. Although many aspects of tropical lakes are similar to those of a summer temperate lake, there are several key differences. The warmer, year-round temperatures support higher productivity, and plankton blooms may occur at any time during the year. Cyanobacteria capable of nitrogen fixation may dominate, reflecting the low N:P ratios frequently found in tropical lakes. Nutrient cycling in temperate lakes is largely driven by physical mixing, whereas biological processes are more important in tropical lakes. In this regard, the microbial loop shown in figure 30.6 can be applied to the epilimnion of larger tropical lakes that receive few allochthonous nutrients.

Lakes vary in nutrient status. Some are oligotrophic and others are eutrophic. Nutrient-poor lakes remain oxic throughout the year, and seasonal temperature shifts usually do not result in distinct oxygen stratification. In contrast, eutrophic lakes usually have bottom sediments that are rich in organic matter. In oligotrophic lakes, cyanobacteria capable of nitrogen fixation may bloom. Members of several genera, notably *Anabaena, Nostoc,* and *Cylindrospermum,* can fix nitrogen under oxic conditions. Those in the genus *Oscillatoria* use hydrogen sulfide as an electron donor for photosynthesis, enabling nitrogen fixation under anoxic conditions. If both nitrogen and phosphorus are present, cyanobacteria compete with algae. Cyanobacteria function more efficiently in alkaline waters (pH 8.5 to 9.5) and higher temperatures (30 to 35°C). By using CO_2 at rapid rates, cyanobacteria also increase the pH, making the environment less suitable for protists. Cyanobacteria have additional competitive advantages. Many produce siderophores to bind iron, making this important trace nutrient less available for protists. Some cyanobacteria also resist predation because they produce toxins. In addition, some synthesize odor-producing compounds that affect the quality of drinking water. These problems become magnified when cyanobacteria or protists (algae) produce massive blooms in strongly eutrophied lakes. ◄◄ *Phylum* Cyanobacteria *(section 21.4)*

Retrieve, Infer, Apply

1. What terms can be used to describe the different parts of a lake?
2. What are some important effects of eutrophication on lakes?
3. Why do cyanobacteria often dominate waters that have been polluted by the addition of phosphorus?

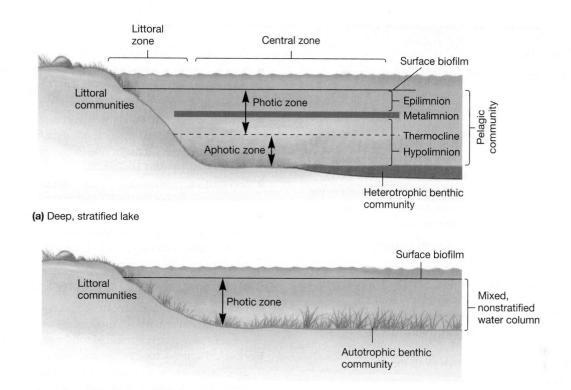

(a) Deep, stratified lake

(b) Shallow, unstratified lake

Figure 30.18 The Depth of a Lake Is a Key Factor in Shaping the Microbial Community. (a) Deep lakes have autotroph-dominated littoral and epilimnion microbial communities. Waters below the thermocline and photic zone form the hypolimnion, which can become oxygen depleted. Here a heterotrophic benthic community thrives. (b) Shallow lakes have little mixing, and the photic zone reaches the lake bottom. Thus autotrophs dominate the water column and benthic regions throughout the lake.

MICRO INQUIRY *How does the contribution of benthic autotrophs differ in deep lakes compared to that in shallow lakes? How does this affect the relative contribution of phytoplankton in these two aquatic systems?*

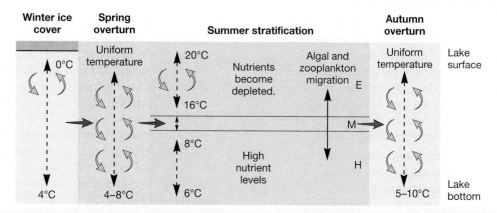

Figure 30.19 Seasonal Changes in a Temperate Lake Are Associated with Blooms. In the spring, the water begins to warm, and lack of thermal stratification ensures ample nutrients for microbes that can grow in the relatively warmer surface waters. During the summer as the surface waters warm, thermal stratification separates the cool, deep, nutrient-rich waters from the warm, illuminated waters. Because microbial growth is favored in the warmer surface water, this water becomes depleted of nutrients. During the autumn, waters cool and storms mix the layers, bringing the bottom nutrients to the surface. This pulse of nutrients can support a bloom of microbes or algae. Water turbulence is indicated by arrows; E, epilimnion; M, metalimnion; H, hypolimnion.

MICRO INQUIRY *Why does water turbulence play only a minor role in mixing the water column in the summer?*

Key Concepts

30.1 Water Is the Largest Microbial Habitat

- Most water on the Earth is marine (97%). The majority of this is cold (2 to 5°C) and at high pressure. Freshwater is a minor but important part of Earth's biosphere.
- The carbonate equilibrium system keeps the oceans buffered at pH 7.6 to 8.2; however, ocean acidification threatens the health of marine ecosystems (**figure 30.1**).
- The penetration of light into surface waters determines the depth of the photic zone. Warming the surface waters can lead to the development of a thermocline.

30.2 Microorganisms in Marine Ecosystems

- Tidal mixing in estuaries causes daily swings in osmolarity. Thus microbes within estuaries have evolved mechanisms to cope with rapid changes in salinity.
- Coastal regions such as estuaries and salt marshes can be the sites of harmful algal blooms (**figure 30.3**).
- Autotrophic microbes in the photic zone within the open ocean account for about one-half of all the carbon fixation on Earth.
- The microbial loop describes the transfer of nutrients between trophic levels while taking into account the multiple contributions of microbes to recycling nutrients. Nutrients are recycled so efficiently, very little organic carbon is buried in the seabed (**figures 30.6–30.8**).
- The carbon and nitrogen budgets of the open ocean are intensely studied because of their implications for the future of climate change.

- Bacteria in the α-proteobacterial clade SAR11 are the most abundant organisms on Earth and demonstrate unique adaptations to life in the oligotrophic open ocean.
- Viruses are present at high concentrations in many waters and occur at ten-fold higher levels than bacteria. In marine systems, they play a major role in nutrient turnover (**figures 30.10–30.12**).
- Sediments deep beneath the ocean's surface are home to perhaps one-half of the world's bacterial and archaeal biomass.
- Methane hydrates, the result of psychrophilic archaeal methanogenesis under extreme atmospheric pressure, may contain more natural gas than is currently found in known reserves.

30.3 Microorganisms in Freshwater Ecosystems

- Glaciers and permanently frozen lakes are sites of active microbial communities. Antarctic lakes lack multicellular organisms and have truncated food webs (**figure 30.15**).
- Nutrient sources for streams and rivers may be autochthonous or allochthonous. Often allochthonous inputs include urban, industrial, and agricultural runoff.
- The depth of a lake determines the microbial communities it will support. Deep lakes have a planktonic, autotrophic community in the epilimnion, and shallow lakes support large autotrophic communities (**figure 30.18**).
- Seasonal mixing of temperate lakes can introduce conditions that support microbial blooms in the spring and fall (**figure 30.19**).

Compare, Hypothesize, Invent

1. The unicellular cyanobacterium *Prochlorococcus* sp. is the most abundant photosynthetic microbe in tropical and subtropical oceans. At least two ecotypes exist: one is adapted to high light and the other to lower light intensities. How does the presence of these two ecotypes contribute to their physiological success and their numerical success? How would you determine the amount of fixed carbon they contribute to these open-ocean ecosystems?

2. How might it be possible to cleanse an aging eutrophic lake? Consider chemical, biological, and physical approaches as you formulate your plan.

3. It is well known that bacterivory (the consumption of bacteria) supports the growth of marine protists that lack plastids. However, measurements of bacterivory among plastid-containing planktonic algae are surprising in showing that small (<5 μm) algae carry out between about 40 to 95% of the bacterivory in the photic zone of the North Atlantic. Discuss how this level of mixotrophy impacts the concept of the microbial loop (figure 30.6).

   Read the original paper: Zubkov, M. V., and Tarran, G. A. 2008. High bacterivory by the smallest phytoplankton in the North Atlantic Ocean. *Nature*, 455:224.

4. Picoplanktonic cyanobacteria belonging to the genera *Synechococcus* and *Prochlorococcus* are estimated to account for up to 25% of primary productivity in certain regions of the open ocean. This makes them profoundly important in local nutrient cycling as well as in the global carbon budget. A group of international scientists developed a model to predict the fate of these microbes as global temperatures continue to rise. After determining that temperature and light are the two most important environmental factors influencing the currently observed variability in cyanobacterial abundance, they project a worldwide 14% and 29% increase in *Synechococcus* and *Prochlorococcus* abundance, respectively. What biological factors might affect the population of these cyanobacteria? How might such a large increase in primary production affect the export of carbon out of the photic zone (refer to figures 30.7 and 30.8)?

   Read the original paper: Flombaum, P. et al., 2013. Present and future global distributions of the marine cyanobacteria *Prochlorococcus* and *Synechococcus*. *Proc. Nat. Acad. Sci., USA.* 110:9824.

31

Microorganisms in Terrestrial Ecosystems

The plant pathogenic fungus Puccinia graminis *grows within the tissues of wheat plants, using plant nutrients and producing rusty streaks of red spores that erupt at the stem and leaf surface, where spores can be dispersed. Fungi are the most important group of plant pathogens.*

A Short History of Rust

Aristotle wrote about them. The Romans had a ceremony to prevent them. Over 2 million Indians starved in the 1940s because of them. They are responsible for the loss of more than a million metric tons of wheat in the United States every year. Norman Borlaug won a Nobel Peace Prize fighting them. What are they? Fungal plant pathogens: the cause of more than 70% of all major crop diseases.

The worst offenders among fungal plant pathogens are the "rusts," which are basidiomycetes (*see section 26.6*). In North America, probably the most important rust is *Puccinia graminis* subsp. *tritici*, the causative agent of wheat stem rust. Other important rusts include barley stem rust, soybean rust, cedar-apple rust, and poplar leaf rust. The introduction of fungi from Europe triggered chestnut blight (*Cryphonectria parasitica*), Dutch elm disease (*Ophiostoma ulmi*), and white pine blister rust (*Cronartium ribicola*). With over 6,000 different rust species known to cause plant diseases, it would seem the only approach to saving millions of dollars of valuable agricultural products would be to spray heavily with a fungicide and hope for the best.

Norman Borlaug knew that expensive fungicides were not an option for inhabitants of developing countries. Shortly after earning his Ph.D. in plant pathology at the University of Minnesota, he moved to Mexico to breed new, disease-resistant wheat strains. He later turned his attention to the Indian subcontinent, where he helped develop hardy strains of rice. In 1970 the Nobel committee noted that "More than any other single person of this age, he has helped provide bread for a hungry world." It is estimated that each day at least half the global population consumes grain that arose through Borlaug's plant breeding efforts. Borlaug was an active scientist until his death at age 91 in 2009.

Most microbes that associate with plants, whether that association is good, bad, or neutral, live in soil. Soil is a hugely underappreciated resource. It is a vast reservoir of nutrients, wastes, energy, and even pharmaceutical products. Although much is known about the geology and chemistry of soil, our understanding of the microbial communities that drive the development and continued health of soils is far from complete. However, the introduction of new, innovative culture techniques combined with culture-independent approaches has made the study of terrestrial microbiology an exciting and dynamic field.

Readiness Check:

Based on what you learned previously, you should be able to:

✔ List environmental factors that influence microbial growth (section 7.4)

✔ Compare and contrast the growth and physiology of obligate aerobes, facultative anaerobes, and strict anaerobes (section 7.4)

✔ Evaluate the importance of intercellular molecular signaling to microbial growth in nature (section 7.5)

✔ Diagram, in general terms, the standard reduction potentials of electron donors as compared to electron acceptors and relate this to standard free energy change (section 10.3)

✔ Recall the basic enzymology of nitrogenase (section 12.5)

✔ Explain why SSU rRNA is used as a taxonomic marker (section 19.3)

✔ Describe members of the *Glomeromycota* (section 26.4)

✔ Identify microorganisms responsible for nitrogen fixation (sections 22.1 and 24.1)

✔ Describe how stable isotope analysis is used to reveal microbial community dynamics (section 29.3)

✔ Explain how the oxidation-reduction (redox) state of an environment impacts the types of microbial metabolic processes that might be found there (section 28.1)

✔ Explain the global impact that microbes have on carbon and nitrogen cycling (section 28.1)

31.1 Soils Are an Important Microbial Habitat

After reading this section, you should be able to:

■ Differentiate between a mineral and an organic soil, and describe the role of microbes in generating soil organic matter

■ Explain why lignin degradation is confined to only a few types of microbes

■ Predict how a specific carbon to nitrogen ratio will influence the microbial community within a given soil

The level of microbial diversity in soil exceeds that of any other habitat on Earth, with estimates of 10,000 to 50,000 species at densities of 10^7 to 10^{10} cells per gram of soil. The soil environment

Figure 31.1 The Soil Habitat.
A typical soil habitat contains a mixture of clay, silt, and sand along with soil organic matter. Roots, animals (e.g., nematodes and mites), and chemoorganotrophic bacteria consume oxygen, which is rapidly replaced by diffusion within the soil pores where the microbes live. Two types of fungi are present: mycorrhizal fungi, which derive their organic carbon from their symbiotic partners, plant roots; and saprophytic fungi, which contribute to the degradation of organic material. In addition to heterotrophs, chemolithotrophic microorganisms are important in elemental cycling within the soil environment.

MICRO INQUIRY *How are filamentous microbes especially well suited to the soil ecosystem?*

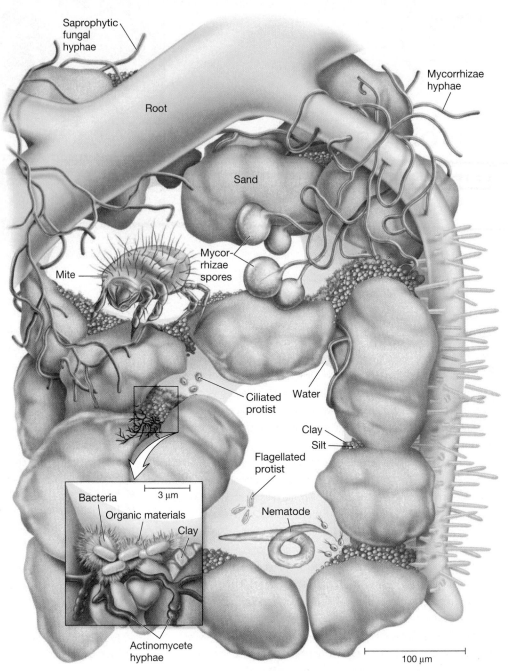

supports active microbial growth that contributes to global elemental cycling and makes nutrients available to plants. Simply put, without soil microbes, there would be no plants, which would eliminate herbivores, and so on up the food chain.

Soil microbial communities are supported by complex physical and chemical parameters, which provide a vast array of microhabitats. These include soil particles and the pore space between them, which is critical for the movement of water and gases (**figure 31.1**). Total pore space, and thus gas diffusion, is determined by the texture of the soil. For instance, sandy soils have larger pore spaces than do clay soils, so sandy soils tend to drain more quickly. Pores are also critical because they provide the optimum environment for microbial growth. Here microbes are within thin water films on particle surfaces where oxygen is present at high levels and can be easily replenished by diffusion.

Soil environments can change rapidly. For instance, rainfall or irrigation may quickly change a well-aerated soil to one with isolated pockets of water, generating miniaquatic habitats. If flooding continues, a waterlogged soil can be created that is more like lake sediment. If oxygen consumption exceeds that of oxygen diffusion, waterlogged soils can become anoxic. Shifts in water content and gas fluxes also affect the concentrations of CO_2 and other gases. These changes are accentuated in the smaller pores,

where many bacteria are found. The roots of plants growing in aerated soils also consume oxygen and release CO_2, influencing the concentrations of these gases in the root environment.

Soils can be divided into two general categories: A **mineral soil** contains less than 20% organic carbon, whereas an **organic soil** possesses at least this amount. By this definition, the vast majority of Earth's soils are mineral. The importance of organic matter within soils cannot be overstated. **Soil organic matter** (**SOM**) helps to retain nutrients, maintain soil structure, and hold water for plant use. SOM is subject to gains and losses, depending on changes in environmental conditions and agricultural practices. Plowing and other disturbances expose SOM to more oxygen, leading to extensive microbiological degradation of organic matter. Irrigation causes periodic wetting and drying, which can also lead to increased degradation of SOM, especially at higher temperatures.

Microbial degradation of plant material results in the release of CO_2 and the incorporation of plant carbon into microbial biomass. However, a small fraction of the decomposed plant material remains in the soil as SOM. When considering this material, it is convenient to divide SOM into humic and nonhumic fractions. Nonhumic SOM is free from significant biochemical degradation. It can represent up to about 20% of the soil organic matter. Humic SOM, or humus, results when the products of microbial metabolism have undergone chemical transformation within the soil. Although it has no defined chemical composition, humus can be described as a complex blend of phenolic compounds, polysaccharides, and proteins. The recalcitrant nature of SOM to degradation is evident by ^{14}C dating: the average age of most SOM ranges from 150 to 1,500 years.

Microbial degradation of plant material and the development of SOM can be thought of as a three-step process. First, easily degraded compounds such as soluble carbohydrates and proteins are broken down by a variety of heterotrophic microorganisms. About half the carbon is respired as CO_2, and the remainder is rapidly incorporated into new biomass. During the second stage, complex carbohydrates, such as the plant structural polysaccharide cellulose, are degraded. Cellulose represents a vast store of carbon, as it is the most abundant organic compound on Earth. Brown rot fungi and filamentous bacteria in the genera *Coniophora* and *Streptomyces,* respectively, and nonfilamentous bacteria such as *Cytophaga* and *Bacillus* spp. produce extracellular cellulase enzymes that break down cellulose into two and three glucose units called cellobiose and cellotriose, respectively. These smaller compounds are readily degraded and assimilated as glucose monomers. Finally, very resistant material, in particular lignin, is attacked.

Lignin is an important structural component of woody plants. While its exact structure differs among plant species, the common building block is the phenylpropene unit. This consists of a hydroxylated benzene ring and a three-carbon linear side chain (**figure 31.2**). A single lignin molecule can consist of up to 600 cross-linked phenylpropene units. It is therefore not surprising that degradation of lignin is much slower than that of cellulose. Basidiomycete fungi commonly termed white rot fungi (e.g., *Phanerochaete* and *Phlebia* spp.) and actinobacteria (e.g., *Streptomyces* spp.) are capable of extracellular lignin degradation. These filamentous microbes produce extracellular phenoloxidase enzymes needed for aerobic lignin degradation. Lignin decomposition is also limited by the physical nature of the material. For example, healthy woody plants are saturated with sap, which limits oxygen diffusion. In addition, high ethylene and CO_2 levels and the presence of phenolic and terpenoid compounds retard the growth of lignin-degrading microbes. It follows that no more than 10% of the carbon found in lignin is recycled into new microbial biomass. Lignin can be degraded anaerobically, but this process is very slow, so lignin accumulates in wet, poorly oxygenated soils, such as peat bogs.

Nitrogen is another important element in soil ecosystems. Nitrogen in soil is often considered in relation to the soil carbon content as the organic **carbon to nitrogen ratio (C/N ratio)**. As a general rule, decomposition is thought to be maximal at a C/N ratio of 30 (i.e., 30 times more carbon than nitrogen). A ratio of less than 30 can result in the loss of soluble nitrogen from the system. Conversely ratios above 30 may cause the microbial community to be nitrogen limited. In fact, many soils are nitrogen limited. This is why each year tons of nitrogen fertilizer are added to agricultural soils. Most fertilizers include ammonium ion because it will bind to the negatively charged clays in a soil until used as a nutrient by the plants. However, when ammonium is in excess, nitrifying microorganisms in a soil can oxidize the ammonium ion to nitrite and nitrate, and these anions can be leached from the plant environment and pollute surface waters and groundwater. ◄◄ *Nitrifying bacteria oxidize ammonium or nitrite to gain energy and electrons (section 22.1); Nitrogen cycle (section 28.1); Global climate change (section 28.2)*

Phosphorus is also commonly in short supply in soils, so it is also added to fertilizers. The binding of anionic phosphate molecules to soils depends on the cation exchange capacity and soil pH. As the soil's phosphorus sorption capacity is reached, the excess, together with phosphorus released by soil erosion, moves to lakes, streams, and estuaries, where it can stimulate the growth of freshwater organisms, particularly nitrogen-fixing cyanobacteria, resulting in eutrophication. ◄◄ *Microorganisms in lakes can cause seasonal blooms (section 30.3)*

Retrieve, Infer, Apply

1. Why are soil pores important?
2. List three reasons why soil organic matter (SOM) is important.
3. Describe the three phases of plant degradation and SOM formation.
4. What microbes commonly degrade cellulose and lignin? Why is their filamentous morphology advantageous?
5. Which of the following soils would a farmer be most likely to fertilize: A soil with a C:N ratio of 23, 28, or 37?
6. Why is most nitrogen fertilizer added as ammonium ion?

31.2 Diverse Microorganisms Inhabit Soil

After reading this section, you should be able to:

- Estimate the relative microbial populations in several soil types
- List the microbial genera that are easily cultured from soil
- Describe the different ecological niches occupied by microbes in soil and how microbes have evolved to take advantage of the niches
- Compare the soil microbial loop with an aquatic microbial loop

An average gram of forest soil contains about 4×10^7 bacteria and archaea, while a gram of grassland or cultivated soil may

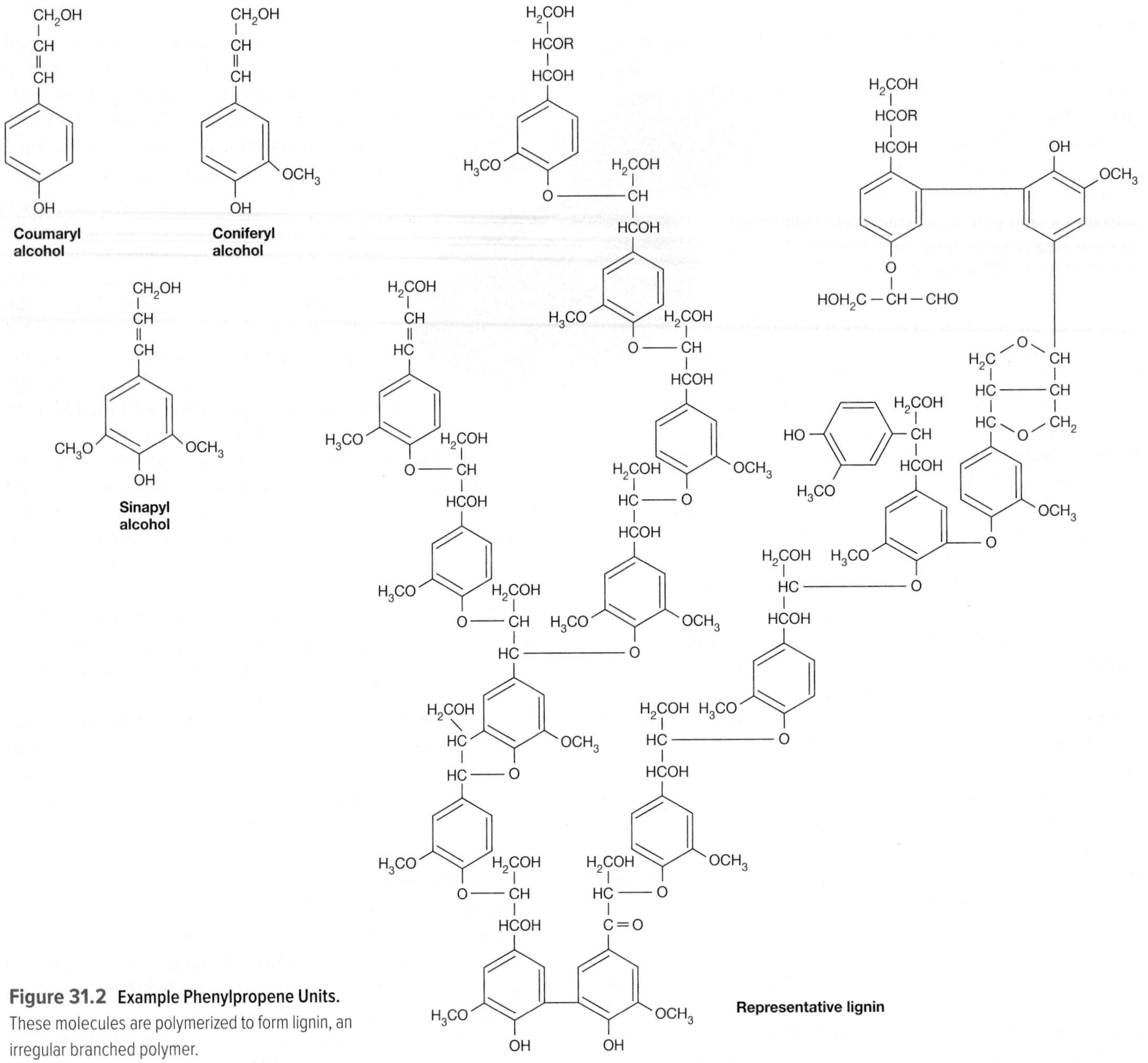

Figure 31.2 **Example Phenylpropene Units.**
These molecules are polymerized to form lignin, an irregular branched polymer.

possess 2×10^9 cells. It is difficult to determine how many different taxa are represented, but estimates based on DNA reassociation range from 2,000 to 18,000 distinct genomes per gram of soil. These numbers suggest that the microbial diversity in just 1 gram of soil may exceed the combined number of known archaeal and bacterial species—just over 16,000. ◄◄ *Genetic methods are used to assess microbial diversity (section 29.2)*

Small subunit rRNA gene surveys from different soils show some major trends. Although most rRNA gene sequences could be assigned to well-studied bacterial phyla, the majority (79 to 89%) could not be placed in any existing genera. Overall, bacteria belonging to at least 32 different phyla have been identified, with over 90% of soil bacterial 16S rRNA assigned to the nine phyla displayed in **figure 31.3**. Interestingly, this includes *Acidobacteria* and *Verrucomicrobia*, phyla about which little is known. Other

important phyla not shown in figure 31.3 include *Chlamydiae, Chlorobi, Cyanobacteria, Deinococcus-Thermus,* and *Nitrospirae.* Coryneforms, nocardioforms, and streptomycetes are easily cultured and therefore well-characterized members of the soil community (**table 31.1**). These actinobacteria play a major role in the degradation of hydrocarbons, older plant materials, and soil humus. In addition, some members of these groups actively degrade pesticides. Filamentous bacteria, primarily of the genus *Streptomyces,* produce an odor-causing compound called geosmin, which gives soils their characteristic earthy odor. ◄◄ Actinobacteria: *High G + C Gram-positive bacteria (chapter 24)*

Of course, bacteria are not the only microbes present in soils. Among archaea, ammonia-oxidizing members of the phylum *Thaumarchaeota* usually prevail, and may constitute up to 10% of the microorganisms identified by SSU rRNA sequencing.

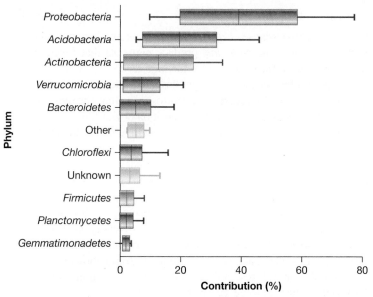

Figure 31.3 Bacterial Phyla Identified from 16S rRNA and 16S rRNA Genes Extracted from Soil. Data from 21 soil metagenomic libraries were analyzed to assess soil bacterial diversity. The horizontal line in the middle of each block represents the mean. The block represents one standard deviation around the mean. The vertical lines extending above and below indicate the maximum and minimum values for each phylum. Gram-negative phyla are shown in red, Gram-positive in violet, unknown and other phyla found in very small numbers in green.

The number of archaea appears to be related to the C/N ratio, with larger archaeal populations occurring in soils with higher concentrations of nitrogen. The relative contribution of bacterial and archaeal nitrification in soils is an active area of research *(see figure 22.15).* ◄◄ *Nitrogen cycle (section 28.1)*

Fungi are also very important members of the soil microbial community. Fungi that grow in the so-called bulk soil—soil not close to plant roots—are saprophytic, helping to degrade complex plant molecules. Fungi can extend up to several hundred meters of

hyphae per gram of soil. Although we tend to think of soil fungi as the mushrooms sprouting on our lawns, the vast majority of fungal biomass is below ground. There filaments bridge open areas between soil particles or aggregates and are exposed to high levels of oxygen. These fungi tend to darken and form oxygen-impermeable structures called sclerotia and hyphal cords. This is particularly important for basidiomycetes, which form these structures as an oxygen-sealing mechanism. Within these structures, the filamentous fungi move nutrients and water over great distances, including across air spaces, a unique part of their functional strategy. Fungal biomass can be enormous; an individual clone of the fungus *Armillaria bulbosa,* which lives associated with tree roots in hardwood forests, covers about 30 acres in the Upper Peninsula of Michigan, USA. It is estimated to weigh a minimum of 100 tons (an adult blue whale weighs about 150 tons) and to be at least 1,500 years old. Thus some fungal mycelia are among the largest and most ancient living organisms on Earth. ◄◄ Basidiomycota *includes mushrooms and plant pathogens (section 26.6)*

The flux of nutrients through soil differs from the microbial loop that operates in aquatic ecosystems. In nearly all terrestrial ecosystems, plants, rather than microbes, account for most primary production. Soil microbes rapidly recycle the organic material derived from plants and animals, including the many nematodes and insects, which function to reduce the size of solid organic matter. In turn, microbes themselves are preyed upon by viruses, other bacteria (e.g., species within the genera *Lysobacter* and *Myxococcus*), and soil protists, whose numbers can reach 100,000 per gram of soil. This makes microbial organic matter available to other trophic levels. Another difference between aquatic and terrestrial nutrient flux reflects the physical and biological properties of soil. Degradative enzymes released by microbes, plants, insects, and other animals do not rapidly diffuse away. Instead, they represent a significant contribution to the biological activity in soil ecosystems by contributing to the many hydrolytic degradation reactions needed for the assimilation and recycling of nutrients.

Retrieve, Infer, Apply

1. What are the differences in preferred soil microhabitats between bacteria and filamentous fungi?
2. What types of archaea have been detected in soils?
3. Compare and contrast the general flux of nutrients in terrestrial and aquatic ecosystems.

31.3 Microbe-Plant Interactions Can Be Positive, Negative, or Neutral

After reading this section, you should be able to:

- Describe mechanisms by which plants and microbes communicate
- Differentiate between the phyllosphere, rhizosphere, and rhizoplane environments and the microbes that dwell there
- Explain the importance of mycorrhizae to the growth of vascular plants
- Contrast and compare the ectomycorrhizal life cycle and that of arbuscular fungi

Table 31.1	Easily Cultured Gram-Positive Bacteria in Soils	
Bacterial Group	**Representative Genera**	**Comments and Characteristics**
Coryneforms	*Arthrobacter*	Rod-coccus cycle
	Cellulomonas	Important in degradation of cellulose
	Corynebacterium	Club-shaped cells
Mycobacteria	*Mycobacterium*	Acid-fast
Nocardioforms	*Nocardia*	Rudimentary branching
Streptomycetes	*Streptomyces*	Aerobic filamentous bacteria
Bacilli	*Thermoactinomyces*	Higher temperature growth

- Diagram the steps that result in the formation of rhizobium nodules on leguminous plants
- Compare the process of stem nodulation with root nodulation
- Discuss other microbes, in addition to rhizobia, that fix nitrogen
- Explain how *Agrobacterium tumefaciens* cause tumors in plants
- List at least five plant pathogens other than *Agrobacterium tumefaciens* and the diseases they cause
- Identify a plant disease that is likely to be found near your home or school

The evolution of aquatic plants into terrestrial organisms roughly 470 million years ago was made possible by cooperation with soil microbes, and many of these microbe-plant interactions remain. Some are commensalistic—no harm is done to the plant, but the microbe gains some advantage. Many others are beneficial to both the microorganism and the plant (e.g., they are mutualistic). Finally, other microbes are plant pathogens and parasitize their plant hosts. In all cases, the microbe and the plant have established the capacity to communicate. The microbe detects and responds to plant-produced chemical signaling molecules. This generally triggers the release of microbial compounds that are in turn recognized by the plant, thereby beginning a two-way "conversation" that employs a molecular lexicon. Once a microbe-plant relationship is initiated, microbes and plants continue to monitor the physiology of their partner and adjust their own activities accordingly. The nature of the signaling molecules and the mechanisms by which both plants and microbes respond have stimulated exciting multidisciplinary research in soil microbiology, ecology, molecular biology, genetics, and biochemistry. ◄◄ *Cell–cell communication within microbial populations (section 7.5)* ►► *Many types of microbial interactions exist (section 32.1)*

Microbes that interact with plants can be broadly divided into two classes: microbes that live on the surface of plants, called **epiphytes,** and those that colonize internal plant tissues, called **endophytes.** We can further classify microbes into those that live in the aboveground, or aerial, surfaces of plants separately and those that inhabit below-ground plant tissues. We begin our discussion of microbe-plant interactions by introducing the microbial communities associated with aerial regions of plants. We then turn our attention to two important microbe-root symbioses—mycorrhizal fungi and nitrogen-fixing rhizobia. Finally, we consider several important microbial plant pathogens.

Phyllosphere Microorganisms Are Stress Resistant

The environment of the aerial portion of a plant (e.g., stems and leaves), called the **phyllosphere,** undergoes frequent and rapid changes in humidity, ultraviolet (UV) light exposure, and temperature. This causes fluctuations in the leaching of organic material (primarily simple sugars) that supports a diverse assortment of microbes, including bacteria, filamentous fungi, yeasts, and photosynthetic and heterotrophic protists. Bacterial densities are estimated to be 10^6 to 10^7 cells per cm^2 of plant tissue. Bacteria include the γ-proteobacteria *Pseudomonas syringae, Erwinia* spp., and *Pantoea* spp. In addition, methylotrophic α-proteobacteria such as *Methylobacterium* spp. are supported by the release of methanol

by pectin methylesterase enzymes. Members of another abundant bacterial genus, *Sphingomonas,* produce pigments that function like sunscreen so they can survive the high levels of UV irradiation striking plant surfaces. The phyllosphere microbial community contributes to global carbon and nitrogen cycling, the removal of airborne pollutants, and the decomposition of leaf litter.

A metagenomic survey of the bacterial diversity of the phyllosphere of the Atlantic Forest in Brazil reveals some intriguing findings. This forest is thought to be the oldest on Earth, with over 20,000 vascular plant species. SSU rRNA analysis of bacteria in the phyllosphere of three plant species shows amazing species richness—from 95 to 671 bacterial species, and all but 3% are members of undescribed species. The diversity is further extended by the observation that each plant species has an almost unique phyllosphere bacterial community. Extrapolation of these results to the entire forest indicates that there are millions of new bacterial species yet to be discovered.

The Rhizoplane and Rhizosphere Support Complex Microbial Communities

Plant roots receive between 30 and 60% of the net photosynthesized carbon. Of this, an estimated 40 to 90% enters the soil as a wide variety of materials, including alcohols, ethylene, sugars, amino and organic acids, vitamins, nucleotides, polysaccharides, and enzymes. The plant root surface, termed the **rhizoplane,** provides an exceptional environment for microorganisms, as these gaseous, soluble, and particulate materials move from plant to soil.

The **rhizosphere** is the region surrounding a plant root rich in plant exudates more easily degraded than lignin and cellulose. This in turn supports bacterial and fungal populations, which can be as much as 100 times greater than surrounding bulk soil that lacks these nutrients. Microbes in these dense rhizosphere communities must compete with one another, and they do so by the secretion of antimicrobial agents, lytic enzymes, and even hydrogen cyanide. Other microbes counteract these inhibitors by efflux pumps, detoxification enzymes, and secretion of molecules that alter competitor gene expression. Importantly, rhizosphere and rhizoplane microorganisms provide nutrients for other organisms, including plants and other microbes, thereby playing critical roles in organic matter synthesis and degradation.

A wide range of microbes in the rhizoplane and rhizosphere promote plant growth, orchestrated by their ability to communicate with plants using chemical signals. Bacteria and fungi enhance plant growth by producing a variety of compounds. These include molecules that regulate plant growth called phytohormones (e.g., auxin and gibberellin) and compounds that stimulate host plant production of phytohormones. The production of volatile organic compounds including acetoin, and phenazines—a diverse class of heterocyclic compounds—stimulates plant responses to stress. In addition, some bacteria secrete compounds that suppress the growth of plant fungal pathogens. The use of these compounds in agriculture is an active area of research.

Associative nitrogen fixation is a critical process performed by nitrogen-fixing microorganisms in the rhizoplane as well as in the rhizosphere. It is carried out by members of a number of genera, including *Azotobacter, Azospirillum,* and *Acetobacter.* These bacteria enhance the growth of agriculturally important crops such as wheat, potato, and barley. Interestingly, some evidence suggests that their major contribution may not be nitrogen fixation but the production of growth-promoting hormones that increase root hair development, thereby enhancing plant nutrient uptake.

Retrieve, Infer, Apply

1. Define rhizosphere, rhizoplane, and associative nitrogen fixation.
2. What unique stresses does a microorganism on a leaf contend with that a microbe in the soil does not?
3. List two ways in which compounds produced by rhizosphere and rhizoplane microbes promote plant growth.
4. What important genera are involved in associative nitrogen fixation?

Mycorrhizae Facilitate the Growth of Most Plants

Mycorrhizae (Greek, "fungus root") are mutualistic relationships that develop between about 80% of all land plants and a limited number of filamentous fungal species. Mycorrhizae are so abundant that up to 100 meters of mycorrhizal filaments can commonly be found in a gram of soil. Unlike most fungi, **mycorrhizal fungi** are not saprophytic; that is, they do not obtain organic carbon from the degradation of organic material. Instead, they use photosynthetically derived carbohydrate from their host. In return, they provide a number of services for their plant hosts. In moist environments, they increase the availability of nutrients, especially phosphorus. In arid environments, where nutrients do not limit plant growth to the same degree, mycorrhizae aid in water uptake, increasing transpiration rates in comparison with nonmycorrhizal plants. ▶▶| *Mutualism (section 32.1)*

Mycorrhizae can be broadly classified as **endomycorrhizae**—those with fungi that enter root cells, or as **ectomycorrhizae**—those that remain extracellular, forming a sheath of interconnecting filaments (hyphae) around roots. Although six types of mycorrhizae are detailed in **table 31.2** and **figure 31.4**, we focus primarily on the two most important types: ectomycorrhizae and arbuscular mycorrhizae. |◀◀ Glomeromycota *are mycorrhizal symbionts (section 26.4)*

Ectomycorrhizae (ECM) are formed by both ascomycete and basidiomycete fungi. ECM colonize almost all trees in cooler climates. Their importance arises from their ability to transfer essential nutrients, especially phosphorus and nitrogen, to the

Table 31.2	Mycorrhizal Associations			
Mycorrhizal Classification	Fungi Involved	Plants Colonized	Fungal Structural Features	Fungal Function
Ectomycorrhizae	Basidiomycetes, including those with large fruiting bodies (e.g., toadstools); some ascomycetes	~90% of trees and woody plants in temperate regions; fungal-plant colonization is often species specific	Hartig net, mantle or sheath; rhizomorphs; root hair development is usually limited	Nutrient (N and P) uptake and transfer
Arbuscular	Glomeromycetes, in particular six genera of glomeromycetes	Wild and crop plants, tropical trees; fungal-plant colonization is not highly specific	Arbuscules: hyphae-filled invaginations of cortical root cell	Nutrient (N and P) uptake and transfer; facilitate soil aggregation; promote seed production; reduce pest and nematode infection; increase drought and disease resistance
Ericaceous	Ascomycetes Basidiomycetes	Low evergreen shrubs, heathers	Some intracellular, some extracellular	Mineralization of organic matter
Orchidaceous	Basidiomycetes	Orchids	Hyphal coils called pelotons within host tissue	Some orchids are non-photosynthetic and others produce chlorophyll when mature; these organisms are almost completely dependent on mycorrhizae for organic carbon and nutrients.
Ectendomycorrhizae	Ascomycetes	Conifers	Hartig net with some intracellular hyphae	Nutrient uptake and mineralization of organic matter
Monotropoid mycorrhizae	Ascomycetes Basidiomycetes	Flowering plants that lack chlorophyll (*Monotropaceae;* e.g., Indian pipe)	Hartig net one cell deep in the root cortex	Nutrient uptake and transfer

Figure 31.4 Mycorrhizae. Fungi can establish mutually beneficial relationships with plant roots, called mycorrhizae. Root cross sections illustrate different mycorrhizal relationships.

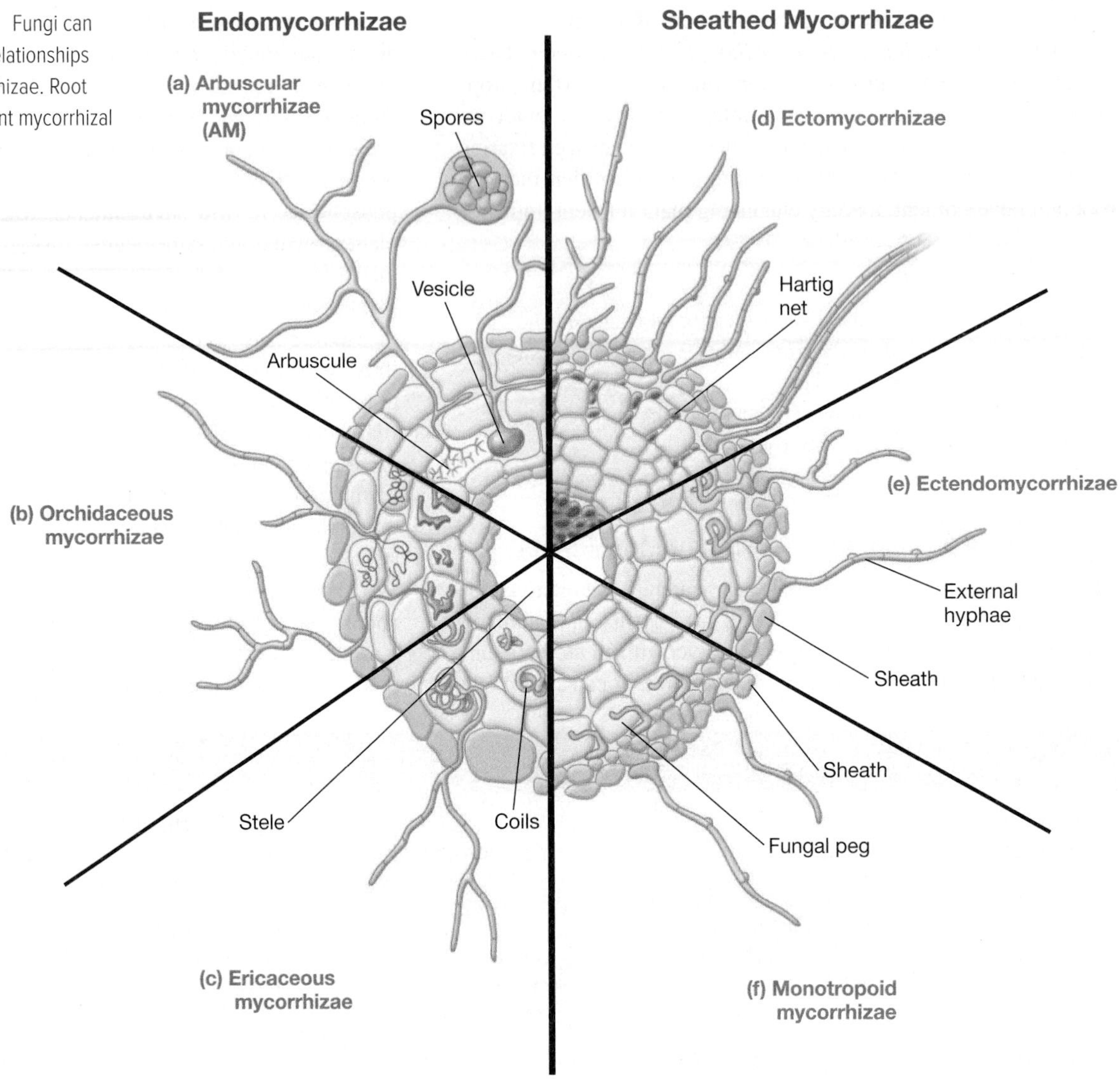

Endomycorrhizae

(a) Arbuscular mycorrhizae (AM)

Spores

Vesicle

Arbuscule

(b) Orchidaceous mycorrhizae

Stele

Coils

(c) Ericaceous mycorrhizae

Sheathed Mycorrhizae

(d) Ectomycorrhizae

Hartig net

(e) Ectendomycorrhizae

External hyphae

Sheath

Sheath

Fungal peg

(f) Monotropoid mycorrhizae

root. The development of ECM starts with the growth of a fungal mycelium around the root. As the mycelium thickens, it forms a sheath or mantle so that the entire root may be covered by the fungal mycelium (**figure 31.5**). Most ECM produce signaling molecules that limit the growth of root hairs; thus ECM-colonized roots often appear blunt and covered in fungi. From the root surface, the fungi extend hyphae into the soil; these filaments may aggregate to form **rhizomorphs**—dense mats of hyphae that are often visible to the naked eye. Hyphae on the inner side of the sheath penetrate between (but not within) the cortical root cells, forming a characteristic meshwork of hyphae called the **Hartig net** (figure 31.4*d*). Soil nutrients taken up by rhizomorphs must first pass through the hyphal sheath and then into the Hartig net filaments, which form numerous contacts with root cells. This results in efficient, two-way transfer of soil nutrients to the plant and carbohydrates to the fungus. This relationship has evolved to the point that some plants synthesize sugars such as mannitol and trehalose that they cannot use but are consumed by their fungal symbionts.

Arbuscular mycorrhizae (AM) are the most common type of mycorrhizae and make up about half the fungal biomass in most soils. They are found in association with many tropical plants and, importantly, with most crop plants. AM fungi belong to the taxon *Glomeromycota* and can be cultured only with roots of their plant hosts. These microbes enter root cells between the

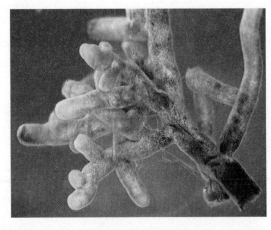

Figure 31.5 Ectomycorrhizae as Found on Roots of a Pine Tree.
Typical irregular branching of the white, smooth mycorrhizae is evident.

plant cell wall and invaginations in the plasma membrane (figure 31.4a). So, although AM are endomycorrhizae, they do not breach the root cell membrane. Instead, treelike hyphal networks called **arbuscules** develop within the folds of the plasma membrane. Individual arbuscules are transient; they last at most 2 weeks. Spores released by nearby root-associated fungi germinate and colonize roots. This requires a number of specific steps, as shown in **figure 31.6**. AM can be vigorous colonizers: a 5 cm segment of root can support the growth of as many as eight species, and hyphae from a single germinated spore can simultaneously colonize multiple roots from unrelated plant species, forming a complex underground highway for nutrient transport.

The AM symbiosis is ancient, having appeared over 400 million years ago. Like the α-proteobacteria belonging to the rhizobium family (discussed next), AM establish symbioses with legumes (among other plants). The secretion of so-called "**Nod factors**" by rhizobia has been well established (p. 677), and analogous "**Myc factors**" have been purified from AM fungi (**figure 31.7**). Both Nod and Myc factors are lipochitooligosaccharides that prompt legume roots to accept the symbiont. Myc factors have a simpler structure than Nod factors. Given that the rhizobium-legume symbiosis is believed to have arisen only about 60 million years ago, it is likely that this signaling mechanism was adopted by rhizobia, who then further modified it.

AM are believed to provide a number of services to their plant hosts, including protection from disease, drought, nematodes, and other pests.

Their capacity to transfer phosphorus to roots has been well documented. AM also provide nitrogen for their plant partners. Interestingly, AM assimilate NO_3^- and NH_4^+ through the glutamine synthetase–glutamate synthase (GOGAT) pathway (*see figure 12.13*), and therefore incorporate nitrogen into glutamine, which they then convert to arginine. But rather than transferring arginine to host cells, intracellular fungal hyphae degrade the amino acids, transferring only the NH_4^+ (**figure 31.8**). Thus while the fungus provides its host with much needed ammonium, it retains the carbon skeleton required for nitrogen uptake. In contrast, little is known about how mycorrhizal fungi take up carbon from their hosts. However, recent studies have revealed

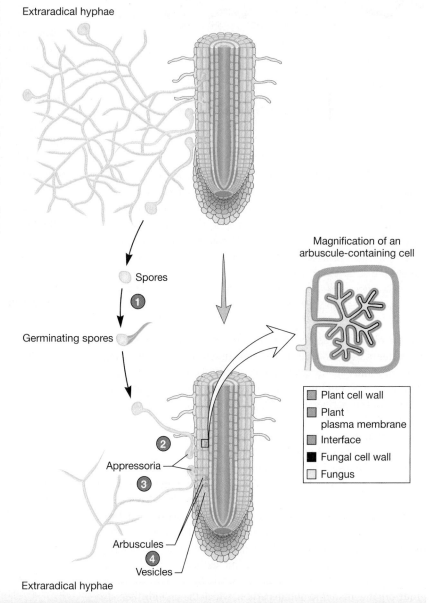

① Fungal spores germinate. Growth is supported by the breakdown of plant carbon storage products.

② Fungi and plant exchange external molecular signals that result in appressorium development.

③ Branching factors released by the host trigger fungal hyphal branching.

④ Fungal hyphae colonize the plant root cells between the cell wall and plasma membrane, forming arbuscules.

Figure 31.6 The Process of Root Colonization by Arbuscular Mycorrhizal Fungi.

MICRO INQUIRY *Which hyphae are growing saprotrophically in this diagram? How does the proportion of the organism growing saprotrophically change during host colonization?*

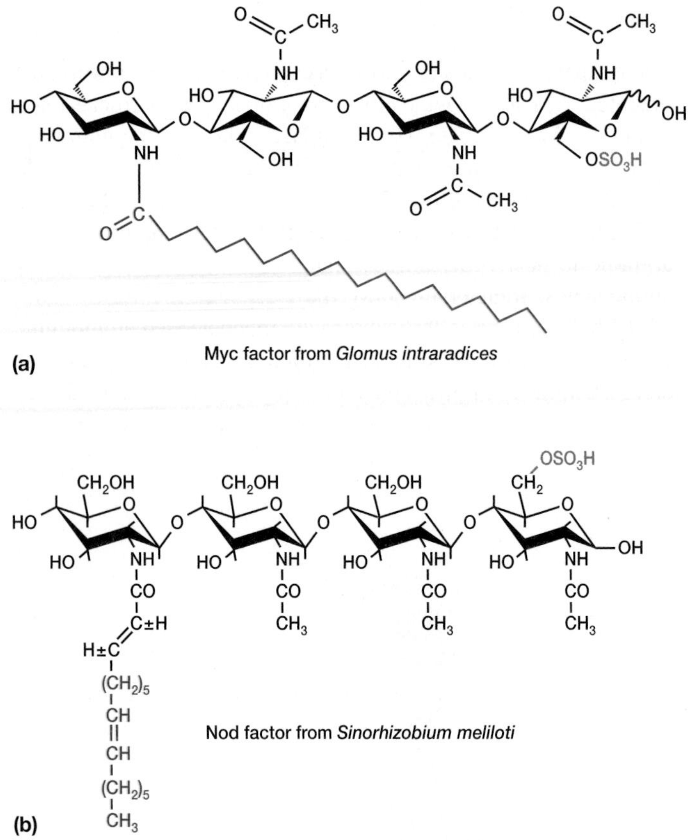

(a) Myc factor from *Glomus intraradices*

(b) Nod factor from *Sinorhizobium meliloti*

Figure 31.7 Myc and Nod Factors.   Both arbuscular mycorrhizae and rhizobia form symbioses with legumes. In both cases, the symbiosis is initiated when the microbe secretes a lipochitooligosaccharide. (a) Myc factors, produced by fungi, have a simpler structure than (b) Nod factors of bacteria. The precise structure of individual Myc and Nod factors depends on the microbial species, but all consist of four to five units of β–1,4 linked *N*-acetyl-*D*-glucosamine bearing a species-specific acyl chain (red) at the nonreducing terminal residue. All Nod factors have a sulfate attached to the reducing end (blue), while not all Myc factors are sulfonated.

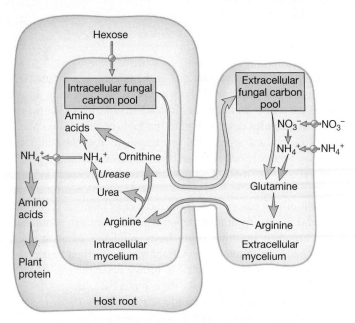

Figure 31.8 Nitrogen and Carbon Exchange Between Arbuscular Mycorrhizal Fungi and Host Plant.   Nitrate and ammonium are taken up by the fungal mycelium that is outside the host plant cell (extracellular mycelium) and converted to arginine. This amino acid is transferred to the mycelium within the host plant cell and broken down so that only the ammonium enters the plant.

that the plants can detect, discriminate, and reward the best fungal partners with more carbohydrates. In return, the fungi transfer more nutrients to the plants that "feed" them best. This tight plant-microbe coupling probably accounts for the evolutionary stability of this partnership. ◄◄ *Inorganic nitrogen assimilation (section 12.5)*

In addition to these two most abundant mycorrhizal types, the orchidaceous mycorrhizae are of particular interest (figure 31.4*b*). Orchids are unusual plants because many never produce chlorophyll, while others only do so after they have matured past the seedling stage. Therefore all orchids have an absolute dependence on their endomycorrhizal partners for at least part of their lives. Indeed, orchid seeds will not germinate unless first colonized by a basidiomycete orchid mycorrhiza. Because orchids cannot produce photosynthetically derived organic carbon (or produce very little), orchid mycorrhizal fungi are saprophytic. They must degrade organic matter to obtain carbon, which the orchids then also consume. In this case, the orchid functions as a parasite.

The nature of some mycorrhizae-host relationships is currently being reevaluated. It is now known that in addition to orchids, some green plants such as serrated wintergreen (*Orthilia*

secunda) also parasitize their mycorrhizal partners. Likewise, the nuisance plant spotted knapweed (*Centaurea maculosa*) in the United States may obtain up to 15% of its organic carbon from mycorrhizal fungi. Finally, mycorrhizae may limit plant species diversity in some ecosystems, as the growth of certain seedlings is promoted by mycorrhizae that will only associate with a particular plant species.

Just as bacteria associate with plant roots, they also interact with mycorrhizal fungi. As the external hyphal network radiates out into the soil, a mycorrhizosphere is formed due to the flow of carbon from the plant into the mycorrhizal hyphal network and then into the surrounding soil. These bacteria are called mycorrhization helper bacteria because they may play a role in the development of mycorrhizal relationships involving ectomycorrhizal fungi; they include species of *Pseudomonas, Bacillus,* and *Burkholderia*. Bacterial symbionts have also been found in the cytoplasm of AM fungi. These organisms appear to form a distinct lineage of β-proteobacteria, most closely related to *Burkholderia, Pandoraea,* and *Ralstonia*. Their role within the fungi remains unclear.

Retrieve, Infer, Apply

1. Describe the two-way relationship between mycorrhizal fungi and the plant host.
2. List three major differences between arbuscular mycorrhizae and ectomycorrhizae.
3. What is the function of the rhizomorph and the Hartig net?
4. Describe the uptake and transfer of ammonium by arbuscular mycorrhizae to the plant host. Why is only the ammonium transferred?
5. Propose two potential functions for mycorrhization helper bacteria.

Nitrogen-Fixing Bacteria Are Vital to Agriculture

The conversion of gaseous nitrogen (N_2) to ammonia (NH_3) often occurs as part of a symbiotic relationship between bacteria and plants. These symbioses produce more than 100 million metric tons of fixed nitrogen annually and are a vital part of the global nitrogen cycle, and without them life on Earth could not have evolved. Currently, symbiotic nitrogen fixation accounts for more than half of the nitrogen used in agriculture (the remainder is applied as fertilizer). The provision of fixed nitrogen enables the growth of host plants in soils that would otherwise be nitrogen limiting. At the same time, it reduces loss of nitrogen by denitrification and leaching. It follows that nitrogen fixation, particularly that by members of the genus *Rhizobium* and related α-proteobacteria in association with their leguminous host plants, has been the subject of intense investigation. ◄◄ *Nitrogen cycle (section 28.1)*

Rhizobia Are Symbionts of Leguminous Plants

Several α-proteobacterial genera contain species that are able to form nitrogen-fixing nodules with legumes. These include *Allorhizobium, Azorhizobium, Bradyrhizobium, Mesorhizobium, Sinorhizobium,* and *Rhizobium.* Collectively these bacteria are often called **rhizobia.** The phylogenetic diversity of rhizobia has been extended by the discovery that the β-proteobacteria *Burkholderia caribensis* and *Cupriavidus* (formerly *Ralstonia*) *taiwanensis* also form nitrogen-fixing nodules on legumes. Here we discuss the general process of nodulation, including some molecular details that have been revealed largely through studies of bacteria in the genus *Rhizobium* ◄◄ *Class* Alphaproteobacteria *includes many oligotrophs (section 22.1); Class* Betaproteobacteria *includes chemoheterotrophs and chemolithotrophs (section 22.2)*

Rhizobia live freely in the soil. In nitrogen-sufficient soils, plants secrete ethylene-mediated compounds that inhibit nodulation. But even in nitrogen-starved soils, when rhizobia approach the plant root, they are assumed to be invaders. The plant responds with an oxidative burst, producing a mixture of compounds that can contain superoxide radicals, hydrogen peroxide, and N_2O. This oxidative burst is critical in determining the fate of the infection process. Rhizobia that become effective colonizers mount antioxidant defenses. Only rhizobia with sufficient antioxidant abilities are able to proceed to the next step in the infection process.

Infection is initiated by the exchange of signaling molecules between the plant and rhizobia in the rhizosphere. Plant roots release flavonoid inducer molecules (2-phenyl-1,4-benzopyrone derivatives) that stimulate rhizobial colonization of the root surfaces (**figure 31.9a**). Flavonoids bind the bacterial protein NodD, which can then function as a transcriptional regulator. NodD activates transcription of *nod* genes, which encode the biosynthetic enzymes needed for the production of Nod factor signaling compounds (figure 31.9b). Conversely flavonoids from nonhost plants inhibit the production of Nod factors. Upon receipt of the Nod factor signal, gene expression in the outer (epidermal) cells of the roots is altered, resulting in oscillations in the levels of intracellular calcium. This "calcium spiking" triggers the root hairs to curl, resembling a shepherd's crook, thereby entrapping bacteria (figure 31.9c,d). Nod factors trigger root cortex cells to initiate cell division, and these cells will eventually form the nodule primordium that accepts the invading rhizobium.

In addition to Nod factors, the bacteria now produce an exopolysaccharide that induces changes in the plant cell wall, and the plant plasma membrane invaginates. This leads to the development of a bacteria-filled, tubelike structure called the **infection thread** (figure 31.9e,f). Plant cell division drives the growth of the infection thread. When bacteria penetrate the base of a root hair cell, they stimulate growth of the infection thread through additional plant cell walls and membranes. This is dependent on the plant hormone cytokinin in concert with Nod factor stimulation of plant cell mitosis. Finally, the infection thread filled with bacteria reaches the inner plant cortex, where each bacterial cell is endocytosed by a plant cell in a discrete, unwalled membrane compartment that arises from the infection thread. This unit, consisting of an individual bacterium and the surrounding endocytic membrane, is called the **symbiosome** (figure 31.9g). It is here that each bacterial cell differentiates into the nitrogen-fixing form called a **bacteroid.** Depending on the plant species, bacteroids may be terminally differentiated: they can neither divide nor revert back to the nondifferentiated state. Terminal differentiation is controlled by the production of nodule-specific cysteine-rich (NCR) peptide signals by the plant host; interestingly, these signals are similar in structure to antimicrobial peptides produced by humans called defensins. ►►| *Antimicrobial peptides and proteins (section 33.3)*

The low levels of oxygen within the symbiosome triggers a regulatory cascade that controls nitrogen fixation and the switch to microaerobic respiration needed to provide ATP and reducing equivalents to the nitrogenase enzyme. These regulatory proteins include an oxygen-sensing, two-component signal transduction system (FixL and FixJ) and an alternative sigma factor, σ^N. Overall, most metabolic processes within bacteroids diminish with the exception of nitrogen fixation and respiration, which are upregulated. The assembly of many symbiosomes gives rise to the **root nodule** (figure 31.9h,i). ◄◄ *Nitrogen fixation (section 12.5); Bacteria combine several regulatory processes to control complex cellular processes (section 14.5)*

The reduction of atmospheric N_2 to ammonia by the differentiated bacteroids depends on the production of a protein called **leghemoglobin.** Recall that the nitrogenase enzyme is disabled by oxygen. To help protect the nitrogenase, leghemoglobin binds oxygen and helps maintain microaerobic conditions within the mature nodule. Leghemoglobin is similar in structure to the hemoglobin found in animals; however, it has a higher affinity for oxygen. Interestingly, the protein moiety is encoded by plant genes, whereas the heme group is a bacterial product.

The transfer of ammonia from the rhizobium to the host plant involves a complex cycling of amino acids. The plant provides certain amino acids to the bacteroids so that they do not need to assimilate ammonia. In return, the bacteroids shuttle amino acids (which bear the newly fixed nitrogen) back to the plant. This creates an interdependent relationship, providing selective pressure for the evolution of mutualism. The bacteria also receive carbon and energy in the form of dicarboxylic acids from their host legume.

The molecular mechanisms by which both the legume host and the rhizobial symbionts establish productive nitrogen-fixing bacteroids within nodules is an intense area of research. A major goal of biotechnology is to enable nitrogen fixation within cereal plants such as rice, wheat, and corn. Three approaches are being pursued: (1) modifying rhizobia to form productive nodules on these nonleguminous plants; (2) creating strains of non-nodulating nitrogen-fixing bacteria that will colonize cereal plants; and (3) introducing nitrogen-fixation genes into plants that are not rhizobium hosts. Although progress has been made using each approach, nitrogen fixation on these important crops has yet to be realized,

and the use of human-made fertilizer, with all its environmental problems, continues. ◀◀ *Global climate change (section 28.2)*

Stem-Nodulating Rhizobia

A few leguminous plants support rhizobia that nodulate stems, rather than roots. Nodules form at the base of roots branching out of the stem just above the soil surface. These plants are generally found in waterlogged soils and riverbanks. Nodulation of the

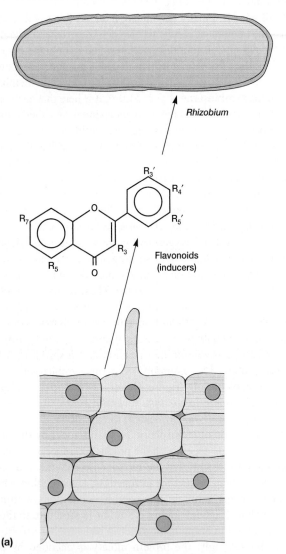

(a)

Figure 31.9 Root Nodule Formation by a *Rhizobium* Species.
Root nodule formation on legumes by rhizobia is a complex process that produces the nitrogen-fixing symbiosis. (a) The plant root releases flavonoids that stimulate the production of various Nod metabolites by bacteria. Many different Nod factors control infection specificity. (b) Attachment of bacterial cells to root hairs involves specific bacterial proteins called rhicadhesins and host plant lectins that affect the pattern of attachment and *nod* gene expression. (c) Structure of a typical Nod factor that promotes root hair curling and plant cortical cell division. The bioactive portion (nonreducing *N*-fatty acyl glucosamine) is highlighted. These Nod factors enter root hairs and migrate to their nuclei. (d) A plant root hair with attached rhizobium undergoing curling.

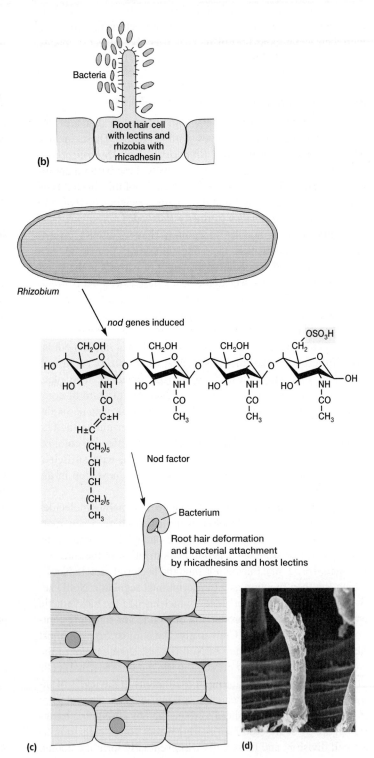

(b)

(c)

(d)

vetch *Aeschynomene sensitiva* by a photoheterotrophic strain of *Bradyrhizobium* has been studied in some detail. *Bradyrhizobium* strain BTAi synthesizes its light-harvesting apparatus only under aerobic day–night cycles of illumination, and optimal nodulation and nitrogen fixation occur only if the microbe is phototrophic. It is thought that light-driven cyclic electron transfer generates a proton gradient used for either ATP synthesis or substrate transfer.

The process of stem nodulation differs from that of root nodulation. Stem nodulation starts with the entry of bacteria through ruptures in stem epidermal cells that occur during emergence of the root tip. Unlike root nodulation, the infection of *Bradyrhizobium* strain BTAi does not involve the production of Nod factors or plant calcium spikes (**figure 31.10**). Nonetheless, as with root nodulation, bacteria within stem nodules are protected from the external environment and can obtain plant-derived energy substrates.

Frankia form Actinorhizae

Another example of symbiotic nitrogen fixation occurs between actinobacteria of the genus *Frankia* and eight nonleguminous host plant families. These bacterial associations with plant roots are called **actinorhizae** or actinorhizal relationships (**figure 31.11**). *Frankia* spp. are important particularly in trees and shrubs. The nodules of some plants (*Alnus, Ceanothus* spp.) can be as large as baseballs. The nodules of *Casuarina* spp. (Australian pine) approach soccer ball size.

Depending on the host plant, the infection process proceeds either by root hair infection or by direct penetration of the matrix between root cells. When infecting through root hairs, the bacteria induce plant cell division, leading to the development of a prenodule. *Frankia* cells then enter the cortex of the root and form a nodule that resembles a lateral root. When infection is independent of root hair invasion, the bacteria are encapsulated in a shell made of pectin that then extends through the plant intercellular spaces. Although *Frankia* nodules vary in morphology, they all bear the general appearance of lateral root clusters that grow by repeated branching of the tips. Some nodules are found just below the surface of the soil, whereas others—particularly those forming on plants that grow in dry, rocky soil—develop very deep below the surface. ◀◀ *Order Frankiales (section 24.1)*

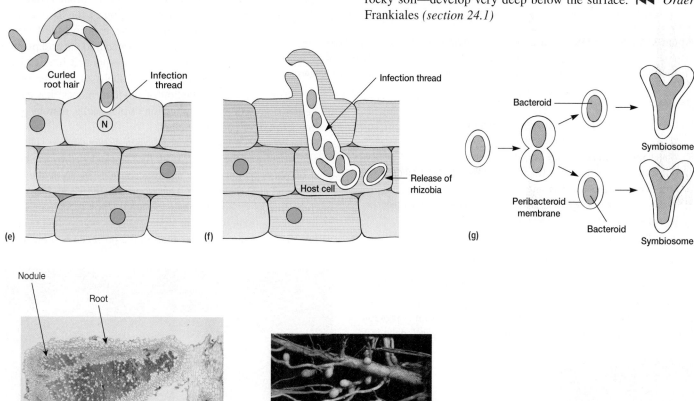

Figure 31.9 Root Nodule Formation by a *Rhizobium* Species (*continued*).   (e) Initiation of bacterial penetration into the root hair cell and infection thread growth coordinated by the plant nucleus "N." (f) Cell-to-cell spread of rhizobia through transcellular infection threads followed by release of rhizobia and infection of host cells. (g) Formation of bacteroids surrounded by plant-derived peribacteroid membranes and differentiation of bacteroids into nitrogen-fixing symbiosomes. The bacteria change morphologically and enlarge around seven to 10 times in volume. The symbiosome contains the nitrogen-fixing bacteroid, a peribacteroid space, and the peribacteroid membrane. (h) Light micrograph of two nodules that develop by cell division (×5). This section is oriented to show the nodules in longitudinal axis and the root in cross section. (i) Nodules on bur clover (*Medicago* sp.).

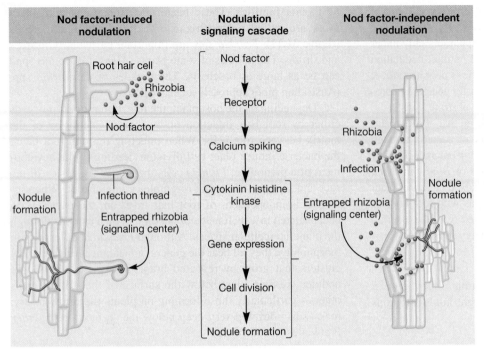

Nod factor-induced nodulation	Nodulation signaling cascade	Nod factor-independent nodulation

Figure 31.10 Root and Stem Nodulation. Root nodulation (left) involves the production of Nod factors by rhizobium and the signaling molecule cytokinin by the host plant. By contrast, Nod factor and plant cytokinin release is not necessary for stem nodulation by at least two rhizobia (right). It is not yet known if this is generally the case.

Retrieve, Infer, Apply

1. List several bacteria that are considered rhizobia.
2. What is the function of the infection thread?
3. What does the term terminally differentiated mean?
4. How does nitrogen transfer between a rhizobium and its host differ from that between AM and its host?
5. What is unusual about leghemoglobin production and what is its function?
6. What are the two general mechanisms by which *Frankia* spp. nodulate their plant hosts? Which types of plants are infected?

Agrobacterium Is a Model Plant Pathogen

Clearly some microbe-plant interactions are beneficial for both partners. However, others involve microbial pathogens that harm or even kill their host. *Agrobacterium tumefaciens* is an α-proteobacterium that has been studied intensely for several decades.

Figure 31.11 Actinorhizae.
Frankia-induced actinorhizal nodules.

MICRO INQUIRY *How do you think Frankia spp. protect nitrogenase from oxygen damage?*

Initially research focused on how this microbe causes crown gall disease, which results in the formation of tumorlike growths in a wide variety of plants (**figure 31.12**). Within the last 20 years or so, however, *A. tumefaciens* has become one of biotechnology's most important tools. The molecular genetics by which this pathogen infects its host are the basis for plant genetic engineering. ◄◄ *Class* Alphaproteobacteria *includes many oligotrophs (section 22.1)* ►►| *Agricultural biotechnology relies on a plant pathogen (section 42.5)*

The genes for plant infection and virulence are encoded on an *A. tumefaciens* plasmid called the **Ti** (**tumor-inducing**) **plasmid.** These genes include 21 *vir* genes (*vir* stands for virulence), found in six separate operons. Two of these genes, *virD1* and *virD2,* encode proteins that excise a specific region of the Ti plasmid, called **T DNA.** After excision, the T DNA fragment is integrated into the host plant's genome. Once incorporated into a plant cell's genome, T DNA directs the overproduction of phytohormones that cause unregulated growth and reproduction of plant cells, thereby generating a tumor or gall in the plant.

The *vir* genes are not expressed when *A. tumefaciens* is living saprophytically in the soil. Instead, they are induced by the presence of plant phenolics and monosaccharides in an acidic (pH 5.2–5.7) and cool (below 30°C) environment (**figure 31.13a**). The microbe infects its host through a wound. Upon reception of the plant signal, a two-component signal transduction system is activated: VirA is a sensor kinase that, in the presence of a phenolic

Figure 31.12 *Agrobacterium.* *Agrobacterium tumefaciens*-caused tumor on a *Kalanchoe* sp. plant.

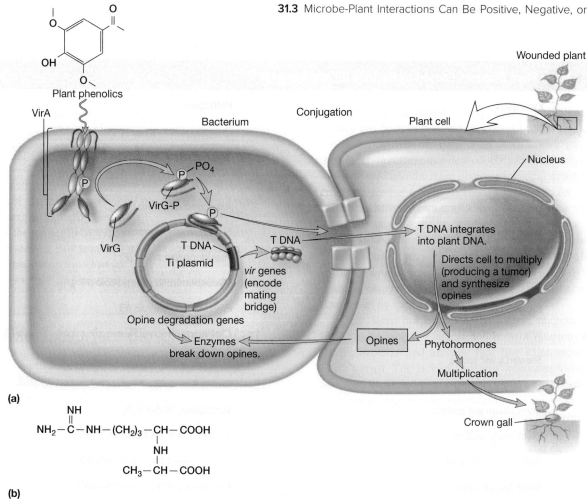

Figure 31.13 Functions of Genes Carried on the *Agrobacterium tumefaciens* Ti Plasmid. (a) Genes carried on the Ti plasmid of *A. tumefaciens* control tumor formation by a two-component regulatory system that stimulates formation of the mating bridge and excision of the T DNA. The T DNA is moved into the plant cell, where it integrates into the plant DNA. T DNA induces the production of plant hormones that cause the plant cells to divide, producing the tumor. (b) The tumor cells produce opines that can serve as a carbon source for the infecting *A. tumefaciens*. Ultimately a crown gall is formed on the stem of the wounded plant above the soil surface.

MICRO INQUIRY *How does the production of opines by the plant accelerate and amplify the* A. tumefaciens *infection process?*

signal, phosphorylates the response regulator VirG. Activated VirG then induces transcription of the other *vir* genes. This enables the bacterial cell to become adequately positioned relative to the plant cell, at which point the *virB* operon expresses the apparatus that will transfer the T DNA. This transfer is similar to bacterial conjugation and involves a type IV secretion system. After VirD1 and VirD2 excise the T DNA from the Ti plasmid, the T DNA, with the VirD2 protein attached to the 5' end, is delivered to the plant cell cytoplasm. The protein VirE2 is also transferred, and together with VirD2, the T DNA is shepherded to the plant cell nucleus, where it is integrated into the host's genome.

Once it is part of the host cell genome, the T DNA has two specific functions. First, it directs the host cell to overproduce phytohormones that cause tumor formation. Second, it stimulates the plant to produce special amino acid and sugar derivatives called opines (figure 31.13b). Opines are not metabolized by the plant, but *A. tumefaciens* is attracted to opines; chemotaxis of bacteria from the surrounding soil population will further

advance the infection because the bacterium can use opines as sources of carbon, energy, nitrogen, and, in some cases, phosphorus. ◀◀ *Bacteria combine several regulatory processes to control complex cellular processes (section 14.5); Bacterial conjugation requires cell-cell contact (section 16.6)*

Microbial Plant Pathogens Are Diverse

In terms of both diversity and economic costs, fungi are the most important plant pathogens, causing about 70% of plant disease (**table 31.3**; also see chapter-opening story). Fungal plant pathogens can be classified by their relationship with their plant hosts: they consume either living or dead plant material. **Biotrophic fungi** assimilate living plant material, and as such this term can also be applied to mutualistic mycorrhizae. Parasitic biotrophic fungi include powdery mildew and rust fungi; these microorganisms have evolved a special feeding apparatus called the haustorium to take up nutrients. Haustoria develop by the expansion of

Key Concepts

31.1 Soils Are an Important Microbial Habitat

■ Soil is a complex mixture of clay, sand, silt, organic matter, and water. This provides many micro-environments that can support a diverse microbial community (**figure 31.1**).

■ Soil organic matter (SOM) helps retain nutrients and water, and maintain soil structure. It can be divided into humic and nonhumic material.

31.2 Diverse Microorganisms Inhabit Soil

■ Soils support the highest levels of microbial diversity measured.

■ Archaea are not as abundant as bacteria in soils; ammonia-oxidizing archaea contribute to nitrification in soils.

■ Saprophytic and mycorrhizal fungi constitute about half the soil microbial biomass in most soils.

31.3 Microbe-Plant Interactions Can Be Positive, Negative, or Neutral

■ The phyllosphere includes all parts of a plant that are aboveground. These structures offer a unique but changeable environment for microbial growth.

■ The rhizosphere is the region around plant roots into which plant exudates are released. The rhizoplane is the plant root surface. A variety of microbes growing in these regions promote plant growth.

■ Mycorrhizal relationships (plant-fungal associations) are varied and complex. Six basic types are observed, including endomycorrhizal and sheathed/ectomycorrhizal types. The hyphal network of the mycobiont can lead to the formation of a mycorrhizosphere (**figures 31.4–31.8; table 31.2**).

■ The rhizobium-legume symbiosis is one of the best-studied examples of plant-microorganism interactions. Rhizobia form nodules on the roots of host plants in which specialized bacterial cells (bacteroids) fix nitrogen (**figure 31.9**).

■ The interaction between rhizobia and host plants is mediated by complex chemicals that serve as communication signals (**figures 31.7–31.10**).

■ *Actinobacteria* in the genus *Frankia* form nitrogen-fixing symbioses with some trees and shrubs.

■ *Agrobacterium tumefaciens* establishes a complex communication system with plant hosts into which they transfer a fragment of DNA. Genes on this DNA encode proteins that promote bacterial growth and result in the formation of plant tumors or galls (**figures 31.12 and 31.13**).

31.4 The Subsurface Biosphere Is Vast

■ The subsurface includes at least four zones: the shallow subsurface; the zone where gas, oil, and coal have accumulated; intermediate depths where oil and methane are found; and the very deep, hot subsurface.

■ Microbial activity in the deep subsurface suggests extraordinarily slow growth fueled by substrates of geological, rather than photosynthetic, origin. These compounds support anaerobic chemolithoautotrophy.

Compare, Hypothesize, Invent

1. Why might vascular plants have developed relationships with so many types of microorganisms? These molecular-level interactions show many similarities when microbe-plant and microbe-human interactions are considered. What does this suggest concerning possible common evolutionary relationships?

2. How might you maintain organisms from the deep hot subsurface under their in situ conditions when trying to culture them? Compare this problem with that of working with microorganisms from deep marine environments.

3. Soil bacteria such as *Streptomyces* spp. produce the bulk of known antibiotics. It has been suggested that in nature antibiotics serve as signaling molecules, allowing members of the same species or strain to communicate. If this is the case, what might these molecular signals be "saying" and how would such a message be transmitted?

4. The Nod factor signals produced by the root nodulating bacterium *Sinorhizobium meliloti* are essential for biofilm formation as well as N_2 fixation. In what ways are nodulation and biofilm formation similar, and why do you think this microbe evolved to use the same signal for the development of both?

5. In 2008 the annotated genome sequence of the ectomycorrhizal basidiomycete *Laccaria bicolor* was published. Surprisingly, *L. bicolor* produces a large number of small, secreted proteins (SSPs), some of which are only expressed in hyphae that colonize the host root. Another interesting finding was the presence of genes encoding enzymes for the degradation of cell wall polysaccharide only found in organisms other than plants. Formulate a hypothesis regarding the role of the SSPs during host colonization, and discuss the metabolic significance of the types of cell walls *L. bicolor* is able to degrade.

   Read the original paper: Martin, F. 2008. The genome of *Laccaria bicolor* provides insights into mycorrhizal symbiosis. *Nature* 452:88.

6. Although the role of viruses and bacteriophages in marine and freshwater systems has been studied for a number of years, little is known about viruses in soils. Using stable isotope probing (see *page 646*), Li and colleagues showed that carbon in rice plants was incorporated into the capsid of a soil bacteriophage. What does this imply about the trophic transfer of carbon in this soil? Suggest two other experimental approaches that might increase our understanding of viruses in soil ecosystems.

   Read the original paper: Li, Y. et al., 2013. Identification of the major capsid gene (*g23*) of T4-type bacteriophages that assimilate substrates from root cap cells under aerobic and anaerobic soil conditions using a DNA-SIP approach. *Soil Biol. Biochem.* 63:97.

the rate and occurrence of blight were decreased. Infection of trees with the less lethal virus strains is the basis of biological control efforts for chestnut blight. The goal is to vaccinate trees against the indigenous lethal strains of *Cryphonectria*.

Retrieve, Infer, Apply

1. What is the difference between the Ti plasmid and T DNA?
2. What functions do the members of the two-component system play in infection of a plant by *A. tumefaciens?* What are the roles of phenolics and opines in this infection process?
3. What are the two general ways by which plant pathogens gain entry into the host plant tissue? What are the advantages and disadvantages of each mechanism?
4. How are plant pathologists attempting to control chestnut blight?

31.4 The Subsurface Biosphere Is Vast

After reading this section, you should be able to:

- Relate the metabolic processes that occur in the subsurface biosphere to the free energy released during the use of each successive compound as a terminal electron acceptor
- Describe the distribution of subsurface methane production
- Explain at least one new discovery that has come from exploring the "deep hot biosphere"

For many years, it was thought that life could exist only in the thin veneer of Earth's surface and that any microbes recovered from sediments hundreds of meters deep were contaminants obtained during sampling. This view was drastically altered in the 1980s when the U.S. Department of Energy started looking for novel ways to store toxic waste. The agency began funding studies that applied modern technologies to sample the deep subsurface biosphere. Subsequent reports of microbes at great depth gained credibility, and international teams of geologists and microbiologists have since recovered microbes from thousands of meters below Earth's surface. The application of culture-independent techniques to quantify the density, diversity, and activity of microbes has revealed that subsurface microbes constitute about one-third of Earth's living biomass. This realization has made deep subsurface microbiology an exciting and active field.

Microbial processes take place in different subsurface regions, including (1) the shallow subsurface where water flowing from the surface moves below the plant root zone; (2) regions where organic matter, originally from the Earth's surface, has been transformed by chemical and biological processes to yield coal (from land plants), kerogens (from marine and freshwater microorganisms), and oil and gas; (3) zones where methane is being synthesized as a result of microbial activity; and (4) the energy-starved deep biosphere.

In the shallow subsurface, surface waters often move through aquifers—porous geological structures below the plant root zone. In a pristine system with an oxic surface zone, the electron acceptors used in catabolism are distributed from the most oxidized and energetically favorable (oxygen) near the surface to the least favorable (in which CO_2 is used in methanogenesis) in lower zones (*see figures 28.1 and 30.13*).

In subsurface regions where organic matter from the Earth's surface has been buried and processed by thermal and biological processes, kerogens and coals break down to yield gas and oil. After their generation, these mobile products, predominantly hydrocarbons, move upward into the more porous geological structures where microorganisms are active. Chemical signature molecules from plant and microbial biomass are present in these petroleum hydrocarbons.

Microorganisms also appear to be growing in intermediate-depth, oil-bearing structures. Active bacterial and archaeal assemblages are present in high-temperature (60 to 90°C) oil reservoirs, including members of such genera as *Thermotoga, Thermoanaerobacter,* and *Thermococcus.* The archaeal species are dominated by methanogens.

The "deep hot biosphere" is a term suggested by Thomas Gold to describe the largest and most mysterious biosphere on Earth: bacteria and archaea that live kilometers beneath the Earth's crust. The lack of apparent substrates to support growth and the calculated generation times in centuries to even 1,000 years seem alien to our understanding of microbial life. Anaerobic chemolithoautotrophy, independent of photosynthetically derived compounds, fuels these microbes; abiotically produced H_2 is believed to the most important energy and electron source. An interesting example of such bacterial activity was discovered in 2.8 km-deep, high-pressure water within a crustal fracture in a South African gold mine. Surprisingly, at least 99% of the microbes belong to a single firmicute species, as determined by 16S rRNA analysis. DNA collected from this unusual habitat was assembled into a complete genome to reveal a novel bacterium provisionally called *Candidatus* Desulforudis audaxviator. It appears to be a motile, spore-forming, sulfate-reducing chemoautotroph that is moderately thermophilic. This microbe most likely infiltrated the fracture zone 3 million to 25 million years ago, and its growth has apparently been sustained by hydrogen as electron donor and sulfate as terminal electron acceptor. ◄◄ *Class* Clostridia *(section 23.1)*

The discovery of deep subsurface microbes has a variety of implications. For example, the presence of primary production in the absence of sunlight suggests that autotrophy might also be possible in the subsurface sediments of Mars. Exploring this possibility, along with investigating the level of metabolic activity among Earth's deep subsurface microbes, remains the next challenge. ◄◄ *Microbial communities in benthic marine sediments are enormous and mysterious (section 30.2)*

Retrieve, Infer, Apply

1. Compare and contrast the metabolism (specifically, electron donors and acceptors) of microbes in the shallow subsurface with that of microbes in the very deep subsurface.
2. What happens in terms of microbiological processes when organic matter leaches from the surface into the subsurface?
3. What microbial genera have been observed in oil field materials?

Key Concepts

31.1 Soils Are an Important Microbial Habitat

- Soil is a complex mixture of clay, sand, silt, organic matter, and water. This provides many micro-environments that can support a diverse microbial community (**figure 31.1**).
- Soil organic matter (SOM) helps retain nutrients and water, and maintain soil structure. It can be divided into humic and nonhumic material.

31.2 Diverse Microorganisms Inhabit Soil

- Soils support the highest levels of microbial diversity measured.
- Archaea are not as abundant as bacteria in soils; ammonia-oxidizing archaea contribute to nitrification in soils.
- Saprophytic and mycorrhizal fungi constitute about half the soil microbial biomass in most soils.

31.3 Microbe-Plant Interactions Can Be Positive, Negative, or Neutral

- The phyllosphere includes all parts of a plant that are aboveground. These structures offer a unique but changeable environment for microbial growth.
- The rhizosphere is the region around plant roots into which plant exudates are released. The rhizoplane is the plant root surface. A variety of microbes growing in these regions promote plant growth.
- Mycorrhizal relationships (plant-fungal associations) are varied and complex. Six basic types are observed, including endomycorrhizal and sheathed/ectomycorrhizal types. The hyphal network of the mycobiont can lead to the formation of a mycorrhizosphere (**figures 31.4–31.8; table 31.2**).
- The rhizobium-legume symbiosis is one of the best-studied examples of plant-microorganism interactions. Rhizobia form nodules on the roots of host plants in which specialized bacterial cells (bacteroids) fix nitrogen (**figure 31.9**).
- The interaction between rhizobia and host plants is mediated by complex chemicals that serve as communication signals (**figures 31.7–31.10**).
- *Actinobacteria* in the genus *Frankia* form nitrogen-fixing symbioses with some trees and shrubs.
- *Agrobacterium tumefaciens* establishes a complex communication system with plant hosts into which they transfer a fragment of DNA. Genes on this DNA encode proteins that promote bacterial growth and result in the formation of plant tumors or galls (**figures 31.12** and **31.13**).

31.4 The Subsurface Biosphere Is Vast

- The subsurface includes at least four zones: the shallow subsurface; the zone where gas, oil, and coal have accumulated; intermediate depths where oil and methane are found; and the very deep, hot subsurface.
- Microbial activity in the deep subsurface suggests extraordinarily slow growth fueled by substrates of geological, rather than photosynthetic, origin. These compounds support anaerobic chemolithoautotrophy.

Compare, Hypothesize, Invent

1. Why might vascular plants have developed relationships with so many types of microorganisms? These molecular-level interactions show many similarities when microbe-plant and microbe-human interactions are considered. What does this suggest concerning possible common evolutionary relationships?

2. How might you maintain organisms from the deep hot subsurface under their in situ conditions when trying to culture them? Compare this problem with that of working with microorganisms from deep marine environments.

3. Soil bacteria such as *Streptomyces* spp. produce the bulk of known antibiotics. It has been suggested that in nature antibiotics serve as signaling molecules, allowing members of the same species or strain to communicate. If this is the case, what might these molecular signals be "saying" and how would such a message be transmitted?

4. The Nod factor signals produced by the root nodulating bacterium *Sinorhizobium meliloti* are essential for biofilm formation as well as N_2 fixation. In what ways are nodulation and biofilm formation similar, and why do you think this microbe evolved to use the same signal for the development of both?

5. In 2008 the annotated genome sequence of the ectomycorrhizal basidiomycete *Laccaria bicolor* was published. Surprisingly, *L. bicolor* produces a large number of small, secreted proteins (SSPs), some of which are only expressed in hyphae that colonize the host root. Another interesting finding was the presence of genes encoding enzymes for the degradation of cell wall polysaccharide only found in organisms other than plants. Formulate a hypothesis regarding the role of the SSPs during host colonization, and discuss the metabolic significance of the types of cell walls *L. bicolor* is able to degrade.

   Read the original paper: Martin, F. 2008. The genome of *Laccaria bicolor* provides insights into mycorrhizal symbiosis. *Nature* 452:88.

6. Although the role of viruses and bacteriophages in marine and freshwater systems has been studied for a number of years, little is known about viruses in soils. Using stable isotope probing (see *page 646*), Li and colleagues showed that carbon in rice plants was incorporated into the capsid of a soil bacteriophage. What does this imply about the trophic transfer of carbon in this soil? Suggest two other experimental approaches that might increase our understanding of viruses in soil ecosystems.

   Read the original paper: Li, Y. et al., 2013. Identification of the major capsid gene (*g23*) of T4-type bacteriophages that assimilate substrates from root cap cells under aerobic and anaerobic soil conditions using a DNA-SIP approach. *Soil Biol. Biochem.* 63:97.

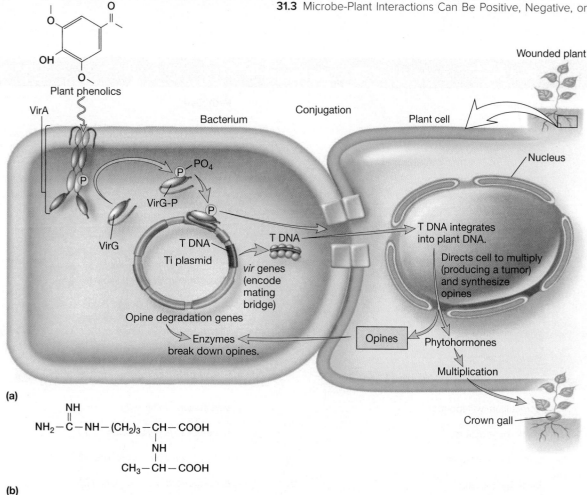

Figure 31.13 Functions of Genes Carried on the *Agrobacterium tumefaciens* Ti Plasmid. (a) Genes carried on the Ti plasmid of *A. tumefaciens* control tumor formation by a two-component regulatory system that stimulates formation of the mating bridge and excision of the T DNA. The T DNA is moved into the plant cell, where it integrates into the plant DNA. T DNA induces the production of plant hormones that cause the plant cells to divide, producing the tumor. (b) The tumor cells produce opines that can serve as a carbon source for the infecting *A. tumefaciens*. Ultimately a crown gall is formed on the stem of the wounded plant above the soil surface.

MICRO INQUIRY *How does the production of opines by the plant accelerate and amplify the* A. tumefaciens *infection process?*

signal, phosphorylates the response regulator VirG. Activated VirG then induces transcription of the other *vir* genes. This enables the bacterial cell to become adequately positioned relative to the plant cell, at which point the *virB* operon expresses the apparatus that will transfer the T DNA. This transfer is similar to bacterial conjugation and involves a type IV secretion system. After VirD1 and VirD2 excise the T DNA from the Ti plasmid, the T DNA, with the VirD2 protein attached to the 5′ end, is delivered to the plant cell cytoplasm. The protein VirE2 is also transferred, and together with VirD2, the T DNA is shepherded to the plant cell nucleus, where it is integrated into the host's genome.

Once it is part of the host cell genome, the T DNA has two specific functions. First, it directs the host cell to overproduce phytohormones that cause tumor formation. Second, it stimulates the plant to produce special amino acid and sugar derivatives called opines (figure 31.13*b*). Opines are not metabolized by the plant, but *A. tumefaciens* is attracted to opines; chemotaxis of bacteria from the surrounding soil population will further

advance the infection because the bacterium can use opines as sources of carbon, energy, nitrogen, and, in some cases, phosphorus. ◄◄ *Bacteria combine several regulatory processes to control complex cellular processes (section 14.5); Bacterial conjugation requires cell-cell contact (section 16.6)*

Microbial Plant Pathogens Are Diverse

In terms of both diversity and economic costs, fungi are the most important plant pathogens, causing about 70% of plant disease (**table 31.3**; also see chapter-opening story). Fungal plant pathogens can be classified by their relationship with their plant hosts: they consume either living or dead plant material. **Biotrophic fungi** assimilate living plant material, and as such this term can also be applied to mutualistic mycorrhizae. Parasitic biotrophic fungi include powdery mildew and rust fungi; these microorganisms have evolved a special feeding apparatus called the haustorium to take up nutrients. Haustoria develop by the expansion of

Table 31.3   Major Plant Diseases Caused by Bacteria (B) and Fungi (F)

Symptoms	Examples	Pathogen
Spots and blights	Wildfire (tobacco)	*Pseudomonas syringae* pv.[1] *tabaci* (B)
	Blight (rice)	*Xanthomonas campestris* pv. *oryzae* (B)
	Citrus blast	*P. syringae* pv. *syringae* (B)
	Leaf spots and blights (many plants)	*Alternaria alternata* (F)
	Head blight (wheat)	*Fusarium graminearum* (F)
	Leaf spot (alfalfa)	*Pseudopeziza trifolii* (F)
	Leaf spot (cereals, grasses)	*Mycosphaerella graminicola* (F)
	Brown spot (corn)	*Physoderma maydis* (F)
Vascular wilts	Wilt (tomato)	*Clavibacter michiganensis* pv. *michiganensis* (B)
	Stewart's wilt (corn)	*Erwinia stewartii* (B)
	Fire blight (apples)	*E. amylovora* (B)
	Verticullium wilt (potato)	*Verticullium dahliae* (F)
Soft rots	Black rot (crucifers)	*X. campestris* pv. *campestris* (B)
	Soft rots (numerous)	*E. carotovora* pv. *carotovora* (B)
	Black leg (potato)	*E. carotovora* pv. *atroseptica* (B)
Smuts	Loose smuts (oats, barley, wheat)	*Ustilago avenae, U. nuda, U. tritici* (F)
	Covered kernel smut (sorghum)	*Sproisorium sorghi* (F)
Cankers	Canker (stone fruit)	*P. syringae* pv. *syringae* (B)
	Hyoxylon canker (oak trees)	*Hyoxylon atropunctatum* (F)
Rusts	Asian soybean rust	*Phakopsora pachyrrhizi* (F)
	Leaf rust (cereals)	*Puccinia recondita* (F)
	White pine blister rust	*Cronartium ribicola* (F)

1 *pv., pathovar, a variety of microorganisms with phytopathogenic properties.*

Source: From Lengler, J. W., et al. 1999. Biology of the prokaryotes. Malden, MA: Blackwell Science, table 34.4.

fungal hyphae that penetrate plant host cells. By contrast, **necrotrophic fungi** kill their host plant cells by releasing toxins or enzymes prior to consuming plant nutrients. Necrotrophic fungi infect (and kill) a broader range of hosts. *Sclerotinia sclerotiorum* (white mold fungus) and *Botrytis cinerea* (grey mold fungus) are two notorious necrotrophic fungi that each parasitize over 200 different plant species and are responsible for the loss of millions of pounds of produce both pre- and post-harvest annually.

Many bacteria are also plant pathogens (table 31.3). Like *A. tumefaciens,* some plant pathogenic bacteria infect through a wound in the plant. Others, like the soft-rot bacteria *Erwinia chrysanthemi* and *E. carotovora,* digest plant tissue by secreting enzymes that degrade pectin, cellulose, and proteins. Another group of important bacterial plant pathogens includes the wallless bacterial phytoplasms that infect vegetable and fruit crops such as sweet potatoes, corn, and citrus.

As discussed in chapters 25 and 27, protists, viruses, and viroids can also be devastating plant pathogens. Examples include the öomycete *Phytophthora infestans,* which was responsible for the Irish potato famine (and continues to extract an economic burden), and tobacco mosaic virus (TMV), the first virus to be characterized. A virus of particular interest in terms of plant-pathogen interactions is a hypovirus (family *Hypoviridae*) that infects the fungus *Cryphonectria parasitica,* the cause of chestnut blight. Based on studies carried out in Europe, workers in the United States noted that if they infected the fungus with this hypovirus,

32

Microbial Interactions

There has been an alarming increase in the number of patients developing Clostridium difficile *infections. This bacterium is normally held in check by other gut microflora, but when antibiotic therapy eliminates these helpful bacteria, antibiotic-resistant* C. difficile *can flourish, causing colitis that can be life-threatening. A small but growing number of physicians are turning to fecal transplants as a last-resort treatment option.*

Embrace Your Gut Flora, for You Know Not What They Do

Samantha D. was desperate. Over the course of 8 months, she had been battling diarrhea so severe she had lost 60 pounds, was confined to a wheelchair, and was wearing adult diapers. For a period of 6 months, Vanessa R. ran high fevers and had diarrhea every 15 minutes on her worst days. Her better days were spent sleeping. Both these women suffered from *Clostridium difficile* infection. Both were eventually cured by a treatment once considered crazy: fecal transplants.

First, the microbe: *C. difficile* (C. diff for short) is a Gram-positive, endospore-forming firmicute (see sections 23.1 and 39.6) that resides in low numbers in the human gut. It is normally outcompeted by nonpathogenic bacteria. Many C. diff infections arise in people undergoing antibiotic therapy. This leads to the inhibition of normal gut microflora, enabling the overgrowth of C. diff and its release of a suite of toxins triggering disabling diarrhea accompanied by sloughing of the gut epithelium. It is most commonly seen in hospitalized patients, but community-acquired C. diff infections are on the rise.

Next the cure: Antibiotic therapy frequently fails because C. diff has acquired resistance to multiple antimicrobials and antibiotic treatment continues to suppress the growth of gut microorganisms needed to outcompete this virulent pathogen. A growing number of physicians are now treating patients with recurrent C. diff infections with fecal microbiome transplants. The patient's fecal donor is often a healthy family member whose stool is first tested to be sure it is free of pathogenic viruses, bacteria, and protozoa. The stool is then mixed with sterile saline and infused into the patient from one end (via a nasogastric tube) or the other (via a colonoscope). About 85 to 90% of patients report a complete cure within 48 hours to a week.

How did such a treatment start? It probably began with farmers who long ago realized that cattle with intestinal illnesses could be cured by ingesting the rumen fluid of another, healthy animal. Current human treatment started in the 1980s, when Australian gastroenterologist Thomas

Borody was caring for a patient who had returned from vacation with incurable colitis caused by an unknown pathogen. After exhausting every conventional treatment, he began looking for alternative therapies. He found one in a 1958 edition of the journal *Surgery,* where he read a report of four patients who were cured by fecal transplants. So Dr. Borody asked the patient's brother to donate his stool, and within days after infusion, the patient's colitis was gone, never to return. But it wasn't until the number of antibiotic-resistant C. diff infections climbed in the 2000s that a handful of doctors around the globe began to treat patients with fecal transplants. Efforts are now under way to encapsulate the fecal microbiota as a more acceptable and standardized treatment.

Apart from the gross factor, what is the message? For a very long time, gut microorganisms have been termed commensals—organisms that live within another without injuring or benefiting the host. Clearly, the severity of C. diff infections and its cure by fecal transplant argue that this term should no longer be considered accurate.

Microbes rarely live alone. They have evolved in specific ways with other organisms, including microbes, plants, and animals. In this chapter, we define types of microbial interactions and present illustrative examples. We then apply these interactions to examine perhaps the most intimate, yet still relatively unexplored, microbial habitat: the human body.

Readiness Check:
Based on what you have learned previously, you should be able to:

✔ List environmental factors that influence microbial growth (section 7.4)
✔ Compare and contrast the growth and physiology of obligate aerobes, facultative anaerobes, and strict anaerobes (section 7.4)
✔ Explain the general morphology of protists and the structure and function of hydrogenosomes (sections 5.6 and 25.2)
✔ Summarize the mechanisms of horizontal gene transfer (section 16.4)
✔ Diagram, in general terms, the standard reduction potentials of electron donors as compared to electron acceptors and relate this to standard free energy change (section 10.3)
✔ Describe the difference between heterotrophy and autotrophy (section 11.1)
✔ Compare anaerobic and aerobic respiration (section 11.2)

✔ Define chemolithotrophy and identify an example electron donor and acceptor (section 11.10)

✔ Compare and contrast oxygenic and anoxygenic phototrophy (section 11.11)

✔ Summarize the fate of pyruvate in fermentation (section 11.8)

✔ Explain carbon and nitrogen fixation (sections 12.3 and 12.5)

✔ Identify substrates used in methanogenesis (section 20.4)

✔ Relate genome reduction to obligate intracellular lifestyles (section 18.7)

✔ Describe the principles and application of metagenomics (sections 18.3 and 29.3)

32.1 Many Types of Microbial Interactions Exist

After reading this section, you should be able to:

- Compare and contrast different types of microbial interactions, including mutualism, cooperation, commensalism, predation, parasitism, amensalism, and competition
- Provide at least one specific example of each of these types of interactions

You need look no further than in a mirror to grasp the importance of microbial-host interactions. Your body possesses about three times as many bacterial cells than it does somatic cells (and that doesn't include the fungi, bacteriophages, and protists you host). Indeed, the vast majority of microbes live in complex communities that include organisms from other domains of life. The stable association of one organism with another is called **symbiosis,** regardless of whether the association is beneficial for both partners, as in mutualism, cooperation, and commensalism, or has a negative impact, described as predation, parasitism, amensalism, and competition (**figure 32.1**).

For many years bacterial symbioses were viewed as the relationship between one bacterial species and one host species. However, the wide application of metagenomics, single-cell sequencing, next-generation sequencing, and new approaches to culturing microbes have revealed that microbial symbioses can range from a single species of microorganism and its host to hosts with thousands of different kinds of microbes. In fact, there have been so many discoveries of new, and never imagined, symbiotic relationships that the vocabulary of microbial interactions is changing. For example, while the smaller organism in a symbiotic relationship has always been called the **symbiont** and the larger organism the host, these organisms are now sometimes regarded as partners. What was once considered a microbe-host relationship is now sometimes called a holobiont or a superorganism to reflect the critical nature of the interaction.

All symbionts must undergo the process of host colonization, reproduction, and persistence within the host, as well as transmission to a new host. Because microbial symbioses are the result of thousands, if not millions, of years of coevolution, these processes have been optimized. However, their specifics differ and make each symbiotic relationship seem like a clever feat of natural history. For example, host specificity can range from very narrow to very broad. Symbionts may be transmitted horizontally from host

to host or vertically during reproduction, or each host may have to acquire the symbiont from the environment. Symbioses can range from those that are very stable (e.g., aphids and their bacterial mutualists, as discussed next), to interactions that are temporary, depending on the life cycle of the host or other variables. Symbionts can live within the host as an **endosymbiont,** or on the surface of the host as an **ectosymbiont.** Finally, symbionts alter not only the health of their hosts, but their behavior, evolution, and reproductive success as well (**Microbial Diversity & Ecology 32.1**) ◄◄ *Metagenomics provides access to uncultured microbes (section 18.3); Microbial biology relies on cultures (section 29.1)*

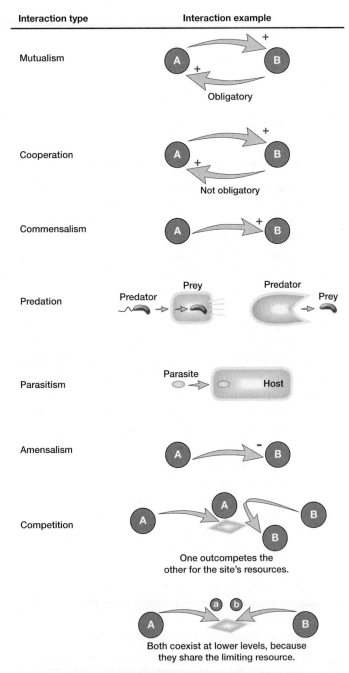

Figure 32.1 Microbial Interactions. Basic characteristics of symbiotic interactions that can occur between different organisms.

MICROBIAL DIVERSITY & ECOLOGY

32.1 *Wolbachia pipientis:* The World's Most Infectious Microbe?

Most people have never heard of the bacterium *Wolbachia pipientis,* but this rickettsia infects more organisms than any other microbe. It is known to infect a variety of crustaceans, spiders, mites, millipedes, and parasitic worms, and may infect more than a million insect species worldwide. To what does *W. pipientis* owe its extraordinary success? Quite simply, this endosymbiont is a master at manipulating its hosts' reproductive biology. In some cases, it can even change the sex of its host.

W. pipientis inhabits the cytoplasm of its host and is transferred from one generation to the next through the eggs of infected females. To survive, *Wolbachia* must ensure the fertilization and viability of infected eggs while decreasing the likelihood that uninfected eggs survive. The mechanism by which this is accomplished depends on the host. In wasps and mosquitoes, *W. pipientis* causes "cytoplasmic incompatibility," which simply means that mating will be abnormal if the only the male is infected. For instance, when infected sperm of the wasp *Nasonia vitripennis* fertilizes uninfected eggs, chromosomes from the *W. pipientis*–laden sperm prematurely try to align with the egg's chromosomes. These eggs then divide as if never fertilized. However, chromosomes behave normally when an infected female mates with an uninfected male. This yields a normal sex distribution, and all progeny are infected with the rickettsia.

In other infected insects, *W. pipientis* may simply kill all the male offspring and induce parthenogenesis of infected females; that is, the mothers simply clone themselves. This limits genetic diversity but allows 100% transmission of rickettsia to the next generation. In still other hosts, the microbe allows the birth of males but then modifies their hormones so that the males become feminized and produce eggs.

The ability of this microbe to manipulate the reproduction of its hosts may play a role in insect speciation. Two genetically similar North American wasp species (*Nasonia giraulti* and *N. longicornis*) each carry a different strain of *W. pipientis.* Predictably, when the two species mate with each other, there are no viable offspring. But when the wasps are first treated with an antibiotic to cure them of their bacterial infections, viable, fertile offspring are produced. It thus appears that *W. pipientis* infection, not genetic differences, forms the reproductive wall that drives speciation. The role of *Wolbachia* bacteria in generating genetic diversity among its hosts is further illustrated by the discovery of horizontal gene transfer (HGT) from the bacterium to its host's genome. This has been confirmed in several insect and nematode species that harbor *W. pipientis* symbionts. In fact, the fruit fly *Drosophila ananassae* contains the entire *W. pipientis* genome within its own.

W. pipientis infection has been implicated in several human diseases, including river blindness, which has a worldwide incidence of about 18 million people. The causative agent, *Onchocerca volvulus,* is a filarial nematode transmitted by black flies in Africa and Latin America. When a fly bites a person, the nematode establishes its home in small nodules beneath the skin, where it can survive for up to 14 years. During that time, it releases millions of larvae, many of which migrate to the eye. Eventually the host mounts an inflammatory response that results in progressive vision loss (**box figure**). It is now recognized that this inflammatory response is principally directed at the bacteria infecting the nematodes, not the worms themselves. Nematode reproduction stops in patients treated with an antibiotic that kills the endosymbiont. While inflammatory damage cannot be reversed, disease progression is halted. Antibacterial treatment is now combined with antiparasitic medications in the treatment of this disabling condition.

Our understanding of the distribution of *W. pipientis* and the mechanistically clever ways it has evolved to ensure its survival will continue to grow. The microbe may ultimately become a valuable tool in investigating the complexities of speciation, as well as the key to eliminating a devastating disease.

Sources: *St. Andre, A., et al. 2002. The role of endosymbiotic* Wolbachia *bacteria in the pathogenesis of river blindness. Science 295:1892–95; Zimmer, C. 2001.* Wolbachia*: A tale of sex and survival. Science 292:1093–95.*

River blindness. This is the second-leading cause of blindness worldwide. Evidence suggests that it is not the nematode but its endosymbiont, *Wolbachia pipientis,* that causes the severe inflammatory response that leaves many, like the woman shown here, blind.

Mutualism: Obligate Positive Interactions Between Host and Microbe

Mutualism (Latin *mutuus,* borrowed or reciprocal) defines the relationship in which some reciprocal benefit accrues to both partners. This is an obligatory relationship in which the **mutualist** and the host are dependent on each other. In many cases, the individual organisms will not survive when separated. Several examples of mutualism are presented next.

Microorganism-Insect Mutualisms

Mutualistic associations are common between microbes and insects partly because insects often consume plant sap or animal fluids lacking essential vitamins and amino acids. These are provided by bacterial symbionts in exchange for a secure habitat and ample nutrients. Aphids and their γ-proteobacterium *Buchnera aphidicola* partner are a model system for the study of mutualistic symbioses. A mature aphid contains literally millions of these bacteria within specialized cells called bacteriocytes. Strains of *B. aphidicola* provide their hosts with ten essential amino acids absent in sap. If the insect is treated with antibiotics, it dies. Likewise, *B. aphidicola* is an obligate mutualistic symbiont. The inability of either partner to grow without the other indicates that the two organisms underwent **coevolution;** that is, they have evolved together. In fact, the pea aphid (*Acyrthosiphon pisum*) and its *B. aphidicola* symbionts share the genes for the biosynthesis of several amino acids such that certain steps occur only in one organism.

It is estimated that the *B. aphidicola*–aphid endosymbiosis was established about 150 million years ago. The genomes of different endosymbiotic *B. aphidicola* strains reveal that about 75% of the ancestral *B. aphidicola* genome has been eliminated. It follows that these genomes are very small: in most cases, only about 0.52 Mb each. Over 90% of the genes are common among these bacterial strains, and the genomes show no evidence of gene duplication, translocation, inversion, or horizontal transfer. This implies that although the initial acquisition of the endosymbiont by ancestral aphids enabled their use of an otherwise deficient food source (sap), the bacteria have not continued to expand the ecological niche of their insect host through the acquisition of new traits that might be advantageous elsewhere. ◄◄ *Comparative genomics (section 18.7)*

Termites are another terrific model system for the study of endosymbiosis. Some termites harbor protist, bacterial, and archaeal endosymbionts and eat only wood. The main structural polysaccharides of wood are cellulose (an unbranched glucose polymer) and hemicellulose (a smaller, branched glucose polymer), which combine with lignin (*see figure 31.2*) to form lignocellulose. A diet of wood poses two problems for the termite: how to degrade polysaccharides that may possess as many as 15,000 glucose monomers and where to get organic nitrogen needed for nucleotide and protein synthesis. Although these termites produce a cellulolytic enzyme, only their mutualistic protists can complete lignocellulose degradation. Nitrogen-fixing bacteria that live in the termite gut solve the problem of obtaining organic nitrogen. ◄◄ *Nitrogen fixation (section 12.5)*

Most termite-associated protists are members of the *Excavata* supergroup and as such are quite ancient, possessing hydrogenosomes rather than mitochondria. These protists ferment cellulose to acetate, CO_2, and H_2. Acetate is the termite's preferred carbon source, and termites rely on bacterial symbionts to convert the CO_2 and H_2 released by the protists to acetate by the reductive acetyl-CoA pathway (*see figure 20.3*). In addition, termites harbor methanogenic archaea that use these substrates to form CH_4, which is excreted. ◄◄ *Mitochondria related organelles and chloroplasts are involved in energy conservation (section 5.6); Supergroup Excavata (section 25.2)*

Some termite protists also harbor endosymbionts (i.e., endosymbionts of endosymbionts). For instance, the protist *Trichonympha* sp. relies on a bacterial endosymbiont in the new genus *Elusimicrobium* to convert glutamine to other amino acids and nitrogenous compounds (**figure 32.2**). In exchange, the protist supplies *Elusimicrobium* sp. with glucose 6-phosphate, which can directly enter glycolysis. In addition, motility is often provided by spirochetes that cover the protist surface. Motility is essential to prevent protist expulsion by the termite gut and to acquire food.

Zooxanthellae

Many marine invertebrates (sponges, jellyfish, sea anemones, corals, ciliated protists) harbor endosymbiotic dinoflagellates called zooxanthellae within their tissues. Because the degree of host dependency on the dinoflagellate is variable, only one well-known example is presented here. ◄◄ *Alveolata (section 25.4)*

The hermatypic (reef-building) corals satisfy most of their energy requirements using their zooxanthellae, which include members of the dinoflagellate genus *Symbiodinium* (**figure 32.3a**). These protists line the coral gastrodermal tissue at densities between 5×10^5 and 5×10^6 cells per square centimeter of coral animal. *Symbiodinium* cells give up to 95% of the carbon they fix in exchange for nitrogenous compounds, phosphates, CO_2, and protection from ultraviolet light from their coral hosts. In addition to dinoflagellates, corals host distinct bacterial communities in three habitats (figure 32.3b). The surface mucus layer supports the growth of nitrogen-fixing and chitin-degrading bacteria, while the animal tissue within the gastrodermal cavity features another population of bacteria (although little is known about these microbes). The coral skeleton hosts yet another group of bacteria. These microbial communities are responsible for extremely efficient nutrient cycling and tight coupling of trophic levels, which accounts for the stunning success of reef-building corals in developing vibrant ecosystems.

During the past several decades, the number of coral bleaching events has increased dramatically. **Coral bleaching** is defined as a loss of either the photosynthetic pigments from the zooxanthellae or the complete expulsion of the dinoflagellates from the coral. Damage to photosystem II of the zooxanthellae generates reactive oxygen species (ROS); these ROS appear to directly damage many corals (recall that photosystem II uses

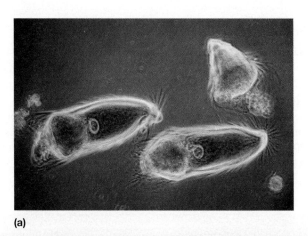

(a)

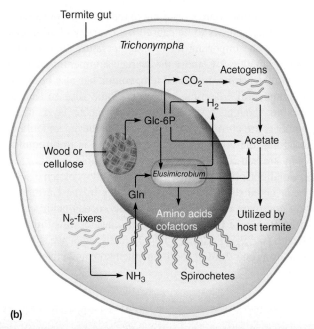

(b)

Figure 32.2 **Mutualism.** (a) Light micrograph of *Trichonympha* sp., a multiflagellated protozoan from the termite's gut (×135). The ability of *Trichonympha* to break down cellulose allows termites to use wood as a food source. (b) Proposed interaction between the termite endosymbiotic protist *Trichonympha* and its bacterial endosymbiont *Elusimicrobium* sp.. Within the termite gut, bacteria convert N_2 to ammonia, which is taken up and converted to glutamine (Gln) by *Trichonympha* sp. The protist relies on *Elusimicrobium* sp. to convert Gln to other amino acids, which are released with other cofactors that the protist cannot synthesize. Both the protist and *Elusimicrobium* sp. ferment glucose to acetate, H_2, and CO_2. These products are exported by the protist and made available to the termite, which also relies on free-swimming gut bacteria to boost acetate levels through reductive acetogenesis (*see figure 20.3*).

MICRO INQUIRY *What methanogenic substrates are also available in the termite gut?*

(a)

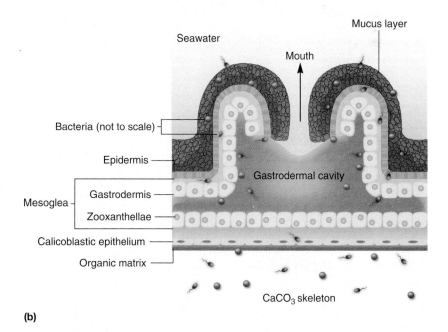

(b)

Figure 32.3 **Zooxanthellae.** (a) The green color of this rose coral (*Manicina* sp.) is due to the abundant zooxanthellae within its tissues. (b) Corals provide three distinct habitats for bacterial growth: the mucus layer at the surface, the gastrodermal cavity, and the coral skeleton. Calicoblastic epithelial cells precipitate calcium carbonate ($CaCO_3$) from the water to build exoskeleton. Zooxanthellae inhabit gastrodermal cells and are shown in green.

water as the electron source, resulting in the evolution of oxygen). Coral bleaching appears to be caused by a variety of additional stressors, but one important factor is temperature. Temperature increases as small as 2°C above the average summer maxima can trigger coral bleaching. Increased temperature has also been correlated with coral infectious diseases. Some disease-causing microbes are thought to reside in the coral mucus layer, where their numbers are normally kept in check. But during and following bleaching, the coral microbial community is in disequilibrium, and stressed corals become more susceptible to pathogenic attack. Conditions such as white plague and yellow blotch disease have been noted to follow bleaching events in the Caribbean, reducing corals by roughly a third. Finally, ocean acidification caused by increased CO₂ dissolution in seawater also threatens coral health (*see figures 30.1 and 30.2*). Sadly, evidence suggests that as many as one-third of corals and their zooxanthellae will be

unable to evolve quickly enough to keep pace with predicted ocean acidification and warming if global climate change continues unchecked. ◄◄ *Light reactions in oxygenic photosynthesis (section 11.11); Global climate change (section 28.2); Water is the largest microbial habitat (section 30.1)*

Sulfide-Based Mutualisms

Several thousand meters below the surface of the ocean, hydrothermal vents are found where the Earth's crustal plates are spreading apart (**figure 32.4**). Vent fluids are anoxic, contain high concentrations of hydrogen sulfide, and can reach a temperature of 350°C. The seawater surrounding these vents has sulfide concentrations around 250 μM and temperatures 10 to 20°C above the ambient seawater temperature of about 2°C. Although the high pressure at such depths prevents the water

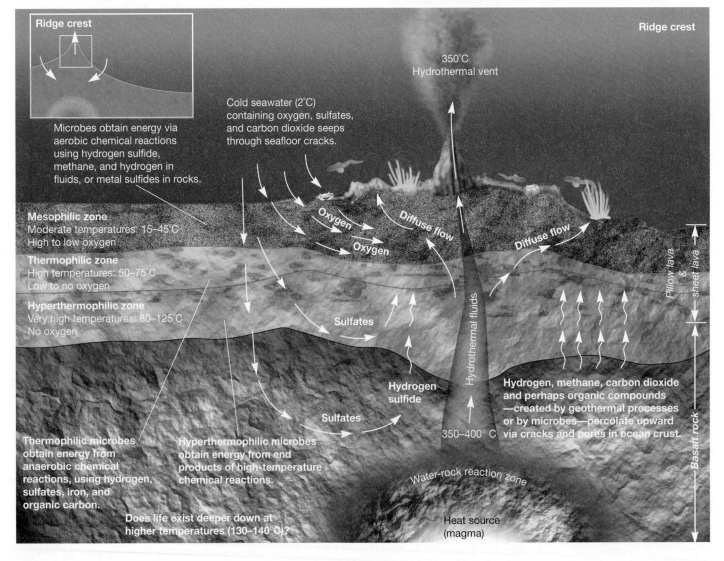

Figure 32.4 Hydrothermal Vents and Related Geological Activity. The chemical reactions between seawater and rocks that occur over a range of temperatures on the seafloor supply the carbon and energy that support a diverse collection of microbial communities in specific niches within the vent system.

from boiling, one might think these conditions are inhospitable to life. But nothing could be further from the truth.

Giant (>1 m in length), red, gutless tube worms (*Riftia* spp., **figure 32.5a**) growing near hydrothermal vents exemplify a remarkably successful mutualism in which bacterial endosymbionts are maintained within specialized cells of the tube worm host (figure 32.5*b,c,d*). *Riftia* worms live at the interface between hot, anoxic, sulfide-containing fluids of the vents and cold, oxygenated seawater. The worm lacks a mouth and gut and relies on endosymbiotic bacteria to provide organic carbon. The bacteria are chemolithotrophic, using sulfide as an electron donor and oxygen as a terminal electron acceptor. In fact, *Rifta* worms have unique hemoglobin that provides both reduced sulfur and oxygen to the endosymbionts. Hydrogen sulfide (H_2S) and O_2 are removed from the seawater by the worm's hemoglobin and delivered to an organ called the trophosome. The trophosome is

packed with chemolithotrophic bacterial endosymbionts that reach densities of up to 10^{11} cells per gram of worm tissue. The endosymbionts use the Calvin-Bensen cycle to fix CO_2 (*see figure 12.4*), which they reduce with electrons provided by H_2S. The CO_2 is carried to the endosymbionts in three ways: (1) freely in the bloodstream, (2) bound to hemoglobin, and (3) as organic acids such as malate and succinate. When these acids are decarboxylated, they release CO_2. This CO_2 is then fixed using the same pathway used by plants and cyanobacteria, but it occurs in the deepest, darkest reaches of the ocean. This mutualism enables *Riftia* worms to grow to an astounding size in densely packed communities.

An interesting, more complex mutualism occurs between the gutless marine oligochaete *Olavius algarvensis* and bacterial endosymbionts that reside just below the outer surface (cuticle) of this worm. Like *Riftia* tube worms, these worms

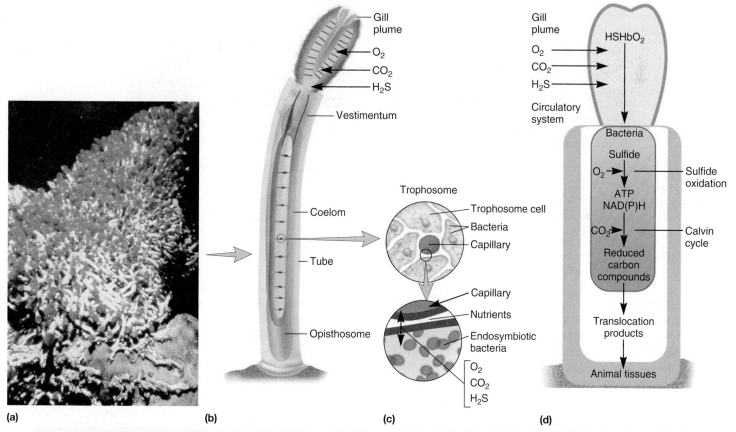

Figure 32.5 The Tube Worm–Bacterial Relationship. (a) A community of tube worms (*Riftia pachyptila*) at the Galápagos Rift hydrothermal vent site (depth 2,550 m). Each worm is more than a meter in length and has a 20-cm gill plume. (b, c) Schematic illustration of the anatomical and physiological organization of the tube worm. The animal's heart and brain is housed within the vestimentum. At its anterior end is a respiratory gill plume. Inside the trunk of the worm is a trophosome consisting primarily of endosymbiotic bacteria, associated cells, and blood vessels. At the posterior end of the animal is the opisthosome, which anchors the worm in its tube. (d) Oxygen, carbon dioxide, and hydrogen sulfide are absorbed through the gill plume and transported to the blood cells of the trophosome. Hydrogen sulfide is bound to the worm's hemoglobin (HSHbO$_2$) and carried to the endosymbiont bacteria. The bacteria oxidize the hydrogen sulfide and use some of the released energy to fix CO_2 in the Calvin cycle. Some of the reduced carbon compounds synthesized by the endosymbiont are translocated to the animal's tissues.

MICRO INQUIRY *What is the nutritional type of the endosymbiotic bacteria?*

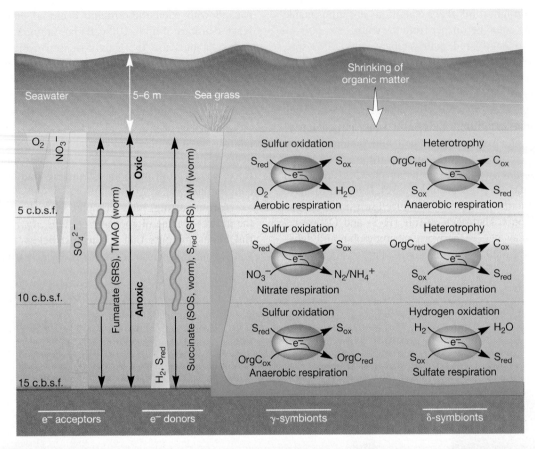

Figure 32.6 **Proposed Energy Metabolism of Bacterial Endosymbionts of the Mediterranean Worm *Olavius algarvensis.*** Electron donors and acceptors change with increasing sediment depth, as measured in centimeters below seafloor (c.b.s.f.). As the worm moves vertically through the sediments, the δ- and γ-proteobacterial endosymbionts switch their metabolism to match the available electron donors and acceptors. OrgC, organic carbon; S_{ox}, oxidized sulfur compounds; S_{red}, reduced sulfur compounds; SOS, sulfur-oxidizing symbionts; SRS, sulfate-reducing symbionts; TMAO, trimethylamine *N*-oxide—an organic electron acceptor produced by the host.

MICRO INQUIRY *Why do the δ-proteobacterial symbionts switch to H_2 as an electron donor only in the deepest sediments inhabited by O. algarvensis?*

lack any digestive tract so they obtain all organic carbon from their endosymbionts. The endosymbiont community is dominated by two δ-proteobacterial and two γ-proteobacterial species. The δ-proteobacteria are metabolically flexible. They use sulfate as their terminal electron acceptor and are chemoorganotrophic, with genes for the uptake and oxidation of a variety of carbohydrates. In addition, they can also function chemolithoautotrophically using H_2 as an electron and energy source and fixing CO_2 using both the acetyl-CoA pathway and reductive TCA cycle *(see figures 20.3 and 12.5, respectively)*. The γ-proteobacteria are chemolithoautotrophs, using reduced sulfur compounds as electron donors and O_2, nitrate, or organic substrates such as fumarate as electron acceptors. ◄◄ *Anaerobic respiration uses the same three steps as aerobic respiration (section 11.7); Chemolithotrophy (section 11.10); Other CO_2-fixation pathways (section 12.3)*

Why does *O. algarvensis* have four endosymbiotic species? Harboring symbionts is energetically expensive, so it can be argued that all four are needed for the worm to survive. Like all mutualisms, the role of each symbiont reflects the environment the host inhabits. In this case, as the worm migrates vertically, three different microhabitats are encountered: oxic, anoxic with nitrate and sulfate available, and anoxic with H_2 and reduced sulfur present **(figure 32.6)**. This requires that the bacterial endosymbionts adjust their metabolism to fit the prevailing electron acceptors and donors. So at the oxic surface, the γ-proteobacteria use O_2 as the terminal electron acceptor and the δ-proteobacteria rely on heterotrophy. Because there is little sulfur at the surface, the bacteria rely on each other to provide reduced sulfur as an electron donor and oxidized sulfur as electron acceptor. In the intermediate zone, the γ-proteobacteria use nitrate as the terminal electron acceptor and may rely on a combination of externally and internally produced reduced sulfur as electron donor. Meanwhile the δ-proteobacteria find plenty of external sulfate to respire. Finally, in the deepest zone, the γ-proteobacteria have access to lots of sulfur in the sediments but must use organics as the terminal electron acceptor. Meanwhile, the δ-proteobacteria may switch to chemolithotrophy and use H_2 as the electron donor and the oxidized sulfur generated by the γ-proteobacteria as the electron acceptor. This complex interplay between bacteria, host, and environment has resulted from millions of years of coevolution.

The Rumen Ecosystem

Ruminants are the most successful and diverse group of mammals on Earth today. Examples include cattle, deer, elk, bison, water buffalo, camels, sheep, goats, giraffes, and caribou. These animals spend vast amounts of time chewing their cud—a small ball of partially digested grasses that the animal has consumed but not yet completely digested. It is thought that ruminants evolved an "eat now, digest later" strategy because their grazing can often be interrupted by predator attacks.

These herbivorous animals have stomachs that are divided into four chambers (**figure 32.7**). The upper part of the stomach is expanded to form a large pouch called the rumen and a smaller, honeycomb-like region, the reticulum. The lower portion is divided into an antechamber, the omasum, followed by the "true" stomach, the abomasum. The rumen is a highly muscular, anaerobic fermentation chamber where grasses eaten by the animal are digested by a diverse microbial community that includes bacteria, archaea, fungi, and protists. This microbial community is large—about 10^{12} organisms per milliliter of digestive fluid. When the animal eats plant material, it is mixed with saliva and swallowed without chewing to enter the rumen. Here, microbial attack and further mixing coats the grass with microbes, reducing it to a pulpy, partially digested mass. At this point, the mass moves into the reticulum, where it is regurgitated as cud, chewed, and reswallowed by the animal. As this process proceeds, the grass becomes progressively more liquefied and flows out of the rumen into the omasum and then the abomasum. Here the nutrient-enriched grass meets the animal's digestive enzymes, and soluble organic and fatty acids are absorbed into the animal's bloodstream.

The microbial community in the rumen is extremely dynamic and, as a whole, forms a mutualistic relationship with the animal host. The rumen is slightly warmer than the rest of the animal, and with a redox potential of about −30 mV, all resident microorganisms must carry out anaerobic metabolism. Very specific interactions occur between microbes. One group of bacteria produces extracellular cellulases that cleave the β(1→4) linkages between successive D-glucose molecules that form plant cellulose. The D-glucose is then fermented to organic acids such as acetate, butyrate, and propionate. These organic acids, as well as fatty acids, are the true carbon and energy sources for the animal. ◀◀ *Fermentation does not involve an electron transport chain (section 11.8)*

In some ruminants, organic matter processing stops with the production of short-chain fatty acids. In others, such as cattle, CO_2, H_2, and to a lesser extent acetate are used by methanogenic archaea to generate methane (CH_4). Methanogens provide two services: they synthesize many of the vitamins needed by their animal host, and they efficiently remove H_2. This is important because the microbial oxidation of a wide range of organic compounds is thermodynamically feasible within rumen and many other anoxic environments only if H_2 is kept at very low levels. The oxidation of these organic substrates, which results in the generation of H_2 quickly consumed by methanogens, is termed **syntrophy** (Greek *syn*, together, and *trophe*, nourishment). Syntrophy is an association in which the growth of one organism either depends on, or is improved by, growth factors, nutrients, or substrates provided by another organism growing nearby.

In addition to the rumen ecosystem, syntrophic relationships occur in anoxic methanogenic ecosystems such as sludge digesters, anoxic freshwater sediments, and flooded soils. Obligate proton-reducing acetogens live in these environments. These microbes oxidize ethanol, butyrate, propionate, and other fermentation end products to H_2, CO_2, and acetate (hence the name "acetogen"). Although the enzymology is not well understood, it is clear that these reactions are thermodynamically unfavorable under standard conditions; that is, they have a positive ΔG. However, because methanogens consume the H_2 as it is generated, the reaction becomes thermodynamically possible. This type of syntrophy is called **interspecies hydrogen transfer.** Thus when fermentative bacteria produce low molecular weight fatty acids (e.g., propionate), anaerobic bacteria such as δ-proteobacteria in the genus *Syntrophobacter* produce H_2 as follows:

$$\text{Propionate}^- + 3H_2O \rightarrow \text{acetate}^- + HCO_3^- + H^+ + 3H_2$$
$$(\Delta G^\circ = +76.1 \text{ kJ/mol})$$

Syntrophobacter spp. use protons ($2H^+ + 2e^- \rightarrow H_2$) as terminal electron acceptors. The products H_2 and CO_2 (HCO_3^-) are then used by methanogenic archaea such as those in the genus *Methanospirillum*:

$$4H_2 + CO_2 \rightarrow CH_4 + 2H_2O \ (\Delta G^\circ = -25.6 \text{ kJ/mol})$$

By synthesizing methane, *Methanospirillum* spp. consume H_2, keeping it at a low concentration in the immediate environment of both microbes. As a consequence, the reaction becomes exergonic, with a ΔG between −20 to −25 kJ/mol (compare to +76.1 kJ/mol under standard H_2 concentrations). Continuous removal of H_2 promotes further fatty acid fermentation and H_2 production. Because increased H_2 production and consumption stimulate the growth of *Syntrophobacter* and *Methanospirillum* spp., all participants in the relationship benefit. The syntrophic nature of this relationship is further illustrated by the observation that the propionate-oxidizing bacterium *Pelotomaculum thermopropionicum* tethers itself to the methanogen *Methanothermobacter*

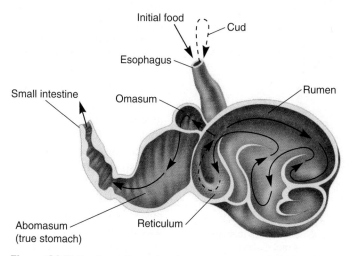

Figure 32.7 Ruminant Stomach.   The stomach compartments of a cow. The microorganisms are active mainly in the rumen. Arrows indicate direction of food movement.

thermautotrophicus. Remarkably, the flagellar tip protein FliD serves as an interspecies signal to increase the rate of methanogenesis by *M. thermautotrophicus*. ◀◀ *Methanogens and methanotrophs (section 20.4)*

Cooperation: Nonobligatory Positive Interactions Between Host and Microbe

For most microbial ecologists, the nonobligatory aspect between host and symbiont differentiates **cooperation** from mutualism (figure 32.1). Unfortunately, it is often difficult to distinguish obligatory from nonobligatory because that which is obligatory in one habitat may not be in another (e.g., the laboratory). Nonetheless, the most useful distinction between cooperation and mutualism is the observation that cooperating organisms can grow independently, although they may not function as well.

So far all our examples have featured symbionts that promote the growth of hosts in exchange for a safe, nutrient-rich home. The example of the Gram-negative bacterium *Xenorhabdus nematophila* and its nematode (worm) host *Steinernema carpocapsae* is remarkable because the bacterium contributes directly to the reproductive success of its host. Juvenile nematodes harboring *X. nematophila* within their guts live in the soil. In order to mature, the juvenile nematode must find an insect to infect and consume; this is when things get really interesting (**figure 32.8**). As the hungry juvenile nematode consumes insect blood (haemolymph), *X. nematophila* is excreted in nematode feces. These are now free-living *X. nematophila* and as these bacteria replicate, they secrete chemicals that kill the already miserable insect. Once the insect is dead, the bacteria switch to the production of different compounds that protect the insect from degradation by other bacteria and from attack by ants, thereby protecting the home of their host nematode. Amazingly, as the process continues, *X. nematophila* produces yet a different set of molecular signals that trigger *S. carpocapsae* development to adulthood. A single insect cadaver may host many adult nematodes that mate, yielding a cadaver that ultimately teems with *S. carpocapsae* eggs. As these eggs mature to the juvenile stage with *X. nematophila* symbionts, they leave to find new insect homes and start the cycle anew. Because the partners can be grown separately in the laboratory, it has been possible to dissect these events at the structural, molecular, and biochemical level, making this an important model system.

Certainly the most provocative cooperative associations to receive recent attention are those between the human host and our gut microbiota. Although the typical human gut hosts 500 to 1,000 different microbial species (most of which have not been cultured), the Gram-negative bacterium *Akkermansia muciniphila* is particularly interesting. This microorganism is critical to maintaining the mucosal barrier that separates microbes within the gut and the underlying, sterile structures. It accomplishes this in several ways, including the stimulation of host mucus and antimicrobial peptide production and promoting tight junction formation between epithelial cells. This is important because recent evidence suggests that the lipopolysaccharide from Gram-negative gut bacteria that enters the host's bloodstream contributes a state of sustained inflammation, called metabolic endotoxemia. This has been linked to conditions such as obesity, type 2 diabetes, and

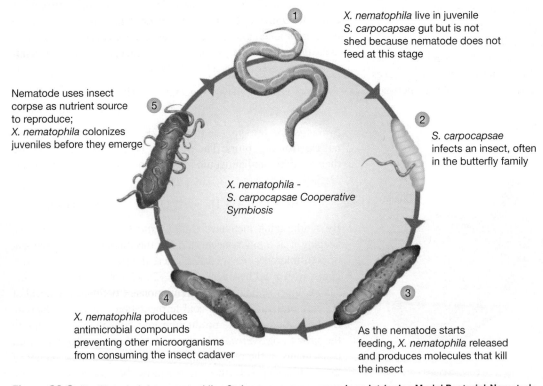

1 *X. nematophila* live in juvenile *S. carpocapsae* gut but is not shed because nematode does not feed at this stage

2 *S. carpocapsae* infects an insect, often in the butterfly family

3 As the nematode starts feeding, *X. nematophila* released and produces molecules that kill the insect

4 *X. nematophila* produces antimicrobial compounds preventing other microorganisms from consuming the insect cadaver

5 Nematode uses insect corpse as nutrient source to reproduce; *X. nematophila* colonizes juveniles before they emerge

X. nematophila - S. carpocapsae Cooperative Symbiosis

Figure 32.8 The *Xenorhabdus nematophila - Steinernema carpocapsae* Is an Intriguing Model Bacterial-Nematode System.

rheumatoid arthritis. ◄◄ *Phylum* Verrucomicrobia *includes human symbionts and methylotrophs (section 21.7)*

Retrieve, Infer, Apply

1. How does cooperation differ from mutualism? What might be some of the evolutionary implications of both types of symbioses?
2. Why is the *X. nematophila–S. carpocapsae* symbiosis considered a cooperation? What other types of microbial interactions are in play during the nematode life cycle?

Commensalism: Microbe Benefits Without Affecting Host

Commensalism (Latin *com,* together, and *mensa,* table) is a relationship in which one symbiont, the **commensal,** benefits, while the host is neither harmed nor helped, as shown in figure 32.1. This is a unidirectional process. The spatial proximity of the two partners permits the commensal to feed on substances captured or ingested by the host, and the commensal often obtains shelter by living either on or in the host. The commensal is not directly dependent on the host's metabolism, so when it is separated from the host experimentally, it can survive without the addition of factors of host origin.

Commensalistic associations also occur when one microbial group modifies the environment to make it better suited for another organism. For instance, the synthesis of acidic waste products during fermentation stimulates the proliferation of more acid-tolerant microorganisms, which may be only a minor part of the microbial community at neutral pH. A good example of this is the succession of microorganisms during milk spoilage (*see figure 41.3*). Biofilm formation provides another example. The colonization of a newly exposed surface by one type of microorganism (an initial colonizer) makes it possible for other microorganisms to attach to the microbially modified surface. This is a good example of the unidirectional interaction that defines commensalism. ◄◄ *Biofilms are common in nature (section 7.5)*

The microorganisms that live in and on humans, known as our microbiome, have traditionally been called commensals. Although this term is still applied broadly to these microbes, it is important to keep in mind that the more we learn about the vast microbial ecology of our bodies, the more we come to understand that we depend on our microbes for a myriad of important functions that we cannot achieve without them. These include the synthesis of vitamins, maturation of our immune systems, maintaining a healthy body weight, and prevention of and recovery from disease. The term commensals does not apply to our microbiome, and it is slowly being replaced by the more appropriate term, mutualists.

Predation: Microbes Harming Other Microbes

As is the case with larger organisms, predation among microbes involves a predator species that attacks and kills its prey. Over the last several decades, microbiologists have discovered a number of fascinating microbes that survive by their ability to prey upon others.

In addition to bdellovibrios, described in chapter 22 *(see figure 22.34),* members of the genera *Vampirococcus* and *Daptobacter* also kill their prey (**figure 32.9**). *Vampirococcus* cells

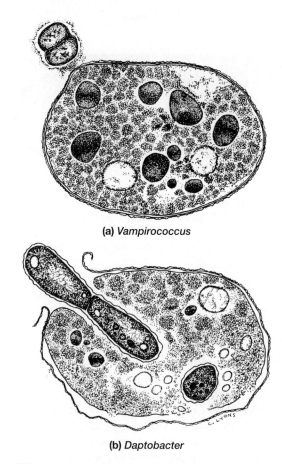

(a) *Vampirococcus*

(b) *Daptobacter*

Figure 32.9 Examples of Predatory Bacteria Found in Nature. (a) *Vampirococcus* has a unique epibiotic mode of attacking a prey bacterium; and (b) *Daptobacter* showing its cytoplasmic location as it attacks a susceptible bacterium.

attach to the outer membrane of prey and secrete degradative enzymes, which result in the release of the prey's cytoplasmic contents. In contrast, a *Daptobacter* cell penetrates the prey cell and consumes the cytoplasmic contents directly. ◄◄ *Order* Bdellovibrionales *includes predators that invade other Gram-negative bacteria (section 22.4)*

In contrast to the predatory bacteria shown in figure 32.9, some bacteria are facultative predators. Sometimes they consume organic matter released from dead organisms, whereas at other times, they actively prey on other microbes. Two good examples are members of the γ-proteobacterial genus *Lysobacter* and the δ-proteobacterial genus *Myxococcus*. Both display what has been called "wolf pack" predation whereby populations of cells use gliding motility to creep toward and cover their prey while releasing an arsenal of degradative enzymes. *M. xanthus* fruiting body formation can be triggered when its prey has been exhausted (**figure 32.10**).

Importantly, predation has many beneficial effects, especially when one considers predators and prey at the population and community levels. Ingestion and assimilation of a prey bacterium can lead to increased rates of nutrient cycling, critical for the functioning of the microbial loop (*see figure 30.6*). Although here we focus on bacterial predators, the impact of viral predation on communities, ecosystems, and even the global carbon budget is vast. As such, it is an active area of research. ◄◄ *Aquatic viruses (section 30.2)*

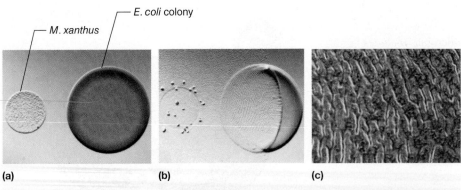

Figure 32.10 Wolf Pack Predation by *Myxococcus xanthus*. (a) In this demonstration, a drop of *M. xanthus cells* is placed to the left of a colony of prey bacteria (*Escherichia coli*). (b) The *E. coli* cells lyse as *M. xanthus* glides over and consumes them; thus the colony becomes clear. (c) As *M. xanthus* cells swarm over the prey bacteria they move in waves. They will go on to form fruiting bodies (not shown; *see figure 22.35*).

MICRO INQUIRY *How does myxobacterial predation differ from that of* Vampirococcus?

Parasitism: Microbes Harm the Host

Parasitism is one of the most complex microbial interactions; the line between parasitism and predation is difficult to define (figure 32.1). In both cases, the relationship between two organisms is beneficial for one and harmful for the other. However, **parasite** and host must coexist, at least temporarily, so the microbe has enough time to reproduce and colonize a new host. Coexistence may involve nutrient acquisition, physical maintenance in or on the host, or both. But what happens if the host-parasite equilibrium is upset? If the balance favors the host (perhaps by a strong immune defense or antimicrobial therapy), the parasite loses its habitat and may be unable to survive. On the other hand, if the equilibrium is shifted to favor the parasite, the host becomes ill and, depending on the specific host-parasite relationship, may die.

On the other hand, a controlled parasite-host relationship can be maintained for long periods of time and the parasitic nature of the relationship may not be obvious. For example, lichens (**figure 32.11**) are the association between specific ascomycetes (fungi) and either green algae or cyanobacteria. The fungal partner is termed the **mycobiont** and the algal or cyanobacterial partner, the **phycobiont.**

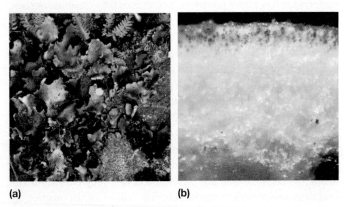

Figure 32.11 Lichens. (a) The flattened leaf-shaped genus *Umbilicaria* is a common foliose lichen. (b) A thin slice of *Umbilicaria* viewed with a compound microscope shows that the green algae occur in a thin upper layer and fungal hyphae make up the rest of the lichen.

The fungus obtains nutrients from its partner by projections of fungal hyphae called haustoria, which penetrate the phycobiont cell wall. It also uses O_2 produced by the phycobiont for respiration. In turn, the fungus protects the phycobiont from high light intensities, provides water and minerals, and creates a firm substratum within which the phycobiont can grow protected from environmental stress. This is an ancient association, having developed prior to the evolution of vascular plants. Based on the above description, one could easily conclude that the relationship is mutualistic. However, those lichens where the cyanobacteria and algae phycobionts grow more quickly when cultured without their mycobiont are considered to be in a relationship in which the mycobiont is the parasite. This is in contrast to positive interactions such as cooperation, in which the symbiont often grows more slowly without its host. ◄◄ Ascomycota *includes yeasts and molds (section 26.5)*; *Phylum* Cyanobacteria *(section 21.4)*

An important aspect of many symbiotic relationships, including parasitism, is that over time, the symbiont will discard excess, unused genomic information, a process called **genomic reduction** (*see figure 18.20*). This occurs when the symbiont has become dependent on the host for specific functions, such as synthesis of key metabolites. It is observed in the aphid endosymbiont *Buchnera aphidicola*; it has also occurred in the human pathogens *Mycobacterium leprae, Mycoplasma genitalia,* and *Encephalitozoon cuniculi,* a microsporidium. These microorganisms now can only survive inside their host cells. ◄◄ *Comparative genomics (section 18.7)*; *Order* Corynebacteriales *includes important human pathogens (section 24.1)*; *Microsporidia are intracellular parasites (section 26.7)*

Retrieve, Infer, Apply

1. How does commensalism differ from cooperation?
2. Why is nitrification a good example of a commensalistic process?
3. Define predation and parasitism. How are these similar and different?
4. Based on your own personal experience, what human diseases would you consider "successful" parasite-host relationships (i.e., long in duration)?
5. What is a lichen? Why is this considered an example of a parasite-host relationship? Why might you also consider it a mutualistic interaction?

Amensalism: Microbial Products Harm Other Microbes

Amensalism describes the adverse effect that one organism has on another organism (figure 32.1). This is a unidirectional process based on the release of a specific compound by one organism that has a negative effect on another. A classic example of amensalism is the microbial production of antibiotics that can inhibit or kill other, susceptible microorganisms (**figure 32.12a**).

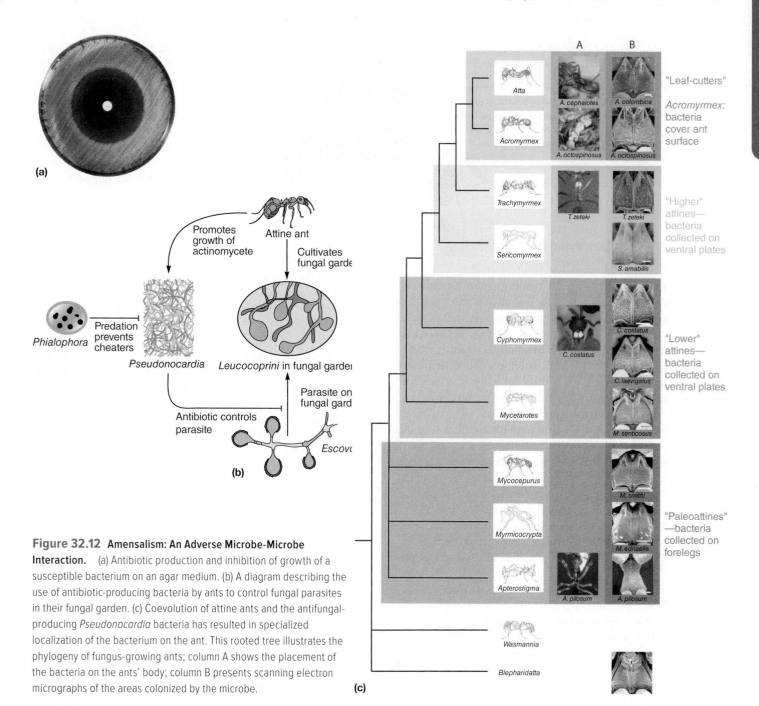

Figure 32.12 Amensalism: An Adverse Microbe-Microbe Interaction. (a) Antibiotic production and inhibition of growth of a susceptible bacterium on an agar medium. (b) A diagram describing the use of antibiotic-producing bacteria by ants to control fungal parasites in their fungal garden. (c) Coevolution of attine ants and the antifungal-producing *Pseudonocardia* bacteria has resulted in specialized localization of the bacterium on the ant. This rooted tree illustrates the phylogeny of fungus-growing ants; column A shows the placement of the bacteria on the ants' body; column B presents scanning electron micrographs of the areas colonized by the microbe.

Amensalisms can be quite complex. Attine ants (ants belonging to a New World tribe) are perhaps the most fascinating example. These ants cultivate for their own nourishment a garden of fungi belonging to the genus *Leucocoprini* (figure 32.12*b*). Their ability to accomplish this advanced task depends upon actinobacteria in the genus *Pseudonocardia*. These bacteria produce inhibitory compounds that prevent the growth of *Escovopsis* spp., a predatory fungus that would otherwise destroy the "crop" *Leucocoprini* fungi.

These unique interactions appear to have evolved 50 million to 65 million years ago in South America. Millions of years of coevolution have resulted in particular groups of ants that cultivate specific strains of *Leucocoprini* fungi, which are then subject to different groups of *Escovopsis* parasites. To promote the amensalistic interaction between *Pseudonocardia* and *Escovopsis*, the ants

have developed intricate crypts within their exoskeletons for the growth of the antifungal-producing *Pseudonocardia*. As shown in figure 32.12*c*, these crypts have been modified throughout the ants' evolutionary history. The most primitive "paleoattine" ants carry the bacterium on their forelegs, "lower" and "higher" attines have evolved special plates on their ventral surfaces, while the entire surface of the most recent attines, leaf-cutter ants of the genus *Acromyrmex*, are covered with the bacterium. Related ants that do not cultivate fungal gardens (e.g., *Atta* spp.) do not host *Pseudonocardia* cells.

Surprisingly, this four-membered symbiosis is even more complicated, as a fifth member of the symbiosis is found in some cases. This member is a black ascomycete yeast of the genus *Phialophora*. It grows in association with the *Pseudonocardia* symbiont, on which it also preys. But why would attine ants evolve the capacity to tolerate a fungus that eats their antifungal-producing bacteria? The answer has to do with cheating. Briefly, it is energetically expensive for *Pseudonocardia* cells to produce an antimicrobial compound, but its production ensures a home on the ant. However, some members of the *Pseudonocardia* population could get a "free ride" by letting their neighbors produce the antifungal agent. It is thought that only antifungal-producing bacteria can keep the *Phialophora* fungus in check; this prevents cheaters from thriving.

Competition: Microbes Vying for Common Goods

Competition arises when different organisms within a population or community try to acquire the same resource, whether this is a physical location or a particular limiting nutrient (figure 32.1). If one of the two competing organisms can dominate the environment, whether by occupying the physical habitat or by consuming a limiting nutrient, it will overtake the other organism. This phenomenon was studied by E. F. Gause, who in 1934 described it as the **competitive exclusion principle.** In chemostats, competition for a limiting nutrient may occur among microorganisms with transport systems of differing affinity. This can lead to the exclusion of the slower-growing population under a particular set of conditions. If the dilution rate in the chemostat is changed, the previously slower-growing population may become dominant. Often two microbial populations that appear to be similar nevertheless coexist. In this case, they share the limiting resource (space, a nutrient) and coexist while surviving at lower population levels. Competitive exclusion most likely accounts, at least in part, for the phenomenon of colonization resistance—the capacity of gut microflora to prevent pathogen growth. ◄◄ *Chemostats (section 7.9)*

Retrieve, Infer, Apply

1. Describe the amensalism observed in attine ant communities. What other types of interactions are present in these communities?

2. The production of antimicrobial agents and their effects on target species have been called a "microbial arms race." Explain what you think this phrase means and its implications for the attine ant system.

3. What is the competitive exclusion principle? List another example where this principle is demonstrated in the natural world.

32.2 The Human-Microbe Ecosystem

After reading this section, you should be able to:

- Explain the role of germfree mice in evaluating how microorganisms regulate health and disease
- Summarize the importance of the human microbiome in maintaining homeostasis
- Infer how data from the Human Microbiome Project might contribute to the development of new disease treatments

The relationship people have with the microbes that live on and in diverse human environments (e.g., skin and gut) is on the forefront of science. Progress in metagenomics and immunology has altered the long-held belief that human-microbe interactions are commensalistic. We are beginning to understand the many ways in which our microbes keep us healthy. After all, the average adult carries about three times more microbial cells (10^{14}) than human somatic cells (average about 4×10^{13}). With a combined microbial contribution of over 8 million protein-encoding genes, we have come to realize that many of these gene products serve valuable functions our genome is not capable of providing.

From the early days of microbiology until the beginning of the twenty-first century, microbiologists relied on the culture of microorganisms for their identification and analysis. For example, the ability to isolate and grow microbes is key to determining the causative agent of infectious disease using Koch's postulates. The observation that the vast majority of microbes associated with healthy humans cannot be grown in pure culture demonstrates that these microorganisms are not the commensals we once thought them to be. The role of metagenomics has been paramount in determining the taxa present; studies involving proteomics, metabolomics, and other omics are used to explore what (and how) microbes affect our health. ◄◄ *Koch's postulates (section 1.3); Metagenomics provides access to uncultured microbes (section 18.3).*

Although our understanding of the microbial contribution to human physiology will take years to tease apart, most studies point to the ecological principle of "niche fulfillment" to define the "who and what" of our microbial partners. This means that microbes colonize specific regions of our bodies in response to the physical and biological parameters found there, and that over time, their activities help shape the structure and function of that environment. However, microbial residency is dynamic, and our microbes must negotiate structural, hormonal, and immunological challenges as humans grow and change. Simply put, we could not live without our microbial flora.

Microorganisms commonly associated with the human body are known as the **normal microbiota** (more traditionally, the "normal flora"). The acquisition of a normal microbiota is a selective process, where a niche may depend on cellular receptors, surface properties, or secreted products. In humans, each microbial niche is also related to age, gender, diet, nutrition, and environment. For instance, many microorganisms revealed by 16S ribosomal RNA sequencing (e.g., members of *Clostridiales* and *Bacteroidales*) can survive only within the anoxic and nutritionally unique environments established in or on the human host, thus preventing their culture in the lab. Furthermore, some organisms long thought to be normal flora turn out to be transient associates or laboratory contaminants.

One method for determining the impact of the normal microbiota is to evaluate the consequences of their absence in and on other organisms. This is accomplished using **germfree** animals, born by cesarean delivery and reared in sterile environments. These animals (usually mice) provide suitable experimental models for investigating the interactions between the vertebrate host and its microbiota. Comparing animals possessing normal microbiota (conventional animals) with germfree animals allows scientists to explore many of the complex relationships between microorganisms, hosts, and specific environments. Germfree experiments also extend and challenge the microbiologist's "pure culture concept" to in vivo research. Data from numerous germfree studies demonstrate that colonization of the vertebrate host intestine with bacteria stimulates the host immune system, assists in digestion of complex dietary components, and triggers anatomical changes of the intestine. Also, association of specific microorganisms with germfree host skin helps explain their role in colonization strategies and some disease states. Overall, association of microorganisms with germfree animals has helped microbiologists understand complex interactions that cannot be readily studied in conventional hosts, or interpreted just from "-omics" data. Together with genomic studies, data from germfree studies have helped to redefine "normal" in terms of microbiota composition and consequence of microbial association.

In December 2007 the National Institutes of Health launched the two-phase Human Microbiome Project (HMP) to better understand the symbiotic relationships that result in human health and disease. The term **microbiome** was coined to define all the microbes associated with the healthy host. Phase 1 of the HMP (FY2007–2012) characterized the composition and diversity of human microbial communities with specific emphasis on identification of the metabolic potential of taxa found in specific body sites. Phase 2 of the HMP (FY2013–2015) employed cohort studies to establish the relationship between the microbiomes of healthy individuals and those of individuals with a disease.

As of April 2014, whole metagenomic sequencing data from ~800 healthy human cohort samples, 16S sequence data from ~5,000 healthy human cohort samples, and associated metadata from these reads have been released. All told, the total amount of HMP data available has surpassed 14 terabytes. Of note are the results of the many HMP projects indicating that there is no single healthy microbiome. Rather, each person harbors a unique and varied collection of microorganisms that assemble over a person's interactive history with their environment, diet, medications, and many other factors. Furthermore, identification of microbial "community types" that are associated with disease risk are clearly emerging. The ongoing task is the association of community types with the myriad of demographic factors (gender, education, being breast-fed, lifestyle attributes, etc.) to identify relationships that will be useful in predicting health and disease, and ultimately to develop therapies (like pre- and probiotics) that would alter community types to obtain healthy outcomes. Importantly, evaluation of human and microbial proteins and metabolites should answer many questions regarding the biochemical processes regulated by the microbial-human symbioses, the diseases resulting from dysfunctional human-microorganism relationships (dysbiosis), and potential biological and therapeutic approaches to treat human disease. ▶▶| *Probiotics (section 41.6)*

Several microbiome-associated health and disease correlations are known. For example, the microbiome (1) assists in the digestion of food and other ingested products, (2) stimulates our immune system to mature and then regulates it, (3) protects against microorganisms that cause disease, and (4) produces the essential vitamins B12, thiamine, and riboflavin, as well as vitamin K. The microbiome also metabolizes bile acids and sterols. Other associations are interesting but require further study. For example

- Type 1 diabetes is an autoimmune disease associated with a less diverse gut microbiome. Bacteria play a role in developing diabetes, in animal studies.
- The presence of a relatively high number of gut firmicutes is strongly correlated with obesity and type 2 diabetes (**Microbial Diversity & Ecology 32.2**).
- Obesity has also been associated with poor microbial diversity in the intestine. Obese twins have a lower diversity of bacteria as compared with lean twins.
- Intestinal disease caused by *Clostridium difficile* infections has been cured using fecal transplants to restore healthy intestinal microbiota.
- Autoimmune disease appears to be passed in families not just by genetic factors but by inheritance of the family microbiome.
- Autoimmune diseases such muscular dystrophy and rheumatoid arthritis are associated with dysfunction of the microbiome.

Clearly our survival depends on an elaborate network integrating microbial gene expression with that of our own. This network extends beyond "metabolism sharing" to the more utilitarian as microbes defend their human ecosystem. This host-microbial interplay is essential in keeping harmful microorganisms and other foreign material from entering the body. Should microorganisms gain access to internal tissues, additional host defenses are summoned to prevent them from establishing another type of relationship, one of parasitism or pathogenicity. **Pathogenicity** (Greek *pathos,* emotion or suffering, and *gennan,* to produce) is the ability to produce pathologic changes or disease. A **pathogen** is any disease-producing microorganism. Although the normal microbiota offer some protection from invading microorganisms, its members may themselves become pathogenic and produce disease under certain circumstances (dysbiosis); they then are termed **opportunistic microorganisms** or **pathogens.** Opportunistic pathogens are adapted to the noninvasive mode of life defined by the environment in which they live. If they are removed from their environment and introduced into different tissues or the bloodstream, disease can result. For example, some streptococci are the most common resident bacteria in the mouth and oropharynx. But if they are introduced into the bloodstream in large numbers (e.g., following tooth extraction), they may settle on deformed or prosthetic heart valves and cause endocarditis. ▶▶| *Pathogenicity and infection (chapter 35)*

Opportunistic microorganisms often cause disease in compromised hosts. A **compromised host** is seriously debilitated and has a lowered resistance to infection. There are many causes of this condition, including malnutrition, alcoholism, illness, trauma from surgery or an injury, an altered normal microbiota

MICROBIAL DIVERSITY & ECOLOGY

32.2 Do Bacteria Make People Fat?

The prevalence of human obesity has increased steadily since 1960. In fact, in the United States, the incidence of obesity increased from 13% to 34% between 1960 and 2006. According to the National Center for Health Statistics, approximately one-third of adults in the United States are obese. As a result, the U.S. Public Health Service has labeled obesity an epidemic. While a sedentary lifestyle coupled with poor nutritional choices probably drives the prevalence of obesity, new evidence suggests that obesity is also related to the bacteria in your gut.

A number of studies show that the relative amounts of two bacterial phyla can determine body weight. Through the use of metagenomic techniques examining intestinal bacteria, higher relative concentrations of members of *Firmicutes* (low G + C Gram-positive bacteria) to *Bacteroidetes* (Gram-negative bacteria) are correlated with obesity. Conversely, lean people have many fewer firmicutes in their intestines as compared to bacteroidetes. Interestingly, obese people who lost at least 25 pounds over 1 year by eating diets low in fat or carbohydrates altered their gut microbiota and subsequently had bacterial profiles that more closely resembled those of lean people.

Critics of this research suggest that this observation is not causal, just coincidence. To address this, investigators used germfree mice (p. 698) as test subjects. They discovered that such mice are inherently resistant to weight gain despite a high fat content in their diet. In fact, in an 8-week study, germfree mice consuming a diet containing 40% fat gained 50% less weight than conventional (having their normal microbiota) mice fed the same amount and type of food. However, when investigators transplanted intestinal bacteria from obese mice into germfree mice, the germfree mice gained more weight than germfree mice controls that received bacteria from lean mice. Thus it appears that unique bacterial profiles occur in obese mice and that specific bacteria are highly correlated with obesity. But what could be the mechanism by which intestinal bacteria drive obesity?

In evaluating the microbiome, investigators have determined that the bacteria from obese mice produced more soluble degradation products from catabolizing complex carbohydrates. Firmicutes thereby release more energy from dietary nutrients than do bacteroidetes. Furthermore, the firmicutes were able to trigger the expression of host genes that slowed fat catabolism and increased fat accrual in various host tissues. It is thus not surprising that their catabolic activity has also been associated with the development of the insulin resistance that characterizes type 2 diabetes.

While these recent data suggest a simplistic view of weight gain, decades of other data support the absolute requirement for aerobic exercise and proper nutrition as part of achieving and maintaining a healthy weight. It seems that our role is to create a selective medium for the right kind of bacterial growth. Remember, a high-fat, high-carbohydrate environment favors firmicute overgrowth, while a low-fat, low-carbohydrate environment favors bacteroidetes. Which bacteria do you want to cultivate?

---

from the prolonged use of antibiotics, and immunosuppression by various factors (e.g., drugs, viruses [HIV], hormones, and genetic deficiencies). For example, *Bacteroides* spp. are one of the most common residents in the large intestine (table 32.2) and are quite harmless in that location. If introduced into the peritoneal cavity or pelvic tissues as a result of trauma, they cause suppuration (the formation of pus) and bacteremia (the presence of bacteria in the blood). Importantly, the normal microbiota is harmless and is often beneficial in its normal location in the host and in the absence of coincident abnormalities. However, these microbes can produce disease if introduced into foreign locations or compromised hosts. Chapters 38–40 discuss the consequences of human interaction with pathogens. We now describe the normal human microbiota that function as mutualistic and commensal symbionts.

Retrieve, Infer, Apply

1. How might a microbial symbiont affect the uptake and metabolism of amino acids needed by its host?

2. Cite three reasons a germfree mouse would be more susceptible to pathogens when compared to a normal mouse.
3. People who suffer from colitis often find that their symptoms improve if they increase the amount of fiber in their diet. Why might this be the case?

32.3 Normal Microbiota of the Human Body Adapt to Many Sites

After reading this section, you should be able to:

- Report the common culturable organisms associated with various body sites
- Predict the environmental host conditions that favor the association of acidophilic, anaerobic, lactic acid fermenting, and halophilic bacteria
- Discuss the mechanism by which the normal microbiota mediates resistance to pathogens

Development of the Normal Microbiota

The human body provides numerous habitats that differ in terms of the environmental conditions they provide. These conditions dictate the microbes that reside at each site. Importantly, the normal microbiota is not static. It begins developing at birth and changes as the human ages. In this section, we begin by describing this development and then discuss the microbiota observed at various body sites of adult humans.

Newborns are initially colonized by microorganisms from the immediate environment. For example, babies born vaginally acquire most of their microorganisms from their mother, whereas babies born by cesarean delivery acquire the microorganisms of their initial caretakers (nurses, doctors, midwives, parents, etc.).

The early colonization of the infant intestinal tract by *E. coli* and streptococci appears to establish a reducing environment for the growth of strict anaerobes such as bifidobacteria and bacteriodetes. Data also suggest that human milk may act as a selective medium for nonpathogenic bacteria, as bottle-fed babies appear to have a much smaller proportion of intestinal bifidobacteria. Furthermore, bifidobacteria appear to be able to synthesize all amino acids and other required growth factors from simple nutrients such as glucose and inorganic sources of nitrogen, sulfur, and phosphorus. Thus bifidobacteria are well adapted for growth in the colonic environment, which has very low concentrations of these nutrients. Bifidobacteria and bacteroidetes have surface proteins that bind sugars, called glycans, present in the diet as well as produced by the host (**figure 32.13**). In the case of bifidobacteria, operons encoding enzymes for the catabolism of starch, pullulan, and amylopectin (necessary for weaning from milk to complex carbohydrates) are also found. Switching to cow's milk or solid food (mostly polysaccharide) appears to result in the loss of bifidobacteria predominance; proteobacteria, firmicutes, and bacteroidetes—specifically enterobacteria, enterococci, lactobacilli, clostridia, and *Bacteroides* spp.—increase and outcompete actinobacteria in a polysaccharide-rich environment.

The initial microbiota changes with its host; age and gender, along with their associated physiological and anatomical differences, substantially contribute to defining the evolving microbiota until a stable resident (normal) microbiota becomes established (**table 32.1**). Each individual cultivates his or her own microbial population that is heavily influenced by genetics, developmental change (e.g., immune system maturity, puberty, and menopause), diet (high protein versus high fat versus high carbohydrate content), personal hygiene (frequency of washing, type of soap, use of cosmetics, use of occlusive bandages, clothing choice, etc.), anatomical site, and "life events" (antimicrobial use, travel, occupation, sexual partners, etc.).

Interestingly, once established, the adult microbiota permits only transient changes in diversity. These changes only occur with substantial host environmental change (like going through puberty or adopting a long-term vegetarian diet) so as to support the new resident taxa. Supporting this observation is the effect that infectious disease and its treatment have on an individual's microbiota. Infection and antimicrobial chemotherapy result in short-term flora shifts (growth of pathogens causing disease and loss of taxa sensitive to the antimicrobial agent) but may not alter the long-term composition of the adult microbiota as long as the drugs eradicate the pathogen and are discontinued. However, long-term shifts in the microbiota can occur with sustained use of antimicrobial agents, as seen with treatment of chronic infectious disease. Thus, most infectious agents are better characterized as transient microorganisms with which the host interacts.

While the adult human microbiota is relatively stable over time, it is highly variable from person to person and at different sites within the same person. In other words, there are relatively few species of microorganisms common to all humans; each human has a relatively unique microbiota. Based on 16S rRNA and whole genome metagenomic data, bacteria common to human skin, the intestinal tract, and the other mucosal surfaces appear to represent five phyla: *Actinobacteria, Bacteroidetes, Firmicutes, Fusobacteria*, and *Proteobacteria* (**figure 32.14**). Additionally, DNA from a number of archaea (euryarchaeotes), fungi, and viruses are consistently recovered. Despite limited diversity at this taxonomic level, each phylum is represented by an enormous number of species.

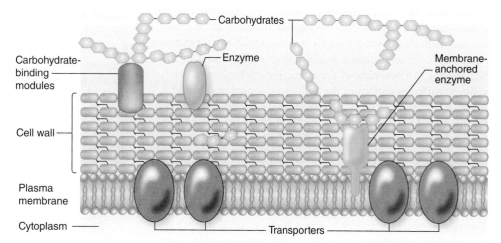

Figure 32.13 Sugar Acquisition Strategy of Bifidobacteria. Carbohydrates are bound to the bifidobacterial cell by use of an enzyme "docking station." Anchored to the cell surface, modular glycanases degrade complex carbohydrates to oligosaccharides, which are then transported across the plasma membrane. Bacteroidetes use a similar strategy to bind glycans to the outer membrane of their Gram-negative cell walls.

Table 32.1	Innate and Environmental Factors That Impact Microbial Diversity and Human Immunity			
Relative Age	**Innate Factors[1]**	**Environmental Factors**	**Microbial Diversity[2]**	**Immunity[3]**
Birth	Genetics	Multiple caregivers, breast milk/formula, immunizations	Mother's vaginal & intestinal flora, transient species from caregivers	Activation of innate systems
Infant	Developing immunity	Solid foods, crawling, mouth contact, first infections, first antibiotics	40–100 species, 7–8 phyla, transient fungi and virus	Induction of acquired systems
Child	Body growth	Changing diet, socialization/contact with peers, play, adventure	40–100 species, 7–8 phyla, rare fungi, rare virus	Challenges of both systems, secondary lymphoid structure development
Teen	Puberty, sexual activity	Travel, sports, fitness	Species change, 7–8 phyla, increasing fungi, increasing virus	Functional tuning with continued antigen exposure
Adult	Weight gain, pregnancy, metabolic disease diagnosis, stress	Stable cohabitation, job changes, shift work, relocations, children, medications	Stable population ~500–1,000 species, 7–8 phyla, 1–10% fungi, 10–40% virus	Colonization resistance, maintenance of lymphocyte reservoir
Senior Citizen	Muscle atrophy, declining immunity, system failures	Day care, diet change, menopause, declining fitness, loss of motility, incontinence, cardiopulmonary disorders, age-related infections, hospital, nursing care	Less stable population, 7–8 phyla, variable fungi, variable virus	Declining responses, increased permissiveness to pathogens

1 Each factor included in subsequent age groups.
2 Type and relative frequency of taxa.
3 Each event may include or yield to subsequent age-group observations.

Sources: Spor, A., O. Koren and R. Ley, 2011. Unravelling the effects of the environment and host genotype on the gut microbiome. Nature Rev Microbiol 9:279–290; and Belkaid, Y., and J. A. Segre, 2014. Dialog between skin microbiota and immunity. Science 346:954–959.

What We Know About Microbiota at Various Body Sites

In a healthy human, regardless of age, the internal tissues (e.g., brain, blood, cerebrospinal fluid, muscles) are normally free of microorganisms. Conversely, the surface tissues (e.g., skin and mucous membranes) are constantly in contact with environmental microorganisms and become readily colonized by various microbial species. An overview of the culturable microbiota typically native to different regions of the body is presented in **table 32.2**. Because bacterial and archaeal species make up most of the normal microbiota, they are emphasized over fungi, protists, viruses (including phages), and arthropods that are also part of the microbial assemblage and about which we are just beginning to learn.

Skin

The adult human is covered with approximately 1.8 m² of skin. The skin surface or epidermis has a slightly acidic pH, a high concentration of sodium chloride, a lack of moisture in many areas, and the oily lubricant sebum and antimicrobial peptides in other areas. It has been estimated that this surface area supports about 10^{10} bacteria dwelling in different microenvironments with distinct moisture, pH, sebum content, temperature, and topography. Microorganisms colonize discrete skin sites dictated by skin

physiology (figure 32.14). Those that are temporarily present are transient microorganisms. Transients usually do not become firmly entrenched and are typically unable to multiply.

The skin surface can be divided into three environmental niches: dry, moist, and sebaceous (containing sebum). In general, bacterial diversity is greatest at dry sites (forearm, buttocks, hands, for example), harboring a mixture of Gram-positive and Gram-negative organisms from the phyla *Actinobacteria*, *Bacteroidetes*, *Firmicutes*, and *Proteobacteria*. Moist areas (umbilicus, underarm, inguinal and gluteal creases, and inside the elbow, for example) exhibit less diversity, supporting mostly members of *Firmicutes* and *Actinobacteria* (*Staphylococcus* and *Corynebacterium* spp., respectively), in addition to low numbers of other bacteria. The skin surfaces with the lowest bacterial diversity are the oily sebaceous sites (forehead, behind the ear, and the back, for example), where members of *Actinobacteria* (*Propionibacterium* spp.) dominate (figure 32.14).

▶▶| *Antimicrobial peptides and proteins (section 33.3)*

Most skin bacteria are found on superficial cells, colonizing dead cells, or closely associated with the oil and sweat glands. Secretions from these glands provide the water, amino acids, urea, electrolytes, and specific fatty acids that serve as nutrients primarily for resident bacteria such as *S. epidermidis*. The oil glands secrete complex lipids that may be partially degraded by the enzymes from certain Gram-positive bacteria, in particular

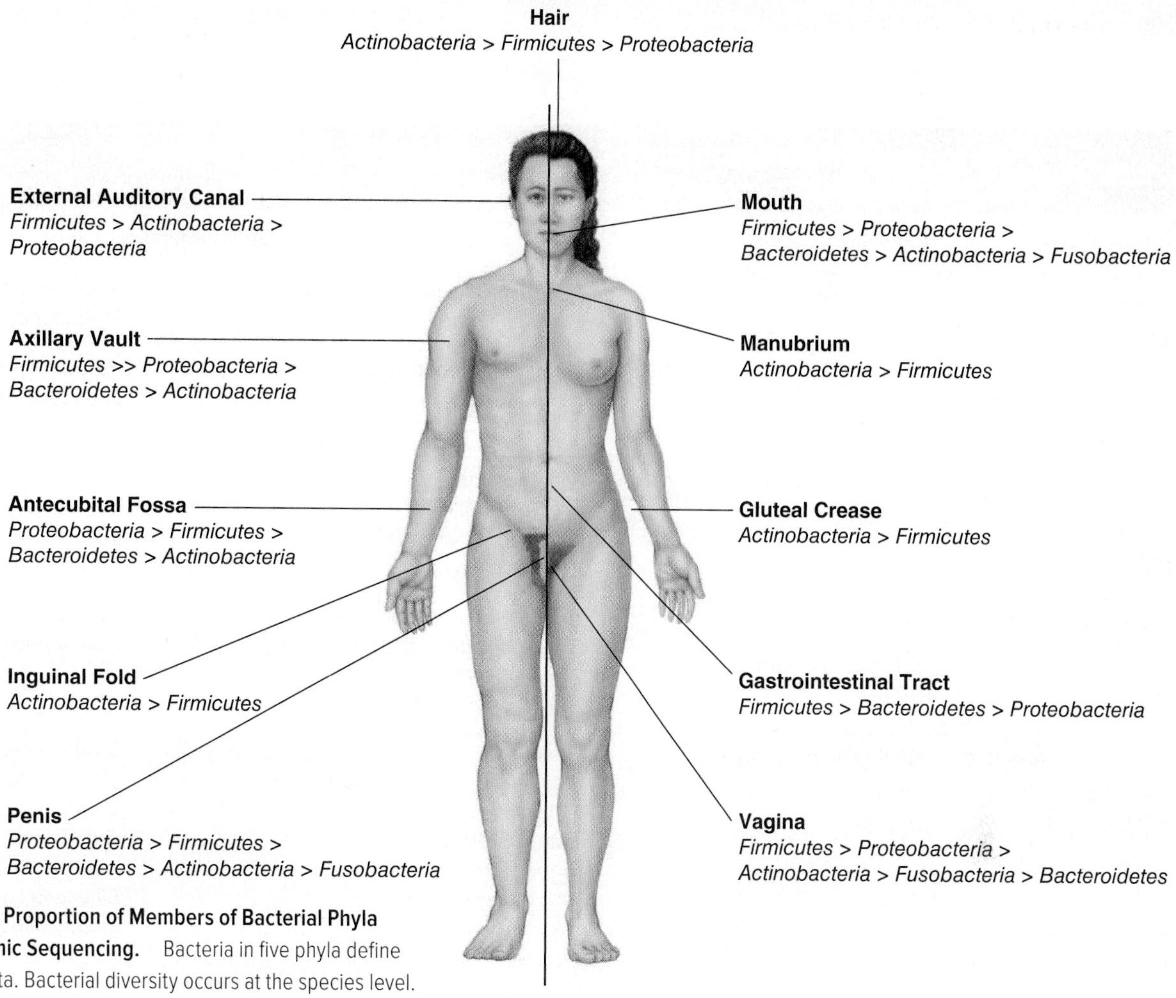

Hair
Actinobacteria > Firmicutes > Proteobacteria

External Auditory Canal
*Firmicutes > Actinobacteria >
Proteobacteria*

Mouth
*Firmicutes > Proteobacteria >
Bacteroidetes > Actinobacteria > Fusobacteria*

Axillary Vault
*Firmicutes >> Proteobacteria >
Bacteroidetes > Actinobacteria*

Manubrium
Actinobacteria > Firmicutes

Antecubital Fossa
*Proteobacteria > Firmicutes >
Bacteroidetes > Actinobacteria*

Gluteal Crease
Actinobacteria > Firmicutes

Inguinal Fold
Actinobacteria > Firmicutes

Gastrointestinal Tract
Firmicutes > Bacteroidetes > Proteobacteria

Penis
*Proteobacteria > Firmicutes >
Bacteroidetes > Actinobacteria > Fusobacteria*

Vagina
*Firmicutes > Proteobacteria >
Actinobacteria > Fusobacteria > Bacteroidetes*

Figure 32.14 Relative Proportion of Members of Bacterial Phyla Identified by Metagenomic Sequencing.   Bacteria in five phyla define the adult human microbiota. Bacterial diversity occurs at the species level.

Propionibacterium acnes. These bacteria usually are harmless; however, they are associated with the skin condition acne vulgaris. They convert lipids secreted by the oil glands to unsaturated fatty acids such as oleic acid that have strong antimicrobial activity against Gram-negative bacteria and some fungi. Some of these fatty acids are volatile and may have a strong odor. This is why many deodorants contain antibacterial substances that act selectively against Gram-positive bacteria to reduce the production of volatile unsaturated fatty acids and body odor.

Retrieve, Infer, Apply

1. Why is it important to understand the normal human microbiota?
2. Why is the skin not always a favorable microenvironment for colonization by bacteria?
3. How do microorganisms contribute to body odor?
4. What physiological role does *P. acnes* play in the establishment of acne vulgaris?

Nose, Pharynx, and Mouth

The normal microbiota of the nose, pharynx, and mouth are similar. Just inside the nostrils, *Staphylococcus aureus* and *S. epidermidis* are the predominant culturable bacteria and are found in approximately the same numbers as on the skin of the face.

The nasopharynx, that part of the pharynx lying above the soft palate, may contain small numbers of potentially pathogenic

bacteria such as *Streptococcus pneumoniae*, *Neisseria meningitidis*, and *Haemophilus influenzae*. Diphtheroids, a large group of nonpathogenic Gram-positive bacteria that resemble *Corynebacterium* spp., are commonly found in both the nose and nasopharynx.

The oropharynx is that division of the pharynx lying between the soft palate and the upper edge of the epiglottis. The most important bacteria found in the oropharynx are various α-hemolytic streptococci (*S. oralis*, *S. milleri*, *S. gordonii*, *S. salivarius*; see figure 23.15); large numbers of diphtheroids; *Branhamella catarrhalis*; and small, Gram-negative cocci related to *N. meningitidis*. The palatine and pharyngeal tonsils harbor a similar microbiota, except within the tonsillar crypts, where there is an increase in *Micrococcus* spp. and anaerobic members of the genera *Porphyromonas*, *Prevotella*, and *Fusobacterium*.

Soon after an infant is born, the mouth is colonized by microorganisms from the surrounding environment. Initially the microbiota belongs mostly to the genera *Streptococcus*, *Neisseria*, *Actinomyces*, *Veillonella*, and *Lactobacillus*, as well as some yeasts. Most microorganisms that initially invade the oral cavity are aerobes and obligate anaerobes. When the first teeth erupt, anaerobes (*Porphyromonas*, *Prevotella*, and *Fusobacterium* spp.) become dominant due to the anoxic nature of the space between the teeth and gums. As the teeth grow, *Streptococcus parasanguis* and *S. mutans* attach to enamel surfaces; *S. salivarius* attaches to the buccal and gingival epithelial surfaces and colonizes the saliva. These streptococci produce a glycocalyx

Table 32.2	Normal Human Microbiota Routinely Cultured from Various Body Sites. Knowing the Identity of These Resident Microorganisms Assists in the Rapid Identification and Treatment of Pathogens.		
Body System	**Body Site**	**Microorganisms**	
Eye	Conjunctiva	Coagulase-negative staphylococci *Hemophilus* spp.	*Staphylococcus aureus* *Streptococcus* spp.
	Outer ear	Coagulase-negative staphylococci Diphtheroids	*Pseudomonas* spp. Enterobacteria (occasionally)
Skin	Nonmucous membrane surfaces	Coagulase-negative staphylococci Diphtheroids *Propionibacterium acnes* *Staphylococcus aureus* *Streptococcus* spp.	*Bacillus* spp. *Malassezia furfur* *Candida* spp. *Mycobacterium* spp. (occasionally)
Respiratory tract	Nose	Coagulase-negative staphylococci *Streptococcus* spp. (including *S. pneumoniae*) *Staphylococcus aureus*	*Neisseria* spp. *Haemophilus* spp.
Gastrointestinal tract	Mouth and oropharynx	*Streptococcus* spp. (including *S. pneumoniae*) Coagulase-negative staphylococci *Veillonella* spp. *Fusobacterium* spp. *Treponema* spp. *Porphyromonas* spp. *Prevotella* spp. *Neisseria* spp.	*Branhamella* spp. *Hemophilus* spp. Diphtheroids *Candida* spp. *Actinomyces* spp. *Eikenella corrodens* *Staphylococcus aureus*
	Stomach	*Streptococcus* spp. *Staphylococcus* spp.	*Lactobacillus* spp. *Peptostreptococcus* spp.
	Small intestine	*Lactobacillus* spp. *Bacteroides* spp. *Clostridium* spp.	*Mycobacterium* spp. *Enterococcus* spp. Enterobacteria
	Large intestine	*Bacteroides* spp. *Fusobacterium* spp. *Clostridium* spp. *Peptostreptococcus* spp. *Escherichia coli* *Klebsiella* spp. *Proteus* spp. *Lactobacillus* spp.	*Enterococcus* spp. *Streptococcus* spp. *Pseudomonas* spp. *Acinetobacter* spp. Coagulase-negative staphylococci *Staphylococcus aureus* *Mycobacterium* spp. *Actinomyces* spp.
Genitourinary tract	Distal urethra	Coagulase-negative staphylococci Diphtheroids *Streptococcus* spp. *Mycobacterium* spp.	*Bacteroides* spp. *Fusobacterium* spp. *Peptostreptococcus* spp.
	Vagina	*Lactobacillus* spp. *Peptostreptococcus* spp. Diphtheroids *Streptococcus* spp.	*Clostridium* spp. *Bacteroides* spp. *Gardnerella vaginalis* *Candida* spp.

and various other adherence factors that enable them to attach to oral surfaces. The presence of these bacteria contributes to the eventual formation of dental plaque, caries, gingivitis, and periodontal disease. ▶▶| *Dental disease (section 39.6)*

In many ways, the mouth provides an ideal habitat for microbes. It has a ready supply of water and nutrients. It also has a neutral pH and moderate temperature. However, it subjects the normal microbiota to mechanical perturbations that can dislodge microbes from the mouth. Thus the normal microbiota of the mouth consists of organisms that resist mechanical removal by adhering to surfaces such as the gums and teeth. These removal mechanisms include flushing of the oral cavity contents to the stomach, where they are destroyed by hydrochloric acid, and the continuous desquamation (shedding) of epithelial cells.

Respiratory Tract

The lower respiratory tract does not have normal microbiota. This is because microorganisms are removed in at least three ways. First, a continuous stream of mucus is generated by the goblet cells. This entraps microorganisms, and the ciliated epithelial cells continually move the entrapped microorganisms out of the respiratory tract (*see figure 33.5*). Second, alveolar macrophages phagocytize and destroy microorganisms. Finally, a bactericidal effect is exerted by the enzyme lysozyme, present in nasal mucus.

Eye and External Ear

At birth and throughout human life, a small number of bacteria are found on the conjunctiva of the eye. The predominant bacterium is *S. epidermidis*, followed by *S. aureus, Haemophilus* spp., and *S. pneumoniae*. By contrast, the normal microbiota of the external ear resembles that of the skin, with coagulase-negative staphylococci and *Corynebacterium* spp. predominating.

Gastrointestinal Tract

The very acidic pH (2 to 3) of the gastric contents kills most microorganisms. As a result, the stomach usually contains less than 10 viable bacteria per milliliter of gastric fluid. These are mainly *Streptococcus, Staphylococcus, Lactobacillus, Peptostreptococcus* spp. and yeasts such as *Candida* spp. Microorganisms may survive if they pass rapidly through the stomach or if the organisms ingested with food are particularly resistant to gastric pH (e.g., mycobacteria).

The small intestine is divided into three anatomical areas: duodenum, jejunum, and ileum. The duodenum (the first 25 cm of the small intestine) contains few microorganisms because of the combined influence of the stomach's acidic juices and the inhibitory action of bile and pancreatic secretions that are added here. Of the bacteria present, Gram-positive cocci and rods comprise most of the microbiota. *Enterococcus faecalis*, lactobacilli, diphtheroids, and the yeast *Candida albicans* are occasionally found in the jejunum. In the distal portion of the small intestine (ileum), the microbiota begins to take on the characteristics of the colon microbiota. It is within the ileum that the pH becomes more alkaline. As a result, anaerobic Gram-negative bacteria and members of the family *Enterobacteriaceae* become established.

The large intestine or colon has the largest microbial community in the body. Microscopic counts of feces approach 10^{12} organisms per gram wet weight. Over 1,000 different bacterial species have been identified in human feces. In fact, some argue that the human colon is home to one of the most diverse microbial communities on Earth. These microorganisms

consist primarily of anaerobic, Gram-negative bacteria and Gram-positive rods (figure 32.14). Not only are the vast majority of microorganisms anaerobic, but many different species are present in large numbers. The metagenomic profile of colonic bacteria in healthy adults includes 60–80% firmicutes, 20–40% bacteriodetes, with the remaining largely proteobacteria and actinobacteria.

Various physiological processes move the microbiota through the colon so an adult eliminates about 3×10^{13} microorganisms daily. These processes include peristalsis, desquamation of the surface epithelial cells to which microorganisms are attached, and continuous flow of mucus that carries adhering microorganisms with it. To maintain homeostasis, the body must continually replace lost microorganisms. The bacterial population in the human colon usually doubles once or twice a day. Under normal conditions, the resident microbial community is self-regulating. Competition and mutualism between different microorganisms and between the microorganisms and their host maintains a status quo.

Genitourinary Tract

The upper urinary tract (kidneys, ureters, and urinary bladder) is usually free of microorganisms. In both males and females, a few common bacteria (*S. epidermidis, E. faecalis*, and *Corynebacterium* spp.) usually are cultured from the distal portion of the urethra. In the male, metagenomic sequencing also suggests several anaerobic Gram-negative bacteria are present.

In contrast, the adult female genital tract, because of its large surface area and mucous secretions, has a complex microbiota that constantly changes with a woman's menstrual cycle. The culturable microorganisms are acid-tolerant lactobacilli, primarily *Lactobacillus acidophilus*, often called Döderlein's bacillus. They ferment glycogen produced by the vaginal epithelium, forming lactic acid. As a result, the pH of the vagina and cervix is maintained between 4.4 and 4.6, inhibiting other microorganisms. In addition to *L. acidophilus*, metagenomic sequencing has identified a variety of anaerobic Gram-negative and Gram-positive bacteria in the female genitourinary tract. Some of these include those that normally inhabit moist skin areas (figure 32.14 and table 32.2).

Retrieve, Infer, Apply

1. What are the most common microorganisms found in the nose? The oropharynx? The nasopharynx? The lower respiratory tract? The mouth? The eye? The external ear? The stomach? The small intestine? The colon? The genitourinary tract?
2. Draw a time line indicating the colonization of the human large intestine. What is the significance of the temporal pattern?
3. What physiological processes move the microbiota through the gastrointestinal tract?
4. What is the role of pH in determining the microbiota of the stomach, and what is the role of microbiota in determining the pH of the vagina?
5. How would you define an opportunistic microorganism or pathogen? A compromised host?

Key Concepts

32.1 Many Types of Microbial Interactions Exist

- Symbiotic interactions include mutualism (mutually beneficial and obligatory), cooperation (mutually beneficial, not obligatory), and commensalism (product of one organism can be used beneficially by another organism). Predation involves one organism (the predator) ingesting or killing a larger or smaller prey, parasitism (a longer-term internal maintenance of another organism or acellular infectious agent), and amensalism (a microbial product inhibiting another organism) (**figure 32.1**).
- Mutual advantage is central to many organism-organism interactions. These interactions can be based on material transfers related to energetics or the creation of physical environmental changes that offer protection. With several important mutualistic interactions, chemolithotrophic microorganisms play a critical role in making organic matter available for use by an associated organism (e.g., endosymbionts in *Riftia* tube worms) (**figure 32.5**).
- The rumen is an excellent example of a mutualistic interaction between an animal and a complex microbial community. In this microbial community, complex plant materials are broken down to simple organic compounds that can be absorbed by the ruminant, as well as forming waste gases such as methane that are released to the environment.
- Commensalism or syntrophism simply means growth together. It does not require physical contact and involves a mutually positive transfer of materials, such as interspecies hydrogen transfer.
- Predation and parasitism are closely related. Predation has many beneficial effects on populations of predators and prey. These include recycling immobilized minerals in organic matter for reuse by chemotrophic and photosynthetic primary producers, and protection of prey from heat and damaging chemicals.

32.2 The Human-Microbe Ecosystem

- The average human adult carries about three times more microbial cells than human cells. Microorganisms that take up residence on a human host are called the normal microbiota.

- The Human Microbiome Project is evaluating the role microbial genes and their products have in maintaining homeostasis in the host.
- Metagenomic analysis of the human microbiome has revealed an enormous diversity of microorganisms living on and in the human body. These microbes are essential in establishing and maintaining the health of the host.
- A pathogen is any disease-producing microorganism.
- Opportunistic pathogens are microorganisms that cause disease if introduced into a sterile site of a healthy host or gain access to an immunologically compromised host.

32.3 Normal Microbiota of the Human Body Adapt to the Human Condition

- Various microbes have adapted to specific niches found on the human host. These niches are uniquely able to support microbial growth by maintaining a relatively constant environment (**figure 32.14**).
- Microorganisms living on or in the skin can be characterized as either transients or residents. The normal microbiota have evolved the ability to tolerate vastly different conditions found on skin; for example, oily, moist, and dry.
- The oral cavity provides a nutrient-rich habitat but subjects microbes to physical processes that can dislodge them. The normal microbiota of the oral cavity is composed of those microorganisms able to resist this mechanical removal.
- The stomach contains very few microorganisms due to its acidic pH.
- The distal portion of the small intestine and the entire large intestine have the largest microbial community in the body. Over 400 species have been cultured, the vast majority of them anaerobic; metagenomic studies suggest that approximately 12,000 different genera inhabit the large intestine (**table 32.2**).
- The upper urinary tract is usually free of microorganisms. The tip of the urethra (a mucus membrane) harbors numerous skin bacteria. In contrast, the adult female genital tract has a more complex microbiota that prevents other microorganisms from colonization.

Compare, Hypothesize, Invent

1. Describe an experimental approach to determine if a plant-associated microbe is a commensal or a mutualist.
2. Some patients who take antibiotics for acne develop yeast infections of the mouth or genitourinary tract. Explain.
3. How does knowing the anatomical location of normal microbiota help clinicians diagnose infection?
4. Compare and contrast the microbial communities that reside in a ruminant with those in the human gut.
5. The fruit fly *Drosophila melanogaster* is an excellent model system. *D. melanogaster* hosts the ubiquitous insect symbiont *Wolbachia pipientis,* but unlike most other insect

hosts, it is not subject to reproductive manipulation by the bacterium. Surprisingly, infection with *W. pipientis* protects the fruit fly from infection by three unrelated viruses that otherwise cause 100% mortality. Discuss the differences in this rickettsial-insect symbiosis with that described in Microbial Diversity & Ecology 32.1. Do you think it is likely that *W. pipientis* also protects wasps against other pathogens, such as viruses? Explain your answer from the point of view of the bacterium and from the insect's point of view.

Read the original paper: Hedges, L. M. 2008. *Wolbachia* and virus protection in insects. *Science* 322:722.

Innate Host Resistance

33

Obesity is an epidemic, rapidly spreading across nations. Excess dietary fat, as often found in fast foods, is implicated in several metabolic diseases. Of great interest is how fat cells produce chemical signals that control inflammatory processes.

Supersize Me!

About one-third of the U.S. population is obese, with 60% considered overweight. This is a staggering statistic, and it has been increasing steadily over the past 50 years. Interestingly, obesity rates correlate to diets high in saturated fats, as found in many fast food products. The fats in foods are not the only problem, though. Increasing portion size has added dramatically to the obesity issue. In many ways, the American diet is supersized. These facts prompted a number of scientists to study the impact of obesity on the immune system. Results of studies have provided additional insight into the mechanisms by which the innate resistance arm of the mammalian immune system functions.

Innate resistance is an elegant system of sensory and response molecules that detect invasion by microorganisms. Capture of microbial pieces and parts by toll-like and NOD-like receptors (TLRs and NLRs, respectively) triggers an ancient gene activation response that upregulates the inflammatory process. This response synthesizes chemical messengers (cytokines) that alert the host to combat the alien invaders by recruiting an army of proteins and cells. An equally elegant system has recently gained attention whereby TLRs and NLRs detect and respond to host cell damage by recognizing damage- or danger-associated molecular patterns (DAMPs), which are typically released from stressed or dying host cells. Recognition of DAMPs likewise stimulates cytokine expression to facilitate cellular repair. Both systems activate the same inflammatory process but usually with very different outcomes. In a series of revealing papers, scientists from Tokyo to Boston studying obesity and its associated problems have pieced together the amazing story of how the intricate system for the detection of MAMPs and DAMPs by TLRs and NLRs is upset when we "supersize."

Originally thought to just be an energy reserve, visceral (belly) fat is a specialized tissue composed of lipid-laden adipocytes (fat cells), preadipocytes, phagocytes called macrophages, innate lymphoid cells, and other cells held in a web of extracellular matrix proteins. Normal adipocytes make pro- and anti-inflammatory cytokines, striking a delicate balance between opposing inflammatory functions to maintain homeostasis needed for tissue repair and regeneration functions.

However, studies now show strong correlations between increasing visceral adipose (obese conditions) and increased serum levels of pro-inflammatory cytokines (adipokines). This is because saturated fatty acids are recognized as DAMPs by visceral adipocytes and their associated macrophages. One of the pro-inflammatory adipokines, interleukin (IL)-18, and a macrophage marker of inflammation, C-reactive protein (CRP), are especially elevated in the obese. These and other data suggest that a low level of homeostatic inflammation is maintained in normal adipose tissue, balancing inflammatory mediators for tissue sustainability. However, shifting from lean to obese conditions causes the visceral adipocytes along with recruited macrophages and lymphocytes, to secrete pro-inflammatory cytokines that alter the structure and function of the adipose tissue. This promotes chronic and systemic, low-grade inflammation with severe consequences for the cardiovascular system. Importantly, return to a leaner mass and increasing exercise reverses the effect. Thus, understanding how tissue-specific inflammation is induced and regulated by MAMPs and DAMPs offers insight into how human disease reflects behavioral choices.

The integrity of any host organism depends on its ability to recognize foreign invading substances and other life forms and provide an appropriate response to prevent invasion by the foreign material. When microorganisms inhabit a multicellular host, competition for resources at the cellular level occurs. In this chapter, we explore the mechanisms by which mammals (principally humans) express and use innate resistance factors to defend themselves against invasion by foreign substances.

Readiness Check:
Based on what you have learned previously, you should be able to:

✔ Explain basic eukaryotic cell biology (sections 5.1–5.7)
✔ Diagram the flow of information in eukaryotes from DNA to protein and regulation of this flow (sections 15.2–15.5)
✔ Explain microbe-human interactions (sections 32.2–32.3)

33.1 Immunity Arises from Innate Resistance and Adaptive Defenses

After reading this section, you should be able to:

■ Identify the major components of the mammalian host immune system
■ Integrate the major immune components and their functions to explain in general terms how the immune system protects the host

To establish an infection, an invading microorganism must first overcome many surface barriers, such as skin, degradative enzymes, and mucus, that have either direct antimicrobial activity or inhibit attachment of the microorganism to the host. Because neither the surface of the skin nor the mucus-lined body cavities are ideal environments for the vast majority of microorganisms, most **pathogens,** or disease-causing microorganisms, must breach these barriers and reach underlying tissues to cause disease. However, any microorganism that penetrates these barriers encounters two levels of host defenses: other innate resistance mechanisms and an adaptive immune response.

Animals (including humans) are continuously exposed to microorganisms that can cause disease. Fortunately animals are equipped with an immune system that usually protects against adverse consequences of this exposure. The **immune system** is composed of widely distributed proteins, cells, tissues, and organs that recognize foreign substances, including microorganisms. Together they act to neutralize or destroy them, maintaining host integrity.

Immunity

The term **immunity** (Latin *immunis,* free of burden) refers to the general ability of a host to resist infection or disease. **Immunology** is the science focused on immune responses and how these responses are used to resist infection. It includes the distinction between "self" and "nonself" and all the biological, chemical, physiological, metabolic, and physical aspects of the immune response.

There are two fundamentally different yet complementary components of the mammalian immune response. The first component is common to all vertebrate animals, which have evolved unique features that inherently protect against invasion by foreign substances. Some of these features are physical and act as barriers. Others are chemical in nature and directly kill or inhibit invaders. Still other features result when specialized cells recognize generic yet highly conserved chemical motifs (e.g., those expressed on bacteria, fungi, and viruses) and initiate processes to engulf and degrade the foreign substance. These features are collectively called the **innate resistance mechanisms.** The second component of the mammalian immune response is much more sophisticated, being directed by highly specialized cells that can respond to specific invaders through receptor-mediated capture events, be programmed to "remember" their encounters with foreign substances, and amplify individual responses, recruiting other components of the host immune system to eliminate or reduce the threat posed by the invader. This component is the **adaptive immune response.** It is the focus of chapter 34.

The innate resistance mechanisms, also known as **nonspecific resistance** and **innate, nonspecific,** or **natural immunity,** are the first line of defense against any foreign material, including microorganisms, encountered by the host. It includes general mechanisms such as skin, mucus, and constitutively produced antimicrobial chemicals such as lysozyme. Innate resistance mechanisms defend against foreign invaders equally and to the same maximal extent each time a foreign invader is encountered.

In contrast, the adaptive immune response, also known as **acquired** or **specific immunity,** defends against a particular foreign agent. The effectiveness of adaptive immune responses increases on repeated exposure to foreign agents such as viruses, bacteria, or toxins; that is to say, adaptive responses have "memory." The adaptive immune response relies on interactions between antigens and antibodies. **Antigens** are substances recognized as foreign. They provoke immune responses and are also called immunogens (immunity generators). **Antibodies** are host proteins produced by certain immune system cells (B cells). Antibodies bind specific antigens and either inactivate them or contribute to their elimination. The binding of antibodies to antigens forms a complex called an antigen-antibody complex. These complexes initiate other immune responses that help eliminate the antigen. These are described in more detail in chapter 34.

The distinction between innate and adaptive systems is somewhat artificial. Although innate systems predominate immediately upon initial exposure to foreign substances, multiple bridges occur between the two immune system components (**figure 33.1**). Importantly, a variety of **white blood cells** (leukocytes) function in both innate and adaptive immunity. Some of these are important because they link the innate arm of the immune system to the adaptive. Thus, leukocytes form the basis for immune responses to invading microbes and foreign substances. Many of these cells reside in specialized tissues and organs. Some tissues and organs provide supportive functions in nurturing the cells so that they can mature and respond correctly.

The bridges that interconnect the innate and adaptive immune responses are numerous and present challenges for anyone studying immunology. However, the amazing complexity of the mammalian immune system and its capacity to prevent infection from the billions of microorganisms encountered every day drive numerous "how" questions, whose answers we explore in this chapter. We begin our discussion with an introduction to the physical and chemical barriers that microbes face when encountering a host. We then present the various cells, tissues, and organs of innate defenses, discussing the mechanism by which they detect and respond to foreign invaders. In this discussion, reference is made to processes and chemicals that are described in detail either later in this chapter or in chapter 34 (adaptive defenses). To help you navigate this information, cross-references to the detailed discussion are provided.

Retrieve, Infer, Apply

1. Define each of the following: immune system, immunity, immunology, antigen, antibody.
2. Compare and contrast the adaptive and innate immune responses.

33.2 Innate Resistance Starts with Barriers

After reading this section, you should be able to:

- Identify the barriers that help prevent microbial invasion of the host
- Explain how physical and mechanical barriers function to prevent microbial invasion of the host
- Relate host anatomy and secretions to the success of innate resistance strategies

Innate and nonspecific	Structure	Effect	Time
Barriers Microbial (commensal organisms) Physical (epithelium, mucus membrane) Chemical (AMPs, APRs, cytokines)[1]	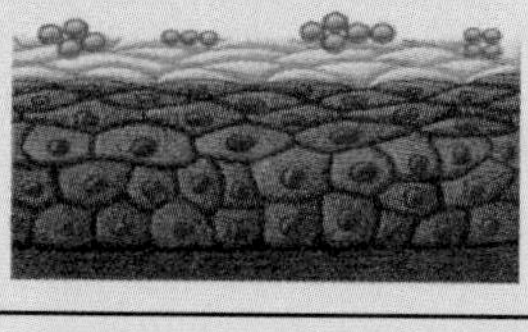	Compete for resources Block entry Neutralize invasion	0–1 hr 0–1 hr 0–6 hr
Cells Dendritic, Macrophage, Granulocytic, NK	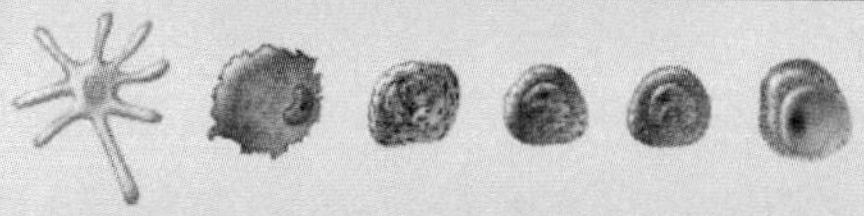	Phagocytosis, apoptosis	6–12 hr
Processes Opsonization, Inflammation, Cell Communication	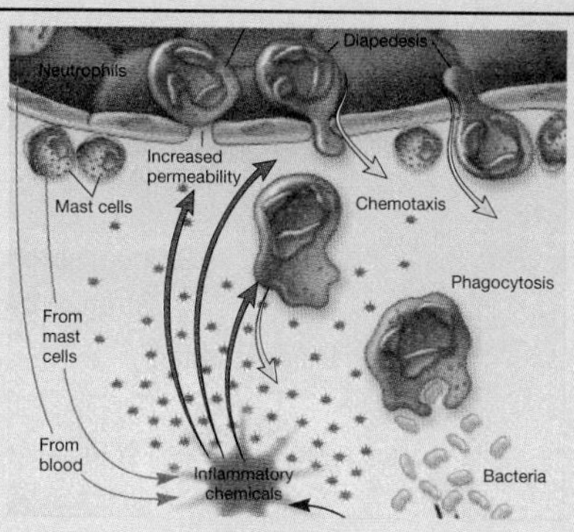	Recruit, combine, integrate responses	6–72 hr

Adaptive and specific	Structure	Effect	Time
Cells T cells, B cells, Dendritic Macrophage	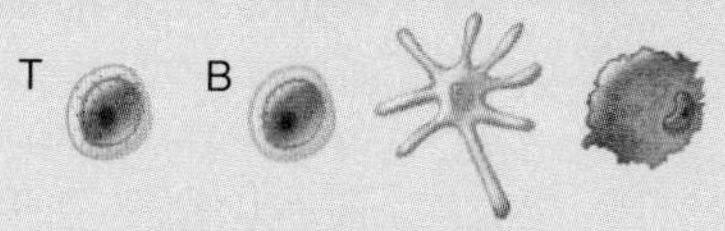	Education, activation, Tissue repair/remodel	2–21 days 21+ days
Chemical Antibody, cytokine	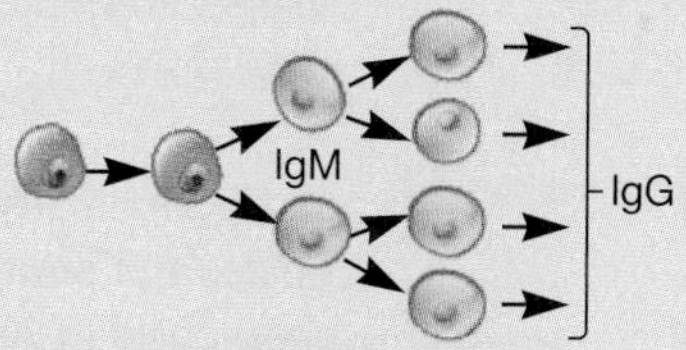	Effector response	21–60 days
Memory T cell, B cell	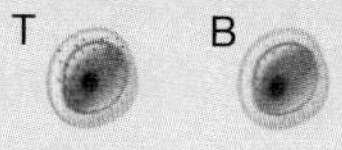	Future response	20+ years

[1]Antimicrobial peptides, (AMPs); Acute Phase Reactants, (APRs)

Figure 33.1 Some Major Components of the Mammalian Immune System. Double-headed arrow indicates potential bridging events that unite innate and acquired forms of immunity. Note the relational timing as innate and acquired aspects of immunity progress.

With few exceptions, a microbial pathogen attempting to invade a human host immediately confronts a vast array of innate defense mechanisms (**figure 33.2**). Although the effectiveness of an individual mechanism may not be great, collectively they are formidable. As we learned in chapter 32, many direct factors (nutrition, physiology, fever, age, genetics) and equally as many indirect factors (personal hygiene, socioeconomic status, living conditions) influence all host-microbe relationships. In addition

to these direct and indirect factors, a vertebrate host has some specific physical and mechanical barriers. These barriers, along with the host's flushing mechanisms (coughing, urinating, etc.), are the first line of defense against chemical, microbiological, and physical assaults. Protection of body surfaces and the mucus-membrane interfaces between the host and its environment is essential in preventing microbial access to the host.

Skin

Skin is stratified, cornified (converted into keratin) epithelium (**figure 33.3a**). It is therefore an effective barrier that prevents unregulated loss of water and solutes and blocks microbial invasion and most other chemical and physical assaults. It is also home to a

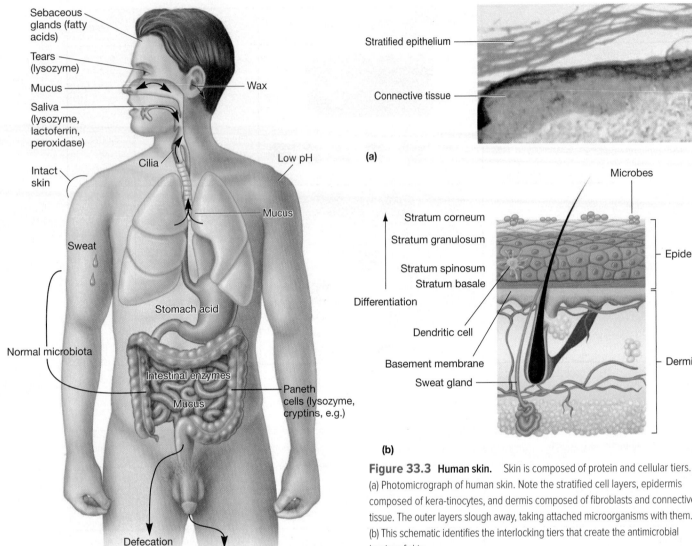

Figure 33.2 Host Defenses. Some nonspecific host defense mechanisms that help prevent entry of microorganisms into the host's tissues.

Figure 33.3 Human skin. Skin is composed of protein and cellular tiers. (a) Photomicrograph of human skin. Note the stratified cell layers, epidermis composed of kera-tinocytes, and dermis composed of fibroblasts and connective tissue. The outer layers slough away, taking attached microorganisms with them. (b) This schematic identifies the interlocking tiers that create the antimicrobial barrier of skin.

myriad of microorganisms that reside on the outside of a human. Skin cells undergo a 4-week differentiation process ending in the cells becoming enucleated, cross-linked shells of proteins (the cornification process) tiled together by extruded lipids (figure 33.3*b*).

Intact skin contributes greatly to innate host resistance because it is a very effective mechanical barrier, and its associated microbiota defend against invaders. Its outer layer consists of thick, closely packed cells called **keratinocytes,** which produce keratins. Keratins are insoluble proteins that are the main components of hair, nails, and the outer skin cells. These outer skin cells shed continuously. Select areas of the skin are bathed in sebum, a lubricating oil produced by the sebaceous glands. The skin surface is topographically, ecologically, and physiologically diverse; there are numerous microenvironments, each with distinct moisture, pH, temperature, and sebum content. These niches influence microbial populations to become resident (figure 32.3). Additionally, the skin surface is cool, acidic (pH 5–6), desiccated,

and periodically bathed in sweat. Skin cells subsist on sebum, lipids, and stratum corneum peptides. The resident microbiota co-evolved to survive these conditions.

Mucous Membranes

The mucous membranes are skin modifications at sites that interface internal structures. Together, they comprise about 4,300 square feet of surface area in the average adult human. Specialized cells at these sites produce mucus (a glycoprotein-rich watery fluid), which bathes the membrane surface, making it slippery for protection from air, food, feces, and so on. The eye (conjunctiva) and the respiratory, digestive, and urogenital systems withstand microbial invasion because the intact stratified epithelium and mucus form a protective covering that resists penetration and traps microorganisms. Many mucosal surfaces are bathed in specific antimicrobial compounds secreted by specialized cells called Paneth cells also found in the mucosa. For example, cervical mucus, prostatic fluid, and tears are toxic to many bacteria. The conjunctiva that lines the interior surface of each eyelid and the exposed surface of the eyeball is a good example of how a mucous membrane

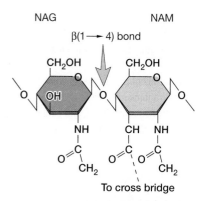

NAG   NAM

$\beta(1 \longrightarrow 4)$ bond

To cross bridge

Figure 33.4 **Action of Lysozyme on Peptidoglycan.** This enzyme degrades the cell wall peptidoglycan by hydrolyzing the $\beta(1{\to}4)$ bonds that connect alternating *N*-acetylglucosamine (NAG) with *N*-acetylmuramic acid (NAM) residues.

functions to provide chemical as well as physical protection from microorganisms. It is kept moist by the continuous flushing action of tears from the lacrimal glands. Tears contain large amounts of lysozyme, lactoferrin, and other antimicrobial chemicals.

Lysozyme (muramidase) is an enzyme that lyses bacteria by hydrolyzing the bond connecting *N*-acetylmuramic acid and *N*-acetylglucosamine of the bacterial cell wall peptidoglycan—especially in Gram-positive bacteria (**figure 33.4**). Tears and other mucous secretions also contain significant amounts of the iron-binding protein lactoferrin. **Lactoferrin** is released by activated phagocytic cells called macrophages and polymorphonuclear leukocytes (PMNs). It sequesters iron from the blood plasma, reducing the amount of iron available to invading microbial pathogens thereby limiting their ability to multiply. Finally, mucous membranes produce lactoperoxidase, an enzyme that catalyzes the production of superoxide radicals, a reactive oxygen species that is toxic to many microorganisms. ◄◄ *Peptidoglycan structure (section 3.4); Oxygen concentration (section 7.4)*

Respiratory System

The average person inhales at least eight microorganisms a minute, or 10,000 each day. Once inhaled, a microorganism must first survive and penetrate the air-filtration system of the upper and lower respiratory tracts. Because the airflow in these tracts is very turbulent, microorganisms are deposited on the moist, sticky mucosal surfaces. Microbes larger than 10 μm usually are trapped by hairs and cilia lining the nasal cavity. The cilia in the nasal cavity beat toward the pharynx, so that mucus with its trapped microorganisms is moved toward the mouth to be expelled (**figure 33.5**). Humidification of the air within the nasal cavity causes many microorganisms to swell, and this aids phagocytosis. Microbes smaller than 10 μm (i.e., most bacterial cells) often pass through the nasal cavity and are trapped by the **mucociliary blanket** that coats the mucosal surfaces of lower portions of the respiratory system. The trapped microbes are transported by ciliary action—the mucociliary escalator—that moves them away from the lungs toward the mouth (figure 33.5). Coughing and sneezing reflexes clear the respiratory system of

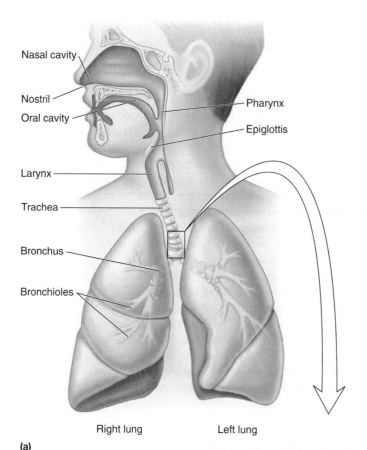

Nasal cavity

Nostril

Oral cavity

Pharynx

Epiglottis

Larynx

Trachea

Bronchus

Bronchioles

Right lung   Left lung

(a)

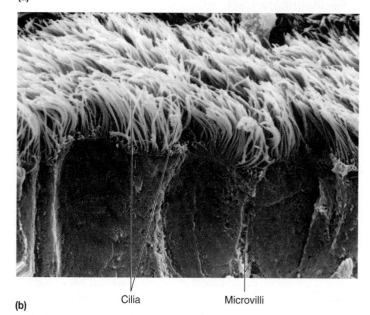

Cilia   Microvilli

(b)

Figure 33.5 **Respiratory Tract.** (a) A schematic of the respiratory tract showing its components. (b) The respiratory tract is lined with a mucous membrane of ciliated epithelial cells. SEM of cillia; these specialized structures sweep particles toward the mouth to be expectorated or swallowed (X5,000).

microorganisms by expelling air forcefully from the lungs through the mouth and nose, respectively. Salivation also washes microorganisms from the mouth and nasopharyngeal areas into the stomach. Microorganisms that succeed in reaching the alveoli of the lungs encounter a population of specialized cells called **alveolar macrophages.** These cells ingest and kill most inhaled microorganisms by phagocytosis (p. 729).

Gastrointestinal Tract

Most microorganisms that reach the stomach are killed by the acidic gastric juice (pH 2–3), which is a mixture of hydrochloric acid, proteolytic enzymes, and mucus. However, some microorganisms and their products (e.g., protozoan cysts, *Helicobacter pylori, Clostridium* spp., and staphylococcal toxins) can survive the stomach acidity. Furthermore, organisms embedded in food particles may be protected from gastric juice and reach the small intestine. There, these microorganisms may be damaged by various pancreatic enzymes, bile, enzymes in intestinal secretions, and the specialized immune cells within the gut-associated lymphoid tissue (GALT) (p. 726). **Peristalsis** (Greek *peri,* around, and *stalsis,* contraction) and the normal shedding of columnar epithelial cells act in concert to purge intestinal microorganisms. As discussed in chapter 32, the microbiota that inhabit the GI tract are tightly linked to human health and homeostasis. These microbes (see table 32.2) stimulate the maturation of the immune system, regulate immunity, and prevent the growth of potential pathogens. Within the small intestine, our microbiota stimulate the release of antimicrobial peptides (AMP) from host Paneth cells (figure 33.2).

Genitourinary Tract

The normal sterility of the kidney, ureters, and urinary bladder is maintained by a complex set of factors. In addition to removing microbes by flushing action, urine kills some bacteria due to its low pH and the presence of urea and other metabolic end products (e.g., uric acid, fatty acids, mucin, enzymes). Portions of the kidney are so hypertonic that few organisms can survive there. In males, the anatomical length of the urethra (20 cm) provides a distance barrier that excludes microorganisms from the urinary bladder. Conversely, the short urethra (5 cm) in females is more readily traversed by microorganisms; this explains why urinary tract infections are 14 times more common in females than in males.

The innate defense of the vagina consists primarily of the normal microbiota combined with the protective measures provided by the mucous membrane. The vaginal microbiota of non-pregnant women is more complex in that there appears to be no single group of organisms that define a normal flora. Rather, 16S rRNA gene-sequence-based surveys suggest that more than 20 vaginal phylotypes of *Lactobacillus* are consistent—forming a number of community state types (CSTs) with *Lactobacillus vaginalis, L. crispatus, L. iners, L. gasseri,* and *L. jensenii* dominating. These lactic-acid-forming bacteria generate an acidic vaginal environment that is inhospitable to pathogenic organisms. Cervical mucus also has antibacterial activity from added defensins, lysozyme, lactoferrin, and other natural AMPs. Adding to the innate defense of the vagina are the phagocytes that migrate through the mucous membrane.

Retrieve, Infer, Apply

1. Why is the skin such a good first line of defense against pathogenic microorganisms?
2. How do intact mucous membranes resist microbial invasion of the host?
3. Describe one specific antimicrobial defense mechanism that operates within the respiratory, gastrointestinal, and genitourinary systems of mammals. Explain how it protects against pathogenic microorganisms.

33.3 Innate Resistance Relies on Chemical Mediators

After reading this section, you should be able to:

- Discuss host mediators that have antimicrobial actions
- Describe in general terms the activation of the host complement system and its three outcomes
- List the three categories of cytokines and discuss their major functions
- Correlate host protection from microbial invasion with specific mediators

Mammalian hosts have a chemical arsenal with which to combat the continuous onslaught of microorganisms. Some of these chemicals (gastric juices, salivary glycoproteins, lysozyme, oleic acid on the skin, urea) have already been discussed with respect to the specific body sites they protect. In addition, blood, lymph, and other body fluids contain a potpourri of antimicrobial chemicals such as defensins and other polypeptides. Importantly, the normal microbiota also assist in host resistance by contributing their own chemicals to fend off competitors. How chemical defenses protect the host is the next topic.

Antimicrobial Peptides and Proteins

Peptides and low molecular weight proteins that exhibit broad-spectrum antimicrobial activity are generally considered to be the most ancient primary defense mechanism of animals. Many are positively charged, and therefore electrostatically attracted to microbial cell surfaces. Antimicrobial peptides are amphipathic; that is, they have hydrophobic and hydrophilic components so they are soluble in both aqueous and lipid-rich environments. Antimicrobial peptides kill target cells through diverse mechanisms after interacting with their membranes. The proteins lysozyme and lactoferrin have broad antimicrobial action, and we discuss these in the context of phagocytosis later in the chapter (p. 727). Another notable antimicrobial protein is granzyme, used by immune system cells called natural killer (NK) cells to lyse their targets (p. 723). Two major types of antimicrobial peptides are cationic peptides and bacteriocins.

Cationic Peptides

Humans produce three generic classes of cationic peptides whose biological activity is related to their ability to damage bacterial plasma membranes. Most form pores or transient gaps, thereby altering membrane permeability, often leading to lysis.

The first group of cationic peptides includes linear, α-helical peptides that lack cysteine amino acid residues. An important example is **cathelicidin** LL37, so named because it is 37 amino acids long, beginning with a pair of leucines. Cathelicidin LL37 is translated from a single transcript but cleaved into a number of antimicrobial fragments that are produced by a variety of cells (e.g., neutrophils, respiratory and urogenital epithelial cells, and alveolar macrophages). ▶▶| *Protein structure (appendix I)*

A second group, called **defensins,** are similarly produced from precursor proteins that mature through protease activity. These peptides are rich in arginine and cysteine, and are thus disulfide linked. Two types of defensins have been reported in humans—α and β. α defensins tend to be slightly smaller than β defensins and are found in the granules of certain immune system cells (neutrophils and intestinal Paneth cells), where they are called cryptidins, and in intestinal and respiratory epithelial cells, where they assist in bacterial degradation. β defensins are larger because they have an additional helical region. They are typically found in epithelial cells and are known to stimulate release of immune system chemical mediators by cells called mast cells (p. 720).

A third group contains larger peptides that exhibit regular structural repeats. **Histatin,** one such peptide isolated from human saliva, has antifungal activity. It translocates to the fungal cytoplasm, where it targets mitochondria.

Other natural antimicrobial products (about 20) are derived from chemokines found in platelets. Still others are fragments from (1) histone proteins, (2) lactoferrin, and (3) immune mediators that recruit cells to specific locations. A number of antibacterial peptides are produced by bacteria as well. The most notable of these are the bacteriocins, as discussed next.

Bacteriocins

Many of the normal bacterial microbiota release toxic proteins called bacteriocins that kill other strains of the same species as well as other closely related bacteria. Bacteriocins are peptides that may give their producers, which are naturally immune to their bacteriocins, an adaptive advantage against other bacteria. Ironically, they sometimes increase bacterial virulence by damaging host immune cells. An example of a bacteriocin-producing bacterium is *Escherichia coli.* It synthesizes bacteriocins called **colicins,** which are encoded on several different Col plasmids (*see table 3.3*). Some colicins bind to specific receptors on the cell envelope of sensitive target bacteria and cause cell lysis; others attack specific intracellular sites such as ribosomes or disrupt energy production. Other examples include the lantibiotics produced by genera such as *Streptococcus, Bacillus, Lactococcus,* and *Staphylococcus.* |◀◀ *Normal microbiota of the human body adapt to the human condition (section 32.3)*

Complement Is a Cascading Enzyme System

Complement refers to a collection of over 30 heat-labile proteins found in human blood plasma. The proteins act in a cascading fashion to lyse cell membranes, augment phagocytosis, and promote inflammation. These activities were said to "complement" the other antibacterial activities of the host; hence the name complement. The **complement system** is complex and has a somewhat

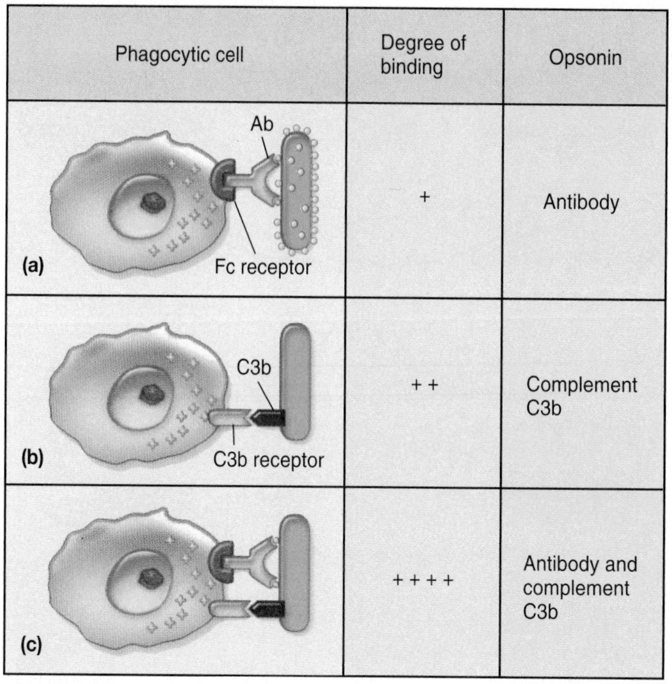

Phagocytic cell	Degree of binding	Opsonin
(a) Ab / Fc receptor	+	Antibody
(b) C3b / C3b receptor	+ +	Complement C3b
(c)	+ + + +	Antibody and complement C3b

Figure 33.6 Opsonization. (a) The intrinsic ability of a phagocyte to bind to a microorganism is enhanced if the microorganism elicits the formation of antibodies (Ab) that act as a bridge to attach the microorganism to the Fc antibody receptor on the phagocytic cell. (b) If the activated complement (C3b) is bound to the microbe, the degree of binding is further increased by the C3b receptor. (c) If both antibody and C3b opsonize, binding is maximal.

confusing nomenclature. We therefore examine each process before considering the complete mechanisms of action. Complement activation can be initiated by three different signals, but all three lead to a common final pathway that (1) defends against bacterial infections by facilitating and enhancing phagocytosis (opsonization), chemotaxis, activation of leukocytes, and lysis of bacteria; (2) bridges innate and specific immune functions by enhancing antibody responses and immunologic memory; and (3) disposes of wastes such as dead host cells and immune complexes, the products of inflammatory injury.

As mentioned earlier, one of the major outcomes of complement activation is enhancement of phagocytosis, a process called opsinization. **Opsonization** (Greek *opson,* to prepare victims for) coats microorganisms or inanimate particles with serum components, thereby preparing them for recognition and ingestion by phagocytic cells. Pathogen clearance by phagocytes increases dramatically when they are coated with opsonins (**figure 33.6**). As the first opsonins to be produced, complement proteins are important in connecting the microorganism with the phagocyte, which bears specific complement receptors.

Complement Activation

Complement proteins are produced in an inactive form; they become active following enzymatic cleavage. There are three pathways of complement activation: the alternative, lectin, and classical pathways (**figure 33.7**). Although each pathway employs

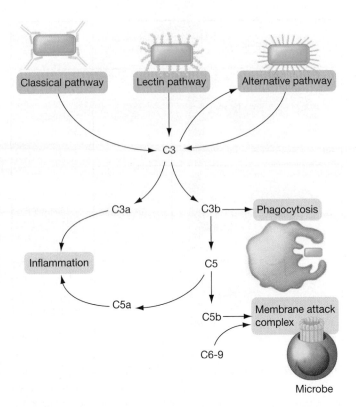

Figure 33.7 Complement Activation and Action. Complement activation involves a series of enzymatic reactions that culminate in the formation of C3 convertase, which cleaves complement component C3 into C3b and C3a. The three pathways converge at C3 convertase production. C3a is a peptide mediator of local inflammation. C3b binds covalently to bacterial cells and opsonizes the bacteria, enhancing phagocytosis. C5a and C5b are generated by the cleavage of C5 by a C5 convertase. C5a is also a powerful peptide mediator of inflammation. C5b promotes assembly of a membrane attack complex from the terminal C6–9 complement components.

similar mechanisms, specific proteins are unique to the first part of each pathway (**table 33.1**). Each complement pathway is activated in a cascade fashion: the activation of one component results in the activation of the next. Thus complement proteins are poised for immediate activity when the host is challenged by an infectious agent.

The **alternative complement pathway** (figure 33.7) is thought to have evolved first. It is initiated in response to bacterial and some fungal molecules with repetitive structures, such as the lipopolysaccharide (LPS) from Gram-negative bacteria. LPS activates complement (C) protein 3 (C3), causing it to cleave into fragments C3a and C3b. C3b becomes stable when it binds to the microbial cell surface, and once bound, C3b acts as an opsonin and enhances phagocytosis (figure 33.6). C3b is also involved in increasing the amount of active complement. This is achieved when C3b interacts with two blood proteins (Factor B and Factor D) to form a complex called C3bBb. This complex functions as the C3 convertase of the alternative complement pathway. It proceeds to cleave more C3 to C3a and C3b, thereby increasing the rate at which C3 is converted and the amount of C3b coating the microbial structure.

The **lectin complement pathway** (also called the mannose-binding lectin pathway) also begins with the activation of C3 convertase. However, in this case, a lectin—a protein that binds to specific carbohydrates—initiates the proteolytic cascade. A lectin of major importance is **mannose-binding protein** (**MBP**). This protein is produced when phagocytic cells called macrophages ingest viruses, bacteria, or other foreign material. MBP, as its name implies, binds mannose, a sugar that is a major component of bacterial cell walls, some virus envelopes, and antigen-antibody complexes. MBP enhances phagocytosis and is therefore an opsonin. MBP forms a complex with another blood protein known as *m*annose-binding lectin-*a*ssociated *s*erine *p*rotease (MASP). When the MBP component of the complex binds to mannose on pathogens, MASP is activated. MASP cleaves the C4 and C2 complement proteins, also circulating in the blood, to produce a C3 convertase. The C3 convertase is the same as that used in the classical complement pathway, discussed next.

Activation of the **classical complement pathway** typically depends on the presence of antibody specific to the pathogen. However, it can also be activated nonspecifically by a few microbial products (e.g., lipid A, staphylococcal protein A). Antibody secretion is part of the adaptive immune response (and is discussed in chapter 34); antibodies are glycoproteins that bind to specific antigens. When the appropriate antibody binds its cognate antigen, the antibody-antigen complex binds C1, a complement component composed of three proteins (q, r, s). This forms a trimolecular complex (C1qrs + antibody + antigen), which has esterase activity. The activated C1s subcomponent attacks and cleaves two other complement proteins (C4 and C2). This leads to binding of a portion of each molecule (C4b and C2a) to the antigen-antibody-C1 complex and the release of C4a and C2b fragments. With the binding of C4b to C2a, an enzyme with trypsinlike proteolytic activity is generated. The natural substrate for this enzyme is C3; thus C4b2a (that is, the antigen-antibody-C1-C4b-C2a complex) is a C3 convertase. Just as we saw with the alternative and lectin pathways, the C3 convertase cleaves C3 into a bound subcomponent C3b and a C3a soluble component. Thus each of the complement pathways is activated by different substrates, all of which signal the presence of a foreign invader.

Complement Action

In each complement activation pathway (alternative, lectin, and classical), the cleavage of C3 into C3a and C3b sets into motion other cascading enzymatic events leading to formation of the membrane attack complex, additional opsonins, and the release of chemical mediators that influence inflammation (figure 33.7).

In each complement pathway, stabilization of C3b and subsequent creation of a C5 convertase lead to the production of C5b. The two proteins, C6 and C7, rapidly bind to C5b, forming a C5b67 complex that is stabilized by binding to a membrane. C8 and C9 then bind, forming the **membrane attack complex** (**MAC**), which creates a pore in the plasma membrane or outer membrane of the target cell (**figure 33.8**). The MAC is very effective against Gram-negative bacteria, less so against Gram-positive

Table 33.1	Key Complement Proteins and Their Functions	
	Protein	**Function**
COMPLEMENT ACTIVATION (initiation through three unique pathways)		
Alternative Pathway	C3	C3 is activated by repeating patterns within microbial structures and then is spontaneously cleaved into C3a and C3b; C3b binds to nearby membrane.
	Factor D	Factor D is activated by C3b to become an active enzyme, which activates Factor B.
	Factor B	Factor B adsorbs to C3b, forming a C3 convertase (protease).
	C3bBb	C3bBb (the active C3 convertase) cleaves additional C3 into its a and b fragments; C3b binds to the nearby membrane.
	Properidin	Properidin stabilizes the C3 convertase.
Lectin Pathway	Mannose-binding protein (MBP)	MBP binds to mannose on microorganisms, then recruits and binds plasma esterase to become mannose-associated serine protease (MASP).
	MASP	MASP (the active C3 convertase) cleaves C3 into C3a and C3b; C3b binds to the nearby membrane.
Classical Pathway	Antibody (Ab) formed by adaptive immune system upon exposure to antigen (Ag)	Ab binds to Ag on microorganism, causing three-dimensional shift in Ab structure, which reveals cryptic amino acids near carboxy-terminus (Fc region); the newly revealed amino acids attract plasma C1 protein.
	C1 (trimer of polypeptides q, r, and s)	C1q binds to Fc region of Ab-Ag complex; C1r,s binds plasma calcium.
	Ag-Ab-C1q,r,s-(Ca^{2+}) complex	Ag-Ab-C1-(Ca^{2+}) complex is an activated enzyme (called the C2/C4 esterase) that cleaves plasma C2 and C4 into their a and b fragments, respectively.
	C2a	C2a binds the Ag-Ab-C1-(Ca^{2+}) complex.
	C4b	C4b binds the Ag-Ab-C1-(Ca^{2+})-C2a complex, forming a C3 convertase (protease).
	Ag-Ab-C1-C2a-C4b	Ag-Ab-C1-C2a-C4b (the active C3 convertase) cleaves plasma C3 into its a and b fragments; C3b binds to the nearby membrane.
COMPLEMENT ACTION (common effector pathway; convergence point for alternative, lectin, and classical activation pathways)	C3b-membrane complex (stabilized C3b)	C3b stabilized in a membrane becomes an active C5 convertase (protease), cleaving plasma C5 into its respective a and b fragments.
	C5b	C5b binds plasma C6 and C7, forming a new, membrane-binding complex.
	C5b-C6-C7-membrane complex	Membrane-bound C5b-C6-C7 recruits plasma C8 and C9, which insert into the membrane adjacent to C5b-C6-C7.
	C5b-C6-C7-C8-C9 complex	C5b-C6-C7-C8-C9 complex forms transmembrane pore known as the membrane attack complex (MAC); MAC formation leads to cell lysis.

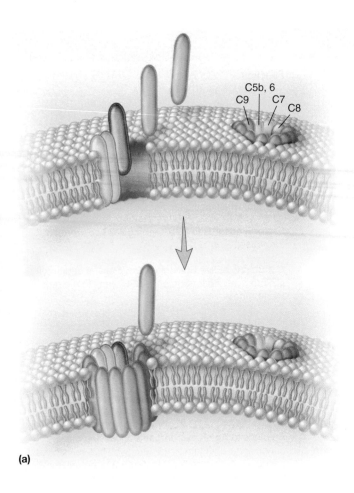

(a)

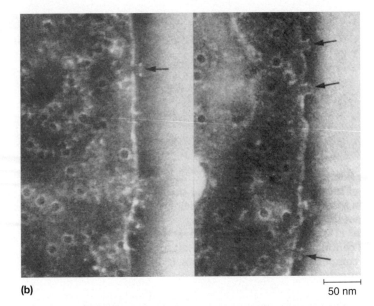

C5b, 6
C9 C7
C8

(b)

|— 50 nm —|

Figure 33.8 The Membrane Attack Complex. The membrane attack complex (MAC) is a tubular structure that forms a transmembrane pore in the target cell's membrane. (a) This representation shows the subunit architecture of the membrane attack complex. The transmembrane channel is formed by a C5b678 complex and 10 to 16 polymerized molecules of C9. (b) MAC pores appear as craters or donuts by electron microscopy.

MICRO INQUIRY *What strategies could a Gram-negative bacterium use to protect itself against lysis by complement?*

bacteria and fungi. If the target cell is eukaryotic, Na^+ and H_2O enter through the pore and the cell lyses.

Importantly, the generation of complement fragments C3a and C5a leads to several pro-inflammatory effects. For example, binding of C3a and C5a to receptors on specific host cells induces the release other biological mediators. These amplify the inflammatory signals by dilating vessels, increasing permeability, stimulating nerves, and recruiting phagocytic cells. C5a also induces chemotactic migration of neutrophils to the site of complement activation, where they are among the first cells to phagocytose invading microorganisms. Other phagocytes in the area synthesize even more complement components to interact with the invading microbes. All of these defensive events promote the ingestion and ultimate destruction of the microbe by dendritic cells, neutrophils, and macrophages.

Our overview of the alternative and lectin complement pathways (figure 33.7) provides a basis for consideration of the function of complement as an integrated system during an animal's defensive effort. Bacteria arriving at a local tissue site for the first time interact with components of the alternative and lectin pathways, resulting in the generation of biologically active fragments, opsonization of the bacteria, and initiation of the lytic sequence. Importantly, and if the bacteria persist or if they invade the host a second time, antibody responses also will activate the classical complement pathway.

Retrieve, Infer, Apply

1. How do cationic peptides function against Gram-positive bacteria?
2. How do bacteriocins function?
3. What effect does the formation of the membrane attack complex have on eukaryotic microbial cells? Bacterial cells? Host cells?
4. Compare and contrast the means by which the alternate, lectin, and classical pathways are activated.
5. What role do complement fragments C3a and C5a play in a host's defense against Gram-negative bacteria?

Cytokines Are Chemical Messages Between Cells

Communication between members of any complex network requires a clear signal and a common language. So it is no surprise to learn that both innate resistance and adaptive defenses use a highly specialized communication system. The "language" used by the mammalian immune system is based on small, soluble proteins whose binding to their receptors initiates a signal transduction pathway that ultimately regulates specific transcription and translation events. The resulting cellular behaviors therefore depend on which commands (signals) are received, how many receptors become occupied, and interference from either external or internal sources.

Cytokine (Greek *cyto*, cell, and *kinesis*, movement) is a generic term for any soluble, low molecular weight protein or glycoprotein released by one cell population that acts as an intercellular (between cells) mediator or signaling molecule. Cytokines are produced by a large number of different cells. Once released, cytokines impact a broad range of cells, resulting in a variety of cellular responses. Cytokine synthesis is inducible, resulting from nonspecific stimuli such as a viral, bacterial, or parasitic infection; cancer; inflammation; or the activation of specific immune cells. Some cytokines also induce the production of other cytokines.

The naming of cytokines has been one of those stories with the left hand not knowing what the right has done. Historically, as new proteins were discovered, a scientist would name it based on its function. As protein biochemistry became more exact, scientists would group proteins based on structure. This resulted in an unfortunate list of proteins with different names yet the same function and proteins with same structures but different names. Thus, cytokines released from mononuclear phagocytes are called **monokines,** cytokines released from T lymphocytes are called **lymphokines,** cytokines that stimulate chemotaxis and chemokinesis (i.e., they direct cell movement) are called **chemokines,** cytokines produced by one leukocyte acting on another leukocyte are called **interleukins,** regulatory cytokines produced by certain eukaryotic cells in response to a viral infection are called **interferons,** and cytokines that stimulate the growth and differentiation of immature leukocytes in the bone marrow are called **colony-stimulating factors** (CSFs).

As noted earlier, cytokines produce biological actions only when bound to specific receptors on the surface of target cells. Most cells have hundreds to a few thousand cytokine receptors, but a maximal cellular response results even when only a small number of these are occupied by cytokine molecules. This is because the affinity of cytokine receptors for their specific cytokines is very high; consequently cytokines are effective at very low concentrations. Because cytokines tend to affect specific aspects of the overall immune response, three categories based on cytokine function have been recently proposed. The three groups are (1) cytokines that regulate innate resistance mechanisms, (2) cytokines that regulate adaptive immunity, and (3) cytokines that stimulate hematopoiesis (**table 33.2**).

The binding of a cytokine to its receptor activates specific intracellular signaling pathways, which results in the transcription of genes encoding proteins essential for a specific cellular function. For example, cytokine binding may result in the cell's production of other cytokines, cell-to-cell adhesion receptors, proteases, lipid-synthesizing enzymes, and nitric oxide synthase (the production of nitric oxide has potent antimicrobial activity). In addition, cytokines can activate cell proliferation or cell differentiation. They also can inhibit cell division and cause **apoptosis** (programmed cell death).

Acute-Phase Proteins

In addition to chemical mediators from leukocytes, a number of biologically active molecules from other somatic cells can cause dramatic physiological changes in the host. Generalized trauma and localized tissue injury result in the release of tissue factors, chemokines, and neurotransmitters. Importantly, macrophages responding to sites of injury are critical in recruiting assistance by releasing pro-inflammatory cytokines that stimulate the liver to rapidly produce **acute-phase proteins.** These mediators assist in the prevention of blood loss and ready the host for microbial invasion. They include C-reactive protein (CRP), mannose-binding protein (MBP; p. 720), and surfactant proteins A (SP-A) and D (SP-D), all of which bind bacterial surfaces and act as opsonins. In addition, CRP interacts with C1q to activate the classical complement pathway and MBP binds to bacteria and fungi to activate complement. SP-A, SP-D, and C1q are also known as collectins—proteins composed of a collagen-like motif connected by α-helices to globular binding sites. These proteins (along with others) police host tissues by binding to ("collecting") foreign materials, thus assisting in the removal of microorganisms and their products.

Retrieve, Infer, Apply

1. Describe the role of cytokines in innate immunity.
2. How might cytokines work to bridge the innate and adaptive immune responses?
3. How might acute-phase reactants assist in pathogen removal? How might collectins do this?

33.4 Cells, Tissues, and Organs Work Collectively to Form an Immune System

After reading this section, you should be able to:

- Recognize the different types of leukocytes involved with innate resistance
- Outline the leukocyte response to microbial invasion
- Integrate leukocyte distribution within the host with host resistance
- Differentiate between primary and secondary lymphoid organs and tissues in terms of structure and function
- Predict connections between innate host resistance and specific immune responses

Blood cell development occurs in the **bone marrow** of mammals during the process of **hematopoiesis.** The immune system is a network of molecules, cells, tissues, and organs, each with a specialized role in defending the host against extracellular microorganisms, cancer cells, and intracellular parasites. While we discuss innate defenses and adaptive immunity separately for clarity, the recognition and removal of damaged host cells and foreign substances is an interconnected and highly regulated symphony of events, connecting both arms of the immune system. We previously discussed some of the protein mediators that signal specific responses. We now examine the cells and tissues that respond to the mediators and to the damaged cells and invading substances. As you read, make note of how host cells function in innate resistance and the linkages to the adaptive defenses.

Table 33.2 Examples of Cytokines Grouped by Function

	Cytokine[1]	Source	Role
Innate Resistance	IL-1	Macrophages, endothelial and epithelial cells	Upregulates inflammatory response, including fever
	IL-6	Macrophages, T cells, endothelial cells, and adipocytes	Upregulates acute phase response, including fever; stimulates neutrophil maturation
	IL-8	Neutrophils, macrophages, dendritic cells, endothelial and epithelial cells	Activates neutrophils, induces chemotaxis, enhances inflammation, stimulates B cells
	IL-15	Monocytes, macrophages	Activates T cells, upregulates NK cell production
	IL-22	Activated dendritic cells, CD4$^+$ helper T_{17} and T_{22} cells, innate lymphoid cells	Links and coordinates innate and adaptive responses, promotes hepatocyte and epithelial cell survival
	IL-23	Macrophages and dendritic cells	Upregulates inflammatory response via IL-17 from T cells; stimulates IL-1, IL-6, TNF, and chemokine production; enhances T-cell activation and memory response
	IL-27	Macrophages and dendritic cells	Enhances antigen recognition by T and B cells
	Type I IFN (IFN α/β)	Somatic cells, especially plasmacytoid dendritic cells, macrophages, and NK cells	Induces broad-spectrum antiviral and anti-tumor activity, enhances other cytolytic function, promotes inflammation
	TNF-α	Monocytes/Macrophages	Upregulates inflammatory response, including fever; stimulates acute-phase protein synthesis; induces tumor regression; mediates septic shock
Adaptive Immunity	IL-2	T cells (autocrine process)	Stimulates growth and differentiation of T cells and NK cells; promotes antibody secretion from B cells
	IL-4	T-cell subset (and putatively basophils)	Induces differentiation of a T-cell subset; stimulates production of antibody, especially antibody associated with allergies; inhibits IL-12 production
	IL-5	T-cell subset and mast cells	Stimulates growth of B cells; enhances antibody secretion; activates eosinophils
	IL-12	Macrophages, dendritic cells, and a T-cell subset	Stimulates growth and function of T-cell subsets, stimulates killing functions of NK cells and cytotoxic lymphocytes
	IL-17	T-cell subset	Monocyte and neutrophil chemokine; induces pro-inflammatory cytokines IL-6, TNF-α, IL-1, chemokines, and prostaglandins from various cells
	Type II IFN (IFN γ)	T-cell subset and NK cells	Enhances phagocytic functions of macrophages; upregulates cytolytic function of NK cells; stimulates antiviral functions
	LT α/β (TNF)	Numerous somatic cells	Upregulates T- and B-cell development; activates neutrophils; lyses tumor cells; upregulates CSF-2 and CSF-3
Hematopoiesis	IL-3	Basophils and activated T cells	Stimulates pluripotent hematopoietic stem cells to become myeloid progenitor cells; stimulates myeloid cell proliferation
	IL-7	Bone marrow and thymic stromal cells, dendritic and epithelial cells, and hepatocytes	Stimulates pluripotent hematopoietic stem cells to become lymphoid progenitor cells; stimulates lymphoid cell proliferation
	CSF-1	Osteoblasts	Induces hematopoietic stem cells to proliferate and differentiate into monocytes/macrophages; promotes monocyte survival
	CSF-2	Macrophages, T cells, endothelial and mast cells, and fibroblasts	Induces hematopoietic stem cells to proliferate and differentiate into granulocytes and monocytes
	CSF-3	Numerous cells and tissues	Induces hematopoietic stem cells to proliferate and differentiate into neutrophils; stimulates neutrophil function and survival

1 Cytokine: IL, interleukin; IFN, interferon; TNF, tumor necrosis factor; CSF, colony-stimulating factor

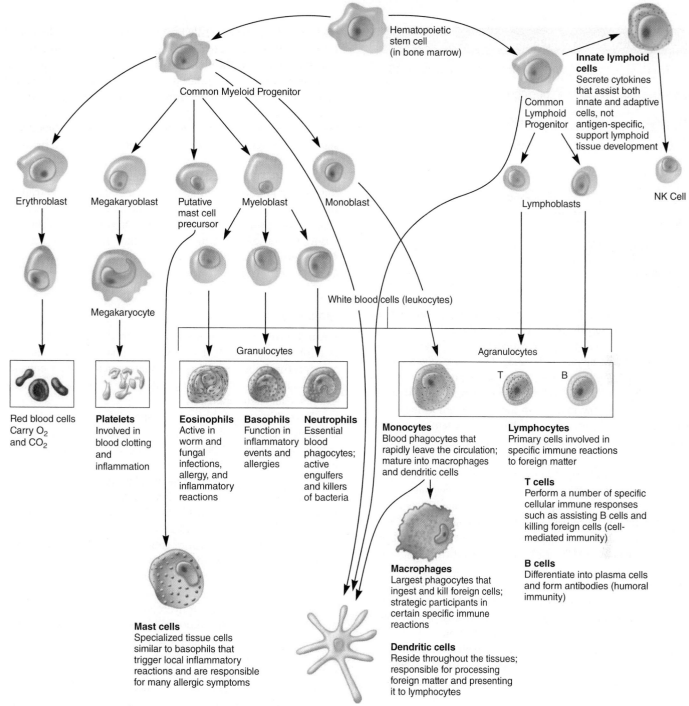

Hematopoietic stem cell (in bone marrow)

Common Myeloid Progenitor

Innate lymphoid cells
Secrete cytokines that assist both innate and adaptive cells, not antigen-specific, support lymphoid tissue development

Common Lymphoid Progenitor

Erythroblast   Megakaryoblast   Putative mast cell precursor   Myeloblast   Monoblast   Lymphoblasts   NK Cell

Megakaryocyte

White blood cells (leukocytes)

Granulocytes   Agranulocytes

T   B

Red blood cells Carry O_2 and CO_2

Platelets
Involved in blood clotting and inflammation

Eosinophils
Active in worm and fungal infections, allergy, and inflammatory reactions

Basophils
Function in inflammatory events and allergies

Neutrophils
Essential blood phagocytes; active engulfers and killers of bacteria

Monocytes
Blood phagocytes that rapidly leave the circulation; mature into macrophages and dendritic cells

Lymphocytes
Primary cells involved in specific immune reactions to foreign matter

T cells
Perform a number of specific cellular immune responses such as assisting B cells and killing foreign cells (cell-mediated immunity)

B cells
Differentiate into plasma cells and form antibodies (humoral immunity)

Macrophages
Largest phagocytes that ingest and kill foreign cells; strategic participants in certain specific immune reactions

Mast cells
Specialized tissue cells similar to basophils that trigger local inflammatory reactions and are responsible for many allergic symptoms

Dendritic cells
Reside throughout the tissues; responsible for processing foreign matter and presenting it to lymphocytes

Figure 33.9 The Different Types of Human Blood Cells.   Pluripotent stem cells in the bone marrow divide to form two blood cell lineages: (1) the common myeloid progenitor cell gives rise to the erythroblast that produces erythrocytes (red blood cells), megakaryocytes that produce platelets, an unknown precursor that gives rise to mast cells, the granulocytes (eosinophils, basophils, and neutrophils), monocytes that give rise to dendritic cells (DCs) and macrophages, and a separate lineage of DCs; and (2) the lymphoid progenitor cell gives rise to innate lymphoid cells that can differentiate into natural killer (NK) cells, a separate lineage of DCs, B cells, and T (regulatory, helper, and cytotoxic) cells. Note that DCs can arise directly from myeloid and lymphoid progenitor cells and from nonproliferating monocytes.

Cells of the Immune System

The cells responsible for both innate defenses and adaptive immunity are the **leukocytes** (Greek *leukos,* white, and *kytos,* cell). All leukocytes originate from pluripotent stem cells in the fetal liver and in the bone marrow (**figure 33.9**). Pluripotent stem cells

have not yet committed to differentiating into one specific cell type. Some differentiate into hematopoietic precursor cells that are destined to become blood cells. When stimulated to undergo further development, some leukocytes become residents within tissues, where they respond to local trauma. These cells may sound the alarm that signals invasion by foreign organisms.

Table 33.3	Normal Adult Blood Count	
Cell Type	**Cells/mm³**	**Percent WBC**
Red blood cells	5,000,000	
Platelets	250,000	
White blood cells	7,400	100
Neutrophils	4,320	60
Lymphocytes	2,160	30
Monocytes	430	6
Eosinophils	215	3
Basophils	70	1

Other leukocytes circulate in body fluids and are recruited to the sites of infection after the alarm has been raised.

The average adult has approximately 7,400 leukocytes per cubic millimeter of blood (**table 33.3**). This average value shifts substantially during an immune response. For example, in most infections the white blood cell count may increase as leukocytes migrate from the bone marrow to the blood, then to the site of invasion. This transient increase in blood-borne leukocytes (leukophilia, also called leukocytosis) may result until the infection subsides.

For leukocytes to respond to pathogens, they must detect their presence. Many leukocytes detect repeating molecular patterns found in microbial macromolecules. These unique microbial signatures (originally called pathogen-associated molecular patterns, or PAMPs) are now also referred to as **microbe-associated molecular patterns** (**MAMPs**). MAMPs are specific regions within common microbial macromolecules, such as lipopolysaccharide [LPS], peptidoglycan, fungal cell wall components, viral nucleic acids, and other microbial structures. MAMPs do not define a specific microorganism, but they do inform the host of a microbial presence, sounding the alarm of potential infection.

The receptors that bind to MAMPs are both soluble and membrane bound; they are called **pattern recognition molecules** (**PRMs**) or pattern recognition receptors (PRRs). Soluble PRMs include the acute-phase proteins, mannose-binding protein (of the lectin complement pathway), and C-reactive protein (CRP). The membrane-bound PRMs are found on phagocytic cells, enabling them to (1) distinguish between potentially harmful microbes and other host molecules, and importantly, (2) respond by ingesting and degrading the source of the MAMP. Additional details of MAMP recognition and removal are discussed in section 33.5. The different types of leukocytes, including those that detect MAMPs, are now briefly examined (in the order presented in figure 33.9).

Mast Cells

Mast cells are bone marrow–derived cells that differentiate in connective tissue. Mast cells (figure 33.9) have an indented nucleus and a cytoplasm filled with specialized organelles called granules. Mast cells are nonphagocytic, but when stimulated, they rapidly release the contents of their granules into the extracellular environment, a process called degranulation. Mast cell granules contain histamine, prostaglandins, serotonin, heparin, dopamine, platelet-activating factor, and leukotrienes. Because these compounds influence the tone and diameter of blood vessels, they are termed **vasoactive mediators.** Additionally, mast cells possess high-affinity receptors for the type of antibody associated with allergic responses. When they become coated with this type of antibody, binding of antigen to the antibody triggers the release of preformed vasoactive mediators. As discussed in chapter 34, vasoactive mediators play a major role in certain allergic responses such as eczema, hay fever, and asthma. ▶▶| *Antibodies are proteins that bind to specific 3-D molecules (section 34.7); Hypersensitivities (section 34.10)*

Granulocytes

Granulocytes also have irregularly shaped nuclei with two to five lobes (figure 33.9). Their cytoplasm has granules that contain reactive substances that kill microorganisms and enhance inflammation. Three types of granulocytes exist: basophils, eosinophils, and neutrophils.

Basophils (Greek *basis*, base, and *philein*, to love) have irregularly shaped nuclei with two lobes (figure 33.9). The granules contain histamine and other substances similar to those in mast cells. However, basophils arise from a different cellular lineage (figure 33.9). Like mast cells, though, basophils tend to infiltrate specific tissue sites rather than circulate through the bloodstream. Basophils are also important in the development of allergies and hypersensitivities, because like mast cells, they possess high-affinity receptors for the type of antibody associated with allergic responses and release preformed vasoactive mediators.

Eosinophils (Greek *eos*, dawn, and *philein*) have a two-lobed nucleus connected by a slender thread of chromatin (figure 33.9). Eosinophil granules contain hydrolytic enzymes (e.g., nucleases and glucuronidases), in addition to peroxidases and a caustic protein called major basic protein. Unlike basophils, eosinophils circulate in low numbers and migrate from the bloodstream into tissue spaces, especially mucous membranes, when recruited by soluble chemotactic mediators. They defend against protozoan and helminth parasites, mainly by releasing enzymes, cationic peptides (p. 712), and reactive oxygen species (p. 729) into the extracellular fluid. These molecules damage the parasite's plasma membrane, killing it. Eosinophils also play a role in allergic reactions, as they have granules that contain histaminase and aryl sulphatase, down-regulators of the inflammatory mediators histamine and leukotrienes, respectively. Thus their numbers in the bloodstream often increase during allergic reactions, especially type 1 hypersensitivities. ▶▶| *Type I hypersensitivity (section 34.10)*

Neutrophils (Latin *neuter*, neither, and *philein*) are highly phagocytic cells with a nucleus that has three to five lobes connected by slender threads of chromatin (figure 33.9). Because of the irregularly shaped nuclei, neutrophils are also called

polymorphonuclear neutrophils, or **PMNs.** Neutrophils have inconspicuous organelles known as primary and secondary granules. Primary granules contain peroxidase, lysozyme, defensins, and various hydrolytic enzymes, whereas the smaller secondary granules have collagenase, lactoferrin, cathelicidin LL37, and lysozyme. These enzymes and other molecules help digest foreign material after it is phagocytosed (section 33.5). Neutrophils also use oxygen-dependent and oxygen-independent pathways to generate additional antimicrobial substances to kill ingested microorganisms.

Mature neutrophils leave the bone marrow and circulate in blood so they can rapidly migrate to a site of tissue damage and infection, where they become the principal phagocytic and microbicidal responders. Neutrophils have toll-like pattern recognition molecules (PRMs), as well as receptors for antibodies and complement proteins, so that MAMPs and opsonized particles, respectively, can be more readily phagocytosed. Neutrophils have a limited life span that shortens upon activation of the phagocytic processes. Neutrophils and their antimicrobial compounds are described in more detail in the context of phagocytosis (section 33.5) and the inflammatory response (section 33.6).

Monocytes, Macrophages, and Dendritic Cells

Monocytes (Greek *monos,* single, and *cyte,* cell) are mononuclear leukocytes with an ovoid- or kidney-shaped nucleus and granules in the cytoplasm (figure 33.9). They are produced in the bone marrow and enter the blood, circulate for about 8 hours, enlarge, migrate into tissues, and mature into macrophages or dendritic cells.

Macrophages (Greek *macros,* large, and *phagein,* to eat) are classified as mononuclear phagocytic leukocytes. However, they are larger than monocytes, contain more organelles that are critical for phagocytosis, and have a plasma membrane covered with microvilli.

Macrophages are similar to dendritic cells in their ability to bridge innate resistance and adaptive immunity. They have PRMs that recognize specific MAMPs, and binding of their PRMs induces phagocytosis. Macrophages spread throughout the host and take up residence in specific tissues, where they are sometimes referred to as "fixed" or resident macrophages. They often serve as sentinel cells in tissues, sounding the chemical alarm when a microbial invader is present. This results in the recruitment of neutrophils to aid in the removal of the pathogen, and preventing its spread to other body sites. Activated macrophages also migrate to lymphoid tissues to present antigen to lymphocytes. Because macrophages are highly phagocytic, their function in innate resistance is discussed in more detail in the context of phagocytosis (section 33.5).

Dendritic cells (DCs) are not a single cell type; they are a heterogeneous group of cells that are crucial in initiating and augmenting immune responses (**figure 33.10**). They arise from various hematopoietic cell lineages; both lymphoid and myeloid DC subtypes are known that differ in location, migration patterns, and specific immunological roles. Immature dendritic cells constitute about 0.2% of peripheral blood leukocytes and

can be stimulated to mature by specific cytokines. Most DCs, however, are tissue-bound, where they play an important role in bridging innate resistance and adaptive immunity. They are present in the skin and mucous membranes of the nose, lungs, and intestines, where they constantly survey their environment. DCs are highly specialized cells that are programmed to detect and phagocytose foreign substances, especially pathogens. Dendritic cells have numerous membrane-bound receptors, including pattern recognition molecules (PRMs). The end product of phagocytosis is the proteolytic degradation of the ingested particles. Once degraded, the resulting remnants of phagocytosis (antigens) are further processed and displayed on the dendritic cell surface. DCs then migrate to lymphoid tissues to "present" the antigens to lymphocytes, sharing vital information about invaders and stimulating the specific immune response. In presenting antigen, DCs are capable of eliciting specific immune responses from naïve T cells (i.e., T cells that have never encountered a specific antigen). DCs can also capture foreign materials with their cell surface receptors that recognize antibodies and complement (figure 33.6). Recall that these opsonins enhance phagocytosis (section 33.5). Importantly, not only do dendritic cells destroy invading pathogens as part of the innate resistance process, they also help trigger specific immune responses. ▶▶ *Molecules that elicit specific immunity are called antigens (section 34.2); T cells oversee and participate in immune functions (section 34.5); Antibodies are proteins that bind to specific 3-D molecules (section 34.7)*

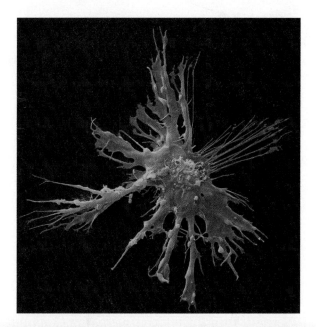

Figure 33.10 The Dendritic Cell. The dendritic cell was named for its cellular extensions, which resemble the dendrites of nerve cells. Dendritic cells reside in most tissue sites, where they survey their local environments for pathogens and altered host cells.

MICRO INQUIRY *For what does the dendritic cell use its cellular extensions?*

1. Describe the structure and function of each of the following blood cells: basophil, eosinophil, neutrophil, monocyte, macrophage, and dendritic cell. Which cells are phagocytic?
2. What is the significance of the respective blood cell percentages in blood?
3. How does a dendritic cell bridge the innate and adaptive arms of immunity?

Lymphocytes

Lymphocytes (Latin *lympha,* water, and *cyte,* cell) are major cells of the adaptive immune system and are divided into three populations: T cells, B cells, and innate lymphoid cells (figure 33.9). B and T lymphocytes differentiate from their respective lymphoid precursor cells in a kind of cellular stasis—not actively replicating until stimulated by foreign substances called antigens (**figure 33.11**). They are the primary responders of the adaptive arm of immunity, and are discussed in chapter 34.

Innate Lymphoid Cells

The **innate lympoid cell (ILC)** is thought to be a pivotal bridge between the innate and adaptive arms of immunity because it amplifies and relays information about infection and tissue damage. There are diverse ILCs, each with specific functions and target cells. Some recognize and kill infected cells while activating T cells to enhance inflammatory activities. Some respond to injured cells to help drive tissue repair and regulate **inflammation.** Yet others assist in homeostasis by regulating antibody production and killing tumor cells.

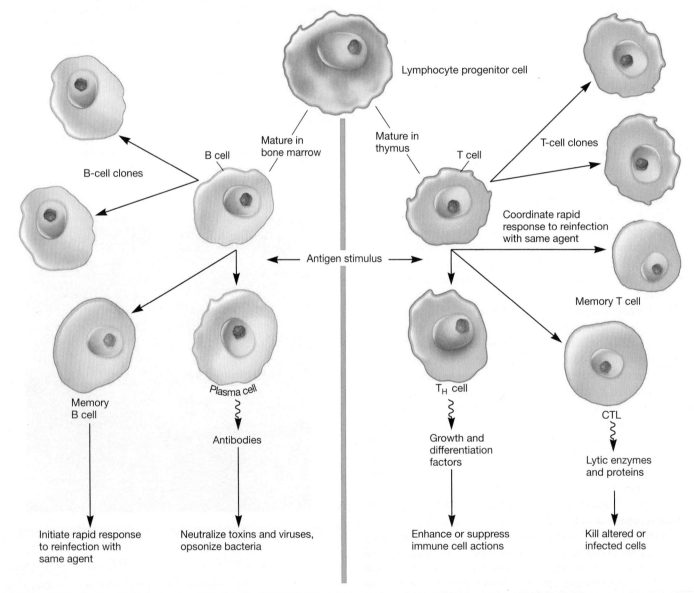

Figure 33.11 The Development and Function of B and T Lymphocytes. B cells and T cells arise from the same cell lineage but diverge into two different functional types. Immature B cells and T cells are indistinguishable by histological staining. However, they express different proteins on their surfaces that can be detected by immunohistochemistry. Additionally, the final secreted products of mature B and T cells can be used to identify the cell type.

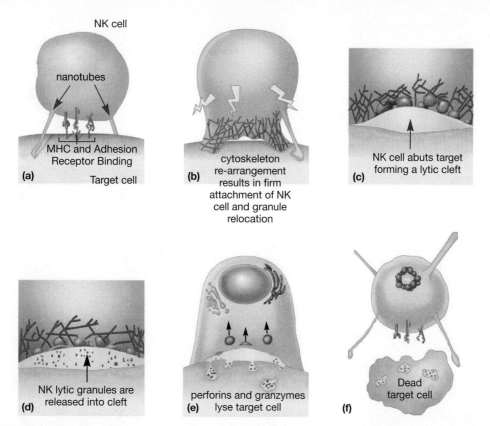

Figure 33.12 **The System Used by Natural Killer (NK) Cells to Recognize and Destroy Abnormal Cells.** (a) NK cell evaluates target cell using membrane nanotubes. (b) Microtubules organize lytic granules toward the target cell. (c) A lytic cleft forms and (d) granules exit the NK cell into the cleft. (e) Lytic perforins and granzymes lyse the target cell. (f) The NK cell releases the dead target cell and migrates toward another cell.

Cytotoxic (killer) ILCs are represented by natural killer cells. **Natural killer (NK) cells** are a small population of large, nonphagocytic granular lymphocytes. The major NK cell function is to detect and destroy stressed, malignant, or virally infected cells (**figure 33.12**). Although the exact mechanism by which NK cells recognize target cells is not known, it is clear that they survey adjacent cells by forming membrane nanotube connections, and by "kissing" targets via several surface receptors (figure 33.12*a*). These interactions provide both kill and don't kill signals to the NK cell. It is the balance of these signals that determines the fate of the target cell. If the target cell is found to be defective or infected, the NK cell kiss becomes deadly.

Once activated to kill the target cell, the NK cell mobilizes its cytoskeletal proteins to form an "immunological synapse," similar to the synapses cytotoxic lymphocytes form with their targets. When kill signals are activated, the NK cell closely adheres to the target cell and releases its cargo of deadly molecules (figure 33.12). These include the pore-forming protein perforin and enzymes called granzymes. Together, these proteins trigger the target cell to commit suicide (apoposis).

NK cells also have another killing mechanism. NK cells have receptors for antibodies, and therefore can attack cells that are opsonized by antibodies. This process is called **antibody-dependent cell-mediated cytotoxicity (ADCC)** (**figure 33.13**), and the

Figure 33.13 **Antibody-Dependent Cell-Mediated Cytotoxicity.** (a) In this mechanism, antibodies bind to a target cell infected with a virus. (b) NK cells have specific antibody receptors on their surface. (c) The antibody bridges the infected cell with the NK cell so that the target is close enough for enzymatic attack.

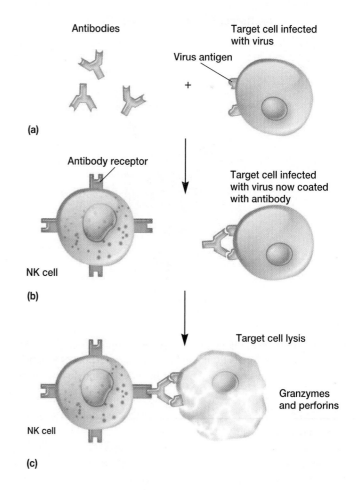

Figure 33.14 Anatomy of the Lymphoid System. Lymph is distributed through a system of lymphatic vessels, passing through many lymph nodes. Lymph enters the lymph node through afferent lymphatic vessels, percolates through and around the follicles in the node, and leaves through the efferent lymphatic vessels. The lymphoid follicles are the sites of cellular interactions and extensive immunologic activity.

MICRO INQUIRY *Why is it necessary to send lymphocytes destined to be T cells out of the bone marrow?*

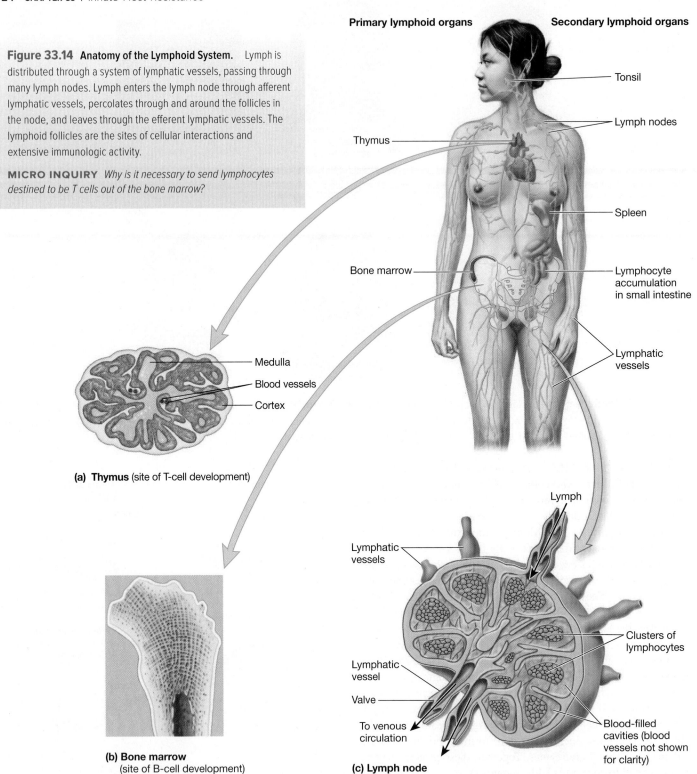

(a) **Thymus** (site of T-cell development)

(b) **Bone marrow**
(site of B-cell development)

(c) **Lymph node**

Primary lymphoid organs   Secondary lymphoid organs

Tonsil

Lymph nodes

Thymus

Spleen

Bone marrow

Lymphocyte accumulation in small intestine

Lymphatic vessels

Medulla

Blood vessels

Cortex

Lymph

Lymphatic vessels

Clusters of lymphocytes

Lymphatic vessel

Valve

To venous circulation

Blood-filled cavities (blood vessels not shown for clarity)

bridging antibodies help facilitate the death of the target cell. Unlike T and B cells, NK cells do not exhibit memory responses to target cells, thus they are considered part of the innate immune response.

Retrieve, Infer, Apply

1. What is a plasma cell?
2. What is the purpose of T-cell secretion of cytokines?
3. Discuss the role of NK cells in protecting the host. What are the two mechanisms NK cells use to detect altered cells?

Organs and Tissues of the Immune System

Based on function, the organs and tissues of the immune system can be divided into primary or secondary lymphoid organs and tissues (**figure 33.14**). The primary organs and tissues are where immature lymphocytes mature and differentiate into antigen-sensitive B and T cells. The thymus is the primary lymphoid organ for T cells, and the bone marrow is the primary lymphoid tissue for B cells. The secondary organs and tissues are areas where lymphocytes may encounter and bind antigen, which triggers their proliferation and differentiation into fully active, antigen-specific effector cells. The spleen is a secondary lymphoid organ, and lymph nodes and mucosal-associated tissues (GALT—gut-associated lymphoid tissue, and SALT—skin-associated lymphoid tissue) are secondary lymphoid tissues. These are now discussed in more detail.

Primary Lymphoid Organs and Tissues

The **thymus** is a highly organized lymphoid organ located above the heart. Precursor cells from the bone marrow migrate into the outer cortex of the thymus, where they proliferate. As they mature, about 98% die. This is due to a process known as **thymic selection** in which T cells that could recognize and respond to the host (self) are destroyed. The remaining 2% move into the medulla of the thymus (figure 33.14*a*), where they mature and subsequently enter the bloodstream. These T cells recognize and respond to the myriad of foreign nonhost substances, collectively referred to as "nonself."

In mammals, the bone marrow (figure 33.14*b*) is the site of B-cell maturation. Like thymic selection during T-cell maturation, a selection process within the bone marrow eliminates B cells that recognize self. Remaining B cells mature in the bone marrow and subsequently enter the bloodstream.

Secondary Lymphoid Organs and Tissues

The **spleen** is the most highly organized secondary lymphoid organ. It is a large organ located in the abdominal cavity that functions to filter the blood and trap blood-borne particles to be assessed for foreignness by phagocytes (figure 33.14). Macrophages and dendritic cells are present in abundance, and once trapped by splenic macrophages or dendritic cells, a pathogen is phagocytosed, killed, and digested. The resulting antigens are presented to lymphocytes, activating a specific immune response.

Lymph nodes lie at the junctions of lymphatic vessels, where macrophages and dendritic cells trap particles that enter the lymphatic system (figure 33.14*c*). If a particle is found to be foreign, it is then phagocytosed and degraded, and the resulting antigens are presented to lymphocytes.

Lymphoid tissues are found throughout the body either as highly organized or as loosely associated cellular complexes (figure 33.14). Some lymphoid cells are closely associated with specific tissues such as skin (skin-associated lymphoid tissue, or SALT) and mucous membranes (mucosal-associated lymphoid tissue, or MALT). SALT and MALT are good examples of highly organized lymphoid tissues that feature macrophages surrounded by specific areas of B and T lymphocytes and sometimes dendritic cells. Loosely associated lymphoid tissue is best represented by the bronchial-associated lymphoid tissue (BALT), because it lacks cellular partitioning. The primary role of these lymphoid tissues is to efficiently organize leukocytes to increase interaction between the innate and the adaptive arms of the immune response. Thus, the lymphoid tissues serve as the interface between the innate resistance mechanisms and adaptive immunity of a host. We now discuss these tissues in more detail.

Despite the skin's defenses, at times pathogenic microorganisms gain access to the tissue under the skin surface. Here they encounter a specialized set of cells called the **skin-associated**

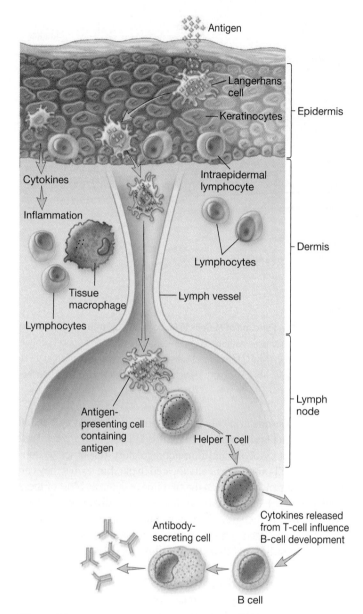

Figure 33.15 Skin-Associated Lymphoid Tissue (SALT).   Keratinocytes make up 90% of the epidermis. They can secrete cytokines that cause an inflammatory response to invading pathogens. Langerhans cells internalize antigen and move to a lymph node, where they present antigen to helper T cells. The intraepidermal lymphocytes may function as T cells that can activate B cells to induce an antibody response.

lymphoid *tissue* (SALT) (figure 33.15). The major function of SALT is to confine microbial invaders to the area immediately underlying the epidermis and to prevent them from gaining access to the bloodstream. One type of SALT cell is the **Langerhans cell,** a dendritic cell that phagocytoses microorganisms that penetrate the skin. Once the Langerhans cell has internalized a foreign particle or microorganism, it migrates from the epidermis to nearby lymph nodes, where it presents antigen to activate nearby lymphocytes, inducing a specific immune response to that antigen. This dendritic cell–lymphocyte interaction illustrates another bridge between innate resistance and adaptive immunity.

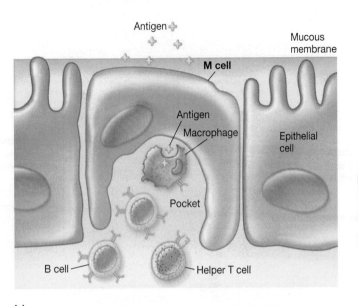

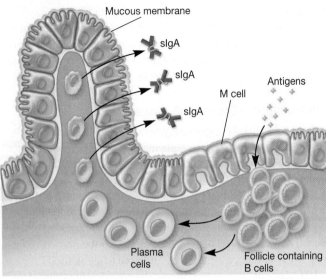

(a)

(b)

Figure 33.16 Function of M Cells in Mucosal-Associated Immunity. (a) Structure of an M cell located between two epithelial cells in a mucous membrane. The M cell endocytoses pathogens and releases them into the pocket containing helper T cells, B cells, and macrophages. It is within the pocket that pathogens often are destroyed. (b) Once in the M cell pocket, antigen enters an organized lymphoid follicle containing B cells. The activated B cells mature into plasma cells, which produce antibodies (sIgA; secretory immunoglobulin A) that are released into the lumen, where they react with the antigen that caused their production.

The epidermis also contains another type of SALT cell called the **intraepidermal lymphocyte** (figure 33.15), a specialized T cell having potent cytolytic and immunoregulatory responses to antigen. These cells are strategically located in the skin so that they can intercept antigens that breach the first line of defense. Most of these specialized SALT cells have limited receptor diversity and have likely evolved to recognize common skin microbe patterns.

The specialized lymphoid tissue in mucous membranes is called *mucosal-associated lymphoid tissue* (**MALT**). There are several types of MALT. The system most studied is the **gut-associated lymphoid tissue** (**GALT**). GALT includes the tonsils, adenoids, diffuse lymphoid areas along the gut, and specialized regions in the intestine called Peyer's patches. Less well-organized MALT also occurs in the respiratory system and is called *bronchial-associated lymphoid tissue* (**BALT**); the diffuse MALT in the genitourinary system does not have a specific name.

MALT operates by two basic mechanisms. First, when a foreign particle arrives at the mucosal surface, it contacts an **M cell** (**figure 33.16a**). They reside above large epithelial pockets containing B cells, T cells, and macrophages. When a foreign particle contacts an M cell, it is endocytosed and released into the pocket. Macrophages engulf the particle or pathogen and degrade it. An M cell also can endocytose particles and release them to a cluster of cells called an organized lymphoid follicle (figure 33.16b). The B cells within this follicle react to the foreign particle and mature into antibody-producing plasma cells. The plasma cells leave the follicle and secrete mucous membrane–associated antibody. In the case of GALT, the antibody is transported into the lumen of the gut, where it interacts with the particles that caused its production. Similar to SALT, GALT intra- and interepithelial lymphocytes are strategically distributed so that the detection of foreign particles

is increased, should the intestinal membrane be breached. ▶▶ *Antibodies are proteins that bind to specific 3-D molecules (section 34.7)*

As seen in our discussion thus far, a microbe attempting to invade a potential host encounters innate mechanisms including physical, chemical, and cellular barriers that are designed to kill the invader, digest the carcass into small antigens, and assist the lymphocytes in formulating long-term protection against the next invasion. We now examine the phagocytic processes in more detail and then consider how the host integrates many of the innate resistance activities into a substantial barrier to microbial invasion, known as the inflammatory response.

Retrieve, Infer, Apply

1. What is the function of the spleen? A lymph node? The thymus? What is the importance of thymic selection?
2. Injury to the spleen can lead to its removal. What impact would this have on host defenses?
3. Describe SALT and MALT functions in innate resistance.

33.5 Phagocytosis: Destroying Invaders and Recycling Their Parts

After reading this section, you should be able to:

- Explain the methods by which pathogens are recognized by phagocytes
- Describe the processes of autophagy and phagocytosis
- Forecast how biochemical activities within the phagocyte result in pathogen destruction

During their lifetimes, humans and other vertebrates encounter many microbial species, but only a few of these can grow and cause serious disease in otherwise healthy hosts. Phagocytic cells (macrophages, dendritic cells, and neutrophils) are an important early defense against invading microorganisms. These cells recognize, ingest, and kill cellular microbes by the process called **phagocytosis** (Greek *phagein*, to eat; *cyte*, cell; and *osis*, a process) and intracellular pathogens by the process of **autophagocytosis** (or autophagy). Phagocytic cells use two basic molecular mechanisms to recognize microorganisms: (1) opsonin-independent (nonopsonic) recognition and (2) opsonin-dependent (opsonic) recognition. They then encase the microorganisms within a "killing chamber." Recall that we discussed opsonic recognition in section 33.3, focusing on complement proteins and collectins. We now discuss nonopsonic recognition of foreignness and provide more detail regarding the result of pattern recognition by phagocytes.

Recognition of Foreignness Is Key for Survival

The ability to discriminate friend from foe affords a clear evolutionary advantage. This is no less true at the cellular level. Recognition of foreignness by mammals evolved in at least two ways. Both involve receptors that recognize molecules bearing microbial-associated molecular patterns (MAMPS). This ancient system is used by cells during innate immune responses. Detection of MAMPS begins with receptors that recognize molecules with repetitive patterns, and ends with common signal transduction mechanisms that produce antimicrobial responses (**figure 33.17**). PRMs are used by sentinel cells (macrophages, mast cells, and dendritic cells) in invader surveillance. They include C-type lectin receptors (CLRs), toll-like receptors (TLRs), nucleotide oligomerization domain (NOD)-like receptors (NLRs), and retinoic acid-inducible gene I (RIG-I)-like receptors (RLRs). We next examine the CLRs and TLRs that are membrane-bound. NLRs and RLRs are found in the cytosol, and are discussed later.

C-Type Lectin Receptors

CLRs are a large group of calcium-dependent membrane-bound proteins having one or more lectin-binding domains characterized by conserved amino acid residues specific for one type of carbohydrate (lectin). Thus a C-type lectin recognizes and binds to a specific carbohydrate found on MAMPs (section 33.4). They generally bind mannose, fucose, or glucan carbohydrates found on some bacteria (*Mycobacterium tuberculosis* and *Helicobacter pylori*), fungi (e.g., *Candida albicans* and *Aspergillus fumigatus*), helminths, and some viruses. CLRs also bind *N*-acetylglucosamine components of a number of different microorganisms, and the capsular polysaccharides of *Streptococcus pneumoniae* and *Klebsiella pneumoniae*. CLRs are concentrated on macrophages and dendritic cells. CLR-MAMP binding results in a signaling cascade that induces cytokine gene expression. This is mediated by the CLR cytoplasmic domain, which is called the immunoreceptor tyrosine-based activation motif (ITAM) in the CLR cytoplasmic domain. Binding also results in receptor-MAMP internalization, its degradation, and subsequent MAMP presentation to lymphocytes, as described later.

Toll-Like Receptors

Introduced in section 33.4, we now continue our discussion of **toll-like receptors** (**TLRs**). TLRs are transmembrane receptors expressed by a variety of cells that participate in innate immunity, especially macrophages and dendritic cells. They are also found on the intracellular organelles of those cells, mainly on membranes of endosomes, lysosomes, endolysosomes, and endoplasmic reticulum, where they detect cytoplasmic MAMPs. These transmembrane receptors have an extracellular (or extra-organelle) leucine-rich repeat domain that binds MAMPs, and a cytoplasmic Toll/interleukin-1 receptor (TIR) domain that activates downstream transcription signaling (figure 33.17). Binding of a MAMP to a TLR causes the TIR to recruit a specific adapter protein, either MyD88 or TRIF, which in turn facilitates transcription of distinct sets of immune response genes. Activation through the MyD88-dependent adapter causes NFκB to activate transcription of pro-inflammatory cytokine genes. All TLRs except TLR3 can use this pathway. Alternatively, activation through the TRIF-dependent adapter causes NFκB and IRF3 pathways to express type I interferon, chemokines, and inflammatory cytokines. TLRs 3 and 4 use TRIF. There are at least 10 distinct human TLRs that recognize and subsequently initiate an immune response.

NOD-Like Receptors

Unlike the transmembranous TLRs, **nucleotide-binding and oligomerization domain (NOD)-like receptors** are cytosolic. They too can bind to MAMPs. Once bound to MAMPs, they recruit and bundle additional proteins in a process known as oligomerization. NOD-like receptors (NLRs) sense intracellular MAMPs (e.g., viral RNA, anthrax toxin, group B streptococci, and *Listeria monocytogenes*), along with host molecules called "damage-associated molecular patterns" (DAMPs), such as uric acid and heat-shock proteins. DAMPs are a diverse group of molecules that arise from cell injury and stress.

NLRs are characterized by leucine-rich repeats (LRRs) within the C-terminus and nucleotide binding domain (NBD). LRRs bind the MAMP or DAMP, stimulating the NBD to bind ribonucleotides. This facilitates oligomerization of several NLRs and recruits co-factors and the enzyme capsase-1. This NLR-capsase complex is called the **inflammasome**, because it enhances inflammation when capsase-1 cleaves and activates pro-inflammatory cytokines, especially pro-IL-1, which is produced in response to signals transduced by other PRMs, such as a TLR. Thus NLRs and TLRs can work together to provide a stronger inflammatory response (**figure 33.18**). Different inflammasomes form in response to the various MAMP or DAMP activators.

RIG-I-Like Receptors

Retinoic-acid-inducible gene I (RIG-I)-like receptors (RLRs) are yet another receptor family that are cytoplasmic. Structurally, RLR proteins have two N-terminal capsase-recruitment domains (CARDS), an RNA helicase domain, and a C-terminal repressor domain (RD). Recognition and binding of viral RNAs are through the helicase and RD. Upon RLR binding of viral RNA,

Figure 33.17 Recognition of Microbe-Associated Molecular Patterns (MAMPs) by Pattern Recognition Molecules (PRMs). MAMPs bind PRMs, especially toll-like receptors (TLRs), C-type lectin receptors (CLRs), and other pattern recognition receptors. PRM binding results in signaling that upregulates cytokine gene expression through common signal transduction pathways, like NF$_\kappa$B.

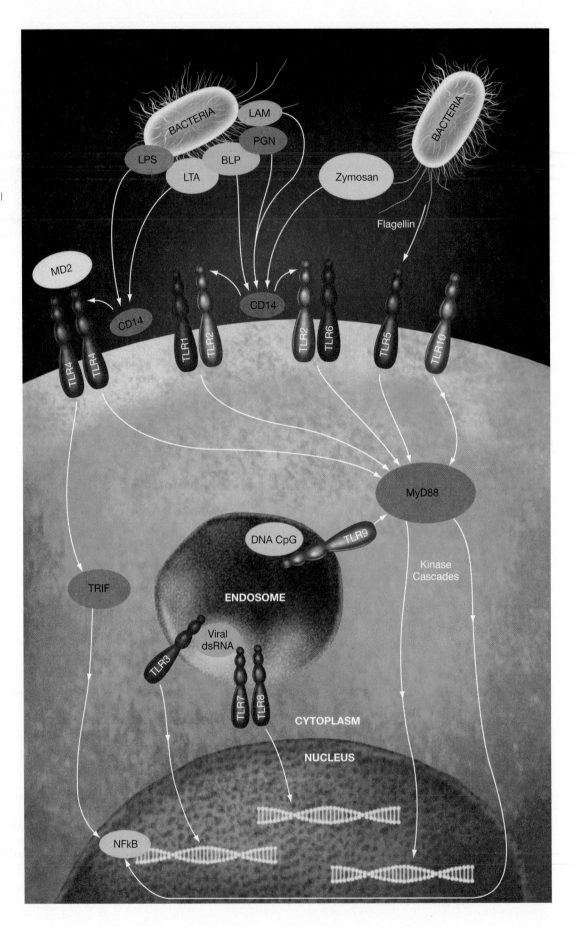

the CARDS domains trigger intracellular signal cascades leading to expression of interferon. Both dsRNA and ssRNA virus replication lead to dsRNA and triphosphate-capped ssRNA during the course of infection. RLRs recognize these forms of viral RNA and stimulate an antiviral cytokine response.

Intracellular Digestion

Ingestion and subsequent digestion of potential nutrient sources is a highly conserved process across species. Importantly, it has evolved into an intricate detection and removal process that extends beyond nutrient acquisition. Two parallel processes, phagocytosis and autophagy, accomplish ingestion and digestion, and are found to be widely distributed in nature. Phagocytosis imports extracellular particles via the extension of the cell membrane (as pseudopodia), to end up with an internal phagosome; this is often receptor-mediated (**figure 33.19*a,b***). Autophagy begins when intracellular microorganisms, or their products, are captured by intracellular membranes through PRMs, or are coated with

ubiquitin, the tagging protein used for recycling by proteasomes (*see figure 5.8*). In this application, ubiquitin "labels" microorganisms for capture by a phagophore (free-floating, open membrane). The phagophore then encircles the microorganism, creating the double-membrane autophagosome (**figure 33.20**). Autophagy is widely recognized as an ancient means of homeostasis whereby cellular components are recycled. ◄◄ *There are several endocytic pathways (section 5.4)*

Once membrane-bound, microorganisms are delivered to and fused with lysosomes, fusion of phagosome and lysosome creates the phagolysosome (figure 33.19*c,d*). Fusion of the autophagosome and lysosome creates the autolysosome (figure 33.20). **Lysosomes** deliver a variety of hydrolases such as lysozyme, phospholipase A2, ribonuclease, deoxyribonuclease, and proteases. The activity of these degradative enzymes is enhanced by the acidic vacuolar pH. Collectively these enzymes participate in the destruction of the entrapped microorganisms. In addition, toxic **reactive oxygen species (ROS)** such as the superoxide radical (O_2^-), hydrogen peroxide (H_2O_2), singlet oxygen ($^1O_2$), and hydroxyl radical ($\cdot OH$) are produced. Neutrophils also contain the heme-protein enzyme, myeloperoxidase, which catalyzes the production of hypochlorous acid. Some reactions that form ROS are shown in **table 33.4**. These reactions result in part from the respiratory, or oxidative, burst that accompanies the increased oxygen consumption and ATP generation needed for phagocytosis. The **oxidative burst** occurs within the phagosome or autophagosome as soon as it is formed. In this way, ROS are immediately effective in killing microorganisms. ◄◄ *Oxygen concentration (section 7.4)*

Macrophages, neutrophils, and mast cells have also been shown to form **reactive nitrogen species** (RNS). These molecules include nitric oxide (NO) and its oxidized forms, nitrite (NO_2^-) and nitrate (NO_3^-). RNS are very potent cytotoxic agents; nitric oxide is probably the most effective RNS. It can block cellular respiration by complexing with iron in electron transport proteins. Macrophages

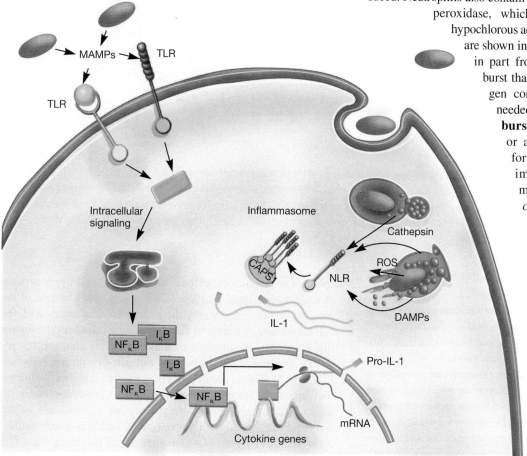

Figure 33.18 **Cooperation Between Two Pattern Recognition Molecules (PRMs).** Binding of toll-like and Nod-like receptors (TLR and NLR, respectively) by microbe-associated molecular patterns (MAMPs) activates two different response mechanisms that cooperate to produce a potent pro-inflammatory cytokine, interleukin (IL)-1. Binding of MAMPs to TLRs activates an ancient signal transduction system that releases the $NF_\kappa B$ transcription factor to upregulate gene expression of pro-IL-1. Importantly, binding of other MAMPs, cathepsin, or reactive oxygen species (from damaged phagolysosomes) to NLRs results in the oligomerization of several NRLs, which then assemble and activate the proteolytic enzyme, capsase (CAPS)-1. CAPS-1 cleaves pro-IL-1 into active IL-1 cytokine, upregulating numerous inflammatory cells and biochemical pathways.

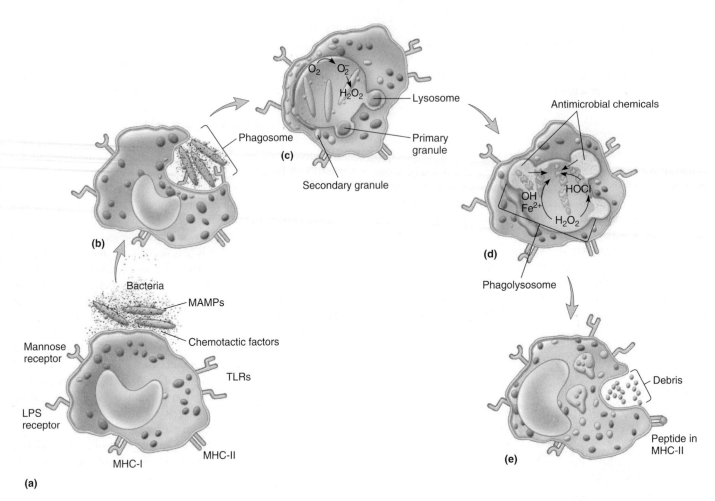

Figure 33.19 Phagocytosis. (a) Receptors on a phagocytic cell (e.g., a macrophage) and the corresponding MAMPs participating in phagocytosis. The process of phagocytosis includes (b) ingestion, (c) participation of granules, and O_2-dependent killing events, (d) intracellular digestion, and (e) exocytosis. As with autophagy, digested fragments can also enter the endoplasmic reticulum to be processed for antigen presentation. MHC proteins are discussed in chapter 34. LPS receptor: lipopolysaccharide receptor; TLRs: toll-like receptors; MHC-I: class I major histocompatibility protein; MHC-II: class II major histocompatibility protein; MAMPs: microbe-associated molecular patterns.

use RNS in the destruction of a variety of infectious agents as well as to kill tumor cells.

Neutrophil granules contain a variety of other microbicidal substances such as several cationic peptides, the bactericidal permeability-increasing protein (BPI), and broad-spectrum antimicrobial peptides, including defensins (section 33.3). These substances are compartmentalized for extracellular secretion or delivery to phagocytic vacuoles. Susceptible microbial targets include a variety of Gram-positive and Gram-negative bacteria, yeasts and molds, and some viruses. Defensins act against bacteria and fungi by permeabilizing cell membranes. This is accomplished by formation of voltage-dependent membrane channels that allow ionic efflux. Antiviral activity involves direct neutralization of enveloped viruses, so they can no longer bind host cell receptors; nonenveloped viruses are not affected by defensins.

Exocytosis

Once the microbial invaders have been killed and digested into small antigenic fragments, the phagocyte may do one of two things. The cell may expel the microbial fragments by the process of **exocytosis** (figure 33.19e). This is essentially a reverse of the phagocytic process whereby the phagolysosome unites with the cell membrane, resulting in the extracellular release of the microbial fragments. Neutrophils tend to do this. By contrast, macrophages and dendritic cells become **antigen-presenting cells** (*see figure 34.5*). This is accomplished by passing some of the microbial fragments from the phagolysosome to the endoplasmic reticulum (figure 33.19e). Here the peptide components of the fragments are united with glycoproteins, called histocompatibility proteins (*see figure 34.4*), destined for the cell membrane. The glycoproteins bind the peptides so that they are presented outwardly from the cell once the glycoprotein is secured in the cell membrane. Antigen presentation is critical because it enables wandering lymphocytes to evaluate killed microbes (as antigens) and be activated. Thus antigen presentation links an innate immune response (phagocytosis and autophagy) to a specific immune response (lymphocyte activation). ▶▶ *Recognition of foreignness is critical for a strong defense (section 34.4)*

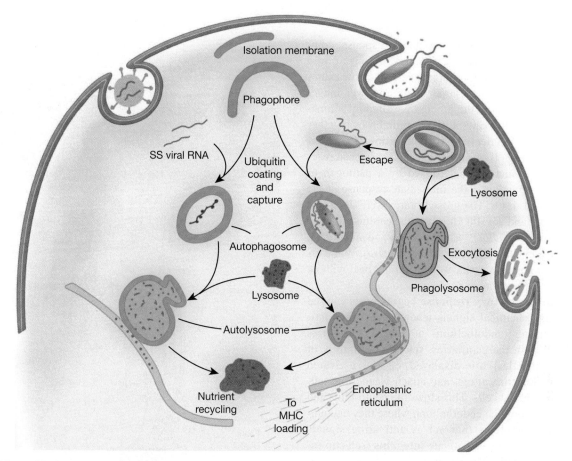

Figure 33.20 Autophagy Is Similar to Phagocytosis. Intracellular parasites such as viruses, and bacteria that escape the phagosome, are captured by free-floating membranes that form a phagophore, often through membrane-bound toll-like receptors. Cytosolic ubiquitin molecules coat the intracellular invaders and their macromolecules (such as viral RNA). The ubiquitinated particles are then surrounded by the phagophore to form an autophagosome. Like the phagosome that becomes the phagolysosome during phagocytosis, the autophagosome (autosome) fuses to lysosomes to digest its contents. Once the lysosome fuses with the autosome, it is called the autolysosome. Both phagolysosome and autolysosome can have some of their debris donated to the endoplasmic reticulum, where antigenic peptides are further processed to bind MHC molecules destined for the phagocyte's cell membrane. The remaining debris is recycled or released to the environment.

Table 33.4	Formation of Reactive Oxygen Intermediates
Oxygen Intermediate	**Reaction**
Superoxide ($O_2^-\cdot$)	$NADPH + 2O_2 \xrightarrow{\text{NADPH oxidase}} 2O_2^-\cdot + H^+ + NADP^+$
Hydrogen peroxide (H_2O_2)	$2O_2^-\cdot + 2H^+ \xrightarrow{\text{Superoxide dismutase}} H_2O_2 + O_2$
Hypochlorous acid (HOCl)	$H_2O_2 + Cl^- \xrightarrow{\text{Myeloperoxidase}} HOCl + OH^+$
Singlet oxygen ($^1O_2$)	$ClO^- + H_2O_2 \xrightarrow{\text{Peroxidase}} {}^1O_2 + Cl^- + H_2O$
Hydroxyl radical ($\cdot OH^-$)	$O_2^-\cdot + H_2O_2 \xrightarrow{\text{Peroxidase}} 2\cdot OH^- + O_2$

Retrieve, Infer, Apply

1. How is the entrapped microorganism destroyed within the phagolysosome and autolysosome?
2. What is the purpose of the respiratory burst that occurs within macrophages and other phagocytic cells? Describe the nature and function of reactive oxygen and nitrogen species.
3. How do macrophages and dendritic cells become antigen-presenting cells?

33.6 Inflammation Unites All the Components of Immunity

After reading this section, you should be able to:

- Outline the sequence of innate host responses that result in inflammation
- Distinguish acute and chronic inflammation in terms of the host responses involved in each
- Construct a concept map relating host cells and processes that remove pathogens

The collective actions of the innate resistance response, combined with responses from adaptive defenses, culminate in a concentrated effort to remove foreign invaders. We now examine how the host responds on a larger scale and over longer time periods. Physiologically, the process is called inflammation, and its function is to bring all the host defenses together in response to injury or infection. **Inflammation** (Latin, *inflammatio,* to set on fire) is an important innate defense reaction to tissue injury, such as that caused by a pathogen or wound. Acute inflammation is the immediate response of the body to injury or cell death. The gross features were described over 2,000 years ago and are still known as the cardinal signs of inflammation: redness (*rubor*), warmth (*calor*), pain (*dolor*), swelling (*tumor*), and altered function (*functio laesa*). While often thought to be a negative event, wounds do not heal without inflammation. 🔁 *Inflammatory Response*

The **acute inflammatory response** begins when injured tissue cells release chemical signals (chemokines) that activate the inner lining (endothelium) of nearby capillaries (**figure 33.21**). Within the capillaries, **selectins** (a family of cell adhesion molecules) are displayed on the activated endothelial cells. Selectins attract and attach wandering neutrophils to the endothelial cells. This slows the neutrophils and causes them to roll along the endothelium, where they encounter inflammatory chemicals that act as activating signals (figure 33.21*b*). These signals activate **integrins** (selectin receptors) on the neutrophils. The integrins then attach tightly to the selectins, causing the neutrophils to stick to the endothelium and stop rolling (margination). The neutrophils now undergo dramatic shape changes, squeeze through the endothelial wall (diapedesis) into the interstitial tissue fluid, migrate to the site of injury (extravasation), and attack the pathogen or other cause of the tissue damage. Neutrophils and other leukocytes are attracted to the infection site by chemotactic factors, which are also called **chemotaxins.** They include substances released by bacteria, endothelial cells, mast cells, and tissue breakdown products. Depending on the severity and nature of tissue damage, other types of leukocytes (e.g., lymphocytes, monocytes, and macrophages) may follow the neutrophils.

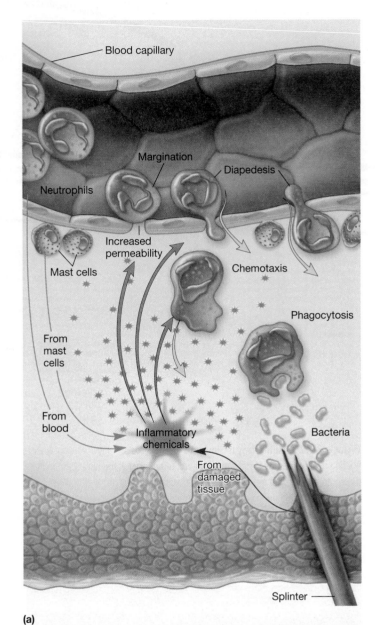

(a)

(b)

Figure 33.21 **Physiological Events of the Acute Inflammatory Response.** (a) At the site of injury (splinter), chemical messengers are released from the damaged tissue, mast cells, and the blood plasma. These inflammatory mediators stimulate neutrophil migration, diapedesis, chemotaxis, and phagocytosis. (b) Neutrophil integrins interact with endothelial selectins (1) to facilitate margination (2) and diapedesis (3).

MICRO INQUIRY *How does the movement of leukocytes in response to chemokines contribute to the cardinal signs of inflammation?*

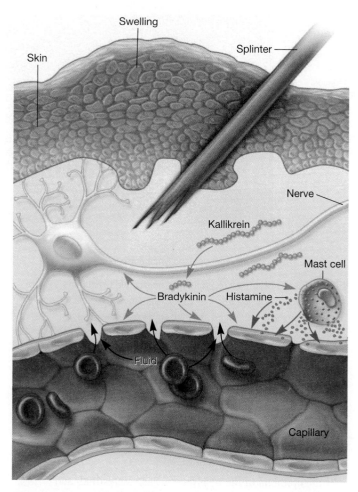

Figure 33.22 Tissue Injury Results in the Recruitment of Kallikrein, from Which Bradykinin Is Released. Bradykinin acts on endothelial and nerve cells, resulting in edema and pain, respectively. It also stimulates mast cells to release histamine. Histamine acts on endothelial cells, further increasing blood leakage into injured tissue sites.

The release of inflammatory mediators from injured tissue cells sets into motion a cascade of events that results in the development of the signs of inflammation. One response that ensues is the stimulation of local macrophages and distant liver cells to release antimicrobial and acute-phase proteins, respectively. In the local response, these mediators increase the acidity in the surrounding extracellular fluid, which activates the extracellular enzyme **kallikrein** (figure 33.22). Cleavage of kallikrein releases the smaller peptide bradykinin. Bradykinin then binds to receptors on the capillary wall, opening the junctions between cells and allowing fluid, red blood cells, and infection-fighting leukocytes to leave the capillary and enter the infected tissue. Simultaneously, bradykinin binds to mast cells in the connective tissue associated with most small blood vessels. This activates mast cells by causing an influx of calcium ions, which leads to degranulation and release of preformed mediators such as histamine. If nerves in the infected area are damaged, they release substance P, which also binds to mast

cells, boosting preformed-mediator release. Histamine in turn makes intercellular junctions in the capillary wall wider so that more fluid, leukocytes, kallikrein, and bradykinin move out, causing swelling or edema. Bradykinin then binds to nearby capillary cells and stimulates the production of prostaglandins (PGE_2 and PGF_2) to promote tissue swelling in the infected area. Prostaglandins also bind free nerve endings, making them fire and start a pain impulse. At the same time, liver cells release additional complement proteins, the iron-binding glycoprotein lactoferrin, and various scavenging proteins called collectins (p. 717).

Activated mast cells also release a small molecule called arachidonic acid, the product of a reaction catalyzed by phospholipase A_2. Arachidonic acid is metabolized by mast cells to form potent mediators, including PGE_2 and PGF_2, thromboxane, slow-reacting substance (SRS), and leukotrienes (LTC_4 and LTD_4). These mediators play specific roles in the inflammatory response. During acute inflammation, the offending pathogen is neutralized and eliminated by a series of important events:

1. The increase in blood flow and capillary dilation bring into the area more antimicrobial factors and leukocytes that destroy the pathogen. Dead host cells also release antimicrobial factors.
2. Blood leakage into tissue spaces increases the temperature at that site and further stimulates the inflammatory response and may inhibit microbial growth.
3. A fibrin clot often forms and may limit the spread of the invaders.
4. Phagocytes collect in the inflamed area and phagocytose the pathogen. In addition, chemicals stimulate the bone marrow to release neutrophils and increase the rate of granulocyte production.

An acute inflammatory response resolves in days to a few weeks. Prolonged stimulation of the inflammatory pathways leads to progressive changes in cellular response and outcome, known as chronic inflammation. Recall that the end result of acute inflammation is tissue healing and repair. However, **chronic inflammation** is detrimental and is characterized by infiltration of lymphocytes and macrophages into the affected site and the formation of new connective tissue. This usually causes permanent tissue damage. As cycles of cellular infiltration and resolution continue, degradative enzymes from macrophages destroy more underlying tissue than is replaced. Furthermore, if innate responses are unable to protect the host from ongoing infection, chronic tissue injury, or the persistence of poorly degradable materials (sutures, implants, etc.), the body attempts to wall off and isolate the site by forming a **granuloma** (Latin, *granulum*, a small particle; Greek, *oma*, to form). A granuloma is a well-organized mass of neutrophils, epithelioid macrophages, eosinophils, multinucleated giant cells (two or more cells fused into one large cell), fibroblasts, and collagen (**figure 33.23**). Together, the cells and extracellular matrix proteins form a spherical mass that can bind calcium and form a recalcitrant nodule, mineralizing the granuloma and making it harder to break down.

Importantly, chronic inflammation can occur as a distinct process without much acute inflammation, as evidenced by the chapter opening story about visceral fat causing chronic inflammation. The persistence of bacteria can also stimulate chronic inflammation. For example, mycobacteria, some of which cause tuberculosis and leprosy, have cell walls with a very high lipid content, making them relatively resistant to phagocytosis and intracellular killing. Granulomas formed around mycobacteria often calcify and appear on chest X rays, indicating potential tuberculosis. Of note is the fact that the mycobacterial granuloma keeps the bacteria from spreading throughout the body. However, if the immune system declines, granuloma integrity fails, releasing the bacteria into the alveoli, where they are coughed up and potentially transmitted to another host, having survived for many years within macrophages. ◀◀ *Order* Corynebacteriales *includes important human pathogens (section 24.1)*

Retrieve, Infer, Apply

1. What major events occur during an inflammatory reaction, and how do they contribute to pathogen destruction?
2. How does chronic inflammation differ from acute inflammation?

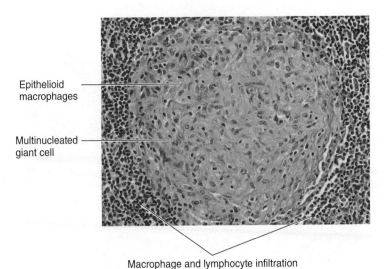

Figure 33.23 Photomicrograph of a Tuberculoid Granuloma. The granuloma is a well-organized mass of cells and proteins resulting from a chronic inflammatory response. Note the epithelioid macrophages in the granuloma as contrasted to the large number of normal macrophages surrounding it. Granuloma formation sequesters the mycobacteria, restricting access to its inflammatory products.

Key Concepts

33.1 Immunity Arises from Innate Resistance and Adaptive Defenses

- There are two interdependent components of the immune response to invading microorganisms and foreign material; innate resistance and adaptive immune responses.
- Innate resistance mechanisms offer substantial host defense against any microorganism or foreign material. They include general mechanisms that are a part of the animal's innate structure and function. The nonspecific system has no immunological memory; that is, nonspecific responses occur to the same extent each time.
- The adaptive immune response resists a particular foreign agent; moreover, adaptive immune responses improve on repeated exposure to the agent.
- Numerous cells and processes bridge the innate and adaptive immune responses (**figure 33.1**).

33.2 Innate Resistance Starts with Barriers

- Many direct factors (age, nutrition) or general barriers contribute in some degree to all host-microbe relationships. At times, they favor the establishment of the microorganism; at other times, they provide the host some measure of general defense (**figure 33.2**).
- Physical and mechanical barriers along with host secretions are the host's first line of defense against pathogens. Examples include the skin and mucous membranes and

the epithelia of the respiratory, gastrointestinal, and genitourinary systems (**figures 33.3–33.5**).

33.3 Innate Resistance Relies on Chemical Mediators

- Mammalian hosts have specific chemical barriers that help combat the continuous onslaught of pathogens. Examples include cationic peptides, bacteriocins, cytokines, acute-phase proteins, and complement.
- The complement system is composed of a large number of serum proteins that play a major role in an animal's defensive immune response. There are three pathways of complement activation: the alternative, lectin, and classical pathways (**figure 33.7** and **table 33.1**).
- Cytokines are required for regulation of both the innate and adaptive immune responses. Cytokines have a broad range of actions on eukaryotic cells (**tables 33.2** and **33.3**).
- Acute-phase proteins are released from liver cells to bind microbial surfaces. They act as opsonins.

33.4 Cells, Tissues, and Organs Work Collectively to Form an Immune System

- The cells responsible for both nonspecific and adaptive immunity are white blood cells, which are also called leukocytes (**figure 33.9**). Examples include monocytes and macrophages, dendritic cells, granulocytes, and mast cells. Phagocytic leukocytes use a receptor system (pattern recognition molecules; PRMs) to detect molecular patterns found on pathogens (microbe-associated molecular patterns; MAMPs).

- Immature undifferentiated lymphocytes are generated in the bone marrow and become committed to a particular antigenic specificity within the primary lymphoid organs and tissues. In mammals, T cells mature in the thymus and B cells in the bone marrow. The thymus is the primary lymphoid organ; the bone marrow is the primary lymphoid tissue (**figure 33.14**).

- Natural killer cells are a small population of large, nonphagocytic lymphocytes that destroy cancer cells and cells infected with microorganisms (**figures 33.12 and 33.13**).

- The secondary lymphoid organs and tissues serve as areas where lymphocytes may encounter and bind antigens, then proliferate and differentiate into fully mature, antigen-specific effector cells. The spleen is a secondary lymphoid organ, and the lymph nodes and mucosal-associated tissues (GALT and SALT) are the secondary lymphoid tissue (**figures 33.14–33.16**).

33.5 Phagocytosis: Destroying Invaders and Recycling Their Parts

- Phagocytic cells use two basic mechanisms for the recognition of extracellular microorganisms:

opsonin-dependent and opsonin-independent. Detection of intracellular microorganisms is with opsonic-independent methods.

- Phagocytes use PRMs to detect MAMPs and damage- or danger-associated molecular patterns (DAMPs) on microorganisms and host cells. Membrane-bound and cytoplasmic PRMs transduce signals via the same molecular pathway to activate the inflammatory response. Toll-like receptors and NOD-like receptors are distinct classes of pathogen recognition receptors (**figures 33.17 and 33.18**).

- Phagocytosis involves the recognition, ingestion, and destruction of pathogens by lysosomal enzymes, reactive oxygen species, defensins, reactive nitrogen species, and metallic ions. A more conserved method of detection and digestion, called autophagy, uses intracellular membranes and lysosomes to kill and degrade intracellular pathogens (**figures 33.19 and 33.20**).

33.6 Inflammation Unites All the Components of Immunity

- Inflammation is one of the host's nonspecific defense mechanisms to tissue injury and infection.

- Inflammation can either be acute or chronic (**figures 33.21–33.23**).

Compare, Hypothesize, Invent

1. Some pathogens invade cells; others invade tissue spaces. Explain how innate responses differ for both types of pathogens.

2. How might the various antimicrobial chemical factors be developed into new methods to control infectious disease, such as for antibiotic resistant bacteria?

3. How might a scientist use selective gene "knockouts" in bacteria to test the role of the toll-like receptor proteins?

4. Conservation of resources is a common theme in biology. However, it is only recently that scientists are piecing together how human responses to intracellular damage, multicellular injury, invading microorganisms, and chronic toxin exposure share survival processes. It seems that

mitochondria may have a prominent role in wide-scale sensing of cellular trouble and alerting ancient evolutionary processes to respond.

Recalling what mitochondria do for the cell, what would you deduce mitochondria do to detect and announce danger to the cell?

What types of experiments could you do to test your hypothesis?

Why would an "ancient" signaling process be used for such an important function?

Read the original paper: Tschopp, Jurg. 2011. Mitochondria: Sovereign of inflammation? *Eur. J. Immunol.* 41:1196.

Learn More

34

Adaptive Immunity

Charles Darwin (pictured here) heavily influenced the writer H.G. Wells, who's book War of the Worlds *ends with the invading Martians dying from infections due to earthly bacteria. His conclusion that adaptation had "won" humans the right to exist reflected Darwin's theory.*

It's in My Genes?

Next time someone wants to know why you never seem to get sick, you might say it's in your genes. A recent paper in *Science* magazine suggests that modern HLA DNA is rooted in an ancestral "family" tree that may include the infusion of DNA from primate relatives of *Homo sapiens*. HLA genes encode proteins important in immune function; these proteins identify "self" to the immune system as well as help activate an immune response when the host is threatened by an infectious agent. Scientists at Stanford University scanned the genomes of several thousand modern humans from around the globe and compared their HLA sequences to HLA sequences from the genomes of three Neanderthal and one Denisovan (the only one known to date) hominoids. They found that modern humans share several long stretches of HLA DNA with the ancient hominids and that these sequences were pervasive in modern genomes. The Stanford scientists believe that because HLA DNA diversifies quickly, the only possible explanation for the shared sequences is interbreeding between the common human ancestors and their archaic "cousins." Importantly, the data demonstrate that the HLA alleles from Neanderthals and Denisovans are more common in Europeans and Asians than Africans, pointing to the possibility that the archaic alleles entered the human gene pool after humans left Africa. All told, the modern HLA DNA has had hundreds of thousands of years to adapt to foreign antigens, and with the potential infusion of DNA from archaic cousins, it is exceedingly robust in protecting modern humans from foreign invasion.

Chapter 33 discusses nonspecific host resistance and the innate mechanisms by which the host is protected from invading microorganisms. Recall that the innate resistance system responds to a foreign substance in the same manner and to the same magnitude each time and that its activation can assist in the formation of adaptive immune responses. We now continue our discussion of the immune response by describing the adaptive responses used to protect the host. Although all vertebrates are born with adaptive immunity, it requires sufficient time to fully develop. Unlike innate immunity, upon subsequent exposure to a substance (antigen),

activation of an adaptive immune response is significantly faster and stronger than the initial response. Additionally, a fully mature immune response involves cooperation between the host's ever-present innate resistance mechanisms and inducible adaptive responses.

Readiness Check:

Based on what you have learned previously, you should be able to:

✔ Explain basic eukaryotic cell biology (sections 5.1–5.7)
✔ Diagram the flow of information in eukaryotes from DNA to protein and regulation of this flow (sections 15.2–15.4)
✔ Discuss the importance of the host complement response (section 33.3)
✔ Recall the structure and function of host leukocytes (section 33.4)
✔ Outline the events that occur during phagocytosis and inflammation (sections 33.5 and 33.6, respectively)

34.1 Adaptive Immunity Relies on Recognition and Memory

After reading this section, you should be able to:

■ Contrast host innate resistance with adaptive immunity
■ Outline the localization of B and T cells during development

Although the separation and discussion of innate and adaptive functions of immunity is artificial, it allows us to distinguish the preprogrammed, automatic responses of the innate side from the more diverse, developmental, and long-lasting aspects of the adaptive side. In reality the two sides of this coin we call immunity represent a single, elegant recognition and response system that is intimately integrated (**figure 34.1**).

The adaptive immune system of animals has three major functions: (1) to recognize anything that is foreign to the body ("nonself"); (2) to respond to this foreign material; and (3) to remember the foreign invader. The recognition response is highly specific. The immune system is able to distinguish one pathogen

they are fully mature. The T-cell progenitors leave the bone marrow and migrate to the thymus, where they develop into at least four subsets of T cells. They also populate secondary lymphoid tissue sites once they mature (*see figure 33.14*). Mature B cells and T cells are induced to elicit an **effector response** when a foreign, nonself material is detected. A successful effector response either eliminates the foreign material or renders it harmless to the host. As a consequence of initial activation, some lymphocytes enter a state of limited activity and are referred to as **memory cells** because they will "remember" the foreign material that initially activated them. If the same substance is encountered at a later time, the B and T memory cells are already programmed to mount a response that is more intense and rapid to eliminate the nonself material to once again protect the host.

Four characteristics distinguish adaptive immunity from innate resistance:

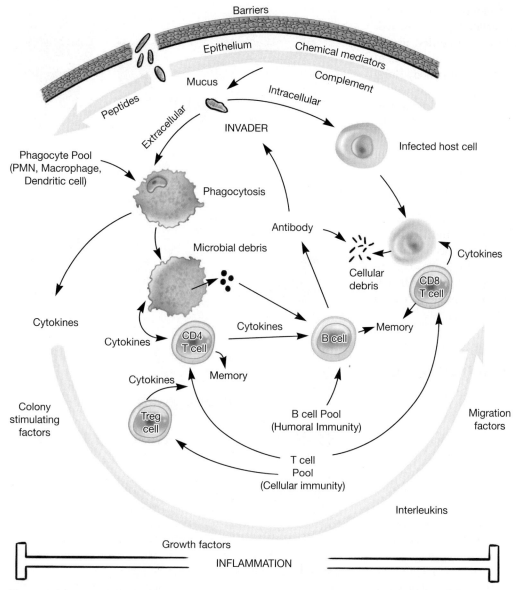

Figure 34.1 Innate and Acquired Components Integrate to Be an Elegant Immune System.   Barriers, cells, chemical mediators, and inflammation of the innate arm of immunity respond maximally to every host insult. B-cell (humoral) and T-cell (cell-mediated) components of the adaptive arm of immunity respond selectively, specifically, and with memory to each insult. Together the two arms of immunity generate a highly effective immune protection system.

from another, to identify cancer cells, and to discriminate the body's own "self" molecules and cells as different from "nonself" molecules and cells. After recognition of nonself, the adaptive system responds by activating and amplifying specific **lymphocytes** to attack it. Recall that we defined lymphocytes in chapter 33 based on their physical appearance and generic duties within the host. In this chapter, we define them based on their specific biological functions.

The lymphocytes of concern are B cells and T cells. To begin, B- and T-cell progenitor cells develop in the embryonic bone marrow and relocate to specialized tissue sites for maturation. The B cells relocate to microenvironments within the bone marrow, leaving to seed secondary tissue sites when

1. **Discrimination between self and nonself.** The adaptive immune system almost always responds selectively to nonself and produces responses tailored to protecting against that type stimulus (e.g., virus vs. bacterium).

2. **Diversity.** The system is able to generate an enormous diversity of molecules such as cellular receptors and soluble proteins, including antibodies, that recognize trillions of different foreign substances.

3. **Specificity.** Immunity is highly selective in that it can be directed against one particular pathogen or foreign substance (among trillions); the immunity to one pathogen or substance usually does not confer immunity to others.

4. **Memory.** When reexposed to the same pathogen or substance, the host reacts so quickly that there is usually no noticeable illness. By contrast, the reaction time for innate defenses is just as long for later exposures to a given antigen as it was for the initial one.

Two branches or arms of adaptive immunity are recognized (figure 34.1): humoral immunity and cellular (cell-mediated) immunity. **Humoral (antibody-mediated) immunity,** named for the fluids or "humors" of the body, is based on the action of soluble glycoproteins called antibodies that occur in body fluids and on the plasma membranes of B lymphocytes. Circulating antibodies bind to microorganisms, toxins, and extracellular viruses, neutralizing them or "tagging" them for destruction by phagocytes and other mechanisms described in section 34.8. **Cellular (cell-mediated) immunity** is based on the action of specific kinds of T lymphocytes that directly attack target cells infected with viruses or parasites, transplanted cells or organs, and cancer cells. T cells release cytokines that can induce or assist with target cell suicide (apoptosis), target cell lysis, B-cell development, antibody production, or other functions that enhance adaptive immunity and innate defenses, such as phagocytosis and inflammation. Because the activity of the acquired immune response is so potent, it is imperative that T and B cells consistently discriminate between self and nonself with great accuracy. How they accomplish this is discussed next.

34.2 Molecules That Elicit Immunity Are Called Antigens

After reading this section, you should be able to:

- Predict the types of molecules that can serve as antigens
- Compare haptens and true antigens

While cells of the innate arm of immunity use a limited library of molecular patterns with which to identify nonself, the adaptive arm of immunity uses a more sophisticated database of biochemical and structural features to recognize foreignness. The term **antigen** refers to any self or nonself substance that elicits an immune response. Antigens include microbe-associated molecular patterns (MAMPS, described in chapter 33) as well as other molecules, including proteins, nucleoproteins, polysaccharides, and some glycolipids. While the term immunogen (*immun*ity *gen*erator) is a more precise term for a substance that elicits an immune response, antigen is used more frequently. Most antigens are large, complex molecules with a molecular mass generally greater than 10,000 daltons (Da). The ability of a molecule to function as an antigen depends on its size, structural complexity, chemical nature, and degree of foreignness to the host. ◄◄ *Lymphocytes (section 33.4)*

Each antigen can have several **antigenic determinant sites,** or **epitopes (figure 34.2).** Epitopes are the regions or sites of the antigen that bind to a specific antibody or T-cell receptor. Chemically, epitopes include sugars, organic acids and bases, amino acid side chains, hydrocarbons, and aromatic groups. The number of epitopes on the surface of an antigen is its **valence.** The valence determines the number of antibody molecules that can combine with the antigen at one time. If one determinant site is present, the antigen is monovalent. Most antigens, however, have more than one copy of the same epitope and so are multivalent. Multivalent

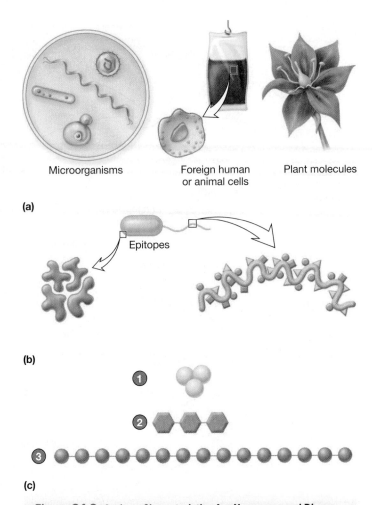

(a)

(b)

(c)

Figure 34.2 Antigen Characteristics Are Numerous and Diverse.
(a) Whole cells, viruses, and some plant molecules are immunogenic, because they (b) have complex, three-dimensional configurations. (c) Molecules that are poorly immunogenic include (1) small molecules (haptens) not attached to a carrier molecule, (2) linear and conserved molecules, and (3) some large but repetitive molecules.

MICRO INQUIRY *What does the term valence mean and how does an antigen's valence influence an immune response?*

antigens generally elicit a stronger immune response than monovalent antigens. **Antibody affinity** relates to the strength with which an antibody binds to its antigen at a given antigen-binding site. Affinity tends to increase during the course of an immune response (section 34.7). The **avidity** of an antibody relates to its overall ability to bind antigen at all antigen-binding sites.

Haptens

Many small organic molecules are not antigenic by themselves but become antigenic if they bond to a larger carrier molecule such as a protein. These small molecules are called **haptens** (Latin *haptein,* to grasp). When lymphocytes are stimulated by combined hapten-carrier molecules, they can react to either the hapten or the larger carrier molecule. One example of a hapten is penicillin. By itself, penicillin is a small molecule that is not antigenic. However, when it

Acquired Immunity

Natural immunity
is acquired through normal life experiences
and is not induced through medical means.

Active immunity
is the consequence of
a person developing his or
her own immune response
to a microbe.

Passive immunity
is the consequence of
one person receiving
preformed immunity
made by another person.

Artificial immunity
is produced purposefully through
medical procedures (also called immunization).

Active immunity
is the consequence of a
person developing his or
her own immune response
to a microbe.

Passive immunity
is the consequence
of one person receiving
preformed immunity
made by another person.

Infection

Maternal antibody

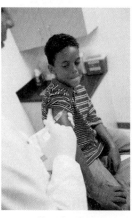
Vaccination

Immune globulin therapy

Figure 34.3 Immunity Can Be Acquired by Various Means.

is combined with certain serum proteins in sensitive individuals, it becomes immunogenic, activates lymphocytes, and initiates a severe and sometimes fatal allergic immune reaction. In these instances, the hapten is acting as an antigenic determinant or epitope.

Retrieve, Infer, Apply

1. Distinguish between self and nonself substances.
2. Define and give several examples of an antigen. What is an antigenic determinant site or epitope?
3. How does a hapten differ from an antigen?

34.3 Adaptive Immunity Can Be Earned or Borrowed

After reading this section, you should be able to:

- Report the methods by which immunity occurs by natural and artificial means
- Distinguish between the active and passive forms of natural and artificial immunity

Adaptive immunity refers to the type of acquired immunity a host develops after exposure to foreign substances or after transfer of antibodies or lymphocytes from an immune donor. Adaptive immunity results in a specific response to foreign substances and can be obtained actively or passively by natural or artificial means (**figure 34.3**).

Naturally Acquired Immunity

Naturally acquired active immunity occurs when an individual's immune system contacts an antigen such as a pathogen that causes an infection. The immune system responds by producing antibodies and activated lymphocytes that inactivate or destroy the pathogen. The immunity produced can be either lifelong, as with measles or chickenpox, or last for only a few years, as with influenza. **Naturally acquired passive immunity** involves the transfer of antibodies from one host to another. For example, some of a pregnant woman's antibodies pass across the placenta to her fetus. If the woman is immune to diseases such as polio or diphtheria, this placental transfer also gives her fetus and newborn temporary immunity to these diseases. Certain other antibodies can pass from a mother to her offspring when the child is fed breast milk. These maternal antibodies are essential for providing immunity to the newborn for the first few weeks or months of life, as the child's own immune system matures. Naturally acquired passive immunity generally lasts only a short time (weeks to months, at most).

Artificially Acquired Immunity

Artificially acquired active immunity results when an animal is vaccinated; that is, intentionally exposed to a foreign material and induced to form antibodies and activated lymphocytes. A **vaccine** may consist of a preparation of killed microorganisms; living, weakened (attenuated) microorganisms; genetically engineered organisms or their products; or inactivated bacterial toxins (toxoids) that are administered to induce immunity artificially. The vaccine material induces production of antibodies and activated lymphocytes that bind to the whole bacterium, virus, toxin, and so forth to inactivate or help remove them from the host. Vaccines and immunizations are discussed in detail in chapter 37.

Artificially acquired passive immunity results when antibodies or lymphocytes produced by one host are introduced into another. Although this type of immunity is immediate, it is short-lived, lasting only a few weeks to a few months. An example of artificially acquired passive immunity would be botulinum antitoxin produced in a horse and given to a human suffering from botulism food poisoning.

Retrieve, Infer, Apply

1. What are the types of adaptive immunity?
2. What distinguishes natural from artificial immunity?
3. What are ways that active immunity is different from passive immunity?
4. Of the four types of acquired immunity, which do you think your immune system has undergone?

34.4 Recognition of Foreignness Is Critical for a Strong Defense

After reading this section, you should be able to:

- Define the method by which a host distinguishes itself from nonself (foreign) materials
- Diagram the host cell receptors that distinguish self from nonself
- Compare the processes by which MHC class I and class II receptors recognize foreignness
- Identify cells that can function as antigen-presenting cells (APCs)
- Explain the need to use "cluster of differentiation" (CD) molecules to name cells

Distinguishing between self and nonself is essential in maintaining host integrity. This distinction must be highly specific and selective so invading pathogens are eliminated but host tissue is not destroyed. An important extension of this exquisite survival mechanism, where self is distinguished from nonself, is seen in the modern use of tissue transplantation, where organs, tissues, and cells from unrelated persons are carefully matched to the self-recognition markers on the recipient's cells. While entire textbooks are devoted to this subject, we only discuss the aspects of foreignness recognition that assist us in understanding why and how lymphocytes respond.

Recall that each cell of a particular host needs to be identified as a member of that host so it can be distinguished from foreign invaders. To accomplish this, each cell must express proteins that mark it as a resident of that host. Furthermore, it is not enough to simply identify resident (self) cells; in addition, effective cooperation between cells must occur so that efficient information sharing and selective responses occur. Such a system has evolved in mammals and is encoded in the major histocompatibility gene complex.

The Major Histocompatibility Complex

The *major histocompatibility complex* (**MHC**) is a collection of genes encoding proteins that enable the host to distinguish between self and nonself. The term histocompatibility is derived from the Greek word for tissue (*histo*) and the ability to get along (*compatibility*). Human MHC is located on chromosome 6 and is called the *human leukocyte antigen* (**HLA**) complex. HLA proteins can be divided into three classes: class I are found on all types of nucleated body cells; class II appear only on cells that can process antigens and present them to other cells (i.e., macrophages, dendritic cells, and B cells); and class III molecules include various secreted proteins that have immune functions. Class III molecules are not required for the discrimination between self and nonself and will not be discussed further.

Class I MHC molecules include HLA types A, B, and C, and serve to identify almost all cells of the body as "self." They consist of a complex of two protein chains, one with a mass of 45,000 daltons, known as the alpha chain, and the other with a mass of 12,000 Da called β_2-microglobulin (**figure 34.4a**). Only the alpha chain spans the plasma membrane, but both chains interact to form an antigen-binding pocket. Because MHC class I proteins are found on all nucleated cells (i.e., only red blood cells lack them), they stimulate an immune response when cells from one host are introduced into another host with different class I molecules. This is the basis for MHC typing when a patient is being prepared for an organ or bone marrow transplant.

HLA proteins differ among individuals; the closer two people are related, the more similar are their HLA molecules. In addition, many forms of each HLA gene exist (i.e., HLA genes are polymorphic). This is because multiple alleles of each gene have arisen by gene mutation, recombination, and other mechanisms. Additionally, because HLA genes are codominant, each person expresses both alleles at each A, B, or C locus. Thus a person produces six different class I MHC proteins.

Class II MHC molecules are produced only by certain white blood cells, such as activated macrophages, dendritic cells, mature B cells, some T cells, and certain cells of other tissues. Importantly, class II molecules are required for T-cell communication with macrophages, dendritic cells, and B cells. Class II MHC molecules are also transmembrane proteins consisting of two distinct (heterodimeric) chains (figure 34.4b). Similar to MHC class I, the protein chains expressed from MHC class II genes combine to form a three-dimensional antigen-binding pocket into which a nonself peptide fragment can be captured for presentation to other cells of the immune system (immunocytes). Although MHC class I and class II molecules are structurally distinct, both fold into similar shapes. Each MHC molecule has a deep groove into which a short peptide derived from a foreign substance can bind (figure 34.4c,d). As discussed in section 34.5, antigen fragments in the MHC groove must be present to activate T cells, which in turn activate other immunocytes.

Class I and class II molecules inform the immune system of the presence of nonself by binding and presenting foreign peptides. These peptides arise in different places within cells as the result of **antigen processing** (**figure 34.5**). Class I molecules bind to peptides that originate in the cytoplasm. Foreign peptides within the cytoplasm of mammalian cells come from viruses or other intracellular pathogens, or are the result of cancerous transformation.

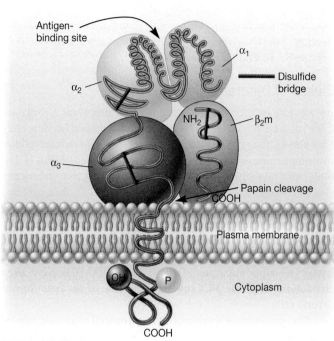

(a) Class I MHC

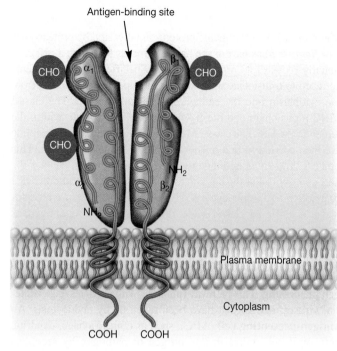

(b) Class II MHC

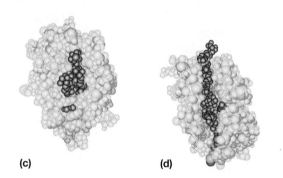

(c) **(d)**

Figure 34.4 **The Membrane-Bound Class I and Class II Major Histocompatibility Complex Molecules.** (a) The class I molecule is a heterodimer composed of the alpha protein, which is divided into three domains: α_1, α_2, and α_3, and the protein β_2 microglobulin (β_2m). (b) The class II molecule is a heterodimer composed of two distinct proteins called alpha and beta. Each is divided into two domains α_1, α_2 and β_1, β_2, respectively. (c) This space-filling model of a class I MHC protein illustrates that it holds shorter peptide antigens (blue) than does a class II MHC. (d) The difference is because the peptide binding site of the class I molecule is closed off, whereas the binding site of the class II molecules is open on both ends.

MICRO INQUIRY *On what types of cells are MHC class I molecules found? MHC class II?*

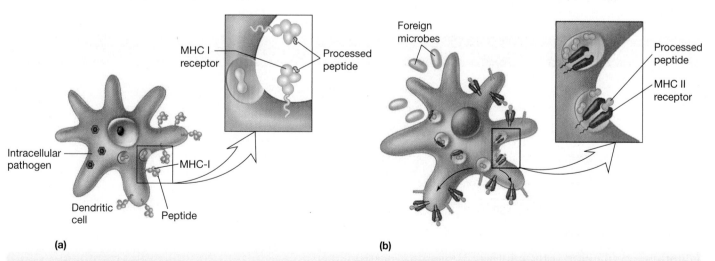

(a) **(b)**

Figure 34.5 **Antigen Presentation.** (a) Antigens arising from within a cell (including intracellular parasites) are degraded by a proteasome and integrated into the antigen-binding site of the MHC class I proteins for presentation on the cell surface. (b) Antigens arising from outside a phagocytic cell are captured by endocytosis and degraded by the lysosomal process and integrated into the antigen-binding site of MHC class II proteins for presentation on the cell surface.

MICRO INQUIRY *What cells recognize and bind to MHC class I molecules that carry antigens? What about MHC class II molecules with antigen?*

These intracellular antigenic proteins are digested by proteasomes (*see figure 5.8*) as part of the natural process by which a cell continually recycles its protein contents. Specific transport proteins are used to translocate the resulting short peptide fragments from the cytoplasm into the endoplasmic reticulum (ER). The class I MHC alpha chain and the β_2-microglobulin associate within the lumen of the ER. The class I MHC molecule and antigenic peptide are then carried to and anchored in the plasma membrane. This process, known as **endogenous antigen processing,** enables the host cell to present the antigen to a subset of T cells called CD8$^+$, or cytotoxic T lymphocytes (CTLs). CD8$^+$ T cells bear a receptor that is specific for class I MHC molecules presenting antigen; as will be discussed in section 34.5, these T cells bind and ultimately kill host cells presenting this foreign protein fragment.

Class II MHC molecules bind to fragments that initially come from antigens outside the cell, thus they undergo **exogenous antigen processing.** This pathway functions with particles (e.g., bacteria, viruses, toxins) that have been taken up by endocytosis. An **antigen-presenting cell** (APC), such as a macrophage, dendritic cell, or B cell, takes in the antigen or pathogen by receptor-mediated endocytosis or phagocytosis and produces antigen fragments by digestion in the phagolysosome. Fragments then combine with preformed class II MHC molecules and are delivered to the cell surface. The peptide within the MHC class II antigen-binding site can now be recognized by CD4$^+$ T-helper cells. Dendritic cells are particularly adept at presenting foreign peptides to T cells and stimulating them to become activated T cells. Unlike CD8$^+$ T cells, CD4$^+$ T cells do not directly kill target cells. Instead, they respond in two distinct ways. One is to proliferate, thereby increasing the number of CD4$^+$ cells that can react to the antigen. Some of these cells will become memory T cells that can respond to subsequent exposures to the same antigen. The second response is to secrete proteins that either directly inhibit the pathogen that produced the antigen or recruit and stimulate other cells to join in the immune response. ◄◄ *Several cytoplasmic membranous organelles function in the secretory and endocytic pathways (section 5.4) Cytokines are chemical messages between cells (section 33.3); Phagocytosis (section 33.5)*

Cluster of Differentiation Molecules

As we note in the previous section, two important T cells are CD4$^+$ and CD8$^+$ T cells. These cells are so named because they bear cell-surface proteins that have specific roles in intercellular communication and are called **cluster of *d*ifferentiation (CDs) molecules** or **antigens** (also known as immunogens). Other cells also bear specific CD molecules. CDs are cell-surface proteins and many are receptors. CDs have both biological and diagnostic significance. They can be measured in situ and from peripheral blood, biopsy samples, or other body fluids. They often are used in a classification system to differentiate between leukocyte subpopulations. To date, over 300 CDs have been characterized. **Table 34.1** summarizes some of the functions of several CDs.

The presence of various CDs on the cell's surface can be used to determine the cell's identity. For example, the CD4 molecule is a cell-surface receptor for the human immunodeficiency virus (HIV; the virus that causes AIDS), and CD34 is found on the surface of stem cells. As we will see, using the CD antigen system to name cells is more efficient than describing all of a cell's functions. We also use this approach in naming specific cell types as we discuss their relative functions in immunity.

Retrieve, Infer, Apply

1. What are MHCs and HLAs? Describe the roles of the three MHC classes.
2. Define CD molecules.
3. Give some examples of the biological significance of cluster of differentiation molecules (CDs).
4. How are foreign peptides processed so as to activate CD4$^+$ T-helper cells and CD8$^+$ cytotoxic T cells, respectively?

Table 34.1	Functions of Some Cluster of Differentiation (CD) Molecules
Molecule	**Function**
CD1 a, b, c	MHC class I–like receptor used for lipid antigen presentation
CD3 δ, ε, γ	T-cell antigen receptor
CD4	MHC class II coreceptor on T cells, monocytes, and macrophages; HIV-1 and HIV-2 (gp120) receptor
CD8	MHC class I coreceptor on cytotoxic T cells
CD11 a, b, c, d	α-subunits of integrin found on various myeloid and lymphoid cells; used for binding to cell adhesion molecules
CD19	B-cell antigen coreceptor
CD34	Stem cell protein that binds to sialic acid residues
CD45	Tyrosine phosphatase common to all hematopoietic cells
CD56	NK cell and neural cell adhesion molecule

34.5 T Cells Oversee and Participate in Immune Functions

After reading this section, you should be able to:

- Categorize T cells based on their CD designation
- Contrast the biological functions of T-cell subsets
- Describe T-cell receptor structure and function
- Illustrate the T-cell developmental process
- Connect antigen presentation within MHC receptors and T-cell subset recognition
- Build a model of the molecular events resulting in T-cell activation

For adaptive immunity to develop, T cells and B cells must be activated. Activation results in proliferation and differentiation of progenitor cells into numerous T- and B-cell subsets, with environmental signals determining the final type of cell that forms. These include T cells ($CD8^+$ and $CD4^+$ cells) that are major players in the cell-mediated immune response and have a major role in B-cell activation. They are immunologically adaptive, can carry a vast repertoire of immunologic memory, and can function in a variety of regulatory and effector ways. Because of their paramount importance, we discuss them first.

T-Cell Receptors

T cells respond to antigen fragments presented by MHC molecules through specific **T-cell receptor (TCR)** complexes on their plasma membrane surface. Each receptor complex is composed of two parts: a heterodimeric polypeptide receptor, and six accessory polypeptides, collectively referred to as CD3 (**figure 34.6a**). The heterodimeric receptor can be one of two possible TCR dimers, the predominant α/β or the γ/δ; thus cells can be designated **α/β T cells** and **γ/δ T cells,** respectively. Each heterodimer forms a transmembrane receptor stabilized by disulfide bonds. The antigen recognition sites of the receptors form when the extracellular amino termini of the α and β or γ and δ; chains interact and create a three-dimensional pocket having terminal variable sections complementary to antigen fragments (similar to MHC antigen-binding sites). The cytoplasmic tails of the receptors associate with the CD3 accessory polypeptides, and transduce antigen-binding events into intracellular signals (figure 34.6b).

T-Cell Activation

During antigen presentation, the antigen held by the MHC of the antigen-presenting cell is bound by the TCR of the T cell, thus forming a bridge between the two (figure 34.6b). This causes a three-dimensional change in the TCR that stimulates the CD3 accessory proteins to send molecular signals to the T-cell nucleus. However, full T-cell activation occurs only when a second signal is communicated along with the antigen. A general discussion of this process in T cells now follows.

All naïve T cells require two signals to be activated into effector cells. **Signal 1** occurs when an antigen fragment, presented in

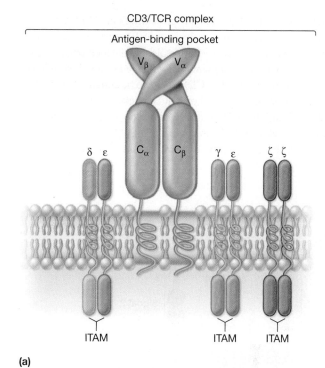

(a)

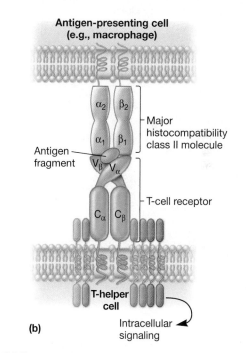

(b)

Figure 34.6 The T-Cell Receptor Protein Complex in T-Helper Cell Activation. (a) The overall structure of the antigen receptor complex on a T-cell plasma membrane. Notice the CD3 accessory proteins that convey the signal of antigen capture intracellularly. Extracellular antigen binding is communicated internally through the immunoreceptor tyrosine-based activation motifs (ITAMs) on the CD3 proteins. (b) An antigen-presenting cell begins the activation process by displaying an antigen fragment (e.g., peptide) within histocompatibility molecules. A T-helper cell is activated after the variable region of its receptor (designated V_α and V_β) reacts with the antigen fragment in a class II MHC molecule on the presenting cell surface.

an MHC molecule of an antigen-presenting cell (APC), fills the appropriate T-cell receptor. In the case of T-helper (T_H) cells, antigen is presented by class II MHC molecules, which triggers CD4 coreceptors on the T_H cell to interact with the antigen-bound MHC molecule (**figure 34.7a**). For $CD8^+$ cells to be activated, an endogenous (cytoplasmic) antigen is presented in class I MHC molecules on the APC, and the CD8 coreceptor on the T cell interacts with the antigen-bound MHC molecule on the APC to induce CTL function (figure 34.7b). In both cases, the T-cell coreceptor assists with signal 1 recognition.

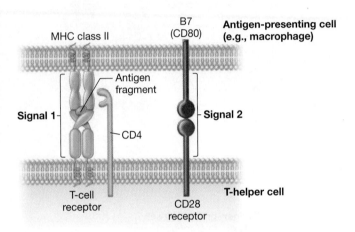

(a)

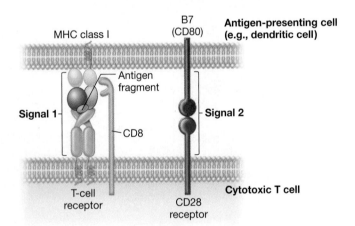

(b)

Figure 34.7 Two Signals (Co-stimulation) Are Essential for T-Helper Cell Activation. The first signal is the presentation of the antigen fragment by a dendritic cell or other antigen-presenting cell (a) in the MHC class II molecule to activate the T-helper cell. (b) The first signal for T_C cells is presentation of the antigen fragment in the MHC class I molecule. The second signal occurs when the dendritic cell presents the B7 (CD80) protein to the T-helper or cytotoxic T cell with its CD28 protein receptor. Note that the CD3 complex is not shown.

MICRO INQUIRY *How can T-helper cells facilitate $CD8^+$ cell activation?*

In addition to signal 1, both naïve $CD4^+$ and $CD8^+$ cells require a second, co-stimulatory **signal 2** to become activated. More than one factor may contribute to signal 2, but the most important seems to be the **B7 (CD80) protein** on the surface of an APC, which binds to the CD28 receptor on the T cell (figure 34.7b). The activated T cell then synthesizes and secretes IL-2 to drive its own proliferation and differentiation. Overall, once T cells have been activated by two signals, they differentiate rapidly to respond to the antigen that activated them. T cells that receive only signal 1 often become **anergic,** or unresponsive.

One type of APC that is particularly good at stimulating naïve T cells is the dendritic cell. This is because it expresses high levels of B7 at all times. Other APCs may not produce sufficient B7 unless stimulated to do so. Poor B7 stimulation of $CD8^+$ cells requires activated $CD4^+$ (also known as T_H) cells to stimulate the APC to generate a different signal 2. This is one way in which T_H cells help $CD8^+$ T cells become activated. ◀◀ *Monocytes, macrophages, and dendritic cells (section 33.4)*

Types of T Cells

T cells originate from $CD34^+$ stem cells in the bone marrow, but T-cell precursors migrate to the thymus for further differentiation. This includes destruction or inactivation of cells that recognize self antigens (so-called self-reactive T cells). T-cell development in the thymus also results in its differentiation into one of four functional lineages. T cells differentiate into progenitor cells destined to become (1) natural T regulatory (nTreg) cells, (2) natural killer T (NKT) cells, (3) γ/δ T cells, and (4) α/β T cells (**figure 34.8**).

The α/β T cells make up the bulk of all T cells. They circulate in the blood in small numbers and are found in the secondary lymphoid tissues. Recall that the α/β designation describes the type of T-cell receptor (TCR) on these cells. In addition, α/β T cells also have either the CD4 or CD8 coreceptor to facilitate TCR recognition of the class II or class I MHC, respectively. Thus, an α/β T cell can be either a helper or cytotoxic T cell. Like all lymphocytes, T cells that have survived the process of development are called mature cells, but they are also considered to be "naïve" cells because they have not yet been activated by a specific MHC-antigen peptide combination. Activation of T cells involves specific molecular signaling events inside the cell, which is discussed in the section on T-cell activation. Antigen stimulation induces α/β T cells to assume their active roles as **effector cells,** as well as memory cells (figure 33.11). Effector T cells carry out specific functions to protect the host against foreign antigen. Effector $CD4^+$ T cells include the T-helper (T_H) cells, and the effector $CD8^+$ T cells mature into CTLs.

T-Helper Cells

Mature, naïve **T-helper (T_H) cells** are also known as **$CD4^+$ T cells.** The activation of specific transcription factors in response to cytokine stimulation drives naive T cells to differentiate into one the following types of $CD4^+$ T-helper cells: T_H type 1 (T_H1) cells, T_H type 2 (T_H2) cells, T_H type 9 (T_H9) cells, T_H type 17

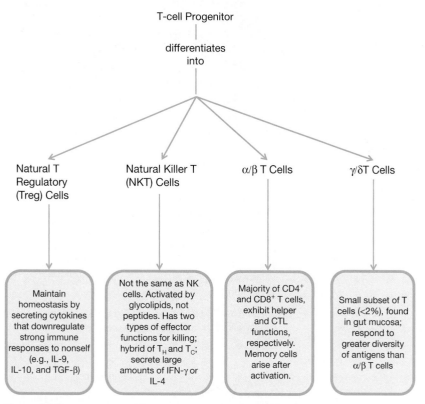

Figure 34.8 T-Cell Populations. T-cell subsets differentiate from their progenitor cells in the thymus. They leave the thymus upon maturation; however, they are still naïve in that they have not been activated by their specific antigen. At least four T-cell subsets leave the thymus and populate secondary lymphoid tissues where they await antigen activation.

$(T_H 17)$ cells, T_H type 22 $(T_H 22)$, T follicular helper type (T_{fh}) cells, $CD4^+$ induced regulatory T cells (iTreg), and T_H memory cells (**figure 34.9a**). $T_H 0$ is often used to designate cells that are simply undifferentiated precursors of these cells.

$T_H 1$ **cells** promote cytotoxic T lymphocyte activity, activate macrophages, and mediate inflammation by producing interleukins, interferon (IFN)-γ, and tumor necrosis factor-β (TNF-β) when the transcription factor T-bet is active. These cytokines are also responsible for delayed-type (type IV) hypersensitivity reactions, in which host cells and tissues are damaged nonspecifically by activated T cells (p. 763).

$T_H 2$ **cells** tend to stimulate antibody responses in general and defend against helminth parasites by producing IL-4, IL-5, IL-6, IL-10, and IL-13. An overabundance of $T_H 2$-type responses may also be involved in promoting allergic reactions.

$T_H 9$ **cells** halt their characteristic $T_H 2$ profile upon exposure to TGF-β and subsequently produce IL-9. IL-9 stimulates hematopoiesis and inhibits programmed cell death (apoptosis). $T_H 9$ cells have been found in inflamed tissue and associated with autoimmune disease.

$T_H 17$ **cells** result from exposure of naïve T_H cells to TGF-β and IL-6. These cells are found predominantly in the skin and intestinal epithelia, where they respond to bacterial invasion by secreting IL-17 and defensins, recruiting neutrophils, and inducing a strong inflammatory response. In addition, $T_H 17$ cells have been implicated in several autoimmune diseases.

$T_H 22$ **cells** differentiate from naïve $CD4^+$ T cells upon exposure to IL-6 and TNF-alpha. They primarily secrete pro-inflammatory cytokines IL-22, IL-13, and TNF-alpha. They are typically found at innate barriers (skin, intestine, lungs), where they act on various cell populations regulating wound healing and repair.

T_{fh} **cells** develop from antigen-stimulated $CD4^+$ T cells. They migrate through secondary lymphoid tissues, coaxing B cells to migrate into regions of lymphoid tissues where they secrete IL-21 and IL-4 to promote B-cell antibody production.

Regulatory T cells (**Tregs**) are derived from approximately 10% of $CD4^+$ T cells and 2% of $CD8^+$ T cells in response to exposure to IL-10. IL-10 activates a transcription factor that leads to production of a protein that binds to B7 (CD80), blocking the second signal required for lymphocyte activation (p. 744). Tregs also exert suppressor/regulatory functions through the secretion of IL-9, IL-10, and TGF-β. These cytokines inhibit $T_H 1$ and $T_H 17$ cells from upregulating inflammatory responses and $T_H 2$ cells from assisting B cells in making antibody and upregulating allergic responses, in addition to a wider range of suppressor activities. The relatively recent reports of $T_H 9$, $T_H 17$, $T_H 22$, and T_{fh} cells suggest that new technologies that assess genome and proteome expression of single cells may continue to reveal a growing list of T-cell subsets. ◀◀ *Single-cell genome sequencing (section 18.2); Proteomics (section 18.4)*

Cytotoxic T Lymphocytes

Naïve $CD8^+$ **T cells** that express antigen-specific T-cell receptors (TCRs) are historically designated as cytotoxic T cells (T_C). Activation by its specific antigen causes a T_C cell to mature into a **cytotoxic T lymphocyte** (**CTL**) that functions to destroy host cells that have been infected by intracellular pathogens, such as a virus, or have altered MHC surface proteins. Recall that to be activated, $CD8^+$ T cells interact with APCs through their class I MHCs (figure 34.9b). This is important because these cells mature into CTLs that must respond only to the same antigen presented in the class I MHC of host cells. Any host cell that presents the same antigen in its class I MHC receptors is thus targeted for destruction. Once activated, CTLs kill target cells in at least two ways: a cytolytic pathway, using **perforins** and granulysins, that is similar to complement-mediated lysis; and apoptotic (programmed cell death) pathways triggered by CD95-Fas-FasL (ligand) or by granzymes and perforins.

Superantigens

Several bacterial and viral proteins can provoke a drastic and harmful response when they are exposed to T cells. These protein antigens are known as **superantigens** because they "trick" a huge

Figure 34.9 T-Cell Development.
(a) Once activated, a CD4⁺ T$_H$0 cell may differentiate into T$_H$1, T$_H$2, T$_H$9, T$_H$17, T$_H$22, or Treg cells. Cytokines near red arrows trigger differentiation of one cell to another while cytokines near blue arrows are produced by the associated cell. Memory T cells are also produced during expansion of the activated T-cell population. Each T$_H$ phenotype secretes specific cytokines to effect the functions of other cells; some cytokines stimulate B-cell growth, replication, and antibody synthesis, others suppress migration and apoptosis, and still others enhance recruitment of phagocytes. (b) CD8⁺ T$_C$ cells differentiate into activated CTLs. CTLs induce apoptosis in target cells. (c) A cytotoxic T cell (left) contacts a target cell (right) (×5,700). (d) The CTL secretes perforin that forms pores in the target cell's plasma membrane. These pores allow the contents of the target cell to leak out and granzymes to enter and induce apoptosis.

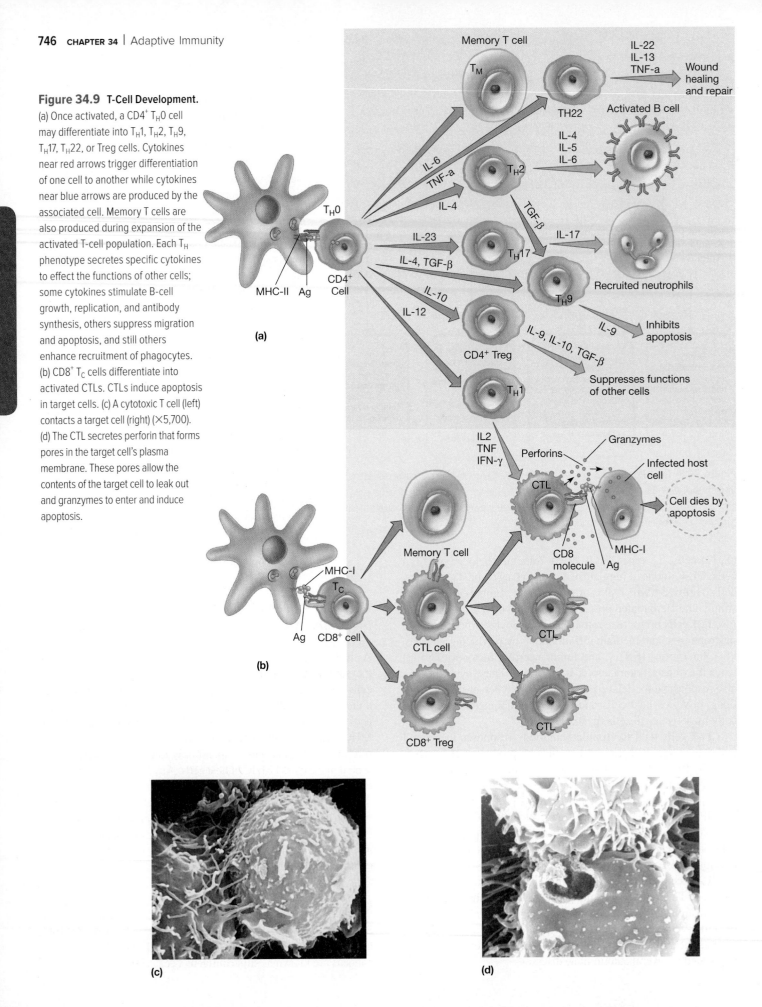

number of T cells into activation when no specific antigen has triggered them. Superantigens accomplish this by bridging class II MHC molecules on APCs to T-cell receptors in the absence of a specific antigen in the MHC-binding site. This interaction results in the activation of many different T cells with different antigen specificities. The consequence of this nonspecific activation is the release of massive quantities of cytokines from $CD4^+$ T cells. Indeed, nearly 30% of T cells can be activated by superantigen to overproduce pro-inflammatory cytokines such as TNF-α and interleukins 1 and 6, resulting in endothelial damage, circulatory shock, and multiorgan failure. Examples of superantigens include the staphylococcal enterotoxins (which can cause food poisoning) and the toxin that causes toxic shock syndrome. Because of these activities, staphylococcal enterotoxin B was originally added to the U.S. government's select agent list as a potential agent of terrorism. The devastating effects of superantigens serve to emphasize the importance of tightly regulating a normal immune response.

Retrieve, Infer, Apply

1. What is the function of an antigen-presenting cell?
2. What is a T-cell receptor and how is it involved in T-cell activation?
3. Describe antigen processing. How does this process differ for endogenous and exogenous antigens?
4. Contrast and compare the activation of $CD4^+$ and $CD8^+$ T cells.
5. Outline the functions of a T-helper cell. How do T_H1, T_H2, T_H9, T_H17, and TH_{22} cells differ in function? Briefly describe how T_H cells are activated by co-stimulation versus superantigen.
6. Briefly describe the two ways in which CTLs destroy target cells.

34.6 B Cells Make Antibodies and Do a Whole Lot More

After reading this section, you should be able to:

- Describe the B-cell receptor structure and function
- Illustrate the B-cell maturation process in response to antigen triggering
- Compare T-dependent- and T-independent B-cell activation
- Build a model of the molecular events resulting in B-cell activation

Stem cells in the bone marrow produce B-cell precursors *(see figure 33.11)*. Similar to T cells, it is becoming clear that unique subsets of B cells form along divergent developmental pathways and that these subsets may have unique functions. It appears that one subset acts more like cells of the innate resistance arm of immunity. Of note is that B cells also appear to express toll-like receptors (TLRs) throughout their developmental stages. Importantly, expression of TLRs on activated B cells may serve as additional (adjuvant-like) switches to regulate antibody production.

Until the science becomes clearer, we focus on the B-cell subset with a well-defined activation process. A specific antigen must activate these B cells so they can proliferate and differentiate into mature, antibody-secreting plasma cells. Prior to antigen stimulation, B cells produce a form of antibody (IgM or IgD) that is attached to their cell membrane and oriented so that the part of the antibody that binds to antigen is facing outward, away from the cell (**figure 34.10**). These cell-surface, transmembrane antibodies (also known as immunoglobulins) act as receptors for the

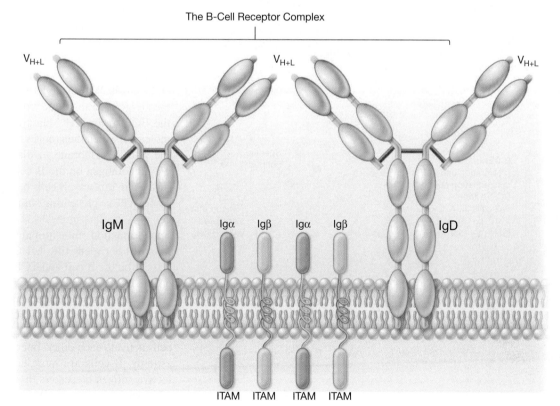

The B-Cell Receptor Complex

V_{H+L} V_{H+L} V_{H+L}

IgM   Igα Igβ Igα Igβ IgD

ITAM   ITAM   ITAM   ITAM

Figure 34.10 The Membrane-Bound B-Cell Receptor (BCR) Complex. The BCR is composed of monomeric IgM and IgD antibody, and the coreceptors Igα and Igβ. A B cell is activated after the variable region of one receptor (designated V_{H+L}) binds an antigen. The antigen is internalized, processed, and presented by the B cell's class II MHC molecule, so as to be recognized by a T_H2 cell (T-dependent activation). Activation can also occur after the variable regions of (two or more) receptors are bridged by an antigen (T-independent activation).

specific antigen that will activate that particular B cell. On a molecular level, the immunoglobulin receptor molecules on the B-cell surface associate with other proteins known as the Ig-a/Ig-b heterodimer proteins (similar to how the T-cell receptor interacts with CD3). Together the transmembrane immunoglobulin and the heterodimer protein complex is called the **B-cell receptor (BCR)**. When an antigen is captured by the immunoglobulin receptor, the receptor communicates this capture to the nucleus through a signal transduction pathway similar to that described for T cells.

Each B cell may have as many as 50,000 BCRs on its surface. While each individual human carries BCRs specific for as many as 10^{13} different antigens, each individual B cell possesses BCRs specific for only one particular epitope on an antigen. Therefore a host produces at least 10^{13} different, undifferentiated B cells. These naïve B cells reside in the host awaiting activation by specific antigenic epitopes. Upon activation, B cells differentiate into antibody-producing **plasma cells** and memory cells (figure 34.1).

So far we have introduced the B-cell receptor and the fact that activated B cells secrete antibody. The fate of antigen bound to the BCR is critical for the transition from naïve B cell to antibody-secreting plasma cell. Once bound, B cells internalize the antigen-bound receptor by receptor-mediated endocytosis. Within the B cell, the antigen is transferred from the BCR to an MHC class II molecule, which then moves to the B-cell membrane to display the antigen on the B-cell surface. In this way, B cells become APCs, presenting antigen to T_H2 cells **(figure 34.11)**. Upon formation of the immune synapse between the MHC class II on the B cell and TCR on the T_H2 cell, the B cell triggers the release of a specific suite of cytokines from the T_H2 cell. Thus the B and T_H2 cells activate each other. Interestingly, B cells are more effective antigen presenters than macrophages, especially at

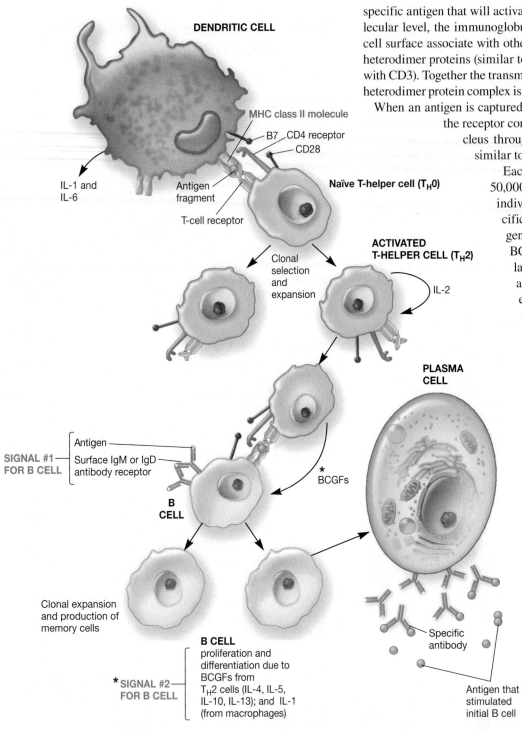

Figure 34.11 T-Dependent Antigen Triggering of a B Cell. Schematic diagram of the events occurring in the interactions between macrophages, T-helper cells, and B cells that produce humoral immunity. Many cytokines (e.g., IL-1, IL-4, IL-5, IL-6, IL-10, IL-13) act as B-cell growth factors (BCGFs) to stimulate B-cell proliferation. Cytokines such as IL-2, IL-4, IL-6, and IL-13 stimulate B-cell differentiation into plasma cells.

MICRO INQUIRY *Which cells are functioning as APCs in this figure?*

low antigen concentrations. In summary, B cells have two immunological roles: (1) they proliferate and differentiate into memory cells and plasma cells, which respond to antigens by making antibodies, and at the same time (2) they can act as antigen-presenting cells, stimulating T cells to also respond.

B-Cell Activation

Similar to T cells, receptor binding is required to initially activate B cells, which may then also require growth and differentiation factors, supplied by other cells, to fully mature. T-helper cells are the primary source of cytokines that assist in this B-cell maturation process. This typical process is known as T-cell-dependent B-cell activation, because after stimulation by the antigen, the B-cell maturation process is dependent on the T_H2 cell for the cytokines to complete maturation. However, some antigens "force" the B cell into antibody production before T-cell growth and differentiation factors have influenced the maturation process. This other B-cell activation process is called T-cell-independent B-cell activation.

T-Dependent B-Cell Activation

In most cases, B cells specific for a given epitope on an antigen (e.g., epitope X) cannot develop into plasma cells that secrete antibody (anti-X) without the collaboration of T-helper cells; this is **T-dependent B-cell activation.** In other words, binding of epitope X to the B cell may be necessary, but it is not usually sufficient for B-cell activation. Antigens that elicit a response with the aid of T-helper cells are called **T-dependent antigens.** Examples include bacteria, foreign red blood cells, certain proteins, and hapten-carrier combinations.

T-dependent antigen triggering of a B cell occurs when an activated T_H2 cell directly associates with the B cell displaying the same antigen-MHC complex presented to the T_H2 on a dendritic cell. B-cell growth factors are then secreted by the activated T_H2 cell (figure 34.11). These interactions cause the B cell to proliferate and differentiate into a plasma cell, which starts producing antibodies. Thus B cells, like T cells, require two signals: antigen-BCR interaction (signal 1) and T-cell cytokines (signal 2). This is a very effective process: one plasma cell can synthesize more than 10 million antibody molecules per hour!

Interestingly, a subset of B cells has been identified that reins in T-cell–dependent inflammatory responses. These cells express the surface receptors CD1d and CD5, and also secrete the anti-inflammatory cytokine IL-10. They appear to be activated by a broad range of antigens and suppress T-dependent antibody responses elicited during inflammation.

T-Independent B-Cell Activation

As mentioned, a few specific antigens can trigger B cells into antibody production without T-cell cooperation. These are called **T-independent antigens,** and their stimulation of B cells is known as **T-independent B-cell activation.** Examples include bacterial lipopolysaccharides, certain tumor-promoting agents, antibodies against other antibodies, and antibodies to certain B-cell differentiation antigens. The T-independent antigens tend to be polymeric; that is, they are composed of repeating sugars or amino acids. The resulting antibody generally has a relatively low affinity for antigen.

The mechanism for activation by T-independent antigens probably depends on their polymeric structure. Large molecules present a wide array of identical epitopes to a B cell specific for that determinant. The repeating epitopes cross-link membrane-bound BCRs such that cell activation occurs and antibody is secreted. However, without T-cell help, the B cell cannot alter its antibody production, and no memory B cells are formed. Thus T-independent B-cell activation is less effective than T-dependent B-cell activation: The antibodies produced have a low affinity for antigen and no immunologic memory is formed.

Microbial Pattern Triggering

Identifying yet another crossover between innate and adaptive arms of the immune system, a unique subset of phenotypically distinct B cells, the innate response activator (IRA) B cells, not only make antibody but also produce granulocyte-macrophage colony-stimulating factor (GM-CSF) and IL-3. IRAs are typically found in the pleural and peritoneal cavities in a steady state of activation. In response to binding of bacterial lipopolysaccharide (LPS) by toll-like receptor-4 (TLR-4), they release large quantities of premade IgM antibody, and directly stimulate another subset of B cells to also release IgM. Additionally, IRAs then migrate to the spleen, where they secrete GM-CSF and IL-3, stimulating the production of phagocytes. Experimental models of sepsis suggest that the IRA B-cells are first-responders that direct emergency IgM release to abate bacteria invading through the lungs or intestines, while they sound an alarm to splenic stem cells, recruiting other defenders (*see figure 35.8*).

> *Retrieve, Infer, Apply*
>
> 1. What are B-cell receptors? How are they involved in B-cell activation?
> 2. Briefly compare and contrast B cells and T cells with respect to their formation, structure, and roles in the immune response.
> 3. How does T-independent antigen triggering of B cells differ from T-dependent triggering?
> 4. How might IRA B cells prevent bacterial sepsis?

34.7 Antibodies Are Proteins That Bind to Specific 3-D Structures

After reading this section, you should be able to:

- Describe the structure of the B-cell receptor that is secreted as antibody
- Compare and contrast the five classes of antibody
- Diagram the antibody changes, induced by antigen binding, that facilitate antigen capture and removal from the host
- Integrate antibody secretion with antigen exposure
- Create a model of genetic diversity that results from recombination, alternative splicing, and somatic hypermutation
- Predict antibody specificity resulting from clonal selection

truly complementary, the antibody will not effectively bind the antigen. Thus a lock-and-key mechanism operates; however, in at least one case, the antigen-binding site changes shape when it complexes with the antigen (an induced-fit mechanism; *see figure 10.16*). In either case, antibody specificity results from the nature of antibody-antigen binding.

Phagocytes have Fc receptors for immunoglobulin on their surface, so bacteria covered with antibodies are better targets for phagocytosis by neutrophils and macrophages. This is termed **opsonization.** Other cells, such as natural killer cells, destroy antibody-coated cells through a process called antibody-dependent cell-mediated cytotoxicity (*see figure 33.12*). Immune destruction also is promoted by antibody-induced activation of the classical complement system.

Immunoglobulin Classes

Immunoglobulin γ, or **IgG,** is the major immunoglobulin in human serum, accounting for 80% of the immunoglobulin pool (**figure 34.13a**). IgG is present in blood plasma and tissue fluids. IgG acts against bacteria and viruses by opsonizing the invaders and neutralizing toxins and viruses (p. 759). It is also one of the two immunoglobulin classes that activate complement by the classical pathway. IgG is the only immunoglobulin molecule able to cross the placenta and provide natural immunity in utero and to the neonate at birth.

There are four human IgG isotypes (IgG1, IgG2, IgG3, and IgG4); these vary chemically in their heavy-chain composition and the number and arrangement of interchain disulfide bonds (figure 34.13b). Differences in biological function have been noted in these isotypes. For example, IgG2 antibodies are opsonic and develop in response to toxins. IgG1 and IgG3, upon recognition of their specific antigens, bind to Fc receptors expressed on neutrophils and macrophages. This increases phagocytosis by these cells. The IgG4 antibodies function as skin-sensitizing immunoglobulins.

Immunoglobulin μ, or **IgM,** accounts for about 5 to 10% of the serum immunoglobulin pool. It is usually a polymer of five monomeric units (pentamer), each composed of two heavy chains and two light chains (**figure 34.14a**). The monomers are arranged in a pinwheel array with the Fc ends in the center, held together by disulfide bonds and a special J (joining) chain. Soluble, pentameric IgM is the immunoglobulin made during B-cell maturation, while individual IgM monomers are expressed on B cells, serving as the antibody component of the BCR (p. 748). Pentameric IgM tends to remain in the bloodstream, where it agglutinates (or clumps) bacteria, activates complement by the classical pathway, and enhances the ingestion of pathogens by phagocytic cells.

Although most IgM appears to be pentameric, around 5% or less of human serum IgM exists in a hexameric form. This molecule contains six monomeric units but seems to lack a J chain. Hexameric IgM activates complement up to 20 times more effectively than does the pentameric form. It has been suggested that bacterial cell-wall antigens such as lipopolysaccharides may

directly stimulate B cells to form hexameric IgM without a J chain. If this is the case, the immunoglobulins formed during primary immune responses are less homogeneous than previously thought.

Immunoglobulin α, or **IgA,** accounts for about 12% of serum immunoglobulin. Some IgA is present in the serum as a monomer. However, IgA is most abundant in mucous secretions, where it is a dimer held together by a J chain (figure 34.14b). IgA, when transported from mucosal-associated lymphoid tissue (MALT) to mucosal surfaces, acquires a protein called the secretory component (*see figure 33.16*). **Secretory IgA (sIgA),** as the modified molecule is called, is the primary immunoglobulin of

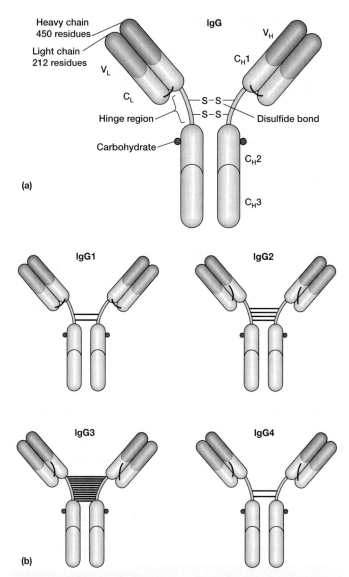

Figure 34.13 Immunoglobulin G. (a) The basic structure of human IgG. (b) The structure of the four human IgG subclasses. Note the arrangement and numbers of disulfide bonds (shown as thin black lines). Carbohydrate side chains are shown in red.

MICRO INQUIRY *What makes this class of immunoglobulin unique?*

Table 34.2	Physicochemical Properties of Human Immunoglobulin Classes				
	IMMUNOGLOBULIN CLASSES				
Property	**IgG[1]**	**IgM**	**IgA[2]**	**IgD**	**IgE**
Heavy chain	γ_1	μ	α_1	δ	ε
Mean serum concentration (mg/ml)	9	1.5	3.0	0.03	0.00005
Percent of total serum antibody	80–85	5–10	5–15	<1	<1
Valency	2	5(10)	2(4)	2	2
Mass of entire molecule (kDa)[3]	146	970	160[3]	184	188
Placental transfer	+	−	−	−	−
Half-life in serum (days)[4]	23	5	6	3	2
Complement activation Classical pathway Alternative pathway	++ −	+++ −	− +	− −	− −
Induction of mast cell degranulation	−	−	−	−	+
% carbohydrate	3	7–10	7	12	11
Major characteristics	Most abundant Ig in body fluids; neutralizes toxins; opsonizes bacteria	First to appear after antigen stimulation; very effective agglutinator; expressed as membrane-bound antibody on B cells	Secretory antibody; protects mucous membranes	Present on B-cell surface; B-cell recognition of antigen	Anaphylactic-mediating antibody; resistance to helminths

1 Properties of IgG subclass 1.
2 Properties of IgA subclass 1.
3 sIgA = 360 − 400 kDa.
4 Time required for half of the antibodies to disappear.

respectively, the five immunoglobulin (Ig) classes: IgG, IgA, IgM, IgD, and IgE (**table 34.2**). Each immunoglobulin class differs in its general properties, half-life, distribution in the body, and interaction with other components of the host's defensive systems.

Retrieve, Infer, Apply

1. What is the variable region of an antibody? The hypervariable or complementarity-determining region? The constant region?
2. What is the function of the Fc region of an antibody? The Fab region?
3. Name the two types of antibody light chains.
4. What determines the class of heavy chain of an antibody? Name the five immunoglobulin classes.

Immunoglobulin Function

Each end of the immunoglobulin molecule has a unique role. The Fab region is concerned with binding antigen, whereas the Fc

region mediates binding to receptors (called Fc receptors) found on various immunocytes or the first component of the classical complement system. The binding of an antibody to an antigen usually does not destroy the antigen, even when the antigen is on a cell or other agent. Rather, the antibody serves to mark and identify the nonself agent as a target for immunological attack and to activate innate immune responses that can destroy the target. ◀◀ *Complement is a cascading enzyme system (section 33.3)*

An antigen binds to an antibody at the antigen-binding site within the Fab region of the antibody. More specifically, a pocket is formed by the folding of the V_H and V_L regions (figure 34.12c). At this site, specific amino acids contact the antigen's epitope and form multiple noncovalent bonds between the antigen and amino acids of the binding site. Because binding is due to weak, noncovalent bonds such as hydrogen bonds and electrostatic attractions, the antigen's shape must exactly match that of the antigen-binding site. If the shapes of the epitope and binding site are not truly complementary, the antibody will not effectively bind the antigen. Thus a lock-and-key mechanism operates; however, in at least one case, the antigen-binding site changes shape when it complexes with the antigen (an induced-fit mechanism; *see figure 10.16*). In either case, antibody specificity results from the nature of antibody-antigen binding.

Phagocytes have Fc receptors for immunoglobulin on their surface, so bacteria covered with antibodies are better targets for phagocytosis by neutrophils and macrophages. This is termed **opsonization.** Other cells, such as natural killer cells, destroy antibody-coated cells through a process called antibody-dependent cell-mediated cytotoxicity (*see figure 33.12*). Immune destruction also is promoted by antibody-induced activation of the classical complement system.

Immunoglobulin Classes

Immunoglobulin γ, or **IgG,** is the major immunoglobulin in human serum, accounting for 80% of the immunoglobulin pool (**figure 34.13a**). IgG is present in blood plasma and tissue fluids. IgG acts against bacteria and viruses by opsonizing the invaders and neutralizing toxins and viruses (p. 759). It is also one of the two immunoglobulin classes that activate complement by the classical pathway. IgG is the only immunoglobulin molecule able to cross the placenta and provide natural immunity in utero and to the neonate at birth.

There are four human IgG isotypes (IgG1, IgG2, IgG3, and IgG4); these vary chemically in their heavy-chain composition and the number and arrangement of interchain disulfide bonds (figure 34.13b). Differences in biological function have been noted in these isotypes. For example, IgG2 antibodies are opsonic and develop in response to toxins. IgG1 and IgG3, upon recognition of their specific antigens, bind to Fc receptors expressed on neutrophils and macrophages. This increases phagocytosis by these cells. The IgG4 antibodies function as skin-sensitizing immunoglobulins.

Immunoglobulin μ, or **IgM,** accounts for about 5 to 10% of the serum immunoglobulin pool. It is usually a polymer of five monomeric units (pentamer), each composed of two heavy chains and two light chains (**figure 34.14a**). The monomers are arranged in a pinwheel array with the Fc ends in the center, held together by disulfide bonds and a special J (joining) chain. Soluble, pentameric IgM is the immunoglobulin made during B-cell maturation, while individual IgM monomers are expressed on B cells, serving as the antibody component of the BCR (p. 748). Pentameric IgM tends to remain in the bloodstream, where it agglutinates (or clumps) bacteria, activates complement by the classical pathway, and enhances the ingestion of pathogens by phagocytic cells.

Although most IgM appears to be pentameric, around 5% or less of human serum IgM exists in a hexameric form. This molecule contains six monomeric units but seems to lack a J chain. Hexameric IgM activates complement up to 20 times more effectively than does the pentameric form. It has been suggested that bacterial cell-wall antigens such as lipopolysaccharides may

directly stimulate B cells to form hexameric IgM without a J chain. If this is the case, the immunoglobulins formed during primary immune responses are less homogeneous than previously thought.

Immunoglobulin α, or **IgA,** accounts for about 12% of serum immunoglobulin. Some IgA is present in the serum as a monomer. However, IgA is most abundant in mucous secretions, where it is a dimer held together by a J chain (figure 34.14b). IgA, when transported from mucosal-associated lymphoid tissue (MALT) to mucosal surfaces, acquires a protein called the secretory component (*see figure 33.16*). **Secretory IgA (sIgA),** as the modified molecule is called, is the primary immunoglobulin of

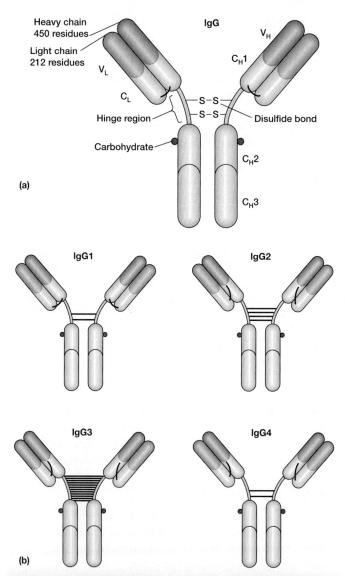

Figure 34.13 Immunoglobulin G.   (a) The basic structure of human IgG. (b) The structure of the four human IgG subclasses. Note the arrangement and numbers of disulfide bonds (shown as thin black lines). Carbohydrate side chains are shown in red.

MICRO INQUIRY *What makes this class of immunoglobulin unique?*

low antigen concentrations. In summary, B cells have two immunological roles: (1) they proliferate and differentiate into memory cells and plasma cells, which respond to antigens by making antibodies, and at the same time (2) they can act as antigen-presenting cells, stimulating T cells to also respond.

B-Cell Activation

Similar to T cells, receptor binding is required to initially activate B cells, which may then also require growth and differentiation factors, supplied by other cells, to fully mature. T-helper cells are the primary source of cytokines that assist in this B-cell maturation process. This typical process is known as T-cell-dependent B-cell activation, because after stimulation by the antigen, the B-cell maturation process is dependent on the T_H2 cell for the cytokines to complete maturation. However, some antigens "force" the B cell into antibody production before T-cell growth and differentiation factors have influenced the maturation process. This other B-cell activation process is called T-cell-independent B-cell activation.

T-Dependent B-Cell Activation

In most cases, B cells specific for a given epitope on an antigen (e.g., epitope X) cannot develop into plasma cells that secrete antibody (anti-X) without the collaboration of T-helper cells; this is **T-dependent B-cell activation.** In other words, binding of epitope X to the B cell may be necessary, but it is not usually sufficient for B-cell activation. Antigens that elicit a response with the aid of T-helper cells are called **T-dependent antigens.** Examples include bacteria, foreign red blood cells, certain proteins, and hapten-carrier combinations.

T-dependent antigen triggering of a B cell occurs when an activated T_H2 cell directly associates with the B cell displaying the same antigen-MHC complex presented to the T_H2 on a dendritic cell. B-cell growth factors are then secreted by the activated T_H2 cell (figure 34.11). These interactions cause the B cell to proliferate and differentiate into a plasma cell, which starts producing antibodies. Thus B cells, like T cells, require two signals: antigen-BCR interaction (signal 1) and T-cell cytokines (signal 2). This is a very effective process: one plasma cell can synthesize more than 10 million antibody molecules per hour!

Interestingly, a subset of B cells has been identified that reins in T-cell–dependent inflammatory responses. These cells express the surface receptors CD1d and CD5, and also secrete the anti-inflammatory cytokine IL-10. They appear to be activated by a broad range of antigens and suppress T-dependent antibody responses elicited during inflammation.

T-Independent B-Cell Activation

As mentioned, a few specific antigens can trigger B cells into antibody production without T-cell cooperation. These are called **T-independent antigens,** and their stimulation of B cells is known as **T-independent B-cell activation.** Examples include bacterial lipopolysaccharides, certain tumor-promoting agents, antibodies against other antibodies, and antibodies to certain B-cell differentiation antigens. The T-independent antigens tend to be polymeric; that is, they are composed of repeating sugars or amino acids. The resulting antibody generally has a relatively low affinity for antigen.

The mechanism for activation by T-independent antigens probably depends on their polymeric structure. Large molecules present a wide array of identical epitopes to a B cell specific for that determinant. The repeating epitopes cross-link membrane-bound BCRs such that cell activation occurs and antibody is secreted. However, without T-cell help, the B cell cannot alter its antibody production, and no memory B cells are formed. Thus T-independent B-cell activation is less effective than T-dependent B-cell activation: The antibodies produced have a low affinity for antigen and no immunologic memory is formed.

Microbial Pattern Triggering

Identifying yet another crossover between innate and adaptive arms of the immune system, a unique subset of phenotypically distinct B cells, the innate response activator (IRA) B cells, not only make antibody but also produce granulocyte-macrophage colony-stimulating factor (GM-CSF) and IL-3. IRAs are typically found in the pleural and peritoneal cavities in a steady state of activation. In response to binding of bacterial lipopolysaccharide (LPS) by toll-like receptor-4 (TLR-4), they release large quantities of premade IgM antibody, and directly stimulate another subset of B cells to also release IgM. Additionally, IRAs then migrate to the spleen, where they secrete GM-CSF and IL-3, stimulating the production of phagocytes. Experimental models of sepsis suggest that the IRA B-cells are first-responders that direct emergency IgM release to abate bacteria invading through the lungs or intestines, while they sound an alarm to splenic stem cells, recruiting other defenders (*see figure 35.8*).

Retrieve, Infer, Apply

1. What are B-cell receptors? How are they involved in B-cell activation?
2. Briefly compare and contrast B cells and T cells with respect to their formation, structure, and roles in the immune response.
3. How does T-independent antigen triggering of B cells differ from T-dependent triggering?
4. How might IRA B cells prevent bacterial sepsis?

34.7 Antibodies Are Proteins That Bind to Specific 3-D Structures

After reading this section, you should be able to:

■ Describe the structure of the B-cell receptor that is secreted as antibody
■ Compare and contrast the five classes of antibody
■ Diagram the antibody changes, induced by antigen binding, that facilitate antigen capture and removal from the host
■ Integrate antibody secretion with antigen exposure
■ Create a model of genetic diversity that results from recombination, alternative splicing, and somatic hypermutation
■ Predict antibody specificity resulting from clonal selection

So far we have discussed antibody function and how B cells are activated to become antibody-secreting plasma cells. We now turn our attention to antibody structure and how a seemingly infinite diversity of antibodies can be synthesized by a single individual. An **antibody** or **immunoglobulin (Ig)** is a glycoprotein. There are five classes of human antibodies; all share a basic structure that varies in order to accomplish specific biological functions.

Immunoglobulin (Antibody) Structure

All immunoglobulin molecules have a basic structure composed of four polypeptide chains: two identical heavy and two identical light chains (**figure 34.12**). Each light-chain polypeptide usually consists of about 220 amino acids and has a mass of approximately 25,000 Da. Each heavy chain consists of about 440 amino acids and has a mass of about 50,000 to 70,000 Da. Heavy and light chains are connected to each other by disulfide bonds (figure 34.12b). The heavy chains are structurally distinct for each immunoglobulin class or subclass. Both light (L) and heavy (H) chains contain two different regions. The **constant (C) regions (C_L and C_H)** have amino acid sequences that do not vary significantly between antibodies of the same class. The **variable (V) regions (V_L and V_H)** have different amino acid sequences, and these regions fold together to form the antigen-binding sites.

The four chains are arranged in the form of a flexible Y with a hinge region. This hinge allows the antibody molecule to be more flexible, adjusting to the different spatial arrangements of epitopes on antigens. The stalk of the Y is termed the **crystallizable fragment (Fc)** and can bind to a host cell by interacting with the cell-surface Fc receptor. The top of the Y consists of two **antigen-binding fragments (Fab)** that bind with compatible epitopes.

The light chain may be either of two distinct forms called kappa (κ) and lambda (λ). These can be distinguished by the amino acid sequence of the constant (C) portion of the chain. In humans, the constant regions of all light chains are identical. Each antibody molecule produced by a sole B cell will contain either κ or λ light chains but never both. The light-chain variable (V) domain contains hypervariable regions, or complementarity-determining regions (CDRs), that differ in amino acid sequence more frequently than the rest of the variable domain. These regions are essential in determining antigen specificity.

Like the light chain, the heavy chain also has a variable region (VH) at its amino terminal domain. The other domains of the heavy chains are termed constant (C) domains. The constant domains of the heavy chain form the constant (C_H) region. The amino acid sequence of this region determines the classes of heavy chains. In humans, there are five classes of heavy chains designated by lowercase Greek letters: gamma (γ), alpha (α), mu (μ), delta (δ), and epsilon (ε), and usually written as G, A, M, D, or E. The properties of these heavy chains determine,

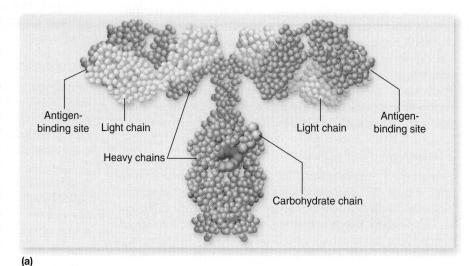

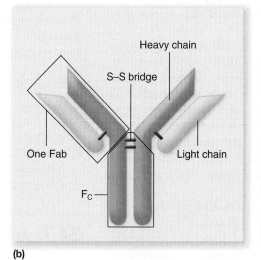

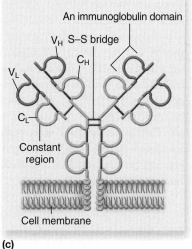

(a)

(b)

(c)

Figure 34.12 Immunoglobulin (Antibody) Structure. (a) A computer generated model of immunoglobulin structure showing the arrangement of the four polypeptide chains. (b) Heavy and light chains are connected to each other by disulfide bridges (bonds). Each heavy and light chain pair form an antigen-binding fragment (Fab). The disulfide-linked heavy chains form the crystalizable fragment (Fc). The Fc fragments are composed only of constant regions, whereas the Fab fragments have both constant and variable regions. (c) Both the heavy and the light chains contain several homologous units of about 100 to 110 amino acids. Within each unit, called a domain, disulfide bonds form a loop of approximately 60 amino acids. Interchain disulfide bonds also link heavy and light chains together. All light chains contain a single variable domain (V_L) and a single constant domain (C_L). Heavy chains contain a variable domain (V_H) and either three or four constant domains (C_H1, C_H2, C_H3, and C_H4). The variable regions (V_H, V_L), when folded together in three-dimensions, form the antigen-binding sites.

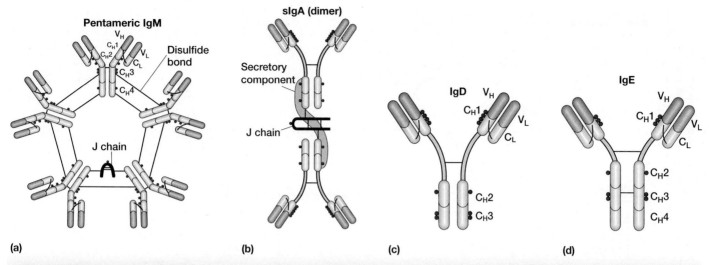

Figure 34.14 **Immunoglobulins M, A, D, and E.** (a) The pentameric structure of human IgM. The disulfide bonds linking peptide chains are shown in black. Note that 10 antigen-binding sites are present. (b) The dimeric structure of human secretory IgA. Notice the secretory component (tan) wound around the IgA dimer and attached to the constant domain of each IgA monomer. (c) The structure of human IgD showing the disulfide bonds that link protein chains (shown in black). (d) The structure of human IgE. Carbohydrate side chains are in red.

MICRO INQUIRY *Where is monomeric IgM commonly found?*

MALT. Secretory IgA is also found in saliva, tears, and breast milk. In these fluids and related body areas, sIgA plays a major role in protecting surface tissues against microorganisms by the formation of an immune barrier. For example, in the intestine, sIgA attaches to viruses, bacteria, and protozoa. This prevents pathogen adherence to mucosal surfaces and invasion of host tissues, a phenomenon known as immune exclusion. In addition, sIgA binds to antigens within the mucosal layer of the small intestine; subsequently the antigen-sIgA complexes are excreted through the adjacent epithelium into the gut lumen. This rids the body of locally formed immune complexes and decreases their access to the circulatory system. Secretory IgA also plays a role in the alternative complement pathway. ◀◀ *Complement is a cascading enzyme system (section 33.3); Secondary lymphoid organs and tissues (section 33.4)*

Immunoglobulin δ, or **IgD,** has a monomeric structure (figure 34.14c) similar to that of IgG. IgD antibodies are abundant in combination with IgM on the surface of B cells and thus are part of the B-cell receptor complex. Therefore their function is to signal the B cell to start antibody production upon initial antigen binding.

Immunoglobulin ε, or **IgE** (figure 34.14d), makes up only a small percent of the total immunoglobulin pool. The skin-sensitizing and anaphylactic antibodies belong to this class. A number of different cell types, especially mast cells, eosinophils, and basophils, can become coated with IgE molecules because these cells have Fc receptors for IgE. When two cell-bound IgE molecules are cross-linked by binding to the same antigen, the cells degranulate. Degranulation releases preformed mediators of inflammation (e.g., histamine). It also stimulates production of an excessive number of eosinophils in the blood (eosinophilia) and

increased rate of movement of the intestinal contents (gut hypermotility), which aid in the elimination of helminthic parasites. Thus although IgE is present in small amounts, this class of antibodies has potent biological capabilities, as is discussed in section 34.10.

Retrieve, Infer, Apply

1. Explain the different functions of IgM when it is bound to B cells versus when it is soluble in serum.
2. Describe the major functions of each immunoglobulin class.
3. Why is the structure of IgG considered the model for all five immunoglobulin classes?
4. Which immunoglobulin can cross the placenta?
5. Which immunoglobulin is most prevalent in the immunoglobulin pool? The least prevalent?

Antibody Kinetics

The synthesis and secretion of antibody can be evaluated with respect to time. Monomeric IgM serves as the B-cell receptor for antigen, and pentameric IgM is the first type of antibody secreted after B-cell activation. Later, under the influence of T-helper cells, IgM-secreting plasma cells may switch and produce another antibody class (e.g., IgG, IgA, or IgE). This is known as antibody class switching (p. 755).

The Primary Antibody Response

When an individual is exposed to an antigen (e.g., an infection or vaccine), there is an initial lag phase, or latent period (up to several weeks), before an antibody response is mounted. During this latent

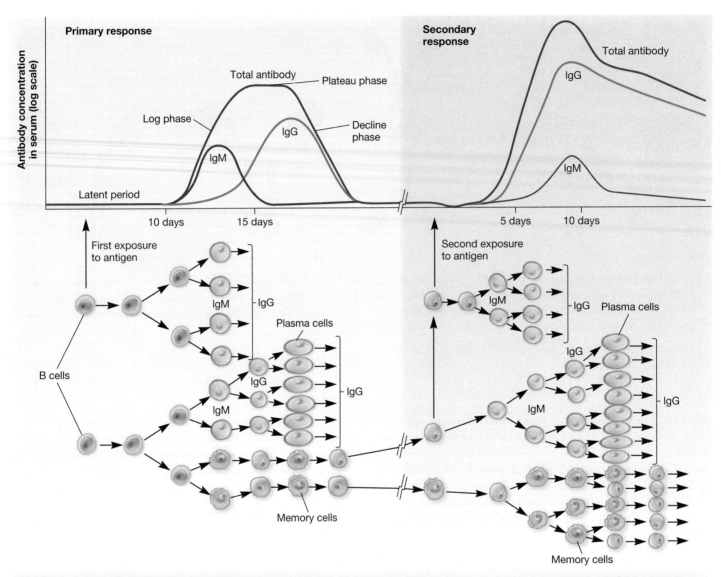

Figure 34.15 Antibody Production and Kinetics.   The four phases of a primary antibody response correlate to the clonal expansion of the activated B cell, differentiation into plasma cells, and secretion of the antibody protein. The secondary response is much more rapid, and total antibody production is nearly 1,000 times greater than that of the primary response.

MICRO INQUIRY   *How does the lag in IgM production compare during a primary and secondary response? Contrast this with the lag seen in IgG production during a primary and secondary response. What does this difference imply about the nature of the secondary response?*

period, no antigen-specific antibody can be detected in the blood (**figure 34.15**). This explains why antibody-based HIV tests, for example, are not accurate until weeks after exposure. Once B cells have differentiated into plasma cells, antibody is secreted and can be detected. During the primary antibody response, IgM appears first, then switches to another antibody class, usually IgG. The affinity of the antibodies for the antigen's determinants is low to moderate during the primary antibody response. The **antibody titer,** which is a measurement of serum antibody concentration (the reciprocal of the highest dilution of an antiserum that gives a positive reaction in the test being used), then rises logarithmically to a plateau during the second, or log, phase. In the plateau phase,

the antibody titer stabilizes. This is followed by a decline phase, during which antibodies are naturally metabolized or bound to the antigen and cleared from circulation.

The Secondary Antibody Response

The primary antibody response primes the immune system so that it possesses specific immunological memory through its clones of memory B cells, each of which responds to different antigen. Upon secondary antigen challenge, as occurs when an individual is reexposed to a pathogen or receives a vaccine booster, B cells mount a heightened secondary response to the same antigen (figure 34.15). Compared to the primary antibody response, the secondary antibody response has a much shorter lag phase and a more rapid log phase, persists for a longer plateau period, attains a higher IgG titer, and produces antibodies with a higher affinity for the antigen.

Diversity of Antibodies

One unique property of antibodies is their remarkable diversity. According to current estimates, each human can synthesize antibodies that can bind to more than 10^{13} (10 trillion) different epitopes. How is this diversity generated? The answer is threefold: (1) rearrangement of antibody gene segments, called combinatorial joining; (2) generation of different codons during antibody gene splicing; and (3) somatic mutations.

Combinatorial joining of immunoglobulin loci occurs because these genes are split into many gene segments. The genes that encode antibody proteins in precursor B cells contain a small number of exons, close together on the same chromosome, that determine the constant (C) region of the light chains. Separated from them but still on the same chromosome is a larger cluster of segments that determines the variable (V) region of the light chains. During B-cell differentiation, exons for the constant region are joined to one segment of the variable region. This occurs by recombination and is mediated by the enzymes called RAG-1 and RAG-2. This splicing process joins the coding regions for constant and variable regions of light chains to produce a gene encoding a complete light chain of an antibody. A similar splicing produces a complete heavy-chain antibody gene.

Because the light-chain genes actually consist of three parts and the heavy-chain genes consist of four, the formation of a finished antibody molecule is slightly more complicated than previously outlined. The germ-line DNA for the light-chain gene contains multiple coding sequences called V and J (joining) regions (**figure 34.16**). During the development of a B cell in the bone marrow, RAG enzymes join one V gene segment with one J segment. This DNA joining process is termed combinatorial joining because it can create many combinations of the V and J regions (**table 34.3**). In addition, an enzyme called *t*erminal *d*eoxynucleotidyl *t*ransferase (tdt) inserts nucleotides at the V-J junction, creating additional diversity. When the light-chain gene is transcribed, transcription continues through the DNA region that encodes the constant portion of the gene. RNA splicing subsequently joins the V, J, and C regions, creating mRNA.

Combinatorial joining in the formation of a heavy-chain gene occurs by means of DNA splicing of the heavy-chain counterparts of V and J along with a third set of D (diversity) sequences (**figure 34.17***a*). Initially, all heavy chains have the µ type of constant region. This corresponds to antibody class IgM (figure 34.17*b*). If a particular B cell is reexposed to its antigen, another DNA splice joins the VDJ region with a different constant

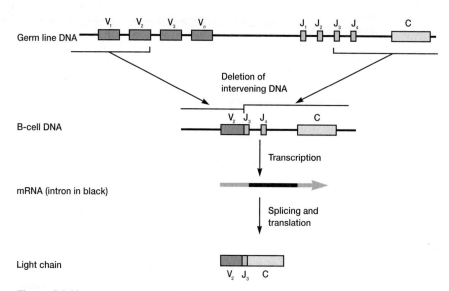

Figure 34.16 Light-Chain Production in a Mouse.   One V segment is randomly joined with one J-C region by deletion of the intervening DNA. The remaining J segments are eliminated from the RNA transcript during RNA processing.

region that can subsequently change the class of antibody produced by the B cell (figure 34.17*c*)—the phenomenon of **antibody class switching.** This process is regulated by antigen-activated helper T cells.

Antibody class switching solves a problem faced by an organism during the immune response. It must be able to bind lots of antigen, but it must also be able to clear the antigen-antibody complexes (immune clearance). Recall that the primary antibody response to antigen is IgM. Because secreted IgM is pentameric, its 10 antigen-binding sites allow it to bind strongly to foreign materials that have multiple repetitive epitopes. However, IgM is not the best antibody class for clearing the antigen from the host. In fact, depending on the

Table 34.3	Theoretical Antibody Diversity Resulting from Combinatorial Joining of Germ-Line Genes[1]
λ light chains	V regions = 2 J regions = 3 Combinations = 2 × 3 = 6
κ light chains	V_κ regions = 250 − 350 J_κ regions = 4 Combinations = 250 × 4 = 1,000 　　　　　　　　= 350 × 4 = 1,400
Heavy chains	V_H = 250 − 1,000 D = 10–30 J_H = 4 Combinations = 250 × 10 × 4 = 10,000 　　　　　　　　= 1,000 × 30 × 4 = 120,000
Diversity of antibodies	κ-containing: 1,000 × 10,000 = 10^7 　　　　　　　1,400 × 120,000 = 2 × 10^8 λ-containing: 6 × 10,000 = 6 × 10^4 　　　　　　　6 × 120,000 = 7 × 10^5

1 Approximate values.

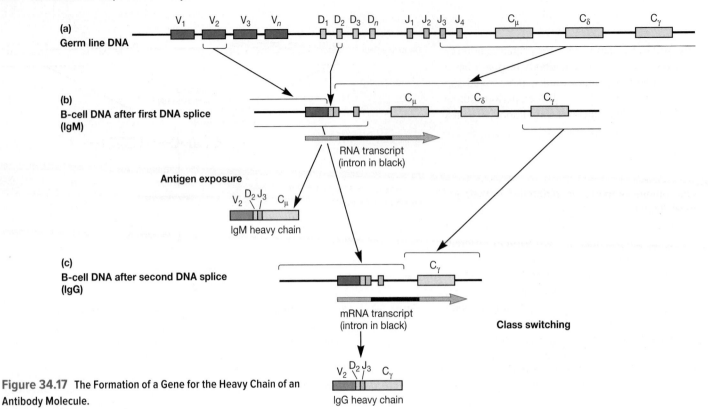

Figure 34.17 The Formation of a Gene for the Heavy Chain of an Antibody Molecule.

antigen and the body site, IgA or IgG provides better effector mechanisms for immune clearance. So how does an antigen-activated B cell change antibody class without altering antigen-binding affinity? Proliferating B cells are able to use combinatorial joining to attach the rearranged coding sequence for the variable region to another heavy-chain gene sequence (figure 34.17).

The switch from IgM to another antibody class results from a unique editing process that occurs when the proliferating B cell synthesizes the enzyme *a*ctivation-*i*nduced cytidine *d*eaminase (AID). AID replaces cytosine residues with uracil when the complementary DNA strands separate during transcription. This causes DNA repair enzymes to remove the inappropriate uracil (now on a DNA backbone). The abasic nucleotide is then excised, nicking one DNA strand. When nicking occurs in the respective μ and γ switch regions, recombination occurs; the μ constant region is excised, and the γ constant region is joined to the VDJ region.

Remember that combinatorial joining is only one of three mechanisms by which antibody diversity is generated. The other two processes are:

1. **Splice-site variability:** The junction for either VJ or VDJ splicing in combinatorial joining can occur between different nucleotides and thus generate different codons in the spliced gene. In addition, the activation of tdt can greatly increase the variability of the nucleotide sequence at the VJ or VDJ junctions during the splicing process. For example, one VJ splicing event can join the V sequence CCTCCC with the J sequence TGGTGG in two ways: CCTCCC + TGGTGG = CCGTGG, which codes for proline and tryptophan. Alternatively, the VJ splicing event can give rise to the sequence CCTCGG, which codes for

proline and arginine. Thus the same VJ joining could produce polypeptides differing in a single amino acid.

2. **Somatic hypermutation of V regions:** The V regions of germ-line DNA are susceptible to a high rate of somatic mutation during B-cell development in response to an antigen challenge. The rearranged variable heavy and light genes encode "hotspots" of four to five nucleotides that are more likely to be altered by point mutation than other areas of the variable regions. The resulting point mutations are recognized by DNA repair enzymes that attempt to correct the altered bases. AID (the same enzyme that mediates antibody class switching) is one of the enzymes that is involved. AID deaminates DNA at cytosine, resulting in base pair mismatches and transition-type mutations. Additionally, repair of the base pair mismatch is error-prone and results in additional sequence diversity. Accumulation of the altered bases in the variable regions of B-cell immunoglobulin genes during antigen challenge promotes significant antibody diversity and thus appropriate antibody affinity. Adding the impact of antibody class switching to the above two processes should begin to explain how it is possible for unique antibodies to respond to the trillions of potential antigens in the environment.

Retrieve, Infer, Apply

1. What is the name of each part of the gene that encodes for the different regions of antibody chains?
2. Describe what is meant by combinatorial joining of V, D, and J gene segments.
3. In addition to combinatorial joining, what other two processes play a role in antibody diversity?

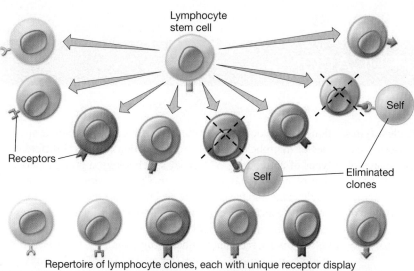

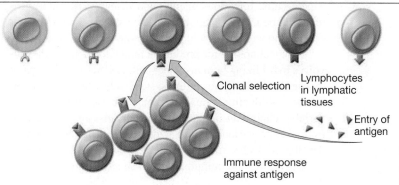

(a) **Antigen-Independent Period**

1 During development of early lymphocytes from stem cells, a given stem cell undergoes rapid cell division to form numerous progeny.

During this period of cell differentiation, random rearrangements of the genes that code for the TCR or BCR occur. The result is a large array of genetically distinct cells, called clones, each clone bearing a different receptor that is specific to react with only a single type of foreign molecule or antigen.

2 At the same time, any lymphocyte clones that have a specificity for self molecules and could be harmful are eliminated from the pool of diversity. This is called immune tolerance.

3 The specificity for a single antigen molecule is programmed into the lymphocyte and is set for the life of a given clone. The end result is an enormous pool of mature but naïve lymphocytes ready to further differentiate under the influence of immune stimuli.

(b) **Antigen-Dependent Period**

4 Lymphocytes will then encounter antigens. These antigens will become the stimulus for the lymphocytes' final activation and immune function. Entry of a specific antigen selects only the lymphocyte clone or clones that carry matching TCR or BCR. This will trigger an immune response, which varies according to the type of lymphocyte involved.

Figure 34.18 Lymphocyte Clonal Expansion. (a) Cell populations expand and are restricted based on their MHC expression. (b) They are further expanded when activated by a specific antigen.

Clonal Selection

As noted, combinatorial joinings, somatic mutations, and variations in the splicing process generate the great variety of antibodies produced by mature B cells. From a large, diverse B-cell pool, specific cells are stimulated by antigens to reproduce and form B-cell clones containing the same genetic information. This is known as **clonal selection.** Both B and T cells undergo clonal selection. It is important because it accounts for immunological specificity and memory.

The clonal selection theory has four components or tenets. The first tenet is that there exists a pool of lymphocytes that can bind to a tremendous range of epitopes (**figure 34.18**). Because some the B and T cells in the pool react with self-epitopes, the second tenet is that these self-reactive cells are eliminated at an early stage of development. Indeed, this has been shown to be true for developing B cells (in the bone marrow) and self-reacting T cells (in the thymus). The third tenet is that once a lymphocyte has been released from its primary development site and is exposed to its specific antigen, it proliferates to form a **clone** (a population of identical cells derived from a single parent cell). Note that this clone has been "selected" by exposure to specific antigen, hence the name of the theory. The final tenet states that all clonal cells react with the same antigenic epitope that stimulated its formation. However, the cells may have somewhat different functions. For instance a B cell forms two different cell populations: antibody-producing plasma cells and memory cells.

Plasma cells are true protein factories that produce about 2,000 antibodies per second in their brief 5- to 7-day life-span. Memory cells circulate more actively from blood to lymph and live much longer (years or even decades) than do plasma cells. Memory cells are responsible for the immune system's rapid secondary antibody response (figure 34.15) to the same antigen. Finally, memory B cells and plasma cells are usually not produced unless the B cell has interacted with and received cytokine signals from activated T-helper cells (figure 34.11). Clonal selection theory has also led to the development of **monoclonal antibody (mAb)** technologies that have been exploited to treat disease (**Techniques & Applications 34.1**).

34.8 Antibody Binding Dooms the Target

After reading this section, you should be able to:

- Explain the consequences of antibody binding of antigen
- Assess the effectiveness of antigen removal by antibody
- Predict which antigens will be most susceptible to antibody action

As we have seen, the antigen-antibody interaction is a bimolecular association that exhibits exquisite specificity. The interactions that occur in animals are absolutely essential in protecting the animal against the continuous onslaught of microorganisms

TECHNIQUES & APPLICATIONS

34.1 Monoclonal Antibody Therapy

The value of antibodies as tools for locating or identifying antigens is well established. For many years, antiserum extracted from human or animal blood was the main source of antibodies for tests and therapy, but most antiserum is problematic. It contains polyclonal antibodies, meaning it is a mixture of different antibodies because it reflects dozens of immune reactions from a wide variety of B-cell clones. This characteristic is to be expected, because several immune reactions may be occurring simultaneously, and even a single species of microbe can stimulate production of several different types of antibodies. With today's technology, a pure preparation of monoclonal antibodies (mAbs) that originate from a single clone and have a single specificity for the same epitope on a single antigen can be made. These monoclonal antibodies can be used therapeutically when they have been modified to deliver toxins, radionuclides, drugs, and even cytokines directly to the antigen. Thus mAbs against specific antigens on cancer cells, infectious agents, and disease-associated proteins can be made to "target" those agents and facilitate their eradication without affecting other proteins, cells, or tissues.

The technology for producing monoclonal antibodies involves hybridizing human cancer cells and activated rabbit or mouse B cells in tissue culture systems. Over the years, methods for producing mouse-human chimeric mAbs (mouse variable chains on human constant chains) and humanized (mouse hypervariable amino acids in human antibodies) were developed. Today transgenic mice containing human antibody genes can be immunized with specific antigens to produce completely human B cells and mAbs. Furthermore, human mAbs are now made in *Escherichia coli* from bacteriophage containing human B cell DNA.

It is estimated that there are over 500 mAbs presently licensed or in development for the specific diagnosis or treatment of human disease. The mAbs target various types of cancer, autoimmune diseases, infection, sepsis, rheumatoid arthritis, Crohn's disease, and Alzheimer's disease, to name a few. Interestingly, immunologists have learned much about B-cell biology from the study of mAbs, including the function of antibody "orphan pieces" (single chain, single domain, Fab, F(ab')$_2$), previously thought to be nonfunctional products of proteolysis.

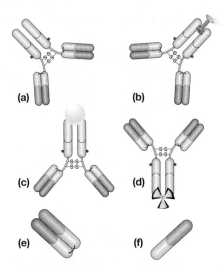

Monoclonal Antibody (mAb) Therapeutics. mAbs that retain their antigen binding specificity can be used (a) as the naked glycoprotein to stimulate antigen recognition and removal, or they can be altered to deliver, (b) cytokines, (c) drugs, or (d) radionuclides to kill their targets. Additionally, antibody pieces such as the (e) single domain and the (f) single-chain fragments can also be used.

and their products, as well as cancer cells. This occurs partly because the antibody coats the invading foreign material, marking it for enhanced recognition by other components of the innate and adaptive immune systems. The mechanisms by which antibodies achieve this are now discussed.

Neutralization

Binding of antibody to biologically active materials (such as bacteria, toxins, or viruses) causes their inactivation or **neutralization** (**figure 34.19**). The capacity of bacteria to colonize the mucosal surfaces of mammalian hosts depends in part on their ability to adhere to mucosal epithelial cells. Secretory IgA (sIgA) antibodies block certain bacterial adherence factors from attaching to host cells. Thus sIgA can protect the host against infection by some pathogenic bacteria and perhaps other microorganisms on mucosal surfaces by neutralizing (preventing) their adherence to host cells, one form of immune exclusion.

Extracellular toxins made by pathogens attach to target host cells before exerting their bioactivity. Antibody binding of toxins prevents toxin-mediated injury to cells: The toxin-antibody complex either is unable to attach to receptor sites on host target cells and is unable to enter the cell, or it is ingested by macrophages. For example, diphtheria toxin inhibits protein synthesis after binding to the cell surface by its B subunit and subsequent passage of the A subunit toxin into the cytoplasm of the target cell. The antibody blocks the toxic effect by binding to the B fragment, thereby preventing the toxin from binding a target host cell. An antibody capable of neutralizing a toxin or

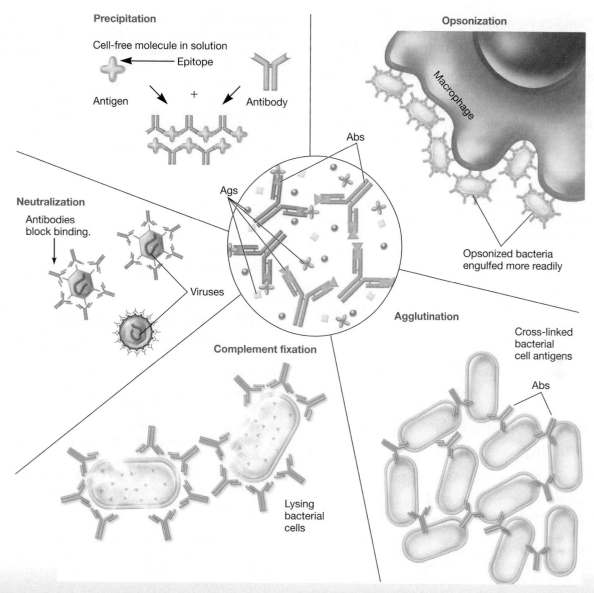

Figure 34.19 Consequences of Antigen-Antibody Binding. Immune complexes form when soluble antigens (Ag) bind soluble antibody (Ab), resulting in precipitation. Opsonization occurs when antibody binds to antigens on larger molecules or cells to be recognized by phagocytic cells. Agglutination results when insoluble antigens (such as viral or bacterial cells) are cross-linked by antibody. The classical complement cascade can be activated by immune complexes; this is called complement fixation. Neutralization results when antibody binds to antigens, preventing the antigen from binding to host cells.

MICRO INQUIRY *What is the difference between a precipitation and an agglutination reaction?*

antiserum containing neutralizing antibody against a toxin is called **antitoxin.** ▶▶| *Virulence defines a pathogen's success (section 35.2)*

In a similar fashion, IgG, IgM, and IgA antibodies can bind to some viruses during their extracellular phase and block their capacity to bind to host cells. This antibody-mediated viral inactivation is called **viral neutralization.**

Opsonization

The phagocytic process (see *figure 33.19*) can be greatly enhanced by opsonization. Phagocytes have an intrinsic ability to bind directly to microorganisms through pattern recognition by

nonspecific cell-surface receptors, engulf the microorganisms, form phagosomes, and digest the microorganisms. As noted in *section 33.3,* opsonization is the process by which microorganisms or other foreign particles are coated with antibody or complement and thus prepared for "recognition" and ingestion by phagocytic cells. Opsonizing antibodies, especially IgG1 and IgG3, bind to Fc receptors on the surface of dendritic cells, macrophages, and neutrophils, after they bind to the target antigen. This process permits effective removal of antigens; the antibody forms a bridge between the phagocyte and the antigen, thereby increasing the likelihood of its phagocytosis (*see figure 33.6*).

Immune Complex Formation

Because antibodies have at least two antigen-binding sites and most antigens have at least two epitopes, cross-linking can occur, producing large aggregates termed **immune complexes** (figure 34.19). If the antigens are soluble molecules and the complex becomes large enough to settle out of solution, a **precipitation** (Latin *praecipitare,* to cast down) or **precipitin reaction** occurs. When the immune complex involves the cross-linking of cells or particles, an **agglutination reaction** occurs. These immune complexes are more rapidly phagocytosed than free antigens. However, immune complex formation can harm the host, as occurs in some types of kidney and vascular pathologies. Finally, formation of immune complexes is the basis of some diagnostic tests. ▶▶| *Immune responses can be measured or exploited to detect infections (section 36.4)*

Retrieve, Infer, Apply

1. How does antigen-antibody binding occur? What is the basis for antibody specificity?
2. How does toxin neutralization occur? Viral neutralization?
3. How does opsonization inhibit microbial adherence to eukaryotic cells?
4. Describe an immune complex. What is the equivalence zone?

34.9 Not Responding Is Also Part of Immunity

After reading this section, you should be able to:

- Evaluate the impact of improper B-cell activation
- Contrast the outcome of B-cell and T-cell exposure to self antigens

Acquired immune tolerance is the body's ability to produce T cells and antibodies against nonself antigens such as microbial antigens, while "tolerating" (not responding to) self antigens. Some of this tolerance arises early in embryonic life, when immunologic competence is being established. Three general tolerance mechanisms have been proposed: (1) negative selection by clonal deletion, (2) the induction of anergy, and (3) inhibition of the immune response by the action of Treg cells.

Negative selection by clonal deletion removes lymphocytes that recognize self antigens. These cells are eliminated during embryogenesis by apoptosis. T-cell tolerance induced in the thymus and B-cell tolerance in the bone marrow is called **central tolerance** (see figure 34.18a). Many autoreactive B cells undergo clonal deletion or become anergic as they mature in the bone marrow. However, another mechanism is needed to prevent immune reactions against self antigens, termed autoimmunity, because many antigens are tissue-specific and are not present in the thymus or bone marrow.

Mechanisms occurring elsewhere in the body are collectively referred to as **peripheral tolerance.** These supplement central tolerance. Peripheral tolerance is thought to be based largely on incomplete activation signals given to the lymphocyte when it encounters self antigens in the periphery of the body. This mechanism leads to a state of unresponsiveness called

anergy, which is associated with impaired intracellular signaling and apoptosis. The deletion of self-reactive B cells takes place in secondary lymphoid tissue, such as the spleen and lymph nodes. Since B cells recognize unprocessed antigen, there is no need for the participation of MHC molecules in these processes. For those self antigens present at relatively low concentrations, immunologic tolerance is often maintained only within the T-cell population. This is sufficient to sustain tolerance because it denies the help essential for antibody production by self-reactive B cells. Regulatory T cells with suppressor activity can inhibit B- and other T-cell activation, as well as modify specific functions (p. 744).

Retrieve, Infer, Apply

1. Describe the three ways acquired immune tolerance develops in the vertebrate host.
2. How would you define anergy?

34.10 Sometimes the Immune System Doesn't Work the Way It Should

After reading this section, you should be able to:

- Illustrate differences between hypersensitivity, autoimmunity, tissue rejection, and immunodeficiency
- Relate types of hypersensitivity to the root biological cause
- Predict disease potential due to the presence in the body of "self-reactive" T and B cells
- Explain tissue rejection as a function of cells responding to the graft
- Propose the impact on immunity when T cells and B cells become deficient

As in any body system, disorders also occur in the immune system. Immune disorders can be categorized as hypersensitivities, autoimmune diseases, transplantation (tissue) rejection, and immunodeficiencies. Each of these is now discussed.

Hypersensitivities

Hypersensitivity is an exaggerated adaptive immune response that results in tissue damage and is manifested in the individual on a second or subsequent contact with an antigen. Hypersensitivity reactions can be classified from immediate to delayed. Obviously, immediate reactions appear faster than delayed ones, but the main difference between them is the nature of the immune response to the antigen. Realizing this fact in 1963, Peter Gell (1914–2001) and Robert Coombs (1921–2006) developed a classification system for reactions responsible for hypersensitivities. Their system correlates clinical symptoms with information about immunologic events that occur during hypersensitivity reactions. The **Gell-Coombs classification** system divides hypersensitivities into four types: I, II, III, and IV.

Type I Hypersensitivity

An **allergy** (Greek *allos,* other, and *ergon,* work) is one of the most common kinds of **type I hypersensitivity** reaction. Allergic reactions occur when an individual who has produced IgE antibody in response to the initial exposure to an antigen (**allergen**) subsequently encounters the same allergen. Upon initial exposure to a soluble allergen, B cells are stimulated to differentiate into plasma cells and produce specific IgE antibodies with the help of T_H cells (**figure 34.20**). This IgE is sometimes called a **reagin,** and the individual has a hereditary predisposition for its production. Once synthesized, IgE binds to the Fc receptors of mast cells (basophils and eosinophils can also be bound) and sensitizes these cells, making the individual sensitized to the allergen. When a subsequent exposure to the allergen occurs, the allergen attaches to the surface-bound IgE on the sensitized mast cells, causing mast cell degranulation.

Degranulation releases physiological mediators such as histamine, leukotrienes, heparin, prostaglandins, PAF (*p*latelet-*a*ctivating *f*actor), ECF-A (*e*osinophil *c*hemotactic *f*actor of *a*naphylaxis), and proteolytic enzymes. These mediators trigger smooth muscle contractions, vasodilation, increased vascular permeability, and mucus secretion (figure 34.20). The inclusive term for these responses is **anaphylaxis** (Greek *ana,* up, back again, and *phylaxis,* protection). Anaphylaxis can be divided into systemic and localized reactions.

Systemic anaphylaxis is a generalized response that is immediate due to a sudden burst of mast cell mediators. Usually there is respiratory impairment caused by smooth muscle constriction in the bronchioles. The arterioles dilate, which greatly

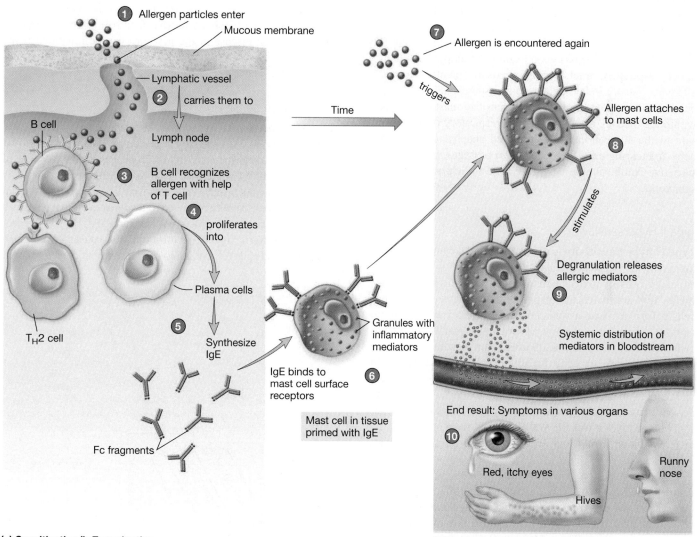

(a) Sensitization/IgE production

(b) Subsequent exposure to allergen

Figure 34.20 Type I Hypersensitivity (Allergic Response).   (a) The initial contact (sensitization) of lymphocytes by small-protein allergens at mucous membranes results in T_H2 cell–assisted antibody class switching; plasma cells secrete IgE antibody. The IgE binds to its receptor on tissue mast cells (1–6). (b) Subsequent exposure to the same allergens results in their capture by the cell-bound IgE (7), triggering mast cell degranulation (8, 9). Characteristic signs and symptoms (hives, swelling, itching, etc.) of allergy ensue (10).

MICRO INQUIRY *What are some of the allergic mediators released during mast cell degranulation?*

reduces arterial blood pressure and increases capillary permeability with rapid loss of fluid into the tissue spaces. These physiological changes can be rapid and severe enough to be fatal within a few minutes from reduced blood flow, asphyxiation, reduced blood pressure, and circulatory shock. Common examples of allergens that can produce systemic anaphylaxis include drugs (penicillin), passively administered antisera, peanuts, and insect venom from wasps, hornets, or bees.

Localized anaphylaxis is called an **atopic** ("out of place") **reaction.** The symptoms that develop depend primarily on the route by which the allergen enters the body. Allergens that enter through the skin stimulate a localized reaction that results in a raised (wheal) and reddened (flare) lesion. Hay fever (allergic rhinitis) is a good example of atopy involving the upper respiratory tract. Initial exposure involves airborne allergens—such as plant pollen, fungal spores, animal hair and dander, and house dust mites—that sensitize mast cells located within the mucous membranes of the respiratory tract. Reexposure to the allergen causes a localized anaphylactic response: itchy and tearing eyes, congested nasal passages, coughing, and sneezing. Antihistamine drugs are used to help alleviate these symptoms.

Type II Hypersensitivity

Type II hypersensitivity is generally called a cytolytic or cytotoxic reaction because it results in the destruction of host cells, either by lysis or toxic mediators. In type II hypersensitivity, IgG or IgM antibodies are inappropriately directed against host cell-surface or tissue-associated antigens. They usually stimulate the classical complement pathway and a variety of effector cells (**figure 34.21**). The antibodies interact with complement (Clq) and the effector cells through their Fc regions. The cellular damage is similar to the lytic events that complement exerts on microbial membranes (*see figure 33.8*). Examples of type II hypersensitivity reactions include the response exhibited by a person who receives a transfusion with blood from a donor with a different blood group and erythroblastosis fetalis, which we now discuss.

Blood transfusion was often fatal prior to the discovery of distinct blood types by Karl Landsteiner in 1904. Landsteiner's observation that sera from one person could agglutinate the blood cells of another person led to his identification of four distinct types of human blood. The red blood cell types were subsequently determined to result from cell-surface glycoproteins, now called the **ABO blood groups** (**figure 34.22a**). The type II hypersensitivity reaction seen in blood transfusion occurs as complement is activated by cross-linking antibodies (figure 34.22b).

Blood typing can be accomplished by a slightly more sophisticated method of Landsteiner's process whereby blood from one host is mixed with antibodies specific for type A or type B

blood. Agglutination of red cells by antibody (figure 34.22c) is used as a diagnostic tool to determine blood type. Another red blood cell antigen often reported with the ABO type was discovered during experiments with rhesus monkeys. The so-called rhesus (Rh) factor (or D antigen) is determined by the expression of two alleles, one dominant (coding for the factor) and one recessive (not coding for the factor). Thus an individual who is homozygous or heterozygous for the dominant Rh allele has the antigen and is indicated as Rh^+. Expression of two recessive alleles results in the designation of Rh^- (no Rh factor). Incompatibility between Rh^- mothers and their Rh^+ fetus can result in maternal anti-Rh antibodies destroying fetal blood cells (**figure 34.23**). This type II hypersensitivity is called **erythroblastosis fetalis.** Control of this potentially fatal hemolytic disease of the newborn can be mitigated if the mother is passively immunized with anti-Rh factor antibodies (RhoGAM).

Type III Hypersensitivity

Type III hypersensitivity involves the formation of immune complexes (**figure 34.24**). Normally these complexes are phagocytosed effectively by dendritic cells and macrophages. In the presence of excess amounts of some soluble antigens, the antigen-antibody complexes may not be efficiently removed. Their accumulation can lead to a hypersensitivity reaction from complement that triggers a variety of inflammatory processes. The antibodies of type III reactions are primarily IgG. The inflammation caused by immune complexes and cells responding to such inflammation can result in

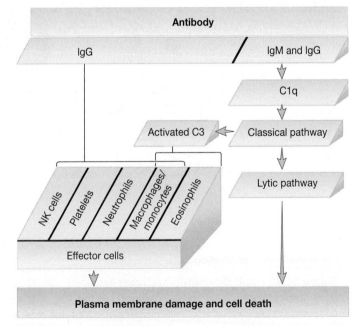

Figure 34.21 Type II Hypersensitivity. The action of antibody occurs through effector cells or the membrane attack complex, which damages target cell plasma membranes, causing cell destruction.

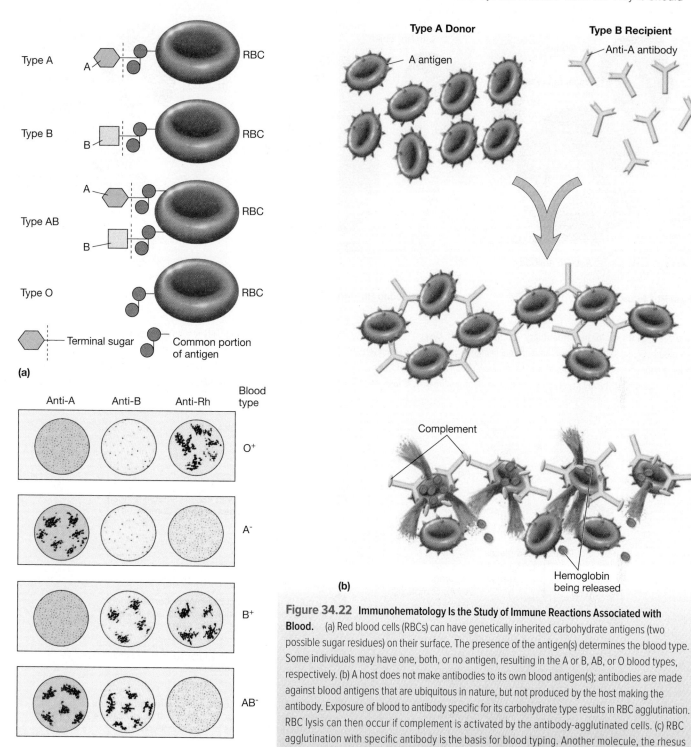

(a)

(b)

(c)

Figure 34.22 Immunohematology Is the Study of Immune Reactions Associated with Blood. (a) Red blood cells (RBCs) can have genetically inherited carbohydrate antigens (two possible sugar residues) on their surface. The presence of the antigen(s) determines the blood type. Some individuals may have one, both, or no antigen, resulting in the A or B, AB, or O blood types, respectively. (b) A host does not make antibodies to its own blood antigen(s); antibodies are made against blood antigens that are ubiquitous in nature, but not produced by the host making the antibody. Exposure of blood to antibody specific for its carbohydrate type results in RBC agglutination. RBC lysis can then occur if complement is activated by the antibody-agglutinated cells. (c) RBC agglutination with specific antibody is the basis for blood typing. Another molecule, the rhesus (Rh) factor, is another major RBC antigen that is typed to determine blood compatibility.

MICRO INQUIRY *What would be the genotype of an individual with type O, Rh⁻ blood?*

significant damage, especially of blood vessels (vasculitis), kidney glomerular basement membranes (glomerulonephritis), joints (arthritis), and skin (systemic lupus erythematosus).

Type IV Hypersensitivity

Type IV hypersensitivity involves delayed, cell-mediated immune reactions. A major factor in the type IV reaction is the time required for T cells to migrate to and accumulate near the antigens. Both T_H and CTL cells can elicit type IV hypersensitivity reactions, depending on the pathway in which the antigen is processed and presented. These events usually take a day or more to plateau. Two important type IV hypersensitivities are tuberculin hypersensitivity and allergic contact dermatitis. Other examples of type IV hypersensitivity reactions include the cell and tissue destruction caused by leprosy, tuberculosis, leishmaniasis, candidiasis, and herpes simplex lesions. These infectious diseases are discussed in chapters 38–40.

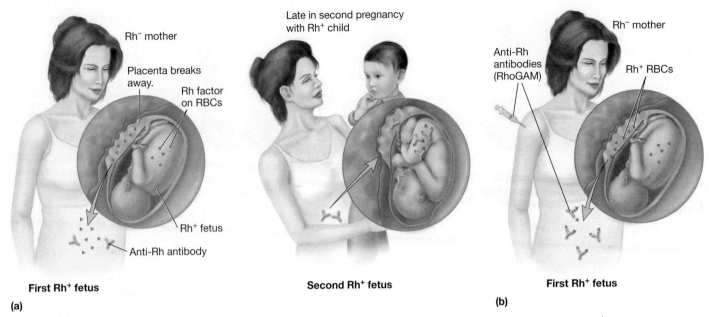

(a)

Figure 34.23 **Rh Factor Incompatibility Can Result in RBC Lysis.** (a) A naturally occurring blood cell incompatibility results when an Rh⁺ fetus develops within an Rh⁻ mother. Initial sensitization of the maternal immune system occurs when fetal blood passes the placental barrier. In most cases, the fetus develops normally. However, a subsequent pregnancy with an Rh⁺ fetus results in a severe, fetal hemolysis. (b) Anti-Rh antibody (RhoGAM) can be administered to Rh⁻ mothers during pregnancy to help bind, inactivate, and remove any Rh factor that may be transferred from the fetus. In some cases, RhoGAM is administered before sensitization occurs.

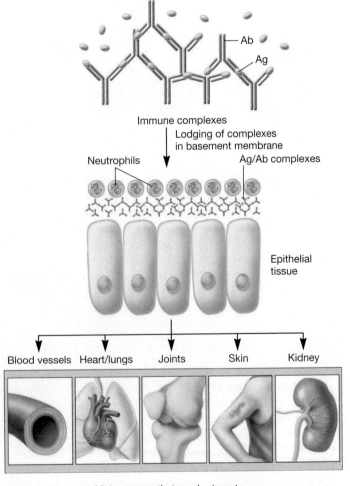

1. Antibody combines with excess soluble antigen, forming large quantities of Ab/Ag complexes.

2. Circulating immune complexes become lodged in the basement membrane of epithelia in sites such as kidney, lungs, joints, skin.

3. Fragments of complement cause release of histamine and other mediator substances.

4. Neutrophils migrate to the site of immune complex deposition and release enzymes that cause severe damage to the tissues and organs involved.

Major organs that can be targets of immune complex deposition

Figure 34.24 **Type III Hypersensitivity.** Circulating immune complexes may lodge at various tissue sites, where they activate complement and subsequently cause tissue cell lysis. Complement activation also results in granulocyte recruitment and release of their mediators, causing further injury to the tissue. Increased vascular permeability, resulting from granulocyte mediators, allows immune complexes to be deposited deeper into tissue sites, where platelet recruitment leads to microthrombi (blood clots), which impair blood flow, resulting in additional tissue damage.

In general, type IV reactions occur when antigens, especially those binding to serum proteins or tissue cells, are processed and presented to T cells. If the antigen is phagocytosed, it will be presented to T_H cells by the class II MHC molecules on the antigen presenting cell. This activates the $T_H 1$ cell, causing it to proliferate and secrete cytokines such as IFN-γ and TNF-α. If the antigen is lipid soluble, it can cross the cell membrane to be processed within the cytosol and be presented to CTL cells by class I MHC molecules. CTLs induce apoptosis to kill the cell that is presenting the antigen. T_H-cell and CTL signals also stimulate the expression of adhesion molecules on the local endothelium to increase vascular permeability, allowing fluid and cells to enter the tissue space. The cytokines attract and recruit lymphocytes, macrophages, and basophils, exacerbating the inflammatory response. Extensive tissue damage may result.

In tuberculin hypersensitivity, partially purified proteins collectively called tuberculin are obtained from the bacterium that causes tuberculosis. After they are injected into the skin of the forearm, the response in a tuberculin-positive individual begins in about 8 hours, and a reddened area surrounding the injection site becomes indurated (firm and hard) within 12 to 24 hours. The $T_H 1$ cells that migrate to the injection site are responsible for the induration. The reaction peaks in 48 hours and then subsides. The size of the induration is directly related to the amount of antigen that was introduced and to the degree of hypersensitivity of the tested individual.

Allergic contact dermatitis is a type IV reaction caused by haptens that combine with proteins in the skin to form the allergen that elicits the immune response (**figure 34.25**). The haptens are the antigenic determinants, and the skin proteins are the carrier molecules for the haptens. Examples of these haptens include cosmetics, plant materials (catechol molecules from poison ivy and poison oak), topical chemotherapeutic agents, metals, and jewelry (especially jewelry containing nickel).

Retrieve, Infer, Apply

1. Discuss the mechanism of type I hypersensitivity reactions and how these can lead to systemic and localized anaphylaxis.
2. What causes the typical raised and reddened skin reaction to antigen?
3. Why are type II hypersensitivity reactions called cytolytic or cytotoxic?
4. What characterizes a type III hypersensitivity reaction? Give an example.
5. Characterize a type IV hypersensitivity reaction.

Autoimmunity and Autoimmune Diseases

As we discuss in sections 34.1 and 34.9, the body is normally able to distinguish its own self antigens from foreign nonself antigens and usually does not mount an immunologic attack against itself. This phenomenon is called immune tolerance. At times the body loses tolerance and mounts an abnormal immune attack, either with antibodies or T cells, against its own self antigens.

It is important to distinguish between autoimmunity and autoimmune disease. Autoimmunity often is benign, whereas autoimmune disease often is fatal. **Autoimmunity** is characterized by the presence of serum antibodies that react with self antigens. These antibodies are called autoantibodies. The formation of autoantibodies is a normal consequence of aging; is readily inducible by infectious agents or drugs; and is potentially reversible (it disappears when the offending "agent" is removed or eradicated). **Autoimmune disease** results from the activation of self-reactive T and B cells that, following stimulation by genetic or environmental triggers, cause actual tissue damage (**table 34.4**). Examples include rheumatoid arthritis and type I diabetes mellitus.

Some investigators believe that the release of abnormally large quantities of self antigens may occur when an infectious agent causes host tissue damage. Others suggest that a change in the normal host-microbiota relationship (dysbiosis) results in

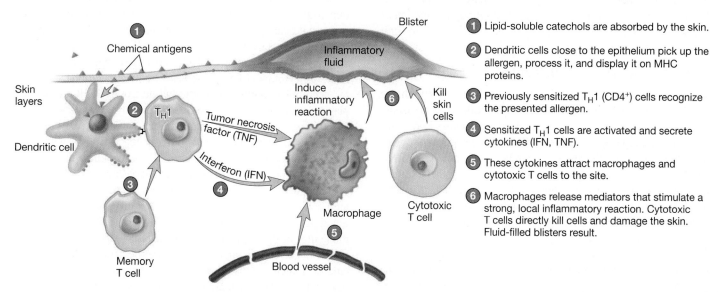

Figure 34.25 Contact Dermatitis. In contact dermatitis to poison ivy, a person initially becomes exposed to the allergen, predominantly 3-n-pentadecyl-catechol found in the resinous sap material urushiol, which is produced by the leaves, fruit, stems, and bark of the poison ivy plant. The catechol molecules, acting as haptens, combine with high molecular weight skin proteins. After 7 to 10 days, sensitized T cells are produced and give rise to memory T cells. Upon second contact, the catechols bind to the same skin proteins, and the memory T cells become activated in only 1 to 2 days, leading to an inflammatory reaction (contact dermatitis).

Table 34.4	Some Autoimmune Diseases in Humans	
Disease	**Autoantigen**	**Pathophysiology**
Acute rheumatic fever	Streptococcal cell wall antigens mimic self antigens and induce antibodies that cross-react with antigens on cardiomyocytes and other cells.	Type II hypersensitivity leads to myocarditis, heart valve scarring, and arthritis.
Autoimmune hemolytic anemia	Rh blood group antigen induces antibody to Rh antigen on red blood cells.	Type II hypersensitivity leads to anemia when red blood cells are destroyed by complement and phagocytosis.
Goodpasture's syndrome	Damage to kidney basement membrane exposes cryptic collagen protein, inducing anti-collagen antibody.	Type II hypersensitivity results in glomerulonephritis and pulmonary hemorrhage.
Graves' disease	Antibody to thyroid-stimulating hormone (TSH) receptor mimics TSH.	Overstimulation of TSH receptor leads to hyperthyroidism.
Multiple sclerosis	Antibody and activated T cells to several nervous system antigens	Types II and IV hypersensitivities alter nerve communications, leading to numbness, weakness, spasm, and loss of motor and cognitive function.
Myasthenia gravis	Antibody to acetycholine receptor in skeletal muscle	Antibody blockade of neurotransmitter receptors results in progressive muscular weakness.
Rheumatoid arthritis	IgG antibody (to synovial joint cartilage antigen) recognized as foreign	Type III hypersensitivity from immune complexes of antibodies to antibodies results in joint inflammation and destruction.
Systemic lupus erythematosus	Antibodies to various cellular (DNA, nucleoprotein, cardiolipin) and blood clotting components	Type III hypersensitivity results in immune complex-induced arthritis, glomerulonephritis, vasculitis, and rash.
Insulin-dependent diabetes mellitus	Antibody and activated T cells to pancreatic beta cell antigens	Types II and IV hypersensitivities destroy beta cells, resulting in insulin deficiency.

immune dysfunction. The same agents also may cause body proteins to change into forms that stimulate antibody production or T-cell activation. Simultaneously, the capacity of T cells to limit this type of reaction seems to be repressed.

Transplantation (Tissue) Rejection

Transplants between genetically different individuals within a species are termed **allografts** (Greek *allos,* other). Some transplanted tissues do not stimulate an immune response. For example, a transplanted cornea is rarely rejected because lymphocytes do not circulate into the anterior chamber of the eye. This site is considered an immunologically privileged site. Another example of a privileged tissue is the heart valve, which in fact can be transplanted from a pig to a human without stimulating an immune response. Such a graft between different species is termed a **xenograft** (Greek *xenos,* strayed).

Transplanting tissue that is not immunologically privileged generates the possibility that the recipient's cells will recognize the donor's tissues as foreign. This triggers the recipient's immune mechanisms, which may destroy the donor tissue. Such a response is called **host-versus-graft disease,** or tissue rejection. Tissue rejection can occur by two different mechanisms. First, foreign MHC molecules on transplanted tissue, or the "graft," are recognized by host T-helper cells, which aid cytotoxic T cells in graft destruction (**figure 34.26**). Cytotoxic T cells then recognize the graft because it bears foreign class I MHC molecules. This response is much like the activation of CTLs by virally infected host cells. A second mechanism involves T-helper cells reacting to the graft and releasing cytokines. The cytokines stimulate macrophages to enter, accumulate within the graft, and destroy it. MHC molecules play a dominant role in tissue rejection reactions because of their unique association with the recognition system of T cells. Unlike antibodies, T cells cannot recognize or react directly with non-MHC molecules (viruses, allergens). They recognize these molecules only when presented by an MHC molecule.

Because class I MHC molecules are present on every nucleated cell in the body, they are important targets of the rejection reaction. The greater the antigenic difference between class I molecules of the recipient and donor tissues, the more rapid and severe the rejection reaction is likely to be. Class II MHC molecule mismatch can also result in rejection and may be even more severe than class I mismatch reactions. However, the reaction can sometimes be minimized if recipient and donor tissues are matched as closely as possible. Most recipients are not 100% matched to their donors, so immunosuppressing drugs are used to prevent host-mediated rejection of the graft.

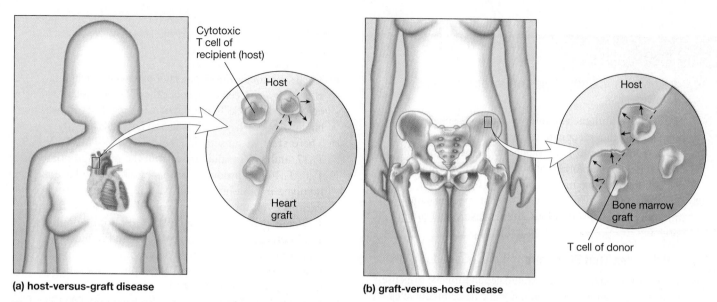

(a) host-versus-graft disease

(b) graft-versus-host disease

Figure 34.26 Potential Transplantation Reactions. (a) Donated tissues that are not from an identical twin contain cellular MHC proteins that are recognized as foreign by the recipient host (host-versus-graft disease). The tissue is then attacked by host CTLs, resulting in its damage and rejection. (b) Donated tissue may also contain immune cells that react against host antigens. Recognition of a foreign host by donor CTLs results in graft-versus-host disease.

Organ transplant recipients also can develop **graft-versus-host disease.** This occurs when the transplanted tissue contains immune cells that recognize host antigens and attack the host. The immunosuppressed recipient cannot control the response of the grafted tissue. Graft-versus-host disease is a common problem in bone marrow transplants. The transplanted bone marrow contains many mature T cells. These cells recognize the host MHC antigens and attack the immunosuppressed recipient's normal tissue cells.

Immunodeficiencies

Defects in one or more components of the immune system can result in its failing to recognize and respond properly to antigens. Such **immunodeficiencies** can make a person more prone to infection than those people capable of a complete and active immune response. Despite the increase in knowledge of functional derangements and cellular abnormalities in the various immunodeficiency disorders, the fundamental biological errors responsible for them remain largely unknown. To date, most genetic errors associated with these immunodeficiencies are located on the X chromosome and produce primary or congenital immunodeficiencies (**table 34.5**). Other immunodeficiencies can be acquired because of infections by immunosuppressive microorganisms, such as HIV.

Retrieve, Infer, Apply

1. What is an autoimmune disease and how might it develop?
2. What is an immunologically privileged site and how is it related to transplantation success?
3. How does a tissue rejection reaction occur? How might a patient about to receive a bone marrow transplant be prepared? Explain why this is necessary.
4. Describe an immunodeficiency. How might immunodeficiencies arise?

Table 34.5	**Some Congenital Immune Deficiencies in Humans**	
Condition	**Symptoms**	**Cause**
Chronic granulomatous disease	Defective monocytes and neutrophils leading to recurrent bacterial and fungal infections	Failure to produce reactive oxygen intermediates due to defective NADPH oxidase
X-linked agammaglobulinemia	Plasma cell or B-cell deficiency and inability to produce adequate specific antibodies	Defective B-cell differentiation due to loss of tyrosine kinase
DiGeorge syndrome	T-cell deficiency and very poor cell-mediated immunity	Lack of thymus or a poorly developed thymus
Severe combined immunodeficiency disease (SCID)	Both antibody production and cell-mediated immunity impaired due to a great reduction of B- and T-cell levels	Various mechanisms (e.g., defective B- and T-cell maturation because of X-linked gene mutation; absence of adenosine deaminase in lymphocytes)

Key Concepts

34.1 Adaptive Immunity Relies of Recognition and Memory

- The adaptive immune response system consists of lymphocytes that recognize foreign molecules (antigens) and respond to them. Two branches or arms of immunity are recognized: humoral (antibody-mediated) immunity and cellular (cell-mediated) immunity (**figure 34.1**).
- Acquired immunity refers to the type of specific (adaptive) immunity that a host develops after exposure to a suitable antigen.

34.2 Molecules That Elicit Immunity Are Called Antigens

- An antigen is a substance that stimulates an immune response. Each antigen can have several antigenic determinant sites called epitopes that stimulate production of and combine with specific antibodies (**figure 34.2**).
- Haptens are small organic molecules that are not antigenic by themselves but can become antigenic if bound to a larger carrier molecule.

34.3 Adaptive Immunity Can Be Earned or Borrowed

- Immunity can be acquired by natural means—actively through infection or passively through receipt of preformed antibodies, as through colostrum (**figure 34.3**).
- Immunity can be acquired by artificial means—actively through immunization or passively through receipt of preformed antibodies, as with antisera.

34.4 Recognition of Foreignness Is Critical for a Strong Defense

- MHC molecules are cell-surface proteins coded by a group of genes termed the major histocompatibility complex. Class I MHC proteins are found on all nucleated cells of mammals. Class II MHC proteins are expressed on macrophages, dendritic cells, and B cells. The human MHC gene products are called the human leukocyte antigens (HLAs) (**figure 34.4**).
- Class I MHC proteins collect foreign peptides processed by the proteasome and present them to cytotoxic T cells (**figure 34.5**).
- Class II MHC proteins collect foreign peptides processed by the phagosome (macrophages and dendritic cells) or endosome (B cells) and present them to T-helper cells.

34.5 T Cells Oversee and Participate in Immune Functions

- T cells are pivotal elements of the immune response. T cells have antigen-specific receptor proteins (**figure 34.6**).
- Antigen-presenting cells—macrophages, dendritic cells, and B cells—take in foreign antigens or pathogens, process them, and present antigenic fragments complexed with MHC molecules to T-helper cells.

- T cells control the development of other cells, including effector B and T cells. T-helper cells (CD4$^+$) regulate cell behavior; cytotoxic T cells (CD8$^+$) also regulate cell behavior, but in addition, they can kill altered host cells directly.
- There are five subsets of T-helper cells—T_H1, T_H2, T_H9, T_H17, and T_H22—that arise from the undifferentiated T_H0. T_H1 cells produce various cytokines and are involved in cellular immunity. T_H2 cells also produce various cytokines but are involved in humoral immunity. T_H9 cells primarily produce IL-9 to inhibit apoptosis. T_H17 cells primarily produce IL-17 and induce inflammation by recruiting neutrophils. T_H22 cells produce IL-22, IL-13, and TNF-alpha to facilitate wound healing (**figure 34.9**).
- Regulatory T (Treg) cells can arise from both CD4$^+$ and CD8$^+$ T cells and can downregulate specific immune reactions, especially those rare reactions attacking self.
- Cytotoxic T lymphocytes recognize target host cells such as virus-infected cells that have foreign antigens and class I MHC molecules on their surface. The CTLs then attack and destroy the target cells using the CD95 pathway, the perforin pathway, or both.

34.6 B Cells Make Antibodies and Do a Whole Lot More

- B cells defend against antigens by differentiating into plasma cells that secrete antibodies into the blood and lymph, providing humoral or antibody-mediated immunity.
- B cells can be stimulated to divide or differentiate to secrete antibody when triggered by the appropriate signals. One type of such a signal is the microbe-associated molecular pattern (MAMP) that binds to B-cell toll-like receptors.
- B cells have receptor immunoglobulins on their plasma membrane surface that are specific for given three-dimensional epitopes. Contact with the epitope is required for the B cell to divide and differentiate into plasma cells and memory cells (**figures 34.10** and **34.11**).

34.7 Antibodies Are Proteins That Bind Specific 3-D Structures

- Antibodies (immunoglobulins) are a group of glycoproteins present in the blood, tissue fluids, and mucous membranes of vertebrates. All immunoglobulins have a basic structure composed of four polypeptide chains (two light and two heavy) connected to each other by disulfide bonds (**figure 34.12**). In humans, five immunoglobulin classes exist: IgG, IgA, IgM, IgD, and IgE (**figures 34.13** and **34.14** and **table 34.2**).
- The primary antibody response in a host occurs following initial exposure to the antigen. This response has lag, log, plateau, and decline phases. Upon secondary antigen challenge, the B cells mount a heightened and accelerated response (**figure 34.15**).

- Antibody diversity results from the rearrangement and splicing of the individual gene segments on the antibody-coding chromosomes, somatic mutations, the generation of different codons during splicing, and the independent assortment of light- and heavy-chain genes (**figures 34.16** and **34.17**).
- Immunologic specificity and memory are partly explained by the clonal selection theory (**figure 34.18**).

34.8 Antibody Binding Dooms the Target

- Various types of antigen-antibody reactions occur in vertebrates and initiate the participation of other body processes that determine the ultimate fate of the antigen.
- The complement system can be activated, leading to cell lysis, phagocytosis, chemotaxis, or stimulation of the inflammatory response.
- Other defensive antigen-antibody interactions include toxin neutralization, viral neutralization, adherence inhibition, opsonization, and immune complex formation (**figure 34.19**).

34.9 Not Responding Is Also Part of Immunity

- Acquired immune tolerance is the ability of a host to react against nonself antigens while tolerating self antigens. It can be induced in several ways.
- Anergy is a state of immune unresponsiveness.

34.10 Sometimes the Immune System Doesn't Work the Way It Should

- When the immune response occurs in an exaggerated form and results in tissue damage to the individual, the term hypersensitivity is applied. There are four types of hypersensitivity reactions, designated as types I through IV (**figures 34.20–34.25**).
- Autoimmune diseases result when self-reactive T and B cells attack the body and cause tissue damage. A variety of factors can influence the development of autoimmune disease (**table 34.4**).
- The immune system can act detrimentally and reject tissue transplants. There are different types of transplants. Xenografts involve transplants of privileged tissue between different species, and allografts are transplants between genetically different individuals of the same species (**figure 34.26**).
- Immunodeficiency diseases are a diverse group of conditions in which an individual's susceptibility to various infections is increased; several severe diseases can arise because of one or more defects in the adaptive or innate immune responses.

Compare, Hypothesize, Invent

1. What properties of proteins make them suitable molecules from which to make antibodies?

2. What other biotechnologies could be invented based upon clonal selection theory (in addition to monoclonal antibody technique)s?

3. Speculate as to how MHC, TCR, and BCR molecules evolved to require accessory proteins or coreceptors for a signal to be sent within the cell?

4. Why do you think two signals are required for B- and T-cell activation but only one signal is required for activation of an APC?

5. Most immunizations require multiple exposures to the vaccine (i.e., boosters). Why is this the case?

6. In an effort to better understandf the mechanisms by which hosts tolerate their normal gut flora, researchers introduced the Gram-negative bacterium *Bacteroides thetaiotaomicron* (a normal inhabitant of the guts of both mice and humans) into gnotobiotic mice. They discovered that the more sIgA the mice produced, the better the microbe was tolerated, as monitored by the level of intestinal inflammation. We usually think of antibody production and inflammation as partners in the fight against infectious agents. Why do you think that in this case, the production of antibody was not coupled with inflammation and, to the contrary, protected the host against inflammation? What does this suggest about the mechanism by which normal flora is "permitted" to make a specific host its home?

Read the original paper: Peterson, D. A., et al. 2007. IgA response to symbiotic bacteria as a mediator of gut homeostasis. *Cell Host Microbe* 2:328–39.

7. While the distribution of memory T cells is well understood, the distribution of memory B cells after an infection has not been as well studied. To address this, Joo and coworkers used a mouse influenza pneumonia model to track the distribution of IgG and IgA memory B cells. The influenza infection was limited to the lung; no virus was detected outside the lung. During and after infection, memory B cells were widely dispersed to secondary lymphoid tissues as well as into the lung itself. Why do you think memory B cells localize to secondary lymphoid tissues if this virus is specific to lung tissue? What might the implications of this study be for the distribution of memory B cells following vaccination, particularly for a subunit vaccine that does not localize to the normal site of infection?

Read the original paper: Joo, H. M. 2008. Broad dispersion and lung localization of virus-specific memory B cells induced by influenza pneumonia. *Proc. Natl. Acad. Sci. USA* 105:3485–90.

Pathogenicity and Infection

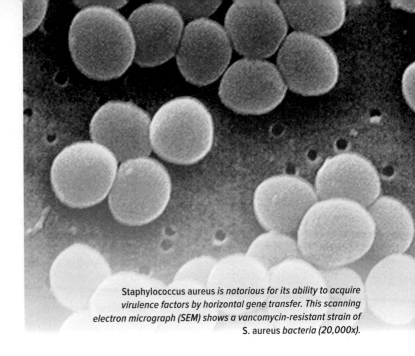

Staphylococcus aureus is notorious for its ability to acquire virulence factors by horizontal gene transfer. This scanning electron micrograph (SEM) shows a vancomycin-resistant strain of S. aureus *bacteria (20,000x).*

Sneaky Little Buggers

Have you ever wondered why some microorganisms seem to be more infectious than others? Scientists from the Pasteur Institute, the University of Lisbon (Portugal), and Harvard University also think about this. In separate studies, these "microbiology detectives" used bioinformatics, genomics, epidemiology, and evolutionary biology techniques to investigate the question of why specific numbers of microorganisms are needed to initiate disease. One measurement is called the infectious dose 50% (ID_{50}), and it refers to the number of microorganisms needed to infect 50% of the test hosts. The French and Portuguese scientists scoured the literature to find 48 different bacterial pathogens whose ID_{50} they could correlate with motility, growth rate, immune cell interactions, and use of quorum sensing. They determined that bacteria with a higher infectivity (a lower ID_{50}) were less motile, slower to reproduce, and had little interaction with others of their kind. By contrast, the faster, more social, and more rapid reproducers had lower infectivity, requiring a greater ID_{50}. One important interpretation of these data is that distantly related bacteria use the same common mechanisms to cause infectious disease; that is, ID_{50} correlates better with acquisition of virulence factors by horizontal gene transfer (HGT) than with taxonomy and phylogenetics.

The U.S. study evaluated the genetic changes that occur in bacteria during their encounters with human hosts. These researchers used large-scale genomic sequencing of hundreds of *Burkholderia* samples that infected cystic fibrosis patients to identify selective pressures (host immunity and medical treatment) on bacterial chromosomal genes over generations. The scientists compared the *Burkholderia* genes that had mutated multiple times to those that had mutated only once. They found that the selective pressures imposed while in the human host leads to mutation in at least 17 genes including genes encoding antibiotic resistance, adhesion molecules, secretion system proteins, proteins for oxygen sensing and regulation, and, unexpectedly, a gene for modifying lipopolysaccharide (LPS).

Both studies suggest that successful pathogens adapt to selective pressures imposed by the host, mutating or acquiring genes by HGT to overcome survival challenges. Infection of subsequent hosts, as in an epidemic, alters the selective pressures (different hosts with different immune responses, medical therapies, etc.), revealing the important challenges for survival as a pathogen. Together, these studies confirm that bacterial genomes are extremely flexible and resilient, enabling many species to live in and adapt to the harshest of environmental conditions.

Chapter 32 introduces the concept of symbiosis and deals with its subordinate categories, including commensalism and mutualism. In this chapter, the process of parasitism is presented along with one of its possible consequences—pathogenicity. The parasitic way of life is so successful, it has evolved independently in nearly all groups of organisms. Understanding host-parasite relationships requires an interdisciplinary approach, drawing on knowledge of cell biology, microbiology, entomology, immunology, ecology, and zoology. This chapter examines the parasitic way of life in terms of health and disease with an emphasis on viral and bacterial disease mechanisms.

Readiness Check:
Based on what you have learned previously, you should be able to:

✔ Explain symbiosis and differentiate the true definition of parasitism from several common uses of the word (sections 32.1 and 32.2)
✔ Generalize the principles of symbiosis to human-microbe interactions (section 32.2)
✔ Identify the major structures and their functions of bacteria, protists, fungi, and viruses, including their replication processes and energy requirements (chapters 3, 5, 6, 7, 10, and 12)

35.1 Pathogenicity Drives Infectious Disease

After reading this section, you should be able to:

■ Compare and contrast competition between microbial species with competition between microbial and human cells
■ Predict the microbial virulence factors and host cell responses that result in disease
■ Relate the infectious disease process to time, identifying events associated with each stage of the process

Microorganisms actively seek host-specific sites that meet their unique growth and replication requirements. These are molecular determinants that optimize microbial survival. Thus, host and tissue specificity offer microorganisms shelter and access to resources. However, relationships between two organisms can be very complex. A larger organism that supports the survival and growth of a smaller organism is called the **host.** Any organism that causes disease is known as a **pathogen** (Greek *patho,* disease, and *gennan,* to produce). Its ability to cause disease is called **pathogenicity.** Even normal microbiota, such as those associated with the gut or skin, can become pathogens. An **opportunistic pathogen** is such an organism; it infects a host away from its typical niche, especially in a host with a weakened immune system (a compromised host).

Infectious disease results from antagonism with microbial agents such as viruses, bacteria, fungi, and protozoa. When a pathogen is growing and multiplying within or on a host, the host is said to have an **infection.** The nature of an infection can vary widely with respect to severity, location, and type of organisms involved. An infection may or may not result in overt disease. An **infectious disease** is any change from a state of health in which part or all of the host body is not capable of carrying on its normal functions due to the presence of a pathogen or its products (e.g., toxins).

The fundamental process of infection is essentially a competition for resources. The host is a source of protection, nutrients, and energy for the pathogen to use. Infectious agents must therefore develop mechanisms to access and exploit their hosts. Furthermore, to continue surviving, pathogens must also devise methods to move on to better environments once the immediate one declines in value. Dissemination to another point in one host or into another host must then occur. The **infectious disease chain** represents these events in the form of an intriguing mystery, where the disease process becomes understood only when all of the links of the chain are known (**figure 35.1**). Here we introduce these links and the concepts that will be discussed more thoroughly later in the chapter.

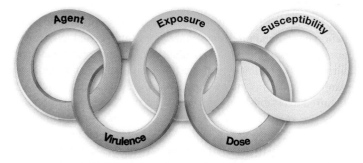

Figure 35.1 The Chain of Infection. The infectious disease process requires that a specific agent of substantial virulence is exposed to a susceptible host in an appropriate dose (number of organisms). All links of the chain are necessary for infectious disease to result. Breaking the chain, by removal of any link, prevents infectious disease.

MICRO INQUIRY *What physical and chemical methods could break the chain of infection?*

Before a microorganism can cause disease in a human, it must contact it. Pathogens come from a source, the location from which the pathogen is immediately transmitted to the host, either directly through the environment or indirectly through an intermediate agent. The source can be either animate (e.g., humans or animals) or inanimate (e.g., water or food). The period of infectivity is the time during which the source—such as a human, or contaminated water—is disseminating the pathogen. If the source of the infection can be eliminated or controlled—by treating the human or disinfecting contaminated water—the infectious disease cycle itself will be interrupted and transmission of the pathogen will be prevented.

A **reservoir** is the natural environmental location in which the pathogen normally resides. It is also the site from which a source acquires the pathogen or where direct infection of the host can occur. Thus a reservoir sometimes functions as a source. Reservoirs also can be animate or inanimate. Examples include birds as reservoir of West Nile virus and contaminated drinking water as the reservoir of *Vibrio cholerae.* The increasing impingement of humans on the environment and increased exposure to antibiotics and mutagens have played significant roles in the substantive change in reservoirs over the last 100 years.

Infectious diseases that can be transmitted from animals to humans are termed **zoonoses** (Greek *zoon,* animal, and *nosos,* disease); thus animals also can serve as reservoirs, as when a rabid dog transmits rabies by biting a human. In addition, being bitten by arthropod **vectors** (organisms that spread disease from one host to another), such as mosquitoes, ticks, fleas, mites, or biting flies, can lead to infection (e.g., West Nile virus, malaria, Lyme disease, Rocky Mountain spotted fever, and plague). These reservoirs are often referred to as vehicles, as they deliver the infectious agent to the host; they are discussed in chapters 38 through 40.

In order for the microorganism to cause disease, it must not just contact the host but be able to survive in it. To survive, the microorganism needs a suitable environment. An overview of the factors leading to infection, from the microorganism's perspective, is presented in **figure 35.2**. We evaluate the transmission of the microorganism from its reservoir to its entry point on the host a little later. But first, we briefly consider the challenges faced when a microorganism initially encounters a human host.

Recall that microorganisms have a range of values—temperature, pH, moisture, oxygen concentration, etc.—in which they can grow and reproduce. Thus, for a pathogen to survive in its human host, the host must provide conditions that fall within those values. When a pathogen is introduced to the human host, it needs to adapt to the realities of its new environment to meet its metabolic and physiological needs. One of those realities is that the microorganism must compete with the cells of its eukaryotic host. The successful pathogen will outcompete host cells and evade host immune defenses through the use of **virulence factors.** We discuss these in the next section, but suffice it to say, virulence factors enable the microbe to counter the challenges of the host environment. ◄◄ *Environmental factors affect microbial growth (section 7.4)*

Survival of the microorganism in the host also depends heavily on its abilities to "find shelter." Some microorganisms survive and replicate inside of host cells; in some cases, they reside in the

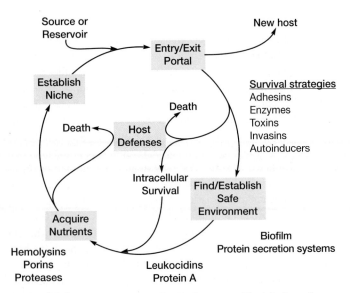

Figure 35.2 Overview of the Infection Process. The infectious disease process is complex and dynamic. To the host, infection is at best an annoyance and at worst lethal, if the microorganism succeeds. To microorganisms, infection is focused on finding a suitable environment, obtaining nutrients, and avoiding host immune responses to establish a niche. Microorganisms use numerous physical and chemical strategies to survive.

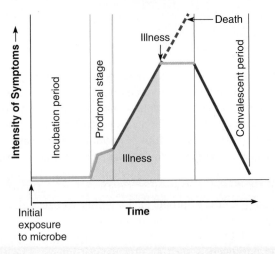

Figure 35.3 The Course of an Infectious Disease. Most infectious diseases occur in four stages. The duration of each stage is a characteristic feature of each disease. The shaded area represents when many diseases are communicable. Should the host be unable to overcome illness, death results.

MICRO INQUIRY *During which stages does the host exhibit signs and symptoms? How does this correlate to when the disease is communicable?*

very cells that are supposed to destroy them. Other organisms attach to and between host cells, make capsules, burrow under mucus, and secrete exopolysaccharides to form communal shelters within biofilms. Some microorganisms produce enzymes that inactivate innate resistance mechanisms. Still other organisms have evolved specialized protein secretion systems to selectively kill host cells. These physical and chemical characteristics reflect microbial virulence; they facilitate microbial survival.

Some microorganisms are pathogenic due to their rapid replication and explosive departure from host cells. Still others are pathogens because they integrate parts of their genome into the host cell DNA, commandeering that cell in the future. Many microorganisms, particularly bacteria, use a strategy that poisons host cells; they use toxins. A **toxin** (Latin *toxicum,* poison) is a substance that alters the normal metabolism of host cells with deleterious effects on the host. The term toxemia is used when toxins have entered the blood of the host. **Toxigenicity** is the pathogen's ability to produce toxins. **Intoxications** are diseases that result from a specific toxin (e.g., botulinum toxin, tetanus toxin) produced by the pathogen. Intoxication diseases do not require the presence of the actively growing pathogen—just its toxin, as in the case of botulism.

Infectious diseases are typically not detected until distinct physical and/or physiological events alert the host of cellular battle. Infectious diseases often have characteristic signs and symptoms. **Signs** are objective changes in the body, such as a fever or rash, that can be directly observed. **Symptoms** are subjective changes, such as pain and loss of appetite, that are experienced by the patient. The term symptom is often used in a broader scope to include the clinical signs. A **disease syndrome** is a set of signs and symptoms that are characteristic of the disease.

Clinically, the course of an infectious disease usually has a characteristic pattern and can be divided into several phases (**figure 35.3**). The **incubation period** is the time between pathogen entry and development of signs and symptoms. The pathogen is reproducing but has not reached a sufficient level to cause clinical manifestations. This period's length varies with disease. Next, the **prodromal stage** occurs with an onset of signs and symptoms that are not yet specific enough to make a clear diagnosis. However, the patient often is contagious. This is followed by the **illness period,** when the disease is most severe and displays characteristic signs and symptoms. The host immune response is typically triggered at this stage. Finally, during the period of decline, the signs and symptoms begin to disappear. The recovery stage often is referred to as **convalescence.**

Infectious Dose

The rate at which an infection proceeds (the timing of the incubation period and prodromal stage, for example) and its severity are functions of the initial inoculum of microorganisms and their virulence (in addition to the host's resistance). Some microorganisms are sufficiently adept at entering a host that very few are required to establish an infection. Conversely, other microorganisms are required in high numbers to establish an infection (see the chapter-opening story). The inocula can be measured experimentally to determine the **infectious dose 50 (ID_{50})**—that is, the number of microorganisms required to cause disease in 50% of the inoculated hosts (**figure 35.4**). One example of this is found in the oral transmission of *Salmonella* spp., which are shed in the feces of infected hosts and often contaminate fingers and hands. During food preparation, the bacteria are transferred to foods, where they can then be ingested. The ID_{50} for *Salmonella* spp. ingested in contaminated food is approximately 1×10^5 bacteria. Suffice it to say, the reason for thorough handwashing as a means of maintaining public health is to decrease the number

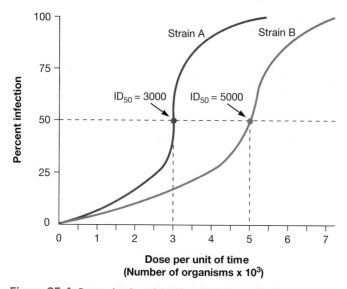

Figure 35.4 Determination of the ID_{50} of a Pathogenic Microorganism. Various doses of a specific pathogen are introduced into experimental host animals. Infections are recorded and a graph constructed. In this example, the graph represents the susceptibility of host animals to two different strains of a pathogen—strain A and strain B. For strain A, the ID_{50} is 3000, and for strain B, it is 5000. Hence strain A is more virulent than strain B.

of pathogens that can be transmitted. For the most part, the lower the infectious dose (the fewer organisms needed for infection), the higher the risk of infection.

Another measure of infectivity, and thus virulence, is the experimental determination of the number of microorganisms required to be lethal to a host species. This is known as the **lethal dose 50 (LD_{50})** and refers to the dose or number of pathogens that kills 50% of an experimental group of hosts within a specified period. Highly virulent pathogens kill at a much reduced dose as compared to less virulent organisms.

Host Susceptibility

The final link in the infectious disease cycle is the host (figure 35.1). The susceptibility of the host to a pathogen depends on the pathogenicity of the organism, the innate host resistance factors, and the adaptive immune mechanisms of the host. Host immunity is the subject of chapters 33 and 34 and will not be discussed here. In addition to host defense mechanisms, nutrition, genetic predisposition, cleanliness, and stress also influence host susceptibility to infection.

Extracellular and Intracellular Pathogens

As noted earlier, a successful pathogen must find an appropriate environment in the host. This can be provided outside host cells or inside host cells, depending on the pathogen. Infectious microorganisms are considered **extracellular pathogens** if, during the course of disease, they remain in tissues and fluids but never enter host cells. For instance, some bacteria and fungi can actively grow and multiply in the blood or tissue spaces. *Y. pestis* and *Aspergillus* (the cause of a variety of fungal infections) are good examples of extremely virulent extracellular pathogens. Microbes that grow and multiply within host cells are called **intracellular pathogens;** they can be further subdivided into two groups. **Facultative**

intracellular pathogens are those organisms that can reside within the cells of the host or in the environment. A bacterial example is *Brucella abortus,* which is capable of growth and replication within macrophages, neutrophils, and trophoblast cells (cells that surround the developing embryo). A protozoal example is *Histoplasma capsulatum,* which grows within phagocytes. However, facultative intracellular pathogens can also be grown in pure culture without host cell support. In contrast, **obligate intracellular pathogens** are incapable of growth and multiplication outside a host cell. By definition, all viruses are obligate intracellular pathogens in that they require a host cell for replication, often to the detriment of that cell. Some bacteria are also obligate intracellular bacterial pathogens, such as *Chlamydia* spp. and rickettsias (which cause diseases such as Rocky Mountain Spotted Fever and typhus). These microbes cannot be grown in the laboratory outside of their host cells. Malarial parasites are examples of protozoa that require host cells for growth and differentiation.

The survival of an invading microorganism depends on its ability to rapidly adapt to the host. Host cues often activate specific microbial genes whose products equip the microbe with specific survival means, such as exotoxins, specialized attachment proteins, tissue-altering enzymes, and biofilm synthesis. These products and the methods by which they are used in the infection process are collectively referred to as virulence factors. They are discussed next.

Retrieve, Infer, Apply

1. Define infection, infectious disease, pathogenicity, virulence, and opportunistic pathogen.
2. What factors determine the outcome of most host-pathogen relationships?
3. Why would a psychrophile not be a human pathogen?
4. What are some important characteristics of a pathogen?
5. Define source and reservoir. How are they related?
6. How would you determine the ID_{50} of a suspected pathogen?
7. What is an obligate intracellular pathogen? How might it survive?
8. Which phase of an infectious disease usually coincides with spread of the disease? Suppose an individual is contagious in the incubation period or in the early prodromal stage. How would that alter the ease with which the pathogen is spread to new hosts?

35.2 Virulence Defines a Pathogen's Success

After reading this section, you should be able to:

■ Identify and describe the features that allow microorganisms to overcome host resistance and immunity
■ Discuss strategies microorganisms have evolved to exploit human cells and tissues as resources for their survival
■ Compare the molecular mechanisms by which microorganisms adhere to and invade human cells and tissues
■ Illustrate the mechanisms by which microbial toxins impact human cells
■ Model disease processes and explain virulence

While pathogenicity is a general term that refers to an organism's potential to cause disease, virulence refers more specifically to the magnitude of harm (pathogenicity) caused. Any structural or soluble product that increases pathogenicity is a virulence factor. For bacteria, protozoa, and fungi virulence also is often directly correlated with a pathogen's ability to survive in the external environment. If a pathogen cannot survive well outside its host and does not use a vector, it depends on host survival and will tend to be less virulent (e.g., rhinoviruses that cause the common cold). When a cellular pathogen can survive for long periods outside its host, it can afford to leave the host and simply wait for a new one to come along. This seems to promote increased virulence. Host health is not critical, but extensive multiplication within the host will increase the efficiency of transmission. Good examples are tuberculosis, diphtheria, and measles caused by *Mycobacterium tuberculosis, Corynebacterium diphtheriae,* and measles viruses, respectively. These microbes survive for a relatively long time, from hours to days and weeks, outside human hosts. Many bacterial virulence factors that facilitate survival are encoded in unique sequences of DNA that are readily swapped between bacteria through conjugation, transduction, or transformation (horizontal gene transfer). We now consider the genes and proteins that encode virulence factors. ◄◄ *Horizontal gene transfer (section 16.4); Microbial evolutionary processes (section 19.5)*

Pathogenicity Islands

Large segments (10 to 200 kilobases) of bacterial chromosomal and plasmid DNA have been found to encode virulence factors. These DNA segments are called **pathogenicity islands,** as they appear to have been inserted into the existing DNA. Many bacteria (e.g., *Yersinia* spp., *Pseudomonas aeruginosa, Shigella flexneri, Salmonella* spp., enteropathogenic *Escherichia coli*) carry at least one pathogenicity island. Pathogenicity islands generally increase bacterial virulence and are absent in nonpathogenic members of the same genus or species (**table 35.1**). For example, the *Staphylococcus aureus* pathogenicity island encodes several superantigen genes, including the gene for the toxin that causes toxic shock syndrome. The detection of pathogenicity islands within a microbe's genome is explained in chapter 18. ◄◄ *Comparative genomics (section 18.7)*

Virulence Factors

Virulence factors can be encoded in the microbial chromosome or as extrachromosomal elements. These factors enable the pathogen to establish an infection (adherence and colonization factors), spread to other tissues (invasion factors), and harm the host (e.g., toxins). Virulence factors are upregulated by specific host triggers, enabling the microorganism to better survive its residency in the host. Examples of their association with the infectious disease process are included in figure 35.2, illustrating how they are used by the microorganism. This is discussed next.

Adherence and Colonization Factors

The first step in the infectious disease process is the entrance and attachment of the microorganism to a susceptible host. Entrance may be accomplished through one of the body surfaces, the **portal of entry:** skin, respiratory system, gastrointestinal system, urogenital system, or the conjunctiva of the eye. Some pathogens enter the host by sexual contact, needle sticks, blood transfusions and organ transplants, or insect vectors.

After gaining access to the host, the pathogen must be able to adhere to and colonize host cells or tissues. In this context, colonization means the establishment of a site of microbial replication on or within a host. It does not necessarily result in tissue

Table 35.1 Examples of Pathogenicity Islands and the Products They Encode

Organism	Pathogenicity Island[1]	Gene Product	Function
Escherichia coli	cagPI	Type IV secretion proteins	Cytotoxin
Helicobacter pylori	PAI-III	Type 1 pili Secreted protein	Attachment Hemolysin, cytotoxic necrosing factor, uropathogenic protein
Legionella pneumophila	icm/dot	Type IV secretion proteins	Intracellular survival
Rhodococcus equi	PAI-vap	Secreted proteins	Intracellular survival
Salmonella enterica	SPI-1, SPI-2	Type III secretion proteins	Cytotoxin
Shigella flexneri	SHI-1, SHI-2	Type III secretion proteins	Cytotoxin
Staphylococcus aureus	SaPI	Secreted proteins	Superantigens
Vibrio cholerae	VPI	Secreted protein	Toxin
Yersinia pestis	HPI-1	Siderophore synthesis	Iron uptake and storage

[1] PI and PAI are both common abbreviations for pathogenicity island.

invasion or damage. Adherence can be thought of as a two-step process. Step 1 is nonspecific and often reversible. It results from hydrophobic, electrostatic, and vibrational forces that encourage "docking." Step 2 is specific and permanent, as complementary molecules (cognate receptors) fulfill a lock-in-key style bonding. This latter event seems to have co-evolved, permitting pathogen exploitation of the characteristics of host cell surfaces. Thus, pathogens adhere with a high degree of specificity to target tissues. Adherence structures, also known as adhesins, such as pili and fimbriae (**table 35.2**), membrane and capsular materials, and specialized adhesion molecules on the invading cell surface bind to host sites (**figure 35.5**). These microbial products and structural components contribute to infectivity, thus they are one type of virulence factor. Examples of pathogen adherence are discussed in the context of specific diseases (*see chapters 38–40*).

Invasion Factors

Pathogens can be described in terms of their infectivity and invasiveness. **Infectivity** is the ability of the organism to establish a discrete focal point of infection. Attachment to specific cell surface proteins permits microbe-mediated infection but may also induce cell-mediated microbe internalization. Viruses are adept at these methods of infectivity. **Invasiveness** is the ability of the organism to spread to adjacent or other tissues. For some pathogens, a localized infection is sufficient to cause disease. However, many pathogens invade other tissues.

Pathogens can either actively or passively penetrate the host's mucous membranes and epithelium. Active penetration may be accomplished through production of lytic substances that alter the host tissue by (1) attacking the extracellular matrix and basement membranes of integuments and intestinal linings, (2) degrading carbohydrate-protein complexes between cells or

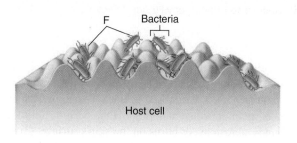

(a) Fimbriae (F)

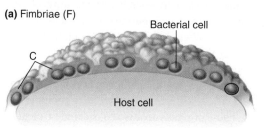

(b) Capsules (C)

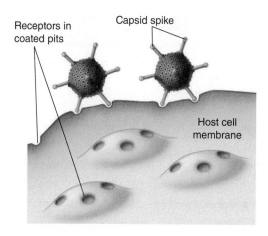

(c) Spikes

Figure 35.5 Microbial Adherence Mechanisms. (a) Bacterial fimbriae. (b) Bacterial capsules. (c) Viral protein spikes.

Table 35.2	Examples of Microbial Attachment Mechanisms		
Microbe	**Disease**	**Adhesion Mechanism**	**Host Receptor**
Neisseria gonorrheae	Gonorrhea	Type IV fimbriae	Mannose on urethral epithelium
Escherichia coli	Diarrhea	Type I fimbriae	Sugar residue on intestinal epithelium
	Hemolytic uremic syndrome	P pili	Sugar residue on kidney cell
	Urinary tract infection	P-pili	Sugar residue on urethral epithelium
Treponema pallidum	Syphilis	Outer membrane protein	Fibronectin on mucosal cell
Mycoplasma pneumoniae	Pneumonia	Membrane protein	Sialic acid on lung cell
Streptococcus pyogenes	Sore throat	Protein F	Protein residue on upper respiratory tract cell
Streptococcus mutans	Dental caries	Sugar residue	Salivary glycoprotein on tooth
Influenza virus	Influenza	Hemagglutinin spike protein	Protein residue on upper respiratory tract cell
HIV-1	AIDS	gp120 protein	CD4 receptor on T cells
Polio virus	Poliomyelitis	Capsid protein VP1	CD 155 protein on intestinal and nerve cells

on the cell surface (the glycocalyx), or (3) disrupting the host cell surface. Passive mechanisms of penetration are not related to the pathogen itself. Examples include (1) small breaks, lesions, or ulcers in a mucous membrane that permit initial entry; (2) wounds, abrasions, or burns on the skin's surface; (3) arthropod vectors that create small wounds while feeding; (4) separation of tight junctions between cells during inflammation, and (5) tissue damage caused by other organisms (e.g., a dog bite).

Once under the mucous membrane, a pathogen may penetrate to deeper tissues and continue disseminating throughout the body of the host. One way the pathogen accomplishes this is by producing specific structures or enzymes that promote spreading (**table 35.3**). These products are also considered virulence factors because they contribute to pathogen success within the host. Bacteria may also enter the small terminal lymphatic capillaries that surround epithelial cells. These capillaries merge into large lymphatic vessels that eventually drain into the circulatory system. Once the circulatory system is reached, the bacteria have access to all organs and systems of the host. The presence of viable bacteria in the bloodstream is called **bacteremia.** The infectious disease process caused by bacterial or fungal toxins in the blood is termed **septicemia** (Greek *septikos,* produced by putrefaction, and *haima,* blood).

Invasiveness varies greatly among pathogens. For example, *Clostridium tetani* (the cause of tetanus) is considered noninvasive because it does not spread from one tissue to another, but its

Table 35.3	Bacterial Virulence Factors Involved in Bacterial Pathogen Invasion and Dissemination	
Product	**Organism Involved**	**Mechanism of Action**
Coagulase	*Staphylococcus aureus*	Coagulates (clots) the fibrinogen in plasma. The clot protects the pathogen from phagocytosis and isolates it from other host defenses.
Collagenase	*Clostridium* spp.	Breaks down collagen that forms the framework of connective tissues; allows the pathogen to spread
Deoxyribonuclease	Group A streptococci, staphylococci, *Clostridium perfringens*	Lowers viscosity of exudates, giving the pathogen more mobility
Elastase and alkaline protease	*Pseudomonas aeruginosa*	Cleaves laminin associated with basement membranes
Hemolysins	Staphylococci, streptococci, *Escherichia coli, Clostridium perfringens*	Lyse erythrocytes; make iron available for microbial growth
Hyaluronidase	Groups A, B, C, and G streptococci, staphylococci, clostridia	Hydrolyzes hyaluronic acid, a constituent of the extracellular matrix that cements cells together and renders the intercellular spaces amenable to passage by the pathogen
Hydrogen peroxide (H_2O_2) and ammonia (NH_3)	*Mycoplasma* spp., *Ureaplasma* spp.	Are produced as metabolic wastes. These are toxic and damage epithelia in respiratory and urogenital systems.
Immunoglobulin A protease	*Streptococcus pneumoniae*	Cleaves immunoglobulin A into Fab and Fc fragments
Lecithinase or phospholipase	*Clostridium* spp.	Destroys the lecithin (phosphatidylcholine) component of plasma membranes, allowing pathogen to spread
Leukocidins	Staphylococci, pneumococci, and other streptococci	Pore-forming exotoxins that kill leukocytes; cause degranulation of lysosomes within leukocytes, which decreases host resistance
Porins	*Salmonella enterica* serovar Typhimurium	Inhibit leukocyte phagocytosis by activating the adenylate cyclase system
Protein A Protein G	*Staphylococcus aureus* *Streptococcus pyogenes*	Located on cell wall. Immunoglobulin G (IgG) binds to either protein A or protein G by its Fc end, thereby preventing complement from interacting with bound IgG.
Pyrogenic exotoxin B (cysteine protease)	Group A streptococci (*Streptococcus pyogenes*)	Degrades proteins
Streptokinase (fibrinolysin, staphylokinase)	Groups A, C, and G streptococci, staphylococci	A protein that binds to plasminogen and activates the production of plasmin, thus digesting fibrin clots; this allows the pathogen to move from the clotted area

toxin becomes blood-borne, thereby causing disease. *Bacillus anthracis* (the cause of anthrax) and *Yersinia pestis* (the cause of plague) also produce toxins but the bacteria are also highly invasive. Members of the genus *Streptococcus* span the spectrum of virulence factors and invasiveness. *Microsporidia* spp. use a novel polar tubule to bore into host cells (*see figure 26.17*), whereas many other fungi use hydrolytic enzymes to invade cells and tissues. These virulence factors are discussed in more detail in chapters 39 and 40.

Exotoxins

Exotoxins are soluble, heat-labile proteins (inactivated at 60 to 80°C) that usually are released into the surroundings as the bacterial pathogen grows. Often exotoxins travel from the site of infection to other body tissues or target cells, where they exert their effects (**figure 35.6**). Exotoxins are often encoded by genes carried on plasmids or prophages carried by specific bacteria. They are associated with specific diseases and often are named for the disease they produce (e.g., the diphtheria toxin). Some are among the most lethal substances known— toxic in nanogram-per-kilogram of body weight concentrations (e.g., botulinum toxin).

Exotoxins exert their biological activity by specific mechanisms and are grouped by either mechanism of action (e.g., a cytotoxin kills cells) or their protein structure. A common type is the **AB toxin,** which gets its name from the fact that it has two distinct toxin subunits, an "A" (or active) component and a "B (or binding) component." The B portion of the toxin binds to a host cell receptor and is separate from the A portion, which enters the cell and has enzyme activity that causes toxicity (figure 35.6a). Thus the B component determines the cell type the toxin will affect, whereas the A component exerts the deleterious effect. AB toxins act on cells by different mechanisms. Many A subunits have ADP ribosylation activity, which catalyzes the transfer of adenosine diphosphate and ribose moieties of host NAD$^+$ to target host molecules. Another type of exotoxin disrupts membranes. Examples of this type of exotoxin are the channel (pore)-forming toxins (figure 35.6b). They destabilize membrane integrity so that the cell lyses and dies. The general properties of several exotoxins are presented in **table 35.4**. As proteins, toxins are easily recognized by the host immune system, which produces antitoxin—antibodies that recognize and bind to their respective toxins—to inactivate and remove them.

Finally, exotoxins called **superantigens** act by stimulating as many as 30% of host T cells to overexpress and release massive amounts of cytokines from other host immune cells in the absence of a specific antigen (**figure 35.7**). The excessive concentration of cytokines causes multiple host organs to fail, giving the pathogen time to disseminate. By triggering this "cytokine storm," superantigens cause life-threatening disease; fever, fluid loss, and low blood pressure result in shock and death. The mechanism by which superantigens provoke the T-cell release of cytokines is discussed in chapter 34. Examples of each specific toxin type are discussed as part of the disease process in

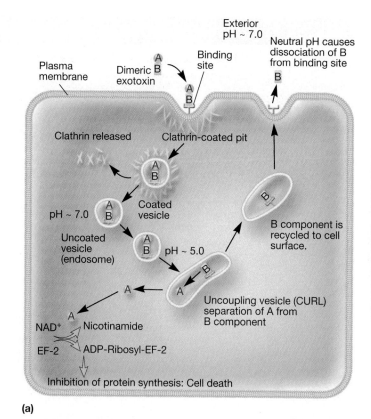

(a)

(b)

Figure 35.6 Two Examples of Exotoxin Mechanisms. (a) This AB toxin binds to the cell receptor within clathrin-coated pits by the B subunit. The intact toxin is endocytosed, and the pH change within the endosome causes the subunits to separate. An endosome in which this separation occurs is sometimes called a compartment of uncoupling of receptor and ligand (CURL). The B subunit is then recycled. The active toxin (A) subunit exerts its effect on its target, in this case the elongation factor-2 (EF-2) of protein synthesis, by enzymatically attaching an ADP-ribose to EF-2. The altered EF-2 is inhibited from assisting in protein synthesis, leading to cell death. (b) This channel-forming (pore-forming) toxin inserts itself into the host cell membrane, forming a channel (or pore). Multiple membrane pores results in an osmolarity shift, as water enters the cell and cytoplasmic contents move out. The resulting effect of this toxin is cell lysis.

Table 35.4 Exotoxins Produced by Human Bacterial Pathogens

Toxin	Organism	Gene Location	Toxin Type	Mechanism of Action
Edema factor (EF) Lethal factor (LF) Protective antigen (PA)	*Bacillus anthracis*	Plasmid	Tripartite AB	EF causes edema. LF is a cytotoxin. PA is a B component.
Pertussis toxin	*Bordetella pertussis*	Chromosome	AB	↓ATP, ↑ cAMP alters cell function, leading to death.
Botulinum toxin	*Clostridium botulinum*	Prophage	AB	Blocks neurotransmitter release, leading to paralysis
CPE enterotoxin	*Clostridium perfringens*	Chromosome	Cytotoxin	Hemolysis
Tetanospasmin	*Clostridium tetani*	Plasmid	AB	Blocks neurotransmitter, leading to spastic paralysis
Diphtheria toxin	*Corynebacterium diphtheriae*	Phage	AB	Alters translation, leading to protein synthesis inhibition
Enterotoxin Shiga-like toxin	*Escherichia coli* *E. coli* O157:H7	Plasmid Phage gene integrated into chromosome	AB AB	↑cAMP, leading to water secretion from cell Inhibits protein synthesis leading to death
Cytolysin	*Salmonella* spp.	Chromosome	Cytotoxin	↑cAMP, leading to water secretion from cell
Shiga toxin	*Shigella dysenteriae*	Chromosome	AB	Inhibits protein synthesis, leading to death
Exfolative toxin Toxic shock syndrome toxin-1 Panton-Valentine leukocidin	*Staphylococcus aureus*	Chromosome Chromosome Phage	Protease Superantigen Cytotoxin	Skin peeling Cytokine-induced shock Necrotizing pneumonia
Streptolysin O Erythrogenic toxin	*Streptococcus pyogenes*	Chromosome Phage	Cytolysin Superantigen	Hemolysis Cytokine-induced shock
Cholera toxin	*Vibrio cholerae*	Phage	AB	↑cAMP, leading to water secretion from cell

chapters 38 through 40. ◄◄ *Cytokines are chemical messages between cells (section 33.3); Superantigens (section 34.5)*

Endotoxins

The lipopolysaccharide (LPS) in the outer membrane of Gram-negative bacteria is toxic to humans. Thus LPS is called an **endotoxin** because it is bound to the bacterium and is released when the microorganism lyses, although some may also be released during cell division. The toxic component of LPS is the lipid portion, called lipid A (*see figure 3.25*). **Lipid A** is not a single macromolecular structure; rather, it is a complex array of lipid residues. Lipid A is heat stable and toxic in nanogram amounts but only weakly immunogenic. ◄◄ *Typical Gram-negative cell walls include additional layers besides peptidoglycan (section 3.4)*

Unlike the structural and functional diversity of exotoxins, the lipid A of various Gram-negative bacteria produces similar systemic effects regardless of the microbe from which it is derived. These include fever (i.e., endotoxin is pyrogenic), shock, blood coagulation, weakness, diarrhea, inflammation, intestinal hemorrhage, and fibrinolysis (enzymatic breakdown of fibrin, the major protein component of blood clots) (**figure 35.8**).

The main biological effect of lipid A is an indirect one, mediated by host molecules and systems, rather than by lipid A itself. For example, endotoxins initially activate a protein called the Hageman factor (blood clotting factor XII), which in turn results in unregulated blood clotting within capillaries (disseminated intravascular coagulation) and multiorgan failure (figure 35.8). Endotoxins also indirectly induce a fever in the host by causing macrophages to release endogenous pyrogens that reset the hypothalamic thermostat. One important endogenous pyrogen is the cytokine interleukin-1 (IL-1). Other cytokines released by macrophages, such as the tumor necrosis factor, also produce fever. The net effect is often called **septic shock** and can also be

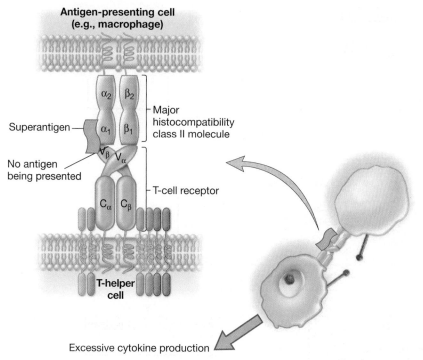

Figure 35.7 Superantigen. Nonspecific activation of T cells by a superantigen, in the absence of antigen, stimulates massive amounts of cytokines that overwhelm the host and result in shock and death.

induced by certain pathogenic fungi and Gram-positive bacteria. All drugs, especially those given intravenously or intramuscularly, must be free of endotoxin for this reason. ◀◀ *Cytokines are chemical messages between cells (section 33.3)*

Mycotoxins

Mycotoxins are toxins produced by fungi. For example, *Aspergillus flavus* and *A. parasiticus* produce aflatoxins, and *Stachybotrys* spp. produce satratoxins. These fungi commonly contaminate food crops and water-damaged buildings, respectively. An estimated 4.5 billion people in developing countries may be exposed chronically to aflatoxins through their diet. Exposure to aflatoxins is known to cause both chronic and acute liver disease and liver cancer. Aflatoxins are extremely carcinogenic, mutagenic, and immunosuppressive. Approximately 18 different types of aflatoxins exist. Aflatoxins are classified in two broad groups according to their chemical structure (*see figure 41.6*). *Stachybotrys trichothecene* mycotoxins are potent inhibitors of DNA, RNA, and protein synthesis. They induce inflammation, disrupt surfactant phospholipids in the lungs, and may lead to pathological changes in tissues.

The fungus *Claviceps purpurea* also produces toxic substances. The products are generically referred to as ergots, reflecting the name of the tuberlike structure of the fungi (*see figure 26.12*). The ergot is a fungal resting stage and is composed of a compact mass of hyphae. The ergots from various *Claviceps* spp. produce alkaloids that have varying physiological effects on humans. One such alkaloid is lysergic acid, the psychotropic hallucinogen LSD.

Biofilm Development

Most bacteria live within stable, sessile communities surrounded by a hydrated polymeric matrix of their own production. These biofilm bacteria are protected from nutrient deprivation, predators, environmental shifts, antimicrobial agents, and host immune cells. In contrast to free-swimming (planktonic) bacteria of the same species, biofilm bacteria are phenotypically different, expressing genes that favor survival when attached and in poor nutrient conditions. However, the role of the biofilm is much more complex, especially for biofilms that cause chronic infections in humans. Using high throughput sequencing, fluorescent in situ hybridization, and global gene analysis, studies have revealed that some pathogenic bacteria within biofilms exchange plasmids, nutrients, and quorum-sensing molecules so as to behave differently than planktonic forms. The differences are striking in that biofilm bacteria coordinate gene expression to upregulate mechanisms that make the biofilm community less sensitive to antibiotics and more resistant to host defense mechanisms. For example, studies on *P. aeruginosa* biofilms reveal that shared communication signals activate global response regulators, which increase the activity of multidrug efflux pumps, increase starvation or stress responses, stimulate exopolysaccharide secretion, increase resistance to host cell attack, suppress host cell killing mechanisms, and change rapidly growing, antibiotic-sensitive cells into slower-growing, antibiotic-resistant persister cells. ◀◀ *Microbial growth in natural environments (section 7.5)*

Importantly, one remarkable activity of biofilms formed by pathogenic *P. aeruginosa* is the ability to suppress host immune responses (**figure 35.9**). This is evident in studies evaluating *P. aeruginosa* mixed with antibiotics, antibodies, and the phagocytic white blood cells macrophages and neutrophils. Antibiotics, antibodies, and the phagocytes readily attack and kill planktonic *P. aeruginosa* using hydrolytic enzymes (lysozyme, lactoferrin, etc.) and free radicals of oxygen and nitrogen. However, antibiotics, antibodies, and phagocytes are ineffective at killing *P. aeruginosa* in biofilms. *P. aeruginosa* in biofilms appears to prevent penetration of antibiotics and antibodies, and to cause phagocytes to downregulate killing mechanisms. In addition, host hydrolytic enzymes are decreased, while enzymes that create super- and nitric oxides are inhibited. In essence, the *P. aeruginosa* biofilms "frustrate" phagocytes by reducing the magnitude of some antibacterial responses and preventing others altogether.

Resisting Host Defenses

Most microorganisms do not cause disease, in part because they are eliminated by the host's normal microbiota or its immune system before infection begins. However, pathogens have evolved

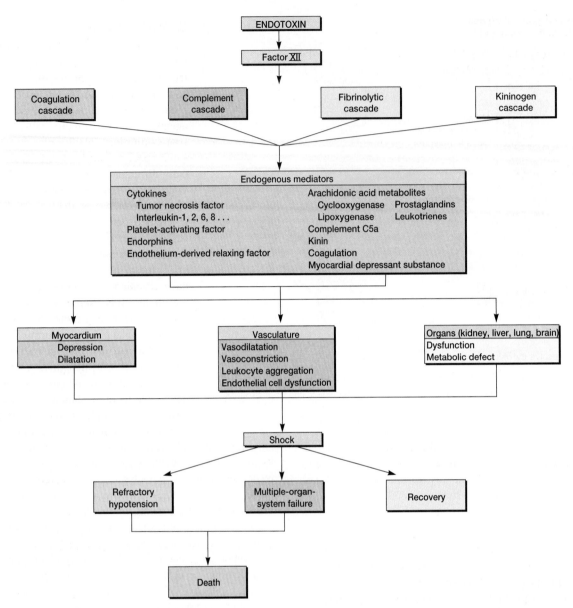

Figure 35.8 The Septic Shock Cascade. Gram-negative bacterial endotoxin triggers biochemical events that lead to such serious complications as shock, acute respiratory distress syndrome, and disseminated intravascular coagulation.

many mechanisms to evade host defenses. As we just discussed, biofilms are one example of how many microorganisms resist host defenses. Although host innate defense mechanisms come into play upon initial host exposure, specific immune responses normally become paramount in preventing further proliferation of infectious agents. Thus successful pathogens have evolved mechanisms that enable them to elude both the initial host responses as well as those governed by the adaptive immune system.

Some viruses, most notably the human immunodeficiency virus (HIV), infect cells of the immune system and diminish their function. HIV as well as measles virus and cytomegalovirus cause fusion of host cells. This allows these viruses to spread from an infected cell to an uninfected cell with minimal exposure to host defenses. Some bacteria such as *Streptococcus pneumoniae*,

Neisseria meningitidis, and *Haemophilus influenzae* can produce a slippery mucoid capsule that prevents the host immune cells from effectively capturing the bacterium. Other bacteria evade phagocytosis by producing specialized surface proteins such as the M protein on *S. pyogenes* (*see figure 39.5*).

Many pathogens have devised methods to prevent detection by host antimicrobial proteins (e.g., defensins) that either directly or indirectly cause membrane (envelope) lysis. Hepatitis B virus–infected cells produce large amounts of proteins not associated with the complete virus particle. These proteins serve as decoys, binding the available antimicrobial proteins so that the virus goes unnoticed. Some Gram-negative bacteria can lengthen the O side chains in their lipopolysaccharide to prevent host detection. Others such as *Neisseria gonorrhoeae* have modified the

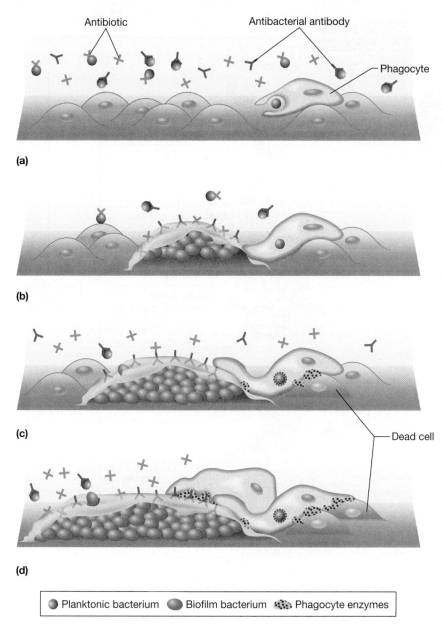

(a)

(b)

(c)

Dead cell

(d)

● Planktonic bacterium ● Biofilm bacterium 🦠 Phagocyte enzymes

Figure 35.9 **Frustrated Phagocytosis.** (a) Planktonic cells are sensitive to antibiotics and are detected by host phagocytes and antibodies. (b) Planktonic cells settle to form a biofilm and become biofilm cells. (c) Biofilm cells resist antibiotics and antibody detection. Host phagocytes detect and attempt to destroy biofilm cells. (d) Unable to capture biofilm cells, phagocytes release antimicrobial products that kill host cells but not biofilm cells.

Actin tail

Figure 35.10 **Formation of Actin Tails by Intracellular Bacterial Pathogens.** Scanning electron micrograph of *Listeria monocytogenes* in a host macrophage. The bacterium has polymerized host actin into a long tail at only one pole that it uses for intracellular propulsion and to move from one host cell to another.

surface, where they push out against the plasma membrane and form protrusions. The protrusions are engulfed by adjacent cells, and the bacteria enter the unsuspecting neighboring host cell. In this way, the infection spreads to adjacent cells.

Pathogens have evolved a variety of ways to suppress or evade the host's immune response. For instance, some change their surface proteins by mutation or recombination (e.g., the influenza virus), or downregulate the level of expression of cell surface proteins. To further evade the specific immune response, some bacteria (e.g., *S. pyogenes*) produce capsules that resemble host tissue components and thus remain covert. *N. gonorrhoeae* can evade the specific immune response by two mechanisms: (1) It makes changes in the gene encoding the pili protein through a process called phase variation, resulting in altered pilin protein sequence and expression. This renders specific host responses useless against the new pili, so that adherence to host tissue is not obstructed. (2) It produces proteases that degrade host proteins that bind the bacteria. This allows adherence. Finally, some bacteria produce special proteins (such as staphylococcal protein A and protein G of *S. pyogenes*) that interfere with the host's ability to detect and remove them (*see figure 23.11b*).

lipooligosaccharides on their surface that interfere with host mechanisms that normally lyse *N. gonorrhoeae*. ◀◀ *Antimicrobial peptides and proteins (section 33.3)*

Some bacteria have evolved the ability to survive inside the host cells that otherwise would destroy bacteria. These microorganisms are very pathogenic because they are impervious to an important innate resistance mechanism of the host (p. 780). One method of evasion is to escape by ejecting themselves from the host cell, as seen with *Listeria monocytogenes, Shigella* spp., and *Rickettsia* spp. These bacteria use actin-based motility to move within mammalian host cells and spread between them. They activate the assembly of an actin tail using host cell actin and other cytoskeletal proteins (**figure 35.10**). The actin tails propel the bacteria through the cytoplasm of the infected cell to its

Retrieve, Infer, Apply

1. Describe several specific adhesins by which bacterial pathogens attach to host cells.
2. What are virulence factors?
3. What are pathogenicity islands and why are they important?
4. Describe some general characteristics of exotoxins.
5. What is a biofilm? How is it a virulence factor?

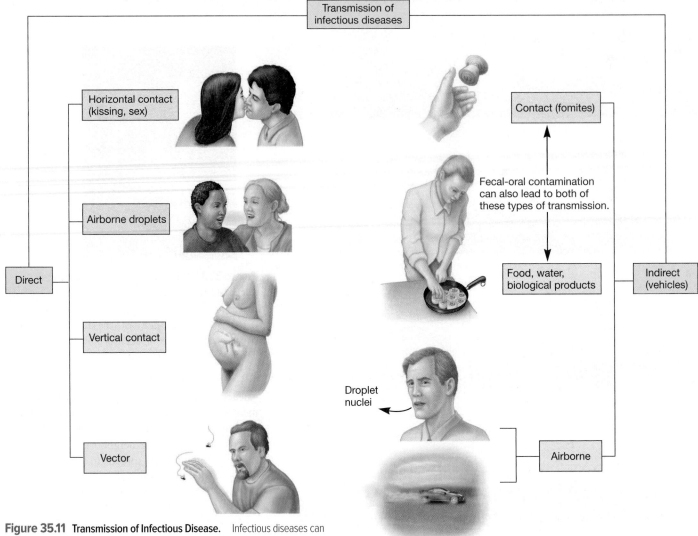

Figure 35.11 Transmission of Infectious Disease. Infectious diseases can be transmitted by various direct and indirect methods.

35.3 Exposure and Transmission Can Lead to Infectious Disease

After reading this section, you should be able to:

- List and describe the means by which microorganisms access humans to cause disease
- Correlate initial microbial numbers and replication rates to infection and lethality
- Synthesize a concept map of the infectious process

An essential feature in the development of an infectious disease is the initial exposure of the host to the pathogen; the pathogen must be transmitted from one host or source to another (figure 35.2). Thus, transmission is a critical component of the infection process and is the next link in the infectious disease chain, occurring directly or indirectly by four main routes: airborne, contact, vehicle, and vector-borne (**figure 35.11**).

As mentioned, a pathogen's virulence may be strongly influenced by its mode of transmission and ability to live outside its host (**Historical Highlights 35.1**). When the pathogen uses a mode of transmission such as direct contact, it cannot afford to

make the host so ill that it will not be transmitted effectively. This is the case with the common cold, which is caused by rhinoviruses and several other respiratory viruses. Cold sufferers must be able to move about and directly contact others. Thus virulence is low and people are not incapacitated by the common cold. If the virus replicates too rapidly and damages its host extensively, the person would be bedridden and not contact others. The efficiency of transmission would drop further because the rhinoviruses would be inactivated by exposure. Cold sufferers must be able to move about and directly contact others. Thus virulence is low and people are not incapacitated by the common cold.

On the other hand, if a pathogen uses a mode of transmission not dependent on host health and mobility, then the person's health will not be a critical matter. The pathogen might be quite successful; that is, it may be transmitted to many new hosts even though it kills its host relatively quickly. Host death means the end of any resident pathogens, but the species as a whole can spread and flourish as long as the increased transmission rate outbalances the loss due to host death. This situation may arise in several ways. In general, such pathogens are highly contagious.

Exposure alone is not sufficient for infection to occur. Rather, the pathogen must also make contact with the appropriate

HISTORICAL HIGHLIGHTS

35.1 The First Indications of Person-to-Person Spread of an Infectious Disease

In 1773 Charles White, an English surgeon and obstetrician, published his "Treatise on the Management of Pregnant and Lying-in Women." In it, he appealed for surgical cleanliness to combat childbed or puerperal fever, an acute febrile condition that can follow childbirth and is caused by streptococcal infection of the uterus or adjacent regions. In 1795 Alexander Gordon, a Scottish obstetrician, published his "Treatise on the Epidemic Puerperal Fever of Aberdeen," which demonstrated for the first time the contagiousness of the disease. In 1843 Oliver Wendell Holmes, a noted physician and anatomist in the United States, published a paper entitled "On the Contagiousness of Puerperal Fever" and also appealed for surgical cleanliness to combat this disease.

However, the first person to realize that a pathogen could be transmitted from one person to another was the Hungarian physician Ignaz Phillip Semmelweis. Between 1847 and 1849, Semmelweis observed that women who had their babies at the hospital with the help of medical students and physicians were four times as likely to contract puerperal fever as those who gave birth with the help of midwives. He concluded that the physicians and students were infecting women with material remaining on their hands after autopsies and other activities (amazingly, these individuals did not wash their hands after such procedures!). Semmelweis thus began washing his hands with a calcium chloride solution before examining patients or delivering babies. This simple procedure led to a dramatic decrease in the number of cases of puerperal fever and saved the lives of many women. As a result, Semmelweis is credited with being the pioneer of antisepsis in obstetrics. Unfortunately, in his own time, most of the medical establishment refused to acknowledge his contribution and adopt his procedures. After years of rejection, Semmelweis had a nervous breakdown in 1865. He died a short time later of a wound infection. It is very probable that it was a streptococcal infection, arising from the same pathogen he had struggled against his whole professional life.

host tissue. For instance, rhinoviruses are spread by airborne transmission from one host to another. Once the virus makes contact with the upper respiratory tract epithelia, the virus invades to cause disease only in the upper respiratory tract; rhinoviruses do not usually infect anywhere else in the host. This specificity is called a **tropism** (Greek *trope,* turning). Many pathogens exhibit cell, tissue, and organ specificities. A tropism by a specific microbe usually reflects the presence of specific cell surface receptors for that microbe on the host cell.

Airborne Transmission

Because air is not a suitable medium for the growth of pathogens, any pathogen that is airborne must have originated from a source such as humans, other animals, plants, soil, food, or water. In **airborne transmission,** the pathogen is suspended in the air in droplets, droplet nuclei, or dust, which travel over a meter or more from the source to the host. Typically, this results from host-to-host interaction (coughing, sneezing).

Droplets form from saliva and mucus. They are 5 μm or more in diameter and form when liquids are placed under force. The route of transmission to humans is through the air for a very short distance, so droplet transmission of a pathogen depends on the proximity of the source and the host. Chickenpox and measles are examples of droplet-spread diseases. **Droplet nuclei** are smaller particles, 1 to 4 μm in diameter, that result from the evaporation of the larger droplets. In contrast to droplets, the much smaller droplet nuclei can remain airborne for hours or days and travel long distances. Contact with oral secretions may result when droplet nuclei contaminate body surfaces that touch mucous membranes (e.g., respiratory secretions on hands that contact eyes).

Dust also is an important route of airborne transmission. Adhering to dust particles, microorganisms contribute to the number of airborne pathogens when the dust is resuspended by some disturbance. A pathogen that can survive for relatively long periods in or on dust creates an epidemiological problem, particularly in hospitals, where dust can be the source of hospital-acquired (nosocomial) infections. Many systemic mycoses (fungal infections) are examples of transmission in dust.

Contact Transmission

Contact transmission implies the coming together or touching of the source or reservoir of the pathogen and the host. Contact can be direct or indirect. Direct contact implies an actual physical interaction with the infectious source (figure 35.11). This route is frequently called person-to-person contact. Person-to-person transmission occurs primarily by touching, kissing, or sexual contact; by contact with oral secretions or body lesions (e.g., herpes and boils); by nursing mothers (e.g., AIDS, staphylococcal infections); and through the placenta (e.g., syphilis). Some pathogens also can be transmitted by direct contact with animals or animal products (e.g., members of the genera *Salmonella* and *Campylobacter*). Indirect contact transmission occurs when something "transfers" the infectious agent between hosts.

Vehicle Transmission

Inanimate materials that indirectly transmit pathogens are called **vehicles (figure 35.12).** In vehicle transmission, a single inanimate vehicle or source serves to spread the pathogen to multiple hosts. Examples include surgical instruments, drinking vessels,

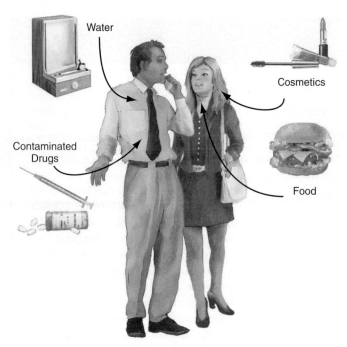

Figure 35.12 **Common Vehicles of Microbial Contamination.** Any inanimate object that transmits infectious agents to humans is referred to as a vehicle.

stethoscopes, bedding, eating utensils, and neckties. These common vehicles are called **fomites** (s., fomes or fomite). A single source containing pathogens (e.g., blood, drugs, intravenous fluids) can contaminate a common vehicle that causes multiple infections. Food, water, and biological materials (fluids and tissues) are important common vehicles for many human diseases. They often support pathogen reproduction.

Vector-Borne Transmission

Direct living transmitters of a pathogen are called vectors. Most vectors are arthropods (e.g., insects, ticks, mites, fleas) or vertebrates (e.g., dogs, cats, skunks, bats). **Vector-borne transmission** often benefits the pathogen by permitting extensive reproduction and spread within the host. If pathogen levels are very high in the host, a vector such as a biting insect has a better chance of picking up the pathogen and transferring it to a new host. Indeed, pathogens transmitted by biting arthropods often are very virulent (e.g., malaria, typhus, sleeping sickness). It is important that such pathogens do not harm their vectors and the vector generally remains healthy, at least long enough for pathogen transmission.

Vertical Transmission

Direct transmission of pathogens is not only mediated by horizontal, person-to-person passage but by vertical passage as well. **Vertical transmission** of pathogens occurs when the unborn child acquires a pathogen from an infected mother. While vertical transmission does not occur as frequently as horizontal transmission, a number of microorganisms exploit this route, extending their host range. Babies born with an infectious disease are said to have a congenital (*Latin:* with birth) infection. Examples of congenital infections include gonorrhea, syphilis, herpes, German measles, and toxoplasmosis. These diseases are discussed in chapters 38–40.

Retrieve, Infer, Apply

1. Describe the four main types of infectious disease transmission methods and give examples of each.
2. Define droplet nuclei, vehicle, fomite, and vector.
3. Why are pathogens that are transmitted person-to-person often less virulent than those transmitted by arthropod vectors?

Key Concepts

35.1 Pathogenicity Drives Infectious Disease

- Infections occur when microorganisms enter a larger organism, the host, and compete for resources.
- The infectious disease process is represented by a chain to demonstrate how each link is required for it to occur (**figure 35.1**).
- Signs of an infectious disease are objective changes in the body. Symptoms are subjective.
- Microorganisms require a suitable environment to survive. They use a number of survival strategies inside the host. These are collectively known as virulence factors (**figure 35.2**).
- The course of an infection follows a characteristic pattern: the incubation period, the prodromal stage, the illness period, and convalescence (**figure 35.3**).

- Pathogens reside in various environments, termed reservoirs. They infect the host from the reservoir and can return to it.
- Infectious diseases passed from animals to humans are called zoonoses.

35.2 Virulence Defines a Pathogen's Success

- Virulence refers to the magnitude of harm that a pathogen can exert on its host.
- Specific gene sequences called pathogenicity islands encode virulence factors (**table 35.1**).
- Pathogens require virulence factors to assist them in accessing a host. These include the ability to adhere to host cells, chemicals that permit cell and tissue degradation, and processes and mediators that overcome host defenses (**tables 35.2 and 35.3**).

- Exotoxins are secreted proteins that have biological activity. Endotoxins are the lipid A components of the lipopolysaccharide found in most Gram-negative bacterial envelopes (**figure 35.6** and **table 35.4**).
- Mycotoxins are metabolites of fungi. Their effects include hallucinations, cancer, and immunosuppression.
- Biofilms provide a unique environment for bacteria to exchange genetic information and collectively resist antimicrobial agents and host defense mechanisms (**figure 35.9**).
- Microorganisms have evolved many strategies for avoiding host defenses.

35.3 Exposure and Transmission Can Lead to Infectious Disease

- Pathogens are transmitted to and between hosts by several common routes, including air, direct contact, arthropod vectors, and vehicles (**figures 35.11** and **35.12**).
- Microorganism growth can be rapid or slow and may occur as extracellular or intracellular.
- Host susceptibility is an additional factor in determining infectious disease.

Compare, Hypothesize, Invent

1. Why does a parasitic organism not have to be a parasite?
2. In general, infectious diseases that are commonly fatal are newly evolved relationships between the parasitic organism and the host. Why is this so?
3. Explain the observation that different pathogens infect different parts of the host.
4. Intracellular bacterial infections present a particular difficulty for the host. Why is it harder to defend against these infections than against viral infections and extracellular bacterial infections?
5. Among the many diseases caused by *Staphylococcus aureus,* necrotizing pneumonia is particularly aggressive, causing lung inflammation, tissue death (necrosis), and hemorrhage. Only *S. aureus* strains that produce the toxin Panton-Valentine leukocidin (PVL) cause necrotizing pneumonia. PVL is a leukocidin because it forms pores in membranes of white blood cells (leukocytes) that would otherwise engulf and phagocytose the pathogen. When *S. aureus* infects the respiratory tract, it relies on adhesins to mediate attachment to airway cells. One of the most important adhesins is called staphylococcal protein A (Spa). Spa also blocks the engulfment activity of host phagocytes. *S. aureus* strains that produce PVL also produce high levels of Spa.

   How could you assess the contributions of each protein independently to pathogenesis and their combined effect (note that a mouse model system has been developed for this disease)? Your answer should state your hypotheses (predictions) about each protein tested alone and together, followed by the approach you would take to test each hypothesis.

Read the original paper: Labandeira-Rey, M., et al., 2007. *Staphylococcus aureus* Panton-Valentine leukocidin causes necrotizing pneumonia. *Science* 315:1130.

6. The Ebola virus outbreak of 2013 was unprecedented in its geographic distribution and number of infections. Yet the incubation period, duration of illness, disease symptomology (i.e., clinical course of infection), and virus transmissibility were similar to those in earlier Ebola outbreaks stemming back to the 1970s. Direct contact with viable virus has been understood to be the route of transmission, especially when contact is made with sick or dead relatives. All individuals who contracted Ebola had direct contact with highly infectious body fluids from other ebola patients. However, speculation regarding the airborne transmission of Ebola virus has been voiced. Experimentally, Ebola virus particles can be aerosolized. Importantly, family members residing in the same apartment as the Dallas index case did not acquire the disease.

   Given that the genomes of several strains of Ebola virus have been sequenced, what experiments could you do to test whether airborne Ebola virus is infectious?

Read the original papers: Team WHOER. Ebola virus disease in West Africa—The first 9 months of the epidemic and forward projections. *New Eng. J. Med.* September 22, 2014; Gire, S. K., et al. Genomic surveillance elucidates Ebola virus origin and transmission during the 2014 outbreak. *Science.* September 12, 2014; **345**:1369–1372.

Learn More

shop.mheducation.com   Enhance your study of this chapter with interactive study tools and practice tests. Also ask your instructor about the resources available through Connect, including adaptive learning tools and animations.

36

Clinical Microbiology and Immunology

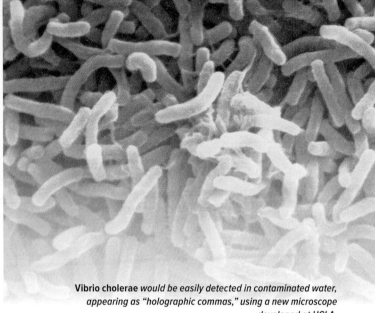

Vibrio cholerae *would be easily detected in contaminated water, appearing as "holographic commas," using a new microscope developed at UCLA.*

Seeing the Next Frontier

The headline could have read *"Star Trek* device now real" when scientists at the University of California–Los Angeles announced their creation of a portable, inexpensive, holographic microscope. The "microscope" does not use glass lenses to focus light the way a traditional microscope does. Rather, it uses a laser and electronic components to "view" specimens. The device is about the size of a small cell phone. It uses inexpensive photosensors to digitally capture specimens and advanced electronics to process the data into high-resolution images. Similar to an electron microscope, it is able to scan solid specimens to provide a three-dimensional hologram of the superstructures, as well as penetrate thin specimens to create a holographic image of intracellular components.

In the "reflection" mode, the laser is divided into two beams using a mirror such that one of the beams illuminates the specimen and is then recombined with the other. Software integrates the information from the two light beams to create a three-dimensional hologram of the specimen. In the "transmission" mode, the laser light penetrates through thin structures to be detected on the other side. Denser structures slow the laser penetration by fractions of nanoseconds compared to the light penetrating less dense areas. The result is also a hologram, but this one is a three-dimensional representation of cellular ultrastructure. To keep the price below $100, there is no display device on the microscope; rather, the data is sent for viewing to a computer, which has more processing power. This technological breakthrough now allows field investigators to rapidly detect bacteria or protozoa in water and blood specimens, identify insect vectors, and even monitor disease states in plants and animals. Microbiologists are now hearing the mandate, "To boldly go...."

Microorganisms are everywhere. Because pathogens, particularly bacteria and yeasts, coexist with harmless microorganisms on or in the host, their rapid detection and identification are essential to prevent or treat disease. The challenge for the fields of clinical microbiology and clinical immunology is to integrate reliable technology and evidence-based practices to achieve these goals. A variety of approaches may be used to detect and identify microorganisms based on morphological, biochemical, immunologic, and molecular procedures. In the absence of a culture, DNA-based and immunologic tests can be used to detect pathogens in specimens such as blood and sputum. Importantly, laboratory safety should always be the highest priority. In the final analysis, the patient's well-being and health can benefit significantly from information provided by the clinical laboratory—the subject of this chapter.

Readiness Check:

Based on what you have learned previously, you should be able to:

✔ Describe the infectious disease process (chapter 35)
✔ Discuss the uses of tools and techniques used to culture and characterize microorganisms (chapters 2, 7, 17, 18, and 29)
✔ Outline important aspects of microbial structure, function, survival requirements, and replication strategies (chapters 3, 4, 5, 6, 7, 13, and 15)
✔ List products of immune cells and their functions (sections 34.2, 34.4, 34.7, and 34.8)
✔ Describe the mechanisms by which microorganisms can be controlled (chapters 8 and 9)

36.1 The Clinical Microbiology Laboratory Is the Front Line for Infectious Disease Detection

After reading this section, you should be able to:

■ Explain the workflow approach to identifying microorganisms in the clinical lab
■ Practice biosafety, especially with clinical specimens
■ Describe the various scientific disciplines involved in clinical microbiology
■ Report the two general methods used to identify pathogens in clinical specimens

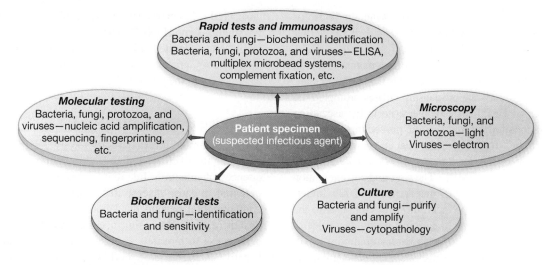

Figure 36.1 The Patient Specimen Is Evaluated by Various Techniques.   The specimen source and patient history guide the decision for use of tests and techniques. Not all tests are performed on all specimens, but all laboratories need to be able to perform these tests or refer the specimen to other laboratories.

The clinical microbiology laboratory is the primary site for the diagnosis of human infectious disease; it is often found in a hospital, although it may also be a stand-alone facility. The clinical microbiology lab also supports public health infrastructure, acting as a sentinel in the rapid detection of naturally occurring outbreaks of infection and bioterrorism. The major goals of the **clinical microbiologist** are (1) rapid and accurate identification of disease-causing microorganisms from clinical specimens, and (2) accurate antimicrobial susceptibility testing of those isolated organisms. Clinical microbiology is interdisciplinary, and the clinical microbiologist must have a working knowledge of microbial morphology, metabolism, growth, reproduction, biochemistry, and physiology. Additionally, the clinical microbiologist needs to understand the principles of immunology, molecular biology, genomics, aseptic technique, sterilization, disinfection, and the dynamics of host-parasite relationships. Importantly, tests developed to exploit (1) the antigen-antibody binding capabilities (the focus of clinical immunology) and (2) nucleic acid sequencing and amplification techniques (genomics) can often detect microorganisms in specimens by identifying microbial antigens and genes or gene products, respectively (**figure 36.1**).

The clinical microbiology laboratory provides both preliminary and definitive identification of microorganisms using various tests and procedures that have the highest probability of rapid identification based on (1) direct recovery of the infectious agent from the clinical specimen or (2) indirect evidence that a specific pathogen was the likely source of the infection. Testing a clinical sample must follow a prescribed process to avoid contamination and waste. Thus, a workflow paradigm is often established in the laboratory to rapidly move the specimen through the testing processes while maintaining its integrity. In hospital laboratories, inpatient specimens arrive barcoded to link them to respective patients and to track their progress. Outpatient samples may be received with barcodes or with paperwork that must be entered into the laboratory information system (LIS). The LIS prints barcodes to be affixed to the outpatient

specimens. A barcode reader attached to the LIS interfaces automated testing machines to request tests. The LIS tracks the testing process and manages the test results. The LIS data are accessible directly to the other hospital departments, reducing the time in which results are delivered to the clinical staff attending the patient.

36.2 Biosafety Practices Protect Lab Workers

After reading this section, you should be able to:

- Describe the risk assessment process leading to a safe laboratory
- Paraphrase the "standard microbiological practices" that guide lab safety
- Differentiate between biosafety levels and provide examples of microorganisms assigned to each level
- Integrate safety practices into laboratory workflow

As a potentially infectious material, the patient specimen is itself a possible hazard to the laboratory professional. Safety for the patient, the hospital or other clinical site, and laboratory staff is of utmost importance. Thus we discuss biosafety and safe working practices before we begin our discussion of clinical microbiology processes.

The Need for Safety Guidance

Although appalling today, not so long ago a visitor could see mouth pipetting, smoking, and eating in the microbiology laboratory. Over the years, several key events have slowly but quite significantly changed laboratory culture to promote safety. Pioneering biosafety techniques developed by Arnold G. Wedum, M.D., Ph.D., at the Army Biological Research Laboratory, Fort Detrick, Maryland, during the 1950s and 1960s set a standard for physical containment strategies and best practices to keep microbiologists from acquiring laboratory infections. Later, the ground-breaking methods for manipulating DNA by Paul Berg in the 1970s sparked an international discussion about the possible hazards of recombinant DNA technology and drove the requirement for the systematic evaluation of safe practices in the laboratory. Subsequently concern over standards needed to safely study cancer-causing viruses prompted the National Institutes of Health to establish criteria for ascending levels of physical containment and specific procedures to prevent human contamination. Together, these best methods and work practices were assembled into the *Biosafety in Microbiological and Biomedical Laboratories,* or BMBL. The fifth and current edition of the BMBL was published in 2007 and updated in 2009. Importantly, the editors insightfully prefaced the fifth edition with the caveat that "the intent was and is to establish a voluntary code of practice, one that all members of a laboratory community will together embrace to safeguard themselves and their colleagues, and to protect the public health and environment."

To help prevent laboratory-acquired infections (LAIs), microbiologists today follow strict procedures, often called standard microbiological practices, based on the thoughtful work by Wedum, Berg, and countless others, and codified in the BMBL (**table 36.1**). In addition, some events have led to the formation of laws affecting how microbiologists think about and practice laboratory safety and security. For example (1) the recognition that the human immunodeficiency virus (HIV) and other viruses are transmitted by blood led to the 1991 Bloodborne Pathogens Act, (2) the use of anthrax as a bioterrorism agent led to the Public Health Security and Bioterrorism Preparedness and Response Act of 2002, and (3) the response to emerging viruses with high infectivity or high lethality (such as SARS and avian influenza) resulted in the 2005 Pandemic and All-Hazards Preparedness Act. Additionally, the accidental *Salmonella* infections from 2010 through 2014 in U.S. teaching labs, along with the 2014 and 2015 events revealing unsecured U.S. laboratory samples of plague and tularemia bacteria, smallpox and H5N1 flu viruses, as well as ricin and botulinum toxins, have reignited efforts to integrate safety into all aspects of microbiology practice. Regardless of the reason, the message is clear: safety in the microbiology lab must be a high priority.

Practicing Biosafety

Safety planning for the microbiology laboratory focuses on assessing the likelihood of an LAI in light of the potential consequences of infection. Inherent in this assessment is not just the potential to cause disease in the user but also the potential for contamination beyond the laboratory. Typically, the laboratory design (e.g., airflow, sink access, decontamination processes) and access to biosafety cabinets are surveyed to minimize the hazard (*see figure 8.6*). Following this initial survey, potential risks associated with manipulation of microorganisms are considered, including: (1) procedure(s) used with each agent, (2) experience of those using the laboratory, and (3) infectious agent risk group. There are four risk groups (RGs) into which pathogens are assigned. These RGs reflect increasing danger to humans (**table 36.2**) and are determined based on a number of criteria, including pathogenicity, virulence factors, mode of transmission, availability of protective vaccines and chemotherapeutic agents, and DNA content beyond wild type, among others.

The rest of the risk assessment determines the biosafety level (BSL). The BSL designation indicates that a particular risk is acknowledged and that a specific set of work practices, safety equipment, and facility engineering controls will be used in an assigned laboratory space to keep risk low. The four biosafety levels reflect an appropriate containment level used to prevent LAI and environmental contamination (**table 36.3**). Each subsequent level builds on the previous. The common component to all

Table 36.1	Standard Microbiological Practices
1.	Persons must wash their hands after working with potentially hazardous materials and before leaving the laboratory.
2.	Eating, drinking, smoking, handling contact lenses, applying cosmetics, and storing food for human consumption must not be permitted in laboratory.
3.	Mouth pipetting is prohibited; mechanical pipetting devices must be used.
4.	Policies for the safe handling of sharps (such as needles, scalpels, pipettes, and broken glassware) must be developed and implemented.
5.	Procedures must be developed and implemented to minimize the creation of splashes and/or aerosols.
6.	Work surfaces must be decontaminated with an appropriate disinfectant after completion of work and after any spill or splash of potentially infectious material.
7.	All cultures, stocks, and other potentially infectious materials must be decontaminated using an effective method before disposal.
8.	A sign incorporating the universal biohazard symbol must be posted at the entrance to the laboratory when infectious agents are present. The sign may include the biosafety level, name of the agent(s) in use, and the name and phone number of the laboratory supervisor or other responsible personnel. Information should be posted according to institutional policy.
9.	Persons must report all injuries incurred in the laboratory to the laboratory supervisor as soon as possible.

Adapted from: Chosewood, L. C., and Wilson, D. E., 2009. Biosafety in Microbiological and Biomedical Laboratories, *5th ed. Washington, D.C.: U.S. Department of Health and Human Services, CDC, National Institutes of Health and Emmert, E. 2013. Biosafety guidelines for handling microorganisms in the teaching laboratory: Development and rationale. J. Microbiol. & Biol. Educ. 14(1). doi:10.1128/jmbe.v14i1.531.*

Table 36.2	Microbiology Risk Group (RG) Characterization with Example Organisms
1.	Not known to consistently cause disease in healthy adults (e.g., *Lactobacillus casei*, *Vibrio fischeri*)
2.	Associated with human disease; potential hazard if percutaneous injury, ingestion, or mucous membrane exposure occurs (e.g., *Salmonella typhi*, *Escherichia coli* O157:H7, *Staphylococcus aureus*)
3.	Indigenous or exotic agents with potential for aerosol transmission; disease may have serious or lethal consequences (e.g., *Coxiella burnetti*, *Yersinia pestis*, herpesviruses)
4.	Dangerous/exotic agents that pose high risk of life-threatening disease, aerosol-transmitted lab infections; or related agents with unknown risk of transmission (e.g., variola major [smallpox virus], Ebola virus, hemorrhagic fever viruses)

Adapted from: Chosewood, L. C., and Wilson, D. E., 2009. Biosafety in Microbiological and Biomedical Laboratories, *5th ed. Washington, D.C.: U.S. Department of Health and Human Services, CDC, National Institutes of Health.*

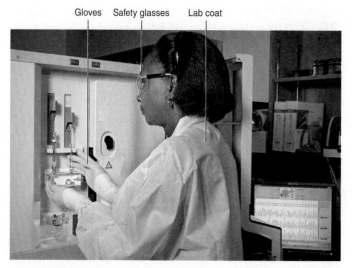

Gloves Safety glasses Lab coat

Figure 36.2 Personal Protective Equipment.   This person is wearing a lab coat, safety glasses, and gloves to prevent contamination of the laboratory and herself from infectious agents.

levels is the foundational standard microbiology practices, encompassing the behaviors directly under the user's control. These represent minimum guidelines for laboratory safety.

In addition to accepted laboratory behaviors, specific pieces of safety equipment are often recommended for each BSL. Personal protective equipment (PPE) includes (1) protective coverings (coats, gowns, or uniforms), (2) protective eyewear, and (3) gloves. PPE should prevent transmission of infectious agents by blocking their access to entry portals (**figure 36.2**). Additional PPE (e.g., face shield, respiratory protection) may be necessary as determined by the risk assessment.

Table 36.3	Biosafety Recommendations for BSL 1-4 Microbiology Laboratories		
Biosafety Level	**Work Practices**	**Safety Equipment**	**Facilities**
1.	Standard microbiology practices	No barriers required. Personal protective equipment (PPE): laboratory coats and gloves; eye, face protection, as needed	Laboratory bench and sink required
2.	BSL-1 practice plus: Limited lab access Biohazard warning signs "Sharps" precautions Biosafety manual defining any needed waste decontamination or medical surveillance policies	Biosafety cabinets (BSCs) or other physical containment devices used for all manipulations of agents that cause splashes or aerosols of infectious materials PPE: Laboratory coats, gloves, face and eye protection, as needed	BSL-1 facilities plus autoclave available
3.	BSL-2 practice plus: Controlled lab access All waste decontaminated Lab clothing decontaminated before laundering Baseline serum testing	BSCs or other physical containment devices used for all open manipulations of agents PPE: Protective lab clothing, gloves, respiratory protection as needed	BSL-2 facilities plus self-closing double door, negative air pressure, room exhaust not recirculated, away from access corridors
4.	BSL-3 practice plus: Different clothing to enter Shower to exit Decontaminate all material on exit	Class III (glove box) cabinet-based work or Class II (laminar flow) BSC with full-body, positive pressure suit	BSL-3 facilities plus separate building or zone, dedicated heat, ventilation, and air conditioning system, vacuum, and decontamination systems

Adapted from: Chosewood, L. C., and Wilson, D. E., 2009. Biosafety in Microbiological and Biomedical Laboratories, *5th ed. Washington, D.C.: U.S. Department of Health and Human Services, CDC, National Institutes of Health.*

36.3 Identification of Microorganisms from Specimens

After reading this section, you should be able to:

- Construct a work plan for the identification of a specific microorganism from a clinical specimen
- Define and provide examples of appropriately collected specimens
- Describe the use of microscopy to identify microorganisms in patient specimens
- Explain common methods for culturing common bacteria, fungi, and viruses from patient specimens
- Interpret the microscopy, culture, biochemical, and molecular assays used to identify microorganisms from patient specimens

The clinical specimen is an unknown, and microbiology laboratory scientists are the detectives who can identify the microorganism(s) in the specimen. In doing so, they determine the cause of a patient's infection. There are two types of clinical specimens: those from normally sterile body sites, such as cerebrospinal fluid or bone, and those from nonsterile sites, such as skin, mucous membranes, or feces. An important consideration for the clinical microbiologist is to remember that any microorganism found in a sterile-site specimen is a potential problem for the patient and thus should be rapidly identified. Analogously, assessment of nonsterile-site specimens requires a working knowledge of the normal microbiota associated with that site *(see table 32.2).* The challenge for the microbiologist is to identify the source of the infection amid the background "noise" of the normal microbiota.

The Clinical Specimen

A clinical specimen represents a portion or quantity of human material that is tested, examined, or studied to determine the presence or absence of particular microorganisms **(figure 36.3)**. In general, the clinical specimen should:

1. Adequately represent the diseased area and may include additional sites (e.g., urine and blood specimens) to isolate and identify potential agents of the particular disease process.

2. Be of sufficient quality and quantity to allow a variety of diagnostic should they be necessary.
3. Avoid contamination from the normal microbiota of the skin, mucous membranes, and environment.
4. Be collected in appropriate containers, kept at an appropriate temperature, and forwarded promptly to the clinical laboratory.
5. Be obtained before antimicrobial agents have been administered to the patient, if possible.
6. Be accompanied by a putative diagnosis permitting laboratory personnel to contact the physician suggesting tests necessary to "rule out" other pathogens.

A number of techniques are used in the modern clinical microbiology laboratory. Choosing the appropriate test or procedure is determined by the specimen and what is generically expected to be in the specimen based on a patient's signs and symptoms. Most laboratories have prescribed algorithms that dictate how each specimen is processed when it enters the laboratory. In general, the specimen is evaluated for microorganisms by direct techniques, including methods that first require replication or growth of the microbe prior to testing. Indirect techniques are used to identify circumstantial evidence of infection. These are often used when direct techniques do not reveal a pathogen. We highlight a number of these, acknowledging that many of the classical microbiological methods previously performed by the medical laboratory technologist at the bench have been integrated into automated mechanical systems.

Direct Identification of Infectious Agents

The direct recovery and identification of a pathogen is the best evidence documenting the source of an infectious disease. To accomplish this task, the clinical laboratory scientist (CLS) can employ several tests to determine the presence of a pathogen in a specimen. First, however, the CLS must ensure the integrity of the specimen; having more than one species of organism in a specimen collected from a normally sterile site could bias test results. Furthermore, adequate numbers of organisms are required for each laboratory test. Thus to rule out contamination, isolate pure cultures, and amplify organisms for testing, the specimen is often cultured.

(a)

(b)

(c)

Figure 36.3 Collection of Patient Specimen for Microorganism Testing. (a) The identification of the microorganism begins with proper instruction for specimen collection. (b) The specimen is protected by transport media and transported in a biohazard containment bag to prevent contamination and spillage. (c) The specimen is processed in a biosafety cabinet. In this case, the specimen is cultured to identify the infectious agent and determine the agent's sensitivity to antimicrobial drugs.

Table 36.4	Media Used for Isolation of Pure Bacterial Cultures from Specimens

Selective Media

A selective medium is prepared by the addition of specific substances to a culture medium that will permit growth of one group of bacteria while inhibiting growth of some other groups. These are examples:

Salmonella-Shigella agar (SS) is used to isolate *Salmonella* and *Shigella* species. Its bile salt mixture inhibits many groups of coliforms. Both *Salmonella* and *Shigella* species produce colorless colonies because they are unable to ferment lactose. Lactose-fermenting bacteria produce pink colonies.

Mannitol salt agar (MS) is used for the isolation of staphylococci. The selectivity is obtained by the high (7.5%) salt concentration that inhibits growth of many groups of bacteria. The mannitol helps differentiate pathogenic from nonpathogenic staphylococci, as the former ferment mannitol to form acid, whereas the latter do not. Thus this medium is also differential.

Bismuth sulfite agar (BS) is used for the isolation of *Salmonella enterica* serovar Typhi, especially from stool and food specimens. *S. enterica* serovar Typhi reduces the sulfite to sulfide, resulting in black colonies with a metallic sheen.

Hektoen enteric agar is used to increase the yield of *Salmonella* and *Shigella* species relative to other microbiota. The high bile salt concentration inhibits the growth of Gram-positive bacteria and retards the growth of many coliform strains.

Differential Media

The incorporation of certain chemicals into a medium may result in diagnostically useful growth or visible change in the medium after incubation. These are examples:

Eosin methylene blue agar (EMB) differentiates between lactose fermenters and nonlactose fermenters. EMB contains lactose, salts, and two dyes—eosin and methylene blue. *E. coli,* which is a lactose fermenter, produces a dark colony or one that has a metallic sheen. *S. enterica* serovar Typhi, a nonlactose fermenter, appears colorless.

MacConkey agar is used for the selection and recovery of members of *Enterobacteriaceae* and related Gram-negative rods. The bile salts and crystal violet in this medium inhibit the growth of Gram-positive bacteria and some fastidious Gram-negative bacteria. Because lactose is the sole carbohydrate, lactose-fermenting bacteria produce colonies that are various shades of red, whereas nonlactose fermenters produce colorless colonies.

Blood agar: addition of citrated blood to tryptic soy agar makes possible variable hemolysis, which permits differentiation of some species of bacteria. Three hemolytic patterns can be observed on blood agar.
1. α-hemolysis—greenish to brownish halo around the colony (e.g., *Streptococcus gordonii* and *S. pneumoniae*)
2. β-hemolysis—complete lysis of blood cells resulting in clearing around the colony (e.g., *Staphylococcus aureus* and *Streptococcus pyogenes*)
3. Nonhemolytic—no change in medium (e.g., *Staphylococcus epidermidis* and *S. saprophyticus*)

Culture-Based Methods

For those bacteria and fungi that can be grown in culture, the identification process begins by isolating them in pure culture. This is often done on certain types of selective or differential media that provide the first clues to the identity of the microbe. The choice of media used depends in part on the source of the specimen and knowledge of likely causes of the infection. Some examples of media used are described in **table 36.4** (*also see table 22.7*). Once isolated in pure culture, growth characteristics of the bacterium are examined and specific biochemical tests are performed. Some of the most common biochemical tests used to identify bacterial isolates are listed in **table 36.5**. Classic dichotomous keys are coupled with the biochemical tests for the identification of bacteria from specimens. Generally, fewer than 20 tests are required to identify clinical bacterial isolates to the species level (**figure 36.4**).

Miniaturization into kits and robotic automation of biochemical tests have accelerated significantly the time for identification of pathogens in specimens. Some smaller laboratories use a "kit approach" biochemical system, such as the API 20E, for the identification of members of the family *Enterobacteriaceae* and other Gram-negative bacteria. The API 20E consists of a plastic strip with 20 microtubes containing dehydrated biochemical substrates

that can detect certain biochemical characteristics (**figure 36.5**). The biochemical substrates are inoculated with a pure culture of bacteria evenly suspended in sterile saline. After 5 to 12 hours of incubation, the 20 test results are converted to a seven- or nine-digit profile. This profile number can be used with a computer or a book called the *API Profile Index* to identify the bacterium.

In addition to traditional techniques and kits, larger laboratories also use semiautomated robotic systems to detect microorganisms. Some of these are culture-independent systems, reporting infectious agents by directly identifying microbial DNA within the patient specimen, with or without PCR amplification. Several of these technologies couple pure culture isolates with electrospray ionization-mass spectroscopy, peptide nucleic acid fluorescence *in situ* hybridization (PNA-FISH), or matrix-assisted laser desorption/ionization–time of flight (MALDI-TOF)-mass spectrometry to provide high-throughput, high-reliability identification capabilities. ◄◄ *Polymerase chain reaction amplifies targeted DNA (section 17.2); Genomic libraries (section 17.5); Molecular characteristics (section 19.3); Molecular techniques (section 29.2)*

One example of an independent identification system is the automated blood culture system. It measures CO_2 concentration

Table 36.5   Some Common Biochemical Tests Used in the Identification and Differentiation of Bacteria

Biochemical Test	Description	Laboratory Application
Carbohydrate fermentation	Acid and/or gas are produced during fermentative growth with sugars or sugar alcohols.	Fermentation of specific sugars used to differentiate enteric bacteria as well as other genera or species
Casein hydrolysis	Detects the presence of caseinase, an enzyme that hydrolyzes the milk protein casein. Bacteria that use casein appear as colonies surrounded by a clear zone.	Used to cultivate and differentiate aerobic actinomycetes based on casein utilization. For example, *Streptomyces* uses casein and *Nocardia* does not.
Catalase	Detects the presence of catalase, which converts hydrogen peroxide to water and O_2	Used to differentiate *Streptococcus* ($-$) from *Staphylococcus* ($+$) and *Bacillus* ($+$) from *Clostridium* ($-$)
Citrate utilization	When citrate is used as the sole carbon source, this results in alkalinization of the medium.	Used in the identification of enteric bacteria. *Klebsiella* ($+$), *Enterobacter* ($+$), *Salmonella* (often $+$); *Escherichia* ($-$), *Edwardsiella* ($-$)
Coagulase	Detects the presence of coagulase. Coagulase causes plasma to clot.	This is an important test to differentiate *Staphylococcus aureus* ($+$) from *S. epidermidis* ($-$).
Decarboxylases (arginine, lysine, ornithine)	The decarboxylation of amino acids releases CO_2 and amine.	Used in the identification of enteric bacteria
Esculin hydrolysis	Tests for the cleavage of a glycoside	Used in the differentiation of *Staphylococcus aureus*, *Streptococcus mitis*, and others ($-$) from *S. bovis*, *S. mutans*, and enterococci ($+$)
β-galactosidase (ONPG) test	Demonstrates the presence of an enzyme that cleaves lactose to glucose and galactose	Used to separate enterics (*Citrobacter* $+$, *Salmonella* $-$) and to identify pseudomonads
Gelatin liquefaction	Detects whether or not a bacterium can produce proteases that hydrolyze gelatin and liquefy solid gelatin medium	Used in the identification of *Clostridium*, *Serratia*, *Pseudomonas*, and *Flavobacterium*
Hydrogen sulfide (H_2S)	Detects the formation of hydrogen sulfide from the amino acid cysteine due to cysteine desulfurase	Important in the identification of *Edwardsiella*, *Proteus*, and *Salmonella*
IMViC (indole; methyl red; Voges-Proskauer; citrate)	The indole test detects the production of indole from the amino acid tryptophan. Methyl red is a pH indicator to determine whether the bacterium carries out mixed acid fermentation. Voges-Proskauer (VP) detects the production of acetoin. The citrate test determines whether or not the bacterium can use sodium citrate as a sole source of carbon.	Used to separate *Escherichia* (MR$+$, VP$-$, indole$+$) from *Enterobacter* (MR$-$, VP$+$, indole$-$) and *Klebsiella pneumoniae* (MR$-$, VP$+$, indole$-$); also used to characterize members of the genus *Bacillus*
Lipid hydrolysis	Detects the presence of lipase, which breaks down lipids into simple fatty acids and glycerol	Used in the separation of clostridia
Nitrate reduction	Detects whether a bacterium can use nitrate as an electron acceptor	Used in the identification of enteric bacteria, which are usually $+$
Oxidase	Detects the presence of cytochrome *c* oxidase which is able to reduce O_2 and artificial electron acceptors	Important in distinguishing *Neisseria* and *Moraxella* spp. ($+$) from *Acinetobacter* ($-$) and enterics (all $-$) from pseudomonads ($+$)
Phenylalanine deaminase	Deamination of phenylalanine produces phenylpyruvic acid, which can be detected colorimetrically.	Used in the characterization of the genera *Proteus* and *Providencia*
Starch hydrolysis	Detects the presence of the enzyme amylase, which hydrolyzes starch	Used to identify typical starch hydrolyzers such as *Bacillus* spp.
SIM (sulfide, indole, motility medium	Detects sulfides, indole (a metabolic product of tryptophan utilization), and motility	Used to differentiate enteric organisms colorimetrically
TSI (triple sugar iron) medium	Detects metabolism of glucose, lactose or sucrose, and sulfides from ammonium sulfate or sodium thiosulfate	Used to identify of enteric organisms colorimetrically
Urease	Detects the enzyme that splits urea to NH_3 and CO_2	Used to distinguish *Proteus*, *Providencia rettgeri*, and *Klebsiella pneumoniae* ($+$) from *Salmonella*, *Shigella* and *Escherichia* ($-$)

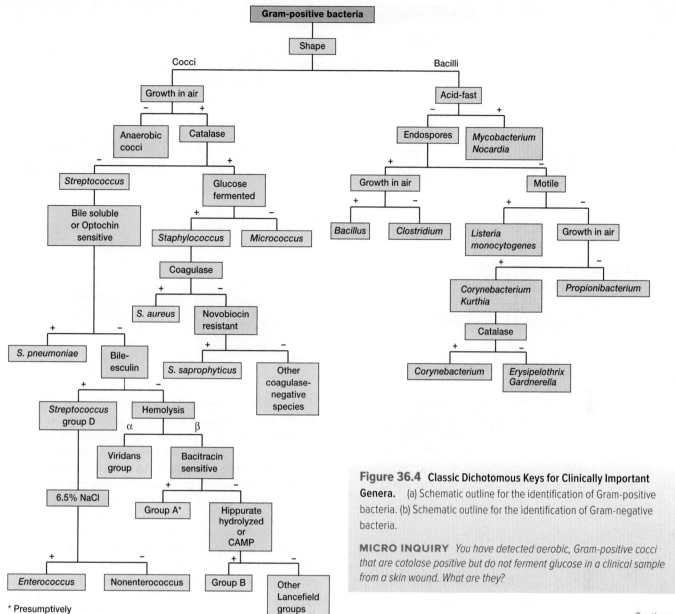

Figure 36.4 Classic Dichotomous Keys for Clinically Important Genera. (a) Schematic outline for the identification of Gram-positive bacteria. (b) Schematic outline for the identification of Gram-negative bacteria.

MICRO INQUIRY *You have detected aerobic, Gram-positive cocci that are catalase positive but do not ferment glucose in a clinical sample from a skin wound. What are they?*

Continued

and gas pressure changes due to glucose utilization as a presumptive indicator of microorganism growth and metabolism. This has reduced the time to detect blood-borne infections to less than 24 hours. Other automated systems not only detect and identify microorganisms (**figure 36.6**), they also determine their antibiotic susceptibility. Bacterial and fungal antimicrobial susceptibility is determined by evaluating organism growth when exposed to various antimicrobial agents and reported by similar means. The "gold standard" for antimicrobial susceptibility testing is still the agar dilution method. However, the most common practice is an automated microbroth dilution susceptibility test, although the manual Etest and Kirby-Bauer tests are still used in many laboratories (*see figures 9.2–9.4*). Together, the organism's identity and its sensitivity to a panel of antibiotics (a profile called its antibiogram) are determined, often within hours.

Cultures remain the standard for the recovery of fungi from patient specimens; however, the time needed to culture fungi varies anywhere from a few days to several weeks, depending on the organism. Cultures of fungi are evaluated for rate and appearance of growth on at least one selective and one nonselective agar medium, with careful examination of colonial morphology, color, and dimorphism. Typically, the isolation of fungi is accomplished by concurrent culture of the specimen on media that is respectively supplemented and unsupplemented with antibiotics and cycloheximide. Antibiotics inhibit bacteria that may be in the specimen, and cycloheximide inhibits saprophytic (living on decaying matter) molds. However, a number of media formulations are routinely used to culture specific fungi (**table 36.6**). Physicians who suspect pathogenic fungal infections will also order serological tests for antibodies to specific fungal pathogens. Culture of viruses and protozoa from clinical specimens is not routine. Their detection and identification is by microscopy, molecular techniques, or both.

Microscopy

Sometimes the specimen or the cultured microorganisms can be visualized by microscopy to aid in the identification process. Wet-mount, heat-fixed, or chemically fixed specimens can be examined

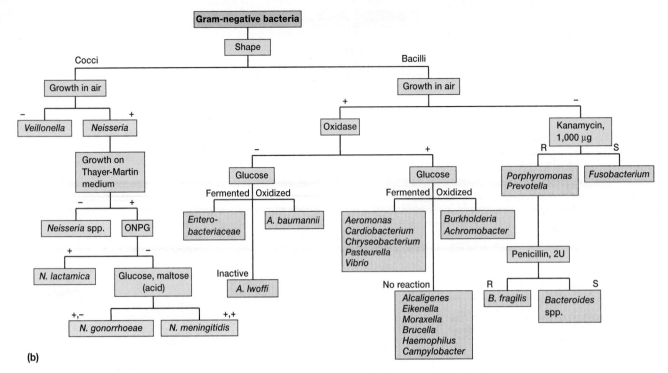

(b)

Figure 36.4 *Continued*

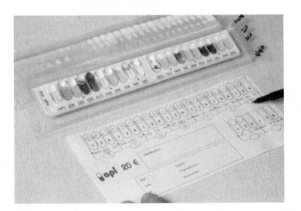

Figure 36.5 A "Kit Approach" to Bacterial Identification. The API 20E manual biochemical system for microbial identification.

Figure 36.6 Automated Clinical Data Collection. Clinical scientists use a luminometer to measure HIV infection of human cells. Specimens of human cells are distributed into 96-well microtiter plates for rapid detection of bioluminescence.

with an ordinary bright-field microscope. Examination can be enhanced with either phase-contrast or dark-field microscopy, found in some laboratories. The latter is the procedure of choice for the detection of spirochetes in skin lesions associated with early syphilis or Lyme disease. Many stains can be used to examine specimens for specific microorganisms. Two of the more widely used bacterial stains are the Gram stain and the fluorescence stain (*see figure 2.19*). Because these stains are based on the chemical composition of cell walls, they are not useful in identifying specific bacteria. However, preliminary reports often indicate the Gram-morphology, such as "Gram-positive cocci in pairs and chains." Depending on the patient symptoms, such information may be helpful in ensuring appropriate antibiotic therapy. Note in figure 36.4 that the Gram reaction and shape of a bacterium are critical pieces of information

Table 36.6	Examples of Media Used to Isolate Fungi
Culture Medium	**Target Fungi**
Antibiotic agar	Fungi in polymicrobial specimens
Caffeine agar	*Aspergillus, Rhizopus,* and others
Chlamydospore agar	Dimorphic fungi
Cornmeal agar	Most fungi, including pathogens
Malt agar	Ascomycetes
Malt extract agar	Basidiomycetes
Potato dextrose-yeast extract agar	Mushrooms
Sabouraud dextrose agar (SAB)	Most fungi
SAB + chloramphenicol + cyclohexamide	Dimorphic fungi and dermatophytes

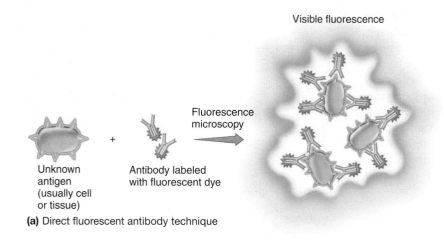

(a) Direct fluorescent antibody technique

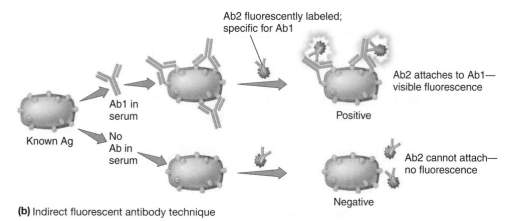

(b) Indirect fluorescent antibody technique

Figure 36.7 Direct and Indirect Immunofluorescence. (a) In the direct fluorescent antibody (DFA) technique, the specimen containing the antigen is fixed to a slide. Fluorescently labeled antibodies that recognize the antigen are then added, and the specimen is examined with a fluorescence microscope for fluorescence. (b) The indirect fluorescent antibody technique (IFA) detects an antibody (Ab) that recognizes a known antigen affixed to a slide. The antigen-antibody complex is located with a fluorescently labeled second or secondary Ab that recognizes other antibodies. (c) A photomicrograph of cells infected with cytomegalovirus (green fluorescence) and adenovirus (yellow fluorescence) using an indirect immunofluorescent staining procedure.

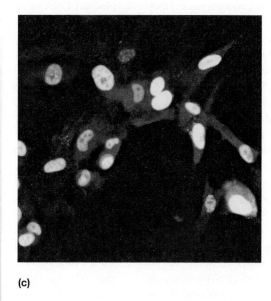

(c)

MICRO INQUIRY *Which type of immunofluorescence would most likely be used to test a clinical sample for the presence of an intracellular pathogen such as hepatitis C virus?*

used to decide which biochemical tests are appropriate to perform.

Definitive identification of most fungi is based on the morphology of reproductive structures (spores). Identification and characterization of ova, trophozoites, and cysts in the specimen result in the definitive diagnosis of protozoa infection. These are typically accomplished by direct microscopic evaluation of the clinical specimen. Typical histological staining of blood is routinely used in the identification of blood parasites. Concentrated wet mounts of stool or urine specimens can be examined microscopically for the presence of eggs, cysts, larvae, or vegetative cells of protozoa and helminths.

The fluorescence microscope can be used to visualize microorganisms after staining with fluorochromes such as acridine orange, which stains nucleic acids. Viruses are too small to visualize with the light microscope, but they can be diagnosed using fluorescence microscopy. ◀◀ *Antibodies are proteins that bind to specific 3-D molecules (section 34.7)*

With the use of sensitive techniques such as fluorescence microscopy, it is possible to perform antibody-based microbial identifications with improved accuracy, speed, and fewer organisms. The sensitivity and specificity of the antigen-antibody reaction have led to their exploitation for various microbial detection and identification techniques. Since antibodies are the focus of immunology, we discuss them in more detail in section 36.4. In general, though, two main kinds of fluorescent antibody techniques are used for microscopy: direct and indirect. **Direct immunofluorescence** involves fixing the specimen containing the antigen of interest onto a slide (**figure 36.7a**). Fluorochrome-labeled antibodies are then added to the slide and incubated. The slide is washed to remove any unbound antibody and examined with the fluorescence microscope (*see figures 2.12 and 2.13*) for fluorescence. The pattern of fluorescence reveals the antigen's location. Direct immunofluorescence is used to identify antigens such as those found on

Labels in figure (a): Visible fluorescence; Fluorescence microscopy; Unknown antigen (usually cell or tissue); Antibody labeled with fluorescent dye

Labels in figure (b): Ab2 fluorescently labeled; specific for Ab1; Ab1 in serum; Known Ag; No Ab in serum; Ab2 attaches to Ab1— visible fluorescence; Positive; Ab2 cannot attach— no fluorescence; Negative

the surface of group A streptococci and to diagnose enteropathogenic *Escherichia coli, Neisseria meningitidis, Salmonella* spp., *Shigella sonnei, Listeria monocytogenes, Mycobacterium tuberculosis,* and *M. avium,* and *Haemophilus influenzae* type b.

Fungi and protozoa can also be identified using fluorescence microscopy. Calcofluor white is typically used to stain fungi from cultures because it binds to the chitan in their cell walls. Fungi in specimens often are diagnosed by direct microscopic examination of specimens using fluorescence. Identification of hyphae in clinical specimens is a presumptive positive result for fungal infection.

Molecular Methods

Probably the most accurate and sensitive methods for the identification of microorganisms are those that report their presence through the analysis of proteins and nucleic acids. Examples include comparison of proteins; physical, kinetic, and regulatory properties of microbial enzymes; nucleic acid–base composition; nucleic acid hybridization; and nucleic acid sequencing (*see figures 17.7 and 18.2–18.3*). Other molecular methods being widely used are nucleic acid probes, real-time PCR amplification of DNA, and DNA fingerprinting. Details of tests that use microbial DNA, RNA, or protein are presented in chapters 17–19. ◄◄ *Microbial taxonomy and phylogeny are largely based on molecular characterization (section 19.3)*

Nucleic acid–based diagnostic methods for the detection and identification of microorganisms have become routine in clinical microbiology laboratories. For example, DNA hybridization technology can identify a microorganism by probing its genetic composition. This type of technology is more sensitive than conventional microbiological techniques, gives results in 2 hours or less, and requires the presence of fewer microorganisms. DNA probe sensitivity can be increased by over 1 million times if the target DNA is first amplified using PCR. PCR can also be monitored in "real time" to detect microbial DNA amplification after each replicative cycle, identifying the microorganisms after only 25 to 30 amplification cycles (**figure 36.8**; also *see figure 17.8*). The most sensitive methods for demonstrating chlamydiae in clinical specimens involves nucleic acid sequencing and PCR-based methods. ◄◄ *Polymerase chain reaction amplifies targeted DNA (section 17.2)* ↺ *DNA Hybridization*

The nucleotide sequence of small subunit ribosomal RNA (rRNA) can be used to identify bacterial genera (*see figure 19.3*). Usually the rRNA gene or gene fragment is amplified by PCR. After nucleotide sequencing, the sequence is compared with those in international databases. This method of bacterial identification, called **ribotyping,** is based on the high level of 16S rRNA conservation among bacteria. As discussed in chapter 19, multiple "housekeeping" genes, instead of 16S rRNA genes, may be sequenced and analyzed in a technique called **multilocus sequence typing** (**MLST**) (*see p. 472*).

Genomic fingerprinting is also used in identifying pathogens. This does not involve nucleotide sequencing; rather, it compares the similarity of specific DNA fragments generated by restriction endonuclease digestion (*see figure 19.4*). Plasmids can

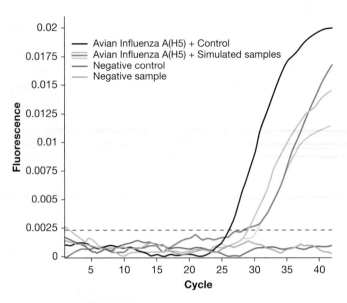

Figure 36.8 PCR Kit Results. Real-time detection of avian influenza is possible using a PCR kit. Note that sufficient DNA amplification occurs after 25 to 30 cycles.

also be used in DNA fingerprinting. **Plasmid fingerprinting** can identify microbial isolates of the same or similar strains because related strains often contain the same plasmids. In contrast, microbial isolates that are phenotypically distinct have different plasmid fingerprints. Plasmid fingerprinting of many *E. coli, Salmonella, Campylobacter,* and *Pseudomonas* strains and species has demonstrated that this method often is more accurate than phenotyping methods such as biotyping, antibiotic-resistance patterns, phage typing, and serotyping. Just as in genomic fingerprinting, isolated plasmid DNA is cut with specific restriction endonucleases (*see table 17.2*). The DNA fragments are then separated by gel electrophoresis to yield a pattern of fragments, which appear as bands in the gel. The molecular weight of each plasmid species can then be determined and patterns of restriction fragments compared. ◄◄ *Gel electrophoresis (section 17.1)*

Indirect Identification of Infectious Agents

When an infectious agent cannot be identified by direct methods, methods that indirectly implicate a specific microorganism are used. Indirect evidence does not reveal a live or replicating microbe; it can, however, suggest the cause of infectious disease, especially when no other evidence can refute it. Several techniques are used in the clinical laboratory for this purpose. They include **serology,** the testing and use of serum for antibody evidence of an infection, and molecular biology methods that find evidence of microbial DNA, RNA, and proteins. We discuss serological techniques in detail as a method of clinical immunology.

Briefly, the presence of antimicrobial antibodies in a patient's serum (the liquid portion of blood devoid of clotting factors) is indirect evidence of an infection. In fact, identification of antibody classes (i.e., IgM or IgG) can also reveal when (relatively speaking) a specific infection occurred. Similarly, antibodies can be used to create tests that detect and identify microbial antigens in clinical specimens. The concept is like that described above for antibody-directed fluorescence microscopy. Modern antibody derivatives, known as monoclonal antibodies, have increased identification sensitivity by several orders of magnitude. ◀◀ *Techniques and Applications 34.1*

Indirect immunofluorescence (figure 36.7*b*) is used to detect the presence of antibodies in serum following an individual's exposure to microorganisms. In this technique, a known antigen (e.g., a virus) is fixed onto a slide. The patient serum suspected of containing (antiviral) antibodies is then added, and if the specific antibody is present, it reacts with antigen to form a complex. When a second, fluorescein-labeled antibody (not from the patient's serum) is added, it reacts with the fixed antibody. After incubation and washing, the slide is examined with the fluorescence microscope. The occurrence of fluorescence shows that antibody specific to the (viral) antigen is present in the serum and that its presence is reported by the fluorescence of the secondary antibody that has attached to the serum antibody (figure 36.7*c*). A common application of indirect immunofluorescence is the identification of *Treponema pallidum* antibodies in the diagnosis of syphilis.

Because serology is so commonly used, numerous microbial detection kits are available to indirectly screen clinical specimens for the presence of specific microorganism antibodies (**table 36.7**). Monoclonal antibodies in these kits have been produced against a wide variety of bacterial, viral, fungal, and protozoal antigens. Coupling sensitive visualization technologies such as fluorescence or scanning tunneling microscopy to mAb detection systems makes it possible to perform microbial identifications with improved accuracy, speed, and fewer organisms.

Table 36.7	Some Common Rapid Immunologic Test Kits for the Detection of Bacteria and Viruses in Clinical Specimens

BioKit HSV-2 (American Screening Corporation, Shreveport, LA) Detects anti-HSV-2 antibody (IgG) in blood or serum.

Directigen (Becton, Dickinson, and Company, Franklin Lakes, NJ) The Directigen Meningitis Combo Test kit is used to detect *S. pneumoniae*, *H. influenzae* type b, *N. meningitidis* groups A, B, C, Y, and W135, Group B Streptococcus, and *E. coli* K1. The Directigen Flu Test kit is used for the direct detection of types A and B influenza.

GonoGen II Test Kit (Becton, Dickinson, and Company, Franklin Lakes, NJ) This test uses monoclonal antibody in a colorimetric assay for the confirmatory testing of *Neisseria gonorrhea* from cultures.

OraQuick detects HIV antibodies in saliva in 10 minutes.

QuickView InLine Strep A Test (Quidel Corporation, San Diego, CA) The test detects group A streptococcal antigen directly from patient throat swab specimens.

Staphaurex (Remel, Thermo Scientific, Lenexa, KS) Staphaurex screens and confirms *Staphylococcus aureus* in 30 seconds.

Retrieve, Infer, Apply

1. Name two specimens for which microscopy would be used in the initial diagnosis of an infectious disease.
2. Describe in general how biochemical tests are used in the API 20E system to identify bacteria.
3. What are the advantages of using monoclonal antibody (mAb) immunofluorescence in the identification of viruses?
4. How can a clinical microbiologist determine the initial identity of a bacterium?
5. Describe a dichotomous key that could be used to identify a bacterium.
6. How can nucleic acid–based detection methods be used by the clinical microbiologist?
7. How can a suspect bacterium be fingerprinted?

36.4 Immune Responses Can Be Measured or Exploited to Detect Infections

After reading this section, you should be able to:

- Diagram the methods by which antibodies are used to identify microorganisms
- Interpret immunologic data used to identify microorganisms from patient specimens
- Design an antibody-based test for the identification of a specific microorganism from a patient specimen

It may not be possible to culture certain microorganisms from clinical specimens because the methodology remains undeveloped (e.g., *Treponema pallidum;* hepatitis A, B, and C viruses; and Epstein-Barr virus), the microbe is unsafe (e.g., rickettsias and HIV), or it is impractical for all but a few microbiology laboratories (e.g., mycobacteria, strict anaerobes, *Borrelia* spp.). Cultures also may be "negative" because of prior antimicrobial therapy. Under these circumstances, detection of antibodies or antigens can be quite valuable.

Immunologic systems for the detection and identification of pathogens from clinical specimens are easy to use, give relatively rapid reaction end points, and are sensitive and specific with a low percentage of false positives and negatives. Some of the more popular immunologic rapid test kits for viruses and bacteria are presented in table 36.7.

Dramatic advances in clinical immunology have given rise to a marked increase in the number, sensitivity, and specificity of serological tests. This reflects a better understanding of (1) immune cell surface antigens (CD antigens; *see table 34.1*), (2) lymphocyte biology, (3) the production of monoclonal antibodies, and (4) the development of sensitive antibody-binding reporter systems. For a number of reasons, the utility of these tests depends on proper test selection and timing of specimen collection. For instance, each individual's immunologic response to a microorganism is quite variable, making the interpretation of immunologic tests potentially difficult. For example, a single, elevated IgG titer does not distinguish between active and past infections. Rather, an elevated IgM titer typically indicates an active infection, especially when subsidence of symptoms correlates with a four-fold (or greater) decrease in antibody titer. Furthermore, a lack of a measurable antibody titer may reflect an organism's lack of immunogenicity or insufficient time for an antibody response to develop following the onset of the infectious disease. Some patients are also immunosuppressed due to other disease processes or treatment procedures (e.g., cancer and AIDS patients) and therefore do not respond. In this section, some of the more common antibody-based techniques employed in the diagnosis of microbial and immunological diseases are discussed. ◄◄ *Immunoglobulin (antibody) structure (section 34.7)*

Serotyping

Serum contains many different components, especially the immunoglobulins (antibodies). **Serotyping** refers to the use of serum or purified antibodies to specifically detect and identify other molecules. Serotyping can be used to identify specific white blood cells or the proteins on cell surfaces. Serotyping can also be used to differentiate strains (serovars or serotypes) of microorganisms that differ in the antigenic composition of a structure or product. Therefore, it is sometimes possible to identify a pathogen serologically by testing for cell antigens. For example, there are 90 different strains of *Streptococcus pneumoniae,* each unique in the nature of its capsular material.

Agglutination

An **agglutination reaction** occurs when an immune complex is formed by cross-linking cells or particles with specific antibodies (*see figure 34.19*). Agglutination reactions usually form visible aggregates or clumps, called **agglutinates,** that can be seen with the

unaided eye. Direct agglutination reactions are very useful in the diagnosis of certain diseases. For example, the Rapid Plasma Reagin test is used to screen for syphilis. Antigenic cardiolipin released from spirochete-damaged host cells stimulates the production of reagin antibodies. These antibodies, collected from syphilis-suspect patients, will agglutinate syphilis bacteria in vitro, yielding a positive test. ◄◄ *Immune complex formation (section 34.8)*

Techniques have also been developed that employ microscopic synthetic latex spheres coated with antigens. These coated microspheres bind antibodies in a patient's serum specimen to identify viral disease rapidly when cultures are not feasible (e.g., HIV). Latex agglutination tests are also used to detect antibodies that develop during certain mycotic (fungal), helminthic (worm), and bacterial infections, as well as in drug testing. Microspheres can also be coated with monoclonal antibodies to capture antigens from patient specimens. Multiplex microbead test systems are increasingly used in larger laboratories where high throughput is needed to balance the workload.

Hemagglutination usually results from antibodies cross-linking red blood cells through attachment to surface antigens and is routinely used in blood typing. In addition, some viruses can accomplish **viral hemagglutination.** For example, if a person has a certain viral disease, such as measles, antibodies will be present in the serum to react with the measles virus particles and neutralize them. Normally hemagglutination occurs when measles virus particles and red blood cells are mixed. However, when a person's serum is mixed first with virions, followed by the addition of red blood cells, the lack of hemagglutination indicates that serum antibodies have neutralized the measles viruses. This is considered a positive test result for the presence of virus-specific antibodies (**figure 36.9**).

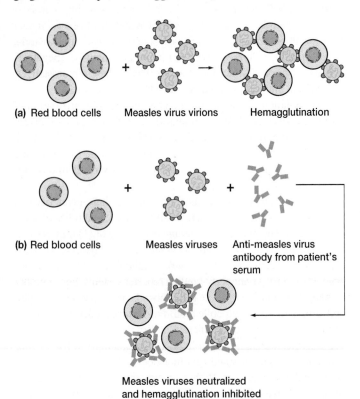

(a) Red blood cells   Measles virus virions   Hemagglutination

(b) Red blood cells   Measles viruses   Anti-measles virus antibody from patient's serum

Measles viruses neutralized and hemagglutination inhibited

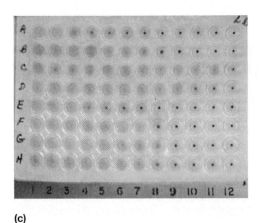

(c)

Figure 36.9 Viral Hemagglutination.   (a) The virions of certain viruses can bind to red blood cells, causing hemagglutination. (b) If patient serum containing specific antibodies to the virus is mixed with the red blood cells, the antibodies will neutralize the virus and inhibit hemagglutination (a positive test). (c) Reovirus hemagglutination test results.

MICRO INQUIRY *Which wells show hemagglutination as characterized by cellular clumping?*

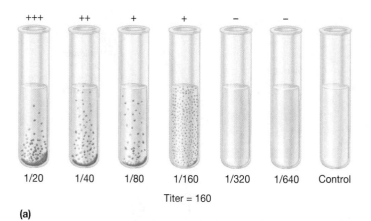

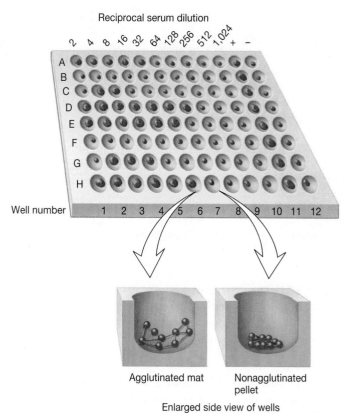

Figure 36.10 Agglutination Tests. (a) Tube agglutination test for determining antibody titer. The titer in this example is 160 because there is no agglutination in the next tube in the dilution series (1/320). The blue in the dilution tubes indicates the presence of the patient's serum. (b) A microtiter plate illustrating hemagglutination. The antibody is placed in the wells (rows 1–10). Positive controls (row 11) and negative controls (row 12) are included. Red blood cells are added to each well. If sufficient antibody is present to agglutinate the cells, they sink as a mat to the bottom of the well. If insufficient antibody is present, they form a pellet at the bottom.

MICRO INQUIRY *What is the titer for the antibody tested in row E?*

Agglutination tests are also used to measure antibody titer. In tube or well agglutination tests, a specific amount of antigen is added to a series of tubes or shallow wells in a microtiter plate (**figure 36.10**). Serial dilutions of serum (1/20, 1/40, 1/80, 1/160, etc.) containing the antibody are then added to each tube or well. The greatest dilution of serum showing an agglutination reaction is determined, and the reciprocal of this dilution is the serum antibody titer.

Complement Fixation

When complement binds to an antigen-antibody complex, it becomes "fixed" and "used up." Complement fixation tests are very sensitive and can be used to detect extremely small amounts of an antibody directed against a suspect microorganism in an individual's serum. A known antigen is mixed with test serum lacking complement (**figure 36.11a**). When immune complexes have had time to form, complement is added (figure 36.11b) to the mixture. If immune complexes are present, they fix and consume complement. Afterward, sensitized indicator cells, usually sheep red blood cells previously coated with complement-fixing antibodies, are added to the mixture. If specific antibodies are present in the test serum and complement is consumed by the immune complexes, insufficient amounts of complement will be available to lyse the indicator cells. On the other hand, in the absence of antibodies, complement remains and lyses the indicator cells (figure 36.11c). Thus the absence of lysis shows that specific antibodies are present in

the test serum. Complement fixation was once used in the diagnosis of syphilis (the Wassermann test) and is still used as a rapid, inexpensive screening method in the diagnosis of certain viral, fungal, rickettsial, chlamydial, and protozoan diseases. *Complement Fixation Test*

Enzyme-Linked Immunosorbent Assay

The ***enzyme-linked immunosorbent assay*** (**ELISA**) has become one of the most widely used serological tests for antibody or antigen detection. The science behind the technique is similar to that described on page 795 for fluorescence microscopy and involves the linking of various "reporting" enzymes to either antigens or antibodies. Two methods are used: the indirect immunosorbent assay and the direct (also called the double antibody sandwich assay).

The indirect immunosorbent assay detects antibodies, rather than antigens. In this assay, antigen in appropriate buffer is incubated in the wells of a microtiter plate (**figure 36.12a**) and is adsorbed onto the walls of the wells. Free antigen is washed away. Test serum is added, and if specific antibody is present, it binds to the antigen. Unbound antibody is washed away. Alternatively, the test sample can be incubated with a suspension of latex beads that have the desired antigen attached to their surface. After allowing time for antibody-antigen complex formation, the beads are trapped on a filter and unbound antibody is washed away. An anti-antibody that has been covalently coupled to an enzyme, such as horseradish peroxidase, is added next. The antibody-enzyme complex (the

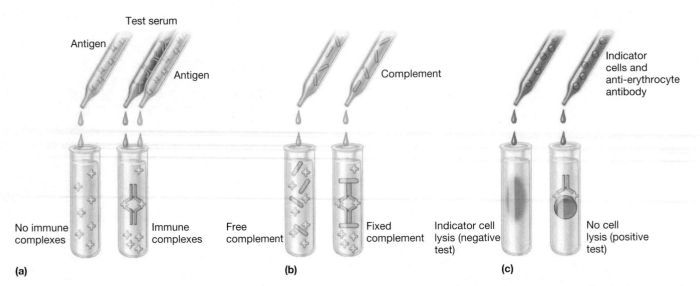

Figure 36.11 Complement Fixation. (a) Test serum is added to one test tube. A fixed amount of antigen is then added to both tubes. If antibody is present in the test serum, immune complexes form. (b) When complement is added, if complexes are present, they fix complement and consume it. (c) Indicator cells and a small amount of anti-erythrocyte antibody are added to the two tubes. If there is complement present, the indicator cells will lyse (a negative test): if the complement has been consumed by the immune complex, no lysis will occur (a positive test).

conjugate) binds to the test antibody, and after unbound conjugate is washed away, the attached indicator antibody is visualized by the addition of a **chromogen.** A chromogen is a colorless substrate acted on by the enzyme portion of the indicator antibody to produce a colored product. The indirect immunosorbent assay currently is used to test for antibodies to HIV and rubella virus (German measles), and to detect certain drugs in serum. For example, antigen-coated nitrocellulose membranes are used in the OraQuick HIV-1 test to detect HIV serum antibodies in about 10 minutes.

The direct ELISA (also known as the double antibody sandwich, or capture, assay) is used for the detection of antigens (figure 36.12c). In this assay, specific antibody is placed in wells of a microtiter plate (or it may be attached to a membrane). The antibody is absorbed onto the walls, coating the plate. A test antigen (in serum, urine, etc.) is then added to each well. If the antigen reacts with the antibody, the antigen is retained when the well is washed to remove unbound antigen. A commercially prepared antibody-enzyme conjugate specific for the antigen is then added to each well. The final complex formed is an outer antibody-enzyme, middle antigen, and inner antibody; that is, it is a layered (Ab-Ag-Ab) sandwich. A substrate that the enzyme will convert to a colored product is then added, and any resulting product is quantitatively measured by optical density scanning of the plate. If the antigen has reacted with the absorbed antibodies in the first step, the ELISA test is positive (i.e., it is colored). If the antigen is not recognized by the absorbed antibody, the ELISA test is negative because the unattached antigen has been washed away and no antibody-enzyme is bound (it is colorless). This assay is currently used for the diagnosis of *Helicobacter pylori* infections, brucellosis, salmonellosis, and cholera. Many other antigens also can be detected by the sandwich method. For example, some ELISA kits on the market test for food allergens. 🔊 *ELISA: Enzyme-Linked Immunosorbent Assay*

Immunoblotting (Western Blotting)

Another immunologic technique used in the clinical microbiology laboratory is immunoblotting, also known as Western blotting. **Immunoblotting** involves polyacrylamide gel electrophoresis of a protein specimen followed by transfer of the separated proteins to sheets of nitrocellulose or polyvinyl difluoride. Protein bands are then visualized by treating the nitrocellulose sheets with solutions of enzyme-tagged antibodies. This procedure demonstrates the presence of common and specific proteins among different strains of microorganisms. Immunoblotting also can be used to show strain-specific immune responses to microorganisms, to serve as an important diagnostic indicator of a recent infection with a particular strain of microorganism, and to allow for prognostic implications with severe infectious diseases (**figure 36.13**). ◀◀ *Gel electrophoresis (section 17.1)*

Immunoprecipitation

The **immunoprecipitation** technique detects soluble antigens that react with antibodies called **precipitins.** The precipitin reaction occurs when bivalent or multivalent antibodies and antigens are mixed in the proper proportions. The antibodies link the antigen to form a large antibody-antigen network or lattice that settles out of solution when it becomes sufficiently large (**figure 36.14a**). Immunoprecipitation reactions occur only at the equivalence zone when there is an optimal ratio of antigen to antibody so that an insoluble lattice forms. If the precipitin reaction takes place in a test tube (figure 36.14b), a precipitation ring forms in the area in which the optimal ratio or equivalence zone develops. ◀◀ *Immune complex formation (section 34.8)*

Indirect ELISA, comparing a positive versus negative reaction. This is the basis for HIV screening tests.

Well A   Well B

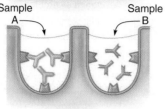

Known antigen is adsorbed to well.

Sample A   Sample B

Serum samples with unknown antibodies

A

B

Well is rinsed to remove unbound (nonreactive) antibodies.

Indicator antibody linked to enzyme attaches to any bound antibody.

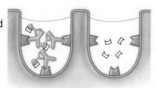

Wells are rinsed to remove unbound indicator antibody. A colorless substrate for enzyme is added.

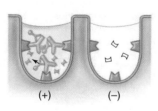

Enzymes linked to indicator Ab hydrolyze the substrate, which releases a dye. Wells that develop color are positive for the antibody; colorless wells are negative.

(+)   (−)

(a)

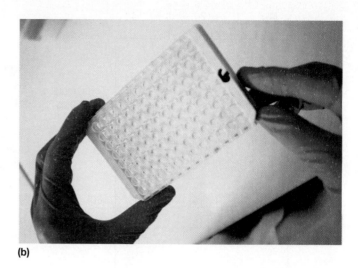

(b)

The direct ELISA method.
Note that an antigen is trapped between two antibodies. This test is used to detect hantavirus and measles virus.

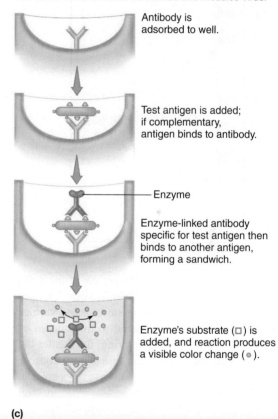

Antibody is adsorbed to well.

Test antigen is added; if complementary, antigen binds to antibody.

Enzyme

Enzyme-linked antibody specific for test antigen then binds to another antigen, forming a sandwich.

Enzyme's substrate (□) is added, and reaction produces a visible color change (•).

(c)

Figure 36.12 The ELISA Test. (a) The indirect immunosorbent assay for detecting antibodies. (b) HIV antibody ELISA test results. (c) The direct or capture or sandwich method for the detection of antigens.

MICRO INQUIRY *Why do you think an indirect ELISA is used to test for HIV?*

Immunodiffusion

Immunodiffusion refers to a precipitation reaction that occurs between an antibody and antigen in an agar gel medium. Two techniques are routinely used: single radial immunodiffusion and double diffusion in agar.

The **single *radial immunodiffusion* (RID) assay** or Mancini technique quantitates antigens. Antibody is added to agar, then the mixture is poured onto slides and allowed to set. Wells are cut in the agar and known amounts of standard antigen added. The unknown test antigen is added to a separate well. The slide is incubated for 24 hours or until equilibrium has been reached,

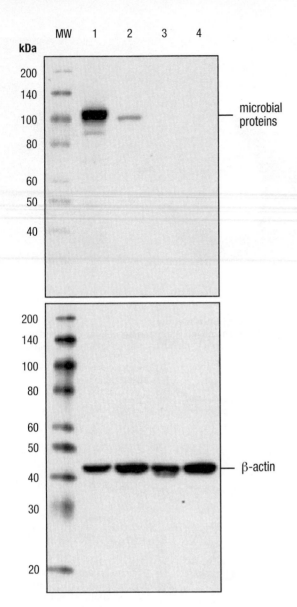

Figure 36.13 The Western Blot. Microbial proteins are separated by gel electrophoresis. For example, lane 1 might have HIV proteins; lane 2, treponemal proteins; lane 3, gonococcal proteins; and lane 4, chlamydial proteins. The proteins in the gel are then blotted onto nitrocellulose filter strips. The nitrocellulose strips are incubated with patient serum samples to permit serum antibody to bind to the proteins. The strips are developed by probing for antibody using radioactive or colorimetric methods. Molecular weight (MW) standards are included to determine antigen mass. To ensure that equal amounts of protein are applied to the gel, a protein control is used (i.e., β-actin in the lower blot).

MICRO INQUIRY *To what microorganism(s) do you think the patient who donated the serum for the above Western blot has been exposed?*

The **double diffusion agar assay (Öuchterlony technique)** is based on the principle that diffusion of both antibody and antigen (hence double diffusion) through agar can form stable and easily observable immune complexes. Test solutions of antigen and antibody are added to the separate wells punched in agar. The solutions diffuse outward, and when antigen and the appropriate antibody meet, they combine and precipitate at the equivalence zone, producing an indicator line (or lines) (**figure 36.15**). The visible line of precipitation permits a comparison of antigens for identity (same antigenic determinants), partial identity, or nonidentity against a given selected antibody.

Immunoelectrophoresis

Some antigen mixtures are too complex to be resolved by simple diffusion and precipitation. Greater resolution is obtained by the technique of **immunoelectrophoresis** in which antigens are first separated based on their electrical charge, then visualized by the

during which time the antigen diffuses out of the wells to form insoluble complexes with antibodies. The size of the resulting precipitation ring surrounding various dilutions of antigen is proportional to the amount of antigen in the well (the wider the ring, the greater the antigen concentration). This is because antigen concentration drops as it diffuses farther out into the agar. The antigen forms a precipitin ring in the agar when its level has decreased sufficiently to reach equivalence and combine with the antibody to produce a large, insoluble network. This method is commonly used to quantitate serum immunoglobulins, complement proteins, and other substances.

Figure 36.14 Immunoprecipitation. (a) Graph showing that a precipitation curve is based on the ratio of antigen to antibody. The zone of equivalence represents the optimal ratio for precipitation. (b) A precipitation ring test. Antibodies and antigens diffuse toward each other in a test tube. A precipitation ring is formed at the zone of equivalence.

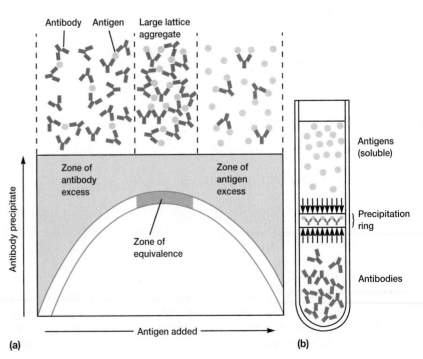

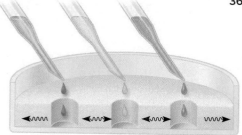

Side view

(a) In one method of setting up a double-diffusion test, wells are punctured in soft agar, and antibodies and antigens are added in a pattern. As the contents of the wells diffuse toward each other, a number of reactions can result, depending on whether antibodies meet and precipitate antigens.

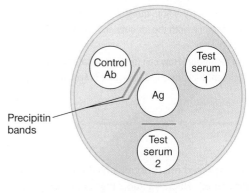

(b) Example of test pattern and results. Antigen (Ag) is placed in the center well and antibody (Ab) samples are placed in outer wells. The control contains known Abs to the test Ag. Note bands that form where Ab and Ag meet. The other wells (1, 2) contain unknown test sera. One is positive and the other is negative. Double bands indicate more than one antigen and antibody that can react.

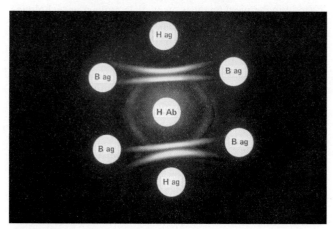

(c) Actual test results for detecting infection with the fungal pathogen *Histoplasma* sp.

Figure 36.15 Immunodiffusion. Double diffusion agar assay showing characteristics of identity (test serum 2) and a reaction of nonidentity (test serum 1).

precipitation reaction. In this procedure, antigens are separated by electrophoresis in an agar gel. Positively charged proteins move to the negative electrode, and negatively charged proteins move to the positive electrode (**figure 36.16a**). A trough is then cut next to the wells (figure 36.16b) and filled with antibody. When the plate is incubated, the antibodies and antigens diffuse and form

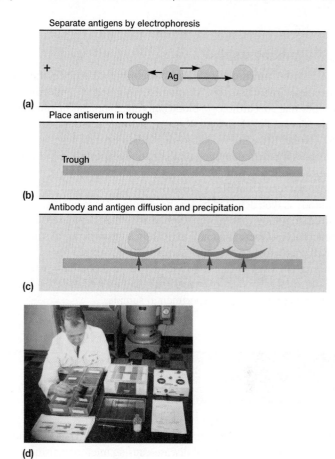

Figure 36.16 Classical Immunoelectrophoresis. (a) Antigens are separated in an agar gel by an electrical charge. (b) Antibody (antiserum) is then placed in a trough cut parallel to the direction of the antigen migration. (c) The antigens and antibodies diffuse through the agar and form precipitin arcs. (d) After staining, better visualization is possible.

precipitation bands or arcs (figure 36.16c) that can be better visualized by staining (figure 36.16d). This assay is used to separate the major blood proteins in serum for certain diagnostic tests. ◀◀ *Gel electrophoresis (section 17.1)*

Flow Cytometry

Flow cytometry allows detection of specific lymphocyte subsets ($CD4^+$ T-cells, for example, in HIV-infected blood) in an easy, reliable, fast way. In flow cytometry, cells are identified on the basis of their unique cytometric parameters or by means of fluorochromes that can be used either independently or bound to specific antibodies. The flow cytometer forces a suspension of cells through a laser beam and measures the light they scatter or the florescence the cells emit as they pass through the beam (*see figure 29.4*). For example, cells can be tagged with a fluorescent antibody directed against a specific surface antigen. As the stream of cells flows past the laser beam, each fluorescent cell can be detected, counted, and even separated from the other cells in the suspension. The cytometer also can measure a cell's shape, size, and content of DNA and RNA. This technique has enabled the development of quantitative methods to assess antimicrobial susceptibility and drug cytotoxicity in a rapid, accurate, and highly reproducible way.

Radioimmunoassay

The *radioimmunoassay* (**RIA**) technique has become an extremely important tool in biomedical research and clinical practice (e.g., in cardiology, blood banking, diagnosis of allergies, and endocrinology). Indeed, Rosalyn Yalow (1921–2011) won the 1977 Nobel Prize in Physiology or Medicine for its development. RIA uses a purified antigen that is radioisotope-labeled. It competes for antibody with unlabeled standard antigen or test antigen in experimental samples. The radioactivity associated with the antibody is then detected by means of radioisotope analyzers and autoradiography (photographic emulsions that show areas of radioactivity). If there is a lot of antigen in an experimental sample, it will compete with the radioisotope-labeled antigen for antigen-binding sites on the antibody, and little radioactivity will be bound. A large amount of bound radioactivity indicates that little antigen is present in the experimental sample.

Retrieve, Infer, Apply

1. What is serology? Why is serology commonly used to detect syphilis?
2. Why does hemagglutination occur and how can it be used in the clinical laboratory?
3. What does a negative complement fixation test show? A positive test?
4. What are the two types of ELISA methods and how do they work? What is a chromogen?
5. Specifically, when do immunoprecipitation reactions occur?
6. Name two types of immunodiffusion tests and describe how they operate.
7. Describe the immunoelectrophoresis technique.
8. Explain how flow cytometry is both qualitative and quantitative (i.e., how it can determine both the identity and the number of a specific cell type).
9. Describe the RIA technique. How can fungi and protozoa be detected in a clinical specimen? Rickettsias? Chlamydiae? Mycoplasmas?

Key Concepts

36.1 The Clinical Microbiology Laboratory Is the Front Line for Infectious Disease Detection

- The major foci of the clinical microbiologist are the rapid and accurate identification of disease-causing microorganisms from clinical specimens and the accurate antimicrobial susceptibility testing of isolated organisms.
- A variety of laboratory tests and techniques are used (**figure 36.1**). Most of the traditional bench tests have been miniaturized or automated. Some have been abandoned, replaced by faster and more sensitive tests.
- Because each laboratory reflects the clinical environment in which it is housed, some laboratories are relatively small and may only perform a few common or rapid tests. Laboratories associated with large hospitals or public health departments typically analyze a wider diversity of specimens.

36.2 Biosafety Practices Protect Lab Workers

- Over 50 years of pioneering work led to protocols and equipment to protect laboratory workers.
- To prevent laboratory-associated infections, scientists follow standard microbiology practices and wear personal protective equipment (**figure 36.2**).

36.3 Identification of Microorganisms from Specimens

- Detection of disease-causing organisms begins with the collection of patient specimens. The correct selection of body fluid or tissue, and subsequent collection, transport, and preanalytical processing of the specimen, are paramount for identifying infecting agents.
- The clinical microbiology laboratory can provide preliminary or definitive identification of microorganisms from sterile site and nonsterile site specimens. A thorough knowledge of the normal human microbiota is necessary to identify pathogens from host organisms (**tables 36.1** and **36.2; figure 36.3**).

- The type of suspected microorganisms dictates which tests are performed. Bacteria and fungi are often grown in culture to purify and amplify the pathogen for identification. Viruses and protozoa are typically identified by immunologic or molecular tests.
- Identification of pathogens is accomplished by (1) direct recovery and (2) indirect or circumstantial evidence of a pathogen. Direct recovery of infectious agents is by culture and biochemical analysis, as well as molecular techniques. The culturing of many types of bacteria and fungi (filamentous and yeast forms) is routine; however, a few require special consideration.
- Identification of bacteria, fungi, and protozoa often can be made by concentrating and staining the organisms. Gram stain for bacteria, Calcofluor for fungi, and Geimsa for protozoa permit detection using the light microscope.
- The initial identity of a bacterial organism may be suggested by (1) the source of the culture specimen; (2) its microscopic appearance; (3) its pattern of growth on selective, differential, enrichment, or characteristic media; and (4) its hemolytic, metabolic, and fermentative properties. A simple dichotomous key is used to facilitate the identification (**figure 36.4**).
- Immunofluorescence is a process in which fluorochromes are irradiated with ultraviolet, violet, or blue light to make them fluoresce. These dyes can be coupled to an antibody. There are two main kinds of fluorescent antibody assays: direct and indirect (**figure 36.7**).
- Various molecular methods are mostly used to identify microorganisms directly from specimens. Complementary methods like real-time PCR, nucleic acid fingerprinting, PNA-FISH, and MALDI-TOF are used on culture isolates to speed identification.

36.4 Immune Responses Can Be Measured or Exploited to Detect Infections

- Serotyping refers to serological procedures used to differentiate strains (serovars or serotypes) of microorganisms that have differences in the antigenic composition of a structure or product.
- In vitro agglutination reactions usually form aggregates or clumps (agglutinates) visible with the naked eye. Tests have been developed, such as the Rapid Plasma Reagin test, latex microsphere agglutination reaction, hemagglutination, and viral hemagglutination, to detect antigen as well as to determine antibody titer (**figures 36.9** and **36.10**).
- The complement fixation test can be used to detect a specific antibody for a suspect microorganism in an individual's serum (**figure 36.11**).
- The enzyme-linked immunosorbent assay (ELISA) involves linking various enzymes to either antigens or antibodies. Two basic methods are involved: the double antibody sandwich method and the indirect immunosorbent assay (**figure 36.12**). The first method detects antigens and the latter, antibodies.
- Immunoblotting involves polyacrylamide gel electrophoresis of a protein specimen followed by transfer of the separated proteins to nitrocellulose sheets and identification of specific bands by labeled antibodies (**figure 36.13**).
- Immunoprecipitation reactions occur only when there is an optimal ratio of antigen and antibody to produce a lattice at the zone of equivalence, which is evidenced by a visible precipitate (**figure 36.14**).
- Immunodiffusion refers to a precipitation reaction that occurs between antibody and antigen in an agar gel medium. Two techniques are routinely used: single radial diffusion and double diffusion in agar (**figure 36.15**).
- In immunoelectrophoresis, antigens are separated based on their electrical charge, then visualized by precipitation and staining (**figure 36.16**).
- Flow cytometry and fluorescence allow detection of specific cell types, such as $CD4^+$ T-cells in an HIV-infected patient, based on their cytometric parameters or by means of certain dyes called fluorochromes.

Compare, Hypothesize, Invent

1. As more ways of identifying the characteristics of microorganisms emerge, the number of distinguishable microbial strains also seems to increase. Why do you think this is the case?

2. Why are miniaturized identification systems used in clinical microbiology? Describe one such system and its advantage over classic dichotomous keys.

3. ELISA tests usually use a primary and secondary antibody. Why? What are the necessary controls one would need to perform to ensure that the antibody specificities are valid (i.e., no false-positive or false-negative reactions)?

4. *Legionella pneumophila* is a bacterium that is often found in water systems (e.g., shower heads, air-cooling towers). The dispersal of these bacteria can lead to the development of Legionnaires' disease, a particularly virulent pneumonia in the elderly and immunocompromised. Water systems are routinely monitored for *L. pneumophila* contamination using GVPC medium, which contains glycine and the antibiotics vancomycin, polymyxin B, and cyclohexamide. However, this method is known to underestimate the number of *L. pneumophila* cells in the environment. Flow cytometry (FC) is one approach to count cells.

   How would you go about determining if FC is a good way to monitor *L. pneumophila* in the environment? Specifically, how would you collect your samples and compare your results to those obtained by the currently accepted approach (i.e., plating on GVPC medium)? What controls would you need to perform? Based on this information, how do you think the results from FC would compare to those of GVPC cultures obtained from the same samples?

   Read the original paper: Allegra, S. 2008. Use of flow cytometry to monitor *Legionella* viability. *Appl. Environ. Microbiol.* 74:7813.

5. Rotaviruses are double-stranded RNA viruses that cause severe diarrhea in children, resulting in over 600,000 deaths annually, most in developing countries. There are seven serotypes of rotaviruses (A through G; A is the most common). Ideally the diagnosis of rotavirus infection includes the serotype. A PCR-ELISA approach may be the most efficient way to accomplish this. PCR-ELISA uses a 96-well microtiter plate format, thereby maximizing efficiency. PCR products are labeled, usually with a chemical known as digoxigenin, during amplification. A special oligonucleotide called a "capture probe" complementary to the PCR product is then added. The capture probe immobilizes the PCR product to the well and ELISA using antibody against digoxigenin is performed.

   Why do you think it is important to report the serotype of a rotavirus infection? How does PCR-ELISA differ from PCR? How does it differ from a regular ELISA? Why do you think PCR-ELISA might be better than the currently used typing PCR protocol? (Hint: Think about how PCR-ELISA could detect more than one PCR product.)

   Read the original paper: Santos, N., et al. 2008. Development of a microtiter plate hybridization-based PCR-enzyme-linked immunosorbent assay for identification of clinically relevant human group A rotavirus G and P genotypes. *J. Clin. Microbiol.* 46:462.

37

Epidemiology and Public Health Microbiology

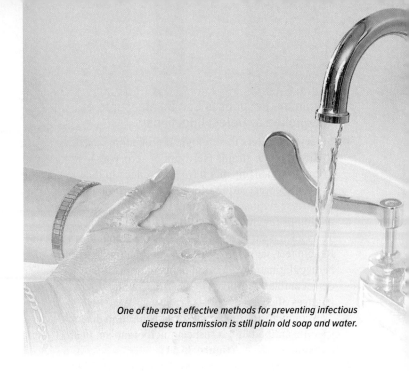

One of the most effective methods for preventing infectious disease transmission is still plain old soap and water.

Practice What You Preach

The years 2011 and 2014 marked a practical opportunity for instructors of microbiology laboratories to practice what we preach. Using DNA fingerprinting techniques, investigators from the Centers for Disease Control and Prevention (CDC) tracked two multistate outbreaks of *Salmonella* infection in a total of 150 people. The two groups were approximately 70% female, 30% male, with a median age of 21.5 years. The original illnesses were identified in employees of clinical microbiology laboratories and students in microbiology teaching laboratories who reported working with *Salmonella* bacteria. The outbreak strain associated with several of the ill persons was indistinguishable from a commercially available strain of *Salmonella enterica* serovar Typhimurium found in several clinical and teaching laboratories associated with ill employees and students. Sadly, several children known to reside in households of persons who work or study in a microbiology laboratory using this "lab strain" became seriously ill. Thus we teachers of microbiology were schooled in infectious disease tracking, case definitions, incidence reporting, laboratory testing, disease control, and disease prevention. Certainly, in addition to all the other lessons learned from this outbreak, proper hand washing upon leaving the laboratory should be a best practice that is never overlooked (ever).

In this chapter, we describe the practical goal of epidemiology: to establish effective disease recognition, control, prevention, and eradication measures for a given population. Because emerging and reemerging diseases and pathogens, health-care–acquired (nosocomial) infections, and bioterrorism are major concerns for public health, these topics are also covered here.

Readiness Check:
Based on what you have learned previously, you should be able to:

✔ Identify specific chemotherapeutic treatments to control infection (chapter 9)
✔ Describe innate human host resistance mechanisms (chapter 33)
✔ Discuss the human adaptive immune response to pathogens (chapter 34)
✔ Explain pathogenicity and the infectious disease process (chapter 35)
✔ Summarize the methods by which infectious disease is detected in humans (chapter 36)

37.1 Epidemiology Is an Evidence-Based Science

After reading this section, you should be able to:

■ Define the agencies that are responsible for disease prevention and control
■ Define the basic vocabulary and processes used in the science of epidemiology

By definition, **epidemiology** (Greek *epi,* upon; *demos,* people or population; and *logy,* study) is the science that evaluates the occurrence, determinants, distribution, and control of health and disease in a defined human population (**figure 37.1**). An individual who practices epidemiology is an **epidemiologist.** Epidemiologists are, in effect, disease detectives. Their major concerns are the discovery of the factors essential to disease occurrence and the development of methods for disease prevention. In the United States, the **Centers for Disease Control and Prevention,** headquartered in Atlanta, Georgia, serves as the national agency for developing and carrying out disease prevention and control, environmental health, and health promotion and education activities (**Historical Highlights 37.1**). Its worldwide counterpart is the **World Health Organization** (WHO), located in Geneva, Switzerland.

The science of epidemiology originated and evolved in response to great epidemic diseases such as cholera, typhoid fever, smallpox, influenza, and yellow fever (**Historical Highlights 37.2**). More recent epidemics of Ebola, HIV/AIDS, cryptosporidiosis, enteropathogenic *Escherichia coli,* SARS, salmonellosis, and novel influenza strains have underscored the importance of epidemiology in preventing global catastrophes caused by infectious diseases

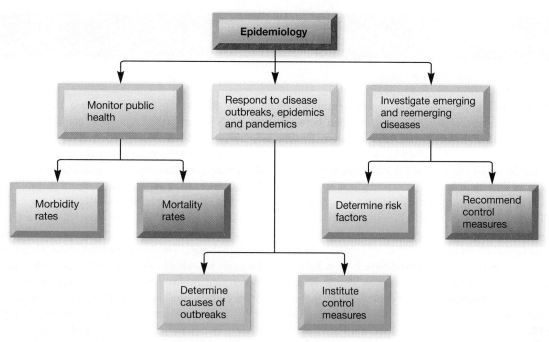

Figure 37.1 **Epidemiology.** Epidemiology is a multifaceted science that investigates diseases to discover their origin, evaluates diseases to assess their risk, and controls diseases to prevent future outbreaks.

HISTORICAL HIGHLIGHTS

37.1 The Birth of Public Health in the United States

The U.S. Public Health Service was born in 1798 out of the need to keep sailors from becoming sick while at sea. As defenders of the new republic, sailors, or seamen as they were called, could not find adequate health care at port cities of the day. Public hospitals were few in number and often did not have the resources to care for the large numbers of sick seamen that could arrive. As the seamen came from all over the country, their health care was determined to be a national responsibility. Thus the Marine Hospital Act was approved in 1798 and established a network of federal hospitals, called the Marine Hospital Service (MHS). The oversight responsibility was assigned to the Revenue Marine Division of the Treasury Department. The act permitted the taxation (20 cents per month) of the seamen, creating the first medical insurance program in the United States.

In 1878 the MHS was reorganized by congressional action from individually operated hospitals into a centrally controlled, national agency headquartered in Washington, D.C. The new federal agency became a separate bureau of the Treasury Department overseen by a central administrator, the "Supervising Surgeon," who was appointed by the Secretary of the Treasury. The administrative title of Supervising Surgeon was changed to Supervising Surgeon General in 1875.

In 1902 the MHS became the Public Health and Marine Hospital Services, and the administrator's title was changed to Surgeon General. A national "hygienic" laboratory was established in 1887 and was located on Staten Island, New York. The Pure Food and Drugs Act was established in 1906. In 1912 the Public Health and Marine Hospital Services was renamed the Public Health Service and authorized to investigate communicable disease.

In 1930 the Hygienic Lab moved to Washington to become the National Institute of Health, but it was not until 1939 that all health, education, and welfare agencies and services created by Congress were combined, under the Federal Security Agency. The Public Health Service opened the Communicable Disease Center in Atlanta, Georgia, in 1946. In 1953 the Federal Security Agency became the Department of Health, Education, and Welfare, and in 1970 the Communicable Disease Center became the Center for Disease Control. Ten years later, the Department of Health, Education, and Welfare became the Department of Health and Human Services, and the Center for Disease Control was renamed the Centers for Disease Control. It was renamed again in 1992 to become the Centers for Disease Control and Prevention (although it is still abbreviated as the CDC).

HISTORICAL HIGHLIGHTS

37.2 John Snow, the First Epidemiologist

Much of what we know today about the epidemiology of cholera is based on the classic studies conducted between 1849 and 1854 by the British physician John Snow. During this period, a series of cholera outbreaks occurred in London, England, and Snow set out to find the source of the disease. Some years earlier when he was still a medical apprentice, Snow was sent to help during a cholera outbreak among coal miners. His observations convinced him that the disease was usually spread by unwashed hands and shared food, not by "bad" air or casual direct contact.

Thus when the outbreak of 1849 occurred, Snow believed that cholera was spread in the same way as among the coal miners. He suspected that water, and not unwashed hands and shared food, was the source of the cholera infection. Snow examined official death records and discovered that most of the victims in the Broad Street area lived close to the Broad Street water pump or were in the habit of drinking from it. He concluded that cholera was spread by drinking water from the Broad Street pump, which was contaminated with raw sewage containing the disease agent. When the pump was disabled by removing its handle, the number of cholera cases dropped dramatically.

In 1854 another cholera outbreak struck London. Part of the city's water supply came from two different suppliers: the Southwark and Vauxhall Company and the Lambeth Company. Snow interviewed cholera patients and found that most of them purchased their drinking water from the Southwark and Vauxhall Company. He also discovered that this company obtained its water from the Thames River below locations at which Londoners discharged their sewage. In contrast, the Lambeth Company took its water from the Thames before the river reached the city. The death rate from cholera was over eight-fold lower in households supplied with Lambeth Company water. Water contaminated by sewage was transmitting the disease. Snow deduced that the cause of the disease must be able to multiply in water. Thus he nearly recognized that cholera was caused by a microorganism, though Robert Koch did not discover the causative bacterium (*Vibrio cholerae*) until 1883.

To commemorate these achievements, the John Snow Pub now stands at the site of the old Broad Street pump. Those who complete the Epidemiologic Intelligence Program at the Centers for Disease Control and Prevention receive an emblem bearing a replica of a barrel of Watney's Ale, the brew dispensed at the John Snow Pub.

(**Historical Highlights 37.3**). Today epidemiology's scope encompasses all public health issues: infectious diseases, genetic abnormalities, metabolic dysfunction, malnutrition, neoplasms, psychiatric disorders, obesity, and aging. This chapter discusses only infectious disease epidemiology.

Epidemiology studies the factors determining and influencing the frequency and distribution of health-related events. When a disease occurs occasionally and at irregular intervals in a human population, it is a sporadic disease (e.g., bacterial meningitis). When it maintains a steady, low-level frequency at a moderately regular interval, it is an **endemic** (Greek *endemos*, dwelling in the same people) **disease** (e.g., the common cold). Hyperendemic diseases gradually increase in frequency beyond the endemic level but not to the epidemic level (e.g., the common cold during winter months). **Incidence** reflects the number of new cases of a disease in a population at risk during a specified time period. An **outbreak** is the sudden, unexpected occurrence of a disease, usually in a limited segment of a population (e.g., the 2014 measles outbreak that occurred at Disneyland California). The **attack rate** is the proportional number of cases that develop in a population that was exposed to an infectious agent. An **epidemic** (Greek *epidemios*, upon the people), on the other hand, is an outbreak affecting many people at once (i.e., there is a sudden increase in the occurrence of a disease above the expected level).

Influenza is an example of a disease that may occur suddenly and unexpectedly in a family and often achieves epidemic status in a community. The influenza virus resides in a **reservoir host** (usually birds or pigs) until it is transmitted to a human host. The first case in an epidemic is called the **index case.** Finally, a **pandemic** (Greek *pan*, all) is an increase in disease occurrence within a large population over at least two countries around the world. Usually pandemic diseases spread among continents. The global H1N1 influenza outbreaks of 1918 and 2009 are good examples.

37.2 Epidemiology Is Rooted in Well-Tested Methods

After reading this section, you should be able to:

- Evaluate the effectiveness of public health surveillance methods
- Deduce the impact of public health methods on the quality of U.S. life over the last 100 years
- Describe the use of geographical information systems to track diseases remotely
- Use standardized surveillance data to measure infectious disease frequency
- Calculate rates of incidence, prevalence, morbidity, and mortality

HISTORICAL HIGHLIGHTS

37.3 A Modern Epidemic Exposed

Guinea, Sierra Leone, and Liberia intersect in the southern Forest Region, at Gueckedou Province. It was somewhere near the village of Meliandou (Gueckedou) in December 2013 that a two-year-old boy unknowingly became infected with one of the deadliest viruses known. Within a week he began to feel ill. His three-year-old sister, their mother, and their grandmother nursed him until he died. As was the tradition at West African funerals, mourners washed, embraced, and caressed the boy's body prior to burial. It wasn't long after the funeral, though, that they fell sick. The village midwife and a health-care worker from a nearby town provided the family with traditional medicines and care. But these did not help. Soon the midwife, health-care worker, and two mourners from the boy's funeral were ill. The symptoms were all the same: sudden fever, muscle pain, fatigue, reddened eyes, headache, and sore throat. These were followed by diarrhea, vomiting, a rash, and bleeding around the gums, nose, eyes, and rectum. The local doctors treated the sick for malaria, with no relief. Treatment did not stop the next wave of death, and four more died. Soon even the doctors succumbed to the disease; three died in the early days of the disease. As the numbers of dead escalated, government officials diagnosed the cause as Ebola virus disease. At the time of this writing the 2014 epidemic was still ongoing—although declining in its toll on life. As of August 18, 2015, there were 28,018 cases and 11,301 dead.

Thought to be sequestered in an undescribed species of fruit bat, Ebola virus is well known in Africa. The virus first erupted in 1976, in Yambuku, Zaire (now the Democratic of the Congo), but it was named for the Ebola River to protect from stigma the local village where many had died. Its gruesome hemorrhages instilled much fear. There were 33 more outbreaks of Ebola virus disease recorded over the next three decades, although curiously no cases were reported between 1979 and 1994. Usually fewer than 500 cases were reported with each outbreak.

The 2013 outbreak differed in both geographical distribution and incidence, spreading to at least ten countries (Guinea, Liberia, Sierra Leone, Nigeria, Senegal, Mali, Spain, Scotland, England, and the United States, Clearly, modern roadways and international travel facilitated the migration of the virus, as the infected went about their business until their symptoms stopped them. This was the story of Thomas Eric Duncan, who contracted the virus in Liberia before traveling to the United States. Mr. Duncan was diagnosed with Ebola virus disease in a Dallas, Texas, hospital on September 30, 2014. Although he did not display symptoms of the disease until several days after his arrival in Texas, by the fifth day he was febrile, hemorrhagic, and barely conscious. He died on October 8, 2014. His hospital room, an active BSL-4 environment, was filled with drums of contaminated medical waste. Somehow while caring for Mr. Duncan, two nurses were exposed and ultimately tested positive for Ebola virus.

This modern epidemic of Ebola virus disease is schooling the best of the public health theorists and practitioners. Unknown still is the exact reservoir and source of the virus. (The sale and consumption of bats and other "bushmeat," delicacies in southern Africa, are now banned by the health ministries, in hopes of preventing additional exposures.) Much of the natural history of Ebola virus disease has been revealed during this epidemic. Infection typically occurs through direct contact with blood, vomit, feces, urine, or other bodily fluids of a symptomatic Ebola patient. The virus gains entry to its new host through the broken skin, mouth, nose, or genitalia of the uninfected. This includes direct contact with virus-contaminated bedding, clothing, or environmental surfaces. Exposed but asymptomatic individuals are not infectious to others. An incubation period of 2 to 21 days, but more typically 8 to 10 days, is needed to generate a viral load that causes symptoms. The Ebola virus of 2015 is from the same lineage as the 1976 virus, but it has mutated and evolved to be less lethal. Today it kills about 40% of its victims as compared to 88% in 1976. Passive immunotherapy is effective in controlling infection as long as it is initiated right after symptoms appear. Thus, sera from Ebola survivors cure the newly infected.

Other important lessons have been learned:

- Most people and countries rally in support of disaster— but nongovernmental organizations, like Doctors without Borders, are able to respond more quickly and with less red tape, often supporting with substantial financial and human resources.
- Some people will always be driven by fear of infectious disease (one of the exposed, asymptomatic nurses traveled, with CDC permission, to Akron Ohio, where her parents were later feared and ridiculed).
- Local public health agencies must have policies and procedures to practice emergency response and quarantine for high-consequence diseases (new CDC guidance has been issued to facilitate protecting cities and states).
- U.S. hospitals must now be prepared for BSL-4, worst-case infections (not for "if" they happen, but for "when" they happen).
- Given the right motivation, new vaccines and drugs can be pushed through otherwise choking red tape.
- Ebola virus is quite sensitive to bleach and chlorine-based disinfectants.

Source: http://www.cdc.gov/vhf/ebola/

Public health is the science of protecting populations and improving the health of human communities through education, promotion of healthy lifestyles, and prevention of disease and injury. Public health practitioners use a methodical approach to identify population health issues to prevent or correct negative outcomes. The methodical approach mirrors the scientific method in that it identifies a population health problem, determines the cause, proposes preventative or corrective action, implements that action, and assesses the outcomes of implementing the action.

Public Health Surveillance

The identification of a population health issue is rooted in public health surveillance. Public health surveillance is the proactive evaluation of genetic background, environmental conditions, human behaviors and lifestyle choices, emerging infectious agents, and microbial responses to chemotherapeutic agents, for example, to monitor the health of a population. In other words, public health practitioners look for cause-and-effect relationships to determine risk. The impact of public health surveillance is obvious when one considers the change in the leading causes of death in the United States from 1900 to 2013 (**table 37.1**). The public health landscape of 1900 was a valley of death due to infectious disease. Surveillance identified infectious disease problems and their associated risks. These were remedied by water treatment, strict sanitation guidelines, and later by the use of antimicrobial agents and vaccines. The result changed the death-by-infection landscape of the early twentieth century to one of very few deaths due to infectious disease by the 1960s. Ironically, the modernization that helped prevent the leading causes of mortality in the 1900s also led to a longer life expectancy and a more sedentary lifestyle in the 2000s. This changed the landscape of 2013 to one of a mountain of metabolic disease.

Public health surveillance really began in the fourteenth century as a means to control bubonic plague. However, it was not until after the germ theory of disease was widely accepted (late 1800s) that a scientifically based monitoring process was used as a means to track communicable disease. This was not the confidential monitoring that we expect today. The first disease tracking practices were intrusive, as people were inspected for signs of clinical disease; those found to be infected were quarantined. The change from inspecting people to monitoring disease occurred around 1940. The emphasis on watching disease progress in a population and disseminating information in advance of the disease radically improved the impact on public health by changing the role of public health practitioners from disease informants to health-care allies (**figure 37.2**). Importantly, disease data come from a number of sources, including mandated clinical reports, laboratory specimens, records and databases, sentinel surveillance, disease registries, and field surveys. These data are used

Table 37.1	Percentage of Total U.S. Citizen Deaths for the 10 Leading Causes of Death in 1900[1] and 2013[2]	
Cause	**Percentage**	
	1900	**2013***
Heart diseases	8.0	23.5
Cancer	3.7	22.5
Pneumonia and influenza	11.8	2.2
Tuberculosis	11.3	ND
Gastrointestinal diseases	8.3	ND
Cerebrovascular diseases	6.2	5.0
Senility (Alzheimer's disease and other dementias)	6.8	3.3
Pulmonary diseases	ND	5.7
Kidney diseases	4.7	1.8
Accidents	4.2	5.0
Diabetes mellitus	ND	2.9
Diphtheria	2.3	ND
Suicide	ND	1.6
All other causes	32.7	26.5
Total	100.0	100.0

1 *Source: National Office of Vital Statistics, 1954.*
2 *Source: Murphy, S. L.; Xu, J.; and Kochanek, K. D. 2012.* Deaths: Preliminary Data for 2010. National Vital Statistics Report; *vol. 60, no. 4. Hyattsville, Md.: National Center for Health Statistics.*
ND: Not determined.
*Latest available data

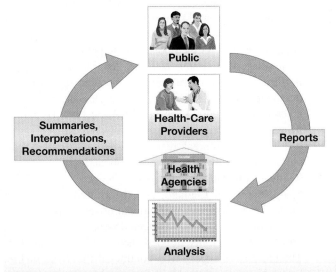

Figure 37.2 Information Loop of Public Health Surveillance. The Centers for Disease Control and Prevention uses a number of different data sources to maintain surveillance of the infectious diseases that may erupt into localized epidemics or more widespread pandemics.

MICRO INQUIRY *How might biased data collection impact the public health recommendation to the public? Explain.*

most often for the generation of reports informing public health officials of (1) population morbidity and mortality, (2) disease effect on school attendance, (3) impact of epidemics on employee absenteeism, and (4) animal and vector control policies.

In 1878 Congress authorized the U.S. Marine Hospital Service (the forerunner of the Public Health Service; Historical Highlights 37.1) to collect illness data associated with cholera, smallpox, plague, and yellow fever. These data were used to quarantine the sick and prevent the spread of these diseases into the United States. The following year, a congressional appropriation directed the collection and publication of these "notifiable diseases." Congress subsequently directed the Surgeon General to compile and publish the data. The CDC assumed this responsibility in 1961. As of 2014, there are 64 infectious diseases on the list of notifiable diseases; it is updated periodically to reflect emerging infectious diseases (e.g., measles, SARS-CoV, H5N1 influenza A, etc.). Importantly, effective reporting of notifiable (and other) diseases is contingent on reliable surveillance data collection systems.

Effective use of public health surveillance data stems from high-integrity data collection (i.e., documented first-hand information of the data origin and the methods used to collect them). Surveillance of health issues is typically accomplished by two methods: population surveys and case reporting. Population surveys tend to be more of a sentinel surveillance method in that they look for trends and outliers within a given population (e.g., immunization records revealing infection control success). Case reports tend to be retrospective comparisons of specific conditions that assign people to risk groups (e.g., influenza test results to rule out SARS).

Remote Sensing and Geographic Information Systems: Charting Infectious Disease Data

Remote sensing and geographic information systems are map-based tools that can be used to study the distribution, dynamics, and environmental correlates of microbial diseases. **Remote sensing (RS)** gathers digital images of Earth's surface from satellites and output from biological sensors, for example, and transforms the data into maps. A **geographic information system (GIS)** is a data management system that organizes and displays digital map data from RS and facilitates the analysis of relationships between mapped features. Statistical relationships often exist between mapped features and diseases in natural host or human populations. Examples include the location of the habitats of the malaria parasite and mosquito vectors in Mexico and Asia, tick that transmit Lyme disease in the United States, and African trypanosome in both humans and livestock in the southeastern United States. RS and GIS may also permit the assessment of human risk from pathogens such as Sin Nombre virus (the virus that causes hantavirus pulmonary syndrome in North America). RS and GIS are most useful if disease dynamics and distributions are clearly related to mapped environmental variables. For example, if a microbial disease is associated with certain vegetation types or physical characteristics (e.g., elevation, precipitation), RS and GIS can identify regions in which risk is relatively high.

Measuring Infectious Disease Frequency

To determine if an outbreak, epidemic, or pandemic is occurring, epidemiologists measure disease frequency at single time points and over time. They then use statistics to analyze the data and determine risk factors and other factors associated with disease.

To determine the frequency or rate of an event, an accurate count of the total population, the population exposed, and the number of affected people needs to be made. Disease surveillance practices enable the tracking of infections. Many infectious diseases (e.g., food- and waterborne diseases) must, by law, be reported within a specific time frame. This permits public health agencies to act swiftly to contain the outbreak and mobilize control measures. Timely notification of "reportable" diseases to public health agencies has stemmed numerous epidemics such as cholera and typhoid.

Measures of frequency usually are expressed as fractions. The numerator is the number of individuals experiencing the event—infection—and the denominator is the number of individuals in whom the event could have occurred, that is, the population at risk. The fraction is a proportion or ratio but is commonly called a rate because a time period is specified. (A rate also can be expressed as a percentage.) In population statistics, rates usually are stated per 1,000 individuals, although other powers of 10 may be used for particular diseases (e.g., per 100 for very common diseases and per 10,000 or 100,000 for uncommon diseases). For example, disease incidence is a measure of the number of diseased people during a defined time period, as compared to the total (healthy) population. The incidence of a disease reports not only the rate of occurrence but the relative risk as well. Since disease reflects a change in health status over time, a **morbidity rate** is used. The rate is commonly determined when the number of new cases of illness in the general population is known from clinical reports. It is calculated as follows:

$$\text{Morbidity rate} = \frac{\substack{\text{Number of new cases of a disease} \\ \text{during a specified period}}}{\text{Number of individuals in the population}}$$

For example, if in 1 month, 700 new cases of influenza per 100,000 individuals occur, the morbidity rate would be expressed as 700 per 100,000, or 0.7%.

The **prevalence rate** refers to the total number of individuals infected in a population at any one time, no matter when the disease began. The prevalence rate depends on both the incidence rate and the duration of the illness. Prevalence is calculated as follows:

$$\text{Prevalence} = \frac{\text{Total number of cases in population}}{\text{Total population}} \times 100$$

The **mortality rate** is the relationship between the number of deaths from a given disease and the total number of cases of the disease. The mortality rate is a simple statement of the proportion of all deaths that are assigned to a single cause. It is calculated as follows:

$$\text{Mortality rate} = \frac{\text{Number of deaths due to a given disease}}{\substack{\text{Size of the total population} \\ \text{with the same disease}}}$$

For example, if 15,000 deaths occurred due to AIDS in a year and the total number of people infected was 30,000, the mortality rate would be 15,000 per 30,000, or 1 per 2, or 50%.

The determination of morbidity, prevalence, and mortality rates helps public health personnel direct health-care efforts to control the spread of infectious diseases. For example, a sudden increase in the morbidity rate of a particular disease may indicate a need for implementation of preventive measures designed to reduce the disease.

Retrieve, Infer, Apply

1. What is epidemiology?
2. What terms are used to describe the occurrence of a disease in a human population?
3. What types of surveillance data are most useful in determining infectious disease penetration into a population?
4. How might remote sensing be used to track Rocky Mountain spotted fever (*see p. 871*)?
5. Define morbidity rate, prevalence rate, and mortality rate.

37.3 Infectious Disease Is Revealed Through Patterns Within a Population

After reading this section, you should be able to:

- Discriminate between a communicable and noncommunicable disease
- Interpret population infection data to define epidemic and pandemic events
- Apply the concept of herd immunity to public health in your community
- Predict potential infectious disease outbreaks from molecular data revealing antigenic drift and antigenic shift events

An **infectious disease** is a disease resulting from an infection by microbial agents such as viruses, bacteria, fungi, protozoa, and helminths. A **communicable disease** is an infectious disease that can be transmitted from person to person (not all infectious diseases are communicable; e.g., rabies is an infectious disease acquired only through contact with a rabid animal). An epidemiologist studying an infectious disease is concerned with the causative agent, the source or reservoir of the disease agent, how it is transmitted, what host and environmental factors can aid development of the disease within a defined population, and how best to control or eliminate the disease. These factors describe the natural history or cycle of an infectious disease. ◄◄ *Pathogenicity drives infectious disease (section 35.1)*

Two major types of epidemics are recognized: common source (noncommunicable) and propagated (communicable). A **common-source epidemic** is characterized as having reached a peak level within a short period of time (1 to 2 weeks), followed by a moderately rapid decline in the number of infected patients (**figure 37.3a**). This type of epidemic usually results from a single, common

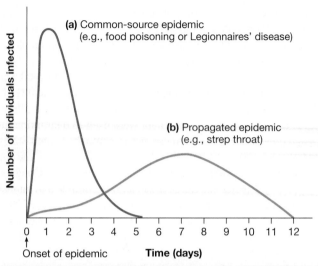

Figure 37.3 Epidemic Curves. (a) In a common-source epidemic, there is a rapid increase up to a peak in the number of individuals infected and then a rapid but more gradual decline. Cases usually are reported for a period that equals approximately one incubation period of the disease. (b) In a propagated epidemic, the curve has a gradual rise and then a gradual decline. Cases usually are reported over a time interval equivalent to several incubation periods of the disease.

MICRO INQUIRY *Is a flu epidemic an example of a common-source or a propagated epidemic? Explain your answer.*

contaminated source such as food (food poisoning) or water (cholera). By contrast, a **propagated epidemic** is characterized by a relatively slow and prolonged rise, and then a gradual decline in the number of individuals infected (figure 37.3b). This type of epidemic usually results from the introduction of a single infected individual into a susceptible population. The initial infection is then propagated to others in a gradual fashion until many individuals within the population are infected. An example is the increase in influenza cases that coincides with the return of college students after their winter break. Droplet and contact transmission of flu virus spreads rapidly among susceptible, and unvaccinated students. Whereas one infected student is sufficient to initiate the epidemic, it is the person-to-person transmission of the flu virus that extends the epidemic. After an infectious disease has been recognized in a population, epidemiologists correlate the disease outbreak with a specific organism; its exact cause must be discovered (**Historical Highlights 37.4**). At this point, the clinical or diagnostic microbiology laboratory enters the investigation. Its purpose is to isolate and identify the organism responsible for the disease. ◄◄ *Clinical microbiology and immunology (chapter 36)*

As mentioned, epidemiologists recognize an infectious disease in a population by using various surveillance methods. Surveillance is a dynamic activity that includes gathering information on the development and occurrence of a disease, collating and analyzing the data, summarizing the findings, and using the information to select control methods (**figure 37.4**). Surveillance may not always require the direct examination of cases. However, to accurately interpret

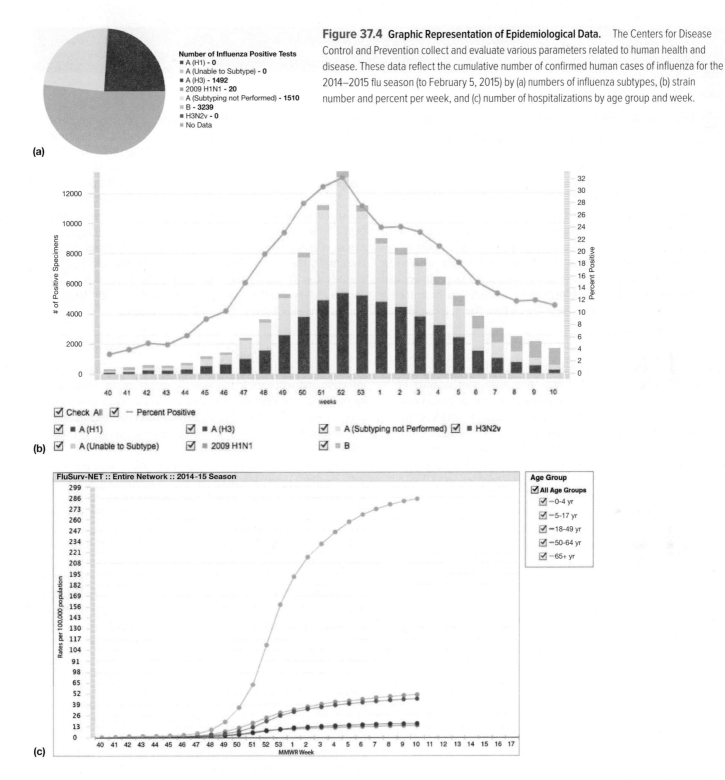

Figure 37.4 Graphic Representation of Epidemiological Data. The Centers for Disease Control and Prevention collect and evaluate various parameters related to human health and disease. These data reflect the cumulative number of confirmed human cases of influenza for the 2014–2015 flu season (to February 5, 2015) by (a) numbers of influenza subtypes, (b) strain number and percent per week, and (c) number of hospitalizations by age group and week.

Number of Influenza Positive Tests
■ A (H1) - 0
■ A (Unable to Subtype) - 0
■ A (H3) - 1492
■ 2009 H1N1 - 20
■ A (Subtyping not Performed) - 1510
■ B - 3239
■ H3N2v - 0
■ No Data

surveillance data and study the course of a disease in individuals, epidemiologists and other medical professionals must be aware of the pattern of infectious diseases that reflects how the disease spreads through a population or at least to one person (*see figure 35.3*). Epidemiologists study disease patterns so they can rapidly identify new outbreaks before substantial morbidity and mortality occur. ◄◄ *Pathogenicity drives infectious disease (section 35.1)*

To understand how epidemics are propagated, consider **figure 37.5**. At time 0, all individuals in this population are susceptible to a hypothetical pathogen. The introduction of an infected individual initiates the epidemic outbreak (lower curve), which spreads to reach a peak by day 15. As individuals recover from the disease, they become immune and no longer transmit the pathogen (upper curve). The number of susceptible individuals

HISTORICAL HIGHLIGHTS

37.4 "Typhoid Mary"

In the early 1900s there were thousands of typhoid fever cases, and many people died of the disease. Most of these cases arose when people drank water contaminated with sewage or ate food handled by or prepared by individuals who were shedding the typhoid fever bacterium (*Salmonella enterica* serovar Typhi). The most famous carrier of the typhoid bacterium was Mary Mallon.

Between 1896 and 1906, Mallon worked as a cook in seven homes in New York City. Twenty-eight cases of typhoid fever occurred in these homes while she worked in them. As a result, the New York City Health Department had Mallon arrested and admitted to an isolation hospital on North Brother Island in New York's East River. Examination of her stools showed that she was shedding large numbers of typhoid bacteria, though she exhibited no external symptoms of the disease. An article published in 1908 in the *Journal of the American Medical Association* referred to Mallon as "Typhoid Mary," an epithet by which she is still known today. She was released when she pledged not to cook for others or serve food to them. Mallon changed her name and began to work as a cook again. For 5 years she managed to avoid capture while continuing to spread typhoid fever. Eventually the authorities tracked her down. She was held in custody for 23 years, until she died in 1938. As a lifetime carrier, Mary Mallon was positively linked with 10 outbreaks of typhoid fever, 53 cases, and 3 deaths.

therefore decreases. The decline in the number of susceptible individuals to the threshold density (the minimum number of individuals necessary to continue propagating the disease) coincides with the peak of the epidemic wave, and the incidence of new cases declines because the pathogen cannot propagate itself.

Herd immunity is the resistance of a population to infection and pathogen spread because of the immunity of a large percentage of the population. The larger the proportion of those immune, the smaller the probability of effective contact between infective and susceptible individuals; that is, many contacts will be immune, and thus the population will exhibit a group resistance. A susceptible member of such an immune population enjoys an immunity that is not of his or her own making but instead arises because of membership in the group.

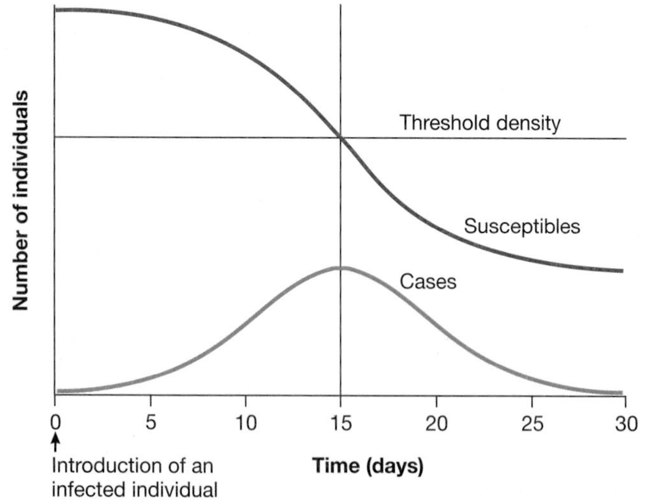

At times public health officials immunize large portions of a susceptible population in an attempt to maintain a high level of herd immunity. Any increase in the number of susceptible individuals may result in an endemic disease becoming epidemic. The proportion of immune to susceptible individuals must be constantly monitored because new susceptible individuals continually enter a population through migration and birth.

Pathogens can cause endemic diseases because infected humans continually transfer them to others (e.g., sexually transmitted diseases) or because the pathogens continually reenter the human population from animal reservoirs (e.g., West Nile virus). Other pathogens continue to evolve and may produce epidemics (e.g., AIDS, influenza virus [A strain], and *Legionella* bacteria). As described in chapter 38, one way in which a pathogen changes is by **antigenic shift,** a major genetically determined change in the antigenic character of a pathogen *(see p. 830 and figure 38.3)*. Genetic changes introduced by an antigenic shift can be so extensive that the pathogen is no longer recognized by the host's immune system. For example, influenza viruses frequently change by recombination from one antigenic type to another. ◀◀ *Minus-strand RNA viruses (section 27.6)* ▶▶| *Influenza (flu) (section 38.1)*

Whenever antigenic shift or drift *(see p. 830)* occurs, the population of susceptible individuals increases because the immune system does not recognize the new or altered strains, respectively. Thus the need for annual flu vaccination. If the percentage of susceptible people is above the threshold density (figure 37.5), the level of protection provided by herd immunity will decrease and the morbidity rate will increase. For example, the morbidity rates

Figure 37.5 The Spread of an Imaginary Propagated Epidemic. The lower curve represents the number of cases, and the upper curve, the number of susceptible individuals. Notice the coincidence of the peak of the epidemic wave with the threshold density of susceptible people.

MICRO INQUIRY *Why does the number of susceptible individuals decline after the epidemic has reached its peak?*

of influenza among schoolchildren may reach epidemic levels if the number of susceptible individuals rises above 30% for the whole population. One of the primary goals of public health is to ensure that the general public is sufficiently protected to withstand epidemics. This has resulted in the recommendation that at least 70% of the population be immunized against common infectious diseases, providing the herd immunity necessary for the protection of those who are not immunized. In other words, by keeping the population sufficiently immunized, the chain of infectious disease transmission is broken, limiting the reach of the disease (*see figure 35.1*). This is especially true for infectious diseases that quickly arise by antigenic shift, the 2014–2015 Influenza season's shift to an H3N2 viral strain being a great example.

Retrieve, Infer, Apply

1. How do epidemiologists recognize an infectious disease in a population?
2. Differentiate between common-source and propagated epidemics.
3. Explain herd immunity. How does this protect the community?
4. What is the significance of antigenic shift and drift in epidemiology?

37.4 Infectious Diseases and Pathogens Are Emerging and Reemerging

After reading this section, you should be able to:

- Report on recently emerged and reemerging global infectious diseases
- Organize geographic data to reveal changes in microbial drug resistance and vector spread of infection

Only a few decades ago, a grateful public trusted that science had triumphed over infectious diseases by building a fortress of health protection. Antibiotics, vaccines, and aggressive public health campaigns had yielded a string of victories over old enemies such as whooping cough, pneumonia, polio, and smallpox. In developed countries, people were lulled into believing that microbial threats were a thing of the past. Trends in the number of deaths caused by infectious diseases in the United States from 1900 through 2013 supported this conclusion (table 37.1). Sadly, despite this overall downward trend, the incidence of infectious disease resulting from emerging and reemerging microbial pathogens has claimed the lives of millions of people. The world has seen the global spread of AIDS, the resurgence of tuberculosis, the H1N1 influenza virus, and the appearance of new enemies such as Sin Nombre virus (also known as hantavirus pulmonary syndrome virus), hepatitis C and E viruses, pandemic emergence of SARS and MERS, pandemic Ebola virus, the Lyme disease spirochete, *Cryptosporidium* spp., and several strains of shiga toxin–producing *Escherichia coli*, not to mention the rise of extended-spectrum drug-resistant bacteria (**table 37.2**). In addition, during this same time period:

1. A "bird flu" virus that had never before attacked humans began to kill people in southeast Asia. ▶▶❙ *Influenza (flu) (section 38.1)*
2. A new variant of a fatal prion disease of the brain, Creutzfeldt-Jakob disease, was identified in the United Kingdom,

Table 37.2	National Institute of Allergy and Infectious Diseases List of Emerging and Reemerging Infectious Diseases
Pathogens Newly Recognized in the Past Two Decades	**Reemerging Pathogens**
BACTERIUM	BACTERIUM
Bartonella henselae	*Clostridium difficile*
Ehrlichia spp.	*Streptococcus pyogenes*
Heliobacter pylori	*Staphylococcus aureus*
Borrelia burgdorferi	
FUNGUS	
Encephalitozoon spp.	
Cryptococcus gattii strains	
PROTIST	
Acanthamoeba spp.	
Babesia spp.	
VIRUS	VIRUS
Australian bat lyssavirus	Ebola virus
Hendra virus	Measles virus
Hepatitis C virus	Mumps virus
Hepatitis E virus	Enterovirus A71
Human herpesvirus 8	Enterovirus D68
Human herpesvirus 6	Polio virus
Parvovirus B19	Dengue virus
MERS-CoV	
Avian influenza A (H7N9)	
Bourbon virus	
Chikungunya virus	

transmitted by beef from animals with "mad cow disease." ▶▶❙ *Prion proteins transmit diseases (section 38.6)*

3. *Staphylococcus* spp. with resistance to methicillin and vancomycin, long the antibiotics of first choice and last resort, respectively, were seen for the first time. ❙◀◀ *Evolution in action (section 16.9)*

4. Several major multistate food-borne outbreaks occurred in the United States and Germany, including those caused by *Salmonella* spp. in peanut butter and sprouts, protists on raspberries, viruses on strawberries, and various bacteria in produce, ground beef, cold cuts, and breakfast cereal. ▶▶❙ *Food-borne disease outbreaks (section 41.3)*

5. A new strain of the tuberculosis bacterium that is resistant to many drugs and occurs most often in people infected with HIV arose in New York and other large cities. ▶▶❙ *Mycobacterium tuberculosis (section 39.1)*

6. The resurgence of vaccine-preventable, childhood diseases, most notably whooping cough, measles, and mumps, erupted in several cities and towns across the United States and Europe.

By the 1990s the idea that infectious diseases no longer posed a serious threat to human health was obsolete. In the twenty-first century, it is clear that humans will continually be faced with both new infectious diseases and the reemergence of older diseases once thought to be conquered. The CDC has defined these diseases as "new, reemerging, or drug-resistant infections whose incidence in humans has increased within the past three decades or whose incidence threatens to increase in the near future." Of note is the fact that while most emerging infectious diseases arise from the "Old World tropics," with today's capacity to travel anywhere in the world within 24 hours, infectious diseases can erupt just about anywhere in the world. In addition, newer data suggest that emergence and reemergence of infectious disease stem from zoonotic sources. Areas that are dense with livestock are highly correlated with disease eruption that then disseminates globally. The increased importance of emerging and reemerging infectious diseases has stimulated the establishment of a field called **systematic epidemiology,** which focuses on the ecological and social factors that influence the development of these diseases. Coupled with systematic epidemiology and remote sensing techniques, the growing field of public health informatics integrates surveillance activities, defines data elements, and disseminates information via the Internet so as to better track emerging and reemerging infectious diseases in real time.

Why are pathogens posing such a problem, despite dramatic advances in medical research, drug discovery, technology development, and sanitation? Many factors characteristic of the modern world undoubtedly favor the development and spread of these microorganisms and their diseases.

As population density increases in cities, the dynamics of microbial exposure and evolution increase in humans. Urbanization often crowds humans and increases exposure to microorganisms. Crowding leads to unsanitary conditions and hinders the effective implementation of adequate medical care, enabling more widespread transmission and propagation of pathogens. In modern societies, crowded workplaces, communal-living settings, day-care centers, large hospitals, and public transportation all facilitate microbial transmission. In this millennium, the speed and volume of international travel are major factors contributing to the global emergence of infectious diseases. The spread of a new disease often used to be limited by the travel time needed to reach a new host population. If the travel time was sufficiently long, as when a ship crossed the ocean, the infected travelers would either recover or die before reaching a new population. Because travel by air has essentially eliminated time between exposure and disease outbreak, a traveler can spread virtually any infectious disease in a matter of hours. The H1N1 influenza pandemic of 2009 is one example of how one virus-infected individual (in Mexico) can lead to a global pandemic as asymptomatic carriers transport the virus to others during normal activities (*see p. 831*).

When changes in climate or ecology occur, it should not be surprising to find changes in both beneficial and detrimental microorganisms. Global warming also affects microorganism survival. Mass migrations of refugees, workers, and displaced persons have led to a steady growth of urban centers at the expense of rural areas.

Furthermore, land development and the exploration and destruction of natural habitats have increased the likelihood of human exposure to new pathogens and may select for pathogens better able to adapt to new hosts and changing environments. The introduction of pathogens to a new environment or host can alter transmission and exposure patterns, leading to sudden proliferation of disease. For example, the spread of Ebola virus disease in 2013–2015 across West Africa clearly resulted when a suspected bat pathogen jumped into the encroaching human population. Migration of infected persons along established roadways facilitated viral transmission. Here, a virus with a week-long incubation period quickly exploited cultural practices to reemerge as an insidious and deadly foe (Historical Highlights 37.3). ◀◀ *Global climate change (section 28.2)*

Emerging and reemerging pathogens and their diseases are therefore the outcome of many different factors. Because the world is now so interconnected, we cannot isolate ourselves from other countries and continents. Changes in the disease status of one part of the world or the misuse of antibiotics in another part may well affect health around the planet. This was clearly understood after the arrival of an Ebola-exposed man in Dallas, Texas, in October 2014 ultimately resulted in the infection of two nurses with Ebola and a country on heightened alert anticipating its own epidemic. As Nobel laureate Joshua Lederberg so eloquently stated, "The microbe that felled one child in a distant continent yesterday can reach yours today and seed a global pandemic tomorrow."

Retrieve, Infer, Apply

1. What factors influence the definitions of emerging or reemerging infectious diseases?
2. What are some of the factors in your community that may lead to the emergence or reemergence of pathogens?
3. Describe how global climate change can lead to new human infectious diseases.

37.5 Health-Care Facilities Harbor Infectious Agents

After reading this section, you should be able to:

- Report the major causes of health-care–associated infection in the United States
- Distinguish community-acquired pathogens from health-care–associated pathogens, and explain why community-acquired pathogens are of concern to hospital personnel
- Recommend measures for preventing and controlling health-care–associated infections

Health-care–associated, sometimes called **health-care–acquired** or **nosocomial infections** (Greek *nosos*, disease, and *komeion*, to take care of), result from infections acquired by patients while in a hospital or other clinical care facility. Besides harming patients, health-care–associated infections (HAI) can affect nurses, physicians, aides, visitors, salespeople, delivery personnel, custodians, and anyone else who has contact with the hospital. Most HAI become clinically apparent while patients are still hospitalized; however, disease onset can occur after patients have been discharged. Infections that are incubating when patients are admitted to a hospital are not health-care associated; they are community acquired. The CDC estimates that about 5% of all hospital patients (up to 2.5 million people) acquire some type of HAI. Thus health-care–associated infections represent a significant proportion of all infectious diseases acquired by humans, averaging approximately 1.87 million infections per year.

Health-care–associated diseases are usually caused by bacteria, most of which are noninvasive and part of the normal human microbiota. **Figure 37.6** summarizes the most common types of health-care–associated infections and pathogens. Interestingly, throughout most of the twentieth century, HAIs were dominated by penicillin-sensitive staphylococci. During subsequent years, the emergence of methicillin- and vancomycin-resistant *Staphylococcus aureus* (MRSA and VRSA, respectively) increased dramatically. First reported in the late 1980s, vancomycin-resistant enterococci (VRE) are now common in U.S. hospitals. Similar patterns are emerging for penicillin-resistant *Streptococcus pneumoniae.*

Today the most common HAIs are (1) catheter-associated urinary tract infections, (2) surgical site infections, (3) central line (catheters)-associated bloodstream infections, and (4) ventilator-associated pneumonias (**figure 37.7**). Recently recognized health-care–associated, Gram-positive species include *Clostridium sordellii, Corynebacterium jeikeium* and *Rhodococcus equi.* The incidence of infections by the Gram-negative pathogens *Pseudomonas aeruginosa, Acinetobacter* spp., *Burkholderia cepacia,* and *Stenotrophomonas maltophilia* have also increased. Many Gram-negative bacilli are resistant to β-lactam antibiotics. These bacteria include *Klebsiella pneumoniae, E. coli,* other *Klebsiella* spp., *Proteus* spp., *Morganella* spp., *Citrobacter* spp., *Salmonella* spp., and *Serratia marcescens,* which are resistant to penicillins; many older cephalosporins; and even some newer cephalosporins such as cefotaxime, ceftriaxone, ceftazidime, and aztreonam. The β-lactamase-resistant members of **Enterobacteriaceae** are collectively referred to as extended spectrum β-lactamase (producing) *Enterobacteriaceae.* ◀◀ *Evolution in action (section 16.9)*

Many potential exogenous sources of infection exist in a hospital. Animate sources are the hospital staff, other patients, and visitors. Some examples of inanimate exogenous sources are food, plants and flowers, computer keyboards, intravenous and respiratory therapy equipment, and water systems (e.g., softeners, dialysis units, and hydrotherapy equipment).

In the United States, HAIs prolong hospital stays by 4 to 14 days, result in an additional $28 billion to $33 billion per year to overall direct health-care costs, and lead to approximately 99,000 deaths annually. The enormity of this problem has led most hospitals to allocate substantial resources to developing methods and programs for the surveillance, prevention, and control of health-care–associated infections.

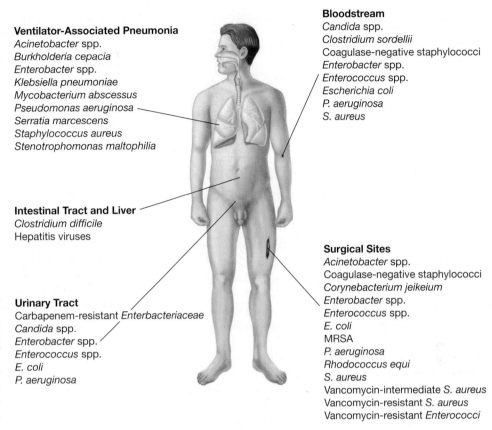

Ventilator-Associated Pneumonia
Acinetobacter spp.
Burkholderia cepacia
Enterobacter spp.
Klebsiella pneumoniae
Mycobacterium abscessus
Pseudomonas aeruginosa
Serratia marcescens
Staphylococcus aureus
Stenotrophomonas maltophilia

Intestinal Tract and Liver
Clostridium difficile
Hepatitis viruses

Urinary Tract
Carbapenem-resistant *Enterbacteriaceae*
Candida spp.
Enterobacter spp.
Enterococcus spp.
E. coli
P. aeruginosa

Bloodstream
Candida spp.
Clostridium sordellii
Coagulase-negative staphylococci
Enterobacter spp.
Enterococcus spp.
Escherichia coli
P. aeruginosa
S. aureus

Surgical Sites
Acinetobacter spp.
Coagulase-negative staphylococci
Corynebacterium jeikeium
Enterobacter spp.
Enterococcus spp.
E. coli
MRSA
P. aeruginosa
Rhodococcus equi
S. aureus
Vancomycin-intermediate *S. aureus*
Vancomycin-resistant *S. aureus*
Vancomycin-resistant *Enterococci*

Figure 37.6 Health-Care–Associated Infections (HAIs). The most common HAIs are caused by bacteria, fungi, and viruses that are transferred to a patient from a health-care worker, contaminated instruments, plants and flowers, fresh foods, and so on.

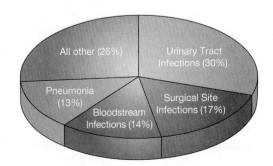

Figure 37.7 Leading Types of Health-Care–Associated Infections in the United States. These data reflect national statistics as of March 2015.

37.6 Coordinated Efforts Are Required to Prevent and Control Epidemics

After reading this section, you should be able to:

- Assemble communicable disease linkages to create a chain of infection
- Reveal the weakest link in the chain of infection
- Advise vaccine use to prevent infectious disease transmission in a population
- Explain the role of disinfection, sanitation, and chemotherapy to control infectious disease transmission in a population

The control of an infectious disease relies heavily on a well-defined network of clinical microbiologists, nurses, physicians, epidemiologists, and infection control personnel who supply epidemiological information to a network of local, state, national, and international organizations. These individuals and organizations comprise the public health system. For example, each state has a public health laboratory that is involved in infection surveillance and control. The communicable disease section of a state laboratory includes specialized laboratory services for the examination of specimens or cultures submitted by physicians, local health departments, hospital laboratories, sanitarians, epidemiologists, and others. These groups share their findings with other health agencies in the state, the Centers for Disease Control and Prevention, and the World Health Organization.

Control Starts with Breaking the Chain of Infection

Just as the development of an infectious disease is a complex process involving many factors, so too is the design of specific epidemiological control measures. Epidemiologists must consider available resources and time constraints, adverse effects of potential control measures, and human activities that might influence the spread of the infection. Often control activities reflect compromises among alternatives. To proceed intelligently, one must identify components of the infectious disease chain that are primarily responsible for a particular epidemic. Control measures should be directed toward that part of the chain that is most susceptible to control—the weakest link in the chain (*see figure 35.1*).

There are three types of control measures. The first type is directed toward reducing or eliminating the source or reservoir of infection: social distancing and isolation of carriers, destruction of an animal reservoir of infection, treatment of water and sewage to reduce contamination (*see chapter 43*), and therapy that reduces or eliminates infectivity of the individual.

The second type of control measure is designed to break the connection between the source of the infection and susceptible individuals. Examples include general sanitation measures: chlorination of water supplies, pasteurization of milk and other beverages, supervision and inspection of food and food handlers, and destruction of vectors (e.g., spraying to eliminate mosquitoes).

The third type of control measure reduces the number of susceptible individuals and raises the general level of herd immunity by immunization. Examples include passive immunization (*see figure 34.3*) to give temporary immunity following exposure to a pathogen or when a disease threatens to become an epidemic, active immunization to protect the host population, and prophylactic treatment to prevent infection (e.g., use of chloroquine when traveling to malaria-endemic countries). Antimicrobial chemotherapy is discussed in chapter 9. We discuss the use of vaccines and the subsequent immunization in the next section.

Vaccines Immunize Susceptible Populations

A **vaccine** (Latin *vacca,* cow) is a preparation of one or more microbial antigens that induces protective immunity in the host. **Immunization** occurs when a vaccine has been successfully delivered. The specific goal of vaccination is to induce antibodies and activated T cells to protect a host from future infection. Many epidemics have been stayed by mass prophylactic immunization (**table 37.3**). Vaccines have eradicated smallpox, pushed polio to the brink of extinction, and spared countless individuals from hepatitis A and B, influenza, measles, rotavirus disease, tetanus, typhus, and other dangerous diseases. **Vaccinomics,** the application of genomics and bioinformatics to vaccine development, is bringing a fresh approach to the Herculean problem of making vaccines against various microorganisms and helminths.
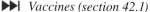 *Vaccines (section 42.1)*

To promote a more efficient immune response, antigens in vaccines can be mixed with an **adjuvant** (Latin *adjuvans,* aiding), which enhances the rate and degree of immunization. Adjuvants can be any nontoxic material that prolongs antigen interaction with immune cells, assists in processing of antigens by antigen-presenting cell (APC), or otherwise nonspecifically stimulates the immune response to the antigen. Several types of adjuvants can be used, including oil in water emulsions (Freund's incomplete adjuvant), aluminum hydroxide salts (alum), beeswax, and various combinations of bacteria (live or killed). In most cases, the adjuvant materials trap the antigen, thereby promoting a sustained release as APCs digest and degrade the preparation. In other cases, the adjuvant activates APCs so that antigen recognition, processing, and presentation are more efficient.

Table 37.3	Examples of Vaccines to Prevent Viral and Bacterial Diseases in Humans		
Disease	**Vaccine**	**Booster***	**Recommendation**
Viral Diseases			
Chickenpox	Attenuated Oka strain	None	Children 12–18 months: older children who have not had chickenpox
Hepatitis A	Inactivated virus	6–12 months	International travelers
Hepatitis B	Viral antigen	1–4 months 6–18 months	High-risk medical personnel: children, birth to 18 months and 11–12 years of age
Human papillomavirus infections	Recombinant protein subunits	2–3 months 6 months	Girls and boys 11–12 years of age
Influenza A/B	Inactivated virus or live attenuated	Yearly	All persons
Measles, Mumps, Rubella	Attenuated viruses (combination MMR vaccine)	None	First dose 12–15 months, 2nd dose 4–6 years
Poliomyelitis	Attenuated (oral poliomyelitis vaccine, OPV) or inactivated virus (inactivated polio vaccine, IPV)	Adults as needed	First dose at 2 months, 2nd at 4 months, 3rd at 16–18 months, 4th at 4–6 years
Rabies	Inactivated virus	None	For individuals in contact with wildlife, animal control personnel, veterinarians
Respiratory disease	Live attenuated adenovirus	None	Military personnel
Smallpox	Live attenuated vaccinia virus	None	Laboratory, health-care, and military personnel
Yellow fever	Attenuated virus	10 years	Military personnel and individuals traveling to endemic areas
Bacterial Diseases			
Anthrax	Extracellular components of unencapsulated *Bacillus anthracis*	None	Agricultural workers, veterinary, and military personnel
Cholera	Fraction of *Vibrio cholerae*	6 months	Individuals in endemic areas, travelers
Diphtheria, Pertussis, Tetanus	Diphtheria and tetanus toxoids, and acellular *Bordetella pertussis* vaccine (DTap) or tetanus toxoid, reduced diphtheria toxoid, and acellular pertussis vaccine (Tdap)	10 years	Children from 2–3 months old to 12 years, and adults; children 10–18 years, at least 5 years after DPT series, should receive Tdap
Haemophilus influenzae type b	Polysaccharide-protein conjugate (HbCV) or bacterial polysaccharide (HbPV)	None	First dose at 2 months, 2nd at 4 months, 3rd at 6 months, 4th at 12–15 months
Meningococcal infections	*Neisseria meningitidis* polysaccharides of serotypes A/C/Y/W-135	None	Military; high-risk individuals; college students living in dormitories; elderly in nursing homes
Plague	Fraction of *Yersinia pestis*	Yearly	Individuals in contact with rodents in endemic areas
Pneumococcal pneumonia	Purified *S. pneumoniae* polysaccharide of 23 pneumococcal types or pneumococcal conjugate of 13 strains	None	Adults over 50 with chronic disease
Q fever	Killed *Coxiella burnetii*	None	Workers in slaughterhouses and meat-processing plants
Tuberculosis	Attenuated *Mycobacterium bovis* (BCG vaccine)	3–4 years	Individuals exposed to TB for prolonged periods of time; used in some countries, not licensed in the U.S.
Typhoid fever	*Salmonella enterica* Typhi Ty21a (live attenuated or polysaccharide)	None	Residents of and travelers to areas of endemic disease
Typhus fever	Killed *Rickettsia prowazekii*	Yearly	Scientists and medical personnel in areas where typhus is endemic

* Subsequent immunization dose(s) after the initial immunization.

HISTORICAL HIGHLIGHTS

37.5 The First Immunizations

Since the time of the ancient Greeks, it has been recognized that people who have recovered from plague, smallpox, yellow fever, and various other infectious diseases rarely contract the diseases again. The first scientific attempts at artificial immunizations were made in the late eighteenth century by Edward Jenner (1749–1823), who was a country doctor from Berkeley, Gloucestershire, England. Jenner investigated the basis for the widespread belief of the English peasants that anyone who had vaccinia (cowpox) never contracted smallpox. Smallpox was often fatal—10 to 40% of the victims died—and those who recovered had disfiguring pockmarks. Yet most English milkmaids, who were readily infected with cowpox, had clear skin because cowpox was a relatively mild infection that left no scars.

On May 14, 1796, Jenner extracted the contents of a pustule from the arm of a cowpox-infected milkmaid, Sarah Nelmes, and injected it into the arm of eight-year-old James Phipps. As Jenner expected, immunization with the cowpox virus caused only mild symptoms in the boy. When he subsequently inoculated the boy with smallpox virus (an act now considered completely unethical), the boy showed no symptoms of the disease. Jenner then inoculated large numbers of his patients with cowpox pus, as did other physicians in England and on the European continent (**box figure**). By 1800 the practice known as variolation had begun in America, and by 1805 Napoleon Bonaparte had ordered all French soldiers to be vaccinated.

Further work on immunization was carried out by Louis Pasteur (1822–1895). Pasteur discovered that if cultures of chicken cholera bacteria were allowed to age for 2 or 3 months, the bacteria produced only a mild attack of cholera when inoculated into chickens. Somehow the old cultures had become less pathogenic (attenuated) for the chickens. He then found that fresh cultures of the bacteria failed to produce cholera in chickens that had been previously inoculated with old, attenuated cultures. To honor Jenner's work with cowpox, Pasteur gave the name *vaccine* to any preparation of a weakened pathogen that was used (as was Jenner's "vaccine virus") to immunize against infectious disease.

LES ŒUVRES PHILANTHROPIQUES du Petit Journal
La vaccination gratuite contre la variole dans le grand hall du Petit Journal

Nineteenth-Century Physicians Performing Vaccinations on Children.

The modern era of vaccines and immunization began in 1798 with Edward Jenner's use of cowpox as a vaccine against smallpox (**Historical Highlights 37.5**) and in 1881 with Louis Pasteur's rabies vaccine. Vaccines for other diseases did not emerge until later in the nineteenth century when, largely through a process of trial and error, methods for inactivating and attenuating microorganisms were improved and vaccines were produced. Vaccines were eventually developed against most of the epidemic diseases that plagued Western Europe and North America (e.g., diphtheria, measles, mumps, pertussis, German measles, and polio). Indeed, by the late twentieth century, it seemed that the combination of vaccines and antibiotics had solved the problem of microbial infections in developed nations. Such optimism was cut short by the emergence of new and previously unrecognized pathogens and antibiotic resistance among existing pathogens. Nevertheless, vaccination is still the most cost-effective weapon for preventing microbial disease (see table 37.2).

Vaccination of most children should begin at birth, initiating protection against the blood-borne hepatitis B virus. Other vaccines

should be given at about age two months; a complete schedule of childhood vaccines is available on the Centers for Disease Control and Prevention (CDC) website. Before that age, children are protected by passive natural immunity from maternal antibodies. Vaccination of teens and most adults depends on their risk for exposure to infectious disease. Individuals living in close quarters (e.g., college students in residence halls, military personnel) and those living with infants, the elderly, and individuals with reduced immunity (e.g., those with chronic and metabolic diseases) should be vaccinated to prevent bacterial meningitis from *Neisseria meningitidis*. Additionally, they should confirm that they have received childhood vaccinations and, where necessary, be revaccinated with newer vaccine formulations. Adults (over 50 years) should also be vaccinated/revaccinated for influenza, measles/mumps/rubella, streptococcal pneumonia, whooping cough, and shingles, as determined by a physician.

Additional or booster vaccines are suggested for many. For instance, international travelers may need to be immunized against cholera, hepatitis A, plague, polio, typhoid, typhus, and yellow

Table 37.4   A Comparison of Vaccine Types[1]

Major Characteristic	Whole Cell Vaccines		Acellular or Subunit Vaccines	Recombinant or DNA Vaccines	
	Live, Attenuated	Inactivated		Recombinant	DNA
Typical Immunogen	Virus	Bacteria	1–20 antigens from microorganism	Virus or bacterial vector	Antigen(s) encoded by DNA
T Response	Strong	Weak	Very strong	Very strong	Very strong
B cell Response	Strong	Strong	Very strong	Very strong	Very Strong
Stability	Decreased without refrigeration	Longer shelf life, freeze-dried	Longer shelf life, freeze-dried	Longer shelf life, freeze-dried	
Potential problems	Reversion to pathogen possible, not given to immuno- compromised	Weaker response, requires booster doses	Identification of antigens is costly and research intensive	Identification of genes is costly and research intensive	

[1]Source: http://www.vaccines.gov/more_info/types/

fever, depending on the public health policies of the country visited. Immunization is the most inexpensive method to prevent some infectious diseases; however, good hand hygiene and food/water cautions are still advised as some countries lack public health basics such as clean water. Veterinarians, forest rangers, and others whose jobs involve contact with animals may be vaccinated against rabies, plague, and anthrax. Health-care workers are typically immunized against hepatitis B virus. Vaccine recommendations for U.S. citizens are updated annually; travelers, veterinarians, and health-care providers should consult the CDC website, which lists vaccine and booster recommendations for all countries outside the United States, as well as recommendations for at-risk individuals. The role of immunization as a protective measure cannot be overstated: immunizations save lives.

Whole-Cell Vaccines

Many of the current vaccines used for humans that are effective against viral and bacterial diseases consist of whole microorganisms, termed **whole-cell vaccines.** These are either inactivated (killed) or attenuated (live but avirulent). The major characteristics of these vaccines are compared in **table 37.4. Inactivated vaccines** are effective, but they are less immunogenic, so they often require several boosters and normally do not adequately stimulate cell-mediated immunity or secretory IgA production. In contrast, **attenuated vaccines** usually are given in a single dose and stimulate both humoral and cell-mediated immunity. ◀◀ *Innate host resistance (chapter 33); Adaptive immunity (chapter 34)*

Even though whole-cell vaccines are considered the "gold standard" of vaccines, they can be problematic. For example, whole-organism vaccines fail to shield against some diseases. Attenuated vaccines can also cause full-blown illness in an individual whose immune system is compromised (e.g., AIDS patients, cancer patients undergoing chemotherapy, the elderly).

These same individuals may also contract the disease from healthy people who have been vaccinated recently. Moreover, attenuated viruses can at times mutate in ways that restore virulence, as happens when the live polio vaccine given in developing nations recombines with Coxsackie virus, leading to vaccine-derived polio. Children in the United States and other developed nations receive the killed polio vaccine, which is incapable of such genetic effects.

Acellular or Subunit Vaccines

A few of the common risks associated with whole-cell vaccines can be avoided by using only specific, purified macromolecules derived from pathogenic microorganisms. There are three general forms of **subunit vaccines:** (1) capsular polysaccharides, (2) recombinant surface antigens, and (3) inactivated exotoxins called **toxoids.** The purified microbial subunits or their secreted products can be prepared as nontoxic antigens to be used in the formulation of vaccines (**table 37.5**).

Recombinant-Vector and DNA Vaccines

Genes isolated from a pathogen that encode major antigens can be inserted into nonvirulent viruses or bacteria. Such recombinant microorganisms serve as vectors, replicating within the host and expressing the gene product of the pathogen-encoded antigenic proteins. The antigens elicit humoral immunity (i.e., antibody production) when they escape from the vector, and they also elicit cellular immunity when they are broken down and properly displayed on the cell surface (just as occurs when host cells harbor an active pathogen). Several microorganisms, such as rotavirus and attenuated *Salmonella enterica*, have been used in the production of these **recombinant-vector vaccines.**

On the other hand, **DNA vaccines** introduce fragments of pathogen DNA directly into the host cell. When injected into muscle cells, the DNA is taken into the nucleus and the pathogen's

Table 37.5	Subunit Vaccines Currently Available for Human Use
Microorganism or Toxin	**Vaccine Subunit**
Capsular polysaccharide	
Haemophilus influenzae type b	Polysaccharide-protein conjugate (HbCV) or bacterial polysaccharide (HbPV)
Neisseria meningitidis	Polysaccharides of serotypes A/C/Y/W-135
Streptococcus pneumoniae	23 distinct capsular polysaccharides
Surface antigen	
Hepatitis B virus	Recombinant surface antigen (HbsAg)
Human papillomavirus	Recombinant protein subunits
Toxoids	
Corynebacterium diphtheriae toxin	Inactivated exotoxin
Clostridium tetani toxin	Inactivated exotoxin

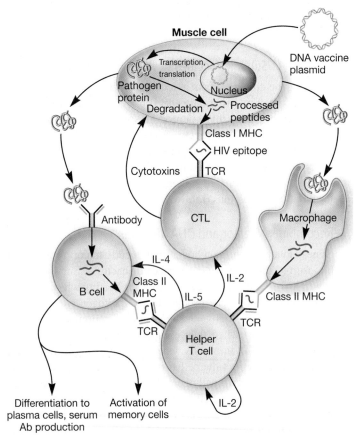

Figure 37.8 DNA Vaccine. DNA coding for a pathogen's proteins is inserted into host cells and expressed. The foreign proteins elicit both cell-mediated and antibody (Ab) responses.

DNA is transiently expressed, generating foreign proteins to which the host's immune system responds (**figure 37.8**). DNA vaccines are very stable; refrigeration is often unnecessary. At present, human trials are underway with several different DNA vaccines against malaria, AIDS, influenza, hepatitis B, and herpes. ⟳ *Constructing Vaccines*

37.7 Bioterrorism Readiness Is an integral Component of Public Health Microbiology

After reading this section, you should be able to:

- Discuss the readiness efforts necessary to be prepared for a bioterrorism attack
- Prioritize microorganisms based on public health threat
- Construct tables of prevention and control information relative to an infectious agent
- Alert appropriate authorities in response to suspected acts of bioterrorism

Bioterrorism is defined as "the intentional or threatened use of viruses, bacteria, fungi, or toxins from living organisms to produce death or disease in humans, animals, and plants." The use of biological agents to effect personal or political outcomes is not new, and the modern use of biological agents is a reality (**Historical Highlights 37.6**). The most notable intentional uses of biological agents for criminal or terror intent are (1) the use of *Salmonella enterica* serovar Typhimurium in 10 restaurant salad bars (by the Rajneeshee religious cult in The Dalles, Oregon, 1984); (2) the intentional release of *Shigella dysenteriae* in a hospital laboratory break room (perpetrator[s] unknown; Texas, 1996); and (3) the use of *Bacillus anthracis* spores delivered through the U.S. postal system (perpetrator[s] unknown, although the FBI named Bruce Ivins, now deceased, as the likely suspect; five eastern U.S. states, 2001). The *S. enterica*–contaminated salads resulted in 751 documented cases and 45 hospitalizations due to salmonellosis. The *S. dysenteriae* release resulted in eight confirmed cases and four hospitalizations for shigellosis. The *B. anthracis* spores infected 22 people (11 cases of inhalation anthrax and 11 cases of cutaneous anthrax) and caused five deaths. The list of biological agents that could pose the greatest public health risk in the event of a bioterrorism attack is relatively short and includes viruses, bacteria, and toxins that, if acquired and properly disseminated, could become a difficult public health challenge in terms of limiting the numbers of casualties and controlling panic (**table 37.6**).

Among weapons of mass destruction, biological weapons can be as destructive as chemical weapons, including nerve gas. In certain circumstances, biological weapons can be as devastating as a nuclear explosion: a few kilograms of anthrax spores could kill as many people as a Hiroshima-size nuclear bomb. Biological agents are likely to be chosen as a means of localized attack (biocrime) or mass casualty (bioterrorism) for several reasons. They are mostly invisible, odorless, tasteless, and difficult

HISTORICAL HIGHLIGHTS

37.6 1346—The First Recorded Biological Warfare Attack

The Black Death, which swept through Europe, Asia, and North Africa in the mid-fourteenth century, was probably the greatest public health disaster in recorded history. Europe, for example, lost an estimated quarter to a third of its population. This is not only of great historical interest but also relevant to current efforts to evaluate the threat of military or terrorist use of biological weapons. ▶▶❘ *Arthropods can transmit bacterial diseases (section 39.2)*

Some believe that evidence for the origin of the Black Death in Europe is found in a memoir by Gabriele de' Mussi of Genoa. According to this fourteenth-century memoir, the Black Death reached Europe from the Crimea (a region of the Ukraine) in 1346 as a result of a biological warfare attack. The Mongol army hurled plague-infected cadavers into the besieged Crimean city of Caffa (now Feodosija, Ukraine), thereby transmitting the disease to the inhabitants; fleeing

survivors then spread the plague from Caffa to the Mediterranean Basin. Such transmission was especially likely at Caffa, where cadavers would have been badly mangled by being hurled and the defenders probably often had cut or abraded hands from coping with the bombardment. Because many cadavers were involved, the opportunity for disease transmission was greatly increased. Disposal of victims' bodies in a major disease outbreak is always a problem, and the Mongol army used their hurling machines as a solution to limited mortuary facilities. It is possible that thousands of cadavers were disposed of this way; de' Mussi's description of "mountains of dead" might have been quite literally true. Indeed, Caffa could be the site of the most spectacular incident of biological warfare ever, with the Black Death as its disastrous consequence. It is a powerful reminder of the horrific consequences that can result when disease is successfully used as a weapon.

to detect. Furthermore, perpetrators may escape undetected as it may take hours to days before signs and symptoms of their use become evident. Additionally, the general public is not likely to be protected immunologically against agents that might be used in bioterrorism. Ultimately the use of biological agents in terrorism results in fear, panic, and chaos.

In 1998 the U.S. government launched the first national effort to create a biological weapons defense. The initiatives included (1) the first-ever procurement of specialized vaccines and medicines for a national civilian protection stockpile;

(2) invigoration of research and development in the science of biodefense; (3) investment of more time and money in genome sequencing, new vaccine research, and new therapeutic research; (4) development of improved detection and diagnostic systems; and (5) preparation of clinical microbiologists and clinical microbiology laboratories as members of first-responder teams to respond in a timely manner to acts of bioterrorism. In 2002 the Congress enacted the Public Health Security and Bioterrorism Preparedness and Response Act, which identified "select" agents whose use is now tightly regulated. A final rule that governs the possession,

Table 37.6	Some CDC-Defined Biological Select Agents and Toxins (BSATs)	
Bacteria	**Virus**	**Toxin**
*Bacillus anthracis**	Crimean-Congo hemorrhagic fever virus	Abrin
Bacillus anthracis Pasteur	Eastern Equine Encephalitis virus	Botulinum toxins*
Brucella abortus	Ebola virus*	Conotoxins
Brucella melitensis	Lassa fever virus	Ricin
Brucella suis	Marburg virus*	Saxitoxin
*Burkholderia mallei**	Monkeypox virus	Staphylococcal enterotoxins A, B, C, D, & E
*Burkholderia pseudomallei**	Nipah virus	T-2 toxin
Coxiella burnetii	Reconstructed 1918 Influenza virus Rift Valley fever virus	Tetrodotoxin
*Clostridium botulinum**	SARS-associated coronavirus	
*Francisella tularensis**	Tick-borne encephalitis viruses	
Rickettsia prowazekii	Variola major virus*	
*Yersinia pestis**	Variola minor virus*	
	Venezuelan equine encephalitis virus	

* Tier 1 agents. The documented risk of causing a high-consequence event is greater for Tier 1 BSATs than for the other BSATs.

Table 37.7	Criteria for Presumptive Identification of Six Bacterial Select Agents		
Pathogen	**Gram Morphology**	**Colonial Morphology**	**Biochemical Results**
Bacillus anthracis	Gram-positive, endospore-forming rod	Gray, flat, "Medusa head" irregularity, nonhemolytic on sheep's blood agar	Catalase positive, oxidase positive, urea negative, Voges-Proskauer (VP) positive, phenylalanine (Phe) deaminase negative, NO_3^- to NO_2^- positive
Brucella suis	Gram-negative rod (tiny)	Nonpigmented, convex-raised, pinpoint after 48 hr, nonhemolytic	Catalase positive, oxidase variable, urea positive, VP negative
Burkholderia mallei	Gram-negative, straight or slightly curved coccobacilli, bundles	Gray, smooth, translucent after 48 hr, nonhemolytic	Catalase positive, oxidase variable, indole negative, arginine dihydrolase positive, NO_3^- to NO_2^- positive
Clostridium botulinum	Gram-positive, endospore-forming rod	Creamy, irregular, rough, broad, nonhemolytic	Catalase negative, urea negative, gelatinase positive, indole negative, VP negative, Phe deaminase negative, NO_3^- to NO_2^- negative
Francisella tularensis	Gram-negative rod (tiny)	Gray-white, shiny, convex, pinpoint after 72 hr, nonhemolytic	Catalase positive (weak), oxidase negative, β-lactamase positive, urea negative
Yersinia pestis	Gram-negative rod (bipolar staining)	Gray-white, "fried-egg" irregularity, nonhemolytic, grows faster and larger at 28°C	Catalase positive, oxidase negative, urea negative, indole negative, VP negative, Phe deaminase negative

use, and transport of select agents was issued in 2005. The select agent definition, however, was revised in 2012 to identify a "Tier 1" subset that (1) could produce a mass casualty event or economic devastation, (2) were highly communicable, (3) had a low infectious dose, and (4) might be readily weaponized (table 37.6).

In 2003 Congress established the Department of Homeland Security to coordinate the defense of the United States against terrorist attacks. As one of many duties, the secretary of Homeland Security is responsible for maintaining the National Incident Management System to monitor large-scale hazardous events. Bioterrorism and other public health incidents are managed within this system. The Department of Health and Human Services will respond and deploy assets as needed within the areas of its statutory responsibility (e.g., the Public Health Service Act and the Federal Food, Drug, and Cosmetic Act), while appraising the secretary of Homeland Security of the nature of the response. The secretary of Health and Human Services directs the CDC to effect the necessary integration of public health activities.

The events of September and October 2001 in the United States changed the world. Global efforts to prevent terrorism, especially using biological agents, are evolving from cautious planning to proactive preparedness. In the United States, the CDC has partnered with academic institutions across the country to educate, train, and drill public health employees, first responders, and numerous environmental and health-care providers. Centers for Public Health Preparedness were established to bolster the overall response capability to bioterrorism. Another CDC-managed program that began in 1999, the Laboratory Response Network (LRN), serves to ensure an effective laboratory response to bioterrorism by helping to improve the nation's public health laboratory infrastructure through its partnership

with the FBI and the Association of Public Health Laboratories (APHL). The LRN maintains an integrated network that links state and local public health, federal, military, and international laboratories so that a rapid and coordinated response to bioterrorism or other public health emergencies (including veterinary, agriculture, military, and water- and food-related) can occur.

In the absence of overt terrorist threats and without the ability to rapidly detect bioterrorism agents, it is likely that a bioterrorism act will be defined by a sudden spike in an unusual (nonendemic) disease reported to the public health system. Also, suddenly increased numbers of zoonoses, diseased animals, or vehicle-borne illnesses may indicate bioterrorism. Important guidelines and standardized protocols prepared for all sentinel (local hospital, contract, clinic, etc.) laboratories to assist in the management of clinical specimens containing select agents have been established by the American Society for Microbiology in coordination with the CDC and the APHL. A summary of the rule-out tests for six Tier 1 bacterial agents is presented in **table 37.7**. The diseases and microbiology associated with specific select agents are discussed in chapters 38–40.

Retrieve, Infer, Apply

1. In what three general ways can epidemics be controlled? Give one or two specific examples of each type of control measure.
2. Name some of the microorganisms that can be used to commit biocrimes. From this list, which pose the greatest risk for causing large numbers of casualties?
3. Why are biological weapons more destructive than chemical weapons?
4. What is the Public Health Security and Bioterrorism Preparedness and Response Act designed to do?

Key Concepts

37.1 Epidemiology Is an Evidence-based Science

- Epidemiology is the science that evaluates the determinants, occurrence, distribution, and control of health and disease in a defined population.
- Specific epidemiological terminology is used to communicate disease incidence in a given population. Frequently used terms include sporadic disease, endemic disease, hyperendemic disease, epidemic, index case, outbreak, and pandemic.

37.2 Epidemiology Is Rooted in Well-Tested Methods

- Public health surveillance is necessary for recognizing a specific infectious disease within a given population. This consists of gathering data on the occurrence of the disease, collating and analyzing the data, summarizing the findings, and applying the information to control measures.
- Population-based and case-based surveillance data are used to track infections within a population.
- Remote sensing and geographic information systems are used to digitally gather data from natural environments.
- Statistics is an important tool used in the study of modern epidemiology.
- Epidemiological data can be obtained from such factors as morbidity, prevalence, and mortality rates.

37.3 Infectious Disease Is Revealed Through Patterns Within a Population

- A common-source epidemic is characterized by a sharp rise to a peak and then a rapid but not as pronounced decline in the number of individuals infected (**figure 37.3**). A propagated epidemic is characterized by a relatively slow and prolonged rise and then a gradual decline in the number of individuals infected.
- Herd immunity is the resistance of a population to infection and pathogen spread because of the immunity of a large percentage of the individuals within the population.

37.4 Infectious Diseases and Pathogens Are Emerging and Reemerging

- Humans will continually be faced with both new infectious diseases and the reemergence of older diseases once thought to be conquered.
- The CDC has defined these diseases as "new, reemerging, or drug-resistant infections whose incidence in humans has increased within the past two decades or whose incidence threatens to increase in the near future."
- Many factors characteristic of the modern world undoubtedly favor the development and spread of these microorganisms and their diseases (such as transportation, commerce, housing development).

37.5 Health-Care Facilities Harbor Infectious Agents

- Health-care–associated (nosocomial) infections are infections acquired within a hospital or other clinical care facility and are produced by a pathogen acquired during a patient's stay. These infections come from either endogenous or exogenous sources (**figures 37.6** and **37.7**).
- Health-care-associated diseases are usually caused by bacteria, most of which are noninvasive and part of the normal human microbiota.

37.6 Coordinated Efforts Are Required to Prevent and Control Epidemics

- The public health system consists of individuals and organizations that function in the control of infectious diseases and epidemics.
- Vaccination is one of the most cost-effective weapons for preventing microbial disease, and vaccines constitute one of the greatest achievements of modern medicine.
- Many of the current vaccines in use for humans (**table 37.3**) consist of whole organisms that are either inactivated (killed) or attenuated (live but avirulent).
- Some of the risks associated with whole-cell vaccines can be avoided by using only specific purified macromolecules derived from pathogenic microorganisms. Currently there are three general forms of subunit or acellular vaccines: capsular polysaccharides, recombinant surface antigens, and inactivated exotoxins (toxoids) (**table 37.5**).
- A number of microorganisms have been used for recombinant-vector vaccines. The attenuated microorganism serves as a vector, replicating within the host and expressing the gene product of the pathogen-encoded antigenic proteins. The proteins can elicit humoral immunity when the proteins escape from the cells and cellular immunity when they are broken down and properly displayed on the cell surface.
- DNA vaccines elicit protective immunity against a pathogen by activating both branches of the immune system: humoral and cellular (**figure 37.8**).
- Epidemiological control measures can be directed toward reducing or eliminating infection sources, breaking the connection between sources and susceptible individuals, or isolating the susceptible individuals and raising the general level of herd immunity by immunization.

37.7 Bioterrorism Readiness Is an integral Component of Public Health Microbiology

- Among weapons of mass destruction, biological weapons can be more destructive than chemical weapons. The list of biological agents that could pose the greatest public health risk in the event of a bioterrorism attack is short and includes viruses, bacteria, parasites, and toxins (**table 37.6**).
- Specific laboratory tests define the process for hospital and other sentinel labs to rule out a select agent.

Compare, Hypothesize, Invent

1. Why is international cooperation a necessity in the field of epidemiology? What specific problems can you envision if there were no such cooperation?

2. What common sources of infectious disease are found in your community? How can the etiologic agents of such infectious diseases spread from their source or reservoir to members of your community?

3. How could you prove that an epidemic of a given infectious disease was occurring?

4. How can changes in herd immunity contribute to an outbreak of a disease on an island?

5. College dormitories are notorious for outbreaks of flu and other infectious diseases. These are particularly prevalent during final exam weeks. Using your knowledge of the immune response and epidemiology, suggest practices that could be adopted to minimize the risks at such a critical time.

6. Why does an inactivated vaccine induce only a humoral response, whereas an attenuated vaccine induces both humoral and cell-mediated responses?

7. Why is a DNA vaccine delivered intramuscularly and not by intravenous or oral routes?

8. Community-associated methicillin-resistant *Staphylococcus aureus* (CA-MRSA) was an unexpected development. In the United States, CA-MRSA isolates are classified as USA300. It was unknown whether USA300 strains are clonal in origin (i.e., arising from a single, common parental strain), represent convergent evolution of multiple strains, or are some combination of the two. To differentiate between these possibilities, the entire genome sequence of 10 USA300 isolates from eight states was compared. There were very few single nucleotide polymorphisms in eight of the 10 isolates (*see figure 19.5 and related discussion*). Furthermore, the same eight were extremely virulent in a mouse model and produced the same exotoxins, whereas the other two were only modestly virulent and produced fewer exotoxins.

Why do you think it was important to determine if USA300 was clonal in origin or if the strains arose from multiple parental strains? What would be the public health implication of each origin?

Based on this evidence, what do you conclude about the origin of these 10 USA300 strains? Explain your answer.

Read the original paper: Kennedy, A. D. 2008. Epidemic community-associated methicillin-resistant *Staphylococcus aureus*: Recent clonal expansion and diversification. *Proc. Natl. Acad. Sci. USA* 105:1327.

Learn More

38

Human Diseases Caused by Viruses and Prions

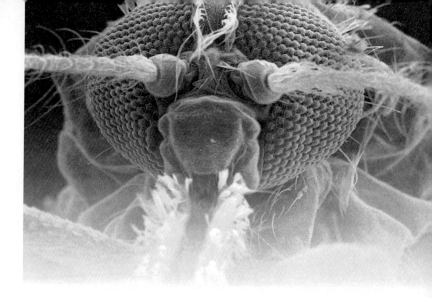

Mosquitoes, like this Anopheles female, transmit a number of viruses to humans.

Honest . . . It Was the Mosquito!

Exactly what happened when they returned from Bandafassi, Senegal, remains a mystery. All they knew for sure is that they had been bitten, a lot, while in Africa. Colorado State University researcher Brian Foy and his graduate assistant Kevin Kobylinski study malaria. They went to Senegal to trap mosquitoes. About 5 days after they returned to Colorado, both men became sick. They had headaches, rash, fatigue, and painfully swollen joints. Foy also had painful urination and bloody semen. Several days later, Foy's wife, Joy, began to complain of many of the same symptoms, along with chills, headache, malaise, and sensitivity to light. Their four children, however, had no signs or symptoms of illness.

About a week later, the symptoms abated. Intrigued by the pattern, Foy was sure he and Kobylinski had been infected in Senegal with one of the many mosquito-borne diseases transmitted through one or more of the bites. But Joy was not bitten; she was nowhere near Senegal. Foy and Kobylinski sought answers, but none were forthcoming. Laboratory tests on blood samples of the husband, wife, and student were finally performed at the U.S. Centers for Disease Control and Prevention's Division of Vector Borne Diseases, which happens to be in nearby Fort Collins, Colorado. As suspected, Foy and Kobylinski tested positive for dengue fever, a mosquito-borne viral disease. Dengue could explain most of the symptoms. But Joy's blood test was negative for dengue.

Kobylinski went back to Senegal a year later where he told his story to a colleague. Without hesitation, the scientist declared that the Americans had been infected with Zika virus. Zika is found primarily in Africa and usually misdiagnosed as dengue. Little is known about Zika other than it is a mosquito-borne flavivirus. Kobylinski relayed the information to Foy, who had the blood samples tested for Zika. Sure enough, all three blood samples had antibodies to Zika.

So how did Joy become infected? A little research revealed that Zika virus infects 50% more women than men. This hinted at sexual transmission. The species of mosquito and the Zika life cycle ruled out the possibility

that Joy was infected by a mosquito in the United States There was no direct evidence of sexual transmission—except, coded in his case report, Foy noted having had vaginal sexual intercourse with his wife upon his return from Senegal and before he had symptoms. Excited by their discovery, the three published a scientific paper describing sexual transmission of Zika virus for the first time. The morals to this story include: Mosquitoes can harbor a number of human pathogens, a fortuitous mosquito bite can lead to a new discovery (e.g., the probable sexual transmission of Zika virus), and being part of a "natural experiment" can get you authorship on a scientific paper.

Chapters 6 and 27 review the general biology of viruses and introduce basic virology. In chapter 38, we continue this coverage by discussing some of the most important viruses that are human pathogens. We group viral diseases according to their mode of acquisition and transmission; viral diseases that occur in the United States are emphasized.

More than 400 different viruses can infect humans. Diseases caused by these viruses are particularly interesting, considering the small amount of viral genetic information introduced into a host cell. This apparent simplicity belies the severe pathological features and clinical consequences that result from many viral diseases. With few exceptions, only prophylactic or supportive treatment is available. Collectively these diseases are some of the most common and yet most puzzling of all infectious diseases. The resulting frustration is compounded when year after year, familiar diseases of unknown etiology or new diseases become linked to viral infections, while vaccines for a range of viral diseases are questioned by some and unavailable to many.

Readiness Check:

Based on what you have learned previously, you should be able to:

✔ Review basic virology (sections 6.1–6.6) and prion biochemistry (section 6.7)
✔ List the major features of each group of viruses in the Baltimore system of viral classification (chapter 27)
✔ Explain pathogenicity and the infection process (chapter 35)

38.1 Viruses Can Be Transmitted by Airborne Routes

After reading this section, you should be able to:

- Report the common viral diseases spread by airborne transmission
- Identify typical signs and symptoms of viral diseases spread by airborne transmission
- Correlate airborne viral infection and disease severity with viral virulence factors

Because air does not support virus multiplication, any virus that is airborne must have originated from a living source. When humans are the source of the airborne virus, it usually is propelled from the respiratory tract by coughing, sneezing, or vocalizing.

Chickenpox (Varicella) and Shingles (Herpes Zoster)

Chickenpox (varicella) is a highly contagious skin disease primarily of children two to seven years of age. Humans are the reservoir and source for this virus, which is acquired by droplet inhalation into the respiratory system (**figure 38.1a**). The virus is highly infectious, with secondary infection rates in susceptible household contacts of 65 to 86%. In the prevaccine era, about 4 million cases of chickenpox occurred annually in the United States, resulting in approximately 11,000 hospitalizations and 100 to 150 deaths.

The causative agent is the enveloped, double-stranded (ds) DNA varicella-zoster virus (VZV; species *Human herpesvirus 3*), a member of the family *Herpesviridae*. The virus produces at least six glycoproteins that play a role in viral attachment to specific receptors on respiratory epithelial cells. Following an incubation period of 10 to 23 days, small vesicles erupt on the face or upper trunk, fill with pus, rupture, and become covered by scabs (figure 38.1b). Healing of the vesicles occurs in about 10 days. During this time, intense itching often occurs.

Laboratory testing for VZV is not normally required, as the diagnosis of chickenpox is typically made by clinical assessment. Laboratory confirmation is recommended, though, to confirm the diagnosis of severe or unusual cases of chickenpox. The introduction of the attenuated varicella vaccine (*see table 37.3*) in the 1990s has drastically reduced the incidence of chickenpox. However because fewer cases are seen, the likelihood of misdiagnosis is increased. Furthermore, in persons who have previously received varicella vaccination, chickenpox is still possible, as immunity wanes. In these individuals, mild or atypical chickenpox symptoms may occur, posing particular challenges for clinical diagnosis. The vaccine is recommended for ages twelve months to twelve years and can be administered separately or as a combined measles, mumps, rubella (MMR) and varicella (MMVR) vaccine.

Individuals who recover from chickenpox are subsequently immune to this disease; however, they are not free of the virus, as viral DNA resides in a dormant (latent) state within the nuclei of cranial nerves and sensory neurons in the dorsal root ganglia. Although viral DNA is maintained in infected cells, virions cannot be detected (**figure 38.2a**). When the infected person becomes immunocompromised by such factors as age, cancers,

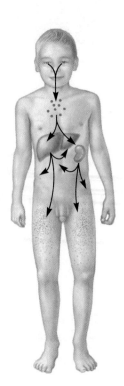

Day 0	Infection of conjunctiva and/or mucosa of upper respiratory tract
	Viral replication in regional lymph nodes
Day 4–6	Primary viremia in bloodstream
	Further viral replication in liver and spleen
	Secondary viremia
Day 10	Infection of skin and appearance of vesicular rash

Incubation period (bracket spanning Day 0 to Day 10)

(a)

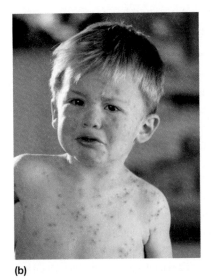

(b)

Figure 38.1 Chickenpox (Varicella). (a) Course of infection. (b) Typical vesicular skin rash. This rash occurs all over the body but is heaviest on the trunk and diminishes in intensity toward the periphery.

organ transplants, AIDS, or psychological or physiological stress, the viruses may become activated (figure 38.2b). They migrate down sensory nerves, initiate viral replication, and produce painful vesicles because of sensory nerve damage (figure 38.2c). In this reactivated form, the pain is called **postherpetic neuralgia.** To manage the intense pain, corticosteroids or the drug gabapentin can be prescribed. The antiviral drug acyclovir is used to shorten the infection in severe cases or in immunocompromised patients who develop chickenpox. Another reactivated form of chickenpox is called **shingles** (**herpes zoster**). Most cases occur in people over fifty years of age. Except for the pain of postherpetic neuralgia, shingles does not require specific therapy; however, in immunocompromised individuals, acyclovir, valacyclovir, or famciclovir are recommended. The U.S. Food and Drug Administration (FDA) has licensed a vaccine for shingles (Zostavax) for people fifty years and older. The Centers for Disease Control and Prevention (CDC) estimates that one in three people will get shingles, although the total number of cases

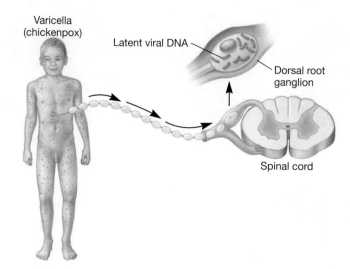

(a) Primary infection—Chickenpox

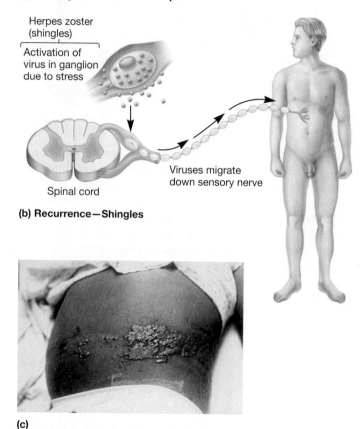

(b) Recurrence—Shingles

(c)

Figure 38.2 Pathogenesis of the Varicella-Zoster Virus. (a) After an initial infection with varicella (chickenpox), the viruses migrate up sensory peripheral nerves to their dorsal root ganglia, producing a latent infection. (b) When a person becomes immunocompromised or is under psychological or physiological stress, the viruses may be activated. (c) They migrate down sensory nerve axons, initiate viral replication, and produce painful vesicles. Since these vesicles usually appear around the trunk of the body, the name *zoster* (Greek for girdle) is used.

MICRO INQUIRY *Where does viral DNA reside during latent infection? Can virions be detected at this time?*

may decline as a shingles vaccine is now available. More than 1 million cases of herpes zoster occur annually in the United States, with 50% of these in people over the age of sixty. There are approximately 96 deaths from herpes zoster annually in the United States.

Influenza (Flu)

Influenza (Italian, *un influenza di freddo*—to be influenced by the cold), or the flu, was described by Hippocrates in 412 BCE. The first well-documented global epidemic of influenza-like disease occurred in 1580. Since then, 31 possible influenza pandemics have been documented, with four occurring in the twentieth century. The worst pandemic on record occurred in 1918 and killed approximately 50 million people around the world. This disaster, traced to the Spanish influenza virus, was followed by pandemics of Asian flu (1957), Hong Kong flu (1968), Russian flu (1977), and the novel swine flu (2009). (The names reflect popular impressions of where or from what the episodes began, although all are now thought to have originated in China.) Another new influenza strain erupted from pigs in 2011. It was the dominant strain of the 2014 flu season, for which that season's vaccine was only 23% effective. ◄◄ *Minus-strand RNA viruses (section 27.7)*

Influenza is a respiratory system disease caused by negative-strand RNA viruses that belong to the family *Orthomyxoviridae* *(see figure 27.28)*. The three genera are *Influenzavirus A*, *Influenzavirus B*, and *Influenzavirus C*; only types A and B cause significant human disease. Influenza viruses are widely distributed in nature, infecting a variety of mammal and bird hosts. All influenza viruses are acquired by inhalation or ingestion of virus-contaminated respiratory secretions (thus the public health recommendation to "cover your cough and wash your hands frequently"). During an incubation period of 1 to 2 days, viral particles adhere to host respiratory epithelium and initiate the replication cycle *(see figure 27.28b)*. Two envelope spikes, hemagglutinin (HA) and neuraminidase (NA) play critical roles. HA is the adherence factor that binds host cells and thereby triggers receptor-mediated endocytosis. This is facilitated by NA, which is thought to hydrolyze the mucus that covers the epithelium. Once within the cytoplasm enclosed in an endosome, HA again functions, this time to bring about release of the virus's 7–8 nucleocapsids into the cytosol. To do this, HA undergoes a dramatic conformational change when the endosomal pH decreases. The hydrophobic ends of the hemagglutinin spring outward and extend toward the endosomal membrane. After they contact the membrane, fusion occurs and the RNA nucleocapsids are released into the cytoplasm. Each influenza virus nucleocapsid consists of a single negative-strand RNA molecule associated with nucleoprotein (NP) and three subunits of the RNA-dependent RNA polymerase (PA, PB1, and PB2; **figure 38.3**). NA also has an additional role in the replication cycle. It aids in the release of newly assembled influenza virus virions from infected cells by cleaving sialic acid residues between virus and host membranes. ◄◄ *Several cytoplasmic membranous organelles function in the secretory and endocytic pathways (section 5.4)*

Influenza viruses are classified into subtypes based on their HA and NA. There are 16 HA (H1–H16) and 9 NA (N1–N9)

Figure 38.3 Influenza Infection.
Influenza A viruses infect birds, pigs, and humans. Reassortment of viral RNA from more than one viral subtype results in novel subtypes, such as the H1N1 cause of the 2009–2010 influenza pandemic.

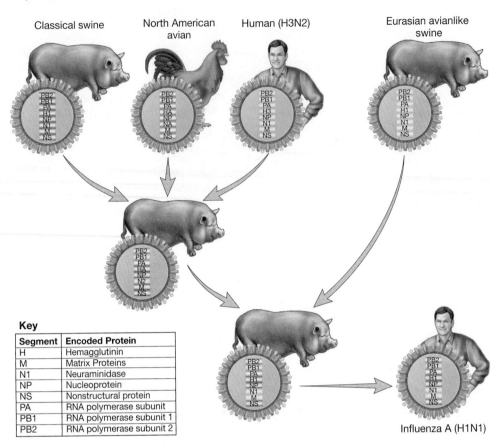

Key

Segment	Encoded Protein
H	Hemagglutinin
M	Matrix Proteins
N1	Neuraminidase
NP	Nucleoprotein
NS	Nonstructural protein
PA	RNA polymerase subunit
PB1	RNA polymerase subunit 1
PB2	RNA polymerase subunit 2

antigenic forms known; they recombine to produce the various HA/NA subtypes. Host specificity can be explained in part by the difference in receptor binding. Human influenza viruses preferentially attach by way of their HA protein to receptors on human tracheal epithelial cells. In contrast, avian influenza viruses preferentially bind receptors found on intestinal epithelial cells of waterfowl (the main viral replication site of avian influenza viruses), for the most part, restricting infection to birds.

Recall that the genomes of influenza viruses are linear RNA segments that change frequently due to point mutations and reassortment (figure 38.3). As RNA viruses, influenza viruses must use an RNA-dependent RNA polymerase (RdRp) to replicate their genomes. This gives rise to a high mutation rate because RdRps lack proofreading capabilities. These smaller genetic changes often result in subtle protein (antigen) changes, especially within the HA and NA structures. This is known as **antigenic drift.** Antigenic drift results from the accumulation of mutations of HA and NA in a single strain of flu virus within a geographic region and was responsible for the 2014–2015 mismatch between one vaccine flu strain and the circulating strain. This usually occurs every 2 to 3 years, causing local increases in the number of flu infections. **Antigenic shift,** however, is a larger protein sequence change made possible by the linear, segmented nature of the RNA genome. It occurs when two different strains of flu viruses (e.g., one from swine and one from humans) infect the same host cell and are incorporated into a single new capsid (figure 38.3). Because the new, recombinant virus contains segments from two different flu strains, there is a greater change with antigenic shift than with antigenic drift. Indeed, antigenic shifts are responsible for major epidemics and pandemics. Antigenic variation occurs almost yearly with influenza A virus, less frequently with the B virus, and thus the need for yearly influenza vaccination. Antigenic shift has not been demonstrated in the C virus. ◄◄ *Infectious disease is revealed through patterns within a population (section 37.3)*

Animal reservoirs are critical to the epidemiology of human influenza. For example, rural China is one region of the world in which chickens, pigs, and humans live in close, crowded conditions. Influenza is widespread in birds, and although birds cannot usually transmit the virus to humans, they can transfer it to pigs. Pigs can transfer it to humans and humans back to pigs. Recombination between human and avian strains thus occurs in pigs, leading to novel HA/NA combinations causing major antigenic shifts. To track the various influenza subtypes, a standard nomenclature system was developed. It includes the following information: group (A, B, or C), host of origin, geographic location, strain number, and year of original isolation. Antigenic descriptions of the HA and NA surface proteins are given in parentheses for type A viruses only. The host of origin is indicated for viruses obtained from nonhuman hosts; for example, A/swine/Iowa/15/30 (H1N1). However, the host origin is not given for human isolates; for example, A/Hong Kong/03/68 (H3N2).

Influenza A viruses having H1, H2, and H3 HA antigens, along with N1 and N2 NA antigens, are predominant in nature, infecting humans since the early 1900s. H1N1 viruses appeared in 1918 and were replaced in 1957 by H2N2 viruses as the predominant subtype. The H2N2 viruses were subsequently replaced by H3N2 as the principle subtype in 1968. The H1N1 subtype reappeared in 1977 and cocirculates today with H2N1, H3N2, H5N2, H7N2, H7N3, H7N7, H9N2, H10N7, and H5N1 viruses. Importantly, the receptor-binding specificity of human and avian influenza viruses just discussed suggests that in order to infect humans, surface HA and NA proteins of avian influenza viruses must gain the capacity to recognize human-type receptors. Indeed, genome sequences of the earliest isolates of the 1918, 1957, and 1968 pandemics demonstrated that otherwise avian influenza A viruses possessed HAs that recognized specific receptors with a characteristic structure usually seen in human influenza strains.

The year 2009 brought another influenza A H1N1 pandemic. Although there was initially much concern, this H1N1 strain was much, much milder than its relative that caused the 1918 pandemic. The novel H1N1 influenza emerged in Veracruz, Mexico, as a new strain when a previous triple reassortment of bird, swine, and human influenza viruses further recombined with a Eurasian pig influenza virus. Within months this new H1N1 strain became pandemic: by the end of the 2010 flu season there were more than 622,000 cases in 122 countries, with over 18,000 deaths. In the United States there were over 113,500 cases and 3,400 deaths.

The novel H1N1 (swine) flu pandemic of 2009 diverted attention from the continuing threat of the highly pathogenic H5N1 (avian) flu. The threat of the H5N1 subtype was first recognized in 1997, when it was responsible for six deaths, although it had been sporadically detected prior to that. Between 2003 to 2014, however, H5N1 was responsible for several bird to human and human to human transmissions in Asia, along with a substantial number of bird infections, resulting in the culling of millions of birds. So far this virus seems to spread rarely from birds to humans and even more rarely between humans. Even so, millions of birds have been destroyed to help prevent transmission of the virus. Unfortunately, highly pathogenic avian influenza H5 viruses H5N8 and H5N2 and a novel H5N1 have been found in both domestic and wild birds in the United States in 2015. This has resulted in the culling of hundreds of thousands of U.S. chickens and turkeys, export restrictions, and soaring egg prices. Bird populations continue to be carefully monitored.

Because of potentially severe consequences of pandemic influenza infection in humans, an international effort coordinated by the World Health Organization (WHO) continues to monitor virus surveillance, epidemiology, and control efforts for each virus subtype, as well as for newly recombinant strains. For example, in 2013 another influenza strain, H7N9, erupted in Asia. As of mid-March 2015 there were 643 human cases of H7N9. H7N9 has caused serious human illness and is fatal in about 30% of its victims. Of note are the additional reports of H5N2, H5N3, H5N8 bird outbreaks in Taiwan, and an H5N1 bird outbreak in Nigeria.

As with many other viral diseases, only the symptoms of influenza usually are treated. The antiviral drugs amantadine and rimantadine are no longer recommended for treatment of influenza, as most strains have developed resistance. Instead the CDC now recommends oseltamivir, zanamivir, and the newly approved peramivir; these can reduce the duration and symptoms of type A influenza if administered during the first two days of illness. These drugs are chemically related antiviral agents known as neuraminidase inhibitors; they have activity against both influenza A and B viruses. Neuraminidase inhibitors prevent the release of newly formed virions from their host cell by blocking the catalytic site of the enzyme neuraminidase. With the neuraminidase blocked, virions do not detach from the host cell membrane and therefore cannot travel to infect another cell. Importantly, aspirin (salicylic acid) should be avoided in children younger than fourteen years to reduce the risk of Reye's syndrome. ◂◂ *Antiviral drugs (section 9.6)*

The mainstay for prevention of influenza since the late 1940s has been inactivated virus vaccines, and the more recently approved live-attenuated, nasal spray vaccines (*see table 37.3*).

Although the flu vaccine is recommended for all, it is particularly important that the chronically ill, individuals under five and over age sixty-five, and residents of nursing homes be immunized, as disease in these patients is most likely to be severe. Because of influenza's high genetic variability, efforts are made each year to incorporate new virus subtypes into the vaccine. Even when no new subtypes are identified in a given year, annual immunization is still recommended because immunity using the inactivated virus vaccine typically lasts only 1 to 2 years.

Measles (Rubeola)

Measles (rubeola: Latin *rubeus,* red) is an extremely contagious disease that is endemic throughout most of the world. There were 145,700 cases of measles reported to WHO in 2013, with about 400 deaths every day. Fortunately, it is no longer endemic in the United States; thanks to an aggressive vaccination program, most U.S. cases of measles today are only those imported from other countries. There are approximately 18 measles-active countries as of 2015. Unfortunately, sporadic cases of measles in teens and young adults have been being reported in the United States since 2011. Notably, in 2014 there were 644 U.S. cases reported to the CDC. More than half of those were traced to an Amish man returning to Ohio from the Philippines. He gave the disease to unvaccinated community members, who passed it on to local college students whose vaccine protection had lapsed. Late in 2014 at Disneyland in California there erupted a second wave of measles cases, also believed to have had its origin in the Philippines. These progressed into the early weeks of 2015 where 141 new cases of measles were reported in 17 states and the District of Columbia. Aggressive public health measures were implemented to contain the epidemic.

Measles virus (MeV) is a negative-strand, enveloped RNA virus in the genus *Morbillivirus* and the family *Paramyxoviridae* (**figure 38.4a**). The MeV genome encodes eight proteins, six of which are structural.

MeV enters the body through the respiratory tract or the conjunctiva of the eyes. The receptors for MeV are found on activated B and T cells, as well as antigen-presenting cells. The hemagglutinin (H) and fusion (F) proteins mediate attachment of the viral envelope to the host cell membrane, facilitating entry by promoting fusion of the viral envelope and the host cell's plasma membrane. Two proteins, L and N, form the nucleocapsid. The remaining proteins (P, V, C, and M) are involved in viral replication.

The incubation period for measles is usually 10 to 14 days, and visible symptoms begin about the tenth day (figure 38.4d). The measles prodrome is characterized by nasal discharge, cough, fever, headache, and conjunctivitis, which intensify several days prior to the onset of rash. Within 3 to 5 days, skin eruptions occur as erythematous (red) maculopapular (discolored area of small bumps) lesions that are at first discrete but gradually become confluent (figure 38.4e). The rash normally lasts about 5 to 7 days. **Koplik's spots,** red lesions with a bluish-white speck in the center of each, are sometimes seen on the mucosal lining of the mouth in the early stages of disease. Recovery from measles begins soon

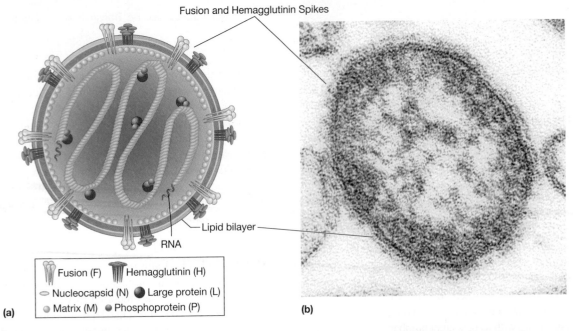

Fusion and Hemagglutinin Spikes

Lipid bilayer

RNA

Fusion (F)   Hemagglutinin (H)
Nucleocapsid (N)   Large protein (L)
Matrix (M)   Phosphoprotein (P)

(a)

(b)

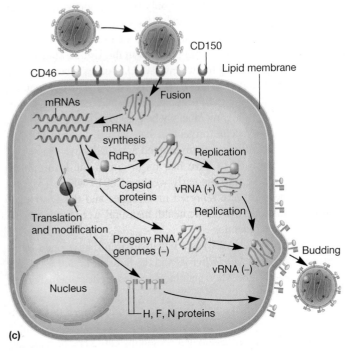

CD150
CD46
Lipid membrane

mRNAs
Fusion
mRNA synthesis
RdRp
Replication
Capsid proteins
vRNA (+)
Replication
Translation and modification
Progeny RNA genomes (−)
vRNA (−)
Budding
Nucleus
H, F, N proteins

(c)

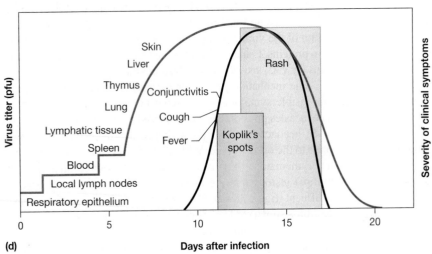

Skin
Liver
Thymus   Conjunctivitis
Lung
Lymphatic tissue   Cough
Spleen
Blood   Fever   Koplik's spots
Local lymph nodes
Respiratory epithelium

Rash

Virus titer (pfu)

Severity of clinical symptoms

0   5   10   15   20

Days after infection

(d)

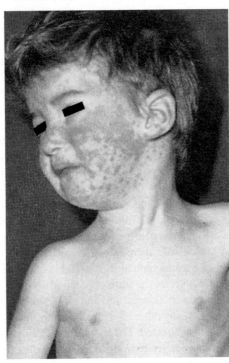

(e)

Figure 38.4 Measles (Rubeola).   (a) Schematic of the spherical, RNA morbillivirus that causes measles. (b) Transmission electron micrograph of a measles virus (MeV) virion (virion is 100 to 200 nm in diameter). (c) Life cycle of MeV. MeV virions exit the host by budding. (d) Severity of measles symptoms (black line) corresponds to the viral load (red line) in the tissue sites. (e) The rash of small, raised spots is typical of measles. The rash usually begins on the face and moves downward to the trunk.

after the rash is reported. Although a progressive degeneration of the central nervous system called **subacute sclerosing panencephalitis** is a rare complication, it is devastating and can result in permanent cognitive deficits.

No specific treatment is available for measles. The use of attenuated measles vaccine in combination with mumps and rubella (MMR vaccine) or as a combined measles, mumps, rubella, and varicella vaccine is recommended for all children (*see table 37.3*). Serious outbreaks of measles are still reported in North America and Europe, especially among the unvaccinated. A measles infection, as well as immunization with measles vaccine, typically provides lifelong immunity against reinfection. Unfortunately, a now-disproven research report had linked (measles) vaccination to autism, convincing many to forgo immunization of their children. While no vaccine is without potential risk, it is very clear that the benefit of vaccination far outweighs that risk.

Mumps

Mumps is an acute, highly contagious disease. The annual mumps incidence in most parts of the unvaccinated world is approximately 1–10 per 10,000 population, with epidemics every 2 to 5 years. U.S. cases range from a total of several hundred to several thousand every year, the low rates largely due to the highly effective vaccine. In 2009–2010 an outbreak occurred in the New York metroplex. The index case was an eleven-year-old boy returning from a trip to the United Kingdom. This ultimately led to about 3,000 cases of mumps. There were also several smaller mumps outbreaks in 2011–2013, mostly on college campuses. However, there were 1,151 cases in 2014, several of which were attributed to members of professional hockey teams. As with measles resurgence, modern mumps cases are typically in those whose immunity has waned or who were never vaccinated.

Mumps virus (MuV) is a member of the genus *Rubulavirus* in the family *Paramyxoviridae*. MuV virions are pleomorphic, enveloped, and contain a helical nucleocapsid containing negative-strand RNA. Like the measles virus, MuV is transmitted in saliva and respiratory droplets, but it is less contagious than measles or chickenpox. The most prominent manifestations of mumps are swelling and tenderness of the salivary (parotid) glands 16 to 18 days after infection of the host by the virus. The swelling usually lasts for 1 to 2 weeks and is accompanied by a low-grade fever. Severe complications of mumps are rare; however, deafness, meningitis, encephalitis, and inflammation of the epididymis and testes (orchitis) leading to sterility can occur, especially in postpubescent males. Therapy is limited to symptomatic and supportive measures. A live, attenuated mumps virus vaccine is available. It usually is given as part of the trivalent or tetravalent MMR vaccine (*see table 37.3*).

Respiratory Syndromes and Viral Pneumonia

Acute viral infections of the respiratory system are among the most common causes of human disease. The infectious agents are called acute respiratory viruses, and they collectively produce a variety of clinical manifestations, including rhinitis (inflammation of the mucous membranes of the nose), tonsillitis, laryngitis, and

bronchitis. Adenoviruses, coronaviruses, picornoviruses (rhinoviruses, coxsackieviruses A and B, echoviruses, and poliovirus), influenza viruses, parainfluenza viruses, respiratory syncytial virus, and human metapneumovirus are thought to be responsible. It should be emphasized that for most of these viruses, there is a lack of specific correlation between the agent and the clinical manifestation; hence the term syndrome. Most of these viruses have many strains, so immunity is not complete, and reinfection is common. The best treatment is rest with symptom management.

Some viral infections of the respiratory system culminate in pneumonia. However, viral pneumonia is typically only diagnosed when other causes, including mycoplasmal pneumonia, (*see p. 866*) have been ruled out. The clinical picture of viral pneumonia is nonspecific. Symptoms may be mild, or there may be severe illness and death. Importantly, secondary (subsequent) bacterial infections may also contribute to respiratory illness initiated by viruses. We highlight two coronaviruses and human respiratory syncytial virus as examples of these pathogens.

Coronaviruses are positive-strand RNA viruses named for the crownlike spikes on their surface. There are four main subtypes of coronaviruses; alpha, beta, gamma, and delta. Coronavirus virions are relatively large (120 to 150 nm), with a helical nucleocapsid surrounded by an envelope. Large spikes protrude from the envelope to aid in attachment and entry into host cells. The protruding envelope spikes extend from the oval-to-spherical virion to give the illusion of a halo, or corona, around the virus (**figure 38.5**). The spike proteins mediate viral attachment and fusion to the host cell membrane. There are many coronaviruses

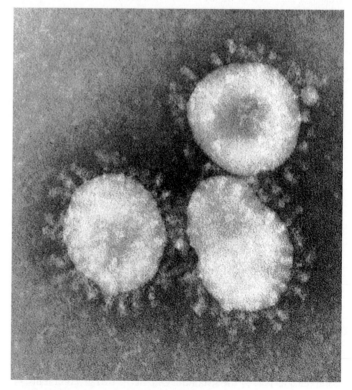

Figure 38.5 SARS Coronavirus (SARS-CoV). Coronaviruses are named for the crownlike (corona) appearance seen by electron microscopy.

that naturally infect animals, but most do not usually infect people. However, two relatively new coronaviruses do infect people: SARS-CoV and MERS-CoV. In addition to humans, SARS-CoV infects monkeys, Himalayan palm civets, raccoon dogs, cats, dogs, and rodents; and MERS-CoV also infects camels and bats.

Severe acute respiratory syndrome (SARS) is a highly contagious viral disease caused by the SARS coronavirus (SARS-CoV). SARS-CoV was first recognized in China in 2002, where it infected 8,098 people; it killed 774 worldwide in about one year. Since 2004 there have been no known cases of SARS-CoV. We report on it here as a great example of how a novel virus can emerge, wreak havoc on society, and burn itself out. The virus caused a febrile (100.4°F, 38°C) lower respiratory tract illness. Sudden, severe illness in otherwise healthy individuals was a hallmark of the disease. Other symptoms included headache, mild flulike discomfort, and body aches. SARS patients developed a dry cough after a few days, and most developed pneumonia. About 10 to 20% of patients had diarrhea. SARS was transmitted by close contact with respiratory secretions (droplet spread). No specific treatment is currently approved and no vaccine is licensed for preventative use. Molecular evidence suggests that substitutions in one or two amino acids within the receptor-binding domain of the human receptor were the likely the reason for reduced virulence of SARS-CoV infections late in 2004, indicating that mutations coding for these proteins may be responsible for the apparent eradication of the SARS virus.

Middle East respiratory syndrome (MERS) is a relatively new viral respiratory illness of humans caused by MERS-CoV. It was first reported in Saudi Arabia in 2012 and has since spread to several other countries. It has caused illness in hundreds of people from several countries in and near the Arabian Peninsula. In 2014 the first confirmed cases of MERS were reported in two travelers to the United States from Saudi Arabia; they were not linked. Most people infected with MERS-CoV developed severe acute respiratory illness, including fever, cough, and shortness of breath four to five days after exposure. Many of the infected also have gastrointestinal distress and diarrhea. MERS has a 30 to 40% mortality rate. There is no currently approved antiviral treatment. Prevention is through diligent respiratory protection and thorough hand hygiene when exposure is possible.

Human respiratory syncytial virus (HRSV) often is described as the most dangerous cause of lower respiratory infections in infants (less than one year of age) and the elderly (age 65 and older). The virus is found worldwide and causes seasonal (November to March) outbreaks lasting several months. In the United States, over 2.1 million children (under five years old) require outpatient visits, and between 100,000 and 126,000 infants are hospitalized each year for HRSV diagnoses. Additionally, there are 177,000 hospitalizations and 14,000 deaths among those 65 years or older.

HRSV is a member of the negative-strand RNA virus family *Paramyxoviridae*. The virion is variable in shape and size (average diameter 120–300 nm). It is enveloped with two virus-specific glycoproteins as part of the structure. One of the two glycoproteins, G, is responsible for binding the virus to the host cell; the other, the fusion or F protein, permits fusion of the viral envelope with the host cell plasma membrane, leading to entry of the virus. The F protein

also induces fusion of the plasma membranes of infected cells. Thus, HRSV gets its name from the resulting formation of a syncytium or multinucleated mass of fused cells. The syncytia are responsible for inflammation, alveolar thickening, and fluid accumulation in alveolar spaces. HRSV is transmitted by direct contact with respiratory secretions of humans. Virions are unstable in the environment (surviving only a few hours on environmental surfaces) and are readily inactivated with soap and water, and disinfectants.

Clinical manifestations of HRSV infection include acute onset of fever, cough, rhinitis, and nasal congestion. In infants and young children, this often progresses to severe bronchitis and viral pneumonia. Diagnosis is by either RSV rapid antigen test kit or RSV detection by reverse transcriptase (rt) PCR. Treatment of uncomplicated cases is with fluid replacement and fever reducers. In more serious cases, oxygen therapy with inhaled ribavirin is used. A series of monoclonal antibody (palivizumab) injections has been shown to reduce the severity of this disease in premature infants and babies who are immunosuppressed. Prevention and control consist of isolation of RSV-infected individuals and strict attention to good hand-washing practices.

Rubella (German Measles)

Rubella (Latin *rubellus*, little red) was first described in Germany in the 1800s as a variant of measles ("first disease"), and then scarlet fever ("second disease"); next it was called "third disease" and then rubella or German measles. It is a moderately contagious disease that occurs primarily in children five to nine years of age. It is caused by rubella virus, an enveloped, positive-strand RNA virus that is a member of the family *Togaviridae*. Rubella is worldwide in distribution, being spread in droplets shed from the respiratory secretions of infected individuals. Once the virus is inside the body, the incubation period ranges from 12 to 23 days. A rash of small red spots (**figure 38.6**),

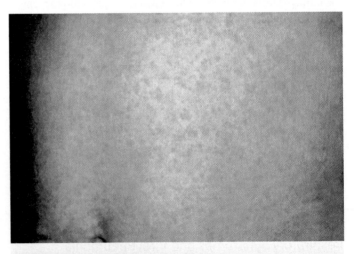

Figure 38.6 German Measles (Rubella). This disease is characterized by a rash of red spots. Spots are not raised above the surrounding skin as in measles (rubeola; see figure 38.4e).

MICRO INQUIRY *Why is it thought that the characteristic rash is not caused by viral infection of skin cells?*

usually lasting no more than 3 days, and a light fever are the normal symptoms. The rash appears as immunity develops and the virus disappears from the blood, suggesting that the rash is immunologically mediated and not caused by the virus infecting skin cells. Because rubella tends to be such a mild infection, usually no treatment is indicated.

However, rubella can be disastrous if contracted by a woman who is in her first trimester of pregnancy. It can lead to fetal death, premature delivery, or a wide array of congenital defects called **congenital rubella syndrome.** These defects include damage to the heart, eyes, and ears, and brain damage that leads to cognitive disabilities. All children and women of childbearing age who have not been previously exposed to rubella should be vaccinated. The United States was declared rubella free in 2009; there have been no endemic cases reported since then. The live attenuated rubella vaccine (part of MMR or MMRV, *see table 37.3*) continues to be recommended.

Smallpox (Variola)

Smallpox (variola) is a highly contagious disease of antiquity. We highlight it for two reasons: it is the only example of vaccine eradication of disease; and it continues to instill concern should it be used in bioterrorism. Both eradication and bioterrorism are linked to the fact that smallpox is uniquely an illness of humans. It is caused by orthopoxviruses belonging to the species *Variola virus.* Variola viruses (VARVs) belong to the family *Poxviridae,* which also includes the species *Monkeypox virus, Molluscum contagiosum virus,* and *Vaccinia virus.* Vaccinia virus (VACV) is the virus that causes cowpox and serves as the antigen in the smallpox vaccine. Poxviruses are large, brick-shaped, and contain a dumbbell-shaped core (*see figure 6.6*). The genome consists of a single, linear molecule of double-stranded DNA that is replicated in the host cell's cytoplasm. The genome is composed of approximately 186 kilobase (kb) pairs with covalently closed ends (**figure 38.7***a*). Variola virus isolates have approximately 200 open reading frames that are closely spaced and nonoverlapping. Interestingly, the size of the virion (300 by 250 to 200 nm) is slightly larger than that of some of the smallest bacteria (e.g., *Chlamydia* spp.).

Smallpox was once one of the most prevalent of all diseases. It was a universally dreaded scourge for more than 3 millennia, with case fatality rates of 20 to 50%; it disabled and disfigured those who survived. The disease begins with a prodrome of systemic aches. Characteristic symptoms of infection also include acute onset of fever 101°F (38.3°C), followed by a rash that features firm, deep-seated vesicles or pustules in the same stage of development (figure 38.7*b*). First subjected to some control in tenth-century India and China, it was gradually suppressed in the industrialized world after Edward Jenner's 1796 landmark discovery that infection with the harmless cowpox (vaccinia) virus renders humans immune to the smallpox virus (*Historical Highlights 37.5*).

Protection from smallpox is through vaccination. The smallpox vaccine contains live vaccinia virus. Since the advent of immunization and because of concerted efforts by WHO, smallpox was eradicated throughout the world—the greatest public

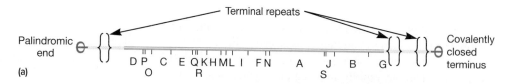

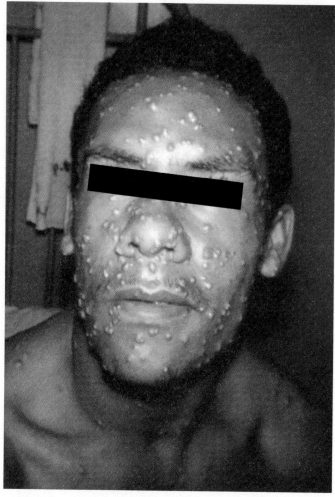

(b)

Figure 38.7 Variola Virus (VARV) Causes Smallpox.
(a) Genome architecture of VARV showing its linear, 186 kb pair, covalently closed DNA. (b) Smallpox rash demonstrating the virus-filled pustules. Pustules erupt at the same time all over the body.

MICRO INQUIRY *What specific clinical and biological features of smallpox made its eradication possible?*

health achievement ever. The last case from a natural infection occurred in Somalia in 1977. The Somali man was part of the vaccination program, but his vaccine did not take. Fortunately he survived the disease and only recently died. Eradication was possible because smallpox has obvious clinical features, very few asymptomatic carriers, only human hosts as reservoirs, and a short period of infectivity (3 to 4 weeks). Furthermore, variola virus is generally transmitted by direct and fairly prolonged face-to-face contact. It also can be spread through direct contact with infected bodily fluids or contaminated objects such as bedding or clothing. Smallpox has been reported to spread through the air in enclosed settings such as buildings, buses, and trains.

Variola virus enters the respiratory tract, seeding the mucous membranes, passing rapidly into regional lymph nodes. The average incubation period is 12 to 14 days but can range from 7 to 17 days. During this time, the virus multiplies in the host, but the host is not contagious. Another brief period of viremia precedes the prodromal phase. The prodromal phase lasts for 2 to 4 days and is characterized by malaise, severe head and body aches, occasional vomiting, and fever (over 104°F, 40°C), all beginning abruptly. During the prodromal phase, the mucous membranes in the mouth and pharynx become infected, resulting in a rash of small, red spots. These spots develop into open sores that spread large amounts of the virus into the mouth and throat. At this time, the person is highly contagious. The virus then invades the capillary epithelium of the skin, leading to the development of the following sequence of lesions in or on the skin: eruptions, papules, vesicles, pustules, crusts, and desquamation (figure 38.7b). The rash appears on the face, spreads to the arms and legs, and then the hands and feet. Usually the rash spreads over the body within 24 hours. By the third day of the rash, it forms raised bumps. By the fourth day, the bumps fill with a thick, opaque fluid and often have a depression in the center that looks like a bellybutton. This is a major distinguishing characteristic of smallpox, as compared to other diseases that exhibit rashes. The fever usually declines as the rash appears but rises and is sustained from the time the vesicles form until they crust over. Oropharyngeal and skin lesions contain abundant virus particles, particularly early in the disease process. Death from smallpox is due to toxemia associated with immune-mediated blood clots and elevated blood pressure.

Routine immunization for smallpox was discontinued in the United States once global eradication was confirmed. An accidental or deliberate release of the virus would be catastrophic in an unimmunized population and could cause a major pandemic. Thus if an outbreak occurred, prompt recognition and institution of control measures would be paramount as there is no FDA-approved treatment for smallpox, even though several antiviral agents have been suggested as adjunct therapies. ◄◄ *Bioterrorism readiness is an integral component of public health microbiology (section 37.7)*

Retrieve, Infer, Apply

1. Why are chickenpox and shingles discussed together?
2. Briefly describe the course of an influenza infection and how the virus causes the symptoms associated with the flu. Why has it been difficult to develop a single flu vaccine?
3. What are some common symptoms of measles? What are Koplik's spots?
4. What is one side effect that mumps can cause in a young postpubescent male?
5. Describe some clinical manifestations caused by acute respiratory viruses.
6. Is viral pneumonia a specific disease? Explain.
7. When is a German measles infection most dangerous and why?
8. What factors inherent in the variola virus and the disease smallpox made it possible to eradicate them?

38.2 Arthropods Can Transmit Viral Diseases

After reading this section, you should be able to:

- Report the two common arthropod-borne viral diseases
- Identify typical signs and symptoms of two arthropod-borne viral diseases
- Correlate arthropod-borne virus infection with geography and time of year

Arthropod-borne viruses (arboviruses) are transmitted by blood-sucking arthropods from one vertebrate host to another. They multiply in the tissues of the arthropod without producing disease, although the arthropod vector acquires a lifelong infection. Approximately 150 of the recognized arboviruses cause illness in humans. Diseases produced by the *arthropod-borne* viruses (arboviruses) can be divided into three clinical syndromes: (1) fevers of an undifferentiated type with or without a rash; (2) encephalitis (inflammation of the brain), often with a high fatality rate; and (3) hemorrhagic fevers, also frequently severe and fatal. No antiviral therapy is effective in treating arbovirus diseases, although supportive treatment is beneficial.

Chikungunya

Chikungunya virus, a member of *Togaviridae*, is a positive-strand, enveloped RNA virus. We highlight it here as an emerging infectious disease. In late 2013 it was found on the Caribbean Islands, and by July 7, 2015, a total of 207 chikungunya virus disease cases had been reported to the CDC from 32 states. All of these cases were in travelers returning from affected areas. At the same time the virus was detected as locally transmitted cases in Puerto Rico and the U.S. Virgin Islands. The virus is transmitted to humans by *Aedes aegypti* and *Aedes albopictus* mosquitoes.

Acute chikungunya virus infection in humans often includes fever and joint pain. Other symptoms may include headache, muscle pain, joint swelling, or rash. These symptoms often resolve in 7 to 10 days. However, many patients report prolonged joint pain persisting for months to years. Rare complications include eye inflammation, myocarditis, hepatitis, nephritis, hemorrhage, encephalitis, and Guillain-Barré syndrome.

Chikungunya virus detection is by tissue culture, viral RNA amplified by RT-PCR within eight days of infection, and antiviral antibodies five days post infection. There is no medicine to treat, or vaccine to prevent, chikungunya virus infection. Travelers to endemic areas should follow guidelines to prevent mosquito-borne infectious disease; they should use insect repellent, wear light-colored clothes with long sleeves and pants, and stay in places with mosquito netting, window and door screens, and fans. ◄◄ *The polymerase chain reaction amplifies targeted DNA (section 17.2)*

Equine Encephalitis

Equine encephalitis is caused by viruses in the genus *Alphavirus*, family *Togaviridae*. They are positive-strand, enveloped RNA viruses. As the name implies, this is a disease of horses, where it can be fatal. However, in humans, the disease can present as a spectrum from fever and headache to (aseptic) meningitis and encephalitis. Human disease can progress to include seizures, paralysis, coma, and death. The virus is transmitted to humans by *Aedes* and *Culex* spp. mosquitoes. Various geographic descriptors are used to define the disease caused by genetically distinct strains of these arboviruses: eastern equine encephalitis (EEE) occurs along the eastern Atlantic coast from Canada to South America; western equine encephalitis (WEE) occurs from Canada to South America along the western coast; and Venezuelan equine encephalitis occurs in central and southern parts of the United States into South America. Reservoir hosts are important in the replication, maintenance, and dissemination of these arboviruses. Treatment consists of the supportive care of symptoms. Currently no vaccine is available to prevent disease. Preventative measures rely on common mosquito precautions.

Heartland Virus

Heartland virus is a member of the genus *Phlebovirus*, in the family *Bunyaviridae*. Heartland virus is yet another emerging virus, this one infecting eight people in Missouri and Tennessee (in the U.S. heartland), between 2012 and 2014. Early data suggest that heartland virus is transmitted by the Lone Star tick, although other ticks may also transmit the virus.

Patients infected with heartland virus presented with fever and overwhelming fatigue. Some also complained of headaches, muscle aches, diarrhea, anorexia, or nausea. They all had laboratory findings found in patients with other tick-borne viral diseases, low leukocytes and platelets. Most patients required hospitalization but fully recovered. One patient died. There is no drug to treat, or vaccine to prevent, the disease caused by heartland virus. Supportive care and treatment of symptoms are recommended. Prevention of heartland virus infection is through standard tick precautions: use of tick repellants, wearing light-colored clothing having long sleeves and pants, and tick checks.

West Nile Fever (Encephalitis)

West Nile fever (encephalitis) is caused by a positive-strand RNA flavivirus (West Nile virus) that occurred primarily in the Middle East, Africa, and Southwest Asia, where the disease was first discovered in 1937 in the West Nile district of Uganda. In 1999 the virus appeared unexpectedly in the United States (New York), causing seven deaths among 62 confirmed human encephalitis cases and extensive mortality in a variety of domestic and exotic birds. Epidemiological data suggest that the virus crossed the Atlantic in an infected bird, mosquito, or human traveler.

By 2015, 47 U.S. states and the District of Columbia reported West Nile virus (WNV) infections in over 41,679 people, resulting in 1,753 deaths. WNV is transmitted to humans predominantly by *Culex* spp. mosquitoes that feed on infected birds (crows and sparrows). Mosquitoes harbor the greatest concentration of virus in the late summer and early fall, and corresponding

to this, there is a peak of disease in late August to early September. The risk of disease then decreases as the mosquitoes die when the weather becomes colder. Although many people are bitten by WNV-infected mosquitoes, most do not know they have been exposed and remain asymptomatic or exhibit only mild, flulike symptoms.

Human-to-human transmission has been reported through blood and organ donation; however, the risk of acquiring WNV infection from donated blood or organs has greatly diminished since the introduction of a PCR-based detection assay in 2003. There are no data to suggest that WNV transmission to humans occurs from handling infected birds (live or dead), but barrier protection is suggested in handling potentially infected animals. The virus can be recovered from *Culex* mosquitoes, birds, and blood taken in the acute stage of a human infection. Diagnosis is by a rise in neutralizing antibody in a patient's serum. Only one antigenic type exists and immunity is presumed permanent. An enzyme-linked immunosorbent assay (ELISA) test for IgM anti-WNV antibody is the FDA-approved diagnostic test. Other immunological and molecular testing is available at state health department laboratories or at the CDC. There is no treatment other than hospitalization and intravenous fluids. There is no human vaccine to prevent WNV. Mosquito abatement and the use of repellents such as DEET are currently the only control measures. ◄◄ *Enzyme-linked immunosorbant assay (section 36.4)*

38.3 Direct Contact Diseases Can Be Caused by Viruses

After reading this section, you should be able to:

- Report the common viral diseases spread by direct contact
- Identify typical signs and symptoms of common viral diseases spread by direct contact
- Correlate direct contact virus infection and disease severity with viral virulence factors

Recall that each microorganism has an optimal portal of entry into the host. In addition to air- and vector-borne transmission, viruses can be exchanged between people. Transmission of disease from one person to another often requires close personal contact. For example, transfer of microorganisms between people occurs by direct contact with the infected person through touching, kissing, sexual contact, contact with body fluids and secretions, or contact with open wounds. We now consider several viral diseases that are transmitted by direct contact.

Acquired Immune Deficiency Syndrome (AIDS)

AIDS (acquired immune deficiency syndrome) was the great pandemic of the second half of the twentieth century. Fortunately the 2011 report issued by the Joint United Nations Programme on HIV/AIDS and the WHO indicates that the number of new HIV infections has decreased 21% since 1997, when the epidemic began to peak globally. Today 35 million people are infected and living with HIV/AIDS worldwide (**figure 38.8**). First described in 1981, AIDS is the result of an infection by the **human immunodeficiency virus** (**HIV**), a positive-strand, enveloped RNA virus within the family *Retroviridae*. In the United States, AIDS is caused primarily by HIV-1 (some cases result from HIV-2 infection); the HIV-2 strain predominantly causes AIDS in West Africa.

HIV is acquired and may be passed from one person to another when infected blood, semen, or vaginal secretions come in contact with an uninfected person's broken skin or mucous membranes. In the developing world, AIDS affects as many women as men, with many women getting AIDS from men who have multiple sex partners. In the United States, the groups most at risk for acquiring AIDS are (in descending order) men who have unprotected sex with other men; intravenous drug users; heterosexuals who have unprotected sex with infected partners; children born of infected mothers, as well as their breast-fed infants; transfusion patients; and transplant recipients. Transmission of HIV in the latter two risk groups is exceedingly rare due to extensive testing of blood products. Transmission through semen increases HIV virulence as much as 100,000-fold relative to transmission by other routes. The mortality rate from AIDS is almost 100% within 9 to 11 years post infection, if it is not treated. The use of combined antiviral medications has significantly reduced the morbidity and mortality of AIDS in developed nations. Modern medications can provide good quality of life and extended life expectancy, making HIV infection essentially a chronic, yet controllable, disease.

HIV-1 virions are approximately 110 nm in diameter, have a cylindrical core capsid, and their membrane envelopes are peppered with viral spike proteins (**figure 38.9a**). The core contains two copies of its RNA genome and several enzymes. Ten virus-specific proteins have been discovered. One of them, the gp120 envelope protein, facilitates attachment to a susceptible host cell (e.g., T-helper cell). Electron cryotomography imaging reveals that the gp120 spikes are approximately 14 nm long and are randomly distributed around the virion. The gp120 proteins form trimeric spikes; that is, three lobes form the final three-dimensional spike structure, which sits on three membrane-bound feet (figure 38.9c). ◀◀ *Electron cryotomography (section 2.4)*

Once inside the body, the virus's gp120 envelope protein (figure 38.9b) binds the CD4 glycoprotein in the plasma membranes of CD4$^+$ T cells, macrophages, dendritic cells, and monocytes. Because dendritic cells are present throughout the body's mucosal surfaces and bear the CD4 protein, it is possible that these are the first cells infected by HIV in sexual transmission. The virus requires a coreceptor in addition to CD4. Macrophage-tropic strains, which seem to predominate early in the disease

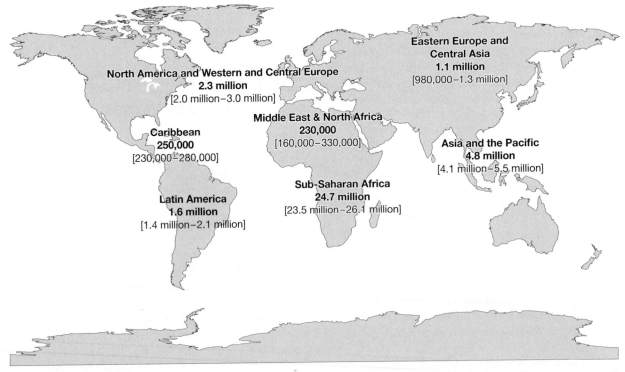

Figure 38.8 The Global HIV/AIDS View. This figure shows data from the 2013 UNAIDS report on the global AIDS epidemic. The report estimates that 35 million people are living with HIV/AIDS. Latest official data available, July 2015.

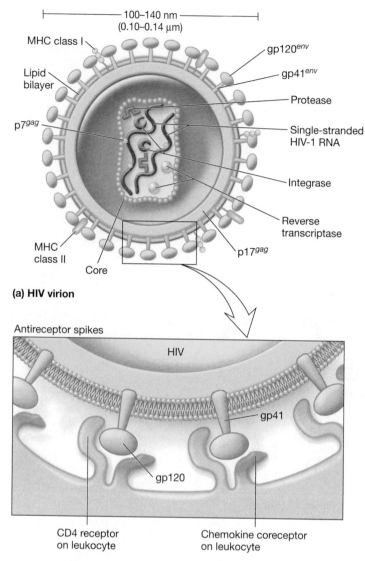

(a) HIV virion

100–140 nm
(0.10–0.14 μm)

MHC class I

gp120^env

gp41^env

Lipid bilayer

Protease

p7^gag

Single-stranded HIV-1 RNA

Integrase

Reverse transcriptase

MHC class II

p17^gag

Core

Antireceptor spikes

HIV

gp41

gp120

CD4 receptor on leukocyte

Chemokine coreceptor on leukocyte

(b) HIV attachment to host cell

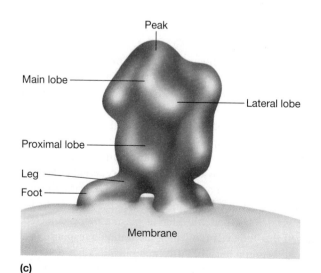

Peak

Main lobe

Lateral lobe

Proximal lobe

Leg

Foot

Membrane

(c)

Figure 38.9 Schematic Diagram of the HIV-1 Virion. (a) The HIV-1 virion is enveloped and contains 72 external spikes. These spikes are formed by the two major viral envelope proteins, gp120 and gp41 (gp stands for glycoprotein—proteins linked to sugars—and the number refers to the mass of the protein, in kilodaltons). The HIV-1 lipid bilayer is also studded with various host proteins, including class I and class II major histocompatibility complex molecules, acquired during virion budding. The cone-shaped core of HIV-1 contains four nucleocapsid proteins (p24, p17, p9, p7), each of which is proteolytically cleaved from a 53 kDa Gag precursor by the HIV-1 protease. The phosphorylated p24 polypeptide forms the chief component of the inner shell of the nucleocapsid, whereas the p17 protein is associated with the inner surface of the lipid bilayer and stabilizes the exterior and interior components of the virion. The p7 protein binds directly to genomic RNA through a zinc finger structural motif and together with p9 forms the core. The retroviral core contains two copies of the single-stranded HIV-1 genomic RNA to which preformed viral reverse transcriptase, integrase, and protease enzymes are attached. (b) The snug attachment of HIV glycoprotein molecules (gp41 and gp120) to their specific receptors on a human cell membrane. These receptors are CD4 and chemokine coreceptors CCR5 and CXCR4 (fusin) that permit docking with the host cell and fusion with the cell membrane. (c) Surface-rendered model of the gp120 ENV protein spike of human and simian immunodeficiency viruses.

and infect both macrophages and T cells, require CCR5 (a chemokine coreceptor) as well as CD4. A second chemokine coreceptor, called CXCR-4, is used by T-cell–tropic strains that are active at later stages of infection. These strains induce the formation of syncytia (multinucleated masses of fused cells). There are rare individuals with two defective copies of the CCR5 gene who do not develop AIDS because the virus cannot infect their macrophages and T cells. People with one good copy of the CCR5 gene do get AIDS but have an altered disease presentation than those with no mutation. ◄◄ *Retroviruses (section 27.8); T-cell receptors (section 34.5)*

The virus can enter host cells by endocytosis. Infection of the host cell begins when the gp120 protein binds to the CD4 and coreceptor. Elegant studies using membrane tracking dyes show that the virus uncoats from within the endocytic compartment as the viral envelope fuses with the endosome membrane, releasing the virion contents into the cytosol (**figure 38.10a**). Another entry method is endocytosis-independent; on the binding of the virus to some host cells, a conformational change in gp120 permits gp41 to extend, coil, and insert into the host cellular membrane. This process pulls the viral and cellular membranes together, allowing them to fuse. Inside the infected cell, the core protein remains associated with the RNA as it is copied into a single strand of DNA by the RNA-dependent DNA polymerase activity of the reverse transcriptase enzyme (*see figure 27.3*). The RNA template is next degraded by another reverse transcriptase component, ribonuclease H, and the newly synthesized DNA strand now serves as template to form a double-stranded DNA

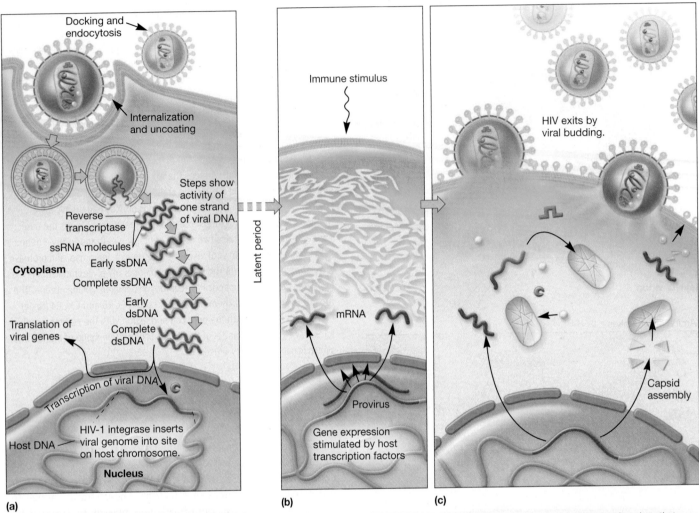

Figure 38.10 HIV Life Cycle. (a) The HIV virion binds to the CD4$^+$ host cell and enters the host cell by endocytosis. Membrane labeling studies show that the viral envelope fuses with the endocytic compartment releasing the virion contents into the cytosol. Reverse transcriptase catalyzes the synthesis of a single complementary strand of DNA from each of the twin RNA molecules. This DNA serves as a template for synthesis of double-stranded (ds) DNA. The dsDNA is inserted into the host chromosome as a provirus (latency). (b) The provirus genes are transcribed. (c) Viral mRNA is translated into virion components (capsid, reverse transcriptase, spike proteins), and the virion is assembled. Mature virus particles bud from the host cell, taking host membrane as their envelope.

MICRO INQUIRY *What is the function of integrase and how is its activity related to latency?*

copy of the original RNA genome. A complex of the double-stranded DNA (provirus) and integrase enzyme moves into the nucleus. Then proviral DNA is integrated into the cell's DNA through a complex sequence of reactions catalyzed by integrase. The integrated provirus can remain latent, giving no clinical sign of its presence. Alternatively the provirus can force the cell to synthesize viral mRNA (figure 38.10b). Some of the RNA is translated to produce viral proteins by host cell ribosomes. Viral proteins and the complete HIV-1 RNA genome are then assembled into new virions that bud from the infected host cell (figure 38.10c). Eventually the host cell lyses. ◀◀ *The major histocompatibility complex (section 34.4)*

Once a person becomes infected with HIV, the course of disease may vary greatly. The acute infection stage occurs 2 to 8 weeks after HIV infection. About 70% of individuals in this stage experience a brief illness referred to as acute retroviral syndrome, with symptoms that may include fever, malaise, headache, macular (small, red, spotty) rash, weight loss, lymph node enlargement (lymphadenopathy), and oral candidiasis. During this stage, the virus multiplies rapidly and disseminates to lymphoid tissues throughout the body, until an acquired immune response (antibodies and cytotoxic T cells) can be generated to bring virus multiplication under control **(figure 38.11)**. During the acute infection stage, levels of HIV increase rapidly, with 10^5 to 10^6 copies of viral RNA per milliliter of plasma being detected. It is believed that the extent to which the immune response is able to control this initial burst of virus multiplication may determine the amount of time required for progression to the next clinical stage.

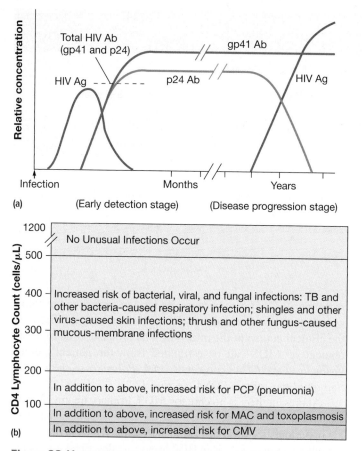

Figure 38.11 The Typical Immunological Pattern in an HIV-1 Infection. (a) HIV-1 antigen (HIV Ag) is detectable as early as 2 weeks after infection and typically occurs before IgG antibodies against gp41 and p24 develop. Seroconversion occurs when HIV-1 antibodies have risen to detectable levels. This usually takes place several weeks to months after infection. The period between HIV-1 infection and seroconversion often is associated with an acute flulike illness. HIV-1 antigen usually disappears following seroconversion but reappears in later stages of the disease. The reappearance of antigen usually indicates impending clinical deterioration. An asymptomatic HIV-1 antigen-positive individual is six times more likely to develop AIDS within 3 years than a similar individual who is HIV-1 antigen negative. Thus testing for the presence of HIV-1 antigens assists clinicians in monitoring progression of the disease. (b) As CD4 lymphocytes are destroyed by HIV, their declining numbers result in increased host susceptibility to specific infections.

The asymptomatic stage of HIV infection may last from 6 months to 10 years (or longer in some individuals). During this stage, the levels of detectable HIV in the blood decrease, but the virus continues to replicate, particularly in lymphoid tissues. Even before any changes in CD4$^+$ T cells can be detected, the virus may affect certain immune functions; for example, memory cell responses to common antigens such as tetanus toxoid or *Candida albicans.*

During the chronic symptomatic stage, which can last for months to years, virus multiplication continues and the number of CD4$^+$ T cells in the blood begins to significantly decrease.

Because these T-helper cells are critically important in the generation of acquired immunity, individuals at this stage develop a variety of symptoms including fever, weight loss, malaise, fatigue, anorexia, abdominal pain, diarrhea, headaches, and lymphadenopathy. Paradoxically, some patients develop increased serum antibody production during this stage, perhaps as a result of generalized immune dysfunction. These antibodies, however, do little to protect the host from infection. As CD4$^+$ T-cell numbers continue to decline, some patients develop opportunistic infections, such as oral candidiasis (thrush), tuberculosis, and infections rarely seen in other patients, including *Mycobacterium avium* complex (MAC), cryptococcal pneumonia, cytomegalovirus (CMV) infections, acute toxoplasmosis, or *Pneumocystis* pneumonia (PCP) (figure 38.11*b*).

In order to simplify a diagnosis of HIV/AIDS disease, the CDC has recently created a revised algorithm on which HIV disease criteria are based. This was necessary to interpret often mixed results of rapid tests. Thus, a confirmed HIV case is now classified in one of five HIV infection stages (0, 1, 2, 3, or "unknown"). Stage 0 defines a diagnosis of acute (primary) HIV, where HIV tests may be equivocal within 6 months of an initial HIV diagnosis (i.e., a sequence of positive and negative HIV tests is reported after an initial positive HIV test, resulting from either the use of an older, less-sensitive test or patient antibody production that is not fully developed). If the criteria for stage 0 are not met, the stage is classified as 1, 2, 3, or unknown, depending on CD4$^+$ T-lymphocyte numbers (**table 38.1**) or whether an opportunistic illness is present (**table 38.2**). If the criteria for a stage of 0, 1, 2, or 3 cannot be met (e.g., because of missing information on CD4 test results), the stage is "unknown."

In addition to its devastating effects on the immune system, HIV infection can also lead to disease of the central nervous system when virus-infected macrophages cross the blood-brain barrier. The classical symptoms of central nervous system disease in AIDS patients are headaches, fevers, subtle cognitive changes, abnormal reflexes, and ataxia (irregularity of muscular action). Dementia and severe sensory and motor changes characterize more advanced stages of the disease. Another potential complication of HIV infection is cancer. Individuals infected with HIV-1 have an increased risk of three types of tumors: (1) the human herpesvirus 8–induced Kaposi's sarcoma (**figure 38.12**), (2) carcinomas of the mouth and rectum, and (3) B-cell lymphomas or lymphoproliferative disorders. It seems likely that the depression of the initial immune response enables other tumor-causing agents to initiate the cancers.

Laboratory diagnosis of HIV infection can be by isolation and culture of the virus or by using assays for viral reverse transcriptase activity or viral antigens. However, diagnosis is most commonly accomplished through the detection of specific anti-HIV antibodies in the blood. A number of newer antibody-based rapid tests have decreased reports of false results. The most sensitive HIV assay employs the polymerase chain reaction (PCR). PCR can be used to amplify and detect tiny amounts of viral RNA and proviral DNA in infected host cells. Quantitative PCR assays provide an estimate of a patient's viral load. This is

Table 38.1	HIV Infection Stages 1-3, Based on Age-Specific CD4⁺ T-Lymphocyte Count or CD4⁺ T-Lymphocyte Percentage of Total Lymphocytes*						
HIV Stage*	**Age at date of CD4⁺ T-lymphocyte test**						
	<1 year		**1–5 years**		**6+ years**	
	Cells/µL	%	Cells/µL	%	Cells/µL	%
1	≥1,500#	≥34#	≥1,000#	≥30#	500–1200#	≥26#
2	750–1,499	26–33	500–999	22–29	200–499	14–25
3	<750	<26	<500	<22	<200	<14

*The stage is based primarily on the CD4⁺ T-lymphocyte count; the CD4⁺ T-lymphocyte count takes precedence over the CD4 T-lymphocyte percentage, and the percentage is considered only if the count is missing (see www.cdc.gov).

#Normal CD4 (cell count and percent) for this age group.

Source of data: *http://www.cdc.gov/hiv/statistics/recommendations/terms.html.*

Table 38.2	Disease Processes Associated with AIDS
Bacterial infection, multiple or recurrent
Candidiasis of bronchi, trachea, or lungs
Candidiasis, esophageal
Cervical cancer, invasive
Coccidioidomycosis, disseminated or extrapulmonary
Cryptococcosis, extrapulmonary
Cryptosporidiosis, chronic intestinal (>1 month's duration)
Cytomegalovirus disease (other than liver, spleen, or lymph nodes)
Cytomegalovirus retinitis (with loss of vision)
Encephalopathy, HIV-related
Herpes: chronic ulcer(s) (>1 month's duration); or bronchitis, pneumonitis, or esophagitis
Histoplasmosis, disseminated or extrapulmonary
Isosporiasis, chronic intestinal (>1 month's duration)
Kaposi's sarcoma
Lymphoma, Burkitt's
Lymphoma, immunoblastic
Lymphoma, primary, of brain
Mycobacterium avium complex or *M. kansasii*
Mycobacterium tuberculosis infection, any site
Mycobacterium infection, other species or unidentified species
Pneumocystis pneumonia
Pneumonia, recurrent
Progressive multifocal leukoencephalopathy
Salmonella septicemia, recurrent
Toxoplasmosis of brain
Wasting syndrome

particularly significant because the level of virions in the blood, as well as the concentration of CD4⁺ cells, is very predictive of the clinical course of the infection. The probable time to development of AIDS can be estimated from the patient's blood virion level and CD4⁺ cell count. ◄◄ *Polymerase chain reaction amplifies targeted DNA (section 17.2)*

At present there is no cure for AIDS. Primary treatment is directed at reducing the viral load and disease symptoms, and treating opportunistic infections and malignancies. The antiviral drugs currently approved for use in HIV disease are described in detail in chapter 9. An overview of their mechanisms of action are presented in figure 9.16*b*. Briefly, there are five types. (1) Nucleoside reverse transcriptase inhibitors (NRTIs) are nucleoside analogues that inhibit the enzyme reverse transcriptase as it synthesizes viral DNA. (2) Nonnucleoside reverse transcriptase inhibitors (NNRTIs) prevent HIV DNA synthesis by selectively binding to and inhibiting the reverse transcriptase enzyme. (3) Protease inhibitors (PIs) work by blocking the activity of the HIV protease, which is responsible for generating mature viral proteins, and thus interferes with

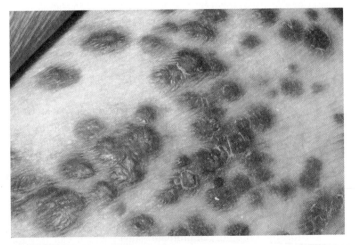

Figure 38.12 Kaposi's Sarcoma on the Arm of an AIDS Patient.   The flat purple tumors can occur in almost any tissue.

virion assembly. (4) Integrase inhibitors (IIs) prevent the incorporation of the HIV genome into the host's chromosomes. (5) Fusion inhibitors (FIs) prevent HIV entry into cells or release of viral RNA from endosomes. The most successful treatment approach in combating HIV/AIDS is to use drug combinations. An effective combination is a cocktail of an NRTI, NNRTI, and a PI or II. The use of such drug combination use is referred to as HAART (highly active antiretroviral therapy).

The development of a vaccine for AIDS has been a long-sought goal. Among the many problems encountered in developing an HIV vaccine is the fact that the viral envelope proteins continually change their antigenic properties; HIV has a high mutation rate because reverse transcriptase lacks proofreading capabilities. One new experimental approach has been to synthesize hybrid antibodies to block the highly conserved gp120 protein from docking with host cells.

Many HIV researchers continue to take great interest in HIV-infected persons who do not develop AIDS; these individuals maintain CD4$^+$ T-cell counts of at least 600 per microliter of blood, have less than 5,000 copies of HIV RNA per milliliter of blood, and have remained this way for more than 10 years after documented infection even in the absence of antiviral agents. Other researchers are actively studying HIV patients who undergo radiation and bone marrow transplants with stem cells that lack the CCR5 co-receptor as a possible means of overcoming the disease.

Prevention and control of AIDS are achieved primarily through education. Understanding risk factors and practicing strategies to reduce risk are essential in the fight against AIDS. Barrier protection from blood and body fluids greatly limits risk of HIV infection. Education to prevent the sharing of intravenous needles and syringes is also very important. Additionally, prevention includes the continued screening of blood and blood products.

Cold Sores

A cold sore, also known as a fever blister, is a fluid-filled lesion typically found on the lips or inside the nostrils. They usually cause pain, itching, or a burning sensation, because they reside on or near nerve endings. They are caused by infection with the double-stranded DNA viruses commonly called herpes simplex viruses (genus *Simplexvirus*). Herpes simplex viruses have icosahedral capsids and are enveloped. The genome encodes approximately 100 proteins. The term herpes is derived from the Greek word meaning "to creep."

Clinical descriptions of herpes lesions date back to the time of Hippocrates (circa 400 BCE). Transmission is through direct contact of epithelial tissue surfaces with the virus. Herpes simplex virus type 1 (HSV-1; species *Human herpesvirus 1*) was initially thought to infect oral mucosal epithelium to cause cold sores or fever blisters, and herpes simplex virus type 2 (HSV-2; species *Human herpesvirus 2*) was thought to infect genital epithelium to cause genital herpes. However, both viruses can infect either tissue site. The herpesvirus genome must first enter an epithelial cell

for the initiation of infection. The initial association is between proteoglycans of the epithelial cell surface and viral glycoproteins. This is followed by a specific interaction with one of several cellular receptors collectively termed HVEMs for *herpesvirus entry mediators*. The capsid, along with some associated proteins, then migrates along the cellular microtubule transport machinery to nuclear envelope pores. This "docking" is thought to result in the viral DNA being injected through nuclear envelope pores while the capsid remains in the cytoplasm.

Active and latent phases have been identified within an infected host. After an incubation period of about a week, the active phase begins. During the active phase, the virus multiplies explosively; between 50,000 and 200,000 new virions are produced from each infected cell. During this replication cycle, herpesvirus inhibits its host cell's metabolism and degrades host DNA, inducing apoptosis. As a result, the cell dies, releasing viral progeny to infect other cells. Such an active infection may be symptom-free, or painful blisters, containing fluid and infectious virions, may occur in the infected tissue. Indeed, apparently healthy people can transmit HSV to other hosts. This is especially true of HSV-2; it can be passed to sexual contacts even when there are no clinical signs of infection. Blisters involving the epidermis and surface mucous membranes of the lips, mouth, and gums (**gingivostomatitis**) are referred to as herpes labialis (**figure 38.13**). Primary and recurring HSV infections also may occur in the eyes, causing **herpetic keratitis** (inflammation of the cornea)—currently a major cause of blindness in the United States. Blisters involving the epidermis and surface mucous membranes of the genitals and perianal region are referred to as genital herpes. In all cases, the blisters are the result of cell lysis and development of a local inflammatory response. Fever, headache, muscle aches and pains, a burning sensation, and general soreness are frequently present during the active phase.

Although blisters usually heal within a week, after a primary infection, a very interesting virulence factor initiates the long-term survival of HSV-1–infected neurons, inducing the

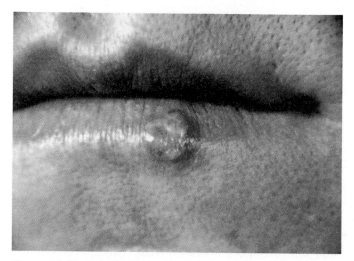

Figure 38.13 Cold Sores. Herpes simplex fever blisters on the lip, caused by herpes simplex type 1 virus.

latent phase of infection. As the active infection is curtailed by a vigorous cellular immune response, viruses released from infected oral epithelia travel to the trigeminal nerve ganglion. The retreat to the nerve ganglion (cells) results in a latent state for the lifetime of the infected host, apparently to prevent further detection by the host immune system. With time, however, HSV-1 occasionally emerges from latency to grow productively, potentially attracting the wrath of the immune system. Stressful stimuli such as excessive sunlight, fever, trauma, chilling, emotional stress, and hormonal changes can reactivate the virus. Once reactivated, the virus moves from the nerve ganglion down a peripheral nerve and back to epithelial cells to produce the active phase. Interestingly, not all infected neurons die from a productive infection. Latently infected neurons and some neurons with productive HSV infection survive. How is this? It turns out that a small HSV gene can protect the infected neuron from host immune cells. The gene encodes the latency-associated transcript (LAT), an 8.5-kb RNA. LAT is spliced out of a transcript similar to that of an mRNA precursor, but no consistently expressed protein has been found to be translated from it. Instead, a long-lived 2-kb microRNA (miR-LAT) remaining in the nucleus appears to promote degradation of two proteins that inhibit cell proliferation and initiate apoptosis. In other words, miR-LAT prevents the infected cell from self-destruction, ensures neuron survival, and provides a hiding place for the virus. ◄◄ *Eukaryotes use several other regulatory mechanisms (section 15.5)*

The drugs acyclovir, valacyclovir, and famciclovir are effective against cold sores. Idoxuridine and trifluridine are used to treat herpes infections of the eye. By adulthood, 70 to 90% of all people in the United States have been infected and have type 1 herpes antibodies. Diagnosis of HSV-1 infection is by ELISA and direct fluorescent antibody screening of tissue (*see figure 36.7*). Diagnosis may also be made through the recovery of viral DNA by PCR. These tests are especially useful in individuals who are particularly susceptible to severe infections.

Common Cold

The **common cold (coryza:** Greek *koryza*, discharge from the nostrils) is one of the most frequent infections experienced by humans. The incidence of infection is greater during the winter months, likely due to increased indoor population density, the effect of dry winter air on mucous membranes, and the decreased immune function that results from the direct effect of cold temperatures. About 50% of the cases are caused by rhinoviruses (Greek *rhinos*, nose), which are nonenveloped, positive-strand RNA viruses in the family *Picornaviridae*. There are over 115 distinct serotypes, and each of these antigenic types has a varying capacity to infect the nasal mucosa and cause a cold. In addition, immunity to many of them is transitory. Several other respiratory viruses are also associated with colds (e.g., coronaviruses and parainfluenza viruses). Thus colds are common because of the diversity of rhinoviruses, the involvement of other respiratory viruses, and the lack of a durable immunity. ◄◄ *Plus-strand RNA viruses (section 27.5)*

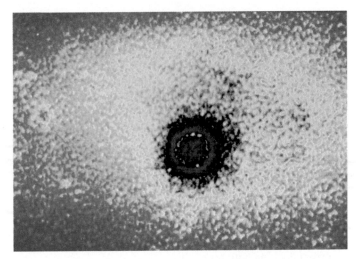

Figure 38.14 Rhinovirus Virion. One of over 100 types of virus that cause the common cold.

Rhinoviruses provide an excellent example of the medical relevance of research on virion morphology (**figure 38.14**). The complete rhinovirus capsid structure has been elucidated with the use of X-ray diffraction techniques. The results help explain rhinovirus resistance to human immune defenses. The capsid protein that recognizes and binds to host cell surface molecules during infection lies at the bottom of a surface cleft (sometimes called a "canyon") about 12 Å deep and 15 Å wide. Thus the binding site is well protected from the immune system while it carries out its functions. Moreover, with greater than 100 serotypes of human rhinoviruses, immunity to one strain may not protect against another strain, making vaccine development problematic. Possibly drugs that could fit in the cleft and interfere with virus attachment can be designed.

Viral invasion of the upper respiratory tract is the basic mechanism in the pathogenesis of a cold. The virus enters the body's cells by binding to cellular adhesion molecules. The clinical manifestations include the familiar nasal stuffiness, sneezing, scratchy throat, and watery discharge from the nose. The discharge becomes thicker and assumes a yellowish appearance over several days. General malaise is commonly present. Fever is usually absent in uncomplicated colds, although a low-grade (100 to 102°F, 37.8 to 38.9°C) fever may occur in infants and children. The disease usually runs its course in about a week. Diagnosis of the common cold is made from observations of clinical symptoms. There are no procedures for direct examination of clinical specimens or for serological diagnosis.

Sources of the cold viruses include infected individuals excreting viruses in nasal secretions, airborne transmission over short distances by way of moisture droplets, and transmission on contaminated hands or fomites. Epidemiological studies of rhinovirus colds have shown that the familiar explosive, noncontained sneeze may not play an important role in virus spread. Rather, hand-to-hand contact between a rhinovirus "donor" and a susceptible "recipient" is more likely. The common cold occurs worldwide with two main seasonal peaks, spring and early autumn. Infection is most common early in life and generally decreases with an increase in age. Treatment for the common cold is mainly rest, extra fluids, and the use of anti-inflammatory agents for alleviating local and systemic discomfort.

Cytomegalovirus Inclusion Disease

Cytomegalovirus inclusion disease is caused by human cytomegalovirus (HCMV), a member of the family *Herpesviridae* (species *Human herpesvirus 5*). HCMV is an enveloped, double-stranded DNA virus with an icosahedral capsid. Most people become infected with this virus at some time during their life; in the United States, as many as 80% of individuals older than thirty-five years have been exposed and carry a lifelong infection. Although most HCMV infections are asymptomatic, certain groups are at risk for serious illness and long-term effects. For example, HCMV remains the leading cause of congenital virus infection in the United States, a significant cause of transfusion-acquired infections, and a frequent contributor to morbidity and mortality among organ transplant recipients and immunocompromised individuals (especially AIDS patients). Because the virus persists in the body, it is shed for several years in saliva, urine, semen, and cervical secretions.

HCMV can infect any cell of the body, where it multiplies slowly and causes the host cell to swell in size; hence the prefix "cytomegalo-," which means "enlarged cell." Cytomegaloviruses are well known for their ability to interfere with many host immune functions, such as antigen presentation, cytokine production, and natural killer cell activity. Infected cells contain unique **intranuclear inclusion bodies** and cytoplasmic inclusions (**figure 38.15**). In fatal cases, cell damage is seen in the gastrointestinal tract, lungs, liver, spleen, and kidneys. In less-severe cases, cytomegalovirus inclusion disease symptoms resemble those of infectious mononucleosis.

Laboratory diagnosis is by virus isolation and culture from urine, blood, semen, and lung or other infected tissue. Serological

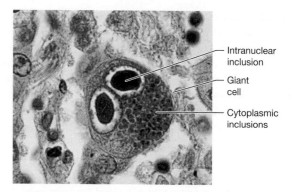

Figure 38.15 Cytomegalovirus Inclusion Disease. Light micrograph of a giant cell in a lung section infected with cytomegalovirus (×480). Cells containing intranuclear inclusion bodies have a typical "owl-eyed" appearance because of the clear halo that surrounds the inclusion. Sometimes inclusions also are visible in the cytoplasm.

tests for anti-HCMV IgM and IgG or by rapid test kits also are available. Detection of HCMV nucleic acid by PCR is also used.

The virus has a worldwide distribution, especially in developing countries, where infection is universal by childhood. The prevalence of this disease increases with poor hygiene or as socioeconomic status declines. The antivirals ganciclovir, valganciclovir, foscavir, and cidofovir are used only for high-risk patients. Infection can be prevented by avoiding close personal contact (including sexual) with an actively infected individual. Transmission by blood transfusion or organ transplantation can be avoided by using blood or organs from seronegative donors.

Genital Herpes

Genital herpes is a lifelong infection caused by herpes simplex virus, predominantly type 2 (HSV-2). (**figure 38.16a**). These viruses have a very short replication cycle. The core DNA is linear

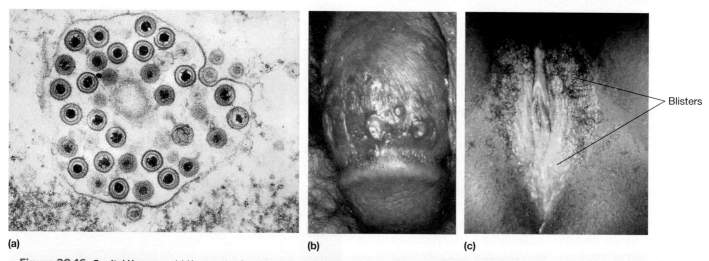

(a) **(b)** **(c)**

Figure 38.16 Genital Herpes. (a) Herpes simplex virus virions inside an infected cell. (b) Herpes vesicles on the penis. The vesicles contain fluid that is infectious. (c) Herpes blistering around the vaginal opening.

MICRO INQUIRY *Where does HSV-2 reside during latency?*

and double stranded. The HSV-2 envelope contains at least eight glycoproteins. HSV-2 is most frequently transmitted by sexual contact. Infection begins when the virus is introduced into a break in the skin or mucous membranes. The virus infects the epithelial cells of the external genitalia, the urethra, and the cervix (figure 38.16*b,c*). Rectal and pharyngeal herpes are also transmitted by sexual contact.

The details of HSV-2 infection are very similar to those of HSV-1. In the case of HSV-2, the viruses retreat to nerve cells in the sacral plexus of the spinal cord, where they remain in a latent form. As described for HSV-1, during the latent phase, the host cell does not die. Because viral genes are not expressed, the infected person is symptom-free. It should be noted that both primary infection and reactivation can occur without any symptoms, and apparently healthy people can transmit HSV-2 to their sexual partners or their newborns during vaginal delivery, the latter leading to **congenital (neonatal) herpes.** Congenital herpes is one of the most life-threatening of all infections in newborns, affecting approximately 30 out of every 100,000 babies per year in the United States. It can result in neurological involvement, as well as blindness. As a result, any pregnant female who has active or new genital herpes should have a cesarian section instead of delivering vaginally. For unknown reasons, HSV-2 is also associated with a higher-than-normal rate of cervical cancer and miscarriages.

Diagnosis of HSV-2 infection is by ELISA screening of blood or serum, direct fluorescent antibody testing of tissue, and/or PCR. Although there is no cure for genital herpes, oral use of the antiviral drugs acyclovir, valacyclovir, or famciclovir has proven to be effective in ameliorating the recurring blister outbreaks. However, famciclovir seems somewhat less effective for suppression of viral shedding. Topical acyclovir is also effective in reducing virus shedding, the time until the crusting of blisters occurs, and new lesion formation.

In the United States, the incidence of genital herpes has increased so much during the past several decades that it is now a very common sexually transmitted disease. It is estimated that 776,000 people in the United States get new herpes infections each year. That's 15.5% of individuals aged fourteen to forty-nine infected with the HSV-2, annually.

Human Herpesvirus 6 Infection

Human herpesvirus 6 (HHV-6) is the etiologic agent of **exanthem subitum** (Greek *exanthema*, rash) in infants. HHV-6 is a unique member of the family *Herpesviridae* that is distinct serologically and genetically from the other herpesviruses. The virus envelope encloses an icosahedral capsid and a core containing double-stranded DNA. The disease caused by HHV-6 was originally termed **roseola infantum** and then given the ordinal designation **sixth disease** to differentiate it from other exanthems and roseolas. Exanthem subitum is a short-lived disease characterized by a high fever of 3 to 4 days' duration, after which the temperature suddenly drops to normal and a macular rash appears on the trunk and then spreads to other areas of the body. HHV-6 infects over 95% of the U.S. infant population, and most children are

seropositive for HHV-6 by three years of age. CD4$^+$ T-cells are the main site of viral replication, whereas monocytes are in an infected, latent state. In adults, HHV-6 is commonly found in peripheral-blood mononuclear cells and saliva, suggesting that the infection is lifelong. Since the salivary glands are the major site of latent infection, transmission is probably by way of saliva.

Because HHV-6 produces latent infections, it occasionally reactivates in immunocompromised hosts, leading to pneumonitis. Furthermore, HHV-6 has been implicated in several other diseases (lymphadenitis and multiple sclerosis) in immunocompetent adults. Diagnosis is by immunofluorescence or enzyme immunoassay, or PCR. To date, there is no antiviral therapy or prevention.

Human Parvovirus B19 Infection

Since its discovery in 1974, **human parvovirus B19** (family *Parvoviridae*, genus *Erythrovirus*) has emerged as a significant human pathogen. B19 virions are uniform, icosahedral, naked particles approximately 23 nm in diameter. Parvoviruses have a genome composed of one ssDNA molecule of about 5,000 bases. Parvoviruses are among the simplest of the DNA viruses. The genome is so small that it must resort to the use of overlapping genes to encode the very few proteins it encodes. The genome does not code for any enzymes, and the virus must use host cell enzymes for all biosynthetic processes. Thus viral DNA can only be replicated in the nucleus during the S phase of the cell cycle, when the host cell replicates its own DNA. Because the viral genome is single stranded and linear, the host DNA polymerase must be tricked into copying it. By using a self-complementary sequence at the ends of the viral DNA, the parvovirus genome folds back on itself to form a primer for replication (*see figure 27.19*). This is recognized by the host DNA polymerase and DNA replication ensues. ◄◄ *Parvoviruses (section 27.3)*

A spectrum of disease is caused by parvovirus B19 infection, ranging from mild symptoms (fever, headache, chills, malaise) in normal persons and **erythema infectiosum (figure 38.17)** in children (also called **fifth disease**), to a joint disease syndrome in

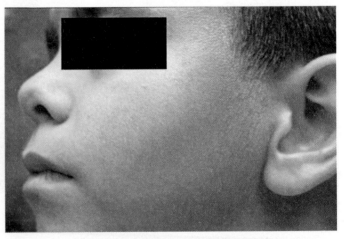

Figure 38.17 Erythema infectiosum. The facial rash of fifth disease, sometimes called "slapped face syndrome."

adults. More serious diseases include aplastic crisis in persons with sickle cell disease and autoimmune hemolytic anemia, and pure red cell aplasia due to persistent B19 virus infection in immunocompromised individuals. The B19 parvovirus can also infect a fetus, resulting in anemia, fetal hydrops (the accumulation of fluid in tissues), and spontaneous abortion. This has prompted a grass-roots awareness of B19 complications among school teachers who may be pregnant. It is assumed that the natural mode of infection is by the respiratory route. The average incubation period is 4 to 14 days. Approximately 20% of infected individuals are asymptomatic, and a smaller percentage of infected individuals have symptoms for up to 3 weeks. Infection typically results in a lifelong immunity to B19.

B19 infection is typically detected by serology (presence of anti-B19 antibodies) and/or PCR. Antiviral antibodies appear to represent the principal means of defense against infection and disease. The treatment of individuals suffering from acute and persistent B19 infections with commercial immunoglobulins containing anti-B19 and human monoclonal antibodies to B19 is an effective therapy. As with other diseases spread by contact with respiratory secretions, frequent hand-washing is the best prevention of the disease.

Mononucleosis (Infectious)

Epstein-Barr virus (EBV; species *Human herpesvirus 4*) is a member of the family *Herpesviridae*. EBV exhibits the characteristic herpesvirus morphology: all herpesviruses consist of an icosahedral capsid (approximately 125 nm in diameter) surrounded by an envelope. The capsid contains double-stranded (ds) DNA. Its dsDNA exists as a linear form in the mature virion and a circular episomal form in latently infected host cells. EBV is the etiologic agent of **infectious mononucleosis** (mono), a disease whose symptoms closely resemble those of cytomegalovirus-induced mononucleosis. Because EBV occurs in oropharyngeal secretions, it can be spread by mouth-to-mouth contact (hence the terminology infectious and kissing disease) or shared drinking bottles and glasses. A person gets infected when the virus from someone else's saliva makes its way into epithelial cells lining the throat. After a brief bout of multiplication in the epithelial cells, the new virions are shed and infect memory B cells. Infected B cells rapidly proliferate and take on an atypical appearance (Downey cells) that is useful in diagnosis (**figure 38.18**). The disease is manifested by enlargement of lymph nodes and spleen, sore throat, headache, nausea, general weakness and tiredness, and a mild fever that usually peaks in the early evening. The disease lasts for 1 to 6 weeks and is self-limited. Like other herpesviruses, EBV becomes latent in its host.

Treatment of infectious mononucleosis is largely supportive and includes plenty of rest. Diagnosis is usually confirmed by demonstration of an increase in circulating mononuclear cells, along with a serological test for nonspecific (cross-reactive) antibodies, specific anti-viral antibodies, or identification of viral nucleic acid. Several rapid tests are available.

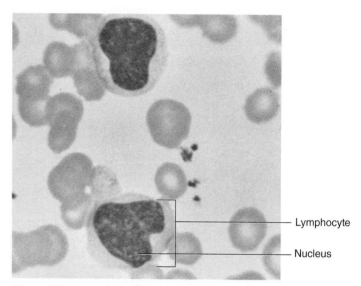

Figure 38.18 **Evidence of Epstein-Barr Virus Infection in the Blood Smear of a Patient with Infectious Mononucleosis.** Note the abnormally large lymphocytes (sometimes called Downey cells) containing indented nuclei with light discolorations.

The peak incidence of mononucleosis occurs in people fifteen to twenty-five years of age. Collegiate populations, particularly those in upper socioeconomic classes, have a high incidence of the disease. About 50% of college students have no immunity, and approximately 15% of these can be expected to contract mononucleosis. People in lower socioeconomic classes tend to acquire immunity to the disease because of early childhood infection. EBV may well be the most common virus in humans, as it infects 80 to 90% of all adults worldwide. EBV infections are associated with the cancers Burkitt lymphoma in tropical Africa and nasopharyngeal carcinoma in Southeast Asia, East and North Africa, and in Inuit populations.

Viral Hepatitides

Inflammation of the liver is called **hepatitis** (pl., hepatitides; Greek *hepaticus*, liver). Currently 11 viruses are recognized as causing hepatitis. Two are herpesviruses (cytomegalovirus [CMV] and Epstein-Barr virus [EBV]) and nine are hepatotropic viruses that specifically target liver hepatocytes. EBV and CMV cause mild, self-resolving forms of hepatitis with no permanent hepatic damage. Both viruses cause the typical infectious mononucleosis syndrome of fatigue, nausea, and malaise. Of the nine human hepatotropic viruses, only five are well characterized; hepatitis G (**table 38.3**) and TTV (transfusion-transmitted virus) were only recently discovered. Hepatitis A (sometimes called infectious hepatitis) and hepatitis E are transmitted by fecal-oral contamination and discussed in the section on food-borne and waterborne diseases (section 38.4). The other major types include hepatitis B (sometimes called serum hepatitis), hepatitis C, and hepatitis D.

Table 38.3	Characteristics of Hepatitides Caused by Hepatotropic Viruses				
Disease	**Genome**	**Classification**	**Transmission**	**Characteristic**	**Prevention**
Hepatitis A	RNA	*Picornaviridae, Hepatovirus*	Fecal-oral	Subclinical, acute infection	Killed HAV (Havrix vaccine)
Hepatitis B	DNA	*Hepadnaviridae, Orthohepadnavirus*	Blood, needles, body secretions, placenta, sexually	Subclinical, acute chronic infection; cirrhosis; primary hepatocarcinoma	Recombinant HBV vaccines
Hepatitis C	RNA	*Flaviviridae, Hepacivirus*	Blood, sexually	Subclinical, acute chronic infection; primary hepatocarcinoma	Routine screening of blood
Hepatitis D	RNA	Satellite	Blood, sexually	Coinfection with HBV required for satellite replication	HBV vaccine
Hepatitis E	RNA	*Hepeviridae, Hepevirus*	Fecal-oral	Subclinical, acute infection (but high mortality in pregnant women)	Improve sanitary conditions

Hepatitis B (serum hepatitis) is caused by **hepatitis B virus** (HBV), an enveloped virus with a partially double-stranded circular DNA genome of complex structure. HBV is classified as an *Orthohepadnavirus* within the family *Hepadnaviridae*. Members of this family are reverse-transcribing DNA viruses, so named because they encode a reverse transcriptase enzyme that is used to generate their genomes. ◀◀ *Reverse transcribing DNA viruses (section 27.8)*

Serum from individuals infected with HBV contains three distinct antigenic particles: a spherical 22 nm particle, a 42 nm spherical particle (containing DNA and a polymerase) called the **Dane particle,** and tubular or filamentous particles that vary in length (**figure 38.19**). The viral genome is 3.2 kb in length, consisting of four partially overlapping, open-reading frames that encode viral proteins. Viral multiplication takes place predominantly in hepatocytes. The infecting virus encases its double-shelled Dane particles within membrane envelopes coated with hepatitis B surface antigen (HBsAg). The inner nucleocapsid core antigen (HBcAg) encloses a single molecule of double-stranded HBV DNA and polymerase. HBsAg in body fluids is (1) an indicator of active hepatitis B infection, (2) used in the large-scale screening of blood for HBV, and (3) the basis for the first vaccine for human use developed by recombinant DNA technology. Diagnosis of HBV is made by detection of HBsAg in unimmunized individuals or HBcAg antibody. Viral load is determined using PCR to measure HBV nucleic acid.

Hepatitis B virus is normally transmitted through blood or other body fluids (saliva, sweat, semen, breast milk, urine, feces) and body-fluid–contaminated equipment (including shared intravenous needles). The virus can also pass through the placenta to the fetus of an infected mother. It is estimated that there have been 2 billion persons worldwide infected with HBV, and more than 400 million persons with chronic, lifelong HBV infection. In the United States, there are currently 40,000 new cases annually and one million chronically infected Americans. Additionally, 5,000 persons die yearly from HBV-related complications.

The clinical signs of hepatitis B vary widely. Most cases are asymptomatic. However, sometimes fever, loss of appetite, abdominal discomfort, nausea, fatigue, and other symptoms gradually appear following an incubation period of 1 to 3 months. The virus infects hepatic cells and causes liver tissue degeneration and the release of liver-associated enzymes (transaminases) into the bloodstream. This is followed by jaundice, the accumulation

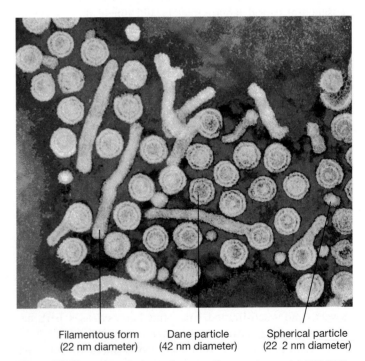

| Filamentous form (22 nm diameter) | Dane particle (42 nm diameter) | Spherical particle (22 2 nm diameter) |

Figure 38.19 Hepatitis B Virus in Serum.   Electron micrograph (×210,000) showing the three distinct types of hepatitis B antigenic particles. The spherical particles and filamentous forms are small spheres or long filaments without an internal structure, and only two of the three characteristic viral envelope proteins appear on their surface. Dane particles are the complete, infectious virion.

of bilirubin (a breakdown product of hemoglobin) in the skin and other tissues with a resulting yellow appearance. Chronic hepatitis B infection also causes the development of primary liver cancer, known as hepatocellular carcinoma. HBV is second only to tobacco as a known cause of human cancer.

General measures for prevention and control involve (1) excluding contact with HBV-infected blood and secretions, and minimizing accidental needlesticks; (2) passive prophylaxis with intramuscular injection of hepatitis B immune globulin within 7 days of exposure; and (3) active prophylaxis with a recombinant vaccines that contains S antigen (envelope protein, HBsAg). The vaccine is widely used and recommended for routine prevention of HBV in infants to eighteen-year-olds and risk groups of all ages (e.g., household contacts of HBV carriers, health-care and public safety professionals, men who have sex with other men, international travelers, hemodialysis patients). Recommended treatments for HBV include pegylated interferon-α (IFN-α), entecavir (ETV), and tenofovir disoproxil fumarate (TDF) as the first-line antiviral chemotherapy.

Hepatitis C is caused by **hepatitis C virus** (HCV), which has a 50-nm enveloped virion that contains a single strand of linear RNA. HCV is a member of the family *Flaviviridae* and is classified into multiple genotypes. It is transmitted by contact with virus-contaminated blood, by the fecal-oral route, by in utero transmission from mother to fetus, sexually, or through organ transplantation. Diagnosis is made by ELISA, which detects serum antibody to a recombinant antigen of HCV, and nucleic acid detection by PCR. HCV is found worldwide. Prior to routine screening, HCV accounted for more than 90% of hepatitis cases developed after a blood transfusion. Worldwide, hepatitis C has reached epidemic proportions, 2 million new cases reported annually. In the United States, HCV is the most common chronic blood-borne infection; approximately 3.2 million persons are chronically infected and 16,500 new cases occur annually. Currently HCV has a 48% annual fatality rate in the United States. Furthermore, HCV is the leading reason for liver transplantation in the United States. According to the U.S. Department of Veterans Affairs, 75 percent of people in the United States infected with HCV have genotype 1. Treatment for genotype 1 disease is now with the 2014 FDA-approved drug combination of sofosbuvir (a nucleotide analog polymerase inhibitor) and ledipasvir (an HCV replication inhibitor). A 12-week treatment with this drug combination results in complete cure of the virus from the host. Other genotypes continue to be treated with various combinations of antiviral drugs, including ribavirin and sofosbuvir for at least 12 weeks. This therapy can rid the virus in up to 80% of those infected with genotype 2. For genotype 3 infection, ribavirin and sofosbuvir are also recommended except for rare occasions where they are combined with pegylated (coupled to polyethylene glycol) recombinant IFN-α to effect cure.

In 1977 a cytopathic hepatitis agent termed the **delta agent** was discovered. Later it was called the hepatitis D virus (HDV) and the disease hepatitis D was designated. It is now named hepatitis delta virus. There are three known genotypes of HDV: genotype 1 has a worldwide distribution; genotype 2 exists in Taiwan, Japan, and northern Asia; and genotype 3 is found in South America. HDV is a satellite virus that is dependent on hepatitis B virus to provide the envelope protein (HBsAg) for its own replication. HDV's dependence on HBV means that HDV only replicates in liver cells coinfected with actively replicating HBV. HDV is spread only to persons who are already infected with HBV (superinfection) or to individuals who get HBV and the satellite at once (coinfection). Thus, the negative-strand RNA of HDV is smaller than the RNA of the smallest picornaviruses, and its circular conformation differs from the linear structure typical of animal negative-strand RNA viruses.

The primary laboratory tools for diagnosis of an HDV infection are serological tests for antidelta antibodies. Treatment of patients with chronic HDV infection remains difficult. There is no FDA-approved treatment for HDV. Some positive results have been obtained with interferon-α treatment for 3 months to 1 year. Liver transplantation is the only alternative to chemotherapy. Worldwide, there are approximately 15 million individuals infected with HDV. Thus because of the propensity of HDV to cause acute as well as chronic liver disease, continued incursion of HDV into areas of the world where persistent hepatitis B infection is endemic has serious implications. Prevention and control involve the widespread use of the hepatitis B vaccine. ◀◀ *Viroids and satellites (section 6.6)*

Warts

Warts, or verrucae (Latin *verruca*, wart), are horny projections on the skin caused by human papillomaviruses. Human papillomavirus (HPV) is the name given to a group of DNA viruses that includes more than 100 different strains, some of which are oncogenic (cancer-associated) (**figure 38.20***a*). They differ in the types of epithelium they infect; some infect cutaneous sites, whereas others infect mucous membranes. More than 30 of these viruses are sexually transmitted; they can cause genital warts— soft, pink cauliflower-like growths that occur on the genital area of women (figure 38.20*b*) and men, as well as on the anal region (figure 38.20*c*). Genital warts are the most common sexually transmitted disease in the United States today.

Papillomaviruses are placed in the family *Papillomaviridae*. These viruses have nonenveloped icosahedral capsids with a double-stranded, supercoiled, circular DNA genome. At least eight distinct genotypes produce benign epithelial tumors that vary in respect to their location, clinical appearance, and histopathologic features. Warts occur principally in children and young adults, and are limited to the skin and mucous membranes. The viruses are spread between people by direct contact; autoinoculation occurs through scratching. Four major kinds of warts are **plantar warts, verrucae vulgaris, flat** or **plane warts,** and **anogenital condylomata (venereal warts).** Treatment includes physical destruction of the wart by electrosurgery, cryosurgery with liquid nitrogen or solid CO_2, laser fulguration (drying), direct application of the drug podophyllum to the wart, or injection of IFN-α.

Anogenital condylomata are sexually transmitted. Once HPV enters the body, the incubation period is 1 to 6 months.

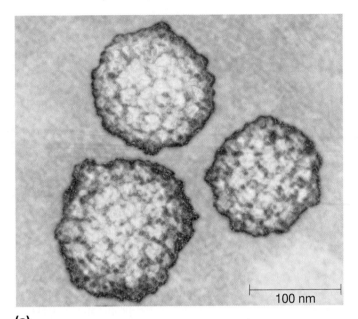

(a)

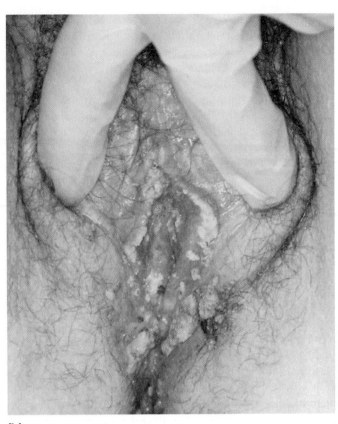

(b)

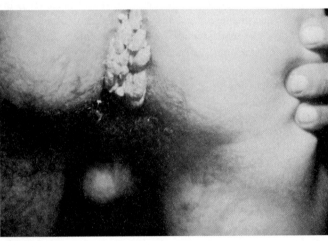

(c)

Figure 38.20 Human Papillomavirus (HPV). (a) Transmission electron micrograph of HPV virions. (b) Genital warts of the vaginal labia. (c) Genital warts in the anal region.

Genital infection with HPV is of considerable importance because specific types of genital HPV play a major role in the pathogenesis of epithelial cancers of the male and female genital tracts. HPV strains are designated by numbers and can be divided into "high-risk" (oncogenic or cancer causing) and "low-risk" (nononcogenic) strains. Four high-risk strains (6, 11, 16, and 18) are responsible for 70% of cervical cancers and 90% of genital warts. Specifically they have been found in association with invasive cancers of the cervix, vulva, penis, or anus (and other sites). Other high-risk strains have been found to be associated with cancers no more than 1% of the time. These other strains cause benign or low-grade cervical cell changes and genital warts but are rarely, if ever, associated with invasive cancer.

About 79 million Americans are currently infected with HPV. About 14 million become newly infected annually. Frequent sexual contact is the most consistent predictor of HPV infection; the number of sexual partners is proportionately linked to the risk of HPV infection. About half of the infected individuals are sexually active teens and young adults (fifteen to twenty-four years of age). HPV is typically transmitted through direct contact, especially during vaginal or anal sex. Other types of genital contact in the absence of penetration (oral-genital, manual-genital, and genital-genital contact) can also lead to infection with HPV, but infection by these routes is less common. Most HPV infections are asymptomatic, transient, and clear on their own without treatment. However, in some individuals, HPV infections result in genital warts weeks or months after infection.

A quadrivalent vaccine against four strains of HPV (6, 11, 16, and 18) was licensed in the United States in 2006. The vaccine is made from noninfectious HPV-like particles and was recommended by the U.S. Advisory Committee on Immunization It is recommended for prophylactic use in females eleven to twenty-six years of age and males eleven to twenty-one years of age. The vaccine has been found to be safe and to cause no serious side effects. Clinical trials have demonstrated 100% efficacy in preventing cervical precancers caused by the four HPV strains and nearly 100% efficacy in preventing vulvar and vaginal precancers and genital warts caused by the same four strains.

Retrieve, Infer, Apply

1. Describe HIV and how it cripples the immune system. What types of pathological changes can result?
2. Why do people periodically get cold sores? Describe the causative agent.
3. Why do people get the common cold so frequently? How are cold viruses spread?
4. Give two major ways in which herpes simplex virus type 2 is spread. Why do herpes infections become active periodically?
5. Describe the causative agent and some symptoms of mononucleosis and exanthem subitum.
6. What are the different causative viruses of hepatitis, and how do they differ from one another? How can one avoid hepatitis? Do you know anyone who is a good candidate for infection with these viruses?
7. What kind of viruses cause the formation of warts? Describe the formation of venereal warts and a method to prevent infection.

38.4 Food and Water Are Vehicles for Viral Diseases

After reading this section, you should be able to:

- Report the common food-borne and waterborne viral diseases
- Identify typical signs and symptoms of common food-borne and waterborne viral diseases
- Correlate food-borne and waterborne virus infection and disease severity with viral virulence factors

Food and water have been recognized as potential carriers (vehicles) of disease since the beginning of recorded history. Collectively more infectious diseases are transmitted by these two routes than any other. A few of the many human viral diseases that are food- and waterborne are now discussed. ▶▶❘ *Purification and sanitary analysis ensure safe drinking water (section 43.1)*

Gastroenteritis (Viral)

Acute viral gastroenteritis, inflammation of the stomach or intestines, is caused by six major categories of viruses: adenoviruses, astroviruses, bocaparvoviruses (bocaviruses and parvoviruses), caliciviruses (noroviruses and sapoviruses), parechoviruses, and rotaviruses (**figure 38.21**). The medical importance of these viruses is summarized in **table 38.4.** Acute viral gastroenteritis is a common illness, with a global incidence of 3 to 5 billion cases and 1.4 million deaths annually. Viral diarrhea is an especially common cause of mortality among children less than five years of age in developing countries, with 1.7 billion episodes

and 700,000 deaths estimated to occur annually. In the United States, acute diarrhea in children accounts for more than 3.5 million cases, resulting in more than 500,000 office visits, 55,000 hospitalizations, and 30 deaths per year. The viruses responsible for gastroenteritis are transmitted by the fecal-oral route.

Noroviruses are estimated to cause about 20 million U.S. cases of acute gastroenteritis (at least 50% of all food-borne outbreaks of gastroenteritis). Furthermore, noroviruses are responsible for 1.9 million outpatient visits, 400,000 emergency department visits (primarily in young children), about 56,000 to 71,000 hospitalizations and 570–800 deaths (mostly of young children and the elderly). At least another 20% of global cases and the majority of severe cases of diarrhea in children under age 5 are due to rotaviruses, which are the leading cause of severe diarrhea in infants and children worldwide. Prior to the widespread use of vaccine to counter rotavirus in the U.S., there were more than 400,000 doctor visits, more than 200,000 emergency room visits, 55,000 to 70,000 hospitalizations, and 20 to 60 deaths, annually. Vaccination has reduced yearly rotavirus hospitalizations to fewer than 20,000, and deaths to single digits. Adenoviruses, sapoviruses, and astroviruses are the other major viral agents responsible for gastroenteritis. Genomic testing of patients with gastroenteritis has revealed two additional viral causes: human bocavirus 2 and parechoviruses (table 38.4).

The average incubation period for most of these viral diseases is 1 to 2 days. Viral gastroenteritis is seen most frequently in infants one to eleven months of age, where the virus attacks the epithelial cells of the upper intestinal villi, causing malabsorption, impairment of sodium transport, and diarrhea. The clinical manifestations typically range from asymptomatic to a relatively mild diarrhea with headache and fever, to a severe, watery, nonbloody diarrhea with abdominal cramps. Fatal dehydration is most common in young children. Vomiting is almost always present, especially in children. Viral gastroenteritis can be self-limiting. Treatment is designed to provide relief through the use

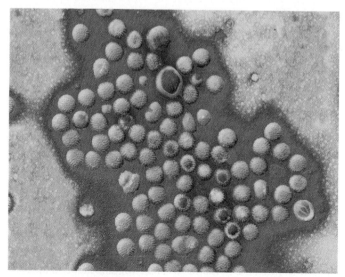

Figure 38.21 Rotavirus Virions. Electron micrograph of rotavirus virions (family *Reoviridae*) in a human gastroenteritis stool filtrate (×90,000).

Family	Virus	Characteristics
Adenoviridae	Adenoviruses (subgroup F, serotypes 40 & 41)	Primarily pediatric cases, outbreaks on military bases, more severe in the immunocompromised
Astroviridae	Human Astroviruses	Primarily pediatric cases and outbreaks in nursing homes, lasts 1–3 days
Caliciviridae	Noroviruses (genotypes I and II)	Pediatric epidemics, outbreaks common in older children and adults (families, communities, social events); often associated with shellfish, food handlers with poor hand hygiene, and cruise ships
	Sapoviruses (genogroups 1, 2, 4, & 5)	Primarily pediatric cases with mild diarrhea, rarely adult cases associated with shellfish and other foods
Parvoviridae	Human bocavirus 2	Primarily pediatric cases; mild to severe diarrhea, rarely with acute flaccid paralysis
Picornaviridae	Human parechoviruses	Primarily pediatric cases; mild diarrhea, rarely with myocarditis and encephalitis complications
Reoviridae	Rotaviruses (Groups A-C)	Primarily pediatric cases, large outbreaks in adults; dehydrating diarrhea (5–7 days) with cramps, fever, nausea, and vomiting in pediatric cases, 3–5 days diarrhea in adult cases

Table 38.4 **Medically Important Gastroenteritis Viruses**

of oral fluid replacement with isotonic liquids, analgesics, and antiperistaltic agents. Symptoms usually last for 1 to 5 days, and recovery often results in protective immunity to subsequent infection.

Hepatitis A

Hepatitis A (infectious hepatitis) usually is transmitted by fecal contamination of food or drink, or shellfish that live in contaminated water. The disease is caused by hepatitis A virus (HAV) of the genus *Hepatovirus* in the family *Picornaviridae*. Recall that while all hepatitis viruses can cause liver disease, not all are taxonomically related (table 38.3). HAV is an icosahedral, linear, positive-strand RNA virus that lacks an envelope. Once in the digestive system, the viruses multiply within the intestinal epithelium. Usually only mild intestinal symptoms result. Occasionally viruses are found in the blood (viremia) and may spread to the liver. The viruses reproduce in the liver, enter the bile, and are released into the small intestine. This explains why feces are so infectious. After about a 4-week incubation period, symptoms develop that include anorexia, general malaise, nausea, diarrhea, fever, and chills. If the liver becomes infected, jaundice ensues. Most cases resolve in 4 to 6 weeks and yield a strong immunity, although some patients relapse, exhibiting symptoms for 6 months or more. Fortunately the mortality rate is low (less than 1%). Approximately 40 to 80% of the U.S. population has antibodies, though few are aware of having had the disease. Control of infection is by simple hygienic measures, the sanitary disposal of excreta, and the HAV vaccine. The number of new cases has been dramatically reduced since the introduction of the hepatitis A vaccine in the 1990s. This vaccine is recommended for travelers (*see table 37.3*) going to regions with high rates of hepatitis A.

Hepatitis E

Hepatitis E is implicated in many epidemics in certain developing countries in Asia, Africa, and Central and South America. It is uncommon in the United States but is occasionally imported by infected travelers. The single, positive-strand RNA viral genome (7,900 nucleotides) is linear. The virion is spherical, nonenveloped, and 32 to 34 nm in diameter.

Infection usually is associated with feces-contaminated drinking water. Presumably HEV enters the blood from the gastrointestinal tract, replicates in the liver, is released from hepatocytes into the bile, and is subsequently excreted in the feces. Like hepatitis A, an HEV infection usually runs a benign course and is self-limiting. The incubation period varies from 15 to 60 days, with an average of 40 days. The disease is most often recorded in patients who are fifteen to forty years of age. Children are typically asymptomatic or present mild signs and symptoms, including abdominal pain, anorexia, dark urine, fever, hepatomegaly, jaundice, malaise, nausea, and vomiting. Case fatality rates are low (1 to 3%), except for pregnant women (15 to 25%), who risk death from fulminant hepatic failure. Diagnosis of HEV infection is by ELISA (IgM or IgG to recombinant HEV) or reverse transcriptase PCR. There are no specific measures for preventing HEV infections, other than those aimed at improving the level of health and sanitation in affected areas.

Poliomyelitis

Poliomyelitis (Greek *polios*, gray, and *myelos*, marrow or spinal cord), **polio**, or **infantile paralysis** is caused by poliovirus (species *Human enterovirus C*) and was first described in England in 1789 as a "leg-wasting" disease of children, although a pictograph from ancient Egypt clearly depicts a man with what appears to be polio (**Historical Highlights 38.1**). The virus is an

HISTORICAL HIGHLIGHTS

38.1 A Brief History of Polio

Like many other infectious diseases, polio is probably of ancient origin. Various Egyptian hieroglyphics dated approximately 2000 BCE depict individuals with wasting, withered legs and arms (**box figure**). In 1840 the German orthopedist Jacob von Heine described the clinical features of poliomyelitis and identified the spinal cord as the problem area. Little further progress was made until 1890, when Oskar Medin, a Swedish pediatrician, portrayed the natural history of the disease as epidemic in form. He also recognized that a systemic phase, characterized by minor symptoms and fever, occurred early and was complicated by paralysis only occasionally. Major progress occurred in 1908, when Karl Landsteiner and William Popper successfully transmitted the disease to monkeys. In the 1930s much public interest in polio occurred because of the polio experienced by U.S. President Franklin D. Roosevelt. This led to the founding of the March of Dimes campaign in 1938; the sole purpose of the March of Dimes was to collect money for research on polio. In 1949 John Enders, Thomas Weller, and Frederick Robbins discovered that poliovirus could be propagated in vitro in cultures of human embryonic tissues of nonneural origin. This was the keystone that later led to the development of vaccines.

In 1952 David Bodian recognized that there were three distinct serotypes of poliovirus. Jonas Salk successfully immunized humans with formalin-inactivated poliovirus in 1952, and this vaccine (IPV) was licensed in 1955. The live attenuated poliovirus vaccine (oral polio vaccine, OPV) developed by Albert Sabin and others had been employed in Europe since 1960 and was licensed for U.S. use in 1962. Both the Salk and Sabin vaccines led to a dramatic decline of paralytic poliomyelitis in most developed countries and, as such, have been rightfully hailed as two of the great accomplishments of medical science.

Ancient Egyptian with Polio. Note the withered leg.

enterovirus—a transient inhabitant of the gastrointestinal tract—and a member of the family *Picornaviridae* and as such is a nonenveloped, positive-strand RNA virus. Three different poliovirus subtypes have been identified. Eradication efforts have eliminated strains 2 and 3, but the vaccine remains tripartite. Like other viruses transmitted by the fecal-oral route, poliovirus virions are very stable, especially at acidic pH, and can remain infectious for relatively long periods in food and water, its main routes of transmission. The incubation period ranges from 6 to 20 days.

Once ingested, the virus multiplies in the mucosa of the throat or small intestine. From these sites, the virus invades the tonsils and lymph nodes of the neck and terminal portion of the small intestine. Generally there are either no symptoms or a brief illness characterized by fever, headache, sore throat, vomiting, and loss of appetite. The virus sometimes enters the bloodstream, causing viremia. In most cases (more than 99%), the viremia is transient and clinical disease does not result. However, in a minority of cases, the viremia persists and the virus enters the central nervous system and causes paralytic polio. The virus has a high affinity for anterior horn motor nerve cells of the spinal cord. Once inside these cells, it multiplies and destroys the cells; this results in motor and muscle paralysis. Since the licensing of the tripartite Salk vaccine (1955) and the tripartite Sabin vaccine (1962), the incidence of polio has decreased markedly. No wild polio viruses exist in the United States. An ongoing global polio eradication effort has been very successful. Excitingly, as of 2014, 80% of the world's population now live in polio-free areas. Three polio-endemic countries remain: Afghanistan, Nigeria, and Pakistan. Substantial efforts are being made to eradicate polio and certify these remaining countries as polio-free. However, religious views and misinformation continue to diminish vaccination efforts and civil wars interrupt public health efforts. Nonetheless, it is likely that polio will be the next human disease to be completely eradicated.

In addition to polio, there are over 100 other enteroviruses that all have the potential to invade the human central nervous system. But like poliovirus, this is very rare. In 2014 there was an outbreak of one such virus, enterovirus D68, that targeted young children with respiratory problems, such as asthma, causing life-threatening respiratory distress and in a few cases paralysis.

Retrieve, Infer, Apply

1. What virus groups are associated with acute viral gastroenteritis? How do they cause the disease's symptoms?
2. Describe some symptoms of hepatitis A.
3. Why is hepatitis A called infectious hepatitis?
4. At what specific sites within the body can the poliomyelitis virus multiply? What is the usual outcome of an infection?

38.5 Zoonotic Diseases Arise from Human-Animal Interactions

After reading this section, you should be able to:

- Report the common viral diseases spread by contact with infected animals
- Identify typical signs and symptoms of zoonotic viral diseases
- Correlate zoonotic virus infection and disease severity with viral attachment factors

The diseases discussed here are caused by viruses that are normally zoonotic (animal-borne). Members of the RNA virus families *Arenaviridae, Bunyaviridae, Flaviviridae, Filoviridae,* and *Picornaviridae* are notable examples of viruses found in animal reservoirs before transmission to and between humans. Some of these viruses are exotic and rare; others are being eradicated by public health efforts. Some are found in relatively small geographic areas; others are distributed across continents. Several cause diseases with substantial morbidity and mortality.

Ebola Virus and Marburg Diseases

Ebola virus disease (**EVD**) is caused by Ebola viruses, first recognized near the Ebola River in the Democratic Republic of the Congo in Africa. They are members of the genus *Ebolavirus* in the family *Filoviridae,* a group of negative-strand RNA viruses. Five Ebola species are known: *Sudan ebolavirus, Tai Forst ebolavirus* (formerly *Côte d'Ivoire ebolavirus*), *Bundibugyo ebolavirus, Zaire ebolavirus,* and *Reston ebolavirus.* Members of the first four species cause disease in humans (**figure 38.22**). The fifth, *Reston ebolavirus,* was first discovered in 1989, and to date Reston viruses have caused disease only in nonhuman primates and pigs. Only Reston virus is spread by aerosol transmission.

All five types of Ebola virus are transmitted by contact with body fluids. Ebola viruses cause varying degrees of hemorrhagic fever.

Viral hemorrhagic fever (**VHF**) is the term used to describe a severe, multisystem syndrome caused by several distinct viruses. Characteristically the overall host vascular system is damaged, resulting in vascular leakage (hemorrhage) and dysfunction (coagulopathy). The incubation period for EVD ranges from 2 to 21 days and is characterized by abrupt fever, headache, joint and muscle aches, sore throat, and weakness, followed by diarrhea, vomiting, and stomach pain. Signs of infection include fever, rash, red eyes, bleeding, and hiccups—symptoms alerting of internal hemorrhage. Minor hemorrhage is symptomatic in all ebolavirus infection. Severe hemorrhage is late stage and can be minimized if patients are resuscitated in time. There is no standard treatment for Ebola infection. Patients receive supportive therapy consisting of balancing patients' fluids and electrolytes, maintaining their oxygen status and blood pressure, and treating them for any complicating infections. Convelscent sera, collected from Ebola survivors, and experimental antibody cocktails have shown promising success in the 2014 treatments of ebolavirus patients. Experimental vaccines are currently being evaluated and show promise in various animal models.

Marburg disease is caused by a genetically unique RNA virus also in the *Filoviridae* family. Marburg virus (MARV) similarly causes a hemorrhagic fever. It is a rare, severe type of hemorrhagic fever that affects both humans and nonhuman primates. MARV was first recognized in 1967, when outbreaks of hemorrhagic fever occurred simultaneously in laboratories in Marburg and Frankfurt, Germany, and in Belgrade, Yugoslavia (now Serbia). The first people infected had been exposed to African green monkeys or their tissues. In Marburg, the monkeys had been imported for research and to prepare polio vaccine. MARV is indigenous to Africa, and the reservoir host of MARV is the African fruit bat, *Rousettus aegyptiacus.* The average incubation period for Marburg hemorrhagic fever is 5 to 10 days. The disease symptoms are abrupt, marked by fever, chills, headache, and myalgia. A maculopapular rash (i.e., discolored with bumps), most prominent on the chest, back, and stomach, typically appears around the fifth day after onset. Nausea, vomiting, chest pain, sore throat,

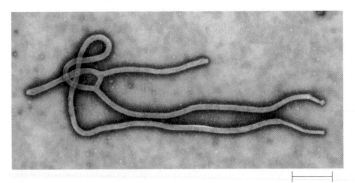

Figure 38.22 Ebola Virus Virions.

abdominal pain, and diarrhea may also occur in infected patients. Symptoms become increasingly severe and may include jaundice, delirium, liver failure, pancreatitis, severe weight loss, shock, and multiorgan dysfunction. A specific treatment for this disease is unknown. However, supportive hospital therapy should be utilized. This includes balancing the patient's fluids and electrolytes, maintaining oxygen status and blood pressure, replacing lost blood and clotting factors, and treating for other complicating infections.

Hantavirus Pulmonary Syndrome

Hantavirus pulmonary syndrome (**HPS**) is a disease caused by a negative-strand RNA virus of the family *Bunyaviridae*. HPS is typically transmitted to humans by inhalation of viral particles shed in urine, feces, or saliva of infected rodents. HPS was first recognized in 1993 and has since been identified throughout the United States. Although rare, HPS is potentially deadly. Rodent control in and around the home remains the primary strategy for preventing hantavirus infection. HPS in the United States is not transmitted from person to person, nor is it known to be transmitted by rodents purchased from pet stores.

Hantaviruses have lipid envelopes that are susceptible to most disinfectants. The length of time hantaviruses can remain infectious in the environment is variable and depends on environmental conditions. Temperature, humidity, exposure to sunlight, and even the rodent's diet, which affects the chemistry of rodent urine, strongly influence viral survival. Viability of dried virus particles has been reported at room temperature for 2 to 3 days. Hantaviruses are shed in body fluids but do not appear to cause disease in their reservoir rodent hosts. Data indicate that viral transfer may occur between rodents through biting, as field studies suggest that viral transmission in rodent populations occurs horizontally and more frequently between males. A specific treatment for HPS is unknown. Supportive therapy is used to treat symptoms, including balancing the patient's fluids and electrolytes, maintaining oxygen status and blood pressure, replacing lost blood and clotting factors, and treating for other complicating infections.

Rabies

Rabies (Latin *rabere*, rage or madness) is caused by a number of different strains of highly neurotropic viruses. Most belong to a single serotype in the genus *Lyssavirus* (Greek *lyssa*, rage or rabies), family *Rhabdoviridae*. The bullet-shaped virion contains a negative-strand RNA genome (**figure 38.23**). Rabies has been the object of human fascination, torment, and fear since the disease was first recognized. Prior to Pasteur's development of a rabies vaccine, few words were more terrifying than the cry of "mad dog!" Improvements in prevention during the past 50 years have led to almost complete elimination of indigenously acquired rabies in the United States, where rabies is primarily a disease of feral animals and domestic cats that contact feral animals. Most wild animals can become infected with rabies, but susceptibility varies according to species. Foxes, coyotes, and wolves are the

most susceptible; intermediate are skunks, raccoons, insectivorous bats, and bobcats, whereas opossums are quite resistant. Worldwide, almost all cases of human rabies are attributed to dog bites. In the more than 150 developing countries and territories, where canine rabies is still endemic, rabies accounts for tens of thousands of deaths per year. In the United States, fewer than 10 cases are reported annually. Feral raccoons, skunks, and bats account for 83% of all animal cases. Occasionally other domestic animals are responsible for transmission of rabies to humans. It should be noted, however, that not all rabid animals exhibit signs of agitation and aggression (known as furious rabies). In fact, paralysis (dumb rabies) is the more common sign exhibited by rabid animals.

The virus multiplies in the salivary glands of an infected host. It is transmitted to humans or other animals by the bite of an infected animal whose saliva contains the virus; by aerosols of the virus that can be spread in caves where bats dwell; or by contamination of scratches, abrasions, open wounds, and mucous membranes with saliva from an infected animal. After inoculation, the virion's envelope spike attaches to the plasma membrane of nearby skeletal muscle cells, which the virus enters. Multiplication of the virus then occurs. When the concentration of the virus in the muscle is sufficient, the virus enters the nervous system through unmyelinated sensory and motor terminals; the binding site is the nicotinic acetylcholine receptor.

The virus spreads by retrograde axonal flow at 8 to 20 mm per day until it reaches the spinal cord, when the first specific symptoms of the disease—pain or paresthesia at the wound site—may occur. A rapidly progressive encephalitis develops as the virus quickly disseminates through the central nervous system. The

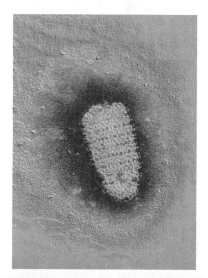

Figure 38.23 Rabies. Electron micrograph of a rabies virus virion (yellow) (×36,700). Note the bullet shape. The external surface of the virion contains spikelike glycoprotein projections that bind specifically to cellular receptors.

MICRO INQUIRY *What type of host cell does the rabies virus initially infect? To where does the virus then travel?*

virus then spreads throughout the body along the peripheral nerves, including those in the salivary glands, where it is shed in the saliva.

Within brain neurons, the virus produces characteristic Negri bodies, masses of virions or unassembled viral subunits that are visible in the light microscope. In the past, diagnosis of rabies consisted solely of examining tissue for the presence of these bodies. Today diagnosis is based on *d*irect immuno*f*luorescent *a*ntibody (DFA) of brain tissue, virus isolation, detection of Negri bodies, a rapid rabies enzyme-mediated immunodiagnosis test and by PCR testing. ▶▶ *Direct identification of infectious agents (section 36.3)*

Symptoms of rabies in humans usually begin 2 to 16 weeks after exposure and include anxiety, irritability, depression, fatigue, loss of appetite, fever, and a sensitivity to light and sound. The disease quickly progresses to paralysis. In about 50% of all cases, intense and painful spasms of the throat and chest muscles occur when the victim swallows liquids. The mere sight, thought, or smell of water can set off spasms. Consequently, rabies has been called hydrophobia (fear of water). Once symptoms of rabies develop in a human, the disease has a 100% mortality rate. Death results from destruction of the regions of the brain that regulate breathing. Safe and effective vaccines (*h*uman *d*iploid-*c*ell rabies *v*accine [HDCV] or *r*abies *v*accine *a*dsorbed [RVA]) against rabies are available; however, to be effective they must be given prior to or soon after the person has been infected. Veterinarians and laboratory personnel, who have a high risk of exposure to rabies, usually are immunized every 2 years and tested for the presence of suitable antibody titer. Prevention and control involve preexposure vaccination of dogs and cats, postexposure vaccination of humans, and preexposure vaccination of humans at special risk, including persons spending a month or more in countries where rabies is common in dogs.

Postexposure prophylaxis—rabies immune globulin for passive immunity and rabies vaccine for active immunity—is initiated to exploit the relatively long incubation period of the disease. Rabies immune globulin is typically administered directly into the wound after cleaning. The vaccine is administered as four doses given on days 3, 7, and 14 after the initial dose. This is usually recommended for anyone bitten by one of the common reservoir species (raccoons, skunks, foxes, and bats), unless it is proven that the animal was uninfected.

Some states and countries (e.g., Hawaii and Great Britain) retain their rabies-free status by imposing quarantine periods on any entering dog or cat. If an asymptomatic, unvaccinated dog or cat bites a human, the animal is typically confined and observed by a veterinarian for at least 10 days. If the animal shows no signs of rabies in that time, it is determined to be uninfected. Animals demonstrating signs of rabies are killed, and brain tissue is submitted for rabies testing.

Retrieve, Infer, Apply

1. Why are Ebola and Marburg hemorrhagic fever diseases so deadly?
2. What precautions can be taken to prevent hantavirus and rabies virus transmission to humans?
3. How does the rabies virus cause death in humans?

38.6 Prion Proteins Transmit Disease

After reading this section, you should be able to:

- Describe diseases caused by prions
- Differentiate prion disease by route of infection

Prion diseases, also called **transmissible spongiform encephalopathies (TSEs)**, are fatal neurodegenerative disorders that have attracted enormous attention not only for their unique biological features but also for their impact on public health. Prions (proteinaceous infectious particles) are thought to consist of abnormally folded proteins (PrP^{Sc}), which can induce normal forms of the protein (PrP^C) to fold abnormally *(see figure 6.24)*. This group of diseases includes scrapie, kuru, Creutzfeldt-Jakob disease (CJD), variant Creutzfeldt-Jakob disease (vCJD), Gerstmann-Sträussler-Scheinker disease (GSD), and fatal familial insomnia (FFI). Scrapie occurs primarily in sheep. The first of these diseases to be studied in humans was kuru, discovered in the Fore tribe of New Guinea. Carlton Gadjusek and others showed that the disease was transmitted by ritual cannibalism (especially where brains and spinal cords were eaten). The primary symptom of the human disorders is dementia, usually accompanied by manifestations of motor dysfunction such as cerebral ataxia (inability to coordinate muscle activity) and myoclonus (shocklike contractions of muscle groups). FFI is also characterized by dysautonomia (abnormal functioning of the autonomic nervous system) and sleep disturbances. These symptoms appear insidiously in middle to late adult life and last from months (CJD, FFI, and kuru) to years (GSD) prior to death. ◀◀ *Prions are composed only of proteins (section 6.7)* 🔁 *Prion Diseases*

Neuropathologically these disorders produce a characteristic spongiform degeneration of the brain, as well as deposition of amyloid plaques (**figure 38.24**). Prion diseases thus share important

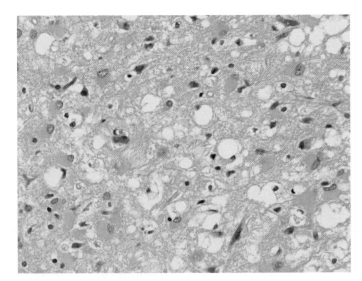

Figure 38.24 Variant Creutzfeldt-Jacob Disease (vCJD). vCJD is a prion-dependent disease that results in prominent spongiotic changes in the cortex, and loss of neurons. Note the spongiotic (holey) appearance of this brain tissue photomicrograph (magnified 100X and stained with hematoxylin and eosin).

clinical, neuropathological, and cell biological features with another, more common cerebral amyloidosis, Alzheimer's disease. Yet they cause a distinct neurological deterioration. Prion diseases are organized into three subcategories: familial (inherited) types, sporadic types, and acquired types. Familial CJD is the result of spontaneous mutation within the prion-encoding gene. Certain residues when mutated are known to give rise to CJD, but it is not clear why these amino acid substitutions confer inducible misfolding to the (PrPSc) conformation. Familial CJD is very rare, accounting for 5 to 15% of all CJD cases. It is rapidly progressive and always fatal, usually within 1 year of onset of illness.

Sporadic CJD (sCJD) has a global incidence of one case per million people per year and accounts for 75% of global CJD cases. It is most common in the 45–75 age group, with the peak age of onset being 60 to 65. There are no known gender, environmental, or lifestyle associations that would induce sCJD. This form of the disease is also rapidly progressive; symptoms move from depression and loss of coordination and walking, to loss of vision and speech, to incontinence and immobility, over an average of 4 to 5 months. Seventy percent of patients die within six months of symptom onset. CJD that is acquired is known as variant (v) CJD, so named as it is clinically and pathologically different from familial or sporadic CJD.

vCJD is acquired when a human consumes meat products from a BSE cow. Such transmission occurred during the 1980s and early 1990s in the UK, resulting in 175 cases of vCJD reported between 1996 and 2011. These data suggest a five-year incubation period. vCJD can also be acquired from a physician or surgeon, a medical treatment, or diagnostic procedures. Known as iatrogenic CJD, it has been transmitted by prion-contaminated human growth hormone, corneal grafts, and grafts of dura mater (tissue surrounding the brain). Donor screening and more thorough testing of grafts have decreased the frequency of prion transmission. Although vCJD symptoms mirror those of other CJD diseases, psychiatric symptoms are more prominent early, with dementia developing later. vCJD represents less than 5% of global CJD cases.

Retrieve, Infer, Apply

1. How are prions different from viruses? How are they similar?
2. In what way are spongiform encephalopathies commonly acquired?

Key Concepts

38.1 Viruses Can Be Transmitted by Airborne Routes

- More than 400 different viruses can infect humans. These viruses can be grouped and discussed according to their mode of transmission and acquisition.
- Most airborne viral diseases involve either directly or indirectly the respiratory system. Examples include chickenpox (varicella; **figure 38.1**), shingles (herpes zoster; **figure 38.2**), influenza (flu; **figure 38.3**), measles (rubeola; **figure 38.4**), rubella (German measles; **figure 38.5**), acute respiratory viruses such as SARS-CoV virus (**figure 38.6**), the eradicated smallpox (variola; **figure 38.7**), and viral pneumonia.

38.2 Arthropods Can Transmit Viral Diseases

- The arthropod-borne viral diseases are transmitted by arthropod vectors from human to human or animal to human.
- Examples of arthropod-borne diseases include eastern, western, and Venezuelan equine encephalitis, Bourbon virus disease, chikungunya disease, Heartland virus disease, and West Nile fever. All these diseases are characterized by fever, headache, nausea, vomiting, and other distinct characteristics.

38.3 Direct Contact Diseases Can Be Caused by Viruses

- Person-to-person contact is another way of acquiring or transmitting a viral disease.
- Examples of direct contact diseases include AIDS (**figures 38.8–38.12**), cold sores (**figure 38.13**), the common cold (rhinovirus; **figure 38.14**), cytomegalovirus inclusion disease (**figure 38.15**), genital herpes (**figure 38.16**), human herpesvirus 6 infections, human parvovirus B19 infection (**figure 38.17**), infectious mononucleosis, hepatitis B (serum hepatitis), hepatitis C, and hepatitis D (delta hepatitis) satellite virus (**table 38.3**).

38.4 Food and Water Are Vehicles for Viral Diseases

- Viruses that are transmitted in food and water usually grow in the intestinal system and leave the body in feces (**table 38.4**). Acquisition is generally by the oral route.
- Examples of diseases caused by food-borne viruses include acute viral gastroenteritis (rotaviruses and others), infectious hepatitis A, hepatitis E, and poliomyelitis.

38.5 Zoonotic Diseases Arise from Human-Animal Interactions

- Diseases transmitted from animals are zoonotic.
- Several animal viruses can cause disease in humans. Examples of viral zoonoses include Ebola and Marburg diseases, hantavirus pulmonary syndrome, and rabies (**figure 38.23**).

38.6 Prion Proteins Transmit Disease

- A prion disease is a pathological process caused by a transmissible agent (a prion) that remains clinically silent for a prolonged period, after which the clinical disease becomes apparent.
- Examples of prion diseases include Creutzfeldt-Jakob disease, variant Creutzfeldt-Jakob disease, kuru, Gerstmann-Sträussler-Scheinker disease, and fatal familial insomnia. These diseases are chronic infections of the central nervous system that result in progressive degenerative changes and eventual death.

Compare, Hypothesize, Invent

1. Explain why antibiotics are ineffective against viral infections. Advise a person about what can be done to relieve symptoms of a viral infection and recover most quickly. Address your advice to (a) someone who has had only a basic course in high school biology and (b) a third-grade student.

2. Several characteristics of AIDS render it particularly difficult to detect, prevent, and treat effectively. Discuss two of them. Contrast the disease with polio and smallpox.

3. From an epidemiological perspective, why are most arthropod-borne viral diseases hard to control?

4. In terms of molecular genetics, why is the common cold such a prevalent viral infection in humans?

5. Will it be possible to eradicate many viral diseases in the same way as smallpox? Why or why not?

6. Prior to the development of detection assay for HIV in the blood supply, HCV was used as a proxy. That is to say, any donated blood found to be contaminated with HBV or HCV was also assumed to be HIV positive as well. What do you think was the rationale behind the use of HCV as an indicator of HIV? Do you think this was a reasonable approach at the time?

7. In 2005 a reconstructed influenza virus containing eight genes from the 1918 Spanish influenza virus was introduced into human lung epithelial cells and mice. This virus was able to rapidly reproduce and cause sudden illness and death in mice. In 2008 these genes were replaced on an individual basis with homologues from a contemporary H1N1 influenza virus and tested for virulence. Substitution of the 1918 gene with the H1N1 hemagglutinin (HA), neuraminidase (NA), or polymerase PB1 subunit severely diminished the virulence of the 1918 virus.

   Explain why each of these genes is so critical for infection and virulence.

   Do you think there would be an additive effect in loss of virulence if more than one gene were replaced? Explain your answer and discuss what this implies with regard to the emergence of another highly virulent influenza virus capable of causing a pandemic.

   Read the original paper: Pappas, C., et al. 2008. Single gene reassortments identify a critical role for PB1, HA, and NA in the high virulence of the 1918 pandemic influenza virus. *Proc. Natl. Acad. Sci. USA* 105:3064–69.

8. Chronic wasting disease (CWD), a prion disease found in cervids (elk, deer, and moose), was first described in the 1960s and is now found in animals in at least 14 states. Despite its long history, it was unclear how the disease spread and if humans were at risk of acquiring the disease. In 2006 researchers designed and executed an experiment to determine if prions could be transmitted between cervids by contact with saliva, blood, feces, and urine. They found that CWD could be transmitted by oral contact with saliva and by blood following transfusion. This finding has raised concern among hunters and others who come in contact with cervid body fluids.

   How would you design a similar experiment to identify the transmission of scrapie between sheep and possible transmission to humans? Be sure to explain the controls you would use and how you would unequivocally document the development of disease.

   Read the original paper: Mathiason, C. K., et al. 2006. Infectious prions in the saliva and blood of deer with chronic wasting disease. *Science* 314:133.

9. In this chapter we have evaluated a number of viruses that cause human disease, suggesting that RNA viruses are the more rapidly evolving human pathogens. One common observation is that although RNA viruses replicate with much lower fidelity, they mutate (resulting in antigenic drift) and recombine (resulting in antigenic shift) with greater frequency and success than DNA viruses. Thus, RNA viruses evolve significantly faster than most DNA viruses. However, there is tremendous variation among the evolutionary rates of different RNA viruses, which is not explained by variation in mutation rates. Scientists at Rutgers University, in New Jersey, surveyed mammalian RNA virus rates of evolution and compared evolutionary rates to different properties of virus genomic architecture (e.g., RNA or DNA, ds or ss, number of segments, etc.) and host cell ecology. Interestingly, they found that the most significant predictor of long-term rates of mammalian RNA virus evolution was the host cell-generated environmental condition influencing virus tropism. In other words, host cell characteristics that entice viruses to infect them seem to drive viral evolution better than any virus-associated factor.

   What aspects of tropism (viral infection of a specific cell type) do you think would drive virus evolution? Why? How would infection of epithelial cells (respiratory or gastrointestinal) be favored over infection of neurons, for example? Can you suggest a mechanism by which RNA viruses would risk extinction in attempting to enhance infectivity?

   Read the original article: Hicks, A. L., and Duffy, S. 2014. Cell tropism predicts long-term nucleotide substitution rates of mammalian RNA viruses. *PLoS Pathog.* 10(1): e1003838. doi:10.1371/journal.ppat.1003838.

39

Human Diseases Caused by Bacteria

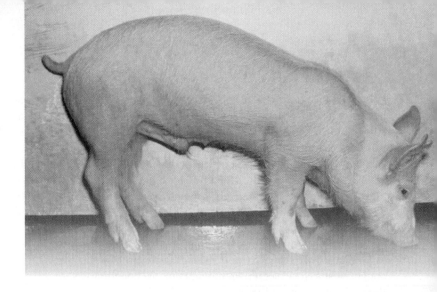

New strains of MRSA have been discovered in pigs on many U.S. and European farms.

"This Little Piggie Stayed Home"

Methicillin-resistant *Staphylococcus aureus* (MRSA) seems to be everywhere. One U.S. lab has tracked a single strain originally from European pigs to the United States. But identification of MRSA is not always easy. A novel strain of methicillin-resistant staphylococci has been identified by two independent groups. In both laboratories, several bacterial isolates were found to be methicillin resistant by culture and sensitivity techniques but negative for the primary MRSA resistance gene, *mecA*, when assayed by PCR amplification. These results suggested the bacterial isolates should not be methicillin resistant. Negative results for the *mecA* gene product, a modified penicillin-binding protein, using a slide agglutination test, also suggested that the isolated bacteria lacked methicillin resistance. To resolve the apparent paradox, scientists at a lab in England (who isolated strains from cattle) and a lab in Ireland (whose strains were from humans) independently used whole genome sequencing to probe the DNA for known drug-resistance genes. The analysis showed that these methicillin-resistant bacteria had a novel β-lactam resistance gene, similar to *mecA*, except this version of the gene was about 60% different from published sequences and the gene used for PCR testing! In a further analysis, about two-thirds of the isolates contained this novel resistance gene (now designated as *mecC*). More concerning are reports from the U.S. lab revealing that this new MRSA has infiltrated horse farms and veterinary hospitals. Only time will tell if, but more likely when, this MRSA strain will commonly be detected as the source of human infections like its *mecA* cousin.

Despite all the press MRSA and other pathogens receive, only a relatively few bacterial species are pathogenic to humans. Some human diseases have been only recently recognized; others have been known since antiquity. In this chapter, we continue our discussion of infectious disease by turning our attention to bacterial pathogens. These include bacteria that cause localized and systemic infections. We present examples of bacteria and the diseases they cause by route of transmission, as well as opportunistic bacteria. Diseases caused by bacteria categorized by the

Centers for Disease Control and Prevention as select agents (potential bioterror agents) are identified within each section.

Readiness Check:
Based on what you have learned previously, you should be able to:

✔ Review basic bacterial cell biology (sections 3.2–3.9)
✔ Explain how key pathogens cause infection (chapter 35)
✔ Compare and contrast the general principles of innate and adaptive immunity (sections 33.1, 34.1–34.3)

39.1 Bacteria Can Be Transmitted by Airborne Routes

After reading this section, you should be able to:

■ Report the common airborne bacterial diseases
■ Identify typical signs and symptoms of airborne bacterial diseases
■ Correlate airborne bacterial infection and disease severity with bacterial virulence factors

Most airborne bacterial diseases infect the respiratory tract. Inhalation of infectious particles or self-inoculation can cause disease of the sinuses, throat, bronchus, or lungs. Some airborne bacteria can cause skin diseases. Some of the better known of these diseases are now discussed.

Chlamydial Pneumonia

Chlamydial pneumonia is caused by *Chlamydophila pneumoniae*. The Gram-negative chlamydiae are obligate intracellular parasites. They must grow and reproduce within host cells. Although their ability to cause human disease is widely recognized, many species grow within protists and animal cells as their natural reservoir. Infection begins when the 0.2 to 0.6 μm diameter elementary bodies (EBs) are inhaled (*see figure 21.17*). EBs are designed exclusively for transmission and infection.

Host cells phagocytose the EBs and they are held in inclusion bodies, where they reorganize to form reticulate bodies (RBs). About 8 to 10 hours after infection, the RBs undergo binary fission, and continue to divide until the host cell dies. Interestingly, a chlamydia-filled inclusion body can become large enough to be visible by light microscopy. RBs differentiate into infectious EBs after 20 to 25 hours. The host cell lyses and releases the EBs 48 to 72 hours after infection. ◄◄ *Phylum* Chlamydiae *(section 21.6)*

Clinically, infections are generally mild; pharyngitis, bronchitis, and sinusitis commonly accompany some lower respiratory tract involvement. Symptoms include fever, a productive cough (respiratory secretion brought up by coughing), sore throat, hoarseness, and pain on swallowing. Infections with *C. pneumoniae* are common but sporadic. Approximately, 5 to 10% of all cases of adult pneumonia and bronchitis are caused by *C. pneumoniae,* and about 50% of adults have antibody to this bacterium. *C. pneumoniae* appears to be primarily a human pathogen directly transmitted from human to human by droplet (respiratory) secretions. Diagnosis of chlamydial pneumonia is based on symptoms and an immunofluorescence test. Macrolides are the first line of treatment, tetracycline and fluoroquinolones are also effective.

In seroepidemiological studies, *C. pneumoniae* infections have been linked with coronary artery disease, as well as vascular disease at other sites. Following a demonstration of *C. pneumoniae*–like particles in atherosclerotic plaque tissue by electron microscopy, *C. pneumoniae* genes and antigens have been detected in artery plaque. However, the microorganism is inconsistently recovered in cultures of artery plaque. As a result of these findings, a possible etiologic role of *C. pneumoniae* infection in coronary artery disease and systemic atherosclerosis is under intense scrutiny.

Diphtheria

Diphtheria (Greek *diphthera,* membrane, and *-ia,* condition) is an acute, contagious disease caused by the Gram-positive bacterium *Corynebacterium diphtheriae* (see figure 24.9). *C. diphtheriae* is well adapted to airborne transmission by way of nasopharyngeal secretions because it is very resistant to drying. Diphtheria mainly affects unvaccinated, poor people living in crowded conditions. Interestingly, the *tox* gene that encodes diphtheria toxin is found on a prophage, so only *C. diphtheriae* lysogens that carry this prophage cause disease. This is an example of phage conversion. Diphtheria toxin causes an inflammatory response and the formation of a grayish pseudomembrane on the pharynx and respiratory mucosa (**figure 39.1**). The pseudomembrane consists of dead host cells and cells of *C. diphtheriae.*

Diphtheria toxin is absorbed into the circulatory system and distributed throughout the body, where it may cause destruction of cardiac, kidney, and nervous tissues by inhibiting protein synthesis. The toxin is an AB toxin encoded by a single gene that gives rise to a protein with two domains, A and B (*see figure 35.6*). The B domain binds to the heparin-binding epidermal growth factor receptor on the surface of various eukaryotic cells. Once

bound, the toxin enters the cytoplasm by endocytosis. The B domain contains a transmembrane region that embeds itself in the endosome membrane. Acidification of the endosome breaks a disulfide bond holding the A and B domains together and releases the A domain into the cytoplasm. The cleaved catalytic domain catalyzes the attachment of ADP-ribose (from NAD$^+$) to translation elongation factor-2 (EF-2). A single enzyme (i.e., catalytic domain) can exhaust the entire supply of cellular EF-2 within hours, resulting in protein synthesis inhibition and cell death. ◄◄ *Order* Corynebacteriales *includes important human pathogens (section 24.1); Exotoxins (section 35.2)*

Typical signs and symptoms of diphtheria include a thick nasal discharge containing both mucus and pus, pharyngitis, fever, cough, and paralysis. Most current diphtheria cases involve people over thirty years of age who have a weakened immunity to the diphtheria toxin and live in tropical areas. Between 1980 and 2013, fewer than six diphtheria cases were reported annually in the United States, and most occurred in nonimmunized individuals. ◄◄ *Vaccines immunize susceptible populations (section 37.6)*

Legionnaires' Disease

In 1976 the term **Legionnaires' disease,** or **legionellosis,** was coined to describe an outbreak of pneumonia that occurred at the Pennsylvania State American Legion Convention in Philadelphia. The bacterium responsible for the outbreak was *Legionella pneumophila*, a nutritionally fastidious, aerobic, Gram-negative rod (**figure 39.2**). This bacterium is part of the natural microbial community of soil and freshwater ecosystems, and it has been found in large numbers in air-conditioning systems and shower stalls. In fact, a variety of free-living amoebae and ciliated protozoa serve as the natural reservoir for *L. pneumophila*. These protists inhabit cooling water found in air-conditioning systems and are therefore an important part of the infection cycle. ◄◄ *Class* Gammaproteobacteria *is the largest bacterial class (section 22.3)*

Legionella spp. multiply intracellularly within these protists, just as they do within human macrophages. This might explain why there is no human-to-human spread of legionellosis, as transmission of host cells containing intracellular bacteria is unlikely. Furthermore, most healthy people infected with *L. pneumophila* display the milder form of the disease known as Pontiac fever. Legionnaires' pneumonia is typically seen in immunocompromised individuals, including the elderly. In these cases, the bacteria reside within the phagosomes of alveolar macrophages, where they multiply and produce localized tissue destruction through export of a cytotoxic exoprotease. Symptoms start 2 to 14 days after exposure and include a high fever, nonproductive cough (respiratory secretions are not brought up during coughing), headache, neurological manifestations, and severe bronchopneumonia. Diagnosis depends on isolation of the bacterium, documentation of a rise in antibody titer over time, or the presence of *Legionella* antigens in the urine as detected by a rapid test kit. Treatment begins with supportive measures and the

Figure 39.1 Diphtheria Pathogenesis. (a) Diphtheria is a well-known, exotoxin-mediated infectious disease caused by *Corynebacterium diphtheriae*. The disease is an acute, contagious, febrile illness characterized by local oropharyngeal inflammation and pseudomembrane formation. If the exotoxin enters the blood, it is disseminated and can damage the peripheral nerves, heart, and kidneys. (b) The clinical appearance includes gross inflammation of the pharynx and tonsils marked by grayish patches (a pseudomembrane) and swelling of the entire area. (c) The B subunit of the diphtheria toxin binds to the host cell membrane facilitating entry of the A subunit, which ADP-ribosylates EF-2 to inhibit protein synthesis.

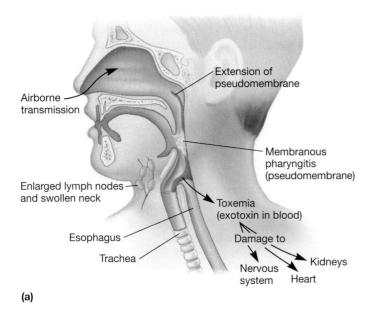

(a)

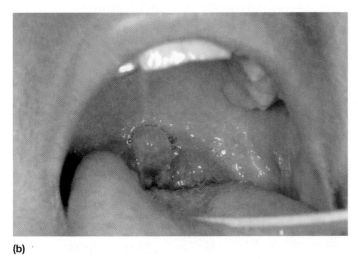

(b)

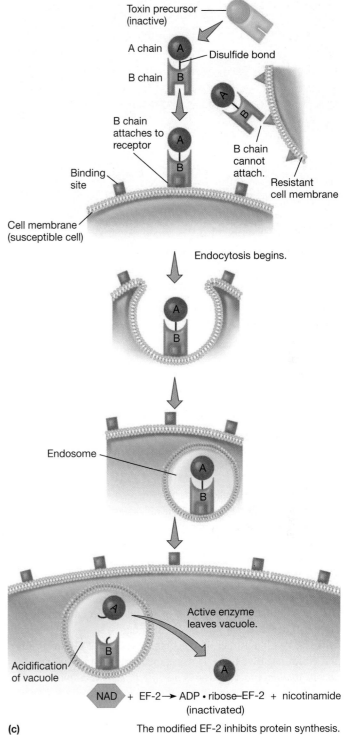

(c)

administration of erythromycin or rifampin. Death occurs in 5 to 30% of symptomatic cases.

Prevention of Legionnaires' disease depends on the identification and elimination of the environmental source of *L. pneumophila*. Chlorination, ozonation, the heating of water, and the cleaning of water-containing devices can help control the multiplication and spread of *L. pneumophila*. A variety of regulatory control measures are now in place to reduce the incidence of Legionnaires' disease in hospitals, hotels, and so on. However, use of private hot tubs, fountains, and other water systems that aerosolize water can be sources of bacteria. Since the initial outbreak of this disease in 1976, many outbreaks during summer months have been recognized in all parts of the United States. About 8,000 to 18,000 people are hospitalized each year with

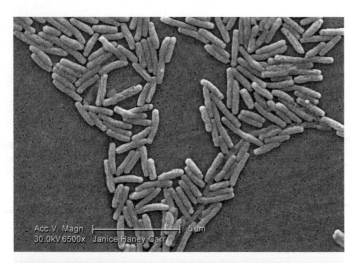

Figure 39.2 *Legionella pneumophila.* This peritrichously flagellated Gram-negative bacterium is the causative agent of Legionnaires' disease; SEM (×10,000).

MICRO INQUIRY *What organisms serve as the environmental reservoir for* L. pneumophila?

Legionnaires' disease, and about another 18,000 or more mild or subclinical cases are thought to occur.

Retrieve, Infer, Apply

1. Why do you think chlamydiae differentiate into specialized cell types for infection and reproduction?
2. Compare and contrast the means by which people contract chlamydial pneumonia with Legionnaires' disease.
3. What causes the typical symptoms of diphtheria? How are individuals protected against this disease?

Meningitis

Meningitis (Greek *meninx*, membrane, and *-itis*, inflammation) is an inflammation of the brain or spinal cord meninges (membranes). There are many causes of meningitis, some of which can be treated with antimicrobial agents. A variety of Gram-positive and Gram-negative bacteria cause meningitis, including *Streptococcus pneumoniae, Neisseria meningitidis, Haemophilus influenzae* serotype b, group B streptococci (*S. agalactiae*), *Listeria monocytogenes, E. coli, Mycobacterium tuberculosis, Nocardia asteroides, Staphylococcus aureus*, and *S. epidermidis*. However, three bacteria tend to be associated with meningitis more frequently than others: *S. pneumoniae, N. meningitidis,* and *H. influenzae* serotype b.

Accurate identification of the causative agent is essential for proper treatment of meningitis. However, this can sometimes be difficult because meningitis symptoms may be present without a microbial agent in Gram-stained specimens of cerebral spinal fluid (CSF), which also yields negative cultures. In such cases, the diagnosis often is called aseptic (meaning lack of an identifiable agent) meningitis syndrome; aseptic meningitis can be caused by a virus or protozoan. Aseptic meningitis is typically more difficult to treat as therapy targeting a specific microorganism is not possible.

The immediate sources of the bacteria responsible for meningitis are respiratory secretions from carriers. *N. meningitidis*, often referred to as meningococcus, is a normal inhabitant of the human nasopharynx (5 to 10% of humans carry the nonpathogenic serotypes). Most disease-causing *N. meningitidis* strains belong to serotypes A, B, C, Y, and W-135. In general, serotype A strains are the cause of epidemic disease in developing countries, whereas serotype B, C, Y, and W-135 strains are responsible for meningitis outbreaks in industrialized countries. Infection results from airborne transmission of the bacteria, typically through close contact with a primary carrier. The disease process is initiated by pili-mediated colonization of the nasopharynx by pathogenic bacteria (**figure 39.3**). The bacteria cross the nasopharyngeal epithelium and the blood-brain barrier (typically through endocytosis) to enter the central nervous system. Often, *N. meningitidis* also invades the bloodstream (meningococcemia), where they proliferate.

Symptoms caused by *N. meningitidis* are variable, depending on whether or not meningococcemia is established. The usual symptoms of meningitis include an initial respiratory illness or sore throat interrupted by one or more of the meningeal syndromes: vomiting, headache, lethargy, confusion, and stiff neck and back. If the bacteria have spread to the bloodstream, a characteristic rash develops along with septic shock. Once bacterial meningitis is suspected, antibiotics (penicillin or ceftriaxone) are administered immediately. In fact, antibiotics (ciprofloxacin or rifampin) are often administered prophylactically to patient contacts; if left untreated, meningococcal meningitis is fatal.

Control of *N. meningitidis* infection is with vaccination and antibiotics. Two vaccine types, meningococcal polysaccharide

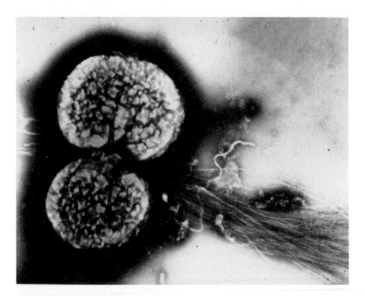

Figure 39.3 *Neisseria meningitidis.* Type IV pili are present on the cell surface as bundled filamentous appendages, facilitating transformation and attachment.

vaccine and meningococcal conjugate vaccine, are available. Both vaccine types are effective against serotypes A, C, Y, and W-135. Two vaccines that use recombinant proteins (antigen) are available for protection against serogroup B. Vaccination is recommended for all college students living in residence halls.

Another agent of meningitis is *H. influenzae* serotype B (Hib), a Gram-negative bacterium. Transmission is by inhalation of droplet nuclei shed by infectious individuals or carriers. Hib infects mucous membranes, resulting in sinusitis, pneumonia, and bronchitis. In addition to gaining access to the CSF, it can disseminate to the bloodstream and cause bacteremia. In the United States, *H. influenzae* disease (including pneumonia and meningitis) is primarily observed in children less than five years of age and thanks to the Hib conjugate vaccine, fewer than one case per 100,000 children occurs annually. Globally, it is estimated that *H. influenzae* serotype b causes at least 3 million cases of serious disease and almost 400,000 deaths each year.

Three to 6 percent of all *H. influenzae* infections are fatal. Furthermore, up to 20% of surviving patients have permanent hearing loss or other long-term sequelae. Therefore it is important that all children begin vaccination with the Hib conjugate vaccine at the age of two months.

Mycobacterium Infections

Mycobacteria are a large group of bacteria that are normal inhabitants of soil, water, and house dust. They cause a number of human infections that are difficult to treat. Recall that mycobacteria make cell walls that have a very high lipid content and contain mycolic acids. These complex fatty acids have 60 to 90 carbon atoms with a shorter β-hydroxy chain and a longer α-alkyl side chain (*see figure 24.11*). The presence of mycolic acids and other lipids outside the peptidoglycan layer makes mycobacteria acid-fast, which means that basic fuchsin dye cannot be removed from the cell by acid alcohol treatment. More importantly, the mycolic acids also make mycobacteria resistant to water-soluble antibiotics. ◄◄ *Differential staining (section 2.3)*

M. avium Complex

Two airborne mycobacteria are noteworthy pathogens in the United States: *Mycobacterium avium* and *Mycobacterium intracellulare*. They are so closely related that they are referred to as the *M. avium* complex (MAC). Globally, *M. tuberculosis* has remained more prevalent in developing countries, whereas MAC has become the most common cause of mycobacterial infections in the United States. ◄◄ *Order Corynebacteriales includes important human pathogens (section 24.1)*

MAC is found worldwide and infects a variety of insects, birds, and animals. Both the respiratory and the gastrointestinal tracts have been proposed as entry portals for MAC; however, person-to-person transmission is not very efficient. MAC causes a pulmonary infection in humans similar to that caused by *M. tuberculosis*. Pulmonary MAC is more common in elderly persons with preexisting pulmonary disease. The gastrointestinal tract is thought to be the most common site of colonization

and dissemination in AIDS patients. It produces disabling symptoms, including fever, malaise, weight loss, night sweats, and diarrhea. Carefully controlled epidemiological studies have shown that MAC shortens survival by 5 to 7 months among persons with AIDS. ◄◄ *Acquired immune deficiency syndrome (AIDS) (section 38.3)*

MAC can be isolated from sputum, blood, and aspirates of bone marrow. Acid-fast stains are of value in making a diagnosis. The most sensitive method for detection is the commercially available blood culture system. No drug is currently approved by the U.S. Food and Drug Administration (FDA) for the therapy of MAC; however, every regimen should contain either azithromycin or clarithromycin and ethambutol as a second drug. One or more of the following can be added: clofazimine, rifabutin, rifampin, ciprofloxacin, and amikacin. Regardless of the specific drugs, MAC infection is treated with two or three antimicrobials for at least 12 months.

Mycobacterium tuberculosis

More than a century ago, Robert Koch identified *Mycobacterium tuberculosis* as the causative agent of **tuberculosis (TB)**. At the time, TB was rampant, causing one-seventh of all deaths in Europe and one-third of deaths among young adults. Today TB remains a global health problem of enormous dimension. It is estimated that one-third of the world's human population is infected (with at least 2 billion people with latent infections). There were 9 million people infected with TB bacilli in 2013, and 1.5 million TB-related deaths. The 2013 incidence of TB in the United States was 9,582 cases. Worldwide, TB is caused by *M. bovis* and *M. africanum*, in addition to *M. tuberculosis* (**figure 39.4a**). After AIDS became pandemic, the number of global TB cases increased dramatically. Because a close association exists between AIDS and TB, further spread of HIV infection among populations with a high prevalence of TB infection is resulting in dramatic increases in TB globally. Importantly, TB is the direct cause of death of over half of all AIDS patients worldwide.

In the United States, TB occurs most commonly among the homeless, elderly, malnourished, alcoholic males, minorities, immigrants, prison populations, and Native Americans. It appears that about one-fourth to one-third of active TB cases in the United States may be due to recent transmission. Thus the majority of active cases result from the reactivation of old, dormant infections.

Infection results when the bacteria are phagocytosed by macrophages in the lungs, where they survive the normal antimicrobial processes. In fact, infected macrophages often die attempting to destroy the bacteria, thus releasing viable bacteria into respiratory spaces. The incubation period is about 4 to 12 weeks, and the disease develops slowly. The symptoms of TB are fever, fatigue, night sweats, and weight loss. A cough, which is characteristic of pulmonary involvement, may result in expectoration of bloody sputum.

M. tuberculosis (Mtb) does not produce classic virulence factors such as exotoxins, capsules, and fimbriae. Instead, Mtb has some rather unique products and properties that contribute to its

Caseous necrosis

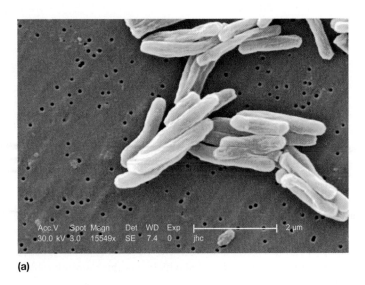

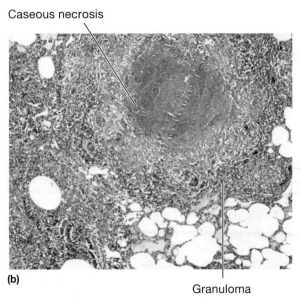

Granuloma

(a)

(b)

Figure 39.4 Tuberculosis. (a) Mycobacteria are rods that are a little larger than 2 μm in length (SEM). (b) In the lungs, tuberculosis is identified by the tubercle, a granuloma of white blood cells, bacteria, fibroblasts, and epithelioid cells. The center of the tubercle contains caseous (cheesy) pus and bacteria.

virulence. The unique lipids and glycolipids in the cell envelope of Mtb (*see figure 24.11*) are directly toxic to eukaryotic cells and create a hydrophobic barrier around the bacterium that facilitates impermeability and resistance to antimicrobial agents. It also protects against killing by acidic and alkaline compounds, osmotic lysis, and lysozyme. Furthermore, cell wall glycolipids associate with mannose, giving Mtb control over entry into macrophages by exploiting the macrophage mannose receptors. Once inside, Mtb inhibits phagosome-lysosome fusion by altering the phagosome membrane. Resistance to oxidative killing, inhibition of phagosome-lysosome fusion, and inhibition of diffusion of lysosomal enzymes are just some of the mechanisms that help explain the survival of Mtb inside macrophages. ◄◄ *Phagocytosis (section 33.5)*

During infection, other immune cells are recruited to the site of infection by cytokines released from the responding macrophages. Together, and in response to several mycobacterial products, a hypersensitivity response results in the formation of small nodules called **tubercles** composed of bacteria, macrophages, T cells, and various human proteins (figure 39.4*b*). Tubercles are characteristic of tuberculosis and give the disease its name. The disease process usually stops at this stage, but the bacteria often remain alive within macrophage phagosomes. In some cases, the disease may become active, even after many years of latency.

In time, a tubercle may change to a cheeselike consistency and is then called a **caseous lesion** or granuloma (figure 39.4*b*). If such lesions calcify, they are called **Ghon complexes,** which show up prominently in a chest X ray. Sometimes the tubercle lesions liquefy and form air-filled **tuberculous cavities.** From these cavities, the bacteria can be disseminated to new foci throughout the body. This spreading is often called **miliary tuberculosis** due to the many tubercles the size of millet seeds that are formed in the infected tissue. It is also called **reactivation tuberculosis** because the bacteria have been reactivated in the initial site of infection. Most cases in the United States are

acquired from other humans through droplet nuclei and the respiratory route (figure 39.4*c*).

TB must be treated with antimicrobial therapy. Several drugs are administered simultaneously (e.g., isoniazid [INH], plus rifampin, ethambutol, and pyrazinamide). These drugs are administered for 6 to 9 months as a way of decreasing the possibility that the bacterium develops drug resistance. However, ***m**ulti**d**rug-**r**esistant strains of **t**uberculosis* (**MDR-TB**) have developed and are spreading. A multidrug-resistant strain is defined as *M. tuberculosis* that is resistant to INH and rifampin, with or without resistance to other drugs. Inadequate therapy is the most common means by which resistant bacteria are acquired, and patients who have previously undergone therapy are presumed to harbor MDR-TB until proven otherwise. Strains of Mtb considered to cause MDR-TB have emerged that are *ex*tensively *d*rug-*r*esistant (XDR); they are resistant to INH, rifampin, and the secondary antibiotics. XDR-TB is a unique challenge to public health as first-line and second-line drug resistance severely limits treatment options. For this reason, MDR-TB and XDR-TB have a higher mortality rate than drug-sensitive TB.

Drug-resistant TB arises because tubercle bacilli have spontaneous, predictable rates of chromosomal mutations that confer resistance to drugs. These mutations are unlinked; hence resistance to one drug is not associated with resistance to an unrelated drug. The observation that these mutations are not linked is the cardinal principle underlying TB chemotherapy. In the circumstances of monotherapy, erratic drug compliance, omission of one or more drugs, suboptimal dosage, poor drug absorption, or an insufficient number of active drugs in a regimen, a susceptible strain of *M. tuberculosis* may become resistant to multiple drugs within a matter of months. This is why many public health organizations worldwide practice *d*irectly *o*bserved *t*reatment *s*hort course (DOTS), in which each dose is taken in the presence of a health-care worker.

Persons infected with Mtb develop cell-mediated immunity because the bacteria stimulate a strong TH1-cell response. These

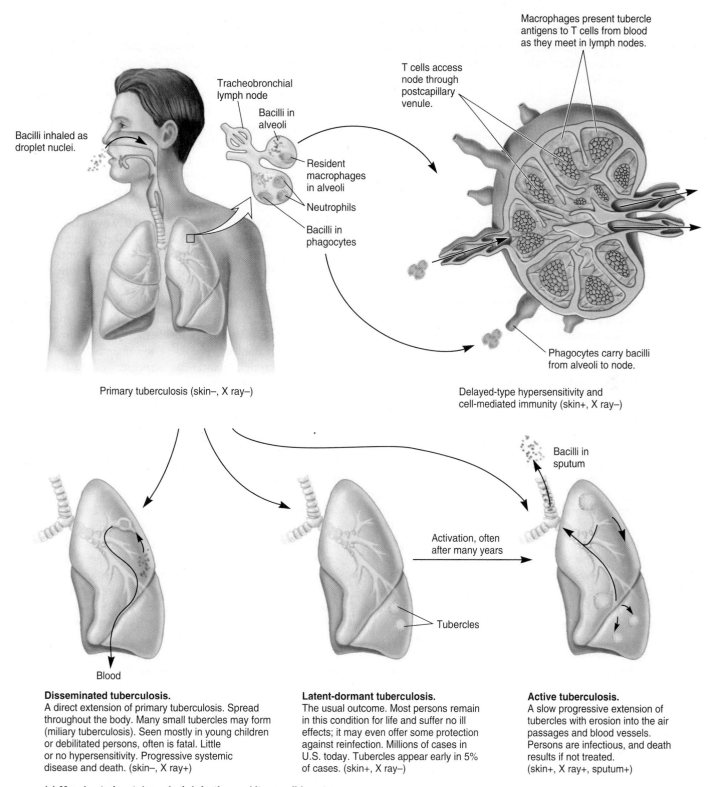

Bacilli inhaled as droplet nuclei.

Tracheobronchial lymph node

Bacilli in alveoli

Resident macrophages in alveoli

Neutrophils

Bacilli in phagocytes

T cells access node through postcapillary venule.

Macrophages present tubercle antigens to T cells from blood as they meet in lymph nodes.

Phagocytes carry bacilli from alveoli to node.

Primary tuberculosis (skin–, X ray–)

Delayed-type hypersensitivity and cell-mediated immunity (skin+, X ray–)

Bacilli in sputum

Activation, often after many years

Tubercles

Blood

Disseminated tuberculosis.
A direct extension of primary tuberculosis. Spread throughout the body. Many small tubercles may form (miliary tuberculosis). Seen mostly in young children or debilitated persons, often is fatal. Little or no hypersensitivity. Progressive systemic disease and death. (skin–, X ray+)

Latent-dormant tuberculosis.
The usual outcome. Most persons remain in this condition for life and suffer no ill effects; it may even offer some protection against reinfection. Millions of cases in U.S. today. Tubercles appear early in 5% of cases. (skin+, X ray–)

Active tuberculosis.
A slow progressive extension of tubercles with erosion into the air passages and blood vessels. Persons are infectious, and death results if not treated. (skin+, X ray+, sputum+)

(c) *Mycobacterium tuberculosis* infection and its possible outcomes.

Figure 39.4 *(Continued)*

sensitized T cells are the basis for the tuberculin skin test. In this test, a *p*urified *p*rotein *d*erivative (PPD) of Mtb is injected intracutaneously (the Mantoux test). If the person has had TB or has been exposed to Mtb, sensitized T cells react with these proteins and a delayed hypersensitivity reaction occurs within 48 hours.

This positive skin reaction appears as an induration (hardening) and reddening of the area around the injection site. In a young person, a positive skin test could indicate active tuberculosis. In older persons, it may result from previous disease, vaccination, or a false-positive test. In these cases, X rays and bacterial isolation are completed to confirm the diagnosis. Other laboratory tests include microscopy of the acid-fast bacterium, immunoassays, and commercially available DNA probes. ◄◄ *Type IV hypersensitivity (section 34.10).*

Retrieve, Infer, Apply

1. What are the three major causes of meningitis? Why is it important to determine which type a person has?
2. How is tuberculosis diagnosed? Describe the various types of tubercular lesions and how they are formed.
3. What is the reason for the complex antibiotic therapy used to treat tuberculosis?
4. How do multidrug-resistant strains of tuberculosis develop?

Mycoplasmal Pneumonia

Typical pneumonia has a bacterial origin (most frequently *Streptococcus pneumoniae*) with fairly consistent signs and symptoms. If the symptoms of pneumonia are different from what is usually observed, the disease is called **atypical pneumonia.** One cause of atypical pneumonia is *Mycoplasma pneumoniae*, a bacterium that lacks a cell wall and has worldwide distribution. Transmission involves close contact and airborne droplets. The disease is fairly common and mild in infants and small children; serious disease is seen principally in older children and young adults. ◄◄ *Class* Mollicutes, *phylum* Tenericutes *(section 21.3)*

Because *M. pneumoniae* lacks a cell wall, cells vary in shape and stain Gram negative. Because they cannot synthesize peptidoglycan precursors, they are resistant to all β-lactam antibiotics (e.g., penicillin). Mycoplasmas typically infect the upper respiratory tract and subsequently move to the lower respiratory tract, where they attach to respiratory mucosal cells. They then produce peroxide, which may be a toxic factor, but the exact mechanism of pathogenesis is uncertain. The manifestations of this disease vary in severity from asymptomatic to a serious pneumonia. The latter is accompanied by death of the surface mucosal cells, lung infiltration, and congestion. Initial symptoms include headache, weakness, a low-grade fever, and a predominant, characteristic cough. The disease and its symptoms usually persist for weeks. The mortality rate is less than 1%.

Several rapid tests using latex-bead agglutination for *M. pneumoniae* antibodies are available for diagnosis of mycoplasmal pneumonia. When isolated from respiratory secretions, some mycoplasmas form distinct colonies with a "fried-egg" appearance on agar (*see figure 21.5*). During the acute stage of disease, diagnosis must be made by clinical observations. Tetracyclines or erythromycin are effective in treatment. There are no preventive measures.

Pertussis

Pertussis (Latin *per*, intensive, and *tussis*, cough), also called "whooping cough," is caused by the Gram-negative bacterium *Bordetella pertussis*. *B. parapertussis* is a closely related species that causes a milder form of the disease. Pertussis bacteria colonize the respiratory epithelium to produce a disease characterized by fever, malaise, uncontrollable cough, and cyanosis (bluish skin color resulting from inadequate tissue oxygenation). Whooping cough gets its name from the characteristically prolonged and paroxysmal coughing that ends in an inspiratory gasp, or whoop. Pertussis is a highly contagious, vaccine-preventable disease that primarily affects children. In 2013 the World Health Organization (WHO) indicated that there are approximately 16 million global cases of pertussis per year, of which 195,000 children, mostly infants less than six months old, die. The overall number of pertussis cases had been increasing in the United States since the 1980s, reaching 48,277 cases (and 20 deaths) in 2012. The number of cases in 2013 decreased to 28,639 (with 13 deaths), but 2014 saw a 30% increase in cases, most likely due to the emergence of new strains of *B. pertussis* not covered by the vaccine. ◄◄ *Order* Burkholderiales *includes recently evolved pathogens (section 22.2)*

Disease transmission occurs by inhalation of the bacterium in droplets released from an infectious person. The incubation period is 7 to 14 days. Once inside the upper respiratory tract, the bacteria colonize the cilia of the mammalian respiratory epithelium through fimbria-like structures called filamentous hem-agglutinins that bind to complement receptors on phagocytes. Additionally, some of the components of the *B. pertussis* toxin assist in adherence to cilia by bridging bacterial and host cells; subunit S2 binds to cilial glycolipids (lactosylceramide), and subunit S3 binds to phagocyte glycoproteins. Thus attachment is an important virulence factor in the initiation of the disease.

B. pertussis produces several toxins, of which the pertussis toxin (PTx) is most important. PTx is an AB exotoxin (*see figure 35.6*). The B portion of the toxin is composed of five polypeptides (S2–S5; there are two S4 subunits) that bind to specific carbohydrates on cell surfaces and then transport the A subunit to the plasma membrane, where it is inserted and released into the cytoplasm. The A subunit (S1) is an ADP-ribosyl transferase, similar to the diphtheria toxin. However, PTx adds ADP-ribose moieties to the regulator of the host cell's adenyl cyclase, locking it in the active state so that ATP is continuously converted to cyclic AMP (cAMP). This results in an increase in intracellular levels of cAMP, which accelerates mucin secretion and alters water transport, affecting electrolyte balance. *B. pertussis* also produces its own adenyl cyclase, which further increases intracellular cAMP, tracheal cytotoxin, and dermonecrotic toxin, which destroy epithelial tissue. Working together, the adenyl cyclase, tracheal cytotoxin, and pertussis toxin provoke secretory cells in the respiratory tract to produce nitric oxide, which kills nearby ciliated cells, inhibiting removal of bacteria and mucus. The secretion of a thick mucus also impedes ciliary action. These actions make breathing difficult and lead to the whooping sound on inhalation.

Pertussis is divided into three stages. (1) The catarrhal stage, so named because of the mucous membrane inflammation, resembles the common cold. (2) The paroxysmal stage is characterized by prolonged coughing sieges. During this stage, the infected person tries to cough up the mucous secretions by making five to 15 rapidly consecutive coughs followed by the characteristic whoop—a hurried, deep inspiration. The catarrhal and paroxysmal stages last about 6 weeks. (3) The convalescent stage, when final recovery occurs, may take several months.

Laboratory diagnosis of pertussis is by culture of the bacterium, fluorescent antibody staining of smears from nasopharyngeal swabs, other antibody-based detection tests, and PCR. The development of a strong, lasting immunity takes place after an initial infection. Treatment is with a macrolide. Trimethoprim-sulfamethoxasole can also be used. Treatment ameliorates clinical illness when begun during the catarrhal phase and may also reduce the severity of the disease when begun within 2 weeks of the onset of the paroxysmal cough. Prevention is with the DTaP vaccine, starting as early as two to three months of age. It is now recommended that all previously vaccinated adults get a booster vaccine.

Retrieve, Infer, Apply

1. Describe the pneumonia caused by *M. pneumoniae*.
2. What is the mechanism by which PTx kills host cells?
3. Describe the three stages of pertussis.

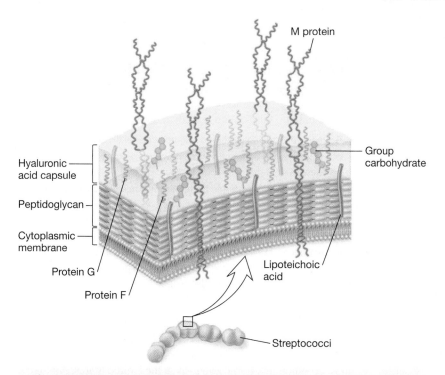

Figure 39.5 Streptococcal Cell Envelope. The M protein is a major virulence factor for group A streptococci. It facilitates bacterial attachment to host cells and has antiphagocytic activity. Protein G prevents attack by antibodies because it binds to the Fc portion of antibody molecules, preventing the antigen-binding site from interacting with the bacterium. Protein F is an epithelial cell attachment factor.

MICRO INQUIRY *How is the M protein also thought to be involved in poststreptococcal diseases?*

Streptococcal Diseases (Group A)

Pathogenic streptococci, commonly called strep, are a heterogeneous group of Gram-positive bacteria. In this group, *Streptococcus pyogenes* (group A, β-hemolytic streptococci) is one of the most important bacterial pathogens. The different serotypes of **group A streptococci** (**GAS**) produce (1) extracellular enzymes that break down host molecules; (2) streptokinases, enzymes that activate a host-blood factor that dissolves blood clots; (3) the cytolysins streptolysin O and streptolysin S, which kill host leukocytes; and (4) capsules and M protein, which help to retard phagocytosis (**figure 39.5**). M protein, a filamentous protein anchored in the streptococcal cell membrane, facilitates attachment to host cells and prevents opsonization by complement protein C3b. It is the major virulence factor of GAS. Although over 50 different *S. pyogenes* M proteins have been identified, types 1, 3, 12, and 28 are commonly found in patients with streptococcal toxic shock and multiorgan failure. ◄◄ *Complement is a cascading enzyme system (section 33.3)*

S. pyogenes is widely distributed among humans; some people become asymptomatic carriers. Individuals with acute infections may spread the pathogen, and transmission can occur through respiratory droplets, as direct or indirect contact. ◄◄ *Order* Lactobacillales *(section 23.3)*

Diagnosis of a streptococcal infection is based on both clinical and laboratory findings. Several rapid tests are available. Treatment is with penicillin or macrolide antibiotics. There is no vaccine for *S. pyogenes*. The best control measure is prevention of transmission. Individuals with a known infection should be isolated and treated. Personnel working with infected patients should follow standard aseptic procedures.

Streptococcal Pharyngitis

Streptococcal pharyngitis is one of the most frequent bacterial infections of humans and is commonly called strep throat (**figure 39.6**). The GAS are spread by droplets of saliva or nasal secretions. The incubation period in humans is 2 to 4 days.

The action of the strep bacteria in the throat (**pharyngitis**) or on the tonsils (**tonsillitis**) stimulates an inflammatory response and the lysis of white and red blood cells. An inflammatory exudate consisting of cells and fluid is released from the blood vessels and deposited in the surrounding tissue in about 50% of patients with strep pharyngitis. This is accompanied by a general feeling of discomfort or malaise, fever, and headache. Prominent physical manifestations include redness, edema, and lymph node enlargement in the throat. Signs and symptoms alone are not diagnostic because they are manifested by several viral infections as well. Several common rapid test kits are available for diagnosing strep throat. In the absence of complications,

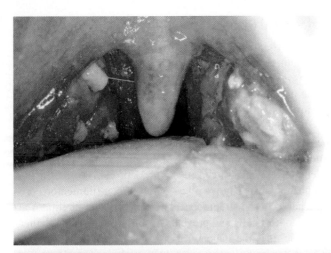

Figure 39.6 Streptococcal Pharyngitis (Strep Throat).

MICRO INQUIRY *What causes the exudate seen in strep throat? Is it seen in all cases?*

the disease can be self-limiting and may disappear within a week. However, antibiotic treatment (penicillin, cephalosporins, or a macrolide antibiotic for penicillin-allergic people) can shorten the infection and clinical syndromes, and is especially important in children for the prevention of complications such as rheumatic fever and glomerulonephritis, as discussed next. Infections in older children and adults tend to be milder and less frequent due in part to the immunity developed against the many serotypes encountered in early childhood.

Poststreptococcal Diseases

The poststreptococcal diseases caused by GAS are glomerulonephritis and rheumatic fever. They occur 1 to 4 weeks after an acute streptococcal infection. These two diseases are the most serious problems associated with streptococcal infections in the United States.

Glomerulonephritis, also called **Bright's disease,** is an inflammatory disease (type II hypersensitivity reaction) of the renal glomeruli—membranous structures within the kidney where blood is filtered. Damage results from the deposition of antigen-antibody complexes, possibly involving the streptococcal M protein, in the glomeruli. The complexes cause destruction of the glomerular membrane, allowing proteins and blood to leak into the urine. Clinically, the affected person exhibits edema, fever, hypertension, and hematuria (blood in the urine). The disease occurs primarily among school-age children. Diagnosis is based on clinical history, physical findings, and confirmatory evidence of prior streptococcal infection. The incidence of glomerulonephritis in the United States is less than 0.5% of streptococcal infections. Penicillin G or erythromycin can be given for any residual streptococci. However, there is no specific therapy once kidney damage has occurred. About 80 to 90% of all cases undergo slow, spontaneous healing of the damaged glomeruli,

whereas others develop a chronic form of the disease. The latter may require a kidney transplant or lifelong renal dialysis. ◄◄ *Type II hypersensitivity (section 34.10)*

Rheumatic fever is an autoimmune disease characterized by inflammatory lesions involving the heart valves, joints, subcutaneous tissues, and central nervous system. The exact mechanism of rheumatic fever development remains unknown. However, it has been associated with prior streptococcal pharyngitis of specific M protein strains. The disease occurs most frequently among children six to fifteen years of age and manifests itself through a variety of signs and symptoms, making diagnosis difficult. In the United States, rheumatic fever has become very rare (less than 0.05% of streptococcal infections). It occurs 100 times more frequently in tropical countries. Therapy is directed at decreasing inflammation and fever, and controlling cardiac failure. Salicylates and corticosteroids are the mainstays of treatment. Although rheumatic fever is rare, it is still the most common cause of permanent heart valve damage in children.

Retrieve, Infer, Apply

1. Describe the streptococcal exotoxins and how they are involved in virulence.
2. Which streptococcal disease is most prevalent? Why do you think this is the case?

39.2 Arthropods Can Transmit Bacterial Diseases

After reading this section, you should be able to:

- Report the common arthropod-borne bacterial diseases
- Identify typical signs and symptoms of arthropod-borne bacterial diseases
- Correlate arthropod-borne bacterial infection and disease severity with bacterial virulence factors

Bacteria cause fewer arthropod-borne diseases than do viruses and protozoa. However, they are of interest either historically (plague, typhus) or because they have emerged into human populations (ehrlichiosis, Lyme disease).

Lyme Disease

Lyme disease (LD, Lyme borreliosis) was first observed among people of Old Lyme, Connecticut, in 1975. It has become the most common tick-borne zoonosis in the United States. Annually, an average of approximately 30,000 confirmed cases of Lyme disease are reported to the CDC, with 94% of these located in only 12 states. The disease is also present in Europe and forested areas of Asia.

At least three spirochete species are responsible for this disease. *Borrelia burgdorferi* is the primary cause of Lyme disease in the United States, whereas *B. garinii* and *B. afzelii* appear to cause the disease in Europe (**figure 39.7a**). Rodents and other

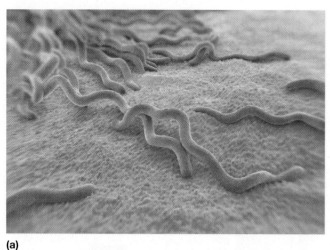
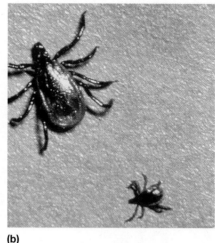
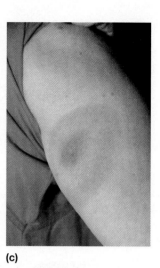

(a) (b) (c)

Figure 39.7 Lyme Disease. (a) The spirochete *Borrelia burgdorferi*. (b) An unengorged adult tick (*Ixodes scapularis*; bottom) is about the size of pinhead, and an engorged adult (top) can reach the size of a jelly bean. (c) The typical rash (erythema migrans) showing concentric rings around the initial site of the tick bite.

MICRO INQUIRY *What are the three stages of Lyme disease? What other spirochete disease does this resemble?*

wild animals are the natural hosts; although adult deer ticks feed on deer, deer do not become infected. In the northeastern United States, the reservoir of *B. burgdorferi* is the white-footed mouse. Spirochetes are transmitted to humans by the bite of infected black-legged (or deer) ticks (*Ixodes scapularis*; figure 39.7*b*). On the Pacific Coast, especially in California, the reservoir is a dusky-footed woodrat and the tick *I. pacificus*. ◄◄ *Phylum Spirochaetes (section 21.8)*

Clinically, Lyme disease is a complex illness with three major stages. The initial, localized stage occurs a week to 10 days after an infectious tick bite. The illness sometimes begins with erythema migrans, an expanding, ring-shaped skin lesion with a red outer border and partial central clearing (figure 39.7*c*). This is accompanied by flulike symptoms (malaise and fatigue, headache, fever, and chills). However, often the tick bite is unnoticed, or the skin lesion is missed due to skin coloration or its obscure location, such as on the scalp. Thus treatment, which is effective at this stage, may not be started because the illness is assumed to be "just the flu."

The second, disseminated stage may appear weeks or months after the initial infection. It consists of several symptoms such as neurological abnormalities, heart inflammation, and bouts of arthritis, usually in major joints such as the elbows or knees. Current research indicates that Lyme arthritis might be an autoimmune response to major histocompatibility (MHC) molecules on cells in the synovium (joint) that are similar to the bacterial antigens. Inflammation that produces organ damage is initiated and possibly perpetuated by the immune response to one or more spirochetal proteins. Finally, like the progression of syphilis—another disease caused by a spirochete—the late stage may appear years later. Infected individuals may develop neuron demyelination with symptoms resembling Alzheimer's disease and multiple sclerosis. Behavioral changes can also occur. ◄◄ *Major histocompatibility complex (section 34.4)*

Laboratory diagnosis of LD is based on (1) serological testing (Lyme ELISA) for IgM or IgG antibodies to the pathogen and Western blot for specific bacterial proteins, (2) detection of *Borrelia* DNA in patient specimens (especially synovial fluid) after amplification by PCR, and (3) recovery of the spirochete from patient specimens, although cultures are laborious with modest success. Treatment with doxycycline early in the illness results in prompt recovery and prevents arthritis and other complications. If nervous or cardiac system involvement is suspected, intravenous penicillin or ceftriaxone is used because it can cross the blood-brain barrier. ◄◄ *Polymerase chain reaction amplifies targeted DNA (section 17.2)*

Prevention and control of LD involve environmental modifications such as clearing and burning tick habitat and the application of acaricidal compounds, which destroy mites and ticks. An individual's risk of acquiring LD may be greatly reduced by education and personal protection.

Plague

Plague (Latin *plaga*, pest) is caused by the Gram-negative bacterium *Yersinia pestis*. It is transmitted from rodent to human by the bite of an infected flea, direct contact with infected animals or their products, or inhalation of contaminated airborne droplets (**figure 39.8***a*). Between 1,000 and 2,000 cases of plague are reported to the WHO each year; U.S. cases average seven annually. Initially spread by contact with flea-infested animals, *Y. pestis* can then spread among people by airborne transmission. Once in the human body, the bacteria multiply in the blood and lymph. Symptoms include subcutaneous hemorrhages, fever, chills, headache, extreme exhaustion, and the appearance of enlarged lymph nodes called **buboes** (hence the old name, **bubonic plague**) (figure 39.8*b*). In 50 to 70% of the untreated cases, death

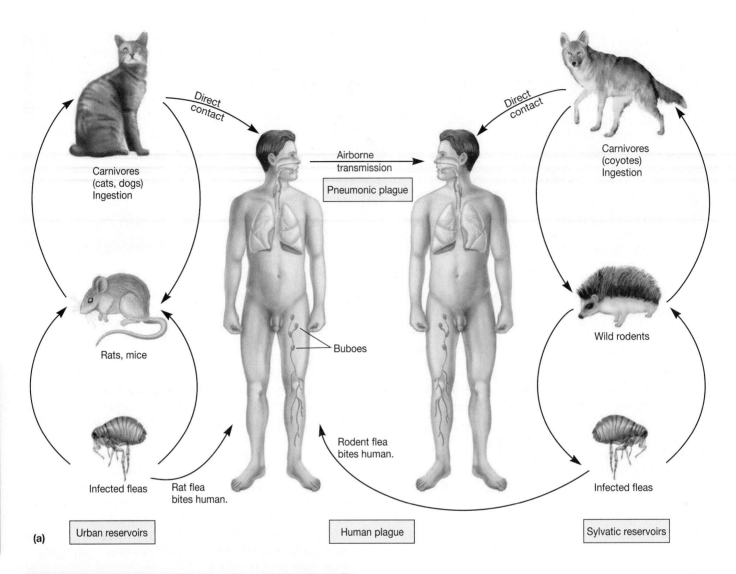

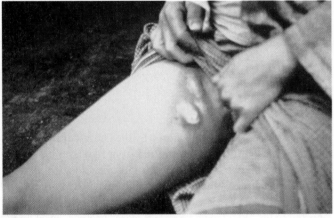

Figure 39.8 Plague. (a) Plague is spread to humans through (1) the urban cycle and rat fleas, (2) the sylvatic cycle, centering on wild rodents and their fleas, or (3) by airborne transmission from an infected person, leading to pneumonic plague. Dogs, cats, and coyotes can also acquire the bacterium by ingestion of infected animals. (b) Enlarged lymph nodes called buboes are characteristic of *Yersinia* infection.

follows in 3 to 5 days from toxic conditions caused by the large number of bacilli in the blood.

An important factor in the virulence of *Y. pestis* is its ability to survive and proliferate inside phagocytic cells, rather than being killed by them. Like many other Gram-negative pathogens, *Y. pestis* uses a type III secretion system (a kind of molecular syringe) to deliver effector proteins. *Y. pestis* secretes plasmid-encoded yersinial outer membrane proteins (YOPS) into phagocytic cells to counteract natural defense mechanisms and help the bacteria multiply and disseminate in the host (**figure 39.9**). ◀◀ *Order* Enterobacteriales *includes pathogens and mutualists (section 22.3); Pathogenicity islands (section 35.2)*

In the southwestern part of the United States, plague occurs primarily in wild ground squirrels, chipmunks, mice, and prairie dogs. However, massive human epidemics occurred in Europe during the Middle Ages, where the disease was known as the Black Death due to black-colored, subcutaneous hemorrhages. Infections now occur in humans only sporadically or in limited outbreaks.

Pneumonic plague arises (1) from primary exposure to infectious respiratory droplets from a person or animal with

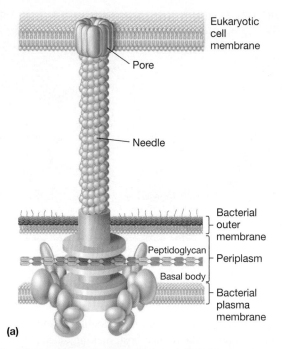

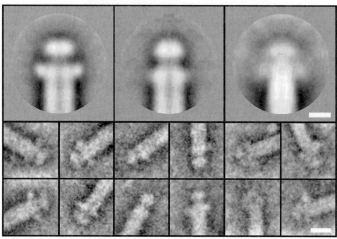

Figure 39.9 Type III Secretion System. (a) X-ray fiber diffraction studies resolve the injectisome as a helical structure. (b) Scanning tunneling electron microscopy reveals the injectisome tip, indicating how it may lock into the translocator pore on the target cell.

respiratory plague or (2) secondary to blood-borne spread in a patient with bubonic or septicemic plague. Pneumonic plague can also arise from accidental inhalation of *Y. pestis* in the laboratory. The mortality rate for this kind of plague is almost 100% if it is not recognized within 12 to 24 hours. Obviously great care must be taken to prevent the spread of airborne infections to personnel who care for pneumonic plague patients.

Because the plague bacillus is listed as a potential bioterrorism agent, sentinel laboratory identification of *Y. pestis* is restricted to initial staining and culture tests. Confirmation of the identity of *Y. pestis* by PCR, advanced serological studies, and other approved practices is restricted to national reference laboratories. Treatment is with streptomycin and gentamycin, although fluoroquinolones, chloramphenicol, or tetracycline can also be used, and recovery from the disease provides immunity.

Prevention and control involve flea and rodent control, isolation of human patients, prophylaxis or abortive therapy of exposed persons, and vaccination (USP Plague vaccine) of persons at high risk. (e.g., park rangers, veterinarians).

Rocky Mountain Spotted Fever

Rocky Mountain spotted fever is caused by the rickettsia, *Rickettsia rickettsii*. Although this disease was originally detected in the Rocky Mountain area, most cases occur east of the Mississippi River. The disease is transmitted by ticks and usually occurs in people who are or have been in tick-infested areas. There are two principal vectors: *Dermacentor andersoni*, the wood tick, which is active during the spring and early summer in the Rocky Mountain states; and *D. variabilis*, the dog tick, which has assumed greater importance and is almost exclusively confined to the eastern half of the United States. Unlike other rickettsias, *R. rickettsii* can pass from generation to generation of ticks through their eggs in a process known as transovarian passage. Therefore, no animals are needed as reservoirs for the continued propagation of this rickettsia in the environment.

The rickettsial bacteria are harbored in the tick intestinal tract and are deposited on humans when the tick defecates after feeding. Humans are typically infected by rubbing or scratching the site where the tick is feeding. Once inside the skin, the rickettsias enter the endothelial cells of small blood vessels, where they multiply and produce a characteristic rash that is caused by inflamed blood vessels. Therefore this rash is considered a vasculitis. The bacteria induce phagocytosis by macrophages responding to the infection, but then escape the phagosome and reproduce in the cytosol (*see figure 22.6*). They continue to reproduce until the host cell bursts, releasing bacteria to infect other host cells. Rickettsias are very different from most other bacteria in physiology and metabolism. Described as an energy parasite, rickettsias have a membrane carrier protein that exchanges internal ADP for external ATP. Thus bacterial growth and reproduction continue at the expense of the host cell. ◄◄ *Rickettsias are obligate intracellular bacteria (section 22.1); Phagocytosis (section 33.5)*

Rocky Mountain spotted fever (RMSF) is characterized by the sudden onset of a headache, high fever, chills, and a skin rash (**figure 39.10**) that initially appears on the ankles and wrists, and then spreads to the trunk of the body. The symptoms reflect the accumulation of lysed host cells and toxic effects of rickettsial cell walls. If the disease is not treated, the rickettsias destroy blood vessels in the heart, lungs, or kidneys and cause death within 8 days after infection. Usually, however, severe pathological changes are avoided by antibiotic therapy (doxycycline), development of immune resistance, and supportive therapy. Diagnosis is made through observation of symptoms and signs such as the characteristic rash and by serological tests. The best means of prevention remains avoidance of tick-infested habitats and animals. RMSF cases are reported year-round, with the greatest incidence between May and August. The CDC reports approximately 2,000 cases per year.

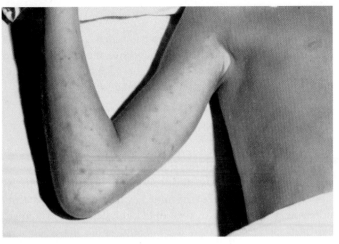

Figure 39.10 Rocky Mountain Spotted Fever. Typical rash occurring on the arms consists of generally distributed, sharply defined macules.

Retrieve, Infer, Apply

1. What strategies are most effective in preventing Lyme disease?
2. What is the causative agent of Lyme disease and how is it transmitted to humans? What preventative measures can an individual take to avoid infection?
3. How is plague transmitted? Distinguish between bubonic and pneumonic plague.
4. Compare the rash seen in Lyme disease with that seen in Rocky Mountain spotted fever.
5. How does transovarian passage occur?

39.3 Direct Contact Diseases Can Be Caused by Bacteria

After reading this section, you should be able to:

- Report the common bacterial diseases spread by direct contact
- Identify typical signs and symptoms of bacterial diseases spread by direct contact
- Correlate direct contact bacterial infection and disease severity with bacterial virulence factors

Most of the direct contact bacterial diseases involve the skin or underlying tissues. Others result from bacterial infection of mucous membranes. Occasionally, bacteria penetrate beyond the skin to cause disseminated disease. We now discuss some of the better-known bacterial contact diseases.

Gas Gangrene (Clostridial Myonecrosis)

Clostridium perfringens, C. novyi, and *C. septicum* are Gram-positive, endospore-forming rods termed the histotoxic clostridia. All can produce a necrotizing infection of skeletal muscle called **gas gangrene** (Greek *gangraina,* an eating sore) or **clostridial myonecrosis** (Greek *myo,* muscles, and *necrosis,* death); however, *C. perfringens* is the most common cause. Analysis of the *C. perfringens* genome sequence reveals that the microbe possesses the genes for fermentation with gas production but lacks genes encoding enzymes for the TCA cycle or a respiratory chain. Nonetheless,

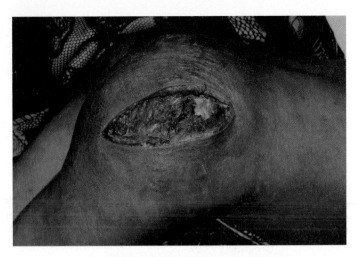

Figure 39.11 Gas Gangrene (Clostridial Myonecrosis). Necrosis of muscle and other tissues results from the numerous toxins produced by *Clostridium perfringens.*

C. perfringens has an extraordinary doubling time of only 8 to 10 minutes when in the human host, making diagnosis and treatment of gangrene critical. ◀◀ *Class* Clostridia *(section 23.1)*

Histotoxic (tissue damaging) clostridia occur in the soil worldwide and also are part of the normal endogenous microflora of the human large intestine. Contamination of injured tissue with spores from soil containing histotoxic clostridia or bowel flora is the usual means of transmission. Infections are commonly associated with wounds resulting from abortions, automobile accidents, military combat, or frostbite. If the spores germinate in anoxic tissue, the bacteria grow and secrete α-toxin (a lecithinase), which breaks down muscle tissue by disrupting cell membranes and causing cell lysis. Growth often results in accumulation of gas (mainly hydrogen as a result of carbohydrate fermentation) and toxic breakdown products of skeletal muscle tissue (**figure 39.11**). ◀◀ *Exotoxins (section 35.2)*

Clinical manifestations of gas gangrene include severe pain, edema, drainage, and muscle necrosis. The pathology arises from progressive skeletal muscle necrosis due to the effects of the toxin. Other enzymes produced by the bacteria degrade collagen and tissue, facilitating spread of the disease.

Gas gangrene is a medical emergency! Laboratory diagnosis is through recovery of the appropriate species of clostridia accompanied by the characteristic disease symptoms. Treatment is extensive surgical removal of all dead tissue, the administration of antitoxin, and antimicrobial therapy with penicillin and tetracycline. Hyperbaric oxygen therapy (the use of high concentrations of oxygen at elevated pressures) is also effective. The oxygen saturates the infected tissue and thereby prevents the growth of the obligately anaerobic clostridia. Prevention and control include debridement of contaminated traumatic wounds plus antimicrobial prophylaxis and prompt treatment of all wound infections. Interestingly, laboratory-raised, sterile maggots are also used to remove dead and infected tissue. Amputation of limbs often is necessary to prevent further spread of the disease.

Group B Streptococcal Disease

Streptococcus agalactiae, or **group B streptococcus** (**GBS**), is a Gram-positive bacterium that causes illness in newborn babies, pregnant women, the elderly, and adults compromised by other severe illness. In fact, GBS is the leading cause of sepsis and meningitis in newborns, a frequent cause of newborn pneumonia, and a common cause of life-threatening neonatal infections. GBS in neonates is more common than rubella and congenital syphilis. Nearly 75% of infected newborns present symptoms in the first week of life; premature babies are more susceptible. Most GBS cases are apparent within hours to less than seven days after birth. This is called "early-onset disease." GBS can also develop in infants who are one week to three months old ("late-onset disease"); however, this is very rare. Meningitis is more common with late-onset disease. Interestingly, about half of the infants with late-onset GBS disease are associated with a mother who is a GBS carrier; the source of infection for the other newborns with late-onset GBS disease is unknown. Infant mortality due to GBS disease is about 5%. Babies who survive GBS disease, particularly those with meningitis, may have permanent disabilities, such as hearing or vision loss or developmental disabilities. The CDC reports approximately 20,000 total cases of invasive GBS annually.

GBS disease is diagnosed when the Gram-positive, β-hemolytic streptococcus is grown from cultures of otherwise sterile body fluids, such as blood or spinal fluid. GBS cultures may take 24 to 72 hours for robust growth to appear. Confirmation of GBS is by latex agglutination immunoassay. An FDA-approved DNA test for GBS is also available if rapid diagnosis of disease is necessary or culture results are equivocal.

GBS is transmitted directly from person to person. Many people are asymptomatic GBS carriers—they are colonized by GBS but do not become ill from it. Adults can carry GBS in the bowel, vagina, bladder, or throat. Twenty-five percent of pregnant women carry GBS in the rectum or vagina. Thus a fetus may be exposed to GBS before or during birth if the mother is a carrier. GBS carriage is temporary; that is, carriers do not harbor the bacteria for life. GBS carriage can be detected during pregnancy by the presence of the bacterium in culture specimens of either the vagina or the rectum. The CDC recommends that pregnant women have vaginal and rectal specimens cultured for GBS in late pregnancy (35 to 37 weeks' gestation). A positive culture result means that the mother carries GBS—not that she or her baby will definitely become ill. Women who carry GBS are not given oral antibiotics before labor because antibiotic treatment at this time does not prevent GBS disease in newborns. An exception to this is when GBS is identified in urine during pregnancy, which indicates an active infection and is treated promptly. Carriage of GBS, in either the vagina or rectum, becomes important at the time of labor and delivery or any time the placental membrane ruptures. At this time, antibiotics are effective in preventing the spread of GBS from mother to baby.

GBS infections in both newborns and adults are usually treated with intravenous penicillin or ampicillin, unless resistance or hypersensitivity is indicated. A GBS vaccine based on genomic analysis has been under development. However, GBS vaccine development is problematic due to antigen-shifting serotypes. ▶▶| *Vaccines (section 42.1)*

Mycobacterial Skin Infections

Many pathogenic species of nontuberculous mycobacteria cause skin or soft-tissue infections. These diseases often are hard to diagnose as they may cause diverse and misleading signs and symptoms. Two skin-associated diseases caused by nontuberculous mycobacteria are presented here. Leprosy has a reputation from history that caused people with the disease to be shunned, rather than helped. We discuss leprosy because it is a complicated disease that still often stigmatizes its victims. In contrast, we also discuss an emerging mycobacterial infection associated with the recent trend of tattoo art.

Leprosy

Leprosy (Greek *lepros*, scaly, scabby, rough), a skin disease caused by *Mycobacterium leprae,* can be severely disfiguring. The only reservoirs of proven significance are humans and armadillos. The disease most often occurs in tropical countries, and the global leprosy incidence is approximately 220,000 cases. The United States reported 82 confirmed cases in 2012 (last available data), with cases in 33 states; 69.5% of the cases were reported by California, Florida, Hawaii, and Texas. Transmission of leprosy is most likely to occur when individuals are exposed for prolonged periods to infected individuals who shed large numbers of *M. leprae*. Nasal secretions probably are the infectious material for family contacts. |◀◀ *Order* Corynebacteriales *includes important human pathogens (section 24.1)*

The incubation period for leprosy is about 3 to 5 years but may be much longer, and the disease progresses slowly. The bacterium invades peripheral nerve and skin cells, and reproduces as an obligately intracellular parasite. It is most frequently found in Schwann cells that surround peripheral nerve axons and in mononuclear phagocytes. The earliest symptom is usually a slightly pigmented skin eruption several centimeters in diameter. Approximately 75% of all individuals with this early, solitary lesion heal spontaneously because they mount an effective cell-mediated immune response. However, in some individuals, this immune response may be so weak that one of two distinct forms of the disease occurs: tuberculoid or lepromatous leprosy.

Tuberculoid (neural) leprosy is a mild, nonprogressive form of leprosy associated with a delayed-type hypersensitivity reaction to antigens on the surface of *M. leprae*. It is characterized by damaged nerves and regions of the skin that have lost sensation and are surrounded by a border of nodules (**figure 39.12**). Afflicted individuals who do not develop hypersensitivity have a relentlessly progressive form of the disease, called **lepromatous (progressive) leprosy,** in which large numbers of *M. leprae* develop in the skin cells. The bacteria kill skin tissue, leading to a progressive loss of facial features, fingers, toes, and other structures. Moreover, disfiguring nodules form all over the body. Nerves are also infected but usually are less damaged than in tuberculoid leprosy. |◀◀ *Type IV hypersensitivity (section 34.10)*

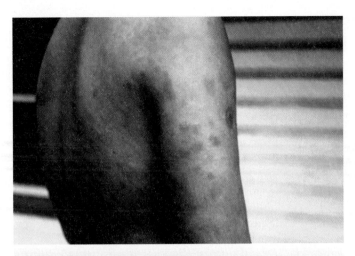

Figure 39.12 Leprosy. In tuberculoid leprosy, the skin manifests shallow, painless lesions.

MICRO INQUIRY *What are the chief differences between tuberculoid and lepromatous leprosy?*

Because the leprosy bacillus cannot be cultured in vitro, laboratory diagnosis is supported by the demonstration of the bacterium in biopsy specimens using direct fluorescent antibody staining. Serodiagnostic methods, such as the fluorescent leprosy antibody absorption test and ELISA, have been developed.

Treatment is accomplished with rifampicin combined with dapsone to treat tuberculous leprosy. Rifampicin with clofazimine is combined with dapsone to treat lepromatous leprosy. Dapsone (diacetyl dapsone) acts by inhibiting the synthesis of dihydrofolic acid through competition with para-aminobenzoic acid (*see figure 9.11*). Identification and treatment of patients with leprosy is the key to control. Children of presumably contagious parents should be given chemoprophylactic drugs until treatment of their parents has made them noninfectious.

Tattoo-Associated Mycobacterial Infections

Sporadic cases and outbreaks of tattoo-associated mycobacterial infections remind us that emerging diseases do not have to be exotic. Rapidly growing mycobacteria (RGM) are a nontuberculous group of mycobacteria commonly found in water, soil, and dust. RGM were found to be the cause of 142 cases, from 11 countries, of tattoo-associated skin infections. *Mycobacterium chelonae* was identified as the elusive agent responsible for these cases. The index case was a 49-year-old man who had over 100 small pimplelike pustules and reddened skin blotches within two-week-old tattoos. Examination of biopsy specimens were negative for bacteria, including acid-fast bacilli. However, culture of the specimen revealed *M. chelonae*. Five additional cases were also reported that had a range of skin lesions, including pink, red, and purple papules; scaly papules; granulomatous papules; lichenoid papules; pustules; and plaques. The common link to all six cases was the inking of tattoos by the same artist; all of the lesions were found to be

concentrated in "gray wash" (areas of shading using gray tones) within the tattoos. Other tattoo-associated infections, caused by *M. haemophilum*, were identified in two otherwise healthy males, who developed painless rashes about three days after receiving tattoos. Their tattoo sites then filled with numerous nodular pustules that failed to resolve with initial antibiotic therapy. In fact, it took nine months of multicourse antibiotic regimens to eradicate the infection in one of the patients. The only common link in these two cases was the use of tap water to dilute the tattoo ink.

Laboratory diagnosis for rapidly growing mycobacterial organisms is by culture and routine bacterial staining. Antibiotic sensitivity testing is also important when weighing treatment options. As with other mycobacterial infections, multiple antibiotics may be used concurrently to control the infection. Treatment of rapidly growing mycobacterial skin infections is successful with macrolide antibiotics (e.g., clarithromycin or azithromycin), orally and topically, but may take many months. Topical steroids may also be indicated to control the lesions.

Retrieve, Infer, Apply

1. How can humans acquire gas gangrene? Group B strep? Leprosy? Describe the major symptoms of each.
2. Why do you think the slow growth rate of *M. leprae* makes leprosy treatment more difficult?

Peptic Ulcer Disease and Gastritis

Peptic ulcer disease is caused by the Gram-negative, spiral bacillus *Helicobacter pylori*. Found in gastric biopsy specimens from patients with gastritis, *H. pylori* is responsible for most cases of chronic **gastritis** not associated with another known primary cause (e.g., autoimmune gastritis or eosinophilic gastritis). Importantly, untreated *H. pylori* infection can lead to gastric cancer and it has been classified as a Class I carcinogen by the World Health Organization (WHO). ◄◄ *Class* Epsilonproteobacteria *ranges from pathogens to deep-sea bacteria (section 22.5)*

H. pylori colonizes only gastric mucus-secreting cells, beneath the gastric mucous layers, and surface fimbriae are believed to be one of the adhesins associated with this process (**figure 39.13**). *H. pylori* binds to Lewis B antigens, which are part of the blood group antigens that determine blood group O. *H. pylori* also binds to the monosaccharide sialic acid, which is found in the glycoproteins on the surface of gastric epithelial cells. The bacterium moves into the mucous layer to attach to mucus-secreting cells. Movement into the mucous layer may be aided by the fact that *H. pylori* is a strong producer of urease. Urease activity creates a localized alkaline environment when hydrolysis of urea produces ammonia. The increased pH may protect the bacterium from gastric acid until it is able to grow under the layer of mucus in the stomach. The potential virulence factors responsible for epithelial cell damage and inflammation probably include proteases, phospholipases, cytokines, and cytotoxins.

Laboratory identification of *H. pylori* is by culture of gastric biopsy specimens, examination of stained biopsies for the presence of bacteria, detection of serum IgG, the urea breath test,

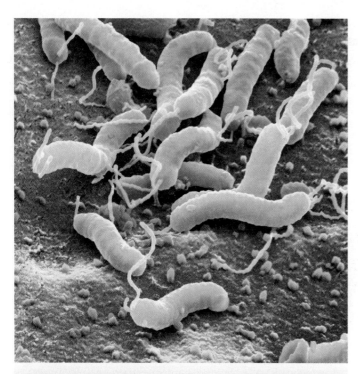

Figure 39.13 *Helicobacter pylori,* **the Causative Agent of Peptic Ulcer Disease.** Scanning electron micrograph (×3,441) of *H. pylori* adhering to gastric cells.

MICRO INQUIRY *How does* H. pylori *increase the local pH in its gastric microenvironment?*

urinary excretion of (^{15}N) ammonia, stool antigen assays, or detection of urease activity in biopsies. The goal of *H. pylori* treatment is the complete elimination of the organism. Treatment is two-pronged: use of drugs to decrease stomach acid (proton pump inhibitors) and antibiotics (metronidazole and tetracycline, amoxicillin, or clarithromycin) to kill the bacteria.

Approximately 50% of the world's population is estimated to be infected with *H. pylori*. It is most likely transmitted from person to person (usually acquired in childhood), although infection from a common exogenous source cannot be completely ruled out, and some think that it is spread by food or water. Support for person-to-person transmission comes from evidence of clustering within families and from reports of higher than expected prevalences in residents of custodial institutions such as nursing homes.

Sexually Transmitted Diseases

Sexually transmitted diseases (STDs) represent a worldwide public health problem. There are over 30 different bacteria, viruses, and parasites that are transmitted between humans through sexual contact. The various viruses that cause STDs are presented in chapter 38, and the yeasts and protozoa in chapter 40. The most common bacterial conditions include chlamydial infections, gonorrhea, and syphilis. These and other sexually transmitted bacteria, and the diseases they cause, are presented in **table 39.1**.

The spread of most STDs is currently out of control. The WHO estimates that the annual global number of new STD cases is 499 million, with most infections occurring in fifteen- to thirty-year-old individuals. Importantly, STDs represent the single most frequent cause of preventable infertility, especially among women. In the United States, STDs remain a major public health challenge. The 2013 sexually transmitted disease *Surveillance Report* published by the Centers for Disease Control and Prevention (CDC) indicated that while substantial progress has been made in preventing, diagnosing, and treating certain STDs, 20 million new infections occur each year, nearly half of them among people aged fifteen to twenty-four. STDs also exact a tremendous economic toll, with direct medical costs estimated at $15.6 billion annually in the United States alone. Many cases of STDs go undiagnosed; others are not reported at all. This is particularly problematic as women with untreated chlamydia or gonorrhea infections risk infertility, and both men and women with papillomavirus infection have a greater risk of cancer.

STDs were formerly called venereal diseases (from Venus, the Roman goddess of love) and may sometimes be referred to as sexually transmitted infections (STIs). Although they occur most frequently in the most sexually active age group, anyone who has sexual contact with an infected individual is at increased risk. In general, the more sexual partners a person has, the more likely the person will acquire an STD.

As noted in previous chapters, some of the microorganisms that cause STDs can also be transmitted by nonsexual means. Examples include transmission by contaminated hypodermic needles and syringes shared among intravenous drug users and transmission from infected mothers to their infants. Some STDs can be cured quite easily, but others are presently difficult or impossible to cure. Because treatments are often inadequate, prevention is essential. Preventive measures are based mainly on better education, barrier protection, and treatment of infected individuals with chemotherapeutic agents.

Chlamydial Diseases

Chlamydia is the most frequently reported sexually transmitted bacterial disease. Over 1.4 million cases of chlamydia were reported to the CDC in 2013. However, this number is thought to be grossly underestimated as most people with chlamydial infections have no symptoms and thus go undiagnosed. The most frequently isolated species causing an STD is *C. trachomatis*. Chlamydia is transmitted through anal, oral, and vaginal sex, and it can be transmitted from mother to child during vaginal childbirth. Symptoms of chlamydial disease vary widely. Males may have few or no manifestations (including a urethral discharge, burning during urination, and itching); however, complications such as inflammation of reproductive structures can occur. Females may be asymptomatic or have a severe infection of the cervix and urethra. Symptoms in women include abnormal vaginal discharge and burning on urination. Bacteria can spread from the vaginal areas to the rectum. The infection may also spread from the cervix to the fallopian tubes, leading to **pelvic inflammatory disease** (**PID**), with symptoms of abdominal

Table 39.1   Summary of the Major Sexually Transmitted Diseases (STDs) Caused by Bacteria

Microorganism	Disease	Comments	Treatment
Chlamydia trachomatis	Nongonococcal urethritis (NGU); cervicitis, pelvic inflammatory disease (PID), lymphogranuloma venereum	Serovars D-K cause most of the STDs in the U.S.; lymphogranuloma venereum rare in the U.S.	Doxycycline and azithromycin
Haemophilus ducreyi	Chancroid ("soft chancre")	Open sores on the genitals can lead to scarring without treatment; on the rise in the U.S.	Azithromycin and ceftriaxone
Helicobacter cinaedi, H. fennelliae	Diarrhea and rectal inflammation in men who have sex with other men	Common in immunocompromised individuals	Metronidazole, macrolides
Mycoplasma genitalium	Implicated in some cases of NGU	Only recently described as an STD	Azithromycin
Mycoplasma hominis	Implicated in some cases of PID	Widespread, often asymptomatic but can cause PID in women	Azithromycin
Neisseria gonorrhoeae	Gonorrhea, PID	Most commonly reported STD in the U.S.; usually symptomatic in men and asymptomatic in women; antibiotic-resistant strains	Doxycycline and azithromycin
Treponema pallidum subsp. *pallidum*	Syphilis, congenital syphilis	Manifests many clinical syndromes	Benzathine penicillin G or doxycycline
Ureaplasma urealyticum	Urethritis	Widespread, often asymptomatic but can cause PID in women and NGU in men; premature birth	Azithromycin or doxycycline

pain, back pain, fever, nausea, and unusual vaginal bleeding (**figure 39.14**). Furthermore, chlamydial infections of the throat can occur in men or women who have oral sex with infected partners. In pregnant females, chlamydial infections are especially serious because they are directly related to miscarriage, stillbirth, inclusion conjunctivitis, and infant pneumonia. Chlamydia is effectively treated with a macrolide or doxycycline; because it is very often co-transmitted with *N. gonorrheae*, patients treated for gonorrhea are also treated for chlamydia.

Gonorrhea

Gonorrhea (Greek *gono*, seed, and *rhein*, to flow) is an acute, infectious, sexually transmitted disease of the mucous membranes of the genitourinary tract, eye, rectum, and throat (table 39.1). It is caused by the Gram-negative, diplococcus *Neisseria gonorrhoeae*. These bacteria are also referred to as **gonococci** and have a worldwide distribution. It is estimated that there are over 700,000 cases of gonorrhea in the United States each year, but fewer than half are reported. For example, in 2013 approximately 301,000 cases were reported to the CDC. ◀◀ *Order* Neisseriales *includes important pathogens (section 22.2)*

Once inside the body, gonococci attach to the microvilli of mucosal cells by means of type IV pili and protein II, which function as adhesins. This attachment prevents the bacteria from

being washed away by normal cervical and vaginal discharges or by the flow of urine. They are often endocytosed by mucosal cells and may even be transported through the cells to the intercellular spaces and subepithelial tissue, a process known as

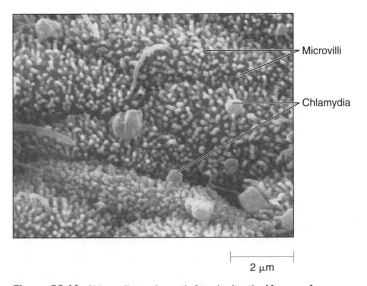

Microvilli

Chlamydia

2 μm

Figure 39.14 *Chlamydia trachomatis* **Attached to the Mucosa of a Fallopian Tube.**

transcytosis. The host tissue becomes infiltrated by immune cells, which phagocytose the bacterium. Neutrophils can be observed that contain gonococci inside phagosomes, where the gonococci can survive because they can withstand the harsh phagosomal environment (**figure 39.15***a*). Because gonococci are intracellular at this time, the host's defenses have little effect on the bacteria. Additionally, gonococci alter their adhesin proteins in response to host recognition (a process known as antigenic or phase variation), forcing immune cells to repeatedly re-identify and re-respond to the bacteria. Neutrophils are later replaced by fibrous tissue that may lead to urethral closing, or stricture, in males. ◄◄ *Cells of the immune system (section 33.4); Phagocytosis (section 33.5)*

In males, the incubation period is 2 to 8 days. The onset consists of a urethral discharge of yellow, creamy pus and frequent, painful urination accompanied by a burning sensation. In females, the cervix is the principal site infected. The disease is more insidious in females, and few women experience any symptoms. However, some symptoms may begin 7 to 21 days after infection. These are generally mild; some vaginal discharge may occur. The gonococci also can infect the Fallopian tubes and surrounding tissues, leading to PID. This occurs in 10 to 20% of infected females. Gonococcal PID is a major cause of sterility and ectopic pregnancies because of scar formation in the Fallopian tubes. Gonococci disseminate most often during menstruation, a time in which there is an increased concentration of free iron available to the bacteria. In both genders, disseminated gonococcal infection with bacteremia may occur. This can lead to involvement of the joints (gonorrheal arthritis), heart (gonorrheal endocarditis), or pharynx (gonorrheal pharyngitis).

Gonorrheal eye infections can occur in newborns as they pass through an infected birth canal. The resulting disease is called **ophthalmia neonatorum,** or **conjunctivitis of the newborn** (figure 39.15*b*). This was once a leading cause of blindness in many parts of the world. To prevent this disease, as well as chlamydial conjunctivitis, erythromycin or silver nitrate in dilute solution is placed in the eyes of newborns. This treatment is required by law in the United States and many other nations.

Laboratory diagnosis of gonorrhea relies on the successful growth of *N. gonorrhoeae* in culture to determine oxidase reaction, Gram-stain reaction, and colony and cell morphology. Because gonococci are very sensitive to adverse environmental conditions and survive poorly outside the body, specimens should be plated directly; when this is not possible, special transport media are necessary. A DNA probe for *N. gonorrhoeae* has been developed and is used to supplement other diagnostic techniques.

Gonorrhea is currently treated with a single injection of cephtriaxone because penicillin- and fluoroquionolone-resistant strains are prevalent worldwide. Alarmingly, some strains are also gaining resistance to cephalosporins. Because *N. gonorrheae* and *C. trachomatis* are often co-transmitted, treatment for gonorrhea also includes a macrolide or tetracycline to treat chlamydia. Later stages of gonorrhea are more difficult to treat and may require much larger drug doses over a longer period of time.

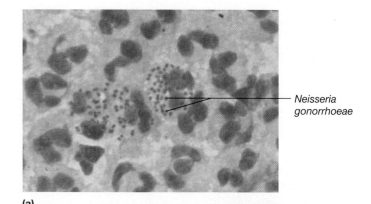

(a)

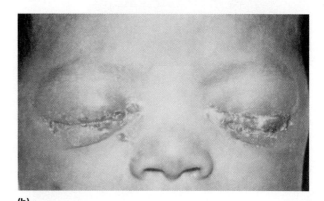

(b)

Figure 39.15 Gonorrhea. (a) Gram stain of male urethral exudate showing *Neisseria gonorrhoeae* (diplococci) inside a neutrophil; light micrograph (×500). Although the presence of Gram-negative diplococci in exudates is a probable indication of gonorrhea, the bacterium should be isolated and identified. (b) Gonococcal ophthalmia neonatorum in a one-week-old infant.

Syphilis

Syphilis (Greek *syn*, together, and *philein*, to love) is a contagious, ulcerative disease (table 39.1) caused by the spirochete *Treponema pallidum* subsp. *pallidum* (*T. pallidum; see figure 21.18*). Infection occurs through intimate contact with the infected lesion of a sexual partner. Syphilis has been called by a number of slang words, but it has also been known as the "great imitator" as its many signs and symptoms mimic those of other diseases. The signs and symptoms of syphilis vary depending on the stage of infection (primary, secondary, latent, and tertiary), and initial signs of infection may be obscured, depending on the site of bacterial entry. **Congenital syphilis** is the disease transmitted in utero from mother to child. ◄◄ *Phylum* Spirochaetes *(section 21.8)*

As a spirochete, *T. pallidum* has a corkscrew morphology allowing it to easily enter the body through mucous membranes and minor breaks or abrasions of the skin. It replicates locally, inducing an ulceration, before it migrates to the regional lymph nodes, where it rapidly spreads throughout the body. Three recognizable stages of syphilis occur in untreated individuals. In the primary stage, after an incubation period of about 10 days to

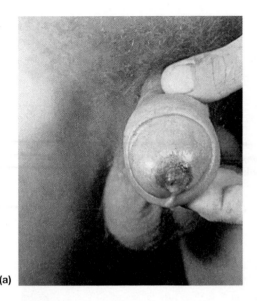

(a)

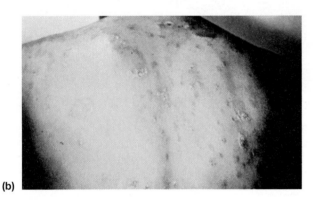

(b)

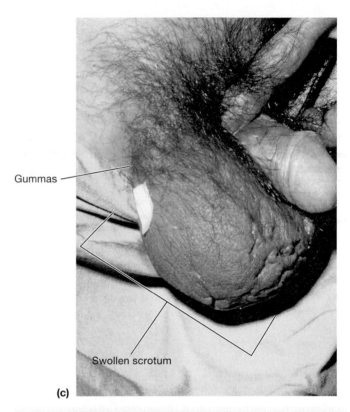

Gummas

Swollen scrotum

(c)

Figure 39.16 Syphilis. (a) Primary syphilitic chancre of the penis. (b) Whole body rash of secondary syphilis. (c) Late-stage tertiary syphilis marked by masses of swollen and dead, fiberlike tissue.

MICRO INQUIRY *Which of these clinical manifestations are associated with syphilis transmission? Explain.*

3 weeks or more, the initial symptom is a small, painless, reddened ulcer, or **chancre** (French, meaning a destructive sore) with a hard ridge that appears at the infection site and contains spirochetes (**figure 39.16a**). Contact with the chancre during sexual contact may result in disease transmission. In about one-third of the cases, the disease does not progress further and the chancre disappears. Serological tests are positive in about 80% of the infected individuals during this stage (**figure 39.17**).

Within 2 to 10 weeks after the primary lesion appears, the disease may enter the secondary stage, which is characterized by a highly variable skin rash (figure 39.16b). By this time, 100% of the individuals are serologically positive. Other symptoms during this stage include the loss of hair in patches, malaise, and fever. Both the chancre and the rash lesions are infectious.

After several weeks, the disease becomes latent. During the latent period, the disease is not normally infectious, except for possible transmission from mother to fetus. After many years, tertiary stage disease develops in about 40% of untreated individuals. During this stage, masses of dead, fibrous tissue called **gummas** (figure 39.16c) form in the skin, bone, and nervous system as the result of hypersensitivity reactions. This stage also

is characterized by a great reduction in the number of spirochetes in the body. Involvement of the central nervous system may result in tissue loss that can lead to cognitive deficits, blindness, a "shuffle" walk (tabes), or insanity. Many of these symptoms have been associated with such well-known people as Al Capone, Francisco Goya, Henry VIII, Adolf Hitler, Scott Joplin, Friedrich Nietzsche, Franz Schubert, Oscar Wilde, and Kaiser Wilhelm (**Disease 39.1**). ◄◄ *Hypersensitivities (section 34.10)*

Diagnosis of syphilis is through clinical history, physical examination, and immunoassay. Because humans respond to *T. pallidum* with the formation of antitreponemal antibody and a complement-fixing reagin, serological tests are very informative. Examples include tests for nontreponemal antigens (VDRL) and treponemal antibodies. PCR testing is also available to amplify and identify specific treponemal genes. ◄◄ *Immune complex formation (section 34.8)*

Treatment in the early stages of the disease is easily accomplished with a single dose of long-acting benzathine penicillin G or a two-week course of doxycycline, for penicillin-allergic individuals. Later stages of syphilis are more difficult to treat and require much larger doses over a longer period. For example, in neurosyphilis cases, treponemes occasionally survive drug treatment. Immunity to syphilis is not complete, and subsequent infections can occur.

Prevention and control of syphilis depend on (1) public education, (2) prompt and adequate treatment of all new cases, (3) follow-up on sources of infection and contact so they can be treated, and (4) prophylaxis (barrier protection) to prevent exposure. At present, the incidence of syphilis, as well as other sexually transmitted diseases, is rising in most parts of the world. An estimated 12 million new cases continue to occur each year. The latest available U.S. data report that there were 56,471 total cases of syphilis in 2013, a 10% increase over the previous year. Most of these cases were in men who have sex with other men, who were also co-infected with HIV.

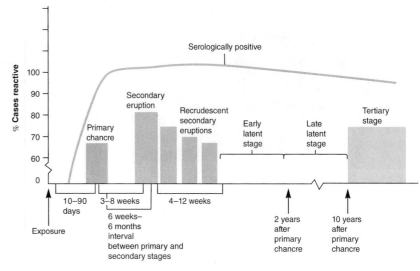

Figure 39.17 The Course of Untreated Syphilis.

Retrieve, Infer, Apply

1. How is *H. pylori* transmitted and what is its global prevalence?
2. What is ophthalmia neonatorum and how is it transmitted?
3. Describe the lesions of syphilis. What do the symptoms indicate about the disease progression?
4. What distinguishes gonorrhea from syphilis?
5. Infection with *N. gonorrhea* does not confer lifelong protection from reinfection. Why do you think this is the case?

Staphylococcal Diseases

Staphylococci are among the most important bacteria that cause disease in humans. They are normal inhabitants of the upper respiratory tract, skin, intestine, and vagina. Staphylococci are members of a group of invasive Gram-positive bacteria known as the **pyogenic** (pus-producing) cocci. These bacteria cause various suppurative (pus-forming) diseases in humans (e.g., boils, carbuncles, abscesses, impetigo contagiosa). The genus *Staphylococcus* consists of Gram-positive cocci, 0.5 to 1.5 μm in diameter (*see figure 23.11*). The cell wall peptidoglycan and teichoic acid contribute to the pathogenicity of staphylococci. ◄◄ *Family Staphylococcaceae (section 23.3)*

Staphylococci can be divided into pathogenic and relatively nonpathogenic strains based on synthesis of the enzyme coagulase. The coagulase-positive species *S. aureus* is the most important human pathogen in this genus. Coagulase-negative staphylococci

---

DISEASE

39.1 A Brief History of Syphilis

Syphilis was first recognized in Europe near the end of the fifteenth century. During this time, the disease reached epidemic proportions in the Mediterranean areas. According to one hypothesis, syphilis is of New World origin and Christopher Columbus (1451–1506) and his crew acquired it in the West Indies and introduced it into Spain after returning from their historic voyage. Another hypothesis is that syphilis had been endemic for centuries in Africa and may have been transported to Europe at the same time that vast migrations of the civilian population were occurring (1500). Others believe that the Vikings, who reached the New World well before Columbus, were the original carriers.

Syphilis was initially called the Italian disease, the French disease, and the great pox as distinguished from smallpox. In 1530 the Italian physician and poet Girolamo Fracastoro wrote the poem "Syphilis sive Morbus Gallicus" ("Syphilis or the French Disease"). In this poem, a Spanish shepherd named Syphilis is punished for being disrespectful to the gods by being cursed with the disease. Several years later, Fracastoro published a series of papers in which he described the possible mode of transmission of the "seeds" of syphilis through sexual contact.

Its venereal transmission was not definitely shown until the eighteenth century. Recognition of the different stages of syphilis was demonstrated in 1838 by Philippe Ricord, who reported his observations on more than 2,500 human inoculations. In 1905 Fritz Schaudinn and Erich Hoffmann discovered the causative bacterium, and in 1906 August von Wassermann introduced the diagnostic test that bears his name. In 1909 Paul Ehrlich introduced an arsenic derivative, arsphenamine or salvarsan, as therapy.

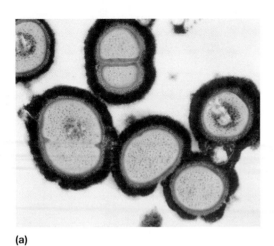

(a)

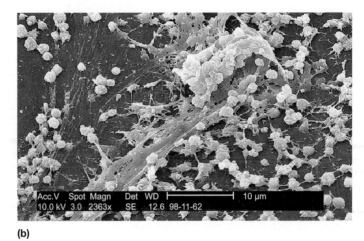

(b)

Figure 39.18 Slime and Biofilm.   (a) Cells of *S. aureus* producing slime (transmission electron microscopy, ×10,000). (b) A biofilm on the inner surface of an intravenous catheter. Extracellular polymeric substances, mostly polysaccharides, surround and encase the staphylococci (scanning electron micrograph, ×2,363).

(CoNS) such as *S. epidermidis* are nonpigmented, and are generally less invasive. However, they have increasingly been associated, as opportunistic pathogens, with serious nosocomial infections.

Staphylococci are further classified into slime producers (SP) and nonslime producers (NSP). The ability to produce slime has been proposed as a marker for pathogenic strains of staphylococci (**figure 39.18***a*). Slime is a viscous, extracellular glycoconjugate that allows these bacteria to adhere to smooth surfaces such as prosthetic medical devices and catheters. Scanning electron microscopy has clearly demonstrated that biofilms (figure 39.18*b*) consisting of staphylococci encased in a slimy matrix are formed in association with biomaterial-related infections (**Disease 39.2**). Slime also appears to inhibit neutrophil chemotaxis, phagocytosis, and antimicrobial

DISEASE

39.2 Biofilms

Biofilms consist of microorganisms immobilized at a substratum surface and typically embedded in an organic polymer matrix of microbial origin. They develop on virtually all surfaces immersed in natural aqueous environments, including both biological (aquatic plants and animals) and abiological (concrete, metal, plastics, stones). Biofilms form particularly rapidly in flowing aqueous systems where a regular nutrient supply is provided to the microorganisms. Extensive microbial growth, accompanied by excretion of copious amounts of extracellular organic polymers, leads to the formation of visible slimy layers (biofilms) on solid surfaces. ◄◄ *Biofilms are common in nature (section 7.5)*

Insertion of a prosthetic device into the human body often leads to the formation of biofilms on the surface of the device. The microorganisms primarily involved are *Staphylococcus epidermidis*, other coagulase-negative staphylococci, and Gram-negative bacteria. These normal skin inhabitants possess the ability to tenaciously adhere to the surfaces of inanimate prosthetic devices. Within the biofilms, they are protected from the body's normal defense mechanisms and also from antibiotics; thus the biofilm also provides a source of infection for other parts of the body as bacteria detach during biofilm sloughing.

Some examples of biofilms of medical importance include:

1. The deaths following massive infections of patients receiving Jarvik 7 artificial hearts
2. Cystic fibrosis patients harboring great numbers of *Pseudomonas aeruginosa* that produce large amounts of alginate polymers, which inhibit the diffusion of antibiotics
3. Teeth, where biofilm forms plaque that leads to tooth decay (figure 39.34)
4. Contact lenses, where bacteria may produce severe eye irritation, inflammation, and infection
5. Air-conditioning and other water retention systems where potentially pathogenic bacteria, such as *Legionella* species, may be protected by biofilms from the effects of chlorination.

Table 39.2	Various Enzymes and Toxins Produced by Staphylococci
Product	**Physiological Action**
β-lactamase	Breaks down β-lactam antibiotics
Catalase	Converts hydrogen peroxide into water and oxygen and reduces killing by phagocytosis
Coagulase	Reacts with prothrombin to form a complex that can cleave fibrinogen and cause the formation of a fibrin clot; fibrin may also be deposited on the surface of staphylococci, which may protect them from destruction by phagocytic cells; coagulase production is synonymous with invasive pathogenic potential
DNase	Destroys DNA
Enterotoxins	Are divided into heat-stable toxins of six known types (A, B, C1, C2, D, E); responsible for the gastrointestinal upset typical of food poisoning
Exfoliative toxins A and B (superantigens)	Cause loss of the surface layers of the skin in scalded-skin syndrome
Hemolysins	Alpha hemolysin destroys erythrocytes and causes skin destruction. Beta hemolysin destroys erythrocytes and sphingomyelin around nerves.
Hyaluronidase	Also known as spreading factor; breaks down hyaluronic acid located between cells, allowing for penetration and spread of bacteria
Panton-Valentine leukocidin	Inhibits phagocytosis by granulocytes and can destroy these cells by forming pores in their phagosomal membranes
Lipases	Break down lipids
Nuclease	Breaks down nucleic acids
Protein A	Is antiphagocytic by competing with neutrophils for the Fc portion of specific opsonins
Proteases	Break down proteins
Toxic shock syndrome toxin-1 (a superantigen)	Is associated with the fever, shock, and multisystem involvement of toxic shock syndrome

agents (*see figure 35.9*). ◄◄ *Biofilms are common in nature (section 7.5)*

Staphylococci are common in nature and harbored by either an asymptomatic carrier or a person with the disease (i.e., an active carrier). They can be spread by the hands, expelled from the respiratory tract, or transported in or on animate and inanimate objects. Staphylococci can produce disease in almost every organ and tissue of the body. For the most part, however, staphylococcal disease occurs in people whose defensive mechanisms have been compromised, such as patients in hospitals.

Staphylococci produce disease through their ability to multiply and spread widely in tissues and through their production of many virulence factors (**table 39.2**). Some of these factors are exotoxins, and others are enzymes thought to be involved in staphylococcal invasiveness (*see figure 23.11b*). Many toxin genes are carried on plasmids; in some cases, genes responsible for pathogenicity reside on both a plasmid and the bacterial chromosome. The pathogenic capacity of a particular *S. aureus* strain is due to the combined effect of extracellular

factors and toxins, together with the invasive properties of the strain. At one end of the disease spectrum is staphylococcal food poisoning, caused solely by the ingestion of preformed enterotoxin (table 39.2). At the other end of the spectrum are staphylococcal bacteremia and disseminated abscesses in most organs of the body.

The classic example of a staphylococcal lesion is the localized abscess of a domed pustule (**figure 39.19**). When *S. aureus* becomes established in a hair follicle, tissue necrosis results. Coagulase is produced and forms a fibrin wall around the lesion that limits the spread. Within the center of the lesion, liquefaction of necrotic tissue occurs, and the abscess spreads in the direction of least resistance. The abscess may be either a furuncle (boil) or a carbuncle. The central necrotic tissue drains, and healing eventually occurs. However, the bacteria may spread from any focus through lymph or blood to other parts of the body.

Newborn infants and children can develop a superficial skin infection characterized by the presence of encrusted pustules. This disease, called impetigo contagiosa, can be caused by

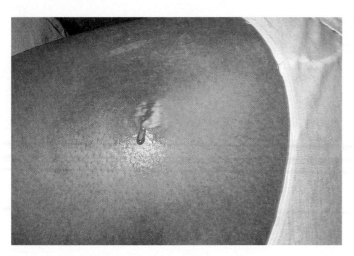

Figure 39.19 Staphylococcal Skin Infections. Draining abscess on the hip.

S. aureus or group A streptococci. It is contagious and can spread rapidly through a nursery, day care, or school. It usually occurs in areas where sanitation and personal hygiene are poor.

Staphylococcal scalded skin syndrome (**SSSS**) is another common staphylococcal disease. SSSS is caused by strains of *S. aureus* that produce **exfoliative toxin** (**exfoliatin**). This protein is usually plasmid encoded, although in some strains, the toxin gene is on the bacterial chromosome. This toxin clips proteins that keep epidermal cells attached to the underlying skin layer, so the epidermis peels off to reveal a red area underneath—thus the name of the disease. SSSS is seen most commonly in infants and children, and neonatal nurseries occasionally suffer large outbreaks of the disease.

Toxic shock syndrome (**TSS**) is a staphylococcal disease with potentially serious consequences. Most cases of this syndrome have occurred in females who used superabsorbent tampons. These tampons can trigger a change in the normal vaginal flora, thereby enabling the growth of toxin-producing *S. aureus*. Toxic shock syndrome is characterized by low blood pressure, fever, diarrhea, an extensive skin rash, and shedding of the skin (**figure 39.20**). TSS results from the massive overproduction of cytokines by T cells induced by the TSST-1 protein (or to staphylococcal enterotoxins B and C1). TSST-1 is a **superantigen** that binds both class II MHC receptors and T-cell receptors, stimulating T-cell responses in the absence of specific antigen. Superantigens activate 5 to 30% of the total T-cell population, whereas specific antigens activate only 0.01 to 0.1% of the T-cell population. The net effect of cytokine overproduction is circulatory collapse leading to shock and multiorgan failure. Tumor necrosis factor α (TNF-α) and interleukins (IL) 1 and 6 are strongly associated with superantigen-induced shock. Mortality rates for TSS are 30 to 70%, and morbidity following surgical debridement and amputation is very high. For these reasons, staphylococcal and streptococcal superantigens are categorized as select agents; their production and use are restricted, as they may be used as bioterror agents.
◀◀ *Superantigens (section 34.5)*

The definitive diagnosis of staphylococcal disease can be made only by isolation and identification of the staphylococcus involved. This requires culture, catalase, coagulase, and other biochemical tests. Commercial rapid test kits also are available. There is no specific prevention for staphylococcal disease. The mainstay of treatment is the administration of specific antibiotics: vancomycin, cefotaxime, ceftriaxone, or rifampin and others. Because of the prevalence of drug-resistant strains (e.g., methicillin-resistant staphylococci), all staphylococcal isolates should be tested for antimicrobial susceptibility. Cleanliness, good hygiene, and aseptic management of lesions are the best means of control.

Methicillin-Resistant *Staphylococcus aureus* (MRSA)

Isolates of *Staphylococcus aureus* resistant to several specific antibiotics are known as **methicillin-resistant S*taphylococcus aureus* (MRSA)**. MRSA is resistant to currently available β-lactam antibiotics, including penicillins (e.g., penicillin, amoxicillin), "antistaphylococcal" penicillins (e.g., methicillin, oxacillin), and cephalosporins (e.g., cephalexin). Antibiotic resistance is driven by the *mecA* (and *mecC*) gene encoding an altered penicillin-binding protein (see chapter-opening story). Otherwise healthy people can acquire MRSA infections, usually on or in the skin, such as abscesses, boils, and other pus-filled lesions. Around 94,000 infections and 18,000 deaths directly due to MRSA occur annually in the United States. When the infection is acquired by people who have not been hospitalized (i.e., in the past year) or had a medical procedure (e.g., dialysis, surgery, catheters) recently, the infection is referred to as a *community-associated* (CA)-MRSA infection. However, MRSA seems to be predominantly a nosocomial infection, with growing evidence that it is widely distributed in livestock populations (cows, pigs, horses). Importantly, CA-MRSA strains are now being reported inside health-care settings. Nearly 300,000 strains of *S. aureus* are known, with a growing number of them resistant to one or more

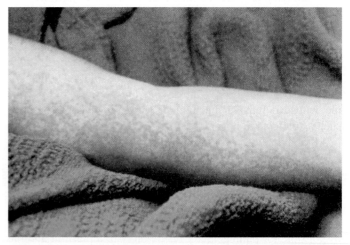

Figure 39.20 Effects of Toxic Shock Toxin. Toxic shock toxin produces a rash, resembling measles, 3 to 5 days into the toxemia.

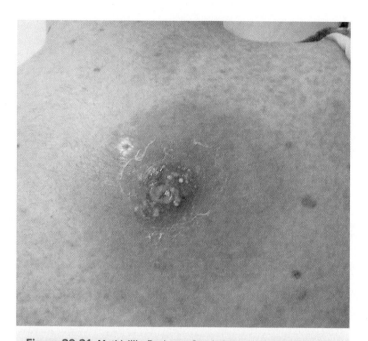

Figure 39.21 Methicillin-Resistant *Staphylococcus aureus* Wound.

MICRO INQUIRY *To what drugs, besides methicillin, are MRSA strains resistant?*

antibiotics. Their movement from farm to grocery, home, and hospital has serious consequences for human health.

MRSA can be diagnosed by culture and sensitivity or molecular testing. However, the CDC suggests that all skin diseases compatible with *S. aureus* infections be treated as MRSA until proven otherwise. Similar to other infections caused by staphylococci, MRSA infections begin as a bump or reddened area on the skin that looks like a large pimple. In other words, MRSA infection would appear as fluid-filled pustules, with a yellow or white center, central point, or "head," that may be draining pus (**figure 39.21**).

Treatment is by incision and drainage, as for other purulent (pus-filled) skin infections. Antimicrobial therapy is prescribed in addition to incision and drainage, while awaiting laboratory results. Doxycycline, minocycline, trimethoprim, and sulfamethoxazole, and rifampin are treatment options. Importantly, vancomycin (an antibiotic thought to be the drug of last resort) is no longer a sure treatment for MRSA as vancomycin-resistant strains occur worldwide. Good aseptic technique is paramount in preventing the transmission of MRSA from patients to healthcare workers.

Streptococcal Diseases

Streptococci that are transmitted through the air are discussed in section 39.1. Superficial cutaneous disease caused by streptococci include cellulitis, impetigo, and erysipelas. Streptococcal disease can also invasively infect the skin, spreading deeper into the tissue, often reaching the underlying muscle. This tissue is a very rich "growth medium" for streptococci, providing the nutrients for rapid replication. We discuss both the superficial and invasive streptococcal diseases next.

Cellulitis, Impetigo, and Erysipelas

Cellulitis is a diffuse, spreading infection of subcutaneous skin tissue. The resulting inflammation is characterized by a defined area of redness (erythema) and the accumulation of fluid (edema). In addition to streptococci, a number of other bacteria can cause cellulitis.

The most frequently diagnosed skin infection caused by *S. pyogenes* is **impetigo** (impetigo also can be caused by *Staphylococcus aureus*; figure 39.19*b*). Impetigo is a superficial cutaneous infection, most commonly seen in children, usually located on the face, and characterized by crusty lesions and vesicles surrounded by a red border. Impetigo is most common in late summer and early fall. Treatment of impetigo is with topical mupirocin. In severe cases, penicillin or a cephalosporin is the drug of choice; erythromycin is prescribed for those individuals who are allergic to β-lactams.

Erysipelas (Greek *erythros*, red, and *pella*, skin) is an acute infection and inflammation of the dermal layer of the skin. It occurs primarily in infants and people over thirty years of age with a history of streptococcal sore throat. The skin often develops painful, reddish patches that enlarge and thicken with a sharply defined edge. Recovery usually takes a week or longer if no treatment is given. The drugs of choice for the treatment of erysipelas are erythromycin and penicillin. Erysipelas may recur periodically at the same body site for years.

Invasive Streptococcal Infections

The development of disease initiated by *Streptococcus pyogenes* (invasive group A strep [GAS, *p. 867*]) appears to depend on the presence of specific virulent strains (e.g., M-1 and M-3 serotypes) and predisposing host factors (surgical or nonsurgical wounds, diabetes, and other underlying medical problems). A life-threatening infection begins when GAS strains penetrate a mucous membrane or take up residence in a wound such as a bruise. This infection can quickly lead either to **necrotizing fasciitis** (Greek *nekrosis*, deadness, Latin *fascis*, band or bandage, and *itis*, inflammation), which destroys the sheath covering skeletal muscles, or to **myositis** (Greek *myos*, muscle, and *-itis*), the inflammation and destruction of skeletal muscle and fat tissue (**figure 39.22**). Because necrotizing fasciitis and myositis arise and spread so quickly, they have been colloquially called "galloping gangrene."

Globally, there are over 10 million cases of strep throat and impetigo annually. It is estimated that approximately 9,000 to 11,500 cases of invasive GAS infections occur annually in the United States (1,000 to 1,800 deaths), and 6 to 7% of these are associated with necrotizing conditions. Rapid treatment is necessary to reduce the risk of death due to necrotizing fasciitis, and penicillin G remains the treatment of choice. In addition,

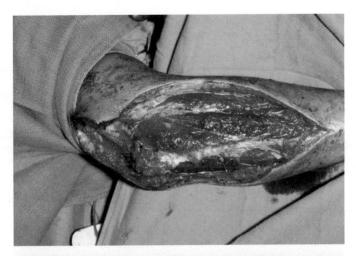

Figure 39.22 Necrotizing Fasciitis. Rapidly advancing streptococcal disease can lead to large necrotic sites, sometimes with blisters that rupture and expose the dying tissue. This is often called flesh-eating disease or necrotizing fasciitis.

MICRO INQUIRY *What virulence factors enable the rapid invasion of certain strep A strains?*

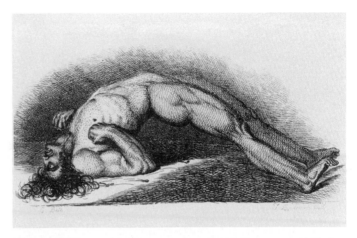

Figure 39.23 The Toll of Tetanus. Sir Charles Bell's portrait (c. 1821) of a soldier wounded in the Peninsular War in Spain shows opisthotonus resulting from tetanus.

surgical removal of dead and dying tissue usually is needed in more advanced cases.

One reason invasive GAS strains are so deadly is that about 85% of them carry the genes for the production of streptococcal pyrogenic exotoxins A and B (Spe exotoxins). Exotoxin A acts as a superantigen, a nonspecific T-cell activator. Like all superantigens, it quickly stimulates T cells to begin producing abnormally large quantities of cytokines. The cytokines damage the endothelial cells that line blood vessels, causing fluid loss and rapid tissue death from a lack of oxygen. Another pathogenic mechanism involves secretion of exotoxin B, a protease that rapidly destroys tissue by breaking down proteins. ◄◄ *Superantigens (section 34.5)*

Since 1986 it has been recognized that GAS infections can also trigger streptococcal toxic shock syndrome (STSS), characterized by a precipitous drop in blood pressure, failure of multiple organs, and a very high fever. STSS is caused by GAS that produce one or more streptococcal pyrogenic exotoxins. STSS has a mortality rate of 40%.

Tetanus

Tetanus (Greek *tetanos*, to stretch) is caused by *Clostridium tetani*, an anaerobic, Gram-positive, endospore-forming rod (*see figure 23.3*). The spores of *C. tetani* are commonly found in hospital environments, in soil and dust, and in the feces of many farm animals and humans. Transmission to humans is associated with skin wounds. Any break in the skin can allow *C. tetani* spores to enter. If the oxygen tension is low enough, the spores germinate and release the neurotoxin tetanospasmin.

Tetanospasmin is an endopeptidase that selectively cleaves the synaptic vesicle membrane protein synaptobrevin. This prevents release of inhibitory neurotransmitters (γ-aminobutyric acid and glycine) at synapses within the spinal cord motor nerves. The result is uncontrolled stimulation of skeletal muscles (spastic paralysis). A second toxin, **tetanolysin,** is a hemolysin that aids in tissue destruction. ◄◄ *Class Clostridia (section 23.1)*

Early in the course of the disease, tetanospasmin causes tension or cramping and twisting in skeletal muscles surrounding the wound and tightness of the jaw muscles. With more advanced disease, there is trismus ("lockjaw"), an inability to open the mouth because of the spasm of the masseter muscles. Facial muscles may go into spasms, producing the characteristic expression known as risus sardonicus. Spasms or contractions of the trunk and extremity muscles may be so severe that there is boardlike rigidity, painful convulsions, and backward bowing of the back so that the heels and back approach each other (opisthotonus; **figure 39.23**). Death results from respiratory failure caused by spasms of the diaphragm and intercostal respiratory muscles.

Prevention of tetanus involves the use of the tetanus toxoid. The toxoid is given routinely with diphtheria toxoid and acellular *B. pertussis* vaccine. Infants receive four doses by 18 months of age and every 10 years afterward. Adults receive boosters every 10 years.

Control measures for tetanus are not possible because of the wide dissemination of the bacterium in the soil and the long survival of its spores. The case fatality rate in generalized tetanus ranges from 30 to 90% because tetanus treatment is not very effective. Therefore prevention is all important and depends on (1) active immunization with toxoid, (2) proper care of wounds contaminated with soil, (3) prophylactic use of antitoxin, and (4) administration of penicillin. Tetanus is rare; fewer than 20 cases of tetanus are now reported annually in the

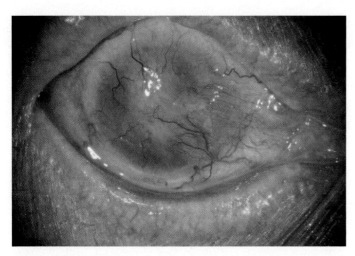

Figure 39.24 Trachoma. An active infection showing marked follicular hypertrophy of the eyelid. The inflammatory nodules cover the thickened conjunctiva of the eye.

United States—the majority of which are in IV drug users and diabetics.

Trachoma

Trachoma (Greek *trachoma*, roughness) is a contagious disease caused by *Chlamydia trachomatis* serotypes A–C. It is one of the oldest known infectious diseases of humans and is the greatest single cause of blindness throughout the world: annually, 2.2. million are infected worldwide, and about half of these develop irreversible blindness. In hyperendemic areas, most children are chronically infected within a few years of birth. Active disease in adults over age 20 is three times as frequent in females as in males because of mother-child contact. Although trachoma is uncommon in the United States, except among Native Americans in the Southwest, it is widespread in Asia, Africa, and South America. *C. trachomatis* can also cause "pink eye," a milder form of eye infection, but also highly contagious.

Trachoma is transmitted by contact with inanimate objects such as soap and towels, by hand-to-hand contact that carries *C. trachomatis* from an infected eye to an uninfected eye, or by flies. The disease begins abruptly with an inflamed conjunctiva. This leads to an inflammatory cell exudate and necrotic eyelash follicles (**figure 39.24**). The disease usually heals spontaneously. However, with reinfection, secondary infections, vascularization of the cornea, pannus (tissue flap) formation, and scarring of the conjunctiva can occur. If scar tissue accumulates over the cornea, blindness results.

Diagnosis of trachoma is by serologic methods. Treatment is surgery to treat the blinding stage and antibiotics (doxycycline, macrolides, or tetracycline) to treat infection. However, prevention and control of trachoma depend more on health education and personal hygiene—such as access to clean water for washing—than on treatment.

39.4 Food and Water Are Vehicles for Bacterial Diseases

After reading this section, you should be able to:

- Report the common food-borne and waterborne bacterial diseases
- Identify typical signs and symptoms of food-borne and waterborne bacterial diseases
- Correlate food-borne and waterborne bacterial infection and disease severity with bacterial virulence factors

Many microorganisms that contaminate food and water can cause acute gastroenteritis—inflammation of the stomach and intestinal lining. When food is the source of the pathogen, the condition is often called **food poisoning.** Gastroenteritis can arise in two ways. Some bacteria cause a food infection. Alternatively, the pathogen may secrete an exotoxin that contaminates the food and is then ingested by the host. This is referred to as a **food intoxication** because the toxin is ingested and the presence of living microorganisms is not required. Because these toxins disrupt the functioning of the intestinal mucosa, they are called **enterotoxins.** Common symptoms of enterotoxin poisoning are nausea, vomiting, and diarrhea.

Worldwide, there are nearly 1.7 billion cases of diarrheal disease annually. In fact, diarrheal diseases are second only to respiratory diseases as a cause of adult death; they are the leading cause of childhood death, and in some parts of the world, they are responsible for more years of potential life lost than all other causes combined. For example, each year around 800,000 children (about 2,200 a day) die from diarrheal diseases. In the United States, estimates exceed 10,000 deaths per year from diarrhea, and an average of 500 childhood deaths are reported.

This section describes several of the more common bacterial gastrointestinal infections, food intoxications, and waterborne diseases. **Table 39.3** summarizes many of the bacterial pathogens responsible for food poisoning, and **table 39.4** lists many important waterborne bacterial pathogens. The protozoa responsible for food-borne and·waterborne diseases are covered in chapter 40. ▶▶ *Food-borne disease outbreaks (section 41.3)*

Table 39.3 Bacteria That Cause Acute Bacterial Diarrhea and Food Poisoning

Organism	Incubation Period (Hours)	Vomiting	Diarrhea	Fever	Common Source
Staphylococcus aureus	1–8 (rarely, up to 18)	+++	+	–	Staphylococci grow in meats, dairy and bakery products and produce enterotoxins.
Bacillus cereus	2–16	+++	++	–	Reheated fried rice causes vomiting or diarrhea.
Clostridium perfringens	8–16	±	+++	–	Clostridia grow in rewarmed meat dishes.
Clostridium botulinum	18–24	±	Rare	–	Clostridia grow in anoxic foods (e.g., canned goods) and produce toxin.
Escherichia coli (enterohemorrhagic)	3–5 days	±	++	±	Generally associated with ingestion of undercooked ground beef and unpasteurized fruit juices and cider.
Escherichia coli (enterotoxigenic strain)	24–72	±	++	–	Raw fruits and vegetables, contaminated water; common cause of traveler's diarrhea.
Vibrio parahaemolyticus	6–96	+	++	±	Organisms grow in seafood and in gut and produce toxin or invade.
Vibrio cholerae	24–72	+	+++	–	Found in fecally contaminated water; grows on exoskeleton of marine invertebrates (e.g., copepods).
Shigella spp. (mild cases)	24–72	±	++	+	Raw fruits and vegetables, contaminated water; causes toxin-mediated disease.
Salmonella spp. (gastroenteritis)	8–48	±	++	+	Poultry is most common source.
Salmonella enterica serovar Typhi (typhoid fever)	10–14 days	±	±	++	Humans are the only source of this serovar; it resides in the gallbladder in carriers.
Clostridium difficile	Days to weeks after antibiotic therapy	–	+++	+	Two reservoirs include humans and environments contaminated with spores.
Campylobacter jejuni	2–10 days	–	+++	++	Infection by oral route from foods, (commonly poultry), pets. Organism grows in small intestine.
Yersinia enterocolitica	4–7 days	±	++	+	Food-borne (most commonly pork).

Adapted from Geo. F. Brooks, et al., Medical Microbiology, 21st ed. Copyright 1998 Appleton & Lange, Norwalk, CT.

Pathogenesis	Clinical Features and Treatment
Enterotoxins act on gut receptors that transmit impulses to medullary centers.	Abrupt onset (2–6 hours after consumption), intense vomiting for up to 24 hours, recovery in 24–48 hours. Occurs in persons eating the same food. No treatment usually necessary except to restore fluids and electrolytes.
Enterotoxins formed in food or in gut from growth of *B. cereus*.	Incubation period of 8–16 hours, mainly vomiting and diarrhea.
Enterotoxins produced during sporulation in gut; causes hypersecretion.	Abrupt onset of profuse diarrhea; vomiting occasionally. Recovery usual without treatment in 1–4 days.
Toxin absorbed from gut and blocks acetylcholine release at neuromuscular junction.	Difficulty speaking or breathing. Treatment requires clearing the airway, ventilation, and intravenous polyvalent antitoxin. Exotoxin present in food and serum. Mortality rate high.
Toxins cause epithelial necrosis in colon; mild to severe complications.	Symptoms vary from mild to severe bloody diarrhea. The toxin can be absorbed, becoming systemic and producing hemolytic uremic syndrome, most frequently in children.
Heat-labile (LT) and heat-stable (ST) enterotoxins cause hypersecretion in small intestine.	Usually abrupt onset of diarrhea; vomiting rare. A serious infection in newborns. In adults, "traveler's diarrhea" is usually self-limited in 1–3 days.
Toxin causes hypersecretion; vibrios invade epithelium; stools may be bloody.	Abrupt onset of diarrhea in groups consuming the same food, especially crabs and other seafood. Recovery is usually complete in 1–3 days. Food and stool cultures are positive.
Toxin causes hypersecretion in small intestine; infective dose $>10^5$ vibrios; cholera stool classically described as "rice-water" consistency.	Abrupt onset of liquid diarrhea in endemic area. Needs prompt replacement of fluids and electrolytes IV or orally. Tetracyclines shorten excretion of vibrios. Stool cultures positive.
Organisms invade epithelial cells; blood, mucus, and neutrophils in stools; infective dose $<10^3$ organisms.	Abrupt onset of diarrhea, often with blood and pus in stools, cramps, and lethargy. Stool cultures are positive. Trimethoprim-sulfamethoxazole, ampicillin, or chloramphenicol given in severe cases. Do not give opiates. Often self-limited. Restore fluids.
Superficial infection of gut, little invasion; infective dose $>10^5$ organisms.	Gradual or abrupt onset of diarrhea and low-grade fever. Nausea, headache, and muscle aches common. Administer no antimicrobials unless systemic dissemination is suspected. Stool cultures are positive. Prolonged carriage is frequent.
Symptoms due to exotoxin; infective dose $\geq 10^7$ organisms.	Initially fever, headache, malaise, anorexia, and muscle pains. Fever may reach 104°F (40°C) by the end of the first week of illness and lasts for 2 or more weeks. Diarrhea often occurs, and abdominal pain, cough, and sore throat may be prominent. Antibiotic therapy shortens duration of the illness.
Toxins causes epithelial necrosis in colon; pseudomembranous colitis.	Especially after abdominal surgery, abrupt bloody diarrhea and fever. Toxins in stool. Oral vancomycin useful in therapy.
Invasion of mucous membrane; toxin production uncertain.	Fever, diarrhea; PMNs and fresh blood in stool, especially in children. Usually self-limited. Special media needed for culture at 43°C. Erythromycin given in severe cases with invasion. Usual recovery in 5–8 days.
Gastroenteritis; occasional bacteremia; toxin produced occasionally.	Severe abdominal pain, diarrhea, fever; PMNs and blood in stool; polyarthritis, erythema nodosum, especially in children. Gentamicin used in severe cases.

Table 39.4 Waterborne Bacterial Pathogens

Organism	Reservoir	Comments
Aeromonas hydrophila	Free-living	Sometimes associated with gastroenteritis, cellulitis, and other diseases
Campylobacter jejuni	Bird and animal reservoirs	Major cause of diarrhea; common in processed poultry; a microaerophile
Helicobacter pylori	Free-living	Can cause gastritis, peptic ulcers, gastric adenocarcinomas
Legionella pneumophila	Free-living and associated with protozoa	Found in cooling towers, evaporators, condensers, showers, and other water sources
Leptospira spp.	Infected animals	Hemorrhagic effects, jaundice
Mycobacterium spp.	Infected animals and free-living	Complex recovery procedure required
Pseudomonas aeruginosa	Free-living	Swimmer's ear and related infections
Salmonella enteriditis	Animal intestinal tracts	Common in many waters
Vibrio cholerae	Free-living	Found in many waters, including estuaries
Vibrio parahaemolyticus	Free-living in coastal waters	Causes diarrhea in shellfish consumers
Yersinia enterocolitica	Frequent in animals and in the environment	Waterborne gastroenteritis

Botulism

Food-borne **botulism** (Latin *botulus*, sausage) is a form of food poisoning where the disease results from ingesting the preformed botulinum toxin, not the bacteria. The most common source of disease is home-canned food that has not been heated sufficiently to kill contaminating *Clostridium botulinum* spores. The spores then germinate, and a toxin is produced during anaerobic vegetative growth. If the food is later eaten without adequate cooking, the active toxin causes disease.

Botulinum toxin is an AB neurotoxin that binds to the same target as tetanus toxin, the synaptobrevin of motor neurons, but with opposite effects (**figure 39.25**). It selectively cleaves the synaptic vesicle membrane protein synaptobrevin, thereby preventing release of the neurotransmitter

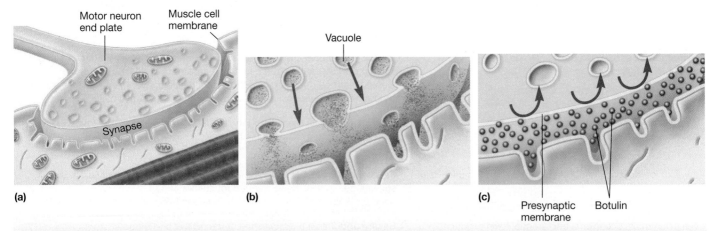

Figure 39.25 The Physiological Effects of *Botulinum* Toxin. (a) The relationship between the motor neuron and the muscle at the neuromuscular junction. (b) In the normal state, acetylcholine released at the synapse crosses to the muscle and creates an impulse that stimulates muscle contraction. (c) In botulism, the toxin enters the motor end plate and attaches to the presynaptic membrane, where it blocks release of the acetylcholine. This prevents impulse transmission and keeps the muscle from contracting.

MICRO INQUIRY *How do the neurological effects of botulinum toxin differ from those of tetanus toxin?*

TECHNIQUES & APPLICATIONS

39.3 Clostridial Toxins as Therapeutic Agents: Benefits of Nature's Most Toxic Proteins

Some toxins are currently used for the treatment of human disease. Specifically, botulinum toxin, the most poisonous biological substance known, is used for the treatment of specific neuromuscular disorders characterized by involuntary muscle contractions. Since approval of type-A botulinum toxin (Botox) by the FDA in 1989 for three disorders (strabismus [crossing of the eyes], blepharospasm [spasmotic contractions of the eye muscles], and hemifacial spasm [contractions of one side of the face]), the number of neuromuscular problems treated has increased to include other tremors, migraine and tension headaches, and other maladies. In 2000 dermatologists and plastic surgeons began using Botox to temporarily eradicate wrinkles caused by repeated muscle contractions

as we laugh, smile, or frown. The remarkable therapeutic utility of botulinum toxin lies in its ability to specifically and potently inhibit involuntary muscle activity for an extended duration. Overall, clostridia (currently one of the largest and most diverse groups of bacteria and includes about 130 species) produce more protein toxins than any other known bacterial genus and are a rich reservoir of toxins for research and medicinal uses. For example, research is underway to use clostridial toxins or toxin domains for drug delivery, prevention of food poisoning, and the treatment of cancer and other diseases. The remarkable success of botulinum toxin as a therapeutic agent has thus created a new field of investigation in microbiology.

---

acetylcholine. As a consequence, muscles do not contract in response to motor neuron activity, and flaccid paralysis results (**Techniques & Applications 39.3**). Symptoms of botulism occur within 12 to 72 hours of toxin ingestion and include blurred vision, difficulty swallowing and speaking, bilateral muscle weakness, nausea, and vomiting. Treatment relies on supportive care and polyvalent antitoxin. If untreated, up to 5% of patients die of either respiratory or cardiac failure within a few days. Approximately 150 cases of botulism occur in the United States annually.

Infant botulism is the most common form of botulism in the United States with about 100 cases reported each year. It appears that ingested spores, which may be naturally present in honey or house dust, germinate in the infant's intestine. *C. botulinum* then multiplies and produces the toxin. The infant becomes constipated, listless, generally weak, and eats poorly. Death may result from respiratory failure. It is therefore recommended that infants not be fed honey.

Campylobacter jejuni Gastroenteritis

Campylobacter jejuni causes an estimated 1.3 million cases of **Campylobacter gastroenteritis** (**campylobacteriosis**) and subsequent diarrhea in the United States each year. Studies with chickens, turkeys, and cattle have shown that as much as 50 to 100% of a flock or herd of these birds or animals excrete *C. jejuni*. It also can be isolated in high numbers from surface waters. It is transmitted to humans by contaminated food and water, contact with infected animals, or anal-oral sexual activity. ◄◄ *Class* Epsilonproteobacteria *ranges from pathogens to deep-sea bacteria (section 22.5)*

The incubation period for campylobacteriosis is 2 to 10 days. *C. jejuni* invades the epithelium of the small intestine, causing

inflammation, and also secretes an exotoxin that is antigenically similar to the cholera toxin. Symptoms include diarrhea, high fever, severe inflammation and ulceration of the intestine, and bloody stools. *C. jejuni* infection has also been linked to Guillain-Barré syndrome, a disorder in which the body's immune system attacks peripheral nerves, resulting in life-threatening paralysis. ◄◄ *Exotoxins (section 35.2)*

Laboratory diagnosis is presumptive using culture-independent tests and microscopy; definitive diagnosis is by culture in a CO_2-enriched atmosphere. The disease is usually self-limited, and treatment is supportive; fluids, electrolyte replacement, and erythromycin may be used in severe cases. Recovery usually takes from 5 to 8 days. Prevention and control involve good personal hygiene and food-handling precautions, including pasteurization of milk and thorough cooking of poultry.

Cholera

Cholera is an acute diarrheal disease caused by infection of the intestine with the Gram-negative, comma-shaped bacterium *Vibrio cholerae* (**figure 39.26**). Infection is usually mild or without symptoms in most healthy adults but can be severe. The disease is characterized by profuse watery diarrhea, vomiting, and leg cramps. These symptoms result from rapid loss of body fluids leading to dehydration and shock. Death can occur within hours if left untreated. Throughout recorded history, cholera (Greek *chole*, bile) has caused seven pandemics in various areas of the world, especially in Asia, the Middle East, and Africa. The disease has been rare in the United States since the 1800s; only a few cases are reported each year (17 cases in 2012), with over 90% of U.S. cases being travel-associated. Unfortunately, 2010 will be remembered for the resurgence of cholera in Haiti. After Haiti

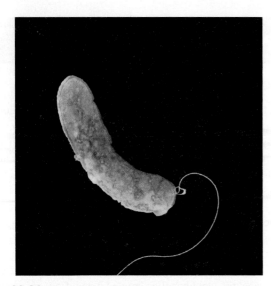

Figure 39.26 *Vibrio cholerae*, the Causative Agent of Cholera.
V. cholerae adhering to intestinal epithelium; scanning electron micrograph
(×12,000). Notice that the bacteria are slightly curved with a single polar flagellum.

had gone almost 100 years without cholera, well-intentioned relief workers responding to a devastating earthquake imported cholera bacteria to Haiti, where they caused 717,203 cholera cases and about 9,000 deaths. Cholera epidemics are a testament to how quickly bacteria can invade and infect.

Although there are many *V. cholerae* serogroups, only the toxigenic O1 and O139 (encapsulated) organisms have caused epidemics. *V. cholerae* O1 is divided into two biotypes, Classical and El Tor. Each biotype is divided into two serotypes, Inaba and Ogawa. Interestingly, the cholera toxin gene (ctxAB) is carried by the temperate CTX filamentous bacteriophage, which exists as a prophage within toxin-producing strains of *V. cholerae*. The receptor for the phage is the toxin coregulated pilus—the same structure used to colonize the host's gut. Thus *V. cholerae* is an excellent example of how horizontal transfer of genes can confer pathogenicity.

Cholera is transmitted by ingesting food or water contaminated by fecal material from infected individuals. The incubation period is 3 to 72 hours and infection with serotypes O1 or O139 results in mild to severe diarrhea. Approximately 20% of those with severe diarrhea have watery diarrhea; 10-20% of those with watery diarrhea produce "rice-water-stool," so named because of the flecks of mucus floating in it. The bacteria adhere to the intestinal mucosa of the small intestine, where they are not invasive but secrete cholera toxin, an AB toxin (*see figure 35.6*). The A subunit enters the intestinal epithelial cells and activates the enzyme adenyl cyclase by the addition of an ADP-ribosyl group in a way similar to that employed by pertussis toxin. This results in hypersecretion of water and chloride ions while inhibiting absorption of sodium ions. The infected person loses massive quantities of fluid and electrolytes, causing abdominal muscle cramps, vomiting, fever, and diarrhea. The diarrhea can be so profuse that a person can lose 10 to 15 liters of fluid during the infection. Death may result from the elevated concentrations of

blood proteins, caused by reduced fluid levels, which leads to circulatory shock and collapse.

Evidence indicates that passage through the human host enhances infectivity, although the exact mechanism is unclear. Before *V. cholerae* exits the body in watery stools, some unknown aspect of the intestinal environment stimulates the activity of certain bacterial genes. These genes, in turn, seem to prepare the bacteria for ever more effective colonization of their next victims, possibly fueling epidemics. *V. cholerae* can also be free-living in warm, alkaline, and saline environments.

Treatment is by oral rehydration therapy with NaCl plus glucose to stimulate water uptake by the intestine; doxycycline or tetracycline antibiotics may also be given. The most reliable control methods are based on proper sanitation, especially of water supplies. The mortality rate without treatment is often over 50%, but with treatment and supportive care, it is less than 1%.

Retrieve, Infer, Apply

1. How can *Clostridium botulinum* cause disease even when the bacteria are dead?
2. Compare the gastroenteritis caused by *Vibrio cholerae* with that of *Campylobacter jejuni*. How are they similar and different?

Escherichia coli Gastroenteritis

Every year, millions of people travel internationally. Unfortunately, a large percentage of these travelers acquire a rapidly acting, dehydrating condition called traveler's diarrhea. This diarrhea results from an encounter with certain viruses, bacteria, or protozoa usually absent from the traveler's normal environment. This is the basis for the popular warnings to international travelers: "Don't drink the local water" and "Boil it, peel it, cook it, or forget it."

One of the major causative agents is *Escherichia coli*. Normal, residential *E. coli* is harmless and provides assistance in the metabolism of the food we eat. *E. coli* strains have become pathogenic by acquiring new genetic information through horizontal gene transfer. Expression of virulence genes often alters the bacterial surface proteins (antigens). Thus, strains of *E. coli* are readily tracked by serotyping. Two of the most commonly used serotyping antigens are the H (flagellar) and the O (LPS) antigens.

Although the vast majority of *E. coli* strains are nonpathogenic members of the normal intestinal microbiota, pathogenic strains cause diarrheal disease by several mechanisms. Six categories (pathotypes) of diarrheagenic *E. coli* are recognized (**figure 39.27**): enterotoxigenic *E. coli* (ETEC), enteroinvasive *E. coli* (EIEC), enteropathogenic *E. coli* (EPEC), enteroaggregative *E. coli* (EAEC), enterohemorrhagic *E. coli* (EHEC), and diffusely adhering *E. coli* (DAEC). ETEC, EIEC, EPEC, and EAEC are all found in human feces; transmission is from poor hand hygiene. ◀◀ *Order* Enterobacteriales *includes pathogens and mutualists (section 22.3)*

The **enterotoxic *E. coli* (ETEC)** strains produce one or both of two distinct enterotoxins responsible for diarrhea. The toxins are distinguished by their heat stability: heat-stable enterotoxin (ST) and heat-labile enterotoxin (LT) (figure 39.27a). The genes for ST

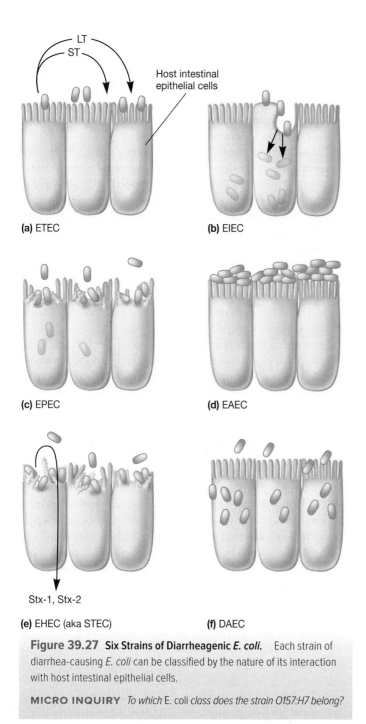

(a) ETEC

(b) EIEC

(c) EPEC

(d) EAEC

Stx-1, Stx-2

(e) EHEC (aka STEC)

(f) DAEC

Figure 39.27 **Six Strains of Diarrheagenic *E. coli*.** Each strain of diarrhea-causing *E. coli* can be classified by the nature of its interaction with host intestinal epithelial cells.

MICRO INQUIRY *To which* E. coli *class does the strain O157:H7 belong?*

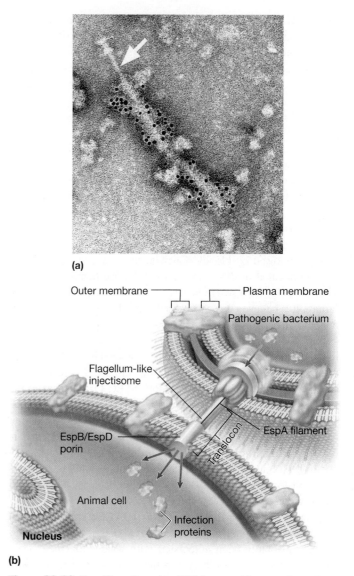

(a)

(b)

Figure 39.28 **Type Three Secretion (TTS) System of Enteropathogenic *E. coli* (EPEC).** (a) Transmission electron micrographs of TTS structures from EPEC cells. TTS sheathlike structures are labeled with 6 nm immunogold particles; the needle complexes, neck, and stem components (white arrows) remain unlabeled. (b) The TTS proteins function like a syringe to inject virulence proteins into a host cell.

and LT production and for colonization factors are usually plasmid-borne and acquired by horizontal gene transfer. ST and LT induce hypersecretion of electrolytes and water into the intestinal lumen, triggering the watery diarrhea characteristic of an ETEC infection.

The **enteroinvasive *E. coli*** (**EIEC**) strains cause diarrhea by penetrating and multiplying within the intestinal epithelial cells (figure 39.27*b*). The ability to invade epithelial cells is similar to that of *Shigella* spp.; it is associated with the presence of a large virulence plasmid. EIEC strains also produce an enterotoxin that stimulates diarrhea and dysentery symptoms.

The **enteropathogenic *E. coli*** (**EPEC**) strains attach to the brush border of intestinal epithelial cells using an adhesin known as intimin and cause a specific type of cell damage called effacing lesions (figure 39.27*c*). Proteins encoded by a LEE (*locus of enterocyte effacement*) region of a pathogenicity island cause effacing lesions or attaching-effacing (AE) lesions. This ultimately results in destruction of brush border microvilli (hemorrhagic colitis) adjacent to adhering bacteria. AE lesions result from the delivery of specific virulence proteins through a LEE-encoded type III secretion (TTS) system, into host cell membranes (**figure 39.28**). The EPEC virulence proteins are homologous to *Yersinia* TTS Yop proteins that lyse red blood cells. Cell destruction leads to the subsequent bloody diarrhea. EPEC is an important cause of infantile diarrhea in underdeveloped countries.

The **enteroaggregative E. coli** (**EAEC**) strains adhere to epithelial cells in localized regions, forming clumps of bacteria with a "stacked brick" appearance (figure 39.27d). EAEC can also acquire the shiga toxin gene(s) to produce toxin that causes prolonged diarrhea in children.

The **enterohemorrhagic E. coli** (**EHEC**) or **Shiga toxin–producing E. coli** (**STEC**) are zoonotic pathogens transferred to humans by water or food. There are 200 to 400 serotypes that carry the bacteriophage-encoded genetic determinants for Shiga toxin (Stx-1 and/or Stx-2 proteins; figure 39.27e). Some EHEC strains also carry the LEE gene sequence to produce AE lesions, causing hemorrhagic colitis (HC) with severe abdominal pain and cramps followed by bloody diarrhea. Stx-1 and Stx-2 have also been implicated in the extraintestinal disease hemolytic uremic syndrome (HUS), a severe hemolytic anemia that leads to kidney failure. It is believed these toxins kill vascular endothelial cells. A major form of STEC (EHEC) in North America is E. coli O157:H7 (STEC O157), which has caused many outbreaks of HC and HUS in the United States since it was first recognized in 1982. A variety of other STEC serotypes cause similar disease. Currently there are an estimated 265,000 STEC (36% STEC O157) cases in the United States each year, resulting in an average of 60 deaths. Despite the variety of STEC serotypes, most laboratories do not test for non-O157 strains, so the actual incidence of EHEC is underreported. Importantly, STECs are the only strains passed in nonhuman (animal) feces. Thus animal manure and farm contamination of foods are major sources of these bacteria. ▶▶| *Food-borne disease outbreaks (section 41.3)*

The **diffusely adhering E. coli** (**DAEC**) strains are not typical human pathogens. They are mentioned here because they have occasionally been associated with watery diarrhea. They have been observed to adhere over the entire surface of epithelial cells and usually cause disease in immunologically naïve or malnourished children (figure 39.27f). It has been suggested that DAEC may have an as-yet undefined virulence factor.

Diagnosis of traveler's diarrhea caused by E. coli is based on travel history and symptoms. Laboratory diagnosis is by isolation of the specific type of E. coli from feces and identification using DNA probes, the determination of virulence factors, and the polymerase chain reaction. Treatment is with fluid and electrolytes plus doxycycline and trimethoprim-sulfamethoxazole. Recovery can be without complications except in EHEC damage to kidneys. Prevention and control involve avoiding contaminated food and water.

Sepsis and Septic Shock

Sepsis is clinically defined as the systemic response to a microbial infection. This response is manifested by two or more of the following conditions: temperature above 102.2°F (39°C) or below 96.8°F (36°C); heart rate above 90 beats per minute; respiratory rate above 20 breaths per minute; and leukocyte count above 12,000 cells per cubic milliliter or below 4,000 cells per cubic milliliter. These lead to **septic shock,** which is sepsis associated with severe hypotension (low blood pressure) despite adequate fluid replacement. Sepsis is a response to an infection that is complicated by the release of cytokines causing body-wide inflammation. Cytokines activate the clotting and complement cascades; they cause blood to clot indiscriminately and to leak from blood vessels.

Gram-negative sepsis is most commonly caused by *E. coli, Salmonella* spp., *Klebsiella* spp., *Enterobacter* spp., or *Pseudomonas aeruginosa*. Endotoxin, or more specifically, the lipid A moiety of lipopolysaccharide (LPS), an integral component of the outer membrane of Gram-negative bacteria, is the primary initiator of septic shock, although Gram-positive bacteria and fungi are also known to do this. We discuss sepsis and septic shock here because (1) some microbial diseases and their effects cannot be categorized under a specific mode of transmission, and (2) we discuss many of the Gram-negative pathogens in this section. |◀◀ *Typical Gram-negative cell walls include additional layers besides peptidoglycan (section 3.4)*

Sepsis is a medical emergency. Septic shock is the most common cause of death in intensive care units and the thirteenth most common cause of death in the United States. Unfortunately, the incidence of sepsis and septic shock continues to rise: more than one million cases per year with an estimated 28–50% of these ending in death.

The pathogenesis of sepsis and septic shock begins with the proliferation of the microorganism at the infection site (*see figure 35.8*). The microorganism may invade the bloodstream directly or may proliferate locally and release various products into the bloodstream. These products include both structural components of the microorganisms (e.g., endotoxin, teichoic acid) along with superantigens (staphylococcal entertoxins) and exotoxins synthesized by the microorganism. All of these products can stimulate the release of the endogenous mediators (e.g., cytokines like TNF-α, IL-1, IL-6, and IL-18) of sepsis from endothelial cells, monocytes, macrophages, neutrophils, and plasma cell precursors. The ensuing "cytokine storm" has profound physiological effects on the heart, vasculature, and other body organs. Other than antimicrobial agents to control infection, there is no drug therapy (despite vigorous research efforts) to stem the septic cascade. Thus, the consequences of septic shock are either recovery or death. Death usually ensues if one or more organ systems fail completely. |◀◀ *Virulence defines a pathogen's success (section 35.2)*

Retrieve, Infer, Apply

1. What are the mechanisms by which E. coli can cause disease?
2. Compare toxigenic E. coli disease with invasive disease. How are they different?
3. Explain how Gram-negative bacteria cause sepsis.

Salmonellosis

Salmonellosis (*Salmonella* gastroenteritis) is caused by over 2,000 *Salmonella* serovars. Based on DNA homology studies, all known *Salmonella* serovars are thought to belong to a single species, *Salmonella enterica*. The most frequently isolated serovars from humans are Typhimurium and Enteritidis. (Serovar names are not italicized and the first letter is capitalized.) An estimated 1.2 million cases of salmonellosis occur annually in the United States; of these, approximately 19,000 result in hospitalization, and 380 result in death.

The initial source of the bacterium is the intestinal tracts of birds and other animals. Humans acquire the bacteria from contaminated water or foods such as beef products, poultry, eggs, or egg products. Once the bacteria are in the body, the incubation time is only about 12 to 72 hours. Disease results when the bacteria multiply and invade the intestinal mucosa. Abdominal pain, cramps, diarrhea, nausea, vomiting, and fever are the most prominent symptoms, which usually persist for 4 to 7 days but can last for several weeks. During the acute phase of the disease, as many as 1 billion *Salmonella* cells can be found per gram of feces. Most adult patients recover, but the loss of fluids can cause problems for children and elderly people. Persons with severe diarrhea may require rehydration therapy with intravenous fluids. Antibiotics are not usually given unless the infection spreads from the intestines.

Other serovars of *Salmonella* are infrequently isolated from humans. We mention a few of these because of their ability to cause occasional epidemics. **Typhoid** (Greek *typhodes*, smoke) fever is caused by *Salmonella enterica* serovar Typhi and is acquired by ingestion of food or water contaminated by feces of infected humans or person-to-person contact. In earlier centuries, the disease occurred in great epidemics. The incubation period for typhoid fever is about 10 to 14 days. The bacteria colonize the small intestine, penetrate the epithelium, and spread to lymphoid tissue, blood, liver, and gallbladder. After approximately 3 months, most individuals stop shedding bacteria in their feces. However, a few individuals continue to shed the pathogen for extended periods but show no symptoms. In these carriers, the bacteria continue to grow in the gallbladder and reach the intestine through the bile duct. This was almost certainly the case for "Typhoid Mary" (*see Historical Highlights 37.4*).

S. Agona, *S.* Heidelberg, *S.* Montevideo, and *S.* Mbandaka, and *S.* Tennessee have also been reported as contaminants of cereal products, chicken, tahini sesame paste, and peanut butter. The ongoing concern over food safety has prompted greater diligence for the detection and reporting of food contamination, so it is not so unusual to learn of foods contaminated with *Salmonella* spp. The CDC and FDA respond quickly with efforts to identify the strain and remove contaminated foods from the supply chain. Food contamination by *Salmonella* spp. demonstrates (1) the continued potential for farmed-food contamination and (2) the speed with which a public health response can occur. Of note, however, is the increasing concern regarding infection resulting from antibiotic-resistant *Salmonella* spp. *S.* Typhimurium is now resistant to at least five antimicrobial agents. Another serovar, *S.* Newport, is resistant to at least seven agents.

Shigellosis

Shigellosis, or bacillary dysentery, is a diarrheal illness resulting from an acute inflammatory reaction of the intestinal tract caused by the four species of the genus *Shigella*. Globally, *Shigella* spp. are estimated to cause up to 165 million cases of disease, with 600,000 deaths, per year. About 14,000 cases a year are reported in the United States. However, real numbers may be 20 times greater because mild cases of disease typically are not reported. The predominant disease-causing species in the United States and other industrialized countries is *S. sonnei*. *S. flexneri* predominates in the developing world. *S. boydii* infections are uncommon, along with *S. dysenteriae*.

The organism is transmitted by the fecal-oral route and is most prevalent among children, especially one- to four-year-olds. The infectious dose is only around 10 to 200 bacteria. In the United States, shigellosis is a particular problem in day-care centers and crowded custodial institutions.

Shigellae are intracellular parasites that multiply within the villus cells of the colonic epithelium. The bacteria induce Peyer's patch cells to phagocytose them. After being ingested by macrophages, the bacteria disrupt the phagosome membrane and are released into the cytosol, where they reproduce. They then invade adjacent mucosal cells. Both endotoxins and exotoxins may participate in disease progression, but the bacteria do not usually spread beyond the colonic epithelium. Watery stools often contain blood, mucus, and pus. In severe cases, the colon can become ulcerated. Virulent shigellae produce a heat-labile AB exotoxin known as Shiga toxin (Stx). The A subunit protein is subsequently released from the B protein units and cleaves host cell 28S rRNA, inhibiting protein synthesis. One target of the B protein seems to be the glomerular endothelium; toxin action on these cells leads to kidney failure. Additionally, like *Y. pestis* and the EPEC and EHEC pathotypes of *E. coli*, shigellae also use a type III secretion system to deliver virulence factors to target epithelial cells. ◄◄ *Phagocytosis (section 33.5); Exotoxins (section 35.2)*

The incubation period for shigellosis usually ranges from 1 to 3 days, and the organisms are shed over a period of 1 to 2 weeks. The disease normally is self-limiting in adults and lasts an average of 4 to 7 days; in infants and young children, it may be fatal. Fluid and electrolyte replacement are usually sufficient to treat disease; antibiotics may not be required in mild cases, although they can shorten the duration of symptoms and limit transmission to family members. Sometimes, particularly in malnourished infants and children, neurological complications and kidney failure occur. Antibiotic-resistant strains are becoming a problem. Prevention is a matter of good personal hygiene and maintenance of a clean water supply.

Staphylococcal Food Poisoning

Staphylococcal food poisoning is the major type of food intoxication in the United States. It is caused by ingestion of improperly stored or cooked food (particularly foods such as ham, processed meats, chicken salad, pastries, ice cream, and hollandaise sauce) in which *S. aureus* has been allowed to reproduce and release enterotoxins. Because these enterotoxins are heat stable, the food can be properly cooked, killing the bacteria without destroying the toxin. Staphylococcal enterotoxins (SE) are primarily produced by *S. aureus,* although other staphylococci produce them. SE are structurally related, single-chain proteins with molecular weights between 26,000 and 29,000; they are antigenically distinct. Thirteen different SE have been identified; SE A, B, C1, C2, C3, D, and E are the most common. Typical symptoms include severe abdominal pain, cramps, diarrhea, explosive vomiting, and nausea. The onset of symptoms is rapid (usually 1 to 8 hours) and of short duration (usually less than 24 hours). The mortality rate of staphylococcal food poisoning is negligible among healthy individuals.

Treatment of staph food poisoning primarily involves keeping hydrated while letting the toxin flush from the body. The toxemia usually resolves without medication. Death due to staph food poisoning is very rare, but it has occurred in infants, the elderly, and individuals who are immunocompromised. Prevention is through thorough hand washing during food preparation and the use of safe cooking and dining practices.

Retrieve, Infer, Apply

1. Compare and contrast *Salmonella* Typhi with other *Salmonella* serovars.
2. Compare the intestinal diseases caused by *Salmonella* and *Shigella* spp. How are they the same? Different?
3. What is the origin of staphylococci that contaminate foods? How is this prevented?

39.5 Zoonotic Diseases Arise from Human-Animal Interactions

After reading this section, you should be able to:

- Report the common bacterial diseases spread by contact with infected animals
- Identify typical signs and symptoms of zoonotic bacterial diseases
- Correlate zoonotic bacterial infection and disease severity with bacterial virulence factors

Diseases transmitted from animals to humans are called zoonotic diseases, or zoonoses. A number of important human pathogens begin as normal flora or parasites of animals and can often adapt to cause disease in humans. Here we highlight a few of the more notable diseases and the agents that cause them.

Anthrax

Anthrax (Greek *anthrax*, coal) is a highly infectious animal disease caused by the Gram-positive, endospore-forming *Bacillus anthracis*. *B. anthracis* is found worldwide and can be transmitted

to humans by direct contact with infected animals (cattle, goats, sheep) or their products, especially hides. *B. anthracis* spores remain viable in soil and animal products for decades (*see figure 23.8a*). Although strains of *B. anthracis* have very little genetic variation, it is now possible to separate all known strains into five categories (providing some clues to their geographic sites of origin) based on the number of tandem repeats in various genes.

There are three forms of anthrax disease. When the bacterium enters through a cut or abrasion of the skin, **cutaneous anthrax** results. Inhaling spores may result in **pulmonary anthrax,** also known as woolsorter's disease. If spores reach the gastrointestinal tract, **gastrointestinal anthrax** may result. *B. anthracis* bacteremia can develop from any form of anthrax. Discounting the 2001 bioterrorism events in the United States, more than 95% of anthrax is the cutaneous form. Spores and vegetative bacteria are detected by skin macrophages and dendritic cells. In pulmonary anthrax, the spores (1 to 2 μm in diameter) are inhaled and lodge in the alveolar spaces, where they are engulfed by alveolar macrophages. Symptoms of gastrointestinal anthrax typically occur after the ingestion of undercooked meat containing spores and include nausea, vomiting, fever, and abdominal pain.

B. anthracis carries two plasmids that encode virulence factors. One plasmid has genes involved in the synthesis of a polyglutamyl capsule that inhibits phagocytosis, and the other bears genes for the synthesis of an exotoxin. *B. anthracis* exotoxin is composed of three proteins: protective antigen (PA), edema factor (EF), and lethal factor (LF). Macrophages have many receptors on their plasma membranes to which PA attaches. Attachment continues until seven PA-receptor complexes gather in a doughnut-shaped ring (**figure 39.29**). The ring acts like a syringe, boring through the plasma membrane

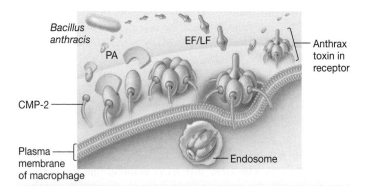

Figure 39.29 Anthrax. A protein called protective antigen (PA) delivers two other proteins, edema factor (EF) and lethal factor (LF), to the capillary morphogenesis protein-2 (CMP-2) receptor on the cell membrane of a target macrophage, where PA, EF, and LF are transported to an endosome. PA then delivers EF and LF from the endosome into the cytoplasm of the macrophage, where they exert their toxic effects.

MICRO INQUIRY *By what mechanisms do EF and LF kill macrophages?*

of the macrophage. Toxin moieties EF and LF are then engulfed by the macrophage and shuttled to an endosome (*see figure 33.19*). Once there, PA molecules form a pore that pierces the endosomal membrane, and EF and LF enter the cytosol. Toxin activity results in fluid release, with the formation of edema. Additionally, LF prevents the transcription factor nuclear factor kappa B (NF$_\kappa$B) from regulating numerous cytokine and other genes needed for an effective immune response. As thousands of macrophages die, they release their lysosomal contents, leading to fever, internal bleeding, septic shock, and rapid death. Without antibiotic treatment and supportive therapy, mortality rates approach 100% for inhalational and gastrointestinal anthrax, and between 20 and 50% for cutaneous anthrax.

Between 2,000 and 20,000 cases of anthrax are estimated to occur worldwide annually, primarily in developing countries; in the United States, the annual incidence was less than one case per year—a rate maintained for 20 years until 2001. The 2001 occurrence of 22 cases of anthrax has spotlighted concern about anthrax as a weapon of bioterrorism.

Treatment of each form of anthrax is the same: ciprofloxacin, penicillin, or doxycycline are successful only if begun before a critical concentration of toxin has accumulated. Although antibiotics may kill the bacterium or suppress its growth, the exotoxin can still eventually kill the patient. Thus, supportive care is essential in preventing death. Vaccination of animals, primarily cattle, is an important control measure. However, people with a high occupational risk, such as those who handle infected animals or their products, including hides and wool, should be immunized. U.S. military personnel also receive the vaccine. ◀◀ *Bioterrorism readiness is an integral component of public health microbiology (section 37.7)*

Brucellosis (Undulant Fever)

The genus *Brucella* contains important human and animal pathogens. **Brucellosis,** also called undulant fever, is caused by faintly staining coccobacilli of the species *B. abortus, B. melitensis, B. suis,* or *B. canis*. These bacteria are commonly transmitted through consumption of contaminated animal products or through abrasions of the skin while handling infected mammals (cattle, sheep, goats, pigs, and, rarely, dogs). The most common route is ingestion of contaminated milk products. Direct person-to-person spread of brucellosis is extremely rare. However, infants may be infected through their mother's breast milk; sexual transmission of brucellosis has also been reported. Although uncommon, transmission of brucellosis may also occur through transplantation of contaminated blood or tissue. A controversy in the northwestern United States over the transmission of brucellosis, endemic in the wild bison and elk populations, to otherwise *Brucella*-free cattle has continued for many years. ◀◀ *Class* Alphaproteobacteria *includes many oligotrophs (section 22.1)*

In the United States, human *Brucella* infections (primarily *B. melitensis*) occur more frequently when individuals ingest unpasteurized milk or dairy products. Brucellosis also has occurred in laboratory workers—culturing the organisms concentrates them and increases the risk of their aerosolization. For the past 10 years, approximately 110 cases of brucellosis have been reported annually. Most of these cases have been in slaughterhouse workers, meat inspectors, animal handlers, veterinarians, and laboratorians. Important for tourists is the potential infection through unpasteurized cheeses, sometimes called "village cheeses."

Brucellosis, in the acute (8 weeks from onset) form, presents as nonspecific, flulike symptoms, including fever, sweats, malaise, anorexia, headache, myalgia (muscle pain), and back pain. In the undulant (rising and falling) form, symptoms of brucellosis include cyclic fevers, arthritis, and testicular inflammation. Neurologic symptoms may occur acutely in up to 5% of the cases. In the chronic form (1 year from onset), brucellosis symptoms may include chronic fatigue, depression, and arthritis. Mortality is low, less than 2%. Treatment is usually with doxycycline and rifampin in combination for 6 weeks to prevent recurring infection. The insidious disease, potential for food supply disruption, and significant morbidity associated with *Brucella* infection have prompted its restriction as a select agent; the public health impact, should it be used as a bioweapon, would be substantial. ◀◀ *Bioterrorism readiness is an integral component of public health microbiology (section 37.7)*

Psittacosis (Ornithosis)

Psittacosis (ornithosis) is a worldwide infectious disease of birds that is transmissible to humans. Since 1996 fewer than 50 confirmed cases have been reported in the United States each year; typically 1 or 2. Psittacosis was first described in association with parrots and parakeets, both of which are psittacine birds. The disease is now recognized in many other birds—among them, pigeons, chickens, ducks, and turkeys—and the general term ornithosis (Latin *ornis,* bird) is used.

Ornithosis is caused by *Chlamydophila psittaci*. Humans contract this disease either by handling infected birds or by inhaling dried bird excreta that contains viable *C. psittaci*. Ornithosis is recognized as an occupational hazard within the poultry industry, particularly to workers in turkey-processing plants. Incubation is typically 5 to 19 days. After entering the respiratory tract, the bacteria are transported to the cells of the liver and spleen. The bacteria multiply within these cells and then invade the lungs, where they cause inflammation, hemorrhaging, and pneumonia. Endocarditis (inflammation of the heart muscle), hepatitis, and neurologic complications may occasionally occur. With doxycycline or tetracycline therapy, the mortality rate is about 2%.

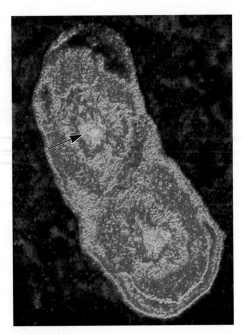

Figure 39.30 *Coxiella burnetii.* Note the unique endospore-like structure (at arrowhead) within the vegetative cell.

Q Fever

The Q in **Q fever** stands for query because the cause of the fever was long unknown. The disease is now known to be caused by the γ-proteobacterium *Coxiella burnetii. C. burnetii* can survive outside host cells by forming resistant, endospore-like bodies (**figure 39.30**). This bacterium infects both wild animals and livestock. Cattle, sheep, and goats are the primary reservoirs. It does not need an arthropod vector for transmission, but in animals, ticks (many species) transmit *C. burnetii.* By contrast, human transmission is primarily by inhalation of dust contaminated with bacteria from dried animal feces or urine, or consumption of unpasteurized milk. Importantly, organisms are shed in high numbers within amniotic fluids and the placenta during birthing. The organisms are resistant to heat, drying, and many common disinfectants. These features enable the bacteria to survive for long periods in the environment. The disease can occur in epidemic form among slaughterhouse workers and sporadically among farmers and veterinarians. Each year, fewer than 30 cases of Q fever are reported in the United States. However, approximately 3% of U.S. adults and an additional 10-20% of high-risk populations (farmers, veterinarians, etc.) have antibodies suggesting prior exposure to *C. burnetii.* ◀◀ *Class* Gammaproteobacteria *is the largest bacterial class (section 22.3)*

In humans, Q fever is an acute illness characterized by the sudden onset of severe headache, malaise, confusion, sore throat, chills, sweats, nausea, chest pain, myalgia, and high fever. It is rarely fatal, but endocarditis occurs in about 10% of the cases. Five to 10 years may elapse between the initial infection and the appearance of endocarditis. During this interval, the bacteria

reside in the liver and often cause hepatitis. The fatality rate is approximately 4%.

Diagnosis of Q fever is made by national reference laboratories using PCR or culture of the bacterium and fluorescent antibody and agglutination tests. Treatment is with doxycycline and is most effective when initiated within the first 3 days of illness. Prevention and control involve public education, protective clothing, and vector control. An attenuated live vaccine is available from the U.S. Army for high-risk laboratory workers. Importantly, *C. burnetii* is a microorganism of concern as a biological threat agent. ◀◀ *Bioterrorism readiness is an integral component of public health microbiology (section 37.7)*

Tularemia

The Gram-negative bacterium *Francisella tularensis* is widely found in animal reservoirs in the United States and causes the disease **tularemia** (from Tulare, a county in California where the disease was first described). It may be transmitted to humans by biting arthropods (ticks and deer flies), direct contact with infected tissue (rabbits), inhalation of aerosolized bacteria, or ingestion of contaminated food or water. However, tularemia is most often transmitted through contact with infected animals; it is called rabbit fever in the central United States because it is often a disease of hunters who skin their catch. After an incubation period of 2 to 10 days, a primary ulcerative lesion appears at the infection site, lymph nodes enlarge, and a high fever develops.

Diagnosis is made by national reference laboratories using PCR or culture of the bacterium and fluorescent antibody and agglutination tests. Treatment is with streptomycin, tetracycline, or aminoglycoside antibiotics for 10 to 21 days. Prevention and control involve public education, protective clothing, and vector control. An attenuated live vaccine is available from the U.S. Army for high-risk laboratory workers. Typically 150 to 200 cases of tularemia are reported annually in the United States. *F. tularensis* is also a microorganism of concern as a biological threat agent. Because public health preparedness efforts in the United States have shifted toward a stronger defense against biological terrorism, public health and medical management protocols following a potential release of tularemia are now in place. ◀◀ *Bioterrorism readiness is an integral component of public health microbiology (section 37.7)*

Retrieve, Infer, Apply

1. How can humans acquire anthrax? Brucellosis?
2. Describe the symptoms of the disease as related to the infection process for anthrax and brucellosis.
3. How is ornithosis transmitted?
4. Describe the disease of tularemia. Why is *F. tularensis* considered a potential agent of bioterrorism?

39.6 Opportunistic Diseases Can Be Caused by Bacteria

After reading this section, you should be able to:

- Explain how disease can result from human microbiota that is normal to another body site
- Cite specific environmental conditions that result from overgrowth of one bacterial species
- Design a model of bacterial biofilm formation to explain its role in human disease

Recall that the normal microbiota of a human can also become pathogenic, especially if they gain access to a tissue site that is not their regular environment or they overgrow other normal flora. Microbes that are otherwise members of the normal microbiota but become pathogens are referred to as opportunists; they cause opportunistic disease. We discuss a few examples of opportunistic disease next.

Antibiotic-Associated Colitis (Pseudomembranous Colitis)

Antibiotic-associated colitis is a spectrum of disease caused by the overgrowth of *Clostridium difficile*. *C. difficile* can be part of the human intestinal microbiota but is kept at low numbers by other bacteria. However, when the other bacteria are inhibited or killed during antibiotic chemotherapy, *C. difficile* grows unchecked. The disease spectrum (*C. difficile* infection, or CDI) includes uncomplicated diarrhea, pseudomembranous colitis (a viscous collection of inflammatory cells, dead cells, necrotic tissue, and fibrin that obstructs the intestine), and toxic megacolon (inflammation that results in intestinal tissue necrosis), which can, in some instances, lead to sepsis and death (**figure 39.31**). *C. difficile* is an anaerobic, spore-forming bacillus that can be found in the intestines of healthy people and often in health-care facilities. The most common antibiotics associated with the development of CDI are amoxicillin, ampicillin, cephalosporins, and clindamycin. When antibiotics reduce the other bacteria, *C. difficile* bacteria multiply, produce toxins, and sporulate. Importantly, even short stays in health-care facilities predispose people to *C. difficile* infection. Three toxins can be expressed by *C. difficile*: enterotoxin (toxin A) and a cytotoxin (toxin B), both of which are responsible for the inflammation and diarrhea, and binary toxin, which alters the host cell cytoskeleton. Toxin genes are located within the pathogenicity locus. The most common symptom of CDI is watery diarrhea, consisting of three or more bowel movements per day for 2 or more days. Other common symptoms include fever, loss of appetite, nausea, and abdominal cramping or tenderness.

C. difficile is the most commonly recognized cause of diarrhea in hospitalized patients, as spores from one patient can be transmitted to others. Antibiotic use is the main modifiable risk factor for CDI. In addition to suppressing other bacteria and

thereby promoting *C. difficile* overgrowth, antibiotics increase the risk of selecting for antimicrobial-resistant strains. It is likely that alteration of the complex colonic ecology provides an environment for *C. difficile* to thrive and produce disease. Ironically, treatment of CDI is by oral metronidazole or oral vancomycin. Newer therapies showing success include (1) the use of a probiotic milk shake containing *Lactobacillus casei, L. bulgaricus,* and *Streptococcus thermophilus* to prevent antibiotic-associated diarrhea caused by *C. difficile*; and (2) the transplantation of fecal material to replenish a healthy colonic microbiota (*see p. 685*).

Bacterial Vaginosis

Bacterial vaginosis is caused by *Gardnerella vaginalis* (a Gram-positive to Gram-variable, pleomorphic, nonmotile rod), *Mobiluncus* spp. (Gram-negative, anaerobic rods), and various other anaerobic bacteria. These microorganisms inhabit the vagina and rectum of 20 to 40% of healthy women, suggesting an opportunistic etiology; however, some consider the disease to be sexually transmitted because the organisms can be shared through intimate contact. Vaginosis of bacterial origin is typically a mild disease, although it is a risk factor for obstetric infections, various adverse outcomes of pregnancy, and pelvic inflammatory disease.

Vaginosis is characterized by a copious, frothy, fishy-smelling discharge with varying degrees of pain or itching. Diagnosis is based on this fishy odor and the microscopic observation of clue

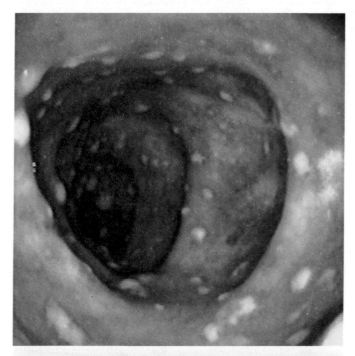

Figure 39.31 Pseudomembranous Colitis. Endoscopic view of *C. difficile*–associated colitis. The white patches represent inflamed tissue.

MICRO INQUIRY *What cells and tissues make up the pseudomembrane?*

cells in the discharge. Clue cells are sloughed-off vaginal epithelial cells covered with bacteria, mostly *G. vaginalis*. Treatment for bacterial vaginosis is with metronidazole, a drug that kills anaerobic streptococci and the *Mobiluncus* spp. that appear to be needed for the continuation of the disease.

Dental Diseases

Only a few of the symbiotic bacteria in the oral cavity can be considered true opportunistic dental pathogens, or odontopathogens. These few odontopathogens are responsible for the most common bacterial diseases in humans: tooth decay and periodontal disease.

Dental Plaque

The human tooth has a natural defense mechanism against bacterial colonization that complements the protective role of saliva. The hard enamel surface selectively absorbs acidic glycoproteins (mucins) from saliva, forming a membranous layer called the acquired enamel pellicle. This pellicle, or organic covering, contains many sulfate (SO_4^{2-}) and carboxylate ($-COO^-$) groups that confer a net negative charge to the tooth surface. Because most bacteria also have a net negative charge, there is a natural repulsion between the tooth surface and bacteria in the oral cavity. Unfortunately, this natural defense mechanism breaks down when dental plaque forms.

Dental plaque is one of the densest collections of bacteria in the body—perhaps the source of the microorganisms seen under a microscope by Antony van Leeuwenhoek in the seventeenth century. Dental plaque formation begins with the initial colonization of the pellicle by *Streptococcus gordonii, S. oralis,* and

S. mitis (**figure 39.32**). These bacteria selectively adhere to the pellicle by specific ionic, hydrophobic, and lectinlike interactions. Once the tooth surface is colonized, subsequent attachment of other bacteria results from a variety of specific coaggregation reactions.

S. mutans and *S. sobrinus* become established on the tooth surface by attaching to these initial colonizers (figure 39.32). *S. mutans* and *S. sobrinus* produce extracellular enzymes that polymerize the glucose moiety of sucrose into a heterogeneous group of extracellular, water-soluble, and water-insoluble polysaccharides. These act like a cement that binds bacterial cells together, forming a plaque ecosystem that is also a biofilm. The fructose by-product can be used in fermentation. Once plaque becomes established, the surface of the tooth becomes anoxic. This leads to the growth of strictly anaerobic bacteria (*Bacteroides melaninogenicus, B. oralis,* and *Veillonella alcalescens*), especially between teeth and in the dental-gingival crevices (**figure 39.33**). In addition, bacteria produce lactic and possibly acetic and formic acids from sucrose and other sugars. Because plaque is not permeable to saliva, the acids are not diluted or neutralized, and they demineralize the enamel to produce a lesion on the tooth. It is this chemical lesion that initiates dental decay. ◄◄ *Biofilms are common in nature (section 7.5); Phylum* Bacteroidetes *includes important gut microbiota (section 21.9)*

Dental Decay (Caries)

As fermentation acids move below the enamel surface, they dissociate and react with the hydroxyapatite of the enamel to form soluble calcium and phosphate ions. As the ions diffuse outward, some reprecipitate as calcium phosphate salts in the tooth's surface

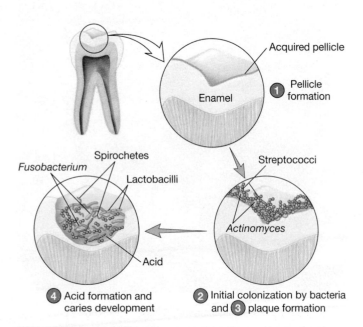

Figure 39.32 Stages in Plaque Development and Cariogenesis. A drawing of a microscopic view of pellicle and plaque formation, acidification, and destruction of tooth enamel.

Figure 39.33 The Microscopic Appearance of Plaque. Scanning electron micrograph of plaque with long filamentous forms and "corn cobs" that are the mixed bacterial aggregates.

MICRO INQUIRY *How is tooth enamel demineralized by bacterial growth?*

layer to create an outer layer overlying a porous subsurface area. Between meals and snacks, the pH returns to neutrality and some calcium phosphate reenters the lesion and crystallizes. The result is a demineralization-remineralization cycle.

When an individual eats foods high in sucrose for prolonged periods, acid production overwhelms the repair process, and demineralization is greater than remineralization. This leads to dental decay or **caries** (Latin, rottenness). Once the hard enamel has been breached, bacteria can invade the dentin and pulp of the tooth and cause its death (figure 39.33).

No drugs are available to prevent dental caries. The main strategies for prevention include minimal ingestion of sucrose; daily brushing, flossing, and rinsing with mouthwashes; and professional cleaning at least twice a year to remove plaque. The use of fluorides in toothpaste, drinking water, and mouthwashes or fluoride and sealants applied professionally to the teeth protects against lactic and acetic acids and reduces tooth decay.

Periodontal Disease

Periodontal disease refers to a diverse group of inflammatory diseases that affect the periodontium and is the most common chronic infection in adults. The **periodontium** is the supporting structure of a tooth and includes the cementum, the periodontal membrane, the bones of the jaw, and the gingivae (gums). The gingiva is dense, fibrous tissue and its overlying mucous membrane that surrounds the necks of the teeth. The gingiva helps to hold the teeth in place.

Disease is initiated by the formation of **subgingival plaque,** the plaque that forms at the dentogingival margin and extends down into the gingival tissue. A number of bacterial species contribute to tissue damage. The result is an initial inflammatory reaction known as **periodontitis,** which is caused by the host's immune response to both the plaque bacteria and the tissue destruction. This leads to swelling of the tissue and formation of periodontal pockets. Bacteria colonize these pockets and cause more inflammation, which leads to formation of a periodontal abscess, bone destruction (periodontosis), inflammation of the gingiva (gingivitis), and general tissue necrosis (**figure 39.34**). If the condition is not treated, the tooth may fall out of its socket.

Periodontal disease can be controlled by frequent plaque removal; by brushing, by flossing, and by rinsing with mouthwashes; and at times, by oral surgery of the gums and antibiotics.

Streptococcal Pneumonia

Streptococcal pneumonia is now considered an opportunistic infection because it is often contracted from one's own normal microbiota. It is caused by Gram-positive *Streptococcus pneumoniae*, normally found in the upper respiratory tract. However, disease usually occurs only in those individuals with predisposing factors such as viral infections of the respiratory tract, physical injury to the tract, alcoholism, or diabetes. In addition to pneumonia, *S. pneumoniae* causes other diseases, including acute middle ear infection, bacteremia, and meningitis.

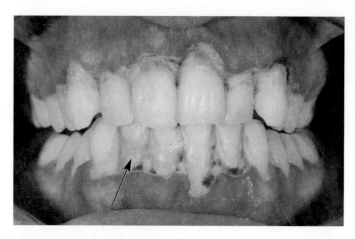

Figure 39.34 Periodontal Disease. Notice the plaque on the teeth (arrow), especially at the gingival (gum) margins, and the inflamed gingiva.

About 60 to 80% of all respiratory diseases known as pneumonia are caused by *S. pneumoniae*. An estimated 4 million pneumococcal illnesses occur in the United States annually, and 22,000 deaths result.

The primary virulence factor of *S. pneumoniae* is its capsular polysaccharide, which is composed of hyaluronic acid. The capsule plays an important role in protecting the organism from ingestion and killing by phagocytes. Pathogenesis is due to the rapid multiplication of the bacteria in alveolar spaces. The bacteria also produce the toxin pneumolysin, which destroys host cells. The alveoli fill with blood cells and fluid, and become inflamed. The sputum is often rust-colored because of blood coughed up from the lungs. The onset of clinical symptoms is usually abrupt, with chills; hard, labored breathing; and chest pain. Diagnosis is by chest X ray, Gram stain, culture, and tests for metabolic products. Treatment needs to be based on antibiotic susceptibility, because many strains are resistant to β-lactam and macrolide antibiotics.

Pneumococcal vaccines (PCV-13 and PPSV23) are available for people who are at risk for exposure (e.g., college students, children, the elderly, and people in chronic-care facilities). The pneumococcal conjugate vaccine (PCV) covers 13 pneumococcal serotypes, while the pneumococcal polysaccharide vaccine (PPSV23) covers 23 types of pneumococcal bacteria. These vaccines are effective because they generate antibodies to the strains that cause 88 to 90% of pneumococcal disease. When these antibodies are deposited on the surface of the bacteria, they become opsonic and enhance phagocytosis (*see figure 33.6*).

Retrieve, Infer, Apply

1. Describe the process by which *C. difficile*–associated disease occurs.
2. In both *C. difficile*–associated disease and bacterial vaginosis, the disease begins when the normal microbiota of the tissue site is altered. Why is this?
3. Name some common odontopathogens that are responsible for dental caries, dental plaque, and periodontal disease.

Key Concepts

39.1 Bacteria Can Be Transmitted by Airborne Routes

- A number of infectious diseases are caused by bacteria as a result of their transmission through air.
- Examples of diseases transmitted by airborne dissemination include chlamydial pneumonia (*Chlamydophila pneumoniae*); diphtheria (*Corynebacterium diphtheriae;* **figure 39.1**); Legionnaires' disease (*Legionella pneumophila;* **figure 39.2**); meningitis (*Neisseria meningitidis* [**figure 39.3**] and *Haemophilus influenzae* type b); mycobacterial infections (*M. avium* and *M. intracellulare* pneumonia and *M. tuberculosis;* **figure 39.4**); mycoplasmal pneumonia (*Mycoplasma pneumoniae*); pertussis (*Bordetella pertussis*); and streptococcal diseases (*Streptococcus* spp.; **figures 39.5** and **39.6**).

39.2 Arthropods Can Transmit Bacterial Diseases

- Bacteria can be transmitted to humans as a result of their interaction with arthropod vectors, such as ticks, lice, and fleas.
- Examples of arthropod-borne diseases include Lyme disease (*Borrelia burgdorferi;* **figure 39.7**); plague (*Yersinia pestis;* **figures 39.8** and **39.9**); and Rocky Mountain spotted fever (*Rickettsia rickettsii;* **figure 39.10**).

39.3 Direct Contact Diseases Can Be Caused by Bacteria

- Direct contact of an uninfected human with sources of bacteria (including infected humans) can result in transmission of bacteria and disease. Direct contact of skin, mucous membranes, open wounds, or body cavities can lead to bacterial colonization and disease.
- Examples of direct contact diseases caused by bacteria include gas gangrene or clostridial myonecrosis (*Clostridium perfringens;* **figure 39.11**); group B streptococcal disease (*Streptococcus agalactiae*); mycobacterial infections including leprosy (*Mycobacterium leprae;* **figure 39.12**) and tattoo-associated infections; peptic ulcer disease (*Helicobacter pylori;* **figure 39.13**); sexually transmitted diseases such as chlamydial diseases (*Chlamydia trachomatis;* **figure 39.14**), gonorrhea (*Neisseria gonorrhea;* **figure 39.15**), and syphilis (*Treponema pallidum;* **figures 39.16** and **39.17**); staphylococcal diseases (*Staphylococcus aureus;* **figures 39.18–39.20**), including methicillin-resistant *Staphylococcus aureus* (**figure 39.21**); streptococcal diseases (*Streptococcus pyogenes;* **figure 39.22**); tetanus (*Clostridium tetani*); and trachoma (*Chlamydia trachomatis;* **figure 39.24**).

39.4 Food and Water Are Vehicles for Bacterial Diseases

- Food and water can serve as vehicles that transport bacteria to humans. Ingestion of contaminated food and water often results in infections of the gastrointestinal tract (**tables 39.3** and **39.4**).

- Some common food-borne and waterborne bacterial infectious diseases of the intestinal tract are botulism (*Clostridium botulinum;* **figure 39.25**), *Campylobacter* gastroenteritis (*Campylobacter jejuni*), cholera (*Vibrio cholerae;* **figure 39.26**), *Escherichia coli* gastroenteritis (**figure 39.27**), salmonellosis (various *Salmonella* serovars), typhoid fever (*Salmonella* serovar Typhi), shigellosis (*Shigella* spp.), and staphylococcal food poisoning (*Staphylococcus aureus* enterotoxins).
- Sepsis and septic shock are most commonly caused by Gram-negative bacteria, mediated by the lipid A component of the outer membrane.
- Some diseases are caused by bacterial action on cells of the intestine. Other diseases result from bacterial toxins.

39.5 Zoonotic Diseases Arise from Human-Animal Interactions

- Diseases of animals that are transmitted to humans are called zoonoses (s., zoonosis).
- Important bacterial zoonoses include anthrax (*Bacillus anthracis;* **figure 39.29**); brucellosis (*Brucella* spp.); psittacosis (*Chlamydophila psittaci*); Q fever (*Coxiella burnetii;* **figure 39.30**); and tularemia (*Francisella tularensis*).

39.6 Opportunistic Diseases Can Be Caused by Bacteria

- Normal microbiota can cause disease especially when they are in tissues that are not the regular environment or they overgrow the regular environment; these are opportunists and cause opportunistic diseases.
- Opportunistic diseases include antibiotic-associated colitis (*Clostridium difficile;* **figure 39.31**); bacterial vaginosis (*Gardnerella vaginalis*); dental infections; and streptococcal pneumonia (*Streptococcus pneumoniae*).
- Dental plaque formation begins on a tooth with the initial colonization of the acquired enamel pellicle by *Streptococcus gordonii, S. oralis,* and *S. mitis*. Other bacteria then become attached and form a plaque ecosystem (**figures 39.32** and **39.33**). The bacteria produce acids that cause a chemical lesion on the tooth and initiate dental decay or caries.
- Periodontal disease is a group of diverse clinical entities that affect the periodontium. Disease is initiated by the formation of subgingival plaque, which leads to tissue inflammation known as periodontitis and to periodontal pockets. Bacteria that colonize these pockets can cause an abscess, periodontosis, gingivitis, and general tissue necrosis (**figure 39.34**).

Compare, Hypothesize, Invent

1. Describe a typhoid carrier. How does one become a carrier?

2. Many consider cholera as the most severe form of gastroenteritis. Why do you think this is so?

3. Compare the three stages of syphilis and Lyme disease. Why do you think both diseases are so hard to treat once they progress beyond the primary stage?

4. While many *Vibrio cholerae* strains are found in aquatic environments, only a small fraction of these cause human disease. When in aquatic ecosystems, *V. cholerae* is frequently found attached to the exoskeleton of zooplankton, which is made of chitin. Indeed, *V. cholerae* produces an extracellular chitinase, so zooplankton-associated growth presumably provides a good source of organic carbon and nitrogen. Growth on zooplankton and in the human gut shares the requirement for attachment proteins. Remarkably, the same protein, called GbpA, has been shown to bind to both the *N*-acetylglucosamine (GlcNAc) of zooplankton chitin and to epithelial cells. The glycoproteins and lipids on the surface of epithelial cells are commonly modified with GlcNAc.

   How would you show whether or not GbpA binds specifically to epithelial GlnNAc? How would you determine if GbpA is needed for virulence? How would you test the hypothesis that GbpA is produced by pathogenic strains but not, or to a lesser extent, by nonpathogenic strains of *V. cholerae*?

   Read the original paper: Kirn, T. J., et al. 2005. A colonization factor links *Vibrio cholerae* environmental survival and human infection. *Nature* 438:863.

5. The spirochete *Borrelia burgdorferi* causes Lyme disease, which is transmitted to humans when its vector, the deer tick, takes a blood meal. When *B. burgdorferi* is in the gut of the tick, its outer surface protein A (OspA) is expressed at high levels. Once it has been transmitted to a human, OspA production diminishes. However, to complete its infection cycle and return to the tick, *B. burgdorferi* cells must increase OspA synthesis. Microbiologists wanted to determine how the spirochete senses the presence of a feeding tick and thus resumes high levels of OspA production. It was discovered that *B. burgdorferi* specifically binds the host neuroendocrine stress hormones epinephrine and norepinephrine. This enables the microbe to detect the presence of a tick, which results in upregulation of OspA.

   Discuss the role of coevolution in the development of *B. burgdorferi*'s ability to sense and respond to these host compounds. How would you show that among all the mediators of inflammation at the site of a tick bite, the spirochete responds specifically to epinephrine and norepinephrine? Your answer should include some carefully considered control experiments.

   Read the original paper: Schekelhoff, M. R., et al. 2007. *Borrelia burgdorferi* intercepts host hormonal signals to regulate expression of outer surface protein A. *Proc. Natl. Acad. Sci. USA* 104:7247–52.

Learn More

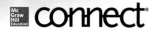

40

Human Diseases Caused by Fungi and Protists

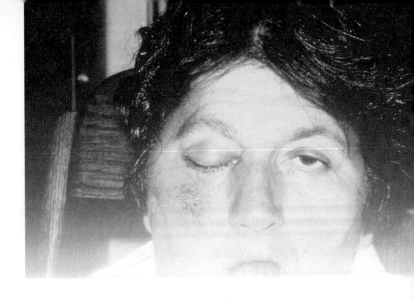

Similar to the patient in the opening story, this woman has mucormycosis. The fungus entered her right eye, causing life-threatening disease.

Death by...Mushroom?

Ask Kentuckian Mark Tatum who would win a battle of man versus mushroom and he'll likely bet on the fungus. In 2000 Tatum lost a portion of his face to a severe mucormycosis infection; such infections are often fatal. In fact, the capacity of certain fungi to degrade animal tissues is so legendary that a research fellow at the Massachusetts Institute of Technology is developing a mushroom strain that will be specifically adept at digesting human tissues. In what is known as the "Infinity Burial Project," Jae Rhim Lee is exploring alternatives to the toxic formaldehyde-based burial process. Growing fungi in Petri dishes filled with human hair, skin, nails, and so forth will facilitate the selective breeding of mushrooms that thrive on this macabre media. She has also designed a "mushroom death suit" into which a suspension of mushroom spores will provide mushroom access to its nutrient source (the dead body), and thus accelerate decomposition. Lee states that she is building an acceptance of natural processes, pointing out that cemetery soil already contains many fungal species. Interestingly, among another of her goals is to determine if mushroom-based decomposition will also eliminate some of the more than 200 toxic chemicals that accumulate in the human body.

Ethicists will debate the religious and spiritual issues, while microbial ecologists consider the impact such a strain might have in the environment. For example, do such strains already exist in cemeteries? How will a "man-eating mushroom" change the microbiota of soil? Will mourners be able to visit the cemeteries without fear of hitchhiking spores coming home to dinner? Regardless of the questions yet unanswered, one thing is certain: Fungi are masters at producing extracellular enzymes to digest complex foods into soluble nutrients. Just ask Mark Tatum.

In this chapter, we describe some of the fungi and protists that are pathogenic to humans and discuss the clinical manifestations, diagnosis, epidemiology, pathogenesis, and treatment of the diseases caused by them. The biology and diversity of these organisms are covered in chapters 5, 25, and 26. This chapter follows the format of the previous two chapters in categorizing their diseases by route of transmission (**figure 40.1**).

Readiness Check:

Based on what you have learned previously, you should be able to:

✔ Review basic eukaryote cell biology (sections 5.1–5.7)
✔ Distinguish between yeast and mold (section 26.1)
✔ Identify the major factors supporting growth and reproduction of protists and fungi (sections 25.1 and 26.1)
✔ Explain pathogenicity and the infection process (chapter 35)
✔ Differentiate different types of vaccines (section 37. 6)

40.1 Relatively Few Fungi and Protists Are Human Pathogens

After reading this section, you should be able to:

■ Describe the system used to categorize fungal diseases
■ Explain why protist infection is becoming a greater challenge

While there are approximately 1.5 million species of fungi on the Earth, only about 300 produce disease in humans. Medical mycology is the discipline that deals with the fungi that cause human disease. These fungal diseases, known as **mycoses** (s., mycosis; Greek *mykes*, fungus), are typically divided into five groups according to the route of infection: superficial, cutaneous, subcutaneous, systemic, and opportunistic mycoses (**table 40.1**). Superficial, cutaneous, and subcutaneous mycoses are direct contact infections of the skin, hair, and nails. Systemic mycoses are fungal infections that have disseminated to visceral tissues. The fungi that cause systemic or deep mycoses are dimorphic; that is, they exhibit a parasitic yeastlike (single-cell) phase (Y) and a saprophytic mold or mycelial (filamentous) phase (M; *see p. 922*). The switch from growth as a mycelial form at external temperatures to growth as a yeast form at human body temperatures is called the **Y-M shift** and is considered a virulence factor. Most systemic mycoses are acquired by inhalation of spores from soil in which the mold phase of the fungus resides. If a susceptible person inhales enough spores, infection begins as a lung

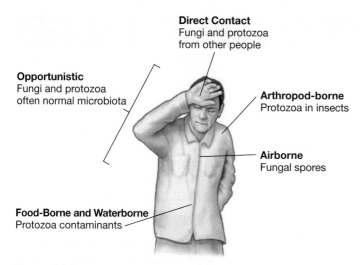

Direct Contact
Fungi and protozoa
from other people

Opportunistic
Fungi and protozoa
often normal microbiota

Arthropod-borne
Protozoa in insects

Airborne
Fungal spores

Food-Borne and Waterborne
Protozoa contaminants

Figure 40.1 **Common Routes of Fungal and Protozoan Disease Transmission.** Examples of disease-causing fungi and protists are presented in this chapter according to route of transmission.

lesion, becomes chronic, and may spread through the bloodstream to other organs (the target organ varies with the species). Some fungi produce toxins (mycotoxins) that are harmful to humans. We discuss them with other toxins in chapter 35 and in chapter 41 (*see figures 41.5 and 41.6*). ◄◄ Fungi *(chapter 26); Mycotoxins (section 35.2)*

Protozoa have become adapted to practically every type of habitat on Earth, including the human body. Many protists are transmitted to humans by arthropod vectors or in contaminated food or water. However, some protozoan diseases are transmitted by direct contact. Although members of fewer than 20 genera of protists cause disease in humans (**table 40.2**), their impact is formidable. For example, approximately 200 million cases of malaria occur in the world each year. In sub-Saharan Africa alone, malaria is responsible for the deaths of more than half a million children under the age of five annually. It is estimated that there are at least 8,000 cases of trypanosomiasis, 1.3 million cases of leishmaniasis, and over 50 million cases of amebiasis yearly.

Table 40.1	Examples of Some Medically Important Fungi		
Disease Category	**Pathogen**	**Location**	**Disease**
Superficial mycoses	*Piedraia hortae*	Scalp	Black piedra
	Trichosporon beigelii	Beard, mustache	White piedra
	Malassezia furfur	Trunk, neck, face, arms	Tinea versicolor
Cutaneous mycoses	*Trichophyton mentagrophytes, T. verrucosum, T. rubrum*	Beard hair	Tinea barbae
	Trichophyton, Microsporum canis	Scalp hair	Tinea capitis
	Trichophyton rubrum	Smooth or bare parts of the skin	Tinea corporis
	Epidermophyton floccosum, T. mentagrophytes, T. rubrum	Groin, buttocks	Tinea cruris (jock itch)
	T. rubrum, T. mentagrophytes, E. floccosum	Feet	Tinea pedis (athlete's foot)
	T. rubrum, T. mentagrophytes, E. floccosum	Nails	Tinea unguium (onychomycosis)
Subcutaneous mycoses	*Phialophora verrucosa, Fonsecaea pedrosoi*	Legs, feet	Chromoblastomycosis
	Madurella mycetomatis	Feet, other areas of body	Maduromycosis
	Sporothrix schenckii	Puncture wounds	Sporotrichosis
Systemic mycoses	*Blastomyces dermatitidis*	Lungs, skin	Blastomycosis
	Coccidioides immitis	Lungs, other parts of body	Coccidioidomycosis
	Cryptococcus neoformans	Lungs, skin, bones, viscera, central nervous system	Cryptococcosis
	Histoplasma capsulatum	Within phagocytes	Histoplasmosis
Opportunistic mycoses	*Aspergillus fumigatus, A. flavus*	Respiratory system	Aspergillosis
	Candida albicans	Skin or mucous membranes	*Candidiasis*
	Pneumocystis jiroveci	Lungs, sometimes brain	*Pneumocystis* pneumonia
	Encephalitozoon, Pleistophora, Enterocytozoon, Microsporidium	Lungs, sometimes brain	Microsporidiosis

Table 40.2	Examples of Medically Important Protozoan Infections	
Initial Infection Site	**Pathogen**	**Disease**
Blood	*Plasmodium* spp.	Malaria
	Trypanosoma brucei	African sleeping sickness
	Trypanosoma cruzi	American trypanosomiasis
Intestinal Tract	*Cryptosporidium* spp.	Cryptosporidiosis
	Cyclospora spp.	Cyclosporidiosis
	Entamoeba histolytica	Amebiasis
	Giardia intestinalis	Giardiasis
	Toxoplasma gondii	Toxoplasmosis
Lungs	*Balamuthia mandrillaris*	*Balamuthia* amebic encephalitis
	Pneumocystis jiroveci	*Pneumocystis* pneumonia
	Trichomonas tenax	Pulmonary trichomoniasis
Muscosal Membranes Ears, Eyes, Nose Genitals	*Acanthamoeba* spp.	Amebic meningoencephalitis
	Naegleria fowleri	Amebic meningoencephalitis
	Trichomonas vaginalis	Trichomoniasis
Skin	*Leishmania braziliensis*	Mucocutaneous leishmaniasis
	Leishmania donovani	Visceral leishmaniasis
	Leishmania tropica	Cutaneous leishmaniasis

Table 40.3	Water-Based Protozoan Reservoirs	
Organism	**Reservoir**	**Comments**
Acanthamoeba spp.	Sewage sludge disposal areas	Can cause granulomatous amebic encephalitis, keratitis, corneal ulcers
Cryptosporidium spp.	Many species of domestic and wild animals	Causes acute enterocolitis; important with immunologically compromised individuals; cysts resistant to chemical disinfection; not antibiotic sensitive
Cyclospora cayetanensis	Waters—does not withstand drying; possibly other reservoirs	Causes long-lasting (43 days average) diarrheal illness; infection self-limiting in immunocompetent hosts; sensitive to prompt treatment with sulfonamide and trimethoprim
Giardia intestinalis	Beavers, sheep, dogs, cats	Major cause of early spring diarrhea; important in cold mountain water
Naegleria fowleri	Warm water (hot tubs), swimming pools, lakes	Inhalation in nasal passages; central nervous system infection; causes primary amebic meningoencephalitis

There is also an increasing problem with *Cryptosporidium* and *Cyclospora* contamination of food and water supplies (**table 40.3**). More of our population is elderly, and a growing number of persons are immunosuppressed due to HIV infection, organ transplantation, or cancer chemotherapy. These populations are at increased risk for protozoan infections. ◄◄ *Protists (chapter 25)*

While seemingly different, fungi and protists share a number of phenotypic features that serve them in their ability to cause infection: microscopic size, eukaryotic physiology, cell walls or wall-like structures, alternative stages for survival outside of the host, degradative enzymes, and others. Some fungi and protists also share transmission routes. We now discuss diseases of fungi and protists based on how they are acquired by the human host.

40.2 Fungi Can Be Transmitted by Airborne Routes

After reading this section, you should be able to:

- Report the common fungal diseases spread by airborne transmission
- Identify typical signs and symptoms of common airborne fungal diseases

A number of eukaryotic pathogens are transmitted to people through air. These infectious agents are small particles that are carried by air currents and are typically inhaled by unsuspecting hosts. While it is unlikely to find human protozoan disease spread this way, protozoan-containing droplet particles can be

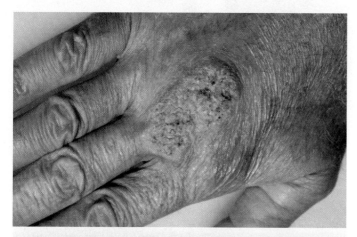

Figure 40.2 Systemic Mycosis: Blastomycosis. Blastomycosis of the hand caused by *Blastomyces dermatitidis*.

MICRO INQUIRY *What is the morphology of* B. dermatitidis *in the environment? In the human host?*

transmitted through the air. Here we present several examples of fungal infectious agents that are acquired through airborne transmission and the diseases they cause.

Blastomycosis

Blastomycosis is the systemic mycosis caused by *Blastomyces dermatitidis*, a fungus that grows as a budding yeast in humans but as a mold on culture media and in the environment. It is found predominately in moist soil enriched with decomposing organic debris, as in the Mississippi and Ohio river basins. *B. dermatitidis* is endemic in parts of the south-central, southeastern, and midwestern United States. Additionally, microfoci have been reported in Central and South America and in parts of Africa. The disease occurs in three clinical forms: cutaneous, pulmonary, and disseminated. The initial infection begins when blastospores are inhaled into the lungs. The fungus can then spread rapidly, especially to the skin, where cutaneous ulcers and abscess formation occur (**figure 40.2**). *B. dermatitidis* can be isolated from pus and biopsy sections. Diagnosis requires the demonstration of thick-walled, yeastlike cells 8 to 15 μm in diameter. Complement fixation, immunodiffusion, and skin tests (blastomycin) are also useful. Amphotericin B, itraconazole, and ketoconazole are the drugs of choice for treatment (*see figure 9.14*). Surgery may be necessary for the drainage of large abscesses. Mortality is about 5%. There are no preventive or control measures. ◄◄ *Antifungal drugs (section 9.5)*

Coccidioidomycosis

Coccidioidomycosis, also known as valley fever, San Joaquin fever, or desert rheumatism because of the geographical distribution of the fungus, is caused by *Coccidioides immitis*. *C. immitis* thrives in the semiarid, highly alkaline soils of southwestern United States and parts of Mexico and South America. In the United States about 30,000 people are infected annually, resulting in approximately 200 deaths. Endemic areas have been defined by massive skin testing with the antigen coccidioidin, where upward of 15% positive cases are reported; endemic areas include Arizona, California, Nevada, New Mexico, and Utah. In these areas coccidiodomycosis causes up to 30% of community-acquired pneumonia. In the soil and on

culture media, this fungus grows as a mold that forms arthroconidia at the tips of hyphae (*see figure 26.4*). Because arthroconidia are so abundant, immunocompromised individuals can acquire the disease by inhalation as they simply move through these endemic areas. Wind turbulence and even construction of outdoor structures have been associated with increased exposure and infection. In humans, the fungus grows as a thick-walled spherule filled with spores (**figure 40.3**). Most cases of coccidioidomycosis are asymptomatic or indistinguishable from ordinary upper respiratory infections. Most cases resolve in a few weeks, and lasting immunity results. A few infections result in progressive chronic pulmonary disease or disseminated infections of other tissues; the fungus can spread throughout the body, involving almost any organ or site.

Diagnosis is accomplished by identification of the large spherules (approximately 80 μm in diameter) in pus, sputum, and aspirates. Culturing clinical samples in the presence of penicillin and streptomycin on the fungal selective medium Sabouraud agar also is diagnostic. Rapid confirmation methods include testing supernatants of liquid cultures for antigens, serology, and skin testing. Miconazole, itraconazole, ketoconazole, and amphotericin B are the drugs of choice for treatment. Prevention involves reducing exposure to dust (soil) in endemic areas. ◄◄ *Antifungal drugs (section 9.5); Identification of microorganisms from specimens (section 36.3)*

Cryptococcosis

Cryptococcosis is a systemic mycosis caused by *Cryptococcus gattii* and *C. neoformans*. These fungi grow as large, budding yeasts in the human host but as smaller yeast cells that undergo monokaryotic or dikaryotic reproduction in the environment, where they are saprophytes with a worldwide distribution. *C. gattii* is responsible for nearly all cryptococcal infections in humans, except where *C. neoformans* infects those with HIV. *C. gattii* disease (primarily pneumonia and meningitis) continues to be an emerging infectious disease in the United States. Worldwide, *C. gatti* kills 13 to 33% of infected individuals. Typically, cryptococci cause disease in the immunocompromised and often cause disease in people with AIDS, cancer, or a recent organ transplant. In fact, cryptococcosis is found in approximately 15% of AIDS patients.

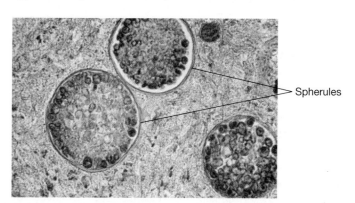

Spherules

Figure 40.3 Systemic Mycosis: Coccidioidomycosis. *Coccidioides immitis* mature spherules filled with endospores within a tissue section; light micrograph (×400).

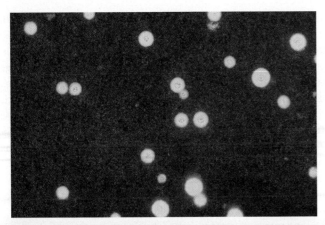

Figure 40.4 Systemic Mycosis: Cryptococcosis. India ink preparation showing *Cryptococcus neoformans*. Note the distinctly outlined capsules surrounding all cells and refractile inclusions in the cytoplasm; light micrograph (×150).

MICRO INQUIRY *Once this microbe is acquired by inhalation, to what other body systems can it spread?*

Aged, dried pigeon droppings are an apparent source of infection. The fungi enter the body by the respiratory tract, causing a minor pulmonary infection in most that is usually transitory. In others who inhale the fungi, a pneumonia-like disease with coughing and fever occurs. Some pulmonary infections spread to the skin, bones, viscera, and central nervous system, causing headaches and mental status changes. Once the nervous system is involved, cryptococcal meningitis usually results. Diagnosis is accomplished by detection of the thick-walled, spherical yeast cells in pus, sputum, or cerebral spinal fluid smears using India ink to define the organism (**figure 40.4**). The fungi can be easily cultured on Sabouraud dextrose agar. Identification of the fungus in body fluids is made by immunologic procedures. Treatment for *C. neoformans* includes amphotericin B often with flucytosine. HIV/AIDS patients with a history of infection are treated prophylactically with fluconazole. Prevention of infection with *C. neoformans* is through avoidance of birds and areas contaminated with bird droppings.

Histoplasmosis

Histoplasmosis is primarily a disease of the lungs caused by *Histoplasma capsulatum* var. *capsulatum*, a facultative, parasitic fungus that grows intracellularly. The incubation period for symptomatic patients is approximately 10 days; symptoms include flulike responses such as fever, cough, headaches, and muscle aches. However, infection is often asymptomatic. *H. capsulatum* is a small, budding yeast in humans and on culture media at 37°C (98.6°F). At 25°C (77°F) it grows as a mold, producing microconidia (1 to 5 μm in diameter) that are borne singly at the tips of short conidiophores (*see figure 26.4*). Large macroconidia (8 to 16 μm in diameter) are also formed on conidiophores (**figure 40.5a**). In humans, the yeastlike form reproduces by budding (figure 40.5b). The mycelial form of the fungus is found in soil and is localized in areas that have been contaminated with bird or bat excrement. The microconidia can become airborne

(a)

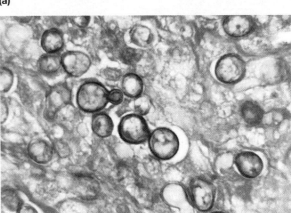

(b)

Figure 40.5 Morphology of *Histoplasma capsulatum* var. *capsulatum*. (a) Mycelia, microconidia, and chlamydospores as found in the soil. These are the infectious particles; light micrograph (×400). (b) Yeastlike form of *H. capsulatum*.

when contaminated soil is disturbed. Microconidia are thus most prevalent where bird droppings—especially from starlings, crows, blackbirds, cowbirds, sea gulls, turkeys, and chickens—have accumulated. It is noteworthy that the birds themselves are not infected because of their high body temperature; their droppings simply provide the nutrients for this fungus. Only bats and humans demonstrate the disease and harbor the fungus.

Infection ensues when microconidia are inhaled. Histoplasmosis is found worldwide, with Africa, Australia, India, and Malaysia being endemic regions. Within the United States, histoplasmosis is endemic within the Mississippi, Kentucky, Tennessee, Ohio, and Rio Grande river basins. As many as 90% of people who reside these areas have antibodies against the fungus. It has been estimated that in endemic areas of the United States, about 500,000 individuals are infected annually: 50,000 to 200,000 become ill; 3,000 require hospitalization; and about 50 die. The total number of infected individuals may be over 40 million in the United States alone. Histoplasmosis is a common disease among poultry farmers, spelunkers (people who explore caves), and bat guano miners (bat guano is used as fertilizer).

Because *H. capsulatum* grows within macrophages, the fungus can be disseminated to many organs (though rarely) as the infected macrophages circulate throughout the body. Lesions may appear in the lungs and show calcification; thus, the disease may resemble tuberculosis. Most infections resolve spontaneously. Laboratory diagnosis is accomplished by complement fixation tests and isolation of the fungus from tissue specimens.

DISEASE

40.1 A Brief History of Malaria

No other single infectious disease has had the impact on humans that malaria has had. The first references to its periodic fever and chills can be found in early Chaldean, Chinese, and Hindu writings. In the late fifth century BCE, Hippocrates described certain aspects of malaria. In the fourth century BCE, the Greeks noted an association between individuals exposed to swamp environments and the subsequent development of periodic fever and enlargement of the spleen (splenomegaly). In the seventeenth century, the Italians named the disease *mal' aria* (bad air) because of its association with ill-smelling vapors from swamps near Rome. At about the same time, the bark of the quinaquina (cinchona) tree of South America was used to treat the intermittent fevers, although it was not until the mid-nineteenth century that quinine was identified as the active alkaloid. The major epidemiological breakthrough came in 1880, when French army surgeon Charles Louis Alphonse Laveran observed gametocytes in fresh blood. Five years later, the Italian histologist Camillo Golgi observed the multiplication of the asexual blood forms. In the late 1890s, Patrick Manson postulated that malaria was transmitted by mosquitoes. Sir Ronald Ross, a British army surgeon in the Indian Medical Service, subsequently observed developing plasmodia in the intestine of mosquitoes, supporting Manson's theory. Using birds as experimental models, Ross definitively established the major features of the life cycle of plasmodia and received the Nobel Prize in 1902.

Human malaria is known to have contributed to the fall of the ancient Greek and Roman empires. Troops in both the U.S. Civil War and the Spanish-American War also were severely incapacitated by the disease. More than 25% of all hospital admissions during these wars were malaria patients. During World War II, malaria epidemics severely threatened both the Japanese and Allied forces in the Pacific. The same can be said for the military conflicts in Korea and Vietnam.

In the twentieth century, efforts were directed toward understanding the biochemistry and physiology of malaria, controlling the mosquito vector, and developing antimalarial drugs. In the 1960s it was demonstrated that resistance to *P. falciparum* among West Africans was associated with the presence of hemoglobin-S (Hb-S) in their erythrocytes. Hb-S differs from normal hemoglobin-A by a single amino acid, valine, in each half of the Hb molecule. Consequently these erythrocytes—responsible for sickle cell disease—have a low binding capacity for oxygen. Because the malarial parasite has a very active aerobic metabolism, it cannot grow and reproduce within these erythrocytes.

In 1955 the World Health Organization began a worldwide malarial eradication program that collapsed by 1976. Among the major reasons for failure were the development of resistance to the insecticide DDT by mosquitoes and the development of resistance to chloroquine by strains of *Plasmodium*. Scientists are exploring new approaches, such as the development of vaccines and more potent drugs. In 2002 the complete DNA sequences of *P. falciparum* and *Anopheles gambiae* (the mosquito that most efficiently transmits this parasite to humans in Africa) were determined. Together with the human genome sequence, researchers now have in hand the genetic blueprints for the parasite, its vector, and its victim. This has made possible a holistic approach to understanding how the parasite interacts with the human host, leading to new antimalarial strategies, including vaccine design. Overall, no greater achievement for microbiology could be imagined than the control of malaria—a disease that has caused untold misery throughout the world since antiquity and remains one of the world's most serious infectious diseases.

Most individuals with this disease exhibit a hypersensitive state that can be demonstrated by the histoplasmin skin test. Currently the most effective treatment is with amphotericin B, ketoconazole, or itraconazole. Prevention and control involve wearing protective clothing and masks before entering or working in infested habitats. Soil decontamination with 3 to 5% formalin is effective where economically and physically feasible.

40.3 Arthropods Can Transmit Protozoal Disease

After reading this section, you should be able to:

- Report the common protist diseases spread by arthropod transmission
- Identify typical signs and symptoms of arthropod-borne protist diseases
- Correlate arthropod-borne protist infection and disease severity with protist virulence factors

Biting insects, fleas, ticks, and mites can harbor microorganisms that can be subsequently passed to humans. In some cases the arthropod vector is a transient or accidental carrier of the organism, as is the case with fungal diseases that result from passage of spores by flying and crawling insects. In other cases, the vector is an intermediate host in which the microorganism undergoes developmental changes. We discuss viral and bacterial diseases transmitted by arthropods in chapters 38 and 39, respectively. This section discusses several examples of protozoan diseases transmitted by insects.

Malaria

The most important human protozoal pathogens belong to the genus *Plasmodium* and are the causative agents of **malaria** (**Disease 40.1**). A 2014 World Health Organization (WHO) report estimates that there are 3.2 billion people (almost half of the world's population) at risk for malaria. In 2013 there were 198 million people infected worldwide (**figure 40.6**). Although

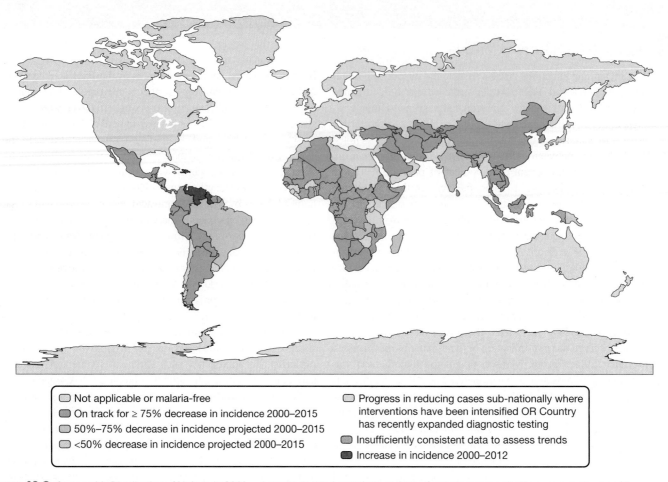

Figure 40.6 **Geographic Distribution of Malaria in 2014.** A country is listed as malaria-endemic if malaria was reported in even a small area of it.

- ○ Not applicable or malaria-free
- ◐ On track for ≥ 75% decrease in incidence 2000–2015
- ◑ 50%–75% decrease in incidence projected 2000–2015
- ○ <50% decrease in incidence projected 2000–2015
- ○ Progress in reducing cases sub-nationally where interventions have been intensified OR Country has recently expanded diagnostic testing
- ◐ Insufficiently consistent data to assess trends
- ● Increase in incidence 2000–2012

mortality rates have fallen by more than 47% since 2000, 584,000 people died of malaria in 2013; 90% of those deaths were children under the age of five. Indeed, 22% of all childhood deaths in Africa are due to malaria, where about one child dies of malaria every minute. About 1,600 cases are reported each year in the United States, divided between returning U.S. travelers and non-U.S. citizens.

Human malaria is caused by five species of *Plasmodium*: *P. falciparum, P. malariae, P. vivax, P. knowlesi,* and *P. ovale.* The life cycle of plasmodial protists is shown in **figure 40.7**. The parasite first enters the bloodstream through the bite of an infected female *Anopheles* mosquito. As she feeds, the mosquito injects a small amount of saliva containing an anticoagulant along with small haploid sporozoites. The sporozoites in the bloodstream immediately enter hepatic cells of the liver. There they undergo multiple asexual fission (schizogony) and produce merozoites. After being released from hepatocytes, merozoites attach to and penetrate erythrocytes. The incubation period is as short as 7 days (frequently due to *P. falciparum*) and as long as 30 days (usually due to *P. malariae*). ◀◀ Alveolata *(section 25.4)*

Once inside the erythrocyte, the plasmodial cell begins to enlarge as a mononucleate cell termed a trophozoite. The trophozoite's nucleus divides asexually to produce a schizont that has 6 to 24 nuclei. The schizont divides and produces mononucleated merozoites that are released when infected erythrocytes lyse. These then infect other erythrocytes. This erythrocytic stage is cyclic and repeats approximately every 48 to 72 hours or longer, depending on the species of *Plasmodium* involved. The sudden release of merozoites, toxins, and erythrocyte debris into the bloodstream triggers an attack of the chills, fever, and sweats characteristic of malaria. Occasionally, merozoites differentiate into macrogametocytes and microgametocytes, which do not rupture the erythrocyte. When these are ingested by a mosquito during a blood meal, they develop into female and male gametes, respectively. In the mosquito's gut, the infected erythrocytes lyse and the gametes fuse to form a diploid zygote called the ookinete. The ookinete migrates to the mosquito's gut wall, penetrates, and forms an oocyst. In a process called sporogony, the oocyst undergoes meiosis and forms sporozoites, which migrate to the salivary glands of the mosquito. The cycle is now complete, and when the mosquito bites another human host, the cycle begins anew.

The pathological changes caused by malaria involve not only erythrocytes but also the spleen and other organs. Classic symptoms first develop with the synchronized release of merozoites and erythrocyte debris into the bloodstream, resulting in

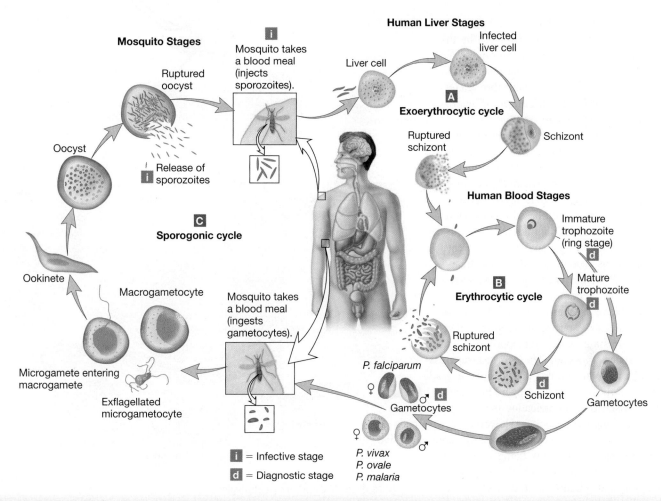

Mosquito Stages

Ruptured oocyst

Oocyst

Release of sporozoites

C Sporogonic cycle

Ookinete

Macrogametocyte

Microgamete entering macrogamete

Exflagellated microgametocyte

i Mosquito takes a blood meal (injects sporozoites).

Mosquito takes a blood meal (ingests gametocytes).

i = Infective stage

d = Diagnostic stage

Human Liver Stages

Liver cell

Infected liver cell

A Exoerythrocytic cycle

Schizont

Ruptured schizont

Human Blood Stages

Immature trophozoite (ring stage)

B Erythrocytic cycle

Mature trophozoite

Ruptured schizont

Schizont

Gametocytes

P. falciparum
♀ ♂ **d**
Gametocytes

♀ ♂
P. vivax
P. ovale
P. malaria

Figure 40.7 Malaria. Life cycle of *Plasmodium vivax*. Note the (A) exoerythrocytic cycle, (B) the erythrocytic cycle, and (C) the sporogonic cycle.

MICRO INQUIRY *How is a schizont formed? What is the cell type that is released from erythrocytes, and where does the protist go next?*

the malarial paroxysms—shaking chills, then burning fever followed by sweating. It may be that the fever and chills are caused partly by a malarial toxin that induces macrophages to release the cytokines TNF-α and interleukin-1; interleukin-1 is a well defined inducer of fever. Several of these paroxysms constitute an attack. If *P. vivax* or *P. ovale* are the causative agents, there is a remission that lasts from a few weeks to several months while dormant protozoa called hypnozooites reside in the liver. This is followed by a relapse. Between paroxysms, the patient feels normal. Anemia can result from the loss of erythrocytes, and the spleen and liver often become enlarged. Children and nonimmune individuals often die of cerebral malaria. ◄◄ *Cytokines are chemical messages between cells (section 33.3)*

Diagnosis of malaria is made by demonstrating the presence of parasites within Wright- or Giemsa-stained erythrocytes (**figure 40.8**). When blood smears are negative, serological testing can establish a diagnosis of malaria. Outside the United States, rapid diagnostic tests using species-specific antibodies

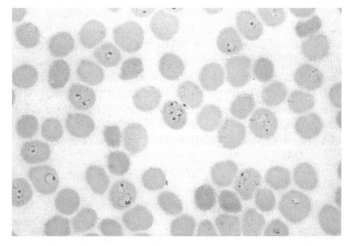

Figure 40.8 Malaria: Erythrocytic Cycle. Trophozoites of *P. falciparum* in circulating erythrocytes; light micrograph (×1,100). The young trophozoites resemble small rings within the erythrocyte cytoplasm.

are commonly used to diagnose malaria; only one of these tests is approved for use by the U.S. Food and Drug Administration (FDA). Also, PCR-based malarial identification and drug resistance testing is available through reference labs. ◄◄ *Identification of micro-organisms from specimens (section 36.3)*

Specific recommendations for treatment are malaria-region-dependent and include administration of chloroquine, amodiaquine, or mefloquine. In the regions where *P. falciparum* predominates, artemisinin is used for treatment. In fact, artemisinin-based combination therapy (ACT)—combining artemisinin with an antimalarial drug having a different mechanism—is recommended for the treatment of all *P. falciparum* malaria. Implementation of the recommendation has been cost-prohibitive in many regions, but most countries are now starting to adopt this regimen. Artemisinin is not yet approved in the United States, however. These drugs suppress protozoan reproduction and are effective in eradicating erythrocytic asexual stages. Primaquine has proved satisfactory in eradicating the exoerythrocytic stages including hypnozooites. However, because resistance to these drugs is occurring rapidly, more expensive drug combinations are now being used. One example is Fansidar, a combination of pyrimethamine and sulfadoxine. It is worth noting that individuals who are traveling to areas where malaria is endemic (figure 40.6) should receive chemoprophylactic treatment with chloroquine, atovaquone/proguanil, or primaquine.

Control of malaria has been a nagging problem for some time. The ideal approach would be to vaccinate all young children in endemic areas. However, to be effective, a vaccine must interfere with a critical malarial component or process (**figure 40.9**). A number of credible vaccines against malaria have been in clinical trials since 2005. A 2014 report of clinical trials indicated that the pre-erythrocytic, recombinant vaccine Mosquiix, reduced the rates of malaria infection in children by 50% and by 25% in infants. However, until a vaccine that can provide 90 to 95% effectiveness is approved for use, malaria prevention is still best attempted with the use of bed netting and insecticides. ◄◄ *Vaccines immunize susceptible populations (section 37.7)*

Leishmaniasis

Leishmanias are flagellated protists that cause a group of several human diseases collectively called **leishmaniasis.** The most common forms of leishmaniasis are cutaneous (skin associated ulcers that heal), diffuse cutaneous (disseminated and chronic skin lesions), mucocutaneous (ulcers of the skin, mouth, nose, and throat), and visceral (affecting liver, spleen, and blood).

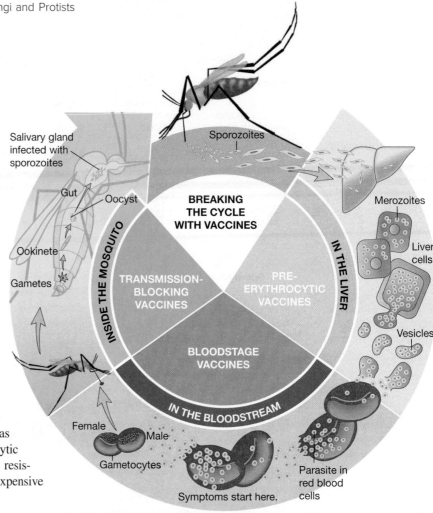

Figure 40.9 Malaria Vaccine Strategies.

MICRO INQUIRY *What is currently the most effective strategy to prevent malaria?*

Co-infection with HIV increases the severity of disease. Worldwide, leishmaniasis is a disease of poverty associated with malnutrition, displacement, poor housing, and discrimination. There are 1.3 million new cases of the cutaneous disease and 400,000 cases of the visceral disease (also called kala-azar) diagnosed around the globe each year. One-tenth of the world's population is at risk of infection; more than 90% of cutaneous disease cases occur in parts of Afghanistan, Algeria, Iran, and the Syrian Arab Republic, as well as Brazil and Colombia, and more than 90% of visceral disease cases occur in parts of Bangladesh, Ethiopia, Sudan, South Sudan, India, and Brazil. About 30,000 people die from leishmaniasis each year, and many more are permanently disfigured (**figure 40.10**).

The primary reservoirs of these parasites are canines and rodents. All species of *Leishmania* use female sand flies such as those of the genera *Lutzomyia* and *Phlebotomus* as intermediate hosts. Sand flies are about one-third the size of a mosquito, so they are hard to see and hear. The leishmanias are transmitted from animals to humans or between humans by sand flies. When an infected sand fly takes a human blood meal, it introduces flagellated promastigotes into the skin of

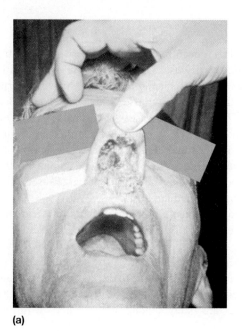

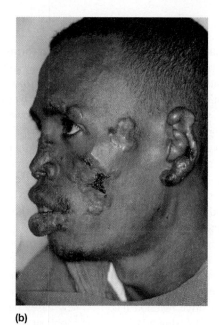

(a)   **(b)**

Figure 40.10 Leishmaniasis.
(a) A person with mucocutaneous leishmaniasis, which has destroyed the nasal septum and deformed the nose and lips. (b) A person with diffuse cutaneous leishmaniasis.

MICRO INQUIRY *What is the primary reservoir of leishmanias? What is the intermediate host?*

the definitive (human) host (**figure 40.11**). Within the skin, the promastigotes are engulfed by macrophages, multiply by binary fission, and form small, nonmotile cells called amastigotes. These destroy the host cell and are engulfed by other macrophages in which they continue to develop and multiply. ◄◄ Euglenozoa *(section 25.2)*

A number of species of *Leishmania* are human pathogens. *L. mexicana* is found in the Yucatan Peninsula (Mexico) and has been reported as far north as Texas. In this disease, a relatively small, red papule forms at the site of each insect bite—the inoculation site. These papules are frequently found on the face and ears. They eventually develop into crustated ulcers (figure 40.10*b*). Healing

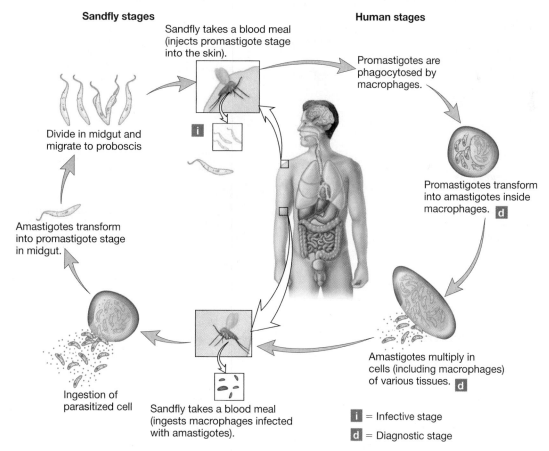

Sandfly stages

Sandfly takes a blood meal (injects promastigote stage into the skin).

Divide in midgut and migrate to proboscis

Amastigotes transform into promastigote stage in midgut.

Ingestion of parasitized cell

Sandfly takes a blood meal (ingests macrophages infected with amastigotes).

Human stages

Promastigotes are phagocytosed by macrophages.

Promastigotes transform into amastigotes inside macrophages. **d**

Amastigotes multiply in cells (including macrophages) of various tissues. **d**

i = Infective stage

d = Diagnostic stage

Figure 40.11 *Leishmania* spp. life cycle.   Note the promastigote-to-amastigote cycle.

occurs with scarring and a permanent immunity. *L. donovani* is endemic in large areas within northern China, eastern India, the Sudan, Mediterranean countries, and Latin America. Infection with *L. donovani* results in visceral disease as infected macrophages lodge in various organs where the ongoing cycle of macrophage capture of promastigotes, amastigote development, macrophage death, and amastigote release damages host organs and tissues.

Laboratory diagnosis of leishmaniasis is based on finding the protist within infected macrophages in stained smears from lesions or infected organs. Blood smears and serological tests are also available for diagnosis. Treatment includes pentavalent antimicrobial compounds (e.g., meglumine antimoniate), paromomycin, and liposomal amphotericin B. Vector and reservoir control, and aggressive epidemiological surveillance are the best options for prevention and containment of this disease. As of 2015, there were no approved vaccines against leishmaniasis, although one experimental vaccine induces a 30% skin-positive reaction that appears to be associated with decreased visceral leishmaniasis.

Trypanosomiasis

Another group of flagellated protists called **trypanosomes** cause the aggregate of diseases termed Human African **Trypanosomiasis** (HAT). HAT is also known as African sleeping sickness. There are two forms of HAT, each referring to the subspecies of trypanosome causing the disease. Gambiense HAT is caused by *Trypanosoma brucei* subspecies *gambiense*, which is distributed over western and central countries and causes over 95% of reported cases. The remaining cases (rhodesiense HAT) are caused by *T. brucei* subspecies *rhodesiense*, found in the upland savannas of eastern and southern Africa. Reservoirs for these trypanosomes are domestic cattle and wild animals, within which they cause severe malnutrition. Both species use tsetse flies (genus *Glossina*) as intermediate hosts.

Trypanosomiasis is such a problem in parts of Africa that millions of hectares are not fit for human habitation. New cases of HAT have been declining since 2009, with approximately 7,000 reported to WHO annually. However, it is likely that the majority of cases are not reported due to the lack of public health infrastructure in endemic regions. As a result, more than 20,000 new cases per year are likely, among an estimated 70 million people at risk.

The parasites are transmitted to humans by the bite of the fly (**figure 40.12**; *also see figure 25.6*). The protists pass through the lymphatic system, enter the bloodstream, and replicate by binary fission as they pass to other body fluids. The tsetse fly bite is painful and can develop into a chancre. Patients may exhibit

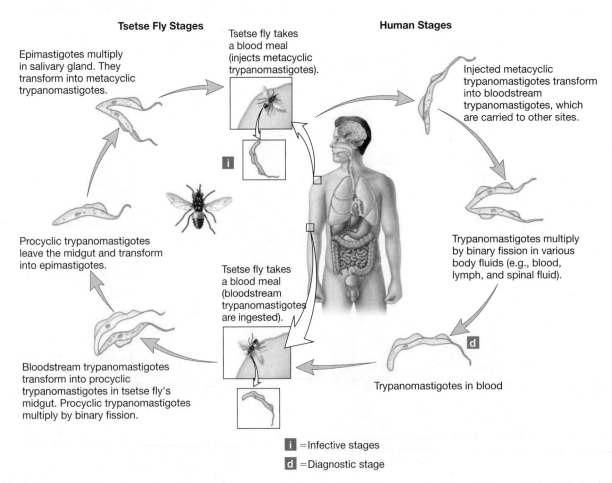

Tsetse Fly Stages

Epimastigotes multiply in salivary gland. They transform into metacyclic trypanomastigotes.

Tsetse fly takes a blood meal (injects metacyclic trypanomastigotes).

Procyclic trypanomastigotes leave the midgut and transform into epimastigotes.

Bloodstream trypanomastigotes transform into procyclic trypanomastigotes in tsetse fly's midgut. Procyclic trypanomastigotes multiply by binary fission.

Tsetse fly takes a blood meal (bloodstream trypanomastigotes are ingested).

Human Stages

Injected metacyclic trypanomastigotes transform into bloodstream trypanomastigotes, which are carried to other sites.

Trypanomastigotes multiply by binary fission in various body fluids (e.g., blood, lymph, and spinal fluid).

Trypanomastigotes in blood

i =Infective stages

d =Diagnostic stage

Figure 40.12 Life Cycle of *Trypanosoma brucei.*

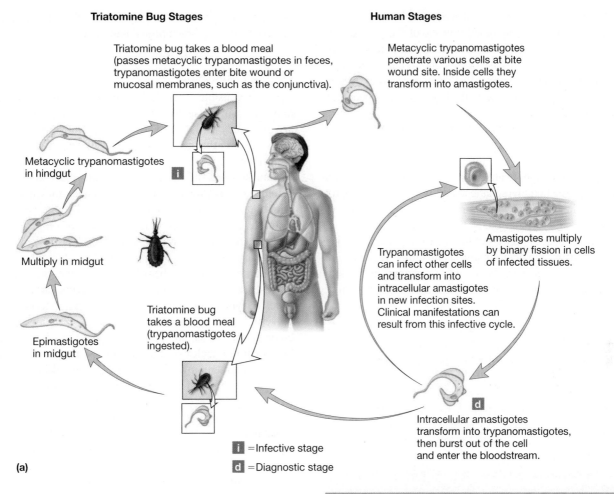

Triatomine Bug Stages

Human Stages

Triatomine bug takes a blood meal (passes metacyclic trypanomastigotes in feces, trypanomastigotes enter bite wound or mucosal membranes, such as the conjunctiva).

Metacyclic trypanomastigotes penetrate various cells at bite wound site. Inside cells they transform into amastigotes.

Metacyclic trypanomastigotes in hindgut

Multiply in midgut

Epimastigotes in midgut

Triatomine bug takes a blood meal (trypanomastigotes ingested).

Amastigotes multiply by binary fission in cells of infected tissues.

Trypanomastigotes can infect other cells and transform into intracellular amastigotes in new infection sites. Clinical manifestations can result from this infective cycle.

Intracellular amastigotes transform into trypanomastigotes, then burst out of the cell and enter the bloodstream.

i = Infective stage

d = Diagnostic stage

(a)

symptoms of fever, severe headaches, extreme fatigue, muscle and joint aches, irritability, and swollen lymph nodes. Some patients develop a skin rash. The disease gets its name from the fact that patients exhibit lethargy—characteristically lying prostrate, drooling from the mouth, and insensitive to pain; they also exhibit progressive confusion, slurred speech, seizures, and personality changes. The protists cause interstitial inflammation and necrosis within the lymph nodes and small blood vessels of the brain and heart. Rhodesiense HAT develops so rapidly that infected individuals often die within a year. *T. brucei* subspecies *gambiense* causes necrotic damage within the central nervous system that leads to a variety of nervous disorders, including the characteristic sleeping sickness. Sleeping sickness is curable with medication; it is otherwise fatal within 2 to 3 years.

T. cruzi causes **American trypanosomiasis (Chagas' disease)**, which occurs in the tropics and subtropics of the continental Americas (southern United States, Central America and northern South America). The parasite uses the triatomine (kissing) bug as a vector (**figure 40.13**). As the triatomine bug takes a blood meal, protists are discharged in the insect's feces, which contain an irritant that stimulates the victim to scratch. This enables trypanosomes to enter the bloodstream and invade the liver, spleen, lymph nodes, and central nervous system; disease

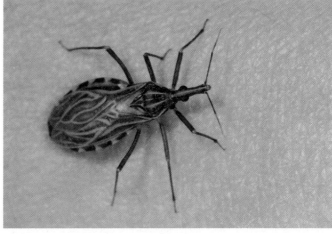

(b)

Figure 40.13 *Trypanosoma cruzi* **Infection.** (a) *T. cruzi* life cycle. (b) The triatomine bug.

occurs immediately after infection, lasting weeks or months (acute phase). Cell invasion stimulates the trypanosome's transformation into amastigotes, resulting in clinical manifestations of infection. The bloodstream trypanomastigotes do not replicate, however, until they enter a cell (becoming amastigotes) or are ingested by the arthropod vector (becoming epimastigotes). A chronic indeterminate (asymptomatic) phase of the disease follows the acute phase. During this phase, a high percentage of heart and gastrointestinal disease occurs due to parasitized cardiac and intestinal cells, respectively. There are approximately 8 million infected people globally, with 28,000 new cases per year and 12,000 annual deaths.

Trypanosomiasis is diagnosed by finding motile protists in fresh blood, spinal fluid, or skin biopsy and by serological testing. Treatment depends on the infecting trypanosome species and the stage of infection. Pentamidine isethionate and suramin are used to treat the hemolymphatic stages of HAT. Melarsoprol is used to treat late disease with central nervous system involvement caused by both trypanosomes. Benznidazole or nifurtimox are used to treat Chagas' disease, but effectiveness is best when administered during the acute phase of infection. Vaccines are not useful because the parasite is able to change its protein coat (**antigenic variation**) to evade the immunologic response. ◀◀ Euglenozoa *(section 25.2)*

Retrieve, Infer, Apply

1. Besides their route of transmission, how else are human fungal diseases categorized?
2. Why are fungal infections of the lungs potentially life-threatening?
3. Why is *Histoplasma capsulatum* found in bird feces but not within bird tissue? What are the public health implications of this?
4. What flagellated protists invade the blood? What diseases do they cause?
5. Describe the malarial life cycle. What stages of the life cycle occur in humans?
6. What malarial stages could vaccines block? What are the challenges to vaccine development?

40.4 Direct Contact Diseases Can Be Caused by Fungi and Protists

After reading this section, you should be able to:

- Report the common fungal and protist diseases spread by direct contact
- Identify typical signs and symptoms of fungal and protist diseases spread by direct contact
- Correlate direct contact fungal infection and disease severity with fungal virulence factors
- Correlate direct contact protist infection and disease severity with protist virulence factors

Transmission of eukaryotic pathogens by direct contact results in a number of infectious diseases of the skin and reproductive organs. The direct contact diseases caused by fungi can be classified by their penetration into the host; superficial, cutaneous, and subcutaneous fungal infections are known. Diseases caused by protists can occur through hand-to-mouth/nose exchange, as well

as their transfer through body fluids. We discuss the fungal pathogens first and then protist pathogens.

Fungal Pathogens

Fungi responsible for infections limited to the outer surface of hair and skin cause superficial mycoses. Superficial mycoses are extremely rare in the United States; most occur in the tropics. The fungi that cause these diseases remain above the basement membrane of skin, consuming keratin as their primary carbon source. Infections of the hair shaft are collectively called **piedras** (Spanish for stone because they are associated with the hard nodules formed by mycelia on the hair shaft). For example, black piedra is caused by *Piedraia hortae* and forms hard, black nodules on the hairs of the scalp. White piedra is caused by the yeast *Trichosporon beigelii* and forms light-colored nodules on the beard and mustache. Some superficial mycoses are called **tineas** (Latin for grub, larva, worm); the specific type is designated by a modifying term. Tineas involve the outer layers of the skin, nails, and hair. Tinea versicolor is caused by the yeast *Malassezia furfur,* which forms brownish-red scales on the skin of the trunk, neck, face, and arms. Treatment involves removal of the skin scales with a cleansing agent and removal of the infected hairs. Good personal hygiene prevents these infections.

Tinea barbae (Latin *barba,* the beard) is an infection of beard hair caused by *Trichophyton mentagrophytes* or *T. verrucosum*. It is predominantly a disease of men who live in rural areas and acquire the fungus from infected animals. Tinea capitis (Latin *capita,* the head) is an infection of the scalp hair (**figure 40.14a**). It is characterized by hair loss, inflammation, and scaling. Tinea capitis is primarily a childhood disease caused by *Trichophyton* or *Microsporum* species. Person-to-person transmission of the fungus occurs frequently when poor hygiene and overcrowded conditions exist. The fungus also occurs in domestic animals, which can transmit it to humans. A Wood's lamp (an ultraviolet [UV] light) can help with the diagnosis of tinea capitis because fungus-infected hair fluoresces when illuminated by UV radiation (figure 40.14b).

Other fungi can infect skin cells; they cause diseases known as cutaneous mycoses. Cutaneous mycoses—also called **dermatomycoses, ringworms,** or tineas—occur worldwide and represent the most common fungal diseases in humans. Members of three genera of cutaneous fungi, or dermatophytes, are involved in these mycoses: *Epidermophyton, Microsporum,* and *Trichophyton*. Diagnosis is by microscopic examination of biopsied areas of the skin cleared with 10% potassium hydroxide and by culture on Sabouraud dextrose agar. Treatment is with topical ointments such as miconazole, tolnaftate, or clotrimazole for 2 to 4 weeks. Griseofulvin and itraconazole are the only agents proven to be effective in treating dermatophytoses. Infections by dermatophytes may be insidious; the desquamated (shed) skin cells can contain viable fungi for months or years. Thus any treatment plan should also include thorough cleaning of bedding, carpets, and any other substrates that could hold desquamated skin cells.

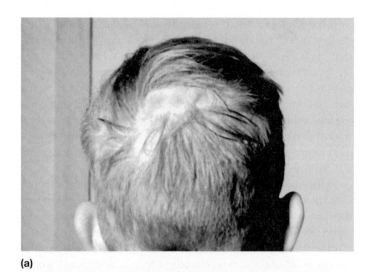

(a)

(b)

Figure 40.14 Cutaneous Mycosis: Tinea Capitis. (a) Fungal infection of the head caused by *Microsporum canis.* (b) Close-up using a Wood's light (a UV lamp).

MICRO INQUIRY *What is the difference between a piedra and a tinea? Which is shown here?*

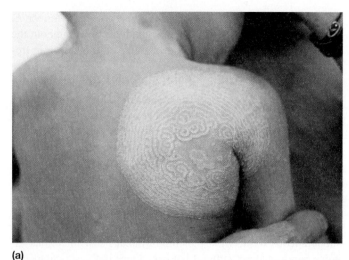

(a)

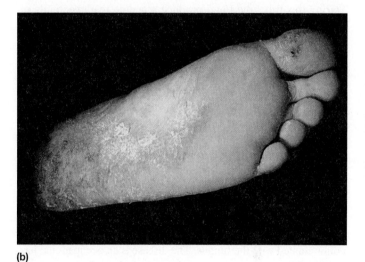

(b)

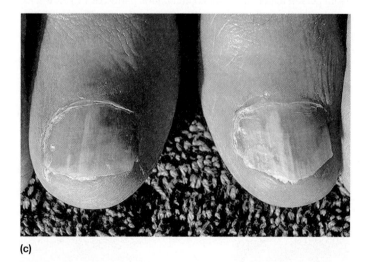

(c)

Figure 40.15 Cutaneous Mycosis. (a) Tinea corporis. Ringworm of the body (shoulder) caused by *Trichophyton concentricum.* (b) Tinea pedis caused by *Trichophyton rubrum.* (c) Tinea unguium of the nails caused by *Trichophyton rubrum.*

Tinea corporis (Latin *corpus,* the body) is a dermatophytic infection that can occur on any part of the skin (**figure 40.15a**). The disease is characterized by circular, red, well-demarcated, scaly, vesiculopustular lesions accompanied by itching. Tinea corporis is caused by *Trichophyton rubrum, T. mentagrophytes, T. concentricum,* or *Microsporum canis.* Fungi are transmitted by direct contact with infected animals and humans or by indirect contact through fomites (inanimate objects).

Tinea pedis (Latin *pes,* the foot), also known as athlete's foot, and tinea manuum (Latin *mannus,* the hand) are dermatophytic infections of the feet (figure 40.15b) and hands, respectively. Clinical symptoms vary from a fine scale to a pimplelike eruption. Itching is frequently present. Warmth, humidity, trauma, and occlusion increase susceptibility to infection. Most infections are caused by *T. rubrum, T. mentagrophytes,* or *E. floccosum.* Tinea pedis and tinea manuum occur throughout the world, are most commonly found in adults, and increase in frequency with age.

Tinea unguium (Latin *unguis,* nail) is a dermatophytic infection of the nail bed (figure 40.15c). In this disease, the nail becomes discolored and then thickens. The nail plate rises and separates from the nail bed. *Trichophyton rubrum* or *T. mentagrophytes* is the causative fungus.

The dermatophytes that infect underneath the skin cause subcutaneous mycoses, such as sporotrichosis, chromoblastomycosis, and maduromycosis. These fungi are normally saprophytic inhabitants of soil and decaying vegetation. Because they are unable to penetrate the skin, they must be introduced into the subcutaneous tissue by a puncture wound. Most infections involve barefooted agricultural workers. Once in the subcutaneous tissue, the disease develops slowly—often over a period of years. During this time, the fungi produce a nodule that eventually ulcerates, and the organisms spread along lymphatic channels, producing more subcutaneous nodules. At times, the nodules drain to the skin surface. Griseofulvin, terbinafine, itraconazole, and fluconazole are common therapeutic options. The administration of oral 5-fluorocytosine, iodides, or amphotericin B, and surgical excision may be used for stubborn infections. Diagnosis is accomplished by culture of the infected tissue.

Sporotrichosis is the subcutaneous mycosis caused by the dimorphic fungus *Sporothrix schenckii.* The disease occurs throughout the world and is the most common subcutaneous mycotic disease in the United States. The fungus can be found in the soil; on living plants, such as barberry shrubs and roses; or in plant debris, such as sphagnum moss, baled hay, and pine-bark mulch. Infection occurs by a puncture wound from a thorn or splinter contaminated with the fungus. The disease is an occupational hazard for florists, gardeners, and forestry workers. It is not spread from person to person. After an incubation period of 1 to 12 weeks, a small, red papule arises and begins to ulcerate (**figure 40.16**).

New lesions appear along lymph channels and can remain localized or spread throughout the body, producing extracutaneous sporotrichosis.

Sporotrichosis is typically treated by oral itraconazole for 3 to 6 months. Systemic or severe disease is treated with a lipid formulation of amphotericin B or ingestion of supersaturated potassium iodide until the lesions are healed, usually over several weeks. Preventative measures include gloves and other protective clothing, as well as avoidance of contaminated landscaping materials, especially sphagnum moss.

Chromoblastomycosis, a disease in which dark brown nodules are formed, is caused by the black molds *Phialophora verrucosa* and *Fonsecaea pedrosoi.* These fungi exist worldwide, especially in tropical and subtropical regions. Most infections involve the legs and feet (**figure 40.17**). **Maduromycosis** is caused by *Madurella mycetomatis,* which is distributed worldwide and is especially prevalent in the tropics. Because the fungus destroys subcutaneous tissue and produces serious deformities, the resulting infection is often called a **eumycotic mycetoma,** or fungal tumor. One form of mycetoma, known as Madura foot, occurs through skin abrasions acquired while walking barefoot on contaminated soil.

Protist Pathogens

Very few human pathogenic protists are transmitted through direct contact. Thus we only discuss one such disease, trichomoniasis.

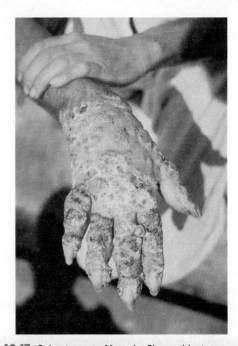

Figure 40.17 Subcutaneous Mycosis: Chromoblastomycosis. Chromoblastomycosis of the hand caused by *Phialophora verrucosa.*

MICRO INQUIRY *How do fungi causing subcutaneous mycoses gain access to tissue below the skin surface?*

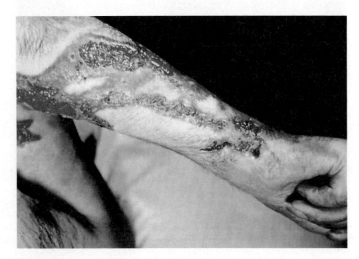

Figure 40.16 Subcutaneous Mycosis: Sporotrichosis. Sporotrichosis of the arm caused by *Sporothrix schenckii.*

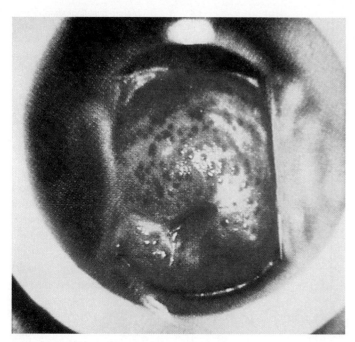

Figure 40.18 Trichomoniasis. *Trichomonas vaginalis* infection of the cervix causing hemorrhages, papules, and vesicles ("strawberry cervix").

Trichomoniasis is a sexually transmitted disease caused by the protozoan flagellate *Trichomonas vaginalis* (*see figure 25.4*). Humans are the only known host of *T. vaginalis.* Infection with this protist is often found with other sexually transmitted infections (STIs) and may increase the risk of acquiring other STIs. It is one of the most common STIs, with an estimated 2.3 million cases (women aged 14–49) annually in the United States and 160 million cases annually worldwide. In response to the protist, the body accumulates leukocytes at the site of infection. In females, this usually results in a profuse, purulent vaginal discharge that is yellowish to light cream in color and characterized by a disagreeable odor. The discharge is accompanied by itching. The cervical mucosa is covered with punctate hemorrhages, papules, and vesicles, referred to as "strawberry cervix" (**figure 40.18**). Males are generally asymptomatic because of the trichomonacidal action of prostatic secretions; however, at times a burning sensation occurs during urination. Diagnosis is made in females by microscopic identification of the protozoan. Infected males demonstrate protozoa in semen or urine. Treatment is by administration of metronidazole to all sexual contacts. ◄◄ Parabasalia (*section 25.2*)

Retrieve, Infer, Apply

1. Describe two piedras that infect humans.
2. Briefly describe the major tineas that occur in humans.
3. Describe the three types of subcutaneous mycoses that affect humans.
4. How would you diagnose trichomoniasis in a female? In a male?

40.5 Food and Water Are Vehicles of Protozoal Diseases

After reading this section, you should be able to:

- Report the common food-borne and waterborne fungal and protist diseases
- Identify typical signs and symptoms of common food-borne and waterborne fungal and protist diseases
- Correlate vehicle-borne protist infection and disease severity with protist virulence factors
- Explain in specific terms why access to clean water is considered a basic human right

With advancing genomic studies, a number of infectious agents transmitted by food and water have been reclassified from fungi to protozoa. Thus we discuss food-borne and waterborne diseases using protist examples.

Amebiasis

Worldwide there are approximately 500 million people infected with *Entamoeba* protozoa. It is now accepted that two species of *Entamoeba* infect humans: *E. dispar* and *E. histolytica*, which causes approximately 10% of invasive amebiasis; both are responsible for **amebiasis** (**amebic dysentery**). These very common parasites are endemic in warm climates where adequate sanitation is lacking. Within the United States, about 3,000 to 5,000 cases are reported annually. However, it is a major cause of parasitic death worldwide; as many as 100,000 die of amebiasis each year. However, this may represent the tip of an iceberg since only 10–20% of infected individuals are symptomatic and thus seek diagnosis and treatment. ◄◄ Entamoebida (*section 25.3*)

Infection occurs by ingestion of mature cysts from fecally contaminated water, food, or hands, or from fecal exposure during sexual contact. After excystation in the lower region of the small intestine, the metacyst divides rapidly to produce eight small trophozoites (**figure 40.19**). These trophozoites move to the large intestine, where they can invade the host tissue, live as commensals in the lumen of the intestine, or undergo encystation. In many hosts, the trophozoites remain in the intestinal lumen, resulting in an asymptomatic carrier state with cysts shed in feces.

If the infective trophozoites invade intestinal tissues, they multiply rapidly and spread laterally, while feeding on erythrocytes, bacteria, and yeasts. The invading trophozoites destroy the epithelial lining of the large intestine by producing a protease that plays a role in intestinal invasion by degrading the extracellular matrix and circumventing the host immune response by cleaving secretory immunoglobulin A (sIgA), IgG, and complement factors. The protease is encoded by at least seven genes, several of which are found in *E. histolytica* but not *E. dispar*. Lesions (ulcers) are characterized by minute points of entry into the mucosa and extensive enlargement of the lesion after penetration into the submucosa. *E. histolytica* also may invade and produce lesions in other tissues, especially the liver, to cause hepatic

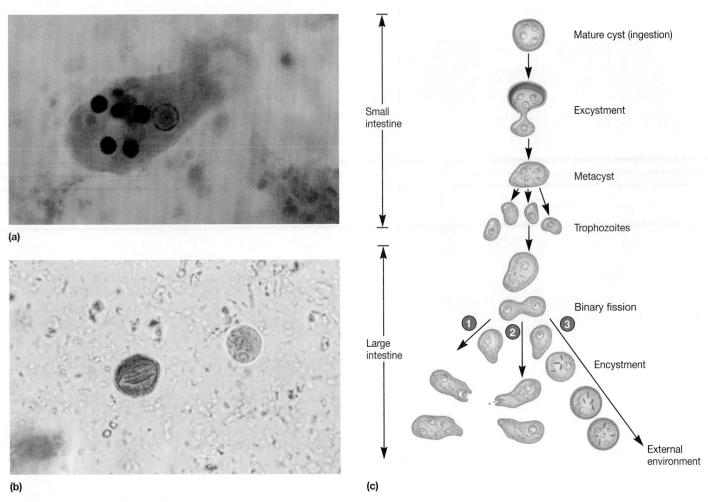

Figure 40.19 Amebiasis Caused by *Entamoeba histolytica.* Light micrographs of (a) a trophozoite (×1,000) and (b) a cyst (×3,100). (c) Life cycle. Infection occurs by ingestion of a mature cyst. Excystment occurs in the lower region of the small intestine, and the metacyst rapidly divides to give rise to eight small trophozoites (only four are shown). These enter the large intestine, undergo binary fission, and may (1) invade host tissues, (2) live in the lumen of the large intestine without invasion, or (3) undergo encystment and pass out of the host in feces.

amebiasis. However, all extraintestinal amebic lesions are secondary to those established in the large intestine. The symptoms of amebiasis are highly variable, ranging from an asymptomatic infection to fulminating dysentery (exhaustive diarrhea accompanied by blood and mucus), appendicitis, and abscesses in the liver, lungs, or brain. ◄◄ *Antibodies are proteins that bind to specific 3-D molecules (section 34.7)*

Laboratory diagnosis of amebiasis can be difficult and is based on finding trophozoites in fresh, warm stools and cysts in ordinary stools. Serological testing for *E. histolytica* is available but often unreliable. The therapy for amebiasis is complex and depends on the location of the infection within the host and the host's condition. Asymptomatic carriers who are passing cysts should always be treated with iodoquinol or paromomycin because they represent the most important reservoir of the protozoan. In symptomatic intestinal amebiasis, paromomycin, metronidazole and iodoquinol are the drugs of choice. Prevention and control of amebiasis are achieved by practicing good personal hygiene and avoiding water or food that might be contaminated with human feces. Viable cysts in water can be destroyed by hyperchlorination or iodination. ◄◄ *Antiprotozoan drugs (section 9.7)*

Amebic Meningoencephalitis and Keratitis

Free-living amoebae of the genera *Naegleria* and *Acanthamoeba* are facultative parasites responsible for causing two clinically distinct central nervous system diseases in humans: primary amoebic meningoencephalitis (PAM) and **keratitis** (inflammation of the cornea), respectively. *Naegleria fowleri* causes PAM, a disease that is typically fatal in 3 to 10 days after infection. They are among the most common protists found in freshwater (including natural and treated pools and hot tubs), drinking water systems, ventilation and air-conditioning systems, and moist soil. In 2011 several deaths were attributed to *Naegleria* infections acquired through the use of neti pots (similar to gravy boats) to "rinse the sinuses." In each case, tap water was reported to be contaminated with the protist. Meningoencephalitis occurred when the protist traversed the thin cell layer between the sinus and the brain.

In addition, several species of *Acanthamoeba* are known to cause encephalitis and to infect the eye, causing a chronically progressive, ulcerative *Acanthamoeba* keratitis that may result in blindness. Wearers of soft contact lenses may be predisposed to this infection and should take care to prevent contamination of their lens-cleaning and soaking solutions. Diagnosis of these infections is by demonstration of amoebae in clinical specimens.

Most freshwater amoebae are resistant to commonly used antimicrobial agents. These amoebae are reported in fewer than 100 human disease cases annually in the United States, although the incidence (especially of *Acanthamoeba keratitis*) is likely higher.

Cryptosporidiosis

A number of cryptosporidial species infect humans and other animals. *Cryptosporidium* ("hidden spore cysts") spp. are found in about 90% of sewage samples, in 75% of river waters, and in 28% of drinking waters. *C. parvum* is a common protist found in the intestine of many birds and mammals. When these animals defecate, oocysts are shed into the environment. If a human ingests food or water that is contaminated with the oocysts, excystment occurs within the small intestine, and sporozoites enter epithelial cells and develop into merozoites. Some of the merozoites subsequently undergo sexual reproduction to produce zygotes, which differentiate into thick-walled oocysts. Oocysts released into the environment begin the life cycle again. ◀◀ Alveolata *(section 25.4)*

Major problems for public health arise from the fact that (1) oocysts are small, only 4 to 6 μm in diameter—much too small to be easily removed by the sand filters used to purify drinking water; (2) they have a low infectious dose, around 10 to 100 oocysts; and (3) *Cryptosporidium* spp. also are extremely resistant to disinfectants such as chlorine and oocysts may remain viable for 2 to 6 months in moist environments.

The incubation period for cryptosporidiosis ranges from 5 to 28 days. Diarrhea, which characteristically may be cholera-like (i.e., very watery, without blood), is the most common symptom. Other symptoms include abdominal pain, nausea, fever, and fatigue. The pathogen is routinely diagnosed by fecal concentration and acid-fast stained smears. Treatment of people with healthy immune systems is with nitazoxanide along with rehydration. Although the disease usually is self-limiting in healthy individuals, patients with late-stage AIDS or who are immunocompromised in other ways may develop prolonged, severe, and life-threatening diarrhea.

Cyclosporiasis

Cyclosporiasis is caused by the unicellular protist *Cyclospora cayetanensis*, first isolated in 1979. The disease is most common in tropical and subtropical environments, although it has been reported in most countries. In Canada and the United States, cyclosporiasis has been responsible for affecting over 3,600 people in at least 11 food-borne outbreaks since 1990, with 304 confirmed cases in 2014. A number of cyclosporiasis outbreaks have been linked to contaminated produce.

The disease presents with frequent, sometimes explosive, diarrhea. Often the patient exhibits loss of appetite, cramps, and bloating due to substantial gas production, nausea, vomiting, fever, fatigue, and substantial weight loss. Patients may report symptoms for days to weeks with decreasing frequency and then relapses. The protozoa infect the small intestine with a mean incubation period of approximately 1 week. Cyclosporan oocysts in freshly passed feces are not infective (**figure 40.20**). Thus direct oral-fecal transmission does not occur. Instead, the oocysts must differentiate into

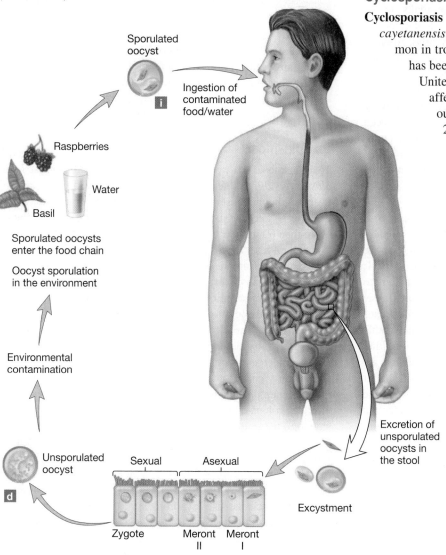

Figure 40.20 Life Cycle of *Cyclospora cayetanensis.* Unsporulated oocysts, shed in the feces of infected animals, sporulate and contaminate food and water. Once ingested, the sporulated oocysts excyst, penetrate host cells, and reproduce by binary fission (merogony). Sexual stages unite to form zygotes, resulting in new oocysts (unsporulated) that are shed in feces.

Some of the elements in this figure were created based on an illustration by Ortega, et al. Cyclospora cayetanensis. In: Advances in Parasitology: Opportunistic protozoa in humans. San Diego: Academic Press; 1998. pp. 399–418.

sporozoites after days or weeks at temperatures between 22 and 32°C (71.6–89.6°F). Sporozoites often enter the food chain when oocyst-contaminated water is used to wash fruits and vegetables prior to their transport to market. Once ingested, the sporozoites are freed from the oocysts and invade intestinal epithelial cells, where they replicate asexually. Sexual development is completed when sporozoites mature into new oocysts and are released into the intestinal lumen to be shed with feces.

Laboratory diagnosis of cyclosporiasis is by the identification of oocysts in feces. Identification may require several specimens over several days. Treatment is with a combination of trimethoprim and sulfamethoxazole, and fluids to restore water lost through diarrhea. Prevention of cyclosporiasis is by avoidance of contaminated food and water. No vaccine is available.

Giardiasis

Giardia intestinalis is a flagellated protist that causes the very common intestinal disease **giardiasis,** which is worldwide in distribution and affects children more seriously than adults. In the United States, this protist is the most common cause of epidemic waterborne diarrheal disease (about 30,000 cases yearly). Approximately 7% of the population are asymptomatic carriers who shed cysts in their feces. *G. intestinalis* is endemic in child day-care centers in the United States, with estimates of 5 to 15% of diapered children being infected. Transmission occurs most frequently by cyst-contaminated water supplies. Epidemic outbreaks have been recorded in wilderness areas, suggesting that humans may be infected from stream water with *G. intestinalis* harbored by rodents, deer, cattle, or household pets. This implies that human infection also can be a zoonosis. As many as 200 million humans may be infected worldwide by ingesting as few as 10 cysts.

Following ingestion, *Giardia* cysts undergo excystment in the duodenum, forming trophozoites. The trophozoites inhabit the upper portions of the small intestine, where they attach to the intestinal mucosa by means of their sucking disks (**figure 40.21**; *also see figure 25.4*). The ability of the trophozoites to adhere to the intestinal epithelium accounts for the fact that they are rarely found in stools. It is thought that the trophozoites feed on mucous secretions and reproduce to become such a large population that they interfere with nutrient absorption by the intestinal epithelium. Giardiasis varies in severity, and asymptomatic carriers are common. Symptoms of giardiasis typically begin 1 week after infection and can last 6 to 8 weeks or longer if untreated. The disease can be acute or chronic. Acute giardiasis is characterized by severe diarrhea, epigastric pain, cramps, voluminous flatulence ("passing gas"), and anorexia. Chronic giardiasis is characterized by intermittent diarrhea, with periodic appearance and remission of symptoms.

Laboratory diagnosis is based on the identification of trophozoites—only in the severest of diarrhea—or cysts in stools. A commercial ELISA test is also available for the detection of *G. intestinalis* antigen in stool specimens. Tinidazole, nitazoxanide, and metronidazole are the drugs of choice for adults, and furazolidone is used for children because it is available in a liquid

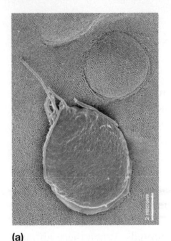
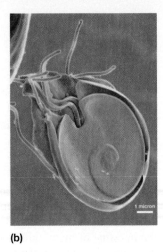

(a)   (b)

Figure 40.21 Giardiasis. (a) Scanning electron micrograph (SEM) depicting a *Giardia muris* protozoan adhering itself to the microvillous border of an intestinal epithelial cell. Each small circular profile under the protozoan represents the rounded tip of a single microvillous, and it is estimated that 2000 to 3000 microvilli cover the surface of a single intestinal epithelial cell. The two circular lesions above the protozoan are impressions made by the ventral adhesive disk of other *G. muris* organisms. (b) This SEM reveals the ultrastructural morphology of a *Giardia* protozoan's ventral adhesive disk. This disk acts like a suction cup, facilitating the organism's attachment to the intestinal surface.

suspension. Prevention and control involve proper treatment of community water supplies, especially the use of slow sand filtration because the cysts are highly resistant to chlorine treatment. ▶▶| *Purification and sanitary analysis ensure safe drinking water (section 43.1); Wastewater treatment (section 43.2)*

Toxoplasmosis

Toxoplasmosis is a worldwide disease caused by the protist *Toxoplasma gondii*. It infects nearly 90% of people over the age of twelve, although most of those infected do not exhibit symptoms of the disease because the immune system usually prevents illness. More than 60 million U.S. citizens are asymptomatically infected. The natural reservoir of *T. gondii* is wild rodents, birds, and other small animals. Cats are an important factor in the transmission of *T. gondii* to humans; cats are the definitive (final) host and are required for completion of the sexual cycle (**figure 40.22**). Kittens especially shed oocysts of the protist in their feces for several weeks after infection. The protist is ubiquitous and finds its way into the food supply; toxoplasmosis can be transmitted by the ingestion of raw or undercooked meat. In fact, the CDC lists toxoplasmosis as the third leading cause of death due to food-borne infection. Oocysts enter the human host by way of the nose or mouth, and the parasites colonize the intestine. Congenital transfer, blood transfusion, or a tissue transplant also can transmit the protist. Originally toxoplasmosis gained public notice when it was discovered that the protist can infect a pregnant woman's fetus, causing serious congenital defects or death. It follows that pregnant women are advised not to empty cat litter boxes. Adults usually complain of an "infectious mononucleosis-like" syndrome.

Acute toxoplasmosis is usually accompanied by lymph node swelling (lymphadenopathy) with white blood cell enlargement. Pulmonary necrosis, myocarditis, and hepatitis caused by tissue necrosis are common. Retinitis (inflammation of the retina of the

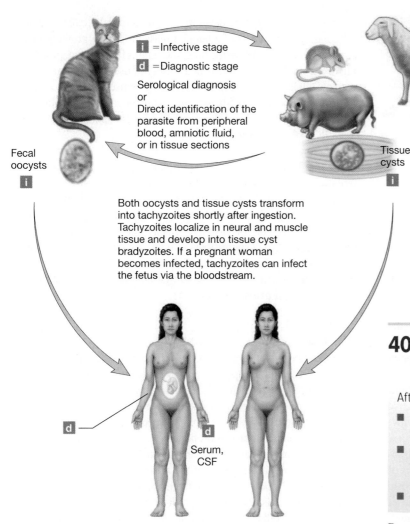

i = Infective stage

d = Diagnostic stage

Serological diagnosis
or
Direct identification of the parasite from peripheral blood, amniotic fluid, or in tissue sections

Fecal oocysts
i

Tissue cysts
i

Both oocysts and tissue cysts transform into tachyzoites shortly after ingestion. Tachyzoites localize in neural and muscle tissue and develop into tissue cyst bradyzoites. If a pregnant woman becomes infected, tachyzoites can infect the fetus via the bloodstream.

d

d

Serum, CSF

Figure 40.22 Life Cycle of *Toxoplasma gondii*. *Toxoplasma gondii* is a protozoan parasite of numerous mammals and birds. However, a cat, the definitive host, is required for completion of the sexual cycle. Oocysts are shed in the feces of infected animals, where they may be ingested by another host. Ingested oocysts transform into tachyzoites, which migrate to various tissue sites via the bloodstream.

MICRO INQUIRY *Upon entry into the human host by the nose or mouth, where do* T. gondii *colonize?*

eye) is associated with necrosis due to proliferation of the parasite within retinal cells. *T. gondii* in the immunocompromised, such as AIDS or transplant patients, can produce a unique encephalitis with necrotizing lesions in the brain accompanied by inflammatory infiltrates. *T. gondii* continues to cause more than 3,000 congenital infections per year in the United States. In immunocompromised or immunosuppressed individuals, infection frequently results in fatal disseminated disease with heavy cerebral involvement. ◀◀ Alveolata *(section 25.5)*

Laboratory diagnosis of toxoplasmosis is by serological tests. Epidemiologically, toxoplasmosis is ubiquitous in all higher animals. Treatment of toxoplasmosis is with a combination of pyrimethamine and sulfadiazine. Prevention and control require minimizing exposure by the following: avoiding eating raw meat and eggs, washing hands after working in soil, cleaning cat litter boxes daily, keeping household cats indoors if possible, and feeding cats commercial food.

40.6 Opportunistic Diseases Can Be Caused by Fungi and Protists

After reading this section, you should be able to:

- Predict the impact to the host resulting when normal fungal microbiota are introduced to a new host environment
- Distinguish growth and reproduction challenges experienced by normal fungal microbiota and ubiquitous environmental fungi when attempting to infect a human
- Cite specific host conditions that promote fungal infection

Recall that an opportunistic microorganism can be a member of the resident microbiota that is generally harmless in its normal environment; it becomes pathogenic when (1) introduced to a new environment, (2) outcompetes the other resident microbes, or (3) the host becomes (immuno-) compromised. A compromised host is seriously debilitated and has lowered resistance to infection. There are many causes of this condition: malnutrition, alcoholism, cancer, diabetes, another infectious disease (e.g., HIV/AIDS), trauma from surgery or injury, an altered microbiota from prolonged use of antibiotics (e.g., in vaginal candidiasis), and immunosuppression (e.g., by drugs, hormones, genetic deficiencies, cancer chemotherapy, and old age). Opportunistic mycoses may start as normal microbiota or ubiquitous environmental contaminants. The importance of opportunistic fungi and protozoa is increasing because of the expansion of the immunocompromised patient population. Here we discuss a few opportunistic fungal pathogens.

Aspergillosis

Of all the fungi that cause disease in human hosts, none is as widely distributed in nature as members of the genus *Aspergillus*. These filamentous fungi are omnipresent: they are found wherever organic debris occurs, especially in soil, decomposing plant

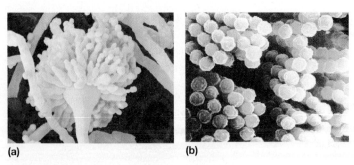

Figure 40.23 Cause of an Opportunistic Mycosis. (a) Conidiophores of *Aspergillus flavus.* (b) SEM of *Aspergillus* spores.

matter, household dust, building materials, some foods, and water. In fact, it is almost impossible to avoid daily inhalation of *Aspergillus* spores (**figure 40.23**). In people with healthy immune systems, *Aspergillus* fungi do not often cause disease. However, infection results when healthy immune functions deteriorate.

Aspergillus fumigatus is apparently more common in air, and is also the usual cause of pulmonary **aspergillosis.** *A. flavus* is the most common cause of both superficial infection and invasive disease of immunosuppressed patients. *A. flavus* is more common in some locations such as hospitals. Invasive disease typically results in pulmonary infection (with fever, chest pain, and cough) that disseminates to the brain, kidney, liver, bone, or skin. Invasive aspergillosis has a mortality rate of 50 to 100%. ◄◄ *Ascomycota includes yeasts and molds (section 26.5)*

The major portal of entry for *Aspergillus* spp. is the respiratory tract. Inhalation of conidiospores can lead to several types of pulmonary aspergillosis. One type is allergic aspergillosis. Infected individuals may develop an immediate allergic response and suffer asthma attacks when exposed to fungal antigens. In bronchopulmonary aspergillosis (another potential response to *Aspergillus* spp.), patients with underlying pulmonary disease (tuberculosis, asthma, cystic fibrosis) suffer the major clinical manifestation of bronchitis resulting from both type I and type III hypersensitivities. Although tissue invasion seldom occurs in bronchopulmonary aspergillosis, the fungus often can be cultured from sputum. Finally, the most common manifestation of pulmonary involvement is the occurrence of colonizing aspergillosis, in which fungal colonies

form within the lungs and develop into "fungus balls" called aspergillomas. These consist of a tangled mass of hyphae growing in a circumscribed area. From the lung, the fungus may spread, producing disseminated aspergillosis in a variety of tissues and organs. In patients whose resistance is severely compromised, invasive aspergillosis may occur and fill the lung with fungal hyphae. Rarely, aspergillomas (or tumor-like growths of fungal spores) can form, as well. ◄◄ *Hypersensitivities (section 34.10)*

Laboratory diagnosis of aspergillosis depends on identification either by direct examination of pathological specimens or by isolation and characterization of the fungus. An enzyme immunoassay that detects galactomannan (an exoantigen of *Aspergillus* spp.) can be used to screen suspected cases of aspergillosis. Treatment for invasive disease is with voriconazole, while allergic-type aspergillosis is treated with itraconazole and corticosteroids. However, successful therapy depends on treatment of the underlying disease so that host resistance increases. ◄◄ *Antifungal drugs (section 9.5)*

Candidiasis

Candidiasis is a mycosis caused by dimorphic fungi of the genus *Candida.* Recall that some fungi can exist as single cells (yeast) or as filaments (molds) adapting to their environmental conditions. Candida spp. use this Y-M shift to disseminate within a human host depending on their need to invade through body fluids (as yeast) or tissues (as molds). There are over 20 species of *Candida* that can cause infection in humans, most notably *C. albicans* (**figure 40.24***a*) and *C. glabrata.* In contrast to other pathogenic fungi, *C. albicans* and *C. glabrata* are members of the normal microbiota within the gastrointestinal tract, respiratory tract, vaginal area, and mouth. In healthy individuals, they do not produce disease because growth is suppressed by other microbiota and host resistance mechanisms. However, if anything upsets the normal microbiota and immune competency, these fungi may multiply rapidly and produce candidiasis. *Candida* species are important nosocomial pathogens. In some hospitals, they may represent almost 10% of nosocomial bloodstream infections. Because these fungi can be transmitted sexually, candidiasis is also listed by the CDC as a sexually transmitted disease.

No other mycotic pathogens produce as diverse a spectrum of disease in humans as do *Candida* spp. Most infections involve the

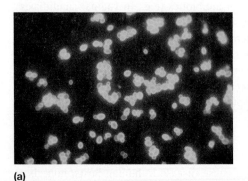
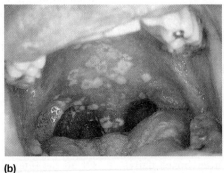
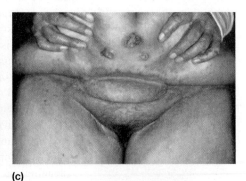

Figure 40.24 Opportunistic Mycoses Caused by *Candida albicans.* (a) Fluorescent antibody (FA) stain revealing the oval budding of the yeast form. (b) Thrush, or oral candidiasis, is characterized by the formation of white patches on the mucous membranes of the tongue and elsewhere in the oropharyngeal area. These patches form a pseudomembrane composed of spherical yeast cells, leukocytes, and cellular debris. (c) Intertriginous candidiasis involves those areas of the body, usually opposed skin surfaces, that cause irritation, inflammation, and areas of cutaneous ulceration.

skin or mucous membranes. This occurs because *Candida* spp. are strict aerobes and find such surfaces very suitable for growth. Cutaneous involvement usually occurs when the skin becomes overly moist or damaged. For instance, paronychia and onychomycosis are associated with infections of the subcutaneous tissues of the digits and nails, respectively. These infections usually result from continued immersion of the appendages in water.

Oropharyngeal candidiasis, or **thrush** (figure 40.24*b*), is a fairly common disease in newborns, persons with dentures, and people who use inhaled steroids. A weakened immune system predisposes people to candidiasis, which appears as many small, white flecks that cover the tongue and mouth. At birth, newborns do not have a normal microbiota in the oropharyngeal area. If the mother's vaginal area is heavily colonized with *Candida* spp., the upper respiratory tract of the newborn becomes colonized during passage through the birth canal. Once the newborn has developed his or her own normal oropharyngeal microbiota, thrush becomes uncommon.

Intertriginous candidiasis involves those areas of the body, usually opposed skin surfaces, that are warm and moist: axillae, groin, skin folds (figure 40.24*c*). Napkin (diaper) candidiasis is typically found in infants whose diapers are not changed frequently and therefore are not kept dry. **Candidal vaginitis** can result as a complication of diabetes, antibiotic therapy, oral contraceptives, pregnancy, or any other factor that compromises the female host. Normally the omnipresent vaginal lactobacilli create a low pH, which limits fungal growth. However, if the numbers of lactobacilli are decreased by any of the aforementioned factors, yeast may proliferate, causing a curdlike, yellow-white discharge from the vagina. *C. albicans* and *C. glabrata* can be transmitted to males during intercourse and lead to **balanitis.** Balanitis is an infection of the male glans penis and occurs primarily in uncircumcised males. The disease begins as vesicles on the penis that develop into patches and are accompanied by severe itching and burning.

Diagnosis of candidiasis is sometimes difficult because (1) these fungi are frequent secondary invaders in diseased hosts, (2) a mixed microbiota is most often found in the diseased tissue, and (3) no completely specific immunologic procedures for the identification of *Candida* spp. currently exist. Mortality is almost 50% when the fungi invade the blood or disseminate to visceral organs, as occasionally seen in immunocompromised patients.

There is no one satisfactory treatment for candidiasis. Cutaneous lesions can be treated with topical agents such as sodium caprylate, sodium propionate, nystatin, miconazole, and trichomycin. Clotrimazole lozenges and nystatin swishing are used to treat oropharyngeal candidiasis. Similar drugs are used to treat genital/vulvovaginal and rectal candidiasis. Ketoconazole, amphotericin B, fluconazole, itraconazole, and flucytosine also can be used for systemic candidiasis.

Microsporidiosis

More than 1,200 species of *Microsporidia* have been catalogued, in 143 genera; at least 14 species are known to be human pathogens. Several domestic and feral animals appear to be reservoirs for several species that infect humans. Increasingly recognized as opportunistic infectious agents, microsporidia infect a wide range of

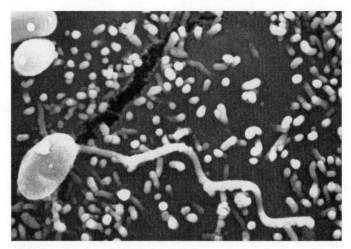

Figure 40.25 Microsporidia. Scanning electron micrograph of a microsporidian spore with an extruded polar tubule inserted into a eukaryotic cell. The spore injects the infective sporoplasms through its polar tubule.

vertebrate and invertebrate hosts. **Microsporidiosis** is an infectious disease found mostly in HIV patients, however. One unifying characteristic of microsporidia is their production of a highly resistant spore, capable of surviving long periods of time in the environment. Spore morphology varies with species, but most superficially resemble enteric bacteria. Microsporidial spores recovered from human infections are oval to rodlike, measuring 1 to 4 μm. Microsporidia also possess a unique organelle known as the polar tubule, which is coiled within the spore (**figure 40.25**; *also see figure 26.17*). ◀◀ Microsporidia (*section 26.7*)

Infection of a host cell results when a microsporidial spore extrudes its polar tubule, which bores through the target cell's plasma membrane. A sudden increase in spore calcium results in the injection of the sporoplasm (cytoplasm-like contents) through the polar tubule into the host cell. Inside the cell, the sporoplasm condenses and undergoes asexual multiplication. Multiplication of the sporoplasm is species-specific, occurring within the host cytoplasm or within a vacuole, and is usually completed through binary (merogony) or multiple (schizogony) fission. Once the sporoplasm has multiplied, it directs the development of new spores (sporogony) by first encapsulating meronts or schizonts (products of asexual multiplication) within a thick spore coating. New spores increase until the host cell membrane ruptures, releasing the progeny.

Microsporidia infections can result in a wide variety of symptoms. These include hepatitis, pneumonia, skin lesions, diarrhea, weight loss, and wasting syndrome. Diagnosis is based on clinical manifestations and identification of microsporidia in fecal smears stained by the "Quick-Hot Gram-Chromotrope technique," also known as the Chromotrope 2R method. Where possible by electron microscopy, identification can be made based on the characteristic polar tubule coiled within a spore. Molecular identification is also possible using PCR and primers to the small subunit ribosomal RNA (*see figure 19.3*). Treatment of microsporidiosis is not well defined. However, some successes have been reported with the use of albendazole (which inhibits tubulin and ATP synthesis in helminths), itraconazole, metronidazole, thalidomide (a toxic, immunosuppressant drug), and fumagillin (an amebicide).

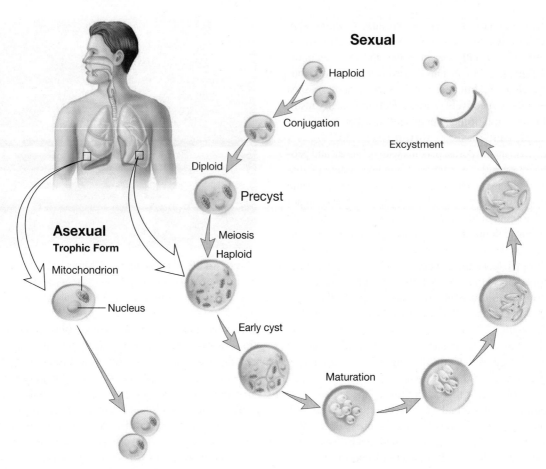

Figure 40.26 Life Cycle of *Pneumocystis jiroveci.* This fungus exhibits both sexual and asexual reproductive stages. Sexual reproduction occurs when haploid cells undergo conjugation followed by meiosis. Asexual reproduction occurs by binary fission. Both forms of reproduction can occur within the human host.

Pneumocystis Pneumonia

Pneumocystis jiroveci is a fungus found in the lungs of a wide variety of mammals. Although it was previously classified as a protist, its rRNA, DNA sequences (from several genes), and biochemical analyses have shown that it is more closely related to fungi than to protists. The life cycle of *P. jiroveci* is presented in **figure 40.26**, although some aspects of its development are not well known. The disease that this fungus causes has been called *Pneumocystis* pneumonia or *Pneumocystis carinii* pneumonia (PCP). Its name was changed to *Pneumocystis jiroveci* in honor of the Czech parasitologist, Otto Jirovec, who first described this pathogen in humans. In recent medical literature, the acronym PCP for the disease has been retained despite the loss of the old species epithet (*carinii*). PCP now stands for *Pneumocystis* pneumonia.

Serological data indicate that most humans are exposed to *P. jiroveci* by age three or four. However, PCP occurs almost exclusively in immunocompromised hosts. Extensive use of immunosuppressive drugs and irradiation for the treatment of cancers and following organ transplants accounts for the formidable prevalence rates for PCP. This pneumonia also occurs in premature, malnourished infants and in more than 80%

of HIV-positive patients who have progressed to AIDS. Both the organism and the disease remain localized in the lungs—even in fatal cases. Within the lungs, *P. jiroveci* causes the alveoli to fill with a frothy exudate.

Laboratory diagnosis of *Pneumocystis* pneumonia can be made definitively only by microscopically demonstrating the presence of the microorganisms in infected lung material or by a PCR analysis. Treatment is by means of oxygen therapy and either a combination of trimethoprim and sulfamethoxazole, atovaquone, or trimetrexate. Prevention and control are through prophylaxis with drugs, such as fluconazole, in susceptible persons.

Retrieve, Infer, Apply

1. How do infections caused by *Entamoeba histolytica* occur?
2. What is the most common cause of epidemic waterborne diarrheal disease?
3. In what two ways does *Toxoplasma* affect human health?
4. Why are some mycotic diseases of humans called opportunistic mycoses?
5. What parts of the human body can be affected by *Candida* infections?
6. Describe the infection process used by microsporidia.
7. When is *Pneumocystis* pneumonia likely to occur in humans?

Key Concepts

40.1 Relatively Few Fungi and Protists Are Human Pathogens

- Human fungal diseases, or mycoses, can be divided into five groups according to the level and mode of entry into the host (**figure 40.1**). These are the superficial, cutaneous, subcutaneous, systemic, and opportunistic mycoses (**table 40.1**).
- Protozoa are responsible for some of the most serious human diseases that affect hundreds of millions of people worldwide (**table 40.2**) and can also be grouped by route of disease transmission.

40.2 Fungi Can Be Transmitted by Airborne Routes

- Most systemic mycoses that occur in humans are acquired by inhaling the spores from the soil where the free-living fungi are found.
- Four types can occur in humans: blastomycosis (**figure 40.2**), coccidioidomycosis (**figure 40.3**), cryptococcosis, (**figure 40.4**), and histoplasmosis (**figure 40.5**).

40.3 Arthropods Can Transmit Protozoal Disease

- The most important human protozoan pathogen is *Plasmodium*, the causative agent of malaria (**figures 40.6–40.9**). Human malaria is caused by five species of *Plasmodium*: *P. falciparum, P. vivax, P. malariae, P. knowlesi,* and *P. ovale*.
- Flagellated protozoa are transmitted by arthropods and infect the blood and tissues of humans. Two major groups occur: the leishmanias, which cause the diseases collectively termed leishmaniasis (**figures 40.10** and **40.11**), and the trypanosomes (**figures 40.12** and **40.13**), which cause trypanosomiasis.

40.4 Direct Contact Diseases Can Be Caused by Fungi and Protists

- Superficial mycoses of the hair shaft are collectively called piedras. Two major types are black piedra and white piedra. Tinea versicolor is a third common superficial mycosis.
- The cutaneous fungi that parasitize the hair, nails, and outer layer of the skin are called dermatophytes, and their infections are termed dermatophytoses, ringworms, or tineas. At least seven types can occur in humans: tinea barbae (ringworm of the beard), tinea capitis (ringworm of the scalp; **figure 40.14**), tinea corporis (ringworm of the body; **figure 40.15a**), tinea pedis (ringworm of the feet; figure 40.15b), and tinea unguium (ringworm of the nails; figure 40.15c).

- The dermatophytes that cause the subcutaneous mycoses are normal saprophytic inhabitants of soil and decaying vegetation. Three types of subcutaneous mycoses can occur in humans: chromoblastomycosis (**figure 40.16a**), maduromycosis, and sporotrichosis (**figure 40.17**).
- Trichomoniasis is a sexually transmitted disease caused by the protozoan flagellate *Trichomonas vaginalis*. *T. vaginalis* causes bleeding and pustules (**figure 40.18**).

40.5 Food and Water Are Vehicles of Protozoal Diseases

- *Entamoeba histolytica* is the amoeboid protozoan responsible for amebiasis. This is a very common disease in warm climates throughout the world. It is acquired when cysts are ingested with contaminated food or water (**figure 40.19**).
- Freshwater parasites such as *Naegleria* and *Acanthamoeba* can cause primary amebic meningoencephalitis.
- *Cryptosporidium parvum* is a common coccidial apicomplexan parasite that causes severe diarrheal disease. It is acquired from contaminated food or water.
- *Cyclospora cayetanensis* is shed in the feces of a current host and can only infect the gastrointestinal tract of a new host after it has developed at 22 to 32°C for several days or weeks. It can be acquired from contaminated water used to rinse fruit or vegetables (**figure 40.20**).
- *Giardia intestinalis* is a flagellated protozoan that causes the common intestinal disease giardiasis (**figure 40.21**). This disease is distributed throughout the world, and in the United States, it is the most common cause of waterborne diarrheal disease.
- Toxoplasmosis is a disease caused by the protozoan *Toxoplasma gondii*. It is one of the major causes of death in AIDS patients (**figure 40.22**).

40.6 Opportunistic Diseases Can Be Caused by Fungi and Protists

- An opportunistic organism is one that is generally harmless in its normal environment but can become pathogenic in a compromised host.
- The most important opportunistic mycoses affecting humans include systemic aspergillosis (**figure 40.23**), candidiasis (**figure 40.24**), and *Pneumocystis* pneumonia (**figure 40.26**).
- An emerging opportunistic fungal disease, especially of the immunocompromised, is caused by the group of unique microbes known as the microsporidia. They live as a spore form and have a unique organelle used for infecting new host cells (**figure 40.25**).

Compare, Hypothesize, Invent

1. What is one distinct feature of fungi that could be exploited for antibiotic therapy?

2. Why do you think most fungal diseases in humans are not contagious?

3. Trypanosomes are notorious for their ability to change their surface antigens frequently. Given the kinetics of a primary immune response (primary antibody production), how often would the surface antigen need to be changed to stay "ahead" of the antibody specificity?

4. The initial, nonsymptomatic liver stage (LS) of *Plasmodium* spp. is considered very promising for prophylactic drug and vaccine development (figure 40.9). However, technical obstacles in obtaining LS protists has hindered analysis of gene and protein expression. Tarun and her colleagues developed a rodent model whereby LS protists could be studied. They identified a set of proteins expressed only in the LS, including enzymes involved in metabolic pathways active only in LS, as well as exported proteins.

    Examine the *P. vivax* life cycle in figure 40.7. Why do you think the metabolic requirements of LS protists are so unique? Having identified these proteins, what would be the next step in identifying drugs that might be effective against LS protozoa? How might this information assist in vaccine development?

Read the original paper: Tarun, A. S., et al. 2008. A combined transcriptome and proteome survey of malaria parasite liver stages. *Proc. Natl. Acad. Sci. USA* 105:305.

5. Infection with the fungal pathogen *Cryptococcus gattii* can be devastating to immunocompromised individuals, yet not much is known about its natural habitat. To explore its association with plants, model systems were developed between *C. gattii* and the mustard plant *Arabidopsis* and *Eucalyptus*. It was discovered that *C. gattii* grows well on these plants and is a plant pathogen only when plants are infected with strains of opposite mating types. In addition, two compounds produced by the plants, myoinositol and indole acetic acid, were found to promote fungal mating.

    Why is it important to understand the fungal niche in nature? List two questions these results raise concerning the transmission of *Cryptococcus* to human hosts. How would you go about addressing these questions?

Read the original paper: Xue, C., et al. 2007. The human fungal pathogen *Cryptococcus* can complete its sexual life cycle during a pathogenic association with plants. *Cell Host Microbe* 1:263.

Learn More

shop.mheducation.com   Enhance your study of this chapter with interactive study tools and practice tests. Also ask your instructor about the resources available through Connect, including adaptive learning tools and animations.

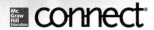

41

Microbiology of Food

Beer brewing started in prehistoric times and has been continuously refined. Now the introduction of genetically modified grain has some brewers and consumers wondering if this is progress.

The Art, Science, and Genetics of Brewing Beer

Beer has a long history. Prehistoric nomads probably learned to make beer from grains and water before they figured out how to bake bread. Beer was on Noah's list of provisions. Six-thousand-year-old Babylonian tablets describe beer recipes. Ancient Chinese, Egyptians, and Romans brewed beer. It is even rumored that the Pilgrims made landfall in Plymouth in 1620 because their beer supply was running low. Apart from water, beer is now considered the most frequently consumed liquid. With a global population of about 7 billion, that's a lot of beer.

There are three essential steps to making beer. First, barley is soaked in warm water to trigger germination. Once germinated, the grain produces the enzymes gluconase, which breaks down plant cell walls, and amylase to degrade the plant starch to a sugar fermentable by yeast. Next the sprouted grain is heated and dried, denaturing the enzymes. This mixture is called malt, and it is flushed with warm water so yeast can begin fermenting the starch to ethanol and CO_2 (section 41.5).

Predictably, many modifications have been made to the art and science of beer brewing. Since the mid-1990s, this has included the development of genetically modified (GM) ingredients. Most breweries find it more cost effective to add gluconases and amylases collected from bacteria that have been genetically engineered to produce these enzymes. Likewise, GM yeasts can be used to increase the brewing efficiency. Herbicide-resistant rice and corn have been used for a number of years by many larger breweries. However, the genetic modification of barley has lagged behind that of other grains.

Australia has become an emerging leader in the production of GM grain, where drought and heat have fostered the development of stress-tolerant crops. Trials are underway to test GM barley with resistance to two species of pathogenic fungi (*Rhizoctonia solani* and *R. oryzae*), as well as GM barley that uses nitrogen more efficiently, and barley that produces heat-stable gluconase.

What is the likely fate of GM barley? Public fear of GM foods is thought to be diminishing as it is realized that for over a decade, 70 to 80% of foods in the United States contain at least one GM ingredient without resulting in public health problems. Other countries are gradually increasing their growth and consumption of GM foods, particularly soy, corn, and cotton. Thus it is likely that GM barley will quietly find its way into much of the world's beer (including yours).

As the history of beer illustrates, humankind long ago figured out how to harness microorganisms to transform raw foods into gastronomic delights, including chocolate, cheeses, pickles, sausages, soy sauce, and wines. It was not until the late nineteenth century, however, that the role microorganisms play in food spoilage was appreciated. This illustrates the two opposing roles of microorganisms in food production and preservation. This chapter begins with an overview of microbial spoilage of food; the second half discusses the importance of other microbes in food and beverage production.

Readiness Check:

Based on what you have learned previously, you should be able to:

✔ List environmental factors that influence microbial growth (section 7.4)

✔ Compare and contrast the growth and physiology of obligate aerobes, facultative anaerobes, and strict anaerobes (section 7.4)

✔ Review physical and chemical agents commonly used to control microbial growth (sections 8.4 and 8.5)

✔ Describe the capacity of a bacterial spore to survive in nature, and list example bacteria that form endospores (sections 3.9, 23.1, and 23.3)

✔ Summarize the possible fates of pyruvate in fermentation (section 11.8)

✔ Explain the principles of a toxin-mediated disease (section 35.2)

✔ Predict methods used to test for food-borne microorganisms (sections 29.2, 36.3, and 36.4)

✔ Discuss the typical bacterial, viral, fungal, and protist diseases transmitted by food (sections 38.4, 39.4, and 40.5)

41.1 Microbial Growth Can Cause Food Spoilage

After reading this section, you should be able to:

- Differentiate between intrinsic and extrinsic factors that influence food spoilage
- Explain how the composition of a food item helps determine the kind of microbial growth that might occur
- List two physical and two biological features of a food that influence spoilage

Just as foods nourish the people who consume them, so too can they support the growth of microorganisms. Microbial growth is controlled by factors related to the food itself, called intrinsic factors; by extrinsic factors, which include the environment in which the food is stored; and by any preservatives that may have been added. Here we review the most important intrinsic and extrinsic factors involved in food spoilage. 🌀 *Food Spoilage*

Intrinsic Factors

Food composition is a critical intrinsic factor that influences microbial growth (**table 41.1**). If a food consists primarily of carbohydrates, microbial spoilage will most likely occur at some point in time. Foods with high water activity (p. 931) including some fruits and vegetables first show spoilage by fungi. For instance, even the slightest bruising of the tomato skin, exposing the interior, will result in rapid fungal growth. This affects the quality of tomato products, including tomato juices and ketchup. Molds also produce enzymes that weaken and penetrate the protective outer skin of fruits and vegetables. Once molds degrade carbohydrates within the skin, bacteria can colonize the food. Colonizing bacteria include those that cause soft rots, such as *Erwinia carotovora,* which produce hydrolytic enzymes such as pectinase. ◄◄ *Solutes affect osmosis and water activity (section 7.4)*

Molds can rapidly grow on grains and corn when these products are stored in moist conditions. The moldy bread pictured in **figure 41.1a** shows extensive fungal hyphal development and sporulation. Contamination of grains by the ascomycete *Claviceps purpurea* causes ergotism, a toxic condition (*see figure 26.12*). Hallucinogenic alkaloids produced by this fungus can lead to altered behavior, abortion, and death if infected grains are eaten.

In contrast to carbohydrates, when foods contain large amounts of proteins or fats (e.g., meat and dairy products), bacterial growth can produce a variety of foul odors. The production of

Table 41.1	Differences in Spoilage Processes in Relation to Food Characteristics		
Substrate	**Food Example**	**Chemical Reactions or Processes**[1]	**Typical Products (and Effects)**
Pectin	Fruits	Pectinolysis	Methanol, uronic acids (loss of fruit structure, soft rots)
Proteins	Meat	Proteolysis, deamination	Amino acids, peptides, amines, H_2S, ammonia, indole (bitterness, souring, bad odor, sliminess)
Carbohydrates	Starchy foods	Hydrolysis, fermentations	Organic acids, CO_2, mixed alcohols (souring, acidification)
Lipids	Butter	Hydrolysis, fatty acid degradation	Glycerol and mixed fatty acids (rancidity, bitterness)

1 Other reactions also occur during the spoilage of these substrates.

(a)

(b)

Figure 41.1 Food Spoilage. Typical examples of fungal spoilage include (a) bread, most likely supporting the growth of *Penicllium* sp. (green hyphae) and other fungi, and (b) ear rot in corn, which can result in major economic losses.

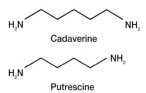

Figure 41.2 Cadaverine and Putrescine. These foul-smelling amines are the products of protein degradation.

short-chained fatty acids from fats renders butter rancid. The anaerobic breakdown of proteins is called **putrefaction.** It yields foul-smelling amine compounds such as cadaverine (imagine the origin of that name) and putrescine (**figure 41.2**). Unpasteurized milk, which contains fats, protein, and carbohydrates, provides a good example of microbial succession during spoilage: acid production by *Lactococcus lactis* subspecies *lactis* is followed by additional acid production associated with the growth of more acid-tolerant organisms such as *Lactobacillus.* Next, yeasts and molds become dominant and degrade the accumulated lactic acid, and the acidity gradually decreases. Eventually protein-digesting bacteria become active, resulting in a putrid odor and bitter flavor. The milk, originally opaque, eventually becomes clear because the proteins and fats have become coagulated (**figure 41.3**).

The pH and oxidation-reduction (redox) potential of a food also are critical intrinsic factors. A low pH favors the growth of yeasts and molds. In neutral or alkaline pH foods, such as meats, bacteria are more dominant in spoilage and putrefaction. Furthermore, when meat products, especially broths, are cooked, they present a reducing environment for microbial growth. These products, with their readily available amino acids, peptides, and growth factors, are ideal media for the growth of anaerobes, including clostridia.

Figure 41.3 Milk Spoilage. Fresh (left) and curdled (right) milk are shown. The spoilage process produces acidic conditions that denature and precipitate the milk casein to yield separated curds and whey.

The physical structure of a food also can affect the course and extent of spoilage. Grinding and mixing of foods such as sausage and hamburger increase the food surface area and distribute contaminating microorganisms throughout the food. This can result in rapid spoilage if such foods are stored improperly.

Many foods contain natural antimicrobial substances, including complex chemical inhibitors and enzymes. Coumarins found in fruits and vegetables have antimicrobial activity. Cow's milk and eggs also contain antimicrobial substances. Eggs are rich in the enzyme lysozyme, which lyses the cell walls of contaminating Gram-positive bacteria (*see figure 33.4*). Herbs and spices, which were once the mainstay of food preservation, often possess significant antimicrobial substances; generally fungi are more sensitive than bacteria to these compounds. Sage and rosemary are two of the most antimicrobial spices. Aldehydic and phenolic compounds that inhibit microbial growth are found in cinnamon, mustard, and oregano. Other important inhibitors are garlic, which contains allicin; cloves, which have eugenol; and basil, which contains rosmarinic acid. Nonetheless, spices can sometimes contain pathogenic and spoilage organisms. Enteric bacteria, *Bacillus cereus, Clostridium perfringens,* and *Salmonella enterica* have been detected in spices. Microorganisms can be eliminated or reduced by ethylene oxide sterilization. This treatment can result in *Salmonella*-free spices and herbs and a 90% reduction in the levels of general spoilage organisms. ◄◄ *Microorganisms are controlled with chemical control agents (section 8.5)*

Extrinsic Factors

Extrinsic factors are more easily controlled than intrinsic factors and are therefore frequently the basis for food preservation. Temperature is an important extrinsic factor in determining whether a food will spoil. For instance, spoilage problems can occur with minimally processed, concentrated frozen citrus products. These were once prepared with little or no heat treatment, and Gram-positive bacteria and yeasts in the genera *Saccharomyces* and *Candida* could grow upon thawing and dilution with water. Today most frozen concentrates are pasteurized (p. 931), filtered, and then concentrated by heating under vacuum. Vitamin C and other flavors are added back to the product before packaging. Ready-to-serve juices are routinely pasteurized, even though this results in some flavor loss.

Humidity is also a key factor in controlling food spoilage. At higher relative humidities, microbial growth is initiated more rapidly, even at lower temperatures (especially when refrigerators are not maintained in a defrosted state). When drier foods are placed in moist environments, moisture absorption can occur on the food surface, promoting microbial growth.

The atmosphere in which food is stored also is important. This is especially true with shrink-wrapped foods because many plastic films allow oxygen diffusion, which results in increased growth of surface-associated microorganisms. Excess CO_2 can decrease the solution pH, inhibiting microbial growth. Storing meat in a high CO_2 atmosphere inhibits Gram-negative bacteria, resulting in a population dominated by the lactobacilli.

41.2 Various Methods Are Used to Control Food Spoilage

After reading this section, you should be able to:

- Review the limitations of refrigeration
- Explain why canning is not always a fail-safe means of preserving food
- Compare and contrast two pasteurization methods
- Describe the principle of water availability and how it can be used to preserve foods
- Explain how commonly used chemicals preserve food
- Defend an informed opinion on the safety and efficacy of food irradiation
- Describe two microbe-based food preservatives
- Examine the packaging of food you frequently consume and assess its capacity to prevent spoilage

Foods can be preserved by a variety of methods. The goal of each method is to eliminate or reduce the populations of spoilage and disease-causing microorganisms while maintaining food quality. We now briefly discuss some of these techniques.

Filtration

Microorganisms can be removed from water, wine, beer, juices, soft drinks, and other liquids by filtration. This can keep bacterial populations low or eliminate them entirely. Removal of large particulates by prefiltration and centrifugation maximizes filter life and effectiveness. Several major brands of beer are filtered rather than pasteurized to better preserve flavor and aroma.

Low Temperature

Refrigeration at 5°C retards microbial growth, although with extended storage, microorganisms eventually grow and produce spoilage. Of particular concern is the growth of *Listeria monocytogenes,* which can grow at temperatures used for refrigeration. It should be kept in mind that refrigeration slows the metabolic activity of most microbes, but it does not lead to significant decreases in existing microbial populations.

High Temperature

Controlling microbial populations in foods by means of high temperatures can significantly limit disease transmission and spoilage. Heating processes, first used by Nicholas Appert in 1809, provide a safe means of preserving foods, particularly when carried out in commercial canning operations (**figure 41.4**). Canned food is heated in special containers called retorts at about 115°C for intervals ranging from 25 to over 100 minutes. The precise time and temperature depend on the nature of the food. Sometimes canning does not kill all microorganisms but only those that will spoil the food, because growth of remaining bacteria is inhibited by an intrinsic property of the food, such as pH. After heat treatment, cans are cooled as rapidly as possible, usually with cold water. ◄◄ *Physical control methods alter microorganisms to make them nonviable (section 8.4)*

Despite efforts to eliminate spoilage microorganisms during canning, canned foods may become spoiled. This may be due to spoilage before canning, underprocessing during canning, and leakage of contaminated water through seams during cooling. Spoiled canned food is often recognized by changes in color, texture, odor, and taste (although sampling is not advisable). Organic acids, sulfides, and gases (particularly CO_2 and H_2S) may be produced. If spoilage microorganisms produce gas, both ends of the can will bulge outward. Sometimes the swollen ends can be moved by thumb pressure (soft swells); in other cases, the gas pressure is so great that the ends cannot be dented by hand (hard swells). However, swelling is not always due to microbial

Figure 41.4 Food Preparation for Canning. Microbial control is important in processing and preserving many foods. Here a worker is pouring peas into a large, clean vat during the preparation of vegetable soup. After preparation, the soup is transferred to cans. Each can is heated for a short period, sealed, processed at temperatures from 110 to 121°C in a canning retort to destroy spoilage microorganisms, and finally cooled.

MICRO INQUIRY *Canned products such as vegetable soup often contain salt. Other than serving as a flavoring, what is the purpose of salt in the canned product?*

Table 41.2	Methods of Pasteurization	
Type of Pasteurization	**Temperature and Duration of Heating**	**Products Pasteurized**
Low-temperature holding (LTH)	68.2°C, 30 minutes	Beer, fruit juice, smaller volumes of milk
High-temperature short-time (HTST)	72°C, 15 seconds	Industry standard for milk and other dairy products in the United States
Ultra-high temperature (UHT)	138°C, 3 seconds	Milk that does not require refrigeration until opened

spoilage; acid in low pH foods may react with iron in the can to release hydrogen and generate a hydrogen swell.

Pasteurization involves heating food to a temperature that kills disease-causing microorganisms and substantially reduces the levels of spoilage organisms. There are three types of pasteurization, which differ by temperature and length of time the product is heated (**table 41.2**). Shorter-term processing results in improved flavor and extended product shelf life. These require stainless steel heat exchangers specifically designed to rapidly heat and cool the liquid being pasteurized. The duration of pasteurization is based on the statistical probability that the number of remaining viable microorganisms will be below a certain level after a particular heating time at a specific temperature. These calculations are discussed in detail in section 8.4. ↺ *Food Pathogens and Temperature*

Water Availability

The presence and availability of water also affect the ability of microorganisms to colonize foods. Simply by drying a food, one can slow or prevent spoilage processes; this is commonly seen in dried fruit, which does not require refrigeration. Water can be made less available by adding solutes such as sugar and salt. Water availability is measured in terms of **water activity** (a_w). This represents the ratio of relative humidity of the air over a test solution compared with that of distilled water, which has an a_w of 1. When large quantities of salt or sugar are added to food, most microorganisms are dehydrated by the hypertonic conditions and cannot grow. Even under these adverse conditions, osmophilic and xerophilic microorganisms may spoil food. In contrast to most spoilage bacteria that thrive at an a_w of about 0.9, these microbes grow best at an a_w of 0.6 to 0.7. Osmophilic (Greek *osmus*, impulse, and *philein*, to love) microbes grow best in or on media with a high osmotic concentration (e.g., jams and jellies), whereas xerophilic (Greek *xerosis*, dry, and *philein*, to love) microorganisms thrive in low a_w environments (e.g., cereals) and may not grow under high a_w conditions. Dehydration, such as lyophilization to produce freeze-dried foods, is a common means of preventing microbial growth. Freeze-drying increases the osmolarity of the product while at the same time reducing the a_w so microbial growth is retarded.

Chemical-Based Preservation

Various chemical agents can be used to preserve foods; these substances are closely regulated by the U.S. Food and Drug Administration (FDA) and are listed as being "*generally recognized as safe*" or **GRAS** (**table 41.3**). They include simple organic acids, sulfite, ethylene oxide as a gas sterilant, sodium nitrite, and ethyl formate. These chemical agents may damage the microbial plasma membrane or denature various cell proteins. Other compounds interfere with the functioning of nucleic acids, thus inhibiting cell reproduction.

Sodium nitrite is an important chemical used to help preserve ham, sausage, bacon, and other cured meats by inhibiting the growth of *Clostridium botulinum* and the germination of its spores. This protects against botulism and reduces the rate of spoilage. Besides increasing meat safety, nitrite decomposes to nitric acid, which reacts with heme pigments to keep the meat red in color. However there is concern about nitrite because it can react with amines to form carcinogenic nitrosamines. ◀◀ *Botulism (section 39.4)*

Low pH can also be used to hinder microbial spoilage. For example, acetic and lactic acids inhibit growth of *Listeria monocytogenes* Organic acids (1–3%) can be used to treat meat carcasses, and poultry can be cleansed with 10% lactic acid/sodium lactate buffer (pH 3) prior to packaging. In addition, low pH can increase the activity of other chemical preservatives. Sodium propionate is most effective at lower pH values, where it is primarily undissociated. Breads, with their low pH values, often contain sodium propionate as a preservative. The use of citric acid is also common. ▶▶ *Organic acids (section 42.1)*

High Hydrostatic Pressure

In the last decade, the use of high hydrostatic pressure (HHP) has grown as an alternative way to preserve food. HHP is the application of pressures from 100 to 800 millipascals (mPa) without significant changes in temperature. Pressures typically range from 350 to 600 mPa for 5 to 15 minutes at temperatures from 5 to 35°C. HHP is most detrimental to cell membranes, thus eukaryotic microbes are more susceptible. Predictably, Gram-positive bacteria, particularly endospores, are harder to kill than Gram-negative bacteria. At present, there is no industry standard for the HHP conditions that are considered effective for various foods and their microbial contaminants.

Radiation

Although the public is wary of food radiation, it is an effective way to prevent food spoilage. Unfortunately, there is the common misperception that food becomes radioactive upon radiation.

Table 41.3	Major Groups of Chemicals Used in Food Preservation		
Preservatives	**Approximate Maximum Use**	**Organisms Affected**	**Foods**
Benzoic acid/benzoates	0.1%	Yeasts and molds	Margarine, pickle relishes, apple cider, soft drinks, tomato ketchup, salad dressings
Dehydroacetic acid	65 ppm	Insects	Pesticide on strawberries, squash
Ethylene/propylene oxides	700 ppm	Yeasts, molds, vermin	Fumigant for spices, nuts
Ethyl formate	15–200 ppm	Yeasts and molds	Dried fruits, nuts
Parabens[1]	0.1%	Yeasts and molds	Bakery products, soft drinks, pickles, salad dressings
Propionic acid/propionates	0.32%	Molds	Bread, cakes, some cheeses; inhibitor of ropy bread dough
Sodium diacetate	0.32%	Molds	Bread
Sodium nitrite	120 ppm	Clostridia	Cold cuts, hot dogs, sausages
Sorbic acid/sorbates	0.2%	Molds	Hard cheeses, figs, syrups, salad dressings, jellies, cakes
SO_2/sulfites	200–300 ppm	Insects and microorganisms	Molasses, dried fruits, wine, lemon juice (not used in meats or other foods recognized as sources of thiamine)

Adapted from James M. Jay. 2000. **Modern Food Microbiology,** *6th ed.*

1 *Methyl, propyl, and heptyl esters of p-hydroxybenzoic acid.*

This would be similar to thinking that you become radioactive following a chest or dental X ray. ◄◄ *Radiation (section 8.4)*

Gamma irradiation of food most often uses a cobalt-60 source; however, cesium-137 is employed in some facilities. Gamma radiation has excellent penetrating power but must be used with moist foods because radiation is effective only if it generates reactive oxygen species from water in the microbial cells, resulting in oxidation of nucleic acids, lipids, and proteins. This process of **radappertization,** named after Nicholas Appert, can extend the shelf life of seafoods, fruits, and vegetables. Unlike canning, radappertization does not heat the food product. To sterilize meat products, 4.5 to 5.6 megarads are commonly used.

Microbial Product–Based Inhibition

Interest is increasing in the use of bacteriocins for the preservation of foods. **Bacteriocins** are proteins secreted by certain bacteria that kill closely related microbes by binding to specific sites on the target cell. They often affect plasma membrane integrity and function. The only bacteriocin currently approved for food preservation is nisin, a small amphiphilic peptide produced by some strains of *Lactococcus lactis.* It is nontoxic to humans and affects Gram-positive bacteria by binding to lipid II during peptidoglycan synthesis and forming pores in the plasma membrane (*see figure 12.10*). Nisin was added to the GRAS list of food additives about 45 years ago. This means that it can be added directly to foods such as packaged meats, cheese, eggs, and canned vegetables. Nisin is effective in low-acid foods,

where it improves inactivation of *Clostridium botulinum* during the canning process and inhibits germination of any surviving spores. ◄◄ *Bacteriocins (section 33.3)*

Interestingly, a preparation of six strains of **bacteriophages** that specifically infect and kill *Listeria monocytogenes* is used to spray ready-to-eat meats such as hot dogs and lunch meats. The spray is applied to the surface of the meats prior to packaging. The phages are present in equal concentration; using multiple phage types significantly reduces the development of phage-resistant strains.

Packaging

Gases present in the atmosphere within a food storage package can have a significant affect on microbial growth. This has led to the development of **modified atmosphere packaging** (**MAP**). Modern shrink-wrap materials and vacuum technology make it possible to package foods with controlled atmospheres. These materials are largely impermeable to oxygen. This prolongs shelf life by a factor of two to five times compared to the same product packaged in air. With a CO_2 content of 60% or greater in the atmosphere surrounding a food, spoilage fungi will not grow, even if low levels of oxygen are present. High-oxygen MAP is also effective because it triggers the formation of superoxide anions within microbial cells. Some products currently packaged using MAP technology include delicatessen meats and cheeses, pizza, grated cheese, some bakery items, and dried products such as coffee.

41.3 Food-Borne Disease Outbreaks

After reading this section, you should be able to:

- Summarize a major food-borne infection that had substantial health and economic consequences
- Describe the origin and consequences of a major food intoxication event in the United States
- Explain why these cases of food-borne disease outbreaks serve as cautionary tales for government regulatory agencies, food producers, and consumers
- Contrast and compare aflatoxins and fumonisins

Most food-borne illnesses are neither recognized nor reported. The source and the identity of a food pathogen are only discovered if it causes an illness dire enough that the victim visits a physician or hospital. That said, the Centers for Disease Control and Prevention (CDC) estimates that are about 48 million cases of food-related illness annually in the United States, 128,000 hospitalizations, and about 3,000 deaths. Of these, only 14 million can be attributed to known pathogens. *Salmonella enterica* serovars causes the most reported illnesses, while noroviruses, *Campylobacter jejuni*, *Escherichia coli* O157:H7, and *Listeria monocytogenes* are other major causes of food-borne diseases. ◀◀ *Food and water are vehicles for viral diseases (section 38.4); Food and water are vehicles for bacterial diseases (section 39.4)*

All food-borne illnesses are transmitted by the fecal-oral route. While the "4-Fs"—feces, fingers, food, and flies—are critical for fecal-oral transmission, fomites, such as faucets, drinking cups, and cutting boards, also play a role in the maintenance of the fecal-oral route of contamination. Here we review several major outbreaks of food-borne diseases that illustrate the personal, economic, and societal impacts of tainted food. We divide our coverage into the two primary types of food-related diseases: food-borne infections and food intoxications. The individual food-borne diseases are covered in sections 38.4, 39.4, and 40.5.

Food-Borne Infection

A **food-borne infection** involves ingestion of the pathogen, followed by its growth in the host, including tissue invasion or the release of toxins. The most common microbes involved are summarized in **table 41.4** (*also see table 39.3*).

Listeriosis, caused by *Listeria monocytogenes*, was responsible for the largest meat recall in U.S. history. In 2002 a seven-state listeriosis outbreak was linked to deli meats and hot dogs produced at a single meat-processing plant in Pennsylvania. Pregnant women, the young and the old, and immunocompromised individuals are especially vulnerable to *L. monocytogenes* infections. In this outbreak, 7 deaths, 3 stillbirths, and 46 illnesses were caused by consumption of contaminated meats. Microbiologists matched the strain of *L. monocytogenes* found in the contaminated food products with samples obtained from floor drains in a Wampler, Inc., packaging plant. This prompted the recall of 27.4 million pounds of meats that had been distributed over a 5-month period to stores, restaurants, and school lunch programs. Following the outbreak, the plant closed for a month and the Wampler brand name was phased out. This episode prompted the U.S. Department of Agriculture (USDA) to step up its environmental testing program for *L. monocytogenes* so that it now tests plants that do not regularly submit data to the USDA. It also performs surprise inspections of those that do. The USDA advises people at risk of contracting listeriosis to avoid eating soft cheeses (e.g., feta, Brie, Camembert), refrigerated smoked meats such as lox, as well as deli meats and undercooked hot dogs. In 2006 the Wampler plant in Pennsylvania was closed, having never recovered from the $100 million cost and the damage to its reputation caused by the 2002 outbreak. Thus a single outbreak resulted in death, illness, bankruptcy, and changes to food safety inspection policy.

Food Intoxication

Microbial growth in food products also can result in **food intoxication** (table 41.4). Intoxication produces symptoms shortly after the food is consumed because growth of the disease-causing microorganism is not required. The classic example is picnickers vomiting 4 to 6 hours after consuming salad contaminated with *Staphylococcus aureus* enterotoxin. While in this case, the consumption of *S. aureus* cells is not necessary, toxins produced in the food can be associated with microbial cells or can be released by microbes growing in the host. Thus the division between food-borne infection and intoxication is not always clear-cut.

For example, the enterohemorrhagic microbe *E. coli* O157:H7 blurs the line between infection and intoxication as the microbe must first replicate in the host before sufficient toxin is produced to cause illness. *E. coli* O157:H7 produces a toxin known as Shiga-like toxin, which is responsible for a potentially life-threatening complication known as hemolytic uremic syndrome, most commonly seen in children and the elderly. The first U.S. outbreaks of *E. coli* O157:H7 (in the 1980s) occurred in individuals who had consumed undercooked hamburgers. This microbe

Table 41.4	Some Food-Borne Bacteria That Cause Acute Bacterial Vomiting and Diarrhea				
Organism	Incubation Period (Hours)	Vomiting	Diarrhea	Fever	Food Source
Staphylococcus aureus	1–6	+++[1]	+	Rare	Meats, cold cuts, and salads (e.g., tuna, egg, pasta), cream-filled bakery goods
Bacillus cereus	1–6 (intoxication); 8–16 (infection)	+++	++	–[2]	Reheated fried rice, sprouts, cucumber
Clostridium perfringens	8–16	±[3]	+++	–	Rewarmed meat dishes
Clostridium botulinum	18–24	±	Rare	–	Canned goods contaminated during processing or packaging, raw honey
Escherichia coli (enterohemorrhagic)	3–5 days	±	++	±	Undercooked hamburger, unpasteurized fruit juices, raw vegetables
Escherichia coli (enterotoxigenic strain)	16–72	±	++	–	Contaminated drinking water; major cause of traveler's diarrhea.
Vibrio parahaemolyticus	6–96	+	++	±	Shellfish, particularly clams and oysters
Vibrio cholerae	24–72	+	+++	–	Contaminated water; shellfish
Shigella spp.	24–72	±	++	+	Salads, sandwiches, produce, contaminated water
Salmonella spp. (gastroenteritis)	8–48	±	++	+	Poultry, eggs, meats, dairy, produce, peanut butter
Salmonella enterica serovar Typhi (typhoid fever)	10–14 days	±	±	++	Spread from a healthy carrier to food via fecal-oral transmission.
Campylobacter jejuni	2–10 days	–	+++	++	Poultry, produce, raw milk
Yersinia enterocolitica	3–14 days	±	+	+	Pork, milk

Adapted from Geo. F. Brooks, et al., Medical Microbiology, 21st ed. Copyright 1998 Appleton & Lange, Norwalk, CT.
1 + *indicates condition is present, number of symbols indicates severity.*
2 − *indicates condition is absent.*
3 ± *indicates condition sometimes occurs.*

inhabits the intestines of a variety of mammals, and ground beef contaminated with cattle feces during slaughter and processing continues to be a common cause of these outbreaks. ◄◄ Escherichia coli *gastroenteritis (section 39.4)*

More recently other types of foods have been found to be contaminated with *E. coli* O157:H7. One example is minimally processed, ready-to-eat fresh salad greens, which have become a popular convenience and represent a multibillion-dollar business in the United States and Europe. The produce is repeatedly washed, then dried and packed into plastic bags. When transported, the produce must be kept cool. Unfortunately, this healthy approach to consuming vegetables is not always without risk. In the past decade, over a dozen outbreaks of enterohemorrhagic *E. coli* O157:H7 were traced to minimally processed produce. The largest and most well publicized was a spinach-associated outbreak in 2006, which sickened more than 200 people. This outbreak prompted intense investigation, and it was determined that contamination occurred while the spinach was still in the field. Surveillance studies determined that the most likely source was feral swine that burrowed

under fences and traveled across the spinach fields. This finding raises more questions than it answers, including: how can thousands of agricultural fields be protected from wild animals; why didn't the postharvest washing procedure cleanse the spinach of the *E. coli*; and what factors were in play that allowed the initial bacterial inoculum on the plants to grow by at least a factor of 10? This single event prompted efforts by both government and industry to ensure that minimally processed produce remains a convenience, not a sporadic source of disease.

Three Gram-positive, endospore-forming rods are important sources of food intoxications and infections: *Clostridium botulinum, C. perfringens,* and *Bacillus cereus. C. botulinum* intoxication is discussed in chapter 39. There are typically 30 or fewer cases of botulism reported each year in the United States, and most are associated with home canning. *C. perfringens* typically causes a 24-hour infection. It has a relatively high infectious dose (10^8 cells) and results in abdominal cramps and diarrhea 8 to 22 hours after consumption. *B. cereus* causes two types of intoxication, each the result of a different toxin. Food poisoning following ingestion of *B. cereus*–contaminated meats, fish, or vegetables can cause a diarrheal illness that is very similar to that of *C. perfringens*. *B. cereus* intoxication, characterized by acute (less than 24 hours) vomiting, is generally associated with consumption of contaminated starchy foods such as rice or pastries.

Some fungus-derived toxins have a more insidious health effect. The most potent are the fungal carcinogens aflatoxins and fumonisins. **Aflatoxins** are produced by *Aspergillis flavus* and *A. parasiticus*. These fungi grow and release toxin in moist grains and nuts. Aflatoxin consumption is most common in developing countries where these products are stored in suboptimal conditions and are consumed with little or no post-harvest processing. Aflatoxins are planar, ringed compounds that wedge between strands of DNA (intercalate), thereby causing frameshift mutations that can cause cancer. This occurs primarily in the liver, where they are converted to unstable derivatives. Currently a total of 18 aflatoxins are known. The most important are shown in **figure 41.5**. Of these, aflatoxin B_1 is the most common and the most potent carcinogen. After ingestion by lactating animals (e.g., dairy cows), aflatoxins B_1 and B_2 are modified in the liver to yield the aflatoxins M_1 and M_2. If cattle consume aflatoxin-contaminated feeds, aflatoxins can contaminate milk and dairy products. Besides their importance in grains, aflatoxins have also been found in beer, cocoa, raisins, peanut butter, and soybean meal.

Diet appears to be related to aflatoxin exposure: the average aflatoxin intake in the typical European-style diet is 19 ng/day, whereas for some Asian diets it is estimated to be 103 ng/day. Aflatoxin sensitivity also can be influenced by prior disease exposure. Individuals who have had hepatitis B have a 30-fold higher risk of liver cancer upon exposure to aflatoxins than individuals who have never had this disease. It has been observed that prevention of hepatitis B infections by vaccination and reduction of carrier populations help control the potential effects of aflatoxins.

The **fumonisins** are fungal contaminants of corn that were first isolated in 1988. They are produced by *Fusarium moniliforme* and cause leukoencephalomalacia in horses (also called "blind staggers"—it is fatal within 2 to 3 days), pulmonary edema in pigs, and esophageal cancer in humans. Fumonisins inhibit ceramide synthase, a key enzyme for the proper use of fatty substances in the cell. This disrupts synthesis and metabolism of sphingolipids, important compounds that influence a wide variety of cell functions. At least 10 different fumonisins exist; the basic structures of fumonisins FB1 and FB2 are shown in **figure 41.6**. Corn and corn-based feeds and foods, including cornmeal and corn grits, can be contaminated. Thus it is extremely important to store corn and corn products under dry conditions, where these fungi cannot develop.

Other eukaryotic microorganisms also synthesize potent toxins. For example, algal toxins contaminate fish and thereby affect the health of marine animals higher in the food chain; they also contaminate shellfish and finfish, which are later consumed by humans. These toxins are produced during harmful algal blooms, as discussed in section 30.2.

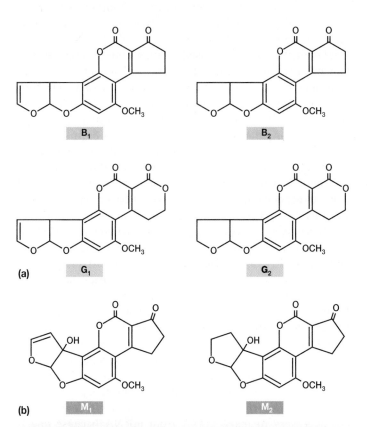

Figure 41.5 Aflatoxins. When *Aspergillus flavus* and related fungi grow on foods, carcinogenic aflatoxins can be formed. These have four basic structures. (a) The letter designations refer to the color of the compounds under ultraviolet light after extraction from the grain and separation by chromatography. The B_1 and B_2 compounds fluoresce with a blue color, and the G_1 and G_2 appear green. (b) The two type M aflatoxins are found in the milk of lactating animals that have ingested type B aflatoxins.

MICRO INQUIRY *What foods are most prone to aflatoxin contamination?*

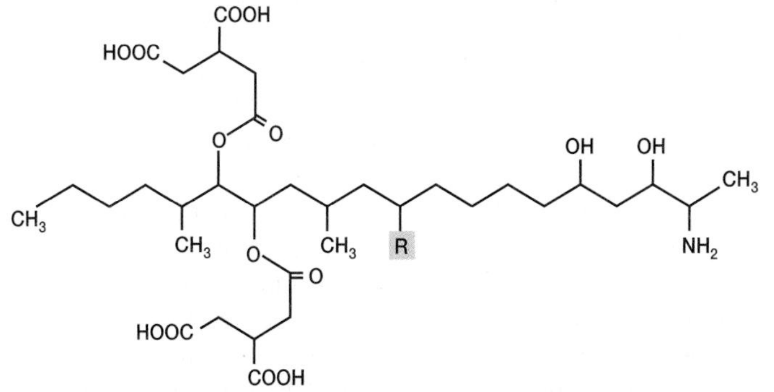

Figure 41.6 Fumonisin Structure. The basic structure of fumonisins FB1 and FB2 produced by *Fusarium moniliforme,* a fungal contaminant that can grow in improperly stored corn. A total of at least 10 different fumonisins have been isolated. These are strongly polar compounds that cause diseases in domestic animals and humans. FB1, R = OH; FB2, R = H.

41.4 Detection of Food-Borne Pathogens Requires Government-Industry Cooperation

After reading this section, you should be able to:

- List the principles that guide the development of food testing strategies
- Describe how a food-borne pathogen consumed by an individual might be traced back to the food source

In 1905 Upton Sinclair's novel *The Jungle* exposed the horrific state of the meat packaging industry. The following year, the first in a series of federal and state laws was enacted in the United States aimed at improving safety of the food supply. In 2001 the FDA Food Safety Modernization Act was passed. This was the first major piece of food safety legislation to be enacted since 1938. It has a number of key provisions; the goal is to change the focus of federal regulators from responding to food contamination to preventing it. Although most look to the government to keep food safe, food safety is a shared responsibility, as precautions must be defined and in place during production, transport, sale, and consumption of any food item. The length and complexity of the chain from producer to consumer has increased dramatically in the last 20 years with the rise in food imports and exports, making ensuring food safety even more complicated.

Here we provide a broad overview of food pathogen detection. Unfortunately it is impossible to test each food item for every potential pathogen. Apart from the logistical implications, such an approach is economically untenable. For example, if the U.S. dairy industry were to test all dairy products daily at each farm with a $5-per-sample assay, it would cost about $150 million annually. Checking each milk tanker truck as it enters a processing plant reduces the cost to about $21 million—a savings of $129 million each year. So the challenge in developing a testing strategy involves identifying which pathogens to detect, as well as when and where to detect them.

Several principles guide the development of the technologies, protocols, and policies used to keep foods safe. These include

(1) specificity and sensitivity for the given pathogen—any given test should not have a high level of false negatives or positives (specificity) and the pathogen in question should be detected even in low concentrations (sensitivity); (2) speed, which is particularly important for fresh produce awaiting distribution; and (3) simplicity: foods should be tested with limited sample preparation. Furthermore, the goal of each detection strategy should be clearly defined. Ideally "testing to prevent," that is, to confirm food is safe before it leaves the farm or processing plant, is the goal. When this is unreasonable due to logistics and cost, then the next level is "testing to protect," which involves analysis before the food is accessible for consumption. These strategies are designed to avoid "testing to recover," when an outbreak has occurred and the origin of the contaminated food must be identified.

The most common approach to detecting food-borne pathogens involves using selective media to culture the microbe followed by the use of biochemical and/or immunological assays for strain identification. Fluorescent antibody tagging and *enzyme-linked immunosorbant assays* (ELISAs; *see figure 36.12*) are frequently employed. A variety of molecular methods are also used. Unlike culture-based assays, molecular techniques can detect viruses that cannot be grown conveniently, microbes that are present in very small numbers, and slow-growing or nonculturable pathogens. In addition, results are usually obtained much faster. For instance, although the important pathogen *E. coli* O157:H7 can be isolated and identified using selective culture media, the application of multilocus sequence typing, serotype-specific probes, and polymerase chain reaction (PCR) enables the rapid detection of a few target cells in large populations of background microorganisms. Several molecular techniques have been specifically developed to ensure food safety; here we mention three. (1) PCR has been combined with restriction fragment length polymorphism (RFLP) analysis. Here, rather than visualizing genomic fragments that have been cleaved, PCR amplifies the variable regions to generate a cleaner genomic fingerprint in a relatively short period of time. (2) Pathogen-specific PCR primers. For instance, different species of *Campylobacter* can be detected by amplifying a region of the gene encoding gyrase B. (3) Microarray technology in which oligonucleotides representing housekeeping and virulence genes are used as probes to which DNA extracted directly from the food item can be hybridized. Note that unlike most microarray procedures, it is not necessary to make cDNA from mRNA because one is looking only for the presence or absence of a given pathogen's genome, not relative levels of gene expression. Microarrays are especially attractive because, unlike some other molecular techniques, there is no need to first enrich for the pathogen by culture-based techniques. ◀◀ *Polymerase chain reaction amplifies targeted DNA (section 17.2); Transcriptome analysis (section 18.5); Microbial taxonomy and phylogeny are largely based on molecular characterization (section 19.3)*

The CDC oversees a program called **PulseNet** for the early detection of food-borne pathogens, particularly during an

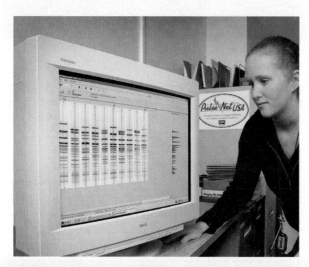

Figure 41.7 PulseNet. This network operates throughout the United States to identify microbes causing food-borne disease using pulsed-field gel electrophoresis of genomes extracted from contaminating microbes. Shown here, a food microbiologist examines the electrophoretic pattern, or fingerprint, generated by a bacterial pathogen.

MICRO INQUIRY *Why is it important that PulseNet coordinate the identities of food-borne pathogens on a national level?*

outbreak. The basis for this program is the use of pulsed-field gel electrophoresis (PFGE) under carefully controlled and duplicated conditions. PFGE generates a strain-specific DNA electrophoresis pattern, or fingerprint, for each bacterial pathogen (**figure 41.7**). PFGE is similar to standard DNA gel electrophoresis but is capable of separating very large fragments of DNA. This requires periodic switching of the voltage supply in three different directions: the central axis of the gel and two that run at a 120° angle on either side. The duration of each pulse is the same but can be increased over the duration of the run. Because PFGE can resolve large (Mb) pieces of DNA, whole genomes can be cut with restriction enzymes that cleave in only a few places to generate a DNA fingerprint. The patterns obtained for each strain of food-borne pathogen are found in a national database maintained by the CDC. This enables strain-specific identification of new isolates through comparison with the database. ◀◀ *Restriction enzymes (section 17.1); Gel electrophoresis (section 17.1)*

PulseNet also fosters real-time communication between community, state, and federal health agencies. With this uniform procedure, it is possible to link pathogens associated with disease outbreaks in different parts of the world to a specific food source. For example, a *Shigella* outbreak in three different areas of North America was traced to Mexican parsley that had been tainted with polluted irrigation water. This program has resulted in the rapid establishment of epidemiological linkages, thereby decreasing additional cases of food-borne disease.

The 2011 outbreak of a novel *E. coli* pathogen in Germany brought the detection and analysis of food-borne pathogens into the twenty-first century. In less than a month, the entire genomes of several *E. coli* O104:H4 strains recovered from patients were sequenced using next-generation sequencing techniques. This outbreak illustrates that the pressing need for food safety combined with the rapid pace of innovation ensures the introduction of newer, faster, and highly accurate techniques will continue. ◀◀ *Next-generation DNA sequencing (section 18.1);* Escherichia coli *gastroenteritis (section 39.4)*

Retrieve, Infer, Apply

1. How are most food-borne pathogens detected?
2. Why is it advantageous to omit the pre-enrichment step in microarray and PCR-based approaches?
3. How is PulseNet used in the surveillance of food-borne diseases?
4. Compare how you might implement the testing of a meat product (e.g., hot dogs) with that of a vegetable (e.g., spinach). Consider the pathogen(s) to be detected, in addition to when and where testing should occur.

41.5 Microbiology of Fermented Foods: Beer, Cheese, and Much More

After reading this section, you should be able to:

- Compare and contrast mesophilic and thermophilic fermentations
- Differentiate between a yeast-lactic and a mold-lactic fermentation
- Outline the process by which cheese is made
- List some fermented meats
- Provide an overview of the production of wine, champagne, beer, and distilled spirits.
- Describe a fermented food item not commonly eaten in the United States

Fermentation has been used to process food for thousands of years; the earliest records date as far back as 6,000 BCE. Microbial growth, either of natural or inoculated populations, causes chemical or textural changes to form a product that can be stored for extended periods. The fermentation process also is used to create pleasing food flavors and odors—such as cheese, yogurt, and chocolate (**Techniques & Applications 41.1**).

The major fermentations used in food microbiology (which are named by their principle end product) are lactic, propionic, and alcoholic fermentations. These fermentations are carried out by a wide range of microbes, many of which have not been characterized. ◀◀ *Fermentation does not involve an electron transport chain (section 11.8)*

Fermented Milks

Although we are most familiar with yogurt, at least 400 different fermented milks are produced around the world. The majority of fermented milk products rely on **lactic acid bacteria (LAB)**,

TECHNIQUES & APPLICATIONS

41.1 Chocolate: The Sweet Side of Fermentation

Chocolate could be characterized as the "world's favorite food," but few people realize that fermentation is an essential part of chocolate production. The Aztecs were the first to develop chocolate fermentation, serving a chocolate drink made from the seeds of the chocolate tree, *Theobroma cacao* (Greek *theos,* god, and *broma,* food, or "food of the gods"). Chocolate trees now grow in Africa as well as South America.

The process of chocolate fermentation has changed very little over the past 500 years. Each tree produces large pods that each hold 30 to 40 seeds in a sticky pulp (**box figure**). Ripe pods are harvested and slashed open to release the pulp and seeds. The sooner the fermentation begins, the better the product, so fermentation occurs on the farm where the trees are grown. The seeds and pulp are placed in "sweat boxes" or in heaps in the ground and covered, usually with banana leaves.

Like most fermentations, this process involves a succession of microbes. First, a community of yeasts, including

(a)

(b)

Cocoa Fermentation. (a) Cocoa pods growing on the cocoa tree. Each pod is 13 to 15 cm in length and contains 30 to 40 seeds in a sticky white pulp. (b) Seeds and pulp are fermented in boxes commonly covered with banana leaves for 5 to 7 days and then dried in the sun. Chocolate cannot be produced without fermentation.

Candida rugosa and *Kluyveromyces marxianus,* hydrolyzes the pectin that covers the seeds and ferments sugars to release ethyl alcohol and CO_2. As the temperature and the alcohol concentration increase, the yeasts are inhibited and lactic acid bacteria increase in number. Lactic acid production drives the pH down, encouraging the growth of bacteria that produce acetic acid as a fermentation end product. Acetic acid is critical to the production of fine chocolate because it kills the sprout inside the seed and releases enzymes that cause further degradation of proteins and carbohydrates, contributing to the overall taste of the chocolate. In addition, acetate esters, derived from acetic acid, are important for the development of good flavor. Fermentation takes 5 to 7 days. An experienced cocoa grower knows when fermentation is complete: if it is stopped too soon, the chocolate will be bitter and astringent; if it lasts too long, microbes start growing on the seeds instead of in the pulp. "Off tastes" arise when Gram-positive bacteria in the genus *Bacillus* and filamentous fungi belonging to the genera *Aspergillus, Penicillium,* and *Mucor* hydrolyze lipids in the seeds to release short-chain fatty acids. As the pH begins to rise, bacteria of the genera *Pseudomonas, Enterobacter,* and *Escherichia* also contribute to bad tastes and odor.

After fermentation, the seeds, now called beans, are spread out to dry. The dried beans are brown and lack pulp. They are bagged and sold to chocolate manufacturers, who first roast the beans to further reduce the bitter taste and kill most of the microbes. The beans are then ground and the nibs—the inner part of each bean—are removed. The nibs are crushed into a thick paste called a chocolate liquor, which contains cocoa solids and cocoa butter but no alcohol. Cocoa solids are brown and have a rich flavor, and cocoa butter has a high fat content and is off-white in color. The two components are separated, and the cocoa solids can be sold as cocoa, while the cocoa butter is used to make white chocolate or sold to cosmetics companies for use in lipsticks and lotions. However, the bulk of these two components will be used to make chocolate. The cocoa solids and butter are reunited in controlled ratios, and sugar, vanilla, and other flavors are added. The better the fermentation, the less sugar needs to be added (and the more expensive the chocolate).

The final product, delicious chocolate, is a combination of over 300 different chemical compounds. This mixture is so complex that no one has yet been able to make synthetic chocolate that can compete with the natural fermented plant. Microbiologists and food scientists are studying the fermentation process to determine the role of each microbe. But like chemists, they have had little luck in replicating the complex, imprecise fermentation that occurs on cocoa farms. In fact, the finest, most expensive chocolate starts as cocoa on farms where the details of fermentation have been handed down through generations. Chocolate production is truly an art as well as a science, while eating it is simply divine.

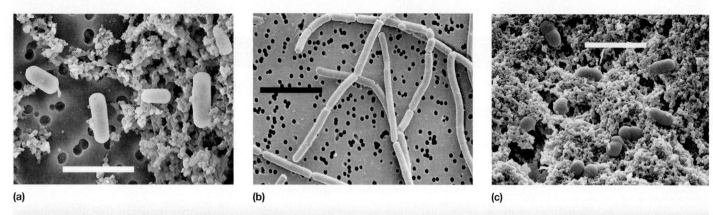

Figure 41.8 Lactic Acid Bacteria (LAB). Colorized scanning electron micrographs of LAB used as starter cultures. (a) *Lactobacillus helveticus*. (b) *Lactobacillus delbrueckii* subspecies *bulgaricus*. (c) *Lactococcus lactis*. The bacteria are supported by filters, seen as holes in the background. Scale bar = 5 µm.

MICRO INQUIRY *Can you name at least two features that make these bacteria well suited for milk fermentations?*

which include species belonging to the genera *Lactobacillus, Lactococcus, Leuconostoc,* and *Streptococcus* (**figure 41.8**). These firmicutes tolerate acidic conditions, are nonsporing, and are aerotolerant with a strictly fermentative metabolism. ◄◄ *Order Lactobacillales (section 23.3)*

Regardless of the type of milk product, **starter cultures** of LAB are added to begin fermentation. For example, *Lactococcus lactis* and *Streptococcus cremoris* are used in cheese production and various *Leuconostoc* species are used to start buttermilk fermentation. A vexing problem for the dairy industry is the presence of bacteriophage that destroy starter cultures. Lactic acid production from a heavily phage-infected starter culture can cease within 30 minutes. Overcoming this problem has revealed that many reliable starter cultures consist of bacteria that prevent phage infection by a variety of mechanisms, such as preventing adsorption or phage DNA entry, the activity of restriction modification, or the use of a CRISPR-Cas system. Nonetheless, various phage control measures are available, and still more are in development. ◄◄ *Restriction enzymes cut DNA at specific sequences (section 17.1); The CRISPR/Cas System (section 27.2)*

Mesophilic Fermentations

Mesophilic milk fermentation results in the production of buttermilk and sour cream. This type of fermentation relies on the acidic products of fermentation to denature milk protein. To carry out the process, milk is typically inoculated with the desired starter culture and is then incubated at an optimum growth temperature (approximately 20 to 30°C). Microbial growth is stopped by cooling, and *Lactobacillus* spp. and *Lactococcus lactis* cultures are used for aroma and acid production. *Lactococcus lactis* subspecies *diacetylactis* converts CO_2 and milk citrate to diacetyl, which gives a richer, buttery flavor to the finished product.

Thermophilic Fermentations

Despite the name, thermophilic fermentations are not carried out at extremely hot temperatures, rather they are conducted around 45°C. An important example is yogurt production.

In commercial production, milk is pasteurized, cooled to 43°C or lower, and inoculated with a 1:1 ratio of *Streptococcus salivarius* subspecies *thermophilus* (*S. thermophilus*) and *Lactobacillus delbrueckii* subspecies *bulgaricus* (*L. bulgaricus*). *S. thermophilus* grows more rapidly at first and renders the milk anoxic and weakly acidic. *L. bulgaricus* then acidifies the milk even more. Acting together, the two species ferment almost all of the lactose to lactic acid and flavor the yogurt with diacetyl (*S. thermophilus*) and acetaldehyde (*L. bulgaricus*). Many yogurts now contain probiotic bacterial strains (p. 944), but these do not necessarily contribute to the fermentation process. Fruits or fruit flavors are pasteurized separately and then combined with the yogurt. Freshly prepared yogurt contains about 10^9 bacteria per gram.

Yeast-Lactic Fermentation

Yeast-lactic fermentations include kefir, a product with an ethanol concentration of up to 2%. This fermented milk originated in the Caucasus Mountains and is produced east into Mongolia. Kefir products tend to be foamy due to active carbon dioxide production. This process is based on the use of kefir "grains" as a starter culture. These are coagulated lumps of casein (milk protein) that contain yeasts, lactic acid bacteria, and acetic acid bacteria, which are recovered at the end of the fermentation. Originally kefir was produced in leather sacks hung by the front door during the day, and passers-by were expected to push and knead the sack to mix and stimulate the fermentation. Fresh milk could be added occasionally to replenish nutrients and maintain activity.

Mold-Lactic Fermentation

Mold-lactic fermentation results in a unique Finnish fermented milk called *viili*. Milk is inoculated with a mixture of the fungus *Geotrichum candidum* and lactic acid bacteria. The cream rises to the surface, and after incubation at 18 to 20°C for 24 hours, lactic acid reaches a concentration of 0.9%. The fungus forms a velvety layer across the top of the final product, which also can be made with a bottom fruit layer.

Table 41.5	Major Types of Cheese and Microorganisms Used in Their Production	
	CONTRIBUTING MICROORGANISMS[1]	
Cheese (Country of Origin)	**Earlier Stages of Production**	**Later Stages of Production**
Soft, unripened Cottage	*Lactococcus lactis*	*Leuconostoc cremoris*
Cream	*Lactococcus cremoris, L. diacetylactis, Streptococcus thermophilus, L. delbrueckii* subspecies *bulgaricus*	
Mozzarella (Italy)	*S. thermophilus, L. bulgaricus*	
Soft, ripened Brie (France)	*Lactococcus lactis, Lactococcus cremoris*	*Penicillium camemberti, P. candidum, Brevibacterium linens*
Camembert (France)	*L. lactis, Lactococcus cremoris*	*Penicillium camemberti, B. linens*
Semisoft Blue, Roquefort (France)	*L. lactis, Lactococcus cremoris*	*P. roqueforti*
Brick, Muenster (United States)	*L. lactis, Lactococcus cremoris*	*B. linens*
Limburger (Belgium)	*L. lactis, Lactococcus cremoris*	*B. linens*
Hard, ripened Cheddar, Colby (Britain)	*L. lactis, Lactococcus cremoris*	*Lactobacillus casei, L. plantarum*
Swiss (Switzerland)	*L. lactis, L. helveticus, S. thermophilus*	*Propionibacterium shermanii, P. freudenreichii*
Very hard, ripened Parmesan (Italy)	*L. lactis, Lactococcus cremoris, S. thermophilus*	*L. delbrueckii* subspecies *bulgaricus*

1 *Lactococcus lactis* stands for *L. lactis* subspecies *lactis. Lactococcus cremoris* is *L. lactis* subspecies *cremoris, Lactococcus diacetylactis* is *L. lactis* subspecies *diacetylactis,* and *Streptococcus thermophilus* is *Streptococcus salivarius* subspecies *thermophilus.*

Cheese Production

Cheese is one of the oldest foods, probably developed roughly 8,000 years ago. About 2,000 distinct varieties of cheese are produced throughout the world, representing approximately 20 general types (**table 41.5**).

The production of cheese varies with the type of cheese, but a general outline includes the following steps. First, the milk may be pasteurized to remove spoilage microorganisms. It must then be cooled to 32°C so that starter cultures can be added. The mixture is held at this temperature for 30 minutes. This ripening step allows fermentation to begin and the pH to decline as acidic fermentation products accumulate. Next, rennet is added; rennet is a mixture of enzymes produced by mammals to aid in the digestion of milk. Rennin is the key enzyme in rennet because it catalyzes a complex reaction that results in hydrolysis of milk protein (casein), converting the milk to solid curds. The curds set for about 30 minutes to allow the formation of a firm coagulum. Further fermentation of curds continues until the pH reaches 6.4, at which point the cheese is cut into small pieces and heated to

38°C, which helps separate remaining liquid (whey) from the curd. After the whey is completely drained from the curd, the process of cheddaring begins (**figure 41.9**). Here curd is stacked and occasionally flipped over; this process helps expel residual whey and allows further fermentation. When the curd mats, now knitted together, reach a pH of 5.1 to 5.5, they are cut into smaller pieces. Depending on what kind of cheese is being made, salt will either be added (e.g., cheddar cheese) or the cheese will be rolled in loaves and placed in a salt solution called brine (e.g., mozzarella). The next step is to take the salted cheese and form it into blocks so it can age for weeks to years. This will, in part, determine the final hardness and texture of the cheese. Finally it will be packaged or waxed (e.g., Jarlsberg cheese).

Lactococcus lactis is used as a starter culture for a number of cheeses. Starter culture density is often over 10^9 colony-forming units (CFU) per gram of cheese curd before ripening. However, the high salt, low pH, and temperatures that characterize the cheese microenvironment quickly reduce these numbers. This enables other bacteria, sometimes called nonstarter lactic

Figure 41.9 Cheddar Cheese Production. Cheddar, a village in England, has given its name to a cheese made in many parts of the world. "Cheddaring" is the process of turning and piling the curd to expel whey and develop the desired cheese texture.

acid bacteria (NSLAB), to grow; their numbers can reach 10^7 to 10^9 CFU/g after several months of aging. Thus both starter and nonstarter LAB contribute to the final taste, texture, aroma, and appearance of the cheese.

In some cases, molds are used to further enhance the cheese. Obvious examples are Roquefort and blue cheese. For these cheeses, *Penicillium roqueforti* spores are added to the curds just before final cheese processing. Sometimes the surface of an already formed cheese is inoculated at the start of ripening; for example, Camembert cheese is inoculated with spores of *Penicillium camemberti*.

Meat and Fish

A variety of meat products can be fermented: sausage, country-cured hams, salami, cervelat, Lebanon bologna, fish sauces (processed by halophilic *Bacillus* species), *izushi,* and *katsuobushi.* *Pediococcus acidilactici* and *Lactobacillus plantarum* are most often involved in sausage fermentations. *Izushi* is based on the fermentation of fresh fish, rice, and vegetables by *Lactobacillus* spp.; *katsuobushi* results from the fermentation of tuna by *Aspergillus glaucus.* These latter two fermentations originated in Japan.

Retrieve, Infer, Apply

1. What are the major types of milk fermentations?
2. Briefly compare how cultured buttermilk and yogurt are made.
3. What major steps are used to produce cheese? How is the cheese curd formed in this process?
4. Which fungal genus is often used in cheese making?

Wines and Champagnes

Wine production, the focus of **enology** (Greek *oinos,* wine, and *ology,* the science of), starts with the collection of grapes, which are then crushed so the liquid, called **must,** can be fermented.

Wine production concludes with a variety of storage and aging steps (**figure 41.10**). All grapes have white juices. To make a red wine, skins of red grapes are allowed to remain in contact with the must before fermentation. Wines can be produced by using the natural grape skin microorganisms, but this natural mixture of bacteria and yeasts gives unpredictable fermentation results. To avoid this, fresh must is treated with a sulfur dioxide fumigant and a desired strain of the yeast *Saccharomyces cerevisiae* or *S. ellipsoideus* is added. After inoculation, the juice is fermented

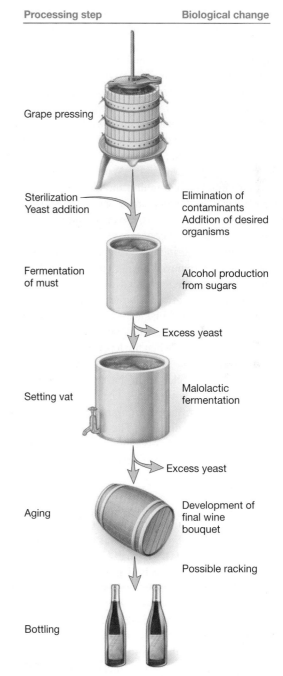

Processing step	Biological change
Grape pressing	
Sterilization Yeast addition	Elimination of contaminants Addition of desired organisms
Fermentation of must	Alcohol production from sugars
	Excess yeast
Setting vat	Malolactic fermentation
	Excess yeast
Aging	Development of final wine bouquet
	Possible racking
Bottling	

Figure 41.10 Wine Making. Once grapes are pressed, the sugars in the juice (the must) can be immediately fermented to produce wine. Must preparation, fermentation, and aging are critical steps.

for 3 to 5 days at temperatures between 20 and 28°C. Depending on the alcohol tolerance of the yeast strain (the alcohol eventually kills the yeast that produces it), the final product may contain 10 to 14% alcohol. Clearing and development of flavor occur during the aging process. Grape juice contains high levels of organic acids, including malic and tartaric acids. If the levels of these acids are not decreased during the fermentation process, the wine will be too acidic and have poor stability and "mouth feel." Malic acid levels are decreased when fermented by the bacteria *Leuconostoc oenos, L. plantarum, L. hilgardii, L. brevis,* and *L. casei.* These microbes transform malic acid (a four-carbon dicarboxylic acid) to lactic acid (a three-carbon monocarboxylic acid) and carbon dioxide. This malolactic fermentation results in deacidification, improvement of flavor stability, and, in some cases, the possible accumulation of bacteriocins in the wines. ◄◄ *Fermentation does not involve an electron transport chain (section 11.8);* Asco-mycota *includes yeasts and molds (section 26.5)*

A critical part of wine making involves the choice of whether to produce a dry (no remaining free sugar) or a sweeter (varying amounts of free sugar) wine. This can be controlled by regulating the initial must sugar concentration. With higher levels of sugar, alcohol will accumulate and inhibit the fermentation before the sugar is completely used, producing a sweeter wine. During final fermentation in the aging process, flavoring compounds accumulate and influence the bouquet of the wine.

Microbial growth during the fermentation process produces sediments, which are removed during **racking.** Racking can be carried out at the time the fermented wine is transferred to bottles or casks for aging or after the wine is placed in bottles.

Champagnes and sparkling wines are essentially the same product, but a champagne must be made in the Champagne region of France. They result when the fermentation is continued in bottles to produce carbon dioxide, which forms bubbles. Sediments that remain are collected in the necks of inverted champagne bottles after the bottles have been carefully turned. The necks of the bottles are then frozen and the corks removed to disgorge accumulated sediments. The bottles are refilled with clear champagne from another disgorged bottle, and the product is ready for final packaging and labeling.

Beers and Ales

Beer and ale production uses cereal grains such as barley, wheat, and rice. The first step in beer production, known as **mashing,** involves germination of the barley grains and activation of their enzymes to produce a **malt (figure 41.11).** The malt is then mixed with water and the desired grains, and the mixture is transferred to the mash tun or cask where the active barley enzymes hydrolyze starch to usable carbohydrates. Once this process is completed, the **mash** is heated with **hops** (dried flowers of the female vine *Humulus lupulus*), which were originally added to the mash to inhibit spoilage microorganisms. The hops also provide flavor and assist in clarification of the liquid portion called the wort **(figure 41.12).** In this heating step, the hydrolytic enzymes are inactivated and the wort can be **pitched**—inoculated—with the desired yeast.

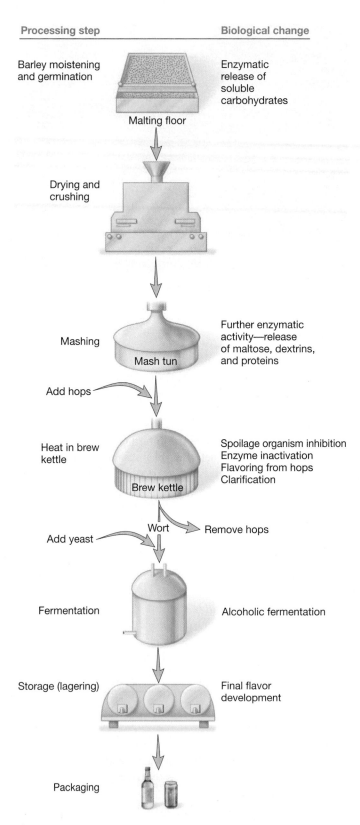

Figure 41.11 Producing Beer. To make beer, the complex carbohydrates in the grain must first be transformed into a fermentable substrate. Beer production thus requires the important steps of malting and the use of hops and boiling for clarification, flavor development, and inactivation of malting enzymes to produce the wort.

Figure 41.12 Brew Kettles Used for Preparation of Wort. In large-scale processes, copper brew kettles can be used for wort preparation, as shown here.

Most beers are fermented with bottom yeasts, related to *Saccharomyces pastorianus*, which settle at the bottom of the fermentation vat. The beer flavor also is influenced by the production of small amounts of glycerol and acetic acid. Bottom yeasts require 7 to 12 days of fermentation to produce beer with a pH of 4.1 to 4.2. With a top yeast, such as *Saccharomyces cerevisiae*, the pH is lowered to 3.8 to produce ales. Freshly fermented (green) beers are aged, and when they are bottled, CO_2 is usually added. Beer can be pasteurized at 40°C or higher or sterilized by passage through membrane filters to minimize flavor changes.

Distilled Spirits

Distilled spirits are produced by an extension of beer production processes. Rather than starting with fresh grain, a **sour mash** is used. This is the liquid fermented grain collected from a previously prepared batch of alcohol. The mash is inoculated with a homolactic (lactic acid is the major fermentation product) bacterium such as *Lactobacillus delbrueckii* subspecies *bulgaricus* (figure 41.8*b*), which can lower the mash pH to around 3.8 in 6 to 10 hours. This limits the development of undesirable organisms. It is boiled, and the volatile components are condensed to yield a product with a higher alcohol content than beer. Rye and bourbon are examples of whiskeys. Rye whiskey must contain at least 51% rye grain, and bourbon must contain at least 51% corn. Scotch whiskey is made primarily of barley. Vodka and grain alcohols are also produced by distillation. Gin is vodka to which resinous flavoring agents—often juniper berries—have been added to provide a unique aroma and flavor.

Production of Breads

Bread is one of the most ancient of human foods. The use of yeasts to leaven bread is carefully depicted in paintings from ancient Egypt, and a bakery at the Giza Pyramid area, from the year 2575 BCE, has been excavated. In bread making, yeast growth is carried out under oxic conditions. This results in increased CO_2 production and minimum alcohol accumulation. The fermentation of bread involves several steps: α- and β-amylases present in the moistened dough release maltose and glucose from starch. Then a baker's strain of the yeast *Saccharomyces cerevisiae*, which produces maltase, invertase, and zymase enzymes, is added. The CO_2 produced by the yeast results in the light texture of many breads, and traces of fermentation products contribute to the final flavor. Usually bakers add sufficient yeast to allow the bread to rise within 2 hours—the longer the rising time, the more additional growth by contaminating bacteria and fungi can occur, making the product less desirable. For instance, bread products can be spoiled by *Bacillus* species that produce ropiness. If the dough is baked after these organisms have grown, stringy and ropy bread results.

Other Fermented Foods

Many other plant products can be fermented, as summarized in **table 41.6**. These include *sufu*, which is produced by the fermentation of tofu, a chemically coagulated soybean milk product. To carry out the fermentation, the tofu curd is cut into small chunks and dipped into a solution of salt and citric acid. After the cubes are heated to pasteurize their surfaces, the fungi *Actinomucor elegans* and some *Mucor* species are added. When a white mycelium develops, the cubes, now called *pehtze*, are aged in salted rice wine. This product has achieved the status of a delicacy in many parts of Asia. Another popular product is tempeh, a soybean mash fermented by *Rhizopus* spp.

Sauerkraut or sour cabbage is produced from shredded cabbage. Usually the mixed microbial community of the cabbage is used. A concentration of 2.2 to 2.8% sodium chloride restricts the growth of Gram-negative bacteria while favoring the development of lactic acid bacteria. The primary microorganisms contributing to this product are *Leuconostoc mesenteroides* and *Lactobacillus plantarum*. A predictable microbial succession occurs in sauerkraut's development. The activities of the lactic acid–producing cocci usually cease when the acid content reaches 0.7 to 1.0%. At this point, *Lactobacillus plantarum* and *Lactobacillus brevis* continue to function. The final acidity is generally about pH 1.7, with lactic acid comprising 1.0 to 1.3% of the total acid in a satisfactory product.

Pickles are produced by placing cucumbers and components such as dill seeds in casks filled with a brine. The sodium chloride concentration begins at 5% and rises to about 16% in 6 to 9 weeks. The salt not only inhibits the growth of undesirable bacteria but also extracts water and water-soluble constituents from the cucumbers. These soluble carbohydrates are fermented to lactic acid. The fermentation, which can require 10 to 12 days, involves the growth of the Gram-positive bacteria *L. mesenteroides, Enterococcus faecalis, Pediococcus acidilactici, Lactobacillus brevis*, and *L. plantarum. L. plantarum* plays the dominant role in this fermentation process. Sometimes, to achieve more uniform pickle quality, natural microorganisms are first destroyed and the cucumbers are fermented using pure cultures of *P. acidilactici* and *L. plantarum*.

When grass, chopped corn, and other fresh animal feeds are stored under moist anoxic conditions, they undergo a mixed acid

Table 41.6	Fermented Foods Produced from Fruits, Vegetables, Beans, and Related Substrates		
Foods	**Raw Ingredients**	**Fermenting Microorganisms**	**Location**
Bacteria-Fermented Foods			
Coffee	Coffee beans	*Erwinia dissolvens, Saccharomyces* spp.	Brazil, Congo, Hawaii, India
Gari	Cassava	*Corynebacterium manihot, Geotrichum* spp.	West Africa
Kimchi	Cabbage and other vegetables	Lactic acid bacteria	Korea
Ogi	Corn	*Lactobacillus plantarum, Lactococcus lactis, Z. rouxii*	Nigeria
Olives	Green olives	*Leuconostoc mesenteroides, Lactobacillus plantarum*	Worldwide
Pickles	Cucumbers	*Pediococcus cerevisiae, L. plantarum, L. brevis*	Worldwide
Poi	Taro roots	Lactic acid bacteria	Hawaii
Sauerkraut	Cabbage	*L. mesenteroides, L. plantarum, Lactobacillus brevis*	Worldwide
Mold-Fermented Foods			
Kenkey	Corn	*Aspergillus* spp., *Penicillium* spp., lactobacilli, yeasts	Ghana, Nigeria
Miso	Soybeans	*Aspergillus oryzae, Zygosaccharomyces rouxii*	Japan
Ontjom	Peanut presscake	*Neurospora sitophila*	Indonesia
Peujeum	Cassava	Molds	Indonesia
Soy sauce	Soybeans	*A. oryzae* or *A. soyae, Z. rouxii, Lactobacillus delbrueckii*	Japan
Sufu	Soybeans	*Actinimucor elegans, Mucor* spp.	China
Tao-si	Soybeans	*A. oryzae*	Philippines
Tempeh	Soybeans	*Rhizopus oligosporus, R. oryzae*	Indonesia, New Guinea, Surinam

Adapted from Jay, James M. 2000. *Modern Food Microbiology, 6th ed.*

fermentation that produces pleasant-smelling **silage.** Trenches or more traditional vertical steel or concrete silos are used to store silage. The accumulation of organic acids in silage can cause rapid deterioration of silos. Older wooden stave silos, if not properly maintained, allow the outer portions of the silage to become oxic, resulting in spoilage of a large portion of the plant material.

Retrieve, Infer, Apply

1. Describe and contrast the processes of wine and beer production.
2. How do champagnes differ from wines?
3. Describe how distilled spirits such as whiskey are produced.
4. What microorganisms are most important in sauerkraut and pickle fermentations?

41.6 Probiotics

After reading this section, you should be able to:

■ Describe how a food is classified as a probiotic and identify two microorganisms used in probiotic foods

The health benefits of fermented foods such as yogurt, which commonly contain bacteria in the genera *Lactobacillus* and *Bifidobacterium,* have been touted for many years. Recent advances in understanding the human gut microbiome show that specific microbial products, often attributed to these bacteria, have a profound influence on immune system function. For example, lactobacilli and bifidobacteria stimulate immune maturation and minimize inflammation. Other health benefits attributed to the consumption of fermented milks involve minimizing lactose intolerance, lowering serum cholesterol, and possibly protecting against colon cancer. Several lactobacilli have antitumor compounds in their cell walls. With only a few exceptions, our increased understanding of the human microbiome has yet to translate into clinically tested probiotic foods or treatments. ◄◄ *The human-microbe* ecosystem *(section 32.2)*

The enormous volume of research currently performed on the human microbiome and the media attention it has received have helped fuel explosive growth in the probiotic food and supplement market. However, most consumers are unaware that the term "probiotic" is not regulated. In an effort to standardize the use of the term **probiotic,** the Food and Agricultural Organization of the United Nations–World Health Organization (FAO-WHO) has defined it as "live microorganisms, which, when administered in adequate amounts, confer a health benefit to the host." FAO-WHO guidelines published in 2002 provide specific requirements that should be met for products to meet this standard (**figure 41.13**).

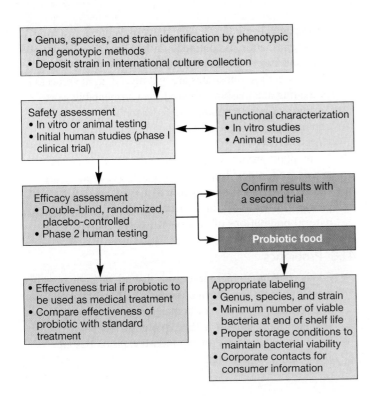

Figure 41.13 The Development of a Probiotic Food.

Probiotic microbes are also used to promote the health of farm animals. Such microbes, primarily *Lactobacillus acidophilus*, are used in beef cattle feed. When the bacteria are sprayed on feed, cattle that eat it appear to have markedly lower (as much as 60%) carriage of the pathogenic *E. coli* strain O157:H7. This can make it easier to produce beef that will meet current standards for microbiological quality at the time of slaughter. Probiotics are also used successfully with poultry. For instance, the USDA has designated a probiotic *Bacillus* strain for use with chickens as GRAS (p. 931). Feeding chickens a strain of *Bacillus subtilis* leads to increased body weight and feed conversion. There is also a reduction in coliforms and *Campylobacter* spp. in the processed carcasses. It has been suggested that this probiotic decreases the need for antibiotics in poultry production and pathogen levels on farms. *Salmonella enterica* can be controlled by spraying a patented blend of 29 bacteria, isolated from the chicken cecum, on day-old chickens. As they preen themselves, the chicks ingest the bacterial mixture, establishing a functional microbial community in the cecum that limits *Salmonella* colonization of the gut in a process called competitive exclusion.

Retrieve, Infer, Apply

1. What bacterial genera are often included in probiotic foods?
2. How are probiotics used in agriculture?

Key Concepts

41.1 Microbial Growth Can Cause Food Spoilage

- Most foods, especially when raw, provide an excellent environment for microbial growth. This growth can lead to spoilage or preservation, depending on the microorganisms present and environmental conditions.
- The course of microbial development in a food is influenced by the intrinsic characteristics of the food itself—pH, salt content, substrates present, water presence, and availability—and extrinsic factors, including temperature, relative humidity, and atmospheric composition.
- Microorganisms can spoil meat, dairy products, fruits, vegetables, and canned goods in several ways. Spices, with their antimicrobial compounds, sometimes protect foods.

41.2 Various Methods Are Used to Control Food Spoilage

- Foods can be preserved in a variety of physical and chemical ways, including filtration, alteration of temperature (cooling, pasteurization, sterilization), drying, addition of chemicals, radiation, and fermentation.
- Modified atmosphere packaging (MAP) is used to control microbial growth in foods and to extend product shelf life. This process involves decreased oxygen and increased carbon dioxide levels in the space between the food surface and the wrapping material.

- Interest is increasing in using bacteriocins for food preservation. Nisin, a product of *Lactococcus lactis*, is a major substance approved for use in foods.

41.3 Food-Borne Disease Outbreaks

- Foods can be contaminated by pathogens at any point in the food production, storage, or preparation processes. Pathogens such as *Salmonella enterica*, *Campylobacter jejuni*, *Listeria monocytogenes*, and *E. coli* can be transmitted to susceptible consumers, where they grow and cause a food-borne infection (**table 41.4**).
- If the pathogen grows in the food before consumption and forms toxins that affect food consumers without further microbial growth, the disease is a food intoxication. Examples are intoxications caused by *Staphylococcus*, *Clostridium*, and *Bacillus* spp. Fungi that grow in foods, especially cereals and grains, can produce important disease-causing chemicals, including the carcinogens aflatoxins (**figure 41.5**) and fumonisins (**figure 41.6**).

41.4 Detection of Food-Borne Pathogens Requires Government-Industry Cooperation

- Detection of food-borne pathogens is a major part of food microbiology. Many assays rely on culture-based approaches.

- The use of immunological and molecular techniques such as DNA and RNA hybridization, PCR, and pulsed-field gel electrophoresis often makes it possible to link disease occurrences to a common infection source. The PulseNet network is used to coordinate these control efforts (**figure 41.7**).

41.5 Microbiology of Fermented Foods: Beer, Cheese, and Much More

- Dairy products can be fermented to yield a wide variety of cultured milk products. These include mesophilic, therapeutic probiotic, thermophilic, yeast-lactic, and mold-lactic products.
- Growth of lactic acid–forming bacteria, often with the additional use of rennin, can coagulate milk solids. These solids can be processed to yield a wide variety of cheeses, including soft unripened, soft ripened, semisoft, hard, and very hard types (**table 41.5**). Both bacteria and fungi are used in these cheese production processes.
- Wines are produced from pressed grapes and can be dry or sweet, depending on the level of free sugar that remains at the end of the alcoholic fermentation (**figure 41.10**).

Champagne is produced when the fermentation, resulting in CO_2 formation, is continued in the bottle.
- Beer and ale are produced from cereals and grains. The starches in these substrates are hydrolyzed, in the processes of malting and mashing, to produce a fermentable wort. *Saccharomyces cerevisiae* and *S. pastorianus* are yeasts commonly used in the production of beer and ale (**figure 41.11**).
- Many plant products can be fermented with bacteria, yeasts, and molds. Important products are breads, soy sauce, *sufu,* and tempeh (**table 41.6**). Sauerkraut and pickles are produced in a fermentation process in which natural populations of lactobacilli play a major role.

41.6 Probiotics

- Foods containing microorganisms that are thought to confer a health benefit have been consumed for hundreds of years. The recent appreciation of human gut microflora has helped engender more interest in these probiotic foods.
- Probiotic foods are a growing industry; probiotics are also used in agriculture.

Compare, Hypothesize, Invent

1. Compare the "sell-by" date of a package of hot dogs with that of a fresh cut of meat. What are the intrinsic and extrinsic factors that contribute to the difference?

2. You are going through a salad line in a cafeteria at the end of the day. Which types of foods would you tend to avoid? Or would you skip the salad altogether? Explain.

3. Why were aflatoxins not discovered before the 1960s? Do you think this was the first time they had grown in a food product to cause disease?

4. What advantage might the Shiga-like toxin give *E. coli* O157:H7? Can we expect to see other "new" pathogens appearing, and what should we do, if anything, to monitor their development?

5. Keep a record of what you eat for a day or two. Determine if the food, beverages, and snacks you ate were produced (at any level) with the aid of microorganisms. Indicate at what levels microorganisms were deliberately used. Be sure to consider ingredients such as citric acid, which is produced at the industrial level by several species of fungi.

6. During cheese production, LAB convert lactose to lactate and casein (milk protein) to amino acids. Lactate and amino acids then become the substrates for further microbial growth, which results in aroma production and deacidification of the cheese. The yeast *Yarrowia lipolytica* grows on the surface of many cheeses; it is capable of both lactate and amino acid catabolism. When grown on a lactate plus amino acid medium, *Y. lipolytica* preferentially consumes amino acids. Amino acid degradation results in the release of ammonia, which increases the pH. Draw a flow chart that shows the LAB fermentation of milk, followed by the growth of *Y. lipolytica.* Indicate which substrates are consumed first and what happens to the pH. Based on this simplified scenario, why do you think most cheeses involve the activity of more than one yeast species?

7. Following recent outbreaks of *E. coli* O157:H7 associated with leafy vegetables, there has been a need to better understand the growth of this pathogen on vegetables. To that end, a study was performed to compare the colonization and growth of *E. coli* O157:H7 on lettuce. In this study, the success of the microbe on intact leaves was compared with regions that had been mechanically damaged by shredding or bruising. Indeed, even within 4 hours after inoculation, there was a marked increase in microbial growth on the injured surfaces of the lettuce.

Why do you think this was the case? If you were to perform this study (or one similar to it), how would you quantify *E. coli* colonization? If you were able to repeat and extend these findings, what kinds of recommendations might you make to the vegetable producers? To consumers?

Read the original paper: Brandl, M. T. 2008. Plant lesions promote the rapid multiplication of *Escherichia coli* O157:H7. *Appl. Environ. Microbiol.* 74:5285.

42

Biotechnology and Industrial Microbiology

There is little, if any, financial incentive for large pharmaceutical companies to develop new antibiotics, despite a pressing need for them.

Where Are the New Antibiotics?

Bill's dad takes a statin drug to lower his cholesterol, and his grandmother takes beta-blocker for her heart. His mom has a steroid inhaler to keep her asthma under control. When Bill comes down with strep throat, his physician prescribes a 10-day course of amoxicillin and he gets better.

Bill's relatives are benefiting from drugs that pharmaceutical companies love to develop. They aren't so wild about helping Bill, though. It's not that they don't want to; they believe they can't afford it. Consider this: it costs over 2 billion U.S. dollars and at least a decade to bring a drug from discovery to market. Then the patent lasts for about another 10 years. Once the patent expires, the cost of the drug (and the profit therein) drops precipitously. Whereas Bill's relatives will very likely take their medicines for the rest of their lives, Bill takes his antibiotic for 10 days. When he's done: no more strep throat, no more profit for the drug manufacturer.

But in the meantime, amoxicillin (like almost all antibiotics) is losing its effectiveness as more and more bacteria develop resistance. So when new antibiotics are developed, they cannot be widely prescribed, lest resistance quickly develop.

Despite these disincentives, the need for new antibiotics grows more urgent with each passing year. Recall that two-thirds of antibiotics are natural microbial products. However, because only about 1% of all bacteria and archaea—and even fewer fungi—have been cultured, the environment holds great untapped resources. But history tells us we need a new strategy to access this potential antibiotic reservoir. Between 1935 and 1962, the seven classes of antibiotics* we still use today were developed. Then from 1962 to 2000, no new drug classes were discovered. Instead, while men

went to the moon and the Internet was born, existing antibiotics were chemically modified, generating multiple generations of existing drugs. Newly introduced drugs all belong to drug classes already in use, subject to the same mechanisms of resistance. It was not until the first decade of the twenty-first century that three new classes were introduced to clinical practice: oxazolidinones, lipopeptides, and mutilins. Currently, drug discovery is becoming synonymous with microbial discovery: looking in new habitats, such as deep-sea trenches and the symbionts of other animals, or inventing new ways to coax microbes from common environments (like soil) to grow in the lab.

Industrial microbiology harnesses the capacity of microbes to synthesize compounds that have important applications in medicine, agriculture, food preparation, and other industrial processes. These compounds are commonly called **natural products.** In this chapter, we discuss these products and how microorganisms (or their genes) are optimized for industrial manufacturing.

Readiness Check:

Based on what you have learned previously, you should be able to:

✔ Compare and contrast at least five different classes of antibiotics (section 9.2)

✔ Describe the difference between batch and continuous culture techniques (sections 7.7 and 7.9)

✔ Recall the principle biochemical pathways involved in central metabolism (chapters 11 and 12)

✔ Explain oxidation-reduction (redox) reactions (section 10.3)

✔ Describe the means by which DNA mutations can be induced (section 16.1)

✔ Explain the mechanisms of horizontal gene transfer (sections 16.4–16.9)

✔ Summarize the process and importance of PCR (section 17.2)

✔ Define metagenomics and discuss its applications (section 18.3)

✔ Discuss the principles by which vaccines confer immunity (section 37.6)

---

* In chronological order of discovery, the seven drug classes include sulfa drugs, β-lactams, chloramphenicol/tetracyclines, aminoglycosides, macrolides, glycopeptides, quinolones, and streptogramins.

42.1 Microbes Are the Source of Many Products of Industrial Importance

After reading this section, you should be able to:

- Describe at least five compounds of industrial importance made by microbes
- Define biocatalysis and explain its advantages over chemical synthesis

Industrial microbiology has profoundly changed our lives and lifespans. Microbial products include antibiotics and other medicines, food additives, enzymes with industrial applications, and biofuels, just to name a few (**table 42.1**). Here we introduce a few of the more common microbial products that are an integral part of modern life.

Antibiotics

More than 65% of all antibiotics are produced by microorganisms. Currently most antibiotics are produced by members of the genus *Streptomyces* and by filamentous fungi (*see table 24.3*). Antibiotics are secreted by these microbes and either used in the native state or chemically modified to produce semisynthetic derivatives. Such derivatives generally are manufactured to address the problem of antibiotic resistance or the desire for a broader-spectrum drug.

The synthesis of penicillin and its derivatives illustrates how an antibiotic is commercially produced. As with any industrially important natural product, the manufacturer will determine how to maximize product yield. In this case, the growth of the fungus *Penicillium chrysogenum* requires careful adjustment of the medium, so that the slowly hydrolyzed disaccharide lactose is used in combination with a limited supply of nitrogen to stimulate a greater accumulation of penicillin after growth has stopped (**figure 42.1**). If a particular penicillin is needed, the specific precursor is added to the medium. For example, phenylacetic acid is added to maximize production of penicillin G, which has a benzyl side chain (*see figure 9.5*). The pH is maintained around neutrality by the addition of sterile alkali, which ensures maximum stability of the newly synthesized penicillin. After about a week, the broth is separated from the fungal hyphae and processed by absorption, precipitation, and crystallization to yield the final product. This basic product can then be modified by chemical procedures to yield a variety of semisynthetic penicillins. Manufacture of semisynthetic β-lactam antibiotics, such as ampicillin and amoxicillin, begins with natural penicillin. The β-lactam ring is preserved, but the side chain is modified to provide a broader spectrum of activity. ◀◀ *Inhibitors of cell wall synthesis (section 9.4)*

Amino Acids

Amino acids such as lysine and glutamic acid are used in the food industry as nutritional supplements in bread products and as flavor-enhancing compounds such as monosodium glutamate (MSG). Amino acid production is typically carried out by regulatory mutants that have a reduced ability to limit synthesis of a specific amino acid or a key intermediate. Wild-type microorganisms avoid overproduction of biochemical intermediates by the careful regulation of cellular metabolism. Production of glutamic acid and several other amino acids is carried out using mutants of *Corynebacterium glutamicum* that lack or have only a limited ability to process the TCA cycle intermediate α-ketoglutarate to succinyl-CoA (*see figure 11.8*). A controlled low biotin level and the addition of fatty acid derivatives increase membrane permeability so that high concentrations of glutamic acid are secreted. The impaired bacteria use the glyoxylate pathway (*see figure 12.22*) to meet their needs for essential biochemical intermediates, especially during the growth phase. By the time nutrients are depleted and growth stops, an

Table 42.1	Major Microbial Products and Processes of Interest in Industrial Microbiology
Substances	**Microorganisms**
Industrial Products	
Ethanol (from glucose)	*Saccharomyces cerevisiae*
Ethanol (from lactose)	*Kluyveromyces fragilis*
Acetone and butanol	*Clostridium acetobutylicum*
2,3-butanediol	*Enterobacter, Serratia*
Enzymes	*Aspergillus, Bacillus, Mucor, Trichoderma*
Agricultural Products	
Gibberellins	*Gibberella fujikuroi*
Food Additives	
Amino acids (e.g., lysine)	*Corynebacterium glutamicum*
Organic acids (citric acid)	*Aspergillus niger*
Nucleotides	*Corynebacterium glutamicum*
Vitamins	*Ashbya, Eremothecium, Blakeslea*
Polysaccharides	*Xanthomonas*
Medical Products	
Antibiotics	*Penicillium, Streptomyces, Bacillus*
Alkaloids	*Claviceps purpurea*
Steroid transformations	*Rhizopus, Arthrobacter*
Insulin, human growth hormone, somatostatin, interferons	*Escherichia coli, Saccharomyces cerevisiae*, and others (recombinant DNA technology)
Biofuels	
Hydrogen	Photosynthetic microorganisms
Methane	*Methanobacterium*
Ethanol	*Zymomonas, Thermoanaerobacter*

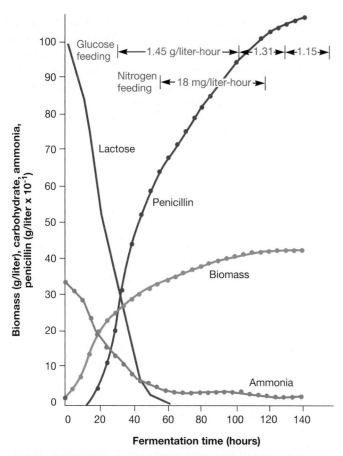

Figure 42.1 Penicillin Fermentation Involves Precise Control of Nutrients. The synthesis of penicillin begins when nitrogen from ammonia becomes limiting. After most of the lactose (a slowly catabolized disaccharide) has been degraded, glucose (a rapidly used monosaccharide) is added along with a low level of nitrogen. This stimulates maximum transformation of the carbon sources to penicillin. The scale factor is presented using the convention recommended by the American Society for Microbiology; that is, a number on the axis should be multiplied by 0.10 to obtain the true value.

MICRO INQUIRY *Why are the cells first fed lactose and then glucose?*

almost complete molar conversion (or 81.7% weight conversion) of isocitrate to glutamate occurs. ◄◄ *Synthesis of amino acids consumes many precursor metabolites (section 12.5)*

Organic Acids

Simply reading the ingredient list on most processed foods will illustrate the widespread use of organic acids such as citric, acetic, and lactic acids. These acids are principally used as preservatives. Organic acid production illustrates how the concentration of trace elements can influence product yield.

Citric acid fermentation involves limiting the amounts of trace metals such as manganese and iron to stop *Aspergillus niger* growth at a specific point. The medium often is treated with ion exchange resins to ensure low and controlled concentrations

of available metals. Generally, high sugar concentrations (15 to 18%) are used, and copper has been found to counteract the inhibition of citric acid production by iron. This reflects the regulation of glycolysis and the tricarboxylic acid cycle (recall that citric acid is a constituent of the TCA cycle). After the active growth phase, when the substrate level is high, citrate synthase activity increases and the activities of aconitase and isocitrate dehydrogenase decrease. This results in citric acid accumulation and excretion by the stressed microorganism.

Biopolymers

Biopolymers are microbially produced polymers, primarily polysaccharides, used to modify the flow characteristics of liquids and to serve as gelling agents. These are employed in many areas of the pharmaceutical and food industries.

Biopolymers include (1) dextrans, which are used as blood expanders and absorbents; (2) *Erwinia* polysaccharides used in paints; (3) polyesters, derived from *Pseudomonas oleovorans,* which are used for specialty plastics; (4) cellulose microfibrils, produced by an *Acetobacter* strain, that serve as a food thickener; (5) polysaccharides such as scleroglucan used by the oil industry as drilling mud additives; and (6) xanthan polymers, which have a variety of applications as food additives as well to enhance oil recovery by thickening drilling mud. This use of xanthan gum, produced by *Xanthomonas campestris,* represents a large market for this microbial product.

Biosurfactants

Biosurfactants are amphiphilic molecules; that is, they possess both hydrophobic and hydrophilic regions. Thus they partition at the interface between fluids that differ in polarity, such as oil and water. For this reason, they are used for emulsification, increasing detergency, wetting, and phase dispersion, as well as solubilization. These properties are especially important in bioremediation, oil spill dispersion, and enhanced oil recovery. The most widely used microbially produced biosurfactants are glycolipids. These are carbohydrates that bear long-chain fatty acids. They can be isolated as extracellular products from a variety of microorganisms, including pseudomonads and yeasts.

Biocatalysts

Biocatalysts are enzymes used in the industrial production of chemicals and pharmaceuticals. The vast majority of such enzymes are of microbial origin and include dehydrogenases, oxygenases, hydrolases, transferases, and lyases. These enzymes may introduce minor changes in molecules, such as the insertion of a hydroxyl or keto function, or the saturation or desaturation of a complex cyclic structure. A typical bioconversion is the hydroxylation of a steroid as shown in **figure 42.2**. In this example, the water-insoluble steroid is dissolved in acetone and then added to the reaction system. The course of the modification is monitored, and the final, soluble product is extracted and purified.

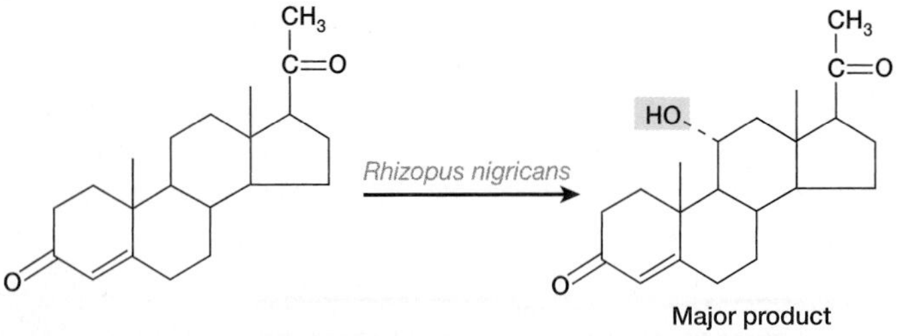

Figure 42.2 Biotransformation to Modify a Steroid. Hydroxylation of progesterone in the 11α position by *Rhizopus nigricans*. The steroid is dissolved in acetone before addition to the pregrown fungal culture.

Biocatalysis generally has the advantage over chemical synthesis in that natural enzymes have stereospecificity so that the product is not a mixture of isomers and is biologically active. Further, microbial enzymes catalyze reactions under mild conditions; some of these reactions would require harsh conditions (e.g., extreme pH, temperature) if performed by chemical synthesis.

Vaccines

Vaccines are the single most important public health weapon. The most common vaccines are live attenuated viruses (e.g., measles vaccine) or killed pathogens (influenza vaccine). However, these traditional forms of vaccines cannot be used for certain pathogens. For instance, if the pathogen cannot be grown in vitro (e.g., hepatitis B and C viruses), neither attenuation or killing the purified microbe is possible. Likewise, different vaccine strategies must be used if the pathogen produces molecules that are too similar to host molecules, as this can trigger autoimmunity, or if there is too much genetic diversity among pathogenic strains. This is the case for group B *Streptococcus,* also called GBS, which causes serious infections in infants, so a vaccine is highly desired. In these cases, **reverse vaccinology** may be a fruitful approach. Reverse vaccinology mines the genomic sequence of the pathogen to assemble a list of antigens that may be good vaccine targets. ◄◄ *Vaccines immunize susceptible populations (section 37.6)*

Recall that an antigen is a foreign molecule targeted by the host immune system. Purified antigens, rather than whole organisms, can be used in vaccines. Such antigens must (1) be expressed by the pathogen during infection; (2) be secreted or found on the surface of the pathogen; (3) be found in all strains of the pathogen; (4) elicit a host immune response; and (5) be essential for the survival of the pathogen, at least while it is in the host.

Considering the thousands of genes that are revealed in each annotated genome, finding genes whose products meet all of these criteria might seem next to impossible. But there are two approaches to generating a list of antigen candidates: Either the genome of a single species is examined for vaccine targets, or a pan-genomic approach is taken so that the genomes of a number of strains are compared. The first approach was used to develop a *Neisseria meningitidis* serogroup B (MenB) vaccine. For many years there were two vaccines made of capsular polysaccharides (CPS) to protect against serogroups A, C, Y, and W-135. Because the CPS of MenB is identical to a human polysaccharide, it could not be included in these vaccines. By using the complete genome sequence of serogroup B *N. meningitidis*, 600 surface proteins were identified and about 300 were tested in mice. Of these about 25 were worthy of further study, and in 2014 the first MenB vaccine was approved for use in the United States.

By contrast, a pan-genomic approach was used to develop a GBS vaccine because there are many genetically variable pathogenic strains. Thus the genomes of eight GBS isolates were compared and 312 surface proteins were identified as potential targets. When tested for their ability to protect mice from infection, four were found to be effective. These were combined into a single vaccine capable of protecting animals against infection by all known clinical strains. This vaccine and others are in clinical trials. Reverse vaccinology is currently being applied in the development of vaccines to protect against a variety of other pathogens, including *Staphylococcus aureus, Chlamydophila pneumoniae,* and *Bacillus anthracis.*

Retrieve, Infer, Apply

1. What critical limiting factors are used in the production of penicillin? Why do you think the microbe must be stressed to produce high levels of drug?
2. Why are regulatory mutants used to increase the production of glutamic acid by *Corynebacterium glutamicum*?
3. Discuss the major uses for biopolymers and biosurfactants.
4. What kinds of molecular changes result from biocatalysis?
5. Why is serogroup B not included in the meningitis vaccine? How was the MenB vaccine developed?

42.2 Biofuel Production Is a Dynamic Field

After reading this section, you should be able to:

- Compare the value of biofuel use with the challenges presented in its manufacture and transport
- Discuss at least two types of biofuels

Global energy consumption is predicted to increase by roughly 40% between 2014 and 2040. Currently the world relies on petroleum to fuel almost all its transportation, while coal and natural gas (methane) are used in abundance to generate power and heat buildings. These fossil fuels need to be phased out of use, not only because they are finite resources but also because they are the major contributors of greenhouse gases and global climate change. Many agree that renewable energy sources based on microbial transformation of organic material are the most viable approach to solve our global energy and climate crises. **Microbial energy conversion** includes the microbial transformation of organic materials into **biofuels,** such as ethanol and hydrogen, that can be burned to fuel cars or other machines. In this section, we discuss the microbial production of the biofuels ethanol (and other alcohols) and hydrogen. ◄◄ *Global climate change (section 28.2)*

The U.S. Department of Energy has set a goal of substituting 30% of gasoline with biofuels by 2030. Currently ethanol is used as an additive to gasoline because it is easy to store and can be burned in cars designed to use pure gasoline. This is why gasoline sold in the United States contains at least 10% ethanol. Corn is currently the substrate most commonly used for generating ethanol in the United States. Its fermentation into ethanol involves the microbial degradation of plant starch using amylases and amyloglucosidases. The resulting sugars are then fermented to ethanol. However, the use of corn to make bioethanol has had unintended consequences, chiefly in the form of higher worldwide food prices. Therefore much research is now devoted to finding other substrates for the production of alcohol biofuels. In addition to the use of cyanobacteria and algae (*see p. 509*), crop residues could significantly boost biofuel yields and ease the cost of corn and corn-based foods. Crop residues are the plant material left in the field after harvest. This material consists of cellulose and hemicellulose—polymers of five different hexoses and pentoses: glucose, xylose, mannose, galactose, and arabinose. Degradation of cellulose and hemicellulose to release these monomers is commonly done by heating the plant material and treating it with acid, which is energy-intensive and corrosive. This is an expensive and potentially hazardous process. Nonetheless, 2014 marked the opening of the first large-scale plant to produce cellulosic ethanol—that is, ethanol made from crop residues. Research continues to eliminate the need for chemical pretreatment. This has led to the development of an *E. coli* strain engineered to express yeast enzymes as well as to the isolation of hyperthermophilic archaea that can degrade cellulose at temperatures in excess of 100°C. These are excellent examples of biocatalysis.

Even if cheaper plant material were found for ethanol production, ethanol as a biofuel has several disadvantages. One is that it absorbs water and cannot be shipped through existing pipelines because they almost always contain some water. So ethanol must be distilled twice to remove the water: once when it is generated and then again after shipping. Furthermore, ethanol contains far less energy than other fuels, including gasoline and higher molecular weight alcohols such as butanol, which are easier to generate. In fact, the entire ethanol manufacturing process consumes more energy than it yields; that is, ethanol is not a carbon-neutral fuel. This has motivated a number of small biotechnology companies to engineer microbes to produce other alcohols and even hydrocarbons. Currently, synthetic biologists have built strains of *E. coli* that produce both short-chain alkanes and alkenes. These products more closely resemble gasoline (petrol), which is a mixture of C4-C12 short-chain hydrocarbons.

In many ways, hydrogen (H_2) as a biofuel compares very favorably to ethanol and other fuels. In fact, H_2 has about three times more potential energy per unit weight than gasoline; it is thus the highest energy-content fuel available. However, until a liquid fuel with a high hydrogen content is developed, the storage and distribution of hydrogen will remain a problem. Also, unlike ethanol, hydrogen cannot simply be mixed with gasoline and used in today's vehicles. Nonetheless, because of its high energy yield, among other factors, there is great interest in developing H_2 as a fuel source, and the concept of a "hydrogen economy" has been promoted, particularly in developing nations.

A diverse group of microbes produce H_2, thus making its manufacture quite flexible. Like ethanol, H_2 can be produced as a product of fermentation. It is also produced by oxygenic photosynthetic microbes (cyanobacteria and algae) and anoxygenic photoheterotrophs. Two different enzymes (thus processes) generate H_2: hydrogenase and nitrogenase. Both are keenly sensitive to oxygen, so H_2 production is an anaerobic process.

Oxygenic photosynthetic microbes produce H_2 at night, when they oxidize cellular storage products such as poly-β-hydroxybutyrate. The use of aquatic phototrophs to generate H_2 holds promise because there is plenty of substrate (light and CO_2). However, hydrogenase is strongly inhibited by its product, so H_2 must be removed from the system as soon as it is created. By contrast, H_2 production by anoxygenic photoheterotrophic bacteria can occur via cellular hydrogenases or nitrogenases. It is the latter that has sparked interest among industrial microbiologists. This is because in an atmosphere that has been artificially purged of all N_2, nitrogenase will form only H_2. This process requires a tremendous amount of ATP and reductant, but unlike hydrogenase, nitrogenase is not inhibited by the accumulation of H_2, and electron conversion to H_2 reaches efficiencies near 100%.

Although we have painted an uncertain picture of which biofuel will eventually dominate, several things are clear. With the global population predicted to expand to 9.6 billion people by 2050, the desire for energy will eventually make fossil fuels rare and prohibitively expensive. At the same time, the need to increase food production will make it untenable to grow feedstocks for biofuel production on scarce arable land. It may be that synthetic biology (p. 956) will save the day, and instead of drilling for oil, we will be harvesting it from bioreactors. Or it may turn out that H_2 will become an important resource, especially in countries that have not yet built an extensive infrastructure to support a hydrocarbon-based lifestyle. But no matter what eventually transpires, these are exciting times for the microbial energy conversion field.

Retrieve, Infer, Apply

1. Why is there interest in developing alternatives to ethanol as the biofuel of choice in the United States?
2. What types of microbes are good candidates for the industrial production of H_2?

42.3 Growing Microbes in Industrial Settings Presents Challenges

After reading this section, you should be able to:

- Explain the use of the term fermenter as it is used in an industrial setting
- Describe two types of fermenters

As our review of natural products and processes points out, the development of appropriate culture media and conditions is critical to the manufacture of the desired product. Culturing microbes in industrial settings is much different than their growth in either nature or the laboratory. However, before describing microbial growth on an industrial scale, we must clarify terminology. The term **fermentation,** used in a physiological sense, is employed differently in the context of industrial microbiology and biotechnology. To industrial microbiologists, fermentation means the mass culture of microorganisms, and plant and animal cells. Industrial fermentations require the development of appropriate culture media and the transfer of small-scale technologies to a much larger scale. In fact, the success of an industrially important microbial product rests on the process of **scale-up.** This is when a procedure developed in a small flask is modified for use in a large fermenter. The microenvironment of the small culture must be maintained despite increases in culture volume. Although the process of transitioning a culture originally developed in a 250-milliliter Erlenmeyer flask to a 100,000-liter reactor seems like it should be relatively straightforward, it is often very difficult to obtain the same yield of the desired product, and a series of troubleshooting experiments must be devised.

Microorganisms are often grown in stirred fermenters or other mass culture systems. Stirred **fermenters** can range in size from 3 liters to over 100,000 liters, depending on production requirements. A typical industrial stirred fermentation unit is illustrated in **figure 42.3**. Upon inoculation of the microbe into sterile growth medium, aeration, pH and temperature adjustments, and sampling are carried out under rigorously controlled conditions. When required, foam control agents are added, especially with high-protein media. Computers are used to monitor outputs from probes that determine microbial biomass, levels of critical metabolic products, pH, input and exhaust gas composition, and other parameters. Environmental conditions can be changed or held constant over time, depending on the requirements of the particular process.

Frequently a critical component in the medium, often the carbon source, is added continuously—a process called **continuous feed**—so that the microorganisms will not have excess substrate available at any given time. This is particularly important with glucose and other carbohydrates. If excess glucose is present at the beginning of a fermentation, it can be catabolized and lost as volatile end products, reducing the final yield of the desired product.

Other culturing approaches exist, among them continuous culture techniques using chemostats. These can sometimes markedly improve rates of substrate use and product yield because microorganisms can be maintained in logarithmic phase. However, for some industrial processes continuous active growth is undesirable, because the product of interest is not made during exponential growth. Microbial products can be classified as primary and secondary metabolites. **Primary metabolites** consist of compounds related to the synthesis of microbial cells during exponential growth. They include amino acids, nucleotides, and fermentation end products such as ethanol and organic acids. In addition, industrially useful enzymes, either associated

with the microbial cells or exoenzymes, often are synthesized by microorganisms during log phase growth. **Secondary metabolites** usually accumulate when nutrients become limited or waste products accumulate following the active growth phase. These compounds are sometimes considered part of a stress response as nutrients become scarce. Most antibiotics fall into this category. ◄◄ *Growth curves consist of five phases (section 7.7); Chemostats (section 7.9)*

Retrieve, Infer, Apply

1. What is the objective of the scale-up process and why is it so critical?
2. What parameters can be monitored in a modern, large-scale industrial fermentation?
3. Which of the following are secondary metabolites: pyruvate, a secreted amylase, an antibiotic, leucine? What kind of system (fermenter vs. chemostat) would most likely be used for the production of the streptomycete antibiotic streptomycin? Explain your answer.

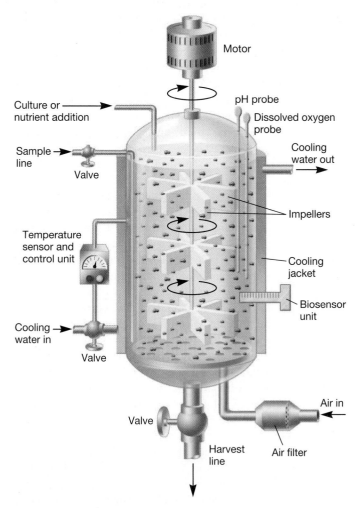

Figure 42.3 An Industrial Fermenter. This unit can be run under oxic or anoxic conditions, and nutrient additions, sampling, and fermentation monitoring can be carried out under aseptic conditions. Biosensors and infrared monitoring can provide real-time information on the course of the fermentation. Specific substrates, metabolic intermediates, and final products can be detected.

42.4 Production Strains Are Developed to Maximize Output of Industrially Important Compounds

After reading this section, you should be able to:

- List three techniques used to optimize microbial output of industrial products
- Compare and contrast three directed evolution technologies
- Describe how metagenomics increases the pool of microbial products available for screening

Most microbes used in industrial processes originate from natural materials (e.g., soil) and are difficult to grow under standard laboratory conditions. Furthermore, even after appropriate growth conditions are determined, the original strains often produce very low amounts of the desired product. Starting with these microbes, microbiologists expend considerable effort to create strains, called **production strains,** that are optimized for industrial purposes. Here we describe some of the approaches they use.

Mutagenesis

Once a microbe is found to produce a promising compound, a variety of mutagenic techniques can be used to increase product yield. For many years, biotechnologists were limited to chemical, ultraviolet (UV) light, and X-ray mutagenesis. For example, the first cultures of *Penicillium notatum* had to be grown in stationary vessels and produced low concentrations of penicillin. In 1943 strain NRRL 1951 of *Penicillium chrysogenum* was isolated and further improved using these methodologies (**figure 42.4**). Today most penicillin is produced with *Penicillium chrysogenum* grown in aerobic, stirred fermenters, yielding 55 times more penicillin than the original static cultures. ◄◄ *Mutations (section 16.1)*

Protoplast Fusion

Protoplast fusion is another technique that can be used to promote genetic variability in microorganisms and plant cells. In this method, protoplasts—cells lacking a cell wall—are prepared by growing microbes in an isotonic solution in the presence of enzymes that degrade the cell wall (e.g., lysozyme for bacteria, chitinase for yeasts). The protoplasts of cells of the same or even different species physically fuse during coincubation. The cell wall is then regenerated during growth on osmotic stabilizers such as sucrose. Protoplast fusion is also inherently mutagenic and is especially useful for prompting recombination. For example, when protoplasts of *Penicillium roqueforti* were fused with those of *P. chrysogenum,* new industrially useful *Penicillium* strains were created.

Genetic Transfer Between Different Organisms

The transfer and expression of genes between different organisms can give rise to novel products, may change the way in which

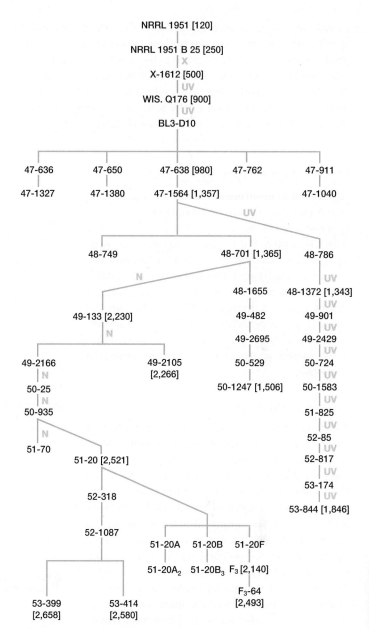

Figure 42.4 Mutation Makes It Possible to Increase Fermentation Yields. The genealogy of the mutational processes used to increase penicillin yields with *Penicillium chrysogenum*. These included X-ray treatment (X), UV treatment (UV), and mustard gas (N). By using these mutational processes, the yield was increased 20-fold. Unmarked transfers were used for mutant growth and isolation. Yields in International Units/mL in brackets.

genes of interest are regulated, or may even change the posttranslational modification of the protein product. Whereas protoplast fusion results in the in vivo recombination between microbial genomes, in vitro genetic transfer of specific genes enables a more targeted approach to strain construction. When genes from one organism are cloned into another and then transcribed and translated into protein, it is called **heterologous gene expression.** An important and early example of heterologous gene expression is the cloning and expression of the human insulin genes in

E. coli. Heterologous gene expression enables the production of specific proteins and peptides without contamination by other products that might be synthesized in the original organism. This approach can decrease the time and cost of recovering and purifying a product. Another major advantage of engineered protein production is that only biologically active stereoisomers are produced. This specificity is required to avoid the possible harmful side effects of inactive stereoisomers.

Modification of Gene Expression

In addition to inserting new genes in organisms, it also is possible to increase product yield by modifying regulatory molecules or the DNA sites to which they bind. For instance, the tight regulation of secondary metabolites such as antibiotics can be diminished by promoter mutations. The goal is to prompt microbes to begin antibiotic synthesis earlier and at a higher level than the parental strain. These approaches make it possible to overproduce a wide variety of products, such as antibiotics, amino acids, and enzymes of industrial importance.

Directed Evolution

As demonstrated in figure 42.4, mutants that overproduce a product or had some other desirable trait were traditionally obtained by "brute force" mutation. That is, mutations that conferred the trait were allowed to accumulate in a microbial strain after repeated rounds of mutagenesis and transfer from culture to culture. Production strains created in this way have a genotype quite different from the parental strain. Indeed, until the advent of genomic sequencing, the number and sites of most mutations remained a mystery. **Directed evolution** takes a more rational approach to the construction of production strains because only genes of interest are targeted for mutagenesis. In most cases, this requires that the gene sequences be altered in vitro, then replaced in the original strain or a heterologous host that may be better suited to produce the compound at high levels.

Several technical advancements were key to the development of direct evolution methodologies. The first is **site-directed mutagenesis.** In this technique, the nucleotide sequence of a specific gene is altered. The steps required are outlined in **figure 42.5.** Biotechnologists have discovered that making even minor amino acid substitutions can lead to unexpected changes in protein function and stability, resulting in higher yields of existing products or new products such as more environmentally resistant enzymes.

Site-directed mutagenesis changes only a few nucleotides at a time. A variety of other directed evolution techniques have been developed that can make larger changes to the gene of interest; some of these are listed in **table 42.2**. Careful inspection of table 42.2 reveals that many natural products are not proteins and therefore cannot be modified by simply mutating a single, structural gene. Instead, altering these nonprotein products requires an understanding of the biosynthetic

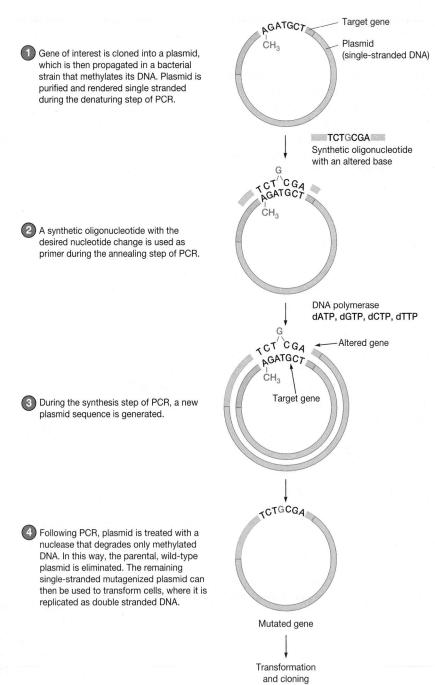

① Gene of interest is cloned into a plasmid, which is then propagated in a bacterial strain that methylates its DNA. Plasmid is purified and rendered single stranded during the denaturing step of PCR.

② A synthetic oligonucleotide with the desired nucleotide change is used as primer during the annealing step of PCR.

③ During the synthesis step of PCR, a new plasmid sequence is generated.

④ Following PCR, plasmid is treated with a nuclease that degrades only methylated DNA. In this way, the parental, wild-type plasmid is eliminated. The remaining single-stranded mutagenized plasmid can then be used to transform cells, where it is replicated as double stranded DNA.

Figure 42.5 Site-Directed Mutagenesis. A synthetic oligonucleotide is used to add a specific mutation to a gene.

Table 42.2   Some Directed Evolution Technologies

Technology	Overview of Approach	Examples of Products Generated
Genome-based strain reconstruction	A new strain is constructed based on the genotype of a production strain previously generated by "brute force." The new strain possesses only the mutations that are required for overproduction of the product.	Lysine
Metabolic pathway engineering	Rational approach to product improvement by either modification of specific biochemical reactions in a pathway, such as by site-directed mutagenesis of biosynthetic enzyme genes or the use of recombinant DNA technology to introduce new genes into the production strain. The goal is to optimize the flux of metabolites so the pathway operates most efficiently and the highest product yields possible are obtained.	Antibiotics: cephamycin C, neomycin, spiramycin, erythromycin L-lysine, aromatic amino acids, ethanol, vitamins
Assembly of designed oligonucleotides (ADO)	If the gene of interest contains a conserved sequence flanking the region to be mutated, oligonucleotides can be designed that include the conserved region and a mixture of point mutations in the region to be mutated. This combination of different (called degenerate) oligonucleotides is used to create variants by in vitro homologous recombination.	Lipases
Error-prone PCR	Variants of a gene are produced by using a DNA polymerase that is particularly error-prone during PCR.	Lycopene
DNA shuffling	Similar genes from different species are randomly fragmented and pooled, and new DNA fragments are generated by in vitro homologous recombination. Progeny sequences encoding desirable traits are identified; these new genes are then shuffled (bred) over and over again, creating new progeny that contain multiple desirable mutations.	Cephalosporinase
Whole genome shuffling	Similar to DNA shuffling described above, except entire genomes are recombined. Protoplast fusion may be used to obtain a single cell with the genomes of two microbial species so that recombination can occur.	Lactic acid

Modified from Adrio, J. L., and Demain, A. L. 2006; Genetic improvement of processes yielding microbial products. FEMS Microbiol. Rev. *30:187–214.*

pathway by which they are catalytically assembled. The genetic manipulation of enzymes with the goal of changing an industrially important product is sometimes called **metabolic engineering.** Metabolic engineering can be used to develop novel biochemical pathways that either improve product yield or generate new products.

The biosynthesis of certain types of natural products makes them particularly amenable to bioengineering. For instance, polyketide and certain peptide antibiotics have become models of bioengineering because the biosynthetic enzymes are grouped into large modular complexes, which function in an assembly-line fashion (**figure 42.6**). Each module within the complex catalyzes a specific type of reaction. For instance, the anticancer agents epothilones C and D are synthesized by a combination of polyketide synthase (PKS) and nonribosomal peptide synthetase (NRPS) enzymes. The first module is responsible for binding a starter unit and transferring it to the first extension module. Each extension module modifies the intermediate and hands it off to the next module in line until it reaches the last module and is released. The mixing and matching of domains encoding extension modules from hundreds of such modular multienzyme complexes can be used to generate new products. Many soil bacteria, such as members of the genera *Streptomyces* and *Bacillus*, possess PKS and NRPS genes. Biotechnologists are mining other microbes and metagenomes for novel PKS and NRPS genes.

The recognition that RNA is a versatile molecule that can have catalytic activity and control gene expression has given rise to a new class of therapeutic agents. Noting that RNA can fold into seemingly an infinite number of complex three-dimensional structures, molecular biologists use in vitro evolution to generate a library of up to 10^{15} different RNA molecules of interest. This procedure is called "systematic evolution of ligands by exponential enrichment" (SELEX) and is outlined in **figure 42.7**. Recall that a molecule that binds to a specific target or receptor is called a ligand, so in this case, the goal is to design an RNA ligand that has biological activity because it fits very tightly into a cleft or groove in a specific target molecule. This class of engineered RNAs have been given the name **aptamers** (Latin *aptus*, means "fitting"). In 2005 the first aptamer-based therapy, for the progressive vision disease macular degeneration, began clinical use. ◄◄ *Ribozymes (section 10.6)*

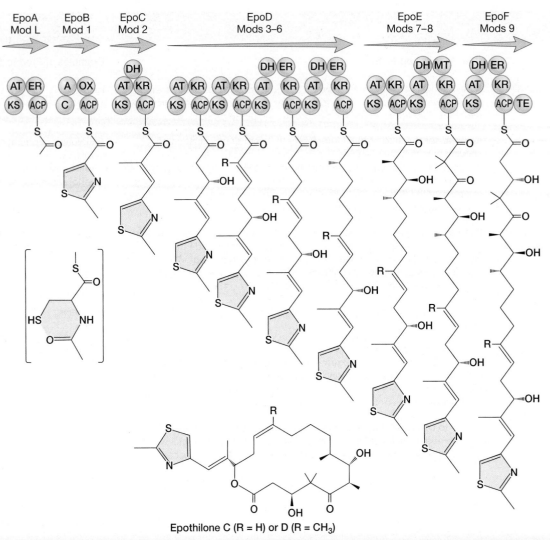

Figure 42.6 Modular Assembly of the Anticancer Agents Epothilone C and D. The genes encoding the enzymes that assemble this polyketide are found on the genome of the myxobacterium *Sorangium cellulosum*. They are arranged on the genome in the order of synthesis. The catalytic domains in the corresponding enzyme are shown as circles. These domains include keto-acyl synthase (KS), acyltransferase (AT), enoyl reductase (ER), acyl carrier protein (ACP), adenylation (A), condensation (C), oxidation (O), dehydratase (DH), ketoreductase (KR), and thioesterase (TE). The progression of synthesis of the molecule is shown, as is the mature structure.

MICRO INQUIRY *Why are modular genes particularly suitable as platforms for the development of new natural products of industrial interest?*

All directed evolution methodologies are designed to produce new genes, proteins, or RNAs that will create hundreds, if not thousands, of variants. Sorting through so many new candidate molecules requires a specific assay that assesses the desired trait, such as stability under reducing conditions or increased ability to bind a specific molecule. Selecting the best variant could represent a tremendous workflow bottleneck; however, the development of **high-throughput screening** (**HTS**) enables the rapid selection of a subset of desirable molecules from tens of thousands of candidates. HTS employs a combination of robotics and computer analysis to screen samples (proteins, natural compounds, and whole cells) for a specific trait, usually in 96-well

microtiter plates. The combination of directed evolution approaches and HTS has propelled industrial microbiology to a new level of exploration and efficiency not previously envisioned.

Synthetic Biology

So far we have discussed the modification of existing regulatory genetic networks and biochemical pathways. **Synthetic biology** looks beyond the constraints of what is already present in an organism and constructs or repurposes entirely new capabilities. Simply put, the new field of synthetic biology involves the engineering of biological systems (plants, microbes) to perform

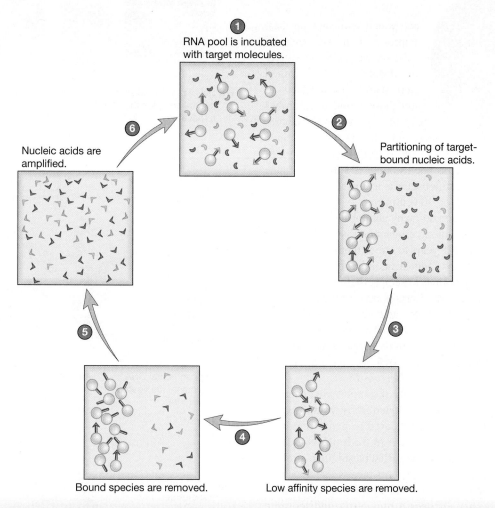

Figure 42.7 Directed Evolution of RNA Molecules. Systematic evolution of ligands by exponential enrichment (SELEX) starts with (1) an assortment of RNA molecules incubated with target molecule. (2) RNAs that have bound the target are separated from those that have not. (3) RNAs that have not bound tightly to the target molecule are removed. (4) RNAs that have bound tightly to the target are removed from the target so that (5) they can be amplified. (6) This pool of tight-binding RNAs is now subject to repeated rounds of SELEX so that a molecule with the highest binding affinity can evolve.

MICRO INQUIRY *Why is SELEX considered an "enrichment" technique?*

completely novel functions. Examples include the design of microorganisms that seek out and kill cancerous tumor cells and the redesign of photosynthesis to produce excess energy that can be fed into existing electrical grids.

There is no set of instructions for something so complex as designing a microbe to perform a completely novel function. Each project requires careful planning followed by the iterative process of design, build, test, redesign, rebuild, retest, and so on. For instance, if one wanted to engineer a bacterium to produce a biofuel, the first step would be to review literature to determine what genes are needed (planning). The next step would be to figure out how these genes are going to be cloned and what bacterial species should host them (design). Once the genes are inserted into the host (build), the amount of biofuel made would be quantified (test). It is safe to say that perfection

is never attained with the first try. However, the results inform the synthetic biologist of what steps need redesign (or if the entire design should be scrapped and fresh approach taken). Synthetic biology has the potential to generate a vast diversity of products, and researchers at the J. Craig Venter Institute have prepared for such versatility by constructing and expressing a completely synthetic genome in mycoplasma cells (*see p. 412*). This genome serves as the platform into which new genes can be added for the production of novel compounds.

Metagenomics

As we have learned, techniques for growing most microbes found in nature have not yet been developed. While having a pure culture of the microorganism that produces an interesting compound is always the best approach, industrial microbiologists would be at a great disadvantage if they limited their search for products to only those found in microbes that can be grown in the laboratory. Indeed, this would exclude over 90% of microbial gene products. The example of antibiotics made by actinobacteria is instructive. Worldwide, the top 10 centimeters of many soils are estimated to contain so many actinobacteria that have never been screened for antibiotic production that at the current rate, it would take another 128 years to screen them all, assuming they can be easily cultured. Industrial microbiologists have therefore turned to metagenomics to sample the genetic diversity from a number of environments. Certain areas of the globe have been dubbed "biodiversity hot spots," and these are of particular interest when exploring genetic diversity. The process of exploring nature for new and potentially useful organisms and their products is called **bioprospecting.** ◀◀ *Metagenomics provides access to uncultured microbes (section 18.3)*

Metagenomic libraries from natural samples are generally screened in two ways. First, all the cloned DNA from the environment can be sequenced and compared to genes that have already been characterized *(see figure 18.8)*. In this way, new versions of known genes can be identified based upon similarity to genes that are already known. This means that genes for which no function can be deduced will be ignored. But what if one of these novel genes encodes a previously undiscovered protein with a highly desirable activity? A functional screen of the metagenomic library can be used to explore this possibility (**figure 42.8**). In a functional screen, the genes are cloned into vectors and expressed in microbes such as *E. coli* or *Bacillus subtilis*, which are then screened for the acquisition of a new, specific function. Of course, these two lines of inquiry are not mutually exclusive. Keep in mind that once interesting genes are identified, they can undergo directed evolution techniques to optimize products for industrial use. ◀◀ *Genomic libraries (section 17.5)*

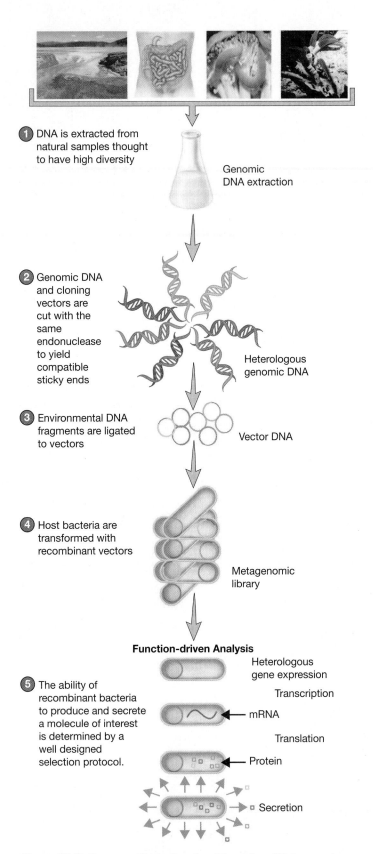

1. DNA is extracted from natural samples thought to have high diversity

Genomic DNA extraction

2. Genomic DNA and cloning vectors are cut with the same endonuclease to yield compatible sticky ends

Heterologous genomic DNA

3. Environmental DNA fragments are ligated to vectors

Vector DNA

4. Host bacteria are transformed with recombinant vectors

Metagenomic library

Function-driven Analysis

Heterologous gene expression

Transcription

5. The ability of recombinant bacteria to produce and secrete a molecule of interest is determined by a well designed selection protocol.

mRNA

Translation

Protein

Secretion

Figure 42.8 Bioprospecting by Functional Screening of Metagenomic Libraries. Rather than trying to culture every organism in a natural sample to look for the production of an important, novel compound, a metagenomic library can be constructed and expressed in host bacteria, which are screened for the desired protein function. This is usually performed in a high-throughput fashion.

42.5 Agricultural Biotechnology Relies on a Plant Pathogen

After reading this section, you should be able to:

- Explain how the natural infection process used by *Agrobacterium tumefaciens* has been leveraged for the genetic modification of plants
- Defend the use of Bt as a pesticide

A plasmid from the plant pathogenic bacterium *Agrobacterium tumefaciens* is chiefly responsible for the successful genetic engineering of plants. In nature, infection of dicotyledonous (dicot) plant cells by this bacterium transforms the cells into tumor cells, and crown gall disease develops. Only strains of *A. tumefaciens* possessing a large conjugative plasmid called the Ti (tumor inducing) plasmid are pathogenic (*see figures 31.12 and 31.13*). This is because the Ti plasmid transfers a region known as T-DNA to the plant's chromosomes at sites where it is stably maintained. ◄◄ *Transposable elements move genes within and between DNA molecules (section 16.5)*

When the molecular nature of crown gall disease was recognized, it became clear that the Ti plasmid and its T-DNA had great potential as a vector for the insertion of recombinant "good" DNA into plant chromosomes. The Ti plasmid has been genetically engineered to improve its utility as a cloning vector. The tumor-inducing genes have been deleted, while elements found in other vectors (e.g., selectable marker and multicloning site; *see figure 17.10*) have been added. Genes required for infection of plant cells by the plasmid are retained. The gene or genes to be expressed by the plant are inserted into the T-DNA region between the direct repeats needed for integration into the plant genome (**figure 42.9**). Then the plasmid is returned to *A. tumefaciens,* plant cells are infected with the bacterium, and transformants are selected by screening for antibiotic resistance (or another trait encoded by T-DNA). Finally, whole plants are regenerated from the transformed cells.

While it is relatively easy to use the *A. tumefaciens* Ti plasmid to modify dicots such as potato, tomato, celery, lettuce, and alfalfa, its use in monocots has only recently been optimized. The creation of new procedures for inserting DNA into monocot plant cells may lead to the use of recombinant DNA techniques with many important crop plants such as corn, wheat, and other grains (*see p. 680*).

Another microbial application to agriculture involves the use of bacteria, fungi, and viruses as **bioinsecticides** and **biopesticides.** These are defined as biological agents; that is, microbes or their components that kill susceptible insects. Bacterial agents include a variety of *Bacillus* species; however, *B. thuringiensis* is most widely used. This bacterium is only weakly toxic to insects as a vegetative cell, but during sporulation, it produces an intracellular protein toxin crystal, the parasporal body, that kills insect groups (*see figure 23.10*). The parasporal crystal, after exposure to alkaline conditions in the insect hindgut, breaks apart to release protoxin. After protoxin reacts with a protease enzyme in the target insect's gut, active toxin is generated. Six of the active toxin units integrate into the plasma membrane to form a hexagonal-shaped pore through the midgut cell, as shown in **figure 42.10**. This leads to the loss of osmotic balance and ATP, and finally to cell lysis. ◄◄ *Order* Bacillales *(section 23.3)*

B. thuringiensis has become an important industrial microbe. The toxin is widely marketed as the insecticide **Bt,** and the toxin genes can be used to genetically modify plants. The manufacture of Bt begins by growing *B. thuringiensis* in fermenters. The spores and crystals are released into the medium when cells lyse. The medium is then centrifuged and converted to a dust or wettable powder for application to plants. This insecticide has been used on a worldwide basis for over 40 years. Unlike chemical insecticides, Bt does not accumulate in the soil or in nontarget animals. Rather, it is readily lost from the environment by microbial and abiotic degradation.

Unlike many other genetically modified organisms, transgenic plants expressing the *B. thuringiensis* toxin gene, *cry* (for *cry*stal), have generally been well accepted. The widespread acceptance of these plants reflects the history of safe application of Bt as an insecticide without adverse environmental or health impacts. In addition, it is well understood that the Cry protein can only be activated in target insects. Long-term studies have shown that Bt is nontoxic to mammals and is not an allergen in humans. One potential problem, the horizontal gene flow of the *cry* gene to weeds and other plants, has not been reliably demonstrated. On the contrary, the decreased need for insecticides when cultivating Bt-modified crops has resulted in the recovery of nontarget insects that were also eliminated by insecticide application. Many of these insects eliminate undesirable pest insects, resulting in enhanced levels of biocontrol where Bt crops have been sown.

Retrieve, Infer, Apply

1. What is the Ti plasmid and how is it modified for the genetic modification of plants?
2. How is Bt toxin produced and why is it so widely accepted?
3. Can you think of other traits that might be useful, either to the farmer or the consumer, that could be introduced into plants?

42.6 Some Microbes Are Products

After reading this section, you should be able to:

- Describe the use of microbes in nanotechnology
- Explain the utility of biosensors

So far, we have discussed the use of microbial products to meet defined goals. However, microbial cells can be marketed as valuable products. Perhaps the most common example is the inoculation of legume seeds with rhizobia to ensure efficient nodulation and nitrogen fixation. Here we introduce several other microbes and microbial structures that are of industrial or agricultural

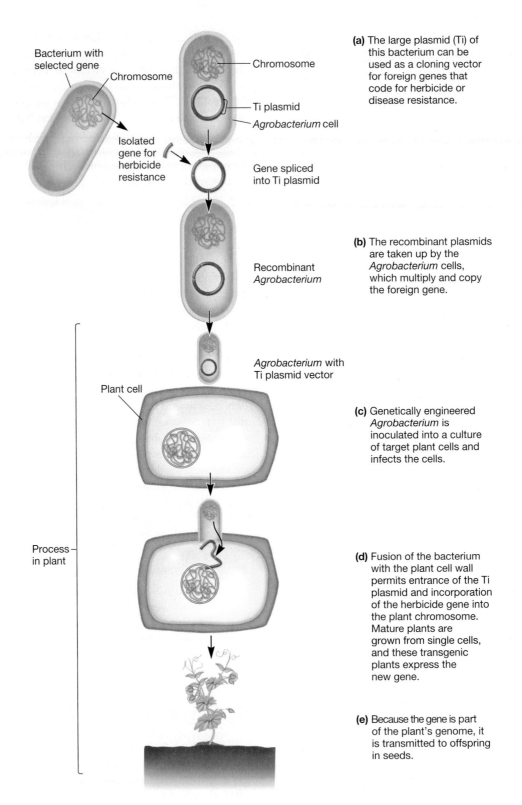

Bacterium with selected gene

Chromosome

Chromosome

Ti plasmid

Agrobacterium cell

Isolated gene for herbicide resistance

Gene spliced into Ti plasmid

(a) The large plasmid (Ti) of this bacterium can be used as a cloning vector for foreign genes that code for herbicide or disease resistance.

Recombinant *Agrobacterium*

(b) The recombinant plasmids are taken up by the *Agrobacterium* cells, which multiply and copy the foreign gene.

Agrobacterium with Ti plasmid vector

Plant cell

(c) Genetically engineered *Agrobacterium* is inoculated into a culture of target plant cells and infects the cells.

Process in plant

(d) Fusion of the bacterium with the plant cell wall permits entrance of the Ti plasmid and incorporation of the herbicide gene into the plant chromosome. Mature plants are grown from single cells, and these transgenic plants express the new gene.

(e) Because the gene is part of the plant's genome, it is transmitted to offspring in seeds.

Figure 42.9 Bioengineering of Plants. Most techniques employ a genetically modified strain of the natural tumor-producing bacterium called *Agrobacterium tumefaciens.*

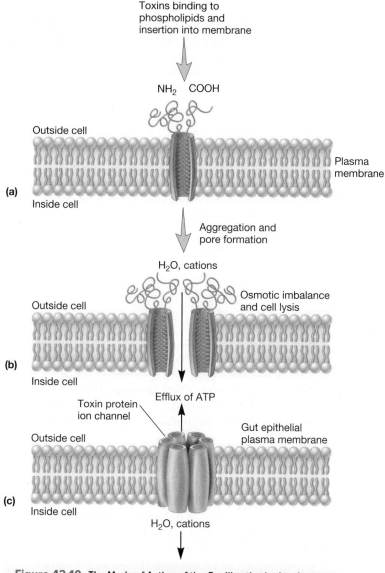

(a)

Toxins binding to
phospholipids and
insertion into membrane

NH$_2$ COOH

Outside cell

Plasma
membrane

Inside cell

Aggregation and
pore formation

H$_2$O, cations

(b)

Outside cell

Osmotic imbalance
and cell lysis

Inside cell

Efflux of ATP

Toxin protein
ion channel

(c)

Outside cell

Gut epithelial
plasma membrane

Inside cell

H$_2$O, cations

**Figure 42.10 The Mode of Action of the *Bacillus thuringiensis*
Toxin.** (a) Insertion of the 68 kDa active toxin molecules into the
membrane. (b) Aggregation and pore formation, showing a cross section
of the pore. (c) Final hexagonal pore, which causes an influx of water
and cations as well as a loss of ATP, resulting in cell imbalance and lysis.

MICRO INQUIRY *Why is it advantageous to have the Bt toxin
inactive until it reaches the insect gut?*

relevance. ◄◄ *Nitrogen-fixing bacteria are vital to agriculture
(section 31.3)*

Diatoms have aroused the interest of nanotechnologists. These
photosynthetic protists produce intricate silica shells that differ according to species (**figure 42.11**). Nanotechnologists are interested
in diatoms because they create precise structures at the micrometer
scale that can be used for making optical, catalytic, or electrical
materials. Three-dimensional structures in nanotechnology are currently built plane by plane, and meticulous care must be taken to
etch each individual structure to its final, exact shape. Diatoms, on
the other hand, build directly in three dimensions and do so while

growing exponentially. The silica in these shells can be
chemically converted to silicon to generate nanoelectronics.
Another application involves the use of diatom shells to coat
solar energy cells. This is attractive because diatom shells
capture three times more electrons than the standard coat-
ings. ◄◄ Stramenopila *(section 25.4)*

Magnetotactic bacteria are also of interest to nanotechnologists. Magnetosomes are formed by certain bacteria that convert iron to magnetite. The sizes and shapes
of the magnetosomes differ among species, but like diatom shells, they are perfectly formed despite the fact that
they are only tens of nanometers in diameter. Although
tiny, magnetic beads can be chemically synthesized, they
are not as precisely formed and lack the membrane that
surrounds magnetosomes. This membrane enables the attachment of useful biological molecules such as enzymes
and antibodies. Potential applications for magnetosomes

(a)

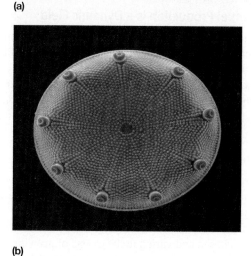

(b)

**Figure 42.11 Marine Diatom Surface Features Are Attractive Platforms
for Nanotechnology.** The precise patterns on the silica shells of these
protists have nanotechnology applications, including biosensing, drug
delivery, and the development of electronic, photonic, and structural
materials. (a) *Actinoptychus* sp. (b) *Aulacodiscus oregonus*. These species
exemplify the precise patterns found on diatom silica shells.

currently under investigation include their use as a contrast medium to improve magnetic resonance tomography (MRI) and as biological probes to detect cancer at early stages. ◄◄ *Inclusions (section 3.6)*

Another rapidly developing area of biotechnology is that of **biosensor** production. In this field of bioelectronics, living microorganisms (or their enzymes or organelles, and most recently diatom frustules) are linked with electrodes, and biological reactions are converted into electrical currents. Biosensors are used to measure specific components in beer, monitor pollutants, detect flavor compounds in food, and study environmental processes such as changes in biofilm concentration gradients. Using biosensors, it is possible to measure the concentration of substances from many

different environments. Biosensors have been developed using immunochemical-based detection systems. These biosensors detect pathogens, herbicides, toxins, proteins, and DNA. Many of these biosensors are based on the use of a streptavidin-biotin recognition system (*see Techniques & Applications 17.1*).

Retrieve, Infer, Apply

1. Why are diatoms and magnetotactic bacteria of interest to nanotechnologists? What do you think might be special challenges when growing these microbes for industrial applications?

2. In what areas are biosensors being used to assist in chemical and biological monitoring efforts?

Key Concepts

42.1 Microbes Are the Source of Many Products of Industrial Importance

- A wide variety of compounds are produced in industrial microbiology that impact our lives in many ways. These include antibiotics, amino acids, organic acids, biopolymers, and biosurfactants.

- Microorganisms also can be used as biocatalysts to carry out specific chemical reactions.

- Vaccine development continues to challenge biotechnologists; reverse vaccinology identifies potential targets and uses their structure and function to inform vaccine production.

42.2 Biofuel Production Is a Dynamic Field

- The field of microbial energy conversion has become extremely relevant. This includes the production of biofuels and the development of microbial sources of H_2.

- Although corn is currently the most common substrate used for the microbial production of ethanol, the use of both corn as substrate and ethanol as product is being challenged.

42.3 Growing Microbes in Industrial Settings Presents Challenges

- Microorganisms can be grown in controlled environments of various types using fermenters and other culture systems (**figure 42.3**).

- If defined constituents are used, growth parameters can be chosen and varied in the course of growing a microorganism. This approach is used particularly for the production of amino acids, organic acids, and antibiotics.

42.4 Production Strains Are Developed to Maximize Output of Industrially Important Compounds

- Historically, microorganisms used in industry were isolated from nature and modified by the use of classic mutational

techniques. Biotechnology involves the use of molecular techniques to modify and improve microorganisms.

- Finding new microorganisms in nature for use in biotechnology is a continuing challenge. Only about 1% of the observable microbial community has been grown, but major advances in growing uncultured microbes are being made.

- Selection and mutation continue to be important approaches for identifying new microorganisms.

- Site-directed mutagenesis and protein engineering are used to modify genes and their expression. These approaches are leading to new and often different products with novel properties (**figures 42.5** and **42.6**).

- Directed evolution includes a suite of techniques designed to target specific molecules for change with the goal of optimizing their function in a particularly useful way (**table 42.2** and **figure 42.7**).

- Metagenomics has become increasingly important in the search for new products, which are detected in a metagenomic functional screen (**figure 42.8**).

42.5 Agricultural Biotechnology Relies on a Plant Pathogen

- Many plants are genetically modified through the use of the Ti plasmid of *Agrobacterium tumefaciens*. The plasmid can be engineered to introduce genes whose products confer new traits to plants (**figure 42.9**).

- *Bacillus thuringiensis* is an important biopesticide, and the Bt gene has been incorporated into several important crop plants (**figure 42.10**).

42.6 Some Microbes Are Products

- Microorganisms are being used in a wide range of biotechnological applications such as nanotechnology and biosensors (**figures 42.11**).

Compare, Hypothesize, Invent

1. You have discovered a secondary metabolite made by an unusual fungus that has great industrial potential. Unfortunately, the fungus makes very little of this product, so you must develop a scheme to increase the yield. Discuss the pros and cons of each of the following strategies: (a) mutagenesis of the fungus followed by repeated passaging until a high-producing strain evolves, (b) heterologous expression of the genes and regulators responsible for producing this product in a bacterial host, (c) heterologous expression of the genes responsible for producing this product in a bacterial host but using a bacterial promoter that will drive constitutive expression of the product.

2. Brewer's yeast, *Saccharomyces cerevisiae,* does not metabolize lactose, but an engineered strain of *S. cerevisiae* able to consume lactose is of industrial interest. This is because such a strain can convert whey (an unusable by-product of cheese production) to ethanol. Such a strain has been engineered by cloning the genes that encode β-galactosidase and lactose permease from another yeast, *Kluyveromyces lactis,* into *S. cerevisiae.* However, the recombinant *S. cerevisiae* expresses the *K. lactis* genes only at a low level, so very little lactose is consumed. In an effort to increase the expression of these heterologous genes, a directed evolution approach was applied.

   Suggest one directed evolution technique that could be applied in vivo (on the entire organism) and one that could be used in vitro; that is, on the two genes from *K. lactis.* Which do you think would be easier to carry out? Which do you think would be more effective?

   Read the original paper: Guimaraes, P. M. R., et al. 2008. Adaptive evolution of a lactose-consuming *Saccharomyces cerevisiae* recombinant. *Appl. Environ. Microbiol.* 74:1748.

3. The detection of unexploded land mines is a horrific problem. Unexploded ordinances emit a small amount of nitroaromatic compounds, affording the opportunity to develop a biosensor. A *Pseudomonas putida* strain has been constructed with a mutated toluene responsive regulator. When activated by bi- and trinitro-substituted toluenes, the regulator triggers the expression of genes transcriptionally fused to GFP (*see figure 17.15*). When spread on soil spotted with nitrotoluenes, there is sufficient activation of GFP-generated light that the location of the chemicals is evident.

   Draw a flow chart of how the signal is emitted, received, and transduced into a light signal that would enable you to explain this biosensor to military personnel.

   Read the original paper: Garmendia, J., et al. 2008. Tracing explosives in soil with transcriptional regulators for *Pseudomonas putida* evolved for responding to nitrotoluenes. *Microbial Biotechnol.* 1:236.

Learn More

shop.mheducation.com   Enhance your study of this chapter with interactive study tools and practice tests. Also ask your instructor about the resources available through Connect, including adaptive learning tools and animations.

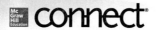   

Applied Environmental Microbiology

The explosion of the Deepwater Horizon offshore oil rig resulted in one of the worst environmental disasters in U.S. history. This photograph was taken while the rig was still burning; it eventually sank.

Deepwater Horizon Oil Consumed by Microbes

One of the planet's largest oil disasters occurred on April 20, 2010, when an explosion aboard BP's Deepwater Horizon oil rig resulted in the death of 11 people and the release of nearly 5 million barrels of oil over 83 days into the Gulf of Mexico. Many may recall the dramatic images of oil spewing from the uncapped well and the resulting oil slicks with booms skimming the oil from the water's surface. But about the fate of the oil between the seafloor and the surface? The release of petroleum at such depth resulted in the dissolution of methane and, to a lesser extent, ethane, propane, and butane. These dissolved gases formed lateral plumes trapped 800 to 1,200 meters below the lighter gas-free seawater.

Mass balance calculations revealed that several years later, 22% of the spilled oil could not be accounted for. Could microbes be responsible for some (or all) of this "missing" oil? Could they have mineralized the oil to CO_2 or immobilized it as biomass? Arriving at a satisfactory answer to this question required a combination of methods, including sequencing 16S rRNA genes, metagenomic and single-cell genome sequencing, measuring the flux of ^{14}C-labeled test hydrocarbons, just to name a few. In the end, the scientific community agreed that microbes present in the sediments degraded a large (but unknown) fraction of the oil. Interestingly, the microbial community changed as the oil was degraded. Early in the spill recovery when concentrations of saturated hydrocarbons were high, single-cell genome sequencing revealed genes for alkane degradation, chemotaxis, and motility. However, as alkane hydrocarbons declined, so did evidence of genes whose products degrade them. The loss of alkanes meant that polyaromatic hydrocarbons (PAHs) were enriched, meaning they represented a larger fraction of the undegraded oil. These large PAHs persist because they have proved to be resistant to microbial attack. These compounds are known to be toxic to eukaryotes, so even though microbial oxidation of many of the contaminants has been a helpful outcome, the persistence of PAHs in the environment could negatively impact ongoing Gulf Coast recovery efforts, and have implications for future bioremediation efforts elsewhere.

The Deepwater Horizon experience vividly illustrates that the natural decontamination of soils, waters, and wastewaters is largely a microbial process. Water cleanliness—whether to decontaminate or to prepare it for drinking—is a multidisciplinary endeavor, involving geology, biochemistry, and physics, among other fields. In this chapter, we address water and how to assess the purity of drinking water and as well as how to treat wastewater before returning it to the environment. We also discuss the means by which microbes can contribute to meeting the world's insatiable demand for energy and their role in degrading environmental contaminants—bioremediation.

Readiness Check:

Based on what you have learned previously, you should be able to:

✔ Identify several waterborne diseases (sections 38.4, 39.4, and 40.5)
✔ List environmental factors that influence microbial growth (section 7.4)
✔ Explain how the term redox potential is used to describe microbial habitats and how this influences the elements found in these habitats (section 28.1)
✔ Distinguish between mineralization and immobilization of an element (section 28.1)
✔ Compare and contrast carbon substrates that are readily degraded with those that are not (section 28.1)
✔ Sketch generalized carbon and nitrogen iron cycles (section 28.1)

43.1 Purification and Sanitary Analysis Ensure Safe Drinking Water

After reading this section, you should be able to:

■ Trace the water purification steps that are often used from source to tap
■ Describe how the U.S. Environmental Protection Agency regulates water purity
■ Discuss the use of indicator organisms and how they are enumerated

Access to clean, drinkable water is a fundamental human right, yet over a billion people worldwide do not enjoy a reliable source of potable water. Water that is not adequately monitored for purity can contain pathogens introduced from human and animal waste. The U.S. Environmental Protection Agency (EPA) has designated certain bacterial, viral, and protozoan pathogens that can survive (or thrive) in water as "microbial containment candidates" because they represent particularly severe health risks (**table 43.1**). When waters are used for recreation or are the source of fish and other food, the possibility for disease transmission exists. In many cases, such waters also are a source of drinking water.

Water purification is therefore a critical link in controlling disease transmission. The choice of purification method depends on the volume, source, and initial quality of the water. Many municipalities draw their water from surface sources, such as reservoirs. Surface water is usually purified by a process that consists of at least three steps (**figure 43.1**). First, the water may be held for a specified duration to allow larger particles to settle out in a **sedimentation basin.** Sometimes coagulants are also added in a procedure called **coagulation** or flocculation, which removes some microorganisms, organic matter, toxic contaminants, and suspended fine particles. The coagulated particles are called flocs. The partially clarified water is then moved to a **settling basin** and mixed with chemicals such as alum (aluminum sulfate) and lime to facilitate further precipitation. The water is further purified by passing it through **rapid sand filters.** Here sand of roughly 1 mm in diameter physically traps fine particles and flocs and removes up to 99% of the bacteria. After filtration, the water is disinfected. This step usually involves chlorination or ozonation.

When chlorination is employed, the chlorine dose must be large enough to leave residual free chlorine at a concentration of 0.2 to 2.0 milligrams per liter. The creation of **disinfection by-products** (**DBPs**) such as trihalomethanes (THMs), formed when chlorine reacts with organic matter, must be monitored because some DBPs are carcinogenic.

These purification processes remove or inactivate disease-causing bacteria and indicator organisms (coliforms). Unfortunately, the use of coagulants, rapid filtration, and chemical disinfection often does not remove *Giardia cysts, Cryptosporidium* oocysts, *Cyclospora* spp., and viruses. *Giardia intestinalis,* a cause of human diarrhea, is now recognized as the most commonly identified water borne pathogen in the United States (*see figure 25.4*). *Cryptosporidium* spp. are also a significant problem. These protists are smaller than *G. intestinalis* and even more difficult to remove from water, in part because they are resistant to chlorine and other disinfectants. A major source of *G. intestinalis* and *Cryptosporidium* spp. contamination is the Canada goose, a migratory bird with an ever-expanding population. Water supplies that receive terrestrial runoff are therefore at particular risk of contamination by these protists. ◄◄ *Food and water are vehicles of fungal and protozoal diseases (section 40.5)*

In the United States, the EPA has developed regulations called the Long Term 2 Enhanced Surface Water Treatment (LT2) Rule as a means of ensuring protection against pathogens. The LT2 rule applies to all public water systems that draw from surface waters or groundwater fed by surface waters. It sets a **maximum containment level goal** (**MCLG**) for specific pathogens. The EPA defines MCLGs as "health goals set at a level at which no known or anticipated adverse effects on health

Table 43.1   Microbial Water Contaminant Candidates Listed by the EPA

Organism	Concern
Caliciviruses	Viruses that cause relatively mild, self-limiting nausea, vomiting, diarrhea; includes norovirus
Campylobacter jejuni	Causes nausea, vomiting, and diarrhea; usually self-limiting
Entamoeba histolytica	Protozoan that causes short-term as well as long-lasting gastrointestinal (GI) illness
E. coli O157:H7	Enterohemorrhagic *E. coli* that causes severe GI illness; can also cause kidney failure
Helicobacter pylori	Bacterium that causes gastric ulcers and cancer
Hepatitis A virus	Causes liver disease leading to jaundice
Legionella pneumophila	Found in hot water systems; causes bacterial pneumonia when inhaled. Usually limited to elderly patients.
Naegleria fowleri	Protozoan found in warm surface and groundwater; causes primary amoebic meningoencephalitis
Salmonella enterica	Causes GI illness; severity depends on subspecies
Shigella sonnei	Causes GI illness, including bloody diarrhea
Vibrio cholerae	Causes cholera

Water purification steps

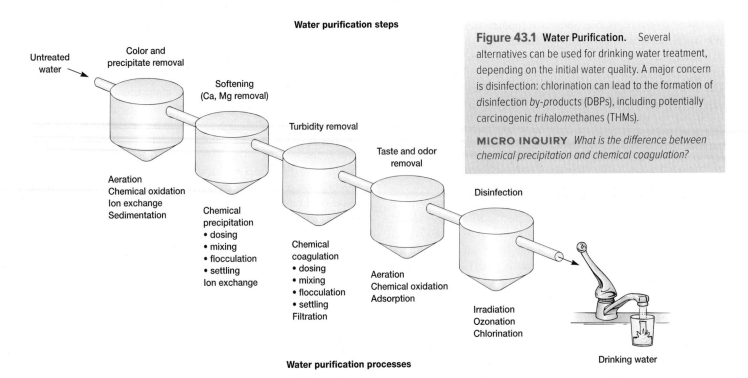

Figure 43.1 **Water Purification.** Several alternatives can be used for drinking water treatment, depending on the initial water quality. A major concern is disinfection: chlorination can lead to the formation of *d*isinfection *b*y-*p*roducts (DBPs), including potentially carcinogenic *tri*halomethanes (THMs).

MICRO INQUIRY *What is the difference between chemical precipitation and chemical coagulation?*

Water purification processes

of persons occur and which allows an adequate margin of safety." MCLGs are set at zero for *Giardia, Cryptosporidium,* and *Legionella* spp., and certain viruses. In addition, the Total Coliform Rule, which applies to all public water systems regardless of supply source, sets an MCLG of zero for total and fecal coliform bacteria. To meet these MCLGs, water systems found to be contaminated must develop additional water treatment strategies to provide 99.9% inactivation of the microbes. The treatment approach depends on the level of contamination but generally involves filtration techniques such as **slow sand filters.** This treatment involves the slow passage of water through a bed of sand in which a microbial layer (biofilm) covers the surface of each sand grain. Waterborne microorganisms are removed by adhesion to the gelatinous surface microbial layer. Coagulation and filtration reduce virus levels about 90 to 99%. Further inactivation of viruses can be achieved by chemical oxidants, high pH, and photooxidation. ◄◄ *Biofilms are common in nature (section 7.5)*

Sanitary Analysis of Waters

Once water has been purified, it must be deemed potable. Although a wide range of viral, bacterial, and protozoan diseases result from the contamination of water with human and other animal fecal wastes (table 43.1), the detection of **indicator organisms** as an index of possible water contamination by human pathogens has long been the standard approach to monitoring drinking water safety. Researchers are still searching for the "ideal" indicator organism to use in sanitary microbiology. Among the suggested criteria for such an indicator microbe are:

1. The indicator bacterium should be suitable for analysis of all types of water: tap, river, ground, impounded, recreational, estuary, sea, and waste.
2. It should be present whenever enteric pathogens are present.

3. It should survive longer than the hardiest enteric pathogen.
4. It should not reproduce in the contaminated water, as this would produce an inflated value.
5. It should be harmless to humans.
6. Its level in contaminated water should have some direct relationship to the degree of fecal pollution.
7. The assay procedure for the indicator should have great specificity; in other words, other bacteria should not give positive results. In addition, the procedure should have high sensitivity and detect low levels of the indicator.
8. The testing method should be easy to perform.

Coliforms, including *Escherichia coli,* are members of the family *Enterobacteriaceae (see p. 527).* These bacteria are commonly found in the intestines of humans and other animals, and are used widely as indicator organisms. They lose viability in freshwater at slower rates than most of the major intestinal bacterial pathogens. When such "foreign" enteric indicator bacteria are not detectable in a specific volume (generally 100 milliliters) of water, the water is considered potable (Latin *potabilis,* fit to drink). ◄◄ *Order* Enterobacteriales *includes pathogens and mutualists (section 22.3)*

The original test for coliforms involved the presumptive, confirmed, and completed tests, as shown in **figure 43.2.** The presumptive step is carried out by means of tubes inoculated with three different sample volumes to give an estimate of the **most probable number (MPN)** of coliforms in the water *(see figure 29.3).* The complete process, including the confirmed and completed tests, requires at least 4 days of incubations and transfers.

The coliforms include a wide range of bacteria whose primary source may not be the intestinal tract. To address this, tests have been developed that assay for the presence of **fecal coliforms.** These are coliforms from the intestines of warm-blooded animals, which grow at the more restrictive temperature of 44.5°C. Other indicator

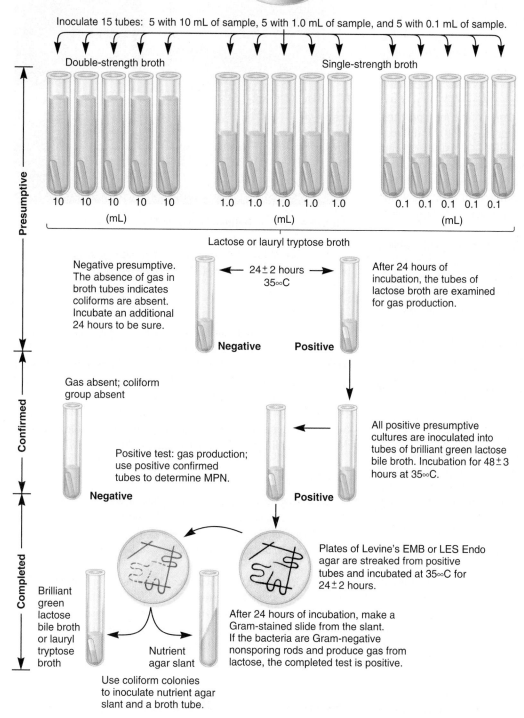

Water sample

Inoculate 15 tubes: 5 with 10 mL of sample, 5 with 1.0 mL of sample, and 5 with 0.1 mL of sample.

Double-strength broth

Single-strength broth

| 10 | 10 | 10 | 10 | 10 |
| (mL) | | | | |

| 1.0 | 1.0 | 1.0 | 1.0 | 1.0 |
| (mL) | | | | |

| 0.1 | 0.1 | 0.1 | 0.1 | 0.1 |
| (mL) | | | | |

Lactose or lauryl tryptose broth

Negative presumptive. The absence of gas in broth tubes indicates coliforms are absent. Incubate an additional 24 hours to be sure.

24±2 hours
35∞C

After 24 hours of incubation, the tubes of lactose broth are examined for gas production.

Negative   **Positive**

Gas absent; coliform group absent

Positive test: gas production; use positive confirmed tubes to determine MPN.

All positive presumptive cultures are inoculated into tubes of brilliant green lactose bile broth. Incubation for 48±3 hours at 35∞C.

Negative   **Positive**

Brilliant green lactose bile broth or lauryl tryptose broth

Nutrient agar slant

Plates of Levine's EMB or LES Endo agar are streaked from positive tubes and incubated at 35∞C for 24±2 hours.

After 24 hours of incubation, make a Gram-stained slide from the slant. If the bacteria are Gram-negative nonsporing rods and produce gas from lactose, the completed test is positive.

Use coliform colonies to inoculate nutrient agar slant and a broth tube.

Presumptive — Confirmed — Completed

Figure 43.2 The Multiple-Tube Fermentation Test. The multiple-tube fermentation technique has been used for many years for the sanitary analysis of water. Lactose broth tubes are inoculated with different water volumes in the presumptive test. Tubes that are positive for gas production are inoculated into brilliant green lactose bile broth in the confirmed test, and positive tubes are used to calculate the most probable number (MPN) value. The completed test is used to establish that coliform bacteria are present.

microorganisms include **fecal enterococci.** These microbes are increasingly being used as an indicator of fecal contamination in brackish and marine water. In saltwater, these bacteria die more slowly than fecal coliforms, providing a more reliable indicator of possible recent pollution. ◄◄ *Order* Lactobacillales *(section 23.3)*

The **membrane filter technique** is a common and preferred method of evaluating the microbiological characteristics of water. The water sample is passed through a membrane filter. The filter with its trapped bacteria is transferred to the surface of a solid medium or to an absorptive pad containing the desired liquid medium. Specific media enable the rapid detection of total coliforms, fecal coliforms, or fecal enterococci by the presence of their characteristic colonies. Samples may first be placed on a less selective resuscitation medium or incubated at a less stressful temperature prior to growth under the final set of selective conditions. An example of such a resuscitation step is the use of a 2-hour incubation on a pad soaked with lauryl sulfate broth, as is carried out in the LES Endo procedure. A resuscitation step often is needed with chlorinated samples, where the microorganisms are especially stressed. Membrane filters have been widely used with water that does not contain high levels of background organisms, sediment, or heavy metals. ◀◀ *Mechanical removal methods rely on barriers (section 8.3)*

More simplified tests for detecting coliforms and fecal coliforms are now available. The **presence-absence (P-A) test** can be used for coliforms. This is a modification of the most probable number procedure in which a larger water sample (100 milliliters) is incubated in a single culture bottle with a triple-strength broth containing lactose broth, lauryl tryptose broth, and bromocresol purple indicator. The P-A test is based on the assumption that no coliforms should be present in 100 milliliters of drinking water. A positive test results in the production of acid (a yellow color), which requires confirmation.

To test for both coliforms and *E. coli,* the related **Colilert defined substrate test** can be used. A water sample of 100 milliliters is added to a specialized medium containing *o*-nitrophenyl-β-D-galactopyranoside (ONPG) and 4-methylumbelliferyl-β-D-glucuronide (MUG) as the only nutrients. If coliforms are present, the medium will turn yellow within 24 hours at 35°C due to the hydrolysis of ONPG, which releases *o*-nitrophenol (**figure 43.3**). To check for *E. coli,* the medium is observed under long-wavelength ultraviolet (UV) light for fluorescence. When *E. coli* is present, MUG is modified to yield a fluorescent product. If the test is negative for the presence of coliforms, the water is considered acceptable for human consumption. Current standards dictate that water be free of coliforms and fecal coliforms. If coliforms are present, tests for fecal coliforms or *E. coli* must be performed.

The use of indicator organisms as a means of detecting all pathogenic microbes is limited. Some pathogens simply do not co-occur with indicator organisms. Another problem is that indicator organisms can be more sensitive to disinfectants; thus when drinking water is tested after such treatment, indicator-based assays may be misleading. In addition, these culture-based methods generally require at least 18 hours to complete, and some assays require almost a week. The consequences of these shortcomings were evident in 1993, when about 400,000 Milwaukee, Wisconsin, residents became ill from drinking water contaminated with *Cryptosporidium parvum.* Currently the Centers for Disease Control and Prevention estimates that in the United States, a million people per year are sickened from drinking impure water and another 1,000 die annually.

It is clear that effective, reliable, rapid, and quantitative methods for determining the presence or absence of specific pathogens are needed. Molecular approaches can be all these things—with the ability to quickly detect multiple, specific pathogens in a single water sample. Some of the methodologies that have been developed include the flow cytometry (*sections 29.2 and 36.3*), fluorescent in situ hybridization (*FISH; section 29.2*), quantitative PCR (*section 17.2*), and microarrays (*sections 18.5 and 29.3*). One example is *Legionella pneumophila* contamination, which is routinely detected by PCR amplification. Results are reported in

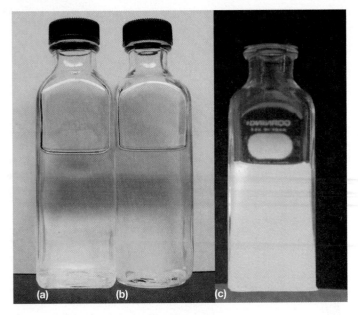

Figure 43.3 The Defined Substrate Test. This test is used to detect coliforms and fecal coliforms in single 100-milliliter water samples. The medium uses ONPG and MUG as defined substrates. (a) Sterile control. (b) Yellow color due to the presence of coliforms. (c) Fluorescence resulting from the presence of fecal coliforms.

genome units per liter, but the equivalence of these units to colony forming units is unclear. Many communities routinely use 16S rRNA amplification to detect coliforms in waters and other environments, including foods. Following a short enrichment step, this approach can detect 1 colony forming unit (CFU) of *E. coli* per 100 milliliters of water. This technique can also be used to differentiate between nonpathogenic and enterotoxigenic strains, including the Shiga-like toxin–producing *E. coli* O157:H7.

Retrieve, Infer, Apply

1. What steps are usually taken to purify drinking water?
2. Why is chlorination, although beneficial in terms of bacterial pathogen control, of environmental concern?
3. How can *Giardia* and *Cryptosporidium* cysts be removed from a water supply?
4. Why is it important that an indicator microorganism survive but not grow in the water to be tested?
5. Why do you think workers who perform these tests must practice excellent personal hygiene and be specifically trained in the implementation of these techniques?

43.2 Wastewater Treatment Maintains Human and Environmental Health

After reading this section, you should be able to:

- Compare and contrast the processes that occur during primary, secondary, and tertiary wastewater treatment
- Explain how, in general, septic systems operate
- Identify at least two challenges for the future of wastewater treatment

Table 43.2	Common Contaminants of Wastewater
Contaminant	**Problems Caused by Contaminant**
Suspended solids	Can cause deposition of sludge leading to anoxic conditions when discharged into aquatic environments
Biodegradable organic compounds	Includes proteins, fats, and carbohydrates that when discharged into streams and rivers cause increased heterotrophic growth leading to anoxic conditions
Pathogenic microorganisms	Cause infectious disease
Priority pollutants	Include organic and inorganic compounds that may be toxic, carcinogenic, mutagenic, or teratogenic
Refractory organic compounds	Organic compounds such as phenols, surfactants, agricultural pesticides, and other toxic compounds that resist conventional wastewater treatment
Heavy metals	Usually discharged by industry; discharge into rivers and streams is toxic to all trophic levels
Dissolved inorganic constituents	Include calcium, sodium, and sulfate that may be added to domestic water supplies and may have to be removed prior to wastewater reuse

Source: Modified from Metcalf and Eddy, Inc. Wastewater Engineering, *4th ed. McGraw-Hill, 2002.*

The flip side of ensuring safe drinking water is wastewater management. In fact, the combination of wastewater management, safe drinking water, vaccines, and antibiotics accounts for most of the gain in longevity accomplished in the last century. Citizens of Western nations take this for granted, yet about 40% of the Earth's population lacks basic sanitation.

Wastewater includes sewage as well as industrial and agricultural effluent and street runoff collected in storm sewers. These waters often contain high levels of organic matter, heavy metals, nutrients, and particulates (**table 43.2**). We begin our discussion of wastewater treatment with an overview of large-scale wastewater treatment processes, followed by the means by which water quality is monitored. We then review home treatment systems.

Wastewater Treatment Processes

An aerial photograph of a modern sewage treatment plant is shown in **figure 43.4**. Wastewater treatment involves at least three spatially segregated steps: primary, secondary, and tertiary treatment (**table 43.3**). At the end of the process, the water is usually chlorinated or treated with ozone before it is released.

Primary treatment prepares wastewater for treatment by physically removing much of the solid material. This can be accomplished in several ways, including screening, precipitation of small particulates, and settling in basins or tanks (**table 43.4**).

Figure 43.4 Aerial View of a Conventional Sewage Treatment Plant. Sewage treatment plants allow natural processes of self-purification that occur in rivers and lakes to be carried out under more intense, managed conditions in large concrete vessels.

MICRO INQUIRY *This plant does not use an anaerobic digester. What would be the advantage of this additional step?*

Table 43.3	Major Steps in Primary, Secondary, and Tertiary Treatment of Wastes
Treatment Step	**Processes**
Primary	Removal of insoluble particulate materials by settling, screening, addition of alum and other coagulation agents, and other physical procedures

Secondary	Biological removal of dissolved organic matter
	    Trickling filters
	    Activated sludge
	    Lagoons
	    Extended aeration systems
	    Anaerobic digesters

Tertiary	Biological removal of inorganic nutrients Chemical removal of inorganic nutrients Virus removal/inactivation Trace chemical removal

Table 43.4   Wastewater Treatment Methods

Nature of Operation	Process	Brief Explanation
Physical operations	Screening	Mechanical removal of large suspended materials with parallel bars, rods, wires, grating, wire mesh, or perforated plates
	Comminution	Comminutors are cutting instruments used to pulverize large floating material before wastewater enters primary settling tank.
	Flow equalization	A basin or basins used to adjust the rate of wastewater entry into secondary and tertiary treatment facilities
	Sedimentation	Also called clarification, this widely used procedure is based on the gravitational separation of large particles. Used in primary settling basins and to separate biological floc in activated sludge settling basins. Also used to precipitate chemically treated particulates when chemical coagulation is employed.
	Flotation	The removal of particles by bubbling air or other gas into the water. Particles float to the surface and are collected by skimming; removes fine particles more efficiently than sedimentation. Chemical additives can be added to improve particle removal.
	Granular-medium filtration	Wastewater is passed through a filter bed composed of granular material such as sand; used to remove suspended solids and chemically precipitated phosphorus.
Chemical operations	Chemical precipitation	Addition of coagulants such as alum, ferric chloride, ferric sulfate, or lime to wastewater prior to sedimentation; facilitates the flocculation of finely divided solids into flocs that will settle
	Adsorption	Soluble material is collected by adsorption of activated carbon; usually performed after biological treatment to remove remaining dissolved organic carbon
	Disinfection	Elimination of pathogens (and other microbes) by a variety of methods including UV light, heat, gamma radiation, and chemical agents such as chlorine, bromine, ozone, phenolic compounds, and others
	Dechlorination	Removal of all chlorine residues before discharge into receiving waters. These compounds are reactive and potentially dangerous.
Biological operations	Activated sludge process	An aerobic, continuous flow system that uses microbes to aerobically degrade organic matter; occurs after primary settling. Material is mechanically mixed to maintain aeration and prevent settling (figure 43.6a).
	Aerated lagoon	A basin 1 to 4 m in depth where wastewater is treated in a process similar to activated sludge treatment. Turbulence created by aeration keeps the contents of the lagoon mixed.
	Trickling filters	Most commonly used aerobic attached-growth biological treatment. A large basin filled with rock or plastic packing material provides a substrate for microbial biofilm growth and increased contact area for wastewater (figure 43.6b).
	Rotating biological contractors	Attached growth process that features large cylinders partially submerged in wastewater. Biofilms form on the cylinders, which are continuously rotated.
	Stabilization ponds	Shallow basin that is mixed naturally or mechanically. Aerobic ponds treat soluble organic wastes and effluents from wastewater treatment facilities; anaerobic ponds stabilize organic wastes. Wastewater retention times from 1 to 3 months
	Anaerobic digestion	Used for treatment of sludge and wastewater with high organic content (figure 43.7)
	Biological nutrient removal	Removal of nitrogen by nitrification-denitrification and anammox reaction (see figure 21.15) and phosphorus, often by proprietary processes (e.g., A/O, PhoStrip, and SBR process)

Modified from Water-Water Treatment Technologies: A General Review. New York: United Nations, 2003.

The resulting solid material is usually called **sludge.** In general, the goal of primary treatment is to generate water (effluent) that can be further purified by biological treatment and produce sludge that can be treated before disposal.

Secondary treatment of the effluent promotes the biological transformation of dissolved organic matter (DOM) into microbial biomass and carbon dioxide. In addition to reducing degradable DOM by about 90 to 95%, many bacterial pathogens are removed by this process. Several approaches can be used in secondary treatment to remove DOM. When microbial growth is completed, under ideal conditions the microorganisms will aggregate and form stable flocs, also called **granular sludge,** that settle. As shown in **figure 43.5**, commercial coagulants are often added to wastewater effluent to increase rate of flocculation and the amount of material that settles. Efficient flocculation is important because a sewage treatment facility that generates stable granular sludge requires up to 75% less space and reduces cost by as much as 25%.

When these processes occur with lower O_2 levels or with a microbial community that is too young or too old, unsatisfactory floc formation and settling can occur. The result is a **bulking sludge,** caused by the massive development of filamentous bacteria such as those in the genera *Sphaerotilus* and *Thiothrix*, together with many poorly characterized filamentous microbes. These filamentous microorganisms form flocs that do not settle well and thus produce effluent quality problems. ◀◀ *Order* Thiotrichales *includes large filamentous bacteria (section 22.3)*

An aerobic **activated sludge** system (**figure 43.6***a*) involves the horizontal flow of materials with recycling of sludge—the active biomass that is formed when organic matter is oxidized and degraded by microorganisms. Activated sludge systems can be designed with variations in mixing. In addition, the ratio of organic matter added to the active microbial biomass can be varied. A low-rate system (low nutrient input per unit of microbial biomass), with slower growing microorganisms, will produce an effluent with low residual levels of dissolved organic matter. A high-rate system (high nutrient input per unit of microbial biomass), with faster growing microorganisms, will remove more dissolved organic carbon per unit time but produce a poorer quality effluent.

Aerobic secondary treatment is often carried out with a **trickling filter** (figure 43.6*b*). The waste effluent is passed over rocks or other solid materials upon which microbial biofilms have developed, and the microbial community degrades the organic waste. A sewage treatment plant can be operated to produce less sludge by employing the **extended aeration** process (figure 43.6*c*). Microorganisms grow on the dissolved organic matter, and the newly formed microbial biomass is eventually consumed to meet maintenance energy requirements. This requires extremely large aeration basins and long aeration times. In addition, with the biological self-utilization of the biomass, minerals originally present in the microorganisms are again released to the water.

All aerobic processes produce excess microbial biomass, or sewage sludge, which contains many recalcitrant organics. Often the sludge from aerobic sewage treatment, together with the materials that settled in primary treatment, are further treated by anaerobic digestion. **Anaerobic digestion** occurs in large tanks

Figure 43.5 **Proper Floc Formation in Activated Sludge.** Microorganisms play a critical role in the functioning of activated sludge systems. Wastewater effluent in the beaker on the far left contains bacteria and other organic materials that resist settling. In this case, a commercial coagulant containing ferric chloride is added (second beaker from left). This accelerates the formation of stable flocs, which settle at the bottom of the beaker. The beaker on the far right shows the effluent after the settled material has been removed.

MICRO INQUIRY *Why is proper floc formation critical to the function of a treatment facility?*

designed to operate with continuous input of untreated sludge and removal of the final, stabilized sludge product (**figure 43.7**). Methane, the major component of natural gas, is vented and often burned for heat and electricity production. Within the anaerobic digester, at least three processes occur: (1) the fermentation of the sludge components to form organic acids, including acetate; (2) production of the methanogenic substrates acetate, CO_2, and hydrogen; and finally, (3) methanogenesis by methane producers. These methanogenic processes involve critical balances between electron acceptors and donors (**table 43.5**). To function most efficiently, the hydrogen concentration must be maintained at a low level (*see p. 693*). If hydrogen and organic acids accumulate, the drop in pH will inhibit methane production, resulting in a stuck digester. Often lime is added to neutralize the pH and restore methanogen function. ◀◀ *Methanogens and methanotrophs (section 20.4)*

Anaerobic digestion has many advantages. Most of the microbial biomass produced in aerobic growth is degraded in the anaerobic digester. Also, because the process of methanogenesis is energetically very inefficient, methanogenic archaea consume about twice as many nutrients to produce an equivalent biomass as that of aerobic systems. Consequently less sludge is produced and it can be easily dried. Dried sludge removed from well-operated anaerobic systems can even be sold as organic garden fertilizer. However, sludge from poorly managed systems may contain high levels of heavy metals and other contaminants, which require removal prior to release into the environment.

Tertiary treatment further purifies wastewaters by removing nitrogen and phosphorus compounds that can promote eutrophication. It also removes heavy metals, biodegradable organics,

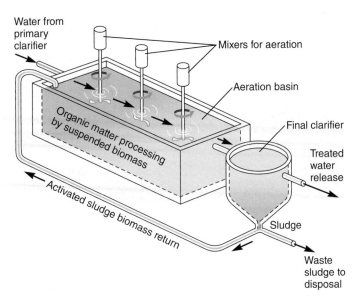

(a) Aerobic activated sludge system

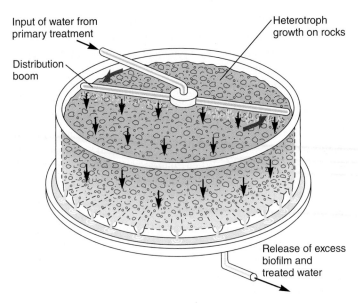

(b) Trickling filter

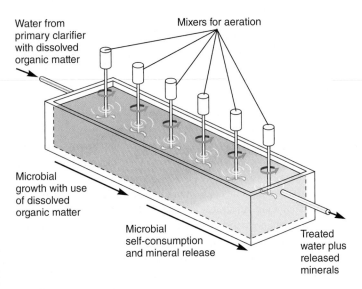

(c) Extended aeration system

Figure 43.6 Aerobic Secondary Sewage Treatment. (a) Activated sludge with microbial biomass recycling. The biomass is maintained in a suspended state to maximize oxygen, nutrient, and waste transfer processes. (b) Trickling filter, where wastewater flows over biofilms attached to rocks or other solid supports, resulting in transformation of dissolved organic matter to new biofilm biomass and carbon dioxide. Excess biomass and treated water flow to a final clarifier. (c) An extended aeration process, where aeration is continued beyond the point of microbial growth, allows the microbial biomass to self-consume due to microbial maintenance energy requirements. (The extended length of the reactor allows this process of biomass self-consumption to occur.) Minerals originally incorporated in the microbial biomass are released to the water as the process occurs.

and many remaining microbes, including viruses. Organic pollutants can be removed with activated carbon filters. Phosphate usually is precipitated as calcium or iron phosphate (e.g., by the addition of lime). To remove phosphorus, oxic and anoxic conditions are used alternately in a series of treatments, and phosphorus accumulates in microbial biomass as polyphosphate. Excess nitrogen may be removed by "stripping," volatilization of NH_3 at high pH. Ammonia itself can be chlorinated to form dichloramine, which is then converted to molecular nitrogen. In some cases, microbial processes can be used to remove nitrogen and phosphorus. A widely used process for nitrogen removal is denitrification, whereby nitrate produced by microbes under aerobic conditions, serves as an electron acceptor during anaerobic respiration. Nitrate reduction yields nitrogen gas (N_2) and nitrous oxide (N_2O) as the major products. In addition to denitrification,

the anammox process is also important. In this reaction, ammonium ion (used as the electron donor) reacts with nitrite (the electron acceptor) to yield N_2. The anammox process can convert up to 80% of the initial ammonium ion to N_2 gas. Tertiary treatment is expensive and is usually not employed except where necessary to prevent obvious ecological disruption. ◄◄ *Phylum Planctomycetes (section 21.5); Nitrogen cycle (section 28.1)*

Measuring Treated Wastewater Quality

The process of wastewater treatment must be monitored to ensure that waters released into the environment do not pose environmental and health risks. Successfully treated water should have very little organic carbon. The amount of carbon during and after wastewater treatment can be measured as **total organic carbon**

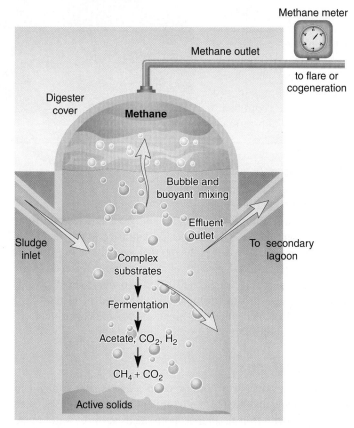

Figure 43.7 Anaerobic Digesters Generate Methane for Local or Municipal Use.   The basic design of an anaerobic digester includes an inlet for the delivery of waste. Fermentation of complex substrates yields acetate, CO_2 and H_2, which archaea convert to CH_4. The effluent outlet functions to optimize the retention time (usually from 22 to 28 days) to maximize CH_4 capture.

(**TOC**), as chemically oxidizable carbon by the **chemical oxygen demand (COD)** test, or as biologically usable carbon by the **biochemical oxygen demand (BOD)** test. The TOC includes all carbon, whether or not it can be used by microorganisms. It is assayed by oxidizing all organic matter in a sample to CO_2 at high temperature in an oxygen stream. The CO_2 produced is measured by infrared or potentiometric techniques. The COD gives a similar measurement, except that lignin (*see figure 31.2*) often does not react with the oxidizing chemical used in this procedure.

The BOD test measures only that portion of total carbon oxidized (i.e., consumed) by microorganisms in a 5-day period at 20°C. It measures the amount of dissolved O_2 needed for microbial oxidation of biodegradable organic material. Ammonia released during organic matter oxidation can also consume O_2 in the BOD test. To prevent this "bottle effect," nitrification is often inhibited by the addition of 2-chloro-6-(trichloromethyl) pyridine (nitrapyrin). Except when treated effluents are analyzed, nitrification is not a major concern.

In terms of speed, the TOC is fastest, but it is less informative with regard to biological processes. The COD involves the use of wet chemicals with higher waste chemical disposal costs. It is critical to note that TOC, COD, and BOD measurements are concerned only with carbon removal. They do not directly address the removal of minerals such as nitrate, phosphate, and sulfate. These impact cyanobacterial and algal growth in lakes, rivers, and oceans by contributing to the process of eutrophication. ◄◄ *Biogeochemical cycling sustains life on earth (section 28.1); Microorganisms in coastal ecosystems are adapted to a changeable environment (section 30.2)*

Retrieve, Infer, Apply

1. Explain how primary, secondary, and tertiary treatments are accomplished.
2. Compare and contrast granular and bulking sludge. Why is granular sludge desirable?
3. What are the steps of organic matter processing that occur in anaerobic digestion?
4. After anaerobic digestion is completed, why is sludge disposal of concern?
5. Compare TOC, COD, and BOD: how are these similar and different?
6. What minerals contribute to eutrophication?

Table 43.5	Sequential Reactions in the Anaerobic Digester		
Process Step	**Substrates**	**Products**	**Major Microorganisms**	
Fermentation	Organic polymers	Butyrate, propionate, lactate, succinate, ethanol, acetate,[1] H_2,[1] CO_2[1]	*Clostridium* *Bacteroides* *Peptostreptococcus*	*Peptococcus* *Eubacterium* *Lactobacillus*
Acetogenic reactions	Butyrate, propionate, lactate, succinate, ethanol	Acetate, H_2, CO_2	*Syntrophomonas* *Syntrophobacter*	
Methanogenic reactions	Acetate	$CH_4 + CO_2$	*Methanosarcina* *Methanothrix* *Methanobrevibacter* *Methanomicrobium*	*Methanogenium* *Methanobacterium* *Methanococcus* *Methanospirillum*
	H_2 and HCO_3^-	CH_4		

1 Methanogenic substrates produced in the initial fermentation step.

Home Treatment Systems

In the absence of municipal sewage systems, a **septic tank** is most commonly used. A conventional septic tank system functions as a simple anaerobic digester, with an anaerobic liquefaction and digestion step. This is followed by organic matter adsorption and entrapment of microorganisms in an aerobic leach field, where biological oxidation occurs. Pathogenic microorganisms and dissolved organic matter are removed from the wastewater in home septic systems during subsurface passage through adsorption and trapping by fine sandy materials, clays, and organic matter (**figure 43.8**). Microorganisms associated with these materials—including predators such as protozoa—can use the trapped pathogens as food. This results in purified water with a lower microbial population.

Because groundwater—the water in gravel beds and fractured rocks below the surface soil—is an important source of potable water, it is imperative that home septic systems be located far above the water table. Thus not all sites are suitable for septic systems. Prior to septic tank installation, a percolation test (or "perc test") must be performed. Perc tests determine if the surrounding soil can absorb the liquid produced by the septic tank, thereby preventing it from contaminating the groundwater beneath. Local governments develop specific protocols for perc tests, which are often witnessed when performed as the outcome of a perc test often determines if a piece of real estate can be developed.

A septic tank may not operate correctly for several reasons. If the retention time of the waste in the septic tank is too short, undigested solids move into the leach field, plugging the system. If the leach field floods and becomes anoxic, biological oxidation does not occur, and effective treatment ceases. Other problems can occur, especially when a suitable soil is not present and the septic tank outflow from a conventional system drains too rapidly to the deeper subsurface. Fractured rocks and coarse gravel materials do not effectively adsorb or filter wastewater. This may result in contamination of well water with pathogens. In addition, nitrogen and phosphorus from waste can pollute groundwater. This leads to nutrient enrichment of ponds, lakes, rivers, and estuaries as the subsurface water enters these environmentally sensitive ecosystems.

Domestic and commercial septic systems can be designed with nitrogen and phosphorus removal steps. Nitrogen is usually removed by nitrification and denitrification, with organic matter provided by sawdust or a similar material. A reductive iron dissolution process can be used for phosphorus removal. Nitrogen and phosphorus releases from septic systems must be controlled to prevent eutrophication of local surface waters.

Subsurface zones also can become contaminated with pollutants from other sources. Land disposal of sewage sludge, illegal dumping of septic tank sludge, improper toxic waste disposal, and runoff from agricultural operations all contribute to groundwater contamination with chemicals and microorganisms. Many pollutants that reach the subsurface will persist and may affect the quality of groundwater for extended periods. Much research is being conducted to find ways to treat groundwater in place—in situ treatment. As discussed in section 43.4, microorganisms and microbial processes are critical in many of these remediation efforts.

Future Challenges

Our planet now has over 7 billion people, and the needs for clean drinking water and wastewater treatment have become ever more pressing. In an effort to address both needs at once, water reuse or recycling has become a reality for many. Water reuse is the purification of wastewater so that it can serve as drinking water. This is not just for the developing world; a southern California municipality began operation of one of the largest water reclamation plants in the United States in 2012.

Not surprisingly, such water reclamation is complex. Not only must the obvious contaminants be removed (i.e., pathogens, heavy metals, nutrients that cause eutrophication), but twenty-first–century sewage also carries alarming levels of pharmaceuticals, introduced by both industry and the households. This has stimulated the development of methods to detect and remove the vast array of drugs that are common in sewage. In addition to recycling water, recent technologies enable the reclamation of biopolymers, plastics, and cellulose fibers in an economically

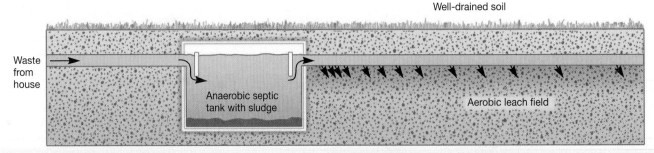

Figure 43.8 The Conventional Septic Tank Home Treatment System. This system combines an anaerobic waste liquefaction unit (the septic tank) with an aerobic leach field. Biological oxidation of the liquefied waste takes place in the leach field, unless the soil becomes flooded.

MICRO INQUIRY *Why would flooding halt biological oxidation of waste in the leach field?*

viable fashion. The current challenge is the widespread application of these technologies.

Population growth also stresses the capacity of municipal systems to carry and treat the ever-increasing volume of wastewater. This leads to sewage overflows during storms. The frequency of sewage overflow and the associated public health risks have resulted in a reexamination of current infrastructure and, in some cases, the construction of overflow basins.

Retrieve, Infer, Apply

1. How, in principle, does a conventional septic tank and leach field system work? What factors can reduce the effectiveness of this system?
2. What type of wastewater treatment system does your hometown use?
3. List two areas of active research in the wastewater treatment field. Which seem more pressing in your community? Explain your answer.

43.3 Microbial Fuel Cells: Batteries Powered by Microbes

After reading this section, you should be able to:

- Diagram the basic design of a microbial fuel cell
- Explain potential applications of microbial fuels cells

An exciting and emerging field in microbiology is the use of bacteria to generate electricity, not through the production of methane (as in anaerobic digesters, figure 43.7) but by directly capturing electrons from the microbe's electron transport chain (ETC). This is possible because when some heterotrophic microorganisms oxidize organic material, they can pass the electrons directly to an electrode. A **microbial fuel cell** (MFC) captures these electrons to generate electricity. To accomplish this, the microbes are continuously fed a rich diet of organic substrates so that very little biosynthesis is required (recall that anabolic reactions require electrons carried by NADPH). In this way, much of the organic substrate is oxidized during catabolism and electrons are donated to the ETC. The basic design of a MFC is relatively simple (**figure 43.9**). It consists two chambers, one oxic and the other anoxic, divided by a membrane through which protons can pass. Biomass is delivered to the bacteria in the anoxic side. The oxidation of this organic matter yields protons and electrons. The protons diffuse across the membrane to the oxic chamber. The electrons are deposited on an anode that connects the two chambers. Electrons flow toward the cathode but can be used to generate electricity while in transit. A catalyst in the oxic chamber combines the protons, electrons, and oxygen to generate water.

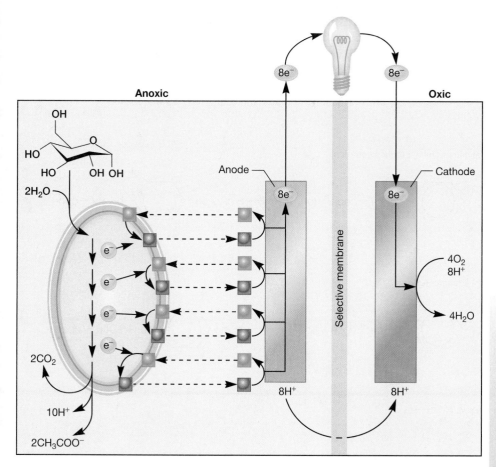

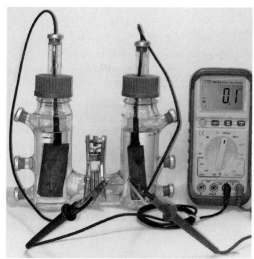

(b)

Figure 43.9 Microbial Fuel Cells. (a) Sugar is catabolized anaerobically by microbes in the anoxic chamber. The resulting protons flow to the oxic chamber, and the electrons are shuttled from the cell membrane to an anode, where they travel to the oxic chamber. During electron transit, they can be used to generate electricity. Green boxes represent oxidized electron shuttles and red boxes represent reduced shuttles. (b) A prototypical two-chambered microbial fuel cell.

MICRO INQUIRY *Why must the membrane between the two chambers be selectively permeable?*

Many different types of heterotrophic microbes can produce power in MFCs. The only requirement is that they oxidize organic matter anaerobically. In fact, communities of microbes can be employed. The largest constraint to obtaining maximum efficiency appears to be delivery of electrons to the electrode. There are several ways this can be accomplished. Often a chemical mediator is used to shuttle electrons from inside the cell, across the cellular membrane, to the anode (figure 43.9). Some microbes, such as the γ-proteobacterium *Shewanella oneidensis,* produce "nanowires" to transmit electrons to the anode (in nature, these nanowires shuttle electrons to external electron acceptors such as Fe^{3+}; *see figure 22.28*).

MFC development is ongoing, and great strides have been made to maximize power output, which has increased orders of magnitude in just a decade. Applications of MFCs include drinking water purification and the treatment of landfill leachate: in both cases, unwanted organic material is used to feed the bacteria in the MFC. MFCs can also be used when energy requirements are small but batteries cannot be used. Environmental monitoring instruments that need to be left in the field for years at a time are good examples. In addition, because MFCs are self-contained, they are attractive for developing countries that lack well-developed power grids. ◄◄ *Order* Alteromonadales *includes anaerobes that use a range of electron acceptors (section 22.3); Order* Desulfuromonales *includes metabolically flexible anaerobes (section 22.4)*

Retrieve, Infer, Apply

1. How does a microbial fuel cell work?
2. Why is it important to supply a high level of organic nutrients to the bacteria in the fuel cell?
3. How might microbial fuel cells be involved in wastewater management?

43.4 Biodegradation and Bioremediation Harness Microbes to Clean the Environment

After reading this section, you should be able to:

- Describe the importance of reductive dehalogenation in bioremediation
- Discuss factors that influence the rate of biodegradation and bioremediation of a compound
- Explain how endogenous microbes can be stimulated to degrade environmental pollutants
- Summarize the rationale on which the addition of microbes and microbial products to contaminated soils and waters is based

The metabolic activities of microbes can also be exploited in complex natural environments for beneficial outcomes. Examples are the use of resident microbial communities to carry out biodegradation, bioremediation, and environmental maintenance processes (see chapter opener), and the addition of microorganisms to soils or plants for the improvement of crop production. We discuss both of these applications, with an emphasis on bioremediation in this section.

Biodegradation Processes

Before discussing biodegradation, it is important to consider its definition. **Biodegradation** has at least three outcomes (**figure 43.10**): (a) a minor change in an organic molecule, leaving the main structure still intact; (b) fragmentation of a complex organic molecule in such a way that the fragments could be reassembled to yield the original structure; and (c) complete mineralization, which is the transformation of organic molecules to inorganic forms.

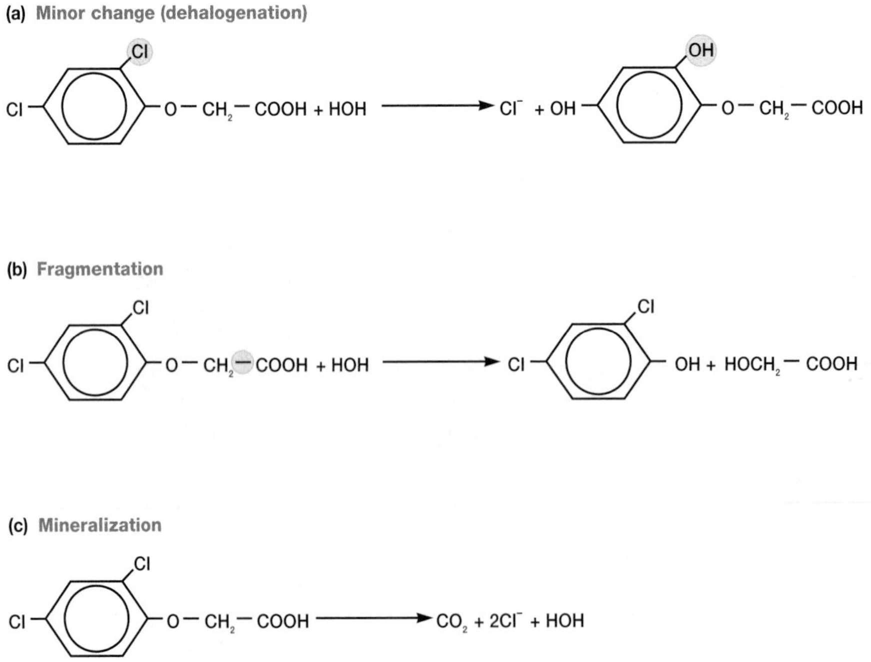

Figure 43.10 Biodegradation Has Several Meanings. Biodegradation is a term that can be used to describe three major types of changes in a molecule: (a) a minor change in the functional groups attached to an organic compound, such as the substitution of a hydroxyl group for a chlorine group; (b) an actual breaking of the organic compound into organic fragments in such a way that the original molecule could be reconstructed; and (c) the complete degradation of an organic compound to minerals.

MICRO INQUIRY *Why does dehalogenation increase the likelihood that the carbon compound will be degraded?*

(a) Minor change (dehalogenation)

(b) Fragmentation

(c) Mineralization

The removal of toxic industrial products in soils and aquatic environments is a daunting and necessary task. Compounds such as perchloroethylene (PCE), trichloroethylene (TCE), and polychlorinated biphenyls (PCBs) are common contaminants. These compounds adsorb onto organic matter in the environment, making decontamination using traditional approaches difficult or ineffective. The use of microbes to transform these contaminants to nontoxic degradation products is called **bioremediation.** To understand how bioremediation takes place at the level of an ecosystem, we first must consider the biochemistry of biodegradation.

Degradation of complex compounds requires several discrete stages, usually performed by different microbes. Initially contaminants are converted to less toxic compounds that are more readily degraded. The first step for many contaminants, including organochloride pesticides, alkyl solvents, and aryl halides, is **reductive dehalogenation.** This is the removal of a halogen substituent (e.g., chlorine, bromine, fluorine) while at the same time adding electrons to the molecule. This can occur in two ways. In hydrogenolysis, the halogen substituent is replaced by a hydrogen atom (**figure 43.11a**). Alternatively dihaloelimination removes two halogen substituents from adjacent carbons while inserting an additional bond between the carbons (figure 43.11b). Both processes require an electron donor. The dehalogenation of PCBs uses electrons derived from water; alternatively hydrogen can be the electron donor for the dehalogenation of different chlorinated compounds. Major genera with species that carry out this process include *Desulfitobacterium, Dehalospirillum,* and *Desulfomonile.*

Reductive dehalogenation usually occurs under anoxic conditions. In fact, humic acids (polymeric residues of lignin decomposition that accumulate in soils and waters) can serve as terminal electron acceptors under what are called humic acid–reducing conditions. The use of humic acids as electron acceptors has been observed with the anaerobic dechlorination of vinyl chloride and dichloroethylene. Once the anaerobic dehalogenation steps are completed, degradation of the main structure of many pesticides and other xenobiotics (foreign compounds) often proceeds more rapidly in the presence of O_2. Thus the degradation of halogenated

toxic compounds generally requires the action of several microbial genera, sometimes referred to as a consortium.

Structure and stereochemistry are critical in predicting the fate of a specific chemical in nature. When a constituent is in the *meta* as opposed to the *ortho* position, the compound will be degraded at a much slower rate. This stereochemical difference is the reason that the common lawn herbicide 2,4-dichlorophenoxyacetic acid (2,4-D), with a chlorine in the *ortho* position, will be largely degraded in a single summer. In contrast, 2,4,5-trichlorophenoxyacetic acid, with a constituent in the *meta* position, will persist in the soils for several years and thus is used for long-term brush control.

Another important factor that influences biodegradation is a compound's chirality or handedness *(see figure AI.8).* Microorganisms often degrade one isomer of a substance but not the other. At least 25% of herbicides are chiral. Thus it is critical to add the herbicide isomer that is effective and also degradable. However, populations that make up a microbial community can change in response to the addition of inorganic or organic substrates. If a particular compound, such as a herbicide, is added repeatedly to a microbial community, the community changes and faster rates of degradation can occur—a process of acclimation. Eventually a microbial community can become so efficient at herbicide degradation that herbicide effectiveness is diminished. To counteract this process, herbicides can be switched to alter the microbial community.

Degradation processes that occur in soils also can be used in large-scale degradation of hydrocarbons or wastes from agricultural operations. The waste material is incorporated into soil or allowed to flow across the soil surface, where degradation occurs. While this is often very effective, sometimes such degradation processes do not reduce toxicity. An example of this problem is the microbial metabolism of 1,1,1-trichloro-2,2-bis-(p-chlorophenyl) ethane (DDT), a pesticide once commonly used in the United States. Degradation removes a chlorine function to give 1,1-dichloro-2,2-bis-(p-chlorophenyl) ethylene (DDE), which is also an environmental toxin (**figure 43.12a**). Another important example is the degradation of trichloroethylene (TCE), a widely used solvent.

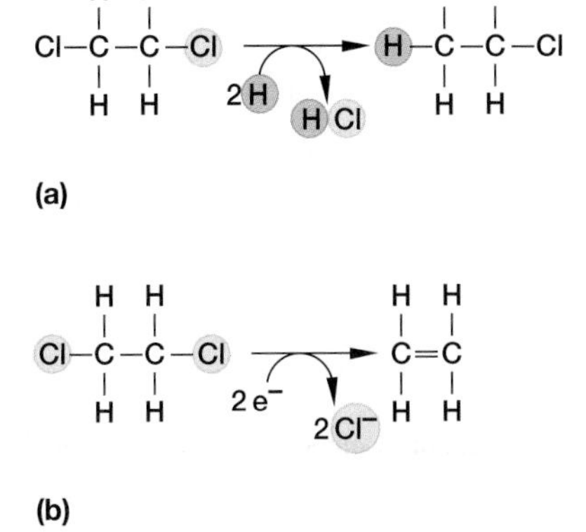

(a)

(b)

Figure 43.11 Reductive Dehalogenation. In these two examples, 1,2 dichlorobenzoate is dehalogenated by (a) alkyl hydrogenolysis and (b) dihaloelimination. Note that although the substrate is the same, these strategies result in different products.

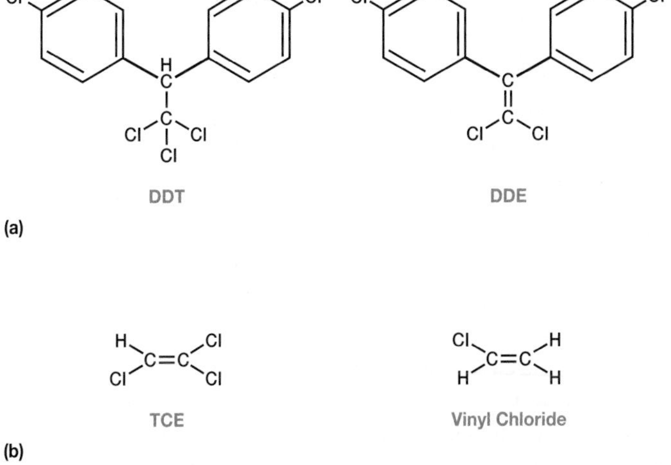

(a)

(b)

Figure 43.12 Toxic Degradation Products. The parent compounds DDT and TCE are metabolized to the toxins DDE and vinyl chloride, respectively.

Figure 43.13 Bioremediation of the *Exxon Valdez* Oil Spill. Workers sprayed the coastline with nitrogen and phosphorus-containing fertilizer to provide limiting nutrients to hydrocarbon-degrading microorganisms. This increased the rate at which oil was degraded.

Stimulating Hydrocarbon Degradation in Waters and Soils

Experience with oil spills in marine environments illustrates how effective stimulating natural microbial communities can be. Hydrocarbons lack phosphorus and nitrogen, so these elements limit their degradation. One of the first demonstrations that simple fertilization accelerates hydrocarbon degradation by natural populations occurred in 1989 following the *Exxon Valdez* oil spill in Prince Williams Sound, Alaska. As seen in **figure 43.13**, the addition of nitrogen and phosphorus-containing fertilizer significantly diminished the oil in coastal sediments. Since that time, it has been recognized that another key variable in promoting bioremediation is the degree of contact between microorganisms and the hydrocarbon substrate. To increase hydrocarbon accessibility by microbes, pellets containing nutrients and an oleophilic (hydrocarbon soluble) preparation are often used. This technique accelerates the degradation of different crude oil slicks by 30 to 40%, in comparison with control oil slicks where additional nutrients are not available. More recent research also focuses on the entire microbial community involved in the bioremediation process. Oil, which is a complex mixture of organic molecules, is degraded by a variety of bacteria (so far, archaea do not appear to degrade oil), but their capacity to do so is influenced by rates of predation by viruses and protists, the rate of nutrient recycling, and even the release of biosurfactants by other microbes (**figure 43.14**). Surfactants

When degraded under anoxic conditions, the dangerous carcinogen vinyl chloride can be formed (figure 43.12*b*).

An emerging approach to the bioremediation of contaminants in soils is the use of microbial fuel cells (figure 43.9). One of the major challenges to the microbial degradation of subsurface organic compounds and heavy metals is the provision of an electron acceptor that will participate in the oxidation of the contaminant. For instance, the biodegradation of petroleum in soils is frequently limited because it is very difficult and expensive to introduce oxygen into sediments. Microbial fuel cells can be inserted in sediments such that the anode can serve as the terminal electron acceptor by bacteria in the community during biodegradation. Several innovative applications of microbial fuel cells have been demonstrated, including the oxidation of toluene by *Geobacter metallireducens*. ◄◄ *Order* Desulfuromonales *includes metabolically flexible anaerobes (section 22.4)*

Bioremediation

Bioremediation usually involves stimulating the degradative activities of microorganisms already present in contaminated waters or soils. These existing microbial communities typically cannot carry out biodegradation processes at the desired rate due to limiting physical or nutritional factors. For example, biodegradation may be limited by low levels of oxygen, nitrogen, phosphorus, or other nutrients. In these cases, it is necessary to determine the limiting factors and supply the needed materials or modify the environment. Often the addition of easily metabolized organic matter such as glucose increases biodegradation of recalcitrant compounds that otherwise would not be used as carbon and energy sources by microorganisms. This process is called **cometabolism.**

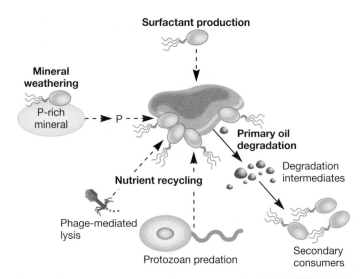

Figure 43.14 Oil Degradation Requires a Microbial Community. Primary oil degradation involves bacteria that begin oil degradation (blue microbes). These bacteria release a variety of degradation products that are further broken down by secondary consumers (yellow microbes). Dashed lines represent indirect interactions that influence the rate of bioremediation, such as phage lysis and protozoan grazing.

serve to disperse the oil; so-called dispersants used during the BP oil spill in the Gulf of Mexico in 2010 became almost as controversial as the oil itself. This has prompted the development of microbial-based oil dispersants, described next.

Bioaugmentation

The acceleration of microbiological processes by the addition of known active microorganisms to soils, waters, or other complex systems is called **bioaugmentation.** This is not a new concept; commercial culture preparations have long been available to facilitate silage formation and to improve septic tank performance. However, the magnitude of contamination coupled with an increased awareness of some of the dangers cleanup can introduce has stimulated the growth of the bioremediation industry. For instance, an oil dispersant consisting of several nonpathogenic *Bacillus* species, wetting agents, and nutrients is now commercially available.

Microorganisms can be added directly to natural communities, where they create their own microhabitats. For example, microorganisms in the water column overlying PCB-contaminated sand-clay soils have been observed to create their own "clay hutches" by binding clays to their outer surfaces with exopolysaccharides. However, in most cases, microbes are added with other components to ensure their viability. A blend of microorganisms can be purchased from commercial bioremediation engineering companies to treat petrochemical contamination of soils and water. These microbes are most often added in a slurry along with nutrients, enzymes, surfactants, and chemicals that release oxygen to promote aerobiosis. In other cases, microbes are introduced with protective inert microhabitats. This helps the foreign microbes establish an effective population in the face of competition from the natural, endogenous microbial community. Thus the application of principles of microbial physiology and ecology can facilitate the successful management of microbial communities in nature.

Metal Bioleaching

Bioleaching is the use of microorganisms that produce acids from reduced sulfur compounds to create acidic environments that solubilize desired metals for recovery. This approach is used to recover metals from ores and mining tailings with metal levels too low for smelting. Bioleaching carried out by natural populations of *Leptospirillum*-like species and thiobacilli, for example, allow recovery of up to 70% of the copper in low-grade ores. This involves the biological oxidation of copper present in these ores to produce soluble copper sulfate. In fact, this process is now used for copper mining.

> ### Retrieve, Infer, Apply
>
> 1. List alternative definitions for the term biodegradation.
> 2. Explain the two mechanisms of reductive dehalogenation. Describe humic acids and the role they can play in anaerobic degradation processes.
> 3. Discuss chirality and isomerization and their importance for understanding biodegradation.
> 4. What components are commonly added to microbes used to treat petrochemical contamination? Why are these added?
> 5. How is bioleaching carried out, and what microbes are involved?

Key Concepts

43.1 Purification and Sanitary Analysis Ensure Safe Drinking Water

- Water purification can involve the use of sedimentation, coagulation, chlorination, and rapid and slow sand filtration. Chlorination may lead to the formation of organic disinfection by-products, including trihalomethane compounds, which are potential carcinogens (**figure 43.1**).
- *Cryptosporidium, Cyclospora, Giardia* spp., and viruses are of concern, as conventional water purification and chlorination will not always ensure their removal and inactivation to acceptable limits.
- Indicator organisms are used to assess the presence of pathogenic microorganisms. Most probable number (MPN) and membrane filtration procedures are employed to estimate the number of indicator organisms present. The presence-absence (P-A) test is used for coliforms and defined substrate tests, for coliforms and *E. coli* (**figures 43.2** and **43.3**).

43.2 Wastewater Treatment Maintains Human and Environmental Health

- Conventional sewage treatment is a controlled intensification of natural self-purification processes, and it can involve primary, secondary, and tertiary treatment steps (**figures 43.4** and **43.5; tables 43.3** and **43.4**).
- The biochemical oxygen demand (BOD) test is an indirect measure of the organic matter that can be oxidized by the aerobic microbial community. The chemical oxygen demand (COD) and total organic carbon (TOC) tests provide information on carbon that is not biodegraded in the 5-day BOD test.
- Home treatment systems operate on general self-purification principles. The conventional septic tank provides anaerobic liquefaction and digestion, whereas the aerobic leach field allows oxidation of the soluble effluent. These systems are now designed to provide nitrogen and phosphorus removal to lessen impacts of on-site sewage treatment systems on vulnerable marine and freshwaters (**figure 43.8**).

43.3 Microbial Fuel Cells: Batteries Powered by Microbes

- Microbial fuel cells are an emerging technology whereby electrons are donated to an anode during heterotrophic bacterial respiration. MFCs have a variety of applications, including the purification of waters while generating power and providing energy for monitoring environmental processes (**figure 43.9**).

43.4 Biodegradation and Bioremediation Harness Microbes to Clean the Environment

- Biodegradation is a critical part of natural systems mediated largely by microorganisms. This can involve minor changes in a molecule, fragmentation, or mineralization (**figure 43.10**).

- Biodegradation can be influenced by many factors, including oxygen levels, humic acids, and the presence of readily usable organic matter. Reductive dehalogenation occurs best under anoxic conditions (**figure 43.11**), and the presence of organic matter can facilitate modification of recalcitrant compounds in the process of cometabolism.

- Degradation management can be carried out in place, whether this be large marine oil spills, soils, or the subsurface. Such large-scale efforts often involve the use of natural microbial communities (**figures 43.13** and **43.14**).

- Bioremediation can also be accomplished by adding microbes directly to contaminated environments. These microbes are usually introduced with other components such as nutrients and dispersants to ensure their success despite competition from endogenous microbial communities.

Compare, Hypothesize, Invent

1. You wish to build a house in a rural community where you will need to install a septic system and drill a well for drinking water. Where will you site the well in relation to the septic system? Following the perc test, you discover that the town will allow you to build your house only if you install a sand filter in addition to your septic tank. Why do you think this might be the case?

2. You are responsible for the bioremediation of soil contaminated by jet fuel. Your coworker suggests you fertilize the soil with nitrogen and phosphorus to promote the growth of naturally occurring hydrocarbon-degrading microbes. However, you just read about a new mixture of microbes sold in a slurry with nutrients and a surfactant. What data do you need to collect to decide which approach would be best? How will you obtain these data?

3. Biofilms are known to form on surfaces in drinking water distribution networks. It is therefore important to understand how such biofilms develop and how long they last. A recent study followed the development of biofilms populated by the protists in the genera *Giardia* and *Cryptosporidium,* as well as a strain of *Poliovirus* and several bacteriophages. Attachment to, growth within, and detachment from model biofilms were measured in tap water flowing through a special reactor designed to simulate the type of turbulence found in a water distribution network.

Why were the investigators especially interested in biofilm stability? Also, why were they interested in bacteriophages? What special considerations do you think need to be addressed when studying protists with relatively complex life cycles? What method do you think would be most appropriate to detect the protists? What about detecting the viruses?

Read the original paper: Helmi, K., et al. 2008. Interactions of *Cryptosporidium parvum, Giardia lamblia,* vaccinal poliovirus type 1, and bacteriophages fX174 and MS2 with a drinking water biofilm and a wastewater biofilm. *Ann. Rev. Microbiol.* 74:2079.

Learn More

shop.mheducation.com Enhance your study of this chapter with interactive study tools and practice tests. Also ask your instructor about the resources available through Connect, including adaptive learning tools and animations.

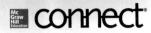

Appendix One

A Review of the Chemistry of Biological Molecules

Appendix I provides a brief summary of the chemistry of organic molecules with particular emphasis on the molecules present in microbial cells. Only basic concepts and terminology are presented; introductory textbooks in biology and chemistry should be consulted for a more extensive treatment of these topics.

Atoms and Molecules

Matter is made of elements that are composed of atoms. An element contains only one kind of atom and cannot be broken down to simpler components by chemical reactions. An atom is the smallest unit characteristic of an element and can exist alone or in combination with other atoms. When atoms combine, they form molecules. Molecules are the smallest particles of a substance. They have all the properties of the substance and are composed of two or more atoms.

Although atoms contain many subatomic particles, three directly influence their chemical behavior: protons, neutrons, and electrons. The atom's nucleus is located at its center and contains varying numbers of protons and neutrons (**figure AI.1**). Protons have a positive charge, and neutrons are uncharged. The mass of these particles and the atoms that they compose is given in terms of the atomic mass unit (AMU), which is 1/12 the mass of the most abundant carbon isotope. Often the term *dalton* (Da) is used to express the mass of molecules. It also is 1/12 the mass of an atom of ^{12}C or 1.661×10^{-24} grams. Both protons and neutrons have a mass of about 1 dalton. The atomic weight is the actual measured weight of an element and is almost identical to the mass number for the element, the total number of protons and neutrons in its nucleus. The mass number is indicated by a superscripted number preceding the element's symbol (e.g., ^{12}C, ^{16}O, and ^{14}N).

Negatively charged particles called electrons circle the atomic nucleus (figure AI.1). The number of electrons in a neu-

tral atom equals the number of its protons and is given by the atomic number, the number of protons in an atomic nucleus. The atomic number is characteristic of a particular type of atom. For example, carbon has an atomic number of six, hydrogen's number is one, and oxygen's is eight (**table AI.1**).

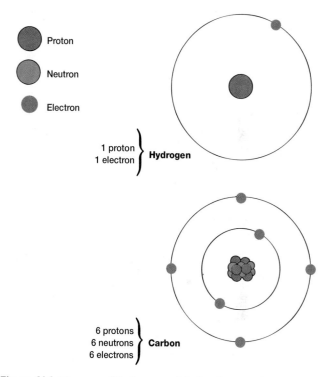

Figure AI.1 Diagrams of Hydrogen and Carbon Atoms. The electron orbitals are represented as concentric circles.

Table AI.1	Atoms Commonly Present in Organic Molecules			
Atom	**Symbol**	**Atomic Number**	**Atomic Weight**	**Number of Chemical Bonds**
Hydrogen	H	1	1.01	1
Carbon	C	6	12.01	4
Nitrogen	N	7	14.01	3
Oxygen	O	8	16.00	2
Phosphorus	P	15	30.97	5
Sulfur	S	16	32.06	2

The electrons move constantly within a volume of space surrounding the nucleus, even though their precise location in this volume cannot be determined accurately. This volume of space in which an electron is located is called its orbital. Each orbital can contain two electrons. Orbitals are grouped into shells of different energy that surround the nucleus. The first shell is closest to the nucleus and has the lowest energy; it contains only one orbital. The second shell contains four orbitals, one circular and three shaped like dumbbells (**figure AI.2a**). It can contain up to eight electrons. The third shell has even higher energy and holds more than eight electrons. Shells are filled beginning with the innermost and moving outward. For example, carbon has six electrons, two in its first shell and four in the second (figures AI.1 and AI.2b). The electrons in the outermost shell are the ones that participate in chemical reactions. The most stable condition is achieved when the outer shell is filled with electrons. Thus the number of bonds an element can form depends on the number of electrons required to fill the outer shell. Since carbon has four electrons in its outer shell and the shell is filled when it contains eight electrons, it can form four covalent bonds (table AI.1).

Chemical Bonds

Molecules are formed when two or more atoms associate through chemical bonding. Chemical bonds are attractive forces that hold together atoms, ions, or groups of atoms in a molecule or other substance. Many types of chemical bonds are present in organic molecules; three of the most important are covalent bonds, ionic bonds, and hydrogen bonds.

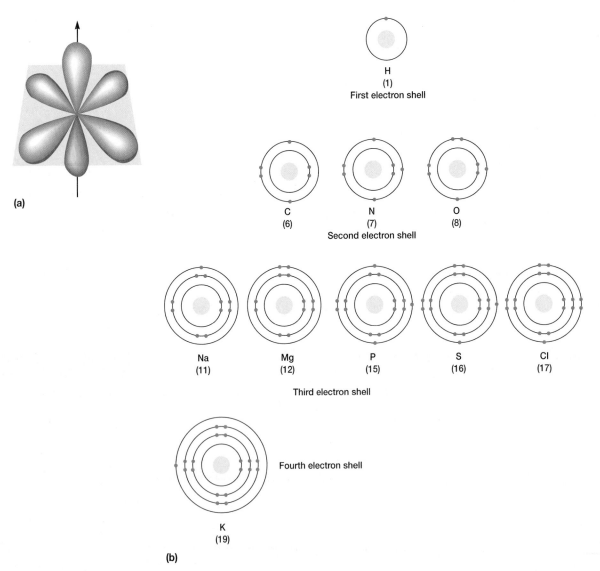

Figure AI.2 Electron Orbitals. (a) The three dumbbell-shaped orbitals of the second shell. The orbitals lie at right angles to each other. (b) The distribution of electrons in some common elements. Atomic numbers are given in parentheses.

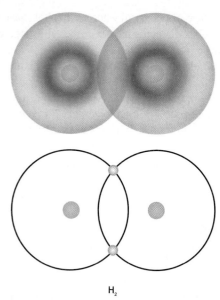

Figure AI.3 The Covalent Bond. A hydrogen molecule is formed when two hydrogen atoms share electrons.

In covalent bonds, atoms are joined together by sharing pairs of electrons (**figure AI.3**). If the electrons are equally shared between identical atoms (e.g., in a carbon-carbon bond), the covalent bond is strong and nonpolar. When two different atoms such as carbon and oxygen share electrons, the covalent bond formed is polar because the electrons are pulled toward the more electronegative atom, the atom that more strongly attracts electrons (the oxygen atom). A single pair of electrons is shared in a single bond; a double bond is formed when two pairs of electrons are shared.

Atoms often contain either more or fewer electrons than the number of protons in their nuclei. When this is the case, they carry a net negative or positive charge and are called ions. Cations carry positive charges and anions have a net negative charge. When a cation and an anion approach each other, they are attracted by their opposite charges. This ionic attraction that holds two groups together is called an ionic bond. Ionic bonds are much weaker than covalent bonds and are easily disrupted by a polar solvent such as water. For example, the Na^+ cation is strongly attracted to the Cl^- anion in a sodium chloride crystal, but sodium chloride dissociates into separate ions (ionizes) when dissolved in water. Ionic bonds are important in the structure and function of proteins and other biological molecules.

When a hydrogen atom is covalently bonded to a more electronegative atom such as oxygen or nitrogen, the electrons are unequally shared and the hydrogen atom carries a partial positive charge. It will be attracted to an electronegative atom such as oxygen or nitrogen, which carries an unshared pair of electrons; this attraction is called a hydrogen bond (**figure AI.4**). Although an individual hydrogen bond is weak, there are so many hydrogen bonds in proteins and nucleic acids that they play a major role in determining protein and nucleic acid structure.

Figure AI.4 Hydrogen Bonds. Representative examples of hydrogen bonds present in biological molecules.

Figure AI.5 Hydrocarbons. Examples of hydrocarbons that are (a) linear, (b) cyclic, and (c) aromatic.

Organic Molecules

Most molecules in cells are organic molecules, molecules that contain carbon. Since carbon has four electrons in its outer shell, it tends to form four covalent bonds in order to fill its outer shell with eight electrons. This property makes it possible to form chains and rings of carbon atoms that also can bond with hydrogen and other atoms (**figure AI.5**). Although adjacent carbons usually are connected by single bonds, they may be joined by double or triple bonds. Rings that have alternating single and double bonds, such as the benzene ring, are called aromatic rings. The hydrocarbon chain or ring provides a chemically inactive skeleton to which more reactive groups of atoms may be attached. These reactive groups with specific properties are known as functional groups. They usually contain atoms of oxygen, nitrogen, phosphorus, or sulfur (**figure AI.6**) and are largely responsible for most characteristic chemical properties of organic molecules.

Organic molecules are often divided into classes based on the nature of their functional groups. Ketones have a carbonyl group within the carbon chain, whereas alcohols have a hydroxyl on the chain. Organic acids have a carboxyl group, and amines have an amino group (**figure AI.7**).

Organic molecules may have the same chemical composition and yet differ in their molecular structure and properties. Such molecules are called isomers. One important class of isomers is the stereoisomers. Stereoisomers have the same atoms arranged in the same nucleus-to-nucleus sequence but differ in the spatial arrangement of their atoms. For example, an amino acid such as alanine can form stereoisomers (**figure AI.8**). L-Alanine and other L-amino acids are the stereoisomer forms normally present in proteins.

Carbohydrates

Carbohydrates are aldehyde or ketone derivatives of polyhydroxy alcohols. The smallest and least complex carbohydrates are the simple sugars or monosaccharides. The most common sugars have five or six carbons. A sugar in its ring form has two isomeric structures, the α and β forms, that differ in the orientation of the hydroxyl on the aldehyde or ketone carbon, which is

Figure AI.6 Functional Groups.   Some common functional groups in organic molecules. The groups are shown in red.

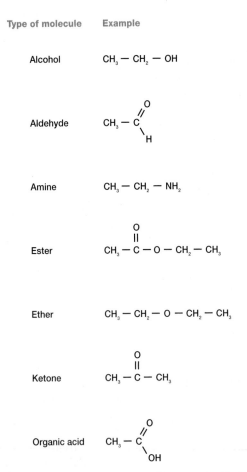

Figure AI.7 Types of Organic Molecules.   These are classified on the basis of their functional groups.

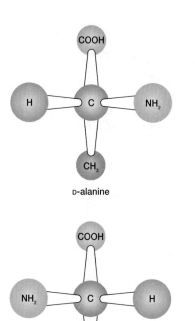

D-alanine

L-alanine

Figure AI.8 The Stereoisomers of Alanine.   The α-carbon is in gray. L-alanine is the form usually present in proteins.

called the anomeric or glycosidic carbon (**figure AI.9**). Microorganisms have many sugar derivatives in which a hydroxyl is replaced by an amino group or some other functional group (e.g., glucosamine).

β-D-glucose   α-D-glucose

Figure AI.9 The Interconversion of Monosaccharide Structures.   The open chain form of glucose and other sugars is in equilibrium with closed ring structures (depicted here with Haworth projections). Aldehyde sugars form cyclic hemiacetals, and keto sugars produce cyclic hemiketals. When the hydroxyl on carbon one of cyclic hemiacetals projects above the ring, the form is known as a β form. The α form has a hydroxyl that lies below the plane of the ring. The same convention is used in showing the α and β forms of hemiketals such as those formed by fructose.

Two monosaccharides can be joined by a bond between the anomeric carbon of one sugar and a hydroxyl or the anomeric carbon of the second (**figure AI.10**). The bond joining sugars is a glycosidic bond and may be either α or β depending on the orientation of the anomeric carbon. Two sugars linked in this way constitute a disaccharide. Some common disaccharides are maltose (two glucose molecules), lactose (glucose and galactose), and sucrose (glucose and fructose). If 10 or more sugars are linked together by glycosidic bonds, a polysaccharide is formed. For example, starch and glycogen are common polymers of glucose that are used as sources of carbon and energy (**figure AI.11**).

(a)   Glucose   Glucose   Maltose $+H_2O$

(b)   Sucrose

(c)   Lactose (α form)

Figure AI.10 Common Disaccharides.
(a) The formation of maltose from two molecules of an α-glucose. The bond connecting the glucose extends between carbons one and four, and involves the α form of the anomeric carbon. Therefore, it is called an α (1→4) glycosidic bond.
(b) Sucrose is composed of a glucose and a fructose joined to each other through their anomeric carbons, and αβ (1→2) bond.
(c) The milk sugar lactose contains galactose and glucose joined by a β (1→4) glycosidic bond.

Figure AI.11 Glycogen and Starch Structure. (a) An overall view of the highly branched structure characteristic of glycogen and most starch. The circles represent glucose residues. (b) A close-up of a small part of the chain (shown in blue in part a) revealing a branch point with its α (1→6) glycosidic bond, which is colored blue.

Lipids

All cells contain a heterogeneous mixture of organic molecules that are relatively insoluble in water but very soluble in nonpolar solvents such as chloroform, ether, and benzene. These molecules are called lipids. Lipids vary greatly in structure and include triacylglycerols, phospholipids, steroids, carotenoids, and many other types. Among other functions, they serve as membrane components, storage forms for carbon and energy, precursors of other cell constituents, and protective barriers against water loss.

Most lipids contain fatty acids, which are monocarboxylic acids that often are straight chained but may be branched. Saturated fatty acids lack double bonds in their carbon chains, whereas unsaturated fatty acids have double bonds. The most common fatty acids are 16 or 18 carbons long.

Two good examples of common lipids are triacylglycerols and phospholipids. Triacylglycerols are composed of glycerol esterified to three fatty acids (**figure AI.12a**). They are used to store carbon and energy. Phospholipids are lipids that contain at least one phosphate group and often have a nitrogenous constituent as well. Phosphatidylethanolamine is an important phospholipid frequently present in bacterial membranes (figure AI.12b).

It is composed of two fatty acids esterified to glycerol. The third glycerol hydroxyl is joined with a phosphate group, and ethanolamine is attached to the phosphate. The resulting lipid is very asymmetric with a hydrophobic nonpolar end contributed by the fatty acids and a polar, hydrophilic end. In cell membranes, the hydrophobic end is buried in the interior of the membrane, while the polar-charged end is at the membrane surface and exposed to water.

Figure AI.12 Examples of Common Lipids. (a) A triacylglycerol or neutral fat. (b) The phospholipid phosphatidylethanolamine. The R groups represent fatty acid side chains.

Proteins

The basic building blocks of proteins are amino acids. An amino acid contains a carboxyl group and an amino group on its alpha carbon (**figure AI.13**). About 20 amino acids are normally found in proteins; they differ from each other with respect to their side chains. In proteins, amino acids are linked together by peptide bonds between their carboxyls and α-amino groups to form linear polymers called peptides. If a peptide contains more than 50 amino acids, it usually is called a polypeptide. Each protein is composed of one or more polypeptide chains and has a molecular weight greater than about 6,000 to 7,000.

Proteins have three or four levels of structural organization and complexity. The primary structure of a protein is the sequence of the amino acids in its polypeptide chain or chains. The structure of the polypeptide chain backbone is also considered part of the primary structure. Each different polypeptide has its own amino acid sequence that is a reflection of the nucleotide sequence in the gene that codes for its synthesis. The polypeptide chain can coil along one axis in space into various shapes such as the α-helix (**figure AI.14**). This arrangement of the polypeptide in space around a single axis is called the secondary structure. Secondary structure is formed and stabilized by the interactions of amino acids that are fairly close to one another on the polypeptide chain. The polypeptide with its primary and secondary structure can be coiled or organized in space along three axes to form a more complex, three-dimensional shape (**figure AI.15**). This level of organization is the tertiary structure (**figure AI.16**). Amino acids more distant from one another on the polypeptide chain contribute to tertiary structure. Secondary and tertiary structures are examples of conformation, molecular shape that can be changed by bond rotation and without breaking covalent bonds. When a protein contains more than one polypeptide chain, each chain with its own primary, secondary, and tertiary structure associates with the other chains to form the final molecule. The way in which polypeptides associate with each other

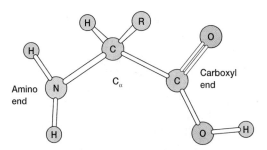

Figure AI.13 **L-Amino Acid Structure.** The uncharged form is shown.

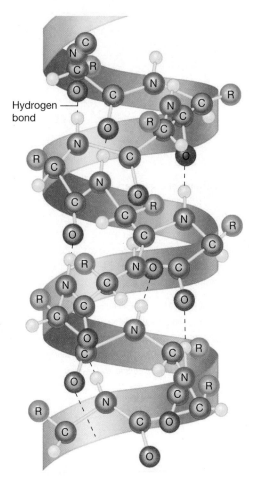

Figure AI.14 **The β-Helix.** A polypeptide twisted into one type of secondary structure, the α-helix. The helix is stabilized by hydrogen bonds joining peptide bonds that are separated by three amino acids.

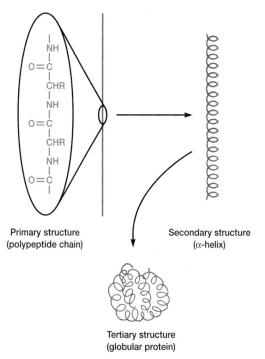

Figure AI.15 **Secondary and Tertiary Protein Structures.** The formation of secondary and tertiary protein structures by folding a polypeptide chain with its primary structure.

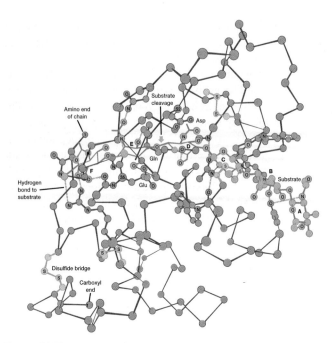

Figure AI.16 **Lysozyme.** The tertiary structure of the enzyme lysozyme. (a) A diagram of the protein's polypeptide backbone with the substrate hexasaccharide shown in blue. The point of substrate cleavage is indicated.

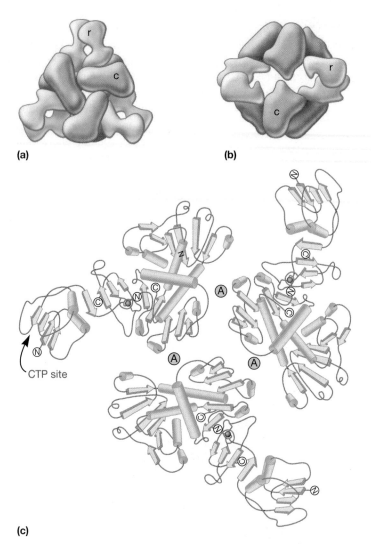

(c)

Figure AI.17 **An Example of Quaternary Structure.** The enzyme aspartate carbamoyltransferase from *Escherichia coli* has two types of subunits, catalytic and regulatory. The association between the two types of subunits is shown: (a) a top view, and (b) a side view of the enzyme. The catalytic (C) and regulator (r) subunits are shown in different colors. (c) The peptide chains shown when viewed from the top as in (a). The active sites of the enzyme are located at the positions indicated by A.

(a and b) Adapted from Krause, et al., in Proceedings of the National Academy of Sciences, *V. 82, 1985, as appeared in* Biochemistry, *3d edition, by Lubert Stryer. Copyright © 1975, 1981, 1988. Reprinted with permission of W. H. Freeman and Company. (c) Adapted from Kantrowitz, et al., in* Trends in Biochemical Science, *V. 5, 1980, as appeared in* Biochemistry, *3d edition, by Lubert Stryer. Copyright © 1975, 1981, 1988. Reprinted with permission of W. H. Freeman and Company.*

in space to form the final protein is called the protein's quaternary structure (**figure AI.17**). The final conformation of a protein is ultimately determined by the amino acid sequence of its polypeptide chains.

Protein secondary, tertiary, and quaternary structure is largely determined and stabilized by many weak noncovalent forces such as hydrogen bonds and ionic bonds. Because of this, protein shape often is very flexible and easily changed. This flexibility is very important in protein function and in the regulation of enzyme activity. Because of their flexibility, however, proteins readily lose their proper shape and activity when exposed to harsh conditions. The only covalent bond commonly involved in the secondary and tertiary structure of proteins is the disulfide bond. The disulfide bond is formed when two cysteines are linked through their sulfhydryl groups. Disulfide bonds generally strengthen or stabilize protein structure.

Nucleic Acids

The nucleic acids, deoxyribonucleic acid (DNA) and ribonucleic acid (RNA), are polymers of deoxyribonucleosides and ribonucleosides joined by phosphate groups. The nucleosides in DNA contain the purines adenine and guanine, and the pyrimidine bases thymine and cytosine. In RNA the pyrimidine uracil is substituted for thymine. Because of their importance for genetics and molecular biology, the chemistry of nucleic acids is introduced earlier in the text. The structure and synthesis of purines and pyrimidines are discussed in chapter 12 (section 12.6). The structures of DNA and RNA are described in chapter 13 (section 13.2).

Appendix Two
Common Metabolic Pathways

This appendix contains a few of the more important pathways discussed in the text, particularly those involved in carbohydrate catabolism. Enzyme names and final end products are given in color. Consult the text for a description of each pathway and its physiologic role.

Figure AII.1 The Embden-Meyerhof Pathway. This pathway converts glucose and other sugars to pyruvate and generates NADH and ATP. In some prokaryotes, glucose is phosphorylated to glucose 6-phosphate during group translocation transport across the plasma membrane.

Figure AII.2 The Entner-Doudoroff Pathway.

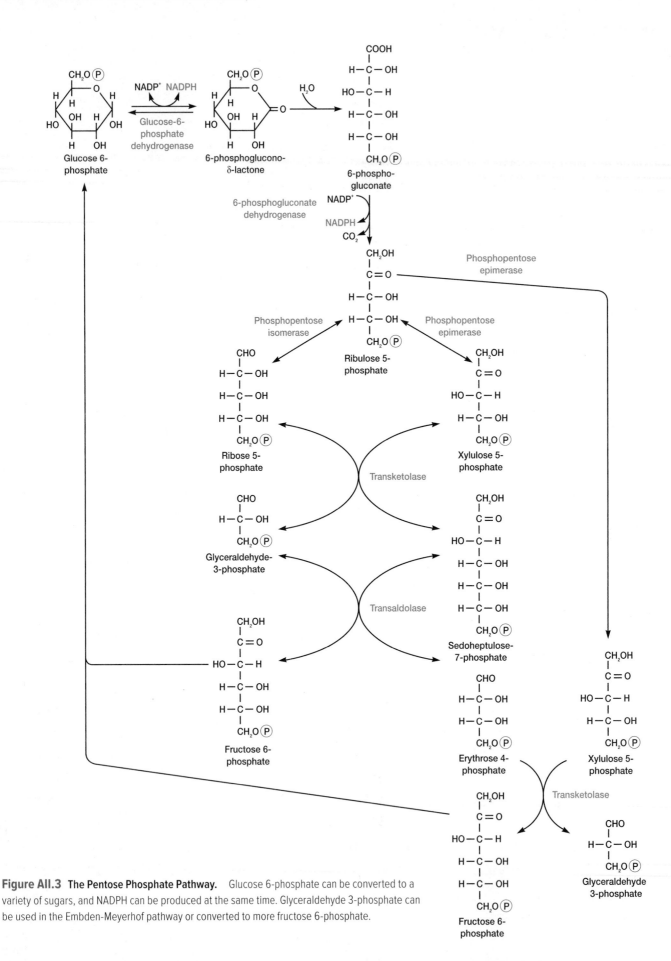

Figure AII.3 The Pentose Phosphate Pathway. Glucose 6-phosphate can be converted to a variety of sugars, and NADPH can be produced at the same time. Glyceraldehyde 3-phosphate can be used in the Embden-Meyerhof pathway or converted to more fructose 6-phosphate.

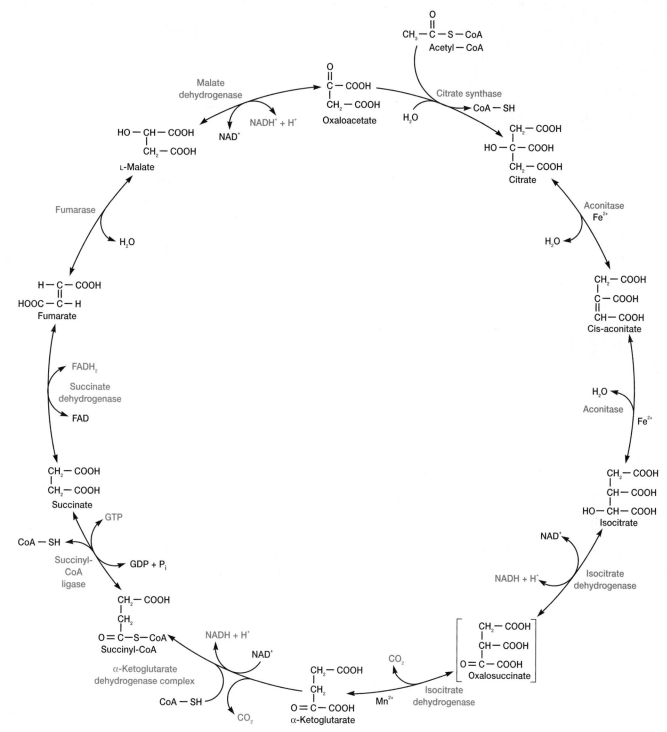

Figure AII.4 The Tricarboxylic Acid Cycle. Cis-aconitate and oxalosuccinate remain bound to aconitase and isocitrate dehydrogenase. Oxalosuccinate has been placed in brackets because it is so unstable.

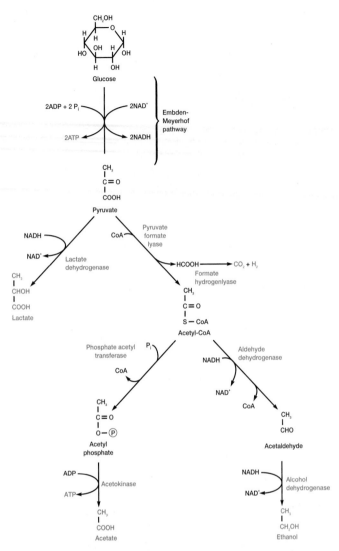

Figure AII.5 The Mixed Acid Fermentation Pathway. This pathway is characteristic of many members of the *Enterobacteriaceae* such as *Escherichia coli.*

Glucose

2ADP + 2 P$_i$ → 2NAD$^+$

2ATP ← 2NADH

Embden-Meyerhof pathway

CH$_3$ | C=O | COOH

Pyruvate

NADH   CoA

NAD$^+$ Pyruvate formate lyase

Lactate dehydrogenase

HCOOH → CO$_2$ + H$_2$

Formate hydrogenlyase

CH$_3$ | CHOH | COOH

Lactate

CH$_3$ | C=O | S—CoA

Acetyl-CoA

Phosphate acetyl transferase   P$_i$

CoA

Aldehyde dehydrogenase   NADH

NAD$^+$

CoA

CH$_3$ | C=O | O—Ⓟ

Acetyl phosphate

CH$_3$ | CHO

Acetaldehyde

ADP   Acetokinase

ATP

NADH   Alcohol dehydrogenase

NAD$^+$

CH$_3$ | COOH

Acetate

CH$_3$ | CH$_2$OH

Ethanol

Glucose

Embden-Meyerhof pathway   2NAD$^+$

2ADP + 2P$_i$ → 2NADH + 2H$^+$

2ATP

CH$_3$ | C=O | COOH

2 pyruvate

CO$_2$ α-Acetolactate synthase

CH$_3$ | C=O | HO—C—COOH | CH$_3$

Acetolactate

CO$_2$ Acetolactate decarboxylase

CH$_3$ | C=O | H—C—OH | CH$_3$

Acetoin

NADH + H$^+$ 2,3-butanediol dehydrogenase

NAD$^+$

CH$_3$ | CHOH | CHOH | CH$_3$

2,3-butanediol

Figure AII.6 The Butanediol Fermentation Pathway. This pathway is characteristic of members of the *Enterobacteriaceae* such as *Enterobacter.* Other products may also be formed during butanediol fermentation.

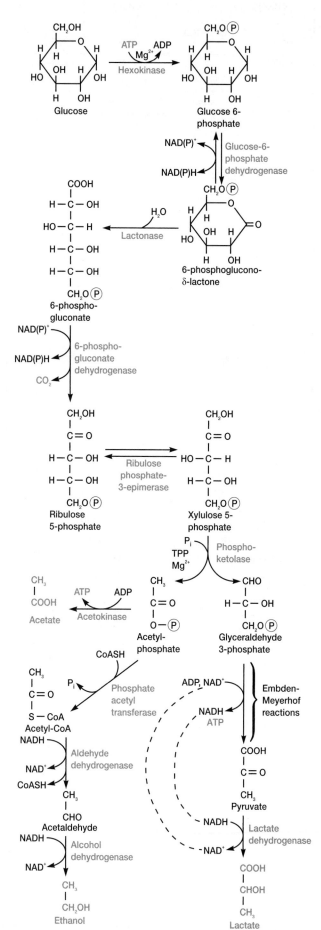

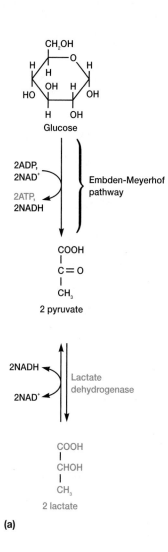

Figure AII.7 Lactic Acid Fermentations. (a) Homolactic fermentation pathway. (b) Heterolactic fermentation pathways.

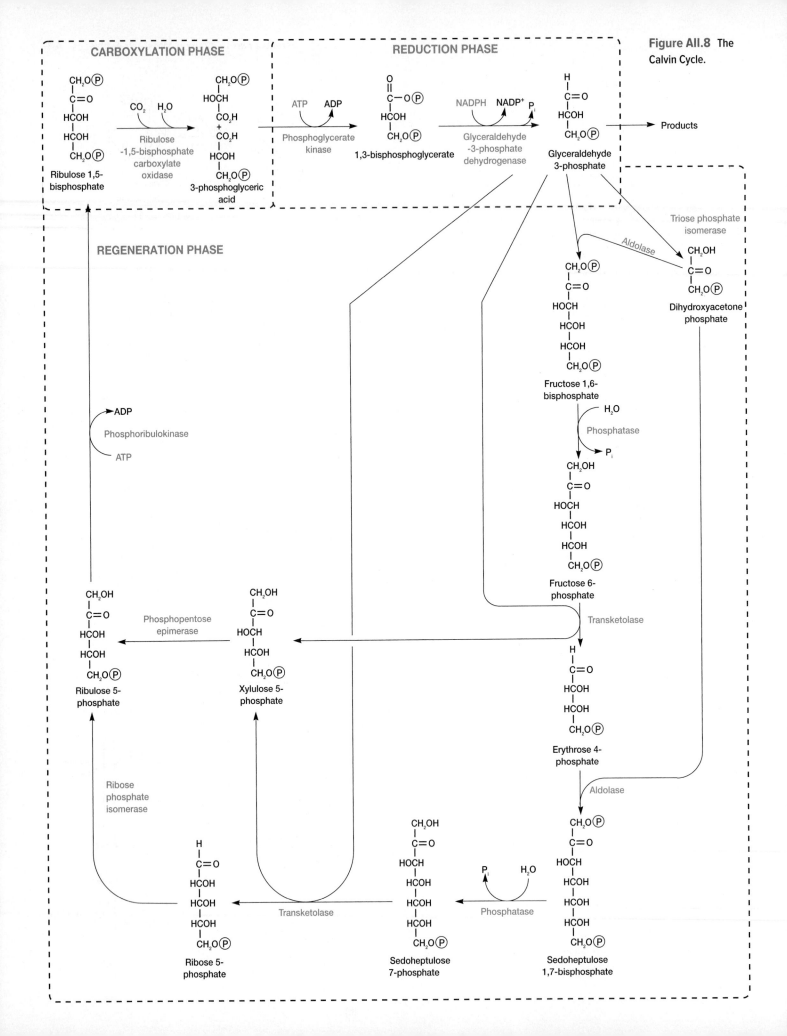

Figure AII.8 The Calvin Cycle.

Figure AII.9 Aromatic Amino Acid Synthesis. The carbons arising from phosphoenolpyruvate (green) and erythrose 4-phosphate (red) are shown. The remaining carbons present in tryptophan are provided by 5-phosphoribosyl-1-pyrophosphate (PRPP) and the amino acid serine. PRPP is also important in purine biosynthesis (figure AII.10).

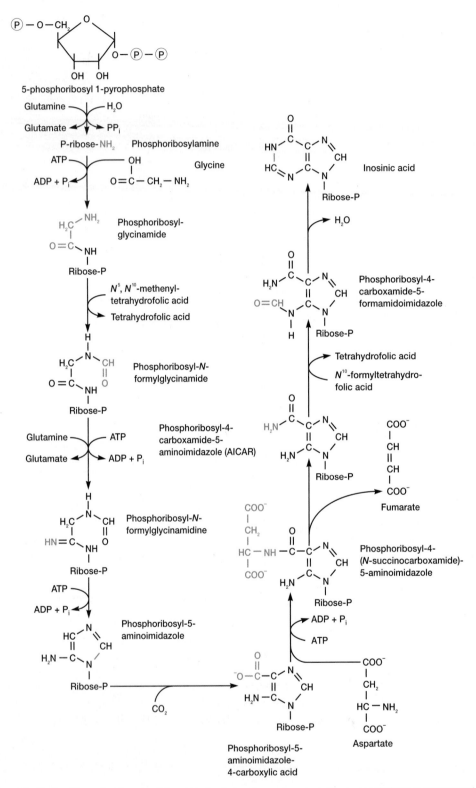

Figure AII.10 The Pathway for Purine Biosynthesis. Inosinic acid is the first purine end product. The purine skeleton is constructed while attached to a ribose phosphate.

Appendix Three
Microorganism Pronunciation Guide

The pronunciation of each name is given in parentheses. The phonetic spelling system is explained at the beginning of the glossary (p. G-1)

Bacteria

Acetobacter (ah-se''to-bak'ter)
Acinetobacter (as''ĭ-net''o-bak'ter)
Actinomyces (ak''tĭ-no-mi'sēz)
Agrobacterium (ag''ro-bak-te're-um)
Alcaligenes (al''kah-lij'ĕ-nēz)
Anabaena (ah-nab'ē-nah)
Arthrobacter (ar''thro-bak'ter)
Bacillus (bah-sil'lus)
Bacteroides (bak''tĕ-roi'dēz)
Bdellovibrio (del''o-vib're-o)
Beggiatoa (bej''je-ah-to'ah)
Beijerinckia (bi''jer-ink'e-ah)
Bifidobacterium (bi''fid-o-bak-te're-um)
Bordetella (bor''dĕ-tel'lah)
Borrelia (bŏ-rel'e ah)
Brucella (broo-sel'lah)
Campylobacter (kam''pi-lo-bak'ter)
Caulobacter (kaw''lo-bak'ter)
Chlamydia (klah-mid'e-ah)
Chlorobium (klo-ro'be-um)
Chromatium (kro-ma'te-um)
Citrobacter (sit''ro-bak'ter)
Clostridium (klo-strid'e-um)
Corynebacterium (ko-ri''ne-bak-te're-um)
Coxiella (kok''se-el'lah)
Cytophaga (si-tof'ah-gah)
Desulfovibrio (de-sul''fo-vib're-o)
Enterobacter (en''ter-o-bak'ter)
Erwinia (er-win'e-ah)
Escherichia (esh''er-i'ke-ah)
Flexibacter (flek''sĭ-bak'ter)
Francisella (fran-si-sel'ah)
Frankia (frank'e-ah)
Gallionella (gal''le-o-nel'ah)
Haemophilus (he-mof'ĭ-lus)
Halobacterium (hal''o-bak-te're-um)
Hydrogenomonas (hi-dro''jĕ-no-mo'nas)
Hyphomicrobium (hi''fo-mi-kro'be-um)
Klebsiella (kleb''se-el'lah)
Lactobacillus (lak''to-bah-sil'lus)
Legionella (le''jun-el'ah)
Leptospira (lep''to-spi'rah)
Leptothrix (lep'to-thriks)

Leuconostoc (loo''ko-nos'tok)
Listeria (lis-te're-ah)
Methanobacterium (meth''ah-no-bak-te're-um)
Methylococcus (meth''il-o-kok'-us)
Methylomonas (meth''il-o-mo'nas)
Micrococcus (mi''kro-kok'us)
Mycobacterium (mi''ko-bak-te're-um)
Mycoplasma (mi''ko-plaz'mah)
Neisseria (nĭs-se're-ah)
Nitrobacter (ni''tro-bak'ter)
Nitrosomonas (ni-tro''so-mo'nas)
Nocardia (no-kar'de-ah)
Pasteurella (pas''tĕ-rel'ah)
Photobacterium (fo''to-bak-te're-um)
Propionibacterium (pro''pe-on''e-bak-te're-um)
Proteus (pro'te-us)
Pseudomonas (soo''do-mo'nas)
Rhizobium (ri-zo'be-um)
Rhodopseudomonas (ro''do-soo''do-mo'nas)
Rhodospirillum (ro''do-spi-ril'um)
Rickettsia (ri-ket'se-ah)
Salmonella (sal''mo-nel'ah)
Sarcina (sar'sĭ-nah)
Serratia (sĕ-ra'she-ah)
Shigella (shi-gel'ah)
Sphaerotilus (sfe-ro'ti-lus)
Spirillum (spi-ril'um)
Spirochaeta (spi''ro-ke'tah)
Spiroplasma (spi''ro-plaz'mah)
Staphylococcus (staf''ĭ-lo-kok'us)
Streptococcus (strep''to-kok'us)
Streptomyces (strep''to-mi'sēz)
Sulfolobus (sul''fo-lo'bus)
Thermoactinomyces (ther''mo-ak''tĭ-no-mi'sēz)
Thermoplasma (ther''mo-plaz'mah)
Thiobacillus (thi''o-bah-sil'us)
Thiothrix (thi'o-thriks)
Treponema (trep''o-ne'mah)
Ureaplasma (u-re'ah-plaz''ma)
Veillonella (va''yon-el'ah)
Vibrio (vib're-o)
Xanthomonas (zan''tho-mo'nas)
Yersinia (yer-sin'e-ah)
Zoogloea (zo''o-gle'ah)

Viruses

Vernacular, nonscientific virus names are not written in italics, and thus the following names are not italicized.

adenovirus (ad″ĕ-no-vi′rus)
arbovirus (ar″bo-vi′rus)
baculovirus (bak″u-lo-vi′rus)
coronavirus (kor″o-nah-vi′rus)
cytomegalovirus (si″to-meg″ah-lo-vi′rus)
Epstein-Barr virus (ep′stĭn-bar′)
hepadnavirus (hep-ad″nə-vi′rus)
hepatitis virus (hep″ah-ti′tis)
herpesvirus (her″pēz-vi′rus)
influenza virus (in″flu-en′zah)
measles virus (me′zelz)
mumps virus (mumps)
orthomyxovirus (or″tho-mik″so-vi′rus)
papillomavirus (pap″ĭ-lo″mah-vi′rus)
paramyxovirus (par″ah-mik″so-vi′rus)
parvovirus (par″vo-vi′rus)
picornavirus (pi-kor″nah-vi′rus)
poliovirus (po″le-o-vi′rus)
polyomavirus (pol″e-o-mah-vi′rus)
poxvirus (poks-vi′rus)
rabies virus (ra′bēz)
reovirus (re″o-vi′rus)
retrovirus (re″tro-vi′rus)
rhabdovirus (rab″do-vi′rus)
rhinovirus (ri″no-vi′rus)
rotavirus (ro′tah-vi″rus)
rubella virus (roo-bel′ah)
togavirus (to″gah-vi′rus)
varicella-zoster virus (var″i-sel′ah zos′ter)
variola virus (vah-ri′o-lah)

Fungi

Agaricus (ah-gar′i-kus)
Amanita (am″ah-ni′tah)
Arthrobotrys (ar″thro-bo′tris)
Aspergillus (as″per-jil′us)
Blastomyces (blas″to-mi′sēz)
Candida (kan′di-dah)
Cephalosporium (sef″ah-lo-spo′re-um)
Claviceps (klav′i-seps)
Coccidioides (kok-sid″e-oi′dēz)

Cryptococcus (krip″to-kok′us)
Epidermophyton (ep″i-der-mof′i-ton)
Fusarium (fu-sa′re-um)
Histoplasma (his″to-plaz′mah)
Microsporum (mi-kros′po-rum)
Mucor (mu′kor)
Neurospora (nu-ros′po-rah)
Penicillium (pen″i-sil′e-um)
Phytophthora (fi-tof′tho-rah)
Pneumocystis (noo″mo-sis′tis)
Rhizopus (ri-zo′pus)
Saccharomyces (sak″ah-ro-mi′sēz)
Saprolegnia (sap″ro-leg′ne-ah)
Sporothrix (spo′ro-thriks)
Trichoderma (trik-o-der′mah)
Trichophyton (tri-kof′i-ton)

Protists

Acanthamoeba (ah-kan″thah-me′bah)
Acetabularia (as″ĕ-tab″u-la′re-ah)
Amoeba (ah-me′bah)
Balantidium (bal″an-tid′e-um)
Chlamydomonas (klah-mid″do-mo′nas)
Chlorella (klo-rel′ah)
Cryptosporidium (krip″to-spo-rid′e-um)
Entamoeba (en″tah-me′bah)
Euglena (u-gle′nah)
Giardia (je-ar′de-ah)
Gonyaulax (gon″e-aw′laks)
Leishmania (lēsh-ma′ne-ah)
Naegleria (na-gle′re-ah)
Paramecium (par″ah-me′she-um)
Plasmodium (plaz-mo′de-um)
Prototheca (pro″to-the′kah)
Spirogyra (spi″ro-ji′rah)
Tetrahymena (tet″rah-hi′mĕ-nah)
Toxoplasma (toks″o-plaz′mah)
Trichomonas (trik″o-mo′nas)
Trypanosoma (tri″pan-o-so′mah)
Volvox (vol′voks)

Glossary

Some of the boldface terms in this glossary are followed by a phonetic spelling in parentheses. These pronunciation aides usually come from *Dorland's Illustrated Medical Dictionary*. The following rules are taken from this dictionary and will help in using its phonetic spelling system.

1. An unmarked vowel ending a syllable (an open syllable) is long; thus, *ma* represents the pronunciation of *may; ne,* that of *knee; ri,* of *wry; so,* of *sew; too,* of *two;* and *vu,* of *view.*
2. An unmarked vowel in a syllable ending with a consonant (a closed syllable) is short; thus, *kat* represents *cat; bed, bed; hit, hit; not, knot; foot, foot;* and *kusp, cusp.*
3. A long vowel in a closed syllable is indicated by a macron; thus, *māt* stands for *mate; sēd,* for *seed; bīl,* for *bile; mōl,* for *mole; fūm,* for *fume;* and *fo͞ol,* for *fool.*

4. A short vowel that ends or itself constitutes a syllable is indicated by a breve; thus, *ĕ-fekt'* for *effect, ĭ-mūn'* for *immune,* and *ŭ-klo͞od'* for *occlude.*

Primary (') and secondary (") accents are shown in polysyllabic words. Unstressed syllables are followed by hyphens.

Some common vowels are pronounced as indicated here.

ə	sofa	ē	met	ŏ	got
ā	mate	ī	bite	ū	fuel
ă	bat	ĭ	bit	ŭ	but
ē	beam	ō	home		

From Dorland's Illustrated Medical Dictionary. *Copyright © 1988 W. B. Saunders, Philadelphia, Pa. Reprinted by permission.*

A

AB toxins Exotoxins composed of two parts (A and B). The B portion is responsible for toxin binding to a cell but does not directly harm it; the A portion enters the cell and disrupts its function.

ABC protein secretion pathway *See* ATP-binding cassette transporters.

accessory pigments Photosynthetic pigments such as carotenoids and phycobiliproteins that aid chlorophyll in trapping light energy.

acellular slime mold *See* plasmodial slime mold.

acetyl-CoA pathway A biochemical pathway used by methanogens to fix CO_2 and by acetogens to generate acetic acid.

acetyl-coenzyme A (acetyl-CoA) A molecule made of acetic acid and coenzyme A that is energy rich; it is produced by many catabolic pathways and is the substrate for the tricarboxylic acid cycle, fatty acid biosynthesis, and other pathways.

acid-fast Refers to bacteria (e.g., mycobacteria) that cannot be easily decolorized with acid alcohol after being stained with dyes such as basic fuchsin.

acid-fast staining A staining procedure that differentiates between bacteria based on their ability to retain a dye when washed with an acid alcohol solution.

acidophile (as'id-o-fīl") A microorganism that has its growth optimum between about pH 0 and 5.5.

acquired immune tolerance The ability to produce antibodies against nonself antigens while "tolerating" (not producing antibodies against) self antigens.

acquired immunity *See* adaptive immunity.

actin filaments One of the proteinaceous components of the eukaryotic cytoskeleton; also called microfilaments.

actinobacteria (ak"tĭ-no-bak-tēr-e-ah) A group of Gram-positive bacteria containing actinomycetes and their high G + C relatives.

actinomycete (ak"tĭ-no-mi'sēt) An aerobic, Gram-positive bacterium that forms branching filaments (hyphae) and asexual spores.

actinorhizae Associations between actinomycetes and plant roots.

activated sludge Solid matter or sediment composed of actively growing microorganisms that participate in the aerobic portion of a biological sewage treatment process.

activation energy The energy required to bring reacting molecules together to reach the transition state in a chemical reaction.

activator-binding site *See* activator protein.

activator protein A transcriptional regulatory protein that binds to a specific site on DNA (activator-binding site) and enhances transcription initiation.

active (catalytic) site The part of an enzyme that binds the substrate to form an enzyme-substrate complex and catalyze the reaction. Also called the catalytic site.

active transport The transport of solute molecules across a membrane against a gradient; it requires a carrier protein and the input of energy. Three major types are primary active transport, which uses

hydrolysis of ATP to power transport; secondary active transport, which uses ion gradients across a membrane to power active transport; and group translocation. *See also* group translocation.

acute-phase proteins Liver proteins that assist in the prevention of blood loss and ready the host for microbial invasion.

N-acylhomoserine lactone *See* quorum sensing.

adaptive immunity Immunity that develops in response to a specific antigen; involves T and B lymphocyte differentiation and antibody production. Immunologic memory is formed to each antigen.

adaptive mutation The notion that mutations will arise when organisms are subjected to prolonged stress that will help the organism survive the stress.

adenine (ad'e-nēn) A purine derivative, 6-aminopurine, found in nucleosides, nucleotides, coenzymes, and nucleic acids.

adenosine diphosphate (ADP) (ah-den'o-sēn) The nucleoside diphosphate usually formed upon the breakdown of ATP when it provides energy for work.

adenosine 5'-triphosphate (ATP) A high energy molecule that serves as the cell's major form of energy currency.

adhesin (ad-he'zin) A molecule on the surface of a microorganism involved in adhesion to a substratum (e.g., host tissue).

adjuvant Material added to an antigen in a vaccine preparation that increases its immunogenicity.

aerial mycelium The mat of hyphae formed by actinomycetes or fungi that grows above the substrate, imparting a fuzzy appearance to colonies.

aerobic anoxygenic phototrophy (AAnP) Phototrophic process by which electron donors such as organic matter or sulfide are used under aerobic conditions.

aerobic respiration A metabolic process in which molecules, often organic, are oxidized with oxygen as the final electron acceptor.

aerotolerant anaerobes Microbes that grow equally well whether or not oxygen is present.

aflatoxin (af″lah-tok′sin) A polyketide secondary fungal metabolite that can cause cancer.

agar (ahg′ar) A complex sulfated polysaccharide, usually from red algae, that is used to solidify culture media.

agglutinates The visible aggregates or clumps formed by an agglutination reaction.

agglutination reaction The formation of an insoluble immune complex by the cross-linking of cells or particles.

airborne transmission The spread of a pathogen as it travels suspended in air over a meter or more from the source to the host.

akinetes (ā-kin′ĕts) Specialized, nonmotile, dormant, thick-walled resting cells formed by some cyanobacteria.

alcoholic fermentation A fermentation process that produces ethanol and CO_2 from sugars.

alga (al′gah) A common term for several unrelated groups of photosynthetic eukaryotic microorganisms. Most are now considered protists.

alignment The comparison of gene, RNA, and protein sequences in a base-by-base (or amino acid by amino acid) fashion; used to calculate the degree of similarity between sequences.

alkaliphile (alkalophile) A microorganism that grows best at pHs from about 8.5 to 11.5.

allele An alternative form of a gene.

allergen An antigen that induces an allergic response.

allergy *See* Type I hypersensitivity.

allochthonous (ăl′ək-thə-nəs) Substances not native to a given environment (e.g., nutrient influx into freshwater ecosystems).

allograft A transplant between genetically different individuals of the same species.

allosteric effector *See* allosteric enzyme.

allosteric enzyme An enzyme whose activity is altered by the noncovalent binding of a small molecule (allosteric effector) at a regulatory site separate from the catalytic site; effector binding causes a conformational change in the enzyme's catalytic site, causing enzyme activation or inhibition.

α/β T cells CD4 or CD8 T cells that possess alpha/beta T-cell receptors.

α-hemolysis *See* hemolysis.

alphaproteobacteria Members of the bacterial class *Alphaproteobacteria*.

alternative complement pathway An antibody-independent pathway of complement activation.

alternate sigma factors Sigma is the subunit of bacterial RNA polymerase required for transcription initiation; alternate sigma factors direct the core RNA polymerase to transcribe distinct sets of inducible genes or operons, and is thereby a mechanism for global regulation.

alternative splicing The use of different exons during RNA splicing to generate different polypeptides from the same gene.

alveolar macrophage A vigorously phagocytic macrophage located on the epithelial surface of the lung alveoli, where it ingests inhaled particulate matter and microorganisms.

amensalism A relationship in which the product of one organism has a negative effect on another organism.

amino acid activation The preparatory step of protein synthesis in which amino acids are attached to transfer RNA molecules. The reaction is catalyzed by aminoacyl-tRNA synthetases.

aminoacyl or **acceptor site (A site)** The ribosomal site that contains an aminoacyl-tRNA at the beginning of the elongation cycle during protein synthesis.

aminoacyl-tRNA synthetases *See* amino acid activation.

aminoglycoside antibiotics A group of antibiotics that contain a cyclohexane ring and amino sugars; all aminoglycoside antibiotics inhibit protein synthesis by binding to the small ribosomal subunit.

amphibolic pathways Metabolic pathways that function both catabolically and anabolically.

amphipathic Term describing a molecule that has both hydrophilic and hydrophobic regions (e.g., phospholipids).

amphitrichous (am-fit′rĕ-kus) A cell with a single flagellum at each end.

anabolism The synthesis of complex molecules from simpler molecules with the input of energy and reducing power.

anaerobic digestion (an″a-er-o′bik) The microbiological treatment of sewage wastes under anoxic conditions to produce methane.

anaerobic respiration An energy-conserving process in which the terminal electron transport chain acceptor is a molecule other than oxygen.

anagenesis Changes in gene frequencies and distribution among species; the accumulation of small genetic changes within a population that introduces genetic variability but is not enough to result in either speciation or extinction.

anammoxosome The compartment within a planctomycete cell in which the anammox reaction occurs.

anammox reaction The coupled use of nitrite as an electron acceptor and ammonium ion as an electron donor under anoxic conditions to yield nitrogen gas.

anaphylaxis (an″ah-fĭ-lak′sis) A severe, immediate type I hypersensitivity reaction following exposure of a sensitized individual to the appropriate antigen.

anaplasia The reversion of an animal cell to a more primitive, undifferentiated state.

anaplerotic reactions (an′ah-plĕ-rot′ik) Reactions that replenish depleted tricarboxylic acid cycle intermediates.

anergy (an′ər-je) A state of immune cell unresponsiveness to antigens.

anoxic Without oxygen present.

anoxygenic photosynthesis Photosynthesis that does not oxidize water to produce oxygen.

antibiotic A microbial product or its derivative that kills susceptible microorganisms or inhibits their growth.

antibody (immunoglobulin) A glycoprotein made by plasma cells in response to the introduction of an antigen.

antibody affinity The strength of binding between an antigen and an antibody.

antibody-dependent cell-mediated cytotoxicity (ADCC) The killing of antibody-coated target cells by immune cells with Fc receptors that recognize the Fc region of the bound antibody.

antibody-mediated immunity *See* humoral immunity.

antibody class switching The process by which plasma cells change their expression from one isotype of immunoglobulin to a second isotype.

antibody titer An approximation of the antibody concentration.

anticodon The three bases on a tRNA that are complementary to the codon on mRNA.

antigen A substance (such as a protein, nucleoprotein, polysaccharide, or sometimes a glycolipid) to which lymphocytes respond.

antigen-binding fragment (Fab) "Fragment antigen binding." A monovalent antigen-binding fragment of an immunoglobulin molecule that consists of one light chain and part of one heavy chain, linked by interchain disulfide bonds.

antigen-presenting cells (APCs) Cells that take in protein antigens, process them, and present peptide fragments bound to MHC molecules to T cells; includes macrophages, dendritic cells, and B cells.

antigen processing The hydrolytic digestion of antigens to produce peptide fragments, which are collected and presented by class I or class II MHC molecules on the surface of an antigen-presenting cell.

antigenic determinant site *See* epitope.

antigenic drift A small change due to mutation in the antigenic character of an organism.

antigenic shift A major change in the antigenic character of an organism that makes it unrecognized by host immune mechanisms. This term is most often used to describe an antigenically distinct virus that results when two different strains of virus recombine genetic material.

antigenic variation The capacity of some microbes to change surface proteins when in a host organism, thereby evading the host immune system.

antimetabolite A compound that blocks metabolic pathway function by competitively inhibiting a key enzyme because it closely resembles the normal enzyme substrate.

antimicrobial agent An agent that kills microorganisms or inhibits their growth.

antiport Coupled transport of two molecules in which one molecule enters the cell as the other leaves the cell.

antisense RNA A single-stranded RNA with a base sequence complementary to a segment of a target RNA molecule; when bound to the target RNA, the target's activity is altered.

antisepsis The prevention of infection or sepsis.

antiseptic Chemical agent applied to tissue to prevent infection by killing or inhibiting pathogens.

antitoxin An antibody to a microbial toxin, usually a bacterial exotoxin, that binds and neutralizes the toxin.

apical complex A set of organelles characteristic of members of the protist subdivision *Apicomplexa*: includes polar rings, subpellicular microtubules, conoid, rhoptries, and micronemes.

apicomplexan (a′pĭ-kom-plek′san) A protist that has an apical complex and a spore-forming stage. It is either an intra- or extracellular parasite of animals; a member of the taxon *Apicomplexa*.

apicoplast (a′pĭ-ko-plast) Organelles in apicomplexan protists derived from cyanobacterial endosymbionts. Apicoplasts, however, are not involved in phototrophy; rather, they are the site of fatty acid, isoprenoid, and heme biosynthesis.

apoenzyme (ap″o-en′zīm) The protein part of an enzyme that also has a nonprotein component.

apoptosis (ap″o-to′sis) Programmed cell death. A physiological suicide mechanism that results in fragmentation of a cell into membrane-bound particles that are eliminated by phagocytosis.

aporepressor The inactive form of a repressor protein; the repressor becomes active when the corepressor binds to it.

appressorium A flattened region of hypha found in some plant-infecting fungi that aids in penetrating the host plant cell wall. Occurs in both pathogenic fungi and nonpathogenic mycorrhizal fungi.

aptamer (ap′tăm-er) A single-stranded DNA or RNA molecule engineered to bind to a target protein to modify the activity of the protein.

arbuscular mycorrhizae (AM) The mycorrhizal fungi in a fungus-root association that penetrate the outer layer of the root, grow intracellularly, and form characteristic much-branched hyphal structures called arbuscules.

arbuscules *See* arbuscular mycorrhizal fungi.

archaea Members of the domain *Archaea*.

archaerhodopsin A transmembranous protein to which retinal is bound; it functions as a light-driven proton pump resulting in photophosphorylation without chlorophyll or bacteriochlorophyll. Found in the purple membrane of halophilic archaea.

arthroconidia (arthrospores) Fungal spores formed by fragmentation.

artificially acquired active immunity The type of immunity that results from immunizing an animal with a vaccine. The immunized animal produces its own antibodies and activated lymphocytes.

artificially acquired passive immunity The temporary immunity that results from introducing into an animal antibodies that have been produced either in another animal or by in vitro methods.

ascocarp A multicellular structure in ascomycete fungi lined with specialized cells called asci.

ascogenous hypha A specialized hypha that gives rise to one or more asci.

ascomycetes (as″ko-mi-se′tēz) A group of fungi that form ascospores.

ascospore (as′ko-spor) A spore contained or produced in an ascus.

ascus (pl., asci) A specialized cell, characteristic of the ascomycetes, in which two haploid nuclei fuse to produce a zygote, which often immediately divides by meiosis; at maturity, an ascus contains ascospores.

aseptate (coenocytic) hypha A multinucleate hypha (filament) that lacks cross walls.

assimilatory reduction (e.g., assimilatory nitrate reduction and assimilatory sulfate reduction) The reduction of an inorganic molecule to incorporate it into organic material. No energy is conserved during this process.

associative nitrogen fixation Nitrogen fixation by bacteria in the plant root zone (rhizosphere).

atomic force microscope *See* scanning probe microscope.

atopic reaction *See* localized anaphylaxis.

ATP *See* adenosine 5′-triphosphate.

ATP-binding cassette transporters (ABC transporters) Transport systems that use ATP hydrolysis to drive translocation across the plasma membrane; can be used for nutrient uptake (ABC importer) or export of substances (ABC exporter), including protein secretion (ABC protein secretion pathway).

ATP synthase An enzyme that catalyzes synthesis of ATP from ADP and P_i, using energy derived from the proton motive force.

attack rate The proportional number of cases that develop in a population exposed to an infectious agent.

attenuated vaccine Live, nonpathogenic organisms used to activate adaptive immunity.

attenuation 1. A mechanism for the regulation of transcription termination of some bacterial operons by aminoacyl-tRNAs. 2. A procedure that reduces or abolishes the virulence of a pathogen without altering its immunogenicity.

attenuator A factor-independent transcription termination site in the leader sequence that is involved in attenuation.

autochthonous (ô-tŏk-thə-nəs) Substances (nutrients) that originate in a given environment.

autoclave An apparatus for sterilizing objects by the use of steam under pressure.

autoimmune disease A disease produced by the immune system attacking self antigens.

autoimmunity A condition characterized by the presence of serum autoantibodies and self-reactive lymphocytes that may be benign or pathogenic.

autoinducer With respect to quorum sensing, a small molecule that induces synthesis of the enzyme responsible for the autoinducer's synthesis. *See also* quorum sensing.

autoinduction In general terms, any molecule that increases its own level of production.

autolysins (aw-tol′ĭ-sins or aw″to-lye′sins) Enzymes that partially digest peptidoglycan in growing bacteria so that the cell wall can be enlarged.

autophagocytosis Ancient means of homeostasis whereby intracellular components are recycled.

autophagosome *See* macroautophagy.

autophagy *See* autophagocytosis and macroautophagy.

autoradiography A procedure that detects radioactively labeled materials using a photographic process.

autotroph (aw′to-trōf) An organism that uses CO_2 as its sole or principal source of carbon.

auxotroph (awk′so-trōf) An organism with a mutation that causes it to lose the ability to synthesize an essential nutrient; because of the mutation, the organism must obtain the nutrient or a precursor from its surroundings.

average nucleotide identity (ANI) A pairwise comparison between the whole or partial genome sequences of two microorganisms; expressed as percent identity.

avidity The combined strength of binding between an antigen and all the antibody-binding sites.

axenic (a-zen′ik) Not contaminated by any foreign organisms; the term is used in reference to pure microbial cultures. *See also* pure culture.

axial fibrils *See* periplasmic flagella.

axial filament *See* periplasmic flagella.

axopodium A thin, needlelike type of pseudopodium with a central core of microtubules.

B

B7 (CD80, CD86) proteins Glycoproteins on the surface of antigen-presenting cells (B cells, macrophages, dendritic cells, as well as T cells); it binds CD28 on T cells, resulting in second signal activation.

bacillus (bah-sil′lus) A rod-shaped bacterium or archaeon.

bacteremia The presence of viable bacteria in the blood.

bacteria Members of the domain *Bacteria*.

bacterial artificial chromosome (BAC) A cloning vector constructed from the *Escherichia coli* F-factor plasmid.

bactericide An agent that kills bacteria.

bacteriochlorophyll (bak-te″re-o-klo′ro-fil) A modified chlorophyll that serves as the primary light-trapping pigment in purple and green photosynthetic bacteria and heliobacteria.

bacteriocin (bak-te′re-o-sin) A secreted bacterial protein that kills other closely related bacteria.

bacteriophage (bak-te′re-o-fāj″) A virus that uses bacteria as its host; often called a phage.

bacteriostatic Inhibiting the growth and reproduction of bacteria.

bacteroid A modified, often pleomorphic, bacterial cell within the root nodule cells of legumes; it carries out nitrogen fixation.

baeocytes Small, spherical, reproductive cells produced by some cyanobacteria through multiple fission.

barophilic (bar″o-fil′ik) or **barophile** *See* piezophilic.

barotolerant Organisms that can grow and reproduce at high pressures but do not require them.

basal body *See* flagellum.

base analogues Molecules that resemble normal DNA nucleotides and can substitute for them during DNA replication, leading to mutations. Also used as antiviral agents.

base excision repair *See* excision repair.

basic dyes Dyes that are cationic, or have positively charged groups, and bind to negatively charged cell structures.

basidiocarp (bah-sid′e-o-karp″) The fruiting body of a basidiomycete that contains the basidia.

basidiomycetes (bah-sid″e-o-mi-se′tēz) A group of fungi in which the spores are borne on club-shaped organs called basidia.

basidiospore (bah-sid′e-ō-spōr) A spore borne on the outside of a basidium.

basidium (bah-sid′e-um; pl., **basidia**) A structure formed by basidiomycetes; it bears on its surface basidiospores (typically four) that are formed following karyogamy and meiosis.

basophil A weakly phagocytic white blood cell in the granulocyte lineage. It synthesizes and stores vasoactive molecules (e.g., histamine) that are released in response to external triggers.

batch culture Growth of microorganisms in a closed culture vessel without adding fresh or removing old (spent) medium.

B cell (B lymphocyte) A type of lymphocyte that, following interaction with antigen, becomes a plasma cell, which synthesizes and secretes antibody molecules involved in humoral immunity.

B-cell receptor (BCR) A transmembrane immunoglobulin complex on the surface of a B cell that binds an antigen and stimulates the B cell. It is composed of a membrane-bound immunoglobulin, usually IgD or a modified IgM, complexed with the Ig-α/Ig-β heterodimer.

benthic Pertaining to the bottom of the sea or other body of water.

β-hemolysis *See* hemolysis.

β-lactam Referring to antibiotics containing a β-lactam ring (e.g., penicillins and cephalosporins).

β-lactam ring The cyclic chemical structure composed of one nitrogen and three carbon atoms. It has antibacterial activity, interfering with bacterial cell wall synthesis.

β-lactamase An enzyme that hydrolyzes the β-lactam ring, rendering the antibiotic inactive. Sometimes called penicillinase.

β-oxidation pathway The major pathway of fatty acid oxidation to produce NADH, FADH$_2$, and acetyl coenzyme A.

betaproteobacteria Members of the bacterial class *Betaproteobacteria*.

binal symmetry The symmetry of some virus capsids (e.g., those of complex phages) that is a combination of icosahedral and helical symmetry.

binary fission Asexual reproduction in which a cell separates into two identical daughter cells.

binomial system The nomenclature system in which an organism is given two names; the first is the capitalized generic name, and the second is the uncapitalized specific epithet.

bioaugmentation Addition of pregrown microbial cultures to an environment to perform a specific task.

biochemical oxygen demand (BOD) The amount of oxygen used by organisms in water under standard conditions; it provides an index of the amount of microbially oxidizable organic matter present.

biochemical pathways Sets of chemical reactions performed by organisms that convert a starting substrate into one or more products.

biocide Chemical or physical agent, usually with a broad spectrum of biological activity, that inactivates microorganisms.

biodegradation The breakdown of a complex chemical through biological processes that results in minor loss of functional groups, fragmentation into smaller constituents, or complete breakdown to CO$_2$ and minerals.

biofilms Organized microbial communities encased in extracellular polymeric substances and associated with surfaces, often with complex structural and functional characteristics.

biofuels Energy-dense compounds (i.e., fuels) produced by microbes, usually by the degradation of plant substrates.

biogeochemical cycling The oxidation and reduction of substances carried out by living organisms and abiotic processes that results in the cycling of elements within and between different parts of the ecosystem and the atmosphere.

bioinformatics The interdisciplinary field that manages and analyzes large biological data sets, including genome and protein sequences.

bioinsecticide A microbe or microbial product used to kill or disable unwanted insect pests.

biological safety cabinets Cabinets that use HEPA filters to project a curtain of sterile air across its opening, preventing microbes from entering or exiting.

biomagnification The increase in concentration of a substance in higher-level consumer organisms.

bioprospecting The collection, cataloging, and analysis of organisms, including microorganisms and plants, with the intent of finding a useful application and/or to document biodiversity.

bioremediation The use of biologically mediated processes to remove or degrade pollutants from specific environments.

biosensor A device for the detection of a particular substance (an analyte) that combines a biological receptor with a physicochemical detector.

biotechnology The process in which living organisms are manipulated, particularly at the molecular genetic level, to form useful products.

bioterrorism The intentional or threatened use of microorganisms or organic toxins to produce death or disease in humans, animals, and plants.

biotransformation (microbial transformation) The use of living organisms to modify substances that are not normally used for growth.

biovar Variant strains of microbes characterized by biological or physiological differences.

BLAST (basic local alignment search tool) A publicly available computer program that identifies regions of similarity between nucleotide or amino acid sequences. It compares input or query sequences with sequence databases and determines the statistical significance of matches. Homologies can be used to infer gene or protein function and evolutionary relationships.

blastospores Fungal spores produced from a vegetative mother cell by budding.

bloom The growth of a single microbial species or a limited number of species in an aquatic environment, usually in response to a sudden pulse of nutrients. *See also* harmful algal bloom.

B lymphocyte *See* B cell.

bright-field microscope A microscope that illuminates the specimen directly with bright light and forms a dark image on a brighter background.

broad-spectrum drugs Chemotherapeutic agents that are effective against many different kinds of pathogens.

bronchial-associated lymphoid tissue (BALT) *See* mucosal-associated lymphoid tissue.

Bt An insecticide that consists of a toxin produced by *Bacillus thuringiensis*.

bubo (bu′bo) A tender, inflamed, enlarged lymph node that results from a variety of infections.

budding A vegetative outgrowth of yeast and some bacteria as a means of asexual reproduction; the daughter cell is smaller than the parent.

bulking sludge Sludge produced in sewage treatment that does not settle properly, usually due to the development of filamentous microorganisms.

butanediol fermentation A type of fermentation most often found in members of the family *Enterobacteriaceae* in which 2,3-butanediol is a major product; acetoin is an intermediate in the pathway and may be detected by the Voges-Proskauer test.

C

calorie The amount of heat needed to raise one gram of water from 14.5 to 15.5°C.

Calvin-Benson cycle The main pathway for the fixation (reduction and incorporation) of CO$_2$ into organic material by photoautotrophs and chemolithoautotrophs.

cancer A malignant tumor that expands locally by invasion of surrounding tissues and systemically by metastasis.

cannulae Tubelike structures extending from the surface of some archaeal cells. They are thought to function in attachment to surfaces.

5′ cap *See* pre-mRNA.

capsid The protein coat or shell that surrounds a virion's nucleic acid.

capsomer The ring-shaped morphological unit of icosahedral capsids.

capsule A layer of well-organized material, not easily washed off, lying outside the cell wall.

carbonate equilibrium system The interchange among CO$_2$, HCO$_3^-$, and CO$_3^{2-}$ that keeps oceans buffered between pH 7.6 to 8.2.

carbon to nitrogen ratio The ratio of carbon to nitrogen in a soil; it can be used to predict rates of organic decomposition and thus the cycling of nutrients in the system.

carboxysomes Polyhedral inclusions where CO$_2$ fixation occurs; contain the CO$_2$-fixation enzyme ribulose 1,5-bisphosphate carboxylase/oxygenase. They are a type of microcompartment.

cardinal temperatures The minimal, maximal, and optimum temperatures for growth.

caseous lesion (ka′se-us) A lesion containing necrotic cells with cheesy appearance and texture; often caused by *Mycobacterium tuberculosis.*

catabolism That part of metabolism in which larger, more complex molecules are broken down into smaller, simpler molecules with the release of energy.

catabolite activator protein (CAP) A protein that regulates the expression of genes encoding catabolite repressible enzymes; also called cyclic AMP receptor protein (CRP).

catabolite repression Inhibition of the synthesis of several catabolic enzymes by a preferred carbon and energy source (e.g., glucose).

catalase An enzyme that catalyzes the destruction of hydrogen peroxide.

catalyst A substance that accelerates a reaction without being permanently changed itself.

catalytic site *See* active site.

catalyzed reporter deposition-fluorescent in situ hybridization (CARD-FISH) A FISH technique in which the signal is amplified so that low levels of fluorescence can be detected. *See also* fluorescent in situ hybridization (FISH).

catenanes (kăt′ə-nāns′) Circular, covalently closed nucleic acid molecules that are locked together like the links of a chain.

cathelicidins Antimicrobial cationic peptides that are produced by a variety of cells (e.g., neutrophils, respiratory epithelial cells, and alveolar macrophages).

caveolin-dependent endocytosis *See* endocytosis.

CD4⁺ cell *See* T-helper cell.

CD8⁺ cell *See* cytotoxic T lymphocyte.

CD95 pathway *See* Fas-FasL pathway.

cell cycle The sequence of events in a cell's growth-division cycle between the end of one division and the end of the next.

cell envelope The plasma membrane plus all other external layers.

cellular (cell-mediated) immunity The type of immunity mediated by T cells.

cellular slime molds Protists with a vegetative phase consisting of amoeboid cells that aggregate to form a multicellular pseudoplasmodium. They belong to the taxon *Dictyostelia;* formerly considered fungi.

cell wall The strong structure that lies outside the plasma membrane; it supports and protects the membrane and helps maintain cell shape.

Centers for Disease Control and Prevention (CDC) Public health agency of the United States, headquartered in Atlanta, GA; serves as the national agency for developing and carrying out disease prevention and control, environmental health, and health promotion and education.

central metabolic pathways Those pathways central to the metabolism of an organism because they function catabolically and anabolically (e.g., glycolytic pathways and tricarboxylic acid cycle).

central tolerance The process by which self-reactive B cells are eliminated in the bone marrow and self-reactive T cells are eliminated in the thymus.

chain of infection *See* infectious disease cycle.

chain termination DNA sequencing method A DNA-sequencing method that uses dideoxynucleotides to cause termination of DNA replication at random sites.

chancre (shang′ker) The primary lesion of syphilis occurring at the site of entry of the pathogen.

chaperone proteins Proteins that assist in the folding and stabilization of other proteins. Some are also involved in directing newly synthesized proteins to protein secretion systems or to other locations in the cell.

chemical fixation *See* fixation.

chemical oxygen demand (COD) The amount of chemical oxidation required to convert organic matter in water and wastewater to CO_2.

chemiosmotic hypothesis (kem″e-o-os-mot′ik) The hypothesis that proton and electrochemical gradients are generated by electron transport and then used to drive ATP synthesis by oxidative phosphorylation or photophosphorylation.

chemoheterotroph (ke″mo-het′er-o-trōf″) *See* chemoorganoheterotroph.

chemokine A type of cytokine that stimulates chemotaxis and chemokinesis of immune cells.

chemolithoautotroph A microorganism that oxidizes reduced inorganic compounds to derive both energy and electrons; CO_2 is the carbon source.

chemolithoheterotroph A microorganism that uses reduced inorganic compounds to derive both energy and electrons; organic molecules are used as the carbon source.

chemolithotroph A microorganism that uses reduced inorganic compounds as a source of energy and electrons.

chemoorganoheterotroph An organism that uses organic compounds as sources of energy, electrons, and carbon for biosynthesis. Also called chemoheterotroph and chemoorganotrophic heterotroph.

chemoreceptors Proteins in the plasma membrane or periplasmic space that bind chemicals and trigger the appropriate chemotactic response.

chemostat A continuous culture apparatus that feeds medium into the culture vessel at the same rate as medium containing microorganisms is removed; the medium in a chemostat contains one essential nutrient in a limiting quantity.

chemotaxins Agents that cause white blood cells to migrate up its concentration gradient.

chemotaxis The pattern of cellular behavior in which the cell moves toward chemical attractants and away from repellents.

chemotherapy The use of chemical agents to treat disease.

chemotrophs Organisms that obtain energy from the oxidation of chemical compounds.

chlamydiae (klə-mid′e-e) Members of the genera *Chlamydia* and *Chlamydiophila:* Gram-negative, obligate intracellular pathogens.

chlorophyll The green photosynthetic pigment that consists of a large tetrapyrrole ring with a magnesium atom in the center.

chloroplast A eukaryotic plastid that contains chlorophyll and is the site of photosynthesis.

chlorosomes Elongated, intramembranous vesicles found in the green sulfur and nonsulfur bacteria; they contain light-harvesting pigments. Sometimes called chlorobium vesicles.

choleragen (kol′er-ah-gen) The cholera toxin.

chromatic adaptation The ability of cyanobacteria to change the relative concentrations of light-harvesting pigments in response to changes in wavelengths of incident light.

chromatin The complex of DNA and proteins, including histones, from which eukaryotic chromosomes are made.

chromatin immunoprecipitation (ChIP) A technique that enables the identification of protein-DNA interactions.

chromatin modification The addition of acetyl, methyl, phosphoryl, and other groups to a histone protein. These modifications change the level of transcription of nearby genes.

chromatin remodeling The alteration of chromatin structure by inserting or removing variant histone proteins from a nucleosome.

chromogen (kro′me-jen) A colorless substrate that is acted on by an enzyme to produce a colored end product.

chromophore group A chemical group with double bonds that absorbs visible light and gives a dye its color.

chromosomes The bodies that have most or all of the cell's DNA and contain most of its genetic information (mitochondria and chloroplasts also contain DNA).

chronic inflammation Inflammation that is characterized by a dense tissue infiltration of lymphocytes and macrophages into the affected site and the formation of new connective tissue, which usually causes permanent tissue damage.

chytrids A term used to describe members of *Chytridiomycota,* fungi that produce motile zoospores with single, posterior, whiplash flagella.

cidal Causing death.

cilia Threadlike appendages extending from the surface of some protists that beat rhythmically to propel them; cilia are membrane-bound cylinders with a complex internal array of microtubules, usually in a 9 + 2 pattern.

ciliates (*Ciliophora*) Protists that move by rapidly beating cilia and belong to *Alveolata.*

citric acid cycle *See* tricarboxylic acid cycle.

class I MHC molecule *See* major histocompatibility complex.

class II MHC molecule *See* major histocompatibility complex.

class switching *See* antibody class switching.

classical complement pathway The antibody-dependent pathway of complement activation; it stimulates lysis of pathogens, phagocytosis, and other host defenses.

clathrin-dependent endocytosis *See* endocytosis.

clonal selection The process by which an antigen binds to the best-fitting B-cell receptor, activating that B cell, resulting in the synthesis of that specific antibody and clonal expansion.

clone (1) A group of genetically identical cells or organisms derived by asexual reproduction from a single parent. (2) A DNA sequence that has been isolated and replicated using a cloning vector.

cloning vector A DNA molecule that can replicate independently of the host chromosome and maintain a piece of inserted foreign DNA, such as a gene, into a recipient cell. It may be a plasmid, phage, cosmid, or artificial chromosome.

club fungi Common name for members of *Basidiomycota*.

cluster of differentiation molecules (CDs) Functional cell surface proteins that can be used to identify leukocyte subpopulations (e.g., interleukin-2 receptor, CD4, and CD8).

coagulase (ko-ag′u-las) An enzyme that induces blood clotting; it is characteristically produced by pathogenic staphylococci.

coagulation The process of adding chemicals to precipitate impurities from water during purification; also called flocculation.

coccolithophores Photosynthetic protists belonging to the group *Stramenopila*; characterized by coccoliths—intricate cell walls made of calcite.

coccus (kok′us, pl., **cocci**, kok′si) A roughly spherical bacterial or archaeal cell.

code degeneracy The presence of more than one codon for each amino acid.

coding sequences (CDS) In genomic analysis, open reading frames presumed to encode proteins but not tRNA or rRNA.

codon (ko′don) A sequence of three nucleotides in mRNA that directs the incorporation of an amino acid during protein synthesis or signals the stop of translation.

coenocytic (se″no-sit′ik) *See* aseptate hypha.

coenzyme A loosely bound cofactor that often dissociates from the enzyme active site after product has been formed.

coenzyme Q (CoQ, ubiquinone) *See* electron transport chain.

coevolution The evolution of two interacting organisms such that they optimize their survival with one another; that is, they develop reciprocal adaptations.

cofactor The nonprotein component of an enzyme required for catalytic activity.

colicin (kol′ĭ-sin) A plasmid-encoded protein produced by enteric bacteria; binds to receptors on the cell envelope of sensitive target bacteria, where it may cause lysis or attack specific intracellular sites such as ribosomes.

coliform A Gram-negative, nonsporing, facultative rod that ferments lactose with gas formation within 48 hours at 35°C.

Colilert defined substrate test An assay designed to detect the presence of coliforms in water.

colonization The establishment of a site of microbial reproduction on an inanimate surface or organism without necessarily resulting in tissue invasion or damage.

colony An assemblage of microorganisms growing on a solid surface; the assemblage often is directly visible.

colony forming units (CFU) The number of microorganisms that form colonies when cultured using spread plates or pour plates, an indication of the number of viable microorganisms in a sample.

colony stimulating factor (CSF) A protein that stimulates the growth and development of specific cell populations (e.g., granulocyte-CSF stimulates granulocytes to be made from precursor stem cells).

colorless sulfur bacteria A diverse group of nonphotosynthetic proteobacteria that can oxidize reduced sulfur compounds such as hydrogen sulfide. Many are chemolithotrophs and derive energy from sulfur oxidation.

cometabolism The modification of a compound not used for growth by a microorganism, which occurs in the presence of another organic material that serves as a carbon and energy source.

commensal Living on or within another organism without injuring or benefiting the other organism.

commensalism A type of symbiosis in which one individual gains from the association (the commensal) and the other is neither harmed nor benefited.

common-source epidemic An epidemic characterized by a sharp rise to a peak and then a rapid but not as pronounced decline in the number of individuals infected; it usually involves a single contaminated source that infects individuals.

communicable disease A disease associated with a pathogen that can be transmitted from one host to another.

comparative genomics Comparison of genomes from different organisms to discern differences and similarities.

compartmentation The differential distribution of enzymes and metabolites to separate cell structures or organelles.

compatible solute A low molecular weight molecule used to protect cells against changes in solute concentrations (osmolarity) in their habitat; it can exist at high concentrations within the cell without hindering metabolism or growth.

competent cell A cell that can take up free DNA fragments and incorporate them into its genome during transformation.

competition An interaction between two organisms attempting to use the same resource (nutrients, space, etc.).

competitive exclusion principle Two competing organisms overlap in resource use, which leads to the exclusion of one of the organisms.

competitive inhibitor A molecule that inhibits enzyme activity by binding the enzyme's active site.

complementary Refers to the matching of two strands of DNA or RNA based on the base-pairing rules.

complementary DNA (cDNA) A DNA copy of an RNA molecule (e.g., a DNA copy of an mRNA).

complement system A group of plasma proteins that plays a major role in innate immunity.

complex medium Culture medium that contains some ingredients of unknown chemical composition.

compromised host A host with lowered resistance to infection and disease.

concatemer A long DNA molecule consisting of several genomes linked together in a row.

conditional mutations Mutations with phenotypes that are expressed only under certain environmental parameters (e.g., temperature).

confocal microscope A light microscope in which monochromatic laser derived light scans across the specimen at a specific level. Stray light from above and below the plane of focus is blocked out to give an image with excellent contrast and resolution.

conidiospore (ko-nid′e-o-spōr) An asexual, thin-walled spore borne on hyphae and not contained within a sporangium; it may be produced singly or in chains.

conidium (ko-nid′e-um; pl., **conidia**) *See* conidiospore.

conjugate redox pair The electron acceptor and electron donor of a redox half reaction.

conjugation (1) The form of gene transfer and recombination in bacteria and archaea that requires direct cell-to-cell contact. (2) A complex form of sexual reproduction commonly employed by protists.

conjugative plasmid A plasmid that carries the genes that enable its transfer to other bacteria during conjugation (e.g., F plasmid).

conjugative transposon A mobile genetic element that is able to transfer itself from one bacterium to another by conjugation.

consensus sequence A commonly occurring sequence of nucleotides within a genetic element.

conserved hypothetical proteins In genomic analysis, proteins of unknown function for which at least one match exists in a database.

consortium A physical association of two or more different organisms, usually beneficial to all of the organisms.

constant region (C_L and C_H) The part of an antibody molecule that does not vary greatly in amino acid sequence among molecules of the same class, subclass, or type.

constitutive gene A gene that is expressed at nearly the same level at all times.

contact transmission Spread of a pathogen by contact of the source or reservoir of the pathogen.

contigs In genomic sequencing, fragments of DNA with identical sequences at the ends; that is, the ends of the fragments overlap in sequence.

continuous culture system A culture system with constant environmental conditions maintained through continual provision of nutrients and removal of wastes. *See also* chemostat and turbidostat.

continuous feed The constant addition of a carbon source during industrial fermentation.

contractile vacuole (vak′u-ōl) In protists and some animals, a clear fluid-filled vacuole that takes up water from within the cell and then contracts, releasing it to the outside through a pore in a cyclical manner. It functions primarily in osmoregulation and excretion.

cooperation A positive but not obligatory interaction between two different organisms.

coral bleaching The loss of photosynthetic pigments by either physiological inhibition or expulsion of the coral photosynthetic endosymbiont, zooxanthellae, a dinoflagellate.

core genome The common set of genes found in all genomes in a species or other taxon.

core polysaccharide *See* lipopolysaccharide.

corepressor A small molecule that binds a transcription repressor protein, thereby activating the repressor and inhibiting the synthesis of a repressible enzyme.

cosmid A plasmid vector with lambda phage *cos* sites that can be packaged in a phage capsid; it is useful for cloning large DNA fragments.

coverage The number of times a nucleotide is sequenced in any given sequencing project. Synonymous with depth of coverage, so "depth" and "coverage" are used interchangeably. Breadth of coverage refers to the percentage of genome sequenced.

cristae (kris′te) Infoldings of the inner mitochondrial membrane.

cryophile *See* psychrophile.

cryptidins Antimicrobial peptides produced by Paneth cells in the intestines.

crystallizable fragment (Fc) The stem of the Y portion of an antibody molecule. Cells such as macrophages bind to the Fc region; it also is involved in complement activation.

culturomics An approach to isolating and identifying microbes that involves cultivation under many different conditions followed by identification by mass spectrometry and SSU rRNA gene sequencing.

curing The loss of a plasmid from a cell.

cut-and-paste transposition *See* simple transposition.

cyanobacteria A large group of Gram-negative, morphologically diverse bacteria that carry out oxygenic photosynthesis.

cyanophycin Inclusions that store nitrogen found in cyanobacteria.

3′, 5′-cyclic adenosine monophosphate (cAMP) A small nucleotide that is involved in regulating a number of cellular processes in bacteria and other organisms. In some bacteria, it functions in catabolite repression.

cyclic dimeric GMP (c-di-GMP) A small nucleotide consisting of two GMP molecules linked together by their phosphates. It functions as a second messenger in many bacteria and some eukaryotes.

cyclic photophosphorylation (fo″to-fos″for-ĭ-la′shun) The formation of ATP when light energy is used to move electrons cyclically through an electron transport chain during photosynthesis.

cyst A general term used for a specialized microbial cell enclosed in a wall. Cysts are formed by protists and a few bacteria. They may be dormant, resistant structures formed in response to adverse conditions or reproductive cysts that are a normal stage in the life cycle.

cytochromes (si′to-krōms) *See* electron transport chain.

cytokine (si′to-kīn) A general term for proteins released by immune cells in response to inducing stimuli; they mediate the activity of other cells.

cytokinesis Processes that apportion the cytoplasm and organelles, synthesize a septum, and divide a cell into two daughter cells during cell division.

cytopathic effect The observable change that occurs in cells as a result of viral replication.

cytoplasm All material in the cell enclosed by the plasma membrane, with the exception of the nucleus in eukaryotic cells.

cytoproct A specific site in certain protists (e.g., ciliates) where digested material is expelled.

cytosine (si′to-sēn) A pyrimidine 2-oxy-4-aminopyrimidine found in nucleosides, nucleotides, and nucleic acids.

cytoskeleton A network of structures made from filamentous proteins (e.g., actin and tubulin) and other components in the cytoplasm of cells.

cytosol Liquid component of the cytoplasm.

cytostome A permanent site in a ciliate protist in which food is ingested.

cytotoxic T lymphocyte (CTL) Effector T cell that can lyse target cells bearing class I MHC molecules complexed with antigenic peptides. CTLs arise from antigen-activated Tc cells.

D

Dane particle A 42-nm spherical particle that is one of three seen in Hepatitis B virus infections. The Dane particle is the complete virion.

dark-field microscopy Microscopy in which the specimen is brightly illuminated while the background is dark.

dark reaction *See* photosynthesis.

deamination The removal of amino groups from amino acids.

decimal reduction time (*D* or *D* value) The time required to kill 90% of the microorganisms or spores in a sample at a specified temperature.

defensins A large family of antimicrobial peptides produced by neutrophils, macrophages, lymphocytes, and a variety of other host cells.

defined (synthetic) medium Culture medium made with components of known composition.

deltaproteobacteria Members of the bacterial class *Deltaproteobacteria*.

denaturation A change in either protein shape or nucleic-acid structure.

denaturing gradient gel electrophoresis (DGGE) A technique by which DNA is rendered single stranded (denatured) while undergoing electrophoresis so that DNA fragments of the same size can be separated according to nucleotide sequence.

dendritic cell An antigen-presenting cell that has long membrane extensions resembling the dendrites of neurons.

denitrification The reduction of nitrate to gaseous products, primarily nitrogen gas, during anaerobic respiration.

deoxyribonucleic acid (DNA) (de-ok″se-ri″bo-nu-kle′ik) A polynucleotide that constitutes the genetic material of all cellular organisms. It is composed of deoxyribonucleotides connected by phosphodiester bonds.

depth filters Filters composed of fibrous or granular materials that are used to decrease microbial load and sometimes to sterilize solutions.

desensitization The process by which a sensitized or hypersensitive individual is made insensitive or nonreactive to a sensitizing agent (e.g., an allergen).

detergent An organic molecule, other than a soap, that serves as a wetting agent and emulsifier; it is normally used as cleanser, but some may be used as antimicrobial agents.

diatoms (*Bacillariophyta*) Photosynthetic protists with siliceous cell walls called frustules belonging to the group *Stramenopila*.

diauxic growth (di-awk′sik) Biphasic growth response in which a microorganism, when exposed to two nutrients, initially uses one of them for growth and then alters its metabolism to make use of the second.

differential interference contrast (DIC) microscope A light microscope that employs two beams of plane polarized light. The beams are combined after passing through the specimen and their interference is used to create the image.

differential media Culture media that distinguish between groups of microorganisms based on differences in their growth and metabolic products.

differential staining Staining procedures that divide bacteria into separate groups based on staining properties.

diglycerol tetraether lipids Archaeal lipids formed when isoprenoid hydrocarbons are linked to two glycerols by ether bonds.

dikaryotic stage (di-kar-e-ot′ik) In fungi, having two separate haploid nuclei, one from each parent.

dilution susceptibility tests A method by which antibiotics are evaluated for their ability to inhibit bacterial growth in vitro. A standardized concentration of bacteria is added to serially diluted antibiotics and incubated. Tubes lacking additional bacterial growth suggest antibiotic concentrations that are bacteriocidal or bacteriostatic.

dinoflagellate (di″no-flaj′e-lāt) A photosynthetic protist characterized by two flagella used in swimming in a spinning pattern; belong to the group *Alveolata*.

diplococcus (dip″lo-kok′us) A pair of cocci.

direct counts Methods for determining population size by examining a sample of the population and counting the cells within the sample.

direct immunofluorescence Technique for imaging of microscopic agents using fluorescently labeled antibodies to bind to the agent.

direct repair A type of DNA repair mechanism in which a damaged nitrogenous base is returned to its normal form (e.g., conversion of a thymine dimer back to two normal thymine bases).

directed evolution A variety of approaches used to mutate a specific gene or genes with the aim of increasing product yield or generating new molecules.

disease syndrome A set of signs and symptoms that are characteristic of a disease.

disinfectant An agent, usually chemical, that disinfects; normally, it is employed only with inanimate objects.

disinfection The killing, inhibition, or removal of microorganisms that may cause disease. It usually refers to the treatment of inanimate objects with chemicals.

disinfection by-products (DBPs) Chlorinated organic compounds such as trihalomethanes formed during chlorine use for water disinfection. Many are carcinogens.

dissimilatory reduction (e.g., dissimilatory nitrate reduction and dissimilatory sulfate reduction) The use of a substance as an electron acceptor for an electron transport chain. The acceptor (e.g., sulfate or nitrate) is reduced but not incorporated into organic matter.

dissolved organic matter (DOM) Nutrients that are available in the soluble, or dissolved, state.

divisome A collection of proteins that aggregate at the region in a dividing microbial cell where a septum will form.

DNA amplification The process of increasing the number of DNA molecules, usually by using the polymerase chain reaction.

DNA-dependent DNA polymerase An enzyme that catalyzes the synthesis of DNA using a DNA molecule as a template.

DNA-dependent RNA polymerase An enzyme that catalyzes the synthesis of RNA using a DNA molecule as a template.

DNA-DNA hybridization (DDH) An assay in which the single-stranded DNA of one organism is hybridized to that of another, and the percent that hybridizes is calculated. There must be at least 70% similarity between two genomes as determined by DDH for microbes to be considered the same species; however, this is only one criterion upon which species assignments are made.

DNA gyrase A topoisomerase enzyme that relieves tension generated by the rapid unwinding of DNA during DNA replication.

DNA ligase An enzyme that joins two DNA fragments together through the formation of a new phosphodiester bond.

DNA microarrays Solid supports that have DNA attached in organized arrays and are used to evaluate gene expression.

DNA polymerase (pol-im′er-ās) An enzyme that synthesizes new DNA using a parental nucleic acid strand (usually DNA) as a template.

double diffusion agar assay (Öuchterlony technique) An immunodiffusion reaction in which both antibody and antigen diffuse through agar to form stable immune complexes, which can be observed visually.

doubling time *See* generation time.

droplet nuclei Small particles (0 to 4 μm in diameter) that represent what is left from the evaporation of larger particles (10 μm or more in diameter) called droplets.

drug inactivation Physical or chemical alteration of chemotherapeutic agent resulting in its loss of biological activity.

D value *See* decimal reduction time.

E

ecotype A population of an organism that is genetically very similar but ecologically distinct from others of the same species.

ectomycorrhizae A mutualistic association between fungi and plant roots in which the fungus surrounds the root with a sheath.

ectoplasm In some protists, the cytoplasm directly under the cell membrane (plasmalemma) is divided into an outer gelatinous region, the ectoplasm, and an inner fluid region, the endoplasm.

efflux pumps Membrane-associated protein assemblies that transport amphiphilic molecules from cytoplasm to the extracellular environment.

electron acceptor A compound that accepts electrons in an oxidation-reduction reaction. Often called an oxidizing agent or oxidant.

electron cryotomography A specialized electron microscopy procedure that involves rapid freezing of intact specimens, maintenance of the specimen in a frozen state while being examined, and imaging from different angles. The information from each angle is used to create a three-dimensional reconstruction.

electron donor An electron donor in an oxidation-reduction reaction. Often called a reducing agent or reductant.

electron shuttle A substance or structure along which electrons are passed from the electron transport chain to an external terminal electron acceptor.

electron transport chain (ETC) A series of electron carriers that operate together to transfer electrons from donors to acceptors such as oxygen. Molecules involved in electron transport include nicotinamide adenine dinucleotide (NAD^+), NAD phosphate ($NADP^+$), cytochromes, heme proteins, nonheme proteins (e.g., iron-sulfur proteins and ferredoxin), coenzyme Q, flavin adenine dinucleotide (FAD), and flavin mononucleotide (FMN). Also called an electron transport system.

electroporation The application of an electric field to create temporary pores in the plasma membrane in order to render a cell temporarily transformation competent.

elementary body (EB) A small, dormant body that serves as the agent of transmission between host cells in the chlamydial life cycle.

elongation factors Proteins that function in the elongation cycle of protein synthesis.

Embden-Meyerhof pathway A pathway that degrades glucose to pyruvate (i.e., a glycolytic pathway).

encystment The formation of a cyst.

endemic disease A disease that exists at a steady, low-level frequency at a moderate, regular interval.

endergonic reaction (end″er-gon′ik) A reaction that does not spontaneously go to completion as written; the standard free energy change is positive, and the equilibrium constant is less than one.

endocytic pathways *See* endocytosis.

endocytosis The process in which a cell takes up solutes or particles by enclosing them in vesicles pinched off from its plasma membrane. It often occurs at regions of the plasma membrane coated by proteins such as clathrin and caveolin. Endocytosis involving these proteins is called clathrin-dependent endocytosis and caveolin-dependent endocytosis, respectively. When the substance endocytosed first binds to a receptor, the process is called receptor-mediated endocytosis.

endogenous antigen processing Antigen degradation by cytoplasmic proteasomes and subsequent complexing with class I MHC molecules for presentation on the processing cell's surface.

endogenous pyrogen A host-derived chemical mediator (e.g., interleukin-1) that acts on the hypothalamus, stimulating a rise in core body temperature (i.e., it stimulates the fever response).

endomycorrhizae Referring to a mutualistic association of fungi and plant roots in which the fungus penetrates into the root cells.

endophyte A microorganism living within a plant but not necessarily parasitic on it.

endoplasm *See* ectoplasm.

endoplasmic reticulum (ER) A system of membranous tubules and flattened sacs (cisternae) in eukaryotic cells. Rough endoplasmic reticulum (RER) bears ribosomes on its surface; smooth endoplasmic reticulum (SER) lacks them.

endosome A membranous vesicle formed by endocytosis. It undergoes a maturational process that starts with early endosomes and proceeds to late endosomes and finally lysosomes.

endospore An extremely heat- and chemical-resistant, dormant, thick-walled spore that develops within some members of the bacterial phylum *Firmicutes*. It has a complex structure that includes (from outermost to innermost) exosporium, spore coat, cortex, spore cell wall, and spore core.

endosymbiosis A type of symbiosis in which one organism is found within another organism.

endosymbiotic hypothesis The hypothesis that mitochondria, and related organelles (e.g., hydrogenosomes), and chloroplasts arose from bacterial endosymbionts of ancestral eukaryotic cells.

endotoxin The lipid A component of lipopolysaccharide (LPS) from Gram-negative bacterial cell walls. Nanogram quantities can induce fever, activate complement and coagulation cascades, act as a mitogen to B cells, and stimulate cytokine release from a variety of cells. Systemic effects of endotoxin are referred to as septic shock.

end product inhibition *See* feedback inhibition.

energy The capacity to do work or cause particular changes.

enhancer A site in DNA to which a eukaryotic activator protein binds.

enology The science of wine making.

enriched media Media that contain nutrients to encourage growth of fastidious microbes.

enrichment culture The growth of specific microbes from natural samples by including media components or manipulating environmental conditions to promote growth of the desired microbes while suppressing the growth of other microbes.

enteric bacteria (enterobacteria) Members of the family *Enterobacteriaceae*; also can refer to bacteria that live in the intestinal tract.

enterotoxin A toxin specifically affecting the cells of the intestinal mucosa, causing vomiting and diarrhea.

enthalpy Heat content of a system.

Entner-Doudoroff pathway A pathway that converts glucose to pyruvate and glyceraldehyde 3-phosphate by producing 6-phosphogluconate and then dehydrating it (i.e., a glycolytic pathway).

entropy A measure of the randomness or disorder of a system; a measure of that part of the total energy in a system that is unavailable for useful work.

envelope In virology, an outer membranous layer that surrounds the nucleocapsid of some viruses.

enveloped virus A virus having virions with a nucleocapsid enclosed within a lipid bilayer.

environmental microbiology The study of microbial processes that occur in soil, water, and other natural habitats. Specific microorganisms are not necessarily investigated; rather, the cumulative impact of the microbial community on the specific biome, as well as global implications, are assessed.

enzyme A protein catalyst with specificity for the reaction catalyzed and its substrates.

enzyme-linked immunosorbent assay (ELISA) A serological assay in which bound antigen or antibody is detected by another antibody that is conjugated to an enzyme. The enzyme converts a colorless substrate to a colored product reporting the antibody capture of the antigen.

eosinophil (e″o-sin′o-fil) A leukocyte that has a two-lobed nucleus and cytoplasmic granules that release chemical mediators in response to parasite infections.

epidemic A disease that suddenly increases in occurrence above the normal level in a given population.

epidemiologist A person who specializes in epidemiology.

epidemiology The study of the factors determining and influencing the frequency and distribution of disease, injury, and other health-related events and their causes in defined human populations.

epifluorescence microscopy The illumination of a microscope sample with a high-intensity light from above. The light is passed through an exciter filter so that when light of a specific wavelength is focused on the specimen, it emits light, thereby making the sample visible as a specific color on a black background.

epilimnion The upper, warmer layer of water in a stratified lake.

epiphyte An organism that grows on the surface of plants.

episome A plasmid that can either exist independently of the host cell's chromosome or be integrated into it.

epitope (ep′i-tōp) An area of an antigen that stimulates the production of, and combines with, specific antibodies; also known as the antigenic determinant site.

epsilonproteobacteria Members of the bacterial class *Epsilonproteobacteria*.

equilibrium The state of a system in which no net change is occurring and free energy is at a minimum; in a chemical reaction at equilibrium,

the rates in the forward and reverse directions exactly balance each other out.

equilibrium constant (K_{eq}) A value that relates the concentration of reactants and products to each other when a reaction is at equilibrium.

ergot The dried sclerotium of *Claviceps purpurea*. Also, an ascomycete that parasitizes rye and other plants; production of an alkaloid causes an animal disease called ergotism.

erythroblastosis fetalis Hemolytic disease of the newborn whereby red blood cells are destroyed by maternal antibody.

eukaryotic cells Cells that have a membrane-delimited nucleus; all plants, fungi, protists and animals are eukaryotic.

eutrophic Nutrient enriched.

eutrophication The enrichment of an aquatic environment with nutrients.

excision repair A type of DNA repair mechanism in which a section of a strand of damaged DNA is excised and replaced, using the complementary strand as a template. Two types are recognized: base excision repair and nucleotide excision repair.

excystment The escape of one or more cells or organisms from a cyst.

exergonic reaction (ek″ser-gon′ik) A reaction that spontaneously goes to completion as written; the standard free energy change is negative, and the equilibrium constant is greater than one.

exfoliative toxin (exfoliatin) An exotoxin produced by *Staphylococcus aureus* that causes the separation of epidermal layers and the loss of skin surface layers; produces scalded skin syndrome.

exit site (E site) The location on a ribosome to which an empty (uncharged) tRNA moves from the P site before it finally leaves during protein synthesis.

exoenzymes Enzymes that are secreted by cells.

exogenous antigen processing Antigen degradation by phagolysosome enzymes and subsequent complexing with class II MHC molecules for presentation on the processing cell's surface.

exon The region in a gene that is retained in mRNA after introns are removed by splicing.

exospore A spore formed outside the mother cell.

exotoxin A heat-labile, toxic protein produced by a bacterium and usually released into the bacterium's surroundings.

exponential (log) phase The phase of the growth curve during which the microbial population is growing at a constant and maximum rate, dividing and doubling at regular intervals.

expression vector A plasmid into which a recombinant gene is cloned; when in host cells, the gene is transcribed and its protein synthesized.

extant organism An organism that exists on Earth today.

exteins Polypeptide sequences of precursor self-splicing proteins that are joined together during formation of the final, functional protein. They are separated from one another by intein sequences.

extended aeration process An approach that can be used during secondary sewage treatment to reduce the amount of sludge produced.

extinction culture technique A method for obtaining an axenic (pure) culture in which natural samples are serial diluted so that the final dilution has between 1 and 10 cells (i.e., diluted to extinction).

extreme environment An environment in which physical factors such as temperature, pH, salinity, and pressure are outside of the normal range for growth of most microorganisms; these conditions allow unique organisms to survive and function.

extreme halophiles *See* haloarchaea.

extremophiles Microorganisms that grow in extreme environments.

F

facilitated diffusion Diffusion across the plasma membrane that is aided by a channel protein or a carrier protein.

facultative anaerobes Microorganisms that do not require oxygen for growth, but grow better in its presence.

Fas-FasL pathway One of two pathways used by cytotoxic T cells to kill target cells. Fas (CD95 receptor) is located on target cells. FasL is the ligand to the CD95 receptor and is found on cytotoxic T cells.

fatty acid synthase The multienzyme complex that makes fatty acids.

fecal coliform Gram-negative, lactose-fermenting bacteria whose normal habitat is the intestinal tract and that can grow at 44.5°C. They are used as indicators of fecal pollution of water.

fecal enterococci (en″ter-o-kok′si) Enterococci found in the intestine of humans and other warm-blooded animals.

feedback inhibition A negative feedback mechanism in which an end product inhibits the activity of an enzyme in the pathway leading to its formation.

fermentation (1) An energy-yielding process in which an organic molecule is oxidized without an exogenous electron acceptor. Usually pyruvate or a pyruvate derivative serves as the electron acceptor. (2) The growth of microbes in very large volumes for the production of industrially important products.

fermenter In industrial microbiology, the large vessel used to culture microorganisms.

ferredoxin *See* electron transport chain.

F factor The fertility factor, a plasmid that carries genes for bacterial conjugation and makes its *Escherichia coli* host the gene donor during conjugation.

filopodia Long, narrow pseudopodia found in certain amoeboid protists.

fimbria (fim′bre-ah; pl., **fimbriae**) A fine, hairlike protein appendage on many bacteria, some archaea, and some fungi. They attach cells to surfaces, and some are involved in twitching motility.

first law of thermodynamics Energy can be neither created nor destroyed (although it can be changed in form or redistributed).

fixation (1) The process in which the internal and external structures of cells and organisms are preserved and fixed in position. (2) The conversion of inorganic, gaseous elements such as carbon (CO_2) and nitrogen (N_2) to organic forms.

flagellin (flaj′ĕ-lin) A family of related proteins in motile bacteria used to construct the filament of a flagellum. An unrelated group of proteins, also called flagellins, are used to construct archaeal flagellar filaments.

flagellum (flah-jel′um; pl., **flagella**) A threadlike appendage on many cells that is responsible for motility. Bacterial flagella are composed of a basal body at the base of the flagellum, which attaches it to the cell, and a hook, which connects the basal body to the filament. The filament is the part of the flagellum that rotates and moves the bacterium.

flavin adenine dinucleotide (FAD) (fla′vin ad′ĕ-nēn) *See* electron transport chain.

flavin mononucleotide *See* electron transport chain.

flow cytometry A method for defining and enumerating cells by sending them in single file past a laser to detect cell size and morphology.

fluid mosaic model The model of cell membranes in which the membrane is a lipid bilayer with integral proteins buried in the lipid and peripheral proteins more loosely attached to the membrane surface.

fluorescence activated cell sorting (FACS) A form of flow cytometry that results in separating cells according to size or other quantifiable trait (e.g., DNA content).

fluorescence microscope A microscope that exposes a specimen to light of a specific wavelength and then forms an image from the fluorescent light produced.

fluorescent in situ hybridization (FISH) A technique for identifying certain genes or organisms in which single-stranded DNA fragments are labeled with fluorescent dye and hybridized to genes of interest.

fluorescent light The light emitted by a substance when it is irradiated with light of a shorter wavelength.

fluorochrome A fluorescent dye.

fomite (pl., **fomites**) Common inanimate materials that transmit pathogens to humans.

food-borne infection Gastrointestinal illness caused by ingestion of microorganisms, followed by their growth within the host.

food intoxication Food poisoning caused by microbial toxins produced in a food prior to consumption. The presence of living bacteria is not required.

food poisoning A general term usually referring to a gastrointestinal disease caused by the ingestion of food contaminated by pathogens or their toxins.

formate dehydrogenase An enzyme system that cleaves formic acid into H_2 and CO_2. It is used by enteric bacteria that perform mixed acid fermentation.

forward mutation A mutation from the wild type to a mutant form.

F′ plasmid An F plasmid that carries some bacterial genes and transmits them to recipient cells when the F′ cell carries out conjugation.

frameshift mutations Mutations arising from the loss or gain of a base or DNA segment, leading to a change in the codon reading frame and thus a change in the amino acids incorporated into protein.

free energy change The total energy change in a system that is available to do useful work as the system goes from its initial state to its final state at constant temperature and pressure.

fruiting body A specialized structure that holds sexually or asexually produced spores; found in fungi, some protists, and some bacteria (e.g., the myxobacteria).

frustule (frus′tūl) The silicified cell wall in the diatoms.

fueling reactions Chemical reactions that supply the ATP, precursor metabolites, and reducing power needed for biosynthesis.

fumonisins A family of toxins produced by molds belonging to the genus *Fusarium*. It primarily affects corn and it is known to be hepato- and nephrotoxic in animals.

functional gene array Microarrays designed to determine potential microbial activities in natural environments. The probes on the array represent genes whose products are involved in specific processes, such as nitrogen fixation.

functional genomics Genomic analysis concerned with determining the way a genome functions.

fungicide An agent that kills fungi.

fungistatic Inhibiting the growth and reproduction of fungi.

fungus (pl., **fungi**) Achlorophyllous, hetero-trophic, spore-bearing eukaryotes with absorptive nutrition and a walled thallus; sometimes called "true fungi" or *Eumycota*.

G

G + C content The percent of guanines and cytosines present in a genome; this percent is used for taxonomic purposes.

gametangium (gam-ĕ-tan′je-um; pl., **gametangia**) A structure that contains gametes or in which gametes are formed.

γ/δ T cells CD4 or CD8 T cells that possess gamma/delta T-cell receptors.

gammaproteobacteria Members of the bacterial class *Gammaproteobacteria*.

gamonts Gametic cells formed by protists when they undergo sexual reproduction.

gas vacuole A gas-filled vacuole found in cyanobacteria and some other aquatic bacteria and archaea that provides flotation. It is composed of gas vesicles, which are made of protein.

gas vesicle *See* gas vacuole.

gel electrophoresis The separation of molecules according to charge and size through a gel matrix.

Gell-Coombs classification System used to codify hypersensitivity reactions.

gene A DNA segment or sequence that codes for a polypeptide, rRNA, or tRNA.

gene ontology A structured vocabulary adopted for annotating functional gene product data.

generalized transduction The transfer of any part of a bacterial or archaeal genome when the DNA fragment is packaged within a virus's capsid by mistake.

generation (doubling) time The time required for a microbial population to double in number.

genetic complementation The capacity of a gene, when introduced into a cell lacking a functional copy of that gene, to restore the wild-type phenotype to the cell. *See also* phenotypic rescue.

genetic drift Changes in the genotype of an organism that collect over time. These changes are generally neutral mutations that occur and are maintained without selection.

genetic engineering The deliberate modification of an organism's genetic information by changing its nucleic acid genome.

genome The full set of genes present in a cell or virus; all the genetic material in an organism.

genome annotation The process of determining the location and potential function of specific genes and genetic elements in a genome sequence.

genomic analysis *See* genomics.

genomic fingerprinting A series of techniques based on restriction enzyme digestion patterns that enable the comparison of microbial species and strains, and is thus useful in taxonomic identification.

genomic island A region in a genome introduced in an ancestral microbe by horizontal gene transfer. Genomic islands often bring new phenotypic traits to the microbe; *see also* pathogenicity island.

genomic library A collection of clones that contains fragments that represent the complete genome of an organism.

genomic reduction The decrease in genomic information that occurs over evolutionary time as an organism or organelle becomes increasingly dependent on another cell or a host organism.

genomics The study of the molecular organization of genomes, their information content, and the gene products they encode.

genotype The specific set of alleles carried in the genome of an organism.

genotypic classification The use of genetic data to construct a classification scheme for the identification of an unknown species or the phylogeny of a group of microbes.

genus A well-defined group of one or more species that is clearly separate from other organisms.

geographic information system (GIS) Use of hardware, software, and data for the capture, management, and display of geographically referenced information.

germfree State of having no microbial flora; describing animals born by cesarean delivery and reared in sterile environments.

germination The stage following spore activation in which the spore breaks its dormant state. Germination is followed by outgrowth.

Ghon complex The initial focus of infection in primary pulmonary tuberculosis.

gliding motility A type of motility in which a microbial cell glides smoothly along a solid surface without the aid of flagella.

global regulatory protein A protein that affects the transcription of numerous operons or genes simultaneously.

global regulatory systems Regulatory systems that simultaneously affect many genes.

glomeromycetes Common name for members of the fungal taxon *Glomeromycota*.

gluconeogenesis (gloo″ko-ne″o-jen′e-sis) The synthesis of glucose from noncarbohydrate precursors such as lactate and amino acids.

glutamine synthetase–glutamate synthase (GS-GOGAT) system A mechanism used by many microbes to incorporate ammonia.

glycerol diether lipids Archaeal lipids formed when isoprenoid hydrocarbons are linked to glycerol by ether bonds.

glycocalyx (gli″ko-kal′iks) A network of polysaccharides extending from the surface of bacteria and other cells.

glycogen (gli′ko-jen) A highly branched polysaccharide containing glucose, which is used to store carbon and energy.

glycolysis (gli-kol′ĭ-sis) The conversion of glucose to pyruvic acid by use of the Embden-Meyerhof pathway, pentose phosphate pathway, or Entner-Doudoroff pathway.

glycolytic pathway A pathway that converts glucose to pyruvic acid.

glycomics Analysis of all the carbohydrates produced by a cell under specific circumstances.

glyoxylate cycle A modified tricarboxylic acid cycle in which the decarboxylation reactions are bypassed by the enzymes isocitrate lyase and malate synthase; it is used to convert acetyl-CoA to succinate and other metabolites.

gnotobiotic (no″to-bi-ot′ik) Animals that are microbe-free or have a known microbiota consisting of one or a few microorganisms.

Golgi apparatus (gol′je) A membranous eukaryotic organelle composed of stacks (dictyosomes) of flattened sacs (cisternae) involved in packaging and modifying materials for secretion and many other processes.

gonococci (gon′o-kok′si) Bacteria of the species *Neisseria gonorrhoeae,* the organism causing gonorrhea.

graft-versus-host disease Pathology resulting from attack of host cells by immunocompetent T cells in a transplanted organ or stem cells.

Gram stain A differential staining procedure that divides bacteria into Gram-positive and Gram-negative groups based on their ability to retain crystal violet when decolorized with an organic solvent such as ethanol.

granular sludge In wastewater management, the stable aggregation of bacteria that readily settle, thereby facilitating recovery of the effluent.

granulocyte A type of white blood cell that stores preformed enzymes and antimicrobial proteins in vacuoles near the cell membrane.

granuloma Term applied to nodular inflammatory lesions containing phagocytic cells.

granzyme Enzyme found in the granules of cytotoxic T cells and natural killer cells that causes targets of these cells to undergo apoptosis.

GRAS An acronym for *g*enerally *r*egarded *a*s *s*afe: a category of food additives recognized in the United States and by the United Nations.

great plate count anomaly (GPCA) The discrepancy between the number of viable microbial cells and the number of colonies that can be cultivated from the same natural sample.

green fluorescent protein (GFP) A protein from the jellyfish *Aequorea victoria* that fluoresces when illuminated at a specific wavelength. The gene for GFP can be fused to gene promoters to determine when a gene is expressed (transcriptional fusion). When fused to the coding sequence of a gene so that a chimeric protein is produced, GFP fluorescence reveals the localization of the protein (translational fusion).

greenhouse gases Gases (e.g., CO_2, CH_4) released from Earth's surface through chemical and biological processes that interact with the chemicals in the stratosphere to decrease radiational cooling of Earth.

green nonsulfur bacteria Anoxygenic photosynthetic bacteria that contain bacteriochlorophylls *a* and *c;* usually photoheterotrophic and display gliding motility. Include members of the phylum *Chloroflexi.*

green sulfur bacteria Anoxygenic photosynthetic bacteria that contain bacteriochlorophylls *a,* plus *c, d,* or *e;* photolithoautotrophic; use H_2, H_2S, or S as electron donor. Include members of the phylum *Chlorobi.*

group A streptococcus (GAS) A Gram-positive, coccus-shaped bacterium often found in the throat and on the skin of humans, having the Lancefield group A of surface carbohydrate.

group B streptococcus (GBS) A Gram-positive, coccus-shaped bacterium found occasionally on mucous membranes of humans, having the Lancefield group B surface carbohydrate.

group translocation An active transport process in which a molecule is moved across a membrane by carrier proteins while being chemically altered at the same time (e.g., phosphoenolpyruvate: sugar phosphotransferase system).

growth factors Organic compounds that must be supplied in the diet for growth because they are essential cell components or precursors of such components and cannot be synthesized by the organism.

growth rate constant (*k*) The rate of microbial population growth expressed in terms of the number of generations per unit time.

guanine (gwan′in) A purine derivative, 2-amino-6-oxypurine found in nucleosides, nucleotides, and nucleic acids.

guanosine tetraphosphate (ppGpp) A small nucleotide that functions as a second messenger for the stringent response. *See also* stringent response.

guild A group of microbes defined by the same physiological activity.

gumma (gum′ah) A soft, gummy, benign tumor occurring in tertiary syphilis.

gut-associated lymphoid tissue (GALT) *See* mucosal-associated lymphoid tissue.

H

haloarchaea (extreme halophiles) A group of archaea that depend on high NaCl concentrations for growth and do not survive at a concentration below about 1.5 M NaCl.

halophile A microorganism that requires high levels of sodium chloride for growth.

halotolerant The ability to withstand large changes in salt concentration.

hami Structures extending from the surface of some archaeal cells that terminate in a grappling hooklike apparatus. They are thought to function in attachment to surfaces.

hapten A molecule not immunogenic by itself that, when coupled to a macromolecular carrier, can elicit antibodies directed against itself.

harmful algal bloom (HAB) In aquatic ecosystems, the growth of a single population of phototroph, either a protist (e.g., diatom, dinoflagellate) or a cyanobacterium, that produces a toxin poisonous to other organisms, sometimes including humans. Alternatively, in the absence of toxin production, the concentration of bloom microbe reaches levels that are inherently harmful to other organisms, such as filter-feeding bivalves.

Hartig net The area of nutrient exchange between ectomycorrhizal fungal hyphae and plant host cells.

health-care–associated infection An infection that is acquired during a patient's stay in a hospital or other type of clinical care facility.

heat fixation *See* fixation.

heat-shock proteins Proteins produced when cells are exposed to high temperatures or other stressful conditions. They protect the cells from damage and often aid in the proper folding of proteins.

helical capsid A viral capsid in the form of a helix.

helicases Enzymes that use ATP energy to unwind DNA ahead of the replication fork.

hemagglutination (hem″ah-gloo″tĭ-na′shun) The clumping of red blood cells by antibodies or components of virus capsids. *See also* viral hemagglutination.

hemagglutination assay A testing procedure based on a hemagglutination reaction.

hemagglutinin (HA) (1) Proteins that cause red blood cells to clump together (hemagglutination). (2) One of the envelope spikes of influenza and some other viruses. They are the basis for identifying different influenza virus strains.

hematopoiesis The process by which blood cells develop into specific lineages from stem cells. Red and white blood cells and platelets develop from this process.

hemolysis The disruption of red blood cells and release of their hemoglobin. In α-hemolysis, a greenish zone of incomplete hemolysis forms around a colony grown on blood agar. A clear zone of complete hemolysis without any obvious color change is formed during β-hemolysis.

herd immunity The resistance of a population to infection and spread of an infectious agent due to the immunity of a high percentage of the population.

heterocysts Specialized cells of cyanobacteria that are the sites of nitrogen fixation.

heterokont flagella A pattern of flagellation found in the members of the protist taxon *Stramenopila,* featuring two flagella, one extending anteriorly and the other posteriorly.

heterolactic fermenters Microorganisms that ferment sugars to form lactate and other products such as ethanol and CO_2.

heterologous gene expression The cloning, transcription, and translation of a gene that has been introduced (cloned) into an organism that normally does not possess the gene.

heterotroph An organism that uses reduced, preformed organic molecules as its principal carbon source.

hexose monophosphate pathway *See* pentose phosphate pathway.

Hfr conjugation Conjugation involving an Hfr strain and an F⁻ strain.

Hfr strain A strain of *Escherichia coli* that donates its chromosomal genes with high frequency to a recipient cell during conjugation because the F factor is integrated into the donor's chromosome.

hierarchical cluster analysis The organization of gene expression data such that induced and repressed genes, or genes of similar function, are grouped separately.

high-efficiency particulate (HEPA) filter Thick, fibrous filter constructed to remove 99.97% of particles that are 0.3 μm or larger.

high phosphate transfer potential A characteristic of a phosphorylated compound such that it readily transfers a phosphoryl group to another molecule concomitant with a large energy release.

high-throughput screening (HTS) A system that combines liquid handling devices, robotics, computers, data processing, and a sensitive detection system to screen thousands of compounds for a single capability.

histatin An antimicrobial peptide composed of 24 to 38 amino acids, heavily enriched with histidine, that targets fungal mitochondria.

histone A small basic protein with large amounts of lysine and arginine that is associated with eukaryotic DNA in chromatin. Related proteins are observed in many archaeal species, where they form archaeal nucleosomes.

holdfast A structure produced by some bacteria (e.g., *Caulobacter*) that attaches them to a solid object.

holoenzyme A complete enzyme consisting of the apoenzyme plus a cofactor. Also refers to a complete enzyme consisting of all protein subunits (e.g., RNA polymerase holoenzyme).

holozoic nutrition Acquisition of nutrients (e.g., bacteria) by endocytosis and the subsequent formation of a food vacuole or phagosome.

homolactic fermenters Organisms that ferment sugars almost completely to lactic acid.

homologous recombination Recombination involving two DNA molecules that are very similar in nucleotide sequence.

hopanoids Lipids found in bacterial membranes that are similar in structure and function to the sterols found in eukaryotic membranes.

hops Dried flowers of the plant *Humulus lupulus* used in the preparation of beer and lager.

horizontal (lateral) gene transfer (HGT/LGT) The process by which genes are transferred from one mature, independent organism to another. In bacteria and archaea, transformation, conjugation, and transduction are the major mechanisms by which HGT can occur.

hormogonia Small motile fragments produced by fragmentation of filamentous cyanobacteria; used for asexual reproduction and dispersal.

hospital-acquired infections *See* health-care-associated infections.

host An organism that harbors another organism.

housekeeping genes Genes that encode proteins that function throughout most of the life cycle of an organism (e.g., enzymes of the Embden-Meyerhof pathway).

human leukocyte antigen complex (HLA) A collection of genes on chromosome 6 that encodes proteins involved in host defenses; also called major histocompatibility complex.

human microbiome The sum of all the microorganisms that live on and in the human body.

humoral (antibody-mediated) immunity The type of immunity that results from the presence of soluble antibodies in blood and lymph.

hydrogen hypothesis A hypothesis that considers the origin of the eukaryotes through the development of the hydrogenosome. It suggests the organelle arose as the result of an endosymbiotic anaerobic bacterium that produced CO_2 and H_2 as the products of fermentation.

hydrogenosome One member of the mitochondrial family of organelles, it is found in some anaerobic protists that produce ATP by fermentation.

hydrophilic A polar substance that has a strong affinity for water (or is readily soluble in water).

hydrophobic A nonpolar substance lacking affinity for water (or which is not readily soluble in water).

3-hydroxypropionate bi-cycle A pathway used by green nonsulfur bacteria to fix CO_2.

hyperthermophile (hi″per-ther′mo-fīl) A microbe that has its growth optimum between 85°C and about 120°C. Hyperthermophiles usually do not grow well below 55°C.

hypha (hi′fah; pl., **hyphae**) The unit of structure of most fungi and some bacteria; a tubular filament.

hypolimnion The colder, bottom layer of water in a stratified lake.

I

icosahedral capsid A viral capsid that has the shape of a regular polyhedron having 20 equilateral triangular faces and 12 corners.

IgA Immunoglobulin A; the class of dimeric immunoglobulins that is present in mucous membranes and in many body secretions.

IgD Immunoglobulin D; the class of immunoglobulins found on the surface of many B lymphocytes, where it serves as an antigen receptor in the stimulation of antibody synthesis.

IgE Immunoglobulin E; the immunoglobulin class that binds to mast cells and basophils, and is responsible for type I or anaphylactic hypersensitivity reactions such as hay fever and asthma. IgE is also involved in resistance to helminth parasites.

IgG Immunoglobulin G; the predominant immunoglobulin class in serum. It serves to

neutralize toxins, opsonize bacteria, activate complement, and cross the placenta to protect the fetus and neonate.

IgM Immunoglobulin M; the class of serum antibody produced first upon infection. It is a large, pentameric molecule that agglutinates pathogens and activates complement. The monomeric form is present on the surface of some B lymphocytes.

immobilization The incorporation of a simple, soluble substance into the body of an organism, making it unavailable for use by other organisms.

immune complex The product of antibody binding to antigen; may also contain components of the complement system.

immune system The defensive system in an animal consisting of the innate and adaptive immune responses. It is composed of widely distributed cells, tissues, and organs that recognize foreign substances and microorganisms and acts to neutralize or destroy them.

immunity Refers to the overall general ability of a host to resist a particular disease; the condition of being immune.

immunization The deliberate introduction of foreign materials into a host to stimulate an adaptive immune response. *See also* vaccine.

immunoblotting The electrophoretic transfer of proteins from polyacrylamide gels to filters to demonstrate the presence of specific proteins through reaction with labeled antibodies.

immunodeficiency The inability to produce a normal innate and/or adaptive immune response.

immunodiffusion A technique involving the diffusion of antigen or antibody within a semisolid gel to produce a precipitin reaction.

immunoelectrophoresis (pl., **immuno-electrophoreses**) The electrophoretic separation of protein antigens followed by diffusion and precipitation in gels using antibodies against the separated proteins.

immunofluorescence A technique used to identify particular antigens microscopically in cells or tissues by binding a fluorescent antibody conjugate.

immunoglobulin (Ig) *See* antibody.

immunology The study of host defenses against invading foreign materials, including pathogenic microorganisms, transformed or cancerous cells, and tissue transplants from other sources.

immunoprecipitation A reaction involving soluble antigens reacting with antibodies to form an aggregate that precipitates out of solution.

inactivated vaccines Nonviable infectious agents prepared to stimulate immunity but not cause infectious disease; also known as killed vaccines.

incidence The number of new cases of a disease in a population at risk.

inclusions (1) Granules, crystals, or globules of organic or inorganic material in the cytosol of cells. (2) Clusters of viral proteins or virions within the nucleus or cytoplasm of virus-infected cells.

incubation period The period after pathogen entry into a host and before signs and symptoms appear.

indel A blending of the words *insertion* and *deletion*; these genetic mutations can be taxonomically useful.

index case The first disease case in an outbreak or epidemic.

indicator organism An organism whose presence indicates the condition of a substance or environment, for example, the potential presence of pathogens. Coliforms are used as indicators of fecal pollution.

indirect immunofluorescence Technique for imaging microscopic agents using antibodies to target the agent and a second, fluorochrome-labeled reagent to bind to the antibodies.

induced mutations Mutations caused by exposure to a mutagen.

inducer A small molecule that stimulates the synthesis of an inducible enzyme.

inducible gene A gene whose expression level can be increased by a regulatory molecule.

induction (1) In virology, the events that trigger a virus to switch from a lysogenic mode to a lytic pathway. (2) In genetics, an increase in gene expression.

infection The invasion of a host by a microorganism with subsequent establishment and multiplication of the agent. An infection may or may not lead to overt disease.

infection thread A tubular structure formed during the infection of a root by nitrogen-fixing bacteria. The bacteria enter the root by way of the infection thread and stimulate the formation of the root nodule.

infectious disease Any change from a state of health in which part or all of the host's body cannot carry on its normal functions because of the presence of an infectious agent or its products.

infectious disease cycle (chain of infection) The linked events that must occur for an infectious disease to be expressed in an individual.

infectious dose 50 (ID$_{50}$) Refers to the number of organisms that infect 50% of an experimental group of hosts within a specified time period.

infectivity Infectiousness; the state or quality of being infectious or communicable.

inflammasomes A complex formed within the cytoplasm of certain immune cells that promotes inflammation; composed of oligomerized NOD-like receptors and formed in response to the capture of microbe-associated molecular patterns and danger-associated molecular patterns.

inflammation A localized protective response to tissue injury or destruction. Acute inflammation is characterized by pain, heat, swelling, and redness in the injured area.

initiation factors Proteins that function during initiation of protein synthesis.

initiator tRNA The tRNA that binds to the start codon during initiation of translation.

innate (natural) immunity Refers to nonspecific defenses that are mounted within minutes to hours after challenge by a microbial or other foreign antigen. Sometimes also called nonspecific immunity.

innate lymphoid cell (ILC) A lymphocyte that functions as a bridge between the innate and adaptive arms of immunity, amplifying and relaying information about infection and tissue damage.

innate resistance mechanisms *See* innate (natural) immunity.

insertion sequence A transposable element that contains genes only for those enzymes, such as transposase, that are required for transposition.

in silico analysis The study of biology through the examination of nucleic acid and amino acid sequence. *See also* bioinformatics.

in situ reverse transcriptase (ISRT)-FISH A method that uses fluorescently labeled probes to identify mRNA transcripts in natural samples. *See also* fluorescent in situ hybridization (FISH).

insulators Sites in eukaryotic DNA that limit the effects of regulatory proteins to only those genes for which they are specified.

integral membrane protein *See* plasma membrane.

integrase An enzyme observed in some viruses; it catalyzes the integration of provirus DNA into the host chromosome.

integrative conjugative elements (ICEs) *See* conjugative transposon.

integrins (in′tə-grinz) Cellular adhesion receptors that mediate cell-cell and cell-substratum interactions, usually by recognizing linear amino acid sequences on protein ligands.

integron A genetic element with an attachment site for site-specific recombination and an integrase gene. It can capture genes and gene cassettes.

inteins Internal intervening sequences of precursor self-splicing proteins that are removed during formation of the final protein.

intercalating agents Molecules that can insert between the stacked bases of a DNA double helix, thereby distorting the DNA and inducing frameshift mutations.

interferon (IFN) A cytokine that stimulates cells to produce antiviral proteins (type 1 IFNs), and regulates growth, differentiation, or function of immune system cells (type 2 IFNs).

interleukin (in″tər-loo′kin) A cytokine that regulates growth and differentiation, particularly of lymphocytes.

intermediate filaments One of the proteinaceous components of the eukaryotic cytoskeleton.

internal transcribed spacer region (ITSR) The region between small subunit rRNA genes found in many microbial genomes that is conserved and transcribed.

interspecies hydrogen transfer The linkage of H$_2$ production by fermentative microorganisms to the use of H$_2$ by archaea during methanogenesis.

intoxication A disease that results from the entrance of a specific toxin into the body of a host. The toxin can induce the disease in the absence of the toxin-producing organism.

intraepidermal lymphocytes T cells found in the epidermis of the skin.

intranuclear inclusion body A structure found within cells infected with cytomegalovirus.

intron A noncoding intervening sequence in a gene that codes for pre-mRNA and is removed (spliced) from the final RNA product.

invasiveness The ability of a microorganism to enter a host, grow and reproduce within the host, and spread throughout its body.

iodophor An antimicrobial agent consisting of an organic compound complexed with iodine.

ionizing radiation Radiation of very short wavelength and high energy that causes atoms to lose electrons (i.e., ionize).

iron-sulfur (Fe-S) protein *See* electron transport chain.

isoelectric focusing An electrophoretic technique in which proteins are separated based on their isoelectric point.

isoelectric point The pH value at which a protein or other molecule no longer carries a net charge.

isoenzyme (isozyme) An enzyme that carries out the same catalytic function but differs in terms of its amino acid sequence, regulatory properties, or other characteristics.

isotope fractionation The preferential use of one stable isotope over another by microbes carrying out metabolic processes.

J

joule The SI (International System of Units) unit of measure for energy or work. One calorie is equivalent to 4.1840 joules.

K

kallikrein An enzyme that acts on kininogen, releasing the active bradykinin protein.

keratinocytes (kĕ-rat′ĭ-no-sīt) Cells of the skin that express keratin.

Kirby-Bauer method A disk diffusion test to determine the susceptibility of a microorganism to chemotherapeutic agents.

Koch's postulates A set of rules for proving that a microorganism causes a particular disease.

Koplik's spots White lesions that appear on mucous membranes in the mouth (buccal mucosa) characteristic of measles infection.

Krebs cycle *See* tricarboxylic acid (TCA) cycle.

L

labyrinthulids A subgroup of stramenopiles characterized by having heterokont flagellated zoospores.

lactic acid bacteria (LAB) Strictly fermentative, Gram-positive bacteria that produce lactic acid as the primary fermentation end product; used in the fermentation of dairy products.

lactic acid fermentation A fermentation that produces lactic acid as the sole or primary product.

lactoferrin An iron-sequestering protein released from macrophages and neutrophils into plasma.

lagging strand The strand of DNA synthesized discontinuously during DNA replication.

lag phase A period following the introduction of microorganisms into fresh culture medium when there is no increase in cell numbers or mass during batch culture.

Lancefield system (group) One of the serologically distinguishable groups (e.g., group A and group B) into which streptococci can be divided.

Langerhans cell Cell found in the skin that internalizes antigen and moves in the lymph to lymph nodes, where it differentiates into a dendritic cell.

lateral gene transfer *See* horizontal gene transfer.

leader A sequence in a gene that lies between the promoter and the start codon. It is transcribed, becoming a nontranslated sequence at the 5′ end of mRNA. It often aids in initiation and regulation of transcription and translation.

leading strand The strand of DNA that is synthesized continuously during DNA replication.

lectin complement pathway An antibody independent pathway of complement activation that is initiated by microbial lectins (proteins that bind carbohydrates).

leghemoglobin A heme-containing pigment produced in leguminous plants. It functions to protect the nitrogenase enzyme from oxygen in nodule-forming, nitrogen-fixing bacteria.

leishmanias (lĕsh″ma′ne-ăs) Trypanosomal protists of the genus *Leishmania*.

lentic An aquatic system that features slow-moving or still waters.

lethal dose 50 (LD_{50}) Refers to the number of organisms that kills 50% of an experimental group of hosts within a specified time period.

leukocyte (loo′ko-sīt) Any white blood cell.

ligand *See* receptor.

light reactions *See* photosynthesis.

lignin An irregularly branched molecule made of phenylpropene units; an important structural component of woody plants.

limnology The study of the biological, chemical, and physical aspects of freshwater systems.

lipid A The lipid component of a lipopolysaccharide; also called endotoxin.

lipidomics Study of all the lipids produced by an organism under specific circumstances.

lipopolysaccharide (LPS) (lip″o-pol″e-sak′ah-rīd) A molecule containing both lipid and polysaccharide, found in the outer membrane of the typical Gram-negative cell wall. In many bacteria, it consists of three components: lipid A, core polysaccharide, and O antigen.

lithoheterotrophy The metabolic strategy characterized by the use of inorganic chemicals as an electron and energy source and organic carbon as a source of carbon. *See also* chemolithoheterotroph.

lithotroph An organism that uses reduced inorganic compounds as its electron source.

littoral zone The coastal region of a lake, stream, or ocean.

lobopodia Rounded pseudopodia found in some amoeboid protists.

localized anaphylaxis (atopic reaction) Type I hypersensitivity that remains contained in one tissue space (e.g., hay fever).

log phase *See* exponential phase.

long-term stationary phase The phase following the death phase of a growth curve in which the population size remains at a more or less constant low level for an extended period.

lophotrichous (lo-fot′rĭ-kus) A cell with a cluster of flagella at one or both ends.

lotic An aquatic system that features fast-moving, free-running waters.

lymph node A small secondary lymphoid organ that contains lymphocytes, macrophages, and dendritic cells. It serves as a site for (1) filtration and removal of foreign antigens and (2) activation and proliferation of lymphocytes.

lymphocyte A nonphagocytic, mononuclear leukocyte that is an immunologically competent cell, or its precursor. Lymphocytes are present in the blood, lymph, and lymphoid tissues. *See also* B cell and T cell.

lymphokine A glycoprotein cytokine (e.g., IL-1) secreted by activated lymphocytes, especially sensitized T cells.

lysis (li′sis) The rupture or physical disintegration of a cell.

lysogenic (li-so-jen′ik) *See* lysogens.

lysogenic conversion A change in the phenotype of a bacterium due to the presence of a prophage.

lysogenic cycle The phase of a temperate virus's life cycle in which it establishes and maintains lysogeny.

lysogens (li′so-jens) Bacterial and archaeal cells that carry a provirus and can produce viruses under the proper conditions.

lysogeny (li-soj′e-ne) The state in which a viral genome remains within a bacterial or archaeal cell after infection and reproduces along with it, rather than taking control of the host cell and destroying it.

lysosome A spherical membranous eukaryotic organelle that contains hydrolytic enzymes and is responsible for the intracellular digestion of substances.

lysozyme An enzyme that degrades peptidoglycan by hydrolyzing the $\beta(1 \rightarrow 4)$ bond that joins *N*-acetylmuramic acid and *N*-acetylglucosamine.

lytic cycle (lit′ik) A viral life cycle that results in the lysis of the host cell.

M

macroautophagy Digestion of cytoplasmic components that involves enclosing the material (e.g., an organelle) in a double-membrane structure called an autophagosome. The autophagosome delivers the material to a lysosome for digestion.

macroelement A nutrient that is required in relatively large amounts (e.g., carbon and nitrogen).

macrolide antibiotic An antibiotic containing a macrolide ring—a large lactone ring with multiple keto and hydroxyl groups—linked to one or more sugars.

macromolecule A large molecule that is a polymer of smaller units joined together.

macronucleus The larger of the two nuclei in ciliate protists. It is normally polyploid and directs the routine activities of the cell.

macrophage The name for a large, mononuclear phagocytic antigen-presenting cell, present in blood, lymph, and other tissues.

madurose The sugar derivative 3-*O*-methyl-D-galactose, which is characteristic of several actinomycete genera that are collectively called maduromycetes.

magnetosomes Magnetic greigite or magnetite particles in magnetotactic bacteria that allow the bacteria to orient themselves in magnetic fields.

maintenance energy The energy a cell requires simply to maintain itself or remain alive and functioning properly. It does not include the energy needed for either growth or reproduction.

major histocompatibility complex (MHC) A chromosome locus encoding the histocompatibility antigens and other components of the immune system. Class I MHC molecules are cell surface glycoproteins present on all nucleated cells; class II MHC glycoproteins are on antigen-presenting cells.

malt Grain soaked in water to soften it, induce germination, and activate its enzymes. The malt is then used in brewing and distilling.

mannose-binding protein (MBP) Also known as mannose-binding lectin, a serum protein that binds to mannose residues in microbial cell walls, thereby initiating complement activation.

marine snow Organic matter that sinks out of the photic zone of the ocean. These flocculent particles are composed of fecal pellets, diatom frustules, and other materials that are not rapidly degraded.

mash The soluble materials released from germinated grains and prepared as a microbial growth medium.

mashing The process in which cereals are mixed with water and incubated to degrade their complex carbohydrates (e.g., starch) to more readily usable forms such as simple sugars.

massively parallel nucleotide sequencing *See* next-generation DNA sequencing.

mass spectrometry A type of spectrometry that determines mass-to-charge ratio of ions formed from the molecule being analyzed. The ratio can be used to identify structures and determine sequences of proteins.

mast cell A white blood cell that produces vasoactive molecules (e.g., histamine) and stores them in vacuoles near the cell membrane where they are released upon cell stimulation by external triggers.

maximum containment level goal The number of pathogens permitted in a water supply to prevent adverse health effects and provide a margin of safety.

M cell Specialized cell of the intestinal mucosa and other sites (e.g., urogenital tract) that delivers antigen from its apical face to lymphocytes clustered within the pocket in its basolateral face.

mediator A complex of proteins in eukaryotic cells that is often involved in transcription initiation and regulation.

meiosis (mi-o′sis) The type of cell division by which a diploid cell divides and forms four haploid cells.

melting temperature (T_m) The temperature at which double-stranded DNA separates into individual strands; used to compare genetic similarity in microbial taxonomy.

membrane attack complex (MAC) The complex of complement proteins (C5b–C9) that creates a pore in the plasma membrane of a target cell and leads to cell lysis.

membrane filter Porous material that retains microorganisms as the suspending liquid passes through the pores.

membrane filter technique The use of a thin porous filter to collect microorganisms from water, air, and food.

memory cell An inactive lymphocyte derived from a sensitized B or T cell capable of a rapid response to subsequent antigen exposure.

mesophile A microorganism with a growth optimum around 20 to 45°C, a minimum of 15 to 20°C, and a maximum about 45°C or lower.

messenger RNA (mRNA) Single-stranded RNA synthesized from a nucleic acid template (DNA in cellular organisms, RNA in some viruses) during transcription; mRNA binds to ribosomes and directs the synthesis of protein.

metabolic channeling The localization of metabolites and enzymes in different parts of a cell.

metabolic engineering The use of molecular techniques to develop novel biochemical pathways that either promote product yield or generate new products.

metabolism The total of all chemical reactions in the cell; almost all are enzyme catalyzed.

metabolite flux The rate of turnover of a metabolite.

metabolites Chemical compounds produced by the metabolic activities of organisms.

metabolomics Analysis of all the small metabolites present in a cell under specific circumstances.

metagenomic library A genomic library constructed of DNA (or cDNA derived from mRNA) extracted directly from the environment.

metagenomics The study of genomes recovered from environmental samples (including the human body) without first isolating members of the microbial community and growing them in cultures.

metalimnion In lakes, the region of the water column between the hypolimnion and the epilimnion.

metaproteomics The emerging field of examining all the proteins produced in a given microbial habitat at a specific time.

metatranscriptomics Large-scale sequencing of mRNAs collected from natural environments as a means to assess microbial activities.

methane hydrate Pools of trapped methane that accumulates in latticelike cages of crystalline water 500 m or more below the sediment surface in many regions of the world's oceans.

methanogenesis The production of methane by certain members of the archaeal phylum *Euryarchaeota*.

methanogens (meth′ə-no-jens″) Strictly anaerobic archaea that derive energy by converting CO_2, H_2, formate, acetate, and other compounds to either methane or methane and CO_2.

methanotroph A microbe that has the ability to grow on methane as its sole carbon and energy source.

methylotroph A bacterium that uses reduced one-carbon compounds such as methane and methanol as its sole source of carbon and energy.

metronidazole Antimicrobial chemotherapeutic agent of the nitromidazole family having biological activity on anaerobic bacteria and protozoa.

Michaelis constant (K_m) A value that represents the substrate concentration required for an enzyme to operate at half maximal velocity.

microaerophile (mi″kro-a′er-o-fīl) A microorganism that requires low levels of oxygen for growth, around 2 to 10%, but is damaged by normal atmospheric oxygen levels.

microautoradiography A technique whereby the uptake of a radioactive substrate by single cells is determined.

microbe-associated molecular pattern (MAMP) Conserved molecular structures that occur in patterns on microbial surfaces. The structures and their patterns are unique to particular types of microorganisms and invariant among members of a given microbial group.

microbial ecology The study of microbial populations and communities in natural habitats, such as soils, water, food, etc.

microbial energy conversion The processes by which microbes are used to generate useful forms of energy, such a biofuel (ethanol, H_2, methanol) production, or electricity through microbial fuel cells.

microbial flora (microbiota) *See* microbiota.

microbial fuel cell An apparatus designed to capture electrons produced during microbial respiration and convert these electrons to electricity.

microbial loop The cycling of organic matter synthesized by photosynthetic microorganisms among other microbes, such as bacteria and protozoa. This process "loops" organic nutrients, minerals, and carbon dioxide back for reuse by the primary producers and makes the organic matter unavailable to higher consumers.

microbial mat A firm structure of layered microorganisms with complementary physiological activities that can develop on surfaces in aquatic environments.

microbial transformation *See* biotransformation.

microbiology The study of organisms that are usually too small to be seen with the naked eye.

microbiome The totality of microorganisms and microbial genomes that constitute a host's normal microbiota.

microbiota The microbes found in a habitat.

microcosms Incubation chambers with conditions that mimic those in a natural setting.

microdroplet culture A technique designed to obtain an axenic culture whereby cells from a microbial community are encapsulated in a gel matrix, which when emulsified generates a porous microdroplet that houses a single cell. The microdroplets can be incubated together, thereby allowing the exchange of small molecules. Following incubation, microdroplets containing colonial growth can be separated by flow cytometry.

microelectrode An electrode with a tip that is sufficiently small to perform nondestructive measurements at intervals of less than 1 mm.

microflora *See* microbiota.

micronucleus The smaller of the two nuclei in ciliate protists. Micronuclei are diploid and involved only in genetic recombination and the regeneration of macronuclei.

micronutrients Nutrients required in very small quantities for growth and reproduction. Also called trace elements.

microorganism An organism that is usually too small to be seen clearly with the naked eye and is often unicellular, or if multicellular, does not exhibit a high degree of differentiation.

micro RNAs (miRNAs) Very small RNAs that inhibit translation of a target mRNA.

microsporidia Primitive obligate intracellular fungi; primarily infect vertebrates.

microtubules (mi″kro-tu′buls) One of the proteinaceous components of the eukaryotic cytoskeleton, flagella, and cilia; composed of tubulin proteins.

mineralization The conversion of organic nutrients into inorganic material during microbial growth and metabolism.

mineral soil Soil that contains less than 20% organic carbon.

minimal inhibitory concentration (MIC) The lowest concentration of a drug that will prevent the growth of a particular microorganism.

minimal lethal concentration (MLC) The lowest concentration of a drug that will kill a particular microorganism.

minus (negative) strand A viral single-stranded nucleic acid that is complementary to the viral mRNA.

mismatch repair A type of DNA repair in which a portion of a newly synthesized strand of DNA containing mismatched base pairs is removed and replaced, using the parental strand as a template.

missense mutation A single base substitution in DNA that changes a codon for one amino acid into a codon for another.

mitochondrion (mi″to-kon′dre-on) The eukaryotic organelle that is the site of electron transport, oxidative phosphorylation, and pathways such as the Krebs cycle; it provides most of a nonphotosynthetic cell's energy under aerobic conditions.

mitosis A process that takes place in the nucleus of a eukaryotic cell and results in the formation of two new nuclei, each with the same number of chromosomes as the parent.

mitosome A member of the mitochondria family of organelles found in certain protists that lack mitochondria or hydrogenosomes.

mixed acid fermentation A type of fermentation carried out by members of the family *Enterobacteriaceae* in which ethanol and a complex mixture of organic acids are produced.

mixotrophy A nutritional strategy that includes different types of metabolism (e.g., phototrophy and chemoorganotrophy).

modified atmosphere packaging (MAP) Addition of gases such as nitrogen and carbon dioxide to packaged foods to inhibit the growth of spoilage organisms.

mold Any of a large group of fungi that exist as multicellular filamentous colonies; typically lack macroscopic fruiting bodies.

molecular chaperones *See* chaperone proteins.

monocistronic mRNA An mRNA containing a single coding region.

monoclonal antibody (mAb) An antibody of a single type that is produced by a population of genetically identical plasma cells; typically produced in cell cultures derived from the fusion of a cancer cell and an antibody-producing cell (i.e., hybridoma).

monocyte A mononuclear phagocytic leukocyte; precursor of macrophages and dendritic cells.

monokine A generic term for a cytokine produced by macrophages or monocytes.

monotrichous (mon-ot′rĭ-kus) Having a single flagellum.

morbidity rate Measures the number of individuals who become ill as a result of a particular disease within a susceptible population during a specific time period.

mordant A substance that helps fix dye on or in a cell.

morphovar A variant strain of a microbe characterized by morphological differences.

mortality rate The ratio of the number of deaths from a given disease to the total number of cases of the disease.

most probable number (MPN) Statistical estimation of the size of a population in a liquid by diluting and determining end points for microbial growth.

mucociliary blanket The layer of cilia and mucus that lines certain portions of the respiratory system; it traps microorganisms and then transports them by ciliary action away from the lungs.

mucosal-associated lymphoid tissue (MALT) Organized and diffuse immune tissues found as part of the mucosal epithelium. It can be specialized to the gut (GALT) or the bronchial system or (BALT).

multicloning site (MCS) A region of DNA on a cloning vector that has a number of restriction enzyme recognition sequences to facilitate the cloning of a gene.

multilocus sequence analysis (MLSA) The application of multilocus sequence typing to more than one species.

multilocus sequence typing (MLST) A method for genotypic classification of bacteria or archaea within a single genus using nucleotide differences among five to seven housekeeping genes.

multiple strand displacement A technique used to amplify very low concentrations of DNA; it can yield enough DNA from a single cell to perform DNA sequencing.

murein *See* peptidoglycan.

must The juices of fruits, including grapes, that can be fermented for the production of alcohol.

mutagen (mu′tah-jen) A chemical or physical agent that causes mutations.

mutation A heritable change in the genetic material.

mutualism A type of symbiosis in which both partners gain from the association and are dependent on each other.

mutualist An organism associated with another in an obligatory relationship that is beneficial to both.

mycelium (mi-se′le-um) A mass of branching hyphae found in fungi and some bacteria.

Myc factor A lipochitooligosaccharide produced by some mycorrhizal fungi that prompts their uptake by the roots of host plants.

mycobiont The fungal partner in a lichen.

mycolic acids Complex 60 to 90 carbon fatty acids with a hydroxyl on the β-carbon and an aliphatic chain on the α-carbon; found in the cell walls of mycobacteria.

mycologist A person specializing in mycology; a student of mycology.

mycology The science and study of fungi.

mycoplasmas Common name for bacteria that are members of the class *Mollicutes,* phylum *Tenericutes.* They lack cell walls and cannot synthesize peptidoglycan precursors.

mycorrhizal fungi Fungi that form stable, mutualistic relationships on (ectomycorrhizal) or in (endomycorrhizal) the roots of vascular plants.

mycosis (mi-ko′sis; pl., **mycoses**) Any disease caused by a fungus.

mycotoxicology (mi-ko′tok″si-kol′o-je) The study of fungal toxins and their effects on various organisms.

myositis (mi″o-si′tis) Inflammation of a striated or voluntary muscle.

myxobacteria A group of Gram-negative, aerobic soil bacteria characterized by gliding motility, a complex life cycle with the production of fruiting bodies, and the formation of myxospores.

myxospores (mik′so-spōrs) Special dormant spores formed by myxobacteria.

N

naked amoeba An amoeboid protist that lacks a cell wall or other supporting structures.

naked viruses *See* nonenveloped virus.

nanowires Threadlike extensions of plasma membrane that transfer electrons from the terminal point in the electron transport chain to an external metal surface.

narrow-spectrum drugs Chemotherapeutic agents effective only against a limited variety of microorganisms.

natural classification A classification system that arranges organisms into groups whose members share many characteristics and reflect as much as possible the biological nature of organisms.

natural immunity *See* innate immunity.

natural killer (NK) cell A type of white blood cell that destroys host cells infected with intracellular pathogens and tumor cells.

naturally acquired active immunity The type of active immunity that develops when an individual's immune system encounters an appropriate antigen during the course of normal activities; it often arises as the result of recovering from an infection and lasts a long time.

naturally acquired passive immunity The type of temporary immunity that involves the transfer of antibodies from one individual to another.

natural products Industrially important compounds, such as antibiotics, produced by microbes.

negative selection In immunology, the process by which lymphocytes that recognize host (self) antigens undergo apoptosis or become anergic (inactive).

negative staining A staining procedure in which a dye is used to make the background dark while the specimen is unstained.

negative-strand DNA *See* minus strand.

negative-strand RNA *See* minus strand.

negative transcriptional control Regulation of transcription by a repressor protein. When bound to the repressor-binding site, transcription is inhibited.

neoplasia Abnormal cell growth and reproduction due to a loss of regulation of the cell cycle; produces a tumor in solid tissues.

neuraminidase An enzyme that cleaves the chemical bond linking neuraminic acids to the sugars present on the surface of animal cells; in virology, one type of envelope spike on influenza viruses has neuraminidase activity and is used to identify different strains.

neutralization Binding of specific immunoglobulins to toxins or viruses, inhibiting their biological activity.

neutral mutation A missense mutation that does not affect the function of the protein (e.g., a polar amino acid is substituted for a polar amino acid).

neutrophil A mature white blood cell in the granulocyte lineage. It has a nucleus with three to five lobes and is very phagocytic.

neutrophile A microorganism that grows best at a neutral pH range between pH 5.5 and 8.0.

next-generation nucleotide sequencing Methodologies that allow rapid genome sequencing without the prior construction of a genomic library. Instead, fragments of genomic DNA are tethered to a solid substrate, and each fragment is sequenced simultaneously or in a massively parallel fashion.

nicotinamide adenine dinucleotide (NAD⁺) (nik″o-tin′ah-mid) An electron-carrying coenzyme; it is particularly important in catabolic processes and usually transfers electrons from an electron source to an electron transport chain.

nicotinamide adenine dinucleotide phosphate (NADP⁺) An electron-carrying coenzyme that most often participates as an electron carrier in biosynthetic metabolism.

nitrification The oxidation of ammonia to nitrate.

nitrifying archaea Mesophilic archaea (thaumarchaeotes) capable of oxidizing ammonium to nitrite.

nitrifying bacteria Chemolithotrophic, Gram-negative bacteria that are members of several families within the phylum *Proteobacteria* that either oxidize ammonia to nitrite or nitrite to nitrate.

nitrogenase (ni′tro-jen-ās) The enzyme that catalyzes biological nitrogen fixation.

nitrogen fixation The metabolic process in which atmospheric molecular nitrogen (N_2) is reduced to ammonia; carried out by nitrogen-fixing bacteria and archaea.

nocardioforms Bacteria that resemble members of the genus *Nocardia;* they develop a substrate mycelium that readily breaks up into rods and coccoid elements (a quality called fugacity).

Nod factors Signaling compounds that alter gene expression in the plant host of a rhizobium.

NOD-like receptors (NLRs) Receptors similar to nucleotide-binding and oligomerization domain (NOD) receptors that sense endogenous metabolites and regulate the production of inflammatory cytokines in response to microbe-associated molecular patterns and danger-associated molecular patterns.

nomenclature The branch of taxonomy concerned with the assignment of names to taxonomic groups in agreement with published rules.

noncompetitive inhibitor A chemical that inhibits enzyme activity by a mechanism that does not involve binding the active site of the enzyme.

noncyclic photophosphorylation The process in which light energy is used to make ATP when electrons are moved from water to $NADP^+$ during oxygenic photosynthesis; both photosystem I and photosystem II are involved.

nonenveloped (naked) virus A virus having virions composed only of a nucleocapsid (i.e., lacking an envelope).

nonheme iron protein *See* electron transport chain.

nonsense (stop) codon A codon that does not code for an amino acid but is a signal to terminate protein synthesis.

nonsense mutation A mutation that converts a sense codon to a nonsense (stop) codon.

nonspecific immune response *See* innate (natural) immunity.

nonspecific resistance *See* innate (natural) immunity.

normal microbiota *See* microbiota.

nosocomial infection (nos″o-ko′me-al) *See* health-care–associated infection.

nuclear envelope The complex double-membrane structure forming the outer boundary of the eukaryotic nucleus. It is covered by nuclear pores through which substances enter and leave the nucleus.

nuclear pore complex The nuclear pore plus about 30 proteins that form the pore and are involved in moving materials across the nuclear envelope.

nucleocapsid (nu″kle-o-kap′sid) The viral nucleic acid and its surrounding capsid; the basic unit of virion structure.

nucleoid An irregularly shaped region in a bacterial or archaeal cell containing genetic material.

nucleoid occlusion A mechanism used by many bacteria to prevent premature formation of a septum during binary fission.

nucleolus (nu-kle′o-lus) The organelle, located within the nucleus and not bounded by a membrane, that is the location of ribosomal RNA synthesis and the assembly of ribosomal subunits.

nucleoside (nu′kle-o-sīd″) A combination of ribose or deoxyribose with a purine or pyrimidine base but lacking phosphate.

nucleosome (nu′kle-o-sōm″) A complex of histones and DNA found in eukaryotic chromatin and some archaea; the DNA is wrapped around the surface of the beadlike histone complex.

nucleotide (nu′kle-o-tīd) A combination of ribose or deoxyribose with phosphate and a purine or pyrimidine base; a nucleoside plus one or more phosphates.

nucleotide excision repair *See* excision repair.

nucleus The eukaryotic organelle enclosed by a double-membrane envelope that contains the cell's chromosomes.

numerical aperture The property of a microscope lens that determines how much light can enter and how great a resolution the lens can provide.

O

O antigen The terminal polysaccharide of the lipopolysaccharide found on some Gram-negative bacterial cell walls.

obligate aerobes Organisms that grow only in the presence of oxygen.

obligate anaerobes Microorganisms that cannot tolerate oxygen and die when exposed to it.

ocean acidification The decline in pH in ocean waters as a result of changes to the carbonate equilibrium system.

oceanography The study of the biological, chemical, and physical aspects of the oceans.

Okazaki fragments Short stretches of polynucleotides produced during discontinuous DNA replication.

oligonucleotide A short fragment of DNA or RNA, usually artificially synthesized, used in a number of molecular genetic techniques such as DNA sequencing and polymerase chain reaction.

oligonucleotide signature sequence Short, conserved nucleotide sequences that are specific for a phylogenetically defined group of organisms. The signature sequences found in small subunit rRNA molecules are most commonly used.

oligotrophic environment (ol″ĭ-go-trof′ik) An environment containing low levels of nutrients that support microbial growth.

oncogene (ong″ko-jēn) A gene whose activity is associated with the conversion of normal cells to cancer cells.

oncovirus A virus known to be associated with the development of cancer.

oocyst (o′o-sist) Cyst formed around a zygote of certain protozoa.

öomycetes (o″o-mi-se′tēz) A collective name for protists also known as water molds. Formerly thought to be fungi.

open reading frame (ORF) A sequence of DNA thought to encode a protein because it is not interrupted by a stop codon and has an apparent promoter and ribosome-binding site at the 5′ end and a terminator at the 3′ end.

operational taxonomic unit (OTU) A term that includes all sources of data used to construct phylogenetic trees; that is, it includes species, strains, and genomic sequences from organisms not yet grown in culture.

operator The segment of DNA in an operon to which the repressor protein binds; it controls the expression of the genes adjacent to it.

operon In bacteria, the sequence of bases in DNA that contains a promoter and one or more structural genes and often an operator or activator-binding site that controls their expression.

opportunistic microorganism or pathogen A microorganism that is usually free living or a part of the host's normal microbiota but may become pathogenic under certain circumstances, such as when the immune system is compromised.

opsonization (op″so-ni-za′shun) The coating of foreign substances by antibody, complement proteins, or fibronectin to make the substances more readily recognized by phagocytic cells.

optical tweezer A focused laser beam that can drag and isolate an individual microorganism from a complex microbial mixture.

organelle A structure within or on a cell that performs specific functions and is related to the cell in a way similar to that of an organ to the body of a multicellular organism.

organic soil A soil that contains at least 20% organic carbon.

organotrophs Organisms that use reduced organic compounds as their electron source.

origin of replication A site on a chromosome or plasmid where DNA replication is initiated.

origin recognition complex (ORC) A complex of six proteins that "marks" the origins of replication in eukaryotic chromosomes.

orthologue A gene found in the genomes of two or more different organisms that share a common ancestry. The products of orthologous genes are presumed to have similar functions.

osmophiles Microorganisms that grow best in or on media of high solute concentration.

osmotolerant Organisms that grow over a fairly wide range of water activity or solute concentration.

osmotrophy A form of nutrition in which soluble nutrients are absorbed through the cytoplasmic membrane; found in bacteria, archaea, fungi, and some protists.

Öuchterlony technique *See* double diffusion agar assay.

outbreak The sudden, unexpected occurrence of a disease in a given population.

outer membrane A membrane located outside the peptidoglycan layer in the cell walls of typical Gram-negative bacteria.

oxidation-reduction (redox) reactions Reactions involving electron transfers; the electron donor (reductant) gives electrons to an electron acceptor (oxidant).

oxidative burst *See* respiratory burst.

oxidative phosphorylation (fos″for-ĭ-la′-shun) The synthesis of ATP from ADP using energy made available during electron transport initiated by the oxidation of a chemical energy source.

oxygenic photosynthesis Photosynthesis that oxidizes water to form oxygen; the form of photosynthesis characteristic of plants, protists, and cyanobacteria.

P

palisade arrangement Angular arrangements of bacteria whereby cells remain partially attached after snapping division.

pandemic An increase in the occurrence of a disease within a large and geographically widespread population (often refers to a worldwide epidemic).

Paneth cell (pah'net) A specialized epithelial cell of the intestine that secretes hydrolytic enzymes and antimicrobial proteins and peptides.

pangenome The collection of genes found in all strains that belong to a single species or other taxonomic group.

paralogue Two or more genes in the genome of a single organism that arose through duplication of a common ancestral gene. The products of paralogous genes may have slightly different functions.

parasite An organism that lives on or within another organism (the host) and benefits from the association while harming its host. Often the parasite obtains nutrients from the host.

parasitism A type of symbiosis in which one organism benefits from the other and the host is usually harmed.

parasporal body An intracellular, solid protein crystal made by the bacterium *Bacillus thuringiensis*. It is the basis of the bacterial insecticide Bt.

parfocal A microscope that retains proper focus when the objectives are changed.

particulate organic matter (POM) Nutrients that are not dissolved or soluble, generally referring to aquatic ecosystems. This includes microorganisms and their cellular debris after senescence or viral lysis.

passive diffusion The movement of molecules from a region of higher concentration to one of lower concentration as a result of random thermal agitation.

pasteurization The process of heating liquids to destroy microorganisms that can cause spoilage or disease.

pathogen (path'o-jən) Any organism that causes disease.

pathogenicity (path″o-je-nis'ĭ-te) The condition or quality of being pathogenic, or the ability to cause disease.

pathogenicity island A segment of DNA in some pathogens that contains genes responsible for virulence; often codes for a type III secretion system that allows the pathogen to secrete virulence proteins and damage host cells.

pattern recognition molecule (PRM) A receptor found on macrophages and other phagocytic cells that binds to microbe-associated molecular patterns on microbial surfaces.

PCR bias Artifactual or misleading results obtained when microbial populations are assessed by amplifying genes or genomes by the polymerase chain reaction (PCR); occurs because not all DNA or RNA is amplified at the same rate.

pelagic zone The central region of a lake or ocean not influenced by events or nutrients found in the coastal regions.

pellicle (pel'ĭ-k'l) A relatively rigid layer of proteinaceous elements just beneath the plasma membrane in many protists. The plasma membrane is sometimes considered part of the pellicle.

penicillin A β-lactam antibiotic. The first true antibiotic to be discovered and used clinically.

penicillinase A β-lactamase enzyme that digests the active moiety of penicillin.

pentose phosphate (hexose monophosphate) pathway A glycolytic pathway that oxidizes glucose 6-phosphate to ribulose 5-phosphate and then converts it to a variety of three to seven carbon sugars.

peplomer (spike) (pep'lo-mer) A protein or protein complex that extends from the viral envelope and often is important in virion attachment to the host cell surface.

peptide interbridge A short peptide chain that connects the tetrapeptide chains in the peptidoglycan of some bacteria.

peptidoglycan (pep″tĭ-do-gli'kan) A large polymer composed of long chains of alternating *N*-acetylglucosamine and *N*-acetylmuramic acid residues. The polysaccharide chains are linked to each other through connections between tetrapeptide chains attached to the *N*-acetylmuramic acids. It provides much of the strength and rigidity possessed by bacterial cell walls; also called murein.

peptidyl or donor site (P site) The site on the ribosome that contains the peptidyl-tRNA at the beginning of the elongation cycle during protein synthesis.

peptidyl transferase The ribozyme that catalyzes the transpeptidation reaction in protein synthesis; in this reaction, an amino acid is added to the growing peptide chain.

perforin Pore-forming proteins secreted along with granzymes by cytotoxic T lymphocytes and NK cells.

peripheral membrane protein *See* plasma membrane.

peripheral tolerance The process by which self-reactive B and T cells are inhibited in body tissues other than bone marrow and thymus, respectively.

periplasm (per'ĭ-plaz-əm) The substance that fills the periplasmic space.

periplasmic flagella The flagella that lie under the outer sheath and extend from both ends of a spirochete cell to overlap in the middle and form the axial filament. Also called axial fibrils, axial filaments, and endoflagella.

periplasmic space The space between the plasma membrane and the outer membrane in typical Gram-negative bacteria, and between the plasma membrane and the cell wall in typical Gram-positive bacteria. A similar space is sometimes observed between the plasma membrane and the cell wall of some archaea.

peristalsis The muscular contractions of the gut that propel digested foods and waste through the intestinal tract.

peritrichous (pĕ-rit'rĭ-kus) A cell with flagella distributed over its surface.

peronosporomycetes Name for öomycetes, a subgroup of *Stramenopila*, distinguished by the formation of large egg cells when they reproduce.

persisters Variants of microbial cells in a population that are resistant to antibiotics.

phage (fāj) *See* bacteriophage.

phagocytic vacuole A membrane-delimited vacuole produced by cells carrying out phagocytosis. It is formed by the invagination of the plasma membrane and contains solid material.

phagocytosis The endocytotic process in which a cell encloses large particles in a phagocytic vacuole (phagosome) and engulfs them.

phagolysosome The vacuole that results from the fusion of a phagosome with a lysosome.

phagosome A membrane-enclosed vacuole formed by the invagination of the cell membrane during endocytosis.

phase-contrast microscope A microscope that converts slight differences in refractive index and cell density into easily observed differences in light intensity.

phenetic system A classification system that groups organisms together based on the similarity of their observable characteristics.

phenol coefficient test A measurement of the effectiveness of disinfectants that compares their activity against test bacteria with that of phenol.

phenotype Observable characteristics of an organism.

phenotypic rescue The identification of a cloned gene based on its ability to complement a genetic deficiency in the host cell.

phosphatase (fos'fah-tās″) An enzyme that catalyzes the hydrolytic removal of phosphate from molecules.

phosphoenolpyruvate:sugar phosphotransferase system (PTS) A group translocation system used by many bacteria. As a sugar is transported into the cell, the hydrolysis of a high-energy phosphate bond fuels its import, and the sugar is modified by the covalent attachment of the phosphoryl group.

phosphorelay system A set of proteins involved in the transfer of phosphate from one protein in the set to another. It is often used to regulate protein activity or transcription.

photic zone The illuminated area of an aquatic habitat from the surface to the depth to which the rate of photosynthesis equals that of respiration.

photoautotroph *See* photolithoautotroph.

photolithoautotroph An organism that uses light energy, an inorganic electron source (e.g., H_2O, H_2, H_2S), and CO_2 as its carbon source. Also called photolithotrophic autotroph or a photoautotroph.

photoorganoheterotroph A microorganism that uses light energy, organic electron sources, and organic molecules as a carbon source. Also called photoorganotrophic heterotroph.

photophosphorylation The synthesis of ATP from ADP using energy made available by the absorption of light.

photoreactivation The process in which blue light is used by an enzyme to repair thymine dimers in DNA.

photosynthate Nutrient material released from phototrophic organisms and contributes to the pool of dissolved organic matter.

photosynthesis The trapping of light energy and its conversion to chemical energy (light reactions), which is then used to reduce CO_2 and incorporate it into organic molecules (dark reactions).

photosystem I The photosystem in eukaryotic cells and cyanobacteria that absorbs longer wavelength light, usually greater than about 680 nm, and transfers the energy to chlorophyll P700 during photosynthesis; it is involved in both cyclic and noncyclic photophosphorylation.

photosystem II The photosystem in eukaryotic cells and cyanobacteria that absorbs shorter wavelength light, usually less than 680 nm, and transfers the energy to chlorophyll P680 during photosynthesis; it participates in noncyclic photophosphorylation.

phototaxis The ability of certain phototrophic organisms to move in response to a light source.

phototrophs Organisms that use light as their energy source.

phycobiliproteins Photosynthetic pigments found in cyanobacteria that are composed of proteins with attached tetrapyrroles.

phycobilisomes Particles on the membranes of cyanobacteria that contain photosynthetic pigments.

phycobiont (fi″ko-bi′ont) The photosynthetic protist or cyanobacterial partner in a lichen.

phycocyanin (fi″ko-si′an-in) A blue phycobiliprotein pigment used to trap light energy during photosynthesis.

phycoerythrin (fi″ko-er′i-thrin) A red photosynthetic phycobiliprotein pigment used to trap light energy.

phycology (fi-kol′o-je) The study of algae.

phyletic classification system *See* phylogenetic classification system.

phyllosphere The surface of plant leaves.

phylochip *See* phylogenetic oligonucleotide arrays.

phylogenetic classification system A classification system based on evolutionary relationships, rather than the general similarity of characteristics.

phylogenetic oligonucleotide arrays Microarrays that contain rRNA sequences as probes and are used as a means to detect complementary rRNA sequences from the environment.

phylogenetic tree A graph made of nodes and branches, much like a tree in shape, that shows phylogenetic relationships between groups of organisms and sometimes also indicates the evolutionary development of groups.

phylogeny The evolutionary development of a species.

phylotype A taxon that is characterized only by its nucleic acid sequence; generally discovered during metagenomic analysis.

physical map In genomics, this refers to a diagram of a chromosome, gene, or other genetic element that shows restriction endonuclease recognition sites. Specific genes and other genetic elements (e.g., origin of replication) may also be noted.

phytoplankton A community of floating photosynthetic organisms, composed of photosynthetic protists and cyanobacteria.

picoplankton Planktonic microbes between 0.2 and 2.0 μm in size, including the cyanobacterial genera *Prochlorococcus* and *Synechococcus*.

piezophilic (pi-e-zo-fil′ic) Describing organisms that prefer or require high pressures for growth and reproduction.

pili (s., pilus) *See* fimbria.

pitching Pertaining to inoculation of a nutrient medium with yeast, for example, in beer brewing.

plaque A clear area in a lawn of host cells that results from their lysis by viruses.

plaque assay A method used to determine the number of infectious virions.

plaque-forming unit The unit of measure of a plaque assay. It usually represents a single infectious virion.

plasma cell A mature, differentiated B lymphocyte that synthesizes and secretes antibody.

plasmalemma The plasma membrane in protists.

plasma membrane The selectively permeable membrane surrounding the cell's cytoplasm; also called the cell membrane, plasmalemma, or cytoplasmic membrane. For most cells it is a lipid bilayer (some archaea have a lipid monolayer) with proteins embedded in it (integral membrane proteins) and associated with the surface (peripheral membrane proteins).

plasmid A double-stranded DNA molecule that can exist and replicate independently of the chromosome. A plasmid is stably inherited but is not required for the host cell's growth and reproduction.

plasmid fingerprinting A technique used to identify microbial isolates as belonging to the same strain because they contain the same number of plasmids with the identical molecular weights and similar phenotypes.

plasmodial (acellular) slime mold A member of the protist division *Amoebozoa (Myxogastria)* that exists as a thin, streaming, multinucleate mass of protoplasm (plasmodium) for some portion of its life cycle.

plasmolysis (plaz-mol′ĭ-sis) The process in which water osmotically leaves a cell, which causes the cytoplasm to shrivel up and pull the plasma membrane away from the cell wall.

plastid A cytoplasmic organelle of algae and higher plants that contains pigments such as chlorophyll, stores food reserves, and often carries out processes such as photosynthesis.

pleomorphic (ple″o-mor′fik) Refers to cells or viruses that are variable in shape and lack a single, characteristic form.

plus (positive) strand A viral nucleic acid strand that is equivalent in base sequence to the viral mRNA.

point mutation A mutation that affects only a single base pair.

polar flagellum A flagellum located at one end of an elongated cell.

3′ poly-A tail *See* pre-mRNA.

poly-β-hydroxybutyrate (PHB) A linear polymer of β-hydroxybutyrate used as a reserve of carbon and energy by many bacteria.

polycistronic mRNA An mRNA that has more than one coding region; formed when an operon is transcribed.

polyhistidine tagging (His-tagging) A method by which a protein is fused to six histidine residues, thereby facilitating the protein's purification through affinity chromatography.

polymerase chain reaction (PCR) An in vitro technique used to synthesize large quantities of specific nucleotide sequences from small amounts of DNA. It employs oligonucleotide primers complementary to specific sequences in the target gene and special heat-stable DNA polymerases (e.g., Taq polymerase).

polymorphonuclear leukocyte (pol″e-mor″fo-noo′kle-ər) A leukocyte that has a variety of nuclear forms.

polyphasic taxonomy An approach in which taxonomic schemes are developed using a wide range of phenotypic and genotypic information.

polyprotein A large polypeptide that contains several proteins, which are released when the polyprotein is cut by proteases. Polyproteins are often produced by viruses having positive-strand RNA genomes.

polyribosome (pol″e-ri′bo-sōm) A complex of several ribosomes translating a single messenger RNA.

population An assemblage of organisms of the same type.

porin proteins Proteins that form channels across the outer membrane of typical Gram-negative bacterial cell walls through which small molecules enter the periplasm.

portal of entry Body surface that is the initial site of pathogen access to the host.

positive transcriptional control Control of transcription by an activator protein. When the activator is bound to the activator-binding site, the level of transcription increases.

postherpetic neuralgia The severe pain after a herpes infection.

posttranslational regulation Alteration of the activity of a protein after it has been synthesized by a structural modification, such as phosphorylation or a conformational change. *See also* allosteric enzyme and reversible covalent modification.

pour plate A Petri dish of solid culture medium with isolated microbial colonies growing both on its surface and within the medium that has been prepared by mixing microorganisms with cooled, still-liquid medium and then allowing the medium to harden.

precipitation (precipitin) reaction The reaction of an antibody with a soluble antigen to form an insoluble precipitate.

precipitin The antibody responsible for a precipitation reaction.

precursor metabolites Intermediates of glycolytic pathways, TCA cycle, and other pathways that serve as starting molecules for biosynthetic pathways that generate monomers and other building blocks needed for synthesis of macromolecules.

pre-mRNA In eukaryotes, the RNA transcript of DNA made by RNA polymerase II prior to processing to form mRNA by the addition of the 5′ cap and 3′ poly-A tail, and the removal of introns.

presence-absence (P-A) test An assay designed to detect the presence of coliforms in water; a positive result must be confirmed by a different assay.

prevalence rate Refers to the total number of individuals infected at any one time in a given population regardless of when the disease began.

Pribnow box A base sequence in bacterial promoters that is recognized by RNA polymerase and is the site of initial polymerase binding.

primary active transport *See* active transport.

primary metabolites Microbial metabolites produced during active growth of an organism.

primary mRNA *See* pre-mRNA.

primary producer Photoautotrophic and chemoautotrophic organisms that incorporate carbon dioxide into organic carbon and thus provide new biomass for the ecosystem.

primary production The incorporation of carbon dioxide into organic matter by photosynthetic organisms and chemoautotrophic organisms.

primary treatment The first step of sewage treatment, in which physical settling and screening are used to remove particulate materials.

primase *See* primosome.

primers *See* polymerase chain reaction and primosome.

primosome In bacteria, a complex of proteins that includes the enzyme primase, responsible for synthesizing the RNA primers needed for DNA replication.

prion (pre′on) An infectious agent consisting only of protein; prions cause a variety of spongiform encephalopathies such as scrapie in sheep.

probe A short, labeled nucleic acid segment complementary in base sequence to part of another nucleic acid that is used to identify or isolate the particular nucleic acid from a mixture through its ability to bind specifically with the target nucleic acid.

probiotic A living microorganism that may provide health benefits beyond its nutritional value when ingested.

prochlorophytes A group of cyanobacteria having both chlorophyll *a* and *b* but lacking phycobilins.

prodromal stage (pro-dro′məl) The period during the course of a disease when signs and symptoms are present, but they are not yet distinctive enough for an accurate diagnosis.

production strains Microbial strains used in industrial processes that have been engineered to generate a high yield of the desired product.

programmed cell death (1) In eukaryotes, apoptosis. (2) In some bacteria, a mechanism proposed to account for the decline in cell numbers during the death phase of a growth curve. *See also* apoptosis.

prokaryotic cells (pro″kar-e-ot′ik) Cells having a type of structure characterized by the lack of a true, membrane-enclosed nucleus. All known members of *Archaea* and most members of *Bacteria* exhibit this type of cell structure; some members of the bacterial phylum *Planctomycetes* have a membrane surrounding their genetic material.

promoter The region on DNA at the start of a gene that RNA polymerase binds to before beginning transcription.

proofreading The ability of enzymes to check their products to ensure that the correct product is made. For instance, DNA polymerase checks newly synthesized DNA and replaces an incorrect nucleotide with the correct one prior further synthesis.

propagated epidemic An epidemic that is characterized by a relatively slow and prolonged rise and then a gradual decline in the number of individuals infected. It usually results from the introduction of an infected individual into a susceptible population and transmission of the pathogen from person to person.

prophage (pro′fāj) **(provirus)** (1) The latent form of a temperate bacterial or archaeal virus that remains within the lysogen, usually integrated into the host chromosome. (2) The form of a eukaryotic virus that remains within the host cell during a latent infection. Also refers to a retroviral genome after it has been integrated into the host's chromosome.

prostheca (pros-the′kah) An extension of a bacterial cell, including the plasma membrane and cell wall, that is narrower than the mature cell.

prosthetic group A tightly bound cofactor that remains at the active site of an enzyme during its catalytic activity.

protease (pro′te-ās) An enzyme that hydrolyzes proteins. Also called a proteinase.

proteasome A large, cylindrical protein complex observed in eukaryotic cells that degrades ubiquitin-labeled proteins to peptides in an ATP-dependent process. Also called 26S proteasome. Similar protein-degrading machinery has been observed in some archaea and bacteria.

protein modeling The process by which the amino acid sequence of a protein is analyzed using software designed to predict the protein's three-dimensional structure.

protein splicing The posttranslational process in which part of a precursor polypeptide is removed before the mature polypeptide folds into its final shape; it is carried out by self-splicing proteins that remove inteins and join the remaining exteins.

proteome The complete collection of proteins that an organism produces.

proteomics The study of the structure and function of cellular proteins.

proteorhodopsin A rhodopsin molecule first identified in marine proteobacteria but since found in a variety of microbes. It enables rhodopsin-based phototrophy. *See also* archaerhodopsin.

protist Unicellular and sometimes colonial eukaryotic organisms that lack cellular differentiation into tissues. Many chemoorganotrophic protists are referred to as protozoa; many phototrophic protists are referred to as algae.

protistology The study of protists.

protomer An individual protein of a viral capsid; a capsomer is made of protomers.

proton motive force (PMF) The potential energy arising from a gradient of protons and a membrane potential that powers ATP synthesis and other processes.

proto-oncogene Normal cellular gene that when mutated or overexpressed contributes to malignant transformation of a cell.

protoplast (1) The plasma membrane and everything within it. (2) A bacterial, archaeal, or fungal cell with its cell wall completely removed. It is spherical in shape and osmotically sensitive.

protoplast fusion The joining of cells that have had their walls weakened or completely removed.

prototroph A microorganism that requires the same nutrients as the majority of naturally occurring members of its species.

protozoa (pro″to-zo′a) A common term for a group of unrelated unicellular, chemoorganotrophic protists.

protozoology The study of protozoa.

provirus *See* prophage.

pseudogene A degraded, nonfunctional gene.

pseudomurein A complex polysaccharide observed in the cell walls of some archaea; it resembles peptidoglycan (murein) in structure and chemical make-up; also called pseudopeptidoglycan.

pseudopodium or pseudopod (soo″do-po′de-um) A nonpermanent cytoplasmic extension of the cell by which amoeboid protists move and feed.

psychrophile (si′kro-fīl) A microorganism that has an optimum growth temperature of 15°C or lower and a temperature maximum around 20°C.

psychrotolerant (psychrotroph) A microorganism that has a growth optimum between 20 and 30°C and a maximum of about 35°C.

PulseNet A program developed and overseen by the Centers for Disease Control and Prevention in which pulse-field gel electrophoresis is used to characterize pathogens found to contaminate food in an effort to detect and track these microbes.

punctuated equilibria The observation based on the fossil record that evolution does not proceed at a slow and linear pace but rather is periodically interrupted by rapid bursts of speciation and extinction driven by abrupt changes in environmental conditions.

pure (axenic) culture A population of cells that are identical because they arise from a single cell.

purine (pu′rēn) A heterocyclic, nitrogen-containing molecule with two joined rings that occurs in nucleic acids and other cell constituents (e.g., adenine and guanine).

purple membrane An area of the plasma membrane of *Halobacterium* that contains archaerhodopsin and is active in trapping light energy.

purple nonsulfur bacteria Anoxygenic photosynthetic α- and β-proteobacteria.

purple sulfur bacteria Anoxygenic photosynthetic γ-proteobacteria.

putrefaction The microbial decomposition of organic matter, especially the anaerobic breakdown of proteins, with the production of foul-smelling compounds such as hydrogen sulfide and amines.

pyogenic Pus-producing.

pyrenoid (pi′rĕ-noid) The differentiated region of the chloroplast that is a center of starch formation in some photosynthetic protists.

pyrimidine (pi-rim′i-dēn) A nitrogen-containing molecule with one ring that occurs in nucleic acids and other cell constituents; the most important pyrimidines are cytosine, thymine, and uracil.

pyruvate formate-lyase (PFL) An enzyme system that cleaves pyruvate into formic acid and acetyl-CoA. It is used by enteric bacteria that perform mixed acid fermentation.

Q

Q cycle The process associated with proton movement across the membrane at complex III of an electron transport chain.

Quellung reaction The increase in visibility or the swelling of the capsule of a microorganism in the presence of antibodies against capsular antigens.

quinine drugs Antimicrobial chemotherapeutic agents derived from the bark of the cinchona tree and used to treat malaria.

quinolones A class of broad-spectrum, bacteriocidal antibiotics, derived from nalidixic acid, that bind to DNA gyrase, inhibiting DNA replication.

quorum sensing The process in which bacteria monitor their own population density or the presence of other species of bacteria by sensing the levels of signal molecules (e.g., *N*-acylhomoserine lactone) released by the microorganisms. When these signal molecules reach a threshold concentration, quorum-dependent genes are expressed.

R

racking The removal of sediments from wine bottles.

radappertization The use of gamma rays from a cobalt source for control of microorganisms in foods.

radioimmunoassay (RIA) (ra″de-o-im″u-no-as′a) A sensitive assay in which a purified radioisotope-labeled antigen or antibody competes for antibody or antigen with unlabeled standard antigen or test antigen in experimental samples to determine the concentration of a substance in the samples.

rapid plasmid reagin A rapid diagnostic test for syphilis.

rapid sand filters A step in drinking water purification that removes particulates by passage over sand.

reaction-center chlorophyll pair The two chlorophyll molecules in a photosystem that are energized by the absorption of light and release electrons to an associated electron transport chain, thus initiating energy conservation by photophosphorylation.

reactive nitrogen intermediate (RNI) Charged nitrogen radicals.

reactive oxygen intermediate (ROI) *See* reactive oxygen species (ROS).

reactive oxygen species (ROS) Superoxide radical (O_2^-.), hydrogen peroxide (H_2O_2), singlet oxygen ($^1O_2$), and hydroxyl radical (OH^-) derivatives of oxygen.

reading frame The way in which nucleotides in DNA and mRNA are grouped into codons for reading the message contained in the nucleotide sequence.

readthrough A phenomenon in which a ribosome ignores a stop codon and continues protein synthesis. This generates a different protein than that produced if the ribosome had stopped at the stop codon.

reagin (re′ah-jin) Antibody that mediates immediate hypersensitivity reactions. IgE is the major reagin in humans.

real-time PCR A type of polymerase chain reaction (PCR) that quantitatively measures the amount of template in a sample as the amount of fluorescently labeled amplified product.

RecA A protein that functions in DNA repair and recombination processes. Homologues of RecA have been found in all three domains of life.

receptor-mediated endocytosis *See* endocytosis.

receptors Proteins that bind signaling molecules (ligands), thereby initiating cellular responses.

recognition sites Specific nucleotide sequences to which DNA-binding proteins (e.g., regulatory proteins, sigma factors, restriction enzymes) bind.

recombinant DNA technology The techniques used in carrying out genetic engineering.

recombinants Organisms produced following a recombination event.

recombinant-vector vaccine A vaccine produced by the introduction of one or more of a pathogen's genes into attenuated viruses or bacteria, which serves as a vector, replicating within the vertebrate host and expressing the gene(s) of the pathogen. The pathogen's antigens induce an immune response.

recombinases Enzymes that catalyze integration and excision of DNA segments during site-specific recombination.

recombination The process in which a new recombinant chromosome is formed by combining genetic material from two different nucleic acids. In some cases, the nucleic acids are from two different organisms.

recombinational repair A DNA repair process that repairs damaged DNA when there is no remaining template; a piece of DNA from a sister molecule is used.

red tides Population blooms of dinoflagellates that release red pigments and toxins, which can lead to paralytic shellfish poisoning. *See also* harmful algal blooms.

redox potential The tendency of a chemical compound or system to accept electrons; more formally called oxidation-reduction potential.

reducing power Molecules such as NADH and NADPH that temporarily store electrons. The stored electrons are used in anabolic reactions such as CO_2 fixation and the synthesis of monomers (e.g., amino acids).

reductive amination pathway A mechanism used by many microbes to incorporate ammonia.

reductive dehalogenation The cleavage of carbon-halogen bonds by anaerobic bacteria that creates a strong electron-donating environment.

reductive TCA cycle A pathway used by some chemolithotrophs and some anoxygenic phototrophs to fix CO_2. It involves the TCA cycle running in the reverse direction; also called reverse citric acid cycle.

refractive index A measure of how much a substance deflects a light ray from a straight path as it passes from one medium (e.g., glass) to another (e.g., air). It is calculated as the ratio of the velocity of light passing through the first medium to that of light passing through the second medium.

regulatory site *See* allosteric enzyme.

regulatory T cells (Tregs) T cells that control the development of effector T cells and B cells.

regulon A collection of genes or operons that is controlled by a common regulatory protein.

remote sensing The gathering of data and transformation of data into maps.

replica plating A technique for isolating mutants by transferring cells from an agar plate with nonselective medium to a piece of velvet and then onto plates with selective media or environmental conditions.

replicase An RNA-dependent RNA polymerase used to replicate the genome of an RNA virus.

replication The process in which an exact copy of parental DNA (or viral RNA) is made with the parental molecule serving as a template.

replication fork The Y-shaped structure where DNA is replicated.

replicative form (RF) A double-stranded nucleic acid that is formed from a single-stranded viral genome and used to synthesize mRNA and new copies of the genome.

replicative transposition A transposition event in which a replicate of the transposon remains in its original site.

replicon A unit of the genome that contains an origin of replication and in which DNA is replicated.

replisome A large protein complex that copies the DNA double helix to form two daughter chromosomes.

repressible gene A gene that encodes a protein whose level drops in the presence of a small molecule. If the gene product is a biosynthetic enzyme, the small molecule is often an end product of the metabolic pathway in which it functions.

repressor protein A protein that can bind to a repressor-binding site and inhibit transcription.

reservoir A site, alternate host, or carrier that normally harbors pathogenic organisms and serves as a source from which other individuals can be infected.

reservoir host An organism other than a human infected with a pathogen that can also infect humans.

residual body A lysosome after digestion of its contents has occurred. It contains undigested material.

resolution The ability of a microscope to distinguish between small objects that are close together.

respiration An energy-yielding process in which the energy substrate is oxidized using an exogenous or externally derived electron acceptor.

respiratory burst An increase in oxygen consumption by an activated phagocytic cell to generate highly toxic oxygen products such as singlet oxygen, superoxide radical, hydrogen peroxide, hydroxyl radical, and hypochlorite within the phagolysosome.

response regulator *See* two-component signal transduction system.

restriction A process used by bacteria to defend the cell against viral infection; accomplished by enzymes that cleave the viral DNA; also called host restriction.

restriction enzymes Enzymes produced by bacterial cells that cleave DNA at specific nucleotide sequences. They evolved to protect bacteria from viral infection and are used in genetic engineering.

restriction fragment length polymorphism (RFLP) Fragments of DNA of different sizes following endonuclease cleavage. The differences can be used to determine genetic similarity among organisms.

reticulate body (RB) The cellular form in the chlamydial life cycle that grows and reproduces within the host cell.

reticulopodia Netlike pseudopodia found in certain amoeboid protists.

retroviruses A group of viruses with RNA genomes that carry the enzyme reverse transcriptase and form a DNA copy of their genome during their life cycle.

reverse electron flow An energy-consuming process used by some chemolithotrophs and phototrophs to generate reducing power.

reverse transcriptase (RT) A multifunctional enzyme used by retroviruses and reverse transcribing DNA viruses during their life cycles. It has RNA-dependent DNA polymerase, DNA-dependent DNA polymerase, and RNAase activity. It synthesizes double-stranded DNA from single-stranded RNA.

reverse vaccinology The use of genomics to identify candidate target molecules on a pathogen for vaccine development.

reversible chain termination sequencing A next-generation DNA sequencing technique that incorporates fluorescently labeled nucleotides that bear a small molecule on the 3′ OH to block addition of the next nucleotide triphosphate. Once the incorporation of the modified base is recorded, the blocking molecule is removed, leaving an unmodified 3′ OH so the next labeled, modified base can be incorporated and recorded; the process continues in an iterative fashion.

reversible covalent modification A mechanism of enzyme regulation in which the enzyme's activity is either increased or decreased by the reversible covalent addition of a group such as phosphate or AMP to the protein.

reversion mutation A mutation in a mutant organism that changes its phenotype back to the wild type; the second mutation is at the same site as the initial mutation.

rhizobia Any one of a number of α- and β-proteobacteria that form symbiotic nitrogen-fixing nodules on the roots of leguminous plants.

rhizomorph A macroscopic, densely packed thread consisting of individual cells formed by some fungi. A rhizomorph can remain dormant or serve as a means of fungal dissemination.

rhizoplane The surface of a plant root.

rhizosphere A region around the plant root where materials released from the root increase the microbial population and its activities.

ribonucleic acid (RNA) A polynucleotide composed of ribonucleotides joined by phosphodiester bonds.

ribosomal frameshifting A phenomenon in which a ribosome shifts the reading frame of the message, thus generating a different protein than it would have if it had stopped at the stop codon.

ribosomal RNA (rRNA) The RNA present in ribosomes; contributes to ribosome structure and also directly involved in the mechanism of protein synthesis.

ribosome The organelle where protein synthesis occurs; the message encoded in mRNA is translated here.

ribosome-binding site (RBS) Nucleotide sequences on mRNA recognized by a ribosome, so that it orients properly on the mRNA and begins translation at the appropriate codon.

riboswitch A site in the leader of an mRNA molecule that interacts with a metabolite or other small molecule, causing the leader to change its folding pattern. In some riboswitches, this change can alter transcription; in others, it affects translation.

ribotyping The use of conserved rRNA sequences to probe chromosomal DNA for classifying bacterial strains.

ribozyme An RNA molecule with catalytic activity.

ribulose 1,5-bisphosphate carboxylase/ oxygenase (RubisCO) The enzyme that catalyzes incorporation of CO_2 in the Calvin-Benson cycle.

RNA-dependent DNA polymerase An enzyme that synthesizes DNA using an RNA template; also called reverse transcriptase.

RNA-dependent RNA polymerase An enzyme that synthesizes RNA using an RNA template.

RNA polymerase An enzyme that catalyzes the synthesis of RNA.

RNA polymerase core enzyme The portion of bacterial RNA polymerase that synthesizes RNA. It consists of all subunits except sigma factor.

RNA polymerase holoenzyme Bacterial RNA polymerase core enzyme plus sigma factor.

RNA-seq Direct sequencing of total cellular mRNA.

RNA silencing A response made by eukaryotic cells that protects them from infections caused by double-stranded RNA viruses.

RNA world The theory that posits that the first self-replicating molecule was RNA and this led to the evolution of the first primitive cell.

rolling-circle replication A mode of DNA replication in which the replication fork moves around a circular DNA molecule, displacing a strand to give a 5′ tail that is also copied to produce new double-stranded DNA.

root nodule Gall-like structures on roots that contain endosymbiotic nitrogen-fixing bacteria.

rough endoplasmic reticulum (RER) *See* endoplasmic reticulum.

R plasmids (R factors) Plasmids bearing one or more drug-resistance genes.

S

sac fungi Common name for the *Ascomycota*.

sanitization Reduction of the microbial population on an inanimate object to levels judged safe by public health standards.

saprophyte (sap′ro-fīt) An organism that takes up nonliving organic nutrients in dissolved form and usually grows on decomposing organic matter.

SAR11 The most abundant group of microbes on Earth, this lineage of α-proteobacteria has been found in almost all marine ecosystems studied.

satellites Subviral infectious agents composed of DNA or RNA encapsidated with the aid of an unrelated helper virus.

scale-up The process of increasing the volume of a culture from a small, experimental system to a large-volume system for the production of an industrially important compound.

scanning electron microscope (SEM) An electron microscope that passes a beam of electrons over the surface of a specimen; an image is formed from the electrons that are emitted by the specimen.

scanning probe microscope A microscope used to study surface features by moving a sharp probe over the object's surface (e.g., atomic force and scanning tunneling microscopes).

scanning tunneling microscope *See* scanning probe microscope.

sclerotia (sklĕ-ro′she-ə) Compact masses of hyphae produced by some filamentous fungi that permit overwintering. In the spring, they germinate to produce either additional hyphae or conidia.

Sec system A system found in all domains of life that transports a protein through or inserts it into a membrane.

secondary active transport *See* active transport.

secondary metabolites Products of metabolism that are synthesized after growth has been completed. Antibiotics are considered secondary metabolites.

secondary treatment The biological degradation of dissolved organic matter in the process of sewage treatment; the organic material is either mineralized or changed to settleable solids.

second law of thermodynamics Physical and chemical processes proceed in such a way that the entropy of the universe (the system and its surroundings) increases to the maximum possible.

second messengers Small molecules that are made in response to an extracellular signal (the first messenger). They "pass" the message to effector molecules that bring about the appropriate response to the signal.

secretory IgA (sIgA) The class of immunoglobulin associated with mucous membranes. *See also* IgA.

secretory pathway The process used by eukaryotic cells to synthesize proteins and lipids, followed by secretion or delivery to organelles or the plasma membrane.

sedimentation basin A containment vessel used during surface water purification. Following the addition of material to facilitate the formation of coagulated impurities, water is left until such material has precipitated, forming sediment at the bottom of the basin.

segmented genome A viral genome that is divided into several parts or fragments, each usually coding for a single polypeptide.

selectable marker A gene conferring a wild-type or mutant phenotype that can be determined by growth on specific media.

selectins A family of cell adhesion molecules displayed on activated endothelial cells; mediate leukocyte binding to the vascular endothelium.

selective media Culture media that favor the growth of specific microorganisms; this may be accomplished by inhibiting the growth of undesired microorganisms.

selective toxicity The ability of a chemotherapeutic agent to kill or inhibit a microbial pathogen while damaging the host as little as possible.

self-assembly The spontaneous formation of a complex structure from its component molecules without the aid of special enzymes or factors.

sense codon A codon that specifies an amino acid.

sensor kinase *See* two-component signal transduction system.

sensory rhodopsin Microbial rhodopsin that senses the spectral quality of light. *See also* archaerhodopsin.

sepsis Systemic response to infection manifested by two or more of the following conditions: temperature >38°C or <36°C; heart rate >90 beats per min; respiratory rate >20 breaths per min, or pCO$_2$ <32 mm Hg; leukocyte count >12,000 cells per ml^3 or >10% immature (band) forms. Sepsis also has been defined as the presence of pathogens or their toxins in blood and other tissues.

septate hyphae Fungal hyphae (filaments) having cross walls.

septation The process of forming a cross wall between two daughter cells during cell division.

septicemia (sep″tĭ-se′me-ah) A disease associated with the presence in the blood of pathogens or bacterial toxins.

septic shock Sepsis associated with severe hypotension despite adequate fluid resuscitation, along with the presence of perfusion abnormalities that may include but are not limited to lactic acidosis, oliguria, or an acute alteration in mental status.

septic tank A tank used to process domestic sewage. Solid material settles out and is partially degraded by anaerobic bacteria as sewage slowly flows through the tank. The outflow is further treated or dispersed in oxic soil.

septum (pl., **septa**) A partition that is used to separate cellular material. Septa are formed during cell division to divide the two daughter cells. They also separate cells in filamentous organisms and a developing endospore from its mother cell.

sequencing by synthesis Any DNA sequencing technology that determines the nucleotide sequence as labeled nucleotides are incorporated.

serology The branch of immunology that is concerned with in vitro reactions involving one or more serum constituents (e.g., antibodies and complement).

serotyping A technique or serological procedure used to differentiate between strains (serovars or serotypes) of microorganisms that have differences in the antigenic composition of a structure or product.

serovar A variant strain of a microbe that has distinctive antigenic properties.

settling basin A basin used during water purification to chemically precipitate fine particles, microorganisms, and organic material by coagulation or flocculation.

sex pilus (pī′lus) A thin protein appendage required for bacterial conjugation. The cell with sex pili donates DNA to recipient cells.

sheath A hollow, tubelike structure surrounding a cell; present in some genera of bacteria.

Shine-Dalgarno sequence A segment in the leader of bacterial mRNA and some archaeal mRNA that binds to a sequence on the 16S rRNA of the small ribosomal subunit. This helps properly orient the mRNA on the ribosome.

shuttle vector A DNA vector (e.g., plasmid, cosmid) that has two origins of replication, each recognized by a different microorganism. Thus the vector can replicate in both microbes.

siderophore (sid′er-o-for″) A small molecule that complexes with ferric iron and supplies it to a cell by aiding in its transport across the plasma membrane.

sigma factor A protein that helps bacterial RNA polymerase core enzyme recognize the promoter at the start of a gene. Thus, it is a transcription factor.

sign An objective change in a diseased body that can be directly observed (e.g., a fever or rash).

signal 1 Primary lymphocyte activation that occurs when antigen is presented to the lymphocyte by its appropriate MHC molecule on the presenting cell.

signal 2 Confirmatory event causing lymphocytes stimulated by signal 1 to complete activation and thus become effector cells.

signal peptide The amino terminal sequence on a protein destined for transport through or into a membrane. It delays protein folding and is recognized by the Sec system.

signal recognition particle A complex of a small RNA and a protein that directs a polypeptide and its translating ribosome to the Sec system during cotranslational translocation.

silage Fermented plant material with increased palatability and nutritional value for animals, which can be stored for extended periods.

silencers Sites on eukaryotic chromosomes to which repressor proteins bind.

silent mutation A mutation that does not change an organism's proteins or phenotype even though the DNA base sequence has been changed.

simple (cut-and-paste) transposition A transposition event in which the transposable element is cut out of one site and ligated into a new site.

single-cell genomic sequencing A method whereby the genome of a single microbial cell is copied many times by the multiple displacement amplification (MDA) technique. The genome is then sequenced by a next-generation technique.

single nucleotide polymorphism (SNP) Difference of one base in genes or genetic elements that are otherwise highly conserved among members of a given taxon.

single radial immunodiffusion (RID) assay An immunodiffusion technique that quantitates antigens by following their diffusion through a gel containing antibodies directed against the test antigens.

site-directed mutagenesis The in vitro process by which a specific nucleotide change is made to a particular gene.

site-specific recombination Recombination between two DNA molecules that does not involve extensive areas of homology between the two molecules. It is catalyzed by enzymes called recombinases distinct from RecA, and it is the mechanism by which transposition occurs.

skin-associated lymphoid tissue (SALT) The lymphoid tissue in the skin that forms a first-line defense as a part of innate immunity.

S-layer A regularly structured layer composed of protein or glycoprotein that lies on the surface of many bacteria and archaea.

slime The viscous extracellular glycoproteins or glycolipids produced by some bacteria. *See also* slime layer.

slime layer A layer of diffuse, unorganized, easily removed material lying outside an archaeal or bacterial cell wall.

slime mold A common term for members of the protist taxa *Myxogastria* and *Dictyostelia*.

slow sand filter A bed of sand through which water slowly flows; the gelatinous microbial layer on the sand grain surface removes waterborne microorganisms, particularly *Giardia*.

sludge A general term for the precipitated solid matter produced during water and sewage treatment; solid particles composed of organic matter and microorganisms that are involved in aerobic sewage treatment (activated sludge).

small interfering RNAs (siRNAs) Small RNA molecules that affect either transcription of target genes or translation of target mRNAs.

small RNAs (sRNAs) Small regulatory RNA molecules that do not function as messenger, ribosomal, or transfer RNAs.

small subunit rRNA (SSU rRNA) The rRNA associated with the small ribosomal subunit: 16S rRNA in bacterial and archaeal cells and 18S rRNA in eukaryotes.

smooth endoplasmic reticulum *See* endoplasmic reticulum.

snapping division A distinctive type of binary fission resulting in an angular or a palisade arrangement of cells; characteristic of members of the genera *Arthrobacter* and *Corynebacterium*.

soil organic matter (SOM) The organic material within a soil; organic material is important because it helps to retain nutrients, maintain soil structure, and hold water for plant use.

SOS response A complex, inducible process that allows bacterial cells with extensive DNA damage to survive, although often in a mutated form; it involves cessation of cell division, upregulation of several DNA repair systems, and induction of translesion DNA synthesis.

sour mash Liquid mixture of fermented grain that is reserved from one batch of distilled spirits to start the fermentation of a new batch; used in the production of whiskey and bourbon.

Southern blotting technique A procedure used to isolate and identify DNA fragments from a complex mixture. The isolated, denatured fragments are transferred from an agarose gel to a nylon filter and identified by hybridization with oligonucleotide probes.

specialized transduction A transduction process in which only a specific set of bacterial or archaeal genes is carried to a recipient cell by a temperate virus; the cell's genes are acquired because of a mistake in the excision of a provirus during the lysogenic life cycle.

species Species of higher organisms are groups of interbreeding or potentially interbreeding natural populations that are reproductively isolated. Archaeal and bacterial species are often defined as collections of strains that have many stable properties in common and differ significantly from other strains. However, there is currently considerable debate about the best definition of archaeal and bacterial species.

specific immune response *See* adaptive immunity.

specific immunity *See* adaptive immunity.

spheroplast (sfēr′o-plast) A spherical cell formed by the weakening or partial removal of the rigid cell wall component.

spike *See* peplomer.

spirillum (spi-ril′um) A rigid, spiral-shaped bacterium.

spirochete (spi′ro-kēt) A flexible, spiral-shaped bacterium with periplasmic flagella.

spleen A secondary lymphoid organ where old erythrocytes are destroyed and blood-borne antigens are trapped and presented to lymphocytes.

spliceosome A complex of proteins and small RNAs that carries out RNA splicing in eukaryotes.

spontaneous generation An early belief that living organisms could develop from nonliving matter.

spontaneous mutations Mutations that occur due to errors in replication or the occurrence of lesions in the DNA (e.g., apurinic sites).

sporangiospore (spo-ran′je-o-spōr) A spore borne within a sporangium.

sporangium (spo-ran′je-um; pl., **sporangia**) A saclike structure or cell, the contents of which are converted into one or more spores.

spore A differentiated, specialized cell that can be used for dissemination, for survival in adverse conditions because of its heat and dessication resistance, and/or for reproduction. Spores are usually unicellular and may develop into vegetative organisms or gametes. They may be produced asexually or sexually.

sporozoite The motile, infective stage of apicomplexan protists.

sporulation The process of spore formation.

spread plate A Petri dish of solid culture medium with isolated microbial colonies growing on its surface that has been prepared by spreading a dilute microbial suspension evenly over the agar surface.

stable isotope analysis The use of stable (nonradioactive) isotopes to determine the fate of a given nutrient or compound in the environment.

stable isotope probing A technique that uses stable isotopes to examine nutrient cycling and identify the microbes involved.

stalk A bacterial appendage produced by the cell and extending from it.

standard free energy change The free energy change of a reaction at 1 atmosphere pressure when all reactants and products are present in their standard states; usually the temperature is 25°C.

standard plate counts *See* viable counting methods.

standard reduction potential A measure of the tendency of an electron donor to lose electrons in an oxidation-reduction (redox) reaction. The more negative the reduction potential of a compound, the better electron donor it is.

start codon The first codon in an mRNA that is translated; it signifies the translation start site.

starter culture An inoculum, consisting of a mixture of microorganisms, used to start a commercial fermentation.

static Inhibiting or retarding growth.

stationary phase The phase of microbial growth in a batch culture when population growth ceases and the growth curve levels off.

stem-nodulating rhizobia Rhizobia that produce nitrogen-fixing structures above the soil surface on plant stems.

sterilization The process by which all living cells, spores, viruses, and viroids are either destroyed or removed from an object or habitat.

Stickland reaction A type of fermentation in which an amino acid serves as the energy source and another amino acid serves as the electron acceptor.

sticky ends The complementary single-stranded ends of double-stranded DNA that result from cleavage with certain restriction endonuclease enzymes. These single-stranded ends can be used to introduce new fragments of DNA to generate a recombinant molecule.

stigma A photosensitive region on the surface of certain protists that is used in phototaxis.

stop codon *See* nonsense codon.

strain A population of organisms that descends from a single organism or pure culture isolate.

streak plate A Petri dish of solid culture medium with isolated microbial colonies growing on its surface that has been prepared by spreading a microbial mixture over the agar surface, using an inoculating loop.

streptomycete Any high G + C Gram-positive bacterium of the genera *Kitasatospora, Streptomyces,* and *Streptoverticillium.*

strict anaerobes *See* obligate anaerobes.

stringent response A decrease in tRNA and rRNA, and an increase in specific amino acid biosynthetic enzymes in response to amino acid starvation.

structural gene A gene that codes for the synthesis of a polypeptide or polynucleotide (i.e., rRNA, tRNA) with a nonregulatory function.

structural proteomics Analysis of the three-dimensional structure of many proteins to predict the structure of other proteins and protein complexes.

subgenomic mRNA Viral mRNA that is smaller than the genomic RNA of an RNA virus.

substrate (1) A reacting molecule in a chemical reaction that is bound by an enzyme. (2) The first molecule of a biochemical pathway. (3) A surface upon which microbes grow.

substrate-level phosphorylation The synthesis of ATP from ADP by phosphorylation coupled with the exergonic breakdown of a high-energy organic substrate molecule.

substrate mycelium In actinobacteria and fungi, a mat of hyphae that penetrates the surface into which the microbes are growing.

subsurface biosphere The region below the plant root zone where microbial populations grow.

subunit vaccines Immunogenic materials derived from whole infectious agents; e.g., bacterial capsular polysaccharide.

sulfate- or sulfur-reducing bacteria (SRB) Anaerobic bacteria that use sulfate or sulfur or both as a terminal electron acceptor during respiration, thereby performing dissimilatory sulfate reduction.

sulfate reduction The use of sulfate as an electron acceptor, which results in the accumulation of reduced forms of sulfur such as sulfide or incorporation of sulfur into organic molecules, usually as sulfhydryl groups.

superantigens Toxic bacterial proteins that overstimulate the immune system.

superoxide dismutase (SOD) (dis-mu′tās) An enzyme found in all aerobes, facultative anaerobes, and microaerophiles that catalyzes the conversion of the radical oxygen species superoxide to hydrogen peroxide and molecular oxygen (O_2).

superphylum A classification of organisms below the level of domain but above phylum; includes several phyla that are united by one or more distinct characteristics.

supportive media Culture media that are able to sustain the growth of many different kinds of microorganisms.

suppressor mutation A mutation at a different site that overcomes the effect of a forward mutation and produces the normal phenotype. The different site can be within the same gene (intragenic suppressor) or different gene (extragenic suppressor) as the forward mutation.

Svedberg unit (sfed′berg) The unit used in expressing the sedimentation coefficient; the greater a particle's Svedberg value, the faster it travels in a centrifuge.

symbiont Any organism that has a specific relationship with another that can be characterized as mutualism, cooperation, commensalism, predation, parasitism, amensalism, or commensalism.

symbiosis The living together or close association of two dissimilar organisms, each of these organisms being known as a symbiont.

symbiosome The organelle-like compartment that houses bacteroids that fix nitrogen during rhizobium nodulation.

symport Linked transport of two substances across a plasma membrane in the same direction.

symptom A change during a disease that a person subjectively experiences (e.g., pain, fatigue, or loss of appetite). Sometimes the term symptom is used more broadly to include any observed signs.

syndrome *See* disease syndrome.

syngamy The fusion of haploid gametes.

synteny Partial or complete conservation of gene order observed when the genomes of two closely related organisms are compared.

synthetic medium *See* defined medium.

syntrophism (sin′trōf-izəm) The association in which the growth of one organism either depends on, or is improved by, the provision of one or more growth factors or nutrients by a neighboring organism. Sometimes both organisms benefit.

systematic epidemiology The field of epidemiology that focuses on the ecological and social factors that influence the development of emerging and reemerging infectious diseases.

systematics The scientific study of organisms with the ultimate objective of characterizing and arranging them in an orderly manner; often considered synonymous with taxonomy.

systemic anaphylaxis Type 1 (immediate) hypersensitivity that involves multiple tissue sites and results in shock. It is often fatal if not treated.

T

tandem mass spectrometry The analysis of molecules using two mass spectrometers sequentially.

Taq polymerase A thermostable DNA polymerase used in the polymerase chain reaction.

Tat system A system used to transport folded proteins across the plasma membrane in bacteria and some archaea; some eukaryotes have homologues of the Tat system in their chloroplasts and mitochondria.

taxon A group into which related organisms are classified.

taxonomy The science of biological classification; it consists of three parts: classification, nomenclature, and identification.

T cell (T lymphocyte) A type of lymphocyte that matures into an immunologically competent cell within the thymus.

T-cell antigen receptor (TCR) The receptor on the T-cell surface consisting of two antigen-binding peptide chains and associated with a several other glycoproteins. Binding of antigen to the TCR in association with MHC activates the T cell.

T-cell receptor complexes *See* T-cell antigen receptor.

T-dependent antigen An antigen that activates B-cell response only with the aid of T-helper cells.

T DNA A portion of the *Agrobacterium tumefaciens* Ti plasmid integrated into the host plant's chromosome.

teichoic acids (ti-ko′ik) Polymers of glycerol or ribitol joined by phosphates; found in the cell walls of most Gram-positive bacteria.

teichoplanin Antibacterial agent that interferes with cell wall synthesis.

teliospores Diploid spores formed by fungi of *Ustilaginomycota*.

telomerase An enzyme in eukaryotes that replicates the ends of chromosomes.

telomeres Complexes of DNA and proteins at the ends of eukaryotic chromosomes.

temperate bacteriophages (viruses) Bacterial and archaeal viruses that can establish a lysogenic relationship, rather than immediately lysing their hosts.

template strand A DNA or RNA strand that specifies the base sequence of a new complementary strand of DNA or RNA.

terminator A sequence that marks the end of a gene and stops transcription.

tertiary treatment The removal from sewage of inorganic nutrients, heavy metals, viruses, and other small or dissolved contaminants by chemical and biological means after microbes have degraded dissolved organic material during secondary sewage treatment.

testate amoeba An amoeba with a covering (test) made either by the protist or from materials collected from the environment.

tetanolysin (tet″ah-nol′ĭ-sin) A hemolysin that aids in tissue destruction; produced by *Clostridium tetani*.

tetanospasmin (tet″ah-no-spaz′min) The neurotoxic component of the tetanus toxin, which causes the muscle spasms of tetanus.

tetracyclines (tet″rah-si′klēns) A family of antibiotics with a common four-ring structure that are isolated from the genus *Streptomyces* or produced semisynthetically.

T$_H$ 0 cell Precursor to T$_H$1, T$_H$17, and T$_H$2 cells.

T$_H$ 1 cell A CD4$^+$ T-helper cell that secretes interferon-gamma, interleukin-2, and tumor necrosis factor.

T$_H$ 2 cell A CD4$^+$ T-helper cell that secretes interleukin (IL)-4, IL-5, IL-6, IL-9, IL-10, and IL-13, influencing growth and differentiation of B cells.

T$_H$ 9 cells T$_H$2 cells that halt their characteristic T$_H$2 profile upon exposure to TGF-β and subsequently produce IL-9.

T$_H$ 17 cells T$_H$ cells found predominantly in the skin and intestinal epithelia, where their release of IL-17 and IL-22 stimulates neutrophils in response to bacterial invasion.

thallus (thal′us) A type of body that is devoid of root, stem, or leaf; characteristic of fungi.

thaumarchaeol Cyclopentane ring-containing lipids unique to members of *Thaumarchaeota*.

T-helper (T$_H$) cell A T lymphocyte that bears the CD4 co-receptor. Subtypes amplify innate immune responses and promote cell-mediated and antibody-mediated immune responses.

therapeutic index The ratio between the toxic dose and the therapeutic dose of a drug, used as a measure of the drug's relative safety.

thermoacidophiles Microbes that grow best at acidic pHs and temperatures 55°C or higher.

thermocline A water column characterized by a steep temperature gradient.

thermodynamics A science that analyzes energy changes in a system.

thermophile (ther′mo-fīl) A microorganism that can grow at temperatures of 55°C or higher; the minimum is usually around 45°C.

thylakoid (thi′lah-koid) A flattened sac in the chloroplast stroma that contains photosynthetic pigments and the photosynthetic electron transport chain. Similar structures are observed in the cytoplasm of cyanobacteria.

thymic selection A process occurring in the thymus in which T-cell precursors are evaluated for functionality.

thymine (thi′min) The pyrimidine 5-methyluracil found in nucleosides, nucleotides, and DNA.

thymus (thi′məs) A primary lymphoid organ in the chest that is necessary in early life for the development of immunological functions, including T-cell maturation.

T-independent antigen An antigen that triggers a B cell to produce immunoglobulin without T-cell cooperation.

Ti plasmid Tumor-inducing plasmid found in plant pathogenic species of the bacterial genus *Agrobacterium*.

T lymphocyte *See* T cell.

toll-like receptor (TLR) A type of pattern recognition molecule (PRM) on phagocytes and other cells that when bound to a microbial ligand, triggers the production of transcription factor NFκB, which stimulates formation of cytokines, chemokines, and other defense molecules.

topoisomerases Enzymes that change the topology of DNA molecules by transiently breaking one or both strands. They play an important role in DNA replication.

toxemia (tok-se′me-ah) The condition caused by toxins in the blood of the host.

toxigenicity (tok″sĭ-jĕ-nis′ĭ-tē) The capacity of an organism to produce a toxin.

toxin neutralization *See* neutralization.

toxoid A bacterial exotoxin that has been inactivated but still stimulates antitoxin formation when injected into a person or animal.

trace elements *See* micronutrients.

trailer That portion of a gene downstream of the coding region. It is transcribed but not translated.

transaminase An enzyme that catalyzes a transamination reaction.

transamination The transfer of an amino group from an amino acid to an α-keto acid acceptor. Transamination is used both for degradation of amino acids and in their synthesis.

transcriptase (trans-krip′tās) An enzyme that catalyzes transcription.

transcription The process in which RNA with a base sequence complementary to the template strand of DNA or RNA is synthesized.

transcriptome All the messenger RNA transcribed from the genome of an organism under a given set of circumstances.

transcriptomics The study of all the mRNA expressed by a cell at the time of sampling.

transduction The transfer of genes between bacterial or archaeal cells by viruses.

transfer RNA (tRNA) A small RNA that binds an amino acid and delivers it to the ribosome for incorporation into a polypeptide chain during protein synthesis.

transformation (1) A mode of gene transfer in bacteria and archaea in which a piece of free DNA is taken up by a cell and integrated into its genome. (2) In recombinant DNA technology, the uptake of free DNA by a cell. (3) The conversion of a cell into a malignant, cancerous cell.

transition mutations Mutations that involve the substitution of a different purine base for the purine present at the site of the mutation or the substitution of a different pyrimidine for the normal pyrimidine.

translation Protein synthesis; the process by which the genetic message carried by mRNA directs the synthesis of polypeptides with the aid of ribosomes and other cell constituents.

translesion DNA synthesis A type of DNA synthesis that occurs during the SOS response. DNA synthesis occurs without an intact template and generates mutations as a result.

translocation (1) During the elongation cycle of protein synthesis, the movement of the ribosome relative to the mRNA that repositions the A site of the ribosome over the next codon to be translated; (2) the movement of a protein across or into a membrane.

transmissible spongiform encephalopathies Degenerative central nervous system diseases in which the brain has a spongy appearance; due to prions.

transmission electron microscope (TEM) A microscope in which an image is formed by passing an electron beam through a specimen and focusing the scattered electrons with magnetic lenses.

transpeptidation (1) The reaction that forms the peptide cross-links during peptidoglycan synthesis. (2) The reaction that forms a peptide bond during the elongation cycle of protein synthesis.

transposable element A small DNA molecule that carries the genes for transposition and thus can move around the genome.

transposase An enzyme that catalyzes the movement of a mobile genetic element.

transposition The movement of a piece of DNA around the chromosome.

transposon (tranz-po′zon) A mobile genetic element that encodes recombinase, which is needed for transposition, and contains other genes that are not required for transposition.

transversion mutations Mutations that result from the substitution of a purine base for the normal pyrimidine or vice versa.

tricarboxylic acid (TCA) cycle The cycle that oxidizes acetyl coenzyme A to CO_2 and generates NADH and $FADH_2$ for oxidation in an electron transport chain; also supplies precursor metabolites for biosynthesis. Also called citric acid cycle and Krebs cycle.

trichocyst A small organelle beneath the pellicle of some protists. It can be extruded to function as either an anchoring device or as a form of defense.

trichome (tri′kōm) A filament of microbial cells that are in close contact with one another over a large area.

trickling filter A bed of rocks covered with a microbial film that aerobically degrades organic waste during secondary sewage treatment.

trimethoprim A synthetic antibiotic that inhibits production of folic acid by binding to dihydrofolate reductase. Trimethoprim has a wide spectrum of activity and is bacteriostatic.

trophozoite (trof″o-zo′īt) The active, motile feeding stage of a protozoan organism.

tropism (1) The movement of living organisms toward or away from a focus of heat, light, or other stimulus. (2) The selective infection of certain organisms or host tissues by a virus due to the presence of the specific receptor to which the virus binds.

trypanosome (tri-pan′o-sōm) A flagellated protozoan of the genus *Trypanosoma;* often live in the blood of humans and other vertebrates and are transmitted by insect bites.

tubercle (too′ber-k′l) A small, rounded nodular lesion produced by *Mycobacterium tuberculosis.*

tuberculous cavity An air-filled cavity that results from a tubercle lesion caused by *M. tuberculosis.*

tumble Random turning or tumbling movements made by flagellated bacteria when they stop swimming in a straight line (a run).

tumor suppressor genes Genes encoding tumor suppressor proteins.

tumor suppressor proteins Proteins that regulate cell cycling or repair DNA. Their inactivation can contribute to the development of cancer.

turbidostat A continuous culture system equipped with a photocell that adjusts the flow of medium through the culture vessel to maintain a constant cell density or turbidity.

twitching motility A type of bacterial motility characterized by short, intermittent jerky motions due to extension and retraction of type IV pili.

two-component signal transduction system A regulatory system that uses the transfer of phosphoryl groups to control gene transcription and protein activity. It has two major components: a sensor kinase and a response regulator protein.

two-dimensional gel electrophoresis An electrophoretic procedure that separates proteins using two procedures: isoelectric focusing and SDS-PAGE.

tyndallization The process of repeated heating and incubation to destroy bacterial spores.

type I hypersensitivity A form of immediate hypersensitivity arising from the binding of antigen to IgE attached to mast cells, which then release anaphylaxis mediators such as histamine. Examples: hay fever, asthma, and food allergies.

type II hypersensitivity A form of immediate hypersensitivity involving the binding of antibodies to antigens on cell surfaces followed by destruction of the target cells (e.g., through complement attack, phagocytosis, or agglutination).

type III hypersensitivity A form of immediate hypersensitivity resulting from the exposure to excessive amounts of antigens to which antibodies bind. These antibody-antigen complexes activate complement and trigger an acute inflammatory response with subsequent tissue damage.

type IV hypersensitivity A delayed hypersensitivity response that results from the binding of antigen to activated T lymphocytes, which then release cytokines to trigger inflammation that damages tissue. Type IV hypersensitivity is seen in contact dermatitis from poison ivy, leprosy, and tertiary syphilis.

type I secretion system *See* ABC protein secretion pathway.

type III secretion system A system in Gram-negative bacteria that secretes virulence factors and injects them into host cells.

type IV secretion system A system in Gram-negative bacteria that secretes proteins; some Gram-negative and Gram-positive bacteria use this secretion system to transfer DNA during bacterial conjugation.

type V secretion system A pathway in Gram-negative bacteria that translocates proteins across the outer membrane after they have been translocated across the plasma membrane by the Sec system.

type VI secretion system A protein secretion system in Gram-negative bacteria that is similar in structure to the machinery used by some bacteriophages to insert their DNA into host cells.

type strain The microbial strain that is the nomenclatural type or holder of the species name. A type strain will remain within that species should nomenclature changes occur.

U

ubiquinone *See* electron transport chain.

ultraviolet (UV) radiation High-energy radiation of short wavelength, about 10 to 400 nm.

uniporters Carrier proteins that move a single solute across a membrane.

universal phylogenetic tree A phylogenetic tree that considers the evolutionary relationships among organisms from all three domains of life: *Bacteria, Archaea,* and *Eukarya.*

upstream activating sequences (UASs) Sequences near eukaryotic genes where activator proteins bind.

uracil (u′rah-sil) The pyrimidine 2,4-dioxy-pyrimidine, which is found in nucleosides, nucleotides, and RNA.

use dilution test Method of evaluating disinfectant effectiveness at a recommended dilution.

V

vaccine A preparation of antigens and adjuvant administered to induce development of an adaptive immune response that will protect the individual against a specific pathogen or a toxin.

vaccinomics The application of genomics to vaccine development.

valence The number of antigenic determinant sites on the surface of an antigen or the number of antigen-binding sites possessed by an antibody molecule.

vancomycin A glycopeptide antibiotic obtained from *Nocardia orientalis* that is effective only against Gram-positive bacteria and inhibits peptidoglycan synthesis.

variable region (V_L and V_H) The region at the N-terminal end of immunoglobulin heavy and light chains whose amino acid sequence varies between antibodies of different specificity. Variable regions form the antigen-binding site.

vector (1) In genetic engineering, a DNA molecule that can replicate and transport a piece of inserted foreign DNA, such as a gene, into a recipient cell. It may be a plasmid, phage, cosmid or artificial chromosome. (2) In epidemiology, it is a living organism, usually an arthropod or other animal, that transfers an infective agent between hosts.

vector-borne transmission The transmission of an infectious pathogen between hosts by means of a vector.

vehicle An inanimate substance or medium that transmits a pathogen.

vertical gene transfer The transfer of genes from a parent organism to its offspring.

vertical transmission Pathogen transfer from mother to unborn child (in utero).

vesicular transport The movement of materials in small membrane-bound vesicles between the endoplasmic reticulum, Golgi apparatus, lysosomes, plasma membrane, and other organelles of the endocytic and secretory pathways in eukaryotic cells.

viable but nonculturable (VBNC) microorganisms Microbes known to be growing in a specific environment but that cannot be cultured under standard laboratory conditions.

viable counting methods Methods for determining population size that depend on a microbe's ability to reproduce when cultured. Also called standard plate counts.

vibrio (vib′re-o) A rod-shaped bacterial cell that is curved to form a commalike shape.

viral hemagglutination The clumping of red blood cells caused by the attachment of some viruses.

viral neutralization An antibody-mediated process in which IgG, IgM, and IgA antibodies bind to some viruses during their extracellular phase and inactivate or neutralize them.

viricide (vir′i-sīd) An agent that inactivates viruses so that they cannot reproduce within host cells.

virion A complete virus particle; at the simplest, it consists of a protein capsid surrounding a single nucleic acid molecule.

virioplankton Viruses that occur in waters; high levels are found in marine and freshwater environments.

viroid An infectious agent that is a single-stranded RNA not associated with any protein; the RNA does not code for any proteins and is not translated.

virology The branch of microbiology that is concerned with viruses and viral diseases.

virulence The degree or intensity of pathogenicity of an organism as indicated by case fatality rates and/or ability to invade host tissues and cause disease.

virulence factor A product, usually a protein or carbohydrate, that contributes to virulence or pathogenicity.

virulent (bacteriophages) viruses Viruses that lyse their host cells at the end of the viral life cycle.

virus An infectious agent having a simple acellular organization with a protein coat and a nucleic acid genome, lacking independent metabolism, and multiplying only within living host cells.

viruslike particles (VLP) Presumptive viruses detected by electron microscopy or fluorescence microscopy; because these approaches do not demonstrate that these entities can infect host cells, they cannot be considered viruses.

W

wastewater treatment The use of physical and biological processes to remove particulate and dissolved material from sewage and to control pathogens.

water activity (a_w) A quantitative measure of water availability in the habitat; the water activity of a solution is one-hundredth its relative humidity.

water molds *See* öomycetes.

white blood cell (WBC) Blood cell having innate or acquired immune function. These cells are named for the white or buffy layer in which they are found when blood is centrifuged.

whole-cell vaccine A vaccine made from complete pathogens, which can be either killed microorganisms or live, attenuated microbes.

whole-genome shotgun sequencing Genome sequencing in which random fragments of a complete genome are individually sequenced. The nucelotide sequences of the fragments are placed in the proper order based on overlapping identical sequences.

wild type Prevalent form of a gene or phenotype.

Winogradsky column A glass column with an anoxic lower zone and an oxic upper zone, which allows growth of microorganisms under conditions similar to those found in a nutrient-rich lake.

wobble The loose base pairing between an anticodon and a codon at the third position of the codon.

World Health Organization An international health agency located in Geneva, Switzerland.

wort The filtrate of malted grains used as the substrate for the production of beer and ale by fermentation.

X

xenograft (zen″o-graft) A tissue graft between animals of different species.

xerophiles Microorganisms that grow best under low water activity (a_w) conditions, and may not be able to grow at high a_w values.

Y

yeast A unicellular, uninuclear fungus that reproduces either asexually by budding or fission, or sexually through spore formation.

yeast artificial chromosome (YAC) Engineered DNA that contains all the elements required to propagate a chromosome in yeast and is used to clone foreign DNA fragments in yeast cells.

Z

zoonosis (zo″o-no′sis; pl., **zoonoses**) A disease that can be transmitted from animals to humans.

zooxanthella (zo″o-zan-thel′ah) A dinoflagellate found living symbiotically within cnidarians (corals) and other invertebrates.

zygomycetes (zi″go-mi-se′tez) A group of fungi that usually has aspetate hyphae. Sexual reproduction normally involves the formation of zygospores.

zygospore A thick-walled, sexual, resting spore characteristic of the zygomycetous fungi.

Credits

Photos

Front Matter
Page iv (left): Courtesy of Joanne M. Willey; p. iv (middle): Courtesy of Linda M. Sherwood; p. iv (right): Courtesy of Christopher J. Woolverton.

Chapter 1
Opener: Tim Pyle/NASA; 1.3 (all): © J. William Schopf; 1.7a: © Dirk Wiersma/SPL/Science Source; 1.7b, 1.11a: © Bettmann/Corbis; 1.11b, 1.11b (inset): © Kathy Park Talaro/Pasadena City College; 1.11c: © Dr. Jeremy Byrgess/SPL/Getty Images; 1.12: © Pixtal/age fotostock; 1.14: © Bettmann/Corbis.

Chapter 2
Opener: © Kenneth Lambert/AP Images; 2.3: © McGraw-Hill Education/James Redfearn, photographer; 2.7a: CDC/Schwartz; 2.7b, 2.9a: © Stephen Durr; 2.9b: © McGraw-Hill Education/James Redfearn, photographer; 2.11: © Stephen Durr; 2.13a: © Dr. Rita B. Moyes; 2.13b: © Evans Roberts; 2.14: © Jeff Errington/Centre for Bacterial Cell Biology/Newcastle University; 2.15 (all): © P. Dirckx/Center for Biofilm Engineering/Montana State University; 2.17 (both): © Dr. Rita B. Moyes; 2.19a, 2.19b: © McGraw-Hill Education/James Redfearn, photographer; 2.19c: Larry Stauffer, Oregon State Public Health Laboratory/CDC; 2.19d: © Steven P. Lynch; 2.19e: © David B. Fankhauser/University of Cincinnati Clermont College; 2.21a: © McGraw-Hill Education/James Redfearn, photographer; 2.21b: © Biology Media/Science Source; 2.22: © McGraw-Hill Education/James Redfearn, photographer; 2.24a: © Ami Images/Science Source; 2.24b: © Dr. Tony Brain/SPL/Science Source; 2.25: © Dr. Kari Lounatmaa/SPL/Science Source; 2.27: Janice Carr/CDC; 2.28a: © AMI Images/SPL/Science Source; 2.28b: © Grant J. Jensen; 2.29: © Driscoll, Yougquist & Baldeschwieler, Cal-tech/SPL/Science Source; 2.31 (both): © Simon Scheuring.

Chapter 3
Opener: © Design Pics/Hammond HSN; 3.1a: © Science Source; 3.1b: Janice Haney Carr/CDC; 3.1c: © McGraw-Hill Education/James Redfearn, photographer; 3.2a: © Media for Medical/Getty Images; 3.2b, 3.2c: Janice Carr/CDC; 3.2d: © Biology Pics/Science Source; 3.2e: © Dr. Amy Gehring; 3.2f: © Phototake, Inc.; 3.4: © Esther R. Angert, Ph.D./Phototake; 3.16 (both): © Egbert Hoiczyk; 3.28: © McGraw-Hill Education/James Redfearn, photographer; 3.29: © Dr. Robert G.E. Murray/University of Western Ontario; 3.30a: © Dr. Joseph Pogliano; 3.30b, 3.30c: © Jeff Errington/Centre for Bacterial Cell Biology/Newcastle University; 3.30d: © Dr. Christine Jacobs-Wagner; 3.31: © Dr. Robert G.E. Murray/University of Western Ontario; 3.34: From J.T. Staley, M.P. Bryant, N. Pfenning and J.G. Holt (Eds), Bergey's Manual of Systematic Bacteriology, Vol. 3. © 1989 Williams and Wilkins Co., Baltimore; 3.35: © Michael Schmid; 3.36: © Daniel Branton/Harvard University; 3.37a: © Dennis Kunkel Microscopy, Inc./Phototake; 3.37b: © Grant J. Jensen; 3.39a: © CNRI/SPL/Science Source; 3.39b: © Dr. Gopal Murti/SPL/Science Source; 3.40: © Thomas Deerinck, NCMIR/Science Source; 3.41a: Dr. William A. Clark/CDC; 3.41b, 3.41c: © McGraw-Hill Education/James Redfearn, photographer; 3.45b: © Gavin Murphy/Nature/Science Source; 3.46: © Dr. Daniel Kearns; 3.47b: © Jacques Izard/Harvard School of Dental Medicine; 3.49b: © CNRI/Science Source.

Chapter 4
Opener: © Ingram Publishing; 4.1 (both): From J.T. Staley, M.P. Bryant, N. Pfenning and J.G. Holt (Eds), Bergey's Manual of Systematic Bacteriology, Vol. 3. © 1989 Williams and Wilkins Co., Baltimore; 4.2: © Prof. Olivier Gros; 4.8: © Dr. Reinhard Rachel; 4.9: © Karl O. Stetter; 4.10 (both): © Dr. Christine Moissl-Eichinger.

Chapter 5
Opener: © Wave Royalty Free/Science Source; 5.1a: © Melba Photo Agency/PunchStock; 5.1b: © PhotoLink/Getty Images; 5.1c: © Jackson Kung'u; 5.1d, 5.1e: © Stephen Durr; 5.1f: © Steven P. Lynch; 5.2: © Johanna Höög/European Molecular Biology Laboratory/Science Source; 5.6: © David C. Amberg/SUNY Upstate Medical University; 5.11a: © Biophoto Associates/Science Source; 5.12: © Don W. Fawcett/Science Source; 5.13a, 5.13b: H. Gao et al, "The structure of the 80S ribosome from Trypanosoma cruzi reveals unique rRNA components" PNAS July 19, 2005 Vol. 102 no. 29, fig. 1, p. 10207. Copyright (2005) National Academy of Sciences, U.S.A.; 5.15b: © Keith R. Porter/Science Source; 5.19: From "Electron microscopic observations of the flagellar hairs of Phytophthora palmivora zoospores" by P. R. Desjardins, G. A. Zentmyer, D. A. Reynolds. Canadian Journal of Botany, 1969, 47(7): 1077-1079, 10.1139/b69-153; 5.21a: © Vincent A. Fischetti, Ph.D/Rockefeller University, www.rockefeller.edu/vaf; 5.21b: © Biophoto Associates/Science Source.

Chapter 6
Opener: © Stockbyte/Getty Images; p. 110: © David Prangishvili; 6.3a: © Robert G. Milne, Plant Virus Institute National Research Council, Italy; 6.3c: © Gerald Stubbs/Vanderbilt University; Keiichi Namba/Osaka University; and Donald Caspar, Florida State University; 6.4b: Dr. F.A. Murphy/CDC; 6.5a: © Biophoto Associates/Science Source; 6.5b: © Division of Computer Research & Technology, NIH/Science Source; 6.6b: © Marek Cyrklaff; 6.7b: © Ami Images/Science Source; 6.8a: © Chris Bjornberg/Science Source; 6.8b: © Dr. Linda Stannard, UCT/Science Source; 6.12: From "SARS-Coronavirus Replication Is Supported by a Reticulovesicular Network of Modified Endoplasmic Reticulum" Knoops K, Kikkert M, Worm SH, Zevenhoven-Dobbe JC, van der Meer Y, Koster AJ, Mommaas AM, Snijder EJ - (2008); 6.14: © Lee D. Simon/Science Source; 6.18: © Carolina Biological Supply Company/Phototake; 6.19: © Norm Thomas/SPL/Science Source.

Chapter 7
Opener: © Brand X Pictures/Stockbyte/Getty Images; 7.2a: © Dr. Kari Lounatmaa/Science Source; 7.2b: © Dr. John B. Waterbury, Woods Hole Oceanographic Inst.; 7.2c: © Nora Ausmees, University of Uppsala; 7.17a: © Dr. Joachim Reitner; 7.17b: © Nicholas G. Sotereanos, M.D.; 7.21a: © Chris Frazee/UW-Madison; 7.21b: © Dr. Margaret Jean McFall-Ngai; Table 7.6: CDC; 7.22 (both): © Kathy Park Talaro/Pasadena City College; 7.23: © Coy Laboratory Products, Inc.; 7.25b, 7.26b: © Kathy Park Talaro/Pasadena City College; 7.28b: © Scimat/Science Source; 7.37: © McGraw-Hill Education/Lisa Burgess, photographer.

Chapter 8

Opener: Megan Mathias and J. Todd Parker/CDC; 8.5: © Callista Images Cultura/Newscom; 8.6a: © ThermoForma of Marietta, Ohio; 8.7a: © BSIP SA/Alamy; 8.8: © McGraw-Hill Education/ James Redfearn, photographer; p. 180: © Brand X Pictures/ Stockbyte/Getty Images; 8.12a: © Andersen Products, www.anpro. com.

Chapter 9

Opener: © McGraw-Hill Education/James Redfearn, photographer; 9.1: © Christine L. Case, Skyline College; 9.2 (both): © McGraw-Hill Education/James Redfearn, photographer; 9.4: © BSIP/Newscom; 9.12b: © Alex MacKerell, Ph.D.

Chapter 10

Opener: © Comstock Images/ Getty Images; p. 210: © California History Collection, California State Library/Getty Images.

Chapter 11

Opener: NASA; 11.16a (top): © Thomas Meier.

Chapter 12

Opener: © Science & Society Picture Library/Superstock.

Chapter 13

Opener: National Archives and Records Administration [NW-CTC-59-INV15E205-86220212(82A)]; p. 285: © Brand X Pictures/PunchStock; 13.9b: From J. Cairns, "The Chromosome of E. coli" in Cold Spring Harbor Symposia on Quantitative Biology, 77, Fig. 2, Pg. 44 © 1963 by Cold Spring Harbor Laboratory Press; 13.23: From "Transcription of the T4 late genes" by Geiduschek EP, Kassavetis GA - (2010); 13.33a: © Steven McKnight and Oscar L. Miller/University of Virginia.

Chapter 14

Opener: © Image Source/ JupiterImages; 14.6b: © Mitchell Lewis/SPL/Science Source; 14.18b: © Laguna Design/SPL/ Science Source; 14.21b: © Janine Maddock.

Chapter 15

Opener: © Comstock Images/ Getty Images; 15.10 (all) © Finn Werner; 15.11: © Stephen D. Bell, MRC Cancer Cell Unit, UK.

Chapter 16

Opener: Tim McCabe/USDA; 16.17: © Dennis Kunkel Microscopy, Inc./Phototake.

Chapter 17

Opener: © Matt Champlin/Getty Images; 17.2: © Dr. A.K. Aggarwal/ Mount Sinai School of Medicine; 17.6b: © Kathy Park Talaro/ Pasadena City College; 17.11a, 17.11b: © Huntington Porter and David Dressler/Time Life Pictures/Getty Images; 17.11c: © Edvotek, Inc. www.edvotek.com; 17.15a: © Mark Buttner; 17.15b: © Klas Flärdh.

Chapter 18

Opener: © Edward Rozzo/Corbis; 18.8a: © Yi Xiang Yeng/iStock/ Getty Images; 18.8b: © McGraw-Hill Education; 18.8c: © Nigel Cattlin/Science Source; 18.8d: © OAR/National Undersea Research Program (NURP)/College of William & Mary/NOAA; 18.14: Liu Zhou, et al. "Transcritomedyamics," PNAS, April 2003,Vol. 100: 4191-4196. Copyright (2003) National Academy of Sciences, U.S.A.; 18.16b: © Tyne/Simon Fraser/Science Source.

Chapter 19

Opener: NASA, ESA, R. O'Connell (University of Virginia), and the Hubble Heritage Team; p. 445: © Pixtal/age Fotostock; 19.13 (Euryarchaeota): © SPL/Science Source; 19.13 (Aquificae): © Karl O. Stetter; 19.13 (Bacilli): © Andre Syred/SPL/ Science Source; 19.13 (Actinobacteria): © Microfield Scientific Ltd/Getty Images; 19.13 (Spirochaetes): © Alfred Paseika/SPL/ Science Source; 19.13 (Cyanobacteria): © McGraw-Hill Education/ Don Rubbelke photographer; 19.13 (Betz): © Dr. Kari Lounatmaa/Science Source; 19.13 (Gamma): Janice Carr/CDC; 19.13 (Delta): © Derek Lovley/Science Source.

Chapter 20

Opener: © Mike Groll/AP Images; 20.8: © DonFink/iStock/ Getty Images; 20.9: © Derek Lovley/Kazem Kashefi/Science Source; 20.10: © Karl O. Stetter; 20.13a: From J.T. Staley, M.P. Bryant, N. Pfenning and J.G. Holt (Eds), Bergey's Manual of Systematic Bacteriology, Vol. 3. © 1989 Williams and Wilkins Co., Baltimore; 20.13b: © DR M. Rohde,GBF/Science Source; 20.16: V. Orphan, "Multiple archaeal groups mediate methane oxidation in anoxic cold seep sediments" PNAS, Vol. 99 no. 11, pages 7663–7668, fig. 1. Copyright (2002) National Academy of Sciences, U.S.A.; 20.17a: © Dr. Mike Dyall-Smith; 20.17b: © Armands Pharyos/Alamy.

Chapter 21

Opener: © Daniel Deitschel/Getty Images; 21.2: © Karl O. Stetter; 21.3a: © Michael J. Daly/Science Source; 21.3b: © Dr. Robert G.E. Murray/University of Western Ontario; 21.4: © Don W. Fawcett/ Science Source; 21.5: From "Isolation and Identification of Acholeplasma sp. from the Mud Crab, Scylla serrata" by Chen JG, Lou D, Yang JF - (2011); 21.11a: © Joanne M. Willey, Ph.D.; 21.12a: © M.I. Walker/Science Source; 21.12b: © Gerd Guenther/SPL/ Science Source; 21.12c: © Sinclair Stammers/SPL/Science Source; 21.12d: © Biophoto Associates/ Science Source; 21.13a: © Jason K. Oyadomari, www.keweenawalgae.mtu.edu; 21.13b: © Biophoto Associates/Science Source; 21.14: © Environmental Protection Agency/National Archives. Photo by Belinda Rain; 21.15 (both): © John A. Fuerst; 21.17a: © Dr. Peter Braun/Max Planck Institute for Infection Biology; 21.18: CDC; 21.19 a2: © CDC/ SPL/Science Source.

Chapter 22

Opener: © Ingram Publishing; 22.3b: © Dr. Andre Kempe/ Sciencefoto.De/Getty Images; 22.3c: © Laguna Design/Science Source; 22.5 (both): Courtesy of James Berleman and Carl Bauer; 22.6: CDC; 22.7: From J.T. Staley, M.P. Bryant, N. Pfenning and J.G. Holt (Eds), Bergey's Manual of Systematic Bacteriology, Vol. 3. © 1989 Williams and Wilkins Co., Baltimore; 22.9a, 22.9b: © Jeanne S. Poindexter/Barnard College; 22.9c: From J.T. Staley, M.P. Bryant, N. Pfenning and J.G. Holt (Eds), Bergey's Manual of Systematic Bacteriology, Vol. 3. © 1989 Williams and Wilkins Co., Baltimore; 22.11b: © John Kaprielian, The National Audubon Society Collection/Science Source; 22.12: © Custom Life Science Images/Alamy; 22.15 (top left): © Dr. John B. Waterbury, Wood Hole Oceanographic Inst.; 22.15 (top right): © Dr. Martin Konneke; 22.15 (bottom): © ISM/Phototake; 22.16 : © Dennis Bazylinski; 22.22: © Johnny Madsen/Alamy; 22.23: © Roger Burks/University of California at Riverside and Mark Schneegurt/Wichita State University, and Cyanosite (www-cyanosite.bio.purdue.edu); 22.24: © Dr. Heide Schulz-Vogt; 22.28b, 22.28c: From "Electrically conductive bacterial nanowires produced by Shewanella oneidensis strain MR-1 and other microorganisms" by Yuri A. Gorby et al. PNAS vol. 103 no. 30, fig 1A, p. 11358–11363. Copyright (2001) National Academy of Sciences, U.S.A.; 22.29a: © Frederick R. McConnaughey/Science Source; 22.29b: © James G. Morin/Cornell University; 22.34b: © Katy Evans, Chi Aizawa and Liz Sockett; 22.35b (all): © Eye of Science/ Science Source; 22.37: © Annette S. Engel, Ph.D.

Chapter 23

Opener: © William Coupon/Corbis; 23.3: © Alfred Pasieka/Getty Images; 23.6: © R. John Parkes/ Cardiff University; 23.8a: © Michael Abbey/SPL/Science Source; 23.8b: © Molecular Probes, Inc.; 23.9 (both): Branda et al., PNAS 25 Dept. 2001, vol. 98, fig 1A, p. 11623 Copyright (2001) National Academy of Sciences, U.S.A.; 23.10: From "Protein crystal structure obtained at 2.9 Å resolution from injecting bacterial cells into an X-ray free-electron laser beam" Michael R. Sawaya, Duilio Cascioa, Mari Gingerya, Jose Rodriguez, Lukasz Goldschmidt, Jacques-Philippe Colletier, Marc M. Messerschmidt, Sébastien Boutet, Jason E. Koglin, Garth J. Williams, Aaron S. Brewster, Karol Nassh, Johan Hattne, Sabine Bothh, R. Bruce Doak, Robert L. Shoeman, Daniel P. DePonte, Hyun-Woo Park, Brian A. Federici, Nicholas K. Sauter, Ilme Schlichting, and David S. Eisenberg, PNAS vol. 111 no. 35, fig 1, p. 12769–12774. Copyright (2014) National Academy of Sciences, U.S.A.; 23.11a: © Science Source; 23.13: © SCIMAT/ Science Source; 23.14a: © James Cavallini/Science Source; 23.14b: © Science Source; 23.15a: © McGraw-Hill Education/James Redfearn, photographer; 23.15b: © Matt Meadows/Getty Images; 23.15c: © McGraw-Hill Education/ James Redfearn, photographer.

Chapter 24

Opener: © Don Hammond/Design Pics/Getty Images; 24.2a: © Dr. Alisa A. Gaskell; 24.3a: © C. L. Jiang, L. H. Xu & S. Suzuki/Society for Actinomycetes Japan; 24.3b: © M. Hayakawa, H. Iino & H. Nonomura/Society for Actinomycetes Japan; 24.3c: Courtesy of Kim Findlay and Mark Buttner; 24.5a: © Biophoto Associates/Science Source; 24.5b: © David M. Phillips/Science Source; 24.6: © Michael Abbey/Science Source; 24.8a: © Dr. Gary Gaugler/Science Source; 24.8b: © M. Wachi/Society for Actinomycetes Japan; 24.9: CDC; 24.12b: © S. Amano & S. Miyadoh/Society for Actinomycetes Japan; 24.12b (inset): © H. Suzuki & A. Seino/Society for Actinomycetes Japan; 24.13b, 24.13c: © Joanne M. Willey, Ph.D.; 24.14a: © Roger Greenwell; 24.14b: © Nigel Cattlin/Alamy; 24.15a: © R.H. Berg/Society for Actinomycetes Japan; 24.15b: © S. Amano, S. Miyadoh & T. Shomura/Society for Actinomycetes Japan; 24.16 (both): © David Benson/University of Connecticut; 24.17: © S. Amano & S. Miyadoh/Society for Actinomycetes Japan.

Chapter 25

Opener: © Dr. Owen Gilbert; 25.1: © Biophoto Associates/ Science Source; 25.2: Dr. Stan Erlandsen/CDC; 25.4a: © Science Source; 25.4b: © David M. Phillips/The Population Council/ Science Source; 25.6a: Dr. Mae Melvin/CDC; 25.6b: © Martin Dohrn/Science Source; 25.8b: © Scott Camazine/Science Source; 25.8c: © Eye of Science/Science Source; 25.8d: © Ray Simons/ Science Source; 25.10b: © Eye of Science/Science Source; 25.11: © David Caron/Science Source; 25.12: © Richard Rowan/Science Source; 25.13a: © Lee W. Wilcox, Ph.D.; 25.13b: © Claude Taylor, III; 25.14 (both): © Eric Grave/ Science Source; 25.19a: © Dr. Anne Smith/SPL/Science Source; 25.19b: © Jim Hinsch/Science Source; 25.21: © Natural History Museum/SPL/Science Source; 25.22a: © M.I. Walker/Science Source; 25.22b: © Micro_photo/ iStock/Getty Images; 25.22c: © Nuridsany et Perennou/Science Source; 25.22d: © De Agostini Picture Library/Science Source.

Chapter 26

Opener: Jim Gathany/CDC; 26.2a, 26.2b: © Carolina Biological Supply Company/Phototake; 26.2c: © Eye of Science/Science Source; 26.2d: © Sinclair Stammers/Science Source; 26.2e: © Robert W. Seagull; 26.2f: Dr. James Becnel, USDA ARS Gainesville, FL and Society of Invertebrate Pathology; 26.6 (both): © Martha Powell; 26.8a: © Juniors Bildarchiv GmbH/Alamy; 26.8b: © Kelly Cline/E+/Getty Images; 26.9a: © SCIMAT/SPL/ Science Source; 26.10: © Microfield Scientific Ltd./SPL/Science Source; 26.12: © David Q. Cavagnaro/Photolibrary/Getty Images; 26.13: © Scimat/Science Source; p. 593: Al Hicks/New York Department of Environmental Conservation/USGS; 26.14a: © Alexandra Lowry/The National Audubon Sociey Collection/ Science Source; 26.15: © J.K. Pataky/University of Illinois.

Chapter 27

Opener: © Patrick Swan/Design Pics; 27.4f: © Lee D. Simon/ Science Source; 27.8: © M. Wurtz/ Science Source; 27.12a: © Mark Young/Montana State University; 27.12b: © Kenneth Stedman; 27.12c: © David Prangishvili; 27.13: Courtesy of Susan Brumfield and Mark Young; 27.15: © AMI Images/SPL/Science Source.

Chapter 28

Opener: © VOISHMEL/AFP/ Getty Images.

Chapter 29

Opener: NOAA/OER; 29.2: © Dr. Rita B. Moyes; p. 641: © age fotostock/Alamy; 29.5: © David C. Gillan/Mons University, Belgium; 29.7: © Seana Davidson; 29.10a: © Reut S. Abramovich/Bar-Ilan University, Israel; 29.11 (both): © Cleber Ouverney.

Chapter 30

Opener: © Bill Bachmann/Science Source; 30.2a: © Natural History Museum/SPL/Science Source; 30.2b: © Dr. M. Debora Iglesias-Rodriguez; 30.3: © Bill Bachmann/ Science Source; 30.4: © Richard Ellis/Alamy; 30.6b: © Jackie Parry; 30.6c: © Greg Antipa/Biophoto Associates/Science Source; 30.8a: Photo by Susumu Honjo © Woods Hole Oceanographic Institution; 30.9: © Image Quest 3-D/NHPA/ Photoshot/Newscom; 30.11: NASA image courtesy Jeff Schmaltz, MODIS Rapid Response Team at NASA GSFC.

Chapter 31

Opener: © Nigel Cattlin/Science Source; p. 669: © Pixtal/AGE Fotostock; 31.5: © R. Henrik Nilsson; 31.9d: © Dr. Jeremy Burgess/ SPL/Science Source; 31.9h: © Biology Pics/Science Source; 31.9i: © Hugh Spencer/Science Source; 31.11: © Biophoto Associates/ Science Source; 31.12: © Dr. Sandor Sule, Plant Protection Institute, Hungary Academy of Sciences.

Chapter 32

Opener: © Ingram Publishing; p. 687: © Andy Crump, TDR, World Health Organization/Science Source; 32.2a: © Michael Abbey/ Science Source; 32.3a: © Mary Beth Angelo/Science Source; 32.5a: OAR/National Undersea Research Program (NURP)/College of William & Mary/NOAA; 32.10 (all): Courtesy of James Berleman and Carl Bauer; 32.11 (both): © Lee W. Wilcox, Ph.D.; 32.12a: © John Durham/SPL/ Science Source; 32.12c: © Cameron R. Currie; p. 698: © Amazon-Images/Alamy; p. 701: © Banana Stock/Punchstock.

Chapter 33

Opener: © McGraw-Hill Education/Mary Reeg, photographer; 33.3: Dr. Steve Kraus/CDC; 33.5b: © SPL/Science Source; 33.8b: © Sucharit Bhakdi; 33.10: © David Scharf/Science Source; 33.23: © Dr. Sanjay Mukhopadhyay.

Chapter 34

Opener: © GraphicaArtis/Getty Images; 34.3 (left), 34.3 (middle left): © EyeWire Collection/Getty Images; 34.3 (middle right): © Photodisc Collection/Getty Images; 34.3 (right): © Creatas/ PictureQuest; 34.9c, 34.9d: © Dr. Gilla Kaplan, PHRI, UMDNJ; p. 762: © Pixtal/age fotostock.

Chapter 35

Opener: Janice Carr/CDC; 35.10: © SPL/Science Source; p. 784: © Ewen Charlton/Getty Images.

Chapter 36

Opener: © Ami Images/Science Source; 36.2: Hsi Liu, Ph.D., MBA; James Gathany/CDC; 36.3 (all): © McGraw-Hill Education/James Redfearn, photographer; 36.5: © John Watney/Science Source; 36.6: Hsi Liu, Ph.D., MBA; James Gathany/CDC; 36.7c: © Millipore Corporation; 36.9: Dorothy Allain/ CDC; 36.12: © Phanie/SuperStock; 36.13: © Cell Signaling Technology, Inc.; 36.15: Dr. Errol Reiss/ CDC; 36.16: Dr. U.P. Kokko/CDC.

Chapter 37

Opener: Amanda Mills/CDC; 37.4 (all): CDC; p. 818: © Justin Gollmer/Getty Images; p. 820: © Ann Ronan Pictures/Print Collector/Getty Images.

Chapter 38

Opener: Paul Howell/CDC; 38.1b: © Brand X Pictures/PunchStock; 38.2c: CDC; 38.4b: Cynthia Goldsmith/CDC; 38.4e: Brian W.J. Mahy, BSc, MA, PhD, ScD, DSc/CDC; 38.5: Dr. Fred Murphy/CDC; 38.6: CDC; 38.7b: Dr. Noble/CDC; p. 837: © Lawrence Manning/Corbis; 38.12: © National Institutes of Health/Stocktrek Images/Getty Images; 38.13: Dr. Herrmann/CDC; 38.14: © Custom Medical Stock Photo; 38.15: © Dan Wiedbrauk, Ph.D/Warke Medical Laboratory Ann Arbor, Michigan; 38.16a: Dr. Fred Murphy & Sylvia Whitfield/CDC; 38.16b: Dr. Fiumara and Dr. Gavin Hart/CDC; 38.16c: Susan Lindsley/CDC; 38.17: CDC; 38.18: © Barbara O'Connor; 38.19: © Dr. Linda M. Stannard, University of Cape Town/Science Source; 38.20a: © James Cavallini/Science Source; 38.20b: Joe Miller/CDC; 38.20c: CDC; 38.21: © Eye of Science/Science Source; p. 853: © Bettmann/Corbis; 38.22: Cynthia Goldsmith/CDC; 38.23: © CNRI/SPL/Science Source; 38.24: Teresa Hammett/CDC.

Chapter 39

Opener: Dr. Jerry J. Callis, PIADC and Dr. Brian W.J. Mahy/CDC; 39.1b: © Mediscan/Alamy; 39.2: Margaret Williams, PhD; Claressa Lucas, PhD; Tatiana Travis, BS/CDC; 39.3: © Tone Tonjum; 39.4a: Janice Carr/CDC; 39.4b: © Biophoto Associates/Science Source; 39.6: © Dr. Allan Harris/Phototake; 39.7a: © Science Picture Co/Getty Images; 39.7b: Michael L. Levin, Ph. D./CDC; 39.7c: James Gathany/CDC; 39.8b: CDC; 39.9b: © Prof. Guy R. Cornelis; 39.10: CDC; 39.11: © Scott Camazine/Phototake; p. 873: CDC; 39.12: Arthur E. Kaye/CDC; 39.13: © Juergen Berger/Science Source; 39.14: © Morris D. Cooper, Ph.D./Southern Illinois University School of Medicine; 39.15a: Bill Schwartz.CDC; 39.15b: J. Pledger/CDC; 39.16a, 39.16b: CDC; 39.16c: Donated by Brian Hill, New Zealand/CDC; 39.18a: © Dr. Kari Lounatmaa/Science Source; 39.18b: Janice Carr/CDC; 39.19, 39.20: CDC; 39.21: Gregory Moran, M.D./CDC; 39.22: © Biomedical Communications/Custom Medical Stock Photo; 39.23: © Everett Collection Historical/Alamy; 39.24: © ISM/Phototake; 39.26: © Ami Images/Science Source; 39.28a: K. Sekiya et al, "Supermolecular structure of the enteropathogenic Escherichia coli type III secretion system and its direct interaction with the EspA-sheath-like structure" PNAS September 25, 2001, Vol. 98, no. 20, p. 11638–11643. Copyright (2001) National Academy of Sciences, U.S.A.; p. 895: © Brand X Pictures/PunchStock; 39.30: © Alfred Pasieka/Science Source; 39.31: © Dr. Paul L. Beck/University of Calgary; 39.33: © David Scharf/Science Source; 39.34: © Daniel Zgombic/E+/Getty Images.

Chapter 40

Opener: Lucille K. Georg/CDC; 40.2: © Scott Camazine/Science Source; 40.3: Lucille K. Georg/CDC; 40.4: CDC; 40.5 (both): Dr. Libero Ajello/CDC; 40.8: CDC; 40.10a: Dr. Mae Melvin/CDC; 40.10b: © Andy Crump, TDR, World Health Organization/Science Source; 40.13b: © Ray Wilson/Alamy; 40.14a: CDC; 40.14b: © Medical-on-Line/Alamy; 40.15a: K. Mae Lennon, Tulane Medical School; Clement Benjamin/CDC; 40.15b, 40.15c: CDC; 40.16, 40.17: Dr. Lucille K. Georg/CDC; 40.18: CDC; 40.19a: Dr. Mae Melvin and Dr. Greene/CDC; 40.19b: CDC; 40.21 (both): Dr. Stan Erlandsen/CDC; 40.23a: © Thierry Berrod, Mona Lisa Production/SPL/Science Source; 40.23b: Robert Simmons/CDC; 40.24a: Maxine Jalbert, Dr. Leo Kaufman/CDC; 40.24b: Sol Silverman, Jr., DDC/CDC; 40.24c: Dr. Hardin/CDC; 40.25: CDC-DPDx.

Chapter 41

Opener: © John A. Rizzo/Getty Images; 41.1a: © imagebroker/Alamy; 41.1b: © Daleen Loest/iStock/Getty Images; 41.3: © Donald Klein; 41.4: © Campbell Soup Company. Photo by Mark Seliger; p. 933: © Noam Armonn/Alamy; 41.7: CDC; p. 938 (top): © Stefano Bertini/iStock/Getty Images; p. 938 (bottom): © Rodrigo Buendia/AFP/Getty Images; 41.8 (all): © Jeffery Broadbent/Utah State University; 41.9: © Joe Munroe/Science Source; 41.12: © William West/AFP/Getty Images.

Chapter 42

Opener: © Brand X Pictures/Punchstock; p. 951: © Stockbyte/Getty Images; p. 958: © T. O'Keefe/PhotoLink/Getty Images; Fig 42.8a: © Yi Xiang Yeng/iStock/Getty Images; Fig 42.8b: © McGraw-Hill Education; Fig. 42.8c: © Nigel Cattlin/Science Source; Fig 42.8d: © OAR/National Undersea Research Program (NURP)/College of William & Mary/NOAA; 42.11a: © Eye of Science/Science Source; 42.11b: © Steve Gschmeissner/Science Source.

Chapter 43

Opener: U.S. Coast Guard photo; p. 965: © 81a/age fotostock; 43.3a,b: © Donald Klein; 43.3c: © Ethan Shelkey and Joanne M. Willey, Ph.D.; 43.4: © John Edwards/The Image Bank/Getty Images; 43.5: © Ecologix Environmental Systems, LLC.; p. 974: © NASA/NOAA/SPL/Getty Images; 43.9b: © Kelly P. Nevin; 43.13: © Accent Alaska.com/Alamy.

Design Elements: (DNA): © Chris Knapton/Getty Images; (test tubes): © Ingram Publishing; (books): © McGraw-Hill Education/Mazer Creative Services; (x-ray): © ER Productions/Corbis; (Yellowstone): © McGraw-Hill Education/Carrie Burger, photographer.

Line Art and Text

Chapter 2

Figure 2.10: Source: MicroscopyU website http://www.microscopyu.com.

Chapter 6

Figure 6.23b: Source: Flores, R., et al., *Annu. Rev. Microbiol.* 68:395–414. © 2014 Annual Reviews.

Chapter 7

Figure 7.4: Source: Lindas, A. C., & Bernander, R. (January 01, 2013). The cell cycle of archaea. *Nature Reviews. Microbiology*, 11, 9, 627-38, p. 633, Fig. 4a-bottom, MacMillan, 2013; **Figure 7.5:** Source: Lenz, P., & Søgaard-Andersen, L. (January 01, 2011). Temporal and spatial oscillations in bacteria. *Nature Reviews. Microbiology*, 9, 8, 565-77, p. 572, Fig. 3a, Macmillan, 2011; **Figure 7.6:** Source: Huang, K.-H., et al. J. Bacteriol. 195(9): 1859. © 2013 ASM.

Chapter 8

Table 8.3, Table 8.4: Source: From Seymour S. Block, DISINFECTION, STERILIZATION AND PRESERVATION. Copyright © 1983 Lea & Febiger, Malvern, PA, 1983. Philadelpha, PA: Lippincott Williams Wilkins.

Chapter 12

Figure 12.31: Source: Bishop, R. E. *Nature* 511:37–38. © 2014 Macmillan.

Chapter 13

Figure 13.13: Source: Kurth, I., and O'Donnell, M. (2013). *Trends in Biochemical Sciences* 38(4):195–203. Elsevier. Fig. 1a. p. 196; **Figure 13.43:** Source: Kim, Y. E., Hipp, M. S., Bracher, A., Hayer-Hartl, M., & Ulrich Hartl, F. (2013). Molecular chaperone functions in protein folding and proteostasis. *Annual Review of Biochemistry*, 82, 323–355.

Chapter 15

Figure 15.1: Source: Kurth, I., and O'Donnell, M. (2013). *Trends in Biochemical Sciences* 38(4):195–203, Fig. 1a, p.196, Elsevier; **Figure 15.5:** Source: Lindas, A.-C. and Bernander R. (2013) *Nature Reviews Microbiology* 11:627. Macmillan. 627–638. Fig. 2. Pg. 629; **Figure 15.14a-b:** Source: Kim, Y. E., Hipp, M. S.,

Bracher, A., Hayer-Hartl, M., & Ulrich Hartl, F. (2013). Molecular chaperone functions in protein folding and proteostasis. *Annual Review of Biochemistry*, 82, 323–355, Fig. 3, p. 329.

Chapter 16
Figure 16.27a-b: Source: Johnston, C., Martin, B., Fichant, G., Polard, P., & Claverys, J. P. (2014). Bacterial transformation: distribution, shared mechanisms and divergent control. *Nature Reviews Microbiology*, 12(3), 181–196, Fig. 2, p. 185.

Chapter 18
Figure 18.4: Source: Illumina, Inc. http://res.illumina.com/documents/products/techspotlights/techspotlight_sequencing.pdf;
Figure 18.7b: Source: Rinke, C., Schwientek, P., Sczyrba, A., Ivanova, N. N., Anderson, I. J., Cheng, J. F., Woyke, T. (2013). Insights into the phylogeny and coding potential of microbial dark matter. *Nature*, 499(7459), 431–437; **Figure 18.15:** Source: Malone and Oliver BMC Biology 2011 9:34 doi:10.1186/1741-7007-9-34; **Figure 18.21a-b:** Source: Guerrero et al., BMC Evolutionary Biol. 2005, 5:55, figure a+d.

Chapter 19
Figure 19.3: Data from Woese, C. P. 1987. Microbiological Rev. 51(2):221–27.

Chapter 20
Figure 20.2: Source: Offre, P., Spang, A., & Schleper, C. (2013). Archaea in biogeochemical cycles. *Annu. Rev. Microbiol.* 67, 437-457. Fig. 1, p. 439; **Figure 20.11:** Source: She et al, *Proc. Nat. Acad. Sci*, July 3, 2001, Vol. 98, Figure 1, p. 7836. Copyright © 2001 National Academy of Sciences, U.S.A.; **Figure 20.19:** "Genome Sequence of Halobacterium species NRC-1" by Wailap Victor Ng et al from *PNAS*, October 3, 2000, Figure 1, p. 12178. Copyright © 2000 National Academy of Sciences, U.S.A.

Chapter 21
Figure 21.6: Adapted from Glass, J. I., et al. 2000. The complete genome of the mucosal pathogen Ureaplasma urealyticum. *Nature.* 407:757–62.

Chapter 22
Figure 22.2: Source: Ferla MP, Thrash JC, Glovannoni SJ, Patrick WM (2013) New rRNA Gene-Based Phylogenies of the Alphaproteobacteria provide prespective on Major Groups, Mitochondrial Ancestry and Phylogenetic Instability. PLoS ONE 8(12): e83383; **Figure 22.17:** Carini et al., 2013, The ISME journal 7: 592-602 doi: 10-1038/ismej.2012.122; **Figure 22.19:** Chain, P., et al., 2003. Complete genome sequence of the ammonia-oxidizing bacterium and obligate chemolithoautotroph Nitrosomonas europaea. *J. Bacteriol.* 185:2759–73; **Figure 22.21:** The Ribosomal Database Project; **Figure 22.26:** Scott et al., *PloS Biology*, December 2006, Vol. 4, Figure 2, p. 2199; **Figure 22.36b:** Source: Wartel M, Ducret A, Thutupalli S, Czerwinski F, et al. (2013) A Versatile Class of Cell Surface Directional Motors Gives Rise to Gliding Motility and Sporulation in Myxococcus xanthus. *PLoS Biol* 11(12): e1001728. doi:10.1371/journal.pbio.1001728.

Chapter 23
Figure 23.2: Brenner, D. J., et al., eds. 2005. *Bergey's Manual of Systemic Bacteriology*, 2nd ed., vol. 2: The Proteobacteria. Garrity, G. M. Ed-in-Chief. New York: Springer; **Figure 23.5:** Adapted from Bruggemann et al., *Proc. National Acad. Sci*, February 2003, Vol. 100:1316-21, figure 4, p. 1320. Copyright © 2003 National Academy of Sciences, U.S.A.; **Figure 23.11b:** Source: Dr TV Bao MD Travancore Medical College, Kollam Kerala.

Chapter 24
Figure 24.10a: Source: http://faculty.ccbcmd.edu/courses/bio141/lecguide/unit1/prostruct/u1fig11. html.

Chapter 26
Figure 26.1: Source: "Reconstructing the early evolution of Fungi using a six-gene phylogeny" by James, T.Y. et al., *NATURE*, Vol. 443, 2006, p. 818–822.

Chapter 27
Figure 27.28a: Source: Tao, Y. J. and Zheng W., (2012). *Science.* Vol. 338 no. 6114, AAAS, pp. 1545–1546.

Chapter 28
Figure 28.1: Adapted from PN, Klinkhammer, GP, Bender, ML., et al., 1979. Geochimica et Cosmochimica Acta 43:1077–1090.

Chapter 31
Figure 31.3: "Identifying the Dominant Soil Bacterial Taxa in Libraries of 16S rRNA and 16S rRNA Genes" by Peter H. Janssen from *Appl. Env. Microbiol.* March 2006, Vol. 72, No.3, Fig. 1, p. 1722.

Chapter 32
Figure 32.14: Sources: Spor, A., O. Koren and R. Ley, 2011. Unravelling the effects of the environment and host genotype on the gut microbiome. *Nature Rev Microbiol* 9:279–290 and Belkaid, Y and J.A. Segre, 2014. Dialog between skin microbiota and immunity. *Science* 346:954–959.

Chapter 33
Figure 33.7: Source: Mitch Leslie (2012). *Science Magazine*, Vol. 337 no. 6098, AAAS, pp. 1035; **Figure 33.12a-f:** Source: Mace et al., *Immunology and Cell Biology* (2014) 92, 245–255; **Figure 33.17:** Source: SA Biosciences.

Chapter 38
Figure 38.7a: Redrawn from Esposito et al, SCIENCE, Vol. 313, 2006, figure 1, p. 808. Copyright © 2006 AAAS; **Figure 38.8:** Source: https://www.aids.gov/hiv-aids-basics/hiv-aids-101/global-statistics/; **Figure 38.11b:** Source of Data: http://www.cdc.gov/hiv/statistics/recommendations/terms.html.

Chapter 40
Figure 40.4: World Health Organization (WHO).

Chapter 42
Figure 42.6: Menzella & Reeves, "Combinatorial Biosynthesis Yields Novel Analogs of Natural Products," *Microbe* vol. 2 no. 9, p. 431. Copyright © 2007 by the American Society for Microbiology. All rights reserved. Used with permission.

Page numbers followed by t and f designate tables and figures, respectively; page numbers in *italics* refer to definitions; page numbers in **boldface** refer to major discussions; pages numbers with A refer to appendix.

A

AAnP (aerobic anoxygenic photosynthesis), *508*

AAnPs (aerobic anoxygenic phototrophs), 658

Abbé, Ernst, 24

Abbé equation, 24–25

ABC (ATP-binding cassette) transporters, *51*, 51f

Abiotic methane, 464

ABO blood groups, *762*

AB toxin, *777*

Acanthamoeba keratitis (cornea inflammation), *918–919*

Acarbose (type 2 diabetes drug), from *Actinoplanetes*, 558

Accessory pigments, in photosynthesis, *254*, 255f

Acellular slime molds, *569*, 569f

Acellular vaccines, *821*

Acetyl-coenzyme A, *236*

Acidaminococcaceae, 544

Acid-fast cell walls, *558*

Acid-fast staining, *33*, 34f

Acidianus convivator, 110

Acidianus two-tailed virus (ATV), 110, 110f

Acidimethylosilex fumarolicum, 498–499

Acidithiobacillus ferrooxidans, 519, 631

Acidithiobacillus species, 251

Acid mine drainage, **518–519**

Acidobacteria, 257, 490, 631

Acidophile microorganisms, *144*

Acne vulgaris, 559, 703

Acquired immunity, *708; See also* Adaptive immunity

Acridine orange stain, 641

ACT (artemisinin-based combination therapy), for malaria, 438, 910

Actin filaments, *94*, 94f

Actinobacteria, **552–562**

Actinomycetales, 555

Bifidobacteriales, 561

characteristics of, 553t

Corynebacteriales, 557–558

Frankiales, 561

Gram positive bacteria in phylum, 56

Micrococcales, 555–556

Micromonosporales, 558

as monoderms, 54

overview, 554–555

Propionibacteriales, 558–559

Streptomycetales, 559–560

Streptosporangiales, 560–561

Actinomycetales, 555

Actinomycetes, *553*, 555f

Actinomycetoma disease, 560

Actinoplanetes, diabetes drug from, 558

Actinorhizae, *679–680*

Actin tails, of intracellular bacterial pathogens, 781f

Activated sludge, in wastewater treatment, *971*, 971f

Activation energy for reactions, *219*, 220f

Activator proteins, *324–325*, **363–366**

Active sites, for reactions, *220*

Active transport

in archaea, 83

in eukaryotic cells, 92–93

primary and secondary, *51–52*, 51f

Acute bacterial diarrhea and food poisoning, 886t–887t

Acute inflammatory response, *732*, 732f

Acute-phase proteins, *717*

Acute viral gastroenteritis, *851–852*, 852t

Acyclovir antiviral drug, 202, 828, 844

Acylhomoserine lactones (AHLs), 172

Adaptive immunity, **736–769**, 737f; *See also* Host resistance, innate

antibodies, **749–757**

binding to targets, 757–760

clonal selection, 757

diversity of, 755–757

immunoglobulin classes, 752–753

immunoglobulin function, 751–752

immunoglobulin structure, 750–751

kinetics of, 753–755

antigens to elicit, 738–739

autoimmunity and autoimmune diseases, 765–766

B cells in, 747–749

earned or borrowed, 739–740, 739f

foreignness, recognition of, 740–742

hypersensitivities, 760–765

not responding as part of, 760

recognition and memory in, 736–738

T cells in, 743–747

transplantation (tissue) rejection, 766–767

Adaptive mutations, *459*

ADCC (antibody-dependent cell-mediated cytotoxicity), 723–724, 723f, 752

Adenine, *276*

Adenine arabinoside antiviral drug, 202

Adenoids, 726

Adenosine 5'-phosphosulfate (APS), 251

Adenosine diphosphate (ADP), *212*

Adenosine monophosphate, 277f

Adenovirus, 113f

Adherence and colonization factors, in pathogenicity, 774–775, 775f, 775t

Adhesion molecules, in *Caulobacter crescentus*, 510

Adipokines, 707

Adjuvants, in vaccines, *818*

ADP (adenosine diphosphate), *212*

Aerial mycelium (hyphal mat), *554*

Aerobes, 147, 157–158

Aerobic anoxygenic photosynthesis (AAnP), *508*

Aerobic anoxygenic phototrophs (AAnPs), 658

Aerobic respiration, **232–243**

electron transport to oxidative phosphorylation, 236–243

ATP yield, 241–243

electron transport chains, 238–240, 239f

overview, 236–237

oxidative phosphorylation process, 240–241

glucose to pyruvate, 232–235

overview, *232*

pyruvate to carbon dioxide, 236